College Algebra

Formulas from Analytical Geometry: $P_1 \to (x_1, y_1)$, $P_2 \to (x_2, y_2)$

Distance between P_1 and P_2

$$d = \sqrt{(x_2 - x_1)^2 + (y_2 - y_1)^2}$$

Slope of Line Containing P_1 and P_2

$$m = \frac{\Delta y}{\Delta x} = \frac{y_2 - y_1}{x_2 - x_1}$$

Equation of Line Containing P_1 and P_2

Point-Slope Form

$$y - y_1 = m(x - x_1)$$

Equation of Line Containing P_1 and P_2

Slope-Intercept Form (slope m, y-intercept b)

$$y = mx + b, \text{ where } b = y_1 - mx_1$$

Parallel Lines

Slopes Are Equal: $m_1 = m_2$

Perpendicular Lines

Slopes Have a Product of -1: $m_1 m_2 = -1$

Intersecting Lines

Slopes Are Unequal: $m_1 \neq m_2$

Dependent (Coincident) Lines

Slopes and y-Intercepts Are Equal: $m_1 = m_2$, $b_1 = b_2$

Logarithms and Logarithmic Properties

$$y = \log_b x \Leftrightarrow b^y = x \qquad \log_b b = 1 \qquad \log_b 1 = 0$$

$$\log_b b^x = x \qquad b^{\log_b x} = x \qquad \log_c x = \frac{\log_b x}{\log_b c}$$

$$\log_b MN = \log_b M + \log_b N \qquad \log_b \frac{M}{N} = \log_b M - \log_b N \qquad \log_b M^P = P \cdot \log_b M$$

Applications of Exponentials and Logarithms

$A \to$ amount accumulated $\qquad P \to$ initial deposit, $p \to$ periodic payment $\qquad n \to$ compounding periods/year

$r \to$ interest rate per year $\qquad R \to$ interest rate per time period$\left(\frac{r}{n}\right)$ $\qquad t \to$ time in years

Interest Compounded n Times per Year

$$A = P\left(1 + \frac{r}{n}\right)^{nt}$$

Interest Compounded Continuously

$$A = Pe^{rt}$$

Accumulated Value of an Annuity

$$A = \frac{p}{R}\left[(1 + R)^{nt} - 1\right]$$

Payments Required to Accumulate Amount A

$$p = \frac{AR}{(1 + R)^{nt} - 1}$$

Sequences and Series:

$a_1 \to$ 1st term, $a_n \to n$th term, $S_n \to$ sum of n terms, $d \to$ common difference, $r \to$ common ratio

Arithmetic Sequences

$$a_1, a_2 = a_1 + d, a_3 = a_1 + 2d, \ldots, a_n = a_1 + (n - 1)d$$

$$S_n = \frac{n}{2}(a_1 + a_n)$$

$$S_n = \frac{n}{2}[2a_1 + (n - 1)d]$$

Geometric Sequences

$$a_1, a_2 = a_1 r, a_3 = a_1 r^2, \ldots, a_n = a_1 r^{n-1}$$

$$S_n = \frac{a_1 - a_1 r^n}{1 - r}$$

$$S_\infty = \frac{a_1}{1 - r}; \ |r| < 1$$

Binomial Theorem

$$(a + b)^n = \binom{n}{0}a^n b^0 + \binom{n}{1}a^{n-1}b^1 + \binom{n}{2}a^{n-2}b^2 + \cdots + \binom{n}{n-1}a^1 b^{n-1} + \binom{n}{n}a^0 b^n$$

$$n! = n(n - 1)(n - 2) \cdots (3)(2)(1) \qquad \binom{n}{k} = \frac{n!}{k!(n - k)!}; \qquad 0! = 1$$

ALEKS® 3.0 Makes the Grade!

Would your students appreciate their own personal math tutor available 24 hours per day, 7 days a week? A tutor who teaches what they're most ready to learn? And a tutor who analyzes their answers to problems and responds with specific advice when they make a mistake? A tutor who provides text specific assets, such as videos, multimedia tutorials and textbook PDF pages when applicable for additional support?

ALEKS® individualizes assessment and learning and is. . .

- Customizable for each course topic with text specific learning assets.
- Provides a robust course management system telling you exactly what your students know and don't know.
- Provides your students with specific course content they are ready to learn, helping your students progress toward mastery of curricular goals.

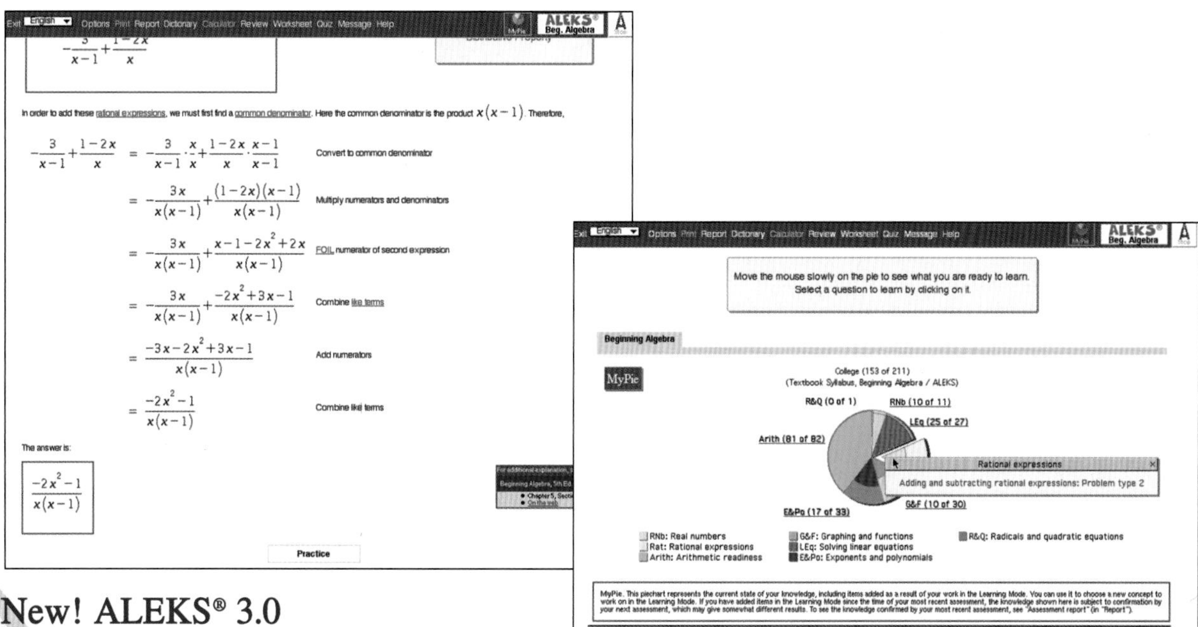

New! ALEKS® 3.0

- Now links to text-specific videos, multimedia tutorials, and textbook pages in PDF format.
- Students and instructors can access their ALEKS® accounts with a single sign-on through recent versions of WebCT and Blackboard and instructors can automatically import results directly into their gradebooks.
- ALEKS® ensures that students can progress toward mastery regardless of their level of preparation.

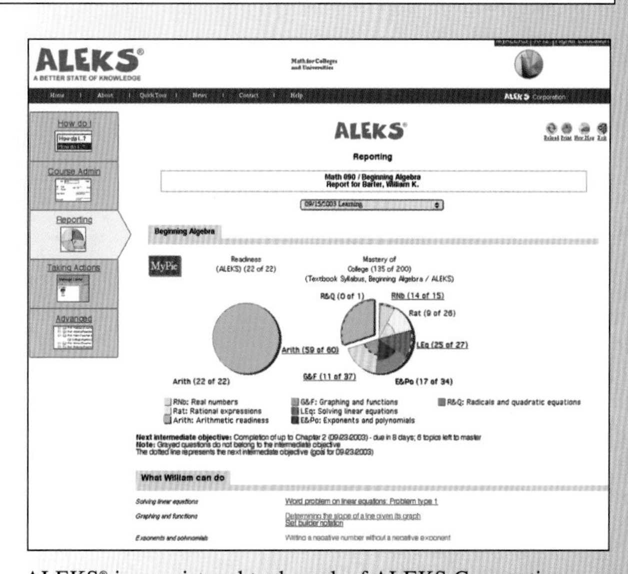

ALEKS® is a registered trademark of ALEKS Corporation.

Be Our Guest!

To request your **FREE 24-hour trial of ALEKS®**, visit our website at www.highed.aleks.com Click on **"Be our Guest"** under the **"For Visitors"** heading.

MathZone™ ■ Assign ■ Assess ■ Administer

MathZone's™ powerful feature set includes book-specific, assignable, algorithmic content, ADA compliant videos and e-Professor animated solutions. MathZone™ provides students virtually unlimited practice through algorithmic quizzing and testing, free live tutoring via NetTutor's™ whiteboard technology, and instructors can track their students' progress in the online gradebook.

Enhance the Learning Experience

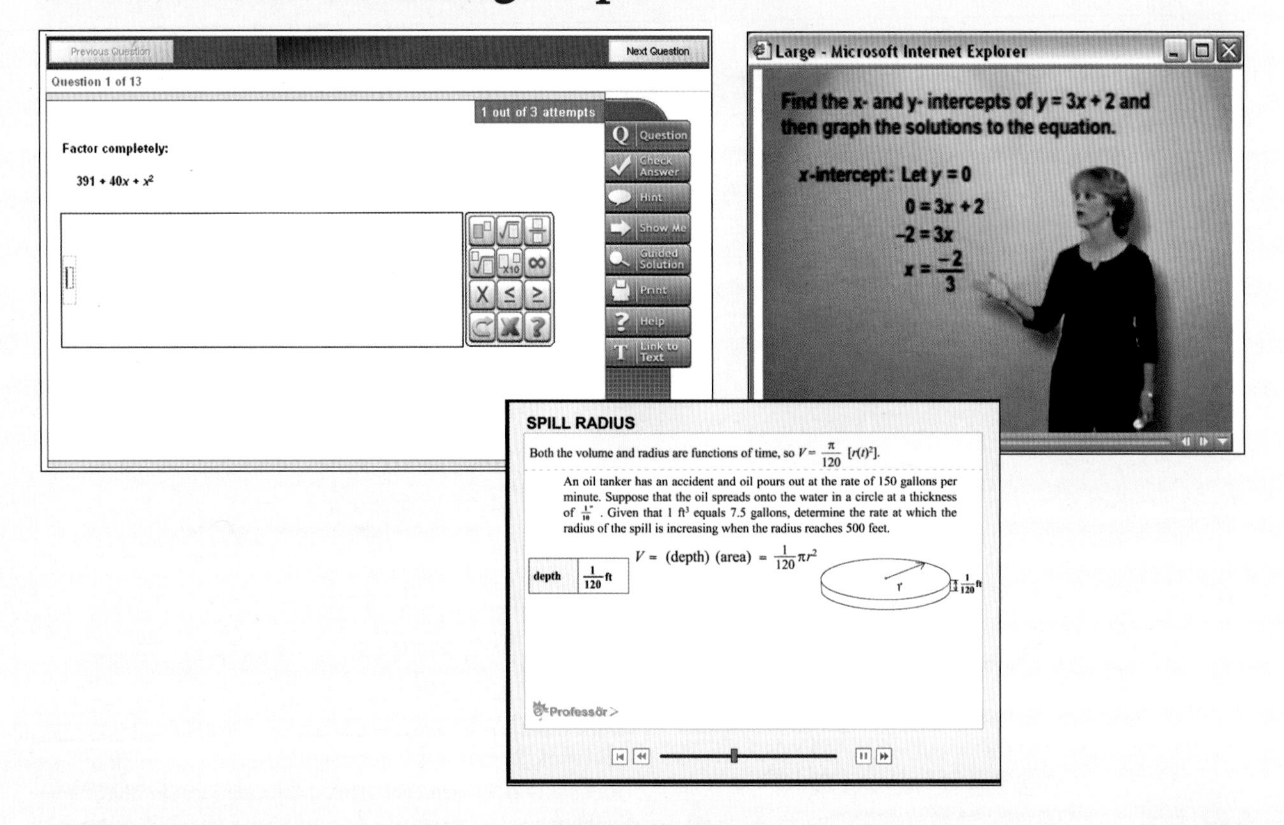

Build Assignments

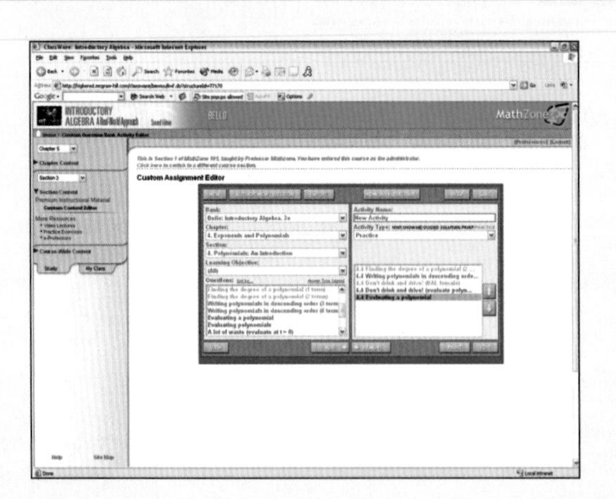

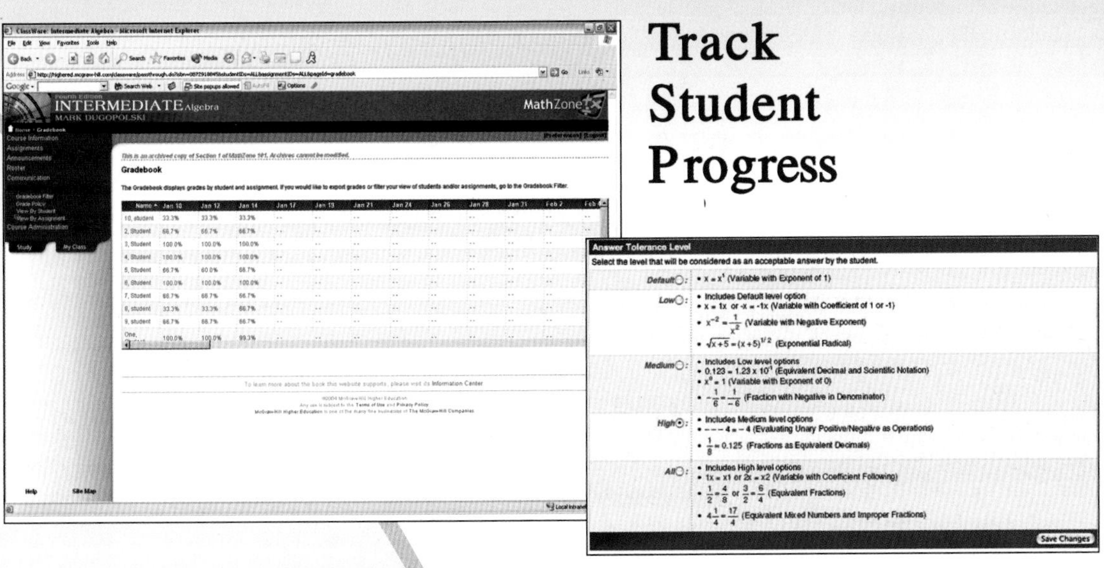

Track Student Progress

Administer Your Course

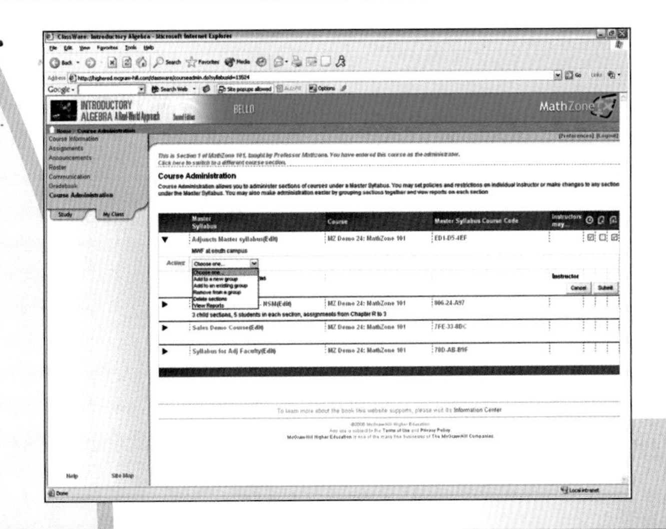

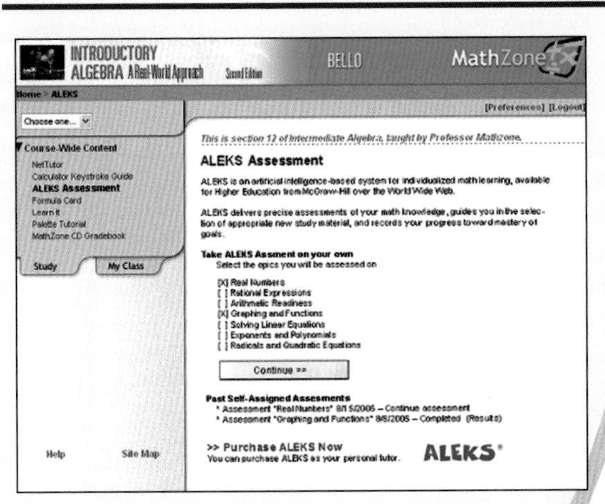

Student Assessment

- MathZone™ 3.0 now incorporates the powerful ALEKS assessment module.
- The ALEKS assessment module will precisely evaluate what each individual student or your entire class is ready to learn next in your syllabus.
- After completing an assessment students and instructors can access individualized assessment reports.
- Students can sign up for a free trial of ALEKS for remediation for those who need extra tutorial assistance.

For More Information

- **www.mathzone.com** click on the technical support tab
- For Instructor Technical Support call **1-800-541-4019**
- Or email: **mathzone@mcgraw-hill.com**

College Algebra

John W. Coburn

St. Louis Community College at Florissant Valley

Boston Burr Ridge, IL Dubuque, IA Madison, WI New York San Francisco St. Louis
Bangkok Bogotá Caracas Kuala Lumpur Lisbon London Madrid Mexico City
Milan Montreal New Delhi Santiago Seoul Singapore Sydney Taipei Toronto

Higher Education

COLLEGE ALGEBRA

Published by McGraw-Hill, a business unit of The McGraw-Hill Companies, Inc., 1221 Avenue of the Americas, New York, NY 10020.

Some ancillaries, including electronic and print components, may not be available to customers outside the United States.

This book is printed on acid-free paper.

1 2 3 4 5 6 7 8 9 0 VNH/VNH 0 9 8 7 6

ISBN-13 978–0–07–290119–1
ISBN-10 0–07–290119–5

ISBN-13 978–0–07–313702–5 (Annotated Instructor's Edition)
ISBN-10 0–07–313702–2

Publisher: *Elizabeth J. Haefele*
Sponsoring Editor: *Steven R. Stembridge*
Director of Development: *David Dietz*
Developmental Editor: *Suzanne Alley*
Senior Marketing Manager: *Dawn R. Bercier*
Senior Project Manager: *Vicki Krug*
Senior Production Supervisor: *Sherry L. Kane*
Lead Media Project Manager: *Stacy A. Patch*
Media Producer: *Amber M. Huebner*
Senior Designer: *David W. Hash*
Cover/Interior Designer: *Maureen McCutcheon*
Cover Photo: *The Gateway Arch, St. Louis, Missouri, © Charles Thatcher/Getty Images*
Senior Photo Research Coordinator: *John C. Leland*
Photo Research: *Emily Tietz*
Supplement Producer: *Melissa M. Leick*
Compositor: *The GTS Companies/York, PA Campus*
Typeface: *10/12 Times Roman*
Printer: *Von Hoffmann Corporation*

Photo Credits
1: © Tom Stewart/CORBIS; **12:** © Ryan McVay/Getty Images; **34:** © Royalty-Free/Corbis
Chapter 1
Opener © Royalty-Free/Corbis; p. 81**:** © Arne Hodalic/Corbis (left); p. 81**:** © AFP/Getty Images (right)
Chapter 2
Opener © Royalty-Free/Corbis; p. 154**:** © Jeffrey Burke/Brand X Pictures/Image State; p. 225**:** © Joe Raedle/Getty Images; p. 228**:** © Leonardo da Vinci/The Bridgeman Art Library/Getty
Chapter 3
Opener © Michael S. Yamashita/CORBIS; p. 271**:** © Keystone, Denis Balibouse, Pool/AP Photo; p. 338**:** © John Coburn
Chapter 4
Opener © Ryan McVay/Getty Images; p. 392**:** © Royalty-Free/CORBIS; p. 436**:** © Royalty-Free/CORBIS; p. 462**:** © Royalty-Free/Corbis
Chapter 5
Opener © Douglas Slone/Corbis; p. 504**:** © Didrik Johnck/Corbis Sygma/Time Inc./Time Life Pictures/Getty Images; p. 557**:** Courtesy of Simon Thomas
Chapter 6
Opener © Alan Schein Photography/CORBIS; p. 576**:** © Royalty-Free/CORBIS
Chapter 7
Opener © RubberBall/SuperStock; p. 704**:** © H. Wiesenhofer/PhotoLink/Getty Images; p. 738**:** © Cindy Charles/Photo Edit
Chapter 8
Opener © Emma Lee/Life File/Getty Images; p. 767**:** © Royalty-Free/CORBIS; p. 805**:** © Galen Rowell/Corbis

Library of Congress Cataloging-in-Publication Data

Coburn, John W.
College algebra / John W. Coburn. — 1st ed.
p. cm.
Includes index.
ISBN 978–0–07–290119–1 — 0–07–290119–5 (acid-free paper)
1. Algebra—Textbooks. I. Title.

QA154.3.C594 2007
512.9—dc22 2005054414
CIP

www.mhhe.com

ABOUT THE AUTHOR

John Coburn grew up in the Hawaiian Islands, the seventh of sixteen children. In 1979 he received a bachelor's degree in education from the University of Hawaii. After being lured into the business world for a number of years, he returned to his first love, accepting a teaching position in high school mathematics where he was recognized as Teacher of the Year in 1987. Soon afterward, John decided to seek a master's degree, which he received two years later from the University of Oklahoma. For the last fifteen years, he has been teaching mathematics at the Florissant Valley campus of St. Louis Community College, where he is now a full professor. During his tenure there he has received numerous nominations as an outstanding teacher by the local chapter of Phi Theta Kappa, and was recognized as Post–Secondary Teacher of the Year in 2004 by the Mathematics Educators of Greater St. Louis (MEGSL). He has made numerous presentations at local, state, and national conferences on a wide variety of topics. His other loves include his family, music, athletics, games, and all things beautiful. We hope this love of life comes through in his writing, and serves to make the learning experience an interesting and engaging one for all students.

DEDICATION

To my wife and best friend Helen, whose love, support, and willingness to sacrifice never faltered.

Contents

Preface xiii
Index of Applications xxvii

CHAPTER R A Review of Basic Concepts and Skills–1

R.1 The Language, Notation, and Numbers of Mathematics 2
R.2 Algebraic Expressions and the Properties of Real Numbers 13
R.3 Exponents, Polynomials, and Operations on Polynomials 22
R.4 Factoring Polynomials 35
R.5 Rational Expressions 44
R.6 Radicals and Rational Exponents 54
Practice Test 68

CHAPTER 1 Equations and Inequalities–71

1.1 Linear Equations, Formulas, and Problem Solving 72
1.2 Linear Inequalities in One Variable with Applications 83
1.3 Solving Polynomial and Other Equations 94
Mid-Chapter Check 106
Reinforcing Basic Concepts: Solving $x^2 + bx + c = 0$ 106
1.4 Complex Numbers 107
1.5 Solving Nonfactorable Quadratic Equations 117
Summary and Concept Review 129
Mixed Review 134
Practice Test 135
Calculator Exploration and Discovery: Evaluating Expressions and Looking for Patterns 136
Strengthening Core Skills: An Alternative Method for Checking Solutions to Quadratic Equations 137

CHAPTER 2 Functions and Graphs–139

2.1 Rectangular Coordinates and the Graph of a Line 140
2.2 Relations, Functions, and Graphs 155
2.3 Linear Functions and Rates of Change 174
Mid-Chapter Check 188
Reinforcing Basic Concepts: The Various Forms of a Linear Equation 189

2.4 Quadratic and Other Toolbox Functions 190
2.5 Functions and Inequalities—A Graphical View 205
2.6 Regression, Technology, and Data Analysis 216
Summary and Concept Review 232
Mixed Review 238
Practice Test 240
Calculator Exploration and Discovery:
Cuts and Bounces: A Closer Look at the Zeroes of a Function 242
Strengthening Core Skills: More on End Behavior 243
Cumulative Review Chapters 1–2 244

CHAPTER 3 Operations on Functions and Analyzing Graphs–247

3.1 The Algebra and Composition of Functions 248
3.2 One-to-One and Inverse Functions 261
3.3 Toolbox Functions and Transformations 274
3.4 Graphing General Quadratic Functions 288
Mid-Chapter Check 299
Reinforcing Basic Concepts: Transformations via Composition 300
3.5 Asymptotes and Simple Rational Functions 301
3.6 Toolbox Applications: Direct and Inverse Variation 312
3.7 Piecewise-Defined Functions 327
3.8 Analyzing the Graph of a Function 341
Summary and Concept Review 358
Mixed Review 365
Practice Test 366
Calculator Exploration and Discovery:
Residuals, Correlation Coefficients, and Goodness of Fit 367
Strengthening Core Skills: Base Functions and Quadratic Graphs 369
Cumulative Review Chapters 1–3 371

CHAPTER 4 Polynomial and Rational Functions–373

4.1 Polynomial Long Division and Synthetic Division 374
4.2 The Remainder and Factor Theorems 383
4.3 Zeroes of Polynomial Functions 393
4.4 Graphing Polynomial Functions 407
Mid-Chapter Check 420
Reinforcing Basic Concepts: Approximating Real Roots 421
4.5 Graphing Rational Functions 422
4.6 Additional Insights into Rational Functions 438
4.7 Polynomial and Rational Inequalities—An Analytic View 451

Summary and Concept Review 464
Mixed Review 469
Practice Test 470
Calculator Exploration and Discovery: Complex Roots, Repeated Roots, and Inequalities 471
Strengthening Core Skills: Solving Inequalities Using the Push Principle 472
Cumulative Review Chapters 1–4 473

CHAPTER 5 Exponential and Logarithmic Functions–475

5.1 Exponential Functions 476
5.2 Logarithms and Logarithmic Functions 485
5.3 *The* Exponential Function and Natural Logarithms 495
Mid-Chapter Check 507
Reinforcing Basic Concepts: Understanding Properties of Logarithms 508
5.4 Exponential/Logarithmic Equations and Applications 509
5.5 Applications from Business, Finance, and Physical Science 521
5.6 Exponential, Logarithmic, and Logistic Regression Models 536
Summary and Concept Review 552
Mixed Review 556
Practice Test 558
Calculator Exploration and Discovery: Investigating Logistic Equations 559
Strengthening Core Skills: More on Solving Exponential and Logarithmic Equations 560
Cumulative Review Chapters 1–5 562

CHAPTER 6 Systems of Equations and Inequalities–565

6.1 Linear Systems in Two Variables with Applications 566
6.2 Linear Systems in Three Variables with Applications 577
6.3 Systems of Linear Inequalities and Linear Programming 590
6.4 Systems and Absolute Value Equations and Inequalities 604
Mid-Chapter Check 613
Reinforcing Basic Concepts: Window Size and Graphing Technology 613
6.5 Solving Linear Systems Using Matrices and Row Operations 616
6.6 The Algebra of Matrices 627
6.7 Solving Linear Systems Using Matrix Equations 640
6.8 Matrix Applications: Cramer's Rule, Partial Fractions, and More 655
Summary and Concept Review 666
Mixed Review 671

Practice Test 673
Calculator Exploration and Discovery:
Optimal Solutions and Linear Programming 674
Strengthening Core Skills: Augmented Matrices and Matrix Inverses 676
Cumulative Review Chapters 1–6 678

CHAPTER 7 Conic Sections and Nonlinear Systems–681

7.1 The Circle and the Ellipse 682
7.2 The Hyperbola 694
Mid-Chapter Check 705
Reinforcing Basic Concepts: More on Completing the Square 706
7.3 Nonlinear Systems of Equations and Inequalities 706
7.4 Foci and the Analytic Ellipse and Hyperbola 716
7.5 The Analytic Parabola 729
Summary and Concept Review 740
Mixed Review 742
Practice Test 743
Calculator Exploration and Discovery: Elongation and Eccentricity 744
Strengthening Core Skills:
Ellipses and Hyperbolas with Rational/Irrational Values of *a* and *b* 745
Cumulative Review Chapters 1–7 746

CHAPTER 8 Additional Topics in Algebra–747

8.1 Sequences and Series 748
8.2 Arithmetic Sequences 759
8.3 Geometric Sequences 768
8.4 Mathematical Induction 781
Mid-Chapter Check 789
Reinforcing Basic Concepts: Applications of Summation 790
8.5 Counting Techniques 792
8.6 Introduction to Probability 807
8.7 The Binomial Theorem 823
Summary and Concept Review 831
Mixed Review 837
Practice Test 838
Calculator Exploration and Discovery: Infinite Series, Finite Result 840
Strengthening Core Skills:
Probability, Quick-Counting, and Card Games 841
Cumulative Review Chapters 1–8 844

Appendix I: U.S. Standard Units and the Metric System A–1
Appendix II: Rational Expressions and the Least Common Denominator A–3
Appendix III: Reduced Row–Echelon Form and More on Matrices A–4
Appendix IV: Deriving the Equation of a Conic A–6
Student Answer Appendix SA-1
Instructor Answer Appendix (Annotated Instructor Edition only) IA-1
Index I-1

Additional Topics Online

(Visit www.mhhe.com/coburn)

R.7 Geometry Review with Unit Conversion
R.8 Expressions, Tables, and Graphing Calculators
7.0 An Introduction to Analytic Geometry
8.8 Conditional Probability and Expected Value
8.9 Probability and the Normal Curve with Applications

Preface

FROM THE AUTHOR

I was raised on the island of Oahu, and was a boy of four when Hawaii celebrated its statehood. From Laie Elementary to my graduation from the University of Hawaii, my educational experience was hugely cosmopolitan. Every day was filled with teachers and fellow students from every race, language, culture, and country imaginable, and this experience made an indelible impression on my view of the world. I can only hope that this exposure to different views and new perspectives contributed to an ability to connect with a diverse audience. It has certainly instilled the desire to communicate effectively with students from all walks of life—students like yours. Even my home experience helped to mold my thinking in this direction, because my education at home was closely connected to my public education. You see, Mom and Dad were both teachers. Mom taught English and Dad, as fate would have it, held advanced degrees in physics, chemistry, and . . . mathematics. But where my father was well known, well respected, and a talented mathematician, I was no prodigy and had to work very hard to see the connections so necessary for success in mathematics. In many ways, my writing is born of this experience, as it seemed to me that many texts offered too scant a framework to build concepts, too terse a development to make connections, and insufficient support in their exercise sets to develop long-term retention or foster a love of mathematics. To this end I've adopted a mantra of sorts, that being, "If you want more students to reach the top, you gotta put a few more rungs on the ladder." These are some of the things that have contributed to the text's unique and engaging style, and I hope in the end, to its widespread appeal.

Chapter Overview

The organization and pedagogy of each chapter support an approach sustained throughout the text, that of laying a firm foundation, building a solid framework, and providing strong connections. In the end, you'll have a beautiful, strong, and lasting structure, designed to support further learning opportunities. Each chapter also offers *Mid-Chapter Checks,* and contains the features *Reinforcing Basic Concepts* and *Strengthening Core Skills,* all designed to support student efforts and build long-term retention. The *Summary and Concept Reviews* offer on-the-spot, structured review exercises, while the *Mixed Review* gives students the opportunity to decide among available solution strategies. All *Practice Tests* have been carefully crafted to match the tone, type, and variety of exercises introduced in the chapter, with the *Cumulative Reviews* closely linked to the *Maintaining Your Skills* feature found in every section. Finally, the *Calculator Exploration and Discovery* feature, well . . . it does just that, offering students the opportunity to go beyond what is possible with paper and pencil alone.

Section Overview

Every section begins by putting some perspective on upcoming material while placing it in the context of the "larger picture." Objectives for the section are clearly laid out. The *Point of Interest* features were carefully researched and help to color the mathematical landscape, or make it more closely connected. The exposition has a smooth and conversational style, and includes helpful hints, mathematical connections, cautions, and opportunities for further exploration. Examples were carefully chosen to weave a tight-knit fabric, and everywhere possible, to link concepts and topics under discussion to real-world experience. A wealth of exercises support the section's main ideas, and due to their range of difficulty, there is very strong support for weaker students, while advanced students are challenged to reach even further. Each exercise set includes the following categories: *Concepts and Vocabulary; Developing Your Skills; Working with Formulas; Applications; Writing, Research, and Decision Making; Extending the Concept;* and *Maintaining Your Skills;* all carefully planned, sequenced, and thought out. The majority of reviewers seemed to think that the applications were first-rate, a staple of this text, and one of its strongest, most appealing features.

Technology Overview

Writing a text that recognizes the diversity that exists among teaching methods and philosophies was a very difficult task. While the majority of the text can in fact be taught with minimal calculator use, there is an abundance of resources for teachers that advocate its total integration into the curriculum. Almost every section contains a detailed *Technology Highlight,* every chapter a *Calculator Exploration and Discovery* feature, and calculator use is demonstrated at appropriate times and in appropriate ways throughout. For the far greater part, an instructor can use graphing and calculating technology where and how they see fit and feel supported by the text. Additionally, there are a number of on-line features and supplements that encourage further mathematical exploration, additional support for the use of graphing and programming technology, with substantive and meaningful student collaborations using the *Mathematics in Action* features available at www.mhhe.com/coburn.

Summary and Conclusion

You have in your hands a powerful tool with numerous features. All of your favorite and familiar features are there, to be used in support of your own unique style, background, and goals. The additional features are closely linked and easily accessible, enabling you to try new ideas and extend others. It is our hope that this textbook and its optional supplements provide all the tools you need to teach the course you've always wanted to teach. Writing these texts was one of the most daunting and challenging experiences of my life, particularly with an 8-year-old daughter often sitting in my lap as I typed, and the twins making off with my calculators so they could draw pretty graphs. But as you might imagine, in undertaking an endeavor of this scope and magnitude, I was blessed to experience the thrill of discovery and rediscovery a thousand times. I'd like to conclude by soliciting your help. As hard as we've worked on this project, and as proud as our McGraw-Hill team is of the result, we know there is room for improvement. Our reviewers have proven many times over there is a wealth of untapped ideas, new perspectives, and alternative approaches that can help bring a new and higher level of clarity to the teaching and learning of mathematics. Please let us know how we can make a good thing better.

ACKNOWLEDGMENTS

I first want to express a deep appreciation for the guidance, comments, and suggestions offered by all reviewers of the manuscript. I found their collegial exchange of ideas and experience very refreshing, instructive, and sometimes chastening, but always helping to create a better learning tool for our students.

Rosalie Abraham
Florida Community College at Jacksonville

Jay Abramson
Arizona State University

Omar Adawi
Parkland College

Carolyn Autrey
University of West Georgia

Jannette Avery
Monroe Community College

Adele Berger
Miami Dade College

Jean Bevis
Georgia State University

Patricia Bezona
Valdosta State University

Patrick Bibby
Miami Dade College

Elaine Bouldin Tenpenny
Middle Tennessee State University

Anna Butler
East Carolina University

Cecil Coone
Southwest Tennessee Community College

Charles Cooper
University of Central Oklahoma

Sally Copeland
Johnson County Community College

Nancy Covey Jenkins
Strayer University

Julane Crabtree
Johnson County Community College

Steve Cunningham
San Antonio College

Tina Deemer
University of Arizona

Jennifer Dollar
Grand Rapids Community College

Patricia Ellington
University of Texas at Arlington

Angela Everett
Chattanooga State Technical Community College

Gerry Fitch
Louisiana State University

James Gilbert
Mississippi Gulf Coast Community College

Ilene Grant
Georgia Perimeter College

Jim Hardman
Sinclair Community College

Brenda Helms
Mississippi Gulf Coast Community College

Laura Hillerbrand
Broward Community College

Linda Hurst
Central Texas College

John Kalliongis
Saint Louis University

Fritz Keinert
Iowa State University

Thomas Keller
Southwest Texas State University

Marlene Kovaly
Florida Community College at Jacksonville

Betty Larson
South Dakota State University

Denise LeGrand
University of Arkansas at Little Rock

Lisa Mantini
Oklahoma State University

Nancy Matthews
University of Oklahoma

Thomas McMillan
University of Arkansas at Little Rock

Owen Mertens
Southwest Missouri State University

James Miller
West Virginia University

Christina Morian
Lincoln University

Jeffrey O'Connell
Ohlone College

Debra Otto
University of Toledo

Luke Papademas
DeVry University–Chicago

Frank Pecchioni
Jefferson Community College

Greg Perkins
Hartnell College

Shahla Peterman
University of Missouri

Jeanne Pirie
Erie Community College

David Platt
Front Range Community College

Evelyn Pupplo-Cody
Marshall University

Lori Pyle
University of Central Florida

Linda Reist
Macomb Community College

Ira Lee Riddle
Pennsylvania State University–Abington

Kathy Rodgers
University of Southern Indiana

Behnaz Rouhani
Georgia Perimeter College

David Schultz
Mesa Community College

John Seims
Mesa Community College–Red Mountain Campus

Delphy Shaulis
University of Colorado

Jean Shutters
Harrisburg Area Community College

Albert Simmons
Ozarks Technical Community College

Mohan Tikoo
Southeast Missouri State University

Diane Trimble
Tulsa Community College–West Campus

Anthony Vance
Austin Community College

Arun Verma
Hampton University

Erin Wall
College of the Redwoods

Anna Wlodarczyk
Florida International University

Kevin Yokoyama
College of the Redwoods

I would also like to thank those who participated in the various college algebra symposia and offered valuable advice.

Robert Anderson
University of Wisconsin–Eau Claire

Rajilakshmi Baradwaj
University of Maryland–Baltimore County

Judy Barclay
Cuesta College

Beverly Broomell
Suffolk County Community College

Donna Densmore
Bossier Parish Community College

Patricia Foard
South Plains College

Nancy Forrester
Northeast State Community College

Steve Grosteffon
Santa Fe Community College

Ali Hajjafar
University of Akron

Ellen Hill
Minnesota State University–Moorhead

Tim Howard
Columbus State University

Miles Hubbard
St. Cloud State University

Tor Kwembe
Jackson State University

Danny Lau
Gainesville College

Kathryn Lavelle
Westchester Community College

Ram Mohapatra
University of Central Florida

Nancy Matthews
University of Oklahoma

Scott Mortensen
Dixie State College

Geoffrey Schulz
Community College of Philadelphia

John Smith
Hawaii Pacific University

Dave Sobecki
Miami University

Anthony Vance
Austin Community College

Additional gratitude goes to Jill Wardynski, Kurt Norlin, Hal Whipple, Teri Lovelace, Tom Smith, Carrie Green, and Sue Schroeder for their superlative work, careful accuracy checking, and helpful suggestions. Thank you to Rosemary Karr and Lesley Seale for authoring the solutions manuals. Rosemary is owed a special debt of gratitude for her tireless attention to detail and her willingness to go above and beyond the call of duty. I would especially like to thank John Leland and Emily Tietz for their efforts in securing just the right photos; Vicki Krug (whose motto is undoubtedly *From Panta Rhei to Fait Accompli*) for her uncanny ability to bring innumerable parts from all directions into a unified whole; Patricia Steele, a copy editor *par excellance* who can tell an en dash from a minus sign at 50 paces; Dawn Bercier for her enthusiasm in marketing the Coburn series; Suzanne Alley for her helpful suggestions, infinite patience, and steady hand in bringing the manuscript to completion; and Steve Stembridge, whose personal warmth, unflappable manner, and down-to-earth approach to problem solving kept us all on time and on target. In truth, my hat is off to all the fine people at McGraw-Hill for their continuing support and belief in this series. A final word of thanks must go to Rick Armstrong, whose depth of knowledge, experience, and mathematical connections seems endless; Anne Marie Mosher for her contributions to various features of the text and to J. D. Herdlick, Richard Pescarino, and the rest of my colleagues at St. Louis Community College whose friendship, encouragement, and love of mathematics makes going to work each day a joy.

A COMMITMENT TO ACCURACY

You have a right to expect an accurate textbook, and McGraw-Hill invests considerable time and effort to make sure that we deliver one. Listed below are the many steps we take to make sure this happens.

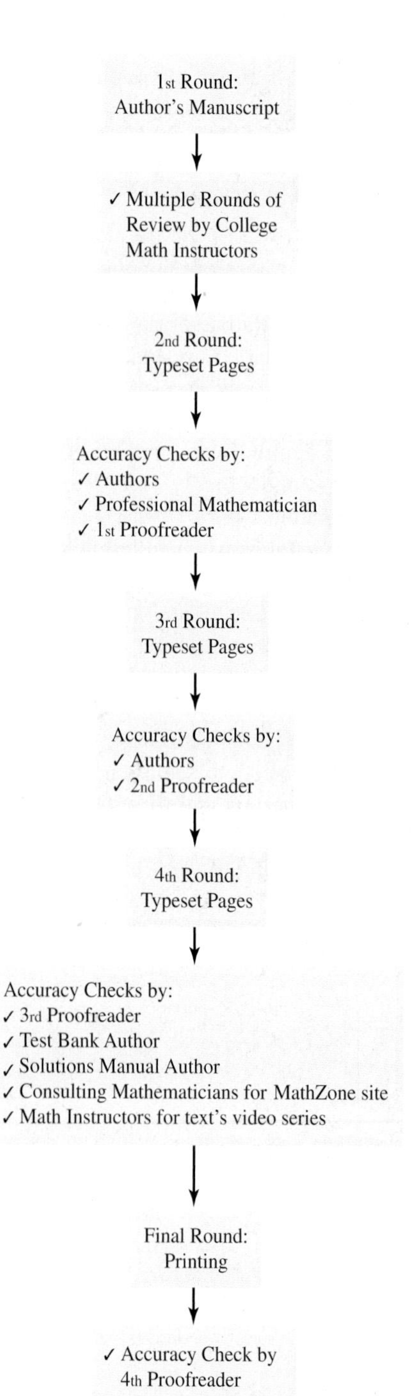

OUR ACCURACY VERIFICATION PROCESS

First Round

Step 1: Numerous **college math instructors** review the manuscript and report on any errors that they may find, and the authors make these corrections in their final manuscript.

Second Round

Step 2: Once the manuscript has been typeset, the **authors** check their manuscript against the first page proofs to ensure that all illustrations, graphs, examples, exercises, solutions, and answers have been correctly laid out on the pages, and that all notation is correctly used.

Step 3: An outside, **professional mathematician** works through every example and exercise in the page proofs to verify the accuracy of the answers.

Step 4: A **proofreader** adds a triple layer of accuracy assurance in the first pages by hunting for errors, then a second, corrected round of page proofs is produced.

Third Round

Step 5: The **author team** reviews the second round of page proofs for two reasons: 1) to make certain that any previous corrections were properly made, and 2) to look for any errors they might have missed on the first round.

Step 6: A **second proofreader** is added to the project to examine the new round of page proofs to double check the author team's work and to lend a fresh, critical eye to the book before the third round of paging.

Fourth Round

Step 7: A **third proofreader** inspects the third round of page proofs to verify that all previous corrections have been properly made and that there are no new or remaining errors.

Step 8: Meanwhile, in partnership with **independent mathematicians,** the text accuracy is verified from a variety of fresh perspectives:

- The **test bank author** checks for consistency and accuracy as they prepare the computerized test item file.
- The **solutions manual author** works every single exercise and verifies their answers, reporting any errors to the publisher.
- A **consulting group of mathematicians,** who write material for the text's MathZone site, notifies the publisher of any errors they encounter in the page proofs.
- A video production company employing **expert math instructors** for the text's videos will alert the publisher of any errors they might find in the page proofs.

Final Round

Step 9: The **project manager,** who has overseen the book from the beginning, performs a **fourth proofread** of the textbook during the printing process, providing a final accuracy review.

⇒ What results is a mathematics textbook that is as accurate and error-free as is humanly possible, and our authors and publishing staff are confident that our many layers of quality assurance have produced textbooks that are the leaders of the industry for their integrity and correctness.

Guided Tour

Laying a Firm Foundation . . .

OUTSTANDING EXAMPLES

EXAMPLE 8 The amount of fuel used by a ship traveling at a uniform speed varies jointly with the distance it travels and the square of the velocity. If 200 barrels of fuel are used to travel 10 mi at 20 nautical miles per hour, how far does the ship travel on 500 barrels of fuel at 30 nautical miles per hour?

Solution:

$F = kdv^2$ — "fuel use *varies jointly* with distance and velocity squared"

$200 = k(10)(20)^2$ — substitute known values

$200 = 4000k$ — simplify and solve for k

$0.05 = k$ — constant of variation

To find the distance traveled when 500 barrels of fuel are used while traveling 30 nautical miles per hour, use $k = 0.05$ in the original formula model and substitute the given values:

$F = kdv^2$ — formula model

$F = 0.05dv^2$ — equation of variation

$500 = 0.05d(30)^2$ — substitute 500 for F and 30 for v

$500 = 45d$ — simplify

$11.\overline{1} = d$ — result

If 500 barrels of fuel are consumed while traveling 30 nautical miles per hour, the ship covers a distance of just over 11 mi.

NOW TRY EXERCISES 41 THROUGH 44

Abundant examples carefully prepare the students for homework and exams. Easily located on the page, Coburn's numerous worked examples expose the learner to more exercise types than most other texts.

Now Try boxes immediately follow most examples to guide the student to specific matched and structured exercises they can try for practice and further understanding.

EXAMPLE 9 Hikers climbing Mt. Everest take a reading of 6.4 cmHg at a temperature of 5°C. How far up the mountain are they?

Solution: For this exercise, $P_0 = 76$, $P = 6.4$, and $T = 5$. The formula yields

$h(T) = (30T + 8000)\ln\frac{P_0}{P}$ — given function

$h(5) = [30(5) + 8000]\ln\frac{76}{6.4}$ — substitute given values

$= 8150 \ln 11.875$ — simplify

$\approx 20{,}167$ — result

The hikers are approximately 20,167 ft above sea level.

NOW TRY EXERCISES 93 AND 94

Annotations located to the right of the solution sequence help the student recognize which property or procedure is being applied.

GRAPHICAL SUPPORT

Graphing the lines from Example 8 as Y1 and Y2 on a graphing calculator, we note the lines do appear to be parallel (they actually *must* be since they have identical slopes). Using the ZOOM **8:ZInteger** feature of the TI-84 Plus (Section 2.1 *Technology Highlight*) we can quickly verify that Y2 indeed contains the point $(-6, -1)$.

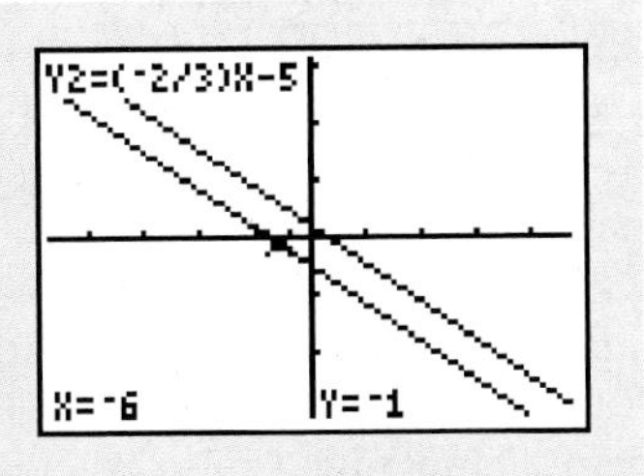

Graphical Support Boxes, located after selected examples, visually reinforce algebraic concepts with a corresponding graphing calculator example.

Building a Solid Framework . . .

SECTION EXERCISES

Concepts and Vocabulary exercises help students recall and retain important mathematical terms, building the solid vocabulary they need to verbalize and understand algebraic concepts.

CONCEPTS AND VOCABULARY

Fill in each blank with the appropriate word or phrase. Carefully reread the section if needed.

1. Points on the grid that have integer coordinates are called ________ points.
2. The graph of a line divides the coordinate grid into two distinct regions, called ________ ________.
3. To find the x-intercept of a line, substitute ________. To find the y-intercept, substitute ________.
4. The notation $\frac{\Delta y}{\Delta x}$ is read ________ y over ________ x and is used to denote a(n) ________ of ________ between the x- and y-variables.
5. What is the slope of the line in Example 9? Discuss/explain the meaning of $m = \frac{\Delta y}{\Delta x}$ in the context of this example.
6. Discuss/explain the relationship between the slope formula, the Pythagorean theorem, and the distance formula. Include several illustrations.

DEVELOPING YOUR SKILLS

Create a table of values for each equation and sketch the graph.

7. $2x + 3y = 6$
8. $-3x + 5y = 10$
9. $y = \frac{3}{2}x + 4$
10. $y = \frac{5}{3}x - 3$

x	y

Developing Your Skills exercises help students reinforce what they have learned by offering plenty of practice with increasing levels of difficulty.

Working with Formulas exercises demonstrate how equations and functions model the real world by providing contextual applications of well-known formulas.

Graphing Calculator icons appear next to examples and exercises where important concepts can be supported by use of graphing technology.

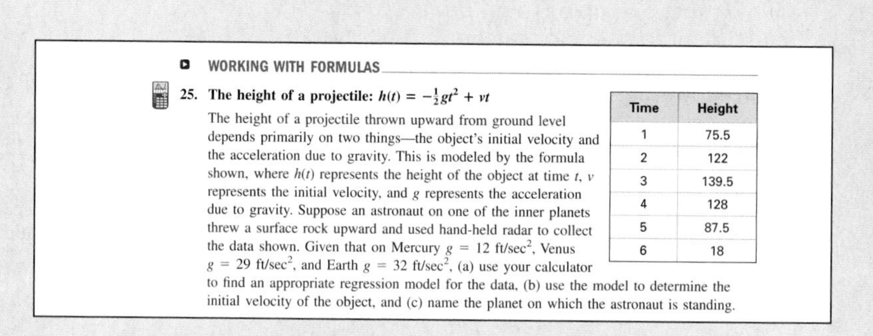

WORKING WITH FORMULAS

25. **The height of a projectile: $h(t) = -\frac{1}{2}gt^2 + vt$**

The height of a projectile thrown upward from ground level depends primarily on two things—the object's initial velocity and the acceleration due to gravity. This is modeled by the formula shown, where $h(t)$ represents the height of the object at time t, v represents the initial velocity, and g represents the acceleration due to gravity. Suppose an astronaut on one of the inner planets threw a surface rock upward and used hand-held radar to collect the data shown. Given that on Mercury $g = 12$ ft/sec², Venus $g = 29$ ft/sec², and Earth $g = 32$ ft/sec², (a) use your calculator to find an appropriate regression model for the data, (b) use the model to determine the initial velocity of the object, and (c) name the planet on which the astronaut is standing.

Time	Height
1	75.5
2	122
3	139.5
4	128
5	87.5
6	18

WRITING, RESEARCH, AND DECISION MAKING

87. Scientists often measure extreme temperatures in degrees Kelvin rather than the more common Fahrenheit or Celsius. Use the Internet, an encyclopedia, or another resource to investigate the linear relationship between these temperature scales. In your research, try to discover the significance of the numbers −273, 0, 32, 100, 212, and 373.

88. In many states, there is a set fine for speeding with an additional amount charged for every mile per hour over the speed limit. For instance, if the set fine is $40 and the additional charge is $12, the *fine for speeding formula* would be $F = 12(S - 65) + 40$, where F is the set fine and S is your speed (assuming a speed limit of 65 mph). (a) What is the slope of this line? (b) Discuss the meaning of the slope in this context and (c) contact your nearest Highway Patrol office and ask about the speeding fines in your area.

Writing, Research, and Decision Making exercises encourage students to communicate their understanding of the topics at hand or explore topics of interest in greater depth.

Wait, There's More!

- **Technology Highlights,** located before most section exercise sets, assist those interested in exploring a section topic with a graphing calculator.
- **Extending the Concept** exercises are designed to be more challenging, requiring synthesis of related concepts or the use of higher-order thinking skills.
- **Maintaining Your Skills** exercises review topics from previous chapters helping students to retain concepts and keep skills sharp.

Mid-Chapter Checks assess student progress before they continue to the second half of the chapter.

Reinforcing Basic Concepts immediately follow the Mid-Chapter Check. This feature extends and explores a chapter topic in greater detail.

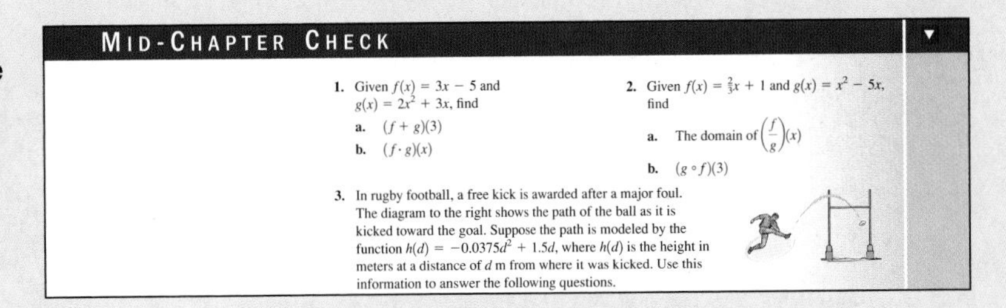

MID-CHAPTER CHECK

1. Given $f(x) = 3x - 5$ and $g(x) = 2x^2 + 3x$, find
   a. $(f + g)(3)$
   b. $(f \cdot g)(x)$
2. Given $f(x) = \frac{2}{3}x + 1$ and $g(x) = x^2 - 5x$, find
   a. The domain of $\left(\frac{f}{g}\right)(x)$
   b. $(g \circ f)(3)$
3. In rugby football, a free kick is awarded after a major foul. The diagram to the right shows the path of the ball as it is kicked toward the goal. Suppose the path is modeled by the function $h(d) = -0.0375d^2 + 1.5d$, where $h(d)$ is the height in meters at a distance of d m from where it was kicked. Use this information to answer the following questions.

END-OF-CHAPTER MATERIAL

The **Summary and Concept Review,** located at the end of Chapters 1–8, lists key concepts and is organized by section. This format provides additional practice exercises and makes it easy for students to review the terms and concepts they will need prior to a quiz or exam.

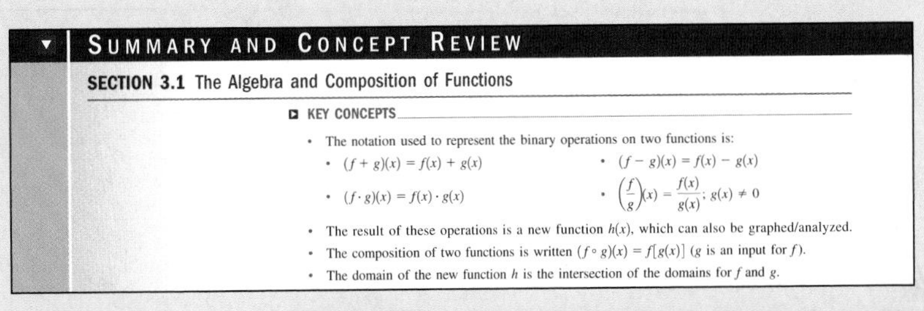

SUMMARY AND CONCEPT REVIEW

SECTION 3.1 The Algebra and Composition of Functions

KEY CONCEPTS

- The notation used to represent the binary operations on two functions is:
  - $(f + g)(x) = f(x) + g(x)$
  - $(f - g)(x) = f(x) - g(x)$
  - $(f \cdot g)(x) = f(x) \cdot g(x)$
  - $\left(\frac{f}{g}\right)(x) = \frac{f(x)}{g(x)};\ g(x) \neq 0$
- The result of these operations is a new function $h(x)$, which can also be graphed/analyzed.
- The composition of two functions is written $(f \circ g)(x) = f[g(x)]$ (g is an input for f).
- The domain of the new function h is the intersection of the domains for f and g.

Mixed Review exercises offer more practice on topics from the entire chapter, are arranged in random order, and require students to identify problem types and solution strategies on their own.

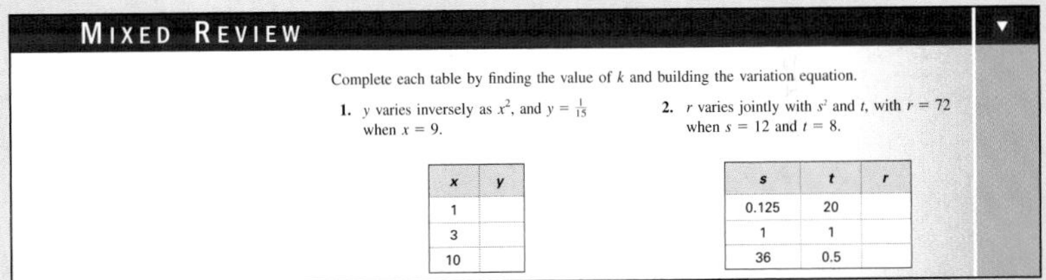

MIXED REVIEW

Complete each table by finding the value of k and building the variation equation.

1. y varies inversely as x^2, and $y = \frac{1}{15}$ when $x = 9$.

x	y
1	
3	
10	

2. r varies jointly with s^2 and t, with $r = 72$ when $s = 12$ and $t = 8$.

s	t	r
0.125	20	
1	1	
36	0.5	

The **Practice Test** gives students the opportunity to check their mastery and prepare for classroom quizzes, tests, and other assessments.

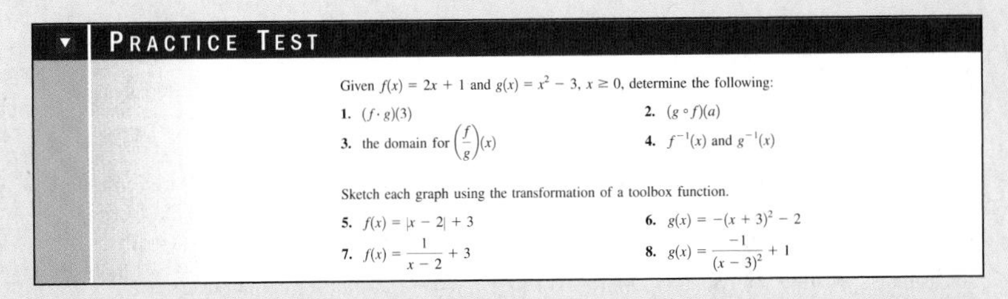

PRACTICE TEST

Given $f(x) = 2x + 1$ and $g(x) = x^2 - 3$, $x \geq 0$, determine the following:

1. $(f \cdot g)(3)$
2. $(g \circ f)(a)$
3. the domain for $\left(\frac{f}{g}\right)(x)$
4. $f^{-1}(x)$ and $g^{-1}(x)$

Sketch each graph using the transformation of a toolbox function.

5. $f(x) = |x - 2| + 3$
6. $g(x) = -(x + 3)^2 - 2$
7. $f(x) = \frac{1}{x - 2} + 3$
8. $g(x) = \frac{-1}{(x - 3)^2} + 1$

Cumulative Reviews help students retain previously learned skills and concepts by revisiting important ideas from earlier chapters.

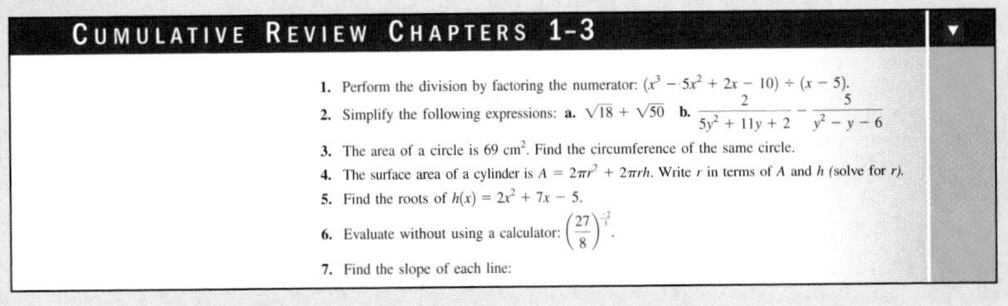

CUMULATIVE REVIEW CHAPTERS 1–3

1. Perform the division by factoring the numerator: $(x^3 - 5x^2 + 2x - 10) \div (x - 5)$.
2. Simplify the following expressions: **a.** $\sqrt{18} + \sqrt{50}$ **b.** $\frac{2}{5y^2 + 11y + 2} - \frac{5}{y^2 - y - 6}$
3. The area of a circle is 69 cm^2. Find the circumference of the same circle.
4. The surface area of a cylinder is $A = 2\pi r^2 + 2\pi rh$. Write r in terms of A and h (solve for r).
5. Find the roots of $h(x) = 2x^2 + 7x - 5$.
6. Evaluate without using a calculator: $\left(\frac{27}{8}\right)^{-\frac{2}{3}}$.
7. Find the slope of each line:

Wait, There's *Still* More!

- The **Calculator Exploration and Discovery** feature is designed to extend the borders of a student's mathematical understanding using the power of graphing and calculating technology.
- **Strengthening Core Skills** exercises help students strengthen skills that form the bedrock of mathematics and lead to continued success.

Providing Strong Connections . . .

THROUGH APPLICATIONS!

interest earnings 152, 324, 531
compound annual 258, 480, 483, 532–533, 555, 557
continuously compounded 523, 532–534
simple 533
mortgage payment, monthly 34, 532
mortgage interest 533
historical data 356
NYSE trading volume 226
savings account balance 766
student loan repayment 779

CHEMISTRY
chemical mixtures 82–83, 130, 155
pH levels 492
froth height 547

prison population 187
speeding fines 154, 271
stopping distance 286

DEMOGRAPHICS
AIDS cases 325
bicycle sales 546
cable television subscriptions 550
centenarians 538
crop allocation 602
eating out 187, 638
females in the work force 229
home-schooling 230
households holding stock 337
internet connections 186
law enforcement 333
lottery numbers 799, 822

faculty salaries 612
GPA 92, 436
grades
average 89, 131
vs. study time 238
home-schooling 230
learning curves 546
library fines 359
memory retention 434, 493
Stooge IQ 625
true/false quizzes 820
working students 838

ENGINEERING
Civil
oil tanker capacity 611
traffic and travel time 474

The **Index of Applications** is located immediately after the Guided Tour and is organized by discipline to help identify applications relevant to a particular field.

Meaningful Applications-over 650 carefully chosen applications explore a wide variety of interests and illustrate how mathematics is connected to other disciplines and the world around us.

82. Baseball card value: After purchasing an autographed baseball card for \$85, its value increases by \$1.50 per year.

a. What is the card's value 7 yr after purchase?

b. How many years will it take for this card's value to reach \$100?

83. Cost of college: For the years 1980 to 2000, the cost of tuition and fees per semester (in constant dollars) at a public 4-yr college can be approximated by the equation $y = 144x + 621$, where y represents the cost in dollars and $x = 0$ represents the year 1980. Use the equation to find: (a) the cost of tuition and fees in 1992 and (b) the year this cost will exceed \$5000.

Source: 2001 *New York Times Almanac*, p. 356

84. Female physicians: In 1960 only about 7% of physicians were female. Soon after, this percentage began to grow dramatically. For the years 1980 to 2002, the percentage of physicians that were female can be approximated by the equation $y = 0.72x + 11$, where ...tage (as a whole number) and $x = 0$ represents the year 1980. Use the ...e percentage of physicians that were female in 1992 and (b) the pro...tage will exceed 30%.

... the 2004 *Statistical Abstract of the United States*, Table 149

87. Find the value of $M(I)$ given
a. $I = 50{,}000I_0$ and **b.** $I = 75{,}000I_0$.

88. Find the intensity I of the earthquake given
a. $M(I) = 3.2$ and **b.** $M(I) = 8.1$.

Intensity of sound: The intensity of sound as perceived by the human ear is measured in units called decibels (dB). The loudest sounds that can be withstood without damage to the eardrum are in the 120- to 130-dB range, while a whisper may measure in the 15- to 20-dB range. Decibel measure is given by the equation $D(I) = 10\log\left(\frac{I}{I_0}\right)$, where I is the actual intensity of the sound and I_0 is the faintest sound perceptible by the human ear—called the reference intensity. The intensity I is often given as a multiple of this reference intensity, but often the constant 10^{-16} (watts per cm^2; W/cm^2) is used as the threshold of audibility.

89. Find the value of $D(I)$ given
a. $I = 10^{-14}$ and **b.** $I = 10^{-4}$.

90. Find the intensity I of the sound given
a. $D(I) = 83$ and **b.** $D(I) = 125$.

Looking for Interactive Applications? Look Online!

The **Mathematics in Action** activities, located at www.mhhe.com/coburn, enable students to work collaboratively as they manipulate applets that apply mathematical concepts in real-world contexts.

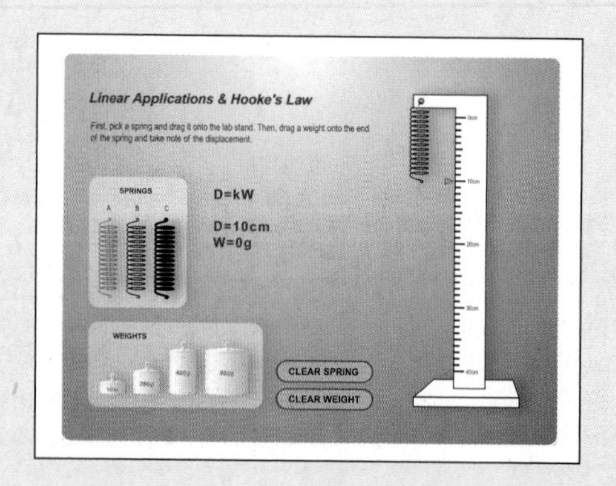

Concepts for Calculus icons identify concepts or skills that a student will likely see in a first semester calculus course.

SUPPLEMENTS FOR INSTRUCTORS

Annotated Instructor's Edition *ISBN-13: 978-0-07-313702-5 (ISBN-10: 0-07-313702-2)*

In the Annotated Instructor's Edition (AIE), **exercise answers appear adjacent to each exercise,** in a color used *only* for annotations. Answers that do not fit on the page appear in the back of the AIE as an appendix.

Instructor's Solutions Manual *ISBN-13: 978-0-07-320066-8 (ISBN-10: 0-07-320066-2)*

Authored by Rosemary Karr and Lesley Seale, the *Instructor's Solutions Manual* contains detailed, **work-out solutions** to all exercises in the text.

Instructor's Testing and Resource CD-ROM *ISBN-13: 978-0-07-320068-2 (ISBN-10: 0-07-320068-9)*

This cross-platform CD-ROM provides a wealth of resources for the instructor. Among the supplements featured on the CD-ROM is a **computerized test bank** utilizing Brownstone Diploma® algorithm-based testing software to quickly create customized exams. This user-friendly program enables instructors to search for questions by topic, format, or difficulty level; to edit existing questions or to add new ones; and to scramble questions and answer keys for multiple versions of a single test. Hundreds of text-specific open-ended and multiple-choice questions are included in the question bank. Sample chapter tests, midterms, and final exams in Microsoft Word® and PDF formats are also provided.

Video Lectures on Digital Video Disk (DVD) *ISBN-13: 978-0-07-320067-5 (ISBN-10: 0-07-320067-0)*

In the videos, qualified teachers work through selected problems from the textbook, following the solution methodology employed in the text. The video series is available on DVD or VHS videocassette, or online as an assignable element of MathZone (see section on MathZone). The DVDs are closed-captioned for the hearing impaired, subtitled in Spanish, and meet the Americans with Disabilities Act Standards for Accessible Design. Instructors can use them as resources in a learning center, for online courses, and/or to provide extra help to students who require extra practice.

MathZone—www.mathzone.com

McGraw-Hill's **MathZone 3.0** is a complete **Web-based tutorial and course management system** for mathematics and statistics, designed for greater ease of use than any other system available. Free upon adoption of a McGraw-Hill textbook, the system enables instructors to **create and share courses and assignments** with colleagues, adjunct faculty members, and teaching assistants with only a few mouse clicks. All **assignments, exercises, e-Professor multimedia tutorials, video lectures, and NetTutor® live tutors** follow the textbook's learning objectives and problem-solving style and notation. Using MathZone's **assignment builder,** instructors can **edit questions and**

algorithms, import their own content, and **create announcements and due dates** for homework and quizzes. MathZone's **automated grading function** reports the results of easy-to-assign algorithmically generated homework, quizzes, and tests. All student activity within MathZone is recorded and available through a **fully integrated gradebook** that can be downloaded to Microsoft Excel®. MathZone also is available on CD-ROM. (See "Supplements for the Student" for descriptions of the elements of MathZone.)

ALEKS

ALEKS (Assessment and **LE**arning in **K**nowledge **S**paces) is an artificial-intelligence-based system for mathematics learning, available over the Web 24/7. Using unique adaptive questioning, ALEKS accurately assesses what topics each student knows and then determines exactly what each student is ready to learn next. ALEKS interacts with the students much as a skilled human tutor would, moving between explanation and practice as needed, correcting and analyzing errors, defining terms and changing topics on request, and helping them master the course content more quickly and easily. Moreover, the new ALEKS 3.0 now links to text-specific videos, multimedia tutorials, and textbook pages in PDF format. ALEKS also offers a robust classroom management system that enables instructors to monitor and direct student progress toward mastery of curricular goals. See www.highed.aleks.com

SUPPLEMENTS FOR STUDENTS

Student's Solutions Manual *ISBN-13: 978-0-07-291761-1 (ISBN-10: 0-07-291761-X)*

Authored by Rosemary Karr and Lesley Seale, the *Student's Solutions Manual* contains detailed, **worked-out solutions** to all the problems in the Mid-Chapter Checks, Reinforcing Basic Concepts, Summary and Concept Review Exercises, Practice Tests, Cumulative Reviews, and Strengthening Core Skills. Also included are **worked-out solutions** for odd-numbered exercises of the section exercises and the mixed reviews. The steps shown in solutions are carefully matched to the style of solved examples in the textbook.

MathZone—www.mathzone.com

McGraw-Hill's MathZone is a powerful Web-based tutorial for homework, quizzing, testing, and multimedia instruction. Also available in CD-ROM format, MathZone offers:

- **Practice exercises** based on the text and generated in an unlimited quantity for as much practice as needed to master any objective.
- **Video** clips of classroom instructors showing how to solve exercises from the text, step-by-step **e-Professor** animations that take the student through step-by-step instructions, delivered on-screen and narrated by a teacher on audio, for solving exercises from the textbook; the user controls the pace of the explanations and can review as needed.
- **NetTutor,** which offers personalized instruction by live tutors familiar with the textbook's objectives and problem-solving methods.

Every assignment, exercise, video lecture, and e-Professor is derived from the textbook.

Video Lectures on Digital Video Disk (DVD) *ISBN-13: 978-0-07-320067-5 (ISBN-10: 0-07-320067-0)*

The video series is based on exercises from the textbook. Each presenter works through selected problems, following the solution methodology employed in the text. The video series is available on DVD or online as part of MathZone. The DVDs are closed-captioned for the hearing impaired, subtitled in Spanish, and meet the Americans with Disabilities Act Standards for Accessible Design.

NetTutor

Available through MathZone, NetTutor is a revolutionary system that enables students to interact with a live tutor over the Web. NetTutor's Web-based, graphical chat capabilities enable students and tutors to use mathematical notation and even to draw graphs as they work through a problem together. Students can also submit questions and receive answers, browse previously answered questions, and view previous sessions. Tutors are familiar with the textbook's objectives and problem-solving styles.

ALEKS

(**A**ssessment and **LE**arning in **K**nowledge **S**paces) is an artificial intelligence-based system for mathematics learning, available online 24/7. ALEKS interacts with the student much as a skilled human tutor would, moving between explanation and practice as needed, helping you master the course content more quickly and easily. NEW! ALEKS 3.0 now links to text-specific videos, multimedia tutorials, and textbook pages in PDF format. See www.highed.aleks.com

Index of Applications

ANATOMY AND PHYSIOLOGY

body proportions 231
height vs. weight 152
height vs. wing span 228
male height vs. shoe size 229

ARCHITECTURE

decorative fireplaces 727
Eiffel Tower 576
elliptical arches 693
pitch of a roof 21
suspension bridges 81
tall buildings 666

ART, FINE ARTS, THEATER

art show lighting 727
arts and crafts 838
candle-making 613
Comedy of Errors 576
concentric rectangles 576
cornucopia composition 839
famous painters 372
graphing and art 437
mathematics and art 557
metal alloys 572
museum collection 621
rare books 625
theater attendance 240
ticket sales 129

BIOLOGY/ZOOLOGY

animal
- birth weight 757, 837
- diets 653
- genus 419
- gestation periods 588
- girth-to-length ratio 831
- length-to-weight models 66, 152
- lifespan 840

bacteria growth 483, 534
fruit fly population 526
predator/prey models 325
species preservation 758
temperature and cricket chirps 154
water-diving birds 215
wildlife population growth 104
yeast culture 543

BUSINESS/ECONOMICS

account balance/service fees 93
advertising and sales 336, 420, 504, 519
annuities 534
balance of payments 418, 470
business loans 584
cell phone charges 103
convenience store sales 652
cost
- car rental 172
- gasoline 614
- manufacturing 435, 443, 447–448, 575, 638
- minimizing 597, 602–603
- packaging material 448
- recycling 311
- repair 21
- running shoes 103
- service call 172

cost/revenue/profit 128, 259
credit card transactions 230
currency conversion 259–260
customer service 819, 839
depreciation 81, 153, 180, 186, 481, 484, 505, 519, 549, 757, 779–780, 839
employee productivity 664
equipment aging 780
fuel consumption 187
gross domestic product 226
households holding stock 337
inflation 484, 757, 779
market demand/consumer interest 311, 324
maximizing profit/revenue 34, 69, 240, 296, 298
maximizing resources 653
mileage rate 165, 172
mixture exercises 82
natural gas prices 338
new product development 556
patent applications 228
personnel decisions 802
phone service charges 339
plant production 637
postage cost history 334
pricing strategies 297
printing and publishing 449
profit/loss 188, 355
real estate sales 226
revenue
- equation models 99, 484, 639
- seasonal 135, 215, 664

salary
- calculations 149, 218, 321
- review 222

sales goals 767
stock purchase 753
supply and demand 575, 626, 714–715
union membership 840
USPS express mail rates 225
USPS package size regulations 81
wage
- hourly 757
- minimum 21
- overtime 339

work per unit time 144–145, 152

CHEMISTRY

chemical mixtures 82–83, 588
pH levels 492
froth height 547

COMMUNICATION

cell phone subscriptions 556
email addresses 806
internet connections 186
parabolic dish 735
phone call volume 325, 547

phone numbers 805
phone service charges 339
radio broadcast range 692, 705
television programming 806

COMPUTERS

animations 767, 780
email addresses, 806
magnetic memory devices 82
ownership 820

CONSTRUCTION

home cost per square foot 151
home improvement 638
lift capacity 92
pitch of a roof 21
suspension bridges 81

CRIMINAL JUSTICE, LEGAL STUDIES

accident investigation 67
law enforcement 227, 333
prison population 187
speeding fines 154, 271
stopping distance 286

DEMOGRAPHICS

AIDS cases 325
bicycle sales 546
cable television subscriptions 550
centenarians 538
crop allocation 602
eating out 187
females in the work force 229
home-schooling 230
households holding stock 337
Internet connections 186
law enforcement 333
lottery numbers 799, 822
lumber imports 245
MDs 240
military
  conflicts, popular support 821
  expenditures 339
  veterans 819
  volunteer service 634
milk production 547
multiple births 338
new books published 223
newspapers published 336
opinion polls 830
Pacific coastal population 273
per capita debt 217
population
  density 433
  doubling time 504, 523
  growth 780
  tripling time 505
post offices 549
raffle tickets 838
smoking 154, 225, 626
tourist population 392
women in politics 225

EDUCATION/TEACHING

campus club membership 638
college costs 154, 779
course scheduling 802
credit hours taught 834
faculty salaries 612
GPA 92, 436
grades
  average 89, 131
  vs. study time 238
home-schooling 230
learning curves 546
library fines 359
memory retention 434, 493
Stooge IQ 625
true/false quizzes 820
working students 838

ENGINEERING

Civil
oil tanker capacity 611
traffic and travel time 474
Electrical
AC circuits 116
impedance calculations 116
resistance 21, 310, 324, 462, 603, 729
resistors in parallel 49
voltage calculations 116
Mechanical
kinetic energy 323
parabolic reflectors 738
pitch diameter 12
solar furnace 738
wind-powered energy 67, 104, 203, 237, 272, 287

ENVIRONMENTAL STUDIES

balance of nature 310
clean up time 324
energy rationing 338
forest fires 260
fuel consumption 318
hazardous waste 602, 705, 790
landfill volume 151
oil spills 254
pollution
  removal 104, 306, 310, 434, 436
recycling cost 311
resource depletion 548
stocking a lake 758
water rationing 338
wildlife population growth 104
wind-powered energy 67, 104, 203, 237, 272, 287

FINANCE

charitable giving 767
debt
  load 392, 507
  per capita 32
federal budget 32
investment
  diversifying 588, 615, 625, 667, 674
  growth 154, 186, 493, 519, 528–529, 576, 589, 839
  return 664
  strategies 523
interest earnings 324, 531
  compound annual 258, 480, 483, 532–533, 555, 557
  continuously compounded 523, 532–534
  simple 533
mortgage payment, monthly 34, 532

mortgage interest 533
historical data 355–356
NYSE trading volume 226
savings account balance 766
student loan repayment 779

GEOGRAPHY, GEOLOGY

cradle of civilization 81
distance between major cities 625
earthquake
epicenter 692
magnitude 490, 493, 557
land area
island nations 673
various states 76
longest rivers 589
natural gas prices 338
temperature of ocean water 319
tidal motion 172

HISTORY

Anthony and Cleopatra 576
child prodigies 613
famous
authors 625, 839
composers 638
Indian Chiefs 536
women 168, 474
major wars 159, 588
mythological deities 234
notable dates 575, 588
postage costs 21, 334
Statue of Liberty 321
Zeno's Paradox 840

INTERNATIONAL STUDIES

countries and languages 168
currency conversion 259–260
shoe sizing 259

MATHEMATICS

arc length
parabolic segment 737
area of
circle 31
ellipse 692
inscribed circle 692
inscribed square 692
inscribed triangle 692
Norman window 663
parabolic segment 715, 737
parallelogram 663
rectangle 601, 637
triangle 601, 663
average rate of change 198, 204
complex numbers 116
absolute value 115, 404
cubes 115
Girolamo Cardano 116
square roots 116, 404
complex polynomials 124, 128
composite solids 94
consecutive integers 82, 88, 105, 132
correlation coefficient 231
curve fitting 653
discriminant of
quadratic 125
distance from point to line 214
equipoise cylinder 272
factorials 804
factoring using the "ac" method 105
focal chords
ellipse 728
hyperbola 703
parabola 733
geometric formulas 80
hailstone sequence 758
nested factoring 43
number puzzles 103
perfect numbers 821
perimeter of
ellipse 726
rectangle 637
polygon angles 768
probability
binomial 830
spinning a spinner 483
Pythagorean Theorem 69, 174
quadratic solutions 137
quartic polynomials 417
radius of a sphere 271
second differences 768
semi-circle equation 214
similar triangles 103
Stirling's Formula 804, 806
sum of
consecutive cubes 463, 779, 789
consecutive fourth powers 789
consecutive squares 766
n integers 296
surface area of
cone 67
cylinder 43, 81, 127–128, 258, 447, 449
frustum 67
sphere 322
USPS package size regulations 81
volume of
cone 272, 602
cube 31, 172
cylinder/cylindrical shells 43, 172, 602
frustum 228
open box 391, 469
pyramid 663, 665
rectangular box 43, 296
spherical cap 381, 450
spherical shells 43, 286

MEDICINE, NURSING, NUTRITION, DIETETICS, HEALTH

AIDS cases 325
appointment scheduling 836
body mass index 92
deaths due to heart disease 366
female physicians 154, 547
hodophobia 821
human life expectancy 153
ideal weight 171
infant growth 540, 563
lithotripsy 727
low birth weight 325
medication in the bloodstream 33, 311, 471, 505
milk fat percentage 576
multiple births 338, 562, 803
number of MDs 240
pediatric dosages/Clark's Rule 12
Poiseuille's Law 44
prescription drugs 186

saline mixtures 78–79
SARS cases 550
smokers 154, 225
weight loss 548

METEOROLOGY

air mass movement 215
atmospheric pressure 227, 504, 519
barometric pressure 500
jet stream altitude 612
rainfall and farm productivity 241
reservoir water levels 418
temperature
  atmospheric 178, 271
  conversions 93, 575
  drop 12, 204, 767
  record high 12
  record low 13
wind speed record 767

MUSIC

famous
  arias 652
  composers 638
notes and frequency 549
rock-n-roll greats 312
Rolling Stones 652

PHYSICS, ASTRONOMY, PLANETARY STUDIES

acceleration 186, 326
atmospheric pressure 227
Beer-Lambert Law 519
Boyle's Law 317
charged particles 324, 704
comet path 723
creating a vacuum 780
deflection of a beam 461
depth and water pressure 372
depth of a dive 298, 612
distance between planets 82
elastic rebound 780, 823
fluid motion 286
gravity
  effects of 323, 326
  free-fall 66, 203–205, 272, 286–287, 322
gravitational attraction 34, 326
Kepler's Third Law 67, 104
light intensity 34, 310, 326
Lorentz transformations 44
metric time 21
mixture exercises 82, 155, 435
Newton's Law of Cooling 515, 518
nuclear power 563, 704
parabolic trajectory 355
pendulums 324, 775, 779, 839
planet
  orbits 693, 727, 743
    aphelion 693
  velocity 551
projected image 271, 322
projectile
  height 103–104, 106, 123, 127–128, 204, 228, 297–298
  range 315
  velocity 198
radio telescopes 739
radioactive
  Carbon-14 dating 505, 519, 534
  decay 484, 505
  half-life 507, 518–519, 527, 534, 554
climb rate, aircraft 152
sound intensity 493–494
spaceship velocity 519
speed of sound 185
spring oscillation 611
star intensity 493
supernova expansion 260
temperature scales 154, 172
uniform motion 77, 82, 104, 105, 575, 576
velocity of a particle 457
volume and pressure 21, 317
weight on other planets 314, 324

POLITICS

dependency on foreign oil 337
electoral votes 576, 589
federal deficit (historical data) 356
flat tax 577
government deficits 405
guns vs. butter 602
military expenditures 339
per capita debt 217
voting tendencies 814
women in politics 225

SOCIAL SCIENCE, HUMAN SERVICES

AIDS cases 325
females in the work force 229
home-schooling 230
law enforcement cost 333
memory retention 434
population density 433
smoking 154, 225

SPORTS AND LEISURE

archery 840
average bowling score 135, 679
basketball
  freethrow percentage 828
  height of players 134
  NBA championship 625
  salaries 549
  stars 168
batting averages 830
bingo 805
butterfly stroke 93
chess tournaments 802
circus clowns 672
Clue 792, 804
darts 820
dice games 811
dominoes 817
eight ball 818
Ellipse Park 722
fitness club membership 672
football field dimensions 576
football player weight 93
horse racing 803
marching formations 767
Olympic
  freestyle records 220
  high jump records 229
ping-pong table dimensions 382
poker probabilities 842
pool table dimensions 382
public park usage 845
playing cards
  Pinochle 816
  standard 808

rugby penalty kick 299
Scrabble 798
seating capacity 763, 767
spelunking 81
stunt pilots 703
team rosters 806, 816
tennis court dimensions 128
tic-tac-toe 806
tourist population 392
training
  diet 653
  regimen 576, 790
travel within US 363
Twister 804
Yahtzee 804

TRANSPORTATION

aircraft N-Numbers 805
flying clubs 704
fuel consumption 187
gasoline cost 614
hydrofoil service 810
radar detection 692, 727–728
  LORAN 704
routing probabilities 817
tire sales 637
tunnel clearance 714

WOMEN'S ISSUES

female physicians 154, 547
females in the work force 229
low birth weight 325
multiple births 338, 562, 803
women in politics 225

Chapter R

A Review of Basic Concepts and Skills

Chapter Outline

R.1 The Language, Notation, and Numbers of Mathematics 2

R.2 Algebraic Expressions and the Properties of Real Numbers 13

R.3 Exponents, Polynomials, and Operations on Polynomials 22

R.4 Factoring Polynomials 35

R.5 Rational Expressions 44

R.6 Radicals and Rational Exponents 54

Preview

This chapter offers a focused review of basic skills that lead to success in college algebra. In fact, college algebra is designed to refine and extend these ideas, enabling us to apply them in new and powerful ways. But regardless of their mathematical sophistication, the power of each new idea can be traced back to the fundamentals reviewed here.[1] In fact, your success in college algebra will likely be measured in direct proportion to how thoroughly you have mastered these skills. As noted mathematician Henri Lebesque (1875–1941) once said, "An idea reaches its maximum level of usefulness only when you understand it so well that it seems like you have always known it. You then become incapable of seeing the idea as anything but a trivial and immediate result."

[1]Note that Section R.7 *Geometry Review* and Section R.8 *Expressions, Tables, and Graphing Calculators* are available online at www.mhhe.com/coburn.

R.1 The Language, Notation, and Numbers of Mathematics

LEARNING OBJECTIVES

In Section R.1 you will review:

A. **Sets of numbers, graphing real numbers, and set notation**

B. **Inequality symbols and order relations**

C. **The absolute value of a real number**

D. **Operations on real numbers and the order of operations**

INTRODUCTION

The most fundamental requirement for learning algebra is mastering the words, symbols, and numbers needed to express mathematical ideas. "Words are the symbols of knowledge, the keys to accurate learning" (author Norman Lewis in *Word Power Made Easy,* Penguin Books).

POINT OF INTEREST

Complete acceptance of the number systems we know today required a long, evolutionary process. For centuries, negative numbers were suspect because they could not be used to describe physical objects. The early Greeks believed the entire universe could be described using only rational numbers—discounting the existence of irrational numbers. Further, it was not until the eighteenth century that the existence of imaginary numbers became widely accepted.

A. Sets of Numbers, Graphing Real Numbers, and Set Notation

To effectively use mathematics as a problem-solving tool, we must first be familiar with the **sets of numbers** used to quantify (give a numeric value to) the things we investigate. Only then can we make comparisons and develop the equation models that lead to informed decisions.

Natural Numbers

The most basic numbers are those used to count physical objects: 1, 2, 3, 4, and so on. These are called **natural numbers** and are represented by the (castellar) capital letter $\mathbb{N}$. We use **set notation** to list or describe a set of numbers. Braces { } are used to group **members** or **elements** of the set, commas separate each member, and three dots ". . ." are used to indicate a pattern that continues indefinitely. The notation $\mathbb{N} = \{1, 2, 3, 4, 5, \ldots\}$ is read, "$\mathbb{N}$ is the set of numbers 1, 2, 3, 4, 5, and so on." To show membership in a set, the symbol $\in$ is used. It is read "is an element of" or "belongs to." The statements $6 \in \mathbb{N}$ and $0 \notin \mathbb{N}$ (0 is not an element of $\mathbb{N}$) are true statements. A set having no elements is called the **empty** or **null set,** and is designated by empty braces { } or the symbol $\varnothing$.

EXAMPLE 1 List the set of natural numbers that are (a) negative, (b) greater than 100, and (c) greater than or equal to 5 and less than or equal to 12.

Solution:

a. { }; all natural numbers are positive.

b. {101, 102, 103, 104, . . .}

c. {5, 6, 7, 8, 9, 10, 11, 12}

NOW TRY EXERCISES 7 AND 8

Whole Numbers

When zero is combined with the natural numbers, a new set is created called the **whole numbers** $\mathbb{W} = \{0, 1, 2, 3, 4, \ldots\}$. We say that the natural numbers are a **subset** of the whole numbers, denoted $\mathbb{N} \subset \mathbb{W}$, since they are contained entirely in this set (every natural number is also a whole number). The symbol $\subset$ means "is a subset of."

EXAMPLE 2 Given set $A = \{1, 2, 3, 4, 5, 6\}$, set $B = \{2, 4\}$, and set $C = \{0, 1, 2, 3, 5, 8\}$, determine whether the following statements are true or false.

a. $B \subset A$ **b.** $B \subset C$ **c.** $C \subset \mathbb{W}$

d. $C \subset \mathbb{N}$ **e.** $104 \in \mathbb{W}$ **f.** $0 \in \mathbb{N}$

g. $2 \notin \mathbb{W}$

Solution:

a. True: Every element of B is in A. **b.** False: $4 \notin C$.

c. True: All elements of C are whole. **d.** False: $0 \notin \mathbb{N}$.

e. True: 104 is a whole number. **f.** False: $0 \notin \mathbb{N}$.

g. False: 2 *is* a whole number

NOW TRY EXERCISES 9 THROUGH 14

Integers

Numbers greater than zero are **positive numbers.** Every positive number has an *opposite* that is a **negative number** (a number less than zero). The set of zero and the natural numbers with their opposites gives the set of **integers** $\mathbb{Z} = \{\ldots, -3, -2, -1, 0, 1, 2, 3, \ldots\}$. We can illustrate the size or **magnitude** of a number (in relation to other numbers) using a **number line** (see Figure R.1).

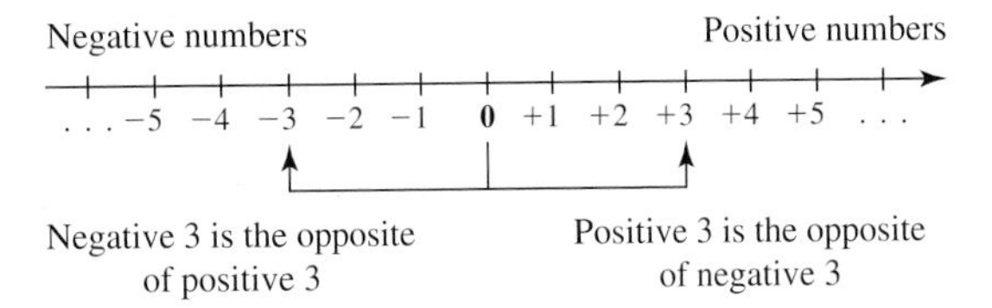

Figure R.1

Any number that corresponds to a point on the number line is called the **coordinate** of that point. When we want to note a specific location on the line, a bold dot "•" is used and a capital letter is assigned to the location. We have then **graphed** the number. Since we need only one coordinate to denote a location on the number line, we call it a **one-dimensional graph.**

Rational Numbers

Fractions and mixed numbers are part of a set called the **rational numbers** $\mathbb{Q}$. A rational number is one that can be written as a fraction with an integer numerator and an integer denominator other than zero. In set notation we write $\mathbb{Q} = \{\frac{a}{b} \mid a, b \in \mathbb{Z}; b \neq 0\}$. The vertical bar "|" is read "such that" and indicates that a description follows. In words, we say, "$\mathbb{Q}$ is the set of numbers of the form a over b, such that a and b are integers and b is not equal to zero."

WORTHY OF NOTE

The integers are a subset of the rational numbers: $\mathbb{Z} \subset \mathbb{Q}$, since any integer can be written as a fraction using a denominator of one: $-2 = \frac{-2}{1}$ and $0 = \frac{0}{1}$.

EXAMPLE 3 Graph the fractions by converting to decimal form and estimating their location between two integers. Use M and N as coordinates: (a) $-2\frac{1}{3}$ and (b) $\frac{7}{2}$.

Solution:

a. $-2\frac{1}{3} = -2.3333333\ldots$ or $-2.\overline{3}$ **b.** $\frac{7}{2} = 3.5$

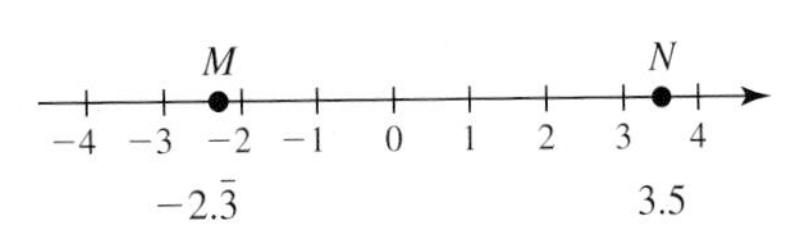

NOW TRY EXERCISES 15 THROUGH 18

In Example 3, note the division $\frac{7}{2}$ **terminated** and the result is a **terminating decimal.** For the mixed number $-2\frac{1}{3}$, the decimal form is **repeating** and **nonterminating.** A repeating decimal is written with a horizontal bar over the digit(s) that repeat. Sometimes the decimal form of a number is **nonrepeating** and **nonterminating.** Such numbers are called irrational.

Irrational Numbers

Although any fraction can be written in decimal form, not all decimal numbers can be written as a fraction. One example is the number represented by the Greek letter π (pi), frequently seen in a study of circular forms. Although we often approximate pi as $\pi \approx 3.14$, its true value has an infinite number of nonrepeating digits and cannot be written as a fraction (the $\approx$ symbol means "approximately equal to," and should be used whenever a value is estimated or rounded). Other numbers of this type can be found by taking **square roots.** The number b is a square root of a only if $(b)(b) = a$. Using the square root symbol $\sqrt{}$ we could also write this as $\sqrt{a} = b$ only if $b^2 = a$. All numbers greater than zero have one positive and one negative square root. The *positive* square root of 9 is 3 since $3^2 = 9$. The positive square root is also called the **principal root.** The *negative* square root of 9 is -3 since $(-3)^2 = 9$. In other words, $\sqrt{9} = 3$ and $-\sqrt{9} = -3$. Unlike the square roots of 9, the two square roots of 10 contain an infinite number of nonrepeating, nonterminating digits and can *never be written as a fraction.* Numbers like π and $\sqrt{10}$ belong to the **irrational numbers** $\mathbb{H}$: $\mathbb{H}$ = {numbers with a nonrepeating and nonterminating decimal form; numbers that cannot be written as a ratio of two integers}. Since the decimal form of $\sqrt{10}$ has an infinite number of digits, we either leave it written as $\sqrt{10}$ called the **exact form,** or obtain an **approximate form** using a calculator and rounding to a specified place value.

THE SQUARE ROOT OF A NUMBER

For any positive real number a:

$\sqrt{a}$ represents the *positive* or *principal* square root of a.

$-\sqrt{a}$ represents the *negative* square root of a.

$\sqrt{a} = b$ only if $b^2 = a$. Note $\sqrt{0} = 0$.

EXAMPLE 4 Use a calculator to approximate the principal square root of each number, then graph them on the number line (round to 100ths): (a) 3, (b) 13, and (c) 36.

Solution: **a.** $\sqrt{3} \approx 1.73$ **b.** $\sqrt{13} \approx 3.61$ **c.** $\sqrt{36} = 6$

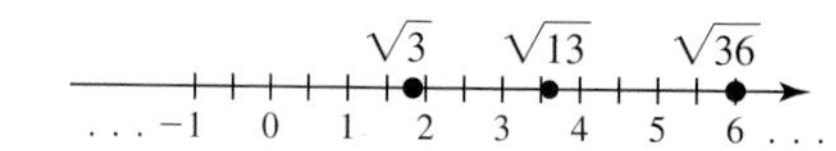

NOW TRY EXERCISES 19 THROUGH 22

Real Numbers

The set of rational numbers with the set of irrational numbers forms the set of **real numbers** $\mathbb{R}$. Figure R.2 helps to illustrate the relationship between the sets of numbers we've discussed so far. Notice how each subset appears "nested" in a larger set.

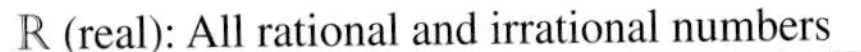

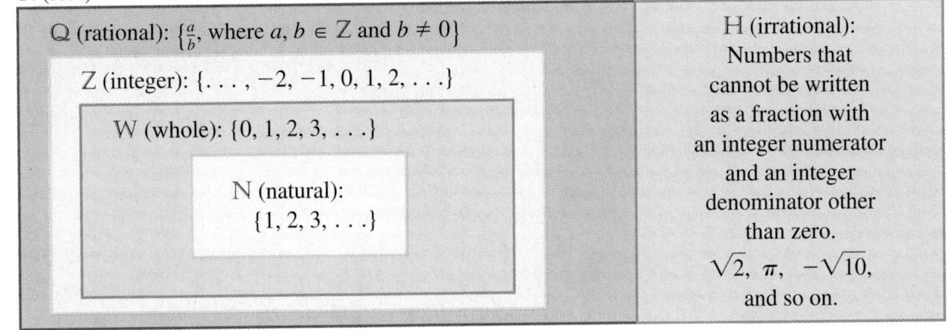

Figure R.2

EXAMPLE 5 List the numbers in set $A = \{-2, 0, 5, \sqrt{7}, 12, \frac{2}{3}, 4.5, \sqrt{21}, \pi, -0.75\}$ that belong to (a) Q, (b) H, (c) W, and (d) Z.

Solution:

a. $-2, 0, 5, 12, \frac{2}{3}, 4.5, -0.75 \in$ Q

b. $\sqrt{7}, \sqrt{21}, \pi \in$ H

c. $0, 5, 12 \in$ W

d. $-2, 0, 5, 12 \in$ Z

NOW TRY EXERCISES 23 THROUGH 26

EXAMPLE 6 Determine whether the statements are true or false.

a. N $\subset$ Q **b.** H $\subset$ Q **c.** W $\subset$ Z **d.** Z $\subset$ R

Solution:

a. True: All natural numbers can be written as a fraction over 1.

b. False: No irrational number can be written in fraction form.

c. True: All whole numbers are integers.

d. True: Every integer is a real number.

NOW TRY EXERCISES 27 THROUGH 38

B. Inequality Symbols and Order Relations

Comparisons between numbers of different size are shown using **inequality notation,** known as the **greater than** ($>$) and **less than** ($<$) symbols. When the numbers -4 and 3 are graphed on the number line, we note that $-4 < 3$ is the same as saying -4 is to the left of 3. In fact, on a number line, a number to the left is smaller than any number to the right of it.

> **ORDER PROPERTY OF REAL NUMBERS**
> Given any two real numbers a and b, $a < b$ if a is to the left of b on the number line. Likewise, $a > b$ if a is to the right of b on the number line.

A **variable** is a symbol, commonly a letter of the alphabet, used to represent an unknown quantity. Over the years x, y, and n have become most common, although any letter (or symbol) can be used. Many times **descriptive variables** are used, or variables that help us remember what they represent. Examples include L for length, D for distance, and so on.

EXAMPLE 7 Use a descriptive variable and an inequality symbol to write a mathematical model for the statement: "To hit a home run in Jacobi Park, the ball must travel over three hundred twenty-five feet."

Solution: Let D represent distance: $D > 325$.

NOW TRY EXERCISES 39 THROUGH 42

In Example 7, note the number 325 itself was not included. If the ball traveled *exactly* 325 ft, it would hit the top of the fence and stay in play (no home run). Numbers that mark the limit or boundary of an inequality are called **endpoints.** If the endpoint(s) are *not* included, we call the relation a **strict inequality.** When the endpoints *are* included, the relation is said to be **nonstrict.** The notation symbols used for nonstrict inequalities include the *less than or equal to symbol* ($\leq$) and the *greater than or equal to symbol* ($\geq$). The decision to *include* or *exclude* an endpoint is often an important one, and many mathematical decisions (and real-life decisions) depend on a clear understanding of the distinction.

C. The Absolute Value of a Real Number

In some applications, our main interest is the size or *magnitude* of a number, rather than its sign. This is called the **absolute value** of a number and can be thought of as its *distance from zero on the number line,* regardless of the direction. Since distance itself is measured in positive units, the absolute value of a number is always positive or zero.

ABSOLUTE VALUE OF A REAL NUMBER

The absolute value of a real number a, denoted $|a|$, is the undirected distance between a and 0 on the number line: $|a| \geq 0$.

EXAMPLE 8 In the table here, the absolute value of a number is given in column 1. Complete the remaining columns.

Solution:

Column 1 (In Symbols)	Column 2 (Spoken)	Column 3 (Result)	Column 4 (Reason)
$\lvert -2 \rvert$	"the absolute value of negative two"	2	the distance between −2 and 0 is 2 units
$\lvert 7.5 \rvert$	"the absolute value of seven and five-tenths"	7.5	the distance between 7.5 and 0 is 7.5 units
$-\lvert -6 \rvert$	"the opposite of the absolute value of negative six"	−6	the distance between −6 and 0 is 6 units, the opposite of 6 is −6

NOW TRY EXERCISES 43 THROUGH 50

Example 8 shows the absolute value of a positive number is the number itself, while the absolute value of a negative number is the *opposite of that number* (also a positive number). For this reason, the definition of absolute value is often given as

DEFINITION OF ABSOLUTE VALUE

$$|x| = \begin{cases} x & \text{if } x \geq 0 \\ -x & \text{if } x < 0 \end{cases}$$

Since "absolute values" involve an undirected distance, the concept can also be used to find the distance between *any* two numbers on a number line. For instance, on the number line we know the distance between 2 and 8 is 6 (by counting). Using absolute values, we can write this as $|8 - 2| = |6| = 6$, or $|2 - 8| = |-6| = 6$. Generally, if a and b are two numbers on the real number line, the distance between them is $|a - b|$ or $|b - a|$.

EXAMPLE 9 Find the distance between -5 and 3 on the number line.

Solution: The distance can be computed as $|-5 - 3| = |-8| = 8$ or $|3 - (-5)| = |8| = 8$.

NOW TRY EXERCISES 51 THROUGH 58

D. Operations on Real Numbers

The operations of addition, subtraction, multiplication, and division are defined for the set of real numbers, and the concept of absolute value plays an important role. However, two ideas involving division and zero deserve special mention. Carefully consider Example 10.

EXAMPLE 10 Determine the result of each quotient *by first writing the related multiplication.*

a. $0 \div 8 = p$ **b.** $\frac{16}{0} = q$ **c.** $\frac{0}{12} = n$

Solution:

a. $0 \div 8 = p$, if $p \cdot 8 = 0 \rightarrow p = 0$.

b. $\frac{16}{0} = q$, if $q \cdot 0 = 16 \rightarrow$ no such number q.

c. $\frac{0}{12} = n$, if $n \cdot 12 = 0 \rightarrow n = 0$.

NOW TRY EXERCISES 59 THROUGH 62

In Example 10(a), a dividend (numerator or first number) of 0 over 8 means we are going to divide zero into eight groups. The related multiplication shows there will be zero in each group. As seen in Example 10(b), an expression with a divisor (denominator or second number) of zero *cannot be computed or checked.* Although it seems trivial, division by zero has many implications in a study of mathematics, so make an effort to know the facts: The quotient of zero and any nonzero number is zero, but *division by zero is undefined.*

DIVISION AND ZERO

The quotient of zero and any real number n is zero $(n \neq 0)$:

$$0 \div n = 0 \quad \text{and} \quad \frac{0}{n} = 0.$$

The expressions $n \div 0$ and $\frac{n}{0}$ are undefined.

Squares, Cubes, and Exponential Form

When a number is repeatedly multiplied by itself as in (10)(10)(10)(10), we write it using **exponential notation** as 10^4. The number used for repeated multiplication (in this case 10) is called the **base,** and the superscript number is called an **exponent.** The exponent tells how many times the base occurs as a factor, and we say 10^4 is written in **exponential**

form. Numbers that result from squaring (exponent of 2) an integer are called **perfect squares,** while numbers that result from cubing (exponent of 3) an integer are called **perfect cubes.** These are often collected into a table, such as Table R.1, and memorized to help complete many common calculations mentally. Only the square and cube of selected positive integers are shown.

Table R.1

Perfect Squares				Perfect Cubes	
N	N^2	N	N^2	N	N^3
1	1	7	49	1	1
2	4	8	64	2	8
3	9	9	81	3	27
4	16	10	100	4	64
5	25	11	121	5	125
6	36	12	144	6	216

EXAMPLE 11 Write the exponential in expanded form, then determine its value.

a. 4^3 b. $(-6)^2$ c. -6^2 d. $\left(\frac{2}{3}\right)^3$

Solution:
a. $4^3 = 4 \cdot 4 \cdot 4 = 64$ b. $(-6)^2 = (-6) \cdot (-6) = 36$

c. $-6^2 = -(6 \cdot 6) = -36$ d. $\left(\frac{2}{3}\right)^3 = \frac{2}{3} \cdot \frac{2}{3} \cdot \frac{2}{3} = \frac{8}{27}$

NOW TRY EXERCISES 63 AND 64

Examples 11(b) and 11(c) illustrate an important distinction. The expression $(-6)^2$ is read, "the square of negative six" and the negative sign is included in both factors. The expression -6^2 is read, "the opposite of six squared," and the square of six is calculated first, then made negative.

Square Roots and Cube Roots

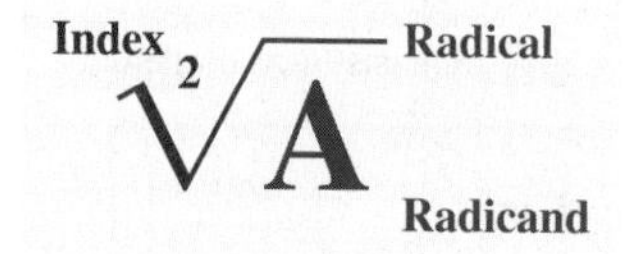

In the computation of square roots, either the $\sqrt{\ }$ or $\sqrt[2]{\ }$ notation can be used. The $\sqrt{\ }$ symbol is called a **radical,** the number under the radical is called the **radicand,** and the small case number 2 is called the **index.** The index tells how many factors are needed to obtain the radicand. For example, $\sqrt[2]{25} = 5$, since $5 \cdot 5 = 25$. The cube root of a number has the form $\sqrt[3]{A} = B$, where $B \cdot B \cdot B = A$. This means $\sqrt[3]{27} = 3$ since $3 \cdot 3 \cdot 3 = 27$.

EXAMPLE 12 Determine the value of each expression.

a. $\sqrt[2]{49}$ b. $\sqrt[3]{125}$ c. $\sqrt{\frac{9}{16}}$ d. $-\sqrt{16}$ e. $\sqrt{-25}$

Solution:
a. 7 since $7 \cdot 7 = 49$ b. 5 since $5 \cdot 5 \cdot 5 = 125$

c. $\frac{3}{4}$ since $\frac{3}{4} \cdot \frac{3}{4} = \frac{9}{16}$ d. -4 since $\sqrt{16} = 4$

e. not a real number since $5 \cdot 5 = (-5)(-5) = 25$

NOW TRY EXERCISES 65 THROUGH 70

In general, we have the following properties:

SQUARE ROOTS	**CUBE ROOTS**
$\sqrt{A} = B$ if $B \cdot B = A$	$\sqrt[3]{A} = B$ if $B \cdot B \cdot B = A$
$(A \geq 0)$	$(A \in \mathbb{R})$
This also means that	This also means that
$\sqrt{A} \cdot \sqrt{A} = A$	$\sqrt[3]{A} \cdot \sqrt[3]{A} \cdot \sqrt[3]{A} = A$
$(\sqrt{A})^2 = A$	$(\sqrt[3]{A})^3 = A$

For square roots, if the radicand is a perfect square or has perfect squares in both the numerator and denominator, the result is a rational number, as in Example 12(c). If the radicand is not a perfect square, the result is an irrational number. Similar statements can be made regarding cube roots.

The Order of Operations

When basic operations are combined into a longer mathematical expression, we use a specified **priority** or **order of operations** to evaluate them. Using a standard order of operations helps prevent getting many different results from the same expression.

THE ORDER OF OPERATIONS

1. Simplify within grouping symbols. If there are "nested" symbols of grouping, begin with the innermost group. If the fraction bar is used as a grouping symbol, simplify the numerator and denominator separately.
2. Evaluate all exponents and roots.
3. Compute all multiplications or divisions *in the order that they occur from left to right.*
4. Compute all additions or subtractions *in the order that they occur from left to right.*

EXAMPLE 13 Simplify using the order of operations:

a. $7500\left(1 + \frac{0.075}{12}\right)^{12 \cdot 15}$ **b.** $\frac{-4.5(8) - 3}{\sqrt[3]{125} + 2^3}$

Solution: **a.** $7500\left(1 + \frac{0.075}{12}\right)^{12 \cdot 15}$ original expression

$= 7500(1.00625)^{12 \cdot 15}$ simplify within the parenthesis (division before addition)

$= 7500(1.00625)^{180}$ simplify the exponent

$= 7500(3.069451727)$ exponents before multiplication

$= 23{,}020.89$ result (rounded to hundredths)

b. $\frac{-4.5(8) - 3}{\sqrt[3]{125} + 2^3}$ original expression

$= \frac{-36 - 3}{5 + 8}$ simplify terms in the numerator and denominator

$= \frac{-39}{13}$ simplify

$= -3$ result

NOW TRY EXERCISES 71 THROUGH 94

R.1 EXERCISES

6. a. $(-5)^2 = (-5)(-5) = 25$
$-5^2 = -(5 \cdot 5) = -25$
b. $-5^3 = -(5)(5)(5) = -125$
$(-5)^3 = (-5)(-5)(-5) = -125$

15. $1.\overline{3}$

16. -0.875

17. $2.\overline{5}$

18. $-1.8\overline{3}$

19. ≈ 2.65

20. ≈ 4.36

21. ≈ 1.73

22. ≈ 6.40

23. a. i. $\{8, 7, 6\}$
ii. $\{8, 7, 6\}$
iii. $\{-1, 8, 7, 6\}$
iv. $\{-1, 8, 0.75, \frac{9}{2}, 5.\overline{6}, 7, \frac{3}{5}, 6\}$
v. { }
vi. $\{-1, 8, 0.75, \frac{9}{2}, 5.\overline{6}, 7, \frac{3}{5}, 6\}$
b. $\{-1, \frac{3}{5}, 0.75, \frac{9}{2}, 5.\overline{6}, 6, 7, 8\}$
c.

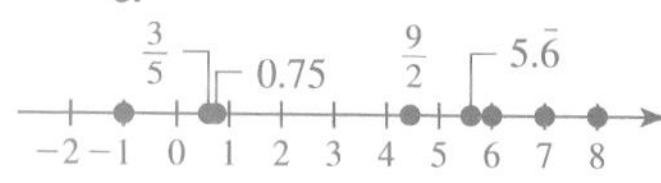

24. a. i. { }
ii. {0}
iii. $\{-7, 0\}$
iv. $\{-7, 2.\overline{1}, 5.73, -3\frac{5}{6}, 0, -1.12, \frac{7}{8}\}$
v. { }
vi. $\{-7, 2.\overline{1}, 5.73, -3\frac{5}{6}, 0, -1.12, \frac{7}{8}\}$
b. $\{-7, -3\frac{5}{6}, -1.12, 0, \frac{7}{8}, 2.\overline{1}, 5.73\}$
c. $-3\frac{5}{6}$ -1.12 $\frac{7}{8}$ $2.\overline{1}$ 5.73

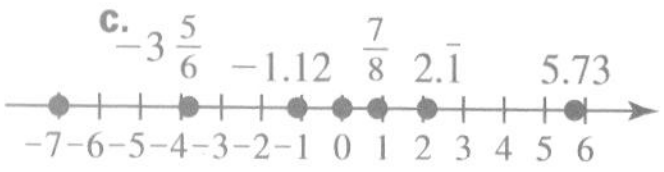

Additional answers can be found in the Instructor Answer Appendix.

CONCEPTS AND VOCABULARY

Fill in each blank with the appropriate word or phrase. Carefully reread the section, if necessary.

1. The symbol $\subset$ means: is a subset of and the symbol $\in$ means: is an element of.

2. A number corresponding to a point on the number line is called the coordinate of that point.

3. Every positive number has two square roots, one positive and one negative. The two square roots of 49 are 7 and −7; $\sqrt{49}$ represents the principal square root of 49.

4. The decimal form of $\sqrt{7}$ contains an infinite number of non repeating and non terminating digits. This means that $\sqrt{7}$ is a(n) irrational number.

5. Discuss/explain why the value of $12 \cdot \frac{1}{3} + \frac{2}{3}$ is $4\frac{2}{3}$ and not 12. Order of operations requires multiplication before addition.

6. Discuss/explain (a) why $(-5)^2 = 25$, while $-5^2 = -25$; and (b) why $-5^3 = (-5)^3 = -125$.

DEVELOPING YOUR SKILLS

7. List the natural numbers that are
a. less than 6. {1, 2, 3, 4, 5}
b. less than 1. { }

8. List the natural numbers that are
a. between 0 and 1. { }
b. greater than 50. {51, 52, 53, 54, . . .}

Identify each of the following statements as either true or false. If false, give an example that shows why.

9. $\mathbb{N} \subset \mathbb{W}$ True

10. $\mathbb{W} \not\subset \mathbb{N}$ True

11. $\{33, 35, 37, 39\} \subset \mathbb{W}$ True

12. $\{2.2, 2.3, 2.4, 2.5\} \subset \mathbb{W}$ False; 2.2 is not a whole number.

13. $6 \in \{0, 1, 2, 3, \ldots\}$ True

14. $1297 \notin \{0, 1, 2, 3, \ldots\}$ False; 1297 is a whole number.

Convert to decimal form and graph by estimating the number's location between two integers.

15. $\frac{4}{3}$

16. $-\frac{7}{8}$

17. $2\frac{5}{9}$

18. $-1\frac{5}{6}$

Use a calculator to find the principal square root of each number (round to hundredths as needed). Then graph each number by estimating its location between two integers.

19. 7

20. 19

21. 3

22. 41

For the sets in Exercises 23 through 26:

a. List all numbers that are elements of (i) $\mathbb{N}$, (ii) $\mathbb{W}$, (iii) $\mathbb{Z}$, (iv) $\mathbb{Q}$, (v) $\mathbb{H}$, and (vi) $\mathbb{R}$.
b. Rewrite the elements of each set in order from smallest to largest.
c. Graph the elements of each set on a number line.

23. $\{-1, 8, 0.75, \frac{9}{2}, 5.\overline{6}, 7, \frac{3}{5}, 6\}$

24. $\{-7, 2.\overline{1}, 5.73, -3\frac{5}{6}, 0, -1.12, \frac{7}{8}\}$

25. $\{-5, \sqrt{49}, 2, -3, 6, -1, \sqrt{3}, 0, 4, \pi\}$

26. $\{-8, 5, -2\frac{3}{5}, 1.75, -\sqrt{2}, -0.6, \pi, \frac{7}{2}, \sqrt{64}\}$

State true or false. If false, state why.

27. $\mathbb{R} \subset \mathbb{H}$ False; not all real numbers are irrational.

28. $\mathbb{N} \subset \mathbb{R}$ True

29. $\mathbb{Q} \subset \mathbb{Z}$ False; not all rational numbers are integers.

30. $\mathbb{Z} \subset \mathbb{Q}$ True

31. $\sqrt{25} \in \mathbb{H}$ False; $\sqrt{25} = 5$ is not irrational.

32. $\sqrt{19} \in \mathbb{H}$ True

Match each set with its correct symbol and description/illustration.

33. c IV Irrational numbers	a.	$\mathbb{R}$	I.	$\{1, 2, 3, 4, \ldots\}$
34. f V Integers	b.	$\mathbb{Q}$	II.	$\{\frac{a}{b}, \vert a, b \in \mathbb{Z}; b \neq 0\}$
35. a VI Real numbers	c.	$\mathbb{H}$	III.	$\{0, 1, 2, 3, 4, \ldots\}$
36. b II Rational numbers	d.	$\mathbb{W}$	IV.	$\{\pi, \sqrt{7}, -\sqrt{13}, \text{etc.}\}$
37. d III Whole numbers	e.	$\mathbb{N}$	V.	$\{\ldots -3, -2, -1, 0, 1, 2, 3, \ldots\}$
38. e I Natural numbers	f.	$\mathbb{Z}$	VI.	$\mathbb{N}, \mathbb{W}, \mathbb{Z}, \mathbb{Q}, \mathbb{H}$

Use a descriptive variable and an inequality symbol ($<$, $>$, $\leq$, $\geq$) to write a model for each statement.

39. To spend the night at a friend's house, Kylie must be at least 6 years old.
40. Monty can spend at most \$2500 on the purchase of a used automobile.
41. If Jerod gets no more than two words incorrect on his spelling test he can play in the soccer game this weekend.
42. Andy must weigh less than 112 lb to be allowed to wrestle in his weight class at the meet.

39. Let a represent Kylie's age: $a \geq 6$.
40. Let a represent the amount: $a \leq 2500$.
41. Let n represent the number of incorrect words: $n \leq 2$.
42. Let w represent Andy's weight: $w < 112$.

Evaluate/simplify each expression.

43. $|-2.75|$ 2.75
44. $|-7.24|$ 7.24
45. $-|-4|$ -4
46. $-|-6|$ -6
47. $\left|\frac{1}{2}\right|$ $\frac{1}{2}$
48. $\left|\frac{2}{5}\right|$ $\frac{2}{5}$
49. $\left|-\frac{3}{4}\right|$ $\frac{3}{4}$
50. $\left|-\frac{3}{7}\right|$ $\frac{3}{7}$

Use the concept of absolute value to complete Exercises 51 to 58.

51. Write the statement two ways, then simplify. "The distance between -7.5 and 2.5 is . . ." 10
52. Write the statement two ways, then simplify. "The distance between $13\frac{2}{5}$ and $-2\frac{3}{5}$ is . . ." 16
53. If n is positive, then $-n$ is negative.
54. If n is negative, then $-n$ is positive.
55. If $n < 0$, then $|n| = -n$.
56. If $n > 0$, then $|n| = n$.
57. What two numbers on the number line are five units from negative three? -8, 2
58. What two numbers on the number line are three units from two? 5, -1

Determine which expressions are equal to zero and which are undefined. Justify your responses by writing the related multiplication.

59. $12 \div 0$ undefined, since $12 \div 0 = k$ implies $k \cdot 0 = 12$
60. $0 \div 12$ $0 \div 12 = 0$, since $0 \cdot 12 = 0$
61. $\frac{7}{0}$ undefined, since $7 \div 0 = k$ implies $k \cdot 0 = 7$
62. $\frac{0}{7}$ $0 \div 7 = 0$, since $0 \cdot 7 = 0$

Without computing the actual answer, state whether the result will be positive or negative. Be careful to note what power is used and whether the negative sign is included in parentheses.

63. a. $(-7)^2$ positive
b. -7^2 negative
c. $(-7)^5$ negative
d. -7^5 negative

64. a. $(-7)^3$ negative
b. -7^3 negative
c. $(-7)^4$ positive
d. -7^4 negative

Evaluate without the aid of a calculator.

65. $-\sqrt{\frac{121}{36}}$ $-\frac{11}{6}$
66. $-\sqrt{\frac{25}{49}}$ $-\frac{5}{7}$
67. $\sqrt[3]{-8}$ -2
68. $\sqrt[3]{-64}$ -4

69. What perfect square is closest to 78? $9^2 = 81$ is closest
70. What perfect cube is closest to -71? $(-4)^3 = -64$ is closest

Perform the operation indicated mentally or using pencil/paper.

71. $-24 - (-31)$ 7 **72.** $-45 - (-54)$ 9 **73.** $7.045 - 9.23$ -2.185 **74.** $0.0762 - 0.9034$ -0.8272

75. $4\frac{5}{6} + (-\frac{1}{2})$ $4\frac{1}{3}$ **76.** $1\frac{1}{8} + (-\frac{3}{4})$ $\frac{3}{8}$ **77.** $(-\frac{2}{3})(3\frac{5}{8})$ $-\frac{29}{12}$ or $-2\frac{5}{12}$ **78.** $(-8)(2\frac{1}{4})$ -18

79. $(12)(-3)(0)$ 0 **80.** $(-1)(0)(-5)$ 0 **81.** $-60 \div 12$ -5 **82.** $75 \div (-15)$ -5

83. $\frac{4}{5} \div (-8)$ $-\frac{1}{10}$ **84.** $-15 \div \frac{1}{2}$ -30 **85.** $-\frac{2}{3} \div \frac{16}{21}$ $-\frac{7}{8}$ **86.** $-\frac{3}{4} \div \frac{7}{8}$ $-\frac{6}{7}$

Evaluate without a calculator, using the order of operations.

87. $-3^2 + 15 - |5 - 15| - \sqrt{169}$ -17

88. $-5^2 + 9 - |7 - 15| - \sqrt{121}$ -35

89. $\sqrt{\frac{9}{16}} - \frac{3}{5} \cdot \left(\frac{5}{3}\right)^2$ $\frac{-11}{12}$

90. $\left(\frac{3}{2}\right)^2 \div \left(\frac{9}{4}\right) - \sqrt{\frac{25}{64}}$ $\frac{3}{8}$

91. $\frac{4(-7) - 6^2}{6 - \sqrt{49}}$ 64

92. $\frac{5(-6) - 3^2}{9 - \sqrt{64}}$ -39

Evaluate using a calculator (round to hundredths).

93. $2475\left(1 + \frac{0.06}{4}\right)^{4 \cdot 10}$ 4489.70

94. $5100\left(1 + \frac{0.078}{52}\right)^{52 \cdot 20}$ 24,241.64

WORKING WITH FORMULAS

95. Pitch diameter: $D = \frac{d \cdot n}{n + 2}$

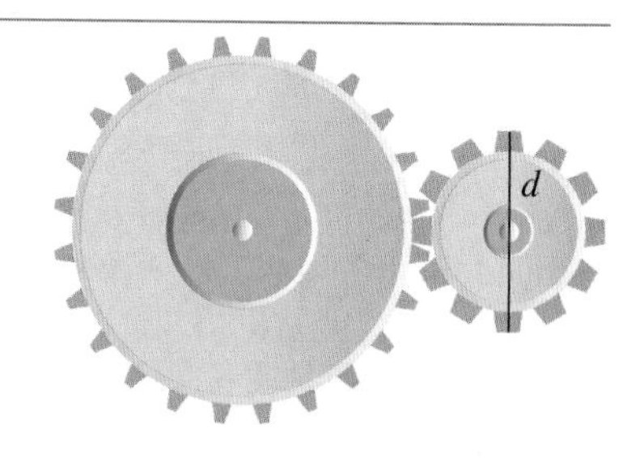

Mesh gears are used to transfer rotary motion and power from one shaft to another. The *pitch diameter* D of a drive gear is given by the formula shown, where d is the outer diameter of the gear and n is the number of teeth on the gear. Find the pitch diameter of a gear with 12 teeth and an outer diameter of 5 cm. $D \approx 4.3$ cm

96. Pediatric dosages and Clark's rule: $D_C = \frac{D_A \cdot W}{150}$

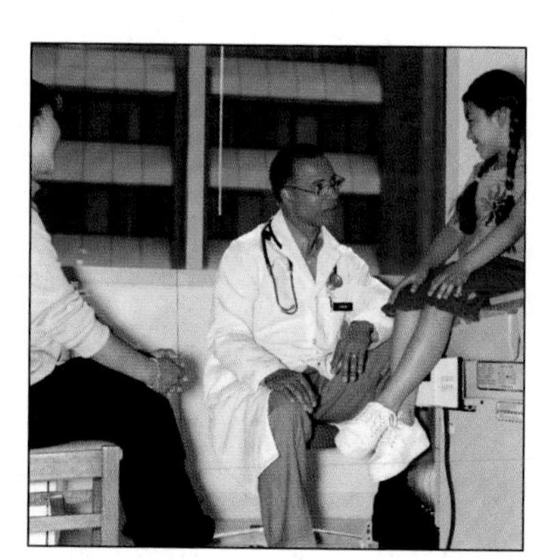

The amount of medication prescribed for young children depends on their weight, height, age, body surface area and other factors. **Clark's rule** is a formula that helps estimate the correct child's dose D_C based on the adult dose D_A and the weight W of the child (an average adult weight of 150 lb is assumed). Compute a child's dose if the adult dose is 50 mg and the child weighs 30 lb. $D_C = 10$ mg

APPLICATIONS

Use positive and negative numbers to model the situation, then compute.

97. At 6:00 P.M., the temperature was 50°F. A cold front moves through that causes the temperature to *drop* 3°F each hour until midnight. What is the temperature at midnight? 32°F

98. Most air conditioning systems are designed to create a 2° *drop* in the air temperature each hour. How long would it take to reduce the air temperature from 86° to 71°? $7\frac{1}{2}$ hr

99. The state of California holds the record for the greatest temperature swing between a record high and a record low. The record high was 134°F and the record low was −45°F. How many degrees *difference* between the record high from the record low? 179°F

100. In Juneau, Alaska, the temperature was 17°F early one morning. A cold front later moved in and the temperature *dropped* 32°F by lunch time. What was the temperature at lunch time? −15°F

EXTENDING THE CONCEPT

Tsu Ch'ung-chih: $\frac{355}{113}$

101. Here are some historical approximations for π. Which one is closest to the true value?

Archimedes: $3\frac{1}{7}$ Tsu Ch'ung-chih: $\frac{355}{113}$ Aryabhata: $\frac{62{,}832}{20{,}000}$ Brahmagupta: $\sqrt{10}$

102. If $A > 0$ and $B < 0$, is the product $A \cdot (-B)$ positive or negative? positive

103. If $A < 0$ and $B < 0$, is the quotient $-(A \div B)$ positive or negative? negative

R.2 Algebraic Expressions and the Properties of Real Numbers

LEARNING OBJECTIVES

In Section R.2 you will review how to:

- **A. Identify variables, coefficients, terms and expressions**
- **B. Create mathematical models**
- **C. Evaluate expressions and use a table of values**
- **D. Identify and use properties of real numbers**
- **E. Simplify algebraic expressions**

INTRODUCTION

To effectively use mathematics as a problem-solving tool, we must develop the ability to translate written or verbal information into a mathematical model. Many times this involves looking for English words that have a *direct* mathematical translation. Other times we look for the *intended* mathematical translation, by mentally visualizing the situation described. After obtaining a model, many applications require working effectively with algebraic terms and expressions. The basic ideas involved are reviewed here.

POINT OF INTEREST

The algebraic notation we use today is also the result of a long, evolutionary process. New ideas often come before the symbols or notation needed to express them clearly, and it took hundreds of years for algebraic symbolism to replace the verbal or prose style called "rhetorical algebra." This example of *rhetorical algebra* is translated from the book *Al-jabr,* written by al-Khowarizmi (c. 825). "What must be the amount of a square, which when one ten is added to it, becomes equal to three roots of that square?" In modern notation, we would simply write $x^2 + 10 = 3x$.

A. Word Phrases and Algebraic Expressions

An **algebraic term** is a *collection of factors* that may include numbers, variables, or parenthesized groups. Here are some examples:

1. 3 2. $-6P$ 3. $5xy$ 4. $-8n^2$ 5. n 6. $2(x + 3)$

If a term consists of a single nonvariable number, it is referred to as a **constant** term. In (1), 3 is a constant term. Any term that contains a variable is called a **variable term.** We call the constant factor of a variable term the **numerical coefficient** or simply the **coefficient.** The coefficients for (2), (3), and (4) are −6, 5, and −8, respectively. In (5), the coefficient of n is 1, since $1 \cdot n = 1n = n$. The term in (6) has two factors, 2 and $(x + 3)$. The coefficient is 2.

An **algebraic expression** is a sum or difference of algebraic terms. To avoid confusion when identifying the coefficient of each term in the expression, it can be rewritten using algebraic addition if desired. Also, it is sometimes helpful to **decompose** a rational term to identify its coefficient, rewriting the term using a unit fraction [see Example 1(b)].

DECOMPOSITION OF RATIONAL TERMS

For any rational term $\frac{A}{B}$ $(B \neq 0)$, $\frac{A}{B} = \frac{1}{B} \cdot \frac{A}{1} = \frac{1}{B} \cdot A$.

e.g. $\frac{2}{3} = \frac{1}{3} \cdot 2$, and $\frac{n-2}{5} = \frac{1}{5}(n-2)$.

WORTHY OF NOTE

Notice how the fraction bar acts as a grouping symbol in Example 1(b), helping us identify $\frac{x+3}{7}$ as a single term with a coefficient of $\frac{1}{7}$. The expression in Example 1(c) consists of a single term whose factors are -1 and $(x-12)$. The coefficient is -1. In Example 1(d), the constant 5 is its own coefficient, since $-2x^2 + (-1x) + 5 = -2x^2 + (-1x) + 5x^0$.

EXAMPLE 1 State the number of terms in each expression and identify the coefficient of each.

a. $2x - 5y$ **b.** $\frac{x+3}{7} - 2x$ **c.** $-(x-12)$ **d.** $-2x^2 - x + 5$

Rewritten:	a. $2x + (-5y)$	b. $\frac{1}{7}(x+3) + (-2x)$	c. $-1(x-12)$	d. $-2x^2 + (-1x) + 5$
Number of terms:	two	two	one	three
Coefficient(s):	2 and -5	$\frac{1}{7}$ and -2	-1	-2, -1, and 5

NOW TRY EXERCISES 7 THROUGH 14

B. Translating Written or Verbal Information into a Mathematical Model

The key to solving many applied problems is finding a mathematical model or algebraic expression that accurately models the situation. This can be done by assigning a variable to an unknown quantity, then building related expressions by noting that many words in the English language suggest a mathematical operation (see Table R.2).

Table R.2

Addition	Subtraction	Multiplication	Division	Equals
and	from	of	into	is
plus	subtract	times	over	equals
more	less	product	divided by	same as
added to	fewer	by	quotient of	makes
together with	minus	percent of	ratio of	leaves
sum	difference	multiplied by	a is to b	yields
total	take away	per		equivalent
increased by	decreased by			results in

twice → 2 times doubled → 2 times tripled → 3 times

Many different phrases from the English language can be translated into a single mathematical phrase using words from this list. Here are several examples.

EXAMPLE 2 The phrases in each group here can be modeled by the same algebraic expression. Assign a variable to the unknown and write the expression.

a. the difference of negative ten and a number, a number subtracted from negative ten, some number less than negative ten, negative ten decreased by a number

b. the quotient of negative twelve and a number, negative twelve divided by a number, the ratio of negative twelve and a number, a number divided into negative twelve

Solution:

a. Let n represent the unknown number: $-10 - n$.

b. Let x represent the unknown number: $-12 \div x$ or $\frac{-12}{x}$.

NOW TRY EXERCISES 15 THROUGH 28

Recall that *descriptive variables* are often used in the modeling process. Capital letters are also used due to their widespread appearance in other fields. In many cases, the algebraic expression will contain more than one operation.

WORTHY OF NOTE

In Example 3(b), note "six less than three times the width" is modeled by $3W - 6$ and not $6 - 3W$. Finding a quantity that is "six less than" some other, requires us to subtract six *from* the original quantity, not the original quantity from six. Remember, we are looking for the *meaning* or *intent* of the phrase, not a word-for-word translation. Also, note the difference between *six is less than 3W*: $6 < 3W$, and *six less than 3W*: $3W - 6$.

EXAMPLE 3 Assign a variable to the unknown number, then translate each phrase into an algebraic expression using descriptive variables.

a. twice a number increased by five

b. six less than three times the width

c. ten less than triple the payment

d. two hundred fifty feet more than double the length

Solution:

a. Let N represent the number. Then $2N$ represents twice the number, and $2N + 5$ represents twice a number increased by five.

b. Let W represent the width. Then $3W$ represents three times the width, and $3W - 6$ represents six less than three times the width.

c. Let P represent the payment. Then $3P$ represents a triple payment, and $3P - 10$ represents 10 less than triple the payment.

d. Let L represent the length. Then $2L$ represents double the length, and $2L + 250$ represents 250 more than double the length.

NOW TRY EXERCISES 29 THROUGH 32

Identifying and translating these phrases *when they occur in context* is an important problem-solving skill. Note how this is done in Example 4.

EXAMPLE 4 *The cost for a rental car is \$35 plus 15 cents per mile.* Express the cost of renting a car in terms of the number of miles driven.

Solution: Let m represent the number of miles driven. Then $0.15m$ represents the cost for each mile and $C = 35 + 0.15m$ represents the total cost for renting the car.

NOW TRY EXERCISES 91 THROUGH 98

C. Evaluating Algebraic Expressions

We often need to **evaluate** expressions to investigate patterns and note relationships. We also use evaluation skills when working with formulas. When evaluating expressions or formulas, it's best to use a **vertical format** with the original expression written first, the substitutions shown next, and the simplified forms and final answer following. The value substituted or "plugged into" an expression is often called the **input value,** and the result is called the **output.**

> **EVALUATING A MATHEMATICAL EXPRESSION**
> 1. Replace each variable with an open parenthesis ( ).
> 2. Substitute the given replacements for each variable.
> 3. Simplify using the order of operations.

If the same expression is evaluated repeatedly, results are often collected and analyzed in a table of values, as shown in Example 5.

EXAMPLE 5 Evaluate $x^2 - 2x - 3$ to complete the table shown. Which input value(s) of x cause the expression to have an output of 0?

Solution:

Input x	$x^2 - 2x - 3$	Output
-2	$(-2)^2 - 2(-2) - 3$	5
-1	$(-1)^2 - 2(-1) - 3$	0
0	$(0)^2 - 2(0) - 3$	-3
1	$(1)^2 - 2(1) - 3$	-4
2	$(2)^2 - 2(2) - 3$	-3
3	$(3)^2 - 2(3) - 3$	0
4	$(4)^2 - 2(4) - 3$	5

The expression has an output of 0 when $x = -1$ and $x = -3$.

NOW TRY EXERCISES 33 THROUGH 58

As a practical matter, the substitution and simplification is often done mentally or on scratch paper, with the table showing only the input and output values that result.

D. Properties of Real Numbers

Consider the product $\frac{1}{3} \cdot (-5) \cdot 9$. If we reorder or *commute* the last two factors, the expression becomes $\frac{1}{3} \cdot 9 \cdot (-5)$ and the result is computed more easily since $\frac{1}{3} \cdot 9 = 3$. When we reorder factors, we are using the **commutative property of multiplication.** A reordering of addends involves the **commutative property of addition.**

> **THE COMMUTATIVE PROPERTIES**
> Given that a and b represent real numbers:
>
> ADDITION: $a + b = b + a$
> Addends can be combined in any order without changing the sum.
>
> MULTIPLICATION: $a \cdot b = b \cdot a$
> Factors can be multiplied in any order without changing the product.

The property can be extended to include any number of addends or factors. While the commutative property implies a *reordering* or *movement* of terms (to commute implies back-and-forth movement), the **associative property** implies a *regrouping* or reassociation of terms. For example the sum $\left(\frac{3}{4} + \frac{3}{5}\right) + \frac{2}{5}$ is easier to compute if we regroup the addends as $\frac{3}{4} + \left(\frac{3}{5} + \frac{2}{5}\right)$. Both give a sum of $1\frac{3}{4}$ but the second can be found more easily. This illustrates the **associative property of addition.** Multiplication is also associative.

THE ASSOCIATIVE PROPERTIES

Given that a, b, and c represent real numbers:

ADDITION:	MULTIPLICATION:
$(a + b) + c = a + (b + c)$	$(a \cdot b) \cdot c = a \cdot (b \cdot c)$
Addends can be regrouped.	Factors can be regrouped.

WORTHY OF NOTE

Is subtraction commutative? Consider a situation involving money. If you had \$100, you could easily buy an item costing \$20: \$100 − \$20 leaves you with \$80. But if you had \$20, could you buy an item costing \$100? Obviously \$100 − \$20 is not the same as \$20 − \$100. Subtraction is *not* commutative.

EXAMPLE 6 Use the commutative and associative properties to simplify each calculation.

a. $\frac{3}{8} - 19 + \frac{5}{8}$ b. $[-2.5 \cdot (-1.2)] \cdot 10$

Solution:

a. $\frac{3}{8} - 19 + \frac{5}{8} = -19 + \left(\frac{3}{8} + \frac{5}{8}\right)$
$= -19 + 1$
$= -18$

b. $[-2.5 \cdot (-1.2)] \cdot 10 = -2.5 \cdot [(-1.2) \cdot 10]$
$= -2.5 \cdot (-12)$
$= 30$

NOW TRY EXERCISES 59 AND 60

An **identity element** "identifies" a given value when combined with a stated operation and the members of a set. For the real numbers, the **additive identity** is zero, since $x + 0 = x$ for any real number x. The **multiplicative identity** is the number 1, since $x \cdot 1 = x$ for any real number x. These properties are used extensively in solving equations.

THE ADDITIVE AND MULTIPLICATIVE IDENTITIES

Given that x is a real number:

$x + 0 = x$ $\quad$ $0 + x = x$	$x \cdot 1 = x$ $\quad$ $1 \cdot x = x$
Zero is the identity for addition.	One is the identity for multiplication.

When combined with a given operation and an element of a set, an **inverse element** yields the related identity. For the real numbers, $-x$ is the **additive inverse** for x, since $-x + x = 0$ for any real number (x and $-x$ are also called **opposites**). The **multiplicative inverse** of any nonzero number x is $\frac{1}{x}$, since $x \cdot \frac{1}{x} = 1$ for any nonzero real number. This property can also be stated as $\frac{a}{b} \cdot \frac{b}{a} = 1$ $(a, b \neq 0)$ for any real number $\frac{a}{b}$. Note that $\frac{a}{b}$ and $\frac{b}{a}$ are **reciprocals.**

THE ADDITIVE AND MULTIPLICATIVE INVERSES

Given that a, b, and x represent real numbers where $a, b \neq 0$:

$-x + x = 0$ $\quad$ $x + (-x) = 0$	$\frac{a}{b} \cdot \frac{b}{a} = 1$ $\quad$ $\frac{b}{a} \cdot \frac{a}{b} = 1$
$-x$ is the additive inverse for any real number x.	$\frac{a}{b}$ is the multiplicative inverse for any real number $\frac{b}{a}$.

EXAMPLE 7 Replace the box to create a true statement:

a. $\square \cdot \frac{-3}{5}x = 1 \cdot x$ **b.** $x + 5.3 + \square = x$

Solution: **a.** $\square = \frac{-5}{3}$, since $\frac{-5}{3} \cdot \frac{-3}{5} = 1$

b. $\square = -5.3$, since $5.3 + (-5.3) = 0$

NOW TRY EXERCISES 61 AND 62

The **distributive property of multiplication over addition** is widely used in a study of algebra, because it enables us to rewrite a product as an equivalent sum and vice versa.

THE DISTRIBUTIVE PROPERTY OF MULTIPLICATION OVER ADDITION

Given that a, b, and c represent real numbers:

$a(b + c) = ab + ac$	$ab + ac = a(b + c)$
A factor outside a sum can be distributed to each addend in the sum.	A factor common to each addend in a sum can be "undistributed" and written outside a group.

EXAMPLE 8 Apply the distributive property as appropriate. Simplify if possible.

a. $7(p + 5.2)$ **b.** $4(2.5 - x)$ **c.** $7x^3 - x^3$ **d.** $\frac{5}{2}n + \frac{1}{2}n$

Solution: **a.** $7p + 36.4$ **b.** $10 - 4x$ **c.** $(7 - 1)x^3 = 6x^3$

d. $\left(\frac{5}{2} + \frac{1}{2}\right)n = 3n$

NOW TRY EXERCISES 63 THROUGH 70

E. Simplifying Algebraic Expressions

Two terms are **like terms** only if they have the *same variable factors* (the coefficient is not used to identify like terms). We simplify expressions by **combining like terms** using the distributive property, along with the commutative and associative properties. An algebraic expression has been **simplified completely** when all like terms have been combined. Many times the distributive property is used to eliminate grouping symbols *and* combine like terms within the same expression.

EXAMPLE 9 Simplify the expression completely: $7(2p^2 + 1) - (p^2 + 3)$.

Solution:

$7(2p^2 + 1) - (p^2 + 3)$	original expression
$= 14p^2 + 7 - 1p^2 - 3$	distributive property
$= (14p^2 - 1p^2) + (7 - 3)$	commutative and associative properties
$= (14 - 1)p^2 + 4$	distributive property
$= 13p^2 + 4$	result

NOW TRY EXERCISES 71 THROUGH 88

The steps for simplifying an algebraic expression are summarized here:

> **TO SIMPLIFY AN EXPRESSION**
> 1. Eliminate parentheses by applying the distributive property (mentally change to algebraic addition if you find it helpful).
> 2. Use the commutative and associative properties to group like terms.
> 3. Simplify using the distributive property to combine like terms.

As you practice with these ideas, many of the steps will become more automatic. At some point, the distributive property, the commutative and associative properties, as well as the use of algebraic addition will be performed mentally.

R.2 EXERCISES

CONCEPTS AND VOCABULARY

Fill in each blank with the appropriate word or phrase. Carefully reread the section, if necessary.

1. A term consisting of a single number is called a(n) constant term.
2. A term containing a variable is called a(n) variable term.
3. The constant factor in a variable term is called the coefficient.
4. When $3 \cdot 14 \cdot \frac{2}{3}$ is written as $3 \cdot \frac{2}{3} \cdot 14$, the commutative property has been used.
5. Discuss/explain why the additive inverse of -5 is 5, while the multiplicative inverse of -5 is $-\frac{1}{5}$. $-5 + 5 = 0$, $-5 \cdot (-\frac{1}{5}) = 1$
6. Discuss/explain how we can rewrite the sum $3x + 6y$ as a product, and the product $2(x + 7)$ as a sum. $3(x + 2y)$, $2x + 14$

DEVELOPING YOUR SKILLS

Identify the number of terms in each expression and the coefficient of each term.

7. $3x - 5y$ two; 3 and -5
8. $-2a - 3b$ two; -2 and -3
9. $2x + \dfrac{x + 3}{4}$ two; 2 and $\frac{1}{4}$
10. $\dfrac{n - 5}{3} + 7n$ two; $\frac{1}{3}$ and 7
11. $-2x^2 + x - 5$ three; -2, 1, and -5
12. $3n^2 + n - 7$ three; 3, 1, and -7
13. $-(x + 5)$ one; -1
14. $-(n - 3)$ one; -1

Translate each phrase into an algebraic expression.

15. seven fewer than a number $n - 7$
16. x decreased by six $x - 6$
17. the sum of a number and four $n + 4$
18. a number increased by nine $n + 9$
19. the difference between a number and five is squared $(n - 5)^2$
20. the sum of a number and two is cubed $(n + 2)^3$
21. thirteen less than twice a number $2n - 13$
22. five less than double a number $2n - 5$
23. a number squared plus the number doubled $n^2 + 2n$
24. a number cubed less the number tripled $n^3 - 3n$
25. five fewer than two-thirds of a number $\frac{2}{3}n - 5$
26. fourteen more than one-half of a number $14 + \frac{1}{2}n$
27. three times the sum of a number and five, decreased by seven $3(n + 5) - 7$
28. five times the difference of a number and two, increased by six $5(n - 2) + 6$

29. Let w represent the width. Then $2w$ represents twice the width and $2w - 3$ represents three meters less than twice the width.
30. Let b represent the base. Then $3b$ represents three times the base and $3b - 6$ represents six centimeters less than three times the base.
31. Let b represent the speed of the bus. Then $b + 15$ represents 15 mph more than the speed of the bus.
32. Let t represent the time for Remus to finish the race. Then $t + 3$ represents three minutes more time than Remus.

Create a mathematical model using descriptive variables.

29. The length of the rectangle is three meters less than twice the width.

30. The height of the triangle is six centimeters less than three times the base.

31. The speed of the car was fifteen miles per hour more than the speed of the bus.

32. It took Romulus three minutes more time than Remus to finish the race.

Evaluate each algebraic expression given $x = 2$ and $y = -3$.

33. $4x - 2y$ 14
34. $5x - 3y$ 19
35. $-2x^2 + 3y^2$ 19
36. $-5x^2 + 4y^2$ 16
37. $2y^2 + 5y - 3$ 0
38. $3x^2 + 2x - 5$ 11
39. $-2(3y + 1)$ 16
40. $-3(2y + 5)$ 3
41. $3x^2y$ -36
42. $6xy^2$ 108
43. $(-3x)^2 - 4xy - y^2$ 51
44. $(-2x)^2 - 5xy - y^2$ 37
45. $\frac{1}{2}x - \frac{1}{3}y$ 2
46. $\frac{2}{3}x - \frac{1}{2}y$ $\frac{17}{6}$
47. $(3x - 2y)^2$ 144
48. $(2x - 3y)^2$ 169
49. $\dfrac{-12y + 5}{-3x + 1}$ $-\frac{41}{5}$
50. $\dfrac{12x + (-3)}{-3y + 1}$ $\frac{21}{10}$
51. $\sqrt{-12y} \cdot 4$ 24
52. $7 \cdot \sqrt{-27y}$ 63

Evaluate each expression for integers from -3 to 3 inclusive. What input(s) give an output of zero?

53. $x^2 - 3x - 4$
54. $x^2 - 2x - 3$
55. $-3(1 - x) - 6$
56. $5(3 - x) - 10$
57. $x^3 - 6x + 4$
58. $x^3 + 5x + 18$

53.

x	Output
−3	14
−2	6
−1	0
0	−4
1	−6
2	−6
3	−4

−1 has an output of 0.

54.

x	Output
−3	12
−2	5
−1	0
0	−3
1	−4
2	−3
3	0

−1 and 3 have outputs of 0.

55.

x	Output
−3	−18
−2	−15
−1	−12
0	−9
1	−6
2	−3
3	0

3 has an output of 0.

56.

x	Output
−3	20
−2	15
−1	10
0	5
1	0
2	−5
3	−10

1 has an output of 0.

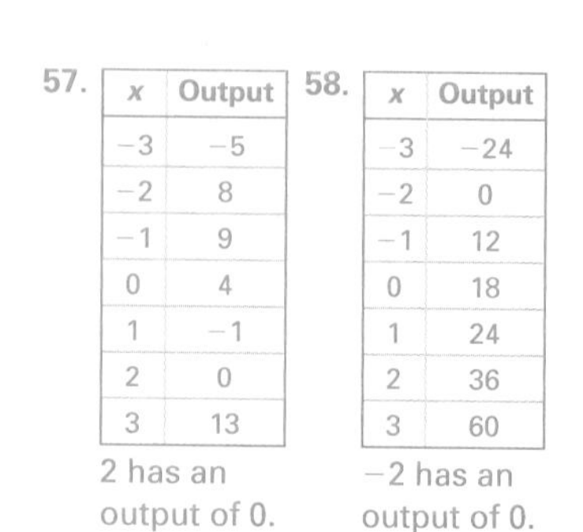

57.

x	Output
−3	−5
−2	8
−1	9
0	4
1	−1
2	0
3	13

2 has an output of 0.

58.

x	Output
−3	−24
−2	0
−1	12
0	18
1	24
2	36
3	60

−2 has an output of 0.

Rewrite each expression using the given property and simplify if possible.

59. Commutative property of addition
a. $-5 + 7$ $7 + (-5) = 2$
b. $-2 + n$ $n + (-2)$
c. $-4.2 + a + 13.6$ $a + (-4.2) + 13.6 = a + 9.4$
d. $7 + x - 7$ $x + 7 - 7 = x$

60. Associative property of multiplication
a. $2 \cdot (3 \cdot 6)$ $(2 \cdot 3) \cdot 6 = 36$
b. $(3a \cdot 4) \cdot b$ $3a \cdot (4 \cdot b) = 12ab$
c. $-1.5 \cdot (6 \cdot a)$ $(-1.5 \cdot 6) \cdot a = -9a$
d. $-6 \cdot (-\frac{5}{6} \cdot x)$ $(-6 \cdot -\frac{5}{6}) \cdot x = 5x$

Replace the box so that a true statement results.

61. a. $x + (-3.2) + \square = x$ 3.2
b. $n - \frac{5}{6} + \boxed{\frac{5}{6}} = n$

62. a. $\boxed{\frac{3}{2}} \cdot \frac{2}{3}x = 1x$
b. $\boxed{\frac{-3}{1}} \cdot \dfrac{n}{-3} = 1n$

Simplify by removing all grouping symbols and combining like terms.

63. $-5(x - 2.6)$ $-5x + 13$
64. $-12(v - 3.2)$ $-12v + 38.4$
65. $\frac{2}{3}(-\frac{1}{5}p + 9)$ $-\frac{2}{15}p + 6$
66. $\frac{5}{6}(-\frac{2}{15}q + 24)$ $-\frac{1}{9}q + 20$
67. $3a + (-5a)$ $-2a$
68. $13m + (-5m)$ $8m$
69. $\frac{2}{3}x + \frac{3}{4}x$ $\frac{17}{12}x$
70. $\frac{5}{12}y - \frac{3}{8}y$ $\frac{1}{24}y$
71. $3(a^2 + 3a) - (5a^2 + 7a)$ $-2a^2 + 2a$
72. $2(b^2 + 5b) - (6b^2 + 9b)$ $-4b^2 + b$
73. $x^2 - (3x - 5x^2)$ $6x^2 - 3x$
74. $n^2 - (5n - 4n^2)$ $5n^2 - 5n$
75. $(3a + 2b - 5c) - (a - b - 7c)$ $2a + 3b + 2c$
76. $(x - 4y + 8z) - (8x - 5y - 2z)$ $-7x + y + 10z$
77. $\frac{3}{5}(5n - 4) + \frac{5}{8}(n + 16)$ $\frac{29}{8}n + \frac{38}{5}$
78. $\frac{2}{3}(2x - 9) + \frac{3}{4}(x + 12)$ $\frac{25}{12}x + 3$
79. $(3a^2 - 5a + 7) + 2(2a^2 - 4a - 6)$ $7a^2 - 13a - 5$
80. $2(3m^2 + 2m - 7) - (m^2 - 5m + 4)$ $5m^2 + 9m - 18$

Simplify by combining like terms.

81. $-4b + 7b - 9b$ $-6b$
82. $6a - 5a + 3a$ $4a$
83. $13g - 4h + 4g + 13h$ $17g + 9h$
84. $-3m + 5n - 8m - 2n$ $-11m + 3n$
85. $5x + 12x^2 - 8x - 3x^2$ $9x^2 - 3x$
86. $3g^2 + 5g - 10g^2 - 5g$ $-7g^2$
87. $6.3y - 11.9x - 7.2y + 0.5x$ $-0.9y - 11.4x$
88. $0.25x + 3.2y - 1.75x - 0.5y$ $-1.5x + 2.7y$

WORKING WITH FORMULAS

89. Electrical resistance: $R = \dfrac{kL}{d^2}$

The electrical resistance in a wire depends on the length and diameter of the wire. This resistance can be modeled by the formula shown, where R is the resistance in ohms, L is the length in feet, and d is the diameter of the wire in inches. Find the resistance if $k = 0.000025$, $d = 0.015$ in., and $L = 90$ ft. 10 ohms

90. Volume and pressure: $P = \dfrac{k}{V}$

If temperature remains constant, the pressure of a gas held in a closed container is related to the volume of gas by the formula shown, where P is the pressure in pounds per square inch, V is the volume of gas in cubic inches, and k is a constant that depends on given conditions. Find the pressure exerted by the gas if $k = 440{,}310$ and $V = 22{,}580\ \text{in}^3$. 19.5 psi

APPLICATIONS

Create the indicated algebraic expression. Use descriptive variables.

91. Cruising speed: A turboprop airliner has a cruising speed that is one-half the cruising speed of a 767 jet aircraft. Express the speed of the turboprop in terms of the speed of the jet. $\frac{1}{2}j$

92. Softball toss: Macklyn can throw a softball two-thirds as far as her father can. Express the distance that Macklyn can throw a softball in terms of the distance her father can throw. $\frac{2}{3}d$

93. Dimensions of a lawn: The length of a rectangular lawn is 3 ft more than twice the width of the lawn. Express the length of the lawn in terms of the width. $2w + 3$

94. Pitch of a roof: To obtain the proper pitch, the crossbeam for a roof truss must be 2 ft less than three-halves of the rafter. Express the length of the cross beam in terms of the rafter. $\frac{3}{2}r - 2$

95. Postage costs: In 2004, a first class stamp cost 22¢ more than it did in 1978. Express the cost of a 2004 stamp in terms of the 1978 cost. If a stamp cost 15¢ in 1978, what was the cost in 2004? $c + 22$; 37¢

96. Minimum wage: In 2004, the federal minimum wage was \$2.85 per hour more than it was in 1976. Express the 2004 wage in terms of the 1976 wage. If the hourly wage in 1976 was \$2.30, what was it in 2004? $w + 2.85$; \$5.15

97. Repair costs: The TV repairman charges a flat fee of \$43.50 to come to your house and \$25 per hour for labor. Express the cost of repairing a TV in terms of the time it takes to repair it. If the repair took 1.5 hr, what was the total cost? $25t + 43.50$; \$81

98. Repair costs: At the local car dealership, shop charges are \$79.50 to diagnose the problem and \$85 per shop hour for labor. Express the cost of a repair in terms of the labor involved. If a repair takes 3.5 hr, how much will it cost? $85t + 79.50$; \$377.00

EXTENDING THE CONCEPT

99. If C must be a positive odd integer and D must be a negative even integer, then $C^2 + D^2$ must be a:

a. positive odd integer. **b.** positive even integer. **c.** negative odd integer.

d. negative even integer. **e.** Cannot be determined. a. positive odd integer

100. Historically, several attempts have been made to create metric time using factors of 10, but our current system won out. If 1 day was 10 metric hours, 1 metric hour was 10 metric minutes, and 1 metric minute was 10 metric seconds, what time would it really be if a metric clock read 4:35 A.M.? 10:26:24 A.M.

R.3 Exponents, Polynomials, and Operations on Polynomials

LEARNING OBJECTIVES

In Section R.3 you will review how to:

A. Apply properties of exponents

B. Perform operations in scientific notation

C. Identify and classify polynomial expressions

D. Add and subtract polynomials

E. Compute the product of two polynomials using F-O-I-L

F. Compute special products: binomial conjugates and binomial squares

INTRODUCTION

In this section we review basic exponential properties and operations on polynomials. Although there are five to eight properties (depending on how you count them), all can be traced back to the basic definition involving repeated multiplication.

POINT OF INTEREST

The triangle of numbers shown in Figure R.3 is known as **Pascal's triangle.** Each entry *within* the triangle is found by adding the two digits that are diagonally above it. Pascal's triangle has proven to be very useful, and entertaining as well—as it contains many unique patterns and relationships. One such pattern involves powers of 2. If you add the entries in each row, the result is always the next power of 2: $1 = 2^0$, $1 + 1 = 2^1$, $1 + 2 + 1 = 2^2$, $1 + 3 + 3 + 1 = 2^3$, and so on.

Figure R.3

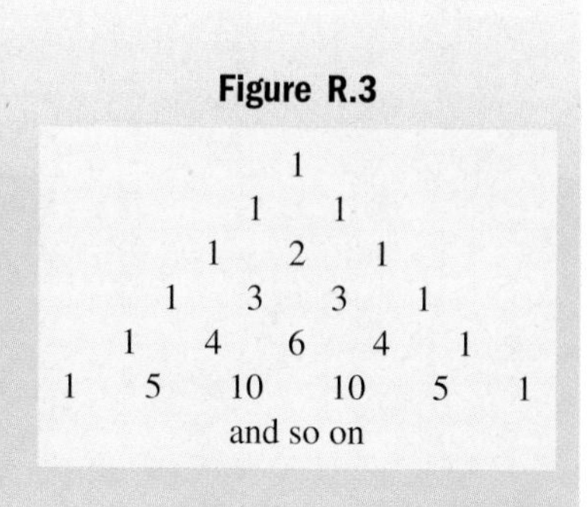

A. The Properties of Exponents

The expression b^3 indicates that b is used as a factor three times: $b^3 = b \cdot b \cdot b$. As noted in Section R.1, the exponent tells how many times the base occurs as a factor, and we say b^3 is written in *exponential form*. In some cases, we may refer to b^3 as an **exponential term.**

EXPONENTIAL NOTATION

An *exponent* tells us how many times the base b is used as a factor.

$$b^n = \underbrace{b \cdot b \cdot b \cdot \ldots \cdot b}_{n \text{ times}} \quad \text{and} \quad \underbrace{b \cdot b \cdot b \cdot \ldots \cdot b}_{n \text{ times}} = b^n$$

The Product and Power Properties

There are two properties that follow immediately from the definition of an exponent. When b^3 is multiplied by b^2, we have an uninterrupted string of five factors: $b^3 \cdot b^2 = (b \cdot b \cdot b) \cdot (b \cdot b)$, which can easily be written as b^5. This is an example of the **product property of exponents.**

PRODUCT PROPERTY OF EXPONENTS

For any base b and positive integers m and n:

$$b^m \cdot b^n = b^{m+n}$$

In words, the property says, *to multiply exponential terms with the* **same base,** *keep the common base and add the exponents.* A special application of the product property uses repeated factors of the *same* exponential term, as in $(x^2)^3$. Using the product property, we have $(x^2)(x^2)(x^2) = x^6$. Notice the same result can be found more quickly by multiplying the inner exponent by the outer exponent: $(x^2)^3 = x^{2 \cdot 3} = x^6$. We can generalize this idea and state the **power property of exponents,** also called the **power to a power property.** In words the property says, *to raise an exponential expression to a power, keep the same base and multiply the exponents.*

POWER PROPERTY OF EXPONENTS
For any base b and positive integers m and n:

$$(b^m)^n = b^{m \cdot n}$$

EXAMPLE 1 Multiply the exponential terms: (a) $-4x^3 \cdot \frac{1}{2}x^2$ and (b) $(p^3)^2 \cdot (p^4)^2$.

Solution:

a. $-4x^3 \cdot \frac{1}{2}x^2 = (-4 \cdot \frac{1}{2})(x^3 \cdot x^2)$ commutative and associative properties
$= (-2)(x^{3+2})$ product property; simplify
$= -2x^5$ result

b. $(p^3)^2 \cdot (p^4)^2 = p^6 \cdot p^8$ power property
$= p^{6+8}$ product property
$= p^{14}$ result

NOW TRY EXERCISES 7 THROUGH 12

The power property can easily be extended to include more than one factor within the parentheses. This application of the power property is sometimes called the **product to a power property.** We can also raise a quotient of exponential terms to a power. The result is called the **quotient to a power property,** and can be extended to include any number of factors. In words the properties say, to raise a product or quotient of exponential expressions to a power, *multiply every exponent inside* the parentheses *by the exponent outside* the parentheses.

PRODUCT TO A POWER PROPERTY
For any bases a and b, and positive integers m, n, and p:

$$(a^m b^n)^p = a^{mp} \cdot b^{np}$$

QUOTIENT TO A POWER PROPERTY
For any bases a and b, and positive integers m, n, and p:

$$\left(\frac{a^m}{b^n}\right)^p = \frac{a^{mp}}{b^{np}}$$

WORTHY OF NOTE
Regarding Examples 2(a) and 2(b), note the difference between the expressions $(-3a)^2 = (-3 \cdot a)^2$ and $-3a^2 = -3 \cdot a^2$. In the first, the exponent acts on both the negative 3 *and* the a; in the second, the exponent acts on only the a.

EXAMPLE 2 Simplify using the power property (if possible): (a) $(-3a)^2$, (b) $-3a^2$, and (c) $\left(\frac{-5a^3}{2b}\right)^2$.

Solution:

a. $(-3a)^2 = (-3)^2 \cdot (a^1)^2$
$= 9a^2$

b. $-3a^2 = -3 \cdot a^2$
$= -3a^2$

c. $\left(\frac{-5a^3}{2b}\right)^2 = \frac{(-5)^2(a^3)^2}{(2b)^2}$
$= \frac{25a^6}{4b^2}$

NOW TRY EXERCISES 13 THROUGH 24

Applications of exponents sometimes involve linking one exponential expression with another using a substitution. The new expression is then simplified using exponential properties.

EXAMPLE 3 The formula for the volume of a cube is $V = S^3$, where S is the length of one edge. If the length of each edge is $2x^2$: (a) find a formula for volume in terms of x and (b) find the volume if $x = 2$.

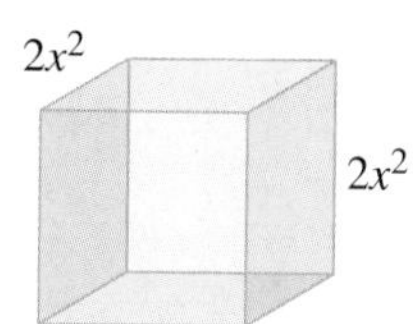

Solution:

a. $V = S^3$ ($S = 2x^2$)

$= (2x^2)^3$

$= 8x^6$

b. For $V = 8x^6$,

$V = 8(2)^6$ substitute 2 for x

$= 8 \cdot 64$ or 512 $(2)^6 = 64$

The volume of the cube would be 512 units^3.

NOW TRY EXERCISES 25 AND 26

The Quotient Property of Exponents

By combining exponential notation and the property $\frac{n}{n} = 1$ for $n \neq 0$, we note a pattern that helps to simplify a quotient of exponential terms. For $\frac{a^4}{a^2} = \frac{\cancel{a} \cdot \cancel{a} \cdot a \cdot a}{\cancel{a} \cdot \cancel{a}}$ or a^2, the exponent of the final result appears to be the *difference between the exponent in the numerator and the exponent in the denominator*. This seems reasonable since the subtraction indicates a removal of the factors that reduce to 1. Regardless of how many factors are used, we can generalize the idea and state the **quotient property of exponents.** In words, the property says, to divide two exponential expressions with the same base, *keep the common base and subtract the exponent of the denominator* from *the exponent of the numerator.*

> **QUOTIENT PROPERTY OF EXPONENTS**
>
> For any base b and integer exponents m and n: $\frac{b^m}{b^n} = b^{m-n}$, $b \neq 0$

Zero and Negative Numbers as Exponents

Considering hat $\frac{a^3}{a^3} = 1$ by division, and $\frac{a^3}{a^3} = a^{3-3} = a^0$ using the quotient property, we conclud that $a^0 = 1$ as long as $a \neq 0$. We can also generalize this observation and state the meaning of zero as an exponent. In words the property says, *any nonzero quantity raised an exponent of zero is equal to 1.*

> **ZERO EXPONENT PROPERTY**
>
> For any base b: $b^0 = 1$, if $b \neq 0$

If the exponent of the denominator is *greater* than the exponent in the numerator, the quotient property yields a negative exponent: $\frac{a^2}{a^5} = a^{2-5} = a^{-3}$. To help understand what a negative exponent *means*, we'll look at the expanded form of the expression: $\frac{a^2}{a^5} = \frac{\cancel{a} \cdot \cancel{a}^1}{\cancel{a} \cdot \cancel{a} \cdot a \cdot a \cdot a} = \frac{1}{a^3}$. A negative exponent can literally be interpreted as "write the factors as a reciprocal." A good way to remember this is:

2^{-3} (three factors of 2, written as a reciprocal) $\frac{2^{-3}}{1} = \frac{1}{2^3} = \frac{1}{8}$

Since the results would be similar regardless of what base is used, we can generalize this idea and state the **property of negative exponents.**

PROPERTY OF NEGATIVE EXPONENTS

For any base $b \neq 0$ and natural number n:

$$\frac{b^{-n}}{1} = \frac{1}{b^n} \qquad \frac{1}{b^{-n}} = \frac{b^n}{1} \qquad \left(\frac{a}{b}\right)^{-n} = \left(\frac{b}{a}\right)^n$$

EXAMPLE 4 Simplify using exponential properties. Answer using positive exponents only.

a. $\left(\frac{2a^3}{b^2}\right)^{-2}$ **b.** $(3hk^{-2})^3(6h^{-2}k^{-3})^{-2}$

c. $(3x^0) + 3x^0 + 3^{-2}$ **d.** $\frac{(-2m^2n^3)^5}{(4mn^2)^3}$

Solution:

a. $\left(\frac{2a^3}{b^2}\right)^{-2} = \left(\frac{b^2}{2a^3}\right)^2$

$= \frac{(b^2)^2}{2^2(a^3)^2}$

$= \frac{b^4}{4a^6}$

b. $(3hk^{-2})^3(6h^{-2}k^{-3})^{-2} = (3^3h^3k^{-6})(6^{-2}h^4k^6)$

$= 3^3 \cdot 6^{-2} \cdot h^{3+4} \cdot k^{-6+6}$

$= \frac{27h^7k^0}{36}$

$= \frac{3h^7}{4}$

c. $(3x)^0 + 3x^0 + 3^{-2} = 1 + 3(1) + \frac{1}{3^2}$

$= 4 + \frac{1}{9}$

$= 4\frac{1}{9}$

d. $\frac{(-2m^2n^3)^5}{(4mn^2)^3} = \frac{(-2)^5(m^2)^5(n^3)^5}{4^3m^3(n^2)^3}$

$= \frac{-32m^{10}n^{15}}{64m^3n^6}$

$= \frac{-1m^7n^9}{2}$ or $-\frac{m^7n^9}{2}$

NOW TRY EXERCISES 27 THROUGH 62

WORTHY OF NOTE

Notice in Example 4(c), we have $(3x)^0 = (3 \cdot x)^0 = 1$, while $3x^0 = 3 \cdot x^0 = 3(1)$. This is another example of operations and grouping symbols working together: $(3x)^0 = 1$ because any *quantity* to the zero power is 1. However, for $3x^0$ there are no grouping symbols, so the exponent 0 acts only on the x and not the 3.

B. Ordinary Notation and Scientific Notation

In many technical and scientific applications, we encounter numbers that are either extremely large or very, very small. For example, the mass of the moon is over 8 sextillion tons (8 followed by 19 zeroes), while the constant for universal gravitation contains 28 zeroes before the first nonzero digit. When computing with numbers of this size, **ordinary notation** (base-10 place values) is inconvenient. **Scientific notation** offers an efficient way to work with these numbers.

WORTHY OF NOTE

Recall that multiplying by 10's (or multiplying by 10^k, where k is positive) shifts the decimal to the right k places, making the number larger. Dividing by 10's (or multiplying by 10^k, where k is negative) shifts the decimal to the left k places, making the number smaller.

SCIENTIFIC NOTATION

A number written in scientific notation has the form

$$N \times 10^k$$

where $1 \le |N| < 10$ and k is an integer.

To convert a number from ordinary notation into scientific notation, we begin by placing the decimal point to the immediate right of the first nonzero digit (creating a number less than 10 but greater than or equal to 1) and multiplying by 10^k. Then we determine the power of 10 (the value of k) needed to ensure that the two forms are equivalent. When writing large or small numbers in scientific notation, we sometimes round the value of N to two or three decimal places.

EXAMPLE 5 Convert from ordinary to scientific notation: The weight of the moon is 80,600,000,000,000,000,000 tons.

Solution: $80{,}600{,}000{,}000{,}000{,}000{,}000 = 8.06 \times 10^k$

Place decimal to the right of first nonzero digit and multiply by 10^k. To get the decimal back to its original position would require 19 shifts to the *right,* so k must be *positive 19.*

$$80{,}600{,}000{,}000{,}000{,}000{,}000 = 8.06 \times 10^{19}$$

The weight of the moon is 8.06×10^{19} tons.

NOW TRY EXERCISES 63 AND 64

Converting a number from scientific notation to ordinary notation is simply an application of multiplication or division and powers of 10.

EXAMPLE 6 Convert to ordinary notation: The constant of gravitation is 9.11×10^{-29}.

Solution: Since the exponent is *negative 29,* shift the decimal *29 places to the left,* using placeholder zeros to maintain correct place value:

$$9.11 \times 10^{-29} = 0.000\,000\,000\,000\,000\,000\,000\,000\,000\,0911$$

NOW TRY EXERCISES 65 THROUGH 68

C. Identifying and Classifying Polynomial Expressions

A **monomial** is a term using *only whole number exponents* on variables, with no variables in the denominator. One important characteristic of a monomial is its **degree.** For a monomial in one variable, the degree is the same as the exponent *on the variable.* The degree of a monomial in two or more variables is the sum of exponents occurring on variable factors. A **polynomial** is a monomial or any sum or difference of monomial terms. For instance, $\frac{1}{2}x^2 - 5x + 6$ is a polynomial, while $3n^{-2} + 2n^1 - 7$ is not (the exponent -2 is not a whole number). Identifying polynomials is an important skill because they represent a very different kind of real-world model than nonpolynomials. In addition, there are different **families of polynomials,** with each family having different applications. We classify polynomials according to their *degree* and *number of terms*. The **degree of a polynomial** in one variable is the largest exponent occurring on any variable. A polynomial with two terms is called a **binomial** (*bi* means two) and a polynomial with three terms is called a **trinomial** (*tri* means three). There are special names for polynomials with four or more terms, but for these, we simply use the general name *polynomial.*

EXAMPLE 7 For each expression: (a) classify as a monomial, binomial, trinomial, or polynomial; (b) state the degree of the polynomial; and (c) name the coefficient of each term.

Solution:

Polynomial	Classification	Degree	Coefficients
$x^2 - 0.81$	binomial	two	$1, -0.81$
$z^3 - 3z^2 + 9z - 27$	polynomial (four terms)	three	$1, -3, 9, -27$
$\frac{-3}{4}x + 5$	binomial	one	$\frac{-3}{4}, 5$
$2x^2 + x - 3$	trinomial	two	$2, 1, -3$

NOW TRY EXERCISES 69 THROUGH 74

A polynomial expression is in **standard form** when the terms of the polynomial are written in *descending order of degree,* beginning with the highest-degree term. The coefficient of the highest-degree term is called the **lead coefficient.**

EXAMPLE 8 Write each polynomial in standard form, then identify the lead coefficient.

Solution:

Polynomial	Standard Form	Lead Coefficient
$9 - x^2$	$-x^2 + 9$	-1
$5z + 7z^2 + 3z^3 - 27$	$3z^3 + 7z^2 + 5z - 27$	3
$2 + (\frac{-3}{4})x$	$\frac{-3}{4}x + 2$	$\frac{-3}{4}$
$-3 + 2x^2 + x$	$2x^2 + x - 3$	2

NOW TRY EXERCISES 75 THROUGH 80

D. Adding and Subtracting Polynomials

Adding polynomials simply involves use of the commutative, associative, and distributive properties. At this point, the properties are usually applied mentally. As with real numbers, the subtraction of polynomials involves adding the opposite of the subtrahend using algebraic addition. For polynomials, this can be viewed as distributing a negative to the second polynomial and combining like terms.

EXAMPLE 9 Combine like terms: $(0.7n^3 + 4n^2 + 8) + (0.5n^3 - n^2 - 6n) - (3n^2 + 7n - 10)$.

Solution:

$(0.7n^3 + 4n^2 + 8) + (0.5n^3 - n^2 - 6n) - (3n^2 + 7n - 10)$ original sum

$= 0.7n^3 + 0.5n^3 + 4n^2 - 1n^2 - 3n^2 - 6n - 7n + 8 + 10$ use real number properties to collect like terms

$= 1.2n^3 - 13n + 18$ combine like terms

NOW TRY EXERCISES 81 THROUGH 86

In Section 4.1, we will review long division and synthetic division of polynomials, which uses subtraction in a vertical format (one polynomial below the other). This is still done by changing the sign of each term in the second polynomial and adding. Note the use of a placeholder zero in Example 10.

EXAMPLE 10 Compute the difference of $x^3 - 5x + 9$ and $x^3 + 3x^2 + 2x - 8$. Use a vertical format.

Solution:

$$\begin{array}{r} x^3 + \mathbf{0}x^2 - 5x + 9 \\ \underline{-(x^3 + 3x^2 + 2x - 8)} \end{array} \longrightarrow \begin{array}{r} x^3 + \mathbf{0}x^2 - 5x + 9 \\ \underline{-x^3 - 3x^2 - 2x + 8} \\ -3x^2 - 7x + 17 \end{array}$$

The difference is $-3x^2 - 7x + 17$.

NOW TRY EXERCISES 87 AND 88

E. The Product of Two Polynomials

The simplest case of polynomial multiplication is monomial $\times$ monomial as seen in Example 1. These were computed using exponential properties along with the properties of real numbers.

Monomial Times Polynomial

To compute the product of monomial $\times$ polynomial we use the distributive property.

EXAMPLE 11 Find the product: $-2a^2(a^2 - 2a + 1)$.

Solution:

$-2a^2(a^2 - 2a + 1) = -2a^2(a^2) - (-2a^2)(2a^1) + (-2a^2)(1)$ distribute

$= -2a^4 + 4a^3 - 2a^2$ simplify

NOW TRY EXERCISES 89 AND 90

Binomial Times Polynomial

For products involving binomials, we still use a version of the distributive property—this time to distribute the entire binomial to each term of the other polynomial factor.

EXAMPLE 12 Multiply as indicated: (a) $(2z + 1)(z - 2)$ and (b) $(2v - 3)(4v^2 + 6v + 9)$.

Solution:

a. $(2z + 1)(z - 2) = 2z(z - 2) + 1(z - 2)$ distribute to every term in the first binomial

$= 2z^2 - 4z + 1z - 2$ eliminate parentheses (distribute again)

$= 2z^2 - 3z - 2$ simplify

b. $(2v - 3)(4v^2 + 6v + 9) = 2v(4v^2 + 6v + 9) - 3(4v^2 + 6v + 9)$ distribute

$= 8v^3 + 12v^2 + 18v - 12v^2 - 18v - 27$ simplify

$= 8v^3 - 27$ combine like terms

NOW TRY EXERCISES 91 THROUGH 96

WORTHY OF NOTE

Sometimes we multiply polynomials in a vertical format, similar to the multiplication of whole numbers. Polynomials have a place value system that mimics that of ordinary numbers, with the "place value" of each term given by the degree of the term.

The F-O-I-L Method

By observing the product of two binomials as in Example 12(a), we note a pattern that can make the process more efficient. We illustrate here using the product $(2x - 1)(3x + 2)$.

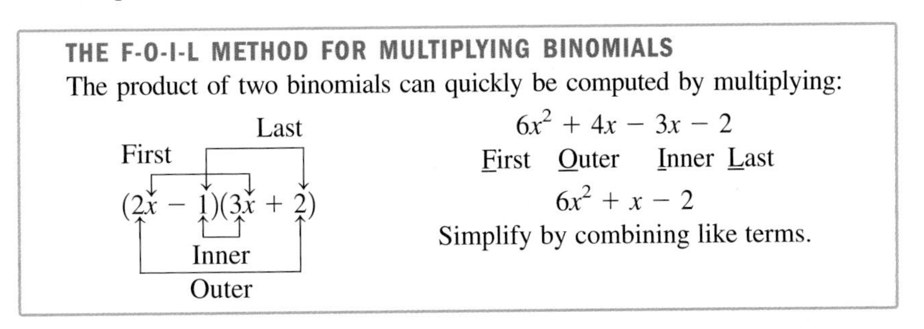

The first term of the result will always be the product of the first terms from each binomial, and the last term of the result is the product of their last terms. We also note that the middle term is found by adding the *outermost product* with the *innermost product*. The result is called the **F-O-I-L method** for multiplying binomials (first-outer-inner-last). These products occur frequently in a study of algebra. As you practice with the F-O-I-L process, much of the work can be done mentally and you can often compute the entire product without writing anything down except the answer.

EXAMPLE 13 Compute the product mentally:
(a) $(5n - 1)(n + 2)$ and (b) $(2b + 3)(5b - 6)$.

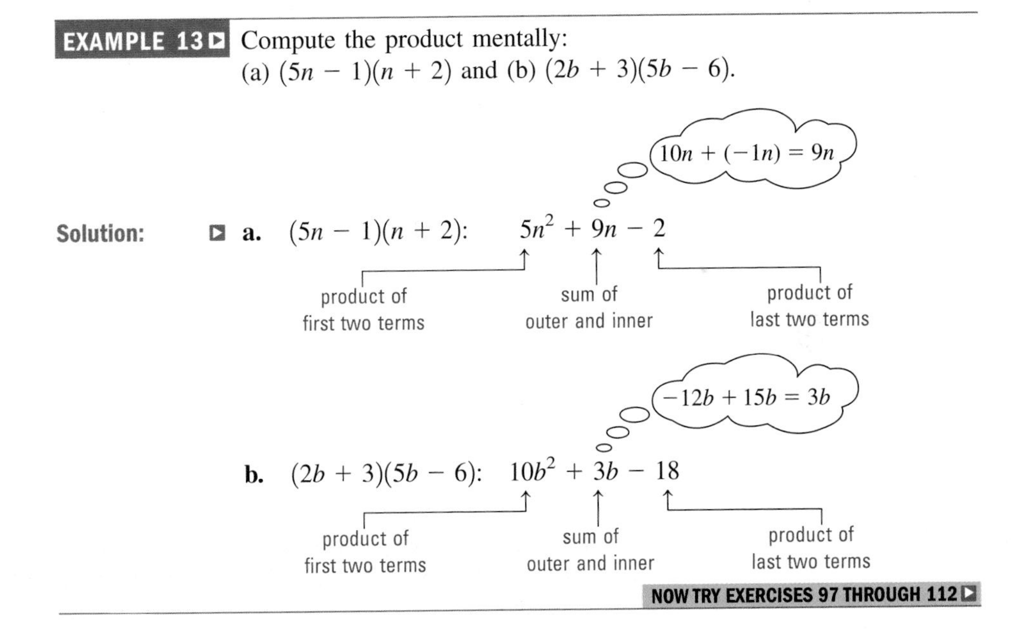

NOW TRY EXERCISES 97 THROUGH 112

F. Special Polynomial Products

Certain polynomial products are considered "special" for two reasons: (1) the product follows a predictable pattern, and (2) the result can be used to simplify expressions, graph functions, solve equations, and/or develop other skills.

Binomial Conjugates

Expressions like $x + 7$ and $x - 7$ are called **binomial conjugates.** For any given binomial, its conjugate is found by using the same two terms with the opposite sign between them. Example 14 shows that when we multiply a binomial and its conjugate, the "outers" and "inners" sum to zero and the result is a **difference of two perfect squares.**

THE PRODUCT OF A BINOMIAL AND ITS CONJUGATE

Given any expression written in the form $A + B$, the conjugate of the expression is $A - B$. The product of a binomial and its conjugate is the difference of two perfect squares:

$$(A + B)(A - B) = A^2 - B^2$$

EXAMPLE 14 Compute each product mentally: (a) $(x + 7)(x - 7)$ and (b) $(2x - 5)(2x + 5)$.

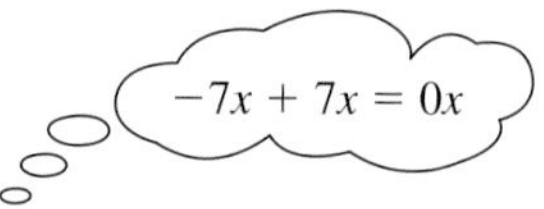

Solution: **a.** $(x + 7)(x - 7) = x^2 - 49$ difference of perfect squares $(x)^2 - (7)^2$

$10xy + (-10xy) = 0xy$

b. $(2x - 5y)(2x + 5y) = 4x^2 - 25y^2$ difference of perfect squares: $(2x)^2 - (5y)^2$

NOW TRY EXERCISES 113 THROUGH 120

Binomial Squares

Expressions like $(x + 7)^2$ are called **binomial squares** and are useful for solving many equations and sketching a number of basic graphs. Note $(x + 7)^2 = (x + 7)(x + 7) = x^2 + 14x + 49$ using the F-O-I-L process. The expression $x^2 + 14x + 49$ is called a **perfect square trinomial** because it is the result of expanding a binomial square. If we write a binomial square in the more general form $(A + B)^2 = (A + B)(A + B)$ and compute the product, we notice a pattern that helps us write the answer in expanded form more quickly.

$$\begin{aligned}(A + B)^2 &= (A + B)(A + B) && \text{repeated multiplication}\\ &= A^2 + AB + AB + B^2 && \text{F-O-I-L}\\ &= A^2 + 2AB + B^2 && \text{simplify (perfect square trinomial)}\end{aligned}$$

The first and last terms of the trinomial are perfect squares coming from the terms A and B of the binomial. Also, the middle term of the trinomial is *twice the product of these two terms*: $AB + AB = 2AB$. The F-O-I-L process clearly shows why. Since the outer and inner products are identical, we always end up with two. A similar result holds for $(A - B)^2$ and the process can be summarized for both cases using the $\pm$ symbol.

THE SQUARE OF A BINOMIAL

Given any expression that can be written in the form $(A \pm B)^2$, the expanded form will be $A^2 \pm 2AB + B^2$.

LOOKING AHEAD

Although a binomial square can always be found using repeated factors and F-O-I-L, learning to expand them using the pattern is a valuable skill. Binomial squares occur often in a study of algebra and it helps to find the expanded form quickly.

EXAMPLE 15 Find each binomial square without using F-O-I-L: (a) $(a + 9)^2$ and (b) $(3x - 5)^2$.

Solution: **a.**

$$\begin{aligned}(a + 9)^2 &= a^2 + 2(a \cdot 9) + 9^2 && \text{special product } A^2 + 2AB + B^2\\ &= a^2 + 18a + 81 && \text{simplify}\end{aligned}$$

b. $(3x - 5)^2 = (3x)^2 - 2(3x \cdot 5) + 5^2$ special product $A^2 - 2AB + B^2$

$= 9x^2 - 30x + 25$ simplify

NOW TRY EXERCISES 121 THROUGH 132

With practice, you will be able to go directly from the binomial square to the resulting trinomial.

R.3 EXERCISES

CONCEPTS AND VOCABULARY

Fill in each blank with the appropriate word or phrase. Carefully reread the section, if necessary.

1. The equation $(3x^2)^3 = 27x^6$ is an example of the __power__ property of exponents.

2. The equation $(2x^3)^{-2} = \frac{1}{4x^6}$ is an example of the property of __negative__ exponents.

3. The sum of the "outers" and "inners" for $(2x + 5)^2$ is __20x__, while the sum of outers and inners for $(2x + 5)(2x - 5)$ is __0__.

4. The expression $2x^2 - 3x - 10$ can be classified as a __trinomial__ of degree __2__, with a lead coefficient of __2__.

5. Discuss/explain why one of the following expressions can be simplified further, while the other cannot: (a) $-7n^4 + 3n^2$; (b) $-7n^4 \times 3n^2$. a. has addition of unlike terms b. is multiplication

6. Discuss/explain why the degree of $2x^2y^3$ is greater than the degree of $2x^2 + y^3$. Include additional examples for contrast and comparison. Answers will vary. Degree 5 and degree 3

DEVELOPING YOUR SKILLS

Determine each product using the product property.

7. $(6p^2q)(p^3q^3)$ $6p^5q^4$

8. $(-1.2vy^2)(6.25v^4y)$ $-7.5v^5y^3$

9. $(-3.2a^2b^2)(5a^3b)$ $-16a^5b^3$

10. $(-0.5c^4d^2)(8.4b^4c)$ $-4.2b^4c^5d^2$

11. $\frac{2}{3}yx^6 \cdot 21xy^6$ $14x^7y^7$

12. $\frac{3}{8}k^8h^3 \cdot \frac{16}{21}g^{10}h$ $\frac{2}{7}g^{10}h^4k^8$

Simplify each expression using the product to a power property.

13. $(6pq^2)^3$ $216p^3q^6$

14. $(-3p^2q)^2$ $9p^4q^2$

15. $(3.2hk^2)^3$ $32.768\,h^3k^6$

16. $(-2.5h^5k)^2$ $6.25h^{10}k^2$

17. $\left(\frac{p}{2q}\right)^2$ $\frac{p^2}{4q^2}$

18. $\left(\frac{b}{3a}\right)^3$ $\frac{b^3}{27a^3}$

19. $(-0.7c^4)^2(10c^3d^2)^2$ $49c^{14}d^4$

20. $(-2.5a^3)^2(3a^2b^2)^3$ $168.75a^{12}b^6$

21. $\left(\frac{3}{4}x^3y\right)^2$ $\frac{9}{16}x^6y^2$

22. $\left(\frac{4}{5}x^3\right)^2$ $\frac{16}{25}x^6$

23. $\left(-\frac{3}{8}x\right)^2\left(16xy^2\right)$ $\frac{9}{4}x^3y^2$

24. $\left(\frac{2}{3}m^2n\right)^2 \cdot \left(\frac{1}{2}mn^2\right)$ $\frac{2}{9}m^5n^4$

25. **Volume of a cube:** The formula for the volume of a cube is $V = S^3$, where S is the length of one edge. If the length of each edge is $3x^2$,

   a. Find a formula for volume in terms of the variable x. $V = 27x^6$

   b. Find the volume of the cube if $x = 2$. 1728 units^3

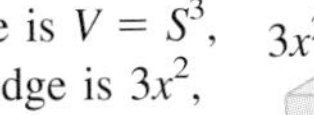

26. **Area of a circle:** The formula for the area of a circle is $A = \pi r^2$, where r is the length of the radius. If the radius is given as $5x^3$,

   a. Find a formula for area in terms of the variable x. $A = 25\pi x^6$

   b. Find the area of the circle if $x = 2$. 1600π units^2

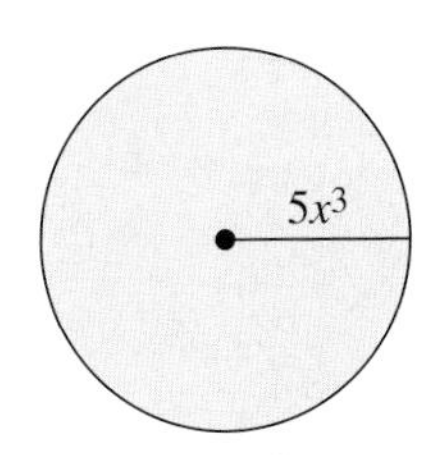

Simplify using the quotient property of exponents. Write answers using positive exponents only.

27. $\dfrac{-6w^5}{-2w^2}$ $3w^3$ **28.** $\dfrac{8z^7}{16z^5}$ $\dfrac{z^2}{2}$ **29.** $\dfrac{-12a^3b^5}{4a^2b^4}$ $-3ab$ **30.** $\dfrac{5m^3n^5}{10mn^2}$ $\dfrac{m^2n^3}{2}$

31. $\left(\frac{2}{3}\right)^{-3}$ $\frac{27}{8}$ **32.** $\left(\frac{5}{6}\right)^{-1}$ $\frac{6}{5}$ **33.** $\dfrac{2}{h^{-3}}$ $2h^3$ **34.** $\dfrac{3}{m^{-2}}$ $3m^2$

Simplify each expression using the quotient to a power property.

35. $\left(\dfrac{2p^4}{q^3}\right)^2$ $\dfrac{4p^8}{q^6}$ **36.** $\left(\dfrac{-5v^4}{7w^3}\right)^2$ $\dfrac{25v^8}{49w^6}$ **37.** $\left(\dfrac{0.2x^2}{0.3y^3}\right)^3$ $\dfrac{8x^6}{27y^9}$ **38.** $\left(\dfrac{-0.5a^3}{0.4b^2}\right)^2$ $\dfrac{25a^6}{16b^4}$

39. $\left(\dfrac{5m^2n^3}{2r^4}\right)^2$ $\dfrac{25m^4n^6}{4r^8}$ **40.** $\left(\dfrac{4p^3}{3x^2y}\right)^3$ $\dfrac{64p^9}{27x^6y^3}$ **41.** $\left(\dfrac{5p^2q^3r^4}{-2pq^2r^4}\right)^2$ $\dfrac{25p^2q^2}{4}$ **42.** $\left(\dfrac{9p^3q^2r^3}{12p^5qr^2}\right)^3$ $\dfrac{27q^3r^3}{64p^6}$

Use properties of exponents to simplify the following. Write the answer using positive exponents only.

43. $\dfrac{9p^6q^4}{-12p^4q^6}$ $\dfrac{3p^2}{-4q^2}$ **44.** $\dfrac{5m^5n^2}{10m^5n}$ $\dfrac{n}{2}$ **45.** $\dfrac{20h^{-2}}{12h^5}$ $\dfrac{5}{3h^7}$ **46.** $\dfrac{5k^3}{20k^{-2}}$ $\dfrac{k^5}{4}$

47. $\dfrac{(a^2)^3}{a^4 \cdot a^5}$ $\dfrac{1}{a^3}$ **48.** $\dfrac{(5^3)^4}{5^9}$ 125 **49.** $\left(\dfrac{a^{-3} \cdot b}{c^{-2}}\right)^{-4}$ $\dfrac{a^{12}}{b^4c^8}$ **50.** $\dfrac{(p^{-4}q^8)^2}{p^5q^{-2}}$ $\dfrac{q^{18}}{p^{13}}$

51. $\dfrac{-6(2x^{-3})^2}{10x^{-2}}$ $\dfrac{-12}{5x^4}$ **52.** $\dfrac{18n^{-3}}{-8(3n^{-2})^3}$ $\dfrac{-n^3}{12}$ **53.** $\dfrac{14a^{-3}bc^0}{-7(3a^2b^{-2}c)^3}$ $\dfrac{-2b^7}{27a^9c^3}$ **54.** $\dfrac{-3(2x^3y^{-4}z)^2}{18x^{-2}yz^0}$ $\dfrac{-2x^8z^2}{3y^9}$

55. $4^0 + 5^0$ 2 **56.** $(-3)^0 + (-7)^0$ 2 **57.** $2^{-1} + 5^{-1}$ $\frac{7}{10}$ **58.** $4^{-1} + 8^{-1}$ $\frac{3}{8}$

59. $3^0 + 3^{-1} + 3^{-2}$ $\frac{13}{9}$ **60.** $2^{-2} + 2^{-1} + 2^0$ $\frac{7}{4}$ **61.** $-5x^0 + (-5x)^0$ -4 **62.** $-2n^0 + (-2n)^0$ -1

Convert the following numbers to scientific notation.

63. In 2004, the value of all \$10 bills in circulation in the United States was approximately \$14,500,000,000. 1.45×10^{10}

Source: *2005 World Almanac and Book of Facts,* p. 118

64. In mid-2004, the U.S. Census Bureau estimated the world population at nearly 6,400,000,000 people. 6.4×10^9

Source: *2005 World Almanac and Book of Facts,* p. 848

Convert the following numbers to ordinary notation.

65. In 2004, the estimated net worth of Bill Gates, the founder of Microsoft, was 4.8×10^9 dollars. \$4,800,000,000

Source: *2005 World Almanac and Book of Facts,* p. 126

66. In 2004, President Bush proposed a U.S. federal budget of nearly 2.25×10^{12} dollars. \$2,250,000,000,000

Source: United States Government Office of Management and Budget

Compute using scientific notation. Show all work.

67. Given $\dfrac{\text{Earth to Jupiter}}{\text{speed of rocket}} = \dfrac{465{,}000{,}000\text{(miles)}}{17{,}500\text{(miles/hour)}}$, how many hours will it take for a rocket ship to get to Jupiter? How many days? Round to tenths. 26571.4 hr; 1107.1 days

68. The 2000 U.S. debt per capita is given here: $\dfrac{\text{2000 U.S. national debt}}{\text{2000 U.S. population}} = \dfrac{4{,}071{,}000{,}000{,}000}{280{,}000{,}000}$. What is the debt-per-capita ratio for 2000? Round to the nearest whole dollar. \$14,539

Identify each expression as a polynomial or nonpolynomial, if a nonpolynomial, state why; classify each as a monomial, binomial, trinomial, or none of these; and state the degree of the polynomial.

69. $-35w^3 + 2w^2 + (-12w) + 14$ polynomial, none of these, degree 3

70. $-2x^3 + \frac{2}{3}x^2 - 12x + 1.2$ polynomial, none of these, degree 3

71. $5n^{-2} + 4n + \sqrt{17}$ nonpolynomial because exponents are not whole numbers, NA, NA

72. $\dfrac{4}{r^3} + 2.7r^2 + r + 1$ nonpolynomial, exponents are not whole numbers, NA, NA

73. $p^3 - \frac{2}{5}$ polynomial, binomial, degree 3

74. $q^3 + 2q^{-2} - 5q$ nonpolynomial, exponents are not whole numbers, NA, NA

75. $-w^3 - 3w^2 + 7w + 8.2; -1$
76. $-2k^2 - k - 12; -2$
77. $c^3 + 2c^2 - 3c + 6; 1$
78. $-3v^3 + 2v^2 - 12v + 14; -3$
79. $\frac{-2}{3}x^2 + 12; \frac{-2}{3}$
80. $2n^2 + 7n + 8; 2$

81. $3p^3 - 3p^2 - 12$
82. $2q^2$
83. $7.85b^2 - 0.6b - 1.9$
84. $\frac{7}{10}n^2 + 3n + \frac{1}{4}$
85. $\frac{1}{4}x^2 - 8x + 6$
86. $-\frac{1}{9}n^2 + 6n - \frac{5}{4}$
87. $q^6 + q^5 - q^4 + 2q^3 - q^2 - 2q$
88. $5x^3 - 3x^2 + 5x$

89. $-3x^3 + 3x^2 + 18x$
90. $-2v^4 - 4v^3 + 30v^2$
91. $3r^2 - 11r + 10$
92. $5s^2 - 11s - 12$
93. $x^3 - 27$
94. $z^3 + 125$
95. $b^3 - b^2 - 34b - 56$
96. $2h^3 - 5h^2 + 11h - 8$
97. $21v^2 - 47v + 20$
98. $12w^2 + 28w - 5$
99. $9 - m^2$
100. $25 - n^2$
101. $p^2 + 1.1p - 9$
102. $q^2 - 3.7q - 5.88$
103. $x^2 + \frac{3}{4}x + \frac{1}{8}$
104. $z^2 + \frac{7}{6}z + \frac{5}{18}$
105. $m^2 - \frac{9}{16}$
106. $n^2 - \frac{4}{25}$
107. $6x^2 + 11xy - 10y^2$
108. $6a^2 + 19ab + 3b^2$
109. $12c^2 + 23cd + 5d^2$
110. $10x^2 - 9xy - 9y^2$
111. $2x^4 - x^2 - 15$
112. $6y^4 - y^2 - 2$

Write the polynomial in standard form and name the lead coefficient.

75. $7w + 8.2 - w^3 - 3w^2$
76. $-2k^2 - 12 - k$
77. $c^3 + 6 + 2c^2 - 3c$
78. $-3v^3 + 14 + 2v^2 + (-12v)$
79. $12 - \frac{2}{3}x^2$
80. $8 + 2n^2 + 7n$

Find the indicated sum or difference.

81. $(3p^3 - 4p^2 + 2p - 7) + (p^2 - 2p - 5)$
82. $(5q^2 - 3q + 4) + (-3q^2 + 3q - 4)$
83. $(5.75b^2 + 2.6b - 1.9) + (2.1b^2 - 3.2b)$
84. $(\frac{2}{5}n^2 + 5n - \frac{1}{2}) + (\frac{3}{10}n^2 - 2n + \frac{3}{4})$
85. $(\frac{3}{4}x^2 - 5x + 2) - (\frac{1}{2}x^2 + 3x - 4)$
86. $(\frac{5}{9}n^2 + 4n - \frac{1}{2}) - (\frac{2}{3}n^2 - 2n + \frac{3}{4})$
87. Subtract $q^5 + 2q^4 + q^2 + 2q$ from $q^6 + 2q^5 + q^4 + 2q^3$ using a vertical format.
88. Find $x^4 + 2x^3 + x^2 + 2x$ decreased by $x^4 - 3x^3 + 4x^2 - 3x$ using a vertical format.

Compute each product.

89. $-3x(x^2 - x - 6)$
90. $-2v^2(v^2 + 2v - 15)$
91. $(3r - 5)(r - 2)$
92. $(s - 3)(5s + 4)$
93. $(x - 3)(x^2 + 3x + 9)$
94. $(z + 5)(z^2 - 5z + 25)$
95. $(b^2 - 3b - 28)(b + 2)$
96. $(2h^2 - 3h + 8)(h - 1)$
97. $(7v - 4)(3v - 5)$
98. $(6w - 1)(2w + 5)$
99. $(3 - m)(3 + m)$
100. $(5 + n)(5 - n)$
101. $(p - 2.5)(p + 3.6)$
102. $(q - 4.9)(q + 1.2)$
103. $(x + \frac{1}{2})(x + \frac{1}{4})$
104. $(z + \frac{1}{3})(z + \frac{5}{6})$
105. $(m + \frac{3}{4})(m - \frac{3}{4})$
106. $(n - \frac{2}{5})(n + \frac{2}{5})$
107. $(3x - 2y)(2x + 5y)$
108. $(6a + b)(a + 3b)$
109. $(4c + d)(3c + 5d)$
110. $(5x + 3y)(2x - 3y)$
111. $(2x^2 + 5)(x^2 - 3)$
112. $(3y^2 - 2)(2y^2 + 1)$

For each binomial, determine its conjugate and then find the product of the binomial with its conjugate.

113. $4m - 3$ — $4m + 3; 16m^2 - 9$
114. $6n + 5$ — $6n - 5; 36n^2 - 25$
115. $7x - 10$ — $7x + 10; 49x^2 - 100$
116. $c + 3$ — $c - 3; c^2 - 9$
117. $6 + 5k$ — $6 - 5k; 36 - 25k^2$
118. $11 - 3r$ — $11 + 3r; 121 - 9r^2$
119. $ab^2 + c$ — $ab^2 - c; a^2b^4 - c^2$
120. $x^2y + z$ — $x^2y - z; x^4y^2 - z^2$

Find each binomial square.

121. $(x + 4)^2$ — $x^2 + 8x + 16$
122. $(a - 3)^2$ — $a^2 - 6a + 9$
123. $(4g + 3)^2$ — $16g^2 + 24g + 9$
124. $(5x - 3)^2$ — $25x^2 - 30x + 9$
125. $(4p - 3q)^2$ — $16p^2 - 24pq + 9q^2$
126. $(5c + 6d)^2$ — $25c^2 + 60cd + 36d^2$
127. $(2m + 3n)^2$ — $4m^2 + 12mn + 9n^2$
128. $(4a - 3b)^2$ — $16a^2 - 24ab + 9b^2$

Compute each product.

129. $(x - 3)(y + 2)$ — $xy + 2x - 3y - 6$
130. $(a + 3)(b - 5)$ — $ab - 5a + 3b - 15$
131. $(k - 5)(k + 6)(k + 2)$ — $k^3 + 3x^2 - 28k - 60$
132. $(a + 6)(a - 1)(a + 5)$ — $a^3 + 10a^2 + 19a - 30$

WORKING WITH FORMULAS

133. Medication in the bloodstream: $M = 0.5t^4 + 3t^3 - 97t^2 + 348t$

If 400 mg of a pain medication are taken orally, the number of milligrams in the bloodstream is modeled by the formula shown, where M is the number of milligrams and t is the time in hours, $0 \le t < 5$. Construct a table of values for $t = 1$ through 5, then answer the following.

a. How many milligrams have reached the bloodstream after 2 hr? 340 mg

b. How many milligrams have reached the bloodstream after 3 hr? 292.5 mg

c. Based on parts a and b, would you expect the number of milligrams in the bloodstream after 4 hr to be less or more than in part b? Why? Less, amount is decreasing

d. Approximately how many hours until the medication wears off (the number of milligrams of the drug in the bloodstream is 0)? after 5 hr

134. Amount of a mortgage payment: $M = \dfrac{A\left(\frac{r}{12}\right)\left(1 + \frac{r}{12}\right)^n}{\left(1 + \frac{r}{12}\right)^n - 1}$

The monthly mortgage payment required to pay off (or amortize) a loan is given by the formula shown, where M is the monthly payment, A is the original amount of the loan, r is the annual interest rate, and n is the term of the loan in months. Find the monthly payment required to purchase a \$198,000 home, if the interest rate is 6.5% and the home is financed over 30 yr. \$1251.49

APPLICATIONS

135. Attraction between particles: In electrical theory, the force of attraction between two particles P and Q with opposite charges is modeled by $F = \frac{kPQ}{d^2}$, where d is the distance between them and k is a constant that depends on certain conditions. This is known as Coulomb's law. Rewrite the formula using a negative exponent. $F = kPQd^{-2}$

136. Intensity of light: The intensity of illumination from a light source depends on the distance from the source according to $I = \frac{k}{d^2}$, where I is the intensity measured in footcandles, d is the distance from the source in feet, and k is a constant that depends on the conditions. Rewrite the formula using a negative exponent. $I = kd^{-2}$

137. Rewriting an expression: In advanced mathematics, negative exponents are widely used because they are easier to work with than rational expressions. Rewrite the expression $\frac{5}{x^3} + \frac{3}{x^2} + \frac{2}{x^1} + 4$ using negative exponents. $5x^{-3} + 3x^{-2} + 2x^{-1} + 4$

138. d.

t	$S(t)$
1	9
2	16
3	21
4	24
5	25
6	24
7	21
8	16
9	9
10	0

138. Swimming pool hours: A swimming pool opens at 8 A.M. and closes at 6 P.M. In summertime, the number of people in the pool at any time can be approximated by the formula $S(t) = -t^2 + 10t$, where S is the number of swimmers and t is the number of hours the pool has been open (8 A.M.: $t = 0$, 9 A.M.: $t = 1$, 10 A.M.: $t = 2$, etc).

a. How many swimmers are in the pool at 6 P.M.? Why? 0, pool closes at 6 P.M.

b. Between what times would you expect the largest number of swimmers? between noon and 2 P.M.

c. Approximately how many swimmers are in the pool at 3 P.M.? 21

d. Create a table of values for $t = 1, 2, 3, 4, \ldots$ and check your answer to part b.

139. Maximizing revenue: A sporting goods store finds that if they price their video games at \$20, they make 200 sales per day. For each decrease of \$1, 20 additional video games are sold. This means the store's revenue can be modeled by the formula $R = (20 - 1x)(200 + 20x)$. Multiply out the binomials and use a table of values to determine what price will give the most revenue. \$15

140. Maximizing revenue: Due to past experience, a jeweler knows that if they price jade rings at \$60, they will sell 120 each day. For each decrease of \$2, five additional sales will be made. This means the jeweler's revenue can be modeled by the formula $R = (60 - 2x)(120 + 5x)$. Multiply out the binomials and use a table of values to determine what price will give the most revenue. \$54

EXTENDING THE CONCEPT

141. If $(3x^2 + kx + 1) - (kx^2 + 5x - 7) + (2x^2 - 4x - k) = -x^2 - 3x + 2$, what is the value of k? 6

142. If $\left(2x + \frac{1}{2x}\right)^2 = 5$, then the expression $4x^2 + \frac{1}{4x^2}$ is equal to what number? 3

R.4 Factoring Polynomials

LEARNING OBJECTIVES

In Section R.4 you will review:

A. **Factoring out the greatest common factor**

B. **Common binomial factors and factoring by grouping**

C. **Factoring quadratic trinomials**

D. **Factoring special forms and quadratic forms**

INTRODUCTION

It is often said that knowing which tool to use is just as important as knowing how to use the tool. In this section, we review the tools needed to factor an expression, an important part of solving polynomial equations. This section will also help us decide which factoring tool is appropriate when many different factorable expressions are presented.

POINT OF INTEREST

In many cases, the process of decision making can be diagrammed in flowchart form. For example, if a car won't start, the mechanic begins the troubleshooting process with a flowchart similar to the one shown in Figure R.4, and follows the appropriate branch to the correct diagnosis. The decision process for factoring can be diagrammed in a similar way.

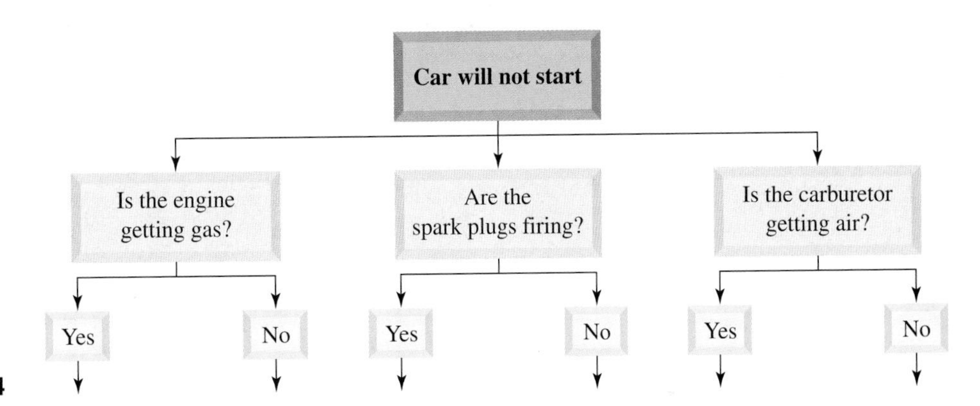

Figure R.4

A. The Greatest Common Factor and Factoring by Grouping

To **factor** an expression means to *rewrite the expression as an equivalent product*. The distributive property is an example of factoring in action. To factor $2x^2 + 6x$, we might first rewrite each term using the common factor $2x$: $2x^2 + 6x = 2x \cdot x + 2x \cdot 3$, then apply the distributive property to obtain $2x(x + 3)$. We commonly say that we have *factored out* $2x$. Recall that the **greatest common factor** (or GCF) is the largest factor common to all terms in the polynomial.

WORTHY OF NOTE

In Example 1(b), the GCF is actually one of the terms in the expression and we use the unit factor "1" to maintain an equivalent expression: $x^2 = x^2 \cdot 1$. We can also view factoring *as removing a factor of* x^2 *by division with* $\frac{x^5}{x^2} = x^3$ *and* $\frac{x^2}{x^2} = 1$

EXAMPLE 1 Factor each polynomial: (a) $12x^2 + 18xy - 30y$ and (b) $x^5 + x^2$.

Solution:

a. Since 6 is common to all three terms, factor using the distributive property.

$$12x^2 + 18xy - 30y \qquad \text{mentally: } 6 \cdot 2x^2 + 6 \cdot 3xy - 6 \cdot 5y$$
$$= 6(2x^2 + 3xy - 5y)$$

b. Since x^2 is common to both terms, factor using the distributive property.

$$x^5 + x^2 \qquad \text{mentally: } x^2 \cdot x^3 + x^2 \cdot 1 = x^5 + x^2$$
$$= x^2(x^3 + 1)$$

NOW TRY EXERCISES 7 AND 8

B. Common Binomial Factors and Factoring by Grouping

If the terms of a polynomial have a **common binomial factor,** it can also be factored out using the distributive property.

EXAMPLE 2 Factor (a) $(x + 3)x^2 + (x + 3)5$ and (b) $x^2(x - 2) - 3(x - 2)$.

Solution:

a. $(x + 3)x^2 + (x + 3)5$
$= (x + 3)(x^2 + 5)$

b. $x^2(x - 2) - 3(x - 2)$
$= (x - 2)(x^2 - 3)$

NOW TRY EXERCISES 9 AND 10

One application of removing a binomial factor involves **factoring by grouping.** Consider the expression $x^3 + 2x^2 + 3x + 6$. At first glance, it appears unfactorable since there are no factors common to all terms. But by grouping the terms (applying the associative property), we can remove a monomial factor from each subgroup, which then reveals a common binomial factor. The factoring process is completed using the ideas shown earlier.

EXAMPLE 3 Factor $3t^3 + 15t^2 - 6t - 30$.

Solution: Notice that all four terms have a common factor of 3. Begin by factoring it out.

$3t^3 + 15t^2 - 6t - 30$ original polynomial
$= 3(t^3 + 5t^2 - 2t - 10)$ factor out 3
$= 3(\underline{t^3 + 5t^2} - \underline{2t - 10})$ group remaining terms
$= 3[t^2(t + 5) - 2(t + 5)]$ factor common *monomial*
$= 3(t + 5)(t^2 - 2)$ factor common *binomial*

NOW TRY EXERCISES 11 AND 12

When asked to factor an expression, the first instinct must be to look for common factors. This will make the resulting expression easier to work with and ensure the final answer is written in **completely factored form** (meaning it cannot be factored further using integers). If a four-term polynomial cannot be factored as written, try rearranging the terms to see if you can find a combination that enables factoring by grouping.

C. Factoring Quadratic Polynomials and Other Expressions

A quadratic polynomial is one that can be written in the form $ax^2 + bx^1 + c$, where a, b, $c \in \mathbb{R}$ and $a \neq 0$. A very common form of factoring involves quadratic trinomials such as $x^2 + 7x + 10$ and $2x^2 - 13x + 15$. While we know $(x + 5)(x + 2) = x^2 + 7x + 10$ and $(2x - 3)(x - 5) = 2x^2 - 13x + 15$ using F-O-I-L, how can we factor these trinomials without seeing the original problem in advance? First, it helps to place the trinomials in two families—those with a lead coefficient of 1 and those with a lead coefficient other than 1.

$ax^2 + bx^1 + c$, where $a = 1$

When $a = 1$, the only factors of x^2 (other than 1) are $x \cdot x$ and the first term in each binomial will be x: $(x \quad)(x \quad)$. The following observation gives the insight needed to complete the factorization. Consider the product $(x + b)(x + a)$:

$$(x + b)(x + a) = x^2 + ax + bx + ab \quad \text{F-O-I-L}$$
$$= x^2 + (a + b)x + ab \quad \text{distributive property}$$

This illustration shows the last term is the product ab (the *lasts*), while the coefficient of the middle term is $a + b$ (the sum of the *outers* and *inners*). We can use these observations to quickly factor trinomials with a lead coefficient of 1. For $x^2 - 8x + 7$, we are seeking two numbers with a product of positive 7 and a sum of negative 8. The numbers are -7 and -1, so the factored form is $(x - 7)(x - 1)$. It is also helpful to note that if the constant term is positive, the binomials will have *like* signs, since only *the product of like signs is positive*. If the constant term is negative, the binomials will have *unlike* signs, since only *the product of unlike signs is negative*. This means we can use the sign of the linear (middle) term to guide our choice of factors.

EXAMPLE 4 Factor these expressions: (a) $x^2 - 3x - 10$ and (b) $x^2 - 11x + 24$.

Solution: **a.** The two numbers with a product of -10 and a sum of -3 are -5 and 2.

$$x^2 - 3x - 10 = (x - 5)(x + 2)$$

b. The two numbers with a product of 24 and a sum of -11 are -8 and -3.

$$x^2 - 11x + 24 = (x - 3)(x - 8)$$

NOW TRY EXERCISES 13 AND 14

Sometimes we encounter **prime polynomials,** or polynomials that cannot be factored. For $x^2 + 9x + 15$, the factor pairs of 15 are $1 \cdot 15$ and $3 \cdot 5$, with neither pair having a sum of $+9$. We conclude that $x^2 + 9x + 15$ is prime.

$ax^2 + bx^1 + c$, where $a \neq 1$

If the lead coefficient is not one, the possible combinations of outers and inners are more numerous and their sum will change depending on where the possible factors are placed. Note that $(2x + 3)(x + 9) = 2x^2 + 21x + 27$ and $(2x + 9)(x + 3) = 2x^2 + 15x + 27$ result in a different middle term, even though identical numbers were used.

To factor $2x^2 - 13x + 15$, note the constant term is positive so the binomials *must have like signs*. The negative linear term indicates these signs will be negative. We then list possibilities for the **first** and last terms of each binomial, then sum the outer and inner products.

Possible First and Last Terms for $2x^2$ and 15	Sum of Outers and Inners
1. $(2x - 1)(x - 15)$	$-30x - 1x = -31x$
2. $(2x - 15)(x - 1)$	$-2x - 15x = -17x$
3. $(2x - 3)(x - 5)$	$-10x - 3x = -13x$ ←
4. $(2x - 5)(x - 3)$	$-6x - 5x = -11x$

As you can see, only possibility 3 yields a linear term of $-13x$, and the correct factorization is then $(2x - 3)(x - 5)$. With practice, this **trial-and-error** process can be completed very quickly. If the constant term is negative, the number of possibilities can be reduced by finding a factor pair with a sum *or* difference equal to the *absolute value* of the linear coefficient. After finding this factor pair, we can arrange the sign of each binomial to obtain the needed coefficients.

EXAMPLE 5 Factor the trinomial $6z^2 - 11z - 35$.

Solution: Two possible first terms are: $(6z \quad)(z \quad)$ and $(3z \quad)(2z \quad)$

$(6z \quad)(z \quad)$	Outers/Inners		$(3z \quad)(2z \quad)$	Outers/Inners	
with Factors of 35	Sum	Diff	with Factors of 35	Sum	Diff
1. $(6z \quad 5)(z \quad 7)$	$47z$	$37z$	3. $(3z \quad 5)(2z \quad 7)$	$31z$	$11z$ ←
2. $(6z \quad 7)(z \quad 5)$	$37z$	$23z$	4. $(3z \quad 7)(2z \quad 5)$	$29z$	$1z$

Since possibility 3 yields the linear term of $11z$, we write the factored form as $6z^2 - 11z - 35 = (3z \quad 5)(2z \quad 7)$ and arrange the signs to obtain a middle term of $-11z$: $(3z + 5)(2z - 7)$

NOW TRY EXERCISES 15 AND 16

D. Factoring Special Forms

Each of the special products reviewed earlier can be factored using the methods shown here.

The Difference of Two Perfect Squares

Multiplying and factoring are reverse processes. Since $(x - 7)(x + 7) = x^2 - 49$, we know that $x^2 - 49 = (x - 7)(x + 7)$. In words, *the difference of two perfect squares will factor into a binomial and its conjugate.* The terms of the factored form can be found by rewriting each term in the original expression as a perfect square: $(\quad)^2$.

FACTORING THE DIFFERENCE OF TWO PERFECT SQUARES

Given any expression that can be written in the form $A^2 - B^2$, the expression can be factored as: $A^2 - B^2 = (A + B)(A - B)$.

Note: The *sum* of two perfect squares $A^2 + B^2$ *cannot be factored* using real numbers (the expression is prime).

As a reminder, always check for a common factor first and be sure to write all results in completely factored form. See Example 6(d).

EXAMPLE 6 Factor each expression completely.

a. $4w^2 - 81$ **b.** $v^2 + 49$ **c.** $-3n^2 + 48$ **d.** $z^4 - \frac{1}{81}$

Solution:

a. $4w^2 - 81 = (2w)^2 - 9^2$
$= (2w + 9)(2w - 9)$

b. $v^2 + 49$ is prime.

c. $-3n^2 + 48 = -3(n^2 - 16)$
$= -3(n - 4)(n + 4)$

d. $z^4 - \frac{1}{81} = (z^2)^2 - (\frac{1}{9})^2$
$= (z^2 - \frac{1}{9})(z^2 + \frac{1}{9})$
$= (z + \frac{1}{3})(z - \frac{1}{3})(z^2 + \frac{1}{9})$

NOW TRY EXERCISES 17 AND 18

Perfect Square Trinomials

Since $(x + 7)^2 = x^2 + 14x + 49$, we know that $x^2 + 14x + 49 = (x + 7)^2$. In words, *a perfect square trinomial will factor into a binomial square.* To use this idea effectively, it is important that we learn to *identify* perfect square trinomials. Note that the first and last terms of $x^2 + 14x + 49$ are *perfect squares* of x and 7, and the middle term is *twice the product of these two terms*: $2(7x) = 14x$. These are the characteristics of a perfect square trinomial.

> **FACTORING PERFECT SQUARE TRINOMIALS**
> Given any expression that can be written in the form $A^2 \pm 2AB + B^2$, the expression will factor as $(A \pm B)^2$.

EXAMPLE 7 Factor the trinomial $12m^3 - 12m^2 + 3m$.

Solution:

$12m^3 - 12m^2 + 3m$ — check for common factors: GCF = $3m$

$= 3m(4m^2 - 4m + 1)$ — factor out $3m$

For the remaining trinomial $4m^2 - 4m + 1 \ldots$

1. Are the first and last terms perfect squares?

$$4m^2 = (2m)^2 \text{ and } 1 = (1)^2 \checkmark \quad \text{Yes.}$$

2. Is the linear term twice the product of $2m$ and 1?

$$2 \cdot 2m \cdot 1 = 4m \checkmark \quad \text{Yes.}$$

Factor as a binomial square: $4m^2 - 4m + 1 = (2m - 1)^2$

This shows $12m^3 - 12m^2 + 3m = 3m(2m - 1)^2$.

NOW TRY EXERCISES 19 AND 20

In actual practice, the tests for a perfect square trinomial are most often performed mentally, with only the factored form being written down.

Sum or Difference of Two Perfect Cubes

Recall that the *difference* of two perfect squares is factorable, but the *sum* of two perfect squares is prime. However, *both the sum and difference of two perfect* cubes *are factorable.*

> **FACTORING THE SUM OR DIFFERENCE OF TWO PERFECT CUBES**
> 1. These will always factor into the product of a binomial and a trinomial.
> 2. The terms of the binomial are the quantities being cubed.
> 3. The terms of the trinomial are the square, product, and square of these two quantities.
> 4. The binomial takes the same sign as what you are factoring.
> 5. The factored form has exactly one negative sign (the constant term of the trinomial is always positive).
>
> $$A^3 + B^3 = (A + B)(A^2 - AB + B^2) \quad \text{and}$$
> $$A^3 - B^3 = (A - B)(A^2 + AB + B^2)$$

EXAMPLE 8 Factor the expression completely: $-5m^3n + 40n^4$.

Solution: $-5m^3n + 40n^4 = -5n(m^3 - 8n^3)$ check for common factors (GCF = $-5n$)

$= -5n[(m)^3 - (2n)^3]$ write terms as perfect cubes

Use the pattern $A^3 - B^3 = (A - B)(A^2 + AB + B^2)$, with $A \to m$ and $B \to 2n$

$$A^3 - B^3 = (A - B)(A^2 + AB + B^2) \quad \text{factoring pattern}$$
$$(m)^3 - (2n)^3 = (m - 2n)[m^2 + m(2n) + (2n)^2] \quad \text{substitute}$$
$$(m)^3 - (2n)^3 = (m - 2n)(m^2 + 2mn + 4n^2) \quad \text{simplify}$$

This shows $-5m^3n + 40n^4 = -5n(m - 2n)(m^2 + 2mn + 4n^2)$.

NOW TRY EXERCISES 21 AND 22

Using *u*-Substitution to Factor Quadratic Forms

For any quadratic expression $ax^2 + bx^1 + c$ in standard form, the degree of the leading term is twice the degree of the middle term. Generally, a trinomial expression is in **quadratic form** if it can be written as $a(\_\_)^2 + b(\_\_)^1 + c$, where the parentheses "hold" the same term. For instance, the equation $x^4 - 13x^2 + 36 = 0$ is in quadratic form since $(x^2)^2 - 13(x^2)^1 + 36 = 0$. In many cases, a **placeholder substitution** helps to factor these expressions, by transforming them into a more recognizable form. In a study of algebra, the letter "u" often plays this role. If we let $u = x^2$, then $u^2 = x^4$ (by squaring both sides), and the expression $(x^2)^2 - 13(x^2)^1 + 36$ becomes $u^2 - 13u^1 + 36$, a quadratic in u that can be factored into $(u - 9)(u - 4)$. After "unsubstituting" (replace u with x^2), we have $(x^2 - 9)(x^2 - 4)$, which gives $(x + 3)(x - 3)(x + 2)(x - 2)$. Note how the technique is used here.

EXAMPLE 9 Write in completely factored form: $(x^2 - 2x)^2 - 2(x^2 - 2x) - 3$.

Solution: Multiplying out the expression would result in a fourth degree polynomial and be very difficult to factor. Instead we note the expression is in *quadratic form*. Letting u represent $x^2 - 2x$ (the variable part of the "middle" term), the expression becomes $u^2 - 2u - 3$.

$$u^2 - 2u - 3 \quad \text{substitute } u \text{ for } x^2 - 2x$$
$$= (u - 3)(u + 1) \quad \text{factor}$$

To finish up, write the expression back in terms of x, substituting $x^2 - 2x$ for u.

$$= (x^2 - 2x - 3)(x^2 - 2x + 1) \quad \text{substitute } x^2 - 2x \text{ for } u$$

The resulting trinomials can now be factored.

$$= (x - 3)(x + 1)(x - 1)^2 \quad \text{result}$$

NOW TRY EXERCISES 23 AND 24

It is well known that information is retained longer and used more effectively when it is placed in an organized form. The process of factoring can easily be put in flowchart form (Figure R.5), similar to the one in the *Point of Interest* at the beginning of this section.

The flowchart is simply a tool that helps to organize our approach to factoring. With some practice the process tends to come more naturally than following a chart, with many of the decisions being made very quickly. There are numerous opportunities to apply these ideas in the exercise set (see Exercises 25 through 52).

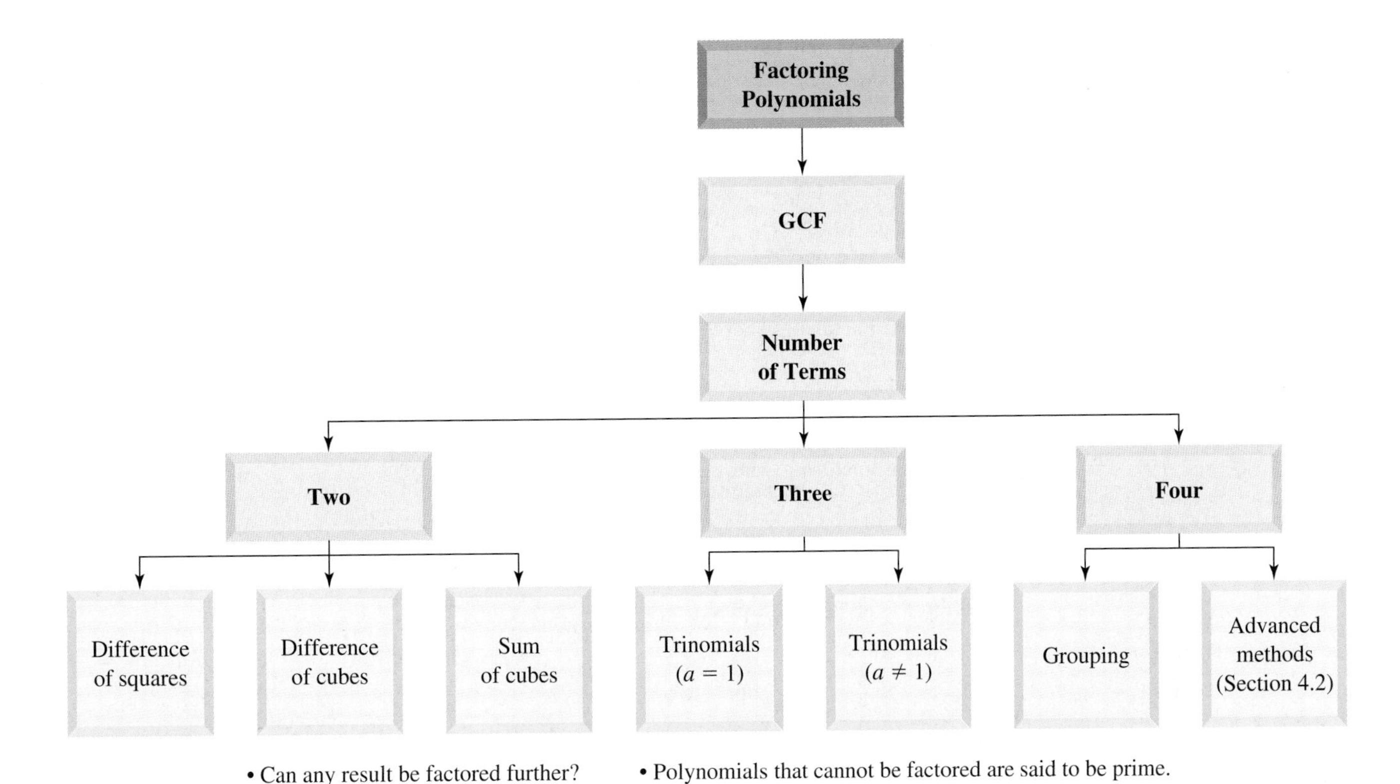

Figure R.5

R.4 EXERCISES

CONCEPTS AND VOCABULARY

Fill in each blank with the appropriate word or phrase. Carefully reread the section, if necessary.

1. To factor an expression means to rewrite the expression as an equivalent _product_.

2. If a polynomial will not factor, it is said to be a(n) _prime_ polynomial.

3. The difference of two perfect squares always factors into a(n) _binomial_ and its _conjugate_.

4. The expression $x^2 + 6x + 9$ is said to be a(n) _perfect square_ trinomial, since its factored form is a perfect (binomial) square.

5. Discuss/explain why $4x^2 - 36 = (2x - 6)(2x + 6)$ is not written in completely factored form, then rewrite it so it is factored completely. Answers will vary.

6. Discuss/explain why $a^3 + b^3$ is factorable, but $a^2 + b^2$ is not. Demonstrate by writing $x^3 + 64$ in factored form, and by exhausting all possibilities for $x^2 + 64$ to show it is prime. Answers will vary.

DEVELOPING YOUR SKILLS

Factor each expression using the method indicated.

Greatest Common Factor

7. a. $-17x^2 + 51$ $-17(x^2 - 3)$ **b.** $21b^3 - 14b^2 + 56b$ $7b(3b^2 - 2b + 8)$ **c.** $-3a^4 + 9a^2 - 6a^3$ $-3a^2(a^2 + 2a - 3)$

8. a. $-13n^2 - 52$ $-13(n^2 + 4)$ **b.** $9p^2 + 27p^3 - 18p^4$ $9p^2(1 + 3p - 2p^2)$ **c.** $-6g^5 + 12g^4 - 9g^3$ $-3g^3(2g^2 - 4g + 3)$

Common Binomial Factor

9. a. $2a(a+2)+3(a+2)$ $(a+2)(2a+3)$ **b.** $(b^2+3)3b+(b^2+3)2$ $(b^2+3)(3b+2)$ **c.** $4m(n+7)-11(n+7)$ $(n+7)(4m-11)$

10. a. $5x(x-3)-2(x-3)$ $(x-3)(5x-2)$ **b.** $(v-5)2v+(v-5)3$ $(v-5)(2v+3)$ **c.** $3p(q^2+5)+7(q^2+5)$ $(q^2+5)(3p+7)$

Grouping

11. a. $9q^3+6q^2+15q+10$ $(3q+2)(3q^2+5)$ **b.** $h^5-12h^4-3h+36$ $(h-12)(h^4-3)$ **c.** $k^5-7k^3-5k^2+35$ $(k^2-7)(k^3-5)$

12. a. $6h^3-9h^2-2h+3$ $(2h-3)(3h^2-1)$ **b.** $4k^3+6k^2-2k-3$ $(2k+3)(2k^2-1)$ **c.** $3x^2-xy-6x+2y$ $(3x-y)(x-2)$

Trinomial Factoring where $a = 1$

13. a. $b^2-5b-14$ $(b-7)(b+2)$ **b.** $a^2-4a-45$ $(a-9)(a+5)$ **c.** $n^2-9n+20$ $(n-4)(n-5)$

14. a. $m^2-13m+42$ $(m-6)(m-7)$ **b.** $x^2+13x+12$ $(x+12)(x+1)$ **c.** $v^2+10v+15$ prime

Trinomial Factoring where $a \neq 1$

15. a. $3p^2-13p-10$ $(3p+2)(p-5)$ **b.** $4q^2+7q-15$ $(4q-5)(q+3)$ **c.** $10u^2-19u-15$ $(5u+3)(2u-5)$

16. a. $6v^2+v-35$ $(2v+5)(3v-7)$ **b.** $20x^2+53x+18$ $(4x+9)(5x+2)$ **c.** $15z^2-22z-48$ $(3z-8)(5z+6)$

Difference of Perfect Squares

17. a. $4s^2-25$ $(2s+5)(2s-5)$ **b.** $9x^2-49$ $(3x+7)(3x-7)$ **c.** $50x^2-72$ $2(5x+6)(5x-6)$ **d.** $121h^2-144$ $(11h+12)(11h-12)$

18. a. $9v^2-\frac{1}{25}$ $(3v+\frac{1}{5})(3v-\frac{1}{5})$ **b.** $25w^2-\frac{1}{49}$ $(5w+\frac{1}{7})(5w-\frac{1}{7})$ **c.** v^4-1 **d.** $16z^4-81$ $(4z^2+9)(2z+3)(2z-3)$

18. c. $(v^2+1)(v+1)(v-1)$

Perfect Square Trinomials

19. a. a^2-6a+9 $(a-3)^2$ **b.** $b^2+10b+25$ $(b+5)^2$ **c.** $4m^2-20m+25$ $(2m-5)^2$ **d.** $9n^2-42n+49$ $(3n-7)^2$

20. a. $x^2+12x+36$ $(x+6)^2$ **b.** $z^2-18z+81$ $(z-9)^2$ **c.** $25p^2-60p+36$ $(5p-6)^2$ **d.** $16q^2+40q+25$ $(4q+5)^2$

Sum/Difference of Perfect Cubes

21. a. $8p^3-27$ **b.** $m^3+\frac{1}{8}$ **c.** $g^3-0.027$ **d.** $-2t^4+54t$

22. a. $27q^3-125$ **b.** $n^3+\frac{8}{27}$ **c.** $b^3-0.125$ **d.** $3r^4-24r$

21. a. $(2p-3)(4p^2+6p+9)$
b. $(m+\frac{1}{2})(m^2-\frac{1}{2}m+\frac{1}{4})$
c. $(g-0.3)(g^2+0.3g+0.09)$
d. $-2t(t-3)(t^2+3t+9)$

22. a. $(3q-5)(9q^2+15q+25)$
b. $(n+\frac{2}{3})(n^2-\frac{2}{3}n+\frac{4}{9})$
c. $(b-0.5)(b^2+0.5b+0.25)$
d. $3r(r-2)(r^2+2r+4)$

***u*-Substitution**

23. a. $9+x^4-10x^2$ $(x+3)(x-3)(x+1)(x-1)$ **b.** $13x^2+x^4+36$ $(x^2+9)(x^2+4)$ **c.** $-8+x^6-7x^3$ $(x-2)(x^2+2x+4)(x+1)(x^2-x+1)$

24. a. $-26x^3+x^6-27$ $(x-3)(x^2+3x+9)(x+1)(x^2-x+1)$ **b.** $3(n+5)^2+2n+10-21$ $(3n+8)(n+8)$ **c.** $2(z+3)^2+3z+9-54$ $(2z-3)(z+9)$

25. Completely factor each of the following (recall that "1" is its own perfect square and perfect cube).

a. n^2-1 $(n+1)(n-1)$ **b.** n^3-1 $(n-1)(n^2+n+1)$ **c.** n^3+1 $(n+1)(n^2-n+1)$ **d.** $28x^3-7x$ $7x(2x+1)(2x-1)$

26. Carefully factor each of the following trinomials, if possible. Note differences and similarities.

a. x^2-x+6 prime **b.** x^2+x-6 $(x+3)(x-2)$ **c.** x^2+x+6 prime **d.** x^2-5x+6 $(x-2)(x-3)$

e. x^2+5x-6 $(x+6)(x-1)$

Factor each expression completely, if possible. Rewrite the expression in standard form and factor out "-1" if needed. If you believe the trinomial will not factor, write "prime."

27. $a^2+7a+10$ **28.** $b^2+9b+20$ **29.** $x^2-12x+20$

30. $z^2-14z+45$ **31.** $64-9m^2$ **32.** $25-16n^2$

33. $-9r+r^2+18$ **34.** $28+s^2-11s$ **35.** $2h^2+7h+6$

36. $3k^2+10k+8$ **37.** $9k^2-24k+16$ **38.** $4p^2-20p+25$

39. $2x^2-13x-21$ **40.** $7z^2-4z-20$ **41.** $12m^2-40m+4m^3$

42. $-30n-4n^2+2n^3$ **43.** $a^2-7a-60$ **44.** $b^2-9b-36$

27. $(a+5)(a+2)$
28. $(b+5)(b+4)$
29. $(x-2)(x-10)$
30. $(z-9)(z-5)$
31. $(8+3m)(8-3m)$
32. $(5+4n)(5-4n)$
33. $(r-3)(r-6)$
34. $(s-7)(s-4)$
35. $(2h+3)(h+2)$
36. $(3k+4)(k+2)$
37. $(3k-4)^2$
38. $(2p-5)^2$
39. prime
40. $(7z+10)(z-2)$
41. $4m(m+5)(m-2)$
42. $2n(n-5)(n+3)$
43. $(a+5)(a-12)$
44. $(b-12)(b+3)$

45. $(2x-5)(4x^2+10x+25)$
46. $(3r+4)(9r^2-12r+16)$
47. $(m-4)(4m-3)$
48. prime
49. $(x-5)(x+3)(x-3)$
50. $(x+3)(x+2)(x-2)$

45. $8x^3 - 125$ **46.** $27r^3 + 64$ **47.** $4m^2 - 19m + 12$

48. $6n^2 - 23n + 11$ **49.** $x^3 - 5x^2 - 9x + 45$ **50.** $x^3 + 3x^2 - 4x - 12$

51. Match each expression with the description that fits *best.*

H **a.** prime polynomial E **b.** standard trinomial $a = 1$

C **c.** perfect square trinomial F **d.** difference of cubes

B **e.** binomial square A **f.** sum of cubes

I **g.** binomial conjugates D **h.** difference of squares

G **i.** standard trinomial $a \neq 1$

A. $x^3 + 27$ **B.** $(x + 3)^2$ **C.** $x^2 - 10x + 25$

D. $x^2 - 144$ **E.** $x^2 - 3x - 10$ **F.** $8s^3 - 125t^3$

G. $2x^2 - x - 3$ **H.** $x^2 + 9$ **I.** $(x - 7)$ and $(x + 7)$

52. Match each polynomial to its factored form. Two of them are prime.

C **a.** $4x^2 - 9$ D **b.** $4x^2 - 28x + 49$ A **c.** $x^3 - 125$

H **d.** $8x^3 + 27$ I **e.** $x^2 - 3x - 10$ E **f.** $x^2 + 3x + 10$

B **g.** $2x^2 - x - 3$ G **h.** $2x^2 + x - 3$ F **i.** $x^2 + 25$

A. $(x - 5)(x^2 + 5x + 25)$ **B.** $(2x - 3)(x + 1)$ **C.** $(2x + 3)(2x - 3)$

D. $(2x - 7)^2$ **E.** prime trinomial **F.** prime binomial

G. $(2x + 3)(x - 1)$ **H.** $(2x + 3)(4x^2 - 6x + 9)$ **I.** $(x - 5)(x + 2)$

WORKING WITH FORMULAS

Nested Factoring

53. As an alternative to evaluating polynomials by direct substitution, **nested factoring** can be used. The method has the advantage of using only products and sums—no powers. For $P = x^3 + 3x^2 + 1x + 5$, we begin by grouping all variable terms and factoring x: $P = [x^3 + 3x^2 + 1x] + 5 = x[x^2 + 3x + 1] + 5$. Then we group the inner terms with x and factor again: $P = x[x^2 + 3x + 1] + 5 = x[x(x + 3) + 1] + 5$. The expression can now be evaluated using any input and the order of operations. If $x = 2$, we quickly find that $P = 27$. Use this method to evaluate $H = x^3 + 2x^2 + 5x - 9$ for $x = -3$. $H = -33$

54. Volume of a cylindrical shell: $\pi R^2h - \pi r^2h$

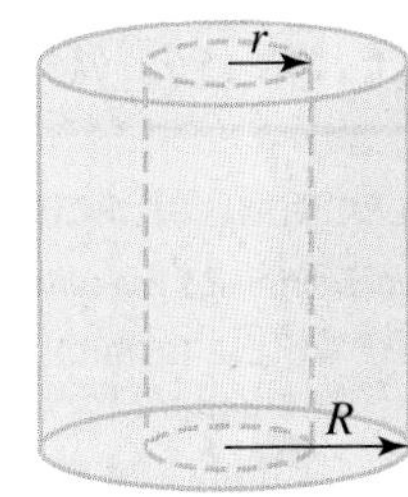

The volume of a cylindrical shell (a larger cylinder with a smaller cylinder removed) can be found using the formula shown, where R is the radius of the larger cylinder and r is the radius of the smaller. Factor out the GCF and use the result to find the volume of a shell where $R = 9$ cm, $r = 3$ cm, and $h = 10$ cm (use $\pi \approx 3.14$).

APPLICATIONS

In many cases, factoring an expression can make it easier to evaluate.

55. $S = \pi r^2(2 + h)$, $S = 75\pi \approx 235.62\text{ cm}^2$
56. $V = \frac{4}{3}\pi(R^3 - r^3) = \frac{4}{3}\pi(R - r)(R^2 + Rr + r^2)$, $V = 3.276\pi\text{ cm}^3 \approx 10.29\text{ cm}^3$

55. The surface area of a cylinder is given by the formula $S = 2\pi r^2 + \pi r^2h$. Write the right-hand side in factored form, then find the surface area if $h = 10$ cm and $r = 2.5$ cm.

56. The volume of a spherical shell (like the outer shell of a cherry cordial) is given by the formula $V = \frac{4}{3}\pi R^3 - \frac{4}{3}\pi r^3$, where R is the outer radius and r is the inner radius of the shell. Write the right-hand side in completely factored form, then find the volume of a shell where $R = 1.8$ cm and $r = 1.5$ cm.

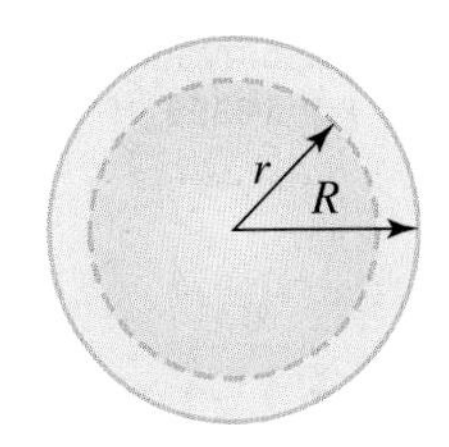

57. The volume of a rectangular box x inches in height is given by the relationship $V = x^3 + 8x^2 + 15x$. Factor the right-hand side to determine: (a) The number of inches that the width exceeds the height, (b) the number of inches the length exceeds the height, and (c) the volume given the height is 2 ft.

58. A publisher ships paperback books stacked x copies high in a box. The total number of books shipped per box is given by the relationship $B = x^3 - 13x^2 + 42x$. Factor the right-hand side to determine (a) how many more or fewer books fit the width of the box (than the height), (b) how many more or fewer books fit the length of the box (than the height), and (c) the number of books shipped per box if they are stacked 10 high in the box.

59. Due to the work of Albert Einstein and other physicists who labored on space-time relationships, it is known that the faster an object moves the shorter it appears to become. This phenomenon is modeled by the **Lorentz transformation** $L = L_0\sqrt{1 - \left(\frac{v}{c}\right)^2}$, where L_0 is the length of the object at rest, L is the relative length when the object is moving at velocity v, and c is the speed of light. Factor the radicand and use the result to determine the relative length of a 12-in. ruler if it is shot past a stationary observer at 0.75 times the speed of light ($v = 0.75c$)

60. As a fluid flows through a tube, it is flowing faster at the center of the tube than at the sides, where the tube exerts a backward drag. **Poiseuille's law** gives the velocity of the flow at any point of the cross section: $v = \frac{G}{4\eta}(R^2 - r^2)$, where R is the inner radius of the tube, r is the distance from the center of the tube to a point in the flow, G represents what is called the pressure gradient, and η is a constant that depends on the viscosity of the fluid. Factor the right-hand side and find v given $R = 0.5$ cm, $r = 0.3$ cm, $G = 15$, and $\eta = 0.25$.

57. a. 3 in.
b. 5 in.
c. $V = 2(5)(7) = 70\ \text{ft}^3$

58. $B = x(x - 6)(x - 7)$
a. 7 fewer
b. 6 fewer
c. $B = 10(4)(3) = 120$ books per box

59. $L = L_0\sqrt{\left(1 + \frac{v}{c}\right)\left(1 - \frac{v}{c}\right)}$
$L = 12\sqrt{(1 + 0.75)(1 - 0.75)}$
$= 3\sqrt{7}$ in. ≈ 7.94 in.

60. $V = \frac{G}{4\eta}(R + r)(R - r)$
$V = \frac{15}{4(0.25)}(0.8)(0.2) = 2.4$

EXTENDING THE CONCEPT

61. Factor out a constant that leaves integer coefficients for each term:

a. $\frac{1}{2}x^4 + \frac{1}{8}x^3 - \frac{3}{4}x^2 + 4$

b. $\frac{2}{3}b^5 - \frac{1}{6}b^3 + \frac{4}{9}b^2 - 1$

62. If $x = 2$ is substituted into $2x^3 + hx + 8$, the result is zero. What is the value of h?

63. Factor the expression: $192x^3 - 164x^2 - 270x$.

61. a. $\frac{1}{8}(4x^4 + x^3 - 6x^2 + 32)$
b. $\frac{1}{18}(12b^5 - 3b^3 + 8b^2 - 18)$

62. -12

63. $2x(16x - 27)(6x + 5)$

R.5 Rational Expressions

LEARNING OBJECTIVES

In Section R.5 you will learn how to:

A. Write a rational expression in simplest form

B. Multiply and divide rational expressions

C. Add and subtract rational expressions

D. Simplify compound rational expressions

E. Simplify formulas and literal equations

INTRODUCTION

A rational number is one that can be written as the quotient of two integers. Similarly, a *rational expression* is one that can be written as the quotient of two polynomials. The skills developed in a study of number fractions (how to reduce, add and subtract, and multiply and divide) will now be applied in a study of **rational expressions,** sometimes called **algebraic fractions.**

POINT OF INTEREST

Robert Boyle (1627–1691) was an Irish chemist and physicist who studied the compression and expansion of gases, and discovered the relation **pressure · volume = *k*,** a constant value. This later became known as Boyle's law. In 1661 he published his observations in a treatise called *The Sceptical Chymist.* Jacques Charles, a French chemist, published his own observations about this relationship in 1787, noting that temperature also played a part in the relationship. The result is known as Charles's law. Boyle's law and Charles's law can be combined to form the *ideal gas law:* $\frac{P_1V_1}{T_1} = \frac{P_2V_2}{T_2}$, where P_1 represents the pressure of the gas when the volume is V_1 and the temperature is T_1, and P_2 represents the pressure of the gas when the volume is V_2 and the temperature is T_2.

A. Writing a Rational Expression in Simplest Form

A rational expression is in **simplest form** or **lowest terms** when the numerator and denominator have no common factors (other than 1). After factoring the numerator and denominator, common factors are reduced using the **fundamental property of rational expressions.**

FUNDAMENTAL PROPERTY OF RATIONAL EXPRESSIONS

If P, Q, and R are polynomials, where Q and $R \neq 0$, then,

$$(1)\ \frac{P \cdot R}{Q \cdot R} = \frac{P}{Q} \quad \text{and} \quad (2)\ \frac{P}{Q} = \frac{P \cdot R}{Q \cdot R}$$

In words, the property says that (1) a rational expression can be simplified by reducing common factors in the numerator and denominator, and (2) an equivalent rational expression can be formed by multiplying numerator and denominator by the same nonzero factor.

WORTHY OF NOTE

When reducing rational expressions, only common *factors* can be reduced. It is incorrect to reduce (or divide out) *addends:* $\frac{x+1}{x-2} \neq \frac{1}{2}$ for all values of x.

EXAMPLE 1 Reduce to lowest terms: $\frac{x^2 - 1}{x^2 - 3x + 2}$.

Solution:

$$\frac{x^2 - 1}{x^2 - 3x + 2} = \frac{(x-1)(x+1)}{(x-1)(x-2)} \quad \text{factor the numerator and denominator}$$

$$= \frac{\cancel{(x-1)}(x+1)}{\cancel{(x-1)}(x-2)} \quad \text{reduce common factors}$$

$$= \frac{x+1}{x-2} \quad \text{simplified form}$$

NOW TRY EXERCISES 7 THROUGH 10

When simplifying rational expressions, we sometimes encounter factors of the form $\frac{a-b}{b-a}$. If we view a and b as two points on the number line, we note that they are the same distance apart, regardless of the order in which they are subtracted. This tells us that the numerator and denominator will have the same absolute value but be opposite in sign, giving a value of -1 (check using a few test values). This can also be seen if we factor -1 from the numerator: $\frac{a-b}{b-a} = \frac{-1\cancel{(b-a)}}{\cancel{b-a}} = -1$. Factors of the form $\frac{a-b}{b-a}$ *can* be simplified, but remember—the result is -1!

EXAMPLE 2 Reduce to lowest terms: $\frac{(6 - 2x)(x^2 + 1)}{x^2 - 9}$.

Solution:

$$\frac{(6-2x)(x^2+1)}{x^2-9} = \frac{2(3-x)(x^2+1)}{(x-3)(x+3)}$$ factor numerator and denominator

$$= \frac{(2)(-1)(x^2+1)}{x+3}$$ reduce: $\frac{(3-x)}{(x-3)} = -1$

$$= \frac{-2(x^2+1)}{x+3}$$ result

NOW TRY EXERCISES 11 THROUGH 16

B. Multiplication and Division of Rational Expressions

Operations on rational expressions use the factoring skills reviewed earlier, along with much of what we know about rational numbers.

MULTIPLYING RATIONAL EXPRESSIONS

Given that P, Q, R, and S are polynomials with Q and $S \neq 0$, then,

$$\frac{P}{Q} \cdot \frac{R}{S} = \frac{PR}{QS}$$

1. Factor all numerators and denominators completely.
2. Reduce common factors.
3. Multiply numerator $\times$ numerator and denominator $\times$ denominator.

EXAMPLE 3 Compute the product: $\frac{2a+2}{3a-3a^2} \cdot \frac{3a^2-a-2}{9a^2-4}$.

Solution:

$$\frac{2a+2}{3a-3a^2} \cdot \frac{3a^2-a-2}{9a^2-4} = \frac{2(a+1)}{3a(1-a)} \cdot \frac{(3a+2)(a-1)}{(3a-2)(3a+2)}$$ factor

$$= \frac{2(a+1)}{3a\cancel{(1-a)}_{1}} \cdot \frac{\cancel{(3a+2)}^{1}\cancel{(a-1)}^{(-1)}}{(3a-2)\cancel{(3a+2)}_{1}}$$ reduce: $\frac{a-1}{1-a} = -1$

$$= \frac{-2(a+1)}{3a(3a-2)}$$ result

The final answer can be left in factored form.

NOW TRY EXERCISES 17 THROUGH 20

Division of Rational Expressions

In the division of fractions, the quotient is found by multiplying the first expression by the *reciprocal of the second*. The quotient of two rational expressions is computed in the same way.

DIVIDING RATIONAL EXPRESSIONS

Given that P, Q, R, and S are polynomials with Q, R, and $S \neq 0$, then,

$$\frac{P}{Q} \div \frac{R}{S} = \frac{P}{Q} \cdot \frac{S}{R} = \frac{PS}{QR}$$

Invert the divisor and multiply as before.

EXAMPLE 4 Compute the quotient of $\frac{m^4n - 9m^2n}{m^2 + 4m - 5}$ and $\frac{m^4n - m^3n - 6m^2n}{m^2 - 1}$.

Solution:

$$\frac{m^4n - 9m^2n}{m^2 + 4m - 5} \div \frac{m^4n - m^3n - 6m^2n}{m^2 - 1} = \frac{m^4n - 9m^2n}{m^2 + 4m - 5} \cdot \frac{m^2 - 1}{m^4n - m^3n - 6m^2n}$$

$$= \frac{m^2n(m^2 - 9)}{m^2 + 4m - 5} \cdot \frac{m^2 - 1}{m^2n(m^2 - m - 6)} \quad \text{factor GCF}$$

$$= \frac{m^2n(m + 3)(m - 3)}{(m + 5)(m - 1)} \cdot \frac{(m - 1)(m + 1)}{m^2n(m - 3)(m + 2)} \quad \text{factor trinomials}$$

$$= \frac{\overset{1}{\cancel{m^2n}}(m + 3)\overset{1}{\cancel{(m - 3)}}}{(m + 5)\underset{1}{\cancel{(m - 1)}}} \cdot \frac{\overset{1}{\cancel{(m - 1)}}(m + 1)}{\underset{1}{\cancel{m^2n}}\underset{1}{\cancel{(m - 3)}}(m + 2)} \quad \text{reduce}$$

$$= \frac{(m + 3)(m + 1)}{(m + 5)(m + 2)} \quad \text{result}$$

NOW TRY EXERCISES 21 THROUGH 40

CAUTION

For products like $\frac{(w + 7)(w - 7)}{(w - 7)(w - 2)} \cdot \frac{(w - 2)}{(w + 7)}$, it is a common mistake to think that all factors "cancel," leaving an answer of zero. Actually, all factors *reduce to 1,* and the result is a value of 1 for all inputs where the product is defined.

C. Addition and Subtraction of Rational Expressions

Recall that the addition and subtraction of *fractions* requires finding the lowest common denominator (LCD) and building equivalent fractions. The sum or difference of the numerators is then placed over this denominator. The procedure for the addition and subtraction of *rational expressions* is very much the same. A complete review is given in Appendix II.

ADDITION AND SUBTRACTION OF RATIONAL EXPRESSIONS

1. Find the LCD of all denominators.
2. Build equivalent expressions.
3. Add or subtract numerators as indicated.
4. Write the result in lowest terms.

EXAMPLE 5 Compute as indicated: (a) $\frac{7}{10x} + \frac{3}{25x^2}$ and (b) $\frac{10x}{x^2 - 9} - \frac{5}{x - 3}$.

Solution: **a.** The LCD of $10x$ and $25x^2$ is $50x^2$. find the LCD

$$\frac{7}{10x} + \frac{3}{25x^2} = \frac{7}{10x} \cdot \frac{(5x)}{(5x)} + \frac{3}{25x^2} \cdot \frac{(2)}{(2)} \quad \text{write equivalent expressions}$$

$$= \frac{35x}{50x^2} + \frac{6}{50x^2} \quad \text{simplify}$$

$$= \frac{35x + 6}{50x^2} \quad \text{add the numerators and write the result over the LCD}$$

The result is in simplest form.

b. The LCD of $x^2 - 9$ and $x - 3$ is $(x - 3)(x + 3)$. — find the LCD

$$\frac{10x}{x^2 - 9} - \frac{5}{x - 3} = \frac{10x}{(x - 3)(x + 3)} \cdot \frac{(1)}{(1)} - \frac{5}{x - 3} \cdot \frac{(x + 3)}{(x + 3)}$$ equivalent expressions

$$= \frac{10x - 5(x + 3)}{(x - 3)(x + 3)}$$ subtract numerators, write the result over the LCD

$$= \frac{5x - 15}{(x - 3)(x + 3)}$$ distribute and simplify

$$= \frac{5\cancel{(x - 3)}^{1}}{\cancel{(x - 3)}_{1}(x + 3)} = \frac{5}{x + 3}$$ factor and reduce

NOW TRY EXERCISES 41 THROUGH 46

Here are two additional examples that have some impact on our later work.

EXAMPLE 6 Perform the operations indicated: (a) $\frac{5}{n - 2} + \frac{3}{2 - n}$ and (b) $\frac{b^2}{4a^2} - \frac{c}{a}$.

Solution: **a.** $$\frac{5}{n - 2} + \frac{3}{2 - n} = \frac{5}{n - 2} \cdot \frac{(1)}{(1)} + \frac{3}{2 - n} \cdot \frac{(-1)}{(-1)}$$ "adjust" second addend

$$= \frac{5}{n - 2} + \frac{(-3)}{n - 2}$$ simplify

$$= \frac{5 + (-3)}{n - 2} = \frac{2}{n - 2}$$ add numerators, write the result over the LCD

b. $$\frac{b^2}{4a^2} - \frac{c}{a} = \frac{b^2}{4a^2} \cdot \frac{(1)}{(1)} - \frac{c}{a} \cdot \frac{(4a)}{(4a)}$$ LCD is $4a^2$

$$= \frac{b^2}{4a^2} - \frac{4ac}{4a^2}$$ simplify

$$= \frac{b^2 - 4ac}{4a^2}$$ subtract numerators, write the result over the LCD

NOW TRY EXERCISES 47 THROUGH 62

WORTHY OF NOTE

Remember that only common *factors* can be reduced. It is incorrect to reduce (or divide out) *addends*.

For the expression $\frac{2}{n - 2}$ in Example 6(a), the 2's do not reduce or cancel out (the expression is in simplest form).

D. Simplifying Compound Rational Expressions

There are two methods used to simplify a compound fraction. Consider this expression:

$$\frac{\frac{3}{4m} - \frac{5}{6}}{\frac{5}{2m} + \frac{1}{3m^2}}$$

numerator

major fraction bar (often longer or darker)

denominator

The expression as a whole is called the **major fraction,** and any fraction occurring in a numerator or denominator is referred to as a **minor fraction.** Method I involves using the LCD to combine terms in the numerator and/or denominator. The result is then simplified further using the "invert and multiply" property from the division of fractions. Method II uses the LCD for the denominators of *all* minor fractions. This simplifies the expression in a single step, rather than two steps as before. We complete the process by factoring and removing any common factors.

SIMPLIFYING COMPOUND FRACTIONS (METHOD I)

1. Add/subtract fractions in the numerator to write as a single expression.
2. Add/subtract fractions in the denominator to write as a single expression.
3. Multiply the numerator by the reciprocal of the denominator and simplify if possible.

SIMPLIFYING COMPOUND FRACTIONS (METHOD II)

1. Find the LCD of all minor fractions in the expression.
2. Multiply all terms in the numerator and denominator by this LCD and simplify.
3. Factor the numerator and the denominator and reduce any common factors.

Method II is illustrated in Example 7.

EXAMPLE 7 Simplify the compound fraction using the LCD:

$$\frac{\frac{2}{3m}-\frac{3}{2}}{\frac{3}{4m}-\frac{1}{3m^2}}.$$

Solution: The LCD for all minor fractions is $12m^2$.

$$\frac{\frac{2}{3m}-\frac{3}{2}}{\frac{3}{4m}-\frac{1}{3m^2}} = \frac{\left(\frac{2}{3m}-\frac{3}{2}\right)\left(\frac{12m^2}{1}\right)}{\left(\frac{3}{4m}-\frac{1}{3m^2}\right)\left(\frac{12m^2}{1}\right)} \quad \text{multiply all minor fractions by } 12m^2$$

$$= \frac{8m - 18m^2}{9m - 4} \quad \text{distribute and simplify}$$

$$= \frac{2m\overset{-1}{\cancel{(4-9m)}}}{\cancel{9m-4}} = -2m \quad \text{factor and write in lowest terms}$$

NOW TRY EXERCISES 63 THROUGH 72

E. Simplifying Formulas and Literal Equations

In many fields of study, formulas and literal equations involve rational expressions and we often need to rewrite them for various reasons.

EXAMPLE 8 In an electrical circuit with two resistors in parallel, the total resistance R is related to resistors R_1 and R_2 by the formula $\frac{1}{R} = \frac{1}{R_1} + \frac{1}{R_2}$. Rewrite the right-hand side as a single term.

Solution: $\frac{1}{R} = \frac{1}{R_1} + \frac{1}{R_2}$ LCD for the right-hand side is R_1R_2

$= \frac{R_2}{R_1R_2} + \frac{R_1}{R_1R_2}$ build equivalent expressions using LCD

$= \frac{R_2 + R_1}{R_1R_2}$ write as a single expression

NOW TRY EXERCISES 75 AND 76

EXAMPLE 9 When studying rational expressions and rates of change, we encounter the expression $\frac{\frac{1}{x+h} - \frac{1}{x}}{h}$. Simplify the compound fraction.

Solution: Using Method I gives:

$$\frac{\frac{1}{x+h} - \frac{1}{x}}{h} = \frac{\frac{x}{x(x+h)} - \frac{x+h}{x(x+h)}}{h}$$ LCD for the numerator is $x(x + h)$

$$= \frac{\frac{x-(x+h)}{x(x+h)}}{h}$$ write numerator as a single expression

$$= \frac{\frac{-h}{x(x+h)}}{h}$$ simplify

$$= \frac{-h}{x(x+h)} \cdot \frac{1}{h}$$ invert and multiply

$$= \frac{-1}{x(x+h)}$$ result

NOW TRY EXERCISES 77 THROUGH 80

R.5 EXERCISES

CONCEPTS AND VOCABULARY

Fill in each blank with the appropriate word or phrase. Carefully reread the section, if necessary.

1. In simplest form, $(a - b)/(a - b)$ is equal to __1__, while $(a - b)/(b - a)$ is equal to __−1__.

2. A rational expression is in __simplest form__ when the numerator and denominator have no common factors, other than __1__.

3. A rational expression is said to be in lowest terms when the numerator and denominator have no __common factor__.

4. Since $x^2 + 9$ is prime, the expression $(x^2 + 9)/(x + 3)$ is already written in __simplest form__. (or lowest terms)

State T or F and discuss/explain your response.

5. $\frac{x}{x+3} - \frac{x+1}{x+3} = \frac{1}{x+3}$ F

6. $\frac{(x+3)(x-2)}{(x-2)(x+3)} = 0$ F

7. a. $-\frac{1}{3}$ b. $\frac{x+3}{2x(x-2)}$

8. a. $-\frac{1}{7}$ b. $\frac{x-6}{2x(x-2)}$

9. a. simplified b. $\frac{a-4}{a-7}$

10. a. $\frac{r+5}{r+3}$ b. simplified

11. a. -1 b. -1

12. a. $\frac{-1(v+4)}{v+7}$ b. $\frac{-1(u-5)}{u+5}$

13. a. $-3ab^9$ b. $\frac{x+3}{9}$ c. $-1(y+3)$ d. $\frac{-1}{m}$

14. a. $\frac{1n^3}{-2m^4}$ b. $\frac{-v+4}{5}$ c. $-1(n+2)$ d. $-w$

15. a. $\frac{2n+3}{n}$ b. $\frac{3x+5}{2x+3}$ c. $x+2$ d. $n-2$

16. a. $\frac{x+5}{x+1}$ b. $\frac{p-3}{p+2}$ c. $\frac{4y-3}{9y^2+3y+1}$ d. $\frac{x^2-3}{x+5}$

17. $\frac{(a-2)(a+1)}{(a+3)(a+2)}$

18. $\frac{b}{(b-3)(b-8)}$ 19. 1

20. $2v+5$ 21. $\frac{(p-4)^2}{p^2}$

22. $\frac{a^2+2a+4}{a^2}$ 23. $\frac{-15}{4}$

24. $\frac{-25}{7}$ 25. $\frac{3}{2}$

26. $p-6$ 27. $\frac{8(a-7)}{a-5}$

28. $m(m-4)$ 29. $\frac{y}{x}$

30. $\frac{7-a}{b-7}$ 31. $\frac{m}{m-4}$

32. $\frac{-3(x-2)}{x+3}$ 33. $\frac{y+3}{3y(y+4)}$

34. $\frac{x-2}{x-7}$ 35. $\frac{x+0.3}{x-0.2}$

36. $\frac{n+\frac{1}{5}}{n+\frac{2}{3}}$

37. $\frac{3(a^2+3a+9)}{2}$

38. $(p+1)(p-7)$

DEVELOPING YOUR SKILLS

Reduce to lowest terms.

7. a. $\frac{a-7}{-3a+21}$ b. $\frac{2x+6}{4x^2-8x}$

8. a. $\frac{x-4}{-7x+28}$ b. $\frac{3x-18}{6x^2-12x}$

9. a. $\frac{x^2-5x-14}{x^2+6x-7}$ b. $\frac{a^2+3a-28}{a^2-49}$

10. a. $\frac{r^2+3r-10}{r^2+r-6}$ b. $\frac{m^2+3m-4}{m^2-4m}$

11. a. $\frac{x-7}{7-x}$ b. $\frac{5-x}{x-5}$

12. a. $\frac{v^2-3v-28}{49-v^2}$ b. $\frac{u^2-10u+25}{25-u^2}$

13. a. $\frac{-12a^3b^5}{4a^2b^{-4}}$ b. $\frac{7x+21}{63}$ c. $\frac{y^2-9}{3-y}$ d. $\frac{m^3n-m^3}{m^4-m^4n}$

14. a. $\frac{5m^{-3}n^5}{-10mn^2}$ b. $\frac{-5v+20}{25}$ c. $\frac{n^2-4}{2-n}$ d. $\frac{w^4-w^4v}{w^3v-w^3}$

15. a. $\frac{2n^3+n^2-3n}{n^3-n^2}$ b. $\frac{6x^2+x-15}{4x^2-9}$ c. $\frac{x^3+8}{x^2-2x+4}$ d. $\frac{mn^2+n^2-4m-4}{mn+n+2m+2}$

16. a. $\frac{x^3+4x^2-5x}{x^3-x}$ b. $\frac{5p^2-14p-3}{5p^2+11p+2}$ c. $\frac{12y^2-13y+3}{27y^3-1}$ d. $\frac{ax^2-5x^2-3a+15}{ax-5x+5a-25}$

Compute as indicated. Write final results in lowest terms.

17. $\frac{a^2-4a+4}{a^2-9}\cdot\frac{a^2-2a-3}{a^2-4}$

18. $\frac{b^2+5b-24}{b^2-6b+9}\cdot\frac{b}{b^2-64}$

19. $\frac{x^2-7x-18}{x^2-6x-27}\cdot\frac{2x^2+7x+3}{2x^2+5x+2}$

20. $\frac{6v^2+23v+21}{4v^2-4v-15}\cdot\frac{4v^2-25}{3v+7}$

21. $\frac{p^3-64}{p^3-p^2}\div\frac{p^2+4p+16}{p^2-5p+4}$

22. $\frac{a^2+3a-28}{a^2+5a-14}\div\frac{a^3-4a^2}{a^3-8}$

23. $\frac{3x-9}{4x+12}\div\frac{3-x}{5x+15}$

24. $\frac{5b-10}{7b-28}\div\frac{2-b}{5b-20}$

25. $\frac{a^2+a}{a^2-3a}\cdot\frac{3a-9}{2a+2}$

26. $\frac{p^2-36}{2p}\cdot\frac{4p^2}{2p^2+12p}$

27. $\frac{8}{a^2-25}\cdot(a^2-2a-35)$

28. $(m^2-16)\cdot\frac{m^2-5m}{m^2-m-20}$

29. $\frac{xy-3x+2y-6}{x^2-3x-10}\div\frac{xy-3x}{xy-5y}$

30. $\frac{2a-ab+7b-14}{b^2-14b+49}\div\frac{ab-2a}{ab-7a}$

31. $\frac{m^2+2m-8}{m^2-2m}\div\frac{m^2-16}{m^2}$

32. $\frac{18-6x}{x^2-25}\div\frac{2x^2-18}{x^3-2x^2-25x+50}$

33. $\frac{y+3}{3y^2+9y}\cdot\frac{y^2+7y+12}{y^2-16}\div\frac{y^2+4y}{y^2-4y}$

34. $\frac{x^2+4x-5}{x^2-5x-14}\div\frac{x^2-1}{x^2-4}\cdot\frac{x+1}{x+5}$

35. $\frac{x^2-0.49}{x^2+0.5x-0.14}\div\frac{x^2-0.10x+0.21}{x^2-0.09}$

36. $\frac{n^2-\frac{4}{9}}{n^2-\frac{13}{15}n+\frac{2}{15}}\div\frac{n^2+\frac{4}{3}n+\frac{4}{9}}{n^2-\frac{1}{25}}$

37. $\frac{3a^3-24a^2-12a+96}{a^2-11a+24}\div\frac{6a^2-24}{3a^3-81}$

38. $\frac{p^3+p^2-49p-49}{p^2+6p-7}\div\frac{p^2+p+1}{p^3-1}$

39. $\dfrac{4n^2-1}{12n^2-5n-3}\cdot\dfrac{6n^2+5n+1}{2n^2+n}\cdot\dfrac{12n^2-17n+6}{6n^2-7n+2}$ $\quad\dfrac{2n+1}{n}$

40. $\left(\dfrac{4x^2-25}{x^2-11x+30}\div\dfrac{2x^2-x-15}{x^2-9x+18}\right)\dfrac{4x^2+25x-21}{12x^2-5x-3}$ $\quad\dfrac{(2x-5)(x+7)}{(x-5)(3x+1)}$

Compute as indicated. Write answers in lowest terms [recall that $a-b=-1(b-a)$].

41. $\dfrac{3}{8x^2}+\dfrac{5}{2x}$

42. $\dfrac{15}{16y}-\dfrac{7}{2y^2}$

43. $\dfrac{7}{4x^2y^3}-\dfrac{1}{8xy^4}$

44. $\dfrac{3}{6a^3b}+\dfrac{5}{9ab^3}$

45. $\dfrac{4p}{p^2-36}-\dfrac{2}{p-6}$

46. $\dfrac{3q}{q^2-49}-\dfrac{3}{2q-14}$

47. $\dfrac{m}{m^2-16}+\dfrac{4}{4-m}$

48. $\dfrac{2}{4-p^2}+\dfrac{p}{p-2}$

49. $\dfrac{2}{m-7}-5$

50. $\dfrac{4}{x-1}-9$

51. $\dfrac{y+1}{y^2+y-30}-\dfrac{2}{y+6}$

52. $\dfrac{4n}{n^2-5n}-\dfrac{3}{4n-20}$

53. $\dfrac{1}{a+4}+\dfrac{a}{a^2-a-20}$

54. $\dfrac{2x-1}{x^2+3x-4}-\dfrac{x-5}{x^2+3x-4}$

55. $\dfrac{3y-4}{y^2+2y+1}-\dfrac{2y-5}{y^2+2y+1}$

56. $\dfrac{-2}{3a+12}-\dfrac{7}{a^2+4a}$

57. $\dfrac{2}{m^2-9}+\dfrac{m-5}{m^2+6m+9}$

58. $\dfrac{m+2}{m^2-25}-\dfrac{m+6}{m^2-10m+25}$

59. $\dfrac{y+2}{5y^2+11y+2}+\dfrac{5}{y^2+y-6}$

60. $\dfrac{m-4}{3m^2-11m+6}+\dfrac{m}{2m^2-m-15}$

Write each term as a rational expression. Then compute the sum or difference indicated.

61. a. $p^{-2}-5p^{-1}$
b. $x^{-2}+2x^{-3}$

62. a. $3a^{-1}+(2a)^{-1}$
b. $2y^{-1}-(3y)^{-1}$

Simplify each compound rational expression. Use either method.

63. $\dfrac{\frac{5}{a}-\frac{1}{4}}{\frac{25}{a^2}-\frac{1}{16}}$

64. $\dfrac{\frac{8}{x^3}-\frac{1}{27}}{\frac{2}{x}-\frac{1}{3}}$

65. $\dfrac{p+\frac{1}{p-2}}{1+\frac{1}{p-2}}$

66. $\dfrac{1+\frac{3}{y-6}}{y+\frac{9}{y-6}}$

67. $\dfrac{\frac{2}{3-x}+\frac{3}{x-3}}{\frac{4}{x}+\frac{5}{x-3}}$

68. $\dfrac{\frac{1}{y-5}-\frac{2}{5-y}}{\frac{3}{y-5}-\frac{2}{y}}$

69. $\dfrac{\frac{2}{y^2-y-20}}{\frac{3}{y+4}-\frac{4}{y-5}}$

70. $\dfrac{\frac{2}{x^2-3x-10}}{\frac{6}{x+2}-\frac{4}{x-5}}$

Rewrite each expression as a compound fraction. Then simplify using either method.

71. a. $\dfrac{1+3m^{-1}}{1-3m^{-1}}$ **b.** $\dfrac{1+2x^{-2}}{1-2x^{-2}}$

72. a. $\dfrac{4-9a^{-2}}{3a^{-2}}$ **b.** $\dfrac{3+2n^{-1}}{5n^{-2}}$

41. $\dfrac{3+20x}{8x^2}$ 42. $\dfrac{15y-56}{16y^2}$
43. $\dfrac{14y-x}{8x^2y^4}$ 44. $\dfrac{9b^2+10a^2}{18a^3b^3}$
45. $\dfrac{2}{p+6}$ 46. $\dfrac{3}{2(q+7)}$
47. $\dfrac{-3m-16}{(m+4)(m-4)}$
48. $\dfrac{2-2p-p^2}{(2+p)(2-p)}$
49. $\dfrac{-5m+37}{m-7}$ 50. $\dfrac{-9x+13}{x-1}$
51. $\dfrac{-y+11}{(y+6)(y-5)}$ 52. $\dfrac{13}{4(n-5)}$
53. $\dfrac{2a-5}{(a+4)(a-5)}$ 54. $\dfrac{1}{x-1}$
55. $\dfrac{1}{y+1}$ 56. $\dfrac{-2a-21}{3a(a+4)}$
57. $\dfrac{m^2-6m+21}{(m+3)^2(m-3)}$
58. $\dfrac{-14m-40}{(m-5)^2(m+5)}$
59. $\dfrac{y^2+26y-1}{(5y+1)(y+3)(y-2)}$
60. $\dfrac{5m^2-5m-20}{(3m-2)(m-3)(2m+5)}$
61. a. $\dfrac{1}{p^2}-\dfrac{5}{p};\ \dfrac{1-5p}{p^2}$
b. $\dfrac{1}{x^2}+\dfrac{2}{x^3};\ \dfrac{x+2}{x^3}$
62. a. $\dfrac{3}{a}+\dfrac{1}{2a};\ \dfrac{7}{2a}$
b. $\dfrac{2}{y}-\dfrac{1}{3y};\ \dfrac{5}{3y}$
63. $\dfrac{4a}{20+a}$ 64. $\dfrac{36+6x+x^2}{9x^2}$
65. $p-1$ 66. $\dfrac{1}{y-3}$
67. $\dfrac{x}{9x-12}$ 68. $\dfrac{3y}{y+10}$
69. $\dfrac{2}{-y-31}$ 70. $\dfrac{1}{x-19}$
71. a. $\dfrac{1+\frac{3}{m}}{1-\frac{3}{m}};\ \dfrac{m+3}{m-3}$
b. $\dfrac{1+\frac{2}{x^2}}{1-\frac{2}{x^2}};\ \dfrac{x^2+2}{x^2-2}$
72. a. $\dfrac{4-\frac{9}{a^2}}{\frac{3}{a^2}};\ \dfrac{4a^2-9}{3}$
b. $\dfrac{3+\frac{2}{n}}{\frac{5}{n^2}};\ \dfrac{3n^2+2n}{5}$

WORKING WITH FORMULAS

73. The cost C, in millions of dollars, for a government to find and seize $P\%$ of a certain illegal drug is modeled by the rational equation $C = \frac{450P}{100 - P}(0 \leq P < 100)$. Complete the table (round to the nearest dollar) and answer the following questions.

a. What is the cost of seizing 40% of the drugs? Estimate the cost at 85%. $300 million $2550 million

b. Why does cost increase dramatically the closer you get to 100%? It would require many resources.

c. Will 100% of the drugs ever be seized? No

P	$\frac{450P}{100 - P}$
40	300
60	675
80	1800
90	4050
93	5979
95	8550
98	22050
100	ERROR

74. Rational equations are often used to model chemical concentrations in the bloodstream. The percent concentration C of a certain drug H hours after injection into muscle tissue can be modeled by $C = \frac{200H^2}{H^3 + 40}$, with $H \geq 0$. Complete the table (round to the nearest tenth of a percent) and answer the following questions.

a. What is the percent concentration of the drug 3 hr after injection? 26.9%

b. Why is the concentration virtually equal at $H = 4$ and $H = 5$? It reaches maximum concentration between these hours.

c. Why does the concentration begin to decrease? It begins to wear off.

d. How long will it take for the concentration to become less than 10%? 20 hr

H	$\frac{200H^2}{H^3 + 40}$
0	0
1	4.9
2	16.7
3	26.9
4	30.8
5	30.3
6	28.1
7	25.6

APPLICATIONS

Rewrite each expression as a single term.

75. $\frac{1}{f_1} + \frac{1}{f_2}$ $\frac{f_2 + f_1}{f_1 f_2}$

76. $\frac{1}{w} + \frac{1}{x} - \frac{1}{y}$

77. $\frac{\frac{a}{x + h} - \frac{a}{x}}{h}$ $\frac{-a}{x(x + h)}$

78. $\frac{\frac{a}{h - x} - \frac{a}{-x}}{h}$ $\frac{-a}{x(x - h)}$

79. $\frac{\frac{1}{2(x + h)^2} - \frac{1}{2x^2}}{h}$

80. $\frac{\frac{a}{(x + h)^2} - \frac{a}{x^2}}{h}$

81. When a hot new stock hits the market, its price will often rise dramatically and then taper off over time. The equation $P = \frac{50(7d^2 + 10)}{d^3 + 50}$ models the price of stock XYZ d days after it has "hit the market." Create a table of values showing the price of the stock for the first 10 days and comment on what you notice. Find the opening price of the stock—does the stock ever return to its original price?

82. The Department of Wildlife introduces 60 elk into a new game reserve. It is projected that the size of the herd will grow according to the equation $N = \frac{10(6 + 3t)}{1 + 0.05t}$, where N is the number of elk and t is the time in years. Approximate the population of elk after 14 yr. $N \approx 282$

83. The number of words per minute that a beginner can type is approximated by the equation $N = \frac{60t - 120}{t}$, where N is the number of words per minute after t weeks, $2 < t < 12$. Use

76. $\frac{xy + wy - wx}{wxy}$

79. $\frac{-(2x + h)}{2x^2(x + h)^2}$

80. $\frac{-a(2x + h)}{x^2(x + h)^2}$

81.

Day	Price
0	10
1	16.67
2	32.76
3	47.40
4	53.51
5	52.86
6	49.25
7	44.91
8	40.75
9	37.04
10	33.81

Price rises rapidly for first four days, then begins a gradual decrease. Yes, on the 35th day of trading.

a table to determine how many weeks it takes for a student to be typing an average of forty-five words per minute. $t = 8$ weeks

84. A group of students is asked to memorize 50 Russian words that are unfamiliar to them. The number N of these words that the average student remembers D days later is modeled by the equation $N = \frac{5D + 35}{D}$ $(D \geq 1)$. How many words are remembered after (a) 1 day? 40 words (b) 5 days? 12 words (c) 12 days? 8 words (d) 35 days? 6 words (e) 100 days? 5 words According to this model, is there a certain number of words that the average student never forgets? How many? yes 5 words

EXTENDING THE CONCEPT

85. One of these expressions is *not* equal to the others. Identify which and explain why.

a. $\frac{20n}{10n}$ **b.** $20 \cdot n \div 10 \cdot n$ **c.** $20n \cdot \frac{1}{10n}$ **d.** $\frac{20}{10} \cdot \frac{n}{n}$

(b); others equal 2

86. The average of A and B is x. The average of C, D, and E is y. The average of A, B, C, D, and E is

a. $\frac{3x + 2y}{5}$ **b.** $\frac{2x + 3y}{5}$ **c.** $\frac{2(x + y)}{5}$ **d.** $\frac{3(x + y)}{5}$

e. None of these (b)

87. Given the rational numbers $\frac{2}{5}$ and $\frac{3}{4}$, what is the reciprocal of the sum of their reciprocals? Given that $\frac{a}{b}$ and $\frac{c}{d}$ are *any* two numbers—what is the reciprocal of the sum of their reciprocals? $\frac{6}{23}$; $\frac{ac}{ad + bc}$

R.6 Radicals and Rational Exponents

LEARNING OBJECTIVES

In Section R.6 you will learn how to:

A. Simplify radical expressions of the form $\sqrt[n]{a^n}$

B. Rewrite and simplify radical expressions using rational exponents

C. Use properties of radicals to simplify radical expressions

D. Identify like radical terms and combine radical expressions

E. Multiply and divide radical expressions; write a radical expression in simplest form

F. Evaluate formulas and simplify literal equations involving radicals

INTRODUCTION

Square roots and cube roots come from a much larger family called **radical expressions.** Expressions containing radicals can be found in virtually every field of mathematical study, and are an invaluable tool for modeling many real world phenomena.

POINT OF INTEREST

Italian physicist and astronomer Galileo Galilei (1564–1642) made numerous contributions to astronomy, physics, and other fields. But perhaps he is best known for his experiments with gravity, in which he dropped objects of different weights from the leaning tower of Pisa. Due in large part to his work, we know the velocity of an object after it has fallen a certain distance is $v = \sqrt{2gs}$, where g is the acceleration due to gravity (32 feet per second/per second), s is the distance in feet the object has fallen, and v is the velocity of the object in feet per second.

A. Simplifying Expressions of the Form $\sqrt[n]{a^n}$

In Section R.1 we noted the square root of a is b only if $b^2 = a$. All numbers greater than zero have two square roots, one positive and one negative. The positive root is also called the principal square root. The expression $\sqrt{-16}$ does not represent a real number because there is no number b such that $b^2 = -16$. This indicates that $\sqrt{a}$ is a real number only if $a \geq 0$. Of particular interest to us now is an inverse operation for a^2. In other words, what operation can be applied to a^2 to return a? Consider the expression $\sqrt{a^2}$ in Example 1.

EXAMPLE 1 Evaluate $\sqrt{a^2}$ for the following values:

a. $a = 3$ **b.** $a = 5$ **c.** $a = -6$

Solution:

a. $\sqrt{3^2} = \sqrt{9}$
$= 3$

b. $\sqrt{5^2} = \sqrt{25}$
$= 5$

c. $\sqrt{(-6)^2} = \sqrt{36}$
$= 6$

NOW TRY EXERCISES 7 AND 8

The pattern seemed to indicate that $\sqrt{a^2} = a$ and that our search for an inverse operation was complete—until Example 1(c), where we found that $\sqrt{(-6)^2} \neq -6$. Using the absolute value concept, we can "fix" this discrepancy and state a general rule for simplifying: $\sqrt{a^2} = |a|$. For expressions like $\sqrt{49x^2}$ and $\sqrt{y^6}$, the radicands can be rewritten as perfect squares with the same idea applied: $\sqrt{49x^2} = \sqrt{(7x)^2}$ or $|7x|$ and $\sqrt{y^6} = \sqrt{(y^3)^2}$ or $|y^3|$.

THE SQUARE ROOT OF a^2: $\sqrt{a^2}$

For any real number a, $\sqrt{a^2} = |a|$.

EXAMPLE 2 Simplify each expression. Assume variables can represent any real number.

a. $\sqrt{169x^2}$ **b.** $\sqrt{x^2 - 10x + 25}$

Solution:

a. $\sqrt{169x^2} = |13x|$
$= 13|x|$ since x could be negative

b. $\sqrt{x^2 - 10x + 25} = \sqrt{(x - 5)^2}$
$= |x - 5|$ since $x - 5$ could be negative

NOW TRY EXERCISES 9 AND 10

To investigate expressions like $\sqrt[3]{x^3}$, note the radicand in both $\sqrt[3]{8}$ and $\sqrt[3]{-64}$ can be written as perfect cubes. From our earlier definition of cube root we know $\sqrt[3]{8} = \sqrt[3]{(2)^3} = 2$, $\sqrt[3]{-64} = \sqrt[3]{(-4)^3} = -4$, and that every real number has only one cube root. The cube root of a positive number is positive, and the cube root of a negative number is negative ($\sqrt[3]{0} = 0$). For this reason, absolute value notation is not used or needed when taking cube roots.

THE CUBE ROOT OF a^3: $\sqrt[3]{a^3}$

For any real number a, $\sqrt[3]{a^3} = a$.

EXAMPLE 3 Simplify the radical expressions. Assume variables can represent any real number.

a. $\sqrt[3]{-27x^3}$ b. $\sqrt[3]{-64n^6}$

Solution: a. $\sqrt[3]{-27x^3} = \sqrt[3]{(-3x)^3}$
$= -3x$

b. $\sqrt[3]{-64n^6} = \sqrt[3]{(-4n^2)^3}$
$= -4n^2$

NOW TRY EXERCISES 11 AND 12

We can extend these ideas to fourth roots, fifth roots, and so on. For example, the fifth root of a number a is b only if $b^5 = a$. In symbols, $\sqrt[5]{a} = b$ implies $b^5 = a$ (the index number indicates how many times a factor must be repeated to obtain the radicand). Since an odd number of negative factors is always negative: $(-2)^5 = -32$, and an even number of negative factors is always positive: $(-2)^4 = 16$, we must take the index into account when evaluating expressions like $\sqrt[n]{a^n}$. If n is even and the radicand is unknown, absolute value notation must be used.

WORTHY OF NOTE

Just as $\sqrt[2]{-16}$ is not a real number, $\sqrt[4]{-16}$ or $\sqrt[6]{-16}$ do not represent real numbers. An even number of repeated factors is always positive!

THE nTH ROOT OF a^n: $\sqrt[n]{a^n}$

For any real number a,

1. $\sqrt[n]{a^n} = |a|$ when n is even.
2. $\sqrt[n]{a^n} = a$ when n is odd.

EXAMPLE 4 Simplify each expression. Assume variables can represent any real number.

a. $\sqrt[4]{81}$ b. $\sqrt[4]{-81}$ c. $\sqrt[5]{32}$ d. $\sqrt[5]{-32}$

e. $\sqrt[4]{16m^4}$ f. $\sqrt[5]{32p^5}$ g. $\sqrt[6]{(m+5)^6}$ h. $\sqrt[7]{(x-2)^7}$

Solution:

a. $\sqrt[4]{81} = 3$

b. $\sqrt[4]{-81}$ is not a real number

c. $\sqrt[5]{32} = 2$

d. $\sqrt[5]{-32} = -2$

e. $\sqrt[4]{16m^4} = \sqrt[4]{(2m)^4}$
$= |2m|$ or $2|m|$

f. $\sqrt[5]{32p^5} = \sqrt[5]{(2p)^5}$
$= 2p$

g. $\sqrt[6]{(m+5)^6} = |m+5|$

h. $\sqrt[7]{(x-2)^7} = x - 2$

NOW TRY EXERCISES 13 AND 14

B. Radical Expressions and Rational Exponents

As an alternative to radical notation, a rational (fractional) exponent is often used, along with the power property of exponents. For $\sqrt[3]{a^3} = a$, notice that an exponent of one-third can replace the cube root notation and accomplish the same job: $\sqrt[3]{a^3} = (a^3)^{\frac{1}{3}} = a^{\frac{3}{3}} = a$. In the same way, an exponent of one-half can replace the square root notation, giving: $\sqrt{a^2} = (a^2)^{\frac{1}{2}} = a^{\frac{2}{2}} = |a|$. In general, the nth root of *any quantity R* can be written using a rational exponent as $\sqrt[n]{R} = \sqrt[n]{R^1} = R^{\frac{1}{n}}$, where n is an integer greater than or equal to two.

RATIONAL EXPONENTS

If $\sqrt[n]{R}$ is a real number with $n \in \mathbb{Z}$ and $n \geq 2$, then $\sqrt[n]{R} = \sqrt[n]{R^1} = R^{\frac{1}{n}}$.

EXAMPLE 5 Simplify by writing the radicand as a perfect nth power, converting to rational exponent notation, then applying the power property.

a. $\sqrt[3]{-125}$ b. $-\sqrt[4]{16x^{20}}$ c. $\sqrt[4]{-81}$ d. $\sqrt[3]{\frac{8w^3}{27}}$

Solution:

a. $\sqrt[3]{-125} = (-5^3)^{\frac{1}{3}} = -5^{\frac{3}{3}} = -5$

b. $-\sqrt[4]{16x^{20}} = -(2^4x^{20})^{\frac{1}{4}} = -\left(2^{\frac{4}{4}}\right)\left(x^{\frac{20}{4}}\right) = -2x^5$

c. $\sqrt[4]{-81} = (-81)^{\frac{1}{4}}$, is not a real number

d. $\sqrt[3]{\frac{8w^3}{27}} = \left[\left(\frac{2w}{3}\right)^3\right]^{\frac{1}{3}} = \left(\frac{2w}{3}\right)^{\frac{3}{3}} = \frac{2w}{3}$

NOW TRY EXERCISES 15 AND 16

WORTHY OF NOTE

Any rational number can be decomposed into the product of a unit fraction and an integer:
$\frac{m}{n} = m \cdot \frac{1}{n}$ or $\frac{m}{n} = \frac{1}{n} \cdot m$.

Note that when a rational exponent is used, as in $\sqrt[n]{R} = \sqrt[n]{R^1} = R^{\frac{1}{n}}$, the denominator of the exponent actually represents the index number, while the numerator of the exponent represents the original power on R. *This is true even when the exponent on R is something other than one!* In other words, the radical expression $\sqrt[4]{16^3}$ can be rewritten as $(16^3)^{\frac{1}{4}} = \left(16^{\frac{3}{1}}\right)^{\frac{1}{4}}$ or $16^{\frac{3}{4}}$. This is further illustrated in Figure R.6. To evaluate the expression $(16^3)^{\frac{1}{4}} = \left(16^{\frac{3}{1}}\right)^{\frac{1}{4}}$ without the aid of a calculator, we can use the commutative property to rewrite $\left(16^{\frac{3}{1}}\right)^{\frac{1}{4}}$ as $\left(16^{\frac{1}{4}}\right)^{\frac{3}{1}}$ and begin with the fourth root of 16: $\left(16^{\frac{1}{4}}\right)^{\frac{3}{1}} = 2^3 = 8$.

Figure R.6

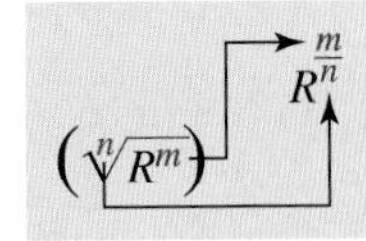

In general, if m and n have no common factor (other than 1) the expression $R^{\frac{m}{n}}$ can be interpreted in two ways. First as the nth root of the quantity R^m: $R^{\frac{m}{n}} = (R^m)^{\frac{1}{n}} = \sqrt[n]{R^m}$, or second, as the nth root of R, raised to the power m: $R^{\frac{m}{n}} = \left(R^{\frac{1}{n}}\right)^m = (\sqrt[n]{R})^m$.

RATIONAL EXPONENTS

For $m, n \in \mathbb{Z}$ with m and n relatively prime and $n \geq 2$,

1. $R^{\frac{m}{n}} = (\sqrt[n]{R})^m$: Simplify $(\sqrt[n]{R})$, then take the mth power.
2. $R^{\frac{m}{n}} = \sqrt[n]{R^m}$: Compute R^m, then take the nth root.

EXAMPLE 6 Find the value of each expression by rewriting the exponent as the product of a unit fraction and an integer, then simplifying the result without a calculator.

a. $27^{\frac{2}{3}}$ b. $(-8)^{\frac{4}{3}}$ c. $\left(\frac{4x^6}{9}\right)^{\frac{5}{2}}$

Solution:

a. $27^{\frac{2}{3}} = 27^{\frac{1}{3} \cdot 2} = \left(27^{\frac{1}{3}}\right)^2 = 3^2$ or 9

b. $(-8)^{\frac{4}{3}} = (-8)^{\frac{1}{3} \cdot 4} = \left[(-8)^{\frac{1}{3}}\right]^4 = [-2]^4$ or 16

c. $\left(\frac{4x^6}{9}\right)^{\frac{5}{2}} = \left(\frac{4x^6}{9}\right)^{\frac{1}{2}\cdot 5}$

$= \left[\left(\frac{4x^6}{9}\right)^{\frac{1}{2}}\right]^5$

$= \left[\frac{2x^3}{3}\right]^5 = \frac{32x^{15}}{243}$

NOW TRY EXERCISES 17 AND 18

As we saw in Example 6, expressions with rational exponents are generally easier to evaluate if we compute the root first, then apply the exponent. Computing the root first also helps us determine whether or not an expression represents a real number.

EXAMPLE 7 Simplify each expression, if possible.

a. $-49^{\frac{3}{2}}$ b. $(-49)^{\frac{3}{2}}$ c. $(-8)^{\frac{2}{3}}$ d. $-8^{-\frac{2}{3}}$

Solution:

a. $-49^{\frac{3}{2}} = -\left(49^{\frac{1}{2}}\right)^3$

$= -(7)^3$ or -343

b. $(-49)^{\frac{3}{2}} = \left[(-49)^{\frac{1}{2}}\right]^3$

$=$ not a real number

c. $(-8)^{\frac{2}{3}} = \left[(-8)^{\frac{1}{3}}\right]^2$

$= (-2)^2$ or 4

d. $-8^{-\frac{2}{3}} = -\left(8^{\frac{1}{3}}\right)^{-2}$

$= -2^{-2}$ or $-\frac{1}{4}$

NOW TRY EXERCISES 19 THROUGH 22

C. Properties of Radicals and Simplifying Radical Expressions

The properties used to simplify radical expressions are closely connected to the properties of exponents. For instance, the product to a power property: $(xy)^n = x^n y^n$ holds true, even when n is a rational number. This means $(xy)^{\frac{1}{2}} = x^{\frac{1}{2}} y^{\frac{1}{2}}$ and $(4 \cdot 25)^{\frac{1}{2}} = 4^{\frac{1}{2}} \cdot 25^{\frac{1}{2}}$. When this statement is expressed in radical form, we have $\sqrt{4 \cdot 25} = \sqrt{4} \cdot \sqrt{25}$, with both having a value of 10. This is called the **product property of radicals,** and can be extended to include cube roots, fourth roots, and so on.

PRODUCT PROPERTY OF RADICALS

If $\sqrt[n]{A}$ and $\sqrt[n]{B}$ represent real-valued expressions,

$$\sqrt[n]{AB} = \sqrt[n]{A} \cdot \sqrt[n]{B} \quad \text{and} \quad \sqrt[n]{A} \cdot \sqrt[n]{B} = \sqrt[n]{AB}.$$

One application of the product property is to simplify radical expressions. In general, the expression $\sqrt[n]{R}$ is in simplified form if R has no factors (other than 1) that are perfect nth roots.

EXAMPLE 8 Write each expression in simplest form using the product property.

a. $5\sqrt[3]{125x^4}$ b. $\frac{-4 + \sqrt{20}}{2}$

Solution:

a. $5\sqrt[3]{125x^4} = 5 \cdot \sqrt[3]{125 \cdot x^4}$

These steps can be done mentally.

$= 5 \cdot \sqrt[3]{125} \cdot \sqrt[3]{x^3} \cdot \sqrt[3]{x^1}$

$= 5 \cdot 5 \cdot x \cdot \sqrt[3]{x}$

$= 25x\sqrt[3]{x}$

b. $\dfrac{-4+\sqrt{20}}{2} = \dfrac{-4+\sqrt{4\cdot 5}}{2}$

$= \dfrac{-4+2\sqrt{5}}{2}$

$= \dfrac{-4}{2} + \dfrac{2\sqrt{5}}{2}$

$= -2 + \sqrt{5}$

NOW TRY EXERCISES 23 AND 24

When radicals are *combined* using the product property, the result may contain a perfect nth root, which can then be simplified. Note that the property indicates that *index numbers must be the same* in order to multiply the expressions.

WORTHY OF NOTE

Rational exponents also could have been used to simplify the expression from Example 9, since $1.2\sqrt[3]{64}\,\sqrt[3]{n^9} = 1.2(4)n^{\frac{9}{3}} = 4.8n^3$. Also see Example 11.

EXAMPLE 9 Combine factors using the product property and simplify: $1.2\sqrt[3]{16n^4}\cdot\sqrt[3]{4n^5}$.

Solution: $1.2\sqrt[3]{16n^4}\cdot\sqrt[3]{4n^5} = 1.2\cdot\sqrt[3]{64\cdot n^9}$ product property

Since the index is 3 we look for perfect cube factors in the radicand.

$= 1.2\cdot\sqrt[3]{64}\cdot\sqrt[3]{n^9}$ product property (to separate)

$= 1.2\cdot\sqrt[3]{64}\cdot\sqrt[3]{(n^3)^3}$ rewrite n^9 as a perfect cube

$= 1.2\cdot 4\cdot n^3$ simplify

$= 4.8n^3$ result

NOW TRY EXERCISES 25 AND 26

The **quotient property of radicals** can also be established using exponential properties, in much the same way as the product property. The fact that $\dfrac{\sqrt{100}}{\sqrt{25}} = \sqrt{\dfrac{100}{25}} = 2$ suggests the following:

QUOTIENT PROPERTY OF RADICALS

If $\sqrt[n]{A}$ and $\sqrt[n]{B}$ represent real numbers, and $B \neq 0$,

$$\sqrt[n]{\frac{A}{B}} = \frac{\sqrt[n]{A}}{\sqrt[n]{B}} \quad\text{and}\quad \frac{\sqrt[n]{A}}{\sqrt[n]{B}} = \sqrt[n]{\frac{A}{B}}.$$

Many times the product and quotient properties must work together to simplify a radical expression, as shown in Example 10.

EXAMPLE 10 Simplify each expression: (a) $\dfrac{\sqrt{18a^5}}{\sqrt{2a}}$, and (b) $\sqrt[3]{\dfrac{81}{125x^3}}$.

Solution: **a.** $\dfrac{\sqrt{18a^5}}{\sqrt{2a}} = \sqrt{\dfrac{18a^5}{2a}}$

$= \sqrt{9a^4}$

$= 3a^2$

b. $\sqrt[3]{\dfrac{81}{125x^3}} = \dfrac{\sqrt[3]{81}}{\sqrt[3]{125x^3}}$

$= \dfrac{\sqrt[3]{27\cdot 3}}{5x}$

$= \dfrac{3\sqrt[3]{3}}{5x}$

NOW TRY EXERCISES 27 AND 28

Radical expressions can also be simplified using rational exponents.

EXAMPLE 11 Simplify using rational exponents: (a) $\sqrt{36p^4q^5}$, (b) $v\sqrt[3]{v^4}$, and (c) $\sqrt[3]{\sqrt{x}}$.

Solution:

a. $\sqrt{36p^4q^5} = (36p^4q^5)^{\frac{1}{2}}$
$= 36^{\frac{1}{2}}p^{\frac{4}{2}}q^{\frac{5}{2}}$
$= 6p^2q^2q^{\frac{1}{2}}$
$= 6p^2q^2\sqrt{q}$

b. $v\sqrt[3]{v^4} = v^1 \cdot v^{\frac{4}{3}}$
$= v^{\frac{3}{3}} \cdot v^{\frac{4}{3}}$
$= v^{\frac{7}{3}}$
$= v^2\sqrt[3]{v}$

c. $\sqrt[3]{\sqrt{x}} = \sqrt[3]{x^{\frac{1}{2}}}$
$= \left(x^{\frac{1}{2}}\right)^{\frac{1}{3}}$
$= x^{\frac{1}{2} \cdot \frac{1}{3}}$
$= x^{\frac{1}{6}}$ or $\sqrt[6]{x}$

NOW TRY EXERCISES 29 AND 30

D. Addition and Subtraction of Radical Expressions

Since $3x$ and $5x$ are like terms, we know $3x + 5x = 8x$, where the variable x can represent *any* real number. Suppose $x = \sqrt[3]{7}$. The relation then becomes $3\sqrt[3]{7} + 5\sqrt[3]{7} = 8\sqrt[3]{7}$, illustrating how like radical expressions can be combined using addition or subtraction. Like radicals are those that have *the same index and the same radicand.* In some cases, we can identify like radicals only after radical terms have been simplified.

EXAMPLE 12 Simplify and add (if possible).

a. $\sqrt{45} + 2\sqrt{20}$ **b.** $\sqrt[3]{16x^5} - x\sqrt[3]{54x^2}$

Solution:

a. $\sqrt{45} + 2\sqrt{20} = 3\sqrt{5} + 2(2\sqrt{5})$ simplify radicals
$= 3\sqrt{5} + 4\sqrt{5}$ like radicals
$= 7\sqrt{5}$ result

b. $\sqrt[3]{16x^5} - x\sqrt[3]{54x^2} = \sqrt[3]{8 \cdot 2 \cdot x^3 \cdot x^2} - x\sqrt[3]{27 \cdot 2 \cdot x^2}$
$= 2x\sqrt[3]{2x^2} - 3x\sqrt[3]{2x^2}$ simplify radicals
$= -x\sqrt[3]{2x^2}$ result

NOW TRY EXERCISES 31 THROUGH 34

E. Multiplication and Division of Radical Expressions

The multiplication of radical expressions is simply an extension of our earlier work with the product property of radicals. The multiplication can take various forms, from the distributive property to any of the special products reviewed in Section R.3: (1) the product of two binomials using F-O-I-L; (2) the product of a binomial and its conjugate: $(A + B)(A - B) = A^2 - B^2$; or (3) the square of a binomial: $(A \pm B)^2 = A^2 \pm 2AB + B^2$. These patterns hold even if A or B is a radical term. As we begin, recall that if $a \geq 0$, $\sqrt[n]{a^n} = a$.

EXAMPLE 13 Compute each product and simplify the result: (a) $5\sqrt{3}(\sqrt{6} - 4\sqrt{3})$, (b) $(2\sqrt{2} + 6\sqrt{3})(3\sqrt{10} + \sqrt{15})$, (c) $(x + \sqrt{7})(x - \sqrt{7})$, and (d) $(3 - \sqrt{2})^2$.

Solution:

a.
$$\begin{aligned} 5\sqrt{3}(\sqrt{6} - 4\sqrt{3}) &= 5\sqrt{18} - 20(\sqrt{3})^2 && \text{distribute; } (\sqrt{3})^2 = 3 \\ &= 5(3)\sqrt{2} - (20)(3) && \text{simplify: } \sqrt{18} = 3\sqrt{2} \\ &= 15\sqrt{2} - 60 && \text{result} \end{aligned}$$

b.
$$\begin{aligned} (2\sqrt{2} + 6\sqrt{3})(3\sqrt{10} + \sqrt{15}) &= 6\sqrt{20} + 2\sqrt{30} + 18\sqrt{30} + 6\sqrt{45} && \text{F-O-I-L} \\ &= 12\sqrt{5} + 20\sqrt{30} + 18\sqrt{5} && \text{extract roots } \textit{and} \text{ simplify} \\ &= 30\sqrt{5} + 20\sqrt{30} && \text{result} \end{aligned}$$

c.
$$\begin{aligned} (x + \sqrt{7})(x - \sqrt{7}) &= x^2 - (\sqrt{7})^2 && (A + B)(A - B) = A^2 - B^2 \\ &= x^2 - 7 && \text{result} \end{aligned}$$

d.
$$\begin{aligned} (3 - \sqrt{2})^2 &= (3)^2 - 2(3)(\sqrt{2}) + (\sqrt{2})^2 && (A + B)^2 = A^2 + 2AB + B^2 \\ &= 9 - 6\sqrt{2} + 2 && \text{simplify each term} \\ &= 11 - 6\sqrt{2} && \text{result} \end{aligned}$$

NOW TRY EXERCISES 35 THROUGH 38

LOOKING AHEAD

Notice that the answer for Example 13(c) contains no radical terms, since the outer and inner products sum to zero. This result will be used to simplify certain radical expressions.

One application of products and powers of radical expressions is to check solutions to certain quadratic equations, as illustrated in Example 14.

EXAMPLE 14 Show that $x = 2 + \sqrt{3}$ is a solution of $x^2 - 4x + 1 = 0$.

Solution:
$$\begin{aligned} x^2 - 4x + 1 &= 0 && \text{original equation} \\ (2 + \sqrt{3})^2 - 4(2 + \sqrt{3}) + 1 &= 0 && \text{substitute } 2 + \sqrt{3} \text{ for } x \\ 4 + 4\sqrt{3} + 3 - 8 - 4\sqrt{3} + 1 &= 0 && \text{multiply} \\ 0 &= 0\checkmark && \text{result} \end{aligned}$$

NOW TRY EXERCISES 39 THROUGH 42

When the quotient property was applied in Example 10, the result was a denominator free of radicals. Sometimes the denominator is not automatically free of radicals, and the need to write radical expressions in *simplest form* comes into play. The procedure used to simplify expressions with a radical in the denominator is called **rationalizing the denominator.** As with other types of simplification, the desired form can be achieved in various ways. If the denominator is a single radical term, we multiply the numerator and denominator by the same radical expression. If the radicand is a rational expression, it is generally easier to build an equivalent fraction *within the radical* having perfect nth root factors in the denominator [see Example 15(b)]. The result each time is a denominator free of radicals.

RADICAL EXPRESSIONS IN SIMPLEST FORM

A radical expression is in simplest form if:

1. The radicand has no perfect nth root factors.
2. The radicand contains no fractions.
3. No radicals occur in a denominator.

EXAMPLE 15 Simplify each expression by rationalizing the denominators. Assume $a \geq 0$.

a. $\dfrac{2}{5\sqrt{3}}$ **b.** $\sqrt[3]{\dfrac{3}{4a^4}}$

Solution:

a. $\dfrac{2}{5\sqrt{3}} = \dfrac{2}{5\sqrt{3}} \cdot \dfrac{\sqrt{3}}{\sqrt{3}}$ multiply numerator and denominator by $\sqrt{3}$

$= \dfrac{2\sqrt{3}}{5(\sqrt{3})^2} = \dfrac{2\sqrt{3}}{15}$ simplify—denominator is now rational

b. $\sqrt[3]{\dfrac{3}{4a^4}} = \sqrt[3]{\dfrac{3}{4a^4} \cdot \dfrac{2a^2}{2a^2}}$ $4 \cdot 2 = 8$ is the smallest perfect cube with 4 as a factor; $a^4 \cdot a^2 = a^6$ is the smallest perfect cube with a^4 as a factor

$= \sqrt[3]{\dfrac{6a^2}{8a^6}}$ the denominator is now a perfect cube—simplify

$= \dfrac{\sqrt[3]{6a^2}}{2a^2}$ result

NOW TRY EXERCISES 43 AND 44

In some applications, we encounter expressions where the numerator or denominator contains a sum or difference of radicals. In this case, the methods used in Example 15 are ineffective, and instead we multiply by a conjugate since $(A + B)(A - B) = A^2 - B^2$. If either A or B are square roots, the result is a denominator free of radicals.

EXAMPLE 16 Simplify the following expressions by rationalizing the denominator. Give the final result in exact form and approximate form (to three decimal places).

$$\frac{2 + \sqrt{3}}{\sqrt{6} - \sqrt{2}}$$

Solution:

$\dfrac{2 + \sqrt{3}}{\sqrt{6} - \sqrt{2}} = \dfrac{2 + \sqrt{3}}{\sqrt{6} - \sqrt{2}} \cdot \dfrac{\sqrt{6} + \sqrt{2}}{\sqrt{6} + \sqrt{2}}$ multiply by the conjugate of the denominator

$= \dfrac{2\sqrt{6} + 2\sqrt{2} + \sqrt{18} + \sqrt{6}}{(\sqrt{6})^2 - (\sqrt{2})^2}$ difference of two perfect squares (in the denominator)

$= \dfrac{3\sqrt{6} + 2\sqrt{2} + 3\sqrt{2}}{6 - 2}$ simplify

$= \dfrac{3\sqrt{6} + 5\sqrt{2}}{4}$ exact form

≈ 3.605 approximate form

NOW TRY EXERCISES 45 THROUGH 48

F. Formulas and Literal Equations

A right triangle is one that has a 90° angle. The longest side (opposite the right angle) is called the **hypotenuse** while the other two sides are simply called "legs." The **Pythagorean theorem** is a formula that says if you add the square of each leg, the result

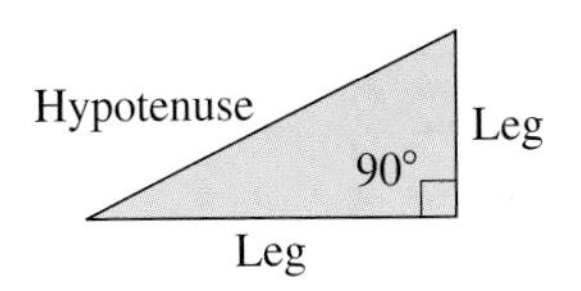

will be equal to the square of the hypotenuse: $\text{leg}^2 + \text{leg}^2 = \text{hyp}^2$. A geometric interpretation of the theorem is given in the figure, which shows $3^2 + 4^2 = 5^2$. The theorem is generally stated as $a^2 + b^2 = c^2$, where c is the hypotenuse.

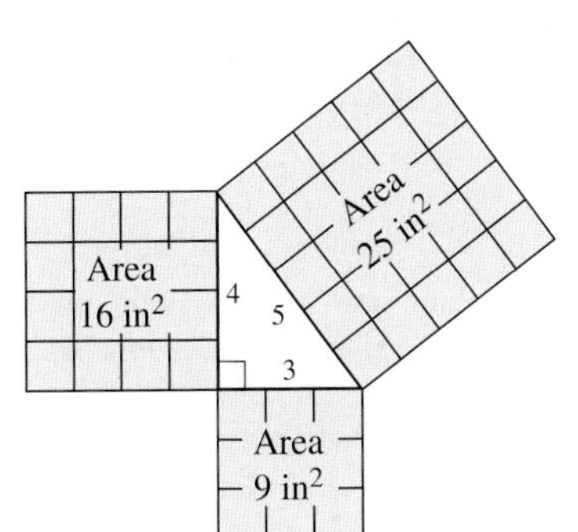

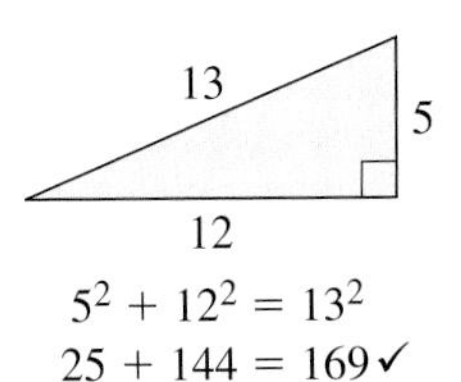

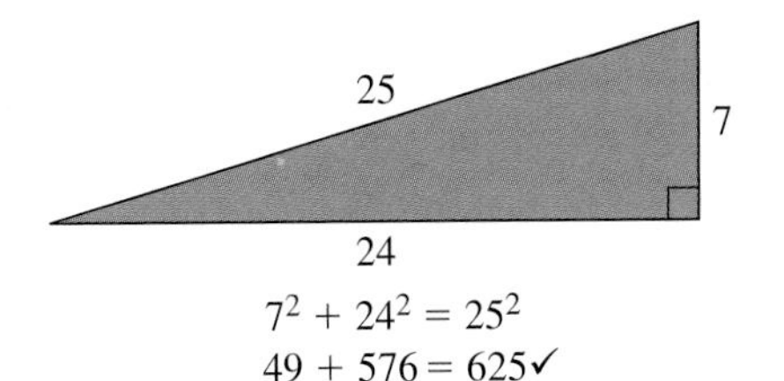

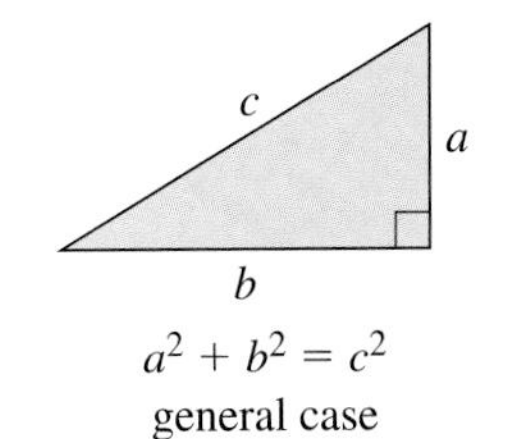

EXAMPLE 17 A 26.5-ft extension ladder is placed 10 ft from the base of a building in an effort to reach a third-story window. Is the ladder long enough to reach a windowsill that is 25 ft high?

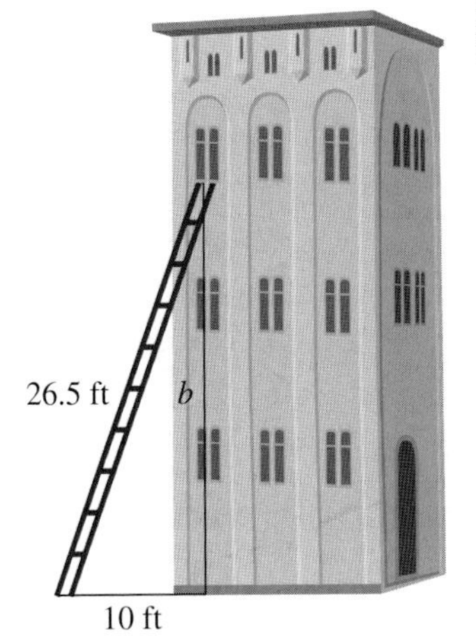

Solution: Let b represent the height of the windowsill.

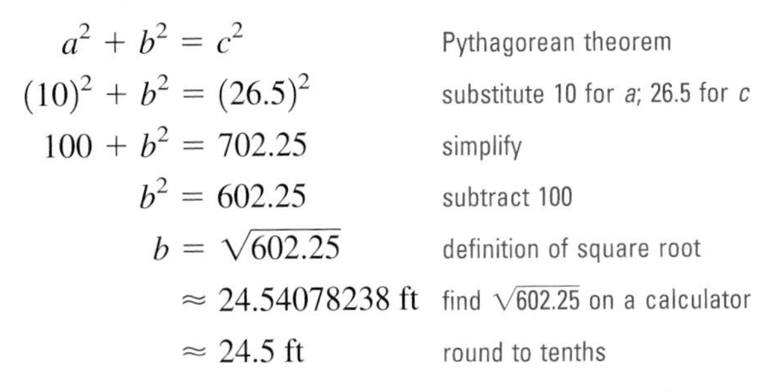

$a^2 + b^2 = c^2$	Pythagorean theorem
$(10)^2 + b^2 = (26.5)^2$	substitute 10 for a; 26.5 for c
$100 + b^2 = 702.25$	simplify
$b^2 = 602.25$	subtract 100
$b = \sqrt{602.25}$	definition of square root
≈ 24.54078238 ft	find $\sqrt{602.25}$ on a calculator
≈ 24.5 ft	round to tenths

The ladder will not quite reach the windowsill $(24.5 < 25)$.

NOW TRY EXERCISES 51 AND 52

As with other algebraic expressions, radical expressions must often be rewritten to make them more convenient to use or to gain needed information.

EXAMPLE 18 Rewrite by rationalizing the *numerator*: $\dfrac{\sqrt{x+h}-\sqrt{x}}{h}$.

Solution:

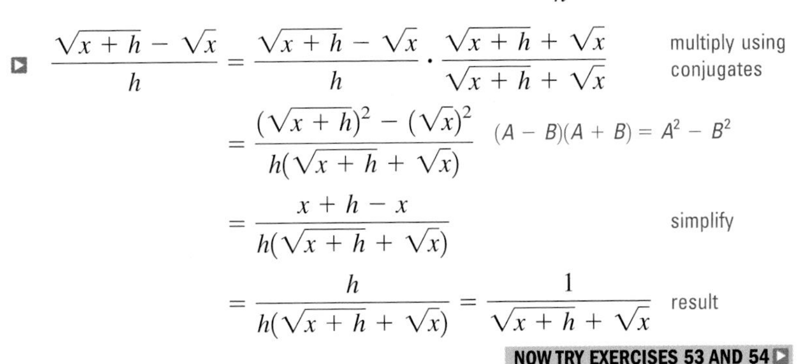

$$\frac{\sqrt{x+h}-\sqrt{x}}{h} = \frac{\sqrt{x+h}-\sqrt{x}}{h} \cdot \frac{\sqrt{x+h}+\sqrt{x}}{\sqrt{x+h}+\sqrt{x}} \quad \text{multiply using conjugates}$$

$$= \frac{(\sqrt{x+h})^2 - (\sqrt{x})^2}{h(\sqrt{x+h}+\sqrt{x})} \quad (A-B)(A+B) = A^2 - B^2$$

$$= \frac{x+h-x}{h(\sqrt{x+h}+\sqrt{x})} \quad \text{simplify}$$

$$= \frac{h}{h(\sqrt{x+h}+\sqrt{x})} = \frac{1}{\sqrt{x+h}+\sqrt{x}} \quad \text{result}$$

NOW TRY EXERCISES 53 AND 54

R.6 EXERCISES

9. a. $7|p|$ b. $|x-3|$ c. $9m^2$ d. $|x-3|$
10. a. $5|n|$ b. $|y+2|$ c. $|v^5|$ d. $|2a+3|$
11. a. 4 b. $-5x$ c. $6z^4$ d. $\frac{v}{-2}$
12. a. -2 b. $-5p$ c. $3q^3$ d. $\frac{w}{-4}$
13. a. 2 b. not a real number c. $3x^2$ d. $-3x$ e. $k-3$ f. $|h+2|$
14. a. simplified b. not a real number c. $4z^3$ d. $-4z^4$ e. $q-9$ f. $|p+4|$
15. a. -5 b. $-3|n^3|$ c. not a real number d. $\frac{7|v^5|}{6}$
16. a. -6 b. $-2m^6$ c. not a real number d. $\frac{5|x^3|}{2}$
17. a. 4 b. $\frac{64}{125}$ c. $\frac{125}{8}$ d. $\frac{9p^4}{4q^2}$
18. a. 27 b. $\frac{8}{27}$ c. $\frac{27}{8}$ d. $\frac{25v^6}{9w^4}$
19. a. -1728 b. not a real number c. $\frac{1}{9}$ d. $\frac{-256}{81x^4}$
20. a. -1000 b. not a real number c. $\frac{1}{25}$ d. $\frac{-16}{x^{12}}$
21. a. $\frac{32n^{10}}{p^2}$ b. $\frac{1}{2y^{\frac{1}{4}}} = \frac{y^{\frac{3}{4}}}{2y}$
22. a. $\frac{36}{x^{\frac{1}{4}}} = \frac{36x^{\frac{3}{4}}}{x}$ b. $\frac{16y^3}{x}$

CONCEPTS AND VOCABULARY

Fill in each blank with the appropriate word or phrase. Carefully reread the section, if necessary.

1. $\sqrt[n]{a^n} = |a|$ if $n > 0$ is a(n) ___even___ integer.

2. The conjugate of $2 - \sqrt{3}$ is ___$2 + \sqrt{3}$___.

3. By decomposing the rational exponent, we can rewrite $16^{\frac{3}{4}}$ as $\left(16^{\frac{?}{?}}\right)^?$. $\left(16^{\frac{1}{4}}\right)^3$

4. $\left(x^{\frac{3}{2}}\right)^{\frac{2}{3}} = x^{\frac{3}{2}\cdot\frac{2}{3}} = x^1$ is an example of the ___power___ property of exponents.

5. Discuss/explain what it means when we say an expression like $\sqrt{A}$ has been written in simplest form. Answers will vary.

6. Discuss/explain how squaring both sides of an equation can introduce extraneous roots. Will cubing both sides yield an "extra" root? Answers will vary; no

DEVELOPING YOUR SKILLS

Evaluate the expression $\sqrt{x^2}$ for the values given.

7. a. $x = 9$ 9 b. $x = -10$ 10

8. a. $x = 7$ 7 b. $x = -8$ 8

Simplify each expression, assuming that variables can represent any real number.

9. a. $\sqrt{49p^2}$ b. $\sqrt{(x-3)^2}$ c. $\sqrt{81m^4}$ d. $\sqrt{x^2 - 6x + 9}$

10. a. $\sqrt{25n^2}$ b. $\sqrt{(y+2)^2}$ c. $\sqrt{v^{10}}$ d. $\sqrt{4a^2 + 12a + 9}$

11. a. $\sqrt[3]{64}$ b. $\sqrt[3]{-125x^3}$ c. $\sqrt[3]{216z^{12}}$ d. $\sqrt[3]{\frac{v^3}{-8}}$

12. a. $\sqrt[3]{-8}$ b. $\sqrt[3]{-125p^3}$ c. $\sqrt[3]{27q^9}$ d. $\sqrt[3]{\frac{w^3}{-64}}$

13. a. $\sqrt[6]{64}$ b. $\sqrt[6]{-64}$ c. $\sqrt[5]{243x^{10}}$ d. $\sqrt[5]{-243x^5}$ e. $\sqrt[5]{(k-3)^5}$ f. $\sqrt[6]{(h+2)^6}$

14. a. $\sqrt[4]{216}$ b. $\sqrt[4]{-216}$ c. $\sqrt[5]{1024z^{15}}$ d. $\sqrt[5]{-1024z^{20}}$ e. $\sqrt[5]{(q-9)^5}$ f. $\sqrt[6]{(p+4)^6}$

15. a. $\sqrt[3]{-125}$ b. $-\sqrt[4]{81n^{12}}$ c. $\sqrt{-36}$ d. $\sqrt{\frac{49v^{10}}{36}}$

16. a. $\sqrt[3]{-216}$ b. $-\sqrt[4]{16m^{24}}$ c. $\sqrt{-121}$ d. $\sqrt{\frac{25x^6}{4}}$

17. a. $8^{\frac{2}{3}}$ b. $\left(\frac{16}{25}\right)^{\frac{3}{2}}$ c. $\left(\frac{4}{25}\right)^{-\frac{3}{2}}$ d. $\left(\frac{-27p^6}{8q^3}\right)^{\frac{2}{3}}$

18. a. $9^{\frac{3}{2}}$ b. $\left(\frac{4}{9}\right)^{\frac{3}{2}}$ c. $\left(\frac{16}{81}\right)^{-\frac{3}{4}}$ d. $\left(\frac{-125v^9}{27w^6}\right)^{\frac{2}{3}}$

19. a. $-144^{\frac{3}{2}}$ b. $\left(-\frac{4}{25}\right)^{\frac{3}{2}}$ c. $(-27)^{-\frac{2}{3}}$ d. $-\left(\frac{27x^3}{64}\right)^{-\frac{4}{3}}$

20. a. $-100^{\frac{3}{2}}$ b. $\left(-\frac{49}{36}\right)^{\frac{3}{2}}$ c. $(-125)^{-\frac{2}{3}}$ d. $-\left(\frac{x^9}{8}\right)^{-\frac{4}{3}}$

Use properties of exponents to simplify. Answer in exponential form without negative exponents.

21. a. $\left(2n^2p^{-\frac{2}{5}}\right)^5$ b. $\left(\frac{8y^{\frac{3}{4}}}{64y^{\frac{3}{2}}}\right)^{\frac{1}{3}}$

22. a. $\left(\frac{24x^{\frac{3}{8}}}{4x^{\frac{1}{2}}}\right)^2$ b. $\left(2x^{-\frac{1}{4}}y^{\frac{3}{4}}\right)^4$

23. a. $3m\sqrt{2}$ b. $10pq^2\sqrt[3]{q}$
c. $\frac{3}{2}mn\sqrt[3]{n^2}$ d. $4pq^3\sqrt{2p}$
e. $-3 + \sqrt{7}$ f. $\frac{9}{2} - \sqrt{2}$

24. a. $2x^3\sqrt{2}$ b. $12a\sqrt[3]{2ab^2}$
c. $\frac{2}{3}b^2\sqrt[3]{a^2}$ d. $3m^3n^4\sqrt{6}$
e. $\frac{3}{2} - \frac{\sqrt{3}}{2}$ f. $-5 + \sqrt{2}$

25. a. $15a^2$ b. $-4b\sqrt{b}$
c. $\frac{x^4\sqrt{y}}{3}$ d. $3u^2v\sqrt[3]{v}$

26. a. $40.8p^3$ b. $-8q^2$
c. $\frac{5}{9}ab^3$ d. $5d\sqrt[3]{c^2}$

27. a. $2m^2$ b. $3n$
c. $\frac{3\sqrt{5}}{4x}$ d. $\frac{18\sqrt[3]{3}}{z^3}$

28. a. $3y^3$ b. $2b\sqrt[3]{3}$
c. $\frac{\sqrt{5}}{x^2}$ d. $\frac{-15}{x^2}$

29. a. $2x^2y^3$ b. $x^2\sqrt[4]{x}$
c. $\sqrt[12]{b}$ d. $\frac{1}{\sqrt[6]{6}} = \frac{\sqrt[6]{6^5}}{6}$

30. a. $3a^3b^4$ b. $a^2\sqrt[5]{a}$
c. $\sqrt[8]{a}$ d. $\sqrt[12]{3}$

31. a. $9\sqrt{2}$ b. $14\sqrt{3}$
c. $16\sqrt{2m}$ d. $-5\sqrt{7p}$

32. a. $-2\sqrt{5}$ b. $16\sqrt{3}$
c. $-19\sqrt{3x}$
d. $15\sqrt{10q}$

33. a. $-x\sqrt[3]{2x}$
b. $2 - \sqrt{3x} + 3\sqrt{5}$
c. $6x\sqrt{2x} + 5\sqrt{2} - \sqrt{7x} + 3\sqrt{3}$

34. a. $15m\sqrt[3]{2} - 4m^2\sqrt[3]{2}$
b. $\sqrt{10b} + 10\sqrt{2b} - 2\sqrt{5} + 2\sqrt{10}$
c. $5r\sqrt{3r} + 4\sqrt{2} - 3\sqrt{3r} + \sqrt{38}$

35. a. 98 b. $\sqrt{15} + \sqrt{21}$
c. $n^2 - 5$ d. $39 - 12\sqrt{3}$

36. a. 0.45 b. $\sqrt{30} - \sqrt{10}$
c. 13 d. $9 + 4\sqrt{5}$

37. a. -19
b. $\sqrt{10} + \sqrt{65} - 2\sqrt{7} - \sqrt{182}$
c. $12\sqrt{5} + 2\sqrt{14} + 36\sqrt{15} + 6\sqrt{42}$

38. a. $-75 - 6\sqrt{10}$
b. $\sqrt{30} + \sqrt{33} + 2\sqrt{5} + \sqrt{22}$
c. $23\sqrt{3} + 7\sqrt{30}$

Simplify each expression. Assume all variables represent non-negative real numbers.

23. **a.** $\sqrt{18m^2}$ **b.** $-2\sqrt[3]{-125p^3q^7}$
c. $\frac{3}{8}\sqrt[3]{64m^3n^5}$ **d.** $\sqrt{32p^3q^6}$
e. $\frac{-6 + \sqrt{28}}{2}$ **f.** $\frac{27 - \sqrt{72}}{6}$

24. **a.** $\sqrt{8x^6}$ **b.** $3\sqrt[3]{128a^4b^2}$
c. $\frac{2}{9}\sqrt[3]{27a^2b^6}$ **d.** $\sqrt{54m^6n^8}$
e. $\frac{12 - \sqrt{48}}{8}$ **f.** $\frac{-20 + \sqrt{32}}{4}$

25. **a.** $2.5\sqrt{18a}\sqrt{2a^3}$ **b.** $-\frac{2}{3}\sqrt{3b}\sqrt{12b^2}$
c. $\sqrt{\frac{x^3y}{3}}\sqrt{\frac{4x^5y}{12y}}$ **d.** $\sqrt[3]{9v^2u}\sqrt[3]{3u^5v^2}$

26. **a.** $5.1\sqrt{2p}\sqrt{32p^5}$ **b.** $-\frac{4}{5}\sqrt{5q}\sqrt{20q^3}$
c. $\sqrt{\frac{ab^2}{3}}\sqrt{\frac{25ab^4}{27}}$ **d.** $\sqrt[3]{5cd^2}\,\sqrt[3]{25cd}$

27. **a.** $\frac{\sqrt{8m^5}}{\sqrt{2m}}$ **b.** $\frac{\sqrt[3]{108n^4}}{\sqrt[3]{4n}}$
c. $\sqrt{\frac{45}{16x^2}}$ **d.** $12\sqrt[3]{\frac{81}{8z^9}}$

28. **a.** $\frac{\sqrt{27y^7}}{\sqrt{3y}}$ **b.** $\frac{\sqrt[3]{72b^5}}{\sqrt[3]{3b^2}}$
c. $\sqrt{\frac{20}{4x^4}}$ **d.** $-9\sqrt[3]{\frac{125}{27x^6}}$

29. **a.** $\sqrt[5]{32x^{10}y^{15}}$ **b.** $x\sqrt[4]{x^5}$
c. $\sqrt[4]{\sqrt[3]{b}}$ **d.** $\frac{\sqrt[3]{6}}{\sqrt{6}}$

30. **a.** $\sqrt[4]{81a^{12}b^{16}}$ **b.** $a\sqrt[5]{a^6}$
c. $\sqrt{\sqrt[4]{a}}$ **d.** $\frac{\sqrt[3]{3}}{\sqrt[4]{3}}$

Simplify and add (if possible).

31. **a.** $12\sqrt{72} - 9\sqrt{98}$
b. $8\sqrt{48} - 3\sqrt{108}$
c. $7\sqrt{18m} - \sqrt{50m}$
d. $2\sqrt{28p} - 3\sqrt{63p}$

32. **a.** $-3\sqrt{80} + 2\sqrt{125}$
b. $5\sqrt{12} + 2\sqrt{27}$
c. $3\sqrt{12x} - 5\sqrt{75x}$
d. $3\sqrt{40q} + 9\sqrt{10q}$

33. **a.** $3x\sqrt[3]{54x} - 5\sqrt[3]{16x^4}$
b. $\sqrt{4} + \sqrt{3x} - \sqrt{12x} + \sqrt{45}$
c. $\sqrt{72x^3} + \sqrt{50} - \sqrt{7x} + \sqrt{27}$

34. **a.** $5\sqrt[3]{54m^3} - 2m\sqrt[3]{16m^3}$
b. $\sqrt{10b} + \sqrt{200b} - \sqrt{20} + \sqrt{40}$
c. $\sqrt{75r^3} + \sqrt{32} - \sqrt{27r} + \sqrt{38}$

Compute each product and simplify the result.

35. **a.** $(7\sqrt{2})^2$
b. $\sqrt{3}(\sqrt{5} + \sqrt{7})$
c. $(n + \sqrt{5})(n - \sqrt{5})$
d. $(6 - \sqrt{3})^2$

36. **a.** $(0.3\sqrt{5})^2$
b. $\sqrt{5}(\sqrt{6} - \sqrt{2})$
c. $(4 + \sqrt{3})(4 - \sqrt{3})$
d. $(2 + \sqrt{5})^2$

37. **a.** $(3 + 2\sqrt{7})(3 - 2\sqrt{7})$
b. $(\sqrt{5} - \sqrt{14})(\sqrt{2} + \sqrt{13})$
c. $(2\sqrt{2} + 6\sqrt{6})(3\sqrt{10} + \sqrt{7})$

38. **a.** $(5 + 4\sqrt{10})(1 - 2\sqrt{10})$
b. $(\sqrt{3} + \sqrt{2})(\sqrt{10} + \sqrt{11})$
c. $(3\sqrt{5} + 4\sqrt{2})(\sqrt{15} + \sqrt{6})$

Use a substitution to verify the solutions to the quadratic equation given.

39. $x^2 - 4x + 1 = 0$ Answers will vary.
a. $x = 2 + \sqrt{3}$ **b.** $x = 2 - \sqrt{3}$

40. $x^2 - 10x + 18 = 0$ Answers will vary.
a. $x = 5 - \sqrt{7}$ **b.** $x = 5 + \sqrt{7}$

41. $x^2 + 2x - 9 = 0$ Answers will vary.
a. $x = -1 + \sqrt{10}$ **b.** $x = -1 - \sqrt{10}$

42. $x^2 - 14x + 29 = 0$ Answers will vary.
a. $x = 7 - 2\sqrt{5}$ **b.** $x = 7 + 2\sqrt{5}$

Rationalize each expression by building perfect nth root factors for each denominator. Assume all variables represent positive quantities.

43. **a.** $\frac{3}{\sqrt{12}}$ $\frac{\sqrt{3}}{2}$ **b.** $\sqrt{\frac{20}{27x^3}}$ $\frac{2\sqrt{15x}}{9x^2}$ **c.** $\sqrt{\frac{27}{50b}}$ $\frac{3\sqrt{6b}}{10b}$ **d.** $\sqrt[3]{\frac{1}{4p}}$ $\frac{\sqrt[3]{2p^2}}{2p}$

44. **a.** $\frac{-4}{\sqrt{20}}$ $\frac{-2\sqrt{5}}{5}$ **b.** $\sqrt{\frac{125}{12n^3}}$ $\frac{5\sqrt{15n}}{6n^2}$ **c.** $\sqrt{\frac{5}{12x}}$ $\frac{\sqrt{15x}}{6x}$ **d.** $\sqrt[3]{\frac{3}{2m^2}}$ $\frac{\sqrt[3]{12m}}{2m}$

Simplify the following expressions by rationalizing the denominators. Where possible, state results in exact form and approximate form, rounded to hundredths.

45. **a.** $\frac{8}{3+\sqrt{11}}$ **b.** $\frac{6}{\sqrt{x}-\sqrt{2}}$

46. **a.** $\frac{7}{\sqrt{7}+3}$ **b.** $\frac{12}{\sqrt{x}+\sqrt{3}}$

47. **a.** $\frac{\sqrt{10}-3}{\sqrt{3}+\sqrt{2}}$ **b.** $\frac{7+\sqrt{6}}{3-3\sqrt{2}}$

48. **a.** $\frac{1+\sqrt{2}}{\sqrt{6}+\sqrt{14}}$ **b.** $\frac{1+\sqrt{6}}{5+2\sqrt{3}}$

45. a. $-12+4\sqrt{11}$; 1.27 b. $\frac{6\sqrt{x}+6\sqrt{2}}{x-2}$

46. a. $\frac{7\sqrt{7}-21}{-2}$; 1.24 b. $\frac{12\sqrt{x}-12\sqrt{3}}{x-3}$

47. a. $\sqrt{30}-2\sqrt{5}-3\sqrt{3}+3\sqrt{2}$; 0.05 b. $\frac{7+7\sqrt{2}+\sqrt{6}+2\sqrt{3}}{-3}$; −7.60

48. a. $\frac{\sqrt{6}-\sqrt{14}+2\sqrt{3}-2\sqrt{7}}{-8}$; 0.39 b. $\frac{5-2\sqrt{3}+5\sqrt{6}-6\sqrt{2}}{13}$; 0.41

WORKING WITH FORMULAS

49. Fish length to weight relationship: $L = 1.13(W)^{\frac{1}{3}}$

The length to weight relationship of a female Pacific halibut can be approximated by the formula shown, where W is the weight in pounds and L is the length in feet. A fisherman lands a halibut that weighs 400 lb. Approximate the length of the fish (round to two decimal places). 8.33 ft

50. Timing a falling object: $t = \frac{\sqrt{s}}{4}$

The time it takes an object to fall a certain distance is given by the formula shown, where t is the time in seconds and s is the distance the object has fallen. Find how long it takes an object to hit the ground, if it is dropped from the top of a building that is 80 ft in height. 2.24 sec

APPLICATIONS

51. Length of a cable: A radio tower is secured by cables that are clamped 21.5 m up the tower and anchored in the ground 9 m from its base. If 30-cm lengths are needed to secure the cable at each end, how long are the cables? Round to the nearest tenth of a meter. 23.9 m

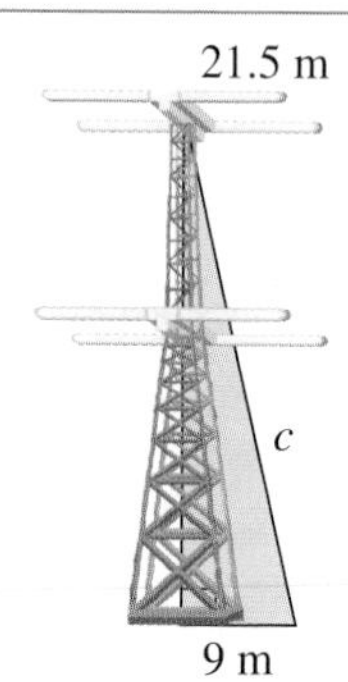

52. Height of a kite: Benjamin Franklin is flying his kite in a storm once again . . . and has let out 200 m of string. John Adams has walked to a position directly under the kite and is 150 m from Ben. How high is the kite to the nearest meter? about 132 m

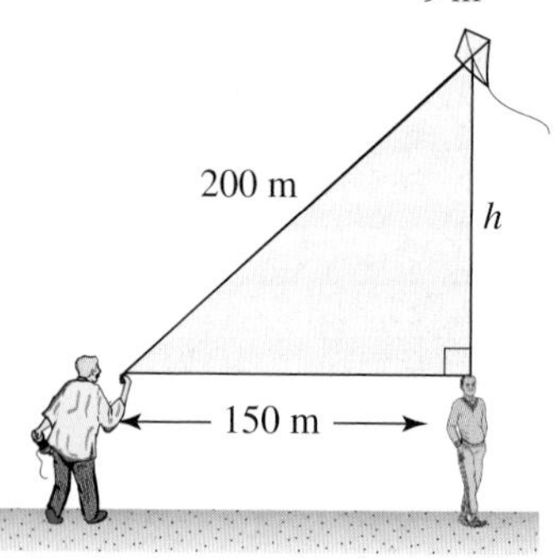

Rewrite each expression by rationalizing the *numerator*.

53. $\frac{\sqrt{x+2}-\sqrt{x}}{2}$ **54.** $\frac{\sqrt{2x+h}-\sqrt{2x}}{h}$

53. $\frac{1}{\sqrt{x+2}+\sqrt{x}}$

54. $\frac{1}{\sqrt{2x+h}+\sqrt{2x}}$

55. a. 365.02 days
b. 688.69 days
c. 87.91 days
56. a. 223.21 days
b. 4280.12 days
c. 10,806.36 days

The time T (in days) required for a planet to make one revolution around the sun is modeled by the function $T = 0.407R^{\frac{3}{2}}$, where R is the maximum radius of the planet's orbit (in millions of miles). This is known as *Kepler's third law of planetary motion.* Use the equation given to approximate the number of days required for one complete orbit of each planet, given its maximum orbital radius.

55. a. Earth: 93 million mi **b.** Mars: 142 million mi **c.** Mercury: 36 million mi

56. a. Venus: 67 million mi **b.** Jupiter: 480 million mi **c.** Saturn: 890 million mi

57. Accident investigation: After an accident, police officers will try to determine the approximate velocity V that a car was traveling using the formula $V = 2\sqrt{6L}$, where L is the length of the skid marks in feet and V is the velocity in miles per hour. (a) If the skid marks were 54 ft long, how fast was the car traveling? (b) Approximate the speed of the car if the skid marks were 90 ft long. a. 36 mph b. 46.5 mph

58. Wind-powered energy: If a wind-powered generator is delivering P units of power, the velocity V of the wind (in miles per hour) can be determined using $V = \sqrt[3]{\frac{P}{k}}$, where k is a constant that depends on the size and efficiency of the generator. Rationalize the radical expression and use the new version to find the velocity of the wind if $k = 0.004$ and the generator is putting out 13.5 units of power. 15 mph

59. Surface area: The lateral surface area (surface area excluding the base) S of a cone is given by the formula $S = \pi r\sqrt{r^2 + h^2}$, where r is the radius of the base and h is the height of the cone. Find the surface area of a cone that has a radius of 6 m and a height of 10 m. Answer in simplest form. $12\pi\sqrt{34} \approx 219.82\ \text{m}^2$

60. Surface area: The lateral surface area S of a frustum (a truncated cone) is given by the formula $S = \pi(a + b)\sqrt{h^2 + (b - a)^2}$, where a is the radius of the upper base, b is the radius of the lower base, and h is the height. Find the surface area of a frustum where $a = 6$ m, $b = 8$ m, and $h = 10$ m. Answer in simplest form. $28\pi\sqrt{26} \approx 448.53\ \text{m}^2$

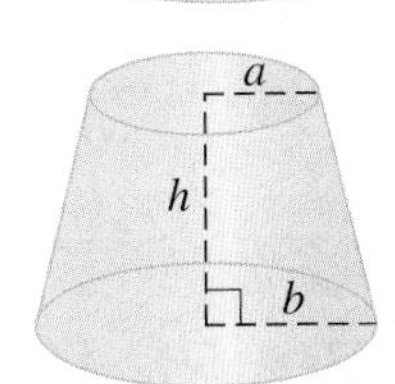

The expression $x^2 - 7$ is not factorable using *integer values*. But the expression *can be written* in the form $x^2 - (\sqrt{7})^2$, enabling us to factor it as a binomial and its conjugate: $(x + \sqrt{7})(x - \sqrt{7})$. Use this idea to factor the following expressions.

61. a. $x^2 - 5$ $(x + \sqrt{5})(x - \sqrt{5})$ **b.** $n^2 - 19$ $(n + \sqrt{19})(n - \sqrt{19})$ **62. a.** $4v^2 - 11$ $(2v + \sqrt{11})(2v - \sqrt{11})$ **b.** $9w^2 - 11$ $(3w + \sqrt{11})(3w - \sqrt{11})$

EXTENDING THE CONCEPT

63. Why is absolute value notation unnecessary when writing the simplified form of $\sqrt{m^8}$, $m \in \mathbb{R}$? Because $m^4 \geq 0$ for $m \in \mathbb{R}$

64. The following three terms—$\sqrt{3x} + \sqrt{9x} + \sqrt{27x} + \cdots$—form a pattern that continues until the sixth term is found. (a) Compute the sum of all six terms; (b) develop a system (investigate the pattern further) that will enable you to find the sum of 12 such terms *without actually writing out the terms.* a. $13\sqrt{3x} + 39\sqrt{x}$ b. Answers will vary.

65. Simplify the expression without the aid of a calculator. 3

$$\left(\left(\left(\left(\left(3^{\frac{5}{6}}\right)^{\frac{3}{2}}\right)^{\frac{4}{5}}\right)^{\frac{3}{4}}\right)^{\frac{2}{5}}\right)^{\frac{10}{3}}$$

66. If $\left(x^{\frac{1}{2}} + x^{-\frac{1}{2}}\right)^2 = \frac{9}{2}$, find the value of $x^{\frac{1}{2}} + x^{-\frac{1}{2}}$. $\frac{3\sqrt{2}}{2}$

PRACTICE TEST

1. State true or false. If false, state why.
   a. $H \subset R$ b. $N \subset Q$
   c. $\sqrt{2} \in Q$ d. $\frac{1}{2} \notin W$

2. State the value of each expression.
   a. $\sqrt{121}$ b. $\sqrt[3]{-125}$
   c. $\sqrt{-36}$ d. $\sqrt{400}$

3. Evaluate each expression:
   a. $\frac{7}{8} - \left(-\frac{1}{4}\right)$ b. $-\frac{1}{3} - \frac{5}{6}$
   c. $-0.7 + 1.2$ d. $1.3 + (-5.9)$

4. Evaluate each expression:
   a. $(-4)\left(-2\frac{1}{3}\right)$ b. $(-0.6)(-1.5)$
   c. $\frac{-2.8}{-0.7}$ d. $4.2 \div (-0.6)$

5. Evaluate using a calculator:
   $2000(1 + \frac{0.08}{12})^{12 \cdot 10}$

6. State the value of each expression, if possible.
   a. $0 \div 6$ b. $6 \div 0$

7. State the number of terms in each expression and identify the coefficient of each.
   a. $-2v^2 + 6v + 5$
   b. $\frac{c+2}{3} + c$

8. Evaluate each expression given $x = -0.5$ and $y = -2$. Round to hundredths as needed.
   a. $2x - 3y^2$
   b. $\sqrt{2} - x(4 - x^2) + \frac{y}{x}$

9. Translate each phrase into an algebraic expression.
   a. Nine less than twice a number is subtracted from the number cubed.
   b. Three times the square of half a number is subtracted from twice the number.

10. Create a mathematical model using descriptive variables.
    a. The radius of the planet Jupiter is approximately 119 mi less than 11 times the radius of the Earth. Express the radius of Jupiter in terms of the Earth's radius.
    b. Last year, Video Venue Inc. earned $1.2 million more than four times what it earned this year. Express last year's earnings of Video Venue Inc. in terms of this year's earnings.

11. Simplify by combining like terms.
    a. $8v^2 + 4v - 7 + v^2 - v$
    b. $-4(3b - 2) + 5b$
    c. $4x - (x - 2x^2) + x(3 - x)$

12. Factor each expression completely.
    a. $9x^2 - 16$
    b. $4v^3 - 12v^2 + 9v$
    c. $x^3 + 5x^2 - 9x - 45$

13. Simplify using the properties of exponents.
    a. $\frac{5}{b^{-3}}$ b. $(-2a^3)^2(a^2b^4)^3$ c. $\left(\frac{m^2}{2n}\right)^3$ d. $\left(\frac{5p^2q^3r^4}{-2pq^2r^4}\right)^2$

14. Simplify using the properties of exponents.
    a. $\frac{-12a^3b^5}{3a^2b^4}$ b. $(3.2 \times 10^{-17}) \times (2.0 \times 10^{15})$ c. $\left(\frac{a^{-3} \cdot b}{c^{-2}}\right)^{-4}$ d. $-7x^0 + (-7x)^0$

15. Compute each product.
    a. $(3x^2 + 5y)(3x^2 - 5y)$
    b. $(2a + 3b)^2$

16. Add or subtract as indicated.
    a. $(-5a^3 + 4a^2 - 3) + (7a^4 + 4a^2 - 3a - 15)$
    b. $(2x^2 + 4x - 9) - (7x^4 - 2x^2 - x - 9)$

Answers

1. a. True
   b. True
   c. False; $\sqrt{2}$ cannot be expressed as a ratio of two integers.
   d. True
2. a. 11
   b. −5
   c. not a real number
   d. 20
3. a. $\frac{9}{8}$ b. $\frac{-7}{6}$ c. 0.5 d. −4.6
4. a. $\frac{28}{3}$ b. 0.9 c. 4 d. −7
5. ≈4439.28
6. a. 0 b. undefined
7. a. 3; −2, 6, 5
   b. 2; $\frac{1}{3}$, 1
8. a. −13 b. ≈7.29
9. a. $x^3 - (2x - 9)$
   b. $2n - 3\left(\frac{n}{2}\right)^2$
10. a. Let r represent Earth's radius. Then $11r - 119$ represents Jupiter's radius.
    b. Let e represent this year's earnings. Then $4e + 1.2$ represents last year's earnings.
11. a. $9v^2 + 3v - 7$
    b. $-7b + 8$
    c. $6x + x^2$
12. a. $(3x + 4)(3x - 4)$
    b. $v(2v - 3)^2$
    c. $(x + 5)(x + 3)(x - 3)$
13. a. $5b^3$ b. $4a^{12}b^{12}$
    c. $\frac{m^6}{8n^3}$ d. $\frac{25}{4}p^2q^2$
14. a. $-4ab$
    b. $6.4 \times 10^{-2} = 0.064$
    c. $\frac{a^{12}}{b^4c^8}$
    d. −6
15. a. $9x^4 - 25y^2$
    b. $4a^2 + 12ab + 9b^2$
16. a. $7a^4 - 5a^3 + 8a^2 - 3a - 18$
    b. $-7x^4 + 4x^2 + 5x$

17. a. -1
b. $\frac{2+n}{2-n}$
c. $x-3$
d. $\frac{x-5}{3x-2}$
e. $\frac{x-5}{3x+1}$
f. $\frac{3(m+7)}{5(m+4)(m-3)}$

18. a. $|x+11|$
b. $\frac{-2}{3v}$
c. $\frac{64}{125}$
d. $-\frac{1}{2}+\frac{\sqrt{2}}{2}$
e. $11\sqrt{10}$
f. x^2-5
g. $\frac{\sqrt{10x}}{5x}$
h. $2(\sqrt{6}+\sqrt{2})$

19. $-0.5x^2+10x+1200$;
a. 10 decreases of 0.50 or \$5.00
b. Maximum revenue is \$1250.

20. 58 cm

Simplify or compute as indicated.

17. **a.** $\frac{x-5}{5-x}$ **b.** $\frac{4-n^2}{n^2-4n+4}$ **c.** $\frac{x^3-27}{x^2+3x+9}$ **d.** $\frac{3x^2-13x-10}{9x^2-4}$

e. $\frac{x^2-25}{3x^2-11x-4} \div \frac{x^2+x-20}{x^2-8x+16}$ **f.** $\frac{m+3}{m^2+m-12}-\frac{2}{5(m+4)}$

18. **a.** $\sqrt{(x+11)^2}$ **b.** $\sqrt[3]{\frac{-8}{27v^3}}$ **c.** $\left(\frac{25}{16}\right)^{\frac{-3}{2}}$ **d.** $\frac{-4+\sqrt{32}}{8}$

e. $7\sqrt{40}-\sqrt{90}$ **f.** $(x+\sqrt{5})(x-\sqrt{5})$ **g.** $\sqrt{\frac{2}{5x}}$ **h.** $\frac{8}{\sqrt{6}-\sqrt{2}}$

19. **Maximizing revenue:** Due to past experience, the manager of a video store knows that if a popular video game is priced at \$30, the store will sell 40 each day. For each decrease of \$0.50, one additional sale will be made. The formula for the store's revenue is then $R = (30 - 0.5x)(40 + x)$. Multiply the binomials and use a table of values to determine (a) the number of 50¢ decreases that will give the most revenue and (b) the maximum amount of revenue.

20. Use the Pythagorean theorem to determine the length of the diagonal of the rectangular prism shown in the figure.

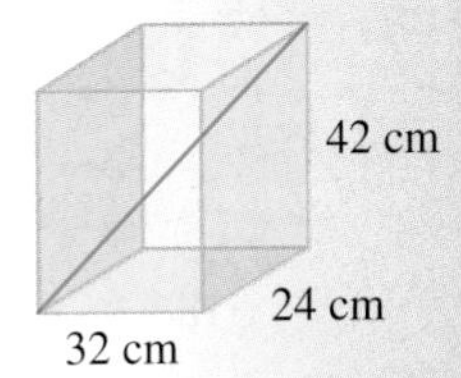

Chapter 1

Equations and Inequalities

Chapter Outline

1.1 Linear Equations, Formulas, and Problem Solving 72

1.2 Linear Inequalities in One Variable with Applications 83

1.3 Solving Polynomial and Other Equations 94

1.4 Complex Numbers 107

1.5 Solving Nonfactorable Quadratic Equations 117

Preview

This chapter is designed to further strengthen basic skills, as we look at numerous extensions of the concepts reviewed in Chapter R or in previous course work. In addition to opening the door to many other applications of mathematics, this material leads directly to a study of linear and quadratic functions—two powerful tools with innumerable applications. Once fundamental concepts and skills are in place, our mathematical journey becomes both fascinating and intriguing as we develop the ability to investigate, explore, model, extend, and apply mathematical ideas.

1.1 Linear Equations, Formulas, and Problem Solving

LEARNING OBJECTIVES

In Section 1.1 you will review how to:

- **A. Solve linear equations using the addition and multiplication properties of equality**
- **B. Recognize and understand equations that are identities or contradictions**
- **C. Solve for a specified variable in a formula or literal equation**
- **D. Use literal equations to find the general solution for a family of linear equations**
- **E. Use the problem-solving guide to solve various problem types**

INTRODUCTION

In a study of algebra, you will encounter many different **families of equations,** or groups of equations that share common characteristics. Of interest to us now is the family of *linear equations.* In addition to *solving linear equations,* we'll use the skills we develop to *solve for a specified variable* in a formula or literal equation, a practice widely used in many academic fields, as well as in business, industry, and research. The techniques we learn are often applied to create forms of an equation that are either more useful or easier to program, and will assist our study of functions in later chapters.

POINT OF INTEREST

The method of *false position* was known to the early Egyptians and used extensively in the Middle Ages to solve many linear equations. To solve the equation $x + \frac{x}{4} = 10$, assume (falsely) that $x = 4$. Although this gives $4 + \frac{4}{4} = 5$, twice 5 gives the desired result (10) and twice 4 gives the correct answer $x = 8$.

A. Solving Equations Using the Addition and Multiplication Properties of Equality

In Section R.2, we learned that an **algebraic expression** is a sum or difference of algebraic terms. Algebraic expressions can be simplified, evaluated, or written in an equivalent form, but they *cannot be solved,* since we are not seeking a specific value of the unknown. An **equation** is a *statement that two expressions are equal.* Our focus now is on **linear equations,** which can be identified using these three tests: (1) the exponent on any variable is a one, (2) no variable is used as a divisor, and (3) no two variables are multiplied together (see Exercises 7 through 12). Alternatively, we can say that a linear equation is any equation that can be written in the **standard form** $Ax + By = C$, where A and B are not simultaneously zero. The equation $2x = 9$ is a linear equation in one variable ($B = 0$), while $2x + 3y = 6$ is a linear equation in two variables ($A = 2$ and $B = 3$). To *solve a linear equation* in one variable, means we determine a specific input that will make the original equation true (left-hand expression equal to the right-hand expression). Inputs that result in a true equation are called **solutions** to the equation. The primary tools used in this process are the **additive and multiplicative properties of equality.**

THE ADDITIVE PROPERTY OF EQUALITY

Like quantities (numbers or terms) can be added to both sides of an equation without affecting the equality. Symbolically, if A and B are algebraic expressions where $A = B$, then $A + C = B + C$ (C can be positive or negative).

THE MULTIPLICATIVE PROPERTY OF EQUALITY

Both sides of an equation can be multiplied by the same nonzero quantity without affecting the equality. Symbolically, if A, B, and C are algebraic expressions where $A = B$, then

$$AC = BC \quad \text{and} \quad \frac{A}{C} = \frac{B}{C}, \; C \neq 0.$$

These fundamental properties apply to all equations, from the very simple to the more complex, and are used to rewrite an equation in **solution form:** *variable = number.* If any coefficients are fractional, we can multiply both sides by the least common multiple or LCM of all denominators to **clear the fractions** and reduce the work needed to solve the equation. The same idea can be applied to decimal coefficients.

EXAMPLE 1 Solve for n: $\frac{1}{4}(n + 8) - 2 = \frac{1}{2}(n - 6)$.

Solution:

$4[\frac{1}{4}(n + 8) - 2] = 4[\frac{1}{2}(n - 6)]$ multiply both sides by LCM = 4

$(n + 8) - 8 = 2(n - 6)$ distribute/simplify

$n = 2n - 12$ simplify

$12 = n$ subtract n and add 12

NOW TRY EXERCISES 13 THROUGH 30

The ideas illustrated in Example 1 can be summarized into a general strategy for solving linear equations. Not all steps are used for every equation, and those stated here are meant only as a guide.

A GENERAL APPROACH TO SOLVING LINEAR EQUATIONS

I. Simplify the equation
- Clear fractions or decimals as needed/desired.
- Eliminate parentheses using the distributive property and combine any like terms.

II. Solve the equation
- Use the additive property of equality to write the equation with all variable terms on one side and constants on the other. Simplify each side.
- Use the multiplicative property of equality to obtain solution form.

Circle your answer. For applications, answer in a complete sentence and be sure to include any units of measure.

B. Identities and Contradictions

The equation in Example 1 is called a **conditional equation,** since the equation is true for $n = 12$, but false for all other values of n. An **identity** is an equation that is *always true* for any real number input. For instance, $2(x + 3) = 2x + 6$ is true for any real number x. **Contradictions** are equations that are *never true,* no matter what real number is used for the variable. The equations $5 = -3$ and $x - 3 = x + 1$ are contradictions. Recognizing these special equations will prevent some surprises and indecision in later chapters.

EXAMPLE 2 Solve for x: $2(x - 4) + 10x = 8 + 4(3x + 1)$.

Solution:

$2(x - 4) + 10x = 8 + 4(3x + 1)$ original equation

$12x - 8 = 12x + 12$ distribute and simplify

$-8 = 12$ subtract $12x$ from both sides

Since -8 is never equal to 12, the original equation is a contradiction.

NOW TRY EXERCISES 31 THROUGH 36

Our attempt to solve for x in Example 2 ended with all variables being eliminated and the result was an equation that is never true—a contradiction (-8 is never equal to 12). There is nothing wrong with the solution process, the result simply tells us the original equation has *no solution.* To state the answer, use the symbol $\varnothing$ or indicate there are no solutions using the empty set "{ }." In other equations, it is possible for all variables to be eliminated but leave an equation that is always true—an identity. This result tells us the original equation has an infinite number of solutions. No matter what value we use for the variable, the result will be a true equation. The solution for an identity is often written in set notation as $\{n | n \in \mathbb{R}\}$.

C. Literal Equations and Solving for a Specified Variable

A **literal equation** is simply one that has two or more unknowns. Formulas are a type of literal equation, but not every literal equation is a formula. A **formula** is an equation that models a known relationship between two or more quantities. For example, $A = P + PRT$ is an equation that models the growth of money in an account earning simple interest, where A represents the total amount accumulated, P represents the initial deposit, R represents the annual interest rate, and T represents the number of years the money is left on deposit. To *describe* $A = P + PRT$, we might say the formula has been "solved for A" or that "A is written in terms of P, R, and T." In some cases, before using a formula it may be more convenient to first solve for one of the other variables, say P. In this case, P is called the **object variable.** Since the object variable occurs in more than one term, we first apply the distributive property, then use the equation-solving skills discussed earlier.

EXAMPLE 3 Given $A = P + PRT$, write P in terms of A, R, and T (solve for P).

Solution:

$A = \mathbf{P} + \mathbf{P}RT$ focus on P—the object variable

$A = \mathbf{P}(1 + RT)$ use distributive property to obtain a single occurrence of P

$\dfrac{A}{1 + RT} = \dfrac{\mathbf{P}(1 + RT)}{(1 + RT)}$ solve for P [divide by $(RT + 1)$]

$\dfrac{A}{1 + RT} = \mathbf{P}$ solution form

NOW TRY EXERCISES 37 THROUGH 48

We solve literal equations for a specified variable using the same methods as for equations and formulas. Remember that it's good practice to *focus on the object variable* to help guide you through the solution process.

WORTHY OF NOTE

In Example 4, notice that in the second step we wrote the subtraction of $2x$ as $-2x + 15$ instead of $15 - 2x$. For reasons that become clear later in this chapter, we generally write variable terms before constant terms.

EXAMPLE 4 Given $2x + 3y = 6$, write y in terms of x (solve for y).

Solution:

$2x + 3\mathbf{y} = 15$	focus on the object variable
$3\mathbf{y} = -2x + 15$	isolate term with y (subtract $2x$)
$\frac{1}{3}(3\mathbf{y}) = \frac{1}{3}(-2x + 15)$	solve for y (multiply by $\frac{1}{3}$)
$\mathbf{y} = \frac{-2}{3}x + 5$	distribute and simplify

NOW TRY EXERCISES 49 THROUGH 54

D. Literal Equations and a General Solution for $ax + b = c$

Solving literal equations for a specified variable can help us develop the general solution for an entire family of equations. This is demonstrated in Example 5 for the family of linear equations written in the form $ax + b = c$. A side-by-side comparison is used with a specific member of this family to illustrate that identical procedures are used.

EXAMPLE 5 Solve $2x + 3 = 15$ *and* the general linear equation $ax + b = c$ for x. Comment on the similarities.

Solution:

Linear Equation		Literal Equation
$2\mathbf{x} + 3 = 15$	focus on object variable	$a\mathbf{x} + b = c$
$2\mathbf{x} = 15 - 3$	subtract constant	$a\mathbf{x} = c - b$
$\mathbf{x} = \dfrac{15 - 3}{2}$	divide by coefficient	$\mathbf{x} = \dfrac{c - b}{a}$

Both equations were solved using the same ideas.

NOW TRY EXERCISES 55 THROUGH 60

In Example 5, we deliberately kept the solution on the left in unsimplified form to show the close relationship between standard equations and literal equations. Of course, the solution would be written as $x = 6$, which should be checked in the original equation. On the right, we now have a formula for all equations of the form $ax + b = c$. For instance, the solution to $4x + 5 = -25$, where $a = 4$, $b = 5$, and $c = -25$, is $x = \frac{-25 - 5}{4}$ or $-\frac{15}{2}$. While this has little practical use here, it does offer practice with identifying the input values and general formula use. In Section 1.5 this idea is used to develop the general solution for the *family of quadratic equations* written in the form $ax^2 + bx + c = 0$, a result with much greater significance.

E. Using the Problem-Solving Guide

Becoming a good problem solver is an evolutionary process. Over a period of time and with continued effort, you will begin to recognize the key fundamentals that make problem solving easier. Most good problem solvers also develop the following characteristics, which all students are encouraged to work on and improve within themselves:

- A positive attitude
- A mastery of basic facts
- Mental arithmetic skills
- Good mental-visual skills
- Estimation skills
- A willingness to persevere

These characteristics form a solid basis for applying what we will call the **Problem-Solving Guide,** which simply organizes the basic elements of good problem solving. Using this guide will help save you from two common pests—indecision and mind block.

PROBLEM-SOLVING GUIDE

- **Gather and organize information.**
  Read the problem several times, forming a mental picture as you read. *Highlight key phrases.* Begin developing ideas about possible approaches and operations to be used. List given information, including any related formulas. *Clearly identify what you are asked to find.*
- **Diagram the problem.**
  Draw and label a chart, table, or diagram as appropriate. This will help you see how different parts of the problem fit together. Label the diagram.
- **Build an equation model and estimate the answer.**
  Assign a descriptive variable to represent what you are asked to find. Build an equation model from the information given in the exercise. *Carefully reread the exercise to double-check your equation model. Determine a reasonable estimate, if possible.*
- **Use the model and given information to solve the problem.**
  Substitute given values, then simplify and solve. See how this result compares to the estimate. State the answer in sentence form, being sure the answer is reasonable and includes any units of measure.

Although every step may not be used each time you solve a problem, these *guidelines* give you a place to start and a sequence to follow. As your problem-solving skills grow, you will tend to use the guide as a road map rather than a formal procedure. Some of the adjustments might include *mentally* noting what information is given and using much less formal diagrams.

Descriptive Translation Exercises

In Section R.2, we learned to translate word phrases into symbols. This skill is used to build equation models from information given in paragraph form. Sometimes the variable *occurs more than once* in our model, because two different items in the same exercise are related. If the relationship involves a comparison of size, we often use line segments or bar graphs in our diagram to model the relative sizes.

EXAMPLE 6 The largest state in the United States is Alaska, which covers an area that is 230 more than 500 times the land area of the smallest state—Rhode Island. If they have a combined area of 616,460 mi^2, how many square miles does each cover?

Solution: Alaska covers 230 more than 500 times the land area of Rhode Island. — highlight key phrase

Let A represent the area of Rhode Island. — assign a variable

Then $500A + 230$ represents Alaska's area. — build related labels

$A + (500A + 230) = 616{,}460$ — equation model

$501A = 616{,}230$ — subtract 230

$A = 1230$ — divide by 501

+230
500 times
615,230
Rhode Island
Alaska

Rhode Island covers an area of approximately 1,230 mi^2, while Alaska covers an area of about $500(1230) + 230 = 615{,}230$ mi^2.

NOW TRY EXERCISES 63 THROUGH 68

Consecutive Integer Exercises

Although they have limited value in the real world, exercises involving **consecutive integers** offer excellent practice in assigning variables to unknown quantities, building related expressions, and the modeling process in general. We sometimes work with consecutive **odd** integers or consecutive **even** integers as well. The number line illustration in Figure 1.1 shows that consecutive odd integers are *two units* apart and labels should be built accordingly: $n, n + 2, n + 4$, and so on. If we know the exercise involves even numbers instead, the same model can be used. For consecutive integers, the labels are $n, n + 1, n + 2$, and so on.

Figure 1.1

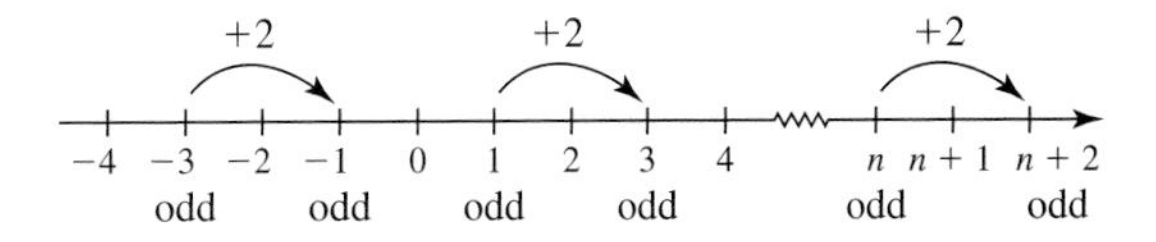

EXAMPLE 7 The sum of three consecutive *odd* integers is 69. What are the integers?

Solution: If n represents the smallest consecutive odd integer, $n + 2$ represents the second odd integer, and $(n + 2) + 2 = n + 4$ represents the third. In words: first + second + third odd integer = 69.

$$n + (n + 2) + (n + 4) = 69 \quad \text{equation model}$$
$$3n + 6 = 69 \quad \text{simplify}$$
$$3n = 63 \quad \text{subtract 6}$$
$$n = 21 \quad \text{solution (divide by 3)}$$

The odd integers are $n = 21$, $n + 2 = 23$, and $n + 4 = 25$.

$$21 + 23 + 25 = 69 ✓$$

NOW TRY EXERCISES 69 THROUGH 72

Uniform Motion (Distance, Rate, Time) Exercises

Uniform motion problems have many variations, and it's always a fun challenge to draw a good diagram, find a close estimate, and complete the exercise. Recall that distance = rate · time.

EXAMPLE 8 I live 260 mi from a popular mountain retreat. On my way there to do some mountain biking, my car had engine trouble—forcing me to bike the rest of the way. If I drove 2 hr longer than I biked, and averaged 60 miles per hour (mph) driving and 10 mph biking, how many hours did I spend peddling to the resort?

Solution: The **rates** are given, the driving **time** is *2 hr more than biking time,* and the sum of the distances travelled must be 260 mi. Here is a diagram and equation model.

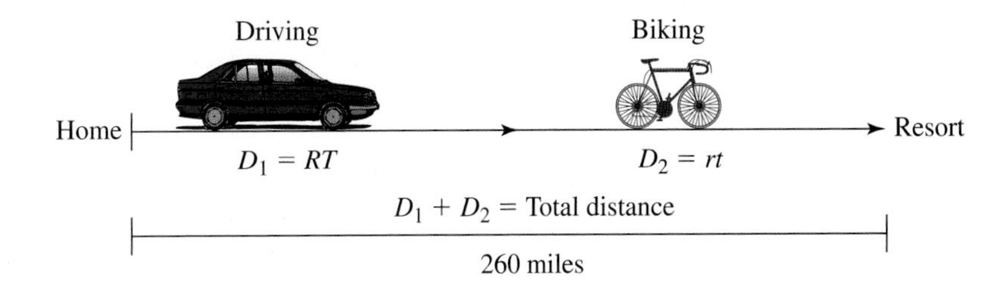

- Verbal model: The total distance is the sum of the driving and biking distances.
- Equation model: $RT + rt = 260$ miles.

Since the driving time is 2 hr more than biking time, $T = t + 2$.

$$RT + rt = 260 \quad \text{equation model: } RT = D_1,\ rt = D_2$$
$$60(t + 2) + 10t = 260 \quad \text{substitute } t + 2 \text{ for } T,\ R = 60, \text{ and } r = 10$$
$$70t + 120 = 260 \quad \text{distribute and simplify}$$
$$70t = 140 \quad \text{subtract 120}$$
$$t = 2 \quad \text{solve for } t$$

I rode my bike for $t = 2$ hr (after I had driven $t + 2 = 4$ hr).

NOW TRY EXERCISES 73 THROUGH 76

Exercises Involving Mixtures

Mixture problems give us another opportunity to refine our problem-solving skills. They lend themselves very nicely to a mental-visual image, allow for use of estimation skills, and have many practical applications in the real world. As with other applications, drawing a diagram or collecting given information in a table often suggests the equation model needed.

EXAMPLE 9 As a nasal decongestant, doctors sometimes prescribe saline solutions with a concentration between 6% and 20%. In "the old days," pharmacists had to create different mixtures, but only needed to stock these concentrations, since any percentage in between could be obtained using a mixture. An order comes in for 50 milliliters (mL) of a 15% solution. How much of each should be used?

Solution: For the estimate, assume we use 25 mL of the 6% solution, and 25 mL of the 20% solution. The final mixture would be 13%: $\frac{6 + 20}{2} = 13\%$. This is too low a concentration (we need a 15% solution), so we estimate that more than 25 mL of the 20% solution will be used. If A represents the amount of 20% solution used, then $50 - A$ represents the amount of 6% solution. Gathering the information in a table yields:

	Percent Concentration	Quantity Used	Amount in the Mixture
First quantity	0.20	A	$0.20A$
Second quantity	0.06	$50 - A$	$0.06(50 - A)$
Total	0.15	50	$0.15(50)$
			(equation column)

(first quantity)(percent) + (second quantity)(percent) = (first + second quantities)(desired percent)

$$0.20A + (0.06)(50 - A) = (50)(0.15) \quad \text{equation model}$$
$$0.14A + 3 = 7.5 \quad \text{distribute and simplify}$$
$$A \approx 32.1 \quad \text{solve for } A \text{ (nearest tenth)}$$

In line with our estimate, about 32.1 mL of the 20% solution and 17.9 mL of the 6% solution are used.

NOW TRY EXERCISES 77 THROUGH 84

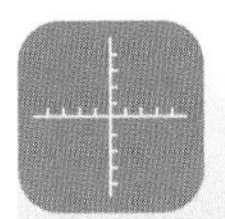

TECHNOLOGY HIGHLIGHT

Using a Graphing Calculator as an Investigative Tool

The keystrokes shown apply to a TI-84 Plus model. Please consult your manual or our Internel site for other models.

Graphing calculators are a wonderful investigative tool that can be used to explore a wide variety of applications. The table shown is an expanded, descriptive version of a numeric table that might be used to help solve applications involving mixtures. It is designed to help you visualize what happens as two different concentrations of saline solution are combined. By initially assuming equal amounts are mixed, we can then estimate whether more of the weaker or more of the stronger solution is required. Using an organized table of values, we can actually determine how much more. This exercise sheds light on how a general mixture equation is set up, and more importantly, why it is set up this way. With some modification, the idea can be extended to cover most mixture applications. Consider the following, How many ounces of a 40% glycerin solution must be added to 10 oz of 80% glycerin so that the resulting solution has a concentration of 56%? Complete the table using your calculator and the pattern shown, extending the table if needed. Then solve analytically and compare results.

Amount I = 10 (first guess)				**Amount II = 10 (given)**				**Percent (Amount I + Amount II)**		
Percent	·	**Amount I**		**Percent**	·	**Amount II**		**Percent**	·	**Total Liquid**
0.40	·	10	+	0.80	·	10	=	P		(10 + 10)
4			+	8			=	$20P$		
$12 = 20P \rightarrow P = 0.60$ (too high), so more of the 40% solution is needed.										
Second guess: Amount I = 13										
0.40	·	13	+	0.80	·	10	=	P		(13 + 10)
5.2			+	8			=	$23P$		
$13.2 = 23P \rightarrow P \approx 0.57$ (still too high), more of the 40% solution is needed.										
Third guess: Amount I = 16										
0.40	·	16	+	0.80	·	10	=	P		(? + ?)
What value goes here?			+	8			=	What value goes here?		
What equation goes here?										

Exercise 1: Should the next guess be more or less than 16 ounces? Why? Less; 0.554 is less than 0.56

Exercise 2: Use this idea to solve Exercises 81 and 82 from the Exercise Set. Answers will vary.

To view these results on the TI-84 Plus, assume x ounces of the 40% solution are used and enter the resulting mixture as Y1, with the result of the mix as Y2 (see Figure 1.2). Then set up a table using 2nd WINDOW (**TBLSET**) using TbStart = 10, ΔTbl = 1 with the calculator set in Indpnt: **AUTO**. The resulting screen is shown in Figure 1.3, where we note that 15 oz of the 40% solution should be used. For help with the TABLE feature, you can go to Section R.8 at www.mhhe.com/coburn.

Figure 1.2

```
Plot1 Plot2 Plot3
\Y1=.4X+.8(10)
\Y2=(10+X).56
\Y3=
\Y4=
\Y5=
\Y6=
\Y7=
```

Figure 1.3

X	Y1	Y2
10	12	11.2
11	12.4	11.76
12	12.8	12.32
13	13.2	12.88
14	13.6	13.44
15	14	14
16	14.4	14.56

X=10

1.1 EXERCISES

CONCEPTS AND VOCABULARY

Fill in each blank with the appropriate word or phrase. Carefully reread the section, if necessary.

1. A(n) identity is an equation that is always true, regardless of the unknown value.

2. A(n) contradiction is an equation that is always false, regardless of the unknown value.

3. A(n) literal equation is an equation having two or more unknowns.

4. For the equation $S = 2\pi r^2 + 2\pi rh$, we can say that S is written in terms of r and h.

5. Discuss/explain the three tests used to identify a linear equation. Give examples and counterexamples in your discussion. Answers will vary.

6. Discuss/explain each of the four basic parts of the *problem-solving guide.* Include a solved example in your discussion. Answers will vary.

DEVELOPING YOUR SKILLS

Identify each equation as linear or nonlinear. If nonlinear, state why. Do not solve.

7. $-2x + 7 = 60$ linear

8. $3m^2 - 5m = 9$ nonlinear; the exponent is two

9. $7 + 9d = 5$ linear

10. $\frac{n}{4} - 8 = 11$ linear

11. $2xy - 3 = 5$ nonlinear; two variables are multiplied together

12. $\frac{5}{x} + 2.5 = 7$ nonlinear; the variable is used as a divisor

Solve each linear equation.

13. $-2(3y + 5) = -7 + 4y - 12$ $\frac{9}{10}$

14. $-3(2x - 5) = x - 5 - 2$ $\frac{22}{7}$

15. $8 - (3n + 5) = -5 + 2(n + 1)$ $\frac{6}{5}$

16. $2a + 4(a - 1) = 3 - (2a + 1)$ $\frac{3}{4}$

17. $2(3m + 5) = 5 - 2(m - 1)$ $\frac{-3}{8}$

18. $7 - 4(x - 2) = 2(3x + 4)$ $\frac{7}{10}$

19. $\frac{1}{2}x + 5 = \frac{1}{3}x + 7$ 12

20. $-4 + \frac{2}{3}y = \frac{1}{2}y + (-5)$ -6

21. $15 = -6 - \frac{3y}{8}$ -56

22. $-15 - \frac{2w}{9} = -21$ 27

23. $\frac{n}{2} + \frac{n}{5} = \frac{2}{3}$ $\frac{20}{21}$

24. $-\frac{x}{3} - 2 = \frac{x}{2}$ $\frac{-12}{5}$

25. $\frac{2}{3}(m + 6) = \frac{-1}{2}$ $\frac{-27}{4}$

26. $\frac{4}{5}(n - 10) = \frac{-8}{9}$ $\frac{80}{9}$

27. $0.2(2.4 - 3.8x) - 5.4 = 0$ $\frac{-123}{19}$

28. $0.4(8.5 - 3.2a) - 9.8 = 0$ -5

29. $-5 - (3n + 4) = -8 - 2n$ -1

30. $-12 - 5y = -9 - (6y + 7)$ -4

Identify the following equations as an identity, a contradiction, or a conditional equation. If conditional, state the solution.

31. $-3(4z + 5) = -15z - 20 + 3z$ contradiction

32. $5x - 9 - 2 = -5(2 - x) - 1$ identity

33. $8 - 8(3n + 5) = -5 + 6(1 + n)$ conditional; $x = -1.1$

34. $2a + 4(a - 1) = 1 + 3(2a + 1)$ contradiction

35. $-4(4x + 5) = -6 - 2(8x + 7)$ identity

36. $-(5x - 3) + 2x = 11 - 4(x + 2)$ conditional; $x = 0$

Solve for the specified variable in each formula or literal equation.

37. $I = PRT$ for R (finance)

38. $V = LWH$ for W (geometry)

39. $C = 2\pi r$ for r (geometry)

40. $C = \pi d$ for d (geometry)

41. $W = I^2R$ for R (circuits)

42. $H = \frac{D^2N}{2.5}$ for N (horsepower)

43. $V = \frac{4}{3}\pi r^2 h$ for h (geometry)

44. $V = \frac{1}{3}\pi r^2 h$ for h (geometry)

45. $\frac{A}{6} = s^2$ for A (geometry)

46. $2A = d^2$ for A (geometry)

37. $R = \frac{I}{PT}$
38. $W = \frac{V}{LH}$
39. $r = \frac{C}{2\pi}$
40. $d = \frac{C}{\pi}$
41. $R = \frac{W}{I^2}$
42. $N = \frac{2.5H}{D^2}$
43. $h = \frac{3V}{4\pi r^2}$
44. $h = \frac{3V}{\pi r^2}$
45. $A = 6s^2$
46. $A = \frac{d^2}{2}$

47. $P = \dfrac{2(S - B)}{S}$

48. $g = \dfrac{2(s - vt)}{t^2}$

49. $y = \dfrac{-Ax}{B} + \dfrac{C}{B}$

50. $y = \dfrac{-2}{3}x + 2$

51. $y = \dfrac{-20}{9}x + \dfrac{16}{3}$

52. $y = \dfrac{6}{7}x - \dfrac{108}{7}$

53. $y = \dfrac{-4}{5}x - 5$

54. $y = \dfrac{-2}{15}x - \dfrac{16}{3}$

47. $S = B + \frac{1}{2}PS$ for P (geometry)

48. $s = \frac{1}{2}gt^2 + vt$ for g (physics)

49. $Ax + By = C$ for y

50. $2x + 3y = 6$ for y

51. $\frac{5}{6}x + \frac{3}{8}y = 2$ for y

52. $\frac{2}{3}x - \frac{7}{9}y = 12$ for y

53. $y - 3 = \frac{-4}{5}(x + 10)$ for y

54. $y + 4 = \frac{-2}{15}(x + 10)$ for y

The following equations are given in $ax + b = c$ form. Solve by identifying the value of a, b, and c, then using the formula $x = \dfrac{c - b}{a}$.

55. $3x + 2 = -19$ $a = 3; b = 2; c = -19; x = -7$

56. $7x + 5 = 47$ $a = 7; b = 5; c = 47; x = 6$

57. $-6x + 1 = 33$ $a = -6; b = 1; c = 33; x = \frac{-16}{3}$

58. $-4x + 9 = 43$ $a = -4; b = 9; c = 43; x = \frac{-17}{2}$

59. $2x - 13 = -27$ $a = 2; b = -13; c = -27; x = -7$

60. $3x - 4 = -25$ $a = 3; b = -4; c = -25; x = -7$

WORKING WITH FORMULAS

61. Surface area of a cylinder: $SA = 2\pi r^2 + 2\pi rh$

The surface area of a cylinder is given by the formula shown, where h is the height of the cylinder and r is the radius of the base. Find the height of a cylinder that has a radius of 8 cm and a surface area of 1256 cm^2. Use $\pi \approx 3.14$. $h = 17$ cm

62. Using the equation-solving process for Exercise 61 as a model, solve the formula $SA = 2\pi r^2 + 2\pi rh$ for h. $h = \dfrac{SA - 2\pi r^2}{2\pi r}$

APPLICATIONS

Solve by building an equation model and using the problem-solving guidelines as needed.

Descriptive Translation Exercises

Exercise 63

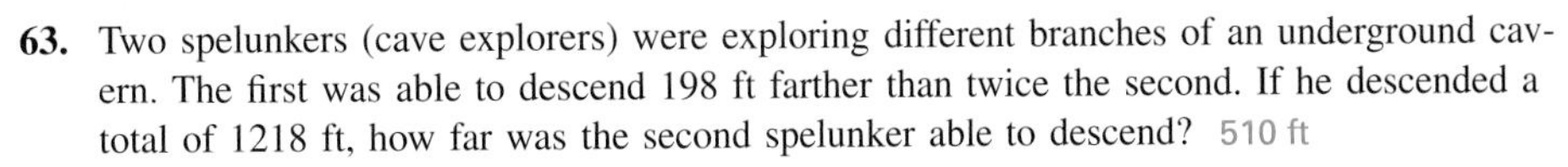

63. Two spelunkers (cave explorers) were exploring different branches of an underground cavern. The first was able to descend 198 ft farther than twice the second. If he descended a total of 1218 ft, how far was the second spelunker able to descend? 510 ft

64. The area near the joining of the Tigris and Euphrates Rivers (in modern Iraq) has often been called the *Cradle of Civilization,* since the area has evidence of many ancient cultures. The length of the Euphrates River exceeds that of the Tigris by 620 mi. If they have a combined length of 2880, how long is each river? Tigris is 1130 mi; Euphrates is 1750 mi long

65. U.S. postal regulations require that a package can have a maximum combined length and girth (distance around) of 108 in. A shipping carton is constructed so that it has a width of 14 in., a height of 12 in., and can be cut or folded to various lengths. What is the maximum length that can be used? 56 in.

Source: www.USPS.com

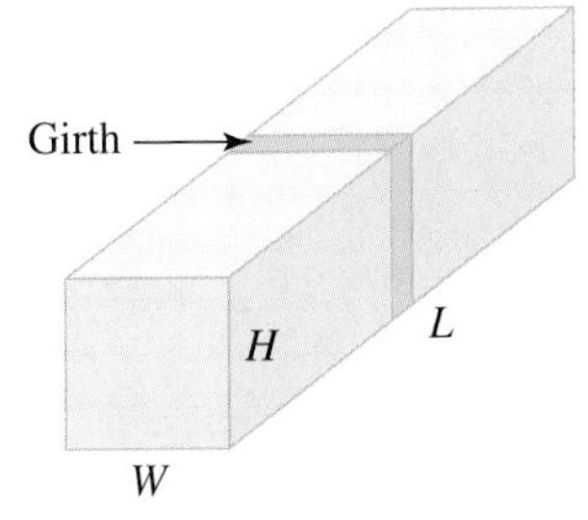

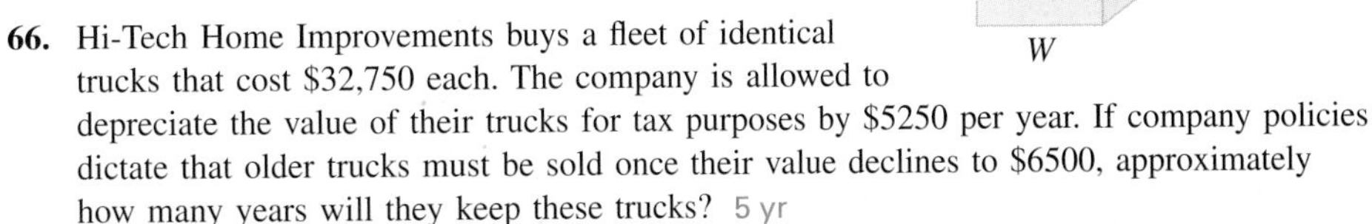

66. Hi-Tech Home Improvements buys a fleet of identical trucks that cost \$32,750 each. The company is allowed to depreciate the value of their trucks for tax purposes by \$5250 per year. If company policies dictate that older trucks must be sold once their value declines to \$6500, approximately how many years will they keep these trucks? 5 yr

67. The longest suspension bridge in the world is the Akashi Kaikyo (Japan) with a length of 6532 feet. Japan is also home to the Shimotsui Straight bridge. The Akashi Kaikyo bridge is three hundred sixty-four feet more than twice the length of the Shimotsui bridge. How long is the Shimotsui bridge? 3084 ft

Source: www.guinnessworldrecords.com

68. The Mars rover *Spirit* landed on January 3, 2004. Just over 1 yr later, on January 14, 2005, the *Huygens* probe landed on Titan (one of Saturn's moons). At their closest approach, the distance from the Earth to Saturn is 29 million mi more than 21 times the distance from the Earth to Mars. If the distance to Saturn is 743 million mi, what is the distance to Mars? 34 million mi

Consecutive Integer Exercises

69. Find two consecutive even integers such that the sum of twice the smaller integer plus the larger integer is one hundred forty-six. 48; 50

70. When the smaller of two consecutive integers is added to three times the larger, the result is fifty-one. Find the smaller integer. 12

71. Seven times the first of two consecutive odd integers is equal to five times the second. Find each integer. 5; 7

72. Find three consecutive even integers where the sum of triple the first and twice the second is eight more than four times the third. 20; 22; 24

Uniform Motion Exercises

73. At 9:00 A.M., Belinda started from home going 30 mph. At 11:00 A.M., Chris started after her on the same road at 45 mph. At what time did Chris overtake Belinda? 3 P.M.

74. A plane flying at 600 mph has a 3-hr head start on a "chase plane," which has a speed of 800 mph. How far from the starting point will the chase plane overtake the first plane? 7200 mi

75. Jeff had a job interview in a nearby city 72 mi away. On the first leg of the trip he drove an average of 30 mph through a long construction zone, but was able to drive 60 mph after passing through this zone. If the trip took $1\frac{1}{2}$ hr, how long was he driving in the construction zone? 36 min

76. At a high-school cross-country meet, Jared jogged 8 mph for the first part of the race, then increased his speed to 12 mph for the second part. If the race was 21 mi long and Jared finished in 2 hr, how far did he jog at the faster pace? 15 mi

Mixture Exercises

Give the total amount of the mix that results and the percent concentration or worth of the mix.

77. Two quarts of 100% orange juice are mixed with 2 quarts of water (0% juice). 4 quarts; 50% O.J.

78. Ten pints of a 40% acid are combined with 10 pints of an 80% acid. 20 pints; 60% acid

79. Eight pounds of premium coffee beans worth \$2.50 per pound are mixed with 8 lb of standard beans worth \$1.10 per pound. 16 lb; \$1.80 lb

80. A rancher mixes 50 lb of a custom feed blend costing \$1.80 per pound, with 50 lb of cheap cottonseed worth \$0.60 per pound. 100 lb; \$1.20 lb

Solve each application of the mixture concept.

81. How much pure antifreeze must be added to 10 gal of 20% antifreeze to make a 50% antifreeze solution? 6 gal

82. How much pure solvent must be added to 600 ounces of a $16\frac{2}{3}\%$ solvent to increase its strength to $37\frac{1}{2}\%$? 200 oz

83. How many pounds of walnuts at 84¢/lb should be mixed with 20 lb of pecans at \$1.20/lb to give a mixture worth \$1.04/lb? 16 lb

84. How many pounds of cheese worth 81¢/lb must be mixed with cheese worth \$1.29/lb to make 16 lb of a mixture worth \$1.11/lb? 6 lb of 81¢/lb cheese

WRITING, RESEARCH, AND DECISION MAKING

85. Whoever developed the concept of magnetic memory devices (tapes, CDs, computer disks, etc.) must have had a sense of humor. The units used to measure memory capacity are bits, nybbles, bytes, big bytes (kilobytes), and great big bytes (megabytes). Research how these units are related. If it takes 8 bits to store one character on a computer disk, how many characters can be stored on a 1,400,000-byte ($3\frac{1}{4}$-in.) disk? about 1.4 million

86. Look up and read the following article. Then turn in a one page summary. "Don't Give Up!," William H. Kraus, *Mathematics Teacher,* Volume 86, Number 2, February 1993: pages 110–112. Answers will vary

EXTENDING THE CONCEPT

87. $106\frac{2}{3}$ oz of 15% acid; $93\frac{1}{3}$ oz of 45% acid
88. 1 way; (18, 19, 20, 21, 22)
89. 69

87. A chemist has four solutions of a very rare and expensive chemical that are 15% acid (cost \$120 per ounce), 20% acid (cost \$180 per ounce), 35% acid (cost \$280 per ounce) and 45% acid (cost \$359 per ounce). She requires 200 ounces of a 29% acid solution. Find the combination of any two of these concentrations that will minimize the total cost of the mix.

88. The sum of at *least* two consecutive positive integers is 100. How many ways can this happen?

89. $P, Q, R, S, T,$ and U represent numbers. The arrows in the figure show the sum of the two or three numbers added in the indicated direction (Example: $Q + T = 23$). Find $P + Q + R + S + T + U$.

Exercise 89

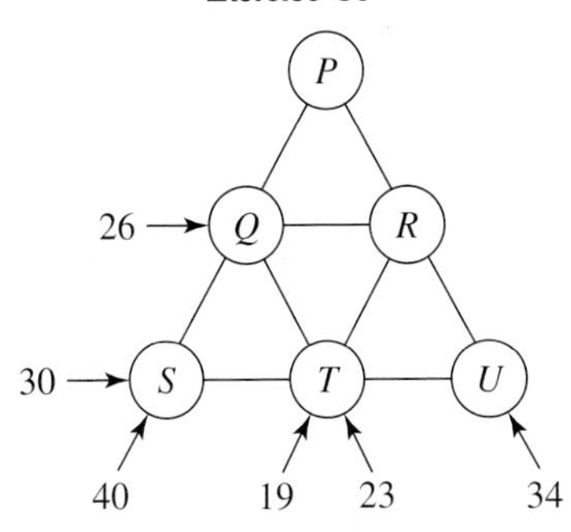

MAINTAINING YOUR SKILLS

90. (R.1) Simplify the expression using the order of operations.

$-2 - 6^2 \div 4 + 8$ -3

91. (R.3) Name the coefficient of each term in the expression:

$-3v^3 + v^2 - \frac{v}{3} + 7$ $-3, 1, -\frac{1}{3}, 7$

92. (R.4) Factor each expression:

a. $4x^2 - 9$ $(2x + 3)(2x - 3)$ **b.** $x^3 - 27$ $(x - 3)(x^2 + 3x + 9)$

93. (R.2) Identify the property illustrated:

$\frac{6}{7} \cdot 5 \cdot 21 = \frac{6}{7} \cdot 21 \cdot 5$ commutative property

94. (R.2) Are the terms $4n$, $-3n$, and n like terms or unlike terms? Why?
$4n$, $-3n$, and n are like terms, since they have like variables with like exponents

95. (R.3) Write the polynomial in standard form:

$2x - 3x^2 + 5x - 9 + x^3$ $x^3 - 3x^2 + 7x - 9$

1.2 Linear Inequalities in One Variable with Applications

LEARNING OBJECTIVES

In Section 1.2 you will review:

A. Inequalities and solution sets
B. Solving linear inequalities
C. Compound inequalities
D. Applications of inequalities

INTRODUCTION

There are many real-world situations where the mathematical model leads to a statement of *inequality* rather than equality. Here are a few examples:

Clarice wants to buy a house costing \$85,000 or less.

To earn a "B," Shantë must score more than 90% on the final exam.

To escape the Earth's gravity, a rocket must travel 25,000 mph or more.

While linear equations have a single solution, linear inequalities often have an *infinite number of solutions*—which means we must develop additional methods for naming a solution set.

POINT OF INTEREST

Thomas Harriot (1560–1621) in his work *Artis Analyticae Praxis (Practice of the Analytic Art)* was the first to denote multiplication using a raised dot as in $2 \cdot 3 = 6$. He also appears responsible for introducing the inequality symbols "<" for less than and ">" for greater than, which were a great improvement over the symbols ⌋ and ⌊ introduced by William Oughtred (1574–1660).

A. Inequalities and Solution Sets

In Section R.1 we introduced the notation used for basic inequalities. In this section, we develop the ability to use inequalities as a modeling and problem-solving tool. The set of numbers that satisfy an inequality is called the **solution set.** Instead of using a simple inequality to write solution sets, we will often use (1) a form of **set notation,** (2) a **number line** graph, or (3) **interval notation.** Interval notation is simply a summary of what is shown on a number line graph. In Section R.1, we marked the *location* of a number on the number line with a bold dot "•." When the number acts as the **boundary point** for an interval (also called an **endpoint**), we use a left bracket "[" or a right bracket "]" to indicate **inclusion** of the endpoint. If the boundary point is **not included,** we use a left parenthesis "(" or right parenthesis ")."

WORTHY OF NOTE

Some texts will use an open dot "○" to mark the location of an endpoint that is not included, and a closed dot "•" for an included endpoint.

EXAMPLE 1 Model the given phrase using the correct symbol. Then state the solution set in set notation, as a number line graph, and in interval notation: "All real numbers greater than or equal to 1."

Solution: Let n represent a real number: $n \geq 1$.

- Set notation: $\{n \mid n \geq 1\}$ "The set of all n such that n is greater than or equal to 1."
- Number line: 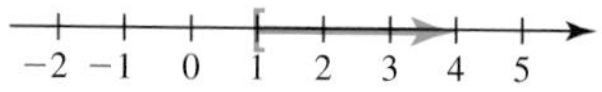

- Interval notation: $n \in [1, \infty)$

NOW TRY EXERCISES 7 THROUGH 18

WORTHY OF NOTE

Since infinity is really a *concept* rather than a number, it is *never included* (using a bracket) as an endpoint for an interval.

Recall the "∈" symbol says the number n is *an element of the set or interval* given. The "∞" symbol represents positive infinity and indicates that the interval continues forever to the right. Note that the endpoints of the interval notation must occur in the same order as on the number line *(smaller value on the left; larger value on the right).*

A summary of various other possibilities is given here

Condition(s)	Set Notation	Number Line Graph	Interval Notation
x is greater than k	$\{x \mid x > k\}$		$x \in (k, \infty)$
x is greater than or equal to k	$\{x \mid x \geq k\}$		$x \in [k, \infty)$
x is less than k	$\{x \mid x < k\}$		$x \in (-\infty, k)$
x is less than or equal to k	$\{x \mid x \leq k\}$	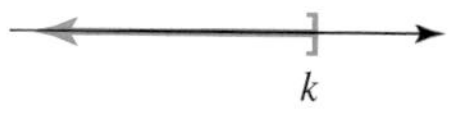	$x \in (-\infty, k]$
x is less than b and greater than a	$\{x \mid a < x < b\}$	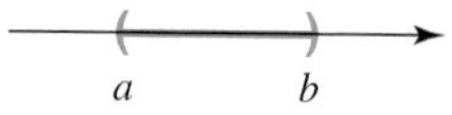	$x \in (a, b)$
x is less than or equal to b and greater than a	$\{x \mid a < x \leq b\}$	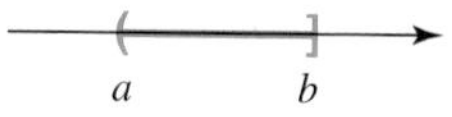	$x \in (a, b]$
x is less than b and greater than or equal to a	$\{x \mid a \leq x < b\}$	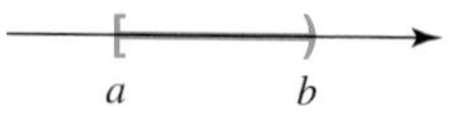	$x \in [a, b)$
x is less than or equal to b and greater than or equal to a	$\{x \mid a \leq x \leq b\}$		$x \in [a, b]$

B. Solving Linear Inequalities

A linear *inequality* resembles a linear *equality* in many respects:

	linear inequality	**related linear equation**
(1)	$x < 3$	$x = 3$
(2)	$\frac{3}{8}p - 2 \geq -12$	$\frac{3}{8}p - 2 = -12$

For a given polynomial inequality, the solutions of the related equation yield the boundary points. For this reason, this section reflects the basic elements of the equation solving seen earlier. A linear inequality in one variable is one that can be written in the form $ax + b < c$, where a, b, and $c \in \mathbb{R}$ and $a \neq 0$. This definition and the following properties also apply when any of the other inequality symbols are used. Solutions for many inequalities are easy to spot. For instance, $x = -2$ is a solution to $x < 3$ since $-2 < 3$. For more involved inequalities we use the **additive property of inequality** (API) and the **multiplicative property of inequality** (MPI). Similar to solving equations, the goal is still to isolate the variable on one side, and obtain a solution form such as *variable* $<$ *number*. The final question we must always ask is, "Are the endpoints included?"

> **THE ADDITIVE PROPERTY OF INEQUALITY**
> Like quantities (numbers or terms) can be added to both sides of an inequality. Symbolically, if A and B are algebraic expressions where $A < B$, then $A + C < B + C$ (C can be positive or negative).

While there is little difference between the additive property of *equality* and the additive property of *inequality,* there is an *important difference* between the multiplicative property of *equality* and the multiplicative property of *inequality.* To illustrate, we begin with the inequality $-2 < 5$. Multiplying by positive three yields $-6 < 15$, also a true inequality. But notice what happens when we **multiply by negative three:**

$$-2 < 5 \quad \text{original inequality}$$
$$-2(-3) < 5(-3) \quad \text{multiply by negative three}$$
$$6 < -15 \quad \text{result}$$

This is a *false* inequality, because 6 is *to the right* of -15 on the number line $(6 > -15)$. Multiplying (or dividing) an inequality by a negative number *changes the order the numbers occur* on the number line, and we must compensate for this by reversing the inequality symbol.

$$6 > -15 \quad \text{change direction of symbol to maintain a true statement}$$

For this reason, the multiplicative property of inequality is stated in two parts.

> **THE MULTIPLICATIVE PROPERTY OF INEQUALITY**
> Assume A and B represent algebraic expressions:
>
> If $A < B$ and C is a ***positive number,*** then $AC < BC$ is a true inequality (the inequality symbol remains the same).
>
> If $A < B$ and C is a ***negative number,*** then $AC > BC$ is a true inequality (the inequality symbol must be reversed).

As before, we attempt to write the equation with the variable terms on one side using the additive property, then simplify and solve for the variable using the multiplicative property.

EXAMPLE 2 Solve the inequality, then graph the solution set and write it in interval notation: $\frac{-2}{3}x + \frac{1}{2} \leq \frac{5}{6}$.

Solution:

$\frac{-2}{3}x + \frac{1}{2} \leq \frac{5}{6}$ original inequality

$6\left(\frac{-2}{3}x + \frac{1}{2}\right) \leq (6)\frac{5}{6}$ clear fractions (multiply by least common multiple)

$-4x + 3 \leq 5$ simplify

$-4x \leq 2$ subtract 3 (API)

$x \geq -\frac{1}{2}$ or -0.5 divide by -4, *reverse inequality sign*

- Number line: 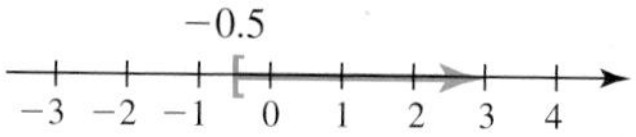

- Interval notation: $x \in \left[-\frac{1}{2}, \infty\right)$

NOW TRY EXERCISES 19 THROUGH 34

To check a linear equation, you have a limited number of choices—the value obtained in the solution process. To check a linear inequality, you often have an infinite number of choices—any number from the solution interval. If the test value results in a true inequality, all numbers in the interval will satisfy the original inequality. For Example 2, $x = 0$ is in the solution interval and sure enough, $\frac{-2}{3}(0) + \frac{1}{2} \leq \frac{5}{6} \rightarrow \frac{1}{2} \leq \frac{5}{6}$✓.

C. Solving Compound Inequalities

In some applications of inequalities, we must consider more than one solution interval. These are called **compound inequalities,** and require us to take a closer look at the operations of **union** "$\cup$" and **intersection** "$\cap$." The intersection of two sets A and B, written $A \cap B$, is the set of all members *common to both sets.* The union of two sets A and B, written $A \cup B$, is the set of all members *that are in either set.* When stating the union of two sets, repetitions are unnecessary.

EXAMPLE 3 For set $A = \{-2, -1, 0, 1, 2, 3\}$ and set $B = \{1, 2, 3, 4, 5\}$, determine $A \cap B$ and $A \cup B$.

Solution: $A \cap B$ is the set of all elements in *both A and B:* $A \cap B = \{1, 2, 3\}$. $A \cup B$ is the set of all elements in *either A or B:* $A \cup B = \{-2, -1, 0, 1, 2, 3, 4, 5\}$.

NOW TRY EXERCISES 35 THROUGH 40

LOOKING AHEAD

These descriptions are used extensively in the solution of various kinds of inequalities, particularly the absolute value inequalities studied in Section 6.4.

Notice the intersection of two sets is described using the word "and," while the union of two sets is described using the word "or." When compound inequalities are formed using these designations, the solution is modeled after the ideas from Example 3. If "and" is used, the solutions must satisfy *both* inequalities. If "or" is used, the solutions can satisfy *either* inequality.

EXAMPLE 4 Solve the compound inequality: $3x + 5 < -1$ *and* $3x + 5 > -13$.

Solution: For the inequality $3x + 5 < -1$ the solution is $x < -2$. For $3x + 5 > -13$ the solution is $x > -6$. The solution for the compound inequality $x < -2$ *and* $x > -6$ can easily be seen by graphing each interval separately, then *noting where they intersect.*

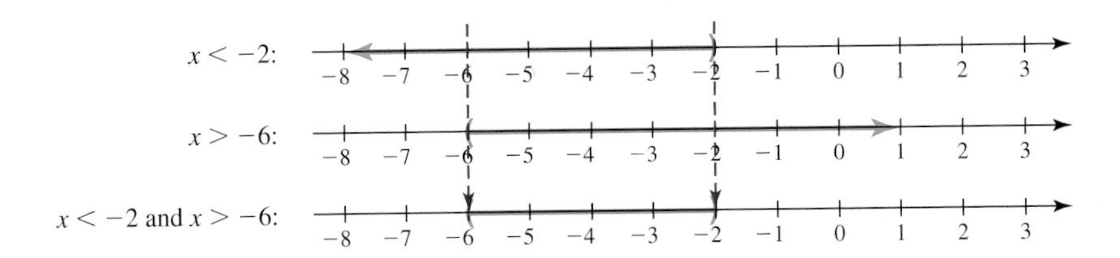

The solution is $x \in (-6, -2)$.

NOW TRY EXERCISES 41 AND 42

EXAMPLE 5 Solve the inequality $-3x - 1 < -4$ *or* $4x + 3 < -6$.

Solution: For $-3x - 1 < -4$, the solution is $x > 1$ (remember to reverse the inequality symbol). For $4x + 3 < -6$ the solution is $x < -\frac{9}{4}$. The solution for $x > 1$ *or* $x < -\frac{9}{4}$ can be seen by graphing each interval separately, *then selecting both intervals (the union).*

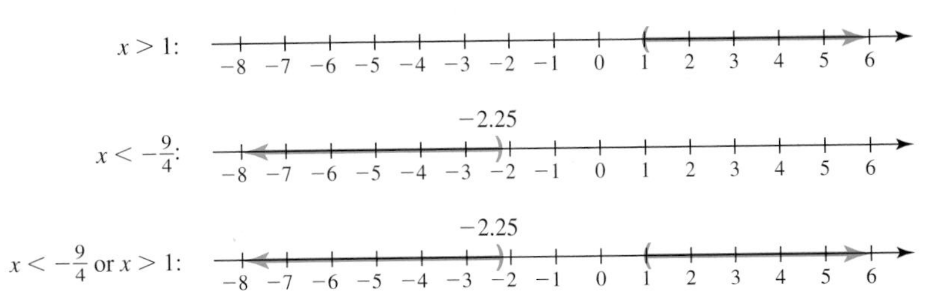

The solution is $x \in \left(-\infty, -\frac{9}{4}\right) \cup (1, \infty)$.

NOW TRY EXERCISES 43 THROUGH 54

The inequality $-12 < x < -6$ is also called a joint inequality, because it "joins" the inequalities $x > -12$ and $x < -6$ (read from middle to left: $-12 < x$, then middle to right $x < -6$). We solve joint inequalities in much the same way as linear inequalities, but must remember that *these have three parts (left-middle-right),* while simple inequalities have just two parts *(left-right).* For joint inequalities, operations must be applied to *all three parts* as you go through the solution process. Our goal is still to isolate the variable, to obtain solution form: *smaller number* $< x <$ *larger number.* The same ideas apply when other inequality symbols are used.

EXAMPLE 6 Solve the compound inequality, then graph the solution set and write it in interval notation: $1 > \frac{2x+5}{-3} \geq -6$.

Solution:

$1 > \frac{2x+5}{-3} \geq -6$ original inequality

$-3 < 2x + 5 \leq 18$ multiply all parts by -3; reverse the inequality symbols

$-8 < 2x \leq 13$ subtract 5 from all parts (API)

$-4 < x \leq 6.5$ divide all parts by 2 (MPI)

- Number line:
- Interval notation: $x \in (-4, 6.5]$

NOW TRY EXERCISES 55 THROUGH 60

As you work your way through the Exercise Set, be aware that some compound inequalities may yield the empty set: { } with no solutions, while others may have all real numbers: $\{x|x \in \mathbb{R}\}$ as the solution.

D. Applications of Inequalities

Table 1.1

x	$\frac{24}{x}$
6	4
−12	−2
$\frac{1}{2}$	48
0	??

Domain and Allowable Values

One application of inequalities involves the concept of **allowable values.** Consider the expression $\frac{24}{x}$. As you see in Table 1.1, we can evaluate this expression using any real number *other than zero,* since the expression $\frac{24}{0}$ is undefined. Using set notation the allowable values are written $\{x|x \in \mathbb{R}, x \neq 0\}$. To graph the solution on a number line, we must be careful to exclude zero, as shown in Figure 1.4.

Figure 1.4

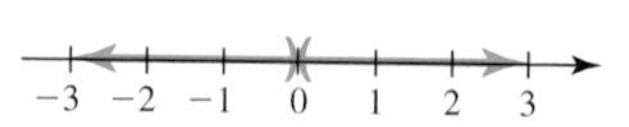

The graph gives us a snapshot of the solution using interval notation, which is written as a union of two intervals so as to exclude zero: $x \in (-\infty, 0) \cup (0, \infty)$. In many cases, the set of allowable values is referred to as the **domain** of the expression. Allowable values are said to be "*in the domain*" of the expression; values that are not allowed are said to be "*outside the domain.*" When the denominator of a fraction contains a variable expression, values of the unknown that make the denominator zero are excluded from the domain.

EXAMPLE 7 Determine the allowable value(s) for the expression $\frac{6}{x-2}$. Give your answer in set notation, as a number line graph, and using interval notation.

Solution: We exclude those numbers that cause the denominator to be zero: $x - 2 \neq 0$ means $x \neq 2$.

- Set notation: $\{x|x \in \mathbb{R}, x \neq 2\}$
- Number line:
- Interval notation: $x \in (-\infty, 2) \cup (2, \infty)$

NOW TRY EXERCISES 61 THROUGH 68

A second area where allowable values are a concern involves the square root operation. Recall that $\sqrt{49} = 7$ since $7 \cdot 7 = 49$. However, the radical $\sqrt{-49}$ cannot be written as the product of two real numbers since $(-7) \cdot (-7) = +49$ and $7 \cdot 7 = +49$. In other words, the square root operation represents a real number only if the radicand is positive or zero. If A represents an algebraic expression, the domain of $\sqrt{A}$ is $\{A | A \geq 0\}$.

EXAMPLE 8 Determine the domain of the expression $\sqrt{x + 3}$. Give your answer in set notation, as a number line graph, and in interval notation.

Solution: To find allowable values, the radicand must represent a nonnegative number. Solving the inequality $x + 3 \geq 0$ gives $x \geq -3$.

- Set notation: $\{x | x \geq -3\}$
- Number line: 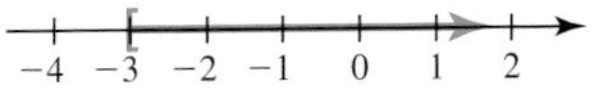

- Interval notation: $x \in [-3, \infty)$

NOW TRY EXERCISES 69 THROUGH 76

Descriptive Translation Exercises

Use the problem-solving guide to solve the application in Example 9.

EXAMPLE 9 Justin earned scores of 78, 72, and 86 on the first three out of four exams. What must he earn on the fourth exam to have an average of at least 80?

Solution:

- **Gather and organize information**
  First the scores: 78, 72, 86. An average of *at least* 80 means $A \geq 80$.

- **Use a chart, table, or diagram**

Test 1	Test 2	Test 3	Test 4		Average
78	72	86	70	$\frac{306}{4}$	76.5
78	72	86	80	$\frac{316}{4}$	79
78	72	86	90	$\frac{326}{4}$	81.5
78	72	86	x	$\frac{\text{total}}{4}$	80

- **Build an equation model, estimate**
  From the table, we estimate that Justin needs about an 85 on test 4.

$$\frac{78 + 72 + 86 + \boldsymbol{x}}{4} \geq 80 \quad \text{compute average}$$

- **Use the model and given information to solve the problem**

$$78 + 72 + 86 + \boldsymbol{x} \geq 320 \quad \text{multiply by 4}$$
$$236 + \boldsymbol{x} \geq 320 \quad \text{simplify}$$
$$\boldsymbol{x} \geq 84 \quad \text{solve for } x$$

Justin must score at least an 84 on the last test to earn an 80 average.

NOW TRY EXERCISES 87 THROUGH 94

TECHNOLOGY HIGHLIGHT

Understanding Unions, Intersections, and Inequalities

The keystrokes shown apply to a TI-84 Plus model. Please consult your manual or our Internet site for other models.

If you're having trouble understanding intersections "∩," unions "∪," as well as the "ands" and "ors," this technology highlight might help. Most graphing calculators are programmed with what are called **logical operators,** that can be used to test a number of different relationships. This feature enables you to set up a relationship, test it, then reason out why certain statements are true (the calculator returns a "1"), while others are false (the calculator returns a "0"). Here we'll set up logical relations on the home screen, and test the relation using numbers entered earlier in a list. For convenience we'll enter the values $-4, -2, 0, 2,$ and 4 in List 1 of the six lists available. To begin, clear out any old entries in L1 by pressing STAT **4:ClrList,** which places **"ClrList"** on the home screen, and clear List 1 by pressing 2nd 1 (**L1**) ENTER. The calculator will notify you that this has been "Done." Data can be entered one-at-a-time on the STAT **1:Edit** screen, or as a set of numbers from the home screen. To use the latter method, enter the set $\{-4, -2, 0, 2, 4\}$ using the braces found above the parentheses keys and tell the calculator to STO→ (store the list in) 2nd 1 (**L1**). The calculator automatically places the entries in List 1 and responds by displaying the list itself (Figure 1.5). First let's test the *and* operator using the relation $-3 < x$ *and* $x < 4$. To be a solution, a number must simultaneously be greater than -3 and less than 4. We can test this relation for all numbers in the list by entering $-3 <$ L1 *and* L1 < 4 on the home screen. Both the inequality symbols and the logical operators are accessed using 2nd MATH, which enables you to choose between the inequality symbols (**TEST**) as well as the relations (**LOGIC**). After pressing ENTER your screen should return the result shown in the first two lines of Figure 1.6. The set displayed is equivalent to {F, T, T, T, F} for each of the numbers in the order they occur in our list. Sure enough, -4 is not greater than -3 (F) and 4 is not less than 4 (4 is *equal to* 4). Now test the relations $-3 < x$ or $x < 4$, then $-3 > x$ and $x < 4$. Did you anticipate the output also shown in Figure 1.6? All numbers in our list satisfy the "or" test since each of them is either greater than -3 or less than 4.

Figure 1.5

```
ClrList L1
                Done
{-4,-2,0,2,4}→L1
      {-4 -2 0 2 4}
```

Figure 1.6

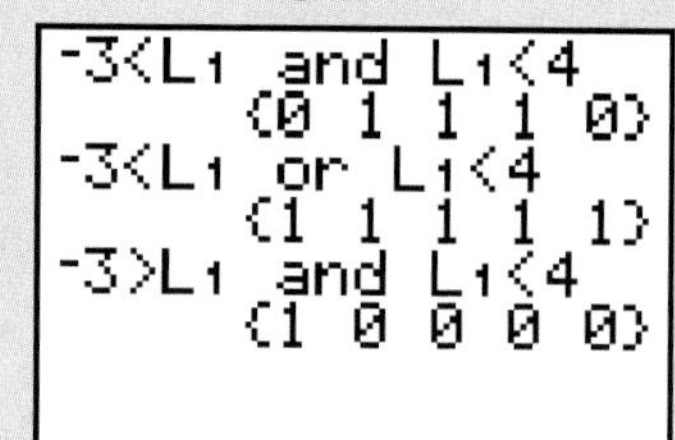

Exercise 1: Repeat these tests after replacing L1 with 0.5L1 − 4 (e.g., for the first test, enter $-3 <$ 0.5L1 − 4 and 0.5L1 − 4 < 4). Analyze the results displayed for each element in the list.

Exercise 2: Repeat these tests using $\{-3, -2, -1, 0, 1, 2, 3, 4\}$. What do you notice about the endpoints?

Exercise 1

1. {0 0 0 0 1}; only 4 satisfies $2 < L_1$ and $L_1 < 16$
   {1 1 1 1 1}; all elements satisfy $2 < L_1$ or $L_1 < 16$
   {1 1 1 0 0}; only $-4, -2, 0$ satisfy $2 > L_1$ and $L_1 < 16$

Exercise 2

2. {0 1 1 1 1 1 1 0}; not included
   {1 1 1 1 1 1 1 1}; included
   {0 0 0 0 0 0 0 0}; not included

1.2 EXERCISES

CONCEPTS AND VOCABULARY

Fill in each blank with the appropriate word or phrase. Carefully reread the section, if necessary.

1. For inequalities, the three ways of writing a solution set are _set_ notation, a number line graph, and _interval_ notation.

2. The mathematical sentence $3x + 5 < 7$ is a(n) _linear_ inequality, while $-2 < 3x + 5 < 7$ is a(n) _compound_ inequality.

11.

12.

13.

14.

15.

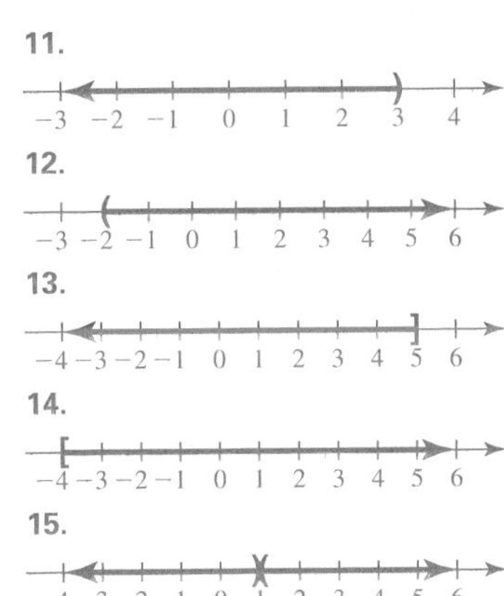

16.

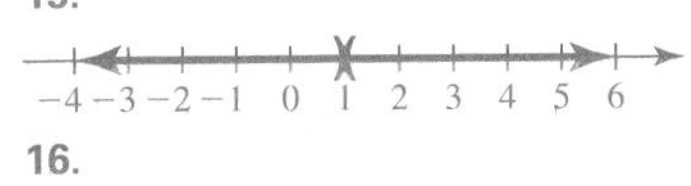

17.

18.

19. $\{a \mid a \geq 2\}$;

$a \in [2, \infty)$

20. $\{n \mid n < \frac{17}{6}\}$;

$n \in (-\infty, \frac{17}{6})$

21. $\{n \mid n \geq 1\}$;

$n \in [1, \infty)$

22. $\{x \mid x > -3\}$;

$x \in (-3, \infty)$

23. $\{x \mid x < \frac{-32}{5}\}$;

$x \in (-\infty, \frac{-32}{5})$

24. $\{y \mid y < -4\}$;

$y \in (-\infty, -4)$

25. $\{y \mid 2 < y < 5\}$;

$y \in (2, 5)$

26. $\{n \mid \frac{-16}{3} < n < \frac{16}{9}\}$;

$n \in (\frac{-16}{3}, \frac{16}{9})$

Additional answers can be found in the Instructor Answer Appendix.

3. The intersection of sets A and B is written $A \cap B$. The union of sets A and B is written $A \cup B$.

4. The intersection of set A with set B is the set of elements in A $\cap$ B. The union of set A with set B is the set of elements in A $\cup$ B.

5. Discuss/explain how the concept of domain and allowable values relates to rational and radical expressions. Include a few examples. Answers will vary.

6. Discuss/explain why the inequality symbol must be reversed when multiplying or dividing by a negative quantity. Include a few examples. Answers will vary.

DEVELOPING YOUR SKILLS

Use an inequality to write a mathematical model for each statement. Use descriptive variables.

7. To qualify for a secretarial position, a person must type at least 45 words per minute. $w \geq 45$

8. The balance in a checking account must remain above $1000 or a fee is charged. $b > 1000$

9. To bake properly, a turkey must be kept between the temperatures of 250° and 450°. $250 < T < 450$

10. To fly effectively, the airliner must cruise at or between altitudes of 30,000 and 35,000 ft. $30{,}000 \leq a \leq 35{,}000$

Graph each inequality on a number line.

11. $y < 3$
12. $x > -2$
13. $m \leq 5$
14. $n \geq -4$
15. $x \neq 1$
16. $x \neq -3$
17. $5 > x > 2$
18. $-3 < y \leq 4$

Solve the inequality, then write the solution set in set notation, number line notation, and interval notation.

19. $5a - 11 \geq 2a - 5$
20. $-8n + 5 > -2n - 12$
21. $2(n + 3) - 4 \leq 5n - 1$
22. $-5(x + 2) - 3 < 3x + 11$
23. $\frac{3x}{8} + \frac{x}{4} < -4$
24. $\frac{2y}{5} + \frac{y}{10} < -2$
25. $1 < \frac{1}{2}y < \frac{5}{2}$
26. $-2 < \frac{3}{8}n < \frac{2}{3}$
27. $-3 < 2m - 7 \leq 5$
28. $-1 < 2x + 5 \leq 10$
29. $3 > 3(2m - 1) - 5 \geq 0$
30. $7 \geq 3(x + 4) - 2 > 0$

Write the solution set illustrated on each graph in set notation and interval notation.

31. $\{x \mid x \geq -2\}; [-2, \infty)$

32. $\{x \mid x < 1\}; (-\infty, 1)$

33. $\{x \mid -2 \leq x \leq 1\}; [-2, 1]$

34. $\{x \mid -2 \leq x < 3\}; [-2, 3]$

Determine the intersection and union of sets A, B, C, and D as indicated, given $A = \{-3, -2, -1, 0, 1, 2, 3\}$, $B = \{2, 4, 6, 8\}$, $C = \{-4, -2, 0, 2, 4\}$, and $D = \{4, 5, 6, 7\}$.

35. $A \cap B$ and $A \cup B$ — $\{2\}; \{-3, -2, -1, 0, 1, 2, 3, 4, 6, 8\}$
36. $A \cap C$ and $A \cup C$ — $\{-2, 0, 2\}; \{-4, -3, -2, -1, 0, 1, 2, 3, 4\}$
37. $A \cap D$ and $A \cup D$ — $\{\,\}; \{-3, -2, -1, 0, 1, 2, 3, 4, 5, 6, 7\}$
38. $B \cap C$ and $B \cup C$ — $\{2, 4\}; \{-4, -2, 0, 2, 4, 6, 8\}$
39. $B \cap D$ and $B \cup D$ — $\{4, 6\}; \{2, 4, 5, 6, 7, 8\}$
40. $C \cap D$ and $C \cup D$ — $\{4\}; \{-4, -2, 0, 2, 4, 5, 6, 7\}$

Express the compound inequalities in number line and interval notation.

41. $x < 5$ and $x \geq -2$
42. $x \geq -4$ and $x < 3$
43. $x < -2$ or $x > 1$
44. $x < -5$ or $x > 5$
45. $x \geq 3$ and $x \leq 1$ no solution
46. $x \geq -5$ and $x \leq -7$ no solution

Solve the compound inequalities and graph the solution set.

47. $4(x - 1) \leq 20$ or $x + 6 > 9$
48. $-3(x + 2) > 15$ or $x - 3 \leq -1$
49. $-2x - 7 \leq 3$ and $2x \leq 0$
50. $-3x + 5 \leq 17$ and $5x \leq 0$

51. $x \in \left(\frac{-1}{3}, \frac{-1}{4}\right)$;

52. $x \in \left(\frac{2}{3}, \frac{5}{4}\right]$;

53. $x \in (-\infty, \infty)$;

54. $x \in (-\infty, -4) \cup (5, \infty)$;

55. $x \in [-4, 1)$;

56. $x \in \left(2, \frac{23}{3}\right]$; $7\frac{2}{3}$

57. $x \in [-1.4, 0.8]$;

58. $x \in (2.3, 9.6)$;

59. $x \in [-16, 8)$;

60. $x \in (3, 45]$;

51. $\frac{3}{5}x + \frac{1}{2} > \frac{3}{10}$ and $-4x > 1$

52. $\frac{2}{3}x - \frac{5}{6} \le 0$ and $-3x < -2$

53. $\frac{3x}{8} + \frac{x}{4} < -3$ or $x + 1 > -5$

54. $\frac{2x}{5} + \frac{x}{10} < -2$ or $x - 3 > 2$

55. $-3 \le 2x + 5 < 7$

56. $2 < 3x - 4 \le 19$

57. $-0.5 \le 0.3 - x \le 1.7$

58. $-8.2 < 1.4 - x < -0.9$

59. $-7 < -\frac{3}{4}x - 1 \le 11$

60. $-21 \le -\frac{2}{3}x + 9 < 7$

Determine the allowable value(s) for each expression. Write your answer in interval notation.

61. $\frac{12}{m}$ $m \in (-\infty, 0) \cup (0, \infty)$

62. $\frac{-6}{n}$ $n \in (-\infty, 0) \cup (0, \infty)$

63. $\frac{5}{y + 7}$ $y \in (-\infty, -7) \cup (-7, \infty)$

64. $\frac{4}{x - 3}$ $x \in (-\infty, 3) \cup (3, \infty)$

65. $\frac{a + 5}{6a - 3}$ $a \in \left(-\infty, \frac{1}{2}\right) \cup \left(\frac{1}{2}, \infty\right)$

66. $\frac{m + 5}{8m + 4}$ $m \in \left(-\infty, \frac{-1}{2}\right) \cup \left(\frac{-1}{2}, \infty\right)$

67. $\frac{15}{3x - 12}$ $x \in (-\infty, 4) \cup (4, \infty)$

68. $\frac{7}{2x + 6}$ $x \in (-\infty, -3) \cup (-3, \infty)$

Determine the domain for each expression. Write your answer in interval notation.

69. $\sqrt{x - 2}$ $x \in [2, \infty)$

70. $\sqrt{y + 7}$ $y \in [-7, \infty)$

71. $\sqrt{3n - 12}$ $n \in [4, \infty)$

72. $\sqrt{2m + 5}$ $m \in \left[\frac{-5}{2}, \infty\right)$

73. $\sqrt{b - \frac{4}{3}}$ $b \in \left[\frac{4}{3}, \infty\right)$

74. $\sqrt{a + \frac{3}{4}}$ $a \in \left[\frac{-3}{4}, \infty\right)$

75. $\sqrt{8 - 4y}$ $y \in (-\infty, 2]$

76. $\sqrt{12 - 2x}$ $x \in (-\infty, 6]$

Place the correct inequality symbol in the blank to make the statement true.

77. If $m > 0$ and $n < 0$, then mn $\underline{<}$ 0.

78. If $m > n$ and $p > 0$, then mp $\underline{>}$ np.

79. If $m < n$ and $p > 0$, then mp $\underline{<}$ np.

80. If $m \le n$ and $p < 0$, then mp $\underline{\ge}$ np.

81. If $m > n$, then $-m$ $\underline{<}$ $-n$.

82. If $m < n$, then $\frac{1}{m}$ $\underline{>}$ $\frac{1}{n}$.

83. If $m > 0$ and $n < 0$, then m^2 $\underline{>}$ n.

84. If $m < 0$, then m^3 $\underline{<}$ 0.

WORKING WITH FORMULAS

85. Body mass index: BMI $= \frac{704W}{H^2}$

The U.S. government publishes a body mass index formula to help people consider the risk of heart disease. An index of 27 means that a person is at risk. Here W represents weight and H represents height in inches. If your height is 5′8″ what could your weight be to remain safe from the risk of heart disease? 177.34 lb or less

Source: www.surgeongeneral.gov/topics.

86. Lift capacity: $75S + 125B \le 750$

The capacity in pounds of the lift used by a roofing company to place roofing shingles and buckets of roofing nails on rooftops is modeled by the formula shown, where S represents packs of shingles and B represents buckets of nails. Use the formula to find (a) the largest number of shingle packs that can be lifted, (b) the largest number of nail buckets that can be lifted, and (c) the largest number of shingle packs that can be lifted along with three nail buckets. 10; 6; 5

APPLICATIONS

Write an inequality to model the given information and solve.

87. Exam scores: Jasmine scored 68% and 75% on two exams. To keep her financial aid, she must bring her average up to at least an 80%. What must she earn on the third exam? $x \ge 97\%$

88. Exam scores: Jacques is going to college on an academic scholarship that requires him to maintain at least a 75% average in all of his classes. So far he has scored 82%, 76%, 65%, and 71% on four exams. What scores are possible on his last exam that will enable him to keep his scholarship? $x \ge 81\%$

89. Temperature conversion: When the outside temperature drops below 45°F or exceeds 85°F, there is concern for the elderly living in the city without air-conditioning and heating subsidies. What would the corresponding Celsius range be? Recall that $F = \frac{9}{5}C + 32$. $7.2°C < T < 29.4°C$

90. Area of a rectangle: Given the rectangle shown, what is the range of values for the width, in order to keep the area less than 150 m^2? $w < 7.5$m

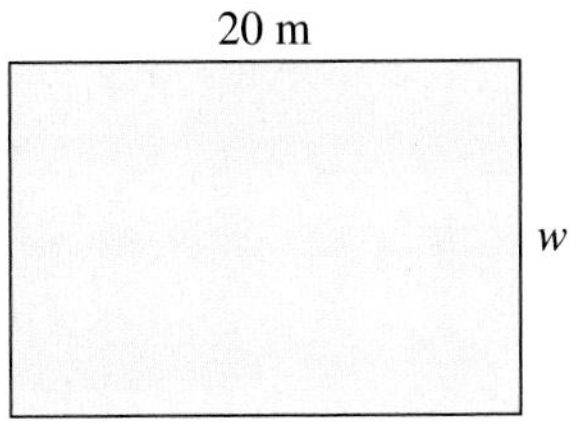

91. Checking account balance: If the average daily balance in a certain checking account drops below $1000, the bank charges the customer a $7.50 service fee. The table gives the daily balance for one customer. What must the daily balance be for Friday to avoid a service charge? $b \geq \$2000$

Weekday	Balance
Monday	$1125
Tuesday	$850
Wednesday	$625
Thursday	$400

92. Average weight: In the National Football League, many consider an offensive line to be "small" if the average weight of the five down linemen is less than 325 lb. Using the table, what must the weight of the right tackle be so that the line will not be considered too small? $w \geq 344$ lb

Lineman	Weight
Left tackle	318 lb
Left guard	322 lb
Center	326 lb
Right guard	315 lb
Right tackle	?

Exercise 93

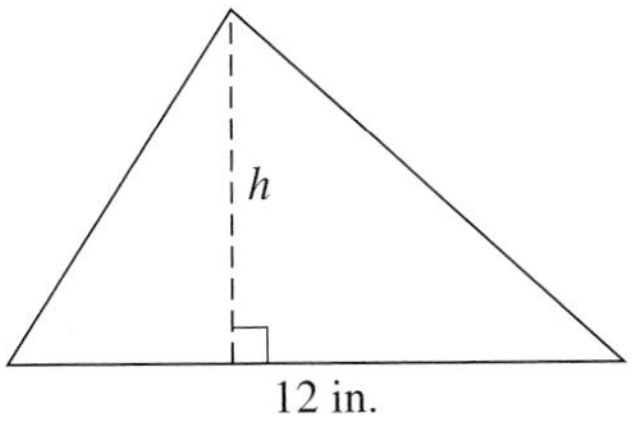

93. Using the triangle shown, find the height that will guarantee an area equal to or greater than 48 in^2. $h \geq 8$ in.

94. In the first three trials of the 100-m butterfly, Johann had times of 50.2, 49.8, and 50.9 sec. How fast must he swim the final timed trial to have an average time of 50 sec? 49.1 sec

WRITING, RESEARCH, AND DECISION MAKING

95. Use a current world almanac or some other source to find the record high and low temperatures for Alaska and Hawaii. Express each temperature range as a compound inequality. Which of the two states has the greatest range (difference between high and low temperatures)? What is the range of temperatures for your home state?

95. Alaska $-80°F \leq T \leq 100°F$; Hawaii $12°F \leq T \leq 100°F$; Alaska; Answers will vary.

96. Use your local library, the Internet, or another resource to find the highest and lowest point on each of the seven continents. Express the range of altitudes for each continent as a compound inequality. Which continent do you consider to be the flattest (having the smallest range)? Answers may vary depending on source

EXTENDING THE CONCEPT

97. Use a table of values or trial and error to find the solution set for the inequality $|x - 2| \geq 4$. Then graph the solution and write it in interval notation.

97.

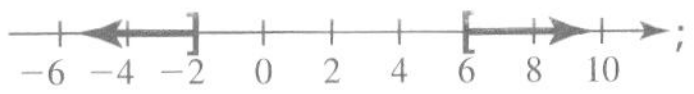

$x \in (-\infty, -2] \cup [6, \infty)$

98. The sum of two consecutive even integers is greater than or equal to 12 and less than or equal to 22. List all possible values for the two integers. 6 and 8; 8 and 10; 10 and 12

MAINTAINING YOUR SKILLS

99. (R.2) Translate into an algebraic expression: eight subtracted from twice a number. $2n - 8$

100. (R.3) Simplify the algebraic expression: $2(\frac{5}{9}x - 1) - (\frac{1}{6}x + 3)$. $\frac{17}{18}x - 5$

101. (R.7) Find the volume of the composite solid. $420 + 30.625\pi$ cm^3

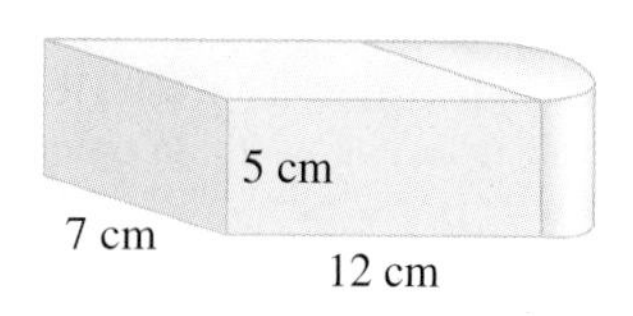

102. (R.6) Find the missing side of the right triangle. 6 yd

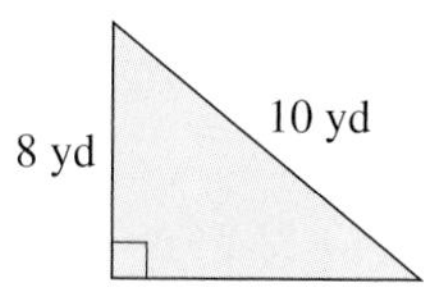

103. (1.1) Solve: $-4(x - 7) - 3 = 2x + 1$ $x = 4$

104. (1.1) Solve: $\frac{4}{5}m + \frac{2}{3} = \frac{1}{2}$ $m = \frac{-5}{24}$

1.3 Solving Polynomial and Other Equations

LEARNING OBJECTIVES

In Section 1.3 you will learn how to:

A. Solve polynomial equations using the zero factor property

B. Solve rational equations

C. Solve radical equations

D. Solve applications using these equation types

INTRODUCTION

The ability to solve linear and quadratic equations is the foundation on which a large percentage of our future studies are built. Both are closely linked to the solution of many other equation types, as well as to the graphs of these equations. In this section we get our first glimpse of these connections, as we learn to solve certain polynomial equations, then use this ability to solve rational and radical equations.

POINT OF INTEREST

While polynomial, rational, and radical equations appear to be very different, all belong to the class of *algebraic functions,* meaning they can be solved using basic algebraic tools (simplifying expressions and properties of equality). Rational and radical equations are often defined in terms of polynomials, making the solution of polynomial equations a key skill. In contrast, logarithmic, exponential, trigonometric, and other equations belong to the class of *transcendental functions,* meaning their solution depends on tools that transcend the algebraic.

A. Polynomial Equations and the Zero Factor Property

A **quadratic equation** is one that can be written as $ax^2 + bx + c = 0$, where $a \neq 0$. In *standard form,* the terms are written in decreasing order of degree and the expression is set equal to zero. The equations $x^2 + 7x + 10 = 0$ and $2x^2 - 18 = 0$ (where $b = 0$) are good examples. With quadratic and other polynomial equations, we cannot isolate the variable on one side using only properties of equality, because the variable is raised to two different powers. Instead, we try to solve the equation by factoring the expression and applying the **zero factor property.** In words, the property says, *If the product of any two (or more) factors is equal to zero, then at least one of the factors must be equal to zero* (later in this chapter we'll study methods for solving equations that cannot be factored).

THE ZERO FACTOR PROPERTY

Given that A and B represent real numbers or real-valued expressions, if $A \cdot B = 0$, then either $A = 0$ or $B = 0$.

EXAMPLE 1 Solve the equation: $2x^3 - 20x = 3x^2$.

Solution:

$2x^3 - 20x = 3x^2$	given equation
$2x^3 - 3x^2 - 20x = 0$	standard form
$x(2x^2 - 3x - 20) = 0$	common factor is x
$x(2x + 5)(x - 4) = 0$	factored form: the product of x, $(2x + 5)$, and $(x - 4)$ is zero
$x = 0$ or $2x + 5 = 0$ or $x - 4 = 0$	set each factor equal to zero
$x = 0$ or $x = \frac{-5}{2}$ or $x = 4$	result

NOW TRY EXERCISES 7 THROUGH 14

The zero factor property can be applied to any polynomial written in factored form. Be sure the equation is in standard form before you begin and remember to first remove any factors common to all terms. For instance, the equations $2x^2 + 6 = 56$ and $d^2 + 37 = 12d + 1$ can be rewritten as $2(x^2 - 25) = 0$ and $d^2 - 12d + 36 = 0$ respectively, and solved by factoring. Verify the solutions are $x = -5$ and $x = 5$ in the first case and $d = 6$ for the second (see Exercises 15–32).

B. Solving Rational Equations

In Section 1.1 we solved linear equations using basic properties of equality. If any equation contained fractional terms, we "cleared the fractions" using the least common multiple (LCM). This idea is also used to solve equations with rational expressions, and the process is summarized here. Since we're working with rational expressions, we must be mindful of values that cause any denominator to become zero and exclude these values. Finally, note the least common denominator and the least common multiple represent the same quantity.

SOLVING RATIONAL EQUATIONS

1. Identify and exclude any values that cause a zero denominator.
2. Multiply both sides by the LCM and simplify (this will eliminate all denominators).
3. Solve the resulting equation using properties of equality.
4. Check the solutions in the original equation.

EXAMPLE 2 Solve for m: $\frac{2}{m} - \frac{1}{m - 1} = \frac{4}{m^2 - m}$.

Solution: Since $m^2 - m = m(m - 1)$, the LCM is $m(m - 1)$, where $m \neq 0$ and $m \neq 1$.

$m(m - 1)\left(\frac{2}{m} - \frac{1}{m - 1}\right) = m(m - 1)\left[\frac{4}{m(m - 1)}\right]$	multiply by LCM
$2(m - 1) - m = 4$	simplify—denominators are eliminated
$2m - 2 - m = 4$	distribute
$m = 6$	solve for m

Check by substituting $m = 6$ into the original equation.

NOW TRY EXERCISES 33 THROUGH 38

It's possible for a rational equation to have more than one solution, or even no solutions, due to domain restrictions. Also, if we solve a rational equation and obtain one of the excluded values as a "solution," that number is called an **extraneous root** and is discarded from the solution set.

EXAMPLE 3 Solve: $x + \frac{12}{x-3} = 1 + \frac{4x}{x-3}$.

Solution: The LCM is $x - 3$, where $x \neq 3$. Multiplying both sides by $x - 3$ gives:

$$(x-3)\left(x + \frac{12}{x-3}\right) = (x-3)\left(1 + \frac{4x}{x-3}\right) \quad \text{multiply both sides by LCM}$$

$$x^2 - 3x + 12 = x - 3 + 4x \quad \text{simplify—denominators are eliminated}$$

$$x^2 - 8x + 15 = 0 \quad \text{simplify and set equal to zero}$$

$$(x-3)(x-5) = 0 \quad \text{factor}$$

$$x = 3 \quad \text{or} \quad x = 5 \quad \text{zero factor property}$$

Checking shows $x = 3$ is an extraneous root, while $x = 5$ is a valid solution.

NOW TRY EXERCISES 39 THROUGH 44

In many fields of study, rational equations and formulas involving rational expressions are used as equation models. There is frequently a need to solve these equations for one variable in terms of others, a skill closely related to our work in Section 1.1.

EXAMPLE 4 Solve for the indicated variable: $S = \frac{a}{1-r}$ for r.

$$S = \frac{a}{1-r} \quad \text{LCM is } 1 - r$$

$$(1-r)S = (1-r)\left(\frac{a}{1-r}\right) \quad \text{multiply both sides by } (1-r)$$

$$S - Sr = a \quad \text{simplify—denominator is eliminated}$$

$$-Sr = a - S \quad \text{isolate term with } r$$

$$r = \frac{a-S}{-S} \quad \text{solve for } r \text{ (divide both sides by } -S)$$

$$r = \frac{S-a}{S} \quad \text{multiply numerator/denominator by } -1$$

NOW TRY EXERCISES 45 THROUGH 52

WORTHY OF NOTE

Generally, we should try to write rational answers with the fewest number of negative signs possible. Multiplying the numerator and denominator in Example 4 by -1 gave $r = \frac{S-a}{S}$, which is a more acceptable answer.

C. Radical Equations and Equations with Rational Exponents

To solve a radical equation, we attempt to isolate a radical term on one side, then apply the appropriate nth power to free up the radicand and solve for the unknown. This is an application of the **power property of equality.**

THE POWER PROPERTY OF EQUALITY

Let $\sqrt[n]{u}$ and v be real numbers or real-valued expressions, where n is an integer and $n \geq 2$.

If $\sqrt[n]{u} = v$, then $(\sqrt[n]{u})^n = v^n$

(recall that if n is even, u must be nonnegative)

Raising both sides of an equation to an *even* power sometimes introduces a "false solution," an extraneous root. For instance, the equation $x = 2$ has 2 as the sole solution, but $x^2 = 4$ has solutions $x = 2$ and $x = -2$. This means we should *check all solutions of an equation where an even power is applied.*

EXAMPLE 5 Solve the radical equation: $\sqrt{x + 1} - 12 = -10$.

Solution:

$\sqrt{x+1} - 12 = -10$ original equation

$\sqrt{x+1} = 2$ isolate radical term (add 12)

$(\sqrt{x+1})^2 = (2)^2$ apply power property (square both sides)

$x + 1 = 4$ simplify $(\sqrt{x+1})^2 = x + 1$

$x = 3$ result

Check: $x = 3$: $\sqrt{3+1} - 12 = \sqrt{4} - 12 = -10$✓

NOW TRY EXERCISES 53 THROUGH 56

Sometimes squaring both sides of an equation still results in an equation with a radical term, but often there is *one fewer* than before. In this case, we simply repeat the solution process, as indicated by the flowchart in Figure 1.7.

EXAMPLE 6 Solve the equation: $\sqrt{x + 15} - \sqrt{x + 3} = 2$.

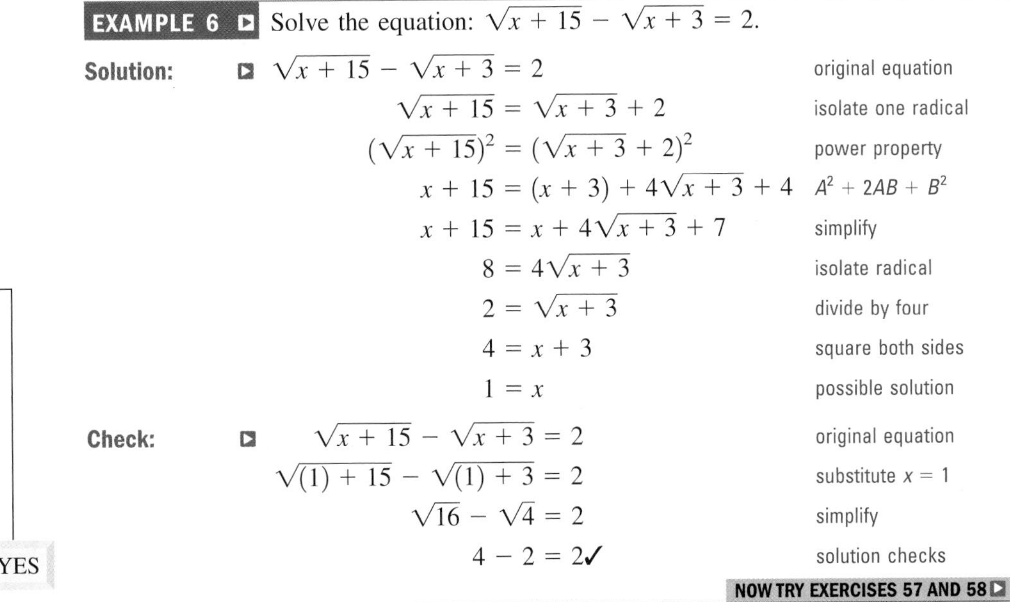

Solution:

$\sqrt{x+15} - \sqrt{x+3} = 2$ original equation

$\sqrt{x+15} = \sqrt{x+3} + 2$ isolate one radical

$(\sqrt{x+15})^2 = (\sqrt{x+3} + 2)^2$ power property

$x + 15 = (x + 3) + 4\sqrt{x+3} + 4$ $A^2 + 2AB + B^2$

$x + 15 = x + 4\sqrt{x+3} + 7$ simplify

$8 = 4\sqrt{x+3}$ isolate radical

$2 = \sqrt{x+3}$ divide by four

$4 = x + 3$ square both sides

$1 = x$ possible solution

Check:

$\sqrt{x+15} - \sqrt{x+3} = 2$ original equation

$\sqrt{(1)+15} - \sqrt{(1)+3} = 2$ substitute $x = 1$

$\sqrt{16} - \sqrt{4} = 2$ simplify

$4 - 2 = 2$✓ solution checks

NOW TRY EXERCISES 57 AND 58

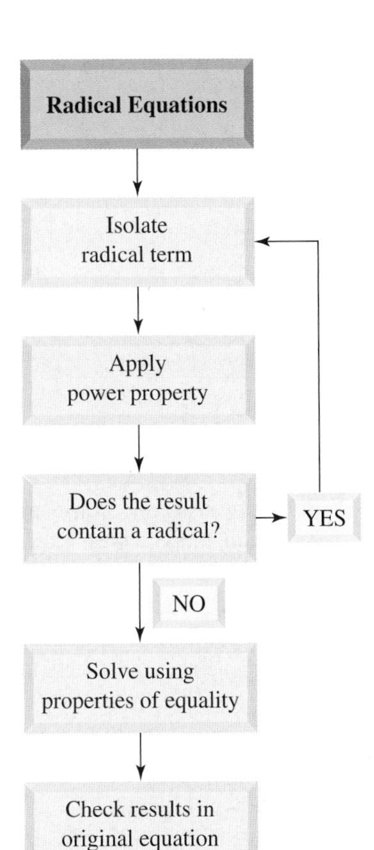

Figure 1.7

Since rational exponents are so closely related to radicals, the solution process for equations with rational exponents is very similar. The goal is still to "undo" the radical (rational exponent) and solve for the unknown.

RATIONAL EXPONENTS AND THE POWER PROPERTY OF EQUALITY

Let u and v be real numbers or real-valued expressions, with m, $n \in \mathbb{Z}$ and $n \neq 0$. If $u = v$, then $u^{\frac{m}{n}} = v^{\frac{m}{n}}$ provided $u^{\frac{1}{n}}$ and $v^{\frac{1}{n}}$ are defined. Both sides of an equation can be raised to a given power.

EXAMPLE 7 Solve the equation: $3(x+1)^{\frac{3}{4}} - 9 = 15$.

Solution:

$(x+1)^{\frac{3}{4}} = 8$ — isolate variable term (add 9, divide by 3)

$\left[(x+1)^{\frac{3}{4}}\right]^{\frac{4}{3}} = 8^{\frac{4}{3}}$ — raise each side to the $\frac{4}{3}$ power: $\frac{3}{4}\cdot\frac{4}{3} = 1$

$x + 1 = 16$ — simplify ($8^{\frac{4}{3}} = 16$)

$x = 15$ — solution

The solution checks.

NOW TRY EXERCISES 59 THROUGH 64

In Section R.4 we used a technique called u-substitution to factor expressions in quadratic form. The following equations are in quadratic form since the degree of the leading term is twice the degree of the middle term: $x^{\frac{2}{3}} - 3x^{\frac{1}{3}} - 10 = 0$, $(x+\frac{1}{2})^2 - 5(x+\frac{1}{2})^1 - 14 = 0$, and $x - 7\sqrt{x+4} + 4 = 0$. [Note: $x^1 - 7(x+4)^{\frac{1}{2}} + 4 = 0$]. A u-substitution will help to solve these equations by factoring. The first equation appears in Example 8, the other two are in the Exercise Set.

EXAMPLE 8 Solve using a u-substitution: $x^{\frac{2}{3}} - 3x^{\frac{1}{3}} - 10 = 0$

Solution: The equation is in quadratic form since we can decompose the fractional exponents and write the equation as $(x^{\frac{1}{3}})^2 - 3(x^{\frac{1}{3}})^1 - 10 = 0$.

$$\text{Let } u = x^{\frac{1}{3}} \longrightarrow u^2 = x^{\frac{2}{3}}$$

The equation becomes: $u^2 - 3u^1 - 10 = 0$, which is factorable in terms of u.

$(u-5)(u+2) = 0$ — factor

$u = 5$ or $u = -2$ — solution in terms of u

$x^{\frac{1}{3}} = 5$ or $x^{\frac{1}{3}} = -2$ — un-substitute $x^{\frac{1}{3}}$ for u

$(x^{\frac{1}{3}})^3 = 5^3$ or $(x^{\frac{1}{3}})^3 = (-2)^3$ — cube both sides: $\frac{1}{3}(3) = 1$

$x = 125$ or $x = -8$ — solve for x

Both solutions check.

NOW TRY EXERCISES 65 THROUGH 78

D. Applications

Polynomial applications come in many different forms. **Number puzzles** and consecutive integer exercises develop the ability to build accurate equation models (see Exercises 81–84). Applications involving **geometry** and **descriptive translation** depend on these models and bring a greater sense of how mathematics is used outside the classroom. Equations involving **revenue models** or **projectile motion** are two of the more significant types, as they are well within reach yet somewhat sophisticated and practical real-world models.

Geometry and Descriptive Translation

EXAMPLE 9 A legal-size sheet of typing paper has a length equal to three inches less than twice its width. If the area of the paper is 119 in^2, find the length and width.

Solution: Let W represent the width of the paper.
Then $2W$ represents twice the width, and $2W - 3$ represents three less than twice the width: $L = 2W - 3$:

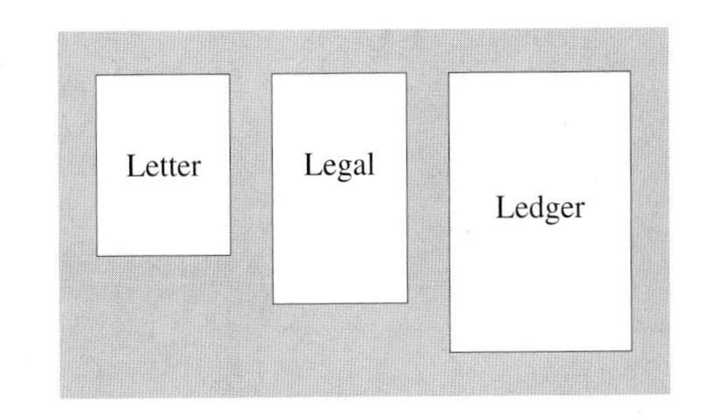

$$(\text{length})(\text{width}) = \text{area} \quad \text{verbal model}$$
$$(2W - 3)(W) = 119 \quad \text{substitute } 2W - 3 \text{ for length}$$

Since the equation is not set equal to zero, multiply and write the equation in standard form.

$$2W^2 - 3W = 119 \quad \text{distribute}$$
$$2W^2 - 3W - 119 = 0 \quad \text{set equal to zero}$$
$$(2W - 17)(W + 7) = 0 \quad \text{factor}$$
$$W = \tfrac{17}{2} \text{ or } W = -7 \quad \text{solve}$$

We ignore $W = -7$, since the width cannot be negative. The width of the paper is $\frac{17}{2} = 8\frac{1}{2}$ in. and the length is $L = 2(\frac{17}{2}) - 3$ or 14 in.

NOW TRY EXERCISES 85 THROUGH 88

Revenue Models

In a consumer-oriented society, we know that if the price of an item is decreased, more people will buy it. This is why stores have sales and bargain days. But if the item is sold too cheaply, revenue starts to decline because less money is coming in—even though more sales are made. This phenomenon is modeled by the formula *revenue = price · number of sales* or $R = P \cdot S$. Note how the formula is used in Example 10.

EXAMPLE 10 When a popular printer is priced at \$300, Compu-Store will sell 15 printers per week. Using a survey, they find that for each decrease of \$8, two additional sales will be made. What price will result in weekly revenue of \$6500?

Solution: Let x represent the number of times the price is decreased by \$8. Then $300 - 8x$ represents the new price and $15 + 2x$ represents the number of additional sales (sales increase by 2 each time the price is decreased).

$$R = P \cdot S \quad \text{revenue model}$$
$$6500 = (300 - 8x)(15 + 2x) \quad R = 6500, P = 300 - 8x, S = 15 + 2x$$
$$6500 = 4500 + 600x - 120x - 16x^2 \quad \text{multiply binomials}$$
$$0 = -16x^2 + 480x - 2000 \quad \text{simplify and write in standard form}$$
$$0 = x^2 - 30x + 125 \quad \text{divide by } -16$$
$$0 = (x - 5)(x - 25) \quad \text{factor}$$
$$x = 5 \text{ or } x = 25 \quad \text{result}$$

Surprisingly, a revenue of \$6500 can be attained after 5 decreases of \$8 each (\$40 total), or 25 price decreases of \$8 each (\$200 total). The related selling prices are $300 - 5(8) = \$260$ and $300 - 25(8) = \$100$. We might assume that due to profitability concerns, the manager of Compu-Store decides to go with the \$260 selling price.

NOW TRY EXERCISES 89 AND 90

Applications of rational equations can also take many forms. Problems using **ratio and proportion** are common, as are those using **descriptive translation.** Applications involving **work, uniform motion,** and **geometry** are also frequently seen (see Exercises 93–96). Example 11 uses a rational equation model.

EXAMPLE 11 In Verano City, the cost C to remove industrial waste from drinking water is given by the equation $C = \frac{80P}{100 - P}$, where P is the percent of total pollutants removed and C is the cost in thousands of dollars. If the City Council budgets \$1,520,000 for the removal of these pollutants, what percentage of the waste will be removed?

Solution:

$$C = \frac{80P}{100 - P} \quad \text{equation model}$$

$$1520 = \frac{80P}{100 - P} \quad \text{substitute 1520 for } C$$

$$1520(100 - P) = 80P \quad \text{multiply by LCM} = (100 - P)$$

$$152{,}000 = 1600P \quad \text{distribute and simplify}$$

$$95 = P \quad \text{result}$$

On a budget of \$1,520,000, 95% of the pollutants will be removed.

NOW TRY EXERCISES 97 AND 98

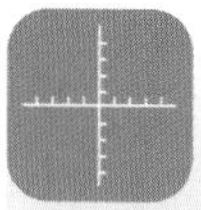

TECHNOLOGY HIGHLIGHT

Graphing Calculators and Rational Exponents

The keystrokes shown apply to a TI-84 Plus model. Please consult your manual or our Internet site for other models.

Expressions with rational exponents are easily evaluated on the home screen of a graphing calculator without using a 2nd function or accessing a submenu. For this reason the preferred method of evaluating a radical expression is to use a rational exponent where possible. Many common rational exponents have a decimal equivalent, which terminates after one, two, or three decimal places: $\frac{1}{2} = 0.5$, $\frac{3}{4} = 0.75$ and $\frac{5}{8} = 0.625$, and so on. Using the decimal form, the expressions $29^{\frac{1}{2}}$, $81^{\frac{3}{4}}$, and $92^{\frac{5}{8}}$ are evaluated in the screen shown in Figure 1.8. If you are unsure of the decimal equivalent, or if the decimal equivalent has a nonterminating form, the rational exponent should be expressed as a division and *grouped within parentheses.* The expressions $53^{\frac{1}{6}}$, $4096^{\frac{5}{12}}$, and $71^{\frac{4}{9}}$ are evaluated in Figure 1.9, as shown. Recall that you must press ENTER to execute the operation. Evaluate each of the following expressions. If the result is an integer or rational number, show why by decomposing the fraction and computing by hand.

Figure 1.8

```
29^.5
        5.385164807
81^.75
                 27
92^.625
        16.87980371
```

Exercise 1: $28^{\frac{4}{5}}$ 14.37892522

Exercise 2: $6561^{\frac{3}{8}}$ 27

Exercise 3: $\left(\frac{125}{343}\right)^{\frac{-2}{3}}$ $\frac{49}{25}$

Exercise 4: $-0.0078125^{\frac{5}{7}}$ $\frac{-1}{32}$

Figure 1.9

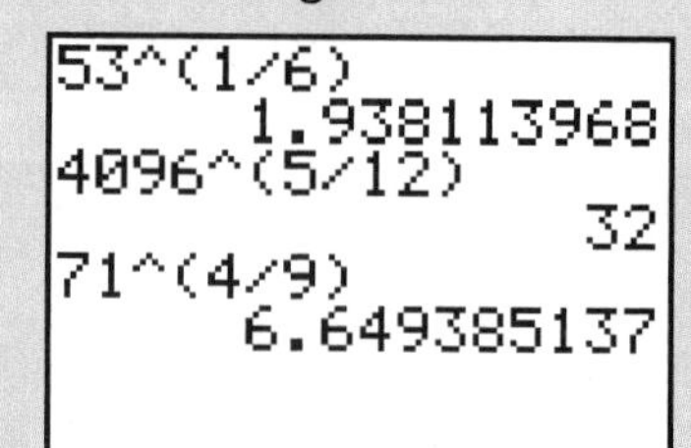

1.3 EXERCISES

CONCEPTS AND VOCABULARY

Fill in each blank with the appropriate word or phrase. Carefully reread the section, if necessary.

1. For rational equations, values that cause a zero denominator must be excluded.

2. The equation or formula for revenue models is revenue = price · sales.

3. "False solutions" to a rational or radical equation are also called extraneous roots.

4. Factorable polynomial equations can be solved using the zero factor property.

5. Explain/discuss the power property of equality as it relates to rational exponents and properties of reciprocals. Use the equation $(x - 2)^{\frac{2}{3}} = 9$ for your discussion. Answers will vary.

6. One factored form of an equation is shown. Discuss/explain why $x = -8$ and $x = 1$ are not solutions to the equation, and what must be done to find the actual solutions: $2(x + 8)(x - 1) = -16$. Equation is not in standard form (set equal to zero).

DEVELOPING YOUR SKILLS

Solve using the zero factor property. Be sure each equation is in standard form and factor out any common factors before attempting to solve. Check all answers in the original equation.

7. $x^2 - 15 = 2x$ **8.** $-21 = z^2 - 10z$ **9.** $m^2 = 8m - 16$

10. $-10n = n^2 + 25$ **11.** $5p^2 - 10p = 0$ **12.** $6q^2 - 18q = 0$

13. $-14h^2 = 7h$ **14.** $9w = -6w^2$ **15.** $a^2 - 17 = -8$

16. $b^2 + 8 = 12$ **17.** $g^2 + 18g + 70 = -11$ **18.** $h^2 + 14h - 2 = -51$

19. $m^3 + 5m^2 - 9m - 45 = 0$ **20.** $n^3 - 3n^2 - 4n + 12 = 0$ **21.** $(c - 12)c - 15 = 30$

22. $(d - 10)d + 10 = -6$ **23.** $9 + (r - 5)r = 33$ **24.** $7 + (s - 4)s = 28$

25. $(t + 4)(t + 7) = 54$ **26.** $(g + 17)(g - 2) = 20$ **27.** $2x^2 - 4x - 30 = 0$

28. $-3z^2 + 12z + 36 = 0$ **29.** $2w^2 - 5w = 3$ **30.** $-3v^2 = -v - 2$

Solve using u-substitution and the zero factor property.

31. $(x^2 - 3x)^2 - 14(x^2 - 3x) + 40 = 0$ $x = 4$ or $x = -1$ or $x = 5$ or $x = -2$

32. $(2x^2 + 3x)^2 - 4(2x^2 + 3x) - 5 = 0$ $x = -\frac{5}{2}$ or $x = 1$ or $x = -\frac{1}{2}$ or $x = -1$

Solve each equation.

33. $\frac{2}{x} + \frac{1}{x + 1} = \frac{5}{x^2 + x}$ $x = 1$

34. $\frac{3}{m + 3} - \frac{5}{m^2 + 3m} = \frac{1}{m}$ $m = 4$

35. $\frac{4}{a + 2} = \frac{3}{a - 1}$ $a = 10$

36. $\frac{4}{2y - 3} = \frac{7}{3y - 5}$ $y = \frac{1}{2}$

37. $\frac{1}{3y} - \frac{1}{4y} = \frac{1}{y^2}$ $y = 12$

38. $\frac{3}{5x} - \frac{1}{2x} = \frac{1}{x^2}$ $x = 10$

39. $x + \frac{14}{x - 7} = 1 + \frac{2x}{x - 7}$ $x = 3$; $x = 7$ is extraneous

40. $\frac{4}{x - 5} + 6 = 1 + \frac{2x}{x - 5}$ $x = 7$

41. $\frac{6}{n + 3} + \frac{20}{n^2 + n - 6} = \frac{5}{n - 2}$ $n = 7$

42. $\frac{7}{p + 2} - \frac{1}{p^2 + 5p + 6} = -\frac{2}{p + 3}$ $p = -\frac{8}{3}$

43. $\frac{a}{2a + 1} - \frac{a^2 + 5}{2a^2 - 5a - 3} = \frac{2}{a - 3}$ $a = -1$

44. $\frac{15}{6n^2 - n - 1} + \frac{3}{2n - 1} = \frac{2}{3n + 1}$ $n = -4$

Solve for the variable indicated.

45. $\frac{1}{f} = \frac{1}{f_1} + \frac{1}{f_2}$; for f

46. $\frac{1}{x} - \frac{1}{y} = \frac{1}{z}$; for z

47. $I = \frac{E}{R + r}$; for r

7. $x = 5$ or $x = -3$
8. $z = 7$ or $z = 3$
9. $m = 4$
10. $n = -5$
11. $p = 0$ or $p = 2$
12. $q = 0$ or $q = 3$
13. $h = 0$ or $h = \frac{-1}{2}$
14. $w = 0$ or $w = \frac{-3}{2}$
15. $a = 3$ or $a = -3$
16. $b = 2$ or $b = -2$
17. $g = -9$
18. $h = -7$
19. $m = -5$ or $m = -3$ or $m = 3$
20. $n = 3$ or $n = 2$ or $n = -2$
21. $c = -3$ or $c = 15$
22. $d = 8$ or $d = 2$
23. $r = 8$ or $r = -3$
24. $s = 7$ or $s = -3$
25. $t = -13$ or $t = 2$
26. $g = 3$ or $g = -18$
27. $x = 5$ or $x = -3$
28. $z = 6$ or $z = -2$
29. $w = -\frac{1}{2}$ or $w = 3$
30. $v = \frac{-2}{3}$ or $v = 1$
45. $f = \frac{f_1 f_2}{f_1 + f_2}$
46. $z = \frac{xy}{y - x}$
47. $r = \frac{E - IR}{I}$

48. $q = \frac{pf}{p - f}$; for p $\quad p = \frac{qf}{q - f}$ **49.** $V = \frac{1}{3}\pi r^2 h$; for h $\quad h = \frac{3V}{\pi r^2}$ **50.** $s = \frac{1}{2}gt^2$; for g $\quad g = \frac{2s}{t^2}$

51. $V = \frac{4}{3}\pi r^3$; for r^3 $\quad r^3 = \frac{3V}{4\pi}$ **52.** $V = \frac{1}{3}\pi r^2 h$; for r^2 $\quad r^2 = \frac{3V}{\pi h}$

Solve each equation and check your solutions by *substitution.* If a solution is extraneous, so state.

53. **a.** $-3\sqrt{3x - 5} = -9$ $\quad x = \frac{14}{3}$

b. $-11 = \frac{\sqrt{3x - 4}}{-2} - 10$ $\quad x = \frac{8}{3}$

c. $2\sqrt{m - 4} = \sqrt{3m + 24}$ $\quad m = 40$

54. **a.** $-2\sqrt{4x - 1} = -10$ $\quad \frac{13}{2}$

b. $-15 = \frac{\sqrt{2x + 5}}{-3} - 12$ $\quad x = 38$

c. $3\sqrt{5p - 4} = \sqrt{2p + 5}$ $\quad p = \frac{41}{43}$

55. **a.** $2 = \sqrt[3]{3m - 1}$ $\quad m = 3$

b. $2\sqrt[3]{7 - 3x} - 3 = -7$ $\quad x = 5$

c. $\frac{\sqrt[3]{2m + 3}}{-5} + 2 = 3$ $\quad m = -64$

d. $\sqrt[3]{2x - 9} = \sqrt[3]{3x + 7}$ $\quad x = -16$

56. **a.** $-3 = \sqrt[3]{5p + 2}$ $\quad p = \frac{-29}{5}$

b. $3\sqrt[3]{3 - 4x} - 7 = -4$ $\quad x = \frac{1}{2}$

c. $\frac{\sqrt[3]{6x - 7}}{4} - 5 = -6$ $\quad x = \frac{-19}{2}$

d. $3\sqrt[3]{x + 3} = 2\sqrt[3]{2x + 17}$ $\quad x = 5$

57. **a.** $\sqrt{x - 2} - \sqrt{2x} = -2$ $\quad x = 2, 18$

b. $\sqrt{x + 5} + \sqrt{x - 10} = 5$ $\quad x = 11$

58. **a.** $\sqrt{12x + 9} - \sqrt{24x} = -3$ $\quad x = 6$; $x = 0$ is extraneous

b. $\sqrt{2x - 3} + \sqrt{3x + 7} = 12$ $\quad x = 1406$ is extraneous

Write the equation in simplified form, then solve. Check all answers by substitution.

59. $x^{\frac{3}{5}} + 17 = 9$ $\quad x = -32$ **60.** $-2x^{\frac{3}{4}} + 47 = -7$ $\quad x = 81$ **61.** $0.\overline{3}x^{\frac{5}{2}} - 39 = 42$ $\quad x = 9$

62. $0.\overline{5}x^{\frac{5}{3}} + 92 = -43$ $\quad x = -27$ **63.** $8x^{-\frac{3}{2}} - 17 = -\frac{11}{8}$ $\quad x = \frac{16}{25}$ **64.** $2x^{-\frac{5}{4}} - 17 = -\frac{29}{16}$ $\quad x = \frac{16}{81}$

Solve each equation using a *u*-substitution. Check all answers.

65. $x^{\frac{2}{3}} - 2x^{\frac{1}{3}} - 15 = 0$ $\quad x = -27, 125$

66. $x^{\frac{2}{3}} + 2x^{\frac{1}{3}} - 8 = 0$ $\quad x = -64, 8$

67. $x^3 - 9x^{\frac{3}{2}} + 8 = 0$ $\quad x = 1, 4$

68. $(x^2 - 3)^2 + (x^2 - 3) - 2 = 0$ $\quad x = \pm 1, \pm 2$

69. $(x^2 + x)^2 - 8(x^2 + x) + 12 = 0$ $\quad x = -3, -2, 1, 2$

70. $(x + \frac{1}{2})^2 - 5(x + \frac{1}{2})^1 - 14 = 0$ $\quad x = \frac{-5}{2}, \frac{13}{2}$

71. $x^{-2} - 3x^{-1} - 4 = 0$ $\quad x = -1, \frac{1}{4}$

72. $x^{-2} - 2x^{-1} - 35 = 0$ $\quad x = -\frac{1}{5}, \frac{1}{7}$

73. $x^{-4} - 13x^{-2} + 36 = 0$ $\quad x = \pm\frac{1}{3}, \pm\frac{1}{2}$

Use a *u*-substitution to solve each radical equation.

74. $3\sqrt{x + 4} = x + 4$ $\quad x = -4, 5$

75. $x + 4 = 7\sqrt{x + 4}$ $\quad x = -4, 45$

76. $2(x + 1) = 5\sqrt{x + 1} - 2$ $\quad x = \frac{-3}{4}, 3$

77. $2\sqrt{x + 10} + 8 = 3(x + 10)$ $\quad x = -6$; $x = \frac{-74}{9}$ is extraneous

78. $4\sqrt{x - 3} = 3(x - 3) - 4$ $\quad x = 7$; $x = \frac{31}{9}$ is extraneous

WORKING WITH FORMULAS

79. Lateral surface area of a cone: $S = \pi r\sqrt{r^2 + h^2}$

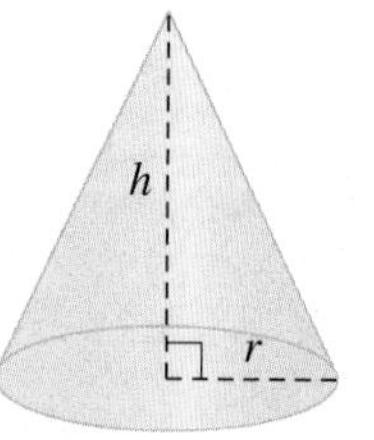

The lateral surface area (surface area excluding the base) S of a cone is given by the formula shown, where r is the radius of the base and h is the height of the cone. Find the surface area of a cone that has a radius of 6 m and a height of 10 m. Answer in simplest form.

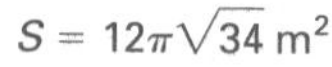
$S = 12\pi\sqrt{34}$ m^2

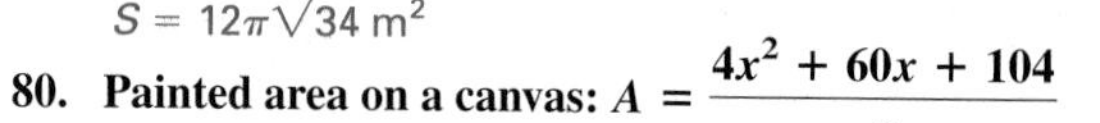
80. Painted area on a canvas: $A = \frac{4x^2 + 60x + 104}{x}$

A rectangular canvas is to contain a small painting with an area of 52 in^2, and requires 2-in. margins on the left and right, with 1-in. margins on the top and bottom for framing.

The total area of such a canvas is given by the formula shown, where x is the height of the *painted* area.

a. What is the area A of the canvas if the height of the painting is $x = 10$ in.? 110.4 in^2

b. If the area of the canvas is $A = 120$ in^2, what are the dimensions of the painted area? 13 in. high, 4 in. wide or 2 in. high, 26 in. wide

MIXED APPLICATIONS

Number puzzles: Find the integers described.

81. Given three consecutive odd integers, the product of the first and third is equal to four less than nine times the second. 7, 9, 11

82. Five less than twice an integer is multiplied by the same integer increased by two. If the result is -9, find the integer. 1

83. When a certain number is added to the numerator and subtracted from the denominator of the fraction $\frac{3}{4}$, the result is -8. Find the number. $n = 5$

84. Three consecutive even integers are chosen so that the ratio of the first and second multiplied by the ratio of the second and third, gives a result of $\frac{9}{10}$. What are the three even integers? 36, 38, 40

85. Envelope sizes: Large mailing envelopes often come in standard sizes, with 5- by 7-in. and 9- by 12-in. envelopes being the most common. The next larger size envelope has an area of 143 in^2, with a length that is 2 in. longer than the width. What are the dimensions of the larger envelope? 11 in. by 13 in.

86. Paper sizes: Letter size paper is 8.5 in. by 11 in. Legal size paper is $8\frac{1}{2}$ in. by 14 in. The next larger (common) size of paper has an area of 187 in^2, with a length that is 6 in. longer than the width. What are the dimensions of the larger size paper? 11 in. by 17 in.

Similar triangles: For each pair of similar triangles, use a proportion to find the length of the missing side (in bold italic).

87.

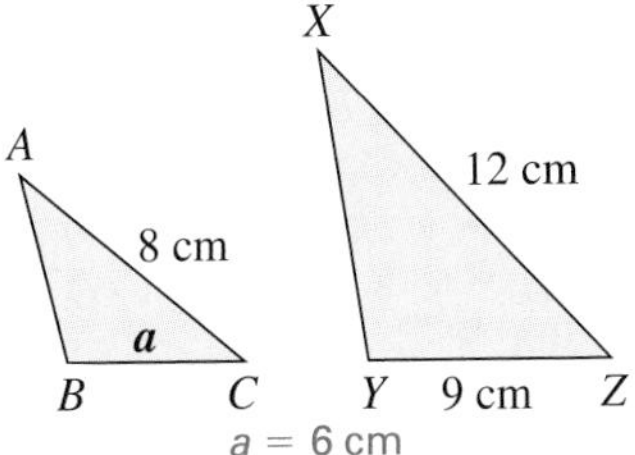

$a = 6$ cm

88.

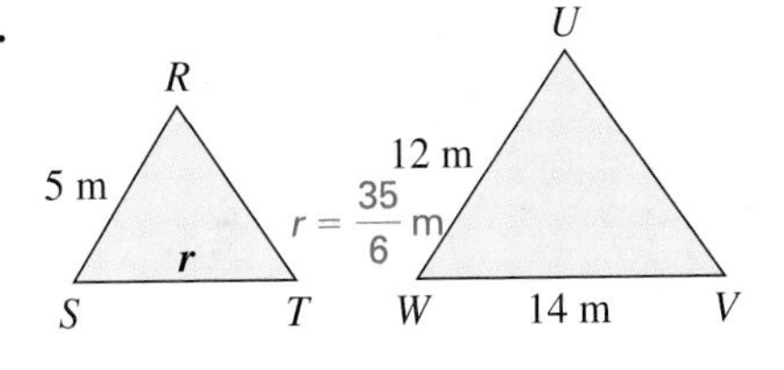

89. Running shoes: When a popular running shoe is priced at \$70, The Shoe House will sell 15 pairs each week. Using a survey, they have determined that for each decrease of \$2 in price, 3 additional pairs will be sold each week. What selling price will give a weekly revenue of \$2250? either \$50 or \$30

90. Cell phone charges: A cell phone service gains 48 new subscribers each month if their monthly fee is \$30. Using a survey, they find that for each decrease of \$1, 6 additional subscribers will join. What charge(s) will result in a monthly revenue of \$2160? \$20 or \$18

Projectile height: In the absence of resistance, the height of an object that is projected upward can be modeled by the equation $h = -16t^2 + vt + k$, where h represents the height of the object (in feet) t sec after it has been thrown, v represents the initial velocity (in feet per second), and k represents the height of the object when $t = 0$ (before it has been thrown). Use this information to complete the following problems.

91. From the base of a canyon that is 480 feet deep (*below* ground level $\rightarrow -480$), a slingshot is used to shoot a pebble upward toward the canyon's rim. If the initial velocity is 176 ft per second:

a. How far is the pebble below the rim after 4 sec? -32 ft

b. How long until the pebble returns to the bottom of the canyon? 11 sec

c. What happens at $t = 5$ and $t = 6$ sec? Discuss and explain. pebble is at canyon's rim

92. A model rocket blasts off. A short time later, at a velocity of 160 ft/sec and a height of 240 ft, it runs out of fuel and becomes a projectile.

- **a.** How high is the rocket three seconds later? Four seconds later? 576 ft, 624 ft
- **b.** How long will it take the rocket to attain a height of 640 ft? 5 sec
- **c.** How many times is a height of 384 ft attained? When do these occur? twice; 1 sec, 9 sec
- **d.** How many seconds until the rocket returns to the ground? 11.32 sec

93. Filling a pool: A swimming pool can be filled by one inlet pipe in 20 hr and by another pipe in 28 hr. How long would it take to fill the pool if both pipes were left open? $11\frac{2}{3}$ hr or 11 hr 40 min

94. Filling a sink: The cold water faucet can fill a sink in 2 min. The drain can empty a full sink in 3 min. If the faucet were left on and the drain was left open, how long would it take to fill the sink? 6 min

95. Uniform motion: On a trip from Bloomington to Chicago, Henri drove at an average speed of 70 mph. On the return trip he had to pull a trailer and was only able to average 50 mph. If the return trip took $\frac{4}{5}$ of an hour longer, how far is it from Bloomington to Chicago? 140 mi

96. Uniform motion: Amy drove 340 miles in $5\frac{1}{2}$ hr. She averaged 70 mph for the first leg, but was later slowed to an average of 55 mph for the rest of the trip. How far did she drive at each speed? 2.5 hr at 70 mph = 175 mi; 3.0 hr at 55 mph = 165 mi

97. Pollution removal: For a steel mill, the cost C (in millions of dollars) to remove toxins from the resulting sludge is given by $C = \dfrac{92P}{100 - P}$, where P is the percent of the toxins removed. What percent can be removed if the mill spends \$100,000,000 on the cleanup? Round to tenths of a percent. $P \approx 52.1\%$

98. Wildlife populations: The Department of Wildlife introduces 60 elk into a new game reserve. It is projected that the size of the herd will grow according to the equation $N = \dfrac{10(6 + 3t)}{1 + 0.05t}$, where N is the number of elk and t is the time in years. If recent counts find 225 elk, approximately how many years have passed? (See Section R.5, Exercise 82.) $t = 8.8$; just less than 9 yr

99. Planetary motion: The time T (in days) for a planet to make one revolution around the sun is modeled by $T = 0.407R^{\frac{3}{2}}$, where R is the maximum radius of the planet's orbit in millions of miles (*Kepler's third law of planetary motion*). Use the equation to approximate the maximum radius of each orbit, given the number of days it takes for one revolution. (See Section R.6, Exercises 55 and 56.)

- **a.** Mercury: 88 days 36 million mi
- **b.** Venus: 225 days 67 million mi
- **c.** Earth: 365 days 93 million mi
- **d.** Mars: 687 days 142 million mi
- **e.** Jupiter: 4333 days 484 million mi
- **f.** Saturn: 10,759 days 887 million mi

100. Wind-powered energy: If a wind-powered generator is delivering P units of power, the velocity V of the wind (in miles per hour) can be determined using $V = \sqrt[3]{\dfrac{P}{k}}$, where k is a constant that depends on the size and efficiency of the generator. Given $k = 0.004$, approximately how many units of power are being delivered if the wind is blowing at 27 miles per hour? (See Section R.6, Exercise 58.) $P \approx 78.7$ units

WRITING, RESEARCH, AND DECISION MAKING

101. To solve the equation $3 - \dfrac{8}{x + 3} = \dfrac{1}{x}$, a student multiplied by the LCM $x(x + 3)$, simplified, and got this result: $3 - 8x = (x + 3)$. Identify and correct the mistake, then find the correct solution(s). The constant "3" was not multiplied by the LCM; $3x(x+3) - 8x = x + 3$; $x = -1, 1$

102. Evaluate $3x^5 + 6x^4 - 2x^3 - 4x^2 + 5x + 10$ for $x = -2$. Then factor by grouping, and evaluate again. Which was easier? Why? Answers will vary.

103. Prior to the widespread use of calculators, square roots were calculated by hand. There is also a "paper-and-pencil" method for calculating *cube roots.* Using an encyclopedia, the Internet, a book on the history of mathematics, or some other resource, locate a discussion of the methods. Report on how each method works and give examples of its application. Answers will vary.

104. The expression $x^2 - 7$ is not factorable using *integer values.* But the expression *can be written* in the form $x^2 - (\sqrt{7})^2$, enabling us to factor it as a binomial and its conjugate: $(x + \sqrt{7})(x - \sqrt{7})$. Use this idea to solve the following equations:

a. $x^2 - 5 = 0$ $x = -\sqrt{5}, \sqrt{5}$ **b.** $n^2 - 19 = 0$ $n = -\sqrt{19}, \sqrt{19}$ **c.** $4v^2 - 11 = 0$ $v = \frac{-\sqrt{11}}{2}, \frac{\sqrt{11}}{2}$ **d.** $9w^2 - 11 = 0$ $w = \frac{-\sqrt{11}}{3}, \frac{\sqrt{11}}{3}$

EXTENDING THE CONCEPT

105. As an alternative to the *guess-and-check* method for factoring $ax^2 + bx + c = 0$, the "***ac* method**" can be used.

a. Compute the product $a \cdot c$ (lead coefficient · constant term).

b. List all factor pairs that give ac and find a pair that *sums* to b.

c. Rewrite b as the sum of these two factors and substitute for b.

d. Factor the result by grouping.

For $5n^2 - 18n - 8 = 0$: $a = 5$, $b = -18$, $c = -8$, and the product ac is -40.

Factor Pairs of −40	Sum of Factor Pairs
$-1 \cdot 40$	39
$1 \cdot (-40)$	−39
$-2 \cdot 20$	18
$2 \cdot (-20)$	−18 ←

Since we've found the right combination, we stop here and rewrite the original trinomial by substituting $2n - 20n$ for $-18n$: $5n^2 - 18n - 8 = 5n^2 + 2n - 20n - 8$. We then factor by grouping: $5n^2 + 2n - 20n - 8 = n(5n + 2) - 4(5n + 2) = (5n + 2)(n - 4) = 0$, and the solutions are $n = \frac{-2}{5}$, $n = 4$. Use the ac method to solve (a) $4x^2 - 23x + 15 = 0$ and (b) $3x^2 + 23x + 14 = 0$. **a.** $x = \frac{3}{4}, 5$ **b.** $x = -7, -\frac{2}{3}$

106. Determine the values of x for which each expression represents a real number.

a. $\frac{\sqrt{x - 1}}{x^2 - 4}$ $x \in [1, 2) \cup (2, \infty)$ **b.** $\frac{x^2 - 4}{\sqrt{x - 1}}$ $x \in (1, \infty)$

MAINTAINING YOUR SKILLS

107. (1.1) Two jets take off on parallel runways going in opposite directions. The first travels at a rate of 250 mph and the second at 325 mph. How long until they are 980 miles apart? 1.7 hr

108. (R.6) Find the missing side.

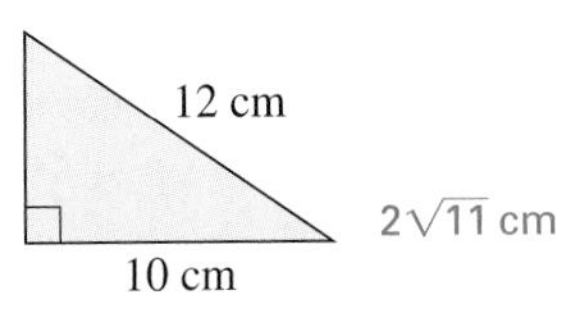

$2\sqrt{11}$ cm

109. (1.1) Of three consecutive odd integers, twice the first decreased by three times the third is equal to the second. What are the integers? −7, −5, −3

110. (1.1) Solve for the specified variable.

a. $A = P + PRT$ for P $P = \frac{A}{1 + RT}$

b. $2x + 3y = 15$ for y $y = \frac{-2}{3}x + 5$

111. (R.3) Simplify using properties of exponents:

$2^{-1} + (2x)^0 + 2x^0$ $3\frac{1}{2}$

112. (1.2) Graph the relation given:

$2x - 3 < 7$ and $x + 2 > 1$

$-1 < x < 5$;

−2 −1 0 1 2 3 4 5 6

MID-CHAPTER CHECK

1. a. $x = \frac{-2}{3}, \frac{2}{3}$
   b. $x = \frac{3}{2}$
   c. $x = -2, 5, 0$
   d. $x = \frac{-1}{3}, \frac{1}{3}$
   e. $x = -16, 3$
   f. $x = \frac{3}{4}, 3$
   g. $x = 2, \frac{-\sqrt{5}}{2}, \frac{\sqrt{5}}{2}$
   h. $x = -1, \frac{3}{2}$

6. $v_0 = \frac{H + 16t^2}{t}$

7. $x = \sqrt{\frac{S}{\pi(2 + y)}}$

8. a. $x \geq 1$ or $x \leq -2$;

   −4 −3 −2 −1 0 1 2 3 4

   b. $16 < x \leq 19$;

   0 15 16 17 18 19 20

1. Solve each polynomial equation by factoring.

a. $9x^2 - 4 = 0$ **b.** $4x^2 - 12x + 9 = 0$ **c.** $3x^3 - 9x^2 - 30x = 0$

d. $x^2 - \frac{1}{9} = 0$ **e.** $x^2 + 13x - 48 = 0$ **f.** $4x^2 - 15x + 9 = 0$

g. $4x^3 - 8x^2 - 5x + 10 = 0$ **h.** $2n^2 - n = 3$

Solve each rational equation and check all solutions by substitution.

2. $\frac{5}{x + 2} + \frac{3}{x} = 2$ $x = 3, -1$

3. $\frac{-2}{x^2 - 4} + \frac{1}{3x - 6} = \frac{3}{x + 2}$ $x = \frac{7}{4}$

Solve each radical equation and check all solutions by substitution.

4. $x + 3 = \sqrt{x^2 - 3}$ $x = -2$

5. $\sqrt{x - 5} + 1 = \sqrt{x}$ $x = 9$

Solve for the variable specified.

6. $H = -16t^2 + v_0t$; for v_0

7. $S = 2\pi x^2 + \pi x^2 y$; for x

8. Solve the inequality and graph the solution set.

a. $-5x + 16 \leq 11$ or $3x + 2 \leq -4$ **b.** $\frac{1}{2} < \frac{1}{12}x - \frac{5}{6} \leq \frac{3}{4}$

9. The ratio of a number and the number decreased by three is equal to one-fourth the number. Find all such numbers. $x = 0, 7$

10. To launch what is called a *Stomp Rocket,* a child jumps or stomps on an air bag that is connected to a model rocket. The compressed air forced through a connecting tube thrusts the rocket upward.

a. If the initial velocity of the rocket is 96 feet per second, how high is the rocket after one second? at $t = 1$, $H = 80$ ft

b. How long until the rocket reaches a height of 140 ft? Use $H = -16t^2 + v_0t$ and assume there are no drag or frictional forces. at $H = 140$, $t = \frac{5}{2}$ or $\frac{7}{2}$ sec

REINFORCING BASIC CONCEPTS

Solving $x^2 + bx + c = 0$

The ability to solve a quadratic equation is a fundamental part of our future course work. We'll use this skill to graph rational equations, solve equations in quadratic form, develop the theory of equations, introduce the conic sections, and in many other areas. Because of these connections, it is important that the most basic components of this skill are developed to a point where they become automatic. This exercise is designed to help accomplish this goal, by identifying three types of quadratic equations that are easily factorable.

I. Solving $x^2 + bx + c = 0$ when $c = 0$: With the constant term missing, the equation becomes $x^2 + bx = 0$, which is easily factorable since x is common to both terms. The factored form is $x(x + b) = 0$, no matter if b is positive or negative. Be careful not to attempt factoring $x^2 + bx$ as the product of two binomials—there is no constant term!

Quickly solve the equations by factoring:

1. $x^2 + 4x = 0$ **2.** $x^2 + 7x = 0$ **3.** $x^2 - 5x = 0$ **4.** $x^2 - 2x = 0$

5. $x^2 + \frac{1}{2}x = 0$ **6.** $x^2 + \frac{2}{5}x = 0$ **7.** $x^2 - \frac{2}{3}x = 0$ **8.** $x^2 - \frac{5}{6}x = 0$

1. $x(x + 4) = 0$; $x = 0, -4$
2. $x(x + 7) = 0$; $x = 0, -7$
3. $x(x - 5) = 0$; $x = 0, 5$
4. $x(x - 2) = 0$; $x = 0, 2$
5. $x(x + \frac{1}{2}) = 0$; $x = 0, -\frac{1}{2}$
6. $x(x + \frac{2}{5}) = 0$; $x = 0, -\frac{2}{5}$
7. $x(x - \frac{2}{3}) = 0$; $x = 0, \frac{2}{3}$
8. $x(x - \frac{5}{6}) = 0$; $x = 0, \frac{5}{6}$

II. Solving $x^2 + bx + c = 0$ when $b = 0$: With the linear term missing and $c < 0$, the equation becomes $x^2 - c = 0$, which can also be factored as the difference of two squares. The factored form is $(x + \sqrt{c})(x - \sqrt{c})$, *no matter if c is a perfect square or a nonperfect square.* We easily recognize that $x^2 - 49 = 0$ factors in this way since we can write it as $x^2 - 7^2 = (x + 7)(x - 7) = 0$. However, for any number $n > 0$, $n = (\sqrt{n})^2$. This means $x^2 - 5 = 0$ can also be written as the difference of two squares and factored in the same way: $x^2 - 5 = x^2 - (\sqrt{5})^2 = (x + \sqrt{5})(x - \sqrt{5}) = 0$.

Quickly solve the equations by factoring:

9. $x^2 - 81 = 0$ **10.** $x^2 - 121 = 0$ **11.** $x^2 - 7 = 0$ **12.** $x^2 - 31 = 0$
13. $x^2 - 49 = 0$ **14.** $x^2 - 13 = 0$ **15.** $x^2 - 21 = 0$ **16.** $x^2 - 16 = 0$

III. Solving $x^2 + bx + c = 0$ when $b \neq 0$ and $c \neq 0$: There is likely no form more common in the algebra sequence. In this case, we are simply looking for two numbers whose product is c and whose sum or difference is b. To aid efficiency, concentrate on the positive factor pairs of c, and mentally determine the sum or difference giving $|b|$. This will yield the correct values, and the needed signs can be then be applied in each binomial factor.

Quickly solve the equations by factoring:

17. $x^2 + 4x - 45 = 0$ **18.** $x^2 - 13x + 36 = 0$ **19.** $x^2 + 10x + 16 = 0$
20. $x^2 - 7x - 44 = 0$ **21.** $x^2 + 6x - 16 = 0$ **22.** $x^2 - 20x + 51 = 0$
23. $x^2 + 8x + 7 = 0$ **24.** $x^2 - 6x - 27 = 0$

9. $(x + 9)(x - 9) = 0$; $x = -9, 9$
10. $(x + 11)(x - 11) = 0$; $x = -11, 11$
11. $(x + \sqrt{7})(x - \sqrt{7}) = 0$; $x = -\sqrt{7}, \sqrt{7}$
12. $(x + \sqrt{31})(x - \sqrt{31}) = 0$; $x = -\sqrt{31}, \sqrt{31}$
13. $(x + 7)(x - 7) = 0$; $x = -7, 7$
14. $(x + \sqrt{13})(x - \sqrt{13}) = 0$; $x = -\sqrt{13}, \sqrt{13}$
15. $(x + \sqrt{21})(x - \sqrt{21}) = 0$; $x = -\sqrt{21}, \sqrt{21}$
16. $(x + 4)(x - 4) = 0$; $x = -4, 4$
17. $(x + 9)(x - 5) = 0$; $x = -9, 5$
18. $(x - 9)(x - 4) = 0$; $x = 9, 4$
19. $(x + 8)(x + 2) = 0$; $x = -8, -2$
20. $(x - 11)(x + 4) = 0$; $x = 11, -4$
21. $(x + 8)(x - 2) = 0$; $x = -8, 2$
22. $(x - 17)(x - 3) = 0$; $x = 17, 3$
23. $(x + 1)(x + 7) = 0$; $x = -1, -7$
24. $(x - 9)(x + 3) = 0$; $x = 9, -3$

1.4 Complex Numbers

LEARNING OBJECTIVES

In Section 1.4 you will learn how to:

A. **Identify and simplify imaginary and complex numbers**

B. **Add and subtract complex numbers**

C. **Multiply complex numbers and find powers of *i***

D. **Divide complex numbers**

INTRODUCTION

For centuries, even the most prominent mathematicians refused to work with equations like $x^2 + 1 = 0$. Using the principal of square roots gave the "solutions" $x = \sqrt{-1}$ and $x = -\sqrt{-1}$, which they found baffling and mysterious, since there is no real number whose square is -1. In this section, we'll see how this "mystery" was finally resolved.

POINT OF INTEREST

Some of the most celebrated names in mathematics can be associated with the history of imaginary numbers and the complex number system. François Viéte (1540–1603) realized their existence, but did not accept them. Girolomo Cardano (1501–1576) found them puzzling, but actually produced solutions to cubic equations that were complex numbers. Albert Girard (1595–1632) was apparently the first to advocate their acceptance, suggesting that this would establish that a polynomial equation has exactly as many roots as its degree. Then in 1799, German mathematician Carl F. Gauss (1777–1855) proved the fundamental theorem of algebra, which states that every polynomial with degree $n \geq 1$ has at least one complex solution. For more information, see Exercise 80.

A. Identifying and Simplifying Imaginary and Complex Numbers

The equation $x^2 = -1$ has no real number solutions, since the square of any real number must be positive. But if we apply the principal of square roots we get $x = \sqrt{-1}$ and $x = -\sqrt{-1}$, which check when we substitute them back into the original equation:

$$x^2 + 1 = 0 \quad \text{original equation}$$

$$(1) \quad (\sqrt{-1})^2 + 1 = 0 \quad \text{substitute } x = \sqrt{-1}$$

$$-1 + 1 = 0 ✓ \quad \text{answer "checks"}$$

$$(2) \quad (-\sqrt{-1})^2 + 1 = 0 \quad \text{substitute } x = -\sqrt{-1}$$

$$-1 + 1 = 0 ✓ \quad \text{answer "checks"}$$

WORTHY OF NOTE

It was René Descartes (in 1637) who first used the term *imaginary* to describe these numbers; Leonhard Euler (in 1777) who introduced the letter i to represent $\sqrt{-1}$; and Carl F. Gauss (in 1831) who first used the phrase *complex number* to describe solutions that had both a real number part and an imaginary part. For more on complex numbers and their story, see www.mhhe.com/coburn.

This is one of many observations that prompted later students of mathematics to accept $x = \sqrt{-1}$ and $x = -\sqrt{-1}$ as valid solutions to $x^2 = -1$, reasoning that although they were not *real* number solutions, *they must be solutions of a different kind.*

One result of this acceptance and evolution of thought was the introduction of the set of **imaginary numbers** and the imaginary unit i. The italicized i represents the number whose square is -1. This means $i^2 = -1$ and $i = \sqrt{-1}$.

IMAGINARY NUMBERS AND THE IMAGINARY UNIT

Imaginary numbers are those of the form $\sqrt{k}$, where $k < 0$.
The imaginary unit i represents the number whose square is -1, yielding $i^2 = -1$ and $i = \sqrt{-1}$.

An imaginary number can be simplified using the product property of square roots and the i notation. For $\sqrt{-12}$ we have: $\sqrt{-12} = \sqrt{-1 \cdot 4 \cdot 3} = i \cdot 2\sqrt{3} = 2i\sqrt{3}$ and we say the expression has been *simplified and written in terms of i.* It's best to write imaginary numbers with the unit "i" *in front of the radical* to prevent it being interpreted as being *under the radical:* $2i\sqrt{3}$ is preferred over $2\sqrt{3}\,i$.

EXAMPLE 1 Rewrite the imaginary numbers in terms of i and simplify.

a. $\sqrt{-81}$ **b.** $\sqrt{-7}$ **c.** $\sqrt{-24}$ **d.** $-3\sqrt{-16}$

Solution:

a. $\sqrt{-81} = \sqrt{-1 \cdot 81} = \sqrt{-1} \cdot \sqrt{81} = i \cdot 9$ or $9i$

b. $\sqrt{-7} = \sqrt{-1 \cdot 7} = \sqrt{-1} \cdot \sqrt{7} = i\sqrt{7}$

c. $\sqrt{-24} = \sqrt{-1 \cdot 24} = \sqrt{-1} \cdot \sqrt{4} \cdot \sqrt{6} = 2i\sqrt{6}$

d. $-3\sqrt{-16} = -3 \cdot \sqrt{-1 \cdot 16} = -3 \cdot \sqrt{-1} \cdot \sqrt{16} = -12i$

NOW TRY EXERCISES 7 THROUGH 12

EXAMPLE 2 The numbers $x = \dfrac{-6 + \sqrt{-16}}{2}$ and $x = \dfrac{6 - \sqrt{-16}}{2}$ are not real, but are known to be solutions of $x^2 + 6x + 13 = 0$. Simplify $\dfrac{-6 + \sqrt{-16}}{2}$.

Solution: Using the i notation and properties of radicals we have:

$$x = \frac{-6 + \sqrt{-1}\sqrt{16}}{2} \quad \text{write in } i \text{ notation}$$

$$x = \frac{-6 + 4i}{2} \quad \text{simplify}$$

$$x = \frac{2(-3 + 2i)}{2} = -3 + 2i \quad \text{factor numerator and reduce}$$

NOW TRY EXERCISES 13 THROUGH 16

The solutions to Example 2 contained both a **real number part** (-3) and an **imaginary part** $(2i)$. Numbers of this type are called **complex numbers.**

> **COMPLEX NUMBERS**
> Complex numbers are those that can be written in the form $a + bi$, where a and b are real numbers and $i = \sqrt{-1}$. The expression $a + bi$ is called the **standard form** of a complex number.

From this definition we note that all real numbers are also complex numbers, since $a + 0i$ is complex with $b = 0$. In addition, *all imaginary numbers are complex numbers,* since $0 + bi$ is a complex number with $a = 0$.

EXAMPLE 3 Write each complex number in the form $a + bi$, and identify the values of a and b.

a. $2 + \sqrt{-49}$ **b.** $\sqrt{-12}$ **c.** 7 **d.** $\frac{4 + 3\sqrt{-25}}{20}$

Solution:

a. $2 + \sqrt{-49}$
$= 2 + \sqrt{-1}\sqrt{49}$
$= 2 + 7i$
$a = 2, b = 7$

b. $\sqrt{-12}$
$= 0 + \sqrt{-1}\sqrt{12}$
$= 0 + 2i\sqrt{3}$
$a = 0, b = 2\sqrt{3}$

c. 7
$= 7 + 0i$
$a = 7, b = 0$

d. $\frac{4}{20} + \frac{3\sqrt{-1 \cdot 25}}{20}$
$= \frac{1}{5} + \frac{3 \cdot 5i}{20}$
$= \frac{1}{5} + \frac{3}{4}i$
$a = \frac{1}{5}, b = \frac{3}{4}$

NOW TRY EXERCISES 17 THROUGH 24

Complex numbers complete the development of our "numerical landscape" for the algebra sequence. Types of numbers and their relationship to each other can be seen in Figure 1.10, which shows how sets of numbers are nested within larger sets.

B. Adding and Subtracting Complex Numbers

The sum and difference of two polynomials is computed by identifying and combining like terms. The sum or difference of two complex numbers is computed in a similar way, by adding the real number parts from each, and the imaginary parts from each. Notice in Example 4 that the commutative, associative, and distributive properties also apply to complex numbers.

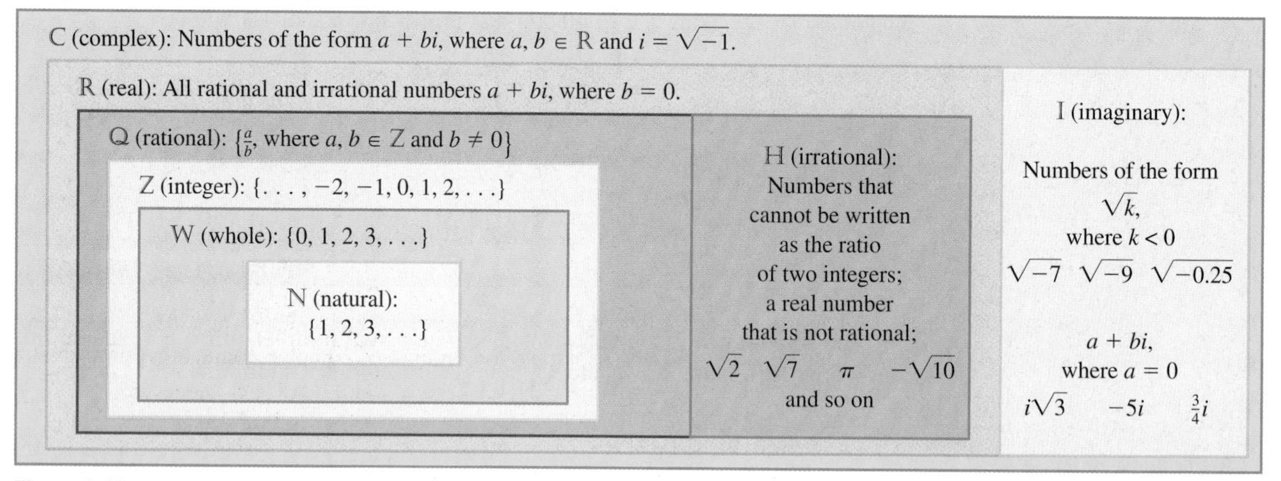

Figure 1.10

EXAMPLE 4 Perform the indicated operation and write the result in the form $a + bi$.

a. $(2 + 3i) + (-5 + 2i)$ **b.** $(-5 - 4i) - (-2 - \sqrt{2}\,i)$

Solution:

a. $(2 + 3i) + (-5 + 2i)$ original sum
$= 2 + 3i + (-5) + 2i$ distribute
$= 2 + (-5) + 3i + 2i$ commute terms
$= [2 + (-5)] + (3i + 2i)$ associate terms
$= -3 + 5i$ result

b. $(-5 - 4i) - (-2 - \sqrt{2}\,i)$ original difference
$= -5 + (-4i) + 2 + \sqrt{2}\,i$ distribute
$= -5 + 2 + (-4i) + \sqrt{2}\,i$ commute terms
$= (-5 + 2) + [(-4i) + \sqrt{2}\,i]$ associate terms
$= -3 + (-4 + \sqrt{2})i$ result

NOW TRY EXERCISES 25 THROUGH 30

C. Multiplying Complex Numbers; Powers of *i*

Recall that expressions like $2x + 5$ and $2x - 5$ are called binomial conjugates. In the same way, $a + bi$ and $a - bi$ are called **complex conjugates.** The product of two complex numbers is computed using the distributive property and the F-O-I-L process in the same way we apply these to binomials. If any result gives a factor of i^2, remember that $i^2 = -1$.

EXAMPLE 5 Find the indicated product and write the answer in $a + bi$ form.

a. $2i(-1 + 3i)$ **b.** $(6 - 5i)(4 + i)$ **c.** $(1 - 2i)^2$ **d.** $(2 + 3i)(2 - 3i)$

Solution:

a. $2i(-1 + 3i)$ monomial · binomial
$= 2i(-1) + 2i(3i)$ distribute
$= -2i + 6i^2$ $i \cdot i = i^2$
$= -2i + 6(-1)$ $i^2 = -1$
$= -6 - 2i$ result

b. $(6 - 5i)(4 + i)$ binomial · binomial
$= (6)(4) + 6i + (-5i)(4) + (-5i)(i)$ F-O-I-L
$= 24 + 6i + (-20i) + (-5)i^2$ $i \cdot i = i^2$
$= 24 + 6i + (-20i) + (-5)(-1)$ $i^2 = -1$
$= 29 - 14i$ result

c. $(1 - 2i)^2$ binomial square
$= (1)^2 - 2(1)(2i) + (-2i)^2$ $A^2 - 2AB + B^2$
$= 1 - 4i + 4i^2$ $i \cdot i = i^2$
$= 1 - 4i + 4(-1)$ $i^2 = -1$
$= -3 - 4i$ result

d. $(2 + 3i)(2 - 3i)$ binomial · conjugate
$= (2)^2 - (3i)^2$ $(A + B)(A - B) = A^2 - B^2$
$= 4 - 9i^2$ $i \cdot i = i^2$
$= 4 - 9(-1)$ $i^2 = -1$
$= 13 + 0i = 13$ result

NOW TRY EXERCISES 31 THROUGH 48

Note from Example 5(d) that the *product* of the complex number $a + bi$ with its complex conjugate $a - bi$ *is a real number.* This relationship is useful when rationalizing expressions with a complex number in the denominator, and we generalize the result as follows:

PRODUCT OF COMPLEX CONJUGATES
Given the complex number $a + bi$ and its complex conjugate $a - bi$, their product is a real number given by $(a + bi)(a - bi) = a^2 + b^2$.

Showing that $(a + bi)(a - bi) = a^2 + b^2$ is left as an exercise, (see Exercise 79) but from here on, when asked to compute the product of complex conjugates, simply refer to the formula as illustrated here: $(-3 + 5i)(-3 - 5i) = (-3)^2 + 5^2$ or 34.

These operations on complex numbers enable us to verify complex solutions by substitution, in the same way we verify solutions for real numbers. In addition to offering contextual practice with these skills, it is fascinating to observe how the complex roots balance or "cancel each other out" to arrive at the solution. In Example 2 we stated that $x = -3 + 2i$ was one solution to $x^2 + 6x + 13 = 0$. This is verified here.

EXAMPLE 6 Verify that $x = -3 + 2i$ is a solution to the equation $x^2 + 6x + 13 = 0$.

Solution:

$$x^2 + 6x + 13 = 0 \quad \text{original equation}$$
$$(-3 + 2i)^2 + 6(-3 + 2i) + 13 = 0 \quad \text{substitute } 3 + 2i \text{ for } x$$
$$(-3)^2 + 2(-3)(2i) + (2i)^2 - 18 + 12i + 13 = 0 \quad \text{square binomial and distribute}$$
$$9 - 12i + 4i^2 + 12i - 5 = 0 \quad \text{simplify}$$
$$9 + (-4) - 5 = 0 \quad \text{combine like terms } (12i - 12i = 0;\ i^2 = -1)$$
$$0 = 0 \checkmark$$

NOW TRY EXERCISES 49 THROUGH 56

EXAMPLE 7 Show that $x = 2 - i\sqrt{3}$ is a solution of $x^2 - 4x = -7$.

Solution:

$$x^2 - 4x = -7 \quad \text{original equation}$$
$$(2 - i\sqrt{3})^2 - 4(2 - i\sqrt{3}) = -7 \quad \text{substitute } 2 - i\sqrt{3} \text{ for } x$$
$$4 - 4i\sqrt{3} + (i\sqrt{3})^2 - 8 + 4i\sqrt{3} = -7 \quad \text{multiply}$$
$$4 - 4i\sqrt{3} - 3 - 8 + 4i\sqrt{3} = -7 \quad (i\sqrt{3})^2 = -3$$
$$-7 = -7 \checkmark \quad \text{solution checks}$$

NOW TRY EXERCISES 57 THROUGH 60

The imaginary unit i has another interesting and useful property. Since $i = \sqrt{-1}$ and $i^2 = -1$, we know that $i^3 = i^2 \cdot i = (-1)i = -i$ and $i^4 = (i^2)^2 = 1$. We can now simplify any **power of *i*** by rewriting the expression in terms of i^4. This is illustrated next:

$$i \text{ or } \sqrt{-1} \qquad i^5 = i^4 \cdot i = 1 \cdot i = i$$
$$i^2 = -1 \qquad i^6 = i^4 \cdot i^2 = -1$$
$$i^3 = i^2 \cdot i = (-1)i = -i \qquad i^7 = i^4 \cdot i^3 = -i$$
$$i^4 = (i^2)^2 = (-1)^2 = 1 \qquad i^8 = (i^4)^2 = 1$$

Notice the powers of i "cycle through" the four values of i, -1, $-i$ and 1. In more advanced classes, powers of complex numbers play a vital role, and here we learn to reduce higher powers using the power property of exponents and $i^4 = 1$.

EXAMPLE 8 Simplify: **a.** i^{22} **b.** i^{28} **c.** i^{57} **d.** i^{75}

Solution: **a.** $i^{22} = (i^4)^5 \cdot (i^2) = -1$ **b.** $i^{28} = (i^4)^7 = 1^7 = 1$

c. $i^{57} = (i^4)^{14} \cdot (i) = i$ **d.** $i^{75} = (i^4)^{18} \cdot (i^3) = 1^{18} \cdot -i = -i$

NOW TRY EXERCISES 61 AND 62

D. Division of Complex Numbers

With $i = \sqrt{-1}$, expressions like $\frac{3 - i}{2 + i}$ actually have a radical in the denominator, which leads to our method for complex number division. We simply apply our earlier method of rationalizing denominators, but this time using a complex conjugate.

EXAMPLE 9 Divide and write each result in the form $a + bi$.

a. $\frac{2}{5 - i}$ **b.** $\frac{3 - i}{2 + i}$ **c.** $\frac{6 + \sqrt{-36}}{3 + \sqrt{-9}}$

Solution: **a.**

$$\begin{aligned}\frac{2}{5 - i} &= \frac{2}{5 - 1i} \cdot \frac{5 + 1i}{5 + 1i} \\ &= \frac{2(5 + i)}{5^2 + 1^2} \\ &= \frac{10 + 2i}{26} \\ &= \frac{10}{26} + \frac{2}{26}i \\ &= \frac{5}{13} + \frac{1}{13}i\end{aligned}$$

b.

$$\begin{aligned}\frac{3 - i}{2 + i} &= \frac{3 - 1i}{2 + 1i} \cdot \frac{2 - 1i}{2 - 1i} \\ &= \frac{6 - 3i - 2i + 1i^2}{2^2 + 1^2} \\ &= \frac{6 - 5i + (-1)}{5} \\ &= \frac{5 - 5i}{5} = \frac{5}{5} - \frac{5i}{5} \\ &= 1 - i\end{aligned}$$

c.

$$\frac{6 + \sqrt{-36}}{3 + \sqrt{-9}} = \frac{6 + \sqrt{-1}\sqrt{36}}{3 + \sqrt{-1}\sqrt{9}} \quad \text{convert to } i \text{ notation}$$

$$= \frac{6 + 6i}{3 + 3i} \quad \text{simplify}$$

The expression can be further simplified by reducing common factors.

$$= \frac{6(1 + 1i)}{3(1 + 1i)} = 2 \quad \text{factor and reduce}$$

NOW TRY EXERCISES 63 THROUGH 68

It is important to note that results from operations in the complex number system can be checked using inverse operations, just as we do for real numbers. From Example 9(b) we found that $(3 - i) \div (2 + i) = 1 - i$. The related multiplication would be $(1 - i)(2 + i)$ giving $2 + 1i - 2i - i^2 = 2 - 1i - (-1) = 3 - i$.✓ Several checks are asked for in the exercises.

TECHNOLOGY HIGHLIGHT

Graphing Calculators and Operations on Complex Numbers

The keystrokes shown apply to a TI-84 Plus model. Please consult your manual or our Internet site for other models.

Virtually all graphing calculators have the ability to find imaginary and complex roots, as well as perform operations on complex numbers.

To use this capability on the TI-84 Plus, we first put the calculator in $a + bi$ mode. Press the MODE key (next to the yellow 2nd key) and the screen shown in Figure 1.11 appears. On the second line from the bottom, note the calculator may be in "Real" mode. To change to "$a + bi$" mode, simply navigate the cursor down to this line using the down arrow, then overlay the "$a + bi$" selection using the right arrow and press the ENTER key. The calculator is now in complex number mode. Press 2nd MODE (QUIT) to return to the home screen. To compute the product $(-2 - 3i)(5 + 4i)$, enter the expression on the home screen exactly as it is written. The number "i" is located above the decimal point on the bottom row. After pressing ENTER the result $2 - 23i$ immediately appears. Compute the product by hand to see if results match.

Figure 1.11

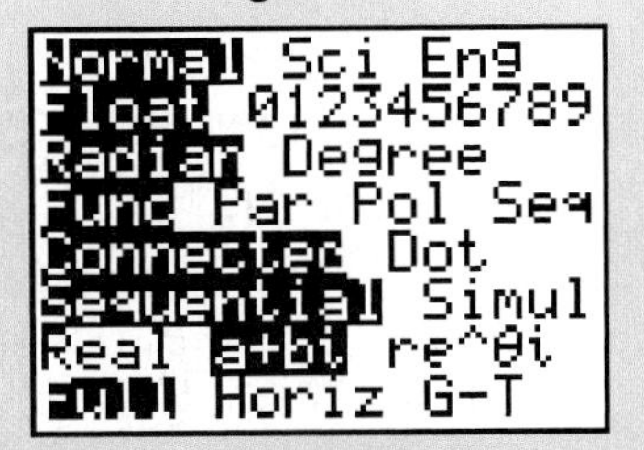

1. GC $3 - 3.464101615i \rightarrow 2i\sqrt{3} \approx 3.464101615i$ Manual $3 - 2i\sqrt{3}$

Exercise 1: Use a graphing calculator to compute the sum $(-2 + \sqrt{-108}) + (5 - \sqrt{-192})$. Note the result is in approximate form. Compute the sum by hand in exact form and compare the results.

Exercise 2: Use a graphing calculator to compute the product $(-3 + 7i)(4 - 5i)$. Then compute the product by hand and compare results. Check your answer using complex number division.

Exercise 3: Use a graphing calculator to compute the quotient $(2i)/(3 + i)$. Then compute the quotient by hand and compare results. Check your answer using multiplication.

2. GC $23 + 43i$ Manual $-12 + 15i + 28i - 35i^2 = 23 + 43i$
Check: $\frac{23 + 43i}{4 - 5i} = -3 + 7i$ ✓

3. GC $0.2 + 0.6i$ Manual $\frac{2i(3 - i)}{(3 + i)(3 - i)} = \frac{6i - 2i^2}{10} = \frac{2}{10} + \frac{6}{10}i = 0.2 + 0.6i$
Check: $(0.2 + 0.6i)(3 + i) = 0.6 + 0.2i + 1.8i + 0.6i^2 = 2i$ ✓

1.4 EXERCISES

CONCEPTS AND VOCABULARY

Fill in each blank with the appropriate word or phrase. Carefully reread the section if needed.

1. Given the complex number $3 + 2i$, its complex conjugate is $\underline{3 - 2i}$.

2. The product $(3 + 2i)(3 - 2i)$ gives the real number $\underline{13}$.

3. If the expression $\frac{4 + 6i\sqrt{2}}{2}$ is written in the standard form $a + bi$, then $a = \underline{2}$ and $b = \underline{3\sqrt{2}}$.

4. For $i = \sqrt{-1}$ (the number whose square root is -1), $i^2 = \underline{-1}$, $i^4 = \underline{1}$, $i^6 = \underline{-1}$, and $i^8 = \underline{1}$, $i^3 = \underline{-i}$, $i^5 = \underline{i}$, $i^7 = \underline{-i}$, and $i^9 = \underline{i}$.

5. Discuss/explain which is correct:

a. $\sqrt{-4} \cdot \sqrt{-9} = \sqrt{(-4)(-9)} = \sqrt{36} = 6$

b. $\sqrt{-4} \cdot \sqrt{-9} = 2i \cdot 3i = 6i^2 = -6$

(b) is correct.

6. Compare/contrast the product $(1 + \sqrt{2})(1 - \sqrt{3})$ with the product $(1 + i\sqrt{2})(1 - i\sqrt{3})$. What is the same? What is different? Answers will vary.

13. a. $1 + i; a = 1, b = 1$
 b. $2 + \sqrt{3}\,i; a = 2, b = \sqrt{3}$
14. a. $8 - \sqrt{2}\,i; a = 8, b = -\sqrt{2}$
 b. $2 + 3\sqrt{5}\,i; a = 2, b = 3\sqrt{5}$
15. a. $4 + 2i; a = 4, b = 2$
 b. $2 - \sqrt{2}\,i; a = 2, b = -\sqrt{2}$
16. a. $\frac{3}{2} - \frac{3\sqrt{2}}{2}i; a = \frac{3}{2}, b = \frac{-3\sqrt{2}}{2}$
 b. $\frac{3}{2} + \frac{5\sqrt{2}}{4}i; a = \frac{3}{2}, b = \frac{5\sqrt{2}}{4}$
17. a. $5 + 0i; a = 5, b = 0$
 b. $0 + 3i; a = 0, b = 3$
18. a. $-2 + 0i; a = -2, b = 0$
 b. $0 - 4i; a = 0, b = -4$
19. a. $18i; a = 0, b = 18$
 b. $\frac{\sqrt{2}}{2}i; a = 0, b = \frac{\sqrt{2}}{2}$
20. a. $-18i; a = 0, b = -18$
 b. $\frac{\sqrt{3}}{3}i; a = 0, b = \frac{\sqrt{3}}{3}$
21. a. $4 + 5\sqrt{2}\,i; a = 4, b = 5\sqrt{2}$
 b. $-5 + 3\sqrt{3}\,i; a = -5, b = 3\sqrt{3}$
22. a. $-2 + 4\sqrt{3}\,i; a = -2, b = 4\sqrt{3}$
 b. $7 + 5\sqrt{3}\,i; a = 7, b = 5\sqrt{3}$
23. a. $\frac{7}{4} + \frac{7\sqrt{2}}{8}i; a = \frac{7}{4}, b = \frac{7\sqrt{2}}{8}$
 b. $\frac{1}{2} + \frac{\sqrt{10}}{2}i; a = \frac{1}{2}, b = \frac{\sqrt{10}}{2}$
24. a. $\frac{7}{4} + \frac{\sqrt{7}}{4}i; a = \frac{7}{4}, b = \frac{\sqrt{7}}{4}$
 b. $\frac{4}{3} + \frac{\sqrt{3}}{2}i; a = \frac{4}{3}, b = \frac{\sqrt{3}}{2}$
25. a. $19 + i$
 b. $2 - 4i$
 c. $9 + 10\sqrt{3}\,i$
26. a. $1 - \sqrt{2}\,i$
 b. $-\sqrt{3} - \sqrt{2}\,i$
 c. $3\sqrt{5} - 3\sqrt{3}\,i$
27. a. $-3 + 2i$
 b. 8
 c. $2 - 8i$
28. a. $1 + 4i$
 b. $5 - i$
 c. $6.8 - 0.7i$
29. a. $2.7 + 0.2i$
 b. $15 + \frac{1}{12}i$
 c. $-2 - \frac{1}{8}i$
30. a. $2.9 - 12.8i$
 b. $14 + \frac{2}{15}i$
 c. $9 - \frac{11}{24}i$

DEVELOPING YOUR SKILLS

Simplify each radical (if possible). If imaginary, rewrite in terms of i and simplify.

7. a. $\sqrt{-16}$ $4i$ b. $\sqrt{-49}$ $7i$ c. $\sqrt{27}$ $3\sqrt{3}$ d. $\sqrt{72}$ $6\sqrt{2}$

8. a. $\sqrt{-81}$ $9i$ b. $\sqrt{-169}$ $13i$ c. $\sqrt{64}$ 8 d. $\sqrt{98}$ $7\sqrt{2}$

9. a. $-\sqrt{-18}$ $-3i\sqrt{2}$ b. $-\sqrt{-50}$ $-5i\sqrt{2}$ c. $3\sqrt{-25}$ $15i$ d. $2\sqrt{-9}$ $6i$

10. a. $-\sqrt{-32}$ $-4i\sqrt{2}$ b. $-\sqrt{-75}$ $-5i\sqrt{3}$ c. $3\sqrt{-144}$ $36i$ d. $2\sqrt{-81}$ $18i$

11. a. $\sqrt{-19}$ $i\sqrt{19}$ b. $\sqrt{-31}$ $i\sqrt{31}$ c. $\sqrt{\frac{-12}{25}}$ $\frac{2\sqrt{3}}{5}i$ d. $\sqrt{\frac{-9}{32}}$ $\frac{3\sqrt{2}}{8}i$

12. a. $\sqrt{-17}$ $i\sqrt{17}$ b. $\sqrt{-53}$ $i\sqrt{53}$ c. $\sqrt{\frac{-45}{36}}$ $\frac{\sqrt{5}}{2}i$ d. $\sqrt{\frac{-49}{75}}$ $\frac{7\sqrt{3}}{15}i$

Write each complex number in the standard form $a + bi$ and clearly identify the values of a and b.

13. a. $\frac{2 + \sqrt{-4}}{2}$ b. $\frac{6 + \sqrt{-27}}{3}$

14. a. $\frac{16 - \sqrt{-8}}{2}$ b. $\frac{4 + 3\sqrt{-20}}{2}$

15. a. $\frac{8 + \sqrt{-16}}{2}$ b. $\frac{10 - \sqrt{-50}}{5}$

16. a. $\frac{6 - \sqrt{-72}}{4}$ b. $\frac{12 + \sqrt{-200}}{8}$

17. a. 5 b. $3i$

18. a. -2 b. $-4i$

19. a. $2\sqrt{-81}$ b. $\frac{\sqrt{-32}}{8}$

20. a. $-3\sqrt{-36}$ b. $\frac{\sqrt{-75}}{15}$

21. a. $4 + \sqrt{-50}$ b. $-5 + \sqrt{-27}$

22. a. $-2 + \sqrt{-48}$ b. $7 + \sqrt{-75}$

23. a. $\frac{14 + \sqrt{-98}}{8}$ b. $\frac{5 + \sqrt{-250}}{10}$

24. a. $\frac{21 + \sqrt{-63}}{12}$ b. $\frac{8 + \sqrt{-27}}{6}$

Perform the addition or subtraction. Write the result in $a + bi$ form.

25. a. $(12 - \sqrt{-4}) + (7 + \sqrt{-9})$
 b. $(3 + \sqrt{-25}) + (-1 - \sqrt{-81})$
 c. $(11 + \sqrt{-108}) - (2 - \sqrt{-48})$

26. a. $(-7 - \sqrt{-72}) + (8 + \sqrt{-50})$
 b. $(\sqrt{3} + \sqrt{-2}) - (\sqrt{12} + \sqrt{-8})$
 c. $(\sqrt{20} - \sqrt{-3}) + (\sqrt{5} - \sqrt{-12})$

27. a. $(2 + 3i) + (-5 - i)$
 b. $(5 - 2i) + (3 + 2i)$
 c. $(6 - 5i) - (4 + 3i)$

28. a. $(-2 + 5i) + (3 - i)$
 b. $(7 - 4i) - (2 - 3i)$
 c. $(2.5 - 3.1i) + (4.3 + 2.4i)$

29. a. $(3.7 + 6.1i) - (1 + 5.9i)$
 b. $\left(8 + \frac{3}{4}i\right) - \left(-7 + \frac{2}{3}i\right)$
 c. $\left(-6 - \frac{5}{8}i\right) + \left(4 + \frac{1}{2}i\right)$

30. a. $(9.4 - 8.7i) - (6.5 + 4.1i)$
 b. $\left(3 + \frac{3}{5}i\right) - \left(-11 + \frac{7}{15}i\right)$
 c. $\left(-4 - \frac{5}{6}i\right) + \left(13 + \frac{3}{8}i\right)$

Multiply and write your answer in $a + bi$ form.

31. a. $5i \cdot (-3i)$ 15
 b. $(4i)(-4i)$ 16

32. a. $3(2 - 3i)$ $6 - 9i$
 b. $-7(3 + 5i)$ $-21 - 35i$

33. a $-7i(5 - 3i)$ $-21 - 35i$
 b. $6i(-3 + 7i)$ $-42 - 18i$

34. a. $(-4 - 2i)(3 + 2i)$ $-8 - 14i$
 b. $(2 - 3i)(-5 + i)$ $-7 + 17i$

35. a. $(-3 + 2i)(2 + 3i)$ $-12 - 5i$
 b. $(3 + 2i)(1 + i)$ $1 + 5i$

36. a. $(5 + 2i)(-7 + 3i)$ $-41 + i$
 b. $(4 - i)(7 + 2i)$ $30 + i$

For each complex number, name the complex conjugate. Then find the product.

37. a. $4 + 5i$ $4 - 5i$; 41 b. $3 - i\sqrt{2}$ $3 + i\sqrt{2}$; 11

38. a. $2 - i$ $2 + i$; 5 b. $-1 + i\sqrt{5}$ $-1 - i\sqrt{5}$; 6

39. a. $7i$ $-7i$; 49 b. $\frac{1}{2} - \frac{2}{3}i$ $\frac{1}{2} + \frac{2}{3}i$; $\frac{25}{36}$

40. a. $-5i$ $5i$; 25 b. $\frac{3}{4} + \frac{1}{5}i$ $\frac{3}{4} - \frac{1}{5}i$; $\frac{241}{400}$

Compute the special products and write your answer in $a + bi$ form.

41. a. $(4 - 5i)(4 + 5i)$ 41
 b. $(7 - 5i)(7 + 5i)$ 74

42. a. $(-2 - 7i)(-2 + 7i)$ 53
 b. $(2 + i)(2 - i)$ 5

43. a. $(3 - i\sqrt{2})(3 + i\sqrt{2})$ 11
 b. $(\frac{1}{6} + \frac{2}{3}i)(\frac{1}{6} - \frac{2}{3}i)$ $\frac{17}{36}$

44. a. $(5 + i\sqrt{3})(5 - i\sqrt{3})$ 28
 b. $(\frac{1}{2} + \frac{3}{4}i)(\frac{1}{2} - \frac{3}{4}i)$ $\frac{13}{16}$

45. a. $(2 + 3i)^2$ $-5 + 12i$ b. $(3 - 4i)^2$ $-7 - 24i$

46. a. $(2 - i)^2$ b. $(3 - i)^2$

46. a. $3 - 4i$
 b. $8 - 6i$

47. a. $(-2 + 5i)^2$ $-21 - 20i$ b. $(3 + i\sqrt{2})^2$ $7 + 6\sqrt{2}\,i$

48. a. $(-2 - 5i)^2$ $-21 + 20i$ b. $(2 - i\sqrt{3})^2$ $1 - 4\sqrt{3}\,i$

Use substitution to determine if the value shown is a solution to the given equation.

49. $x^2 + 36 = 0$; $x = -6$ no

50. $x^2 + 16 = 0$; $x = -4$ no

51. $x^2 + 49 = 0$; $x = -7i$ yes

52. $x^2 + 25 = 0$; $x = -5i$ yes

53. $(x - 3)^2 = -9$; $x = 3 - 3i$ yes

54. $(x + 1)^2 = -4$; $x = -1 + 2i$ yes

55. $x^2 - 2x + 5 = 0$; $x = 1 - 2i$ yes

56. $x^2 + 6x + 13 = 0$; $x = -3 + 2i$ yes

57. $x^2 - 4x + 9 = 0$; $x = 2 + i\sqrt{5}$ yes

58. $x^2 - 2x + 4 = 0$; $x = 1 - \sqrt{3}\,i$ yes

59. Show that $x = 1 + 4i$ is a solution to $x^2 - 2x + 17 = 0$. Then show its complex conjugate $1 - 4i$ is also a solution. Answers will vary.

60. Show that $x = 2 - 3\sqrt{2}\,i$ is a solution to $x^2 - 4x + 22 = 0$. Then show its complex conjugate $2 + 3\sqrt{2}\,i$ is also a solution. Answers will vary.

Simplify using powers of i.

61. a. i^{48} 1 b. i^{26} -1 c. i^{39} $-i$ d. i^{53} i

62. a. i^{36} 1 b. i^{50} -1 c. i^{19} $-i$ d. i^{65} i

Divide and write your answer in $a + bi$ form. Check your answer using multiplication.

63. a. $\dfrac{-2}{\sqrt{-49}}$ b. $\dfrac{4}{\sqrt{-25}}$

64. a. $\dfrac{2}{1 - \sqrt{-4}}$ b. $\dfrac{3}{2 + \sqrt{-9}}$

65. a. $\dfrac{7}{3 + 2i}$ b. $\dfrac{-5}{2 - 3i}$

66. a. $\dfrac{6}{1 + 3i}$ b. $\dfrac{7}{7 - 2i}$

67. a. $\dfrac{3 + 4i}{4i}$ b. $\dfrac{2 - 3i}{3i}$

68. a. $\dfrac{-4 + 8i}{2 - 4i}$ b. $\dfrac{3 - 2i}{-6 + 4i}$

63. a. $\frac{2}{7}i$ b. $\frac{-4}{5}i$

64. a. $\frac{2}{5} + \frac{4}{5}i$ b. $\frac{6}{13} - \frac{9}{13}i$

65. a. $\frac{21}{13} - \frac{14}{13}i$ b. $\frac{-10}{13} - \frac{15}{13}i$

66. a. $\frac{3}{5} - \frac{9}{5}i$ b. $\frac{49}{53} + \frac{14}{53}i$

67. a. $1 - \frac{3}{4}i$ b. $-1 - \frac{2}{3}i$

68. a. -2 b. $-\frac{1}{2}$

WORKING WITH FORMULAS

69. Absolute value of a complex number: $|a + bi| = \sqrt{a^2 + b^2}$

The absolute value of any complex number $a + bi$ (sometimes called the *modulus* of the number) is computed by taking the square root of the sums of the squares of a and b. Find the absolute value of the given complex numbers.

a. $|2 + 3i|$ $\sqrt{13}$ b. $|4 - 5i|$ $\sqrt{41}$ c. $|3 + \sqrt{2}\,i|$ $\sqrt{11}$

70. Binomial cubes: $(A + B)^3 = A^3 + 3A^2B + 3AB^2 + B^3$

The cube of any binomial can be found using this formula, where A and B are the terms of the binomial. Use the formula to compute the cube of $1 - 2i$ (note $A = 1$ and $B = -2i$). $-11 + 2i$

APPLICATIONS

71. In a day when imaginary numbers were imperfectly understood, Girolamo Cardano (1501–1576) once posed the problem, "Find two numbers that have a sum of 10 and whose product is 40." In other words, $A + B = 10$ and $AB = 40$. Although the solution is routine today, at the time the problem posed an enormous challenge. Verify that $A = 5 + \sqrt{15}\,i$ and $B = 5 - \sqrt{15}\,i$ satisfy these conditions. $A + B = 10$ $AB = 40$

72. Suppose Cardano had said, "Find two numbers that have a sum of 4 and a product of 7" (see Exercise 71). Verify that $A = 2 + \sqrt{3}\,i$ and $B = 2 - \sqrt{3}\,i$ satisfy these conditions.

Although it may seem odd, imaginary numbers have several applications in the real world. Many of these involve a study of electrical circuits, in particular *alternating current* or AC circuits. Briefly, the components of an AC circuit are current I (in amperes), voltage V (in volts), and the impedance Z (in ohms). The impedance of an electrical circuit is a measure of the total opposition to the flow of current through the circuit and is calculated as $Z = R + iX_L - iX_C$ where R represents a pure resistance, X_C represents the capacitance, and X_L represents the inductance. Each of these is also measured in ohms (symbolized by Ω).

73. Find the impedance Z if $R = 7\ \Omega$, $X_L = 6\ \Omega$, and $X_C = 11\ \Omega$. $7 - 5i\ \Omega$

74. Find the impedance Z if $R = 9.2\ \Omega$, $X_L = 5.6\ \Omega$, and $X_C = 8.3\ \Omega$. $9.2 - 2.7i\ \Omega$

The voltage V (in volts) across any element in an AC circuit is calculated as a product of the current I and the impedance Z: $V = IZ$.

75. Find the voltage in a circuit with a current of $3 - 2i$ amperes and an impedance of $Z = 5 + 5i\ \Omega$. $25 + 5i$ V

76. Find the voltage in a circuit with a current of $2 - 3i$ amperes and an impedance of $Z = 4 + 2i\ \Omega$. $14 - 8i$ V

In an AC circuit, the total impedance (in ohms) is given by the formula $Z = \frac{Z_1Z_2}{Z_1 + Z_2}$, where Z represents the total impedance of a circuit that has Z_1 and Z_2 wired in parallel.

77. Find the total impedance Z if $Z_1 = 1 + 2i$ and $Z_2 = 3 - 2i$. $\frac{7}{4} + i\ \Omega$

78. Find the total impedance Z if $Z_1 = 3 - i$ and $Z_2 = 2 + i$. $\frac{7}{5} + \frac{1}{5}i\ \Omega$

WRITING, RESEARCH, AND DECISION MAKING

79. (a) Up to this point, we have said that $x^2 - 9$ is factorable and $x^2 + 9$ is prime. Actually we mean that $x^2 + 9$ is nonfactorable *using real numbers,* but it can be factored using complex numbers. Do some research and exploration and see if you can accomplish the task. $(x + 3i)(x - 3i)$ (b) Verify that $(a + bi)(a - bi) = a^2 + b^2$. verified

80. Locate and read the following article. Then turn in a one page summary. "Thinking the Unthinkable: The Story of Complex Numbers," Israel Kleiner, *Mathematics Teacher,* Volume 81, Number 7, October 1988: pages 583–592. Answers will vary.

EXTENDING THE CONCEPT

81. Use the formula from Exercise 70 to compute the cube of $-\frac{1}{2} + \frac{\sqrt{3}}{2}i$. 1

82. If $a = 1$ and $b = 4$, the expression $\sqrt{b^2 - 4ac}$ represents an imaginary number for what values of c? $c > 4$

83. While it is a simple concept for real numbers, the square root of a complex number is much more involved due to the interplay between its real and imaginary parts. For $z = a + bi$ the square root of z can be found using the formula: $\sqrt{z} = \frac{\sqrt{2}}{2}(\sqrt{|z| + a} \pm i\sqrt{|z| - a})$, where the sign

84. a. $P = 4s$; $A = s^2$
b. $P = 2L + 2W$; $A = LW$
c. $P = a + b + c$; $A = \frac{bh}{2}$
d. $C = \pi d$; $A = \pi r^2$

85. a. Six is not a rational number—False.
b. The rational numbers are a subset of the reals—True.
c. 103 is an element of the set {3, 4, 5 . . .}—True.
d. The real numbers are not a subset of the complex—False.

86. $-7 \le x < 15$;
(number line from −7 to 15)
$x \in [-7, 15)$

87. $\frac{x-2}{x-5}$; $x \neq -5, 2, 5$

88. John

89. a. $(x + 2)(x - 2)(x^2 + 4)$
b. $(n - 3)(n^2 + 3n + 9)$
c. $(x - 1)^2(x + 1)$
d. $m(2n - 3m)^2$

is chosen to match the sign of b (see Exercise 69). Use the formula to find the square root of each complex number, then check by squaring.

a. $z = -7 + 24i$ $3 + 4i$ **b.** $z = 5 - 12i$ $3 - 2i$ **c.** $z = 4 + 3i$ $\frac{3\sqrt{2}}{2} + \frac{\sqrt{2}}{2}i$

MAINTAINING YOUR SKILLS

84. (R.7) State the perimeter and area formulas for: (a) squares, (b) rectangles, (c) triangles, and (d) circles.

85. (R.1) Write the symbols in words and state True/False.

a. $6 \notin \mathbb{Q}$ **b.** $\mathbb{Q} \subset \mathbb{R}$

c. $103 \in \{3, 4, 5 \ldots\}$ **d.** $\mathbb{R} \not\subset \mathbb{C}$

86. (1.2) Solve and graph the solution set on a number line: $-6 < \frac{3 - x}{2} \le 5$. Also state the solution in interval notation.

87. (R.4) Multiply:

$$\frac{x^2 - 4x + 4}{x^2 + 3x - 10} \cdot \frac{x^2 - 25}{x^2 - 10x + 25}.$$

State any domain restrictions that exist.

88. (1.1) John can run 10 m/sec, while Rick can only run 9 m/sec. If Rick gets a 2-sec head start, who will hit the 200-m finish line first?

89. (1.3) Factor the following expressions completely.

a. $x^4 - 16$ **b.** $n^3 - 27$

c. $x^3 - x^2 - x + 1$

d. $4n^2m - 12nm^2 + 9m^3$

1.5 Solving Nonfactorable Quadratic Equations

LEARNING OBJECTIVES

In Section 1.5 you will learn how to:

A. Write a quadratic equation in the standard form $ax^2 + bx + c = 0$ and identify the value of a, b, and c

B. Solve quadratic equations using the square root property of equality

C. Solve quadratic equations by completing the square

D. Solve quadratic equations using the quadratic formula

E. Use the discriminant to identify the number of solutions

F. Solve additional applications of quadratic equations

INTRODUCTION

In Section 1.1 we solved the literal equation $ax + b = c$ for x to establish a general solution for all linear equations of this form. In this section, we'll establish a general solution for the quadratic equation $ax^2 + bx + c = 0$, using a process known as *completing the square*. Other applications of completing the square abound in the algebra sequence and include the graphing of parabolas, circles, and other relations from the family of *conic sections*.

POINT OF INTEREST

The process of *completing the square* has an interesting visual interpretation. Consider a square with dimensions x by x and the corresponding area x^2 as shown in Figure 1.12. If we place two rectangles that are x units long by 3 units wide along each side of the square, the new figure has an area of $x^2 + 3x + 3x = x^2 + 6x$—but the figure is no longer square. To *complete the square* would require an additional 3 by 3 piece with an area of 9 square units. The area of the newly completed square would be $x^2 + 6x + 9$, a perfect square trinomial that factors into $(x + 3)^2$.

Figure 1.12

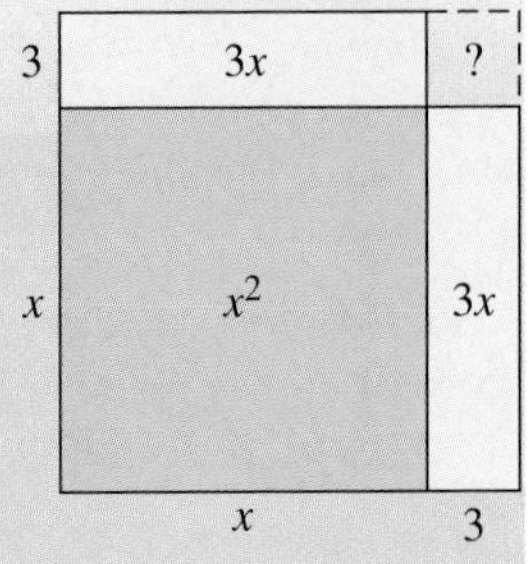

	x	3
3	$3x$	?
x	x^2	$3x$

WORTHY OF NOTE

The word *quadratic* comes from the Latin word *quadratum,* meaning square. The word historically refers to the "four sidedness" of a square, but mathematically to the *area* of a square. Hence its application to polynomials of the form $ax^2 + bx + c$—the variable of the leading term is *squared.*

A. Standard Form and Identifying Coefficients

A polynomial equation is in standard form when its terms are written in descending order of degree and set equal to zero. For quadratic polynomials, this is $ax^2 + bx^1 + c = 0$, where a, b, and c are real numbers and $a \neq 0$. Notice that a is the coefficient of the squared term as well as the lead coefficient, b is the coefficient of the linear (first degree) term, and c is a constant. All quadratic equations have a degree of two, but can have one, two, or three terms. The equation $n^2 - 81 = 0$ is a quadratic equation with two terms, where $a = 1, b = 0$, and $c = -81$: $n^2 + 0n + (-81)$.

EXAMPLE 1 State whether the given equation is quadratic. If yes, identify coefficients a, b, and c.

a. $2x^2 - 18 = 0$ **b.** $z - 12 - 3z^2 = 0$ **c.** $\frac{-3}{4}x + 5 = 0$

d. $z^3 - 2z^2 + 7z = 8$ **e.** $0.8x^2 = 0$

Solution:

	Equation	Quadratic	Coefficients
a.	$2x^2 - 18 = 0$	yes	$a = 2$ $b = 0$ $c = -18$
b.	In standard form, $-3z^2 + z - 12 = 0$	yes	$a = -3$ $b = 1$ $c = -12$
c.	$\frac{-3}{4}x + 5 = 0$	no	(linear equation)
d.	$z^3 - 2z^2 + 7z - 8 = 0$	no	(cubic equation)
e.	$0.8x^2 = 0$	yes	$a = 0.8$ $b = 0$ $c = 0$

NOW TRY EXERCISES 7 THROUGH 18

B. Quadratic Equations and the Square Root Property of Equality

The equation $x^2 - 9 = 0$ can be solved by factoring the left-hand expression and applying the zero factor property: $(x - 3)(x + 3) = 0$ gives solutions $x = 3$ and $x = -3$. Noting these solutions are the positive and negative square roots of 9 ($\sqrt{9} = 3$ or $-\sqrt{9} = -3$) enables us to introduce an alternative method for solving equations of the form $P^2 - k = 0$, known as the **square root (SQR) property of equality.** Since every positive quantity has two square roots, this method can be applied more generally than solutions by factoring.

SQUARE ROOT PROPERTY OF EQUALITY

For an equation of the form $P^2 = k$, where P is any algebraic expression and $k \geq 0$, the solutions are given by $P = \sqrt{k}$ or $P = -\sqrt{k}$, also written as $P = \pm\sqrt{k}$.

EXAMPLE 2 Use the square root property of equality to solve each equation.

a. $-3x^2 + 28 = -23$ **b.** $(x - 5)^2 = 24$

Solution:

a. $-3x^2 + 28 = -23$ original equation

$x^2 = 17$ isolate squared term

$x = \sqrt{17}$ or $x = -\sqrt{17}$ SQR property of equality

b. $(x - 5)^2 = 24$ original equation

$x - 5 = \sqrt{24}$ or $x - 5 = -\sqrt{24}$ SQR property of equality

$x = 5 + 2\sqrt{6}$ $\quad$ $x = 5 - 2\sqrt{6}$ solve for x and simplify radicals

NOW TRY EXERCISES 19 THROUGH 34

CAUTION

For equations of the form $P^2 = k$, where P represents a binomial [see Example 2(b)], students should resist the temptation to expand the binomial square in an attempt to solve by factoring (many times the result is nonfactorable). *Any* equation of the form $P^2 = k$ can quickly be solved using the SQR property of equality.

Answers written using radicals are called **exact** or **closed form** solutions. Results given in decimal form are called **approximate solutions** and must be written using the *approximately equal to* sign "≈." In this section, we will round all approximate solutions to hundredths, yielding approximate solutions of $x \approx 9.90$ and $x \approx 0.10$ for Example 2(b). Actually checking the exact solutions is a nice application of fundamental skills and the check for $x = 5 + 2\sqrt{6}$ is shown here.

Check: $(x - 5)^2 = 24$ original equation

$(5 + 2\sqrt{6} - 5)^2 = 24$ substitute $5 + 2\sqrt{6}$ for x

$(2\sqrt{6})^2 = 24$ simplify

$4(6) = 24$ ✓ result ($x = 5 - 2\sqrt{6}$ also checks)

C. Solving Quadratic Equations by Completing the Square

Consider again the equation $(x - 5)^2 = 24$ from Example 2(b). If we first expand the binomial square we obtain $x^2 - 10x + 25 = 24$, then $x^2 - 10x + 1 = 0$ after simplifying, which *cannot be solved by factoring,* while solutions to $(x - 5)^2 = 24$ were easily found. This observation leads to a strategy for solving nonfactorable quadratic equations, in which we attempt to *create a perfect square trinomial* from the quadratic and linear terms. This process is known as **completing the square.** To transform $x^2 - 10x + 1 = 0$ into $x^2 - 10x + 25 = 24$ [which returns us to $(x - 5)^2 = 24$], it appears we should first subtract 1 from both sides, to "make room" for the addition of 25. Note that with a lead coefficient of 1, *the value that completes the square is* $[(\frac{1}{2})(\textit{linear coefficient})]^2$: $[\frac{1}{2}(-10)]^2 = (-5)^2 = 25$. Adding this to both sides of $x^2 - 10x + \_\_\_\_ = -1 + \_\_\_\_$ gives $x^2 - 10x + 25 = -1 + 25$ and the square is complete since we now have $(x - 5)^2 = 24$. See Exercises 35 through 40 for additional practice.

COMPLETING THE SQUARE TO SOLVE A QUADRATIC EQUATION

To solve the equation $ax^2 + bx + c = 0$ by completing the square:

1. Subtract the constant c from both sides.
2. Divide both sides by the leading coefficient a.
3. Take $[(\frac{1}{2})(\textit{linear coefficient})]^2$ and add the result to both sides.
4. Factor the left-hand side as a binomial square; simplify the right-hand side.
5. Solve using the SQR property of equality.

EXAMPLE 3 Solve by completing the square: $x^2 - 19 = 6x$

Solution:

$x^2 - 19 = 6x$	original equation (note $a = 1$)
$x^2 - 6x - 19 = 0$	standard form—equation is nonfactorable
$x^2 - 6x + \_\_ = 19 + \_\_$	add 19 to both sides; $\left[\left(\frac{1}{2}\right)(\textit{linear coefficient})\right]^2 = \left[\left(\frac{1}{2}\right)(-6)\right]^2 = 9$
$x^2 - 6x + 9 = 19 + 9$	add 9 to both sides (completing the square)
$(x - 3)^2 = 28$	factor and simplify
$x - 3 = \sqrt{28}$ or $x - 3 = -\sqrt{28}$	SQR property of equality
$x = 3 + 2\sqrt{7}$ $\quad$ $x = 3 - 2\sqrt{7}$	simplify radicals and solve for x (exact form)
$x \approx 8.29$ $\quad$ $x \approx -2.29$	approximate form

NOW TRY EXERCISES 41 THROUGH 50

The process of completing the square can be applied to any quadratic equation with a lead coefficient of $a = 1$. If the lead coefficient is not 1, we simply divide through by a before we begin.

EXAMPLE 4 Solve by completing the square: $-5x^2 + 2 = 8x$

Solution:

$-5x^2 + 2 = 8x$	original equation (note $a \neq 1$)
$-5x^2 - 8x + 2 = 0$	standard form (nonfactorable)
$x^2 + \frac{8}{5}x - \frac{2}{5} = 0$	divide through by -5
$x^2 + \frac{8}{5}x + \_\_ = \frac{2}{5} + \_\_$	add $\frac{2}{5}$ to both sides; $\left[\left(\frac{1}{2}\right)(\textit{linear coefficient})\right]^2 = \left[\left(\frac{1}{2}\right)\left(\frac{8}{5}\right)\right]^2 = \frac{16}{25}$
$x^2 + \frac{8}{5}x + \frac{16}{25} = \frac{2}{5} + \frac{16}{25}$	add $\frac{16}{25}$ to both sides (completing the square)
$\left(x + \frac{4}{5}\right)^2 = \frac{26}{25}$	factor and simplify $\left(\frac{2}{5} = \frac{10}{25}\right)$
$x + \frac{4}{5} = \sqrt{\frac{26}{25}}$ or $x + \frac{4}{5} = -\sqrt{\frac{26}{25}}$	SQR property of equality
$x = -\frac{4}{5} + \frac{\sqrt{26}}{5}$ or $x = -\frac{4}{5} - \frac{\sqrt{26}}{5}$	simplify radicals and solve for x (exact form)
$x \approx 0.22$ or $x \approx -1.82$	approximate form

NOW TRY EXERCISES 51 THROUGH 58

D. Solving Quadratic Equations Using the Quadratic Formula

In Section 1.1 we found a general solution to the linear equation $ax + b = c$ by comparing it to $2x + 3 = 15$. We'll use a similar idea to find a general solution for quadratic equations. In a side-by-side format, we solve the equation $2x^2 + 5x + 3 = 0$ and the general equation $ax^2 + bx + c = 0$ by completing the square. Note the similarities.

$2x^2 + 5x + 3 = 0$	given equations	$ax^2 + bx + c = 0$
$x^2 + \frac{5}{2}x + \frac{3}{2} = 0$	divide by leading coefficient	$x^2 + \frac{b}{a}x + \frac{c}{a} = 0$
$x^2 + \frac{5}{2}x + \_\_\_ = -\frac{3}{2}$	subtract constant term	$x^2 + \frac{b}{a}x + \_\_\_ = -\frac{c}{a}$
$\left[\frac{1}{2}\left(\frac{5}{2}\right)\right]^2 = \frac{25}{16}$	$\left[\frac{1}{2}(\text{linear coefficient})\right]^2$	$\left[\frac{1}{2}\left(\frac{b}{a}\right)\right]^2 = \frac{b^2}{4a^2}$
$x^2 + \frac{5}{2}x + \frac{25}{16} = \frac{25}{16} - \frac{3}{2}$	add to each side	$x^2 + \frac{b}{a}x + \frac{b^2}{4a^2} = \frac{b^2}{4a^2} - \frac{c}{a}$
$\left(x + \frac{5}{4}\right)^2 = \frac{25}{16} - \frac{3}{2}$	left side factors as a binomial square	$\left(x + \frac{b}{2a}\right)^2 = \frac{b^2}{4a^2} - \frac{c}{a}$
$\left(x + \frac{5}{4}\right)^2 = \frac{25}{16} - \frac{24}{16}$	determine LCDs	$\left(x + \frac{b}{2a}\right)^2 = \frac{b^2}{4a^2} - \frac{4ac}{4a^2}$
$\left(x + \frac{5}{4}\right)^2 = \frac{1}{16}$	simplify on right	$\left(x + \frac{b}{2a}\right)^2 = \frac{b^2 - 4ac}{4a^2}$
$x + \frac{5}{4} = \pm\sqrt{\frac{1}{16}}$	SQR property of equality	$x + \frac{b}{2a} = \pm\sqrt{\frac{b^2 - 4ac}{4a^2}}$
$x + \frac{5}{4} = \pm\frac{1}{4}$	simplify radicals	$x + \frac{b}{2a} = \pm\frac{\sqrt{b^2 - 4ac}}{2a}$
$x = -\frac{5}{4} \pm \frac{1}{4}$	solve for x	$x = -\frac{b}{2a} \pm \frac{\sqrt{b^2 - 4ac}}{2a}$
$x = \frac{-5 \pm 1}{4}$	combine terms	$x = \frac{-b \pm \sqrt{b^2 - 4ac}}{2a}$
$x = \frac{-5 + 1}{4}$ or $x = \frac{-5 - 1}{4}$	solutions	$x = \frac{-b + \sqrt{b^2 - 4ac}}{2a}$ or $x = \frac{-b - \sqrt{b^2 - 4ac}}{2a}$

On the left, our final solutions are $x = -1$ or $x = -\frac{3}{2}$. The result on the right is called the **quadratic formula,** which can be used to solve *any equation belonging to the quadratic family.*

THE QUADRATIC FORMULA

For the general quadratic equation $ax^2 + bx + c = 0$, where a, b, and c are real numbers and $a \neq 0$, the solutions are given by

$$x = \frac{-b \pm \sqrt{b^2 - 4ac}}{2a},$$

where the "$\pm$" notation indicates the solutions are

$$x = \frac{-b + \sqrt{b^2 - 4ac}}{2a} \quad \text{or} \quad x = \frac{-b - \sqrt{b^2 - 4ac}}{2a}.$$

EXAMPLE 5 Solve $\frac{1}{4}x^2 - \frac{1}{2}x = 1$ using the quadratic formula. State the solution(s) in both exact and approximate form. Check one of the exact solutions in the original equation.

Solution: Although we *could* apply the quadratic formula using fractional values, our work is greatly simplified if we first eliminate the fractions using the LCM.

$\frac{1}{4}x^2 - \frac{1}{2}x = 1$ original equation

$x^2 - 2x - 4 = 0$ multiply by 4; write in standard form

$x = \frac{-(-2) \pm \sqrt{(-2)^2 - 4(1)(-4)}}{2(1)}$ substitute 1 for a, -2 for b, and -4 for c

$x = \frac{2 \pm \sqrt{4 + 16}}{2} = \frac{2 \pm \sqrt{20}}{2}$ simplify

$x = \frac{2 \pm 2\sqrt{5}}{2}$ or $1 \pm \sqrt{5}$ (see following Caution)

$x = 1 - \sqrt{5}$ or $x = 1 + \sqrt{5}$ exact solutions

$x \approx -1.24$ or $x \approx 3.24$ approximate solutions

Check: $\frac{1}{4}x^2 - \frac{1}{2}x = 1$ original equation

$\frac{1}{4}(1 + \sqrt{5})^2 - \frac{1}{2}(1 + \sqrt{5}) = 1$ substitute $1 + \sqrt{5}$ for x

$1 + 2\sqrt{5} + 5 - 2 - 2\sqrt{5} = 4$ multiply by 4, square binomial and distribute

$4 = 4$ ✓ solution checks

NOW TRY EXERCISES 59 THROUGH 94

CAUTION

For $\frac{2 + 2\sqrt{5}}{2}$, be careful not to incorrectly "cancel the twos" as in $\frac{\overset{1}{\cancel{2}} + 2\sqrt{5}}{\underset{1}{\cancel{2}}} = 1 \pm 2\sqrt{5}$. *No!* Use a calculator to verify that the results are not equivalent. Both terms in the numerator are being divided by two and we must factor the numerator (if possible) to see if the expression simplifies further: $\frac{2 \pm 2\sqrt{5}}{2} = \frac{2(1 \pm \sqrt{5})}{2} = (1 \pm \sqrt{5})$. *Yes!*

E. The Discriminant of the Quadratic Formula

Earlier we noted $\sqrt{A}$ represents a real number only when $A \geq 0$. If $A < 0$, the result is an imaginary number. Since the quadratic formula contains the radical $\sqrt{b^2 - 4ac}$, the expression $b^2 - 4ac$, called the **discriminant,** will determine the nature (real or complex) and the number of roots. As shown in the box, there are three possibilities:

WORTHY OF NOTE

Further analysis reveals even more concerning the nature of these roots. If the discriminant is a perfect square, there will be two *rational* roots. If the discriminant is not a perfect square, there will be two *irrational* roots. If the discriminant is zero there is one rational root.

THE DISCRIMINANT OF THE QUADRATIC FORMULA

For the equation $ax^2 + bx + c = 0$, the discriminant is $b^2 - 4ac$.

1. if $b^2 - 4ac = 0$, the equation has one real root.
2. if $b^2 - 4ac > 0$, the equation has two real roots.
3. if $b^2 - 4ac < 0$, the equation has two complex roots.

EXAMPLE 6 Use the discriminant to determine if the equation given has any real root(s). If so, state whether the roots are rational or irrational.

a. $2x^2 + 5x + 2 = 0$ **b.** $x^2 - 4x + 7 = 0$ **c.** $4x^2 - 20x + 25 = 0$

Solution: **a.** $a = 2, b = 5, c = 2$

$b^2 - 4ac = (5)^2 - 4(2)(2)$
$= 9$

Since $9 > 0$,
$\rightarrow$ two rational roots

b. $a = 1, b = -4, c = 7$

$b^2 - 4ac = (-4)^2 - 4(1)(7)$
$= -12$

Since $-12 < 0$,
$\rightarrow$ two complex roots

c. $a = 4, b = -20, c = 25$

$b^2 - 4ac = (-20)^2 - 4(4)(25)$
$= 0$

Since $b^2 - 4ac = 0$,
$\rightarrow$ one rational root

NOW TRY EXERCISES 95 THROUGH 106

A closer look at the quadratic formula also reveals that when $b^2 - 4ac < 0$, the two complex solutions *will be complex conjugates.*

COMPLEX SOLUTIONS

Given $ax^2 + bx + c = 0$ with $a, b, c \in \mathbb{R}$, the complex solutions will occur in conjugate pairs.

EXAMPLE 7 Find the roots of $2x^2 - 4x + 5 = 0$. Simplify and write the result in standard form.

Solution: Evaluate the quadratic formula $x = \dfrac{-b \pm \sqrt{b^2 - 4ac}}{2a}$ for $a = 2$, $b = -4$, and $c = 5$.

$$x = \frac{-(-4) \pm \sqrt{(-4)^2 - 4(2)(5)}}{2(2)} \quad \text{substitute 2 for } a, -4 \text{ for } b, \text{ and 5 for } c$$

$$x = \frac{4 \pm \sqrt{-24}}{4} \quad \text{simplify, note } b^2 - 4ac < 0$$

$$x = \frac{4 \pm 2i\sqrt{6}}{4} \longrightarrow \frac{2(2 \pm i\sqrt{6})}{4} \quad \text{write in } i \text{ form and simplify}$$

$$x = \frac{2 \pm i\sqrt{6}}{2} \longrightarrow 1 \pm \frac{i\sqrt{6}}{2} \quad \text{result}$$

The solutions are the complex conjugates $1 + \frac{i\sqrt{6}}{2}$ and $1 - \frac{i\sqrt{6}}{2}$.

NOW TRY EXERCISES 107 THROUGH 112

F. Applications of the Quadratic Formula

A projectile is any object that is thrown, shot, or *projected* upward. The height of the projectile at any time t is modeled by the equation $h = -16t^2 + vt + k$, where h is the height of the object in feet, t is the elapsed time in seconds, and v is the initial velocity in feet per second. The constant k represents the initial height of the object above ground level, as when a person releases an object 5 ft above the ground in a throwing motion. If the person were on a hill 60 ft high, k would be 65 ft.

EXAMPLE 8 A person standing on a hill 60 ft high, throws a ball upward with an initial velocity of 102 ft/sec (assume the ball is released 5 ft above where the person is standing). Find (a) the height of the object after 3 sec and (b) how many seconds until the ball hits the ground at the bottom of the hill.

Solution: Using the given information, we have $h = -16t^2 + 102t + 65$. To find the height after 3 sec, substitute $t = 3$.

a. $h = -16t^2 + 102t + 65$ original equation
$= -16(3)^2 + 102(3) + 65$ substitute 3 for t
$= 227$ result

After 3 sec, the ball is 227 ft above ground.

b. When the ball hits the ground at the base of the hill, it has a height of zero. Substitute $h = 0$ and solve using the quadratic formula.

$0 = -16t^2 + 102t + 65$ $\quad a = -16, b = 102, c = 65$

$$t = \frac{-b \pm \sqrt{b^2 - 4ac}}{2a}$$ quadratic formula

$$t = \frac{-(102) \pm \sqrt{(102)^2 - 4(-16)(65)}}{2(-16)}$$ substitute $a = -16$, $b = 102$, $c = 65$

$$t = \frac{-102 \pm \sqrt{14{,}564}}{-32}$$ simplify

Since we're trying to find the time in seconds, go directly to the approximate form of the answer.

$t \approx -0.58$ or $t \approx 6.96$ approximate solutions

The ball will strike the ground about 7 sec later. Since t represents time, the solution $t \approx -0.58$ does not apply.

NOW TRY EXERCISES 115 THROUGH 124

Virtually all techniques that are applied in order to solve polynomial equations with real coefficients can still be applied when the coefficients and/or solutions are complex numbers. This means the quadratic formula can be applied to solve *any* quadratic equation, *even those whose coefficients are complex!* We will initially apply this idea to examples that are carefully chosen, as a more general application must wait until a future course, when the square root of a complex number is fully developed.

EXAMPLE 9 Use the quadratic formula to solve the complex quadratic equation given. Check one of the roots by substitution: $0.5z^2 + (5 - 3i)z - 15i = 0$.

Solution: For the equation given we have $a = 0.5$, $b = 5 - 3i$, and $c = -15i$. The quadratic formula gives:

$$z = \frac{-(5 - 3i) \pm \sqrt{(5 - 3i)^2 - 4(0.5)(-15i)}}{2(0.5)}$$

$$= \frac{(-5 + 3i) \pm \sqrt{(25 - 30i - 9) + 30i}}{1} = (-5 + 3i) \pm \sqrt{16}$$

$$= (-5 + 3i) \pm 4$$

The solutions are $z = -1 + 3i$ and $-9 + 3i$. Checking $z = -1 + 3i$ yields

$0.5(-1 + 3i)^2 + (5 - 3i)(-1 + 3i) - 15i = 0$ substitute $-1 + 3i$ for z
$0.5(-8 - 6i) + (4 + 18i) - 15i = 0$ simplify
$-4 - 3i + 4 + 18i - 15i = 0$✓ solution checks

NOW TRY EXERCISES 125 THROUGH 130

TECHNOLOGY HIGHLIGHT

Programs, Quadratic Equations, and the Discriminant

The keystrokes shown apply to a TI-84 Plus model. Please consult your manual or our Internet site for other models.

Quadratic equations play an important role in a study of college algebra, forming a sort of bridge between the elementary equations studied earlier and the more advanced equations to come. As seen in this section, the discriminant of the quadratic formula $b^2 - 4ac$ gives us information on the type and number of roots, and it will often be helpful to have this information in advance of trying to solve or graph the equation. This means the discriminant will be evaluated repeatedly as we work with each new equation, making it a prime candidate for a short program. To begin a new program press PRGM ▶ ▶ (NEW) ENTER. The calculator will prompt you to name the program using the green ALPHA letters (eight letters max), then allow you to start entering program lines. In PRGM mode, pressing PRGM once again will bring up menus that contain all needed commands. For very basic programs these commands will be in the **I/O** (Input/Output) sub-menu, with the most common options being **1:Input, 3:Disp,** and **8:CLRHOME.** As you can see, we have named our program *DISCRMNT.*

PROGRAM: DISCRMNT	
:CLRHOME	Clears the home screen, places cursor in upper left position
:DISP "THIS PRGM WILL"	Displays the words *THIS PRGM WILL* on the screen
:DISP "EVALUATE THE"	Displays the words *EVALUATE THE* on the screen
:DISP "DISCRIMINANT OF"	Displays the words *DISCRIMINANT OF* on the screen
:DISP "AX^2 + BX + C"	Displays the words $AX^2 + BX + C$ on the screen
:PAUSE: CLRHOME	Pauses the program until the user presses ENTER, then clears the screen
:DISP "AX^2 + BX + C"	Displays the words $AX^2 + BX + C$ on the screen
:DISP ""	Displays a blank line (for formatting purposes)
:DISP "ENTER A"	Displays the words *ENTER A* on the screen
:INPUT A	Accepts the input and places it in memory location A
:DISP "ENTER B"	Displays the words *ENTER B* on the screen
:INPUT B	Accepts the input and places it in memory location B
:DISP "ENTER C"	Displays the words *ENTER C* on the screen
:INPUT C	Accepts the input and places it in memory location C
(B^2 – 4AC)→D	Computes $B^2 - 4AC$ using stored values and places result in location D
:CLRHOME	Clears the home screen, places cursor in upper left position
:DISP "B^2 – 4AC IS"	Displays the words $B^2 - 4AC$ *IS* on the screen
:DISP D	Displays the computed value of D

Exercise 1: Run the program for $y = x^2 - 3x - 10$ and $y = x^2 + 5x - 14$, then check to see if each expression is factorable. What do you notice? *D* is a perfect square.

Exercise 2: Run the program for $y = 25x^2 - 90x + 81$ and $y = 4x^2 + 20x + 25$, then check to see if each is a perfect square trinomial. What do you notice? $D = 0$

Exercise 3: Run the program for $y = x^2 + 2x + 10$ and $y = x^2 - 2x + 5$. Do these equations have real number solutions? no

Exercise 4: Run the program for $y = x^2 - 4x + 2$ and $y = x^2 + 4x + 1$. Do these equations have real number solutions? yes What are they? $2 \pm \sqrt{2}; -2 \pm \sqrt{3}$ or approximately 3.41, 0.59; −0.27, −3.73

1.5 EXERCISES

CONCEPTS AND VOCABULARY

1. A polynomial equation is in standard form when written in __descending__ order of degree and set equal to __0__.

2. The solution $x = 2 + \sqrt{3}$ is called an __exact__ form of the solution. Using a calculator, we find the __approximate__ form is $x \approx 3.732$.

3. To solve a quadratic equation by completing the square, the coefficient of the __quadratic__ term must be a __1__.

4. The quantity $b^2 - 4ac$ is called the __discriminant__ of the quadratic equation. If $b^2 - 4ac > 0$, there are __2__ real roots.

5. Discuss/explain why the quadratic formula need not be used to solve $4x^2 - 5 = 0$, then solve the equation using some other method. $x = \pm\frac{\sqrt{5}}{2}$
The square root property is easier.

6. Discuss/explain why this version of the quadratic formula: $x = -b \pm \dfrac{\sqrt{b^2 - 4ac}}{2a}$ is incorrect.
$-b$ must be divided by $2a$.

DEVELOPING YOUR SKILLS

Determine whether each equation is quadratic. If so, identify the coefficients a, b, and c.

7. $2x - 15 - x^2 = 0$
8. $21 + x^2 - 4x = 0$
9. $\frac{2}{3}x - 7 = 0$
10. $12 - 4x = 9$
11. $\frac{1}{4}x^2 = 6x$
12. $0.5x = 0.25x^2$
13. $2x^2 + 7 = 0$
14. $5 = -4x^2$
15. $-3x^2 + 9x - 5 + 2x^3 = 0$
16. $z^2 - 6z + 9 - z^3 = 0$
17. $(x - 1)^2 + (x - 1) + 4 = 9$
18. $(x + 5)^2 - (x + 5) + 4 = 17$

Solve the following equations using the square root property of equality. Write answers in exact form and approximate form rounded to hundredths. If there are no real solutions, so state.

19. $m^2 = 16$
20. $p^2 = 49$
21. $y^2 - 28 = 0$
22. $m^2 - 20 = 0$
23. $p^2 + 36 = 0$
24. $n^2 + 5 = 0$
25. $x^2 = \frac{21}{16}$
26. $y^2 = \frac{13}{9}$
27. $(n - 3)^2 = 36$
28. $(p + 5)^2 = 49$
29. $(w + 5)^2 = 3$
30. $(m - 4)^2 = 5$
31. $(x - 3)^2 + 7 = 2$
32. $(m + 11)^2 + 5 = 3$
33. $(m - 2)^2 = \frac{18}{49}$
34. $(x - 5)^2 = \frac{12}{25}$

Fill in the blank so the result is a perfect square trinomial, then factor into a binomial square.

35. $x^2 + 6x +$ ___ $9; (x + 3)^2$
36. $y^2 + 10y +$ ___ $25; (y + 5)^2$
37. $n^2 + 3n +$ ___ $\frac{9}{4}; (n + \frac{3}{2})^2$
38. $x^2 - 5x +$ ___ $\frac{25}{4}; (x - \frac{5}{2})^2$
39. $p^2 + \frac{2}{3}p +$ ___ $\frac{1}{9}; (p + \frac{1}{3})^2$
40. $x^2 - \frac{3}{2}x +$ ___ $\frac{9}{16}; (x - \frac{3}{4})^2$

Solve by completing the square. Write your answers in both exact form and approximate form rounded to the hundredths place. If there are no real solutions, so state.

41. $x^2 + 6x = -5$
42. $m^2 + 8m = -12$
43. $p^2 - 6p + 3 = 0$
44. $n^2 = 4n + 10$
45. $p^2 + 6p = -4$
46. $x^2 - 8x - 1 = 0$
47. $m^2 + 3m = 1$
48. $n^2 + 5n - 2 = 0$
49. $n^2 = 5n + 5$
50. $w^2 - 7w + 3 = 0$
51. $2x^2 = -7x + 4$
52. $3w^2 - 8w + 4 = 0$

7. $a = -1; b = 2; c = -15$
8. $a = 1; b = -4; c = 21$
9. not quadratic
10. not quadratic
11. $a = \frac{1}{4}; b = -6; c = 0$
12. $a = 0.25; b = -0.5; c = 0$
13. $a = 2; b = 0; c = 7$
14. $a = -4; b = 0; c = -5$
15. not quadratic
16. not quadratic
17. $a = 1; b = -1; c = -5$
18. $a = 1; b = 9; c = 7$
19. $m = \pm 4$
20. $p = \pm 7$
21. $y = \pm 2\sqrt{7}; y \approx \pm 5.29$
22. $m = \pm 2\sqrt{5}; m \approx \pm 4.47$
23. no real solutions
24. no real solutions
25. $x = \pm\frac{\sqrt{21}}{4}; x \approx \pm 1.15$
26. $y = \pm\frac{\sqrt{13}}{3}; y \approx \pm 1.20$
27. $n = 9; n = -3$
28. $p = 2; p = -12$
29. $w = -5 \pm \sqrt{3}; w \approx -3.27$ or $w \approx -6.73$
30. $m = 4 \pm \sqrt{5}; m \approx 6.24$ or $m \approx 1.76$
31. no real solutions
32. no real solutions
33. $m = 2 \pm \frac{3\sqrt{2}}{7}; m \approx 2.61$ or $m \approx 1.39$
34. $x = 5 \pm \frac{2\sqrt{3}}{5}; x \approx 5.69$ or $x \approx 4.31$
41. $x = -1; x = -5$
42. $m = -2; m = -6$
43. $p = 3 \pm \sqrt{6}; p \approx 5.45$ or $p \approx 0.55$
44. $n = 2 \pm \sqrt{14}; n \approx 5.74$ or $n \approx -1.74$
45. $p = -3 \pm \sqrt{5}; p \approx -0.76$ or $p \approx -5.24$
46. $x = 4 \pm \sqrt{17}; x \approx 8.12$ or $x \approx -0.12$
47. $m = \frac{-3}{2} \pm \frac{\sqrt{13}}{2}; m \approx 0.30$ or $m \approx -3.30$
48. $n = \frac{-5}{2} \pm \frac{\sqrt{33}}{2}; n \approx 0.37$ or $n \approx -5.37$
49. $n = \frac{5}{2} \pm \frac{3\sqrt{5}}{2}; n \approx 5.85$ or $n \approx -0.85$
50. $w = \frac{7}{2} \pm \frac{\sqrt{37}}{2}; w \approx 6.54$ or $w \approx 0.46$
51. $x = \frac{1}{2}$ or $x = -4$
52. $w = 2$ or $w = \frac{2}{3}$

54. $p = \frac{5}{4} \pm \frac{\sqrt{33}}{4}$; $p \approx 2.69$ or $p \approx -0.19$

55. $p = \frac{3}{8} \pm \frac{\sqrt{41}}{8}$; $p \approx 1.18$ or $p \approx -0.43$

56. $x = \frac{-5}{6} \pm \frac{\sqrt{97}}{6}$; $x \approx 0.81$ or $x \approx -2.47$

57. $m = \frac{7}{2} \pm \frac{\sqrt{33}}{2}$; $m \approx 6.37$ or $m \approx 0.63$

58. $a = 2 \pm \sqrt{19}$; $a \approx 6.36$ or $a \approx -2.36$

59. $x = 6$ or $x = -3$

60. $w = -3 \pm \sqrt{10}$; $w \approx 0.16$ or $w \approx -6.16$

61. $m = \pm\frac{5}{2}$

62. $a = \frac{1 \pm \sqrt{2}}{2}$; $a \approx 1.21$ or $a \approx -0.21$

63. $n = \frac{2 \pm \sqrt{5}}{2}$; $n \approx 2.12$ or $n \approx -0.12$

64. $x = 1 \pm \frac{\sqrt{6}}{2}i$; $x \approx 1 \pm 1.22i$

65. $w = \frac{2}{3}$ or $w = \frac{-1}{2}$

66. $a = \frac{5}{6} \pm \frac{\sqrt{47}}{6}i$; $a \approx 0.8\overline{3} \pm 1.14i$

67. $m = \frac{3}{2} \pm \frac{\sqrt{6}}{2}i$; $m \approx 1.5 \pm 1.22i$

68. $p = 0$ or $p = \frac{-1}{3}$ 69. $n = \pm\frac{3}{2}$

70. $x = \frac{-3}{4}$ or $x = 1$

71. $w = \frac{-4}{5}$ or $w = 2$

72. $m = \frac{-2}{3}$ or $m = 3$

73. $a = \frac{1}{6} \pm \frac{\sqrt{23}}{6}i$; $a \approx 0.1\overline{6} \pm 0.80i$

74. $n = \frac{1 \pm \sqrt{10}}{3}$; $n \approx 1.39$ or $n \approx -0.72$

75. $p = \frac{3 \pm 2\sqrt{6}}{5}$; $p \approx 1.58$ or $p \approx -0.38$

76. $x = -\frac{1}{4} \pm \frac{\sqrt{23}}{4}i$; $x \approx -0.25 \pm 1.20i$

77. $w = \frac{1 \pm \sqrt{21}}{10}$; $w \approx 0.56$ or $w \approx -0.36$

78. $m = -\frac{1}{3}$ or $m = 2$

79. $a = \frac{3}{4} \pm \frac{\sqrt{31}}{4}i$; $a \approx 0.75 \pm 1.39i$

80. $n = -2 \pm 2\sqrt{3}$; $n \approx 1.46$ or $n \approx -5.46$

81. $p = 1 \pm \frac{3\sqrt{2}}{2}i$; $p \approx 1 \pm 2.12i$

82. $x = \frac{5 \pm \sqrt{57}}{16}$; $x \approx 0.78$ or $x \approx -0.16$

83. $w = \frac{-1}{3} \pm \frac{\sqrt{2}}{3}$; $w \approx 0.14$ or $w \approx -0.80$

84. $m = \frac{16 \pm \sqrt{226}}{15}$; $m \approx 2.07$ or $m \approx 0.06$

85. $a = \frac{-6 \pm 3\sqrt{2}}{2}$; $a \approx -0.88$ or $a \approx -5.12$

86. $n = \frac{9 \pm \sqrt{321}}{12}$; $n \approx 2.24$ or $n \approx -0.74$

87. $p = \frac{4 \pm \sqrt{394}}{6}$; $p \approx 3.97$ or $p \approx -2.64$

88. $x = \frac{48 \pm 3\sqrt{1006}}{50}$; $x \approx 2.86$ or $x \approx -0.94$

89. $w = -5$ or $w = 2$

90. $m = 7$ or $m = -2$

91. $a = \frac{1 \pm \sqrt{57}}{4}$; $a \approx 2.14$ or $a \approx -1.64$

Additional answers can be found in the Instructor Answer Appendix.

53. $2n^2 - 3n - 9 = 0$
$n = 3$ or $n = \frac{-3}{2}$

54. $2p^2 - 5p = 1$

55. $4p^2 - 3p - 2 = 0$

56. $3x^2 + 5x - 6 = 0$

57. $\frac{m}{2} = \frac{7}{2} - \frac{2}{m}$

58. $\frac{a}{5} - \frac{3}{a} = \frac{4}{5}$

Solve each equation using the most efficient method: factoring, SQR property of equality, or the quadratic formula. Write your answer in both exact and approximate form (rounded to hundredths). Check one of the exact solutions in the original equation.

59. $x^2 - 3x = 18$

60. $w^2 + 6w - 1 = 0$

61. $4m^2 - 25 = 0$

62. $4a^2 - 4a = 1$

63. $4n^2 - 8n - 1 = 0$

64. $2x^2 - 4x + 5 = 0$

65. $6w^2 - w = 2$

66. $3a^2 - 5a + 6 = 0$

67. $4m^2 = 12m - 15$

68. $3p^2 + p = 0$

69. $4n^2 - 9 = 0$

70. $4x^2 - x = 3$

71. $5w^2 = 6w + 8$

72. $3m^2 - 7m - 6 = 0$

73. $3a^2 - a + 2 = 0$

74. $3n^2 - 2n - 3 = 0$

75. $5p^2 = 6p + 3$

76. $2x^2 + x + 3 = 0$

77. $5w^2 - w = 1$

78. $3m^2 - 2 = 5m$

79. $2a^2 + 5 = 3a$

80. $n^2 + 4n - 8 = 0$

81. $2p^2 - 4p + 11 = 0$

82. $8x^2 - 5x - 1 = 0$

83. $w^2 + \frac{2}{3}w = \frac{1}{9}$

84. $\frac{5}{4}m^2 - \frac{8}{3}m + \frac{1}{6} = 0$

85. $0.2a^2 + 1.2a + 0.9 = 0$

86. $-5.4n^2 + 8.1n + 9 = 0$

87. $\frac{2}{7}p^2 - 3 = \frac{8}{21}p$

88. $\frac{5}{9}x^2 - \frac{16}{15}x = \frac{3}{2}$

89. $w + 3 - \frac{10}{w} = 0$

90. $m - 5 - \frac{14}{m} = 0$

91. $a - \frac{7}{2a} = \frac{1}{2}$

92. $z - \frac{1}{5z} = -\frac{1}{2}$

93. $n - \frac{4}{n - 3} = 0$
$n = 4, n = -1$

94. $x = \frac{5}{x + 4}$
$x = -5, x = 1$

Use the discriminant to determine whether the given equation has two irrational roots, two rational roots, one repeated root, or two complex roots. Do not solve.

95. $-3x^2 + 2x + 1 = 0$

96. $2x^2 - 5x - 3 = 0$

97. $-4x + x^2 + 13 = 0$

98. $-10x + x^2 + 41 = 0$

99. $15x^2 - x - 6 = 0$

100. $10x^2 - 11x - 35 = 0$

101. $-4x^2 + 6x - 5 = 0$

102. $-5x^2 - 3 = 2x$

103. $2x^2 + 8 = -9x$

104. $x^2 + 4 = -7x$

105. $4x^2 + 12x = -9$

106. $9x^2 + 4 = 12x$

Solve the quadratic equations given. Simplify each result.

107. $-6x + 2x^2 + 5 = 0$

108. $17 + 2x^2 = 10x$

109. $5x^2 + 5 = -5x$

110. $x^2 = -2x - 19$

111. $-2x^2 = -5x + 11$

112. $4x - 3 = 5x^2$

WORKING WITH FORMULAS

113. Height of a projectile: $h = -16t^2 + vt$ $t = \frac{v \pm \sqrt{v^2 - 64h}}{32}$

If an object is projected vertically upward from ground level with no continuing source of propulsion, the height of the object (in feet) is modeled by the equation shown, where v is the initial velocity, and t is the time in seconds. Use the quadratic formula to solve for t in terms of v and h. (*Hint:* Set the equation equal to zero and identify the coefficients as before.)

114. Surface area of a cylinder: $A = 2\pi r^2 + 2\pi rh$

The surface area of a cylinder is given by the formula shown, where h is the height and r is the radius of the base. The equation can be considered a quadratic in the variable r. Use the quadratic formula to solve for r in terms of h. (*Hint:* Rewrite the equation in standard form and identify the coefficients as before.) $r = \frac{-2\pi h \pm \sqrt{4\pi^2h^2 + 8\pi A}}{4\pi}$

APPLICATIONS

115. $t = \frac{6 + \sqrt{138}}{2}$ sec, $t \approx 8.87$ sec

116. $t = \frac{15}{4} - \frac{\sqrt{119}}{4}$ sec, $t \approx 1.02$ sec

115. Height of a projectile: The height of an object thrown upward from the roof of a building 408 ft tall, with an initial velocity of 96 ft/sec, is given by the equation $h = -16t^2 + 96t + 408$, where h represents the height of the object after t seconds. How long will it take the object to hit the ground? Answer in exact form and decimal form rounded to the nearest thousandth.

116. Height of a projectile: The height of an object thrown upward from the floor of a canyon 106 ft deep, with an initial velocity of 120 ft/sec, is given by the equation $h = -16t^2 + 120t - 106$, where h represents the height of the object after t seconds. How long will it take the object to rise to the height of the canyon wall? Answer in exact form and decimal form rounded to hundredths.

117. Cost, revenue, and profit: The revenue for a manufacturer of microwave ovens is given by the equation $R = x(40 - \frac{1}{3}x)$, where revenue is in thousands of dollars and x thousand ovens are manufactured and sold. What is the minimum number of microwave ovens that must be sold to bring in a revenue of \$900,000? 30,000 ovens

118. Cost, revenue, and profit: The revenue for a manufacturer of computer printers is given by the equation $R = x(30 - 0.4x)$, where revenue is in thousands of dollars and x thousand printers are manufactured and sold. What is the minimum number of printers that must be sold to bring in a revenue of \$440,000? 20,000 printers

Exercises 119 and 120

Singles

Doubles

119. Tennis court dimensions: A regulation tennis court for a doubles match is laid out so that its length is 6 ft more than two times its width. The area of the doubles court is 2808 ft^2. What is the length and width of the doubles court? 36 ft, 78 ft

120. Tennis court dimensions: A regulation tennis court for a singles match is laid out so that its length is 3 ft less than three times its width. The area of the singles court is 2106 ft^2. What is the length and width of the singles court? 27 ft, 78 ft

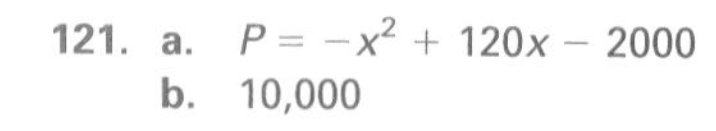

121. a. $P = -x^2 + 120x - 2000$
b. 10,000

121. Cost, revenue, and profit: The cost of raw materials to produce plastic toys is given by the cost equation $C = 2x + 35$, where x is the number of toys in hundreds. The total income (revenue) from the sale of these toys is given by $R = -x^2 + 122x - 1965$. (a) Determine the profit equation (profit = revenue − cost). During the Christmas season, the owners of the company decide to manufacture and donate as many toys as they can, without taking a loss (i.e., they break even: profit or $P = 0$). (b) How many toys will they produce for charity?

122. a. $P = -x^2 + 310x - 18{,}526$
b. 229,000

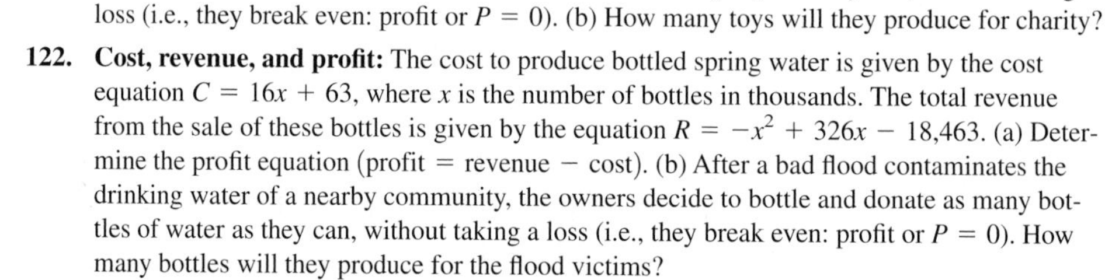

122. Cost, revenue, and profit: The cost to produce bottled spring water is given by the cost equation $C = 16x + 63$, where x is the number of bottles in thousands. The total revenue from the sale of these bottles is given by the equation $R = -x^2 + 326x - 18{,}463$. (a) Determine the profit equation (profit = revenue − cost). (b) After a bad flood contaminates the drinking water of a nearby community, the owners decide to bottle and donate as many bottles of water as they can, without taking a loss (i.e., they break even: profit or $P = 0$). How many bottles will they produce for the flood victims?

123. Height of an arrow: If an object is projected vertically upward from ground level with no continuing source of propulsion, its height (in feet) is modeled by the equation $h = -16t^2 + vt$, where v is the initial velocity and t is the time in seconds. Use the quadratic formula to solve for t, given an arrow is shot into the air with $v = 144$ ft/sec and $h = 260$ ft. See Exercise 113. $t = 2.5$ sec, 6.5 sec

124. Surface area of a cylinder: The surface area of a cylinder is given by $A = 2\pi r^2 + 2\pi rh$, where h is the height and r is the radius of the base. The equation can be considered a quadratic in the variable r. Use the quadratic formula to solve for r, given $A = 4710$ cm^2 and $h = 35$ cm. See Exercise 114. $r \approx 15$ cm

Solve using the quadratic formula. Verify one complex solution by substitution (note that $\frac{1}{i} = -i$).

125. $z^2 - 3iz = -10$

126. $z^2 - 9iz = -22$

127. $4iz^2 + 5z + 6i = 0$

128. $2iz^2 - 9z + 26i = 0$

129. $0.5z^2 + (7 + i)z + (6 + 7i) = 0$

130. $0.5z^2 + (4 - 3i)z + (-9 - 12i) = 0$

125. $x = -2i; x = 5i$
126. $x = -2i; x = 11i$
127. $x = \frac{-3}{4}i; x = 2i$
128. $x = \frac{-13i}{2}; x = 2i$
129. $x = -1 - i; x = -13 - i$
130. $x = 1 + 3i; x = -9 + 3i$

WRITING, RESEARCH, AND DECISION MAKING

131. Given $2x^2 + 3kx + 18 = 0$, use the discriminant to find the value of k that will yield one real root. $k = \pm 4$

132. Locate and read the following article. Then turn in a one-page summary.
"Complete the Square Earlier," Thoddi C. T. Kotiah, *Mathematics Teacher,* Volume 84, Number 9, December 1991: pages 730–731. Answers will vary.

133. Go to your local video store and rent the movie *October Sky* (Jake Gyllenhaal, Laura Dern, Chris Cooper; Universal Studios, 1999). View the movie carefully (have fun with some classmates), especially the episode where Homer Hickam, Jr., and his group of rocketeers are wrongfully blamed for the fire. How did the rocketeers finally exonerate themselves? Write a one-page summary of the movie, paying special attention to the role that mathematics plays in the plot.
Answers will vary.

EXTENDING THE CONCEPT

134. Solve by completing the square:
$2x^2 - 392x - 12{,}544 = 0.$
$x = -28$ or $x = 224$

135. Solve in less than 30 sec:
$(x - 3)(x^2 + 3x - 10) +$
$(x - 2)(x^2 + 2x - 15) = 0.$
$x = 2$ or $x = 3$ or $x = -5$

MAINTAINING YOUR SKILLS

136. (R.7) State the formula for the perimeter and area of each figure illustrated.

a. L, W — $P = 2L + 2W, A = LW$

b. r — $P = 2\pi r, A = \pi r^2$

c. b_1, c, h, b_2

d. a, h, c, b

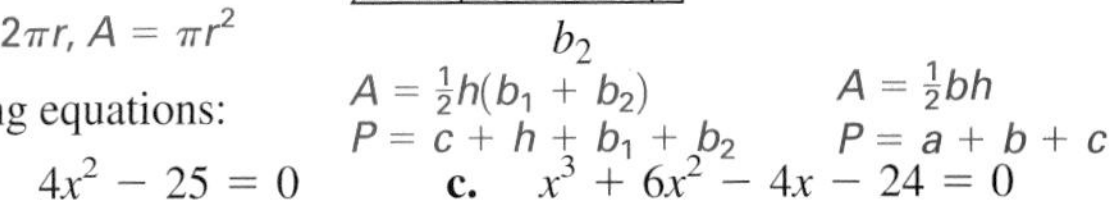

c. $A = \frac{1}{2}h(b_1 + b_2)$, $P = c + h + b_1 + b_2$

d. $A = \frac{1}{2}bh$, $P = a + b + c$

137. (1.3) Factor and solve the following equations:

a. $x^2 - 5x - 36 = 0$ **b.** $4x^2 - 25 = 0$ **c.** $x^3 + 6x^2 - 4x - 24 = 0$

138. (1.2) Solve the inequality and give the answer in set notation, number line notation, and interval notation.

$$-\frac{2}{7}x + 5 \geq 3$$

139. (1.1) A total of 900 tickets were sold for a recent concert and \$25,000 was collected. If good seats were \$30 and cheap seats were \$20, how many of each type were sold?

140. (1.1) Solve for C: $P = C + Ct$.

141. (1.4) Simplify the expression: $\dfrac{6 - \sqrt{-18}}{3}$.

137. a. $x = 9$ or $x = -4$
b. $x = \pm\frac{5}{2}$
c. $x = -6, -2, 2$

138. $\{x \mid x \in \mathbb{R}; x \leq 7\}$;

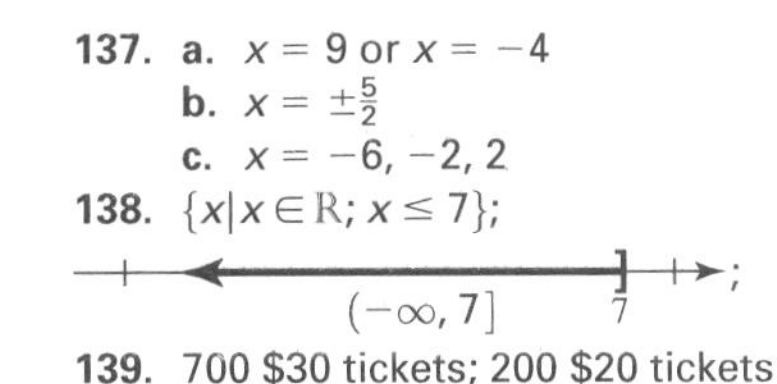

$(-\infty, 7]$

139. 700 \$30 tickets; 200 \$20 tickets

140. $C = \dfrac{P}{1 + t}$

141. $2 - i\sqrt{2}$

SUMMARY AND CONCEPT REVIEW

SECTION 1.1 Linear Equations, Formulas, and Literal Equations

KEY CONCEPTS

- A linear equation can be identified using these three tests: (1) the exponent on any variable is one, (2) no variables are used as a divisor, and (3) no two variables are multiplied together. Alternatively, a linear equation is one that can be written in the form $Ax + By = C$, where A and B are not both zero.
- When solving basic linear equations, our goal is to isolate the term containing the variable using the additive property, then isolate (solve for) the variable using the multiplicative property.
- To solve a literal equation or formula, focus on the object variable and apply properties of equality to write the object variable in terms of the other variables.
- An equation can be an identity (always true), a contradiction (never true), or conditional [true or false depending on the input value(s)].

- If an equation contains fractions, multiplying both sides of the equation by the least common multiple of all denominators is used to clear denominators and reduce the amount of work required to solve.

EXERCISES

1. Identify each equation as linear or nonlinear, and justify your answer. Do not solve.

   a. $4x + 7 = 5$ linear
   b. $\frac{3}{m} + 2.3 = 9.7$ nonlinear, variable as divisor
   c. $3g(g - 4) = 12$ nonlinear, exponent on g will be 2

Solve each equation.

2. $-2b + 7 = -5$ $b = 6$
3. $3(2n - 6) + 1 = 7$ $n = 4$
4. $4m - 5 = 11m + 2$ $m = -1$
5. $\frac{1}{2}x + \frac{2}{3} = \frac{3}{4}$ $x = \frac{1}{6}$
6. $6p - (3p + 5) - 9 = 3(p - 3)$ no solution
7. $-\frac{g}{6} = 3 - \frac{1}{2} - \frac{5g}{12}$ $g = 10$

Solve for the specified variable in each formula or literal equation.

8. $V = \pi r^2 h$ for h $h = \frac{V}{\pi r^2}$
9. $P = 2L + 2W$ for L $L = \frac{P - 2W}{2}$
10. $ax + b = c$ for x $x = \frac{c - b}{a}$
11. $2x - 3y = 6$ for y $y = \frac{2}{3}x - 2$

Use the problem-solving guidelines to solve each of the following applications.

12. At a large family reunion, two kegs of lemonade are available. One is 20% sugar (too sour) and the second is 50% sugar (too sweet). Twelve gallons are needed for the reunion and a 40% sugar mix is deemed just right. How much of each keg should be used? 4 gal of 20% sugar, 8 gal of 50% sugar
13. A rectangular window with a width of 3 ft and a height of 4 ft is topped by a semicircle. Find the area of the window. $12 + \frac{9}{8}\pi$ ft^2
14. Two cyclists start from the same location and ride in opposite directions, one riding at 7 mph and the other at 9 mph. How long until they are 12 mi apart? $\frac{3}{4}$ hr

SECTION 1.2 Linear Inequalities in One Variable with Applications

KEY CONCEPTS

- The solution of an inequality can be given using inequality, set, or interval notation or can be graphed on a number line.
- When solving inequalities, always check whether the endpoint(s) are included or excluded, and use the appropriate notation in the solution set.
- Some applications involve compound inequalities such as $-3 \le x < 5$, also called joint inequalities.
- Given any two sets A and B, the intersection of A and B, written $A \cap B$, is the set of all members *common to both sets.* The union of A and B, written $A \cup B$, is the set of all members *in either set.*
- Inequalities are solved using properties similar to those for solving equalities. The exception is the multiplicative property of inequality, since the truth of the resulting statement depends on whether we multiply/divide by a positive or negative quantity.

EXERCISES

Use inequality symbols to write a mathematical model for each statement.

15. You must be 35 yr old or older to run for president of the United States. $a \ge 35$
16. A child must be under 2 yr of age to be admitted free. $a < 2$
17. The speed limit on many interstate highways is 65 mph. $s \le 65$

18. Our caloric intake should not be less than 1200 calories per day. $c \geq 1200$

Solve the inequality and write the solution using interval notation.

19. $7x > 35$ $(5, \infty)$

20. $-\frac{3}{5}m < 6$ $(-10, \infty)$

21. $2(3m - 2) \leq 8$ $(-\infty, 2]$

22. $-1 < \frac{1}{3}x + 2 \leq 5$ $(-9, 9]$

23. $-4 < 2b + 8$ and $3b - 5 > -32$ $(-6, \infty)$

24. $-5(x + 3) > -7$ or $x - 5.2 > -2.9$ $(-\infty, \frac{-8}{5}) \cup (2.3, \infty)$

25. Find the allowable values for each of the following. Write your answer in interval notation.

a. $\frac{7}{n - 3}$ b. $\frac{5}{2x - 3}$ c. $\sqrt{x + 5}$ d. $\sqrt{-3n + 18}$

25. a. $(-\infty, 3) \cup (3, \infty)$
b. $(-\infty, \frac{3}{2}) \cup (\frac{3}{2}, \infty)$
c. $[-5, \infty)$
d. $(-\infty, 6]$

26. Latoya has earned grades of 72%, 95%, 83%, and 79% on her first four exams. What grade must she make on her fifth and last exam so that her average is 85% or more? $x \geq 96\%$

SECTION 1.3 Solving Polynomial and Other Equations

KEY CONCEPTS

- A quadratic equation in standard form is written in decreasing order of degree and set equal to zero.
- If a quadratic equation is factorable, we solve it using the zero factor property: If the product of two (or more) factors is zero, then at least one of the factors must be equal to zero.
- To solve a rational equation, clear denominators using the least common multiple, noting any values that must be excluded. Solve by factoring or using properties of equality, and check results in the original equation.
- "Solutions" that cause any denominator to be zero are called extraneous roots.
- To solve a radical equation, we use the power property of equality. Isolate the radical on one side, then apply the appropriate "*n*th power" to free up the radicand and solve for the unknown. If there is more than one radical, repeat the process after isolating the remaining radical. See the flowchart on page 97.
- Equations with rational exponents are treated similar to radical equations. For the rational exponent $\frac{a}{b}$, raising both sides to the $\frac{b}{a}$ power will give an exponent of 1 and enable you to solve for the unknown.

EXERCISES

27. Solve using the zero factor property.

a. $(x + 3)(x - 5)(x + 1)(x - 4) = 0$
$x = -3$ or $x = 5$ or $x = -1$ or $x = 4$

b. $3x\left(x + \frac{5}{2}\right)(x - 9)\left(x - \frac{1}{2}\right) = 0$
$x = 0$ or $x = \frac{-5}{2}$ or $x = 9$ or $x = \frac{1}{2}$

28. Solve by factoring.

a. $x^2 + 7x - 18 = 0$ b. $n^2 = -12n - 27$

c. $2z^2 - 3 = z$ d. $-7r^3 + 21r^2 + 28r = 0$

e. $-3b^3 + 27b = 0$ f. $4a^3 - 6a^2 - 16a + 24 = 0$

28. a. $x = -9$ or $x = 2$
b. $n = -9$ or $n = -3$
c. $z = \frac{3}{2}$ or $z = -1$
d. $r = 0$ or $r = 4$ or $r = -1$
e. $b = 0$ or $b = -3$ or $b = 3$
f. $a = \frac{3}{2}$ or $a = -2$ or $a = 2$

Solve each equation.

29. $\frac{3}{5x} + \frac{7}{10} = \frac{1}{4x}$

30. $\frac{3}{h + 3} - \frac{5}{h^2 + 3h} = \frac{1}{h}$

31. $\frac{n}{n + 2} - \frac{3}{n - 4} = \frac{n^2 + 1}{n^2 - 2n - 8}$

32. $\frac{\sqrt{x^2 + 7}}{2} + 3 = 5$

33. $3\sqrt{x + 4} = x + 4$

34. $\sqrt{3x + 4} = 2 - \sqrt{x + 2}$

29. $x = \frac{-1}{2}$
30. $h = 4$
31. $n = -1$
32. $x = -3; x = 3$
33. $x = -4; x = 5$
34. $x = -1$

Solve using the problem-solving guidelines.

35. Given two consecutive integers, the square of the second is equal to one more than seven times the first. Find the integers. 0 and 1; 5 and 6

36. The area of a common stenographer's tablet, commonly called a *steno book,* is 54 in^2. The length of the tablet is 3 in. more than the width. Model the situation with a quadratic equation and find the dimensions of the tablet. width, 6 in.; length, 9 in.

37. A batter has just flied out to the catcher, who catches the ball while standing on home plate. If the batter made contact with the ball at a height of 4 ft and the ball left the bat with an initial velocity of 128 ft/sec, how long will it take the ball to reach a height of 116 ft? 1 sec How high is the ball 5 sec after contact? 244 ft If the catcher catches the ball at a height of 4 ft, how long was it airborne? 8 sec

38. Using a survey, a fire wood distributor finds that if they charge $50 per load they will sell 40 loads each winter month. For each decrease of $2, five additional loads will be sold. What selling price(s) will result in new monthly revenue of $2520? $24 per load; $42 per load

SECTION 1.4 Complex Numbers

KEY CONCEPTS

- The italicized i represents the number whose square is -1. This means $i^2 = -1$ and $i = \sqrt{-1}$.
- Since $i^4 = i^2 \cdot i^2 = (-1) \cdot (-1)$ or 1, larger powers of i can be simplified by writing them in terms of i^4.
- The square root of a negative number can be rewritten using "i" notation: $\sqrt{-4} = \sqrt{-1 \cdot 4} = \sqrt{-1}\sqrt{4}$ or $2i$. We say the expression has been *written in terms of i and simplified.*
- The standard form of a *complex number* is $a + bi$, where a is the *real number part* and bi is the *imaginary number part.*
- To add or subtract complex numbers, combine the like terms.
- For any complex number $a + bi$, its *complex conjugate* is $a - bi$.
- The *product* of a complex number and its conjugate is a real number.
- The *discriminant* of the quadratic formula $b^2 - 4ac$ gives the number and nature of the roots.
- The commutative, associative, and distributive properties also apply to complex numbers and are used to perform basic operations.
- To multiply complex numbers, use the F-O-I-L method and combine like terms.
- The *sum* of a complex number and its complex conjugate is a real number.
- To find a *quotient* of complex numbers, multiply the numerator and denominator by the conjugate of the denominator.
- If $a + bi$ is one solution to a polynomial equation, then its complex conjugate $a - bi$ is also a solution.

EXERCISES

Simplify each expression and write the result in standard form.

39. $\sqrt{-72}$ $6\sqrt{2}\,i$

40. $6\sqrt{-48}$ $24\sqrt{3}\,i$

41. $\dfrac{-10 + \sqrt{-50}}{5}$ $-2 + \sqrt{2}\,i$

42. $\sqrt{3}\sqrt{-6}$ $3\sqrt{2}\,i$

43. i^{57} i

Perform the operation indicated and write the result in standard form.

44. $(5 + 2i)^2$ $21 + 20i$

45. $\dfrac{5i}{1 - 2i}$ $-2 + i$

46. $(-3 + 5i) - (2 - 2i)$ $-5 + 7i$

47. $(2 + 3i)(2 - 3i)$ 13

48. $4i(-3 + 5i)$ $-20 - 12i$

Use substitution to show the given complex number and its conjugate are solutions to the equation shown.

49. $x^2 - 9 = -34; x = 5i$

50. $x^2 - 4x + 9 = 0; x = 2 + i\sqrt{5}$

49.
$(5i)^2 - 9 = -34$ $(-5i)^2 - 9 = -34$
$25i^2 - 9 = -34$ $25i^2 - 9 = -34$
$-25 - 9 = -34$✓ $-25 - 9 = -34$✓

50. $(2 + i\sqrt{5})^2 - 4(2 + i\sqrt{5}) + 9 = 0$
$4 + 4i\sqrt{5} + 5i^2 - 8 - 4i\sqrt{5} + 9 = 0$
$5 + (-5) = 0$✓

$(2 - i\sqrt{5})^2 - 4(2 - i\sqrt{5}) + 9 = 0$
$4 - 4i\sqrt{5} + 5i^2 - 8 + 4i\sqrt{5} + 9 = 0$
$5 + (-5) = 0$✓

SECTION 1.5 Solving Nonfactorable Quadratic Equations

KEY CONCEPTS

- The standard form of a *quadratic equation* is $ax^2 + bx^1 + c = 0$, where a, b, and c are real numbers and $a \neq 0$. The coefficient of the squared term is called the lead coefficient.
- The square root property of equality states that if $P^2 = k$, where $k \geq 0$, then $P = \sqrt{k}$ or $P = -\sqrt{k}$.
- A general quadratic equation $ax^2 + bx + c = 0$, where a, b, and c are real numbers and $a \neq 0$, can be solved by *completing the square* or by using the *quadratic formula.*
- It is often important to distinguish between the *exact* form of an answer—given with radicals—and the *approximate* form—given as a decimal approximation rounded to a specified place value.
- A quadratic equation may have two real roots, one real root, or no real roots, depending on the value of the discriminant $b^2 - 4ac$: (1) if $b^2 - 4ac = 0$, the equation has one real root; (2) if $b^2 - 4ac > 0$, the equation has two real roots; and (3) if $b^2 - 4ac < 0$, the equation has two complex roots.

EXERCISES

51. Determine whether the given equation is quadratic. If so, write the equation in standard form and identify the values of a, b, and c.

a. $-3 = 2x^2$ **b.** $7 = -2x + 11$ **c.** $99 = x^2 - 8x$ **d.** $20 = 4 - x^2$

52. Solve using the square root property of equality.

a. $x^2 - 9 = 0$ **b.** $2(x - 2)^2 + 1 = 11$ **c.** $3x^2 + 15 = 0$ **d.** $-2x^2 + 4 = -46$

53. Solve by completing the square. Give real number solutions in both exact and approximate form.

a. $x^2 + 2x = 15$ **b.** $x^2 + 6x = 16$ **c.** $-4x + 2x^2 = 3$ **d.** $3x^2 - 7x = -2$

54. Solve using the quadratic formula. Give solutions in both exact and approximate form.

a. $x^2 - 4x = -9$ **b.** $4x^2 + 7 = 12x$ **c.** $2x^2 - 6x + 5 = 0$

51. a. $2x^2 + 3 = 0; a = 2, b = 0, c = 3$
b. not quadratic
c. $x^2 - 8x - 99 = 0; a = 1, b = -8, c = -99$
d. $x^2 + 16 = 0; a = 1, b = 0, c = 16$

52. a. $x = \pm 3$
b. $x = 2 \pm \sqrt{5}$
c. $x = \pm\sqrt{5}\, i$
d. $x = \pm 5$

53. a. $x = 3$ or $x = -5$
b. $x = -8$ or $x = 2$
c. $x = 1 \pm \frac{\sqrt{10}}{2}; x \approx 2.58$ or $x \approx -0.58$
d. $x = 2$ or $x = \frac{1}{3}$

54. a. $x = 2 \pm \sqrt{5}\, i; x \approx 2 \pm 2.24i$
b. $x = \frac{3 \pm \sqrt{2}}{2}; x \approx 2.21$ or $x \approx 0.79$
c. $x = \frac{3}{2} \pm \frac{1}{2}i$

Solve the following quadratic applications. Recall the height of a projectile is modeled by $h = -16t^2 + v_0t + k$.

55. A projectile is fired upward from ground level with an initial velocity of 96 ft/sec. (a) To the nearest tenth of a second, how long until the object first reaches a height of 100 ft? 1.3 sec (b) How long until the object is again at 100 ft? 4.66 sec (c) How many seconds until it returns to the ground? 6 sec

56. A person throws a rock upward from the top of an 80-ft cliff with an initial velocity of 64 ft/sec. (a) To the nearest tenth of a second, how long until the object is 120 ft high? 0.8 sec (b) How long until the object is again at 120 ft? 3.2 sec (c) How many seconds until the object hits the ground at the base of the cliff? 5 sec

57. The manager of a large, 14-screen movie theater finds that if he charges $2.50 per person for the matinee, the average daily attendance is 4000 people. With every increase of 25 cents the attendance drops an average of 200 people. (a) What admission price will bring in a revenue of $11,250? (b) How many people will purchase tickets at this price? $3.75; 3000

58. After a storm, the Johnson's basement flooded and the water needed to be pumped out. A cleanup crew is sent out with two powerful pumps to do the job. Working alone (if one of the pumps were needed at another job), the larger pump would be able to clear the basement in 3 hr less time than the smaller pump alone. Working together, the two pumps can clear the basement in 2 hr. How long would it take the smaller pump alone? 6 hr

MIXED REVIEW

1. Find the allowable values for each expression. Write your response in interval notation.

a. $\dfrac{10}{\sqrt{x-8}}$ $x \in (8, \infty)$ **b.** $\dfrac{-5}{3x+4}$ $x \in (-\infty, \frac{-4}{3}) \cup (\frac{-4}{3}, \infty)$

2. Perform the operations indicated.

a. $\sqrt{-18} + \sqrt{-50}$ **b.** $(1 - 2i)^2$

c. $\dfrac{3i}{1+i}$ **d.** $(2 + i\sqrt{3})(2 - i\sqrt{3})$

3. Factor each expression completely.

a. $x^3 + 10x^2 + 16x$ $x(x+2)(x+8)$ **b.** $-2m^2 + 12m + 54$ $-2(m-9)(m+3)$

c. $18z^2 - 50$ $2(3z+5)(3z-5)$ **d.** $v^3 + 2v^2 - 9v - 18$ $(v+2)(v+3)(v-3)$

Solve for the variable indicated.

4. $V = \frac{1}{3}\pi r^2 h + \frac{2}{3}\pi r^3$; for h $h = \dfrac{3V - 2\pi r^3}{\pi r^2}$ **5.** $3x + 4y = -12$; for y $y = \frac{-3}{4}x - 3$

Solve as indicated, using the method of your choice.

6. a. $-20 \le 4x + 8 < 56$ $-7 \le x < 12$ **b.** $-2x + 7 \le 12$ and $3 - 4x > -5$ $x \in [-\frac{5}{2}, 2)$

7. a. $5x - (2x - 3) + 3x = -4(5 + x) + 3$ $x = -2$ **b.** $\frac{n}{5} - 2 = 2 - \frac{5}{3} - \frac{4}{15}n$ $n = 5$

8. $5x(x - 10)(x + 1) = 0$ **9.** $x^2 - 18x + 77 = 0$ **10.** $3x^2 - 10 = 5 - x + x^2$

11. $4x^2 - 5 = 19$ **12.** $3(x + 5)^2 - 3 = 30$ **13.** $25x^2 + 16 = 40x$

14. $3x^2 - 7x + 3 = 0$ **15.** $2x^4 - 50 = 0$

16. a. $\dfrac{2}{x} - \dfrac{x}{5x+12} = 0$ **b.** $\dfrac{1}{n-1} - \dfrac{2}{n^2-1} = -\dfrac{1}{2}$

17. a. $\sqrt{2v - 3} + 3 = v$ **b.** $\sqrt[3]{x^2 - 9} + \sqrt[3]{x - 11} = 0$

18. The local Lion's Club rents out two banquet halls for large meetings and other events. The records show that when they charge $250 per day for use of the halls, there are an average of 156 bookings per year. For every increase of $20 per day, there will be three less bookings. (a) What price per day will bring in $61,950 for the year? (b) How many bookings will there be at the price from part (a)?

19. The Jefferson College basketball team has two guards who are 6′3″ tall and two forwards who are 6′7″ tall. How tall must their center be to ensure the "starting five" will have an average height of at least 6′6″? 6′10″

Exercise 20

20. Two friends are passing out flyers along an oval-shaped boulevard by starting at the same spot and walking in opposite directions. The total distance of the route is 4 mi. If one friend distributes the flyers at a rate of 4 mph, while the other distributes them at 2.4 mph, how long until they meet? Answer in minutes. 37.5 min

2. a. $8i\sqrt{2}$
b. $-3 - 4i$
c. $\frac{3}{2} + \frac{3}{2}i$
d. 7

8. $x = -1, 0, 10$
9. $x = 7, 11$
10. $x = -3, \frac{5}{2}$
11. $x = -\sqrt{6}, \sqrt{6}$
12. $x = -5 \pm \sqrt{11}$
13. $x = \frac{4}{5}$
14. $x = \frac{7 - \sqrt{13}}{6}; x = \frac{7 + \sqrt{13}}{6}$
15. $x = \pm\sqrt{5}, \pm i\sqrt{5}$
16. a. $x = -2, 12$
b. $n = -3$
17. a. $v = 6$
b. $x = -5; x = 4$
18. (a) 17 increases of $20; 250 + (17)(20) = $590/booking
(b) 105 bookings

PRACTICE TEST

Solve each equation.

1. $-\frac{2}{3}x - 5 = 7 - (x + 3)$ $x = 27$
2. $P = C + kC$; for C $C = \frac{P}{1 + k}$
3. $-5.7 + 3.1x = 14.5 - 4(x + 1.5)$ $x = 2$
4. How much water that is 102°F, must be mixed with 25 gal of water at 91°F, so that the resulting temperature of the water will be 97°F? 30 gal

Solve each inequality.

5. $-\frac{2}{5}x + 7 > 19$ $x < -30$
6. $-1 < 3 - x \leq 8$ $-5 \leq x < 4$
7. $\frac{1}{2}x + 3 < 9$ or $\frac{2}{3}x - 1 \geq 3$ $x \in \mathbb{R}$
8. To make the bowling team, Jacques needs a three-game average of 160. If he bowled 141 and 162 for the first two games, what must be bowled in the third game so that his average is at least 160? Jacques needs at least a 177

Solve each equation by factoring, if possible.

9. $z^2 - 7z - 30 = 0$ $z = -3, 10$
10. $4x^2 - 25 = 0$ $x = \pm\frac{5}{2}$
11. $3x^2 - 20x = -12$ $x = \frac{2}{3}, 6$
12. $4x^3 + 8x^2 - 9x - 18 = 0$ $x = -2, \frac{-3}{2}, \frac{3}{2}$
13. The Spanish Club at Rock Hill Community College has decided to sell tins of gourmet popcorn as a fundraiser. The suggested selling price is \$3.00 per tin, but Maria, who also belongs to the Math Club, decides to take a survey to see if they can increase "the fruits of their labor." The survey shows it's likely that 120 tins will be sold on campus at the \$3.00 price, and for each price increase of \$0.10, 2 fewer tins will be sold. (a) What price per tin will bring in a revenue of \$405? (b) How many tins will be sold at the price from part (a)? \$4.50/tin 90 tins

Simplify each expression.

14. $\frac{-8 + \sqrt{-20}}{6}$ $\frac{-4}{3} + \frac{i\sqrt{5}}{3}$
15. i^{39} $-i$
16. Given the complex numbers $\frac{1}{2} + \frac{\sqrt{3}}{2}i$ and $\frac{1}{2} - \frac{\sqrt{3}}{2}i$, find
    a. the sum. 1
    b. the difference. $i\sqrt{3}$
    c. the product. 1
17. Compute the quotient: $\frac{3i}{2 - i}$.
18. Find the product: $(3i + 5)(5 - 3i)$.
19. Solve the equation: $(x - 1)^2 + 3 = 0$.
20. Show that $x = 2 - 3i$ is a solution of $x^2 - 4x + 13 = 0$

17. $-\frac{3}{5} + \frac{6}{5}i$
18. 34
19. $x = 1 \pm \sqrt{3}\,i$
20. $(2 - 3i)^2 - 4(2 - 3i) + 13 = 0$
    $-5 - 12i - 8 + 12i + 13 = 0$
    $0 = 0$ ✓

Solve by completing the square.

21. $2x^2 - 20x + 49 = 0$ $x = 5 \pm \frac{\sqrt{2}}{2}$
22. $2x^2 - 5x + 4 = 0$ $x = \frac{5}{4} \pm \frac{i\sqrt{7}}{4}$

Solve using the quadratic formula.

23. $3x^2 + 2 = 6x$ $x = 1 \pm \frac{\sqrt{3}}{3}$
24. $x^2 - 2x + 10 = 0$ $x = 1 \pm 3i$
25. Due to the seasonal nature of the business, the revenue of Wet Willey's Water World can be modeled by the equation $r = -3t^2 + 42t - 135$, where t is a month of the year ($t = 1$ corresponds to January) and r is the amount of revenue in thousands of dollars. (a) What month does Wet Willey's open? (b) What month does Wet Willey's close? (c) Does Wet Willey's bring in more revenue in July or August? How much more?

25. (a) $t = 5$ (May)
    (b) $t = 9$ (September)
    (c) July; \$3000 more

Calculator Exploration and Discovery

Evaluating Expressions and Looking for Patterns

The keystrokes shown apply to a TI-84 Plus model. Please consult your manual or our Internet site for other models.

These "explorations" are designed to explore the full potential of a graphing calculator, as well as to use this potential to investigate patterns and discover connections that might otherwise be overlooked. In this *exploration and discovery,* we point out the various ways an expression can be evaluated on a graphing calculator. Some ways seem easier, faster, and/or better than others, but each has advantages and disadvantages, depending on the task at hand, and it will help to be aware of them all for future reference and use.

One way to evaluate an expression is to use the TABLE feature of a graphing calculator, with the expression entered as Y1 on the Y= screen. If you want the calculator to generate inputs, use the 2nd WINDOW (**TBLSET**), screen to indicate a starting value (**TblStart=**) and an increment value (**ΔTbl=**), and set the calculator in **Indpnt:** AUTO **ASK** mode (to input specific values, the calculator should be in **Indpnt: AUTO** ASK mode). After pressing 2nd GRAPH (**TABLE**), the calculator shows the corresponding input and output values. For help with the basic TABLE feature of the TI-84 Plus, you can visit Section R.7 at www.mhhe.com/coburn.

Expressions can also be evaluated on the home screen for a single value or a series of values. To illustrate how, we'll use the linear expression $-\frac{3}{4}x + 5$. Enter this expression on the Y= screen (see Screen I) and use 2nd MODE (**QUIT**) to get back to the home screen. To evaluate this expression, we access Y1 using VARS ▶ (**Y-VARS**), and use the first option **1:Function** ENTER . This brings us to a submenu where any of the equations Y1 through Y0 (actually Y10) can be accessed. Since the default setting is the one we need **1:Y1,** simply press ENTER and Y1 now appears on the home screen. To evaluate the expression for a single input, simply enclose it in parentheses. To evaluate the expression for more than one input, enter the numbers as a set of values with the set enclosed in parentheses. In Screen II, Y1 has been evaluated for $x = -4$, then simultaneously for $x = -4, -2, 0,$ and 2.

Screen I

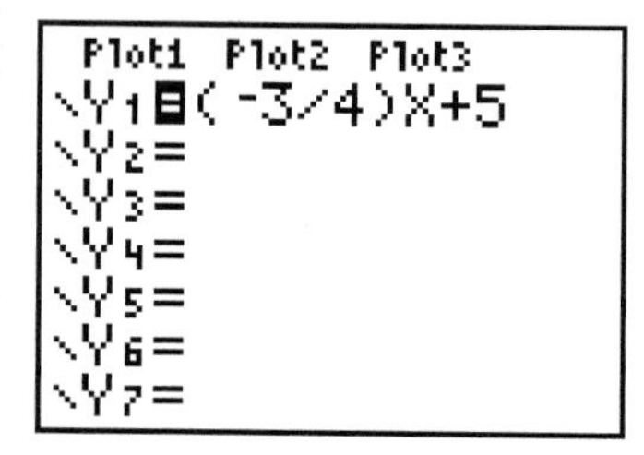

Screen II

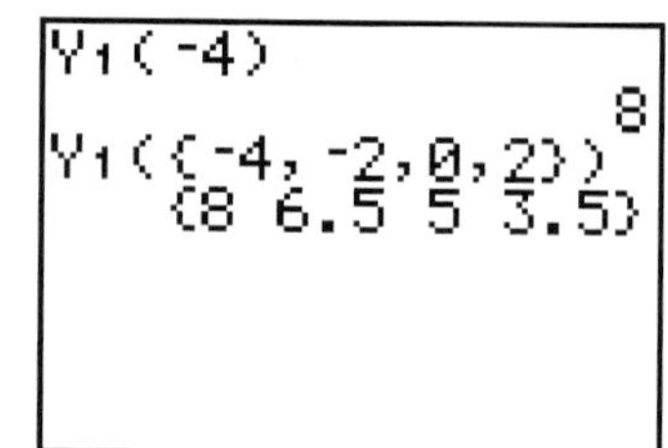

A third way to evaluate an equation is using a list, with the desired inputs entered in List 1 (L1), and List 2 (L2) defined in terms of L1. For example L2 = $-\frac{3}{4}$L1 + 5 will return the same values for inputs of −4, −2, 0, and 2 seen previously on the home screen (remember to clear the lists first). From the *Technology Highlight* in Section 1.2, lists are accessed by pressing STAT **1:Edit.** Enter the numbers −4, −2, 0 and 2 in L1, then use the right arrow ▶ to move to L2. It is important to note that you *next press the up arrow key* ▲ so that the cursor overlies L2. The bottom of the screen now reads **L2=** and the calculator is waiting for us to define L2 in some way. After entering L2 = $-\frac{3}{4}$L1 + 5 your screen should look like Screen III as shown, and after pressing ENTER we obtain the same outputs as before. The advantage of using the "list" method is that we can *further explore or experiment with the output values* in a search for patterns. We already know that the **inputs** differ by two. Now carefully look at the outputs—can you detect a pattern? It appears that the outputs all differ by 1.5! To be sure, we can use an operation called "delta list" and defined as ΔList, which automatically calculates the differences between the output values in a list. If the input values have a constant difference, the result of the ΔList

Screen III

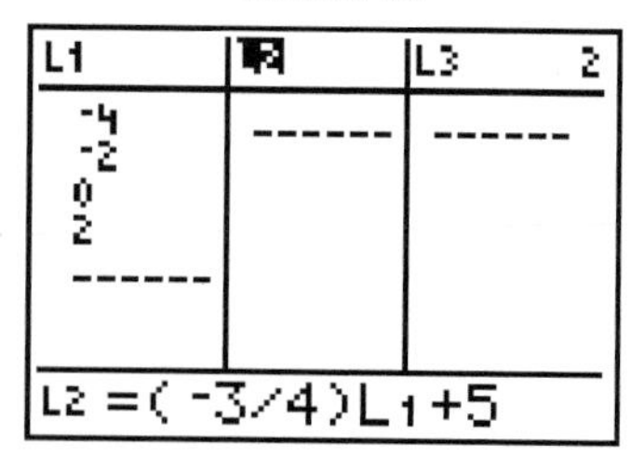

Screen IV

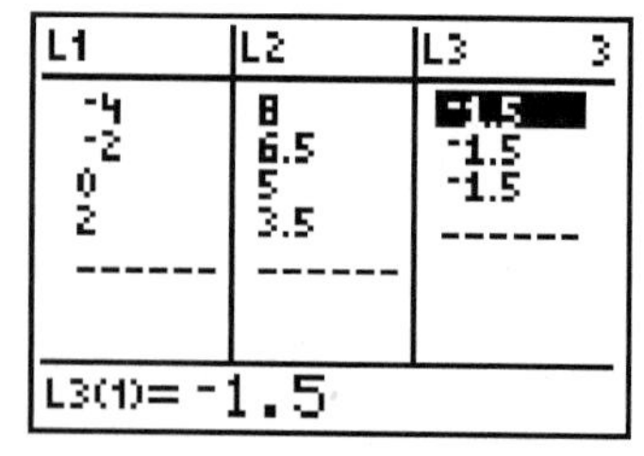

operation is called a **list of first differences.** One of the defining characteristics of linear data is that first differences are constant. There are similar ways of identifying data that are quadratic, cubic, and exponential, which will be explored in later chapters. To use the ΔList operation, use the arrow keys to have the cursor overlay L3, then access the operation using 2nd STAT ▶ (OPS). Note **7:ΔList** gives the desired operation and you can push the number 7 directly or move the cursor to this option and press ENTER. **"L3 = ΔList("** now appears on the last line of the list screen, waiting for you to tell the calculator which list you want to use. After entering 2nd 2 (L2) and pressing ENTER, the result is Screen IV, as shown, which demonstrates unmistakably that the first differences are constant.

Exercise 1: Evaluate the expression $-5\text{L1} + 7$ on the list screen, using consecutive integer inputs from -6 to 6 inclusive. What do you notice about the outputs? They differ by 5.

Exercise 2: Evaluate the expression $0.2\text{L1} + 3$ on the list screen, using consecutive integer inputs from -6 to 6 inclusive. What do you notice about the outputs? They differ by 0.2.

Exercise 3: Evaluate the expression $\sqrt{2}\,\text{L1} - \sqrt{9.1}$ on the list screen, using consecutive integer inputs from -6 to 6 inclusive. We suspect there is a pattern to the output values, but this time the pattern is very difficult to see. Compute a list of first difference in L3. What do you notice? They differ by $\sqrt{2}$ or ≈ 1.41.

STRENGTHENING CORE SKILLS

An Alternative Method for Checking Solutions to Quadratic Equations

To solve $x^2 - 2x - 15 = 0$ by factoring, students will often begin by looking for two numbers whose product is -15 (the constant term) and whose sum is -2 (the linear coefficient). The two numbers are -5 and 3 since $(-5)(3) = -15$ and $-5 + 3 = -2$. In factored form, we have $(x - 5)(x + 3) = 0$ with solutions $x_1 = 5$ and $x_2 = -3$. When these solutions are compared *to the original coefficients,* we can still see the sum/product relationship, but note that while $(5)(-3) = -15$ still gives the constant term, $5 + (-3) = 2$ gives the linear coefficient *with opposite sign.* Although more difficult to accomplish, this method can be applied to *any* factorable quadratic equation $ax^2 + bx + c = 0$ if we divide through by a, giving $x^2 + \frac{b}{a}x + \frac{c}{a} = 0$. For $2x^2 - x - 3 = 0$, we divide both sides by 2 and obtain $x^2 - \frac{1}{2}x - \frac{3}{2} = 0$, then look for two numbers whose product is $-\frac{3}{2}$ and whose sum is $-\frac{1}{2}$. The factors are $\left(x - \frac{3}{2}\right)$ and $(x + 1)$ since $\left(-\frac{3}{2}\right)(1) = -\frac{3}{2}$ and $-\frac{3}{2} + 1 = -\frac{1}{2}$, showing the solutions are $x_1 = \frac{3}{2}$ and $x_2 = -1$. We again note the product of the solutions is the constant $-\frac{3}{2} = \frac{c}{a}$, and the sum of the solutions is the linear coefficient *with opposite sign:* $\frac{1}{2} = -\frac{b}{a}$. No one actually promotes this method for solving trinomials, where $a \neq 1$, but it does illustrate an important and useful concept:

If x_1 and x_2 are the two roots of $x^2 + \frac{b}{a}x + \frac{c}{a} = 0$,

then $x_1x_2 = \frac{c}{a}$ and $x_1 + x_2 = -\frac{b}{a}$

Justification for this can be found by taking the product and sum of the general solutions $x_1 = \frac{-b}{2a} + \frac{\sqrt{b^2 - 4ac}}{2a}$ and $x_2 = \frac{-b}{2a} - \frac{\sqrt{b^2 - 4ac}}{2a}$. Although the computation looks impressive, the product can be computed as a binomial times its conjugate, and the radical parts add to zero for the sum, each yielding the results as already stated.

This observation provides a useful technique for checking solutions to a quadratic equation, *even those having irrational roots!* To test whether $x_1 = 2 + \sqrt{7}$ and $x_2 = 2 - \sqrt{7}$ are solutions to $x^2 - 4x - 3 = 0$, we note $c = -3$ and $-b = 4$. Computing the product gives $(2 + \sqrt{7})(2 - \sqrt{7}) = 2^2 - (\sqrt{7})^2 = 4 - 7 = -3✓$, with a sum of $4✓$ (by inspection). If someone claims the solutions to $4x^2 - 8x + 1 = 0$ are $x_1 = \frac{2 + \sqrt{3}}{2}$ and $x_2 = \frac{2 - \sqrt{3}}{2}$, we can check without actually having to substitute or re-solve the equation. For this equation $-\frac{b}{a} = 2$ and $\frac{c}{a} = \frac{1}{4}$ and checking the product and sum verifies the solutions are correct (try it!). This method of checking solutions can even be applied when the solutions are *complex numbers.* Check the solutions shown in these exercises.

Exercise 1: $2x^2 - 5x - 7 = 0$ $\quad \frac{7}{2} + (-1) = \frac{5}{2} = -\frac{b}{a}✓$

$x_1 = \frac{7}{2}$ $\quad \frac{7}{2} \cdot (-1) = \frac{-7}{2} = \frac{c}{a}✓$

$x_2 = -1$

Exercise 2: $2x^2 - 4x - 7 = 0$ $\quad \frac{2 + 3\sqrt{2}}{2} + \frac{2 - 3\sqrt{2}}{2} = \frac{4}{2} = \frac{-b}{a}✓$

$x_1 = \frac{2 + 3\sqrt{2}}{2}$ $\quad \frac{2 + 3\sqrt{2}}{2} \cdot \frac{2 - 3\sqrt{2}}{2} = \frac{-14}{4} = \frac{-7}{2} = \frac{c}{a}✓$

$x_2 = \frac{2 - 3\sqrt{2}}{2}$

Exercise 3: $x^2 - 10x + 37 = 0$ $\quad (5 + 2\sqrt{3}i) + (5 - 2\sqrt{3}i) = 10 = \frac{-b}{a}✓$

$x_1 = 5 + 2\sqrt{3}\,i$ $\quad (5 + 2\sqrt{3}i)(5 - 2\sqrt{3}i) = 25 + 12 = 37 = \frac{c}{a}✓$

$x_2 = 5 - 2\sqrt{3}\,i$

Exercise 4: Verify this sum/product check by computing the sum and product of the general solutions.

Chapter 2

Functions and Graphs

Chapter Outline

2.1 Rectangular Coordinates and the Graph of a Line 140

2.2 Relations, Functions, and Graphs 155

2.3 Linear Functions and Rates of Change 174

2.4 Quadratic and Other Toolbox Functions 190

2.5 Functions and Inequalities—A Graphical View 205

2.6 Regression, Technology, and Data Analysis 216

Preview

In a study of mathematics, we often place equations with similar characteristics into the same category or family. This type of organization makes each group easier to study and enables us to make comparisons and distinctions between groups. In this section, we introduce the concept of a function and work with some of the related ideas, while discussing some important distinctions between the family of *functions* and the family of *nonfunctions.* Although linear and quadratic functions will play the lead role, there are actually eight basic "toolbox functions" commonly used as elementary mathematical models.

2.1 Rectangular Coordinates and the Graph of a Line

LEARNING OBJECTIVES

In Section 2.1 you will learn how to:

A. Use a table of values to graph linear equations

B. Graph linear equations using the intercept method

C. Use the slope formula to find rates of change

D. Determine when lines are parallel or perpendicular

E. Apply the midpoint and distance formulas

F. Use linear graphs in an applied context

INTRODUCTION

In Section 1.1, we learned that a linear equation has these characteristics: (1) the exponent on any variable is 1, (2) no variable is used as a divisor, and (3) no two variables are multiplied together. In this section, we extend this definition to a study of **linear equations in two variables.** In standard form, these can be written $Ax + By = C$, where A, B, and C are constants, with A and B not simultaneously zero. Although they are fairly simple models, these linear equations and their related graphs have applications in almost all fields of study. In addition, they help introduce one of the most central ideas in mathematics—the concept of a function.

POINT OF INTEREST

The use of graphing to illustrate the solution of certain algebraic equations is very ancient. But the general application of graphical methods had to wait until 1637 when René Descartes (1596–1650) published his work *Discours de la Méthode,* to which he appended *La Géométrie*—which offered examples of how to apply *la Méthode.* The Cartesian coordinate system is named in his honor.

A. The Graph of a Linear Equation

The solution to a linear equation in x is any value of x that creates a true equation. The **solution to a linear equation in two variables** x and y, is any pair of substitutions for x and y that result in a true equation. For example, $x = -2$ and $y = 5$ form a solution to $2x + y = 1$, since $2(-2) + 5 = 1$ is true. When more than one variable forms a solution, the numbers are usually placed in parentheses and separated by a comma, as in $(-2, 5)$. The result is called an **ordered pair,** since the variables are listed in a specific order with the x-value always listed first. Linear equations in two variables often have many solutions, which we organize into an **input/output table.** After substituting a chosen value for x (the **input**), we solve for the corresponding value of y (the **output**). If the choice of inputs is left to you, select them from the context of the situation or simply choose integer values between -10 and 10 for convenience.

EXAMPLE 1 Create a table of values for $3x + 2y = 4$.

Solution: Select $x = -2, x = 0, x = 1$, and $x = 4$ as inputs. The resulting outputs are found and entered in the table (only calculations for $x = -2$ and $x = 1$ are shown).

Equation:	$3x + 2y = 4$	$3x + 2y = 4$
Substitute:	$3(-2) + 2y = 4$	$3(1) + 2y = 4$
Simplify:	$-6 + 2y = 4$	$3 + 2y = 4$
Result:	$y = 5$	$y = \frac{1}{2}$
	ordered pair $(-2, 5)$	ordered pair $(1, \frac{1}{2})$

x Inputs	y Outputs	(x, y) Ordered Pairs
-2	5	$(-2, 5)$
0	2	$(0, 2)$
1	$\frac{1}{2}$	$(1, \frac{1}{2})$
4	-4	$(4, -4)$

NOW TRY EXERCISES 7 THROUGH 10

While the solution to a linear equation in one variable is graphed on a number line, solutions to linear equations in two variables are graphed on a **rectangular coordinate system.** It consists of a horizontal number line and a vertical number line intersecting at zero. The point of intersection is called the **origin.** We refer to the horizontal number line as the ***x*-axis** and the vertical number line as the ***y*-axis,** which together divide the **coordinate plane** into four regions called **quadrants.** These are labeled using Roman numerals from I to IV, beginning in the upper right and moving counterclockwise. The **grid lines** or **tick marks** placed along each axis denote the integer values on each axis and further divide the plane into a **coordinate grid,** which we use to name the location of a point using an ordered pair. Since a point at the origin has not moved along either axis, it has coordinates (0, 0). To *plot the ordered pair* $(-2, 5)$, begin at (0, 0) and first move 2 units in the negative direction along the x-axis, then 5 units in the positive direction parallel to the y-axis (Figure 2.1). After graphing the remaining ordered pairs from Example 1, a noticeable pattern emerges—the points seem to lie along a straight line (Figure 2.2). Equations of the form $Ax + By = C$ might also be called *linear* because after plotting solutions, a straight line can be drawn through them. We have then *graphed the line* or *drawn the graph* of the equation (Figure 2.3).

Figure 2.1

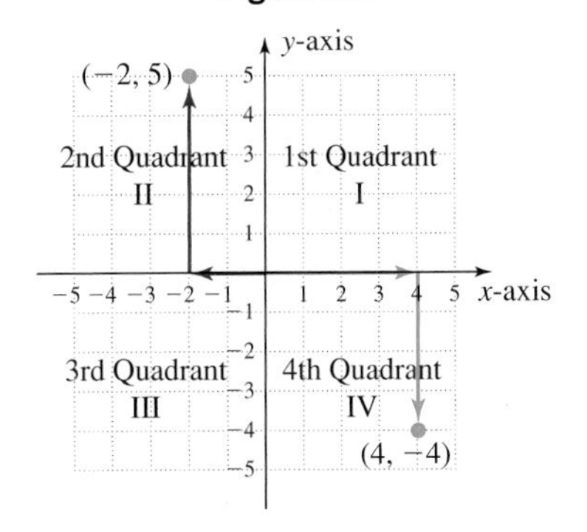

Figure 2.2

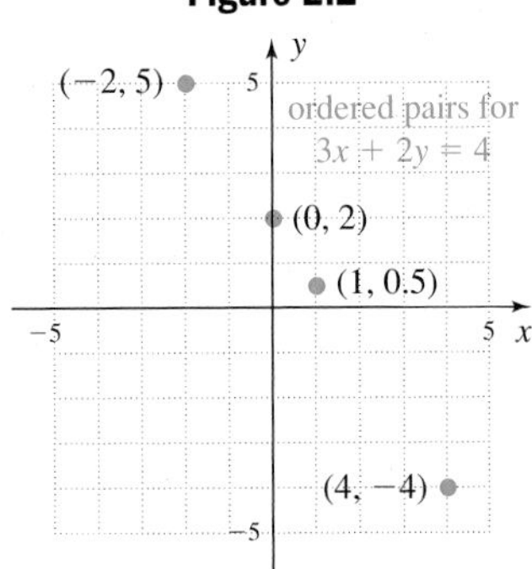

Figure 2.3

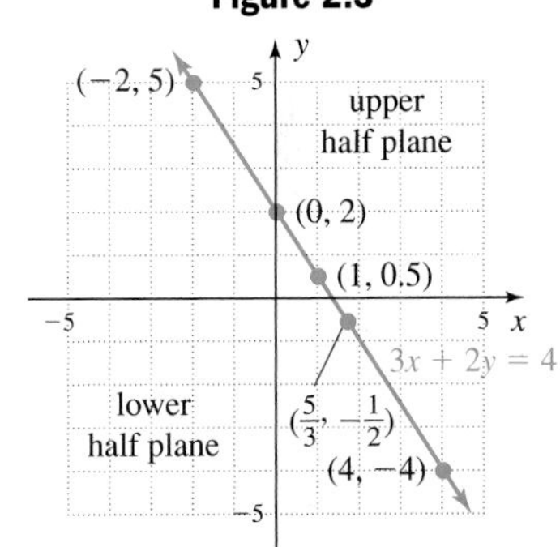

Points where both x and y have integer values are called **lattice points.** It can be shown that *every* ordered pair solution for $3x + 2y = 4$, including those with noninteger coordinates, correspond to a point on this line. For instance, $(\frac{5}{3}, -\frac{1}{2})$ is on the line since $3(\frac{5}{3}) + 2(-\frac{1}{2}) = 4$✓. In other words, although we use only a few specific points to graph the line, it is actually made up of *an infinite number of ordered pairs* that satisfy the equation. All of these points together make the graph *continuous,* which for our purposes means you can draw the entire graph without lifting your pencil from the paper. The arrowheads used on both ends of the line indicate the infinite extension of the graph. For future reference, notice how the line divides the grid into two distinct regions, called **half planes.** For more on this idea, see Exercises 11 and 12.

CAUTION

While $(\frac{5}{3}, -\frac{1}{2})$ was read from the graph and checked out (satisfied the equation), keep in mind that "solutions" read from a graph may only *approximate* the location of an actual solution.

B. Graphing Lines Using *x*- and *y*-Intercepts

Regardless of the coefficients or how the equation is written, any linear equation can be graphed by plotting two or more points. The graph in Figure 2.3 cuts through or **intercepts** the y-axis at (0, 2), and is called the ***y*-intercept** of the line. In general,

y-intercepts have the form $(0, y)$. Although more difficult to see graphically, the line also intercepts the x-axis at $(\frac{4}{3}, 0)$, and this point is called the **x-intercept.** In general, x-intercepts have the form $(x, 0)$. The x- and y-intercepts are usually easier to calculate than other points (since $y = 0$ or $x = 0$, respectively) and we often graph linear equations using only these two points, with one additional point to check our work. This is called the **intercept method** for graphing linear equations.

GRAPHING LINES USING THE INTERCEPT METHOD

1. Substitute $x = 0$ and solve for y. This will give you the y-intercept $(0, y)$.
2. Substitute $y = 0$ and solve for x. This will give you the x-intercept $(x, 0)$.
3. Select any additional input x, substitute, and solve for y. This will give a third point (x, y).
4. Connect the points with a straight line. If you cannot draw a straight line through the points, an error has been made and you should go back and check your calculations.

EXAMPLE 2 Graph the equation $2x + 5y = 6$ using the intercept method.

Solution:

substitute $x = 0$ (y-intercept)	substitute $y = 0$ (x-intercept)	substitute $x = -2$
$2(0) + 5y = 6$	$2x + 5(0) = 6$	$2(-2) + 5y = 6$
$5y = 6$	$2x = 6$	$-4 + 5y = 6$
$y = 1.2$	$x = 3$	$5y = 10$
$(0, 1.2)$	$(3, 0)$	$y = 2$
		$(-2, 2)$

x	y
0	1.2
3	0
−2	2

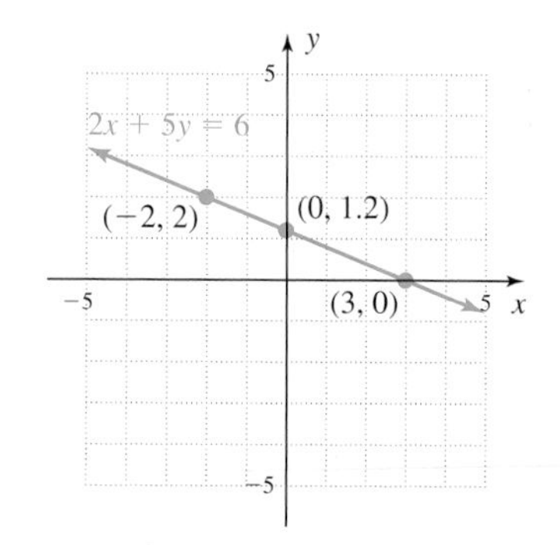

NOW TRY EXERCISES 13 THROUGH 32

Horizontal Lines and Vertical Lines

Horizontal and vertical lines have a number of important applications. They are sometimes used to find the **boundaries of a graph** or to perform certain tests on nonlinear graphs. To better understand them, consider that in *one dimension* the graph of $x = 2$ is a single point (Figure 2.4), indicating a location on the line 2 units from zero in the positive direction. In *two dimensions*, the equation $x = 2$ represents *all points* with an x-coordinate of positive two (Figure 2.5). Since there are an infinite number of these points, we end up with a solid *vertical line* with equation $x = 2$ (Figure 2.6).

Figure 2.4

−2 −1 0 1 2 3 4 5 x

Figure 2.5

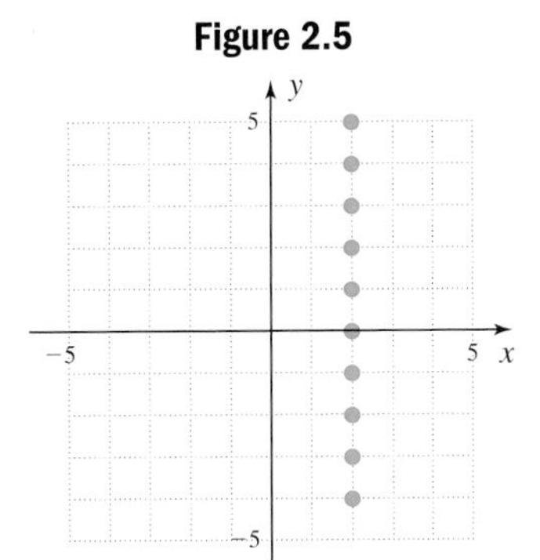

Figure 2.6

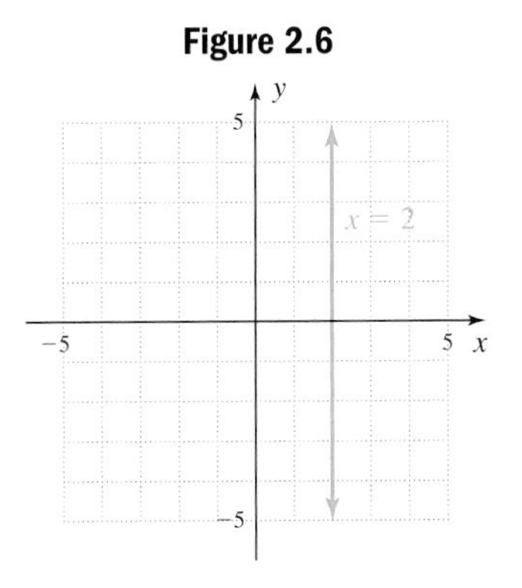

The same idea can be applied to horizontal lines. In *two dimensions,* the equation $y = 4$ represents *all points* with a y-coordinate of positive four, and there are an infinite number of these points as well. The result is a solid *horizontal line* whose equation is $y = 4$.

VERTICAL LINES

The equation of a vertical line has the form $x = h$, where $(h, 0)$ is a point on the x-axis. The x-intercept is $(h, 0)$.

HORIZONTAL LINES

The equation of a horizontal line has the form $y = k$, where $(0, k)$ is a point on the y-axis. The y-intercept is $(0, k)$.

EXAMPLE 3 Graph the lines $y = -3$ and $x = 2$ on the same grid. Where do they intersect?

Solution: The graph of $y = -3$ is a horizontal line through the point $(0, -3)$. The graph of $x = 2$ is a vertical line through the point $(2, 0)$. They intersect at the point $(2, -3)$.

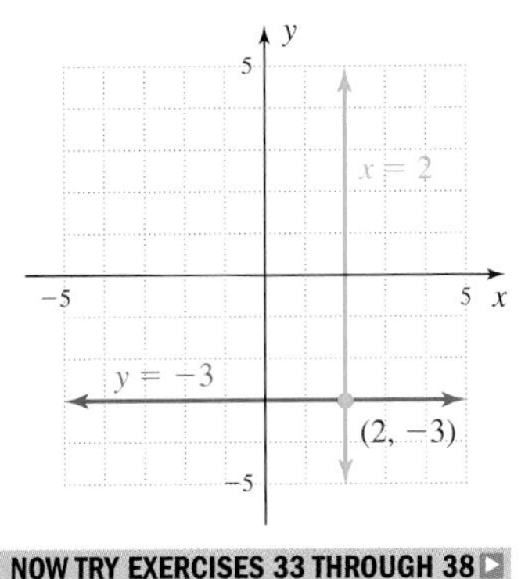

NOW TRY EXERCISES 33 THROUGH 38

Figure 2.7

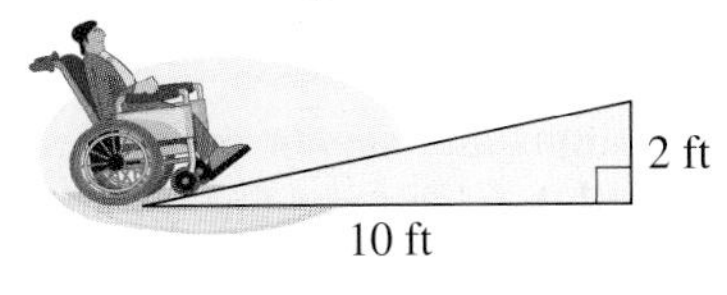

Figure 2.8

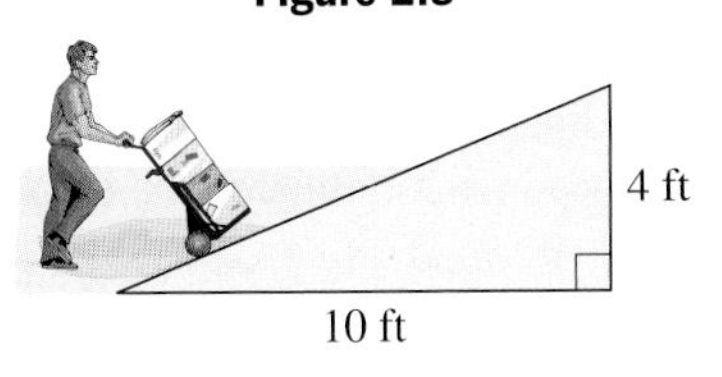

C. The Slope Formula and Rates of Change

Consider the two ramps shown in Figures 2.7 and 2.8. The first is used to make buildings more wheelchair accessible. The second is used to move merchandise up to a loading dock. Both ramps have the same horizontal length, but by simple observation we see the second ramp is *steeper,* since it rises a greater vertical distance. In practical applications, this steepness is referred to as the **slope** (of the ramp), and is measured using the ratio $\frac{\text{ramp height}}{\text{ramp length}} = \frac{\text{vertical change}}{\text{horizontal change}}$. The first ramp has a slope of $\frac{2 \text{ ft}}{10 \text{ ft}} = \frac{1}{5}$ and the second ramp has a slope of $\frac{4}{10} = \frac{2}{5}$. Notice that $\frac{2}{5} > \frac{1}{5}$, and the steeper ramp has the larger slope. To apply the concept of slope to the graph of a line, we can actually draw one of these ramps using a right triangle, called the **slope triangle,** with any segment of the line as the hypotenuse. For the line in Figure 2.9 we select the lattice points $(-5, 1)$ and $(5, 5)$ for the line segment, and draw the slope triangle as shown. By simple counting, the horizontal change is 10 units and the vertical change is 4 units, giving a slope of $\frac{4}{10} = \frac{2}{5}$. It's worth noting the reduced ratio $\frac{2}{5}$ can still be interpreted as $\frac{\text{vertical change}}{\text{horizontal change}}$. In fact, from the point $(-5, 1)$ a vertical change of 2 units followed by a horizontal change of 5 units puts you at $(0, 3)$—*which is another point on the line!* In other words, the slope of a line is constant.

Figure 2.9

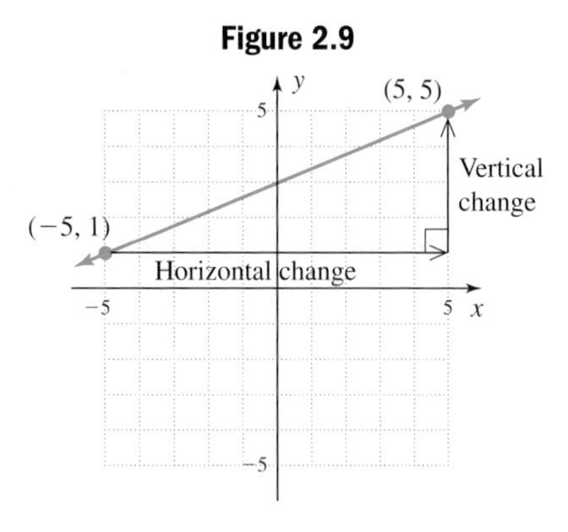

The slope value actually does much more than quantify the slope of a line, it expresses a **rate of change** between the two quantities y and x. In many real-world applications, the ratio $\frac{\text{change in } y}{\text{change in } x}$ is symbolized as $\frac{\Delta y}{\Delta x}$. The symbol Δ is the Greek letter *delta* and has come to represent a change in some quantity. In algebra, we typically use the letter m to represent slope, and $m = \frac{\Delta y}{\Delta x}$ is read, "slope is equal to the change in y over the change in x." Interpreting slope as a rate of change has many significant applications in college algebra and beyond.

EXAMPLE 4 The graph shown models the relationship between the number of crates unloaded from a container ship and the time required. Determine the slope of the line and interpret what the slope ratio means in this context.

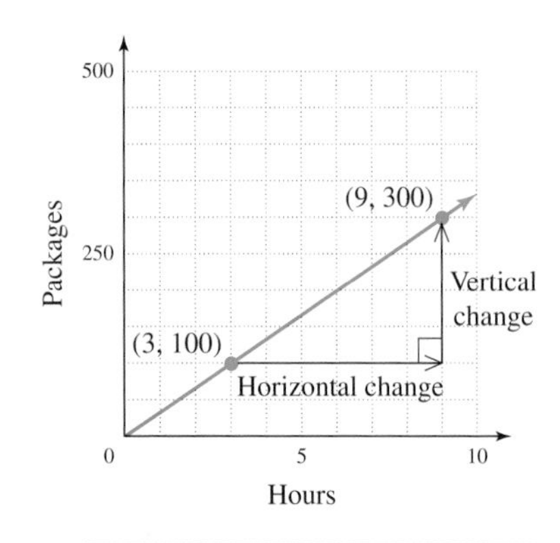

Solution: Selecting lattice points $(3, 100)$ and $(9, 300)$ for the line segment, we draw the slope triangle as shown. The horizontal change is positive 6 and the vertical change is positive 200: $m = \frac{200}{6} = \frac{100}{3}$. This rate of change compares $\frac{\text{packages unloaded}}{\text{time in hours}}$, and indicates that 100 crates are unloaded every 3 hr. As a unit rate, $\frac{100}{3} = 33\frac{1}{3}$ crates per hour are being unloaded.

NOW TRY EXERCISES 39 THROUGH 44

Figure 2.10

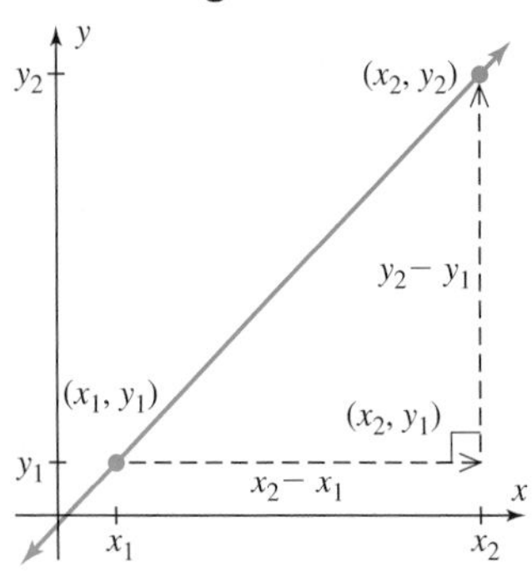

Using a slope triangle is too crude for practical applications and cannot be applied unless the graph is given. Instead, we generalize the idea to find the slope of a line through *any two points* P_1 and P_2. To distinguish one point from the other, subscripts are used as in $P_1 = (x_1, y_1)$ and $P_2 = (x_2, y_2)$ (see Figure 2.10). To help develop a formula for $\frac{\Delta y}{\Delta x}$, we again consider a line segment between these points as a hypotenuse, and draw a slope triangle with the point (x_2, y_1) at the vertex of the right angle. The **vertical change,** also called the **rise,** can be calculated as the difference in y-coordinates of the two points, or $y_2 - y_1$. The **horizontal change,** called the **run,** is the difference in x-coordinates: $x_2 - x_1$. The result is called the **slope formula,** written $m = \frac{y_2 - y_1}{x_2 - x_1}$.

THE SLOPE FORMULA

Given any two points $P_1 = (x_1, y_1)$ and $P_2 = (x_2, y_2)$, the slope of any nonvertical line through P_1 and P_2 is given by

$$m = \frac{\Delta y}{\Delta x} = \frac{\text{rise}}{\text{run}} = \frac{\text{vertical change}}{\text{horizontal change}} = \frac{y_2 - y_1}{x_2 - x_1}, \text{ where } x_2 \neq x_1.$$

EXAMPLE 5 Julia works on an assembly line for an auto parts re-manufacturing company. By 9:00 A.M. her group has assembled 29 carburetors. By 12:00 noon, they have 87 carburetors complete. Assuming the relationship is linear, find the slope of the line and discuss its meaning in this context.

Solution: Let c represent carburetors and t represent time. This gives

$$(t_1, c_1) = (9, 29) \text{ and } (t_2, c_2) = (12, 87).$$

$$\frac{\Delta c}{\Delta t} = \frac{c_2 - c_1}{t_2 - t_1} = \frac{87 - 29}{12 - 9}$$

$$= \frac{58}{3} = 19.\overline{3}.$$

Julia's group can assemble 58 carburetors every 3 hrs, or about $19\frac{1}{3}$ carburetors per hour.

NOW TRY EXERCISES 45 THROUGH 52

Actually, the assignment of (t_1, c_1) to $(9, 29)$ and (t_2, c_2) to $(12, 87)$ is arbitrary. The slope ratio is the same *as long as the order of subtraction is the same*. In other words, for $(t_1, c_1) = (12, 87)$ and $(t_2, c_2) = (9, 29)$, we have $m = \frac{29 - 87}{9 - 12} = \frac{-58}{-3}$ which is equivalent to the previous result.

Positive and Negative Slope

In Example 4, the slope m was a positive number and the line *sloped upward* as you moved from left to right. In general, if $m > 0$ (positive slope), the line slopes upward from left to right since y-values are increasing. If $m < 0$ (negative slope), the line slopes downward as you move left to right since y-values are decreasing.

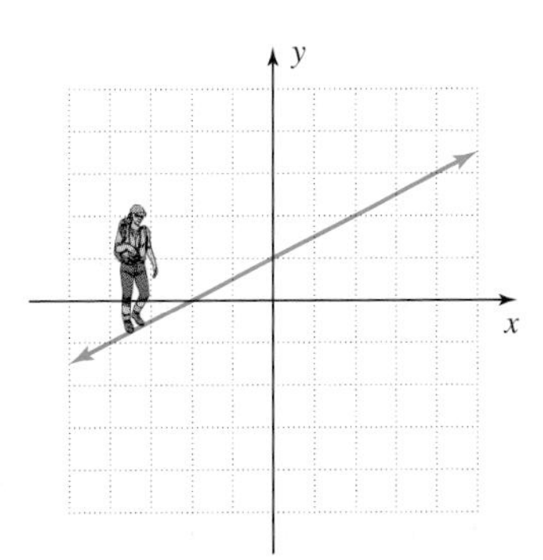

$m > 0$, positive slope
y-values *increase* from left to right

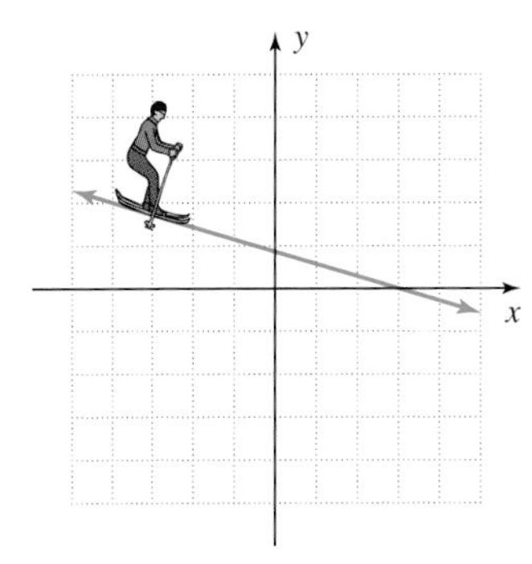

$m < 0$, negative slope
y-values *decrease* from left to right

The Slope of Horizontal and Vertical Lines

So far, the slope formula has been applied only to lines that were nonhorizontal and nonvertical. So what *is* the slope of a horizontal line? On an intuitive level, you expect that a perfectly level highway would have a slope or incline of zero. In general, for any two points on a horizontal line, $y_2 = y_1$ and $y_2 - y_1 = 0$, giving a slope of $m = \frac{0}{x_2 - x_1} = 0$. For vertical lines, any two distinct points give $x_2 = x_1$ and $y_2 \neq y_1$. This makes $x_2 - x_1 = 0$ and the slope ratio $m = \frac{y_2 - y_1}{0}$ is undefined.

THE SLOPE OF HORIZONTAL AND VERTICAL LINES

Given (x_1, y_1) and (x_2, y_2) are two distinct points

(a) on a horizontal line:

$$y_2 = y_1 \text{ and } x_2 \neq x_1: \frac{y_2 - y_1}{x_2 - x_1} = \frac{0}{x_2 - x_1}.$$

The slope of a horizontal line is zero.

(b) on a vertical line:

$$y_2 \neq y_1 \text{ and } x_2 = x_1: \frac{y_2 - y_1}{x_2 - x_1} = \frac{y_2 - y_1}{0}.$$

The slope of a vertical line is undefined.

D. Parallel and Perpendicular Lines

Two lines in the same plane that never intersect are called **parallel lines.** When we place these lines on the coordinate grid, we find that "never intersect" is equivalent to saying "the lines have equal slopes but different y-intercepts." In Figure 2.11, notice that the same slope triangle fits both L_1 and L_2 exactly, and that by counting $\frac{\Delta y}{\Delta x}$ both lines have slope $m = \frac{3}{4}$.

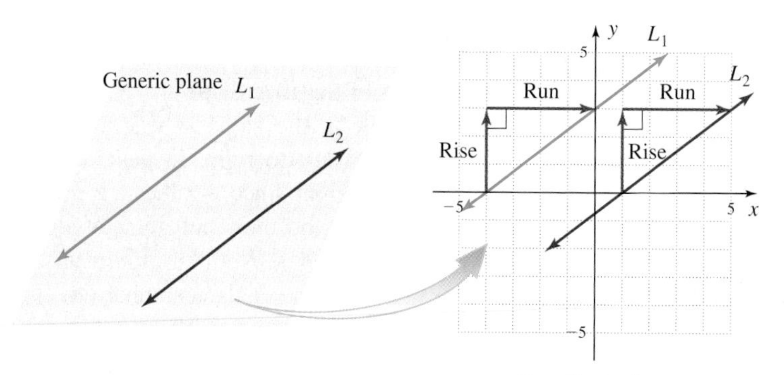

Figure 2.11

PARALLEL LINES

Given m_1 is the slope for line 1 and m_2 is the slope for line 2. If $m_1 = m_2$, then line 1 is parallel to line 2. In symbols we write $L_1 \| L_2$.

Two lines in the same plane that intersect at right angles are called **perpendicular lines.** Using the coordinate grid, we note that "intersect at right angles" is equivalent to saying "their slopes have a product of -1." In Figure 2.12, note the slope triangle for L_1 gives $m_1 = \frac{4}{3}$, the slope triangle for L_2 gives $m_2 = \frac{-3}{4}$, and that $m_1 \cdot m_2 = \frac{4}{3} \cdot \frac{-3}{4} = -1$.

Alternatively, we can say their slopes are **negative reciprocals,** since the negative reciprocal of $\frac{4}{3}$ is $\frac{-3}{4}$ $\left(m_1 \cdot m_2 = -1 \text{ implies } m_1 = -\frac{1}{m_2}\right)$.

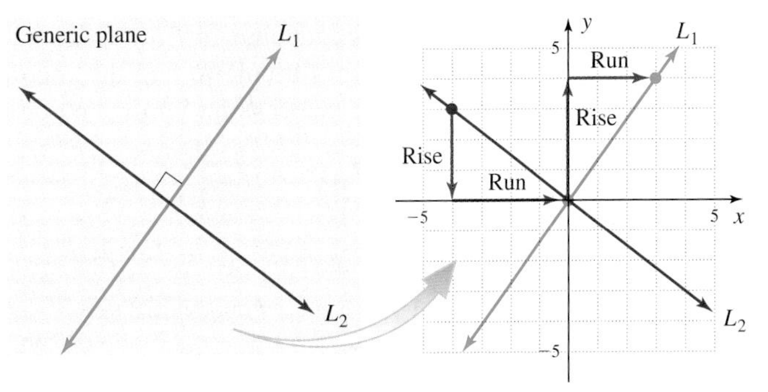

Figure 2.12

> **PERPENDICULAR LINES**
> Given m_1 is the slope for line 1 and m_2 is the slope for line 2.
> If $m_1 \cdot m_2 = -1$, then line 1 is perpendicular to line 2.
> In symbols we write $L_1 \perp L_2$.

EXAMPLE 6 The three points $P_1 = (5, 1)$, $P_2 = (3, -2)$, and $P_3 = (-3, 2)$ form the vertices of a triangle. Use the slope formula to determine if they form a *right* triangle.

Solution: Because a right triangle must have two sides that are perpendicular (forming the right angle), we look for slopes that have a product of -1.

Using P_1 and P_2

$$m_1 = \frac{-2 - 1}{3 - 5} = \frac{-3}{-2} = \frac{3}{2}$$

Using P_1 and P_3

$$m_2 = \frac{2 - 1}{-3 - 5} = \frac{1}{-8}$$

Using P_2 and P_3

$$m_3 = \frac{2 - (-2)}{-3 - 3} = \frac{4}{-6} = \frac{2}{-3}$$

Since $m_1 \cdot m_3 = -1$, the triangle has a right angle and must be a right triangle.

NOW TRY EXERCISES 53 THROUGH 64

E. The Midpoint and Distance Formulas

As the name suggests, the **midpoint of a line segment** is a point located halfway between two points. On a standard number line, the midpoint of the line segment with endpoints 1 and 5 is 3, but more important, note that 3 is the **average distance** (from zero) of 1 unit and 5 units: $\frac{1 + 5}{2} = \frac{6}{2} = 3$. This observation can be extended to find the midpoint between two points (x_1, y_1) and (x_2, y_2) *on the coordinate plane*. We simply find the average distance between the x-coordinates and the average distance between the y-coordinates.

> **THE MIDPOINT FORMULA**
> Given any line segment with endpoints $P_1 = (x_1, y_1)$ and $P_2 = (x_2, y_2)$, the coordinates of the midpoint M can be found by calculating the average value of the given x- and y-coordinates.
>
> $$M: \left(\frac{x_1 + x_2}{2}, \frac{y_1 + y_2}{2}\right)$$

The midpoint formula can be used in many different ways. Here we'll use it to find the coordinates of the center of a circle.

EXAMPLE 7 The diameter of a circle has endpoints at $P_1 = (-3, -2)$ and $P_2 = (5, 4)$. Use the midpoint formula to find the coordinates of the center, then graph the center point on the coordinate grid.

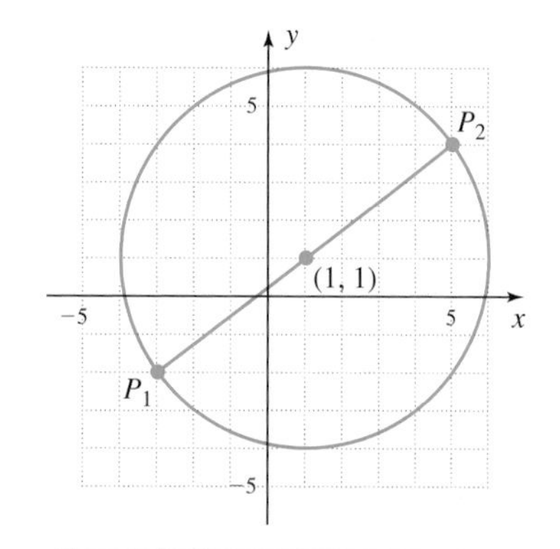

Solution: Midpoint: $\left(\frac{x_1 + x_2}{2}, \frac{y_1 + y_2}{2}\right)$

$$M: \left(\frac{-3 + 5}{2}, \frac{-2 + 4}{2}\right)$$

$$M: \left(\frac{2}{2}, \frac{2}{2}\right)$$

The center is located at (1, 1).

NOW TRY EXERCISES 65 THROUGH 74

Figure 2.13

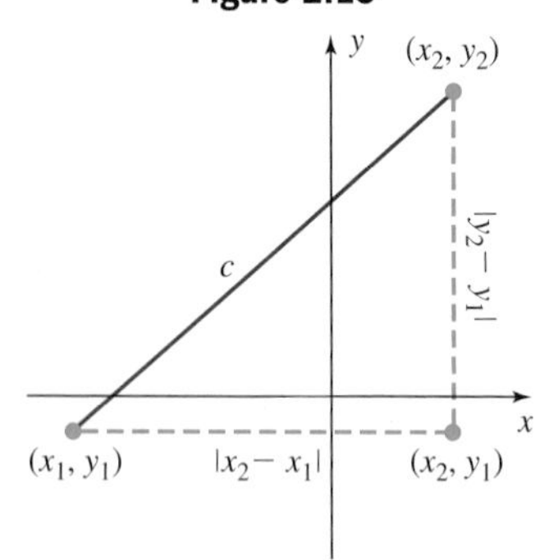

The Distance Formula

Using the "slope triangle" for (x_1, y_1) and (x_2, y_2) introduced earlier, we note the base of the triangle is $|x_2 - x_1|$ units long and the height (vertical distance) is $|y_2 - y_1|$ units (Figure 2.13). From the Pythagorean theorem (Section R.6) we see that $c^2 = a^2 + b^2$ corresponds to $c^2 = (|x_2 - x_1|)^2 + (|y_2 - y_1|)^2$, and taking the square root of both sides yields the **distance formula:** $c = \sqrt{(x_2 - x_1)^2 + (y_2 - y_1)^2}$, although it is most often written using d for distance, rather than c. Note the absolute value bars are dropped since the square of any quantity is always positive.

THE DISTANCE FORMULA

Given any two points $P_1 = (x_1, y_1)$ and $P_2 = (x_2, y_2)$, the straight line distance between them can be found using the Pythagorean theorem.

$c^2 = a^2 + b^2$ becomes

$$d^2 = (x_2 - x_1)^2 + (y_2 - y_1)^2$$

or

$$d = \sqrt{(x_2 - x_1)^2 + (y_2 - y_1)^2}$$

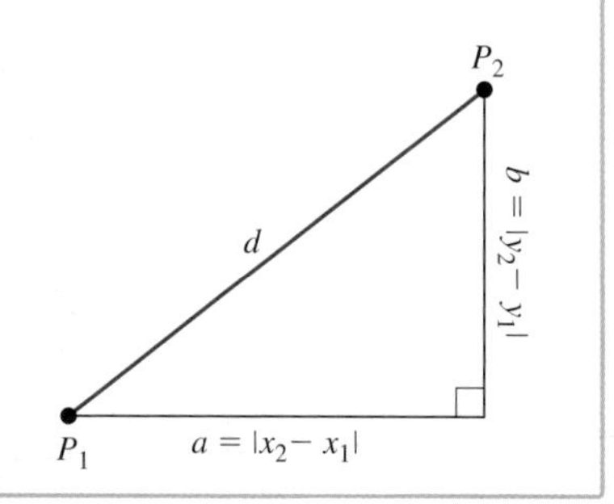

EXAMPLE 8 Use the distance formula to find the diameter of the circle from Example 7.

Solution: For $(x_1, y_1) = (-3, -2)$ and $(x_2, y_2) = (5, 4)$, the distance formula gives

$$\begin{aligned} d &= \sqrt{(x_2 - x_1)^2 + (y_2 - y_1)^2} \\ &= \sqrt{[5 - (-3)]^2 + [4 - (-2)]^2} \\ &= \sqrt{8^2 + 6^2} \text{ or } \sqrt{100} = 10 \end{aligned}$$

The diameter of the circle is 10 units long.

NOW TRY EXERCISES 75 THROUGH 78

F. Applications of Linear Graphs

The graph of a linear equation can be used to help solve many applied problems. If the numbers you're working with are either very small or very large, *scale the axes* appropriately. This can be done by letting each tic mark represent a smaller or larger unit so the points located will fit on the grid. Also, many applications use only nonnegative values and although points with negative coordinates may be used to graph a line, only ordered pairs in Quadrant I (QI) can be meaningfully interpreted.

EXAMPLE 9 Use the information given to create a linear equation model in two variables, then graph the line and use the graph to answer this question: *A salesperson gets a daily \$20 meal allowance plus \$7.50 for every item she sells. How many sales are needed for a daily income of \$125?*

Solution: Let x represent sales and y represent income.

Verbal model: Income (y) equals \$7.50 per sale (x) + \$20 for meals

Equation model: $y = 7.5x + 20$

Using $x = 0$ and $x = 10$, we find (0, 20) and (10, 95) are points on this graph. From the graph, we estimate that 14 sales are needed to generate a daily income of \$125.00. Substituting $x = 14$ into the equation verifies that (14, 125) is indeed on the graph.

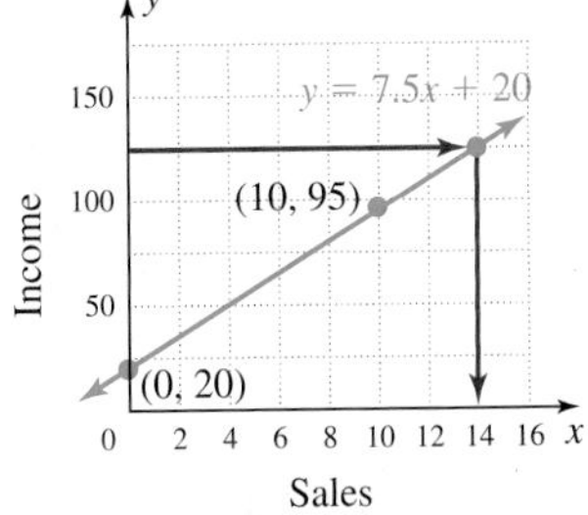

NOW TRY EXERCISES 81 THROUGH 86

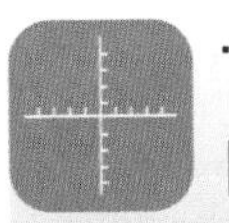

TECHNOLOGY HIGHLIGHT

Linear Equations, Graphing Calculators, and Window Size

The keystrokes shown apply to a TI-84 Plus model. Please consult your manual or our Internet site for other models.

To graph linear equations on the TI-84 Plus, we (1) solve the equation for the variable y as before, (2) enter the equation on the Y= screen, and (3) GRAPH the equation and adjust the WINDOW if necessary.

1. Solve the equation for y.
   For the equation $2x - 3y = -3$, we have

$2x - 3y = -3$ given equation

$-3y = -2x - 3$ subtract $2x$ from each side

$y = \frac{2}{3}x + 1$ divide both sides by -3

2. Enter the equation on the Y= screen.
   On the Y= screen, enter $\frac{2}{3}x + 1$. Note that for some calculators parentheses are needed to group $(2 \div 3)x$, to prevent the calculator from interpreting this term as $2 \div (3x)$.

3. GRAPH the equation, adjust the WINDOW.
   Since much of our work is centered at (0, 0) on the coordinate grid, the calculator's default settings have a domain of $x \in [-10, 10]$ and a range of $y \in [-10, 10]$, as shown in Figure 2.14. This is referred to as the WINDOW size. To graph the line in this window, it is easiest to use the ZOOM key and select **6:ZStandard,** which resets the window to these default settings and graphs the line automatically. The graph is shown in Figure 2.15. The Xscl and Yscl entries give the

Figure 2.14

WINDOW
Xmin=-10
Xmax=10
Xscl=1
Ymin=-10
Ymax=10
Yscl=1
Xres=1

Figure 2.15

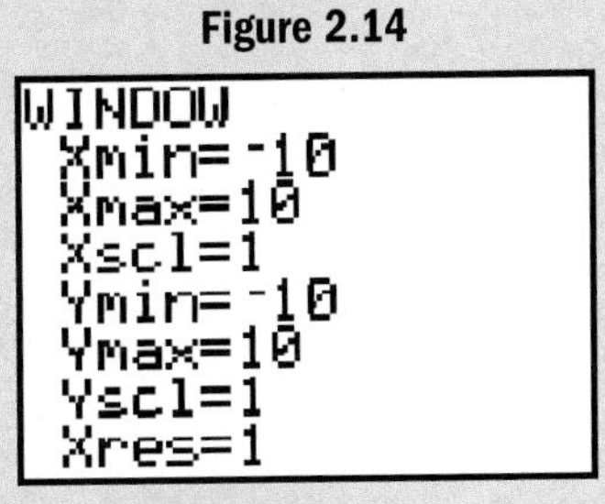

scale used on each axis, indicating that each "tic mark" represents 1 unit. Graphing calculators have many features that enable us to find ordered pairs on a line. One is the (2nd GRAPH) (**TABLE**) feature we have seen previously. We can also use the calculator's TRACE feature. As the name implies, this feature enables us to trace along the line by moving a blinking cursor using the left ◀ and right ▶ arrow keys. The calculator simultaneously displays the coordinates of the current location of the cursor. After pressing the TRACE button, the cursor appears automatically—usually at the y-intercept. Moving the cursor left and right, note the coordinates changing at the bottom of the screen. The point (3.4042553, 3.2695035) is on the line and satisfies the equation of the line (check this using the **TABLE** feature). The calculator is displaying decimal values because the screen is exactly 94 pixels wide, 47 pixels to the left of the y-axis, and 47 pixels to the right. This means that each time you press the left or right arrow, the x-value changes by 1/47—which is *not* a nice round number. To have the calculator TRACE through "friendlier" values, we can use the ZOOM **4:ZDecimal** feature, which sets Xmin = −4.7 and Xmax = 4.7, or **8:ZInteger,** which sets Xmin = −47 and Xmax = 47. Press ZOOM **4:ZDecimal** and the calculator will automatically regraph the line. Then press the **TRACE** key once again and move the cursor. Notice that more "friendly" decimal values are displayed. More will be said about friendly windows in future *Technology Highlights.*

Exercise 1: Use the ZOOM **4:ZDecimal** and **TRACE** features to identify the x- and y-intercepts for $Y_1 = \frac{2}{3}x + 1$. (−1.5, 0), (0, 1)

Exercise 2: Use the ZOOM **8:ZInteger** and **TRACE** features to graph the line $79x - 55y = 869$, then identify the x- and y-intercepts. (11, 0), (0, −15.8)

2.1 EXERCISES

CONCEPTS AND VOCABULARY

Fill in each blank with the appropriate word or phrase. Carefully reread the section if needed.

1. Points on the grid that have integer coordinates are called __lattice__ points.

2. The graph of a line divides the coordinate grid into two distinct regions, called __half__ __planes__.

3. To find the x-intercept of a line, substitute __$y = 0$__. To find the y-intercept, substitute __$x = 0$__.

4. The notation $\frac{\Delta y}{\Delta x}$ is read __change in__ y over __change in__ x and is used to denote a(n) __rate__ of __change__ between the x- and y-variables.

5. What is the slope of the line in Example 9? $m = \frac{15}{2}$ Discuss/explain the meaning of $m = \frac{\Delta y}{\Delta x}$ in the context of this example. Answers will vary.

6. Discuss/explain the relationship between the slope formula, the Pythagorean theorem, and the distance formula. Include several illustrations. Answers will vary.

DEVELOPING YOUR SKILLS

Create a table of values for each equation and sketch the graph.

7. $2x + 3y = 6$

x	y
−6	6
−3	4
0	2
3	0

8. $-3x + 5y = 10$

x	y
−5	−1
0	2
2	3.2
5	5

9. $y = \frac{3}{2}x + 4$

x	y
−2	1
0	4
2	7
4	10

10. $y = \frac{5}{3}x - 3$

x	y
−6	−13
−3	−8
0	−3
3	2

7\.

(−6, 6) (−3, 4) (0, 2) (3, 0)

8\.

(2, 3.2) (5, 5) (0, 2) (−5, −1)

9\.

(4, 10) (2, 7) (0, 4) (−2, 1)

10\.

(3, 2) (0, −3) (−3, −8) (−6, −13)

11. $-0.5 = \frac{3}{2}(-3) + 4$
$-0.5 = -\frac{9}{2} + 4$
$-0.5 = -0.5$ ✓
$\frac{19}{4} = \frac{3}{2}\left(\frac{1}{2}\right) + 4$
$\frac{19}{4} = \frac{3}{4} + 4$
$\frac{19}{4} = \frac{19}{4}$ ✓

12. $-5.5 = \frac{5}{3}(-1.5) - 3$
$-5.5 = -2.5 - 3$
$-5.5 = -5.5$ ✓
$\frac{37}{6} = \frac{5}{3}\left(\frac{11}{2}\right) - 3$
$\frac{37}{6} = \frac{55}{6} - 3$
$\frac{37}{6} = \frac{37}{6}$ ✓

13. 14.

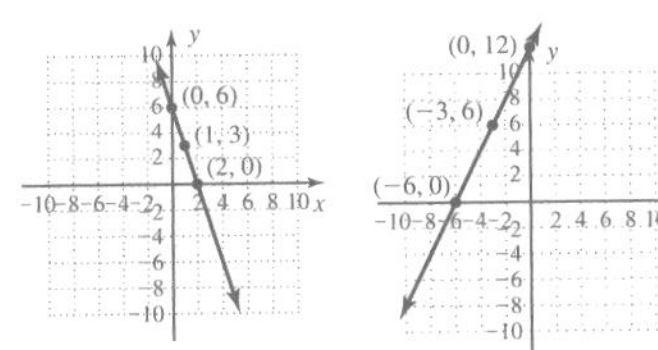

15. 16.

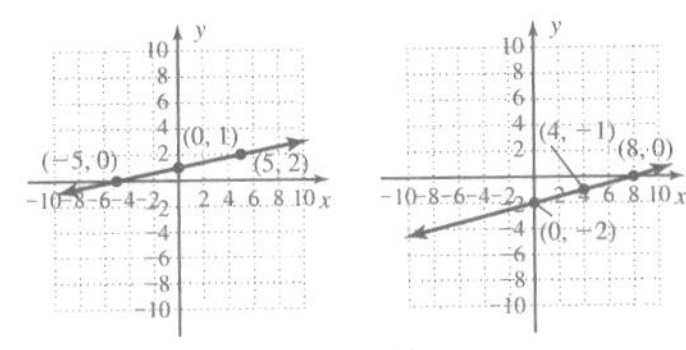

17. 18.

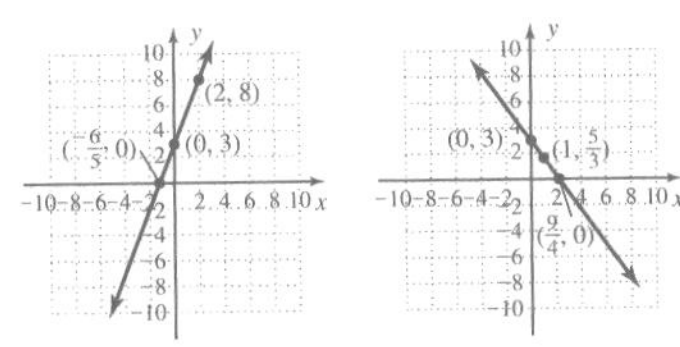

19. 20.

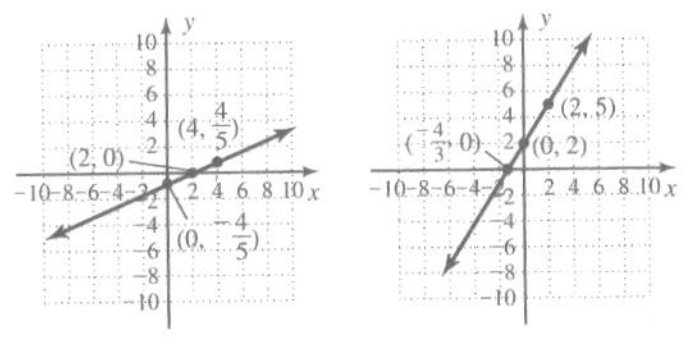

21. 22.

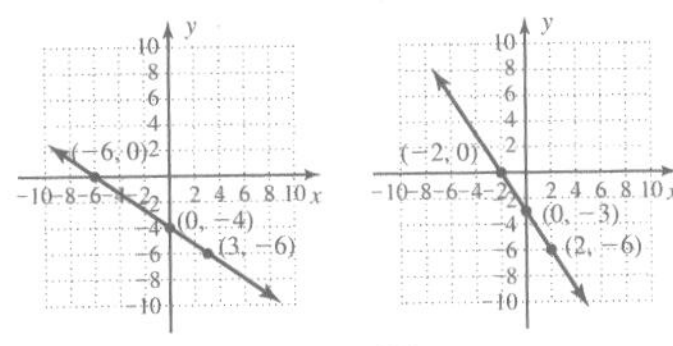

23. 24.

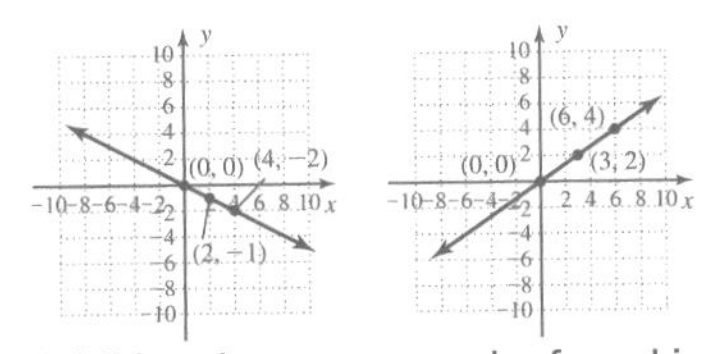

Additional answers can be found in the Instructor Answer Appendix.

11. From Exercise 9, verify the graph of the line appears to go through $(-3, -0.5)$ and $\left(\frac{1}{2}, \frac{19}{4}\right)$, then show they satisfy the equation.

12. From Exercise 10, verify the graph of the line appears to go through $(-1.5, -5.5)$ and $\left(\frac{11}{2}, \frac{37}{6}\right)$, then show they satisfy the equation.

Graph the following equations using the intercept method. Plot a third point as a check.

13. $3x + y = 6$
14. $-2x + y = 12$
15. $5y - x = 5$
16. $-4y + x = 8$
17. $-5x + 2y = 6$
18. $3y + 4x = 9$
19. $2x - 5y = 4$
20. $-6x + 4y = 8$

Graph by plotting points or using the intercept method. Plot at least three points for each graph. If the coefficient is a fraction, choose inputs that will help simplify the calculation.

21. $2x + 3y = -12$
22. $-3x - 2y = 6$
23. $y = -\frac{1}{2}x$
24. $y = \frac{2}{3}x$
25. $y - 25 = 50x$
26. $y + 30 = 60x$
27. $y = -\frac{2}{5}x - 2$
28. $y = \frac{3}{4}x + 2$
29. $2y - 3x = 0$
30. $y + 3x = 0$
31. $3y + 4x = 12$
32. $-2x + 5y = 8$
33. $x = -3$
34. $y = 4$
35. $x = 2$
36. $y = -2$

Write the equation for each line L_1 and L_2 shown. Specifically state their point of intersection.

37.

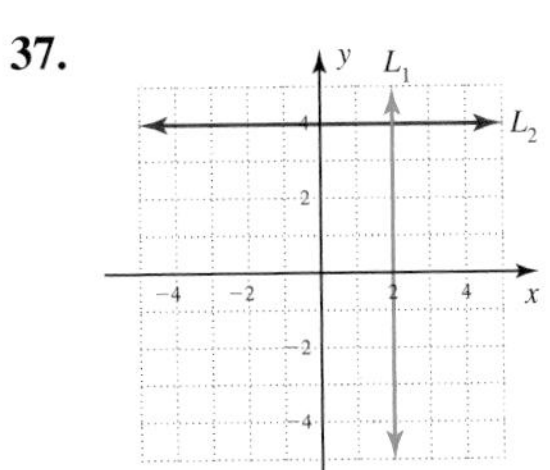

L_1: $x = 2$; L_2: $y = 4$; point of intersection (2, 4)

38.

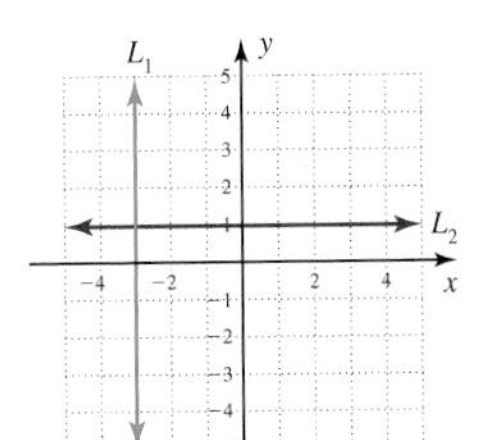

L_1: $x = -3$; L_2: $y = 1$; point of intersection $(-3, 1)$

39. The graph shown models the relationship between the cost of a new home and the size of the home in square feet. (a) Determine the slope of the line using any two lattice points and interpret what the slope ratio means in this context and (b) estimate the cost of a 3000 ft^2 home. **a.** $m = 125$, cost increased \$125,000 per 1000 sq ft **b.** \$375,000

Exercise 39

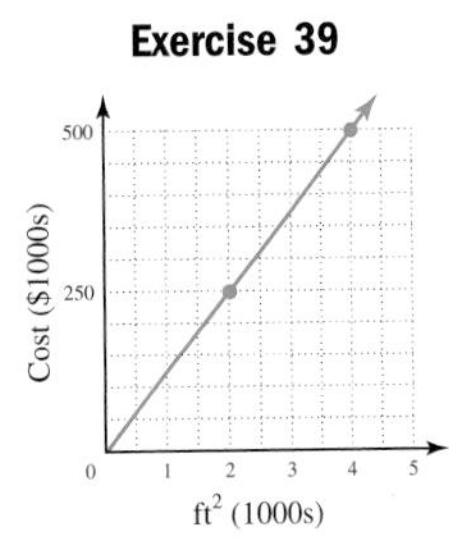

Exercise 40

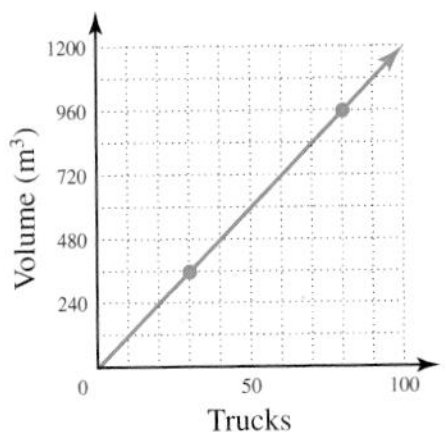

Volume (m³)
Trucks

40. The graph shown models the relationship between the volume of garbage that is dumped in a landfill and the number of commercial garbage trucks that enter the site. (a) Determine the slope of the line and interpret what the slope ratio means in this context and (b) estimate the number of trucks entering the site daily if 1000 m^3 of garbage is dumped per day. **a.** $m = 12$, 12 m^3 dumped per garbage truck **b.** 83 trucks

41. The graph shown models the relationship between the distance of an aircraft carrier from its home port and the number of hours since departure. (a) Determine the slope of the line

and interpret what the slope ratio means in this context and (b) estimate the distance from port after 8.25 hours. **a.** $m = 22.5$, distance increases 22.5 mph **b.** about 186 mi

Exercise 41

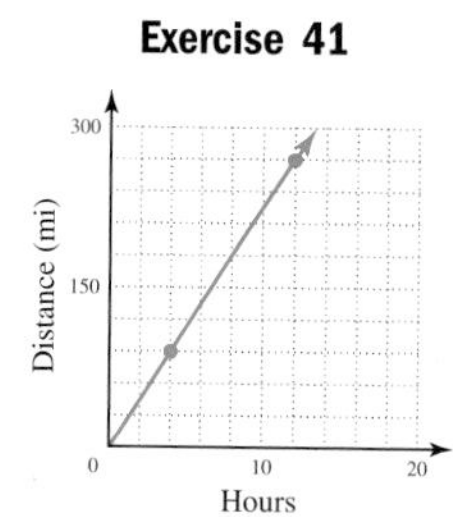

Exercise 42

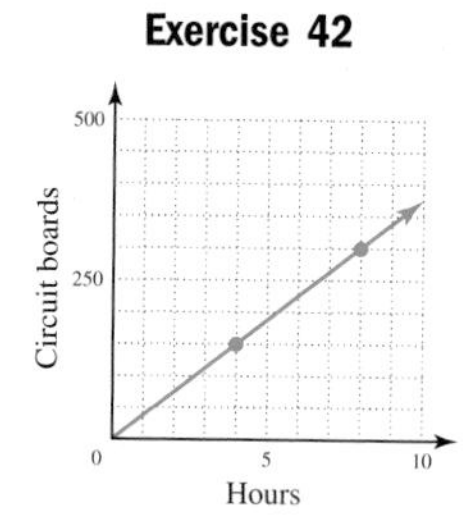

42. The graph shown models the relationship between the number of circuit boards that have been assembled at a factory and the number of hours since starting time. (a) Determine the slope of the line and interpret what the slope ratio means in this context and (b) estimate how many hours the factory has been running if 225 circuit boards have been assembled. **a.** $m = 37.5$, 37.5 circuit boards are assembled per hour **b.** 6 hr

43. Height and weight: While there are many exceptions, numerous studies have shown a close relationship between an average height and average weight. Suppose a person 70 in. tall weighs 165 lb, while a person 64 in. tall weighs 142 lb. Assuming the relationship is linear, (a) find the slope of the line and discuss its meaning in this context and (b) determine how many pounds are added for each inch of height.

44. Rate of climb: Shortly after takeoff, a plane increases altitude at a constant (linear) rate. In 5 min the altitude is 10,000 feet. Fifteen minutes after takeoff, the plane has reached its cruising altitude of 32,000 ft. (a) Find the slope of the line and discuss its meaning in this context and (b) determine how long it takes the plane to climb from 12,200 feet to 25,400 feet.

Compute the slope of the line through the given points, then graph the line and use $m = \frac{\Delta y}{\Delta x}$ to find two additional points on the line. Answers may vary.

45. (3, 5), (4, 6) **46.** (−2, 3), (5, 8) **47.** (10, 3), (4, −5) **48.** (−3, −1), (0, 7)

49. (1, −8), (3, 7) **50.** (0, 5), (0, −5) **51.** (−3, 6), (6, 6) **52.** (−2, −4), (−3, −1)

Two points on L_1 and two points on L_2 are given. Use the slope formula to determine if lines L_1 and L_2 are parallel, perpendicular, or neither.

53. L_1: (−2, [illegible]) and (0, 6)
L_2: (1, 8 [illegible] and (0, 5)

54. L_1: (1, 10) and (−1, 7)
L_2: (0, 3) and (1, 5)

55. L_1: (−3, −4) and (0, 1)
L_2: (0, 0) and (−4, 4)

56. L_1: (6, 2[illegible] and (8, −2)
L_2: (5, [illegible] and (3, 0)

57. L_1: (6, 3) and (8, 7)
L_2: (7, 2) and (6, 0)

58. L_1: (−5, −1) and (4, 4)
L_2: (4, −7) and (8, 10)

In Exercises 59 to 64, three points that form the vertices of a triangle are given. Determine if any of the triangles are right triangles.

59. (5, 2), (0, −3), (4, −4) **60.** (7, 0), (−1, 0), (7, 4) **61.** (−4, 3), (−7, −1), (3, −2)

62. (−3, 7), (2, 2), (5, 5) **63.** (−3, 2), (−1, 5), (−6, 4) **64.** (0, 0), (−5, 2), (2, −5)

Find the midpoint of each segment with the given endpoints.

65. (1, 8), (5, −6) (3,1)

66. (5, 6), (6, −8) $(\frac{11}{2}, -1)$

67. (−4.5, 9.2), (3.1, −9.8) (−0.7, −0.3)

68. (5.2, 7.1), (6.3, −7.1) (5.75, 0)

69. $\left(\frac{1}{5}, -\frac{2}{3}\right), \left(-\frac{1}{10}, \frac{3}{4}\right)$ $\left(\frac{1}{20}, \frac{1}{24}\right)$

70. $\left(-\frac{3}{4}, -\frac{1}{3}\right), \left(\frac{3}{8}, \frac{5}{6}\right)$ $\left(\frac{-3}{16}, \frac{1}{4}\right)$

43. **a.** $m = \frac{23}{6}$, a person weighs 23 lb more for each additional 6 in. in height
b. 3.8

44. **a.** $m = 2200$; the plane climbs 2200 ft/min
b. 6 min

45. $m = 1$; (2, 4) and (1, 3)

46. $m = \frac{5}{7}$; (−9, −2) and (13, 12)

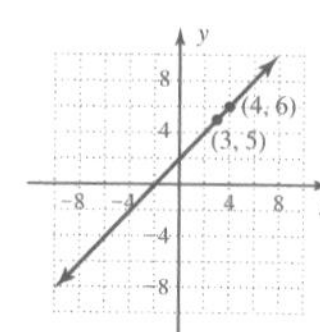

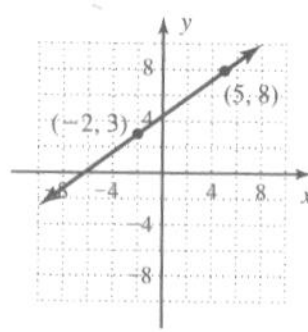

47. $m = \frac{4}{3}$; (5, −1) and (1, −9)

48. $m = \frac{8}{3}$; (−9, −6) and (3, 16)

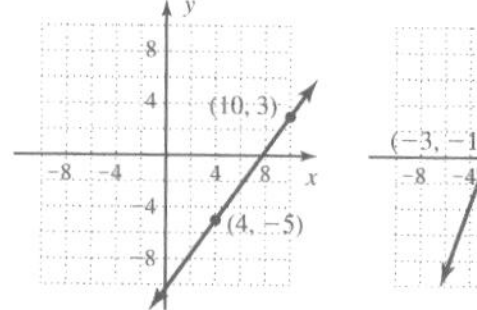

49. $m = \frac{15}{2} = \frac{7.5}{1}$; (2, −0.5) and (4, 14.5)

50. slope is undefined; (0, −6) and (0, 3)

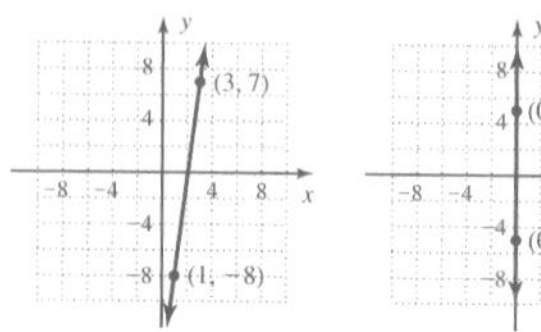

51. $m = 0$; (6, −8) and (6, 3)

52. $m = -3$; (−4, 2) and (−1, −7)

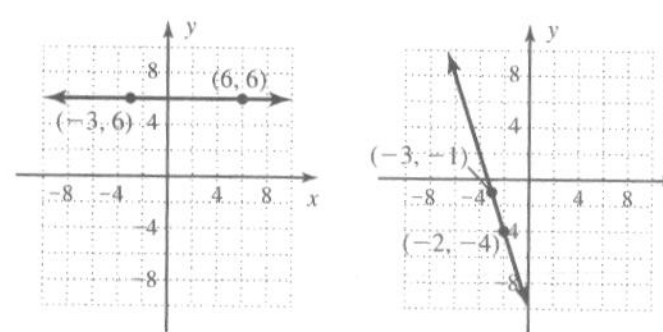

53. parallel
54. neither
55. neither
56. perpendicular
57. parallel
58. neither
59. not a right triangle
60. right triangle
61. not a right triangle
62. right triangle
63. right triangle
64. not a right triangle

Find the midpoint of each segment.

71. 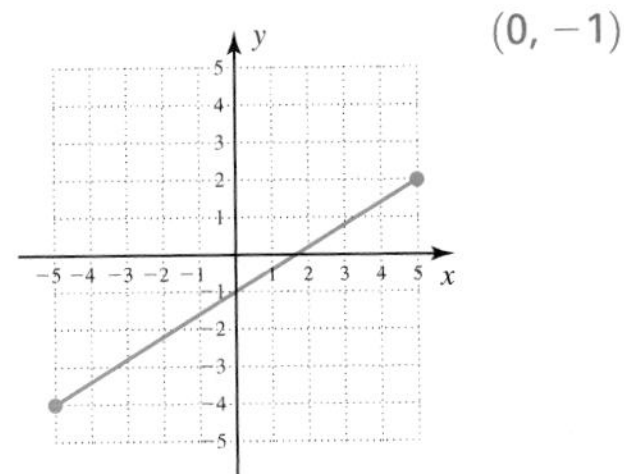 (0, −1)

72. 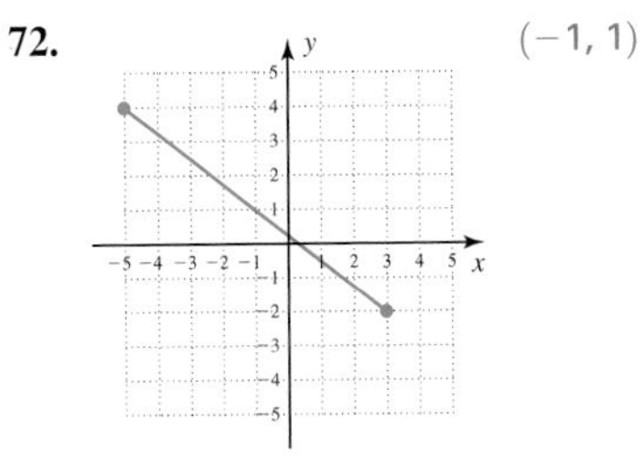 (−1, 1)

Find the center of each circle. Assume the endpoints of the diameter are lattice points.

73. 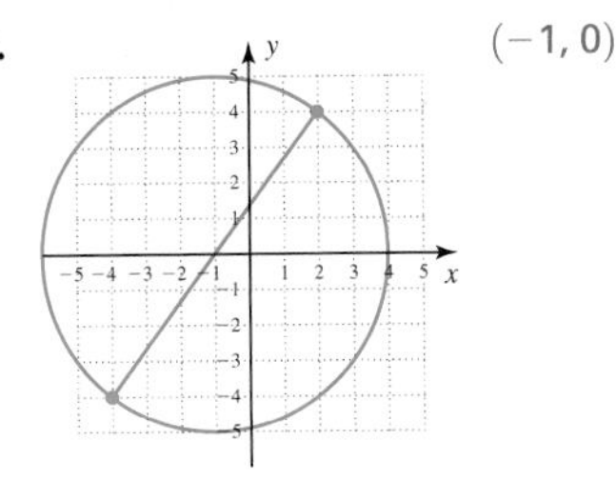 (−1, 0)

74. 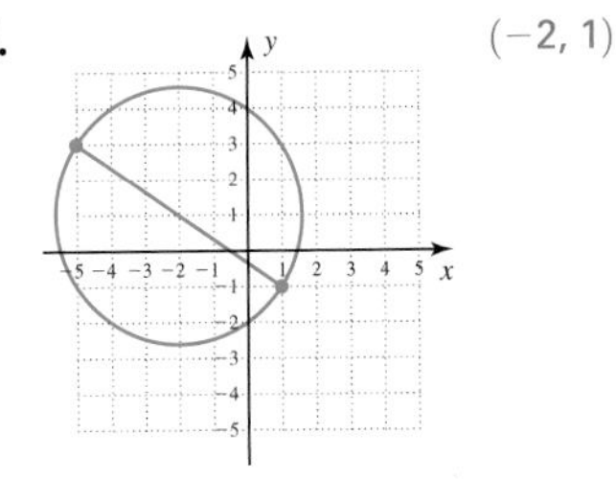 (−2, 1)

75. Use the distance formula to find the length of the line segment in Exercise 71. $2\sqrt{34}$

76. Use the distance formula to find the length of the line segment in Exercise 72. 10

77. Use the distance formula to find the length of the diameter for the circle in Exercise 73. 10

78. Use the distance formula to find the length of the diameter for the circle in Exercise 74. $2\sqrt{13}$

WORKING WITH FORMULAS

79. Human Life Expectancy: $L = 0.11T + 74.2$

The average number of years that human beings live has been steadily increasing over the years due to better living conditions and improved medical care. This relationship is modeled by the formula shown, where L is the average life expectancy and T is number of years since 1980. (a) What was the life expectancy in the year 2000? (b) In what year will average life expectancy reach 77.5 yr? **a.** 76.4 yr **b.** 2010

80. Interest earnings: $I = \left(\frac{7}{100}\right)(5000)T$

If $5000 dollars is invested in an account paying 7% simple interest, the amount of interest earned is given by the formula shown, where I is the interest and T is the time in years. (a) How much interest is earned in 5 yr? (b) How much is earned in 10 yr? (c) Use the two points (5 yr, interest) and (10 yr, interest) to calculate the slope of this line. What do you notice? **a.** $1750 **b.** $3500 **c.** $350 Interest increases $350 per year

APPLICATIONS

81. **a.** $3500
b. 5 yr

81. Business depreciation: A business purchases a copier for $8500 and anticipates it will depreciate in value $1250 per year.

a. What is the copier's value after 4 yr of use?

b. How many years will it take for this copier's value to decrease to $2250?

82. a. $95.50
b. 10 yr

82. Baseball card value: After purchasing an autographed baseball card for $85, its value increases by $1.50 per year.

a. What is the card's value 7 yr after purchase?

b. How many years will it take for this card's value to reach $100?

83. a. $2349
b. 2011

83. Cost of college: For the years 1980 to 2000, the cost of tuition and fees per semester (in constant dollars) at a public 4-yr college can be approximated by the equation $y = 144x + 621$, where y represents the cost in dollars and $x = 0$ represents the year 1980. Use the equation to find: (a) the cost of tuition and fees in 1992 and (b) the year this cost will exceed $5000.

Source: 2001 *New York Times Almanac*, p. 356

84. Female physicians: In 1960 only about 7% of physicians were female. Soon after, this percentage began to grow dramatically. For the years 1980 to 2002, the percentage of physicians that were female can be approximated by the equation $y = 0.72x + 11$, where y represents the percentage (as a whole number) and $x = 0$ represents the year 1980. Use the equation to find: (a) the percentage of physicians that were female in 1992 and (b) the projected year this percentage will exceed 30%. 2006 20%

Source: Data from the 2004 *Statistical Abstract of the United States*, Table 149

85. Decrease in smokers: For the years 1980 to 2002, the percentage of the U.S. adult population who were smokers can be approximated by the equation $y = -\frac{7}{15}x + 32$, where y represents the percentage of smokers (as a whole number) and $x = 0$ represents 1980. Use the equation to find: (a) the percentage of adults who smoked in the year 2000 and 23% (b) the year the percentage of smokers is projected to fall below 20%. 2005

Source: Statistical Abstract of the United States, various years

86. Temperature and cricket chirps: Biologists have found a strong relationship between temperature and the number of times a cricket chirps. This is modeled by the equation $T = \frac{N}{4} + 40$, where N is the number of times the cricket chirps per minute and T is the temperature in Fahrenheit. Use the equation to find: (a) the outdoor temperature if the cricket is chirping 48 times per minute and (b) the number of times a cricket chirps 52°F if the temperature is 70°. 120 chirps per minute

WRITING, RESEARCH, AND DECISION MAKING

87. Scientists often measure extreme temperatures in degrees Kelvin rather than the more common Fahrenheit or Celsius. Use the Internet, an encyclopedia, or another resource to investigate the linear relationship between these temperature scales. In your research, try to discover the significance of the numbers −273, 0, 32, 100, 212, and 373. Answers will vary.

88. a. $m = 12$
b. the additional charge for every mile per hr over 65 mph
c. Answers will vary.

88. In many states, there is a set fine for speeding with an additional amount charged for every mile per hour over the speed limit. For instance, if the set fine is $40 and the additional charge is $12, the *fine for speeding formula* would be $F = 12(S - 65) + 40$, where F is the set fine and S is your speed (assuming a speed limit of 65 mph). (a) What is the slope of this line? (b) Discuss the meaning of the slope in this context and (c) contact your nearest Highway Patrol office and ask about the speeding fines in your area.

EXTENDING THE CONCEPT

89. Let m_1, m_2, m_3, and m_4 be the slopes of lines L_1, L_2, L_3, and L_4, respectively. Which of the following statements is true?

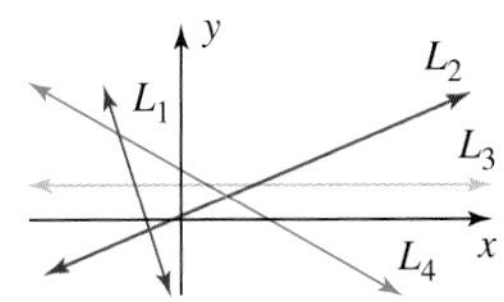

a. $m_4 < m_1 < m_3 < m_2$

b. $m_3 < m_2 < m_4 < m_1$

c. $m_3 < m_4 < m_2 < m_1$

d. $m_1 < m_3 < m_4 < m_2$

e. $m_1 < m_4 < m_3 < m_2$ e

90. Given the lines $4y + 2x = -5$ and $3y + ax = -2$ are perpendicular, what is the value of a? $a = -6$

MAINTAINING YOUR SKILLS

91. (1.1) Simplify the equation, then solve. Check your answer by substitution:
$3x^2 - 3 + 4x + 6 = 4x^2 - 3(x + 5)$

92. (R.7) Identify the following formulas:
$P = 2L + 2W$ $V = LWH$
$V = \pi r^2 h$ $C = 2\pi r$

93. (1.1) How many gallons of a 35% brine solution must be mixed with 12 gal of a 55% brine solution in order to get a 45% solution?

94. (1.1) Two boats leave the harbor at Lahaina, Maui, going in opposite directions. One travels at 15 mph and the other at 20 mph. How long until they are 70 mi apart?

95. (1.3) Solve the rational equation and state all excluded values.
$$\frac{1}{x - 5} = \frac{-10}{x^2 - 2x - 15} + 1$$

96. (1.5) Solve using the quadratic formula. Give answers in both exact and approximate form.
$3x^2 + 5x = 9$

91. $x = 9, x = -2$
92. perimeter of a rectangle
volume of a rectangular prism
volume of a cylinder
circumference of a circle
93. 12 gal
94. 2 hr
95. $x \neq 5, x \neq -3; x = 7$ or $x = -4$
96. $x = \frac{-5 \pm \sqrt{133}}{6}; x \approx 1.09$ or $x \approx -2.76$

2.2 Relations, Functions, and Graphs

LEARNING OBJECTIVES

In Section 2.2 you will learn how to:

A. State a relation in mapping notation and ordered pair form
B. Graph a relation
C. Identify functions and state their domain and range
D. Use function notation

INTRODUCTION

In this section we introduce one of the most central ideas in mathematics—the concept of a function. A study of functions helps to establish the cause-and-effect relationship that is so important to using mathematics as a modeling and decision-making tool. In addition, the study will help to unify and expand on many ideas that are already familiar to you.

POINT OF INTEREST

The definition of a function has gone through a long, evolutionary process. Although the ancient Babylonians might be credited with the first "definition by illustration" in their use of mathematical tables, more definitive ideas seem to have originated around the time of René Descartes (~1638). Some of the most famous names in mathematics are associated with further refinement of the concept, including Gottfried von Leibniz, Johann Bernoulli, Leonhard Euler, Joseph-Louis Lagrange, Lejeune Dirichlet, and Georg Cantor.

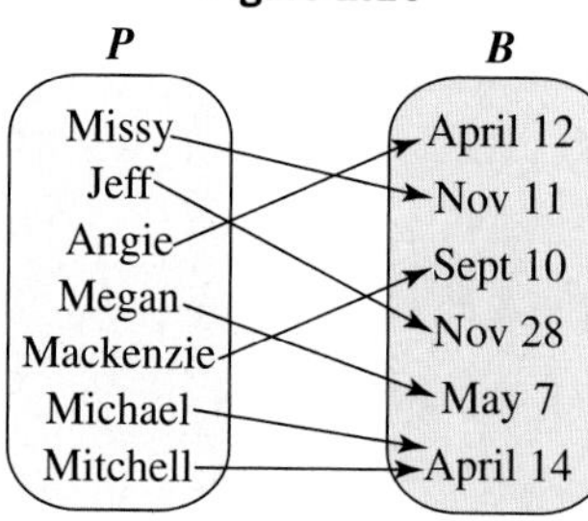

Figure 2.16

A. Relations and Mapping Notation

In the most general sense, a **relation** is simply a correspondence between two sets. Relations can be represented in many different ways and may even be very "unmathematical," like the relation shown here between a set of people and the set of their corresponding birthdays. If P represents the set of people p, and B represents the set of birthdays b, the following four statements are equivalent: (1) the birthday relation maps set P to set B, (2) P corresponds to B, (3) $P \rightarrow B$, and (4) (p, b). Statement 3 and the diagram in Figure 2.16 are given in **mapping notation,** while statement 4 is written in **ordered pair form.** From a purely practical standpoint, we note that while it is possible for two different people to share the same birthday, it is quite impossible for the same person to have two different birthdays. This observation will help mark the difference between a relation and a function.

The bar graph in Figure 2.17 is also an example of a relation. The graph relates the birth *weight* of five children to their *length* at birth. In ordered pair form the relation is (6, 19), (7.5, 20), (6.8, 18), (7.2, 19.5), and (5.5, 17). In this form, the set of all first coordinates (in this case the weights), is called the **domain** of the relation. The set of all second coordinates (corresponding to domain members) is called the **range.**

Figure 2.17

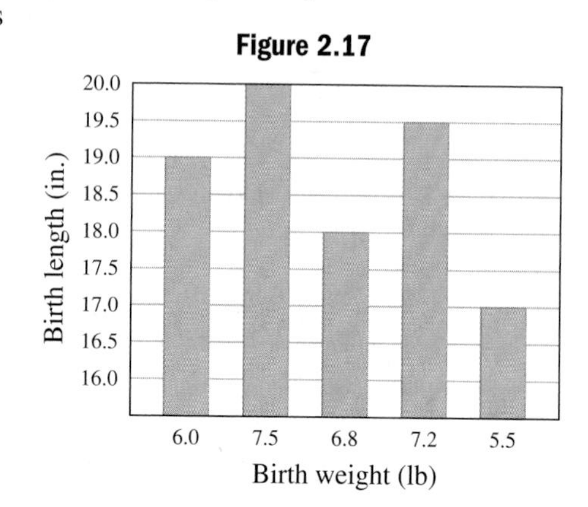

EXAMPLE 1 Represent the relation from Figure 2.17 in mapping notation, then state its domain and range.

Solution: Let W represent weight and L represent length. The mapping $W \rightarrow L$ gives the diagram shown here. The domain (in order of appearance) is the set {6, 7.5, 6.8, 7.2, 5.5}, and the range is {19, 20, 18, 19.5, 17}.

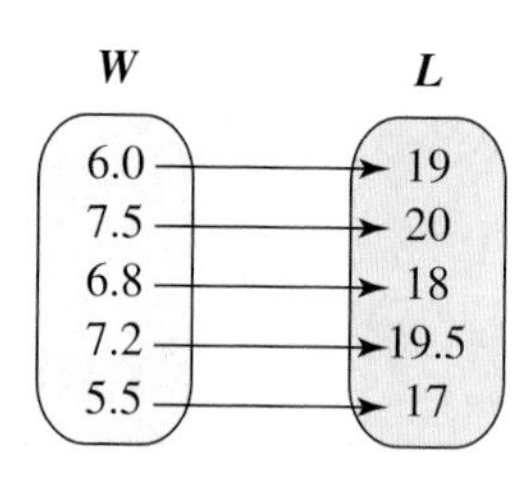

NOW TRY EXERCISES 7 THROUGH 12

A relation can also be given in **equation form.** The equation $y = x - 1$ gives a relation where each y-value is one less than the corresponding x-value. The equation $x = |y|$ states a relation where the absolute value of y gives a corresponding value of x. Using the domain values {0, 1, 2, 3, 4} for illustration, each can also be written in ordered pair form. For $y = x - 1$ we have $\{(0, -1), (1, 0), (2, 1), (3, 2), (4, 3)\}$. For $x = |y|$, the result is $\{(0, 0), (1, -1), (1, 1), (2, -2), (2, 2), (3, -3), (3, 3), (4, -4), (4, 4)\}$.

B. The Graph of a Relation

If relations are defined by a set of ordered pairs, the graph of each relation is simply the plotted points. The graph of a relation in equation form is the set of *all* ordered pairs (x, y) that satisfy the equation. While we often use a few select points to determine the general shape of a graph, the complete graph consists of all ordered pairs that satisfy the equation—including any that are rational or irrational.

EXAMPLE 2 Graph the relations $y = x - 1$ and $x = |y|$ using the ordered pairs given earlier.

Solution: For $y = x - 1$, we plot the points then connect them with a straight line, with the result seen in Figure 2.18. For $x = |y|$, the plotted points form a V-shaped graph made up of two directed line segments, opening to the right (see Figure 2.19).

Figure 2.18

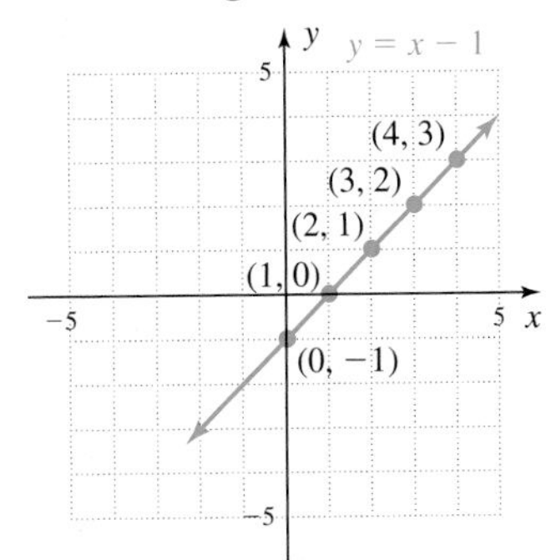

Figure 2.19

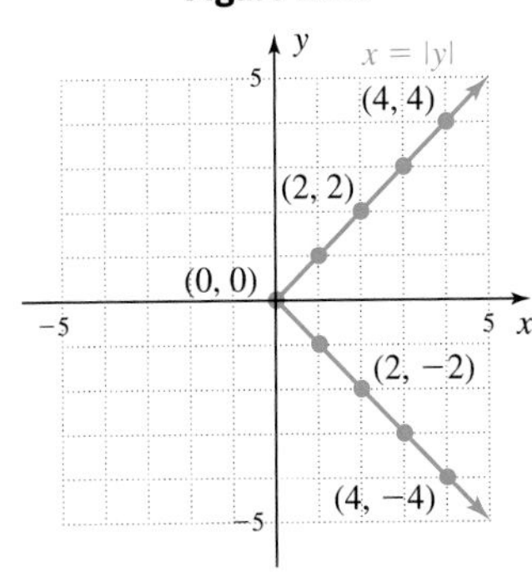

NOW TRY EXERCISES 13 THROUGH 16

Actually, a majority of graphs cannot be drawn using only a straight line or directed line segments. In these cases, we rely on a "sufficient number" of points to outline the basic shape of the graph, then connect the points *with a line or smooth curve,* as indicated by any patterns formed. As your experience with graphing increases, this "sufficient number of points" tends to get very small as you learn to anticipate what the graph of a given relation should look like.

EXAMPLE 3 Graph the following relations by completing the tables shown for $-4 \le x \le 4$.

a. $y = x^2 - 2x$ **b.** $y = \sqrt{9 - x^2}$ **c.** $x = y^2$

Solution: For each relation, we input each x-value in turn and determine the related y-output(s), if they exist. Results are entered in the table and used to draw the graph. Remember to scale the axes (if needed) to comfortably fit the ordered pairs.

a. $y = x^2 - 2x$

x	y	(x, y) Ordered Pairs
−4	24	(−4, 24)
−3	15	(−3, 15)
−2	8	(−2, 8)
−1	3	(−1, 3)
0	0	(0, 0)
1	−1	(1, −1)
2	0	(2, 0)
3	3	(3, 3)
4	8	(4, 8)

Figure 2.20

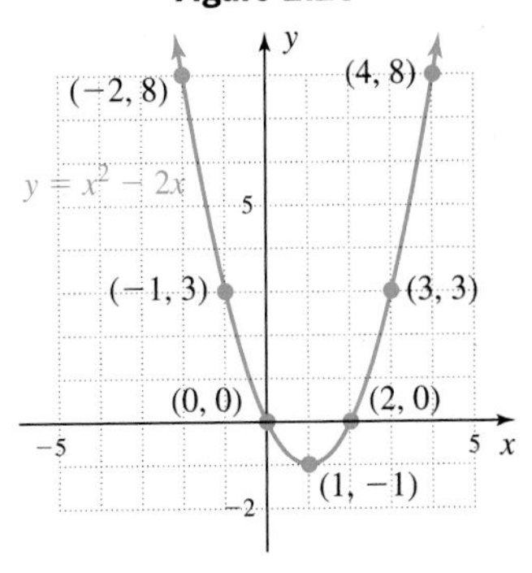

This is a fairly common graph (Figure 2.20), called a **vertical parabola.** Although (−4, 24) and (−3, 15) cannot be plotted on this grid, the arrowheads indicate an infinite extension of the graph, which will include these points.

b. $y = \sqrt{9 - x^2}$

x	y	(x, y) Ordered Pairs
−4	not real	—
−3	0	(−3, 0)
−2	$\sqrt{5}$	$(-2, \sqrt{5})$
−1	$2\sqrt{2}$	$(-1, 2\sqrt{2})$
0	3	(0, 3)
1	$2\sqrt{2}$	$(1, 2\sqrt{2})$
2	$\sqrt{5}$	$(2, \sqrt{5})$
3	0	(3, 0)
4	not real	—

Figure 2.21

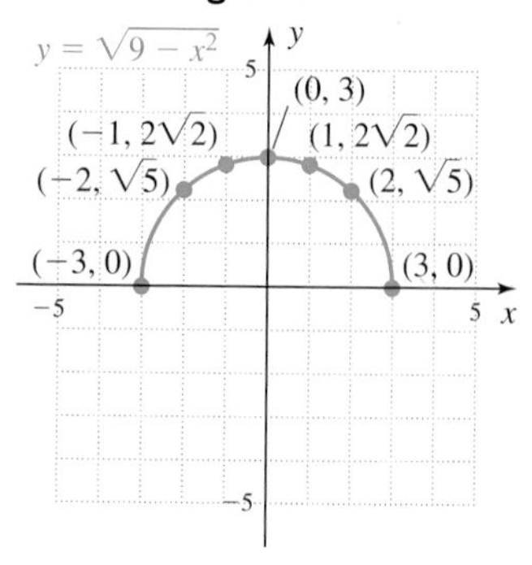

This is the graph of a **semicircle** (Figure 2.21). The points with irrational coordinates were graphed by estimating their location. Note that when $x < -3$ or $x > 3$, the relation $y = \sqrt{9 - x^2}$ does not represent a real number. For example, when $x = -4$ we have $\sqrt{9 - (-4)^2} = \sqrt{-7}$, which does not represent a point in the (real valued) xy-plane and cannot be a part of the graph.

c. Similar to $x = |y|$, the relation $x = y^2$ is defined only for $x \geq 0$ since y^2 is always positive ($-1 = y^2$ has no real solutions). In addition, we reason that each nonzero x-value will likewise correspond to two y-values. For example given $x = 4$, $(4, -2)$ and $(4, 2)$ are both solutions.

$y^2 = x$

x	y	(x, y) Ordered Pairs
−4	not real	—
−3	not real	—
−2	not real	—
−1	not real	—
0	0	(0, 0)
1	−1, 1	(1, −1) and (1, 1)
2	$-\sqrt{2}, \sqrt{2}$	$(2, -\sqrt{2})$ and $(2, \sqrt{2})$
3	$-\sqrt{3}, \sqrt{3}$	$(3, -\sqrt{3})$ and $(3, \sqrt{3})$
4	−2, 2	(4, −2) and (4, 2)

Figure 2.22

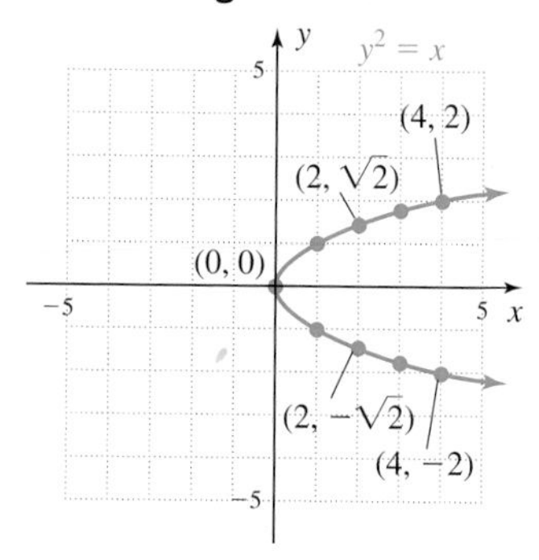

This is the graph of a **horizontal parabola** (Figure 2.22).

NOW TRY EXERCISES 17 THROUGH 24

C. Functions, Graphs, Domain, and Range

There is a certain type of relation that merits further attention. A **function** is a relation where each element of the domain corresponds to exactly one element of the range. In other words, for each first coordinate or input value, there is only one possible second coordinate or output. This definite and certain assignment of one input to a unique output is what makes functions invaluable as a mathematical tool.

FUNCTIONS

A *function* is a relation, rule, or equation that pairs each element from the *domain* with exactly one element of the *range*.

WORTHY OF NOTE

Although Michael and Mitchell share the same birthday (Figure 2.16) this does not violate the definition of a function since each of them has *only one* birthday. A good way to view the distinction is to consider a mail carrier. It is possible for the carrier to put more than one letter into the same mailbox (more than one x going to the same y), but quite impossible for the carrier to place the same letter in two different boxes (one x going to two y's).

If the relation is defined by a mapping, we need only check that each element of the domain is mapped to exactly one element of the range. This is indeed that case for the mapping $P \to B$ from Figure 2.16, where we saw that each person corresponded to only one birthday, and that it was impossible for one person to be born on two different days. For the relation $x = |y|$ shown in Figure 2.19, each element of the domain except zero is paired with *more than one* element of the range. The relation $x = |y|$ is *not* a function.

EXAMPLE 4 Three different relations are given in mapping notation below. Determine whether each relation is a function.

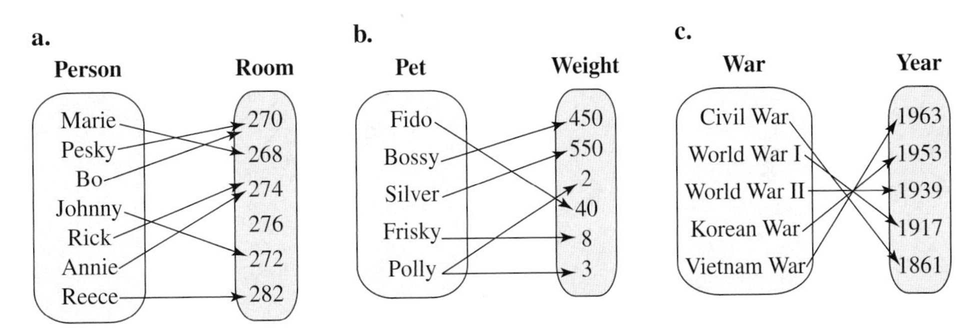

Solution: Relation (a) is a function, since each person corresponds to exactly one room. This relation pairs math professors with their respective office numbers. Notice that while two people can be in one office, it is impossible for one person to physically be in two different offices. Relation (b) is not a function, since we cannot tell whether Polly the Parrot weighs 2 lb or 3 lb (one element of the domain is paired with two elements of the range). Relation (c) is a function, where each major war is paired with the year it began.

NOW TRY EXERCISES 25 THROUGH 28

If the relation is defined by a set of ordered pairs or a set of individual and distinct plotted points, we need only check that no two points have the same first coordinate with a different second coordinate.

EXAMPLE 5 Two relations named f and g are given; f is stated as a set of ordered pairs, while g is given as a set of plotted points. Determine whether each is a function.

f: $(-3, 0), (1, 4), (2, -5), (4, 2), (-3, -2), (3, 6), (0, -1),$
$(4, -5)$, and $(6, 1)$

Solution: The relation f is not a function, since -3 is paired with two different outputs. The relation g shown in the figure *is* a function. Each input corresponds to exactly one output, otherwise one point would be directly above the other and have the same first coordinate.

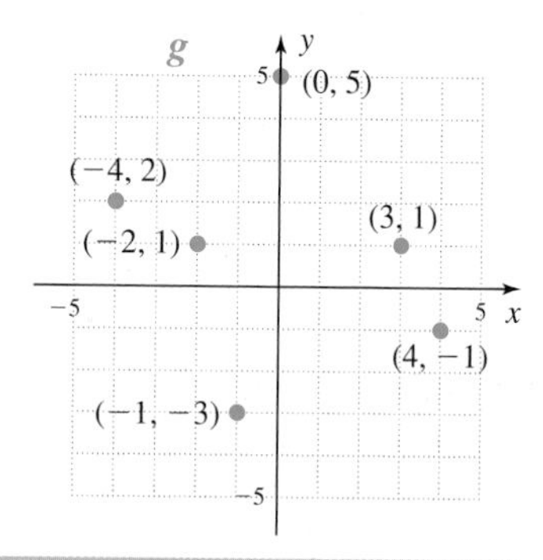

NOW TRY EXERCISES 29 THROUGH 36

The graphs from Example 2 also offer a great deal of insight into the definition of a function. Figure 2.23 shows the line $y = x - 1$ with emphasis on the plotted points (4, 3) and $(-3, -4)$. The vertical movement shown from the x-axis to a point on the graph illustrates *the pairing of a given x value with one related y value*. Note the vertical line shows *only one related y value*. Figure 2.24 gives the graph of $x = |y|$, highlighting the points (4, 4) and $(4, -4)$. The vertical movement shown here branches in two directions, associating one x-value with more than one y-value. The relation $y = x - 1$ is a function, the relation $x = |y|$ is not.

Figure 2.23

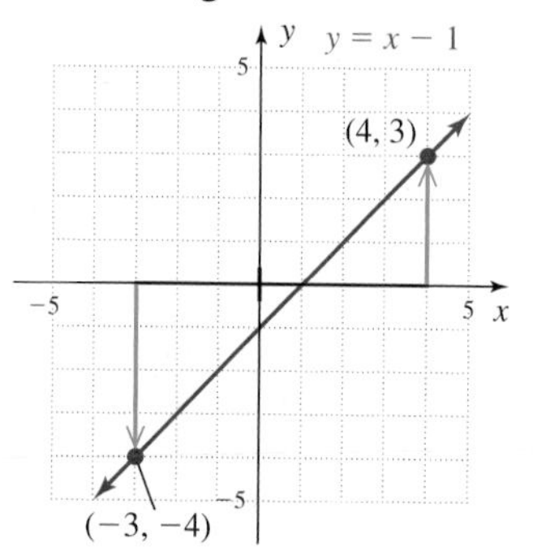

Figure 2.24

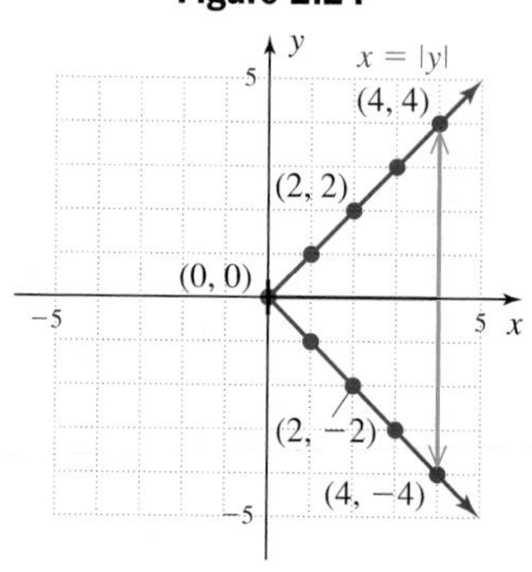

This "vertical connection" of a location on the x-axis to a point on the graph can be generalized into a **vertical line test for functions.**

> **VERTICAL LINE TEST**
> If every vertical line intersects the graph of a relation in at most one point, the relation is a function.

EXAMPLE 6 Use the vertical line test to determine if any of the relations graphed in Example 3 are functions.

Solution: Draw a vertical line on each coordinate grid (shown in blue), then mentally shift the line to the left and right as shown in Figures 2.25, 2.26, and 2.27. In Figures 2.25 and 2.26, every vertical line intersects the graph only once, indicating both $y = x^2 - 2x$ and $y = \sqrt{9 - x^2}$ are functions. In Figure 2.27, a vertical line intersects the graph twice for any $x > 0$. The relation $x = y^2$ is not a function.

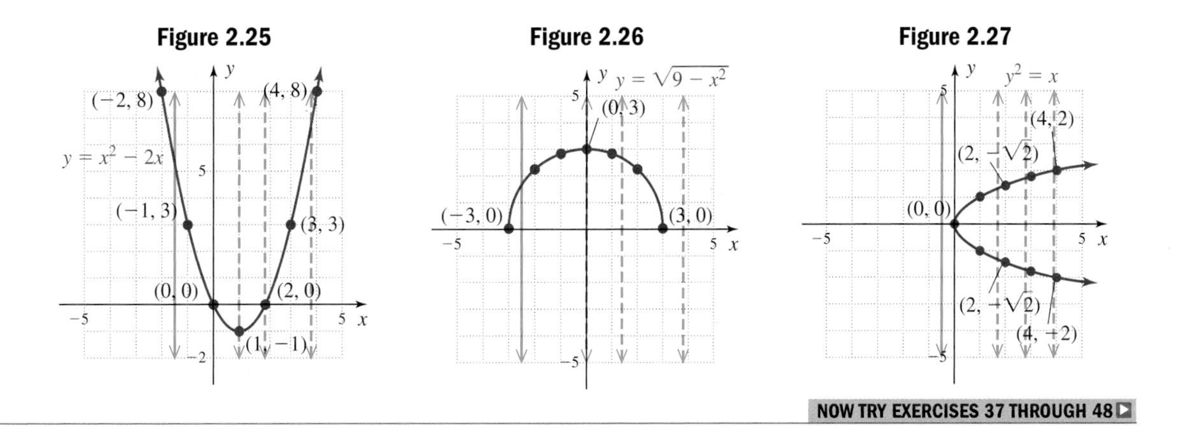

NOW TRY EXERCISES 37 THROUGH 48

Another interesting relation is the **absolute value relation,** defined by the equation $y = |x|$. It is "close cousin" to linear relations, because the two branches of the graph are actually linear. A table of values (Table 2.1) for $y = |x|$ and the corresponding graph are shown (Figure 2.28). Note the result is a V-shaped graph opening upward, with branches formed by the positive portion of $y = x$ on the right and $y = -x$ on the left. The "nose" of the graph (at the origin) is called the **vertex.**

Table 2.1

x Inputs	*y* Outputs	(*x*, *y*)
−4	$y = \lvert -4 \rvert = 4$	(−4, 4)
−3	$y = \lvert -3 \rvert = 3$	(−3, 3)
−2	$y = \lvert -2 \rvert = 2$	(−2, 2)
	and so on	

Figure 2.28

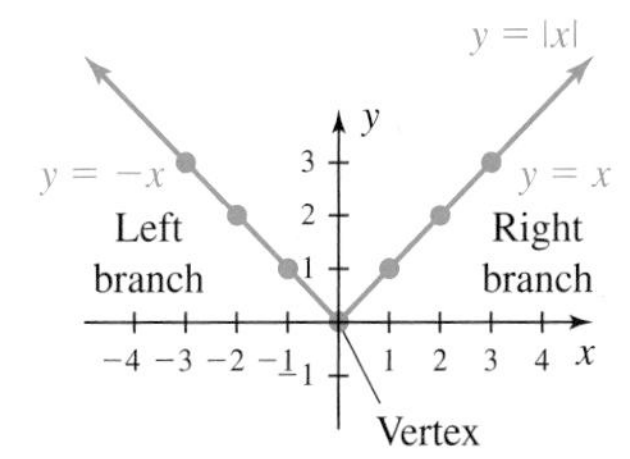

Domain and Vertical Boundary Lines

With practice, the domain and range of a relation can be determined from its graph. In addition to its use as a function versus nonfunction test, a vertical line can help. Consider the graph of $y = \sqrt{x}$ shown in Figure 2.29.

From left to right, a vertical line will not intersect the graph until $x = 0$, and then intersects the graph for all values $x \geq 0$. These **vertical boundary lines** show the domain is $x \in [0, \infty)$ and that the relation is a function. For the graph of $y = |x|$ shown in Figure 2.28, a vertical line will always intersect the graph or its infinite extension. The domain is $x \in (-\infty, \infty)$. Using

Figure 2.29

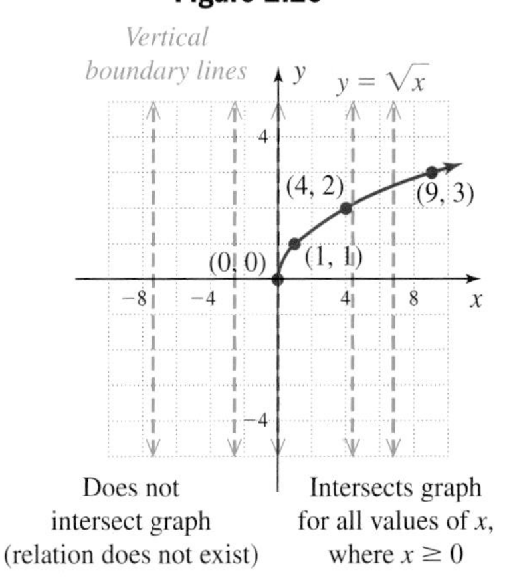

vertical boundary lines in this way shows the domain of $x = |y|$ (Figure 2.24) is $y \in [0, \infty)$, while the domain of $y = x - 1$ (Figure 2.23) is $x \in (-\infty, \infty)$.

Range and Horizontal Boundary Lines

The range of a relation can be found using a **horizontal boundary line,** since it will associate a value on the y-axis with a point on the graph (if it exists). Simply move the line upward or downward until you determine the graph will extend infinitely and always intersect the line, or will no longer intersect the line. This will give you the boundaries of the range. Mentally applying this idea to the graph of $y = \sqrt{x}$ (Figure 2.29) shows the range is $y \in [0, \infty)$. Although shaped very differently, a horizontal boundary line shows the range of $y = |x|$ (Figure 2.28) is also $y \in [0, \infty)$.

EXAMPLE 7 Determine the domain and range of the functions from Examples 3(a) and 3(b).

Solution: For $y = x^2 - 2x$, Figure 2.30 shows a vertical line will intersect the graph or its extension anywhere it is placed. The domain is $x \in (-\infty, \infty)$. Figure 2.31 shows a horizontal line will intersect the graph only for values of y that are greater than or equal to -1. The range is $y \in [-1, \infty)$. Recall that in interval notation, the smaller endpoint is always written first (-1 before positive infinity).

Figure 2.30

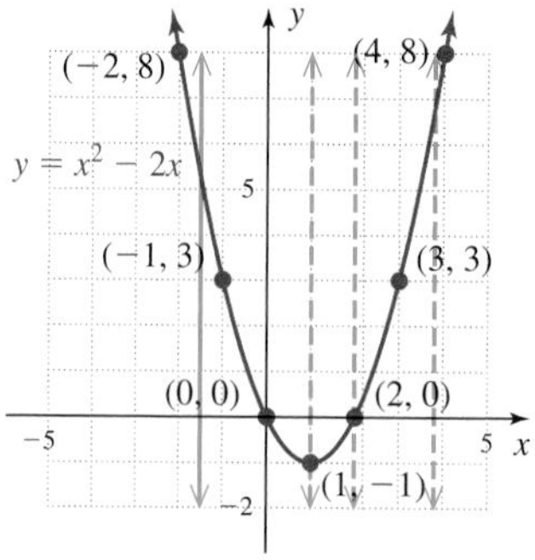

Figure 2.31

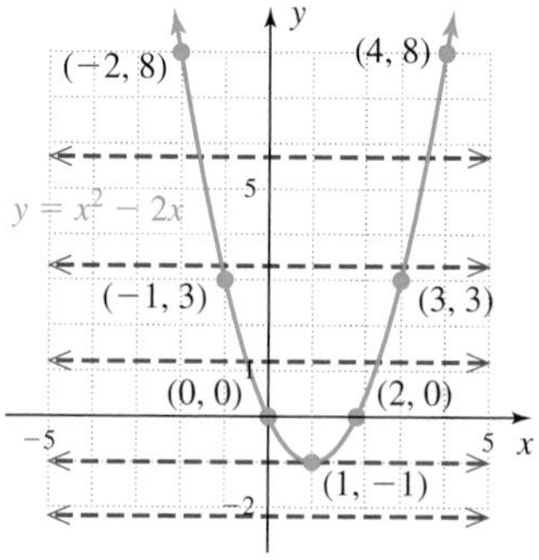

When the same technique is applied to $y = \sqrt{9 - x^2}$, we find the domain is $x \in [-3, 3]$ and the range is $y \in [0, 3]$. See Figures 2.32 and 2.33, respectively.

Figure 2.32

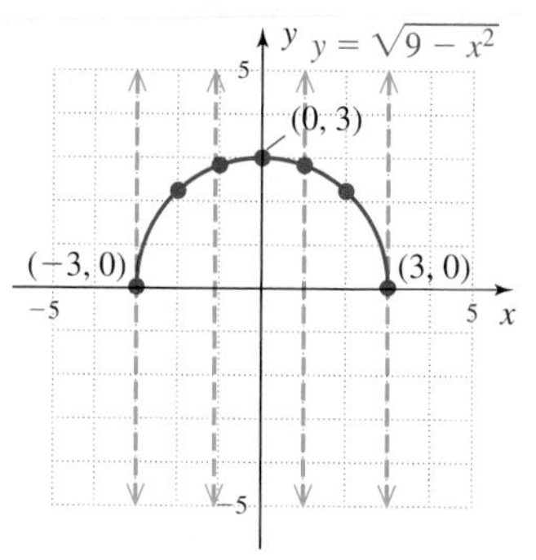

Figure 2.33

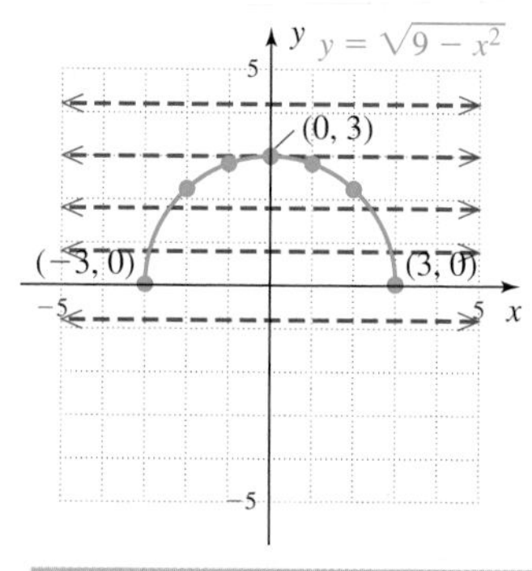

NOW TRY EXERCISES 49 THROUGH 60

Implied Domains

When stated in equation form, the domain of a relation is implicitly given by the expression used to define it. This **implied domain** is the set of all real numbers x for which the relation is defined. For $y = x - 1$, the domain is $x \in \mathbb{R}$. Since the absolute value of a number is always positive, the domain of $x = |y|$ is the set of nonnegative real numbers: $x \in [0, \infty)$.

If the function involves a rational expression, the domain will exclude any input that causes a denominator of zero. If the function involves a square root expression, the domain will exclude inputs that create a negative radicand. In the latter case, it is probably more correct to say that the domain is implicitly determined by the set of real numbers x for which the output y is also real.

EXAMPLE 8 Determine the domain of the functions given.

a. $y = \frac{3}{x + 2}$ **b.** $y = \sqrt{2x + 3}$

c. $y = \frac{x - 5}{x^2 - 9}$ **d.** $y = x^2 - 5x + 7$

Solution:

a. By inspection, we note that an x-value of -2 gives a zero denominator and must be excluded. The domain is $x \in (-\infty, -2) \cup (-2, \infty)$.

b. Since the radicand must be nonnegative, we solve the inequality $2x + 3 \geq 0$, giving $x \geq \frac{-3}{2}$. The domain is $x \in [\frac{-3}{2}, \infty)$.

c. To prevent division by zero, inputs of -3 and 3 must be excluded (set $x^2 - 9 = 0$ and solve by factoring). The domain is $\{x|x \in \mathbb{R}, x \neq -3, 3\}$. Note that $x = 5$ *is in the domain* since $\frac{0}{16} = 0$ is defined.

d. Since squaring a number and multiplying a number by a constant are defined for all reals, the domain is $x \in (-\infty, \infty)$.

NOW TRY EXERCISES 61 THROUGH 78

D. Function Notation

When studying functions, the relationship between input and output values is an important one. Think of a function as a simple machine, which can *process inputs* using a stated sequence of operations, then deliver a single output. The inputs are x-values, the program f performs the operations on x, and y is the resulting output (see Figure 2.34 In this case we say, "the value of y depends on the value of x," or "y is a function of x." Notationally, we write "y is a function of x" as $y = f(x)$ using **function notation.** You are already familiar with letting a variable represent a number. Here we do something quite different, as the letter f is used to represent *a sequence of operations to be performed on x*. This notation is a good model of how the function machine operates, enabling us to evaluate functions while keeping track of the related input and output values. Consider the function $y = \frac{x}{2} + 1$, which we will now write as $f(x) = \frac{x}{2} + 1$

Figure 2.34

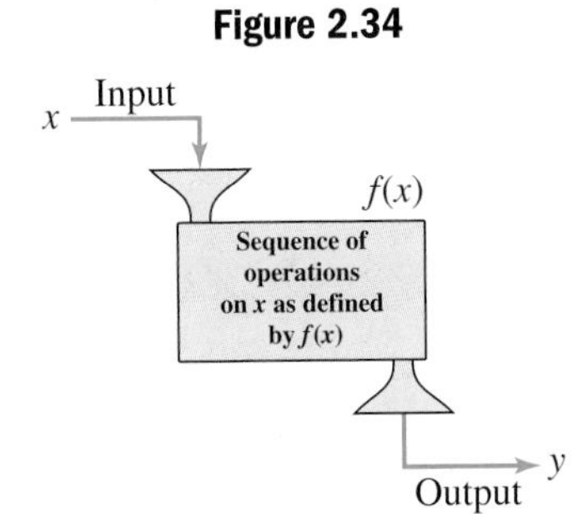

[since $y = f(x)$]. In words the function says, "divide inputs by 2, then add 1." To evaluate the function at $x = 4$ (Figure 2.35) we have:

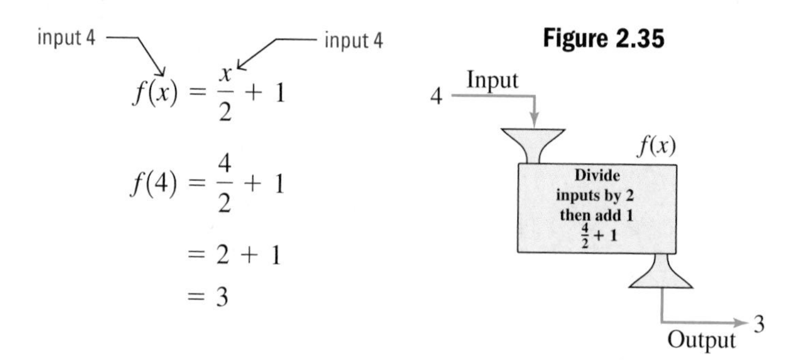

$$f(x) = \frac{x}{2} + 1$$
$$f(4) = \frac{4}{2} + 1$$
$$= 2 + 1$$
$$= 3$$

Figure 2.35

Instead of saying, ". . . when $x = 4$, the value of the function is 3," we simply say "f of 4 is 3," or write $f(4) = 3$. Note that the ordered pair (4, 3) is equivalent to $(4, f(4))$.

CAUTION

Although $f(x)$ is the favored notation for a "function of x," other letters can also be used. For example, $g(x)$ and $h(x)$ also denote functions of x, where g and h might represent a different sequence of operations. It is also important to remember that these represent *function values* and not the product of two variables: $f(x) \neq f \cdot (x)$.

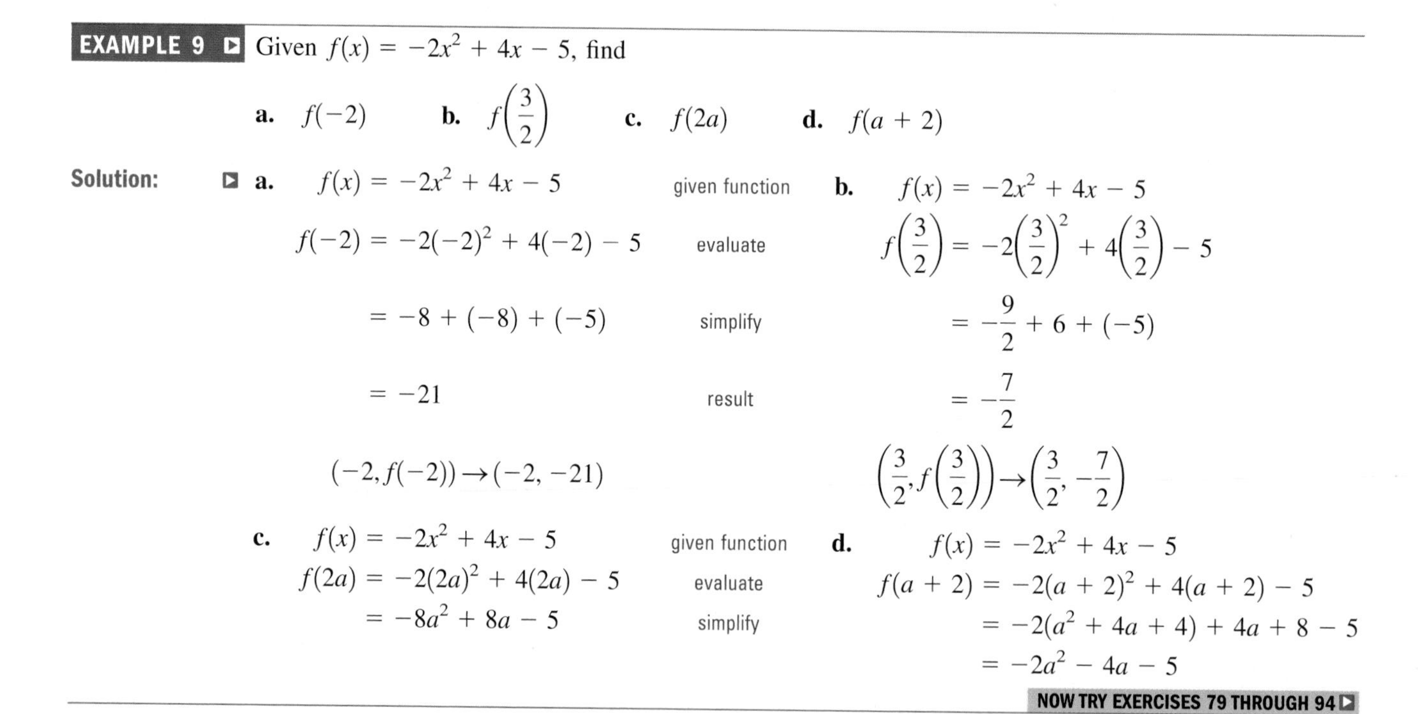

EXAMPLE 9 Given $f(x) = -2x^2 + 4x - 5$, find

a. $f(-2)$ **b.** $f\left(\frac{3}{2}\right)$ **c.** $f(2a)$ **d.** $f(a + 2)$

Solution:

a.
$$f(x) = -2x^2 + 4x - 5 \quad \text{given function}$$
$$f(-2) = -2(-2)^2 + 4(-2) - 5 \quad \text{evaluate}$$
$$= -8 + (-8) + (-5) \quad \text{simplify}$$
$$= -21 \quad \text{result}$$
$$(-2, f(-2)) \rightarrow (-2, -21)$$

b.
$$f(x) = -2x^2 + 4x - 5$$
$$f\left(\frac{3}{2}\right) = -2\left(\frac{3}{2}\right)^2 + 4\left(\frac{3}{2}\right) - 5$$
$$= -\frac{9}{2} + 6 + (-5)$$
$$= -\frac{7}{2}$$
$$\left(\frac{3}{2}, f\left(\frac{3}{2}\right)\right) \rightarrow \left(\frac{3}{2}, -\frac{7}{2}\right)$$

c.
$$f(x) = -2x^2 + 4x - 5 \quad \text{given function}$$
$$f(2a) = -2(2a)^2 + 4(2a) - 5 \quad \text{evaluate}$$
$$= -8a^2 + 8a - 5 \quad \text{simplify}$$

d.
$$f(x) = -2x^2 + 4x - 5$$
$$f(a + 2) = -2(a + 2)^2 + 4(a + 2) - 5$$
$$= -2(a^2 + 4a + 4) + 4a + 8 - 5$$
$$= -2a^2 - 4a - 5$$

NOW TRY EXERCISES 79 THROUGH 94

Graphs are an important part of studying functions, and learning to read and interpret them correctly is a high priority. Part of the reason is this highlights and emphasizes the all-important input/output relationship that defines a function. In addition, reading graphs promotes a better understanding of function notation. Here we hope to firmly establish that statements like $f(-2) = 5$, $(-2, 5)$, and $f(x) = 5$ when $x = -2$ are synonymous.

EXAMPLE 10 For the functions $f(x)$ and $g(x)$ whose graphs are shown in Figures 2.36 and 2.37, (a) state the domain of the function, (b) evaluate the function at $x = 2$, (c) determine the value(s) of x for which the function value is 3, and (d) state the range of the function.

Figure 2.36

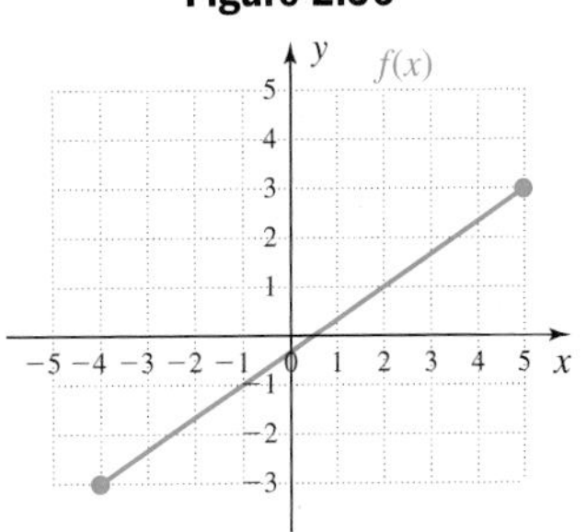

Figure 2.37

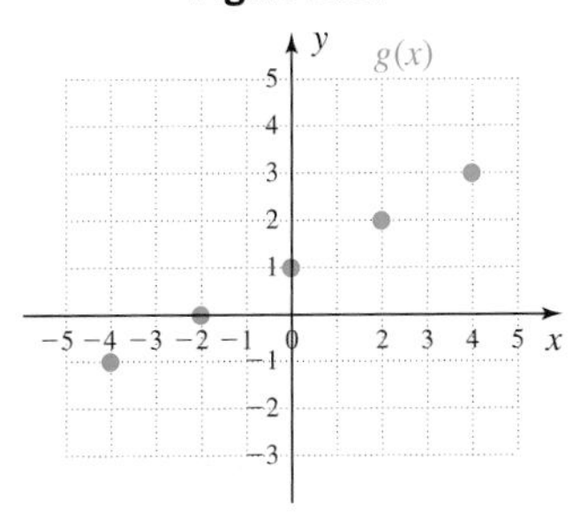

Solution: For $f(x)$,

a. The graph is a continuous line segment with endpoints at $(-4, -3)$ and $(5, 3)$, so we state the domain in interval notation. Using a vertical boundary line we note the smallest input is -4 and the largest is 5. The domain is $x \in [-4, 5]$.

b. The graph shows an input of $x = 2$ corresponds to $y = 1$: $f(2) = 1$ since $(2, 1)$ is a point on the graph.

c. For $f(x) = 3$ (or $y = 3$) the input value must be $x = 5$ since $(5, 3)$ is the point on the graph.

d. Using a horizontal boundary line, the smallest output value is -3 and the largest is 3. The range is $y \in [-3, 3]$.

For $g(x)$,

a. Since the graph is pointwise defined, we state the domain as the set of first coordinates: $D = \{-4, -2, 0, 2, 4\}$.

b. An input of $x = 2$ corresponds to $y = 2$: $g(2) = 2$ since $(2, 2)$ is on the graph.

c. For $g(x) = 3$ (or $y = 3$) the input value must be $x = 4$, since $(4, 3)$ is the point on the graph.

d. The range is the set of all second coordinates: $R = \{-1, 0, 1, 2, 3\}$.

NOW TRY EXERCISES 95 THROUGH 100

In many applications involving functions, the domain and range can be determined by the context or situation given.

EXAMPLE 11 Paul's 1993 Voyager has a 20-gal tank and gets 18 mpg. The number of miles he can drive (his range) depends on how much gas is in the tank. As a function we have $M(g) = 18g$, where $M(g)$ represents the total distance in miles and g represents the gallons of gas in the tank. Find the domain and range.

Solution: Begin evaluating at $x = 0$, since the tank cannot hold less than zero gallons. On a full tank the maximum range of the van is $20 \cdot 18 = 360$ miles or $M(g) \in [0, 360]$. Because of the tank's size, the domain is $g \in [0, 20]$.

NOW TRY EXERCISES 103 THROUGH 110

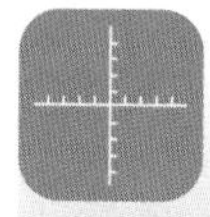

TECHNOLOGY HIGHLIGHT

Using LISTs and STATPLOT Features

The keystrokes shown apply to a TI-84 Plus model. Please consult your manual or our Internet site for other models.

The TI-84 Plus has the ability to plot individual points on the coordinate grid. Later in the text, we will make extensive use of this ability to look at applications involving real data. Consider the ordered pairs (−5, −7), (−2, −1), (0, 3), and (2, 7) from the equation $-2x + y = 3$. To graph these as individual points, we enter the domain values $\{-5, -2, 0, 2\}$ (the first coordinates of each point) in an ordered data list, and the corresponding range values $\{-7, -1, 3, 7\}$ (the second coordinates of each point) in a second data list.

1. *Clear old data:* We begin by making sure the data lists are *clear,* allowing for the input of new data. Press the STAT key and select option 4 "**ClrList.**" This places the **ClrList** command on the home screen. We tell the calculator which lists to clear by pressing 2nd 1 to indicate List1 (L1), then enter a comma using the , key and continue entering other lists we wish to clear: 2nd 2 , 2nd 3 ENTER will clear List1 (L1), List2 (L2), and List3 (L3).
2. *Enter new data:* We can now enter the domain values $\{-5, -2, 0, 2\}$ of the ordered pairs into list L1. Press the STAT key and select option **1:Edit.** This places the cursor in the first position of List1, where we simply enter the values in order: −5 ENTER −2 ENTER 0 ENTER 2 ENTER . Use the right arrow key ▶ to navigate over to List2, and enter the range values $\{-7, -1, 3, 7\}$ in sequence: −7 ENTER −1 ENTER 3 ENTER 7 ENTER .
3. *Display the data:* With the ordered pairs held in these lists, we can now display them on the coordinate grid. First press the Y= key and clear any old equations that might be there (navigate the cursor to any existing equation and press CLEAR). Then press 2nd Y= to access the **"STATPLOTS"** screen. With the cursor over option 1, press ENTER and be sure the options shown are highlighted. If you need to make any changes, navigate the cursor to the desired option and press ENTER . Since the ordered pairs will all "fit" on the standard screen, graph them by pressing ZOOM **6:ZStandard.** Notice that the points seem to lie on an imaginary line. You can double-check the address or location of each plotted point by pressing the TRACE button. A cursor will appear on one of the points and the coordinates of the point are given at the bottom of the screen. Walk the cursor to the other points by pressing the left ◀ and right ▶ arrow keys. The following exercises will give some useful practice with these keystrokes and ideas. Be sure to clear the old lists each time, or to overwrite each old entry with the new data and delete any old data that remains.

Exercise 1: Plot the points (−8, −7), (−3, −4.5), (4, −1), (8, 1) on your graphing calculator and name the quadrant of each point. Then state whether the points seem to lie on an imaginary line. QIII, QIII, QIV, QI, yes

Exercise 2: Plot these points on your graphing calculator. Name the quadrant each point is in and state whether all points seem to lie on an imaginary straight line. (−6, −3), (−1, −2), (3, 1), (6, 8). QIII, QIII, QI, QI, no

Exercise 3: Plot the points (−5, −7), (−2, −1), (0, 3), and (2, 7) (these are the original four points from this page). Then enter $Y_1 = 2x + 3$ on the Y= screen and press Zoom 6. What do you notice? The points are all on the graph of Y_1.

2.2 EXERCISES

CONCEPTS AND VOCABULARY

Fill in each blank with the appropriate word or phrase. Carefully reread the section if needed.

1. If a relation is given in ordered pair form, we state the domain by listing all of the __first__ coordinates in a set.

2. A relation is a function if each element of the __domain__ is paired with __exactly one__ element of the range.

3. The set of output values for a function is called the __range__ of the function.

4. Write using function notation: The function f evaluated at 3 is negative 5: __$f(3) = -5$__

5. Discuss/explain why the relation $y = x^2$ is a function, while the relation $x = y^2$ is not. Justify your response using graphs, ordered pairs, and so on. Answers will vary.

6. Discuss/explain the process of finding the domain and range of a function given its graph, using vertical and horizontal boundary lines. Include a few illustrative examples. Answers will vary.

DEVELOPING YOUR SKILLS

Represent each relation in mapping notation, then state the domain and range.

7.

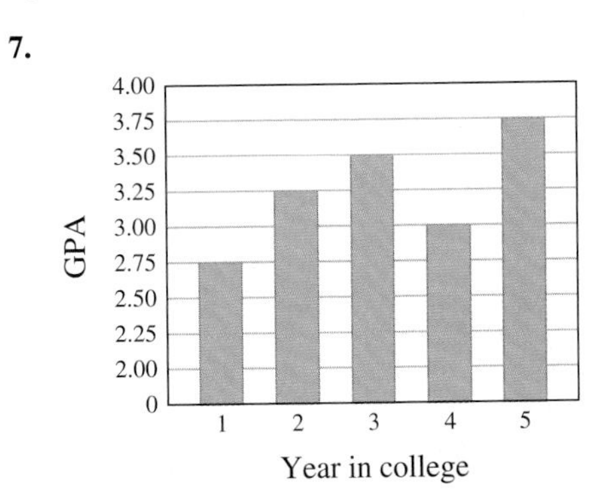

8.

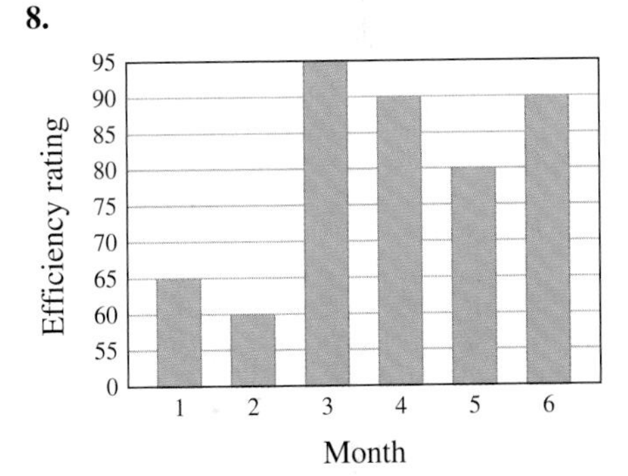

State the domain and range of each relation.

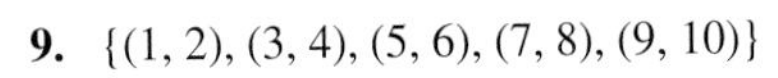

9. $\{(1, 2), (3, 4), (5, 6), (7, 8), (9, 10)\}$

10. $\{(-2, 4), (-3, -5), (-1, 3), (4, -5), (2, -3)\}$

11. $\{(4, 0), (-1, 5), (2, 4), (4, 2), (-3, 3)\}$

12. $\{(-1, 1), (0, 4), (2, -5), (-3, 4), (2, 3)\}$

Complete each table for the values given to find ordered pair solutions for the related equation. For Exercises 15 and 16, each x input corresponds to two possible y outputs. Be sure to find both.

13. $y = -\frac{2}{3}x + 1$

x	y
−6	5
−3	3
0	1
3	−1
6	−3
8	$\frac{-13}{3}$

14. $y = -\frac{5}{4}x + 3$

x	y
−8	13
−4	8
0	3
4	−2
8	−7
10	$\frac{-19}{2}$

15. $x + 2 = |y|$

x	y
−2	0
0	2, −2
1	3, −3
3	5, −5
6	8, −8
7	9, −9

7. Year in college → GPA
$D = \{1, 2, 3, 4, 5\}$
$R = \{2.75, 3.00, 3.25, 3.50, 3.75\}$

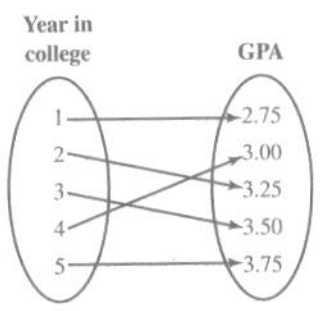

8. Month → Efficiency rating
$D = \{1, 2, 3, 4, 5, 6\}$
$R = \{60, 65, 80, 90, 95\}$

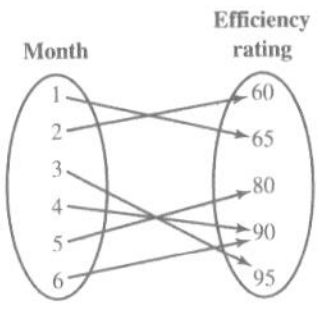

9. $D = \{1, 3, 5, 7, 9\}$; $R = \{2, 4, 6, 8, 10\}$

10. $D = \{-2, -3, -1, 4, 2\}$; $R = \{4, -5, 3, -3\}$

11. $D = \{4, -1, 2, -3\}$; $R = \{0, 5, 4, 2, 3\}$

12. $D = \{-1, 0, 2, -3\}$; $R = \{1, 4, -5, 3\}$

16. $|y + 1| = x$

x	y
0	−1
1	0, −2
3	2, −4
5	4, −6
6	5, −7
7	6, −8

17. $y = x^2 - 1$

x	y
−3	8
−2	3
0	−1
2	3
3	8
4	15

18. $y = -x^2 + 3$

x	y
−2	−1
−1	2
0	3
1	2
2	−1
3	−6

19. $y = \sqrt{25 - x^2}$

x	y
−4	3
−3	4
0	5
2	$\sqrt{21}$
3	4
4	3

20. $y = \sqrt{169 - x^2}$

x	y
−12	5
−5	12
0	13
3	$4\sqrt{10}$
5	12
12	5

21. $x - 1 = y^2$

x	y
10	3, −3
5	2, −2
4	$\sqrt{3}, -\sqrt{3}$
2	1, −1
1.25	0.5, −0.5
1	0

22. $y^2 + 2 = x$

x	y
2	0
3	1, −1
4	$\sqrt{2}, -\sqrt{2}$
5	$\sqrt{3}, -\sqrt{3}$
6	2, −2
11	3, −3

23. $y = \sqrt[3]{x + 1}$

x	y
−9	−2
−2	−1
−1	0
0	1
4	$\sqrt[3]{5}$
7	2

24. $y = (x - 1)^3$

x	y
−2	−27
−1	−8
0	−1
1	0
2	1
3	8

Determine whether the mappings shown represent functions or nonfunctions. If a nonfunction, explain how the definition of a function is violated.

25. function
26. function
27. Not a function. The Shaq is paired with two heights.
28. Not a function. Canada is paired with 2 languages and Brazil is paired with 2.

25.

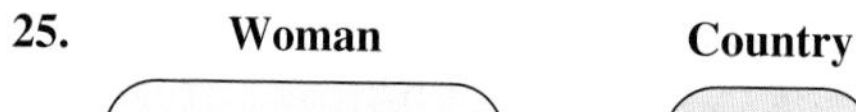

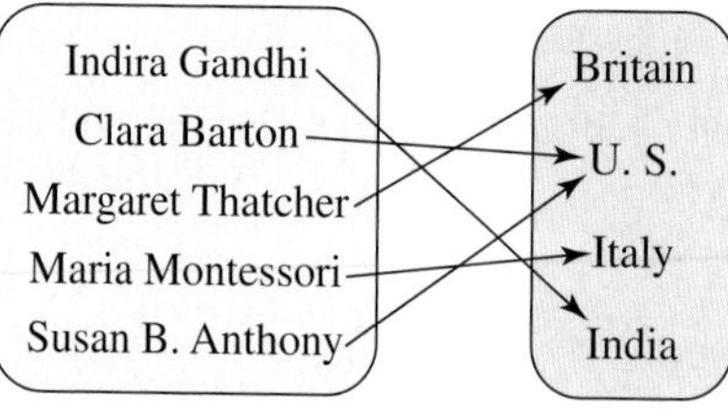

26.

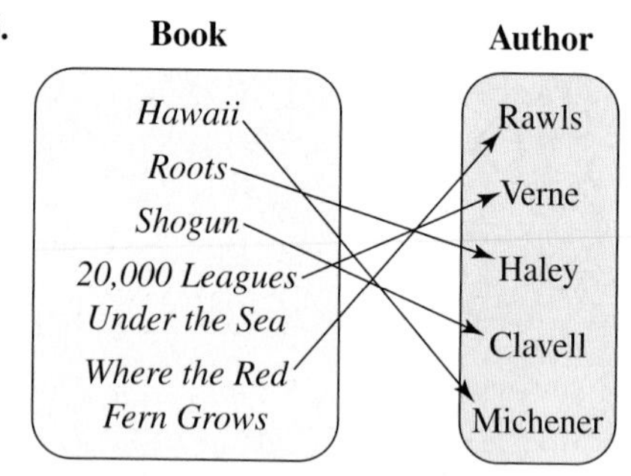

27.

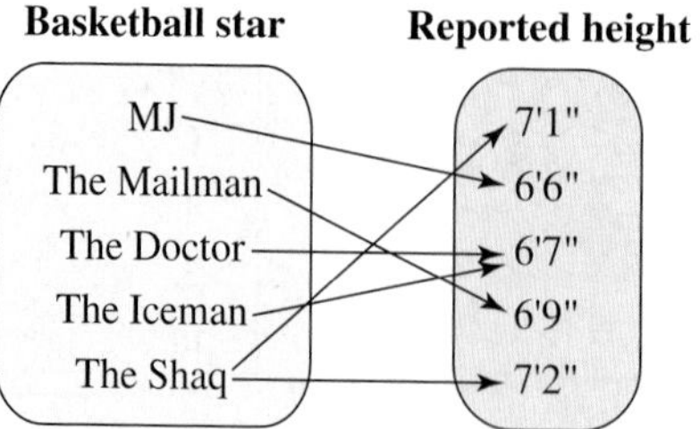

28.

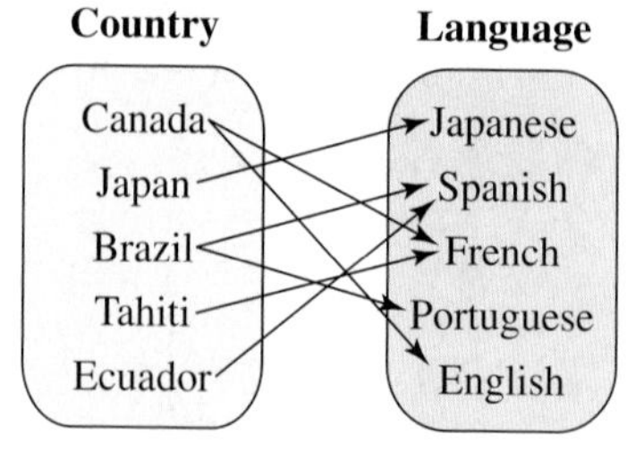

29. Not a function; 4 is paired with 2 and -5.
30. Not a function; -5 is paired with 3 and 7.
31. function
32. function

Determine whether the relations indicated represent functions or nonfunctions. If the relation is a nonfunction, explain how the definition of a function is violated.

29. $(-3, 0), (1, 4), (2, -5), (4, 2), (-5, 6), (3, 6), (0, -1), (4, -5)$, and $(6, 1)$

30. $(-7, -5), (-5, 3), (4, 0), (-3, -5), (1, -6), (0, 9), (2, -8), (3, -2)$, and $(-5, 7)$

31. $(9, -10), (-7, 6), (6, -10), (4, -1), (2, -2), (1, 8), (0, -2), (-2, -7)$, and $(-6, 4)$

32. $(1, -81), (-2, 64), (-3, 49), (5, -36), (-8, 25), (13, -16), (-21, 9), (34, -4)$, and $(-55, 1)$

33.

(−3, 4) (2, 4) (−1, 1) (4, 2) (−5, 0) (1, −4)

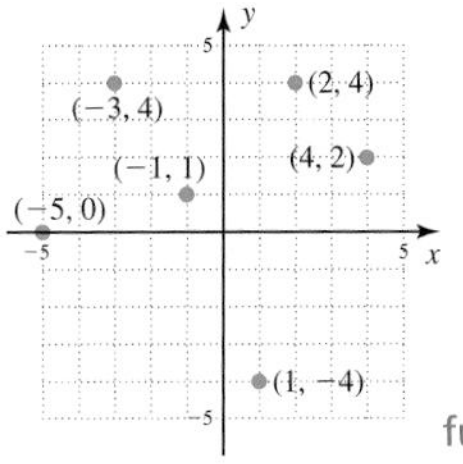

function

34.

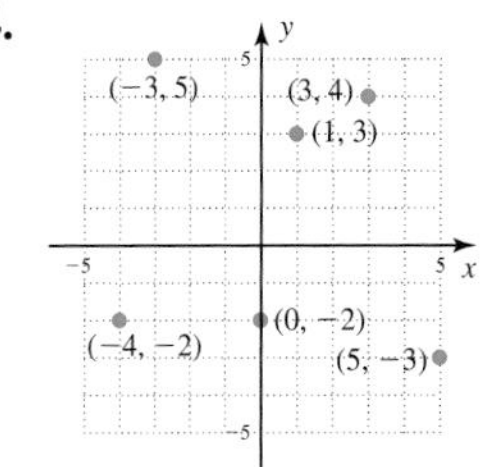

function

35.

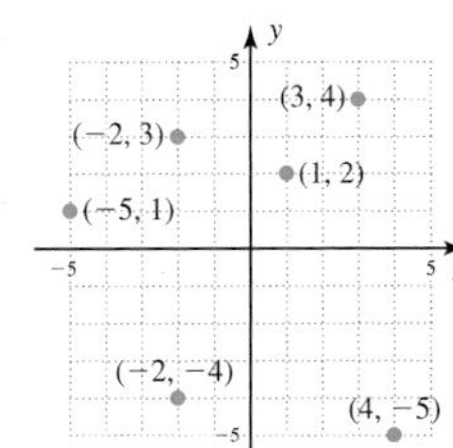

Not a function; -2 is paired with 3 and -4.

36.

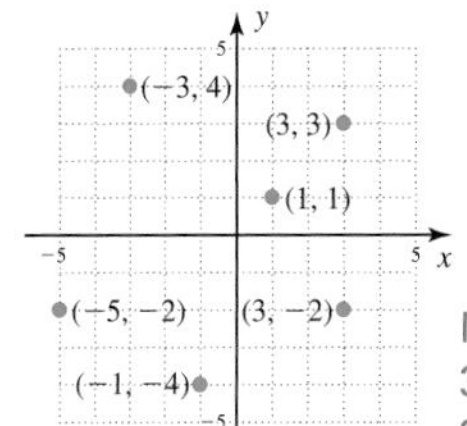

Not a function; 3 is paired with 3 and -2.

37. function
38. function
39. function
40. function
41. Not a function; 0 is paired with 4 and -4.
42. Not a function; 2 is paired with -2.3 and 2.3
43. function
44. function
45. Not a function; -3 is paired with -2 and 2.

Determine whether the relations indicated represent functions or nonfunctions. If a nonfunction, explain how the definition of a function is violated.

37.

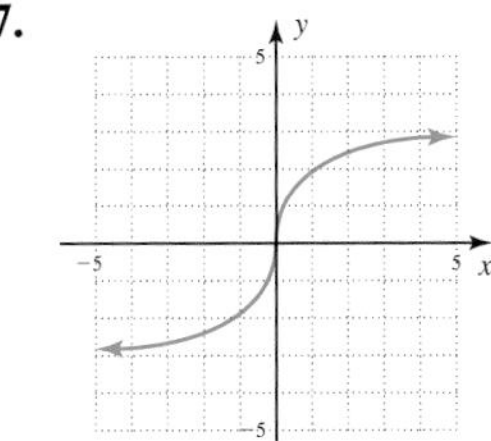

38.

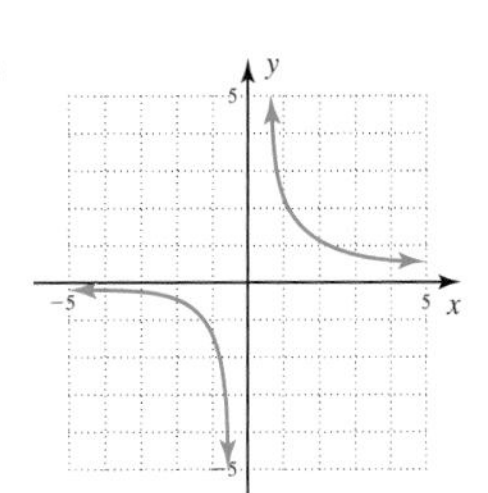

39.

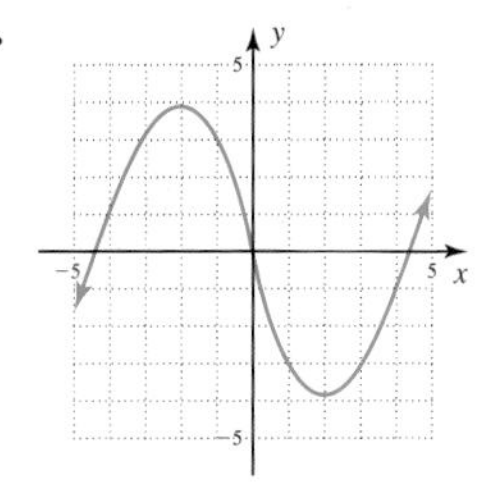

40.

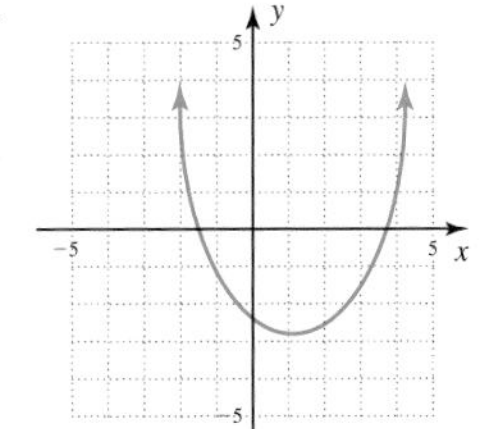

41.

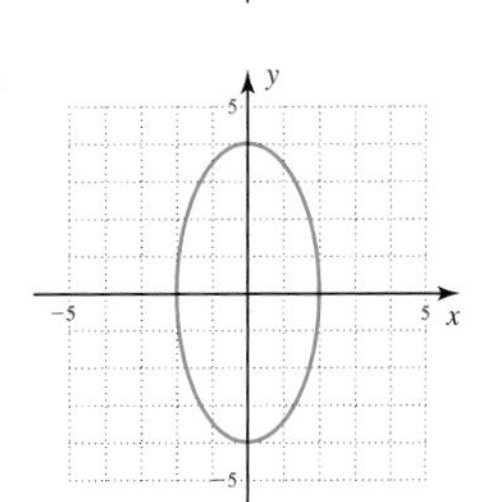

42.

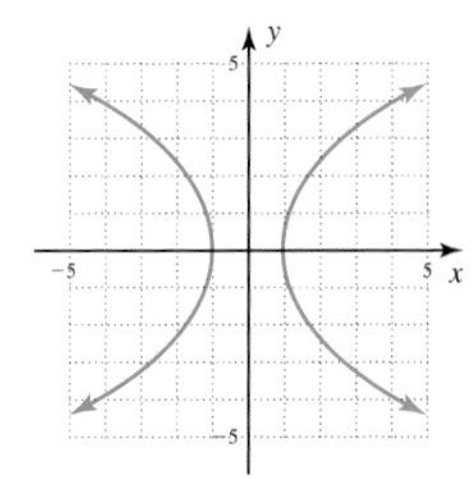

43.

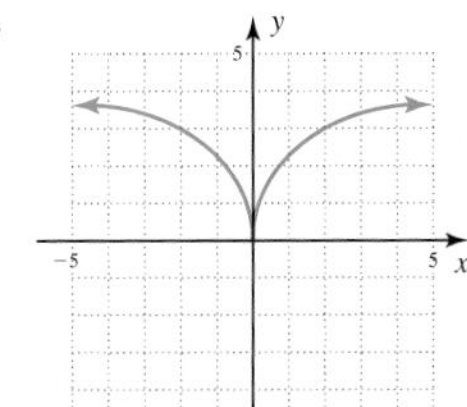

44.

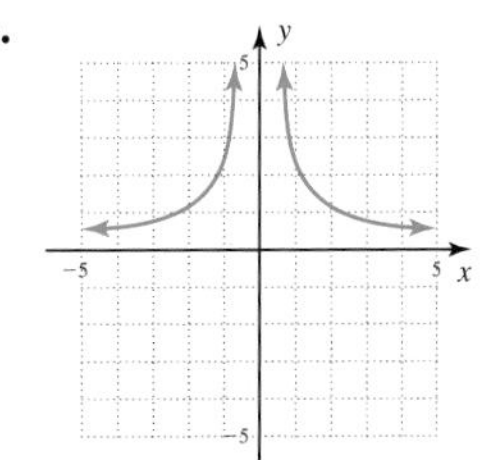

45.

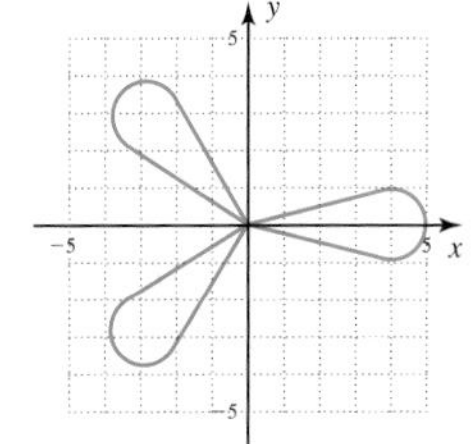

46. Not a function; 0 is paired with 4 and −4.
47. function
48. function

46.

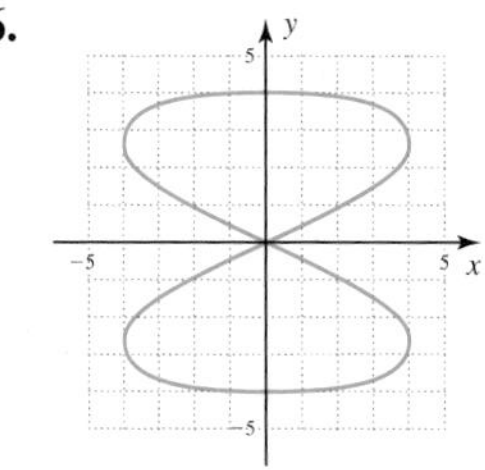

47.

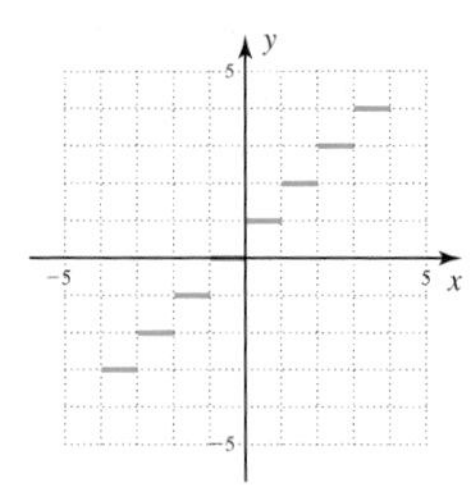

48.

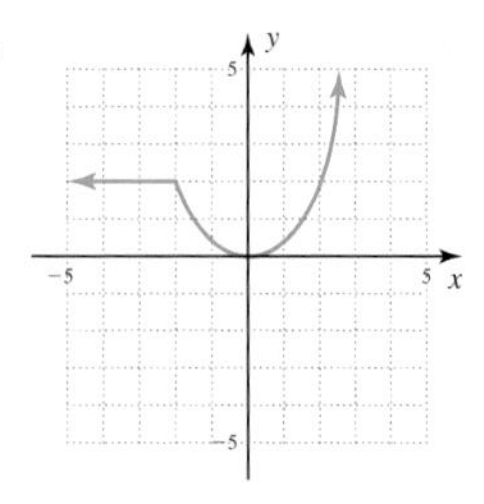

Determine whether the relations indicated represent functions or nonfunctions, then determine the domain and range of the relation using vertical and horizontal boundary lines. Assume the endpoints of all intervals (if they exist) are integer values.

49. function, $x \in [-4, -5]$ $y \in [-2, 3]$
50. function, $x \in [-4, \infty)$ $y \in (-\infty, 5]$
51. function, $x \in [-4, \infty)$ $y \in [-4, \infty)$

49.

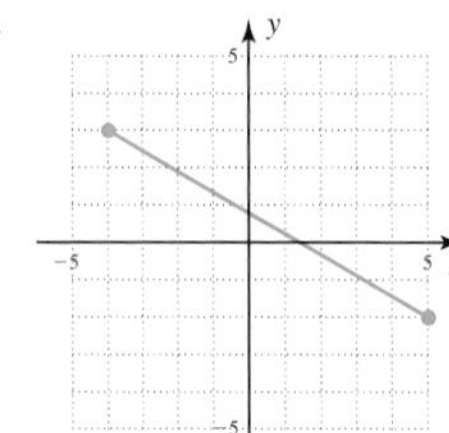

50.

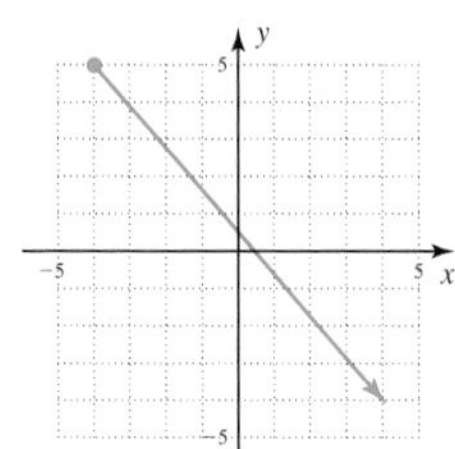

51.

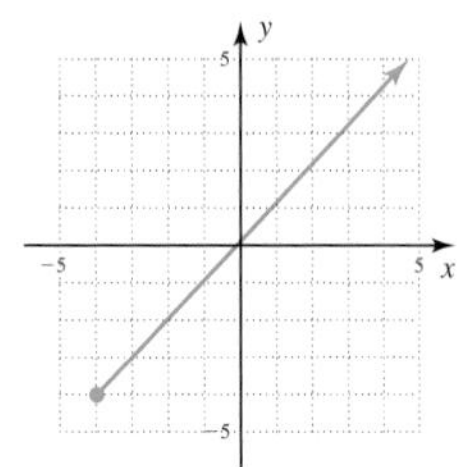

52. Not a function, $x \in [0, 3]$ $y \in [-4, 4]$
53. function, $x \in [-4, 4]$ $y \in [-5, -1]$
54. function, $x \in (-\infty, \infty)$ $y \in (-\infty, \infty)$

52.

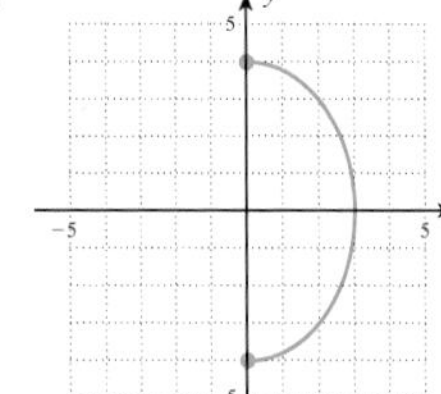

53.

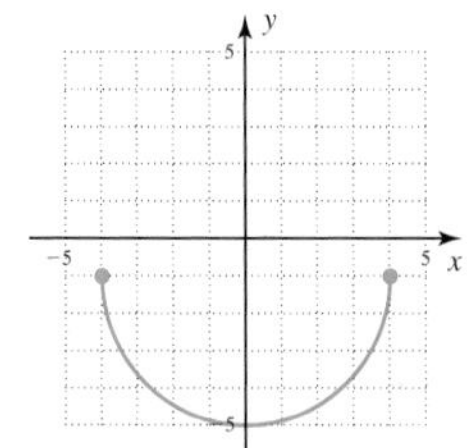

54.

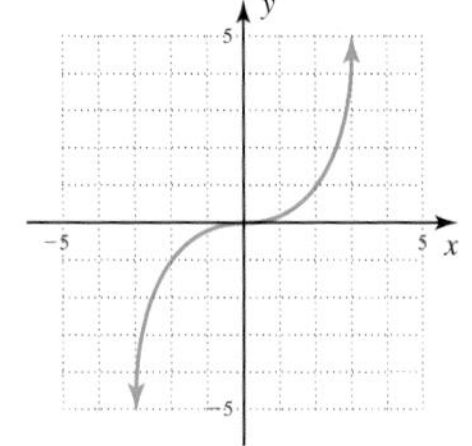

55. function, $x \in (-\infty, \infty)$ $y \in (-\infty, \infty)$
56. Not a function, $x \in [-3, 4]$ $y \in [-3, 4]$
57. Not a function, $x \in [-3, 5]$ $y \in [-3, 3]$

55.

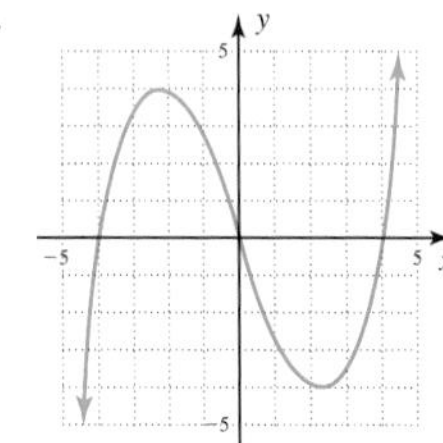

56.

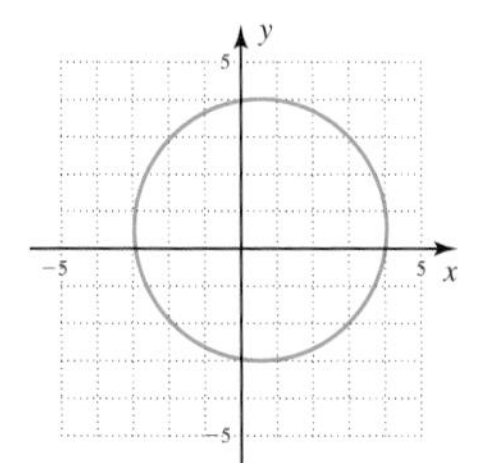

57.

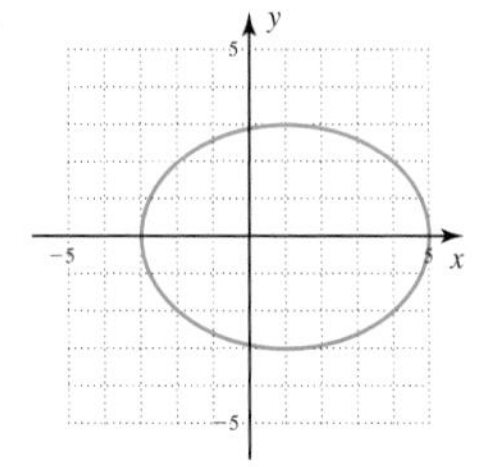

58. function, $x \in (-\infty, \infty)$ $y \in [-3, \infty)$
59. Not a function, $x \in (-\infty, 3]$ $y \in (-\infty, \infty)$
60. function, $x \in [-4, \infty)$ $y \in [0, \infty)$

58.

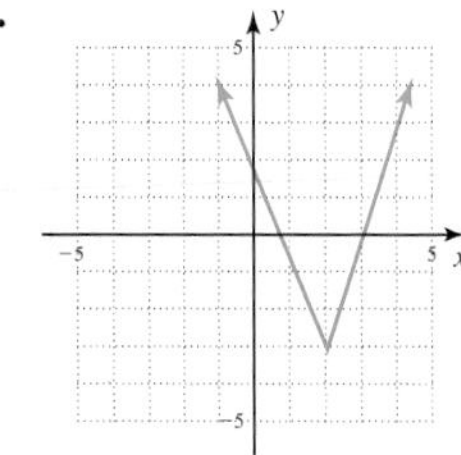

59.

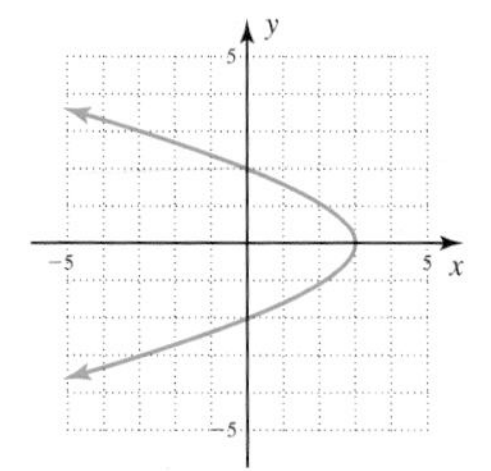

60.

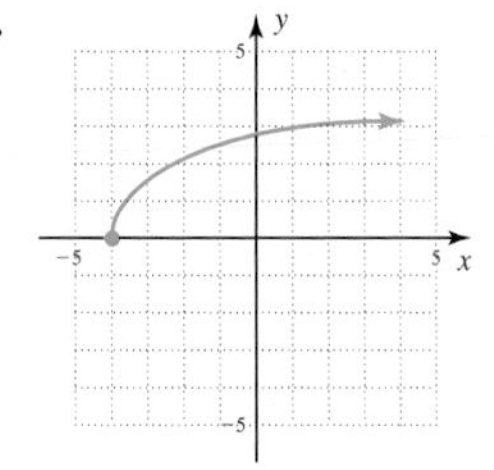

61. $x \in (-\infty, 5) \cup (5, \infty)$
62. $x \in (-\infty, -3) \cup (-3, \infty)$
63. $x \in [\frac{-5}{3}, \infty)$
64. $x \in [\frac{2}{5}, \infty)$
65. $x \in (-\infty, -5) \cup (-5, 5) \cup (5, \infty)$
66. $x \in (-\infty, -7) \cup (-7, 7) \cup (7, \infty)$

Determine the domain of the following functions.

61. $y = \dfrac{3}{x - 5}$

62. $y = \dfrac{-2}{3 + x}$

63. $b = \sqrt{3a + 5}$

64. $b = \sqrt{5a - 2}$

65. $y_1 = \dfrac{x + 2}{x^2 - 25}$

66. $y_2 = \dfrac{x - 4}{x^2 - 49}$

67. $x \in (-\infty, -3\sqrt{2}) \cup (-3\sqrt{2}, 3\sqrt{2}) \cup (3\sqrt{2}, \infty)$
68. $x \in (-\infty, -2\sqrt{3}) \cup (-2\sqrt{3}, 2\sqrt{3}) \cup (2\sqrt{3}, \infty)$
69. $x \in (-\infty, \infty)$
70. $x \in (-\infty, \infty)$
71. $x \in (-\infty, \infty)$
72. $x \in (-\infty, \infty)$
73. $x \in (-\infty, \infty)$
74. $x \in (-\infty, \infty)$
75. $x \in (-\infty, -2) \cup (-2, 5) \cup (5, \infty)$
76. $x \in (-\infty, -5) \cup (-5, 3) \cup (3, \infty)$
77. $x \in [2, \frac{5}{2}) \cup (\frac{5}{2}, \infty)$
78. $x \in (-1, \frac{-2}{3}) \cup (\frac{-2}{3}, \infty)$
79. $f(-6) = 0$, $f(\frac{3}{2}) = \frac{15}{4}$, $f(2c) = c + 3$, $f(c + 2) = \frac{1}{2}c + 4$
80. $f(-6) = -9$, $f(\frac{3}{2}) = -4$, $f(2c) = \frac{4c}{3} - 5$, $f(c + 2) = \frac{2}{3}c - \frac{11}{3}$
81. $f(-6) = 132$, $f(\frac{3}{2}) = \frac{3}{4}$, $f(2c) = 12c^2 - 8c$, $f(c + 2) = 3c^2 + 8c + 4$
82. $f(-6) = 54$, $f(\frac{3}{2}) = 9$, $f(2c) = 8c^2 + 6c$, $f(c + 2) = 2c^2 + 11c + 14$
83. $h(3) = 1$, $h(\frac{-2}{3}) = \frac{-9}{2}$, $h(3a) = \frac{1}{a}$, $h(a - 1) = \frac{3}{a - 1}$
84. $h(3) = \frac{2}{9}$, $h(\frac{-2}{3}) = \frac{9}{2}$, $h(3a) = \frac{2}{9a^2}$, $h(a - 1) = \frac{2}{(a - 1)^2}$
85. $h(3) = 5$, $h(\frac{-2}{3}) = -5$, $h(3a) = -5$ if $a < 0$ or 5 if $a > 0$, $h(a - 1) = -5$ if $a < 1$ or 5 if $a > 1$
86. $h(3) = 4$, $h(\frac{-2}{3}) = -4$, $h(3a) = -4$ if $a < 0$ or 4 if $a > 0$, $h(a - 1) = -4$ if $a < 1$ or 4 if $a > 1$
87. $g(0.4) = 0.8\pi$, $g(\frac{9}{4}) = \frac{9}{2}\pi$, $g(h) = 2\pi h$, $g(h + 3) = 2\pi h + 6\pi$
88. $g(0.4) = 0.8\pi h$, $g(\frac{9}{4}) = \frac{9}{2}\pi h$, $g(h) = 2\pi h^2$, $g(h + 3) = 2\pi h^2 + 6\pi h$
89. $g(0.4) = 0.16\pi$, $g(\frac{9}{4}) = \frac{81}{16}\pi$, $g(h) = \pi h^2$, $g(h + 3) = \pi(h + 3)^2$
90. $g(0.4) = 0.16\pi h$, $g(\frac{9}{4}) = \frac{81}{16}\pi h$, $g(h) = \pi h^3$, $g(h + 3) = \pi h^3 + 6\pi h^2 + 9\pi h$
91. $p(0.5) = 2$, $p(\frac{9}{4}) = \frac{\sqrt{30}}{2}$, $p(a) = \sqrt{2a + 3}$, $p(a + 3) = \sqrt{2a + 9}$
92. $p(0.5) = 1$, $p(\frac{9}{4}) = 2\sqrt{2}$, $p(a) = \sqrt{4a - 1}$, $p(a + 3) = \sqrt{4a + 11}$
93. $p(0.5) = -17$, $p(\frac{9}{4}) = \frac{163}{81}$, $p(a) = \frac{3a^2 - 5}{a^2}$, $p(a + 3) = \frac{3a^2 + 18a + 22}{a^2 + 6a + 9}$

Additional answers can be found in the Instructor Answer Appendix.

67. $u = \dfrac{v - 5}{v^2 - 18}$ **68.** $p = \dfrac{q + 7}{q^2 - 12}$ **69.** $y = \dfrac{17}{25}x + 123$

70. $y = \dfrac{11}{19}x - 89$ **71.** $m = n^2 - 3n - 10$ **72.** $s = t^2 - 3t - 10$

73. $y = 2|x| + 1$ **74.** $y = |x - 2| + 3$ **75.** $y_1 = \dfrac{x}{x^2 - 3x - 10}$

76. $y_2 = \dfrac{x - 4}{x^2 + 2x - 15}$ **77.** $y = \dfrac{\sqrt{x - 2}}{2x - 5}$ **78.** $y = \dfrac{\sqrt{x + 1}}{3x + 2}$

Determine the value of $f(-6)$, $f(\frac{3}{2})$, $f(2c)$, and $f(c + 2)$, then simplify as much as possible.

79. $f(x) = \frac{1}{2}x + 3$ **80.** $f(x) = \frac{2}{3}x - 5$ **81.** $f(x) = 3x^2 - 4x$ **82.** $f(x) = 2x^2 + 3x$

Determine the value of $h(3)$, $h(-\frac{2}{3})$, $h(3a)$, and $h(a - 1)$, then simplify as much as possible.

83. $h(x) = \dfrac{3}{x}$ **84.** $h(x) = \dfrac{2}{x^2}$ **85.** $h(x) = \dfrac{5|x|}{x}$ **86.** $h(x) = \dfrac{4|x|}{x}$

Determine the value of $g(0.4)$, $g(\frac{9}{4})$, $g(h)$, and $g(h + 3)$, then simplify as much as possible.

87. $g(r) = 2\pi r$ **88.** $g(r) = 2\pi rh$ **89.** $g(r) = \pi r^2$ **90.** $g(r) = \pi r^2 h$

Determine the value of $p(0.5)$, $p(\frac{9}{4})$, $p(a)$, and $p(a + 3)$, then simplify as much as possible.

91. $p(x) = \sqrt{2x + 3}$ **92.** $p(x) = \sqrt{4x - 1}$ **93.** $p(x) = \dfrac{3x^2 - 5}{x^2}$ **94.** $p(x) = \dfrac{2x^2 + 3}{x^2}$

Use the graph of each function given to (a) state the domain, (b) evaluate $f(2)$, (c) determine the value x for which $f(x) = 4$, and (d) state the range. Assume all results are integer-valued.

95.

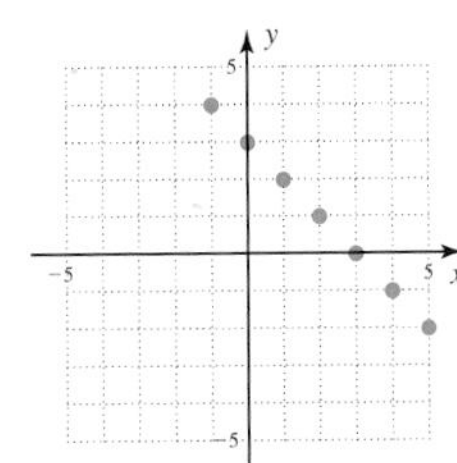

96.

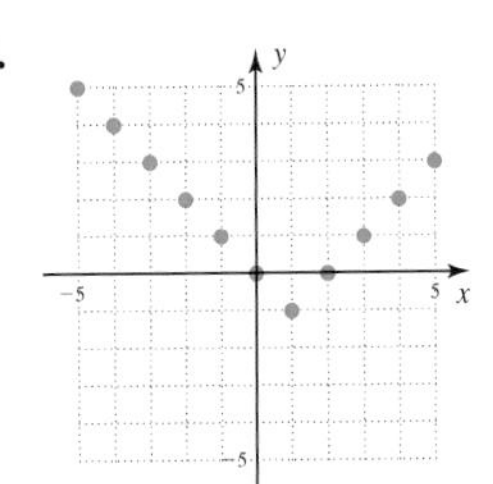

97.

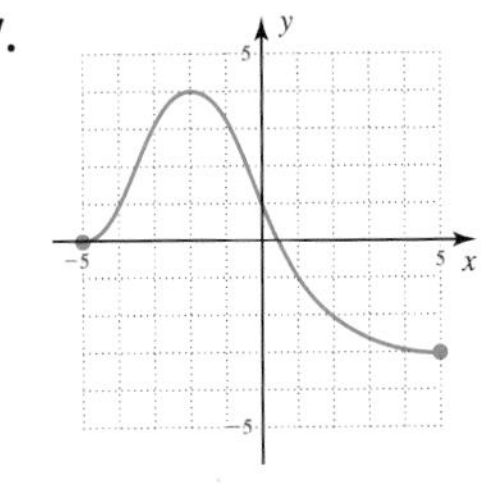

98.

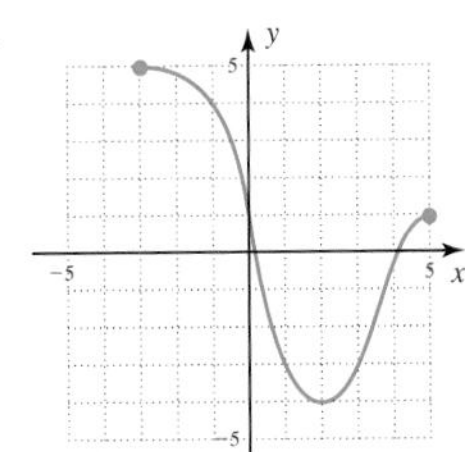

99.

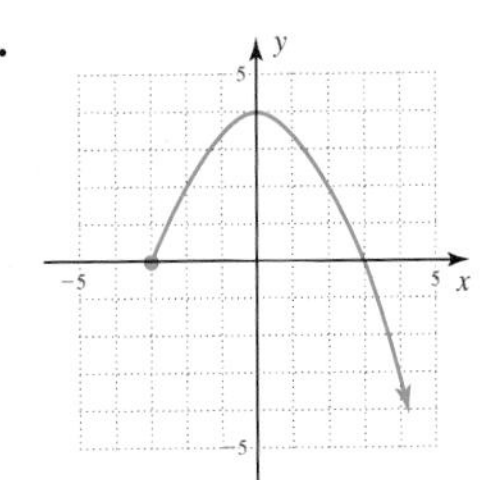

100.

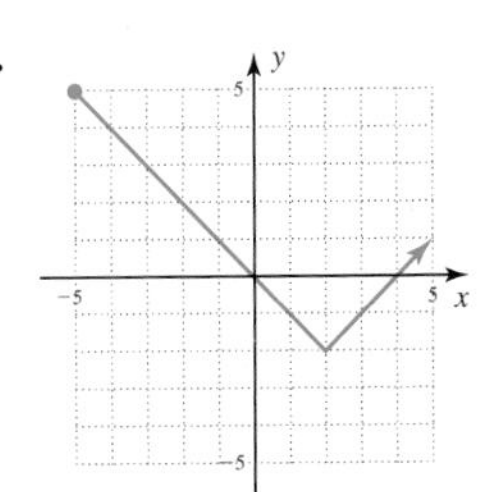

WORKING WITH FORMULAS

101. Ideal weight for males: $W(H) = \frac{9}{2}H - 151$

The ideal weight for an adult male can be modeled by the function shown, where W is his weight in pounds and H is his height in inches. (a) Find the ideal weight for a male who is 75 in. tall. (b) If I am 72 in. tall and weigh 210 lb, how much weight should I lose?
a. 186.5 lb **b.** 37 lb

102. Celsius to Fahrenheit conversions: $C = \frac{5}{9}(F - 32)$
The relationship between Fahrenheit degrees and degrees Celsius is modeled by the function shown. (a) What is the Celsius temperature if °F = 41? (b) Use the formula to solve for F in terms of C, then substitute the result from part (a). What do you notice?
a. 5°C **b.** They give the same result.

APPLICATIONS

103. Gas mileage: John's old '87 LeBaron has a 15-gal gas tank and gets 23 mpg. The number of miles he can drive is a function of how much gas is in the tank. (a) Write this relationship in equation form and (b) determine the domain and range of the function in this context. **a.** $N(g) = 23g$ **b.** $g \in [0, 15]$; $N \in [0, 345]$

104. Gas mileage: Jackie has a gas-powered model boat with a 5-oz gas tank. The boat will run for 2.5 min on each ounce. The number of minutes she can operate the boat is a function of how much gas is in the tank. (a) Write this relationship in equation form and (b) determine the domain and range of the function in this context.
a. $N(g) = 2.5g$ **b.** $g \in [0, 5]$; $N \in [0, 12.5]$

105. Volume of a cube: The volume of a cube depends on the length of the sides. In other words, volume is a function of the sides: $V(s) = s^3$. (a) In practical terms, what is the domain of this function? (b) Evaluate $V(6.25)$ and (c) evaluate the function for $s = 2x^2$.
a. $[0, \infty)$ **b.** ≈ 244 units^3 **c.** $8x^6$

106. Volume of a cylinder: For a fixed radius of 10 cm, the volume of a cylinder depends on its height. In other words, volume is a function of height: $V(h) = 100\pi h$. (a) In practical terms, what is the domain of this function? (b) Evaluate $V(7.5)$ and (c) evaluate the function for $h = \frac{8}{\pi}$. **a.** $[0, \infty)$ **b.** 750π **c.** 800

107. Rental charges: Temporary Transportation Inc. rents cars (local rentals only) for a flat fee of $19.50 and an hourly charge of $12.50. This means that cost is a function of the hours the car is rented plus the flat fee. (a) Write this relationship in equation form; (b) find the cost if the car is rented for 3.5 hr; (c) determine how long the car was rented if the bill came to $119.75; and (d) determine the domain and range of the function in this context, if your budget limits you to paying a maximum of $150 for the rental. **a.** $c(t) = 12.50t + 19.50$ **b.** $63.25 **c.** ≈ 8 hr **d.** $t \in [0, 10.44]$; $c \in [0, 150]$

108. Cost of a service call: Paul's Plumbing charges a flat fee of $50 per service call plus an hourly rate of $42.50. This means that cost is a function of the hours the job takes to complete plus the flat fee. (a) Write this relationship in equation form; (b) find the cost of a service call that takes $2\frac{1}{2}$ hr; (c) find the number of hours the job took if the charge came to $262.50; and (d) determine the domain and range of the function in this context, if your insurance company has agreed to pay for all charges over $500 for the service call. **a.** $c(t) = 42.50t + 50$ **b.** $156.25 **c.** 5 hr **d.** $t \in [0, 10.6]$; $c \in [0, 500]$

109. Predicting tides: The graph shown approximates the height of the tides at Fair Haven, New Brunswick, for a 12-hr period. (a) Is this the graph of a function? Why? (b) Approximately what time did high tide occur? (c) How high is the tide at 6 P.M.? (d) What time(s) will the tide be 2.5 m?

Exercise 109

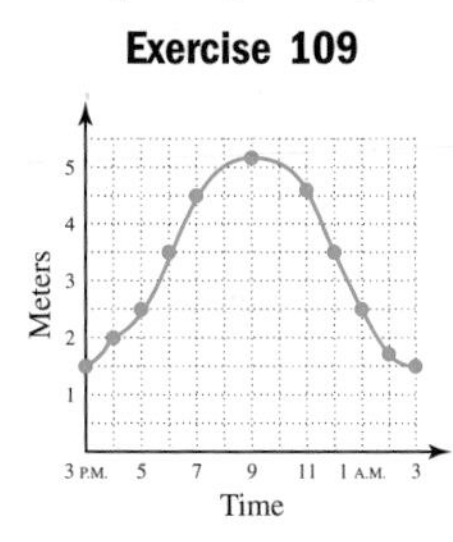

110. Predicting tides: The graph shown approximates the height of the tides at Apia, Western Samoa, for a 12-hr period. (a) Is this the graph of a function? Why? (b) Approximately what time did low tide occur? (c) How high is the tide at 2 A.M.? (d) What time(s) will the tide be 0.7 m?

109. **a.** Yes. Each x is paired with exactly one y.
b. 9 P.M.
c. $3\frac{1}{2}$ m
d. 5 P.M. and 1 A.M.

110. **a.** Yes. Each x is paired with exactly one y.
b. 10 P.M.
c. 0.9 m
d. 7 P.M. and 1 A.M.

Exercise 110

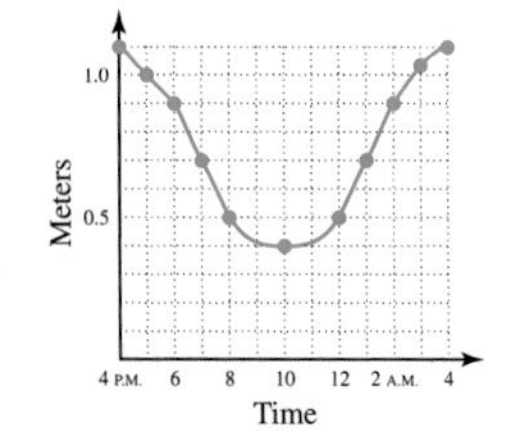

WRITING, RESEARCH, AND DECISION MAKING

111. Outside of a mathematical context, the word "domain" is still commonly used in everyday language. Using a college-level dictionary, look up and write out the various meanings of the word, noting how closely the definitions given (there are several) tie in with its mathematical application. Answers will vary.

112. Complete the following statements, then try to create two additional, like statements.

a. If you work at McDonalds, wages are a function of _time_.

b. Placing a warm can of soda-pop in the fridge, its temperature is a function of _time_.

c. The area of a circle is a function of _the radius_.

EXTENDING THE CONCEPT

113. A father challenges his son to a 400-m race, depicted in the graph shown here.

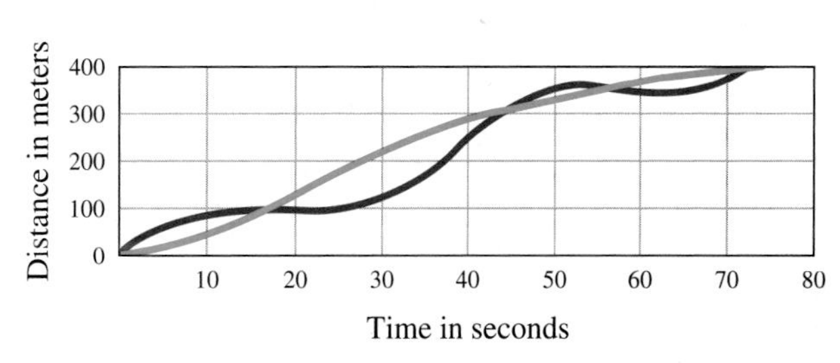

a. Who won and what was the approximate winning time? Son, 72.5 sec

b. Approximately how many meters behind was the second place finisher? 10 m

c. Estimate the number of seconds the father was in the lead in this race. 45 sec

d. How many times during the race were the father and son tied? 3

114.

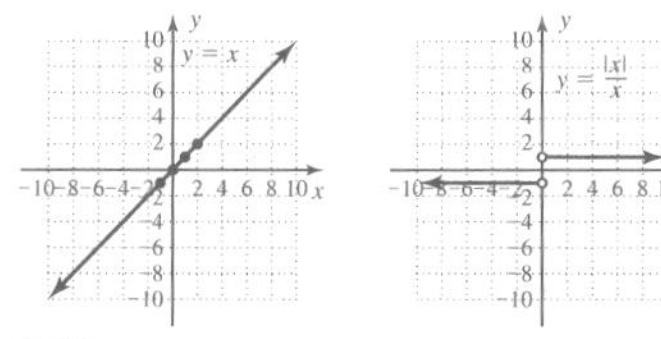

115.

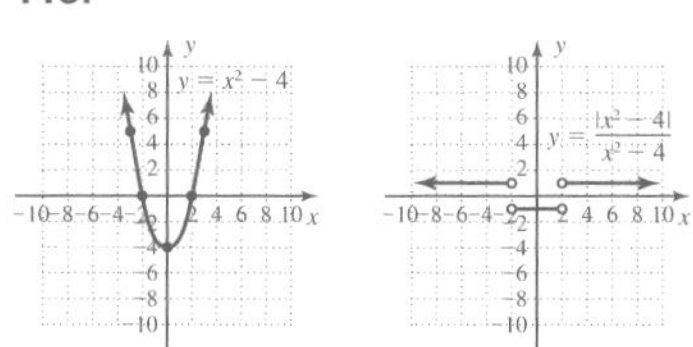

114. Sketch the graph of $f(x) = x$, then discuss how you could use this graph to obtain the graph of $F(x) = |x|$ without computing additional points. What would the graph of $g(x) = \frac{|x|}{x}$ look like?

115. Sketch the graph of $f(x) = x^2 - 4$, then discuss how you could use this graph to obtain the graph of $F(x) = |x^2 - 4|$ without computing additional points. Determine what the graph of $g(x) = \frac{|x^2 - 4|}{x^2 - 4}$ would look like.

MAINTAINING YOUR SKILLS

116. (2.1) Which line has a steeper slope, the line through $(-5, 3)$ and $(2, 6)$, or the line through $(0, -4)$ and $(9, 4)$? line through $(0, -4)$ and $(9, 4)$

117. (R.6) Compute the sum and product indicated:

a. $\sqrt{24} + 6\sqrt{54} - \sqrt{6}$ $19\sqrt{6}$

b. $(2 + \sqrt{3})(2 - \sqrt{3})$ 1

118. (1.5) Solve the equation using the quadratic formula, then check the result(s) using substitution:

$x^2 - 4x + 1 = 0$ $2 \pm \sqrt{3}$

119. (R.4) Factor the following polynomials completely:

a. $x^3 - 3x^2 - 25x + 75$ $(x - 3)(x - 5)(x + 5)$

b. $2x^2 - 13x - 24$ $(2x + 3)(x - 8)$

c. $8x^3 - 125$ $(2x - 5)(4x^2 + 10x + 25)$

120. (R.6) Use the Pythagorean theorem to help you find the perimeter and area of the triangle shown. If you recognize a Pythagorean triple, use it.
$P = 30$ cm, $A = 30$ cm^2

13 cm
5 cm

121. (R.3) Simplify using properties of exponents:

a. $\left(\frac{x^3y}{z^{-2}}\right)^2$ $x^6y^2z^4$ **b.** $\left(\frac{2}{3}\right)^{-3}$ $\frac{27}{8}$

2.3 Linear Functions and Rates of Change

LEARNING OBJECTIVES

In Section 2.3 you will learn how to:

A. **Write a linear equation in function form**

B. **Use function form to identify the slope**

C. **Use slope-intercept form to graph linear functions**

D. **Write a linear equation in point-slope form**

E. **Use the point-slope form to solve applications**

INTRODUCTION

The concept of slope is an important part of mathematics, because it gives us a way to measure and compare change. The value of an automobile changes with time, the circumference of a circle increases as the radius increases, and the tension in a spring grows the more it is stretched. The real world is filled with examples of how one change affects another, and slope helps us understand how these changes are related.

POINT OF INTEREST

Although we've given the word "function" a somewhat formal definition, we should never divorce it from its more intuitive meaning. Life is filled with functions, where one "thing" depends on another. For example,

The amount of a paycheck is a function of (depends on) hours worked.

Water pressure is a function of (depends on) the depth of a dive.

The radius of a balloon is a function of (depends on) the surrounding temperature.

A. Linear Equations and Function Form

In Section 1.1, formulas and literal equations were written in an alternate form by solving for an object variable. The new form made evaluating the formula more efficient, in that we could gain information on the object variable without having to repeatedly solve the original equation. Solving for y in the equation $ax + by = c$ offers similar advantages to linear graphing and applications.

EXAMPLE 1 Solve $2y - 6x = 4$ for y, then evaluate at $x = 4$, $x = 0$, and $x = -\frac{1}{3}$.

Solution:

$2y - 6x = 4$ given equation

$2y = 6x + 4$ add $6x$ to both sides

$y = 3x + 2$ divide by 2

Since the coefficients are integers, evaluate the function mentally. Inputs are multiplied by 3, then increased by 2, yielding the ordered pairs $(4, 14)$, $(0, 2)$, and $(-\frac{1}{3}, 1)$.

NOW TRY EXERCISES 7 THROUGH 12

This form of the equation (where y has been written in terms of x) is sometimes called **function form.** This is likely due to the fact that we can immediately identify what operations the function performs on x in order to obtain y—they appear as the coefficient of x and the constant term. Using function notation we write the result above as $f(x) = 3x + 2$, and note again how this particular function is "programmed": *multiply inputs by 3, then add 2.*

WORTHY OF NOTE

In Example 2, the final form can be written $f(x) = \frac{2}{3}x + 2$ as shown (inputs are multiplied by two-thirds, then increased by 2), or written as $f(x) = \frac{2x}{3} + 2$ (inputs are multiplied by two, the result divided by 3 and this amount increased by 2). The two forms are equivalent.

EXAMPLE 2 Write the linear equation $3y - 2x = 6$ in function form (solve for y), then identify the new coefficient of x and the constant term.

Solution:

$3y - 2x = 6$ given equation

$3y = 2x + 6$ add $2x$

$y = \frac{2}{3}x + 2$ divide by 3

$f(x) = \frac{2}{3}x + 2$ function form

The new coefficient of x is $\frac{2}{3}$ and the constant term is 2.

NOW TRY EXERCISES 13 THROUGH 18

When the coefficient of x is rational, it's helpful to select inputs that are multiples of the denominator to evaluate the function. This enables us to perform the operations mentally and quickly locate the two or three points needed to graph the function. For $f(x) = \frac{2}{3}x + 2$, possible inputs might be $x = -9, -6, 0, 3, 6$, and so on. See Exercises 19 through 24.

B. Function Form and the Slope of a Line

In Section 2.1, linear equations were graphed using the intercept method. When a linear equation is written in function form, we notice a powerful connection between the graph of the function and its equation.

EXAMPLE 3 Find the intercepts of $4x + 5y = -20$ and use them to graph the line. Then,

a. Use these points to calculate the slope of the line.

b. Write the equation in function form and compare the calculated slope and y-intercept to the equation in function form.

c. Comment on what you notice.

Solution: Substituting 0 for x in $4x + 5y = -20$, we find the y-intercept is $(0, -4)$. Substituting $y = 0$ gives an x-intercept of $(-5, 0)$. The graph is displayed here.

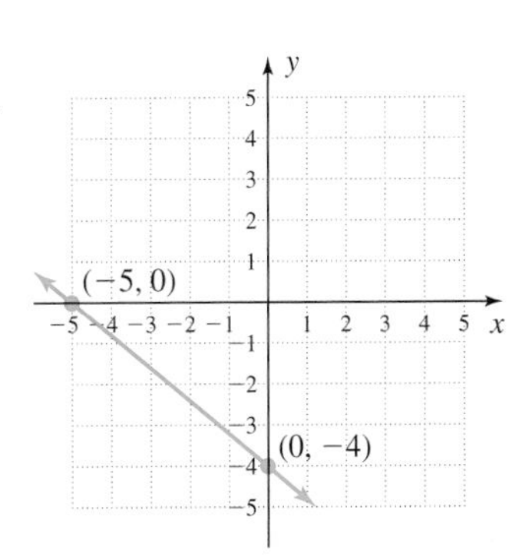

a. By calculation or counting $\frac{\Delta y}{\Delta x}$, the slope is $m = -\frac{4}{5}$.

b. Solving for y gives $y = -\frac{4}{5}x - 4$.

c. The slope value seems to be the coefficient of x while the y-intercept is the constant term.

NOW TRY EXERCISES 25 THROUGH 30

With the equation in function form, an input of $x = 0$ causes the "x term" to become zero, so the y-intercept is *automatically the constant term*. This is true for all functions where y is written as a function of x. As Example 4 illustrates, the function form also enables us to immediately identify the slope of the line—it will be the coefficient of x. In general, a linear equation of the form $y = mx + b$ is said to be in **slope-intercept form** since the slope of the line is m and the y-intercept is $(0, b)$.

> **SLOPE-INTERCEPT FORM**
> For a linear equation written as $y = mx + b$ or $f(x) = mx + b$,
> the slope of the line is m and the y-intercept is $(0, b)$.

EXAMPLE 4 Write each equation in slope-intercept form and clearly identify the slope and y-intercept of each line.

a. $3x - 2y = 9$ **b.** $y + x = 5$ **c.** $x = 2y$

Solution:

a.
$$3x - 2y = 9$$
$$-2y = -3x + 9$$
$$y = \frac{3}{2}x - \frac{9}{2}$$
$$m = \frac{3}{2}$$
y-intercept $\left(0, -\frac{9}{2}\right)$

b.
$$y + x = 5$$
$$y = -x + 5$$
$$y = -1x + 5$$
$$m = -1$$
y-intercept $(0, 5)$

c.
$$2y = x$$
$$y = \frac{x}{2}$$
$$y = \frac{1}{2}x + 0$$
$$m = \frac{1}{2}$$
y-intercept $(0, 0)$

NOW TRY EXERCISES 31 THROUGH 38

C. Slope-Intercept Form and the Graph of a Line

If the slope and y-intercept of a linear function are known or can be found, we're able to construct the equation using slope-intercept form $y = mx + b$ by substituting these values directly.

EXAMPLE 5 Find the slope-intercept form of the linear function shown.

Solution: By calculation using $(-3, -2)$ and $(-1, 2)$, or by simply counting, the slope is $m = \frac{4}{2}$ or $\frac{2}{1}$. By inspection we see the y-intercept is $(0, 4)$, which can be verified by counting $\frac{\Delta y}{\Delta x} = \frac{2}{1}$ [up two $(\Delta y = 2)$ and right one $(\Delta x = 1)$] from $(-1, 2)$. Substituting $\frac{2}{1}$ for m and 4 for b in the slope-intercept form we obtain the equation $y = 2x + 4$.

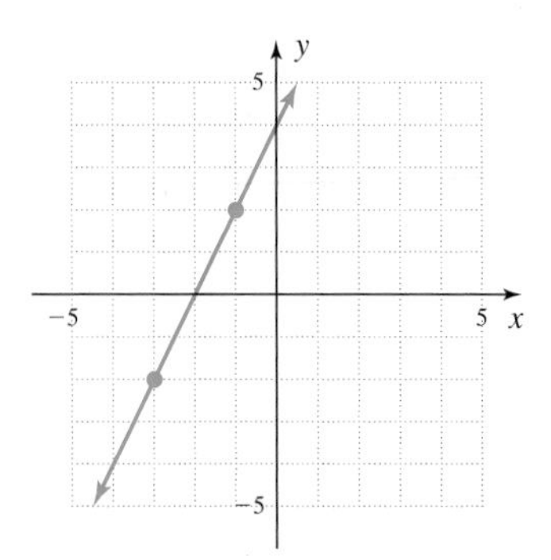

NOW TRY EXERCISES 39 THROUGH 44

Actually, if the slope is known and we have *any* point (x, y) on the line, we can still construct the equation, since the given point *must satisfy the equation of the line*. In this case, we're treating $y = mx + b$ as a simple formula, solving for b after substituting known values of m, x, and y.

EXAMPLE 6 Find the equation of a line that has slope $m = \frac{4}{5}$ and goes through $(-5, 2)$.

Solution: Using the "formula" $y = mx + b$, we have $m = \frac{4}{5}$, $x = -5$, and $y = 2$.

$y = mx + b$ function form

$2 = \frac{4}{5}(-5) + b$ substitute $\frac{4}{5}$ for m, -5 for x, and 2 for y

$2 = -4 + b$ simplify

$6 = b$ solve for b

The equation of this line is $y = \frac{4}{5}x + 6$.

NOW TRY EXERCISES 45 THROUGH 50

Writing a linear function in slope-intercept form enables us to draw its graph with a minimum of effort, since we can easily locate the y-intercept and a second point using $m = \frac{\Delta y}{\Delta x}$.

EXAMPLE 7 Write the equation $3y - 5x = 9$ in slope-intercept form, then graph the line using the y-intercept and slope $m = \frac{\Delta y}{\Delta x}$.

Solution: $3y - 5x = 9$ given equation

$3y = 5x + 9$ isolate y term

$y = \frac{5}{3}x + 3$ divide by 3

The slope is $m = \frac{5}{3}$ and the y-intercept is $(0, 3)$. Plot the y-intercept, then use $\frac{\Delta y}{\Delta x} = \frac{5}{3}$ (up 5 and right 3) to find another point on the line (shown in red). Finish by drawing a line through these points.

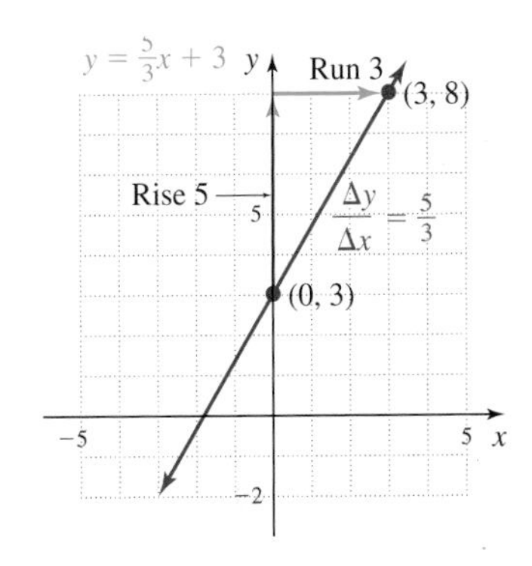

NOW TRY EXERCISES 51 THROUGH 62

Parallel and Perpendicular Lines

From Section 2.1 we know parallel lines have equal slopes: $m_1 = m_2$, and perpendicular lines have slopes with a product of -1: $m_1 \cdot m_2 = -1$ or $m_1 = -\frac{1}{m_2}$. In some applications, we need to find the equation of a second line parallel or perpendicular to a given line, through a known point. Using the slope-intercept form makes this a three-step process: (1) find the slope m_1 of the given line, (2) find the slope m_2 using the parallel or perpendicular relationship, and (3) use the given point with m_2 to find the required equation.

EXAMPLE 8 Find the equation of a line that goes through $(-6, -1)$ and is parallel to $2x + 3y = 6$.

Solution: Begin by writing the equation in function form to identify the slope.

$2x + 3y = 6$ given line

$3y = -2x + 6$ isolate y term

$y = \frac{-2}{3}x + 2$ result

The original line has slope $m_1 = \frac{-2}{3}$ and this will also be the slope of any line parallel to it. Using $m_2 = \frac{-2}{3}$ with $(x, y) \rightarrow (-6, -1)$ we have

$$y = mx + b \quad \text{function form}$$
$$-1 = \frac{-2}{3}(-6) + b \quad \text{substitute } \tfrac{-2}{3} \text{ for } m, -6 \text{ for } x, \text{ and } -1 \text{ for } y$$
$$-5 = b \quad \text{simplify}$$

The equation of the new line is $y = \frac{-2}{3}x - 5$.

NOW TRY EXERCISES 63 THROUGH 74 ▶

GRAPHICAL SUPPORT

Graphing the lines from Example 8 as Y1 and Y2 on a graphing calculator, we note the lines do appear to be parallel (they actually *must* be since they have identical slopes). Using the ZOOM **8:ZInteger** feature of the TI-84 Plus (Section 2.1 *Technology Highlight*) we can quickly verify that Y2 indeed contains the point $(-6, -1)$.

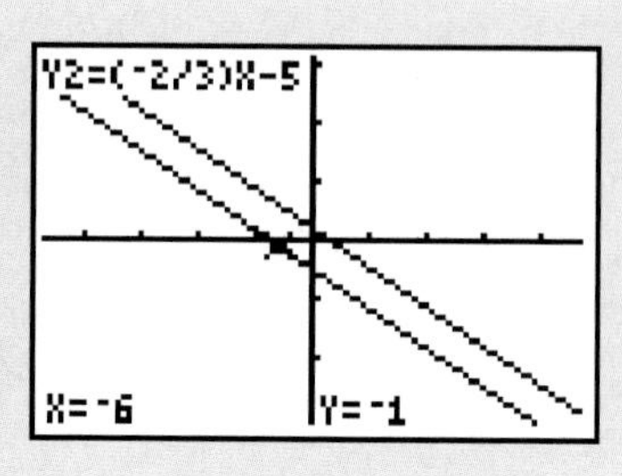

There are a large variety of applications that use linear models. In many cases, the coefficients are noninteger and descriptive variables are used.

Throughout the remainder of this chapter, it will be important to remember that slope represents a *rate of change*. The notation $m = \frac{\Delta y}{\Delta x}$ literally means the quantity y is changing with respect to changes in x.

EXAMPLE 9 ▶ In meteorological studies, atmospheric temperature depends on the altitude according to the function $T(h) = -3.5h + 58.6$, where $T(h)$ represents the approximate Fahrenheit temperature at height h (in thousands of feet).

a. Interpret the meaning of the slope in this context.

b. Determine the temperature at an altitude of 12,000 ft.

c. If temperature is $-10°$F what is the approximate altitude?

Solution: ▶ **a.** Notice that h is the input variable and T is the output. This shows $\frac{\Delta T}{\Delta h} = \frac{-3.5}{1}$, meaning for every 1000-ft increase in altitude, the temperature drops 3.5°.

b. Since height is in thousands, use $h = 12$.

$$T(h) = -3.5h + 58.6 \quad \text{original function}$$
$$T(12) = -3.5(12) + 58.6 \quad \text{substitute } h = 12$$
$$= 16.6 \quad \text{result}$$

At a height of 12,000 ft, the temperature is about 17°F.

WORTHY OF NOTE

When using function notation, it is important to remember that $y = f(x)$ literally means y can be substituted for $f(x)$, and $f(x)$ can be substituted for y. In other words, the notation "$f(x)$" represents a single quantity. In the same way, $T(h)$ represents a *single quantity*—the *temperature* at height h. This is why we replaced $T(h)$ with -10 in Example 9(c).

c. Replacing $T(h)$ with -10 and solving gives

$$\begin{aligned} T(h) &= -3.5h + 58.6 && \text{original function} \\ -10 &= -3.5h + 58.6 && \text{substitute } -10 \text{ for } T(h) \\ -68.6 &= -3.5h && \text{simplify} \\ 19.6 &= h && \text{result} \end{aligned}$$

The temperature is $-10°$F at a height of $19.6 \times 1000 = 19{,}600$ ft.

NOW TRY EXERCISES 103 AND 104

D. Linear Equations in Point-Slope Form

As an alternative to using $y = mx + b$, we can find the equation of the line using the slope formula $\frac{y_2 - y_1}{x_2 - x_1} = m$, and the fact that *the slope of a line is constant*. If we let (x_1, y_1) represent a *given* point on the line and (x, y) represent *any other point* on the line, the formula becomes $\frac{y - y_1}{x - x_1} = m$. Isolating the "$y$" terms on one side gives a new formula for the equation of a line, called the **point-slope form:**

$$\begin{aligned} \frac{y - y_1}{x - x_1} &= m && \text{slope form} \\ \frac{(x - x_1)}{1}\left(\frac{y - y_1}{x - x_1}\right) &= m(x - x_1) && \text{multiply both sides by } (x - x_1) \\ y - y_1 &= m(x - x_1) && \text{simplify} \rightarrow \text{point-slope form} \end{aligned}$$

> **THE POINT-SLOPE FORM OF A LINEAR EQUATION**
> Given a line with slope m and any point (x_1, y_1) on this line, the equation of the line is $y - y_1 = m(x - x_1)$.

While using $y = mx + b$ as in Example 6 may appear to be easier, both the y-intercept form and point-slope form each have their own advantages and it will help to be familiar with both.

WORTHY OF NOTE

It is helpful to note we can write the slope $\frac{\Delta y}{\Delta x} = \frac{2}{3}$ in the equivalent form $\frac{\Delta y}{\Delta x} = \frac{-2}{-3}$, and from a known point, count 2 units down and 3 units left to arrive at a second point on the line as well. For any negative slope $\frac{\Delta y}{\Delta x} = -\frac{a}{b}$, note $-\frac{a}{b} = \frac{-a}{b} = \frac{a}{-b}$.

EXAMPLE 10 Find the equation of the line in point-slope form, given $m = \frac{2}{3}$ and $(-3, -3)$ is on the line. Then write the equation in function form.

Solution:

$$\begin{aligned} y - y_1 &= m(x - x_1) && \text{point-slope form} \\ y - (-3) &= \frac{2}{3}[x - (-3)] && \text{substitute } \tfrac{2}{3} \text{ for } m; (-3, -3) \text{ for } (x_1, y_1)\text{: point-slope form} \\ y + 3 &= \frac{2}{3}x + 2 && \text{distribute and simplify} \\ y &= \frac{2}{3}x - 1 && \text{solve for } y\text{: function form} \end{aligned}$$

NOW TRY EXERCISES 75 THROUGH 92

E. Applications of Point-Slope Form

As a mathematical tool, equation models from the **family of linear functions** rank among the most powerful, common, and versatile. With many applications, particularly when working with real data, the slope of the line modeling the data is unknown, but we can easily find the coordinates of *two points on the line*. After calculating the slope with these points, we then use the point-slope form to find the equation and answer any related questions. One such application is *linear depreciation,* as when a government allows businesses to depreciate vehicles and equipment over time (the less a piece of equipment is worth, the less you pay in taxes).

WORTHY OF NOTE

Actually, it doesn't matter which of the two points are used in Example 11. Once the point (0, 60,000) is plotted, a constant slope of $m = -6000$ will "drive" the line through (3, 42,000). If we first graph (3, 42,000), the same slope would "drive" the line through (0, 60,000). Convince yourself by reworking the problem using the other point.

EXAMPLE 11 In 2002, a newspaper company bought a new printing press for \$60,000. By 2005 the value of the press had depreciated to \$42,000. Find a linear function that models this depreciation and discuss the slope and y-intercept in context.

Solution: Since the value of the press depends on its age, the ordered pairs have the form (age, value) where *age* is the input, *value* is the output. This means our first ordered pair is (0, 60,000), while the second is (3, 42,000).

$$m = \frac{y_2 - y_1}{x_2 - x_1} \quad \text{slope formula}$$

$$= \frac{42{,}000 - 60{,}000}{3 - 0} \quad (x_2, y_2) = (3, 42{,}000),\ (x_1, y_1) = (0, 60{,}000)$$

$$= \frac{-18{,}000}{3} \text{ or } \frac{-6000}{1} \quad \text{simplify and reduce}$$

The slope of the line is $\frac{\Delta\text{Value}}{\Delta\text{Years}} = \frac{-6000}{1}$. In this context it indicates the printing press loses \$6000 in value with each passing year.

$$y - y_1 = m(x - x_1) \quad \text{point-slope form}$$

$$y - 60{,}000 = -6000(x - 0) \quad \text{substitute } -6000 \text{ for } m; (0, 60{,}000) \text{ for } (x_1, y_1)$$

$$y - 60{,}000 = -6000x \quad \text{simplify}$$

$$y = -6000x + 60{,}000 \quad \text{solve for } y$$

The depreciation equation is $y = -6000x + 60{,}000$. Here the y-intercept (0, 60,000) simply indicates the original price of the equipment.

NOW TRY EXERCISES 105 THROUGH 108

Once the depreciation equation is found, it represents the (age, value) relationship for all future (and intermediate) ages of the press. In other words, we can now predict the value of the press for any given year. If the input age is *between* known data points, we're using the equation to **interpolate** information. If the input age is *beyond* or outside the known data, we're using the equation to **extrapolate** information. Care should be taken when extrapolating information from an equation, since some equation models are valid for only a set period of time.

EXAMPLE 12 Referring to Example 11,

a. How much will the press be worth in 2009?

b. How many years until the value of the equipment is less than \$3000?

c. Is this equation model valid for $t = 15$ yr (why or why not)?

Solution:

a. In the year 2009, the press will be 2009 − 2002 = 7 yr old. Evaluating the function at $t = 7$ gives

$$\begin{aligned} V(t) &= -6000t + 60{,}000 && \text{value at time } t \\ V(7) &= -6000(7) + 60{,}000 && \text{substitute 7 for } t \\ &= 18{,}000 && \text{result (7, 18,000)} \end{aligned}$$

In the year 2009, the printing press will only be worth \$18,000 dollars.

b. "Value is less than \$3000" means $V(t) < 3000$

$$\begin{aligned} V(t) &< 3000 && \text{value at time } t \\ -6000t + 60{,}000 &< 3000 && \text{substitute } -6000t + 60{,}000 \text{ for } V(t) \\ -6000t &< -57{,}000 && \text{solve for } t \\ t &> 9.5 && \text{divide and } \textit{reverse inequality symbol} \end{aligned}$$

In the year 2011, 9.5 yr after 2002, the printing press will be worth less than \$3000 dollars.

c. The equation model is not valid for $t = 15$, since $V(15)$ yields a negative quantity. In the current context, the model is only valid while $V(t) \geq 0$. In this case the domain of the function is $t \in [0, 10]$.

NOW TRY EXERCISES 109 THROUGH 114

TECHNOLOGY HIGHLIGHT

Graphing Calculators and the Equation of a Line

The keystrokes shown apply to a TI-84 Plus model. Please consult your manual or our Internet site for other models.

In Section 2.2, we used the **LIST** feature of the TI-84 Plus to plot points in the coordinate plane. In this *Highlight,* we learn how to use **LISTs** and the regression capabilities of a graphing calculator to find the *equation of a line.* Given any two points on a line, we can find its equation by computing the slope and using the point-slope formula. The TI-84 Plus can find the equation of the line using two points, by (1) entering the points in an ordered list and (2) calculating an equation, called the **regression equation,** using these points. We'll illustrate using the points (−6, −5) and (2, 7).

1. Enter the points in a List

As in Section 2.2 we enter the domain (x) values in list L1 and the range (y) values in L2. After clearing List1 and List2, press the STAT ENTER to select option **1:EDIT**. This places the cursor in the first position of List1, where we simply enter the domain values in order: −6 ENTER 2 ENTER. Use the right arrow key ▶ to navigate over to List2, and enter the range values in sequence: −5 ENTER 7 ENTER.

2. Find the equation of the line

To have the calculator compute the equation of this line, we press the STAT ▶ to highlight the **CALC** submenu. The fourth option is **4:LinReg(ax + b)** (see Figure 2.38). This option tells the calculator to use a technique called *linear regression* to find the equation of the line in the form $y = ax + b$ (the calculator uses "a" for slope instead of "m"). Pressing the number 4 on the number pad or navigating to **4:LinReg(ax + b)** and pressing ENTER will place **LinReg(ax + b)** on the Home Screen. The default lists are L1 for the x inputs and L2 for the y outputs, but we can also tell the calculator

Figure 2.38

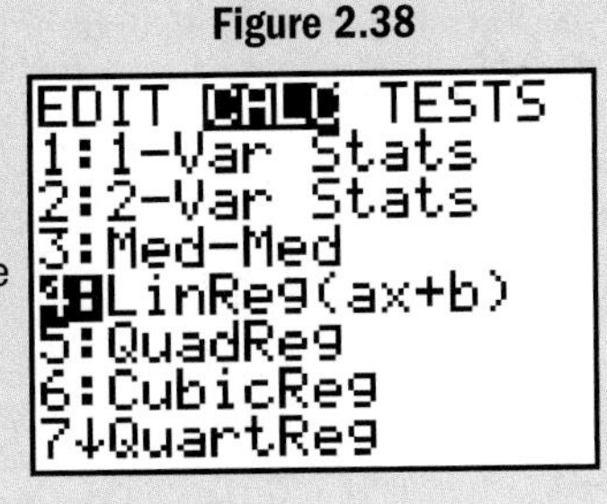

to use any two lists by entering the list number directly after the LinReg(*ax* + *b*) command. In this case we enter L1 and L2 (using 2nd 1 , 2nd 2). The screen should now read **LinReg(ax + b) L1, L2.** Press ENTER once again and the calculator gives the equation of the line in the form shown (see Figure 2.39). The equation of this line is $y = \frac{3}{2}x + 4$. As always, we should *use our estimation skills* as a double check on any result given by a calculator, in case some of the information was entered incorrectly. Given two points, mentally estimate whether the slope is positive or negative, greater than or less than one, and so on. Also, the true value of the TI-84's regression abilities will be seen in later sections, when we find a *line of best fit* for a large number of points.

Figure 2.39

```
LinReg
 y=ax+b
 a=1.5
 b=4
 r²=1
 r=1
```

Use the regression abilities of the TI-84 Plus to find the equation of the line through each pair of points in Exercises 1 through 4, enter this equation on the Y= screen, then use the **TABLE** feature to verify that the point midway between each pair is also on the line (use the midpoint formula).

Exercise 1:	(quantity purchased, cost each)	(97, \$49)	(122, \$44)
Exercise 2:	(laborers assigned, days to complete roadway)	(24, 108)	(40, 84)
Exercise 3:	(years since 1995, median cost of a house):	(0, \$97,500)	(9, \$118,200)
Exercise 4:	(years since 1995, college professor's average salary)	(0, \$42,165)	(9, \$49,419)

Exercise 5: Use the equation from Exercise 3 to predict the median cost of a house in the year 2008.

Exercise 1. $y = -0.2x + 68.4$; (109.5, 46.5) is on the line.✓
Exercise 2. $y = -1.5x + 144$; (32, 96) is on the line.✓
Exercise 3. $y = 2300x + 97{,}500$; (4.5, 107,850) is on the line.✓
Exercise 4. $y = 806x + 42{,}165$; (4.5, 45,792) is on the line.✓
Exercise 5. In 2008 ($x = 13$), the median price will be \$127,400.

2.3 EXERCISES

CONCEPTS AND VOCABULARY

Fill in each blank with the appropriate word or phrase. Carefully reread the section if needed.

1. For the equation $y = -\frac{7}{4}x + 3$, the slope is $\frac{-7}{4}$ and the y-intercept is (0, 3).
2. The notation $\frac{\Delta\text{cost}}{\Delta\text{time}}$ indicates the cost is changing in response to changes in time.
3. Line 1 has a slope of -0.4. The slope of any line perpendicular to line 1 is 2.5.
4. The equation $y - y_1 = m(x - x_1)$ is called the point-slope form of a line.
5. Discuss/explain how to graph a line using only the slope and a point on the line (no equations). Answers will vary.
6. Given $m = -\frac{3}{5}$ and $(-5, 6)$ is on the line. Compare and contrast finding the equation of the line using $y = mx + b$ versus $y - y_1 = m(x - x_1)$. Answers will vary.

DEVELOPING YOUR SKILLS

Solve each equation for y and evaluate the result using $x = -5, x = -2, x = 0, x = 1$, and $x = 3$.

7. $4x + 5y = 10$
8. $3y - 2x = 9$
9. $-0.4x + 0.2y = 1.4$
10. $-0.2x + 0.7y = -2.1$
11. $\frac{1}{3}x + \frac{1}{5}y = -1$
12. $\frac{1}{7}y - \frac{1}{3}x = 2$

7. $y = \frac{-4}{5}x + 2$

x	y
−5	6
−2	$\frac{18}{5}$
0	2
1	$\frac{6}{5}$
3	$\frac{-2}{5}$

8. $y = \frac{2}{3}x + 3$

x	y
−5	$\frac{-1}{3}$
−2	$\frac{5}{3}$
0	3
1	$\frac{11}{3}$
3	5

9. $y = 2x + 7$

x	y
−5	−3
−2	3
0	7
1	9
3	13

10. $y = \frac{2}{7}x - 3$

x	y
−5	$\frac{-31}{7}$
−2	$\frac{-25}{7}$
0	−3
1	$\frac{-19}{7}$
3	$\frac{-15}{7}$

11. $y = \frac{-5}{3}x - 5$

x	y
−5	$\frac{10}{3}$
−2	$\frac{-5}{3}$
0	−5
1	$\frac{-20}{3}$
3	−10

12. $y = \frac{7}{3}x + 14$

x	y
−5	$\frac{7}{3}$
−2	$\frac{28}{3}$
0	14
1	$\frac{49}{3}$
3	21

Write each equation in function form and identify the new coefficient of x and new constant term.

13. $6x - 3y = 9$ — $f(x) = 2x - 3$, new coeff. 2, constant −3
14. $9y - 4x = 18$ — $f(x) = \frac{4}{9}x + 2$, new coeff. $\frac{4}{9}$, constant 2
15. $-0.5x - 0.3y = 2.1$ — $f(x) = \frac{-5}{3}x - 7$, new coeff. $\frac{-5}{3}$, constant −7
16. $-0.7x + 0.6y = -2.4$ — $f(x) = \frac{7}{6}x - 4$, new coeff. $\frac{7}{6}$, constant −4
17. $\frac{5}{6}x + \frac{1}{7}y = -\frac{4}{7}$ — $f(x) = \frac{-35}{6}x - 4$, new coeff. $\frac{-35}{6}$, constant −4
18. $\frac{7}{12}y - \frac{4}{15}x = \frac{7}{6}$ — $f(x) = \frac{16}{35}x + 2$, new coeff. $\frac{16}{35}$, constant 2

19.

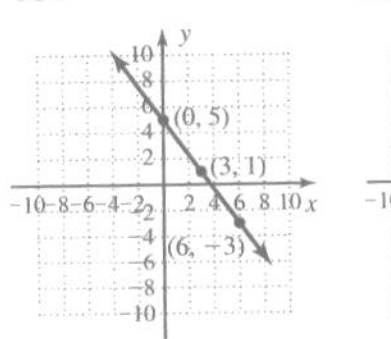

20.

21.

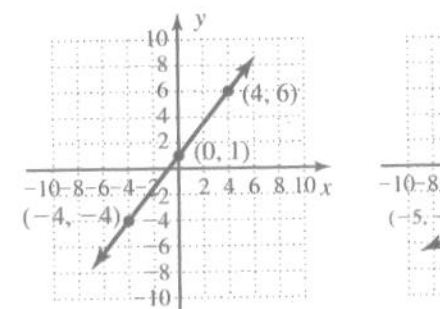

22.

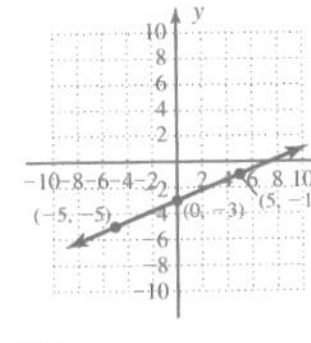

23. 24.

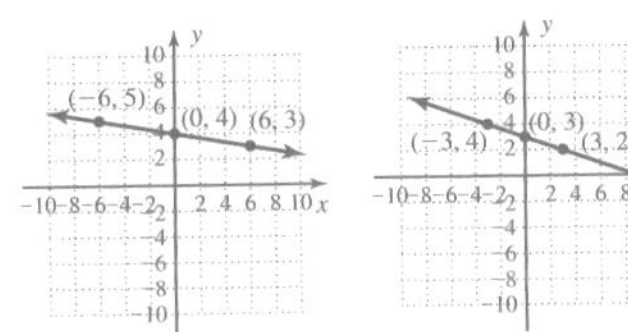

25. a. $\frac{-3}{4}$
b. $f(x) = \frac{-3}{4}x + 3$
c. The coeff. of x is the slope and the constant is the y-intercept.

26. a. $\frac{2}{3}$
b. $f(x) = \frac{2}{3}x - 2$
c. The coeff. of x is the slope and the constant is the y-intercept.

27. a. $\frac{2}{5}$
b. $f(x) = \frac{2}{5}x - 2$
c. The coeff. of x is the slope and the constant is the y-intercept.

28. a. $\frac{-2}{3}$
b. $f(x) = \frac{-2}{3}x + 3$
c. The coeff. of x is the slope and the constant is the y-intercept.

29. a. $\frac{4}{5}$
b. $f(x) = \frac{4}{5}x + 3$
c. The coeff. of x is the slope and the constant is the y-intercept.

30. a. $\frac{-6}{5}$
b. $f(x) = \frac{-6}{5}x - 5$
c. The coeff. of x is the slope and the constant is the y-intercept.

31. $y = \frac{-2}{3}x + 2$, $m = \frac{-2}{3}$, y-intercept (0, 2)

32. $y = \frac{3}{4}x + 3$, $m = \frac{3}{4}$, y-intercept (0, 3)

33. $y = \frac{-5}{4}x + 5$, $m = \frac{-5}{4}$, y-intercept (0, 5)

34. $y = -2x + 4$, $m = -2$, y-intercept (0, 4)

Additional answers can be found in the Instructor Answer Appendix.

Evaluate each function by selecting three inputs that will result in integer values. Then graph each line.

19. $y = -\frac{4}{3}x + 5$ **20.** $y = -\frac{3}{2}x - 2$ **21.** $y = \frac{5}{4}x + 1$

22. $y = \frac{2}{5}x - 3$ **23.** $y = -\frac{1}{6}x + 4$ **24.** $y = -\frac{1}{3}x + 3$

Find the x- and y-intercepts for each line, then (a) use these two points to calculate the slope of the line, (b) write the equation in function form (solve for y) and compare the calculated slope and y-intercept to the equation in function form, and (c) comment on what you notice.

25. $3x + 4y = 12$ **26.** $3y - 2x = -6$ **27.** $2x - 5y = 10$

28. $2x + 3y = 9$ **29.** $4x - 5y = -15$ **30.** $5y + 6x = -25$

Write each equation in slope-intercept form, then identify the slope and y-intercept.

31. $2x + 3y = 6$ **32.** $4y - 3x = 12$ **33.** $5x + 4y = 20$

34. $y + 2x = 4$ **35.** $x = 3y$ **36.** $2x = -5y$

37. $3x + 4y - 12 = 0$ **38.** $5y - 3x + 20 = 0$

For Exercises 39 to 50, use the slope-intercept formula to find the equation of each line.

39. $y = \frac{2}{3}x + 1$

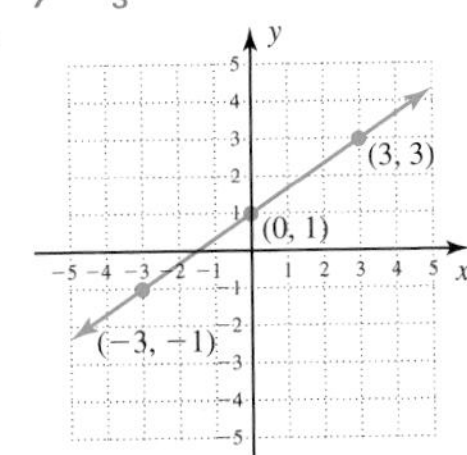

40. $y = \frac{-2}{5}x + 3$

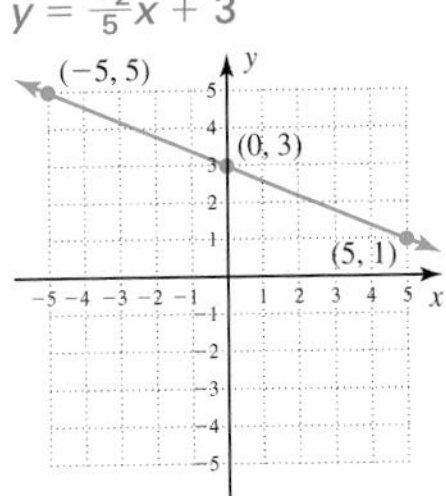

41. $y = 3x + 3$

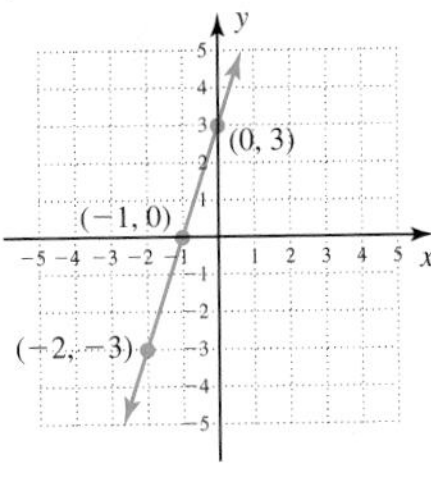

42. $m = -2$; y-intercept $(0, -3)$ $y = -2x - 3$

43. $m = 3$; y-intercept $(0, 2)$ $y = 3x + 2$

44. $m = -\frac{3}{2}$; y-intercept $(0, -4)$ $y = \frac{-3}{2}x - 4$

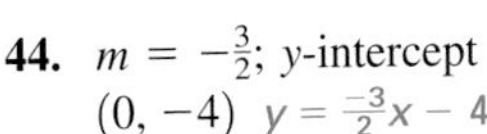

45.

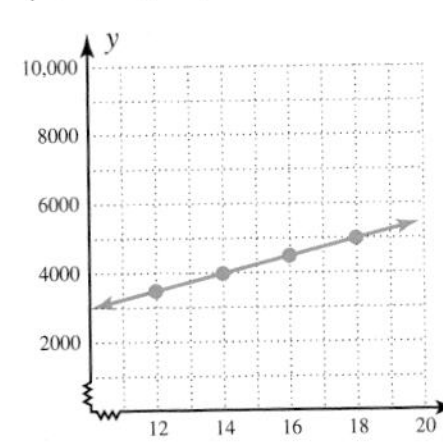

46.

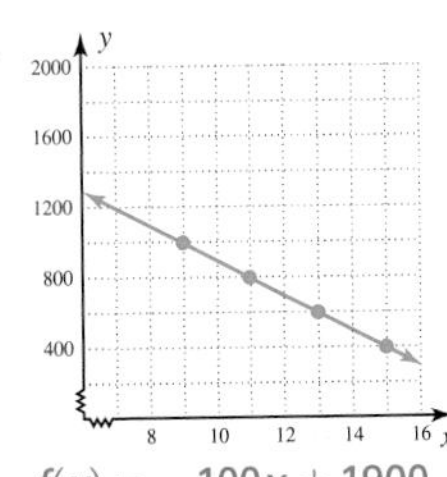

47.

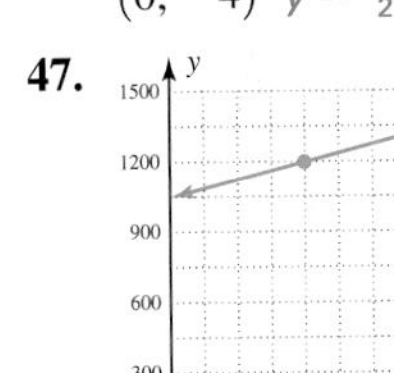

$f(x) = 250x + 500$ $f(x) = -100x + 1900$ $f(x) = \frac{75}{2}x + 150$

48. $m = -4$; $(-3, 2)$ is on the line $f(x) = -4x - 10$

49. $m = 2$; $(5, -3)$ is on the line $f(x) = 2x - 13$

50. $m = -\frac{3}{2}$; $(-4, 7)$ is on the line $f(x) = \frac{-3}{2}x + 1$

Write each equation in slope-intercept form, then use the slope and intercept to graph the line.

51. $4x + 5y = 20$ **52.** $2y - x = 4$ **53.** $5x - 3y = 15$ **54.** $2x + 5y = 10$

Graph each linear equation using the y-intercept and the slope $m = \frac{\Delta y}{\Delta x}$.

55. $y = \frac{2}{3}x + 3$ **56.** $y = \frac{5}{2}x - 1$ **57.** $y = \frac{-1}{3}x + 2$

58. $y = \frac{-4}{5}x + 2$ **59.** $y = 2x - 5$ **60.** $y = -3x + 4$

61. $f(x) = \frac{1}{2}x - 3$ **62.** $f(x) = \frac{-3}{2}x + 2$

Find the equation of the line using the information given. Write answers in slope-intercept form.

63. parallel to $2x - 5y = 10$, through the point $(3, -2)$ $y = \frac{2}{5}x - \frac{16}{5}$

64. parallel to $6x + 9y = 27$, through the point $(-3, -5)$ $y = \frac{-2}{3}x - 7$

65. perpendicular to $5y - 3x = 9$, through the point $(2, 4)$ $y = \frac{-5}{3}x + \frac{22}{3}$

66. perpendicular to $x - 4y = 7$, through the point $(-5, 3)$ $y = -4x - 17$

67. parallel to $12x + 5y = 65$, through the point $(-2, -1)$ $y = \frac{-12}{5}x - \frac{29}{5}$

68. parallel to $15y - 8x = 50$, through the point $(3, -4)$ $y = \frac{8}{15}x - \frac{28}{5}$

Write the lines in slope-intercept form and state whether they are parallel, perpendicular, or neither.

69. $4y - 5x = 8$; $5y + 4x = -15$ perpendicular

70. $3y - 2x = 6$; $-2x + 3y = -3$ parallel

71. $2x - 5y = 20$; $4x - 3y = 18$ neither

72. $5y = 11x + 135$; $11y + 5x = -77$ perpendicular

73. $-4x + 6y = 12$; $2x + 3y = 6$ neither

74. $3x + 4y = 12$; $6x + 8y = 2$ parallel

Find the equation of the line in point-slope form, then write the equation in function form and graph the line.

75. $m = 2;\ P_1 = (2, -5)$

76. $m = -1;\ P_1 = (2, -3)$

77. $m = \frac{3}{8};\ P_1 = (3, -4)$

78. $m = \frac{-5}{6};\ P_1 = (-1, 6)$

79. $m = 0.5;\ P_1 = (1.8, -3.1)$

80. $m = 1.5;\ P_1 = (-0.75, -0.125)$

A *secant line* is one that intersects a graph at two or more points. For the graph of each function given, find the equation of the line (a) parallel and (b) perpendicular to the secant line, through the point indicated.

81.

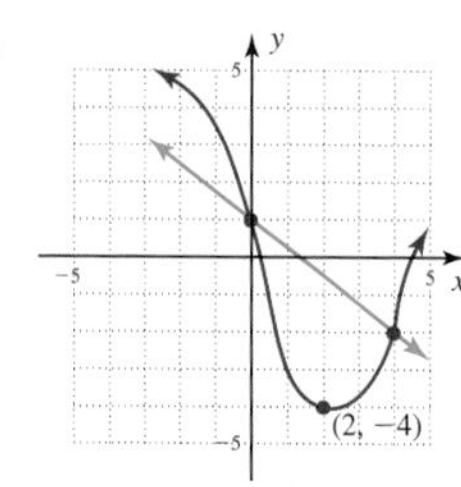

82.

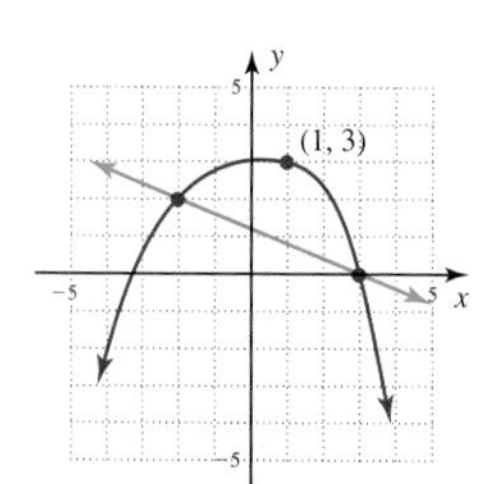

83.

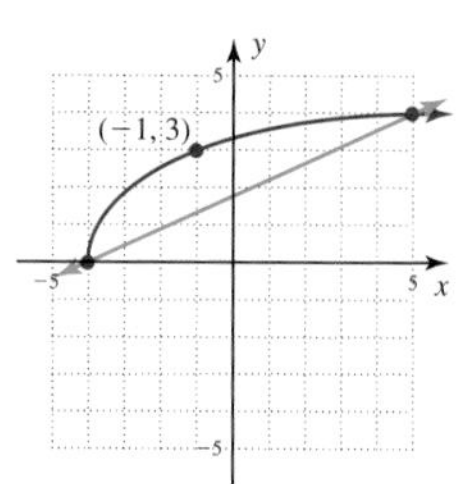

84.

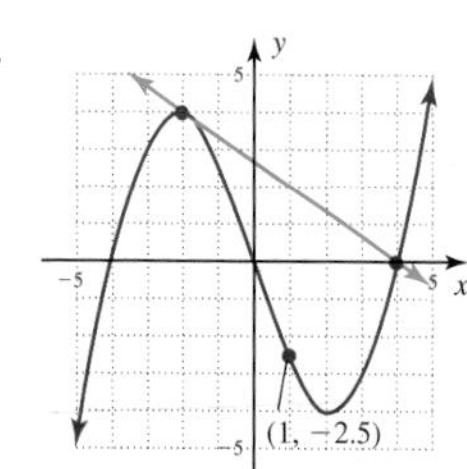

85.

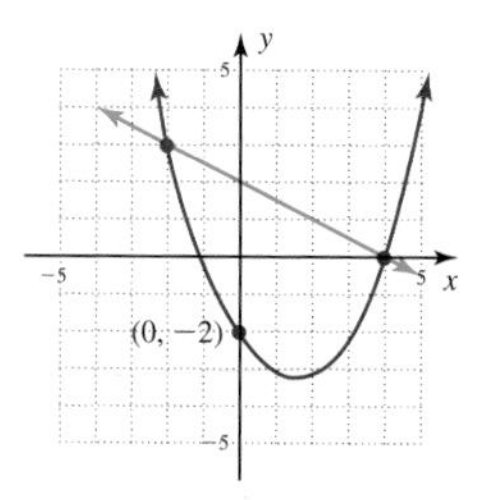

86.

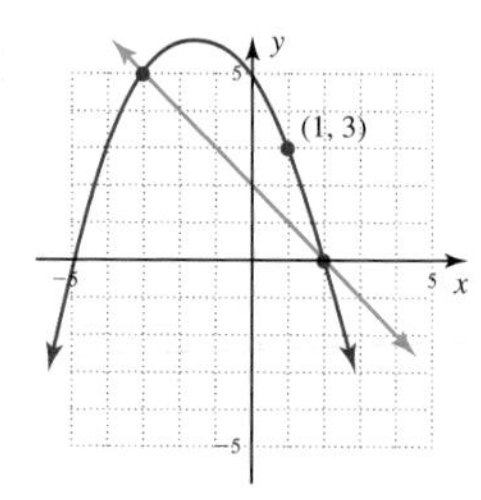

Find the equation of the line in point-slope form, then write the equation in slope-intercept form and state the meaning of the slope in context—what information is the slope giving us?

87.

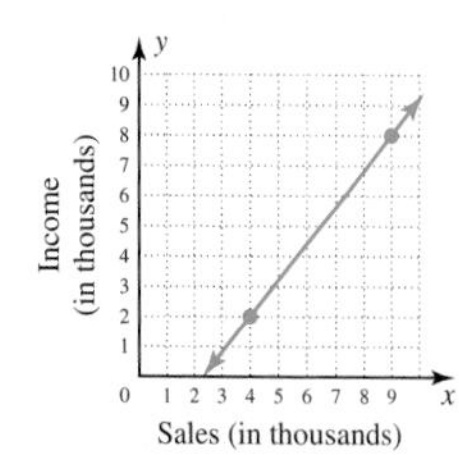

88.

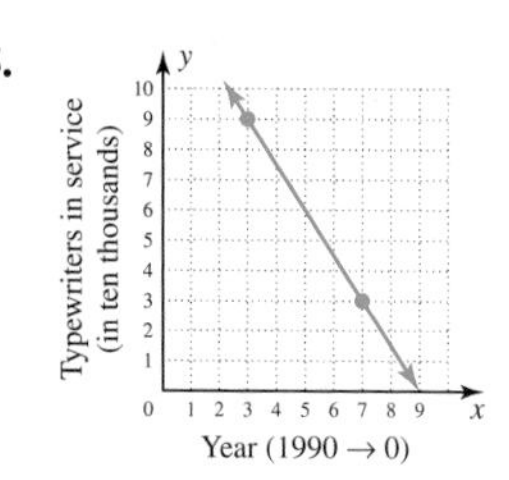

89.

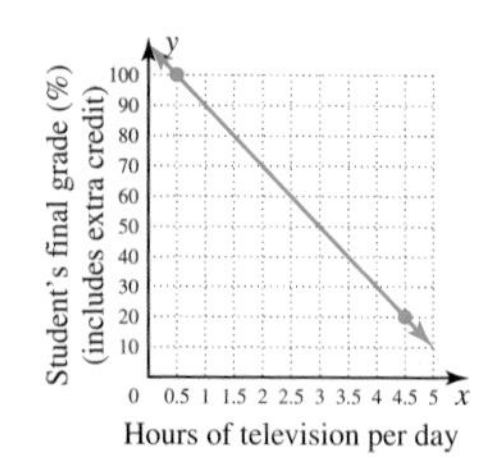

75. $f(x) = 2x - 9$

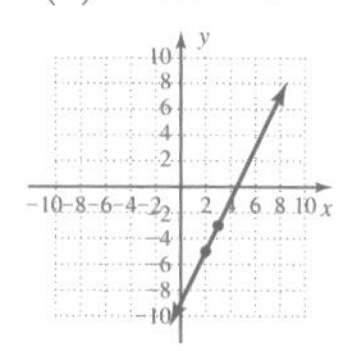

76. $f(x) = -x - 1$

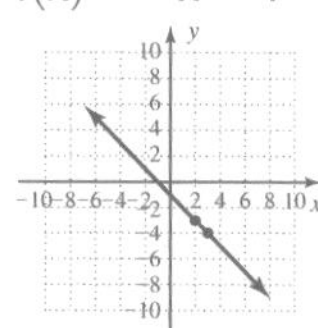

77. $f(x) = \frac{3}{8}x - \frac{41}{8}$

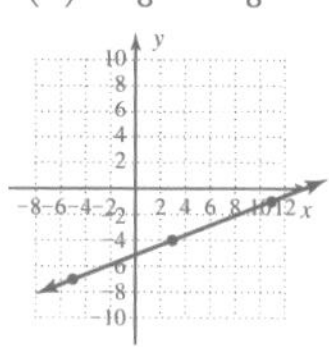

78. $f(x) = \frac{-5}{6}x + \frac{31}{6}$

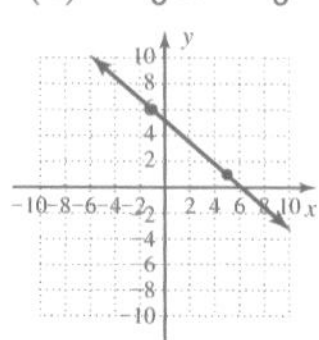

79. $f(x) = 0.5x - 4$

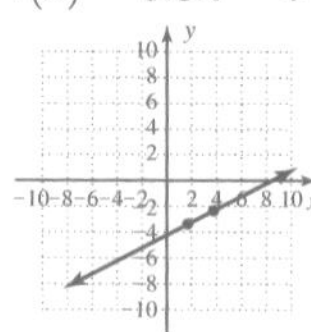

80. $f(x) = 1.5x + 1$

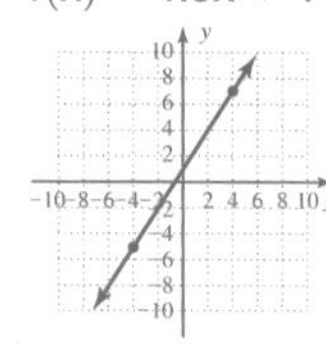

81. a. $y = \frac{-3}{4}x - \frac{5}{2}$
b. $y = \frac{4}{3}x - \frac{20}{3}$

82. a. $y = \frac{-2}{5}x + \frac{17}{5}$
b. $y = \frac{5}{2}x + \frac{1}{2}$

83. a. $y = \frac{4}{9}x + \frac{31}{9}$
b. $y = \frac{-9}{4}x + \frac{3}{4}$

84. a. $y = \frac{-2}{3}x - \frac{11}{6}$
b. $y = \frac{3}{2}x - 4$

85. a. $y = \frac{-1}{2}x - 2$
b. $y = 2x - 2$

86. a. $y = -x + 4$
b. $y = x + 2$

Additional answers can be found in the Instructor Answer Appendix.

90. $y = \frac{3}{7}x + \frac{15}{7}$; Every 7000 investors increases the number of online brokerage houses by 3.
91. $y = \frac{35}{2}x + \frac{5}{4}$; Every 2 in. of rainfall increases the number of cattle raised per acre by 35.
92. $y = \frac{2}{5}x - 22$; For every 5°F rise in temperature there are 2 additional eggs per hen per week.

90.

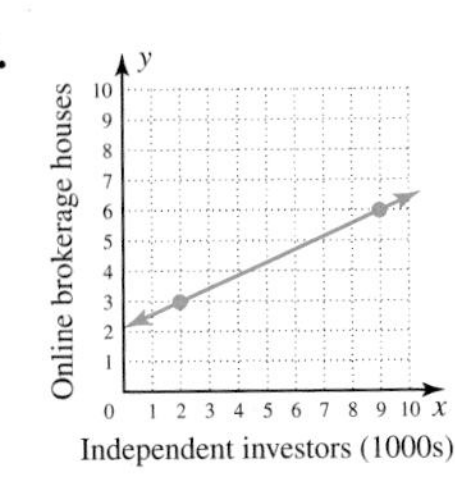

91.

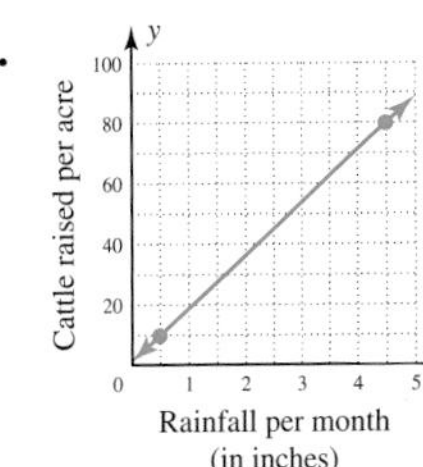

92.

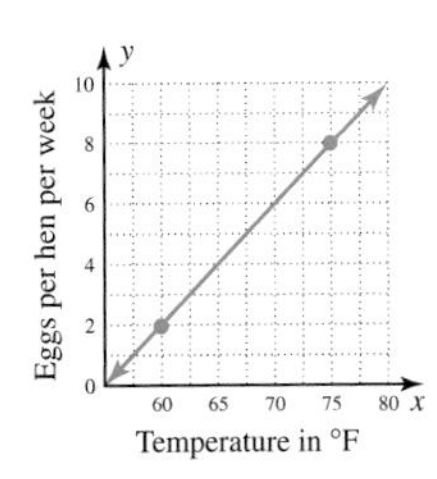

Using slope $= \frac{\Delta y}{\Delta x}$, match each description with the graph that best illustrates it. Assume time is scaled on the horizontal axes, and height, speed, or distance from the origin (as the case may be) is scaled on the vertical axis.

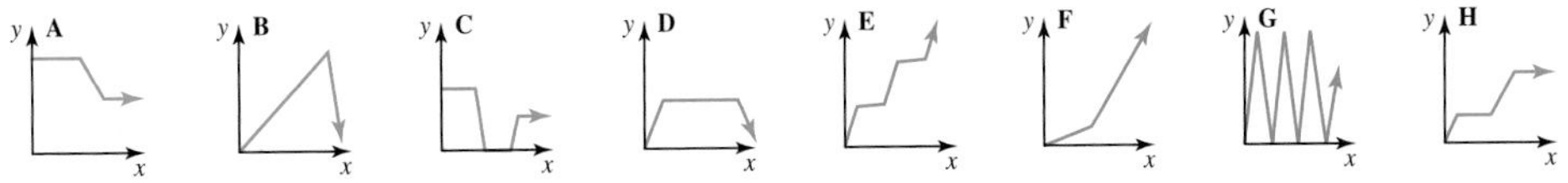

A B C D E F G H

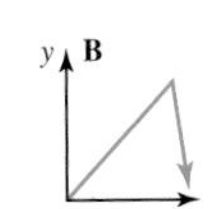
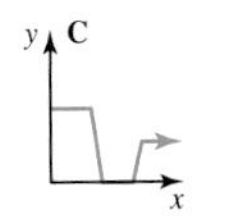
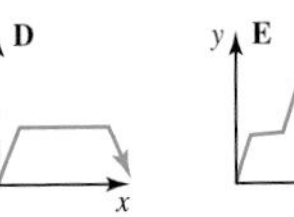
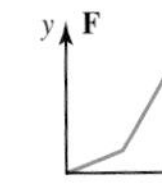
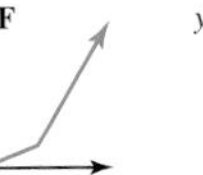
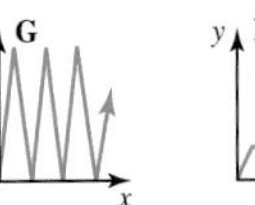

93. While driving today, I got stopped by a state trooper. After she warned me to slow down, I continued on my way. C

94. After hitting the ball, I began trotting around the bases shouting, "Ooh, ooh, ooh!" When I saw it wasn't a home run, I began sprinting. H

95. At first I ran at a steady pace, then I got tired and walked the rest of the way. A

96. While on my daily walk, I had to run for a while when I was chased by a stray dog. F

97. I climbed up a tree, then I jumped out. B

98. I steadily swam laps at the pool yesterday. G

99. I walked toward the candy machine, stared at it for a while then changed my mind and walked back. D

100. For practice, the girl's track team did a series of 50-m sprints, with a brief rest in between. E

WORKING WITH FORMULAS

101. a. $m = \frac{-3}{4}$, y-intercept $(0, 2)$
b. $m = \frac{-2}{5}$, y-intercept $(0, -3)$
c. $m = \frac{5}{6}$, y-intercept $(0, 2)$
d. $m = \frac{5}{3}$, y-intercept $(0, 3)$
102. a. $(5, 0)$ and $(0, 2)$
b. $(-4, 0)$ $(0, 3)$
c. $(\frac{8}{5}, 0)$ $(0, 2)$
Slope m is always equal to $\frac{-k}{h}$.
103. a. As the temperature increases 5°C, the velocity of sound waves increases 3 m/s. At a temperature of 0°C, the velocity is 331 m/s.
b. 343 m/s
c. 50°C

101. General linear equation: $ax + by = c$

The general equation of a line is shown here, where a, b, and c are real numbers, with a and b not simultaneously zero. Solve the equation for y and note the slope (coefficient of x) and y-intercept (constant term). Use these in their "formula form" to find the slope and y-intercept of the following lines, without solving for y or computing points.

a. $3x + 4y = 8$ **b.** $2x + 5y = -15$ **c.** $5x - 6y = -12$ **d.** $3y - 5x = 9$

102. Intercept/Intercept form of a linear equation: $\frac{x}{h} + \frac{y}{k} = 1$

The x- and y-intercepts of a line can also be found by writing the equation in the form shown (with the equation set equal to 1). The x-intercept will be $(h, 0)$ and the y-intercept will be $(0, k)$. Find the x- and y-intercepts of the following lines using this method: (a) $2x + 5y = 10$, (b) $3x - 4y = -12$, and (c) $5x + 4y = 8$. How is the slope of each line related to the values of h and k?

APPLICATIONS

103. Speed of sound: The speed of sound as it travels through the air depends on the temperature of the air according to the function $V(C) = \frac{3}{5}C + 331$, where $V(C)$ represents the velocity of the sound waves in meters per second (m/s), at a temperature of $C°$ Celsius.

a. Interpret the meaning of the slope and y-intercept in this context.

b. Determine the speed of sound at a temperature of 20°C.

c. If the speed of sound is measured at 361 m/s, what is the temperature of the air?

104. a. Every 5 seconds the velocity is increasing 26 ft/sec. The initial velocity is 60 ft/sec.
b. 108.88 ft/sec
c. $t \approx 7.7$ sec

105. a. $V(t) = \frac{20}{3}t + 150$
b. Every 3 yr the value of the coin increases by $20; the initial value was $150.

106. a. $V(t) = -3500t + 18{,}500$
b. Every 1 yr the equipment decreases in value by $3500; the initial value was $18,500.

107. a. $N(t) = 7x + 9$
b. Every 1 yr the number of homes with Internet access increases by 7 million.
c. 1993

108. a. $S(t) = 14.8t + 72$
b. Every 1 yr sales increase by $14.8 billion dollars.
c. 2007

109. a. $223.33
b. 15 years, in 2013
c. 3 yr

110. a. $4500
b. 3.6 yr
c. 5 yr

111. a. 86 million
b. 13 yr
c. 2010

104. Acceleration: A driver going down a straight highway is traveling 60 ft/sec (about 41 mph) on cruise control, when he begins accelerating at a rate of 5.2 ft/sec^2. The final velocity of the car is given by the function $V(t) = \frac{26}{5}t + 60$, where $V(t)$ is the velocity at time t. (a) Interpret the meaning of the slope and y-intercept in this context. (b) Determine the velocity of the car after 9.4 seconds. (c) If the car is traveling at 100 ft/sec, for how long did it accelerate?

105. Investing in coins: The purchase of a "collector's item" is often made in hopes the item will increase in value. In 1998, Mark purchased a 1909-S VDB Lincoln Cent (in fair condition) for $150. By the year 2004, its value had grown to $190.

a. Use the relation (time since purchase, value) with $t = 0$ corresponding to 1998 to find a linear equation modeling the value of the coin.

b. Discuss what the slope and y-intercept indicate in this context.

106. Depreciation: Once a piece of equipment is put into service, its value begins to depreciate. A business purchases some computer equipment for $18,500. At the end of a two-year period, the value of the equipment has decreased to $11,500.

a. Use the relation (time since purchase, value) to find a linear equation modeling the value of the equipment.

b. Discuss what the slope and y-intercept indicate in this context.

107. Internet connections: The number of households that are hooked up to the Internet (homes that are online) has been increasing steadily in recent years. In 1995, approximately 9 million homes were online. By 2001, this figure had climbed to about 51 million.

Source: 2004 *Statistical Abstract of the United States,* Table 965

a. Use the relation (year, homes online) with $t = 0$ corresponding to 1995 to find an equation model for the number of homes online.

b. Discuss what the slope indicates in this context.

c. According to this model, in what year did the first homes begin to come online?

108. Prescription drugs: Retail sales of prescription drugs has been increasing steadily in recent years. In 1995, retail sales hit 72 billion dollars. By the year 2000, sales had grown to about 146 billion dollars.

Source: 2004 *Statistical Abstract of the United States,* Table 965

a. Use the relation (year, retail sales of prescription drugs) with $t = 0$ corresponding to 1995 to find a linear equation modeling the growth of retail sales.

b. Discuss what the slope indicates in this context.

c. According to this model, in what year will sales reach 250 billion dollars?

109. Investing in coins: Referring to Exercise 105: (a) How much will the penny be worth in 2009? (b) How many years after purchase will the penny's value exceed $250? (c) If the penny is now worth $170, how many years has Mark owned the penny?

110. Depreciation: Referring to Exercise 106: (a) What is the equipment's value after 4 yr? (b) How many years after purchase will the value decrease to $6000? (c) Generally, companies will sell used equipment while it still has some value and use the funds to help purchase new equipment. According to the function, how many years will it take this equipment to depreciate in value to $1000?

111. Internet connections: Referring to Exercise 107,

a. If the rate of change stays constant, how many households will be on the Internet in 2006?

b. How many years after 1995 will there be over 100 million households connected?

c. If there are 115 million households connected, what year is it?

112. a. $220 billion
b. 14 yr
c. 2010

113. a. $P(t) = 58{,}000t + 740{,}000$
b. Each year, the prison population increases by 58,000.
c. 1,726,000

114. a. $M(t) = 2.7t + 143$
b. Each year, the number of restaurant meals increases by about 3 meals.
c. about 186 meals/yr

112. Prescription drug sales: Referring to Exercise 108,

a. According to the model, what was the value of retail prescription drug sales in 2005?

b. How many years after 1995 will retail sales exceed $279 billion?

c. If yearly sales totaled $294 billion, what year is it?

113. Prison population: In 1990, the number of persons sentenced and serving time in state and federal institutions was approximately 740,000. By the year 2000, this figure had grown to nearly 1,320,000. (a) Find a linear function with $t = 0$ corresponding to 1990 that models this data, (b) discuss the slope ratio in context, and (c) use the equation to estimate the prison population in 2007 if this trend continues.

Source: Bureau of Justice Statistics at www.ojp.usdoj.gov/bjs

114. Eating out: In 1990, Americans bought an average of 143 meals per year at restaurants. This phenomenon continued to grow in popularity and in the year 2000, the average reached 170 meals per year. (a) Find a linear function with $t = 0$ corresponding to 1990 that models this growth, (b) discuss the slope ratio in context, and (c) use the equation to estimate the average number of times an American will eat at a restaurant in 2006 if the trend continues.

Source: The NPD Group, Inc., National Eating Trends, 2002

WRITING, RESEARCH, AND DECISION MAKING

115. Locate and read the following article. Then turn in a one-page summary.
"Linear Function Saves Carpenter's Time," Richard Crouse,
Mathematics Teacher, Volume 83, Number 5, May 1990: pp. 400–401. Answers will vary.

116. Is there a relationship between the number of passengers an airliner can carry and the amount of fuel it uses? Presumably more passengers mean more weight, more baggage, a larger aircraft, and so on. The Boeing B767-300 can carry about 215 passengers and uses about 1583 gal of fuel per hour. The Boeing B717-200 carries about 110 passengers and uses close to 575 gal of fuel per hour. Assuming the relationship is linear, find an equation model using these two data points, then use the model to complete the third column of the table here. Finally, use an almanac, encyclopedia, the Internet, or other research tools to complete the fourth column. Comment on what you find. $F(p) = 9.6p - 481$

Aircraft Type	Approximate Number of Passengers	Approximate Fuel Consumption (gal/hr) from Linear Function	Approximate Fuel Consumption (gal/hr) from Research
B747-400	380	3167	3350
L1011-100	325	2639	2560
DC10-10	285	2255	2310
B767-300	**215**	**1583**	**1583**
B757-200	180	1247	1063
MD-80	140	863	940
B717-200	**110**	**575**	**575**

(Answers may vary.)

EXTENDING THE CONCEPT

117. Match the correct graph to the conditions stated for m and b. There are more choices than graphs.

a. $m < 0, b < 0$ **b.** $m > 0, b < 0$ **c.** $m < 0, b > 0$ **d.** $m > 0, b > 0$

e. $m = 0, b > 0$ **f.** $m < 0, b = 0$ **g.** $m > 0, b = 0$ **h.** $m = 0, b < 0$

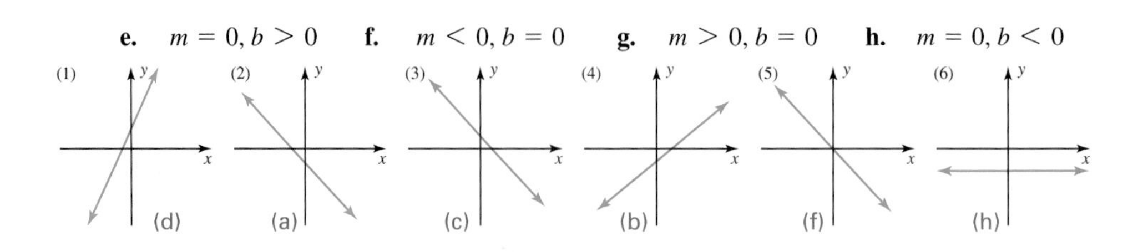

MAINTAINING YOUR SKILLS

118. (2.2) Determine the domain of the functions:

a. $f(x) = \sqrt{2x - 5}$ $x \in [\frac{5}{2}, \infty)$

b. $g(x) = \dfrac{5}{2x^2 + 3x - 2}$ $x \in (-\infty, -2) \cup (-2, \frac{1}{2}) \cup (\frac{1}{2}, \infty)$

119. (1.5) Solve using the quadratic formula. Answer in exact and approximate form: $3x^2 - 10x = 9$.
$x = \frac{5 \pm 2\sqrt{13}}{3}$; $x \approx -0.74$ or $x \approx 4.07$

120. (1.1) Three equations follow. One is an identity, another is a contradiction, and a third has a solution. State which is which.

$2(x - 5) + 13 - 1 = 9 - 7 + 2x$ identity

$2(x - 4) + 13 - 1 = 9 + 7 - 2x$ has a solution ($x = 3$)

$2(x - 5) + 13 - 1 = 9 + 7 + 2x$ contradiction

121. (R.7) Compute the area of the circular sidewalk shown here. Use your calculator's value of π and round the answer (only) to hundredths. 113.10 yd^2

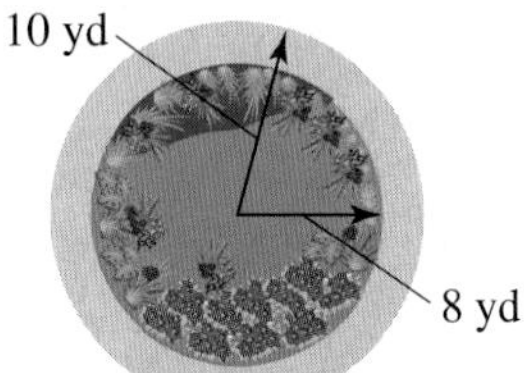

122. (2.2) Does this set of ordered pairs represent a function? Why/Why not? $\{(-7, 4), (5, -4), (-3, -2), (-7, -3), (0, 6)\}$ No, -7 is paired with 4 and -3.

123. (1.2) Solve the following inequality and state the solution set using interval notation: $-3 < 2x + 5$ and $x - 2 < 3$ $(-4, 5)$

MID-CHAPTER CHECK

1. 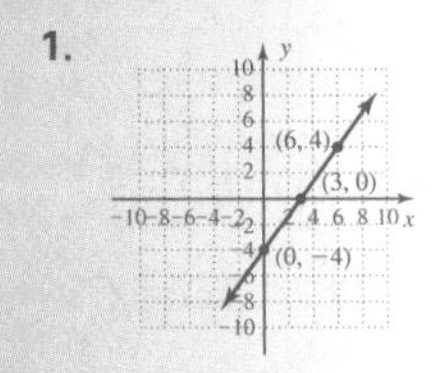

3. positive, loss is decreasing (profit is increasing); $m = \frac{3}{2}$, yes; $\frac{1.5}{1}$, each year Data.com's loss decreases by 1.5 million.

4. 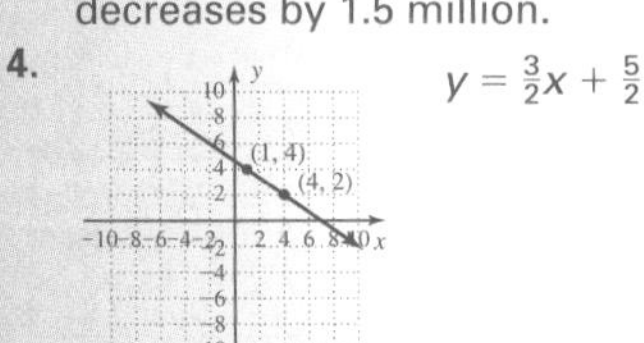

$y = \frac{3}{2}x + \frac{5}{2}$

5. $x = -3$; no; Input -3 is paired with more than one output.

Additional answers can be found in the Instructor Answer Appendix.

1. Sketch the graph of the line. Plot and label at least three points: $4x - 3y = 12$.
2. Find the slope of the line passing through the given points: $(-3, 8)$ and $(4, -10)$. $\frac{-18}{7}$
3. In 2002, Data.com lost \$2 million. In 2003, they lost \$0.5 million. Will the slope of the line through these points be positive or negative? Why? Calculate the slope. Were you correct? Write the slope as a unit rate and explain what it means in this context.
4. Sketch the line passing through (1, 4) with slope $m = \frac{-2}{3}$ (plot and label at least two points). Then find the equation of the line *perpendicular to this line* through (1, 4).
5. Write the equation for line L_1 shown to the right. Is this the graph of a function? Discuss why or why not.
6. Write the equation for line L_2 shown to the right. Is this the graph of a function? Discuss why or why not.
7. For the graph of function $h(x)$ shown, (a) determine the value of $h(2)$; (b) state the domain; (c) determine the value of x for which $h(x) = 4$; and (d) state the range.

Exercises 5 and 6

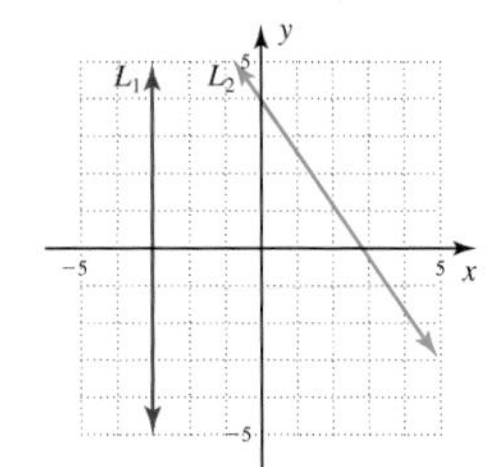

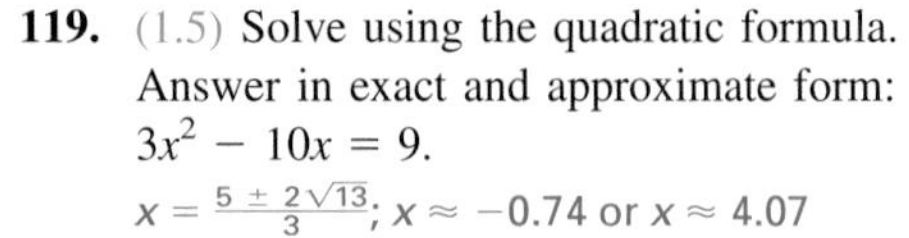

Exercises 7 and 8

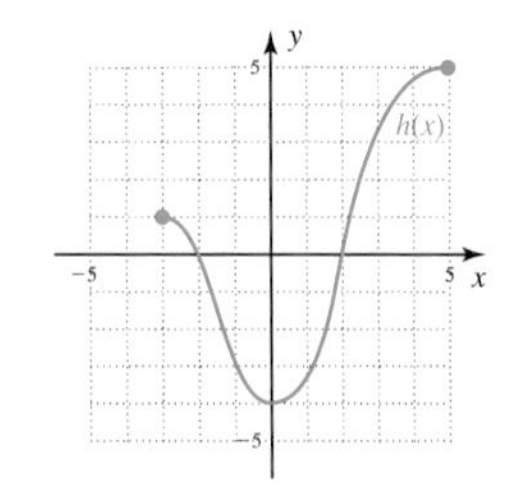

8. from $x = 1$ to $x = 2$; steeper line $\rightarrow$ greater slope
9. $F(p) = \frac{3}{4}p + \frac{5}{4}$; For every 4000 pheasants, the fox population increases by 300; 1625.
10. a. $x \in \{-3, -2, -1, 0, 1, 2, 3, 4\}$ $y \in \{-3, -2, -1, 0, 1, 2, 3, 4\}$
    b. $x \in [-3, 4]$ $y \in [-3, 4]$
    c. $x \in (-\infty, \infty)$ $y \in (-\infty, \infty)$

8. Judging from the appearance of the graph alone, compare the average rate of change from $x = 1$ to $x = 2$ to the rate of change from $x = 4$ to $x = 5$. Which rate of change is larger? How is that demonstrated graphically?

9. Find a linear function that models the graph of $F(p)$ given. Explain the slope of the line in this context, then use your model to predict the fox population when the pheasant population is 20,000.

Exercise 9

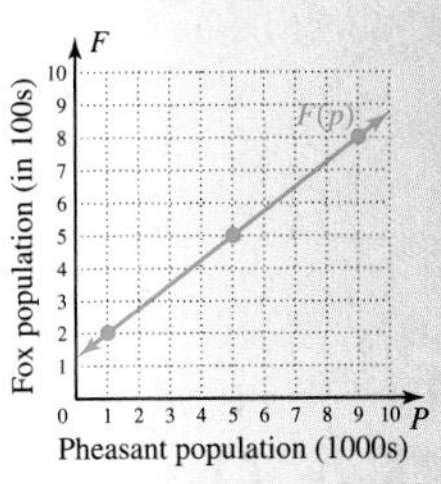

10. State the domain and range for each function below.

a.

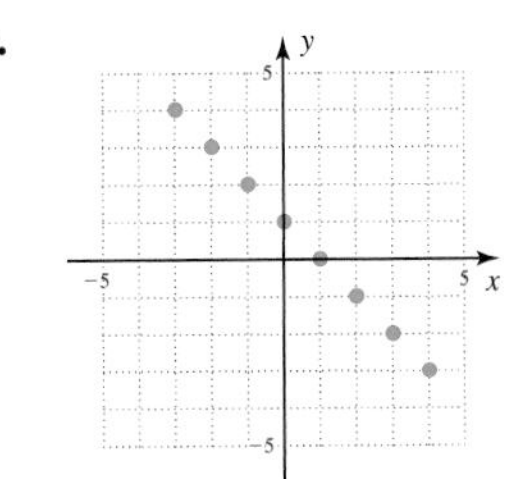

b.

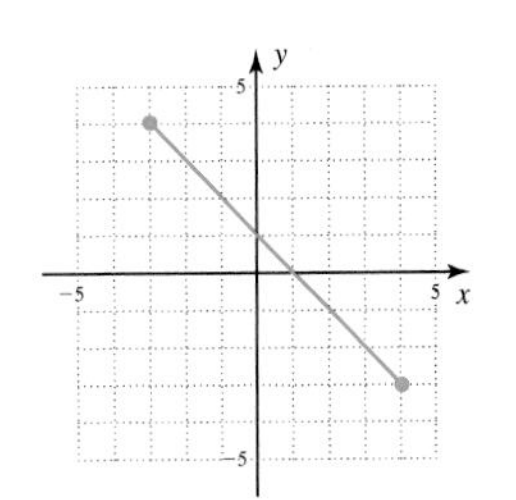

c.

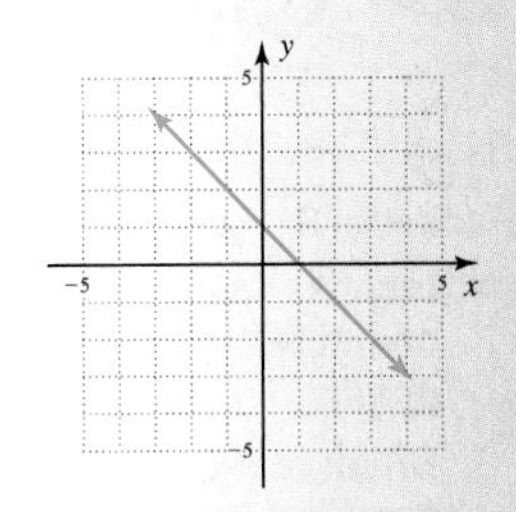

REINFORCING BASIC CONCEPTS

The Various Forms of a Linear Equation

In a study of mathematics, getting a glimpse of the "big picture" can be an enormous help. Learning mathematics is like building a skyscraper: The final height of the skyscraper ultimately depends on the strength of the foundation and quality of the frame supporting each new floor as it is built. Our work with linear functions and their graphs, while having a number of useful applications, is actually the foundation on which *much of your future work will be built*. The study of quadratic and polynomial functions and their applications all have their roots in linear equations. For this reason, it's important that you gain a certain fluency with linear functions—even to a point where things come to you effortlessly and automatically. This level of performance requires a strong desire and a sustained effort. We begin by reviewing the basic facts a student MUST know to reach this level. MUST is an acronym for memorize, understand, synthesize, and teach others. Don't be satisfied until you've done all four. Given points (x_1, y_1) and (x_2, y_2):

Forms and Formulas

slope formula	point-slope form	slope-intercept form	standard form
$m = \dfrac{y_2 - y_1}{x_2 - x_1}$	$y - y_1 = m(x - x_1)$	$y = mx + b$	$Ax + By = C$
given any two points on the line	given slope m and any point (x_1, y_1)	given slope m and y-intercept $(0, b)$	also used in linear systems (Chapter 6)

Characteristics of Lines

y-intercept	x-intercept	increasing	decreasing
$(0, y)$	$(x, 0)$	$m > 0$	$m < 0$
let $x = 0$, solve for y	let $y = 0$, solve for x	line slants upward from left to right	line slants downward from left to right

Practice for Speed and Accuracy

For the two points given, (a) compute the slope of the line and state whether the line is increasing or decreasing; (b) find the equation of the line using point-slope form;

Additional answers can be found in the Instructor Answer Appendix.

(c) write the equation in slope-intercept form; (d) write the equation in standard form; and (e) find the x- and y-intercepts and graph the line.

1. $P_1(0, 5); P_2(6, 7)$ **2.** $P_1(3, 2); P_2(0, 9)$ **3.** $P_1(3, 2); P_2(9, 5)$
4. $P_1(-5, -4); P_2(3, 2)$ **5.** $P_1(-2, 5); P_2(6, -1)$ **6.** $P_1(2, -7); P_2(-8, -2)$

2.4 Quadratic and Other Toolbox Functions

LEARNING OBJECTIVES

In Section 2.4 you will learn how to:

A. Identify basic characteristics of quadratic graphs

B. Graph factorable quadratic functions

C. Graph other toolbox functions

D. Compute the average rate of change for toolbox functions

INTRODUCTION

For many applications of mathematics, our first objective is to build or select a function model appropriate to the situation, and use the model to answer questions or make decisions. So far we've looked extensively at linear functions and briefly at the absolute value function. These are two of the eight **toolbox functions,** so called because they give us a variety of "tools" (equation models) to model the world around us. In this section, we introduce the quadratic, square root, cubic, and cube root functions. In the same way that a study of arithmetic depends heavily on the multiplication table, a study of algebra and mathematical modeling depends (in large part) on a solid working knowledge of these functions.

POINT OF INTEREST

The marriage of geometry and algebra was consummated in 1637, when Descartes published *La Géométrie,* cementing the connection between algebra (the function's equation) and geometry (the function's graph). Looking back at the history of mathematics, it appears that Descartes's work may have been the dividing line between medieval and modern mathematics.

A. Characteristics of Quadratic Graphs

Before we can effectively apply the toolbox functions as problem-solving tools, we need to know more about their graphs. While we can accurately graph a line using only two (or three) points, graphs of most toolbox functions usually require more points to show all of the graph's important features. However, our work is greatly simplified by the fact that each function belongs to a **function family,** in which all graphs from a particular family share like characteristics. This means the number of points required quickly decreases as we start anticipating what the graph of a given function should look like. Knowledge of a graph's important features, along with an awareness of the related domain and range, are critical components of problem solving and mathematical modeling.

The Squaring Function

The squaring function $f(x) = x^2$ is a quadratic function where $a = 1, b = 0$, and $c = 0$. Although it is the most basic quadratic, its graph is sufficient to illustrate all of the features that distinguish it from a linear graph and the graphs of other function families.

EXAMPLE 1 Graph the squaring function $f(x) = x^2$ by plotting points, using integer values between $x = -3$ and $x = 3$.

Solution:

x	f(x)	Ordered Pair
−3	f(−3) = 9	(−3, 9)
−2	f(−2) = 4	(−2, 4)
−1	f(−1) = 1	(−1, 1)
0	f(0) = 0	(0, 0)
1	f(1) = 1	(1, 1)
2	f(2) = 4	(2, 4)
3	f(3) = 9	(3, 9)

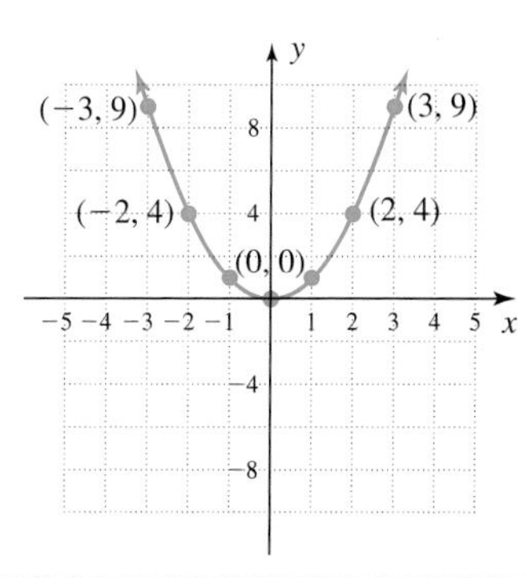

NOW TRY EXERCISES 7 THROUGH 10

The resulting graph is called a **parabola.** Parabolas have three special features that make them excellent real-world models, with applications in ballistics, astronomy, manufacturing, and many other fields. These three features follow and are illustrated in Figure 2.40.

Figure 2.40

(1) Concavity; (2) Axis of symmetry; $x = h$; y-intercept $(0, c)$; $(x_1, 0)$; $(x_2, 0)$; x-intercept(s); (h, k) (3) Vertex

1. **Concavity:** When we say a quadratic graph is **concave up,** we mean that the branches of the graph point in the positive y-direction. This "branching" is also referred to as **end behavior,** and we can describe the end behavior here as, "up on the left, up on the right," or simply, "up/up."
2. **Axis of symmetry:** Parabolas also have a feature called the *line* or *axis of symmetry,* an imaginary line that cuts the graph in half, with each half an exact reflection of the other.
3. **Vertex:** Unlike the graph of a line, a parabola will always have a highest or lowest point called the *vertex*. If the parabola is concave up, the y-value of the vertex is the **minimum value** of the function—the smallest possible y-value. If the parabola is concave down, the y-value of the vertex is the **maximum value.**

All quadratic graphs share these characteristics. Due to the symmetry of the graph, the axis of symmetry will *always go through the vertex* of the parabola. Traditionally, the coordinates of the vertex are written (h, k), meaning the axis of symmetry will be $x = h$, which we know is a vertical line through $(h, 0)$. As with all graphs, the y-intercept is found by substituting $x = 0$. For the x-intercepts (if they exist) substitute $f(x) = 0$ and solve for x. As drawn, the graph in Figure 2.40 has two x-intercepts.

EXAMPLE 2 Given the graph of $f(x) = -x^2 + 6x - 5$ shown in the figure, (a) describe or identify the features indicated and (b) use boundary lines to state the domain and range of the function. Assume noted features are lattice points.

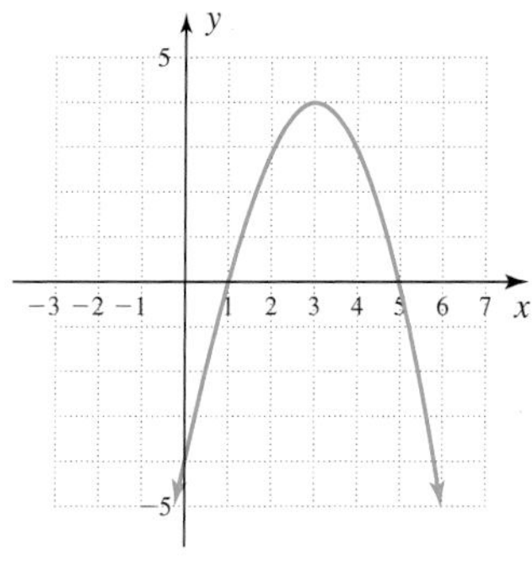

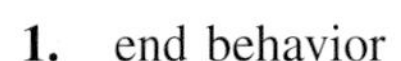

1. end behavior **2.** vertex

3. axis of symmetry **4.** y-intercept

5. x-intercept(s)

Solution:

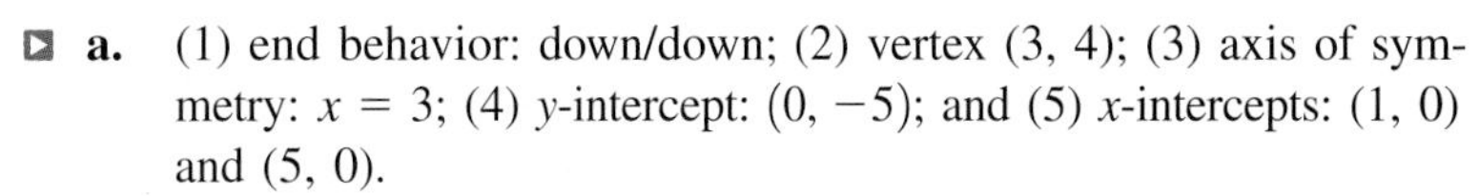

a. (1) end behavior: down/down; (2) vertex (3, 4); (3) axis of symmetry: $x = 3$; (4) y-intercept: $(0, -5)$; and (5) x-intercepts: (1, 0) and (5, 0).

b. A vertical line will intersect the graph or its extension anywhere it is placed, so the domain is $x \in (-\infty, \infty)$. A horizontal line will intersect the graph for values of y that are less than or equal to 4. The range is $y \in (-\infty, 4]$.

NOW TRY EXERCISES 11 THROUGH 16

As a double check on Example 2, we substitute $f(x) = 0$ to verify the x-intercepts. This gives $0 = -(x - 1)(x - 5)$ after factoring, with solutions $x = 1$ and $x = 5$—the same intercepts already noted.

B. Graphing Factorable Quadratic Functions

For the general quadratic $f(x) = ax^2 + bx + c$, the y-intercept is always $(0, c)$ since any term with a variable becomes zero. As before, we substitute $f(x) = 0$ for the x-intercepts, giving $0 = a^2 + bx + c$, which (for now) we attempt to solve by factoring. The graph in Example 2 had two x-intercepts and "cut" through the x-axis twice. It's possible for a quadratic function to have only one x-intercept, if the graph "bounces" off the x-axis as in Example 1; or no x-intercepts, if the graph is entirely above or below the x-axis. These functions will be investigated in Section 3.4. Our earlier work with linear functions and our observations here, suggest that real-number solutions to the equation $f(x) = 0$ *appear graphically as x-intercepts.* This is a powerful connection between a function and its graph, and one that is used throughout our study.

Just as only two points uniquely determine the graph of a line, it can be shown that three points are sufficient to determine a unique parabola. But to effectively graph parabolas with a limited number of plotted points, we must be very familiar with how these unique features can be determined *from the equation.* The basic ideas are discussed here in greater detail.

1. *End behavior (concavity):* The function in Example 1 had a positive lead coefficient ($a = 1$) and its graph was concave up. The function in Example 2 had a negative lead coefficient ($a = -1$) and its graph was concave down. This observation can be extended to all quadratic functions.

   For $f(x) = ax^2 + bx + c$, the graph will be concave up if $a > 0$ (graph is smiling ☺) and concave down if $a < 0$ (graph is frowning ☹).

WORTHY OF NOTE

As we will see later in our study, this method can still be applied, even when the roots are irrational or complex: $h = \frac{x_1 + x_2}{2}$

2. *Axis of symmetry:* For factorable quadratics, the graph will have either one or two x-intercepts, as seen in Examples 1 and 2, respectively. If there is only one x-intercept, this point will also be the vertex due to the parabola's shape, and the axis of symmetry will go through this point. In the case of two x-intercepts, the axis of symmetry will go through a point *halfway between them*—simply compute their average value.

   The axis of symmetry can be found using the average value of the x-intercepts: $h = \frac{x_1 + x_2}{2}$.

   In Example 2, the x-intercepts were $(1, 0)$ and $(5, 0)$. The "halfway point" is $h = \frac{x_1 + x_2}{2} = \frac{1 + 5}{2}$ or $h = 3$. Note this vertical line indeed cuts the graph into symmetric halves.

3. *Vertex:* Again due to the parabola's symmetry, the axis of symmetry will always go through the vertex of the parabola. The x-value for the axis is also the x-coordinate of the vertex, with the y-coordinate found by substitution. For

Example 2, the axis of symmetry was $h = 3$. Evaluating $f(3)$ gives 4, and the point (3, 4) is indeed the vertex of the parabola.

The vertex of a quadratic function is (h, k), where $h = \frac{x_1 + x_2}{2}$ and $k = f(h)$ (the function evaluated at h).

Here is a summary of the procedure for graphing factorable quadratic functions. As mentioned, graphs of nonfactorable equations are studied in Section 3.4.

GRAPHING FACTORABLE QUADRATIC FUNCTIONS

For the quadratic function $f(x) = ax^2 + bx + c$,

1. Determine end behavior: concave up if $a > 0$, concave down if $a < 0$.
2. Find the y-intercept by substituting 0 for x: $f(0) = c$.
3. Find the x-intercept(s) by substituting 0 for $f(x)$ and solving for x.
4. Find the axis of symmetry: $h = \frac{x_1 + x_2}{2}$.
5. Find the vertex $(h, f(h)) = (h, k)$.
6. Use these features to help sketch a parabolic graph.

EXAMPLE 3 Graph the function $f(x) = -2x^2 - 5x + 3$.

Solution: Using the features just discussed, we have:

End behavior: Since $a < 0$, concave down

y-intercept: $f(0) = 3$, the y-intercept is (0, 3).

x-intercept(s): Setting $f(x) = 0$ gives: $0 = -2x^2 - 5x + 3$.

$$0 = -(2x - 1)(x + 3) \longrightarrow x = \frac{1}{2} \quad \text{or} \quad x = -3$$

The x-intercepts are $(-3, 0)$ and $(\frac{1}{2}, 0)$.

Axis of symmetry: The average value of the x-intercepts is

$$x = \frac{-3 + 0.5}{2} \text{ or } -1.25.$$

Vertex: The x-coordinate of the vertex is -1.25, and substituting -1.25 for x gives

$$f(-1.25) = -2(-1.25)^2 - 5(-1.25) + 3$$
$$= 6.125$$

The vertex is at $(-1.25, 6.125)$.

The graph is shown here. Note we've also graphed the point $(-2.5, 3)$, which we obtained using the graph's symmetry. The point (0, 3) is 1.25 units *to the right* of the axis of symmetry, and there must be a point 1.25 units *to the left* of the axis: $(1.25)(2) = 2.5$.

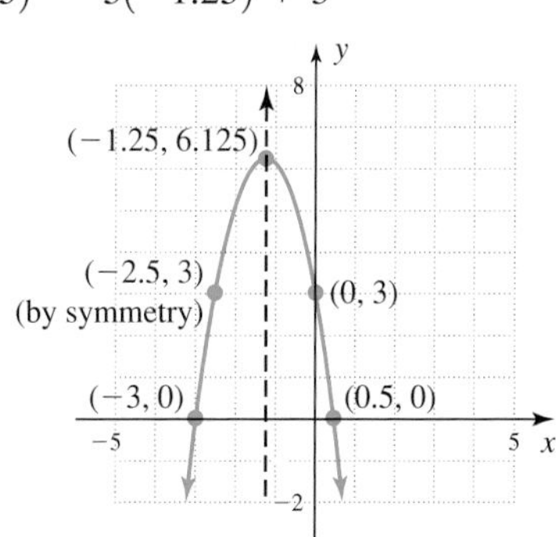

NOW TRY EXERCISES 17 THROUGH 28

C. The Square Root, Cubic, and Cube Root Functions

While linear and quadratic functions tend to appear more frequently as mathematical models, there are many instances when other models must be applied. In these cases, the other toolbox functions play a vital role, since they have features that lines and parabolas do not. Each new function is given an intuitive and descriptive name to help us recall its basic shape. We'll now look at the square root function $f(x) = \sqrt{x}$ and make some observations about its graph.

The Square Root Function

Recall the expression $\sqrt{x}$ represents a real number only for $x \geq 0$, indicating the domain of the square root function is $x \in [0, \infty)$. For graphing, we will select inputs that yield integer outputs.

EXAMPLE 4 Graph the square root function $f(x) = \sqrt{x}$ by plotting points.

Solution: List perfect squares for x in the first column, and the corresponding outputs for y in the second column. Only the ordered pairs given in color are displayed on the graph.

x	f(x) = √x
0	0
1	1
4	2
9	3
16	4
25	5
36	6

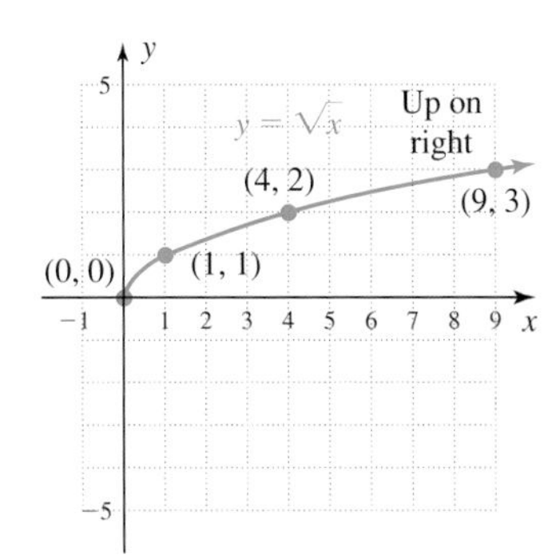

NOW TRY EXERCISES 29 THROUGH 32

Because of domain restrictions, graphs of square root functions always begin at a specific point called the **node** and extend from this point. This can happen in a variety of ways, as illustrated in Figure 2.41.

Figure 2.41

We'll refer to the graph of a square root function as a **one-wing graph.** The end behavior of the graph in Example 4 is "up on right."

EXAMPLE 5 Graph the function $r(x) = -\sqrt{2x + 13} + 2$.

Solution: First locate the node by checking the domain. For $\sqrt{2x + 13}$, we have $2x + 13 \geq 0$, giving $x \in [-\frac{13}{2}, \infty)$. The node is at $(-\frac{13}{2}, 2)$.

y-intercept: The y-intercept is $r(0) = -\sqrt{13} + 2 \approx -1.6$.

x-intercept(s): Set $r(x) = 0$ and solve to locate the x-intercept.

$$0 = -\sqrt{2x + 13} + 2 \quad \text{original equation}$$
$$-2 = -\sqrt{2x + 13} \quad \text{subtract 2}$$
$$4 = 2x + 13 \quad \text{square both sides}$$
$$-\frac{9}{2} = x \quad \text{solve for } x$$

The x-intercept is $(-\frac{9}{2}, 0)$. Having located the node and the intercepts, we determine the graph will decrease from left to right and can now complete the graph. The additional point $(6, -3)$ was computed to assist in sketching the graph.

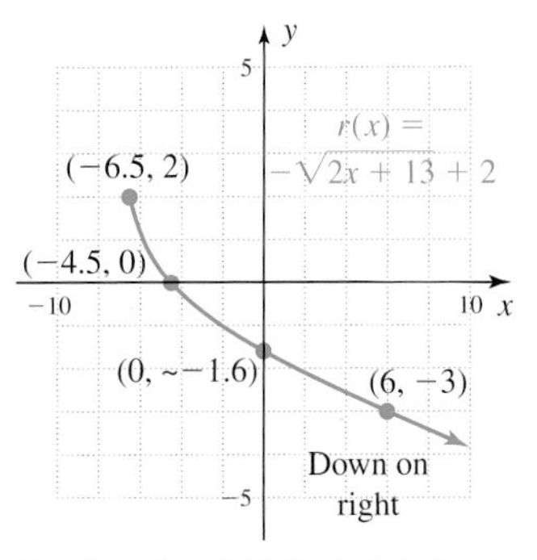

NOW TRY EXERCISES 33 THROUGH 38

The Cubing Function

From our study of exponents in Section R.3, we realize that $y = x^2$ will always be nonnegative, while $y = x^3$ will have the same sign as the value input. This difference shows up graphically in that the branches of x^2 both point upward (in the positive direction), while the branches of x^3 point in opposite directions.

EXAMPLE 6 Use a table of values to graph $f(x) = x^3$ by plotting points.

Solution: Create a table of values using $x \in [-3, 3]$ since output values are very large when $x < -3$ or $x > 3$. Only the ordered pairs shown in color are displayed on the graph.

x	$f(x) = x^3$
−3	−27
−2	−8
−1	−1
0	0
1	1
2	8
3	27

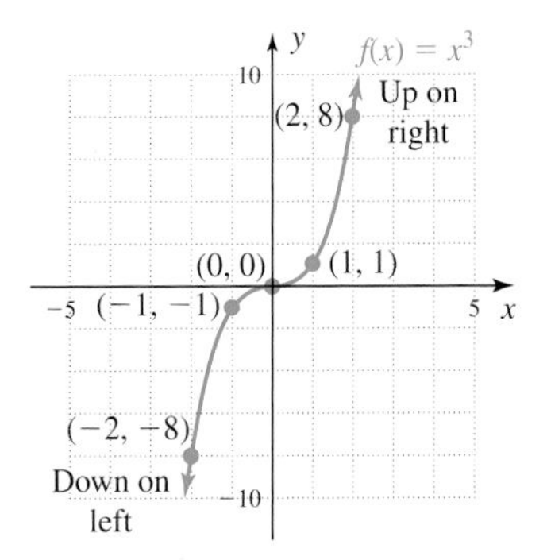

NOW TRY EXERCISES 39 THROUGH 42

Similar to the graph of a line, the "ends" of a cubic graph will always point in opposite directions. More correctly, we say that the end behavior for $y = x^3$ is *down on the left, up on the right*. As seen in the table and graph in Example 6, x^3 is negative if $x < 0$

and positive if $x > 0$. We also note the graph has a central pivot point, called a **point of inflection.** In actual practice, points of inflection are more difficult to locate than intercepts, and we only estimate their location here. All cubic graphs have a point of inflection and opposing end behavior, but may have one, two, or three x-intercepts. Because of the obvious similarity to propellers found on aircraft and ships, we refer to the basic cubic graph as a **vertical propeller.**

EXAMPLE 7 Given the graph of $f(x) = -x^3 + 6x^2 - 12x + 8$ in the figure, describe or identify the features indicated and use boundary lines to state the domain and range of the function. Assume the noted features are lattice points (note how the graph is scaled).

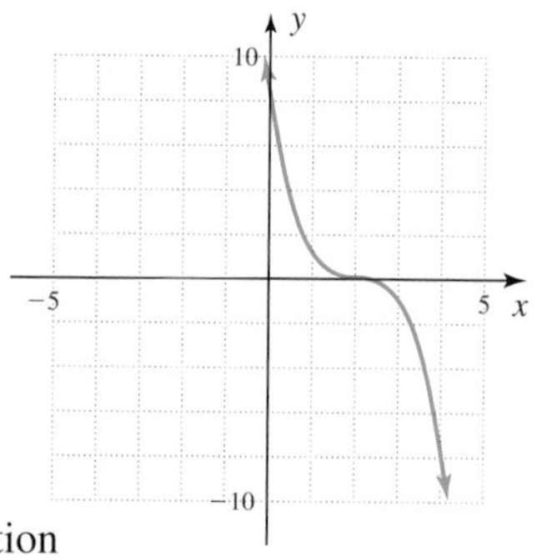

y 10 −5 5 x −10

1. end behavior
2. y-intercept
3. x-intercept(s)
4. point of inflection

Solution:

1. end behavior: up/down
2. y-intercept: $(0, 8)$
3. x-intercept: $(2, 0)$
4. point of inflection: appears to be $(2, 0)$

A vertical line will intersect the graph or its extension anywhere it is placed, as will a horizontal line. The domain is $x \in \mathbb{R}$, the range is $y \in \mathbb{R}$.

NOW TRY EXERCISES 43 THROUGH 48

The cubic function in Example 6 had a positive lead coefficient and the end behavior of the graph was down on the left, up on the right. In Example 7 the lead coefficient was negative and the end behavior was up on the left, down on the right. This indication of end behavior can be extended to all cubic functions.

EXAMPLE 8 Graph the function $f(x) = x^3 - 4x$ using end behavior and intercepts.

Solution: As in our preceding discussion:

End behavior: The lead coefficient is positive, so the end behavior must be down on the left, up on the right (or simply down, up).

y-intercept: The y-intercept is $f(0) = 0$ and the graph goes through the origin.

x-intercept(s): Set $f(x) = 0$ and solve to locate the x-intercepts.

$$\begin{aligned} 0 &= x^3 - 4x \\ &= x(x^2 - 4) \\ &= x(x - 2)(x + 2), \text{ giving } x = 0,\ x = 2, \text{ or } x = -2 \end{aligned}$$

The x-intercepts are $(0, 0)$, $(2, 0)$, and $(-2, 0)$.

In graph (a) shown, we've begun the graph using the information we have so far.

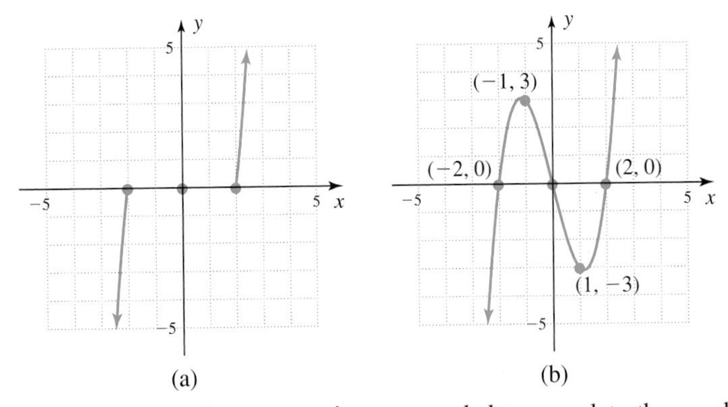

WORTHY OF NOTE

It is important to realize that mid-interval points are simply a device to help us "round-out" the graph and get an idea of its general shape. While it may appear so on the graph, they are not necessarily the maximum or minimum value in the interval. In fact, the maximum and minimum values for this function are irrational numbers very close to 3 and −3.

At this point we realize more points are needed to complete the graph, since we don't know how the graph behaves between each pair of x-intercepts (only that the graph must go through them). In other words, our "sufficient number of points" must be increased. Selecting midinterval points between the x-intercepts, we evaluate the function at $x = -1$ and $x = 1$ to obtain $(-1, 3)$ and $(1, -3)$, then use them to complete the graph, as shown in graph (b). The point of inflection appears to be the origin $(0, 0)$.

NOW TRY EXERCISES 49 THROUGH 54

GRAPHICAL SUPPORT

Graphing $f(x) = x^3 - 4x$ on a graphing calculator supports the information we obtained from the equation from Example 8 regarding end behavior, intercepts, midinterval points, and the point of inflection.

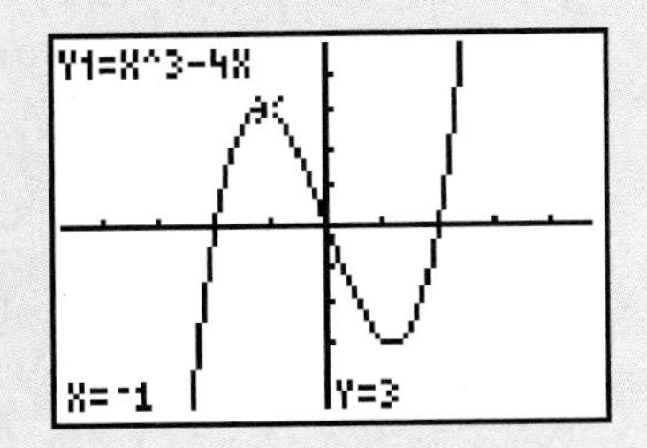

The Cube Root Function

From Section R.6, the cube root function $f(x) = \sqrt[3]{x}$ is defined for all real numbers. As with the square root function, we often select inputs that yield integer-value outputs when graphing.

EXAMPLE 9 Graph $y = \sqrt[3]{x}$ by plotting points.

Solution: List perfect cubes for x in the first column and the corresponding outputs for y in the second column. Only the ordered pairs given in color appear on the graph.

x	$f(x) = \sqrt[3]{x}$
−27	−3
−8	−2
−1	−1
0	0
−1	1
−8	2
27	3

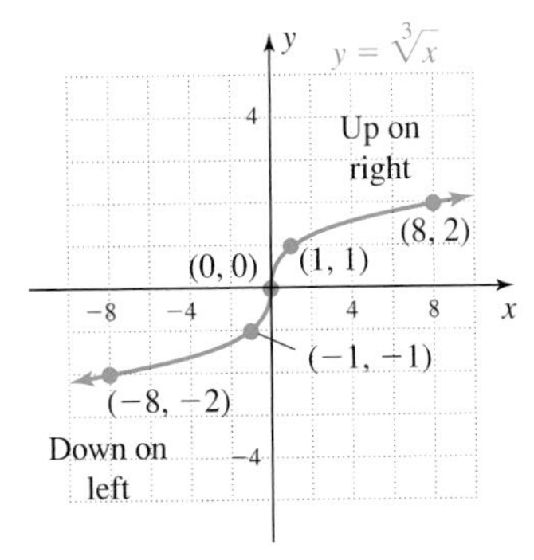

NOW TRY EXERCISES 55 THROUGH 64

The "ends" of a cube root graph also point in opposite directions, and we say the end behavior here is *down on the left, up on the right*. This distinction is harder to see in the case of the cube root function, and we will sometimes say the graph *increases* as you move left-to-right instead, much like a line whose slope is positive (the terms **increasing function** and **decreasing function** are more fully developed in Chapter 3). We note this graph also has a point of inflection, and refer to the graph of $f(x) = \sqrt[3]{x}$ as a **horizontal propeller.** All functions from the cube root family have the same basic characteristics. Similar to the previous functions studied, a negative lead coefficient reverses the end behavior of the graph.

WORTHY OF NOTE

As noted earlier, these terms are defined more formally in advanced courses, but for our purposes **smooth** means there are no jagged edges or sharp turns, and **continuous** means you can draw the entire graph without lifting your pencil.

With the exception of the absolute value function, the graphs of these toolbox functions are both *smooth* and *continuous*. The graph of $f(x) = |x|$ is continuous but not smooth, and in Section 3.5 we'll introduce the last two toolbox functions, which are smooth but not continuous. See Exercises 65 through 76.

D. The Toolbox Functions and Average Rate of Change

In Section 2.3 and other places, we noted that linear functions have a constant rate of change. Regardless of what two points on the line are chosen, the slope formula

$\frac{\Delta y}{\Delta x} = \frac{y_2 - y_1}{x_2 - x_1}$ will yield the same value. For nonlinear functions, the rate of change is not constant, but we can still use a version of the slope formula to investigate the **average rate of change** between two points. Since y_2 corresponds to the function value at x_2, we have $y_2 = f(x_2)$ in function notation. Likewise, $y_1 = f(x_1)$. Making the obvious substitutions in the slope formula gives $\frac{\Delta y}{\Delta x} = \frac{f(x_2) - f(x_1)}{x_2 - x_1}$, or the average rate of change between x_1 and x_2 for *any function f*.

THE AVERAGE RATE OF CHANGE FOR A FUNCTION

Given that f is continuous on the interval containing x_1 and x_2, the average rate of change of f between x_1 and x_2 is given by

$$\frac{\Delta y}{\Delta x} = \frac{f(x_2) - f(x_1)}{x_2 - x_1}$$

Average Rate of Change Applied to Projectile Velocity

A projectile is any object that is thrown or shot upward, with no continuing source of propulsion. A baseball, an arrow, and a water rocket are good examples. On Earth, if the object is thrown directly upward, the height of the object (in feet) after t sec is modeled by the function $h(t) = -16t^2 + vt + k$, where v is the initial velocity, and k is the height at release. For instance, if a football is punted upward with an initial speed of 80 ft/sec, the height of the ball at time t is $h(t) = -16t^2 + 80t$, where t is the time in seconds. This is a quadratic function whose graph is **concave down** because the lead coefficient is negative. To complete the graph modeling the ball's height, we find the x-intercepts, axis of symmetry, and vertex as before:

$$0 = -16t^2 + 80t \quad \text{original function, set equal to zero}$$
$$= -16t(t - 5) \quad \text{factored form}$$
$$t = 0 \text{ or } t = 5 \quad \text{result}$$

Figure 2.42

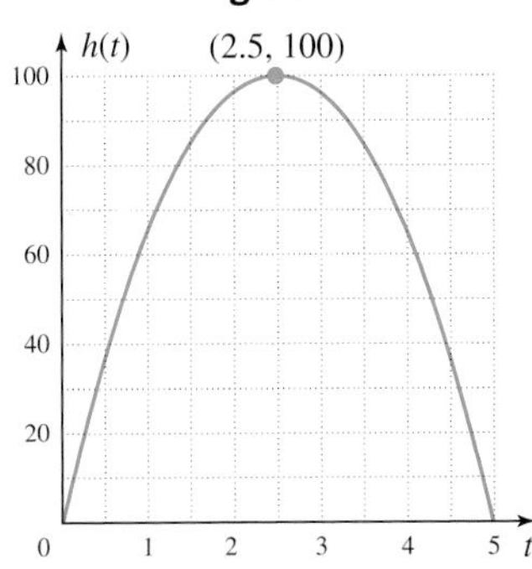

The x-intercepts are (0, 0) and (5, 0). For the vertex, we note $\frac{0+5}{2} = 2.5$ and find $h(2.5) = 100$ (verify this—see Figure 2.42). Common sense or personal experience tells us the football is traveling at a faster rate immediately after it is kicked, as compared to when it nears its maximum height, where it slows down, momentarily stops, then begins its descent. In other words, the rate of change $\frac{\Delta h}{\Delta t}$ has a larger value any time prior to reaching the maximum height. To quantify this, let's find the average rate of change between $t = 0.5$ and $t = 1$, and compare it to the average rate of change between $t = 2$ and $t = 2.5$.

EXAMPLE 10 Given $h(t) = -16t^2 + 80t$: (a) find the average rate of change for $0.5 \le t \le 1$, (b) find the average rate of change for $2 \le t \le 2.5$, and (c) sketch the graph of the function along with the lines representing the average rate of change.

Solution: Use the points given for parts (a) and (b) in the formula $\frac{\Delta h}{\Delta t} = \frac{h(t_2) - h(t_1)}{t_2 - t_1}$.

a.
$$\frac{\Delta h}{\Delta t} = \frac{h(1) - h(0.5)}{1 - 0.5} = \frac{64 - 36}{0.5} = 56$$

b.
$$\frac{\Delta h}{\Delta t} = \frac{h(2.5) - h(2)}{2.5 - 2} = \frac{100 - 96}{0.5} = 8$$

For $0.5 \le t \le 1$, $\frac{\Delta h}{\Delta t} = \frac{56}{1}$, meaning the height of the football is increasing at an average rate of 56 ft/sec. For $2 \le t \le 2.5$, $\frac{\Delta h}{\Delta t} = \frac{8}{1}$, and the height of the football is increasing at a rate of only 8 ft/sec.

c. The points (0.5, 36) and (1, 64) form the line representing the average rate of change shortly after the ball is kicked (in red). The points (2, 96) and (2.5, 100) form the line representing the average rate of change as the ball approaches the vertex (in blue). Note the red line has a much steeper slope than the blue line—a nice graphical illustration of what the average rate of change means. From Section 2.3, recall that a line containing two or more points on the graph of a function is called a **secant line.**

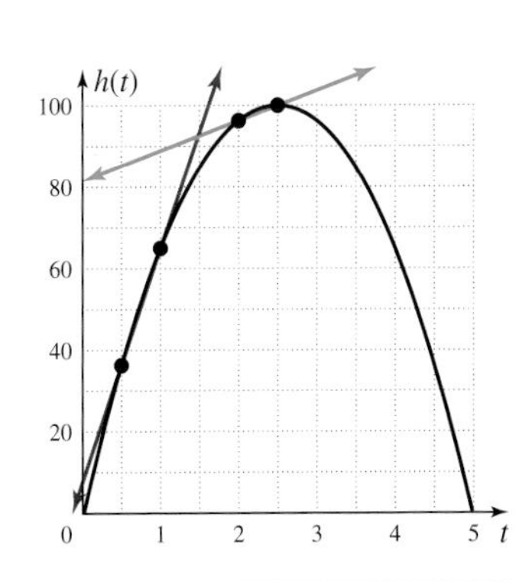

NOW TRY EXERCISES 79 THROUGH 84

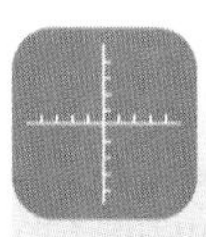

TECHNOLOGY HIGHLIGHT

Graphing Quadratic Functions on a Graphing Calculator

The keystrokes shown apply to a TI-84 Plus model. Please consult your manual or our Internet site for other models.

Many calculators can easily graph quadratic functions, but if you aren't careful, you may think your calculator is broken. To understand why, enter the function $y = x^2 - 14x - 15$ on the Y= screen and press ZOOM **6:ZStandard** to graph this function in the standard window. The result is shown in Figure 2.43 and looks like the graph of a line. The reason is that both the *vertex* and *y-intercept* are located outside the viewing window! This illustrates the value of having some advance information about the function you're graphing, so that you can *set an appropriate viewing window.* This is a crucial part of using technology in the study of mathematics. For this function, we know the graph is concave up since $a > 0$ and that the *y*-intercept is $(0, -15)$. The *x*-intercepts are found by setting $f(x) = 0$ and solving for *x*, showing that $(-1, 0)$ and $(15, 0)$ are the *x*-intercepts (verify this for yourself). This means we must set the **Xmin** and **Xmax** values so that $x = -1$ and $x = 15$ are included. As a general rule, we try to include a "frame" around the *x*-intercepts, and we select Xmin = −5 and Xmax = 20 (Figure 2.44).

Figure 2.43

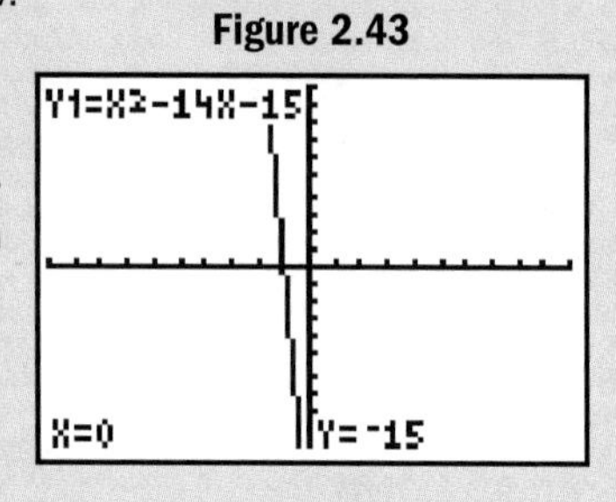

Figure 2.44

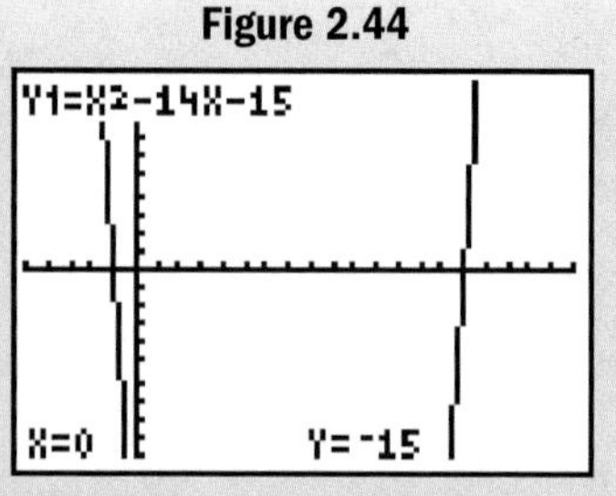

At this point we can compute the average of the *x*-intercepts to locate the vertex and set our **Ymin** and **Ymax** values accordingly, or we can use a trial-and-error process to find a good viewing window. Opting for trial-and-error and noting the Ymin value should be much, much smaller than −10, we try Ymin = −100 (remember to adjust Yscl as well—Yscl = 5 or so). This produces an acceptable graph—but it leaves too much "wasted space" below the graph. Resetting this value to Ymin = −70 gives us a better window, where we can investigate properties of the graph, **TRACE** through values, or make any decisions that the situation or application might require (see Figure 2.45). Use these ideas to find an optimum viewing window for the following functions. Answers will vary.

Figure 2.45

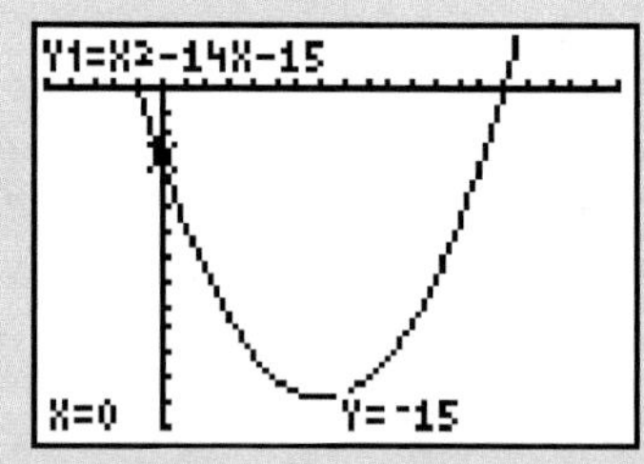

Exercise 1: $f(x) = x^2 - 18x + 45$ $[-2, 20, 2; -50, 50, 10]$

Exercise 2: $f(x) = x^2 + 18x - 54$ $[-30, 10, 2; -140, 20, 10]$

Exercise 3: $f(x) = -x^2 - 12x + 28$ $[-20, 10, 2; -40, 80, 10]$

Exercise 4: $f(x) = -x^2 + 10x + 11$ $[-5, 15, 2; -20, 40, 5]$

2.4 EXERCISES

CONCEPTS AND VOCABULARY

Fill in each blank with the appropriate word or phrase. Carefully reread the section if needed.

1. The graph of a quadratic function is called a(n) ___parabola___.

2. If the graph of a quadratic is concave down, the *y*-coordinate of the vertex is the ___maximum___ value.

3. All cubic graphs have a "pivot point," called the ___point___ of ___inflection___.

4. Graphs of ___square___ ___root___ functions are "one-wing" graphs, that begin at a point called the ___node___.

7. 8.

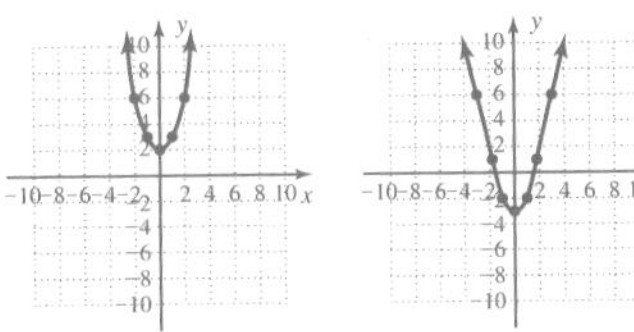

9. 10.

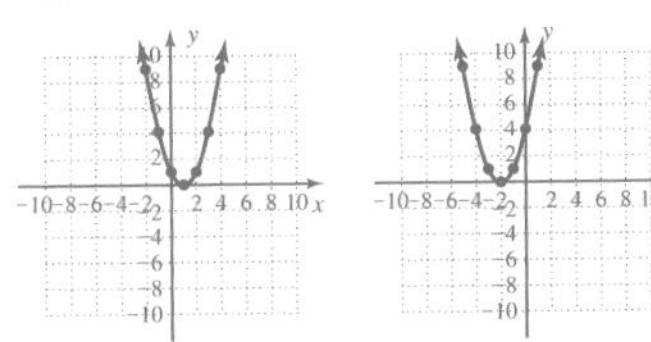

11. a. up/up, $(-2, -4)$, $x = -2$, $(0, 0)$, $(-4, 0)$, $(0, 0)$
 b. $x \in (-\infty, \infty)$, $y \in [-4, \infty)$

12. a. down/down, $(1, 1)$, $x = 1$, $(0, 0)$, $(2, 0)$, $(0, 0)$
 b. $x \in (-\infty, \infty)$, $y \in (-\infty, 1]$

13. a. up/up, $(1, -4)$, $x = 1$, $(-1, 0)$, $(3, 0)$, $(0, -3)$
 b. $x \in (-\infty, \infty)$, $y \in [-4, \infty)$

14. a. down/down, $(1, 9)$, $x = 1$, $(-2, 0)$, $(4, 0)$, $(0, 8)$
 b. $x \in (-\infty, \infty)$, $y \in (-\infty, 9]$

15. a. up/up, $(2, -9)$, $x = 2$, $(-1, 0)$, $(5, 0)$, $(0, -5)$
 b. $x \in (-\infty, \infty)$, $y \in [-9, \infty)$

16. a. up/up, $(-3, 4)$, $x = -3$, $(-5, 0)$ $(-1, 0)$, $(0, 5)$
 b. $x \in (-\infty, \infty)$, $y \in [-4, \infty)$

17. 18.

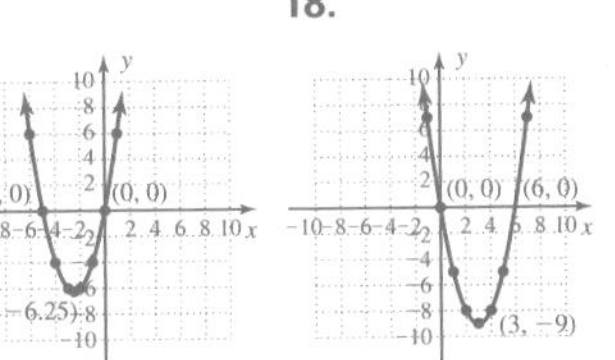

19. 20.

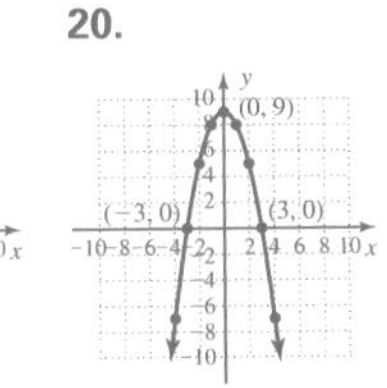

21. 22.

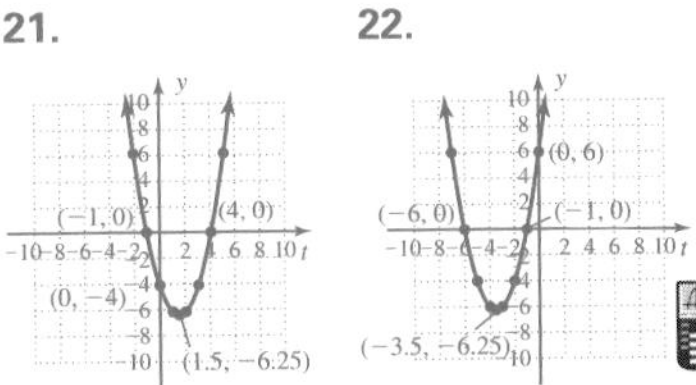

Additional answers can be found in the Instructor Answer Appendix.

5. Compare/contrast the end behavior of quadratic, cubic, and square root graphs. Include a discussion of *why* the behavior differs. Answers will vary.

6. Discuss/explain how the concept of the average rate of change can be applied to a race car that goes from 0 to 175 mph in 10 sec. Answers will vary.

DEVELOPING YOUR SKILLS

Graph each squaring function by *plotting points* using integers values from $x = -5$ to $x = 5$.

7. $f(x) = x^2 + 2$ 8. $g(x) = x^2 - 3$ 9. $p(x) = (x - 1)^2$ 10. $q(x) = (x + 2)^2$

For each quadratic graph given, (a) describe or identify the end behavior, vertex, axis of symmetry, and x- and y-intercepts; and (b) determine the domain and range. Assume noted features are lattice points.

11. $f(x) = x^2 + 4x$

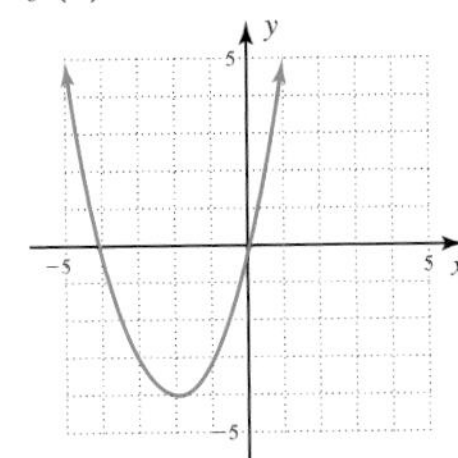

12. $g(x) = -x^2 + 2x$

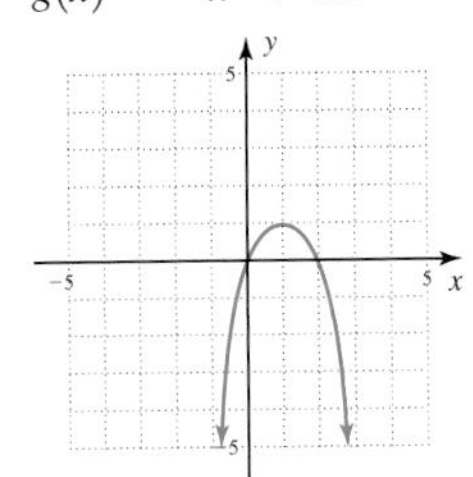

13. $p(x) = x^2 - 2x - 3$

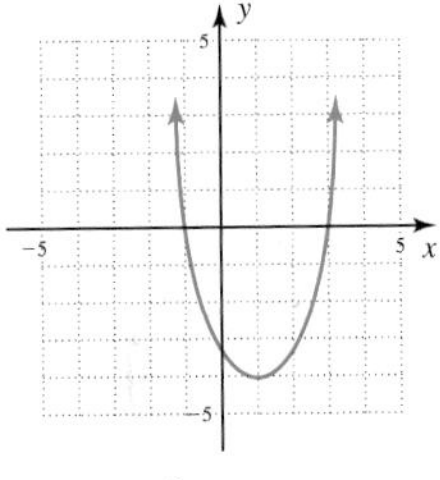

14. $q(x) = -x^2 + 2x + 8$

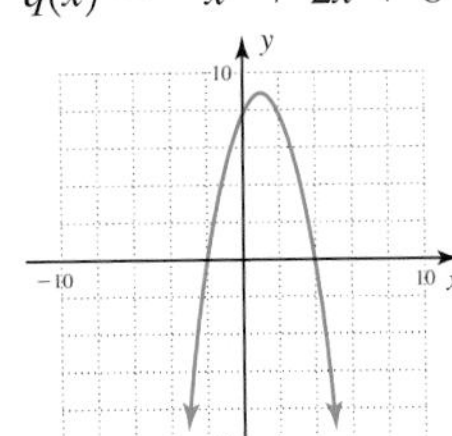

15. $f(x) = x^2 - 4x - 5$

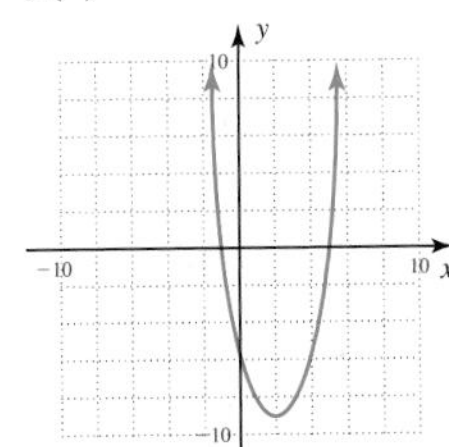

16. $g(x) = x^2 + 6x + 5$

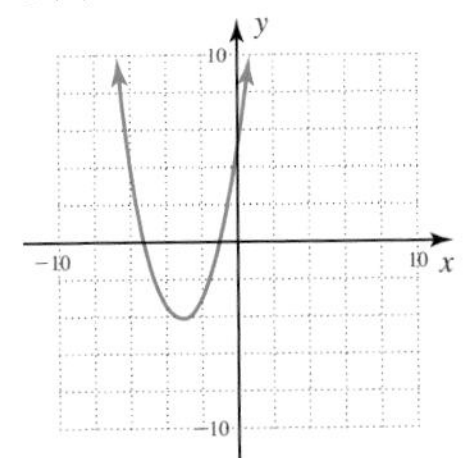

Draw a complete graph of each function by first identifying the concavity, x- and y-intercepts, axis of symmetry, and vertex.

17. $f(x) = x^2 + 5x$ 18. $g(x) = x^2 - 6x$ 19. $p(x) = 4 - x^2$
20. $q(x) = 9 - x^2$ 21. $r(t) = t^2 - 3t - 4$ 22. $s(t) = t^2 + 7t + 6$
23. $f(x) = -x^2 + 4x - 4$ 24. $g(x) = -x^2 + 6x + 9$ 25. $y = 2x^2 + 7x - 4$
26. $y = 3x^2 + 8x - 3$ 27. $p(t) = 12 - t^2 - 4t$ 28. $q(t) = 10 - 3t - t^2$

Graph each square root function by *plotting points* using inputs that result in integer outputs.

29. $f(x) = \sqrt{x} + 1$ 30. $g(x) = \sqrt{x} - 2$ 31. $p(x) = 2\sqrt{x + 3}$
32. $f(x) = -2\sqrt{x - 2}$

Draw a complete graph of each function by first identifying the node, end behavior, and x- and y-intercepts, then using any additional points needed to complete the graph.

33. $f(x) = \sqrt{x + 3} - 2$ 34. $g(x) = \sqrt{x - 2} + 1$ 35. $r(x) = -\sqrt{x + 4} - 1$
36. $s(x) = -\sqrt{x - 1} + 2$ 37. $p(x) = 2\sqrt{x + 1} - 3$ 38. $q(x) = -2\sqrt{x + 1} + 4$

Graph each cubing function by *plotting points* using integers values from $x = -5$ to $x = 5$.

39. $f(x) = x^3 + 1$ 40. $g(x) = x^3 - 2$ 41. $p(x) = (x + 2)^3$
42. $q(x) = (x - 3)^3$

49. **50.**

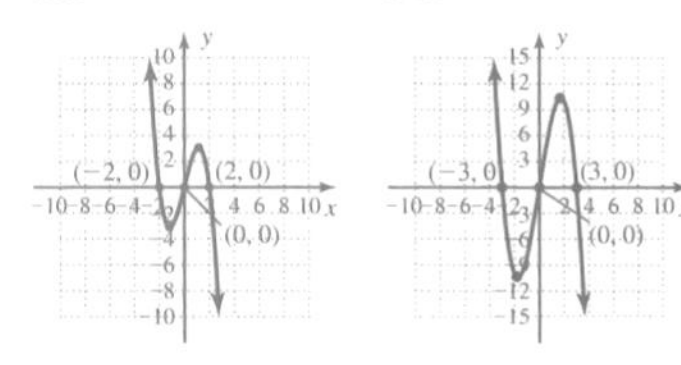

51. **52.**

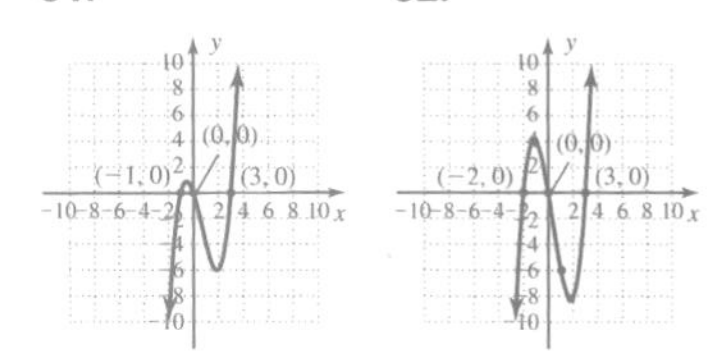

53. **54.**

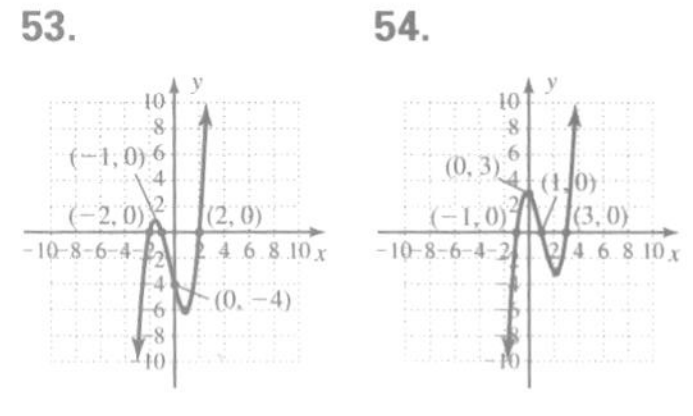

55. **56.**

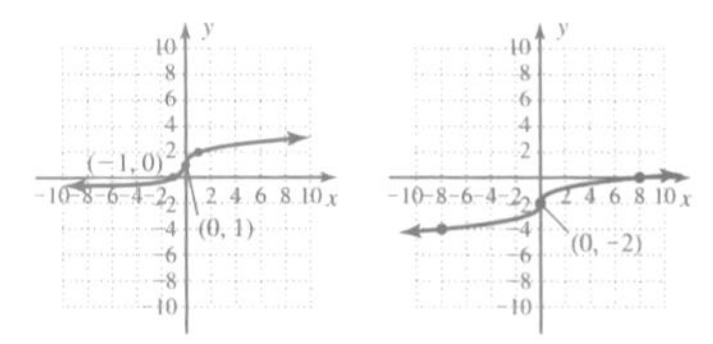

57. **58.**

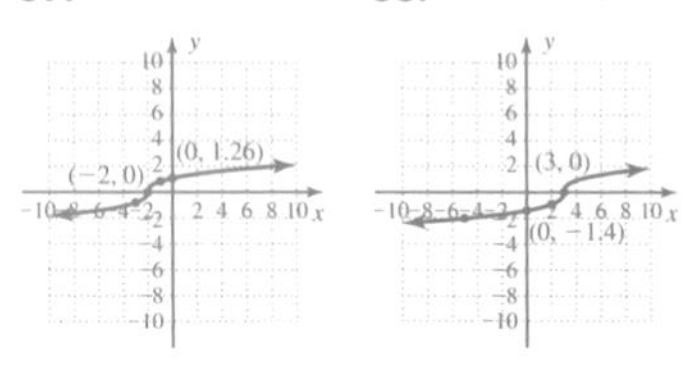

59. **60.**

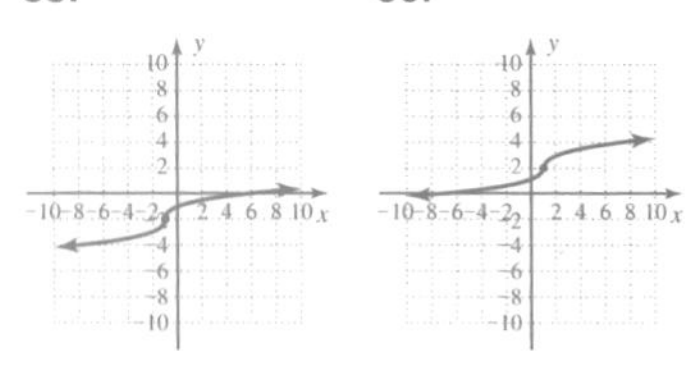

61. **62.**

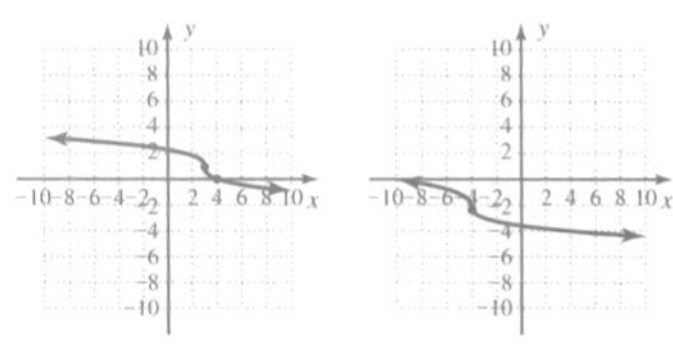

63. **64.**

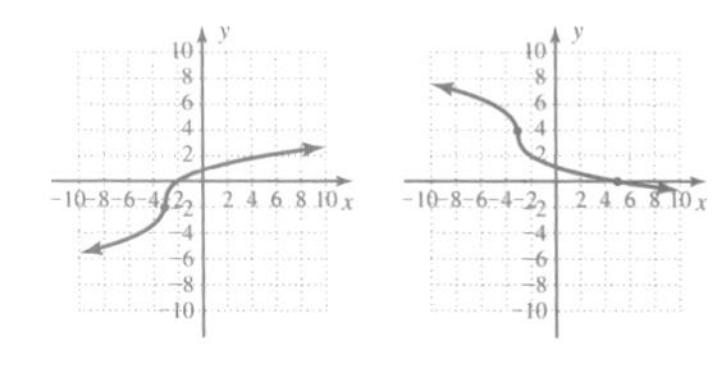

For each cubic graph given, (a) describe or identify the end behavior and x- and y-intercepts (assume the intercepts are lattice points), (b) determine the domain and range, and (c) estimate the coordinates of the point of inflection to the nearest tenth (Note scaling of axis).

43. $f(x) = -x^3 + 3x^2 - 3x + 1$

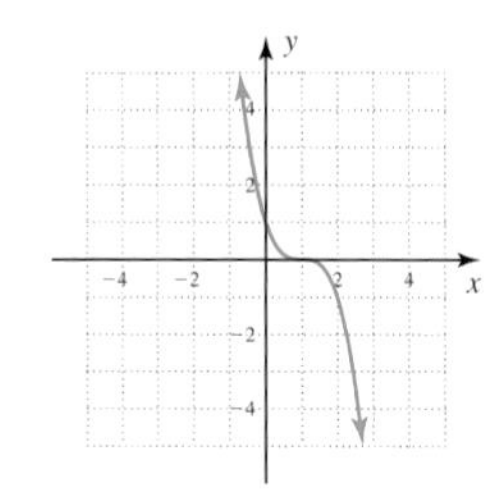

a. up on left, down on right; (1, 0), (0, 1)
b. $x \in (-\infty, \infty)$
$y \in (-\infty, \infty)$
c. (1, 0)

44. $g(x) = x^3 + 6x^2 + 12x + 8$

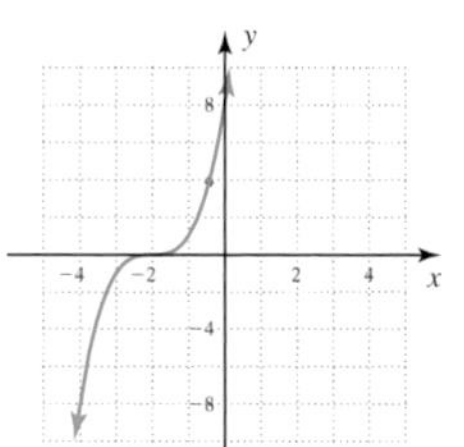

a. down on left, up on right; (−2, 0), (0, 8)
b. $x \in (-\infty, \infty)$
$y \in (-\infty, \infty)$
c. (−2, 0)

45. $p(x) = x^3 + 4x^2 - x - 4$

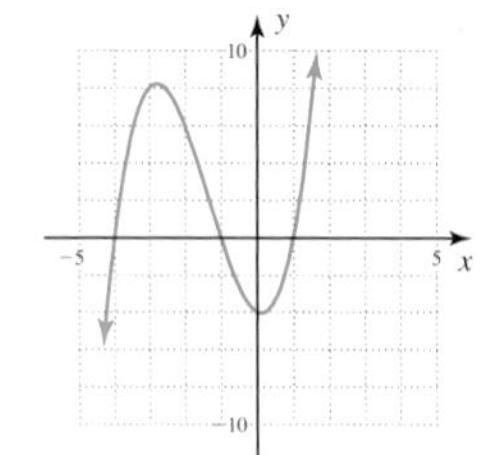

a. down on left, up on right; (−4, 0), (−1, 0), (1, 0), (0, −4)
b. $x \in (-\infty, \infty)$
$y \in (-\infty, \infty)$
c. (−1.3, 2.1)

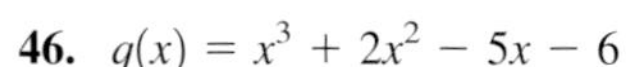

46. $q(x) = x^3 + 2x^2 - 5x - 6$

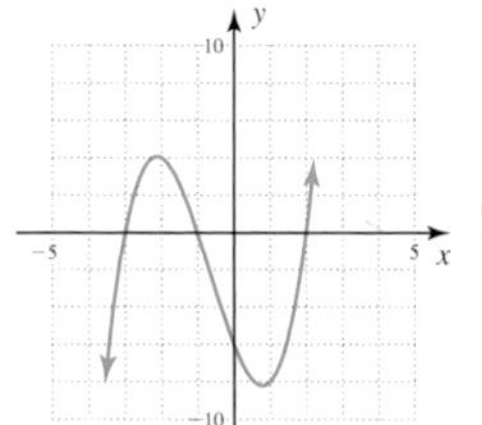

a. down on left, up on right; (−3, 0), (−1, 0), (2, 0), (0, −6)
b. $x \in (-\infty, \infty)$
$y \in (-\infty, \infty)$
c. (−0.7, −2.1)

47. $v(x) = -x^3 + 5x^2 - 2x - 8$

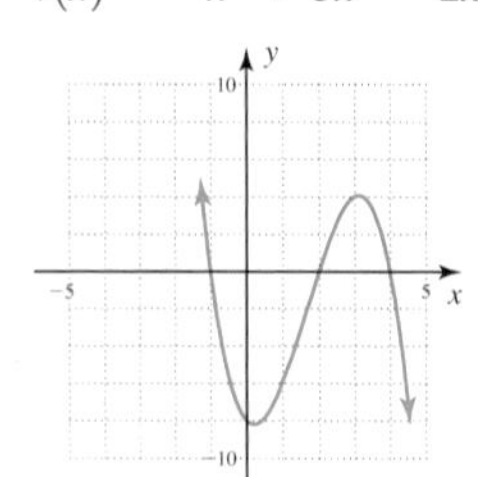

a. up on left, down on right; (−1, 0), (2, 0), (4, 0), (0, −8)
b. $x \in (-\infty, \infty)$
$y \in (-\infty, \infty)$
c. (1.7, −2.1)

48. $w(x) = -x^3 - 5x^2 - 2x + 8$

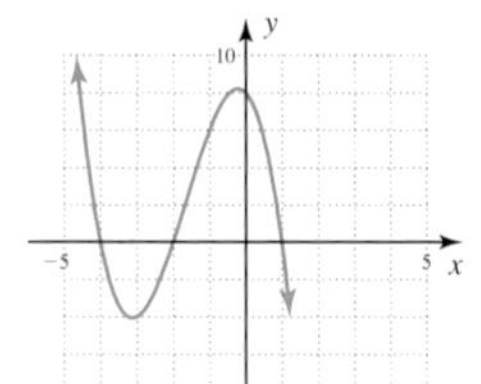

a. up on left, down on right; (−4, 0), (−2, 0), (1, 0), (0, 8)
b. $x \in (-\infty, \infty)$
$y \in (-\infty, \infty)$
c. (−1.7, 2.1)

Draw a complete graph of each function by first identifying the end behavior and x- and y-intercepts, then using any midinterval points needed to complete the graph.

49. $f(x) = 4x - x^3$ **50.** $g(x) = 9x - x^3$ **51.** $v(x) = x^3 - 2x^2 - 3x$
52. $w(x) = x^3 - x^2 - 6x$ **53.** $r(x) = x^3 + x^2 - 4x - 4$ **54.** $g(x) = x^3 - 3x^2 - x + 3$

Graph each cube root function by *plotting points* using inputs that result in integer outputs.

55. $f(x) = \sqrt[3]{x} + 1$ **56.** $g(x) = \sqrt[3]{x} - 2$ **57.** $p(x) = \sqrt[3]{x + 2}$
58. $q(x) = \sqrt[3]{x - 3}$

Draw a complete graph of each function by first identifying the end behavior and x- and y-intercepts, then using any additional points needed to complete the graph.

59. $f(x) = \sqrt[3]{x + 1} - 2$ **60.** $g(x) = \sqrt[3]{x - 1} + 2$ **61.** $r(x) = -\sqrt[3]{x - 3} + 1$
62. $s(x) = -\sqrt[3]{x + 4} - 2$ **63.** $p(x) = 2\sqrt[3]{x + 3} - 2$ **64.** $q(x) = -2\sqrt[3]{x + 3} + 4$

Use the graphical characteristics of each toolbox function family, along with x- and y-intercepts, to match each equation to its graph. Justify your choices.

65. $f(x) = \sqrt{x + 3} - 1$ c **66.** $g(x) = x^2 + 2x - 3$ g **67.** $p(x) = |x + 1| - 2$ a
68. $q(x) = 4x - x^2$ b **69.** $r(x) = -|x - 1| + 1$ k **70.** $s(x) = -\frac{2}{3}x + 2$ h
71. $Y_1 = \sqrt[3]{x} + 1$ d **72.** $Y_2 = x^3 - 3x^2 + 3x - 1$ j **73.** $f(x) = -\sqrt[3]{x + 1}$ i

74. $g(x) = -\sqrt{x+2} + 1$ e **75.** $p(x) = \frac{3}{2}x - 2$ l **76.** $q(x) = 4x - x^3$ f

a.

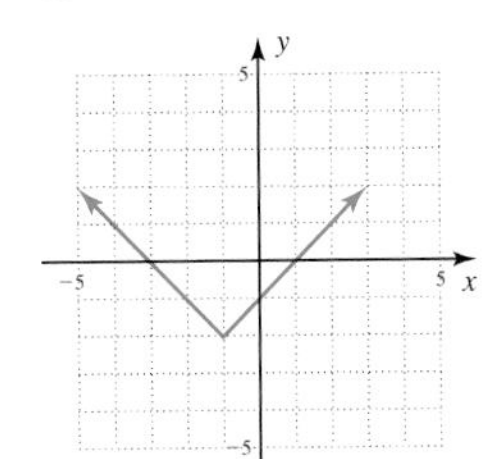

b.

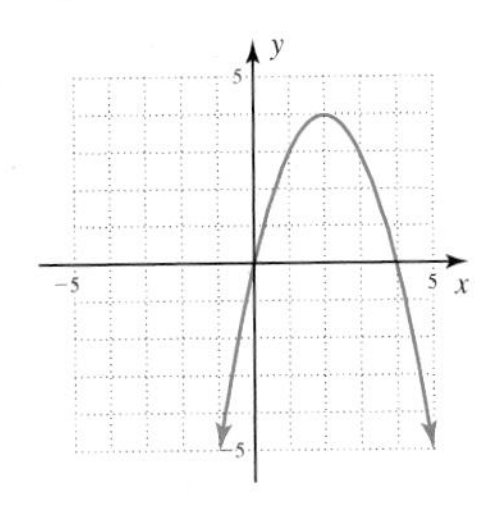

c.

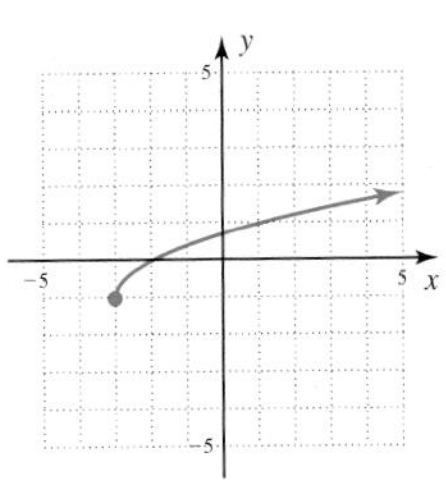

d.

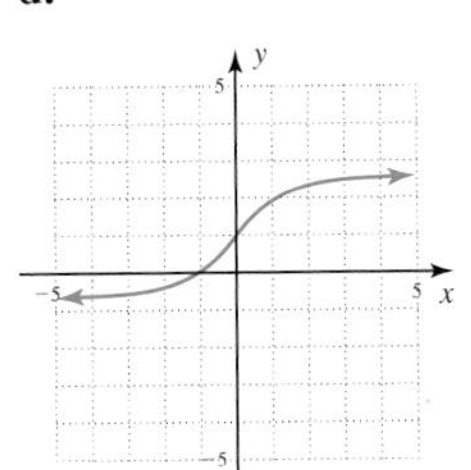

e.

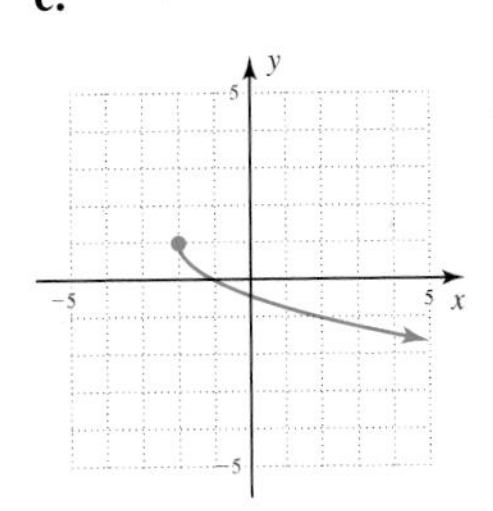

f.

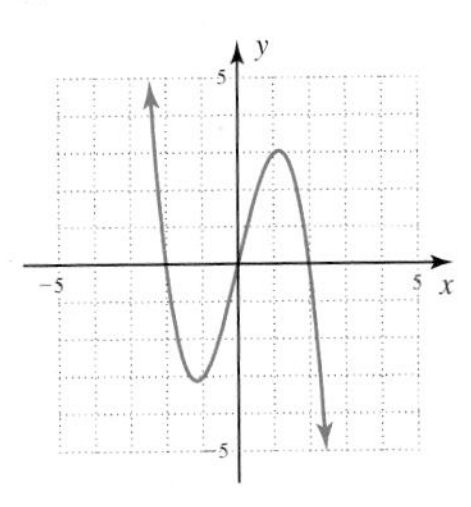

g.

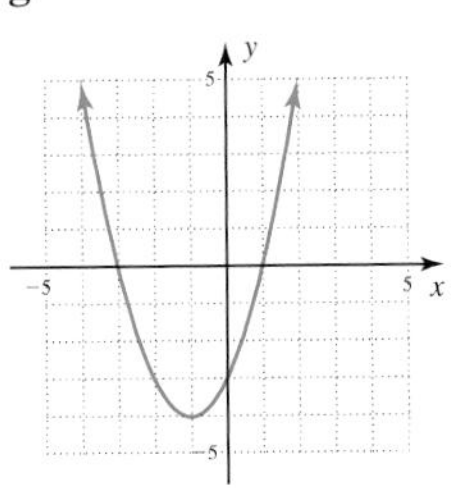

h.

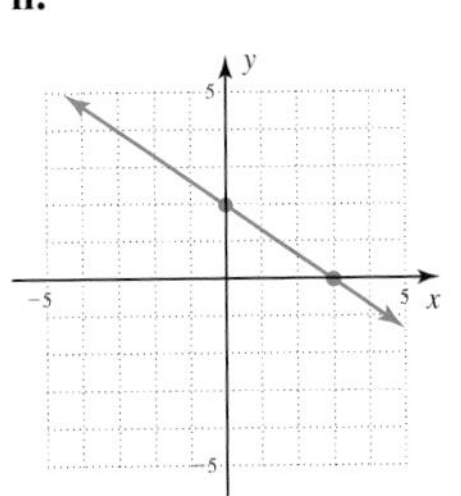

i.

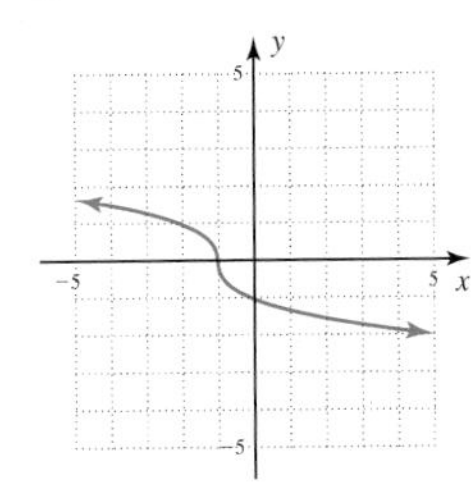

j.

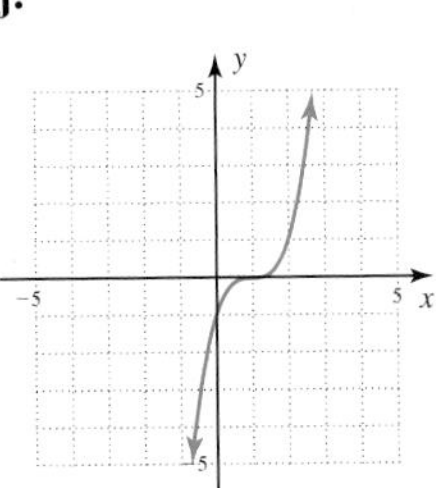

k.

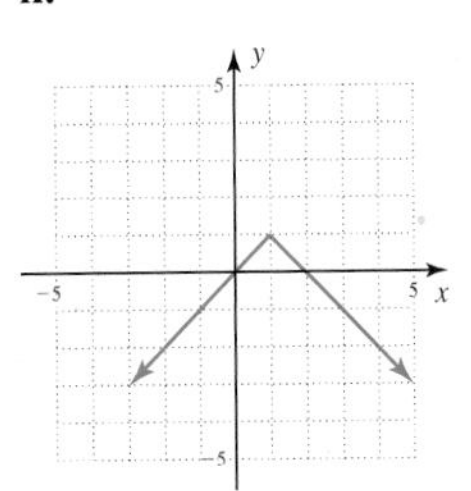

l.

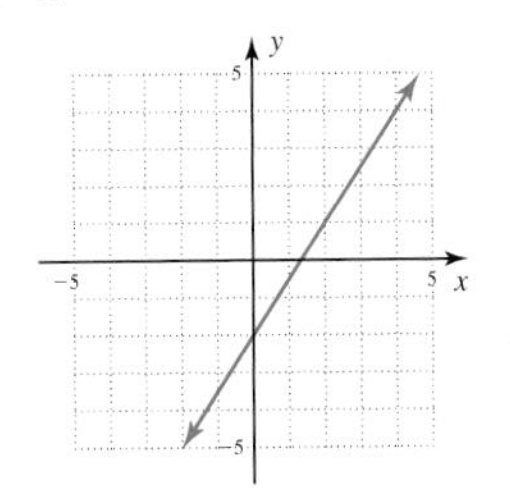

WORKING WITH FORMULAS

77. Velocity of a falling body: $v(s) = \sqrt{2gs}$

The impact velocity of an object dropped from a height is modeled by the formula shown, where v is the velocity in feet per second (ignoring air resistance), g is the acceleration due to gravity (32 ft/sec^2 near the Earth's surface), and s is the height from which the object is dropped. (a) Find the velocity of a wrench as it hits the ground if it was dropped by a telephone lineman from a height of 25 ft, and (b) solve for s in terms of v and find the height the wrench fell from if it strikes the ground at $v = 84$ ft/sec^2. 40 ft/sec, 110.25 ft

78. Power of a wind-driven generator: $P(w) = 0.0004w^3$

The amount of horsepower (hp) delivered by a wind-powered generator can be modeled by the formula shown, where $P(w)$ represents the horsepower delivered at wind speed w in miles per hour. (a) Find the power delivered at a wind speed of 25 mph and (b) solve the formula for w [use P for $P(w)$] and use the result to find the wind speed if 36.45 hp is being generated. 6.25 hp, 45 mph

APPLICATIONS

79. a. 7
b. 7
c. They are the same.
d. Slopes are equal.

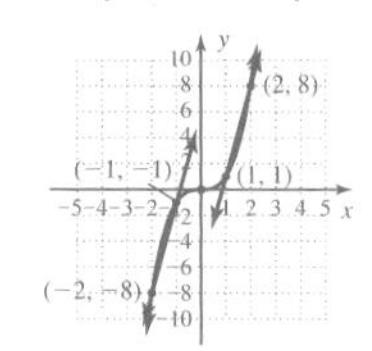

80. a. Between $x = 0$ and $x = 1$
b. 1, 0.09
c. 11

79. Average rate of change: For $f(x) = x^3$, (a) calculate the average rate of change for the interval $x = -2$ and $x = -1$ and (b) calculate the average rate of change for the interval $x = 1$ and $x = 2$. (c) What do you notice about the answers from parts (a) and (b)? (d) Sketch the graph of this function along with the lines representing these average rates of change and comment on what you notice.

80. Average rate of change: Knowing the general shape of the graph for $f(x) = \sqrt[3]{x}$, (a) is the average rate of change greater between $x = 0$ and $x = 1$ or between $x = 7$ and $x = 8$? Why? (b) Calculate the rate of change for these intervals and verify your response. (c) Approximately how many times greater is the rate of change?

81. Height of an arrow: If an arrow is shot from a bow with an initial speed of 192 ft/sec, the height of the arrow can be modeled by the function $h(t) = -16t^2 + 192t$, where $h(t)$ represents the height of the arrow after t sec (assume the arrow was shot from ground level).

a. What is the arrow's height at $t = 1$ sec? 176 ft

b. What is the arrow's height at $t = 2$ sec? 320 ft

c. What is the average rate of change from $t = 1$ to $t = 2$? 144 ft/sec

d. What is the rate of change from $t = 10$ to $t = 11$? Why is it the same as (c) except for the sign? −144 ft/sec; The arrow is going down.

82. Height of a water rocket: Although they have been around for decades, water rockets continue to be a popular toy. A plastic rocket is filled with water and then pressurized using a handheld pump. The rocket is then released and off it goes! If the rocket has an initial velocity of 96 ft/sec, the height of the rocket can be modeled by the function $h(t) = -16t^2 + 96t$, where $h(t)$ represents the height of the rocket after t sec (assume the rocket was shot from ground level).

a. Find the rocket's height at $t = 1$ and $t = 2$ sec. 80 ft, 128 ft

b. Find the rocket's height at $t = 3$ sec. 144 ft

c. Would you expect the average rate of change to be greater between $t = 1$ and $t = 2$, or between $t = 2$ and $t = 3$? Why? Between $t = 1$ and $t = 2$; the rocket is decelerating.

d. Calculate each rate of change and discuss your answer. 48 ft/sec, 16 ft/sec

83. Velocity of a falling object: From Exercise 77, the impact velocity of an object dropped from a height is modeled by $v = \sqrt{2gs}$, where v is the velocity in feet per second (ignoring air resistance), g is the acceleration due to gravity (32 ft/sec^2 near the Earth's surface), and s is the height from which the object is dropped.

a. Find the velocity at $s = 5$ ft and $s = 10$ ft. 17.89 ft/sec; 25.30 ft/sec

b. Find the velocity at $s = 15$ ft and $s = 20$ ft. 30.98 ft/sec; 35.78 ft/sec

c. Would you expect the average rate of change to be greater between $s = 5$ and $s = 10$, or between $s = 15$ and $s = 20$? Between 5 and 10.

d. Calculate each rate of change and discuss your answer. 1.482 ft/sec, 0.96 ft/sec

84. One day in November, the town of Coldwater was hit by a sudden winter storm that caused temperatures to plummet. During the storm, the temperature T (in degrees Fahrenheit) could be modeled by the function $T(h) = 0.8h^2 - 16h + 60$, where h is the number of hours since the storm began. Graph the function and use this information to answer the following questions.

a. What was the temperature as the storm began? 60°

b. How many hours until the temperature dropped below zero degrees? 5 hr

c. How many hours did the temperature remain below zero? 10 hr

d. What was the coldest temperature recorded during this storm? −20°F

85. a. They are the same.
 b. Slope ($\frac{2}{3}$) is constant for a line.
 c. $\frac{2}{3}$, Every change of 1 in x results in $\frac{2}{3}$ unit change in y.

86. a. Mars
 b. 1500 ft
 c. Earth $\frac{\Delta h}{\Delta t} = \frac{192 \text{ ft}}{1 \text{ sec}}$; Mars $\frac{\Delta h}{\Delta t} = \frac{222 \text{ ft}}{1 \text{ sec}}$; The object is rising faster on Mars.

89.

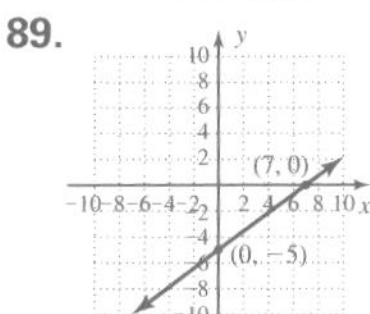

90. $y = \frac{5}{7}x - 5$, $m = \frac{5}{7}$, y-intercept $(0, -5)$

91. $x \in (-6, +\infty)$

92. $\frac{3}{2(x-2)}$; $x \neq -5, 2, 5, 0$

93. No, $x = 2$ is paired with $y = 2$ and $y = -2$.

94. 24 ft and 51 ft

WRITING, RESEARCH, AND DECISION MAKING

85. For the function $f(x) = \frac{2}{3}x - 3$, find the average rate of change between $x = 3$ and $x = 4$, then again between $x = 8$ and $x = 9$. (a) Comment on what do you notice. (b) Why do you think the average rate of change doesn't change? (c) Without any calculations, what is the average rate of change for this function between $x = 20$ and $x = 21$? Why?

86. From Example 10, the function $h(t) = -16t^2 + vt$ models the height of an object thrown upward from the Earth's surface, with a velocity of v ft/sec. The function $h(t) = -6t^2 + vt$ models the same phenomena on the surface of Mars. If a given object is projected upward at 240 ft/sec, (a) on which planet will the object reach a higher point? (b) How many feet higher will it go? (*Hint:* Substitute $v = 240$ in the equations and find each vertex.) (c) Find the average rate of change between $t = 1$ and $t = 2$ for each planet and interpret the result in light of your answer to (a).

EXTENDING THE CONCEPT

87. The area of a rectangle with a fixed perimeter is given by the formula $A(L) = -L^2 + L\left(\frac{P}{2}\right)$, where L is the length of the rectangle and $A(L)$ represents the area for a fixed perimeter P. The City of Carlton wants to resod their city park one section at a time. What is the largest area they can fence off with 1000 ft of barrier fencing? 62,500 ft^2

88. A bridge spans a narrow canyon. The support frame under the bridge forms the shape of a parabola. The height of the frame above the ground can be determined using the function $h(x) = -0.1x^2 + 4x$, where $h(x)$ represents the height of the frame x ft from the canyon's edge. (a) How long is the bridge? (b) How deep is the canyon?
40 ft 40 ft

MAINTAINING YOUR SKILLS

89. (2.1) Graph the line $5x - 7y = 35$ using the intercept method.

90. (2.1) Write the equation from Exercise 89 in slope-intercept form and verify the slope m and y-intercept $(0, b)$.

91. (1.2) Solve the inequality. Write the result in interval notation: $-\frac{2}{3}x + 7 < 11$.

92. (R.5) Find the quotient: $\frac{3x}{x^2 + 3x - 10} \div \frac{2x^2 - 10x}{x^2 - 25}$. What are the domain restrictions?

93. (2.2) Is the relation $x + 2 = y^2$ also a function? Explain why or why not.

94. (1.1) The 75-ft rope is cut so that one piece is 3 ft more than twice the other. How long is each piece?

2.5 Functions and Inequalities—A Graphical View

LEARNING OBJECTIVES

In Section 2.5 you will learn how to:

A. Solve linear function inequalities

B. Solve quadratic function inequalities

C. Solve function inequalities using interval tests

D. Solve applications involving function inequalities

INTRODUCTION

Equations have a finite number of solutions, since we're looking for *specific* value(s) that make an equation true. On the other hand, *inequalities* can have an infinite number of solutions, since the solution set may include an entire region of the plane or interval(s) of the real number line. In this section, we investigate inequalities of the form $f(x) > 0$ and $f(x) < 0$, which play an important role in future sections.

POINT OF INTEREST

In the movie *The Flight of the Navigator* (1986—Joey Cramer, Cliff DeYoung, and Veronica Cartwright), a young boy is captured by some "friendly" aliens in a futuristic spaceship that is capable of traveling in space, in the atmosphere, and under the ocean. According to the diagram in Figure 2.46, between what x-coordinates was this spaceship under water?

Figure 2.46

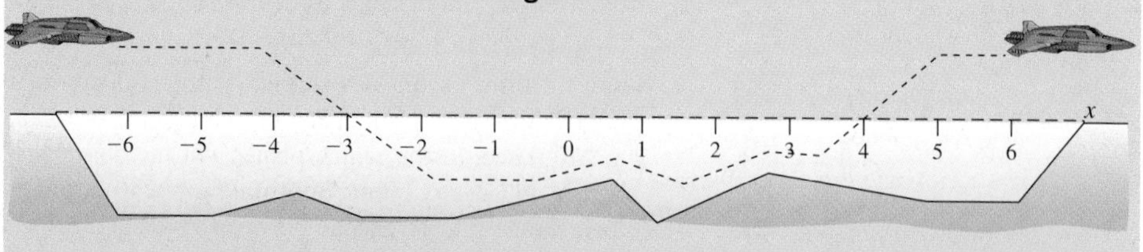

The figure clearly shows the spaceship was below sea level between $x = -3$ and $x = 4$, and was above sea level when $x < -3$ and $x > 4$. This is exactly the idea we use when solving function inequalities, as we determine when the graph of a function is below or above the x-axis.

A. Inequalities and Linear Functions

In any study of algebra, you'll be asked to solve many different kinds of inequalities. For those of the form $f(x) > 0$ or $f(x) < 0$, our focus is again on the input/output nature of the function, as we seek all inputs that cause outputs to be either positive or negative, as indicated by the inequality. In this case the solution set will be an interval of the number line, rather than a region of the xy-plane. As a simplistic illustration, consider the inequalities related to $f(x) = 2x - 1$.

$$f(x) = 2x - 1$$

$f(x) > 0$	inequalities	$f(x) < 0$
$\downarrow$		$\downarrow$
$2x - 1 > 0$	replace $f(x)$ with $2x - 1$	$2x - 1 < 0$

We'll use the "greater than" example to make our observations, but everything said can be applied just as well to $f(x) < 0$, or the inequalities $f(x) \leq 0$ and $f(x) \geq 0$. The key idea is to recognize that the following interpretations of $f(x) > 0$ all mean the same thing:

1. For what inputs are function values greater than zero?
2. For what inputs are the outputs positive?
3. For what inputs is the graph *above the x-axis*?
   Note $y = f(x)$ is positive in Quadrants I and II.

EXAMPLE 1 For $f(x) = 2x - 1$, solve $f(x) > 0$. Respond to all three of the preceding questions to justify your answer.

Solution: **1.** Solve the resulting inequality:

$$\begin{aligned} f(x) &> 0 && \text{given} \\ 2x - 1 &> 0 && \text{replace } f(x) \text{ with } 2x - 1 \\ x &> 0.5 && \text{solve for } x \end{aligned}$$

Function values are greater than zero when $x > 0.5$.

2. Although the answer will be identical, we could use a table of values.

Table 2.2

Input x	Output y $2x - 1$	Input x	Output y $2x - 1$
−2	−5	0.5	0
−1.5	−4	**1**	1
−1	−3	**1.5**	2
−0.5	−2	**2**	3
0	−1	**2.5**	4

From the table, it appears inputs *greater than* 0.5 produce outputs that are positive. The solution is $x > 0.5$, or $x \in (0.5, \infty)$ in interval notation.

3. Graph $f(x) = 2x - 1$. Note from the graph and table of values that $(0.5, 0)$ is the x-intercept. The graph is above the x-axis for $x > 0.5$, again verifying that the solution is $x \in (0.5, \infty)$.

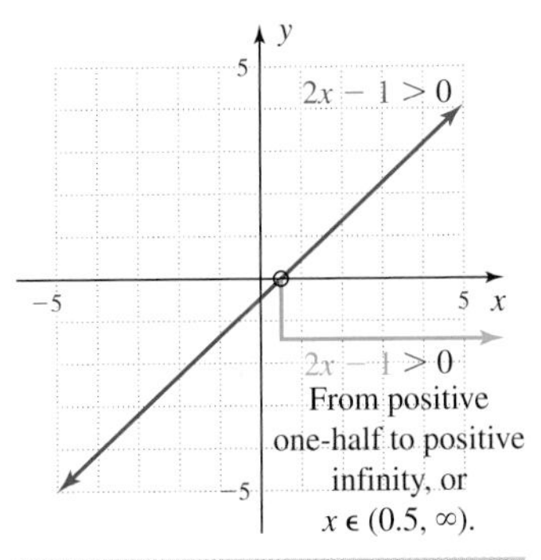

NOW TRY EXERCISES 7 THROUGH 12

Example 1 illustrates some very important concepts related to inequalities. First, since an x-intercept is the input value that gives an output of zero, x-intercepts are also referred to as the **zeroes of a function.** Just as zero on the number line separates the positive numbers from the negative numbers, the zeroes of a linear function separate *intervals where a function is positive from intervals where the function is negative*. This can be seen in Example 1, where the graph is above the x-axis (outputs are positive) when $x > 0.5$, and the graph is below the x-axis (outputs are negative) for $x < 0.5$. Although the idea must be modified to hold for all functions, it definitely applies here and to functions where the x-intercept(s) come from linear factors. This observation makes solving linear inequalities a matter of locating the x-intercept *and simply observing the slope of the line*. This is illustrated in Example 2.

EXAMPLE 2 For $g(x) = -\frac{1}{2}x - \frac{3}{2}$, solve the inequality $g(x) \geq 0$.

Solution: Note this is a greater than *or equal to* inequality. The slope of the line is negative and the zero of the function is $x = -3$ [verify by solving $g(x) = 0$]. Plot $(-3, 0)$ on the x-axis and *sketch* a line through $(-3, 0)$ with negative slope.

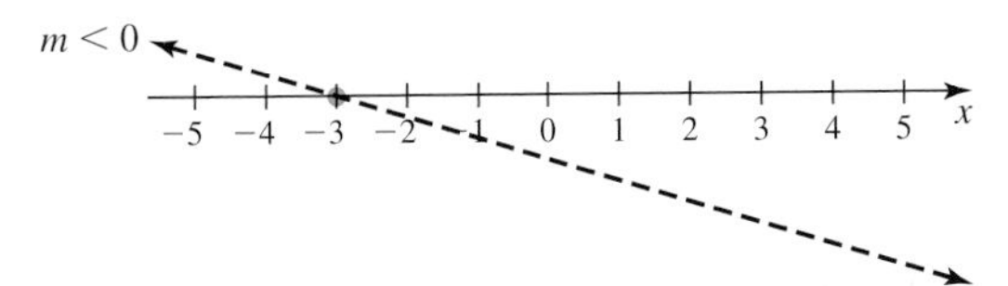

The figure clearly shows the graph is *above or on* the x-axis (outputs are nonnegative) when $x \leq -3$. The solution is $x \in (-\infty, -3]$.

NOW TRY EXERCISES 13 THROUGH 20

Although linear inequalities are easily solved without the slope/intercept analysis, the concept creates a strong bridge to nonlinear inequalities that are not so easily solved. These ideas will also be used in other parts of this text, as well as in future course work. Students should make every effort to understand the fundamentals illustrated. In particular, when solving function inequalities, the statements—(1) function values are greater than zero, (2) outputs are positive, and (3) the graph is above the x-axis—are virtually synonymous.

B. Solving Quadratic Inequalities

In a manner very similar to that used for linear functions, we now solve inequalities that involve *quadratic functions*. We need only (a) locate the zeroes and (b) *observe the concavity of the graph*. If there are no x-intercepts, the graph is entirely above the x-axis (all y-values positive), or entirely below the x-axis (all y-values negative), making the solution either all real numbers or the empty set (see Examples 4 and 5).

EXAMPLE 3 For $f(x) = x^2 + x - 6$, solve $f(x) > 0$.

Solution: The graph is concave up since $a > 0$. After factoring we find the zeroes are $x = -3$ and $x = 2$ (verify). Since both factors are linear, output values will change sign at these zeroes. Using the x-axis alone, we plot $(-3, 0)$ and $(2, 0)$ and sketch a parabola through them that is concave up.

Figure 2.47

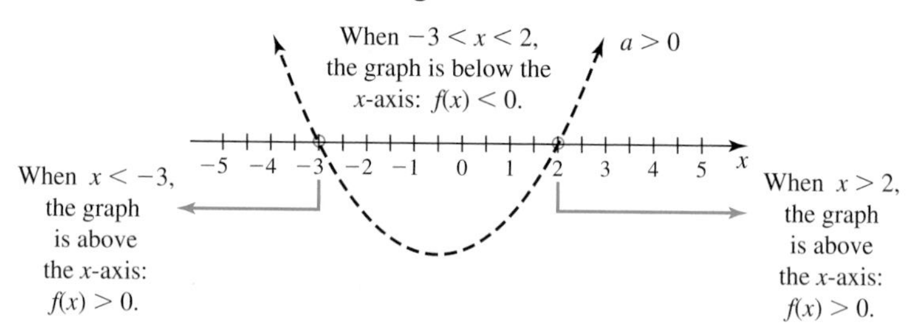

Figure 2.47 clearly shows that the graph is *above* the x-axis (outputs are positive) when $x < -3$ or when $x > 2$. The solution is $x \in (-\infty, -3) \cup (2, \infty)$.

For reference, the complete graph is given in Figure 2.48.

Figure 2.48

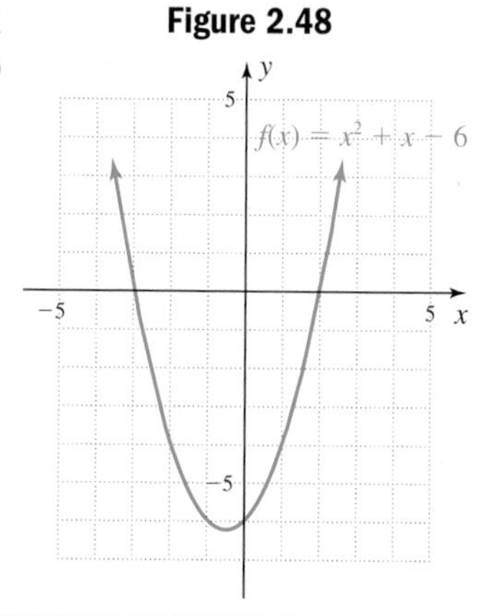

NOW TRY EXERCISES 21 THROUGH 56

EXAMPLE 4 For $f(x) = -x^2 + 6x - 9$, solve the inequality $f(x) \leq 0$.

Solution: Since $a < 0$, the graph will be concave down. The factored form is $0 = -(x - 3)^2$, which is *not a linear factor* (it has degree two) and gives the sole zero $x = 3$. Using the x-axis, we graph the point $(3, 0)$ and sketch a parabola through this point that is concave down (see Figure 2.49).

Figure 2.49

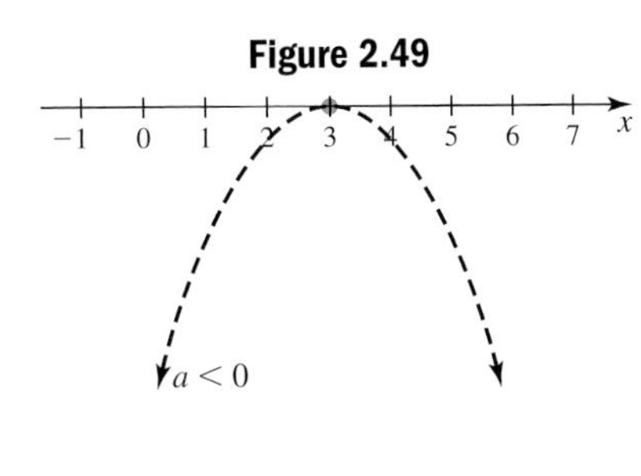

Figure 2.50

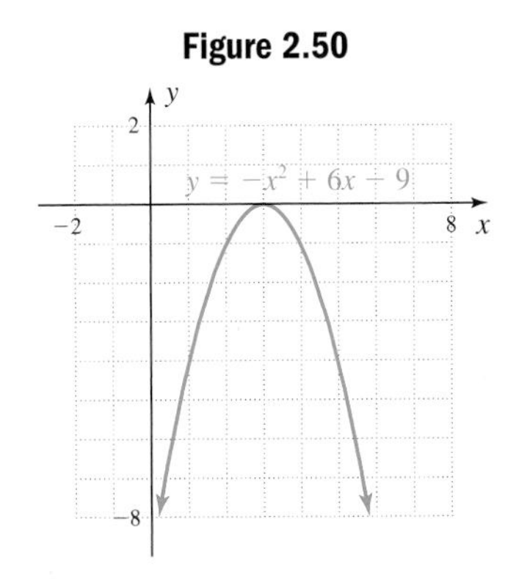

WORTHY OF NOTE

Consider again the factored form $f(x) = -(x - 3)^2$ from Example 4. Because $(x - 3)$ is squared, the outputs will be positive or zero regardless of the input value. The negative sign preceding the factor then makes all outputs negative, resulting in a graph entirely below the x-axis, except at its vertex (3, 0).

Figure 2.49 shows the graph will be *below* the x-axis for *all values* of x except $x = 3$. But since this is a less than *or equal to* inequality, the solution is $x \in \mathbb{R}$. The complete graph is given in Figure 2.50.

NOW TRY EXERCISES 57 THROUGH 62

EXAMPLE 5 For $f(x) = 2x^2 - 5x + 5$, solve $f(x) < 0$.

Solution: Since $a > 0$, the graph will be concave up. The expression will not factor, so we use the quadratic formula to find the x-intercepts.

$$x = \frac{-b \pm \sqrt{b^2 - 4ac}}{2a} \quad \text{quadratic formula}$$

$$= \frac{-(-5) \pm \sqrt{(-5)^2 - 4(2)(5)}}{2(2)} \quad \text{substitute 2 for } a, -5 \text{ for } b, 5 \text{ for } c$$

$$= \frac{5 \pm \sqrt{-15}}{4} \quad \text{simplify}$$

Because the radicand is negative, there are no real roots and the graph has no x-intercepts. Since the graph is concave up, we reason it must be entirely above the x-axis and output values for this function are always positive. The solution for $f(x) < 0$ is the empty set { }.

NOW TRY EXERCISES 63 THROUGH 68

C. Solving Function Inequalities Using Interval Tests

Although somewhat less conceptual, an **interval test method** can also be used to solve quadratic inequalities. The x-intercepts of the function are plotted on the x-axis, then a test number is selected from each interval between and beyond these intercepts. This gives an indication of the function's sign in each interval.

EXAMPLE 6 For $f(x) = x^2 - 3x - 10$, solve the inequality $f(x) \leq 0$.

Solution: The function is factorable and we find the x-intercepts are $(-2, 0)$ and $(5, 0)$. Using the x-axis alone, we graph these intercepts, noting this creates three intervals on the x-axis as shown.

Figure 2.51

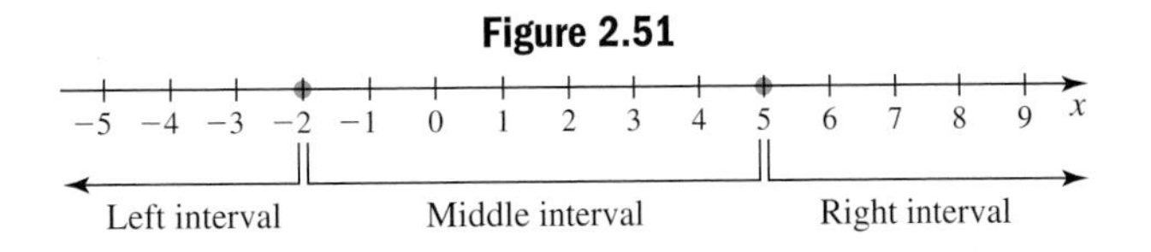

Substituting a test value from each interval in the original function will give the information needed to solve the inequality. Select $x = -3$ from the left interval, $x = 0$ from the middle, and $x = 6$ from the interval on the right.

Chosen Value	Test in Function	Result	Conclusion
$x = -3$	$f(-3) = 8$	positive	$f(x) > 0$ in left interval
$x = 0$	$f(0) = -10$	negative	$f(x) < 0$ in middle interval
$x = 6$	$f(6) = 8$	positive	$f(x) > 0$ in right interval

The original inequality was $f(x) \leq 0$, and we note that outputs were negative only in the middle interval. The solution set is $x \in [-2, 5]$. This is supported by the graph of $f(x)$, shown in Figure 2.52.

Figure 2.52

$f(x) = x^2 - 3x - 10$

NOW TRY EXERCISES 69 THROUGH 74

The ideas presented here are easily extended to any of the toolbox functions. By locating the zeroes and noting end behavior, we're able to solve many inequalities very quickly—sometimes even mentally.

D. Applications of Function Inequalities

One application of function inequalities involves the domain of certain radical functions. As we've seen, functions of the form $f(A) = \sqrt[n]{A}$, where n is an even number, have real number outputs only when $A \geq 0$. When A represents a linear or quadratic function, the ideas just presented can be used to determine the domain. This will be particularly helpful in our study of the composition of functions (Section 3.1), conic sections (Chapter 7), and in other areas.

EXAMPLE 7 What is the domain of $r(x) = \sqrt{4 - x^2}$?

Solution: Here the radicand is nonnegative when $4 - x^2 \geq 0$. Graphically $y = 4 - x^2$ represents a parabola that is concave down, with x-intercepts $(-2, 0)$ and $(2, 0)$. Graphing these zeroes and using the concavity gives the diagram in Figure 2.53, where we see the outputs are nonnegative for $x \in [-2, 2]$. The domain of $r(x) = \sqrt{4 - x^2}$ is $x \in [-2, 2]$. The graph of $r(x)$ is a semicircle (Figure 2.54).

Figure 2.53

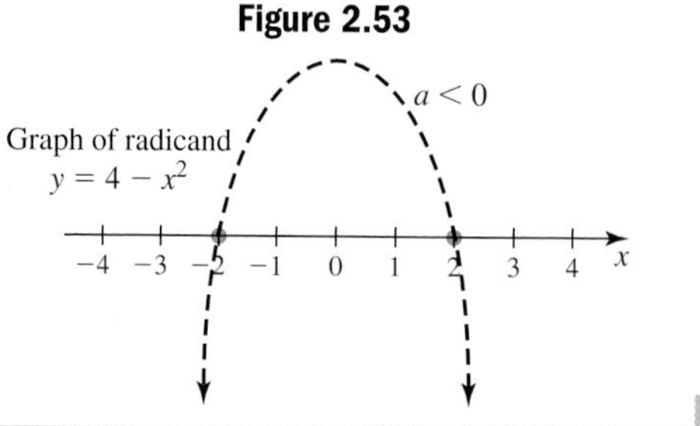

Figure 2.54

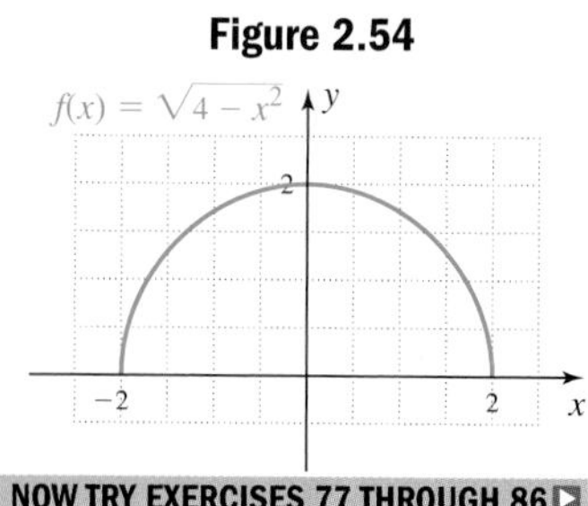

NOW TRY EXERCISES 77 THROUGH 86

CAUTION

Try not to confuse the sketch used to find the domain, with the graph of the actual function. The function inequality $4 - x^2 \geq 0$ is simply a tool to help us understand the domain and graph of $r(x) = \sqrt{4 - x^2}$.

These ideas can be applied to any polynomial function whose zeroes can be determined. In Example 8 they are applied to a factorable cubic equation.

EXAMPLE 8 Given $p(x) = x^3 + x^2 - 9x - 9$, solve $p(x) \geq 0$.

Solution: The lead coefficient is positive, so the end behavior will be down, up. The y-intercept is $(0, -9)$. For the x-intercepts we set $p(x) = 0$ and factor:

$$\begin{aligned} 0 &= x^3 + x^2 - 9x - 9 \\ &= x^2(x + 1) - 9(x + 1) \longrightarrow (x + 1)(x^2 - 9) \\ &= (x + 1)(x - 3)(x + 3) \end{aligned}$$

The x-intercepts are $(-3, 0)$, $(-1, 0)$, and $(3, 0)$. Using these intercepts along with the end behavior produces a general version of the graph as shown in the figure, where we note the solutions to $p(x) \geq 0$ are: $x \in [-3, -1] \cup [3, \infty)$.

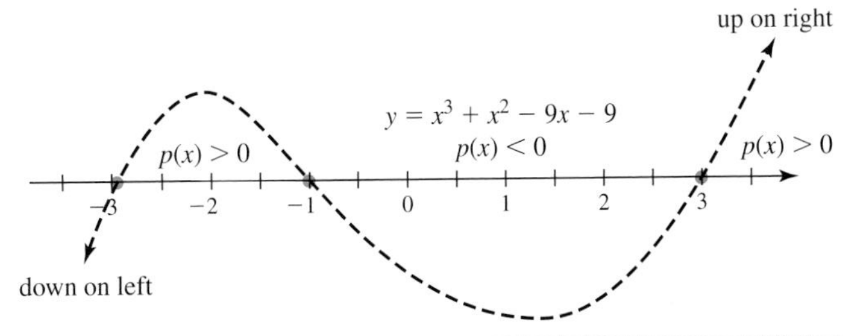

NOW TRY EXERCISES 87 THROUGH 92

TECHNOLOGY HIGHLIGHT

Using the TRACE Feature on a Graphing Calculator

The keystrokes shown apply to a TI-84 Plus model. Please consult your manual or our Internet site for other models.

The TRACE feature of the TI-84 Plus is a wonderful tool for understanding the various characteristics of a graph. We'll illustrate using $f(x) = 0.375x^2 - 0.75x - 5.625$. Enter it as Y_1 on the Y= screen, then graph the function on the standard window (use ZOOM 6). After pressing the TRACE key, the cursor appears on the graph at the y-intercept $(0, -5.625)$ and its location is displayed at the bottom of the screen (Figure 2.55). As we have seen in other *Technology Highlights,* we can walk the cursor along the curve in either direction using the left arrow ◄ and right arrow ► keys. Because the location of the cursor is constantly displayed as we move, finding intervals where the function is positive or negative is *simply a matter of watching the sign of y!* Walk the cursor to the right until the sign of y changes from negative to

Figure 2.55

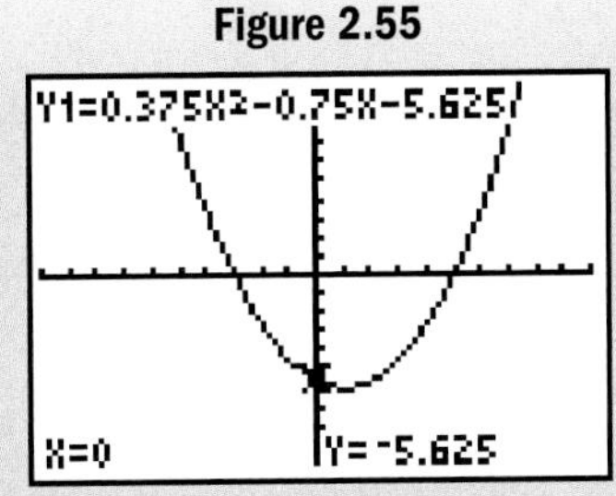

positive. Notice that this occurs about where $x = 5$, but we can't tell exactly. If the function has x-intercepts that are integers, we can locate them using a friendly window. So far we've used ZOOM **4:ZDecimal** and ZOOM **8:ZInteger.** However, sometimes these produce screens that are too small or too large, and we now introduce a third alternative. Since the screen is 94 pixels wide and 62 pixels tall, setting the window as shown in Figure 2.56 will enable us to TRACE through "friendly" values. Enter the window shown and investigate. As it turns out, the x-intercept is (5, 0) and walking the cursor to the left and right of $x = 5$ shows the function is positive for values greater than 5 and negative for values less than 5. Now walk the cursor over to the other intercept near $x = -3$. As we might suspect, this intercept is $(-3, 0)$ and the function is positive for values less than $x = -3$. This demonstrates that $f(x) < 0$ for $x \in (-3, 5)$ and $f(x) \geq 0$ for $x \in (-\infty, -3] \cup [5, \infty)$. Use these ideas to solve $f(x) \geq 0$ and $f(x) \leq 0$ for these functions:

Figure 2.56

```
WINDOW
 Xmin=-9.4
 Xmax=9.4
 Xscl=1
 Ymin=-6.2
 Ymax=6.2
 Yscl=1
 Xres=1
```

Exercise 1: $y = 0.2x^2 + 0.8x - 4.2$

Exercise 2: $y = -0.16x^2 + 0.96x + 2.56$

1. $f(x) \geq 0$ for $x \in (-\infty, -7] \cup [3, \infty)$; $f(x) \leq 0$ for $x \in [-7, 3]$
2. $f(x) \geq 0$ for $x \in [-2, 8]$; $f(x) \leq 0$ for $x \in (-\infty, -2] \cup [8, \infty)$

2.5 EXERCISES

CONCEPTS AND VOCABULARY

Fill in each blank with the appropriate word or phrase. Carefully reread the section if needed.

1. The x-intercepts of a polynomial graph are also called the zeroes of the function.

2. To solve a quadratic inequality, we need only determine the zeroes of the function and the concavity of the graph.

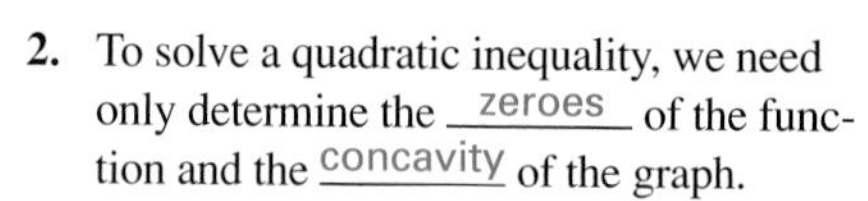

3. If the graph of an absolute value function $f(x)$ opens upward with a vertex at (5, 1), the solution set for $f(x) > 0$ is $x \in$ R.

4. If the graph of a quadratic function $g(x)$ is concave down with a vertex at $(5, -1)$, the solution set for $g(x) > 0$ is empty.

5. State the interval(s) where $f(x) > 0$ and discuss/explain your response.

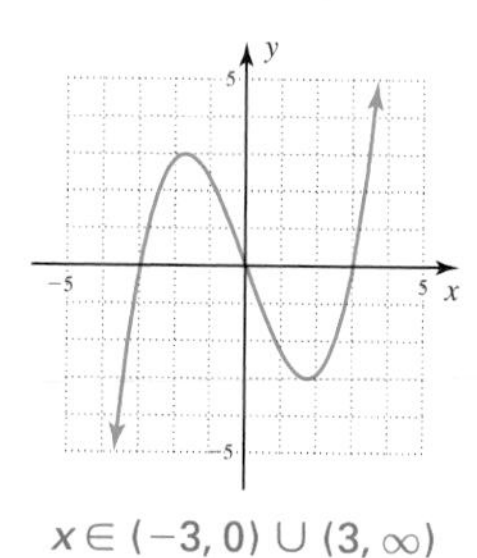

$x \in (-3, 0) \cup (3, \infty)$

6. State the interval(s) where $f(x) \leq 0$ and discuss/explain your response.

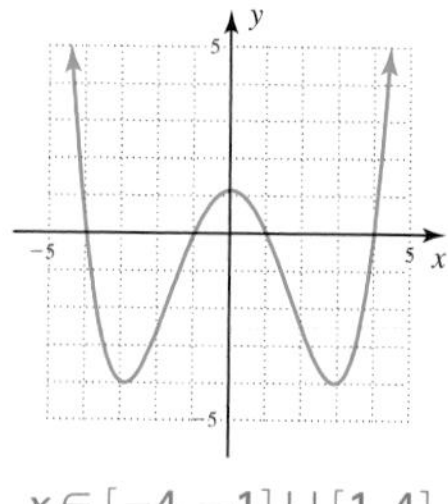

$x \in [-4, -1] \cup [1, 4]$

DEVELOPING YOUR SKILLS

For each function given, solve the inequality indicated using a table of values, sketching a graph, and noting where the graph is above or below the x-axis.

7. $f(x) = 3x - 2; f(x) > 0$ $x \in (\frac{2}{3}, \infty)$

8. $g(x) = 4x - 3; g(x) > 0$ $x \in (\frac{3}{4}, \infty)$

9. $h(x) = -\frac{1}{2}x + 4; h(x) \leq 0$ $x \in [8, \infty)$

10. $p(x) = -\frac{2}{3}x + 1; p(x) \leq 0$ $x \in [\frac{3}{2}, \infty)$

11. $q(x) = 5; q(x) < 0$ no solution

12. $r(x) = -2; r(x) > 0$ no solution

Solve the inequality indicated by locating the x-intercept and noting the slope of the line.

13. $Y_1 = 2x - 5; Y_1 < 0$ $x \in (-\infty, \frac{5}{2})$

14. $Y_2 = 3x - 4; Y_2 < 0$ $x \in (-\infty, \frac{4}{3})$

15. $r(x) = -\frac{3}{2}x + 2; r(x) > 0$ $x \in (-\infty, \frac{4}{3})$

16. $s(x) = -\frac{4}{3}x - 1; s(x) > 0$ $x \in (-\infty, \frac{-3}{4})$

17. $v(x) = -0.5x + 4; v(x) \le 0$ $x \in [8, \infty)$

18. $w(x) = -0.4x + 2; w(x) \le 0$ $x \in [5, \infty)$

19. $f(x) = -x + 4; f(x) \le 0$ $x \in [4, \infty)$

20. $g(x) = 5 - x; g(x) \le 0$ $x \in [5, \infty)$

Graph each function by plotting points, then use the graph to solve the inequality indicated. See page 161 for reference.

21. $f(x) = |x| - 3; f(x) \le 0$ $x \in [-3, 3]$

22. $g(x) = |x| - 5; g(x) \ge 0$ $x \in (-\infty, -5] \cup [5, \infty)$

23. $h(x) = |x - 2| + 1; h(x) < 0$ no solution

24. $p(x) = -|x + 1| - 2; p(x) > 0$ no solution

25. $q(x) = -|x| - 3; q(x) < 0$ $x \in (-\infty, \infty)$

26. $r(x) = |x| + 2; r(x) > 0$ $x \in (-\infty, \infty)$

Solve the indicated inequality using the graph given. Assume all intercepts are lattice points.

27. $f(x) > 0$ $x \in (-\infty, 3)$

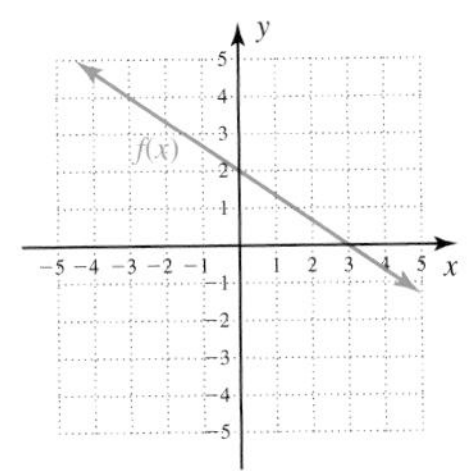

28. $g(x) < 0$ $x \in (-\infty, 1)$

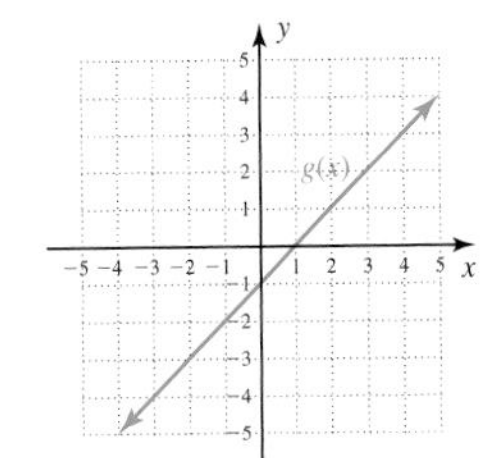

29. $h(x) \ge 0$ $x \in (-\infty, -3]$

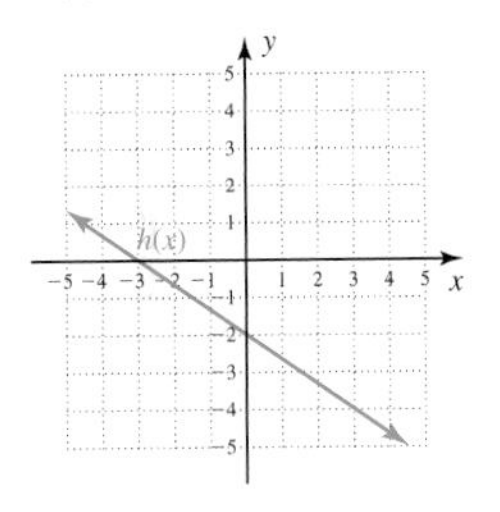

30. $f(x) > 0$ $x \in (0, \infty)$

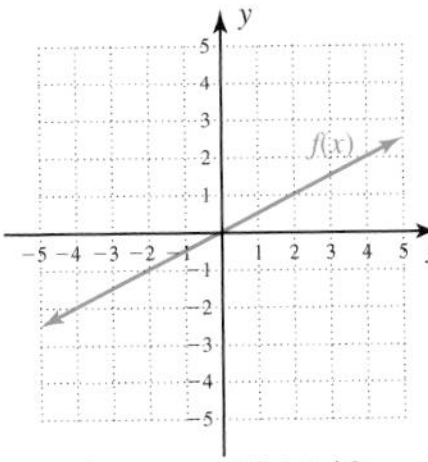

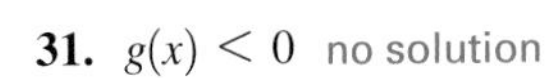

31. $g(x) < 0$ no solution

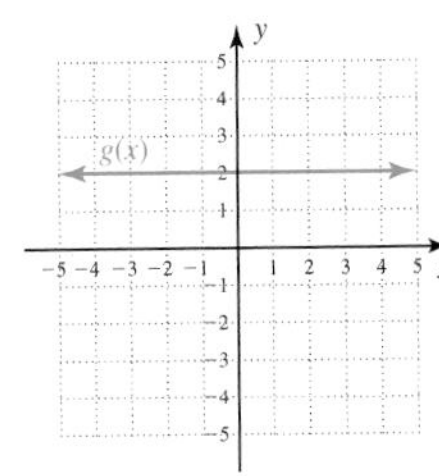

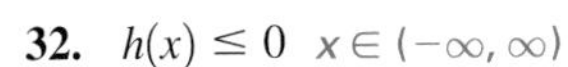

32. $h(x) \le 0$ $x \in (-\infty, \infty)$

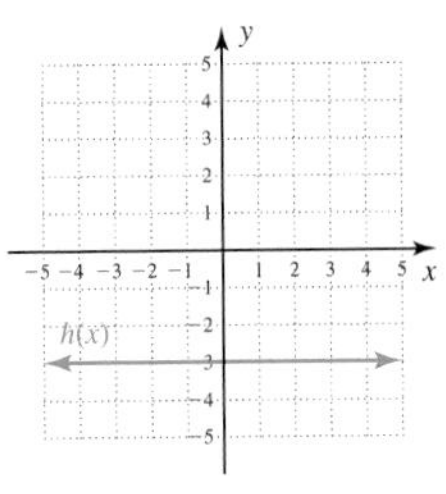

33. $f(x) > 0$ $x \in (-\infty, -2) \cup (3, \infty)$

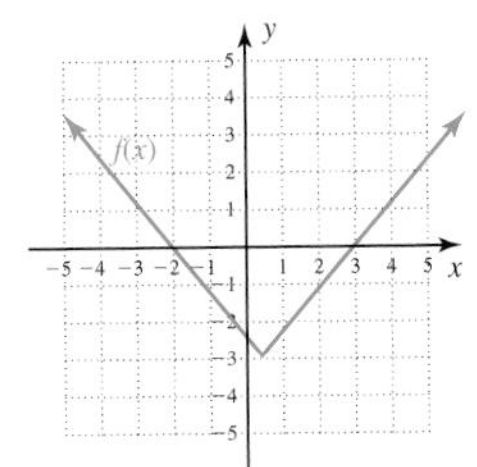

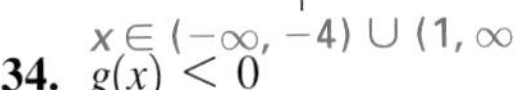

34. $g(x) < 0$ $x \in (-\infty, -4) \cup (1, \infty)$

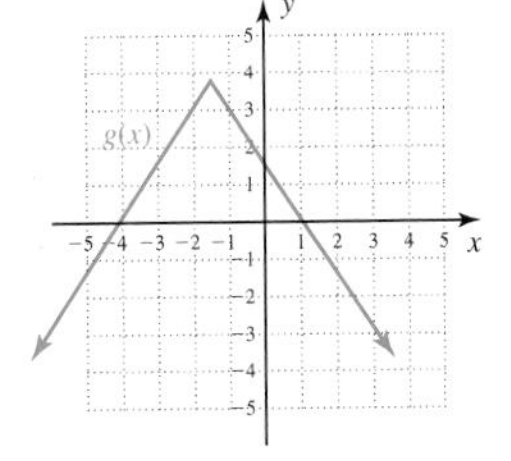

35. $h(x) \ge 0$ $x \in [-4, 3]$

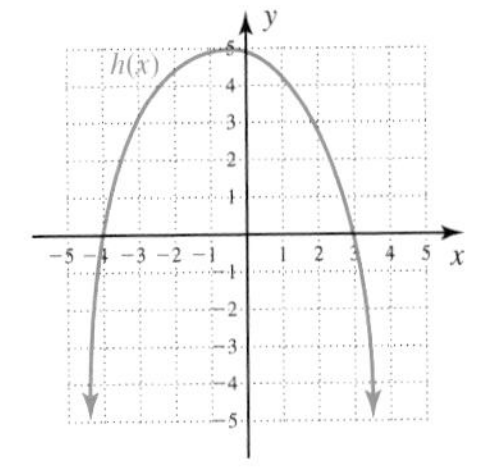

36. $f(x) > 0$ $x \in (-\infty, -5) \cup (2, \infty)$

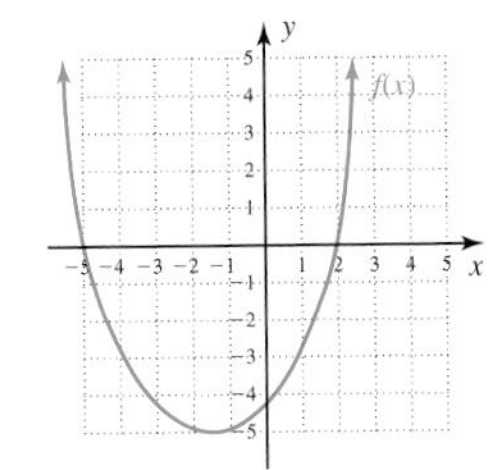

37. $g(x) < 0$ no solution

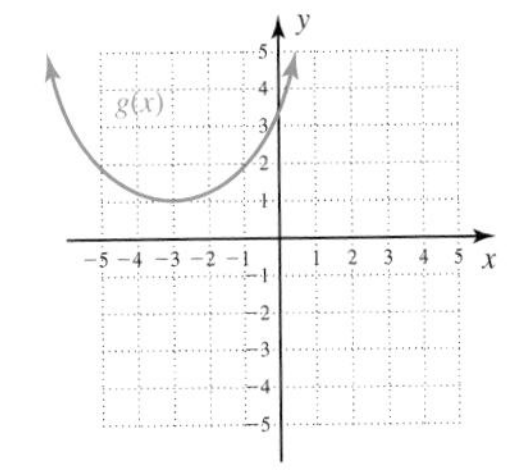

38. $h(x) \le 0$ $x \in (-\infty, \infty)$

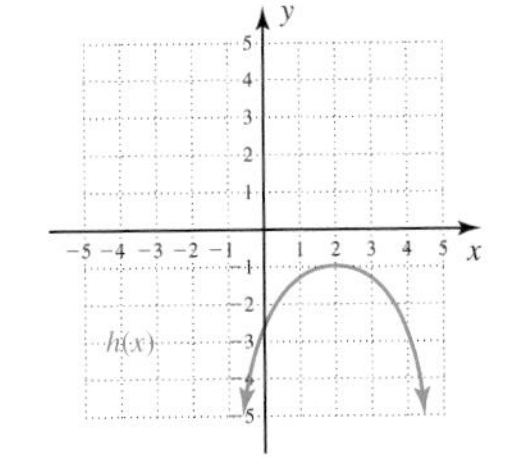

39. $f(x) > 0$ $x \in (-\infty, -4) \cup (-1, 2)$

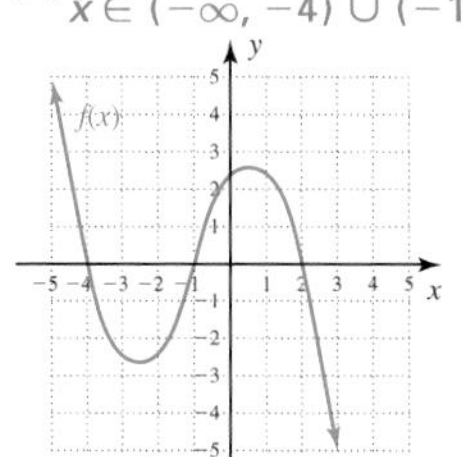

40. $g(x) < 0$ $x \in (-\infty, -4) \cup (0, 3)$

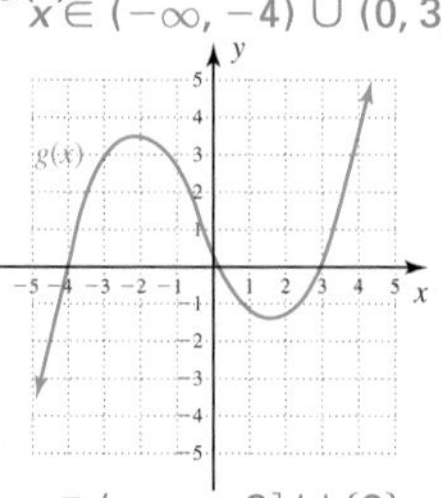

41. $h(x) \leq 0$ $x \in [-4, -3]$

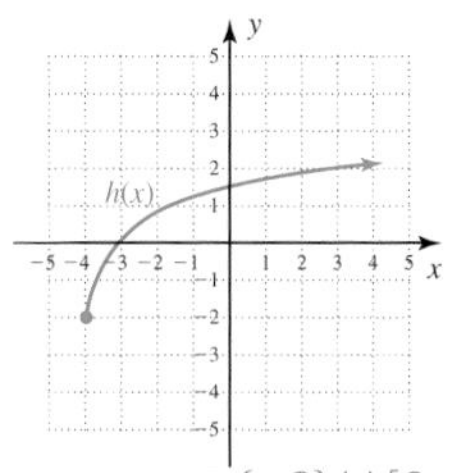

42. $p(x) > 0$ $x \in (-\infty, 3)$

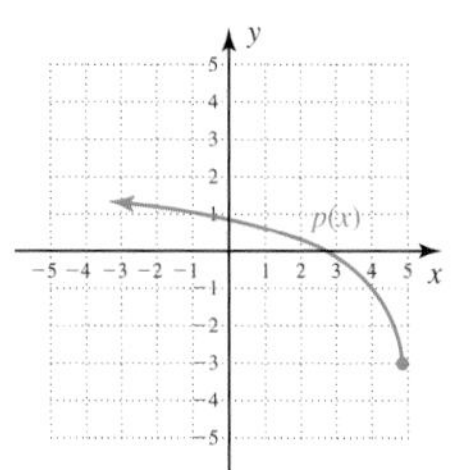

43. $q(x) \geq 0$

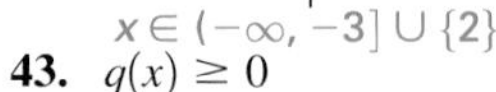
$x \in (-\infty, -3] \cup \{2\}$

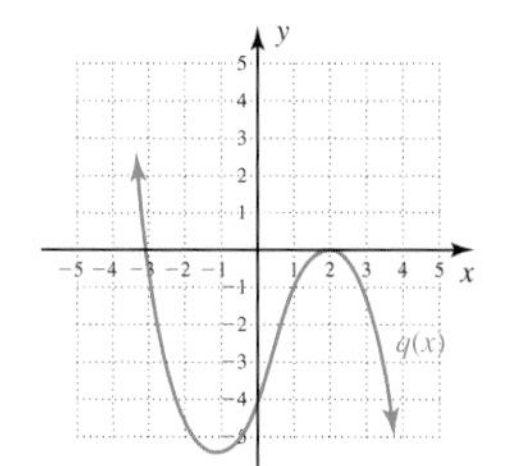

44. $r(x) \leq 0$ $x \in \{-3\} \cup [2, \infty)$

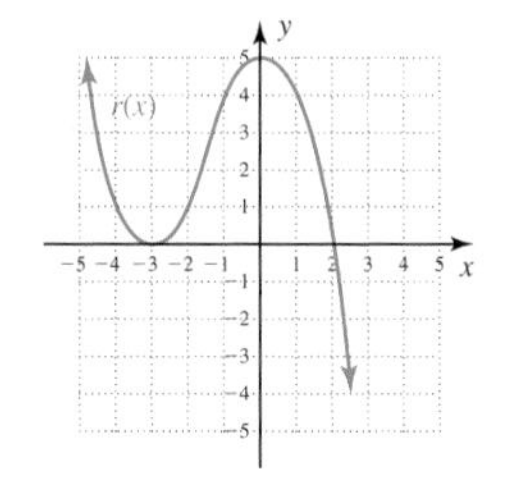

Solve each inequality by locating the x-intercept(s) and noting the concavity of the graph.

45. $f(x) = -x^2 + 4x; f(x) > 0$

46. $g(x) = x^2 - 5x; g(x) < 0$

47. $h(x) = x^2 + 4x - 5; h(x) \geq 0$

48. $p(x) = -x^2 + 3x + 10; p(x) \leq 0$

49. $q(x) = 2x^2 - 5x - 7; q(x) < 0$

50. $r(x) = -2x^2 - 3x + 5; r(x) > 0$

51. $s(x) = 7 - x^2; s(x) \geq 0$

52. $t(x) = 13 - x^2; t(x) \geq 0$

53. $Y_1 = x^2 + 3x - 6; Y_1 \geq 0$

54. $Y_2 = x^2 - 5x - 2; Y_2 \geq 0$

55. $Y_3 = 3x^2 - 2x - 5; Y_3 \geq 0$

56. $Y_4 = 4x^2 - 3x - 7; Y_4 \geq 0$

57. $s(x) = x^2 - 8x + 16; s(x) \geq 0$

58. $t(x) = x^2 - 6x + 9; t(x) \geq 0$

59. $r(x) = 4x^2 + 12x + 9; r(x) < 0$

60. $f(x) = 9x^2 - 6x + 1; f(x) > 0$

61. $g(x) = -x^2 + 10x - 25; g(x) \leq 0$

62. $h(x) = -x^2 + 14x - 49; h(x) \geq 0$

63. $Y_1 = -x^2 - 2; Y_1 > 0$

64. $Y_2 = x^2 + 4; Y_2 < 0$

65. $f(x) = x^2 - 2x + 5; f(x) > 0$

66. $g(x) = -x^2 + 3x - 3; g(x) > 0$

67. $p(x) = 2x^2 - 6x + 9; p(x) \geq 0$

68. $q(x) = 5x^2 - 4x + 4; q(x) \geq 0$

Solve the following inequalities using the interval test method.

69. $f(x) = x^2 - 2x - 15; f(x) \geq 0$

70. $g(x) = x^2 + 3x - 18; g(x) \geq 0$

71. $h(x) = 2x^2 - 7x - 15; h(x) < 0$

72. $p(x) = 3x^2 - 11x - 20; p(x) < 0$

73. $Y_1 = x^3 - 8x; Y_1 \geq 0$

74. $Y_2 = x^3 - 12x; Y_2 \leq 0$

45. $x \in (0, 4)$
46. $x \in (0, 5)$
47. $x \in (-\infty, -5] \cup [1, \infty)$
48. $x \in (-\infty, -2] \cup [5, \infty)$
49. $x \in (-1, \frac{7}{2})$
50. $x \in (\frac{-5}{2}, 1)$
51. $x \in [-\sqrt{7}, \sqrt{7}]$
52. $x \in [-\sqrt{13}, \sqrt{13}]$
53. $x \in (-\infty, \frac{-3-\sqrt{33}}{2}] \cup [\frac{-3+\sqrt{33}}{2}, \infty)$
54. $x \in (-\infty, \frac{5-\sqrt{33}}{2}] \cup [\frac{5+\sqrt{33}}{2}, \infty)$
55. $x \in (-\infty, -1] \cup [\frac{5}{3}, \infty)$
56. $x \in (-\infty, -1] \cup [\frac{7}{4}, \infty)$
57. $x \in (-\infty, \infty)$
58. $x \in (-\infty, \infty)$
59. no solution
60. $x \in (-\infty, \frac{1}{3}) \cup (\frac{1}{3}, \infty)$
61. $x \in (-\infty, \infty)$
62. $\{7\}$
63. no solution
64. no solution
65. $x \in (-\infty, \infty)$
66. no solution
67. $x \in (-\infty, \infty)$
68. $x \in (-\infty, \infty)$
69. $x \in (-\infty, -3] \cup [5, \infty)$
70. $x \in (-\infty, -6] \cup [3, \infty)$
71. $x \in (\frac{-3}{2}, 5)$
72. $x \in (\frac{-4}{3}, 5)$
73. $x \in [-2\sqrt{2}, 0] \cup [2\sqrt{2}, \infty)$
74. $x \in (-\infty, -2\sqrt{3}] \cup [0, 2\sqrt{3}]$

WORKING WITH FORMULAS

75. The equation of a central semicircle: $f(x) = \sqrt{r^2 - x^2}$

The equation of a semicircle is given by the function shown, where r represents the radius. Use the techniques of this section to find the domain of the function if the semicircle has a *diameter* of 8 ft. $x \in [-4, 4]$

76. The perpendicular distance from a point to a line: $D = \dfrac{Ax + By + C}{\pm\sqrt{A^2 + B^2}}$

The perpendicular distance from a given point (x, y) to the line $Ax + By + C = 0$ is given by the formula shown, where the sign is chosen so as to ensure the expression is positive.

(a) Discuss why this formula must hold for all points (x, y) and real coefficients A, B, and C. (b) Find the distance between the point $(12, -1)$ and the line $-2x + 3y - 12 = 0$. (c) Verify that this is the *perpendicular* distance. **a.** Answers will vary. **b.** $3\sqrt{13}$ **c.** Given line: $y = \frac{2}{3}x + 4$. Perpendicular line through $(12, -1)$: $y = \frac{-3}{2}x + 17$. These intersect at $(6, 8)$ and the distance between these points is also $3\sqrt{13}$.

APPLICATIONS

Determine the domain of the following functions.

77. $f(x) = \sqrt{-3x + 4}$ **78.** $g(x) = \sqrt{5 - 2x}$ **79.** $h(x) = \sqrt{x^2 - 25}$

80. $p(x) = \sqrt{25 - x^2}$ **81.** $q(x) = \sqrt{x^2 - 5x}$ **82.** $r(x) = \sqrt{6x - x^2}$

83. $s(x) = \sqrt{x^2 - 2x - 15}$ **84.** $t(x) = \sqrt{-x^2 + 3x - 4}$

85. $Y_1 = \sqrt{x^2 - 6x + 9}$ **86.** $Y_2 = \sqrt{x^2 - 10x + 25}$

87. $f(x) = \sqrt{x^3 - 2x^2 - 15x}$ **88.** $g(x) = \sqrt{x^3 - 2x^2 - 9x + 18}$

77. $x \in (-\infty, \frac{4}{3}]$
78. $x \in (-\infty, \frac{5}{2}]$
79. $x \in (-\infty, -5] \cup [5, \infty)$
80. $x \in [-5, 5]$
81. $x \in (-\infty, 0] \cup [5, \infty)$
82. $x \in [0, 6]$
83. $x \in (-\infty, -3] \cup [5, \infty)$
84. no solution
85. $x \in (-\infty, \infty)$
86. $x \in (-\infty, \infty)$
87. $x \in [-3, 0] \cup [5, \infty)$
88. $x \in [-3, 2] \cup [3, \infty)$
89. April–September; December–March, October–December

89. Seasonal profits: Due to the seasonal nature of the business, the profit earned by Scotty's Water Sports Equipment fluctuates predictably over a 12-month period. The profit graph is shown, where $0 \rightarrow$ December 31, $1 \rightarrow$ January 31, $2 \rightarrow$ February 28, and so on. During what months is the company making a profit? During what months is the company losing money?

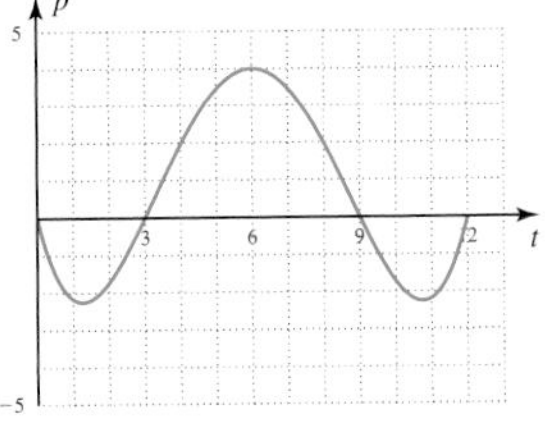

Exercise 90

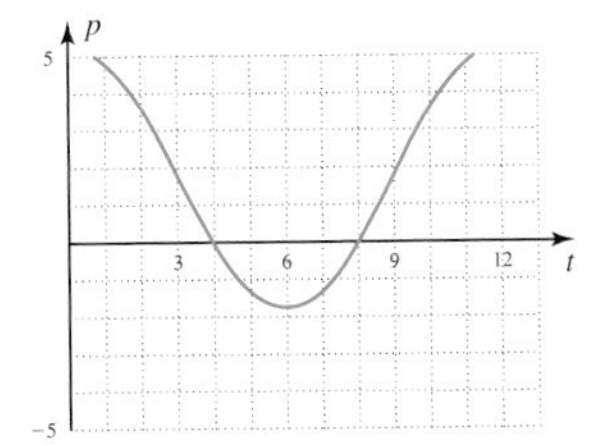

90. Seasonal profits: Due to the seasonal nature of the business, the profit earned by Sally's Ski Shop fluctuates predictably over a 12-month period. The profit graph is shown to the left, where $0 \rightarrow$ December 31, $1 \rightarrow$ January 31, $2 \rightarrow$ February 28, and so on. During what months is the company making a profit? What months are they losing money? December–April, September–December; May–August

91. Birds gone fishing: There are a number of birds who feed by diving into a body of water to capture a fish, then swim to the surface and fly away with their prey. Suppose the function $h(t) = 5|t - 3| - 5$ models the height (in feet) of the bird as it begins to dive. Construct a table of values using inputs from $t = 0$ to $t = 6$ sec, graph the function, and answer the following questions.

a. How many seconds after the bird begins its dive does it hit the water? 2 sec

b. How many seconds after the bird hits the water does it surface? 2 sec

c. How deep does the bird dive? 5 ft

d. What was the bird's height above the water when the dive began? 10 ft

92. Cold air mass movement: One cold, winter evening at 12 o'clock midnight, a freezing arctic air mass swept over Montana, causing the temperature to drop precipitously. The temperature was already a chilly 30°F and began falling from there. Suppose the temperature t (in degrees Fahrenheit) was modeled by the function $F(t) = \frac{5}{2}t^2 - 20t + 30$. Use this function to answer the following questions.

a. How many hours until the temperature dropped below 0°F? 2 hr

b. How many hours until the temperature rose above 0°F? 4 hr

c. How cold did it get? −10°F

d. How many hours was the temperature below zero? 4 hr

Exercise 94

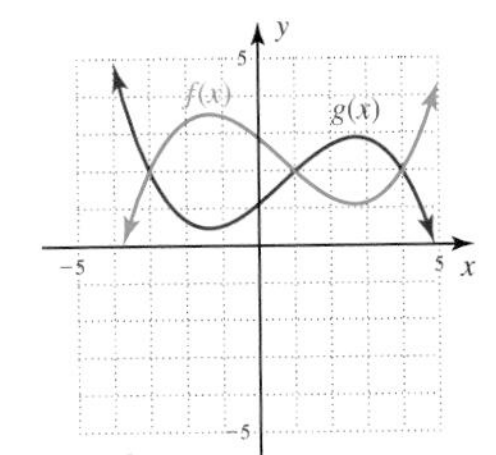

WRITING, RESEARCH, AND DECISION MAKING

93. How would you write the solution set for $f(x) \geq 0$, whose graph is shown to the right? Give a complete description of the processes and concepts involved, as though you were trying to explain the ideas to a friend who was absent on the day these ideas were explored. $x \in \{-2\} \cup [4, \infty)$

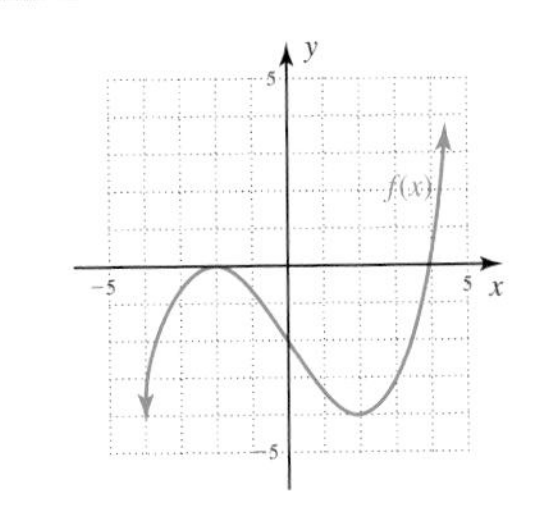

94. Using the graphs of $f(x)$ and $g(x)$ given, name the points or intervals where: (a) $f(x) = g(x)$, (b) $f(x) < g(x)$, and

(c) $f(x) > g(x)$. Then, (d) estimate the area bounded between the two curves using the grid (count squares and partial squares). In a calculus class, techniques are introduced that enable us to find the exact area.
a. $x = -3, 1, 4$ b. $x \in (-\infty, -3) \cup (1, 4)$ c. $x \in (-3, 1) \cup (4, \infty)$
d. Answers will vary. $A \approx 10$ units^2

EXTENDING THE CONCEPT

Determine the domain of each function.

95. $f(x) = \sqrt{x^3 - 9x}$
$x \in [-3, 0] \cup [3, \infty)$

96. $\dfrac{\sqrt{4 - x^2}}{x^2 - 1}$
$x \in [-2, -1) \cup (-1, 1) \cup (1, 2]$

MAINTAINING YOUR SKILLS

97. (R.3) Simplify each expression:
a. $(7x)^0 + 5x^0 + 2^{-1}$ $6\frac{1}{2}$
b. $\dfrac{(3m^{-3}n^4)^{-2}}{3mn}$ $\dfrac{m^5}{27n^9}$

98. (2.3) For the interval $x = 1$ to $x = 2$, which function has the greater rate of change: $f(x) = \frac{3}{2}x + 2$ or $g(x) = x^2 - 1$?
$g(x) = x^2 - 1$

99. (2.3) Find the slope of the line $3x - 4y = 7$. $\frac{3}{4}$

100. (R.6) Solve the equation: $-2\sqrt{x + 1} + 7 = 1$ $x = 8$

101. (1.4) Find the sum, difference, product, and quotient of $2 + 3i$ and $2 - 3i$.
$4, 6i, 13, -\frac{5}{13} + \frac{12}{13}i$

102. (R.7) Find the area and perimeter of the triangle shown.

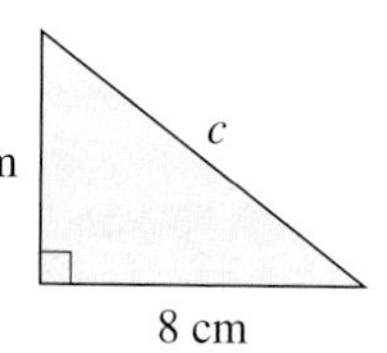

$A = 24$ cm^2, $P = 24$ cm

2.6 Regression, Technology, and Data Analysis

LEARNING OBJECTIVES

In Section 2.6 you will learn how to:

A. Draw a scatter-plot and identify positive and negative associations
B. Use a scatter-plot to identify linear and nonlinear associations
C. Use a scatter-plot to identify strong and weak associations
D. Estimate a line of best fit for a set of data
E. Use linear regression to find the line of best fit
F. Use quadratic regression to find the parabola of best fit

INTRODUCTION

Collecting and analyzing data is a tremendously important mathematical endeavor, having applications throughout business, industry, science, government, and a score of other fields. When it comes to linear and quadratic applications, the link between classroom mathematics and real-world mathematics is called a **regression,** in which we attempt to find an equation that will act as a model for the raw data. In this section, we focus on linear and quadratic equation models.

POINT OF INTEREST

The collection and use of data seems to have originally been motivated by two unrelated investigations. The first was the processing of statistical data for insurance rates and mortality tables, the second was to answer questions related to gambling and games of chance. In 1662, a London merchant named John Gaunt (1620–1674) wrote *Natural and Political Observations Made upon Bills of Mortality.* It is widely held that this work helped launch a more formal study of statistics and data collection.

A. Scatter-Plots and Positive/Negative Association

In this section we continue our study of ordered pairs and functions, but this time using data collected from various sources or from observed real-world relationships. You can hardly pick up a newspaper or magazine without noticing it contains a large volume of data—graphs, charts, and tables seem to appear throughout the pages. In addition, there are many simple experiments or activities that enable you to collect your own data. After it's been collected, we begin analyzing the data using a **scatter-plot,** which is simply a graph of all of the ordered pairs in a data set. Much of the time, real data (sometimes called **raw data**) is not very "well behaved" and the points may be somewhat scattered—which is the reason for the name.

Positive and Negative Associations

Earlier we noted that lines with positive slope rise from left to right, while lines with negative slope fall from left to right. We can extend this idea to the data from a scatter-plot. The data points in Example 1 seem to *rise* as you move from left to right, with larger input values resulting in larger outputs. In this case, we say there is a **positive association** between the variables. If the data seems to decrease or fall as you move left to right, we say there is a **negative association.**

EXAMPLE 1 The ratio of the federal debt to the total population is known as the *per capita debt*. The per capita debt of the United States is shown in the table for the odd-numbered years from 1995 to 2003. Draw a scatter-plot of the data and state whether the association is positive or negative.

Data from the Bureau of Public Debt at www.publicdebt.treas.gov

Year	Per Capita Debt (1000s)
1995	18.9
1997	20.0
1999	20.7
2001	20.5
2003	23.3

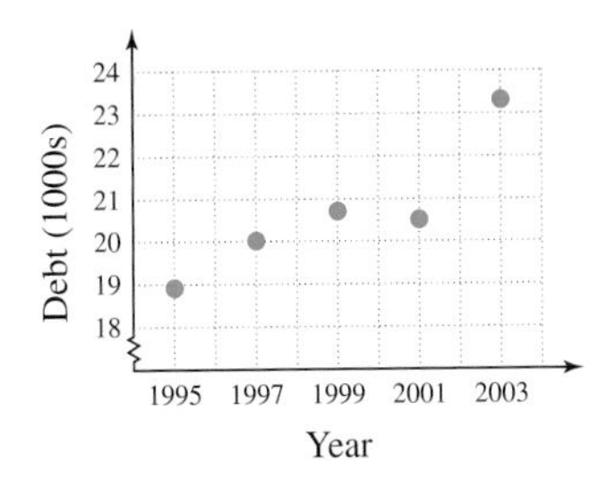

Solution: Since the amount of debt depends on the year, *year* is the input x and *per capita debt* is the output y. Scale the x-axis from 1995 to 2003 and the y-axis from 18 to 23 to comfortably fit the data (the "squiggly line" near the 18 in the graph is used to show that some initial values have been skipped). The graph indicates there is a positive association between the variables, meaning the debt is generally *increasing* as time goes on.

NOW TRY EXERCISES 7 AND 8

EXAMPLE 2 A cup of coffee is placed on a table and allowed to cool. The temperature of the coffee is measured every 10 min and the data are shown in the table. Draw the scatter-plot and state whether the association is positive or negative.

Elapsed Time (minutes)	Temperature (°F)
0	110
10	89
20	76
30	72
40	71

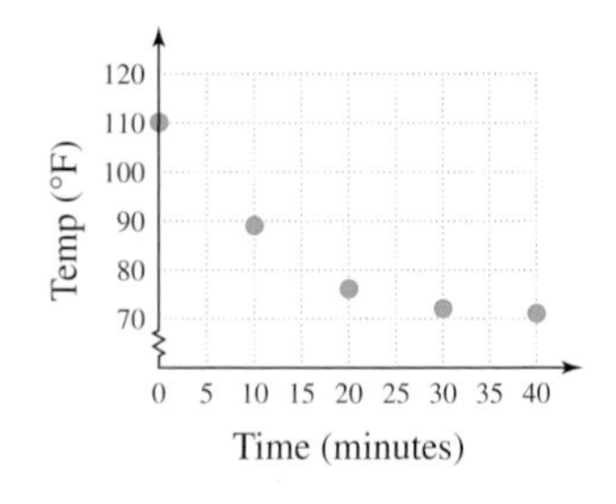

Solution: Since temperature depends on cooling time, *time* is the input x and *temperature* is the output y. Scale the x-axis from 0 to 40 and the y-axis from 70 to 110 to comfortably fit the data. As you see in the figure, there is a negative association between the variables, meaning the temperature *decreases* over time.

NOW TRY EXERCISES 9 THROUGH 12

B. Scatter-Plots and Linear/Nonlinear Associations

The data in Example 1 had a positive association, while the association in Example 2 was negative. But the data from these examples differ in another important way. In Example 1, the data seem to cluster about an imaginary line. This indicates a linear equation model might be a good approximation for the data, and we say there is a **linear association** between the variables. The data in Example 2 could not accurately be modeled using a straight line, and we say the variables *time* and *cooling temperature* exhibit a **nonlinear association.**

EXAMPLE 3 A college professor tracked her annual salary for 1997 to 2004 and the data are shown in the table. Draw the scatter-plot and determine if there is a linear or nonlinear association between the variables. Also state whether the association is positive, negative, or cannot be determined.

Year	Salary (1000s)
1997	30.5
1998	31
1999	32
2000	33.2
2001	35.5
2002	39.5
2003	45.5
2004	52

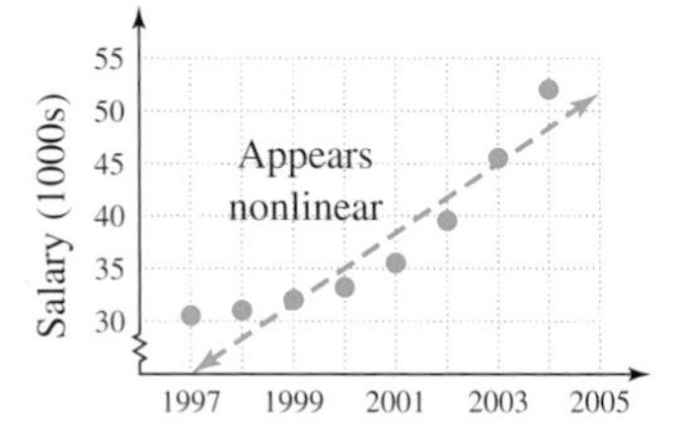

Solution: Since salary earned depends on a given year, *year* is the input x and *salary* is the output y. Scale the x-axis from 1996 to 2005, and the y-axis from 30 to 55 to comfortably fit the data. A line doesn't seem to model the data very well, and the association appears to be nonlinear. The data rises from left to right, indicating a positive association between the variables. This makes good sense, since we expect our salaries to increase over time.

NOW TRY EXERCISES 13 AND 14

C. Strong and Weak Associations

Using Figures 2.57 and 2.58, we can make one additional observation regarding the data in a scatter-plot. While both associations appear linear, the data in Figure 2.57 seems to cluster more tightly about an imaginary straight line than the data in Figure 2.58.

Figure 2.57

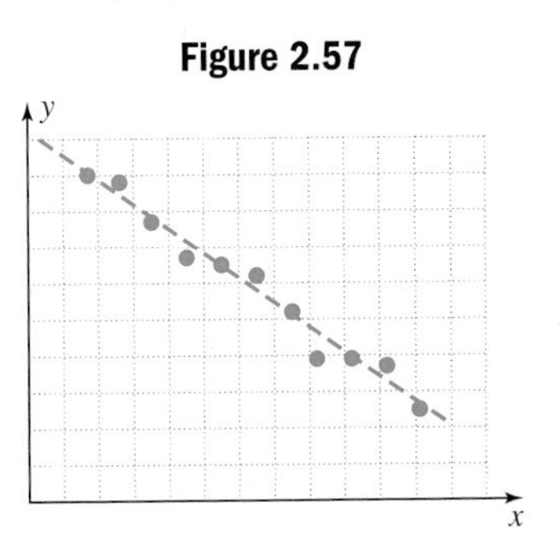

Figure 2.58

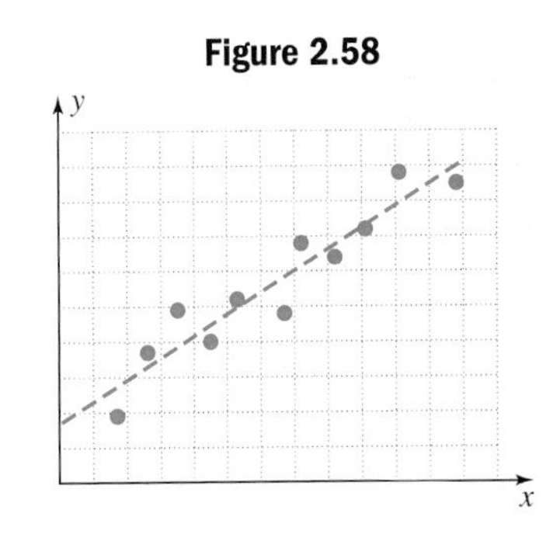

We refer to this "clustering" as the **tightness of fit** or in statistical terms, the **strength of the correlation.** To quantify this fit we use a measure called the **correlation coefficient *r*,** which tells whether the association is positive or negative—$r > 0$ or $r < 0$, and quantifies the strength of the association: $|r| \leq 100\%$. Actually, the coefficient is given in decimal form, making it a number from -1.0 to $+1.0$, depending on the association. If the data points form a perfectly straight line, we say the strength of the correlation is either -1 or 1. If the data points appear clustered about the line, but are scattered on either side of it, the strength of the correlation falls somewhere between -1 and 1, depending on how tightly or loosely they're scattered. This is summarized in Figure 2.59.

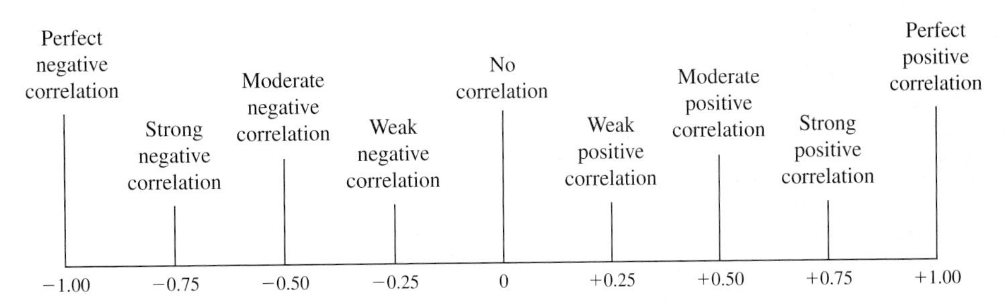

Figure 2.59

The following scatter-plots help to further illustrate this idea. Figure 2.60 shows a linear and negative association between the value of a car and the age of a car, with a strong correlation. Figure 2.61 shows there is no apparent association between family income and the number of children, and Figure 2.62 shows a linear and positive association between a man's height and weight, with a moderate correlation.

Figure 2.60

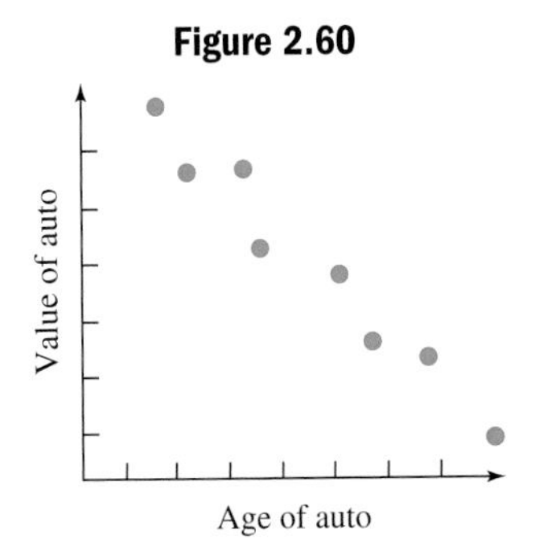

Figure 2.61

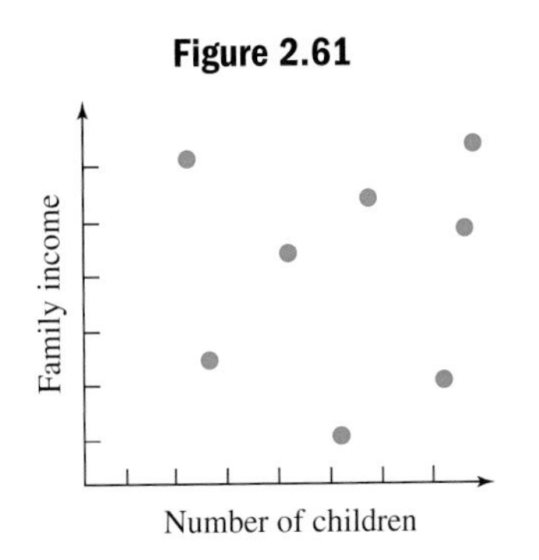

Figure 2.62

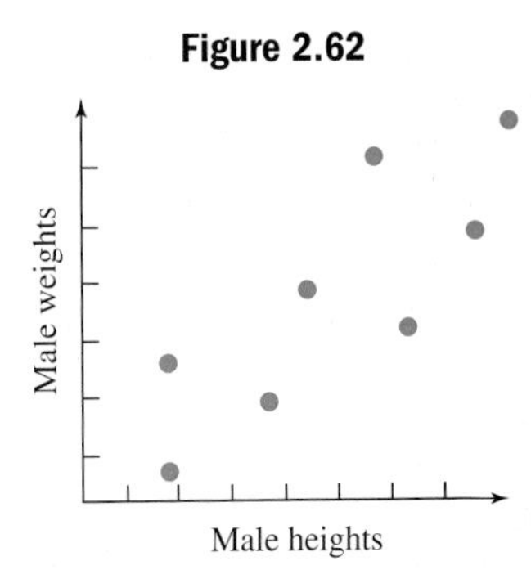

Use these ideas to complete Exercises 15 and 16.

D. Calculating a Linear Equation Model for a Set of Data

Calculating a **linear equation model** for a set of data involves visually estimating and sketching a line that appears to fit the data. This means answers will vary slightly, but a good, usable equation model can often be obtained. To find the equation, we select two points on this imaginary line and use the point-slope formula to construct the equation. Note that points on this estimated line *but not in the data set* can still be used to help determine the equation model.

EXAMPLE 4 The men's 400-m freestyle times (to the nearest second) for the 1960 through 2000 Olympics are given in the table shown. Let the year be the input x, and race time be the output y. Based on the data, draw a scatter-plot and answer the following:

a. Does the association appear linear or nonlinear?

b. Classify the correlation as weak, moderate, or strong.

c. Is the association positive or negative?

d. Find the equation of an estimated line of best fit and use it to predict the winning time for the 2004 Olympics.

Source: www.athens2004.com

Year (x) $(1900 \to 0)$	Time (y)
60	258
64	252
68	249
72	240
76	232
80	231
84	231
88	227
92	225
96	228
100	221

WORTHY OF NOTE

Sometimes it helps to draw a straight line on an overhead transparency, then lay it over the scatter-plot. By shifting the transparency up and down, and rotating it left and right, the line can more accurately be placed so that it's centered among and through the data.

Solution: Begin by choosing an appropriate scale for the axes. The x-axis (year) is scaled from 60 to 100, and the y-axis (time) should only be scaled from 210 to 260 so the data will not be too crowded or hard to read. After plotting the points we obtain the scatter-plot shown in the figure.

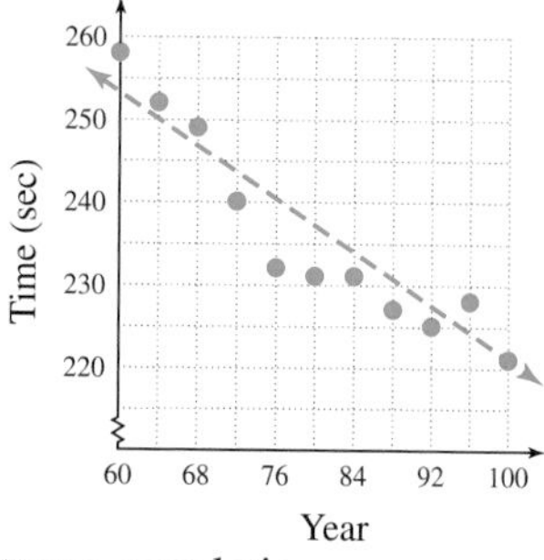

a. The association appears to be linear.

b. There is a moderate or moderate-to-strong correlation.

c. The association is negative, showing that finishing times tend to decrease over the years.

d. The points (100, 221) and (64, 252) were chosen to develop the equation.

$$m = \frac{y_2 - y_1}{x_2 - x_1} = \frac{252 - 221}{64 - 100}$$

$$\approx -0.86 \quad \text{simplify}$$

$$y - y_1 = m(x - x_1) \quad \text{point-slope formula}$$

$$y - 221 = -0.86(x - 100) \quad \text{use } x \approx -0.86 \text{ and } (100, 221)$$

$$y = -0.86x + 307 \quad \text{simplify and solve for } y$$

One equation model of this data is $y = -0.86x + 307$. Once again, slightly different equations may be obtained, depending on the points chosen. Based on this model, the predicted time for the 2004 Olympics would be

$$f(x) = -0.86x + 307 \quad \text{estimated line of best fit}$$
$$f(104) = -0.86(104) + 307 \quad \text{substitute 104 for } x \text{ (year 2004)}$$
$$\approx 217.6 \quad \text{result}$$

For 2004, the winning time was projected to be about 217.6 sec. The actual time was 223 sec, swum by Ian Thorpe of Australia.

NOW TRY EXERCISES 17 THROUGH 24

Once again, great care should be taken to use equation models obtained from data wisely. It would be foolish to assume that in the year 2257, the swim times for the 400-m freestyle would be near 0 seconds—even though that's what the equation model gives for $x = 357$. Most equation models are limited by numerous constraining factors.

E. Linear Regression and the Line of Best Fit

There is actually a sophisticated method for calculating the equation of a line that best fits a data set, called the **regression line.** The method minimizes the vertical distance between all data points and the line itself, making it the unique **line of best fit.** Most graphing calculators have the ability to perform this calculation quickly, and we'll illustrate using the TI-84 Plus. The process involves these steps: (1) clearing old data, (2) entering new data; (3) displaying the data; (4) calculating the regression line; and (5) displaying and using the regression line. We'll illustrate by finding the regression line for the data from Example 4.

Step 1: Clear Old Data, Step 2: Enter New Data, and Step 3: Display the Data

Instructions for completing steps 1, 2, and 3 were given in the *Technology Highlight* in Section 2.2. Carefully review these steps to input and display the data. When finished, you should obtain the screen shown in Figure 2.63. To set an appropriate window, refer to the *Technology Highlights* from Sections 2.1, 2.4, or 2.5. The data in L1 (the Xlist) ranges from 60 to 100, and the data in L2 (the Ylist) ranges from 221 to 258, so we set the display window on the calculator accordingly, allowing for a **frame around the window** to comfortably display all points. For instance, we'll use [50, 110] and [210, 270] for the Xlist and Ylist respectively.

Figure 2.63

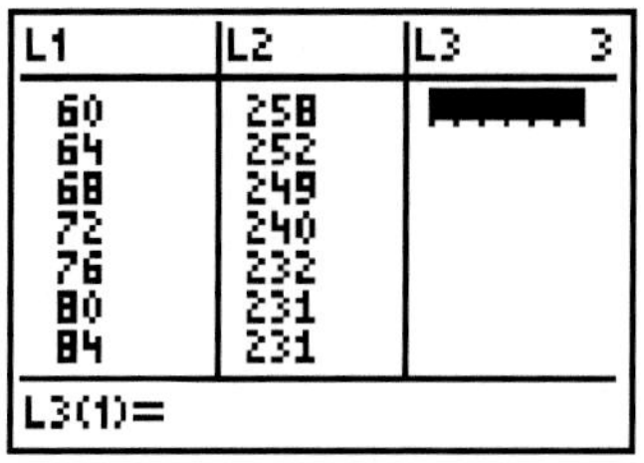

L1	L2	L3 3
60	258	
64	252	
68	249	
72	240	
76	232	
80	231	
84	231	

L3(1)=

WORTHY OF NOTE

As a rule of thumb, the tick marks for Xscl can be set by mentally calculating $\frac{|Xmax| + |Xmin|}{10}$ and using a convenient number in the neighborhood of the result. The same goes for Yscl.

Step 4: Calculate the Regression Equation

To have the calculator compute the regression equation, press the STAT and ▶ keys to move the cursor over to the **CALC** options (see Figure 2.64). Note that the fourth option reads **4:LinReg (ax + b).** Pressing the number 4 places **LinReg(ax + b)** on the home screen, and pressing ENTER computes the value of a, b, and the correlation coefficient r (the calculator

Figure 2.64

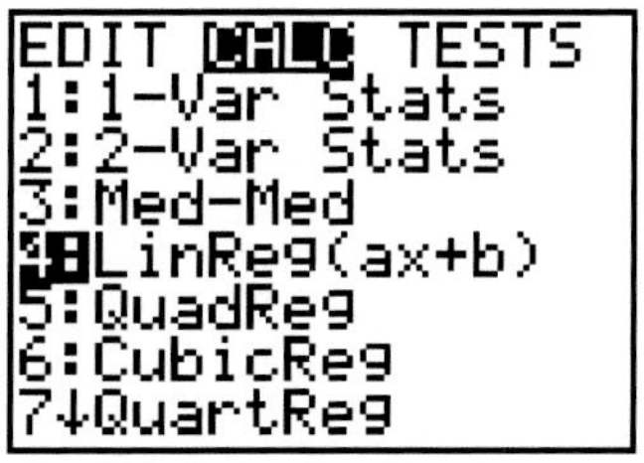

EDIT CALC TESTS
1:1-Var Stats
2:2-Var Stats
3:Med-Med
4:LinReg(ax+b)
5:QuadReg
6:CubicReg
7↓QuartReg

WORTHY OF NOTE

The correlation coefficient is a *diagnostic feature* of the regression calculation and can sometimes be misleading (see Exercise 41). Other indications of a good or poor correlation include a study of **residuals,** which we investigate after Chapter 3. To turn the diagnostic feature on or off, use 2nd 0 (CATALOG) and scroll down to the "D's" (all options offered on the TI-84 Plus are listed in the CATALOG) until you reach **DiagnosticOff** and **DiagnosticOn.** Highlight your choice and press ENTER.

automatically uses the data in L1 and L2 unless instructed otherwise). Rounded to hundredths the equation is $y = -0.86x + 304.91$ (Figure 2.65), which is very close to the estimated equation. An ***r*-value** (correlation coefficient) of -0.94 tells us the association is *negative* and *very strong*.

Figure 2.65

```
LinReg
 y=ax+b
 a=-.8636363636
 b=304.9090909
 r²=.885998282
 r=-.9412748175
```

Step 5: Displaying and Using the Results

Although the TI-84 Plus can paste the regression equation directly into Y_1 on the Y= screen, for now we'll enter $y = -0.86x + 304.91$ by hand. Afterward, pressing the GRAPH key will plot the data points (if Plot1 is still active) and graph the line. Your display screen should now look like the one in Figure 2.66. The regression line is the best estimator for the set of data as a whole, but there will still be some difference between the values it generates and the values from the set of raw data.

Figure 2.66

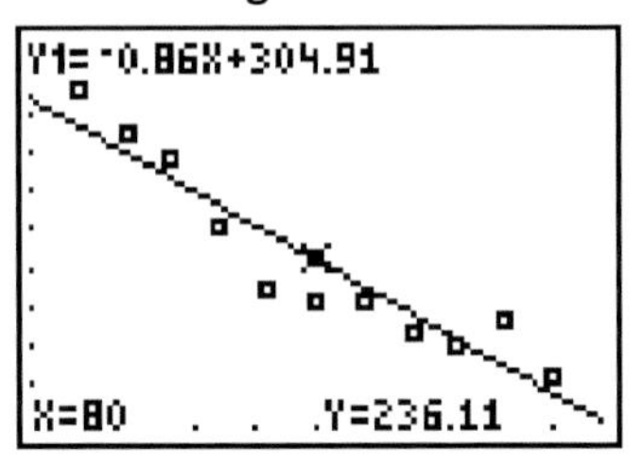

EXAMPLE 5 Riverside Electronics reviews employee performance semiannually, and awards increases in their hourly rate of pay based on the review. The table shows Thomas's hourly wage for the last 4 yr (eight reviews). Find the regression equation for the data and use it to project his hourly wage for the year 2007, after his fourteenth review.

Year (*x*)	Wage (*y*)
(2001) 1	9.58
2	9.75
(2002) 3	10.54
4	11.41
(2003) 5	11.60
6	11.91
(2004) 7	12.11
8	13.02

Solution: Following the prescribed sequence produces the equation $y \approx 0.48x + 9.09$. For $x = 14$ we obtain $y = 0.48(14) + 9.09$ or a wage of \$15.81. According to this model, Thomas will be earning \$15.81 per hour in 2007.

NOW TRY EXERCISES 27 THROUGH 32

If the input variable is a unit of time, particularly the time in years, we often **scale the data** to avoid working with large numbers. For instance, if the data involved the cost of attending a major sporting event for the years 1980 to 2000, we would say 1980 corresponds to 0 and use input values of 0 to 20 (subtracting the smallest value from itself and all other values has the effect of scaling down the data). This is easily done on a graphing calculator. Simply enter the four-digit years in L1, then with the cursor in the header of L1—use the keystrokes 2nd 1 (**L1**) – 1980 ENTER and the data in this list automatically adjusts.

F. Quadratic Regression and the Parabola of Best Fit

Once the data have been entered, graphing calculators have the ability to find many different regression equations. The choice of regression depends on the context of the data, patterns formed by the scatter-plot, and/or some foreknowledge of how the data are related. Earlier we focused on linear regression equations. We now turn our attention to quadratic regression equations.

EXAMPLE 6A Since 1990, the number of *new* books published each year has been growing at a rate that can be approximated by a quadratic function. The table shows the number of books published in the United States for selected years. Draw a scatter-plot and sketch an estimated parabola of best fit by hand.

Source: 1998, 2000, 2002, and 2004 *Statistical Abstract of the United States.*

Year (1990→0)	Books Published (1000s)
0	46.7
2	49.2
3	49.8
4	51.7
5	62.0
6	68.2
7	65.8
9	102.0
10	122.1

Solution: Begin by drawing the scatter-plot, being sure to scale the axes appropriately. The data appears to form a quadratic pattern, and we sketch a parabola that seems to best fit the data (see graph).

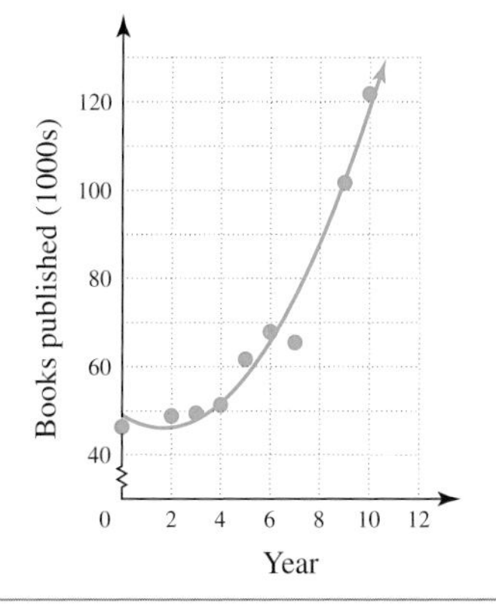

The regression abilities of a graphing calculator can be used to find a **parabola of best fit** and the steps are identical to those for linear regression.

EXAMPLE 6B Use the data from Example 6A to calculate a quadratic regression equation, then display the data and graph. How well does the equation match the data?

Solution: Begin by entering the data in L1 and L2 as shown in Figure 2.67. Press 2nd Y= to be sure that Plot 1 is still active and is using L1 and L2 with the desired point type. Set the window size to comfortably fit the data. Finally, press STAT and the right arrow ▶ to overlay the CALC option. The quadratic regression option is number **5:QuadReg.** Pressing 5 places this option directly on the home screen. Lists L1 and L2 are the default lists, so pressing ENTER will have the calculator compute the regression equation for the data in L1 and L2. After "chewing on the data" for a short while, the calculator returns the regression equation in the form shown in Figure 2.68. To maintain a higher degree of accuracy,

Figure 2.67

L1	L2	L3 2
3	49.8	
4	51.7	
5	62	
6	68.2	
7	65.8	
9	102	
10	[illegible]	

L2(9) =122.1

WORTHY OF NOTE

The TI-84 Plus can round all coefficients and the correlation coefficient to any desired number of decimal places. For three decimal places, press MODE and change the **Float** setting to "3." Also, be aware that there are additional methods for pasting the equation in Y1.

we can actually paste the entire regression equation in Y_1. Recall the last operation using 2nd ENTER, and **QuadReg** should (re)appear. Then enter the function Y_1 after the QuadReg option by pressing VARS ▶ **(Y-Vars)** and ENTER **(1:Function)** and ENTER (Y_1). After pressing ENTER once again, the full equation is automatically pasted in Y_1. To compare this equation model with the data, simply press GRAPH and both the graph and plotted data will appear. The graph and data seem to match very well (Figure 2.69).

Figure 2.68

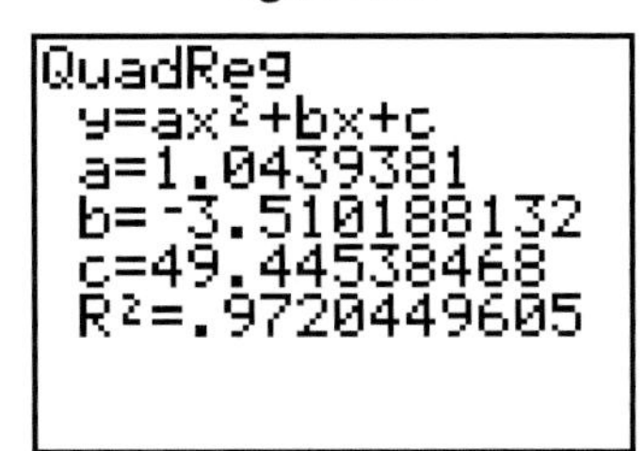

Figure 2.69

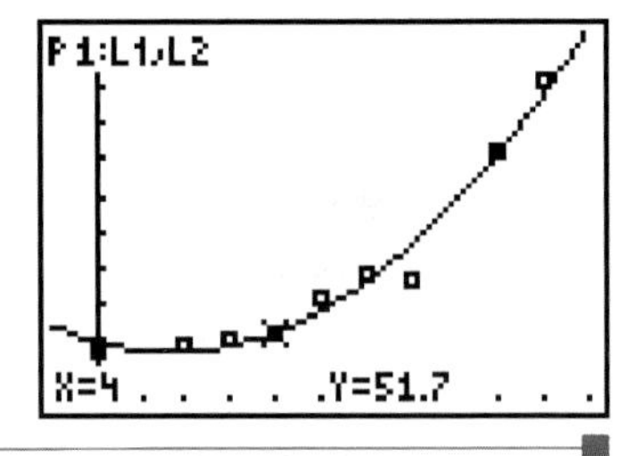

EXAMPLE 6C Use the equation from Example 6B to answer the following questions: According to the function model, how many new books were published in 1991? If this trend continues, how many new books will be published in 2007?

Solution: Since the year 1990 corresponds to 0 in this data set, we use an input value of 1 for 1991, and an input of 15 for 2007. Accessing the table (2nd GRAPH) feature and inputting 1 and 17 gives the screen shown. Approximately 47,000 new books were published in 1991, and about 291,600 will be published in the year 2007.

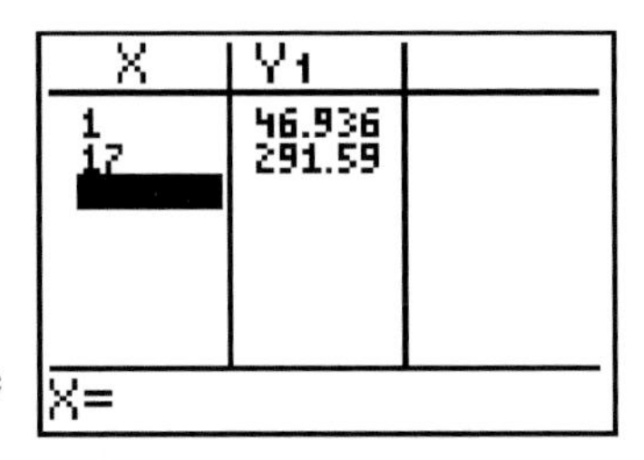

NOW TRY EXERCISES 33 THROUGH 38

2.6 EXERCISES

CONCEPTS AND VOCABULARY

Fill in each blank with the appropriate word or phrase. Carefully reread the section if needed.

1. When the ordered pairs from a set of data are plotted on a coordinate grid, the result is called a(n) scatter-plot.

2. If the data points seem to form a curved pattern or if no pattern is apparent, the data is said to have a(n) nonlinear association.

3. If the data points seems to cluster along an imaginary line, the data is said to have a(n) linear association.

4. If the pattern of data points seems to increase as they are viewed left to right, the data is said to have a(n) positive association.

5. Compare/contrast: One scatter-plot is nonlinear, with a strong and positive association. Another is linear, with a strong and negative association. Give a written description of each. Answers will vary.

6. Discuss/explain how this is possible: Working from the same scatter-plot, Demetrius obtained the equation $y = -0.64x + 44$ as a model, while Jessie got the equation $y = -0.59 + 42$. Answers will vary.

7. positive
8. negative
9. cannot be determined
10. positive
11. positive
12. negative
13. a. linear
    b. positive

DEVELOPING YOUR SKILLS

Draw a scatter-plot for the following data sets, then decide if the association between tl.e input and output variables is positive, negative, or cannot be determined.

7.

x	y
0	2
2	3
3	5
5	8
7	9
8	11

8.

x	y
10	11
12	8
15	7
17	4
20	3
21	0

9.

x	y	x	y
0	10	14	15
2	200	16	60
4	25	18	75
6	85	20	190
8	24	22	176
10	170	24	89
12	60	26	225

10.

x	y	x	y
0	1	21	27
3	2	24	34
6	4	27	42
9	7	30	55
12	11	33	79
15	16	36	120
18	21	39	181

11. For mail with a high priority, "Express Mail" offers next day delivery by 12:00 noon to most destinations, 365 days of the year. The service was first offered by the U.S. Postal Service in the early 1980s and has been growing in use ever since. The cost of the service (in cents) for selected years is shown in the table. Draw a scatter-plot of the data, then decide if the association is positive, negative, or cannot be determined.

Source: 2004 *Statistical Abstract of the United States*

x	y
1981	935
1985	1075
1988	1200
1991	1395
1995	1500
1999	1575
2002	1785

12. After the Surgeon General's first warning in 1964, cigarette consumption began a steady decline as advertising was banned from television and radio, and public awareness to the dangers of cigarette smoking grew. The percentage of the U.S. adult population who considered themselves smokers is shown in the table for selected years. Draw a scatter-plot of the data, then decide if the association is positive, negative, or cannot be determined.

Source: 1998 *Wall Street Journal Almanac* and 2004 *Statistical Abstract of the United States,* Table 188

Exercise 12

x	y
1965	42.4
1974	37.1
1979	33.5
1985	29.9
1990	25.3
1995	24.6
2000	23.1
2002	22.4

13. Since the 1970s women have made tremendous gains in the political arena, with more and more female candidates running and winning seats in the U.S. Senate and U.S. Congress. The number of women candidates for the U.S. Congress is shown in the table for selected years. Draw a scatter-plot of the data and then decide (a) if the association is linear or nonlinear and (b) if the association is positive or negative.

Source: Center for American Women and Politics at www.cawp.rutgers.edu/Facts3.html

Exercise 13

x	y
1972	32
1978	46
1984	65
1992	106
1998	121
2004	141

14. a. nonlinear
b. positive

14. The number of shares traded on the New York Stock Exchange experienced dramatic change in the 1990s as more and more individual investors gained access to the stock market via the Internet and online brokerage houses. The volume is shown in the table for 2002, and the odd numbered years from 1991 to 2001 (in billions of shares). Draw a scatter-plot of the data then decide (a) if the association is linear or nonlinear; and (b) if the association is positive, negative, or cannot be determined.

Source: 2000 and 2004 *Statistical Abstract of the United States,* Table 1202

x	y
1991	46
1993	67
1995	88
1997	134
1999	206
2001	311
2002	369

For the scatter-plots given, arrange them in order from the weakest correlation to the strongest correlation and state whether the correlation is positive, negative, or cannot be determined.

15. a, d, c, b
16. a, d, b, c
17. a.

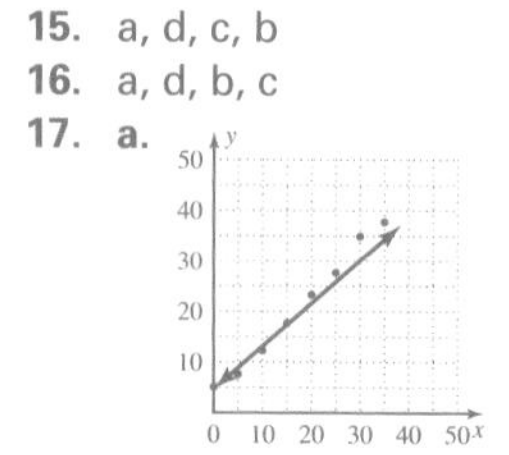

b. positive c. strong

18. a. (graph: Sales vs. Price; y: 21, 45, 69, 93, 117, 141; x: 130 160 190 220 250 280)

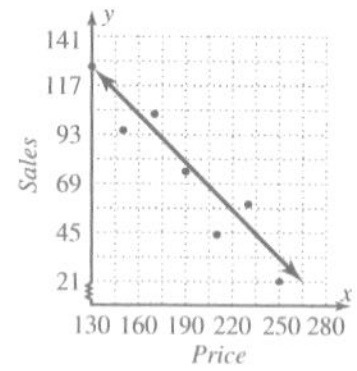

b. negative c. moderate

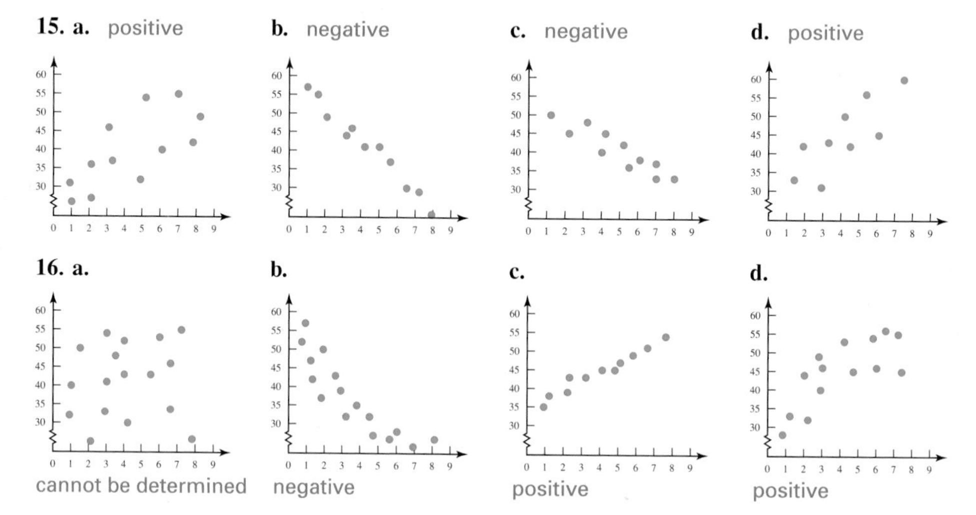

The data sets in Exercises 17 and 18 are known to be linear.

17. The total value of the goods and services produced by a nation is called its gross domestic product or GDP. The *GDP per capita* is the ratio of the GDP for a given year with the population that year, and is one of many indicators of economic health. The GDP per capita (in \$1000s) for the United States is shown in the table for selected years. (a) Draw a scatter-plot using a scale that appropriately fits the data; (b) sketch an estimated line of best fit and decide if the association is positive or negative; then (c) comment on the strength of the correlation (weak, moderate, strong, or in between).

Source: 2004 *Statistical Abstract of the United States,* Tables 2 and 641

Exercise 17

x 1970 → 0	y
0	5.1
5	7.6
10	12.3
15	17.7
20	23.3
25	27.7
30	35.0
33	37.8

18. Real estate brokers carefully track sales of new homes looking for trends in location, price, size, and other factors. The table relates the average selling price within a price range (homes in the \$120,000 to \$140,000 range are represented by the \$130,000 figure), to the number of new homes sold by Homestead Realty in 2004. (a) Draw a scatter-plot using a scale that appropriately fits the data; (b) sketch an estimated line of best fit and decide if the association is positive or negative; then (c) comment on the strength of the correlation (weak, moderate, strong, or in between).

Price	Sales
130's	126
150's	95
170's	103
190's	75
210's	44
230's	59
250's	21

The data sets in Exercises 19 through 24 are known to be linear. Use the data to

a. Draw a scatter-plot using a scale that appropriately fits the data.

b. Sketch an estimated line of best fit and decide if the association is positive or negative, then comment on the strength of the correlation (weak, moderate, strong, or in between).

c. Find an equation for the estimated line of best fit (answers may vary).

19.

x	y
65	61
70	54
75	47
80	50
85	53
90	46
95	40
100	48
105	40

20.

x	y
13	47
16	65
23	83
25	92
33	99
35	115
44	145
47	158
51	167

21.

x	y
2	30
21	40
33	49
67	63
89	74
115	87
139	100
167	112
193	126

22.

x	y
5	134
23	153
38	110
78	135
104	60
135	130
163	74
196	40
227	85

23. In most areas of the country, law enforcement has become a major concern. The number of law enforcement officers employed by the federal government and having the authority to carry firearms and make arrests is shown in the table for selected years.

x	y (1000s)
1993	68.8
1996	74.5
1998	83.1
2000	88.5
2004	93.4

a. Draw a scatter-plot using a scale that appropriately fits the data.

b. Sketch an estimated line of best fit, decide if the association is positive or negative, and comment on the strength of the correlation (weak, moderate, strong, or in between).

c. Find an equation for the estimated line of best fit and use it to predict the number of federal law enforcement officers in 1995 and the projected number for 2007. Answers may vary.

Source: U.S. Bureau of Justice, Statistics at www.ojp.usdoj.gov/bjs/fedle.htm

24. Due to atmospheric pressure, the temperature at which water will boil varies predictably with the altitude. Using special equipment designed to duplicate atmospheric pressure, a lab experiment is set up to study this relationship for altitudes up to 8000 ft. The data collected is shown to the right, with the boiling temperature y in degrees Fahrenheit, depending on the altitude x in feet.

Exercise 24

x	y
−1000	213.8
0	212.0
1000	210.2
2000	208.4
3000	206.5
4000	204.7
5000	202.9
6000	201.0
7000	199.2
8000	197.4

a. Draw a scatter-plot using a scale that appropriately fits the data.

b. Sketch an estimated line of best fit and decide if the association is positive or negative, then comment on the strength of the correlation (weak, moderate, strong, or in between).

c. Find an equation for the estimated line of best fit and use it to predict the boiling point of water on the summit of Mt. Hood in Washington State (11,239 ft height) and along the shore of the Dead Sea (approximately 1,312 ft below sea level. Answers may vary.

19. a.

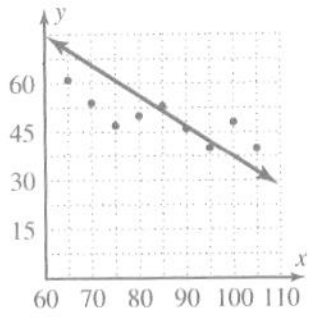

b. negative, moderate

c. $y = -0.4x + 82.8$

20. a.

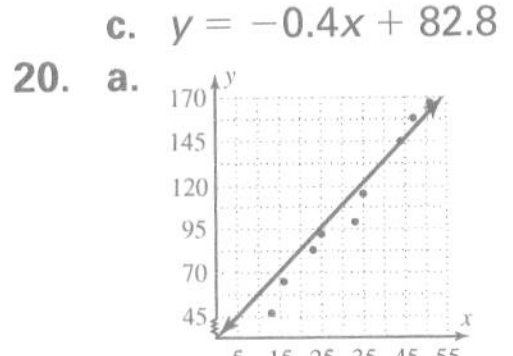

b. positive, strong

c. $x = 3x + 11.1$

21. a.

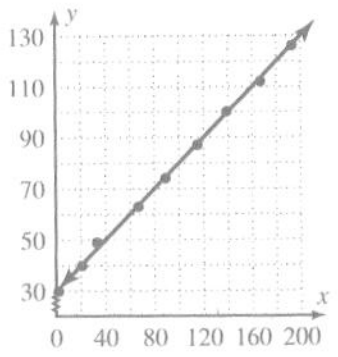

b. positive, strong

c. $y = 0.5x + 30.2$

22. a.

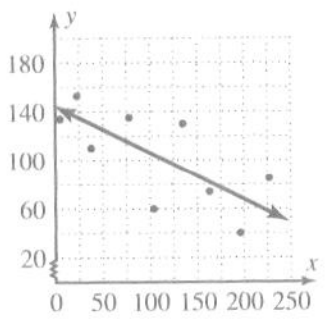

b. negative, weak

c. $y = -0.4x + 140.1$

23. a.

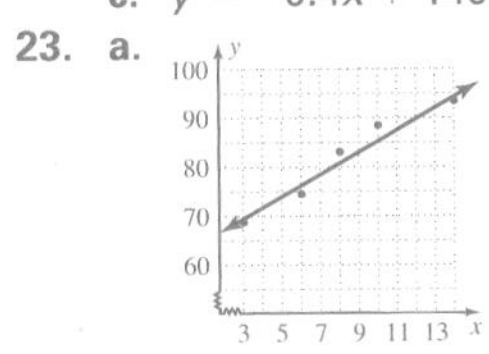

b. positive, strong

c. $y = 2.4x + 69.4$, 74,200, 103,000

24. a.

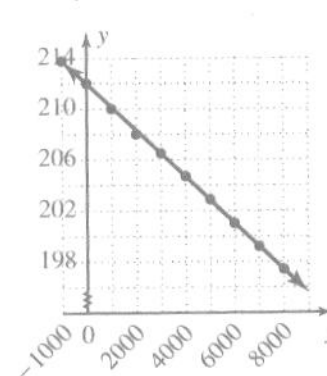

b. negative, strong

c. $y = -0.002x + 212.005$; 189.5°F; 214.6°F

25. a. $h(t) = -14.5t^2 + 90t$
b. $v = 90$ ft/sec
c. Venus

26. a. $V = \pi ha^2$
b. when $h = a - b$
c. ≈ 24.99 cm

27. a.
b. linear c. positive
d. $y = 0.96x + 1.55$, 63.95 in.

28. a.
b. linear
c. positive
d. $y = 17.18x + 135.94$; about 428,000

Exercise 27

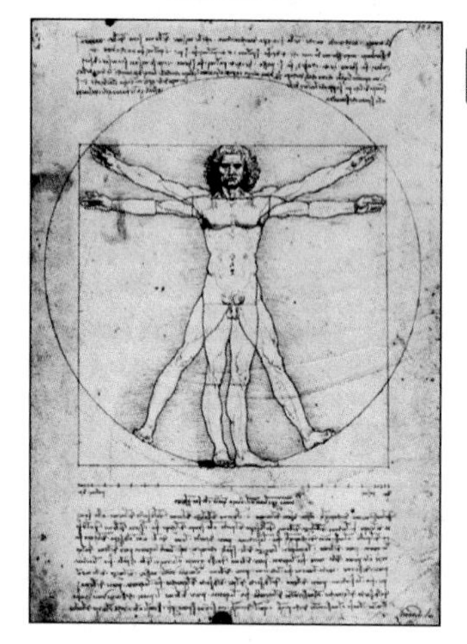

29. a.
b. linear
c. positive
d. $y = 9.55x + 70.42$; about 232,800
The number of applications, since the line has a greater slope.

WORKING WITH FORMULAS

25. The height of a projectile: $h(t) = -\frac{1}{2}gt^2 + vt$

The height of a projectile thrown upward from ground level depends primarily on two things—the object's initial velocity and the acceleration due to gravity. This is modeled by the formula shown, where $h(t)$ represents the height of the object at time t, v represents the initial velocity, and g represents the acceleration due to gravity. Suppose an astronaut on one of the inner planets threw a surface rock upward and used hand-held radar to collect the data shown. Given that on Mercury $g = 12$ ft/sec^2, Venus $g = 29$ ft/sec^2, and Earth $g = 32$ ft/sec^2, (a) use your calculator to find an appropriate regression model for the data, (b) use the model to determine the initial velocity of the object, and (c) name the planet on which the astronaut is standing.

Time	Height
1	75.5
2	122
3	139.5
4	128
5	87.5
6	18

26. Volume of a frustum: $V = \frac{1}{3}\pi h(a^2 + ab + b^2)$

The volume of the frustum of a right circular cone is given by the formula, where h is the height, and a and b are the smaller and larger radii, respectively. (a) What happens to the formula if $a = b$? (b) Under what conditions can the formula be rewritten using a difference of cubes? (c) Solve the formula for h and use the result to find the height of a frustum with radii $a = 5$ cm, $b = 8$ cm, and a volume of 3375.5 cm^3.

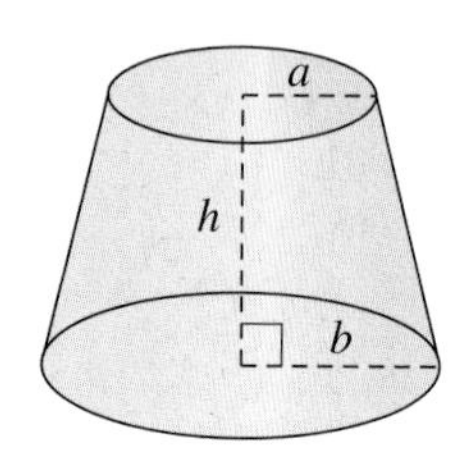

APPLICATIONS

Use the regression capabilities of a graphing calculator to complete Exercises 27 through 32.

27. Height versus wingspan: Leonardo da Vinci's famous diagram is an illustration of how the human body comes in predictable proportions. One such comparison is a person's wingspan to their height. Careful measurements were taken on eight students and the data is shown here. Using the data: (a) draw the scatter-plot; (b) determine whether the association is linear or nonlinear; (c) determine whether the association is positive or negative; and (d) find the regression equation and use it to predict the height of a student with a wingspan of 65 in.

Height (x)	Wingspan (y)
61	60.5
61.5	62.5
54.5	54.5
73	71.5
67.5	66
51	50.75
57.5	54
52	51.5

28. Patent applications: Every year the United States Patent and Trademark Office (USPTO) receives thousands of applications from scientists and inventors. The table given shows the number of appplications received for the odd years from 1993 to 2003 ($1990 \rightarrow 0$). Use the data to: (a) draw the scatter-plot; (b) determine whether the association is linear or non-linear; (c) determine whether the association is positive or negative; and (d) find the regression equation and use it to predict the number of applications that will be received in 2007.

Year ($1990 \rightarrow 0$)	Applications (1000's)
3	188.0
5	236.7
7	237.0
9	278.3
11	344.7
13	355.4

Source: United States Patent and Trademark Office at www.uspto.gov/web

29. Patents issued: An increase in the number of patent applications (see Exercise 28), typically brings an increase in the number of patents issued, though many applications are denied due to improper filing, lack of scientific support, and other reasons. The table given shows the number of patents issued for the odd years from 1993 to 2003 ($1999 \rightarrow 0$). Use the data to:

Exercise 29

Year (1990 → 0)	Patents (1000's)
3	107.3
5	114.2
7	122.9
9	159.2
11	187.8
13	189.6

(a) draw the scatter-plot; (b) determine whether the association is linear or non-linear; (c) determine whether the association is positive or negative; and (d) find the regression equation and use it to predict the number of applications that will be approved in 2007. Which is increasing faster, the number of patent applications or the number of patents issued? How can you tell for sure?

Source: United States Patent and Trademark Office at www.uspto.gov/web

30. High jump records: In the sport of track and field, the high jumper is an unusual athlete. They seem to defy gravity as they launch their bodies over the high bar. The winning height at the summer Olympics (to the nearest unit) has steadily increased over time, as shown in the table for selected years. Using the data: (a) draw the scatter-plot, (b) determine whether the association is linear or nonlinear, (c) determine whether the association is positive or negative, and (d) find the regression equation using $t = 0$ corresponding to 1900 and predict the winning height for the 2000 and 2004 Olympics.

Source: athens2004.com

Year (x)	Height (y)
00	75
12	76
24	78
36	80
56	84
68	88
80	93
88	94
92	92
96	94
100	95.44
104	96.32

30. a.

b. linear c. positive
d. $y = 0.22x + 73.44$

31. a.

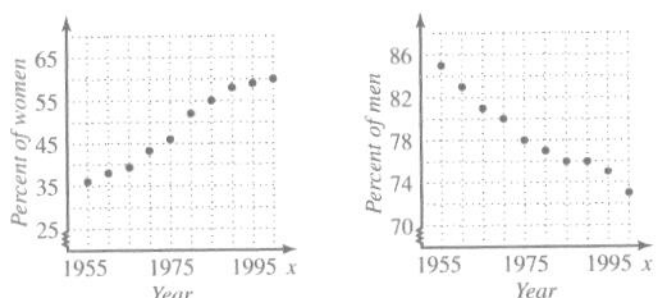

31. Females/males in the workforce: Over the last 4 decades, the percentage of the female population in the workforce has been increasing at a fairly steady rate. At the same time, the percentage of the male population in the workforce has been declining. The data is shown in the tables. Using the data; (a) draw scatter-plots for both data sets, (b) determine whether the associations are linear or nonlinear, (c) determine whether the associations are positive or negative, and (d) determine if the percentage of females in the workforce is increasing faster than the percentage of males is decreasing? Discuss/explain how you can tell for sure.

Source: 1998 *Wall Street Journal Almanac,* p. 316

b. women: linear
c. positive

Exercise 31 (women)

Year (x)	Percent
1955	36
1960	38
1965	39
1970	43
1975	46
1980	52
1985	55
1990	58
1995	59
2000	60

Exercise 31 (men)

Year (x)	Percent
1955	85
1960	83
1965	81
1970	80
1975	78
1980	77
1985	76
1990	76
1995	75
2000	73

b. men: linear
c. negative
d. yes, |slope| is greater

32. a.

b. linear, moderate
c. positive
d. $y = 0.47x - 22.58$, 15.02(15), 5.62(5.5)

32. Height versus male shoe size: While it seems reasonable that taller people should have larger feet, there is actually a wide variation in the relationship between height and shoe size. The data in the table show the height (in inches) compared to the shoe size worn for a random sample of 12 male chemistry students. Using the data: (a) draw the scatter-plot, (b) determine whether the association is linear or nonlinear and comment on the strength of the correlation (weak, moderate, strong, or in between), (c) determine whether the association is positive or negative, and (d) find the regression equation and use it to predict the shoe size of a man 80 inches tall and another that is 60 inches tall.

Exercise 32

Height	Shoe Size
66	8
69	10
72	9
75	14
74	12
73	10.5
71	10
69.5	11.5
66.5	8.5
73	11
75	14
65.5	9

The data sets in Exercises 33 through 36 are known to be nonlinear. Use a graphing calculator to

a. Draw a scatter-plot using a scale that appropriately fits the data.

b. Sketch an estimated parabola of best fit, and comment on the strength of the correlation (weak, moderate, strong, or in between).

c. Find a quadratic regression model for the data and compare the input/output values of the model with the actual data. What do you notice?

33.

x	y
0	20
5	16
12	8
15	6
20	6
24	12
30	22

34.

x	y
3	12
10	20
16	24
24	26
32	25
42	18
50	10

35.

x	y
50	339
75	204
100	96
125	45
150	50
175	90
190	180

36.

x	y
0	130
20	105
45	90
80	100
100	130
140	190
165	300

37. Plastic Money: The total amount of business transacted using credit cards has been changing rapidly over the last 15 to 20 yr. The total volume (in billions of dollars) is shown in the table for selected years.

a. Use a graphing calculator to draw a scatter-plot of the data and decide on an appropriate form of regression.

b. Calculate a regression equation with $x = 1$ corresponding to 1991 and display the scatter-plot and graph on the same screen. Comment on correlation (weak, medium, strong).

c. According to the equation model, how many billions of dollars was transacted in 2003? How much will be transacted in the year 2007?

x	y
1991	481
1992	539
1994	731
1997	1080
1998	1157
1999	1291
2000	1458
2002	1638

Source: Statistical Abstract of the United States, various years

38. Homeschool education: Since the early 1990s the number of parents electing to homeschool their children has been steadily increasing. Estimates for the number of children homeschooled (in 1000s) are given in the table for selected years.

a. Use a graphing calculator to draw a scatter-plot of the data and decide on an appropriate form of regression.

b. Calculate a regression equation with $x = 0$ corresponding to 1985 and display the scatter-plot and graph on the same screen. Comment on correlation (weak, medium, strong, or in between).

c. According to the equation model, how many children were homeschooled in 1991? If growth continues at the same rate, home many children will be homeschooled in 2006?

x	y
1985	183
1988	225
1990	301
1992	470
1993	588
1994	735
1995	800
1996	920
1997	1100

Source: National Home Education Research Institute

33. a. 34. a.

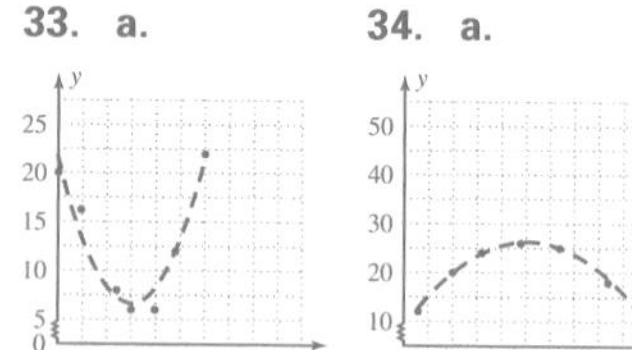

b. strong
c. $y = 0.07x^2 - 2.02x + 21.77$

b. strong
c. $y = -0.03x^2 + 1.39x + 8.44$

35. a.

b. strong
c. $y = 0.04x^2 - 11.32x + 807.88$

36. a.

b. strong
c. $y = 0.016x^2 - 1.71x + 132.32$

37. a. linear

b. $y = 108.2x + 330.2$, strong
c. \$1736.8 billion; about \$2170 billion

38. a. quadratic

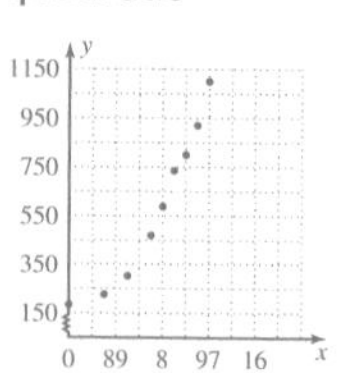

b. $y = 6.68x^2 - 3.48x + 176.30$, strong
c. about 396,000 about 3,049,000

WRITING, RESEARCH, AND DECISION MAKING

39. (b), since there is a recognizable and fixed correspondence between the independent and dependent variables
40. Answers will vary.

39. One of the scatter-plots shown here was drawn using data collected from a formula. The other was drawn from data collected during a survey that compared a person's shoe size with their grade on a final exam. Which was drawn from the data collected from a formula? Discuss why.

a.

b.

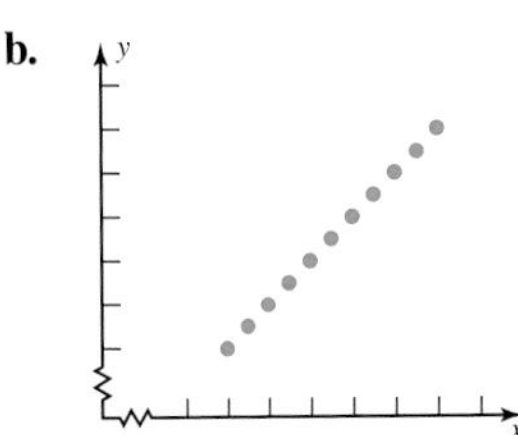

40. In his book *Gulliver's Travels,* Jonathan Swift describes how the Lilliputians were able to measure Gulliver for new clothes, even though he was a giant compared to them. According to the text, "Then they measured my right thumb, and desired no more . . . for by mathematical computation, once around the thumb is twice around the wrist, and so on to the neck and waist." Is it true that once around the neck is twice around the waist? Find at least 10 willing subjects and take measurements of their necks and waists in millimeters. Arrange the data in ordered pair form (circumference of neck, circumference of waist). Draw the scatter-plot for this data. Does the association appear to be linear? Find the equation of the best fit line for this data. What is the slope of this line? Is the slope near $m = 2$? Is there a moderate to strong correlation?

41. Very strong; virtually equal; context and goodness of fit.

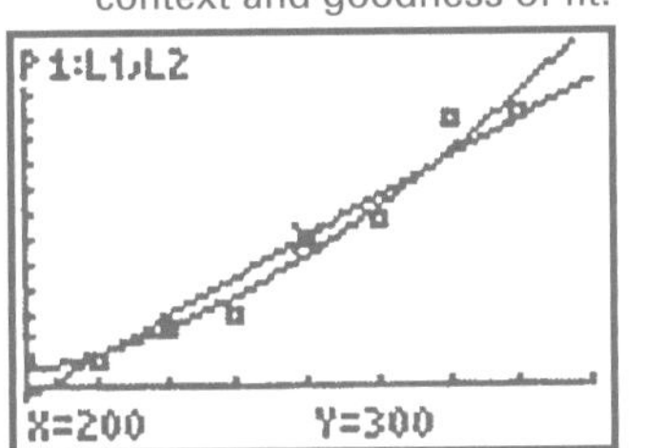

41. It can be very misleading to rely on the correlation coefficient alone when selecting a regression model. To illustrate, run a linear regression on the data set given, without doing a scatter-plot. What do you notice about the strength of the correlation (the correlation coefficient)? Now run a quadratic regression and comment on what you see. Finally, graph the scatter-plot and both regression equations. What factors besides the correlation coefficient should you take into account when choosing a form of regression?

Exercise 41

x	y
0	50
50	60
100	120
150	140
200	300
250	340
300	540
350	559

42. about 23 yrs; 95 yrs
43. Answers will vary.

EXTENDING THE CONCEPT

42. The average age of the first seven people to arrive at Gramps's birthday was 21. When Frank (29) arrived, the mean age increased to 22. Margo (also 29) arrived next. What was the mean age after Margo's arrival? The tenth and last person to arrive at Gramps's party was Gramps himself, and the mean age increased to 30 years old. How old is Gramps on this birthday?

43. Most graphing calculators offer numerous forms of regression. Using the data given in the table to the right, explore some additional forms of regression and find one that appropriately fits this data. Do you recognize the pattern of the scatter-plot from our studies in Section 2.4?

x	y	x	y
0	−31	28	57
4	20	32	45
8	56	36	39
12	75	40	42
16	81	44	57
20	77	48	88
24	68	52	120

44. No, it does not pass the vertical line test.
45. $66 + 9\pi$ cm, $40.5\pi + 432$ cm^2
46. $r = \frac{A - P}{Pt}$
47. $w = \frac{-7}{10} \pm \frac{\sqrt{51}}{10}i$
48. yes
49. $\frac{3}{2}, \pm\sqrt{7}i$

MAINTAINING YOUR SKILLS

44. (2.2) Is the graph of a function shown here? Discuss why or why not.

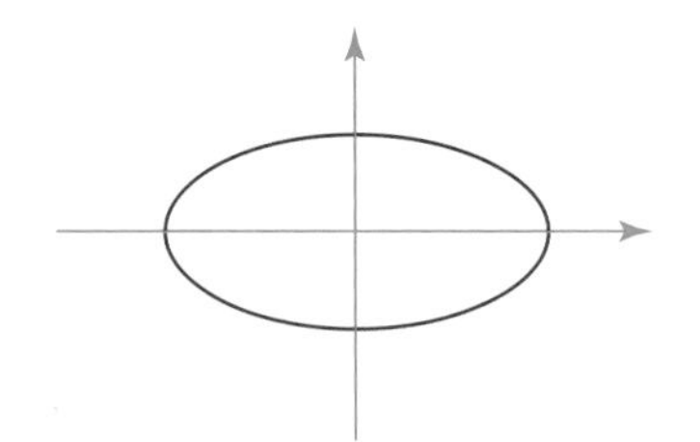

45. (R.7) Determine the perimeter and area of the figure shown.

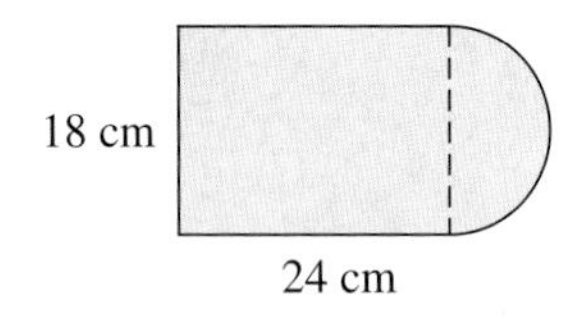

46. (1.1) Solve for r: $A = P + Prt$

47. (1.5) Solve for w: $-2(3w^2 + 5) + 4 = 7w - (w^2 + 1)$

48. (1.1) John and Rick are out orienteering. Rick finds the last marker first and is heading for the finish line, 1275 yd away. John is just seconds behind, and after locating the last marker tries to overtake Rick, who by now has a 250-yd lead. If Rick runs at 4 yd/sec and John runs at 5 yd/sec, will John catch Rick before they reach the finish line?

49. (R.4/1.5) Use factoring to find all zeroes, real and complex, of the function $g(x) = 2x^3 - 3x^2 + 14x - 21$.

SUMMARY AND CONCEPT REVIEW

SECTION 2.1 Rectangular Coordinates and the Graph of a Line

KEY CONCEPTS

- The solution to a linear equation in two variables is an ordered pair (x, y) that makes the equation true.
- Points on the grid where both x and y have integer values are called lattice points.
- The x- and y-axes divide the plane into four quadrants I to IV, with quadrant I in the upper right.
- The graph of a linear equation is a straight line, which can be graphed using the intercept method.
- For the intercept method: $x = 0$ gives the y-intercept and $y = 0$ gives the x-intercept $(x, 0)$. Draw a straight line through these points. If the line goes through $(0, 0)$, an additional point must be found.
- Given any two points on a line, the slope of the line is the ratio $\frac{\text{vertical change}}{\text{horizontal change}}$ as you move from one point to the other.
- Other designations for slope are $m = \frac{\text{rise}}{\text{run}} = \frac{\text{change in } y}{\text{change in } x} = \frac{\Delta y}{\Delta x}$.
- The slope formula is $m = \frac{y_2 - y_1}{x_2 - x_1}$, where $x_2 \neq x_1$.
- In applications, the slope of the line gives a rate of change, indicating how fast the quantity measured on the vertical axis is changing with respect to that measured on the horizontal axis. This change is denoted $\frac{\Delta y}{\Delta x}$.
- Lines with positive slope $(m > 0)$ rise from left to right; lines with negative slope $(m < 0)$ fall from left to right.

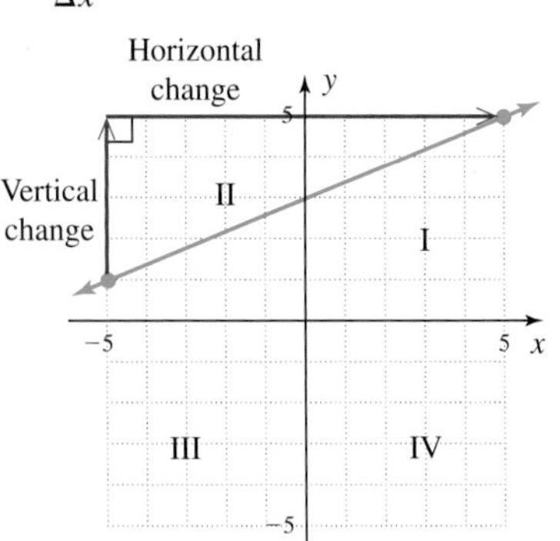

1. a. b.

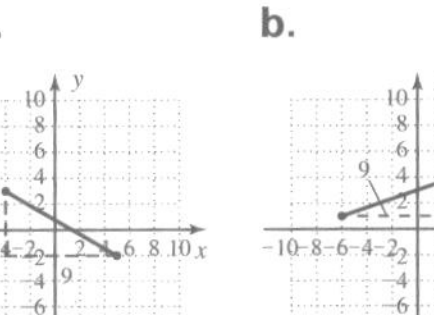

$\frac{-5}{9}$, (14, −7) $\frac{1}{3}$, (0,3)

2. a. parallel
   b. perpendicular
3. a. b.

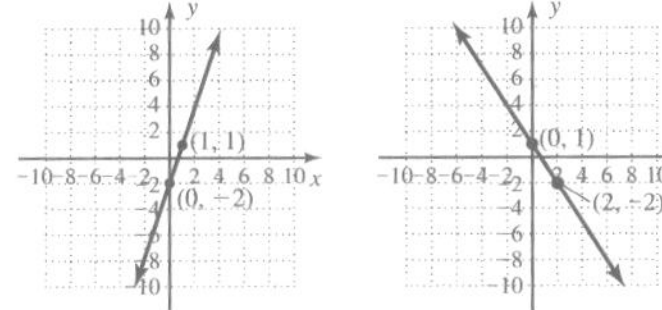

4. a. b.

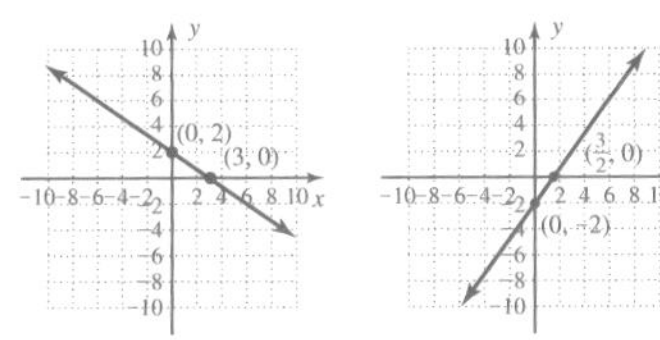

5. a. vertical
   b. horizontal
   c. neither

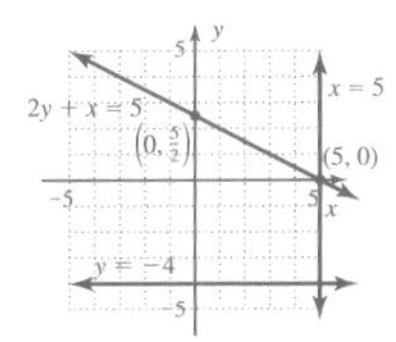

6. yes
7. $m = \frac{2}{3}$, y-intercept (0, 2), when the rodent population increases by 2000, the hawk population increases by 300.
8. $(\frac{1}{2}, 0)$, $\sqrt{149}$

- The slope of a horizontal line is zero ($m = 0$); the slope of a vertical line is undefined.
- Parallel lines have equal slopes ($m_1 = m_2$); perpendicular lines have slopes where $m_1 \cdot m_2 = -1$.
- The midpoint of a line segment with endpoints (x_1, y_1) and (x_2, y_2) is $\left(\frac{x_1 + x_2}{2}, \frac{y_1 + y_2}{2}\right)$.
- The distance between the points (x_1, y_1) and (x_2, y_2) is $d = \sqrt{(x_2 - x_1)^2 + (y_2 - y_1)^2}$.

EXERCISES

1. Plot the points, determine the slope using a slope triangle, then use $\frac{\Delta y}{\Delta x}$ to find an additional point on the line: (a) $(-4, 3)$ and $(5, -2)$ and (b) $(3, 4)$ and $(-6, 1)$.
2. Use the slope formula to determine if the lines L_1 and L_2 are parallel, perpendicular, or neither:
   a. L_1: $(-2, 0)$ and $(0, 6)$; L_2: $(1, 8)$ and $(0, 5)$
   b. L_1: $(1, 10)$ and $(-1, 7)$: L_2: $(-2, -1)$ and $(1, -3)$
3. Graph each equation by plotting points: (a) $y = 3x - 2$ and (b) $y = -\frac{3}{2}x + 1$.
4. Find the intercepts for each line and sketch the graph: (a) $2x + 3y = 6$ and (b) $y = \frac{4}{3}x - 2$.
5. Identify each line as either horizontal, vertical, or neither, and graph each line.
   a. $x = 5$ b. $y = -4$ c. $2y + x = 5$
6. Determine if the triangle with the vertices given is a right triangle: $(-5, -4)$, $(7, 2)$, $(0, 16)$.
7. Find the slope and y-intercept of the line shown and discuss the slope ratio in this context.

Exercise 7

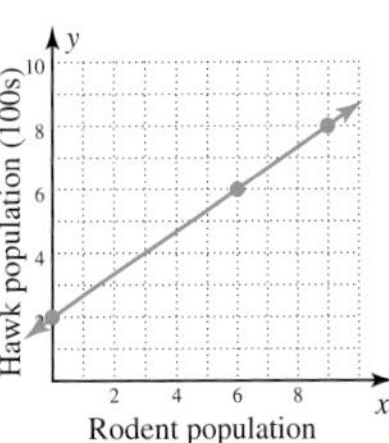

Exercise 8

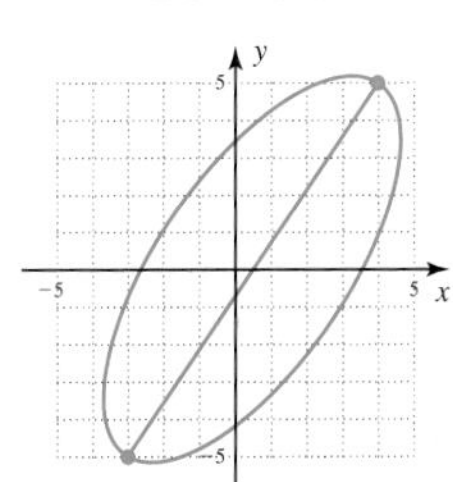

8. Find the center and diameter of the ellipse shown. Assume the endpoints are lattice points.

SECTION 2.2 Relations, Functions, and Graphs

KEY CONCEPTS

- A relation is a collection of ordered pairs (x, y) and can be given in set or equation form.
- As a set of ordered pairs, the domain of the relation is the set of all first coordinates and the range is the set of all corresponding second coordinates.
- A relation can be expressed in mapping notation $x \to y$, indicating an element from the domain is mapped to (corresponds to or is associated with) an element from the range.
- To graph a relation we plot a sufficient number of points and connect them with a straight line or smooth curve, depending on the pattern formed.
- A function is a relation where each x-value from the domain corresponds to only one y-value in the range.
- The domain of a function is the set of allowable inputs (x-values) and can be determined by analyzing restrictions on the input variable, by the context of a problem, or from the graph.

- The range of a function is the set of outputs y generated by the domain. For a linear function the range is $y \in (-\infty, \infty)$.
- The phrase, "y is a function of x" is written as $y = f(x)$. The notation enables us to evaluate functions while tracking corresponding x- and y-values.
- Vertical lines ($x = h$) and horizontal lines ($y = k$) can help name the boundaries of the domain and range.
- If any vertical line intersects the graph of a relation only once, the relation is a function.
- The absolute value function is determined by $y = |x|$ and gives a graph that is V shaped.

EXERCISES

9. Represent the relation in mapping notation, then state the domain and range. $\{(-7, 3), (-4, -2), (5, 1), (-7, 0), (3, -2), (0, 8)\}$

10. Graph the relation from Exercise 9. Is this relation a function? Justify your response.

11. State the implied domain of each function:

a. $f(x) = \sqrt{4x + 5}$

b. $g(x) = \dfrac{x - 4}{x^2 - x - 6}$

12. Determine $h(-\frac{2}{3})$, $h(3a)$, and $h(a - 1)$ for $h(x) = 2x^2 - 3x$.

13. Graph the relation $y = \sqrt{36 - x^2}$ by completing the table, then state the domain and range of the relation. Is this relation also a function? Why or why not?

Exercise 13

x	y
−6	0
−4	4.47
$-\sqrt{11}$	5
0	6
$\sqrt{11}$	5
4	4.47
6	0

Exercise 14

Mythological deities	Primary concern
Apollo	messenger
Jupiter	war
Ares	craftsman
Neptune	love and beauty
Mercury	music and healing
Venus	oceans
Ceres	all things
Mars	agriculture

14. Determine if the mapping given represents a function. If not, explain how the definition of a function is violated.

15. For the graph of each function shown: (a) state the domain and range; (b) find the value of $f(2)$; and (c) determine the value(s) of x for which $f(x) = 1$. Assume all values are lattice points.

I.

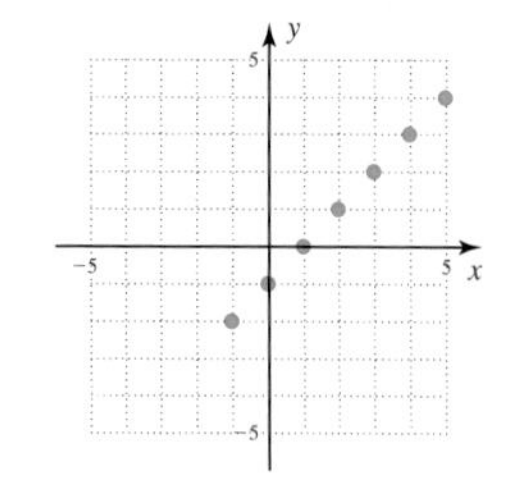

II.

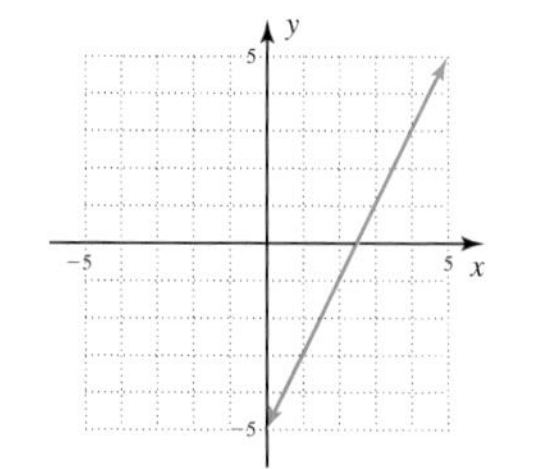

III.

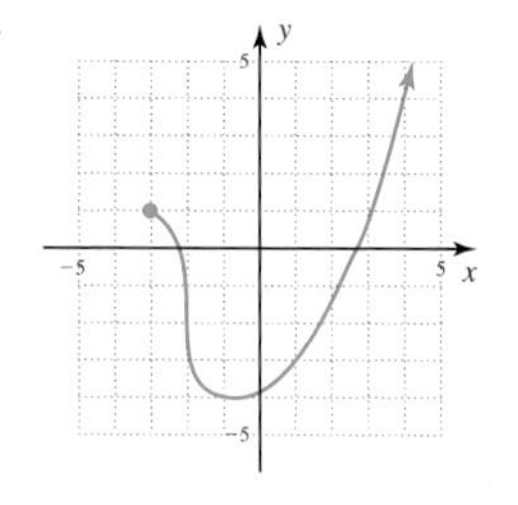

9. $x \in \{-7, -4, 0, 3, 5\}$
$y \in \{-2, 0, 1, 3, 8\}$

10. No, −7 is paired with 3 and 0.

11. **a.** $x \in [-\frac{5}{4}, \infty)$
b. $x \in (-\infty, -2) \cup (-2, 3) \cup (3, \infty)$

12. $\frac{26}{9}$; $18a^2 - 9a$; $2a^2 - 7a + 5$

13. $x \in [-6, 6]$
$y \in [0, 6]$
Yes, passes the vertical line test

14. yes

15. **I. a.** $D = \{-1, 0, 1, 2, 3, 4, 5\}$
$R = \{-2, -1, 0, 1, 2, 3, 4\}$
b. 1
c. 2
II. a. $x \in [-5, 4]$
$y \in [-5, 4]$
b. −3
c. −2
III. a. $x \in [-3, \infty)$
$y \in [-4, \infty)$
b. −1
c. −3 or 3

SECTION 2.3 Linear Functions and Rates of Change

KEY CONCEPTS

- When $2y - 4x = -6$ is rewritten as $y = 2x - 3$, the equation is said to be in function form, since we can immediately see the operations performed on x (the input) in order to obtain y (the output).
- The function form of a linear equation is also called the slope-intercept form and is denoted $y = mx + b$. In slope-intercept form the point $(0, b)$ is the y-intercept and the slope of the line is m (the coefficient of x).
- To graph a line using the slope-intercept method, begin at y-intercept $(0, b)$ and count off the slope ratio $\frac{\Delta y}{\Delta x}$. Plot this point and use a straightedge to draw a line through both points.
- The notation $m = \frac{\Delta y}{\Delta x}$ literally says the quantity y is changing with respect to changes in x.
- If the slope m and a point (x_1, y_1) on the line are known, the equation of the line can be written in point-slope form: $y - y_1 = m(x - x_1)$.

EXERCISES

16. Write each equation in slope-intercept form, then identify the slope and y-intercept.

a. $4x + 3y - 12 = 0$

b. $5x - 3y = 15$

17. Graph functions using the y-intercept and $\frac{\Delta y}{\Delta x}$. Then comment on the slope (does it "rise or fall").

a. $f(x) = -\frac{2}{3}x + 1$ **b.** $h(x) = \frac{5}{2}x - 3$

18. Use a slope triangle to graph a line with the given slope through the given point.

a. $m = \frac{2}{3}$; $(1, 4)$ **b.** $m = -\frac{1}{2}$; $(-2, 3)$

19. What is the equation of the horizontal line and the vertical line passing through the point $(-2, 5)$? Which line is the point $(7, 5)$ on?

20. Find the equation of the line passing through the points $(1, 2)$ and $(-3, 5)$. Write your final answer in slope-intercept form.

21. Find the equation for the line that is parallel to $4x - 3y = 12$ and passes through the point $(3, 4)$. Write your final answer in function form.

22. Determine the slope and y-intercept of the line shown. Then write the equation of the line in slope-intercept form and interpret the slope ratio $m = \frac{\Delta W}{\Delta R}$ in the context of this exercise.

23. Use the point-slope form to (a) find the equation for the line shown, (b) use the equation to predict the x- and y-intercepts, (c) write the equation in function form, and (d) find $f(20)$ and the value of x for which $f(x) = 15$.

Exercise 22

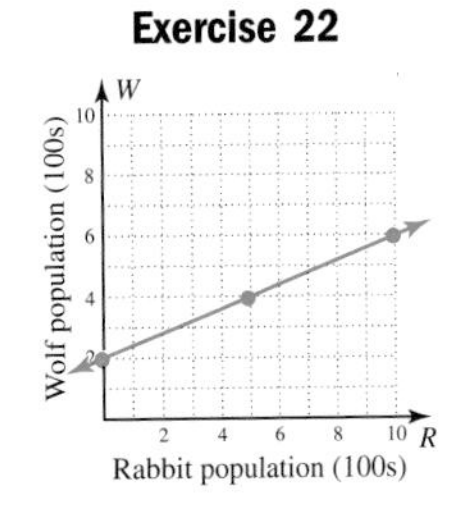

Exercise 23

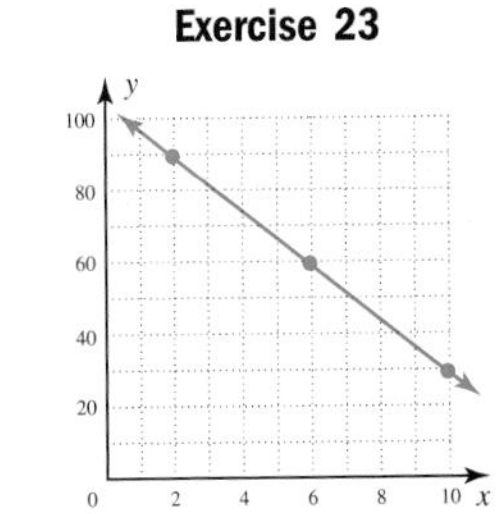

16. a. $y = \frac{-4}{3}x + 4$, $m = \frac{-4}{3}$, y-intercept $(0, 4)$
b. $y = \frac{5}{3}x - 5$, $m = \frac{5}{3}$, y-intercept $(0, -5)$

17. a. falls b. rises

18. a. b.

19. $y = 5$, $x = -2$; $y = 5$

20. $y = \frac{-3}{4}x + \frac{11}{4}$

21. $f(x) = \frac{4}{3}x$

22. $m = \frac{2}{5}$, y-intercept $(0, 2)$, $y = \frac{2}{5}x + 2$. When the rabbit population increases by 500, the wolf population increases by 200.

23. a. $y = \frac{-15}{2}x + 105$
b. $(14, 0)$, $(0, 105)$
c. $f(x) = \frac{-15}{2}x + 105$
d. $f(20) = -45$, $x = 12$

SECTION 2.4 Quadratic and Other Toolbox Functions

KEY CONCEPTS

- A quadratic function is any function that can be written in the form $f(x) = ax^2 + bx + c$; $a \neq 0$. The simplest quadratic is the squaring function $f(x) = x^2$, where $a = 1$ and b and $c = 0$.
- The graph of any quadratic function is called a parabola. A parabola has three distinctive features that we use to graph the function along with the x- and y-intercepts:
  - concavity
  - line of symmetry
  - vertex
- For a quadratic function in the standard form $f(x) = ax^2 + bx + c$; $a \neq 0$:
  - concavity: The graph will be concave up if $a > 0$ and concave down if $a < 0$.
  - y-intercept: The y-intercept is $(0, c)$, found by substituting $x = 0$ [evaluate $f(0)$].
  - x-intercepts: The x-intercept(s) (if they exist) can be found by substituting $f(x) = 0$ and solving the equation for x.
  - line of symmetry: The line of symmetry for factorable quadratic functions can be found by computing the average value of the x-intercepts: $h = \frac{x_1 + x_2}{2}$.
  - vertex: The vertex has coordinates (h, k), where $f(h) = k$.
- If the parabola is concave down, the y-coordinate of the vertex is the maximum value of $f(x)$.
- If the parabola is concave up, the y-coordinate of the vertex is the minimum value of $f(x)$.
- The *toolbox functions* commonly used as mathematical models and bridges to advanced topics are:
  - linear: $f(x) = mx + b$, straight line
  - absolute value: $f(x) = |x|$, V function
  - quadratic: $f(x) = x^2$, parabola
  - cubic: $f(x) = x^3$, vertical propeller
  - cube root: $f(x) = \sqrt[3]{x}$, horizontal propeller
  - square root: $f(x) = \sqrt{x}$, one-wing graph
- Each toolbox function has a certain domain, range, and end behavior associated with it.
- For nonlinear functions, we use the slope formula with function notation to calculate an average rate of change between two points (x_1, y_1) and (x_2, y_2) on the graph: $m = \frac{\Delta y}{\Delta x} = \frac{f(x_2) - f(x_1)}{x_2 - x_1}$.

EXERCISES

Graph each function by plotting points, using input values from $x = -5$ to $x = 5$ as needed.

24. $f(x) = (x - 2)^2$ **25.** $g(x) = x^3 - 4x$ **26.** $p(x) = \sqrt[3]{x - 1}$ **27.** $q(x) = \sqrt{x} + 2$

For each graph given: (a) State the x and y-intercepts (if they exist); (b) describe the end behavior; and (c) state the location of the vertex, node, or point of inflection as applicable.

28.

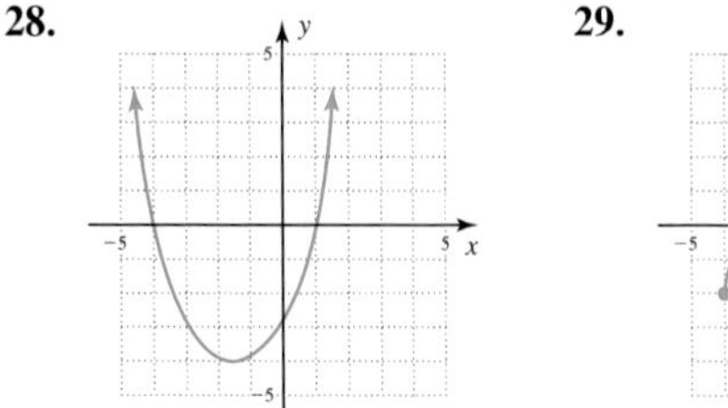

29.

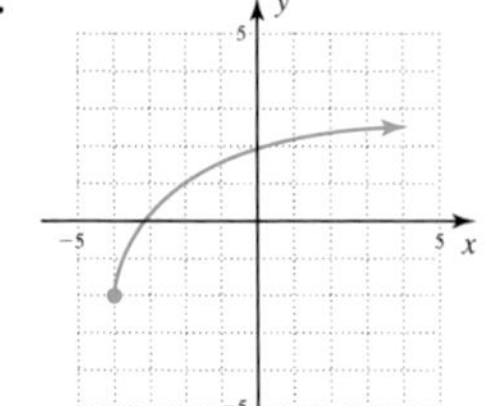

30.

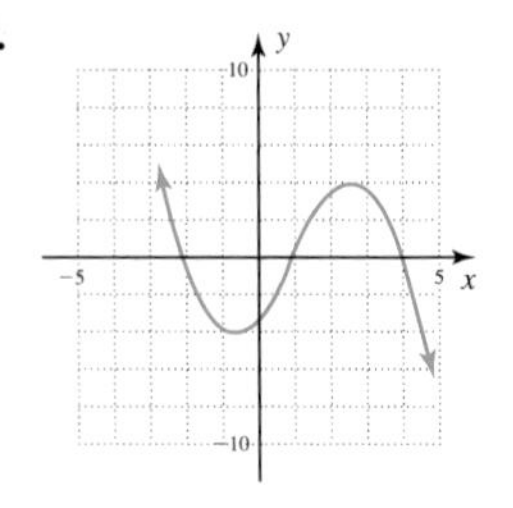

24.

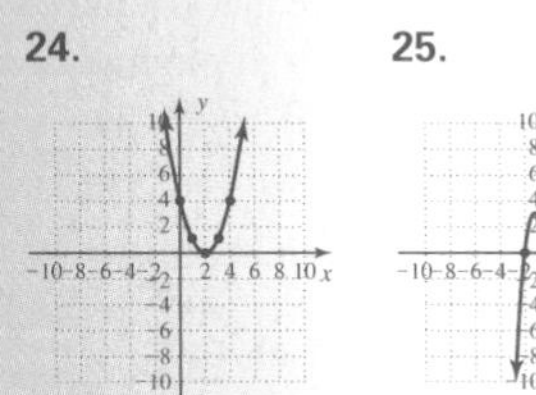

25.

26. **27.**

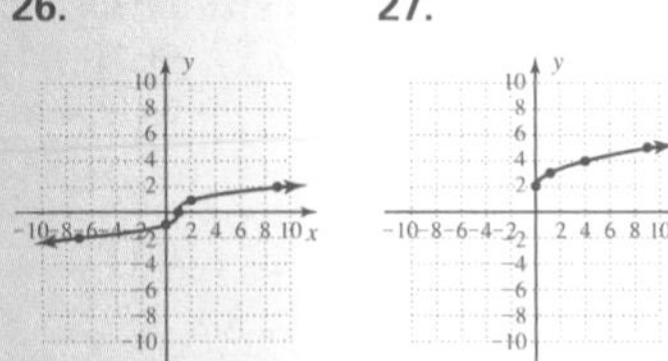

28. a. $(-4, 0)$, $(1, 0)$, $(0, -3)$
b. up/up
c. $(-1.5, -4)$

29. a. $(-3, 0)$, $(0, 2)$
b. up on the right
c. $(-4, -2)$

30. a. $(-2, 0)$, $(1, 0)$, $(4, 0)$, $(0, -3)$
b. up on left, down on right
c. $(1, 0)$

31. a. $(-1.5, 0)$, $(2, 0)$, $(0, -1.5)$
b. up/up
c. $(\frac{1}{2}, -2)$

32. a. $(-2, 0)$, $(0, -2)$
b. up on left, down on right
c. $(-1, -1)$

33. a. $(1, 0)$, $(0, -2)$
b. down on left, up on right
c. not applicable

34. 35.

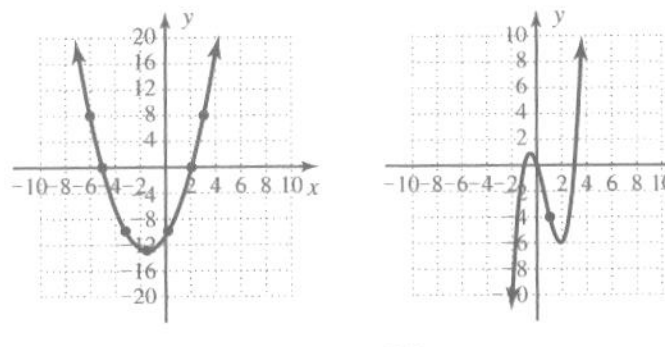

36. 37.

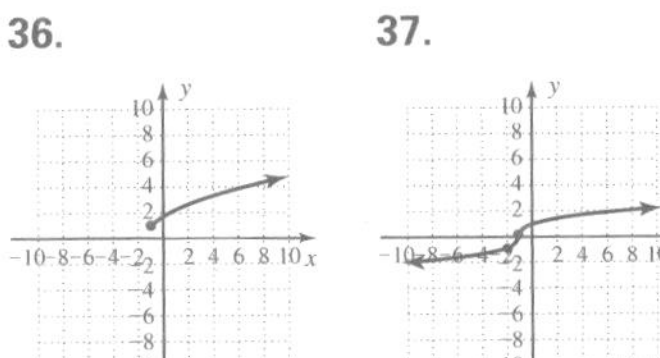

38. $w = 10$ to $w = 15$; $\frac{7}{100}$; $\frac{19}{100}$

31.

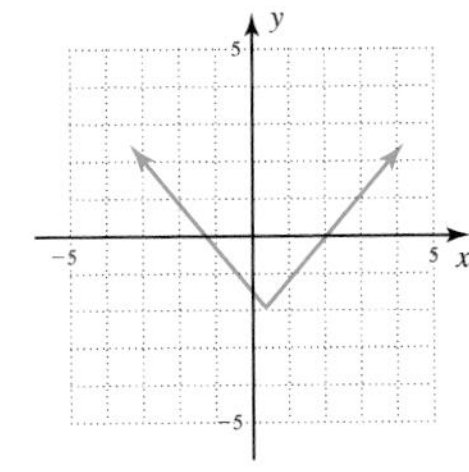

32.

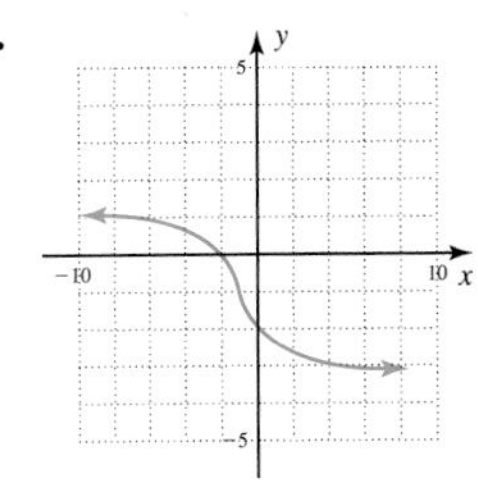

33.

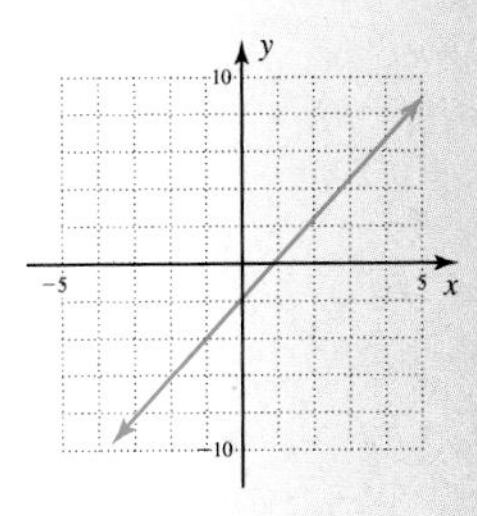

Graph each function using the distinctive features of its function family (not by simply plotting points).

34. $p(x) = x^2 + 3x - 10$

35. $q(x) = x^3 - 2x^2 - 3x$

36. $f(x) = \sqrt{2x + 3}$

37. $g(x) = \sqrt[3]{x + 1}$

38. The amount of horsepower delivered by a wind-powered generator can be modeled by the formula $P = 0.0004w^3$, where P is the horsepower and w is the wind speed in miles per hour. Based on the shape of a cubic graph, (a) would you expect the rate of change $\frac{\Delta P}{\Delta W}$ to be greater in the interval from $w = 5$ to $w = 10$, or greater in the interval from $w = 10$ to $w = 15$? (b) Calculate the rate of change in these intervals and justify your response.

SECTION 2.5 Functions and Inequalities—A Graphical View

KEY CONCEPTS

- The zeroes of a function appear graphically as x-intercepts and divide the x-axis into intervals.
- For linear equations and polynomial equations with linear factors, intervals where outputs are positive are separated from intervals where outputs are negative by zeroes of the function (the x-intercepts).
- The following questions are synonymous: (1) For what inputs are function values greater than zero? (2) For what inputs are the outputs positive? (3) For what inputs is the graph *above* the x-axis?
- To solve a linear inequality, find the x-intercept (if it exists) and note the slope of the line.
- To solve a quadratic inequality, find the x-intercepts (if they exist) and note concavity of the graph.
- To solve a functional inequality using interval tests: (1) find zeroes of the function; (2) plot the zeroes on the x-axis; (3) use a test number for each interval; and (4) state the appropriate solution set.
- If the graph of a quadratic function is concave up with no x-intercepts, then $f(x) > 0$ for all x. If the graph of a quadratic function is concave down with no x-intercepts, then $f(x) < 0$ for all x.

EXERCISES

State the solution set for each function inequality indicated.

39. $x \in (-4, 1)$
40. $x \in (-\infty, -4) \cup (3, \infty)$
41. $x \in (-\infty, -2]$
42. $x \in (-\infty, \infty)$

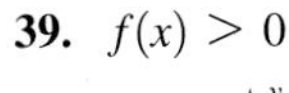

39. $f(x) > 0$

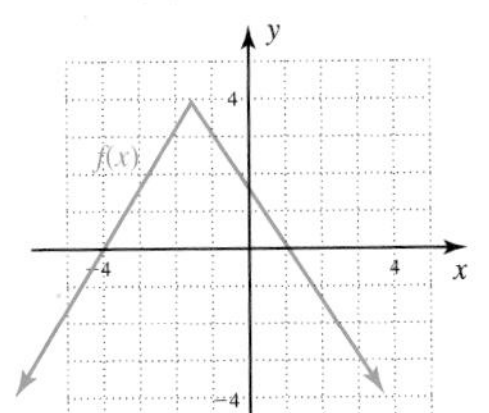

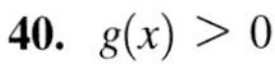

40. $g(x) > 0$

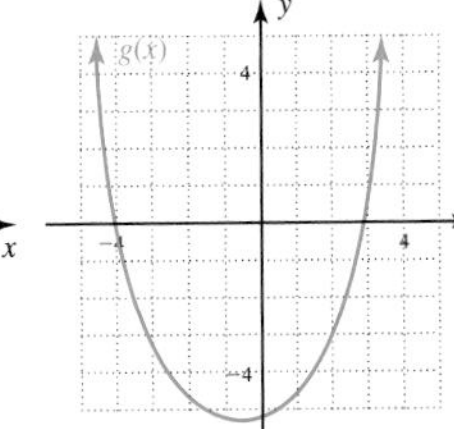

41. $h(x) \geq 0$

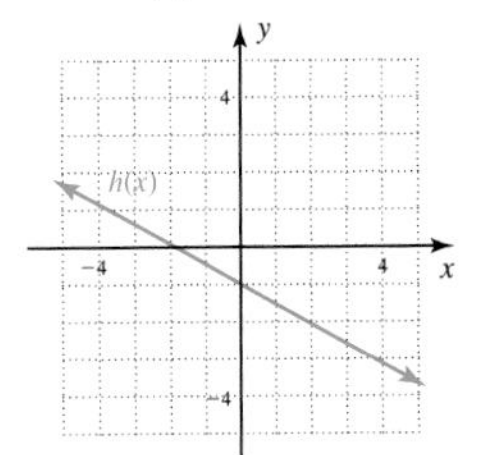

42. $f(x) \geq 0$

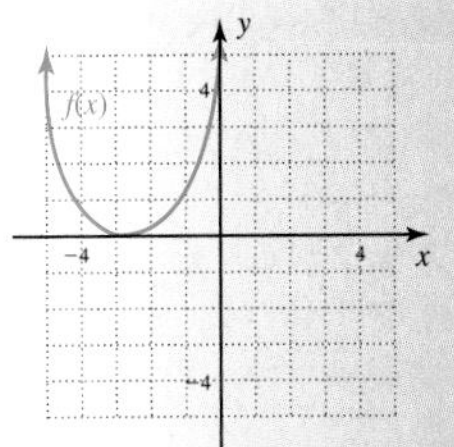

43. $x \in (\frac{2}{3}, \infty)$
44. no solution
45. $x \in (0, 5)$
46. $x \in [-5, 1]$
47. $x \in (-\infty, -2) \cup (2, \infty)$
48. $x \in (-\infty, -1) \cup (0, 1)$
49. $x \in (-\infty, 0] \cup [5, \infty)$
50. $x \in [-1, 0] \cup [1, \infty)$

Solve each function inequality.

43. $f(x) = 3x - 2; f(x) > 0$

44. $g(x) = \sqrt{x} - 3; g(x) < 0$

45. $h(x) = x^2 - 5x; h(x) < 0$

46. $p(x) = x^2 + 4x - 5; p(x) \leq 0$

47. $q(x) = -|x| + 2; q(x) < 0$

48. $r(x) = x^3 - x; r(x) < 0$

49. Use your response to Exercise 45 to help find the domain of $f(x) = \sqrt{x^2 - 5x}$.

50. Use your response to Exercise 48 to help find the domain of $g(x) = \sqrt{x^3 - x}$.

SECTION 2.6 Regression, Technology, and Data Analysis

KEY CONCEPTS

- A scatter-plot is the graph of all the ordered pairs in a real data set.
- When drawing a scatter-plot, we must be sure to scale the axes to comfortably fit the data.
- If larger inputs tend to produce larger output values, we say there is a positive association.
- If larger inputs tend to produce smaller output values, we say there is a negative association.
- If the data seem to cluster around an imaginary line, we say there is a linear association between the variables and attempt to model the data using an estimated line of best fit or a linear regression equation.
- If the data clearly cannot be approximated by a straight line, we say the variables exhibit a nonlinear association (or sometimes no association).
- If the data seem to cluster around an imaginary parabola, we say that there is a quadratic association between the variables and attempt to model the data using a quadratic regression equation.
- The correlation coefficient r measures how tightly a set of data points cluster about an imaginary curve. The strength of the correlation is given as a value between -1 and 1. Measures close to -1 or 1 indicate a very strong correlation. Measures close to 0 indicate a very weak correlation.
- The regression equation minimizes the vertical distance between all data points and the graph itself, making it the unique line or parabola of best fit.

EXERCISES

51. a.

b. linear
c. yes
d. positive

52. $y = 0.35x + 56.10$
53. 98

51. To determine the value of doing homework, a student in a math class records the time spent by classmates on their homework in preparation for a quiz the next day. Then she records their scores. The data are entered in the table. (a) Draw a scatter-plot. (b) Is the association linear or nonlinear? (c) Is there a strong correlation? (d) Is the association positive or negative?

x (min study)	y (score)
45	70
30	63
10	59
20	67
60	73
70	85
90	82
75	90

52. If the association is linear, draw an estimated line of best fit and find its equation using the point-slope form.

53. According to the equation model, what grade can I expect if I study for 120 min?

MIXED REVIEW

1. $y = \frac{-5}{3}x - 3$
2. $y = \frac{-1}{2}x + \frac{7}{2}$
3. a. $x \in (-\infty, -5) \cup (-5, 5) \cup (5, \infty)$
   b. $x \in [\frac{5}{3}, \infty)$
4. a. 7
   b. $8v^2 - 6v + 5$
   c. $2v^2 - 15v + 32$

1. Write the given equation in slope-intercept form: $5x + 3y = -9$.

2. Find the equation of the line perpendicular to $2x - y = 3$ that goes through the point $(-1, 4)$.

3. Find the implied domain of the functions:

a. $f(x) = \dfrac{x + 1}{x^2 - 25}$ **b.** $g(x) = \sqrt{3x - 5}$

4. Given $h(x) = 2x^2 - 3x + 5$, find

a. $h\left(-\dfrac{1}{2}\right)$ **b.** $h(2v)$ **c.** $h(v - 3)$

5. $y = \frac{-3}{4}x + 1$

6. a. $x \in [-4, \infty)$
   b. $q(4) = 3$
   c. $k = -2$

5. Give the equation of the line shown. Write it in slope-intercept form.

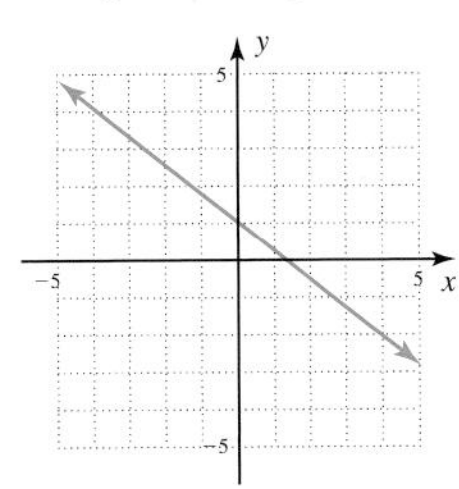

6. For the function q whose graph is given, find (a) domain, (b) $q(4)$, and (c) k if $q(k) = -3$.

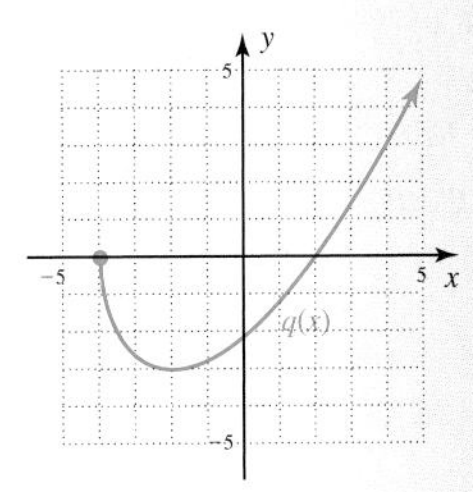

7. Find the length of any diagonal and the coordinates of the center.

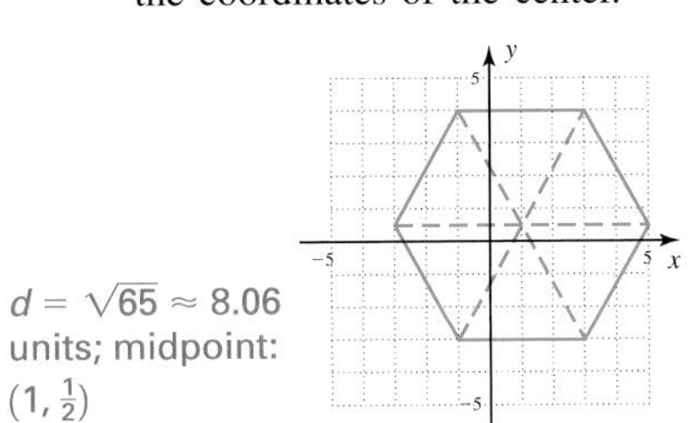

$d = \sqrt{65} \approx 8.06$ units; midpoint: $(1, \frac{1}{2})$

8. Discuss the end behavior of $T(x)$ and name the vertex, axis of symmetry, and all intercepts.

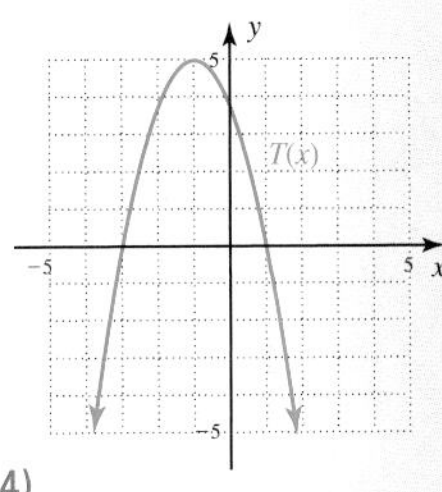

down/down;
vertex: $(-1, 5)$
axis: $x = -1$
y-intercept: $(0, 4)$
x-intercepts: $(-3, 0)$ and $(1, 0)$

9. 10.

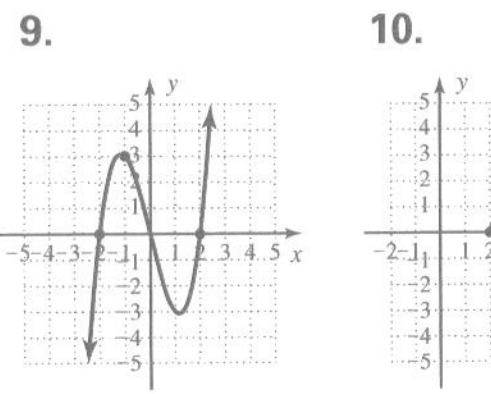

11. 12.

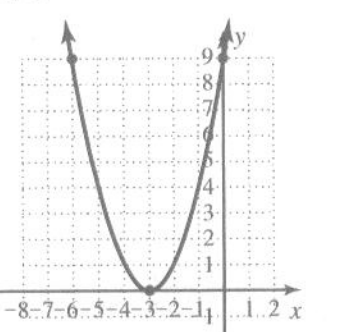

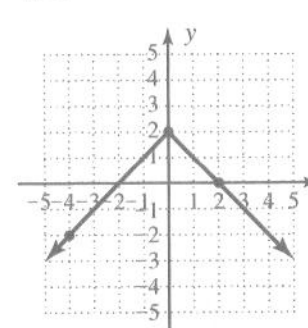

13. 14.

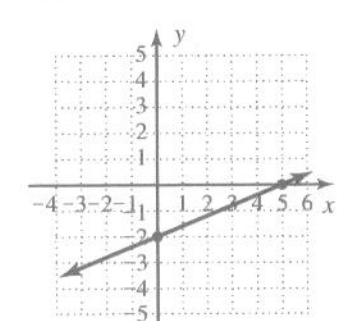

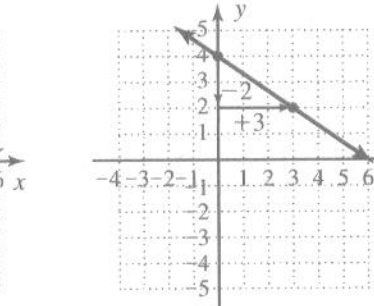

Graph each function by plotting a few points and using known features of the related function family.

9. $f(x) = x^3 - 4x$

10. $g(x) = \sqrt{x - 2}$

11. $h(x) = (x + 3)^2$

12. $p(x) = -|x| + 2$

13. Graph using the intercept method: $2x - 5y = 10$.

14. Graph by plotting the y-intercept, then counting $m = \frac{\Delta y}{\Delta x}$ to find additional points: $y = \frac{-2}{3}x + 4$.

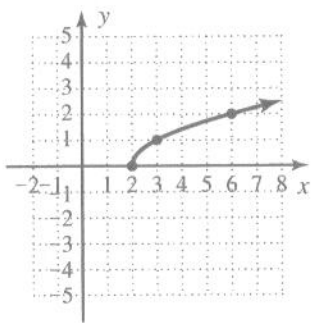

Solve each inequality using the graph provided.

15. $f(x) = 2\sqrt{x + 4} - 2$; $f(x) \leq 0$

$x \in [-4, -3]$

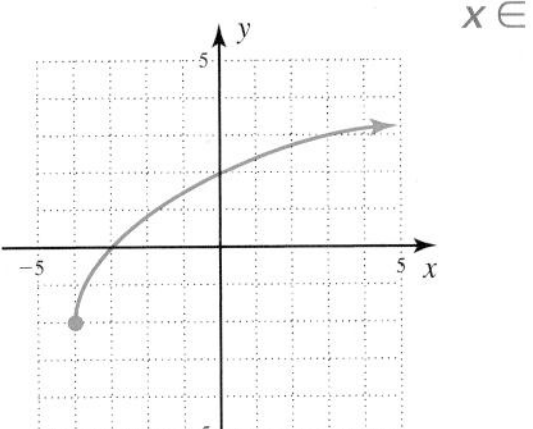

16. $g(x) = x^2 - 2x - 3$; $g(x) > 0$

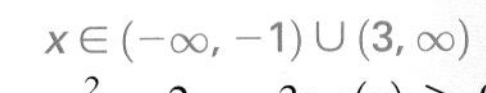

$x \in (-\infty, -1) \cup (3, \infty)$

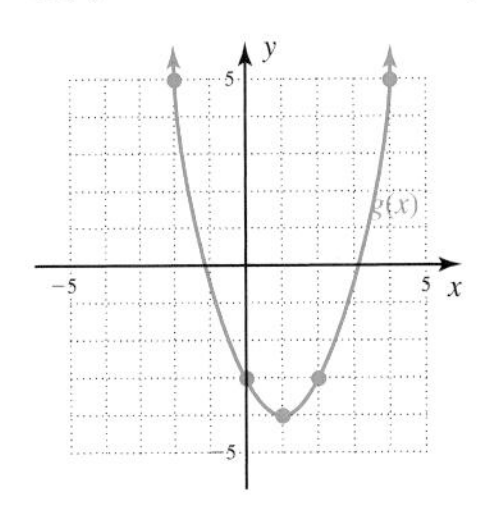

17. Determine the domain of $q(x) = \sqrt{x^2 - 9}$. $x \in (-\infty, -3] \cup [3, \infty)$

18. Graph the function $p(x) = -2x^2 + 8x$. By observing the graph, is the average rate of change positive or negative in the interval $[-2, -1]$? Why? Do you expect the rate of change in $[1, 2]$ to be greater or less than the rate of change in $[-2, -1]$? Calculate the average rate of change in each interval and comment.

18.

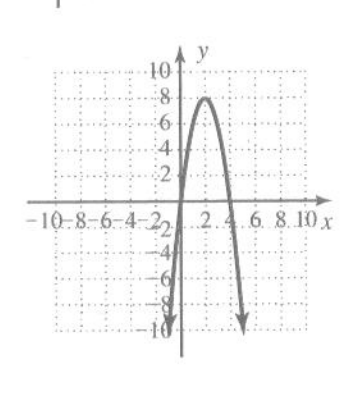

rate of change is positive in $[-2, -1]$ since p is increasing in $(-\infty, 2)$; less; $\frac{\Delta y}{\Delta x} = \frac{14}{1}$ in $[-2, -1]$; $\frac{\Delta y}{\Delta x} = \frac{2}{1}$ in $[1, 2]$

19. a. 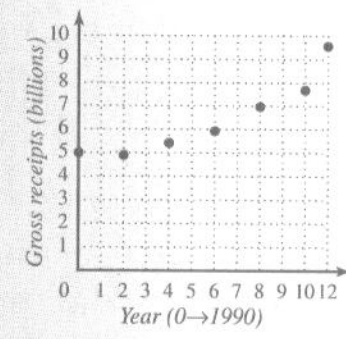

quadratic

b. $g(t) = 0.0357t^2 - 0.0602t + 4.9795$

c. $g(15) \approx 12.1$ billion (2005); $g(18) \approx 15.5$ billion (2008)

20. a. 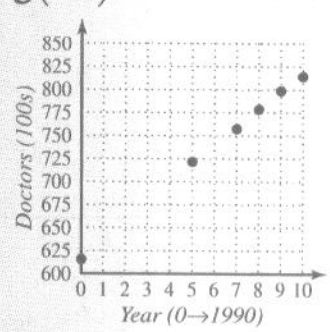

linear

b. $D(t) = 19.953t + 617.253$

c. $D(15) = 916.55 \times 1000$ (2005) $D(20) = 1016.3 \times 1000$ (2010)

19. Since 1990, the total gross receipts of movie theaters in the United States has been increasing. The data for even numbered years is given in the table, with 1990 corresponding to year 0 and gross receipts in billions of dollars. Use the data to (a) draw a scatter-plot and decide on an appropriate form of regression, (b) find the regression equation, and (c) use the equation to find the projected total gross receipts for the years 2005 and 2008.

Source: National Association of Theater Owners at www.natoonline.org

Exercise 19

Year	Gross (billions)
0	5.02
2	4.87
4	5.40
6	5.91
8	6.95
10	7.67
12	9.52

Exercise 20

Year	Doctors (1000s)
0	615.4
5	720.3
7	756.7
8	777.9
9	797.6
10	813.8

20. Since 1990, the total number of doctors of medicine in the United States has been growing. The data for selected years from 1990 to 2000 is given in the table, with 1990 corresponding to year 0 and the total number of M.D.s in thousands. Use the data to (a) draw a scatter-plot and decide on an appropriate form of regression, (b) find the regression equation, and (c) use the equation to find the projected number of M.D.s in the United States in the years 2005 and 2010.

Source: 2002 *Statistical Abstract of the United States,* Table 146

PRACTICE TEST

1. a. a and c are nonfunctions, they do not pass the vertical line test

2. 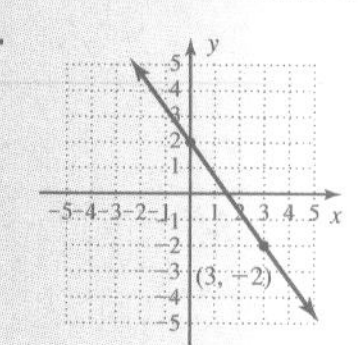

3. neither
4. $y = \frac{-2}{3}x$
5. a. yes b. no
6. $a(t) = \frac{20}{3}t + \frac{20}{3}$; 80 mph
7. a. (7.5, 1.5), b. $\approx$61.27 mi

1. Two relations here are functions and two are not. Identify the nonfunctions (justify your response).

a. $x = y^2 + 2y$ **b.** $y = \sqrt{5 - 2x}$ **c.** $|y| + 1 = x$ **d.** $y = x^2 + 2x$

2. Graph the line using the slope and y-intercept: $y = -\frac{4}{3}x + 2$.

3. Determine if the lines are parallel, perpendicular, or neither: L_1: $2x + 5y = -15$ and L_2: $y = \frac{2}{5}x + 7$.

4. Find the equation of the line parallel to $2x + 3y = 6$, going through the origin.

5. a. Is $(-2, 5)$ on the graph of $2x + 7y = 31$?

b. Is $(\frac{-31}{2}, 0)$ on the graph of $2x + 7y = 31$?

6. After 2 sec, a car is traveling 20 mph. After 5 sec its speed is 40 mph. Assuming the acceleration is constant, find the velocity equation and use it to determine the speed of the car after 11 sec.

7. My partner and I are at coordinates $(-20, 15)$ on a map. If our destination is at coordinates $(35, -12)$, (a) what are the coordinates of the rest station located halfway to our destination? (b) How far away is our destination? Assume that each unit is 1 mi.

8. L_1: $x = -3$
   L_2: $y = 4$

9. a. $x \in \{-4, -2, 0, 2, 4, 6\}$
   $y \in \{-2, -1, 0, 1, 2, 3\}$
   b. $x \in [-2, 6]$
   $y \in [1, 4]$

14. $x \in (-\infty, \frac{10}{3})$

15. $x \in (-\infty, -7] \cup [5, \infty)$

16. I. a. square root
   b. $x \in [-4, \infty)$, $y \in [-3, \infty)$
   c. $(-2, 0)$, $(0, 1)$
   d. up on right
   e. $x \in (-2, \infty)$
   f. $x \in [-4, -2)$

   II. a. cubic
   b. $x \in (-\infty, \infty)$ $y \in (-\infty, \infty)$
   c. $(2, 0)$, $(0, -1)$
   d. down on left, up on right
   e. $x \in (2, \infty)$
   f. $x \in (-\infty, 2)$

   III. a. absolute value
   b. $x \in (-\infty, \infty)$ $y \in (-\infty, 4]$
   c. $(-1, 0)$, $(3, 0)$, $(0, 2)$
   d. down/down
   e. $x \in (-1, 3)$
   f. $x \in (-\infty, -1) \cup (3, \infty)$

   IV. a. quadratic
   b. $x \in (-\infty, \infty)$; $y \in [-5.5, \infty)$
   c. $(0, 0)$, $(5, 0)$
   d. up/up
   e. $x \in (-\infty, 0) \cup (5, \infty)$
   f. $x \in (0, 5)$

17. 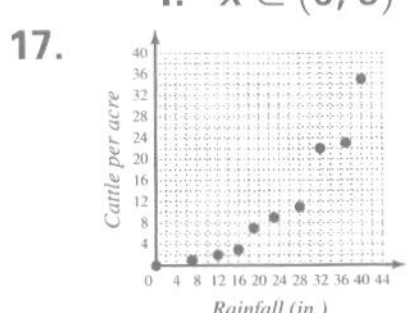

18. yes; positive

19. ≈ 53 cattle per acre

20. a. no; graph is less steep
   b. $\frac{\Delta S}{\Delta t} = 25$ for $[5, 6]$
   $\frac{\Delta S}{\Delta t} = 29$ for $[6, 7]$

8. Write the equations for lines L_1 and L_2 shown on the grid here.

9. State the domain and range for the relations shown on graphs 9(a) and 9(b).

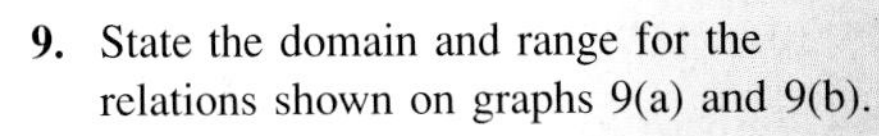

Exercise 8

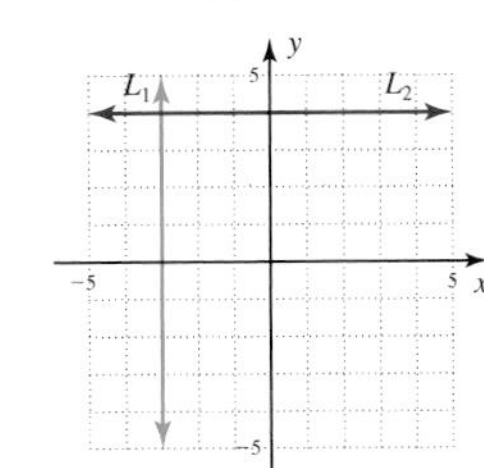

Exercise 9(a)

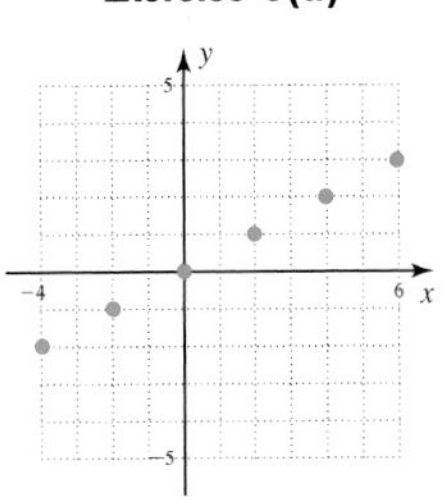

Exercise 9(b)

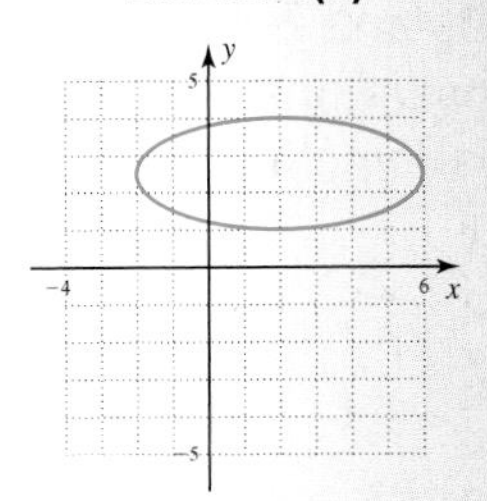

10. For the linear function shown here,

a. Determine the value of $W(24)$ from the graph. 300

b. What input h will give an output value of $W(h) = 375$? 30

c. Find a linear function that models the graph. $W(h) = \frac{25}{2}h$

d. What does the slope indicate in this context? wages are $12.50 per hr

e. State the domain and range of the function. $h \in [0, 40]$; $w \in [0, 500]$

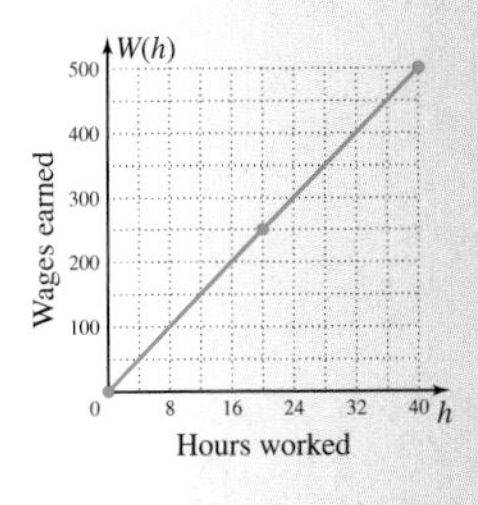

11. Given $f(x) = -2x^2 - 3x + 4$, compute the following: $f(\frac{-3}{2})$, $f(1 + \sqrt{2})$, and $f(2 + 3i)$. 4; $-5 - 7\sqrt{2}$; $8 - 33i$

12. Solve by completing the square: $18x = 3x^2 + 29$. $x = 3 \pm \frac{\sqrt{6}}{3}i$

13. Solve using the quadratic formula: $-2x^2 + 7x = 3$. $x = 3, \frac{1}{2}$

14. For $f(x) = \frac{-3}{2}x + 5$; solve $f(x) > 0$.

15. For $g(x) = x^2 + 2x - 35$; solve $f(x) \geq 0$.

16. Each function graphed here is from a toolbox function family. For each graph, (a) identify the function family, (b) state the domain and range, (c) identify x- and y-intercepts, (d) discuss the end behavior, and (e) solve the inequality $f(x) > 0$, and (f) solve $f(x) < 0$.

I.

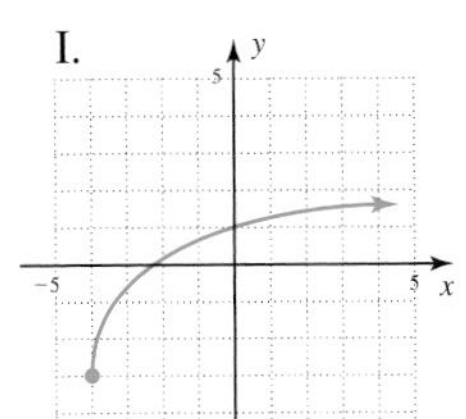

II.

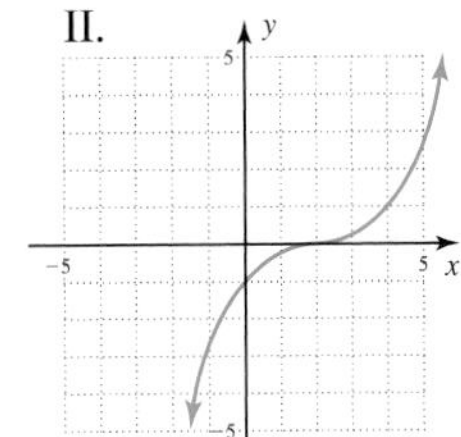

III.

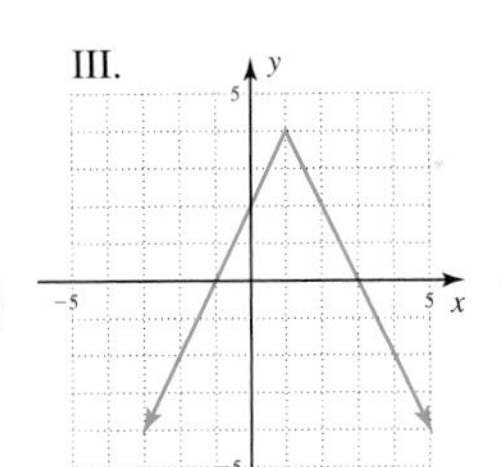

IV.

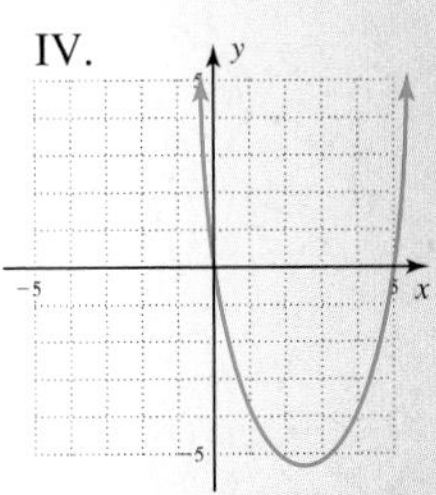

17. To study how annual rainfall affects livestock production, a local university collects data on the average annual rainfall for a particular area and compares this to the average number of free-ranging cattle per acre for ranchers in that area. The data collected are shown in the table. After scaling the axes appropriately, draw a scatter-plot for the data.

Rainfall (in.)	Cattle per Acre
0	0
7	1
12	2
16	3
19	7
23	9
28	11
32	22
37	23
40	35

18. Does the association from Exercise 17 appear nonlinear? Is the association positive or negative?

19. Use a graphing calculator to find the regression equation, then use the equation to predict the number of cattle per acre for an area receiving 50 in. of rainfall per year.

20. Monthly sales volume for a new company is modeled by $S(t) = 2x^2 + 3x$, where $S(t)$ represents sales volume in thousands in month t ($t = 0$ corresponds to January 1). (a) Would you expect the average rate of change from May to June to be greater than that from June to July? Why? (b) Calculate the rates of change in these intervals to verify your answer.

Calculator Exploration and Discovery

Cuts and Bounces: A Closer Look at the Zeroes of a Function

It is said that the most simple truths often lead to the most elegant results. This exploration brings together and connects many of the "simple truths" we've encountered thus far, and from them we hope to gain "elegant" results that we can build on in future chapters. Consider the function $Y_1 = x - 2$. Of a certainty, this is the graph of a *line* that intersects or "cuts" the x-axis at $x = 2$, with function values positive on one side of 2 and negative on the other (Figure 2.70). What we find interesting is that the function maintains these characteristics, even when the factor $(x - 2)$ occurs with other factors. For example, the graph of $Y_2 = (x - 2)(x + 1)$ still cuts the x-axis at $x = 2$, with function values positive on one side and negative on the other (Figure 2.71). A parabolic shape is formed because *both* linear factors must "cut the x-axis" to form the zeroes, and have function values with opposite signs on either side.

Figure 2.70

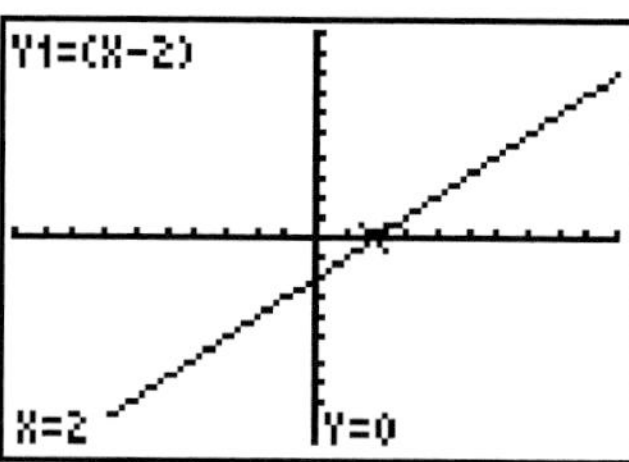

Figure 2.71

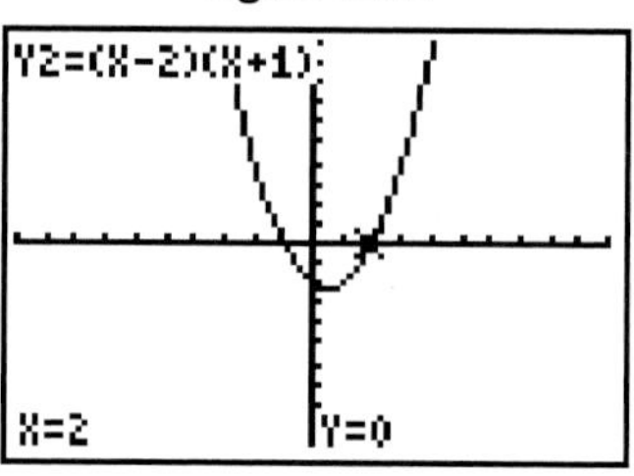

Now consider the function of $Y_3 = (x - 2)^2$, a basic parabola with vertex at (2, 0). There is a zero at $x = 2$ but due to the nature of the graph, it "bounces" off the x-axis with $f(x) > 0$ on *both sides* of 2 (Figure 2.72). However, just as with the linear function, this function again *maintains these characteristics* when combined with other factors. Notice the graph of $Y_4 = (x - 2)^2(x + 1)$ still bounces at $x = 2$, even while the graph cuts back through the x-axis to form the zero at $x = -1$ (Figure 2.73). At the same time, note this is a cubic function with a positive lead coefficient, and the graph exhibits the down, up end behavior we expect!

Figure 2.72

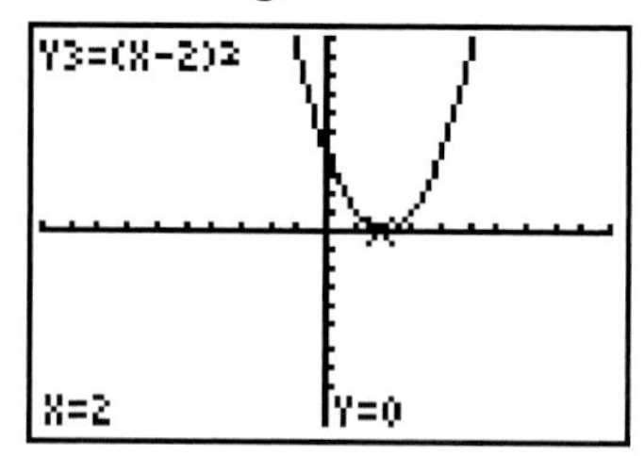

Figure 2.73

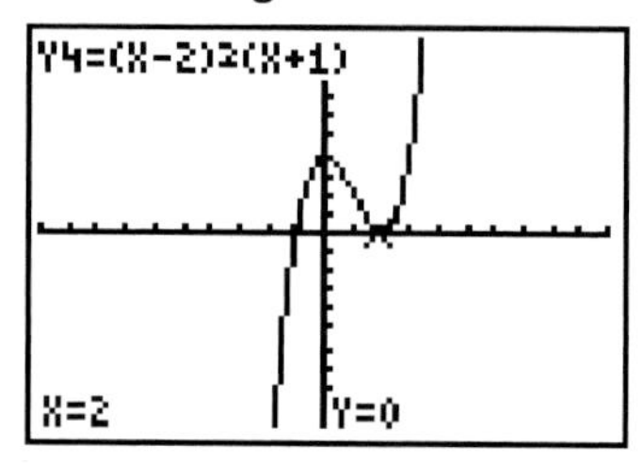

Finally, suppose we wanted to construct a function with all these features, but that also contained the point (3, 2) instead of (3, 4) as it currently does: $Y_4(3) = 4$. As it stands, the function Y_4 implicitly shows a lead coefficient of $a = 1$. To transform the graph so that it contains (3, 2) use the "formula" $Y_4 = a(x - 2)^2(x + 1)$ with $x = 3$ and $y = 2$, then solve for the new value of a. This gives $a = \frac{1}{2}$ and the function $Y_5 = \frac{1}{2}(x - 2)^2(x + 1)$. Note the graph does everything we expect that it "should" (Figure 2.74).

Figure 2.74

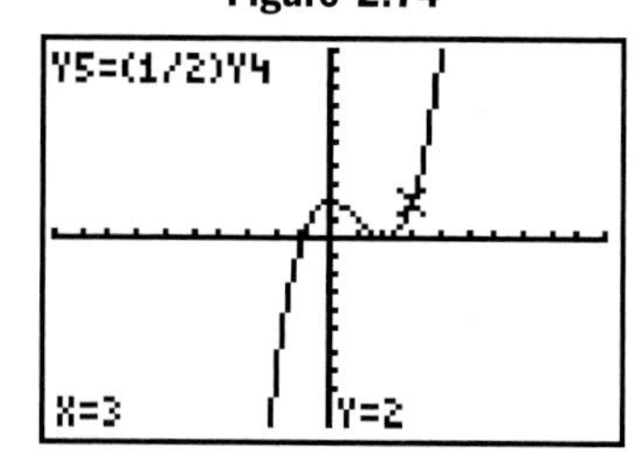

Now for the elegant result—these results hold true for all polynomials! In Chapter 4 the ideas will be combined with others that will enable us to graph almost any polynomial from its factored form. Here we use the ideas to create polynomials with stated characteristics, then check the result on a graphing calculator.

$y = \frac{7}{9}(x - 1)^2(x + 3)$

Exercise 1: Write the equation (in factored form) of a polynomial that bounces off the x-axis at $x = 1$, cuts the x-axis at $x = -3$, has down, up end behavior and contains the point $(-2, 7)$.

Exercise 2: Write the equation (in factored form) of a polynomial that cuts the x-axis at $x = 4$, bounces off the x-axis at $x = -3$, has up, down end behavior with a y-intercept of (0, 6). $y = -\frac{1}{6}(x - 4)(x + 3)^2$

Exercise 3: Create your own stipulations, build the equation, and check the result on a graphing calculator. Answers will vary.

Strengthening Core Skills

More on End Behavior

For quadratic functions, the graph is concave up if $a > 0$, concave down if $a < 0$. For cubic functions (and linear functions), the end behavior is *down, up* if the lead coefficient is positive, and *up, down* otherwise. Up to this point we've determined the end behavior of the toolbox functions by *observation,* noting that the lead coefficient plays a critical role. Here we seek to understand *why* this is true. The reason is that for large values of x, the leading term is much more "powerful" than the remaining terms. As the value of x gets larger, terms of higher degree (larger exponents) will dominate the other terms in an expression, so the degree and coefficient of the leading term will dictate the *end behavior* of the graph. Consider $f(x) = x^2 - 5x - 6$, which should be concave up since $a > 0$, and the table shown. For values of $x \in [0, 6]$ the linear and constant terms "gang up" on the squared term, causing negative or zero outputs. But for larger input values, the squared term easily "gobbles the others up" and dictates that eventually outputs will be positive. This phenomenon is responsible for the end behavior of a graph. For $g(x) = -x^3 + 5x^2 + 4x + 12$ the end behavior should be *up, down* since the lead coefficient is negative. Once again a table of values shows why—for values of $x \in [0, 6]$ the linear, constant, and squared terms "gang up" on the cubic term, causing positive or zero outputs. But for larger input values, the cubic term will eventually dominate.

x	x^2	$-5x$	-6	$x^2 - 5x - 6$
0	0	0	−6	−6
1	1	−5	−6	−10
2	4	−10	−6	−12
3	9	−15	−6	−12
4	16	−20	−6	−10
5	25	−25	−6	−6
6	36	−30	−6	**0**
7	49	−35	−6	**8**
8	64	−40	−6	**18**

x	$-x^3$	$5x^2$	$4x$	12	$-x^3 + 5x^2 + 4x + 12$
0	0	0	0	12	12
1	−1	5	4	12	20
2	−8	20	8	12	32
3	−27	45	12	12	42
4	−64	80	16	12	44
5	−125	125	20	12	32
6	−216	180	24	12	**0**
7	−343	245	28	12	**−58**
8	−512	320	32	12	**−148**

These ideas will play an important role in our study of general polynomial and rational functions in Chapter 4. Use them to complete the following exercises.

Exercise 1: Construct a table of values and do a similar investigation for $f(x) = -x^2 + 3x + 24$ using $x \in [0, 8]$. At what x-value does the squared term begin to "gobble up" the other terms? between 6 and 7

Exercise 1.

x	$-x^2$	$3x$	24	$-x^2 + 3x + 24$
0	0	0	24	24
1	−1	3	24	26
2	−4	6	24	26
3	−9	9	24	24
4	−16	12	24	20
5	−25	15	24	14
6	−36	18	24	6
7	−49	21	24	−4
8	−64	24	24	−16

Additional answers can be found in the Instructor Answer Appendix.

Exercise 2: Using the function $f(x) = x^2 - 9$, can you *anticipate* when the squared term will overcome (gobble up) the -9? How is this related to our study of intervals where a function is positive or negative? Use your conclusions to state the end behavior of $f(x) = 9 - x^2$, and name the interval where $f(x) > 0$.
at $x > 3$; answers will vary; down/down, $x \in (-3, 3)$

Exercise 3: Use a calculator to explore the end behavior of the fourth-degree and fifth-degree polynomials given. Can you detect a pattern emerging regarding end behavior and the degree of the polynomial? Comment and discuss.

a. $p(x) = x^4 - 3x^3 - 5x^2 - 6x - 15$

b. $q(x) = -x^5 + 3x^4 + 5x^3 + 2x^2 + 7x + 9$
Answers will vary.

Cumulative Review Chapters 1–2

1. Translate from words into a mathematical phrase: "Five less than twice a number is equal to three more than the number." $2n - 5 = n + 3$

2. Perform the operations indicated:

a. $2x^3 + 3x - (x - 1) + x(1 + x^2)$ $3x^3 + 3x + 1$

b. $(2x - 3)(2x + 3)$ $4x^2 - 9$

3. Simplify using properties of exponents:

a. $\dfrac{-15n^3m^4}{10nm^3}$ $\dfrac{-3n^2m}{2}$

b. $(5.1 \times 10^{-9}) \times (3 \times 10^6)$ 15.3×10^{-3}

c. $\left(\dfrac{2ab^{-2}}{c^2}\right)^{-3}$ $\dfrac{b^6c^6}{8a^3}$

d. $2x^0 + (2x)^0 + 2^{-1}$ $3\frac{1}{2}$

4. Determine which of the following statements are true:

a. $\mathbb{N} \subset \mathbb{Z} \subset \mathbb{W} \subset \mathbb{Q} \subset \mathbb{R}$ False

b. $\mathbb{W} \subset \mathbb{N} \subset \mathbb{Z} \subset \mathbb{Q} \subset \mathbb{R}$ False

c. $\mathbb{N} \subset \mathbb{W} \subset \mathbb{Z} \subset \mathbb{Q} \subset \mathbb{R}$ True

d. $\mathbb{N} \subset \mathbb{R} \subset \mathbb{Z} \subset \mathbb{Q} \subset \mathbb{W}$ False

5. Add the rational expressions:

a. $\dfrac{-2}{x^2 - 3x - 10} + \dfrac{1}{x + 2}$ $\dfrac{x - 7}{(x - 5)(x + 2)}$

b. $\dfrac{b^2}{4a^2} - \dfrac{c}{a}$ $\dfrac{b^2 - 4ac}{4a^2}$

6. Simplify the radical expressions:

a. $\dfrac{-10 + \sqrt{72}}{4}$ $\dfrac{-5}{2} + \dfrac{3\sqrt{2}}{2}$

b. $\dfrac{1}{\sqrt{2}}$ $\dfrac{\sqrt{2}}{2}$

7. Solve for x: $-2(3 - x) + 5x = 4(x + 1) - 7$. $x = 1$

8. Solve for t: $rt + Rt = D$ $t = \dfrac{D}{r + R}$

9. Find the solution set: $2 - x < 5$ and $3x + 2 < 8$. $-3 < x < 2$

10. Show that $x = 1 + 5i$ is a solution to $x^2 - 2x + 26 = 0$.

11. Compute as indicated:

a. $(2 + 5i)^2$

b. $\dfrac{1 - 2i}{1 + 2i}$

12. Solve by factoring:

a. $6x^2 - 7x = 20$

b. $x^3 + 5x^2 - 15 = 3x$

13. Solve by completing the square. Answer in both exact and approximate form: $2x^2 + 49 = -20x$.

14. Solve using the quadratic formula. If solutions are complex, write them in $a + bi$ form. $2x^2 + 20x = -51$

15. The *National Geographic Atlas of the World* is a very large, rectangular book with an almost inexhaustible panoply of information about the world we live in. The length of the front cover is 16 cm more than its width, and the area of the cover is 1457 cm^2. Use this information to write an equation model, then use the quadratic formula to determine the length and width of the *Atlas*.

10. $(1 + 5i)^2 - 2(1 + 5i) + 26 = 0$; $-24 + 10i - 2 - 10i + 26 = 0$; $0 = 0$ ✓

11. **a.** $-21 + 20i$ **b.** $\frac{-3}{5} - \frac{4}{5}i$

12. **a.** $x = \frac{-4}{3}, \frac{5}{2}$ **b.** $x = -5, -\sqrt{3}, \sqrt{3}$

13. $x = -5 \pm \dfrac{\sqrt{2}}{2}$; $x \approx -5.707$; $x \approx -4.293$

14. $x = -5 \pm \dfrac{i\sqrt{2}}{2}$

15. $W = 31$ cm, $L = 47$ cm

16. Given $f(x) = x^3 - 4x$, find the solution interval(s) for $f(x) \le 0$. $x \in (-\infty, -2] \cup [0, 2]$

17. Graph by plotting the y-intercept, then counting $m = \frac{\Delta y}{\Delta x}$ to find additional points: $y = \frac{1}{3}x - 2$.

18. Since 1990, lumber imports from Canada have grown at a fairly steady rate. The data for selected years is given in the table, with 1990 corresponding to year 0 and lumber imports in billions of board feet. (a) Draw a scatter-plot and decide on an appropriate form of regression; (b) find the regression equation, and (c) use the equation to find projected lumber imports from Canada for the years 2005 and 2008. (d) Using the equation, in what year were 22 billion board feet imported?

Source: 2002 *Statistical Abstract of the United States,* Table 839 (figures have been rounded)

Exercise 18

Year	Imports
0	13.1
3	15.4
4	16.5
5	17.5
6	18.4
7	18.2
8	19.0
9	19.9
10	20.2

19. A theorem from elementary geometry states, "A line tangent to a circle is perpendicular to the radius at the point of tangency." Find the equation of the tangent line for the circle and radius shown.

20. A triangle has its vertices at $(-4, 5)$, $(4, -1)$, and $(0, 8)$. Find the perimeter of the triangle and determine whether or not it is a *right* triangle.

17.

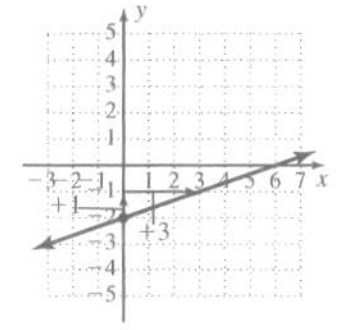

18. a. Linear

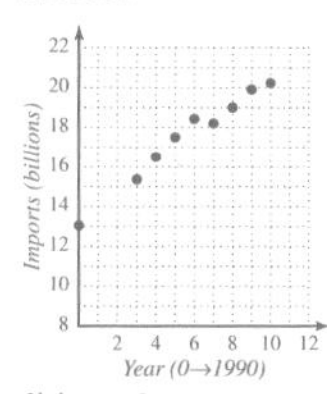

b. $I(t) = 0.711t + 13.470t$

c. $I(15) \approx 24.1$
$I(18) \approx 26.3$

d. year $t = 12$; 2002

19. $m_1 = \frac{1}{2}$
$m_2 = -2$
$\Rightarrow y = -2x + 4$

20. $p = 15 + \sqrt{97}$ units
≈ 24.8 units
No, it is not a right triangle.
$5^2 + (\sqrt{97})^2 \neq 10^2$

Chapter 3

Operations on Functions and Analyzing Graphs

Chapter Outline

3.1 The Algebra and Composition of Functions 248

3.2 One-to-One and Inverse Functions 261

3.3 Toolbox Functions and Transformations 274

3.4 Graphing General Quadratic Functions 288

3.5 Asymptotes and Simple Rational Functions 301

3.6 Toolbox Applications: Direct and Inverse Variation 312

3.7 Piecewise-Defined Functions 327

3.8 Analyzing the Graph of a Function 341

Preview

The foundation, ideas, and structures developed in previous chapters were designed to support your studies through the remainder of college algebra. In Chapter 3, we continue building on these ideas, while making connections and illustrating consistent themes that help relate new ideas to those introduced earlier. Each section develops new concepts and components that contribute to a better overall understanding of functions and graphs—dominant themes in college algebra and the cornerstones of mathematical modeling.

3.1 The Algebra and Composition of Functions

LEARNING OBJECTIVES

In Section 3.1 you will learn how to:

A. Compute a sum or difference of functions and determine the domain of the result

B. Compute a product or quotient of functions and determine the domain of the result

C. Compose two functions and find the domain

D. Apply a composition of functions in context

E. Decompose a function *h* into functions *f* and *g*

INTRODUCTION

In previous course work, you likely learned how to find the sum, difference, product, and quotient of polynomials. In this section, we note the result of these operations is also a function, which can be evaluated, graphed, and analyzed. We call this combining of functions with the basic operations the **algebra of functions,** and use them in many real-world applications.

POINT OF INTEREST

Have you ever tried to multiply or divide using Roman numerals? Although it's possible (various methods have been developed), it's difficult because the notation and numerals used interfere rather than aid the process. The history of mathematics is filled with similar situations, where a wonderful and useful idea was hindered by the notation used to express it. In contrast, the function notation we use today gives us an effective way to write and study operations on functions.

A. Sums and Differences of Functions

This section introduces the notation used for basic operations on functions. We'll further note the result is also a function whose domain depends on the functions involved. In general, if f and g are functions *with overlapping domains,* $f(x) \pm g(x) = (f \pm g)(x)$.

SUMS AND DIFFERENCES OF FUNCTIONS

For functions f and g with domains P and Q respectively, the sum and difference of f and g are defined by:

	Domain of result
$(f + g)(x) = f(x) + g(x)$	$P \cap Q$
$(f - g)(x) = f(x) - g(x)$	$P \cap Q$

EXAMPLE 1 Given $f(x) = x^2 - 5x$ and $g(x) = 2x - 10$, find $h(x) = (f - g)(x)$ and state the domain of h.

Solution:

$$\begin{aligned} h(x) &= (f - g)(x) && \text{given difference} \\ &= f(x) - g(x) && \text{by definition} \\ &= (x^2 - 5x) - (2x - 10) && \text{replace } f(x) \text{ with } (x^2 - 5x) \text{ and } g(x) \text{ with } (2x - 10) \\ &= x^2 - 7x + 10 && \text{distribute and combine like terms} \end{aligned}$$

Since the domain of both f and g is the set of real numbers $\mathbb{R}$, the domain of h is also $\mathbb{R}$.

NOW TRY EXERCISES 7 THROUGH 12

CAUTION

From Example 1, note the importance of using grouping symbols with the algebra of functions. Without them, we could easily confuse the signs of $g(x)$ when computing the difference. Also, although the new function $h(x)$ can be factored, we were only asked to compute the difference of f and g, so we stop there.

When two functions are combined to create a new function, we often need to evaluate the result. Let's again consider the difference $h(x) = f(x) - g(x)$ from Example 1. To find $h(3)$, we could first compute $f(3)$ and $g(3)$, then subtract: $h(3) = f(3) - g(3)$. With $f(3) = -6$ and $g(3) = -4$, we have $h(3) = -6 - (-4) = -2$. As an alternative, we could first *subtract g from f, then evaluate the result:* $h(3) = (f - g)(3)$. For $h(x) = x^2 - 7x + 10$ from Example 1, we have $h(3) = (3)^2 - 7(3) + 10 = -2$✓.

When one of the functions is constant, the sum or difference yields a predictable and useful result, as illustrated in Example 2.

EXAMPLE 2 For $f(x) = x^2 - 4$ and $g(x) = 5$, find $h(x) = (f - g)(x)$, then graph f and h on the same grid and comment on what you notice.

Solution:

$h(x) = (f - g)(x)$ given difference

$= f(x) - g(x)$ by definition

$= (x^2 - 4) - (5)$ replace $f(x)$ with $(x^2 - 4)$ and $g(x)$ with (5)

$= x^2 - 9$ combine like terms

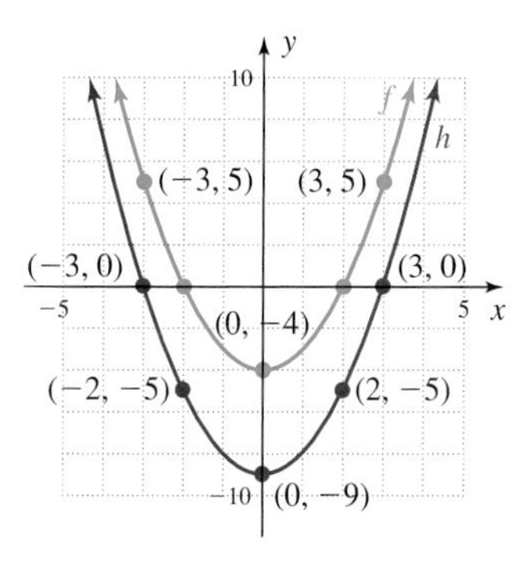

The graph of f is a parabola with vertex and y-intercept at $(0, -4)$, and x-intercepts of $(-2, 0)$ and $(2, 0)$. The graph of h is a parabola with vertex and y-intercept at $(0, -9)$ and x-intercepts of $(-3, 0)$ and $(3, 0)$.

Observe that the graphs of f and h are identical, but h has been "shifted down" 5 units. This is no coincidence and this "downward shift" can be seen numerically in Table 3.1. For any input, the outputs of h are 5 less than the outputs for f: $h(x) = f(x) - 5$. This and similar observations are connected to a number of concepts we'll see in Section 3.3.

Table 3.1

x	$f(x) = x^2 - 4$	$h(x) = x^2 - 9$	x	$f(x) = x^2 - 4$	$h(x) = x^2 - 9$
−5	21	16	2	0	−5
−4	12	7	3	5	0
−3	5	0	4	12	7
−2	0	−5	5	21	16
0	−4	−9			

NOW TRY EXERCISES 13 AND 14

B. Products and Quotients of Functions

The product and quotient of two functions is defined in a manner similar to that for sums and differences. For example, if f and g are functions *with overlapping domains,* $(f \cdot g)(x) = f(x) \cdot g(x)$ and $\left(\frac{f}{g}\right)(x) = \frac{f(x)}{g(x)}$. As you might expect, for quotients we must stipulate $g(x) \neq 0$.

PRODUCTS AND QUOTIENTS OF FUNCTIONS

For functions f and g with domains P and Q respectively, the product and quotient of f and g are defined by:

	Domain of result
$(f \cdot g)(x) = f(x) \cdot g(x)$	$P \cap Q$
$\left(\dfrac{f}{g}\right)(x) = \dfrac{f(x)}{g(x)}$	$P \cap Q$ for all $g(x) \neq 0$

EXAMPLE 3 Given $f(x) = \sqrt{1 + x}$ and $g(x) = \sqrt{3 - x}$: (a) find $h(x) = (f \cdot g)(x)$, (b) evaluate $h(2)$ and $h(4)$, and (c) state the domain of h.

Solution:

a. $h(x) = (f \cdot g)(x)$ given product

$= f(x) \cdot g(x)$ by definition

$= \sqrt{1 + x} \cdot \sqrt{3 - x}$ substitute $\sqrt{1 + x}$ for f and $\sqrt{3 - x}$ for g

$= \sqrt{3 + 2x - x^2}$ combine using properties of radicals

b. $h(2) = \sqrt{3 + 2(2) - (2)^2}$ substitute 2 for x

$= \sqrt{3} \approx 1.732$ result

$h(4) = \sqrt{3 + 2(4) - (4)^2}$ substitute 4 for x

$= \sqrt{-5}$ *not a real number*

c. To see why $h(4)$ is not a real number, consider that the domain of f is $x \in [-1, \infty)$ while the domain of g is $x \in (-\infty, 3]$. The intersection of domains gives $[-1, 3]$, which is the domain for h and shows that h is not defined for $x = 4$.

NOW TRY EXERCISES 15 THROUGH 18

In future sections, we use polynomial division as a tool for factoring, an aid to graphing, and to determine whether two expressions are equivalent. Understanding the notation and domain issues related to division will strengthen our ability to use division in these ways.

EXAMPLE 4 Given $f(x) = x^3 - 3x^2 + 2x - 6$ and $g(x) = x - 3$, find the function $h(x)$, where $h(x) = \left(\dfrac{f}{g}\right)(x)$. Then state the domain of h.

Solution:

$h(x) = \left(\dfrac{f}{g}\right)(x)$ given quotient

$= \dfrac{f(x)}{g(x)}$ by definition

$= \dfrac{x^3 - 3x^2 + 2x - 6}{x - 3}$ replace f with $x^3 - 3x^2 + 2x - 6$ and g with $x - 3$

$= \dfrac{x^2(x - 3) + 2(x - 3)}{x - 3}$ factor the numerator by grouping

$= \dfrac{(x^2 + 2)\cancel{(x - 3)}}{\cancel{x - 3}}$ common factor of $(x - 3)$

$= x^2 + 2;\ x \neq 3$ simplify

While the domain of both f and g is $\mathbb{R}$ and hence their intersection is also $\mathbb{R}$, we must remember $g(x)$ cannot be equal to 0, *even if the result is a polynomial*. The domain of h is $x \in (-\infty, 3) \cup (3, \infty)$.

NOW TRY EXERCISES 19 THROUGH 46

For additional practice with the algebra of functions, see Exercises 35 to 46.

C. Composition of Functions

The composition of functions is best understood by studying the "input/output" nature of a function. Consider $g(x) = x^2 - 3$. To describe how this function operates on input values, we might say, "inputs are squared, then decreased by three." Using a function box, we could "program" the box to perform these operations and in diagram form we have:

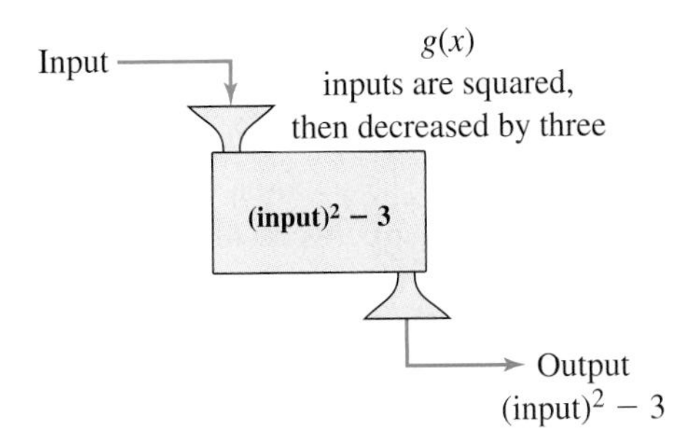

In many respects, a function box can be regarded as a very simple machine, running a simple program. It doesn't matter what the input is, this machine is going to *square the input then subtract three.*

WORTHY OF NOTE

It's important to note that t and $t - 4$ are two different, distinct values—the number represented by t, and a number four less than t. Examples would be 7 and 3, 12 and 8, as well as -10 and -14. There should be nothing awkward or unusual about evaluating $f(t)$ versus evaluating $f(t - 4)$.

EXAMPLE 5 For $g(x) = x^2 - 3$, find (a) $g(-5)$, (b) $g(t)$, and (c) $g(t - 4)$.

Solution:

a. $g(x) = x^2 - 3$ — original function

input -5

$g(-5) = (-5)^2 - 3$ — square input, then subtract 3

$= 25 - 3$ — simplify

$= 22$ — result

b. $g(x) = x^2 - 3$ — original function

input t

$g(t) = (t)^2 - 3$ — square input, then subtract 3

$= t^2 - 3$ — result

c. $g(x) = x^2 - 3$ — original function

input $t - 4$

$g(t - 4) = (t - 4)^2 - 3$ — square input, then subtract 3

$= t^2 - 8t + 16 - 3$ — expand binomial

$= t^2 - 8t + 13$ — result

NOW TRY EXERCISES 47 AND 48

When the input value is itself a function (rather than a single number or variable), this process is called the **composition of functions.** The evaluation method is exactly the same, we are simply using a function input. Using a general function $g(x)$ and a function box as before, the process is illustrated in Figure 3.1.

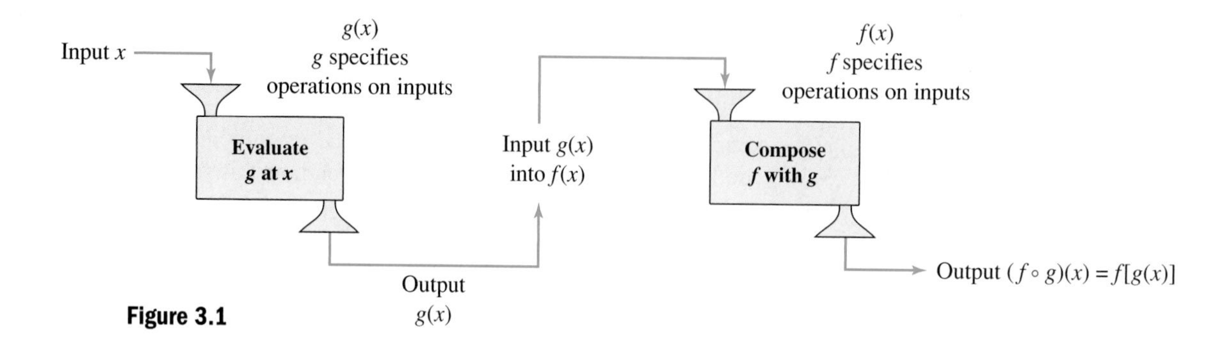

Figure 3.1

The notation used for the composition of functions f and g is an open circle "$\circ$" placed between them, and indicates we will use the second function as an input for the first. In other words, $(f \circ g)(x)$ indicates that $g(x)$ is an input for f: $(f \circ g)(x) = f[g(x)]$. If the order is reversed as in $(g \circ f)(x)$, $f(x)$ becomes the input for g: $(g \circ f)(x) = g[f(x)]$. The diagram in Figure 3.1 also helps us determine the domain of a composite function, in that the first function box can operate only if x is a valid input for g, and the second function box can operate only if $g(x)$ is a valid input for f. In other words, $(f \circ g)(x)$ is defined for *all x in the domain of g, such that $g(x)$ is in the domain of f.*

THE COMPOSITION OF FUNCTIONS

Given two functions f and g, the composition of f with g is defined by

$$(f \circ g)(x) = f[g(x)],$$

for all x in the domain of g such that $g(x)$ is in the domain of f.

In Figure 3.2 the ideas are displayed using the mapping notation from Section 2.2, which can sometimes help clarify concepts related to the domain. The diagram shows that not all elements in the domain of g are automatically in the domain of $(f \circ g)$, since $g(x)$ may represent inputs unsuitable for f. This means the range of g and the domain of f will intersect, while the domain of $(f \circ g)$ is a subset of the domain of g.

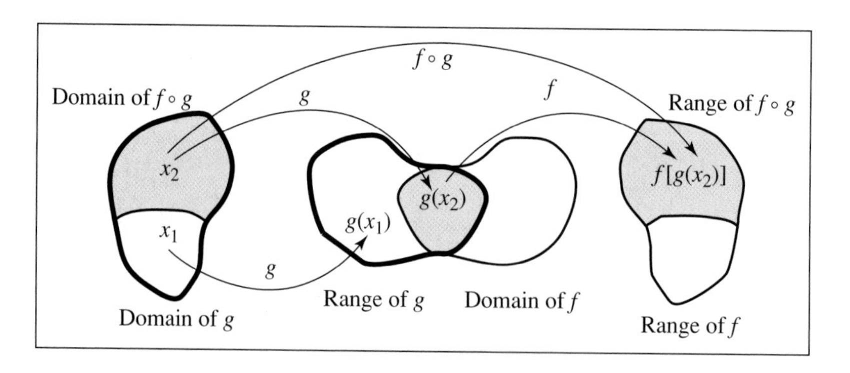

Figure 3.2

EXAMPLE 6 Given $f(x) = \sqrt{x - 4}$ and $g(x) = 3x + 2$, find (a) $(f \circ g)(x)$ and (b) $(g \circ f)(x)$. Also determine the domain for each.

Solution: **a.** Begin by describing what the function f does to inputs: $f(x) = \sqrt{x - 4}$ says "decrease inputs by 4, and take the square root of the result."

$$\begin{aligned}(f \circ g)(x) &= f[g(x)] && g(x) \text{ is an input for } f\\ &= \sqrt{g(x) - 4} && \text{decrease input by 4, and take the square root of the result}\\ &= \sqrt{(3x + 2) - 4} && \text{substitute } 3x + 2 \text{ for } g(x)\\ &= \sqrt{3x - 2} && \text{result}\end{aligned}$$

For the domain of $f[g(x)]$, we note g is defined for all real numbers x, but we must have $g(x) \geq 4$ (in blue in the preceding) or $f[g(x)]$ will not represent a real number. This gives $3x + 2 \geq 4$ so $x \geq \frac{2}{3}$. In interval notation, the domain of $(f \circ g)(x)$ is $x \in [\frac{2}{3}, \infty)$.

b. The function g says "inputs are multiplied by 3, then increased by 2."

$$\begin{aligned}(g \circ f)(x) &= g[f(x)] && f(x) \text{ is an input for } g\\ &= 3f(x) + 2 && \text{multiply input by 3, then increase by 2}\\ &= 3\sqrt{x - 4} + 2 && \text{substitute } \sqrt{x-4} \text{ for } f(x)\end{aligned}$$

For the domain of $g[f(x)]$, although g can accept any real number input, f can supply only those where $x \geq 4$. The domain of $(g \circ f)(x)$ is $x \in [4, \infty)$.

NOW TRY EXERCISES 49 THROUGH 64

Example 6 shows $(f \circ g)(x)$ is not generally equal to $(g \circ f)(x)$. On those occasions when they *are* equal, the functions have a unique relationship that we'll study in Section 3.2.

TECHNOLOGY HIGHLIGHT

Using a Graphing Calculator to Study Composite Functions

The keystrokes shown apply to a TI-84 Plus model. Please consult your manual or our Internet site for other models.

The graphing calculator is truly an amazing tool when it comes to studying composite functions. Using this powerful tool, composite functions can be graphed, evaluated, and investigated with ease. To begin, enter the functions $y = x^2$ and $y = x - 5$ as Y_1 and Y_2 on the Y= screen. Enter the composition $(Y_1 \circ Y_2)(X)$ as $Y_3 = Y_1(Y_2(X))$, as shown in Figure 3.3 [in our standard notation we have $f(x) = x^2$, $g(x) = x - 5$, and $h(x) = (f \circ g)(x) = f[g(x)]$. On the TI-84 *Plus*, we access the function variables Y_1, Y_2, Y_3, and so on by pressing VARS ▶ ENTER and selecting the function desired. Pressing ZOOM **6:ZStandard** will graph all three functions in the standard window. Although there are many relationships we could investigate, let's concentrate on the relationship between Y_1 and Y_3. Deactivate Y_2 and regraph Y_1 and Y_3. What do you notice about the graphs? Y_3 is the same as the graph of Y_1, but shifted 5 units to the

Figure 3.3

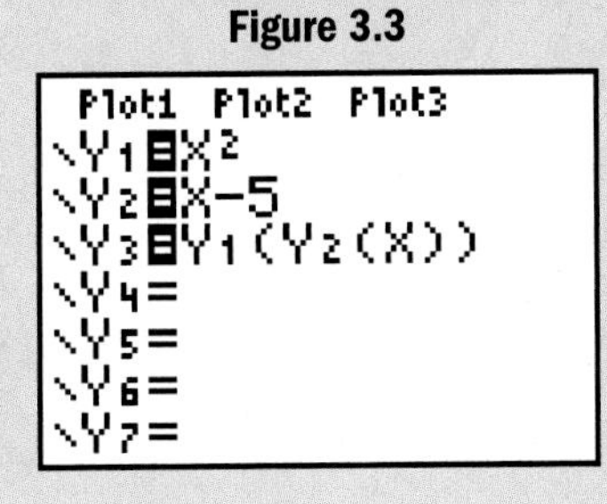

right! Does this have any connection to $Y_2 = x - 5$? Try changing Y_2 to $Y_2 = x + 4$, then regraph Y_1 and Y_3. Use what you notice to complete the following exercises and continue the exploration.

Exercise 1: Change Y_1 to $Y_1 = \sqrt{x}$, then experiment by changing Y_2 to $x + 3$, then to $x - 6$. Did you notice anything similar? What would happen if we changed Y_2 to $Y_2 = x + 7$? yes; graph shifts 7 units to the left.

Exercise 2: Change Y_1 to $Y_1 = x^3$, then experiment by changing Y_2 to $x + 5$, then to $x - 1$. Did the same "shift" occur? What would happen if we changed Y_1 to $Y_1 = |x|$? yes; the basic function $f(x) = |x|$ shifts 1 unit right.

D. Applications of Composition

Consider this hypothetical situation. Due to a collision, an oil tanker is spewing oil into the open ocean. The oil is spreading outward in a shape that is roughly circular, with the radius of the circle modeled by the function $r(t) = 2\sqrt{t}$, where t is the time in minutes and r is measured in feet. How could we determine the *area* of the oil slick in terms of t? As you can see, the radius depends on the time and the area depends on the radius. In diagram form we have:

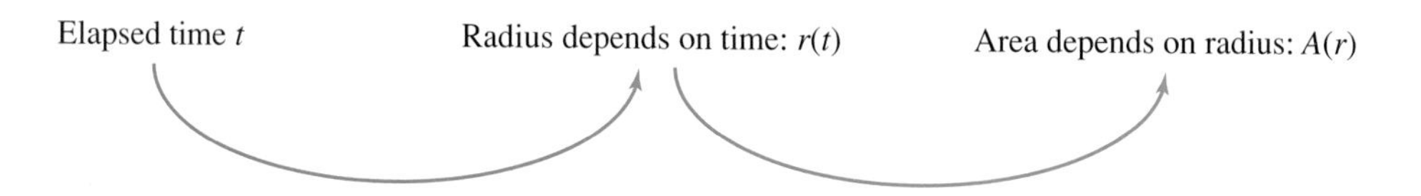

It is possible to create a direct relationship between the elapsed time and the area of the circular spill using a composition of functions.

EXAMPLE 7 Given $r(t) = 2\sqrt{t}$ and $A(r) = \pi r^2$, (a) write A directly as a function of t by computing $(A \circ r)(t)$; and (b) find the area of the oil spill after 30 min.

Solution: **a.** The function A squares inputs, then multiplies by π.

$$\begin{aligned} (A \circ r)(t) &= A[r(t)] && r(t) \text{ is the input for } A \\ &= [r(t)]^2 \cdot \pi && \text{square input, multiply by } \pi \\ &= [2\sqrt{t}]^2 \cdot \pi && \text{substitute } 2\sqrt{t} \text{ for } r(t) \\ &= 4\pi t && \text{result} \end{aligned}$$

Since the result contains no variable r, we can now compute the area of the spill directly, given the elapsed time t: $A(t) = 4\pi t$.

b. To find the area after 30 min, use $t = 30$.

$$\begin{aligned} A(t) &= 4\pi t && \text{composite function} \\ A(30) &= 4\pi(30) && \text{substitute 30 for } t \\ &= 120\pi && \text{simplify} \\ &\approx 377 && \text{result (rounded to the nearest unit)} \end{aligned}$$

After 30 min, the area of the spill is approximately 377 ft^2.

NOW TRY EXERCISES 69 THROUGH 76

E. Decomposing a Composite Function

Based on the diagram in Figure 3.4, would you say that the circle is inside the square or the square is inside the circle? The decomposition of a composite function is related to a similar question, as we ask ourselves what function (of the composition) is on the "inside"—the input quantity—and what function is on the "outside." For instance, consider $h(x) = \sqrt{x - 4}$, where we see that $x - 4$ is "inside" the radical. Letting $g(x) = x - 4$ and $f(x) = \sqrt{x}$, we have $h(x) = (f \circ g)(x)$ or $f[g(x)]$.

Figure 3.4

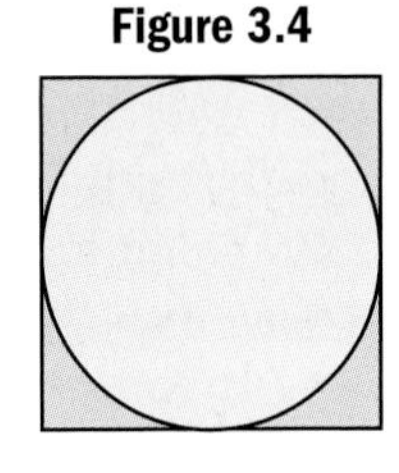

EXAMPLE 8 Given $h(x) = (\sqrt[3]{x} + 1)^2 - 3$, identify two functions f and g so that $(f \circ g)(x) = h(x)$, then check by composing the functions to obtain $h(x)$.

Solution: Noting that $\sqrt[3]{x} + 1$ is inside the squaring function, we assign $g(x)$ as this inner function: $g(x) = \sqrt[3]{x} + 1$. The outer function is the squaring function decreased by 3, so $f(x) = x^2 - 3$.

$$(f \circ g)(x) = f[g(x)] \quad \text{g(x) is an input for f}$$
$$= [g(x)]^2 - 3 \quad \text{f squares inputs, then decreases the result by 3}$$
$$= [\sqrt[3]{x} + 1]^2 - 3 \quad g(x) = \sqrt[3]{x} + 1$$
$$= h(x) \checkmark$$

NOW TRY EXERCISES 77 AND 78

The decomposition of a function is not unique and can often be done in many different ways.

TECHNOLOGY HIGHLIGHT

Graphing Calculators and the Domain of a Function

The keystrokes shown apply to a TI-84 Plus model. Please consult your manual or our Internet site for other models.

The TRACE feature of a graphing calculator is a wonderful tool for understanding the characteristics of a function. We'll illustrate using the function $h(x) = \frac{f(x)}{g(x)}$, where $f(x) = x^3 - 2x^2 - 3x + 6$ and $g(x) = x - 2$ (similar to Example 4). Enter $\frac{f(x)}{g(x)}$ on the Y= screen as Y_1 (Figure 3.5), then graph the function using ZOOM **4:ZDecimal**. Recall this will allow the calculator to trace through Δx intervals of 0.1. After pressing the TRACE key, the cursor appears on the graph at the *y*-intercept (0, −3) and its location is displayed at the bottom of the screen. Note that there is a "hole" in the graph in the first quadrant (Figure 3.6). As we've seen in other *Technology Highlight* boxes, we can walk the cursor along the curve in either direction using the left arrow ◄ and right arrow ► keys. Because the location of the cursor is constantly displayed as we move, we can determine exactly where this hole occurs. Walking the cursor to the right, we note *that no output is displayed for x* = 2.

Figure 3.5

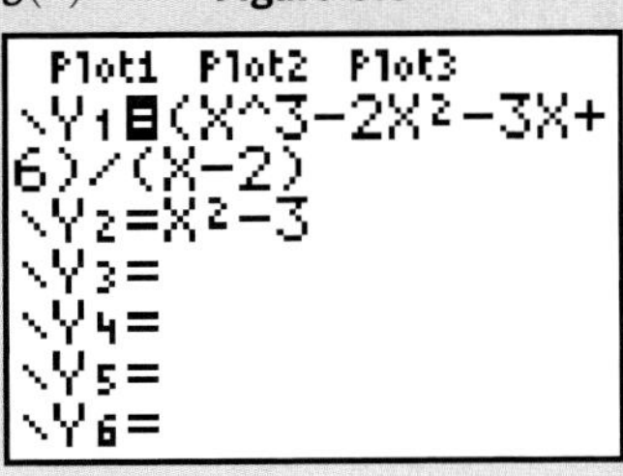

Figure 3.6

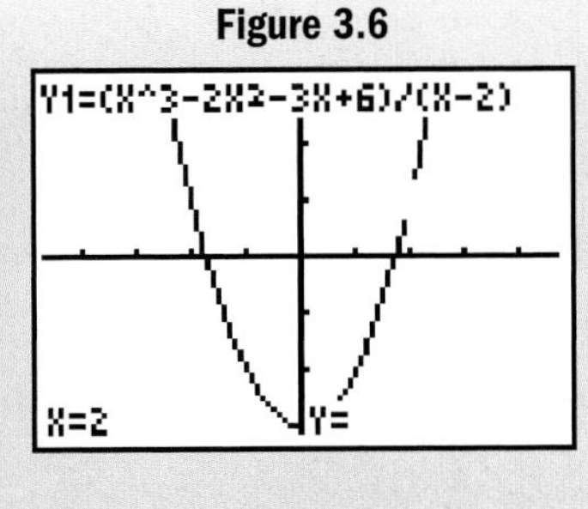

Now enter the result of $\frac{f(x)}{g(x)}$ after simplification as Y_2. After factoring the numerator by grouping and reducing the common factors, we find $h(x) = x^2 - 3$. Graphing both functions reveals that they are identical, except that $Y_2 = x^2 - 3$ *covers the hole left by* Y_1. In other words, Y_1 is equivalent to Y_2 except at $x = 2$. This can also be seen using the TABLE feature of a calculator, which displays an error message for Y_1 when $x = 2$ is input, but shows an output of 1 for Y_2 (Figure 3.7). The bottom line is—the domain of h is all real numbers except $x = 2$.

Figure 3.7

X	Y1	Y2
-2	1	1
-1	-2	-2
0	-3	-3
1	-2	-2
2	ERROR	1
3	6	6
4	13	13

X= -2

For $f(x)$ and $g(x)$ as given, determine the function $h(x)$, where $h(x) = \frac{f(x)}{g(x)}$. Then state the domain of h.

Exercise 1: $f(x) = x^2 - 9$ and $g(x) = x + 3$

Exercise 2: $f(x) = x^3 - 3x^2 - 4x + 12$ and $g(x) = x - 3$

Exercise 3: $f(x) = x^2 + x - 6$ and $g(x) = \sqrt{x + 2}$

Exercise 4: $f(x) = |x|$ and $g(x) = \sqrt{x + 5}$

Exercise 1 $h(x) = x - 3$; $x \in (-\infty, -3) \cup (-3, \infty)$

Exercise 2 $h(x) = x^2 - 4$; $x \in (-\infty, 3) \cup (3, \infty)$

Exercise 3 $h(x) = \frac{x^2 + x - 6}{\sqrt{x + 2}}$; $x \in (-2, \infty)$

Exercise 4 $h(x) = \frac{|x|}{\sqrt{x + 5}}$; $x \in (-5, \infty)$

3.1 EXERCISES

CONCEPTS AND VOCABULARY

Fill in each blank with the appropriate word or phrase. Carefully reread the section, if necessary.

1. Given function f with domain A and function g with domain B, the sum $f(x) + g(x)$ can also be written $\underline{(f + g)(x)}$. The domain of the result is $\underline{A \cap B}$.

2. For the product $h(x) = f(x) \cdot g(x)$, $h(5)$ can be found by evaluating f and g then multiplying the result, or multiplying $f \cdot g$ and evaluating the result. Notationally these are written $\underline{(f \cdot g)(5)}$ and $\underline{f(5) \cdot g(5)}$.

3. When combining functions f and g using basic operations, the domain of the result is the __intersection__ of the domains of f and g. For division, we further stipulate that $\underline{g(x)}$ cannot equal zero.

4. When evaluating functions, if the input value is a function itself, the process is called the __composition__ of functions. The notation $(f \circ g)(x)$ indicates that $\underline{g(x)}$ is the input value for $\underline{f(x)}$, which we can also write as $\underline{f[g(x)]}$.

5. For $f(x) = 2x^3 - 50x$ and $g(x) = x - 5$, discuss/explain why the domain of $h(x) = \left(\frac{f}{g}\right)(x)$ must exclude $x = 5$, even though the resulting quotient is the polynomial $2x^2 + 10x$. Answers will vary.

6. For $f(x) = \sqrt{2x + 7}$ and $g(x) = \frac{2}{x - 1}$, discuss/explain how the domain of $h(x) = (f \circ g)(x)$ is determined. In particular, why is $h(1)$ not defined even though $f(1) = 3$? Answers will vary.

DEVELOPING YOUR SKILLS

Find $h(x)$ as indicated and state the domain of the result.

7. $h(x) = f(x) + g(x)$, where $f(x) = 2x^2 - x - 3$ and $g(x) = x^2 + 5x$
$h(x) = 3x^2 + 4x - 3$; $x \in (-\infty, \infty)$

8. $h(x) = f(x) - g(x)$, where $f(x) = 2x^2 - 18$ and $g(x) = -3x - 7$
$h(x) = 2x^2 + 3x - 11$; $x \in (-\infty, \infty)$

For the functions p and q given and $h(x) = p(x) - q(x)$, find $h(-3)$ two ways: (a) $h(-3) = p(-3) - q(-3)$ and (b) $h(-3) = (p - q)(-3)$. Verify that you obtain the same result each time.

9. $p(x) = 2x^3 + 4x^2 - 7$; $q(x) = 5x + 4x^2$
-46

10. $p(x) = x^2 + 4x - 21$; $q(x) = -2x + 6$
-36

For the functions f and g given and $H(x) = f(x) - g(x)$, find (a) $H(\frac{3}{2})$ and (b) $H(-\frac{1}{2})$; and (c) state the domain of H.

11. $f(x) = \sqrt{4x - 5}$ and $g(x) = 8x^3 + 125$

12. $f(x) = 4x^2 - 2x + 3$ and $g(x) = \sqrt{1 - 4x}$

13. For $f(x) = \sqrt{x}$ and $g(x) = -2$, find $h(x) = f(x) + g(x)$, then graph both f and h on the same grid using a table of values. Comment on what you notice.

14. For $f(x) = |x|$ and $g(x) = 3$, find $h(x) = f(x) - g(x)$, then graph both f and h on the same grid using a table of values. Comment on what you notice.

For the functions f and g given, compute the product $h(x) = (f \cdot g)(x)$ and determine the domain of h.

15. $f(x) = 3x^2 - 2x - 4$ and $g(x) = 2x + 1$

16. $f(x) = x^2 - 2x + 4$ and $g(x) = x + 2$

For the functions p and q given, (a) compute the product $H(x) = (p \cdot q)(x)$, (b) evaluate $H(-2)$ and $H(3)$, and (c) determine the domain of H.

17. $p(x) = \sqrt{x + 5}$ and $q(x) = \sqrt{2 - x}$

18. $p(x) = \sqrt{x + 5}$ and $q(x) = \sqrt{x - 2}$

For the functions f and g given, compute the quotient $h(x) = \left(\frac{f}{g}\right)(x)$ and determine the domain of h.

19. $f(x) = x^3 - 7x^2 + 6x$ and $g(x) = x - 1$

20. $f(x) = x^3 - 1$ and $g(x) = x - 1$

For the functions p and q given, (a) compute the quotient $H(x) = \left(\frac{p}{q}\right)(x)$, (b) evaluate $H(-2)$ and $H(5)$, and (c) determine the domain of H.

21. $p(x) = 2x - 3$ and $q(x) = \sqrt{x^2 - x - 6}$

22. $p(x) = x^2 - 1$ and $q(x) = \sqrt{16 - x^2}$

Find $h(x) = \left(\frac{f}{g}\right)(x)$ and determine the domain of h.

23. $f(x) = x + 1$ and $g(x) = x - 5$

24. $f(x) = x + 3$ and $g(x) = 2x - 7$

25. $f(x) = x - 5$ and $g(x) = \sqrt{x - 2}$

26. $f(x) = x + 1$ and $g(x) = \sqrt{x + 3}$

27. $f(x) = x^2 - 9$ and $g(x) = \sqrt{x + 1}$

28. $f(x) = x^2 - 1$ and $g(x) = \sqrt{x - 3}$

29. $f(x) = x^2 - 16$ and $g(x) = x + 4$

30. $f(x) = x^2 - 49$ and $g(x) = x - 7$

31. $f(x) = x^3 + 4x^2 - 2x - 8$ and $g(x) = x + 4$

32. $f(x) = x^3 - 5x^2 + 2x - 10$ and $g(x) = x - 5$

33. $f(x) = \dfrac{6}{x - 3}$ and $g(x) = \dfrac{2}{x + 2}$

34. $f(x) = \dfrac{4x}{x + 1}$ and $g(x) = \dfrac{2x}{x - 2}$

For each pair of functions f and g given, find the sum, difference, product, and quotient, then determine the domain of each result.

35. $f(x) = 2x + 3$ and $g(x) = x - 2$

36. $f(x) = x - 5$ and $g(x) = 2x - 3$

37. $f(x) = x^2 + 7$ and $g(x) = 3x - 2$

38. $f(x) = x^2 - 3x$ and $g(x) = x + 4$

39. $f(x) = x^2 + 2x - 3$ and $g(x) = x - 1$

40. $f(x) = x^2 - 2x - 15$ and $g(x) = x + 3$

41. $f(x) = 3x + 1$ and $g(x) = \sqrt{x - 3}$

42. $f(x) = x + 2$ and $g(x) = \sqrt{x + 6}$

43. $f(x) = 2x^2$ and $g(x) = \sqrt{x + 1}$

44. $f(x) = x^2 + 2$ and $g(x) = \sqrt{x - 5}$

45. $f(x) = \dfrac{2}{x - 3}$ and $g(x) = \dfrac{5}{x + 2}$

46. $f(x) = \dfrac{4}{x - 3}$ and $g(x) = \dfrac{1}{x + 5}$

---

11. a. -151 b. not defined
c. $x \in [\frac{5}{4}, \infty)$

12. a. not defined b. $5 - \sqrt{3}$
c. $x \in (-\infty, \frac{1}{4}]$

13. $h(x) = \sqrt{x} - 2$; $h(x)$ shifts $f(x)$ down 2 units.

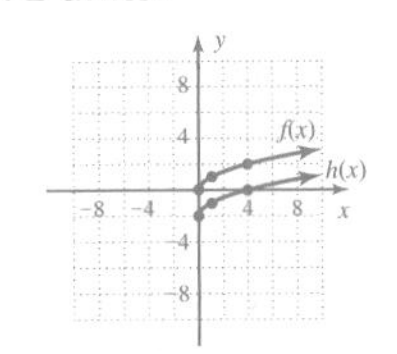

14. $h(x) = |x| - 3$; $h(x)$ shifts $f(x)$ down 3 units.

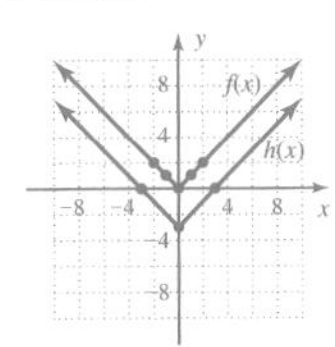

15. $h(x) = 6x^3 - x^2 - 10x - 4$;
$x \in (-\infty, \infty)$

16. $h(x) = x^3 + 8$; $x \in (-\infty, \infty)$

17. a. $H(x) = \sqrt{(x + 5)(2 - x)}$,
b. $2\sqrt{3}$; $H(3)$ is not a real number
c. $[-5, 2]$

18. a. $H(x) = \sqrt{(x + 5)(x - 2)}$,
b. $H(-2)$ is not a real number; $2\sqrt{2}$
c. $x \in [2, \infty)$

19. $h(x) = x^2 - 6x$;
$x \in (-\infty, 1) \cup (1, \infty)$

20. $h(x) = x^2 + x + 1$;
$x \in (-\infty, 1) \cup (1, \infty)$

21. a. $H(x) = \dfrac{2x - 3}{\sqrt{x^2 - x - 6}}$
b. $H(-2)$ is not a real number;
$H(5) = \dfrac{\sqrt{14}}{2}$
c. $x \in (-\infty, -2) \cup (3, \infty)$

22. a. $H(x) = \dfrac{x^2 - 1}{\sqrt{16 - x^2}}$
b. $H(-2) = \dfrac{\sqrt{3}}{2}$; $H(5)$ is not a real number
c. $x \in (-4, 4)$

23. $h(x) = \dfrac{x + 1}{x - 5}$;
$x \in (-\infty, 5) \cup (5, \infty)$

24. $h(x) = \dfrac{x + 3}{2x - 7}$;
$x \in (-\infty, \frac{7}{2}) \cup (\frac{7}{2}, \infty)$

25. $h(x) = \dfrac{x - 5}{\sqrt{x - 2}}$; $x \in (2, \infty)$

Additional answers can be found in the Instructor Answer Appendix.

47. 0; 0; $a^2 - 5a - 14$; $a^2 - 9a$
48. 0; −10; $t^3 - 9t$; $t^3 + 3t^2 - 6t - 8$
49. a. $h(x) = \sqrt{2x - 2}$
b. $H(x) = 2\sqrt{x + 3} - 5$
c. D of $h(x)$: $x \in [1, \infty)$; D of $H(x)$: $x \in [-3, \infty)$
50. a. $h(x) = \sqrt{9 - x^2} + 3$
b. $H(x) = \sqrt{-x^2 - 6x}$
c. D of $h(x)$: $x \in [-3, 3]$; D of $H(x)$: $x \in [-6, 0]$
51. a. $h(x) = \frac{10}{5 + 3x}$ b. $H(x) = \frac{5x + 15}{2x}$
c. D of $h(x)$: $\{x | x \in \mathbb{R}, x \neq 0, x \neq \frac{-5}{3}\}$; D of $H(x)$: $\{x | x \in \mathbb{R}, x \neq -3, x \neq 0\}$
52. a. $h(x) = \frac{-3x + 6}{x}$ b. $H(x) = \frac{3}{2x + 3}$
c. D of $h(x)$: $\{x | x \in \mathbb{R}, x \neq 0, x \neq 2\}$; D of $H(x)$: $\{x | x \in \mathbb{R}, x \neq 0, x \neq \frac{-3}{2}\}$
53. a. $h(x) = x^2 + x - 2$
b. $H(x) = x^2 - 3x + 2$
c. D of $h(x)$: $x \in (-\infty, \infty)$; D of $H(x)$: $x \in (-\infty, \infty)$
54. a. $h(x) = 18x^2 + 24x + 7$
b. $H(x) = 6x^2 - 1$
c. D of $h(x)$: $x \in (-\infty, \infty)$; D of $H(x)$: $x \in (-\infty, \infty)$
55. a. $h(x) = x^2 + 7x + 8$
b. $H(x) = x^2 + x - 1$
c. D of $h(x)$: $x \in (-\infty, \infty)$; D of $H(x)$: $x \in (-\infty, \infty)$
56. a. $h(x) = x^2 - 8x + 14$
b. $H(x) = x^2 - 4x$
c. D of $h(x)$: $x \in (-\infty, \infty)$; D of $H(x)$: $x \in (-\infty, \infty)$
57. a. $h(x) = \sqrt{3x + 1}$
b. $H(x) = 3\sqrt{x - 3} + 4$
c. D of $h(x)$: $x \in [-\frac{1}{3}, \infty)$; D of $H(x)$: $x \in [3, \infty)$
58. a. $h(x) = 2\sqrt{x + 1}$
b. $H(x) = 4\sqrt{x + 5} - 1$
c. D of $h(x)$: $x \in [-1, \infty)$; D of $H(x)$: $x \in [-5, \infty)$
59. a. $h(x) = |-3x + 1| - 5$
b. $H(x) = -3|x| + 16$
c. D of $h(x)$: $x \in (-\infty, \infty)$; D of $H(x)$: $x \in (-\infty, \infty)$
60. a. $h(x) = |3x - 7|$
b. $H(x) = 3|x - 2| - 5$
c. D of $h(x)$: $x \in (-\infty, \infty)$; D of $H(x)$: $x \in (-\infty, \infty)$
61. a. $h(x) = 4x - 20$ b. $H(x) = \frac{x}{4 - 5x}$
c. D of $h(x)$: $\{x | x \in \mathbb{R}, x \neq 5\}$; D of $H(x)$: $\{x | x \in \mathbb{R}, x \neq 0, x \neq \frac{4}{5}\}$
62. a. $h(x) = 3x - 6$ b. $H(x) = \frac{x}{3 - 2x}$
c. D of $h(x)$: $\{x | x \in \mathbb{R}, x \neq 2\}$; D of $H(x)$: $\{x | x \in \mathbb{R}, x \neq 0, x \neq \frac{3}{2}\}$

47. Given $f(x) = x^2 - 5x - 14$, find $f(-2)$, $f(7)$, $f(a)$, and $f(a - 2)$.

48. Given $g(x) = x^3 - 9x$, find $g(-3)$, $g(2)$, $g(t)$, and $g(t + 1)$.

For each pair of functions below find (a) $h(x) = (f \circ g)(x)$ and (b) $H(x) = (g \circ f)(x)$, and (c) determine the domain of each result.

49. $f(x) = \sqrt{x + 3}$ and $g(x) = 2x - 5$

50. $f(x) = x + 3$ and $g(x) = \sqrt{9 - x^2}$

51. $f(x) = \frac{2x}{x + 3}$ and $g(x) = \frac{5}{x}$

52. $f(x) = \frac{-3}{x}$ and $g(x) = \frac{x}{x - 2}$

53. $f(x) = x^2 - 3x$ and $g(x) = x + 2$

54. $f(x) = 2x^2 - 1$ and $g(x) = 3x + 2$

55. $f(x) = x^2 + x - 4$ and $g(x) = x + 3$

56. $f(x) = x^2 - 4x + 2$ and $g(x) = x - 2$

57. $f(x) = \sqrt{x - 3}$ and $g(x) = 3x + 4$

58. $f(x) = \sqrt{x + 5}$ and $g(x) = 4x - 1$

59. $f(x) = |x| - 5$ and $g(x) = -3x + 1$

60. $f(x) = |x - 2|$ and $g(x) = 3x - 5$

61. $f(x) = \frac{4}{x}$ and $g(x) = \frac{1}{x - 5}$

62. $f(x) = \frac{3}{x}$ and $g(x) = \frac{1}{x - 2}$

63. For $f(x) = x^2 - 8$, $g(x) = x + 2$, and $h(x) = (f \circ g)(x)$, find $h(5)$ in two ways:
a. $(f \circ g)(5)$ 41 **b.** $f[g(5)]$ 41

64. For $p(x) = x^2 - 8$, $q(x) = x + 2$, and $H(x) = (p \circ q)(x)$, find $H(-2)$ in two ways:
a. $(p \circ q)(-2)$ −8 **b.** $p[q(-2)]$ −8

WORKING WITH FORMULAS

65. Surface area of a cylinder: $A = 2\pi rh + 2\pi r^2$

If the height of a cylinder is fixed at 20 cm, the formula becomes $A = 40\pi r + 2\pi r^2$. Write this formula in factored form and find two functions $f(r)$ and $g(r)$ such that $A(r) = (f \cdot g)(r)$. Then find $A(5)$ by direct calculation and also by computing the product of $f(5)$ and $g(5)$, then comment on the results. $A = 2\pi r(20 + r)$; $f(r) = 2\pi r$, $g(r) = 20 + r$; $A(5) = 250\pi$ units^2

66. Compound annual growth: $A(r) = P(1 + r)^t$

The amount of money A in a savings account t yr after an initial investment of P dollars depends on the interest rate r. If \$1000 is invested for 5 yr, find $f(r)$ and $g(r)$ such that $A(r) = (f \circ g)(r)$. $g(r) = 1 + r$, $f(r) = 1000r^t$; other answers possible

APPLICATIONS

67. Reading a graph: Use the given graph to find the result of the operations indicated. [*Hint:* Note $f(-4) = 5$ and $g(-4) = -1$.]

a. $(f + g)(-4)$ 4 **b.** $(f \cdot g)(1)$ 0 **c.** $(f - g)(4)$ 2
d. $(f + g)(0)$ 3 **e.** $\left(\frac{f}{g}\right)(3)$ $\frac{2}{3}$ **f.** $(f \cdot g)(-2)$ 6
g. $(g \cdot f)(2)$ −3 **h.** $(f - g)(-1)$ 1 **i.** $(f + g)(8)$ 1
j. $\left(\frac{f}{g}\right)(7)$ undefined **k.** $(f + g)(4)$ 8 **l.** $(f \cdot g)(6)$ −6

68. Reading a graph: The graph given shows the number of sales of cars and trucks from Ullery Used Autos for the years 1994 to 2004. Use the graph to estimate the number of (a) cars sold in 2003; (b) trucks sold in 2003; (c) vehicles sold in 2003, Total $= C(t) + T(t)$; and (d) the difference between the number of cars and trucks sold in 2003, Total $= C(t) - T(t)$. a. 6000 b. 2000 c. 8000 d. 4000

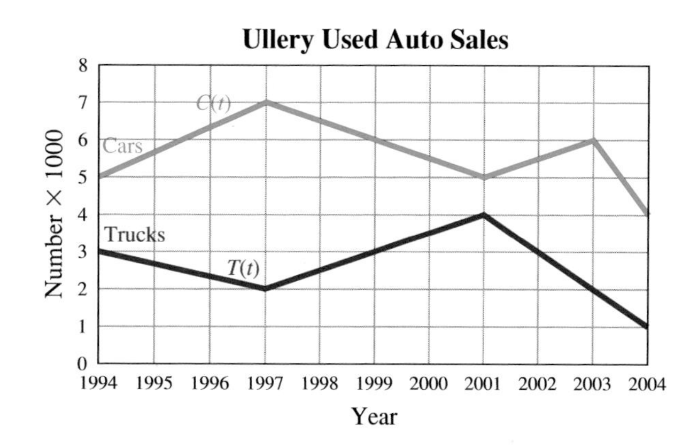

69. Cost, revenue, and profit: Suppose the total cost of manufacturing a certain computer component can be modeled by the function $C(n) = 0.1n^2$, where n is the number of components made and $C(n)$ is in dollars. If each component is sold at a price of \$11.45, the revenue is modeled by $R(n) = 11.45n$. Use this information to complete the following.

a. Find the function that represents the total profit made from sales of the components. $p(n) = 11.45n - 0.1n^2$

b. How much profit is earned if 12 components are made and sold? \$123

c. How much profit is earned if 60 components are made and sold? \$327

d. Explain why the company is making a "negative profit" after the 114th component is made and sold. $C(115) > R(115)$

70. Cost, revenue, and profit: For a certain manufacturer, revenue has been increasing but so has the cost of materials and the cost of employee benefits. Suppose revenue can be modeled by $R(t) = 10\sqrt{t}$, the cost of materials by $M(t) = 2t + 1$, and the cost of benefits by $C(t) = 0.1t^2 + 2$, where t represents the number of months since operations began and outputs are in thousands of dollars. Use this information to complete the following.

a. Find the function that represents the total manufacturing costs. $T(t) = 0.1t^2 + 2t + 3$

b. Find the function that represents how much more the operating costs are than the cost of materials. $D(t) = 0.1t^2 - 2t + 1$

c. What was the total cost of operations in the 10th month after operations began? \$12,000

d. How much less were the operating costs than the cost of materials in the 10th month? \$9,000

e. Find the function that represents the profit earned by this company. $P(t) = 10\sqrt{t} - 0.1t^2 - 2t - 3$

f. Find the amount of profit earned in the 5th month and 10th month. Discuss each result. approximately \$6861 (profit); −\$1377(loss)

71. International shoe sizes: Peering inside her athletic shoes, Morgan notes the following shoe sizes: *US 8.5, UK 6, EUR 40*. The function that relates the U.S. sizes to the European (EUR) sizes is $g(x) = 2x + 23$, while the function that relates EURopean sizes to sizes in the United Kingdom (UK) is $f(x) = 0.5x - 14$. Find the function $h(x)$ that relates the U.S. measurement directly to the UK measurement by finding $h(x) = (f \circ g)(x)$. Find The UK size for a shoe that has a U.S. size of 13. $h(x) = x - 2.5$; 10.5

72. Currency conversion: On a trip to Europe, Megan had to convert American dollars to euros using the function $E(x) = 1.12x$, where x represents the number of dollars and $E(x)$ is the equivalent number of euros. Later, she converts her euros to Japanese yen using the function $Y(x) = 1061x$, where x represents the number of euros and $Y(x)$ represents the equivalent number of yen. (a) Convert 100 U.S. dollars to euros. (b) Convert the answer from part (a) into Japanese yen. (c) Express yen as a function of dollars by finding $M(x) = (Y \circ E)(x)$, then use $M(x)$ to convert 100 dollars directly to yen. Do parts (b) and (c) agree?

Source: 2005 *World Almanac*, p. 231

a. 112 euros **b.** 118,832 yen **c.** $M(x) = 1188.32x$; yes

73. Currency conversion: While traveling in the Far East, Timi must convert U.S. dollars to Thai baht using the function $T(x) = 41.6x$, where x represents the number of dollars and $T(x)$ is the equivalent number of baht. Later she needs to convert her baht to Malaysian ringgit using the function $R(x) = 10.9x$. (a) Convert 100 dollars to baht. (b) Convert the result from part (a) to ringgit. (c) Express ringgit as a function of dollars using $M(x) = (R \circ T)(x)$, then use $M(x)$ to convert 100 dollars to ringgit directly. Do parts (b) and (c) agree?

Source: 2005 *World Almanac*, p. 231 a. 4160 b. 45,344 c. $M(x) = 453.44x$; yes

74. Spread of a fire: Due to a lightning strike, a forest fire begins to burn and is spreading outward in a shape that is roughly circular. The radius of the circle is modeled by the function $r(t) = 2t$, where t is the time in minutes and r is measured in meters. (a) Write a function for the area burned by the fire directly as a function of t by computing $(A \circ r)(t)$. (b) Find the area of the circular burn after 60 min. a. $A(t) = 4\pi t^2$ b. $14{,}400\pi$ m^2

75. a. 6 ft
b. 36π ft^2
c. $A(t) = 9\pi t^2$; yes

76. a. 2.1 Gm
b. 17.64π Gm^2
c. $h(t) = 4.41\pi t^2$; yes

75. Radius of a ripple: As Mark drops firecrackers into a lake one 4th of July, each "pop" caused a circular ripple that expanded with time. The radius of the circle is a function of time t. Suppose the function is $r(t) = 3t$, where t is in seconds and r is in feet. (a) Find the radius of the circle after 2 sec. (b) Find the area of the circle after 2 sec. (c) Express the area as a function of time by finding $A(t) = (A \circ r)(t)$ and use $A(t)$ to find the area of the circle after 2 sec. Do the answers agree?

76. Expanding supernova: The surface area of a star goes through an expansion phase prior to going *supernova*. As the star begins expanding, the radius becomes a function of time. Suppose this function is $r(t) = 1.05t$, where t is in days and $r(t)$ is in gigameters (Gm). (a) Find the radius of the star two days after the expansion phase begins. (b) Find the surface area after two days. (c) Express the surface area as a function of time by finding $h(t) = (S \circ r)(t)$, then use $h(t)$ to compute the surface area after two days directly. Do the answers agree?

77. For $h(x) = (\sqrt{x - 2} + 1)^3 - 5$, find two functions f and g such that $(f \circ g)(x) = h(x)$. Answers may vary.

78. For $H(x) = \sqrt[3]{x^2 - 5} + 2$, find two functions p and q such that $(p \circ q)(x) = h(x)$. Answers may vary.

WRITING, RESEARCH, AND DECISION MAKING

79. a. 1995 to 1996; 1999 to 2004
b. 30; 1995
c. 20 seats; 1997
d. The total number in the senate (50); the number of additional seats held by the majority

79. In a certain country, the function $C(x) = 0.0345x^4 - 0.8996x^3 + 7.5383x^2 - 21.7215x + 40$ approximates the number of Conservatives in the senate for the years 1995 to 2007, where $x = 0$ corresponds to 1995. The function $L(x) = -0.0345x^4 + 0.8996x^3 - 7.5383x^2 + 21.7215x + 10$ gives the number of Liberals for these years. Use this information to answer the following. (a) During what years did the Conservatives control the senate? (b) What was the greatest difference between the number of seats held by each faction in any one year? In what year did this occur? (c) What was the minimum number of seats held by the Conservatives? In what year? (d) Assuming no independent or third-party candidates are elected, what information does the function $T(x) = C(x) + L(x)$ give us? What information does $t(x) = |C(x) - L(x)|$ give us?

80. Given $f(x) = x^3 + 2$ and $g(x) = \sqrt[3]{x - 2}$, graph each function on the same axes by plotting the points that correspond to integer inputs for $x \in [-3, 3]$. Do you notice anything? Next, find $h(x) = (f \circ g)(x)$ and $H(x) = (g \circ f)(x)$. What happened? Look closely at the functions f and g to see how they are related. Can you come up with two additional functions where the same thing occurs? Answers will vary.

81. Given $f(x) = \sqrt{1 - x}$ and $g(x) = \sqrt{x - 2}$, what can you say about the domain of $(f + g)(x)$? Enter the functions as Y_1 and Y_2 on a graphing calculator, then enter $Y_3 = Y_1 + Y_2$. See if you can determine why the calculator gives an error message for Y_3, regardless of the input. Answers will vary.

EXTENDING THE CONCEPT

82. If $f(x) = 1 - \dfrac{1}{x}$, then $f(-n)$ is equal to

a. $f(n)$ **b.** $\dfrac{1}{f(n)}$ **c.** $f\left(\dfrac{-1}{n}\right)$

d. $f\left(\dfrac{1}{n}\right)$ **e.** none of these e

83. Given $f(x) = 2x^2 + 3x + 1$ and $g(x) = 3x - 5$, find

a. $[(f \circ g) \circ f](2)$ 3321

b. $[(f + g) \circ (f \cdot g)](1)$ 212

85. a. b. c.

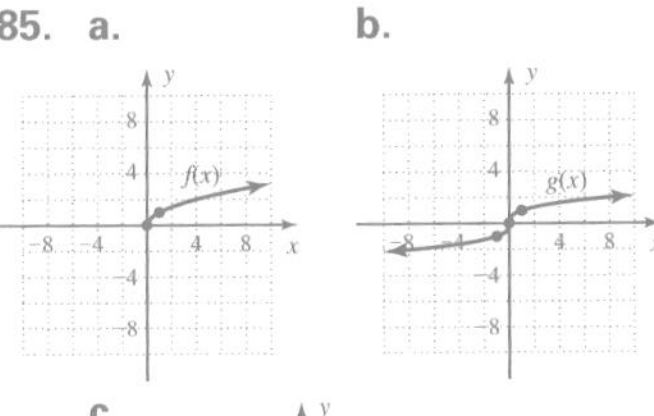

MAINTAINING YOUR SKILLS

84. (1.4) Find the sum and product of the complex numbers $2 + 3i$ and $2 - 3i$. sum 4, product 13

85. (2.4) Draw a sketch of the functions (a) $f(x) = \sqrt{x}$, (b) $g(x) = \sqrt[3]{x}$, and (c) $h(x) = |x|$ *from memory.*

86. (1.5) Use the quadratic formula to solve $2x^2 - 3x + 4 = 0$. $\frac{3}{4} \pm \frac{\sqrt{23}}{4}i$

87. (2.5) Find the domain of the functions $f(x) = \sqrt{4 - x^2}$ and $g(x) = \sqrt{x^2 - 4}$.

88. (R.6) Simplify the following expressions without a calculator: **a.** $27^{-\frac{2}{3}}$ and **b.** $81^{\frac{5}{4}}$. a. $\frac{1}{9}$ b. 243

89. (R.7) Identify the following formulas: **a.** $V = \frac{1}{3}\pi r^2 h$ **b.** $V = \frac{4}{3}\pi r^3$
a. volume of a cone b. volume of a sphere

87. D of $f(x)$: $x \in [-2, 2]$
D of $g(x)$: $x \in (-\infty, -2] \cup [2, \infty)$

3.2 One-to-One and Inverse Functions

LEARNING OBJECTIVES

In Section 3.2 you will learn how to:

A. Identify one-to-one functions

B. Investigate inverse functions using ordered pairs

C. Find inverse functions using an algebraic method

D. Graph a function and its inverse on the same grid

INTRODUCTION

Throughout the algebra sequence, inverse operations are used to solve basic equations. To solve the equation $2x - 3 = -8$ we add 3 to both sides, then divide by 2 since subtraction and addition are inverse operations, as are division and multiplication. In this section, we introduce the idea of an *inverse function,* or one function that "undoes" the operations of another.

POINT OF INTEREST

In the old children's bedtime story, Hansel and Gretel lay out a trail of bread crumbs as they walk into the forest, hoping to eventually follow the trail back home. Although they were foiled in this attempt (birds ate the crumbs), the idea of retracing your steps to get home is a familiar theme. In a related way, an inverse function helps to "find our way back" to the variable, and thereby solve equations.

A. Identifying One-to-One Functions

From our earlier work we know that if every vertical line crosses the graph of a relation in at most one point, the relation is a function. In other words, each first coordinate x must correspond to only one second coordinate y. Consider the graphs of $y = 2x + 3$ and $y = x^2$ given in Figures 3.8 and 3.9, respectively.

Figure 3.8

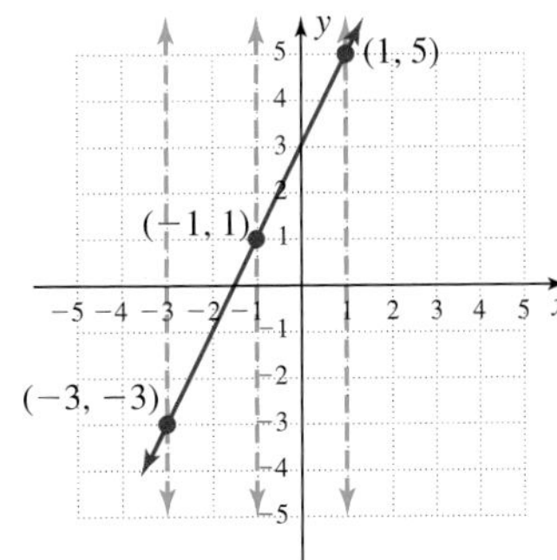

Figure 3.9

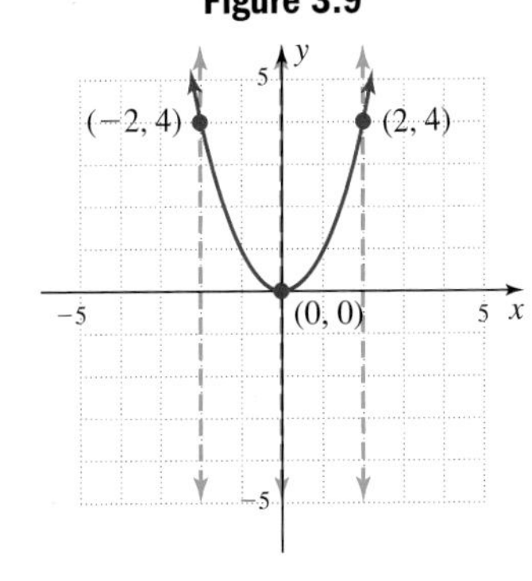

Figure 3.10

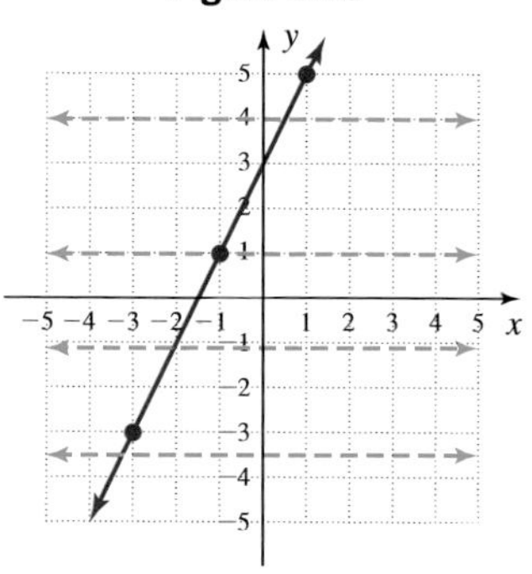

The dashed, vertical lines indicated on each graph clearly show that each x corresponds to only one y. For $y = 2x + 3$, the points $(-3, -3)$, $(-1, 1)$, and $(1, 5)$ are indicated, while for $y = x^2$ we have $(-2, 4)$, $(0, 0)$, and $(2, 4)$. Both are functions, but the points from $y = 2x + 3$ have one characteristic those from $y = x^2$ do not—*each second coordinate y corresponds to a unique first coordinate x.* Note the output "4" from the range of $y = x^2$ corresponds to both -2 and 2 from the domain. If each element from the range of a function corresponds to a unique element of the domain, the function is said to be **one-to-one.** Identifying one-to-one functions is an important part of finding inverse functions.

> **ONE-TO-ONE FUNCTIONS**
> A function f with domain D and range R is said to be one-to-one if no two elements in D correspond to the same element in R:
> If $f(x_1) = f(x_2)$, then $x_1 = x_2$. If $f(x_1) \neq f(x_2)$, then $x_1 \neq x_2$.

Figure 3.11

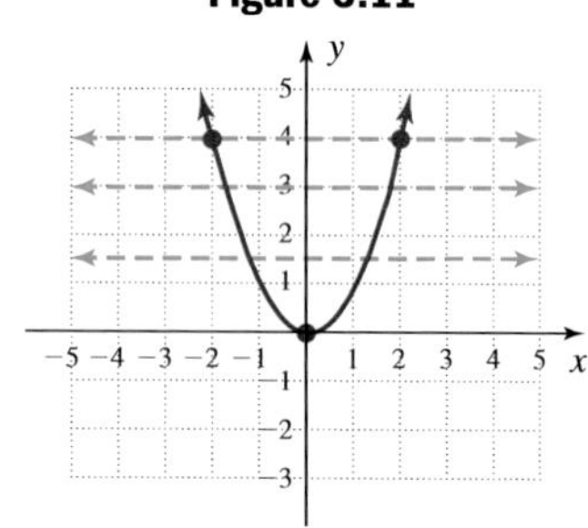

From this definition we conclude the graph of a one-to-one function must not only pass the vertical line test to show that each x corresponds to only one y, it must also pass a **horizontal line test,** to show that each y also corresponds to only one x.

> **HORIZONTAL LINE TEST**
> If every horizontal line intersects the graph of a function in at most one point, the function is one-to-one.

Notice the graph of $y = 2x + 3$ (Figure 3.10) passes the horizontal line test, while the graph of $y = x^2$ (Figure 3.11) does not.

EXAMPLE 1 Use the horizontal line test to determine whether each graph given here is the graph of a one-to-one function.

a.

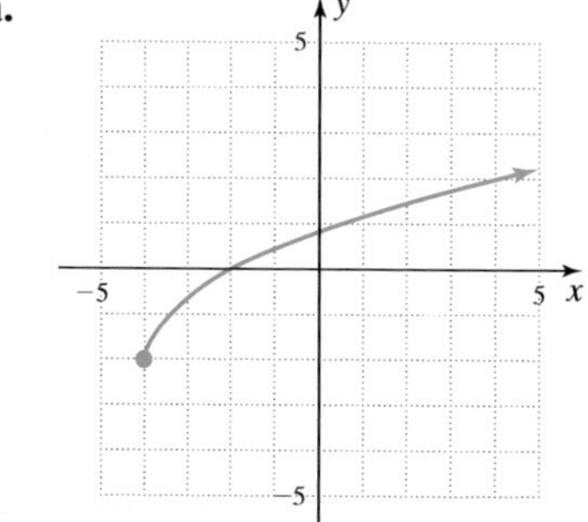

b.

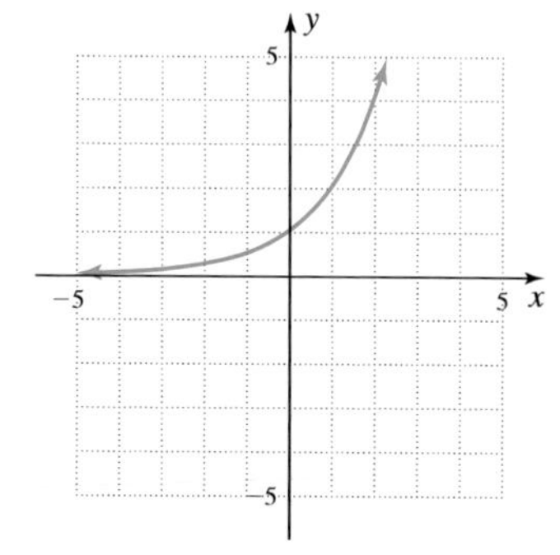

c.

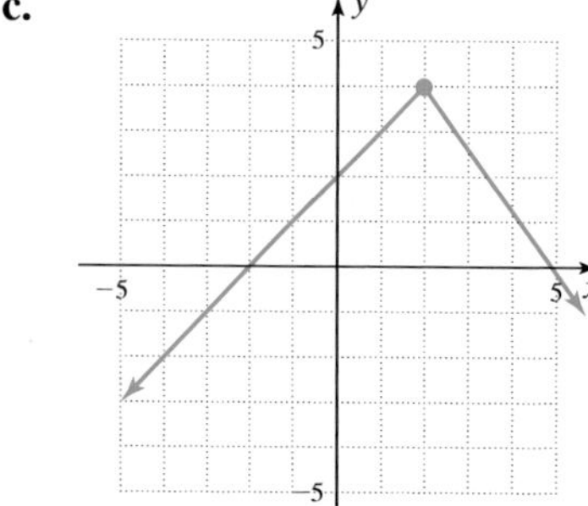

d.

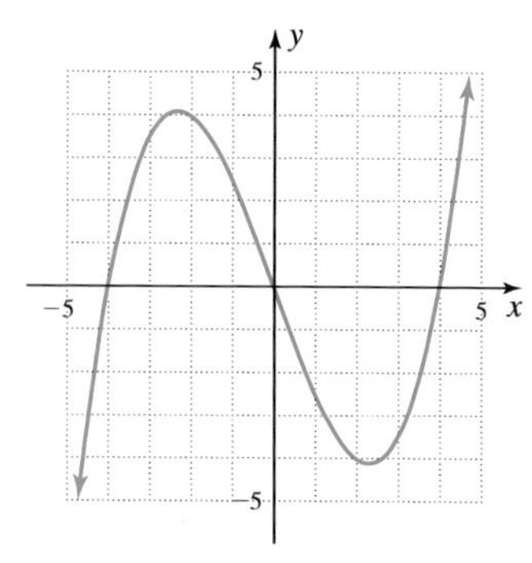

Solution: A careful inspection shows all four graphs depict a function, since each passes the vertical line test. Only (a) and (b) pass the horizontal line test and are *one-to-one* functions.

NOW TRY EXERCISES 7 THROUGH 28

If the function is given in ordered pair form, we simply check to see that no given second coordinate is paired with more than one first coordinate.

B. Inverse Functions and Ordered Pairs

Consider the linear function $f(x) = x + 3$. This function simply "adds 3 to the input value," and some of the ordered pairs generated are $(-7, -4), (-5, -2), (0, 3)$, and $(2, 5)$. On an intuitive level, we might say the inverse of this function would have to "undo" the addition of 3, and $g(x) = x - 3$ is a likely candidate. Some ordered pairs for g are $(-4, -7), (-2, -5), (3, 0)$, and $(5, 2)$. Note that if you interchange the x- and y-coordinates of f, you get exactly the coordinates of the points from g! This shows how the second function "undoes" the operations of the first and vice versa, an observation that will help lead us to the general definition of an **inverse function.** For now, if f is a one-to-one function with ordered pairs (a, b), then the inverse of f is the one-to-one function with ordered pairs of the form (b, a). The inverse function is denoted $f^{-1}(x)$ and is read "f inverse," or "the inverse of f."

CAUTION

The notation $f^{-1}(x)$ is simply a way of denoting an inverse function and has nothing to do with exponential properties. In particular, $f^{-1}(x)$ does *not* mean $\frac{1}{f(x)}$.

It's important to note that if a function is not one-to-one, no inverse function exists since the interchange of x- and y-coordinates will result in a nonfunction. For example, interchanging the coordinates of $(-2, 4)$, $(0, 0)$, and $(2, 4)$ from $y = x^2$ results in $(4, -2)$, $(0, 0)$, and $(4, 2)$, and we have one x-value being mapped to two y-values, in violation of the function definition.

EXAMPLE 2 Find the inverse of each one-to-one function given:

a. $f(x) = \{(-4, 13), (-1, 7), (0, 5), (2, 1), (5, -5), (8, -11)\}$

b. $p(x) = \frac{2}{3}x$

Solution:

a. The inverse function for part (a) can be found by simply interchanging the x- and y-coordinates: $f^{-1}(x) = \{(13, -4), (7, -1), (5, 0), (1, 2), (-5, 5), (-11, 8)\}$.

b. For part (b), we reason the inverse function for p must undo the multiplication of $\frac{2}{3}$ and $q(x) = \frac{3}{2}x$ is a good possibility. Creating a few ordered pairs for p yields $(-3, -2), (-1, -\frac{2}{3}), (0, 0), (2, \frac{4}{3})$, and $(6, 4)$. After interchanging the x- and y-coordinates, we check to see if $(-2, -3), (-\frac{2}{3}, -1), (0, 0), (\frac{4}{3}, 2)$, and $(4, 6)$ satisfy q. Since this is the case, we assume $q(x) = \frac{3}{2}x$ is a likely candidate for $p^{-1}(x)$, deferring a formal proof until later in the section.

NOW TRY EXERCISES 29 THROUGH 40

One important consequence of the relationship between a function and its inverse is that the *domain* of the function becomes the range of the inverse, and the *range* of the function becomes the domain of the inverse (since all input and output values are interchanged). This fact plays an important role in our development of the exponential and logarithmic functions in Chapter 5 and is closely tied to the following definition.

INVERSE FUNCTIONS

Given f is a one-to-one function with domain D and range R, the inverse function $f^{-1}(x)$ has domain R and range D, where $f(x) = y$ implies $f^{-1}(y) = x$, and $f^{-1}(y) = x$ implies $f(x) = y$ for all y in R.

Using this definition, we more clearly see that if $f(-4) = 13$, as in Example 2(a), $f^{-1}(13) = -4$.

C. Finding Inverse Functions Using an Algebraic Method

The fact that interchanging x- and y-values helps determine an inverse function can be generalized to develop an **algebraic method** for finding inverses. Instead of interchanging specific x- and y-values, we actually *interchange the x and y variables*, then solve the equation for y. The process is summarized here.

ALGEBRAIC METHOD FOR FINDING THE INVERSE OF A ONE-TO-ONE FUNCTION

1. Use y instead of $f(x)$.
2. Interchange x and y.
3. Solve the equation for y.
4. The result gives the inverse function: substitute $f^{-1}(x)$ for y.

EXAMPLE 3 Use the algebraic method to find the inverse function for $f(x) = \sqrt[3]{x + 5}$.

Solution:

$f(x) = \sqrt[3]{x+5}$ given function

$y = \sqrt[3]{x+5}$ use y instead of $f(x)$

$x = \sqrt[3]{y+5}$ interchange x and y

$x^3 = y + 5$ cube both sides

$x^3 - 5 = y$ solve for y

$x^3 - 5 = f^{-1}(x)$ the result is $f^{-1}(x)$

For $f(x) = \sqrt[3]{x + 5}$, $f^{-1}(x) = x^3 - 5$.

NOW TRY EXERCISES 41 THROUGH 48

Actually, there is a conclusive way to *prove* that one function is the inverse of another. Just as we generalized the interchange of x- and y-coordinates to develop a method to *find* an inverse function, we can generalize the method we've used to *verify* that we found the inverse function. Consider the result of Example 3, where we saw that for $f(x) = \sqrt[3]{x + 5}$, the inverse function was $f^{-1}(x) = x^3 - 5$. Substituting 3 into f gives 2, and substituting 2 into f^{-1} returns us to 3. Along these same lines, substituting an arbitrary value v in f will yield $\sqrt[3]{v + 5}$, and substituting $\sqrt[3]{v + 5}$ into f^{-1} should return us to v. This indicates that by **composing** f and f^{-1}, we can conclusively verify whether or not one function is the inverse of another.

INVERSE FUNCTIONS

If f is a one-to-one function, then the inverse of f is the function f^{-1} such that $(f \circ f^{-1})(x) = x$ and $(f^{-1} \circ f)(x) = x$.
Note the composition must be verified both ways.

EXAMPLE 4 Use the algebraic method to find the inverse function for $f(x) = \sqrt{x + 2}$. Then verify that you've found the correct inverse.

Solution: From Section 2.4 we know the graph of f is a "one-wing" (square root) function, with domain $x \in [-2, \infty)$ and range $y \in [0, \infty)$. This is important since the *domain and range values will be interchanged for the inverse function*. The domain of f^{-1} will be $x \in [0, \infty)$ and its range $y \in [-2, \infty)$.

$f(x) = \sqrt{x + 2}$	given function; $x \geq -2$
$y = \sqrt{x + 2}$	use y instead of $f(x)$
$x = \sqrt{y + 2}$	interchange x and y
$x^2 = y + 2$	solve for y (square both sides)
$x^2 - 2 = y$	subtract 2
$f^{-1}(x) = x^2 - 2$	the result is $f^{-1}(x)$; D: $x \in [0, \infty)$, R: $y \in [-2, \infty)$

Verify:

$(f \circ f^{-1})(x) = f[f^{-1}(x)]$	$f^{-1}(x)$ is an input for f
$= \sqrt{f^{-1}(x) + 2}$	f adds 2 to inputs, then takes the square root
$= \sqrt{(x^2 - 2) + 2}$	substitute $x^2 - 2$ for $f^{-1}(x)$
$= \sqrt{x^2}$	simplify
$= x$ ✓	since the domain of $f^{-1}(x)$ is $x \in [0, \infty)$

Verify:

$(f^{-1} \circ f)(x) = f^{-1}[f(x)]$	$f(x)$ is an input for f^{-1}
$= [f(x)]^2 - 2$	f^{-1} squares inputs, then subtracts 2
$= [\sqrt{x + 2}]^2 - 2$	$f(x) = \sqrt{x + 2}$
$= x + 2 - 2$	simplify
$= x$ ✓	result

NOW TRY EXERCISES 49 THROUGH 74

D. The Graph of a Function and Its Inverse

When a function and its inverse are graphed on the same grid, an interesting and useful relationship is noted—they are reflections across the line $y = x$ (the identity function). Consider the function $f(x) = 2x + 3$, and its inverse function $f^{-1}(x) = \frac{x - 3}{2} = \frac{1}{2}x - \frac{3}{2}$. The intercepts of f are $(0, 3)$ and $(-\frac{3}{2}, 0)$ and the points $(-4, -5)$ and $(1, 5)$ are on the graph. The intercepts for f^{-1} are $(0, -\frac{3}{2})$ and $(3, 0)$ with both $(-5, -4)$ and $(5, 1)$ on its graph (note the interchange of coordinates once again). When these points are plotted on a coordinate grid (Figure 3.12), we see they are symmetric to the line $y = x$. When both graphs are drawn (Figure 3.13), this relationship is seen even more clearly, with the graphs intersecting on the line of symmetry at $(-3, -3)$.

Figure 3.12

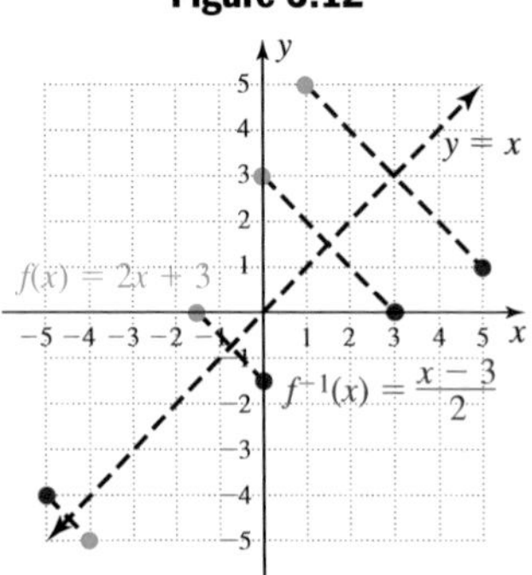

Figure 3.13

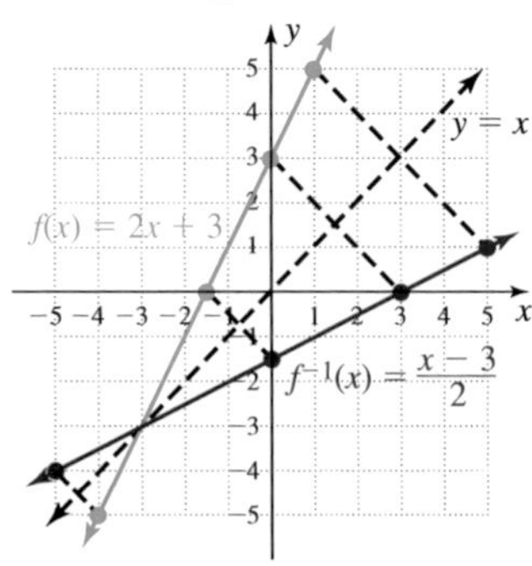

EXAMPLE 5 In Example 4, we found the inverse function for $f(x) = \sqrt{x + 2}$ was $f^{-1}(x) = x^2 - 2, x \geq 0$. Plot these functions on the same grid and comment on how the graphs are related. Also state where they intersect.

Solution: The graph of f is a square root function with the node at $(-2, 0)$, a y-intercept of $(0, \sqrt{2})$, and an x-intercept of $(-2, 0)$ (Figures 3.14 and 3.15 in blue). The graph of $x^2 - 2, x \geq 0$ is the right-hand branch of a parabola, with y-intercept at $(0, -2)$ and an x-intercept at $(\sqrt{2}, 0)$ (Figures 3.14 and 3.15 in red).

Figure 3.14

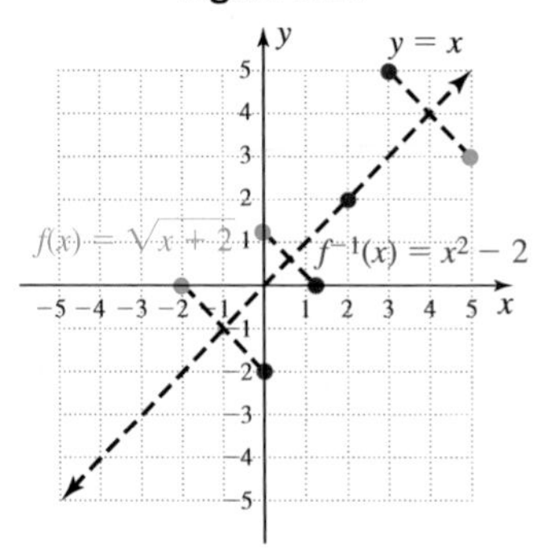

Figure 3.15

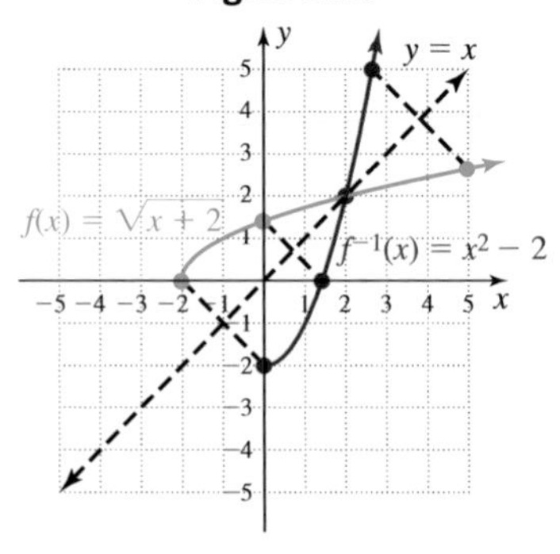

Their graphs are symmetric to the line $y = x$ and intersect on the line of symmetry at $(2, 2)$.

NOW TRY EXERCISES 75 THROUGH 82

EXAMPLE 6 Given the graph shown in Figure 3.16, use the grid in Figure 3.17 to draw a graph of the inverse function.

Figure 3.16

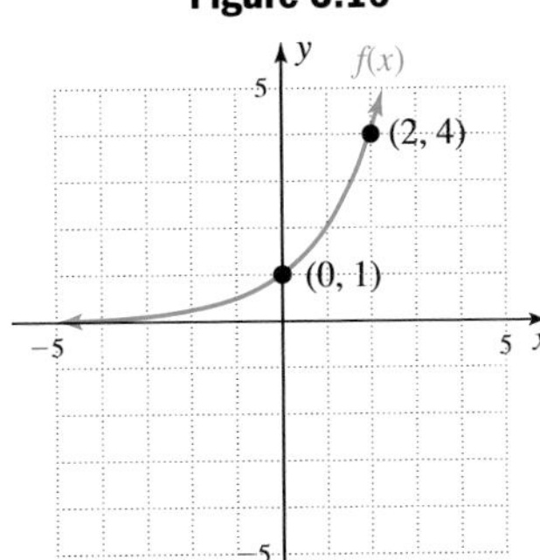

Figure 3.17

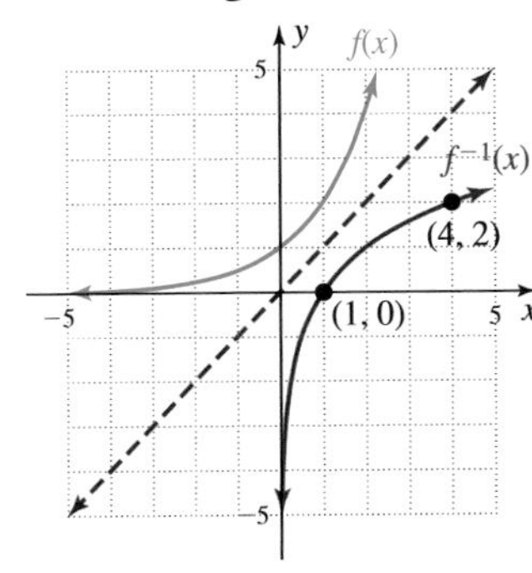

Solution: From the graph, the domain of f appears to be $x \in \mathbb{R}$ and the range is $y \in (0, \infty)$. This means the domain of f^{-1} will be $x \in (0, \infty)$ and the range will be $y \in \mathbb{R}$. The points (0, 1), (1, 2), and (2, 4) seem to be on the graph of f. To sketch f^{-1}, draw the line $y = x$, interchange the x- and y-coordinates of the selected points, and use the domain and range boundaries as a guide. The resulting graph is that of f^{-1} (shown in red).

NOW TRY EXERCISES 83 THROUGH 88

A summary of important points is given here.

FUNCTIONS AND INVERSE FUNCTIONS

1. If a function passes the horizontal line test, it is a one-to-one function.
2. If a function f is one-to-one, the inverse function f^{-1} exists.
3. The domain of f is the range of f^{-1}, and the range of f is the domain of f^{-1}.
4. For a one-to-one function f and its inverse function f^{-1}, $(f \circ f^{-1})(x) = x$ and $(f^{-1} \circ f)(x) = x$.
5. The graphs of f and f^{-1} are symmetric with respect to the line $y = x$.

TECHNOLOGY HIGHLIGHT

Using a Graphing Calculator to Investigate Inverse Functions

The keystrokes shown apply to a TI-84 Plus model. Please consult your manual or our Internet site for other models.

Many of the important points from this section can be illustrated and verified using a graphing calculator. To begin, enter $Y_1 = x^3$ and $Y_2 = \sqrt[3]{x}$ (which are clearly inverse functions) on the Y= screen, then press ZOOM **4:ZDecimal** to graph these equations on a friendly window. The vertical and horizontal propeller functions appear on the screen and seem to be reflections across the line $y = x$ as expected. To verify, use the TABLE feature with the inputs $x = 2$ and

$x = 8$. As illustrated in Figure 3.18, the calculator shows the point (2, 8) is on the graph of Y_1, and the point (8, 2) is on the graph of Y_2. As another check, we can have the calculator locate points of intersection. Recall this is done using the keystrokes 2nd TRACE (CALC) 5, moving the cursor to a location near the desired point of intersection, then pressing ENTER ENTER ENTER. As shown in Figure 3.19, the graphs intersect at (1, 1), which is clearly on the line $y = x$. Finally, we could just return to the Y= screen, enter $Y_3 = x$ and GRAPH all three functions. The line $y = x$ shows a beautiful symmetry between the graphs. Work through the following exercises, then use a graphing calculator to check your results as illustrated in this *Technology Highlight.*

Figure 3.18

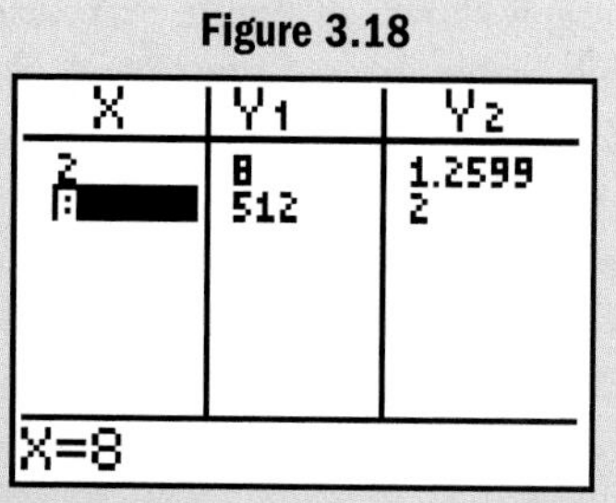

X	Y1	Y2
2	8	1.2599
8	512	2

X=8

Figure 3.19

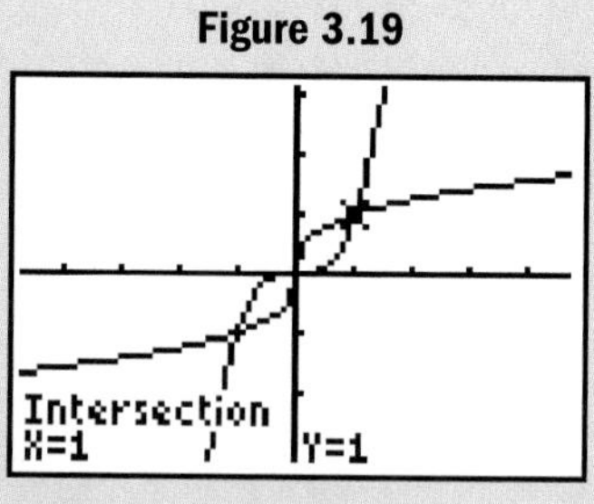

Exercise 1: Given $f(x) = 2x + 1$, find the inverse function $f^{-1}(x)$, then verify they are inverses by (a) using ordered pairs and (b) showing the point of intersection is on the line $y = x$. $f^{-1}(x) = \frac{x-1}{2}$

Exercise 2: Given $f(x) = x^2 + 1$; $x \geq 0$, find the inverse function $f^{-1}(x)$, then verify they are inverses by (a) using ordered pairs and (b) showing each is a reflection of the other across the line $y = x$. $f^{-1}(x) = \sqrt{x-1}$

3.2 EXERCISES

CONCEPTS AND VOCABULARY

Fill in each blank with the appropriate word or phrase. Carefully reread the section if needed.

1. A function is one-to-one if each second coordinate corresponds to exactly one first coordinate.

2. If every horizontal line intersects the graph of a function in at most 1 point, the function is one-to-one.

3. A certain function is defined by the ordered pairs $(-2, -11)$, $(0, -5)$, $(2, 1)$, and $(4, 19)$. The inverse function is $(-11, -2)$, $(-5, 0)$, $(1, 2)$, $(19, 4)$.

4. To find f^{-1} using the algebraic method, we (1) use y instead of $f(x)$, (2) interchange x and y, (3) solve for y and replace y with $f^{-1}(x)$.

5. State true or false and explain why: *To show that g is the inverse function for f, simply show that* $(f \circ g)(x) = x$. Include an example in your response. False, answers will vary.

6. Discuss/explain why no inverse function exists for $f(x) = (x + 3)^2$ and $g(x) = \sqrt{4 - x^2}$. How would the domain of each function have to be restricted to allow for an inverse function? Neither function is one-to-one; one possibility is $f(x) = (x + 3)^2$, $x \geq 0$, and $g(x) = \sqrt{4 - x^2}$, $x \geq 0$.

DEVELOPING YOUR SKILLS

Determine whether each graph given is the graph of a one-to-one function. If not, give examples of how the definition of one-to-oneness is violated.

7. one-to-one
8. not one-to-one, fails horizontal line test, $x = 0$ and $x = -4$ are each paired with $y = 1$
9. one-to-one

7.

8.

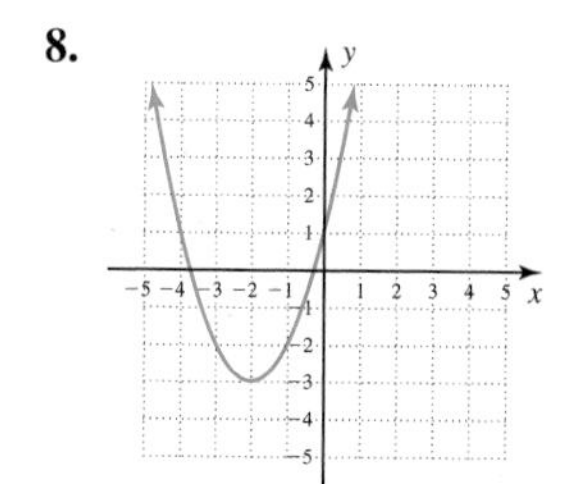

9.

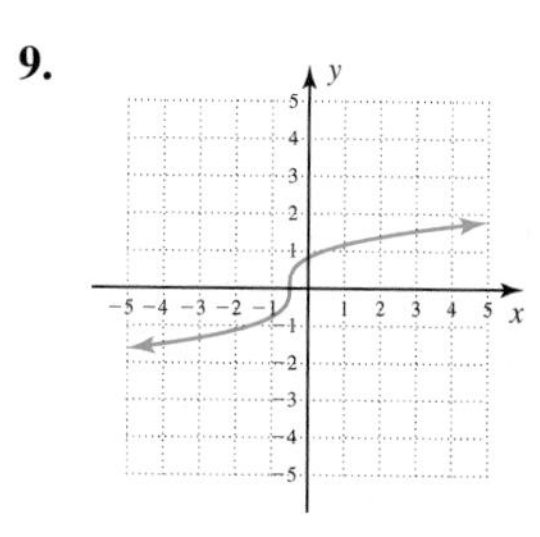

10. one-to-one
11. not a function (cannot be a one-to-one *function*)
12. not one-to-one, fails horizontal line test: $x = -3$, $x = -0.5$ and $x = 2$ are paired with $y = 0$
13. one-to-one
14. not one-to-one, fails horizontal line test: $x = -4$ and $x = 4$ are paired with $y = 0$
15. one-to-one
16. one-to-one
17. not one-to-one, $y = 7$ is paired with $x = -2$ and $x = 2$
18. not one-to-one, $y = 1$ is paired with $x = -6$ and $x = 8$
19. one-to-one
20. one-to-one
21. one-to-one
22. not one-to-one; $h(x) < 3$, corresponds to two x-values
23. not one-to-one; $p(t) > 5$, corresponds to two x-values
24. one-to-one
25. one-to-one
26. not one-to-one; $y = 3$ corresponds to more than one x-value
27. one-to-one
28. one-to-one
29. $f^{-1}(x) = \{(1, -2), (4, -1), (5, 0), (9, 2), (15, 5)\}$
30. $g^{-1}(x) = \{(30, -2), (11, -1), (4, 0), (3, 1), (2, 2)\}$
31. $v^{-1}(x) = \{(3, -4), (2, -3), (1, 0), (0, 5), (-1, 12), (-2, 21), (-3, 32)\}$
32. $w^{-1}(x) = \{(4, -6), (2.5, -5), (-2, -2), (-5, 0), (-9.5, 3), (-11, 4), (-15.5, 7)\}$

33. $f^{-1}(x) = x - 5$
34. $g^{-1}(x) = x + 4$
35. $p^{-1}(x) = \frac{-5}{4}x$
36. $r^{-1}(x) = \frac{4}{3}x$
37. $f^{-1}(x) = \frac{x - 3}{4}$
38. $g^{-1}(x) = \frac{x + 2}{5}$
39. $Y_1^{-1} = x^3 + 4$
40. $Y_2^{-1} = x^3 - 2$

10.

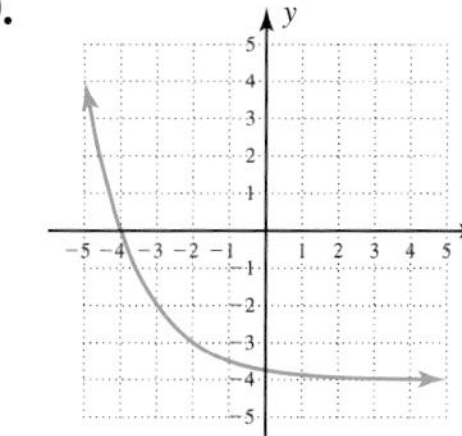

11.

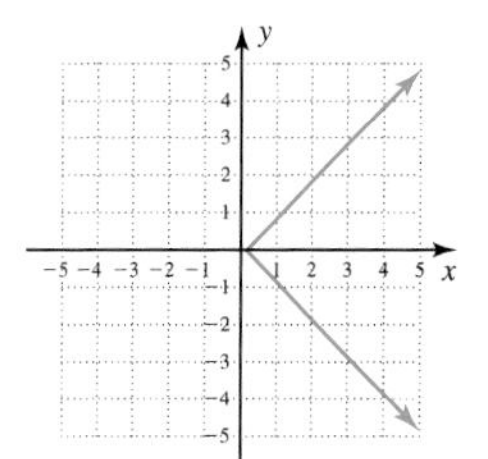

12.

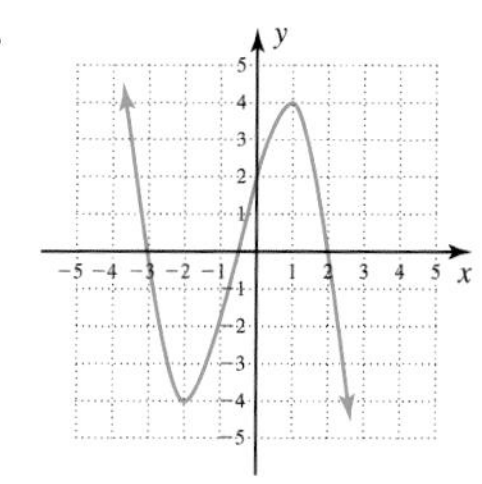

13.

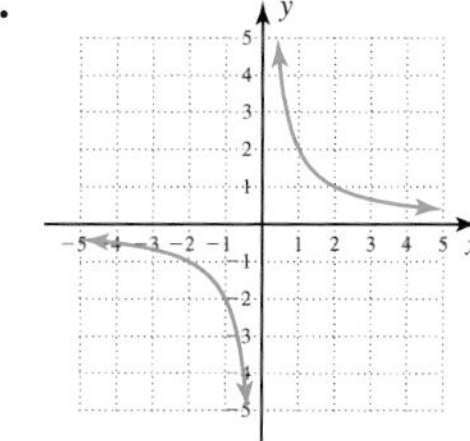

14.

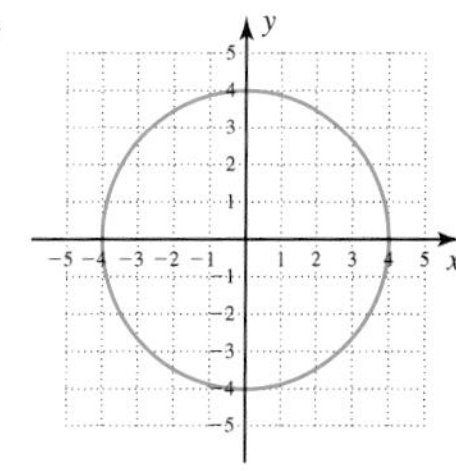

15.

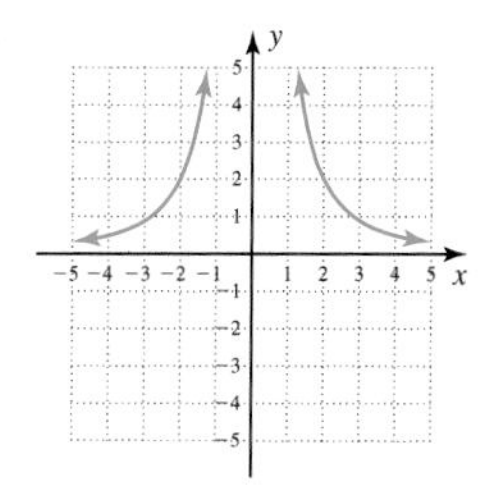

Determine whether the functions given are one-to-one. If not, state why.

16. $\{(-7, 4), (-1, 9), (0, 5), (-2, 1), (5, -5)\}$

17. $\{(9, 1), (-2, 7), (7, 4), (3, 9), (2, 7)\}$

18. $\{(-6, 1), (4, -9), (0, 11), (-2, 7), (-4, 5), (8, 1)\}$

19. $\{(-6, 2), (-3, 7), (8, 0), (12, -1), (2, -3), (1, 3)\}$

Determine if the functions given are one-to-one by noting the function family to which each belongs and mentally picturing the shape of the graph. If a function is not one-to-one, discuss how the definition of one-to-oneness is violated.

20. $f(x) = 3x - 5$

21. $g(x) = (x + 2)^3 - 1$

22. $h(x) = -|x - 4| + 3$

23. $p(t) = 3t^2 + 5$

24. $s(t) = \sqrt{2t - 1} + 5$

25. $r(t) = \sqrt[3]{t + 1} - 2$

26. $y = 3$

27. $y = -2x$

28. $y = x$

For Exercises 29 to 32, find the inverse function of the one-to-one functions given.

29. $f(x) = \{(-2, 1), (-1, 4), (0, 5), (2, 9), (5, 15)\}$

30. $g(x) = \{(-2, 30), (-1, 11), (0, 4), (1, 3), (2, 2)\}$

31. $v(x)$ is defined by the ordered pairs shown.

32. $w(x)$ is defined by the ordered pairs shown.

X	Y_1	
-4	3	
-3	2	
0	1	
5	0	
12	-1	
21	-2	
32	-3	

X=32

X	Y_1	
-6	4	
-5	2.5	
-2	-2	
0	-5	
3	-9.5	
4	-11	
7	-15.5	

X=7

Determine a likely candidate for the inverse function by reasoning and test points.

33. $f(x) = x + 5$

34. $g(x) = x - 4$

35. $p(x) = -\frac{4}{5}x$

36. $r(x) = \frac{3}{4}x$

37. $f(x) = 4x + 3$

38. $g(x) = 5x - 2$

39. $Y_1 = \sqrt[3]{x - 4}$

40. $Y_2 = \sqrt[3]{x + 2}$

Find the inverse of each function given, then compute at least five ordered pairs and check the result. Note that choices of ordered pairs will vary.

41. $f(x) = 2x + 7$ **42.** $f(x) = 5x - 4$ **43.** $f(x) = \sqrt{x - 2}$ **44.** $f(x) = \sqrt{x + 3}$

45. $f(x) = x^2 + 3;\ x \ge 0$ **46.** $f(x) = x^2 - 4;\ x \ge 0$ **47.** $f(x) = x^3 + 1$ **48.** $f(x) = x^3 - 2$

For each function $f(x)$ given, prove (using a composition) that $g(x) = f^{-1}(x)$.

49. $f(x) = -2x + 5,\ g(x) = \dfrac{x - 5}{-2}$

50. $f(x) = 3x - 4,\ g(x) = \dfrac{x + 4}{3}$

51. $f(x) = \sqrt[3]{x + 5},\ g(x) = x^3 - 5$

52. $f(x) = \sqrt[3]{x - 4},\ g(x) = x^3 + 4$

53. $f(x) = \frac{2}{3}x - 6,\ g(x) = \frac{3}{2}x + 9$

54. $f(x) = \frac{4}{5}x + 6,\ g(x) = \frac{5}{4}x - \frac{15}{2}$

55. $f(x) = x^2 - 3;\ x \ge 0,\ g(x) = \sqrt{x + 3}$

56. $f(x) = x^2 + 8;\ x \ge 0,\ g(x) = \sqrt{x - 8}$

Find the inverse of each function $f(x)$ given, then prove (by composition) your inverse function is correct. Note the domain of f is all real numbers.

57. $f(x) = 3x - 5$ **58.** $f(x) = 5x + 4$ **59.** $f(x) = \dfrac{x - 5}{2}$ **60.** $f(x) = \dfrac{x + 4}{3}$

61. $f(x) = \frac{1}{2}x - 3$ **62.** $f(x) = \frac{2}{3}x + 1$ **63.** $f(x) = x^3 + 3$ **64.** $f(x) = x^3 - 4$

65. $f(x) = \sqrt[3]{2x + 1}$ **66.** $f(x) = \sqrt[3]{3x - 2}$ **67.** $f(x) = \dfrac{(x - 1)^3}{8}$ **68.** $f(x) = \dfrac{(x + 3)^3}{-27}$

Find the inverse of each function given, then prove (by composition) your inverse function is correct. Note the implied domain of each function and use it to state any necessary restrictions on the inverse.

69. $f(x) = \sqrt{3x + 2}$ **70.** $g(x) = \sqrt{2x - 5}$ **71.** $p(x) = 2\sqrt{x - 3}$

72. $q(x) = 4\sqrt{x + 1}$ **73.** $v(x) = x^2 + 3;\ x \ge 0$ **74.** $w(x) = x^2 - 1;\ x \ge 0$

Plot each function $f(x)$ and its inverse $f^{-1}(x)$ on the same grid and "dash-in" the line $y = x$. Note how the graphs are related. Then verify the "inverse function" relationship using a composition.

75. $f(x) = 4x + 1;\ f^{-1}(x) = \dfrac{x - 1}{4}$

76. $f(x) = 2x - 7;\ f^{-1}(x) = \dfrac{x + 7}{2}$

77. $f(x) = \sqrt[3]{x + 2};\ f^{-1}(x) = x^3 - 2$

78. $f(x) = \sqrt[3]{x - 7};\ f^{-1}(x) = x^3 + 7$

79. $f(x) = 0.2x + 1;\ f^{-1}(x) = 5x - 5$

80. $f(x) = \dfrac{2}{9}x + 4;\ f^{-1}(x) = \dfrac{9}{2}x - 18$

81. $f(x) = (x + 2)^2;\ x \ge -2;$ $f^{-1}(x) = \sqrt{x} - 2$

82. $f(x) = (x - 3)^2;\ x \ge 3;$ $f^{-1}(x) = \sqrt{x} + 3$

Determine the domain and range for each function whose graph is given, and use this information to state the domain and range of the inverse function. Then sketch in the line $y = x$, estimate the location of two or more points on the graph, and use these to graph $f^{-1}(x)$ on the same grid.

83.

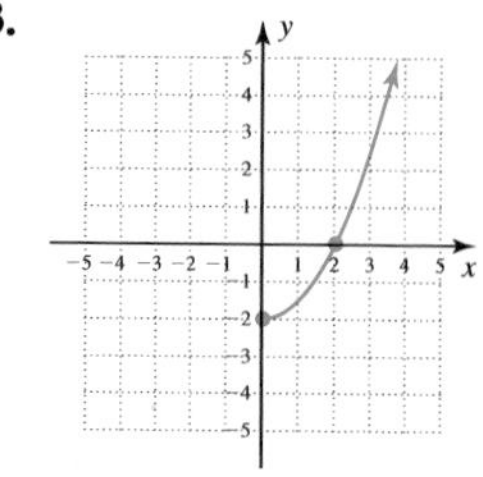

84.

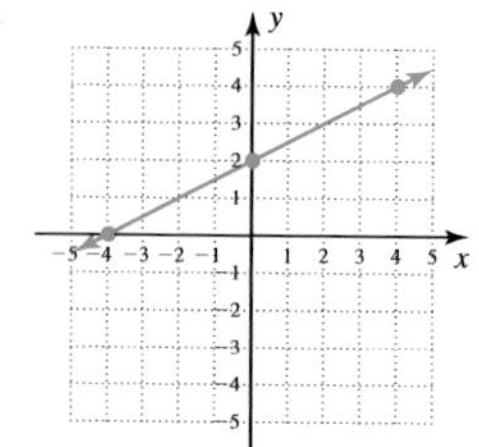

85.

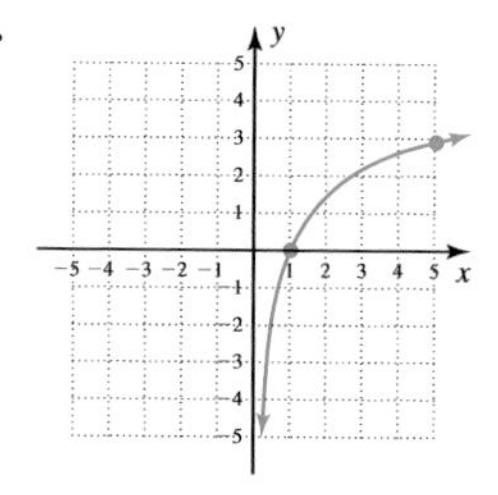

41. $f^{-1}(x) = \dfrac{x - 7}{2}$
42. $f^{-1}(x) = \dfrac{x + 4}{5}$
43. $f^{-1}(x) = x^2 + 2;\ x \ge 0$
44. $f^{-1}(x) = x^2 - 3;\ x \ge 0$
45. $f^{-1}(x) = \sqrt{x - 3};\ x \ge 3$
46. $f^{-1}(x) = \sqrt{x + 4};\ x \ge -4$
47. $f^{-1}(x) = \sqrt[3]{x - 1}$
48. $f^{-1}(x) = \sqrt[3]{x + 2}$
49. $(f \circ g)(x) = x,\ (g \circ f)(x) = x$
50. $(f \circ g)(x) = x,\ (g \circ f)(x) = x$
51. $(f \circ g)(x) = x,\ (g \circ f)(x) = x$
52. $(f \circ g)(x) = x,\ (g \circ f)(x) = x$
53. $(f \circ g)(x) = x,\ (g \circ f)(x) = x$
54. $(f \circ g)(x) = x,\ (g \circ f)(x) = x$
55. $(f \circ g)(x) = x,\ (g \circ f)(x) = x$
56. $(f \circ g)(x) = x,\ (g \circ f)(x) = x$
57. $f^{-1}(x) = \dfrac{x + 5}{3}$
58. $f^{-1}(x) = \dfrac{x - 4}{5}$
59. $f^{-1}(x) = 2x + 5$
60. $f^{-1}(x) = 3x - 4$
61. $f^{-1}(x) = 2x + 6$
62. $f^{-1}(x) = \frac{3}{2}(x - 1)$
63. $f^{-1}(x) = \sqrt[3]{x - 3}$
64. $f^{-1}(x) = \sqrt[3]{x + 4}$
65. $f^{-1}(x) = \dfrac{x^3 - 1}{2}$
66. $f^{-1}(x) = \dfrac{x^3 + 2}{3}$
67. $f^{-1}(x) = 2\sqrt[3]{x} + 1$
68. $f^{-1}(x) = -3\sqrt[3]{x} - 3$
69. $f^{-1}(x) = \dfrac{x^2 - 2}{3},\ x \ge 0$
70. $g^{-1}(x) = \dfrac{x^2 + 5}{2};\ x \ge 0$
71. $p^{-1}(x) = \dfrac{x^2}{4} + 3;\ x \ge 0$
72. $q^{-1}(x) = \dfrac{x^2}{16} - 1;\ x \ge 0$
73. $v^{-1}(x) = \sqrt{x - 3}$
74. $w^{-1}(x) = \sqrt{x + 1}$
75. 76.

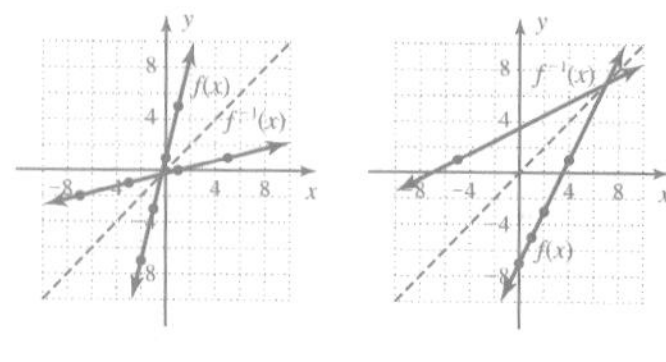

Additional answers can be found in the Instructor Answer Appendix.

86. $D: x \in [-4, 4]$, $R: y \in [-5, 5]$; $D: x \in [-5, 5]$, $R: y \in [-4, 4]$

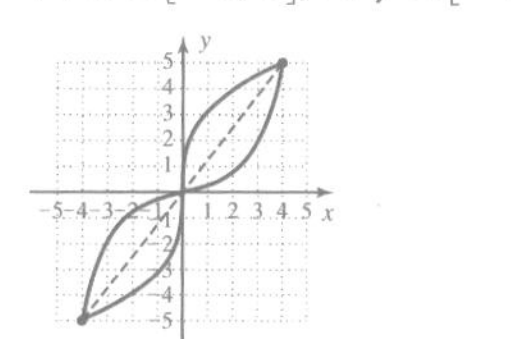

87. $D: x \in (-\infty, 4]$, $R: y \in (-\infty, 4]$; $D: x \in (-\infty, 4]$, $R: y \in (-\infty, 4]$

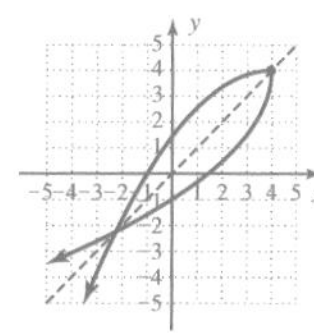

88. $D: x \in (-\infty, \infty)$, $R: y \in (-\infty, \infty)$; $D: x \in (-\infty, \infty)$, $R: y \in (-\infty, \infty)$

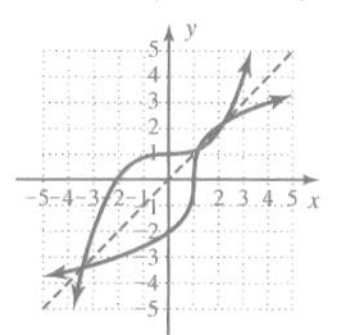

89. a. 31.5 cm b. The result is 80 cm. It gives the distance of the projector from the screen.

90. a. 15 in. b. The result is 14,130 in^3. It gives the volume of the sphere.

91. a. -63.5°F
b. $f^{-1}(x) = \frac{-2}{7}(x - 59)$; it is 35
c. 22,000 ft

92. a. \$220
b. $f^{-1}(x) = \dfrac{x + 560}{12}$
c. 61 mph

86.

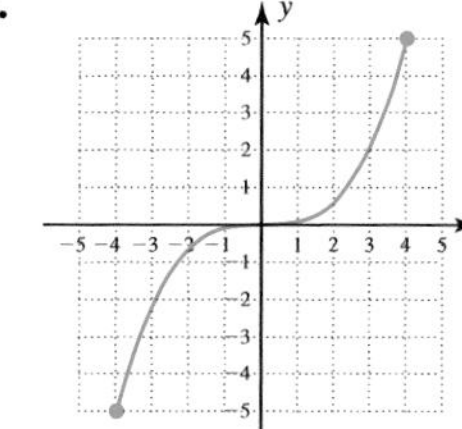

87.

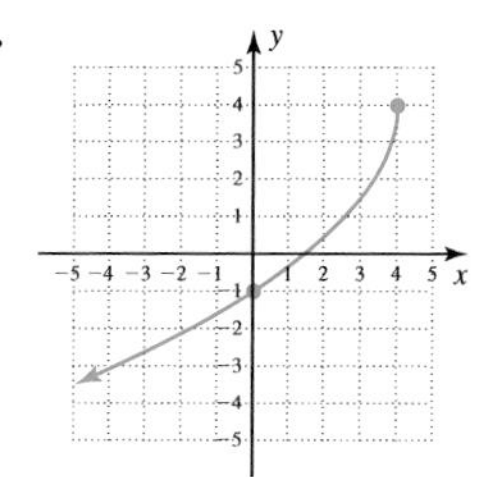

88.

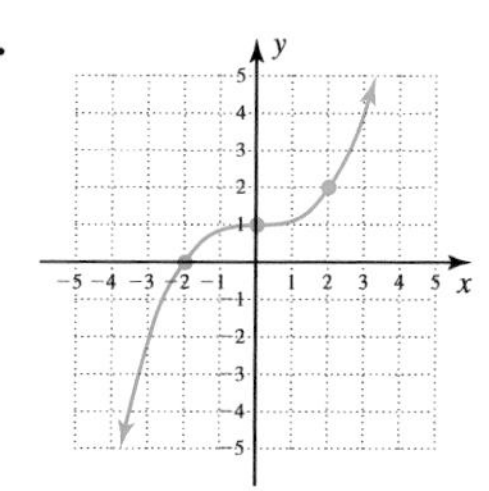

WORKING WITH FORMULAS

89. The height of a projected image: $f(x) = \frac{1}{2}x - 8.5$

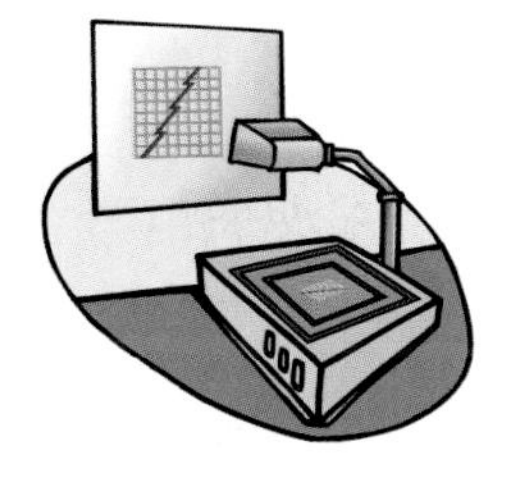

The height of an image projected on a screen by an overhead projector is given by the formula shown, where $f(x)$ represents the actual height of the image on the projector (in centimeters) and x is the distance of the projector from the screen (in centimeters). (a) When the projector is 80 cm from the screen, how large is the image? (b) Show that the inverse function is $f^{-1}(x) = 2x + 17$, then input your answer from part (a) and comment on the result. What information does the inverse function give?

90. The radius of a sphere: $r(x) = \sqrt[3]{\dfrac{3V}{4\pi}}$

In generic form, the radius of a sphere is given by the formula shown, where $r(x)$ represents the radius and V represents the volume of the sphere in cubic units. (a) If a weather balloon that is roughly spherical holds 14,130 in^3 of air, what is the radius of the balloon (use $\pi \approx 3.14$)? (b) Show that the inverse function is $f^{-1}(x) = \frac{4}{3}\pi r^3$, then input your answer from part (a) and comment on the result. What information does the inverse function give?

APPLICATIONS

91. Temperature and altitude: The temperature (in degrees Fahrenheit) at a given altitude can be approximated by the function $f(x) = -\frac{7}{2}x + 59$, where $f(x)$ represents the temperature and x represents the altitude in thousands of feet. (a) What is the approximate temperature at an altitude of 35,000 ft (normal cruising altitude for commercial airliners)? (b) Find $f^{-1}(x)$, then input your answer from part (a) and comment on the result. (c) If the temperature outside a weather balloon is -18°F, what is the approximate altitude of the balloon?

92. Fines for speeding: In some localities, there is a set formula to determine the amount of a fine for exceeding posted speed limits. Suppose the amount of the fine for exceeding a 50 mph speed limit was given by the function $f(x) = 12x - 560$ where $f(x)$ represents the fine in dollars for a speed of x mph. (a) What is the fine for traveling 65 mph through this speed zone? (b) Find $f^{-1}(x)$, then input your answer from part (a) and comment on the result. (c) If a fine of \$172 were assessed, how fast was the driver going through this speed zone?

93. a. 144 ft
b. $f^{-1}(x) = \frac{\sqrt{x}}{4}$, 3 sec, the original input for $f(x)$
c. 7 sec
94. a. 314 ft^2
b. $f^{-1}(r) = \sqrt{\frac{r}{\pi}}$, 10 ft, the original input for $f(x)$
c. 20 ft
95. a. 28,260 ft^3
b. $f^{-1}(h) = \sqrt[3]{\frac{3h}{\pi}}$, 30 ft, the original input for $f(h)$
c. 9 ft
96. a. 10.8 hp
b. $f^{-1}(x) = \sqrt[3]{2500x}$, it is 30 mph, the original input for $f(x)$
c. 40 mph
97. a. 5 cm
b. $f^{-1}(x) = \sqrt[3]{\frac{x}{\pi}}$, $f^{-1}(392.5) \approx 5$; same as original input for $f(x)$
c. $f^{-1}(x)$
98. Answers will vary.
99. a. verified
b.

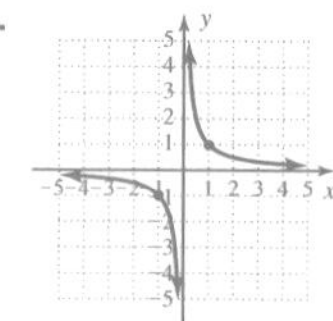

c. (1, 1) and (−1, −1); x and y coordinates are identical on $f(x) = x$

93. Effect of gravity: Due to the effect of gravity, the distance an object has fallen after being dropped is given by the function $f(x) = 16x^2$; $x \geq 0$, where $f(x)$ represents the distance in feet after x sec. (a) How far has the object fallen 3 sec after it has been dropped? (b) Find $f^{-1}(x)$, then input your answer from part (a) and comment on the result. (c) If the object is dropped from a height of 784 ft, how many seconds until it hits the ground (stops falling)?

94. Area and radius: In generic form, the area of a circle is given by $f(r) = \pi r^2$, where $f(r)$ represents the area in square units for a circle with radius r. (a) A pet dog is tethered to a stake in the backyard. If the tether is 10 ft long, how much area does the dog have to roam (use $\pi \approx 3.14$)? (b) Find $f^{-1}(r)$, then input your answer from part (a) and comment on the result. (c) If the owners want to allow the dog 1256 ft^2 of area to live and roam, how long a tether should be used?

95. Volume of a cone: In generic form, the volume of an equipoise cone (height equal to radius) is given by $f(h) = \frac{1}{3}\pi h^3$, where $f(h)$ represents the volume in units^3 and h represents the height of the cone. (a) Find the volume of such a cone if $r = 30$ ft (use $\pi \approx 3.14$). (b) Find $f^{-1}(h)$, then input your answer from part (a) and comment on the result. (c) If the volume of water in the cone is 763.02 ft^3, how deep is the water at its deepest point?

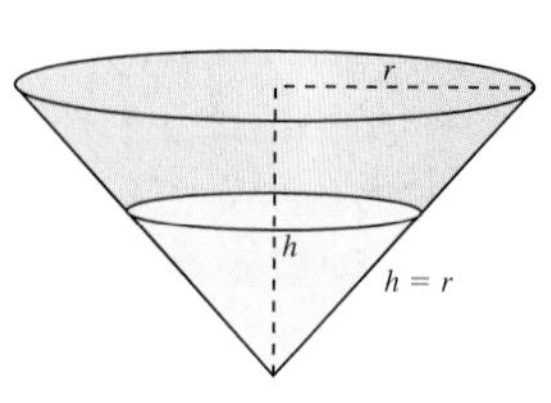

96. Wind power: The power delivered by a certain wind-powered generator can be modeled by the function $f(x) = \frac{x^3}{2500}$, where $f(x)$ is the horsepower (hp) delivered by the generator and x represents the speed of the wind in miles per hour. (a) Use the model to determine how much horsepower is generated by a 30 mph wind. (b) The person monitoring the output of the generators (wind generators are usually erected in large numbers) would like a function that gives the wind speed based on the horsepower readings on the gauges in the monitoring station. For this purpose, find $f^{-1}(x)$. Check your work by using your answer from part (a) as an input in $f^{-1}(x)$ and comment on the result. (c) If gauges show 25.6 hp is being generated, how fast is the wind blowing?

WRITING, RESEARCH, AND DECISION MAKING

97. The volume of an equipoise cylinder (height equal to radius) is given by $f(x) = \pi x^3$, where $f(x)$ represents the volume and x represents the height (or radius) of the cylinder. (a) What radius is needed to produce cans with volume $V = 392.5$ cm^3 (use $\pi \approx 3.14$)? (b) If a can manufacturer will be producing many different sizes based on customer need, would it make more sense to give the metalworkers a formula for the radius required based on required volume? Find $f^{-1}(x)$ to produce this formula, then input $V = 392.5$ cm^3 and comment on the result. (c) Which formula found the radius more efficiently?

98. Inverse functions can be illustrated in very practical ways by retracing sequences we use everyday. Consider this sequence: (a) leave house, (b) take keys from pocket, (c) open car door, (d) get into driver's seat, (e) insert keys in ignition, (f) turn car on, and (g) drive to work. To get back into your home after work would require that you "undo" each step in this sequence and in reverse order. Think of another everyday sequence that has at least three steps, and list what must be done to undo each step. Comment on how this might relate to finding the inverse function for $f(x) = 2x^2 - 3$.

99. The function $f(x) = \frac{1}{x}$ is one of the few functions that is its own inverse. This means the ordered pairs (a, b) and (b, a) must satisfy both f and f^{-1}. (a) Find f^{-1} using the algebraic method to verify that $f(x) = f^{-1}(x) = \frac{1}{x}$. (b) Graph the function $f(x) = \frac{1}{x}$ using a table of integers from -4 to 4. Note that for any ordered pair (a, b) on f, the ordered pair (b, a) is also on f. (c) State where the graph of $y = x$ will intersect the graph of this function and discuss why.

EXTENDING THE CONCEPT

100. Which of the following is the inverse function for $f(x) = \frac{2}{3}\left(x - \frac{1}{2}\right)^5 + \frac{4}{5}$?

a. $\sqrt[5]{\frac{1}{2}\left(x - \frac{2}{3}\right)} - \frac{4}{5}$ **b.** $\frac{3}{2}\sqrt[5]{(x - 2)} - \frac{5}{4}$

c. $\frac{3}{2}\sqrt[5]{\left(x + \frac{1}{2}\right)} - \frac{5}{4}$ **d.** $\sqrt[5]{\frac{3}{2}\left(x - \frac{4}{5}\right)} + \frac{1}{2}$

101. Suppose a function is defined as $f(x) =$ *the exponent that goes on 9 to obtain x*. For example, $f(81) = 2$ since 2 is the exponent that goes on 9 to obtain 81, and $f(3) = \frac{1}{2}$ since $\frac{1}{2}$ is the exponent that goes on 9 to obtain 3. Determine the value of each of the following:

a. $f(1)$ **b.** $f(729)$ **c.** $f^{-1}(2)$ **d.** $f^{-1}\left(\frac{1}{2}\right)$

MAINTAINING YOUR SKILLS

102. (2.5) Given $f(x) = x^2 - x - 2$, solve the inequality $f(x) \leq 0$ using the x-intercepts and concavity of the graph.

103. (2.4) For the function $y = 2\sqrt{x + 3}$, find the average rate of change between $x = 1$ and $x = 2$, and between $x = 4$ and $x = 5$. Which is greater? Why?

104. (R.7) Write as many of the following formulas as you can from memory:

a. perimeter of a rectangle **b.** area of a circle
c. volume of a cylinder **d.** volume of a cone
e. circumference of a circle **f.** area of a triangle
g. area of a trapezoid **h.** volume of a sphere
i. Pythagorean theorem

105. (1.5) Find the x-intercepts using the quadratic formula. Give results in both exact and approximate form: $f(x) = x^2 - 4x - 41$.

106. (1.3) Solve the following cubic equations by factoring:

a. $x^3 - 5x = 0$
b. $x^3 - 7x^2 - 4x + 28 = 0$
c. $x^3 - 3x^2 = 0$
d. $x^3 - 3x^2 - 4x = 0$

107. (2.6) The percentage of the U.S. population that can be categorized as living in *Pacific coastal areas* has been growing steadily for decades, as indicated by the data given for selected years. Using the data with $t = 0$ corresponding to 1970,

a. Draw the scatterplot, scaling the axes to comfortably fit the data.

b. Decide on an appropriate form of regression, find the regression equation, and comment on the strength of the correlation.

c. Use the equation model to predict the percentage of the population living in Pacific coastal areas in 2005 and 2010.

Source: 2004 *Statistical Abstract of the United States,* Table 23

Year	%
1970	22.8
1980	27.0
1990	33.2
1995	35.2
2000	37.8
2001	38.5
2002	38.9
2003	39.4

100. d
101. a. 0 b. 3 c. 81 d. 3
102. $x \in [-1, 2]$
103. ≈ 0.472, ≈ 0.365; rate of change is greater in [1, 2] due to shape of the one-wing function.
104. a. $P = 2l + 2w$
b. $A = \pi r^2$
c. $V = \pi r^2 h$
d. $V = \frac{1}{3}\pi r^2 h$
e. $C = 2\pi r$
f. $A = \frac{1}{2}bh$
g. $A = \frac{1}{2}(b_1 + b_2)h$
h. $V = \frac{4}{3}\pi r^3$
i. $a^2 + b^2 = c^2$
105. $x = 2 \pm 3\sqrt{5}$, $x = 8.71$, $x = -4.71$
106. a. $x = 0, x = \pm\sqrt{5}$
b. $x = 2, x = -2, x = 7$
c. $x = 0, x = 3$
d. $x = 0, x = 4, x = -1$
107. a.

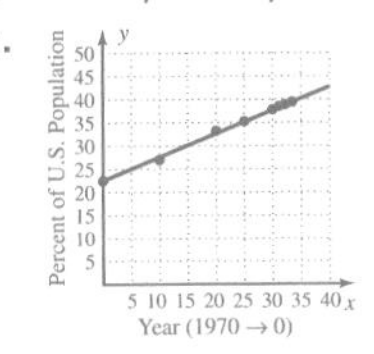

b. $P(t) = 0.51t + 22.51$, very strong
c. 2005: 40.4%, 2010: 43%

3.3 Toolbox Functions and Transformations

LEARNING OBJECTIVES

In Section 3.3 you will learn how to:

A. Perform vertical and horizontal shifts of a basic graph

B. Perform vertical and horizontal reflections of a basic graph

C. Perform stretches and compressions on a basic graph

D. Transform the graph of a general function $f(x)$

E. Compute the area bounded by a basic graph

INTRODUCTION

In our study of functions in Chapter 2, we introduced the following toolbox functions: linear, absolute value, quadratic, square root, cubic, and cube root. In this section, we'll explore these functions further in an effort to effectively apply them in context and to acquire additional skills for use in future chapters. The basic vehicle for this will be the **transformation** of a graph, in which the graph retains all of its characteristic features, while being transposed (moved) or transfigured (morphed) in various ways. Earlier, the concept of *average rate of change* was applied to these functions, where we noted the rate of change varied widely among them—just as rates of change vary widely in the real world. Here we'll study the *area bounded by these graphs* (which also varies widely), an important concept linked to many applications of mathematics.

POINT OF INTEREST

The shift of a basic graph is also called a **translation** of the graph. The word *translate* is of Latin origin with the prefix *trans* meaning "to travel across or beyond." The second syllable is from the word *latus,* meaning "to be carried." In language, you carry over the meaning of words from one language to another. In graphing, you carry over the graph from one position to another.

A. Vertical and Horizontal Shifts

In preparation for the new concepts in this section, basic facts related to each toolbox function should be reviewed carefully. Central to this review is the graph of each function, along with its domain and range. See section 2.4 and the inside back cover of the text.

Previously we've noted the graph of any function from a given family maintains the same general shape. The graphs of $y = -2x^2 - 5x + 3$ and $y = x^2$ are both parabolas, the graphs of $y = \sqrt[3]{x}$ and $y = -\sqrt[3]{x-2} + 1$ are both "horizontal propellers," and so on for the other functions. Once you're aware of the main features of a basic function, you can graph any function from that family using far fewer points, and analyze the graph more efficiently. As we study specific transformations of a graph, it's important to develop a *global view of the transformations,* as they can be applied to virtually any function (see Example 8).

Vertical Translations

We've already glimpsed a vertical translation in our study of the *algebra of functions* (Section 3.1, Example 2). Here we'll investigate the idea more thoroughly using the absolute value function family.

EXAMPLE 1 Construct a table of values for $f(x) = |x|$, $g(x) = |x| + 1$, and $h(x) = |x| - 3$ and graph the functions on the same coordinate grid. Discuss what you observe.

Solution: A table of values for all three functions is shown here, with the corresponding graphs shown in the figure.

x	f(x) = |x|	g(x) = |x| + 1	h(x) = |x| − 3
−3	3	4	0
−2	2	3	−1
−1	1	2	−2
0	0	1	−3
1	1	2	−2
2	2	3	−1
3	3	4	0

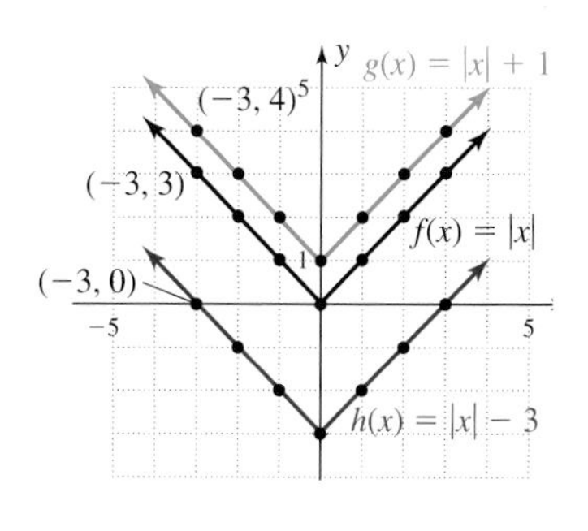

From the table we note that outputs of $g(x)$ are one more than the outputs for $f(x)$, and that each point on the graph of f has been shifted *upward 1 unit* to form the graph of g. Similarly, each point on the graph of f has been shifted *downward 3 units* to form the graph of h: $h(x) = f(x) - 3$.

NOW TRY EXERCISES 7 THROUGH 18

We describe the transformations in Example 1 as **vertical shifts** or the **vertical translation** of a basic graph. The graph of g is the same as the graph of f but *shifted up 1 unit,* and the graph of h is the same as f but *shifted down 3 units*. In general, we have the following:

VERTICAL TRANSLATIONS OF A BASIC GRAPH

Given any function whose graph is determined by $y = f(x)$ and $k > 0$,

1. The graph of $y = f(x) + k$ is the graph of $f(x)$ shifted upward k units.
2. The graph of $y = f(x) - k$ is the graph of $f(x)$ shifted downward k units.

Horizontal Translations

The graph of a parent function can also be shifted left or right. This happens when we *alter the inputs to the basic function,* as opposed to adding or subtracting something to the basic function. For $Y_1 = x^2 + 2$ it's clear that we first square inputs, then add 2, which results in a vertical shift. For $Y_2 = (x + 2)^2$, we add 2 to x *prior to squaring* and since the input values are affected, we might anticipate the graph will shift along the x-axis—horizontally.

EXAMPLE 2 Construct a table of values for $f(x) = x^2$ and $g(x) = (x + 2)^2$, then graph the functions on the same grid and discuss what you observe.

Solution: Both f and g belong to the quadratic family and their graphs will be parabolas. A table of values is shown along with the corresponding graphs.

x	f(x) = x²	g(x) = (x + 2)²
−3	9	1
−2	4	0
−1	1	1
0	0	4
1	1	9
2	4	16
3	9	25

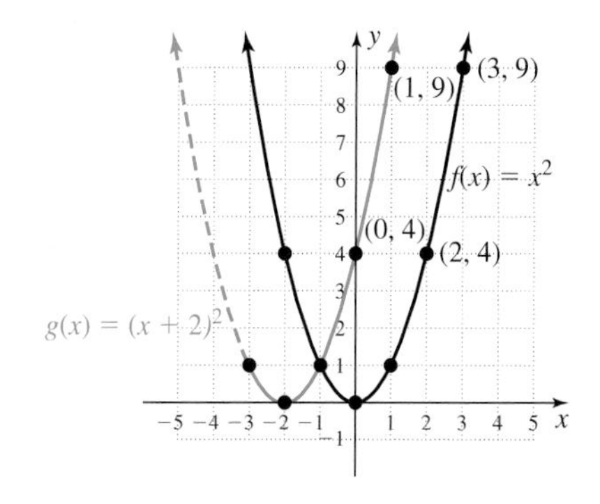

It is apparent the graphs of g and f are identical, but the graph of g has been shifted horizontally 2 units left (the left branch of g can be completed using additional inputs or by simply completing the parabola).

NOW TRY EXERCISES 19 THROUGH 22 ▶

We describe the transformation in Example 2 as a **horizontal shift** or **horizontal translation** of a basic graph. The graph of g is the same as that of f, but *shifted 2 units to the left*. Once again it seems reasonable that since *input* values were altered, the shift must be horizontal rather than vertical. From this example, we also learn the direction of the shift is **opposite the sign:** $y = (x + 2)^2$ is 2 units *to the left* of $y = x^2$. Although it may seem counterintuitive, the shift *opposite the sign* can be "seen" by locating the new x-intercept, which in this case is also the vertex. Substituting 0 for y gives $0 = (x + 2)^2$ with $x = -2$, as shown in the graph in Example 2. In general, we have

HORIZONTAL TRANSLATIONS OF A BASIC GRAPH

Given any function whose graph is determined by $y = f(x)$ and $h > 0$,

1. The graph of $y = f(x + h)$ is the graph of $f(x)$ shifted *to the left h* units.
2. The graph of $y = f(x - h)$ is the graph of $f(x)$ shifted *to the right h* units.

EXAMPLE 3 ▶ Sketch the graphs of $g(x) = |x - 2|$ and $h(x) = \sqrt{x + 3}$ using a horizontal shift of the parent function and a few characteristic points (not a table of values).

Solution: ▶ The graph of $g(x) = |x - 2|$ (Figure 3.20) is a basic "V" function shifted 2 units to the right (shift the vertex and two other points from $y = |x|$). The graph of $h(x) = \sqrt{x + 3}$ (Figure 3.21) is a "one-wing" function, shifted 3 units to the left (shift the node and one or two points from $y = \sqrt{x}$).

Figure 3.20

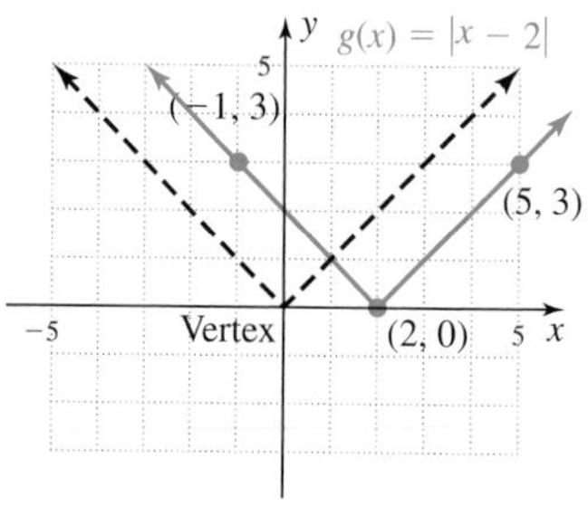

Figure 3.21

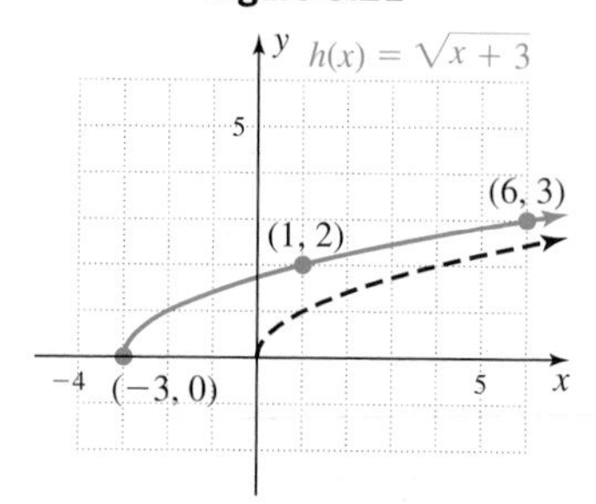

NOW TRY EXERCISES 23 THROUGH 26 ▶

B. Vertical and Horizontal Reflections

The next transformation we investigate is called a **vertical reflection,** in which we compare the function $Y_1 = f(x)$ with the negative of the function: $Y_2 = -f(x)$.

Vertical Reflections

EXAMPLE 4 ▶ Construct a table of values for $Y_1 = x^2$ and $Y_2 = -x^2$, then graph the functions on the same grid and discuss what you observe.

Solution: A table of values is given for both functions, along with the corresponding graphs.

x	$Y_1 = x^2$	$Y_2 = -x^2$
−2	4	−4
−1	1	−1
0	0	0
1	1	−1
2	4	−4

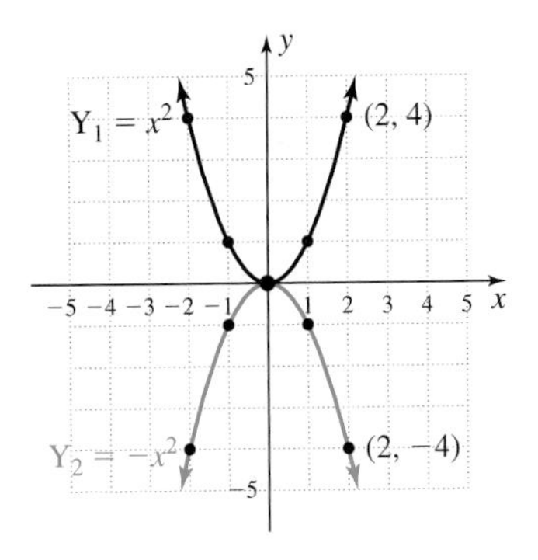

As you might have anticipated, the outputs for f and g differ only in sign. Each output is a **reflection** of the other, being an equal distance from the x-axis but on opposite sides. **NOW TRY EXERCISES 27 AND 28**

The vertical reflection in Example 4 is sometimes called a **reflection across the *x*-axis** or a **north/south reflection.** In general,

VERTICAL REFLECTIONS OF A BASIC GRAPH
Given any function whose graph is determined by $y = f(x)$, the graph of $y = -f(x)$ is the graph of $f(x)$ reflected across the x-axis.

It's also possible for a graph to be reflected horizontally *across the y-axis.* Just as we noted that $f(x)$ versus $-f(x)$ resulted in a vertical reflection, $f(x)$ versus $f(-x)$ results in a horizontal reflection.

Horizontal Reflections

EXAMPLE 5 Construct a table of values for $f(x) = \sqrt{x}$ and $g(x) = \sqrt{-x}$, then graph the functions on the same coordinate grid and discuss what you observe.

Solution: A table of values is given here, along with the corresponding graphs.

x	$f(x) = \sqrt{x}$	$g(x) = \sqrt{-x}$
−4	not real	2
−2	not real	$\sqrt{2} \approx 1.41$
−1	not real	1
0	0	0
1	1	not real
2	$\sqrt{2} \approx 1.41$	not real
4	2	not real

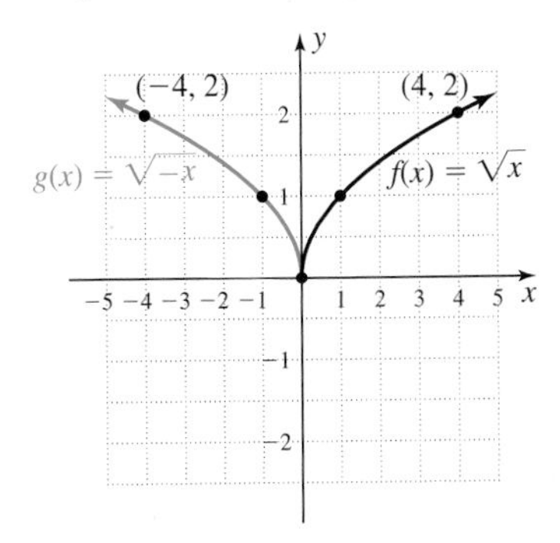

The graph of g is the same as the graph of f, but it has been reflected across the y-axis. A study of the domain shows why—f represents a real number only for nonnegative inputs, so its graph occurs to the right of the y-axis, while g represents a real number for nonpositive inputs, so its graph occurs to the left. **NOW TRY EXERCISES 29 AND 30**

The transformation in Example 5 is called a **horizontal reflection** (or an **east/west reflection**) of a basic graph. In general,

> **HORIZONTAL REFLECTIONS OF A BASIC GRAPH**
> Given any function whose graph is determined by $y = f(x)$,
> the graph of $y = f(-x)$ is the graph of $f(x)$ reflected across the y-axis.

Since the actual shape of a graph remains unchanged after the previous transformations are applied, they are often referred to as **rigid transformations.**

C. Stretching/Compressing a Basic Graph

Stretches and compressions of a basic graph are called **nonrigid transformations.** As the name implies, the shape of a graph is changed or transformed when these are applied. However, the transformation doesn't actually "deform" the graph, and we can still identify the function family as well as all important characteristics.

EXAMPLE 6 Construct a table of values for $f(x) = x^2$, $g(x) = 3x^2$, and $h(x) = \frac{1}{3}x^2$, then graph the functions on the same grid and discuss what you observe.

Solution: A table of values is given for all three functions with the corresponding graphs.

x	$f(x) = x^2$	$g(x) = 3x^2$	$h(x) = \frac{1}{3}x^2$
−3	9	27	3
−2	4	12	$\frac{4}{3}$
−1	1	3	$\frac{1}{3}$
0	0	0	0
1	1	3	$\frac{1}{3}$
2	4	12	$\frac{4}{3}$
3	9	27	3

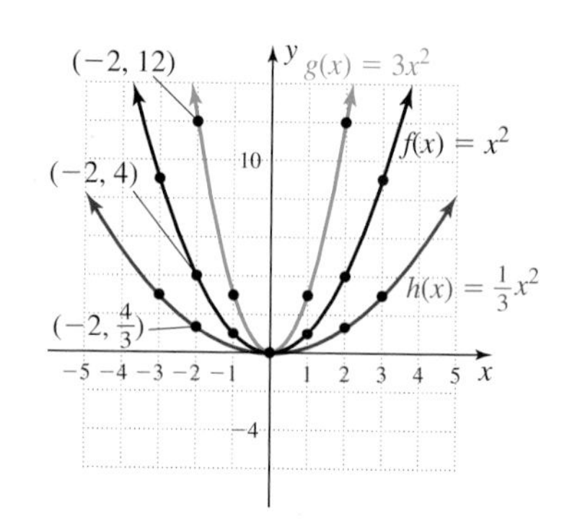

The outputs of g are triple those of f, *stretching* g upward and causing its branches to hug the vertical axis (making it more narrow). The outputs of h are one-third those of f and the graph of h is *compressed* downward, with its branches farther away from the vertical axis (making it wider).

NOW TRY EXAMPLES 31 THROUGH 38

The transformations in Example 6 are called **vertical stretches** or **compressions.** In general, we have,

> **STRETCHES AND COMPRESSIONS OF A BASIC GRAPH**
> Given $a > 0$ and any function whose graph is determined by $y = f(x)$,
> the graph of $y = af(x)$ is the graph of $f(x)$ stretched vertically
> if $|a| > 1$ and compressed vertically if $0 < |a| < 1$.

D. Transformations of a General Function $y = f(x)$

Often more than one transformation acts on the same function at the same time. Although the transformations can be applied in almost any order, it's helpful to use an organized sequence when graphing them.

> **GENERAL TRANSFORMATIONS OF A BASIC GRAPH**
>
> Given any transformation of a function whose graph is defined by $y = f(x)$, the graph of the transformed function can be found by:
>
> 1. Applying the stretch or compression.
> 2. Reflecting the result.
> 3. Applying the horizontal and/or vertical shifts.
>
> These are usually applied to a few characteristic points, with the new graph drawn through these points.

EXAMPLE 7 Sketch the graphs of $g(x) = -(x + 2)^2 + 3$ and $h(x) = 4\sqrt[3]{x - 2} - 1$ using transformations of a parent function and a few characteristic points.

Solution: The graph of $g(x) = -(x + 2)^2 + 3$ is a basic parabola reflected across the x-axis, shifted left 2 and up 3. This sequence of transformations in shown in Figures 3.22 through 3.24. The graph of $h(x) = 4\sqrt[3]{x - 2} - 1$ is a horizontal propeller, stretched by a factor of 4, then shifted right 2 and down 1. This sequence is shown in Figures 3.25 through 3.27.

Figure 3.22

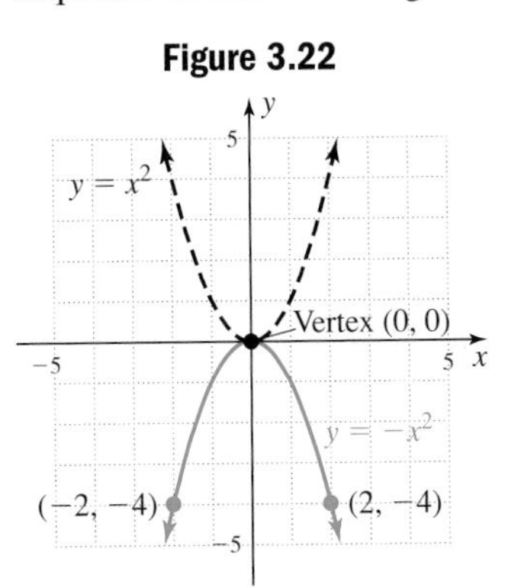

Reflected across x-axis

Figure 3.23

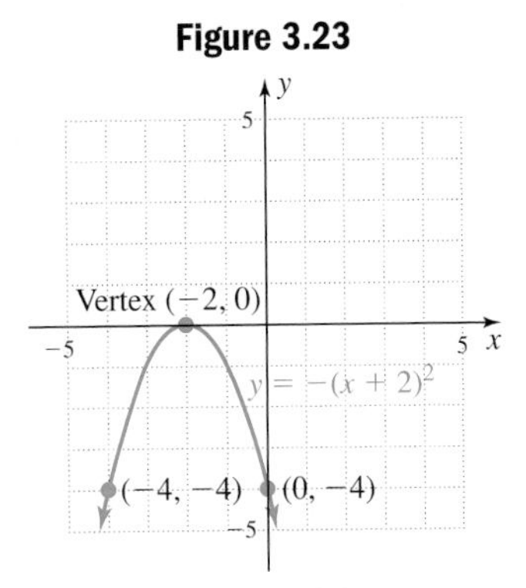

Shifted left 2

Figure 3.24

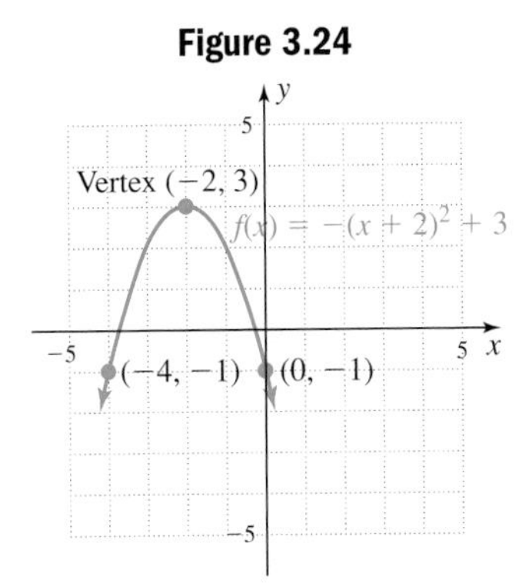

Shifted up 3

Figure 3.25

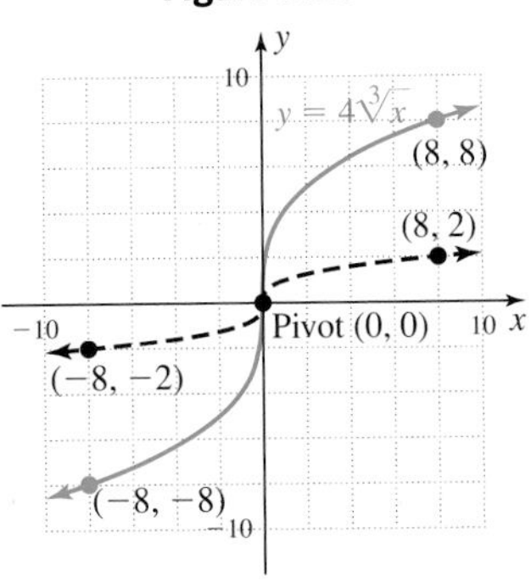

Stretched by a factor of 4

Figure 3.26

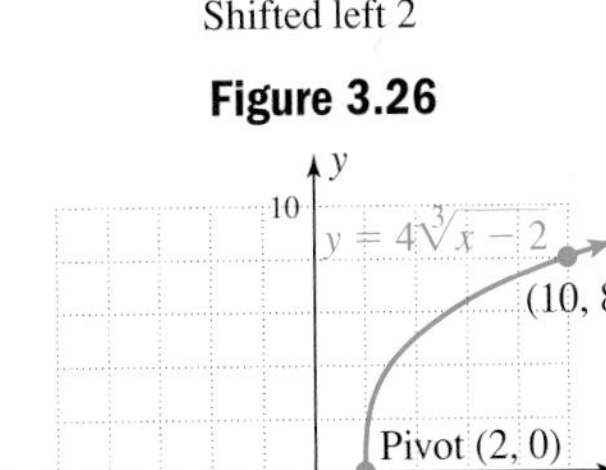

Shifted right 2

Figure 3.27

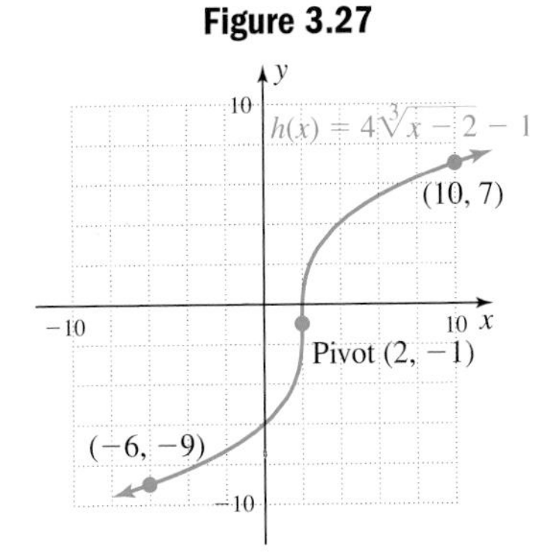

Shifted down 1

NOW TRY EXERCISES 39 THROUGH 68

As mentioned, it's important to realize that these transformations can actually be applied to *any function,* even those that are new and unfamiliar. Consider the following pattern:

Parent Function		Transformation of Parent Function
quadratic:	$y = x^2$	$y = -2(x - 3)^2 + 1$
cubic:	$y = x^3$	$y = -2(x - 3)^3 + 1$
absolute value:	$y = \|x\|$	$y = -2\|x - 3\| + 1$
square root:	$y = \sqrt{x}$	$y = -2\sqrt{x - 3} + 1$
cube root:	$y = \sqrt[3]{x}$	$y = -2\sqrt[3]{x - 3} + 1$
general:	$y = f(x)$	$y = -2f(x - 3) + 1$

In each case, the transformation involves a vertical stretch, then a vertical reflection with the result shifted right 3 and up 1. Since the shifts are the same regardless of the initial function, we can extend and globalize these results to a general function $y = f(x)$.

Parent Function	Transformation of Parent Function
$y = f(x)$	$y = af(x \pm h) \pm k$

- a: north/south reflections, vertical stretches and compressions
- h: horizontal shift h units, opposite direction of sign
- k: vertical shift k units, same direction as sign

Use this illustration to complete Exercise 8. Remember—if the graph of a function is shifted, the *individual points* on the graph are likewise shifted.

EXAMPLE 8 Given the graph of $f(x)$ shown in Figure 3.28, graph $g(x) = -f(x + 1) - 2$.

Solution: For g, the graph of f is reflected across the x-axis, then shifted horizontally 1 unit left and vertically 2 units down. The result is shown in Figure 3.29.

Figure 3.28

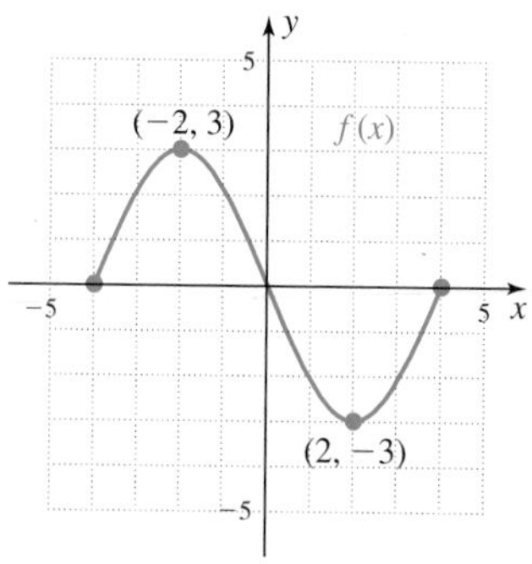

Figure 3.29

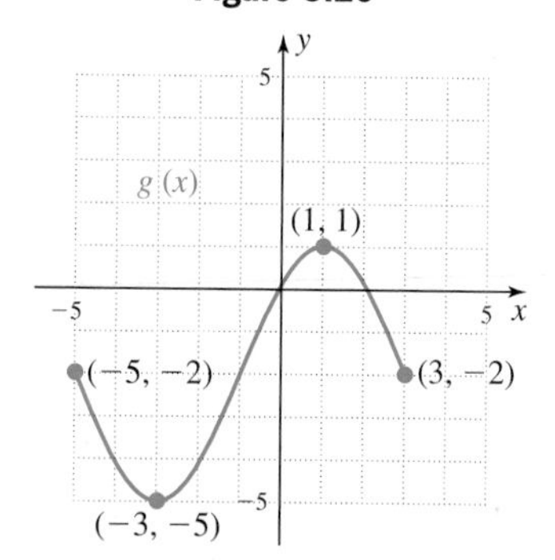

NOW TRY EXERCISES 69 THROUGH 72

E. Transformations and the Area Under a Curve

The transformations studied here are linked to numerous topics in future courses. Surprisingly, one "link" involves a simple computing of the area beneath the graph of a basic function after it's been transformed. Such areas have a number of important real-world applications. Consider a jogger who is running at a steady pace of 600 ft/min (about

Figure 3.30

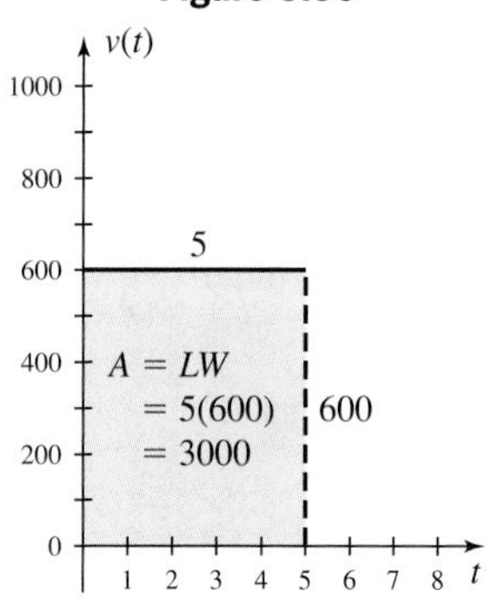

Figure 3.31

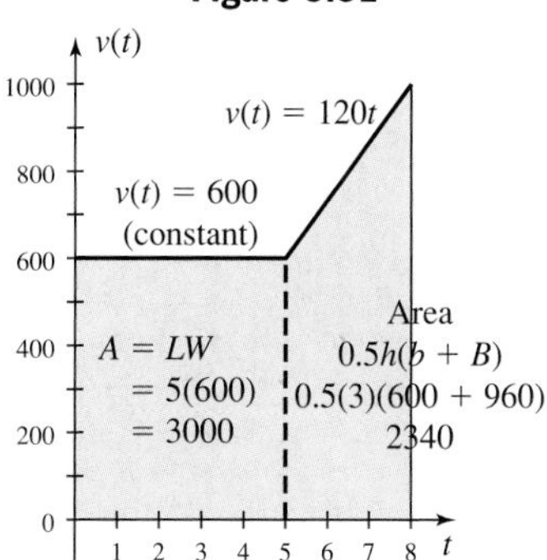

7 mph). If she continues this pace for 5 min, she'll run 3000 ft ($D = RT$). Graphically her running speed is represented by the horizontal line $v(t) = 600$, where $v(t)$ represents the velocity at time t in minutes, with minutes scaled on the horizontal axes. Note this creates a rectangular shape with an area that is numerically the same as the distance run (see Figure 3.30). While this may seem coincidental, it is actually an accurate depiction of the relationship between velocity and time. In other words, $A = LW$ bears a strong relationship to $D = RT$ as well as to other phenomena that result from the product of two factors (force equals mass times acceleration: $F = MA$; cost equals price times quantity: $C = PQ$; and many others). Of great interest to us (and even greater intrigue to early mathematicians), is that this relationship holds *even when the velocity is not constant.* Suppose our jogger decides to "finish strong" and steadily increases her velocity for the next 3 min of the run. It seems reasonable that she'll cover a greater distance than if she continued at 600 ft/min, and to no one's surprise, the area under the graph also grows. If her velocity becomes $V(t) = 120t$ between the fifth and eighth minutes, the graph takes on the shape shown in Figure 3.31, where a trapezoidal area is formed. Recall the area of a trapezoid is $A = \frac{h}{2}(B + b)$, where b and B represent the lengths of the parallel sides (the bases), and h is the height. The "height" here can be read along the t-axis ($t = 3$) and the shorter base b is 600 (same as the rectangle). The longer base B can be found by evaluating $V(t)$ at $t = 8$, giving $V(8) = 120(8)$ or 960. The distance run in the last three minutes was then $\frac{3}{2}(600 + 960)$ or 2340 ft.

Finally, what if the jogger steadily increased her pace from the beginning to the fourth minute, then steadily decreased her pace at the same rate from the fourth to the eighth minute. The area representing the distance covered could possibly resemble one of those in Figures 3.32 to 3.35.

Figure 3.32

Triangle on a rectangle

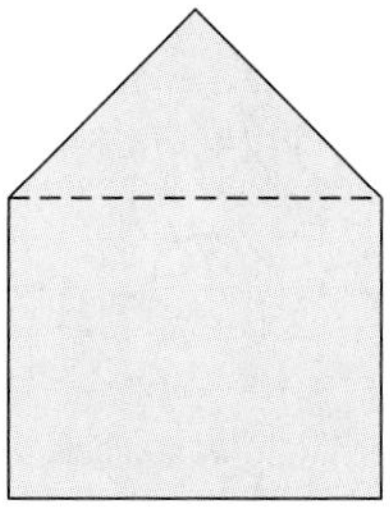

Figure 3.33

Semicircle on a rectangle

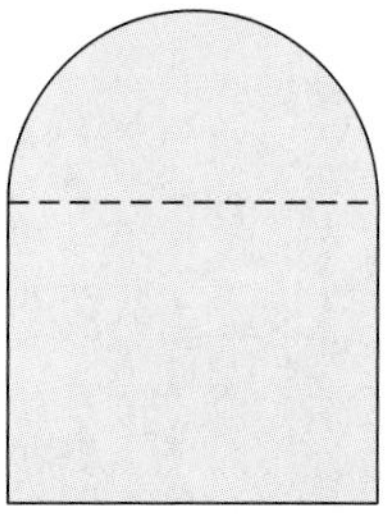

Figure 3.34

Semi-ellipse on a rectangle

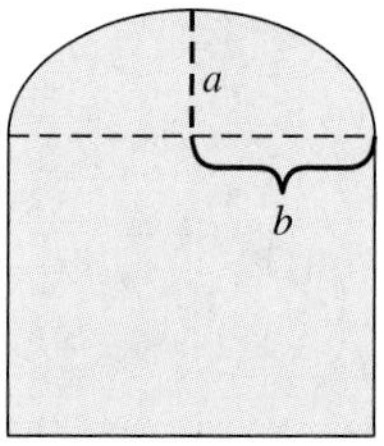

a

b

Figure 3.35

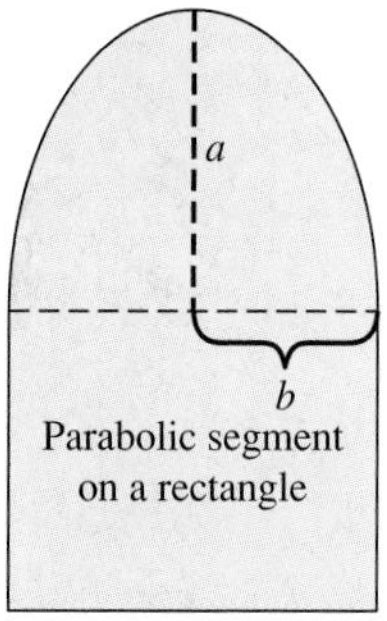

a

b

Parabolic segment on a rectangle

To compute the area of each rectangular portion, we use $A = LW$. For any triangular area (Figure 3.32) we know $A = \frac{1}{2}bh$, while for the area of the semicircle (Figure 3.33), we simply calculate the area of half a circle: $A = \frac{\pi r^2}{2}$. The area of a semi-ellipse (Figure 3.34) is $A = \frac{\pi ab}{2}$, where a represents the "height" of the semi-ellipse and b represents one-half the "base" (note that if a and b are equal, the result is the same formula as that of a semicircle). The area of a parabolic segment (Figure 3.35) is given by $A = \frac{4}{3}ab$, with a and b as shown. Example 9 uses these ideas in connection with our study of transformations.

EXAMPLE 9 For $f(x) = -\frac{1}{2}(x - 4)^2 + 15$, graph the function using transformations of $y = x^2$. Using the x- and y-axes as sides, sketch a rectangle *beneath the graph* using the y-intercept as the height, then shade in the resulting area and find the area of the shaded region.

Solution: The graph of $f(x) = -\frac{1}{2}(x - 4)^2 + 15$ is a basic parabola compressed by a factor of $\frac{1}{2}$, reflected across the x-axis, shifted right 4 and up 15. This means the vertex will be (4, 15). The y-intercept is $f(0) = 7$ and the point on the parabola an equal distance from the axis of symmetry is (8, 7). The resulting graph is shown in Figure 3.36. After sketching in the rectangle described, the shaded portion directly under the parabola is shown in Figure 3.37.

Figure 3.36

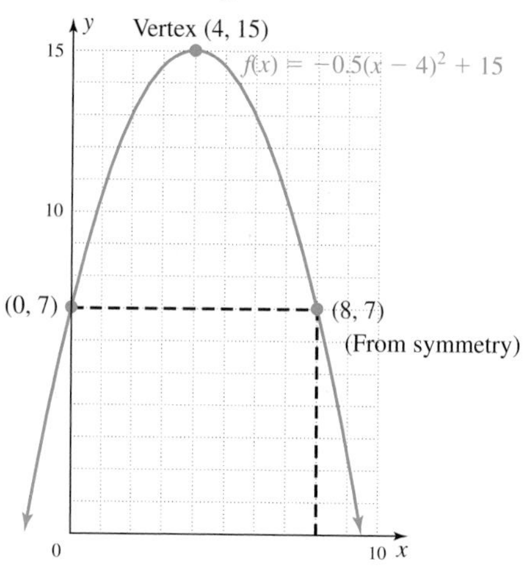

Figure 3.37

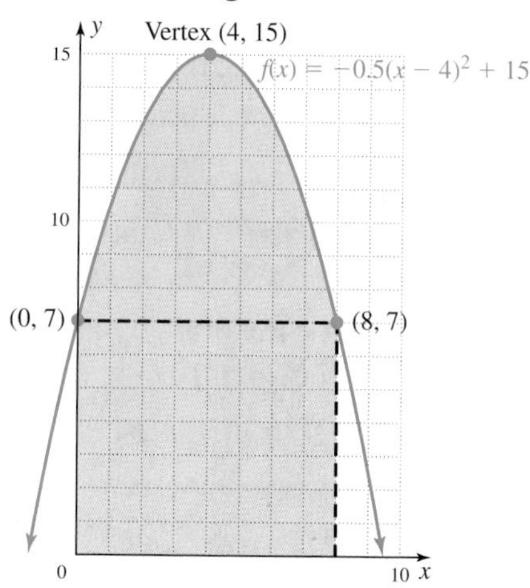

The shaded area is the area of the rectangle plus the area of the parabolic segment, given by $A = LW + \frac{4}{3}ab$, where $W = 7$, $L = 8$, $b = 4$, and $a = 8$. The result is

$$A = LW + \frac{4}{3}ab \quad \text{formula for total area}$$

$$= (8)(7) + \frac{4}{3}(8)(4) \quad \text{substitute 7 for } W\text{, 8 for } L\text{, 8 for } a\text{, and 4 for } b$$

$$= 56 + \frac{128}{3} \quad \text{multiply}$$

$$= 98\frac{2}{3} \quad \text{result}$$

The shaded area is $98\frac{2}{3}$ units^2.

NOW TRY EXERCISES 73 THROUGH 78

TECHNOLOGY HIGHLIGHT

Using a Graphing Calculator to Study Function Families

The keystrokes shown apply to a TI-84 Plus model. Please consult your manual or our Internet site for other models.

Graphing calculators are able to display a number of graphs simultaneously, making them a wonderful tool for studying families of functions. Our main purpose here is to demonstrate that all functions in a particular family have the same basic shape, making them easier to understand and analyze. Let's begin by entering the function $y = |x|$ [actually $y = \text{abs}(x)$] as Y_1 on the Y= screen. Next, we enter different variations of the function, but always in terms of its variable name "Y_1." This enables us to simply change the basic function, and observe how the changes affect the new family. Recall that to access the function name Y_1 we use this sequence of keystrokes: VARS ▶ (to access the Y-VARS menu) ENTER (to access the function variables menu) and ENTER (to select Y_1). Your screen should look like Figure 3.38 when finished. Enter the functions $Y_2 = Y_1 + 3$ and $Y_3 = Y_1 - 6$, then graph all three functions in the ZOOM **6:ZStandard** window. The calculator draws each graph in the order they were entered and you can always identify the functions by pressing the TRACE key and then the up arrow ▲ or down arrow ▼ keys. In the upper left corner of the window shown in Figure 3.39, the calculator identifies which function the cursor is currently on. Most importantly, note that all functions in this family maintain the same "V" shape. Now change Y_1 to $Y_1 = \text{abs}(x - 3)$, leaving Y_2 and Y_3 as is. What do you notice when these are graphed again?

Figure 3.38

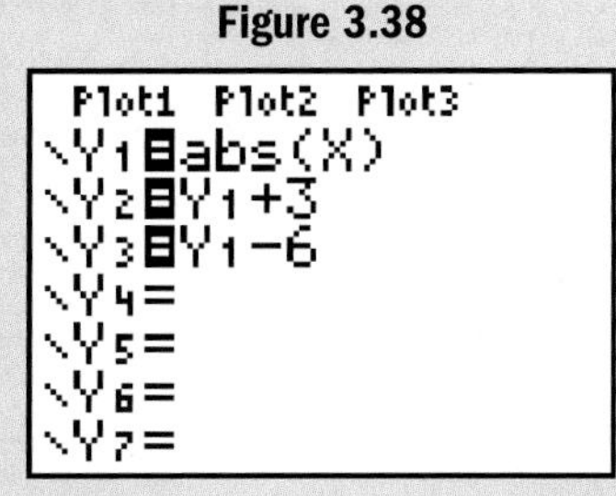

Figure 3.39

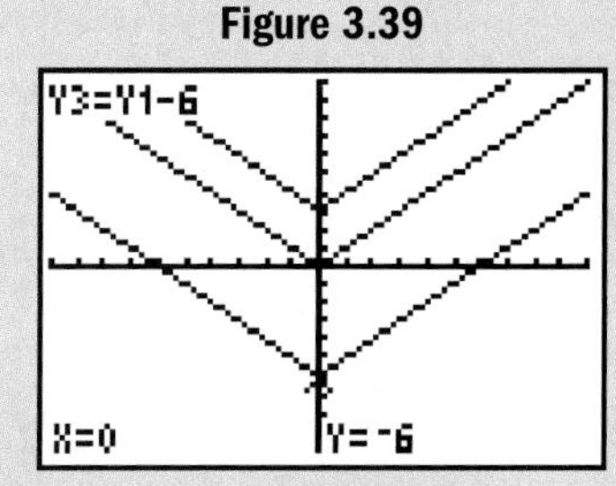

Exercise 1: Change Y_1 to read $Y_1 = \sqrt{x}$ and graph, then enter $Y_1 = \sqrt{x - 3}$ and graph once again. What do you observe? What comparisons can be made with the translations of $Y_1 = \text{abs}(x)$? shifted right 3 units; answers will vary

Exercise 2: Change Y_1 to read $Y_1 = x^2$ and graph, then enter $Y_1 = (x - 3)^2$ and graph once again. What do you observe? What comparisons can be made with the translations of $Y_1 = \text{abs}(x)$ and $Y_1 = \sqrt{x}$? shifted right 3 units; answers will vary

3.3 EXERCISES

CONCEPTS AND VOCABULARY

Fill in each blank with the appropriate word or phrase. Carefully reread the section if needed.

1. After a vertical stretch, points on the graph are closer to the y-axis. After a vertical compression, points on the graph are farther from the y-axis.

2. If the transformation applied changes only the location of a graph and not its shape or form, it is called a rigid transformation. These include translations and reflections.

3. The vertex of $h(x) = 3(x + 5)^2 - 9$ is at $(-5, -9)$ and the graph is concave up.

4. The pivot point of $f(x) = -2(x - 4)^3 + 11$ is at $(4, 11)$ and the end behavior is up, down.

5. Given the graph of a general function $f(x)$, discuss/explain how the graph of $F(x) = -2f(x + 1) - 3$ can be obtained. If $(0, 5)$, $(6, 7)$, and $(-9, -4)$ are on the graph of f, where do they end up on the graph of F? Answers will vary.

6. Discuss/explain why the shift of $f(x) = x^2 + 3$ is a *vertical shift* of 3 units in the *positive* direction, while the shift of $g(x) = (x + 3)^2$ is a *horizontal shift* 3 units in the *negative* direction. Include several examples linked to a table of values. Answers will vary.

DEVELOPING YOUR SKILLS

Identify and discuss the characteristic features of each graph, including the function family, intercepts, vertex, node, pivot point, and end behavior.

11. 12.

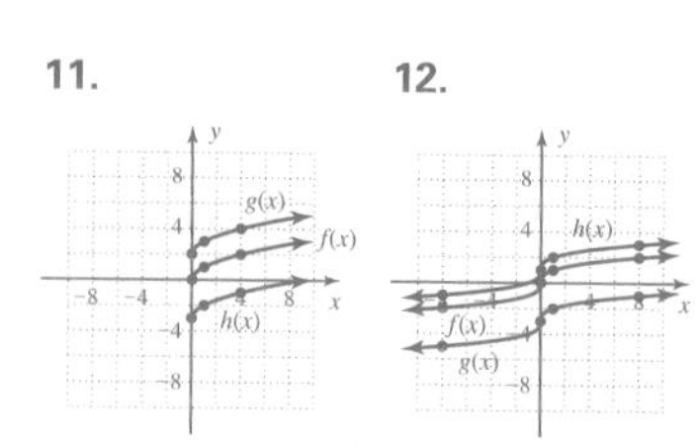

13. 14.

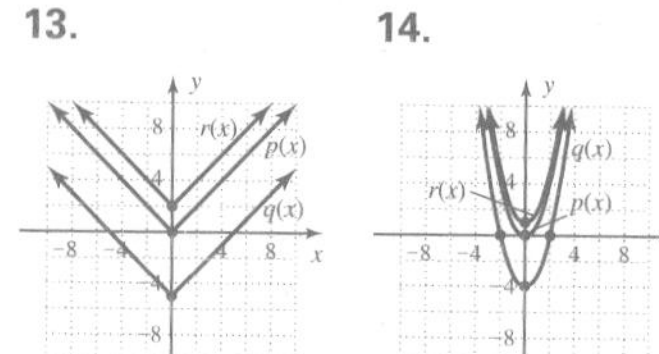

15. 16.

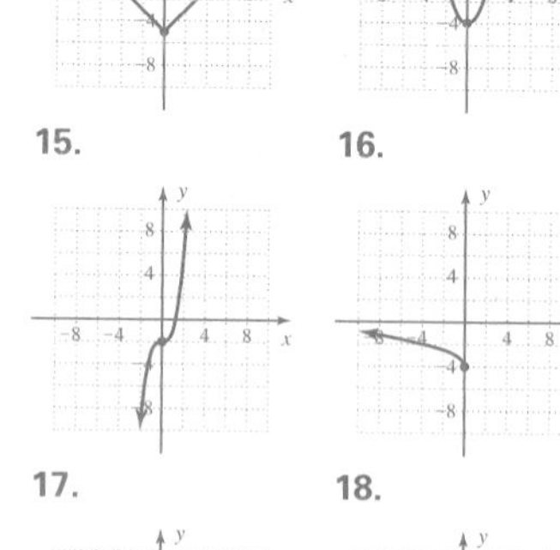

17. 18.

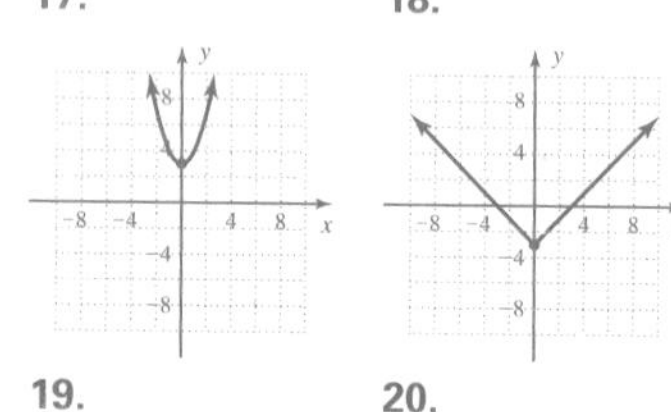

19. 20.

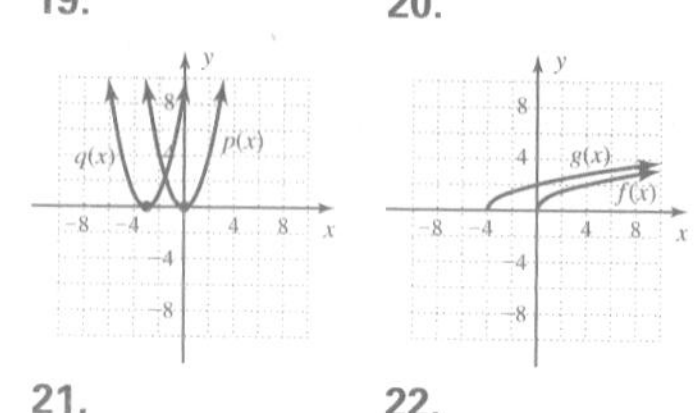

21. 22.

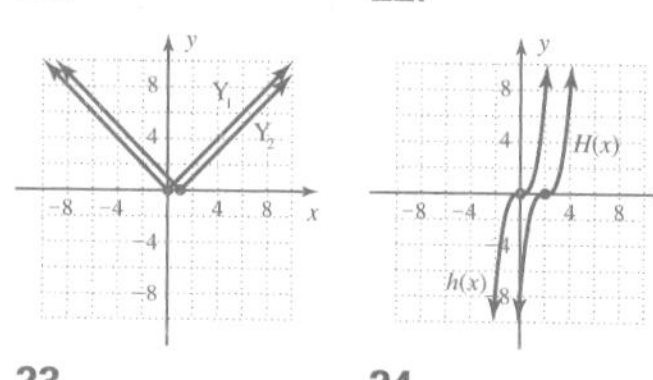

23. 24.

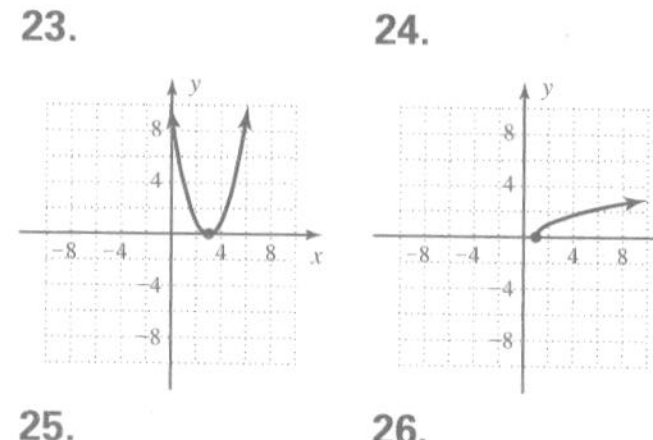

25. 26.

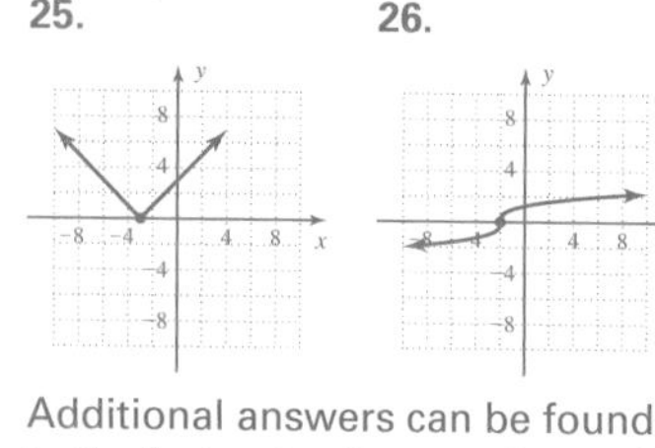

Additional answers can be found in the Instructor Answer Appendix.

7.

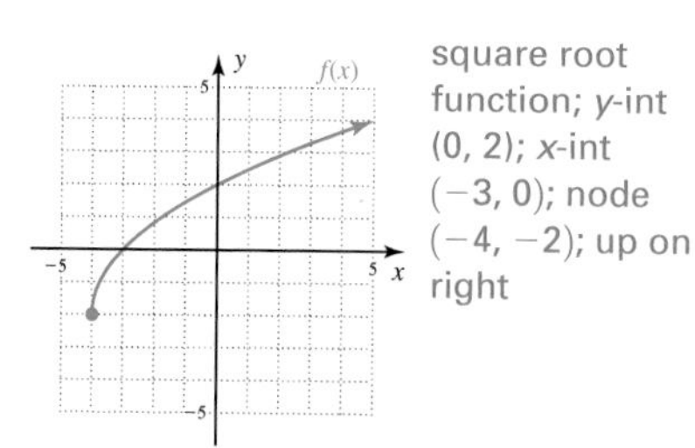

square root function; y-int (0, 2); x-int (−3, 0); node (−4, −2); up on right

8.

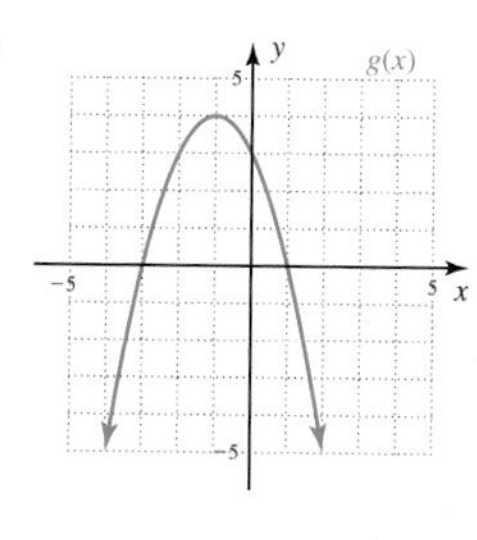

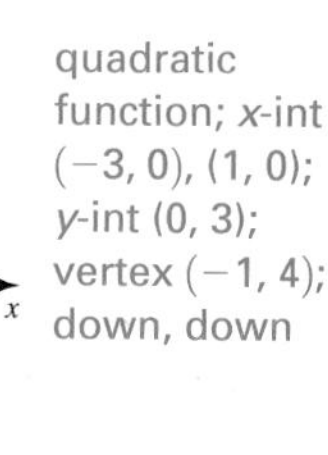

quadratic function; x-int (−3, 0), (1, 0); y-int (0, 3); vertex (−1, 4); down, down

9.

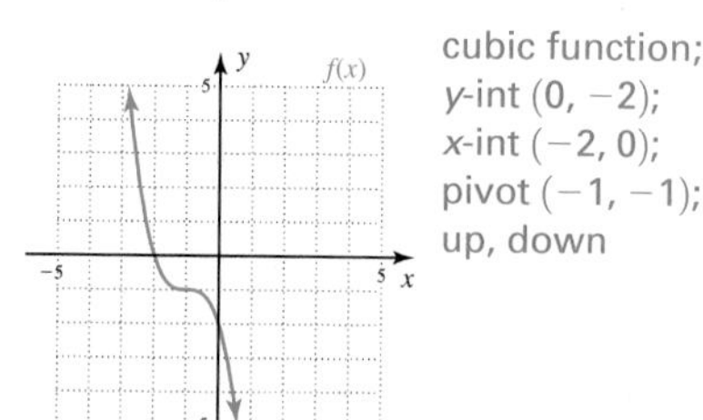

cubic function; y-int (0, −2); x-int (−2, 0); pivot (−1, −1); up, down

10.

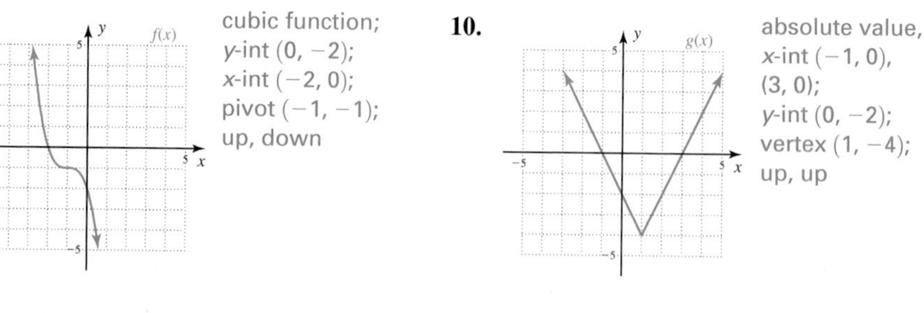

absolute value, x-int (−1, 0), (3, 0); y-int (0, −2); vertex (1, −4); up, up

Use a table of values to graph the functions given on the same grid. Comment on what you observe.

11. $f(x) = \sqrt{x}, \quad g(x) = \sqrt{x} + 2,$ $h(x) = \sqrt{x} - 3$

12. $f(x) = \sqrt[3]{x}, \quad g(x) = \sqrt[3]{x} - 3,$ $h(x) = \sqrt[3]{x} + 1$

13. $p(x) = |x|, \quad q(x) = |x| - 5,$ $r(x) = |x| + 2$

14. $p(x) = x^2, \quad q(x) = x^2 - 4,$ $r(x) = x^2 + 1$

Sketch each graph using transformations of a parent function (without a table of values).

15. $f(x) = x^3 - 2$ **16.** $g(x) = \sqrt{-x} - 4$ **17.** $h(x) = x^2 + 3$ **18.** $Y_1 = |x| - 3$

Use a table of values to graph the functions given on the same grid. Comment on what you observe.

19. $p(x) = x^2, \quad q(x) = (x + 3)^2$

20. $f(x) = \sqrt{x}, \quad g(x) = \sqrt{x + 4}$

21. $Y_1 = |x|, \quad Y_2 = |x - 1|$

22. $h(x) = x^3, \quad H(x) = (x - 2)^3$

Sketch each graph using transformations of a parent function (without a table of values).

23. $p(x) = (x - 3)^2$

24. $Y_1 = \sqrt{x - 1}$

25. $h(x) = |x + 3|$

26. $f(x) = \sqrt[3]{x + 2}$

27. $g(x) = -|x|$

28. $Y_2 = -\sqrt{x}$

29. $f(x) = \sqrt[3]{-x}$

30. $g(x) = (-x)^3$

Use a table of values to graph the functions given on the same grid. Comment on what you observe.

31. $p(x) = x^2, \quad q(x) = 2x^2, \quad r(x) = \frac{1}{2}x^2$

32. $f(x) = \sqrt{-x}, \quad g(x) = 4\sqrt{-x},$ $h(x) = \frac{1}{4}\sqrt{-x}$

33. $Y_1 = |x|, \quad Y_2 = 3|x|, \quad Y_3 = \frac{1}{3}|x|$

34. $u(x) = x^3, \quad v(x) = 2x^3, \quad w(x) = \frac{1}{5}x^3$

Sketch each graph using transformations of a parent function (without a table of values).

35. $f(x) = 4\sqrt[3]{x}$ **36.** $g(x) = -2|x|$ **37.** $p(x) = \frac{1}{3}x^3$ **38.** $q(x) = \frac{3}{4}\sqrt{x}$

Use the characteristics of each function family to match a given function to its corresponding graph. The graphs are not scaled—make your selection based on a careful comparison.

39. $f(x) = \frac{1}{2}x^3$ g

40. $f(x) = \frac{-2}{3}x + 2$ h

41. $f(x) = -(x - 3)^2 + 2$ i

42. $f(x) = -\sqrt[3]{x - 1} - 1$ d

43. $f(x) = |x + 4| + 1$ e

44. $f(x) = -\sqrt{x + 6}$ f

45. $f(x) = -\sqrt{x + 6} - 1$ j

46. $f(x) = x + 1$ k

47. $f(x) = (x - 4)^2 - 3$ l

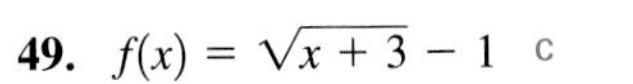

48. $f(x) = |x - 2| - 5$ b **49.** $f(x) = \sqrt{x + 3} - 1$ c **50.** $f(x) = -(x + 3)^2 + 5$ a

a.

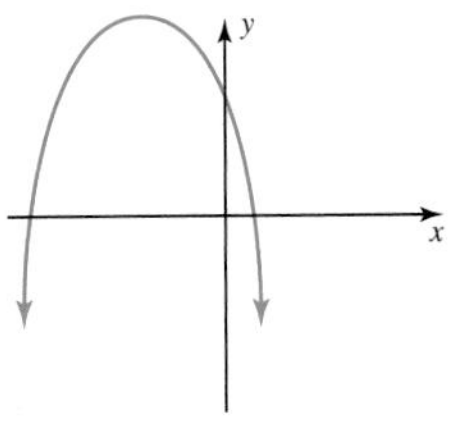

b.

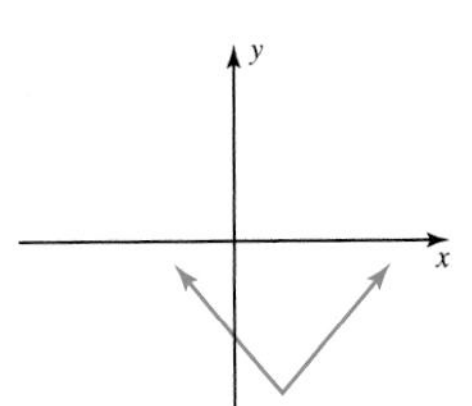

c.

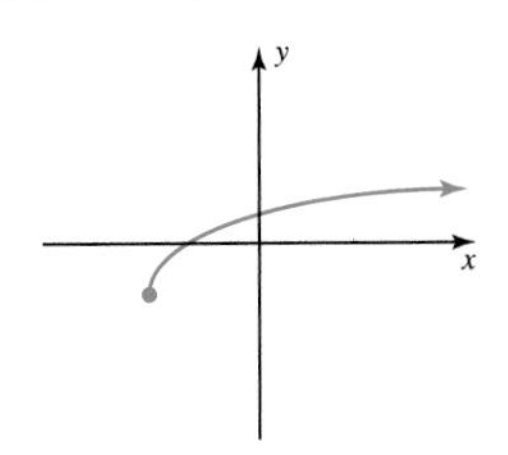

d.

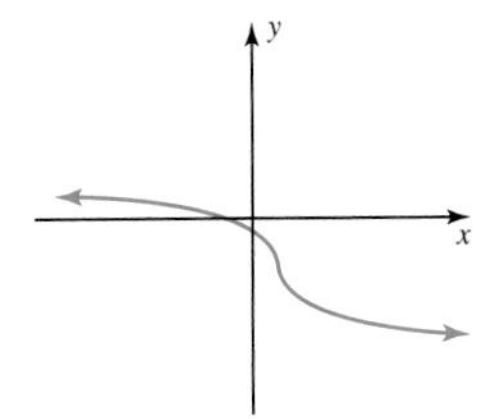

e.

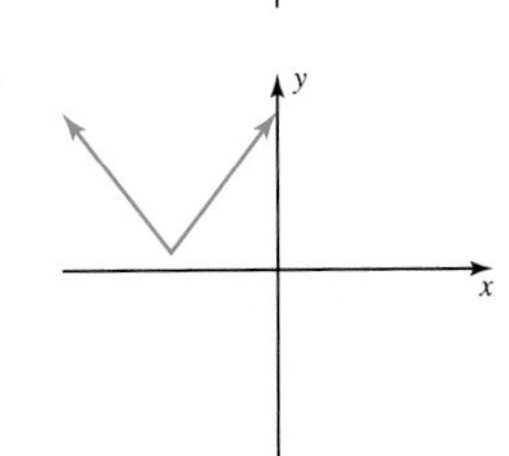

f.

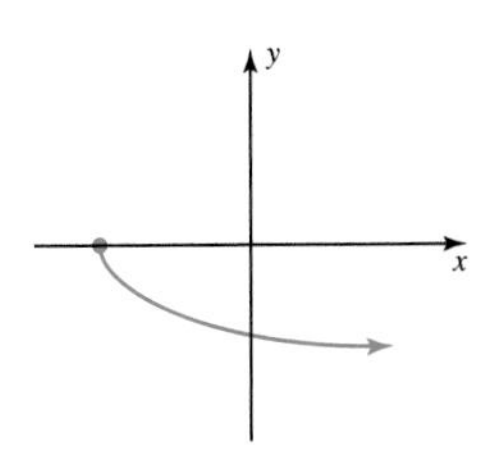

g.

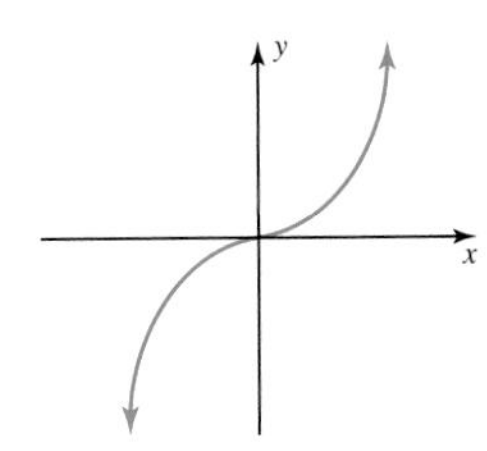

h.

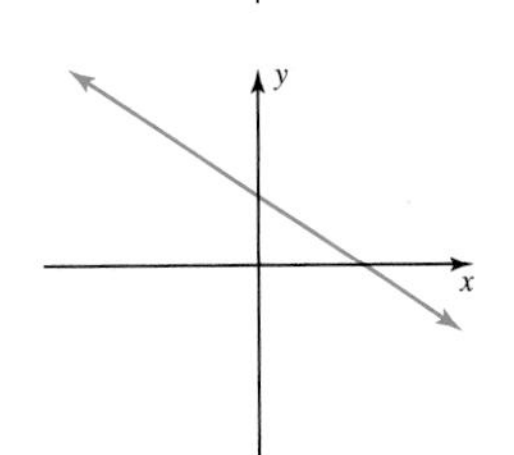

i.

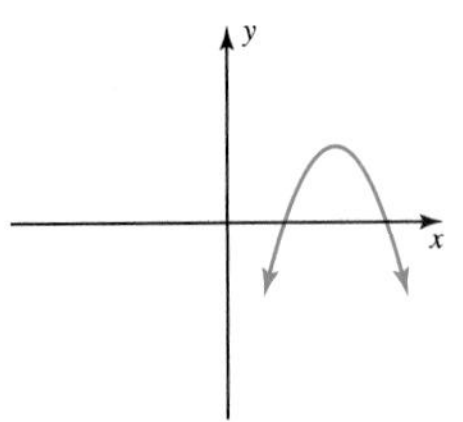

j.

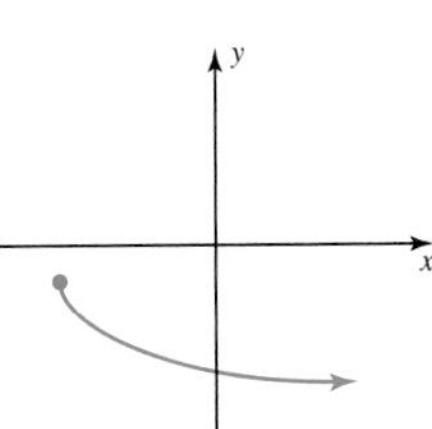

k.

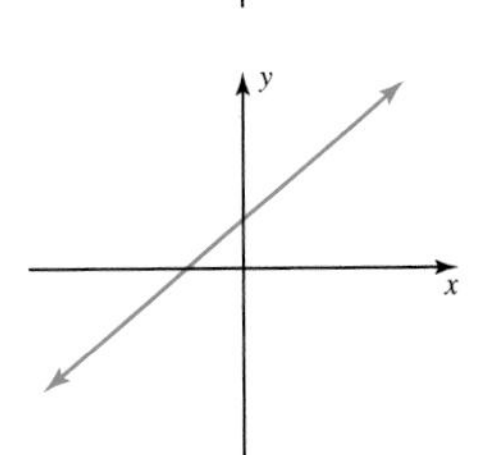

l.

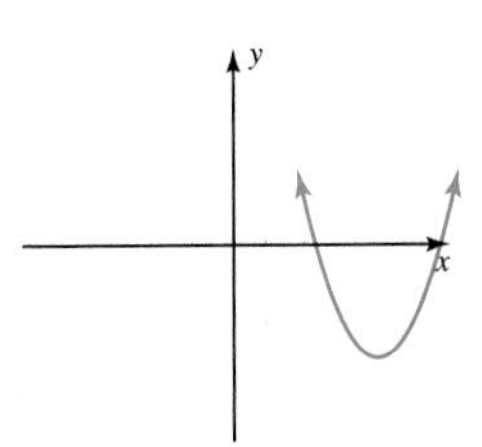

Graph each function using shifts of a parent function and a few characteristic points. *Clearly state and indicate the transformations used* and identify the location of all vertices, nodes, and/or pivot points.

51. $f(x) = \sqrt{x + 2} - 1$ **52.** $g(x) = \sqrt{x - 3} + 2$ **53.** $h(x) = -(x + 3)^2 - 2$

54. $H(x) = -(x - 2)^2 + 5$ **55.** $p(x) = (x + 3)^3 - 1$ **56.** $q(x) = (x - 2)^3 + 1$

57. $Y_1 = \sqrt[3]{x + 1} - 2$ **58.** $Y_2 = \sqrt[3]{x - 3} + 1$ **59.** $f(x) = -|x + 3| - 2$

60. $g(x) = -|x - 4| - 2$ **61.** $h(x) = -2(x + 1)^2 - 3$ **62.** $H(x) = \frac{1}{2}|x + 2| - 3$

63. $p(x) = -\frac{1}{3}(x + 2)^3 - 1$ **64.** $q(x) = 5\sqrt[3]{x + 1} + 2$ **65.** $Y_1 = -2\sqrt{x - 4} - 3$

66. $Y_2 = 3\sqrt{-x + 2} - 1$ **67.** $h(x) = \frac{1}{5}(x - 3)^2 + 1$ **68.** $H(x) = -2|x - 3| + 4$

Apply the transformations indicated for the graph of the general functions given.

69.

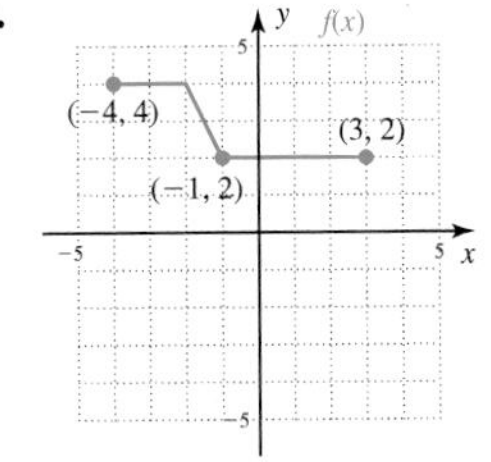

a. $f(x - 2)$ **b.** $-f(x) - 3$

c. $\frac{1}{2}f(x + 1)$ **d.** $f(-x) + 1$

70.

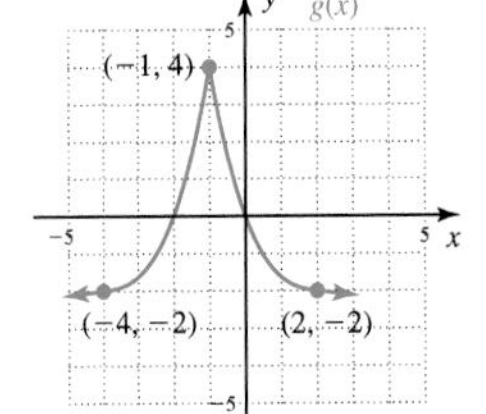

a. $g(x) - 2$ **b.** $-g(x) + 3$

c. $2g(x + 1)$ **d.** $\frac{1}{2}g(x - 1) + 2$

51. left 2, down 1

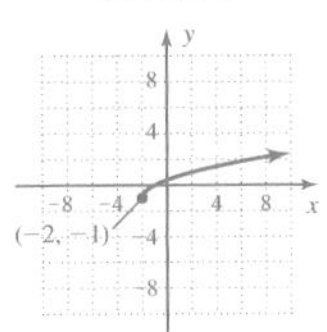

52. right 3, up 2

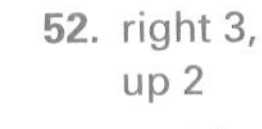

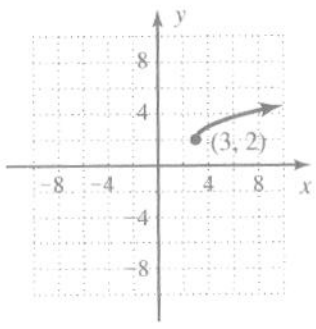

53. reflected across x-axis, left 3, down 2

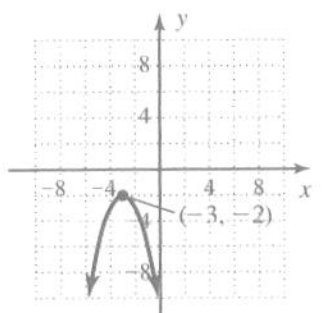

54. reflected across x-axis, right 2, up 5

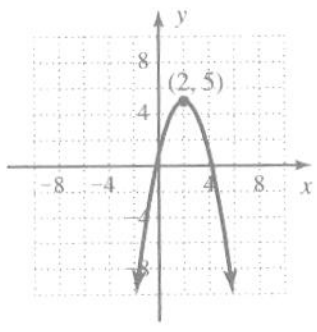

55. left 3, down 1

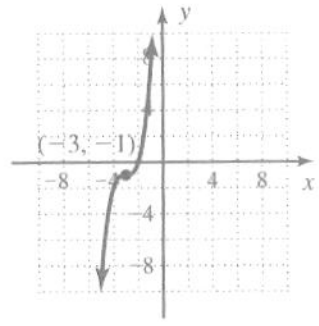

56. right 2, up 1

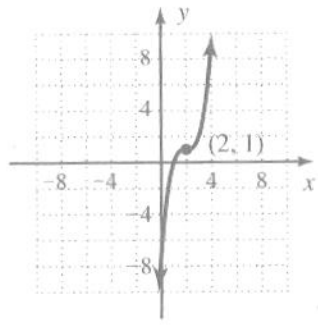

57. left 1, down 2

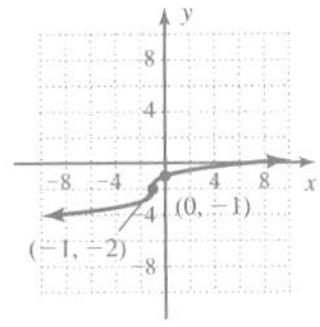

58. right 3, up 1

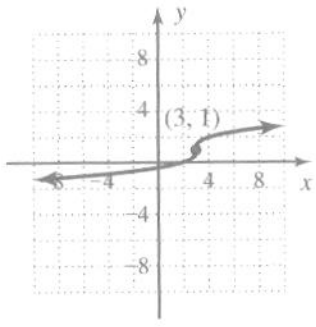

59. reflected across x-axis, left 3, down 2

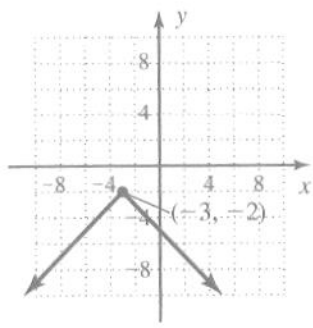

60. reflected across x-axis, right 4, down 2

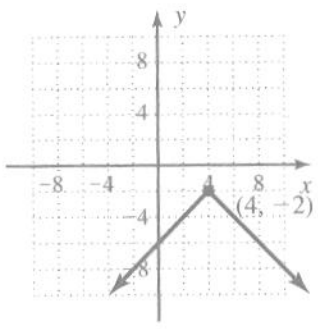

61. stretched vertically, reflected across x-axis, left 1, down 3

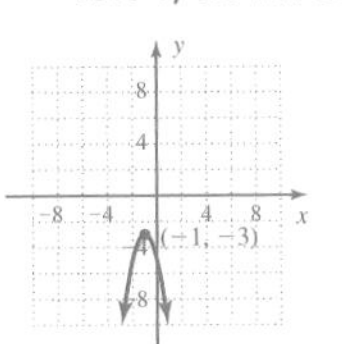

62. compressed vertically, left 2, down 3

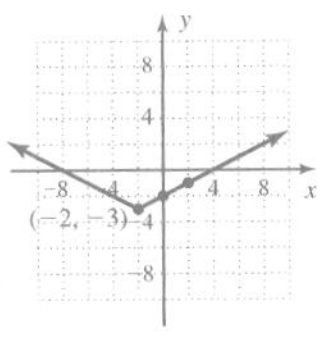

Additional answers can be found in the Instructor Answer Appendix.

71. a. b.

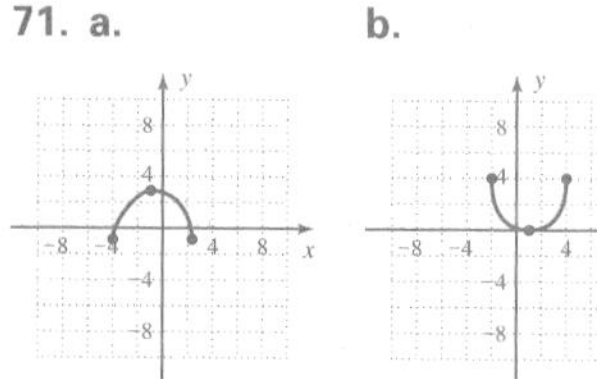

c. d.

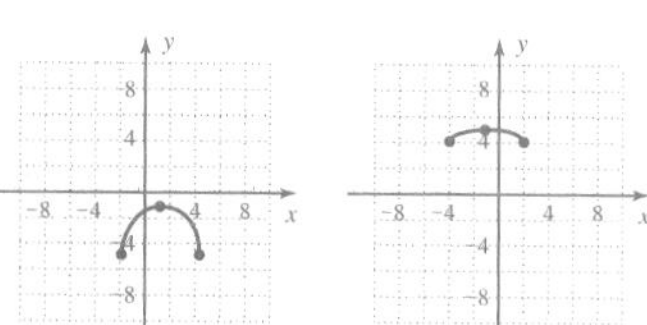

72. a. b.

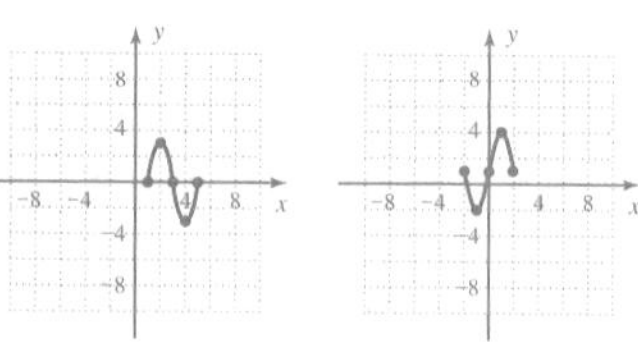

c. d.

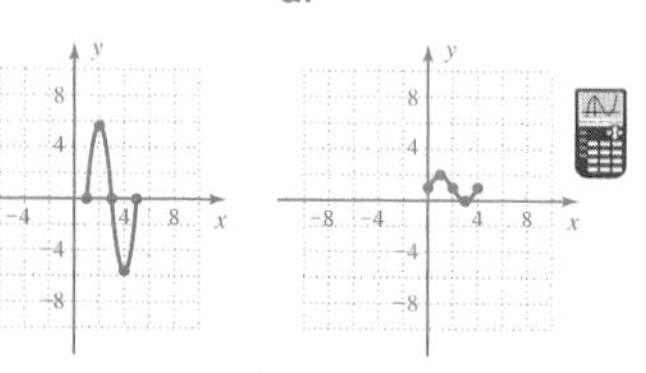

73. $A = 27$ units^2 74. $A = 32$ units^2

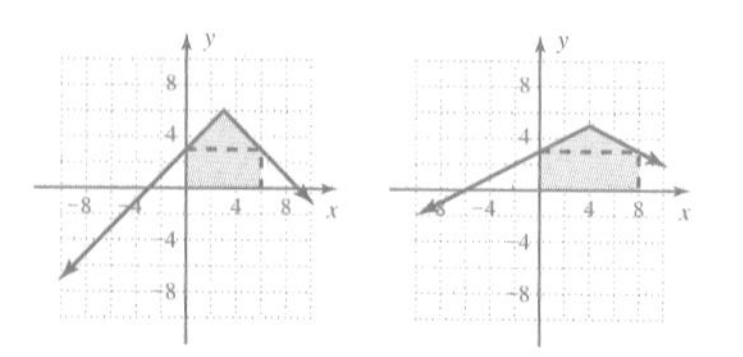

75. $A = 22\frac{2}{3}$ units^2 76. $A = 61\frac{1}{3}$ units^2

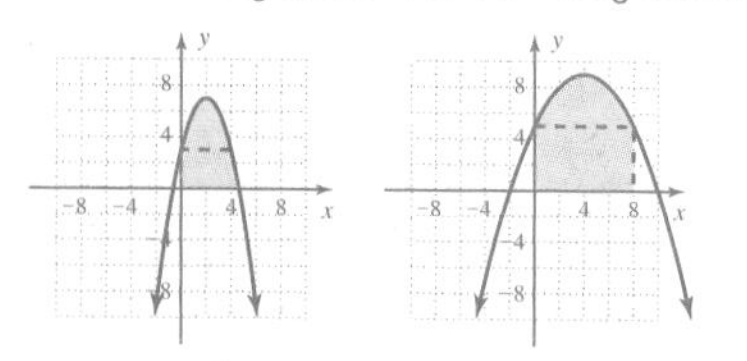

77. $A = \frac{9}{2}\pi + 12$ units^2 78. $A = 2\pi + 20$ units^2

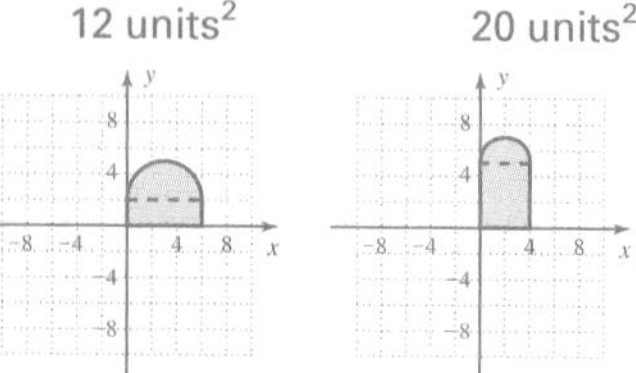

Additional answers can be found in the Instructor Answer Appendix.

71.

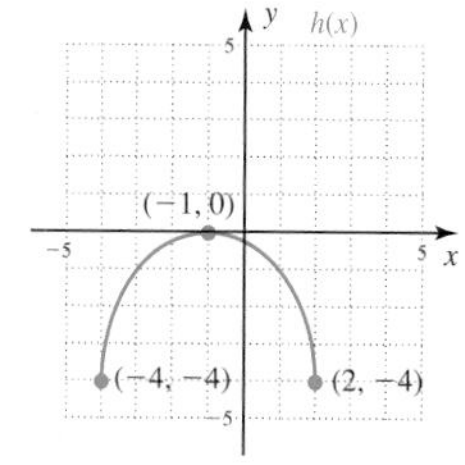

a. $h(x) + 3$ b. $-h(x - 2)$
c. $h(x - 2) - 1$ d. $\frac{1}{4}h(x) + 5$

72.

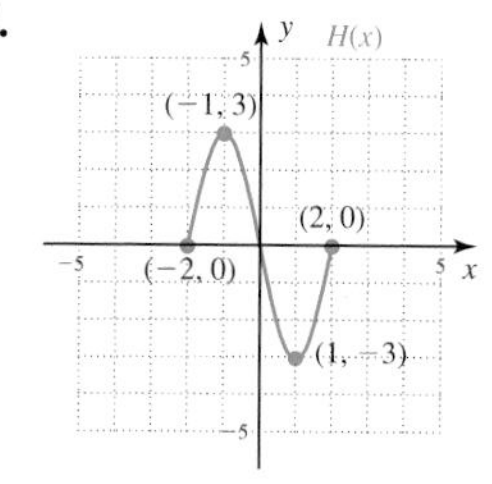

a. $H(x - 3)$ b. $-H(x) + 1$
c. $2H(x - 3)$ d. $\frac{1}{3}H(x - 2) + 1$

Sketch each graph using transformations of a basic function or semicircle ($y = \sqrt{r^2 - x^2}$). Then use the x- and y-axes as sides to sketch a rectangle *beneath the graph* using the y-intercept as the height. Finally, shade in the area directly beneath the graph and find the area of the shaded region.

73. $f(x) = -|x - 3| + 6$
74. $r(x) = -\frac{1}{2}|x - 4| + 5$
75. $g(x) = -(x - 2)^2 + 7$
76. $p(x) = -\frac{1}{4}(x - 4)^2 + 9$
77. $q(x) = \sqrt{9 - (x - 3)^2} + 2$
78. $h(x) = \sqrt{4 - (x - 2)^2} + 5$

WORKING WITH FORMULAS

79. Volume of a sphere: $V(r) = \frac{4}{3}\pi r^3$

The volume of a sphere is given by the function shown, where $V(r)$ is the volume in cubic units and r is the radius. Note this function belongs to the *cubic family* of functions. Approximate the value of $\frac{4}{3}\pi$ to one decimal place, then graph the function on the interval [0, 3] using one-half unit increments (0, 0.5, 1, 1.5, and so on). From your *graph,* estimate the volume of a sphere with radius 2.5 in. Then compute the actual volume. Are the results close? ~4.2, 70 units^3, 65.4 units^3, yes

80. Fluid motion: $V(h) = -4\sqrt{h} + 20$

Suppose the velocity of a fluid flowing from an open tank (no top) through an opening in its side is given by the function shown, where $V(h)$ is the velocity of the fluid (in feet per second) at water height h (in feet). Note this function belongs to the *square root family* of functions. An open tank is 25 ft deep and filled to the brim with fluid. Use a table of values to graph the function on the interval [0, 25]. From your graph, estimate the velocity of the fluid when the water level is 7 ft, then find the actual velocity. Are the answers close? If the fluid velocity is 5 ft/sec, how high is the water in the tank? about 9 ft/sec, ~9.4 ft/sec, yes; about 14 ft

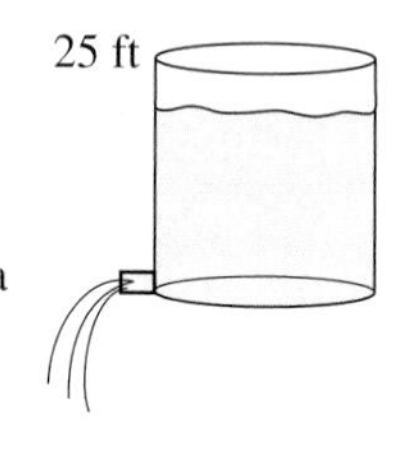

APPLICATIONS

81. Gravity, distance, time: After being released, the time it takes an object to fall x ft is given by the function $T(x) = \frac{1}{4}\sqrt{x}$, where $T(x)$ is in seconds. Describe the transformation applied to obtain the graph of T from the graph of $y = \sqrt{x}$, then sketch the graph of T for $x \in [0, 100]$. How long would it take an object to hit the ground if it were dropped from a height of 81 ft? compressed vertically, 2.25 sec

82. Stopping distance: In certain weather conditions, accident investigators will use the function $v(x) = 4.9\sqrt{x}$ to estimate the speed of a car (in miles per hour) that has been involved in an accident, based on the length of the skid marks x (in feet). Describe the transformation applied to obtain the graph of v from the graph of $y = \sqrt{x}$, then sketch the graph of v for $x \in [0, 400]$. If the skid marks were 225 ft long, how fast was the car traveling? Is this point on your graph? stretched vertically, 73.5 mph; yes

83. 84.

85. 86.

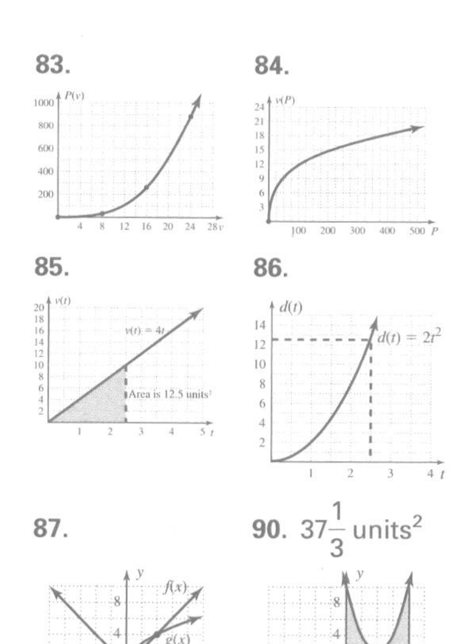

87. 90. $37\frac{1}{3}$ units^2

91. Any points in Quadrants III and IV will reflect in x-axis and move to Quadrants I and II

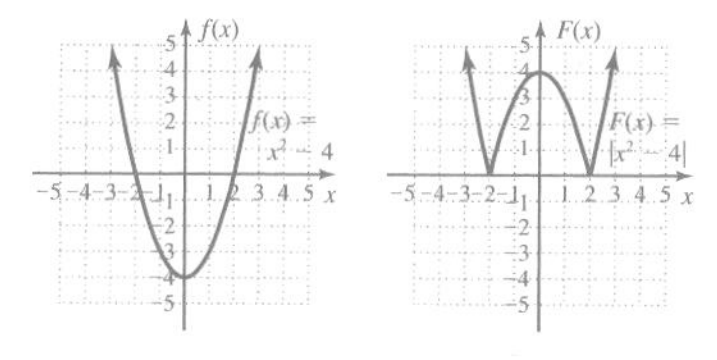

92. Graph in Quadrants III and IV will reflect across x-axis and move to Quadrants I and II

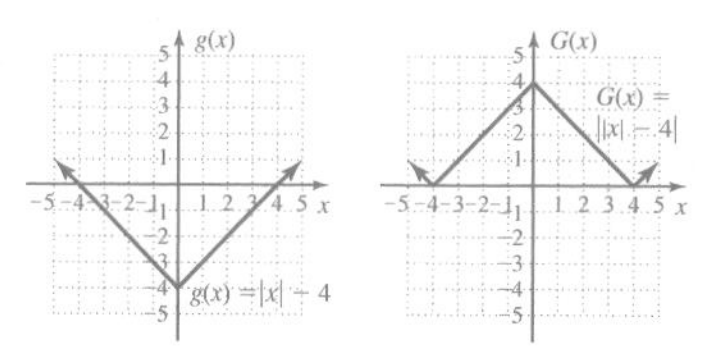

93. $x = 2, x = 2 - 1 \pm \sqrt{3}i$

94. $y = \frac{-2}{3}x + 5$

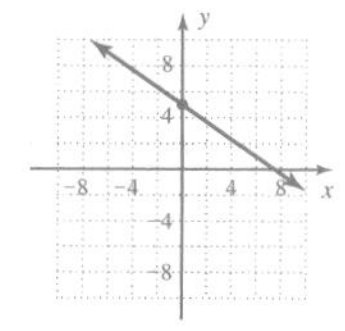

83. Wind power: The power P generated by a certain wind turbine is given by the function $P(v) = \frac{8}{125}v^3$ where $P(v)$ is the power in watts at wind velocity v (in miles per hour). Describe the transformation applied to obtain the graph of P from the graph of $y = v^3$, then sketch the graph of P for $v \in [0, 25]$ (scale the axes appropriately). How much power is being generated when the wind is blowing at 15 mph? compressed vertically, 216 W

84. Wind power: If the power P (in watts) being generated by a wind turbine is known, the velocity of the wind can be determined using the function $v(P) = (\frac{5}{2})\sqrt[3]{P}$. Describe the transformation applied to obtain the graph of v from the graph of $y = \sqrt[3]{P}$, then sketch the graph of v for $P \in [0, 512]$ (scale the axes appropriately). How fast is the wind blowing if 343 W of power is being generated? stretched vertically, 17.5 mph

85. Acceleration due to gravity: The *velocity* of a steel ball bearing as it rolls down an inclined plane is given by the function $v(t) = 4t$, where $v(t)$ represents the velocity in feet per second after t sec. Describe the transformation applied to obtain the graph of v from the graph of $y = t$, then sketch the graph of v for $t \in [0, 3]$. What is the velocity of the ball bearing after 2.5 sec? How far has the ball bearing rolled after 2.5 sec? (*Hint*: See Example 9.) vertical stretch by a factor of 4, 10 ft/sec, 12.5 ft

86. Acceleration due to gravity: The *distance* a ball rolls down an inclined plane is given by the function $d(t) = 2t^2$, where $d(t)$ represents the distance in feet after t sec. Describe the transformation applied to obtain the graph of d from the graph of $y = t^2$, then sketch the graph of d for $t \in [0, 3]$. How far has the ball rolled after 2.5 sec? How does this compare with the result from Exercise 85? vertical stretch by a factor of 2, 12.5 ft, results are identical

WRITING, RESEARCH, AND DECISION MAKING

87. Carefully graph the functions $f(x) = |x|$ and $g(x) = 2\sqrt{x}$ on the same coordinate grid. From the graph, in what interval is the graph of $g(x)$ *above* the graph of $f(x)$? Pick a number (call it h) from this interval and substitute it in both functions. Is $g(h) > f(h)$? In what interval is the graph of $g(x)$ below the graph of $f(x)$? Pick a number from this interval (call it k) and substitute it in both functions. Is $g(k) < f(k)$? $x \in (0, 4)$; yes, $x \in (4, \infty)$; yes

88. For any function $f(x)$, the graph of $f(-x)$ is a horizontal reflection (across the y-axis), while $-f(x)$ is a vertical reflection (across the x-axis). Given $f(x) = \sqrt[3]{x}$, compare the graph of $f(-x)$ with the graph of $-f(x)$. What do you observe? For $f(x) = x^3$, does the comparison of $f(-x)$ with $-f(x)$ yield similar results? Can you find another function that exhibits the same relationship? The reflections produce identical results; yes; any odd function

89. The transformations studied in this section can also be applied to linear functions, with surprising results. For $f(x) = 2x$, compare the graph of $y = 2(x - 3)$ [shifts graph of f 3 units right] with the graph of $y = 2x - 6$ [shifts graph of f 6 units down]. What do you notice? What is the connection? The result are identical; $2(x - 3) = 2x - 6$ via the distributive property

EXTENDING THE CONCEPT

90. Sketch the graph of $f(x) = \frac{1}{2}(x - 4)^2 + 2$ using transformations of the parent function, then determine the area of the region in quadrant I that is beneath the graph and bounded by the vertical line $x = 8$.

91. Sketch the graph of $f(x) = x^2 - 4$, then sketch the graph of $F(x) = |x^2 - 4|$ using your intuition and the meaning of absolute value (not a table of values). What happens to the graph?

92. Sketch the graph of $g(x) = |x| - 4$, then sketch the graph of $G(x) = ||x| - 4|$ using your intuition and the meaning of absolute value (not a table of values). What happens to the graph? Discuss the similarities between Exercises 91 and 92.

MAINTAINING YOUR SKILLS

93. (1.3) Solve the equation $x^3 - 8 = 0$. Find all zeroes, real and complex.

94. (2.3) Solve the equation for y, then sketch its graph using the slope/intercept method: $2x + 3y = 15$.

95. $d = 29$, $m = \frac{-21}{20}$
96. $p = 140$ in. $A = 1168$ in^2
97. $x = -5$
98. $h(x) = 2x^2 + 11x + 14$

95. (2.1) Find the distance between the points $(-13, 9)$ and $(7, -12)$, and the slope of the line containing these points.

96. (R.7) Find the perimeter and area of the figure shown (note the units).

32 in.
32 in.
2 ft
38 in.

97. (1.1) Solve for x: $\frac{2}{3}x + \frac{1}{4} = \frac{1}{2}x - \frac{7}{12}$.

98. (3.1) Given $f(x) = 2x^2 + 3x$ and $g(x) = x + 2$, find $h(x) = (f \circ g)(x)$.

3.4 Graphing General Quadratic Functions

LEARNING OBJECTIVES

In Section 3.4 you will learn how to:

A. Graph quadratic functions by completing the square and transforming $y = x^2$

B. Graph a general quadratic function using the vertex formula

C. Determine the equation of a function from its graph

D. Solve applications involving extreme values of quadratic functions

INTRODUCTION

In Section 3.3, we graphed variations of the basic toolbox functions by transforming the graph of a parent function. In this section, we focus on a useful connection between the "shifted form" of a quadratic function and the general quadratic function $y = ax^2 + bx + c$. In addition, an alternative to the quadratic formula is introduced that greatly simplifies the work required to find x-intercepts, once the vertex of the parabola is known.

POINT OF INTEREST

Of the entire family of polynomial equations, perhaps no other has received more attention than *quadratic equations*, represented by $ax^2 + bx + c = 0$, $a \neq 0$. Methods to solve this equation were developed independently by almost every major civilization, including the Arabs, Babylonians, Hindus, Greeks, and others. Most were developed from a geometric viewpoint. In the early 1600s, René Descartes formalized the connection between a function and its graphical representation in the coordinate plane. Over time, additional ideas were introduced to help sketch and understand all aspects of a quadratic graph, leading to a better understanding of its many applications.

A. Graphing a Quadratic Function by Completing the Square

In Example 7 from Section 3.3 we graphed $y = -(x + 2)^2 + 3$ using a vertical reflection and shifting the parent function 2 units to the left and 3 units up. Since the original vertex also shifts by these amounts, the new vertex was $(-2, 3)$. For obvious reasons, this is called the **shifted form** of a quadratic function and is generally written $y = a(x - h)^2 + k$, with the horizontal shift given by the value of h directly, rather than being considered "opposite the sign." A useful connection between the shifted form and the general quadratic function can be established by completing the square. When completing the square on a *quadratic equation* (as in Section 1.5), we apply the standard properties of equality to both sides of the equation. When completing the square on a *quadratic function,* the process is altered slightly so that we operate on only one side. For instance, instead of "adding $[\frac{1}{2}(\text{linear coefficient})]^2$ to *both* sides," we simultaneously add and subtract this term, then regroup as illustrated in Example 1.

EXAMPLE 1 Graph $y = -2x^2 - 8x - 3$ by completing the square.

Solution:

$y = -2x^2 - 8x - 3$	given function
$= (-2x^2 - 8x + \_\_\_) - 3$	group variable terms
$= -2(x^2 + 4x + \_\_\_) - 3$	factor out "a"
$= -2[(x^2 + 4x + 4) - 4] - 3$	$[(\frac{1}{2})(4)]^2 = 4$; add 4 then subtract 4 and regroup
$= -2[(x + 2)^2 - 4] - 3$	factor trinomial
$= -2(x + 2)^2 + 5$	distribute and simplify

The parabola is stretched vertically, concave down ($a < 0$), shifted 2 units left and 5 up with the vertex at $(-2, 5)$. The y-intercept is $(0, -3)$. Since the graph is concave down with a vertex *above the x-axis*, there are two x-intercepts, which we find using the quadratic formula (the expression does not factor).

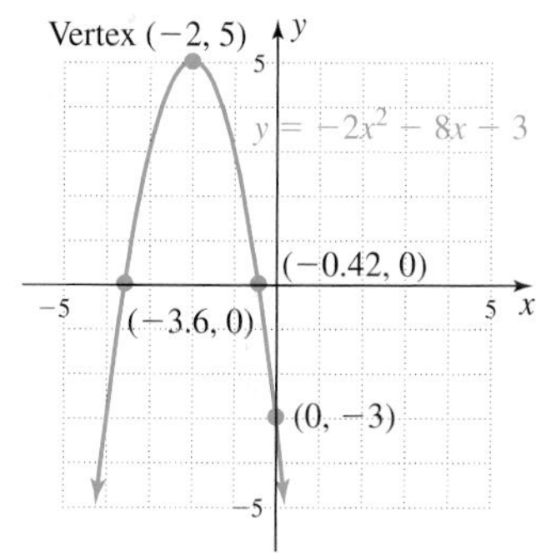

$x = \dfrac{-b \pm \sqrt{b^2 - 4ac}}{2a}$	quadratic formula
$= \dfrac{-(-8) \pm \sqrt{(-8)^2 - 4(-2)(-3)}}{2(-2)}$	substitute
$= \dfrac{8 \pm \sqrt{40}}{-4} = \dfrac{8 \pm 2\sqrt{10}}{-4}$	simplify
$= \dfrac{-4 \pm \sqrt{10}}{2}$	exact form
$x \approx -3.6 \qquad x \approx -0.42$	approximate form

NOW TRY EXERCISES 7 THROUGH 18

The main ideas are highlighted here:

GRAPHING QUADRATIC FUNCTIONS BY COMPLETING THE SQUARE

1. Group the variable terms apart from the constant "c".
2. Factor out the lead coefficient "a."
3. Compute $\left[\frac{1}{2}\left(\frac{b}{a}\right)\right]^2$, then add and subtract the result to the variable terms and regroup to form a factorable trinomial.
4. Factor the grouped terms as a binomial square; then distribute and combine constant terms.
5. Graph using transformations of $y = x^2$.

In cases like $y = 3x^2 - 10x + 3$ where the linear coefficient has no integer factors of a, we simultaneously factor out 3 and divide by 3 to begin the process. This yields $y = 3\left(x^2 - \frac{10}{3}x + \_\_\_\right) + 3$, with the process continuing as before. See Exercises 19 and 20.

If the lead coefficient is positive ($a > 0$) and the vertex (h, k) is below the x-axis, the graph will have two x-intercepts (Figure 3.40). If $a > 0$ and the vertex is above the x-axis, the graph will not intersect the x-axis (Figure 3.41). Similar statements can be made for the case when a is negative.

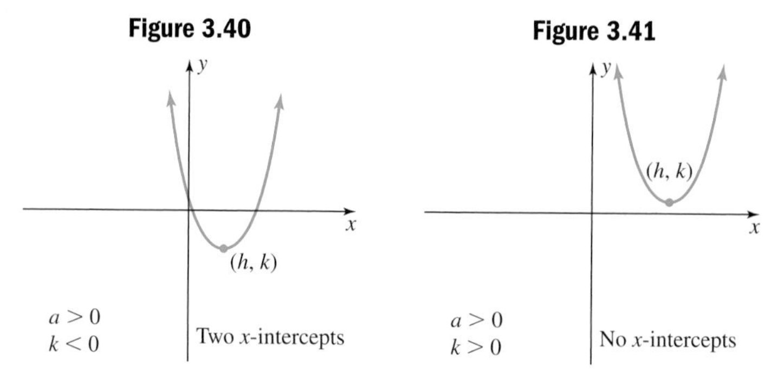

B. Graphing General Quadratic Functions

When this process is applied to the general function $ax^2 + bx + c$ we obtain

$$f(x) = ax^2 + bx + c \qquad \text{general quadratic function}$$

$$= (ax^2 + bx + \_\_\_) + c \qquad \text{group variable terms apart from the constant "c"}$$

$$= a\left(x^2 + \frac{b}{a}x + \_\_\_\right) + c \qquad \text{factor out "a"}$$

We next compute $\left[\frac{1}{2}\left(\frac{b}{a}\right)\right]^2 = \frac{b^2}{4a^2}$, then add and subtract the result to the terms within parentheses. By regrouping these terms, we simultaneously create a factorable trinomial while maintaining an equivalent expression. After factoring, use the distributive property to simplify the final expression as shown:

$$y = a\left[\left(x^2 + \frac{b}{a}x + \frac{b^2}{4a^2}\right) - \frac{b^2}{4a^2}\right] + c \qquad \text{add and subtract } \frac{b^2}{4a^2} \text{ and regroup}$$

$$= a\left[\left(x + \frac{b}{2a}\right)^2 - \frac{b^2}{4a^2}\right] + c \qquad \text{factor the trinomial}$$

$$= a\left(x + \frac{b}{2a}\right)^2 - \frac{b^2}{4a} + c \qquad \text{distribute: } a \cdot \left(\frac{b^2}{4a^2}\right) = \frac{b^2}{4a}$$

$$= a\left(x + \frac{b}{2a}\right)^2 + \frac{4ac - b^2}{4a} \qquad \text{write constants as a single term}$$

By comparing this result with the transformations from Section 3.3, we note the x-coordinate of the vertex (h, k) is $h = \frac{-b}{2a}$ (since the graph shifts horizontally "opposite the sign" of the binomial). Instead of using the expression $\frac{4ac - b^2}{4a}$ to find k, we substitute $\frac{-b}{2a}$ back into the function: $k = f\left(\frac{-b}{2a}\right)$. The result is called the **vertex formula.**

VERTEX FORMULA

For a quadratic function written in standard form $f(x) = ax^2 + bx + c$, the coordinates of the vertex are given by

$$(h, k) = \left(\frac{-b}{2a}, f\left(\frac{-b}{2a}\right)\right).$$

Graphing quadratic functions by completing the square was primarily a vehicle to lead us to the vertex formula. Since all characteristic features of a quadratic graph (concavity, vertex, axis of symmetry, x-intercepts, and y-intercept) can now be determined from the original equation, we'll rely on these features to sketch quadratic graphs, rather than to continue completing the square.

EXAMPLE 2 Graph $f(x) = 2x^2 + 10x + 7$ and locate its zeroes (if they exist).

Solution: The graph will be concave up since $a > 0$, with the y-intercept at $(0, 7)$. The vertex formula $\left(\frac{-b}{2a}, f\left(\frac{-b}{2a}\right)\right)$ yields $\left(-\frac{5}{2}, -\frac{11}{2}\right)$. For the x-intercepts, the quadratic formula gives

$$x = \frac{-b \pm \sqrt{b^2 - 4ac}}{2a} \qquad \text{quadratic formula}$$

$$= \frac{-10 \pm \sqrt{(10)^2 - 4(2)(7)}}{2(2)} \qquad \text{substitute}$$

$$= \frac{-10 \pm \sqrt{44}}{4} = \frac{-10 \pm 2\sqrt{11}}{4} \qquad \text{simplify}$$

$$= \frac{-5}{2} \pm \frac{\sqrt{11}}{2} \qquad \text{exact form}$$

$$x \approx -4.16 \qquad x \approx -0.84 \qquad \text{approximate form}$$

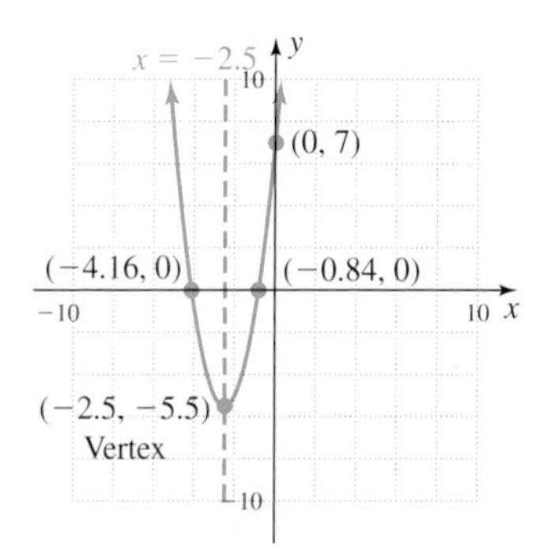

The graph is shown in the figure.

NOW TRY EXERCISES 21 THROUGH 24

As in Example 2, the shifted form easily gives us the vertex of the parabola, but unless the original equation is factorable, finding x-intercepts requires the quadratic formula. However, since the vertex (h, k) of the parabola is known, an alternative formula for finding x-intercepts can be developed using the *general* shifted form:

$$a(x - h)^2 + k = y \qquad \text{shifted form of the general quadratic}$$

$$a(x - h)^2 + k = 0 \qquad \text{set equal to zero to find } x\text{-intercepts}$$

$$a(x - h)^2 = -k \qquad \text{subtract } k \text{ from both sides}$$

$$(x - h)^2 = -\frac{k}{a} \qquad \text{divide by "}a\text{"}$$

$$x - h = \pm\sqrt{-\frac{k}{a}} \qquad \text{take the square root of both sides}$$

$$x = h \pm \sqrt{-\frac{k}{a}} \qquad \text{solve for } x \text{ (add } h \text{ to both sides)}$$

The result is called the **vertex/intercept formula** and as with the quadratic formula, will yield both solutions, *even when roots are irrational or complex.*

THE VERTEX/INTERCEPT FORMULA

Given a quadratic function with lead coefficient a and vertex at (h, k), the zeroes of the function are given by $x = h \pm \sqrt{-\frac{k}{a}}$.

- If a and k have unlike signs, there are two real roots.
- If a and k have like signs, there are two complex roots.
- If the ratio $-\frac{k}{a}$ is positive and a perfect square, roots are rational.

EXAMPLE 3 Graph $h(x) = x^2 - 4x + 7$ and locate its zeroes (if they exist).

Solution: The graph will be concave up since $a > 0$, with the y-intercept at $(0, 7)$. The vertex formula $\left(\frac{-b}{2a}, f\left(\frac{-b}{2a}\right)\right)$ yields the vertex $(2, 3)$. Since the graph is concave up with a vertex above the x-axis, there are no x-intercepts. Using the vertex/intercept formula with $a = 1$, the complex zeroes are: $x = 2 \pm \sqrt{-\frac{3}{1}} = 2 \pm i\sqrt{3}$.

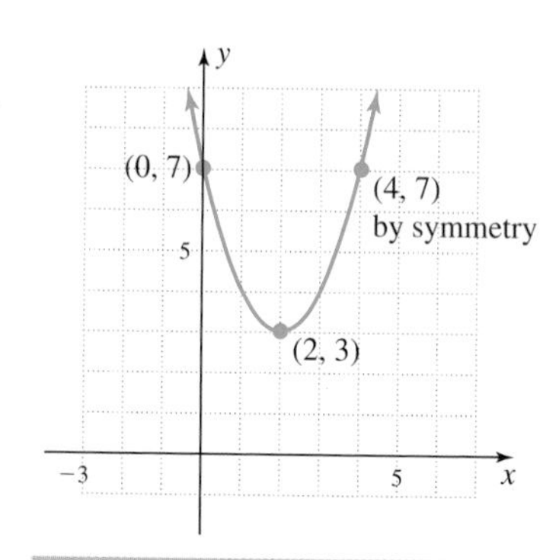

NOW TRY EXERCISES 25 THROUGH 42

C. Finding the Equation of a Function from Its Graph

In shifted form, we can identify the vertex, node, or pivot point of any toolbox function that has been transformed in some way. Using this identification process "in reverse" enables us to determine the original equation of a function, given its graph. The method is similar to that used in Section 2.3, where we found the equation of a line from its graph using the slope and a point on the line (recall how we used $y = mx + b$ as a "formula"). Given the graph of a toolbox function, we note the coordinates of the node, vertex, or pivot point, along with another point on the graph, then use the shifted form $y = af(x - h) + k$ as a formula. After substituting the characteristic points h and k, along with the x- and y-values of the point, we solve for the value of a to complete the equation.

EXAMPLE 4 Find the equation of the toolbox function $f(x)$ shown in Figure 3.42.

Figure 3.42

y
5
f(x)
(1, 2)
(−3, 0)
(5, 0)
−5
5
x
−5

Solution: The function f belongs to the absolute value family. The vertex (h, k) is at $(1, 2)$. For an additional point, choose the x-intercept $(-3, 0)$ and work as follows:

$y = a|x - h| + k$ shifted form

$0 = a|(-3) - 1| + 2$ substitute 1 for h and 2 for k from vertex $(h, k) = (1, 2)$; substitute -3 for x and 0 for y from intercept $(-3, 0)$

$0 = 4a + 2$ simplify

$-\frac{1}{2} = a$ solve for a

The equation for f is $y = -\frac{1}{2}|x - 1| + 2$.

NOW TRY EXERCISES 43 THROUGH 48

D. Quadratic Functions and Extreme Values

If $a > 0$, the parabola is concave up and the y-coordinate of the vertex is a minimum value—the smallest value attained by the function anywhere in its domain. Conversely, if $a < 0$, the parabola is concave down and the vertex yields a maximum value. These greatest and least points are known as **extreme values** and have a number of significant applications.

EXAMPLE 5 An airplane manufacturer can produce up to 15 planes per month. The profit made from the sale of these planes can be modeled by $P(x) = -0.2x^2 + 4x - 3$, where $P(x)$ is the profit in hundred-thousands of dollars per month and x is the number of planes made and sold. Based on this model,

a. Find the y-intercept and explain what it means in this context.

b. How many planes should be made and sold to maximize profit?

c. What is the maximum profit?

d. Find the x-intercepts and explain what they mean in this context.

Solution: **a.** $P(0) = -3$, which means the manufacturer loses \$300,000 each month if the company produces no planes.

b. The phrase "maximize profit" indicates we're seeking the maximum value of the function. Using the vertex formula with $a = -0.2$ and $b = 4$ gives

$x = \frac{-b}{2a}$ vertex formula

$= \frac{-4}{2(-0.2)}$ substitute -0.2 for a and 4 for b

$= 10$ result

The graph will be concave down $(a < 0)$, with the maximum value occurring at $x = 10$. This shows that 10 planes should be made and sold each month.

c. Note that while $x = 10$ tells *when* maximum profit occurs, it's the y-coordinate of the vertex that actually names the extreme value. Evaluating $P(10)$ we find that 17 "hundred thousand dollars" in profit will be earned ($1,700,000).

d. Using the vertex/intercept formula with $a = -0.2$, the x-intercepts are $x = 10 \pm \sqrt{-\frac{17}{-0.2}} = 10 \pm \sqrt{85}$. In approximate form: (0.78, 0) and (19.22, 0). The intercepts tell us the company just about breaks even (has zero profit) if 1 plane is sold or if 19 planes are made and sold.

NOW TRY EXERCISES 51 THROUGH 60

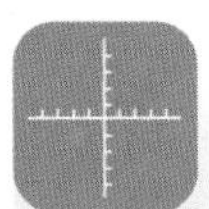

TECHNOLOGY HIGHLIGHT

Estimating Irrational Roots Using a Graphing Calculator

The keystrokes shown apply to a TI-84 Plus model. Please consult your manual or our Internet site for other models.

The solutions of an equation are called the "roots" of the equation. Graphically, we refer to these roots as x-intercepts. With relation to functions, these solutions are called **zeroes** of the function, because they are the *input* values that produce an *output of zero.* Once a function is graphed on the TI-84 Plus, an estimate for irrational zeroes can easily be found. Enter the function $y = x^2 - 8x + 9$ on the Y= screen and graph using the standard window (ZOOM 6). We access the option for finding zeroes by pressing 2nd TRACE (CALC), which displays the screen in Figure 3.43. Pressing the number "2" selects the **2:zero** option and returns you to the graph, where you are asked to enter a "Left Bound." The calculator is asking you to narrow down the area it has to search for the x-intercept. Select any number that is conveniently to the left of the x-intercept you're interested in. For this graph, we entered a left bound of "0" (press ENTER). The calculator marks this choice at the top of the screen with a "▶" marker (pointing to the right), then asks you to enter a "Right Bound." Select any value to the right of this x-intercept, but be sure the value you enter *bounds only one intercept* (see Figure 3.44). For this graph, a choice of 10 would include both x-intercepts, while a choice of 3 would bound only the x-intercept on the left. After entering 3, the calculator asks for a "guess." This option is used only when there are many different zeroes close by or if you entered a large interval. Most of the time we'll simply bypass this option by pressing ENTER . The cursor will be located at the zero you chose, with the coordinates displayed at the bottom of the screen in Figure 3.45. The x-value is an approximation of the irrational root, and the y-value is zero. Find the zeroes of these functions using the 2nd TRACE (CALC) **2:Zero** feature.

Figure 3.43

CALCULATE
1:value
2:zero
3:minimum
4:maximum
5:intersect
6:dy/dx
7:∫f(x)dx

Figure 3.44

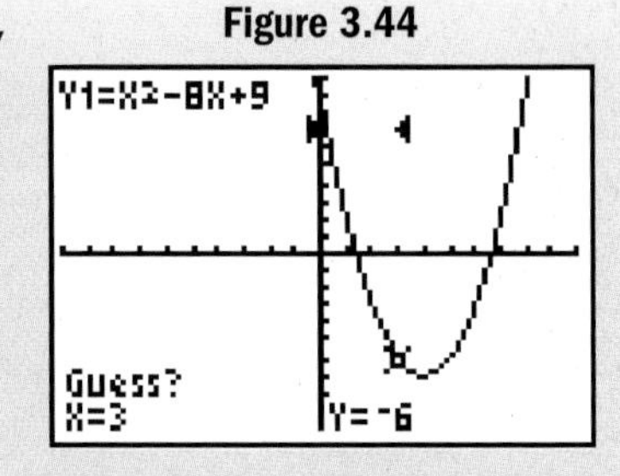

Figure 3.45

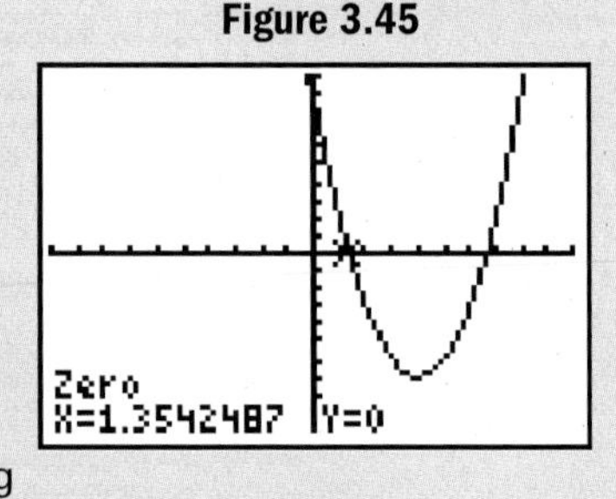

Exercise 1: $y = x^2 - 8x + 9$ 1.35, 6.65

Exercise 2: $y = 3a^2 - 5a - 6$ −0.81, 2.47

Exercise 3: $y = 2x^2 + 4x - 5$ −2.87, 0.87

Exercise 4: $y = 9w^2 + 6w - 1$ −0.80, 0.14

3.4 EXERCISES

7.

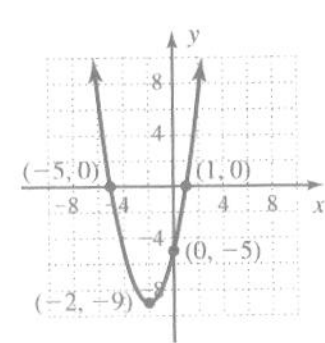

left 2, down 9

8.

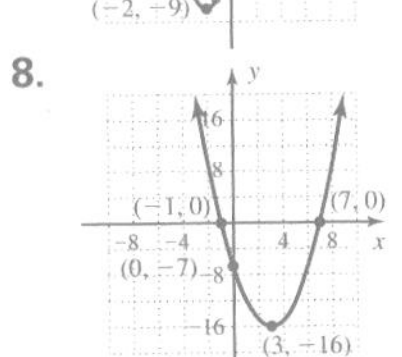

right 3, down16

9.

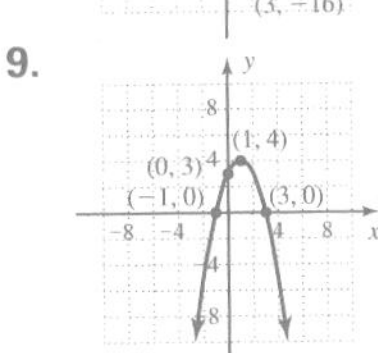

reflected across *x*-axis; right 1, up 4

10.

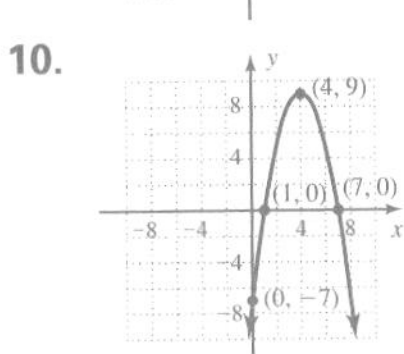

reflected across *x*-axis; right 4, up 9

11.

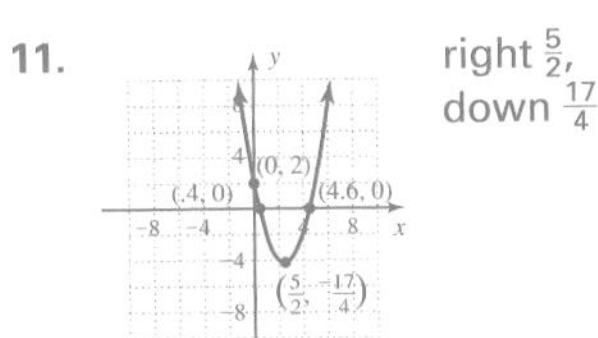

right $\frac{5}{2}$, down $\frac{17}{4}$

12. left $\frac{7}{2}$, down $\frac{33}{4}$

13.

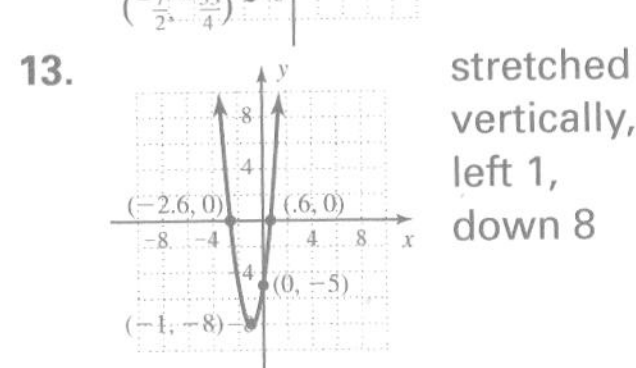

stretched vertically, left 1, down 8

14. stretched vertically, right 3, down 21

15.

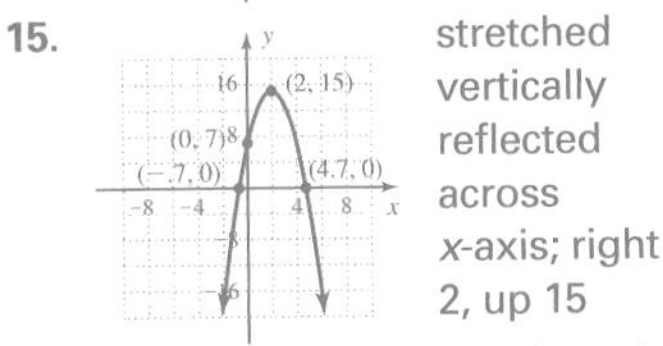

stretched vertically reflected across *x*-axis; right 2, up 15

Additional answers can be found in the Instructor Answer Appendix.

CONCEPTS AND VOCABULARY

Fill in each blank with the appropriate word or phrase. Carefully reread the section if needed.

1. Fill in the blank to complete the square given $f(x) = -2x^2 - 10x - 7$: $f(x) = -2(x^2 + 5x + \frac{25}{4}) - 7 +$ __$\frac{25}{2}$__.

2. The maximum and minimum values are called __extreme__ values and can be found using the __shifted__ form or the __vertex__ formula.

3. To find the x-intercepts of $f(x) = ax^2 + bx + c$, we use the __vertex__ formula.

4. To find the x-intercepts of $f(x) = a(x - h)^2 + k$, we use the __vertex/intercept__ formula.

5. Compare/contrast how to complete the square on an *equation*, versus how to complete the square on a function. Use the equation $2x^2 + 6x - 3 = 0$ and the function $f(x) = 2x^2 + 6x - 3 = 0$ to illustrate. Answers will vary.

6. Discuss/explain why the graph of a quadratic function has no x-intercepts if a and k have like signs. What happens to the radicand of the vertex-intercept formula when they do? Under what conditions will the function have a single real root? No real roots, it's negative if $k = 0$

DEVELOPING YOUR SKILLS

Graph each function by completing the square to write the function in shifted form. Clearly state the transformations used to obtain the graph, and label the vertex and all intercepts (if they exist). Use the quadratic formula to find the x-intercepts.

7. $f(x) = x^2 + 4x - 5$	**8.** $g(x) = x^2 - 6x - 7$	**9.** $h(x) = -x^2 + 2x + 3$
10. $H(x) = -x^2 + 8x - 7$	**11.** $p(x) = x^2 - 5x + 2$	**12.** $q(x) = x^2 + 7x + 4$
13. $Y_1 = 3x^2 + 6x - 5$	**14.** $Y_2 = 4x^2 - 24x + 15$	**15.** $f(x) = -2x^2 + 8x + 7$
16. $g(x) = -3x^2 + 12x - 7$	**17.** $h(x) = -\frac{1}{2}x^2 + 5x - 7$	**18.** $H(x) = -\frac{1}{3}x^2 + 2x + 5$
19. $p(x) = 2x^2 - 7x + 3$	**20.** $q(x) = 4x^2 - 9x + 2$	**21.** $f(x) = -3x^2 - 7x + 6$
22. $g(x) = -2x^2 + 9x - 7$	**23.** $Y_1 = 3x^2 + 5x - 1$	**24.** $Y_2 = 2x^2 + 5x + 1$

Find the zeroes of each function (real or complex) using the vertex/intercept formula.

25. $y = (x + 3)^2 - 5$ $x = -3 \pm \sqrt{5}$	**26.** $y = -(x - 4)^2 + 3$ $x = 4 \pm \sqrt{3}$	**27.** $y = 2(x + 4)^2 - 7$ $x = -4 \pm \frac{\sqrt{14}}{2}$
28. $y = -3(x - 2)^2 + 6$ $x = 2 \pm \sqrt{2}$	**29.** $s(t) = 0.2(t + 0.7)^2 - 0.8$ $x = -2.7,\ x = 1.3$	**30.** $r(t) = -0.5(t - 0.6)^2 + 2$ $x = -1.4,\ x = 2.6$

Graph each function using the concavity, y-intercept, x-intercept(s), vertex, and symmetry. Label the vertex and all intercepts (if they exist). Use the vertex-intercept formula to find the x-intercepts (round to tenths as needed).

31. $f(x) = x^2 + 2x - 6$	**32.** $g(x) = x^2 + 8x + 11$	**33.** $h(x) = -x^2 + 4x + 2$
34. $H(x) = -x^2 + 10x - 19$	**35.** $Y_1 = 0.5x^2 + 3x + 7$	**36.** $Y_2 = 0.2x^2 - 2x + 8$
37. $Y_1 = -2x^2 + 10x - 7$	**38.** $Y_2 = -2x^2 + 8x - 3$	**39.** $f(x) = 4x^2 - 12x + 3$
40. $g(x) = 3x^2 + 12x + 5$	**41.** $p(x) = \frac{1}{2}x^2 + 3x - 5$	**42.** $q(x) = \frac{1}{3}x^2 - 2x - 4$

Use the graph given and the points indicated to determine the equation of the function shown.

43. $f(x) = -(x - 2)^2$

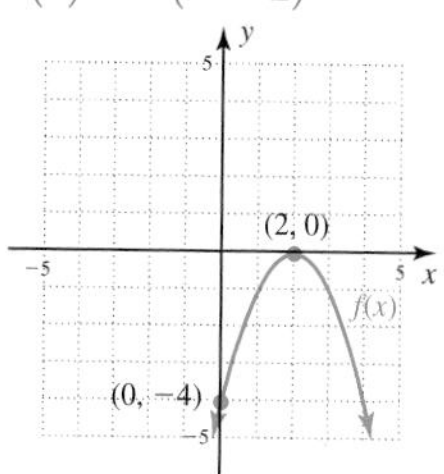

44. $g(x) = \frac{2}{5}x^2 - 4$

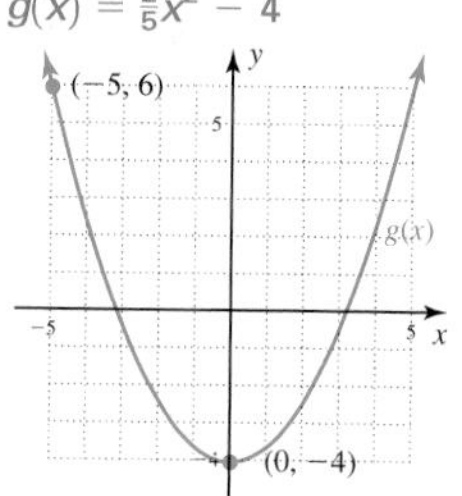

45. $p(x) = 1.5\sqrt{x + 3}$

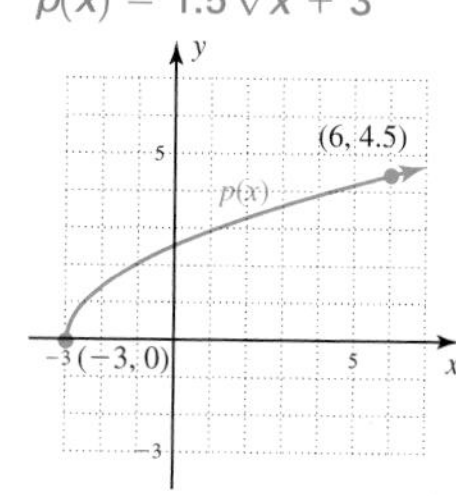

46.

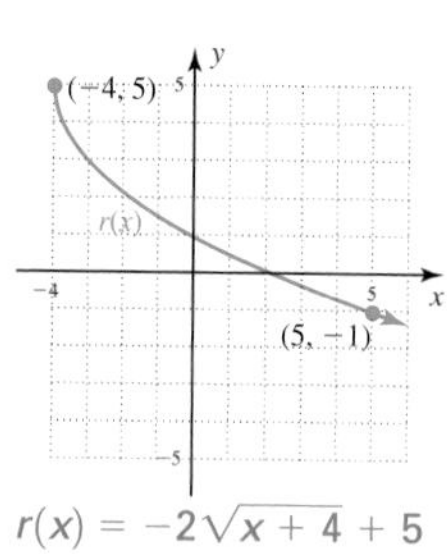

$r(x) = -2\sqrt{x + 4} + 5$

47.

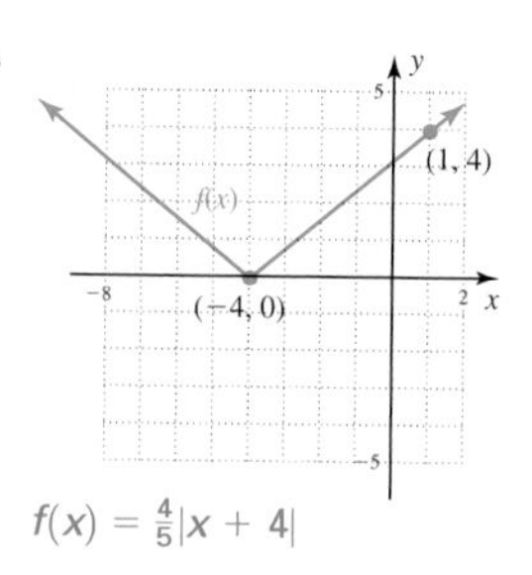

$f(x) = \frac{4}{5}|x + 4|$

48.

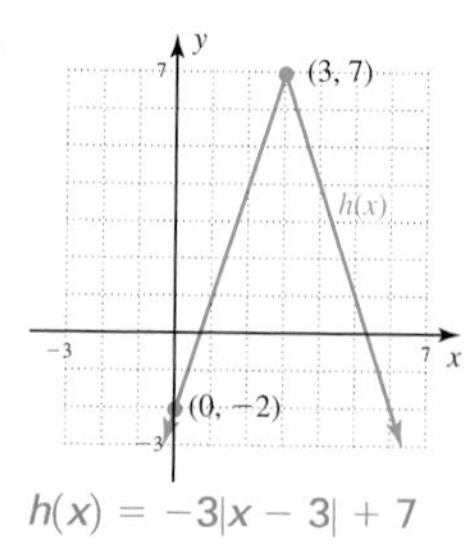

$h(x) = -3|x - 3| + 7$

WORKING WITH FORMULAS

49. Sum of the first n positive integers: $S = \frac{1}{2}n^2 + \frac{1}{2}n$

The sum of the first n positive integers is given by the formula shown, where n is the desired number of integers in the sum. Use the formula to compute the sum of the first 10 positive integers, then verify the result by computing the sum by hand. 55

50. Surface area of a rectangular box with square ends: $S = 2h^2 + 4Lh$

The surface area of a rectangular box with square ends is given by the formula shown, where h is the height and width of the square ends and L is the length of the box. If L is 3 ft and the box must have a surface area of 32 ft^2, find the dimensions of the square ends. 2 ft × 2 ft

APPLICATIONS

51. Maximum profit: An automobile manufacturer can produce up to 300 cars per day. The profit made from the sale of these vehicles can be modeled by the function $P(x) = -10x^2 + 3500x - 66{,}000$, where $P(x)$ is the profit in dollars and x is the number of automobiles made and sold. Based on this model:

a. Find the y-intercept and explain what it means in this context.

b. Find the x-intercepts and explain what they mean in this context.

c. How many cars should be made and sold to maximize profit?

d. What is the maximum profit?

51. a. (0, −66,000); when no cars are produced, there is a loss of $66,000.
b. (20, 0), (330, 0); no profit will be made if less than 20 or more than 330 cars are produced.
c. 175
d. $240,250

52. Maximum profit: The profit for a manufacturer of collectible grandfather clocks is given by the function shown here, where $P(x)$ is the profit in dollars and x is the number of clocks made and sold. Answer the following questions based on this model: $P(x) = -1.6x^2 + 240x - 375$.

a. Find the y-intercept and explain what it means in this context.

b. Find the x-intercepts and explain what they mean in this context.

a. (0, −375); when no clocks are made, there is a loss of $375. b. (1.58, 0), (148.4, 0); no profit will be made if you produce less than 2 or more than 148 clocks. c. 75 d. $8625

c. How many clocks should be made and sold to maximize profit?

d. What is the maximum profit?

53. Depth of a dive: As it leaves its support harness, a minisub takes a deep dive toward an underwater exploration site. The dive path is modeled by the function $d(x) = x^2 - 12x$, where $d(x)$ represents the depth of the minisub in hundreds of feet at a distance of x mi from the surface ship.

a. How far from the mother ship did the minisub reach its deepest point? 6 mi

b. How far underwater was the submarine at its deepest point? 3600 ft

c. At $x = 4$ mi, how deep was the minisub explorer? 3200 ft

d. How far from its entry point did the minisub resurface? 12 mi

54. Optimal pricing strategy: The director of the Ferguson Valley drama club must decide what to charge for a ticket to the club's performance of *The Music Man.* If the price is set too low, the club will lose money; and if the price is too high, people won't come. From past experience she estimates that the income I from sales (in hundreds) can be approximated by $I(x) = -x^2 + 46x - 88$, where x is the cost of a ticket and $0 \le x \le 50$.

a. Find the lowest cost of a ticket that would allow the club to break even. \$2

b. What is the highest cost that the club can charge to break even? \$44

c. If the theater were to close down before any tickets are sold, how much money would the club lose? \$8800

d. How much should the club charge to maximize their profits? What is the maximum profit? \$23; \$44,100

55. Maximum profit: A kitchen appliance manufacturer can produce up to 200 appliances per day. The profit made from the sale of these machines can be modeled by the function $P(x) = -0.5x^2 + 175x - 3300$, where $P(x)$ is the profit in dollars and x is the number of appliances made and sold. Based on this model,

a. Find the y-intercept and explain what it means in this context.

b. Find the x-intercepts and explain what they mean in this context.

c. Determine the domain of the function and explain its significance.

d. How many should be sold to maximize profit? What is the maximum profit?

a. (0, −3300); if no appliances are sold, the loss will be \$3300. **b.** (20, 0), (330, 0); if less than 20 or more than 330 appliances are made and sold, there will be no profit. **c.** $0 \le x \le 200$ **d.** 175, \$12,012.50

By now you are familiar with the function that models the height of a projectile: $h(t) = -16t^2 + vt + k$. The function applies to any object projected upward with an initial velocity v, but not to objects under propulsion (such as a rocket). Consider this situation and answer the questions that follow.

56. Model rocketry: The member of the local rocketry club launches her latest rocket from a large field. At the moment its fuel is exhausted, the rocket has a velocity of 240 ft/sec and an altitude of 544 ft (t is in seconds).

a. Write the function that models the height of the rocket. $h(t) = -16t^2 + 240t + 544$

b. How high is the rocket at $t = 0$? If it took off from the ground, why is it this high at $t = 0$? 544 ft; that is when the fuel is exhausted.

c. How high is the rocket after 5 sec? 1344 ft

d. How high is the rocket after 10 sec? 1344 ft

e. How could it stay at the same height for 5 sec? It is coming back down.

f. What is the maximum height attained by the rocket? 1444 ft

57. b.

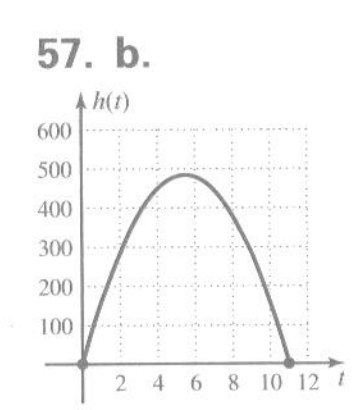

57. Height of a projectile: A projectile is thrown upward with an initial velocity of 176 ft/sec. After t sec, its height $h(t)$ above the ground is given by the function $h(t) = -16t^2 + 176t$.

a. Find the projectile's height above the ground after 2 sec. 288 ft

b. Sketch the graph modeling the projectile's height.

c. What is the projectile's maximum height? What is the value of t at this height? 484 ft; 5.5 sec

d. How many seconds after it is thrown will the projectile strike the ground? 11 sec

58. Height of a projectile: In the movie *The Court Jester* (1956; Danny Kaye, Basil Rathbone, Angela Lansbury, and Glynis Johns), a catapult is used to toss the nefarious adviser to the king into a river. Suppose the path flown by the king's adviser is modeled by the function $h(d) = -0.02d^2 + 1.64d + 14.4$, where $h(d)$ is the height of the adviser in feet at a distance of d ft from the base of the catapult. **a.** 14.4 ft **b.** 41 ft **c.** 48.02 ft **d.** 90 ft

a. How high was the release point of this catapult?

b. How far from the catapult did the adviser reach a maximum altitude?

c. What was this maximum altitude attained by the adviser?

d. How far from the catapult did the adviser splash into the river?

59. Cost of production: The cost of producing a plastic toy is given by the function $C(x) = 2x + 35$, where x is the number of hundreds of toys. The revenue from toy sales is given by $R(x) = -x^2 + 122x - 365$. Since profit = revenue − cost, the profit function must be $P(x) = -x^2 + 120x - 400$ (verify). How many toys sold will produce the maximum profit? What is the maximum profit? 6000; \$3200

60. Cost of production: The cost to produce bottled spring water is given by $C(x) = 16x + 63$, where x is the number of thousands of bottles. The total income (revenue) from the sale of these bottles is given by the function $R(x) = -x^2 + 326x - 7463$. Since profit = revenue − cost, the profit function must be $P(x) = -x^2 + 310x - 7400$ (verify). How many bottles sold will produce the maximum profit? What is the maximum profit?
155,000; \$16,625

WRITING, RESEARCH, AND DECISION MAKING

61. Catapults have a long and fascinating history. Although very ancient devices, catapults were used for hundreds of years, even in modern times. Do some research on catapults, reporting on various kinds that have been built and the ways they were used. Based on your report, create an interesting or innovative application problem of your own, using a quadratic equation to model the path of the object thrown by your catapult. Answers will vary.

62. Use the general solutions from the quadratic formula to show that the average value of the x-intercepts is $\frac{-b}{2a}$. Explain/discuss why the result is valid even if the roots are complex.

$$x_1 = \frac{-b + \sqrt{b^2 - 4ac}}{2a} \qquad x_2 = \frac{-b - \sqrt{b^2 - 4ac}}{2a}$$

Answers will vary.

EXTENDING THE CONCEPT

63. Write the equation of a quadratic function whose x-intercepts are given by $x = 2 \pm 3i$. $f(x) = x^2 - 4x + 13$

64. Write the equation for the parabola given.

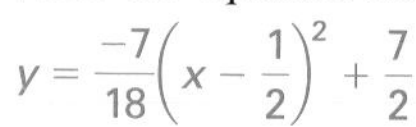

$y = \frac{-7}{18}\left(x - \frac{1}{2}\right)^2 + \frac{7}{2}$

Exercise 64

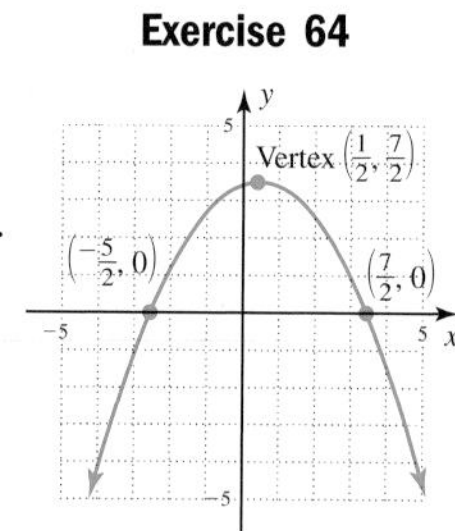

MAINTAINING YOUR SKILLS

65. (1.3) Solve the power equation: $x^{\frac{3}{4}} + 7 = 34$ $x = 81$

66. (1.4) Determine the quotient: $\frac{2 + 3i}{3 - 2i}$. i

67. (2.3) Identify the slope and y-intercept for $-4x + 3y = 9$. Do not graph.
$m = \frac{4}{3}$, y-intercept $(0, 3)$

68. (R.5) Multiply:
$\frac{x^2 - 4x + 4}{x^2 + 3x - 10} \cdot \frac{x^2 - 25}{x^2 - 10x + 25}$ $\frac{x - 2}{x - 5}$

69. (3.1) Given $f(x) = \sqrt[3]{x+3}$ and $g(x) = x^3 - 3$, find $(f \circ g)(x)$ and $(g \circ f)(x)$. $(f \circ g)(x) = x$, $(g \circ f)(x) = x$

70. (2.5) Given $f(x) = 3x^2 + 7x - 6$, solve $f(x) \le 0$ using the x-intercepts and concavity of f. $x \in \left[-3, \frac{2}{3}\right]$

MID-CHAPTER CHECK

5. a. **b.**

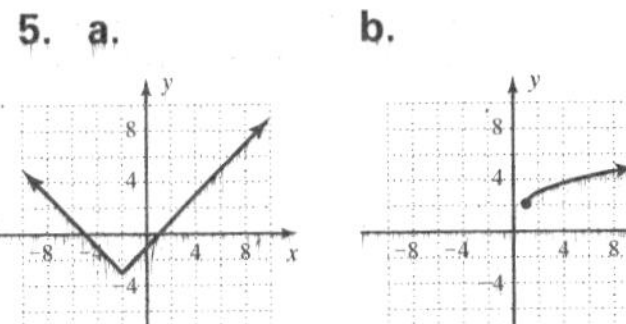

c. **d.**

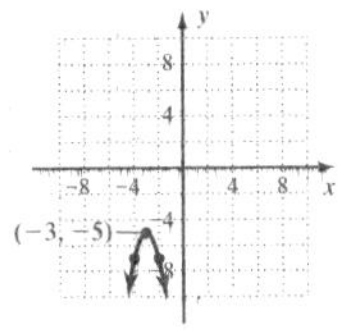

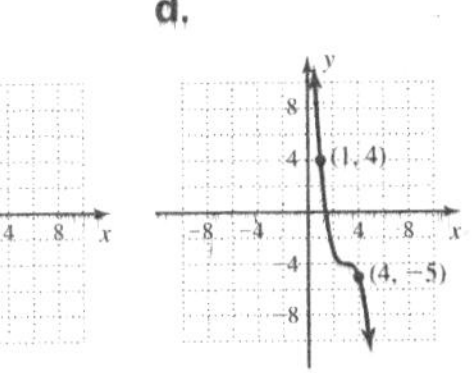

7.

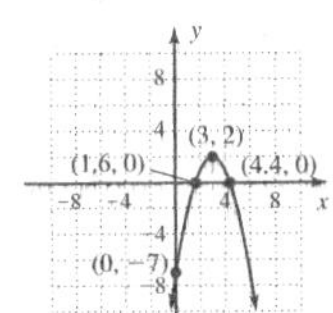

8.

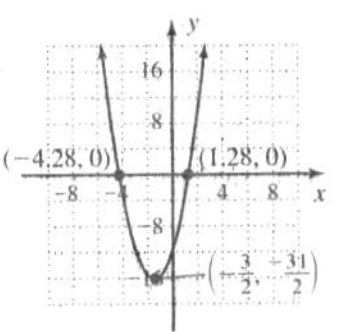

9.

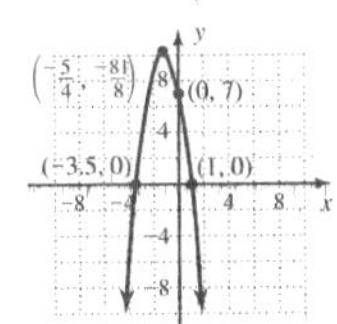

10. Answers will vary.

1. Given $f(x) = 3x - 5$ and $g(x) = 2x^2 + 3x$, find

a. $(f + g)(3)$ 31

b. $(f \cdot g)(x)$ $6x^3 - x^2 - 15x$

2. Given $f(x) = \frac{2}{3}x + 1$ and $g(x) = x^2 - 5x$, find

a. The domain of $\left(\frac{f}{g}\right)(x)$ $x \in (-\infty, 0) \cup (0, 5) \cup (5, \infty)$

b. $(g \circ f)(3)$ -6

3. In rugby football, a free kick is awarded after a major foul. The diagram to the right shows the path of the ball as it is kicked toward the goal. Suppose the path is modeled by the function $h(d) = -0.0375d^2 + 1.5d$, where $h(d)$ is the height in meters at a distance of d m from where it was kicked. Use this information to answer the following questions.

a. How long was the kick? 40 m

b. What is the maximum height of the kick? 15 m

c. How high was the ball at a horizontal distance of 10 m from the point where is was kicked? 20 m away? 11.25 m 15 m

d. Assume the kick was "true" and kicked from 37 m out. If regulation crossbars are 4 m high, will the kick be good? yes

4. Use the graph shown here to find the values indicated. Assume outputs are integer valued.

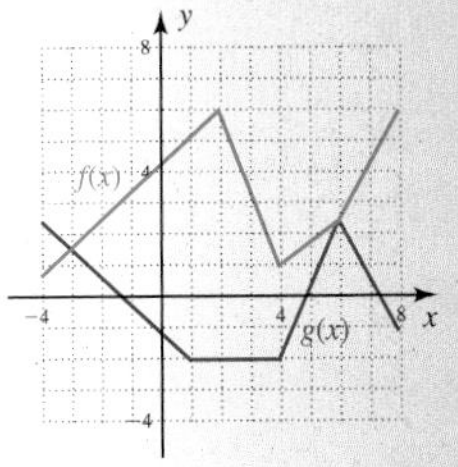

a. $(f + g)(-3)$ 3 b. $(f \cdot g)(1)$ -10

c. $(f - g)(4)$ 3 d. $\left(\frac{f}{g}\right)(2)$ -3

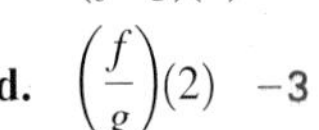

5. Graph the following functions by shifting the parent function and performing the appropriate transformation(s). Plot only a minimal number of points.

a. $f(x) = |x + 2| - 3$ b. $g(x) = \sqrt{x - 1} + 2$

c. $f(x) = -2(x + 3)^2 - 5$ d. $f(x) = -(x - 3)^3 - 4$

6. Given $f(x) = \sqrt{x - 3}$, find the inverse function and state the domain and range of $f^{-1}(x)$. Then verify they are inverse functions using a composition.

$f^{-1}(x) = x^2 + 3$ D: $x \in [0, \infty)$; R: $y \in [3, \infty)$; verified

Graph each quadratic function by completing the square. Be sure to find the vertex, y-intercept, and x-intercepts (if they exist). Plot additional points only as necessary.

7. $f(x) = -x^2 + 6x - 7$ **8.** $g(x) = 2x^2 + 6x - 11$ **9.** $h(x) = -2x^2 - 5x + 7$

10. Define, discuss, and/or explain the following terms. Include examples and counterexamples as part of your discussion: relation, function, one-to-one, end-behavior, zeroes, and axis of symmetry.

Reinforcing Basic Concepts

Transformations via Composition

Historically, many of the transformations studied in this chapter played a critical and fundamental role in the development of modern algebra. These were the transformations that opened the door to the solution of cubic and quartic equations in the sixteenth century, although ancient mathematicians used them centuries earlier to solve quadratic equations. To make the connection, we note that many transformations can be viewed as a composition of functions. For instance, for $f(x) = x^2 + 2$ (a parabola shifted two units up) and $g(x) = (x - 3)$, the composition $h(x) = f[g(x)]$ yields $(x - 3)^2 + 2$, a parabola shifted 2 units up *and* 3 units right. Actually any quadratic function $f(x)$ can be made to shift 3 units right using $h(x) = f[g(x)]$, where $g(x) = x - 3$. Consider $f(x) = x^2 - 4x - 5$. To find the equation of *this* parabola shifted three units right, we apply the same composition $g(x) = x - 3$, noting that this is the *identity function* shifted 3 units right. The composition $h(x) = f[g(x)]$ gives $h(x) = (x - 3)^2 - 4(x - 3) - 5$, or $h(x) = x^2 - 10x + 16$ after simplifying. Enter $f(x)$ as Y_1 and $h(x)$ as Y_2 on your graphing calculator, then graph and inspect the results. As you see, we do obtain the same parabola shifted 3 units to the right. But now, notice what happens when we compose using $g(x) = x + 2$: $h(x) = f[g(x)] = (x + 2)^2 - 4(x + 2) - 5$ [shifts $f(x)$ 2 units *left*]. After simplification, the result is $h(x) = x^2 - 9$ or *a quadratic function whose zeroes can easily be solved by taking square roots,* since the linear term is eliminated. The zeroes of h (the shifted quadratic) are $x = -3$ and $x = 3$, which means the zeroes of f (the original function) can be found by shifting two units *right,* returning them to their original position. The zeroes are $x = -3 + 2 = -1$ and $x = 3 + 2 = 5$ (which we can verify by factoring the original equation). Transformations of this type are especially insightful when the zeroes of a quadratic equation are irrational, since it enables us to find the radical portion of the root by taking square roots, and the rational portion by addition. The key is to shift the quadratic function using $x - \frac{b}{2a}$. Let's find the zeroes of $f(x) = x^2 + 6x - 11$ in this way. We find that $\frac{b}{2a} = 3$ and the necessary transformation is $g(x) = x - 3$. This gives $h(x) = f[g(x)] = f(x - 3)$ or $(x - 3)^2 + 6(x - 3) - 11$, which simplifies to $h(x) = x^2 - 20$. The zeroes of h are $x = -2\sqrt{5}$ and $x = 2\sqrt{5}$, so the solutions to the original equation must be $x = -2\sqrt{5} - 3$ and $x = 2\sqrt{5} - 3$, which can be verified using the quadratic formula or your calculator. For Exercises 1–3, use this method to: a. find such functions $h(x)$, and b. use the zeroes of h to find the zeroes of f. Verify each solution using a calculator.

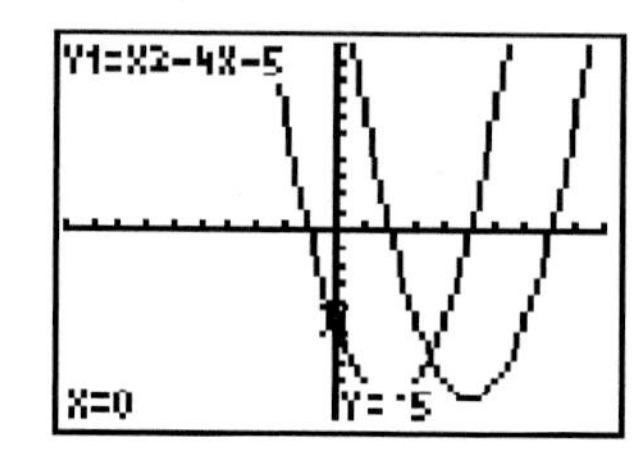

$h(x) = x^2 - 28;\ x = 4 \pm 2\sqrt{7}$

Exercise 1: $f(x) = x^2 - 8x - 12$

$h(x) = x^2 + 1;\ x = -2 \pm i$

Exercise 2: $f(x) = x^2 + 4x + 5$

Exercise 3: $f(x) = 2x^2 - 10x + 11$ $h(x) = 2x^2 - \frac{3}{2};\ x = \frac{5}{2} \pm \frac{\sqrt{3}}{2}$

As a more challenging exercise involving composition, the quadratic term of a *cubic equation* $y = ax^3 + bx^2 + cx + d$ can be eliminated using the transformation $g(x) = x - \frac{b}{3a}$, even though the eventual solution still remains out of reach for now. Perform the transformation to find $h(x)$ for the cubics given.

$h(x) = x^3 - 9x - 3$

Exercise 4: $f(x) = x^3 + 3x^2 - 6x - 11$

$h(x) = x^3 - 10x - 5$

Exercise 5: $f(x) = x^3 - 6x^2 + 2x + 7$

3.5 Asymptotes and Simple Rational Functions

LEARNING OBJECTIVES

In Section 3.5 you will learn how to:

A. Graph the reciprocal function $f(x) = \frac{1}{x}$, the reciprocal quadratic function $g(x) = \frac{1}{x^2}$ and use *direction/approach* notation to discuss the behavior of their graphs

B. Identify horizontal and vertical asymptotes, then use them to graph transformations and determine the equation of a rational function from its graph

C. Solve applications of simple rational functions

INTRODUCTION

In this section, we introduce an entirely new kind of relation called a **rational function.** Even though the functions studied thus far give us great ability to model and explore the real world, they do not have the characteristics necessary to model a number of other important and relevant situations. For example, the functions that model the amount of medication that remains in the bloodstream over a period of time, the relationship between altitude and weight or weightlessness, and the relationship between predator and prey populations are all rational functions.

POINT OF INTEREST

Questions about infinity and eternity have been objects of consideration for both theologians and mathematicians for thousands of years. Queries like, "Is there a largest possible real number?" and "What is the result of taking one-half of one and then half once again, and repeating the process without end?" have intrigued many mathematicians. In this section, we have our first "flirtation with infinity,"as we observe the graphs of two simple rational functions.

A. Simple Rational Functions: $f(x) = \frac{1}{x}$ and $g(x) = \frac{1}{x^2}$

WORTHY OF NOTE

The operation $1 \div 0$ or $\frac{1}{0}$ has no inverse since $0 \cdot$ (any number) is 0, never 1. As a result, expressions like $\frac{1}{0}$ cannot be evaluated, simplified, or written as a known number.

Several times in our study of mathematics, division by zero was discussed and labeled "not possible" or "undefined." But why is it that fractions like $\frac{1}{6}, \frac{1}{5}, \frac{1}{4}, \frac{1}{3}, \frac{1}{2}$, and $\frac{1}{1}$ all make perfect sense (we can draw pictures and diagrams of them), yet the fraction $\frac{1}{0}$ makes no sense at all? One way to explore this question is to study the graph of $f(x) = \frac{1}{x}$.

The Reciprocal Function: $f(x) = \frac{1}{x}$

The reciprocal function takes any input (other than zero) and gives its reciprocal as the output. This means that large inputs produce very small outputs. Figure 3.46 shows a graph of the reciprocal function, with the related table of values presented in Table 3.2.

Table 3.2

Input x	Output y	Input x	Output y
-1000	$-\frac{1}{1000}$	$\frac{1}{1000}$	1000
-5	$-\frac{1}{5}$	$\frac{1}{3}$	3
-4	$-\frac{1}{4}$	$\frac{1}{2}$	2
-3	$-\frac{1}{3}$	1	1
-2	$-\frac{1}{2}$	2	$\frac{1}{2}$
-1	-1	3	$\frac{1}{3}$
$-\frac{1}{2}$	-2	4	$\frac{1}{4}$
$-\frac{1}{3}$	-3	5	$\frac{1}{5}$
$-\frac{1}{1000}$	-1000	1000	$\frac{1}{1000}$
0	undefined	10,000	$\frac{1}{10,000}$

Figure 3.46

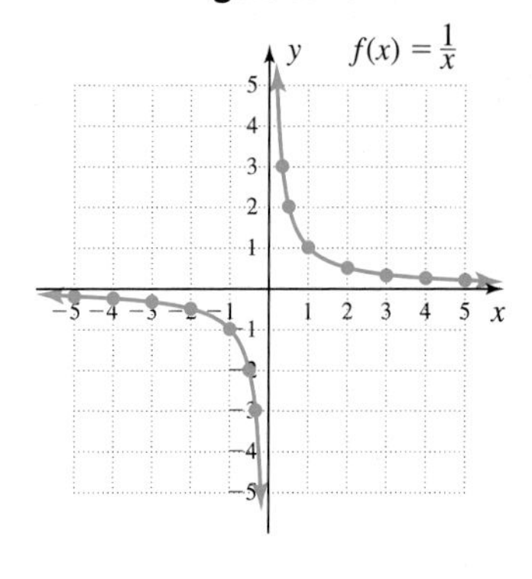

The table and graph reveal some interesting features. First, note the graph passes the vertical line test, verifying that $y = \frac{1}{x}$ is indeed a function. Second, since division by zero is undefined, there can be no corresponding point on the graph, *creating a break at* $x = 0$. This helps render the domain of the reciprocal function, which is defined at every point other than zero. The domain is $x \in (-\infty, 0) \cup (0, \infty)$. Third, since the reciprocal of a positive number is positive, and the reciprocal of a negative number is negative, one "branch" of the graph occurs in the first quadrant and another in the third quadrant, giving the graph its unique shape. Finally, we note that as x becomes an infinitely large positive number, y becomes a very small positive number. To describe this **end behavior** and other characteristics of rational graphs, we introduce a new tool we'll call **direction/approach notation.** Using this notation, the phrase "as x becomes an infinitely large positive number, y becomes a very small positive number" is written: as $x \to \infty$, $y \to 0^+$. This is shown graphically as the plotted points and the related graph become very close to, or *approach, the x-axis*. We alternatively say that "y approaches zero from above." The superscript plus and minus signs are used to indicate the *direction of the approach,* meaning *from the positive side* (right or above) or *from the negative side* (left or below), respectively. The approach notation can be applied to either the input or output variable. For instance, the phrase "as x approaches zero from the right, y becomes an infinitely large positive number" is written: as $x \to 0^+$, $y \to \infty$.

EXAMPLE 1 For $f(x) = \frac{1}{x}$, verbally describe the end behavior of the graph in Quadrant III, then use the direction/approach notation.

Solution: Similar to the graph's behavior in Quadrant I, we have

1. In words: "As x becomes an infinitely large negative number, y approaches zero from below." Using notation: as $x \to -\infty$, $y \to 0^-$.
2. In words: "As x approaches zero from the left, y becomes an infinitely large negative number." Using the notation: as $x \to 0^-$, $y \to -\infty$.

NOW TRY EXERCISES 7 AND 8

If you have trouble reading the graphs in this way or using the direction/approach notation, try this technique. Use a rectangle with its length L along the x-axis, its width W along the y-axis, and one corner of the rectangle on the graph of the function, as shown in Figures 3.47 through 3.49.

Figure 3.47

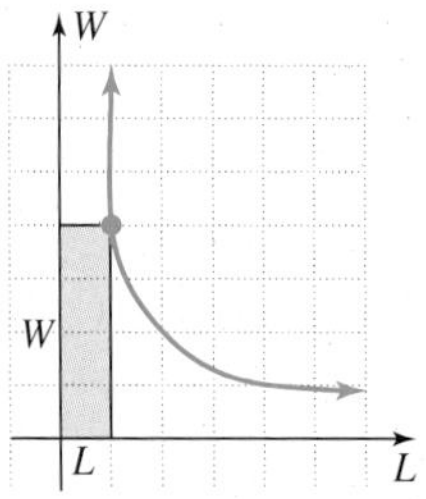

Figure 3.48

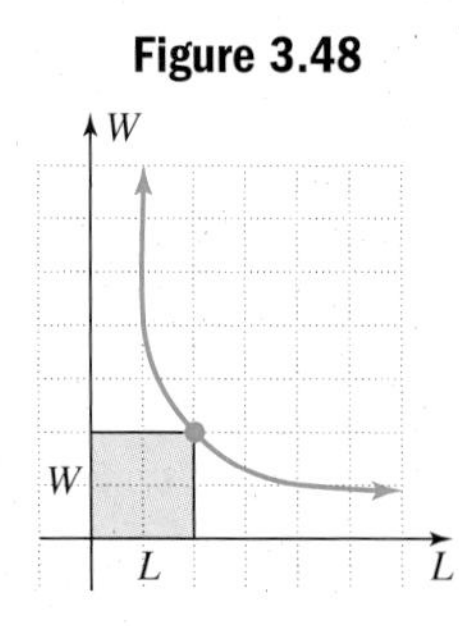

Figure 3.49

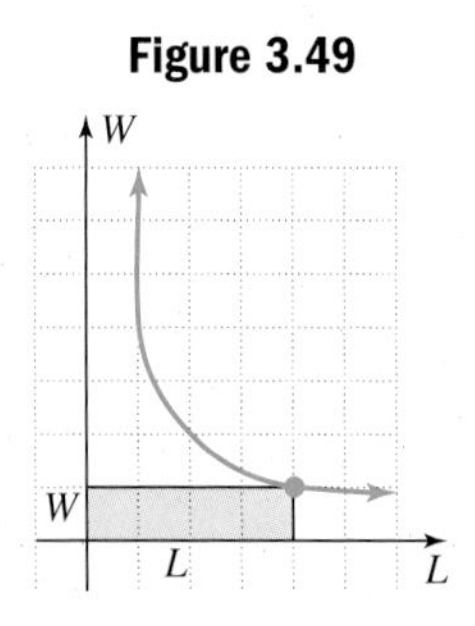

This makes it somewhat easier to see that as the length (x) gets larger, the width (y) gets smaller from above. Using the notation: as $L \to \infty$, $W \to 0^+$.

The Reciprocal Quadratic Function $g(x) = \frac{1}{x^2}$

We anticipate the graph of g will also have a break at $x = 0$, since no valid output can be obtained. This again creates a graph with two branches. But since the square of any negative number is positive, the branches of the **reciprocal quadratic function** are both *above the x-axis,* in Quadrants I and II. Figure 3.50 shows a graph of this function, with the related table of values presented in Table 3.3.

Table 3.3

Input x	Output y	Input x	Output y
-1000	$\frac{1}{1{,}000{,}000}$	$\frac{1}{1000}$	1,000,000
-5	$\frac{1}{25}$	$\frac{1}{3}$	9
-4	$\frac{1}{16}$	$\frac{1}{2}$	4
-3	$\frac{1}{9}$	1	1
-2	$\frac{1}{4}$	2	$\frac{1}{4}$
-1	1	3	$\frac{1}{9}$
$-\frac{1}{2}$	4	4	$\frac{1}{16}$
$-\frac{1}{3}$	9	5	$\frac{1}{25}$
$-\frac{1}{1000}$	1,000,000	1000	$\frac{1}{1{,}000{,}000}$
0	undefined	10,000	$\frac{1}{100{,}000{,}000}$

Figure 3.50

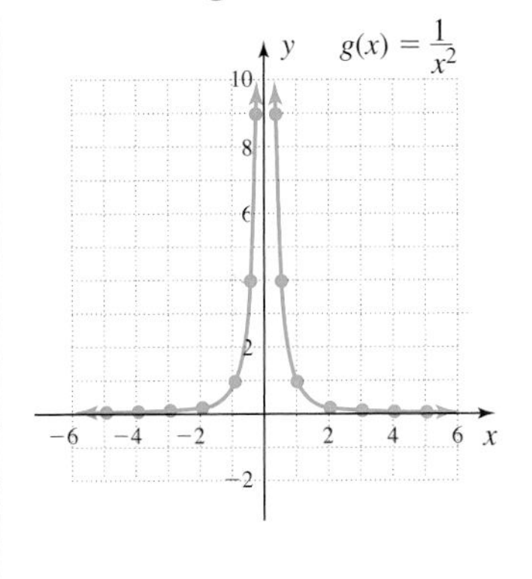

WORTHY OF NOTE

Notice how these statements relate directly to the table of values. As x approaches zero from the left, y does become infinitely large. For $x = -0.00001$, y is 10,000,000,000!

The graph of this function is similar to the reciprocal function, in that large positive inputs have very small (positive) outputs—causing these points and the related graph to *approach the x-axis*. The notation is: as $x \to \infty$, $y \to 0^+$. This "approach" is magnified for $y = \frac{1}{x^2}$, since the square of any proper fraction is an even smaller fraction $[(\frac{1}{10})^2 = \frac{1}{100}$ is much smaller than $\frac{1}{10}]$.

EXAMPLE 2 For $g(x) = \frac{1}{x^2}$, verbally describe the end behavior of the graph in Quadrant II, then use the direction/approach notation.

Solution: Similar to the graph's behavior in Quadrant I, we have

1. In words: "As x becomes an infinitely large negative number, y approaches zero from above." Using the notation: as $x \to -\infty$, $y \to 0^+$.
2. In words: "As x approaches zero from the left, y becomes an infinitely large positive number." Using the notation: as $x \to 0^-$, $y \to \infty$.

NOW TRY EXERCISES 9 AND 10

B. Horizontal and Vertical Asymptotes

Carefully consider our previous observations as we focus on the branch of $g(x) = \frac{1}{x^2}$ in the first quadrant. In Figure 3.50, note that as $x \to \infty, y \to 0^+$. Although defined more formally in future classes, we will consider a **horizontal asymptote** to be any horizontal line that the graph of a function approaches as $|x|$ becomes very large. For both $g(x) = \frac{1}{x^2}$ and $f(x) = \frac{1}{x}$, the line $y = 0$ (the x-axis) is a horizontal asymptote. The asymptote is not actually part of the graph, but can act as a guide when graphing functions from these families. As shown in Figures 3.51 and 3.52, they appear as a dashed line "guiding" the branches of the graph. Figure 3.51 shows a horizontal asymptote at $y = 1$, which means the graph of $f(x) = \frac{1}{x}$ has been shifted upward 1 unit. Figure 3.52 shows a horizontal asymptote at $y = -2$, which means the graph of $g(x) = \frac{1}{x^2}$ has been shifted downward 2 units.

Figure 3.51

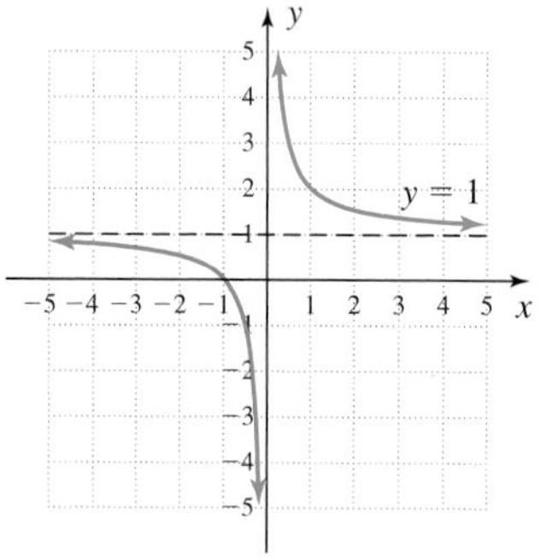

Figure 3.52

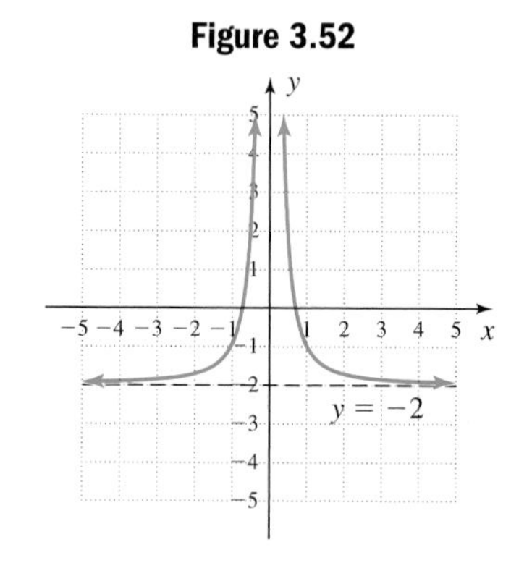

Using direction/approach notation, Figure 3.51 shows (1) as $x \to \infty, y \to 1^+$ and (2) as $x \to -\infty, y \to 1^-$. Figure 3.52 shows (1) as $x \to \infty, y \to -2^+$ and (2) as $x \to -\infty$, $y \to -2^+$.

Further observation in Quadrant I of Figure 3.50 also reveals that as x becomes *smaller and close to zero,* y increases without bound: as $x \to 0^+, y \to \infty$. This is an indication of asymptotic behavior *in the vertical direction* or a **vertical asymptote.** The line $x = 0$ is a vertical asymptote for both $y = \frac{1}{x^2}$ and $y = \frac{1}{x}$, because the values of y become extremely large as $|x|$ becomes very close to zero. Once again the vertical asymptote is not actually part of the graph, but acts as a guide when graphing the function. It is important to note that for rational functions, *vertical asymptotes occur at the zeroes of the denominator.* This is a fact we will use repeatedly in Chapter 4 and elsewhere.

HORIZONTAL AND VERTICAL ASYMPTOTES

Given constants h and k:

- the line $y = k$ is a horizontal asymptote if, as x increases or decreases without bound, y approaches k: as $|x| \to \infty, y \to k$.
- the line $x = h$ is a vertical asymptote if, as x approaches h, $|y|$ increases or decreases without bound: as $x \to h, |y| \to \infty$.

Identifying vertical and horizontal asymptotes is important, because the functions $y = \frac{1}{x}$ and $y = \frac{1}{x^2}$ can be transformed *in exactly the same way as the toolbox functions.* When their graphs shift, the vertical and horizontal asymptotes shift with them and can be used as guides to redraw the graph in its new location. These two functions are, in fact, the seventh and eighth members of our toolbox function family. In shifted form we have $F(x) = \frac{a}{x - h} + k$ for the reciprocal function and $G(x) = \frac{a}{(x - h)^2} + k$ for the reciprocal quadratic function. Here is a brief summary.

Reciprocal Function	**Reciprocal Quadratic Function**
$f(x) = \frac{1}{x}$	$g(x) = \frac{1}{x^2}$
Domain: $x \in (-\infty, 0) \cup (0, \infty)$	Domain: $x \in (-\infty, 0) \cup (0, \infty)$
Range: $y \in (-\infty, 0) \cup (0, \infty)$	Range: $y \in (0, \infty)$
Descriptive name: *butterfly function*	Descriptive name: *volcano function*

EXAMPLE 3 State the equation of the horizontal and vertical asymptotes of the function shown, then name the parent function, discuss the shifts involved, and give the equation related to the graph. Assume $a = 1$.

Solution: The equation of the vertical asymptote is $x = 1$, while the horizontal asymptote has equation $y = -1$. The graph is a member of the reciprocal function family, with parent function $f(x) = \frac{1}{x}$. Since the graph has been shifted 1 unit right and 1 unit downward, its equation is $y = \frac{1}{x - 1} - 1$.

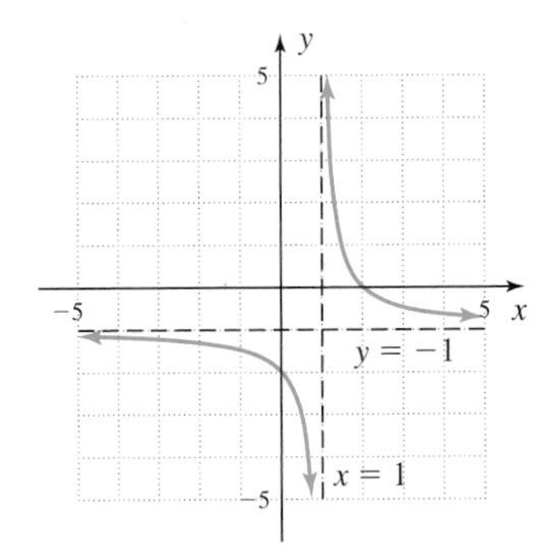

NOW TRY EXERCISES 11 THROUGH 22

When horizontal and/or vertical shifts are applied to simple rational functions, we first apply them to the asymptotes, then calculate the x- and y-intercepts as before. An additional point or two can be computed as needed to round out the graph.

EXAMPLE 4 Sketch the graph of $g(x) = \frac{1}{x - 2} + 1$ using transformations of the parent function.

Solution: The graph of g is the same as that of $y = \frac{1}{x}$, but shifted 2 units right and 1 unit upward. This means the vertical asymptote is also shifted

2 units right, and the horizontal asymptote is shifted 1 unit up. The y-intercept is $g(0) = \frac{1}{2}$. For the x-intercept:

$$0 = \frac{1}{x-2} + 1 \quad \text{substitute 0 for } g(x)$$

$$-1 = \frac{1}{x-2} \quad \text{subtract 1}$$

$$-1(x-2) = 1 \quad \text{multiply by } (x-2)$$

$$x = 1 \quad \text{solve}$$

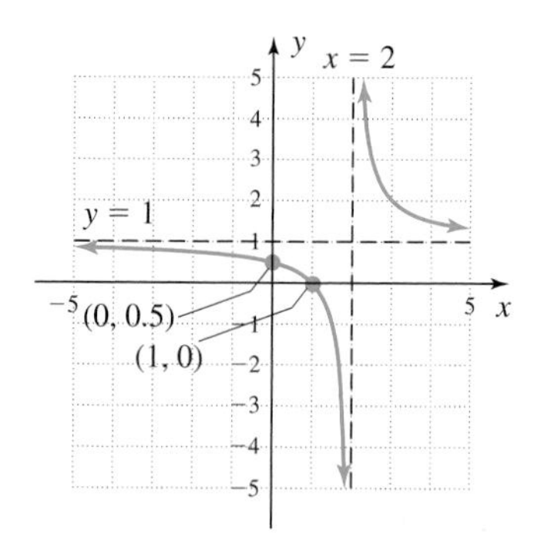

The x-intercept is (1, 0). Knowing the graph will be a butterfly function and using the asymptotes and intercepts yields the graph shown.

NOW TRY EXERCISES 23 THROUGH 46

When numerous transformations are used, it helps to follow the sequence outlined previously for the toolbox functions: (1) stretch/compress, (2) reflect, and (3) apply horizontal/vertical shifts.

C. Applications of Rational Functions

Applications of rational functions are numerous and very diverse. In many situations the coefficients can be rather large and the graph should be scaled appropriately to accommodate the larger numbers, as in Example 5.

EXAMPLE 5 For a large urban-centered county, the cost to remove chemical waste and other pollutants from a local river is given by the function $C(p) = \frac{-18{,}000}{p - 100} - 180$, where $C(p)$ represents the cost (in thousands of dollars) to remove p percent of the pollutants. (a) Find the cost to remove 25%, 50%, and 75% of the pollutants and comment on the results; (b) graph the function using an appropriate scale; and (c) use the direction/approach notation to state what happens as the county attempts to remove 100% of the pollutants.

Solution:

a. We evaluate the function as indicated, finding that $C(25) = 60$, $C(50) = 180$, and $C(75) = 540$. The cost is escalating rapidly. The change from 25% to 50% brought a $120,000 increase, but the change from 50% to 75% brought *a $360,000 increase!*

b. $C(p)$ is a reciprocal function where $a = -18{,}000$, giving a butterfly graph that has been reflected across the x-axis, shifted 100 units right and 180 units down. There is a vertical asymptote at $x = 100$ and a horizontal asymptote at $y = -180$. For the C-intercept we substitute $p = 0$ and find $C(0) = 0$, which seems

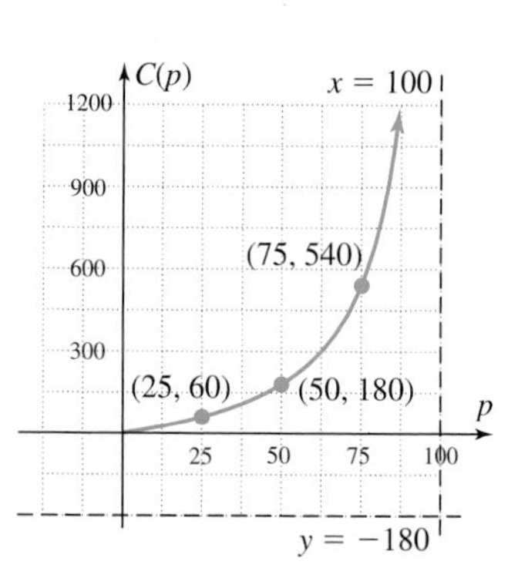

reasonable since 0% is removed if \$0 dollars are invested. From the context, we need only graph the portion from $0 \le p < 100$, or the upper-left wing of the butterfly. This produces the graph shown.

c. As the percentage of pollutants removed approaches 100%, the cost of the cleanup skyrockets. Using the notation: as $p \to 100^-$, $C \to \infty$. See the *Technology Highlight* that follows.

NOW TRY EXERCISES 49 THROUGH 56

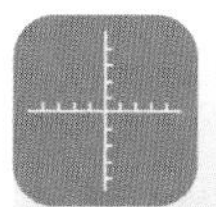

TECHNOLOGY HIGHLIGHT

Rational Functions and Appropriate Domains

The keystrokes shown apply to a TI-84 Plus model. Please consult your manual or our Internet site for other models.

As noted, the function $C(p) = \frac{-18{,}000}{p - 100} - 180$ from Example 5 is actually a butterfly function, but portions of the graph are ignored due to the context of the application. To see the full graph, we use the fact that the second branch occurs on the opposite side of the vertical and horizontal asymptotes, and set a window size like the one shown in Figure 3.53. After entering $C(p)$ as Y_1 on the Y= screen and pressing GRAPH, the full graph shown on Figure 3.54 appears (the horizontal asymptote was drawn using $Y_2 = -180$). Note the result really is a "butterfly graph."

Figure 3.53

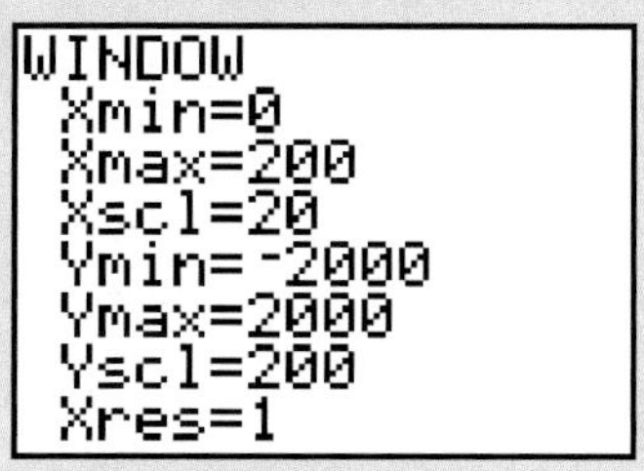

WINDOW
Xmin=0
Xmax=200
Xscl=20
Ymin=-2000
Ymax=2000
Yscl=200
Xres=1

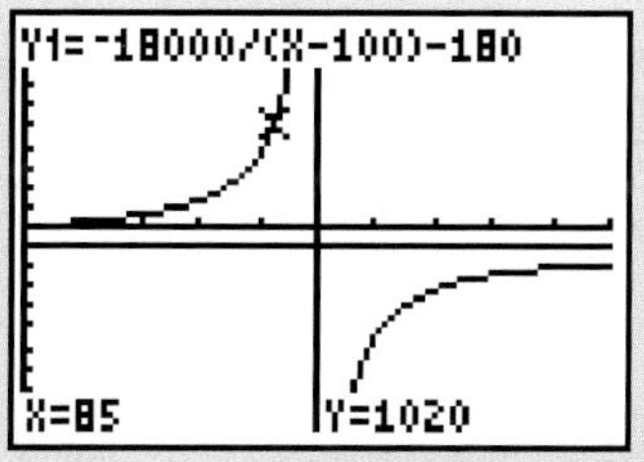

Use your calculator to respond to the following.

Figure 3.54

Y1=-18000/(X-100)-180
X=85 Y=1020

Exercise 1: Use the TRACE feature to verify that as $p \to 100^-$, $C \to \infty$. Approximately how much money must be spent to remove 95% of the pollutants? What happens when you TRACE to 100% Past 100%?

Exercise 2: Calculate the rate of change $\frac{\Delta C}{\Delta p}$ for the intervals [60, 65], [85, 90], and [90, 95] (use the *Technology Extension* from Chapter 2 at www.mhhe.com/coburn if desired). Comment on what you notice.

Exercise 3: Reset the window size changing only Xmax to 100 and Ymin to 0 for a more relevant graph. How closely does it resemble the graph from Example 5?

Exercise 1: \$3,420,000; undefined since cost becomes negative **Exercise 2:** The rate of change is much greater in the interval from 90 to 95. **Exercise 3:** very closely

3.5 EXERCISES

CONCEPTS AND VOCABULARY

Fill in each blank with the appropriate word or phrase. Carefully re-read the section if needed.

1. Write the following in direction/approach notation. *As x becomes an infinitely large negative number, y approaches 2 from below:* $x \to -\infty, y \to 2^-$

2. For any constant k, the notation as $|x| \to \infty$, $y \to k$ is an indication of a horizontal asymptote, while $x \to k$, $|y| \to \infty$ indicates a vertical asymptote.

3. Given the function $g(x) = \frac{1}{(x-3)^2} + 2$, a vertical asymptote occurs at $x = 3$ and a horizontal asymptote at $y = 2$.

4. The graph of $Y_1 = \frac{1}{x}$ has branches in Quadrants I and III. The graph of $Y_2 = -\frac{1}{x}$ has branches in Quadrants II and IV.

5. Discuss/explain how and why the range of the reciprocal function differs from the range of the reciprocal quadratic function. In the reciprocal quadratic function, all range values are positive.

6. If the graphs of $Y_1 = \frac{1}{x}$ and $Y_2 = \frac{1}{x^2}$ were drawn on the same grid, where would they intersect? In what interval(s) is $Y_1 > Y_2$? (1, 1); $(1, \infty)$

■ DEVELOPING YOUR SKILLS

Use the direction/approach notation to describe the end behavior of the graphs given (each exercise should produce four statements). Then state the domain and range of each function.

7.

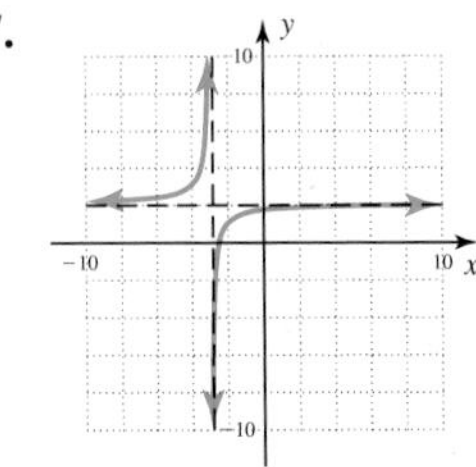

as $x \to \infty$, $y \to 2^-$,
as $x \to -\infty$, $y \to 2^+$,
as $x \to -3^-$, $y \to \infty$,
as $x \to -3^+$, $y \to -\infty$,
$D: x \in (-\infty, -3) \cup (-3, \infty)$,
$R: y \in (-\infty, 2) \cup (2, \infty)$

8.

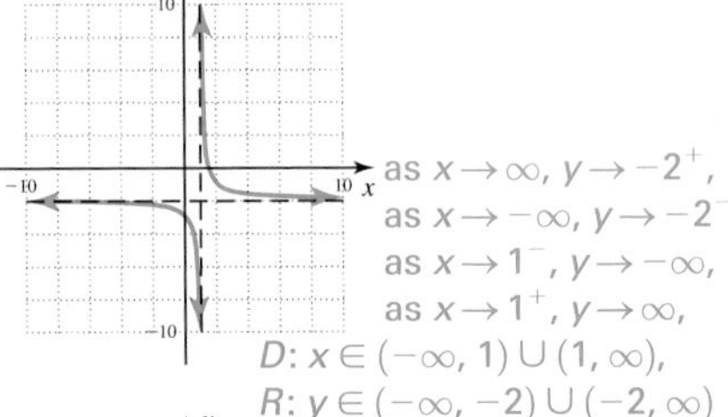

as $x \to \infty$, $y \to -2^+$,
as $x \to -\infty$, $y \to -2^-$,
as $x \to 1^-$, $y \to -\infty$,
as $x \to 1^+$, $y \to \infty$,
$D: x \in (-\infty, 1) \cup (1, \infty)$,
$R: y \in (-\infty, -2) \cup (-2, \infty)$

9.

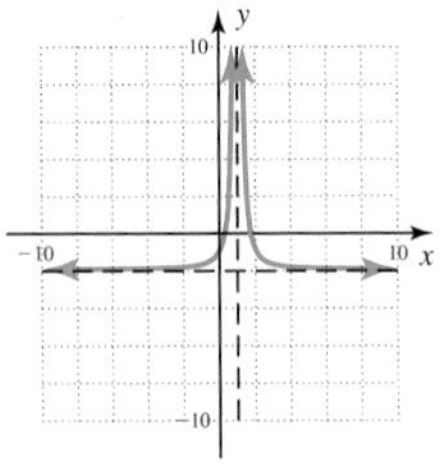

as $x \to \infty$, $y \to -2^+$,
as $x \to -\infty$, $y \to -2^+$,
as $x \to 1^-$, $y \to \infty$,
as $x \to 1^+$, $y \to \infty$,
$D: x \in (-\infty, 1) \cup (1, \infty)$,
$R: y \in (-2, \infty)$

10.

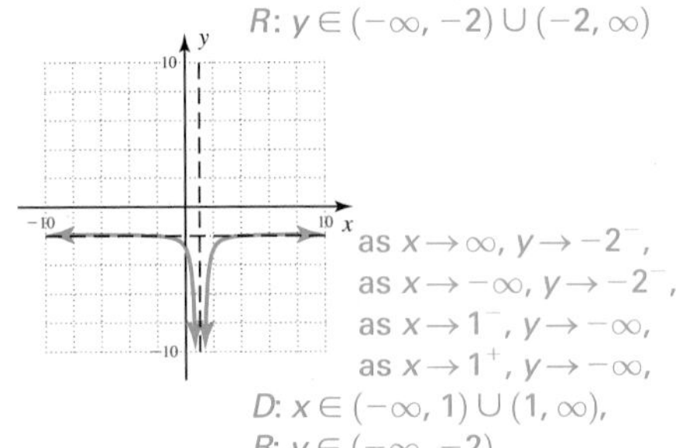

as $x \to \infty$, $y \to -2^-$,
as $x \to -\infty$, $y \to -2^-$,
as $x \to 1^-$, $y \to -\infty$,
as $x \to 1^+$, $y \to -\infty$,
$D: x \in (-\infty, 1) \cup (1, \infty)$,
$R: y \in (-\infty, -2)$

State the equation of the horizontal and vertical asymptotes for the toolbox function shown, and give the equation related to the graph. Assume $|a| = 1$.

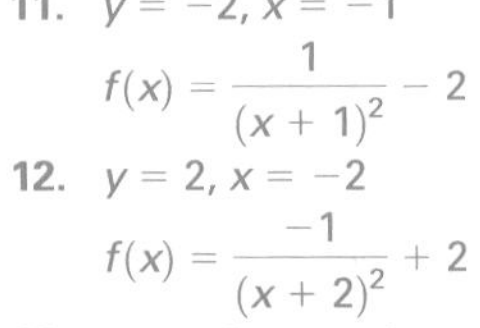

11. $y = -2, x = -1$
$f(x) = \frac{1}{(x+1)^2} - 2$

12. $y = 2, x = -2$
$f(x) = \frac{-1}{(x+2)^2} + 2$

13. $y = -2, x = -1$
$f(x) = \frac{1}{x+1} - 2$

14. $y = -2, x = 1$
$f(x) = \frac{-1}{x-1} - 2$

15. $y = -5, x = -2$
$f(x) = \frac{1}{(x+2)^2} - 5$

16. $y = -1, x = 1$
$f(x) = \frac{1}{x-1} - 1$

11.

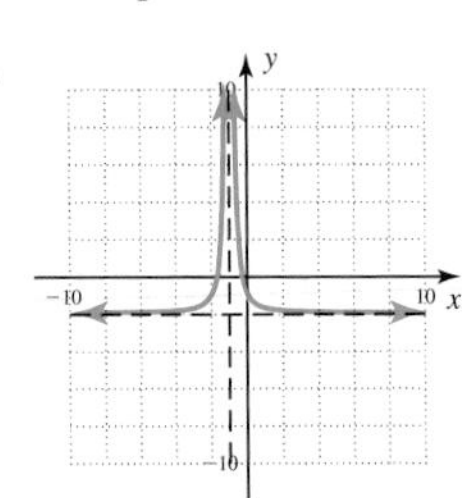

12.

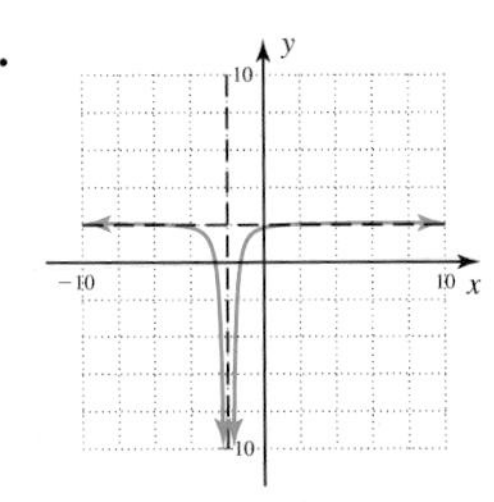

13.

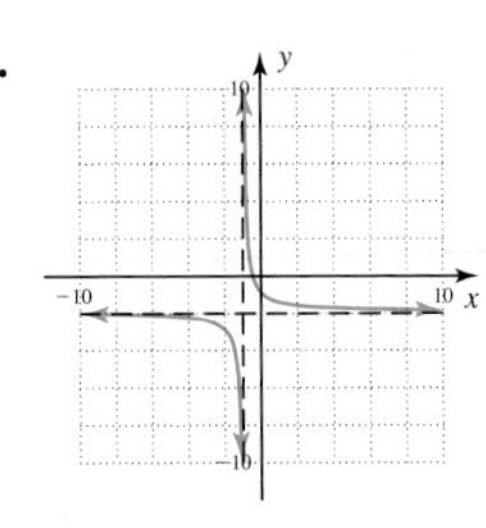

14.

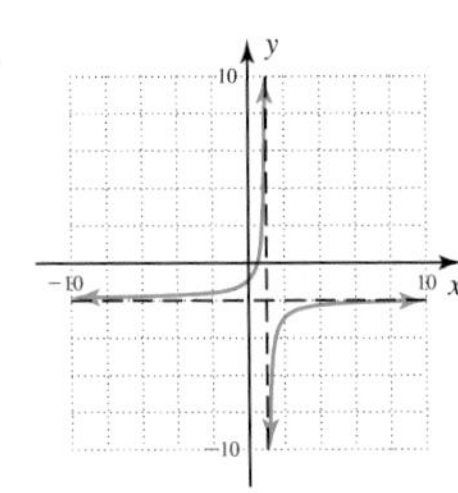

15.

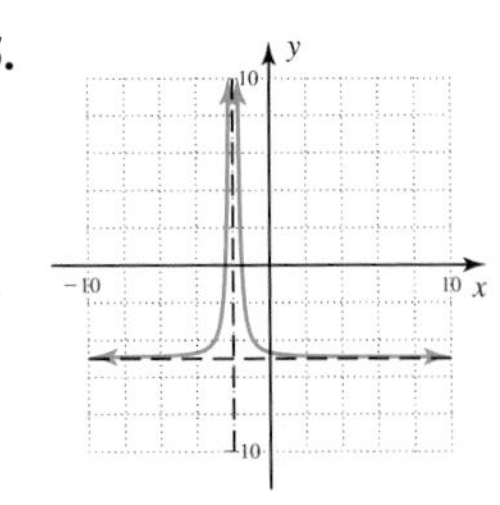

16.

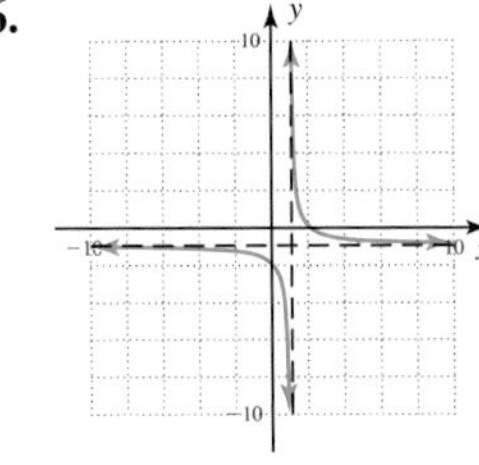

Use the graph shown to complete each statement using the direction/approach notation.

Exercises 17 through 22

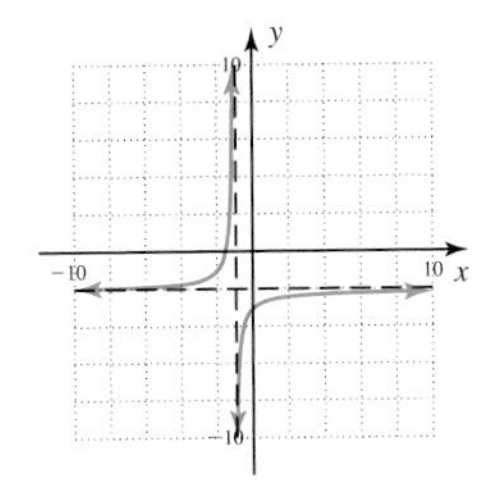

17. As $x \to -\infty$, y $\underline{\to -2^+}$.

18. As $x \to \infty$, y $\underline{\to -2^-}$.

19. As $x \to -1^+$, y $\underline{\to -\infty}$.

20. As $x \to -1^-$, y $\underline{\to \infty}$.

21. The line $x = -1$ is a vertical asymptote, since: as $x \to$ $\underline{-1}$, $y \to$ $\underline{\pm\infty}$.

22. The line $y = -2$ is a horizontal asymptote, since: as $x \to$ $\underline{\pm\infty}$, $y \to$ $\underline{-2}$.

Sketch the graph of each function using transformations of the parent function (not by plotting points). Clearly state the transformations used, and label the horizontal and vertical asymptotes as well as the x- and y-intercepts (if they exist). Also state the domain and range of each function.

23. $f(x) = \frac{1}{x} - 1$

24. $g(x) = \frac{1}{x} + 2$

25. $h(x) = \frac{1}{x + 2}$

26. $f(x) = \frac{1}{x - 3}$

27. $g(x) = \frac{-1}{x - 2}$

28. $h(x) = \frac{-1}{x} - 2$

29. $f(x) = \frac{1}{x + 2} - 1$

30. $g(x) = \frac{1}{x - 3} + 2$

31. $h(x) = \frac{1}{(x - 1)^2}$

32. $f(x) = \frac{1}{(x + 5)^2}$

33. $g(x) = \frac{-1}{(x + 2)^2}$

34. $h(x) = \frac{-1}{x^2} - 2$

35. $f(x) = \frac{1}{x^2} - 2$

36. $g(x) = \frac{1}{x^2} + 3$

37. $h(x) = 1 + \frac{1}{(x + 2)^2}$

38. $g(x) = -2 + \frac{1}{(x - 1)^2}$

39. $f(x) = 3 + \frac{1}{x + 4}$

40. $f(x) = -2 + \frac{1}{x + 2}$

41. $f(x) = \frac{-1}{x - 1} - 3$

42. $f(x) = \frac{-1}{x + 3} - 1$

43. $h(x) = \frac{-1}{(x - 2)^2} + 3$

44. $g(x) = \frac{-1}{(x + 1)^2} - 2$

45. $h(x) = 3 - \frac{1}{(x + 2)^2}$

46. $g(x) = -2 - \frac{1}{(x - 5)^2}$

WORKING WITH FORMULAS

47. Gravitational attraction: $F = \frac{km_1m_2}{d^2}$

The gravitational force F between two objects with masses m_1 and m_2 depends on the distance d between them and some constant k. If the mass of the two objects is constant while the distance between them gets larger and larger, what happens to F? Let m_1 and m_2 equal 1 mass unit with $k = 1$ as well, and investigate using a table of values. What family does this function belong to? F becomes very small; $y = \frac{1}{x^2}$

48. Area of a circular sector: $A = \frac{\theta}{360°}\pi r^2$

The area of a circular sector (a "pie slice") is given by the formula shown, where r is the radius of the circle and θ is the measure of the central angle in degrees. If the angle measure is held constant at 45°, the formula becomes a function of r: $A(r) = \frac{1}{8}\pi r^2$. (a) To what family of toolbox functions does $A(r)$ belong? (b) Solve for r in terms of A and determine the radius of a circle that has an area of 319 cm^2. **a.** $y = x^2$ **b.** $r = \sqrt{\frac{8A}{\pi}}$, $r = 28.5$ cm

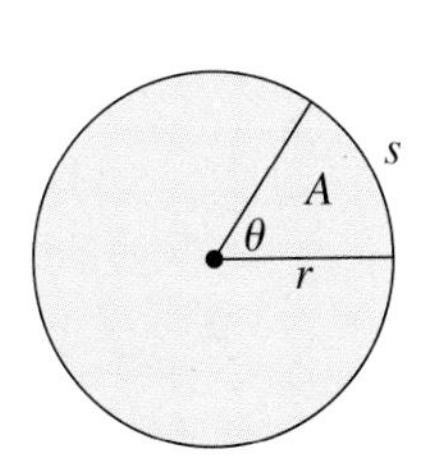

23. down 1, $x \in (-\infty, 0) \cup (0, \infty)$, $y \in (-\infty, -1) \cup (-1, \infty)$

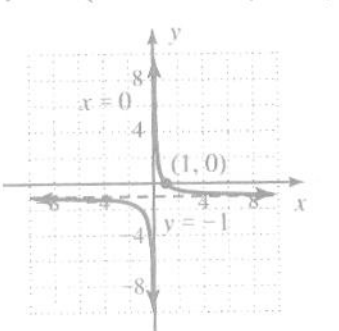

24. up 2, $x \in (-\infty, 0) \cup (0, \infty)$, $y \in (-\infty, 2) \cup (2, \infty)$

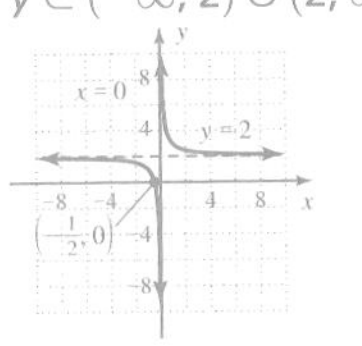

25. left 2, $x \in (-\infty, -2) \cup (-2, \infty)$, $y \in (-\infty, 0) \cup (0, \infty)$

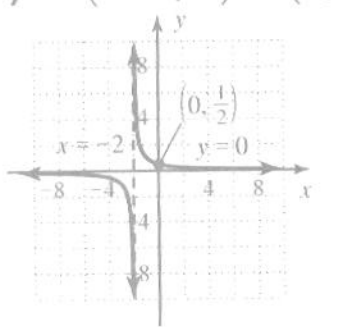

26. right 3, $x \in (-\infty, 3) \cup (3, \infty)$, $y \in (-\infty, 0) \cup (0, \infty)$

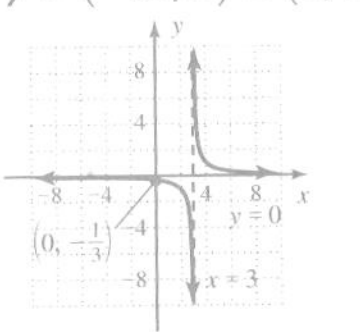

27. reflected across x-axis, right 2, $x \in (-\infty, 2) \cup (2, \infty)$, $y \in (-\infty, 0) \cup (0, \infty)$

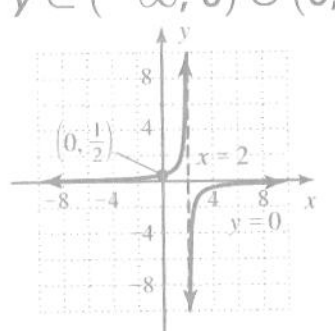

28. reflected across x-axis, down 2, $x \in (-\infty, 0) \cup (0, \infty)$, $y \in (-\infty, -2) \cup (-2, \infty)$

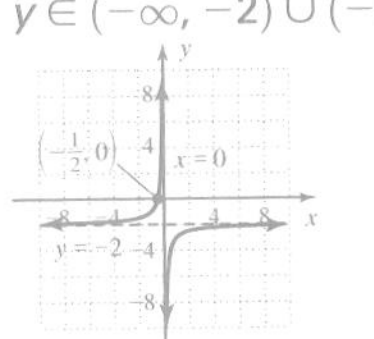

29. left 2, down 1, $x \in (-\infty, -2) \cup (-2, \infty)$, $y \in (-\infty, -1) \cup (-1, \infty)$

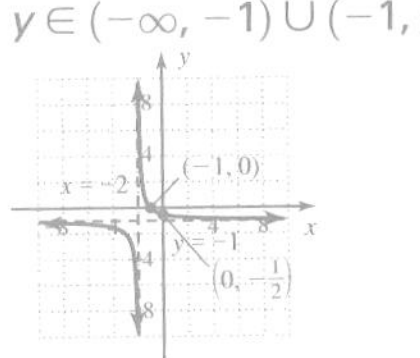

Additional answers can be found in the Instructor Answer Appendix.

49. c. 50. c. 51. c. 52. c.

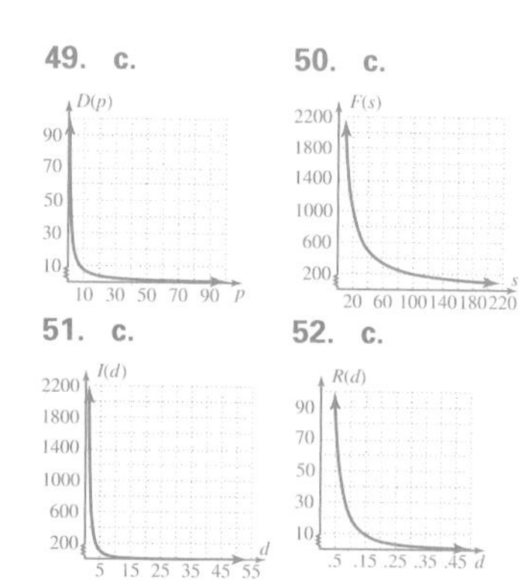

APPLICATIONS

49. Balance of nature: By banding deer over a period of 10 yr, a capture-and-release project determines the number of deer per square mile in the Mark Twain National Forest can be modeled by the function $D(p) = \frac{75}{p}$, where p is the number of predators present and D is the number of deer. Use this model to answer the following.

a. As the number of predators increases, what will happen to the population of deer? Evaluate the function at $D(1)$, $D(3)$, and $D(5)$ to verify. It decreases; 75, 25, 15

b. What happens to the deer population if the number of predators becomes very large? It approaches 0.

c. Graph the function using an appropriate scale. Judging from the graph, what happens to the deer population if the number of predators becomes very small (less than 1 per square mile)? Write this relationship using the direction/approach notation. as p decreases, D becomes very large; as $p \to 0$, $D \to \infty$

50. Balance of nature: A marine biology research group finds that in a certain reef area, the number of fish present depends on the number of sharks in the area. The relationship can be modeled by the function $F(s) = \frac{20{,}000}{s}$, where $F(s)$ is the fish population when s sharks are present.

a. As the number of sharks increases, what will happen to the population of fish? Evaluate the function at $F(10)$, $F(50)$, and $F(200)$ to verify. It decreases; 2000, 400, 100

b. What happens to the fish population if the number of sharks becomes very large? It approaches 0.

c. Graph the function using an appropriate scale. Judging from the graph, what happens to the fish population if the number of sharks becomes very small? Write this relationship using the direction/approach notation. As s decreases, F increases; as $s \to 0$, $F \to \infty$

51. Intensity of light: The intensity I of a light source depends on the distance of the observer from the source. If the intensity is 100 W/m^2 at a distance of 5 m, the relationship can be modeled by the function $I(d) = \frac{2500}{d^2}$. Use the model to answer the following.

a. As the distance from the lightbulb increases, what happens to the intensity of the light? Evaluate the function at $I(5)$, $I(10)$, and $I(15)$ to verify. It decreases; 100, 25, 11.1.

b. If the intensity is increasing, is the observer moving away or toward the light source? toward the light source

c. Graph the function using an appropriate scale. Judging from the graph, what happens to the intensity if the distance from the lightbulb becomes very small? Write this relationship using the direction/approach notation. Intensity becomes large; as $d \to 0$, $I \to \infty$

52. Electrical resistance: The resistance R (in ohms) to the flow of electricity is related to the length of the wire and its gauge (diameter in fractions of an inch). For a certain wire with fixed length, this relationship can be modeled by the function $R(d) = \frac{0.2}{d^2}$, where $R(d)$ represents the resistance in a wire with diameter d.

a. As the diameter of the wire increases, what happens to the resistance? Evaluate the function at $R(0.05)$, $R(0.25)$, and $R(0.5)$ to verify. It decreases; 80, 3.2, 0.8.

b. If the resistance is increasing, is the diameter of the wire getting larger or smaller? smaller

c. Graph the function using an appropriate scale. Judging from the graph, what happens to the resistance in the wire as the diameter gets larger and larger? Write this relationship using the direction/approach notation. It decreases; as $d \to \infty$, $R \to 0^+$.

53. For a certain coal-burning power plant, the cost to remove pollutants from plant emissions can be modeled by $C(p) = \frac{-8000}{p - 100} - 80$, where $C(p)$ represents the cost (in thousands of dollars) to remove p percent of the pollutants. (a) Find the cost to remove 20%, 50%, and 80% of the pollutants, then comment on the results; (b) graph the function using an appropriate

53. a. \$20,000, \$80,000, \$320,000; cost increases dramatically

b.

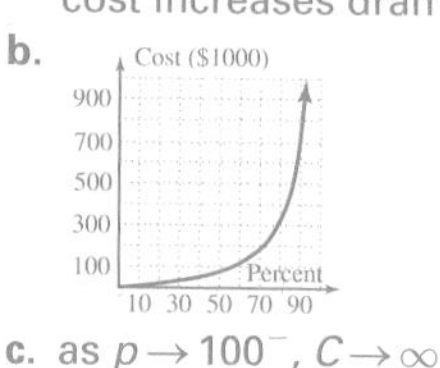

c. as $p \to 100^-$, $C \to \infty$

scale; and (c) use the direction/approach notation to state what happens if the power company attempts to remove 100% of the pollutants.

54. a. \$733,333, \$2,200,000, \$6,600,000; cost increases dramatically
b.

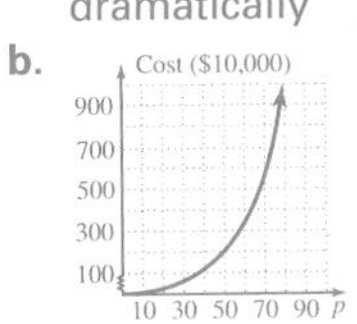

c. as $p \to 100^-$, $C \to \infty$

54. A large city has initiated a new recycling effort, and wants to distribute recycling bins for use in separating various recyclable materials. City planners anticipate the cost of the program can be modeled by the function $C(p) = \frac{-22{,}000}{p - 100} - 220$, where $C(p)$ represents the cost (in \$10,000) to distribute the bins to p percent of the population. (a) Find the cost to distribute bins to 25%, 50%, and 75% of the population, then comment on the results; (b) graph the function using an appropriate scale; and (c) use the direction/approach notation to state what happens if the city attempts to give recycling bins to 100% of the population.

55. The concentration C of a certain medicine in the bloodstream h hours after being injected into the shoulder is given by the function: $C(h) = \frac{2h^2 + h}{h^3 + 70}$. Use the given graph of the function to answer the following questions.

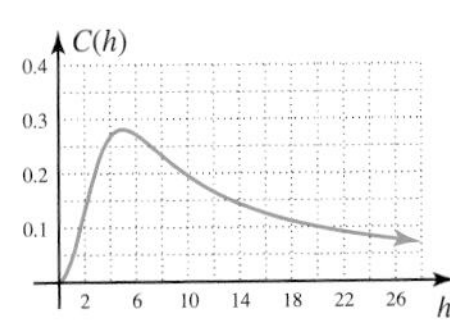

a. Approximately how many hours after injection did the maximum concentration occur? What was the maximum concentration? 5 hr; about 0.28

b. Use $C(h)$ to *compute* the rate of change for the intervals $h = 8$ to $h = 10$ and $h = 20$ to $h = 22$. What do you notice? -0.019, -0.005; As the number of hours increases, the rate of change decreases.

c. Use the direction/approach notation to state what happens to the concentration C as the number of hours becomes infinitely large. What role does the h-axis play for this function? $h \to \infty$, $C \to 0^+$; horizontal asymptote

56. In response to certain market demands, manufacturers will quickly get a product out on the market to take advantage of consumer interest. Once the product is released, it is not uncommon for sales to initially skyrocket, taper off and then gradually decrease as consumer interest wanes. For a certain product, sales can be modeled by the function $S(t) = \frac{250t}{t^2 + 150}$, where $S(t)$ represents the daily sales (in \$10,000) t days after the product has debuted. Use the given graph of the function to answer the following questions.

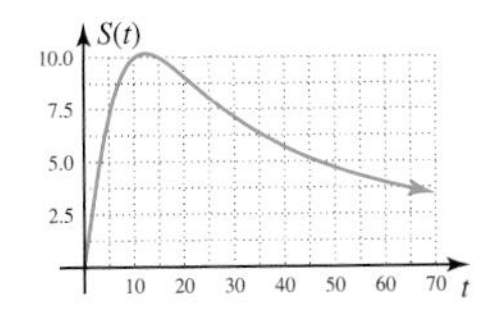

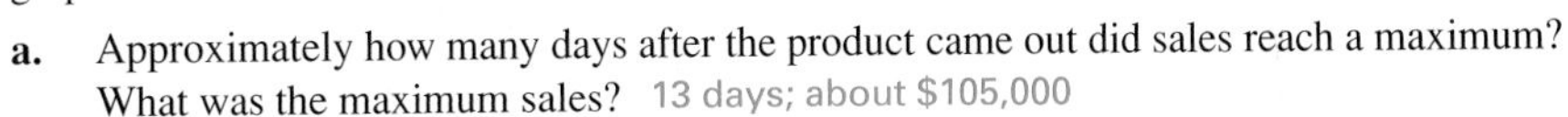

a. Approximately how many days after the product came out did sales reach a maximum? What was the maximum sales? 13 days; about \$105,000

b. Use $S(t)$ to compute the rate of change for the intervals $t = 7$ to $t = 8$ and $t = 60$ to $t = 62$. What do you notice? 0.55, -0.06; rate of change is positive and large in first interval; negative and small in second

c. Use the direction/approach notation to state what happens to the daily sales S as the number of days becomes infinitely large. What role does the t-axis play for this function? $t \to \infty$, $S \to 0^+$; horizontal asymptote

WRITING, RESEARCH, AND DECISION MAKING

57. $f^{-1}(x) = f(x)$ Answers will vary.

57. Find the inverse function for $f(x) = \frac{1}{x}$ using the algebraic method. What do you notice? Compose the functions to show your inverse is correct. Can you think of any other functions where $f(x) = f^{-1}(x)$? (*Hint:* One of them is a toolbox function.)

58. Consider the graph of $f(x) = \frac{1}{x}$ once again, and the x by $f(x)$ rectangles used on page 302 to help with the direction/approach notation. Calculate the area of each rectangle formed for $x \in \{1, 2, 3, 4, 5, 6\}$. What do you notice? Repeat the exercise for $g(x) = \frac{1}{x^2}$ and the x by $g(x)$ rectangles. Can you detect the pattern formed here? Answers will vary.

EXTENDING THE CONCEPT

59. Referring to Exercise 58, there are also interesting patterns formed when calculating the area of the x by y rectangles formed by points on the graphs of $h(x) = \frac{1}{x^3}$ and $H(x) = \frac{1}{x^4}$. Calculate enough areas to determine the pattern for each. Can you generalize the pattern so that it will hold for all x by y rectangles with a point on the graph of $y = \frac{1}{x^n}$ as a corner?
Answers will vary.

MAINTAINING YOUR SKILLS

60. (2.3) Find the equation of the line that is perpendicular to $3x - 4y = 12$ and has a y-intercept of $(0, 7)$. $y = \frac{-4}{3}x + 7$

61. (1.5) Solve the following equation using the quadratic formula, then write the equation in factored form: $12x^2 + 55x - 48 = 0$. $\frac{-16}{3}, \frac{3}{4}$ $(3x + 16)(4x - 3) = 0$

62. (3.4) Sketch the graph of $f(x) = x^2 - 5x + 3$ by completing the square. State the location of the vertex and all intercepts.

62. (0, 3), (.7, 0), (4.3, 0), $\left(\frac{5}{2}, \frac{-13}{4}\right)$

63. (1.4) Verify that $x = -5 + i\sqrt{3}$ is a solution to $x^2 + 10x + 28 = 0$. State the other solution without using the quadratic formula. $-5 - i\sqrt{3}$

64. (2.2) Using a scale from 1 (lousy) to 10 (great), Charlie gave the following ratings: {(The Beatles, 9.5), (The Stones, 9.6), (The Who, 9.5), (Queen, 9.2), (The Monkees, 6.1), (CCR, 9.5), (Aerosmith, 9.2), (Lynyrd Skynyrd, 9.0), (The Eagles, 9.3), (Led Zepplin, 9.4), (The Stones, 9.8)} Is the relation (group, rating) as given, also a function? State why or why not.
No, The Stones are paired with two different ratings.

65. (3.2) Verify that $f(x) = \frac{1}{x - 2} + 3$ and $g(x) = \frac{1}{x - 3} + 2$ are inverse functions.

3.6 Toolbox Applications: Direct and Inverse Variation

LEARNING OBJECTIVES

In Section 3.6 you will learn how to:

- **A. Solve direct variations using toolbox functions**
- **B. Use toolbox functions to solve inverse variations**
- **C. Solve applications of joint variations**
- **D. Use properties of toolbox functions to aid data analysis**

INTRODUCTION

One area where the toolbox functions find numerous and meaningful applications is that of direct and inverse variations. Variations are at the heart of studying how one quantity "varies" with relation to another, a study that can involve careful observations, analysis of gathered data, and the application of an appropriate equation model. In addition, many familiar formulas are actually toolbox functions that can be understood as direct or inverse variations.

POINT OF INTEREST

The study of functions is actually the study of how one variable changes (or varies) with respect to another. This can be seen in some of the names we give x and y as we evaluate a function: (input, output), (cause, effect), (independent variable, dependent variable). In more advanced studies involving data, certain tests are applied to the output values of the data list, to see if a linear, quadratic, rational, radical, or some other function model is most appropriate for the data. This study of how outputs vary with respect to an ordered list of inputs gave rise to the **(abscissa, ordinate)** designation sometimes seen for ordered pairs. The word "ordinate" comes form the Latin *ordinatus,* which means to set in order, arrange, or regulate.

A. Toolbox Functions and Direct Variation

If a car gets 17 miles per gallon of gas, we could model the total miles traveled as $M(g) = 17g$, where $M(g)$ represents the number miles traveled on g gallons.

$$\begin{aligned} &\text{1 gallon of gas} \rightarrow \text{17 miles:} && M(1) = 17 \cdot 1 \\ &\text{2 gallons of gas} \rightarrow \text{34 miles:} && M(2) = 17 \cdot 2 \\ &\text{3 gallons of gas} \rightarrow \text{51 miles:} && M(3) = 17 \cdot 3 \end{aligned}$$

The miles traveled by the car changes in **direct** or **constant proportion** to the number of gallons used. This may remind you of the $m = \frac{\Delta y}{\Delta x}$ notation used earlier for the slope of a line, since the slope of a line (its rate of change) is also constant. When working with variations, the constant k is preferred over m, and in this case we have $\frac{\Delta Miles}{\Delta gallons} = \frac{\Delta M}{\Delta g} = k$. For this application, we note a linear model is most appropriate and we say, "the number of miles traveled *varies directly* with the number of gallons used." The equation $M = 17g$ is called a **direct variation,** and the coefficient 17 is called the **constant of variation.** It's easy to see why, since for each case stated above, $\frac{\Delta M}{\Delta g} = \frac{17}{1} = \frac{34}{2} = \frac{51}{3} = 17$. We further note that graphically, the constant k would represent the slope of the line. In general, we have the following:

> **DIRECT VARIATIONS**
> Given two quantities x and y, if there is a nonzero constant k such that
> $$y = kx$$
> we say that *y varies directly with x or y is directly proportional to x.*
> k is called the *constant of variation.*

EXAMPLE 1 Use descriptive variables to write the variation equation for these statements:

a. Wages earned varies directly with the number of hours worked.

b. The circumference of a circle varies directly with the length of the diameter.

Solution:

a. *W*ages varies directly with *h*ours worked: $W = kh$.

b. *C*ircumference varies directly with *d*iameter: $C = kd$.

NOW TRY EXERCISES 7 THROUGH 10

Once we determine the relationship between two variables is a direct variation, we try to find the value of k and develop an equation model. We then use the equation to study further relationships between the variables involved. Note that "varies directly" indicates that one value is a constant multiple of the other. In Example 1(b), your instincts may have told you that for $C = kd$, $k = \pi$ since the formula for a circle's circumference is $C = \pi d$. This insight helps illustrate the procedure for finding k. If k is a "constant relationship" between C and d—*only one known relation is needed to solve for k!* The basic ideas for solving applications of variation are summarized here:

SOLVING VARIATION PROBLEMS

1. Use descriptive variables to translate the situation given into a formula model.
2. Substitute the first set of values given and solve for k.
3. Use this value of k in the original formula model.
4. Substitute the remaining information to answer the question posed.

EXAMPLE 2 The weight of an astronaut on the surface of Mars *varies directly* with her weight on Earth. A woman weighing 120 lb on Earth weighs only 45.6 lb on Mars. How much would a 135-lb astronaut weigh on Mars?

Solution:

$M = kE$ "*M*ars weight *varies directly* with *E*arth weight"

$45.6 = k(120)$ substitute first set of values

$k = 0.38$ solve for k (constant of variation)

Use this value of k in the original equation model, then use it to find the weight of a 135-lb astronaut if she were transported to Mars.

$M = 0.38E$ use $k = 0.38$ in the original equation

$= 0.38(135)$ substitute $E = 135$

$= 51.3$ result

An astronaut weighing 135 lb on Earth weighs only 51.3 lb on Mars.

NOW TRY EXERCISES 11 THROUGH 14

As we apply the toolbox functions in this way, it is important not to separate the application from its corresponding graph. This will enormously assist our ability to select an appropriate model and apply it correctly.

EXAMPLE 3 Graph the variation equation from Example 2, then *use the graph* to estimate the weight of an astronaut on Mars, if he weighed 180 lb on Earth. Be sure to use an appropriate scale.

Solution: Since we expect the weight of an astronaut to be positive, we use only Quadrant I for the graph. Somewhat arbitrarily selecting Earth weights from 100 to 200 lb, we have $M = 0.38(100) = 38$ lb as a "minimum" weight on Mars, and $M = 0.38(200) = 76$ lb as a "maximum." Scaling the axes accordingly and using the points (100, 38) and (200, 76) produces the graph shown, where we note the astronaut will weigh about 68 to 69 lb on Mars.

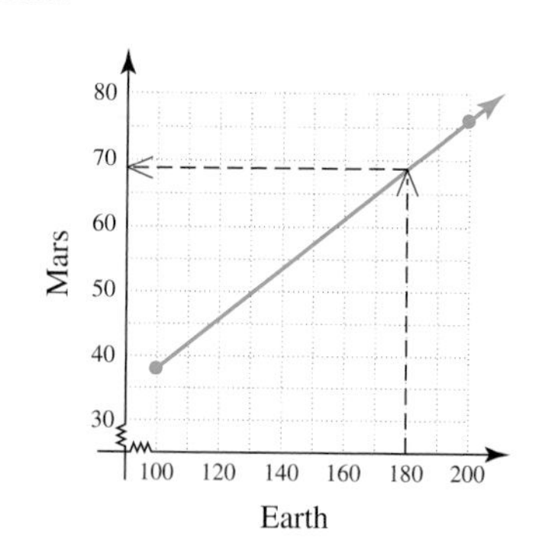

NOW TRY EXERCISES 15 AND 16

Each of the toolbox functions provides a link to new and interesting applications of mathematics. Our knowledge of their graphs and defining characteristics strengthens a contextual understanding of their application as a function model.

EXAMPLE 4 Use descriptive variables to write the variation equation for these statements:

a. In free fall, the distance traveled by an object varies directly with the square of the time.

b. The area of a circle varies directly with the square of its radius.

Solution: **a.** *D*istance varies directly with the square of the *t*ime: $D = kt^2$.

b. *A*rea varies directly with the square of the *r*adius: $A = kr^2$.

NOW TRY EXERCISES 17 THROUGH 20

Each situation in Example 4 uses a quadratic function model. In this context, notice that k represents the concavity and the amount of stretch or compression on the function. Regardless of the model used, the solution process for variations remains the same.

EXAMPLE 5 The range of a projectile varies directly with the square of its initial velocity. As part of a circus act, Bailey the Human Bullet is shot out of a cannon with an initial velocity of 80 ft/sec (fps), into a net 200 ft away.

a. Find the constant of variation and write the variation equation.

b. Graph the equation and *use the graph* to estimate how far away the net should be placed if initial velocity is increased to 95 fps.

c. Determine the accuracy of the estimate from (b) using the variation equation.

Solution **a.**

$R = kv^2$ "*R*ange varies directly with the square of the *v*elocity"

$200 = k(80)^2$ substitute first set of values

$k = 0.03125$ solve for k (constant of variation)

$R = 0.03125v^2$ variation equation (replace k with 0.03125)

b. Since velocity and distance are positive, we again use only Quadrant I. The graph is a parabola, concave up, with vertex at (0, 0). Selecting velocities from 50 to 100 fps, we have $R = 0.03125(50)^2 \approx 78$ ft, and $R = 0.03125(100)^2 \approx 313$ ft as additional points to help draw the graph. Scaling the axes accordingly and using the points (50, 78) and (100, 313) produces the graph shown, where we note that at 95 fps, the net should be placed about 280 ft away.

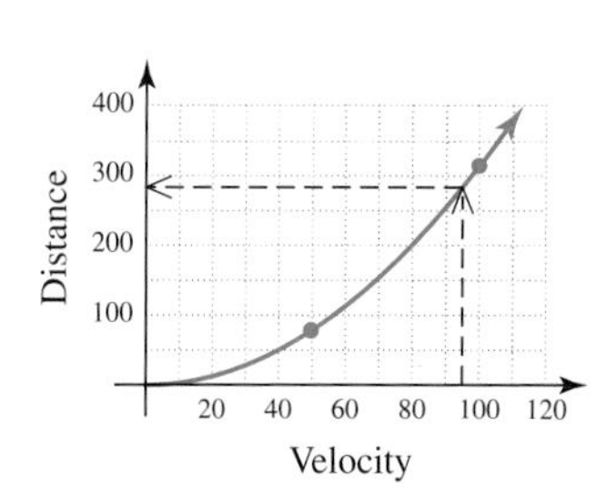

c. Using the variation equation we have

$$R = 0.03125v^2 \quad \text{use } k = 0.03125 \text{ in the original equation}$$
$$= 0.03125(95)^2 \quad \text{substitute } v = 95$$
$$= 282.03125 \quad \text{result}$$

As indicated by the graph, the net should be placed about 282 ft away.

NOW TRY EXERCISES 21 THROUGH 26

Table 3.4

Price (dollars)	Demand (1000s)
8	288
9	144
10	96
11	72
12	57.6

B. Inverse Variation

In a consumer-oriented society, numerous studies have been done relating the price of a commodity to the demand—the willingness of a consumer to pay that price. For instance, if there is a sudden increase in the price of a popular tool, hardware stores know there will be a corresponding decrease in the demand for that tool. The question remains, "What is this rate of decrease?" Can it be modeled by a linear function with a negative slope? A parabola that is concave down? To help answer this question, consider Table 3.4, which shows some (simulated) data regarding price versus demand. It appears that a linear function is not appropriate because the rate of change in the number of tools sold (the demand) is not constant. Likewise a quadratic model seems inappropriate, since we don't expect demand to suddenly start rising again as the price continues to rise. This phenomenon is actually an example of an **inverse variation,** which is modeled by the reciprocal function $y = \frac{k}{x}$. Using descriptive variables and decomposing the fraction to clearly see the inverse relationship, gives $D = k\left(\frac{1}{P}\right)$, where k is the constant of variation, D represents the demand for the product, and P the price of the product. In words, we say that "demand *varies inversely* as the price." In other applications of inverse variation, one quantity may vary inversely *as the* square *of another,* and, in general, we have

INVERSE VARIATIONS

Given two quantities x and y,

- if there is a nonzero constant k such that
$$y = k\left(\frac{1}{x}\right)$$
we say that *y varies inversely with x* or *y is inversely proportional to x.*

- if there is a nonzero constant k such that
$$y = k\left(\frac{1}{x^2}\right)$$
we say that *y varies inversely with x^2* or *y is inversely proportional to x^2.*

EXAMPLE 6 Use descriptive variables to write the variation equation for these statements:

a. In a closed container, pressure varies inversely with the volume of gas.

b. The intensity of light varies inversely with the square of the distance from the source.

Solution: **a.** *P*ressure varies inversely with the *V*olume of gas: $P = k\left(\frac{1}{V}\right)$.

b. *I*ntensity of light varies inversely with the square of the *d*istance: $I = k\frac{1}{d^2}$.

NOW TRY EXERCISES 27 THROUGH 30

EXAMPLE 7 Boyle's law tells us that in a closed container with constant temperature, the pressure of a gas varies inversely with its volume. Suppose the steam in the cylinder of an antique locomotive exerts a pressure of 210 pounds per square inch (psi) when the volume of the cylinder is 50 in^3.

a. Find the constant of variation and write the variation equation.

b. Use the equation to determine the pressure when the return stroke of the piston increases the volume to 150 in^3.

Solution: **a.**

$$P = k\left(\frac{1}{V}\right) \quad \text{"Pressure varies inversely with the volume"}$$

$$210 = k\left(\frac{1}{50}\right) \quad \text{substitute first set of values}$$

$$k = 10{,}500 \quad \text{constant of variation}$$

$$P = 10{,}500\left(\frac{1}{V}\right) \quad \text{variation equation (replace } k \text{ with 10,500)}$$

b. Using the variation equation we have:

$$P = 10{,}500\left(\frac{1}{V}\right) \quad \text{use } k = 10{,}500 \text{ in the original equation}$$

$$= 10{,}500\frac{1}{150} \quad \text{substitute } v = 150$$

$$= 70 \quad \text{result}$$

When volume is 150 in^3, the pressure is about 70 psi.

NOW TRY EXERCISES 31 THROUGH 34

C. Joint or Combined Variations

Just as many of your decisions might be based on more than one consideration, many times the relationship between two variables depends on a combination of factors. Imagine a wooden plank laid across the banks of a stream for hikers to cross the streambed (see Figure 3.55). The amount of weight the plank will support depends on the type of wood, the width of the plank, the thickness of the plank, and the distance between the supported ends. This is an example of a **joint variation,** which can combine any number of variables in many different ways. Two of the many possibilities include: (1) *y varies jointly with*

Figure 3.55

the product of x and p: $y = kxp$; and (2) *y varies jointly with the product of x and p, and inversely with the square of q:* $y = kxp\left(\frac{1}{q^2}\right)$. For practice writing joint variations as an equation model, see Exercises 35 through 40.

EXAMPLE 8 The amount of fuel used by a ship traveling at a uniform speed varies jointly with the distance it travels and the square of the velocity. If 200 barrels of fuel are used to travel 10 mi at 20 nautical miles per hour, how far does the ship travel on 500 barrels of fuel at 30 nautical miles per hour?

Solution:

$F = kdv^2$ "fuel use *varies jointly* with distance and velocity squared"

$200 = k(10)(20)^2$ substitute known values

$200 = 4000k$ simplify and solve for k

$0.05 = k$ constant of variation

To find the distance traveled when 500 barrels of fuel are used while traveling 30 nautical miles per hour, use $k = 0.05$ in the original formula model and substitute the given values:

$F = kdv^2$ formula model

$F = 0.05dv^2$ equation of variation

$500 = 0.05d(30)^2$ substitute 500 for F and 30 for v

$500 = 45d$ simplify

$11.\overline{1} = d$ result

If 500 barrels of fuel are consumed while traveling 30 nautical miles per hour, the ship covers a distance of just over 11 mi.

NOW TRY EXERCISES 41 THROUGH 44

It is interesting to note that the ship in Example 8 covers just over one additional mile, but consumes over 2.5 times the amount of fuel. The additional speed requires a great deal more fuel (30 nautical miles per hour is approximately 34.5 mph—very fast for a ship).

There is a variety of interesting and relevant applications in the Exercise Set. See Exercises 47 through 55.

D. Toolbox Functions and Data Analysis

WORTHY OF NOTE

When working with data, certain tests can be applied to the output values to indicate which model might be most appropriate. See the *Strengthening Core Skills* feature for this chapter.

The linear equation model in Example 3 had a positive slope, but if we were given only raw data, how would we know a linear function was most appropriate? After all, several of the other toolbox functions are also increasing on a significant part of their domain. The answer lies in their **rate of increase.** The output values of a linear function increase at a constant rate, the output values of the other functions do not. In Examples 1 through 8 the equation models were built by modeling the words used to describe them. When equation models are used to model raw data, we rely on a combination of the scatter-plot, intuition, experience, known relationships, and/or a process of elimination. In Example 9 the rate of increase is *not* constant.

In Section 2.6 we introduced applications of data analysis that involved polynomial functions, primarily the linear and quadratic "toolbox" functions. There we learned that real data was not nearly so "well-behaved" as data from formulas. The data from a formula most often resulted in coefficients that were integers or small rational numbers, while the real data sometimes produced "dreadful" decimal coefficients. In this section we'll employ **power regressions** to illustrate how toolbox functions are used to model real data. A power function is one that can be written in the form $y = ax^r$, where a and r are constants. The reciprocal function $f(x) = \frac{a}{x}$, and the reciprocal quadratic function $g(x) = \frac{a}{x^2}$, are power functions, since each can be written in "power" form: $f(x) = ax^{-1}$ and $g(x) = ax^{-2}$. As with other regression models, we hope to obtain a model that accurately reflects the trends indicated by the data, while knowing the coefficients, and in this case, the exponents may appear as decimal values.

EXAMPLE 9 The temperature of ocean water depends on several factors, including salinity, latitude, depth, and density. However, between depths of 125 m and 2000 m, ocean temperatures are relatively predictable as indicated by the data shown for tropical oceans. Use the data shown in the table and a graphing calculator to draw a scatter plot of the data and respond to the following.

Depth (meters)	Temperature (°C)
125	13.0
250	9.0
500	6.0
750	5.0
1000	4.4
1250	3.8
1500	3.1
1750	2.8
2000	2.5

a. Draw a curve through the scatterplot that models the data. Why can we rule out linear and quadratic models as possibilities?

b. As water depth increases, what happens to the temperature? Given that ocean water freezes near 0°C, and the deepest point in the ocean is roughly 11,000 m, write this relationship using direction/approach notation.

c. Run a power regression on the data and identify the equation model. According to the model, what is the ocean temperature at the bottom of the Marianas Trench, some 10,900 m below sea level?

Solution: **a.** After viewing the scatterplot (Figure 3.56), a linear function is ruled out since the data are obviously nonlinear. A quadratic model isn't feasible since it's unreasonable to assume that temperatures will begin increasing again, as depth increases.

Figure 3.56

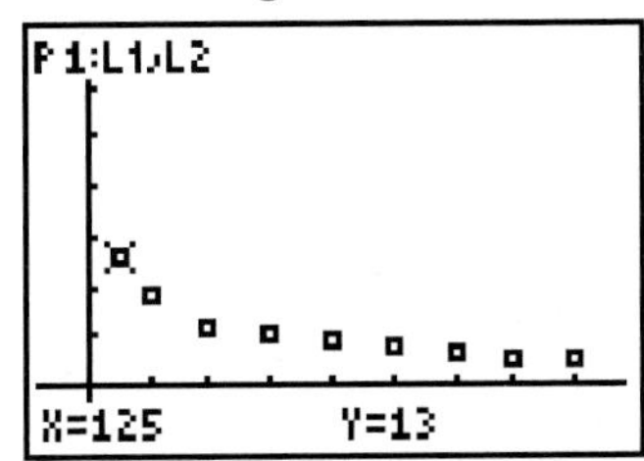

b. As water depth approaches 11,000 m, the temperature of the water approaches 0. In notation: as $d \to 11{,}000^-$, $t \to 0^+$.

Figure 3.57

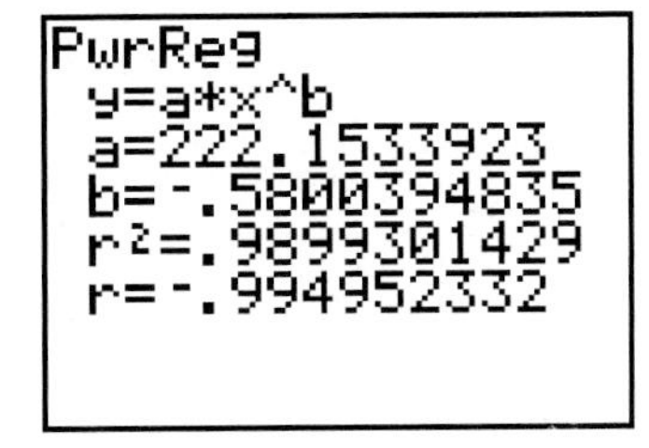

c. Using a power regression (Figure 3.57), we obtain the equation model $f(x) \approx \frac{222}{x^{0.58}}$

from the power function family. For the temperature at the bottom of the Marianas Trench, the model gives $f(10{,}900) \approx 1°\text{C}$.

Source: University of California at Los Angeles at www.msc.ucla.oceanglobe/pdf/thermo_plot_lab

NOW TRY EXERCISES 56 THROUGH 58

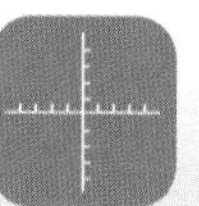

TECHNOLOGY HIGHLIGHT

Studying Joint Variations Using a Graphing Calculator

The keystrokes shown apply to a TI-84 Plus model. Please consult your manual or our Internet site for other models.

Although a graphing calculator is limited to displaying the relationship between only two variables (for the most part), it has a feature that enables us to see how these two are related with respect to a third. Consider the variation equation from Example 8: $F = 0.05dv^2$. If we want to investigate the relationship between fuel consumption and velocity, we can have the calculator display multiple versions of the relationship *simultaneously for different values of d*. This is accomplished using the { and } symbols, which are 2nd functions to the parentheses. When the calculator sees values between these grouping symbols and separated by commas, it is programmed to use each value independently of the others, graphing or evaluating the relation for each value in the set. We illustrate by graphing the relationship $f = 0.05dv^2$ for three different values of d. Enter the equation on the Y= screen as $Y_1 = 0.05\{10, 20, 30\}x^2$, which tells the calculator to graph the equations $Y_1 = 0.05(10)x^2$, $Y_1 = 0.05(20)x^2$, and $Y_1 = 0.05(30)x^2$ on the same grid. Note that since d is constant, each graph is a parabola. Set the viewing window using the values given in Example 8 as a guide. The result is the graph shown in Figure 3.58, where we can study the relationship between these three variables using the up ▲ and down ▼ arrows. From our work with parabolas and the toolbox functions, we know the widest parabola used the coefficient "10," while the narrowest parabola used the coefficient "30." As shown, the graph tells us that at a speed of 15 nautical miles per hour (X = 15), it will take 112.5 barrels of fuel to travel 10 mi (the first number in the list). After pressing the ▲ key, the cursor jumps to the second curve, which shows values of X = 15 and Y = 225. This means at 15 nautical miles per hour, it would take 225 barrels of fuel to travel 20 mi. Use these ideas to complete the following exercises:

Figure 3.58

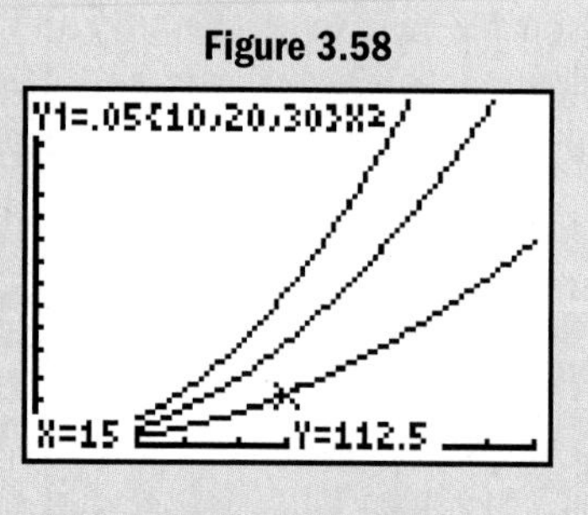

Exercise 1: The comparison of distance covered versus fuel consumption at different speeds also makes an interesting study. This time velocities are constant values and the distance varies. On the Y= screen, enter $Y_1 = 0.05x\{10, 20, 30\}^2$. What family of equations results? Use the up/down arrow keys for $x = 15$ (a distance of 15 mi) to find how many barrels of fuel it takes to travel 15 mi at 10 mph, 15 mi at 20 mph, and 15 mi at 30 mph. Comment on what you notice.

Exercise 2: The maximum safe load S for a wooden, horizontal plank supported at both ends varies jointly with the width W of the beam, the square of its thickness T, and inversely with its length L. A plank 10 ft long, 12 in. wide, and 1 in. thick will safely support 450 lb. Find the value of k and write the variation equation, then use the equation to explore: $k = 375$, $S = \frac{375WT^2}{L}$

a. Safe load versus thickness for a constant width and given lengths (quadratic function). Use $w = 8$ in. and {8, 12, 16} for L. $S = 375T^2$, $S = 250T^2$, $S = 187.5T^2$

b. Safe load versus length for a constant width and given thicknesses (reciprocal function). Use $w = 8$ in. and $\{\frac{1}{4}, \frac{1}{2}, \frac{3}{4}\}$ for thickness. $S = \frac{187.5}{L}$, $S = \frac{750}{L}$, $S = \frac{1687.5}{L}$

Exercise 1. linear; 75; 300; 675; Fuel required increases dramatically.

Exercise 2.

3.6 EXERCISES

CONCEPTS AND VOCABULARY

Fill in each blank with the appropriate word or phrase. Carefully reread the section if needed.

1. The phrase "y varies directly with x" is written $y = kx$, where k is called the constant of variation.

2. If more than two quantities are related in a variation equation, the result is called a joint variation.

3. The statement "y varies inversely with the square of x" is written $y = \frac{k}{x^2}$. This is actually a reciprocal quadratic function, whose graph resembles a volcano.

4. For $y = kx$, $y = kx^2$, $y = kx^3$, and $y = k\sqrt{x}$, it is true that as $x \to \infty$, $y \to \infty$ (functions increase). One important difference among the functions is $\frac{\Delta y}{\Delta x}$, or their rate of increase.

5. Discuss/explain the general procedure for solving applications of variation. Include references to keywords, and illustrate using an example. Answers will vary.

6. The basic percent formula is *amount equals percent times base,* or $A = PB$. In words, write this out as a direct variation with B as the constant of variation, then as an inverse variation with the amount A as the constant of variation. Answers will vary.

DEVELOPING YOUR SKILLS

Use descriptive variables to write the variation equation for each statement.

7. distance traveled varies directly with rate of speed $d = kr$

8. cost varies directly with the quantity purchased $c = kq$

9. force varies directly with acceleration $F = ka$

10. length of a spring varies directly with attached weight $l = kw$

For Exercises 11 through 14, find the constant of variation and write the variation equation. Then use the equation to complete the table or solve the application.

11. y varies directly with x; $y = 0.6$ when $x = 24$. $y = 0.025x$

x	y
500	12.5
650	16.25
750	18.75

12. w varies directly with v; $w = \frac{1}{3}$ when $v = 5$. $w = \frac{1}{15}v$

v	w
291	19.4
327	21.8
339	22.6

13. Wages earned varies directly with the number of hours worked. Last week I worked 37.5 hr and my gross pay was \$344.25. How much will I gross this week if I work 35 hr? What does the value of k represent in this case? \$321.30; the hourly wage; $k = \$9.18$/hr

14. The thickness of a paperback book varies directly as the number of pages. A book 3.2 cm thick has 750 pages. Approximate the thickness of *Roget's 21st Century Thesaurus* (paperback—2nd edition), which has 957 pages. $T = \frac{8}{1875}P$; 4.1 cm

15. b.

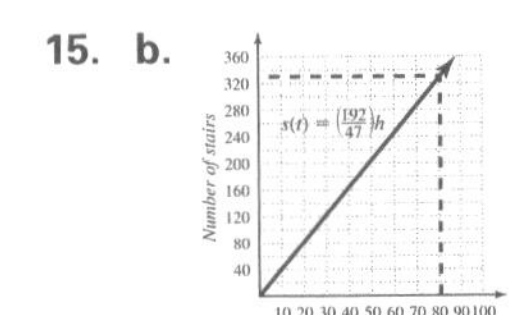

15. The number of stairs in the stairwell of tall buildings and other structures varies directly as the height of the structure. The base and pedestal for the Statue of Liberty are 47 m tall, with 192 stairs from ground level to the observation deck at the top of the pedestal (at the statue's feet). (a) Find the constant of variation and write the variation equation, (b) graph the variation equation, (c) use the graph to estimate the number of stairs from ground level to the
a. $k = \frac{192}{47}$ $S = \frac{192}{47}h$ **c.** 330 stairs **d.** $S = 331$; yes

observation deck in the statue's crown 81 m above ground level, and (d) use the equation to check this estimate. Was it close?

16. The height of a projected image varies directly as the distance of the projector from the screen. At a distance of 48 in., the image on the screen is 16 in. high. (a) Find the constant of variation and write the variation equation, (b) graph the variation equation, (c) use the graph to estimate the height of the image if the projector is placed at a distance of 5 ft 3 in., and (d) use the equation to check this estimate. Was it close?

a. $k = \frac{1}{3}$; $h = \frac{1}{3}d$ c. 21 in. d. $h = 21$; yes

16. b.

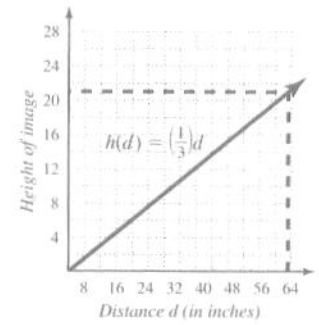

17. $A = kS^2$
18. $E = kd^2$
19. $P = kc^2$
20. $C = km^2$

Use descriptive variables to write the variation equation for each statement.

17. surface area of a cube varies directly with the square of a side

18. potential energy in a spring varies directly with the square of the distance the spring is compressed

19. electric power varies directly with the square of the current (amperes)

20. manufacturing cost varies directly as the square of the number of items made.

For Exercises 21 through 26, find the constant of variation and write the variation equation. Then use the equation to complete the table or solve the application.

21. p varies directly with the square of q; $p = 280$ when $q = 50$

q	p
45	226.8
55	338.8
70	548.8

$k = 0.112$; $p = 0.112\,q^2$

22. n varies directly with m squared; $n = 24.75$ when $m = 30$

m	n
40	44
60	99
88	212.96

$k = 0.0275$; $n = 0.0275\,m^2$

23. The surface area of a cube varies directly as the square of one edge. A cube with edges of $14\sqrt{3}$ cm has a surface area of 3528 cm^2. Find the surface area in square meters of the spaceships used by the Borg Collective in *Star Trek—The Next Generation,* cubical spacecraft with edges of 3036 m.

24. The area of an equilateral triangle varies directly as the square of one side. A triangle with sides of 50 yd has an area of 1082.5 yd^2. Find the area in mi^2 of the region bounded by straight lines connecting the cities of Cincinnati, Ohio, Washington, D.C., and Columbia, South Carolina, which are each approximately 400 mi apart (1 mi = 1760 yd).

25. The distance an object falls varies directly as the square of the time it has been falling. The cannonballs dropped by Galileo from the Leaning Tower of Pisa fell about 169 ft in 3.25 sec. (a) Find the constant of variation and write the variation equation, (b) graph the variation equation, (c) use the graph to estimate how long it would take a hammer, accidentally dropped from a height of 196 ft by a bridge repair crew, to splash into the water below, and (d) use the equation to check this estimate. Was it close? (e) According to the equation, if a camera accidentally fell out of the *News 4 Eye-in-the-Sky* helicopter from a height of 121 ft, how long until it strikes the ground?

26. When a child blows small soap bubbles, they come out in the form of a sphere because the surface tension in the soap seeks to minimize the surface area. The surface area of any sphere varies directly with the square of its radius. A soap bubble with a $\frac{3}{4}$ in. radius has a surface area of approximately 7.07 in^2. (a) Find the constant of variation and write the variation equation, (b) graph the variation equation, (c) use the graph to estimate the radius of a seventeenth-century cannonball that has a surface area of 113.1 in^2, and (d) use the equation to check this estimate. Was it close? (e) According to the equation, what is the surface area of an orange with a radius of $1\frac{1}{2}$ in?

23. $k = 6$, $A = 6s^2$; about 55,303,776 m^2
24. $k = 0.433$, $A = 0.433s^2$, 69,280 mi^2
25. a. $k = 16$ $d = 16t^2$
    b.
    c. about 3.5 sec
    d. 3.5 sec; yes
    e. 2.75 sec
26. a. $k = \frac{2828}{225}$ $A = \frac{2828}{225}r^2$
    b.
    c. about 3 in.
    d. 3 in.; yes
    e. 28.28 in^2

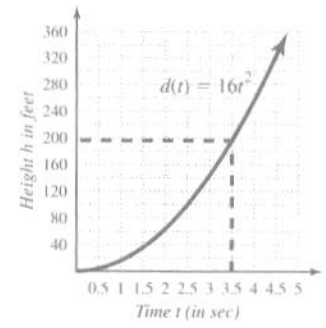

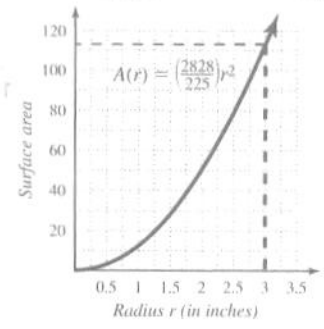

27. $F = \frac{k}{d^2}$ 28. $P = \frac{k}{A}$
29. $S = \frac{k}{L}$ 30. $I = \frac{k}{d^2}$

31. $Y = \frac{12{,}321}{Z^2}$
32. $A = \frac{1960}{B}$

33. $w = \frac{3{,}072{,}000{,}000}{r^2}$; 48 kg
34. $d = \frac{247{,}500}{C}$; 4500 orders

35. $I = krt$
36. $H = kcd^2$
37. $A = kh(B + b)$
38. $A = kbh$
39. $V = ktr^2$
40. $R = \frac{kL}{A}$
41. $C = \frac{6.75R}{S^2}$
42. $J = \frac{23.75P}{\sqrt{Q}}$
43. $E = 0.5\,mv^2$; 612.50 J

Use descriptive variables to write the variation equation for each statement.

27. the force of gravity varies inversely as the square of the distance between objects.

28. pressure varies inversely as the area over which it is applied.

29. the safe load of a beam supported at both ends varies inversely as its length.

30. the intensity of sound varies inversely as the square of its distance from the source.

For Exercises 31 through 34, find the constant of variation and write the variation equation. Then use the equation to complete the table or solve the application.

31. Y varies inversely as the square of Z; $Y = 1369$ when $Z = 3$

Z	Y
37	9
74	2.25
111	1

32. A varies inversely with B; $A = 2450$ when $B = 0.8$

B	A
140	14
320	6.125
560	3.5

33. The effect of Earth's gravity on an object (its weight) varies inversely as the square of its distance from the center of the planet (assume the Earth's radius is 6400 km). If the weight of an astronaut is 75 kg on Earth (when $r = 6400$), what would this weight be at an altitude of 1600 km *above the surface* of the Earth?

34. The demand for a popular new running shoe varies inversely with the cost of the shoes. When the wholesale price is set at $45, the manufacturer ships 5500 orders per week to retail outlets. Based on this information, how many orders would be shipped if the wholesale price rose to $55?

Use descriptive variables to write the variation equation for each statement.

35. Interest earned varies jointly with the rate of interest and the length of time on deposit.

36. Horsepower varies jointly as the number of cylinders in the engine and the square of the cylinder's diameter.

37. The area of a trapezoid varies jointly with its height and the sum of the bases.

38. The area of a triangle varies jointly with its height and the length of the base.

39. The volume of metal in a circular coin varies directly with the thickness of the coin and the square of its radius.

40. The electrical resistance in a wire varies directly with its length and inversely as the cross-sectional area of the wire.

Find the constant of variation and write the related model. Then use the model to complete the table.

41. C varies directly with R and inversely with S squared, and $C = 21$ when $R = 7$ and $S = 1.5$.

R	S	C
120	6	22.5
200	12.5	8.64
350	15	10.5

42. J varies directly with P and inversely with the square root of Q, and $J = 19$ when $P = 4$ and $Q = 25$.

P	Q	J
47.5	90.25	118.75
112	31.36	475
18.76	44.89	66.5

43. Kinetic energy: Kinetic energy (energy attributed to motion) varies jointly with the mass of the object and the square of its velocity. Assuming a unit mass of $m = 1$, an object with a velocity of 20 m per sec (m/s) has kinetic energy of 200 J. How much energy is produced if the velocity is increased to 35 m/s?

44. $S = \frac{3wh^2}{64l}$; 1.5625 tons

44. Safe load: The load that a horizontal beam can support varies jointly as the width of the beam, the square of its height, and inversely as the length of the beam. A beam 4 in. wide and 8 in. tall can safely support a load of 1 ton when the beam has a length of 12 ft. How much could a similar beam 10 in. tall safely support?

WORKING WITH FORMULAS

45. cube root family; answers will vary; 0.054 or 5.4%

45. Required interest rate: $R(A) = \sqrt[3]{A} - 1$

To determine the simple interest rate R that would be required for each dollar ($1) left on deposit for 3 yr to grow to an amount A, the formula $R(A) = \sqrt[3]{A} - 1$ can be applied. To what function family does this formula belong? Complete the table using a calculator, then use the table to estimate the interest rate required for each $1 to grow to $1.17. Compare your estimate to the value you get by evaluating $R(1.17)$.

Amount *A*	Rate *R*
1.0	0.000
1.05	0.016
1.10	0.032
1.15	0.048
1.20	0.063
1.25	0.077

46. a. Force varies jointly with the product of the charges and inversely as the square of the distance between them.
b. 1.8×10^9

46. Force between charged particles: $F = k\frac{Q_1Q_2}{d^2}$

The force between two charged particles is given by the formula shown, where F is the force (in joules—J), Q_1 and Q_2 represent the electrical charge on each particle (in coulombs—C), and d is the distance between them (in meters). If the particles have a like charge, the force is repulsive; if the charges are unlike, the force is attractive. (a) Write the variation equation in words. (b) Solve for k and use the formula to find the electrical constant k, given $F = 0.36$ J, $Q_1 = 2 \times 10^{-6}$ C, $Q_2 = 4 \times 10^{-6}$ C, and $d = 0.2$ m. Express the result in scientific notation.

APPLICATIONS

Find the constant of variation "k" and write the variation equation, then use the equation to solve.

47. $T = \frac{48}{V}$; 32 volunteers

47. Cleanup time: The time required to pick up the trash along a stretch of highway varies inversely as the number of volunteers who are working. If 12 volunteers can do the cleanup in 4 hr, how many volunteers are needed to complete the cleanup in just 1.5 hr?

48. $P = 0.064S^3$; 2744 *W*

48. Wind power: The wind farms in southern California contain wind generators whose power production varies directly with the cube of the wind's speed. If one such generator produces 1000 W of power in a 25 mph wind, find the power it generates in a 35 mph wind.

49. $M = \frac{1}{6}E$; 41.7 kg

49. Pull of gravity: The weight of an object on the moon varies directly with the weight of the object on Earth. A 96-kg object on Earth would weigh only 16 kg on the moon. How much would a 250-kg astronaut weigh on the moon?

50. $T = \frac{\sqrt{5}}{2}\sqrt{L}$; 6.12 sec

50. Period of a pendulum: The time that it takes for a simple pendulum to complete one period (swing over and back) varies directly as the square root of its length. If a pendulum 20 ft long has a period of 5 sec, find the period of a pendulum 30 ft long.

51. $D = 21.6\sqrt{S}$; 144.9 ft

51. Stopping distance: The stopping distance of an automobile varies directly as the square root of its speed when the brakes are applied. If a car requires 108 ft to stop from a speed of 25 mph, estimate the stopping distance if the brakes were applied when the car was traveling 45 mph.

52. $d = \frac{850{,}000}{p}$; 12,000 units

52. Supply and demand: A chain of hardware stores finds that the demand for a special power tool varies inversely with the advertised price of the tool. If the price is advertised at $85, there is a monthly demand for 10,000 units at all participating stores. Find the projected demand if the price were lowered to $70.83.

53. $C = 8.5LD$; $76.50

54. $R = \frac{7.5L}{D^2}$; 24.49 Ω

53. Cost of copper tubing: The cost of copper tubing varies jointly with the length and the diameter of the tube. If a 36-ft spool of $\frac{1}{4}$-in.-diameter tubing costs $76.50, how much does a 24-ft spool of $\frac{3}{8}$-in.-diameter tubing cost?

54. Electrical resistance: The electrical resistance of a copper wire varies directly with its length and inversely with the square of the diameter of the wire. If a wire 30 m long with

55. $C = (4.4 \times 10^{-4})\frac{p_1p_2}{d^2}$; about 223 calls

56. a. Scatter-plot shows data are obviously nonlinear; no sudden increase is reasonable/expected

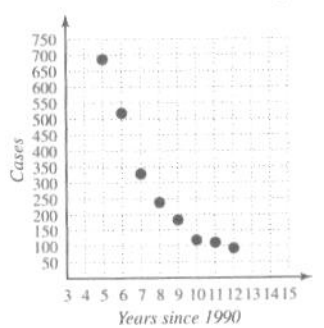

b. $C(t) = \frac{36{,}579}{x^{2.428}}$; 51; 25

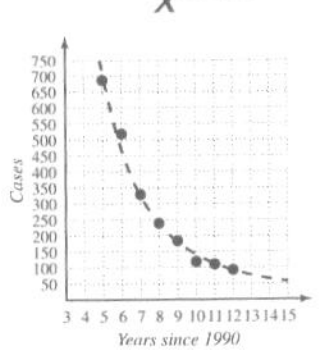

57. a. Scatter-plot shows data are obviously nonlinear; decreasing to increasing pattern rules out a power function.

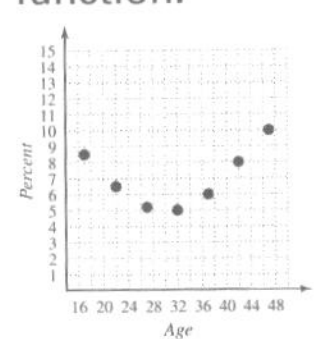

b. $p(t) = 0.0148t^2 - 0.9175t + 19.5601$; 6.5%; 6.3%

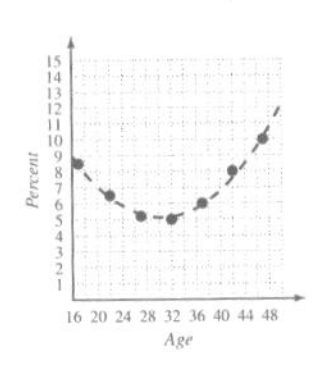

58. a. Scatter-plot shows data are obviously nonlinear; no sudden increase in rodent population is reasonable/expected.

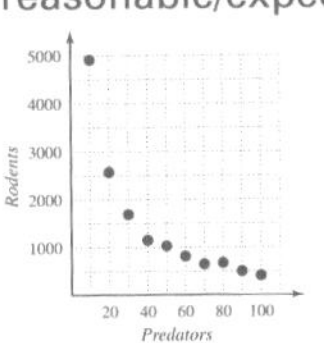

b. as $p \to \infty$, $r \to 0$

a diameter of 3 mm has a resistance of 25 Ω, find the resistance of a wire 40 m long with a diameter of 3.5 mm.

55. **Volume of phone calls:** The number of phone calls per day between two cities varies directly as the product of their populations and inversely as the square of the distance between them. The city of Tampa, Florida (pop. 300,000), is 430 mi from the city of Atlanta, Georgia (pop. 420,000). Telecommunications experts estimate there are about 300 calls per day between the two cities. Use this information to estimate the number of daily phone calls between Amarillo, Texas (pop. 170,000), and Denver, Colorado (pop. 550,000), which are also separated by a distance of about 430 mi. Note: Population figures are for the year 2000 and rounded to the nearest ten-thousand.

Source: 2005 World Almanac, p. 626.

56. Largely due to research, education, prevention, and better health care, estimates of the number of AIDS (Acquired Immune Deficiency Syndrome) cases diagnosed in children less than 13 yr of age has been declining. Data for the years 1995 through 2002 are given in the table.

Years Since 1990	Cases
5	686
6	518
7	328
8	238
9	183
10	118
11	110
12	92

a. Use the data to draw a scatter-plot and decide on an appropriate form of regression. Discuss why linear and quadratic models can be ruled out.

b. Find a regression equation that accurately models the data and graph the scatter-plot and equation on the same screen. According to the model, how many diagnosed cases of AIDS in children are projected for 2005? For 2010?

Source: National Center for Disease Control and Prevention

57. For many years, the association between low birth weight (less than 2500 g or about 5.5 lb) and a mother's age has been well documented. The data given in the table are grouped by age and give the percent of total births with low birth weight.

Ages	Percent
15–19	8.5
20–24	6.5
25–29	5.2
30–34	5
35–39	6
40–44	8
45–54	10

a. Using the median age of each group, draw a scatter-plot and decide on an appropriate form of regression. Discuss why linear and power models can be ruled out.

b. Find a regression equation that accurately models the data and graph the scatter-plot and equation on the same screen. According to the model, what percent of total births will have a low birth weight for mothers 22 yr old? 39 yr old?

Source: National Vital Statistics Report, vol. 50, No. 5, February 12, 2002

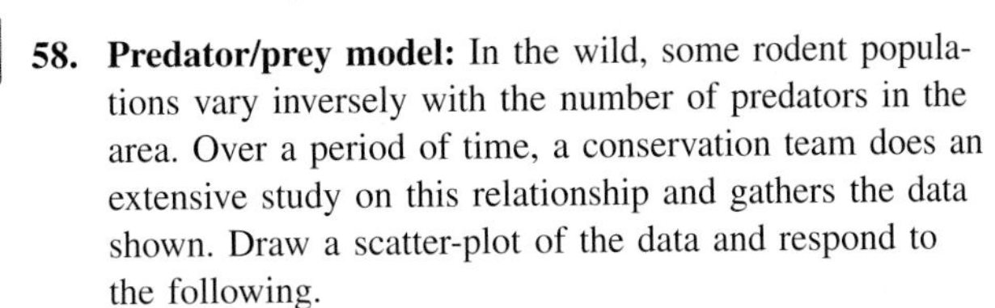

58. **Predator/prey model:** In the wild, some rodent populations vary inversely with the number of predators in the area. Over a period of time, a conservation team does an extensive study on this relationship and gathers the data shown. Draw a scatter-plot of the data and respond to the following.

Predators	Rodents
10	4910
20	2570
30	1690
40	1150
50	1030
60	815
70	650
80	675
90	500
100	410

a. Draw a curve through the scatter-plot that models the data. Discuss why linear and quadratic models should not be used.

b. As the number of predators increases, what happens to the number of rodents? Write this relationship using direction/approach notation.

c. $r = \frac{57{,}319}{p^{1.046}}$; about 25 predators

c. Run a power regression, identify the equation model, then graph the scatter-plot and equation on the same screen. According to the model, how many predators are in the area if studies show a rodent population of 2000 animals?

WRITING, RESEARCH, AND DECISION MAKING

59. On the Earth, the height of a projectile that is thrown from the surface is modeled by the variation $g(t) = \frac{1}{2}gt^2$, where g is acceleration due to gravity (32 ft/sec) and t is the time in seconds. The value of g is different on other planets, due to their various sizes. Use the Internet, an encyclopedia, or some other resource to find the value of g on some of the other planets. Then calculate how far an object would fall in 10 sec if it were dropped from a height of 1000 ft, for each planet researched. Does the object hit the surface in less than 10 sec on any planet? Answers will vary.

60. For f: $\frac{\Delta y}{\Delta x} = \frac{-10}{3}$
For g: $\frac{\Delta y}{\Delta x} = \frac{-110}{9}$; less; for both f and g, as $x \to \infty$, $y \to 0^+$

60. In function form, the variations $Y_1 = k\frac{1}{x}$ and $Y_2 = k\frac{1}{x^2}$ become $f(x) = k\frac{1}{x}$ and $g(x) = k\frac{1}{x^2}$. Both graphs appear similar in Quadrant I and both may "fit" a scatter-plot fairly well, but there is a big difference between them—they decrease as x gets larger, but *they decrease at very different rates*. Use the ideas from Section 2.4 to compute the rate of change for f and g for the interval from $x = 0.5$ to $x = 0.6$. Were you surprised? In the interval $x = 0.7$ to $x = 0.8$, will the rate of decrease for each function be greater or less than in the interval $x = 0.5$ to $x = 0.6$? Why?

EXTENDING THE CONCEPT

61. 6.67×10^{-7}

61. The gravitational force F between two celestial bodies varies jointly as the product of their masses and inversely as the square of the distance d between them. The relationship is modeled by Newton's law of universal gravitation: $F = k\frac{m_1m_2}{d^2}$. Given that $k = 6.67 \times 10^{-11}$, what is the gravitational force exerted by a 1000-kg sphere on another identical sphere that is 10 m away?

62. a. about 3.5 ft
b. about 6.9 ft

62. The intensity of light and sound both vary inversely as the square of their distance from the source.

a. Suppose you're relaxing one evening with a copy of *Twelfth Night* (Shakespeare), and the reading light is placed 5 ft from the surface of the book. At what distance would the intensity of the light be twice as great?

b. *Tamino's Aria* (*The Magic Flute*—Mozart) is playing in the background, with the speakers 12 ft away. At what distance from the speakers would the intensity of sound be three times as great?

MAINTAINING YOUR SKILLS

63. $\frac{9y^2}{4x^2}$
64. $x = 0, x = -2 \pm 2i$
66. a. $x \in (-\infty, -4) \cup (-4, 4) \cup (4, \infty)$
b. $x \in (-\infty, -4) \cup (4, \infty)$
67. 60 gal
68.

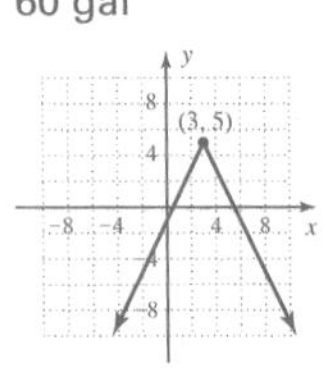

63. (R.3) Evaluate: $\left(\frac{2x^4}{3x^3y}\right)^{-2}$

64. (1.5) Find all roots, real and complex: $x^3 + 4x^2 + 8x = 0$.

65. (1.4) Check by substitution: Is $x = 1 + 2i$ a solution to $x^2 - 2x + 5 = 0$? yes

66. (2.2) State the domains of f and g given here:

a. $f(x) = \frac{x-3}{x^2-16}$ **b.** $g(x) = \frac{x-3}{\sqrt{x^2-16}}$

67. (1.1) You have 40 gal of a 25% acid solution. How much water must be added to reduce the strength of the acid solution to 10%?

68. (3.3) Graph by using transformations of the parent function and plotting a minimum of points: $f(x) = -2|x - 3| + 5$.

3.7 Piecewise-Defined Functions

LEARNING OBJECTIVES

In Section 3.7 you will learn how to:

A. State the domain of a piecewise-defined function

B. Evaluate piecewise-defined functions

C. Graph functions that are piecewise-defined

D. Solve applications involving piecewise-defined functions

INTRODUCTION

Most of the functions we've studied thus far have been smooth and continuous. Although "smooth" and "continuous" are defined more formally in advanced courses, for our purposes *smooth* simply means the graph has no sharp turns or jagged edges, and *continuous* means you can draw the entire graph without lifting your pencil. In this section, we study a special class of functions, called **piecewise-defined functions,** whose graphs may be various combinations of smooth/not smooth and continuous/not continuous. Such functions have a tremendous number of applications in the real world.

POINT OF INTEREST

Although the word "piecewise" is somewhat self-descriptive, the etymology (history and evolution) of the word is still worthy of mention. The word "piece," of course, indicates a portion or fragment of a whole, while the word "wise" means *"able to see, in the manner of, or with regard to."* A piecewise-defined function is one where we are "able to see the pieces," and therefore we are able to evaluate, graph, and analyze each piece, just as we do with other functions.

A. The Effective Domain of a Piecewise-Defined Function

From 1995 to 1998, theater admissions in the United States grew at a rate that was very close to linear. After peaking in 1998, theater attendance dropped for the years 1999 and 2000, then began increasing once again. This decline and return to growth followed a pattern that was nearly parabolic. From Table 3.5 and the related scatter-plot (Figure 3.59),

Table 3.5

Year	Admissions (billions)	Year	Admissions (billions)
5	1.30	9	1.43
6	1.37	10	1.44
7	1.43	11	1.55
8	1.51	12	1.74

(1990 corresponds to year 0)

Source: National Association of Theater Owners at www.natoonline.org/statisticsadmissions.html

Figure 3.59

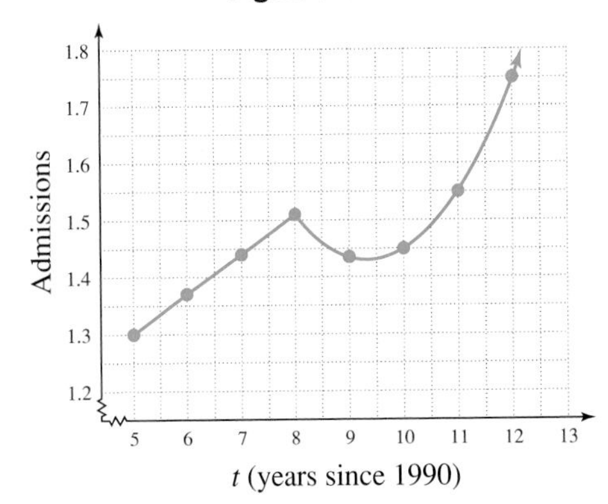

we note a linear model fits the data very nicely from 1995 to 1998, but a quadratic model would be needed from 1999 to 2003 and (perhaps) beyond. The result would be a single function designed to model admissions to U.S. theaters, but it would be *defined in two pieces*. This is an example of a piecewise-defined function.

The notation we use for such functions is a large "left brace," that indicates the equations it groups are part of a single function model. Using the techniques from Section 2.6, the equation for the linear portion of the graph is approximated by

$A(t) = 0.069t + 0.954$, while the quadratic portion can be approximated by $A(t) = 0.046t^2 - 0.856t + 5.434$, where t is the time in years since 1990 and $A(t)$ is the attendance in billions for year t. To write these as a single function we (1) name the function, (2) state the pieces of the function, and (3) list the **effective domain** for each piece. The effective domain consists of input values for which each piece of the function is defined. Since the number of admissions is a function of time, we have:

function name function pieces effective domain

$$A(t) = \begin{cases} 0.069t + 0.954 & 5 \le t < 8 \\ 0.046t^2 - 0.856t + 5.434 & t \ge 8 \end{cases}$$

These ideas are further illustrated in Example 1.

EXAMPLE 1 The function shown has one linear piece modeled by $y = -2x + 10$, and one quadratic piece modeled by $y = -x^2 + 9x - 14$. Use the correct notation to write the piecewise function and state the effective domain for each piece by inspecting the graph.

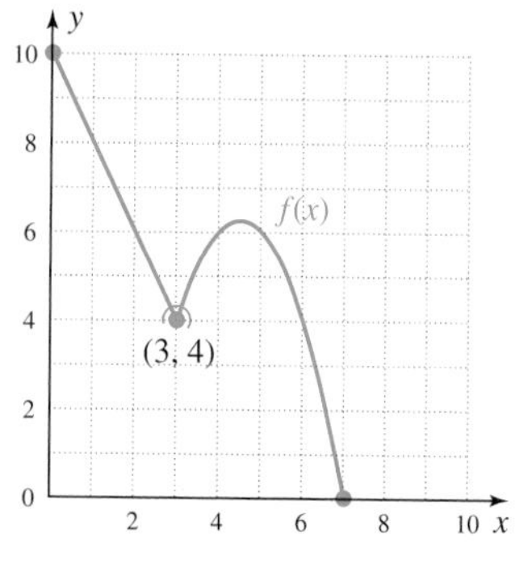

Solution: function name function pieces effective domain

$$f(x) = \begin{cases} -2x + 10 & 0 \le x \le 3 \\ -x^2 + 9x - 14 & 3 < x \le 7 \end{cases}$$

Note that we indicated the exclusion of $x = 3$ from the second piece of the function using an open half-circle.

NOW TRY EXERCISES 7 AND 8

Piecewise-defined functions can be composed of more than two pieces, and can be a combination of any two or more functions. Example 2 shows a function where both pieces are linear, and illustrates a method for graphing the endpoints of a piecewise-defined function to clearly portray the effective domain.

EXAMPLE 2 Suppose a drought in Jenkins County forced the county to ration water. For the first 300 gal used, the cost C might be \$0.10/gal, or $C(g) = 0.10g$. After 300 gal, assume the cost increases to \$0.20/gal, and the cost function becomes $C(g) = 0.20g - 30$. Use the correct notation to write a piecewise function and state the effective domain for each piece.

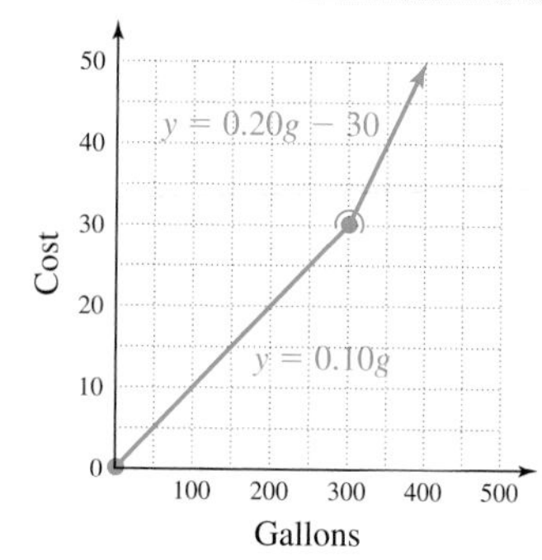

Solution: function name function pieces effective domain

$$C(g) = \begin{cases} 0.10g & 0 \le g \le 300 \\ 0.20g - 30 & g > 300 \end{cases}$$

NOW TRY EXERCISES 9 AND 10

B. Evaluating a Piecewise-Defined Function

The main reason we need to identify the effective domain for each piece of the function is to determine which piece or branch to use when evaluating it. In Example 2, if we used the first piece $C(g) = 0.10g$ to calculate our water bill after using 500 gallons, the cost would come to only $C(500) = 0.10(500) = \$50$, while in reality the water company is expecting a check for $C(g) = 0.20(500) - 30 = \$70$! They could cut off your water supply until you paid the difference. Always check the effective domain of a piecewise-defined function before evaluating it, and be sure you evaluate the piece that is "effective" for the indicated input.

EXAMPLE 3 For the function in Example 2, determine a customer's water bill for the use of 179, 300, 301, and 450 gal.

Solution: The function from Example 2 is $C(g) = \begin{cases} 0.10g & 0 \le g \le 300 \\ 0.20g - 30 & g > 300 \end{cases}$

The inputs $g = 179$ and $g = 300$ are in the effective domain of the first piece, yielding $C(179) = 17.9$ and $C(300) = 30$.

The cost is \$17.90 for 179 gal and \$30.00 for 300 gal.

The inputs $g = 301$ and $g = 450$ are in the effective domain of the second piece, and $C(301) = 30.2$ and $C(450) = 60$ shows the cost is \$30.20 for 301 gal and \$60.00 for 450 gal.

NOW TRY EXERCISES 11 THROUGH 16

C. Graphing a Piecewise-Defined Function

The easiest way to graph a piecewise-defined function is to graph each function in its entirety, then erase those parts that are outside of the corresponding effective domain. Repeat this procedure for each piece of the function. One interesting and highly instructive aspect of these functions is the opportunity to investigate restrictions on their domain and the ranges that result.

Piecewise and Continuous Functions

EXAMPLE 4 Graph the function and state the domain and range:

$$f(x) = \begin{cases} -x^2 + 6x + 3 & 0 < x \le 6 \\ 3 & x > 6 \end{cases}$$

Solution: The first piece of f is a parabola, concave down. Graphing the function by completing the square we have:

$$\begin{aligned} f(x) &= -1(x^2 - 6x + \underline{\quad}) + 3 && \text{group variable terms} \\ &= -1[(x^2 - 6x + 9) - 9] + 3 && \text{add 9, subtract 9 and regroup} \\ &= -1(x - 3)^2 + 12 && \text{distribute and simplify} \end{aligned}$$

Using the vertex/intercept formula with $a = -1$, the x-intercepts are $x = 3 \pm \sqrt{12}$, or approximately $(-0.46, 0)$ and $(6.46, 0)$.

1. Graph entire function (Figure 3.60).

2. Erase portion outside domain of $0 < x \leq 6$ (Figure 3.61).

Figure 3.60

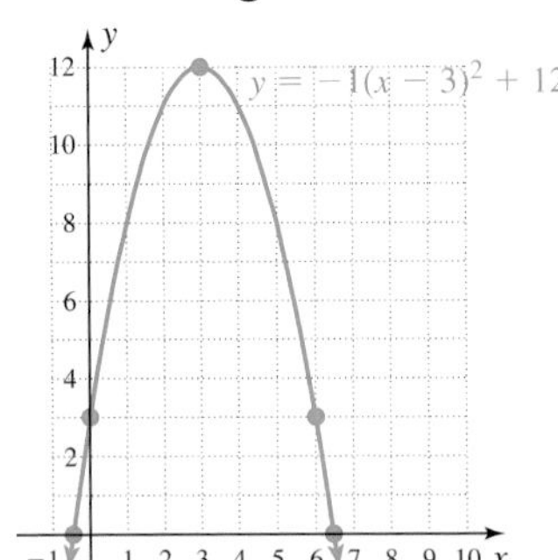

Figure 3.61

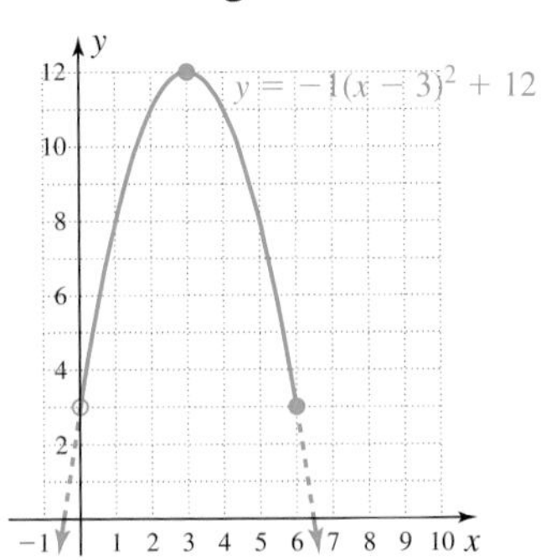

The second function is simply a horizontal line through (0, 3).

3. Graph entire function (Figure 3.62).

4. Erase portion outside domain of $x > 6$ (Figure 3.63).

Figure 3.62

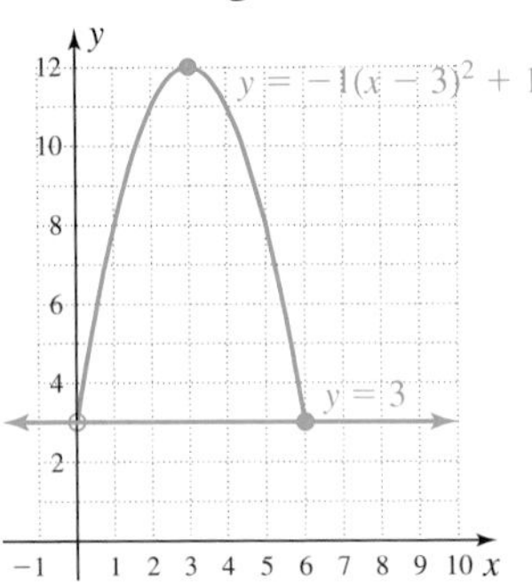

Figure 3.63

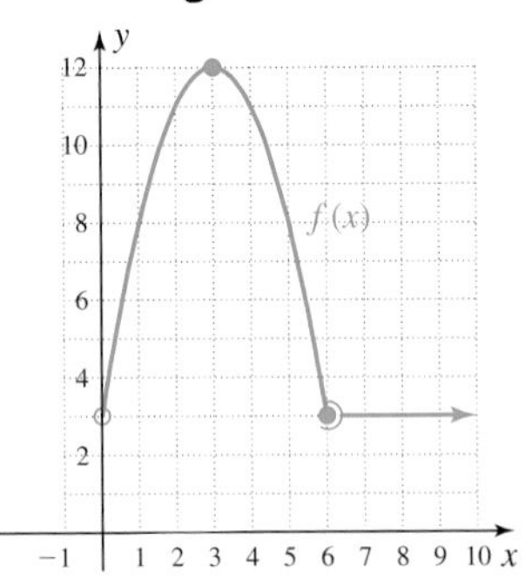

The domain of f is $x \in (0, \infty)$, the corresponding range is $y \in [3, 12]$.

NOW TRY EXERCISES 17 THROUGH 20

Piecewise and Discontinuous Functions

Notice that although the function in Example 4 was piecewise-defined, the graph of f was actually continuous—we could draw the entire graph without lifting our pencil. Piecewise graphs also come in the *discontinuous* variety, which makes the domain and range issues all the more important. We've already noted one type of discontinuity, the **asymptotic discontinuities** from our study of simple rational functions. Here we'll explore two additional types.

EXAMPLE 5 Graph $g(x)$ and state the domain and range:

$$g(x) = \begin{cases} -\frac{1}{2}x + 6 & 0 \leq x \leq 4 \\ -|x - 6| + 10 & 4 < x \leq 9. \end{cases}$$

Solution: The first piece of g is a line, with y-intercept (0, 6) and slope of $\frac{\Delta y}{\Delta x} = -\frac{1}{2}$.

1. Graph entire function (Figure 3.64).

2. Erase portion outside domain of $0 \le x \le 4$ (Figure 3.65).

Figure 3.64

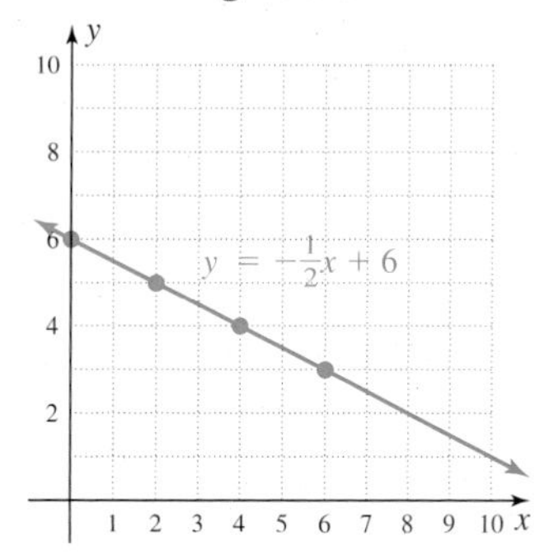

Figure 3.65

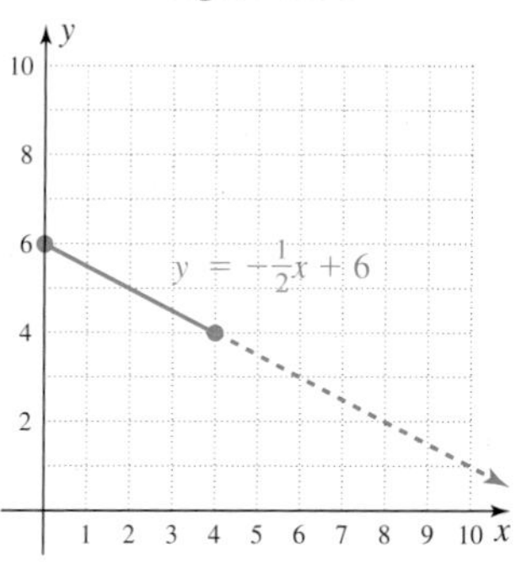

The second is an absolute value "V" function, reflected across the x-axis, then shifted right 6 and up 10.

3. Graph entire function (Figure 3.66).

4. Erase portion outside domain of $4 < x \le 9$ (Figure 3.67).

Figure 3.66

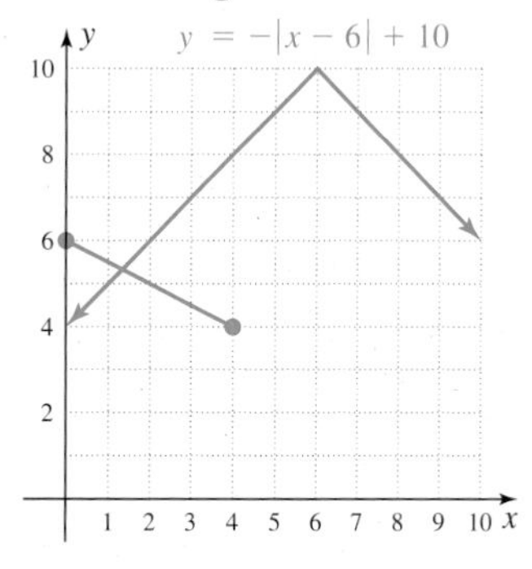

Figure 3.67

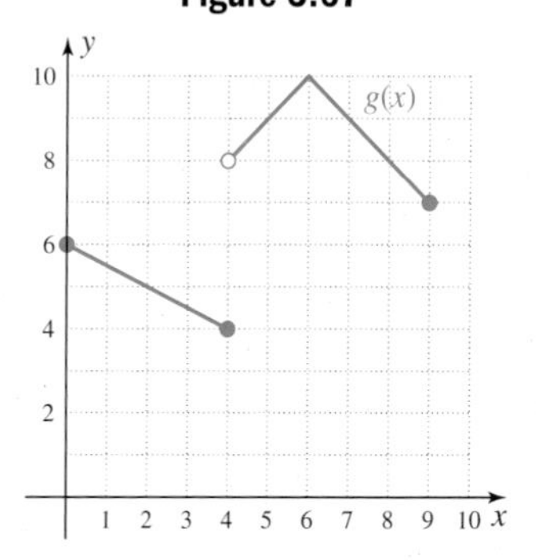

This time the result is a discontinuous graph, having what is called a **jump discontinuity** or a **nonremovable discontinuity,** as there is no way to draw the graph other than by jumping the pencil from where one piece ends to where the next begins. Using a vertical boundary line, we note the domain of g includes all values between 0 and 9 inclusive: $x \in [0, 9]$. Using a horizontal boundary line shows the smallest y-value is 4 and the largest is 10, but no range values exist between 6 and 7. The range is $y \in [4, 6] \cup [7, 10]$.

NOW TRY EXERCISES 21 THROUGH 24

EXAMPLE 6 Graph the function and state the domain and range:

$$h(x) = \begin{cases} \dfrac{x^2 - 4}{x - 2} & x \neq 2 \\ 1 & x = 2. \end{cases}$$

Solution: The first piece of h is not defined at $x = 2$, but this time the result is not asymptotic because the denominator is a factor of the numerator: $y = \dfrac{x^2 - 4}{x - 2} = \dfrac{(x + 2)(x - 2)}{x - 2} = x + 2$. It turns out to be a line, with no corresponding y-value for $x = 2$. This leaves a hole at (2, 4) designated with an open dot.

1. Graph entire function (Figure 3.68).
2. Erase portion outside of the domain (Figure 3.69)

Figure 3.68

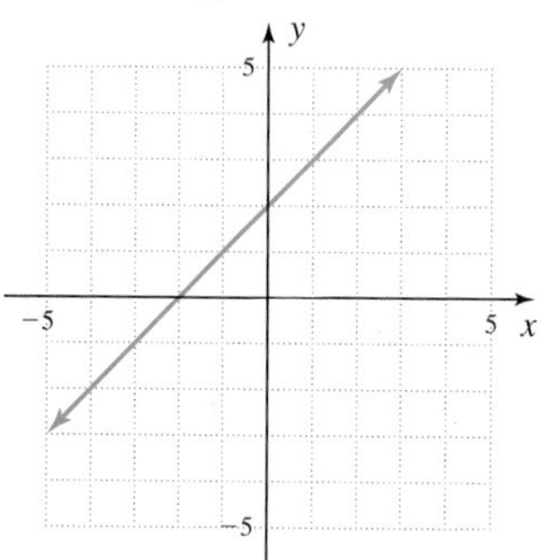

Figure 3.69

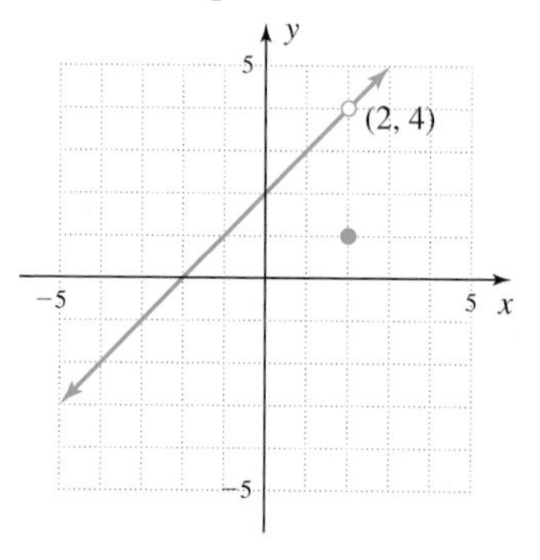

The second piece is pointwise-defined, much as the functions we worked with in Section 2.2. Its graph is simply the point at (2, 1) shown (it cannot be a horizontal line since we defined it only at $x = 2$). It is interesting to note that while the domain is $x \in \mathbb{R}$ (the function *is* defined at all points), the range is $y \in (-\infty, 4) \cup (4, \infty)$ as the function never takes on the value $y = 4$. The discontinuity illustrated here is called a **removable** (or fixable) **discontinuity,** because we could "repair" the hole by redefining the second piece as $y = 4$ when $x = 2$. There is no way to repair jump or asymptotic discontinuities.

NOW TRY EXERCISES 25 THROUGH 28

D. Applications of Piecewise-Defined Functions

The number of applications for piecewise-defined functions is practically limitless. It is actually fairly rare for a single function to accurately model a situation over a long period of time. Laws change, spending habits change, and technology can bring abrupt alterations in many areas of our lives. To accurately model these changes often requires a piecewise-defined function. For all of these reasons and more, time spent studying these functions is time well spent.

EXAMPLE 7 For the first half of the twentieth century, per capita spending on police protection followed a pattern that was nearly linear. This expenditure can be modeled by $S(t) = 0.54t + 12$, where $S(t)$ represents per capita spending on police protection in year t (1900 corresponds to year 0). After 1950, perhaps due to the growth of American cities, this spending began growing at a rate that was still linear—but with a greatly increased slope: $S(t) = 3.65t - 144$. Write these as a single piecewise-defined function $S(t)$, state the effective domain for each piece, then graph the function. According to this model, how much was spent (per capita) on police protection in 2000 and 2005?

Source: Data taken from the *Statistical Abstract of the United States* for various years.

Solution: function name function pieces effective domain

$$S(t) = \begin{cases} 0.54t + 12 & 0 \le t \le 50 \\ 3.65t - 144 & t > 50 \end{cases}$$

Since both pieces are linear, we can graph each part using two points. For the first function, $S(0) = 12$ and $S(50) = 39$. For the second function $S(50) \approx 39$ and $S(80) = 148$. The information indicates that our t (input) axis should be scaled from 0 to 110 and our $S(t)$ (output) axis from 0 to perhaps 240. From the graph, we estimate that in 2000 ($t = 100$) about \$220 per capita was spent on police protection. Evaluating S at $t = 100$ shows our graph is fairly accurate.

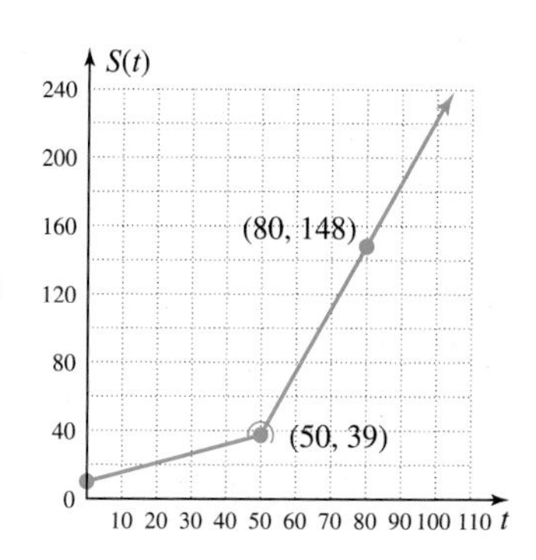

$$\begin{aligned} S(t) &= 3.65t - 144 \\ S(100) &= 3.65(100) - 144 \\ &= 365 - 144 \\ &= 221 \end{aligned}$$

About \$221 per capita was spent on police protection in the year 2000. For 2005, the model shows just over \$239 per capita was spent.

NOW TRY EXERCISES 31 THROUGH 38

Step Functions

The last function we'll explore is referred to as a **step function,** so called because the pieces of the function form a series of horizontal steps. These functions find frequent application in the way consumers are charged for services.

EXAMPLE 8 As of November 2004, the first-class postage rate for U.S. mail was 37¢ for the first ounce, then an additional 23¢ per ounce thereafter, up to 13 ounces. Graph the function and state its domain and range. Use the graph to state the cost of mailing a report weighing (a) 7.5 oz, (b) 8 oz, and (c) 8.1 oz.

Solution: The 37¢ charge applies to letters weighing between 0 oz and 1 oz. Zero is not included since we have to mail *something,* but 1 is included since a letter weighing exactly one ounce still costs 37¢. The graph will be a horizontal line segment.

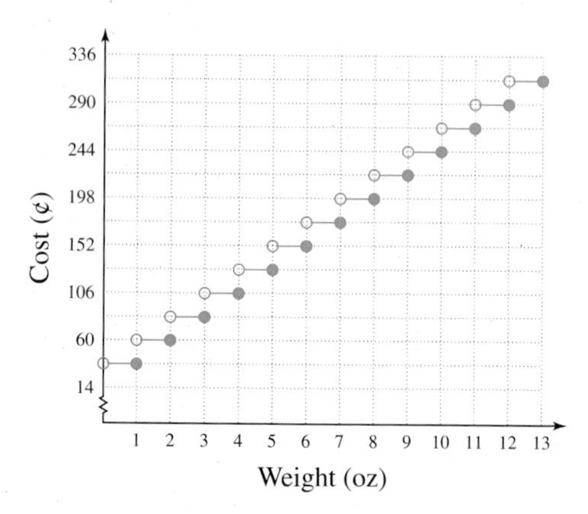

The function is defined for all weights between 0 and 13 oz, excluding zero and including 13: $x \in (0, 13]$. The range consists of single outputs corresponding to the step intervals: $R \in \{37, 60, 83, 106, \ldots, 290, 313\}$.

a. The cost of mailing an 7.5-oz report would be 198¢.

b. The cost of mailing an 8.0-oz report would still be 198¢.

c. The cost of mailing an 8.1-oz report would be $198 + 23 = 221$¢, since this brings you up to the next step.

NOW TRY EXERCISES 39 AND 40

There is a variety of additional applications in the exercise set. Enjoy.

TECHNOLOGY HIGHLIGHT

Peicewise-Defined Functions and Graphing Calculators

The keystrokes shown apply to a TI-84 Plus model. Please consult your manual or our Internet site for other models.

Most graphing calculators are able to graph piecewise-defined functions in support of further investigations. In addition, when properly set, the **TRACE** and **2nd** **GRAPH** (**TABLE**) features of the calculator offer numerous opportunities for exploring these functions. Consider the function f shown here:

$$f(x) = \begin{cases} x + 2 & x < 2 \\ (x - 4)^2 + 3 & x \geq 2 \end{cases}$$

Both "pieces" are well known—the first is a line with slope $m = 1$ and y-intercept (0, 2). The second is a parabola, concave up, shifted 4 units to the right and 3 units up. If we attempt to graph $f(x)$ using $Y_1 = x + 2$ and $Y_2 = (x - 4)^2 + 3$ as they stand, the resulting graph may be difficult to analyze because the pieces conflict and intersect (Figure 3.70) Ideally, we would have the calculator graph Y_1 only for values less than 2, then Y_2 for all other values ($x \geq 2$). This is done by indicating the effective domain for each function, separated by a forward slash and enclosed in parentheses. For instance, for the first piece we enter $Y_1 = x + 2/(x < 2)$, and for the second, $Y_2 = (x - 4)^2 + 3/(x \geq 2)$ (Figure 3.71). The forward slash looks like (is) the division symbol, but in this context, the calculator interprets it as a means of separating the function from the domain stipulations. In other words, it is programmed to "know" that an effective domain is coming. The inequality symbols are accessed using the **2nd** **MATH** (**TEST**) keys. The final result is shown on Figure 3.72, where we see the function is linear for $x \in (-\infty, 2)$ and quadratic for $x \in [2, \infty)$. How does the calculator remind us the function is defined only for $x = 2$ on the second piece? Using the **2nd** **GRAPH** (**TABLE**) feature reveals the calculator will give an **ERR:** (ERROR) message if you attempt to evaluate a function outside of its effective domain (Figure 3.73).

Figure 3.70

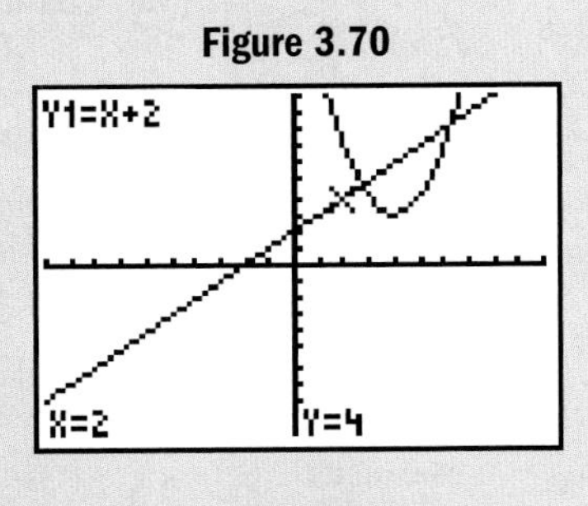

Figure 3.71

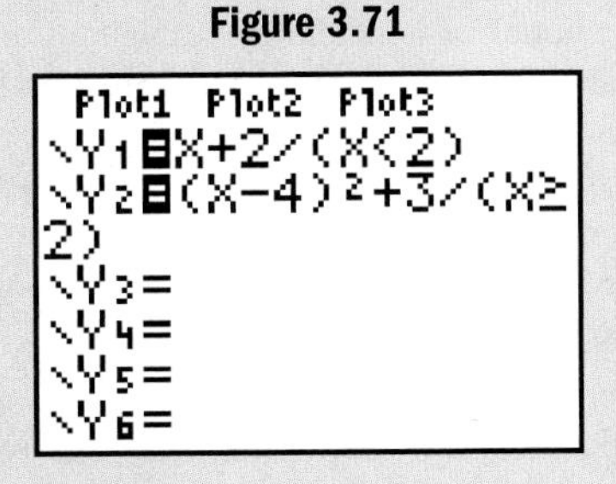

Figure 3.72

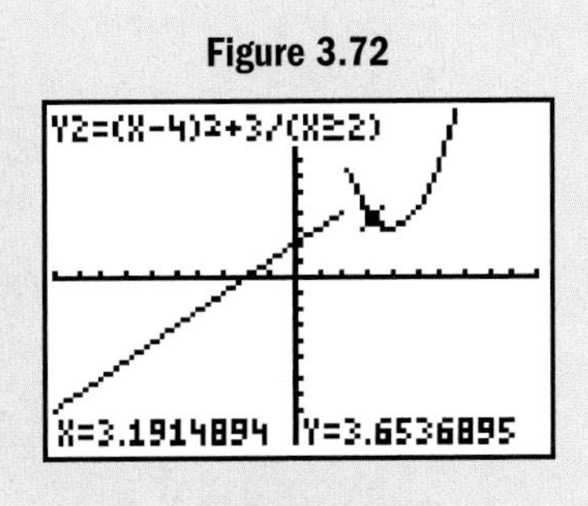

Figure 3.73

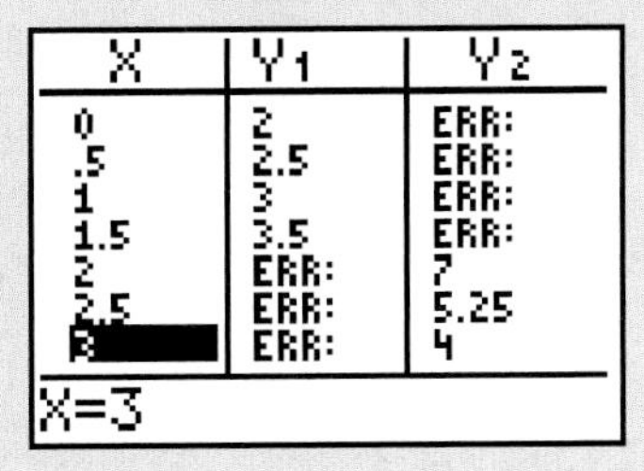

X	Y_1	Y_2
0	2	ERR:
.5	2.5	ERR:
1	3	ERR:
1.5	3.5	ERR:
2	ERR:	7
2.5	ERR:	5.25
3	ERR:	4

X=3

One of the more interesting explorations we can perform on a graphing calculator is to investigate an endpoint of the effective domain. For instance, we know that $Y_1 = x + 2$ is not defined for $x = 2$, but what about numbers very close to 2? Go to **2nd** **WINDOW** (**TBLSET**) and place the calculator in the Indpnt: Auto **Ask** mode. With both Y_1 and Y_2 enabled, use the **2nd** **GRAPH** (**TABLE**) feature to investigate what happens when you evaluate the functions at numbers very near 2. Use $x = 1.9, 1.99, 1.999$, and so on.

Exercise 1: What appears to be happening to the output values for Y_1? What about Y_2?

Exercise 2: What do you notice about the output values when 1.99999 (five nines) is entered? Use the right arrow key ▶ to move the cursor into columns Y_1 and Y_2. Comment on what you think the calculator is doing. Will Y_1 ever really have an output equal to 4?

Exercise 1. They are approaching 4; not defined

Exercise 2. $Y_1 = 4$, Y_2 has an error; calculator is *rounding* to 4; No.

3.7 EXERCISES

CONCEPTS AND VOCABULARY

Fill in each blank with the appropriate word or phrase. Carefully reread the section if needed.

1. A function whose entire graph can be drawn without lifting your pencil is called a continuous function.

2. The input values for which each part of a piecewise function is defined, is called the effective domain.

3. A graph is called smooth if it has no sharp turns or jagged edges.

4. When graphing $2x + 3$ over an effective domain of $x > 0$, we leave an open dot at (0, 3).

5. Discuss/explain the differences between a removable (repairable) discontinuity, a jump discontinuity, and an asymptotic discontinuity. Include examples of each type. Answers will vary.

6. Discuss/explain how it is possible for the domain of a function to be defined for all real numbers, but have a range that is defined on more than one interval. Construct an illustrative example. Answers will vary.

DEVELOPING YOUR SKILLS

For the functions given in Exercises 7 to 10, (a) use the correct notation to write them as a single piecewise-defined function, state the effective domain for each piece by inspecting the graph, and (b) state the range of the function.

7. a. $f(x) = \begin{cases} x^2 - 6x + 10 & 0 \le x \le 5 \\ \frac{3}{2}x - \frac{5}{2} & 5 < x \le 9 \end{cases}$
b. $y \in [1, 11]$

8. a. $f(x) = \begin{cases} -1.5|x - 5| + 10 & 1 \le x < 7 \\ -\sqrt{x - 7} + 5 & x \ge 7 \end{cases}$
b. $y \in (-\infty, 10]$

9. a. $f(t) = \begin{cases} -t^2 + 6t & 0 \le t \le 5 \\ 5 & t > 5 \end{cases}$
b. $y \in [0, 9]$

10. a. $f(t) = \begin{cases} -0.13t^2 + 8.1t + 208 & 4 \le t \le 38 \\ -5.75|t - 46| + 374 & 38 < t < 54 \\ -2.45t + 460 & t \ge 54 \end{cases}$
b. $y \in (-\infty, 374]$

7. $Y_1 = x^2 - 6x + 10$; $Y_2 = \frac{3}{2}x - \frac{5}{2}$

8. $Y_1 = -1.5|x - 5| + 10$;
$Y_2 = -\sqrt{x - 7} + 5$

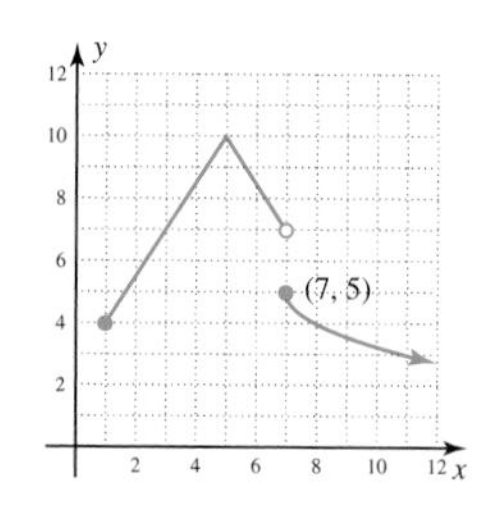

9. Due to heavy advertising, initial sales of the Lynx Digital Camera grew very rapidly, but started to decline once the advertising blitz was over. During the advertising campaign, sales were modeled by the function $S(t) = -t^2 + 6t$, where $S(t)$ represents hundreds of sales in month t. However, as Lynx Inc. had hoped, the new product secured a foothold in the market and sales leveled out at a steady 500 sales per month.

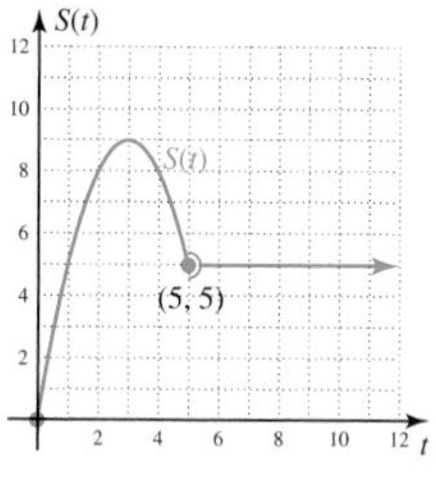

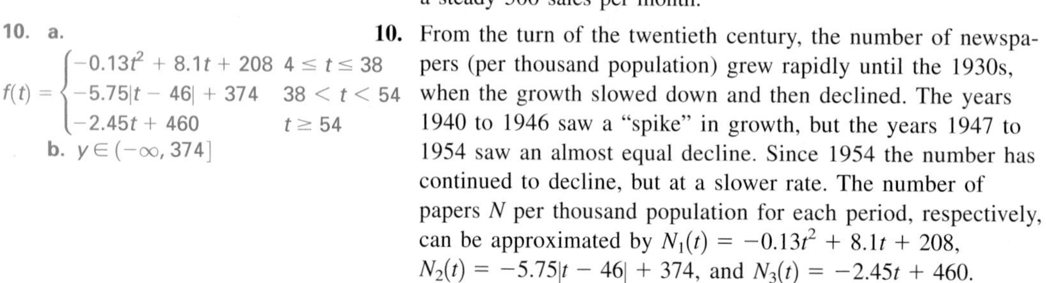

10. From the turn of the twentieth century, the number of newspapers (per thousand population) grew rapidly until the 1930s, when the growth slowed down and then declined. The years 1940 to 1946 saw a "spike" in growth, but the years 1947 to 1954 saw an almost equal decline. Since 1954 the number has continued to decline, but at a slower rate. The number of papers N per thousand population for each period, respectively, can be approximated by $N_1(t) = -0.13t^2 + 8.1t + 208$, $N_2(t) = -5.75|t - 46| + 374$, and $N_3(t) = -2.45t + 460$.

Source: Data from the *Statistical Abstract of the United States,* various years; data from *The First Measured Century, The AEI Press,* Caplow, Hicks, and Wattenberg, 2001.

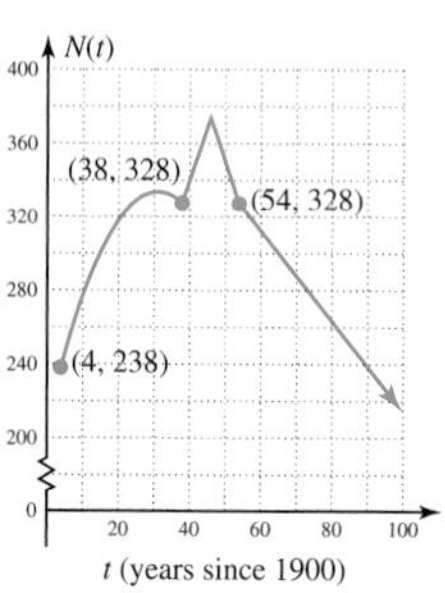

Evaluate each piecewise-defined function as indicated (if possible).

11. $-2, 2, 0, \frac{1}{2}, 2.999, 5$
12. $-3, 0, 2.998, 2$, undefined, 5
13. $5, 5, 0, -4, 5, 11$
14. $0, 2, 2, 2, 2, 2$

11. $h(-5), h(-2), h(-\frac{1}{2}), h(0), h(2.999)$, and $h(3)$

$$h(x) = \begin{cases} -2 & x < -2 \\ |x| & -2 \le x < 3 \\ 5 & x \ge 3 \end{cases}$$

12. $H(-3), H(-\frac{3}{2}), H(-0.001), H(1), H(2)$, and $H(3)$

$$H(x) = \begin{cases} 2x + 3 & x < 0 \\ x^2 + 1 & 0 \le x < 2 \\ 5 & x > 2 \end{cases}$$

13. $p(-5), p(-3), p(-2), p(0), p(3)$, and $p(5)$

$$p(x) = \begin{cases} 5 & x < -3 \\ x^2 - 4 & -3 \le x \le 3 \\ 2x + 1 & x > 3 \end{cases}$$

14. $q(-3), q(-1), q(0), q(1.999), q(2)$, and $q(4)$

$$q(x) = \begin{cases} -x - 3 & x < -1 \\ 2 & -1 \le x < 2 \\ -\frac{1}{2}x^2 + 3x - 2 & x \ge 2 \end{cases}$$

15. a.

Year (0 → 1950)	Percent
5	7.33
15	14.13
25	14.93
35	22.65
45	41.55
55	60.45

b. Each piece gives a slightly different value due to rounding of coefficients in each model. At $t = 30$, we use the "first" piece: $P(30) = 13.08$.

16. a.

Year (0 → 1980)	Barrels (billions)
3	1.183
9	2.115
15	2.68
25	4.01

b. About 1.1 billion barrels

17. $x \in (-\infty, \infty)$; $y \in (-\infty, \infty)$

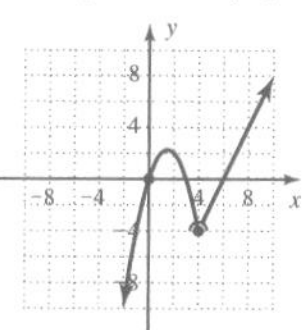

18. $x \in (-\infty, 5]$; $y \in (-\infty, 6]$

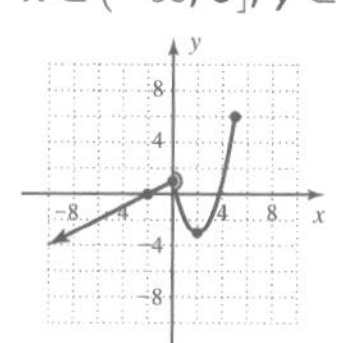

19. $x \in (-\infty, \infty)$; $y \in (-\infty, 0]$

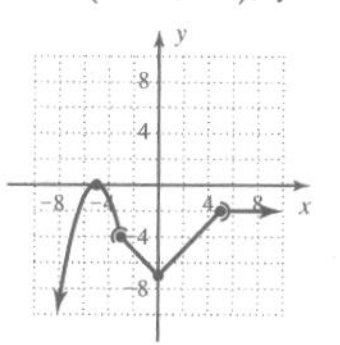

20. $x \in (-\infty, \infty)$; $y \in (-\infty, 8)$

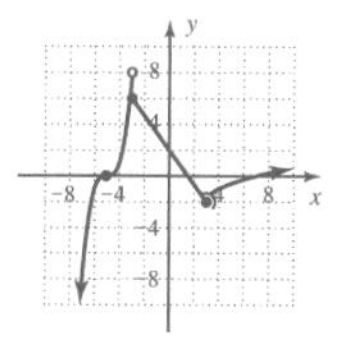

Additional answers can be found in the Instructor Answer Appendix.

15. The percentage of American households that own publicly traded stocks began rising in the early 1950s, peaked in 1970, then began to decline until 1980 when there was a dramatic increase due to easy access over the Internet, an improved economy, and other factors. This phenomenon is modeled by the function $P(t)$, where $P(t)$ represents the percentage of households owning stock in year t, with 1950 corresponding to year 0.

$$P(t) = \begin{cases} -0.03t^2 + 1.28t + 1.68 & 0 \le t \le 30 \\ 1.89t - 43.5 & t > 30 \end{cases}$$

a. According to this model, what percentage of American households held stock in the years 1955, 1965, 1975, 1985, and 1995? If this pattern continues, what percentage will hold stock in 2005?

b. Why is there a discrepancy in the outputs of each piece of the function for the year 1980 ($t = 30$)? According to how the function is defined, which output should be used?

Source: 2004 *Statistical Abstract of the United States,* Table 1204; various other years

16. America's dependency on foreign oil has always been a "hot" political topic, with the amount of imported oil fluctuating over the years due to political climate, public awareness, the economy, and other factors. The amount of crude oil imported can be approximated by the function given, where $A(t)$ represents the number of barrels imported in year t (in billions), with 1980 corresponding to year 0.

$$A(t) = \begin{cases} 0.047t^2 - 0.38t + 1.9 & 0 \le t < 8 \\ -0.075t^2 + 1.495t - 5.265 & 8 \le t \le 11 \\ 0.133t + 0.685 & t > 11 \end{cases}$$

a. Use $A(t)$ to estimate the number of barrels imported in the years 1983, 1989, 1995, and 2005.

b. What was the minimum number of barrels imported between 1980 and 1988?

Source: 2004 *Statistical Abstract of the United States,* Table 897; various other years

Graph each piecewise-defined function and state its domain and range. Use transformations of the toolbox functions where possible.

17. $g(x) = \begin{cases} 3x - x^2 & x \le 4 \\ 2x - 12 & x > 4 \end{cases}$

18. $h(x) = \begin{cases} \frac{1}{2}x + 1 & x \le 0 \\ x^2 - 4x + 1 & 0 < x \le 5 \end{cases}$

19. $p(x) = \begin{cases} -(x+5)^2 & x < -3 \\ |x| - 7 & -3 \le x \le 5 \\ -2 & x > 5 \end{cases}$

20. $q(x) = \begin{cases} (x+5)^3 & x < -3 \\ -\frac{4}{3}x + 2 & -3 \le x \le 3 \\ \sqrt{x-3} - 2 & x > 3 \end{cases}$

21. $H(x) = \begin{cases} -x + 3 & x < 1 \\ -|x - 5| + 6 & 1 \le x < 9 \end{cases}$

22. $w(x) = \begin{cases} \sqrt[3]{x+1} & x < 1 \\ (x-3)^2 - 2 & 1 \le x \le 6 \end{cases}$

23. $f(x) = \begin{cases} -x - 3 & x < -3 \\ 9 - x^2 & -3 \le x < 2 \\ 4 & x \ge 2 \end{cases}$

24. $h(x) = \begin{cases} -\frac{1}{2}x - 1 & x < -3 \\ -|x| + 5 & -3 \le x \le 5 \\ 3\sqrt{x-5} & x > 5 \end{cases}$

Use a table of values as needed to graph each function, then state its domain and range. If the function has a pointwise discontinuity, state how the second piece could be redefined so that a continuous function results.

25. $f(x) = \begin{cases} \dfrac{x^2 - 9}{x + 3} & x \ne -3 \\ -4 & x = -3 \end{cases}$

26. $f(x) = \begin{cases} \dfrac{x^2 - 3x - 10}{x - 5} & x \ne 5 \\ 4 & x = 5 \end{cases}$

27. $f(x) = \begin{cases} \dfrac{x^3 - 1}{x - 1} & x \ne 1 \\ -4 & x = 1 \end{cases}$

28. $f(x) = \begin{cases} \dfrac{4x - x^3}{x + 2} & x \ne -2 \\ 4 & x = -2 \end{cases}$

29. Graph is discontinuous at $x = 0$; $f(x) = 1$ for $x > 0$; $f(x) = -1$ for $x < 0$.

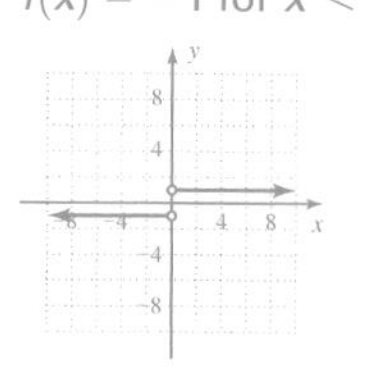

30. Answers will vary.

31. $C(p) = \begin{cases} 0.09p & 0 \le p \le 1000 \\ 0.18p - 90 & p > 1000 \end{cases}$;

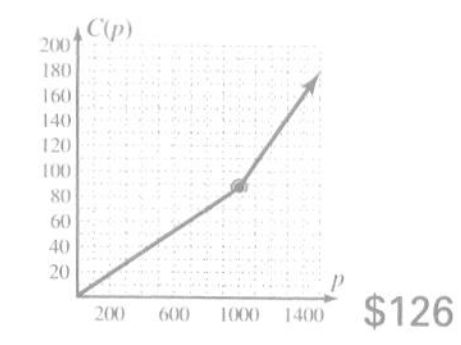

\$126

32. $C(g) = \begin{cases} 0.05g & 0 \le g \le 5000 \\ 0.10g - 250 & g > 5000 \end{cases}$;

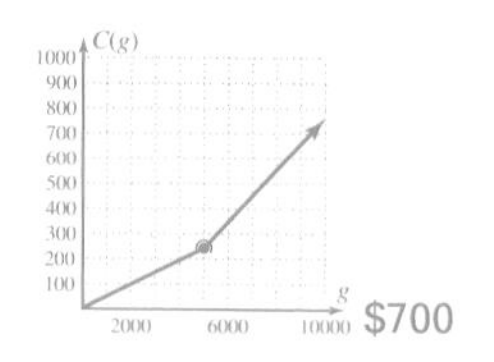

\$700

33. $C(t) = \begin{cases} 0.75t & 0 \le t \le 25 \\ 1.5t - 18.75 & t > 25 \end{cases}$;

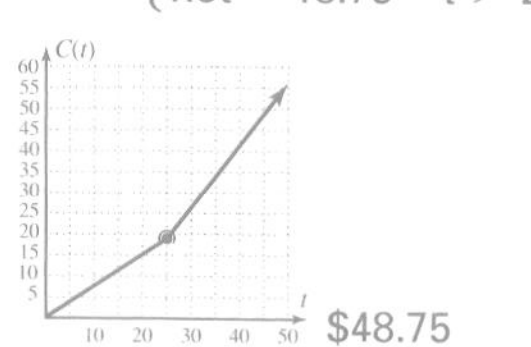

\$48.75

34. $T(x) = \begin{cases} -0.21x^2 + 6.1x + 52 & 5 \le x \le 15 \\ 4.53x + 28.3 & x > 15 \end{cases}$

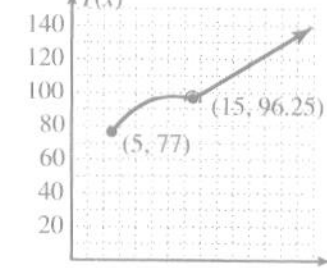

92,000 119,000 142,000 164,000

WORKING WITH FORMULAS

29. Definition of absolute value: $|x| = \begin{cases} -x & x < 0 \\ x & x \ge 0 \end{cases}$

The absolute value function can be stated as a piecewise-defined function, a technique that is sometimes useful in graphing variations of the function or solving absolute value equations and inequalities (Section 6.4). How does this definition ensure that the absolute value of a number is always positive? Use this definition to help sketch the graph of $f(x) = \frac{|x|}{x}$. Discuss what you notice.

30. Sand dune function: $f(x) = \begin{cases} -|x - 2| + 1 & 1 \le x < 3 \\ -|x - 4| + 1 & 3 \le x < 5 \\ -|x - 2k| + 1 & 2k - 1 \le x < 2k + 1, \text{ for } k \in \mathbb{N} \end{cases}$

There are a number of interesting graphs that can be created using piecewise-defined functions, and these functions have been the basis for more than one piece of modern art. (a) Use the descriptive name and the pieces given to graph the function f. Is the function accurately named? (b) Use any combination of the toolbox functions to explore your own creativity by creating a piecewise-defined function with some interesting or appealing characteristics.

APPLICATIONS

31. Energy rationing: In certain areas of the United States, power blackouts have forced some counties to ration electricity. Suppose the cost is \$0.09 per kilowatt (kW) for the first 1000 kW a household uses: $C(p) = 0.09p$. After 1000 kW, the cost increases to 0.18 per kW: $C(p) = 0.18p - 90$. Write these charges for electricity in the form of a piecewise-defined function and state the effective domain for each piece. Then sketch the graph and determine the cost for 1200 kW.

32. Water rationing: Many southwestern states have a limited water supply, and some state governments try to control consumption by manipulating the cost of water usage. Suppose for the first 5000 gal a household uses per month, the charge is \$0.05 per gallon: $C(g) = 0.05g$. Once 5000 gal is used the charge doubles to \$0.10 per gallon: $C(g) = 0.10g - 250$. Write these charges for water usage in the form of a piecewise-defined function and state the effective domain for each piece. Then sketch the graph and determine the cost to a household that used 9500 gal of water during a very hot summer month.

33. Pricing for natural gas: A local gas company charges \$0.75 per therm for natural gas, up to 25 therms. Once the 25 therms has been exceeded, the charge doubles to \$1.50 per therm due to limited supply and great demand. Consumer costs can be modeled by the equation $C(t) = 0.75t$ for the first 25 therms and $C(t) = 1.5t - 18.75$ if more than 25 therms are consumed. Write these charges for natural gas consumption in the form of a piecewise-defined function and state the effective domain for each piece. Then sketch the graph and determine the cost to a household that used 45 therms during a very cold winter month.

34. Multiple births: The number of multiple births has steadily increased in the United States during the twentieth century and beyond. Between 1985 and 1995 the number of twin births could be modeled by the function $T(x) = -0.21x^2 + 6.1x + 52$, where x is the number of years since 1980 and T is in thousands. After 1995, the incidence of twins becomes more linear, with $T(x) = 4.53x + 28.3$ serving as a better model. Write the piecewise-defined function modeling the incidence of twins for these years, including the effective domain of

35.

$$S(t) = \begin{cases} -1.35t^2 + 31.9t + 152; & 0 \le t \le 12 \\ 2.5t^2 - 80.6t + 950; & 12 < t \le 22 \end{cases}$$

$498 billion, $653 billion, $782 billion

36.

$$T(x) = \begin{cases} \dfrac{x}{10} & 0 \le x \le 200 \\ 0.001x^2 - 0.3x + 40 & x > 200 \end{cases}$$

75 tickets

37.

$$c(m) = \begin{cases} 3.3m & 0 \le m \le 30 \\ 7m - 111 & m > 30 \end{cases};$$

$2.11

38. $$w(h) = \begin{cases} 9.50h & 0 \le h \le 40 \\ 14.25h - 190 & 40 < h \le 48; \\ 19h - 418 & 48 < h \le 84 \end{cases}$$

$608

39. $$C(a) = \begin{cases} 0 & a < 2 \\ 2 & 2 \le a < 13 \\ 5 & 13 \le a < 20 \\ 7 & 20 \le a < 65 \\ 5 & a \ge 65 \end{cases}$$

$38

40. −3, −3, −2, 0, 1

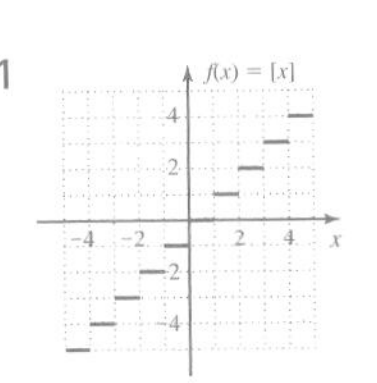

each piece. Then sketch the graph and use the function to estimate the incidence of twins in 1990, 2000, and 2005. If this trend continues, how many sets of twins will be born in 2010?

Source: National Vital Statistics Report, Vol. 50, No. 5, February 12, 2002

35. U.S. military expenditures: Except for the year 1991 when military spending was cut drastically, the amount spent by the U.S. government on national defense and veterans' benefits rose steadily from 1980 to 1992. These expenditures can be modeled by the function $S(t) = -1.35t^2 + 31.9t + 152$, where $S(t)$ is in billions of dollars and 1980 corresponds to $t = 0$.

Source: 1992 *Statistical Abstract of the United States,* Table 525

From 1992 to 1996 this spending declined, then began to rise in the following years. From 1993 to 2002, military-related spending can be modeled by $S(t) = 2.5t^2 - 80.6t + 950$.

Source: 2004 *Statistical Abstract of the United States,* Table 492

Write $S(t)$ as a single piecewise-defined function, stating the effective domain for each piece. Then sketch the graph and use the function to find the projected amount the United States will spend on its military in 2005, 2008, and 2010.

36. Amusement arcades: At a local amusement center, the owner has the SkeeBall machines programmed to reward very high scores. For scores of 200 or less, the function $T(x) = \frac{x}{10}$ models the number of tickets awarded (rounded to the nearest whole). For scores over 200, the number of tickets is modeled by $T(x) = 0.001x^2 - 0.3x + 40$. Write these equation models of the number of tickets awarded in the form of a piecewise-defined function and state the effective domain for each piece. Then sketch the graph and find the number of tickets awarded to a person who scores 390 points.

37. Phone service charges: When it comes to phone service, a large number of calling plans are available. Under one plan, the first 30 min of any phone call costs only 3.3¢ per minute. The charge increases to 7¢ per minute thereafter. Write this information in the form of a piecewise-defined function and state the effective domain for each piece. Then sketch the graph and find the cost of a 46-min phone call.

38. Overtime wages: Tara works on an assembly line, putting together computer monitors. She is paid $9.50 per hour for regular time (0 to 40 hr), $14.25 for overtime (41 to 48 hr), and when demand for computers is high, $19.00 for double-overtime (49 to 84 hr). Write this information in the form of a simplified piecewise-defined function, and state the effective domain for each piece. Then sketch the graph and find the gross amount of Tara's check for the week she put in 54 hr.

39. Admission prices: At Wet Willy's Water World, infants under 2 are free, then admission is charged according to age. Children 2 and older but less than 13 pay $2, teenagers 13 and older but less than 20 pay $5, adults 20 and older but less than 65 pay $7, and senior citizens 65 and older get in at the teenage rate. Write this information in the form of a piecewise-defined function and state the effective domain for each piece. Then sketch the graph and find the cost of admission for a family of nine which includes: one grandparent (70), two adults (44/45), 3 teenagers, 2 children, and one infant.

40. Greatest integer function: The greatest integer function $y = [\![x]\!]$ is a close relative of the postal charge function from Example 8. If fact, you may think of it as a charge of $1 per ounce, except the post office charges $1 per ounce up to and including the next integer ounce, while the greatest integer function charges you $1 per ounce up to but excluding the next integer—at the next integer it jumps to the next higher price. Actually the function is defined as *the greatest integer less than or equal to x,* so for example, $[\![-1.1]\!] = -2$, $[\![-\frac{1}{3}]\!] = -1$, $[\![3]\!] = 3$, $[\![3.7]\!] = 3$, $[\![3.9]\!] = 3$, $[\![3.999]\!] = 3$, $[\![4]\!] = 4$. Sketch the graph of this function and find the value of $[\![-2.5]\!]$, $[\![-2.1]\!]$, $[\![-1.9]\!]$, $[\![0.5]\!]$, and $[\![1]\!]$.

41. yes; $h(x) = \begin{cases} 5 & x \le -3 \\ -2x - 1 & -3 < x < 2 \\ -5 & x \ge 2 \end{cases}$

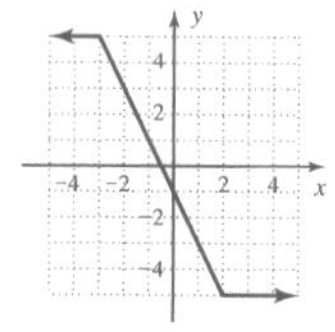

42. yes; $H(x) = \begin{cases} -2x - 1 & x < -3 \\ 5 & -3 \le x \le 2 \\ 2x + 1 & x > 2 \end{cases}$

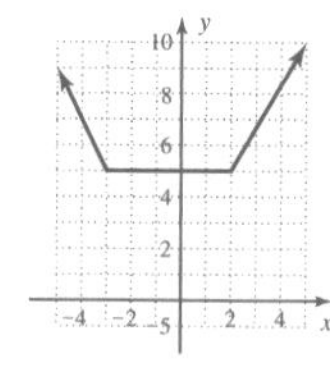

43. Answers will vary.
44. Y_1 has a removable discontinuity at $x = -2$; Y_2 has a nonremovable (jump) discontinuity at $x = -2$; Y_3 has an nonremovable (asymptotic) discontinuity at $x = -2$
45. $h(x) = 4x - 3,\ 1 \le x \le 3$
46. f has a removable discontinuity at $x = 2$;
$F(x) = \begin{cases} \dfrac{x^3 - 8}{x - 2} & x \ne 2 \\ 12 & x = 2 \end{cases}$

47. $x = -7, x = 4$
48. a. {3} b. {3} c. $x \in [-1, \frac{3}{2}]$
49. $y = 2\sqrt{x + 4} - 1$
50. a. $4\sqrt{5}$ cm b. $16\sqrt{5}$ cm^2 c. $V = 320\sqrt{5}$ cm^3
51. a. 25 b. $6x^2 + 1$
52. a. $-21 - 20i$ b. 29

41. Combined absolute value graphs: Carefully graph the function $h(x) = |x - 2| - |x + 3|$ using a table of values over the interval $x \in [-5, 5]$. Is the function continuous? Write this function in piecewise-defined form and state the effective domain for each piece.

42. Combined absolute value graphs: Carefully graph the function $H(x) = |x - 2| + |x + 3|$ using a table of values over the interval $x \in [-5, 5]$. Is the function continuous? Write this function in piecewise-defined form and state the effective domain for each piece.

WRITING, RESEARCH, AND DECISION MAKING

43. The figure shown gives a graph of speed versus time for a cyclist who is out riding for 90 min. Give the graph a story line, writing a story about the cyclist that corresponds to the graph.

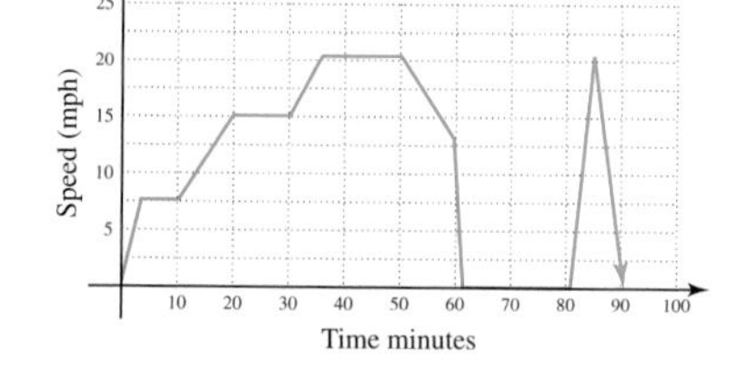

44. You've heard it said, "*any number divided by itself is one.*" Consider the functions $Y_1 = \dfrac{x + 2}{x + 2}$, $Y_2 = \dfrac{|x + 2|}{x + 2}$, and $Y_3 = \dfrac{x - 2}{x + 2}$. Are these functions continuous? If not, what kind of discontinuities does each exhibit? Which of the discontinuities, if any, are removable (repairable)?

EXTENDING THE CONCEPT

45. Find a linear function $h(x)$ that will make the function shown a *continuous* function. Be sure to include its effective domain.

$$f(x) = \begin{cases} x^2 & x < 1 \\ h(x) & \\ 2x + 3 & x > 3 \end{cases}$$

46. Graph the function $f(x) = \dfrac{x^3 - 8}{x - 2}$ by plotting points from $x = -5$ to 5. What do you notice? Define (construct, build) a *continuous* piecewise-defined function $F(x)$, with $\dfrac{x^3 - 8}{x - 2}$ as the first piece.

MAINTAINING YOUR SKILLS

47. (1.3) Solve: $\dfrac{3}{x - 2} + 1 = \dfrac{30}{x^2 - 4}$.

48. (2.5) Solve the inequalities: (a) $|x - 3| \le 0$, (b) $\sqrt{x - 3} \le 0$, and (c) $f(x) = 2x^2 - x - 3;\ f(x) \le 0$.

49. (3.2/3.3) Write the equation for the function given $[a \cdot f(x - h) + k]$, assuming $a = 2$. Then sketch its inverse function.

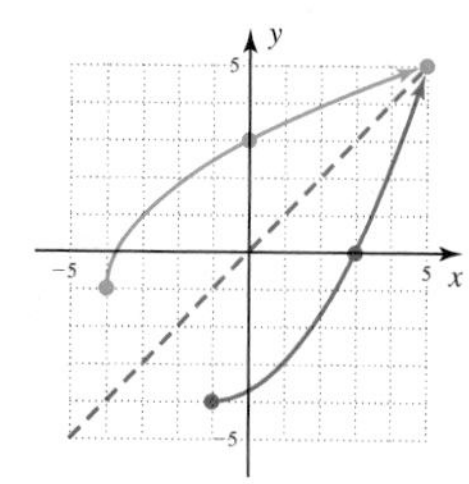

50. (R.7) For the figure shown, (a) find the length of the missing side, (b) state the area of the triangular base, and (c) compute the volume of the prism.

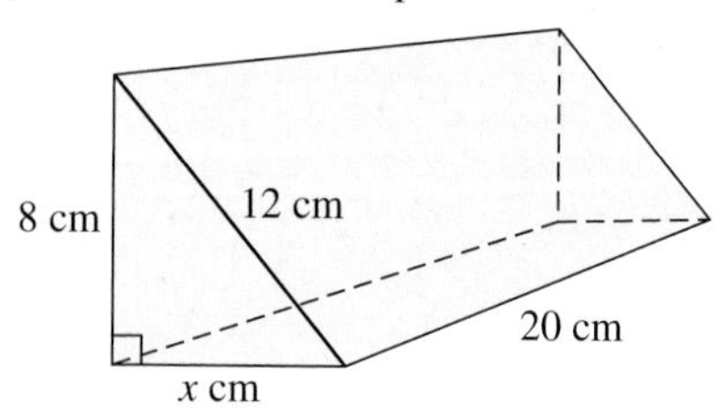

51. (3.1) Given $f(x) = 2x + 1$ and $g(x) = 3x^2$, find: (a) $(f \circ g)(-2)$ and (b) $(f \circ g)(x)$.

52. (1.4) Perform the operations indicated: (a) $(2 - 5i)^2$ and (b) $(2 - 5i)(2 + 5i)$.

3.8 Analyzing the Graph of a Function

LEARNING OBJECTIVES

In Section 3.8 you will learn how to:

A. Determine whether a function is even, odd, or neither

B. Determine intervals where a function is positive or negative

C. Determine where a function is increasing or decreasing

D. Identify the maximum and minimum values of a function

E. Develop a formula to calculate rates of change

INTRODUCTION

Here is a summary of the many things we've touched on related to functions:

Definition of a function	Domain and range	Input/output nature of functions
Function notation	The graph of a function	Simple max/min values
Zeroes of a function	Function inequalities	Increasing/decreasing functions

This section is designed to generalize and refine these ideas, as they are an important part of applying functions as real-world models.

POINT OF INTEREST

In 218 B.C., Hannibal took his army and his elephants from Spain to Italy to attack Rome. During this journey, his army crossed plains, valleys, and mountains (including the Alps). With this image in mind, consider the highly simplified version of a similar trek shown in Figure 3.74. We analyze the journey as follows: The total journey covered a horizontal distance (domain) of 6800 ft. The lowest

Figure 3.74

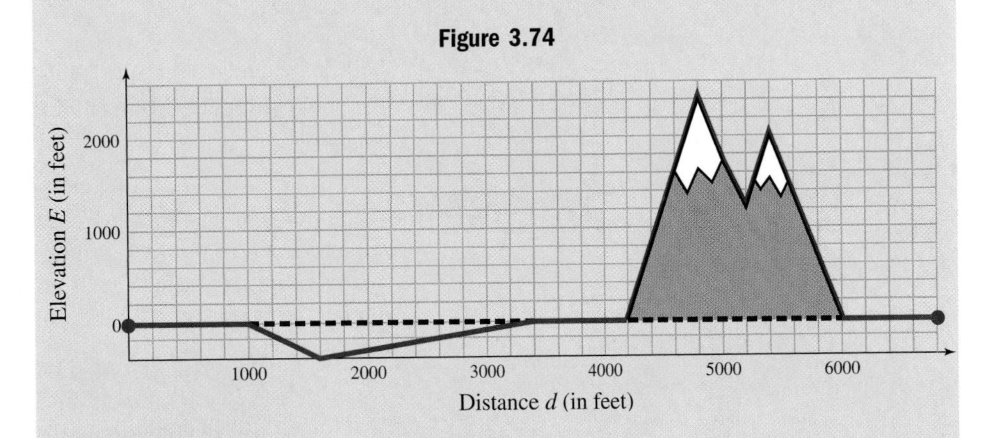

elevation (minimum value) of -400 ft occurred at $d = 1600$, and the highest elevation (maximum value) of 2400 ft occurred at $d = 4800$. The army marched on level ground from $d = 0$ to 1000, 3400 to 4200, and 6000 to 6800. Their march was below sea level ($E < 0$) from $d = 1000$ to 3400, and above sea level ($E > 0$) from $d = 4200$ to 6000. The army was moving downhill from $d = 1000$ to 1600, 4800 to 5200, and 5400 to 6000, and uphill from $d = 1600$ to 3400, 4200 to 4800, and 5200 to 5400. The result is

D: $d \in [0, 6800]$	min. at $(1600, -400)$	$E(d) < 0$ for $d \in (1000, 3400)$
R: $E \in [-400, 2400]$	max. at $(4800, 2400)$	$E(d) > 0$ for $d \in (4200, 6000)$

Function is decreasing for $x \in (1000, 1600) \cup (4800, 5200) \cup (5400, 6000)$

Function is increasing for $x \in (1600, 3400) \cup (4200, 4800) \cup (5200, 5400)$

A. Graphs and Symmetry

While the domain and range of a function will remain dominant themes in our study, for the moment we turn our attention to other characteristics of a function's graph. We begin with the concept of symmetry.

Symmetry with Respect to the y-Axis

In Section 2.4 we used *a vertical axis of symmetry* to aid our graphing of quadratic functions. Here we seek to extend and generalize the idea. Consider the graph of $f(x) = x^4 - 4x^2$ shown in Figure 3.75. A function is **symmetric to the y-axis** if, given a point (x, y) on the graph, the point $(-x, y)$ is also on the graph. We note that $(-1, -3)$ is on the graph, as is $(1, -3)$, and that $(-2, 0)$ is a zero of the function, as is $(2, 0)$. Functions that are symmetric to the y-axis are also known as **even functions** and in general we have:

Figure 3.75

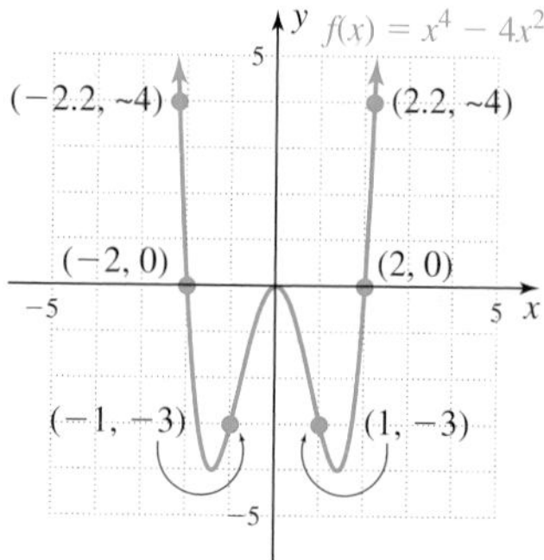

EVEN FUNCTIONS: SYMMETRY TO THE y-AXIS

If (x, y) is on the graph of f, then $(-x, y)$ is also on the graph.
In function notation: $f(-x) = f(x)$.

Symmetry can be a tremendous help in graphing new functions, enabling us to plot just a few points, then completing the graph using the symmetric "reflection."

EXAMPLE 1 (a) The function $g(x)$ in Figure 3.76 is known to be even. Draw the complete graph (only the left half is shown). (b) Show that $h(x) = x^{\frac{2}{3}}$ is an even function using the arbitrary value $x = k$, then sketch the complete graph using $h(1)$, $h(8)$, and y-axis symmetry.

Solution: **a.** To complete the graph of g (see Figure 3.76) use the points $(-4, 1)$, $(-2, -3)$, and $(-1, 2)$ with the definition of y-axis symmetry to find additional points. The corresponding ordered pairs are $(4, 1)$, $(2, -3)$, and $(1, 2)$, which we use to help draw a "mirror image" of the partial graph given.

Figure 3.76

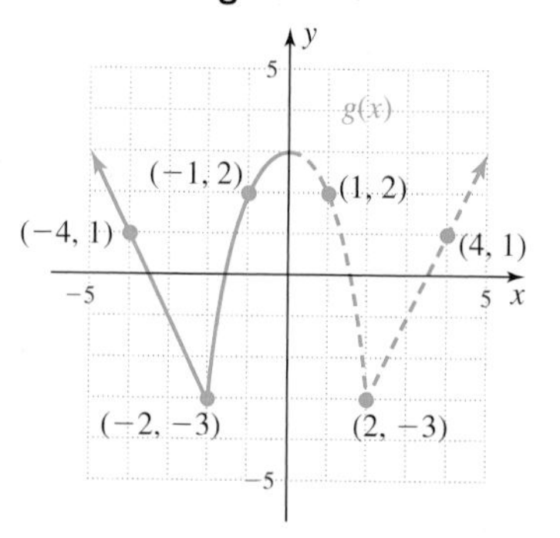

Figure 3.77

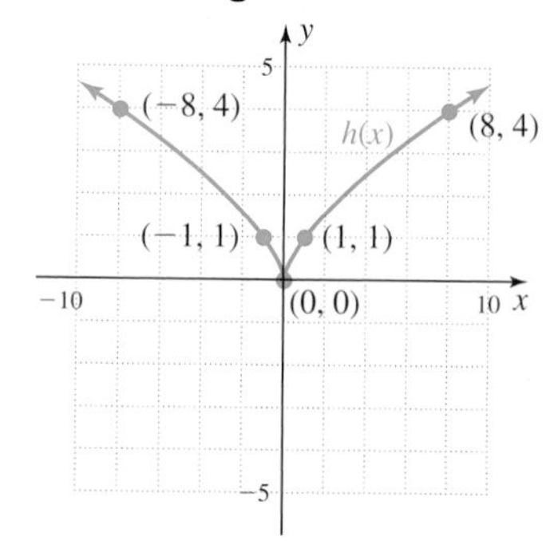

b. To prove that $h(x) = x^{\frac{2}{3}} = (x^{\frac{1}{3}})^2$ is an even function, we must show $h(-k) = h(k)$.

$$h(-k) \stackrel{?}{=} h(k)$$
$$[(-k)^{\frac{1}{3}}]^2 \stackrel{?}{=} (k^{\frac{1}{3}})^2$$
$$(\sqrt[3]{-k})^2 = (\sqrt[3]{k})^2 \checkmark$$

Using $h(0) = 0$, $h(1) = 1$, and $h(8) = 4$ with y-axis symmetry produces the graph shown in Figure 3.77.

NOW TRY EXERCISES 7 THROUGH 18

Symmetry with Respect to the Origin

Another common form of symmetry is known as **symmetry to the origin.** As the name implies, the graph is somehow "centered" at (0, 0). This form of symmetry is easy to see for closed figures with their center at (0, 0), like certain polygons, circles, and ellipses (these will exhibit both y-axis symmetry *and* symmetry to the origin). Note the relation graphed in Figure 3.78 contains the points $(-3, 3)$ and $(3, -3)$, along with $(-1, -4)$ and $(1, 4)$. But notice that function $f(x)$ in Figure 3.79 also contains these points and is, in the same sense, symmetric to the origin (like a pinwheel).

Figure 3.78

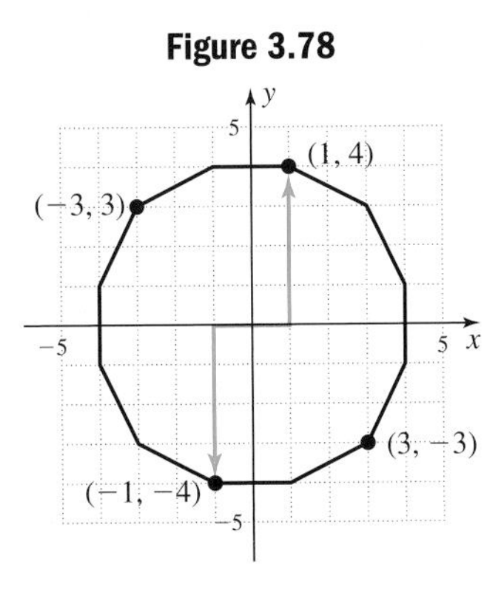

Figure 3.79

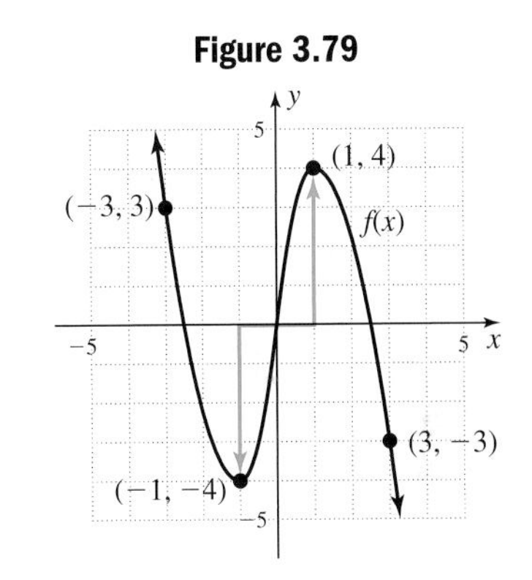

Functions symmetric to the origin are known as **odd functions** and in general we have:

> **ODD FUNCTIONS: SYMMETRY TO THE ORIGIN (0,0)**
> If (x, y) is on the graph, then $(-x, -y)$ is on the graph.
> In function notation: $f(-x) = -f(x)$.

The graph of an odd function can also be identified by noting that a vertical reflection followed by a horizontal reflection *will result in the same graph.*

EXAMPLE 2 (a) In Figure 3.80, the function $q(x)$ given is known to be *odd*. Draw the complete graph (only the left half is shown). (b) Show that $h(x) = x^3 - 4x$ is an odd function using the arbitrary value $x = k$, then sketch the graph using $h(-2)$, $h(-1)$, $h(0)$, and odd symmetry.

Solution:

a. To complete the graph of q, use the points $(-6, 3)$, $(-4, 0)$, and $(-2, 2)$ with the definition of odd symmetry to find additional points. The corresponding ordered pairs are $(6, -3)$, $(4, 0)$, and $(2, -2)$, which we use to help draw a "mirror image" of the partial graph given (see Figure 3.80).

Figure 3.80

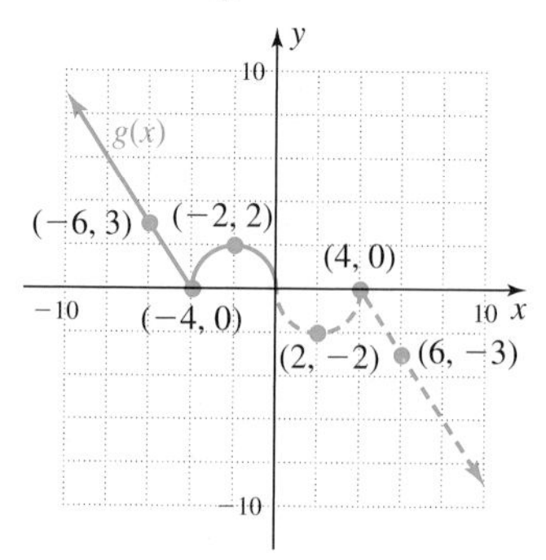

Figure 3.81

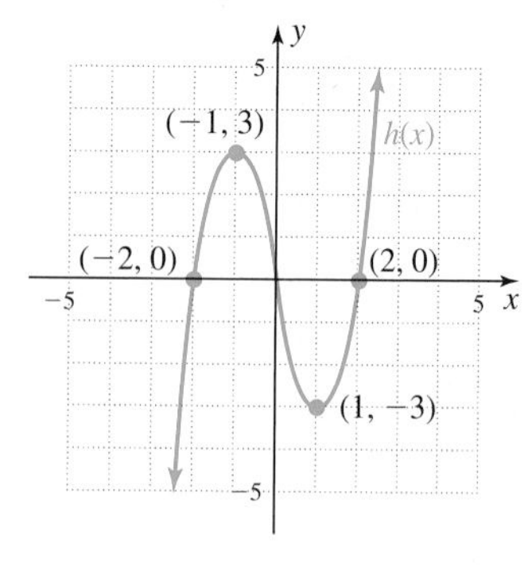

b. To prove that $h(x) = x^3 - 4x$ is an odd function, we must show that $h(-k) = -h(k)$.

$$h(-k) \stackrel{?}{=} -h(k)$$
$$(-k)^3 - 4(-k) \stackrel{?}{=} -[k^3 - 4k]$$
$$-k^3 + 4k = -k^3 + 4k \checkmark$$

Using $h(-2) = 0$, $h(-1) = 3$, and $h(0) = 0$ with symmetry about the origin produces the graph shown in Figure 3.81.

NOW TRY EXERCISES 19 THROUGH 24

B. Intervals Where a Function Is Positive or Negative

Earlier we noted that since y-values are positive in Quadrants I and II, the function is positive $[f(x) > 0]$ when the graph is above the x-axis. We also noted the zeroes of a function that come from *linear factors* cause the graph to "cut" the x-axis, separating intervals where the function is negative from intervals where the function is positive (also see the Chapter 2 *Calculator Exploration and Discovery*). The zeroes that come from "*quadratic factors*" cause the gra'h to "bounce" off the x-axis and the function does not change sign. Compare the graph of $f(x) = x^2 - 4$ with that of $g(x) = (x - 4)^2$ (see Figures 3.82 and 3.83).

Figure 3.82

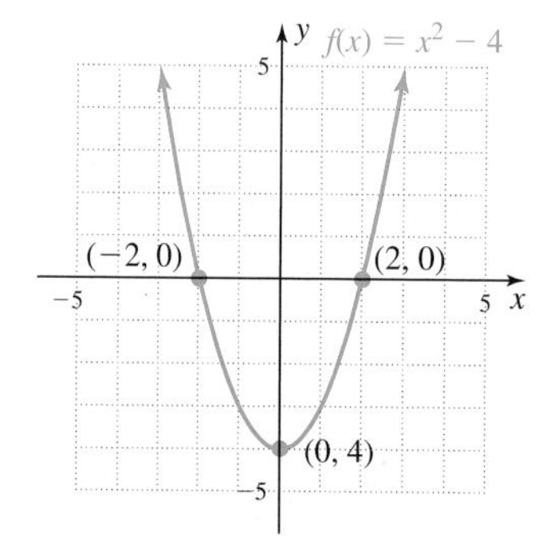

Figure 3.83

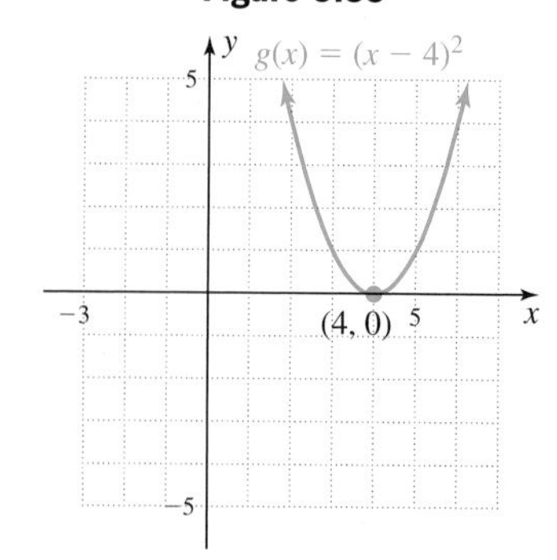

The first is a parabola, concave up, shifted 4 units down, with the x-intercepts at $(-2, 0)$ and $(2, 0)$ coming from the *linear* factors $(x + 2)(x - 2) = 0$. Since the function will alternate sign at these zeroes, $f(x) \geq 0$ for $x \in (-\infty, -2] \cup [2, \infty)$ and $f(x) < 0$ for $x \in (-2, 2)$. The second graph is also a parabola, concave up but *shifted 4 units right.* For $(x - 4)^2 = 0$, $x - 4$ is a quadratic factor with $x = 4$ as the only x-intercept. The graph "bounces" off the x-axis at this point showing $f(x) \geq 0$ for $x \in \mathbb{R}$. These observations form the basis for studying polynomials of higher degree, where we extend the idea of "cuts" and "bounces" to factors of the form $(x - r)^n$ in a study of **roots of multiplicity.** To prevent any confusion with future concepts, we'll refer to factors of the form $(x - r)^2$ as *linear factors of multiplicity two,* rather than as quadratic factors.

EXAMPLE 3 The graph of $g(x) = x^3 - 2x^2 - 4x + 8$ is shown. Solve the inequality $g(x) \geq 0$.

Solution: From the graph, x-intercepts occur at $x = -2$ and $x = 2$ (verify using factoring by grouping). For $g(x) \geq 0$, the graph must be above or on the x-axis giving the solution $x \in [-2, \infty)$.

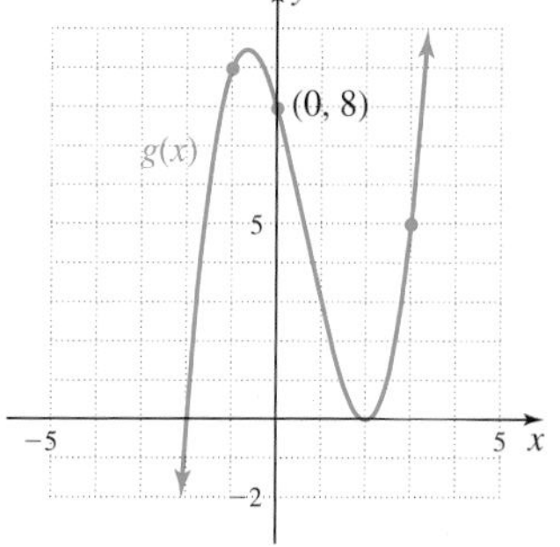

NOW TRY EXERCISES 25 THROUGH 28

C. Intervals Where a Function Is Increasing or Decreasing

Up to now we've relied on an intuitive look at increasing/decreasing functions, saying the graph of a function is increasing if it "rises" as you view it from left to right. More exactly, the function is increasing *on a given interval* if larger and larger x (input) values produce larger and larger y (output) values. Although we'll rely almost exclusively on the graph of a function to find increasing and decreasing intervals, the tests for each are stated here in more formal terms.

INCREASING AND DECREASING FUNCTIONS

Given an interval I that is a subset of the domain, with x_1 and x_2 in this interval and $x_2 > x_1$,

1. A function is increasing on I if $f(x_2) > f(x_1)$ for all x_1 and x_2 in I (larger inputs produce larger outputs).
2. A function is decreasing on I if $f(x_2) < f(x_1)$ for all x_1 and x_2 in I (larger inputs produce smaller outputs).
3. A function is constant on I if $f(x_2) = f(x_1)$ for all x_1 and x_2 in I (larger inputs produce identical outputs).

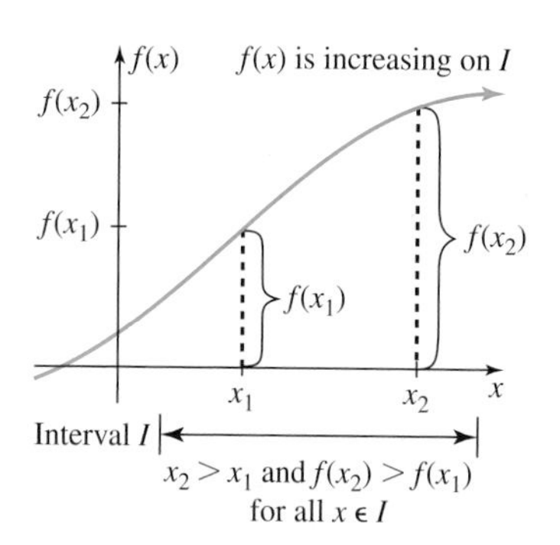

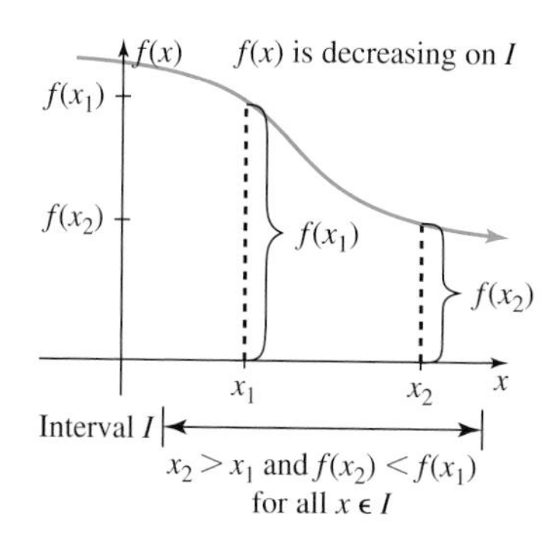

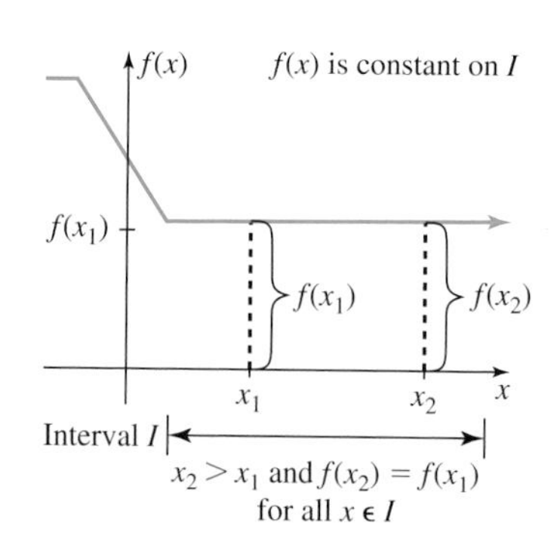

WORTHY OF NOTE

Questions about the behavior of a function are asked with respect to the *y* outputs: where is the *function* positive, where is the *function* increasing, etc. Due to the input/output, cause/effect nature of functions, the response is given in terms of *x*, that is, what is *causing* outputs to be negative, or to be decreasing.

Consider the graph of $f(x) = -x^2 + 4x + 5$ in Figure 3.84. Since the graph is concave down with the vertex at (2, 9), the function must increase until it reaches this maximum value at $x = 2$, and decrease thereafter. Notationally we'll write this as $f(x)\uparrow$ for $x \in (-\infty, 2)$ and $f(x)\downarrow$ for $x \in (2, \infty)$. Using the interval $(-3, 2)$ shown, we see that any larger input value from the interval will indeed produce a larger output value, and $f(x)\uparrow$ on the interval. For instance,

Figure 3.84

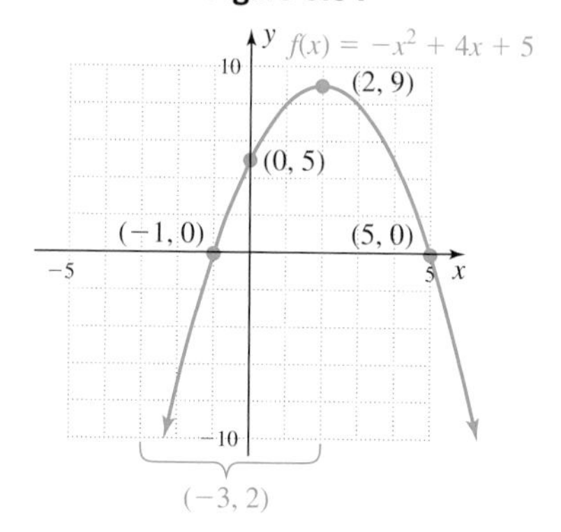

$$1 > -2 \qquad\qquad x_2 > x_1$$

$$\text{and} \qquad\qquad \text{and}$$

$$f(1) = 8 > f(-2) = -7 \qquad\qquad f(x_2) > f(x_1)$$

When identifying intervals where a function is increasing or decreasing, it helps to keep the general concept of a maximum/minimum value in mind, meaning some point at which the function reaches a highest or lowest point, then changes direction from increasing to decreasing or vice versa.

EXAMPLE 4 The graph of a function $v(x)$ is given in the figure. Use the graph to name the interval(s) where v is increasing, decreasing or constant.

Solution: From left to right, the graph of v increases until leveling off at $(-2, 2)$, then it remains constant until reaching $(1, 2)$. The graph then increases once again until reaching a peak at $(3, 5)$ and decreases thereafter. The result is $v(x)\uparrow$ for $x \in (-\infty, -2) \cup (1, 3)$, $v(x)\downarrow$ for $x \in (3, \infty)$, and $v(x)$ is constant for $x \in (-2, 1)$.

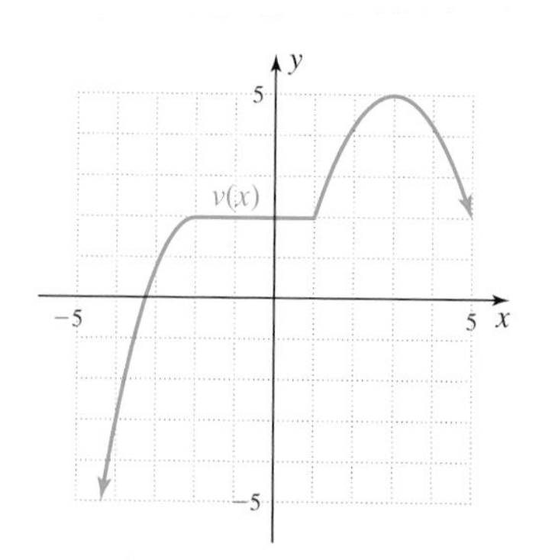

NOW TRY EXERCISES 29 THROUGH 32

D. More on Maximum and Minimum Values

The y-coordinate of the vertex of a quadratic graph and the y-coordinate of "peaks" and "valleys" from other graphs are called **maximum** and **minimum values.** As we work with more diverse function models, it helps to be aware that maximum values can appear in different forms. While these may be discussed more formally in future course work, a **global maximum** names the largest range value over the entire domain. **Local maximums** give the largest range value in a specified interval; and **endpoint maximums** can occur at endpoints of the domain. The same can be said for the corresponding minimum values. Consider the illustration given in the *Point of Interest,* where the maximum value occurs at $d = 4800$ when the mountain peak is 2400 ft high. This is the *global* maximum. But at $d = 5400$, the army had to climb another peak, though not as high—only 2000 ft. This is an example of a *local* maximum. In addition, although the global minimum of this journey was -400 at $d = 1600$, the mountain gorge between these two peaks (1200 ft at $d = 5200$) still represents a *local* minimum, as it's the smallest range value in the "neighborhood."

We already have the ability to locate maximum and minimum values for a quadratic function using the vertex formula. In future courses, methods are developed to help locate maximum and minimum values for almost *any* function. For now, our work will rely chiefly on the graph of a function.

EXAMPLE 5 Analyze the graph of function f shown in Figure 3.85. Include specific mention of (a) domain and range, (b) maximum (max) and minimum (min) values, (c) intervals where f is increasing or decreasing, (d) intervals where $f(x) \geq 0$ and $f(x) < 0$, and (e) any symmetry noted.

Figure 3.85

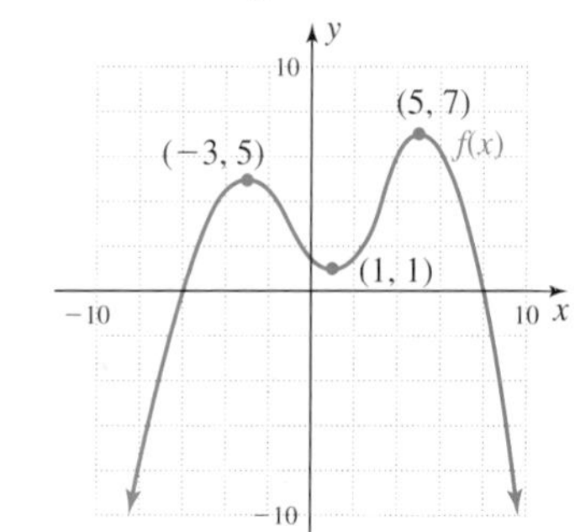

Solution:

a. domain: $x \in \mathbb{R}$;
range: $y \in (-\infty, 7]$.

b. max: $(-3, 5)$ and $(5, 7)$
min: $(1, 1)$

c. $f(x)\uparrow$ for $x \in (-\infty, -3) \cup (1, 5)$ shown in blue in Figure 3.86 and $f(x)\downarrow$ for $x \in (-3, 1) \cup (5, \infty)$ shown in red in Figure 3.86.

d. $f(x) \geq 0$ for $x \in [-6, 8]$ and $f(x) < 0$ for $x \in (-\infty, -6) \cup (8, \infty)$.

e. The function is neither even nor odd.

Figure 3.86

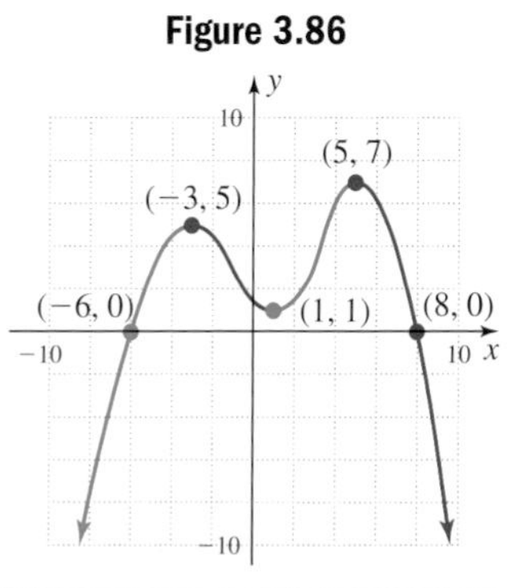

NOW TRY EXERCISES 33 THROUGH 41

The ideas just presented can be applied to functions of all kinds, including rational functions, piecewise-defined functions, step functions, and so on.

EXAMPLE 6 Analyze the function $p(x) = \begin{cases} -\frac{1}{2}x - 1 & x < -2 \\ -2|x| + 6 & -2 \le x \le 3. \\ 3\sqrt{x - 3} & x > 3 \end{cases}$

Include specific mention of (a) domain and range, (b) maximum and minimum values, (c) intervals where f is increasing or decreasing, (d) intervals where $f(x) \ge 0$ and $f(x) < 0$, and (e) any symmetry noted. Begin by graphing the function.

Solution: The first piece of this function is a line with slope $m = -\frac{1}{2}$ and a y-intercept of $(0, -1)$, but it is only defined on $(-\infty, -2)$. The second piece is a "V" function stretched by a factor of 2 reflected across the x-axis, and shifted upward 6 units, giving it x-intercepts of $(-3, 0)$ and $(3, 0)$, but this function is defined only for $[-2, 3]$. The remaining piece is a one-wing function stretched by a factor of 3 and shifted 3 units right. The resulting graph is shown.

a. domain: $x \in (-\infty, \infty)$;
range: $y \in [0, \infty)$.

b. max: $(0, 6)$
min: $(3, 0)$
Note that $(-2, 2)$ is not an endpoint minimum since it is not an endpoint of the domain.

c. $f(x)\downarrow$ for $x \in (-\infty, -2) \cup (0, 3)$;
$f(x)\uparrow$ for $x \in (-2, 0) \cup [3, \infty)$.

d. $f(x) \ge 0$ for $x \in \mathbb{R}$.

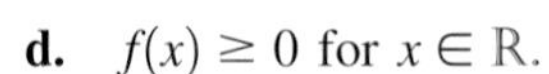

e. There are no symmetries.

NOW TRY EXERCISES 42 THROUGH 45

E. Rates of Change and the Difference Quotient

We complete our study of analyzing graphs by revisiting the concept of average rates of change. In many business, scientific, and economic applications, this is the attribute of a function that draws the most attention. In Section 2.4 we computed average rates of change for a function $f(x)$ by selecting two points on the graph that were fairly close, and using the function form of the slope formula: $\frac{\Delta y}{\Delta x} = \frac{f(x_2) - f(x_1)}{x_2 - x_1}$ (see also Chapter 2: *Technology Extension—Average Rates of Change*). This approach works very well, but requires us to recalculate $\frac{\Delta y}{\Delta x}$ over each new interval. By using a slightly different approach, we can use the same slope calculation to develop a new *formula* that allows us to find rates of change with greater efficiency. This is done by selecting a point $x_1 = x$ in the domain and a point $x_2 = x + h$ that is close to x. Here, h is assumed to be some very small arbitrary number. Making these substitutions in the given formula gives $\frac{\Delta y}{\Delta x} = \frac{f(x + h) - f(x)}{(x + h) - x}$ or $\frac{f(x + h) - f(x)}{h}$. The advantage is that this version, called the **difference quotient,** can be simplified using basic algebra skills with the result being a *formula* for average rates of change.

EXAMPLE 7 Use the difference quotient to find an *average rate of change* formula for the reciprocal function $f(x) = \frac{1}{x}$. Then use the formula to find the average rate of change on the intervals [0.5, 0.51] and [3, 3.01]. Note $h = 0.01$ for both intervals.

Solution: For $f(x) = \frac{1}{x}$ we have $f(x + h) = \frac{1}{x + h}$ and we compute as follows:

$$\frac{\Delta y}{\Delta x} = \frac{f(x + h) - f(x)}{h} \quad \text{difference quotient}$$

$$= \frac{\frac{1}{(x + h)} - \frac{1}{x}}{h} \quad \text{substitute } \frac{1}{x + h} \text{ for } f(x + h) \text{ and } \frac{1}{x} \text{ for } f(x)$$

$$= \frac{\frac{x - (x + h)}{(x + h)x}}{h} \quad \text{common denominator}$$

$$= \frac{\frac{-h}{(x + h)x}}{h} \quad \text{simplify numerator}$$

$$= \frac{-1}{(x + h)x} \quad \text{invert and multiply}$$

This is the formula for computing rates of change given $f(x) = \frac{1}{x}$. To find $\frac{\Delta y}{\Delta x}$ on the interval [0.5, 0.51] we have:

$$\frac{\Delta y}{\Delta x} = \frac{-1}{(0.5 + 0.01)(0.5)} \quad \text{substitute } 0.5 + 0.01 \text{ for } x + h\text{; } 0.5 \text{ for } x$$

$$\approx \frac{-4}{1} \quad \text{result (approximate)}$$

On this interval y is decreasing by about 4 units for every 1 unit x is increasing. The slope of the secant line (in blue) through these points is $m \approx -4$. For the interval [3, 3.01] we have

$$\frac{\Delta y}{\Delta x} = \frac{-1}{(3 + 0.01)(3)} \quad \text{substitute } 3 + 0.01 \text{ for } x + h \text{; substitute } 3 \text{ for } x$$

$$\approx -\frac{1}{9} \quad \text{result (approximate)}$$

On this interval, y is decreasing 1 unit for every 9 units x is increasing. The slope of the secant line (in red) through these points is $m \approx -\frac{1}{9}$.

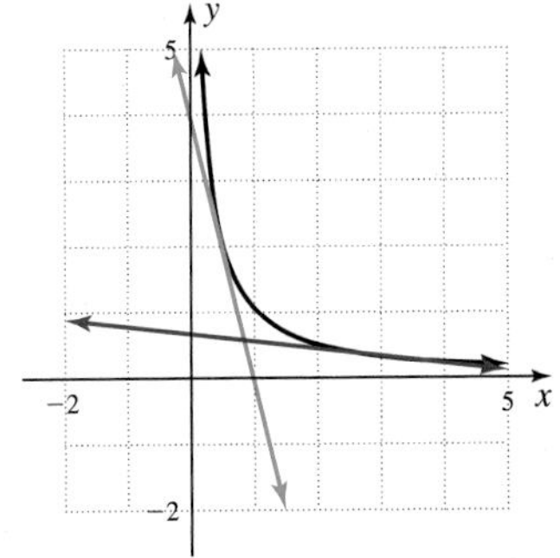

NOW TRY EXERCISES 46 THROUGH 49

TECHNOLOGY HIGHLIGHT

Locating Maximums and Minimums Using a Graphing Calculator

The keystrokes shown apply to a TI-84 Plus model. Please consult your manual or our Internet site for other models.

In the *Technology Highlight* from Section 3.4, we used the 2nd TRACE (**CALC**) **2:zero** option to locate the x-intercepts of a function. The maximum and minimum values of a function are located in much the same way. To begin, enter the function $y = x^3 - 3x - 2$ on the Y= screen and graph the function in a convenient window. As seen in Figure 3.87, it appears a local maximum occurs at $x = -1$ and a local minimum at $x = 1$. To check, we access the **CALC 4:maximum** option, which returns you to the graph and asks you for a *Left Bound,* a *Right Bound,* and a *Guess,* as before. We entered a left bound of "-3," a right bound of "0," and bypassed the Guess option by pressing ENTER (the calculator again sets the "►" and "◄" markers as each bound is set). The calculator will locate the maximum value in this interval, with the coordinates displayed at the bottom of the screen. Due to the algorithm the calculator uses to find these values, a decimal number is sometimes displayed, even if the actual value is an integer (see Figure 3.88). To be certain, we evaluate $f(-1)$ and find the local maximum is indeed 0 $[f(-1) = 0]$. Find all maximum and minimum values for the following functions using these ideas. Round to the nearest hundredth.

Figure 3.87

Y1=X^3-3X-2
X=0 Y=-2

Figure 3.88

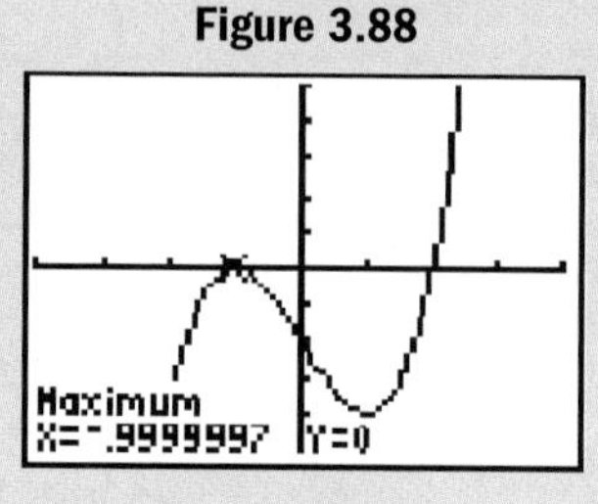

Exercise 1: $y = x^2 - 8x + 9$ (4, −7)

Exercise 2: $y = a^3 - 2a^2 - 4a + 8$ (−.67, 9.48), (2, 0)

Exercise 3: $y = x^4 - 5x^2 - 2x$ (−1.47, −3.20), (−0.20, 0.20) (1.67, −9.51)

Exercise 4: $y = x\sqrt{x + 4}$ (−2.67, −3.08)

3.8 EXERCISES

CONCEPTS AND VOCABULARY

Fill in each blank with the appropriate word or phrase. Carefully reread the section if needed.

1. The graph of a polynomial will ___cut___ through the x-axis at zeroes of ___linear___ factors, and ___bounce___ off the x-axis at the zeroes from quadratic factors.

2. If $f(-x) = f(x)$ for all x in the domain, we say that f is an ___even___ function and symmetric to the ___y___ axis. If $f(-x) = -f(x)$, the function is ___odd___ and symmetric to the ___origin___.

3. If $f(x_2) > f(x_1)$ for $x_1 < x_2$ for all x in a given interval, the function is ___increasing___ in the interval.

4. If $f(c) \geq f(x)$ for all x in a specified interval, we say that $f(c)$ is a local ___maximum___ for this interval.

5. Discuss/explain the following statement and give an example of the conclusion it makes. "If a function f is decreasing to the left of $(c, f(c))$ and increasing to the right of $(c, f(c))$, then $f(c)$ is either a local or a global minimum." Answers will vary.

6. Without referring to notes or textbook, list as many features/attributes as you can that are related to analyzing the graph of a function. Include details on how to locate or determine each attribute. Answers will vary.

DEVELOPING YOUR SKILLS

Use vertical and horizontal boundary lines to help state the domain and range of the functions given.

7. $f(x) = \dfrac{1}{x - 3} - 4$ D: $x \in (-\infty, 3) \cup (3, \infty)$ R: $y \in (-\infty, -4) \cup (-4, \infty)$

8. $g(x) = \dfrac{1}{(x + 3)^2} - 2$

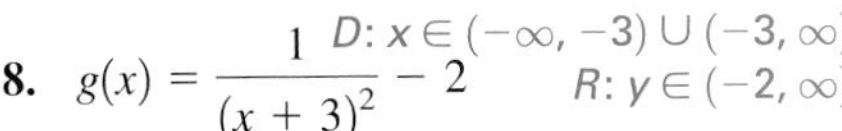

D: $x \in (-\infty, -3) \cup (-3, \infty)$ R: $y \in (-2, \infty)$

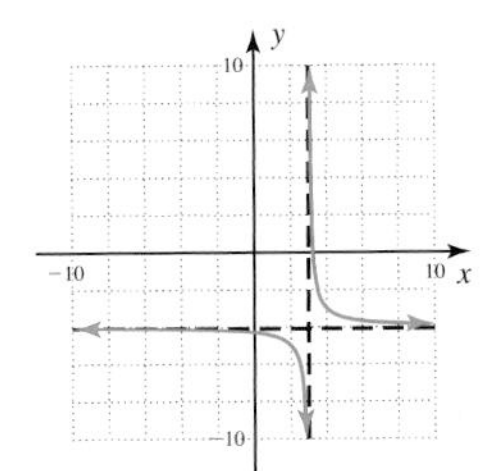

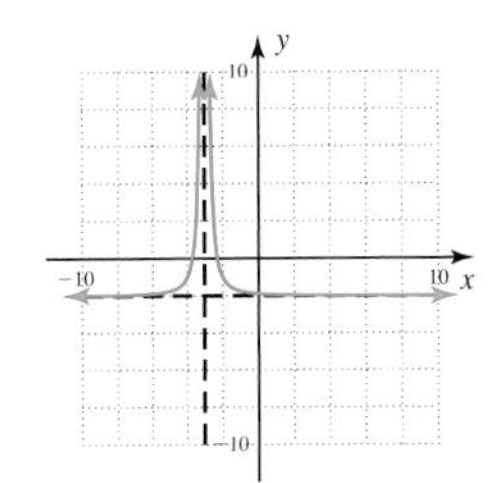

9. $p(x) = \begin{cases} (x + 3)^3 + 3 & x < -3 \\ -1.5x + 0.5 & -3 \le x \le 3 \\ -4 & x > 3 \end{cases}$

D: $x \in (-\infty, \infty)$ R: $y \in (-\infty, 5]$

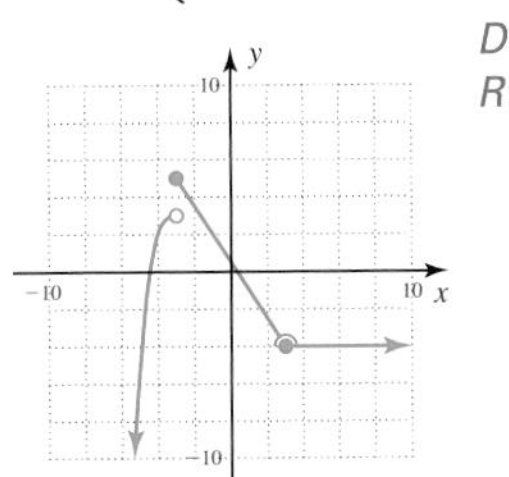

10. $q(x) = \begin{cases} |x - 5| + 3 & 1 \le x < 7 \\ \sqrt{x - 7} + 8 & x \ge 7 \end{cases}$

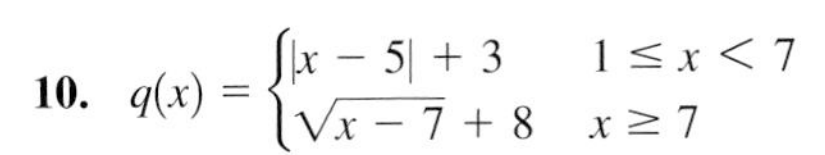

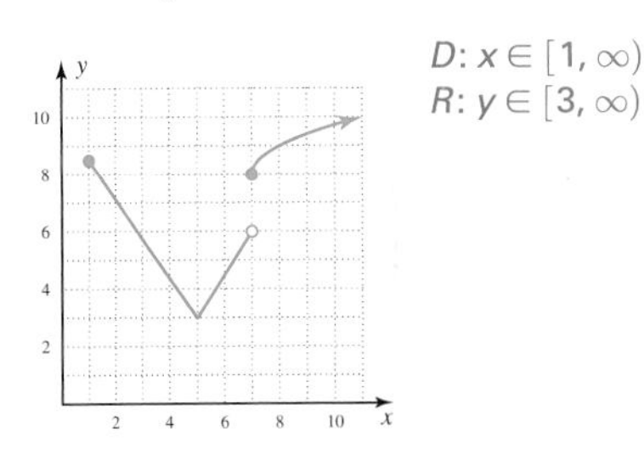

D: $x \in [1, \infty)$ R: $y \in [3, \infty)$

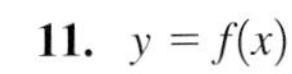

11. $y = f(x)$

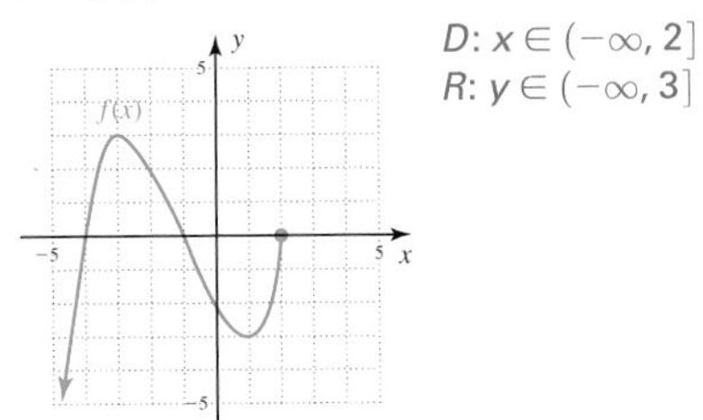

D: $x \in (-\infty, 2]$ R: $y \in (-\infty, 3]$

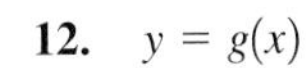

12. $y = g(x)$

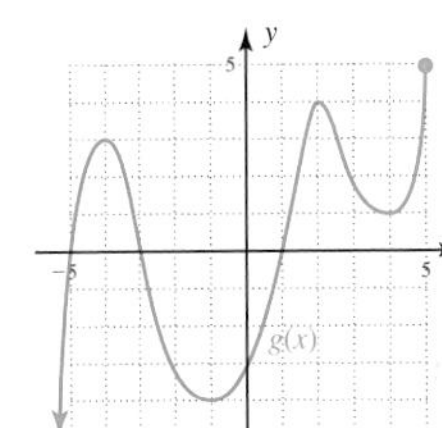

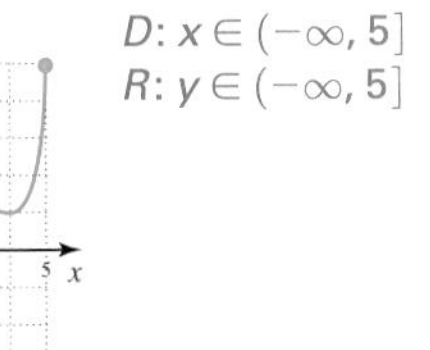

D: $x \in (-\infty, 5]$ R: $y \in (-\infty, 5]$

The following functions are known to be even. Complete each graph using symmetry.

13.

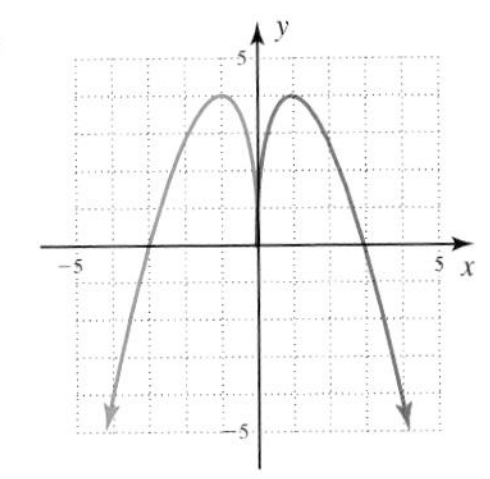

14.

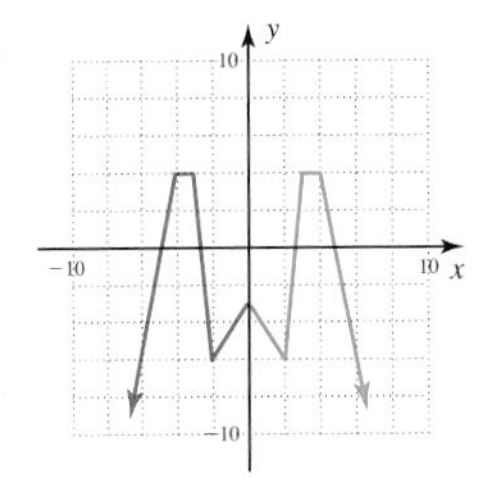

Determine whether the following functions are even using $x = k$.

15. $f(x) = -7|x| + 3x^2 + 5$ even

16. $g(x) = \frac{1}{3}x^4 - 5x^2 + 1$ even

17. $p(x) = 2x^4 - 6x + 1$ not even

18. $q(x) = \frac{1}{x^2} - |x|$ even

The following functions are known to be odd. Complete each graph using symmetry.

19.

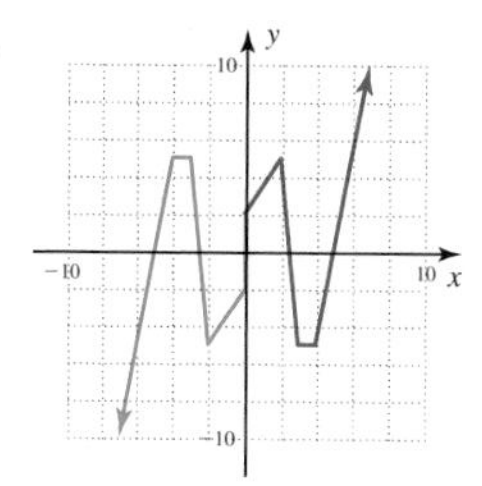

20.

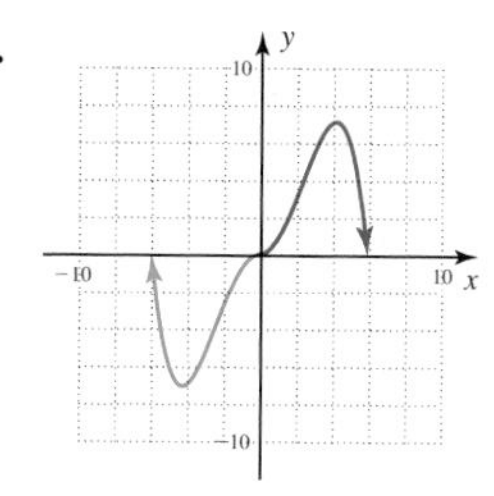

Determine whether the following functions are odd using $x = k$.

21. $f(x) = 4\sqrt[3]{x} - x$ odd

22. $g(x) = \frac{1}{2}x^3 - 6x$ odd

23. $p(x) = 3x^3 - 5x^2 + 1$ not odd

24. $q(x) = \frac{1}{x} - x$ odd

Use the graphs given to solve the inequalities indicated. Write all answers in interval notation.

25. $f(x) = x^3 - 3x^2 - x + 3$; $f(x) \geq 0$

$x \in [-1, 1] \cup [3, \infty)$

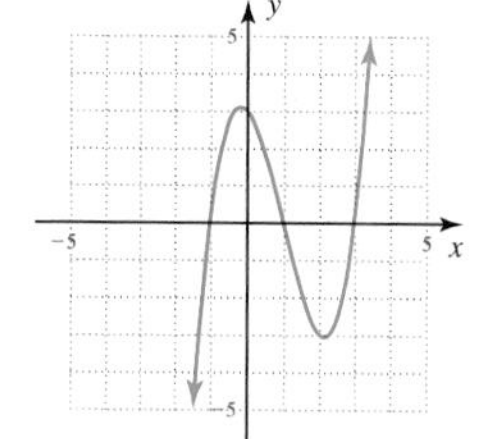

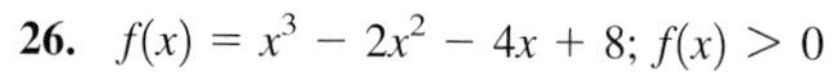

26. $f(x) = x^3 - 2x^2 - 4x + 8$; $f(x) > 0$

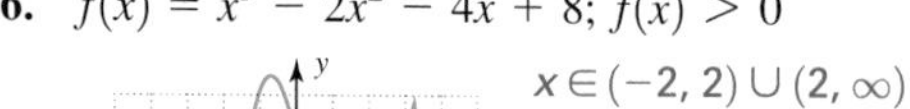

$x \in (-2, 2) \cup (2, \infty)$

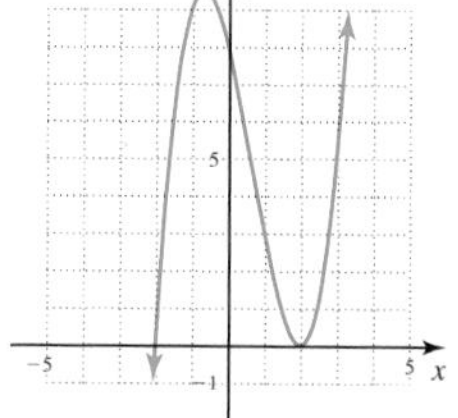

27. $f(x) = x^4 - 2x^2 + 1$; $f(x) > 0$

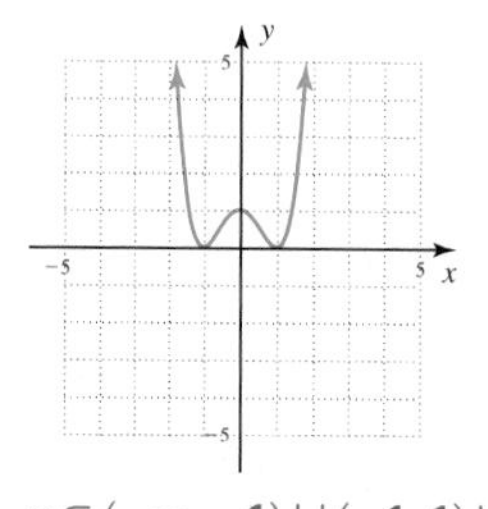

$x \in (-\infty, -1) \cup (-1, 1) \cup (1, \infty)$

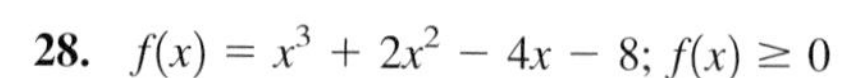

28. $f(x) = x^3 + 2x^2 - 4x - 8$; $f(x) \geq 0$

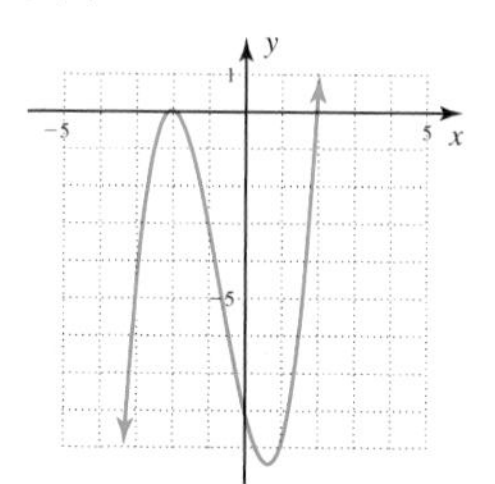

$x \in [2, \infty) \cup \{-2\}$

Name the interval(s) where the following functions are increasing, decreasing, or constant. Write answers using interval notation. Assume all endpoints have integer values.

29. $y = f(x)$

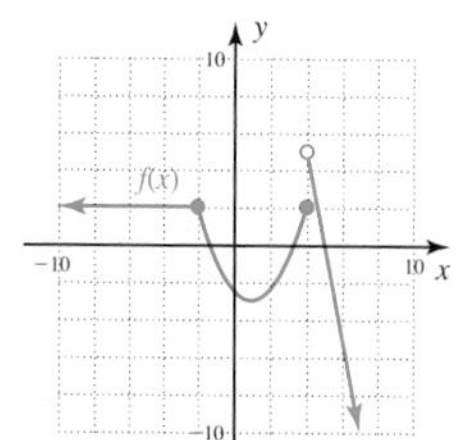

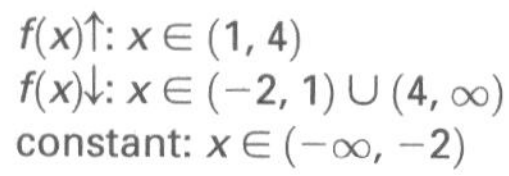

$f(x)\uparrow$: $x \in (1, 4)$
$f(x)\downarrow$: $x \in (-2, 1) \cup (4, \infty)$
constant: $x \in (-\infty, -2)$

30. $y = g(x)$

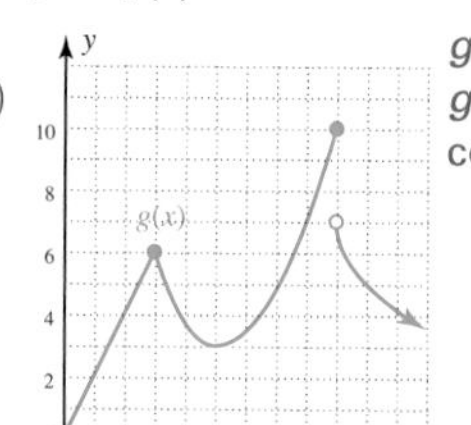

$g(x)\uparrow$: $x \in (0, 3) \cup (5, 9)$
$g(x)\downarrow$: $x \in (3, 5) \cup (9, \infty)$
constant: none

33. a. $x \in (-\infty, \infty)$, $y \in (-\infty, 5]$
b. $(1, 0)$, $(3, 0)$
c. $H(x) \geq 0$: $x \in [1, 3]$
$H(x) \leq 0$: $x \in (-\infty, 1] \cup [3, \infty)$
d. $H(x)\uparrow$: $x \in (-\infty, 2)$
$H(x)\downarrow$: $x \in (2, \infty)$
e. max: $(2, 5)$
f. none

34. a. $x \in (-\infty, \infty)$, $y \in (-\infty, \infty)$
b. $(-2, 0)$
c. $p(x) \geq 0$: $x \in [-2, \infty)$
$p(x) \leq 0$: $x \in (-\infty, -2]$
d. $p(x)\uparrow$: $x \in (-\infty, \infty)$
$p(x)\downarrow$: none
e. none f. none

35. a. $x \in (-\infty, \infty)$, $y \in (-\infty, \infty)$
b. $(-1, 0)$
c. $q(x) \geq 0$: $x \in (-\infty, -1]$
$q(x) \leq 0$: $x \in [-1, \infty)$
d. $q(x)\uparrow$: none
$q(x)\downarrow$: $x \in (-\infty, \infty)$
e. none f. none

36. a. $x \in (-\infty, \infty)$, $y \in (-\infty, \infty)$
b. $(-3.5, 0)$, $(0, 0)$, $(3.5, 0)$
c. $f(x) \geq 0$: $x \in [-3.5, 0] \cup [3.5, \infty)$
$f(x) \leq 0$: $x \in (-\infty, -3.5] \cup [0, 3.5]$
d. $f(x)\uparrow$: $x \in (-\infty, -2) \cup (2, \infty)$
$f(x)\downarrow$: $x \in (-2, 2)$
e. max: $(-2, 3)$,
min: $(2, -3)$
f. none

37. a. $x \in (-\infty, \infty)$, $y \in (-\infty, \infty)$
b. $(-1, 0)$, $(5, 0)$
c. $g(x) \geq 0$: $x \in [-1, \infty)$
$g(x) \leq 0$: $x \in (-\infty, -1] \cup \{5\}$
d. $g(x)\uparrow$: $x \in (-\infty, 1) \cup (5, \infty)$
$g(x)\downarrow$: $x \in (1, 5)$
e. max: $(1, 6)$, min: $(5, 0)$
f. none

38. a. $x \in (-\infty, \infty)$, $y \in [-3, \infty)$
b. $(-3, 0)$, $(2, 0)$, $(4, 0)$
c. $p(x) \geq 0$: $x \in (-\infty, 2] \cup [4, \infty)$
$p(x) \leq 0$: $x \in [2, 4] \cup \{-3\}$
d. $p(x)\uparrow$: $x \in (-3, 0) \cup (3, \infty)$
$p(x)\downarrow$: $x \in (-\infty, -3) \cup (0, 3)$
e. max: $(0, 6)$
min: $(-3, 0)$, $(3, -3)$
f. none

39. a. $x \in (-\infty, \infty)$, $y \in (-\infty, \infty)$
b. $(-2, 0)$, $(4, 0)$, $(8, 0)$
c. $q(x) \geq 0$: $x \in \{-2\} \cup [4, \infty)$
$q(x) \leq 0$: $x \in (-\infty, 4) \cup \{8\}$
d. $q(x)\uparrow$: $x \in (-\infty, -2) \cup (1, 6) \cup (8, \infty)$
$q(x)\downarrow$: $x \in (-2, 1) \cup (6, 8)$
e. max: $(-2, 0)$, $(6, 2)$
min: $(1, -5)$, $(8, 0)$
f. none

Additional answers can be found in the Instructor Answer Appendix.

31. $y = V(x)$

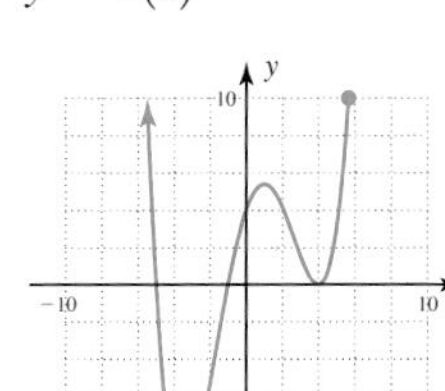

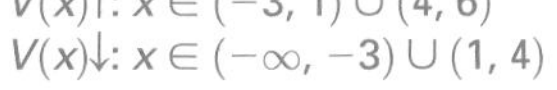

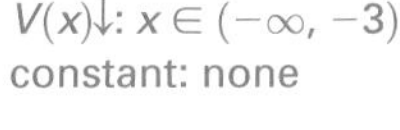

32. $H(x) = \dfrac{-1}{(x - 1)} - 2$

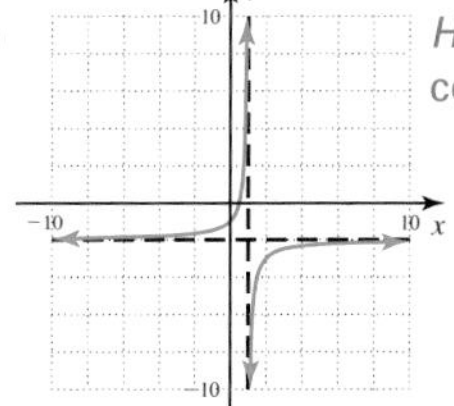

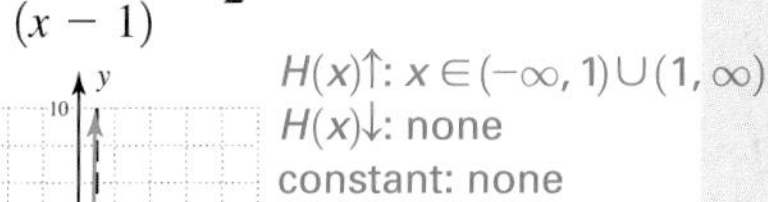

For Exercises 33 through 45, determine the following (answer in interval notation as appropriate): (a) domain and range of the function; (b) zeroes of the function; (c) interval(s) where the function is greater than or equal to zero or less than or equal to zero; (d) interval(s) where the function is increasing, decreasing, or constant; (e) location of any max or min value(s); and (f) equations of asymptotes (if any).

33. $H(x) = -5|x - 2| + 5$

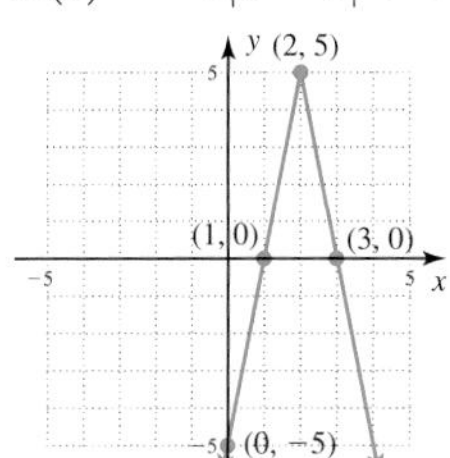

34. $p(x) = 0.5(x + 2)^3$

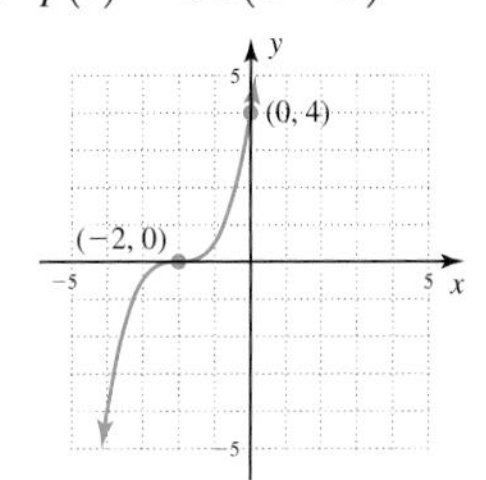

35. $q(x) = -\sqrt[3]{x + 1}$

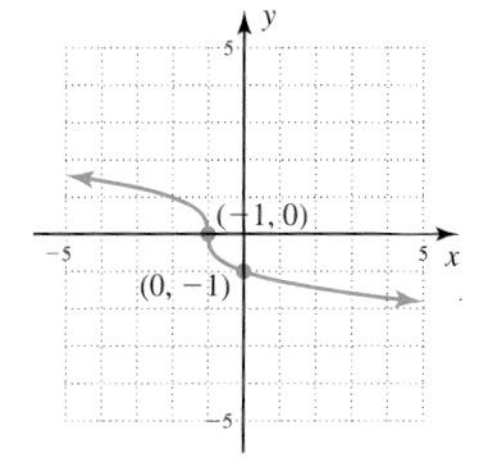

36. $y = f(x)$

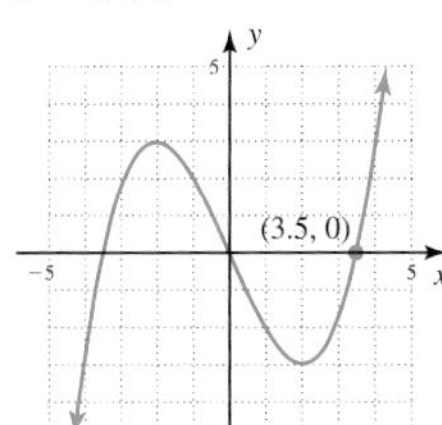

37. $y = g(x)$

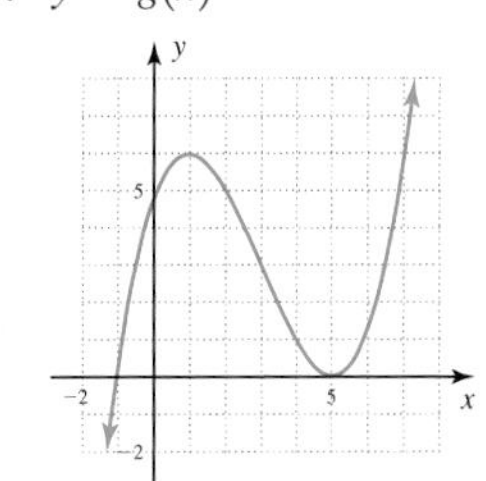

38. $y = p(x)$

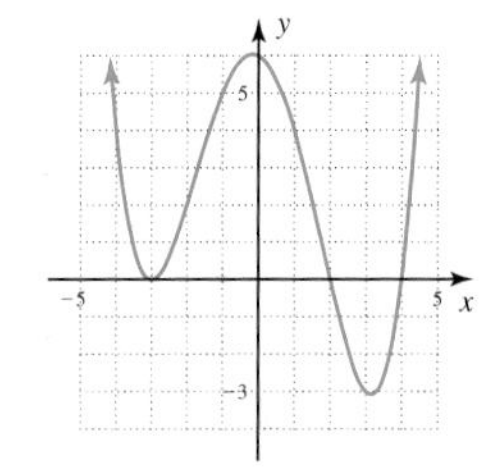

39. $y = q(x)$

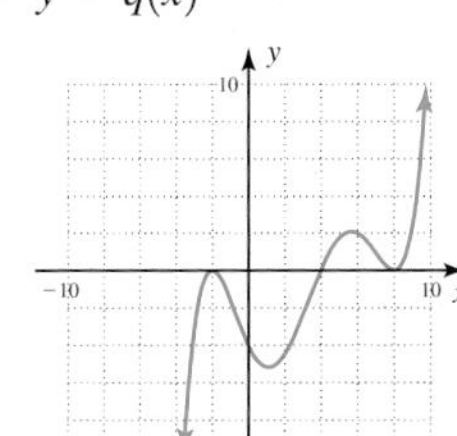

40. $y = Y_1$

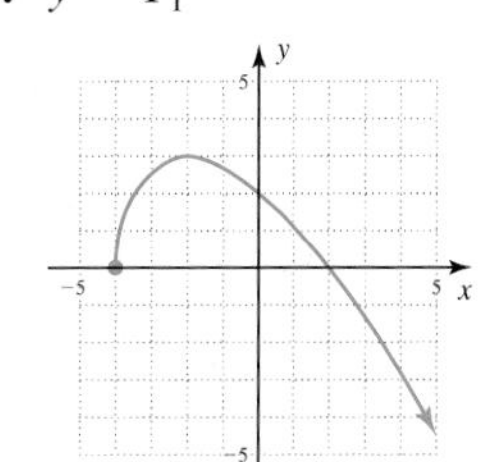

41. $y = Y_2$

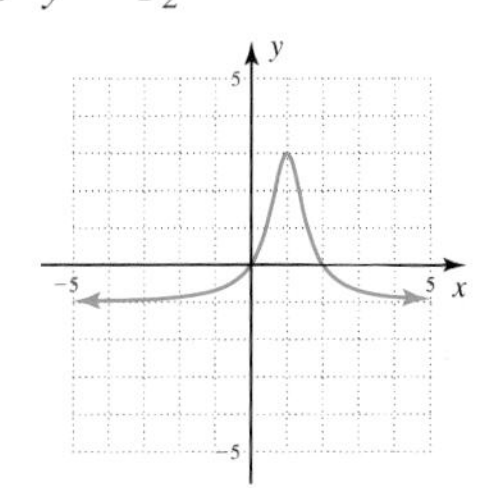

42. $Y_1 = \dfrac{1}{x - 2} - 1$

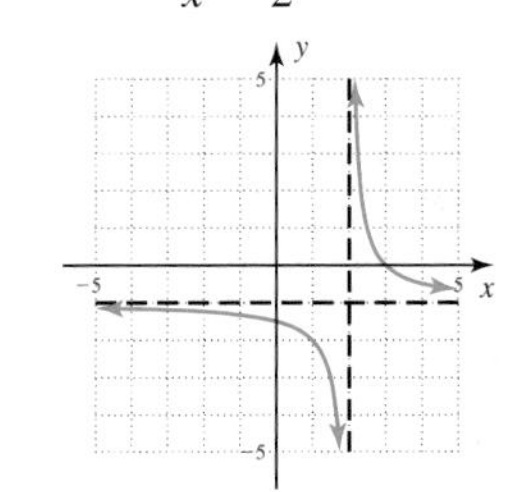

43. $Y_2 = \dfrac{-4}{(x - 2)^2} + 4$

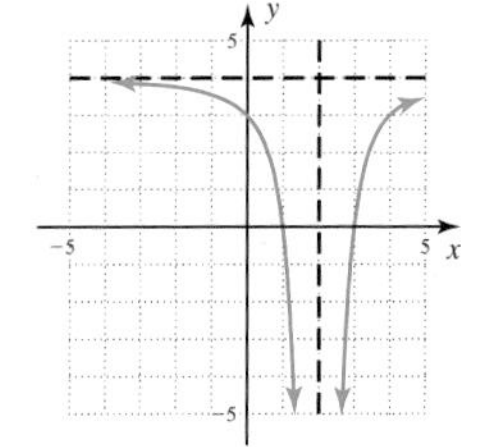

44. a. $x \in (-\infty, \infty)$, $y \in [-4, \infty)$
b. $(-1, 0)$, $(2, 0)$
c. $f(x) \geq 0$: $x \in (-\infty, -1] \cup [2, \infty)$
$f(x) \leq 0$: $x \in [-1, 2]$
d. $f(x)\uparrow$: $x \in (0, \infty)$
$f(x)\downarrow$: $x \in (-\infty, 0)$
e. min: $(0, -4)$
f. none

45. a. $x \in (-\infty, \infty)$, $y \in \{-4\} \cup [-3, \infty)$
b. $(-4, 0)$
c. $H(x) \geq 0$: $x \in (-\infty, -4] \cup (-2, 0) \cup (0, 2)$
$H(x) \leq 0$: $x \in [-4, -2] \cup [3, \infty)$
d. $H(x)\uparrow$: $x \in (-2, 0)$
$H(x)\downarrow$: $x \in (-\infty, -2) \cup (0, 2)$
$H(x)$ constant: $x \in (3, \infty)$
e. none
f. $x = 0$

46. $\frac{\Delta y}{\Delta x} = \frac{-2x - h}{x^2(x+h)^2}$;
$[0.50, 0.51]$: $\frac{\Delta y}{\Delta x} \approx -15.5$;
$[1.50, 1.51]$: $\frac{\Delta y}{\Delta x} \approx -0.6$;
Answers will vary.

47. $\frac{\Delta y}{\Delta x} = 2x - 4 + h$;
$[0.00, 0.01]$: $\frac{\Delta y}{\Delta x} = -3.9$;
$[3.00, 3.01]$: $\frac{\Delta y}{\Delta x} \approx 2.0$;
Answers will vary.

48. $\frac{\Delta y}{\Delta x} = 3x^2 + 3xh + h^2$;
$[-2.01, -2.00]$: $\frac{\Delta y}{\Delta x} \approx 12.1$;
$[0.40, 0.41]$: $\frac{\Delta y}{\Delta x} \approx 0.5$;
Answers will vary.

49. $\frac{\Delta y}{\Delta x} = \frac{1}{\sqrt{x+h} + \sqrt{x}}$
$[1.00, 1.01]$: $\frac{\Delta y}{\Delta x} \approx 0.5$;
$[4.00, 4.01]$: $\frac{\Delta y}{\Delta x} \approx .025$;
Answers will vary.

50. a. $x \in (-\infty, -3] \cup [3, \infty)$, $y \in [0, \infty)$
b. $(-3, 0)$, $(3, 0)$
c. $f(x)\uparrow$: $x \in (3, \infty)$
$f(x)\downarrow$: $x \in (-\infty, -3)$
d. even

Additional answers can be found in the Instructor Answer Appendix.

44.

$$f(x) = \begin{cases} \sqrt{-1-x} & x \leq -1 \\ x^2 - 4 & -1 < x \leq 2 \\ x - 2 & x > 2 \end{cases}$$

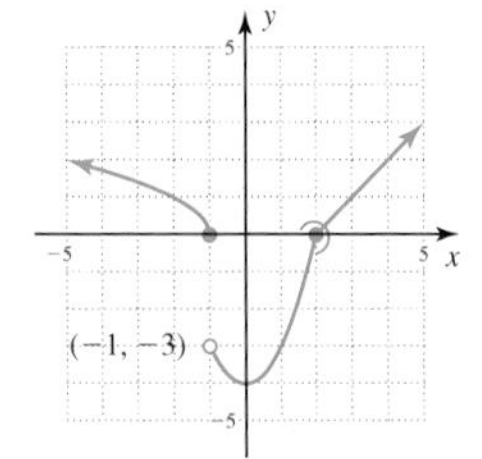

45.

$$H(x) = \begin{cases} -1.5x - 6 & x \leq -2 \\ \frac{4}{x^2} - 1 & -2 < x < 2; x \neq 0 \\ -4 & x \geq 3 \end{cases}$$

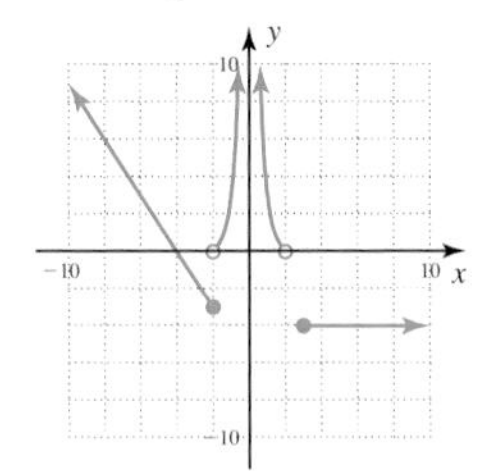

Use the difference quotient to find a rate of change formula for the functions given, then calculate the rate of change for the intervals indicated. Comment on how the rate of change in each interval corresponds to the graph of the function.

46. $h(x) = \frac{1}{x^2}$
$[0.50, 0.51]$, $[1.50, 1.51]$

47. $f(x) = x^2 - 4x$
$[0.00, 0.01]$, $[3.00, 3.01]$

48. $g(x) = x^3 + 1$
$[-2.01, -2.00]$, $[0.40, 0.41]$

49. $r(x) = \sqrt{x}$
$[1.00, 1.01]$, $[4.00, 4.01]$

WORKING WITH FORMULAS

50. Conic sections—hyperbola: $f(x) = \frac{1}{3}\sqrt{4x^2 - 36}$

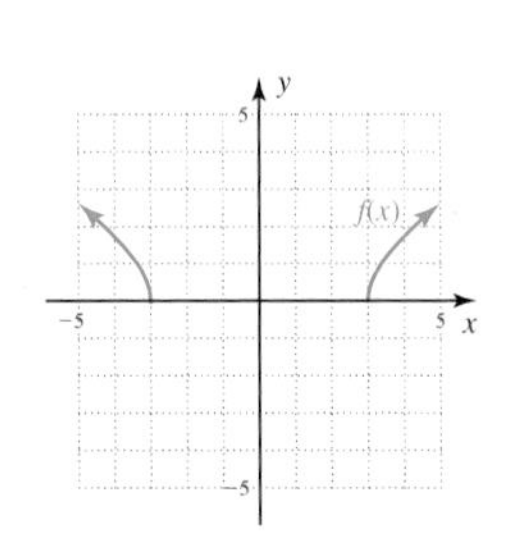

Even though the conic sections are not covered in detail until Chapter 7, we've already developed a number of tools that will help us understand these relations and their graphs. The equation here gives the "upper branches" of a hyperbola, as shown in the figure. Find the following by analyzing the equation: (a) the domain and range; (b) the zeroes of the relation; (c) interval(s) where y is increasing or decreasing; and (d) whether the relation is even, odd, or neither.

51. Trigonometric graphs: $y = \sin(x)$ and $y = \cos(x)$

The trigonometric functions are also studied at some future time, but we can apply the same tools to analyze the graphs of these functions as well. The graphs of $y = \sin(x)$ and $y = \cos(x)$ are given, graphed over the interval $x \in [-180, 360]$. Use them to find (a) the range of the functions; (b) the zeroes of the functions; (c) interval(s) where y is increasing/decreasing; (d) location of minimum/maximum values; and (e) whether each relation is even, odd, or neither.

$y = \sin x$

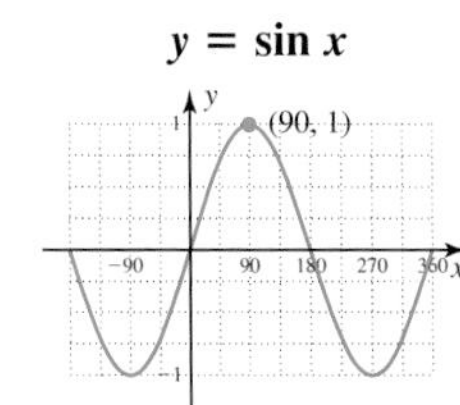

$y = \cos x$

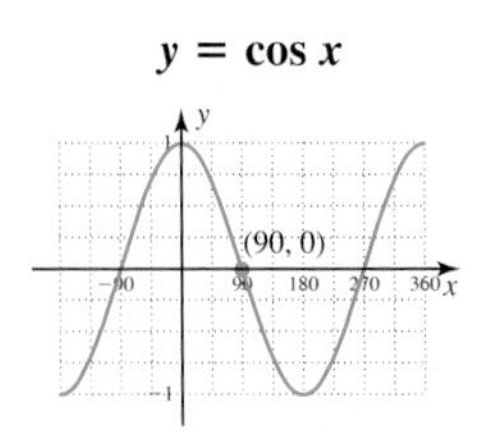

52. a. $x \in [0, 260]$, $y \in [0, 80]$
b. 80 ft
c. 120 ft
d. yes
e. (0, 120)
f. (120, 260)

55. d. $h(x) = |x^2 - 4| - 5$
$h(x)\uparrow$: $x \in (-2, 0) \cup (2, \infty)$;
$h(x)\downarrow$: $x \in (-\infty, -2) \cup (0, 2)$

56.
a. $t \in [72, 96]$, $I \in [7.25, 16]$
b. $I(t)\uparrow$: $t \in (72, 74) \cup (77, 81) \cup (83, 84) \cup (93, 94)$
$I(t)\downarrow$: $t \in (74, 75) \cup (81, 83) \cup (84, 86) \cup (90, 93) \cup (94, 95)$
$I(t)$ constant: $t \in (75, 77) \cup (86, 90) \cup (95, 96)$
c. max: (74, 9.25), (81, 16) (global max), (84, 13), (94, 8.5), min: (72,7.5), (83, 12.75), (93, 7.5)
d. Increase: 80 to 81; Decrease: 82 to 83 or 85 to 86

APPLICATIONS

52. **Catapults and projectiles:** Catapults have a long and interesting history that dates back to ancient times, when they were used to launch javelins, rocks, and other projectiles. The diagram given illustrates the path of the projectile after release, which follows a parabolic arc. Use the graph to determine the following:

a. State the domain and range of the projectile.
b. What is the maximum height of the projectile?
c. How far from the catapult did the projectile reach its maximum height?
d. Did the projectile clear the castle wall, which was 40 ft high and 210 ft away?
e. On what interval was the height of the projectile increasing?
f. On what interval was the height of the projectile decreasing?

53. **Profit and loss:** The profit of DeBartolo Construction Inc. is illustrated by the graph shown. Use the graph to estimate the point(s) or the interval(s) for which the profit P was:

a. increasing $t \in (0, 1) \cup (3, 4) \cup (7, 10)$
b. decreasing $t \in (4, 7)$
c. constant $t \in (1, 3)$
d. a maximum (4, 12), (10, 16)
e. a minimum (7, −4)
f. positive $t \in (0, 6) \cup (8, 10)$
g. negative $t \in (6, 8)$
h. zero (6, 0), (8, 0)

54. **Functions and rational exponents:** The graph of $f(x) = x^{\frac{2}{3}} - 1$ is shown. Use the graph to find:

a. domain and range of the function $x \in (-\infty, \infty)$; $y \in [-1, \infty)$
b. zeroes of the function (−1, 0), (1, 0)
c. interval(s) where $f(x) \geq 0$ or $f(x) < 0$
$f(x) \geq 0$: $x \in (-\infty, -1] \cup [1, \infty)$;
$f(x) < 0$: $x \in (-1, 1)$;
d. interval(s) where $f(x)$ is increasing, decreasing, or constant
$f(x)\uparrow$: $x \in (0, \infty)$, $f(x)\downarrow$: $x \in (-\infty, 0)$
e. location of any max or min value(s) min: (0, −1)
f. equations of asymptotes (if any) none

Exercise 54

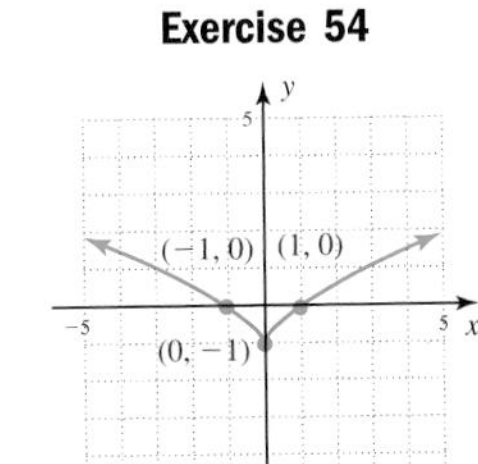

Exercise 55

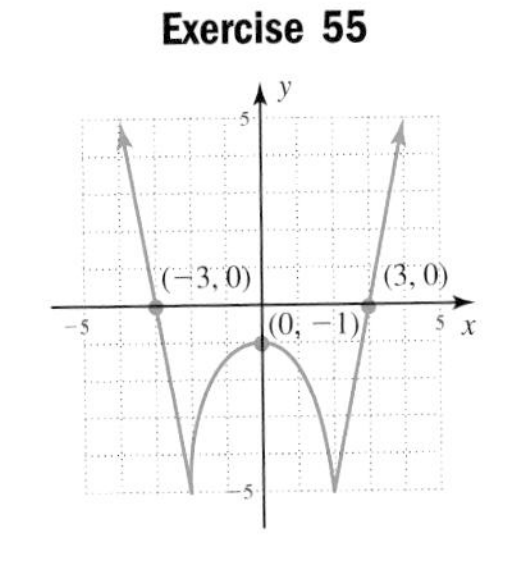

55. **Analyzing a composition:** Given $f(x) = |x| - 5$ and $g(x) = x^2 - 4$. Their composition produces the function $h(x) = (f \circ g)(x)$, whose graph is shown. Write out the composition and use the graph to find: $h(x) = |x^2 - 4| - 5$

a. domain and range of the function $x \in (-\infty, \infty)$, $y \in [-5, \infty)$
b. zeroes of the function (−3, 0), (3, 0)
c. interval(s) where $h(x) \geq 0$ or $h(x) \leq 0$
$h(x) \geq 0$: $x \in (-\infty, -3] \cup [3, \infty)$;
$h(x) \leq 0$: $x \in [-3, 3]$
d. interval(s) where $f(x)$ is increasing, decreasing or constant
e. location of any max or min value(s) max: (0, −1); min: (−2, −5), (2, −5)
f. equations of asymptotes (if any) none

56. **Analyzing interest rates:** The graph shown approximates the average annual interest rates on 30-year fixed mortgages, rounded to the nearest $\frac{1}{4}\%$. Use the graph to estimate the following (write all answers in interval notation).

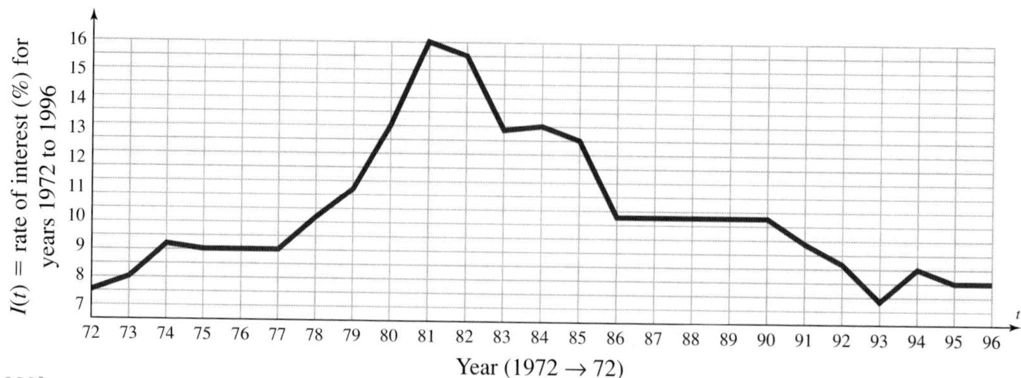

a. domain and range

b. interval(s) where $I(t)$ is increasing, decreasing, or constant

c. location of the maximum and minimum values.

d. the one-year period with the greatest rate of increase and the one-year period with the greatest rate of decrease

Source: 1998 *Wall Street Journal Almanac,* p. 446; 2004 *Statistical Abstract of the United States,* Table 1178

57.
a. $t \in [75, 102]$, $D \in [-300, 230]$
b. $D(t)\uparrow$: $t \in (76, 77) \cup (83, 84) \cup (86, 87) \cup (92, 100)$; $D(t)\downarrow$: $t \in (75, 76) \cup (77, 83) \cup (84, 86) \cup (89, 92) \cup (100, 102)$
$D(t)$ constant: $t \in (87, 89)$
c. max: $(75, -40)$, $(77, -50)$, $(84, -170)$, $(100, 240)$; min: $(76, -70)$, $(83, -210)$, $(86, -220)$, $(92, -300)$, $(102, -140)$
d. increase: 96 to 97 or 99 to 100; decrease: 101 to 102

57. Analyzing the deficit: The following graph approximates the Federal Deficit of the United States. Use the graph to estimate the following (write answers in interval notation).

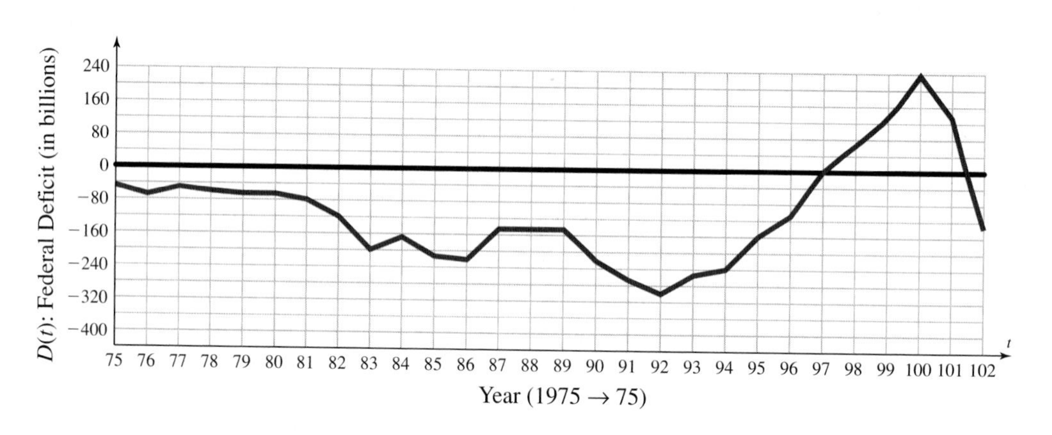

a. the domain and range

b. interval(s) where $D(t)$ is increasing, decreasing or constant

c. the location of the maximum and minimum values

d. the one-year period with the greatest rate of increase, and the one-year period with the greatest rate of decrease

Source: 2005 *Statistical Abstract of the United States,* Table 461

58. zeroes: $(-8, 0)$, $(-4, 0)$, $(0, 0)$, $(4, 0)$; min: $(-2, -1)$, $(4, 0)$; max: $(-6, 2)$, $(2, 2)$

58. Constructing a graph: Draw the function f that has the following characteristics, then state the zeroes and the location of all maximum and minimum values. [*Hint:* Write them as $(c, f(c))$.]

a. Domain: $x \in (-10, \infty)$

b. Range: $y \in (-6, \infty)$

c. $f(0) = 0; f(4) = 0$

d. $f(x)\uparrow$ for $x \in (-10, -6) \cup (-2, 2) \cup (4, \infty)$

e. $f(x)\downarrow$ for $x \in (-6, -2) \cup (2, 4)$

f. $f(x) \geq 0$ for $x \in [-8, -4] \cup [0, \infty)$

g. $f(x) < 0$ for $x \in (-\infty, -8) \cup (-4, 0)$

59. zeroes: $(-9, 0)$, $(-3, 0)$, $(6, 0)$; min: $(-6, -6)$, $(6, 0)$; max: $(3, 6)$

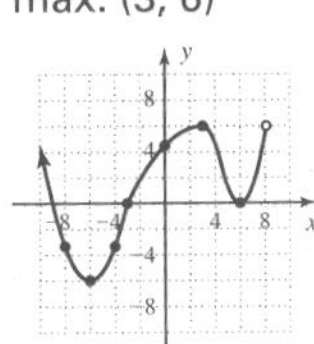

59. Constructing a graph: Draw the function g that has the following characteristics, then state the zeroes and the location of all maximum and minimum values. [*Hint:* Write them as $(c, g(c))$.]

a. Domain: $x \in (-\infty, 8)$
b. Range: $y \in [-6, \infty)$
c. $g(0) = 4.5$; $g(6) = 0$
d. $g(x)\uparrow$ for $x \in (-6, 3) \cup (6, 8)$
e. $g(x)\downarrow$ for $x \in (-\infty, -6) \cup (3, 6)$
f. $g(x) \geq 0$ for $x \in (-\infty, -9] \cup [-3, 8]$
g. $g(x) < 0$ for $x \in (-9, -3)$

WRITING, RESEARCH, AND DECISION MAKING

60. Does the function shown have a maximum value? Does it have a minimum value? Discuss/explain/justify why or why not. no; no; Answers will vary.

61. Look up the word *analyze* in a full-sized college-level dictionary. Consider the etymology (roots or origins) of the word, and the various meanings and examples the dictionary gives. Is the title of this section (Section 3.8) appropriate? Answers will vary.

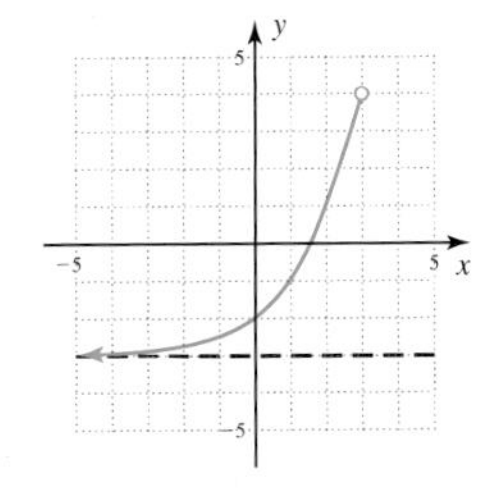

EXTENDING THE CONCEPT

62. The graph drawn here depicts a 400-m race between a mother and her daughter. Analyze the graph to answer questions (a) through (f).

a. Who wins the race, the mother or daughter? daughter
b. By approximately how many meters? 20 m
c. By approximately how many seconds? 10 sec
d. Who was leading at $t = 40$ seconds? mother
e. During the race, how many seconds was the daughter in the lead? about 38 sec
f. During the race, how many seconds was the mother in the lead? about 27 sec

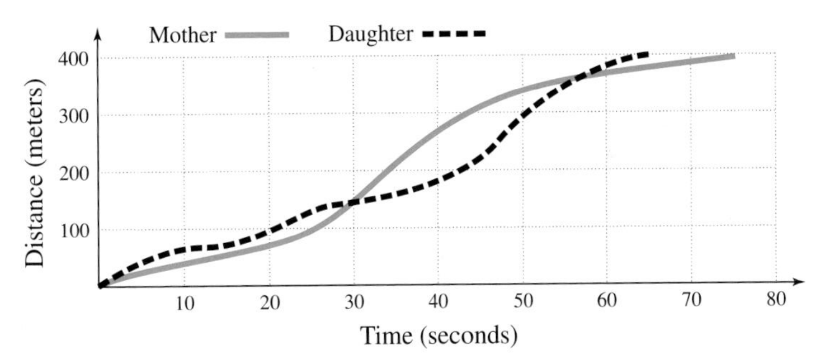

63. $f(x)$ local max local min min value > max value

63. Draw a general function $f(x)$ that has a local *maximum* at $(a, f(a))$ and a local *minimum* at $(b, f(b))$ but with $f(a) < f(b)$. Answers will vary.

64. In the *Point of Interest*, the army marched a horizontal distance of 6800 ft. Use the graph to determine the *actual number of feet* it took to cover the terrain (uphill, downhill, etc). about 11,817 ft

MAINTAINING YOUR SKILLS

65. (1.5) Solve the given quadratic equation three different ways: (a) factoring, (b) completing the square, and (c) using the quadratic formula: $x^2 - 8x - 20 = 0$ $x = -2, x = 10$

66. (3.6) If the flow volume is constant, the velocity of a fluid flowing through a pipe varies inversely with the area of the pipe's cross section. Write this relationship as a variation. $V = \frac{k}{A}$

67. (R.6) Simplify each expression.

a. $16^{\frac{-3}{2}}$ $\frac{1}{64}$ **b.** $\left(\frac{27}{8}\right)^{\frac{-2}{3}}$ $\frac{4}{9}$

68. (R.5) Multiply and write the answer in simplest form

$$\frac{n^2 - 9}{n^2 - 4n - 21} \cdot \frac{n^2 - 2n - 35}{n^2 + 2n - 15} \quad 1$$

69. (3.3) Write the equation for the function illustrated below (assume $a = 1$).

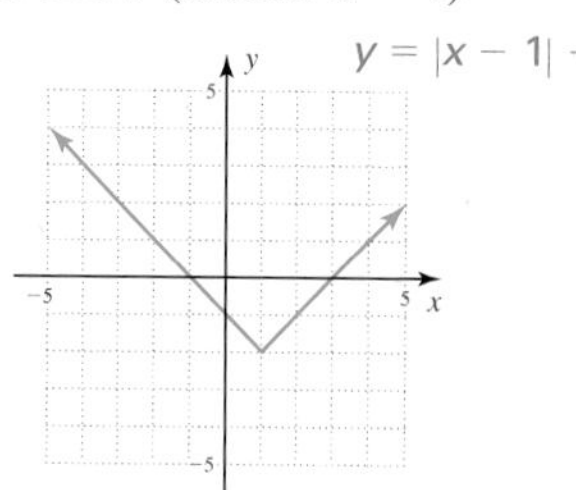

70. (R.7) Find the surface area and volume of the cylinder shown.

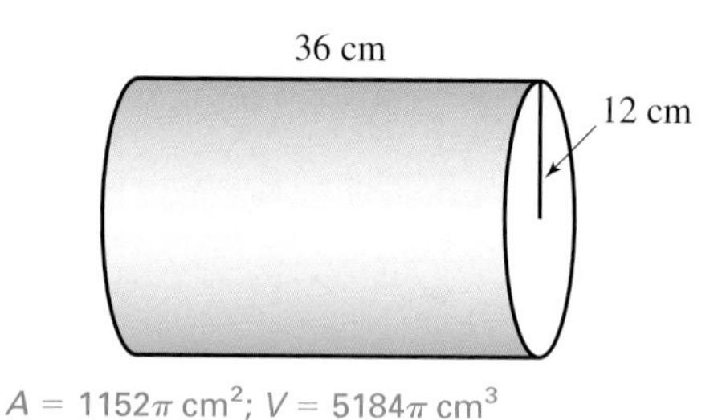

$A = 1152\pi\ \text{cm}^2$; $V = 5184\pi\ \text{cm}^3$

Summary and Concept Review

SECTION 3.1 The Algebra and Composition of Functions

KEY CONCEPTS

- The notation used to represent the binary operations on two functions is:
  - $(f + g)(x) = f(x) + g(x)$
  - $(f - g)(x) = f(x) - g(x)$
  - $(f \cdot g)(x) = f(x) \cdot g(x)$
  - $\left(\frac{f}{g}\right)(x) = \frac{f(x)}{g(x)}$; $g(x) \neq 0$
- The result of these operations is a new function $h(x)$, which can also be graphed/analyzed.
- The composition of two functions is written $(f \circ g)(x) = f[g(x)]$ (g is an input for f).
- The domain of the new function h is the intersection of the domains for f and g.
- To evaluate $(f \circ g)(2)$, we find $(f \circ g)(x)$ then substitute $x = 2$, or find $g(2) = k$ then find $f(k)$.
- A composite function $h(x) = (f \circ g)(x)$ can be "decomposed" into individual functions by identifying functions f and g such that $(f \circ g)(x) = h(x)$. The decomposition is not unique.

EXERCISES

For $f(x) = x^2 + 4x$ and $g(x) = 3x - 2$, find the following:

1. $(f + g)(a)$ $a^2 + 7a - 2$ **2.** $(f \cdot g)(3)$ 147 **3.** the domain of $\left(\frac{f}{g}\right)(x)$ $x \in \left(-\infty, \frac{2}{3}\right) \cup \left(\frac{2}{3}, \infty\right)$

Given $p(x) = 4x - 3$, $q(x) = x^2 + 2x$, and $r(x) = \frac{x + 3}{4}$ find:

4. $(p \circ q)(x)$ **5.** $(q \circ p)(3)$ **6.** $(p \circ r)(x)$ and $(r \circ p)(x)$

For each function here, find functions $f(x)$ and $g(x)$ such that $h(x) = f[g(x)]$:

7. $h(x) = \sqrt{3x - 2} + 1$ **8.** $h(x) = 3 - |x^2 - 1|$ **9.** $h(x) = x^{\frac{2}{3}} - 3x^{\frac{1}{3}} - 10$

10. A stone is thrown into a pond causing a circular ripple to move outward from the point of entry. The radius of the circle is modeled by $r(t) = 2t + 3$, where t is the time in seconds. Find a function that will give the area of the circle directly as a function of time. In other words, find $A(t)$. $A(t) = \pi(2t + 3)^2$

4. $4x^2 + 8x - 3$
5. 99
6. x; x
7. $f(x) = \sqrt{x} + 1$; $g(x) = 3x - 2$
8. $f(x) = 3 - |x|$; $g(x) = x^2 - 1$
9. $f(x) = x^2 - 3x - 10$; $g(x) = x^{\frac{1}{3}}$

SECTION 3.2 One-to-One and Inverse Functions

KEY CONCEPTS

- A function is one-to-one if each element of the range corresponds to a unique element of the domain.
- If every horizontal line intersects the graph of a function in at most one point, the function is one-to-one.
- If f is a one-to-one function with ordered pairs (a, b), then the inverse of f is that one-to-one function f^{-1} with ordered pairs of the form (b, a).
- To find f^{-1} using the algebraic method, use the following four-step process:
  1. Use y instead of $f(x)$.
  2. Interchange x and y.
  3. Solve the equation for y.
  4. Substitute $f^{-1}(x)$ for y
- If f is a one-to-one function, the inverse of f is the function f^{-1} such that $(f \circ f^{-1})(x) = x$ and $(f^{-1} \circ f)(x) = x$.
- The graphs of $f(x)$ and $f^{-1}(x)$ are symmetric with respect to the identity function $y = x$.

EXERCISES

Determine whether the functions given are one-to-one by noting the function family to which each belongs and mentally picturing the shape of the graph.

11. $h(x) = -|x - 2| + 3$ no

12. $p(x) = 2x^2 + 7$ no

13. $s(x) = \sqrt{x - 1} + 5$ yes

Find the inverse of each function given. Then show graphically and using composition that your inverse function is correct. State any necessary restrictions.

14. $f(x) = -3x + 2$

15. $f(x) = x^2 - 2, x \geq 0$

16. $f(x) = \sqrt{x - 1}$

Determine the domain and range for each function whose graph is given, and use this information to state the domain and range of the inverse function. Then sketch in the line $y = x$, estimate the location of three points on the graph, and use these to graph $f^{-1}(x)$ on the same grid.

17.

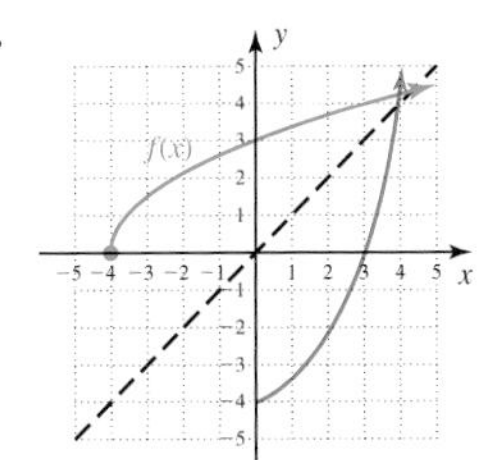

18.

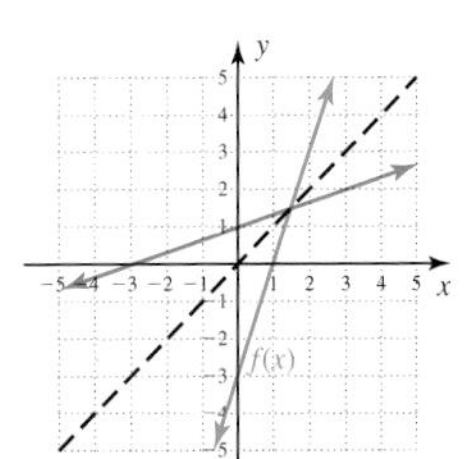

19.

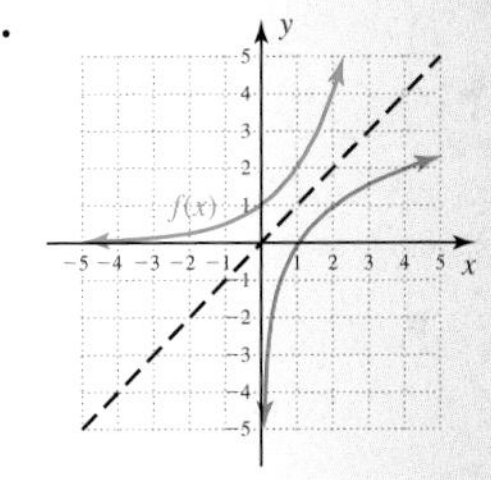

20. Fines for overdue material: Some libraries have set fees and penalties to discourage patrons from holding borrowed materials for an extended period. Suppose the fine for overdue DVDs is given by the function $f(t) = 0.15t + 2$, where $f(t)$ is the amount of the fine t days after it is due. (a) What is the fine for keeping a DVD seven (7) extra days? (b) Find $f^{-1}(t)$, then input your answer from part (a) and comment on the result. (c) If a fine of \$3.80 was assessed, how many days was the DVD overdue?

14. $f^{-1}(x) = \frac{x - 2}{-3}$
15. $f^{-1}(x) = \sqrt{x + 2}$
16. $f^{-1}(x) = x^2 + 1; x \geq 0$
17. $f(x)$: $D: x \in [-4, \infty)$, $R: y \in [0, \infty)$; $f^{-1}(x)$: $D: x \in [0, \infty)$, $R: y \in [-4, \infty)$
18. $f(x)$: $D: x \in (-\infty, \infty)$, $R: y \in (-\infty, \infty)$; $f^{-1}(x)$: $D: (-\infty, \infty)$, $R: y \in (-\infty, \infty)$
19. $f(x)$: $D: x \in (-\infty, \infty)$, $R: y \in (0, \infty)$; $f^{-1}(x)$: $D: x \in (0, \infty)$, $R: y \in (-\infty, \infty)$
20. a. \$3.05
    b. $f^{-1}(t) = \frac{t - 2}{0.15}$
    $f^{-1}(3.05) = 7$
    c. 12 days

SECTION 3.3 Toolbox Functions and Transformations

KEY CONCEPTS

- The following are six of the eight "toolbox" functions used to build mathematical models and build bridges to other functions (for a complete review of their characteristics and graphs, see Section 2.4).

21. quadratic

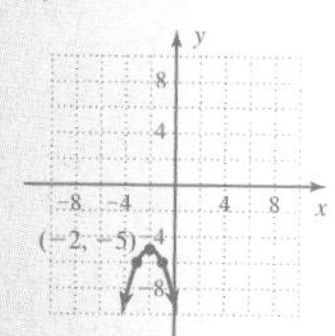

22. absolute value

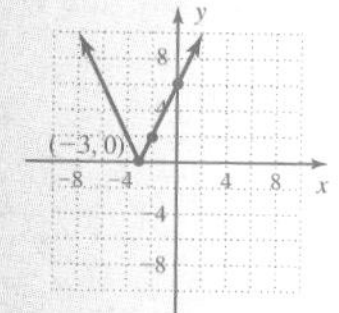

23. cubic

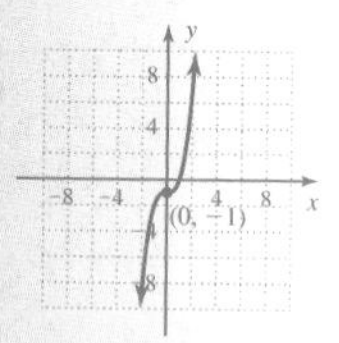

24. square root

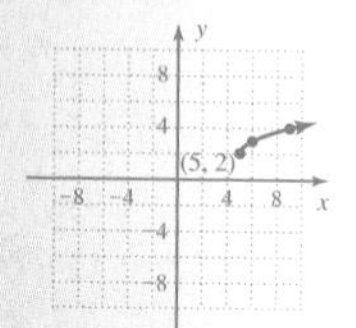

25. cube root

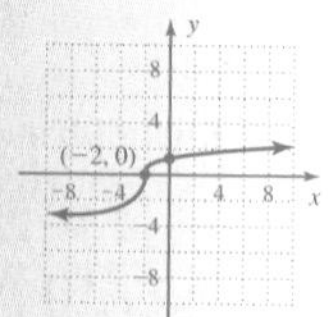

26. linear

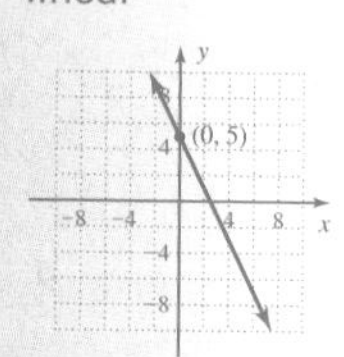

27.

a. (−2, 3) (3, 3) (0.5, −1)

b.

c. $(-4, \frac{3}{2})$ $(1, \frac{3}{2})$

28. $g(x) = -2|x + 1| + 4$

- Identity: $f(x) = x$
- Cubic: $f(x) = x^3$
- Absolute value: $f(x) = |x|$
- Square root function: $f(x) = \sqrt{x}$
- Quadratic: $f(x) = x^2$
- Cube root function: $f(x) = \sqrt[3]{x}$

- A rigid transformation does not change the form or shape of a graph—only its location.

RIGID TRANSFORMATIONS OF A BASIC GRAPH

Given a function whose graph is determined by $y = f(x)$ with $h, k > 0$:

Vertical Shifts

The graph of $y = f(x) \pm k$ is the graph of $f(x)$ *shifted vertically k units in the same direction as the sign.*

Horizontal Shifts

The graph of $y = f(x \pm h)$ is the graph of $f(x)$ *shifted horizontally h units in a direction opposite the sign.*

Vertical Reflections

The graph of $y = -f(x)$ is the graph of $f(x)$ *reflected vertically across the x-axis.*

- A nonrigid transformation stretches/compresses a graph, though the function family can still be noted.

NONRIGID TRANSFORMATIONS OF A BASIC GRAPH

Given any function whose graph is determined by $y = f(x)$:

Stretches

The graph of $y = af(x)$ is the graph of $f(x)$: stretched vertically upward if $|a| > 1$. (Graph is closer to y-axis.)

Compressions

The graph of $y = af(x)$ is the graph of $f(x)$: compressed vertically downward if $0 < |a| < 1$. (Graph is farther from y-axis.)

- To graph the transformation of a given or known graph: (1) apply any stretches or compressions, (2) reflect the graph if indicated, and (3) apply any vertical and/or horizontal shifts. These are usually applied to a few characteristics points, with the new graph drawn through these points.

EXERCISES

Identify each function as belonging to the linear, quadratic, square root, cubic, cube root, or absolute value family. Then sketch the graph using shifts of a parent function and a few characteristic points.

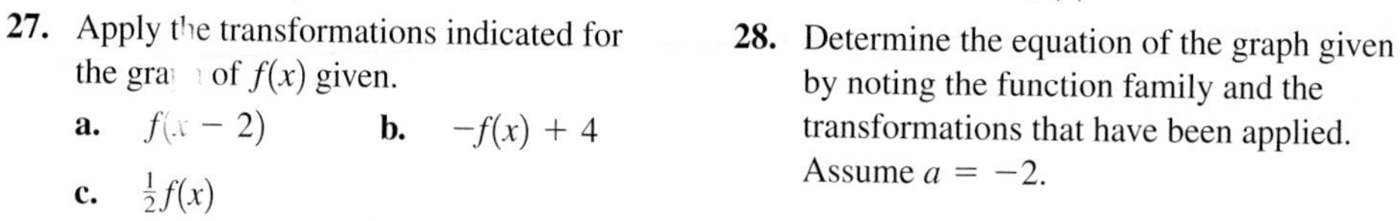

21. $f(x) = -(x + 2)^2 - 5$ **22.** $f(x) = 2|x + 3|$ **23.** $f(x) = x^3 - 1$

24. $f(x) = \sqrt{x - 5} + 2$ **25.** $f(x) = \sqrt[3]{x + 2}$ **26.** $f(x) = -2x + 5$

27. Apply the transformations indicated for the graph of $f(x)$ given.

a. $f(x - 2)$ **b.** $-f(x) + 4$

c. $\frac{1}{2}f(x)$

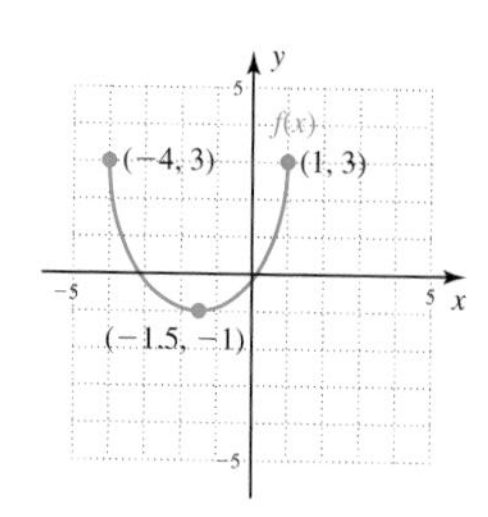

28. Determine the equation of the graph given by noting the function family and the transformations that have been applied. Assume $a = -2$.

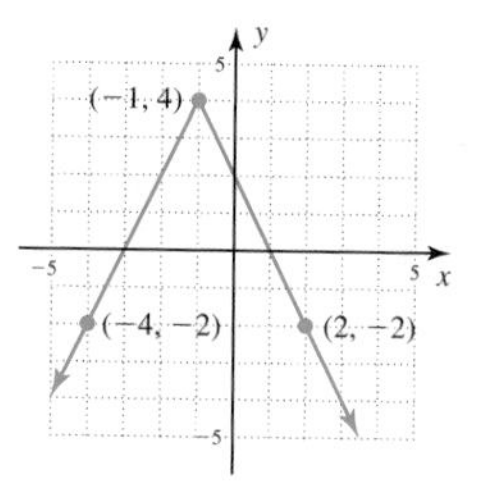

SECTION 3.4 Graphing General Quadratic Functions

KEY CONCEPTS

- The function $f(x) = ax^2 + bx + c$ can be written in *shifted form* $f(x) = a(x - h)^2 + k$ by completing the square.
  - Group the variable terms apart from the constant "c."
  - Factor out the lead coefficient "a."
  - Compute $\left[\frac{1}{2}\left(\frac{b}{a}\right)\right]^2$ and add, then subtract the result and regroup to form a factorable trinomial.
  - Factor the grouped terms as a binomial square; distribute, then combine the constant terms.
  - Graph using transformations of $y = x^2$
- If the vertex (h, k) is known, the x-intercepts can be found using the vertex/intercept formula: $x = h \pm \sqrt{-\frac{k}{a}}$
- Given the graph of a toolbox function, its equation can be found by using the characteristic points and the formula $y = af(x - h) + k$, where f is the corresponding parent function.

EXERCISES

Graph each quadratic by completing the square and using the shifts of the parent function. Find the x-intercepts (if they exist) using the vertex/intercept formula.

29. $f(x) = x^2 + 8x + 15$ **30.** $f(x) = -x^2 + 4x - 5$ **31.** $f(x) = 4x^2 - 12x + 3$

Determine the equation of each function shown (note scaling).

32.

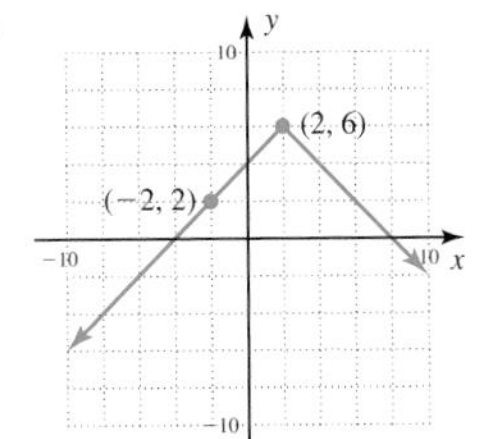

33.

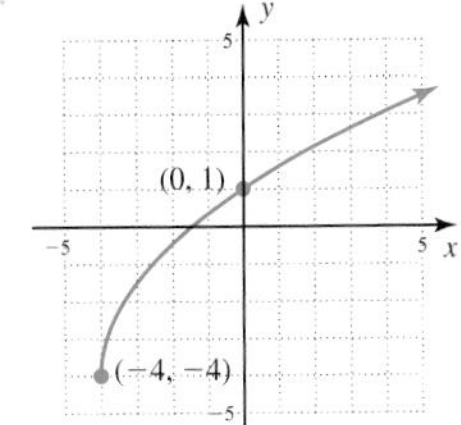

34.

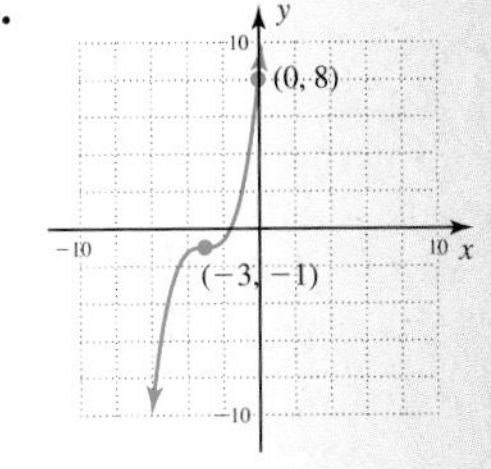

35. Height of a punt: A punter gets in several practice kicks by kicking the ball straight upward, catching it, and kicking it again. The height of the ball (in feet) at time t (in seconds) is given by $h(t) = -16t^2 + 96t + 2$. (a) How high is the ball the moment it is kicked? (b) How high is the ball at $t = 2$ sec? (c) How many seconds until the ball is again at the height found in part (b)? (d) What is the maximum height attained by the ball? At what time t did this occur?

29. (−5, 0), (−3, 0), (−4, −1)

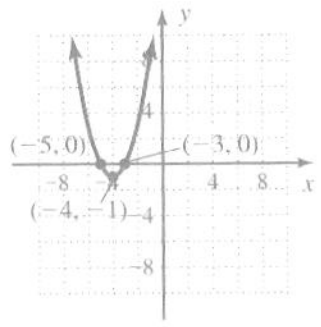

30. (2, −1), (0, −5)

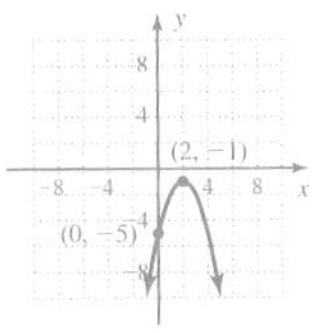

31. (0, 3), (0.3, 0), (2.7, 0), $\left(\frac{3}{2}, -6\right)$

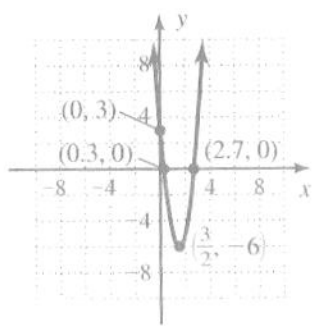

32. $y = -|x - 2| + 6$

33. $y = \frac{5}{2}\sqrt{x + 4} - 4$

34. $y = \frac{1}{9}(x + 3)^3 - 1$

35. a. 2 ft b. 130 ft c. 4 sec d. 146 ft; 3 sec

SECTION 3.5 Asymptotes and Simple Rational Functions

KEY CONCEPTS

- The last two toolbox functions are the reciprocal function $f(x) = \frac{1}{x}$ (butterfly function) and the reciprocal quadratic function $f(x) = \frac{1}{x^2}$ (volcano function).
- The line $y = k$ is a *horizontal asymptote* if, as x increases or decreases without bound, y approaches k: as $|x| \to \infty$, $y \to k$.

36. 37.

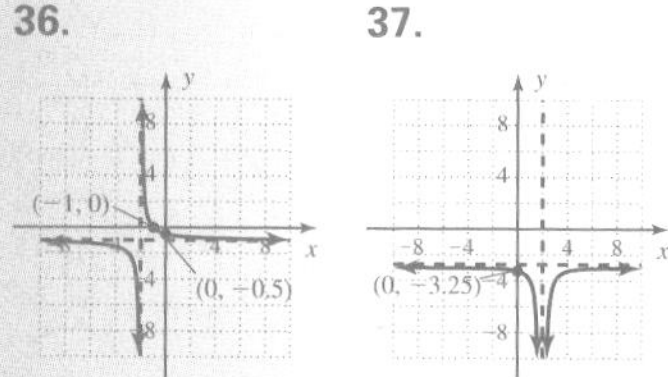

38. a. ≈\$32,143; \$75,000; \$175,000; \$675,000; cost increases dramatically

b.

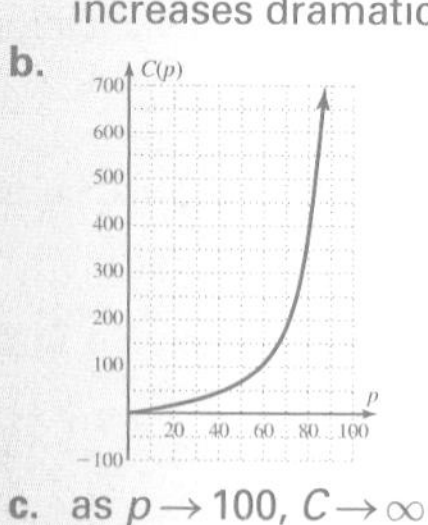

c. as $p \rightarrow 100$, $C \rightarrow \infty$

- The line $x = h$ is a *vertical asymptote* if, as x approaches h, $|y|$ increases or decreases without bound.
- The reciprocal and reciprocal quadratic functions can be transformed using the same shifts, stretches, and reflections as applied to other basic functions, with the asymptotes also shifted to help redraw the graph.

EXERCISES

Sketch the graph of each function using shifts of the parent function (not by using a table of values). Find and label the x- and y-intercepts (if they exist).

36. $f(x) = \dfrac{1}{x + 2} - 1$ **37.** $h(x) = \dfrac{-1}{(x - 2)^2} - 3$

38. In a certain county, the cost to keep public roads free of trash is given by $C(p) = \dfrac{-7500}{p - 100} - 75$, where $C(p)$ represents the cost (thousands of dollars) to keep p percent of the trash picked up. (a) Find the cost to pick up 30%, 50%, 70%, and 90% of the trash, and comment on the results. (b) Sketch the graph using the transformation of a toolbox function. (c) Use the direction/approach notation to state what happens if the county tries to keep 100% of the trash picked up.

SECTION 3.6 Toolbox Applications: Direct and Inverse Variation

KEY CONCEPTS

- *Direct variation:* Given two quantities x and y, if there is a nonzero constant k such that $y = kx$, we say, "y varies directly with x" or "y is directly proportional to x" (k is called the constant of variation).
- *Inverse variation:* Given two quantities x and y, if there is a nonzero constant k such that $y = \dfrac{k}{x}$ we say, "y varies inversely with x" or y is inversely proportional to x.
- In some cases, direct and inverse variations work simultaneously to form a *joint variation.*

SOLVING VARIATION PROBLEMS

1. Use descriptive variables to translate the situation given into a formula model.
2. Substitute the first set of values given and solve for k.
3. Use this value of k in the original formula model.
4. Substitute the remaining information to answer the question posed.

EXERCISES

Find the constant of variation and write the equation model, then use this model to complete the table.

39. y varies directly as the cube root of x; $y = 52.5$ when $x = 27$.
$k = 17.5$; $y = 17.5\sqrt[3]{x}$

x	y
216	105
0.343	12.25
729	157.5

40. z varies directly as v and inversely as the square of w; $z = 1.62$ when $w = 8$ and $v = 144$. $k = 0.72$; $z = \dfrac{0.72v}{w^2}$

v	w	z
196	7	2.88
38.75	1.25	17.856
24	0.6	48

41. The time that it takes for a simple pendulum to complete one period (swing over and back) is directly proportional to the square root of its length. If a pendulum 16 ft long has a period of 3 sec, find the time it take for a 36-ft pendulum to complete one period. 4.5 sec

SECTION 3.7 Piecewise-Defined Functions

KEY CONCEPTS

- The pieces of a piecewise-defined function each have a domain over which that piece is defined.
- To evaluate a piecewise-defined function, each input is used in its corresponding effective domain.
- To graph a piecewise-defined function, graph each piece in its entirety, then erase those portions of the graph outside the effective domain of each piece.
- The graph of a smooth function is one having no sharp turns or "jagged" edges.
- The graph of a continuous function can be drawn without lifting your pencil from the paper.
- Asymptotic and "jump" discontinuities are called nonremovable (nonrepairable) since there is no way to redefine the function so that it is smooth, continuous, and still a function.
- A pointwise discontinuity is said to be removable because we can redefine the function to "fill the hole."
- Step functions are discontinuous and formed by a series of horizontal steps.

EXERCISES

42. a.
$$f(x) = \begin{cases} 5 & x \le -3 \\ -x + 1 & -3 < x \le 3 \\ 3\sqrt{x-3} - 1 & x > 3 \end{cases}$$

b. $R: y \in [-2, \infty)$

43.

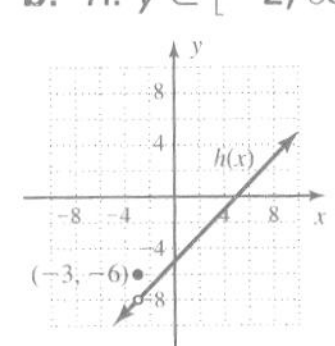

$D: x \in (-\infty, \infty)$,
$R: y \in (-\infty, -8) \cup (-8, \infty)$,

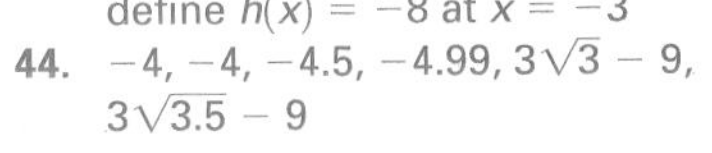

discontinuity at $x = -3$;
define $h(x) = -8$ at $x = -3$

44. $-4, -4, -4.5, -4.99, 3\sqrt{3} - 9, 3\sqrt{3.5} - 9$

45. $D: x \in (-\infty, \infty)$ $R: y \in [-4, \infty)$

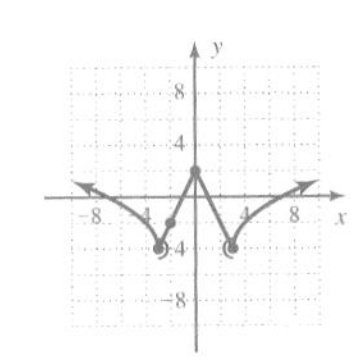

46.

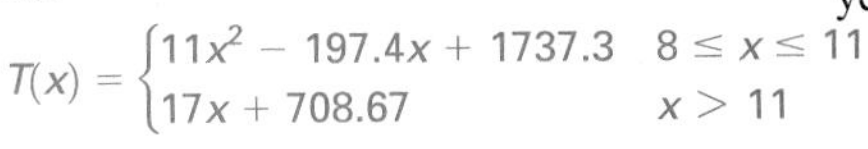

$$T(x) = \begin{cases} 11x^2 - 197.4x + 1737.3 & 8 \le x \le 11 \\ 17x + 708.67 & x > 11 \end{cases}$$

42. For the graph and functions given, (a) use the correct notation to write the relation as a single piecewise-defined function, stating the effective domain for each piece by inspecting the graph; and (b) state the range of the function: $Y_1 = 5$, $Y_2 = -x + 1$, $Y_3 = 3\sqrt{x-3} - 1$.

43. Use a table of values as needed to graph $h(x)$, then state its domain and range. If the function has a pointwise discontinuity, state how the second piece could be redefined so that a continuous function results.

$$h(x) = \begin{cases} \dfrac{x^2 - 2x - 15}{x + 3} & x \ne -3 \\ -6 & x = -3 \end{cases}$$

44. Evaluate the piecewise-defined function $p(x)$: $p(-4)$, $p(-2)$, $p(2.5)$, $p(2.99)$, $p(3)$, and $p(3.5)$

$$p(x) = \begin{cases} -4 & x < -2 \\ -|x| - 2 & -2 \le x < 3 \\ 3\sqrt{x} - 9 & x \ge 3 \end{cases}$$

45. Sketch the graph of the function and state its domain and range. Use transformations of the toolbox functions where possible.

$$q(x) = \begin{cases} 2\sqrt{-x-3} - 4 & x \le -3 \\ -2|x| + 2 & -3 < x < 3 \\ 2\sqrt{x-3} - 4 & x \ge 3 \end{cases}$$

46. For the years 1998 to 2003, leisure travel within the United States (trips of over 50 mi) can be modeled by the table and graph given, where 1990 corresponds to year 0. Assume the model is parabolic from 1998 to 2001, and linear from 2001 to 2003. Find the regression equation modeling each piece and write the single piecewise function $T(x)$ modeling leisure travel for these years.

Year	Trips (millions)	Year	Admissions (billions)
8	863	11	896
9	849	12	912
10	866	13	930

Source: 2005 *World Almanac*, p. 743

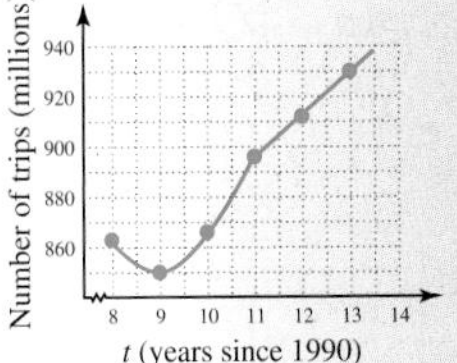

SECTION 3.8 Analyzing the Graph of a Function

47. a. D: $x \in (-\infty, \infty)$, R: $y \in [2, \infty)$
b. 38, 18, 6, 3, 2
c. none
d. $f(x) < 0$: none
$f(x) > 0$: $x \in (-\infty, \infty)$
e. min: (3, 2)
f. $f(x)\uparrow$: $x \in (3, \infty)$
$f(x)\downarrow$: $x \in (-\infty, 3)$

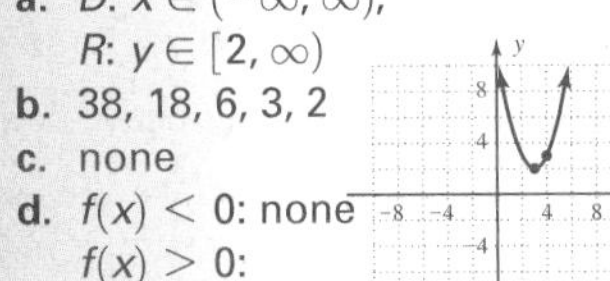

48. a. D: $x \in (-\infty, \infty)$, R: $y \in (-\infty, 3]$
b. $-2, 0, 2, 3, 2$
c. $(-1, 0)$, $(5, 0)$
d. $f(x) < 0$: $x \in (-\infty, -1) \cup (5, \infty)$;
$f(x) > 0$: $x \in (-1, 5)$
e. max: (2, 3)
f. $f(x)\uparrow$: $x \in (-\infty, 2)$
$f(x)\downarrow$: $x \in (2, \infty)$

49.

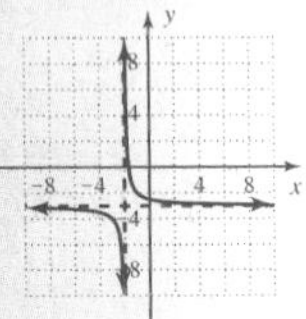

a. D: $x \in (-\infty, -2) \cup (-2, \infty)$, R: $y \in (-\infty, -3) \cup (-3, \infty)$
b. $-4, -2, -2.\overline{6}, -2.75, -2.8$
c. $(\frac{-5}{3}, 0)$
d. $f(x) < 0$: $x \in (-\infty, -2) \cup (\frac{-5}{3}, \infty)$
$f(x) > 0$: $x \in (-2, \frac{-5}{3})$
e. none
f. $f(x)\uparrow$: none
$f(x)\downarrow$: $x \in (-\infty, -2) \cup (-2, \infty)$

50. D: $x \in (-\infty, \infty)$
R: $y \in [-5, \infty)$
$f(x)\uparrow$: $x \in (2, \infty)$
$f(x)\downarrow$: $x \in (-\infty, 2)$
$f(x) > 0$: $x \in (-\infty, -1) \cup (5, \infty)$
$f(x) < 0$: $x \in (-1, 5)$

51. D: $x \in [-3, \infty)$
R: $y \in (-\infty, 0)$
$f(x)\uparrow$: none
$f(x)\downarrow$: $x \in (-3, \infty)$
$f(x) > 0$: none
$f(x) < 0$: $x \in (-3, \infty)$

52. D: $x \in (-\infty, \infty)$, R: $y \in (-\infty, \infty)$
$f(x)\uparrow$: $x \in (-\infty, -3) \cup (1, \infty)$
$f(x)\downarrow$: $x \in (-3, 1)$
$f(x) > 0$: $x \in (-5, -1) \cup (4, \infty)$
$f(x) < 0$: $x \in (-\infty, -5) \cup (-1, 4)$

53.

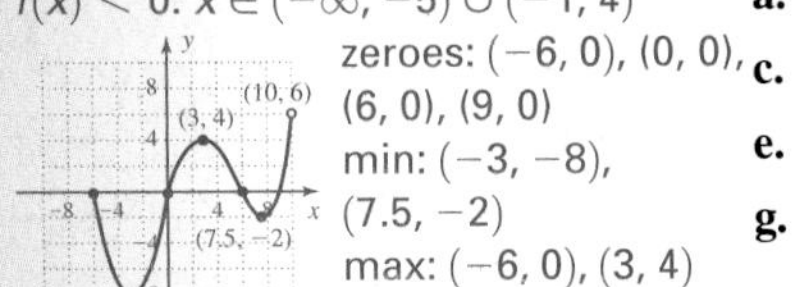

zeroes: $(-6, 0)$, $(0, 0)$, $(6, 0)$, $(9, 0)$
min: $(-3, -8)$, $(7.5, -2)$
max: $(-6, 0)$, $(3, 4)$

KEY CONCEPTS

- A graph is symmetric to the y-axis if given (x, y) is on the graph, then $(-x, y)$ is on the graph. Graphs with y-axis symmetry are said to be even. In function notation: $f(-x) = f(x)$.
- A graph is symmetric to the origin if given (x, y) is on the graph, then $(-x, -y)$ is also on the graph. Graphs symmetric to the origin are said to be odd. In function notation: $f(-x) = -f(x)$.
- *Intuitive descriptions of the characteristics of a graph are given here. The formal definitions can be found within Section 3.8.*
  - A function is *increasing* in an interval if the graph rises from left to right in the interval (larger inputs produce larger outputs).
  - A function is *decreasing* in an interval if the graph falls from left to right in the interval (larger inputs produce smaller outputs).
  - A function is *constant* if the graph is parallel to the x-axis.
  - A function is *positive* in an interval if the graph is above the x-axis in that interval.
  - A function is *negative* in an interval if the graph is below the x-axis in that interval.
  - A maximum value can be an *endpoint* maximum, *local* maximum, or *global* maximum. A similar statement can be made for minimum values.

EXERCISES

Graph the functions in Exercises 47 to 49 using a shift of the parent function. Then use the graph to find the following: (a) the domain and range; (b) the value of $f(-3)$, $f(-1)$, $f(1)$, $f(2)$, and $f(3)$; (c) the zeroes of the function; (d) interval(s) where $f(x)$ is negative or positive; (e) location of any maximum or minimum values; and (f) interval(s) where $f(x)$ is increasing, decreasing, or constant.

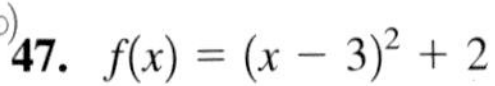

47. $f(x) = (x - 3)^2 + 2$ **48.** $f(x) = -|x - 2| + 3$ **49.** $f(x) = \dfrac{1}{x + 2} - 3$

State the domain and range for each function $f(x)$ given. Then state the intervals where f is increasing or decreasing and intervals where f is positive or negative. Assume all endpoints have integer values.

50.

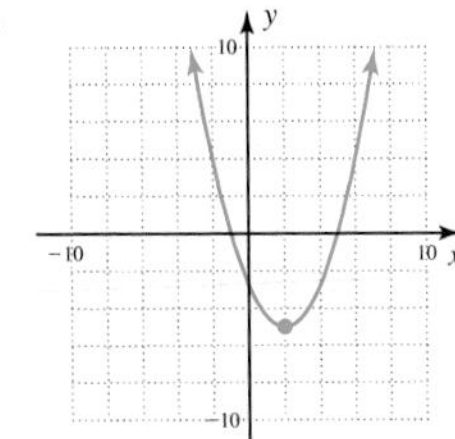

51.

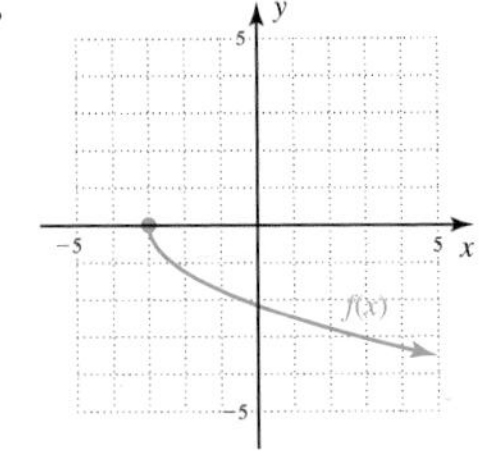

52.

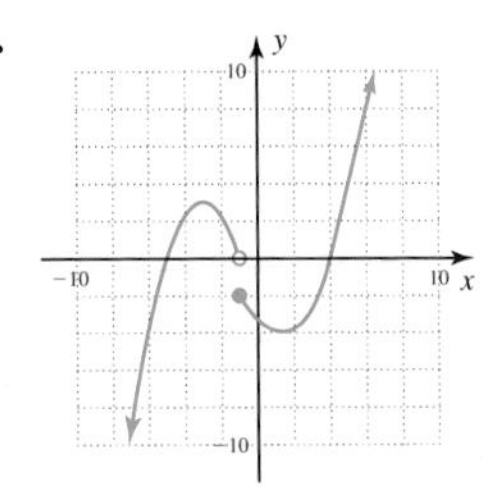

53. Draw the function f that has the following characteristics, then name the zeroes of the function and the location of all maximum and minimum values.
[*Hint:* Write them in the form $(c, f(c))$.]

a. Domain: $x \in [-6, 10)$
b. Range: $y \in (-8, 6)$
c. $f(0) = 0$
d. $f(x)\downarrow$ for $x \in (-6, -3) \cup (3, 7.5)$
e. $f(x)\uparrow$ for $x \in (-3, 3) \cup (7.5, 10)$
f. $f(x) < 0$ for $x \in (-6, 0) \cup (6, 9)$
g. $f(x) > 0$ for $x \in (0, 6) \cup (9, 10)$

MIXED REVIEW

10.

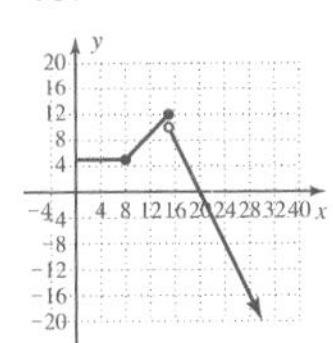

11. cube root family

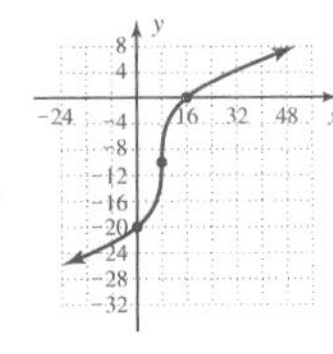

12. quadratic family

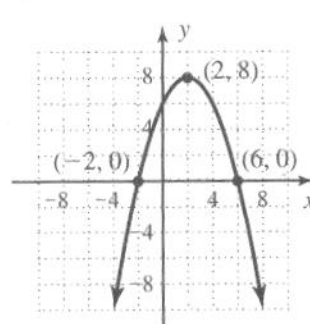

13.
D: $x \in (-\infty, \infty)$; R: $y \in (-\infty, 7]$
$f(x)\uparrow$: $x \in (-\infty, -3) \cup (0, 3)$
$f(x)\downarrow$: $x \in (-3, 0) \cup (3, \infty)$
$f(x)$ constant: none
$f(x) > 0$: $x \in (-4, -1) \cup (1, 5)$
$f(x) < 0$: $x \in (-\infty, -4) \cup (-1, 1) \cup (5, \infty)$
max: $(-3, 4)$ and $(3, 7)$
min: $(0, -2)$

14.
D: $x \in (-\infty, 6)$; R: $y \in (-\infty, 3)$
$g(x)\uparrow$: $x \in (-\infty, -6) \cup (3, 6)$
$g(x)\downarrow$: $x \in (-3, 3)$
$g(x)$ constant: $x \in (-6, -3)$
$g(x) > 0$: $x \in (-7, -1)$
$g(x) < 0$: $x \in (-\infty, -7) \cup (-1, 6)$
max: $y = 3$ for $x \in (-6, -3)$; $(6, 0)$
min: $(3, -3)$

15. **16.**

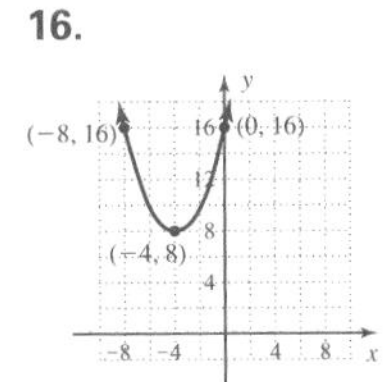

17. zeroes: $(2, 0)$, $(10, 0)$
max: $(15, 10)$
min: $(5, -10)$

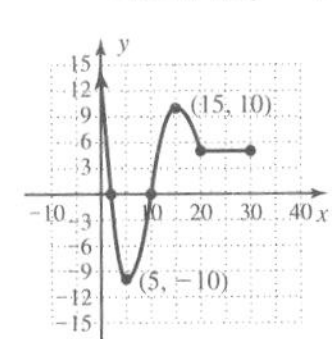

Complete each table by finding the value of k and building the variation equation.

1. y varies inversely as x^2, and $y = \frac{1}{15}$ when $x = 9$.
$k = 5.4$; $y = \dfrac{5.4}{x^2}$

x	y
1	5.4
3	0.6
10	0.054

2. r varies jointly with s^2 and t, with $r = 72$ when $s = 12$ and $t = 8$.
$k = 0.0625$, $r = 0.0625s^2t$

s	t	r
0.125	20	0.02
1	1	0.0625
36	0.5	40.5

Given $f(x) = \dfrac{1}{x-2}$ and $g(x) = x^2 - 2x$ find

3. $(f - g)(-1)$ $\frac{-10}{3}$

4. $(f \cdot g)(x)$ $x, x \neq 2$

5. $\dfrac{g}{f}\left(\dfrac{1}{2}\right)$ $\frac{9}{8}$

6. $(f \circ g)(x)$ $\dfrac{1}{x^2 - 2x - 2}$

7. domain of $(f \circ g)$ $\{x | x \in \mathbb{R}, x \neq 1 \pm \sqrt{3}\}$

8. $f^{-1}(x)$ $\dfrac{1}{x} + 2$

9. Given $h(x) = \sqrt[3]{x^2 + 5}$, find two functions f and g such that $h(x) = (f \circ g)(x)$.
$f(x) = \sqrt[3]{x}$; $g(x) = x^2 + 5$ (others possible)

10. Sketch the function h as defined.

$$h(x) = \begin{cases} 5 & 0 \le x < 8 \\ x - 3 & 8 \le x \le 15 \\ -2x + 40 & x > 15 \end{cases}$$

Identify each function as belonging to the linear, quadratic, square root, cubic, cube root, or absolute value family. Then sketch the graph using shifts of a parent function and a few characteristic points.

11. $f(x) = 5\sqrt[3]{x - 8} - 10$

12. $g(x) = -\frac{1}{2}(x - 2)^2 + 8$

State the domain and range for the graphs of $f(x)$ and $g(x)$ given here. Then state intervals where the functions are increasing or decreasing, and intervals where the functions are positive or negative. Also name any maximum or minimum values. Assume all endpoints have integer values.

13.

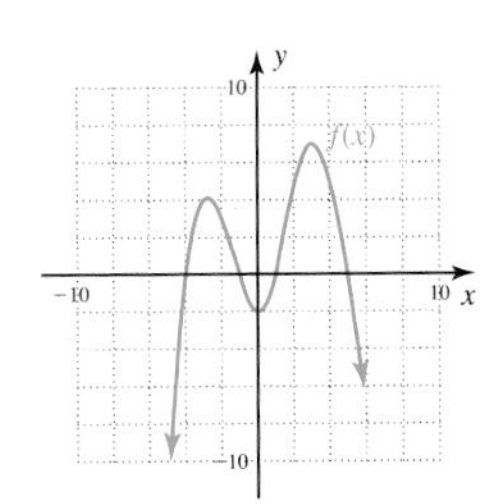

14.

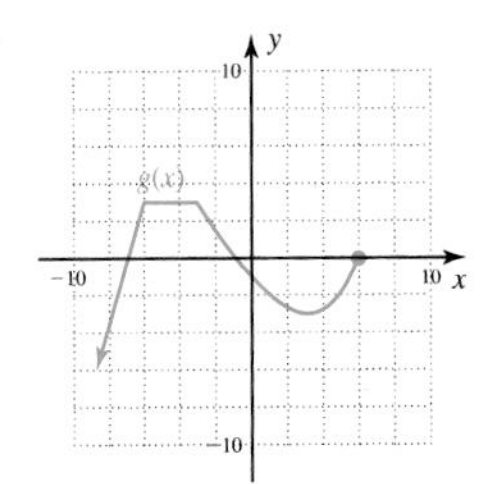

Graph the quadratic functions using the method indicated. Label all characteristic features.

15. Use zeroes and symmetry.
$p(x) = -x^2 + 10x - 16$

16. Complete the square.
$p(x) = \frac{1}{2}x^2 + 4x + 16$

17. Draw a function f that has the following characteristics, then write the zeroes of the function and the location of all maximum and minimum values.

- **a.** domain: $x \in (0, 30]$
- **b.** range: $y \in [-10, \infty)$
- **c.** $f(2) = f(10) = 0$
- **d.** $f(x)\downarrow$ from $x \in (0, 5) \cup (15, 20)$
- **e.** $f(x)\uparrow$ for $x \in (5, 15)$
- **f.** $f(x) < 0$ for $x \in (2, 10)$
- **g.** $f(x) > 0$ for $x \in (0, 2) \cup (10, 30)$
- **h.** $f(x) = 5$ for $x \in [20, 30]$

18. $c(x) = 0.05(6x^2 + 8x)$; \$128
19. $f(x) = -2x^2 + x + 3$
20. $f(x) = -6.96x + 431$; 201,320 in 2008

18. For a rectangular trunk with square ends, the materials needed to cover the exterior cost \$0.05 per square inch. The length of the trunk is 2 in. more than the length of the square ends. Find the cost function for covering the exterior of the trunk and the cost of covering the trunk if $x = 20$ in.

19. Find the equation of the function $f(x)$ whose graph is given.

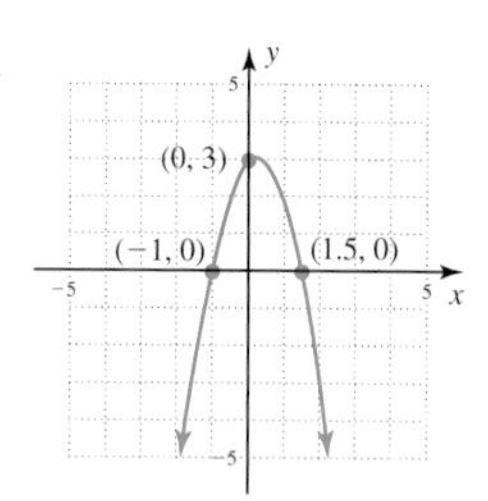

20. Since 1975, the number of deaths in the United States due to heart disease has been declining at a rate that is close to linear. Find an equation model if there were 431 thousand deaths in 1975 and 257 thousand deaths in 2000 (let x represent years since 1975 and $f(x)$ deaths in thousands). How many deaths due to heart disease does the model predict for 2008?

Source: 2004 *Statistical Abstract of the United States*, Table 102

PRACTICE TEST

5.
6.
7.
8.
9.
10.
11. 1; −3; 1
12.

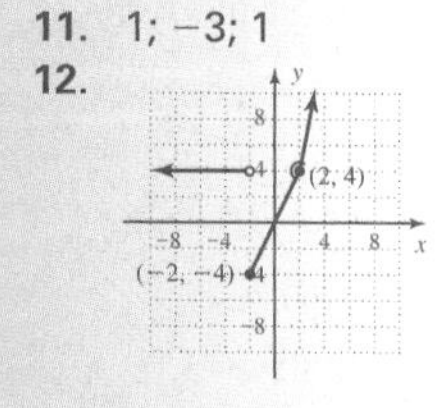

Given $f(x) = 2x + 1$ and $g(x) = x^2 - 3$, $x \geq 0$, determine the following:

1. $(f \cdot g)(3)$ 42

2. $(g \circ f)(a)$ $4a^2 + 4a - 2$

3. the domain for $\left(\frac{f}{g}\right)(x)$ $x \in [0, \sqrt{3}) \cup (\sqrt{3}, \infty)$

4. $f^{-1}(x)$ and $g^{-1}(x)$ $f^{-1}(x) = \frac{x-1}{2}$; $g^{-1}(x) = \sqrt{x+3}$

Sketch each graph using the transformation of a toolbox function.

5. $f(x) = |x - 2| + 3$

6. $g(x) = -(x + 3)^2 - 2$

7. $f(x) = \dfrac{1}{x - 2} + 3$

8. $g(x) = \dfrac{-1}{(x - 3)^2} + 1$

Graph each quadratic function using the indicated method. Label the vertex and intercepts. Round to tenths as needed.

9. Completing the square. $f(x) = 2x^2 + 8x + 3$

10. Using zeroes and symmetry. $f(x) = -x^2 - 4x$

11. Find $f(1), f(-3)$, and $f(5)$ for the piecewise-defined function.

$$f(x) = \begin{cases} 2x + 3 & x < 0 \\ x^2 & 0 \leq x < 2 \\ 1 & x \geq 2 \end{cases}$$

12. Sketch the graph of the piecewise-defined function. Label important points.

$$h(x) = \begin{cases} 4 & x < -2 \\ 2x & -2 \leq x \leq 2 \\ x^2 & x > 2 \end{cases}$$

For Exercises 13 and 14, determine the following for each graph.

a. the domain and range

b. estimate the value of $f(-1)$

c. the zeroes of the function

d. interval(s) where $f(x)$ is negative or positive

e. location of any maximum or minimum values

f. interval(s) where $f(x)$ is increasing, decreasing, or constant

g. the equation for $f(x)$

13.
a. $D: x \in (-\infty, \infty)$; $R: y \in (-\infty, 4]$
b. $f(-1) = 4$
c. $(-4, 0)$ and $(2, 0)$
d. $f(x) < 0: x \in (-\infty, -4) \cup (2, \infty)$
$f(x) > 0: x \in (-4, 2)$
e. max: $(-1, 4)$; min: none
f. $f(x)\uparrow: x \in (-\infty, -1)$
$f(x)\downarrow: x \in (-1, \infty)$
g. $f(x) = -\frac{4}{9}(x + 1)^2 + 4$

14. a. $D: x \in (-4, \infty)$;
$R: y \in (-3, \infty)$
b. $f(-1) \approx 2.2$
c. $(-3, 0)$
d. $f(x) < 0: x \in (-4, -3)$
$f(x) > 0: x \in (-3, \infty)$
e. max: none; min: $(-4, -3)$
f. $f(x)\uparrow: x \in (-3, \infty)$
$f(x)\downarrow$: none
g. $f(x) = 3\sqrt{x + 4} - 3$

15. a. $V(t) = \frac{4}{3}\pi(\sqrt{t})^3$ b. 36π in^3
16. a. b.

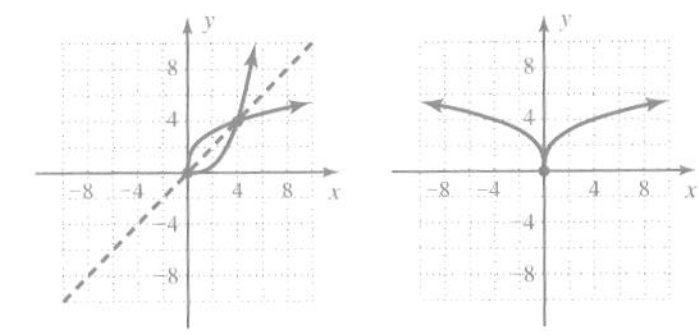

16. c.

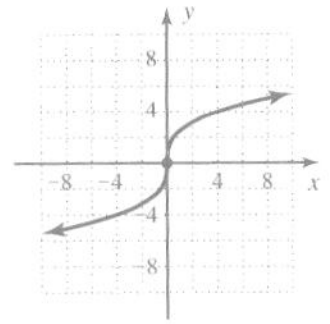

17. a. Yes, passes vertical line test.
b. Yes, passes horizontal line test.
c. odd
d. (6, 4)
$(-4, -2)$
18. a. $g(x)$
b. $g(x)$
19. a. 40 ft, 48 ft
b. 49 ft
c. 14 sec
20. 520 lb

13.

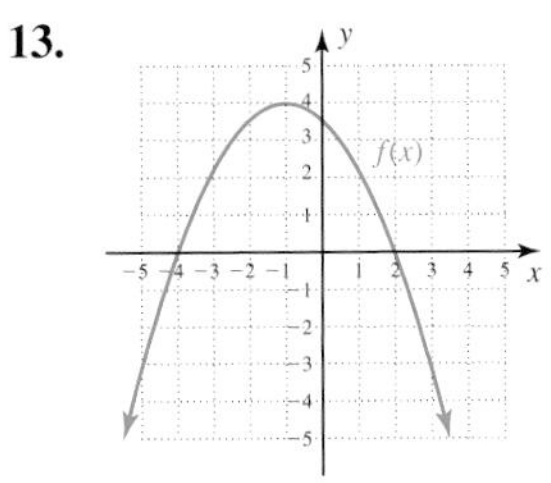

14.

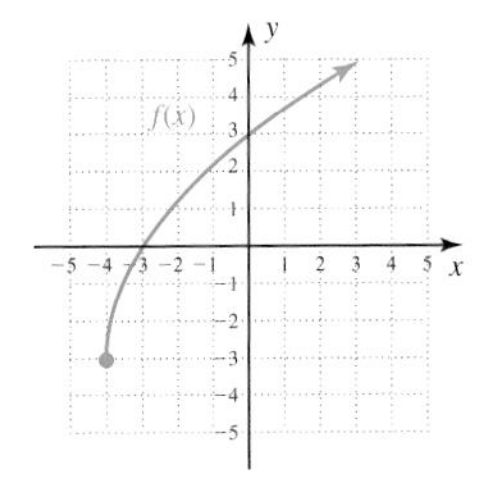

15. A snowball increases in size as it rolls downhill. The snowball is roughly spherical with a radius that can be modeled by the function $r(t) = \sqrt{t}$, where t is time in seconds and r is measured in inches. The volume of the snowball is given by the function $V(r) = \frac{4}{3}\pi r^3$. Use a composition to (a) write V directly as a function of t and (b) find the volume of the snowball after 9 sec.

16. For the function $h(x)$ whose graph is given, (a) draw the inverse function, (b) complete the graph if h is known to be even and only half the graph is given, and (c) complete the graph if h is known to be odd and only half the graph is given.

Exercise 16

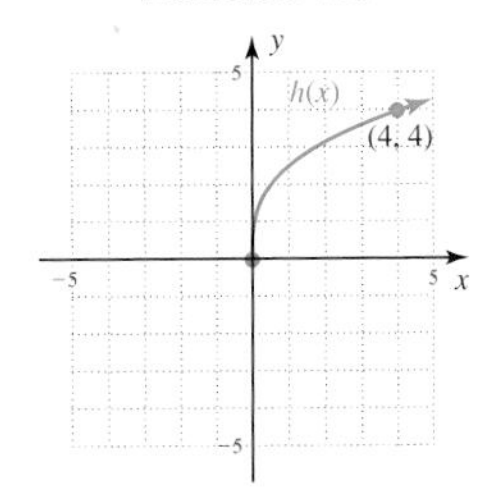

Exercise 17

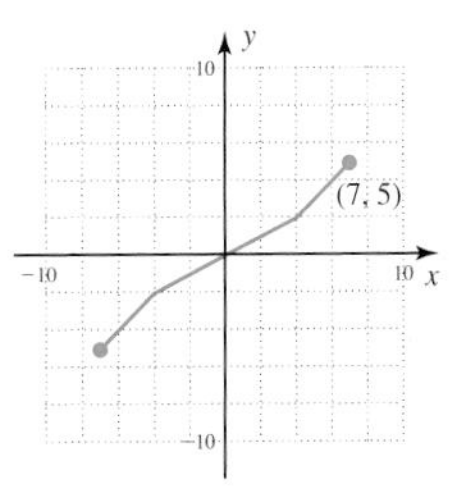

17. Answer the following given the graph of the general relation shown.

a. Is the relation shown a function? Discuss why or why not.

b. Is the relation shown one-to-one? Discuss why or why not.

c. Is the relation even, odd, or neither?

d. Complete the ordered pairs: (___, 4) and $(-4$, ___).

18. Given $f(x) = \frac{1}{x}$ and $g(x) = \frac{1}{x^2}$, use the formula for average rate of change to determine which of these functions is decreasing faster in the intervals: (a) [0.5, 0.6] and (b) [1.5, 1.6].

19. Suppose the function $d(t) = t^2 - 14t$ models the depth of a scuba diver at time t, as she dives underwater from a steep shoreline, reaches a certain depth, and swims back to the surface. Use the function to answer the following questions.

a. What is her depth after 4 sec? After 6 sec?

b. What was the maximum depth of this dive?

c. How many seconds was the diver beneath the surface?

20. The maximum load that can be supported by a rectangular beam varies jointly with its width and its height squared, and inversely with its length. If a beam 10 ft long, 3 in. wide, and 4 in. high can support 624 lb, how many pounds could a beam support with the same dimensions but 12 ft long?

CALCULATOR EXPLORATION AND DISCOVERY

Residuals, Correlation Coefficients, and Goodness of Fit

When using technology to calculate a regression equation, we must avoid relying on the correlation coefficient as the sole indicator of how well a model fits the data. It is actually possible for a regression to have a high r-value (correlation coefficient) but fit the data very poorly. In addition, regression models are often used to predict future values, extrapolating well beyond the given set of data. Sometimes the model fails miserably when extended—even when it fits the data on the specified interval very well. This fact highlights (1) the importance of studying the behavior of the toolbox functions and other graphs, as we often need to choose between two models that seem to fit the given data; (2) the need to consider the context of the data; and (3) the need for an additional means to evaluate the "goodness of fit." For the third item, we investigate something called a **residual.** As the name implies, we are interested in the difference between the outputs generated by the equation model, and the actual data: equation value − data value = residual. Residuals that are fairly random and scattered indicate the equation model has done a good job of capturing the curvature of the data. If the residuals exhibit a detectable pattern of some sort, this often indicates trends in the data that the equation model did not account for. As a simplistic illustration, enter the data from Table 3.6 in L1 and L2, then graph the scatter-plot. The TI-84 Plus offers a window option specifically designed for scatter-plots that will plot the points in an "ideal" window: ZOOM **9:ZoomStat** (Figure 3.89). Upon inspection, it appears the data could be linear with positive slope, quadratic with $a > 0$, or some other toolbox function with increasing behavior. Many of these forms of regression will *give a correlation coefficient in the high 90s*. This is where an awareness of the context is important. (1) Do we expect the data to increase steadily over time (linear data)? (2) Do we expect the rate of change (the growth rate) to increase over time (quadratic data)? At what rate will values increase? (3) Do we expect the growth to increase dramatically with time (power or exponential regression)? Running a linear regression **(LinReg L1, L2, Y1)** gives the equation in Figure 3.90, with a very high r-value. This model appears to fit the data very well, and will reasonably approximate the data points within the interval. But could we use this model to accurately predict future values (do we expect the outputs to grow indefinitely at a linear rate)? In some cases, the answer will be clear from the context. Other times the decision is more difficult and a study of the residuals can help. The TI-84 Plus provides a residual function [2nd STAT (LIST)], but to help you understand more exactly what residuals are, for now we'll calculate them via function values. For this calculation, we go to the STAT ENTER (EDIT) screen. In the header of L3 (List3), input **Y1(L1) − L2** ENTER (Figure 3.91), which will evaluate the function at the input values listed in L1, compute the difference between these outputs and the data in L2, and place the results in L3. Scrolling through the residuals reveals a distinct lack of randomness, as there is a large interval where outputs are continuously positive (Figure 3.92).

Table 3.6

x	y
5	19
7.5	75
10	140
12.5	215
15	297
17.5	387
20	490

Figure 3.89

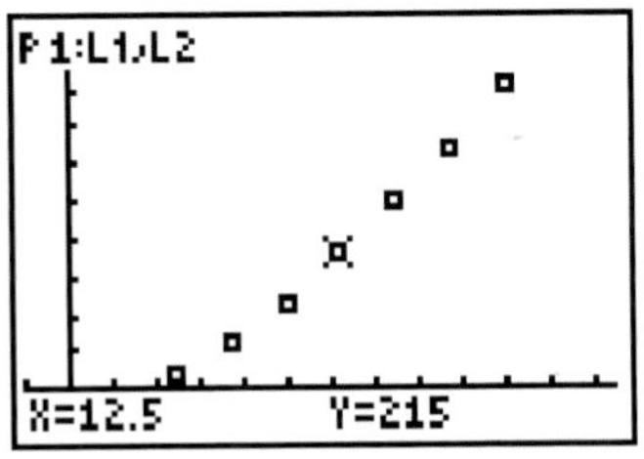

P1:L1,L2
X=12.5 Y=215

Figure 3.90

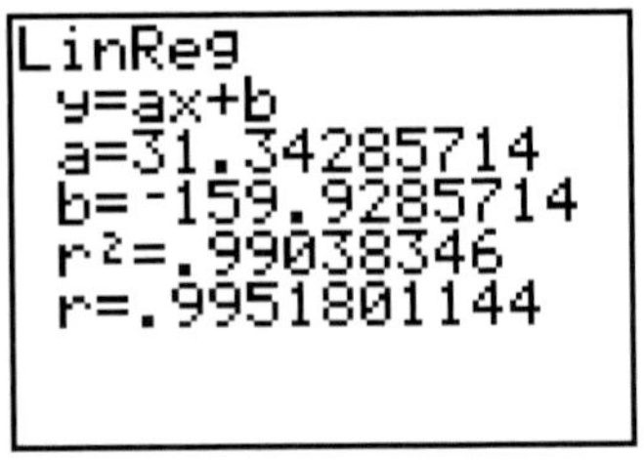

LinReg
y=ax+b
a=31.34285714
b=−159.9285714
r²=.99038346
r=.9951801144

Figure 3.91

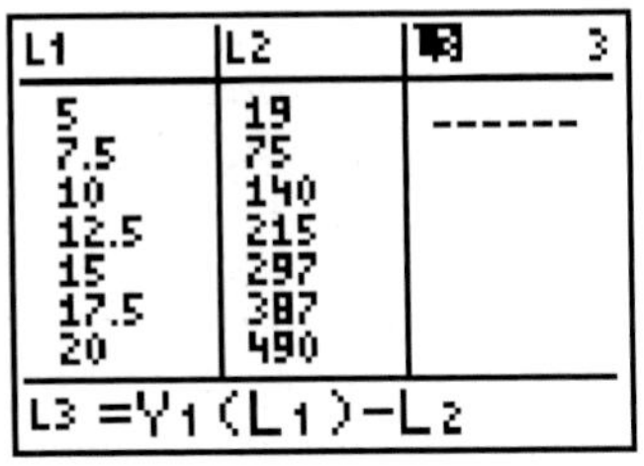

L1	L2	L3
5	19	------
7.5	75	
10	140	
12.5	215	
15	297	
17.5	387	
20	490	

L3 =Y1(L1)−L2

Figure 3.92

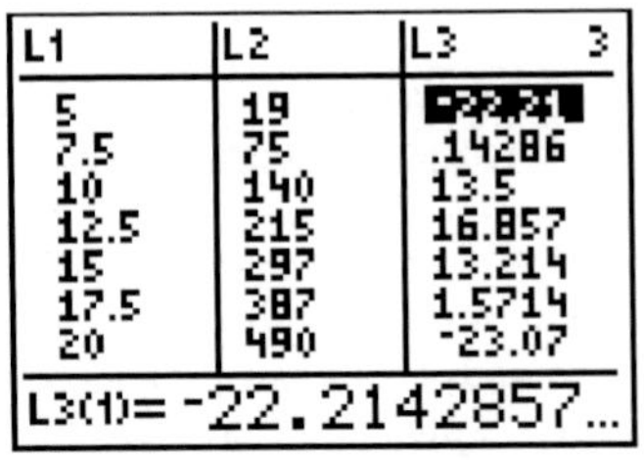

L1	L2	L3
5	19	−22.21
7.5	75	.14286
10	140	13.5
12.5	215	16.857
15	297	13.214
17.5	387	1.5714
20	490	−23.07

L3(1)=−22.2142857...

Figure 3.93

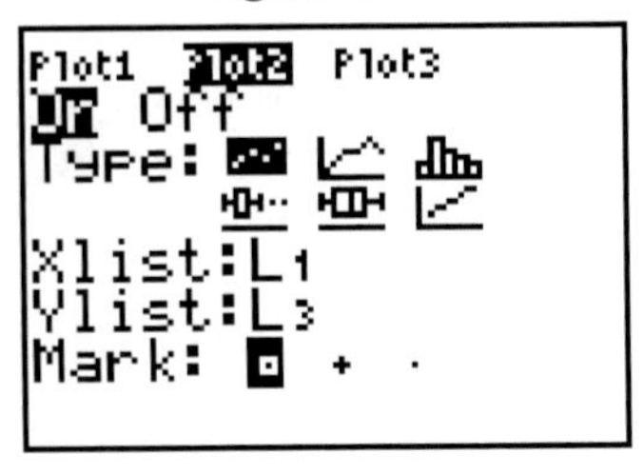

We can also analyze the residuals *graphically* by going to the 2nd Y= (STATPLOT) screen to activate **2:PLOT2,** setting it up to recognize L1 and L3 as the **XList** and **YList** respectively (Figure 3.93). After deactivating all other plots and functions, pressing ZOOM 9:**ZoomStat** gives Figure 3.94 shown, with the residuals following a definite (quadratic) pattern. Performing the same sequence of steps using quadratic regression results in a higher correlation coefficient and an increased randomness in residuals (Figures 3.95 and 3.96), with the residuals appearing to increase over time.

Figure 3.94

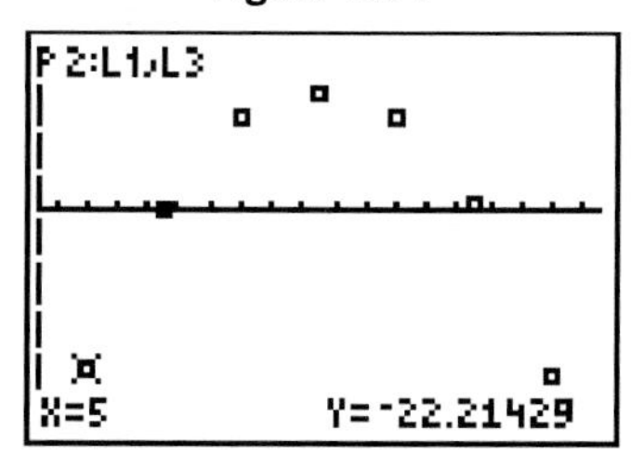

Figure 3.95

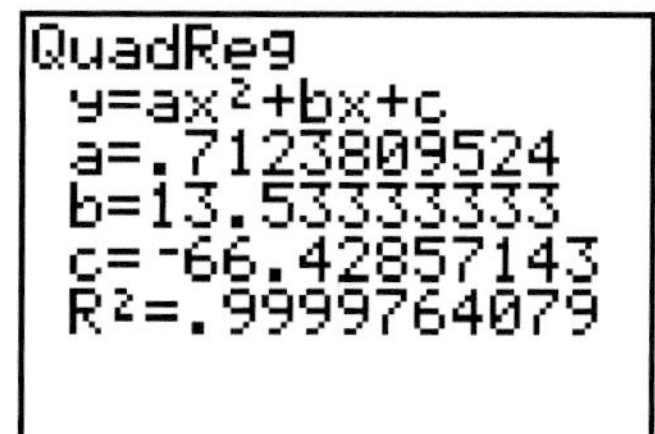

Figure 3.96

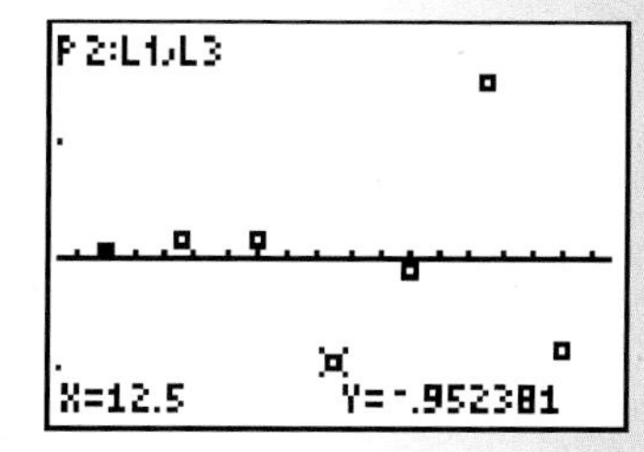

Exercise 1. a. linear: $r \approx 0.99$; residuals form a quadratic pattern (not random); quadratic: $r \approx 0.997$, residuals appear random, we do not expect time to begin decreasing; power: $r \approx 0.999$, residuals appear random, context suggests a power function.
b. 2.01 sec, 2.63 sec
c. 59 cm, 89 cm

Exercise 1: As part of a science lab, students are asked to determine the relationship between the length of a pendulum and the time it takes to complete one back-and-forth cycle, called its *period.* They tie a 500-g weight to 10 different lengths of string, suspend them from a doorway, and collect the data shown in Table 3.7.

a. Use a combination of the context, the correlation coefficient, and an analysis of the residuals to determine whether a linear, quadratic, or power model is most appropriate for the data. State the r-value of each regression and justify your final choice of equation model.

b. According to the data, what would be the period of a pendulum with a 90-cm length? A 150-cm length?

c. If the period was 1.6 sec, how long was the pendulum? If the period were 2 sec, how long was the pendulum?

Table 3.7

Length (cm)	Time (sec)
12	0.7
20	0.9
28	1.09
36	1.24
44	1.35
52	1.55
60	1.64
68	1.73
76	1.80
84	1.95

STRENGTHENING CORE SKILLS

Base Functions and Quadratic Graphs

Certain transformations of quadratic graphs offer an intriguing alternative to graphing these functions by completing the square. In many cases, the process is less time consuming and ties together a number of basic concepts. To begin, we note that for $f(x) = ax^2 + bx + c$, $F(x) = ax^2 + bx$ is called the **base function** or *the original function less the constant term.* By comparing $f(x)$ with $F(x)$, four things are immediately apparent: (1) F and f share the same axis of symmetry since one is a vertical shift of the other; (2) the x-intercepts of F can be found by factoring; (3) the axis of symmetry is simply the

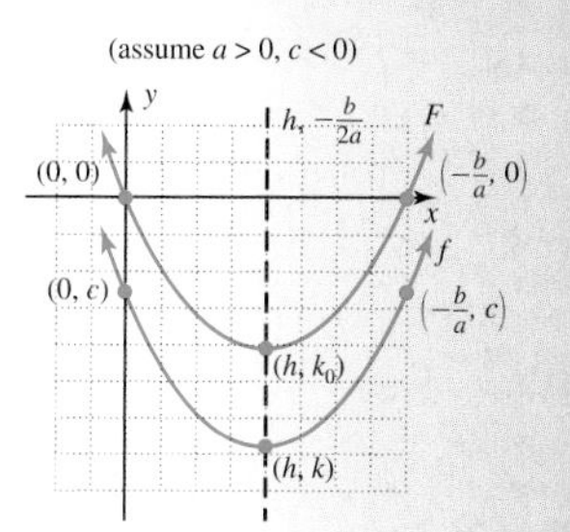

average value of the x-intercepts; $h = \frac{x_1 + x_2}{2}$; and (4) the vertices of F and f differ only by the constant c. Consider these vertices to be (h, k_0) and (h, k), respectively, with $k = k_0 + c$. Knowing the vertex of *any* parabola is $\left(\frac{-b}{2a}, f\left(\frac{-b}{2b}\right)\right)$, we evaluate the base function at $h = \frac{x_1 + x_2}{2} = \frac{-b}{2a}$ and find that for the base function, $F\left(-\frac{b}{2a}\right) = -ah^2$:

$$F(x) = ax^2 + bx \quad \text{original function}$$

$$F\left(-\frac{b}{2a}\right) = a\left(-\frac{b}{2a}\right)^2 + b\left(-\frac{b}{2a}\right) \quad \text{substitute } -\frac{b}{2a} \text{ for } x$$

$$= \frac{b^2}{4a} - \frac{b^2}{2a} = \frac{b^2 - 2b^2}{4a} \text{ or } \frac{-b^2}{4a} \quad \text{multiply and combine terms}$$

$$= \frac{-b^2}{4a} \cdot \frac{a}{a} \quad \text{multiply by } \frac{a}{a}$$

$$= -a\left(\frac{b}{2a}\right)^2 \quad \text{rearrange factors}$$

From $h = -\frac{b}{2a}$, we have $-h = \frac{b}{2a}$ and it follows that

$$F\left(-\frac{b}{2a}\right) = -a(-h)^2 \quad \text{substitute } -h \text{ for } \frac{b}{2a}$$

$$= -ah^2 \quad (-h)^2 = h^2$$

This verifies the vertex of F is (h, k_0), where $k_0 = -ah^2$.

It's significant to note that the vertex of both $F(x)$ and $f(x)$ can now be determined *using only elementary operations on the single value h*, since $k_0 = -ah^2$ and $k = k_0 + c$. And since the vertex of f is known, the vertex/intercept formula (Section 3.4) can be used to find theroots of f with no further calculations: $x = h \pm \sqrt{-\frac{k}{a}}$. Finally, this approach enables easy access to the exact form of the roots, even when they happen to be irrational or complex (no quadratic formula needed). Several examples follow, with the actual graphs left to the student—only the process is illustrated here.

ILLUSTRATION 1 Graph $f(x) = x^2 - 10x + 17$ and locate its zeroes (if they exist).

Solution: For $F(x) = x^2 - 10x$, the zeroes/x-intercepts are (0, 0) and (10, 0) by inspection, with $h = 5$ (halfway point) as the axis of symmetry. Noting $a = 1$ and $c = 17$, the vertex of F is at $(h, -ah^2)$ or $(5, -25)$. After adding 17 units to the y-coordinates of the points from F, we find the y-intercept for f is (0, 17), its "symmetric point" is (10, 17), and the vertex is at $(5, -8)$. The x-intercepts of f are $(h \pm \sqrt{-k}, 0)$ or $(5 \pm \sqrt{8}, 0)$.✓

If "b" is an odd number, we compute $\left(\frac{|b|}{2}\right)^2$ and convert the result to decimal form. For instance, if $b = -7$, $\left(\frac{7}{2}\right)^2 = \frac{49}{4}$ or 12.25.

ILLUSTRATION 2 Graph $f(x) = x^2 + 13x - 15$ and locate its zeroes (if they exist).

Solution: For $F(x) = x^2 + 13x$, the zeroes are (0, 0) and $(-13, 0)$ by inspection, with $h = \frac{-13}{2}$ as the axis of symmetry. Noting $a = 1$, $c = -15$, and $(\frac{13}{2})^2 = \frac{169}{4}$ or 42.25, the vertex of F is $(-6.5, -42.25)$. After subtracting 15 units from the y-coordinates of the points from F, we find the y-intercept for f is

1.

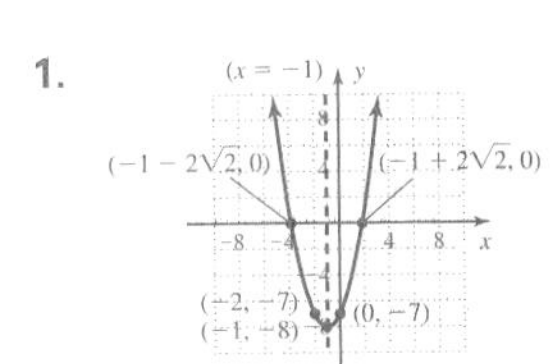

2.

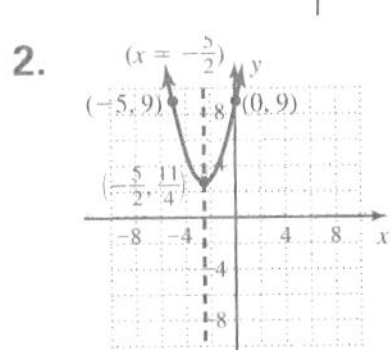

3.

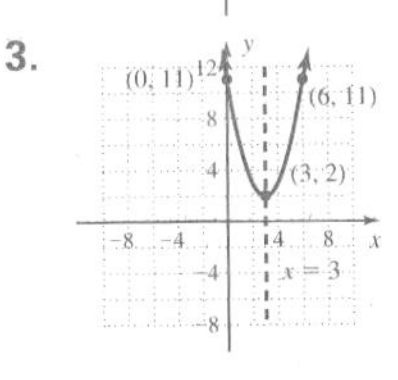

Additional answers can be found in the Instructor Answer Appendix.

(0, −15), its "symmetric point" is (−13, −15), and the vertex is at (−6.5, −57.25). The x-intercepts of f are $(-6.5 \pm \sqrt{57.25}, 0)$.✓

Even when $a \neq 1$ the method lends a measure of efficiency to graphing quadratic functions, as shown in Illustration 3.

ILLUSTRATION 3 Graph $f(x) = -2x^2 + 5x - 4$ and locate its zeroes (if they exist).

Solution: For $F(x) = -2x^2 + 5x$, the zeroes are (0, 0) and $(\frac{5}{2}, 0)$ by inspection, with $h = \frac{5}{4}$ as the halfway point and axis of symmetry. Noting $a = -2$ and $c = -4$, the vertex of F is at $(\frac{5}{4}, \frac{25}{8})$. After subtracting $4 = \frac{32}{8}$ units from the y-coordinates of the points from F, we find the y-intercept for f is $(0, -4)$, its "symmetric point" is $(\frac{5}{2}, -4)$, and the vertex is at $(\frac{5}{4}, \frac{-7}{8})$. The roots of f are $x = \frac{5}{4} \pm \sqrt{\frac{-7}{16}} = \frac{5}{4} \pm \frac{\sqrt{7}}{4}i$✓, showing the graph has no x-intercepts.

Use this method for graphing quadratic functions to sketch a complete graph of the following functions. Find and clearly indicate the axis of symmetry, vertex, x-intercept(s), and the y-intercepts along with its "symmetric point."

Exercise 1: $f(x) = x^2 + 2x - 7$

Exercise 2: $g(x) = x^2 + 5x + 9$

Exercise 3: $h(x) = x^2 - 6x + 11$

Exercise 4: $H(x) = -x^2 + 10x - 17$

Exercise 5: $p(x) = 2x^2 + 12x + 21$

Exercise 6: $q(x) = 2x^2 - 7x + 8$

CUMULATIVE REVIEW CHAPTERS 1-3

1. $x^2 + 2$
2. a. $8\sqrt{2}$ b. $\frac{-23y - 11}{(5y + 1)(y + 2)(y - 3)}$
3. 29.45 cm
4. $r = \frac{-\pi h \pm \sqrt{\pi^2 h^2 + 2\pi A}}{2\pi}$
5. $x = \frac{-7 \pm \sqrt{89}}{4}$
6. $\frac{4}{9}$
7. a. $\frac{-1}{3}$ b. $\frac{3}{5}$
8. a. b.

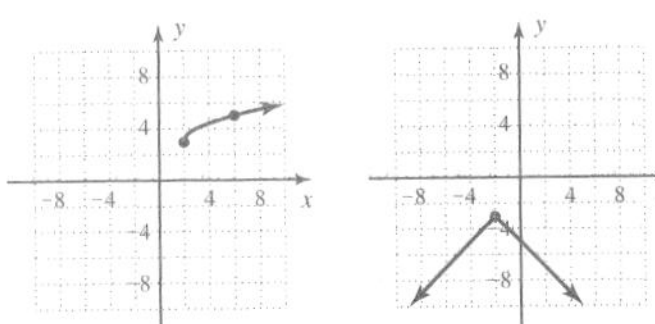

Additional answers can be found in the Instructor Answer Appendix.

1. Perform the division by factoring the numerator: $(x^3 - 5x^2 + 2x - 10) \div (x - 5)$.
2. Simplify the following expressions: **a.** $\sqrt{18} + \sqrt{50}$ **b.** $\frac{2}{5y^2 + 11y + 2} - \frac{5}{y^2 - y - 6}$
3. The area of a circle is 69 cm². Find the circumference of the same circle.
4. The surface area of a cylinder is $A = 2\pi r^2 + 2\pi rh$. Write r in terms of A and h (solve for r).
5. Find the roots of $h(x) = 2x^2 + 7x - 5$.
6. Evaluate without using a calculator: $\left(\frac{27}{8}\right)^{\frac{-2}{3}}$.
7. Find the slope of each line:
   a. through the points: (−4, 7) and (2, 5)
   b. line with equation $3x - 5y = 20$
8. Graph using transformations of a parent function:
   a. $f(x) = \sqrt{x - 2} + 3$
   b. $f(x) = -|x + 2| - 3$
9. Graph the line passing through (−3, 2) with a slope of $m = \frac{1}{2}$, then state its equation.
10. Use a substitution to determine if $x = 2 + 3i$ is a solution to $x^2 + 4x + 13 = 0$.
11. Graph the quadratic by completing the square: $g(x) = x^2 - 4x - 5$.
12. Solve the quadratic inequality: $x^2 - 7x + 6 \leq 0$.
13. Given $f(x) = 3x^2 - 6x$ and $g(x) = x - 2$ find: $(f \cdot g)(x)$, $(f \div g)(x)$, and $(g \circ f)(-2)$.
14. Given $f(x) = \frac{3}{5}x - 4$, find the inverse function $f^{-1}(x)$ and graph them both on the same grid.
15. Given $g(x) = \frac{1}{x - 2} + 1$:
   a. Find the x- and y-intercepts for g
   b. Sketch the graph of the function

16. a. D: $x \in (-\infty, 8]$, R: $y \in [-4, \infty)$
b. 5, −3, −3, 1, 2
c. (−2, 0)
d. $f(x) < 0$: $x \in (-2, 2)$
$f(x) > 0$: $x \in (-\infty, -2) \cup [2, 8]$
e. min: (0, −4), max: (8, 7)
f. $f(x)\uparrow$: $x \in (0, 8)$
$f(x)\downarrow$: $x \in (-\infty, 0)$

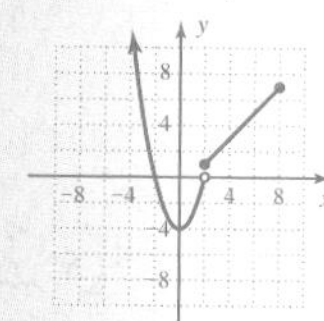

17. a. $f(x)$
b. $g(x)$
18. \$0.83
19. $y = 0.42x + 0.81$, about 43 ppsi
20. No, Raphael corresponds to the School of Athens and Parnassus; Michelangelo corresponds to no element of the second set.

16. Graph the piecewise-defined function $f(x) = \begin{cases} x^2 - 4 & x < 2 \\ x - 1 & 2 \le x \le 8 \end{cases}$ and determine the following:

a. the domain and range

b. the value of $f(-3), f(-1), f(1), f(2)$, and $f(3)$

c. the zeroes of the function

d. interval(s) where $f(x)$ is negative or positive

e. location of any maximum or minimum values

f. interval(s) where $f(x)$ is increasing, decreasing, or constant

17. Given $f(x) = x^2$ and $g(x) = x^3$, use the formula for average rate of change to determine which of these functions is increasing faster in the intervals: (a) [0.5, 0.6], (b) [1.5, 1.6].

18. The cost of PCV piping varies directly with the length and diameter of the pipe. A pipe 6 cm in diameter and 3 m long costs \$4.98. How much would a 2-m length of 1.5-cm-diameter pipe cost?

19. Find an appropriate regression equation for the data shown in the table.

Depth (ft)	Water Pressure (ppsi)
15	6.94
25	11.85
35	15.64
45	19.58
55	24.35
65	28.27
75	32.68

What water pressure can be expected at a depth of 100 ft?

20. Determine if the following relation is a function. If not, how is the definition of a function violated?

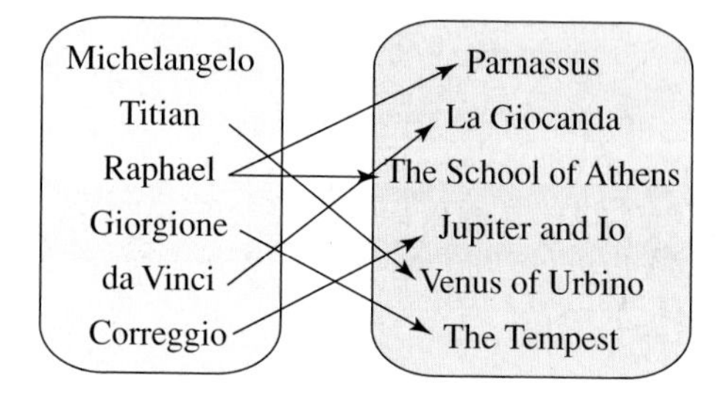

Chapter 4

Polynomial and Rational Functions

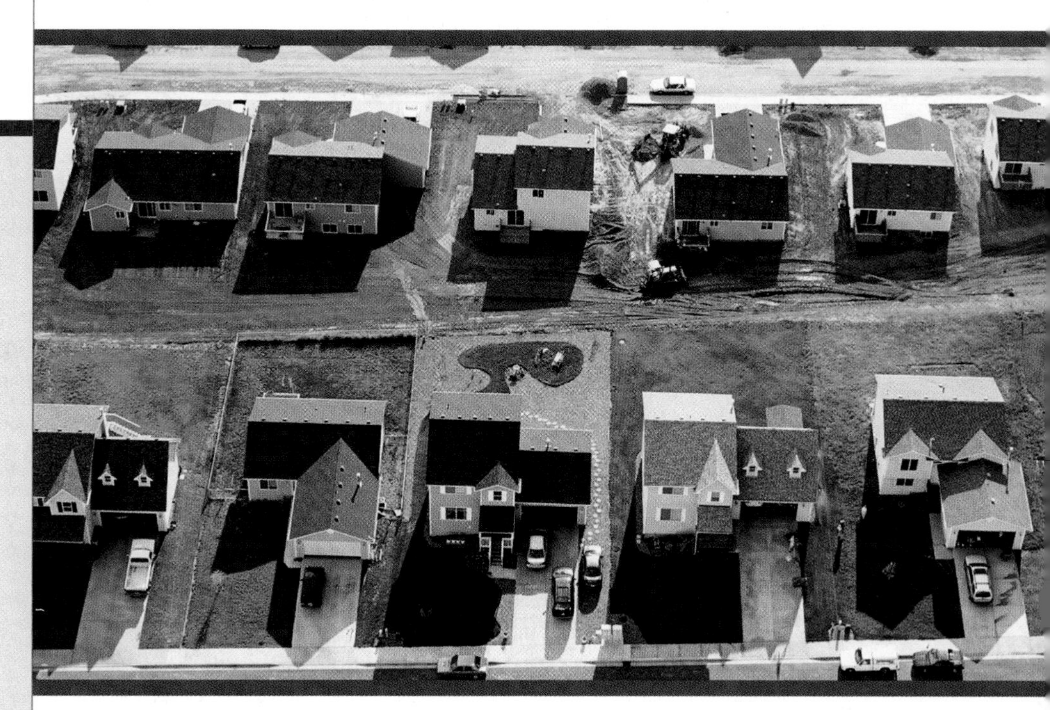

Chapter Outline

4.1 Polynomial Long Division and Synthetic Division 374

4.2 The Remainder and Factor Theorems 383

4.3 The Zeroes of Polynomial Functions 393

4.4 Graphing Polynomial Functions 407

4.5 Graphing Rational Functions 422

4.6 Additional Insights into Rational Functions 438

4.7 Polynomial and Rational Inequalities—An Analytical View 451

Preview

This chapter will bring together many of the concepts that were developed in previous chapters. The connections are numerous and will lead to additional discoveries as we strengthen the ability to use mathematics as a problem-solving tool. As these abilities grow, mathematics becomes a true resource for modeling innumerable relationships in the real world, and for understanding how and why they exist.

4.1 Polynomial Long Division and Synthetic Division

LEARNING OBJECTIVES

In Section 4.1 you will review/learn how to:

A. Divide polynomials using long division

B. Divide polynomials using synthetic division

C. Use synthetic division to factor certain polynomials

INTRODUCTION

The quotient of two polynomials has a number of surprising applications in a study of algebra, and plays a significant role in the introduction to polynomial and rational graphs. In this section we'll review polynomial long division and introduce a technique called synthetic division, which greatly decreases the time needed to complete the division algorithm (algorithm: a well-defined sequence of steps that can be repeated until a task is complete). This enables us to use division more effectively as a mathematical tool.

POINT OF INTEREST

One of the earliest known records of synthetic division is from the work *Libro de Algebra en Arithmetica y Geometria,* written by Pedro Nuñez in 1567. It's interesting to note that in a time when negative numbers were still suspect, Nuñez demonstrated the need for them to complete the division algorithm.

A. The Quotient of a Polynomial and a Binomial

Recall from whole number division that for $\frac{56}{17} = 3$ remainder 5, 56 is called the **dividend,** 17 is the **divisor,** 3 is the **quotient,** and 5 is the **remainder.** The result is written in one of two ways: (1) $\frac{56}{17} = 3 + \frac{5}{17} = 3\frac{5}{17}$ or (2) $56 = 17(3) + 5$. The quotient $\frac{x^2 - 3x - 10}{x - 5}$ can be computed by factoring and yields $\frac{x^2 - 3x - 10}{x - 5} = x + 2$ remainder 0. The expression $x^2 - 3x - 10$ is the dividend, $x - 5$ is the divisor, $x + 2$ is the quotient, and 0 is the remainder. This result can be written in the same two ways:

$$1.\quad \frac{x^2 - 3x - 10}{x - 5} = (x + 2) + \frac{0}{x - 5}$$

$$= (\text{quotient}) + \frac{\text{remainder}}{\text{divisor}}$$

$$2.\quad x^2 - 3x - 10 = (x + 2)(x - 5) + 0$$

$$\text{dividend} = (\text{quotient})(\text{divisor}) + \text{remainder}$$

In general, the division algorithm for polynomials says:

DIVISION OF POLYNOMIALS

Given $p(x)$ and $d(x) \neq 0$ are polynomials, there exist unique polynomials $q(x)$ and $r(x)$ such that

$$\frac{p(x)}{d(x)} = q(x) + \frac{r(x)}{d(x)} \quad \text{or} \quad p(x) = d(x)q(x) + r(x),$$

where either $r(x) = 0$ or the degree of $r(x)$ is less than the degree of $d(x)$.

When the division cannot be completed by factoring, polynomial long division is used and closely resembles whole number division. The main difference is that *we group each partial product* in parentheses to prevent errors in subtraction.

In the division process, zero "place holders" are sometimes used to ensure that like place values will "line up" as we carry out the algorithm.

EXAMPLE 1 Find the quotient of $8x^3 - 27$ and $2x - 3$.

Solution: Write the dividend as $8x^3 + 0x^2 + 0x - 27$. To find our first multiplier, we use the quotient of leading terms from each expression: $\frac{8x^3 \text{ from dividend}}{2x \text{ from divisor}} = 4x^2$. This shows $4x^2$ will be our first multiplier.

$$\begin{array}{rrl}
 & 4x^2 + 6x + 9 & \leftarrow \text{quotient} \\
\text{divisor} \rightarrow 2x - 3 & \overline{)8x^3 + 0x^2 + 0x - 27} & \leftarrow \text{dividend} \\
 & -(8x^3 - 12x^2) \qquad\qquad & 4x^2(2x - 3) = 8x^3 - 12x^2 \\
 & 12x^2 + 0x \qquad & \text{subtract, bring down next term} \\
\text{find next multiplier: } \frac{12x^2}{2x} = 6x & -(12x^2 - 18x) \qquad & 6x(2x - 3) = 12x^2 - 18x \\
 & 18x - 27 & \text{subtract, bring down next term} \\
\text{find next multiplier: } \frac{18x}{2x} = 9 & -(18x - 27) & 9(2x - 3) = 18x - 27 \\
 & 0 & \leftarrow \text{remainder}
\end{array}$$

This shows $(8x^3 - 27) \div (2x - 3) = (4x^2 + 6x + 9) + 0$. A check by multiplication verifies the result, as does a check by factoring $8x^3 - 27$ as the difference of two perfect cubes.

NOW TRY EXERCISES 7 THROUGH 24

It's helpful to note that when a higher degree polynomial is divided by a linear (degree one) polynomial, the degree of the quotient will always be one less than the degree of the dividend. We see this in Example 1, where the dividend had a degree of three $(8x^3 - 27)$, the divisor had a degree of one $(2x^1 - 3)$, and the quotient had a degree of two $(4x^2 + 6x + 9)$. In general, if the dividend has degree n and the divisor is linear, the quotient will have degree $n - 1$. In the most simplistic and illustrative case: $\frac{x^n}{x^1} = x^{n-1}$ using the quotient rule. For division by higher degree polynomials, the process is markedly similar.

B. Synthetic Division

Recall that if one number divides evenly into another (no remainder), it must be a factor of the original number. Since $\frac{51}{17} = 3$, we know that $3 \cdot 17 = 51$. The same idea holds for polynomials. In Example 1, $(8x^3 - 27) \div (2x - 3) = 4x^2 + 6x + 9$ (no remainder), showing $(2x - 3)(4x^2 + 6x + 9) = 8x^3 - 27$. *This means division can be used as a tool to factor larger polynomials*. But to make it an *effective* tool, we need two things: (1) a more efficient method of dividing polynomials and (2) a way to find divisors that give a remainder of zero. The second need is addressed in Part C. For the first we investigate a process called **synthetic division.** Consider the quotient $(x^3 - 2x^2 - 13x - 17) \div (x - 5)$ and the following comparison. On the left, the long division algorithm is fully illustrated. The same division is shown to the right, but in abbreviated form. Note the terms in red are identical to those directly above them, and have been eliminated on the right.

$$\begin{array}{r}
x^2 + 3x + 2 \\
x - 5\overline{)x^3 - 2x^2 - 13x - 17} \\
\underline{-(x^3 - 5x^2)} \downarrow \quad \\
3x^2 - 13x \quad \\
\underline{-(3x^2 - 15x)} \downarrow \\
2x - 17 \\
\underline{-(2x - 10)} \\
-7 \text{ remainder}
\end{array}
\qquad
\begin{array}{r}
x^2 + 3x + 2 \\
x - 5\overline{)x^3 - 2x^2 - 13x - 17} \\
\underline{-(\quad - 5x^2)} \quad \\
3x^2 \quad \\
\underline{-(\quad - 15x)} \\
2x \\
\underline{-(\quad - 10)} \\
-7 \text{ remainder}
\end{array}$$

Also, since the dividend must be written in standard form (decreasing order of degree with place holder zeroes as needed), only the coefficients are actually required to perform the calculations since like place values are automatically aligned (see the next illustration on the left). We can further simplify the process by applying the "double negative rule" or "distributing the negative." In effect, we're using algebraic addition—adding opposites instead of subtracting.

$$\begin{array}{r}
1 \quad 3 \quad 2 \\
x - 5\overline{)1 \quad -2 \quad -13 \quad -17} \\
\underline{-(\quad - 5)} \qquad\qquad \\
3 \qquad\qquad \\
\underline{-(\quad - 15)} \qquad \\
2 \qquad \\
\underline{-(\quad - 10)} \\
-7 \text{ remainder}
\end{array}
\qquad
\begin{array}{r}
1 \quad 3 \quad 2 \\
x - 5\overline{)1 \quad -2 \quad -13 \quad -17} \\
\underline{5} \qquad\qquad \\
3 \qquad\qquad \\
\underline{15} \qquad \\
2 \qquad \\
\underline{10} \\
-7 \text{ remainder}
\end{array}$$

The entire process can be condensed by vertically compressing the rows of the division so that a minimum of space is used (see below and left).

$$\begin{array}{rrrrrl}
 & & 1 & 3 & 2 & \text{quotient} \\
x - 5) & 1 & -2 & -13 & -17 & \text{dividend} \\
 & & 5 & 15 & 10 & \text{products} \\
\hline
 & & 3 & 2 & 7 & \text{sums}
\end{array}
\qquad
\begin{array}{rrrrrl}
 & & 1 & 3 & 2 & \\
x - 5) & 1 & -2 & -13 & -17 & \text{dividend} \\
 & \downarrow & 5 & 15 & 10 & \text{products} \\
\hline
 & 1 & 3 & 2 & -7 & \text{quotient and remainder}
\end{array}$$

We end up by making two final observations. If we include the lead coefficient in the bottom row (see above and right), the coefficients in the top or quotient row are duplicated and no longer necessary. Finally, note all entries in the product row (in red) are five times the prior sum. This allows us to simply use the constant 5, rather than the entire divisor $x - 5$. These connections and replacements preserve all information held in the original algorithm, and the result is called synthetic division. A simple change in format makes the compressed form easier to use, and is illustrated next for the same exercise. Note the process begins by dropping the lead coefficient into place on the new quotient line (shown in blue).

$$\begin{array}{rrrrrrl}
\text{divisor (use 5, not } -5) \rightarrow & 5\rfloor & 1 & -2 & -13 & -17 & \leftarrow \text{coefficients of dividend} \\
 & & \downarrow & & & & \\
\hline
\text{drop lead coefficient into place} \rightarrow & & 1 & & & \llcorner & \leftarrow \text{quotient and remainder appear in this row}
\end{array}$$

We then multiply this coefficient by the divisor, place the result in the next column, and add. In a sense, we "multiply in the diagonal direction" and "add in the vertical direction." Continue the process until the division is complete.

	5⌋	1	−2	−13	−17	
			5			multiply divisor by lead coefficient,
multiply 5 · 1		1	3		⌊	place result in next column and add
	5⌋	1	−2	−13	−17	
			5	15		multiply divisor by previous sum: (5)(3)
multiply 5 · 3		1	3	2	⌊	place result in next column and add
	5⌋	1	−2	−13	−17	
			5	15	10	multiply divisor by previous sum: (5)(2)
multiply 5 · 2		1	3	2	⌊−7	place result in next column and add

Both the quotient and remainder are found in the last row. Since the last position (on the right) is reserved for the remainder (-7), we conclude the quotient polynomial is made up of the constant 2, the linear term $3x$ and the quadratic term $1x^2$: $(x^3 - 2x^2 - 13x - 17) \div (x - 5) = (1x^2 + 3x + 2) + \frac{-7}{x - 5}$. This is the same result as before, but the new process is much more efficient than long division, especially since all stages are actually computed on a single template as shown here:

5⌋	1	−2	−13	−17
		5	15	10
	1	3	2	⌊−7

Note that synthetic division is only employed when the divisor is a linear polynomial and of the form $x - r$ (the coefficient of x must be 1). Also note that for $x - r$, r is used for the division, while for $x + r$, $-r$ is used.

EXAMPLE 2 Use synthetic division to show that $(x + 2)$ is a factor of $(x^3 + 3x^2 - 4x - 12)$.

Solution:

use −2 as a "divisor"	−2⌋	1	3	−4	−12	
		↓	−2	−2	12	multiply divisor by previous sum,
		1	1	−6	0	place result in next column and add

Since the remainder is zero, $x + 2$ must be a factor of $x^3 + 3x^2 - 4x - 12$.

NOW TRY EXERCISES 25 THROUGH 28

EXAMPLE 3 Use synthetic division to compute the quotient: $(2x^3 - 9x - 29) \div (x - 3)$, then write the result as: (a) $\frac{\text{dividend}}{\text{divisor}} = (\text{quotient}) + \frac{\text{remainder}}{\text{divisor}}$ and (b) dividend = (quotient)(divisor) + remainder.

Solution:

use 3 as a "divisor"	3⌋	2	0	−9	−29	note placeholder 0 for x^2 term
		↓	6	18	27	
		2	6	9	−2	

a. $\frac{2x^3 - 9x - 29}{x - 3} = (2x^2 + 6x + 9) + \frac{-2}{x - 3}$

$\frac{\text{dividend}}{\text{divisor}} = (\text{quotient}) + \frac{\text{remainder}}{\text{divisor}}$

b. $2x^3 - 9x - 29 = (2x^2 + 6x + 9)(x - 3) + (-2)$

dividend = (quotient)(divisor) + remainder

NOW TRY EXERCISES 29 THROUGH 32

C. Synthetic Division and Factorable Polynomials

Synthetic division provides the *method* we need for efficient division, making it easier to use division as a mathematical tool. To help find *the divisors that give a remainder of zero,* we make the following observation concerning any polynomial with a lead coefficient of 1 or -1.

Factorable Polynomials

To factor $x^2 - 5x - 24$ by trial and error, a beginner might write out all possible binomial pairs where the *first* terms in the F-O-I-L process give x^2 with the *last* terms multiplying to 24. These are shown here:

$$(x \quad 1)(x \quad 24) \qquad (x \quad 2)(x \quad 12)$$
$$(x \quad 3)(x \quad 8) \qquad (x \quad 4)(x \quad 6)$$

Since these are all the possibilities, if $x^2 - 5x - 24$ is factorable (using integers) the correct factors must be somewhere in this list. We specifically note that the last term in each binomial *must be a factor of 24*. This observation can be extended to a polynomial of any degree, and used to factor certain polynomials.

PRINCIPLE OF FACTORABLE POLYNOMIALS

Given a polynomial of degree $n > 1$ with integer coefficients and a lead coefficent of 1 or -1, the linear factors of the polynomial must be of the form $(x - p)$, where p is a factor of the constant term.

EXAMPLE 4 Use synthetic division to help factor $Q(x) = x^4 + x^3 - 10x^2 - 4x + 24$ completely.

Solution: If Q is factorable using integers, factors must have the form $(x - p)$, where p is a factor of 24. The possibilities are ± 1, ± 24, ± 2, ± 12, ± 3, ± 8, ± 4, and ± 6. Using 1 and -1 does not produce a remainder of zero, so we try 2 and continue the search.

use 2 as a "divisor"

$$\begin{array}{r|rrrrr} 2 & 1 & 1 & -10 & -4 & 24 \\ & \downarrow & 2 & 6 & -8 & -24 \\ \hline & 1 & 3 & -4 & -12 & \underline{|0} \end{array}$$

Since the remainder is zero, $(x - 2)$ is a factor and Q can be written as

$$x^4 + x^3 - 10x^2 - 4x + 24 = (x^3 + 3x^2 - 4x - 12)(x - 2) + 0$$
$$\text{dividend} = \text{(quotient)(divisor)} \quad + \quad \text{remainder}$$

Using factoring by grouping on the quotient polynomial, Q can be factored further and written in completely factored form:

$$\begin{aligned} x^4 + x^3 - 10x^2 - 4x + 24 &= (\underline{x^3 + 3x^2} - \underline{4x - 12})(x - 2) \\ &= [x^2(x + 3) - 4(x + 3)](x - 2) \\ &= [(x + 3)(x^2 - 4)](x - 2) \\ \text{completely factored form} &= (x + 3)(x + 2)(x - 2)^2 \end{aligned}$$

NOW TRY EXERCISES 33 THROUGH 46

WORTHY OF NOTE

In support of our work in Section 3.8, note the factored form contains two linear factors (the graph of Q will cut through the x-axis at $x = -3$ and $x = -2$), and a linear factor of degree two (the graph will bounce off the x-axis at $x = 2$).

GRAPHICAL SUPPORT

The results from Example 4 are easily verified on a graphing calculator. Enter $Q(x) = x^4 + x^3 - 10x^2 - 4x + 24$ as Y_1 on the Y= screen. Noting the y-intercept is 24 and the zeroes of Q are close to (0, 0), we set the window size at $x \in [-5, 5]$ and $y \in [-10, 30]$, using Yscl = 3.

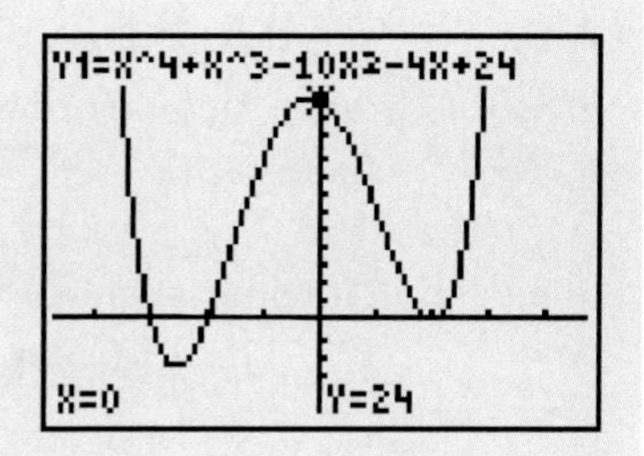

Another application of synthetic division involves the *creation* of factorable polynomials. For instance, what value(s) of k will make $x - 5$ a factor of $x^2 - kx - 10$? After some study and perhaps a trial-and-error process, you might notice that $x - 5$ is a factor of $x^2 - 3x - 10$, meaning $k = 3$ is a possibility. But using trial and error on larger polynomials becomes virtually impossible and a more systematic approach can be appreciated. The approach simply incorporates the unknown coefficient into the synthetic division process, adding each column as before while combining any like terms.

LOOKING AHEAD

In Example 5 where $f(x) = x^2 + kx - 14$, note that $f(-2) = -2k - 10$ (substitute $x = -2$ and check). This "coincidence" is actually no coincidence at all, and in Section 4.2, it will lead us to an even more substantial use of synthetic division.

EXAMPLE 5 What value(s) of k will make $x + 2$ a factor of $f(x) = x^2 + kx - 14$?

Solution: If $x + 2$ is to be a factor of $x^2 + kx - 14$, their quotient must give a remainder of zero.

use −2 as a "divisor"

$$\begin{array}{r|rrr} -2 & 1 & k & -14 \\ & \downarrow & -2 & -2k+4 \\ \hline & 1 & k-2 & -2k-10 \end{array} \quad \text{remainder}$$

The remainder is $-2k - 10$, yielding the equation $-2k - 10 = 0$, with solution $k = -5$. Sure enough, $x^2 - 5x - 14 = (x + 2)(x - 7)$.

NOW TRY EXERCISES 49 THROUGH 56

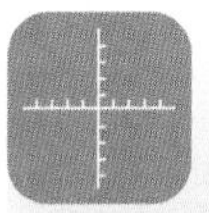

TECHNOLOGY HIGHLIGHT

Graphical Tests for Factors of a Polynomial

The keystrokes shown apply to a TI-84 Plus model. Please consult your manual or our Internet site for other models.

Our study of pointwise and asymptotic discontinuities offers us a *graphical test* of whether one polynomial is a factor of the other. In our study of basic rational functions (Section 3.5), we noted that vertical asymptotes occur at the zeroes of the denominator. However, from Example 6 in Section 3.7, we saw that $h(x) = \frac{x^2 - 4}{x - 2}$ did not have an asymptote at $x = 2$, since $x - 2$ was a factor of $x^2 - 4 = (x - 2)(x + 2)$. The common factors reduced to 1, leaving the linear function $H(x) = x + 2$, with a "hole" at (2, 4) (where $x = 2$). This basic idea can be extended to other polynomials that we're testing for divisibility. If the graph results in an asymptote, the divisor is not a factor of the given polynomial (it didn't reduce to 1). If no asymptotic behavior is evident, the divisor is likely a factor. For verification, we consider $f(x) = x^2 - 3x$ and $g(x) = x^2 - 4x$, and graphically test to see if $x - 3$ is a factor of either polynomial (meaning it "divides evenly"). On the **Y=** screen, enter $Y_1 = (x^2 - 3x)/(x - 3)$ and $Y_2 = (x^2 - 4x)/(x - 3)$. First **GRAPH** Y_1, which turns out to be linear—no asymptotic behavior (Figure 4.1). By inspection we see $x - 3$ is indeed a factor of $x^2 - 3x$. The graph of Y_2 has a vertical asymptote at $x = 3$, showing $x - 3$ is not a factor of $x^2 - 4x$ (Figure 4.2).

Figure 4.1

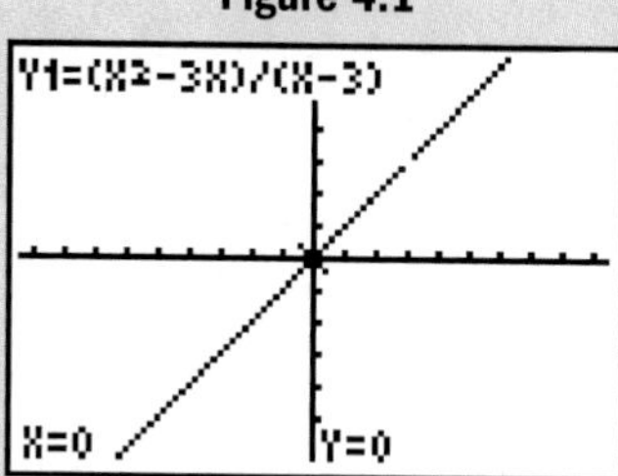

Figure 4.2

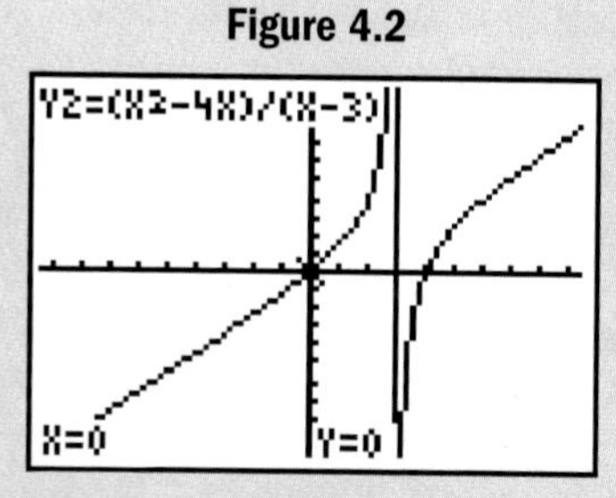

Use this method of testing for factors to complete this exercise.

Exercise 1: Given $x + 3$ is a factor of exactly two of the following polynomials, determine which two.

a. $f(x) = x^2 + 2x + 1$

b. $g(x) = x^3 + x^2 - 14x - 24$

c. $p(x) = x^3 + 6x^2 - x - 30$

d. $q(x) = x^3 + 7x^2 + 4x - 12$ b and c

4.1 EXERCISES

CONCEPTS AND VOCABULARY

Fill in each blank with the appropriate expression, word, or phrase. Carefully reread the section if needed. For Exercises 1–6, consider the division $\frac{x^3 + 2x^2 + 3x - 4}{x - 1} = x^2 + 3x + 6 + \frac{2}{x - 1}$.

1. The *dividend* is $\underline{x^3 + 2x^2 + 3x - 4}$.

2. The *divisor* is $\underline{\quad x - 1 \quad}$.

3. The *quotient* is $\underline{\quad x^2 + 3x + 6 \quad}$.

4. The *remainder* is $\underline{\quad 2 \quad}$.

5. Write the result in the alternative form *dividend = (quotient)(divisor) + remainder.* Compute the product of the quotient and divisor, then combine like terms, and verify you get a true statement.
$x^3 + 2x^2 + 3x - 4 = (x^2 + 3x + 6)(x - 1) + 2$

6. Discuss/explain the principle of factorable polynomials by writing the following polynomials in standard form:
(a) $f(x) = (x + 2)(x - 3)(x + 4)$ and
(b) $h(x) = (x - a)(x + b)(x - c)$.
$f(x) = x^3 + 3x^2 - 10x - 24$; $h(x) = x^3 - ax^2 + bx^2 - cx^2 - abx - bcx + acx + abc$

DEVELOPING YOUR SKILLS

Compute each quotient using long division and check all results using the related multiplication.

7. $\frac{x^2 - 2x - 35}{x + 5}$

8. $\frac{x^2 + 4x - 21}{x - 3}$

9. $\frac{6r^2 - r - 15}{3r - 5}$

10. $\frac{4s^2 + 16s + 15}{2s + 3}$

11. $\frac{x^3 - 5x^2 - 4x + 20}{x - 2}$

12. $\frac{x^3 + 12x^2 - 4x - 15}{x + 1}$

13. $(8n^3 + 1) \div (2n + 1)$

14. $(27m^3 - 8) \div (3m - 2)$

15. $(3b^3 - 5b - 2) \div (b + 1)$

16. $(2x^3 - 7x + 2) \div (x + 2)$

Compute the quotient using long division. Write all answers in two ways: (1) *dividend = (quotient)(divisor) + remainder* and (2) $\frac{dividend}{divisor} = (quotient) + \frac{remainder}{divisor}$.

17. $(9b^2 - 24b + 16) \div (3b - 4)$

18. $(4c^2 + 20c + 25) \div (2c + 5)$

7. $x - 7$
8. $x + 7$
9. $2r + 3$
10. $2s + 5$
11. $x^2 - 3x - 10$
12. $x^2 + 11x - 15$
13. $4n^2 - 2n + 1$
14. $9m^2 + 6m + 4$
15. $3b^2 - 3b - 2$
16. $2x^2 - 4x + 1$
17. (1) $9b^2 - 24b + 16 = (3b - 4)(3b - 4)$
(2) $\frac{9b^2 - 24b + 16}{3b - 4} = 3b - 4$
18. (1) $4c^2 + 20c + 25 = (2c + 5)(2c + 5)$
(2) $\frac{4c^2 + 20c + 25}{2c + 5} = 2c + 5$

19. (1) $2n^3 - n^2 - 19n + 4 = (2n^2 - 7n + 2)(n + 3) - 2$
(2) $\dfrac{2n^3 - n^2 - 19n + 4}{n + 3} = 2n^2 - 7n + 2 + \dfrac{-2}{n + 3}$

20. (1) $h^3 + 8h^2 + 14h + 13 = (h^2 + 5h - 1)(h + 3) + 16$
(2) $\dfrac{h^3 + 8h^2 + 14h + 13}{h + 3} = h^2 + 5h - 1 + \dfrac{16}{h + 3}$

21. (1) $g^4 - 15g^2 + 10g + 24 = (g^3 - 4g^2 + g + 6)(g + 4)$
(2) $\dfrac{g^4 - 15g^2 + 10g + 24}{g + 4} = g^3 - 4g^2 + g + 6$

22. (1) $d^4 - 3d^3 + 4d^2 - 36 = (d^3 + 4d + 12)(d - 3)$
(2) $\dfrac{d^4 - 3d^3 + 4d^2 - 36}{d - 3} = d^3 + 4d + 12$

23. (1) $(x^4 - 16x^2 - 5x - 24) = (x^3 - 4x^2 - 5)(x + 4) - 4$
(2) $\dfrac{x^4 - 16x^2 - 5x - 24}{x + 4} = x^3 - 4x^2 - 5 - \dfrac{4}{x + 4}$

24. (1) $(v^4 - 2v^3 - v + 10) = (v^3 - 1)(v - 2) + 8$
(2) $\dfrac{v^4 - 2v^3 - v + 10}{v - 2} = v^3 - 1 + \dfrac{8}{v - 2}$

25.

-2	1	5	-1	-14
		-2	-6	14
	1	3	-7	0

26.

5	1	-8	5	50
		5	-15	-50
	1	-3	-10	0

27.

-7	1	12	34	-7
		-7	-35	7
	1	5	-1	0

28.

3	1	-10	23	-6
		3	-21	6
	1	-7	2	0

29. $x^3 + 3x^2 - 8x - 13 = (x^2 + 2x - 10)(x + 1) - 3$; $x^2 + 2x - 10 + \frac{-3}{x + 1}$

36. $(x + 3)(x + 1)(x - 2)$
37. $(x - 1)(x - 2)(x + 6)$
38. $(x + 5)(x + 3)(x - 3)$
39. $(x + 3)(x - 1)(x - 2)$
40. $(x + 4)(x - 1)(x - 3)$
41. $(x + 2)(x - 3)(x - 6)$
42. $(x + 3)(x - 6)^2$
43. $(x + 5)(x + 4)(x - 6)$
44. $(x + 7)(x - 3)(x - 5)$
45. $(x + 2)(x + 1)(x - 1)(x - 3)$

19. $(-n^2 - 19n + 2n^3 + 4) \div (n + 3)$

20. $(13 + 8h^2 + h^3 + 14h) \div (3 + h)$

21. $(g^4 - 15g^2 + 10g + 24) \div (g + 4)$

22. $(d^4 - 3d^3 + 4d^2 - 36) \div (d - 3)$

23. $(x^4 - 16x^2 - 5x - 24) \div (x + 4)$

24. $(v^4 - 2v^3 - v + 10) \div (v - 2)$

Use synthetic division to show:

25. $(x + 2)$ is a factor of $x^3 + 5x^2 - x - 14$.

26. $(x - 5)$ is a factor of $x^3 - 8x^2 + 5x + 50$.

27. $\dfrac{x^3 + 12x^2 + 34x - 7}{x + 7} = x^2 + 5x - 1$.
Verify the result using a test value.

28. $\dfrac{x^3 - 10x^2 + 23x - 6}{x - 3} = x^2 - 7x + 2$.
Verify the result using a test value.

Use synthetic division to compute each quotient. Write the result in the two forms shown in this section.

29. $(x^3 + 3x^2 - 8x - 13) \div (x + 1)$

30. $(x^3 - 6x^2 + 11x - 5) \div (x - 2)$
$x^3 - 6x^2 + 11x - 5 = (x^2 - 4x + 3)(x - 2) + 1$; $x^2 - 4x + 3 + \frac{1}{x - 2}$

31. $\dfrac{x^3 - 15x + 12}{x - 3}$
$x^3 - 15x + 12 = (x^2 + 3x - 6)(x - 3) - 6$; $x^2 + 3x - 6 + \frac{-6}{x - 3}$

32. $\dfrac{x^3 + 9x^2 - 91}{x + 5}$
$x^3 + 9x^2 - 91 = (x^2 + 4x - 20)(x + 5) + 9$; $x^2 + 4x - 20 + \frac{9}{x + 5}$

Use synthetic division and the statement on factorable polynomials to factor each polynomial completely.

33. $x^2 - 2x - 63$ $(x - 9)(x + 7)$

34. $x^2 - 7x - 18$ $(x - 9)(x + 2)$

35. $x^3 + 4x^2 + x - 6$ $(x + 2)(x - 1)(x + 3)$

36. $x^3 + 2x^2 - 5x - 6$

37. $x^3 + 3x^2 - 16x + 12$

38. $x^3 + 5x^2 - 9x - 45$

39. $x^3 - 7x + 6$

40. $x^3 - 13x + 12$

41. $x^3 - 7x^2 + 36$

42. $x^3 - 9x^2 + 108$

43. $x^3 + 3x^2 - 34x - 120$

44. $x^3 - x^2 - 41x + 105$

45. Use synthetic division to show that $x + 2$ is a factor of $x^4 - x^3 - 7x^2 + x + 6$. Then use the result to write the polynomial in completely factored form (use factoring by grouping).

46. Use synthetic division to show that $x - 3$ is a factor of $x^4 - 8x^3 + 14x^2 + 8x - 15$. Then use the result to write the polynomial in completely factored form (use factoring by grouping).
$(x - 3)(x - 1)(x - 5)(x + 1)$

WORKING WITH FORMULAS

47. Area of a trapezoid: $A = \dfrac{h(B + b)}{2}$

The area A of a trapezoid is given by the formula shown, where h represents the height of the rectangle, B represents the longer base, and b represents the shorter base. When asked to solve for the variable B, some students submitted $B = \dfrac{2A - hb}{h}$ as the answer, while other students gave $B = \dfrac{2A}{h} - b$. Use term-by-term division to verify that the solutions are identical. $B = \frac{2A}{h} - \frac{hb}{h} = \frac{2A}{h} - b$

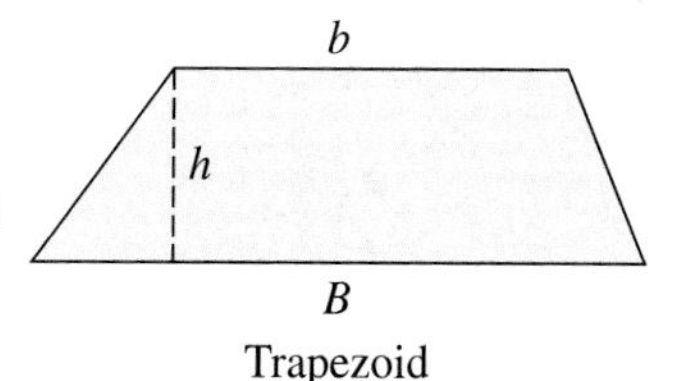

Trapezoid

48. Volume of a spherical cap: $V = \dfrac{\pi h^2}{3}(3r - h)$

The volume of a spherical cap is given by the formula shown, where r is the radius of the sphere and h is the height of the spherical cap (see diagram). When asked to solve for the variable r, some students submitted $r = \dfrac{3V + h^3\pi}{3\pi h^2}$ as the answer, while other students gave $r = \dfrac{V}{\pi h^2} + \dfrac{h}{3}$. Use term-by-term division to verify that the solutions are identical. $r = \frac{3V}{3\pi h^2} + \frac{h^3\pi}{3\pi h^2} = \frac{V}{\pi h^2} + \frac{h}{3}$

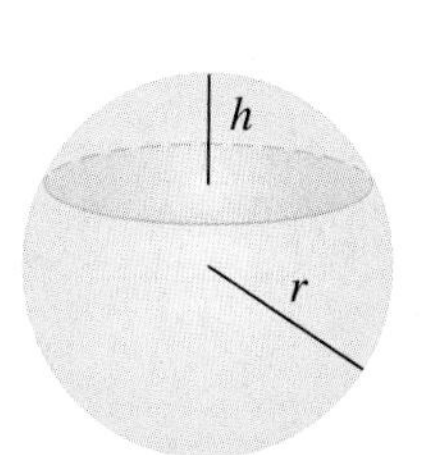

Spherical cap

APPLICATIONS

Division as a mathematical tool.

49. What value(s) of k will make $x - 3$ a factor of $(x^2 - kx - 27)$? $k = -6$

50. What value(s) of k will make $x + 4$ a factor of $(kx^2 - 7x + 20)$? $k = -3$

51. Divide using synthetic division until you get to the last step, then replace k with a number that will ensure $x + 2$ is a factor of the dividend (note the constant term is "k"): $(5x^2 + 2x + k) \div (x + 2)$. $k = -16$

52. Divide using synthetic division until you get to the last step, then replace k with a number that will ensure $x - 3$ is a factor of the dividend (note the constant term is "k"): $(2x^2 + 5x + k) \div (x - 3)$. $k = -33$

53. What value(s) of k will make $x + 3$ a factor of $(x^3 + kx^2 - 7x + 15)$? $k = -1$

54. What value(s) of k will make $x - 5$ a factor of $(x^4 + kx^3 - 125x + 375)$? $k = -3$

Divide using synthetic division until you get to the last step, then replace k with a number that will ensure $x + 2$ is a factor of the dividend. Verify results using a graphing calculator.

55. $(x^3 + 5x^2 + 2x + k) \div (x + 2)$ $k = -8$

56. $(2x^3 + 2x + k) \div (x + 2)$ $k = 20$

57. Area of a ping-pong table: Given that the area of a regulation ping-pong table (in square inches) is represented by the polynomial $5x^2 + 228x + 756$.

a. Use synthetic division to find the length if the width is given as $x + 42$. $5x + 18$

b. Find the value of x if the ratio of length to width is $\frac{\text{length}}{\text{width}} = \frac{9}{5}$. $x = 18$

c. Find the actual dimensions of a ping-pong table. 60×108

58. Area of a pool table: Given that the area of a regulation pool table (in square inches) is represented by the polynomial $7x^2 + 172x + 253$.

a. Use synthetic division to find the length if the width is given as $x + 23$. $7x + 11$

b. Find the value of x if the ratio of length to width is $\frac{\text{length}}{\text{width}} = \frac{2}{1}$. $x = 7$

c. Find the actual dimensions of a pool table. 30×60

WRITING, RESEARCH, AND DECISION MAKING

59. A good friend of yours was absent on the day the polynomial long division was discussed. Using the exercise $(2n^3 + 3n^2 - 3n + 4) \div (n + 2)$, prepare a list of step-by-step instructions that will guide her through the process. Answers will vary.

60. Create *any* four-term cubic polynomial with a lead coefficient of 1 and a constant term of k (one example would be $x^3 - 5x^2 + 7x + k$). With this cubic as the dividend, use synthetic division to compute the quotient $(x^3 - 5x^2 + 7x + k) \div (x - c)$, where c is any integer between -5 and 5 (for convenience). When you get to the remainder, discuss/explain how the result helps to support the *principle of factorable polynomials*. In other words, how does the remainder support the contention that c must be a factor of the constant term? Answers will vary.

EXTENDING THE CONCEPT

61. Find a value of k that ensures the quotient $(n^3 + 3n^2 - 3kn - 10) \div (n + 2)$ has a remainder of -5. $k = \frac{1}{6}$

62. The synthetic division algorithm also applies to polynomials with complex roots. Use the algorithm to show $x - (2 + 3i)$ is a factor of $x^3 - 5x^2 + 17x - 13$. verified

63. Find the value(s) of k that will ensure $x - k$ is a factor of $x^3 + kx - 12$. $k = 2$

64. Find the value(s) of k that will ensure $x - 3$ is a factor of $x^3 - 5x^2 + k^2x - 25k$. $k = 9$, $k = \frac{-2}{3}$

MAINTAINING YOUR SKILLS

65. (R.3) Compute the quotient using properties of exponents: $\frac{6.48 \times 10^6}{1.62 \times 10^{-2}}$. 4×10^8

66. (1.3/1.4) Solve by factoring. Find all roots, real and complex: $3x^4 - 75 = 0$.
$x = \pm\sqrt{5},\ x = \pm\sqrt{5}i$

67. (1.3/3.5) Solve for x: $\frac{12}{(x-4)^2} - 3 = 0$.
$x = 6,\ x = 2$

68. (3.3) Sketch the graph of $h(x) = \sqrt{x+3} - 5$ using transformations of the parent function.

68.

69. (2.5) Find the domain of each function given. $x \in (-\infty, -4] \cup [7, \infty)$

a. $f(x) = \sqrt{x^2 - 3x - 28}$

b. $g(x) = \sqrt{2x + 3}$ $x \in [\frac{-3}{2}, \infty)$

70. (2.4) Calculate the average rate of change for $f(x) = x^2 + 2x - 5$ in the interval $[1, 1.2]$. 4.2

4.2 The Remainder and Factor Theorems

LEARNING OBJECTIVES

In Section 4.2 you will learn how to:

A. Use the remainder theorem to evaluate polynomials

B. Use the factor theorem to factor and build polynomials

C. Apply the remainder theorem, the factor theorem, and synthetic division to complex factors and roots

D. Construct polynomials using complex roots and roots of multiplicity

INTRODUCTION

The history of science and mathematics is full of examples where a casual observation leads to a stunning result. One can just picture Archimedes sitting in his bath and watching the water rise, or Madam Curie coming out to her workshop and noting the mysterious discoloration on a photoplate accidently left near some radioactive material. To a smaller degree, it seems reasonable that a similar thing happened in the discovery of what we call the *remainder theorem.*

POINT OF INTEREST

In previous Points of Interest, we've noted that while early students of mathematics found negative numbers puzzling, imaginary numbers were thought to be "manifestly impossible" (Girolamo Cardano 1501–1576). In fact, it was not until 1629 that a mathematician named Albert Girard (1593–1632) began advocating a greater respect for and use of complex numbers, asserting that their acceptance would verify that a polynomial equation has exactly as many roots as its degree. He further suggested that their acceptance would enable one to make more general observations between the coefficients of a polynomial and its roots, helping to pave the way for a proof of the Fundamental Theorem of Algebra.

A. The Remainder Theorem

In Section 4.1, we used synthetic division to determine if expressions of the form $x - p$ were factors of some larger polynomial, and how to write the result in two different forms. For $x + 3$ and $P(x) = x^3 + 5x^2 + 2x - 8$, synthetic division gives

use −3 as a "divisor"

$$\begin{array}{r|rrrr} -3 & 1 & 5 & 2 & -8 \\ & \downarrow & -3 & -6 & 12 \\ \hline & 1 & 2 & -4 & \lfloor 4 \end{array}$$

with a quotient of $x^2 + 2x - 4$ and a remainder of 4. This shows $x + 3$ is *not* a factor of $P(x)$, since it didn't "divide evenly." When written in the form $P(x) =$ (quotient)(divisor) + remainder, we have $P(x) = (x^2 + 2x - 4)(x + 3) + 4$ and the stage is set for a remarkable "casual observation." Observe that if we evaluate $P(x)$ at $x = -3$, *the divisor portion of our answer becomes zero,* leaving $P(-3) = 4$:

$$\begin{aligned} P(-3) &= [(-3)^2 + 2(-3) - 4](-3 + 3) + 4 \\ &= (-1)(0) + 4 \\ &= 4 \end{aligned}$$

This can be verified by evaluating $P(x)$ in its original (polynomial) form:

$$\begin{aligned} P(x) &= x^3 + 5x^2 + 2x - 8 \\ P(-3) &= (-3)^3 + 5(-3)^2 + 2(-3) - 8 \\ &= -27 + 45 + (-6) - 8 \\ &= 4 \end{aligned}$$

The result is no coincidence, and illustrates what is called the **remainder theorem.**

THE REMAINDER THEOREM
If a polynomial $P(x)$ is divided by a linear factor $(x - r)$, the remainder is identical to $P(r)$—the original function evaluated at r.

Proof of the Remainder Theorem

From our previous work, any number r used in synthetic division will occur as the factor $(x - r)$ when written as: (quotient)(divisor) + remainder: $P(x) = (x - r)q(x) + R$. Here, $q(x)$ represents the quotient polynomial and R is a constant. Evaluating $P(r)$ gives

$$\begin{aligned} P(r) &= (r - r)q(x) + R \\ &= 0 \cdot q(x) + R \\ &= R✓ \end{aligned}$$

One useful consequence of the theorem is the ability to evaluate polynomials more efficiently, which is a great asset to applications of polynomials and polynomial graphing.

EXAMPLE 1 Use the remainder theorem to find the value of $H(-5)$ for $H(x) = x^4 + 3x^3 - 8x^2 + 5x - 6$, and check via substitution.

Solution: use -5 as a "divisor"

$$\begin{array}{r|rrrrr} -5 & 1 & 3 & -8 & 5 & -6 \\ & & -5 & 10 & -10 & 25 \\ \hline & 1 & -2 & 2 & -5 & 19 \end{array}$$

$H(-5) = 19$. To check:

$$\begin{aligned} H(-5) &= (-5)^4 + 3(-5)^3 - 8(-5)^2 + 5(-5) - 6 \\ &= 625 - 375 - 200 - 25 - 6 \\ &= 19✓ \end{aligned}$$

NOW TRY EXERCISES 7 THROUGH 24

Note the direct evaluation of $H(-5)$ involves very large numbers and a long series of calculations. Synthetic division reduces the process to simple products and sums.

B. The Factor Theorem

A second useful consequence of the remainder theorem is known as the **factor theorem.**

THE FACTOR THEOREM
Given $P(x)$ is a polynomial,
1. If $P(r) = 0$, then $x - r$ is a factor of $P(x)$.
2. If $x - r$ is a factor of $P(x)$, then $P(r) = 0$.

The factor theorem is closely related to the zero factor property from Section 1.3. The remainder theorem helps *prove* the factor theorem in a more general sense, enabling us to apply it in other ways.

Proof of the Factor Theorem

1. Consider a polynomial P written in the form $P(x) = (x - r)q(x) + R$. From the remainder theorem we know $P(r) = R$, and substituting $P(r)$ for R in the preceding equation gives

$$P(x) = (x - r)q(x) + P(r)$$

This shows that if $P(r) = 0$, $x - r$ is a factor of $P(x)$:

$$P(x) = (x - r)q(x)\checkmark$$

2. Further, if $(x - r)$ is a factor of $P(x)$, we have $P(x) = (x - r)q(x)$. Evaluating at $x = r$ produces a result of zero.

$$\begin{aligned} P(r) &= (r - r)q(r) \\ &= 0\checkmark \end{aligned}$$

WORTHY OF NOTE

The result obtained in Example 2 is not unique, since any polynomial of the form $a(x^3 - 3x^2 - 2x + 6)$ for $a \in \mathbb{R}$ will also have the same three roots.

EXAMPLE 2 A cubic polynomial has solutions $x = 3$, $x = \sqrt{2}$, and $x = -\sqrt{2}$. Use the factor theorem to find a polynomial P with these three roots.

Solution: By the factor theorem, the factors of $P(x)$ must be $(x - 3)$, $(x - \sqrt{2})$, and $(x + \sqrt{2})$. Computing the product yields the polynomial P.

$$\begin{aligned} P(x) &= (x - 3)(x - \sqrt{2})(x + \sqrt{2}) \\ &= (x - 3)(x^2 - 2) \\ &= x^3 - 3x^2 - 2x + 6 \end{aligned}$$

NOW TRY EXERCISES 25 THROUGH 32

C. Complex Numbers, Coefficients, and the Remainder and Factor Theorems

A polynomial equation with real coefficients sometimes has complex roots. For example, the equation $x^3 - 3x^2 + 4x - 12 = 0$ can be factored by grouping into $(x - 3)(x^2 + 4) = 0$, which has a real solution of $x = 3$, and the complex solutions $x = -2i$ and $x = 2i$. Using the factor theorem, the equation can be written $(x - 3)(x + 2i)(x - 2i) = 0$ and expanded to give back the polynomial form. It's important to note the remainder and factor theorems, as well as synthetic division, *can also be applied when complex numbers are involved*. This will form an important link to our study of complex polynomials (polynomials with complex coefficients) in Section 4.3.

EXAMPLE 3 Use the remainder theorem to show $x = 2i$ is a zero of $P(x) = x^3 - 3x^2 + 4x - 12$.

Solution: use $2i$ as a "divisor"

$$\begin{array}{r|rrrr} 2i & 1 & -3 & 4 & -12 \\ & \downarrow & 2i & -6i-4 & 12 \\ \hline & 1 & -3+2i & -6i & \underline{|0} \end{array}$$

$(2i)(-6i) = -12i^2 = 12$

Since the remainder is zero, $x = 2i$ is a zero of P.

NOW TRY EXERCISES 33 THROUGH 46

D. Complex Conjugates and Roots of Multiplicity

Another link to our pending study of polynomial functions involves the properties of complex conjugates. From our study of quadratics in Section 1.5, if the discriminant $D = b^2 - 4ac$ is not a perfect square, there will be two irrational roots that are conjugates: $-\frac{b}{2a} + \frac{\sqrt{D}}{2a}$ and $-\frac{b}{2a} - \frac{\sqrt{D}}{2a}$. If $b^2 - 4ac < 0$, the roots will be complex conjugates. In fact, if any polynomial with real coefficients has complex roots, they will *always* occur in conjugate pairs.

COMPLEX CONJUGATES THEOREM

Given polynomial $P(x)$ with real number coefficients, complex solutions will occur in conjugate pairs.

If $a + bi$, $b \neq 0$, is a solution, then $a - bi$ must also be a solution.

To prove this for polynomials of degree $n > 2$, we let $z_1 = a + bi$ and $z_2 = c + di$ be complex numbers and $\bar{z}_1 = a - bi$ and $\bar{z}_2 = c - di$ represent their conjugates. Note the following properties:

1. The conjugate of a sum is equal to the sum of the conjugates.

sum: $z_1 + z_2$		**sum of conjugates: $\bar{z}_1 + \bar{z}_2$**
$(a + bi) + (c + di)$		$(a - bi) + (c - di)$
$(a + c) + (b + d)i$	$\rightarrow$ *conjugate of sum* $\rightarrow$	$(a + c) - (b + d)i$ ✓

2. The conjugate of a product is equal to the product of the conjugates.

product: $z_1 \cdot z_2$		**product of conjugates: $\bar{z}_1 \cdot \bar{z}_2$**
$(a + bi) \cdot (c + di)$		$(a - bi) \cdot (c - di)$
$ac + adi + bci + bdi^2$		$ac - adi - cbi + bdi^2$
$(ac - bd) + (ad + bc)i$	$\rightarrow$ *conjugate of product* $\rightarrow$	$(ac - bd) - (ad + bc)i$ ✓

Since polynomials involve only sums and products, and the complex conjugate of any real number is the number itself, we have the following.

Proof of Complex Conjugates Theorem

Given polynomial $P(x) = a_nx^n + a_{n-1}x^{n-1} + \cdots + a_1x^1 + a_0$, where $a_n, a_{n-1}, \ldots, a_1, a_0$ are real numbers and $z = a + bi$ is a zero of P, we must show that $\bar{z} = a - bi$ is also a zero.

$a_nz^n + a_{n-1}z^{n-1} + \cdots + a_1z^1 + a_0 = P(z)$ evaluate $P(x)$ at z

$a_nz^n + a_{n-1}z^{n-1} + \cdots + a_1z^1 + a_0 = 0$ $P(z) = 0$ given

$\overline{a_nz^n + a_{n-1}z^{n-1} + \cdots + a_1z^1 + a_0} = \bar{0}$ conjugate both sides

$\overline{a_nz^n} + \overline{a_{n-1}z^{n-1}} + \ldots + \overline{a_1z^1} + \overline{a_0} = \bar{0}$ property 1

$\bar{a}_n(\bar{z}^n) + \bar{a}_{n-1}(\bar{z}^{n-1}) + \cdots + \bar{a}_1(\bar{z}^1) + \bar{a}_0 = \bar{0}$ property 2

$a_n(\bar{z}^n) + a_{n-1}(\bar{z}^{n-1}) + \cdots + a_1(\bar{z}^1) + a_0 = 0$ conjugate of a real number is the number

$P(\bar{z}) = 0$ ✓ result

An immediate and useful result of this theorem is that any polynomial of odd degree must have at least one real root, an idea we'll explore further in Section 4.3.

EXAMPLE 4 A cubic polynomial P with real coefficients has solutions $x = -1$ and $x = 2 + \sqrt{3}i$. Find the polynomial (assume a lead coefficient of 1).

Solution: By the factor theorem, two factors of P are $(x + 1)$ and $x - (2 + \sqrt{3}i)$. From the complex conjugates theorem, $x = 2 - \sqrt{3}i$ must also be a solution and $x - (2 - \sqrt{3}i)$ is also a factor. This gives

$$P(x) = (x + 1)[x - (2 + \sqrt{3}i)][x - (2 - \sqrt{3}i)]$$

$$
\begin{aligned}
P(x) &= (x + 1)[(x - 2) - \sqrt{3}i][(x - 2) + \sqrt{3}i] && \text{distribute and regroup} \\
&= (x + 1)[(x - 2)^2 - (\sqrt{3}i)^2] && (A + B)(A - B) = A^2 - B^2 \\
&= (x + 1)[(x^2 - 4x + 4) + 3] && \text{expand binomial; } (\sqrt{3}i)^2 = -3 \\
&= (x + 1)(x^2 - 4x + 7) && \text{simplify} \\
&= x^3 - 3x^2 + 3x + 7 && \text{result}
\end{aligned}
$$

The polynomial is $P(x) = x^3 - 3x^2 + 3x + 7$, which can be verified using synthetic division and any of the original roots.

NOW TRY EXERCISES 47 THROUGH 58

The remainder and factor theorems, along with the complex conjugates theorem, enable us to make some additional observations regarding polynomials with real coefficients. We will develop the idea here, then provide a proof in Section 4.3. Consider illustrations A through J:

	Equation	Degree of Polynomial	Solution Method	Solution(s)
A.	$3x - 21 = 0$	1	inverse operations $3x = 21$	7
B.	$x^2 - 3x - 40 = 0$	2	factoring $(x - 8)(x + 5) = 0$	-5 and 8
C.	$x^2 - 6x + 9 = 0$	2	factoring $(x - 3)(x - 3) = (x - 3)^2 = 0$	3 and 3
D.	$x^2 - 2x - 1 = 0$	2	completing the square $(x - 1)^2 = 2$	$1 + \sqrt{2}$ and $1 - \sqrt{2}$
E.	$x^2 - 2x + 5 = 0$	2	completing the square $(x - 1)^2 = -4$	$1 + 2i$ and $1 - 2i$
F.	$x^3 - 2x^2 - 25x + 50 = 0$	3	factor by grouping $(x - 2)(x^2 - 25) = 0$	2, 5, and -5
G.	$x^3 - 3x^2 - 9x + 27 = 0$	3	factor by grouping $(x - 3)(x^2 - 9) = (x - 3)^2(x + 3) = 0$	3, 3, and -3
H.	$x^3 - 6x^2 + 12x - 8 = 0$	3	rearrange terms and factor by grouping $(x - 2)(x^2 - 4x + 4) = (x - 2)^3 = 0$	2, 2, and 2
I.	$x^3 - 8 = 0$	3	factoring/quadratic formula $(x - 2)(x^2 + 2x + 4) = 0$	$2, -1 + \sqrt{3}i$, and $-1 - \sqrt{3}i$
J.	$x^4 - 81 = 0$	4	factoring $(x^2 - 9)(x^2 + 9) = 0$	$-3, 3, 3i$, and $-3i$

LOOKING AHEAD

In Section 3.8 we saw the graph of a function crosses the x-axis at linear roots (multiplicity one), and bounces off the x-axis at roots of multiplicity two. In Section 4.4 this idea will be extended to include roots of higher multiplicity as well.

First, observe that some equations produce repeated roots, called **roots of multiplicity** (C: 3 occurs twice; G: 3 occurs twice; H: 2 occurs three times). If a root repeats, the factor it came from must also repeat and we combine all such factors into the form $(x - r)^m$. If m is *even* (C, G), we have a **root of even multiplicity.** If m is *odd* (H), it's called a **root of odd multiplicity.** For any factor of the form $(x - r)^m$, $x = r$ is called a root of multiplicity m.

EXAMPLE 5 Solve each equation using factoring by grouping, then state the multiplicity of each root:
$x^3 - 2x^2 - 4x + 8 = 0$.

Solution: **a.**

$$\begin{aligned} x^3 - 2x^2 - 4x + 8 &= 0 && \text{original exercise} \\ x^2(x-2) - 4(x-2) &= 0 && \text{factor by grouping} \\ (x-2)(x^2-4) &= 0 && \text{common factor } (x-2) \\ (x-2)(x-2)(x+2) &= 0 && \text{factored form} \\ (x-2)^2(x+2) &= 0 && \text{result—three roots} \end{aligned}$$

$x = 2$ is a root of multiplicity two (even), $x = -2$ is a root of multiplicity one (odd).

NOW TRY EXERCISES 59 THROUGH 66

Examples A through J seem to indicate that if roots of multiplicity m are counted m times, a real polynomial of degree n *will have exactly n roots.* Note that examples F, G, H and I are all degree 3 polynomials with exactly three solutions, although G and H have repeated roots.

POLYNOMIAL ZEROES THEOREM

A polynomial equation of degree n has exactly n roots, (real and complex) where roots of multiplicity m are counted m times.

The ideas discussed here have a number of applications in both applied and theoretical mathematics and can be used to construct polynomials meeting specified criteria, as well as a tool for graphing polynomials. Other more practical applications will come later in this chapter.

EXAMPLE 6 Find a fourth-degree polynomial P with real coefficients, if $x = 3$ is the only real root and $x = 2i$ is also a root of P.

Solution: The complex roots must occur in conjugate pairs so $x = -2i$ is also a root, but this accounts for only three roots. Since P has degree 4, $x = 3$ must be a *repeated* root. The factors of P are $(x - 3)(x - 3)(x - 2i)(x + 2i)$.

$$\begin{aligned} P(x) &= (x^2 - 6x + 9)(x^2 - 4i^2) && \text{multiply binomials} \\ &= (x^2 - 6x + 9)(x^2 + 4) && \text{simplify} \\ &= x^4 - 6x^3 + 13x^2 - 24x + 36 && \text{multiply polynomials} \end{aligned}$$

NOW TRY EXERCISES 69 THROUGH 78

TECHNOLOGY HIGHLIGHT

Polynomials with Complex Roots

The keystrokes shown apply to a TI-84 Plus model. Please consult your manual or our Internet site for other models.

In the *Technology Highlight* from Section 1.4, we learned how to check results of operations on complex numbers using a graphing calculator. The calculator is also quite capable of evaluating a polynomial for any complex number *z*, enabling us to check the complex zeroes of polynomial functions. Due to the way the TI-84 Plus is programmed, this evaluation must be done using the home screen alone [without the **2nd** **GRAPH** **(TABLE)** and **Y=** screens]. To illustrate, we'll use the known solutions to $x^3 - 3x^2 + 3x + 7 = 0$ from Example 4. To verify that $2 + \sqrt{3}i$ is a solution, **CLEAR** the home screen and **STO➡** $2 + \sqrt{3}i$ in memory location **X,T,θ,n** as shown (Figure 4.3). Next write the expression $x^3 - 3x^2 + 3x + 7$ directly on the home screen using the variable **X,T,θ,n** and press **ENTER**. The calculator immediately evaluates the polynomial using the value stored and displays the result (Figure 4.4—top). As you see, $x = 2 + \sqrt{3}i$ is a solution to $x^3 - 3x^2 + 3x + 7 = 0$, as is $x = 2 - \sqrt{3}i$ (Figure 4.4—bottom) and $x = -1$ (not shown).

Use a calculator to complete the following exercise.

Figure 4.3

Figure 4.4

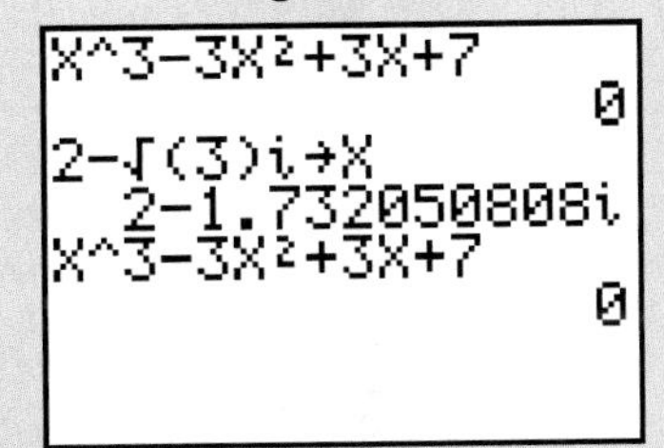

Exercise 1: All four zeroes of $P(x) = x^4 - 5x^3 + 5x^2 + 17x - 42 = 0$ are among the possible solutions listed. Find the zeroes using the ideas presented in this *Technology Highlight.* $x = -2$, $x = 3$, $x = 2 \pm i\sqrt{3}$

$x = -3$ $\quad x = -1$ $\quad x = 1$ $\quad x = 2$

$x = 3 + i\sqrt{2}$ $\quad x = 1 + i\sqrt{3}$ $\quad x = 3 - i\sqrt{2}$ $\quad x = 1 - i\sqrt{3}$

4.2 EXERCISES

CONCEPTS AND VOCABULARY

Fill in each blank with the appropriate word or phrase. Carefully reread the section if needed.

1. If the polynomial $P(x)$ is divided by the linear factor $x - k$, the remainder is identical to $P(k)$. This is a statement of the remainder theorem.

2. If $P(k) = 0$, then $x - k$ must be a factor of the polynomial $P(x)$. Conversely, if $x - k$ is a factor of $P(x)$, then $P(k) = 0$. These are statements from the factor theorem.

3. If a polynomial $P(x)$ has real coefficients and $z = a + bi$ is a zero, then $a - bi$ must also be a zero. This is a statement of the complex conjugate theorem.

4. A polynomial equation of degree n has exactly n roots (real and complex), where roots of multiplicity k are counted k times. This is a statement of the polynomial zeroes theorem.

5. Discuss/explain how the factor theorem and the product $(A + B)(A - B) = A^2 - B^2$ can be used to find the quadratic equation whose roots are $x = 2 + 3i$ and $x = 2 - 3i$. Answers will vary.

6. Discuss/explain how to find the fourth-degree polynomial having a complex root of $x = 1 - 2i$ and whose only real root is $x = 3$. How many x-intercepts does this polynomial have? Answers will vary; 1

DEVELOPING YOUR SKILLS

Use synthetic division and the remainder theorem to show the given value is a zero of $P(x)$.

7. $P(x) = x^3 + 2x^2 - 5x - 6$
$x = -3$

8. $P(x) = x^3 + 3x^2 - 16x + 12$
$x = -6$

9. $P(x) = x^3 - 7x + 6$
$x = 2$

10. $P(x) = x^3 - 13x + 12$
$x = -4$

11. $P(x) = x^3 - 6x^2 + 32$
$x = 4$

12. $P(x) = x^3 - 7x^2 + 36$
$x = 6$

Use synthetic division and the remainder theorem to evaluate $P(x)$ as given. Verify using a second method.

13. $P(x) = x^3 - 6x^2 + 5x + 12$
a. $P(-2)$ −30 **b.** $P(5)$ 12

14. $P(x) = x^3 + 4x^2 - 8x - 15$
a. $P(-2)$ 9 **b.** $P(3)$ 24

15. $P(x) = 2x^3 - x^2 - 19x + 4$
a. $P(-3)$ −2 **b.** $P(2)$ −22

16. $P(x) = 3x^3 - 8x^2 - 14x + 9$
a. $P(-2)$ −19 **b.** $P(4)$ 17

17. $P(x) = x^4 - 4x^2 + x + 1$
a. $P(-2)$ −1 **b.** $P(2)$ 3

18. $P(x) = x^4 + 3x^3 - 2x - 4$
a. $P(-2)$ −8 **b.** $P(2)$ 32

19. $P(x) = 2x^3 - 6x^2 - 7x + 21$
a. $P(-2)$ −5 **b.** $P(3)$ 0

20. $P(x) = -2x^3 + 9x^2 - x - 12$
a. $P(-2)$ 42 **b.** $P(4)$ 0

21. $P(x) = 2x^3 + 3x^2 - 9x - 10$
a. $P(\frac{3}{2})$ −10 **b.** $P(-\frac{5}{2})$ 0

22. $P(x) = 3x^3 + 11x^2 + 2x - 16$
a. $P(\frac{1}{3})$ −14 **b.** $P(-\frac{8}{3})$ 0

23. $P(x) = x^3 - 2x^2 + 3x - 2$
a. $P(\frac{1}{2})$ $\frac{-7}{8}$ **b.** $P(\frac{1}{3})$ $\frac{-32}{27}$

24. $P(x) = x^3 + x^2 - 2x + 3$
a. $P(\frac{2}{3})$ $\frac{65}{27}$ **b.** $P(-\frac{3}{2})$ $\frac{39}{8}$

A polynomial P with integer coefficients has the zeroes and degree indicated. Use the factor theorem to write the function in factored form and standard form.

25. $x = -2, x = 3, x = -5$; degree 3

26. $x = 1, x = -4, x = 2$; degree 3

27. $x = -2, x = \sqrt{3}, x = -\sqrt{3}$; degree 3

28. $x = \sqrt{5}, x = -\sqrt{5}, x = 4$; degree 3

29. $x = -5, x = 2\sqrt{3}, x = -2\sqrt{3}$; degree 3

30. $x = 4, x = 3\sqrt{2}, x = -3\sqrt{2}$; degree 3

31. $x = 1, x = -2, x = \sqrt{10}, x = -\sqrt{10}$; degree 4

32. $x = \sqrt{7}, x = -\sqrt{7}, x = 3, x = -1$; degree 4

Use the remainder theorem to show the value given is a zero of P. Recall that $(a + bi)(a - bi) = a^2 + b^2$.

33. $P(x) = x^3 - 4x^2 + 9x - 36; x = 3i$

34. $P(x) = x^3 + 2x^2 + 16x + 32; x = -4i$

35. $P(x) = x^4 + x^3 + 2x^2 + 4x - 8; x = -2i$

36. $P(x) = x^4 + x^3 - 5x^2 + x - 6; x = i$

37. $P(x) = -x^3 + x^2 - 3x - 5; x = 1 + 2i$

38. $P(x) = -x^3 + x^2 - 8x - 10; x = 1 + 3i$

The degree and zeroes of a polynomial with integer coefficients are given. Use the factor theorem to find the polynomial. Then check one of the real roots using the remainder theorem.

39. $x = -1, x = 1 + 3i, x = 1 - 3i$; degree 3

40. $x = -2, x = 2 + i, x = 2 - i$; degree 3

41. $x = \sqrt{3}, x = -\sqrt{3}, x = 1 + 2i$, $x = 1 - 2i$; degree 4

42. $x = \sqrt{2}, x = -\sqrt{2}, x = 1 + 5i$, $x = 1 - 5i$; degree 4

43. $x = 1, x = -3, x = 1 + \sqrt{2}i$, $x = 1 - \sqrt{2}i$; degree 4

44. $x = -2, x = -1, x = 1 + \sqrt{3}i$, $x = 1 - \sqrt{3}i$; degree 4

---

7. −3| 1 2 −5 −6
−3 3 6
1 −1 −2 0

8. −6| 1 3 −16 12
−6 18 −12
1 −3 2 0

9. 2| 1 0 −7 6
2 4 −6
1 2 −3 0

10. −4| 1 0 −13 12
−4 16 −12
1 −4 3 0

11. 4| 1 −6 0 32
4 −8 −32
1 −2 −8 0

12. 6| 1 −7 0 36
6 −6 −36
1 −1 −6 0

25. $P(x) = (x + 2)(x - 3)(x + 5)$, $P(x) = x^3 + 4x^2 - 11x - 30$
26. $P(x) = (x - 1)(x + 4)(x - 2)$, $P(x) = x^3 + x^2 - 10x + 8$
27. $P(x) = (x + 2)(x - \sqrt{3})(x + \sqrt{3})$, $P(x) = x^3 + 2x^2 - 3x - 6$
28. $P(x) = (x - \sqrt{5})(x + \sqrt{5})(x - 4)$, $P(x) = x^3 - 4x^2 - 5x + 20$
29. $P(x) = (x + 5)(x - 2\sqrt{3})(x + 2\sqrt{3})$, $P(x) = x^3 + 5x^2 - 12x - 60$
30. $P(x) = (x - 4)(x - 3\sqrt{2})(x + 3\sqrt{2})$, $P(x) = x^3 - 4x^2 - 18x + 72$
31. $P(x) = (x - 1)(x + 2)(x - \sqrt{10})(x + \sqrt{10})$, $P(x) = x^4 + x^3 - 12x^2 - 10x + 20$
32. $P(x) = (x - \sqrt{7})(x + \sqrt{7})(x - 3)(x + 1)$, $P(x) = x^4 - 2x^3 - 10x^2 + 14x + 21$
33. $P(3i) = 0$
34. $P(-4i) = 0$
35. $P(-2i) = 0$
36. $P(i) = 0$
37. $P(1 + 2i) = 0$
38. $P(1 + 3i) = 0$
39. $P(x) = x^3 - x^2 + 8x + 10$
40. $P(x) = x^3 - 2x^2 - 3x + 10$
41. $P(x) = x^4 - 2x^3 + 2x^2 + 6x - 15$
42. $P(x) = x^4 - 2x^3 + 24x^2 + 4x - 52$
43. $P(x) = x^4 - 4x^2 + 12x - 9$
44. $P(x) = x^4 + x^3 + 8x + 8$

45. $P(x) = x^4 - 4x^3 + 2x^2 + 4x - 8$
46. $P(x) = x^4 - 10x^3 + 38x^2 - 62x + 33$
47. $P(x) = x^3 - 4x^2 + 9x - 36$
48. $P(x) = x^3 + 3x^2 + 25x + 75$
49. $P(x) = x^3 - 9x^2 + 33x - 65$
50. $P(x) = x^3 + x^2 + 20x + 78$
51. $P(x) = x^3 - 8x^2 + 23x - 22$
52. $P(x) = x^3 + x^2 + 4x - 6$
53. $P(x) = x^4 + 7x^2 + 12$
54. $P(x) = x^4 + 11x^2 + 18$
55. $P(x) = x^4 - x^3 - 3x^2 + 17x - 30$
56. $P(x) = x^4 - 16x^2 + 40x - 25$
57. $P(x) = x^4 - 6x^3 + 15x^2 - 26x + 6$
58. $P(x) = x^4 - 6x^3 + 14x^2 - 10x - 7$
59. $x = -3$; multiplicity one; $x = 3$; multiplicity two; degree 3
60. $x = -4$; multiplicity one; $x = 4$; multiplicity two; degree 3
61. $x = 2$; multiplicity three; degree 3
62. $x = 5$; multiplicity three; degree 3
63. $x = 3$; multiplicity three; $x = -3$; multiplicity one; degree 4
64. $x = 1$; multiplicity three; $x = -1$; multiplicity one; degree 4
65. $x = -4$; multiplicity two; $x = 3$; multiplicity two; $x = -3$; multiplicity one; degree 5
66. $x = 1$; multiplicity four; $x = 2$; multiplicity one; degree 5
67. 4-in. squares; 16 in. × 10 in. × 4 in.
68. 6.5-in. squares
69. $P(x) = x^3 - 2x^2 + 9x - 18$
70. $P(x) = x^3 + 5x^2 + 16x + 80$
71. $P(x) = x^4 - 4x^3 + 13x^2 - 36x + 36$
72. $P(x) = x^4 + 10x^3 + 41x^2 + 160x + 400$
73. $P(x) = x^3 - 3x^2 + 7x - 5$
74. $P(x) = x^3 - 3x^2 + x + 5$
75. $P(x) = (x - 1)^3(x - 1 - 2i)(x - 1 + 2i)$
76. $P(x) = (x + 1)^3(x - 2 + i)(x - 2 - i)$
77. $P(x) = (x + 3)^2(x - 1 - \sqrt{2})(x - 1 + \sqrt{2})(x - 5i)(x + 5i)$
78. $P(x) = (x + 2)^2(x - 2 + \sqrt{3})(x - 2 - \sqrt{3})(x - 4i)(x + 4i)$

45. $x = 1 + \sqrt{5}, x = 1 - \sqrt{5}, x = 1 + i, x = 1 - i$; degree 4

46. $x = 3, x = 1, x = 3 + \sqrt{2}i, x = 3 - \sqrt{2}i$; degree 4

Find a cubic polynomial with real coefficients having the roots specified. Assume a lead coefficient of 1.

47. $x = 4, x = -3i$

48. $x = -3, x = 5i$

49. $x = 5, x = 2 + 3i$

50. $x = -3, x = 1 - 5i$

51. $x = 2, x = 3 + \sqrt{2}i$

52. $x = 1, x = -1 + \sqrt{5}i$

Find a quartic polynomial (degree 4) with real coefficients having the roots specified. Assume a lead coefficient of 1.

53. $x = 2i, x = \sqrt{3}i$

54. $x = -3i, x = \sqrt{2}i$

55. $x = 2, x = -3, x = 1 + 2i$

56. $x = 1, x = -5, x = 2 - i$

57. $x = 2 + \sqrt{3}, x = 1 + \sqrt{5}i$

58. $x = 1 - \sqrt{2}, x = 2 + \sqrt{3}i$

Factor each polynomial completely, then state the multiplicity of its roots and the degree of P.

59. $P(x) = x^3 - 3x^2 - 9x + 27$

60. $P(x) = x^3 - 4x^2 - 16x + 64$

61. $P(x) = x^3 - 6x^2 + 12x - 8$

62. $P(x) = x^3 - 15x^2 + 75x - 125$

63. $P(x) = (x^2 - 6x + 9)(x^2 - 9)$

64. $P(x) = (x^2 - 1)(x^2 - 2x + 1)$

65. $P(x) = (x^3 + 4x^2 - 9x - 36)(x^2 + x - 12)$

66. $P(x) = (x^3 - 3x^2 + 3x - 1)(x^2 - 3x + 2)$

WORKING WITH FORMULAS

Volume of an open box: $V = 4x^3 - 84x^2 + 432x$

An open box is constructed by cutting square corners from a 24 in. by 18 in. sheet of cardboard and folding up the sides. Its volume is given by the formula shown, where x represents the size of the square cut.

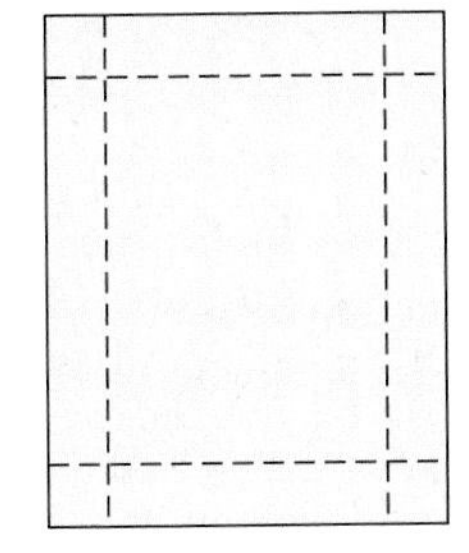

67. Given a volume of 640 in^3, use synthetic division and the remainder theorem to determine if the squares were 2-, 3-, 4-, or 5-in. squares and state the dimensions of the box. (*Hint:* Set $640 = 4x^3 - 84x^2 + 432x$ equal to zero and divide.)

68. Given the volume is 357.5 in^3, use synthetic division and the remainder theorem to determine if the squares were 5.5-, 6.5-, or 7.5-in. squares. (*Hint:* Set $357.5 = 4x^3 - 84x^2 + 432x$ equal to zero and divide.)

APPLICATIONS

Find a polynomial P with real coefficients having the degree specified and only one real, rational root and the indicated zeroes. Assume a lead coefficient of 1. Leave 75 to 78 in factored form.

69. $x = 2, x = -3i$; degree 3

70. $x = -5, x = 4i$; degree 3

71. $x = 2, x = -3i$; degree 4

72. $x = -5, x = 4i$; degree 4

73. $x = 1, x = 1 + 2i$; degree 3

74. $x = -1, x = 2 - i$; degree 3

75. $x = 1, x = 1 + 2i$; degree 5

76. $x = -1, x = 2 - i$; degree 5

77. $x = -3, x = 1 + \sqrt{2}, x = 5i$; degree 6

78. $x = -2, x = 2 - \sqrt{3}, x = 4i$; degree 6

79.

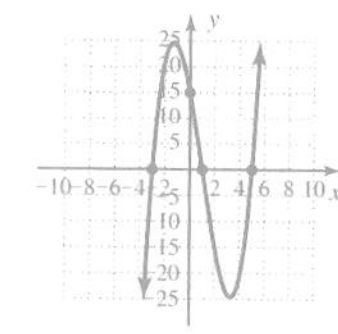

80.

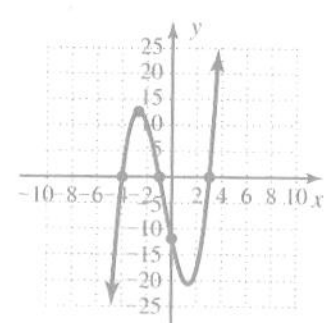

81.

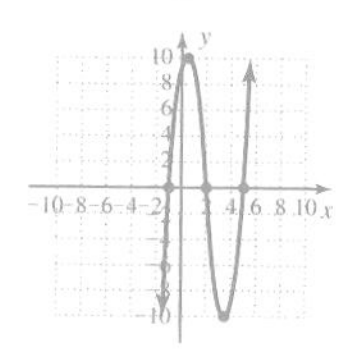

82.

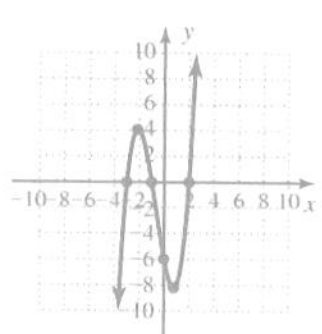

83.

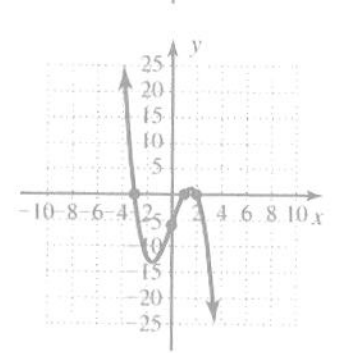

84.

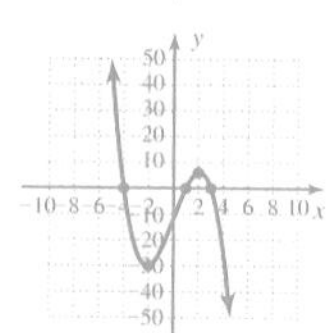

85. a. week 10, 22.5 thousand
b. one week before closing, 36 thousand
c. week 9

86. a. month 10, \$7.5 million
b. after the eleventh month, \$36 million
c. month 8

87. a. $P(x) = x^2 - 4x + 13$
b. $P(x) = x^2 - 2x + 3$

88. Answers will vary.

89. $k = 22$

90. $k = 2$

Use the remainder theorem to verify the values given are zeroes of f and to find additional "midinterval" points. Use these points along with the end behavior and y-intercept to graph the functions.

79. $f(x) = x^3 - 3x^2 - 13x + 15$; $x = -3, x = 1, x = 5$

80. $f(x) = x^3 + 2x^2 - 11x - 12$; $x = -4, x = -1, x = 3$

81. $f(x) = x^3 - 6x^2 + 3x + 10$; $x = -1, x = 2, x = 5$

82. $f(x) = x^3 + 2x^2 - 5x - 6$; $x = -3, x = -1, x = 2$

83. $f(x) = -x^3 + 7x - 6$; $x = -3, x = 1, x = 2$

84. $f(x) = -x^3 + 13x - 12$; $x = -4, x = 1, x = 3$

85. Tourist population: During the 12 weeks of summer, the population of tourists at a popular beach resort is modeled by the polynomial $P(w) = -0.1w^4 + 2w^3 - 14w^2 + 52w + 5$, where $P(w)$ is the tourist population (in 1000s) during week w. Use the remainder theorem to help answer the following questions.

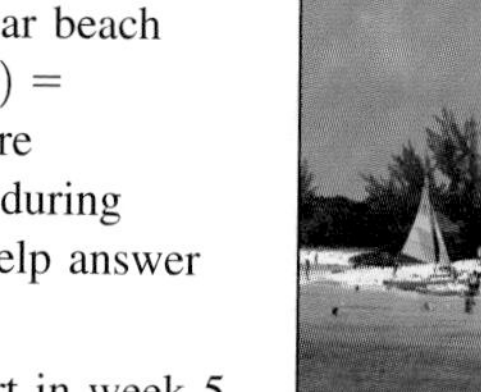

a. Were there more tourists at the resort in week 5 ($w = 5$) or week 10? How many more?

b. Were more tourists at the resort one week after opening ($w = 1$) or one week before closing ($w = 11$). How many more?

c. The tourist population peaked (reached its highest) between weeks 7 and 10. Use the remainder theorem to determine the peak week.

86. Debt load: Due to a fluctuation in tax revenues, a county government is projecting a deficit for the next 12 months, followed by a quick recovery and the repayment of all debt near the end of this period. The projected debt can be modeled by the polynomial $D(m) = 0.1x^4 - 2x^3 + 15x^2 - 64x - 3$, where $D(m)$ represents the amount of debt (in millions of dollars) in month m. Use the remainder theorem to help answer the following questions.

a. Was the debt higher in month 5 ($m = 5$) or month 10 of this period? How much higher?

b. Was the debt higher in the first month of this period (one month into the deficit) or after the eleventh month (one month before the expected recovery)? How much higher?

c. The total debt reached its maximum between months 7 and 10. Use the remainder theorem to determine which month.

WRITING, RESEARCH, AND DECISION MAKING

87. Recall th product of the complex conjugates $(a + bi)(a - bi)$ is $a^2 + b^2$. Show that any quadrati having the zeroes $x = a + bi$ and $x = a - bi$ can be written directly as $x^2 - 2ax + (a^2 + b^2)$. For $x = 1 + 2i$ and $x = 1 - 2i$: $a = 1, b = \pm 2, a^2 + b^2 = 5$, and we have $x^2 - 2x + 5$ directly. Use the new method to write quadratic polynomials with these roots: ($x = 2 + 3i$ and $x = 2 - 3i$ and (b) $x = 1 + \sqrt{2}i$ and $x = 1 - \sqrt{2}i$.

88. Given that $x = -2, x = -3$, and $x = 1 + 2i$ are zeroes of a fourth-degree polynomial $P(x)$, discuss how you might find the value of $P(1)$ without having to write $P(x)$ in polynomial form.

EXTENDING THE CONCEPT

89. For what value of k will $x = -2, x = 1 - 5i$, and $x = 1 + 5i$ be zeroes of $P(x) = x^3 + kx + 52$?

90. Find a value of k so that the following quotient has a remainder of 6: $(x^3 + 3x^2 - 3kx - 10) \div (x + 2)$.

91. $S_3 = 36$; $S_5 = 225$
92. $f^{-1}(x) = \sqrt{x} + 3$; D: $[0, \infty)$, R: $[3, \infty)$
93.

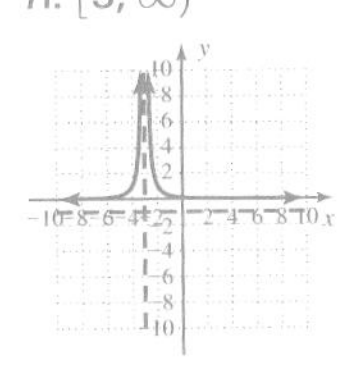

94. $P(x) = 1400x + 5000$, where x is the number of years since 2000
95. $x = 0$, $x = \frac{3}{2}$, $x = -1$
96. $\frac{25}{4}$
97. -1.9

91. The sum of the first n perfect cubes is given by the formula $S = \frac{1}{4}(n^4 + 2n^3 + n^2)$. Use the remainder theorem on S to find the sum of (a) the first three perfect cubes (divide by $n - 3$) and (b) the first five perfect cubes (divide by $n - 5$). Check results by adding the perfect cubes manually. To avoid working with fractions you can initially ignore the $\frac{1}{4}$ (use $n^4 + 2n^3 + n^2 + 0n + 0$), as long as you divide the remainder by 4.

MAINTAINING YOUR SKILLS

92. (3.2) Given $f(x) = (x - 3)^2$ for $x \geq 3$, find $f^{-1}(x)$ and state its domain and range.

93. (3.5) Graph $g(x) = \dfrac{1}{(x + 3)^2} - 1$ using transformations of the parent function.

94. (2.3) The profit of a small business increased linearly from \$5000 in 2000 to \$12,000 in 2005. Find a linear function modeling the growth of the company's profit.

95. (1.3) Solve by factoring: $2x^3 - x^2 - 3x = 0$.

96. (3.4) Create a perfect square trinomial: $x^2 - 5x + \_\_\_$.

97. (2.4) Given $f(x) = x^2 - 4x$, use the average rate of change formula to find $\dfrac{\Delta y}{\Delta x}$ in the interval $x \in [1.0, 1.1]$.

4.3 The Zeroes of Polynomial Functions

LEARNING OBJECTIVES

In Section 4.3 you will learn how to:

A. **Work with complex polynomials and apply the fundamental theorem of algebra**

B. **Locate the roots of a real polynomial using the intermediate value theorem**

C. **Find the rational roots of a polynomial using the rational roots theorem**

D. **Find bounds on zeroes of real polynomials using Descartes's rule of signs and the upper and lower bounds rule**

INTRODUCTION

This section represents one of the highlights in the college algebra curriculum, because it offers a look at what many call *the big picture*. The ideas presented are the result of a cumulative knowledge base developed over a long period of time, and give a fairly comprehensive view of the study of polynomial functions.

POINT OF INTEREST

It is well known that the great mathematician Carl Friedreich Gauss (1777–1855) gave the first proof of the fundamental theorem of algebra at the age of 20, assuming the coefficients of the polynomial were real numbers. It is far less known that he offered a fourth and last proof at the age of 70, in which he assumed the polynomial coefficients were complex numbers. Because complex numbers are bound by the same algebraic properties as real numbers, our work with polynomial operations and properties, synthetic division, as well as the remainder and factor theorems can also be extended to complex polynomials. Differences exist, however, as complex polynomials cannot be graphed on the real plane and complex solutions of a complex polynomial need not occur in conjugate pairs.

A. The Fundamental Theorem of Algebra

From Section 1.4, we know that real numbers are a subset of the complex numbers: $\mathbb{R} \subset \mathbb{C}$. Because complex numbers are the "larger" set (containing all other number sets), properties and theorems about complex numbers are more powerful and far reaching than theorems about real numbers. In the same way, real polynomials are a subset of the complex polynomials, and the same principle applies. Complex polynomials are still written in the form $P(x) = a_n x^n + a_{n-1}x^{n-1} + \cdots + a_1 x^1 + a_0$, but now $a_n, a_{n-1}, \ldots, a_1, a_0$ *represent complex numbers.* For hundreds of years, the greatest mathematical minds sought to bring unification and finality to the study of polynomial functions by proving that solutions existed for all polynomial equations, and that there were exactly as many solutions as the degree of the polynomial. But it was not until 1797 that Carl Friedreich Gauss offered the first satisfactory proof in a dissertation entitled, *A New Proof that Every Rational Integral Function of One Variable Can Be Resolved into Real Factors of the First or Second Degree.* The proof of this statement is based on a theorem that is the foundational bedrock for a complete study of polynomial functions, and has come to be known as the **fundamental theorem of algebra.**

THE FUNDAMENTAL THEOREM OF ALGEBRA

Every complex polynomial of degree $n \geq 1$ has at least one complex root.

Although the statement may seem trivial, it enables us to draw two important conclusions. The first is that our search for a solution will not be fruitless or wasted—solutions for *all* polynomial equations exist. Second, the fundamental theorem combined with the factor theorem enables us to state the **linear factorization theorem,** which is stated here in both informal and formal terms.

THE LINEAR FACTORIZATION THEOREM

Less formal: Every complex polynomial of degree $n \geq 1$ can be rewritten as the product of a nonzero constant and exactly n linear factors.

More formal: If $P(x)$ is a complex polynomial of degree $n \geq 1$, then P has exactly n linear factors and can be written in the form $P(x) = a(x - c_1)(x - c_2)\cdot \ldots \cdot(x - c_n)$ where $a \neq 0$ and $c_1, c_2, \ldots, c_n$ are (not necessarily distinct) complex numbers. Note that some factors may have multiplicities greater than 1.

Proof of the Linear Factorization Theorem

Given $P(x) = a_n x^n + a_{n-1}x^{n-1} + \cdots + a_1 x^1 + a_0$ is a complex polynomial, the fundamental theorem of algebra establishes that P has at least one complex zero, which we will call c_1. The factor theorem stipulates $(x - c_1)$ must be a factor of P, giving

$$P(x) = (x - c_1)q_1(x)$$

where $q_1(x)$ is a complex polynomial of degree $n - 1$.

Since $q_1(x)$ is a complex polynomial in its own right, it too must have a complex zero, call it c_2. Then $(x - c_2)$ must be a factor of $q_1(x)$, giving

$$P(x) = (x - c_1)(x - c_2)q_2(x)$$

where $q_2(x)$ is a complex polynomial of degree $n - 2$.

Repeating this rationale n times will cause $P(x)$ to be rewritten in the form

$$P(x) = (x - c_1)(x - c_2)\cdot\ldots\cdot(x - c_n)q_n(x)$$

where $q_n(x)$ has a degree of $n - n = 0$, resulting in a nonzero constant typically called a_n.
The result is $P(x) = a_n(x - c_1)(x - c_2)\cdot\ldots\cdot(x - c_n)$, and the proof is complete.

The impact of the theorem was hinted at in Section 4.2 (polynomial zeroes theorem) but is now more conclusive: *Every* polynomial equation, real or complex, has exactly n roots, counting roots of multiplicity. Example 1 is designed to help you see the "big picture," as we apply the ideas to select complex polynomials. The ability to solve more general complex equations must wait until a future course, when the square root of a complex number is developed (see *Working with Formulas,* Exercises 95 and 96).

EXAMPLE 1 Find all zeroes of the complex polynomial C, given $x = 1 - i$ is a zero. Then write C in completely factored form:
$C(x) = x^3 + (-1 + 2i)x^2 + (5 - i)x + (-6 + 6i)$.

Solution: Begin by using $x = 1 - i$ and synthetic division to find the quotient polynomial.

$$\begin{array}{r|rrrr} 1-i & 1 & -1+2i & 5-i & -6+6i \\ & \downarrow & 1-i & 1+i & 6-6i \\ \hline & 1 & i & 6 & \lfloor 0 \end{array}$$

Since the remainder is zero, $1 - i$ is indeed a root and the quotient polynomial is $x^2 + ix + 6$. This gives $C(x) = (x^2 + ix + 6)[x - (1 - i)]$ in factored form. To find the remaining zeroes (there must be a total of three), we apply the quadratic formula to the quotient polynomial with $a = 1$, $b = i$, and $c = 6$.

$$x = \frac{-b \pm \sqrt{b^2 - 4ac}}{2a}$$

$$= \frac{-i \pm \sqrt{(i)^2 - 4(1)(6)}}{2(1)} = \frac{-i \pm \sqrt{-1 - 24}}{2}$$

$$= \frac{-i \pm \sqrt{-25}}{2} = \frac{-i \pm 5i}{2}$$

$$x = \frac{-i + 5i}{2} = 2i \text{ and } x = \frac{-i - 5i}{2} = -3i$$

The completely factored form is $C(x) = (x - 2i)(x + 3i)(x - 1 + i)$

NOW TRY EXERCISES 7 THROUGH 14

WORTHY OF NOTE

Specifically note that in Example 1, the zeroes of C were *not complex conjugates*, because the coefficients were complex numbers (not real numbers). The complex conjugates theorem applies only to real polynomials.

If we limit ourselves to polynomials with real coefficients, complex roots must occur in conjugate pairs and we can state this corollary to the linear factorization theorem. Note that a polynomial with no real roots is said to be **irreducible.**

COROLLARY TO THE LINEAR FACTORIZATION THEOREM
Given f is a polynomial with real coefficients, it can be factored into a product of linear factors (which are not necessarily distinct) and irreducible quadratic factors having real coefficients.

For further confirmation, see Example 5.

B. Real Polynomials and the Intermediate Value Theorem

The fundamental theorem of algebra is called an **existence theorem,** as it affirms the *existence* of the zeroes but does not tell us where to locate or how to find them. Because polynomial graphs are continuous (there are no holes or breaks in the graph), the **intermediate value theorem (IVT)** can be used for this purpose (see Figure 4.5).

Figure 4.5

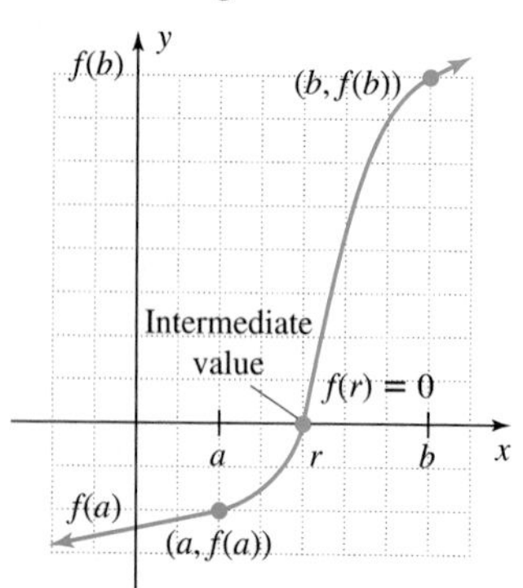

THE INTERMEDIATE VALUE THEOREM (IVT)
Given f is a polynomial with real coefficients, if $f(a)$ and $f(b)$ have opposite signs, there is *at least* one value r between a and b such that $f(r) = 0$.

You might recall a similar idea was used in our solution of quadratic inequalities (Section 2.5), where we noted the zeroes from linear factors of a polynomial cut the domain of the function into intervals where function values change sign from positive to negative or from negative to positive.

EXAMPLE 2 Use the IVT to show $P(x) = x^3 - 9x + 6$ has a zero in the interval given:

a. $[-4, -3]$ **b.** $[0, 1]$

Solution: Begin by evaluating P at $x = -3$ and $x = -4$.

a. $P(-3) = x^3 - 9x + 6$ $= -27 + 27 + 6$ $= 6$

$P(-4) = (-4)^3 - 9(-4) + 6$ $= -64 + 36 + 6$ $= -22$

Since $P(-3) > 0$ and $P(-4) < 0$, there must be a number r_1 between -4 and -3 where $P(r_1) = 0$. The graph must cross the x-axis at r_1.

b. Evaluate P at $x = 0$ and $x = 1$.

$P(0) = (0)^3 - 9(0) + 6$ $= 0 - 0 + 6$ $= 6$

$P(1) = (1)^3 - 9(1) + 6$ $= 1 - 9 + 6$ $= -2$

Since $P(0) > 0$ and $P(1) < 0$, there must be a number r_2 between 0 and 1 where $P(r_2) = 0$. The graph must cross the x-axis at r_2 (see the *Technology Highlight* that follows).

NOW TRY EXERCISES 15 THROUGH 18

Note the remainder theorem can also be used to evaluate the endpoints of an interval.

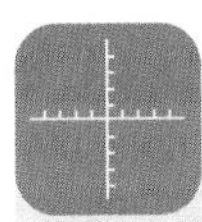

TECHNOLOGY HIGHLIGHT

The Intermediate Value Theorem and Split Screen Viewing

The keystrokes shown apply to a TI-84 Plus model. Please consult your manual or our Internet site for other models.

Graphical support for the results of Example 2 is shown in Figure 4.6 using the window $x \in [-5, 5]$ and $y \in [-10, 20]$. The zero of P between 0 and 1 is highlighted, and the zero between $x = -4$ and $x = -3$ is clearly seen. Note there is also a third zero between 2 and 3.

The TI 84 Plus (and other models) offer a useful feature called *split screen viewing,* that enables us to view a table of values and the graph of a function at the same time. To illustrate, enter the function $y = x^3 - 9x + 6$ for Y_1 on the Y= screen. Press the ZOOM **4:ZDecimal** keys to view the graph, then adjust the viewing window as needed to get a comprehensive view. Set up your table in **AUTO** mode with ΔTbl = 1 [use 2nd WINDOW **(TBLSET)**]. Use the table of values (2nd GRAPH) to locate any real zeroes of f [look for where $f(x)$ changes in sign]. To support this concept we can view *both the graph and table at the same time.* Press the MODE key and notice the second-to-last entry on this screen reads: **Full** (for full screen viewing), **Horiz** for splitting the screen horizontally with the graph above a reduced table of values, and **G-T,** which represents **Graph-Table** and splits the screen vertically. In the **G-T** mode, the graph appears on the left and the table of values on the right. Navigate the cursor to the **G-T** mode and press ENTER . Pressing the GRAPH key at this point should give you a screen similar to Figure 4.7. Use this feature to complete the following exercises.

Figure 4.6

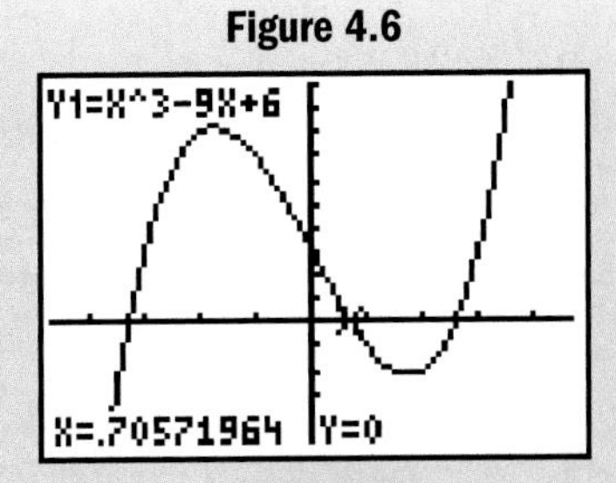

Figure 4.7

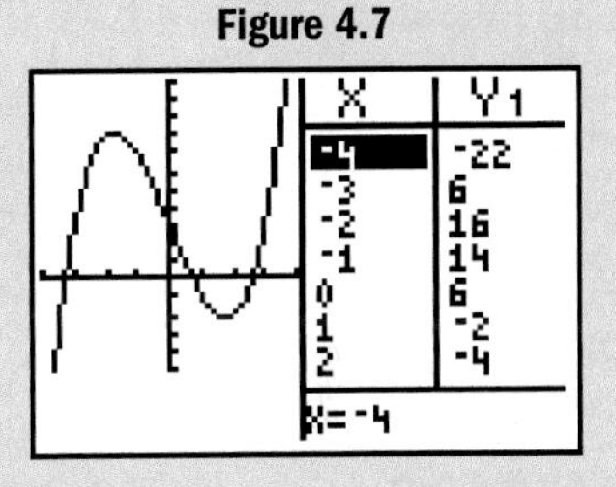

Exercise 1: What do the graph, table and the IVT tell you about the zeroes of this function?

Exercise 2: Go to TBLSET and reset TblStart = −4 and ΔTbl = 0.1. Use 2nd GRAPH to walk through the table values. Does this give you a better idea about where the zeroes are located?

Exercise 3: Press the TRACE key. What happens to the table as you trace through the points on Y_1?

C. The Rational Roots Theorem

The fundamental theorem of algebra tells us that roots of a polynomial equation *exist*. The intermediate value theorem tells us how to *locate* the roots. Our next theorem gives us the information we need to actually *solve* for certain roots of a polynomial.

In Section 4.1 we noted two things were needed to make division an effective mathematical tool: (1) a more efficient method of dividing polynomials and (2) a way to find divisors that give a remainder of zero. These would make it possible to factor larger polynomials. Synthetic division provides the efficient *method* for division. To find *divisors that give a remainder of zero,* we make the following observations. To solve $3x^2 - 11x - 20 = 0$ by factoring, a beginner might write out all possible binomial pairs where the *first* term in the F-O-I-L process multiplies to $3x^2$ and the *last* term multiplies to 20. The six possibilities are shown here:

$(3x \quad 1)(x \quad 20)$ $\quad$ $(3x \quad 20)(x \quad 1)$ $\quad$ $(3x \quad 2)(x \quad 10)$ $\quad$ $(3x \quad 10)(x \quad 2)$

$(3x \quad 4)(x \quad 5)$ $\quad$ $(3x \quad 5)(x \quad 4)$

If $3x^2 - 11x - 20$ is factorable using integers, the factors *must be somewhere in this list*. Also, the first term in each binomial must be a factor of the leading coefficient (3) and the second term must be a factor of the constant term (20). This means that regardless of which factored form is correct, the solution must be a rational number whose numerator comes from the factors of 20, and whose denominator comes from the factors of 3. The correct factored form is shown here, along with the solution:

$$3x^2 - 11x - 20 = 0$$
$$(3x + 4)(x - 5) = 0$$
$$3x + 4 = 0 \qquad x - 5 = 0$$
$$x = \frac{-4}{3} \begin{matrix} \leftarrow \text{from the factors of 20} \\ \leftarrow \text{from the factors of 3} \end{matrix} \qquad x = \frac{5}{1} \begin{matrix} \leftarrow \text{from the factors of 20} \\ \leftarrow \text{from the factors of 3} \end{matrix}$$

This same principle also applies to polynomials of higher degree, with the result of these observations stated as the **rational roots theorem (RRT).**

THE RATIONAL ROOTS THEOREM (RRT)

Given a real polynomial $P(x)$ with degree $n \geq 1$ and integer coefficients, the rational roots of P (if they exist) must be of the form $\frac{p}{q}$, where p is a factor of the constant term a_0 and q is a factor of the lead coefficient a_n $\left(\frac{p}{q} \text{ must be written in lowest terms}\right)$.

Note that if the lead coefficient is 1 ($a_n = 1$), the RRT reaffirms the *principle of factorable polynomials* from Section 4.1. If the lead coefficient is not a "1" and the constant term has a large number of factors, the list of possible rational roots becomes rather large. In this case it helps to begin with all of the factor *pairs* of the constant a_0, then divide each of these by the factors of a_n to see what possibilities are not already in the list and must be added.

EXAMPLE 3 List possible rational roots for $3x^4 + 14x^3 - x^2 - 42x - 24 = 0$, but do not solve.

Solution: Using the RRT, rational roots must be of the form $\frac{p}{q}$, where p is a factor of $a_0 = -24$ and q is a factor of $a_n = 3$. The factor pairs of -24 are: ±1, ±24, ±2, ±12, ±3, ±8, ±4, and ±6. Dividing each by ±1 and ±3 (the factor pairs of 3), we note division by ±1 will not change any of the previous values, while division by ±3 gives $\pm\frac{1}{3}$, $\pm\frac{2}{3}$, $\pm\frac{8}{3}$, $\pm\frac{4}{3}$ as additional possibilities. Any rational roots must be in the set $\{\pm1, \pm24, \pm2, \pm12, \pm3, \pm8, \pm4, \pm6, \pm\frac{1}{3}, \pm\frac{2}{3}, \pm\frac{8}{3}, \pm\frac{4}{3}\}$.

NOW TRY EXERCISES 19 THROUGH 26

The actual solutions to the equation in Example 3 are $x = \sqrt{3}$, $x = -\sqrt{3}$, $x = -\frac{2}{3}$, and $x = -4$. Although the *rational* roots are indeed in the set noted, it's very apparent we need a way to narrow down the number of possibilities (we don't want to try all 24 possible roots). If we're able to find even one factor easily, we can rewrite the polynomial using this factor and the quotient, with the hope of factoring further using trinomial factoring or factoring by grouping. Many times testing to see if 1 or -1 are zeroes will help. To see if $x = 1$ is a zero, simply *add the coefficients of the polynomial,* since 1 to any power is still 1. To test -1, first change the sign of all terms with odd degree, then add the coefficients [since $(-1)^{\text{even}} = 1$, while $(-1)^{\text{odd}} = -1$]. In summary, for any real polynomial equation $f(x) = 0$:

1. *If the sum of all coefficients is zero, $x = 1$ is a root and $(x - 1)$ is a factor.*
2. *After changing the sign of all terms with odd degree, if the sum of the coefficients is zero, then $x = -1$ is a root and $(x + 1)$ is a factor.*

EXAMPLE 4 Find all rational zeroes of $f(x) = 3x^4 - x^3 - 8x^2 + 2x + 4$, and use them to write the function in factored form. Use the factored form to state *all* zeroes of f.

Solution: Instead of listing all possibilities using the RRT, let's first test for 1 and -1, then see if we're fortunate enough to complete the factorization using other means. The sum of the coefficients is: $3 - 1 - 8 + 2 + 4 = 0$, which means $x = 1$ is a root and $x - 1$ is a factor. By changing the sign on terms of odd degree, we have $3x^4 + x^3 - 8x^2 - 2x + 4$ and $3 + 1 - 8 - 2 + 4 = -2$, showing $x = -1$ is *not* a root. Using $x = 1$ and synthetic division gives:

use 1 as a "divisor"

$$\begin{array}{r|rrrrr} 1 & 3 & -1 & -8 & 2 & 4 \\ & & 3 & 2 & -6 & -4 \\ \hline & 3 & 2 & -6 & -4 & 0, \end{array}$$

enabling us to rewrite f as $f(x) = (x - 1)(3x^3 + 2x^2 - 6x - 4)$. Noting that $q(x)$ can be factored by grouping, we need not employ the RRT or continue with synthetic division.

$$\begin{aligned} f(x) &= (x - 1)(3x^3 + 2x^2 - 6x - 4) && f(x) = (x-1)[q(x)] \\ &= (x - 1)[x^2(3x + 2) - 2(3x + 2)] && \text{factor by grouping} \\ &= (x - 1)(3x + 2)(x^2 - 2) && \text{linear and quadratic factors} \\ &= (x - 1)(3x + 2)(x + \sqrt{2})(x - \sqrt{2}) && \text{completely factored form} \end{aligned}$$

The zeroes of f are $x = 1$, $x = \frac{-2}{3}$, and $x = \pm\sqrt{2}$.

NOW TRY EXERCISES 27 THROUGH 48

In Example 4, we completed the factorization by factoring the quotient polynomial. In cases where $q_1(x)$ cannot be factored easily, we continue with synthetic division and other possible roots, until the remaining zeroes can be determined. See Exercises 49 through 68.

D. Descartes's Rule of Signs and Upper/Lower Bounds

Testing $x = 1$ and $x = -1$ is one way to reduce the number of possible rational roots, but unless we're very lucky, factoring the polynomial can still be a challenge. **Descartes's rule of signs** and the **upper and lower bounds property** offer additional assistance.

WORTHY OF NOTE

It's useful to note that if $f(x)$ is a *complete polynomial* (with no zero coefficients), the equation will have as many negative, real roots as the number of times there are two positive terms or two negative terms in succession, or an even number less.

DESCARTES'S RULE OF SIGNS

Given the real polynomial equation $f(x) = 0$,

1. The number of positive real roots is equal to the number of variations in sign for $f(x)$, or an even number less.
2. The number of negative real roots is equal to the number of variations in sign for $f(-x)$, or an even number less.

EXAMPLE 5 For $f(x) = 2x^5 - 5x^4 + x^3 + x^2 - x + 6$, (a) use the RRT to list all possible rational roots; (b) apply Descartes's rule to count the possible number of positive, negative, and complex roots; and (c) use this information, the tests for -1 and 1, synthetic division, and the factor theorem to factor f completely.

Solution:

a. Applying the RRT, the possible rational roots are $\left\{\pm 1, \pm 6, \pm 2, \pm 3, \pm\frac{1}{2}, \pm\frac{3}{2}\right\}$.

b. When using Descartes's rule, it helps to organize your work in a table. Begin by finding the possible number of positive roots, then decrease this number by two and then by two again, if possible (the resulting number must be greater than or equal to zero). Repeat for the negative roots. Since a polynomial of degree n must have n roots, the total possibilities for each *row* must sum to n. To illustrate Descartes's rule, we have positive terms in blue and negative terms in red: $f(x) = 2x^5 - 5x^4 + x^3 + x^2 - x + 6$. The terms change sign a total of four times, meaning there are four, two, or zero positive roots. For the negative roots, we look for two positive or two negative terms in succession (since f is a complete polynomial). There are two blue (positive) terms in succession, meaning there is exactly one negative root. Knowing there's a total of five roots (the degree of f is 5) gives the table shown.

Possible Positive Roots	Possible Negative Roots	Possible Complex Roots	Total Number of Roots
4	1	0	5
2	1	2	5
0	1	4	5

c. Testing 1 and -1 shows $x = 1$ is not a root, but $x = -1$ *is*. After changing the sign on terms of odd degree, the coefficients sum to zero: $-2 - 5 - 1 + 1 + 1 + 6 = 0$. Using $x = -1$ and synthetic division gives

$$\begin{array}{r|rrrrrrl} -1 & 2 & -5 & 1 & 1 & -1 & 6 & \text{coefficients of } f(x) \\ & & -2 & 7 & -8 & 7 & -6 & \\ \hline & 2 & -7 & 8 & -7 & 6 & 0 & q_1(x) \text{ is not easily factored} \end{array}$$

(use -1 as a "divisor")

Since there is only one negative root, we've cut the number of possibilities in half and we need only check the remaining positive roots given by the RRT. The quotient $q_1(x)$ is not easily factored, so we continue with synthetic division using the next larger possible root, $x = 2$.

$$\begin{array}{r|rrrrrl} 2 & 2 & -7 & 8 & -7 & 6 & \text{coefficients of } q_1(x) \\ & & 4 & -6 & 4 & -6 & \\ \hline & 2 & -3 & 2 & -3 & 0 & q_2(x) \text{ is easily factored} \end{array}$$

(use 2 as a "divisor")

The partial factored form is $f(x) = (x + 1)(x - 2)(2x^3 - 3x^2 + 2x - 3)$, which we can complete using factoring by grouping. The factored form is

$$\begin{aligned} f(x) &= (x + 1)(x - 2)(2x^3 - 3x^2 + 2x - 3) && f(x) = (x+1)(x-2)[q_2(x)] \\ &= (x + 1)(x - 2)[x^2(2x - 3) + 1(2x - 3)] && \text{factor by grouping} \\ &= (x + 1)(x - 2)(2x - 3)(x^2 + 1) && \text{linear and quadratic factors} \\ &= (x + 1)(x - 2)(2x - 3)(x + i)(x - i) && \text{completely factored form} \end{aligned}$$

As you can see, the middle row of our table held the correct combination of two positive roots, one negative root, and two complex roots.

NOW TRY EXERCISES 61 THROUGH 82

Another idea that helps cut back a long list of possible roots is the **upper and lower bounds property.** A number b is an **upper bound** on the positive roots of a function if no positive root is greater than b. In the same way, a number a is a **lower bound** on the negative roots of a function if no negative root is less than a.

WORTHY OF NOTE

Many other tests for upper and lower bounds are available. Most involve a study of the coefficients of the given polynomial. Some tests give bounds that are too large for practical use, while the time required to apply other tests decreases their utility. See the *Writing, Research, and Decision Making* exercises.

UPPER AND LOWER BOUNDS PROPERTY

Given $f(x)$ is a real polynomial,

1. If $f(x)$ is divided by $x - b$ $(b > 0)$ using synthetic division and all coefficients in the quotient row are either positive or zero, then b is an upper bound on the zeroes of f.
2. If $f(x)$ is divided by $x - a$ $(a < 0)$ using synthetic division and all coefficients in the quotient row alternate in sign, then a is a lower bound on the zeroes of f. Note that zero coefficients can be either positive or negative as needed.

While this test certainly helps narrow the possibilities, we gain the additional benefit of knowing the property actually places boundaries on *all* roots of the polynomial, both rational and irrational.

In Example 5, part (c), the last row of the first synthetic division alternates in sign, showing $x = -1$ is both a zero and a lower bound.

TECHNOLOGY HIGHLIGHT
Complex Polynomials and the Big Picture

The keystrokes shown apply to a TI-84 Plus model. Please consult your manual or our Internet site for other models.

Although a study of complex polynomials is rarely developed to its fullest extent in a college algebra course, it is possible to "extend the boundaries" using the capabilities of a graphing calculator to catch a greater glimpse of the "big picture." In the *Technology Highlight* from Section 4.2, we used the TI-84 Plus to evaluate a real polynomial using complex numbers. The same ideas can be used to evaluate a complex polynomial, enabling us to determine whether an indicated number is a root, find substantial uses for operations on complex numbers, and verify several of the connections we used in our study of real polynomials. For example, consider the complex polynomial $C(x) = 2ix^2 + (4 + 2i)x + 8 - 4i$. To verify that $x = 1 + 2i$ is a zero, we store $1 + 2i$ in memory location x, then simply enter the complex polynomial expression on the home screen and press ENTER (Figure 4.8.) Since the result is zero, $x = 1 + 2i$ is a root. To find the remaining root, we can use synthetic division and the quotient polynomial, along with our calculator, to carry out the needed operations (Figure 4.9.)

Figure 4.8

```
1+2i→X
                    1+2i
2iX²+(4+2i)X+8-4
i
                       0
```

$$\begin{array}{r|rrr} 1+2i & 2i & 4+2i & 8-4i \\ & & -4+2i & -8+4i \\ \hline & 2i & 4i & \underline{|0} \end{array} \quad q_1(x)$$

1 + 2i as "divisor"

This gives $C(x) = [x - (1 + 2i)](2ix + 4i)$, and we can find the second zero of $C(x)$ by setting the quotient polynomial equal to zero and solving for x. The equation

Figure 4.9

```
(1+2i)*2i
                   -4+2i
(1+2i)*4i
                   -8+4i
```

$2ix + 4i = 0$ gives $x = \frac{-4i}{2i}$ or $x = -2$. This zero can also be verified on the home screen using -2 STO→ X,T,θ,n and recalling (2nd ENTER) the complex expression.

Finally, it seems intriguing that the same properties we've observed in our previous study of real polynomials are actually properties of the entire family of complex polynomials. For instance, given the complex quadratic polynomial $ax^2 + bx + c = 0$, the sum of the roots must always be equal to $-\frac{b}{a}$, while the product of the roots is $\frac{c}{a}$. This is verified in Figure 4.10. Use these ideas to complete the following exercise.

Figure 4.10

```
-(4+2i)/(2i)
                   -1+2i
-2+(1+2i)
                   -1+2i
(8-4i)/(2i)
                   -2-4i
-2*(1+2i)
```

Exercise 1: Use your calculator to verify that $x = 1 - 5i$ is one zero of $C(x) = x^2 - (3 - 2i)x + 17 - 7i$, then use synthetic division and your calculator to find the second root. Finally, verify the product of the two roots is $\frac{c}{a}$, and the sum of the two roots is $-\frac{b}{a}$. $2 + 3i$, verified

4.3 EXERCISES

CONCEPTS AND VOCABULARY

Fill in each blank with the appropriate word or phrase. Carefully reread the section if needed.

1. A complex polynomial is one where one or more <u>coefficients</u> are complex numbers.

2. The number c is called an <u>upper bound</u> if there is no positive root greater than c.

3. If $x = 1$ is a root, all <u>coefficients</u> of the polynomial <u>sum</u> to <u>0</u> and $(x - 1)$ is a <u>root</u> of the polynomial.

4. According to Descartes's rule of signs, there are as many <u>positive</u> real roots as changes in sign from term to term, or an <u>even</u> number less.

5. Which of the following values is *not* a possible root of $f(x) = 6x^3 - 2x^2 + 5x - 12$:

a. $x = \frac{4}{3}$ **b.** $x = \frac{3}{4}$ **c.** $x = \frac{1}{2}$

Discuss/Explain why. b; 4 is not a factor of 6

6. Discuss/explain the two conclusions drawn from *the fundamental theorem of algebra* (page 394). In particular, discuss the role of the theorem in the statement of the *linear factorization theorem.* Answers will vary.

7. $C(x) = (x - 4i)(x + 3)(x - 2)$, $x = -3, x = 2$
8. $C(x) = (x - 9i)(x + 4)(x + 1)$, $x = -4, x = -1$
9. $C(x) = (x - 3i)(x - 1 - 2i)(x - 1 + 2i)$, $x = 1 \pm 2i$
10. $C(x) = (x - i)(x - 2 - 5i)(x - 2 + 5i)$, $x = 2 \pm 5i$
11. $C(x) = (x - 6i)(x - 1 - \sqrt{3}\,i)(x - 1 + \sqrt{3}\,i)$, $x = 1 \pm \sqrt{3}\,i$
12. $C(x) = (x + 4i)(x - 3 - \sqrt{2}\,i)(x - 3 + \sqrt{2}\,i)$, $x = 3 \pm \sqrt{2}\,i$
13. $C(x) = (x - 2 + i)(x - 3i)(x + i)$, $x = 3i, x = -i$
14. $C(x) = (x - 2 + 3i)(x - 5i)(x + 2i)$, $x = 5i, x = -2i$
19. $\frac{\{\pm 1, \pm 15, \pm 3, \pm 5\}}{\{\pm 1, \pm 4, \pm 2\}}$
20. $\frac{\{\pm 1, \pm 20, \pm 2, \pm 10, \pm 4, \pm 5\}}{\{\pm 1, \pm 3\}}$
21. $\frac{\{\pm 1, \pm 15, \pm 3, \pm 5\}}{\{\pm 1, \pm 2\}}$
22. $\frac{\{\pm 1, \pm 14, \pm 2, \pm 7\}}{\{\pm 1, \pm 2\}}$
23. $\frac{\{\pm 1, \pm 28, \pm 2, \pm 14, \pm 4, \pm 7\}}{\{\pm 1, \pm 6, \pm 2, \pm 3\}}$
24. $\frac{\{\pm 1, \pm 36, \pm 2, \pm 18, \pm 3, \pm 12, \pm 4, \pm 9, \pm 6\}}{\{\pm 1, \pm 7\}}$
25. $\frac{\{\pm 1, \pm 3\}}{\{\pm 1, \pm 32, \pm 2, \pm 16, \pm 4, \pm 8\}}$
26. $\frac{\{\pm 1, \pm 6, \pm 2, \pm 3\}}{\{\pm 1, \pm 24, \pm 2, \pm 12, \pm 3, \pm 8, \pm 4, \pm 6\}}$
27. $(x + 4)(x - 1)(x - 3)$, $x = -4, 1, 3$
28. $(x - 1)(x - 4)(x + 5)$, $x = 1, 4, -5$
29. $(x + 3)(x + 2)(x - 5)$, $x = -3, -2, 5$
30. $(x + 4)(x + 2)(x - 6)$, $x = -4, -2, 6$
31. $(x + 3)(x - 1)(x - 4)$, $x = -3, 1, 4$
32. $(x + 2)(x - 1)(x - 5)$, $x = -2, 1, 5$
33. $(x + 2)(x - 3)(x - 5)$, $x = -2, 3, 5$
34. $(x + 4)(x - 2)(x - 6)$, $x = -4, 2, 6$

Additional answers can be found in the Instructor Answer Appendix.

DEVELOPING YOUR SKILLS

For each complex polynomial, one of its complex zeroes is given. Use the given zero to write the polynomial in completely factored form and find all other zeroes.

7. $C(x) = x^3 + (1 - 4i)x^2 + (-6 - 4i)x + 24i$; $x = 4i$
8. $C(x) = x^3 + (5 - 9i)x^2 + (4 - 45i)x - 36i$; $x = 9i$
9. $C(x) = x^3 + (-2 - 3i)x^2 + (5 + 6i)x - 15i$; $x = 3i$
10. $C(x) = x^3 + (-4 - i)x^2 + (29 + 4i)x - 29i$; $x = i$
11. $C(x) = x^3 + (-2 - 6i)x^2 + (4 + 12i)x - 24i$; $x = 6i$
12. $C(x) = x^3 + (-6 + 4i)x^2 + (11 - 24i)x + 44i$; $x = -4i$
13. $C(x) = x^3 + (-2 - i)x^2 + (5 + 4i)x + (-6 + 3i)$; $x = 2 - i$
14. $C(x) = x^3 - 2x^2 + (19 + 6i)x + (-20 + 30i)$; $x = 2 - 3i$

Use the intermediate value theorem to determine if the given polynomial has a zero "c_i" in the intervals specified (state yes or no). Do not find the zeroes.

15. $f(x) = x^3 + 2x^2 - 8x - 5$
 a. $[-4, -3]$ yes b. $[-2, -1]$ no c. $[2, 3]$ yes

16. $g(x) = x^4 - 2x^2 + 6x - 3$
 a. $[-3, -2]$ yes b. $[-1, 0]$ no c. $[0, 1]$ yes

17. $h(x) = 2x^3 + 13x^2 + 3x - 36$
 a. $[-5, -4]$ no b. $[-3, -2]$ yes c. $[1, 2]$ yes

18. $H(x) = 2x^4 + 3x^3 - 14x^2 - 9x + 8$
 a. $[-4, -3]$ yes b. $[-2, -1]$ yes c. $[1, 2]$ no

List all possible rational roots for the polynomials given, but do not solve.

19. $f(x) = 4x^3 - 19x - 15$
20. $g(x) = 3x^3 - 2x + 20$
21. $h(x) = 2x^3 - 5x^2 - 28x + 15$
22. $H(x) = 2x^3 - 19x^2 + 37x - 14$
23. $p(x) = 6x^4 - 2x^3 + 5x^2 - 28$
24. $q(x) = 7x^4 + 6x^3 - 49x^2 + 36$
25. $Y_1 = 32t^3 - 52t^2 + 17t + 3$
26. $Y_2 = 24t^3 + 17t^2 - 13t - 6$

Use the rational roots theorem to write each function in factored form and find all zeroes. Note $a = 1$.

27. $f(x) = x^3 - 13x + 12$
28. $g(x) = x^3 - 21x + 20$
29. $h(x) = x^3 - 19x - 30$
30. $H(x) = x^3 - 28x - 48$
31. $p(x) = x^3 - 2x^2 - 11x + 12$
32. $q(x) = x^3 - 4x^2 - 7x + 10$
33. $Y_1 = x^3 - 6x^2 - x + 30$
34. $Y_2 = x^3 - 4x^2 - 20x + 48$
35. $Y_3 = x^4 - 15x^2 + 10x + 24$
36. $Y_4 = x^4 - 23x^2 - 18x + 40$
37. $f(x) = x^4 + 7x^3 - 7x^2 - 55x - 42$
38. $g(x) = x^4 + 4x^3 - 17x^2 - 24x + 36$

Find all rational zeroes of the functions given and use them to write the function in factored form. Use the factored form to state *all* zeroes of f. Begin by applying the tests for 1 and -1.

39. $f(x) = 4x^3 - 7x + 3$
40. $g(x) = 9x^3 - 7x - 2$
41. $h(x) = 4x^3 + 8x^2 - 3x - 9$
42. $H(x) = 9x^3 + 3x^2 - 8x - 4$
43. $Y_1 = 2x^3 - 3x^2 - 9x + 10$
44. $Y_2 = 3x^3 - 14x^2 + 17x - 6$
45. $p(x) = 2x^4 + 3x^3 - 9x^2 - 15x - 5$
46. $q(x) = 3x^4 + x^3 - 11x^2 - 3x + 6$
47. $r(x) = 3x^4 - 5x^3 + 14x^2 - 20x + 8$
48. $s(x) = 2x^4 - x^3 + 17x^2 - 9x - 9$

Find the zeroes of the polynomials given using any combination of the RRT, synthetic division, testing for 1 and -1, and/or the remainder and factor theorems.

49. $f(x) = 2x^4 - 9x^3 + 4x^2 + 21x - 18$
 $x = 1, 2, 3, \frac{-3}{2}$
50. $g(x) = 3x^4 + 4x^3 - 21x^2 - 10x + 24$
 $x = 1, 2, -3, \frac{-4}{3}$

51. $h(x) = 3x^4 + 2x^3 - 9x^2 + 4$

52. $H(x) = 7x^4 + 6x^3 - 49x^2 + 36$

53. $p(x) = 2x^4 - 3x^3 - 21x^2 - 2x + 24$

54. $q(x) = 3x^4 - 10x^3 - 15x^2 + 30x - 8$

55. $r(x) = 3x^4 - 17x^3 + 23x^2 + 13x - 30$

56. $s(x) = 4x^4 - 15x^3 + 9x^2 + 16x - 12$

57. $Y_1 = x^5 + 6x^2 - 49x + 42$

58. $Y_2 = x^5 + 2x^2 - 9x + 6$

59. $P(x) = 3x^5 + x^4 + x^3 + 7x^2 - 24x + 12$

60. $P(x) = 2x^5 - x^4 - 3x^3 + 4x^2 - 14x + 12$

61. $Y_1 = x^4 - 5x^3 + 20x - 16$

62. $Y_2 = x^4 - 10x^3 + 90x - 81$

63. $r(x) = x^4 + 2x^3 - 5x^2 - 4x + 6$

64. $s(x) = x^4 + x^3 - 5x^2 - 3x + 6$

65. $p(x) = 2x^4 - x^3 + 3x^2 - 3x - 9$

66. $q(x) = 3x^4 + x^3 + 13x^2 + 5x - 10$

67. $f(x) = 2x^5 - 7x^4 + 13x^3 - 23x^2 + 21x - 6$

68. $g(x) = 4x^5 + 3x^4 + 3x^3 + 11x^2 - 27x + 6$

Gather information on each polynomial using the rational roots theorem, testing for 1 and −1, applying Descartes's rule of signs, and using the upper and lower bounds property. Respond explicitly to each.

69. $f(x) = x^4 - 2x^3 + 4x - 8$

70. $g(x) = x^4 + 3x^3 - 7x - 6$

71. $h(x) = x^5 + x^4 - 3x^3 + 5x + 2$

72. $H(x) = x^5 + x^4 - 2x^3 + 4x - 4$

73. $p(x) = x^5 - 3x^4 + 3x^3 - 9x^2 - 4x + 12$

74. $q(x) = x^5 - 2x^4 - 8x^3 + 16x^2 + 7x - 14$

75. $r(x) = 2x^4 + 7x^2 + 11x - 20$

76. $s(x) = 3x^4 - 8x^3 - 13x - 24$

Use Descartes's rule of signs to determine the possible combinations of real and complex roots for each polynomial. Then graph the function on the standard window of a graphing calculator and adjust it as needed until you're certain all real roots are in clear view. Use this screen and a list of the possible rational zeroes (RRT) to factor the polynomial and find all zeroes (real and complex).

77. $f(x) = 4x^3 - 16x^2 - 9x + 36$

78. $g(x) = 6x^3 - 41x^2 + 26x + 24$

79. $h(x) = 6x^3 - 73x^2 + 10x + 24$

80. $H(x) = 4x^3 + 60x^2 + 53x - 42$

81. $p(x) = 4x^4 + 40x^3 - 97x^2 - 10x + 24$

82. $q(x) = 4x^4 - 42x^3 - 70x^2 - 21x - 36$

Find all zeroes (real and complex) of the polynomials given.

83. $f(x) = 9x^3 + 6x^2 - 5x - 2$

84. $g(x) = 8x^3 - 18x^2 + 7x + 3$

85. $q(x) = x^4 - 3x^3 - 11x^2 + 3x + 10$

86. $r(x) = -2x^4 + x^3 + 17x^2 - x - 15$

87. $r(x) = 2x^3 + 11x^2 - x - 30$

88. $s(x) = 3x^3 - 4x^2 - 59x + 20$

89. $Y_1 = 4x^3 - 41x^2 + 94x - 21$

90. $Y_2 = 6x^3 - 49x^2 + 98x - 15$

91. $f(x) = 3x^4 + 2x^3 - 9x^2 + 4$

92. $g(x) = 5x^4 - 18x^3 + 17x^2 - 4$

93. $h(t) = 32t^3 - 52t^2 + 17t + 3$

94. $p(n) = 24n^3 + 17n^2 - 13n - 6$

WORKING WITH FORMULAS

95. The absolute value of a complex number $z = a + bi$: $|z| = \sqrt{a^2 + b^2}$

The absolute value of a complex number z, denoted $|z|$, represents the distance between the origin and the point (a, b) in the complex plane. Use the formula to find $|z|$ for the complex numbers given (also see Section 1.4, Exercise 69): (a) $3 + 4i$, (b) $-5 + 12i$, and (c) $1 + \sqrt{3}\,i$.

96. The square root of $z = a + bi$: $\sqrt{z} = \frac{\sqrt{2}}{2}\left(\sqrt{|z| + a} \pm i\sqrt{|z| - a}\right)$

The square roots of a complex number are given by the relations shown, where $|z|$ represents the absolute value of z and the sign is chosen to match the sign of b. Use the formula to find the square root of each complex number from Exercise 95, then check your answer by squaring the result (also see Section 1.4, Exercise 83).

51. $x = -2, 1, \frac{-2}{3}$
52. $x = -3, 2, 1, \frac{-6}{7}$
53. $x = -2, \frac{-3}{2}, 1, 4$
54. $x = 1, 4, -2, \frac{1}{3}$
55. $x = -1, 2, 3, \frac{5}{3}$
56. $x = -1, 2, \frac{3}{4}$
57. $x = 1, 2, -3, \pm\sqrt{7}i$
58. $x = -2, 1, \pm\sqrt{3}\,i$
59. $x = -2, \frac{2}{3}, 1, \pm\sqrt{3}\,i$
60. $x = -2, 1, \frac{3}{2}, \pm\sqrt{2}\,i$
61. $x = 1, 2, 4, -2$
62. $x = -3, 1, 3, 9$
63. $x = -3, 1, \pm\sqrt{2}$
64. $x = -2, 1, \pm\sqrt{3}$
65. $x = -1, \frac{3}{2}, \pm\sqrt{3}\,i$
66. $x = -1, \frac{2}{3}, \pm\sqrt{5}\,i$
67. $x = \frac{1}{2}, 1, 2, \pm\sqrt{3}\,i$
68. $x = -2, 1, \frac{1}{4}, \pm\sqrt{3}\,i$
69. possible roots: $\{\pm1, \pm8, \pm2, \pm4\}$; neither −1 nor 1 is a root; 3 or 1 positive roots, 1 negative root; roots must lie between −2 and 2
70. possible roots: $\{\pm1, \pm6, \pm2, \pm3\}$ neither −1 nor 1 is a root; 1 positive root, 3 or 1 negative roots; roots must lie between −3 and 2
71. possible roots: $\{\pm1, \pm2\}$; −1 is a root; 2 or 0 positive roots, 3 or 1 negative roots; roots must lie between −3 and 2
72. possible roots: $\{\pm1, \pm4, \pm2\}$; $x = 1$ is a root; 3 or 1 positive roots, 2 or 0 negative roots; roots must lie between −2 and 1
73. possible roots: $\{\pm1, \pm12, \pm2, \pm6, \pm3, \pm4\}$; $x = 1$ and $x = -1$ are roots; 4, 2, or 0 positive roots, 1 negative root; roots must lie between −1 and 4
74. possible roots: $\{\pm1, \pm14, \pm2, \pm7\}$; $x = 1$ and $x = -1$ are roots; 3 or 1 positive roots, 2 or 0 negative roots; roots must lie between −3 and 4
75. possible roots: $\frac{\{\pm1, \pm20, \pm2, \pm10, \pm4, \pm5\}}{\{\pm1, \pm2\}}$; $x = 1$ is a root; 1 positive root, 1 negative root; roots must lie between −2 and 1
76. possible roots: $\frac{\{\pm1, \pm24, \pm2, \pm12, \pm3, \pm8, \pm4, \pm6\}}{\{\pm1, \pm3\}}$; $x = -1$ is a root; 1 positive root, 1 negative root; roots must lie between −1 and 4

Additional answers can be found in the Instructor Answer Appendix.

APPLICATIONS

97. yes

97. Maximum and minimum values: To locate the maximum and minimum values of $F(x) = x^4 - 4x^3 - 12x^2 + 32x + 15$ requires finding the zeroes of $f(x) = 4x^3 - 12x^2 - 24x + 32$. Use the RRT and synthetic division to find the zeroes of f, then graph $F(x)$ on a calculator and see if the graph tends to support your calculations—do the maximum and minimum values occur at the zeroes of f?

98. $x = -3, 5, 1 \pm \sqrt{2}$

98. Graphical analysis: Use the RRT and synthetic division to find the zeroes of $F(x) = x^4 - 4x^3 - 12x^2 + 32x + 15$ (see Exercise 97).

99. yes

99. Maximum and minimum values: To locate the maximum and minimum values of $G(x) = x^4 - 6x^3 + x^2 + 24x - 20$ requires finding the zeroes of $g(x) = 4x^3 - 18x^2 + 2x + 24$. Use the RRT and synthetic division to find the zeroes of g, then graph $G(x)$ on a calculator and see if the graph tends to support your calculations—do the maximum and minimum values occur at the zeroes of g?

100. $x = -2, 1, 2, 5$

100. Graphical analysis: Use the RRT and synthetic division to find the zeroes of $G(x) = x^4 - 6x^3 + x^2 + 24x - 20$ (see Exercise 99).

Geometry: The volume of a cube is $V = x \cdot x \cdot x = x^3$, where x represents the length of the edges. If a slice 1 unit thick is removed from the cube, the remaining volume is $v = x \cdot x \cdot (x - 1) = x^3 - x^2$. Use this information for Exercises 101 and 102.

101. a. 4 cm × 4 cm × 4 cm
b. 5 cm × 5 cm × 5 cm

101. A slice 1 unit in thickness is removed from one side of a cube. Use the RRT and synthetic division to find the original dimensions of the cube, if the remaining volume is (a) 48 cm^3 and (b) 100 cm^3.

102. a. 4 cm × 4 cm × 4 cm
b. 5 cm × 5 cm × 5 cm

102. A slice 1 unit in thickness is removed from one side of a cube, then a second slice of the same thickness is removed from a different side (not the opposite side). Use the RRT and synthetic division to find the original dimensions of the cube, if the remaining volume is (a) 36 cm^3 and (b) 80 cm^3.

Geometry: The volume of a rectangular box is $V = LWH$. For the box to satisfy certain requirements, its length must be twice the width, and its height must be two inches less than the width. Use this information for Exercises 103 and 104.

103. length 10 in., width 5 in., height 3 in.

103. Use the rational roots theorem and synthetic division to find the dimensions of the box if it must have a volume of 150 in^3.

104. length 8 in., width 4 in., height 2 in.

104. Suppose the box must have a volume of 64 in^3. Use the rational roots theorem and synthetic division to find the dimensions required.

Government deficits: Over a 14-yr period, the balance of payments (deficit versus surplus) for a certain county government was modeled by the function $f(x) = \frac{1}{4}x^4 - 6x^3 + 42x^2 - 72x - 64$, where $x = 0$ corresponds to 1990 and $f(x)$ is the deficit or surplus in tens of thousands of dollars. Use this information for Exercises 105 and 106.

105. 1994, 1998, 2002, 5 yr

105. Use the rational roots theorem and synthetic division to find the years the county "broke even" (debt = surplus = 0) from 1990 to 2004. How many years did the county run a surplus during this period?

106. 1990, 2001

106. The deficit was at the \$84,000 level $[f(x) = -84]$, four times from 1990 to 2004. Given this occurred in 1992 and 2000 ($x = 2$ and $x = 10$), use the rational roots theorem, synthetic division, and the remainder theorem to find the other two years the deficit was at \$84,000.

WRITING, RESEARCH, AND DECISION MAKING

107. $x = -4, 1, 3, \pm i$

107. Over the years a number of methods have been employed to place bounds on the roots of a polynomial. A few of them, while mathematically correct, seem rather fanciful. For example, given $f(x)$ is a real polynomial with a lead coefficient of 1, if the first negative coefficient follows k coefficients that are positive or zero, and C represents the *largest* negative coefficient, then an upper bound for the zeroes of f is given by $B = 1 + \sqrt[k]{|C|}$ (assume $\sqrt[1]{|C|} = C$). A lower bound for the negative roots of a function can likewise be found by applying this test to $f(-x)$ (if this results in a negative lead coefficient, multiply both sides by -1). Use this method to find an upper bound for the roots of $f(x) = x^5 - 12x^3 + 12x^2 - 13x + 12$ (remember to count the missing fourth-degree term). Then use the RRT to find all zeroes of the function.

108. Answers will vary.

108. The fundamental theorem of arithmetic is related to the fundamental theorem of algebra studied in this section. Do some research to find exactly what this statement guarantees, and comment on the impact of the theorem on the study of arithmetic.

EXTENDING THE CONCEPT

109. a. $w = 1, 1 \pm \sqrt{3}i$
b. $w = 2, 2 \pm 3i$

109. Every general cubic equation $aw^3 + bw^2 + cw + d = 0$ can be written in the form $x^3 + px + q = 0$ (where the squared term has been "depressed"), using the transformation $w = x - \frac{b}{3}$. Use this transformation to solve the following equations.

a. $w^3 - 3w^2 + 6w - 4 = 0$ **b.** $w^3 - 6w^2 + 21w - 26 = 0$

110. a. $x = 3$
b. $\sqrt[3]{2} + \sqrt[3]{-4}$

110. From Exercise 109, it is actually very rare that the transformation produces a value of $q = 0$ for the "depressed" cubic $x^3 + px + q = 0$, and general solutions must be found using what has become known as *Cardano's formula.* For a complete treatment of cubic equations and their solutions, visit our website at www.mhhe.com/coburn. Here we'll focus on the primary root of selected cubics. Cardano's formula tells us that one solution of $x^3 + px + q = 0$ is given by $z_1 = \sqrt[3]{A} + \sqrt[3]{B}$ where $A = \frac{-q}{2} + \sqrt{\frac{q^2}{4} + \frac{p^3}{27}}$ and $B = \frac{-q}{2} - \sqrt{\frac{q^2}{4} + \frac{p^3}{27}}$. Use these relationships to find the primary solution of the following equations, then verify the solution using a calculator.

a. $x^3 - 6x - 9 = 0$ **b.** $x^3 + 6x + 2 = 0$

111. 2, 1, 4

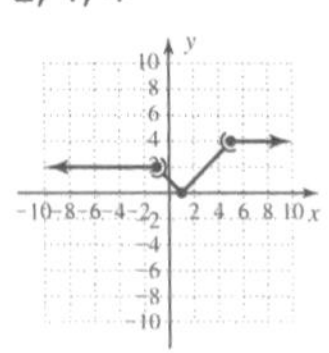

112. $S = \frac{kwd^2}{l}$

113.
a. $f(x) \geq 0$: $x \in [-4, -1] \cup [1, \infty)$
b. max: $(-3, 4)$; min: $(0, -2)$
c. $f(x)\uparrow$: $x \in (-\infty, -3) \cup (0, \infty)$
$f(x)\downarrow$: $x \in (-3, 0)$

114. $f(x) = \frac{1}{(x-2)^2} - 3$

115. $\{(22, -3), (13, -2), (6, -1), (1, 0), (-2, 1), (-3, 2)\}$;
D: $\{22, 13, 6, 1, -2, -3\}$;
R: $\{-3, -2, -1, 0, 1, 2\}$

116. $f(x) = 2(x - 3)^2 - 9$
vertex: $(3, -9)$
axis of symmetry: $x = 3$
y-intercept: $(0, 9)$
x-intercepts: $(3 + 1.5\sqrt{2}, 0)$ and $(3 - 1.5\sqrt{2}, 0)$

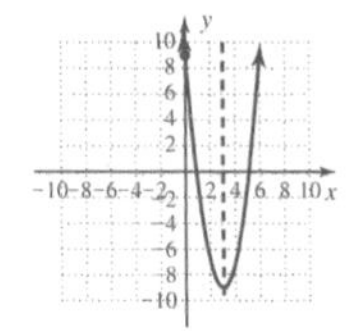

MAINTAINING YOUR SKILLS

111. (3.7) Graph the piecewise-defined function and

$$f(x) = \begin{cases} 2 & x \leq -1 \\ |x - 1| & -1 < x < 5 \\ 4 & x \geq 5 \end{cases}$$

find the value of $f(-3)$, $f(2)$, and $f(5)$.

112. (3.6) The safe load of a rectangular beam supported at both ends varies jointly with its width, square of its depth, and inversely as its length. Write the variation equation.

113. (2.6/3.8) Use the graph given to (a) state intervals where $f(x) \geq 0$, (b) locate maximum and minimum values, and (c) state intervals where $f(x)\uparrow$ and $f(x)\downarrow$.

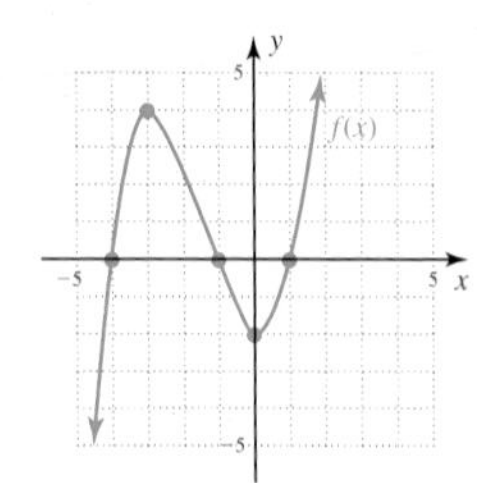

114. (3.3/3.5) Write the equation of the function given on the grid shown. Assume $a = 1$.

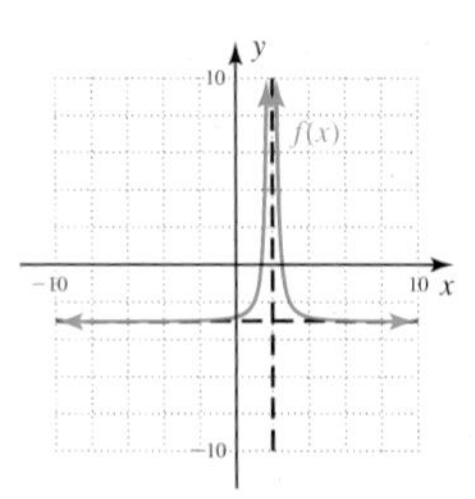

115. (3.2) A pointwise function is defined by the set $\{(-3, 22), (-2, 13), (-1, 6), (0, 1), (1, -2), (2, -3)\}$. Name the inverse function and state its domain and range.

116. (3.4) Graph the quadratic function by completing the square. Name the location of the vertex, axis of symmetry, and x- and y-intercepts.
$f(x) = 2x^2 - 12x + 9$

4.4 Graphing Polynomial Functions

LEARNING OBJECTIVES

In Section 4.4 you will learn how to:

A. Describe the end behavior of a polynomial graph

B. Describe the attributes of a polynomial graph with zeroes of odd and even multiplicity

C. Graph polynomial functions

INTRODUCTION

A great many of the concepts developed in previous sections are brought together in this study of polynomial graphing. Virtually all of the learning objectives have a familiar ring to them, and with the connections made here, they provide a strong foundation for understanding and sketching polynomial graphs, as well as the rational graphs in Section 4.5.

POINT OF INTEREST

The word **graph** is likely a derivative of the Indo-European root word *gerbh,* meaning "to scratch." In the early history of civilization, words and symbols were originally scratched on wood, stone, clay tablets, or some other material. Ancient students and teachers of geometry scratched pictures on the ground, which may have helped influence the meaning of the word as we know it today. A graph is a picture of the relationship between two (or more) variables.

A. The End Behavior of a Polynomial Graph

We were first introduced to the **end behavior** of a graph while studying quadratic functions in Section 2.4. For functions of degree 2 with a positive lead coefficient ($a > 0$), the end behavior is *up on the left* and *up on the right* (up/up). If the lead coefficient is negative ($a < 0$), end behavior is *down on the left* and *down on the right* (down/down). These descriptions were also applied to the graph of a linear function (degree 1), whose ends always point in *opposite* directions. A positive lead coefficient ($m > 0$) indicates positive slope, and the graph is *down on the left, up on the right* (down/up). All polynomial graphs exhibit some form of end behavior.

EXAMPLE 1 Describe the end behavior of each graph shown:

a. $f(x) = x^3 - 4x + 1$ **b.** $g(x) = -2x^5 + 7x^3 - 4x$ **c.** $h(x) = -2x^4 + 5x^2 + x - 1$

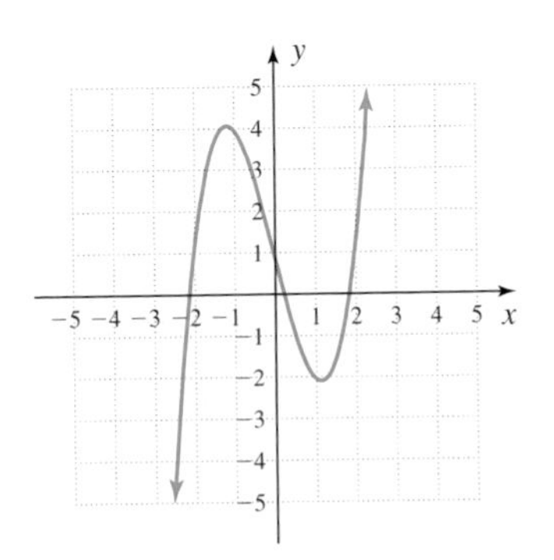

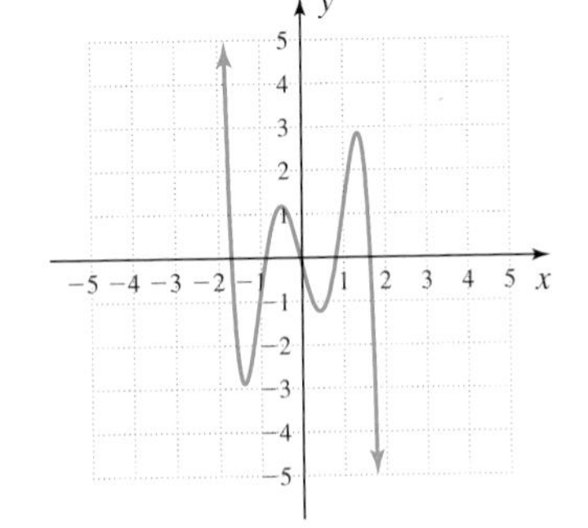

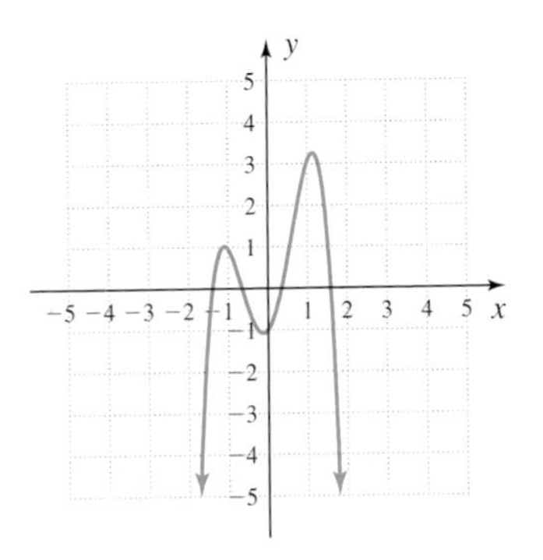

Solution: **a.** down/up **b.** up/down **c.** down/down

NOW TRY EXERCISES 7 THROUGH 10

Basically, end behavior is a way to describe what happens to a polynomial graph as $|x|$ becomes very large. As shown in the tables from *Strengthening Core Skills* in Chapter 2, the term of highest degree tends to *dominate* the other terms in an expression as $|x|$ becomes large. Consider the polynomial $f(x) = x^4 - 4x^3 - 10x^2 - 9x - 18$. For $x \in [0, 6]$ the lower degree terms (which are all negative) "gang up" on the **quartic** (degree 4) term causing negative or zero outputs (see Table 4.1). But for larger input values, the quartic term easily "gobbles the others up" and dictates that all future outputs will be positive. Additionally, for any quartic function $y = ax^4 + bx^3 + cx^2 + dx + e$, x^4 is always positive so both ends of the graph must point in the *same* direction, and it's actually the lead coefficient that determines whether they point up/up or down/down [see the figure in Example 1(c), where $a < 0$]. This is likewise true for all polynomials with leading term ax^n, where n is an *even* integer.

Table 4.1

x	$f(x)$
0	−18
1	−40
2	−92
3	−162
4	−214
5	−188
6	0
7	458
8	1318
9	2736

Similar reasoning can be applied to functions with the leading term ax^n, *where n is odd.* The ends of the graph will point in opposite directions since x^n is positive for $x > 0$ and negative for $x < 0$. Since this term will "dominate" any lesser degree terms as $|x|$ becomes large, the lead coefficient will govern whether they point down/up or up/down [see the figures in Example 1(a) $(a > 0)$ and Example 1(b) $(a < 0)$]. From this we are confident that the end behavior of a polynomial graph depends on two things: *the degree of the function (even/odd)* and *the sign of the lead coefficient (positive/negative).*

THE END BEHAVIOR OF A POLYNOMIAL GRAPH

If the *degree* of the polynomial is *odd,* the ends will point in *opposite directions:*

1. positive lead coefficient: down on left, up on right (like $y = x^3$).
2. negative lead coefficient: up on left, down on right (like $y = -x^3$).

If the *degree* of the polynomial is *even,* the ends will point in *the same direction:*

1. positive lead coefficient: up on left, up on right (like $y = x^2$).
2. negative lead coefficient: down on left, down on right (like $y = -x^2$).

EXAMPLE 2 State the end behavior and y-intercept of each function, without actually graphing.

a. $f(x) = 0.5x^4 + 3x^3 - 5x + 6$ **b.** $g(x) = -2x^5 - 5x^3 - 3$

Solution:

a. The function has degree 4 (even), and the ends will point in the same direction. The lead coefficient is positive, so end behavior is up/up. The y-intercept is (0, 6).

b. The function has degree 5 (odd), and the ends will point in opposite directions. The lead coefficient is negative, so the end behavior is up/down. The y-intercept is $(0, -3)$.

NOW TRY EXERCISES 11 THROUGH 16

B. Attributes of Polynomial Graphs with Roots of Multiplicity

One of the finer points of polynomial graphing involves the behavior of the graph at or near its zeroes. To explore this phenomenon, we'll consider functions of the form $y = ax^n$ **(power functions)** separately for even powers and odd powers, even though both cases depend on the same principle: *For $a > 0$ and x between -1 and 1, $ax^{n+2} \leq ax^n$.* In words, the graph of ax^{n+2} is *below the graph* of ax^n in this interval, becoming very flat near $x = 0$.

For $y = x^4$ versus $y = x^2$, note that both graphs exhibit the same end behavior, "bounce" off the x-axis at (0, 0), and intersect again at $(-1, 1)$ and (1, 1) (see Figure 4.11). But the graph of $y = x^4$ (in blue) is *below* that of $y = x^2$ for $-1 < x < 1$, and *"flattens out" near the zero.* A similar thing occurs with $y = x^5$ and $y = x^3$ (see Figure 4.12). Both have the same end behavior, "cross" through the x-axis at (0, 0), and intersect at $(-1, -1)$ and (1, 1), with the graph of $y = x^5$ (in blue) also "flattening out" near the zero.

Figure 4.11

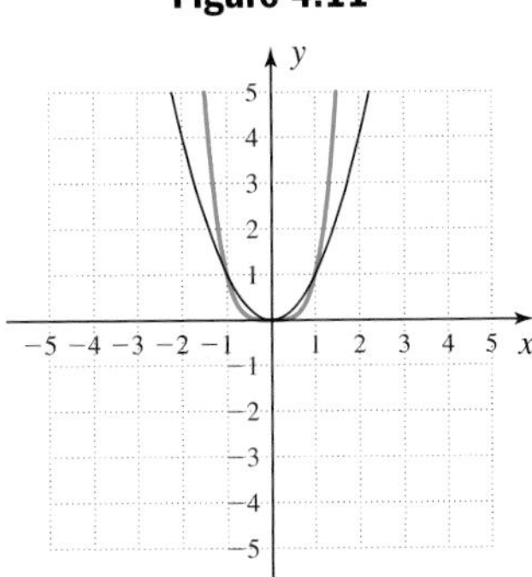

Figure 4.12

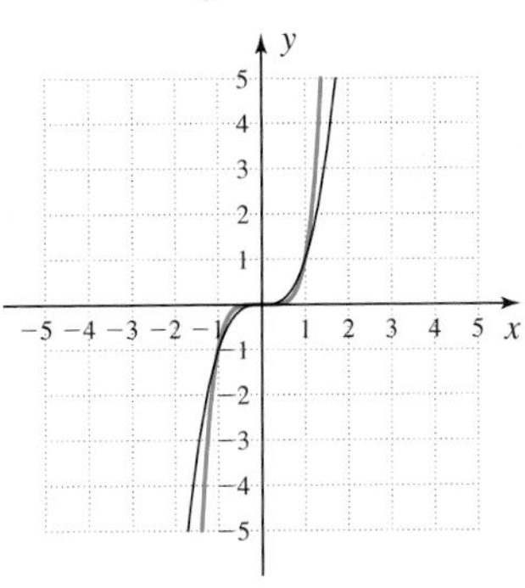

In the final analysis, this "flattening out" will occur at all zeroes of a function with multiplicity greater than 1. In other words, the graph of $y = (x - 2)^4$ (even multiplicity) will bounce off the x-axis and be very flat at $x = 2$, just as $y = x^4$ was at $x = 0$. The graph of $y = (x - 2)^5$ (odd multiplicity) will cross the x-axis and be very flat at $x = 2$, just as $y = x^5$ was at $x = 0$.

These ideas are important because the graph of a polynomial is built around its degree and its x- and y-intercepts. As a prelude to graphing, we'll first look at how to determine the degree and y-intercept of a function from its factored form, knowing the x-intercepts come directly from the factors. In completely factored form, the degree of any irreducible quadratic factor is 2, and the multiplicity of each linear factor is simply its exponent [recall $(x - c) = (x - c)^1$]. To find the degree of the polynomial, add the exponents on the linear factors and the degree of each irreducible quadratic factor. For the y-intercept, we apply the exponent on each factor to its constant term (since $x = 0$), then compute the product of all such factors as shown in Example 3.

EXAMPLE 3 State the degree, y-intercept, and end behavior of each function.

a. $f(x) = (x + 2)^3(x - 3)$ **b.** $g(x) = -(x + 1)^2(x^2 + 3)(x - 6)$

Solution: **a.** The degree of f is $3 + 1 = 4$, and the y-intercept is $(2)^3(-3) = -24$. With even degree and positive lead coefficient, end behavior is up/up.

b. The degree of g is $2 + 2 + 1 = 5$, and the y-intercept is $-1(1)^2(3)(-6) = 18$. With odd degree and negative lead coefficient, end behavior is up/down.

NOW TRY EXERCISES 17 THROUGH 30 ▶

Figure 4.13
$y = (x - 2)(x + 1)^2$

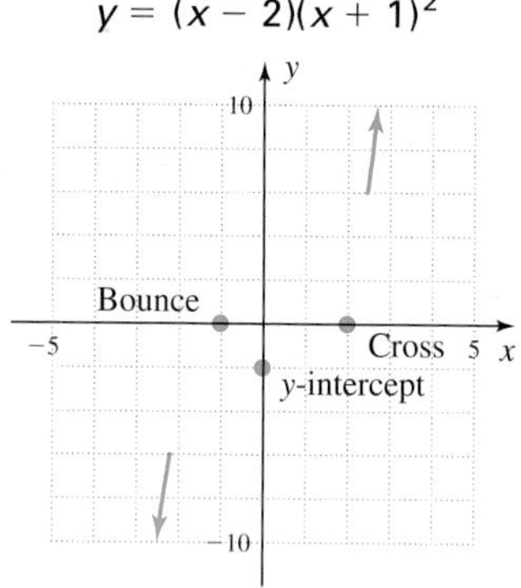

As noted earlier, the graph of a function will *cross through* the x-axis at zeroes of odd multiplicity and *bounce off* the x-axis at real zeroes of even multiplicity. The graph of $y = x - 2$ is a line that crosses the x-axis at $x = 2$, and the graph of $y = (x + 1)^2$ is a parabola that bounces at $x = -1$. The graph of $f(x) = (x - 2)(x + 1)^2$, made up of these same factors, is a cubic function whose graph *exhibits the same behavior at these zeroes!* Since the degree of the function is odd and the lead coefficient is positive, its end behavior is down/up. The y-intercept is $(0, -2)$. This information is displayed in Figure 4.13. The graph must contain the points plotted, yet still follow the prescribed end behavior. These conditions are satisfied by the graph shown in Figure 4.14 [$f(1) = -4$ was used to check the behavior of the graph between 0 and 2]. As noted, the higher the multiplicity, the flatter the graph will be near each zero. For comparison, note the graph of $y = (x - 2)^3(x + 1)^4$ in Figure 4.15 behaves very much like the graph of $f(x) = (x - 2)(x + 1)^2$ from Figure 4.14, except it's flatter at $x = -1$ and $x = 2$. We lose sight of the graph between 0 and 2 since a fourth-degree factor produces larger values than the original grid size could accommodate.

Figure 4.14
$y = (x - 2)(x + 1)^2$

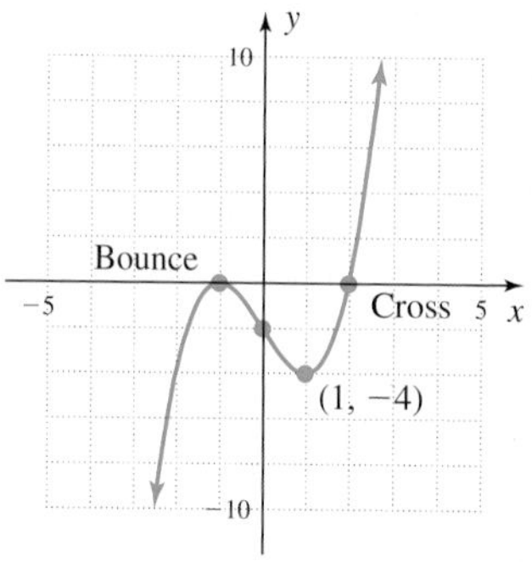

Figure 4.15
$y = (x - 2)^3(x + 1)^4$

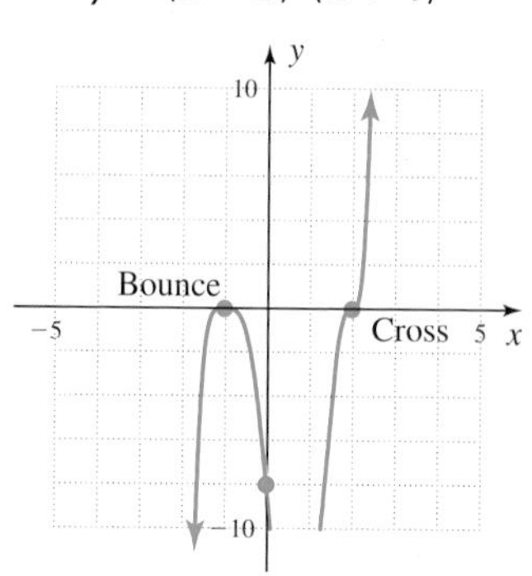

POLYNOMIAL GRAPHS AND ZEROES OF MULTIPLICITY

Given $P(x)$ is a completely factored real polynomial, with linear factors of the form $(x - r)^m$:

- *If m is odd,*
  r is a zero of odd multiplicity and the graph will *cross through* the x-axis at r.
- *If m is even,*
  r is a zero of even multiplicity and the graph will *bounce off* the x-axis at r.

The larger the value of m, the flatter (more compressed) the graph becomes near r.

The complex zeroes from irreducible quadratic factors cannot be graphed in the real plane.

EXAMPLE 4 The graph of a polynomial $f(x)$ is shown. (a) State whether the degree of f is even or odd. (b) Use the graph to estimate the zeroes of f, then state whether their multiplicity is even or odd. (c) State the minimum possible degree of f.

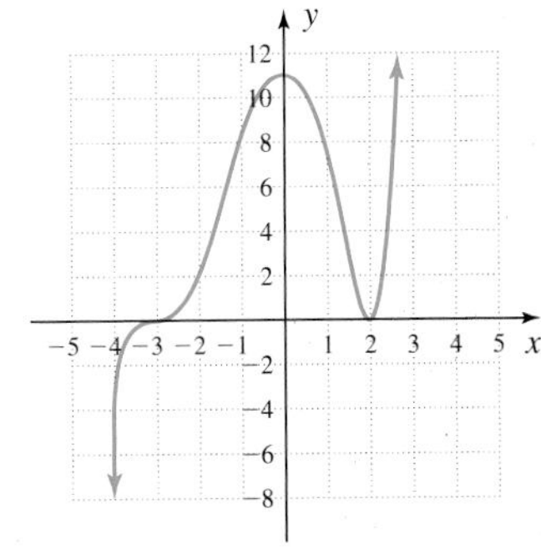

Solution:

a. Since the ends of the graph point in opposite directions, the degree of the function must be odd.

b. The graph cuts through the x-axis at $x = -3$ and is somewhat compressed, meaning it must have odd multiplicity. The graph bounces off the x-axis at $x = 2$, and 2 must be a zero of even multiplicity.

c. The minimum possible degree of f is 5, as in $f(x) = a(x - 2)^2(x + 3)^3$.

NOW TRY EXERCISES 31 THROUGH 36

EXAMPLE 5 The following functions all have zeroes at $x = -2, -1$, and 1. Match each function to the corresponding graph *using its degree and the multiplicity of each zero.*

a. $y = (x + 2)(x + 1)^2(x - 1)^3$ **b.** $y = (x + 2)(x + 1)(x - 1)^3$

c. $y = (x + 2)^2(x + 1)^2(x - 1)^3$ **d.** $y = (x + 2)^2(x + 1)(x - 1)^3$

Solution: The functions in Figures 4.16 and 4.18 must have even degree due to end behavior, so each corresponds to (a) or (d). Their intercepts at $x = 1$ appear identical, but at $x = -1$ the graph in Figure 4.16 "crosses," while the graph in Figure 4.18 "bounces." This indicates Figure 4.16 matches equation (d), while Figure 4.18 matches equation (a). The graphs in Figures 4.17 and 4.19 must have odd degree due to end behavior, so each corresponds to (b) or (c). Their intercepts at $x = 1$ also appear identical, but one graph "bounces" at $x = -2$, while the other "crosses." The graph in Figure 4.17 matches equation (c), the graph in Figure 4.19 matches equation (b).

Figure 4.16

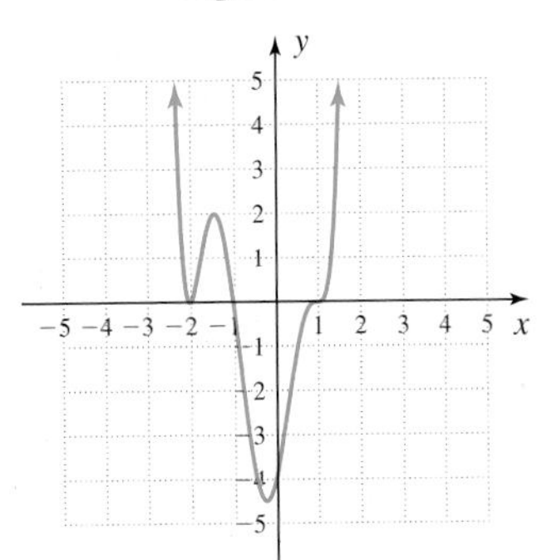

Figure 4.17

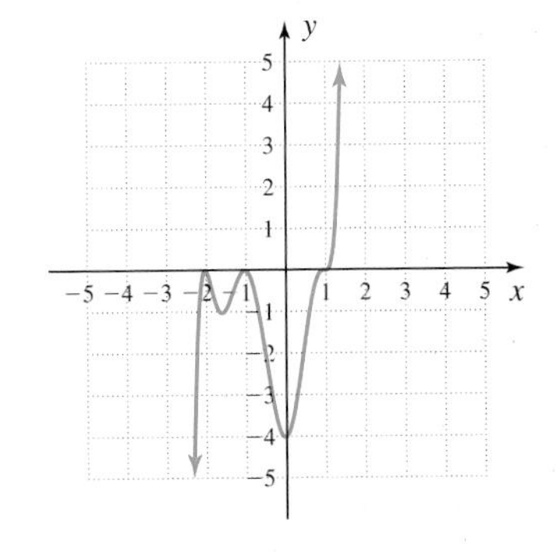

Figure 4.18

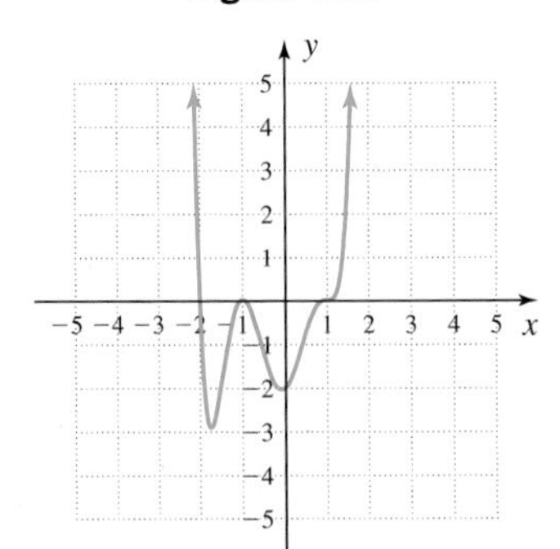

Figure 4.19

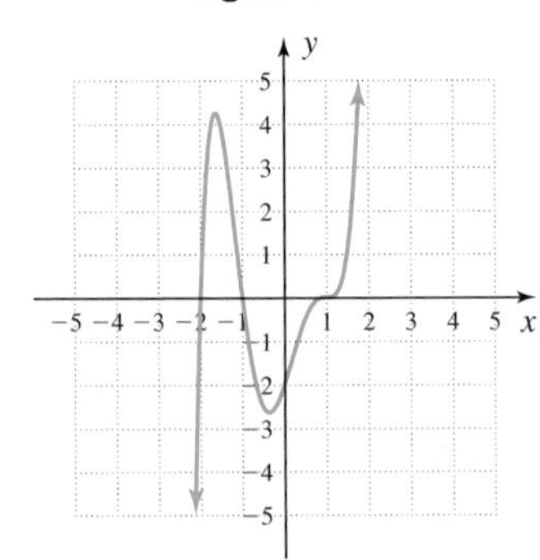

NOW TRY EXERCISES 37 THROUGH 42

EXAMPLE 6 Sketch the graph of $f(x) = (x + 2)^2(x - 1)^3(x + 1)$ using end behavior; the x- and y-intercepts; zeroes of multiplicity; and a smooth, continuous curve.

Solution: Adding the multiplicities of each zero, we find that f is a function of degree 6 with a positive lead coefficient, so end behavior will be up/up. The y-intercept is $f(0) = -4$. The graph will bounce off the x-axis at -2 (even multiplicity), and cross through the axis at -1 and 1 (odd multiplicities). The graph will "flatten out" at $x = 1$ because of its higher multiplicity. To help "round-out" the graph we evaluate f at $x = -1.5$, giving $(0.5)^2(-2.5)^3(-0.5) \approx 1.95$ (note scaling of the x- and y-axes).

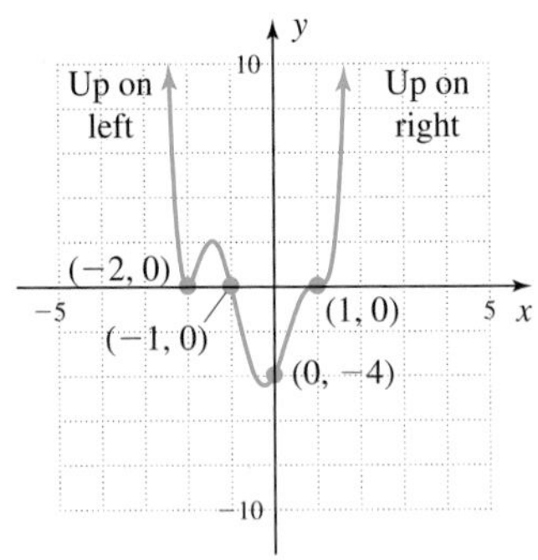

NOW TRY EXERCISES 43 THROUGH 56

C. The Graph of a Polynomial Function

In Example 6, note that as the graph continues from one x-intercept to the next, it must change from increasing to decreasing or vice versa. In other words, since the graph of a polynomial function is continuous, it must, at some point, turn around and head toward the next x-intercept. The precise points at which the graph changes direction are the local maximums and local minimums (also called **turning points** or **extreme values**) studied in Section 3.8. Although not proven until a first course in calculus, it seems reasonable that a polynomial of degree n can have at most $n - 1$ turning points. A linear graph has no turning points, a quadratic function has one, etc.

POLYNOMIAL GRAPHS AND TURNING POINTS

If $f(x)$ is a polynomial function of degree n, the graph of f can have at most $n - 1$ turning points.

Using the cumulative observations from this and previous sections, a general strategy emerges for the graphing of polynomial functions.

WORTHY OF NOTE

Although of somewhat limited value, symmetry (item f in the guidelines) can sometimes aid in the graphing of polynomial functions. If all terms of the function have even degree, the graph will be symmetric to the y-axis (even). If all terms have odd degree, the graph will be symmetric about the origin. Recall that a constant term has degree zero, an even number.

GUIDELINES FOR GRAPHING POLYNOMIAL FUNCTIONS

1. Determine the end behavior of the graph.
2. Find the y-intercept $f(0) = a_0$.
3. Find the x-intercepts using any combination of the rational roots theorem, the factor and remainder theorems, factoring, and the quadratic formula.
4. Use the y-intercept, end behavior, tests for 1 and -1, the multiplicity of each zero, and a few midinterval points to sketch a smooth, continuous curve.

Additional tools include (a) polynomial zeroes theorem, (b) complex conjugates theorem, (c) number of turning points, (d) Descartes's rule of signs, (e) upper and lower bounds, and (f) symmetry.

EXAMPLE 7 Sketch the graph of $g(x) = -x^4 + 9x^2 - 4x - 12$.

Solution: The function has degree 4 (even) with a negative lead coefficient, so end behavior is *down/down*. The y-intercept is $(0, -12)$. Checking to see if 1 is a zero (adding the coefficients) gives $-1 + 9 - 4 - 12 = -8$, meaning $x = 1$ is not a zero but $(1, -8)$ is a point on the graph. If we change the sign of all terms with odd degree, the sum is $-1 + 9 + 4 - 12 = 0$, so $(-1, 0)$ is an x-intercept and $(x + 1)$ is a factor. Using $x = -1$ with synthetic division gives:

use -1 as a "divisor"

$$\begin{array}{r|rrrrr} -1 & -1 & 0 & 9 & -4 & -12 \\ & & 1 & -1 & -8 & 12 \\ \hline & -1 & 1 & 8 & -12 & 0 \end{array}$$

The quotient polynomial is not easily factored so we continue with synthetic division on the quotient polynomial. Using the rational roots theorem, the possible rational roots are $\{\pm1, \pm12, \pm2, \pm6, \pm3, \pm4\}$, so we try $x = 2$.

use 2 as a "divisor" on *the quotient polynomial*

$$\begin{array}{r|rrrr} 2 & -1 & 1 & 8 & -12 \\ & & -2 & -2 & 12 \\ \hline & -1 & -1 & 6 & 0 \end{array}$$

This shows $x = 2$ is a root, $x - 2$ is a factor, and the function can now be written as $f(x) = (x + 1)(x - 2)(-x^2 - x + 6)$. Factoring -1 from the trinomial gives

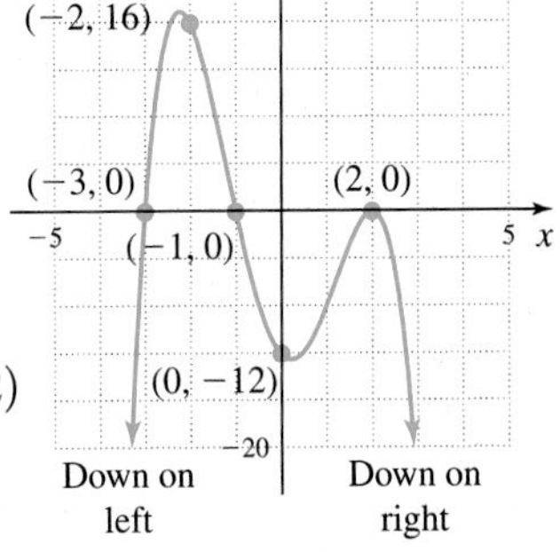

$$\begin{aligned} f(x) &= -1(x + 1)(x - 2)(x^2 + x - 6) \\ &= -1(x + 1)(x - 2)(x + 3)(x - 2) \\ &= -1(x + 1)(x - 2)^2(x + 3) \end{aligned}$$

The x-intercepts are $x = -1$ and $x = -3$, both with multiplicity 1, and $x = 2$ with multiplicity 2. To help "round-out" the graph we evaluate the midinterval point

$x = -2$ using the remainder theorem, which shows that $(-2, 16)$ is also a point on the graph.

use −2 as a "divisor"	$\underline{-2}\rfloor$	−1	0	9	−4	−12
			2	−4	−10	28
		−1	2	5	−14	16

The final result is the graph shown.

NOW TRY EXERCISES 59 THROUGH 78

CAUTION

Sometimes using a midinterval point to help draw a graph will give the illusion that a maximum or minimum value has been located. This is rarely the case, as demonstrated in the figure in Example 7, where the maximum value in Quadrant II is actually closer to $(-2.22, 16.95)$.

TECHNOLOGY HIGHLIGHT

Using a Graphing Calculator to Explore Polynomial Graphs

The keystrokes shown apply to a TI-84 Plus model. Please consult your manual or our Internet site for other models.

One of the greatest advantages a graphing calculator brings is the ability to "play" at mathematics. Polynomial graphing and a study of complex polynomials are two very rich areas of mathematics, and the TI-84 Plus (and other models) offers the ability to explore, discover, and make connections in ways simply unheard of just 15 years ago or so. We'll illustrate by playfully creating polynomial functions in a game called *What's My Equation?* We begin by describing certain attributes of a graph, then form an equation to fit the description, and end by checking what the calculator displays against what we've anticipated. As you might guess, our biggest concern will be window size. Note that the equations derived will not necessarily be unique.

EXAMPLE 1: I am a real polynomial of odd degree. I bounce at $x = -2$, cut very gradually (flat) at $x = 1$, and have a y-intercept of -8 and two complex roots. What's my equation? State the window size that might be used for optimal viewing.

Solution: Call the polynomial $P(x)$. The root that bounces at $x = -2$ could be $(x + 2)^2$ and the root that cuts at $x = 1$ might be $(x - 1)^3$. So far this gives $P(x) = (x + 2)^2(x - 1)^3$, which has an odd degree, but a y-intercept of only $P(0) = -4$. For the required y-intercept of -8 we need a constant of 2, and since the complex roots must be conjugates, the last factor could be $(x^2 + 2)$. This gives $P(x) = (x + 2)^2(x - 1)^3(x^2 + 2)$, which fits the specified requirements. Using the window shown in Figure 4.20 gives the graph of P shown in Figure 4.21.

Figure 4.20

WINDOW
Xmin=-3
Xmax=3
Xscl=1
Ymin=-30
Ymax=30
Yscl=4
Xres=1

Use these ideas to complete the following exercise.

Exercise 1: I am a real polynomial equation of even degree. I am flat near $x = 1$ and cut through the x-axis here. I bounce at $x = -1$ and again at $x = -2$. I have an additional zero somewhere and a y-intercept at $(0, 8)$. What's my equation? Suggest a window size that might be used for optimal viewing of the result.

Figure 4.21

Y1=(X+2)²(X-1)^3(X²+2)
X=0 Y=-8

$P(x) = (x + 1)^2(x + 2)^2(x - 1)^3(x - 2);\ x \in [-3, 3],\ y \in [-12, 12]$

4.4 EXERCISES

CONCEPTS AND VOCABULARY

Fill in each blank with the appropriate word or phrase. Carefully reread the section if needed.

1. A polynomial equation of degree four is called a quartic equation.

2. A polynomial function of degree n has n zeroes and at most $n - 1$ "turning points."

3. The graphs of $Y_1 = (x - 2)^2$ and $Y_2 = (x - 2)^4$ both bounce at $x = 2$, but the graph of Y_2 is flatter than the graph of Y_1 at this point.

4. Since $x^4 > 0$ for all x, the ends will always point in the same direction. Since $x^3 > 0$ when $x > 0$ and $x^3 < 0$ when $x < 0$, the ends will always point in the opposite direction.

5. In your own words, explain/discuss how to find the degree and y-intercept of a function that is given in factored form. Use $f(x) = (x + 1)^3(x - 2)(x + 4)^2$ to illustrate. Answers will vary.

6. Name all of the "tools" at your disposal that play a role in the graphing of polynomial functions. Which tools are indispensable and always used? Which tools are used only as the situation merits? Answers will vary.

DEVELOPING YOUR SKILLS

State the end behavior of the functions given.

7. $f(x) = -x^3 + 6x - 1$ up/down

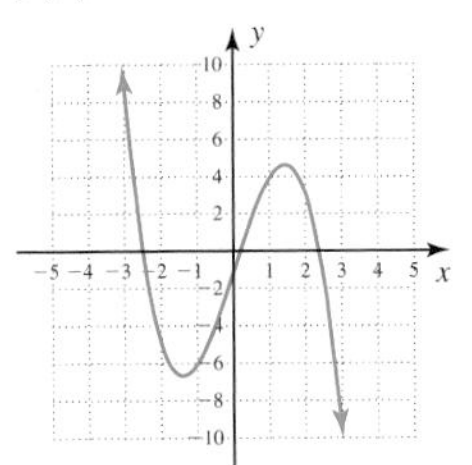

8. $g(x) = x^4 + x^3 - 8x^2 + 2x + 4$ up/up

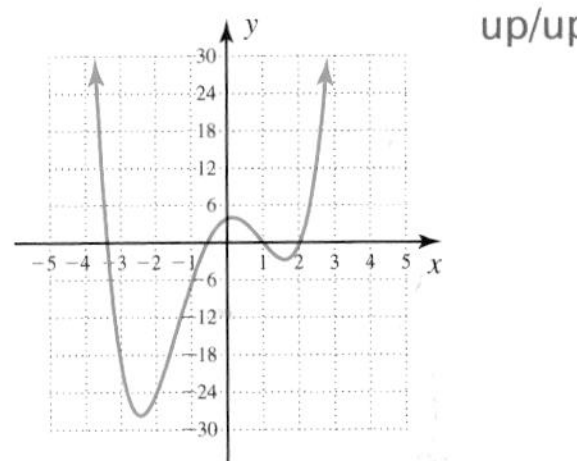

9. $H(x) = -x^6 - 4x^5 + 2x^4 + 20x^3 + 11x^2 - 16x - 12$ down/down

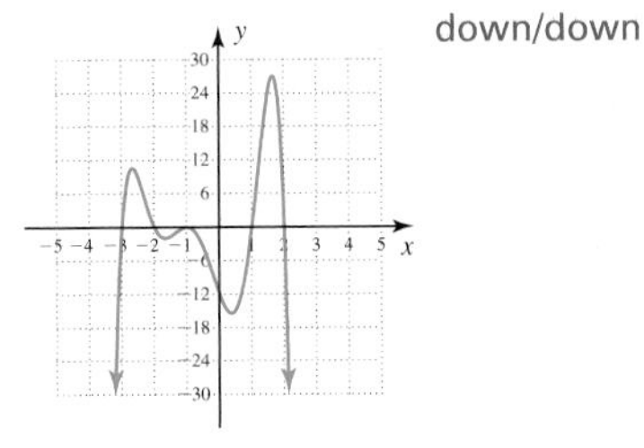

10. $h(x) = x^5 - 2x^4 - 21x^3 + 22x^2 + 40x$ down/up

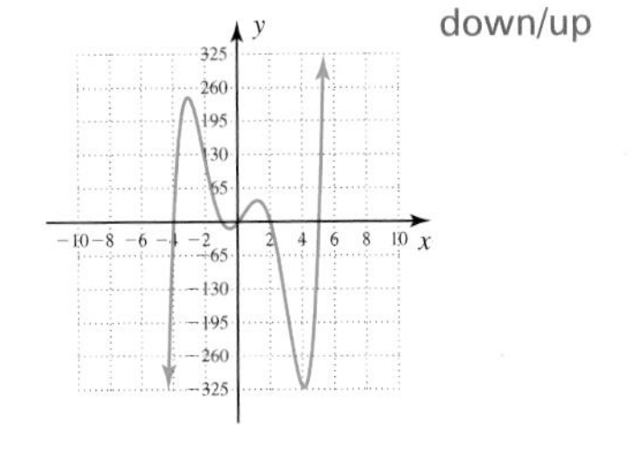

State the end behavior and y-intercept of the functions given. Do not graph.

11. $f(x) = x^3 + 6x^2 - 5x - 2$
12. $g(x) = x^4 - 4x^3 - 2x^2 + 16x - 12$
13. $p(x) = -2x^4 + x^3 + 7x^2 - x - 6$
14. $q(x) = -2x^3 - 18x^2 + 7x + 3$
15. $Y_1 = -3x^5 + x^3 + 7x^2 - 6$
16. $Y_2 = -x^6 - 4x^5 + 4x^3 + 16x - 12$

11. down/up; (0, −2)
12. up/up; (0, −12)
13. down/down; (0, −6)
14. up/down; (0, 3)
15. up/down; (0, −6)
16. down/down; (0, −12)

Use the behavior of each graph at $x = 2$ to match each equation to its corresponding graph.

17. $y = x - 2$ c
18. $y = (x - 2)^4$ b
19. $y = (x - 2)^5$ e

20. $y = (x - 2)^3$ a **21.** $y = (x - 2)^6$ f **22.** $y = (x - 2)^2$ d

a.

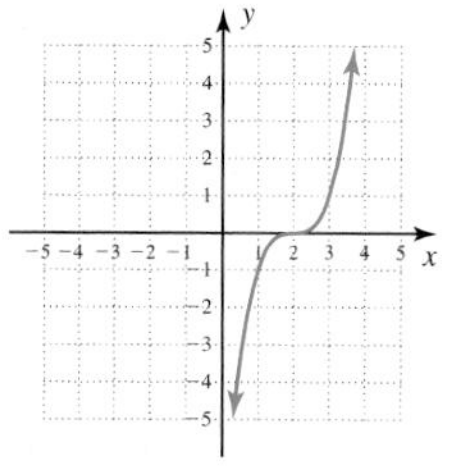

b.

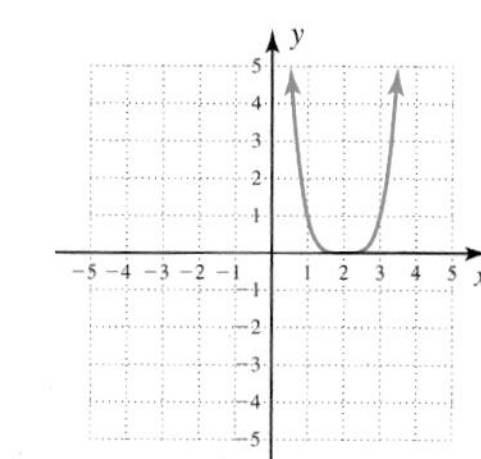

c.

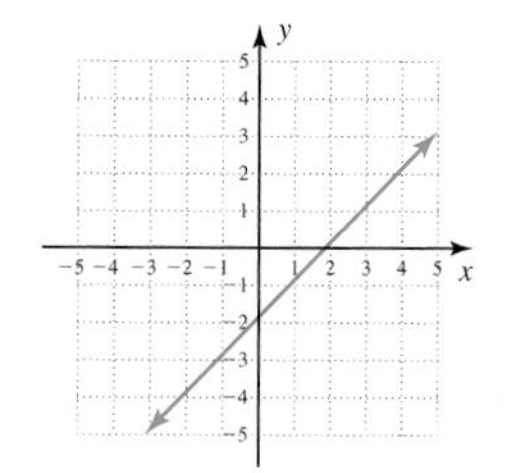

d.

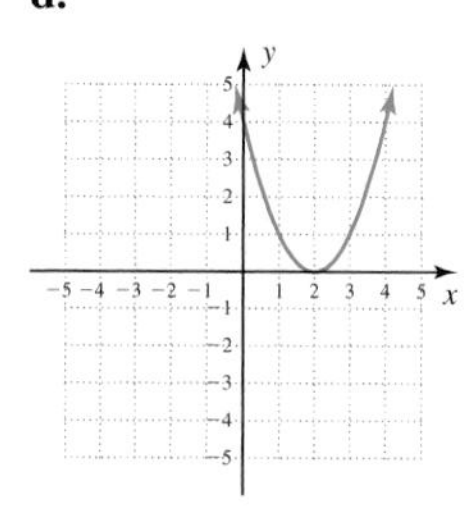

e.

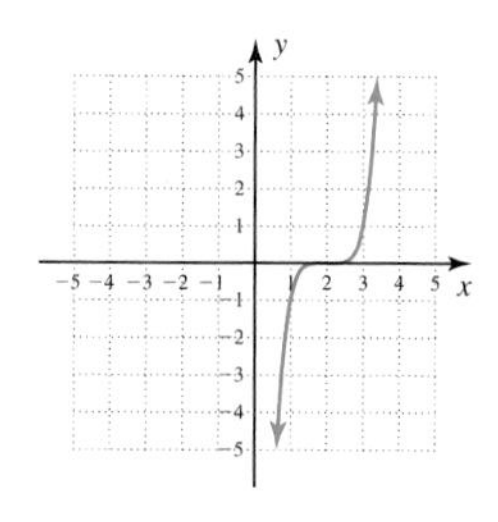

f.

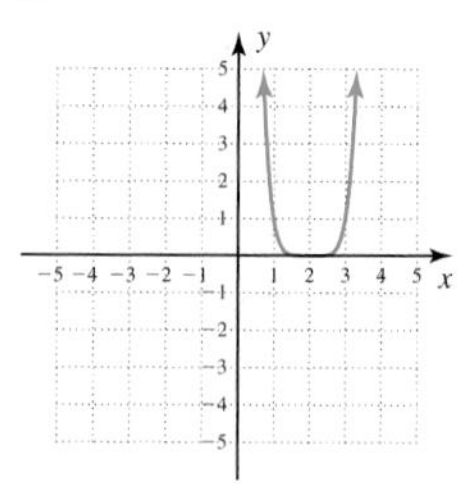

23. degree 6; up/up; (0, −12)
24. degree 4; up/up; (0, −16)
25. degree 5; up/down; (0, −24)
26. degree 5; up/down; (0, −24)
27. degree 6; up/up; (0, −192)
28. degree 6; up/up; (0, 20)
29. degree 5; up/down; (0, 2)
30. degree 5; up/down; (0, 32)
31. **a.** even
b. −3 odd, −1 even, 3 odd
c. $f(x) = (x + 3)(x + 1)^2(x - 3)$
32. **a.** odd
b. −3 odd, −1, even, 2 even
c. $f(x) = (x + 3)(x + 1)^2(x - 2)^2$
33. **a.** even
b. −3 odd, −1 odd, 2 odd, 4 odd
c. $f(x) = -(x + 3)(x + 1)(x - 2)(x - 4)$
34. **a.** even
b. −3 odd, −2 even, 1 odd, 2 even
c. $f(x) = (x + 3)(x + 2)^2(x - 1)(x - 2)^2$
35. **a.** odd
b. −1 even, 3 odd
c. $f(x) = -(x + 1)^2(x - 3)$
36. **a.** odd
b. −3 odd, −1 even, 1 odd, 2 odd
c. $f(x) = -(x + 3)(x + 1)^2(x - 1)(x - 2)$

State the degree, end behavior, and y-intercept of each function.

23. $f(x) = (x - 3)(x + 1)^3(x - 2)^2$
24. $g(x) = (x + 2)^2(x - 4)(x + 1)$
25. $Y_1 = -(x + 1)^2(x - 2)(2x - 3)(x + 4)$
26. $Y_2 = -(x + 1)(x - 2)^3(5x - 3)$
27. $r(x) = (x^2 + 3)(x + 4)^3(x - 1)$
28. $s(x) = (x + 2)^2(x - 1)^2(x^2 + 5)$
29. $h(x) = (x^2 + 2)(x - 1)^2(1 - x)$
30. $H(x) = (x + 2)^2(2 - x)(x^2 + 4)$

For each polynomial graph, (a) state whether the degree of the function is even or odd; (b) use the graph to estimate the zeroes of f, then state whether their multiplicity is even or odd; and (c) find the minimum possible degree of f and write it in factored form. Assume all zeroes are real.

31.

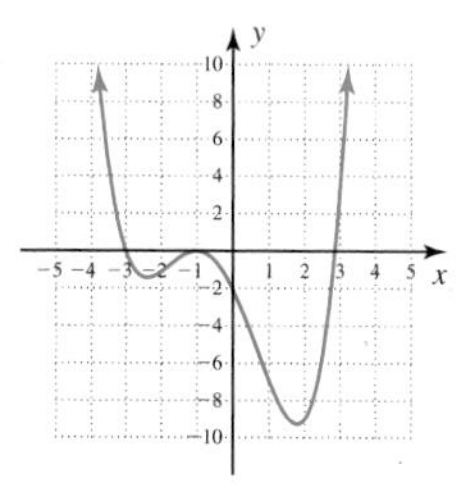

32.

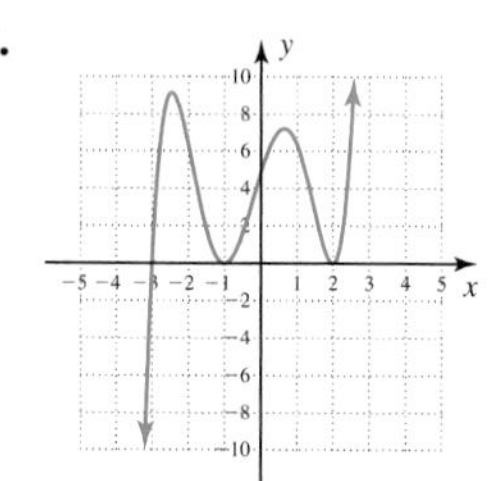

33.

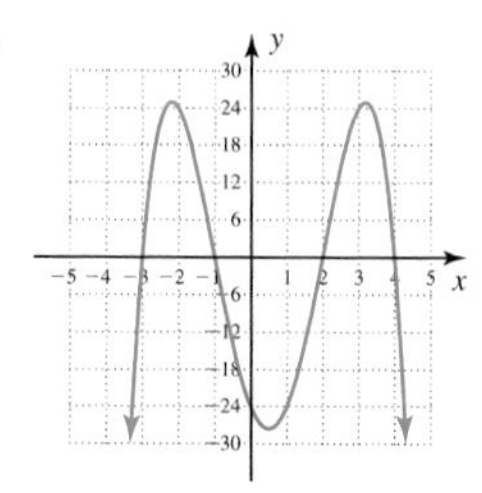

34.

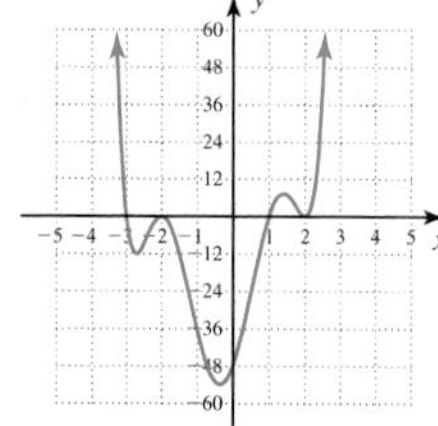

35.

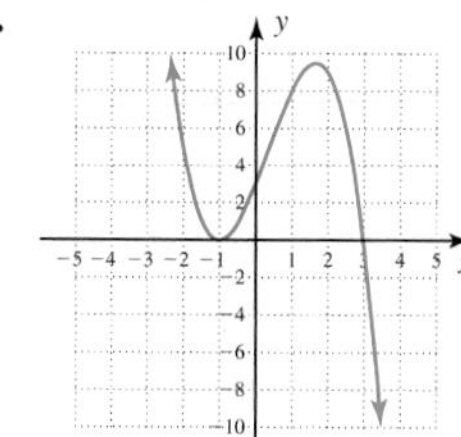

36.

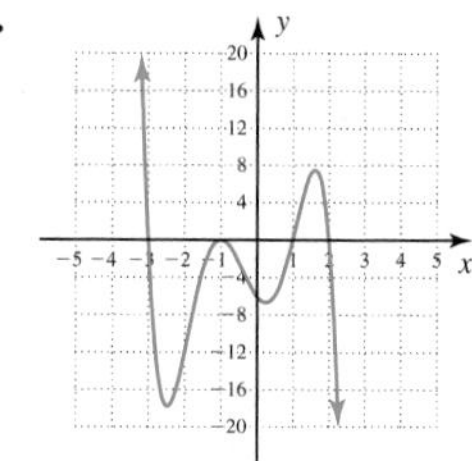

Every function in Exercises 37 through 42 has the zeroes $x = -1$, $x = -3$, and $x = 2$. Match each to its corresponding graph using degree, end behavior, and the multiplicity of each zero.

37. $f(x) = (x + 1)^2(x + 3)(x - 2)$ b
38. $F(x) = (x + 1)(x + 3)^2(x - 2)$ d
39. $g(x) = (x + 1)(x + 3)(x - 2)^3$ e
40. $G(x) = (x + 1)^3(x + 3)(x - 2)$ a

43. **44.**

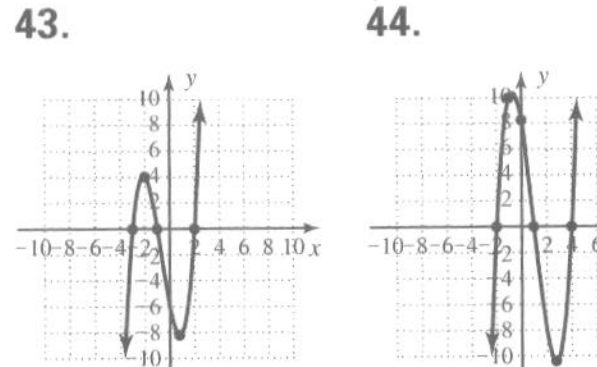

45. **46.**

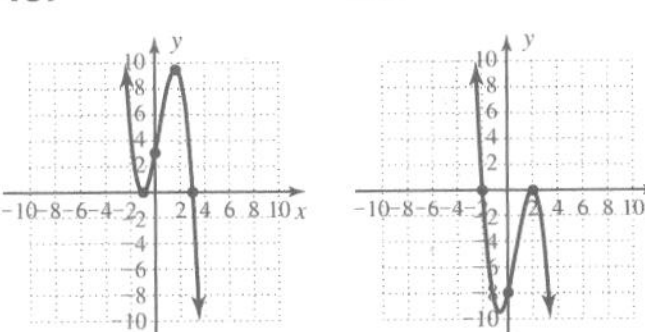

47. **48.**

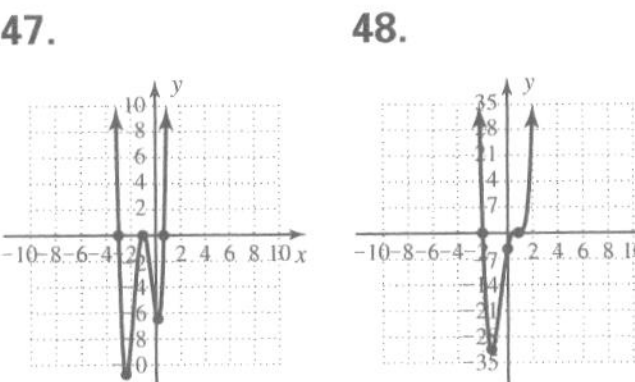

49. **50.**

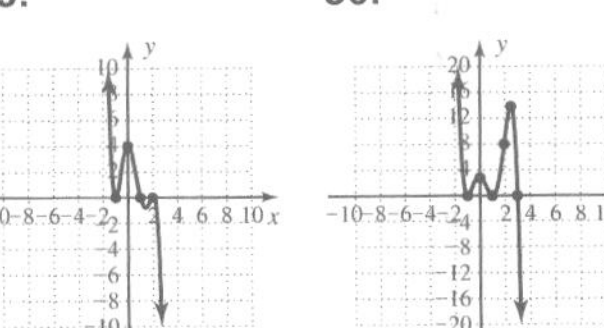

51. **52.**

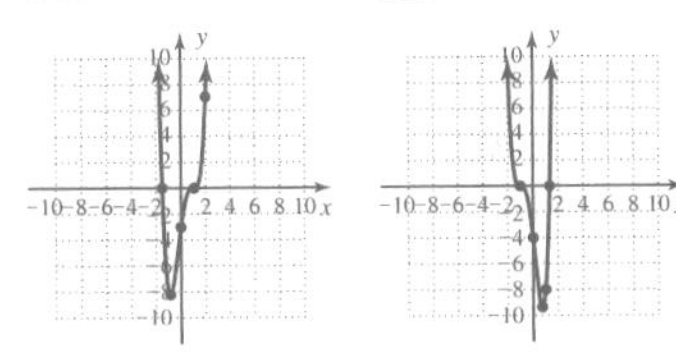

53. **54.**

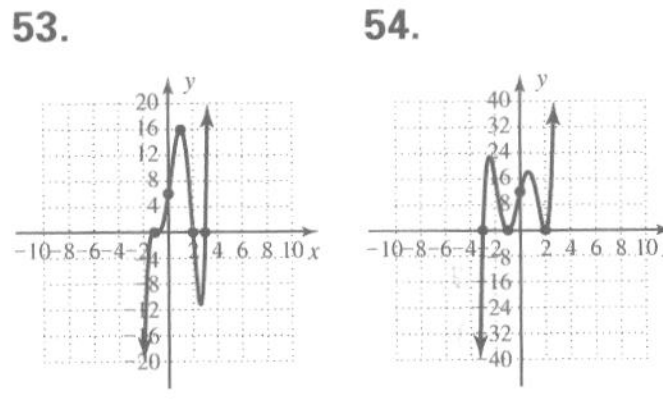

55. **56.**

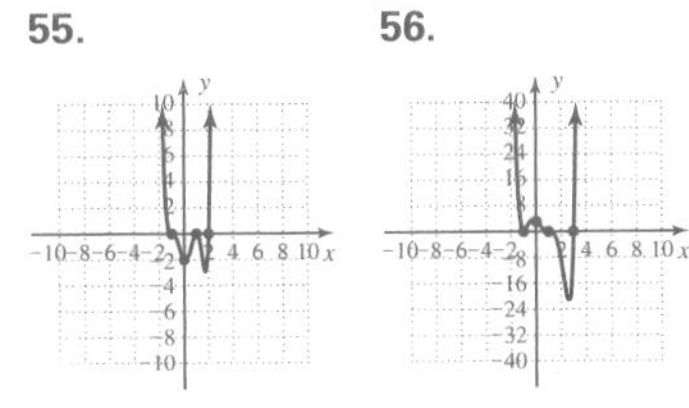

57. $P(x) = x^4 - 2x^3 - 13x^2 + 14x + 24$

58. $P(x) = x^4 - 2x^3 + 2x^2 + 6x - 15$

Additional answers can be found in the Instructor Answer Appendix.

41. $Y_1 = (x + 1)^2(x + 3)(x - 2)^2$ c

42. $Y_2 = (x + 1)^3(x + 3)(x - 2)^2$ f

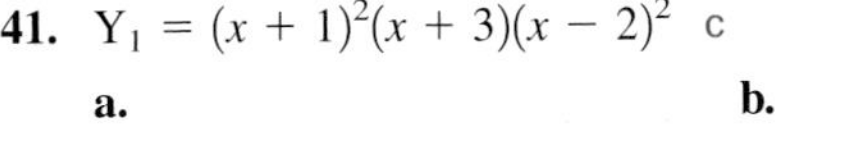

a.

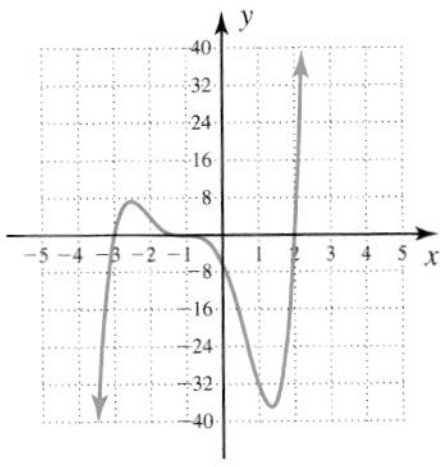

b.

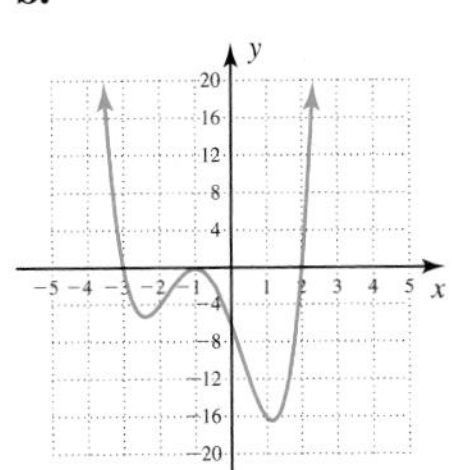

c.

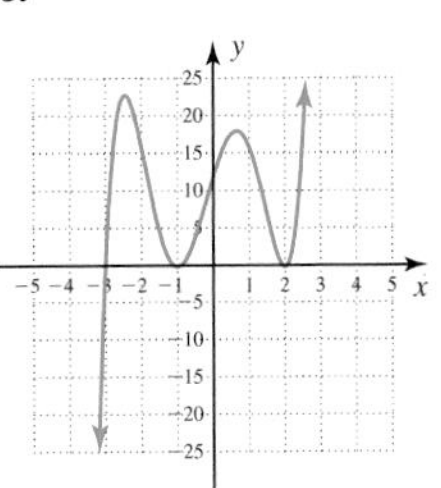

d.

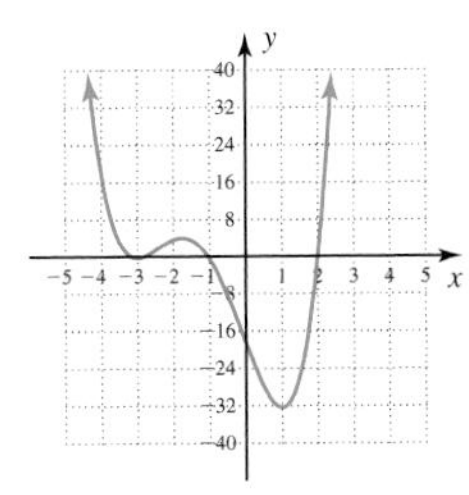

e.

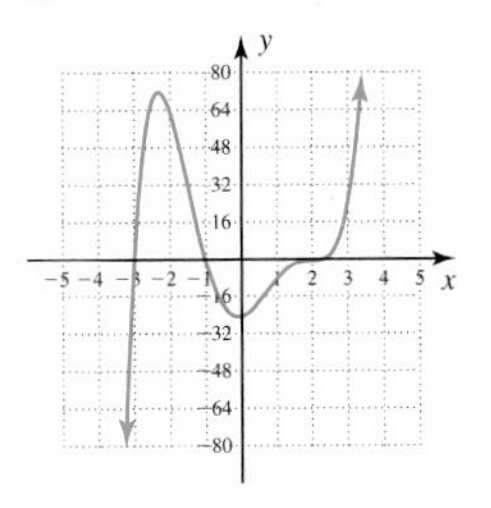

f.

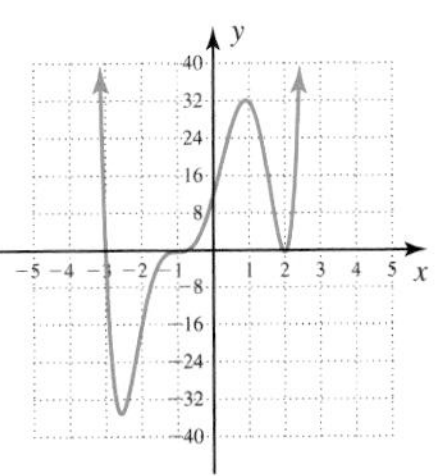

Sketch the graph of each function using the degree, end behavior, x- and y-intercepts, zeroes of multiplicity, and a few midinterval points to round-out the graph. Connect all points with a smooth, continuous curve.

43. $f(x) = (x + 3)(x + 1)(x - 2)$

44. $g(x) = (x + 2)(x - 4)(x - 1)$

45. $p(x) = -(x + 1)^2(x - 3)$

46. $q(x) = -(x + 2)(x - 2)^2$

47. $Y_1 = (x + 1)^2(3x - 2)(x + 3)$

48. $Y_2 = (x + 2)(x - 1)^2(5x - 2)$

49. $r(x) = -(x + 1)^2(x - 2)^2(x - 1)$

50. $s(x) = -(x - 3)(x - 1)^2(x + 1)^2$

51. $f(x) = (2x + 3)(x - 1)^3$

52. $g(x) = (3x - 4)(x + 1)^3$

53. $h(x) = (x + 1)^3(x - 3)(x - 2)$

54. $H(x) = (x + 3)(x + 1)^2(x - 2)^2$

55. $Y_3 = (x + 1)^3(x - 1)^2(x - 2)$

56. $Y_4 = (x - 3)(x - 1)^3(x + 1)^2$

WORKING WITH FORMULAS

57. Root tests for quartic polynomials: $ax^4 + bx^3 + cx^2 + dx + e = 0$

If u, v, w, and x represent the roots of a quartic polynomial, then the following relationships are true: (a) $u + v + w + x = -b$, (b) $u(v + x) + v(w + x) + w(u + x) = c$, (c) $u(vw + wx) + v(ux + wx) = -d$, and (d) $u \cdot v \cdot w \cdot x = e$. Use these tests to verify that $x = -3, -1, 2, 4$ are the solutions to $x^4 - 2x^3 - 13x^2 + 14x + 24 = 0$, then use these zeroes and the factored form to write the equation in polynomial form to confirm results.

58. It is worth noting that the root tests in Exercise 57 still apply when the roots are irrational and/or complex. Use these tests to verify that $x = -\sqrt{3}$, $\sqrt{3}$, $1 + 2i$, and $1 - 2i$ are the solutions to $x^4 - 2x^3 + 2x^2 + 6x - 15 = 0$, then use these zeroes and the factored form to write the equation in polynomial form to confirm results.

APPLICATIONS

Use the *Guidelines for Graphing Polynomial Functions* to graph the polynomials.

59. $y = x^3 + 3x^2 - 4$

60. $y = x^3 - 13x + 12$

61. $f(x) = x^3 - 3x^2 - 6x + 8$

62. $g(x) = x^3 + 2x^2 - 5x - 6$

63. $h(x) = x^3 + x^2 - 5x + 3$

64. $H(x) = x^3 + x^2 - 8x - 12$

65.

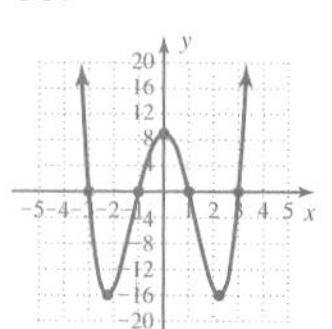

66.

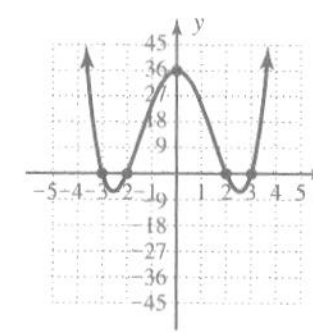

67.

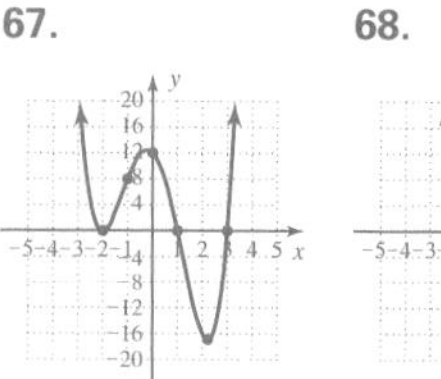

68.

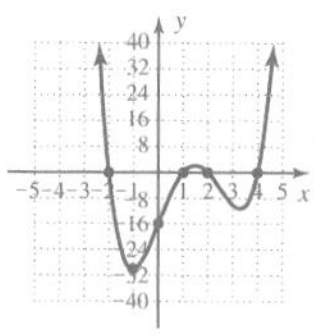

69.

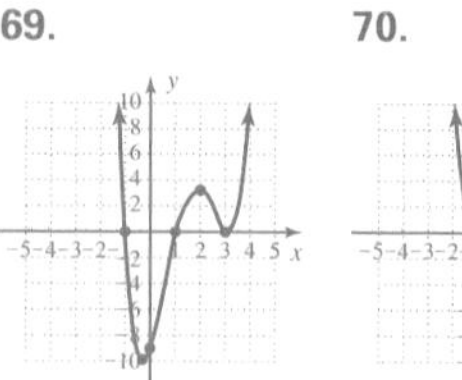

70.

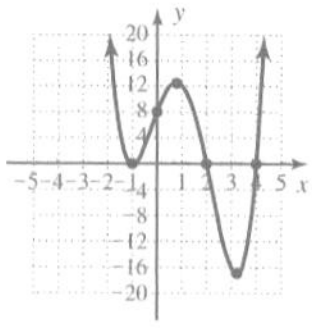

71. 72.

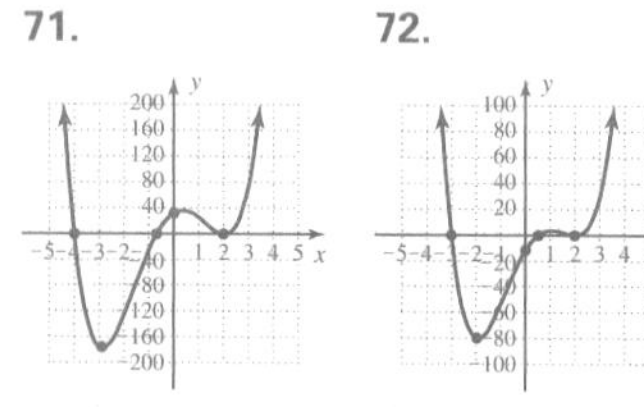

73. 74.

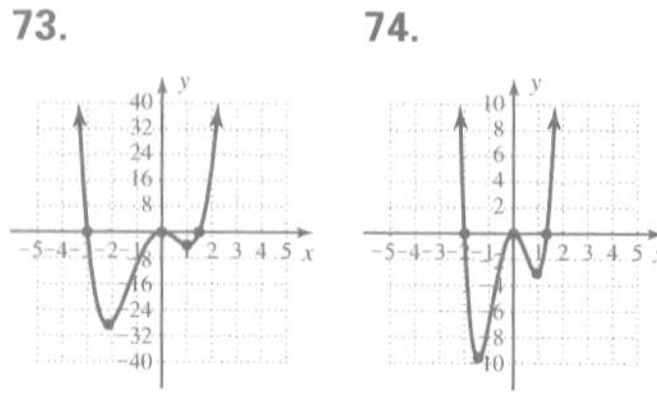

75. 76.

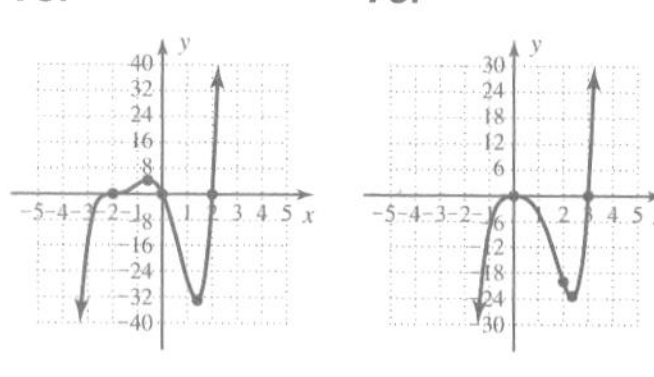

77. 78.

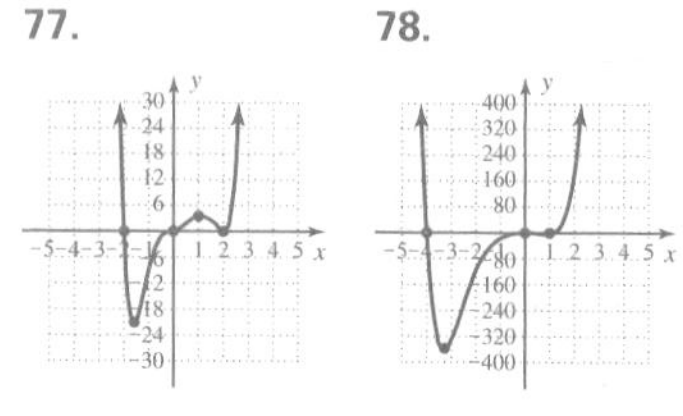

79. $h(x) = (x + 4)(x - \sqrt{3})$
$(x + \sqrt{3})(x - \sqrt{3}i)(x + \sqrt{3}i)$

80. $H(x) = (x + 5)(x - \sqrt{2})$
$(x + \sqrt{2})(x - \sqrt{2}i)(x + \sqrt{2}i)$

Additional answers can be found in the Instructor Answer Appendix.

65. $p(x) = x^4 - 10x^2 + 9$

66. $q(x) = x^4 - 13x^2 + 36$

67. $r(x) = x^4 - 9x^2 - 4x + 12$

68. $s(x) = x^4 - 5x^3 + 20x - 16$

69. $Y_1 = x^4 - 6x^3 + 8x^2 + 6x - 9$

70. $Y_2 = x^4 - 4x^3 - 3x^2 + 10x + 8$

71. $Y_3 = 3x^4 + 2x^3 - 36x^2 + 24x + 32$

72. $Y_4 = 2x^4 - 3x^3 - 15x^2 + 32x - 12$

73. $F(x) = 2x^4 + 3x^3 - 9x^2$

74. $G(x) = 3x^4 + 2x^3 - 8x^2$

75. $f(x) = x^5 + 4x^4 - 16x^2 - 16x$

76. $g(x) = x^5 - 3x^4 + x^3 - 3x^2$

77. $h(x) = x^6 - 2x^5 - 4x^4 + 8x^3$

78. $H(x) = x^6 + 4x^5 + x^4 + 2x^3 - 12x^2$

In preparation for future course work, it becomes helpful to recognize the most common square roots in mathematics: $\sqrt{2} \approx 1.414$, $\sqrt{3} \approx 1.732$, and $\sqrt{6} \approx 2.449$. Graph the following polynomials *on a graphing calculator,* and use the calculator to locate the maximum/minimum values and all zeroes. Use the zeroes to write the polynomial in factored form, then verify the y-intercept from the factored form and polynomial form.

79. $h(x) = x^5 + 4x^4 - 9x - 36$

80. $H(x) = x^5 + 5x^4 - 4x - 20$

81. $f(x) = 2x^5 + 5x^4 - 10x^3 - 25x^2 + 12x + 30$

82. $g(x) = 3x^5 + 2x^4 - 24x^3 - 16x^2 + 36x + 24$

Use the graph of each function to construct its equation in factored form and in polynomial form. Be sure to check the y-intercept and adjust the lead coefficient if necessary. Assume each tic mark is 1 unit.

83.

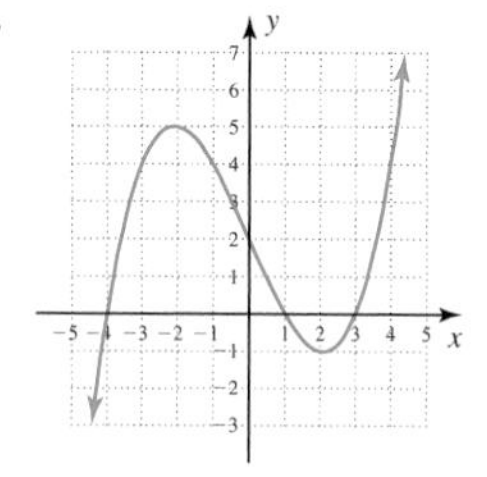

84.

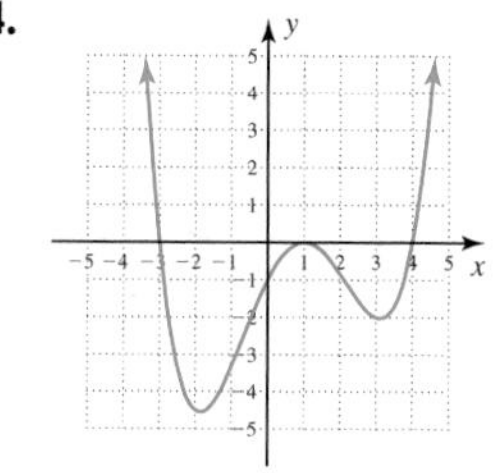

85. The graph shown represents the balance of payments (surplus versus deficit) for a large county over a 9-yr period. Use it to answer the following:

a. What is the minimum possible degree polynomial that can model this graph?

b. How many years did this county run a deficit?

c. Construct an equation model in factored form and in polynomial form, adjusting the lead coefficient as needed. How large was the deficit in year 8?

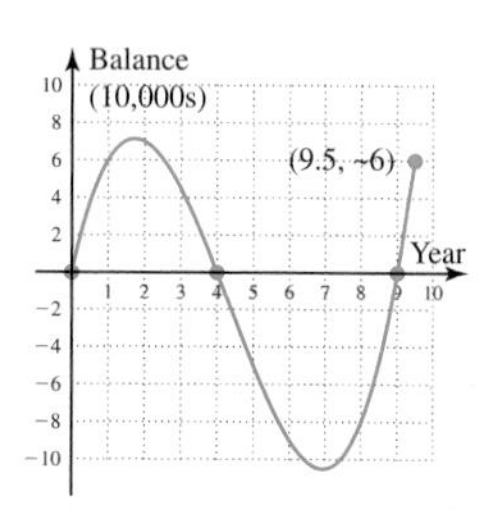

86. The graph shown represents the water level in a reservoir (above and below normal) that supplies water to a metropolitan area, over a 6-month period. Use it to answer the following:

a. What is the minimum possible degree polynomial that can model this graph?

b. How many months was the water level below normal in this 6-month period?

c. At the beginning of this period ($m = 0$), the water level was 36 in. above normal, due to a long period of rain. Use this fact to help construct an equation model in factored form and in polynomial form, adjusting the lead coefficient as needed. Use the equation to determine the water level in months three and five.

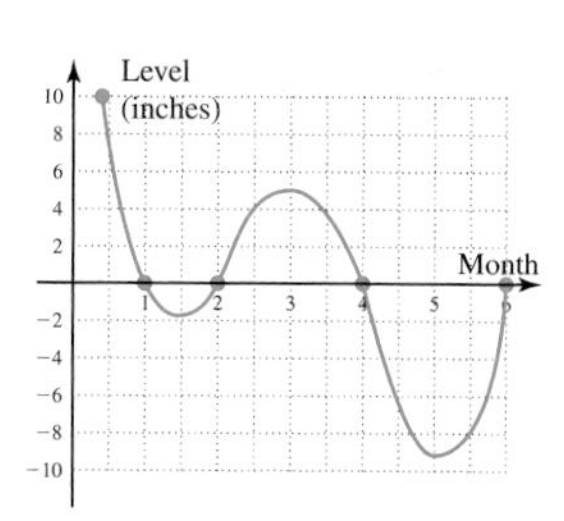

WRITING, RESEARCH, AND DECISION MAKING

87. End behavior precludes extended use.

88. $f(2\sqrt{3}) = (2\sqrt{3})^2 - 7\sqrt{3}(2\sqrt{3}) + 30 = 12 - 42 + 30 = 0$; $f(5\sqrt{3}) = (5\sqrt{3})^2 - 7\sqrt{3}(5\sqrt{3}) + 30 = 75 - 105 + 30 = 0$; The polynomial must have complex coefficients $a + bi$ where $b \neq 0$.

87. Polynomials can be used to model the fluctuations in many real-world phenomena, but only for a predetermined interval or period of time. Why is it that polynomial models fail over an extended time period, and cannot be used for an indefinite period?

88. Is it possible for a quadratic equation to have irrational roots that are *not* conjugates? Surprisingly, the answer is yes, if we allow the polynomial to have irrational coefficients. Verify that $x = 2\sqrt{3}$ and $x = 5\sqrt{3}$ are solutions to $x^2 - 7\sqrt{3}x + 30$. Under what conditions is it possible for a polynomial to have complex roots that are not conjugates? Refer to Section 4.3 if needed.

EXTENDING THE CONCEPT

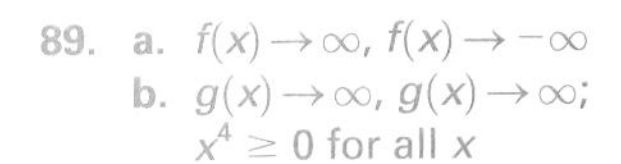

89. a. $f(x) \to \infty$, $f(x) \to -\infty$
b. $g(x) \to \infty$, $g(x) \to \infty$; $x^4 \geq 0$ for all x

89. As discussed in this section, the study of end behavior looks at what happens to the graph of a function as $|x| \to \infty$. From our study of asymptotes in Section 3.5, we know that as $|x| \to \infty$, both $\frac{1}{x}$ and $\frac{1}{x^2}$ approach zero. This general idea can be used to study the end behavior of polynomial graphs.

a. For $f(x) = x^3 + x^2 - 3x + 6$, factoring out x^3 gives the expression $f(x) = x^3\left(1 + \frac{1}{x} - \frac{3}{x^2} + \frac{6}{x^3}\right)$. What happens to the value of the expression as $x \to \infty$? As $x \to -\infty$?

b. Factor out x^4 from $g(x) = x^4 + 3x^3 - 4x^2 + 5x - 1$. What happens to the value of the expression as $x \to \infty$? As $x \to -\infty$? How does this affirm the end behavior must be up/up?

90. $c = -18$

90. For what value of c will three of the four real roots of $x^4 + 5x^3 + x^2 - 21x + c = 0$ be shared by the polynomial $x^3 + 2x^2 - 5x - 6 = 0$?

Show that the following equations have no rational roots.

91. $x^5 - x^4 - x^3 + x^2 - 2x + 3 = 0$

92. $x^5 - 2x^4 - x^3 + 2x^2 - 3x + 4 = 0$

91. verified
92. verified

MAINTAINING YOUR SKILLS

93. $h(x) = \frac{1 - 2x}{x^2}$; $D: x \in \{x \mid x \neq 0\}$; $R: y \in \{y \geq -1\}$; $H(x) = \frac{1}{x^2 - 2x}$; $D: x \in \{x \mid x \neq 0, x \neq 2\}$; $R: y \in \{y \mid y \neq 0\}$
94. a. $x = 2$
b. $x = 8$
c. $x = 4, x = -6$
95. yes
96. $\approx 11{,}519.17\ \text{m}^3$
97. $\approx 2827.43\ \text{m}^2$

93. (3.1) Given $f(x) = x^2 - 2x$ and $g(x) = \frac{1}{x}$, find the compositions $h(x) = (f \circ g)(x)$ and $H(x) = (g \circ f)(x)$, then state the domain and range of each.

94. (1.1/1.4/1.5) Solve each of the following equations.

a. $-(2x + 5) - (6 - x) + 3 = x - 3(x + 2)$

b. $\sqrt{x + 1} + 3 = \sqrt{2x} + 2$

c. $\frac{2}{x - 3} + 5 = \frac{21}{x^2 - 9} + 4$

95. (2.2) Determine if the relation shown is a function. If not, explain how the definition of a function is violated.

feline, canine, dromedary, equestrian, bovine → cow, horse, cat, camel, dog

96. (R.7) Find the volume of the composite figure shown.

97. (R.7) Find the surface area of the composite figure shown.

Exercises 96 and 97

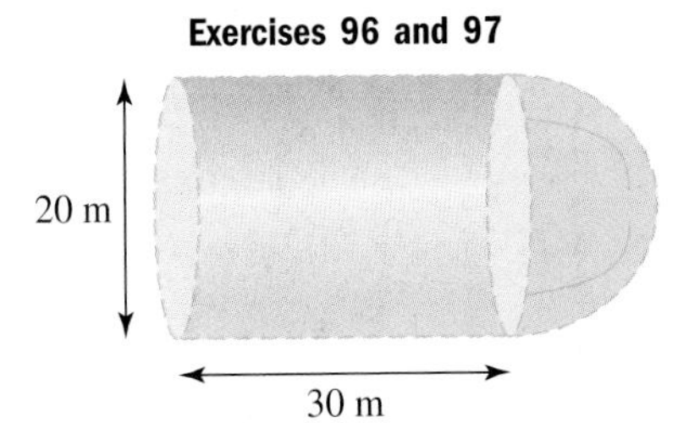

98. a. $y \approx 10x + 120$
b. \$270 billion
c. year 18 $\rightarrow$ 2008

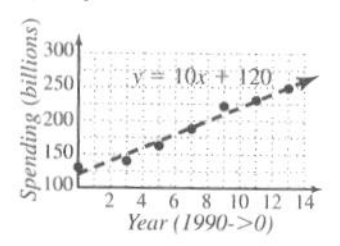

98. (2.6) The data shown indicates the amount spent on advertising (in billions) for selected years, by corporations anxious to keep their name in the public eye. Draw a scatter-plot and decide on an appropriate form of regression, then complete the following.

a. Find the regression equation.

b. Use the equation to project spending on advertising in 2005.

c. According to the model, in what year will advertising expenditures exceed 300 billion dollars?

Source: 2004 Statistical Abstract of the United States, Table 1274; various other years

Year 1990 $\rightarrow$ 0	Amount (billions)
0	130
3	140
5	163
7	188
9	222
11	231
13	249

MID-CHAPTER CHECK

1. (1) $x^3 + 8x^2 + 7x - 14 = (x^2 + 6x - 5)(x + 2) - 4$
(2) $\frac{x^3 + 8x^2 + 7x - 14}{x + 2} = x^2 + 6x - 5 - \frac{4}{x + 2}$
2. $f(x) = (2x + 3)(x + 1)(x - 1)(x - 2)$
3. $f(-2) = 7$
4. $f(x) = x^3 - 2x + 4$
5. $g(2) = -8$ and $g(3) = 5$ have opposite signs
6. $f(x) = (x - 2)(x + 1)(x + 2)(x + 4)$
7. $x = -2, x = 1, x = -1 \pm 3i$
8. 9.

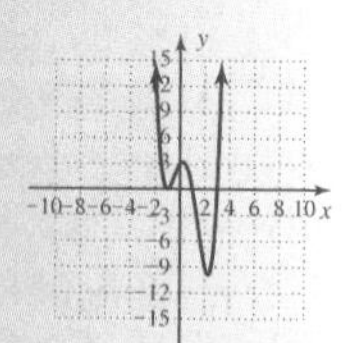

10. a. degree 4; three turning points
b. 2 sec
c. $A(t) = t^4 - 10t^3 + 32t^2 - 38t + 15$
$A(2) = 3$; altitude is 300 ft above hard-deck
$A(4) = -9$; altitude is 900 ft below hard-deck

1. Compute $(x^3 + 8x^2 + 7x - 14) \div (x + 2)$ using long division and write the result in two ways: (1) dividend = (quotient)(divisor) + remainder and (2) $\frac{\text{dividend}}{\text{divisor}} = (\text{quotient}) + \frac{\text{remainder}}{\text{divisor}}$.

2. Given that $x - 2$ is a factor of $f(x) = 2x^4 - x^3 - 8x^2 + x + 6$, use synthetic division and the quotient polynomial to write $f(x)$ in completely factored form.

3. Use synthetic division and the remainder theorem to evaluate $f(-2)$, given $f(x) = -3x^4 + 7x^2 - 8x + 11$.

4. Use the factor theorem to find a third-degree polynomial having $x = -2$ and $x = 1 + i$ as roots.

5. Use the intermediate value theorem to show that $g(x) = x^3 - 6x - 4$ has a root in the interval (2, 3).

6. Use the rational roots theorem, tests for -1 and 1, synthetic division, and the remainder theorem to write $f(x) = x^4 + 5x^3 - 20x - 16$ in completely factored form.

7. Find all the zeroes of h, real and complex: $h(x) = x^4 + 3x^3 + 10x^2 + 6x - 20$.

8. Sketch the graph of p using its degree, end behavior, y-intercept, zeroes of multiplicity, and any midinterval points needed, given $p(x) = (x + 1)^2(x - 1)(x - 3)$.

9. Use the *Guidelines for Graphing* to draw the graph of $q(x) = x^3 + 5x^2 + 2x - 8$.

10. When fighter pilots train for dogfighting, a "hard-deck" is usually established below which no competitive activity can take place. The polynomial graph given shows Maverick's altitude above and below this hard-deck during a 5-sec interval.

a. What is the minimum possible degree polynomial that could form this graph? Why?

b. How many seconds (total) was Maverick below the hard-deck for these 5 sec of the exercise?

c. At the beginning of this time interval ($t = 0$), Maverick's altitude was 1500 ft above the hard-deck. Use this fact and the graph given to help construct an equation model in factored form and in polynomial form, adjusting the lead coefficient if needed. Use the equation to determine Maverick's altitude in relation to the hard-deck at $t = 2$ and $t = 4$.

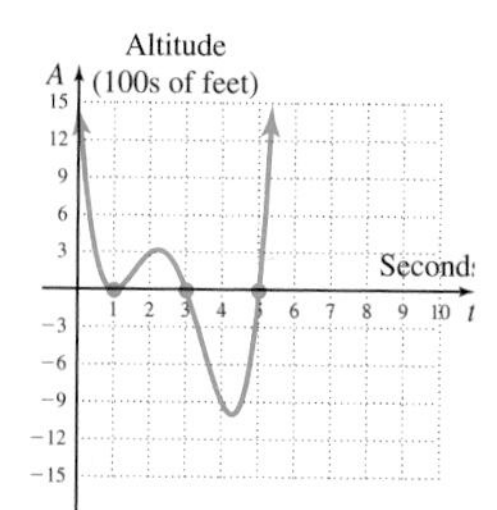

REINFORCING BASIC CONCEPTS

Approximating Real Roots

Consider the equation $x^4 + x^3 + x - 6 = 0$. Using the RRT, the possible rational zeroes are $\{\pm1, \pm6, \pm2, \pm3\}$. The tests for 1 and -1 indicate that neither is a root: $f(1) = -3$ and $f(-1) = -7$. Descartes's rule of signs reveals there must be one positive real root since the coefficients of $f(x)$ change sign one time: $f(x) = x^4 + x^3 + x - 6$, and one negative real root since $f(-x)$ also changes sign one time: $f(-x) = x^4 - x^3 - x - 6$. The remaining two roots must be complex. Using $x = 2$ with synthetic division shows 2 is not a root, but the coefficients in the quotient row are all positive, so 2 is an upper bound.

$$\begin{array}{r|rrrrrl} 2 & 1 & 1 & 0 & 1 & -6 & \text{coefficients of } f(x) \\ & & 2 & 6 & 12 & 26 & \\ \hline & 1 & 3 & 6 & 13 & 20 & q(x) \end{array}$$

Using $x = -2$ shows that -2 is a root *and a lower bound* for all other roots (quotient row alternates in sign):

$$\begin{array}{r|rrrrrl} -2 & 1 & 1 & 0 & 1 & -6 & \text{coefficients of } f(x) \\ & & -2 & 2 & -4 & 6 & \\ \hline & 1 & -1 & 2 & -3 & 0 & q_1(x) \end{array}$$

This means the remaining real zero must be a positive irrational number less than 2 (all other possible rational zeroes were eliminated). The quotient polynomial $q_1(x) = x^3 - x^2 + 2x - 3$ is not factorable, yet we're left with the challenge of finding this final root. While there are many advanced techniques available for approximating irrational roots, at this level either technology or a technique called **bisection** is commonly used. The bisection method combines the intermediate value theorem with successively smaller intervals of the input variable, to narrow down the location of the irrational root. Although "bisection" implies halving the interval each time, any number within the interval will do. The bisection method may be most efficient using a succession of short input/output tables as shown, with the number of tables increased if greater accuracy is desired. Since $f(1) = -3$ and $f(2) = 20$, the intermediate value theorem tells us the root must be in the interval [1, 2]. We begin our search here, rounding non-integer outputs to the nearest 100th. As a visual aid, positive outputs are in blue, negative outputs in red.

x	$f(x)$	Conclusion
1	−3	← Root is here, use $x = 1.25$ next
1.5	3.94	
2	20	

x	$f(x)$	Conclusion
1	−3	
1.25	−0.36	← Root is here, use $x = 1.30$ next
1.5	3.94	

x	$f(x)$	Conclusion
1.25	−0.36	← Root is here, use $x = 1.275$ next
1.30	0.35	
1.5	3.94	

A reasonable estimate for the root appears to be $x = 1.275$. Evaluating the function at this point gives $f(1.275) \approx 0.0098$, which is very close to zero.

Naturally, a closer approximation is obtained using the capabilities of a graphing calculator (see *Technology Highlight,* Section 3.4). To seven decimal places the root is $x \approx 1.2756822$.

Exercise 1: Use the intermediate value theorem to show that $f(x) = x^3 - 3x + 1$ has a zero in the interval [1, 2], then use bisection to locate the zero to three decimal place accuracy. 1.532

Exercise 2: The function $f(x) = x^4 + 3x - 15$ has two real zeroes in the interval $[-5, 5]$. Use the intermediate value theorem to locate the roots, then use bisection to find the zeroes accurate to three decimal places. −2.152, 1.765

4.5 Graphing Rational Functions

LEARNING OBJECTIVES

In Section 4.5 you will learn how to:

A. Find the domain of a rational function

B. Apply the concept of "multiplicity" to rational graphs

C. Find x- and y-intercepts of a rational function

D. Find horizontal asymptotes of a rational function

E. Graph general rational functions

INTRODUCTION

A rational function is simply the ratio of two polynomials. As such, many connections can be made with our study of polynomials, particularly in finding zeroes and in the behavior of the graph at these zeroes. For a rational function, the roots of multiplicity in the numerator and denominator show up in interesting and intriguing ways, but always consistent with what we've learned about even and odd multiplicities.

POINT OF INTEREST

The sale of cardboard puzzles is a big business in the United States, as many people are drawn to the simple satisfaction of having completed a task by putting together its diverse pieces. The most common strategy for completing a puzzle is to first complete the border using the "straight" pieces, then separate pieces having the same color scheme and piece these together first, which breaks the task into smaller and related parts. Near the end, these are joined with the border and the remaining pieces are puzzled into place. The graph of a rational function shares the same puzzle-like nature; and in exactly the same way, we will first "frame-out" the graph, then piece together smaller and related parts, and finish by "puzzling" the graph into place.

Figure 4.22

$r(x) = \dfrac{1}{x+2}$

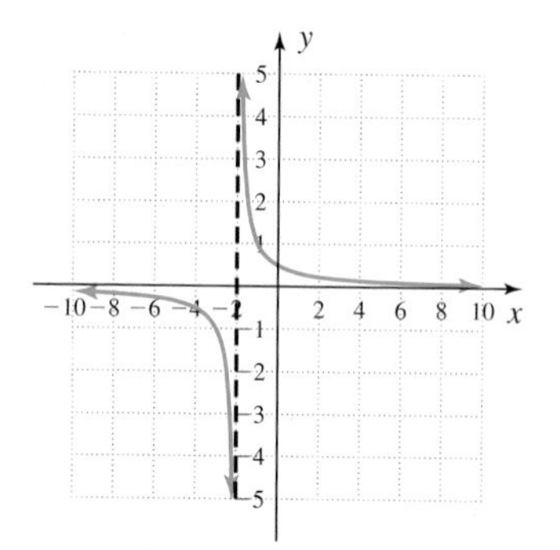

A. Vertical Asymptotes and the Domain

The basic rational functions from Section 3.5 have numerous connections with more general forms. To note and strengthen these connections, we'll look at four rational functions of various types. Figures 4.22 through 4.25 show that rational graphs come in many shapes and sizes, and often in "pieces." Note the first graph is simply the reciprocal function shifted two units left: $r(x) = \dfrac{1}{x+2}$, with a vertical asymptote at $x = -2$, and a horizontal asymptote at $y = 0$ (the x-axis). Asymptotic behaviors can also be seen in the other graphs.

Figure 4.23

$R(x) = \dfrac{2x}{x^2 - 1}$

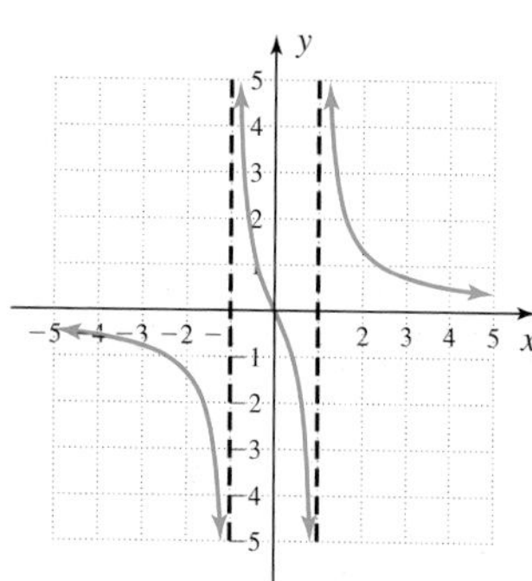

Figure 4.24

$h(x) = \dfrac{3}{x^2 + 1}$

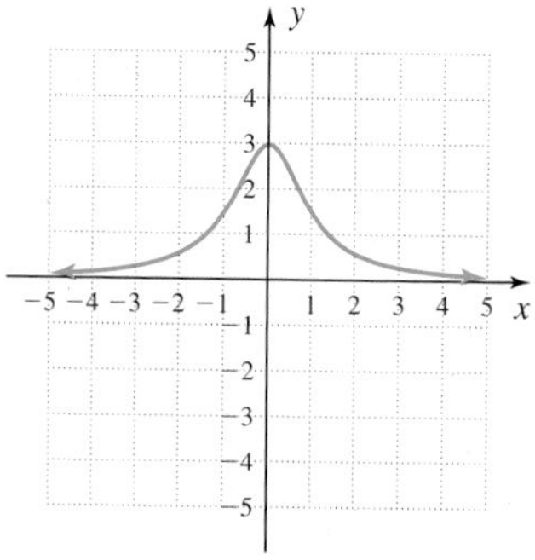

Figure 4.25

$H(x) = \dfrac{x^2}{x^2 - 2x - 3}$

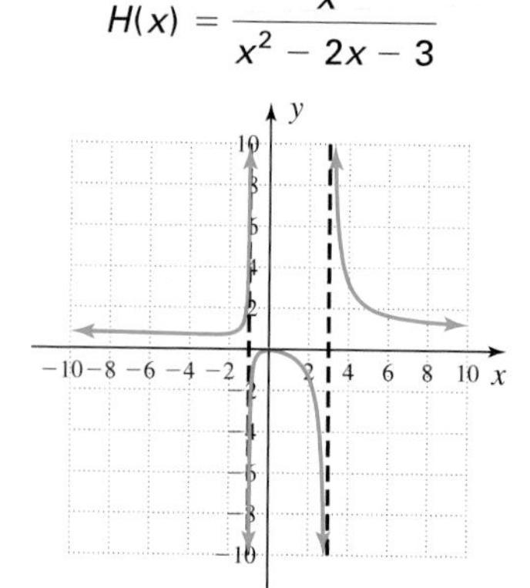

WORTHY OF NOTE

For a review of asymptotes, refer to Section 3.5. In Section 3.7, we studied special cases where f and g shared a common factor. In this section we'll assume that $r(x)$ is given in simplest form (the numerator and denominator have no common factors).

In general, a **rational function** is any function of the form $r(x) = \frac{f(x)}{g(x)}$, where f and g are polynomials and $g(x) \neq 0$. This immediately tells us the domain of a rational function is all real numbers, *except the zeroes of the denominator.*

THE DOMAIN OF A RATIONAL FUNCTION

Given the rational function $r(x) = \frac{f(x)}{g(x)}$,
the domain is $D: \{x | x \in \mathbb{R}; g(x) \neq 0\}$.

In our previous work with $y = \frac{1}{x}$ and $y = \frac{1}{x^2}$, we noted that vertical asymptotes occur at zeroes of the denominator. This idea actually applies to all rational functions in simplified form. For $r(x) = \frac{f(x)}{g(x)}$, if $x = h$ is a zero of g, the function can be evaluated at every point near h, but not *at* h. This creates a **break** or **discontinuity** in the graph, resulting in the asymptotic behavior.

VERTICAL ASYMPTOTES OF A RATIONAL FUNCTION

Given $r(x) = \frac{f(x)}{g(x)}$ is a rational function in lowest terms,
vertical asymptotes will occur at the real zeroes of g.

EXAMPLE 1 Locate the vertical asymptote(s) of each function and state the domain in set notation.

a. $r(x) = \frac{1}{x + 2}$ **b.** $R(x) = \frac{2x}{x^2 - 1}$

c. $h(x) = \frac{3}{x^2 + 1}$ **d.** $H(x) = \frac{x^2}{x^2 + 3x - 10}$

Solution:

a. The denominator is zero when $x = -2$, which is the equation of the vertical asymptote. The domain is $\{x | x \in \mathbb{R}; x \neq -2\}$.

b. Setting the denominator equal to zero gives $x^2 - 1 = 0$, so vertical asymptotes will occur at $x = -1$ and $x = 1$. The domain is $\{x | x \in \mathbb{R}; x \neq 1, x \neq -1\}$.

c. Since the equation $x^2 + 1 = 0$ has no real zeroes, there are no vertical asymptotes. The domain is $x \in \mathbb{R}$.

d. Solving $x^2 + 3x - 10 = 0$ gives $(x + 5)(x - 2) = 0$, with solutions $x = -5$ and $x = 2$. The domain is $\{x | x \in \mathbb{R}; x \neq -5, x \neq 2\}$ with vertical asymptotes at $x = -5$ and 2.

NOW TRY EXERCISES 7 THROUGH 14

B. Vertical Asymptotes and Multiplicities

Figure 4.26

$y = \frac{1}{(x-1)^2}$

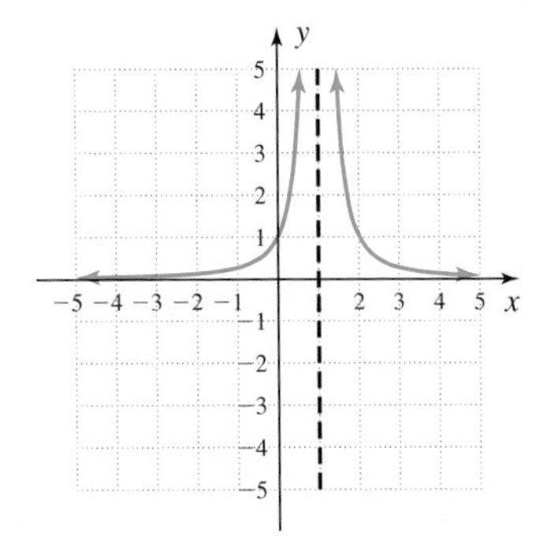

Figure 4.27

$y = \frac{1}{(x-1)^2(x+2)}$

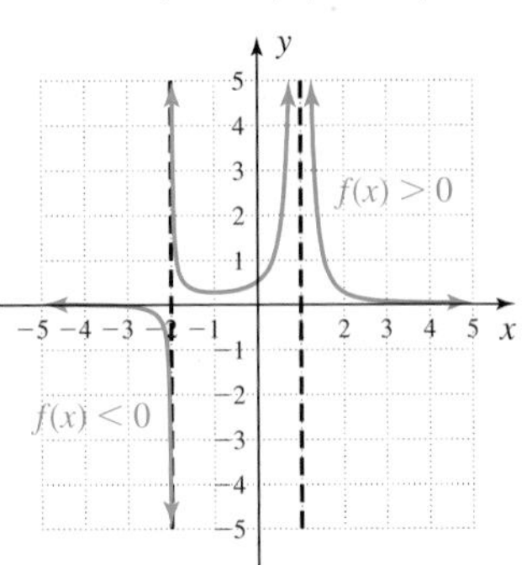

The "cross" and "bounce" concept used for polynomial graphs can also be applied to rational graphs, particularly when viewed in terms of sign changes in output values. As seen in Figure 4.22, the graph of $r(x) = \frac{1}{x+2}$ changes sign at the asymptote $x = -2$, and we note the denominator has multiplicity 1 (odd). The graph of $y = \frac{1}{(x-1)^2}$ is a "volcano function" shifted one unit to the right (Figure 4.26), and does not change sign from one side of the asymptote to the other. Its denominator has multiplicity 2 (even) and is positive for all $x \neq 1$. As with our earlier study of multiplicities, when these two factors are combined into a single rational function, $y = \frac{1}{(x-1)^2(x+2)}$, the same behaviors are exhibited (Figure 4.27).

EXAMPLE 2 Locate the vertical asymptotes of each function and state whether the function will change sign at the asymptote(s).

a. $p(x) = \frac{x^2 - 4x + 4}{x^2 - 2x - 3}$ **b.** $q(x) = \frac{x^2 + 2}{x^2 + 2x + 1}$

Solution:

a. Setting $x^2 - 2x - 3$ equal to zero gives $(x+1)(x-3) = 0$ after factoring, with zeroes $x = -1$ and $x = 3$ (both multiplicity 1). The function will change sign at each asymptote (Figure 4.28).

b. From $x^2 + 2x + 1 = 0$, we have $(x+1)^2 = 0$. There is a vertical asymptote at $x = -1$, but the function will not change sign since it's a zero of even multiplicity (Figure 4.29).

Figure 4.28

$p(x) = \frac{x^2 - 4x + 4}{x^2 - 2x - 3}$

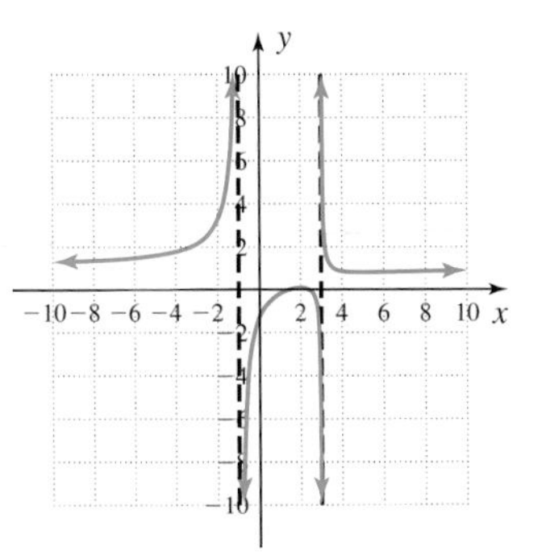

Figure 4.29

$q(x) = \frac{x^2 + 2}{x^2 + 2x + 1}$

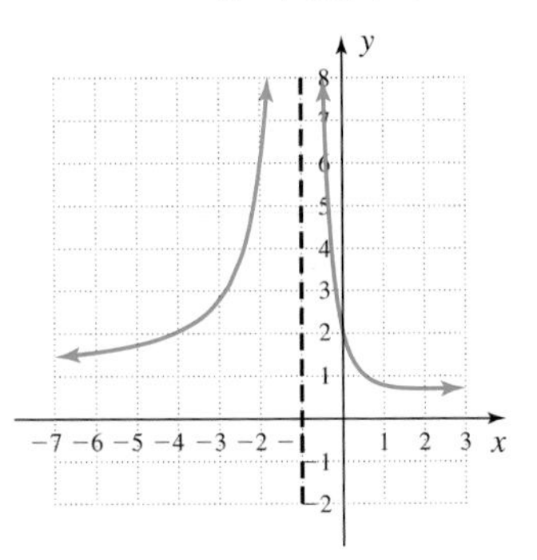

NOW TRY EXERCISES 15 THROUGH 20

C. Locating the *x*- and *y*-Intercepts of a Rational Function

The basic ideas for finding *x*- and *y*-intercepts are consistent for all function families. We locate the *y*-intercept by substituting 0 for *x* (if the expression is defined at $x = 0$).

To find the x-intercept(s), if they exist, substitute 0 for $r(x)$. Note that setting $r(x)$ equal to zero produces the following sequence:

$$r(x) = \frac{f(x)}{g(x)} \quad \text{rational function}$$

$$0 = \frac{f(x)}{g(x)} \quad \text{substitute 0 for } r(x)$$

$$g(x) \cdot 0 = \frac{f(x)}{\cancel{g(x)}} \cdot \overset{1}{\cancel{g(x)}} \quad \text{multiply by } g(x)$$

$$0 = f(x) \quad \text{result}$$

In words, the x-intercepts of a rational function (if they exist) are given by *the zeroes of the numerator.* It's worthwhile to note that if $x = h$ is a zero of even multiplicity, the graph of r will bounce off the x-axis as before and $r(x)$ will not change sign at h (see Figure 4.25). If $x = h$ has odd multiplicity, the graph will cross the x-axis and $r(x)$ changes sign at h.

x- AND y-INTERCEPTS OF A RATIONAL FUNCTION

Given $r(x) = \frac{f(x)}{g(x)}$ is in lowest terms, and $x = 0$ in the domain of r,

1. To find the y-intercept, substitute 0 for x and simplify. If 0 is not in the domain, the function has no y-intercept.
2. To find the x-intercept(s), substitute 0 for $f(x)$ and solve. If the equation has no real zeroes, there are no x-intercepts.

EXAMPLE 3 Determine the x- and y-intercepts for the functions from Example 1, and state the behavior of the graph at the x-intercepts. Confirm by inspecting the graphs.

a. $r(x) = \frac{1}{x + 2}$ **b.** $R(x) = \frac{2x}{x^2 - 1}$

c. $h(x) = \frac{3}{x^2 + 1}$ **d.** $H(x) = \frac{x^2}{x^2 + 3x - 10}$

Solution:

a. y-intercept $(0, \frac{1}{2})$: $r(0) = \frac{1}{2}$.
x-intercept(s): Numerator is constant, graph has no x-intercepts.

b. y-intercept $(0, 0)$: $R(0) = 0$.
x-intercept(s): Setting $2x$ equal to zero shows $x = 0$ is a root of multiplicity 1. The x-intercept is $(0, 0)$ and the graph will cross through the origin.

c. y-intercept $(0, 3)$: $h(0) = 3$.
x-intercept(s): Numerator is constant, has no x-intercepts.

d. y-intercept $(0, 0)$: $H(0) = 0$.
x-intercept(s): Setting x^2 equal to zero shows $x = 0$ is a root of multiplicity 2. The graph will bounce at the origin.

Check results using a graphing calculator.

NOW TRY EXERCISES 21 THROUGH 26

D. Horizontal Asymptotes

A study of horizontal asymptotes is closely related to our study of "dominant terms" in the previous section. Recall the highest degree term in a polynomial tends to dominate all other terms as $|x| \to \infty$. For the function $h(x) = \frac{2x^2 + 4x + 3}{x^2 + 2x + 1}$, both polynomials have the same degree so $\frac{2x^2 + 4x + 3}{x^2 + 2x + 1} \approx \frac{2x^2}{x^2} = 2$ for large values of x. For instance $h(4) = 2.04$, but $h(4000) \approx 2.00000006$.

From Section 3.5, when the degree of the denominator is *larger* than the degree of the numerator we found that as $|x| \to \infty$, $y \to 0$. The result is a horizontal asymptote of $y = 0$ (the x-axis) as seen in the graphs of $y = \frac{1}{x}$ and $y = \frac{1}{x^2}$ (also see Exercise 89, Section 4.4). In summary:

LOOKING AHEAD

In Section 4.6 we will explore two additional kinds of asymptotic behavior: (1) oblique (slant) asymptotes and (2) asymptotes that are nonlinear.

HORIZONTAL ASYMPTOTES

Given $h(x) = \frac{f(x)}{g(x)}$ is a rational function in lowest terms, where the lead term of f is ax^n and the lead term of g is bx^m (polynomial f has degree n, polynomial g has degree m),

1. If $n < m$, the graph of h has a horizontal asymptote at $y = 0$ (the x-axis).
2. If $n = m$, the graph of h has a horizontal asymptote at $y = \frac{a}{b}$ (the ratio of lead coefficients).
3. If $n > m$, the graph of h has no horizontal asymptote.

Finally, while the graph of a rational function can never "cross" the vertical asymptote $x = h$ (since the function simply cannot be evaluated at h), it is *possible* for a graph to cross the horizontal asymptote $y = k$ (some do, others do not). To find out which is the case, we set the function equal to k and solve.

EXAMPLE 4 Locate the horizontal asymptote for each function, if one exists. Then determine if the graph will cross its horizontal asymptote.

a. $r(x) = \frac{3x}{x^2 + 2}$ **b.** $R(x) = \frac{x^2 - 4}{x^2 - 1}$

c. $T(x) = \frac{3x^2 - x - 6}{x^2 + x - 6}$

Solution:

a. The degree of numerator < degree of denominator: horizontal asymptote at $y = 0$. Since the equation $\frac{3x}{x^2 + 2} = 0$ has solution $x = 0$, r will cross the horizontal asymptote here (see Figure 4.30).

Figure 4.30 $r(x) = \frac{3x}{x^2 + 2}$

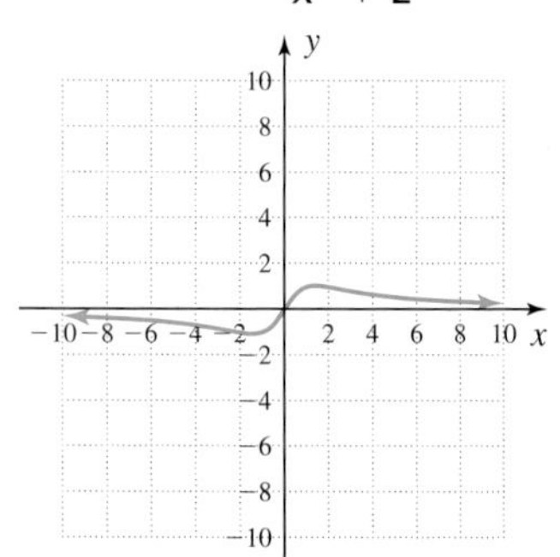

b. The degree of numerator = degree of denominator: horizontal asymptote at $y = \frac{1}{1} = 1$.

Solve the equation $\frac{x^2-4}{x^2-1} = 1$:

$$\frac{x^2-4}{x^2-1} = 1 \qquad r(x) = 1 \rightarrow \text{horizontal asymptote}$$
$$x^2 - 4 = x^2 - 1 \qquad \text{cross multiply}$$
$$-4 = -1 \qquad \text{no solution}$$

The graph will not cross the asymptote (see Figure 4.31).

c. The degree of numerator = degree of denominator: horizontal asymptote at $y = \frac{3}{1} = 3$.

Solve the equation $\frac{3x^2 - x - 6}{x^2 + x - 6} = 3$:

$$\frac{3x^2 - x - 6}{x^2 + x - 6} = 3 \qquad r(x) = 3 \rightarrow \text{horizontal asymptote}$$
$$3x^2 - x - 6 = 3(x^2 + x - 6) \qquad \text{cross multiply}$$
$$3x^2 - x - 6 = 3x^2 + 3x - 18 \qquad \text{distribute}$$
$$-4x + 12 = 0 \qquad \text{simplify}$$
$$x = 3 \qquad \text{solve}$$

The graph will cross the asymptote at $x = 3$ (see Figure 4.32).

The corresponding graphs are given here, with the horizontal asymptotes shown.

Figure 4.31

$R(x) = \frac{x^2 - 4}{x^2 - 1}$

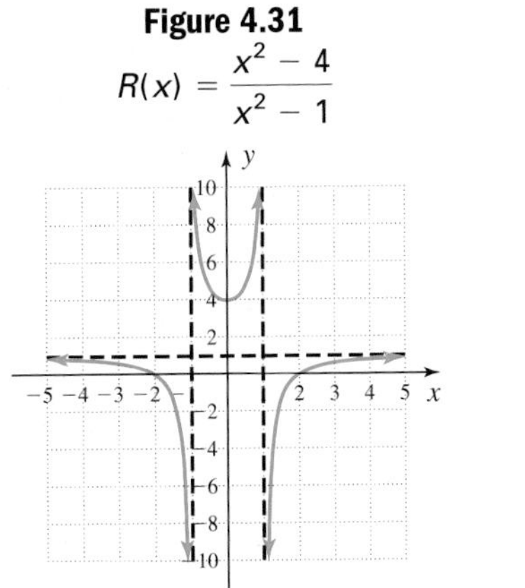

Figure 4.32

$T(x) = \frac{3x^2 - x - 6}{x^2 + x - 6}$

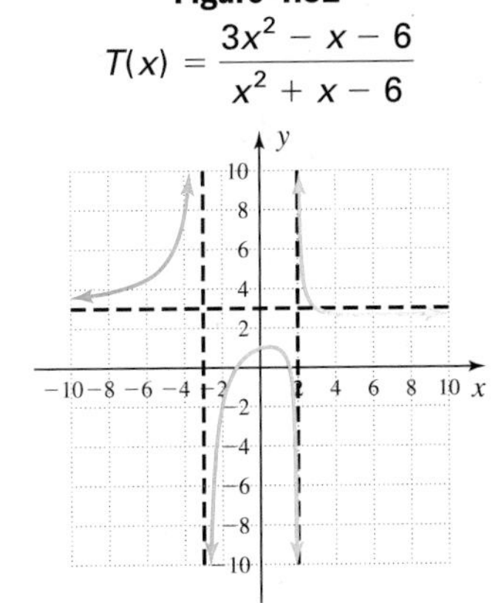

NOW TRY EXERCISES 27 THROUGH 32

E. The Graph of a Rational Function

In some instances, polynomial division enables us to rewrite a rational function in "shiftable" form (see Exercises 33 through 38). However, in most cases, all previous observations are used to formulate a more general strategy for graphing rational functions. Not all components are used every time, but together they provide an effective approach.

GUIDELINES FOR GRAPHING RATIONAL FUNCTIONS

Given $r(x) = \frac{f(x)}{g(x)}$ is a rational function in lowest terms, with $g(x) \neq 0$.

1. Find the y-intercept [evaluate $r(0)$].
2. Locate vertical asymptotes $x = h$ [solve $g(x) = 0$].
3. Find the x-intercepts (if any) [solve $f(x) = 0$].
4. Locate the horizontal asymptote $y = k$ (check degree of numerator and denominator).
5. Determine if the graph will cross the horizontal asymptote [solve $r(x) = k$ from step 4].
6. If needed, compute the value of any "midinterval" points needed to round-out the graph.
7. Draw the asymptotes, plot the intercepts and additional points, and use intervals where $r(x)$ changes sign to complete the graph. The graph *must* go through plotted points and "approach" the asymptotes.

It's useful to note that the number of "pieces" forming a rational graph will always be one more than the number of vertical asymptotes. The graph of $y = \frac{3}{x^2 + 1}$ (Figure 4.24) has no vertical asymptotes and one piece, $y = \frac{1}{x}$ has one vertical asymptote and two pieces, $y = \frac{x^2 - 4}{x^2 - 1}$ (Figure 4.31) has two vertical asymptotes and three pieces, and so on.

EXAMPLE 5 Graph each rational function using the Guidelines.

a. $s(x) = \frac{x^2 - x - 6}{x^2 + x - 6}$ **b.** $S(x) = \frac{3x^2 - 6x + 3}{x^2 - 7}$

Solution: **a.** Writing $s(x)$ in factored form gives $s(x) = \frac{(x + 2)(x - 3)}{(x + 3)(x - 2)}$.

1. y-intercept: $s(0) = 1$

2. vertical asymptote(s): $x = -3$ and $x = 2$.

3. x-intercepts: $(-2, 0)$ and $(3, 0)$

4. horizontal asymptote:
degree of numerator = degree of denominator, $y = 1$ is a horizontal asymptote (ratio of lead coefficients).

5. Solve

$$\frac{x^2 - x - 6}{x^2 + x - 6} = 1 \quad s(x) = 1 \rightarrow \text{horizontal asymptote}$$

$$x^2 - x - 6 = x^2 + x - 6 \quad \text{cross multiply}$$

$$x = 0 \quad \text{solve}$$

The graph will cross its horizontal asymptote at $(0, 1)$.

The information from steps 1 through 5 produces Figure 4.33, and indicates additional points would be helpful.

6. Compute $s(-4) = 2.\overline{3}$ and $s(1) = 1.5$: $(-4, 2.\overline{3})$ and $(1, 1.5)$.
7. All factors of $s(x)$ are linear, so function values will alternate sign in the intervals created by x-intercepts and vertical asymptotes. The y-intercept $(0, 1)$ shows the function is positive in the middle interval and will be negative in the intervals to either side (then alternate in sign thereafter). This information and that of steps 1 through 6 are shown in Figure 4.33. To meet all necessary conditions, we complete the graph as shown in Figure 4.34.

Figure 4.33

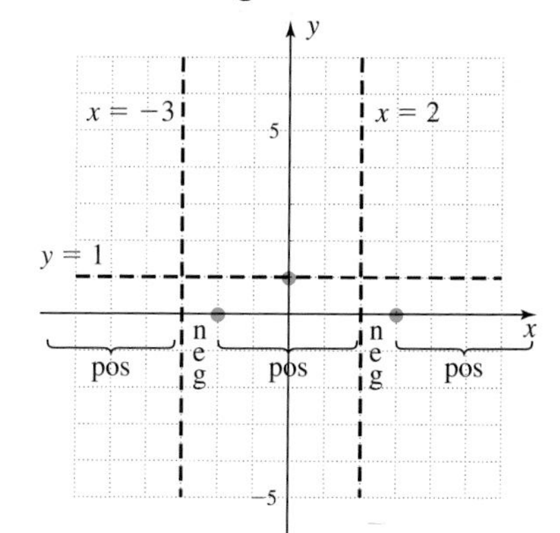

Figure 4.34

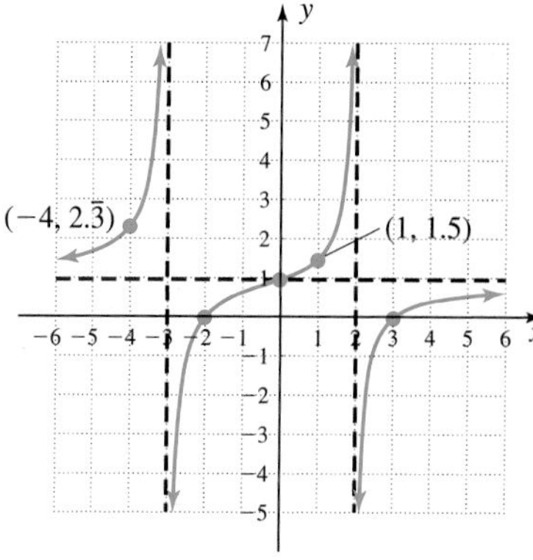

b. Writing $S(x) = \dfrac{3x^2 - 6x + 3}{x^2 - 7}$ in factored form

gives $S(x) = \dfrac{3(x - 1)^2}{(x + \sqrt{7})(x - \sqrt{7})}$.

1. y-intercept: $S(0) = -\frac{3}{7}$
2. vertical asymptote(s): $x = \pm\sqrt{7} \approx \pm 2.6$.
3. x-intercept(s): $(1, 0)$ multiplicity 2
4. horizontal asymptote:
   degree of numerator = degree of denominator, $y = 3$ is a horizontal asymptote (ratio of lead coefficients)
5. Solve

$$\frac{3x^2 - 6x + 3}{x^2 - 7} = 3 \qquad S(x) = 3 \rightarrow \text{horizontal asymptote}$$

$$3x^2 - 6x + 3 = 3x^2 - 21 \qquad \text{cross multiply}$$

$$x = 4 \qquad \text{solve}$$

The graph will cross its horizontal asymptote at $(4, 3)$.

The information from steps 1 to 5 is shown in Figure 4.35, and again indicates that additional points would be helpful (there are no points to the left of $x = -\sqrt{7}$).

6. Compute $S(-4) = 8.\overline{3}$, $S(-1) = -2$, $S(2) = -1$, and $S(3) = 6$ gives

$$(-4, 8.\overline{3}), \quad (-1, -2) \quad (2, -1) \quad (3, 6)$$

7. Since factors of the denominator have odd multiplicity, function values will alternate sign on either side of the asymptotes. The factor in the numerator has even multiplicity, so the graph will bounce off the x-axis at $x = 1$ (function values will not change sign). The y-intercept $(0, -\frac{3}{7})$ shows the function is negative in the interval containing zero. This information and the completed graph are shown in Figure 4.36.

Figure 4.35

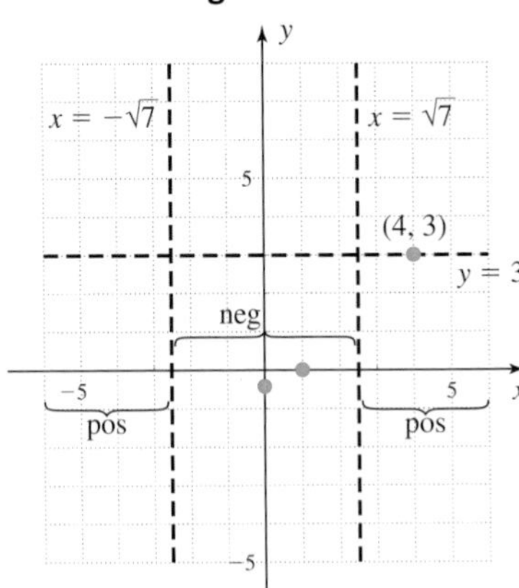

Figure 4.36

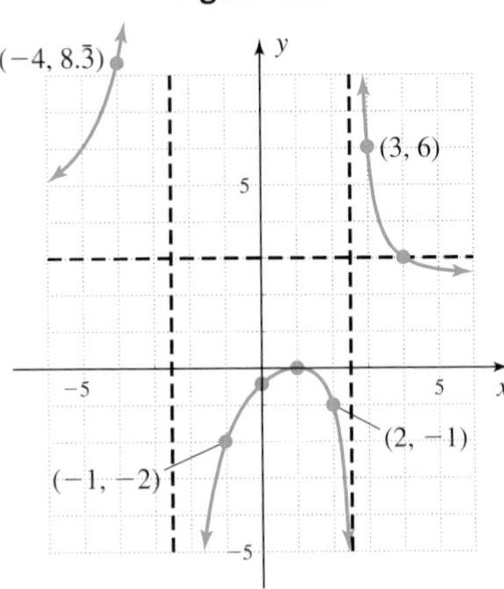

NOW TRY EXERCISES 39 THROUGH 56

Example 5 further demonstrates that graphs of rational functions come in a large variety of shapes and sizes. Once the components of the graph have been found, completing the graph offers an intriguing and puzzle-like challenge, since all conditions must be met. See if you can reverse the process. Given the graph of a rational function, can you construct a likely equation?

EXAMPLE 6 Use the graph of the $r(x)$ shown in the figure to construct its equation.

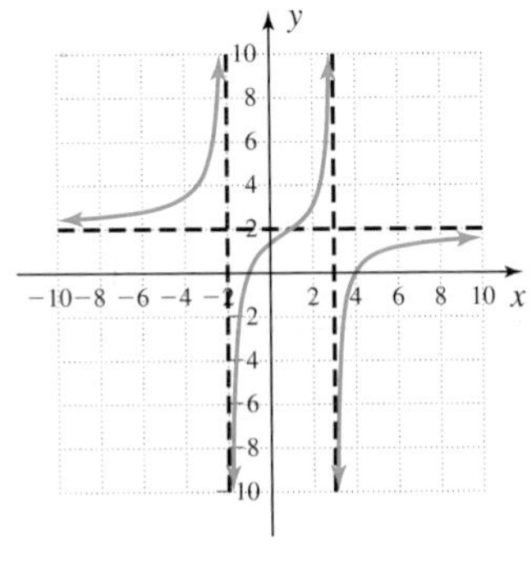

Solution: The x-intercepts are $(-1, 0)$ and $(4, 0)$, so the numerator must contain the factors $(x + 1)$ and $(x - 4)$. The vertical asymptotes are at $x = -2$ and $x = 3$, so the denominator must have the factors $(x + 2)(x - 3)$. So far we have

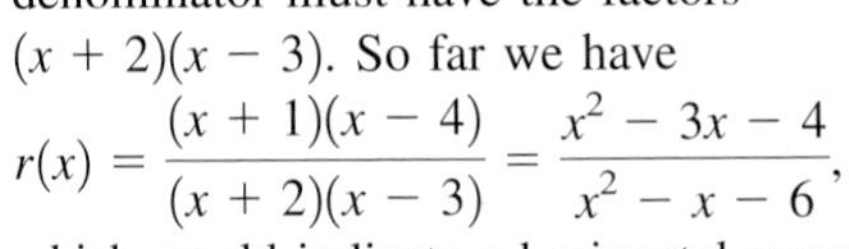

$$r(x) = \frac{(x + 1)(x - 4)}{(x + 2)(x - 3)} = \frac{x^2 - 3x - 4}{x^2 - x - 6},$$

which would indicate a horizontal asymptote at $y = 1$. Since the actual asymptote is at $y = 2$, we're missing a factor of 2 in the numerator. One possible function is $r(x) = \dfrac{2(x^2 - 3x - 4)}{x^2 - x - 6}$, giving a y-intercept of $r(0) = \frac{4}{3}$, which seems to fit the graph very well.

NOW TRY EXERCISES 57 THROUGH 60

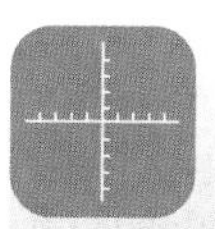

TECHNOLOGY HIGHLIGHT

Graphing Calculators and Rational Graphs

The keystrokes shown apply to a TI-84 Plus model. Please consult your manual or our Internet site for other models.

Consider the graph of $p(x) = \dfrac{x^2 - 4x + 4}{x^2 - 2x - 3}$ from Example 2, shown in Figure 4.37. Calculator-generated graphs of rational functions sometimes appear disfigured or incomplete. This is primarily due to the limited number of pixels the calculator has to display the image, and tends to heighten the importance of an analytical study of these functions. At a vertical asymptote, the calculator tries to display asymptotic behavior to both the left and right of the asymptote—and there simply aren't enough pixels to do it well in every instance. There are a few things we can do to improve the display, *but in the end it will be our conceptual understanding of these graphs that carries the day.* First of all, the screen of the TI-84 Plus is 94 pixels wide and 62 pixels tall. Sometimes we can create a *friendly window* using Xmin and Xmax settings that are multiples of 4.7, and this will clear things up a little (Figure 4.38). As an alternative, we can place the calculator in "DOT" MODE, having the calculator plot only the point computed at each pixel, rather than having it try to connect these points with a smooth curve (Figure 4.39).

Figure 4.37

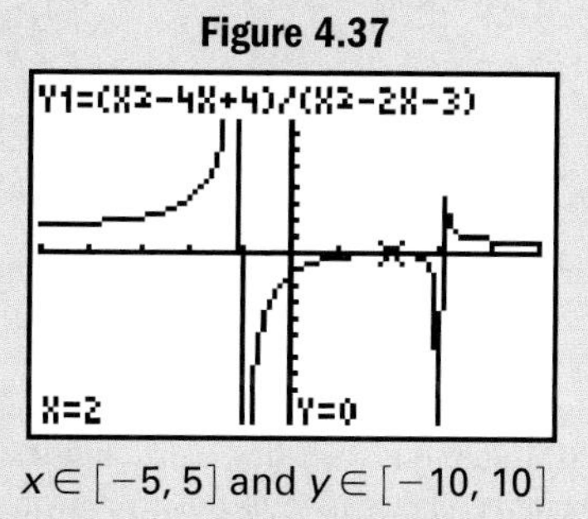

$x \in [-5, 5]$ and $y \in [-10, 10]$

Figure 4.38

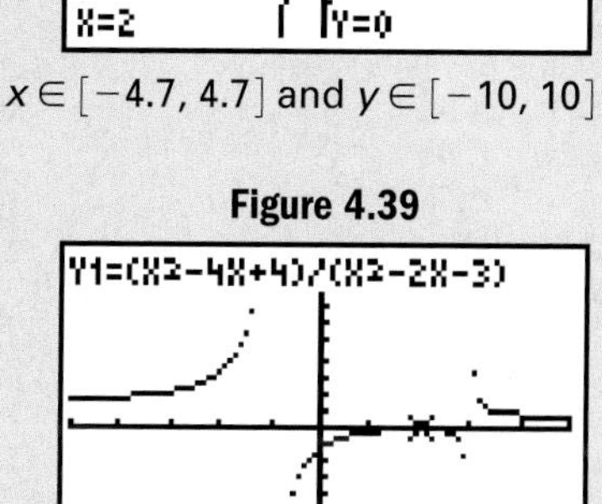

$x \in [-4.7, 4.7]$ and $y \in [-10, 10]$

Figure 4.39

$x \in [-5, 5]$ and $y \in [-10, 10]$

As a final alternative, some times playing with different window settings can improve the display. Try these ideas on the following functions to see what options give the best graphical display.

Exercise 1: $h(x) = \dfrac{2x^2 - x - 3}{x^2 - 2x - 4}$ Answers will vary.

Exercise 2: $h(x) = \dfrac{x^2 - 2x}{x^2 - 3}$ Answers will vary.

4.5 EXERCISES

CONCEPTS AND VOCABULARY

Fill in each blank with the appropriate word or phrase. Carefully reread the section if needed.

1. The domain of a rational function is __all__ __real__ numbers, except the __zeroes__ of the denominator.

2. The graph of a rational function will always have one more "piece" than the number of __vertical__ __asymptotes__.

3. Vertical asymptotes are found by setting the __denominator__ equal to zero. The x-intercepts are found by setting the __numerator__ equal to zero.

4. If the degree of the numerator is equal to the degree of the denominator, a horizontal asymptote occurs at $y = \frac{a}{b}$, where $\frac{a}{b}$ represents the ratio of the __lead__ __coefficients__.

7. $x = 3$, $\{x \mid x \in \mathbb{R}, x \neq 3\}$
8. $x = \frac{3}{2}$, $\{x \mid x \in \mathbb{R}, x \neq \frac{3}{2}\}$
9. $x = 3$, $x = -3$, $\{x \mid x \in \mathbb{R}, x \neq 3, x \neq -3\}$
10. $x = \frac{2}{3}$, $x = \frac{-2}{3}$, $\{x \mid x \in \mathbb{R}, x \neq \frac{2}{3}, x \neq \frac{-2}{3}\}$
11. $x = \frac{-5}{2}$, $x = 1$, $\{x \mid x \in \mathbb{R}, x \neq \frac{-5}{2}, x \neq 1\}$
12. $x = \frac{3}{2}$, $x = -1$, $\{x \mid x \in \mathbb{R} \neq \frac{3}{2}, x \neq -1\}$
13. No V.A., $\{x \mid x \in \mathbb{R}\}$
14. No V.A., $\{x \mid x \in \mathbb{R}\}$
15. $x = 3$, yes; $x = -2$, yes
16. $x = 5$, yes; $x = -4$, yes
17. $x = 3$, no
18. $x = 2$, no
19. $x = 2$, yes; $x = -2$, no
20. $x = -1$, no; $x = 1$, yes
21. (0, 0) cut, (3, 0) cut
22. (0, 0) cut, (2, 0) cut
23. (−4, 0) cut, (0, 4)
24. (−6, 0) cut, (−1, 0) cut, (0, −3)
25. (0, 0) cut, (3, 0) bounce
26. (−2, 0) bounce, (0, 0) cut
27. $y = 0$, crosses at $(\frac{3}{2}, 0)$
28. $y = 0$, crosses at $(-\frac{3}{4}, 0)$
29. $y = 4$, crosses at $(-\frac{21}{4}, 4)$
30. $y = 2$, crosses at $(-20, 2)$
31. $y = 3$
32. $y = 3$
33. $\frac{2}{x-1} + 3$ 34. $\frac{-1}{x+2} + 2$

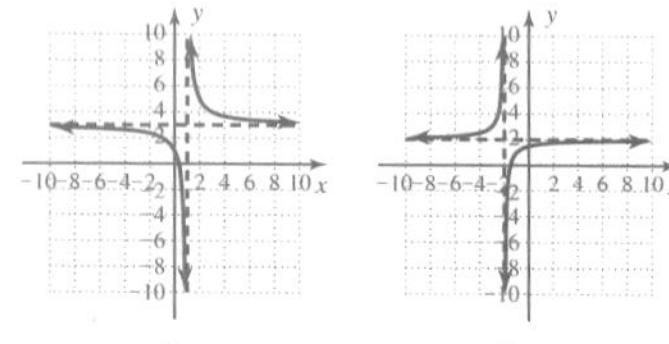

35. $\frac{2}{x+3} + 1$ 36. $\frac{2}{x-2} + 1$

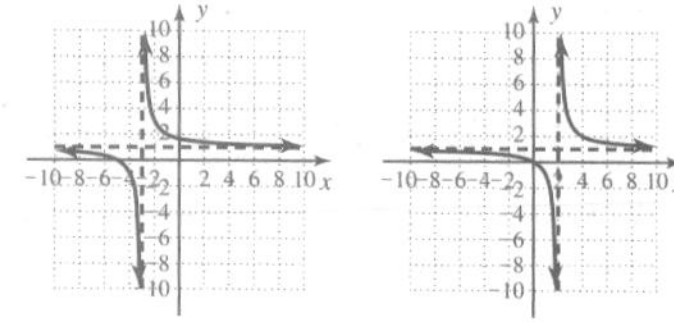

37. $\frac{-1}{(x-3)^2} + 1$ 38. $\frac{-1}{(x-2)^2} + 3$

5. Use a table of integer values and $g(x) = \frac{3x^2 - 5}{x^2 - 1}$ to discuss the concept of horizontal asymptotes. At what positive value of x is the graph of g within 0.01 of its horizontal asymptote? $x = 15$

6. Name all of the "tools" at your disposal that play a role in the graphing of rational functions. Which tools are indispensable and always used? Which are used only as the situation merits? Answers will vary.

DEVELOPING YOUR SKILLS

Give the location of the vertical asymptote(s) if they exist, and state the function's domain in set notation.

7. $f(x) = \frac{x+2}{x-3}$
8. $F(x) = \frac{4x}{2x-3}$
9. $g(x) = \frac{3x^2}{x^2-9}$
10. $G(x) = \frac{x+1}{9x^2-4}$
11. $h(x) = \frac{x^2-1}{2x^2+3x-5}$
12. $H(x) = \frac{x-5}{2x^2-x-3}$
13. $p(x) = \frac{2x+3}{x^2+x+1}$
14. $q(x) = \frac{2x^3}{x^2+4}$

Give the location of the vertical asymptote(s) if they exist, and state whether function values will change sign (positive to negative or negative to positive) from one side of the asymptote to the other.

15. $Y_1 = \frac{x+1}{x^2-x-6}$
16. $Y_2 = \frac{2x+3}{x^2-x-20}$
17. $r(x) = \frac{x^2+3x-10}{x^2-6x+9}$
18. $R(x) = \frac{x^2-2x-15}{x^2-4x+4}$
19. $Y_1 = \frac{x}{x^3+2x^2-4x-8}$
20. $Y_2 = \frac{-2x}{x^3+x^2-x-1}$

Give the location of the x- and y-intercepts (if they exist), and discuss the behavior of the function (bounce or cut) at each x-intercept.

21. $f(x) = \frac{x^2-3x}{x^2-5}$
22. $F(x) = \frac{2x-x^2}{x^2+2x-3}$
23. $g(x) = \frac{x^2+3x-4}{x^2-1}$
24. $G(x) = \frac{x^2+7x+6}{x^2-2}$
25. $h(x) = \frac{x^3-6x^2+9x}{4-x^2}$
26. $H(x) = \frac{4x+4x^2+x^3}{x^2-1}$

For the functions given, (a) determine if a horizontal asymptote exists and (b) determine if the graph will cross the asymptote, and if so, where it crosses.

27. $Y_1 = \frac{2x-3}{x^2+1}$
28. $Y_2 = \frac{4x+3}{2x^2+5}$
29. $r(x) = \frac{4x^2-9}{x^2-3x-18}$
30. $R(x) = \frac{2x^2-x-10}{x^2+5}$
31. $p(x) = \frac{3x^2-5}{x^2-1}$
32. $P(x) = \frac{3x^2-5x-2}{x^2-4}$

Use polynomial division or synthetic division to rewrite the function, then sketch the graph using transformations of a parent function and the x- and y-intercepts (if they exist).

33. $p(x) = \frac{3x-1}{x-1}$
34. $P(x) = \frac{2x+3}{x+2}$
35. $q(x) = \frac{x+5}{x+3}$
36. $Q(x) = \frac{x}{x-2}$
37. $h(x) = \frac{x^2-6x+8}{x^2-6x+9}$
38. $H(x) = \frac{3x^2-12x+11}{x^2-4x+4}$

39. 40.

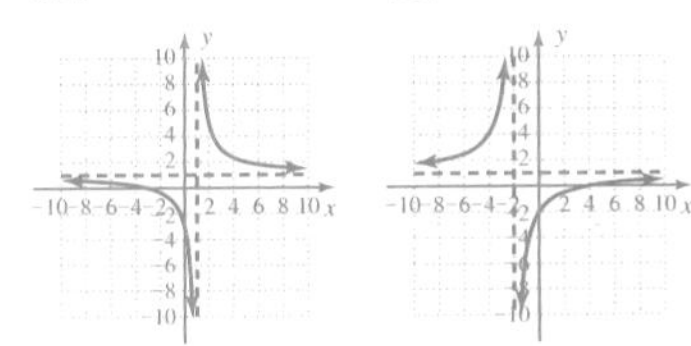

41. 42.

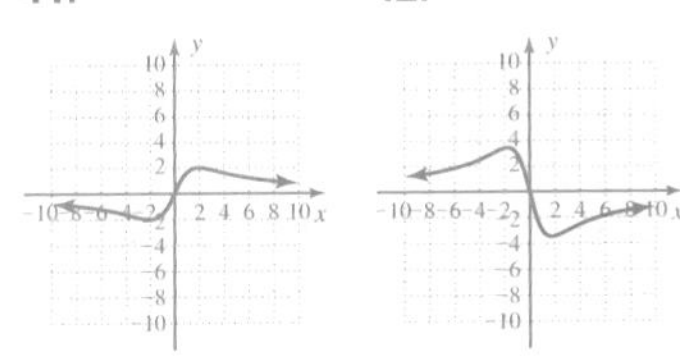

43. 44.

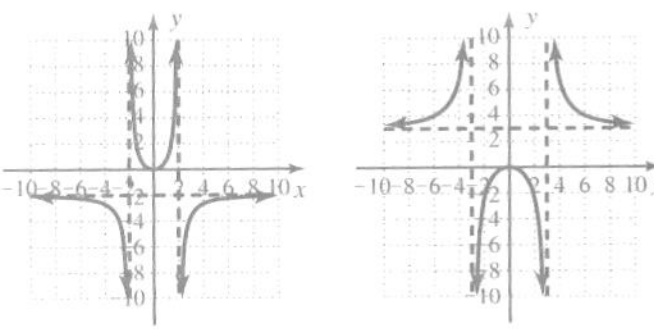

45. 46.

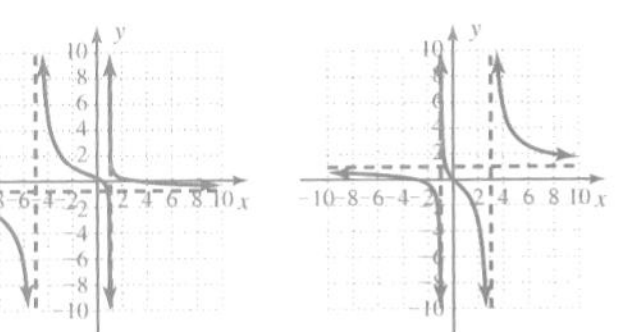

47. 48.

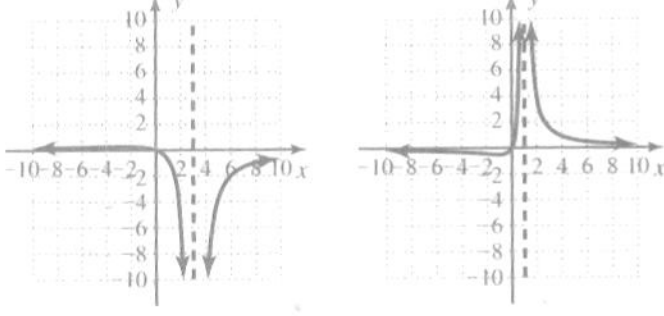

49. 50.

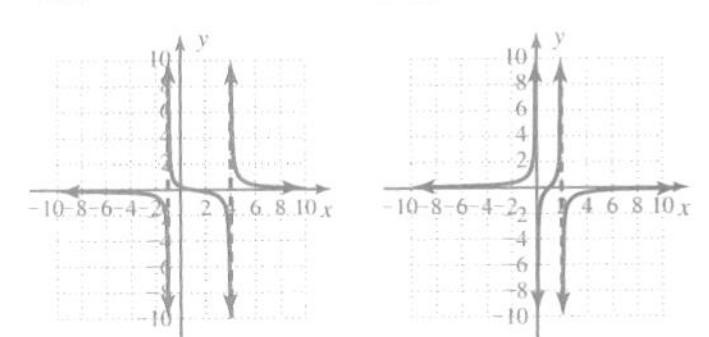

51. 52.

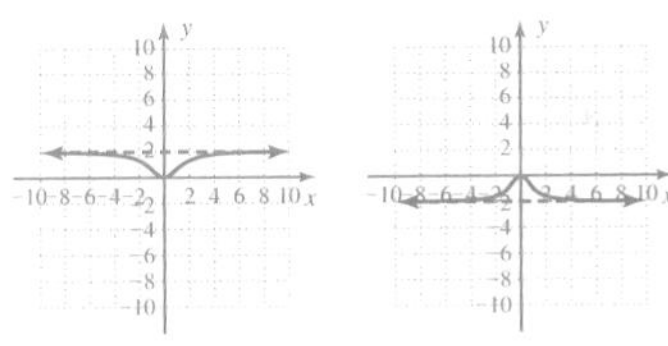

53. 54.

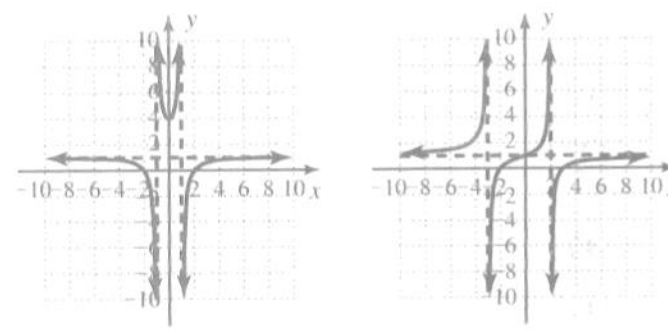

Additional answers can be found in the Instructor Answer Appendix.

Use the *Guidelines for Graphing Rational Functions* to graph the functions given.

39. $f(x) = \dfrac{x+3}{x-1}$ **40.** $g(x) = \dfrac{x-4}{x+2}$ **41.** $F(x) = \dfrac{8x}{x^2+4}$

42. $G(x) = \dfrac{-12x}{x^2+3}$ **43.** $p(x) = \dfrac{-2x^2}{x^2-4}$ **44.** $P(x) = \dfrac{3x^2}{x^2-9}$

45. $q(x) = \dfrac{2x-x^2}{x^2+4x-5}$ **46.** $Q(x) = \dfrac{x^2+3x}{x^2-2x-3}$ **47.** $h(x) = \dfrac{-3x}{x^2-6x+9}$

48. $H(x) = \dfrac{2x}{x^2-2x+1}$ **49.** $Y_1 = \dfrac{x-1}{x^2-3x-4}$ **50.** $Y_2 = \dfrac{1-x}{x^2-2x}$

51. $s(x) = \dfrac{4x^2}{2x^2+4}$ **52.** $S(x) = \dfrac{-2x^2}{x^2+1}$ **53.** $Y_1 = \dfrac{x^2-4}{x^2-1}$

54. $Y_2 = \dfrac{x^2-x-6}{x^2+x-6}$ **55.** $v(x) = \dfrac{-2x}{x^3+2x^2-4x-8}$ **56.** $V(x) = \dfrac{3x}{x^3+x^2-x-1}$

Use the vertical asymptotes, x-intercepts, and their multiplicities to construct an equation that corresponds to each graph. Be sure the y-intercept on the graph matches the value given by your equation. Check work on a graphing calculator.

57.

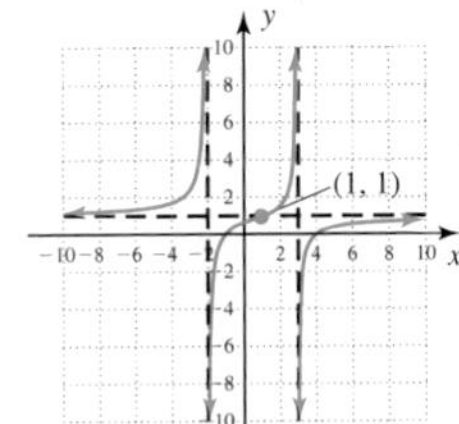

58.

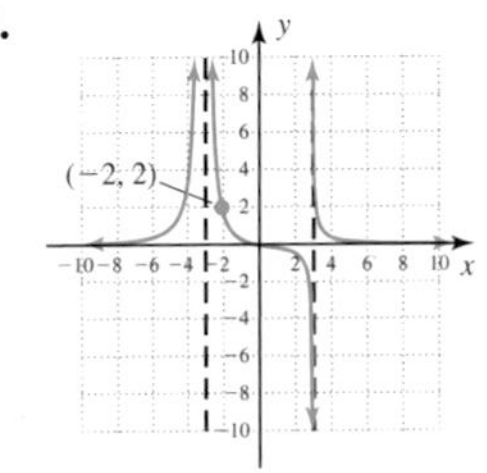

59.

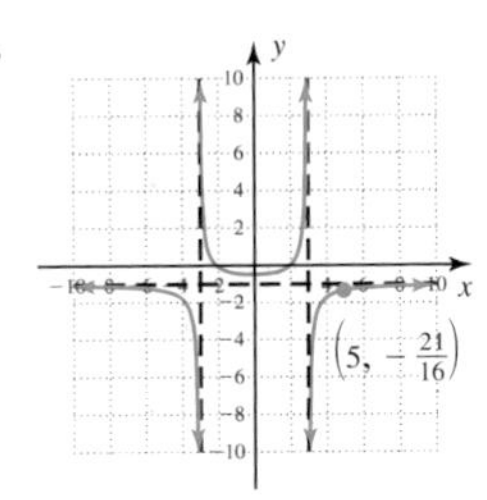

60.

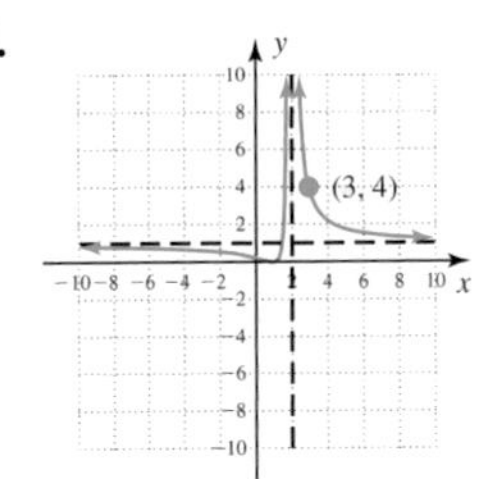

WORKING WITH FORMULAS

Population density: $D(x) = \dfrac{ax}{x^2+b}$

The population density of urban areas (in people per square mile) can be modeled by the formula shown, where a and b are constants related to the overall population and sprawl of the area under study, and $D(x)$ is the population density (in hundreds), x mi from the center of downtown.

61. Graph the function for $a = 63$ and $b = 20$ over the interval $x \in [0, 25]$, and then use the graph to answer the following questions.

a. What is the significance of the *horizontal asymptote* (what does it mean in this context)? Population density approaches zero far from town

b. How far from downtown does the population density fall below 525 people per square mile? How far until the density falls below 300 people per square mile? 10 mi, 20 mi

c. Use the graph and a table to determine how far from downtown the population density reaches a maximum? What is this maximum? 4.5 mi, 704

Cost of removing pollutants: $C(x) = \dfrac{kx}{100 - x}$

Some industries resist cleaner air standards because the cost of removing pollutants rises dramatically as higher standards are set. This phenomenon can be modeled by the formula given, where $C(x)$ is the cost (in thousands of dollars) of removing $x\%$ of the pollutant and k is a constant that depends on the type of pollutant and other factors.

62.

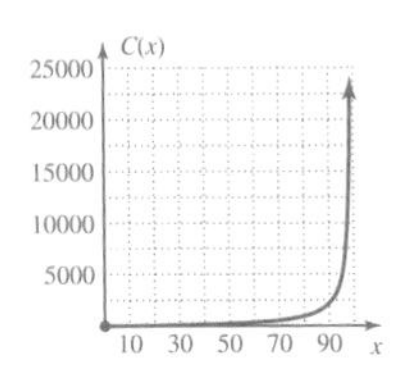

62. Graph the function for $k = 250$ over the interval $x \in [0, 100]$, and then use the graph to answer the following questions.

a. What is the significance of the *vertical asymptote* (what does it mean in this context)? It is impossible to remove 100% of the pollutants.

b. If new laws are passed that require 80% of a pollutant to be removed, while the existing law requires only 75%, how much will the new legislation cost the company? Compare the cost of the 5% increase from 75% to 80% with the cost of the 1% increase from 90% to 91%. $250 thousand, $277.8 thousand

c. What percent of the pollutants can be removed if the company budgets 2250 thousand dollars? 90%

APPLICATIONS

Memory retention: Due to their asymptotic behavior, rational functions are often used to model the mind's ability to retain information over a long period of time—the "use it or lose it" phenomenon.

63.

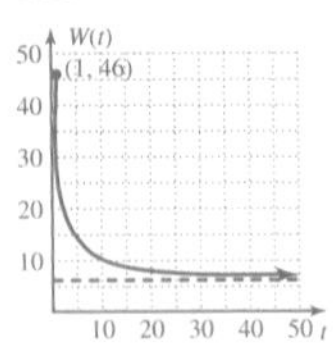

63. A large group of students is asked to memorize a list of 50 Italian words, a language that is unfamiliar to them. The group is then tested regularly to see how many of the words are retained over a period of time. The average number of words retained is modeled by the function $W(t) = \dfrac{6t + 40}{t}$, where $W(t)$ represents the number of words remembered after t days.

a. Graph the function over the interval $t \in [0, 40]$. How many days until only half the words are remembered? How many days until only one-fifth of the words are remembered? 2; 10

b. After 10 days, what is the average numbered of words retained? How many days until only 8 words can be recalled? 10; 20

c. What is the significance of the horizontal asymptote (what does it mean in this context)? On average, 6 words will be remembered for life.

64.

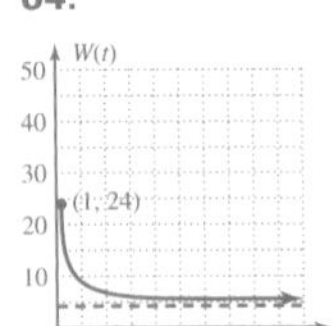

64. A similar study asked students to memorize 50 Hawaiian words, a language that is both unfamiliar and phonetically foreign to them (see Exercise 63). The average number of words retained is modeled by the function $W(t) = \dfrac{4t + 20}{t}$, where $W(t)$ represents the number of words after t days.

a. Graph the function over the interval $t \in [0, 40]$. How many days until only half the words are remembered? How does this compare to Exercise 63? How many days until only one-fifth of the words are remembered? 1; lower retention; 3 days

b. After 7 days, what is the average numbered of words retained? How many days until only 5 words can be recalled? about 7; 20 days

c. What is the significance of the horizontal asymptote (what does it mean in this context)? On average, 4 words will be remembered for life.

Concentration and dilution: When antifreeze is mixed with water, it becomes diluted—less than 100% antifreeze. The more water added, the less concentrated the antifreeze becomes, with this process continuing until a desired concentration is met. This application and many similar to it can be modeled by rational functions.

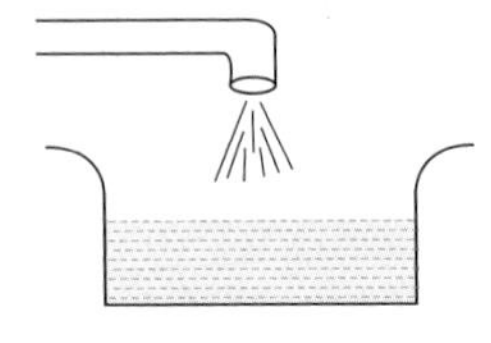

65. a.

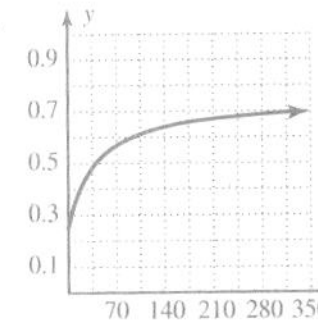

b. 35%; 62.5%; 160 gal
c. 160 gal; 200 gal
d. 70%; 75%

66. a. $A(t) = 40 + 10t$ $S(t) = 2t$
b. $C(6) = 0.12$ oz/gal;
$C(28) = 0.175$ oz/gal;
320 gal

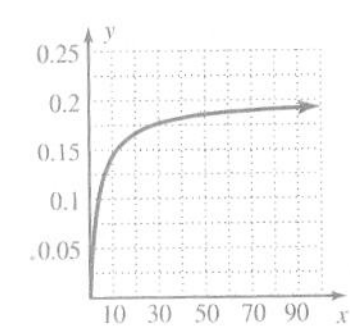

c. 46 min; 500 gal
d. 0.19 oz/gal; 0.20 oz/gal

65. A 400-gal tank currently holds 40 gal of a 25% antifreeze solution. To raise the concentration of the antifreeze in the tank, x gal of a 75% antifreeze solution is pumped in.

a. Show the formula for the resulting concentration is $C(x) = \dfrac{40 + 3x}{160 + 4x}$ after simplifying, and graph the function over the interval $x \in [0, 360]$.

b. What is the concentration of the antifreeze in the tank after 10 gal of the new solution are added? After 120 gal have been added? How much liquid is now in the tank?

c. If the concentration level is now at 65%, how many gallons of the 75% solution have been added? How many gallons of liquid are in the tank now?

d. What is the maximum antifreeze concentration that can be attained in a tank of this size? What is the maximum concentration that can be attained in a tank of "unlimited" size?

66. A sodium chloride solution has a concentration of 0.2 oz (weight) per gallon. The solution is pumped into a 800-gal tank currently holding 40 gal of pure water, at a rate of 10 gal/min.

a. Find a function $A(t)$ modeling the amount of liquid in the tank after t min, and a function $S(t)$ for the amount of sodium chloride in the tank after t min.

b. The concentration $C(t)$ in ounces per gallon is measured by the ratio $\dfrac{S(t)}{A(t)}$, a rational function. Graph the function on the interval $t \in [0, 100]$. What is the concentration level (in ounces per gallon) after 6 min? After 28 min? How many gallons of liquid are in the tank at this time?

c. If the concentration level is now 0.184 oz/gal, how long have the pumps been running? How many gallons of liquid are in the tank now?

d. What is the maximum concentration that can be attained in a tank of this size? What is the maximum concentration that can be attained in a tank of "unlimited" size?

Average cost of manufacturing an item: The cost "C" to manufacture an item depends on the relatively fixed costs "K" for remaining in business (utilities, maintenance, transportation, etc.) and the actual cost "c" of manufacturing the item (labor and materials). For x items the cost is $C(x) = K + cx$. The average cost "A" of manufacturing an item is then $A(x) = \dfrac{C(x)}{x}$.

67. a. \$225; \$175
b. 2000 heaters
c. 4000
d. The horizontal asymptote at $y = 125$ means the average cost approaches \$125 as monthly production gets very large. Due to limitations on production (maximum of 5000 heaters) the average cost will never fall below $A(5000) = 135$.

67. A company that manufactures water heaters finds their fixed costs are normally \$50,000 per month, while the cost to manufacture each heater is \$125. Due to factory size and the current equipment, the company can produce a maximum of 5000 water heaters per month during a good month.

a. Use the average cost function to find the average cost if 500 water heaters are manufactured each month. What is the average cost if 1000 heaters are made?

b. What level of production will bring the average cost down to \$150 per water heater?

c. If the average cost is currently \$137.50, how many water heaters are being produced that month?

d. What's the significance of the horizontal asymptote for the average cost function (what does it mean in this context)? Will the company ever break the \$130 average cost level? Why or why not?

68. An enterprising company has finally developed a disposable diaper that is biodegradable. The brand becomes wildly popular and production is soaring. The fixed cost of production is \$20,000 per month, while the cost of manufacturing is \$6.00 per case (48 diapers). Even while working three shifts around-the-clock, the maximum production level is 16,000 cases per month. The company figures it will be profitable if it can bring costs down to an average of \$7 per case.

68. a. \$16; \$11
b. 10,000
c. 5000
d. $y = 6$ means the average cost will never be less than \$6; no; 20,000

a. Use the average cost function to find the average cost if 2000 cases are produced each month. What is the average cost if 4000 cases are made?

b. What level of production will bring the average cost down to \$8 per case?

c. If the average cost is currently \$10 per case, how many cases are being produced?

d. What's the significance of the horizontal asymptote for the average cost function (what does it mean in this context)? Will the company ever reach its goal of \$7/case at its maximum production? What level of production would help them meet their goal?

Test averages and grade point averages: To calculate a test average simply requires the sum of all test points P divided by the number of tests N: $\frac{P}{N}$. To compute the score or scores needed on future tests to raise the average grade to a desired grade G, we add the number of additional tests n to the denominator, and the number of additional tests *times* the projected grade g on each test to the numerator: $G(n) = \frac{P + ng}{N + n}$. The result is a rational function with some "eye-opening" results.

69. a. 5
b. 18
c. The horizontal asymptote at $y = 95$ means her average grade will approach 95 as the number of tests taken increases; no
d. 6

70. a. 12
b. 28 hr
c. The horizontal asymptote at $y = 4$ means his GPA will approach 4 as the number of 4 point credit hours earned increases; no

69. After four tests, Bobby Lou's test average was an 84. [*Hint:* $P = 4(84) = 336$.]

a. Assume that she gets a 95 on all remaining tests ($g = 95$). Graph the resulting function on a calculator using the window $n \in [0, 20]$ and $G(n) \in [80$ to $100]$. Use the calculator to determine how many tests are required to lift her grade to a 90 under these conditions.

b. At some colleges, the range for an "A" grade is 93–100. How many tests would Bobby Lou have to score a 95 on, to raise her average to higher than 93? Were you surprised?

c. Describe the significance of the horizontal asymptote of the average grade function. Is a test average of 95 possible for her under these conditions?

d. Assume now that Bobby Lou scores 100 on all remaining tests ($g = 100$). Approximately how many more tests are required to lift her grade average to higher than 93?

70. At most colleges, $A \rightarrow 4$ grade points, $B \rightarrow 3$, $C \rightarrow 2$, and $D \rightarrow 1$. After taking 56 credit hours, Aurelio's GPA is 2.5. [*Hint:* In the formula given, $P = 2.5(56) = 140$.]

a. Assume Aurelio is determined to get A's (4 grade points or $g = 4$), for all remaining credit hours. Graph the resulting function on a calculator using the window $n \in [0, 60]$ and $A(n) \in [2, 4]$. Use the calculator to determine the number of credit hours required to lift his GPA to over 2.75 under these conditions.

b. At some colleges, scholarship money is available only to students with a 3.0 average or higher. How many (perfect 4.0) credit hours would Aurelio have to earn, to raise his GPA to 3.0 or higher? Were you surprised?

c. Describe the significance of the horizontal asymptote of the GPA function. Is a GPA of 4.0 possible for him under these conditions?

WRITING, RESEARCH, AND DECISION MAKING

71. In addition to determining *if* a function has a vertical asymptote, we are often interested in *how fast* the graph approaches the asymptote. As in previous investigations, this involves the function's rate of change over a small interval. Exercise 62 describes the rising cost of removing pollutants from the air. As noted there, the rate of increase in the cost changes as higher requirements are set. To quantify this change, we'll compute the rate of change $\frac{\Delta C}{\Delta x} = \frac{C(x_2) - C(x_1)}{x_2 - x_1}$ for $C(x) = \frac{250x}{100 - x}$.

a. Find the rate of change of the function in the following intervals:

$$x \in [60, 61] \quad x \in [70, 71] \quad x \in [80, 81] \quad x \in [90, 91].$$

b. What do you notice? How much did the rate increase from the first interval to the second? From the second to the third? From the third to the fourth?

c. Recompute parts (a) and (b) using the function $C(x) = \frac{350x}{100 - x}$. Comment on what you notice.

72. Consider the function $f(x) = \frac{ax^2 + k}{bx^2 + h}$, where a, b, k, and h are constants and $a, b > 0$.

a. What can you say about asymptotes and intercepts of this function if $h, k > 0$?

b. Now assume $k < 0$ and $h > 0$. How does this affect the asymptotes? The intercepts?

c. If $b = 1$ and $a > 1$, how does this affect the results from part (b)?

d. How is the graph affected if $k > 0$ and $h < 0$?

e. Find values of a, b, h, and k that create a function with a horizontal asymptote at $y = \frac{3}{2}$, x-intercepts at $(-2, 0)$ and $(2, 0)$, a y-intercept of $(0, -4)$, and no vertical asymptotes.

EXTENDING THE CONCEPT

73. Use grouping or the rational roots theorem to factor the numerator and denominator and graph the function $f(x) = \frac{x^3 - 5x^2 - 4x + 20}{x^3 - 5x^2 - 9x + 45}$. Clearly indicate all discontinuities. If any discontinuity is removable, repair the hole with a piecewise-defined function.

74. Graphing and art: Construct the equation of a rational function with zeroes at $(3, 0)$ and $(-3, 0)$, vertical asymptotes at $x = \pm 2$ and $x = \pm 4$, and a horizontal asymptote at $y = 0$. Enter the function on a graphing calculator and graph. What do you notice? Change some of the intercepts and asymptotes to see what other interesting "art forms" you can create using rational functions.

MAINTAINING YOUR SKILLS

75. (R.1) Describe/define each set of numbers: complex $\mathbb{C}$, rational $\mathbb{Q}$, and integers $\mathbb{Z}$.

76. (2.1) Find the equation of a line that is perpendicular to $3x - 4y = 12$ and contains the point $(2, -3)$.

77. (3.2) The graph of a function is given here. Does this function have an inverse? Explain why or why not.

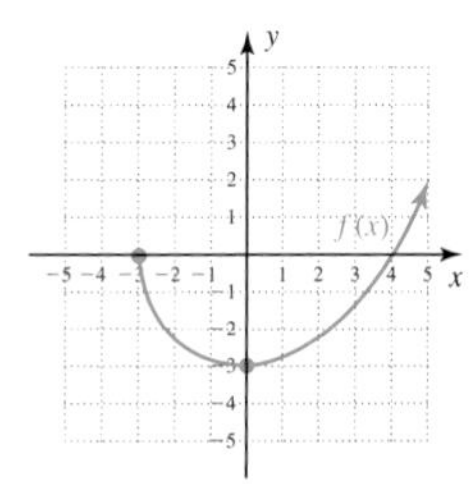

78. (R.7/2.1) Find the perimeter and area of triangle ABC shown, and the length of segment $\overline{CD}$, given $(\overline{CB})^2 = \overline{AB} \cdot \overline{BD}$.

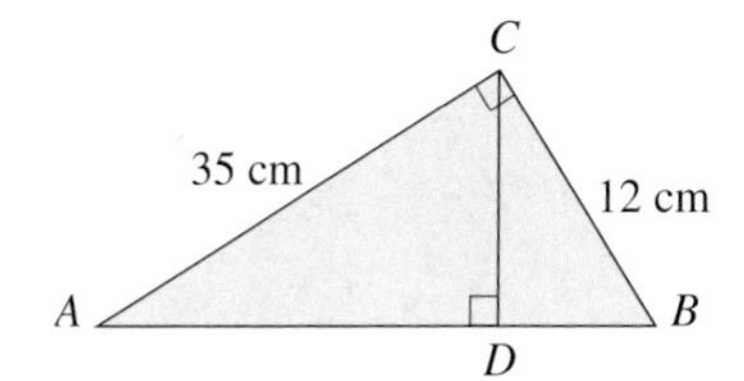

79. (4.2) Use synthetic division and the remainder theorem to find the value of $f(4)$, $f(\frac{3}{2})$, and $f(2)$:
$f(x) = 2x^3 - 7x^2 + 5x + 3$.

80. (1.4/4.3) By direct substitution, show that $x = 1 - 2i$ is a solution to $x^2 - 2x + 5 = 0$. Then name a second solution.

---

71. a. 16.0 28.7 65.8 277.8
b. 12.7, 37.1, 212.0
c. (A) 22.4, 40.2, 92.1, 388.9
(B) 17.8, 51.9, 296.8; answers will vary

72. a. horizontal asymptote at $y = \frac{a}{b}$; no vertical asymptote; y-intercept at $\left(0, \frac{k}{h}\right)$; no x-intercepts
b. no change in asymptotes; y-intercept at $\left(0, \frac{k}{h}\right)$; x-intercepts at $\left(\pm\sqrt{\frac{k}{a}}, 0\right)$
c. no change except in horizontal asymptote
d. horizontal asymptote at $y = \frac{a}{b}$; vertical asymptotes at $x = \pm\sqrt{\frac{h}{b}}$; y-intercept at $\left(0, \frac{k}{h}\right)$; no x-intercepts
e. $a = 3$, $b = 2$
$h = 3$, $k = -12$
$f(x) = \frac{3x^2 - 12}{2x^2 + 3}$

73. $f(x) = \frac{(x - 5)(x - 2)(x + 2)}{(x - 5)(x - 3)(x + 3)}$; vertical asymptotes at $x = -3$, $x = 3$, hole at $\left(5, \frac{21}{16}\right)$;
$$f(x) = \begin{cases} f(x) & x \neq 5 \\ \frac{21}{16} & x = 5 \end{cases}$$

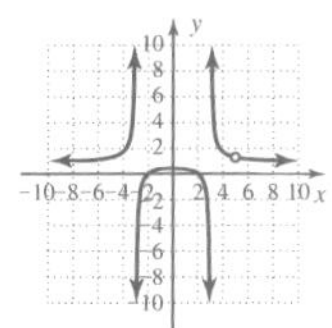

74. Answers will vary.

75. $\mathbb{C}: a + bi$, where $a, b \in \mathbb{R}$, $i = \sqrt{-1}$
$\mathbb{Q}: \left\{\frac{a}{b} \middle| a \in \mathbb{Z}, b \in \mathbb{Z}; b \neq 0\right\}$
$\mathbb{Z}: \{...-2, -1, 0, 1, 2...\}$

76. $y = \frac{-4}{3}x - \frac{1}{3}$

77. no, $f(x)$ is not one-to-one

78. $p = 84$ cm $A = 210$ cm^2
$CD \approx 11.35$

79. $39, \frac{3}{2}, 1$

80. $1 + 2i$

4.6 Additional Insights into Rational Functions

LEARNING OBJECTIVES

In Section 4.6 you will learn how to:

A. Graph rational functions having a common factor in the numerator and denominator

B. Graph rational functions where degree of numerator is greater than degree of denominator

C. Solve applications involving rational functions

INTRODUCTION

Although we've already studied a wide variety of polynomial and rational functions to this point, certain real-world phenomena exhibit characteristics that cannot be accurately modeled without expanding our knowledge base a little further. In this section we focus on functions having asymptotes that are neither vertical nor horizontal, called **oblique asymptotes.**

POINT OF INTEREST

Linear asymptotes that are neither vertical nor horizontal are called oblique (or slant) asymptotes. The word comes from the Latin *obliquus,* which was used to describe an object that was set at an incline or otherwise slanted. In geometry, all triangles except for right triangles can be described as oblique—with reference to a horizontal base the other two sides are "slanted." Over time the word has also come to describe answers that are indirect or evasive, as when a politician offers an oblique answer to a direct question.

WORTHY OF NOTE

As a reminder, the graph of $f(x) = \frac{1}{(x-2)^2}$ also has a break at $x = 2$ (a vertical asymptote). However, this break is a **nonremovable discontinuity,** since no function can be found to repair the break. A vertical line is not a function, and since $f(x)$ is defined for all values *except* $x = 2$, the branches of the graph could never "connect" in any case.

A. Rational Functions and Common Factors

In Example 6 of Section 3.7, we graphed the piecewise-defined function $r(x) = \begin{cases} \frac{x^2-4}{x-2} & x \neq 2 \\ 1 & x = 2 \end{cases}$, noting the first piece was actually a line with a hole at $x = 2$ [the second piece was simply the point (2, 1)]. Using the TABLE feature of a graphing calculator, we confirm the first piece of r can be evaluated at any value close to 2, but not *at* 2 (see Figure 4.40). Further, we note that as $x \to 2$, $Y_1 = r(x) \to 4$, and that this break could be repaired by redefining the second piece of r as $y = 4$, when $x = 2$. This would create a new and continuous function that we'll name $R(x) = \begin{cases} \frac{x^2-4}{x-2} & x \neq 2 \\ 4 & x = 2 \end{cases}$ (see Figure 4.41).

Figure 4.40

X	Y1
1.997	3.997
1.998	3.998
1.999	3.999
2	ERR:
2.001	4.001
2.002	4.002
2.003	4.003

X=1.997

Figure 4.41

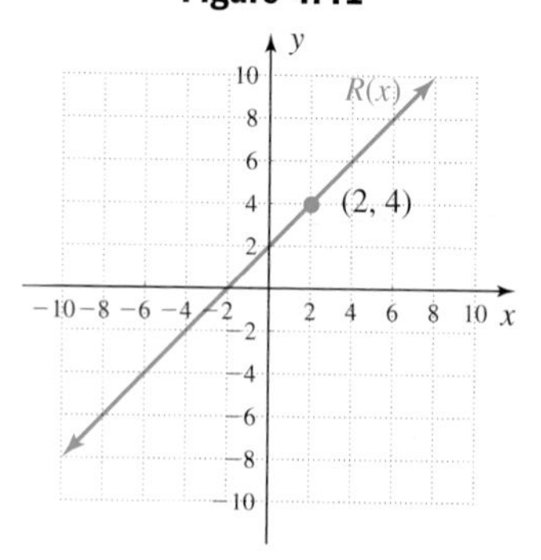

It's possible for a rational graph to have more than one removable discontinuity, or to be nonlinear with a removable discontinuity. For these functions, we will adopt the convention of using the corresponding uppercase letter to name the redefined function, as done in this discussion.

EXAMPLE 1 Graph the function $r(x) = \frac{x^3+1}{x+1}$. If there is a removable discontinuity, repair the break by redefining the given function using an appropriate piecewise-defined function.

Solution: Factor the numerator and denominator to write the expression in lowest terms: $r(x) = \frac{x^3 + 1}{x + 1} = \frac{\cancel{(x + 1)}(x^2 - x + 1)}{\cancel{x + 1}} = x^2 - x + 1;$ $x \neq -1$. The graph of r is the same as $y = x^2 - x + 1$, except there is a removable discontinuity (a hole) at $x = -1$ (Figure 4.42). The graph of y is a parabola, concave up, y-interceptat (0, 1), no x-intercepts (the discriminant is negative), and a vertex at (0.5, 0.75). Evaluating y at $x = -1$ gives $y = (-1)^2 - (-1) + 1 = 3$, and shows that $(-1, 3)$ is the point "missing" from the graph of r. We can repair the discontinuity by redefining r as

Figure 4.42

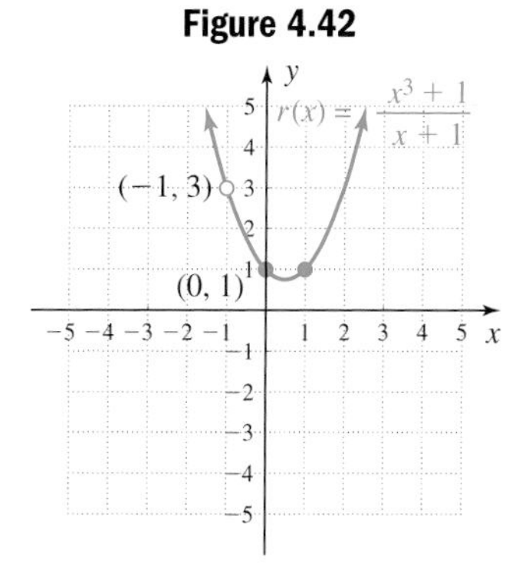

$$R(x) = \begin{cases} \frac{x^3 + 1}{x + 1} & x \neq -1 \\ 3 & x = -1 \end{cases}$$

NOW TRY EXERCISES 7 THROUGH 18

For more on removable discontinuities, see the *Technology Highlight* feature just prior to the Exercise set.

B. Rational Functions with Oblique and Nonlinear Asymptotes

In the last section we saw that when the degree of the numerator is less than the degree of the denominator, the graph has a horizontal asymptote at $y = 0$. If the degree of the numerator is equal to that of the denominator, the ratio of lead coefficients gives the equation of the asymptote. But what happens if the degree of the numerator is *greater than* the degree of the denominator? To explore this question, consider the functions p, q, and r shown in Figures 4.43, 4.44, and 4.45, whose only difference is the degree of the numerator.

Figure 4.43

$p(x) = \frac{2x}{x^2 + 1}$

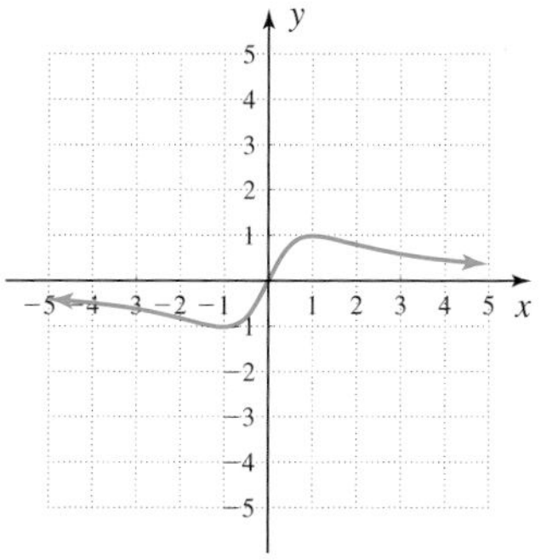

Figure 4.44

$q(x) = \frac{2x^2}{x^2 + 1}$

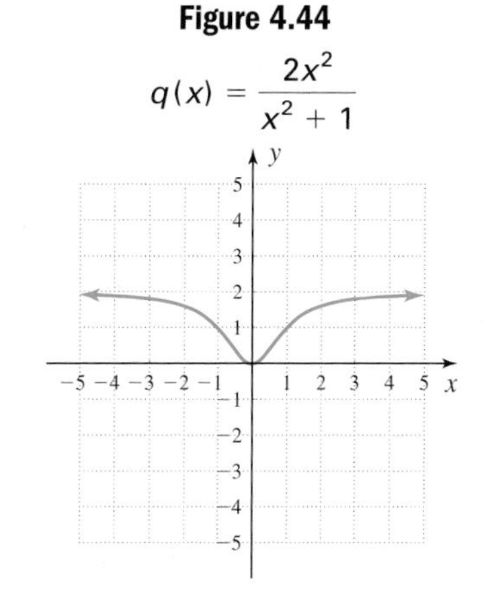

Figure 4.45

$r(x) = \frac{2x^3}{x^2 + 1}$

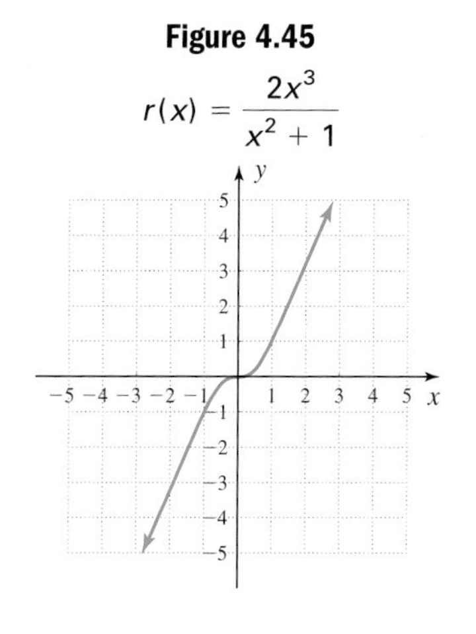

The graph of p has a horizontal asymptote at $y = 0$ (as $|x| \to \infty$, $y \to 0$) since the denominator is of larger degree. As we might have anticipated, the horizontal asymptote for q is $y = 2$, the ratio of lead coefficients (as $|x| \to \infty, y \to 2$). The graph of r has no horizontal asymptote, yet appears to be asymptotic to some slanted line. To investigate its end behavior we use a table of values (Figure 4.46) and after careful inspection, it appears that *as* $|x| \to \infty, y \to 2x = Y_2$. Judging from the numerator and denominator of r, we sense that a comparison of their degrees continues to play a part in determining asymptotic behavior. In fact, these "ratios" have *everything* to do with end behavior, when interpreted to mean the *computed quotient* of the polynomials forming a rational function. For $r(x)$, this division gives $r(x) = 2x - \frac{2x}{x^2 + 1}$ (verify), where we note for $|x| \to \infty$, the term $\frac{2x}{x^2 + 1}$ becomes very small and $r(x) \approx 2x$ for large x.

Figure 4.46

X	Y1	Y2
0	0	0
25	49.92	50
50	99.96	100
75	149.97	150
100	199.98	200
125	249.98	250
150	299.99	300

X=0

OBLIQUE AND NONLINEAR ASYMPTOTES

Given $r(x) = \left(\frac{f}{g}\right)(x)$ is a rational function in lowest terms, where the degree of f is greater than the degree of g. The graph will have an oblique or nonlinear asymptote as determined by $q(x)$, where $q(x)$ is the quotient polynomial of $\left(\frac{f}{g}\right)(x)$.

WORTHY OF NOTE

If the denominator is a monomial, term-by-term division is the most efficient means of computing the quotient. If the denominator is not a monomial, either synthetic division or long division must be used.

Based on our study of polynomial division, we conclude an oblique (linear) asymptote occurs when the degree of the numerator is one more than the degree of the denominator, and a nonlinear asymptote occurs when its degree is larger by two or more.

EXAMPLE 2 Graph the function $r(x) = \frac{x^2 - 1}{x}$.

Solution: Using the guidelines given previously, we find $r(x) = \frac{(x + 1)(x - 1)}{x}$ and proceed:

1. no y-intercept: (zero is not in the domain)
2. vertical asymptote: $x = 0$; multiplicity one. The function will change sign at $x = 0$.
3. x-intercepts: $(-1, 0)$ and $(1, 0)$, both "cross" the x-axis and the function will change sign at these intercepts. Since there is no y-intercept, we select -4 and 4 as test points to find the sign of r in the intervals $(-\infty, -1)$ and $(1, \infty)$ respectively: $r(-4) = -3.75$ and $r(4) = 3.75$.
4. The degree of numerator > degree of denominator so we compute the quotient using term-by-term division: $r(x) = \frac{x^2 - 1}{x} = \frac{x^2}{x} - \frac{1}{x} = x - \frac{1}{x}$. The quotient is $q(x) = x$ and the graph has an oblique asymptote at $y = x$.

5. To determine if the function will cross the asymptote, we solve

$$\frac{x^2 - 1}{x} = x \quad q(x) = x \text{ is the slant asymptote}$$

$$x^2 - 1 = x^2 \quad \text{cross multiply}$$

$$-1 = 0 \quad \text{no solutions possible}$$

The graph will not cross the oblique asymptote.

The information from steps 1 through 5 is displayed in Figure 4.47. Since the graph must alternate sign from interval to interval as stipulated above, we complete the graph as shown in Figure 4.48. Additional points can be computed and graphed for confirmation.

Figure 4.47

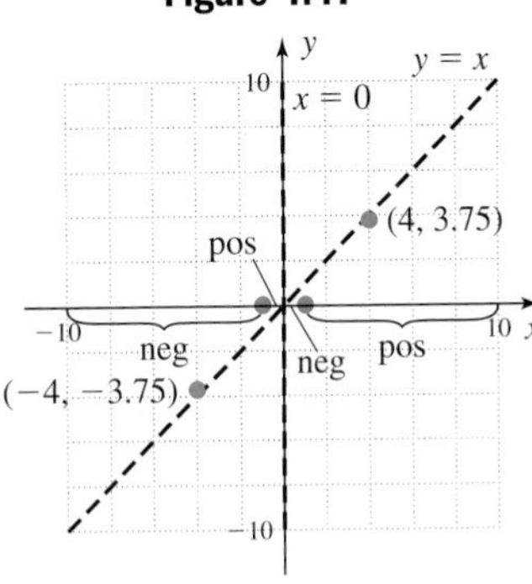

Figure 4.48

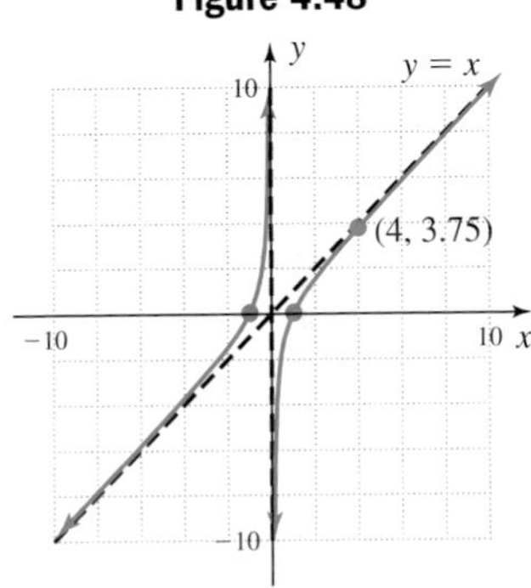

NOW TRY EXERCISES 19 THROUGH 24

EXAMPLE 3 Graph the function: $h(x) = \dfrac{x^2}{x - 1}$

Solution: The function is already in factored form.

1. y-intercept: $(0, 0)$
2. vertical asymptote: $x = 1$; multiplicity one—the function will change sign at $x = 1$.
3. x-intercept: $(0, 0)$; multiplicity two—the function will not change sign at $x = 0$. To find the sign of h in the interval $(-\infty, 0)$ (formed by the x-intercept), and the interval $(1, \infty)$ (formed by the vertical asymptote), we select $x = -3$ and $x = 2$: $h(-3) = -2.25$ and $h(2) = 4$.
4. The degree of numerator > degree of denominator so we compute the quotient using synthetic division (the denominator is not a monomial):

use 1 as a "divisor"	1	1	0	0	coefficients of numerator
		↓	1	1	
		1	1	1	quotient and remainder

The line $q(x) = x + 1$ is a slant asymptote (the remainder plays no part).

5. Solve

$$\frac{x^2}{x-1} = x + 1 \quad q(x) = x + 1 \text{ is the slant asymptote}$$
$$x^2 = x^2 - 1 \quad \text{cross multiply}$$
$$0 = -1 \quad \text{no solutions possible}$$

The graph will not cross the slant asymptote.

The information gathered in steps 1 through 5 is shown in Figure 4.49, and is actually sufficient to complete the graph. If you feel a little unsure about how to "puzzle" out the graph, additional points can be found in the first and fourth quadrants: $h(\frac{1}{2}) = -\frac{1}{2}$ and $h(\frac{3}{2}) = \frac{9}{2}$. Since output values will alternate in sign as stipulated, all conditions are met with the graph shown in Figure 4.50.

Figure 4.49

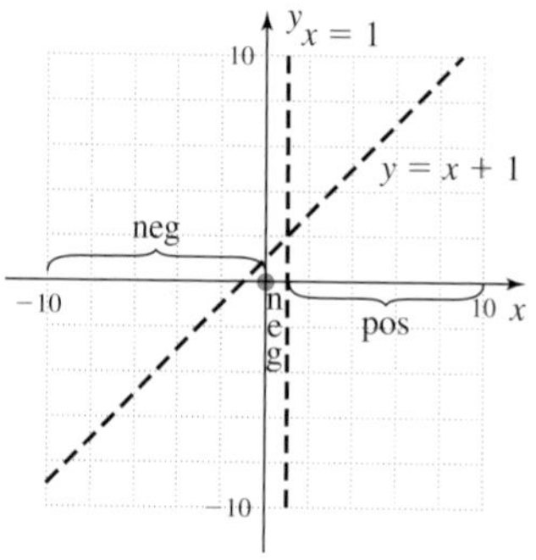

Figure 4.50

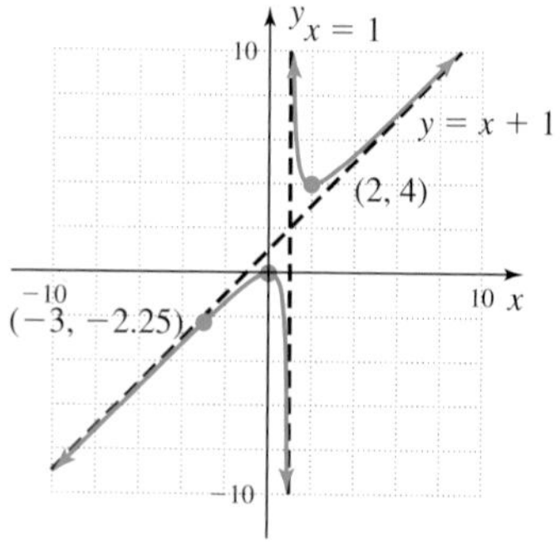

NOW TRY EXERCISES 25 THROUGH 46

Finally, it would be a mistake to think that all asymptotes are linear. In fact, when the degree of the numerator is two more than the degree of the denominator, a parabolic asymptote results. Functions of this type often occur in applications of rational functions, and are used to minimize cost, materials, distances, or other considerations of great importance to business and industry. For $r(x) = \frac{x^4 + 1}{x^2}$, term-by-term division gives $x^2 + \frac{1}{x^2}$ and the quotient $q(x) = x^2$ is a nonlinear, parabolic asymptote (see Figure 4.51). For more on nonlinear asymptotes, see Exercises 47 through 50.

Figure 4.51

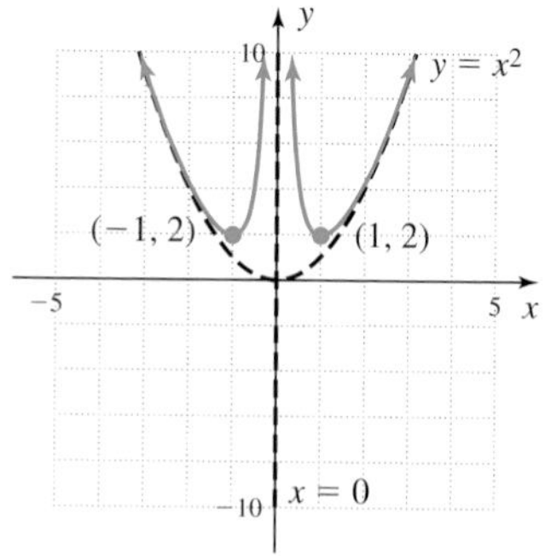

C. Applications of Rational Functions

Rational functions have applications in a wide variety of fields, including environmental studies, manufacturing, and various branches of medicine. In most practical applications, only the values from Quadrant I have meaning since the inputs and outputs must often be positive. Here we investigate an application involving manufacturing and average cost.

EXAMPLE 4 Suppose the cost (in hundreds of dollars) of manufacturing x thousand of a given item is modeled by the function $C(x) = x^2 + 4x + 3$. The *average cost* of each item would then be expressed by $A(x) = \dfrac{x^2 + 4x + 3}{x}$.

a. Graph the function $A(x)$.

b. Find how many thousand items are manufactured when the average cost is \$8.

c. Find how many thousand items should be manufactured to obtain the *minimum average cost* (use the graph to estimate this minimum average cost).

Solution: **a.** The function is already in simplest form.

1. y-intercept: none $[A(0)$ is undefined$]$
2. vertical asymptote: $x = 0$, multiplicity one; the function will change sign at $x = 0$.
3. x-intercept(s): After factoring, the zeroes of the numerator are $x = -1$ and $x = -3$, both with multiplicity one. The graph will cross the x-axis at each intercept.
4. The degree of numerator > degree of denominator, so we divide using term-by-term division:

$$\frac{x^2 + 4x + 3}{x} = \frac{x^2}{x} + \frac{4x}{x} + \frac{3}{x}$$
$$= x + 4 + \frac{3}{x}$$

The line $q(x) = x + 4$ is an oblique asymptote.

5. Solve

$$\frac{x^2 + 4x + 3}{x} = x + 4 \quad q(x) = x + 4 \text{ is a slant asymptote}$$
$$x^2 + 4x + 3 = x^2 + 4x \quad \text{cross multiply}$$
$$3 = 0 \quad \text{no solutions possible}$$

The graph will not cross the slant asymptote.

The function changes sign at both x-intercepts and at the asymptote $x = 0$. The information from steps 1 through 5 is shown in Figure 4.52 and perhaps one additional point in Quadrant I would help: $A(1) = 8$ and the point $(1, 8)$ is on the graph, showing A is positive in the interval containing 1. Since output values will alternate in sign as stipulated above, all conditions are met with the graph shown in Figure 4.53.

Figure 4.52

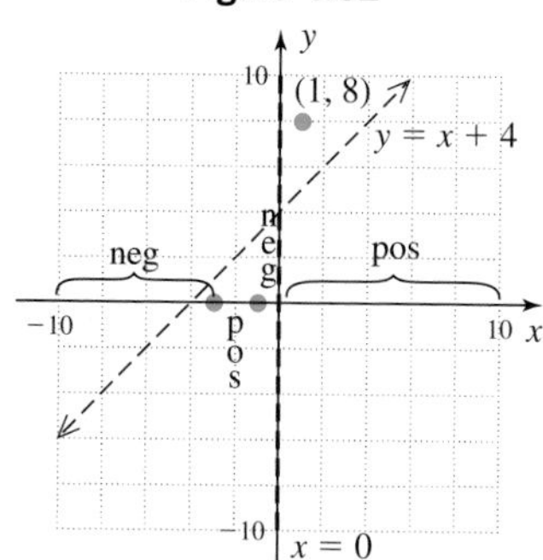

Figure 4.53

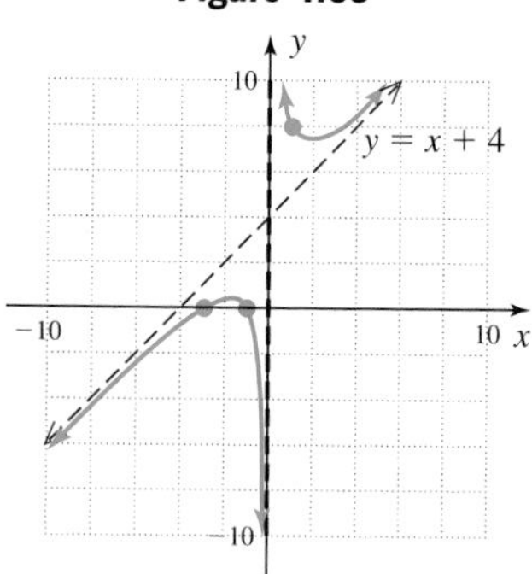

b. To find the number of items manufactured when average cost is \$8, we replace $A(x)$ with 8 and solve: $\frac{x^2 + 4x + 3}{x} = 8$:

$$\begin{aligned} x^2 + 4x + 3 &= 8x \\ x^2 - 4x + 3 &= 0 \\ (x - 1)(x - 3) &= 0 \\ x = 1 \quad \text{or} \quad x &= 3 \end{aligned}$$

The average cost is \$8 when 1000 items or 3000 items are manufactured.

c. From the graph, it appears that the minimum average cost is close to \$7.50, when approximately 1500 to 1800 items are manufactured.

NOW TRY EXERCISES 55 THROUGH 60

GRAPHICAL SUPPORT

In the *Technology Highlight* from Section 3.8, we saw how a graphing calculator can be used to locate the extreme values of a function. Applying this technology to the graph from Example 5 we find that the minimum average cost is approximately \$7.46, when about 1732 items are manufactured.

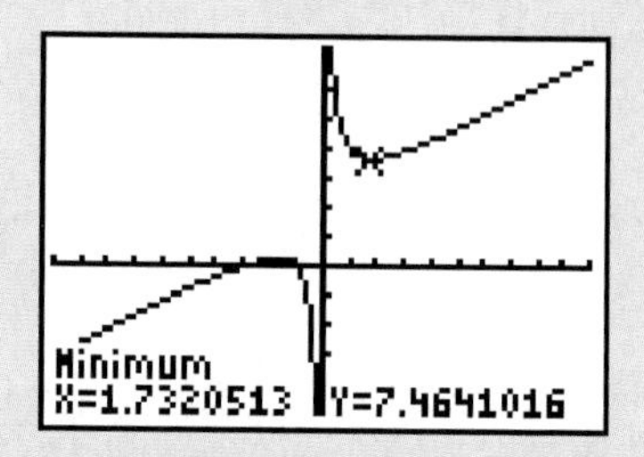

TECHNOLOGY HIGHLIGHT

Removable Discontinuities and Graphing Technology

The keystrokes shown apply to a TI-84 Plus model. Please consult your manual or our Internet site for other models.

When it comes to studying removable discontinuities, graphing calculators are a wonderful asset as they can vividly demonstrate that the given function is defined for *all other values.* Enter the function $r(x) = \frac{x^2 - 4x + 3}{x - 1}$ on the Y= screen, then use the TBLSET feature to set up the table as shown in Figure 4.54. Pressing 2nd GRAPH

Figure 4.54

displays the expected table, which shows the function cannot be evaluated at $x = 1$ (see Figure 4.55). Now change the TBLSET screen so that Δ**Tbl = 0.01** and return to the table screen. Note once again that the function is defined for all values *near* $x = 1$, just not precisely *at* $x = 1$. Reset the table once again so that Δ**Tbl = 0.001** and investigate further.

Figure 4.55

X	Y1
.5	-2.5
.6	-2.4
.7	-2.3
.8	-2.2
.9	-2.1
1	ERROR
1.1	-1.9

X=.5

We can actually see the gap or hole in the graph using a "friendly window". Since the screen of the TI-84 Plus is 94 pixels wide and 62 pixels high, using multiples of 4.7 for the Xmin and Xmax values will enable us to "see what happens" at integer (and other) values (see Figure 4.56). Pressing GRAPH gives the last display shown (Figure 4.57), which shows a noticeable gap at (1, −2). With the TRACE feature, move the cursor over to the gap and notice what happens.

Figure 4.56

```
WINDOW
 Xmin=-9.4
 Xmax=9.4
 Xscl=1
 Ymin=-6.2
 Ymax=6.2
 Yscl=1
 Xres=1
```

Figure 4.57

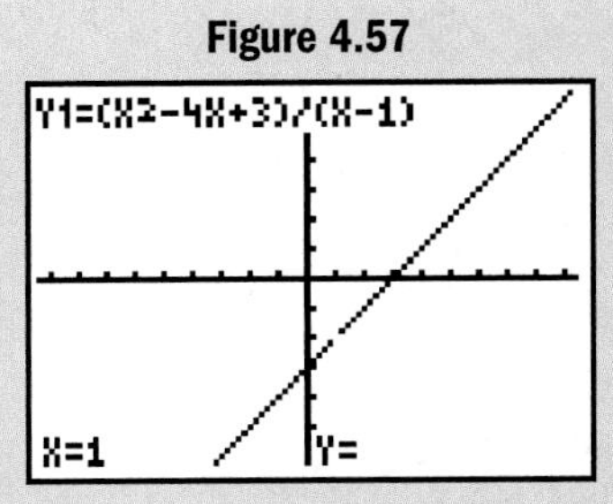

Use these ideas to investigate discontinuities in the following rational functions on your graphing calculator. Discuss what you find.

Exercise 1: $r(x) = \dfrac{x^2 - 4}{x + 2}$ Answers will vary.

Exercise 2: $f(x) = \dfrac{x^2 - 2x - 3}{x + 1}$ Answers will vary.

Exercise 3: $r(x) = \dfrac{x^3 + 1}{x + 1}$ Answers will vary.

Exercise 4: $f(x) = \dfrac{x^3 - 7x + 6}{x^2 + x - 6}$ Answers will vary.

4.6 EXERCISES

CONCEPTS AND VOCABULARY

Fill in each blank with the appropriate word or phrase. Carefully reread the section if needed.

1. The discontinuity in the graph of $y = \dfrac{1}{(x + 3)^2}$ is called a <u>nonremovable</u> discontinuity, since it cannot be "repaired."

2. If the degree of the numerator is greater than the degree of the denominator, the graph will have an <u>oblique</u> or <u>nonlinear</u> asymptote.

3. If the degree of the numerator is <u>two</u> more than the degree of the denominator, the graph will have a parabolic asymptote.

4. If the denominator is a monomial, use <u>term</u> by <u>term</u> division to find the quotient. Otherwise <u>synthetic</u> or <u>long</u> division must be used.

5. Discuss/explain how you would create a function with a parabolic asymptote and two vertical asymptotes. Answers will vary.

6. Complete Exercise 7 in expository form. That is, work this exercise out completely, discussing each step of the process as you go. Answers will vary.

DEVELOPING YOUR SKILLS

Graph each function. If there is a removable discontinuity, repair the break by redefining the function using an appropriate piecewise-defined function.

7. $f(x) = \dfrac{x^2 - 4}{x + 2}$

8. $f(x) = \dfrac{x^2 - 9}{x + 3}$

9. $g(x) = \dfrac{x^2 - 2x - 3}{x + 1}$

10. $g(x) = \dfrac{x^2 - 3x - 10}{x - 5}$

11. $h(x) = \dfrac{3x - 2x^2}{2x - 3}$

12. $h(x) = \dfrac{4x - 5x^2}{5x - 4}$

13. $p(x) = \dfrac{x^3 - 8}{x - 2}$

14. $p(x) = \dfrac{8x^3 - 1}{2x - 1}$

15. $q(x) = \dfrac{x^3 - 7x - 6}{x + 1}$

16. $q(x) = \dfrac{x^3 - 3x + 2}{x + 2}$

17. $r(x) = \dfrac{x^3 + 3x^2 - x - 3}{x^2 + 2x - 3}$

18. $r(x) = \dfrac{x^3 - 2x^2 - 4x + 8}{x^2 - 4}$

Graph each function using the *Guidelines for Graphing Rational Functions,* which is simply modified to include nonlinear asymptotes. Clearly label all intercepts and asymptotes and any additional points used to sketch the graph.

19. $Y_1 = \dfrac{x^2 - 4}{x}$

20. $Y_2 = \dfrac{x^2 - x - 6}{x}$

21. $v(x) = \dfrac{3 - x^2}{x}$

22. $V(x) = \dfrac{7 - x^2}{x}$

23. $w(x) = \dfrac{x^2 + 1}{x}$

24. $W(x) = \dfrac{x^2 + 4}{2x}$

25. $h(x) = \dfrac{x^3 - 2x^2 + 3}{x^2}$

26. $H(x) = \dfrac{x^3 + x^2 - 2}{x^2}$

27. $Y_1 = \dfrac{x^3 + 3x^2 - 4}{x^2}$

28. $Y_2 = \dfrac{x^3 - 3x^2 + 4}{x^2}$

29. $f(x) = \dfrac{x^3 - 3x + 2}{x^2}$

30. $F(x) = \dfrac{x^3 - 12x - 16}{x^2}$

31. $Y_3 = \dfrac{x^3 - 5x^2 + 4}{x^2}$

32. $Y_4 = \dfrac{x^3 + 5x^2 - 6}{x^2}$

33. $r(x) = \dfrac{x^3 - x^2 - 4x + 4}{x^2}$

34. $R(x) = \dfrac{x^3 - 2x^2 - 9x + 18}{x^2}$

35. $g(x) = \dfrac{x^2 + 4x + 4}{x + 3}$

36. $G(x) = \dfrac{x^2 - 2x + 1}{x - 2}$

37. $f(x) = \dfrac{x^2 + 1}{x + 1}$

38. $F(x) = \dfrac{x^2 + x + 1}{x - 1}$

39. $Y_3 = \dfrac{x^2 - 4}{x + 1}$

40. $Y_4 = \dfrac{x^2 - x - 6}{x - 1}$

41. $v(x) = \dfrac{x^3 - 4x}{x^2 - 1}$

42. $V(x) = \dfrac{9x - x^3}{x^2 - 4}$

43. $w(x) = \dfrac{16x - x^3}{x^2 + 4}$

44. $W(x) = \dfrac{x^3 - 7x + 6}{2 + x^2}$

45. $Y_1 = \dfrac{x^3 - 3x + 2}{x^2 - 9}$

46. $Y_2 = \dfrac{x^3 - x^2 - 12x}{x^2 - 7}$

47. $p(x) = \dfrac{x^4 + 4}{x^2 + 1}$

48. $P(x) = \dfrac{x^4 - 5x^2 + 4}{x^2 + 2}$

49. $q(x) = \dfrac{10 + 9x^2 - x^4}{x^2 + 5}$

50. $Q(x) = \dfrac{x^4 - 2x^2 + 3}{x^2}$

Graph each function and its nonlinear asymptote on the same screen, using the window specified. Then locate the minimum value in the first quadrant.

51. $f(x) = \dfrac{x^3 + 500}{x}$; $x \in [-24, 24], y \in [-500, 500]$ 119.1

52. $f(x) = \dfrac{2\pi x^3 + 750}{x}$; $x \in [-12, 12], y \in [-750, 750]$ 287.9

7. $F(x) = \begin{cases} \dfrac{x^2 - 4}{x + 2} & x \neq -2 \\ -4 & x = -2 \end{cases}$

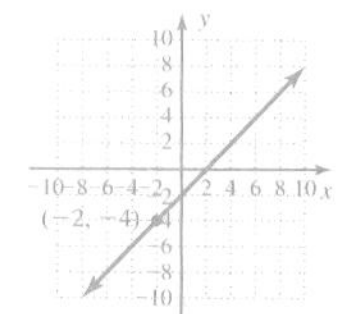

8. $F(x) = \begin{cases} \dfrac{x^2 - 9}{x + 3} & x \neq -3 \\ -6 & x = -3 \end{cases}$

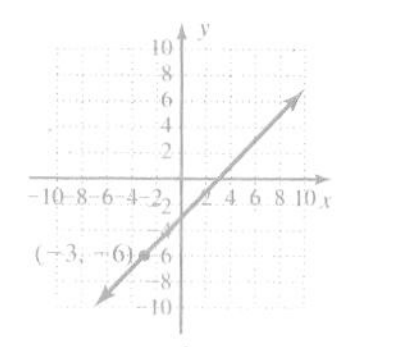

9. $G(x) = \begin{cases} \dfrac{x^2 - 2x - 3}{x + 1} & x \neq -1 \\ -4 & x = -1 \end{cases}$

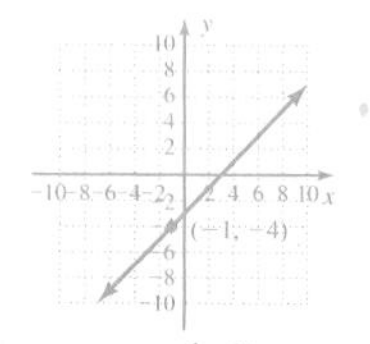

10. $G(x) = \begin{cases} \dfrac{x^2 - 3x - 10}{x - 5} & x \neq 5 \\ 7 & x = 5 \end{cases}$

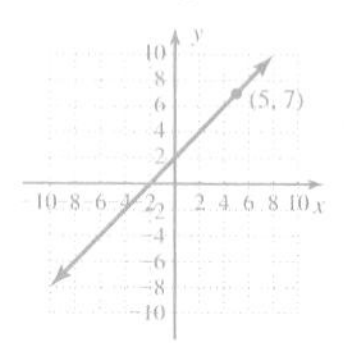

11. $H(x) = \begin{cases} \dfrac{3x - 2x^2}{2x - 3} & x \neq \dfrac{3}{2} \\ \dfrac{-3}{2} & x = \dfrac{3}{2} \end{cases}$

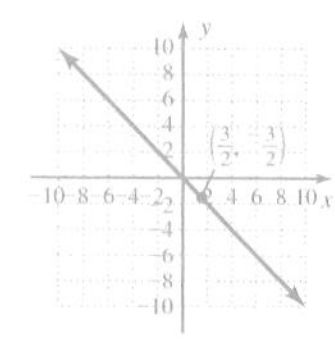

12. $H(x) = \begin{cases} \dfrac{4x - 5x^2}{5x - 4} & x \neq \dfrac{4}{5} \\ \dfrac{-4}{5} & x = \dfrac{4}{5} \end{cases}$

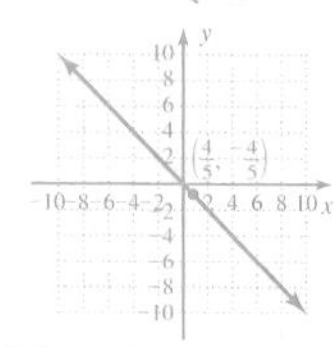

Additional answers can be found in the Instructor Answer Appendix.

WORKING WITH FORMULAS

53. Area of a first quadrant triangle: $A(a) = \frac{1}{2}\left(\frac{ka^2}{a - h}\right)$

The area of a triangle in the first quadrant, formed by a line with negative slope through the point (h, k) is given by the formula shown, where a represents the x-intercept of the resulting line $(h < a)$. The area of the triangle varies with the slope of the line. Assume the line contains the point (5, 6).

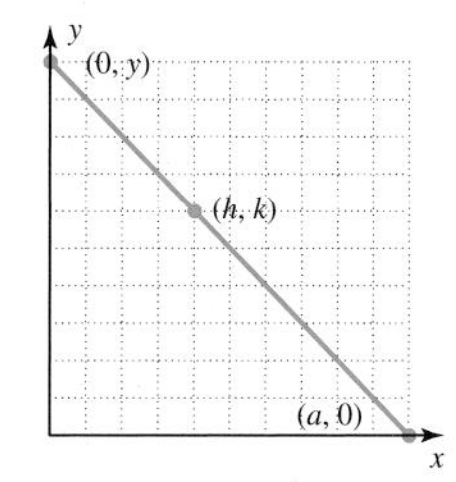

a. Find the equation of the vertical and slant asymptotes. $a = 5,\ y = 3a$

b. Find the area of the triangle if it has an x-intercept of (11, 0). 60.5

c. Use a graphing calculator to graph the function on an appropriate window. Does the shape of the graph look familiar? Use the calculator to find the value of a that minimizes $A(a)$. That is, find the x-intercept that results in a triangle with the smallest possible area. 10

54. Surface area of a cylinder with fixed volume: $S = \frac{2\pi r^3 + 2V}{r}$

It's possible to construct many different cylinders that will hold a specified volume, by changing the radius and height. This is critically important to producers who want to minimize the cost of packing canned goods and marketers who want to present an attractive product. The surface area of the cylinder can be found using the formula shown, where the radius is r and $V = \pi r^2 h$ is known. Assume the fixed volume is 750 cm^3.

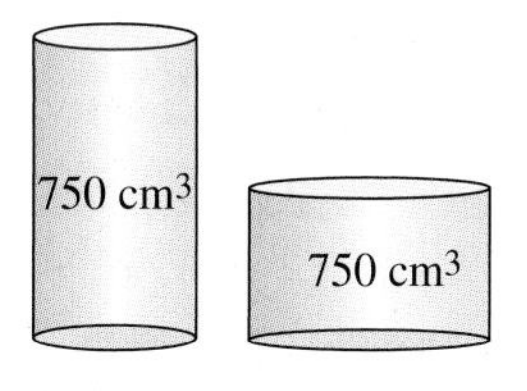

a. Find the equation of the vertical asymptote. How would you describe the nonlinear asymptote? $r = 0$, parabolic

b. If the radius of the cylinder is 2 cm, what is its surface area? ≈ 775.1 cm^2

c. Use a graphing calculator to graph the function on an appropriate window, and use it to find the value of x that minimizes $S(r)$. That is, find the radius that results in a cylinder with the smallest possible area, while still holding a volume of 750 cm^3. $r \approx 4.9$ cm

APPLICATIONS

Costs of manufacturing: As in Example 4, the cost $C(x)$ of manufacturing is sometimes nonlinear and can increase dramatically with each item. For the average cost function $A(x) = \frac{C(x)}{x}$, consider the following.

55. Assume the monthly cost of manufacturing custom-crafted storage sheds is modeled by the function $C(x) = 4x^2 + 53x + 250$.

a. Write the average cost function and state the equation of the vertical and oblique asymptotes.

b. Enter the cost function $C(x)$ as Y_1 on a graphing calculator, and the average cost function $A(x)$ as Y_2. Using the TABLE feature, find the cost and average cost of making 1, 2, and 3 sheds.

c. Scroll down the table to where it appears that average cost is a minimum. According to the table, how many sheds should be made each month to minimize costs? What is the minimum cost?

55. a. $A(x) = \frac{4x^2 + 53x + 250}{x}$; $x = 0$, $g(x) = 4x + 53$
b. cost: \$307, \$372, \$445, Avg. cost: \$307, \$186, \$148.33
c. 8, \$116.25
d.

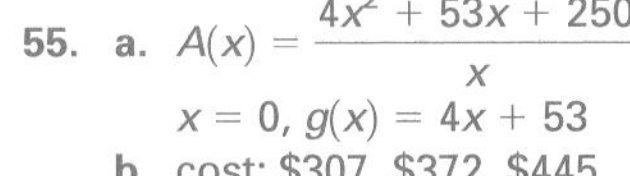

d. Graph the average cost function and its asymptotes, using a window that shows the entire function. Use the graph to confirm the result from part (c).

56.
a. $A(x) = \dfrac{5x^2 + 94x + 576}{x}$; $x = 0$, $g(x) = 5x + 94$
b. cost: \$675, \$784, \$903; avg. cost: \$675, \$392, \$301
c. 11; \$201.36; no
d.

56. Assume the monthly cost of manufacturing playground equipment that combines a play house, slides, and swings is modeled by the function $C(x) = 5x^2 + 94x + 576$. The company has projected that they will be profitable if they can bring their average cost down to \$200 per set of playground equipment.

a. Write the average cost function and state the equation of the vertical and oblique asymptotes.

b. Enter the cost function $C(x)$ as Y_1 on a graphing calculator, and the average cost function $A(x)$ as Y_2. Using the TABLE feature, find the cost and average cost of making 1, 2, and 3 playground equipment combinations. Why would the average cost fall so dramatically early on?

c. Scroll down the table to where it appears that average cost is a minimum. According to the table, how many sets of equipment should be made each month to minimize costs? What is the minimum cost? Will the company be profitable under these conditions?

d. Graph the average cost function and its asymptotes, using a window that shows the entire function. Use the graph to confirm the result from part (c).

Minimum cost of packaging: Similar to Exercise 54, manufacturers can minimize their costs by shipping merchandise in packages that use a minimum amount of material. After all, rectangular boxes come in different sizes and there are many combinations of length, width, and height that will hold a specified volume.

57. a. $S(x, y) = 2x^2 + 4xy$; $V(x, y) = x^2y$
b. $S(x) = \dfrac{2x^3 + 48}{x}$
c. $S(x)$ is asymptotic to $y = 2x^2$.
d. $x = 2$ ft 3.5 in.; $y = 2$ ft 3.5 in.

57. A clothing manufacturer wishes to ship lots of 12 ft^3 of clothing in boxes with square ends and rectangular sides.

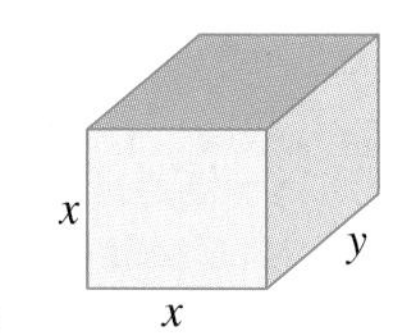

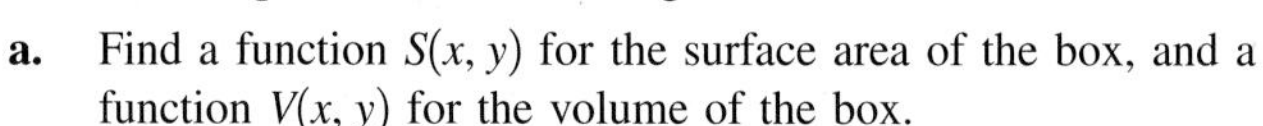

a. Find a function $S(x, y)$ for the surface area of the box, and a function $V(x, y)$ for the volume of the box.

b. Solve for y in $V(x, y) = 12$ (volume is 12 ft^3) and use the result to write the surface area as a function $S(x)$ in terms of x alone (simplify the result).

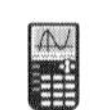

c. On a graphing calculator, graph the function $S(x)$ using the window $x \in [-8, 8]$; $y \in [-100, 100]$. Then graph $y = 2x^2$ on the same screen. How are these two graphs related?

d. Use the graph of $S(x)$ in Quadrant I to determine the dimensions that will minimize the surface area of the box, yet still hold 12 ft^3 of clothing. Clearly state the values of x and y, *in terms of feet and inches*, rounded to the nearest $\frac{1}{2}$ in.

58. a. $S(x, y) = 2x(x + 2) + 2y(x + 2) + 2xy$; $V(x, y) = (x^2 + 2x)y$
b. $y = \dfrac{36}{x^2 + 2x}$
$S(x) = \dfrac{2x^4 + 8x^3 + 8x^2 + 144x + 144}{x(x + 2)}$
c. $S(x)$ is asymptotic to $y = 2x^2 + 4x$.
d. $x \approx 2$ ft 7 in.; $y \approx 3$ ft $\frac{1}{2}$ in.

58. A maker of packaging materials needs to ship 36 ft^3 of foam "peanuts" to his customers across the country, using boxes with the dimensions shown.

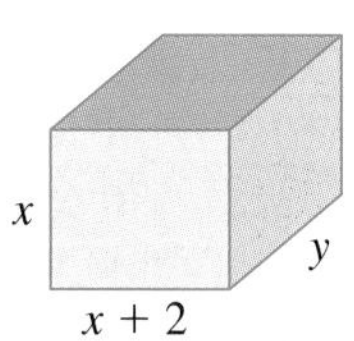

a. Find a function $S(x, y)$ for the surface area of the box, and a function $V(x, y)$ for the volume of the box.

b. Solve for y in $V(x, y) = 36$ (volume is 36 ft^3) and use the result to write the surface area as a function $S(x)$ in terms of x alone (simplify the result).

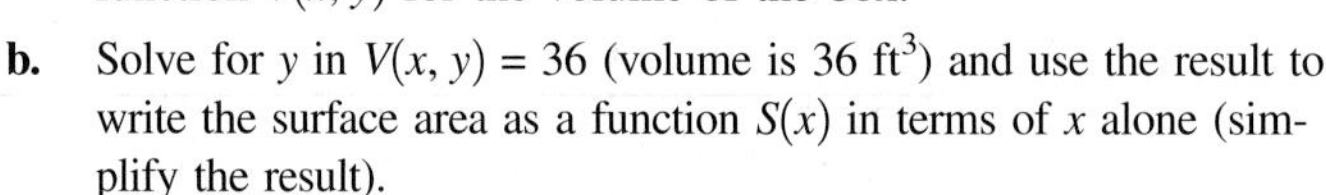

c. On a graphing calculator, graph the function $S(x)$ using the window $x \in [-10, 10]$; $y \in [-200, 200]$. Then graph $y = 2x^2 + 4x$ on the same screen. How are these two graphs related?

d. Use the graph of $S(x)$ in Quadrant I to determine the dimensions that will minimize the surface area of the box, yet still hold the foam peanuts. Clearly state the values of x and y, *in terms of feet and inches*, rounded to the nearest $\frac{1}{2}$ in.

Printing and publishing: In the design of magazine pages, posters, and other published materials, an effort is made to maximize the usable area of the page while maintaining an attractive border, or minimizing the page size that will hold a certain amount of print or art work.

59. a. $A(x, y) = xy$; $R(x, y) = (x - 2.5)(y - 2)$
b. $y = \dfrac{2x + 55}{x - 2.5}$; $A(x) = \dfrac{2x^2 + 55x}{x - 2.5}$
c. $A(x)$ is asymptotic to $y = 2x + 55$
d. $x \approx 11.16$ in.; $y = 8.93$ in.

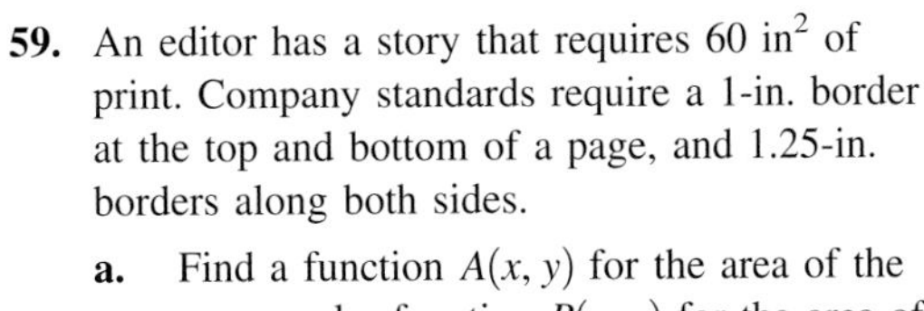

59. An editor has a story that requires 60 in^2 of print. Company standards require a 1-in. border at the top and bottom of a page, and 1.25-in. borders along both sides.

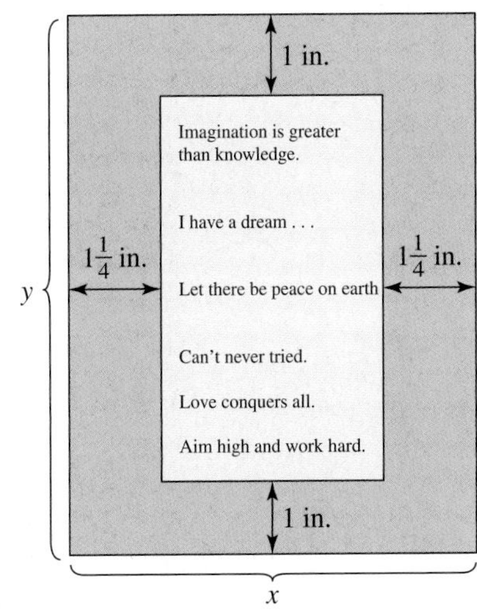

a. Find a function $A(x, y)$ for the area of the page, and a function $R(x, y)$ for the area of the inner rectangle (the printed portion).

b. Solve for y in $R(x, y) = 60$, and use the result to write the area from part (a) as a function $A(x)$ in terms of x alone (simplify the result).

c. On a graphing calculator, graph the function $A(x)$ using the window $x \in [-30, 30]$; $y \in [-100, 200]$. Then graph $y = 2x + 55$ on the same screen. How are these two graphs related?

d. Use the graph of $A(x)$ in Quadrant I to determine the page of minimum size that satisfies these border requirements and holds the necessary print. Clearly state the values of x and y, rounded to the nearest hundredth of an inch.

60. a. $A(x, y) = xy$; $R(x, y) = (x - 5)(y - 5)$
b. $y = \dfrac{5x + 475}{x - 5}$; $A(x) = \dfrac{5x^2 + 475x}{x - 5}$
c. $A(x)$ is asymptotic to $y = 5x + 475$.
d. $x \approx 27.36$ in.; $y \approx 27.36$ in.

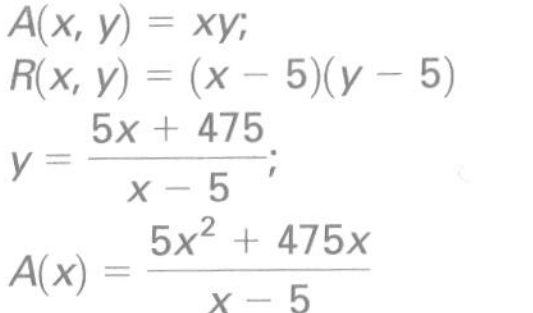

60. *The Poster Shoppe* creates posters, handbills, billboards, and other advertising for business customers. An order comes in for a poster with 500 in^2 of usable area, with margins of 2 in. across the top, 3 in. across the bottom, and 2.5 in. on each side.

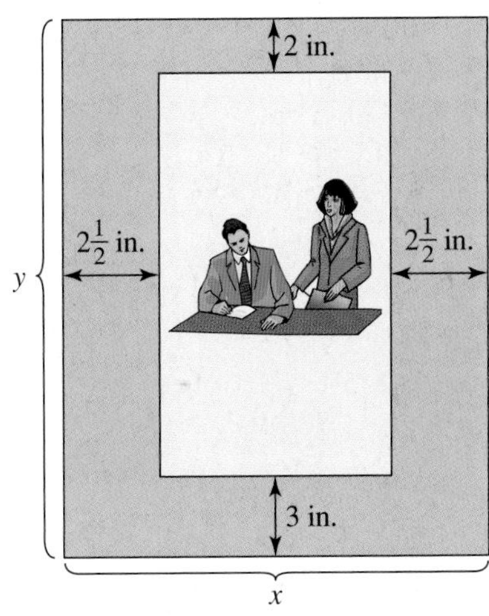

a. Find a function $A(x, y)$ for the area of the page, and a function $R(x, y)$ for the area of the inner rectangle (the usable area).

b. Solve for y in $R(x, y) = 500$, and use the result to write the area from part (a) as a function $A(x)$ in terms of x alone (simplify the result).

c. On a graphing calculator, graph $A(x)$ using the window $x \in [-100, 100]$; $y \in [-800, 1600]$. Then graph $y = 5x + 475$ on the same screen. How are these two graphs related?

d. Use the graph of $A(x)$ in Quadrant I to determine the poster of minimum size that satisfies these border requirements and has the necessary usable area. Clearly state the values of x and y, rounded to the nearest hundredth of an inch.

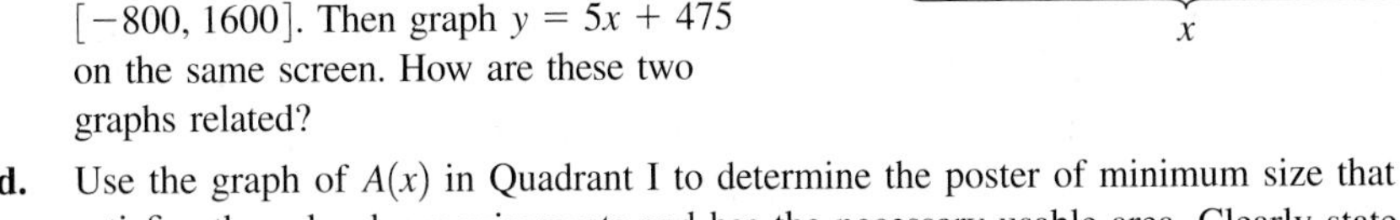

61. a. $h = \dfrac{V}{\pi r^2}$
b. $S = 2\pi r^2 + \dfrac{2V}{r}$
c. $S = \dfrac{2\pi r^3 + 2V}{r}$
d. $r \approx 5.76$ cm, $h \approx 11.51$ cm; $S \approx 625.13$ cm^2

61. The formula from Exercise 54 has an interesting derivation. The volume of a cylinder is $V = \pi r^2 h$, while the surface area is given by $S = 2\pi r^2 + 2\pi rh$ (the circular top and bottom + the area of the side).

a. Solve the volume formula for the variable h.

b. Substitute the resulting expression for h into the surface area formula and simplify.

c. Combine the resulting two terms using the least common denominator, and the result is the formula from Exercise 54.

d. Assume the volume of a can must be 1200 cm^3. Use a calculator to graph the function on an appropriate window, then use it to find the radius r and height h that will result in a cylinder with the smallest possible area, while still holding a volume of 1200 cm^3. Also see Exercise 66.

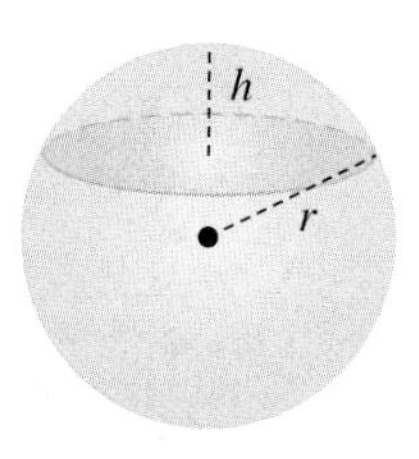

62. The surface area of a spherical cap is given by $S = 2\pi rh$, where r is the radius of the sphere and h is the distance from the circumference to the plane intersecting the sphere, forming the cap. The volume of the cap is $V = \frac{1}{3}\pi h^2(3r - h)$. Similar to Exercise 61, a formula can be found that will minimize the area of a cap that holds a specified volume.

a. Solve the volume formula for the variable r.

b. Substitute the resulting expression for r into the surface area formula and simplify. The result is a formula for surface area given solely in terms of the volume V and the height h.

c. Assume the volume of the spherical cap is 500 cm^3. Use a graphing calculator to graph the resulting function on an appropriate window, and use the graph to find the height h that will result in a spherical cap with the smallest possible area, while still holding a volume of 500 cm^3.

d. Use this value of h and $V = 500$ cm^3 to find the radius of the sphere.

62. a. $r = \frac{3V + \pi h^3}{3\pi h^2}$ b. $S = \frac{2(3V + \pi h^3)}{3h}$ c. $S(h) = \frac{2(1500 + \pi h^3)}{3h}$, $h = 6.20$ d. $r \approx 6.20$ cm

WRITING, RESEARCH, AND DECISION MAKING

63. Consider rational functions of the form $f(x) = \frac{x^2 - a}{x - b}$. Use a graphing calculator to explore cases where $a = b^2 + 1$, $a = b^2$, and $a = b^2 - 1$. What do you notice? Explain/discuss why the graphs differ. It's helpful to note that when graphing functions of this form, the "center" of the graph will be at $(b, b^2 - a)$, and the window size can be set accordingly for an optimal view. Do some investigation on this function and determine/explain *why* the "center" of the graph is at $(b, b^2 - a)$. Answers will vary.

64. We've already discussed and graphed rational functions that have linear and parabolic asymptotes. Do you suppose that some rational graphs have cubic (propeller-shaped) asymptotes? Do some investigation using simple functions of the form $f(x) = \frac{x^4 + a}{x + b}$ for small values of a and b (start with $a = -1$ and $b = 0$). What happens when the function has no real zeroes? What happens when the function has two real zeroes?
Answers will vary.

EXTENDING THE CONCEPT

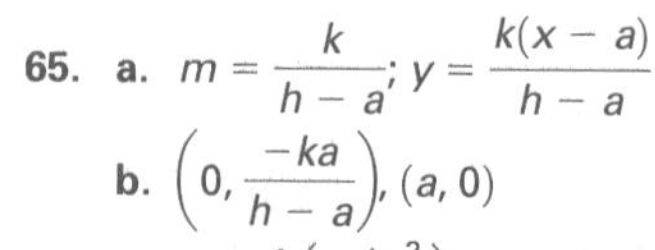

65. a. $m = \frac{k}{h - a}$; $y = \frac{k(x - a)}{h - a}$
b. $\left(0, \frac{-ka}{h - a}\right)$, $(a, 0)$

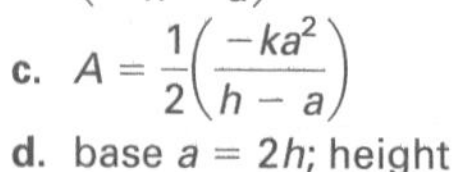

c. $A = \frac{1}{2}\left(\frac{-ka^2}{h - a}\right)$

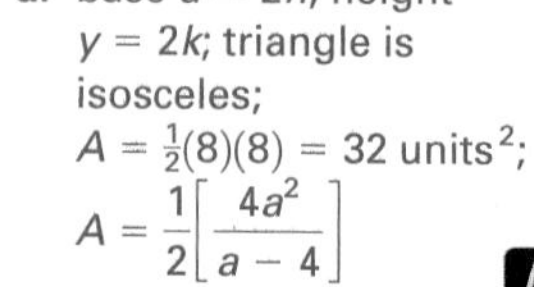

d. base $a = 2h$; height $y = 2k$; triangle is isosceles; $A = \frac{1}{2}(8)(8) = 32$ units^2; $A = \frac{1}{2}\left[\frac{4a^2}{a - 4}\right]$ has a minimum at (8, 32)

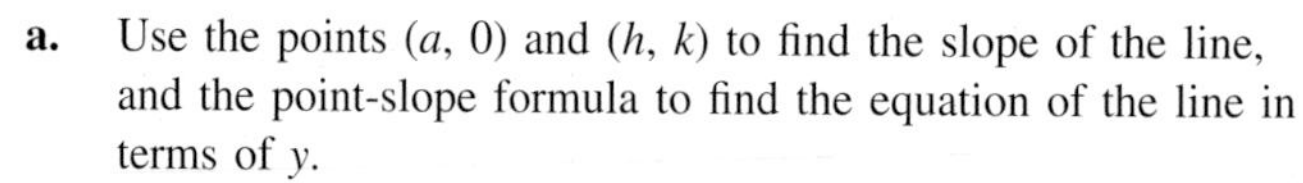

65. The formula from Exercise 53 also has an interesting derivation, and the process involves this sequence:

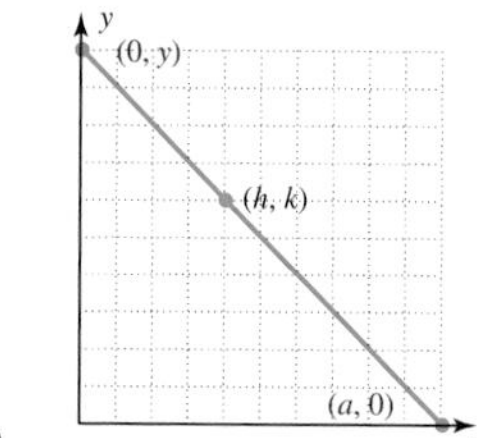

a. Use the points $(a, 0)$ and (h, k) to find the slope of the line, and the point-slope formula to find the equation of the line in terms of y.

b. Use this equation to find the x- and y-intercepts of the line in terms of a, k, and h.

c. Complete the derivation using these intercepts and the triangle formula $A = \frac{1}{2}BH$.

d. Find the dimensions of a triangle with minimum area through (h, k) where $h = k$. What do you notice? Verify using the points (4, 4), (5, 5), and (6, 6).

66. Referring to Exercises 54 and 61, suppose that instead of a closed cylinder, with both a top and bottom, we needed to manufacture *open cylinders,* like tennis ball cans that use a lid made from a different material. Derive the formula that will minimize the surface area of an open cylinder, and use it to find the cylinder with minimum surface area that will hold 90 in^3 of material. $r = 3.1$ in., $h \approx 3$ in.

MAINTAINING YOUR SKILLS

67. (3.4/3.8) Given $g(x) = 2x^2 - 8x + 3$, name intervals where: (a) $g(x) > 0$, (b) $g(x)\downarrow$
$x \in (-\infty, \frac{4-\sqrt{10}}{2}) \cup (\frac{4+\sqrt{10}}{2}, \infty)$; $x \in (-\infty, 2)$

68. (4.3) Use the rational roots theorem and synthetic division to write P in completely factored form: $P(x) = x^4 - 2x^3 - 7x^2 + 8x + 12$. $= (x+1)(x+2)(x-2)(x-3)$

69. (2.5) Compute the quotient $\frac{5i}{1+2i}$, then check your answer using multiplication. $2 + i$

70. (2.3) Write the equation of the line in slope intercept form and state the slope and y-intercept: $-3x + 4y = -16$.

71. (1.6) Given $f(x) = ax^2 + bx + c$, for what values of a, b, and c will the function have: (a) two, real/rational roots, (b) two, real/irrational roots, (c) one real and rational root, (d) one real/irrational root, (e) one complex root, and (f) two complex roots

70. $y = \frac{3}{4}x - 4$, $m = \frac{3}{4}$, $(0, -4)$
71. a. $b^2 - 4ac > 0$, with $b^2 - 4ac$ a perfect square
b. $b^2 - 4ac > 0$, but not a perfect square
c. $b^2 - 4ac = 0$
d. none
e. none
f. $b^2 - 4ac < 0$

72. (2.7) The average monthly cost of cable TV has been rising steadily since it become very popular in the early 1980s. The data given shows the average monthly rate for selected years (year 0 corresponds to 1980).

Year	Monthly Charge
0	7.69
5	9.73
10	16.78
15	23.07
20	30.70

a. Use the data to draw a scatter-plot and decide if a linear or quadratic model is more appropriate. quadratic

b. Find the regression equation using a calculator, then use it to estimate the monthly cable charge for 2008 and 2010.
$y = 0.03x^2 + 0.59x + 7.21$; \$47.12, \$51.76

Source: 2004–2005 Statistical Abstract of the United States (page 725, Table 1138)

4.7 Polynomial and Rational Inequalities—An Analytical View

LEARNING OBJECTIVES

In Section 4.7 you will learn how to:

A. Solve inequalities involving polynomial functions

B. Solve inequalities involving rational functions

C. Solve applications involving polynomial and rational inequalities

INTRODUCTION

The study of polynomial and rational inequalities is simply an extension of our earlier work in analyzing functions (Section 3.8). The main difference is that we've expanded our study of how functions behave with regard to end behavior and the multiplicities of their zeroes. In the end, the solution set to an inequality can be determined from the graph of the function or using a simple number line showing its zeroes and vertical asymptotes (in the case of rational functions), along with an analysis of the end behavior and how the graph behaves at each zero (cross or bounce).

POINT OF INTEREST

While the functions $f(x) = x - 1$ and $F(x) = \frac{1}{x-1}$ certainly have many differences, they are alike in one important respect: when $x > 1$, both functions are positive, and when $x < 1$, both are negative. This observation enables us to continue our analysis of function inequalities in much the same way as we've done previously—by locating the zeroes of a function, observing their multiplicity, and using the concept of "alternating intervals." In the case of rational functions, we simply include the zeroes of the denominator in our analysis.

A. Polynomial Inequalities

Figure 4.58

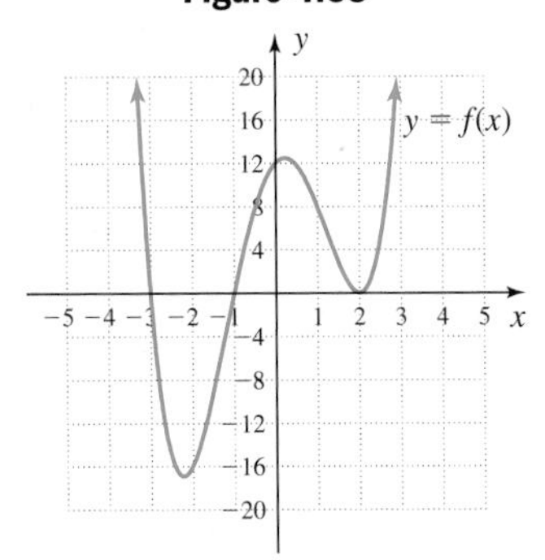

The graph of $f(x) = x^4 - 9x^2 + 4x + 12$ is shown in Figure 4.58. By carefully reviewing the solution sets for $f(x) \geq 0$ and $f(x) > 0$, we can develop a process for solving inequalities without having to use numerous test values or actually graphing the function. Observe the zeroes of f are $x = -3$, $x = -1$, and $x = 2$, indicating that $(x + 3)$, $(x + 1)$, and $(x - 2)$ are factors. Since the graph crosses the x-axis at -3 and -1, the related factors must have odd multiplicity and function values will change sign at these zeroes. The graph bounces at $x = 2$ so $(x - 2)$ must have even multiplicity and the function does not change sign here. The polynomial f has degree 4, giving the factored form $f(x) = (x + 3)(x + 1)(x - 2)^2$. Noting the y-intercept is $f(0) = 12$ helps confirm this result. For $f(x) \geq 0$ the solution set is $x \in (-\infty, -3] \cup [-1, \infty)$ where the graph is above or touching the x-axis. The solution set for $f(x) > 0$ includes only those points where the graph is strictly above the x-axis, giving $x \in (-\infty, -3) \cup (-1, 2) \cup (2, \infty)$, with $x = 2$ excluded. These observations help to affirm and extend our earlier approach to solving inequalities (Section 2.5).

SOLVING POLYNOMIAL INEQUALITIES

Given $f(x)$ is a polynomial in standard form,

1. Use any combination of factoring, tests for 1 and -1, the RRT, and synthetic division to write P in factored form, noting the multiplicity of each zero.
2. Plot the zeroes on a number line (x-axis) and determine if the graph crosses (odd multiplicity) or bounces (even multiplicity) at each zero. Recall that complex zeroes from irreducible quadratic factors can be ignored.
3. Use end behavior, the y-intercept, or a test point to determine the sign of the function in a given interval, then label all other intervals as $P(x) < 0$ or $P(x) > 0$ by analyzing the multiplicity of neighboring zeroes.
4. State the solution using interval notation, noting strict/nonstrict inequalities.

EXAMPLE 1 Given $f(x) = x^3 + 4x^2 - 3x - 18$, solve $f(x) < 0$.

Solution: The polynomial cannot be factored by grouping, and testing 1 and -1 shows neither is a zero. Using $x = 2$ and synthetic division gives a remainder of zero with a quotient of $x^2 + 6x + 9$ (verify this).

1. The factored form is $(x - 2)(x^2 + 6x + 9) = (x - 2)(x + 3)^2$.

2. $x = 2$ has odd multiplicity (cross) and $x = -3$ has even multiplicity (bounce).

An open dot is used at each zero due to the strict inequality (Figure 4.59).

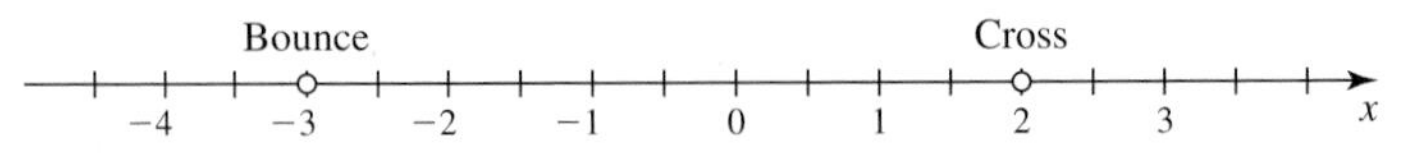

Figure 4.59

3. The polynomial has odd degree with a positive lead coefficient, so end behavior is down/up. The y-intercept is $(0, -18)$ indicating that function values will be negative in the interval containing zero. The solution diagram is shown in Figure 4.60 with negative intervals in red and positive intervals in blue.

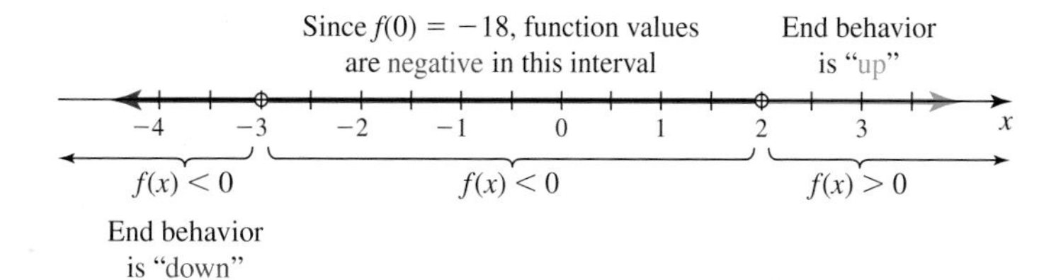

Figure 4.60

For $f(x) < 0$, the solution is $x \in (-\infty, -3) \cup (-3, 2)$.

NOW TRY EXERCISES 7 THROUGH 18

GRAPHICAL SUPPORT

The results from Example 1 can easily be verified using a graphing calculator. The graph shown here is displayed using a window of $x \in [-4.7, 4.7]$ and $y \in [-30, 30]$, and definitely shows the graph is below the x-axis $[f(x) < 0]$ for $x \in (-\infty, 2)$, except at $x = -3$ where the graph bounces off the x-axis.

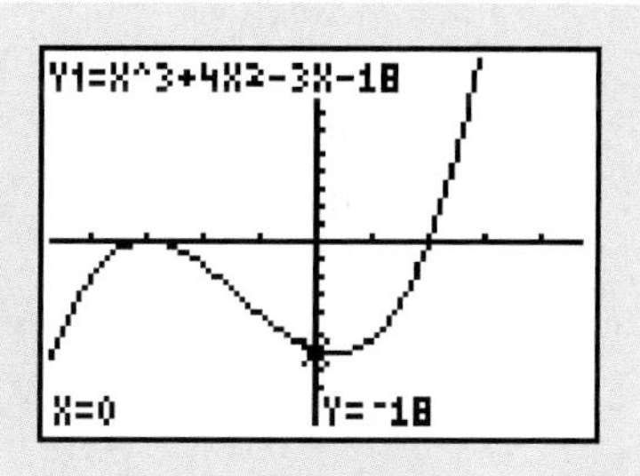

If the inequality is not given in function form, begin by writing the polynomial in standard form with zero on one side.

EXAMPLE 2 Solve the inequality: $x^4 + 4x \geq 9x^2 - 12$.

Solution: First write the polynomial in standard form: $x^4 - 9x^2 + 4x + 12 \geq 0$. The equivalent inequality is $f(x) \geq 0$. Testing 1 and -1 shows $x = 1$ is not a zero, but $x = -1$ is, and synthetic division gives $q_1(x) = x^3 - x^2 - 8x + 12$. Using $x = 2$ and synthetic division with $q_1(x)$ gives $q_2(x) = x^2 + x - 6$, which is easily factored.

1. factored form: $(x + 1)(x - 2)(x^2 + x - 6)$ or
$(x + 3)(x + 1)(x - 2)^2 \geq 0$.

2. The graph will "cross" at $x = -3$ and -1 and bounce at $x = 2$ (why?). See Figure 4.61.

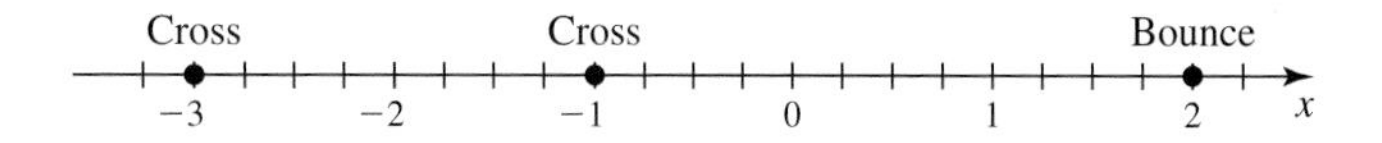

Figure 4.61

3. With even degree and positive lead coefficient, the end behavior is up/up. The y-intercept is $(0, 12)$ indicating that function values

will be positive in the interval containing zero. The solution diagram is shown in Figure 4.62.

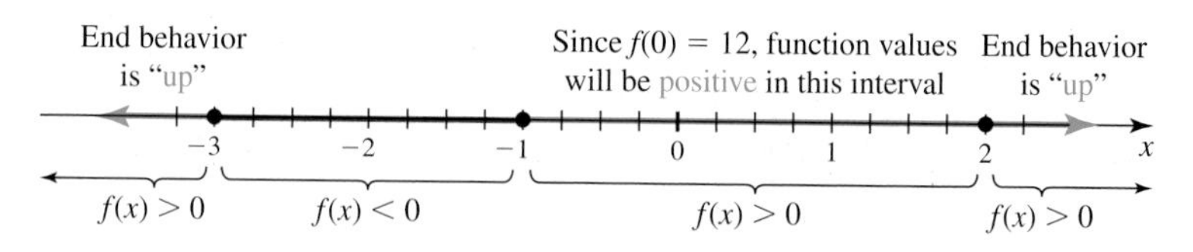

Figure 4.62

For $f(x) \geq 0$, the solution is $x \in (-\infty, -3] \cup [-1, \infty)$.

NOW TRY EXERCISES 19 THROUGH 24

GRAPHICAL SUPPORT

As with Example 1, the results from Example 2 can be confirmed using a graphing calculator. The graph shown here is displayed using $x \in [-4.7, 4.7]$ and $y \in [-20, 20]$. The graph is above or touching the x-axis $[f(x) \geq 0]$ for $x \in (-\infty, -3] \cup [-1, \infty)$.

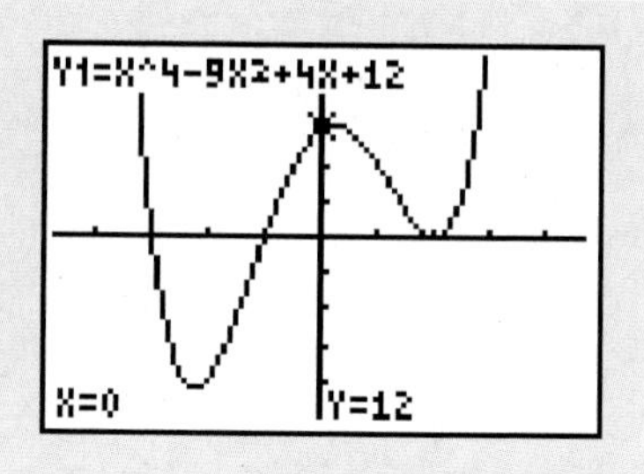

Figure 4.63

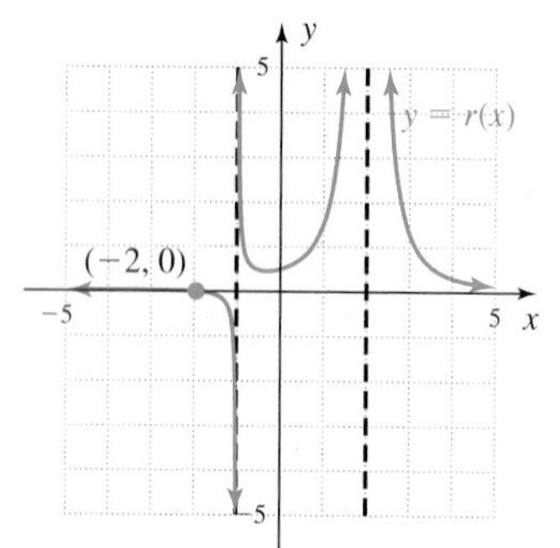

B. Rational Inequalities

The graph of $r(x) = \frac{x + 2}{x^3 - 3x^2 + 4}$ is shown in Figure 4.63. Much as with polynomial inequalities, we can use the solution set for $r(x) \geq 0$ or $r(x) > 0$ to generalize the ideas needed to solve *any* rational inequality. We observe the only x-intercept is $x = -2$, while the vertical asymptotes at $x = -1$ and $x = 2$ must be zeroes of the denominator. Since function values change sign at $x = -2$ and -1, they must have odd multiplicity. Function values do not change sign at $x = 2$, indicating even multiplicity. In factored form $r(x) = \frac{x + 2}{(x + 1)(x - 2)^2}$. For $r(x) \geq 0$, the graph must be above or touching the x-axis, but *exclude the zeroes of the denominator:* $x \in (-\infty, -2] \cup (-1, 2) \cup (2, \infty)$. In the most general sense, the solution process for polynomial and rational inequalities is virtually identical, once we recognize that vertical asymptotes also break the x-axis into intervals where function values may change sign, depending on multiplicity. For this reason, solution diagrams will label x-intercepts and the location of vertical asymptotes as, "function changes sign (*change*)" or, "function does not change sign (*no change*)," from this point forward.

EXAMPLE 3 Solve $\frac{x^2 - 9}{x^3 - x^2 - x + 1} \leq 0$.

Solution: The numerator and denominator are in standard form. The numerator factors easily, and the denominator can be factored by grouping.

1. The factored form is $h(x) = \frac{(x - 3)(x + 3)}{(x - 1)^2(x + 1)}$.
2. The graph will change sign at $x = 3, -3$, and -1 (odd multiplicity) but will not change sign at $x = 1$ (even multiplicity). Note

WORTHY OF NOTE

End behavior can also be used to analyze rational inequalities, although using the y-intercept may be more efficient. For the function $h(x)$ from Example 3 we have $\frac{x^2 - 9}{x^3 - x^2 - x + 1} \approx \frac{x^2}{x^3} = \frac{1}{x}$ for large values of x, indicating $h(x) > 0$ to the far right and $h(x) < 0$ to the far left. The analysis of each interval can then begin from either side.

that zeroes of the denominator will always be indicated by open dots as they are excluded from any solution set (see Figure 4.64).

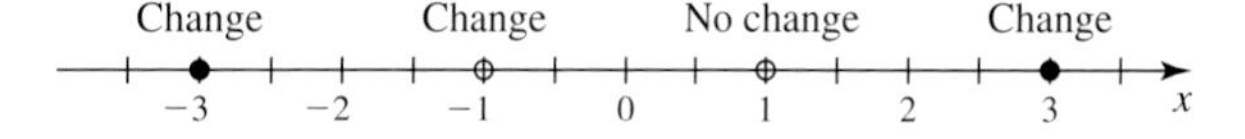

Figure 4.64

3. The y-intercept is $(0, -9)$, indicating function values will be negative in the interval containing zero. Using the "crosses/changes sign and bounces/no sign change" approach, gives the solution indicated in Figure 4.65.

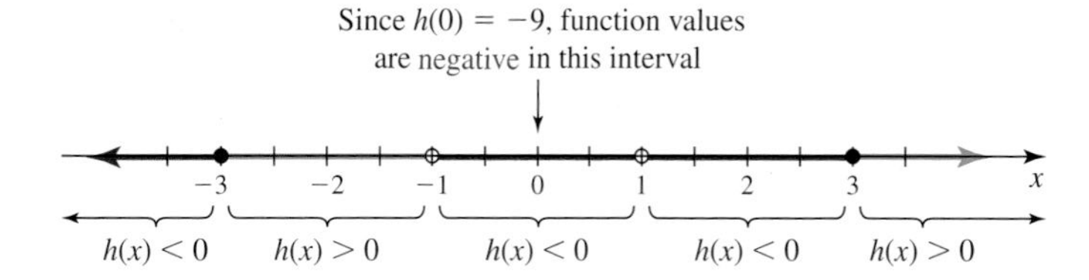

Figure 4.65

For $h(x) \leq 0$, the solution is $x \in (-\infty, -3] \cup (-1, 1) \cup (1, 3]$.

NOW TRY EXERCISES 25 THROUGH 36

GRAPHICAL SUPPORT

Sometimes finding a window that clearly displays all features of a rational function can be difficult. In these cases, we can investigate each piece separately to confirm solutions. For Example 3, most of the features of h can be seen using a window of $x \in [-5, 5]$ and $y \in [-20, 10]$, and we note the graph displayed strongly tends to support the stated solution.

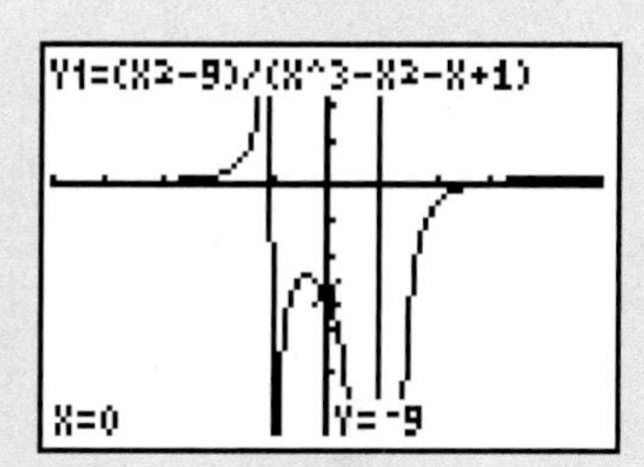

If the rational inequality is not given in function form or is composed of more than one term, start by writing the inequality with zero on one side, then combine terms into a single expression.

EXAMPLE 4 Solve $\frac{x-2}{x-3} \leq \frac{1}{x+3}$.

Solution: Rewrite the inequality with zero on one side: $\frac{x-2}{x-3} - \frac{1}{x+3} \leq 0$.

Combining terms on the left-hand side gives

$$\frac{(x-2)(x+3) - 1(x-3)}{(x+3)(x-3)} \leq 0 \quad \text{combine terms using LCD}$$

$$\frac{x^2 - 3}{(x+3)(x-3)} \leq 0 \quad \text{multiply and combine like terms}$$

1. The factored form is $\dfrac{(x - \sqrt{3})(x + \sqrt{3})}{(x + 3)(x - 3)} \leq 0$.

2. Function values change sign at $x = -3, 3, -\sqrt{3}$, and $\sqrt{3}$, as all have odd multiplicity. See Figure 4.66.

Figure 4.66

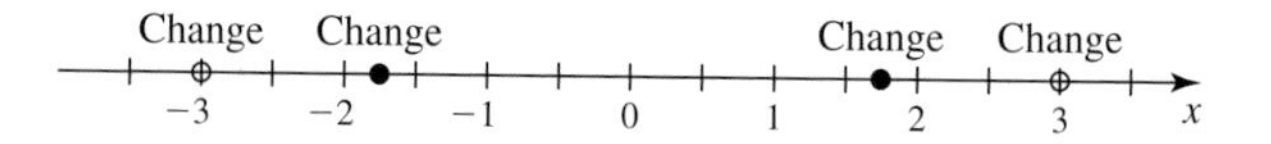

3. The y-intercept is $h(0) = \frac{1}{3}$, indicating function values will be positive in the interval containing zero. This produces the diagram shown in Figure 4.67.

Since $h(0) = \frac{1}{3}$, function values are positive in this interval

−3 −2 −1 0 1 2 3

$h(x) > 0$ $h(x) < 0$ $h(x) > 0$ $h(x) < 0$ $h(x) > 0$

Figure 4.67

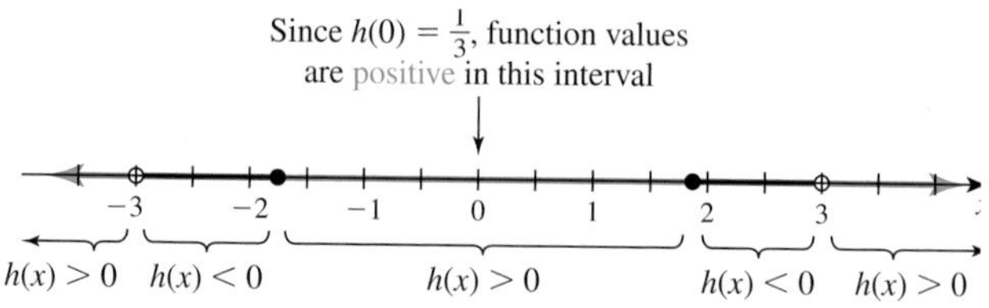

The solution for $\dfrac{x - 2}{x - 3} \leq \dfrac{1}{x + 3}$ is $x \in (-3, -\sqrt{3}] \cup [\sqrt{3}, 3)$.

NOW TRY EXERCISES 37 THROUGH 42

GRAPHICAL SUPPORT

Although checking answers using the original inequality is a high priority, checking solutions to $\dfrac{x - 2}{x - 3} \leq \dfrac{1}{x + 3}$ proves difficult and $\dfrac{x - 2}{x - 3} - \dfrac{1}{x + 3} \leq 0$ is used as an alternative. The graph shown is displayed using the window $x \in [-5, 5]$ and $y \in [-10, 10]$, and we note the graph lends strong support to the stated solution.

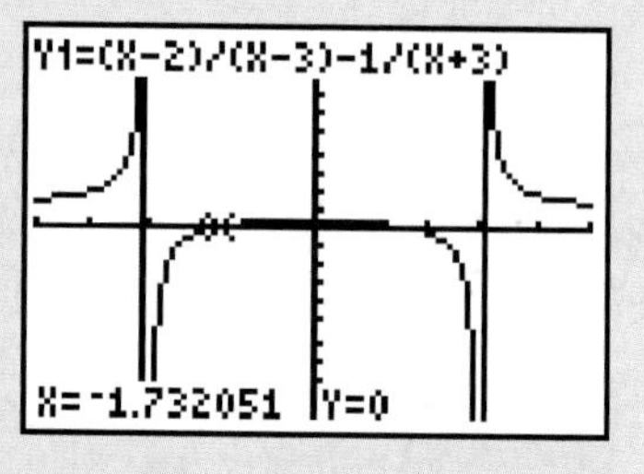

Since complex zeroes of a polynomial or rational function can never represent x-intercepts or vertical asymptotes, they play no part in the solution to real inequalities and can be ignored when determining the solution set. For $\dfrac{x^4 - 16}{x^3 - 1} \geq 0$, the factored form $\dfrac{(x - 2)(x + 2)(x^2 + 4)}{(x - 1)(x^2 + x + 1)} \geq 0$ will produce the same solution set as $\dfrac{(x - 2)(x + 2)}{(x - 1)} \geq 0$ (with the irreducible quadratic factors removed). For more on this idea, see Exercises 43 through 48, and the *Calculator Exploration and Discovery* feature for Chapter 4.

C. Applications of Polynomial and Rational Inequalities

Applications of inequalities are numerous and varied. In addition to further developing algebraic skills, these exercises compel us to consider the context of each application as we state the solution set.

EXAMPLE 5 The velocity of a particle (in feet/sec) as it floats through turbulence is given by $V(t) = t^5 - 10t^4 + 35t^3 - 50t^2 + 24t$, where t is the time in seconds and $0 < t < 4.5$. During what times is the particle moving in the positive direction $[V(t) > 0]$?

Solution: Begin by writing V in factored form. Testing 1 and -1 shows $t = 1$ is a root and $t = -1$ is not. Factoring out t and using $t = 1$ with synthetic division gives a quotient polynomial of $q_1(t) = t^3 - 9t^2 + 26t - 24$. Using $t = 2$ with synthetic division and $q_1(t)$ also gives a remainder of zero, allowing us to write $V(t) = t(t - 1)(t - 2)(t^2 - 7t + 12)$.

1. The completely factored form is $V(t) = t(t - 1)(t - 2)(t - 3)(t - 4)$.
2. All zeroes have multiplicity one and the function will change sign.
3. With odd degree and positive lead coefficient, the end behavior is down/up, and function values will be negative in the interval to the left of $x = 0$ (even though the interval is outside the context of the application) and must alternate thereafter. The solution diagram is shown in Figure 4.68

Figure 4.68

Since end behavior is down/up, function values are negative in this interval, and alternate thereafter.

neg pos neg pos neg pos

-1 0 1 2 3 4 t

For $V(t) > 0$, the solution is $t \in (0, 1) \cup (2, 3) \cup (4, 4.5)$. The velocity of the particle is positive in these intervals of time.

NOW TRY EXERCISES 55 THROUGH 64

GRAPHICAL SUPPORT

To verify our analysis in Example 5, we graph $V(t)$ using the window $x \in [-1, 5]$ and $y \in [-5, 5]$. As the graph clearly shows, function values are positive (graph is above the x-axis) when $t \in (0, 1) \cup (2, 3) \cup (4, 4.5)$. Also see the *Technology Highlight* on page 458.

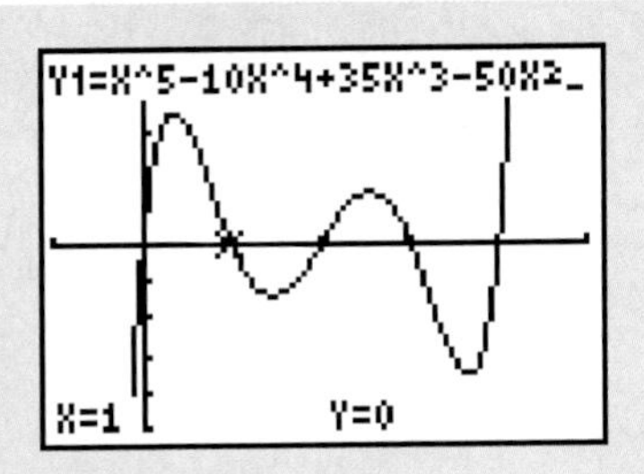

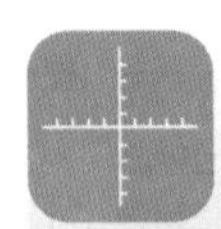

TECHNOLOGY HIGHLIGHT

Polynomial and Rational Inequalities

The keystrokes shown apply to a TI-84 Plus model. Please consult your manual or our Internet site for other models.

Most graphing calculators offer enhancements that enable a clearer view of certain graphical characteristics. While these add little to the "nuts and bolts" of a solution process, they do enable a visual confirmation of the processes and solutions under study. Consider the results from Example 5, where we solved the inequality $V(t) > 0$ for $V(t) = t^5 - 10t^4 + 35t^3 - 50t^2 + 24t$. To emphasize that we are seeking intervals where the function is above the x-axis (the horizontal line $y = 0$), we can have the calculator *shade these areas* of the function for us. Begin by entering $V(t)$ as Y_1 on the **Y=** screen, and the line $y = 0$ as Y_2. Using $x \in [0, 4.7]$ (a "friendly" window) and **GRAPH**ing the functions produces Figure 4.69. For emphasis, comparison, and contrast, we'll have the calculator seek out and shade all portions of the graph that are above the x-axis, since we're interested in the inequality $V(t) \geq 0$. This is done on the home screen, using the **2nd** **PRGM** **(DRAW) 7:Shade** feature. This feature requires six arguments, all separated by commas. These are (in order) lower function, upper function, left endpoint, right endpoint, pattern choice, and density. The calculator will then shade the area between the lower and upper functions, between the left and right endpoints, using the pattern and density given. The patterns are (1) vertical lines, (2) horizontal lines, (3) lines with negative slope, and (4) lines with positive slope. There are eight density settings, which instruct the calculator to shade anywhere from every pixel (1), to every eight pixels (8). Figure 4.70 shows the options we've selected, with the resulting graph shown in Figure 4.71. The friendly window makes it very easy to investigate the inequality further using the **TRACE** feature.

Figure 4.69

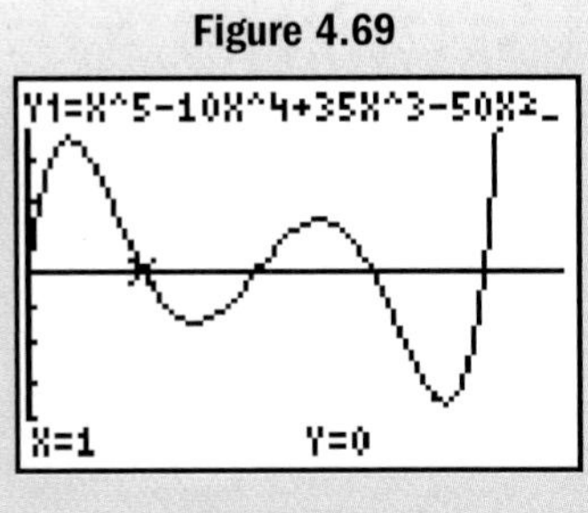

Figure 4.70

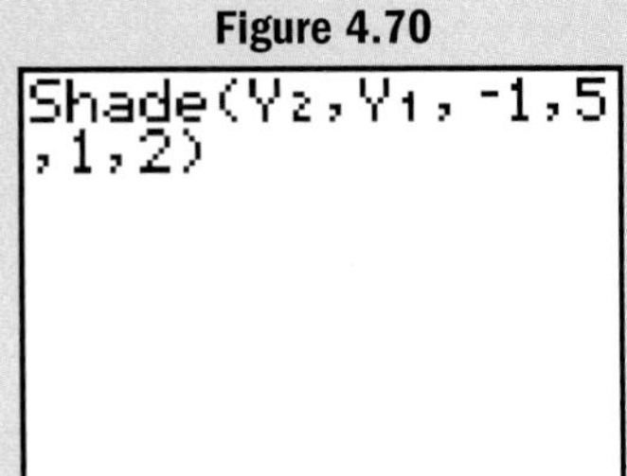

Figure 4.71

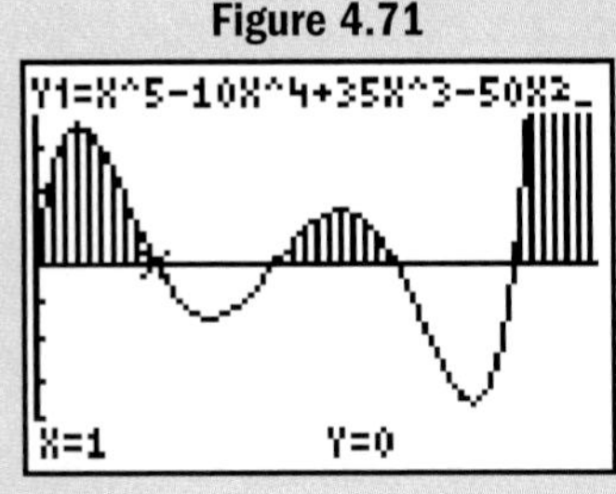

Use these ideas to visually study and explore the solution to the following inequality.

Exercise 1: Use window size $x \in [-4.7, 4.7]$; $y \in [-10, 20]$, **(DRAW) 7:Shade,** and **TRACE** to solve $P(x) < 0$ for $P(x) = x^4 + 1.1x^3 - 9.37x^2 - 4.523x + 16.4424$. $P(x) < 0$: $x \in (-3.1, -1.7) \cup (1.3, 2.4)$

4.7 EXERCISES

CONCEPTS AND VOCABULARY

Fill in each blank with the appropriate word or phrase. Carefully reread the section if needed.

1. To solve a polynomial or rational inequality, begin by plotting the location of all zeroes and __vertical__ asymptotes (if they exist), then consider the __multiplicity__ of each.

2. For strict inequalities, the zeroes are __excluded__ from the solution set. For non-strict inequalities, zeroes are __included__. The values at which a vertical asymptote occurs are always __excluded__.

3. If the inequality is not in function form, rewrite the expression with __0__ on one side. For rational expressions, combine the terms. Then consider $f(x) < 0$ or $f(x) > 0$ as indicated.

4. To solve a polynomial/rational inequality, it helps to find the sign of f in some interval. This can quickly be done using the __end behavior__ or __y-intercept__ of the function.

5. Compare/contrast the process for solving $x^2 - 3x - 4 \geq 0$ with $\frac{1}{x^2 - 3x - 4} \geq 0$. Are there similarities? What are the differences? Answers will vary.

6. Compare/contrast the process for solving $(x + 1)(x - 3)(x^2 + 1) > 0$ with $(x + 1)(x - 3) > 0$. Are there similarities? What are the differences? Answers will vary.

DEVELOPING YOUR SKILLS

Solve the inequality indicated using a number line and the behavior of the graph at each zero. Write all answers in interval notation.

7. $(x + 3)(x - 5) < 0$
8. $(x - 2)(x + 7) < 0$
9. $(x + 1)^2(x - 4) \geq 0$
10. $(x + 6)(x - 1)^2 \leq 0$
11. $(x + 2)^3(x - 2)^2(x - 4) \leq 0$
12. $(x - 1)^3(x + 2)^2(x - 3) \geq 0$
13. $x^2 + 4x + 1 < 0$
14. $x^2 - 6x + 4 > 0$
15. $x^3 + x^2 - 5x + 3 \leq 0$
16. $x^3 + x^2 - 8x - 12 \leq 0$
17. $x^3 - 7x + 6 > 0$
18. $x^3 - 13x + 12 > 0$
19. $x^4 - 10x^2 > -9$
20. $x^4 + 36 < 13x^2$
21. $x^4 - 9x^2 > 4x - 12$
22. $x^4 - 16 > 5x^3 - 20x$
23. $x^4 - 6x^3 \leq -8x^2 - 6x + 9$
24. $x^4 - 3x^2 + 8 \leq 4x^3 - 10x$
25. $f(x) = \frac{x + 3}{x - 2}$; $f(x) \leq 0$
26. $F(x) = \frac{x - 4}{x + 1}$; $F(x) \geq 0$
27. $g(x) = \frac{x + 1}{x^2 + 4x + 4}$; $g(x) < 0$
28. $G(x) = \frac{x - 3}{x^2 - 2x + 1}$; $G(x) > 0$
29. $\frac{2 - x}{x^2 - x - 6} \geq 0$
30. $\frac{1 - x}{x^2 - 2x - 8} \leq 0$
31. $\frac{2x - x^2}{x^2 + 4x - 5} < 0$
32. $\frac{x^2 + 3x}{x^2 - 2x - 3} > 0$
33. $\frac{x^2 - 4}{x^3 - 13x + 12} \geq 0$
34. $\frac{x^2 + x - 6}{x^3 - 7x + 6} \leq 0$
35. $\frac{x^2 + 5x - 14}{x^3 + x^2 - 5x + 3} > 0$
36. $\frac{x^2 + 2x - 8}{x^3 + 5x^2 + 3x - 9} < 0$
37. $\frac{2}{x - 2} \leq \frac{1}{x}$
38. $\frac{5}{x + 3} \geq \frac{3}{x}$
39. $\frac{x - 3}{x + 17} > \frac{1}{x - 1}$
40. $\frac{1}{x + 5} < \frac{x - 2}{x - 7}$
41. $\frac{x + 1}{x - 2} \geq \frac{x + 2}{x + 3}$
42. $\frac{x - 3}{x - 6} \leq \frac{x + 1}{x + 4}$
43. $\frac{x + 2}{x^2 + 9} > 0$
44. $\frac{x^2 + 4}{x - 3} < 0$
45. $\frac{x^3 + 1}{x^2 + 1} > 0$
46. $\frac{x^2 + 4}{x^3 - 8} < 0$

7. $x \in (-3, 5)$
8. $x \in (-7, 2)$
9. $x \in [4, \infty)$
10. $x \in (-\infty, -6] \cup \{1\}$
11. $x \in [-2, 4]$
12. $x \in (-\infty, 1] \cup [3, \infty)$
13. $x \in (-2 - \sqrt{3}, -2 + \sqrt{3})$
14. $x \in (-\infty, 3 - \sqrt{5}) \cup (3 + \sqrt{5}, \infty)$
15. $x \in (-\infty, -3] \cup \{1\}$
16. $x \in (-\infty, 3]$
17. $x \in (-3, 1) \cup (2, \infty)$
18. $x \in (-4, 1) \cup (3, \infty)$
19. $x \in (-\infty, -3) \cup (-1, 1) \cup (3, \infty)$
20. $x \in (-3, -2) \cup (2, 3)$
21. $x \in (-\infty, -2) \cup (-2, 1) \cup (3, \infty)$
22. $x \in (-\infty, -2) \cup (1, 2) \cup (4, \infty)$
23. $x \in [-1, 1] \cup \{3\}$
24. $x \in \{-1\} \cup [2, 4]$
25. $x \in [-3, 2)$
26. $x \in (-\infty, -1) \cup [4, \infty)$
27. $x \in (-\infty, -2) \cup (-2, -1)$
28. $x \in (3, \infty)$
29. $x \in (-\infty, -2) \cup [2, 3)$
30. $x \in (-2, 1] \cup (4, \infty)$
31. $x \in (-\infty, -5) \cup (0, 1) \cup (2, \infty)$
32. $x \in (-\infty, -3) \cup (-1, 0) \cup (3, \infty)$
33. $x \in (-4, -2] \cup (1, 2] \cup (3, \infty)$
34. $x \in (-\infty, 1)$
35. $x \in (-7, -3) \cup (2, \infty)$
36. $x \in (-\infty, -4) \cup (1, 2)$
37. $x \in (-\infty, -2] \cup (0, 2)$
38. $x \in (-3, 0) \cup [\frac{9}{2}, \infty)$
39. $x \in (-\infty, -17) \cup (-2, 1) \cup (7, \infty)$
40. $x \in (-\infty, -5) \cup (-3, 1) \cup (7, \infty)$
41. $x \in (-3, \frac{-7}{4}] \cup (2, \infty)$
42. $x \in (-\infty, -4) \cup [1, 6)$
43. $x \in (-2, \infty)$
44. $x \in (-\infty, 3)$
45. $x \in (-1, \infty)$
46. $x \in (-\infty, 2)$

47. $(-\infty, -3) \cup (3, \infty)$
48. $x \in (-4, -2) \cup (2, 5)$

47. $\dfrac{x^4 - 5x^2 - 36}{x^2 - 2x + 1} > 0$

48. $\dfrac{x^4 - 3x^2 - 4}{x^2 - x - 20} < 0$

Match the correct solution with the inequality and graph given.

49. $f(x) < 0$ b

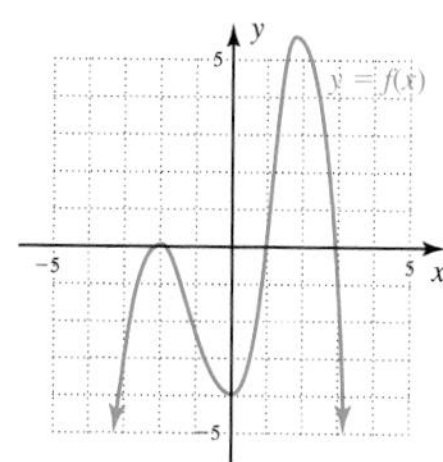

a. $x \in (-5, -2) \cup (3, 5)$

b. $x \in (-\infty, -2) \cup (-2, 1) \cup (3, \infty)$

c. $x \in (-\infty, -2) \cup (3, \infty)$

d. $x \in (-\infty, -2) \cup (-2, 1] \cup [3, \infty)$

e. none of these

50. $g(x) \geq 0$ d

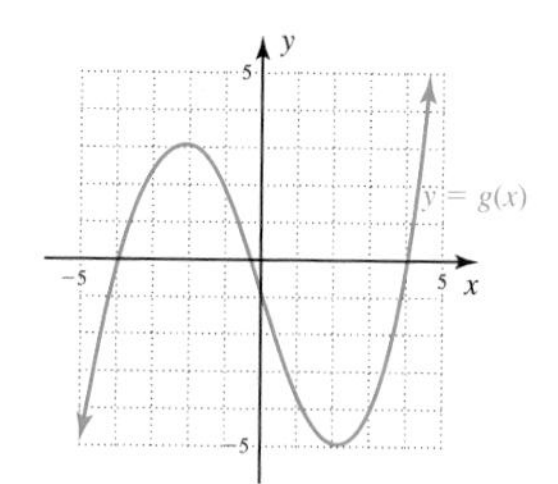

a. $x \in (-4, -0.5) \cup (4, \infty)$

b. $x \in [-0.5, 4] \cup [4, 5]$

c. $x \in (-\infty, -4) \cup (-0.5, 4)$

d. $x \in [-4, -0.5] \cup [4, \infty)$

e. none of these

51. $r(x) \geq 0$ b

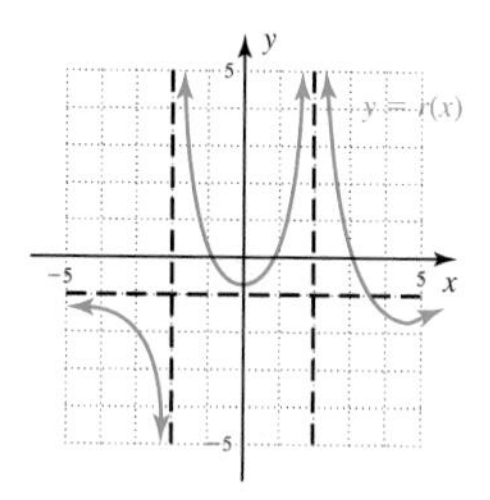

a. $x \in (-\infty, -2) \cup [-1, 1] \cup [3, \infty)$

b. $x \in (-2, -1] \cup [1, 2) \cup (2, 3]$

c. $x \in (-\infty, -2) \cup (2, \infty)$

d. $x \in (-2, -1) \cup (1, 2) \cup (2, 3]$

e. none of these

52. $R(x) \leq 0$ d

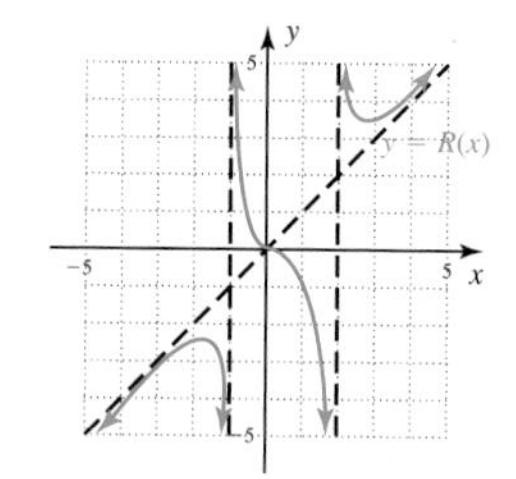

a. $x \in (-\infty, -1) \cup (0, 2)$

b. $x \in [1, 0] \cup (2, \infty]$

c. $x \in [-5, -1] \cup [2, 5]$

d. $x \in (-\infty, -1) \cup [0, 2)$

e. none of these

WORKING WITH FORMULAS

53. Discriminant of the reduced cubic $x^3 + px + q = 0$: $D = -(4p^3 + 27q^2)$

The discriminant of a cubic equation is less well known than that of the quadratic, but serves the same purpose. The discriminant of the reduced cubic is given by the formula shown, where p is the linear coefficient and q is the constant term. If $D > 0$, there will be three real and distinct roots. If $D = 0$, there are still three real roots, but one is a repeated root (multiplicity two). If $D < 0$, there are one real and two complex roots. Suppose we wish to study the family of cubic equations where $q = p + 1$.

a. Verify the resulting discriminant is $D = -(4p^3 + 27p^2 + 54p + 27)$.

b. Determine the values of p and q for which this family of equations has a repeated real root. In other words, solve the equation $-(4p^3 + 27p^2 + 54p + 27) = 0$ using the RRT and synthetic division to write D in completely factored form.

c. Use the factored form from part (b) to determine the values of p and q for which this family of equations has three real and distinct roots. In other words, solve $D > 0$.

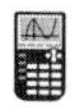

d. Verify the results of parts (b) and (c) on a graphing calculator.

53. a. verified
b. $D = -(p + 3)^2(p + \frac{3}{4})$, $p = -3, q = -2$; $p = \frac{-3}{4}, q = \frac{1}{4}$
c. $(-\infty, -3) \cup (-3, \frac{-3}{4})$
d. verified

54.

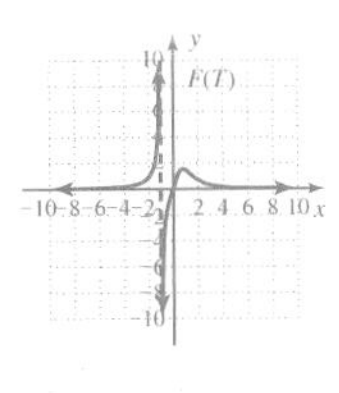

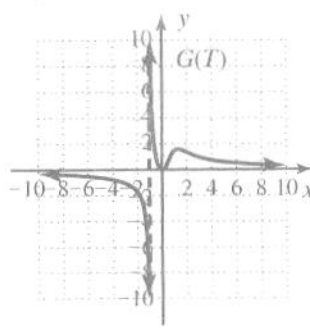

b. $(-\infty, -1) \cup (0, \infty)$
c. $(-1, 0) \cup (0, \infty)$
d. yes; $T = 0$ and $T = 1$

54. Coordinates for the folium of Descartes: $\begin{cases} x = \dfrac{3kT}{1 + T^3} \\ y = \dfrac{3kT^2}{1 + T^3} \end{cases}$

Folium of Descartes

The interesting relation shown here is called the folium (leaf) of Descartes. The folium is most often graphed using what are called *parametric equations*, in which the coordinates x and y are expressed in terms of the parameter T ("k" is a constant that affects the size of the leaf). Since each is an individual function, the x- and y-coordinates can be investigated individually, using $F(T) = \dfrac{3T}{1 + T^3}$ and $G(T) = \dfrac{3T^2}{1 + T^3}$ (assume $k = 1$ for now).

a. Graph each function using the techniques from this section.

b. According to your graph, for what values of T will the *x-coordinate* of the folium be positive? In other words, solve $F(T) = \dfrac{3T}{1 + T^3} > 0$.

c. For what values of T will the *y-coordinate* of the folium be positive? Solve $G(T) = \dfrac{3T^2}{1 + T^3} > 0$.

d. Will $F(T)$ ever be equal to $G(T)$? If so, for what values of T?

APPLICATIONS

Deflection of a beam: The amount of deflection in a rectangular wooden beam of length L ft can be approximated by $d(x) = k(x^3 - 3L^2x + 2L^3)$, where k is a constant that depends on the characteristics of the wood and the force applied, and x is the *distance from the unsupported end* of the beam ($x < L$).

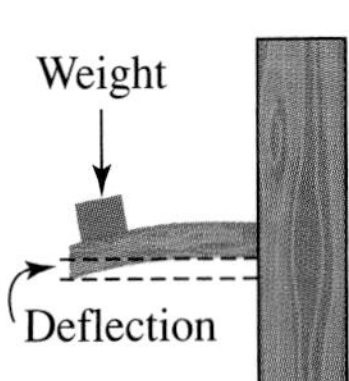

55. $d(x) = k(x^3 - 192x + 1024)$
a. $x \in (5, 8]$
b. 320 units
c. $x \in [0, 3)$
d. 2 ft

56. $d(x) = k(x^3 - 243x + 1458)$
a. $x \in (6, 9]$
b. 550 units
c. $x \in [0, 4)$
d. the longer beam gives greater deflection

57. $x \in [-3, 1] \cup [\frac{5}{2}, \infty)$
58. $x \in [-4, \frac{3}{2}] \cup [2, \infty)$
59. $x \in (-5, -2] \cup (7, \infty)$
60. $x \in (-\infty, -2) \cup [-1, 1] \cup (3, \infty)$

55. Find the equation for a beam 8 ft long and use it for the following:

a. For what distances x is the quantity $\dfrac{d(x)}{k}$ less than 189 units?

b. What is the amount of deflection 4 ft from the unsupported end ($x = 4$)?

c. For what distances x is the quantity $\dfrac{d(x)}{k}$ greater than 475 units?

d. If safety concerns prohibit a deflection of more than 648 units, how far from one end of the beam can the force be applied?

56. Find the equation for a beam 9 ft long and use it for the following:

a. For what distances x is the quantity $\dfrac{d(x)}{k}$ less than 216 units?

b. What is the amount of deflection 4 ft from the unsupported end ($x = 4$)?

c. For what distances x is the quantity $\dfrac{d(x)}{k}$ greater than 550 units?

d. Compare the answer to 55b with the answer to 56b. What can you conclude?

The domain of radical functions: The concepts in this chapter and section are often applied to find the domain of certain radical expressions. Recall that if n is an even number, the expression $\sqrt[n]{A}$ represents a real number only if $A \geq 0$. Use this idea to find the domain of the following functions.

57. $f(x) = \sqrt{2x^3 - x^2 - 16x + 15}$

58. $g(x) = \sqrt[4]{2x^3 + x^2 - 22x + 24}$

59. $p(x) = \sqrt[4]{\dfrac{x + 2}{x^2 - 2x - 35}}$

60. $q(x) = \sqrt{\dfrac{x^2 - 1}{x^2 - x - 6}}$

Average speed for a round-trip: Surprisingly, the average speed of a round-trip is *not* the sum of the average speed in each direction divided by two. For a fixed distance D, consider rate r_1 in time t_1 for one direction, and rate r_2 in time t_2 for the other, giving $r_1 = \frac{D}{t_1}$ and $r_2 = \frac{D}{t_2}$. The average speed for the round-trip is $R = \frac{2D}{t_1 + t_2}$.

61. The distance from St. Louis, Missouri, to Wentzville, Missouri, is approximately 80 mi. Suppose that Sione, due to the age of his vehicle, made the round-trip with an average speed of 40 mph.

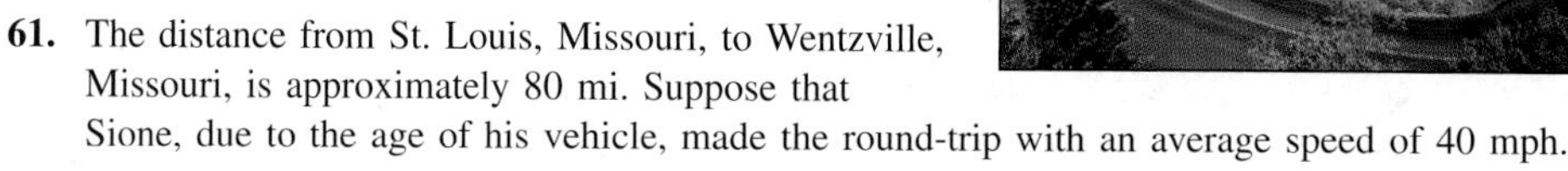

- **a.** Use the relationships above to verify that $r_2 = \frac{20r_1}{r_1 - 20}$.
- **b.** Discuss the meaning of the horizontal and vertical asymptotes in this context.
- **c.** Verify algebraically the speed returning would be greater than the speed going for $20 < r_1 < 40$. In other words, solve the inequality $\frac{20r_1}{r_1 - 20} > r_1$ using the ideas from this section.

62. The distance from Boston, Massachusetts, to Hartford, Connecticut, is approximately 100 mi. Suppose that Stella, due to excellent driving conditions, made the round-trip with an average speed of 60 mph.

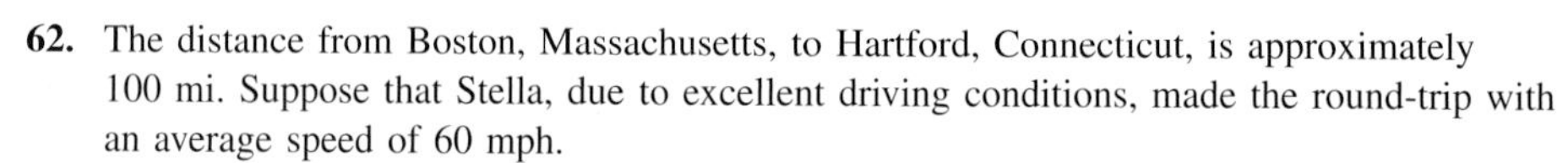

- **a.** Use the relationships above to verify that $r_2 = \frac{30r_1}{r_1 - 30}$.
- **b.** Discuss the meaning of the horizontal and vertical asymptotes in this context.
- **c.** Verify algebraically the speed returning would be greater than the speed going for $30 < r_1 < 60$. In other words, solve the inequality $\frac{30r_1}{r_1 - 30} > r_1$ using the ideas from this section.

Electrical resistance and temperature: The amount of electrical resistance R in a medium depends on the temperature, and for certain materials can be modeled by the equation $R(t) = 0.01t^2 + 0.1t + k$, where $R(t)$ is the resistance (in ohms Ω) at temperature t in degrees Celsius, and k is the resistance at $t = 0°C$.

63. Suppose $k = 30$ for a certain medium. Write the resistance equation and use it to answer the following.

- **a.** For what temperatures is the resistance less than 42 Ω?
- **b.** For what temperatures is the resistance greater than 36 Ω?
- **c.** If it becomes uneconomical to run electricity through the medium for resistances greater than 60 Ω, for what temperatures should the electricity generator be shut down?

64. Suppose $k = 20$. Write the resistance equation and solve the following.

- **a.** For what temperatures is the resistance less than 26 Ω?
- **b.** For what temperatures is the resistance greater than 40 Ω?
- **c.** If it becomes uneconomical to run electricity through the medium for resistances greater than 50 Ω, for what temperatures should the electricity generator be shut down?

65. Sum of consecutive squares: The sum of the first n squares $1^2 + 2^2 + 3^2 + \cdots + n^2$ is given by the formula $S(n) = \frac{2n^3 + 3n^2 + n}{6}$. Use the equation to solve the following inequalities.

- **a.** For what number of consecutive squares is $S(n) \geq 30$?

61. a. verified
b. horizontal: $r_2 = 20$, as r_1 increases, r_2 decreases to maintain $R = 40$
vertical: $r_1 = 20$, as r_1 decreases, r_2 increases to maintain $R = 40$
c. $r_1 \in (20, 40)$

62. a. verified
b. horizontal: $r_2 = 30$, as r_1 increases, r_2 decreases to maintain $R = 60$
vertical: $r_1 = 30$, as r_1 decreases, r_2 increases to maintain $R = 60$
c. $r_1 \in (30, 60)$

63. a. $(0°, 30°)$
b. $(20°, \infty)$
c. $(50°, \infty)$

64. a. $(0°, 20°)$
b. $(40°, \infty)$
c. $(50°, \infty)$

65. a. $n \geq 4$
b. $n \leq 9$
c. 13

66. a. $n \geq 4$
b. $n \leq 7$
c. 7
67. a. yes
b. the method of this section
c. lose critical values
68. a. yes, $x^2 \geq 0$
b. yes, $\frac{x^2}{x^2 + 1} \geq 0$

b. For what number of consecutive squares is $S(n) \leq 285$?

c. What is the maximum number of consecutive squares that can be summed without the result exceeding three digits?

66. Sum of consecutive cubes: The sum of the first n cubes $1^3 + 2^3 + 3^3 + \cdots + n^3$ is given by the formula $S(n) = \frac{n^4 + 2n^3 + n^2}{4}$. Use the equation to solve the following inequalities.

a. For what number of consecutive cubes is $S(n) \geq 100$?

b. For what number of consecutive cubes is $S(n) \leq 784$?

c. What is the maximum number of consecutive cubes that can be summed without the result exceeding three digits?

WRITING, RESEARCH, AND DECISION MAKING

67. Consider the inequality $\frac{x}{x - 2} < 3$. Solve the inequality in two ways: (1) using the techniques of this section, and (2) by multiplying both sides by $(x - 2)$ to clear denominators.

a. Do the solution sets differ?

b. Which solution is correct?

c. Discuss/explain why multiplying both sides of an inequality by a variable quantity can affect the solution set.

68. (a) Is it possible for the solution set of a polynomial inequality to be all real numbers? If not, discuss why. If so, provide an example. (b) Is it possible for the solution set of a rational inequality to be all real numbers? If not, discuss why. If so, provide an example.

EXTENDING THE CONCEPT

69. $x(x + 2)(x - 1)^2 > 0$; $\frac{x(x + 2)}{(x - 1)^2} > 0$
70. $x \in (-2, -1] \cup (2, \infty)$
71. $f'(x) > 0$ for $x \in (-2, 1) \cup (4, \infty)$
72. $r'(x) < 0$ for $x \in (2, 8) \cup (8, 14)$

69. Find one polynomial inequality and one rational inequality that have the solution $x \in (-\infty, -2) \cup (0, 1) \cup (1, \infty)$.

70. Without graphing the function (by analysis only), state the solution set for $\frac{x\sqrt{x^2 - 1}}{x^2 - 4} \geq 0$.

71. Using the tools of calculus, it can be shown that $f(x) = x^4 - 4x^3 - 12x^2 + 32x + 39$ is increasing in the intervals where $f'(x) = x^3 - 3x^2 - 6x + 8$ is positive. Solve the inequality $f'(x) > 0$ using the ideas from this section, then verify $f(x)\uparrow$ in these intervals by graphing f on a graphing calculator and using the **TRACE** feature.

72. Using the tools of calculus, it can be shown that $r(x) = \frac{x^2 - 3x - 4}{x - 8}$ is decreasing in the intervals where $r'(x) = \frac{x^2 - 16x + 28}{(x - 8)^2}$ is negative. Solve the inequality $r'(x) < 0$ using the ideas from this section, then verify $r(x)\downarrow$ in these intervals by graphing r on a graphing calculator and using the **TRACE** feature.

MAINTAINING YOUR SKILLS

74. yes, passes the horizontal line test
75.

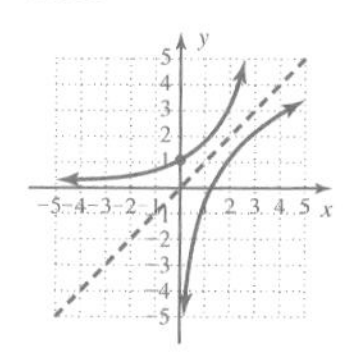

73. (3.3) Use the graph of $f(x)$ given to sketch the graph of $y = f(x + 2) - 3$.

74. (3.2) Is the function $f(x)$ given in Exercise 73 one-to-one? Discuss why or why not.

75. (3.2) Sketch the graph of $f^{-1}(x)$ for $f(x)$ as given in Exercise 73.

Exercise 73

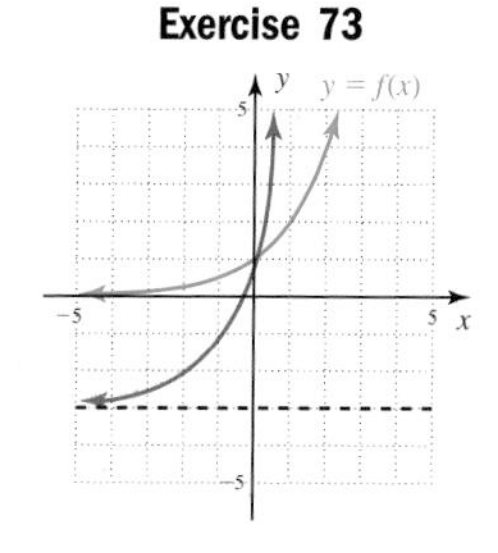

76. $g^{-1}(x) = \frac{x-2}{3}$

77. $x = 0$

78.

76. (3.2) Given $g(x) = 3x + 2$, find $g^{-1}(x)$ using the algebraic method.

77. (1.3) Solve the equation $\frac{1}{2}\sqrt{16 - x} - \frac{x}{2} = 2$. Check solutions in the original equation.

78. (1.2/2.5) Graph the solution set for the relation: $3x + 1 < 10$ *and* $x^2 - 3 < 1$.

▼ SUMMARY AND CONCEPT REVIEW

SECTION 4.1 Polynomial Long Division and Synthetic Division

KEY CONCEPTS

- For long division, write the dividend and divisor in standard form using placeholder zeroes as needed.
- At each stage of the division, the new multiplier is found using a ratio of leading terms.
- Group each partial product in parentheses, being sure to "distribute the negative" as you subtract.
- Synthetic division is an abbreviated form of long division, used when the divisor is of the form $x - r$. Only the coefficients of the dividend are used, since "standard form" ensures like place values are aligned.
- To divide by $x - r$, use r in the synthetic division; to divide by $x + r$, use $-r$. Drop the lead coefficient of the dividend into place, then multiply in the diagonal direction, place the product in the next column, and add in the vertical direction, continuing until the last column is reached.
- The sum in the right-most column will be the remainder, the numbers preceding it are the coefficients of the quotient polynomial. For $\frac{2x^3 + 5x^2 - 6x - 9}{x + 3}$:

−3⌋	2	5	−6	−9
		−6	3	9
	2	−1	−3	⌊0

$2x^3 + 5x^2 - 6x - 9 = (x + 3)(2x^2 - x - 3) + 0.$

- If a polynomial has a leading coefficient of −1 or 1, its linear factors must be of the form $x - p$, where p is a factor of the constant term.
- Synthetic division can also be used with an artificial coefficient "k" to help build polynomials with specific characteristics or values (end behavior, zeroes, y-intercept, and so on).

EXERCISES

Divide using long division and clearly identify the quotient and remainder:

1. $\frac{x^3 + 4x^2 - 5x - 6}{x - 2}$ **2.** $\frac{x^3 + 2x - 4}{x^2 - x}$

3. Use synthetic division to show that $x + 7$ is a factor of $2x^4 + 13x^3 - 6x^2 + 9x + 14$.

4. Compute the division and write the result in the two forms shown in Section 4.1: $\frac{x^3 - 4x + 5}{x - 2}$.

5. Use synthetic division and the principle of factorable polynomials to factor $P(x) = x^3 + 2x^2 - 11x - 12$.

6. Use synthetic division to find a value of k that will make $x + 4$ a factor of $x^3 + 3x^2 + k$.

1. $q(x) = x^2 + 6x + 7$; $R = 8$
2. $q(x) = x + 1$; $R = 3x - 4$
3.

−7⌋	2	13	−6	9	14
		−14	7	−7	−14
	2	−1	1	2	⌊0

Since $R = 0$, −7 is a root and $x + 7$ is a factor.

4. $x^3 - 4x + 5 = (x - 2)(x^2 + 2x) + 5$; $\frac{x^3 - 4x + 5}{x - 2} = x^2 + 2x + \frac{5}{x - 2}$
5. $(x + 4)(x + 1)(x - 3)$
6. $k = 16$

SECTION 4.2 The Remainder and Factor Theorems

KEY CONCEPTS

- The remainder theorem states: If a polynomial $P(x)$ is divided by $x - r$, the remainder will be identical to $P(r)$. The theorem can be used to evaluate polynomials at $x = r$.
- The factor theorem states: For a polynomial $P(x)$, if $P(r) = 0$, then $x = r$ is a zero of P and $(x - r)$ is a factor. Conversely, if $(x - r)$ is a factor of P, then $P(r) = 0$. The theorem can be used to build a polynomial P from its known zeroes.
- The remainder and factor theorems still apply when r is a complex number.
- Complex roots of a real polynomial must occur in conjugate pairs. If $a + bi$ is a root, $a - bi$ is also a root.
- When a polynomial is written in completely factored form with like factors combined and written as $(x - r)^m$, r is called a root of multiplicity m. If m is odd, r is a root of odd multiplicity; if m is even, r is a root of even multiplicity.
- A real polynomial of degree n will have exactly n roots (real and complex), counting roots of multiplicity.
- These relationships can be used as tools to help factor and graph polynomial functions.

EXERCISES

Use the remainder theorem.

7. Show that $x = \frac{1}{2}$ is a zero of $P(x) = 4x^3 + 8x^2 - 3x - 1$.
8. Show that $x = 3i$ is a zero of $P(x) = x^3 - 2x^2 + 9x - 18$.
9. Find $P(-7)$ given $P(x) = x^3 + 9x^2 + 13x - 10$.

Use the factor theorem.

10. Find a cubic polynomial with zeroes $x = 1$, $x = -\sqrt{5}$, and $x = \sqrt{5}$.
11. Find a fourth-degree polynomial with one real root, given $x = 1$ and $x = -2i$ are roots.
12. Use synthetic division and the remainder theorem to answer: At a busy shopping mall, customers are constantly coming and going. One summer afternoon during the hours from 12 o'clock noon to 6 in the evening, the number of customers in the mall could be modeled by $C(t) = 3t^3 - 28t^2 + 66t + 35$, where $C(t)$ is the number of customers (in tens), t hr after 12 noon.
    a. How many customers were in the mall at noon? Were more customers in the mall at 2 o'clock or at 3 o'clock P.M.? How many more?
    b. Was the mall busier at 1 o'clock (after lunch) or 6 o'clock (around dinner time)?

7.
$$\begin{array}{r|rrrr} \frac{1}{2} & 4 & 8 & -3 & -1 \\ & & 2 & 5 & 1 \\ \hline & 4 & 10 & 2 & 0 \end{array}$$
Since $R = 0$, $\frac{1}{2}$ is a root and $\left(x - \frac{1}{2}\right)$ is a factor.

8.
$$\begin{array}{r|rrrr} 3i & 1 & -2 & 9 & -18 \\ & & 3i & -9 - 6i & 18 \\ \hline & 1 & -2 + 3i & -6i & 0 \end{array}$$
Since $R = 0$, $3i$ is a root and $(x - 3i)$ is a factor.

9.
$$\begin{array}{r|rrrr} -7 & 1 & 9 & 13 & -10 \\ & & -7 & -14 & 7 \\ \hline & 1 & 2 & -1 & -3 \end{array}$$
$P(-7) = -3$

10. $P(x) = x^3 - x^2 - 5x + 5$
11. $C(x) = x^4 - 2x^3 + 5x^2 - 8x + 4$
12. a. $C(0) = 350$ customers, more at 2 P.M., 170
    b. Busier at 1 P.M. $760 > 710$

SECTION 4.3 The Zeroes of Polynomial Functions

KEY CONCEPTS

- The fundamental theorem of algebra states: Every complex polynomial of degree $n \geq 1$ has at least one complex root. This guarantees the existence of a solution and leads to other important results.
- The linear factorization theorem states: Every complex polynomial of degree $n \geq 1$ has exactly n linear factors, and can be written in the form $P(x) = a(x - c_1)(x - c_2)\cdots(x - c_n)$, where $a \neq 0$ and $c_1, c_2, \ldots, c_n$ are (not necessarily distinct) complex numbers.
- A corollary to the preceding theorem states: If f is a polynomial with real coefficients, it can be factored into linear factors (not necessarily distinct) and irreducible quadratic factors having real coefficients.

- The intermediate value theorem states: Given f is a polynomial with real coefficients, if $f(a)$ and $f(b)$ have opposite signs, then there is at least one r between a and b such that $f(r) = 0$. In other words, for $f(a) < 0$ the graph is below the x-axis, for $f(b) > 0$ the graph is above the x-axis, and because polynomials are continuous the graph must *cross* the x-axis at some point r, where $a < r < b$.
- The rational roots theorem states: If a real polynomial $f(x)$ has integer coefficients, rational roots must be of the form $\frac{p}{q}$, where p is a factor of the constant term and q is a factor of the lead coefficient.
- Descartes's rule of signs, the upper and lower bounds property, the tests for -1 and 1, synthetic division, and graphing technology can all be used in conjunction with the rational roots theorem to factor, solve, and graph polynomial equations.

EXERCISES

13. Given $x = 3i$ is a root of $C(x)$, find the other two roots: $C(x) = x^3 - 8ix^2 - 19x + 12i$.

14. Use the intermediate value theorem, along with synthetic division and the remainder theorem, to identify two intervals from the following list that contain a root of $f(x) = x^4 - 3x^3 - 8x^2 + 12x + 6$: $[-2, -1]$, $[1, 2]$, $[2, 3]$, and $[4, 5]$.

15. Is $x = -2$ a lower bound on the zeroes of $f(x) = x^3 - 7x + 2$? Discuss why or why not. Is $x = -3$ a lower bound?

16. Use Descartes's rule of signs to discuss the possible number of positive, negative, and complex roots of $g(x) = x^4 + 3x^3 - 2x^2 - x - 30$. Then identify which combination is correct using a graphing calculator.

17. Use any of the mathematical tools discussed in this section to write $P(x) = 2x^3 - 3x^2 - 17x - 12$ in completely factored form, then state all zeroes of P.

18. Use the rational roots theorem and synthetic division to show that $h(x) = x^4 - 7x^2 - 2x + 3$ has no rational roots.

13. $x = i, x = 4i$

14. Zeroes are in $[1, 2]$ and $[4, 5]$.

15. No, -2 is not since last row does not alternate in sign. Yes, -3 is a lower bound.

16. $g(x)$ has one variation in sign $\longrightarrow$ 1 pos root; $g(-x)$ has three variations in sign $\longrightarrow$ 3 or 1 neg root

Pos	Neg	Complex
1	3	0
1	1	2

A grapher shows the second row is correct.

17. $P(x) = (x - 4)(x + 1)(2x + 3)$; $x = 4, -1, \frac{-3}{2}$

18. The possibilities are ± 1 and ± 3, none give a zero remainder. Therefore, h has no rational roots.

SECTION 4.4 Graphing Polynomial Functions

KEY CONCEPTS

- If the degree of a polynomial is odd, the ends of its graph will point in opposite directions (as $|x| \to \infty$). If the degree is even, the ends will point in the same direction. The sign of the lead coefficient determines the actual behavior.
- The "behavior" of a polynomial graph near its zeroes is determined by the multiplicity of the zero. For any factor $(x - r)^m$, the graph will "cross through" the x-axis if m is odd and "bounce off" the x-axis if m is even. The larger the value of m, the flatter (more compressed) the graph near the zero.
- A polynomial of degree n has *at most* $n - 1$ turning points. The precise points at which these turning points occur are the local maximums or local minimums of the function.
- To "round-out" a graph, locate additional *midinterval points* using values between the known x-intercepts.
- These ideas help to establish the *Guidelines for Graphing Polynomial Functions*. See page 413.

EXERCISES

State the degree, end behavior, and y-intercept, but do not graph.

19. $f(x) = -3x^5 + 2x^4 + 9x - 4$

20. $g(x) = (x - 1)(x + 2)^2(x - 2)$

19. degree 5; up/down; $(0, -4)$

20. degree 4; up/up; $(0, 8)$

21. 22.

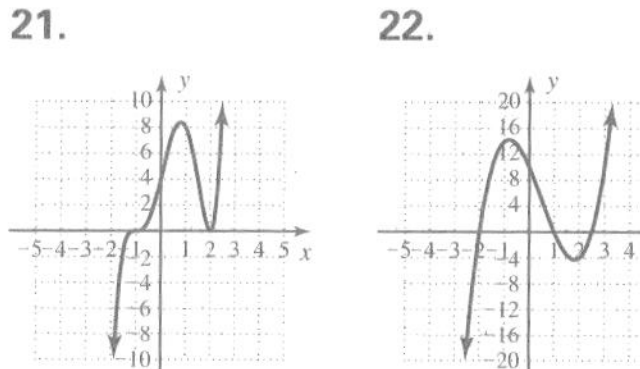

23.

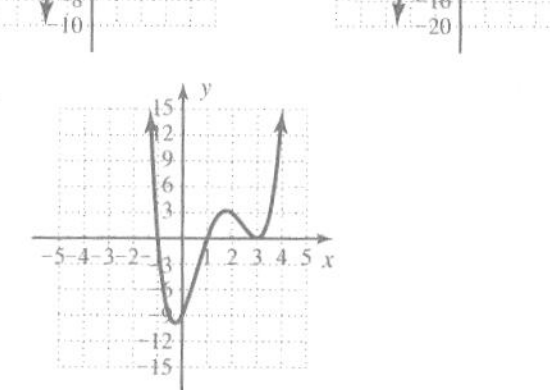

Graph using the *Guidelines for Graphing Polynomials.*

21. $p(x) = (x + 1)^3(x - 2)^2$ **22.** $q(x) = 2x^3 - 3x^2 - 9x + 10$

23. $h(x) = x^4 - 6x^3 + 8x^2 + 6x - 9$

24. For the graph of $P(x)$ shown: (a) state whether the degree of P is even or odd, (b) use the graph to locate the zeroes of P and state whether their multiplicity is even or odd, and (c) find the minimum possible degree of P and write it in factored form. Assume all zeroes are real.

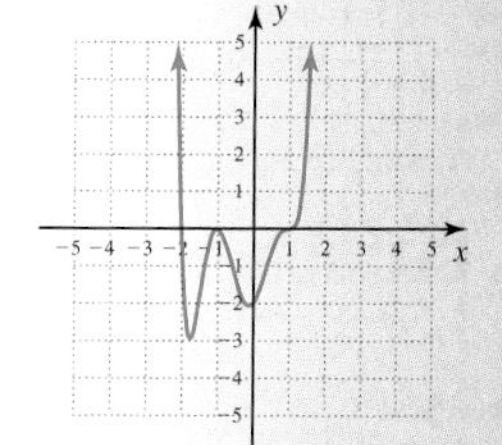

24. a. even
b. $x = -2$, odd; $x = -1$, even; $x = 1$, odd
c. deg 6: $P(x) = (x + 2)(x + 1)^2(x - 1)^3$

SECTION 4.5 Graphing Rational Functions

KEY CONCEPTS

- A rational function is one of the form $r(x) = \frac{f(x)}{g(x)}$, where f and g are polynomials and $g(x) \neq 0$.
- The domain of r is all real numbers, except the zeroes of g.
- A vertical asymptote occurs at the zeroes of g, creating a nonremovable break in the graph.
- If zero is in the domain of r, substitute $x = 0$ to find the y-intercept. The x-intercepts are the zeroes of f (if they exist).
- In certain cases, applying polynomial division to a rational function enables us to rewrite a rational function in "shiftable form," meaning we can graph the function using transformations of $y = \frac{1}{x}$ or $y = \frac{1}{x^2}$.
- If the degree of the numerator is *less than* the degree of the denominator, the line $y = 0$ (x-axis) is a horizontal asymptote. If their degrees *are equal,* a horizontal asymptote is found at $y = \frac{a}{b}$, where a and b are the lead coefficients of the numerator and denominator respectively.
- These ideas help to establish the *Guidelines for Graphing Rational Functions.* See page 428.

25. a. $\{x | x \in \mathbb{R}; x \neq -1, 4\}$
b. HA: $y = 1$;
VA: $x = -1, x = 4$
c. $r(0) = \frac{9}{4}$ (y-intercept);
$x = -3, 3$ (x-intercepts)
d. $r(1) = \frac{4}{3}$

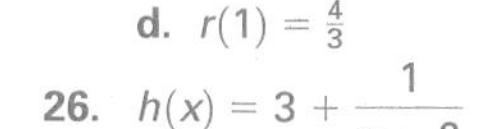

26. $h(x) = 3 + \frac{1}{x - 2}$

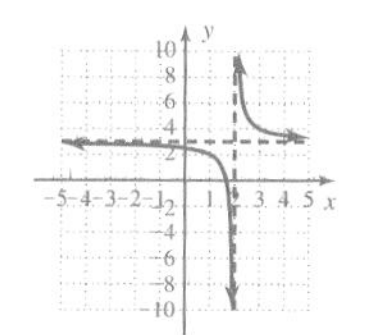

27.

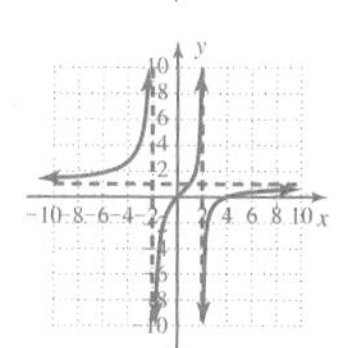

28.

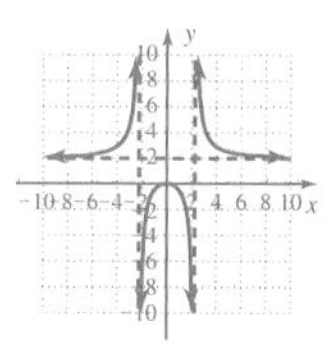

29. $r(x) = \frac{x^2 - x - 12}{x^2 - x - 6}$;
$r(0) = 2$

EXERCISES

25. For the function $r(x) = \frac{x^2 - 9}{x^2 - 3x - 4}$, state the following but do not graph: (a) domain, (b) equations of the horizontal and vertical asymptotes, (c) the x- and y-intercept(s), and (d) the value of $r(1)$.

26. Use synthetic division to rewrite $h(x) = \frac{3x - 5}{x - 2}$, then graph the result using a transformation of $y = \frac{1}{x}$.

Graph using the *Guidelines for Graphing Rational Functions.*

27. $r(x) = \frac{x^2 - 4x}{x^2 - 4}$ **28.** $t(x) = \frac{2x^2}{x^2 - 5}$

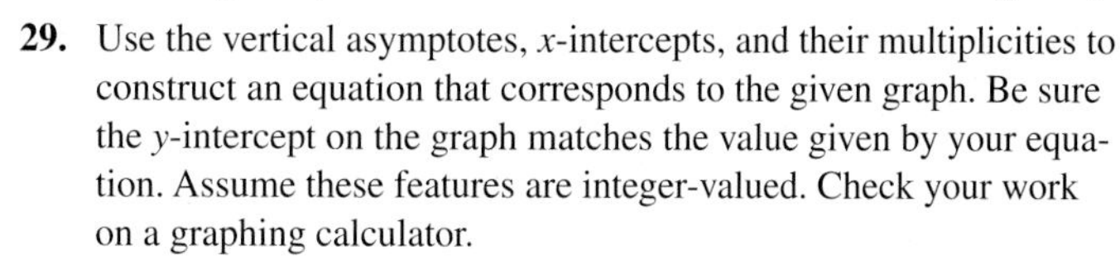

29. Use the vertical asymptotes, x-intercepts, and their multiplicities to construct an equation that corresponds to the given graph. Be sure the y-intercept on the graph matches the value given by your equation. Assume these features are integer-valued. Check your work on a graphing calculator.

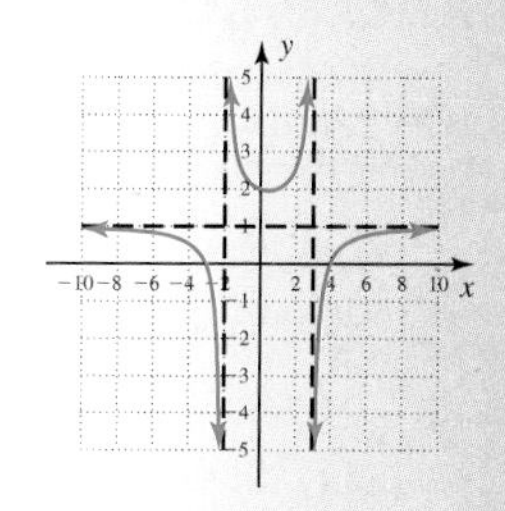

30. a. $y = 15$; as $|x| \to \infty$ $A(x) \to 15^+$. As production increases, average cost decreases and approaches 15.
b. $x > 2000$

30. The average cost of producing a popular board game is given by the function $A(x) = \dfrac{5000 + 15x}{x}$; $x \geq 1000$. (a) Identify the horizontal asymptote of the function and explain its meaning in this context. (b) To be profitable, management believes the average cost must be below \$17.50. What levels of production will make the company profitable?

SECTION 4.6 Additional Insights into Rational Functions

31.

$$H(x) = \begin{cases} \dfrac{x^2 - 3x - 4}{x + 1} & x \neq -1 \\ -5 & x = -1 \end{cases}$$

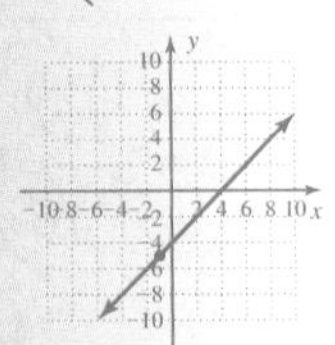

32. **33.**

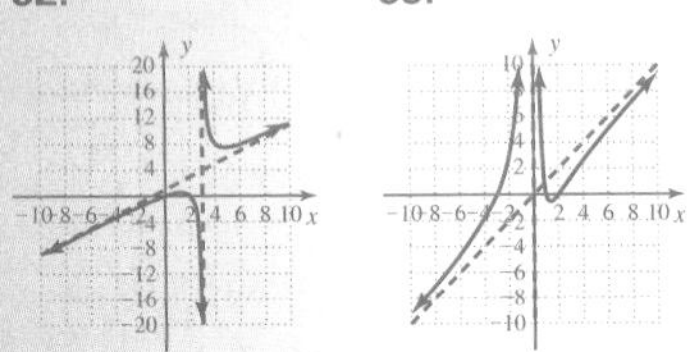

34. factored form
$(x + 4)(x - 1)(x - 2) > 0$

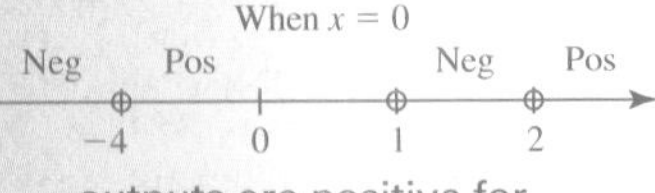

outputs are positive for $x \in (-4, 1) \cup (2, \infty)$

KEY CONCEPTS

- When the numerator and denominator of a rational function h share the common factor $(x - r)$, the graph will have a removable discontinuity (a hole or gap) at $x = r$. The discontinuity can be "removed" (repaired) by redefining h using a piecewise-defined function.
- If no common factors exist and the numerator's degree is greater than the denominator's, the result is an oblique or nonlinear asymptote, as determined by the quotient polynomial $q(x)$.
- If no common factors exist and the numerator's degree is greater by 1, the result is a linear, oblique asymptote. If the numerator's degree is greater by 2, the result is a parabolic asymptote.
- The *Guidelines for Graphing Rational Functions* still apply.

EXERCISES

31. Sketch the graph of $h(x) = \dfrac{x^2 - 3x - 4}{x + 1}$. If there is a removable discontinuity, repair the break by redefining h using an appropriate piecewise-defined function.

Graph the functions using the *Guidelines for Graphing Rational Functions.*

32. $h(x) = \dfrac{x^2 - 2x}{x - 3}$ **33.** $t(x) = \dfrac{x^3 - 7x + 6}{x^2}$

SECTION 4.7 Polynomial and Rational Inequalities—An Analytical View

35.
$\dfrac{x^2 - 3x - 10}{x - 2} = \dfrac{(x - 5)(x + 2)}{x - 2} \geq 0$

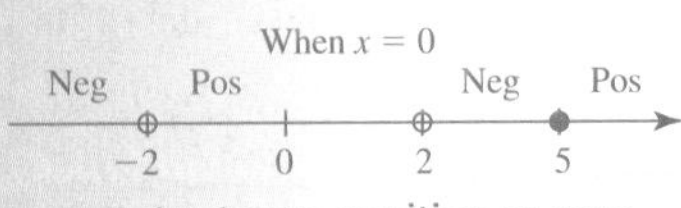

outputs are positive or zero for $x \in [-2, 2) \cup [5, \infty)$

36. $\dfrac{(x + 2)(x - 1)}{x(x - 2)} \leq 0$

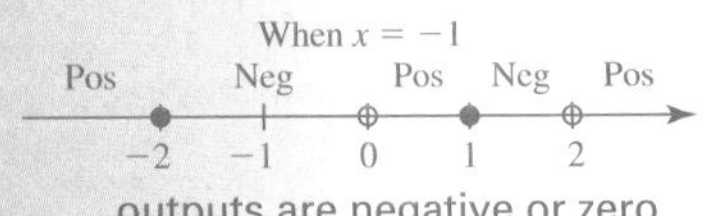

outputs are negative or zero for $x \in [-2, 0) \cup [1, 2)$

KEY CONCEPTS

- To solve polynomial inequalities, write $P(x)$ in factored form and note the multiplicity of real zeroes.
- Plot real zeroes on a number line. The graph will cross the x-axis at zeroes of odd multiplicity, and bounce off the axis at zeroes of even multiplicity.
- Use the end behavior, the y-intercept, or a test point to determine the sign of P in a given interval, then label all other intervals as $P(x) > 0$ or $P(x) < 0$ by analyzing the multiplicity of neighboring zeroes.
- The solution process for rational inequalities and polynomial inequalities is virtually identical, considering that vertical asymptotes also create intervals where function values may change sign, depending on their multiplicity

EXERCISES

Solve each inequality indicated using a number line and the behavior of the graph at each zero.

34. $x^3 + x^2 > 10x - 8$ **35.** $\dfrac{x^2 - 3x - 10}{x - 2} \geq 0$ **36.** $\dfrac{x}{x - 2} \leq \dfrac{-1}{x}$

MIXED REVIEW

1. $q(x) = x^2 - 5$; $R = 8$
2. $q(x) = x^3 - 2x^2 + x + 3$; $R = -7$
3. a, c, d
4.

$3i$	1	$-2i$	$3 + 2i$	6
		$3i$	-3	-6
	1	$1i$	$2i$	0

Since $R = 0$, $x = 3i$ is a root.

5. $k = 6$
6. a. $P(-1) = 42$
   b. $P(1) = -26$
   c. $P(5) = 6$
7. $x^3 - 3x^2 + 25x - 75$
8. $P(x) = x^2 - 4x + 13$
9. $x = 9$; $x = \frac{8}{3}$
10. $P(x) = (x - 2)(x + 1)(x^2 + 9)$; $x = 2, x = -1, x = -3i, x = 3i$

11.

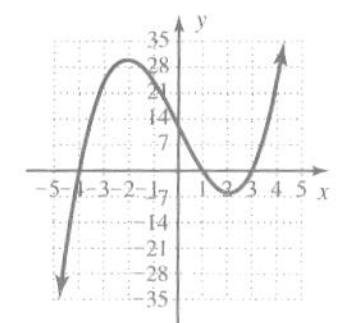

12.

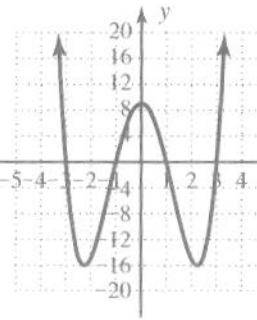
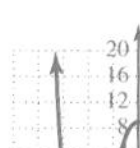

13.

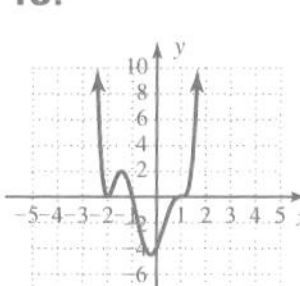

14.

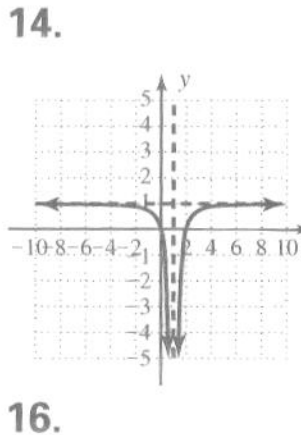

15.

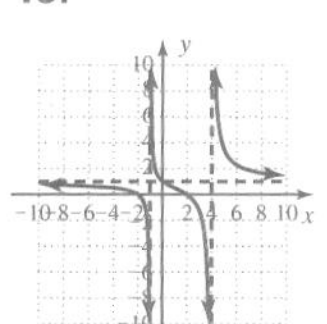

16.

17.

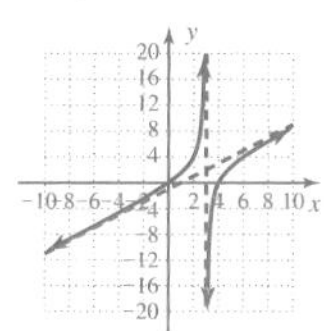

18. $x \in (-\infty, 3) \cup (-2, 2)$
19. $x \in (-2, 0) \cup [6, \infty)$
20. a. $V(x) = (24 - 2x)(16 - 2x)(x)$ $= 4x^3 - 80x^2 + 384x$
    b. $512 = 4x^3 - 80x^2 + 384x$; $0 = x^3 - 20x^2 + 96x - 128$
    c. for $0 < x < 16$, possible rational zeroes are 1, 2, 4, and 8
    d. $x = 4$
    e. $x = 8 - 4\sqrt{2} \approx 2.34$ in.

1. Divide using long division and name the quotient and remainder: $\dfrac{x^3 + 3x^2 - 5x - 7}{x + 3}$.

2. Divide using synthetic division and name the quotient and remainder: $\dfrac{x^4 - 3x^2 + 5x - 1}{x + 2}$.

Use synthetic division and the remainder theorem to complete Exercises 3 to 6.

3. State which of the following *are not factors* of $x^3 - 9x^2 + 2x + 48$: (a) $(x + 6)$, (b) $(x - 8)$, (c) $(x - 12)$, (d) $(x - 4)$, (e) $(x + 2)$.

4. Show that $x = 3i$ is a zero of $C(x) = x^3 - 2ix^2 + (3 + 2i)x + 6$.

5. Find the value of k that makes $(x + 2)$ a factor of $x^3 + 4x^2 + 7x + k$.

6. Given $P(x) = 6x^3 - 23x^2 - 40x + 31$, find (a) $P(-1)$, (b) $P(1)$, and (c) $P(5)$.

Use the factor theorem to complete Exercises 7 and 8.

7. Find a polynomial of degree 3 with roots $x = 3$ and $x = -5i$.

8. Find a polynomial of degree 2 with $x = 2 - 3i$ as one of the roots.

9. According to the rational roots theorem, which of the following *cannot be* roots of $6x^3 + x^2 - 20x - 12 = 0$? $x = 9$ $x = -3$ $x = \frac{3}{2}$ $x = \frac{8}{3}$ $x = -\frac{2}{3}$

10. Use the rational roots theorem to write P in completely factored form. Then state all zeroes of P, real and complex. $P(x) = x^4 - x^3 + 7x^2 - 9x - 18$.

Use the *Guidelines for Graphing Polynomials* to complete Exercises 11 to 13.

11. $f(x) = x^3 - 13x + 12$

12. $g(x) = x^4 - 10x^2 + 9$

13. $h(x) = (x - 1)^3(x + 2)^2(x + 1)$

Use the *Guidelines for Graphing Rational Functions* to complete Exercises 14 to 17.

14. $p(x) = \dfrac{x^2 - 2x}{x^2 - 2x + 1}$

15. $q(x) = \dfrac{x^2 - 4}{x^2 - 3x - 4}$

16. $r(x) = \dfrac{x^3 - 13x + 12}{x^2}$ (see Exercise 11)

17. $y = \dfrac{x^2 - 4x}{x - 3}$

Solve each inequality.

18. $x^3 - 4x < 12 - 3x^2$

19. $\dfrac{4}{x + 2} \geq \dfrac{3}{x}$

20. An open, rectangular box is to be made from a 24-in. by 16-in. piece of sheet metal, by cutting a square from each corner and folding up the sides.

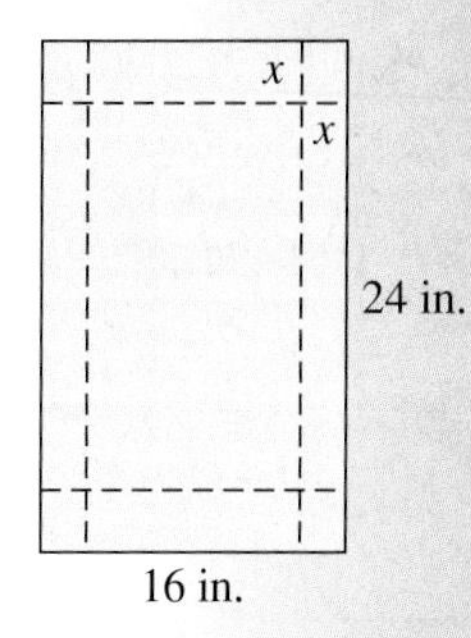

a. Show that the resulting volume is given by $V(x) = 4x^3 - 80x^2 + 384x$.

b. Show that for a desired volume of 512 in^3, the height "x" of the box can be found by solving $x^3 - 20x^2 + 96x - 128 = 0$.

c. According to the rational roots theorem *and the context of this application,* what are the possible rational zeroes for this equation?

d. Find the *rational zero* x (the height) that gives the box a volume of 512 in^3.

e. Use the zero from part (d) and synthetic division to help find the (positive) *irrational zero* x that also gives the box a volume of 512 in^3. Round the solution to hundredths.

PRACTICE TEST

1. Compute the quotient using long division: $\dfrac{x^3 - 3x^2 + 5x - 2}{x^2 + 2x + 1}$.

2. Find the quotient and remainder using synthetic division: $\dfrac{x^3 + 4x^2 - 5x - 20}{x + 2}$.

3. Use synthetic division to find a value of k that gives a remainder of 3: $\dfrac{x^4 + 4x^2 - 3x + k}{x + 2}$.

4. Use the remainder theorem to show $(x + 3)$ is a factor of $x^4 - 15x^2 - 10x + 24$.

5. Given $f(x) = 2x^3 + 4x^2 - 5x + 2$, find the value of $f(-3)$ using synthetic division and the remainder theorem.

6. Given $x = 2$ and $x = 3i$ are two roots of a real polynomial $P(x)$ with degree 3. Use the factor theorem to find $P(x)$.

7. Factor the polynomial and state the multiplicity of each root: $Q(x) = (x^2 - 3x + 2)(x^3 - 2x^2 - x + 2)$.

8. Given $x = 3i$ is a zero of $C(x)$, use synthetic division to write C in completely factored form: $C(x) = x^3 + (6 - 3i)x^2 + (8 - 18i)x - 24i$.

9. Given $C(x) = x^4 + x^3 + 7x^2 + 9x - 18$, (a) use the rational roots theorem to list all possible rational roots; (b) apply Descartes's rule of signs to count the number of possible positive, negative, and complex roots; and (c) use this information along with the tests for 1 and -1, synthetic division, and the factor theorem to factor C completely.

10. Over a 10-year period, the balance of payments (deficit versus surplus) for a small county was modeled by the function $f(x) = \frac{1}{2}x^3 - 7x^2 + 28x - 32$, where $x = 0$ corresponds to 1990 and $f(x)$ is the deficit or surplus in millions of dollars. (a) Use the rational roots theorem and synthetic division to find the years the county "broke even" (debt = surplus = 0) from 1990 to 2000. (b) How many years did the county run a surplus during this period? (c) What was the surplus/deficit in 1993?

11. Sketch the graph of $f(x) = (x - 3)(x + 1)^3(x + 2)^2$ using the degree, end behavior, x- and y-intercepts, zeroes of multiplicity, and a few "midinterval" points.

12. Use the *Guidelines for Graphing Polynomials* to graph $g(x) = x^4 - 9x^2 - 4x + 12$.

13. Use the *Guidelines for Graphing Rational Functions* to graph $h(x) = \dfrac{x - 2}{x^2 - 3x - 4}$.

14. Suppose the cost of cleaning contaminated soil from a dump site is modeled by $C(x) = \dfrac{300x}{100 - x}$, where $C(x)$ is the cost (in \$1000s) to remove x% of the contaminants. Graph using $x \in [0, 100]$, and use the graph to answer the following questions.
    a. What is the significance of the *vertical asymptote* (what does it mean in this context)?
    b. If EPA regulations are changed so that 85% of the contaminants must be removed, instead of the 80% previously required, how much will the new regulations cost the company? Compare the cost of the 5% increase from 80% to 85% with the cost of the 5% increase from 90% to 95%. What do you notice?
    c. What percent of the pollutants can be removed if the company budgets \$2,200,000?

Graph using the *Guidelines for Graphing Rational Functions.*

15. $r(x) = \dfrac{x^3 - x^2 - 9x + 9}{x^2}$

16. $R(x) = \dfrac{x^3 + 7x - 6}{x^2 - 4}$

17. Find the level of production that will minimize the average cost of an item, if production costs are modeled by $C(x) = 2x^2 + 25x + 128$, where $C(x)$ is the cost to manufacture x hundred items. 800

---

1. $x - 5 + \dfrac{14x + 3}{x^2 + 2x + 1}$
2. $x^2 + 2x - 9 + \dfrac{-2}{x + 2}$
3. $k = -35$
4.

$\underline{-3}|$	1	0	−15	−10	24	
		−3	9	18	−24	
	1	−3	−6	8	0	$R = 0$✓

5. -1
6. $P(x) = x^3 - 2x^2 + 9x - 18$
7. $Q(x) = (x - 2)^2(x - 1)^2(x + 1)$, 2 mult 2, 1 mult 2, −1 mult 1
8. $C(x) = (x - 3i)(x + 4)(x + 2)$
9. a. $\dfrac{\pm 1, \pm 18, \pm 2, \pm 9, \pm 3, \pm 6}{\pm 1}$
   b. 1 positive root, 3 or 1 negative roots; 2 or 0 complex roots
   c. $C(x) = (x + 2)(x - 1)(x - 3i)(x + 3i)$
10. a. 1992, 1994, 1998
    b. 4 yr
    c. surplus of \$2.5 million
11.

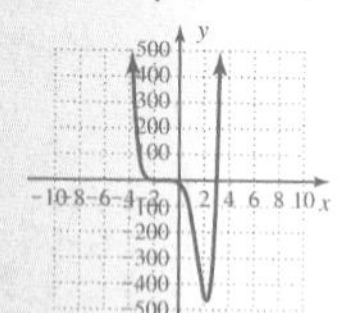

12.

13.

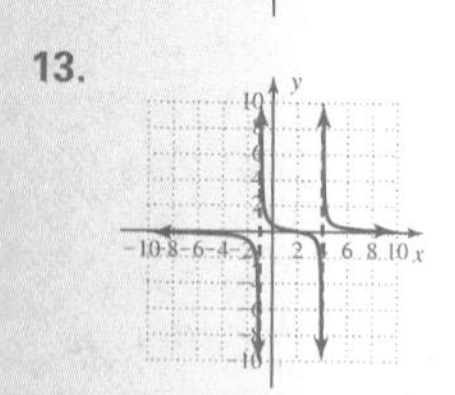

14. a. removal of 100% of the contaminants
    b. \$1,700,000 \$500,000 \$3,000,000
    c. 88%
15. 16.

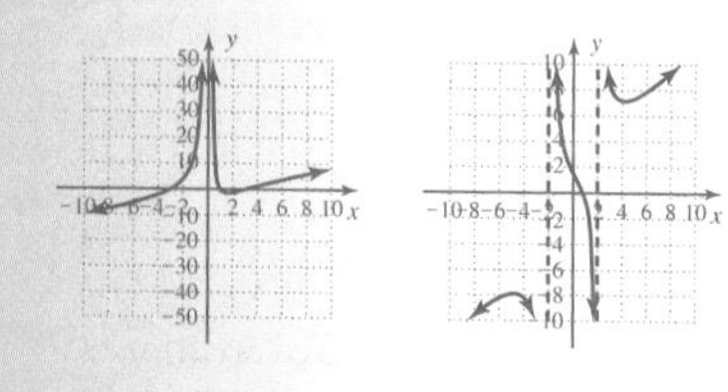

18. $x \in (-\infty, -3] \cup [-1, 4]$
19. $x \in (-\infty, -4) \cup (0, 2)$
20. a.

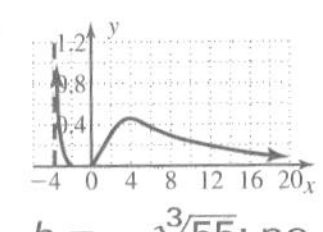

b. $h = -\sqrt[3]{55}$; no
c. 28.6% 29.6%
d. 12 hr
e. 4 hr 43.7%
f. The chemical will eventually disappear from the bloodstream.

Solve the inequalities using a number line and the behavior of the graph at the zeroes and asymptotes (in the case of rational functions).

18. $x^3 - 13x \le 12$

19. $\dfrac{3}{x-2} < \dfrac{2}{x}$

20. Suppose the concentration of a chemical in the bloodstream of a large animal h hours after injection into muscle tissue is modeled by the formula $C(h) = \dfrac{2h^2 + 5h}{h^3 + 55}$.

- **a.** Sketch a graph of the function for the intervals $x \in [-5, 20]$, $y \in [0, 1]$.
- **b.** Where is the vertical asymptote? Does it play a role in this context?
- **c.** What is the concentration after 2 hr? After 8 hr?
- **d.** How long does it take the concentration to fall below 20% $[C(h) < 0.2]$?
- **e.** When does the maximum concentration of the chemical occur? What is this maximum?
- **f.** Describe the significance of the horizontal asymptote in this context.

CALCULATOR EXPLORATION AND DISCOVERY

Complex Roots, Repeated Roots, and Inequalities

The keystrokes discussed apply to a TI-84 Plus model. Please consult your manual or our Internet site for other models.

This *Calculator Exploration and Discovery* will explore the relationship between the solution of a polynomial (or rational) inequality and the complex roots and repeated roots of the related function. After all, if complex roots can never create an x-intercept, how do they affect the function? And if a root of even multiplicity never crosses the x-axis (always bounces), can it still affect a nonstrict (*less than or equal to* or *greater than or equal to*) inequality? These are interesting and important questions, with numerous avenues of exploration. To begin, consider the function $Y_1 = (x + 3)^2(x^3 - 1)$. In completely factored form $Y_1 = (x + 3)^2(x - 1)(x^2 + x + 1)$, a polynomial function of degree 5 with two real roots (one repeated), two complex roots (the quadratic factor is irreducible), and after viewing the graph on Figure 4.72, four turning points. From the graph (or by analysis), we have $Y_1 \le 0$ for $x \le 1$. Now let's consider $Y_2 = (x + 3)^2(x - 1)$, the same function as Y_1, less the quadratic factor. Since complex roots never "cross the x-axis" anyway, the removal of this factor *cannot affect the solution set of the inequality!* But how does it affect the function? Y_2 is now a function of degree three, with three real roots (one repeated) and only two turning points (Figure 4.73). But even so, the solution to $Y_2 \le 0$ is the same as for $Y_1 \le 0$: $x \le 1$. Finally, let's look at $Y_3 = x - 1$, the same function as Y_2 but with the repeated root removed. The key here is to notice that since $(x - 3)^2$ will be nonnegative for any value of x, it too does not change the solution set of the "less than or equal to inequality," only the shape of the graph. Y_3 is a function of degree 1, with one real root and no turning points, *but with the same solution interval as Y_2 and Y_3*: $x \le 1$ (see Figure 4.74).

Figure 4.72

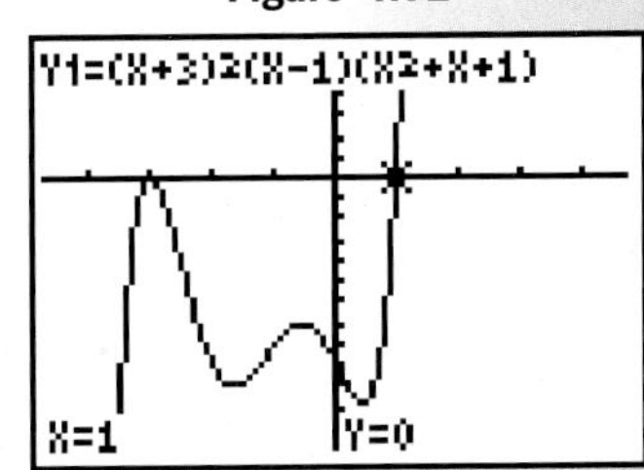

Figure 4.73

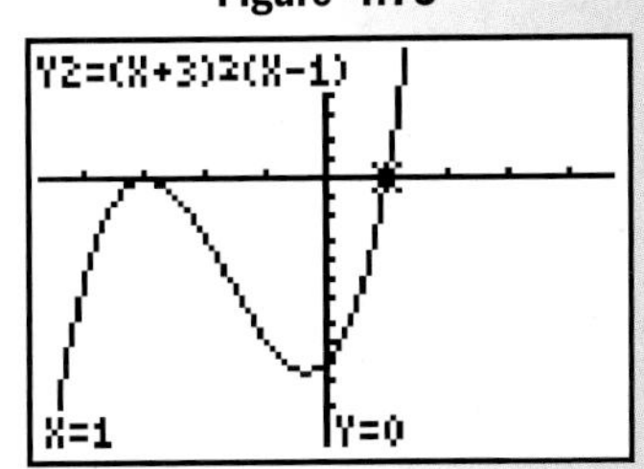

Figure 4.74

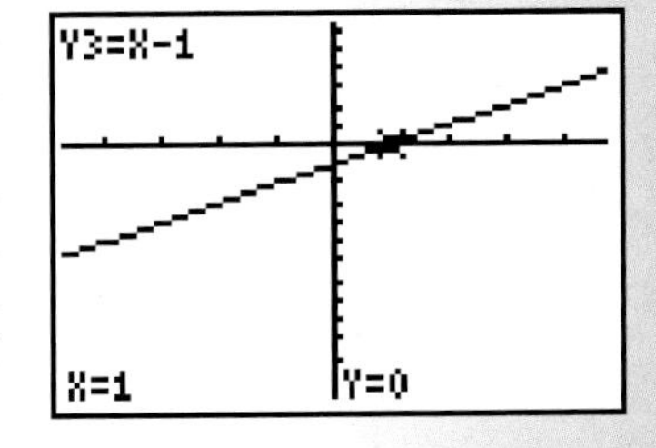

Explore these relationships further using the following exercises and a "greater than or equal to" inequality. Begin by writing Y_1 in completely factored form.

Exercise 1: $Y_1 = (x^3 - 6x^2 + 32)(x^2 + 1)$
$Y_2 = x^3 - 6x^2 + 32$
$Y_3 = x + 2$

Exercise 2: $Y_1 = (x + 3)^2(x^3 - 2x^2 + x - 2)$
$Y_2 = (x + 3)^2(x - 2)$
$Y_3 = x - 2$

Exercise 3: Based on what you've noticed, comment on how the irreducible factors of a polynomial affect its graph. What role do they play in the solution of inequalities? They do not affect the solution.

Exercise 4: How do roots of even multiplicity affect the solution set of non-strict inequalities (less/greater than or equal to)? Can you make a general statement that would include solutions to strict inequalities? They would not affect the answer.

For more on these ideas, see the *Strengthening Core Skills* feature from this chapter.

STRENGTHENING CORE SKILLS

Solving Inequalities Using the Push Principle

The most common method for solving polynomial inequalities involves finding the zeroes of the function and checking the sign of the function in the intervals between these zeroes. In Section 4.7, we relied on the end behavior of the graph, the sign of the function at the y-intercept, and the multiplicity of the zeroes to determine the solution. There is a third method that is more conceptual in nature, but in many cases highly efficient. It is based on two very simple ideas, the first involving only order relations and the number line:

A. Given any number x and constant $k > 0$: $x > x - k$ and $x < x + k$.

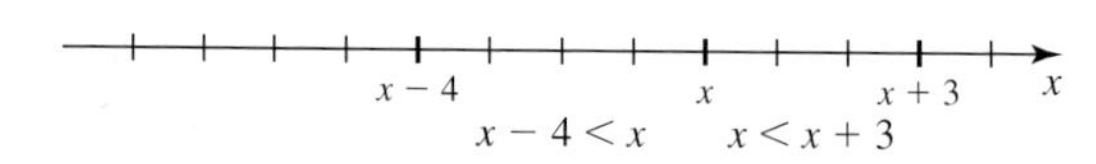

This statement simply reinforces the idea that if a is left of b on the number line, then $a < b$. As shown in the diagram, $x - 4 < x$ and $x < x + 3$, from which $x - 4 < x + 3$ for any x.

B. The second idea reiterates well-known ideas regarding the multiplication of signed numbers. For any number of factors:

if there are an even number of negative factors, the result is positive;
if there are an odd number of negative factors, the result is negative.

These two ideas work together to solve inequalities using what we'll call the *Push Principle*. Consider the inequality $x^2 - x - 12 > 0$. The factored form is $(x - 4)(x + 3) > 0$ and we want the product of these two factors to be positive. From (A), both factors will be positive if $(x - 4)$ is positive, since it's the smaller of the two, and both factors will be negative if $x + 3 < 0$, since it's the larger. The solution set is found by solving these two simple inequalities: $x - 4 > 0$ gives $x > 4$ and $x + 3 < 0$ gives $x < -3$. If the inequality were $(x - 4)(x + 3) < 0$ instead, we require one negative factor and one positive factor. Due to order relations and the number line, the larger factor must be the positive one: $x + 3 > 0$ so $x > -3$. The smaller factor must be the negative one: $x - 4 < 0$ and $x < 4$. This gives the solution $-3 < x < 4$ as can be verified using any alternative method. Solutions to all other polynomial and rational inequalities are an extension of these two cases.

ILLUSTRATION 1 Solve $x^3 - 7x + 6 < 0$ using the Push Principle.

Solution: The polynomial can be factored using the tests for 1 and -1 and synthetic division. The factors are $(x - 2)(x - 1)(x + 3) < 0$, which we've conveniently written in increasing order. For the product of three factors to be negative we require: (1) three negative factors or (2) one negative and two positive factors. The first condition is met by simply making the largest factor negative, as it will ensure the smaller factors are also negative: $x + 3 < 0$ so $x < -3$. The second condition is met by making the smaller factor negative and the "middle" factor positive: $x - 2 < 0$ *and* $x - 1 > 0$. The second solution interval is $x < 2$ and $x > 1$, or $1 < x < 2$.

Note the Push Principle does not require the testing of intervals between the zeroes, nor the "cross/bounce" analysis at the zeroes and vertical asymptotes (of rational functions). In addition, irreducible quadratic factors can still be ignored as they contribute nothing to the solution of real inequalities, and factors of even multiplicity can be overlooked precisely because there is no sign change at these roots.

ILLUSTRATION 2 Solve $(x^2 + 1)(x - 2)^2(x + 3) \geq 0$ using the Push Principle.

Solution: Since the factor $(x^2 + 1)$ does not affect the solution set, this inequality will have the same solution as $(x - 2)^2(x + 3) \geq 0$. Further, since $(x - 2)^2$ will be nonnegative for all x, the original inequality *has the same solution set as* $(x + 3) \geq 0$*!* The solution is $x \geq -3$.

ILLUSTRATION 3 Solve $\dfrac{x^3 - x^2 + 4x - 4}{x^3 - 3x^2 - 9x + 27} < 0$ using the Push Principle.

Solution: In factored form the inequality is $\dfrac{(x^2 + 4)(x - 1)}{(x - 3)^2(x + 3)} < 0$. The factor $(x^2 + 4)$ does not affect the solution set and can be ignored. The factor $(x - 3)^2$ will be nonnegative everywhere the expression is defined and can also be ignored (or overlooked). This indicates that $\dfrac{(x - 1)}{(x + 3)} < 0$, $x \neq -3$, has the same solution set as the original inequality. For the ratio of two factors to be negative requires one negative and one positive factor, giving the solution $x - 1 < 0$ and $x + 3 > 0$ (the smaller factor must be negative and the larger factor positive). The solution is $-3 < x < 1$.

With some practice, the Push Principle can be an effective tool. Use it to solve the following Exercises. Check all solutions by graphing the function on a graphing calculator.

Exercise 1: $x^3 - 3x - 18 \leq 0$ Exercise 2: $\dfrac{x + 1}{x^2 - 4} > 0$

Exercise 3: $x^3 - 13x + 12 < 0$ Exercise 4: $x^3 - 3x + 2 \geq 0$

Exercise 5: $x^4 - x^2 - 12 > 0$ Exercise 6: $(x^2 + 5)(x^2 - 9)(x + 2)^2(x - 1) \geq 0$

1. $x \in (-\infty, 3]$
2. $x \in (-2, -1) \cup (2, \infty)$
3. $x \in (-\infty, -4) \cup (1, 3)$
4. $x \in [-2, \infty)$
5. $x \in (-\infty, -2) \cup (2, \infty)$
6. $x \in [-3, 1] \cup [3, \infty)$

CUMULATIVE REVIEW CHAPTERS 1-4

1. Solve for R: $\dfrac{1}{R} = \dfrac{1}{R_1} + \dfrac{1}{R_2}$

2. Solve for x: $\dfrac{2}{x + 1} + 1 = \dfrac{5}{x^2 - 1}$

3. Factor the expressions:
   a. $x^3 - 1$
   b. $x^3 - 3x^2 - 4x + 12$

4. Solve using the quadratic formula. Write answers in both exact and approximate form: $2x^2 + 4x + 1 = 0$.

5. Solve the inequality and graph the solution on a number line: $x + 3 < 5$ *or* $5 - x < 4$.

6. Name the eight toolbox functions, give their descriptive names, then draw a sketch of each.

7. Use substitution to show that $x = 2 - 3i$ is a solution to $x^2 - 4x + 13$.
   verified

8. Solve the rational inequality: $\dfrac{x + 4}{x - 2} < 3$. $x \in (-\infty, 2) \cup (5, \infty)$

1. $R = \dfrac{R_1R_2}{R_1 + R_2}$
2. $x = -4$ or $x = 2$
3. a. $(x - 1)(x^2 + x + 1)$
   b. $(x - 3)(x + 2)(x - 2)$
4. $x = \dfrac{-2 \pm \sqrt{2}}{2}$, $x \approx -0.29, -1.71$
5. all reals

Additional answers can be found in the Instructor Answer Appendix.

9. $y = \frac{11}{60}x + \frac{1009}{60}$; 39 min, driving time increases 11 min every 60 days
10. No, Herschel is paired with mathematics and astronomy.
11. $y = 1.18x^2 - 10.99x + 4.6$; 9
12.

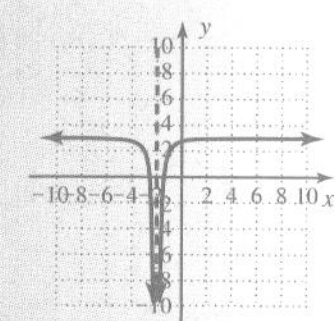

13. $f^{-1}(x) = \frac{x^3 + 3}{2}$
14. a. $x \in (-\infty, \infty)$
    b. $x \in (2, \infty)$

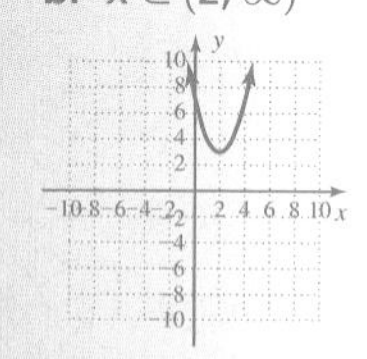

15. 16.

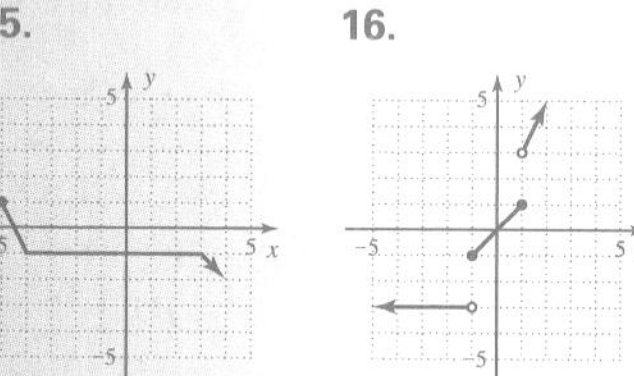

17. $X = 63$
18. $x = 1, -3, 1 \pm 2i$
19. 20.

$x \in (-2, 1] \cup (2, \infty)$

9. As part of a study on traffic conditions, the mayor of a small city tracks her driving time to work each day for six months and finds a linear and increasing relationship. On day 1, her drive time was 17 min. By day 61 the drive time had increased to 28 min. Find a linear function that models the drive time and use it to estimate the drive time on day 121, if the trend continues. Explain what the slope of the line means in this context.

10. Does the relation shown represent a function? If not, discuss/explain why not.

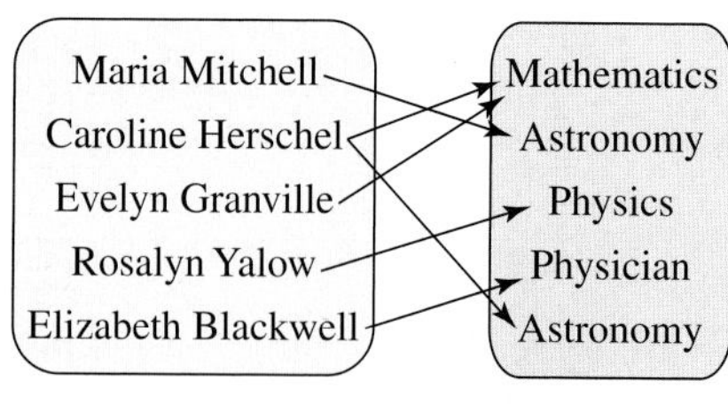

11. The data given shows the profit of a new company for the first 6 months of business. Draw a scatter-plot of the data and decide on the best form of regression. Use the equation model to find the first month a profit is earned.

Exercise 11

Month	Profit (1000s)
1	−5
2	−13
3	−18
4	−20
5	−21
6	−19

12. Graph the function $g(x) = \frac{-1}{(x + 2)^2} + 3$ using transformations of a basic function.

13. Find $f^{-1}(x)$, given $f(x) = \sqrt[3]{2x - 3}$, then use composition to verify your inverse is correct.

14. Graph $f(x) = x^2 - 4x + 7$ by completing the square, then state intervals where:

a. $f(x) \geq 0$

b. $f(x)\uparrow$

15. Given the graph of a general function $f(x)$, graph $F(x) = -f(x + 1) + 2$.

Exercise 15

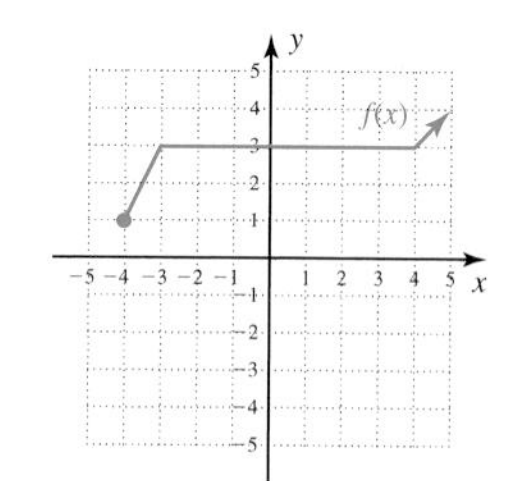

16. Graph the piecewise-defined function given:

$$f(x) = \begin{cases} -3 & x < -1 \\ x & -1 \leq x \leq 1 \\ 3x & x > 1 \end{cases}$$

17. Y varies directly with X and inversely with the square of Z. If $Y = 10$ when $X = 32$ and $Z = 4$, find X when $Z = 15$ and $Y = 1.4$.

18. Use the rational roots theorem and synthetic division to find all roots (real and complex) of $f(x) = x^4 - 2x^2 + 16x - 15$.

19. Sketch the graph of $f(x) = x^3 - 3x^2 - 6x + 8$.

20. Sketch the graph of $h(x) = \frac{x - 1}{x^2 - 4}$ and use the zeroes and vertical asymptotes to solve $h(x) \geq 0$.

Chapter 5

Exponential and Logarithmic Functions

Chapter Outline

5.1 Exponential Functions 476

5.2 Logarithms and Logarithmic Functions 485

5.3 The Exponential Function and Natural Logarithms 495

5.4 Exponential/ Logarithmic Equations and Applications 509

5.5 Applications from Business, Finance, and Physical Science 521

5.6 Exponential, Logarithmic, and Logistic Regression Models 536

Preview

So far we've investigated a sizable number of functions with a large variety of applications. Still, many situations arising in business, science, and industry require the development of two additional tools—the exponential and logarithmic functions. Their applications are virtually limitless, and offer an additional glimpse of the true power and potential of mathematics. This chapter marks another significant advance in your mathematical studies, as we begin using the ideas underlying an inverse function to lay the groundwork needed for the introduction of these new function families.

5.1 Exponential Functions

LEARNING OBJECTIVES

In Section 5.1 you will learn how to:

A. Evaluate an exponential function

B. Graph exponential functions

C. Solve certain exponential equations

D. Solve applications of exponential functions

INTRODUCTION

Demographics is the statistical study of human populations. Perhaps surprisingly, an accurate study of how populations grow or decline cannot be achieved using only the functions discussed previously. In this section, we introduce the family of *exponential functions,* which are widely used to model population growth, with additional applications in finance, science, and engineering. As with other functions, we begin with a study of the graph and its characteristics.

POINT OF INTEREST

When a fair coin is flipped, we expect that roughly half the time it will turn up heads. In other words, the probability of a head showing on the first flip is one out of two or $\frac{1}{2}$. The probability that heads shows up two times in succession is $\frac{1}{2} \cdot \frac{1}{2} = \frac{1}{4}$, and for three heads in a row the probability is $\frac{1}{2} \cdot \frac{1}{2} \cdot \frac{1}{2} = \frac{1}{8}$. As you can see, the probability of the coin continuing to turn up heads gets smaller and smaller and can be modeled by the function $f(x) = \left(\frac{1}{2}\right)^x$, where x represents the number of times the coin is flipped. This is one example from the family of exponential functions, which are used extensively in many different areas of scientific endeavor.

A. Evaluating Exponential Functions

In the boomtowns of the Old West, it was not uncommon for a town to double in size every year (at least for a time) as the lure of gold drew more and more people westward. When this type of growth is modeled using mathematics, exponents play a leading role. Suppose the town of Goldsboro had 1000 residents just before gold was discovered. After 1 yr the population doubled and the town had **2**000 residents. The next year it doubled again to **4**000, then again to **8**000, then to **16**,000 and so on. You probably recognize the digits in bold as powers of two (indicating the population is doubling), with each one multiplied by 1000 (the initial population). This suggests we can model the relationship using $P(x) = 1000 \cdot 2^x$, where $P(x)$ is the total population after x years. Sure enough, $P(4) = 1000 \cdot 2^4 = 16{,}000$. Further, we can also evaluate this function, called an **exponential function,** for *fractional parts of a year* using rational exponents. Using the ideas from Section R.6, the population of Goldsboro one-and-a-half years after the gold rush $(t = \frac{3}{2})$ was: $P(\frac{3}{2}) = 1000 \cdot 2^{\frac{3}{2}} = 1000 \cdot (\sqrt{2})^3 \approx 2828$ people. To actually *graph the function* using real numbers requires that we define the expression 2^x when x is irrational. For example, what does $2^{\sqrt{5}}$ mean? We suspect it represents a real number since $2 < \sqrt{5} < 3$ seems to imply that $2^2 < 2^{\sqrt{5}} < 2^3$. While the technical details require calculus, it can be shown that successive approximations of $2^{\sqrt{5}}$ as in $2^{2.2360}$, $2^{2.23606}$ $2^{2.236067}$, . . . approach a unique real number. This means we can approximate its value to any desired level of accuracy: $2^{\sqrt{5}} \approx 4.71111$ to five decimal places. In general, as long as b is a positive real number, b^x *is a real number for all real numbers* x and we have the following:

EXPONENTIAL FUNCTIONS

For $b > 0$, $b \neq 1$, $f(x) = b^x$ defines the base b exponential function. The domain of f is all real numbers.

Limiting b to positive values ensures outputs will be real numbers [if $b = -4$ and $x = \frac{1}{2}$, we have $(-4)^{\frac{1}{2}} = \sqrt{-4}$, which is not a real number]. The restriction $b \neq 1$ is needed because $y = 1^x$ is a constant function (1 raised to *any* power is still 1).

WORTHY OF NOTE

Exponential functions are very different from the power functions studied earlier. For power functions, the base is variable and the exponent is constant: $y = x^b$, while for exponential functions the *exponent is a variable* and the *base is constant:* $y = b^x$.

EXAMPLE 1 Evaluate the expressions given, rounding each to five decimal places.

a. $2^{3.14}$ **b.** $2^{3.1416}$ **c.** $2^{3.141592}$ **d.** 2^{π}

Solution: **a.** $2^{3.14} \approx 8.81524$ **b.** $2^{3.1416} \approx 8.82502$

c. $2^{3.141592} \approx 8.82497$ **d.** $2^{\pi} \approx 8.82498$

NOW TRY EXERCISES 7 THROUGH 12

Note the domain of the exponential function includes negative numbers as well, and expressions such as 2^{-3} and $2^{-\sqrt{5}}$ can easily be calculated: $2^{-3} = \frac{1}{2^3} = \frac{1}{8}$, and $2^{-\sqrt{5}} = \frac{1}{2^{\sqrt{5}}} \approx 0.21226$. In fact, all of the familiar properties of exponents continue to hold for irrational exponents.

EXPONENTIAL PROPERTIES

Given a, b, x, and t are real numbers, with $b, c > 0$,

$$b^x b^t = b^{x+t} \qquad \frac{b^x}{b^t} = b^{x-t} \qquad (b^x)^t = b^{xt}$$

$$(bc)^x = b^x c^x \qquad b^{-x} = \frac{1}{b^x} \qquad \left(\frac{b}{a}\right)^{-x} = \left(\frac{a}{b}\right)^x$$

B. Graphing Exponential Functions

To gain a better understanding of exponential functions, we'll graph $y = b^x$ on a coordinate grid and note some of its important features. Since the base b cannot be equal to 1, it seems reasonable that we graph one exponential function where $b > 1$ and one where $0 < b < 1$.

EXAMPLE 2 Graph $y = 2^x$ using a table of values.

Solution: To get an idea of the graph's shape we'll use integer values from -3 to 3 in our table, then draw the graph as a continuous curve, knowing the function is defined for all real numbers.

x	$y = 2^x$
-3	$2^{-3} = \frac{1}{8}$
-2	$2^{-2} = \frac{1}{4}$
-1	$2^{-1} = \frac{1}{2}$
0	$2^0 = 1$
1	$2^1 = 2$
2	$2^2 = 4$
3	$2^3 = 8$

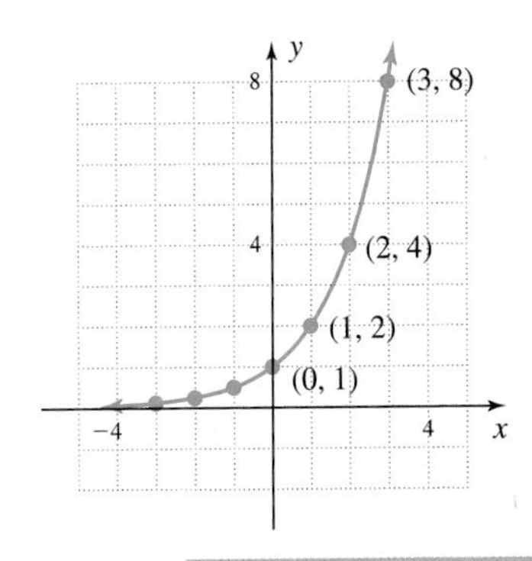

NOW TRY EXERCISES 13 AND 14

WORTHY OF NOTE

Functions that are increasing for all $x \in D$ are said to be **monotonically increasing** or simply **monotonic functions.**

Several important observations can now be made. First note the x-axis (the line $y = 0$) is a horizontal asymptote for the function, because as $x \to -\infty$, $y \to 0$. Second, it is evident that the function is increasing over its entire domain, giving the function a range of $y \in (0, \infty)$.

EXAMPLE 3 Graph $y = \left(\frac{1}{2}\right)^x$ using a table of values.

Solution: Using properties of exponents, we can write $\left(\frac{1}{2}\right)^x$ as $\left(\frac{2}{1}\right)^{-x} = 2^{-x}$. Again using integers from -3 to 3, we plot the ordered pairs and draw a continuous curve.

x	$y = 2^{-x}$
-3	$2^{-(-3)} = 2^3 = 8$
-2	$2^{-(-2)} = 2^2 = 4$
-1	$2^{-(-1)} = 2^1 = 2$
0	$2^0 = 1$
1	$2^{-1} = \frac{1}{2}$
2	$2^{-2} = \frac{1}{4}$
3	$2^{-3} = \frac{1}{8}$

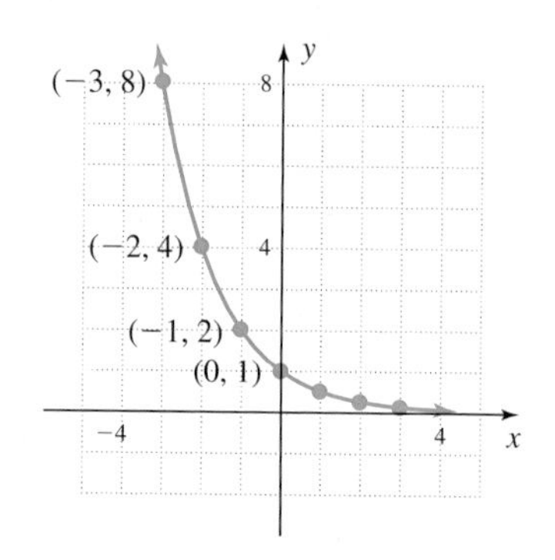

NOW TRY EXERCISES 15 AND 16

We note this graph is also asymptotic to the x-axis, but *decreasing on its domain.* In addition, we see that both $y = 2^x$ and $y = 2^{-x} = \left(\frac{1}{2}\right)^x$ are one-to-one, and have a y-intercept of (0, 1)—which we expect since any base to the zero power is 1. Finally, observe that $y = b^{-x}$ is *a reflection of $y = b^x$ across the y-axis,* a property that suggests these basic graphs might also be transformed in other ways, as the toolbox functions were in Section 3.3. The most important characteristics of exponential functions are summarized here:

$$f(x) = b^x,\ b > 0 \text{ and } b \neq 1$$

- one-to-one function
- domain: $x \in \mathbb{R}$
- increasing if $b > 1$
- asymptotic to the x-axis
- y-intercept (0, 1)
- range: $y \in (0, \infty)$
- decreasing if $0 < b < 1$

WORTHY OF NOTE

When an exponential function is increasing, it can be referred to as a "growth function." When decreasing, it is often called a "decay function." Each of the graphs shown in Figures 5.1 and 5.2 should now be added to your repertoire of basic functions, to be sketched from memory and analyzed or used as needed.

Figure 5.1

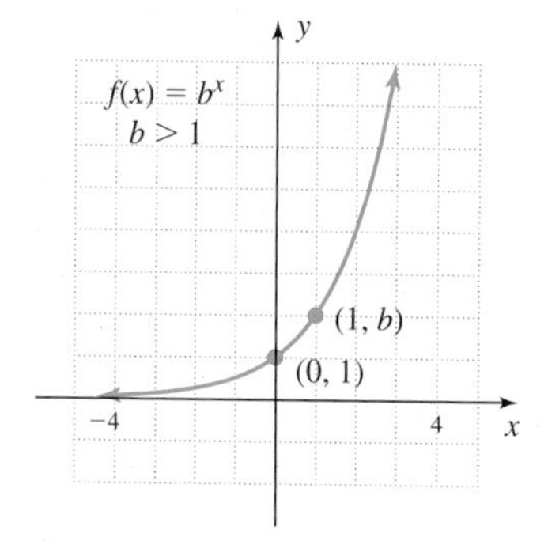

Figure 5.2

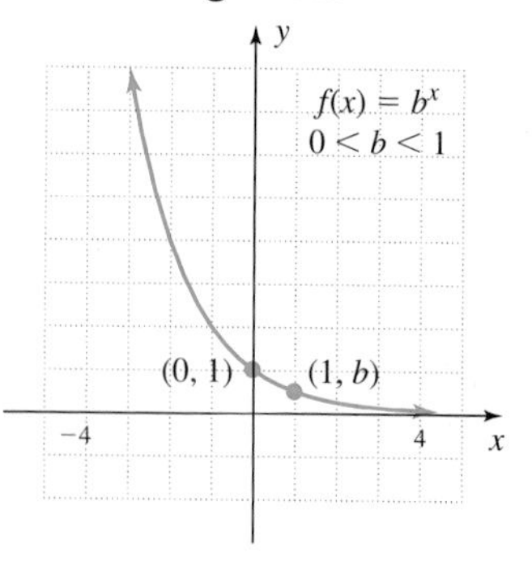

Just as a quadratic function maintains its parabolic shape regardless of the transformations applied, exponential functions will also maintain their general shape and features. Any sum or difference applied to the basic function will cause a vertical shift in the same direction as the sign, and any change to input values ($y = b^{x+h}$ versus $y = b^x$) will cause a horizontal shift in a direction opposite the sign.

EXAMPLE 4 Graph $F(x) = 2^{x-1} + 2$ using transformations of the basic function $f(x) = 2^x$ (not by simply plotting points). Clearly state what transformations are applied.

Solution: The graph of F is that of the basic function shifted 1 unit right and 2 units up. With this in mind we can sketch the horizontal asymptote at $y = 2$ and plot the point (1, 3). The y-intercept of F is at (0, 2.5) and to help sketch a fairly accurate graph, the additional point (3, 6) is used: $F(3) = 6$.

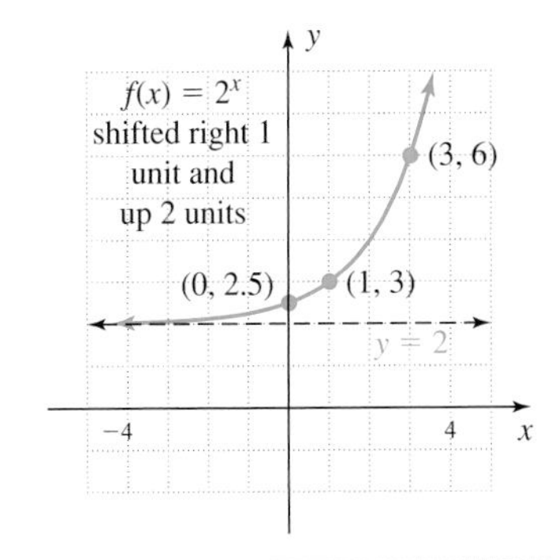

NOW TRY EXERCISES 17 THROUGH 38

C. Solving Exponential Equations

The fact that exponential functions are one-to-one enables us to solve equations where each side is an exponential term with the identical base. This is because one-to-oneness guarantees a unique solution to the equation.

EXPONENTIAL EQUATIONS WITH LIKE BASES: THE UNIQUENESS PROPERTY

For all real numbers b, m, and n, where $b > 0$ and $b \neq 1$,

If $b^m = b^n$, then $m = n$.	If $m = n$, then $b^m = b^n$.

Equal bases imply exponents are equal.

The equation $2^x = 32$ can be written as $2^x = 2^5$, and we note $x = 5$ is a solution. Although $3^x = 32$ can be written as $3^x = 2^5$, the bases are not alike and the solution to this equation must wait until additional tools are developed in Section 5.4.

EXAMPLE 5 Solve the exponential equations using the uniqueness property.

a. $3^{2x-1} = 81$ **b.** $25^{-2x} = 125^{x+7}$ **c.** $\left(\frac{1}{6}\right)^{-3x-2} = 36^{x+1}$

Solution: **a.**

$$3^{2x-1} = 81 \quad \text{given}$$
$$3^{2x-1} = 3^4 \quad \text{rewrite using base 3}$$
$$\Rightarrow 2x - 1 = 4 \quad \text{uniqueness property}$$
$$x = \frac{5}{2} \quad \text{solve for } x$$

Check:

$$3^{2x-1} = 81 \quad \text{given}$$
$$3^{2\left(\frac{5}{2}\right)-1} = 81 \quad \text{substitute } \tfrac{5}{2} \text{ for } x$$
$$3^{5-1} = 81 \quad \text{simplify}$$
$$3^4 = 81\checkmark \quad \text{result checks}$$

The remaining checks are left to the student.

b.

$$25^{-2x} = 125^{x+7} \quad \text{given}$$
$$(5^2)^{-2x} = (5^3)^{x+7} \quad \text{rewrite using base 5}$$
$$5^{-4x} = 5^{3x+21} \quad \text{power property of exponents}$$
$$\Rightarrow -4x = 3x + 21 \quad \text{uniqueness property}$$
$$x = -3 \quad \text{solve for } x$$

c.

$$\left(\frac{1}{6}\right)^{-3x-2} = 36^{x+1} \quad \text{given}$$
$$(6^{-1})^{-3x-2} = (6^2)^{x+1} \quad \text{rewrite using base 5}$$
$$6^{3x+2} = 6^{2x+2} \quad \text{power property of exponents}$$
$$\Rightarrow 3x + 2 = 2x + 2 \quad \text{uniqueness property}$$
$$x = 0 \quad \text{solve for } x$$

NOW TRY EXERCISES 39 THROUGH 54

D. Applications of Exponential Functions

One application of exponential functions involves money invested at a fixed rate of return. If interest is compounded n times per year, the amount of money in an account after t years is modeled by the function $A(t) = P\left(1 + \frac{r}{n}\right)^{nt}$, where P is the principal amount invested, r is the annual interest rate, and $A(t)$ represents the amount of money in the account after t years. The result is called the **compound interest formula** (for a full development of this formula, see Section 5.5). After substituting given values, this formula simplifies to a more recognizable form, $y = ab^x$.

EXAMPLE 6 When she was 8 yr old, Marcy invested \$1000 of the money she earned working a paper route in an account paying 8% interest compounded quarterly (four times per year). At 18 yr old, Marcy now wants to withdraw the money to help with college expenses. Determine how much money is in the account.

Solution: For this exercise, $P = 1000$, $r = 8\%$, $n = 4$, and $t = 10$. The formula yields

$$A(t) = P\left(1 + \frac{r}{n}\right)^{nt} \quad \text{given}$$
$$= 1000\left(1 + \frac{0.08}{4}\right)^{4 \cdot 10} \quad \text{substitute given values}$$
$$= 1000(1.02)^{40} \quad \text{simplify, note } ab^x \text{ form}$$
$$\approx 1000(2.20804) \quad \text{evaluate: } 1.02^{40} \approx 2.20804$$
$$\approx 2208.04 \quad \text{result}$$

After 10 yr, there is approximately \$2208.04 in the account.

NOW TRY EXERCISES 57 AND 58

EXAMPLE 7 For insurance purposes, it is estimated that large household appliances lose $\frac{1}{4}$ of their value each year. The current value can then be modeled by the function $V(t) = V_0\left(\frac{3}{4}\right)^t$, where V_0 is the initial value and $V(t)$ represents the value after t years. How many years does it take a washing machine that cost \$256 new, to depreciate to a value of \$81?

Solution: For this exercise, $V_0 = \$256$ and $V(t) = \$81$. The formula yields

$$V(t) = V_0\left(\frac{3}{4}\right)^t \quad \text{given}$$

$$81 = 256\left(\frac{3}{4}\right)^t \quad \text{substitute known values}$$

$$\frac{81}{256} = \left(\frac{3}{4}\right)^t \quad \text{divide by 256}$$

$$\left(\frac{3}{4}\right)^4 = \left(\frac{3}{4}\right)^t \quad \text{equate bases } (3^4 = 81 \text{ and } 4^4 = 256)$$

$$\Rightarrow 4 = t \quad \text{Uniqueness Property}$$

After 4 yr, the washing machine's value has dropped to \$81.

NOW TRY EXERCISES 59 THROUGH 64

TECHNOLOGY HIGHLIGHT

Solving Exponential Equations Graphically

The keystrokes shown apply to a TI-84 Plus model. Please consult our Internet site or your manual for other models.

In this section, we showed that the exponential function $f(x) = b^x$ was defined for all real numbers. This is important because it establishes that equations like $2^x = 7$ must have a solution, even if x is not rational. In fact, since $2^2 = 4$ and $2^3 = 8$, the solution must be between 2 and 3:

$$2^2 = 4 < 2^x = 7 < 2^3 = 8$$
$$\text{and}$$
$$2 < x < 3$$

Until we develop an inverse for exponential functions, we are unable to solve many of these equations in exact form. We can, however, get a very close approximation using a graphing calculator. For the equation $2^x = 7$, enter $Y_1 = 2^x$ and $Y_2 = 7$ on the Y= screen. Then press ZOOM 6 to graph both functions (see Figure 5.3). To have the calculator compute the point of intersection, press 2nd CALC and select option **5: intersect** and press ENTER *three* times. The x- and y-coordinates of the point of intersection will appear at the bottom of the screen, with the x-coordinate being the solution. As you can see, x is indeed between 2 and 3. Solve the following equations. First estimate the answer by bounding it between two integers, then solve the equation graphically. Adjust the viewing window as needed.

Figure 5.3

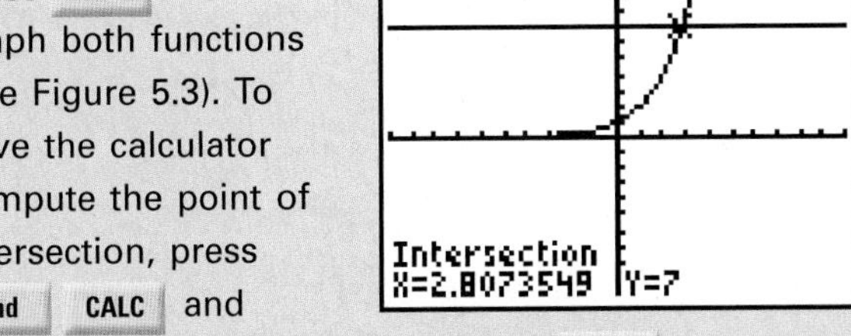

Exercise 1: $3^x = 22$ $x = 2.8$ Exercise 2: $2^x = 0.125$ $x = -3$

Exercise 3: $5^{x-1} = 61$ $x = 3.6$

5.1 EXERCISES

CONCEPTS AND VOCABULARY

Fill in each blank with the appropriate word or phrase. Carefully reread the section if needed.

1. An exponential function is one of the form $y =$ b^x, where b > 0, b $\neq 1$, and x is any real number.

2. The domain of $y = b^x$ is all real numbers, and the range is $y \in$ $(0, \infty)$. Further, as $x \to -\infty$, y $\to 0$.

3. For exponential functions of the form $y = ab^x$, the y-intercept is $(0,$ a $)$, since $b^0 =$ 1 for any real number b.

4. If each side of an equation can be written as an exponential term with the same base, the equation can be solved using the uniqueness property.

5. State true or false and explain why: $y = b^x$ is always increasing if $0 < b < 1$. False; for $|b| < 1$ and $x_2 > x_1$, $b^{x_2} < b^{x_1}$ so function is decreasing

6. Discuss/explain the statement, "For $k > 0$, the y intercept of $y = ab^x + k$ is $(0, a + k)$." Answers will vary.

DEVELOPING YOUR SKILLS

Use a calculator (as needed) to evaluate each function as indicated. Round answers to thousandths.

7. $P(t) = 2500 \cdot 4^t$; $t = 2$, $t = \frac{1}{2}$, $t = \frac{3}{2}$, $t = \sqrt{3}$

8. $Q(t) = 5000 \cdot 8^t$; $t = 2$, $t = \frac{1}{3}$, $t = \frac{5}{3}$, $t = 5$

9. $f(x) = 0.5 \cdot 10^x$; $x = 3$, $x = \frac{1}{2}$, $x = \frac{2}{3}$, $x = \sqrt{7}$

10. $g(x) = 0.8 \cdot 5^x$; $x = 4$, $x = \frac{1}{4}$, $x = \frac{4}{5}$, $x = \pi$

11. $V(n) = 10{,}000\left(\frac{2}{3}\right)^n$; $n = 0$, $n = 4$, $n = 4.7$, $n = 5$

12. $W(m) = 3300\left(\frac{4}{5}\right)^m$; $m = 0$, $m = 5$, $m = 7.2$, $m = 10$

Graph each function using a table of values and integer inputs between -3 and 3. Clearly label the y-intercept and one additional point, then indicate whether the function is increasing or decreasing.

13. $y = 3^x$ increasing
14. $y = 4^x$ increasing
15. $y = \left(\frac{1}{3}\right)^x$ decreasing
16. $y = \left(\frac{1}{4}\right)^x$ decreasing

Graph each of the following functions by *translating the basic function* $y = b^x$ and strategically plotting a few points to round out the graph. Clearly state what shifts are applied.

17. $y = 3^x + 2$
18. $y = 3^x - 3$
19. $y = 3^{x+3}$
20. $y = 3^{x-2}$
21. $y = 2^{-x}$
22. $y = 3^{-x}$
23. $y = 2^{-x} + 3$
24. $y = 3^{-x} - 2$
25. $y = 2^{x+1} - 3$
26. $y = 3^{x-2} + 1$
27. $y = \left(\frac{1}{3}\right)^x + 1$
28. $y = \left(\frac{1}{3}\right)^x - 4$
29. $y = \left(\frac{1}{3}\right)^{x-2}$
30. $y = \left(\frac{1}{3}\right)^{x+2}$
31. $f(x) = \left(\frac{1}{3}\right)^x - 2$
32. $g(x) = \left(\frac{1}{3}\right)^x + 2$

Match each graph to the correct exponential equation.

33. $y = 5^{-x}$ e
34. $y = 4^{-x}$ c
35. $y = 3^{-x+1}$ a
36. $y = 3^{-x} + 1$ f
37. $y = 2^{x+1} - 2$ b
38. $y = 2^{x+2} - 1$ d

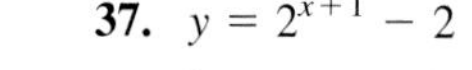
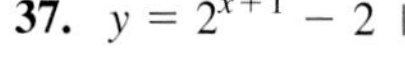

a.

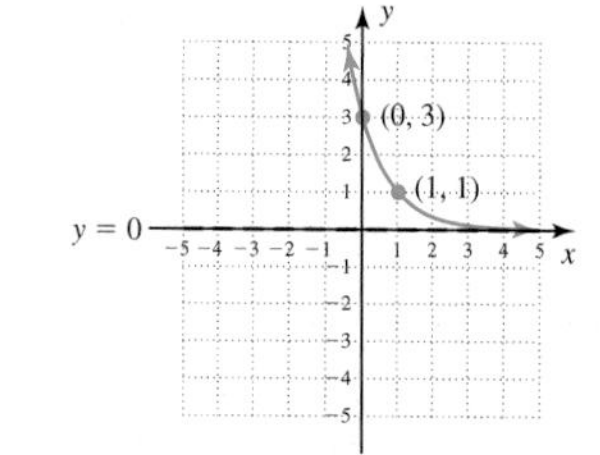

b.

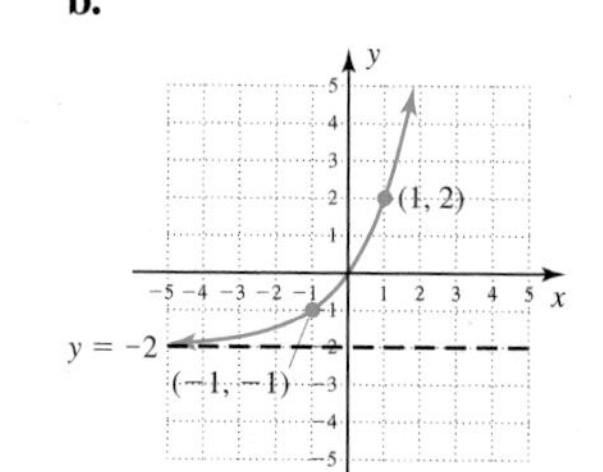

c.

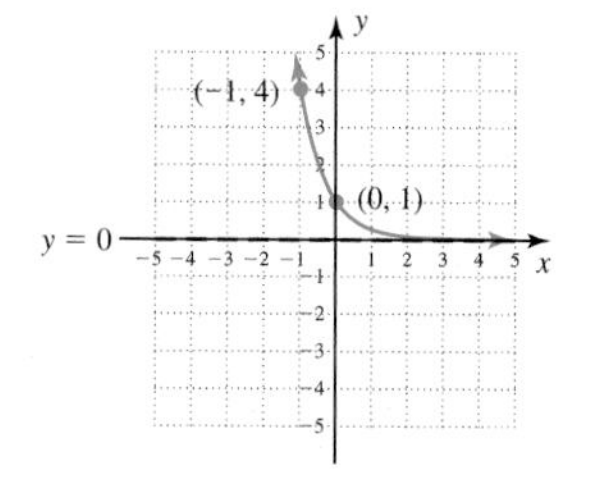

7. 40,000; 5000; 20,000; 27,589.162
8. 320,000; 10,000; 160,000; 163,840,000
9. 500; 1.581; 2.321; 221.168
10. 500; 1.196; 2.899; 125.594
11. 10,000; 1975.309; 1487.206; 1316.872
12. 3300; 1081.344; 661.853; 354.335

13. (2, 9); (0, 1); $y = 0$; increasing
14. (1, 4); (0, 1); $y = 0$; increasing
15. (−2, 9); (0, 1); $y = 0$; decreasing
16. (−1, 4); (0, 1); $y = 0$; decreasing
17. up 2 — (1, 5); (0, 3); $y = 2$
18. down 3 — (2, 6); (0, −2); $y = -3$
19. left 3
20. right 2

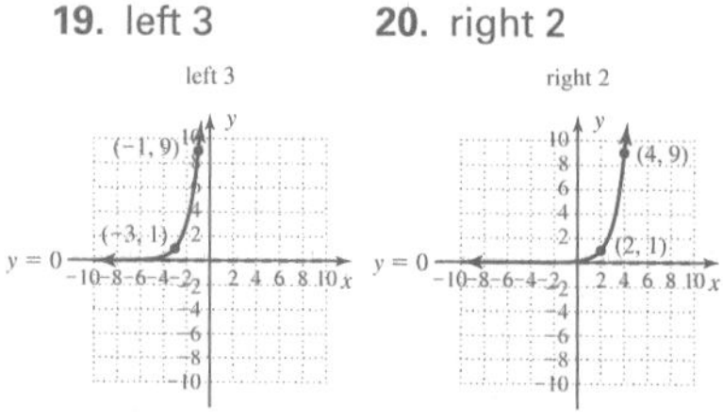

Additional answers can be found in the Instructor Answer Appendix.

d.

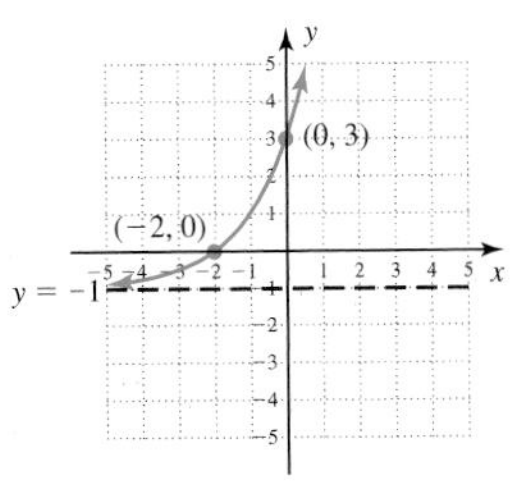

e.

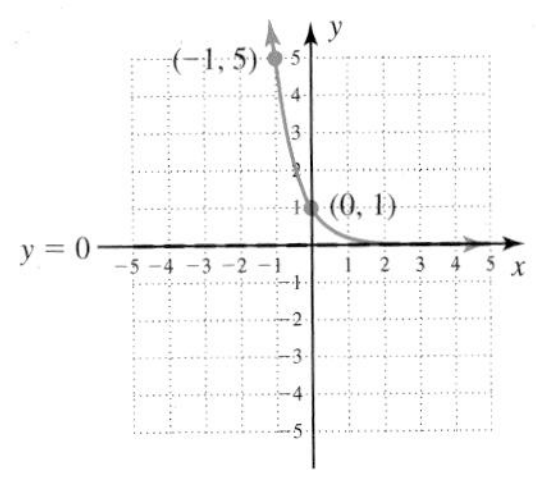

f.

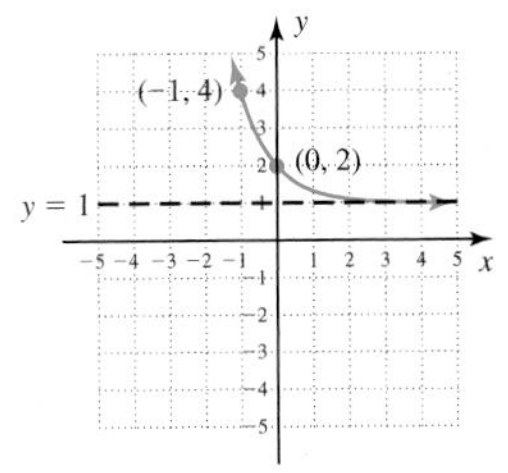

Solve each exponential equation and check your answer by substituting into the original equation.

39. $10^x = 1000$ 3 **40.** $144 = 12^x$ 2 **41.** $25^x = 125$ $\frac{3}{2}$ **42.** $81 = 27^x$ $\frac{4}{3}$

43. $8^{x+2} = 32$ $\frac{-1}{3}$ **44.** $9^{x-1} = 27$ $\frac{5}{2}$ **45.** $32^x = 16^{x+1}$ 4 **46.** $100^{x+2} = 1000^x$ 4

47. $\left(\frac{1}{5}\right)^x = 125$ −3 **48.** $\left(\frac{1}{4}\right)^x = 64$ −3 **49.** $\left(\frac{1}{3}\right)^{2x} = 9^{x-6}$ 3 **50.** $\left(\frac{1}{2}\right)^{3x} = 8^{x-2}$ 1

51. $\left(\frac{1}{9}\right)^{x-5} = 3^{3x}$ 2 **52.** $2^{-2x} = \left(\frac{1}{32}\right)^{x-3}$ 5 **53.** $25^{3x} = 125^{x-2}$ −2 **54.** $27^{2x+4} = 9^{4x}$ 6

WORKING WITH FORMULAS

55. a. 1732, 3000, 5196, 9000
b. yes
c. as $t \to \infty$, $P \to \infty$
d.

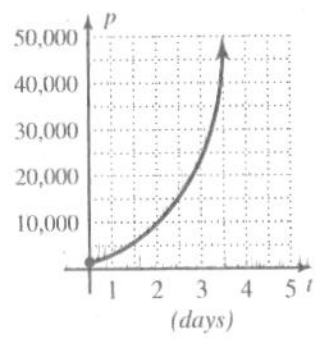

56. a. $\frac{1}{4}$
b. $\frac{1}{256}$
c. as $x \to \infty$, $P \to 0$

55. The growth of a bacteria population: $P(t) = 1000 \cdot 3^t$

If the initial population of a common bacterium is 1000 and the population triples every day, its population is given by the formula shown, where $P(t)$ is the total population after t days. (a) Find the total population 12 hours, 1 day, $1\frac{1}{2}$ days, and 2 days later. (b) Do the outputs show the population is tripling every 24 hr (1 day)? (c) Explain why this is an increasing function. (d) Graph the function using an appropriate scale.

56. Games involving a spinner with numbers 1 through 4: $P(x) = \left(\frac{1}{4}\right)^x$

Games that involve moving pieces around a board using a fair spinner are fairly common. If the spinner has the numbers 1 through 4, the probability that any one number is spun repeatedly is given by the formula shown, where x represents the number of spins and $P(x)$ represents the probability the same number results x times. (a) What is the probability that the first player spins a 2? (b) What is the probability that all four players spin a 2? (c) Explain why this is a decreasing function.

APPLICATIONS

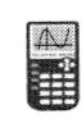

Round all dollar figures to the nearest whole dollar. Use the compound interest formula for Exercises 57 and 58: $A = P\left(1 + \frac{r}{n}\right)^{nt}$.

57. a. ≈\$12,875
b. ≈\$152

57. Investment savings: Janielle decides to begin saving for her 2-yr-old son's college education by depositing her \$5000 bonus check in an account paying 6% interest compounded semiannually. (a) How much will the account be worth when her son enters college (16 yr later)? (b) How much more would be in the account if the interest were compounded monthly?

58. a. ≈\$17,046
b. ≈\$113

58. Investment savings: To be more competitive, a local bank notifies its customers that all accounts earning 9% interest compounded *quarterly* will now earn 9% interest compounded *monthly.* (a) If \$7000 is invested for 10 yr in the account that paid 9% compounded quarterly, what would the account be worth? (b) How much additional interest will the bank have to pay if the same \$7000 were invested for 10 yr in an account that paid 9% interest compounded monthly?

59. a. \$100,000
b. 3 yr

59. Depreciation: The financial analyst for a large construction firm estimates that its heavy equipment loses one-fifth of its value each year. The current value of the equipment is then modeled by the function $V(t) = V_0\left(\frac{4}{5}\right)^t$, where V_0 represents the initial value, t is in years, and $V(t)$ represents the value after t years. (a) How much is a large earthmover worth after 1 yr if it cost \$125 thousand new? (b) How many years does it take for the earthmover to depreciate to a value of \$64 thousand?

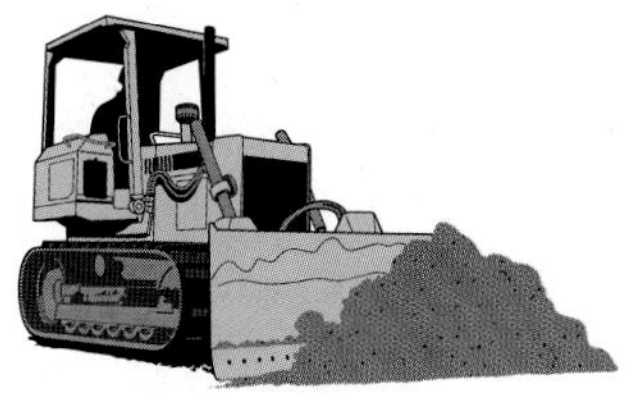

60. a. $40,000
b. 2 yr

61. a. ≈$86,806
b. 3 yr

62. a. 12.8 cm
b. 2 mo

63. a. $40 million
b. 7 yr

64. a. $864 thousand
b. 5 yr

65. ≈$32,578

66. ≈$4.81

67. a. 8 g
b. 48 min

68. a. 16 g
b. 112 hr

69. 9.5×10^{-7}; answers will vary

60. Depreciation: Photocopiers have become a critical part of the operation of many businesses, and due to their heavy use they can depreciate in value very quickly. If a copier loses $\frac{3}{8}$ of its value each year, the current value of the copier can be modeled by the function $V(t) = V_0\left(\frac{5}{8}\right)^t$, where V_0 represents the initial value, t is in years, and $V(t)$ represents the value after t years. (a) How much is this copier worth after one year if it cost $64 thousand new? (b) How many years does it take for the copier to depreciate to a value of $25 thousand?

61. Depreciation: Margaret Madison, DDS, estimates that her dental equipment loses one-sixth of its value each year. (a) Determine the value of an x-ray machine after 5 yr if it cost $216 thousand new, and (b) determine how long until the machine is worth less than $125 thousand.

62. Exponential decay: The groundskeeper of a local high school estimates that due to heavy usage by the baseball and softball teams, the pitcher's mound loses one-fifth of its height every month. (a) Determine the height of the mound after 3 months if it was 25 cm to begin, and (b) determine how long until the pitcher's mound is less than 16 cm high (meaning it must be rebuilt).

63. Exponential growth: Similar to a small town doubling in size after a discovery of gold, a business that develops a product in high demand has the potential for doubling its revenue each year for a number of years. The revenue would be modeled by the function $R(t) = R_0 2^t$, where R_0 represents the initial revenue, and $R(t)$ represents the revenue after t years. (a) How much revenue is being generated after 4 yr, if the company's initial revenue was $2.5 million? (b) How many years does it take for the business to be generating $320 million in revenue?

64. Exponential growth: If a company's revenue grows at a rate of 150% per year (rather than doubling as in Exercise 63), the revenue would be modeled by the function $R(t) = R_0\left(\frac{3}{2}\right)^t$, where R_0 represents the initial revenue, and $R(t)$ represents the revenue after t years. (a) How much revenue is being generated after 3 yr, if the company's initial revenue was $256 thousand? (b) How long until the business is generating $1944 thousand in revenue? (*Hint:* Reduce the fraction.)

Modeling inflation: Assuming the rate of inflation is 5% per year, the predicted price of an item can be modeled by the function $P(t) = P_0(1.05)^t$, where P_0 represents is the initial price of the item and t is in years. Use this information to solve Exercises 65 and 66.

65. What will the price of a new car be in the year 2010, if it cost $20,000 in the year 2000?

66. What will the price of a gallon of milk be in the year 2010, if it cost $2.95 in the year 2000? Round to the nearest cent.

Modeling radioactive decay: The half-life of a radioactive substance is the time required for half an initial amount of the substance to disappear through decay. The amount of the substance remaining is given by the formula $Q(t) = Q_0\left(\frac{1}{2}\right)^{\frac{t}{h}}$, where h is the half-life, t represents the elapsed time, and $Q(t)$ represents the amount that remains (t and h must have the *same unit of time*). Use this information to solve Exercises 67 and 68.

67. Some isotopes of the substance known as thorium have a half-life of only 8 min. (a) If 64 grams are initially present, how many gram (g) of the substance remain after 24 min? (b) How many minutes until only 1 gram (g) of the substance remains?

68. Some isotopes of sodium have a half-life of about 16 hr. (a) If 128 g are initially present, how many grams of the substance remain after 2 days (48 hr)? (b) How many hours until only 1 g of the substance remains?

WRITING, RESEARCH, AND DECISION MAKING

69. In the *Point of Interest* of this section, the formula $f(x) = \left(\frac{1}{2}\right)^x$ was given to determine the probability that "x" number of flips all resulted in heads (or tails). First determine the probability that 20 flips results in *20 heads in a row.* Then use the Internet or some other resource to determine the probability of winning a state lottery (expressed as a decimal). Which has the greater probability? Were you surprised?

70. Answers will vary.

70. In mathematics, it is generally true that for any function $f(x)$, we solve the equation $af(x + h) + k = c$ by isolating the function $f(x + h)$ on one side before the inverse function or procedure can be applied. Solve the equations $118 = 12 \cdot \sqrt{x} + 10$ and $118 = 12 \cdot 3^x + 10$ side-by-side to see how this idea applies to each. Carefully state which operations are used in each step. Comment on how this might apply to a new function or the general function $f(x)$: $118 = 12 \cdot f(x) + 10$.

EXTENDING THE CONCEPT

71. $\frac{1}{5}$

72. 2.4 hr

71. If $10^{2x} = 25$, what is the value of 10^{-x}?

72. Two candles have the same height, but different diameters. Both are lit at the same time and burn at a constant rate. If the first is consumed in 4 hr and the second in 3 hr, how long after being lit was the first candle twice the height of the second?

MAINTAINING YOUR SKILLS

73. 5
-7
9
$2a^2 - 3a$
$2a^2 + 4ah + 2h^2 - 3a - 3h$

74. $x \in [-2, \infty)$, $y \in [-1, \infty)$

73. (2.2) Given $f(x) = 2x^2 - 3x$, determine: $f(-1)$, $f\left(\frac{1}{3}\right)$, $f(a)$, $f(a + h)$

74. (3.3) Graph $g(x) = \sqrt{x + 2} - 1$ using a shift of the parent function. Then state the domain and range of g.

75. (3.3) Given the parent function is $y = |x|$, what is the equation of the function shown? $y = 1|x - 2| + 1$

Exercise 75

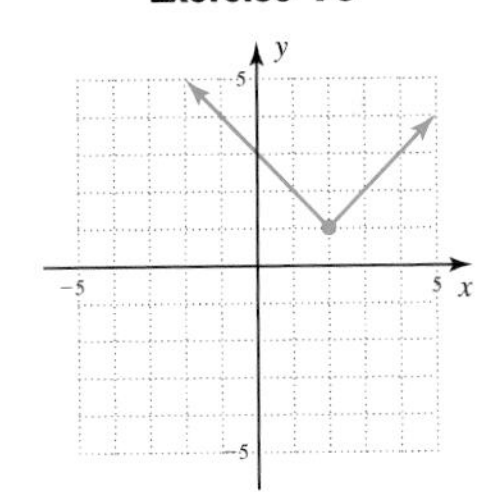

76. (4.3) Factor $P(x)$ completely using the RRT and synthetic division:
$P(x) = 3x^4 - 19x^3 + 15x^2 + 27x - 10$.
$P(x) = (x - 2)(x - 5)(x + 1)(3x - 1)$

77. (1.3) Solve the following equations:

a. $-2\sqrt{x - 3} + 7 = 21$ $\emptyset$

b. $\dfrac{9}{x + 3} + 3 = \dfrac{12}{x - 3}$ $\{-5, 6\}$

78. a. volume of a sphere
b. area of a triangle
c. volume of a rectangular prism
d. Pythagorean theorem

78. (R.7) Identify each formula:

a. $\frac{4}{3}\pi r^3$ b. $\frac{1}{2}bh$

c. lwh d. $a^2 + b^2 = c^2$

5.2 Logarithms and Logarithmic Functions

LEARNING OBJECTIVES

In Section 5.2 you will learn how to:

A. Write exponential equations in logarithmic form

B. Graph logarithmic functions and find their domains

C. Use a calculator to find common logarithms

D. Solve applications of logarithmic functions

INTRODUCTION

A **transcendental function** is one whose solutions are beyond or *transcend* the methods applied to polynomial functions. The exponential function and its inverse, called the logarithmic function, are of this type. In this section, we'll use the concept of an inverse to develop an understanding of the logarithmic function, which has numerous applications that include measuring pH levels, sound and earthquake intensities, barometric pressure, and other natural phenomena.

POINT OF INTEREST

It appears that logarithms were invented by the Scottish mathematician John Napier (1550–1617). The word **logarithm** has its origins in the Greek work *logos,* which means "to reckon," and *arithmos,* which simply means "number." Until calculators and computers became commonplace, logarithms had been used for centuries to manually "reckon" or calculate with numbers, particularly if products, quotients, or powers were involved (see Exercise 96 in *Writing, Research, and Decision Making*). Still, technology has by no means diminished the usefulness of logarithms, since the logarithmic function is required to solve applications involving exponential functions and both exist in abundance.

A. Exponential Equations and Logarithmic Form

While exponential functions have a large number of significant applications, we can't appreciate their full value until we develop the inverse function. Without it, we're unable to solve all but the simplest equations, of the type encountered in Section 5.1. Using the fact that $f(x) = b^x$ is one-to-one, we have the following:

1. The inverse function $f^{-1}(x)$ must exist.
2. We can graph $f^{-1}(x)$ by interchanging the x- and y-coordinates of points from $f(x)$.
3. The domain of $f(x)$ will become the range of $f^{-1}(x)$.
4. The range of $f(x)$ will become the domain of $f^{-1}(x)$.
5. The graph of $f^{-1}(x)$ will be a reflection of $f(x)$ across the line $y = x$.

A table of selected values for $f(x) = 2^x$ is shown in Table 5.1. The points for $f^{-1}(x)$ in Table 5.2 were found by interchanging x- and y-coordinates. Both functions were then graphed using these points.

Table 5.1

$f(x)$: $y = 2^x$

x	y
−3	$\frac{1}{8}$
−2	$\frac{1}{4}$
−1	$\frac{1}{2}$
0	1
1	2
2	4
3	8

Table 5.2

$f^{-1}(x)$: $x = 2^y$

x	$y = f^{-1}(x)$
$\frac{1}{8}$	−3
$\frac{1}{4}$	−2
$\frac{1}{2}$	−1
1	0
2	1
4	2
8	3

Figure 5.4

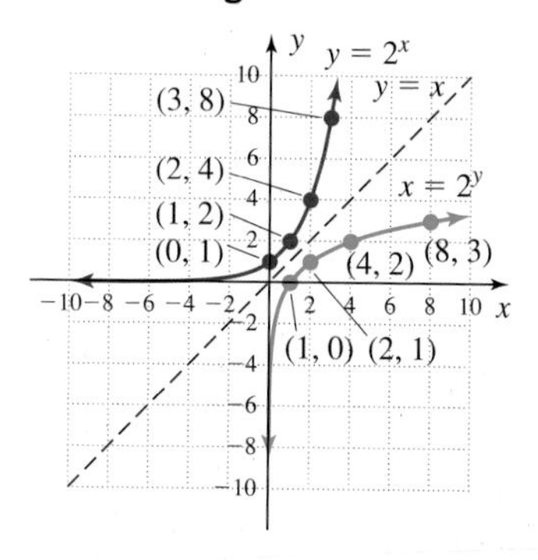

The interchange of x and y and the graphs in Figure 5.4 show that $f^{-1}(x)$ has a y-intercept of (1, 0), a vertical asymptote at $x = 0$, a domain of $x \in (0, \infty)$, and a range of $y \in (-\infty, \infty)$, giving us a great deal of information about the function. To find *an equation* for $f^{-1}(x)$, we'll attempt to use the algebraic approach employed previously. For $f(x) = 2^x$,

1. use y instead of $f(x)$: $y = 2^x$.
2. interchange x and y: $x = 2^y$.

At this point we have an implicit equation for the inverse function, but no algebraic operations that enable us to solve explicitly for y in terms of x. Instead, we write $x = 2^y$ in function form by noting that "y is the exponent that goes on base 2 to obtain x."

In the language of mathematics, this phrase is represented by $y = \log_2 x$ and is called a base 2 **logarithmic function.** For $y = b^x$, $y = \log_b x$ is the inverse function and is read "y is the base-b logarithm of x." For this new function, always keep in mind what y represents—y is an exponent. In fact, *y is the exponent that goes on base b to obtain x: $y = \log_b x$.*

LOGARITHMIC FUNCTIONS

For $b > 0$, $b \neq 1$, the base-b logarithmic function is defined as

$y = \log_b x$ if and only if $x = b^y$

Domain: $x \in (0, \infty)$ Range: $y \in \mathbb{R}$

EXAMPLE 1 Write each equation in exponential form.

a. $3 = \log_2 8$ **b.** $1 = \log_2 2$ **c.** $0 = \log_2 1$ **d.** $-2 = \log_2 \frac{1}{4}$

Solution: **a.** $2^3 = 8$ **b.** $2^1 = 2$ **c.** $2^0 = 1$ **d.** $2^{-2} = \frac{1}{4}$

NOW TRY EXERCISES 7 THROUGH 22

This relationship can be applied to *any* base b, where $b > 0$ and $b \neq 1$. When given the exponential form, as in $5^3 = 125$, note the exponent on the base and begin there: 3 is the exponent that goes on base 5 to obtain 125, or more exactly, *3 is the base-5 logarithm of 125*: $3 = \log_5 125$.

EXAMPLE 2 Write each equation in logarithmic form.

a. $6^3 = 216$ **b.** $2^{-1} = \frac{1}{2}$ **c.** $b^0 = 1$ **d.** $9^{\frac{3}{2}} = 27$

Solution:

a. $6^3 = 216$
3 is the base-6 logarithm of 216
$3 = \log_6 216$

b. $2^{-1} = \frac{1}{2}$
-1 is the base-2 logarithm of $\frac{1}{2}$
$-1 = \log_2 \frac{1}{2}$

c. $b^0 = 1$
0 is the base-b logarithm of 1
$0 = \log_b 1$

d. $9^{\frac{3}{2}} = 27$
$\frac{3}{2}$ is the base-9 logarithm of 27
$\frac{3}{2} = \log_9 27$

NOW TRY EXERCISES 23 THROUGH 38

EXAMPLE 3 Determine the value of each expression without using a calculator:

a. $\log_2 8$ **b.** $\log_5 \frac{1}{25}$ **c.** $\log_b b$ **d.** $\log_{10} 100$

Solution:

a. $\log_2 8 = 3$, since $2^3 = 8$
b. $\log_5 \frac{1}{25} = -2$, since $5^{-2} = \frac{1}{25}$
c. $\log_b b = 1$, since $b^1 = b$
d. $\log_{10} 100 = 2$, since $10^2 = 100$

NOW TRY EXERCISES 39 THROUGH 62

Alternating between logarithmic form and exponential form reveals the following four properties, which we'll use in Section 5.3 to simplify expressions and in Section 5.4 to solve equations involving logarithms. Also see the *Technology Highlight* on page 491.

LOGARITHMIC PROPERTIES

For any base b, where $b > 0$ and $b \neq 1$,

I. $\log_b b = 1$, since $b^1 = b$

II. $\log_b 1 = 0$, since $b^0 = 1$

III. $\log_b b^x = x$, since the exponential form gives $b^x = b^x$

IV. $b^{\log_b x} = x$, since $\log_b x$ is "the exponent on b" for x

Figure 5.5

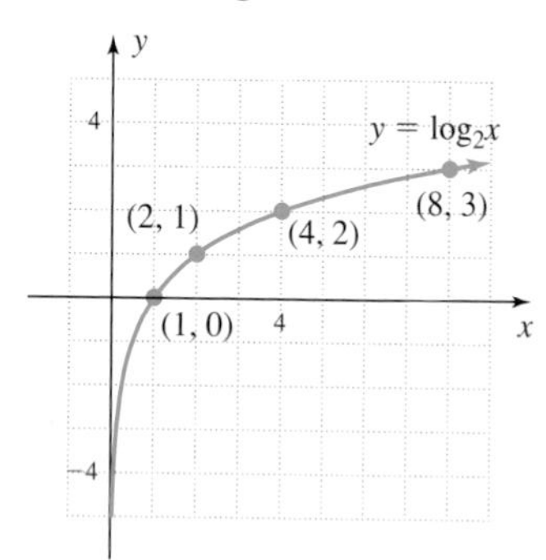

B. Graphing Logarithmic Functions

As with the other basic graphs we've encountered, logarithmic graphs maintain the same characteristics when transformations are applied, and its graph *should be added to your collection of basic functions,* ready for recall and analysis as the situation requires. We've already graphed the function $y = \log_2 x$ $(b > 1)$ earlier in this section, using $x = 2^y$ as the inverse function for $y = 2^x$. The graph is repeated in Figure 5.5.

EXAMPLE 4 Graph $Y = \log_2(x - 1) + 1$ using *transformations of* $y = \log_2 x$ (not by simply plotting points). Clearly state what transformations are applied.

Solution: The graph of Y is the same as that of $y = \log_2 x$, shifted 1 unit right and 1 unit up. With this in mind we can sketch the vertical asymptote at $x = 1$ and plot the point (2, 1). Knowing the general shape of the graph, we need only one or two additional points to complete the graph. For $x = 5$ and $x = 9$ we find $Y = \log_2 4 + 1 = 2 + 1 = 3$, and $Y = \log_2 8 + 1 = 3 + 1 = 4$, respectively. The graph is shown in the figure. Note the domain of this function is $x \in (1, \infty)$.

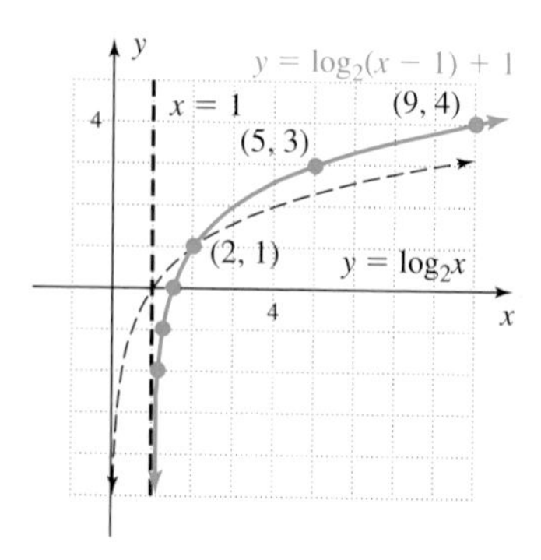

NOW TRY EXERCISES 63 THROUGH 70

Figure 5.6

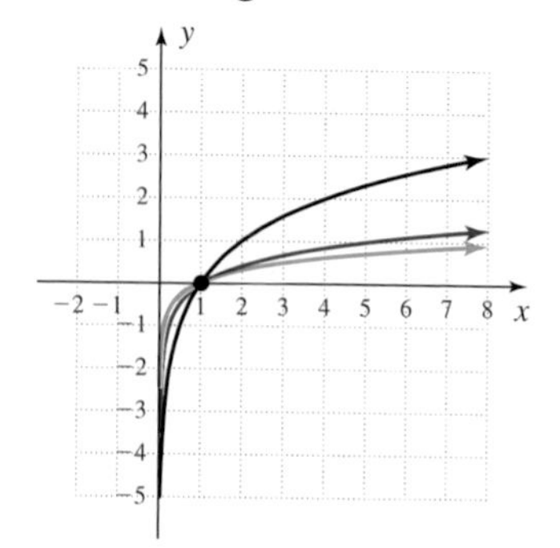

For convenience and ease of calculation, examples of graphing have been done using base-2 logarithms. However, the basic shape of a logarithmic graph remains unchanged regardless of the base used and transformations can be applied to $y = \log_b x$ for any value of b. The illustration in Figure 5.6 shows representative graphs for $y = \log_2 x$, $y = \log_5 x$, and $y = \log_{10} x$.

Example 5 illustrates how the domain of a logarithmic function can change when certain transformations are applied. Since the domain consists of *positive* real numbers, the argument of a logarithmic function must be greater than zero. This means finding the domain often consists of solving various inequalities, which can be done using the skills acquired in Sections 2.5 and 4.7.

EXAMPLE 5 Determine the domain of each function.

a. $p(x) = \log_2(2x + 3)$

b. $q(x) = \log_5(x^2 - 2x)$

c. $r(x) = \log_{10}\left(\dfrac{3 - x}{x + 3}\right)$

Solution:

a. Solving $2x + 3 > 0$ gives $x > -\frac{3}{2}$. $D: x \in [-\frac{3}{2}, \infty)$.

b. For $x^2 - 2x > 0$, we note $y = x^2 - 2x$ is a parabola, concave up, with roots at $x = 0$ and $x = 2$. This means $x^2 - 2x$ will be positive for $x < 0$ and $x > 2$. $D: x \in (-\infty, 0) \cup (2, \infty)$.

c. For $\frac{3 - x}{x + 3} > 0$, $y = \frac{3 - x}{x + 3}$ has a zero at 3 and a vertical asymptote at $x = -3$. Outputs are positive in the interval containing $x = 0$, so y is positive in the interval $(-3, 3)$. $D: x \in (-3, 3)$.

NOW TRY EXERCISES 71 THROUGH 76

GRAPHICAL SUPPORT

The domain for $r(x) = \log_{10}\left(\frac{3 - x}{x + 3}\right)$ from Example 5(c) can be confirmed using the LOG key on a graphing calculator. Use the key to enter the equation as Y_1 on the Y= screen, then graph the function using the ZOOM **4:ZDecimal** option of the TI-84 Plus. Both the graph and TABLE feature help to confirm the domain is $x \in (-3, 3)$.

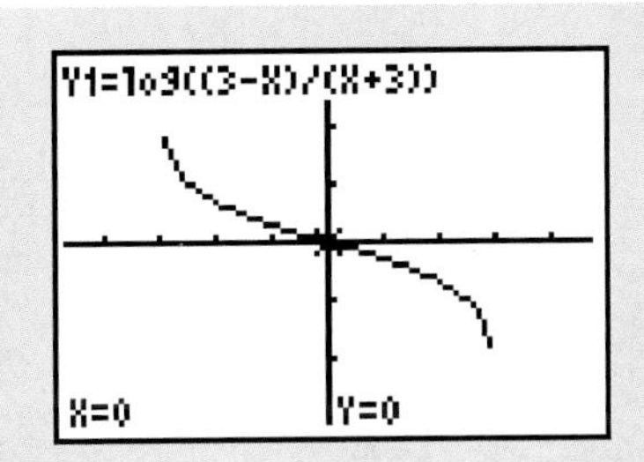

WORTHY OF NOTE

We do something similar with square roots. Technically, the "square root of x" should be written $\sqrt[2]{x}$. However, square roots are so common we often omit the 2, assuming that if no index is written, an index of 2 is intended.

WORTHY OF NOTE

After their invention in about 1614, logarithms became widely used as a tool for performing difficult computations. In 1624, Henry Briggs (1561–1630), a professor of geometry at Gresham College, London, published tables that included the base-10 logarithm (the exponent on base 10) for the numbers 1 to 20,000 to *14 decimal places.* This made it possible to compute many products, quotients, and powers by doing little more than looking up the needed values in a table and using the properties of exponents. Logarithmic tables remained in common use for nearly 400 years, right up until the invention of modern calculating technology.

C. Calculators and Common Logarithms

Of all possible bases for $\log_b x$, one of the most common is base 10, presumably due to our base-10 numeration system. In fact, $\log_{10} x$ is called a **common logarithm** and we most often do not write the "10." Instead, we simply write $\log x$ for $\log_{10} x$.

Some base-10 logarithms are easy to evaluate: $\log 1000 = 3$ since $10^3 = 1000$; and $\log \frac{1}{100} = -2$ since $10^{-2} = \frac{1}{100}$. But what about expressions like log 850 and $\log \frac{1}{4}$? Due to the relationship between exponential and logarithmic functions, values must exist for these expressions, as well as for expressions like $\log \sqrt{5}$. Further, the inequality $100 < 850 < 1000$ implies $\log 100 = 2 < \log 850 < \log 1000 = 3$, telling us that log 850 is **bounded** between 2 and 3. A similar inequality can be constructed for $\log \frac{1}{4}$: $\frac{1}{10} < \frac{1}{4} < 1$ implies that $\log \frac{1}{10} < \log \frac{1}{4} < \log 1$, and the value of $\log \frac{1}{4}$ is bounded between -1 and 0. Fortunately, modern calculators have been programmed to compute base-10 logarithms with great speed, and often with 10-decimal-place accuracy. For log 850, press the LOG key, input 850 and press ENTER. The display should read 2.929418926. For $\log \frac{1}{4}$ we can use the decimal form: LOG 0.25 ENTER $\approx -.6020599913$.

EXAMPLE 6 Estimate the value of each expression by writing an inequality that bounds it between two integers. Then use a calculator to find an approximate value rounded to four decimal places. Note all examples use base 10.

a. log 492 **b.** log 2.7 **c.** log 0.009

Solution:

a. Since $100 < 492 < 1000$, log 492 is bounded between 2 and 3: $\log 100 = 2 < \log 492 < \log 1000 = 3$. Using a calculator, $\log 492 \approx 2.6920$.

b. Since $1 < 2.7 < 10$, $\log 2.7$ is bounded between 0 and 1: $\log 1 = 0 < \log 2.7 < \log 10 = 1$. Using a calculator, $\log 2.7 \approx .4314$.

c. Since $0.001 < 0.009 < 0.010$, $\log 0.009$ is bounded between -3 and -2: $\log 0.001 = -3 < \log 0.009 < \log 0.010 = -2$. Using a calculator we find $\log 0.009 \approx -2.0458$.

NOW TRY EXERCISES 77 THROUGH 82

D. Applications of Common Logarithms

One application of common logarithms involves the measurement of earthquake intensities, in units called **magnitudes** (also called **Richter values**). The most damaging earthquakes have magnitudes of 6.5 and higher, while the slightest earthquakes are of magnitude 1 and are barely perceptible. The magnitude of the intensity $M(I)$ is given by $M(I) = \log\left(\frac{I}{I_0}\right)$, where I is the measured intensity and I_0 represents the smallest earth movement that can be recorded on a seismograph, called the **reference intensity.** The value of I is often given as a multiple of this reference intensity.

EXAMPLE 7A Find the magnitude of an earthquake (round to hundredths), given (a) $I = 4000I_0$ and (b) $I = 8{,}252{,}000I_0$.

Solution: **a.**

$$M(I) = \log\left(\frac{I}{I_0}\right) \quad \text{given}$$

$$M(4000I_0) = \log\left(\frac{4000I_0}{I_0}\right) \quad \text{substitute } 4000I_0 \text{ for } I$$

$$= \log 4000 \quad \text{simplify}$$

$$\approx 3.60 \quad \text{result}$$

The earthquake had a magnitude of 3.6.

b.

$$M(I) = \log\left(\frac{I}{I_0}\right) \quad \text{given}$$

$$M(8{,}252{,}000I_0) = \log\left(\frac{8{,}252{,}000I_0}{I_0}\right) \quad \text{substitute } 8{,}252{,}000I_0 \text{ for } I$$

$$= \log 8{,}252{,}000 \quad \text{simplify}$$

$$\approx 6.92 \quad \text{result}$$

The earthquake had a magnitude of 6.92.

EXAMPLE 7B How many times more intense than the reference intensity I_0 is an earthquake with a magnitude of 6.7?

Solution:

$$M(I) = \log\left(\frac{I}{I_0}\right) \quad \text{given}$$

$$6.7 = \log\left(\frac{I}{I_0}\right) \quad \text{substitute } M(I) = 6.7$$

$$10^{6.7} = \frac{I}{I_0} \quad \text{write in exponential form}$$

$$I = 10^{6.7}I_0 \quad \text{solve for } I$$
$$= 5{,}011{,}872I_0 \quad 10^{6.7} \approx 5{,}011{,}872$$

An earthquake of magnitude 6.7 is over 5 million times more intense than the reference intensity.

NOW TRY EXERCISES 85 THROUGH 90

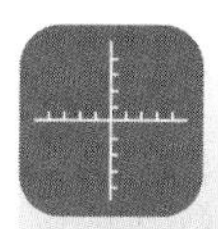

TECHNOLOGY HIGHLIGHT

Logarithms and Exponents

The keystrokes shown apply to a TI-84 Plus model. Please consult our Internet site or your manual for other models.

Although a calculator can rarely be the final word in a mathematical demonstration, it can be used to verify certain ideas. In particular, we'll look at the properties given on page 488, which show an inverse function relationship exists between $y = \log_b x$ and $y = b^x$. Here, we'll simply demonstrate the inverse relation using a graphing calculator, although we'll limit ourselves to base-10 logarithms. Figure 5.7 shows how the base-10 logarithm "undoes" the base-10 exponential. Figure 5.8 shows how the base-10 exponential "undoes" the base-10 logarithm. Use these examples to work the following exercises.

Figure 5.7

```
log(10^3)
                3
log(10^-2)
               -2
log(10)
                1
log(10^-π)
```

Figure 5.8

```
10^log(50)
               50
10^log(.01)
              .01
10^log(1)
                1
10^log(π)
```

Exercise 1: As you can see, the display screen in Figure 5.8 could not display the result of the final calculation without scrolling. What will the display show after ENTER is pressed? $-3.141592654 = -\pi$

Exercise 2: As in Exercise 1, the display screen in Figure 5.8 did not have enough room to display the result of the final calculation. What will the result be? $3.141592654 = \pi$

Exercise 3: First simplify each expression by hand, then use a graphing calculator to verify each result: (a) $\log_{10}10^{2.5}$, (b) $\log_{10}10^{\pi}$, (c) $10^{\log_{10}2.5}$, (d) $10^{\log_{10}\pi}$
2.5 $\quad \pi \quad$ 2.5 $\quad \pi$

5.2 EXERCISES

CONCEPTS AND VOCABULARY

Fill in each blank with the appropriate word or phrase. Carefully reread the section if needed.

1. A logarithmic function is of the form $y =$ __$\log_b x$__, where __b__ > 0, __b__ $\neq 1$ and inputs are __greater__ than zero.

2. The range of $y = \log_b x$ is all __real numbers__ and the domain is $x \in$ __$(0, \infty)$__. Further, as $x \to 0$, $y \to$ __$-\infty$__.

3. For logarithmic functions of the form $y = \log_b x$, the x-intercept is __(1, 0)__, since $\log_b 1 =$ __0__.

4. The function $y = \log_b x$ is an increasing function if __$b > 1$__, and a decreasing function if __$0 < b < 1$__.

5. What number does the expression $\log_2 32$ represent? Discuss/explain how $\log_2 32 = \log_2 2^5$ justifies this fact.
5; answers will vary

6. Explain how the graph of $Y = \log_b(x - 3)$ can be obtained from $y = \log_b x$. Where is the "new" x-intercept? Where is the new asymptote? shifts right 3; (4, 0); $x = 3$

29. $\log_{\frac{1}{3}} 27 = -3$
30. $\log_{\frac{1}{5}}(25) = -2$
31. $\log 1000 = 3$
32. $\log 10 = 1$
33. $\log \frac{1}{100} = -2$
34. $\log \frac{1}{100{,}000} = -5$

63. shift up 3

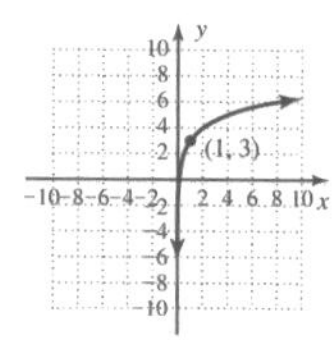
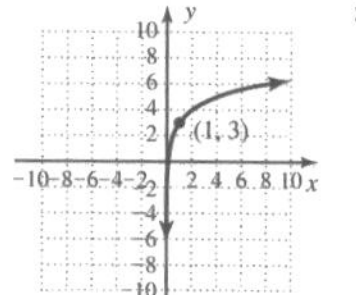

64. shift right 2

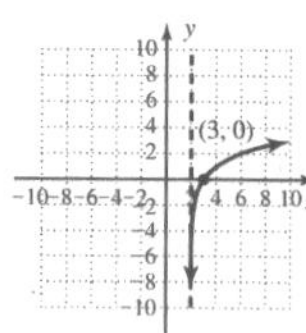

65. shift right 2, up 3

66. shift down 2

67. shift left 1

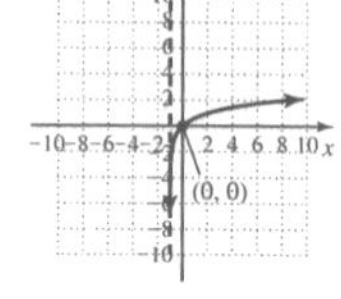

68. shift left 1, down 2

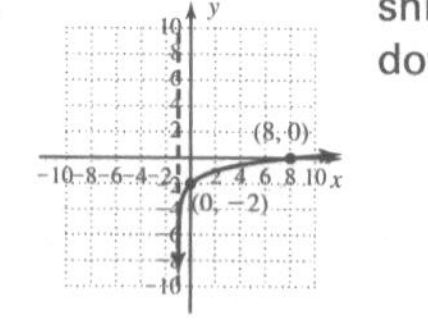

69. reflect across x-axis, shift left 1

70. reflect across x-axis, shift up 2

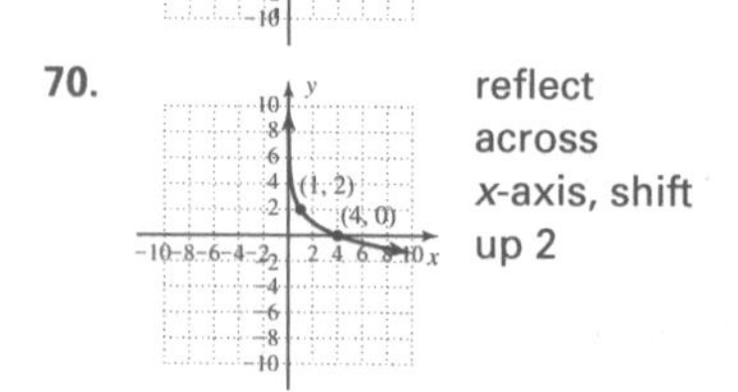

71. $x \in (-\infty, -1) \cup (3, \infty)$
72. $x \in (-\infty, -3) \cup (2, \infty)$
73. $x \in \left(\frac{3}{2}, \infty\right)$

DEVELOPING YOUR SKILLS

Write each equation in exponential form.

7. $3 = \log_2 8$ $\quad 2^3 = 8$
8. $2 = \log_3 9$ $\quad 3^2 = 9$
9. $-1 = \log_7 \frac{1}{7}$ $\quad 7^{-1} = \frac{1}{7}$
10. $-3 = \log_5 \frac{1}{125}$ $\quad 5^{-3} = \frac{1}{125}$
11. $0 = \log_9 1$ $\quad 9^0 = 1$
12. $0 = \log_4 1$ $\quad 4^0 = 1$
13. $\frac{1}{3} = \log_8 2$ $\quad 8^{\frac{1}{3}} = 2$
14. $\frac{1}{2} = \log_{81} 9$ $\quad 81^{\frac{1}{2}} = 9$
15. $1 = \log_2 2$ $\quad 2^1 = 2$
16. $1 = \log_{20} 20$ $\quad 20^1 = 20$
17. $\log_7 49 = 2$ $\quad 7^2 = 49$
18. $\log_4 16 = 2$ $\quad 4^2 = 16$
19. $\log_{10} 100 = 2$ $\quad 10^2 = 100$
20. $\log_{10} 10{,}000 = 4$ $\quad 10^4 = 10{,}000$
21. $\log_{10} 0.1 = -1$ $\quad 10^{-1} = 0.1$
22. $\log_{10} 0.001 = -3$ $\quad 10^{-3} = 0.001$

Write each equation in logarithmic form.

23. $4^3 = 64$ $\quad \log_4 64 = 3$
24. $2^5 = 32$ $\quad \log_2 32 = 5$
25. $3^{-2} = \frac{1}{9}$ $\quad \log_3 \frac{1}{9} = -2$
26. $2^{-3} = \frac{1}{8}$ $\quad \log_2 \frac{1}{8} = -3$
27. $9^0 = 1$ $\quad \log_9 1 = 0$
28. $8^0 = 1$ $\quad \log_8 1 = 0$
29. $\left(\frac{1}{3}\right)^{-3} = 27$
30. $\left(\frac{1}{5}\right)^{-2} = 25$
31. $10^3 = 1000$
32. $10^1 = 10$
33. $10^{-2} = \frac{1}{100}$
34. $10^{-5} = \frac{1}{100{,}000}$
35. $4^{\frac{3}{2}} = 8$ $\quad \log_4 8 = \frac{3}{2}$
36. $216^{\frac{2}{3}} = 36$ $\quad \log_{216} 36 = \frac{2}{3}$
37. $4^{\frac{-3}{2}} = \frac{1}{8}$ $\quad \log_4 \frac{1}{8} = \frac{-3}{2}$
38. $27^{\frac{-2}{3}} = \frac{1}{9}$ $\quad \log_{27} \frac{1}{9} = \frac{-2}{3}$

Determine the value of each expression without using a calculator.

39. $\log_{11} 121$ $\quad 2$
40. $\log_{12} 144$ $\quad 2$
41. $\log_3 243$ $\quad 5$
42. $\log_6 216$ $\quad 3$
43. $\log_7 \frac{1}{49}$ $\quad -2$
44. $\log_9 \frac{1}{81}$ $\quad -2$
45. $\log_4 4$ $\quad 1$
46. $\log_9 9$ $\quad 1$
47. $\log_{10} 10$ $\quad 1$
48. $\log_8 8$ $\quad 1$
49. $\log_4 2$ $\quad \frac{1}{2}$
50. $\log_{81} 9$ $\quad \frac{1}{2}$

Determine the value of x by writing the equation in exponential form.

51. $\log_5 x = 2$ $\quad 25$
52. $\log_3 x = 3$ $\quad 27$
53. $\log_{36} x = \frac{1}{2}$ $\quad 6$
54. $\log_{64} x = \frac{1}{2}$ $\quad 8$
55. $\log_x 36 = 2$ $\quad 6$
56. $\log_x 64 = 2$ $\quad 8$
57. $\log_x \frac{1}{4} = -2$ $\quad 2$
58. $\log_x \frac{1}{3} = -1$ $\quad 3$
59. $\log_{25} x = -\frac{3}{2}$ $\quad \frac{1}{125}$
60. $\log_{16} x = -\frac{5}{4}$ $\quad \frac{1}{32}$
61. $\log_8 32 = x$ $\quad \frac{5}{3}$
62. $\log_9 27 = x$ $\quad \frac{3}{2}$

Graph each function *using transformations* of $y = \log_b x$ and strategically plotting a few points. Clearly state the transformations applied.

63. $f(x) = \log_2 x + 3$
64. $g(x) = \log_2(x - 2)$
65. $h(x) = \log_2(x - 2) + 3$
66. $p(x) = \log_3 x - 2$
67. $q(x) = \log_3(x + 1)$
68. $r(x) = \log_3(x + 1) - 2$
69. $Y_1 = -\log_2(x + 1)$
70. $Y_2 = -\log_2 x + 2$

Determine the domain of the following functions.

71. $y = \log_6\left(\frac{x + 1}{x - 3}\right)$
72. $y = \log_3\left(\frac{x - 2}{x + 3}\right)$
73. $y = \log_5 \sqrt{2x - 3}$
74. $y = \log_4 \sqrt{5 - 3x}$ $\quad x \in \left(-\infty, \frac{5}{3}\right)$
75. $y = \log(9 - x^2)$ $\quad x \in (-3, 3)$
76. $y = \log(9x - x^2)$ $\quad x \in (0, 9)$

Without using a calculator, write an inequality that bounds each expression between two integers. Then use a calculator to find an approximate value, rounded to four decimal places.

77. $\log 175$ $\quad \approx 2.2430$
78. $\log 8.2$ $\quad \approx 0.9138$
79. $\log 127{,}962$ $\quad \approx 5.1071$
80. $\log 9871$ $\quad \approx 3.9944$
81. $\log \frac{1}{5}$ $\quad \approx -0.6990$
82. $\log 0.075$ $\quad \approx -1.1249$

WORKING WITH FORMULAS

83. **pH level: $f(x) = -\log_{10} x$**

The pH level of a solution indicates the concentration of hydrogen (H^+) ions in a unit called *moles per liter*. The pH level $f(x)$ is given by the formula shown, where x is the ion concentration (given in scientific notation). A solution with pH < 7 is called an acid (lemon juice: pH ≈ 2), and a solution with pH > 7 is called a base (household ammonia:

pH ≈ 11). Use the formula to determine the pH level of tomato juice if $x = 7.94 \times 10^{-5}$ moles per liter. Is this an acid or base solution? pH ≈ 4.1; acid

84. Time required for an investment to double: $T(r) = \dfrac{\log 2}{\log(1 + r)}$

The time required for an investment to double in value is given by the formula shown, where r represents the interest rate (expressed as a decimal) and $T(r)$ gives the years required. How long would it take an investment to double if the interest rate were (a) 5%, (b) 8%, (c) 12%? ≈14.2 yr ≈9.0 yr ≈6.1 yr

APPLICATIONS

Brightness of a star: The brightness or intensity I of a star as perceived by the naked eye is measured in units called *magnitudes.* The brightest stars have magnitude 1 $[M(I) = 1]$ and the dimmest have magnitude 6 $[M(I) = 6]$. The magnitude of a star is given by the equation $M(I) = 6 - 2.5 \cdot \log\left(\frac{I}{I_0}\right)$, where I is the actual intensity of light from the star and I_0 is the faintest light visible to the human eye, called the reference intensity. The intensity I is often given as a multiple of this reference intensity.

85. Find the value of $M(I)$ given **a.** $I = 27I_0$ and **b.** $I = 85I_0$. ≈2.4 ≈1.2

86. Find the intensity I of a star given **a.** $M(I) = 1.6$ and **b.** $M(I) = 5.2$. ≈$57.5I_0$ ≈$2.1I_0$

Earthquake intensity: The intensity of an earthquake is also measured in units called *magnitudes.* The most damaging quakes have magnitudes of 6.5 or greater $[M(I) > 6.5]$ and the slightest are of magnitude 1 $[M(I) = 1]$ and are barely felt. The magnitude of an earthquake is given by the equation $M(I) = \log\left(\frac{I}{I_0}\right)$, where I is the actual intensity of the earthquake and I_0 is the smallest earth movement that can be recorded on a seismograph—called the reference intensity. The intensity I is often given as a multiple of I_0.

87. Find the value of $M(I)$ given **a.** $I = 50{,}000I_0$ and **b.** $I = 75{,}000I_0$. ≈4.7 ≈4.9

88. Find the intensity I of the earthquake given **a.** $M(I) = 3.2$ and **b.** $M(I) = 8.1$. ≈$1584.9I_0$ ≈$125{,}892{,}541.2I_0$

Intensity of sound: The intensity of sound as perceived by the human ear is measured in units called decibels (dB). The loudest sounds that can be withstood without damage to the eardrum are in the 120- to 130-dB range, while a whisper may measure in the 15- to 20-dB range. Decibel measure is given by the equation $D(I) = 10 \log\left(\frac{I}{I_0}\right)$, where I is the actual intensity of the sound and I_0 is the faintest sound perceptible by the human ear—called the reference intensity. The intensity I is often given as a multiple of this reference intensity, but often the constant 10^{-16} (watts per cm^2; W/cm^2) is used as the threshold of audibility.

89. Find the value of $D(I)$ given **a.** $I = 10^{-14}$ and **b.** $I = 10^{-4}$. 20 dB 120 dB

90. Find the intensity I of the sound given **a.** $D(I) = 83$ and **b.** $D(I) = 125$. $199{,}526{,}231.5I_0$ ≈$3.2 \times 10^{12}I_0$

Memory retention: Under certain conditions, a person's retention of random facts can be modeled by the equation $P(x) = 95 - 14 \log_2 x$, where $P(x)$ is the percentage of those facts retained after x number of days. Find the percentage of facts a person might retain after:

91. a. 1 day **b.** 4 days **c.** 16 days a. 95% b. 67% c. 39%

92. a. 32 days **b.** 64 days **c.** 78 days a. 25% b. ≈11% c. ≈7%

93. Use the formula given in Exercise 83 to determine the pH level of black coffee if $x = 5.1 \times 10^{-5}$ moles per liter. Is black coffee considered an acid or base solution? ≈4.3; acid

94. The length of time required for an amount of money to *triple* is given by the formula $T(r) = \frac{\log 3}{\log(1 + r)}$ (refer to Exercise 84). Construct a table of values to help estimate what interest rate is needed for an investment to triple in nine years. $\approx 13\%$

WRITING, RESEARCH, AND DECISION MAKING

95. The decibel is a unit based on the faintest sound a person can hear, called the threshold of audibility. It is a base-10 logarithmic scale, meaning a sound 10 times more intense is 1-dB louder. Many texts and reference books give estimates of the noise level (in decibels) of common sounds. Through reading and research, try to locate or approximate where the following sounds would fall along this scale. In addition, determine at what point pain or ear damage begins to occur. Answers will vary.

a. threshold of audibility 0 dB **b.** lawn mower 90 dB **c.** whisper 15 dB

d. loud rock concert 120 dB **e.** lively party 100 dB **f.** jet engine 120 dB

96. Tables of base-10 logarithms are still readily available over the Internet, at a library, and in other places. Locate such a table and take an excursion back in time. After reading the example that follows, compute the value of $\frac{2843}{981}$ *using the table* and properties of exponents (verify the result on a calculator). Consider the product 93×207. If each number is rewritten using a base-10 exponential the calculation becomes $10^{1.968482949} \times 10^{2.315970345}$. Using the properties of exponents yields $10^{1.96848294 + 2.31597034} = 10^{4.284453294}$. From the entries within the log table, we find that the number whose log is 4.28445329 is 19,251—which *is* the product of 93×207. 2.898063201

97. Base-10 logarithms are sometimes called *Briggsian* logarithms due to the work of Henry Briggs. See the *Worthy of Note* preceding Example 6. Using resources available to you, locate some additional information on Henry Briggs and write up a short summary. Include information on his contributions to other areas of mathematics. Answers will vary.

EXTENDING THE CONCEPT

98. Determine the value of x that makes the equation true: $\log_3[\log_3(\log_3 x)] = 0$. 27

99. Use properties of exponents *and* logarithms to show $y = \log_{\frac{1}{2}} x$ is equivalent to $y = -\log_2 x$. $\left(\frac{1}{2}\right)^y = x$, $2^{-y} = x$, $-y = \log_2 x$, $y = -\log_2 x$

100. Find the value of each expression.

a. $\log_{64}\frac{1}{16}$ $\frac{-2}{3}$ **b.** $\log_{\frac{4}{9}}\frac{27}{8}$ $\frac{-3}{2}$ **c.** $\log_{0.25} 32$ $\frac{-5}{2}$

MAINTAINING YOUR SKILLS

101. (1.2) Graph the solution set for (a) $x < 3$ and $x > -1$ and (b) $x < 3$ or $x > -1$.

102. (3.3) Graph $g(x) = \sqrt[3]{x + 2} - 1$ by shifting the parent function. Then state the domain and range of g.

103. (R.4) Factor the following expressions:

a. $x^3 - 8$ **b.** $a^2 - 49$ **c.** $n^2 - 10n + 25$ **d.** $2b^2 - 7b + 6$

104. (1.4) Find the sum, difference, product, and quotient of the complex numbers $1 + 3i$ and $1 - 3i$. 2; $6i$; 10; $\frac{-4}{5} + \frac{3}{5}i$

105. (4.4) For the graph shown, write the solution set for $f(x) < 0$. Then write the equation of the graph in factored form and in polynomial form. $x \in (-\infty, -5)$; $f(x) = (x + 5)(x - 4)^2 = x^3 - 3x^2 - 24x + 80$

Exercise 105

106. (2.2) A function $f(x)$ is defined by the ordered pairs shown in the table. Is the function (a) linear? (b) increasing? Justify your answers. a. No b. No, answers will vary.

101. a.

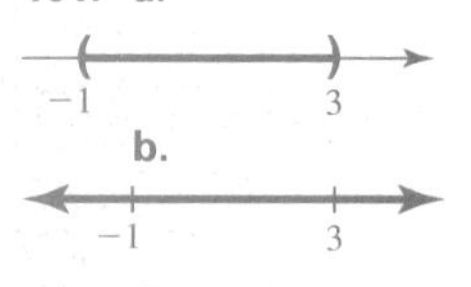

102. $D: x \in \mathbb{R}$
$R: y \in \mathbb{R}$

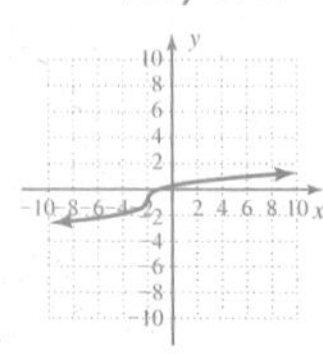

103. a. $(x - 2)(x^2 + 2x + 4)$
b. $(a + 7)(a - 7)$
c. $(n - 5)(n - 5)$
d. $(2b - 3)(b - 2)$

Exercise 106

x	y
−10	0
−9	−2
−8	−8
−6	−18
−5	−50
−4	−72

5.3 The Exponential Function and Natural Logarithms

LEARNING OBJECTIVES

In Section 5.3 you will learn how to:

- **A. Evaluate and graph base e exponential functions**
- **B. Evaluate and graph base e logarithmic functions**
- **C. Apply the properties of logarithms**
- **D. Use the change-of-base formula**
- **E. Solve applications of natural logarithms**
- **F. Compute average rates of change for $y = e^x$ and $y = \ln(x)$**

INTRODUCTION

Up to this point we've seen a large number of base-2 exponential and logarithmic functions. This is because they're convenient for most calculations and enable a study of the basic graphs without the outputs getting too large. We've also used a number of base-10 functions, primarily for traditional reasons and their connections with our base-10 numeration system. However, in virtually all future course work, base $e \approx 2.718$ will be more common by far. So much so, the base-e exponential function is referred to as *the* exponential function, as in the title of this section. We explore the reasons why here.

POINT OF INTEREST

In addition to the "rate of growth" advantage it offers, using base *e* simplifies a number of important calculations in future courses (some bases are easier to work with than others); adds a great deal of understanding to the study of complex numbers; and is extensively used in science, engineering, business, and finance applications.

A. *The* Exponential Function $y = e^x$

In Section 5.1, we discussed the city of Goldsboro—a hypothetical boomtown from the Old West whose population at time t in years was $P(t) = 1000 \cdot 2^t$. Suppose we were more interested in the *rate of growth* of this town, or more specifically, its rate of growth expressed as a percentage of the current population. In Section 5.6 we'll show that $P(t)$ can be equivalently written as $P(t) = 1000e^{kt}$, where k is a constant that gives this growth rate *exactly,* and e is an irrational number whose approximate value is 2.718281828 (to nine decimal places). Knowing this rate of growth offers a huge advantage in applications of the exponential function $y = e^x$, also known as the **natural exponential function.**

In Section 5.5 the value of e is developed in the context of compound interest. For now we'll define it as the number that $\left(1 + \frac{1}{x}\right)^x$ approaches as x becomes very large. It can be shown that as $x \to \infty$, $\left(1 + \frac{1}{x}\right)^x$ approaches the unique, irrational number e we approximated earlier. Table 5.3 gives values of the expression for selected inputs x. Just as we use the symbol π to represent the irrational number 3.141592654 . . . (the ratio of a circle's circumference to its diameter), we use the symbol e to represent the irrational number 2.718281828 . . . . This symbol is used in honor of the famous mathematician Leonhard Euler (1707–1783), who studied the number extensively.

WORTHY OF NOTE

For the Goldsboro example, k is approximately 0.693147. Using a graphing calculator, compare the functions $Y_1 = 1000 \cdot 2^t$ and $Y_2 = 1000(2.7182818)^{0.693147t}$ using the TABLE feature. What do you notice? Soon we'll introduce a much more convenient way to write this exponential base.

Table 5.3

x	$\left(1 + \frac{1}{x}\right)^x$
1	2
10	2.59
100	2.705
1000	2.7169
10,000	2.71815
100,000	2.718268
1,000,000	2.7182805
10,000,000	2.71828169

Instead of having to enter a decimal approximation when computing with e, most calculators have an e^x key, usually as the 2nd function for the key marked LN. To find the value of e^2, use the keystrokes 2nd LN 2 ENTER, and the calculator display should read 7.389056099.

EXAMPLE 1 Use a calculator to evaluate each expression, rounded to six decimal places.

a. e^3 **b.** e^1 **c.** e^0 **d.** $e^{\frac{1}{2}}$

Solution: **a.** $e^3 \approx 20.085537$ **b.** $e^1 \approx 2.718282$

c. $e^0 = 1$ (exactly) **d.** $e^{\frac{1}{2}} \approx 1.648721$

NOW TRY EXERCISES 7 THROUGH 14

Although e is an irrational number, the graph of $y = e^x$ behaves in exactly the same way and has the same characteristics as other exponential graphs. Figure 5.9 shows the graph on the same grid as $y = 2^x$ and $y = 3^x$. As we might expect, all three graphs are increasing, have an asymptote at $y = 0$, and contain the point (0, 1), with the graph of $y = e^x$ "between" the other two. The domain for all three functions, as with all basic exponential functions, is $x \in (-\infty, \infty)$ with a range of $y \in (0, \infty)$. The same transformations applied earlier can also be applied to the graph of $y = e^x$. See Exercises 15 through 20.

Figure 5.9

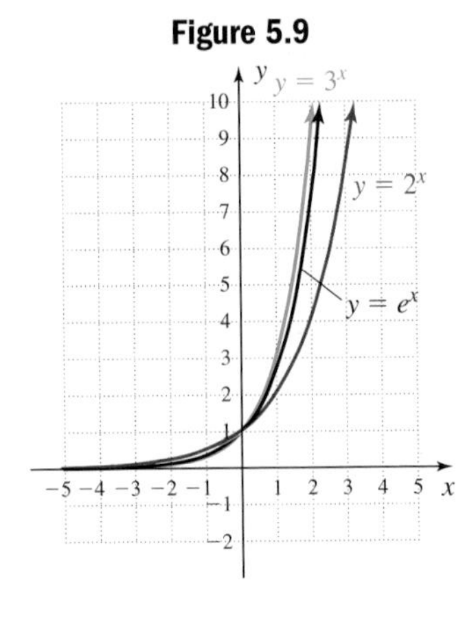

B. The Natural Log Function $y = \ln x$

In Section 5.2 we introduced the *common* logarithmic function or logarithms that use base 10: $y = \log x$. The **natural logarithmic function** uses base e, the irrational number just introduced: $y = \log_e x$. Partly due to its widespread use, the notation $y = \log_e x$ is abbreviated $y = \ln x$, and read "y is equal to the natural log of x." At this point you might realize that $y = \ln x$ is the inverse function for $y = e^x$, just as $y = \log_2 x$ was the inverse function for $y = 2^x$. Also, it's important to remember that *regardless of the base used,* a logarithm represents the exponent that goes on base b to obtain x. In other words, $y = \ln x$ can be written in exponential form as $e^y = x$. The graph of $y = \ln x$ is shown in Figure 5.10. Note it has the same shape and characteristics as other logarithmic graphs.

Figure 5.10

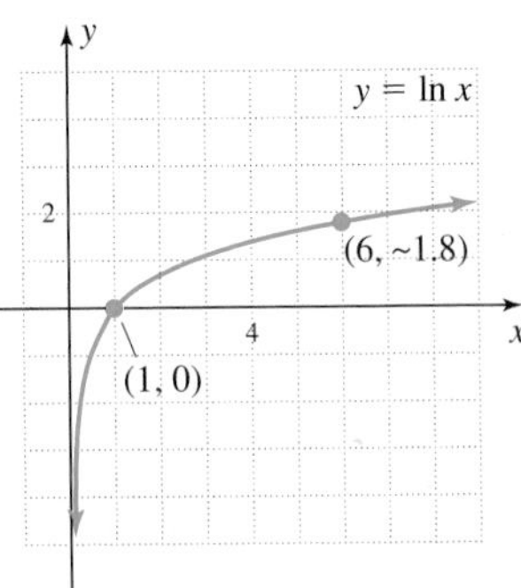

NATURAL LOGARITHMIC FUNCTION

$$y = \log_e x = \ln x$$

- domain: $x \in (0, \infty)$
- range: $y \in \mathbb{R}$
- one-to-one function
- x-intercept (1, 0)
- increasing
- asymptotic to the y-axis

To evaluate $y = \ln x$, the LN key is used in exactly the same way the LOG key was used for $y = \log x$. Refer to the *Technology Highlight* on page 491 if needed.

EXAMPLE 2 Use a calculator to evaluate each expression, rounded to six decimal places.

a. $\ln 22$ **b.** $\ln 0.28$ **c.** $\ln 7$ **d.** $\ln\left(-\frac{5}{2}\right)$

Solution: **a.** $\ln 22 \approx 3.091042$ **b.** $\ln 0.28 \approx -1.272966$

c. $\ln 7 \approx 1.945910$ **d.** $\ln\left(-\frac{5}{2}\right)$ is not a real number ($-\frac{5}{2}$ is not in the domain)

NOW TRY EXERCISES 21 THROUGH 28

EXAMPLE 3 Solve each equation by writing it in exponential form. Answer in exact form and approximate form using a calculator (round to thousandths).

a. $\ln x = 2$ **b.** $\ln x = -2.8$

Solution: **a.** $\ln x = 2$
$e^2 = x$
$x \approx 7.389$

b. $\ln x = -2.8$
$e^{-2.8} = x$
$x \approx 0.061$

NOW TRY EXERCISES 29 THROUGH 36

EXAMPLE 4 Solve each equation by writing it in logarithmic form. Answer in exact form and approximate form using a calculator (round to five decimal places).

a. $e^x = 120$ **b.** $e^x = 0.043214$

Solution: **a.** $e^x = 120$
$x = \ln 120$
≈ 4.78749

b. $e^x = 0.043214$
$x = \ln 0.043214$
≈ -3.14159

NOW TRY EXERCISES 37 THROUGH 44

C. The Product, Quotient, and Power Properties

To solve the equation $2\sqrt{x} + 5\sqrt{x} = 35$, we first combine like terms on the left and isolate the radical, before squaring both sides: $7\sqrt{x} = 35 \rightarrow \sqrt{x} = 5$, giving the solution $x = 25$. The same principle holds for equations that involve logarithms. Before we can solve $\log_4 x + \log_4(x + 6) = 2$, we must find a way to combine terms on the left. Due to the close connection between exponents and logarithms, their properties are very similar. For the product of exponential terms with like bases, we add the exponents: $x^m x^n = x^{m+n}$. This is reflected in the product property of logarithms, where the logarithm of a product also results in a sum of exponents, *which is exactly what "$\log_b M + \log_b N$" represents.*

PROPERTIES OF LOGARITHMS

Given M, N, and b are *positive* real numbers, where $b \neq 1$, and *any* real number x.

Product Property: $\log_b(MN) = \log_b M + \log_b N$

In words: "The log of a product is equal to a sum of logarithms."

Quotient Property: $\log_b \frac{M}{N} = \log_b M - \log_b N$

In words: "The log of a quotient is equal to a difference of logarithms."

Power Property: $\log_b(M)^x = x \log_b M$

In words: "The log of a number to a power is equal to the power times the log of the number."

Proof of the Product Property:
Given $P = \log_b M$ and $Q = \log_b N$, we have $b^P = M$ and $b^Q = N$ in exponential form. It follows that

$$\begin{aligned} \log_b(MN) &= \log_b(b^P b^Q) && \text{substitute } b^P \text{ for } M \text{ and } b^Q \text{ for } N \\ &= \log_b(b^{P+Q}) && \text{properties of exponents} \\ &= P + Q && \text{log property III (page 488)} \\ &= \log_b M + \log_b N && \text{substitute } \log_b M \text{ for } P \text{ and } \log_b N \text{ for } Q \end{aligned}$$

The proof of the quotient property is very similar to that of the product property, and is left as an exercise (see Exercise 110).

Proof of the Power Property:
Given $P = \log_b M$, we have $b^P = M$ in exponential form. It follows that

$$\begin{aligned} \log_b(M)^x &= \log_b(b^P)^x && \text{substitute } b^P \text{ for } M \\ &= \log_b(b^{Px}) && \text{properties of exponents} \\ &= Px && \text{log property III} \\ &= x \log_b M && \text{substitute } \log_b M \text{ for } P \end{aligned}$$

EXAMPLE 5 Use properties of logarithms to write each expression as a single term.

a. $\log_2 7 + \log_2 5$ **b.** $\ln x + \ln(x + 6)$

c. $\log_6 30 - \log_6 10$ **d.** $\ln(x + 2) - \ln x$

Solution:

a. $$\begin{aligned} \log_2 7 + \log_2 5 &= \log_2(7 \cdot 5) && \text{product property} \\ &= \log_2 35 && \text{simplify} \end{aligned}$$

b. $$\begin{aligned} \ln x + \ln(x + 6) &= \ln[x(x + 6)] && \text{product property} \\ &= \ln[x^2 + 6x] && \text{simplify} \end{aligned}$$

c. $$\begin{aligned} \log_6 30 - \log_6 10 &= \log_6 \tfrac{30}{10} && \text{quotient property} \\ &= \log_6 3 && \text{simplify} \end{aligned}$$

d. $$\ln(x + 2) - \ln x = \ln\left(\frac{x + 2}{x}\right) \quad \text{quotient property}$$

NOW TRY EXERCISES 45 THROUGH 60

EXAMPLE 6 Use the power property to rewrite each term as a product.

a. $\ln 5^x$ **b.** $\log 32^{x+2}$ **c.** $\log \sqrt{x}$ **d.** $\log_2 125$

Solution:

a. $$\ln 5^x = x \ln 5 \quad \text{power property}$$

b. $$\log 32^{x+2} = (x + 2)\log 32 \quad \text{power property (note the use of parentheses)}$$

c. $$\begin{aligned} \log \sqrt{x} &= \log x^{\frac{1}{2}} && \text{rewrite argument using a rational exponent} \\ &= \tfrac{1}{2}\log x && \text{power property} \end{aligned}$$

d. $$\begin{aligned} \log_2 125 &= \log_2 5^3 && \text{rewrite argument as an exponential term} \\ &= 3 \log_2 5 && \text{power property} \end{aligned}$$

NOW TRY EXERCISES 61 THROUGH 68

CAUTION

Note from Example 6(b) that parentheses *must be used* whenever the exponent is a sum or difference. There is a huge difference between $(x + 2)\log 32$ and $x + 2\log 32$.

In some cases, applying these properties can help to rewrite an expression in a form that enables certain techniques to be applied more easily. Example 7 actually lays the foundation for more advanced work.

EXAMPLE 7 Use the properties of logarithms to write the following expressions as a sum or difference of simple logarithmic terms.

a. $\log(x^2y)$ **b.** $\log\left(\sqrt{\dfrac{x}{x+5}}\right)$

Solution:

a.
$$\log(x^2y) = \log x^2 + \log y \quad \text{product property}$$
$$= 2\log x + \log y \quad \text{power property}$$

b.
$$\log\left(\sqrt{\frac{x}{x+5}}\right) = \log\left(\frac{x}{x+5}\right)^{\frac{1}{2}} \quad \text{write radicals in exponential form}$$
$$= \frac{1}{2}\log\left(\frac{x}{x+5}\right) \quad \text{power property}$$
$$= \frac{1}{2}[\log x - \log(x+5)] \quad \text{quotient property}$$

NOW TRY EXERCISES 69 THROUGH 78

D. The Change-of-Base Formula

Although base-10 and base-e logarithms dominate the mathematical landscape, there are many practical applications that use other bases. Fortunately, a formula exists that will convert any given base into either base 10 or base e. It's called the **change-of-base formula.**

CHANGE-OF-BASE FORMULA

Given the positive real numbers M, b, and d, where $b \neq 1$ and $d \neq 1$,

$$\log_b M = \frac{\log M}{\log b} \qquad \log_b M = \frac{\ln M}{\ln b} \qquad \log_b M = \frac{\log_d M}{\log_d b}$$

base 10 base e arbitrary base d

Proof of the Change-of-Base Formula:

Given $P = \log_b M$, we have $b^P = M$ in exponential form. It follows that

$$\log_d(b^P) = \log_d M \quad \text{take the logarithm of both sides}$$
$$P\log_d b = \log_d M \quad \text{apply power property of logarithms}$$
$$P = \frac{\log_d M}{\log_d b} \quad \text{divide by } \log_d b$$
$$\log_b M = \frac{\log_d M}{\log_d b} \quad \text{substitute } \log_b M \text{ for } P$$

EXAMPLE 8 Find the value of each expression using the change-of-base formula. Answer in exact form and approximate form using nine digits, *then verify the result* using the original base.

a. $\log_3 29$ **b.** $\log_5 3.6$

Solution: **a.**

$$\log_3 29 = \frac{\log 29}{\log 3} = 3.065044752$$

b.

$$\log_5 3.6 = \frac{\log 3.6}{\log 5} = 0.795888947$$

Check: $3^{3.065044752} = 29$✓ **Check:** $5^{0.795888947} = 3.6$✓

NOW TRY EXERCISES 79 THROUGH 86

The change-of-base formula can also be used to study and graph logarithmic functions of *any* base. For $f(x) = \log_b x$, the right-hand expression is simply rewritten using the formula and the equivalent function is $f(x) = \frac{\log x}{\log b}$. The new function can then be evaluated in this form, or used to study the graph as in the *Technology Highlight* following Example 11. Also see Exercises 87 through 90.

E. Applications of Natural Logarithms

One application of the natural log function involves the relationship between altitude and barometric pressure. The altitude or height above sea level can be determined by the formula $h(T) = (30T + 8000)\ln\frac{P_0}{P}$, where $h(T)$ represents the height in feet at temperature T in degrees Celsius, P is the barometric pressure in **centimeters of mercury** (cmHg), and P_0 is barometric pressure at sea level (76 cmHg).

EXAMPLE 9 Hikers climbing Mt. Everest take a reading of 6.4 cmHg at a temperature of 5°C. How far up the mountain are they?

Solution: For this exercise, $P_0 = 76$, $P = 6.4$, and $T = 5$. The formula yields

$$h(T) = (30T + 8000)\ln\frac{P_0}{P} \quad \text{given function}$$

$$h(5) = [30(5) + 8000]\ln\frac{76}{6.4} \quad \text{substitute given values}$$

$$= 8150 \ln 11.875 \quad \text{simplify}$$

$$\approx 20{,}167 \quad \text{result}$$

The hikers are approximately 20,167 ft above sea level.

NOW TRY EXERCISES 93 AND 94

A second application of natural logarithms involves interest compounded continuously. The formula $T = \frac{1}{r}\ln\frac{A}{P}$ gives the length of time T (in years) required for an initial principal P to grow to an amount A at a given interest rate.

EXAMPLE 10 If Shelley deposits \$2000 at 6% compounded continuously, how many years will it take the money to grow to \$3500?

Solution: For this exercise, $A = 3500$, $P = 2000$, and $r = 6\%$.

$$T = \frac{1}{r}\ln\frac{A}{P} \quad \text{given formula}$$
$$= \frac{1}{0.06}\ln\frac{3500}{2000} \quad \text{substitute given values}$$
$$= \frac{\ln 1.75}{0.06} \quad \text{simplify}$$
$$\approx \frac{0.55961579}{0.06} \quad \text{find value of ln 9.6}$$
$$\approx 9.33 \quad \text{result}$$

Under these conditions, \$2000 grows to \$3500 in about 9 yr and 4 months.

NOW TRY EXERCISES 95 AND 96

F. Rates of Change

As with the functions previously introduced, we are very interested in the concept of average rates of change due to the important role it plays in applications of mathematics. From the graph of $y = \ln x$, we note the function is "very steep" (increases very quickly) for $x \in (0, 0.25)$, with the secant line having a large and positive slope. The secant lines then become much less steep as $x \to \infty$, with very small (but always positive/increasing) slopes. We can quantify these descriptions using the rate of change formula from Section 2.4.

EXAMPLE 11 Use the rate-of-change formula to find the average rate of change of $y = \ln x$ in these intervals:

a. [0.20, 0.21], **b.** [0.99, 1.00], and **c.** [4.99, 5.00].

Solution: Apply the formula for average rates of change: $\frac{\Delta y}{\Delta x} = \frac{f(x_2) - f(x_1)}{x_2 - x_1}$.

a. $\frac{\Delta y}{\Delta x} = \frac{\ln 0.21 - \ln 0.20}{0.21 - 0.20}$
≈ 4.9

b. $\frac{\Delta y}{\Delta x} = \frac{\ln 1.00 - \ln 0.99}{1.00 - 0.99}$
≈ 1

c. $\frac{\Delta y}{\Delta x} = \frac{\ln 5.00 - \ln 4.99}{5.00 - 4.99}$
≈ 0.2

The corresponding secant lines are drawn on the graph shown here. As you can see, the secant line through [0.20, 0.21] (in blue) is much steeper than the secant line through [4.99, 5.00] (in black).

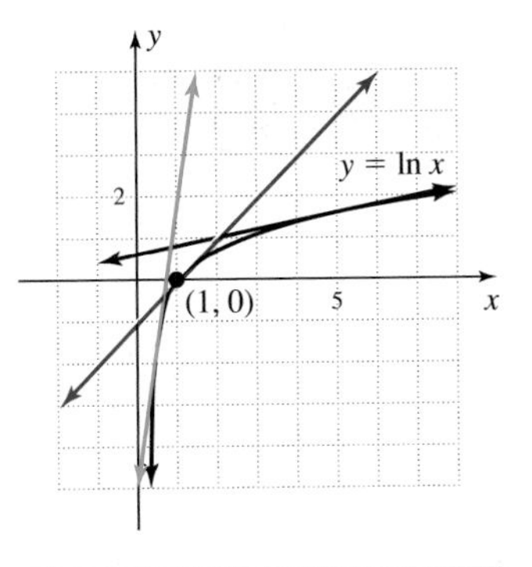

NOW TRY EXERCISES 103 AND 104

One of the many fascinating things about the exponential function involves the relationship between its rate of change in a small interval, and the value of the function in that interval. You are asked to explore this relationship in Exercises 105 and 106.

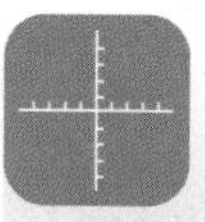

TECHNOLOGY HIGHLIGHT

Using the Change-of-Base Formula to Study Logarithms

The keystrokes shown apply to a TI-84 Plus model. Please consult your manual or our Internet site for other models.

Using the change-of-base formula, we can study logarithmic functions of any base. Many times we find base 2 or $y = \log_2 x$ more convenient to study than $y = \log_{10} x$, since the related exponential function $y = 10^x$ grows very large, very fast; or $y = \ln x$, where so many of the results are irrational numbers. Let's verify many of the things we've learned about logarithms using $y = \log_2 x$ (actually its equivalent equation $y = \frac{\log x}{\log 2}$) by using the change-of-base formula. Enter this expression as Y_1 on the Y= screen as shown in Figure 5.11. Since we know the general shape of the function and that the domain is $x \in (0, \infty)$, we can preset the viewing window before pressing GRAPH (see WINDOW screen in Figure 5.12).

Figure 5.11

```
Plot1 Plot2 Plot3
\Y1=log(X)/log(2
)
\Y2=
\Y3=
\Y4=
\Y5=
\Y6=
```

Figure 5.12

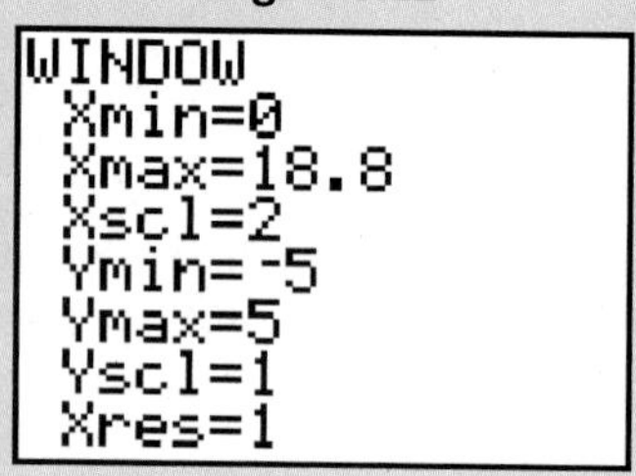

Set Xmax = 18.8 to ensure a friendly window as we TRACE through values on the graph. Now press GRAPH and the graph of $y = \log_2 x$ appears as seen in Figure 5.13. Use GRAPH, TABLE, TRACE or 2nd TRACE (CALC) features of your calculator to work the following exercises.

Figure 5.13

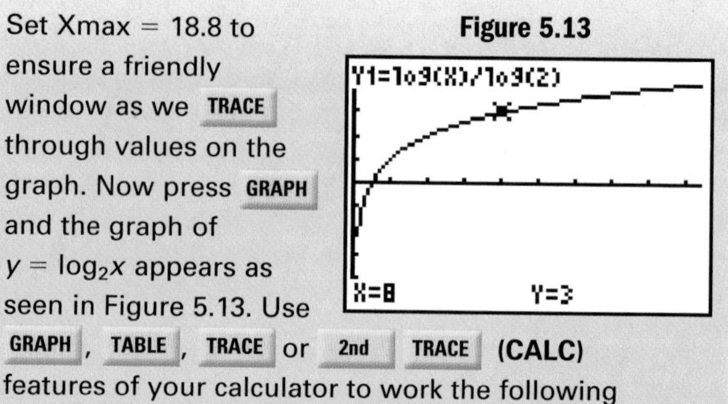

Exercise 1: Verify that the x-intercept of the function is (1, 0). $\frac{\log 1}{\log 2} = \frac{0}{\log 2} = 0$

Exercise 2: Solve by inspection, then verify using TRACE or TABLE: $\log_2 16 = k$. 4

Exercise 3: Find the value of $\sqrt{2}$ on your home screen, then find the value of $\log_2\sqrt{2}$. 0.5

Exercise 4: Solve by inspection, then verify using TABLE: $\log_2 x = -1$. What is the output for an input of $x = 0.25$? -2

Exercise 5: How would you find the value of $\log_2\sqrt{3}$? $\frac{\log\sqrt{3}}{\log 2}$

Exercise 6: How would you solve the equation $\log_2 x = \sqrt{3}$? $2^{\sqrt{3}}$

5.3 EXERCISES

CONCEPTS AND VOCABULARY

Fill in each blank with the appropriate word or phrase. Carefully reread the section if needed.

1. The number e is defined as the value $\underline{(1 + \frac{1}{x})^x}$ approaches as x $\underline{\to \infty}$.

2. The statement $\log_e 10 = \frac{\log 10}{\log e}$ is an example of the $\underline{\text{change}}$-of-$\underline{\text{base}}$ property.

3. The expression $\log_e(x)$ is commonly written $\underline{\ln x}$, but still represents the $\underline{\text{exponent}}$ that goes on base $\underline{e}$ to obtain x.

4. If $\ln 12 \approx 2.485$, then $e^{2.485} \approx \underline{12}$. If (1.4, 4.055) is a point on the graph of $y = e^x$, then ($\underline{4.055}$, $\underline{1.4}$) is a point on the graph of $y = \ln x$.

5. Without using the change-of-base formula, which of the following represents a larger number: $\log_2 9$ or $\log_3 26$? Explain your reasons and justify your response. $\log_2 9 > 3$; $\log_3 26 < 3$

6. Compare/contrast the graphs of each function including a discussion of their domains, intercepts, and whether each is increasing or decreasing: $f(x) = \ln x$, $g(x) = -\ln x$, $p(x) = \ln(-x)$, and $q(x) = -\ln(-x)$. Answers will vary.

DEVELOPING YOUR SKILLS

Use a calculator to evaluate each expression, rounded to six decimal places.

7. e^1 2.718282 **8.** e^0 1 **9.** e^2 7.389056 **10.** e^5 148.413159
11. $e^{1.5}$ 4.481689 **12.** $e^{-3.2}$ 0.040762 **13.** $e^{\sqrt{2}}$ 4.113250 **14.** e^{π} 23.140693

Graph each exponential function.

15. $f(x) = e^{x+3} - 2$ **16.** $g(x) = e^{x-2} + 1$ **17.** $r(t) = -e^t + 2$
18. $s(t) = -e^{t+2}$ **19.** $p(x) = e^{-x+2} - 1$ **20.** $q(x) = e^{-x-1} + 2$

Use a calculator to evaluate each expression, rounded to six decimal places.

21. $\ln 50$ 3.912023 **22.** $\ln 28$ 3.332205 **23.** $\ln 0.5$ −0.693147 **24.** $\ln 0.75$ −0.287682
25. $\ln 225$ 5.416100 **26.** $\ln 382$ 5.945421 **27.** $\ln \sqrt{2}$ 0.346574 **28.** $\ln \pi$ 1.144730

Solve each equation by writing it in exponential form. Answer in exact form and approximate form using a calculator (round to thousandths).

29. $\ln x = 1$ $x = e$; $x \approx 2.718$ **30.** $\ln x = 0$ $x = 1$; $x = 1$ **31.** $\ln x = -1.961$ $x = e^{-1.961}$; $x \approx 0.141$

32. $\ln x = 2.485$ $x = e^{2.485}$; $x \approx 12.001$ **33.** $-2.4 = \ln\left(\frac{1}{x^2}\right)$ $x = \pm\sqrt{e^{2.4}}$; $x \approx \pm 3.320$ **34.** $-0.345 = \ln\left(\frac{1}{x^3}\right)$ $x = \sqrt[3]{e^{0.345}}$; $x \approx 1.122$

35. $\ln e^{2x} = -8.4$ $x = -4.2$; $x = -4.2$ **36.** $\ln e^{3x} = -9.6$ $x = -3.2$; $x = -3.2$

Solve each equation by writing it in logarithmic form. Answer in exact form and approximate form using a calculator (round to five decimal places).

37. $e^x = 1$ **38.** $e^{3x} = \sqrt{2}$ **39.** $e^x = 7.389$ **40.** $e^x = 54.598$
41. $e^{\frac{2x}{5}} = 1.396$ **42.** $e^{\frac{3x}{2}} = 4.482$ **43.** $e^x = -0.30103$ **44.** $e^x = -23.14069$

Use properties of logarithms to write each expression as a single term.

45. $\ln(2x) + \ln(x - 7)$ **46.** $\ln(x + 2) + \ln(3x)$ **47.** $\log(x + 1) + \log(x - 1)$
48. $\log(x - 3) + \log(x + 3)$ **49.** $\log_3 28 - \log_3 7$ **50.** $\log_6 30 - \log_6 10$
51. $\log x - \log(x + 1)$ **52.** $\log(x - 2) - \log x$ **53.** $\ln(x - 5) - \ln x$
54. $\ln(x + 3) - \ln(x - 1)$ **55.** $\ln(x^2 - 4) - \ln(x + 2)$ **56.** $\ln(x^2 - 25) - \ln(x + 5)$
57. $\log_2 7 + \log_2 6$ **58.** $\log_9 2 + \log_9 15$ **59.** $\log_5(x^2 - 2x) + \log_5 x^{-1}$
60. $\log_3(3x^2 + 5x) - \log_3 x$

Use the power property of logarithms to rewrite each term as the product of a constant and a logarithmic term.

61. $\log 8^{x+2}$ $(x + 2)\log 8$ **62.** $\log 15^{x-3}$ $(x - 3)\log 15$ **63.** $\ln 5^{2x-1}$ $(2x - 1)\ln 5$ **64.** $\ln 10^{3x+2}$ $(3x + 2)\ln 10$
65. $\log \sqrt{22}$ $\frac{1}{2}\log 22$ **66.** $\log \sqrt[3]{34}$ $\frac{1}{3}\log 34$ **67.** $\log_5 81$ $4\log_5 3$ **68.** $\log_7 121$ $2\log_7 11$

---

15. **16.**

17. **18.**

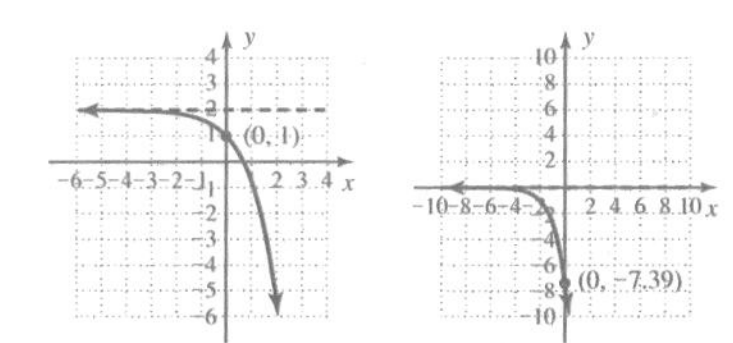

19. **20.**

37. $x = 0$; $x = 0$
38. $x = \frac{\ln\sqrt{2}}{3}$; $x \approx 0.11552$
39. $x = \ln 7.389$; $x \approx 1.99999$
40. $x = \ln 54.598$; $x \approx 4.00000$
41. $x = \frac{5\ln 1.396}{2}$; $x \approx 0.83403$
42. $x = \frac{2\ln 4.482}{3}$; $x \approx 1.00005$
43. not a real number
44. not a real number
45. $\ln(2x^2 - 14x)$
46. $\ln(3x^2 + 6x)$
47. $\log(x^2 - 1)$
48. $\log(x^2 - 9)$
49. $\log_3 4$

Additional answers can be found in the Instructor Answer Appendix.

79. $\frac{\ln 60}{\ln 7}$; 2.104076884
80. $\frac{\ln 92}{\ln 8}$; 2.174520652
81. $\frac{\ln 152}{\ln 5}$; 3.121512475
82. $\frac{\ln 200}{\ln 6}$; 2.957047225
83. $\frac{\log 1.73205}{\log 3}$; 0.499999576
84. $\frac{\log 1.41421}{\log 2}$; 0.499996366
85. $\frac{\log 0.125}{\log 0.5}$; 3
86. $\frac{\log 0.008}{\log 0.2}$; 3
87. $f(x) = \frac{\log(x)}{\log(3)}$; $f(5) \approx 1.4650$; $f(15) \approx 2.4650$; $f(45) \approx 3.4650$; outputs increase by 1; $f(3^3 \cdot 5) = 4.465$
88. $g(x) = \frac{\log(x)}{\log(2)}$; $g(5) \approx 2.3219$; $g(10) \approx 3.3219$; $g(20) \approx 4.3219$; outputs increase by 1; $g(2^3 \cdot 5) \approx 5.3219$
89. $h(x) = \frac{\log(x)}{\log(9)}$; $h(2) \approx 0.3155$; $h(4) \approx 0.6309$; $h(8) \approx 0.9464$; outputs are multiples of 0.3155; $h(2^4) = 4(0.3155) \approx 1.2619$
90. $H(x) = \frac{\log(x)}{\log(\pi)}$; $H(\sqrt{2}) \approx 0.3028$; $H(2) \approx 0.6055$; $H(\sqrt{2^3}) \approx 0.9083$; outputs are multiples of 0.3028; $H(\sqrt{2^4}) = H(4) \approx 1.2110$

Exercise 93

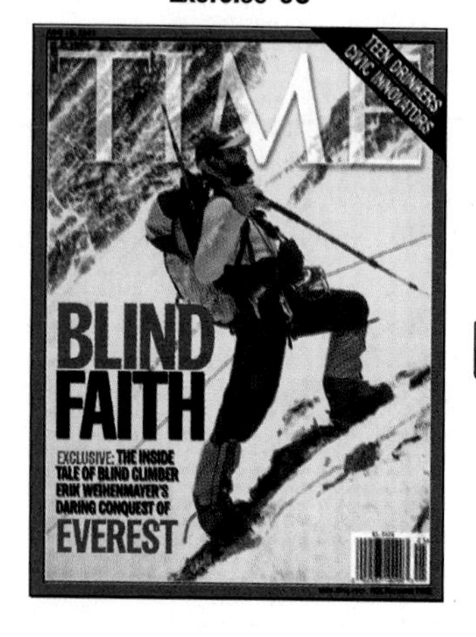

Use the properties of logarithms to write the following expressions as a sum or difference of simple logarithmic terms.

69. $\log(a^3b)$ $3\log a + \log b$

70. $\log(m^2n)$ $2\log m + \log n$

71. $\ln(x\sqrt[4]{y})$ $\ln x + \frac{1}{4}\ln y$

72. $\ln(\sqrt[3]{pq})$ $\frac{1}{3}\ln p + \ln q$

73. $\ln\left(\frac{x^2}{y}\right)$ $2\ln x - \ln y$

74. $\ln\left(\frac{m^2}{n^3}\right)$ $2\ln m - 3\ln n$

75. $\log\left(\sqrt{\frac{x-2}{x}}\right)$ $\frac{1}{2}[\log(x-2) - \log x]$

76. $\log\left(\sqrt[3]{\frac{3-v}{2v}}\right)$ $\frac{1}{3}[\log(3-v) - \log 2 - \log v]$

77. $\ln\left(\frac{7x\sqrt{3-4x}}{2(x-1)^3}\right)$ $\ln 7 + \ln x + \frac{1}{2}\ln(3-4x) - \ln 2 - 3\ln(x-1)$

78. $\ln\left(\frac{x^4\sqrt{x^2-4}}{\sqrt[3]{x^2+5}}\right)$ $4\ln x + \frac{1}{2}\ln(x^2-4) - \frac{1}{3}\ln(x^2+5)$

Evaluate each expression using the change-of-base formula and either base 10 or base *e*. Answer in exact form and in approximate form using nine digits, then verify the result using the original base.

79. $\log_7 60$ **80.** $\log_8 92$ **81.** $\log_5 152$ **82.** $\log_6 200$

83. $\log_3 1.73205$ **84.** $\log_2 1.41421$ **85.** $\log_{0.5} 0.125$ **86.** $\log_{0.2} 0.008$

Use the change-of-base formula to write an equivalent function, then evaluate the function as indicated (round to four decimal places). Investigate and discuss any patterns you notice in the output values, then determine the next input that will continue the pattern.

87. $f(x) = \log_3 x$; $f(5)$, $f(15)$, $f(45)$

88. $g(x) = \log_2 x$; $g(5)$, $g(10)$, $g(20)$

89. $h(x) = \log_9 x$; $h(2)$, $h(4)$, $h(8)$

90. $H(x) = \log_\pi x$; $H(\sqrt{2})$, $H(2)$, $H(\sqrt{2^3})$

WORKING WITH FORMULAS

91. The altitude of an airplane: $A(P) = -4.762 \ln(0.068P)$

The altitude of an airplane is given by the formula shown, where P represents the air pressure (in pounds per square inch) and $A(P)$ represents the altitude in miles. As a plane is flying, the following pressure readings are obtained: (a) 3.25 lb/in^2, (b) 7.12 lb/in^2, and (c) 10.24 lb/in^2. Use the formula to determine the altitude of the plane at each reading (to the nearest quarter mile). Is the plane gaining or losing altitude? a. 7.19 mi b. 3.45 mi c. 1.72 mi; losing altitude

92. Time required for a population to double: $T(r) = \frac{\ln(2)}{r}$

The time required for a population to double is given by the formula shown, where r represents the growth rate of the population (expressed as a decimal) and $T(r)$ gives the years required. How long would it take a population to double (rounded to the nearest whole number of years) if the growth rate were (a) 5%; (b) 10%; and (c) 23%? a. 14 yr b. 7 yr c. 3 yr

APPLICATIONS

93. Altitude: Refering to Example 9, hikers on Mt. Everest take successive readings of 42 cm of mercury at 5°C, 30 cm of mercury at 2°C, and 12 cm of mercury at −6°C. (a) How far up the mountain are they at each reading? (b) Approximate the height of Mt. Everest if the temperature at the summit is −12°C and the barometric pressure is 1.7 cm of mercury. a. 4833.5 ft, 7492.1 ft 14,434.4 ft b. 29,032.8 ft

94. Business: An advertising agency has determined that the number of items sold is related to the amount A spent on advertising by the equation $N(A) = 1500 + 315\ \ln(A)$, where A represents the amount spent on advertising and $N(A)$ gives the number of sales. Determine the approximate number of items that will be sold if (a) \$10,000 is spent on advertising and (b) \$50,000 is spent on advertising. (c) Use the TABLE feature of a calculator to estimate how large an advertising budget is needed (to the nearest \$500) to sell 5000 items. a. 4401 b. 4908 c. \$67,000

Use the formula from Example 10 $\left[T = \frac{1}{r}\ln\left(\frac{A}{P}\right)\right]$ for the length of time T (in years) required for an initial principal P to grow to an amount A at a given interest rate r.

95. **Investment growth:** A small business is planning to build a new \$350,000 facility in 8 yr. If they deposit \$200,000 in an account that pays 5% interest compounded continuously, will they have enough for the new facility in 8 yr? If not, what amount should be invested on these terms to meet the goal? No; \$234,612.01

96. **Investment growth:** After the twins were born, Sasan deposited \$25,000 in an account paying 7.5% compounded continuously, with the goal of having \$120,000 available for their college education 20 yr later. Will Sasan meet the 20-yr goal? If not, what amount should be invested on these terms to meet the goal? No; \$26,775.62

97. **Population:** The time required for a population to *triple* is given by $T(r) = \frac{\ln 3}{r}$, where r represents the growth rate (expressed as a decimal) and $T(r)$ gives the years required. How long would it take a population to triple if the growth rate were (a) 3%, (b) 5.5%, and (c) 8%? (d) Use the TABLE feature of a calculator to estimate what growth rate will cause a population to triple in 10 years. a. $\approx$36.6 yr b. $\approx$20 yr c. $\approx$13.7 yr d. $\approx$11%

98. **Radioactive decay:** The rate of decay for radioactive material is related to the half-life of the substance by the formula $R(h) = \frac{\ln 2}{h}$, where h represents the half-life of the material and $R(h)$ is the rate of decay expressed as a decimal. An element known as potassium-42 is often used in biological studies and has a half-life of approximately 12.5 hr. (a) Find its rate of decay to the nearest hundredth of a percent. (b) Find the half-life of a given substance (to the nearest whole number) whose rate of decay is 2.89% per hour. a. 5.55% b. 24 hr

99. **Drug absorption:** The time required for a certain percentage of a drug to be absorbed by the body depends on the absorption rate of the drug. This can be modeled by the function $T(p) = \frac{-\ln p}{k}$, where p represents the percent of the drug that *remains* (expressed as a decimal), k is the absorption rate of the drug, and $T(p)$ represents the elapsed time. (a) Find the time required (to the nearest hour) for the body to *absorb* 35% of a drug that has an absorption rate of 7.2%. (b) Use the TABLE feature of a calculator to estimate the percent of this drug (to the nearest whole percent) that remains unabsorbed after 24 hr. a. 6 hr b. 18%

100. **Depreciation:** As time passes, the value of an automobile tends to depreciate. The amount of time required for a certain new car to depreciate to a given value can be determined using the formula $T(v_c) = 5\ln\left(\frac{25{,}000}{v_c}\right)$, where v_c represents the current value of the car and $T(v_c)$ gives the elapsed time in years. (a) Determine how many years (to the nearest one-half year) it will take for this car's value to drop to \$10,000. (b) Use the TABLE feature of a calculator to estimate the value of the car after 2 yr (to the nearest \$250). a. $4\frac{1}{2}$ yr b. \$16,750

Carbon-14 dating: All living organisms (plants, animals, humans, and so on) contain trace amounts of the radioactive element known as carbon-14. Through normal metabolic activity, the ratio of carbon-14 to non-radioactive carbon remains constant throughout the life of the organism. After death, the carbon-14 begins disintegrating at a known rate, and no further replenishment of the element can take place. By measuring the percentage p that remains, as compared to other stable elements, the formula $T = -8266\ln p$ can be used to estimate the number of years since the organism died, where p is the percentage of carbon-14 that remains (expressed as a decimal) and T is the time in years since the organism died.

101. Bits of charcoal from Lascaux Cave (home of the prehistoric Lascaux Cave Paintings) were found to contain approximately 12.4% of their original amount of carbon-14. Approximately how many years ago did the fire burn in Lascaux Cave? $\approx$17,255 yr

102. Organic fragments found near Stonehenge were found to contain approximately 62.2% of their original amount of carbon-14. Approximately how many years ago did the organism live? $\approx$3925 yr

AVERAGE RATES OF CHANGE

103. a. ~ 1 b. ~ 0.5 c. ~ 0.33 d. ~ 0.25
104. a. ~ 10 b. ~ 5 c. ~ 3.33 d. ~ 2.5
105. $\frac{\Delta y}{\Delta x} \approx 20$, $f(3) \approx 20$; $\frac{\Delta y}{\Delta x} \approx 7.39$; $f(2) \approx 7.39$; apparently $\frac{\Delta y}{\Delta x} = f(x)$; $f(4) \approx 54.6$; $\frac{\Delta y}{\Delta x} \approx 54.6$
106. $f(x) = 2.5^x$, $\frac{\Delta y}{\Delta x} \approx 0.916$; $g(x) = 3^x$, $\frac{\Delta y}{\Delta x} \approx 1.099$; $y = e^x$, $\frac{\Delta y}{\Delta x} \approx 1.000$

103. Compute the average rate of change of $y = \ln x$ in the intervals (a) [1, 1.001], (b) [2, 2.001], and (c) [3, 3.001]. Based on what you observe, (d) estimate the rate of change for $y = \ln(x)$ in the interval [4.0, 4.001], then check your answer using the formula.

104. Compute the average rate of change of $y = \ln x$ in the intervals (a) [0.1, 0.1001], (b) [0.2, 0.2001], and (c) [0.3, 0.3001]. Based on what you observe, (d) estimate the rate of change for $y = \ln(x)$ in the interval [0.4, 0.4001], then check your answer using the formula.

105. Compute the average rate of change of $f(x) = e^x$ in the interval [3, 3.0001], then evaluate the function at $f(3)$. Repeat for the interval [2, 2.0001] and the value $f(2)$. What do you notice? Based on this observation, estimate the rate of change in the interval [4, 4.0001], then check your estimate with the value given by the formula for this interval.

106. Compute the average rate of change of $f(x) = 2.5^x$ and $g(x) = 3^x$ in the interval [0.0, 0.0001]. Based on the fact that $2 < e < 3$, make a conjecture about the average rate of change for $y = e^x$ in this interval, then check your estimate with the value given by the formula for this interval.

WRITING, RESEARCH, AND DECISION MAKING

107. a. D_f: $x \in (0, \infty)$; R_f: $y \in \mathbb{R}$; D_g: $x \in [0, \infty)$; R_y: $y \in [-1, \infty)$
b. (1, 0)
c. $-0.16, -30.42, -374.38$; as $x \to \infty$, the difference between the two functions increases
108. Answers will vary.
109. Answers will vary.
110. Answers will vary.

107. Although the graphs of $f(x) = \ln x$ and $g(x) = \sqrt{x} - 1$ appear similar in many respects, each function serves a very different purpose and is used to model a very different phenomenon. Research and investigate why by carefully graphing both functions for $x \in (0, 20)$. (a) State the domain and range of each function. (b) Find or estimate the location of any points of intersection. (c) Compute the value of $f(x) - g(x)$ for $x = 15$, $x = 1500$, and $x = 150{,}000$, and discuss why they cannot be used interchangeably as mathematical models.

108. Until calculators and computers became commonplace, logarithms had been used for centuries to manually "reckon" or calculate with numbers, particularly if products, quotients, or powers were involved. Do some reading and research into the history of logarithms and how they were used to do difficult computations without the aid of a calculator. Prepare a short report that includes some sample computations.

109. Use test values to demonstrate that the following relationships are *false*.

$$\ln(p \cdot q) = \ln p \ln q \qquad \ln\left(\frac{p}{q}\right) = \frac{\ln(p)}{\ln(q)} \qquad \ln p + \ln q = \ln(p + q)$$

110. Prove the quotient property of logarithms using the proof of the product property as a model.

EXTENDING THE CONCEPT

111. Verify that $\ln(x) = \ln(10) \cdot \log(x)$, then use the relationship to find the value of $\ln(e)$, $\ln(10)$, and $\ln(2)$ to three decimal places. verified; 1.000, 2.303, 0.693

112. Suppose you and I represent two different numbers. Is the following cryptogram true or false? *The log of me to base me is one and the log of you to base you is one, but the log of you to base me is equal to the log of me to base you turned upside down.* True

113. a. $\log_3 4 + \log_3 5 = 2.7269$ b. $\log_3 4 - \log_3 5 = -0.2031$ c. $2\log_3 5 = 2.9300$
114. a. $2\log_5 3 - \log_5 2 = 0.9345$ b. $3\log_5 3 + 3\log_5 2 = 3.3399$ c. $\frac{1}{3}[\log_5 3 + \log_5 2] = 0.3711$

Use prime factors, properties of logs, and the values given to evaluate each expression without a calculator. Check each result using the change-of-base formula.

113. $\log_3 4 \approx 1.2619$ and $\log_3 5 \approx 1.4650$: (a) $\log_3 20$, (b) $\log_3 \frac{4}{5}$, and (c) $\log_3 25$.

114. $\log_5 2 \approx 0.4307$ and $\log_5 3 \approx 0.6826$: (a) $\log_5 \frac{9}{2}$, (b) $\log_5 216$, and (c) $\log_5 \sqrt[3]{6}$.

MAINTAINING YOUR SKILLS

115. (2.4/3.5) Name all eight basic toolbox functions and draw a quick sketch of each.

116. (3.4) Graph the function by completing the square and label all important features: $f(x) = -2x^2 + 12x - 9$.

117. (4.5) Use polynomial long division to sketch the graph using shifts of a basic toolbox function: $r(x) = \dfrac{x^2 + 4x + 3}{x^2 + 4x + 4}$. Label all asymptotes and intercepts.

118. (3.7) Graph the piecewise-defined function and state its domain and range.

$$p(x) = \begin{cases} -2 & -4 \le x < 0 \\ 4x - x^2 & 0 \le x < 4 \\ 8 - 2|x - 8| & 4 \le x \le 10 \end{cases}$$

119. (R.7/3.3) Sketch the graph using transformations of a basic function, then use basic geometry to compute the area in Quadrant I that is under the graph: $y = -|x - 3| + 6$. 31.5 units^2

120. (2.6) The data given tracks the total amount of debt carried by a family over a six-month period. Draw a scatter-plot of the data, decide on an appropriate form of regression, and use a graphing calculator to determine a regression, equation. (a) How fast is the debt load growing each month? (b) How much debt will the family have accumulated at the end of 12 months?

Exercise 120

Month (x)	Debt (y)
(start)	$0
Jan.	471
Feb.	1105
March	1513
April	1921
May	2498
June	3129

Additional answers can be found in the Instructor Answer Appendix.

MID-CHAPTER CHECK

1. Write the following in logarithmic form.

a. $27^{\frac{2}{3}} = 9$ $\quad \frac{2}{3} = \log_{27} 9$

b. $81^{\frac{5}{4}} = 243$ $\quad \frac{5}{4} = \log_{81} 243$

2. Write the following in exponential form.

a. $\log_8 32 = \frac{5}{3}$ $\quad 8^{\frac{5}{3}} = 32$

b. $\log_{1296} 6 = 0.25$ $\quad 1296^{0.25} = 6$

3. Solve each equation for the unknown.

a. $4^{2x} = 32^{x-1}$ $\quad x = 5$

b. $(\frac{1}{3})^{4b} = 9^{2b-5}$ $\quad b = \frac{5}{4}$

4. Solve each equation for the unknown.

a. $\log_{27} x = \frac{1}{3}$ $\quad x = 3$

b. $\log_b 125 = 3$ $\quad b = 5$

5. The homes in a popular neighborhood are growing in value according to the formula $V(t) = V_0(\frac{9}{8})^t$, where t is the time in years, V_0 is the purchase price of the home, and $V(t)$ is the current value of the home. (a) In 3 yr, how much will a $50,000 home be worth? (b) Use the TABLE feature of your calculator to estimate how many years (to the nearest year) until the home doubles in value. a. $71,191.41 b. 6 yr

6. Estimate the value of each expression by bounding it between two integers. Then use a calculator to find the exact value.

a. $\log 243$ $\quad 2.385606274$

b. $\log 85{,}678$ $\quad 4.93286932$

7. Estimate the value of each expression by bounding it between two integers. Then use the change-of-base formula to find the exact value.

a. $\log_3 19$ $\quad \dfrac{\log 19}{\log 3} \approx 2.68$

b. $\log_2 60$ $\quad \dfrac{\log 60}{\log 2} \approx 5.91$

8. a. 2.48%
b. 12 yr

8. The rate of decay for radioactive material is related to the half-life of the substance by the formula $R(h) = \dfrac{\ln(2)}{h}$, where h represents the half-life of the material and $R(h)$ is the rate of decay expressed as a decimal. The strontium isotope $^{90}_{38}$Sr-14 has a half-life of approximately 28 yr. (a) Find its rate of decay to the nearest hundredth of a percent. (b) Find the half-life of a given substance (to the nearest whole number) whose rate of decay is 5.78% per year.

9. Use properties of logarithms to write each expression as a single logarithmic term:

a. $\log(2x - 3) + \log(2x + 3)$ $\quad \log(4x^2 - 9)$

b. $\log(x + 5) - \log(x^2 - 25)$ $\quad \log\left(\dfrac{1}{x-5}\right)$

10. Given $\log_7 5 \approx 0.8271$ and $\log_7 10 \approx 1.1833$, use properties of logarithms to estimate the value of

a. $\log_7 50$ b. $\log_7 2$ c. $\log_7 25$ d. $\log_7 500$

a. 2.0104 b. 0.3562 c. 1.6542 d. 3.1937

REINFORCING BASIC CONCEPTS

Understanding Properties of Logarithms

To effectively use the properties of logarithms as a mathematical tool, a student must attain some degree of comfort and fluency in their application. Otherwise we are resigned to using them as a template or formula, leaving little room for growth or insight. This *Reinforcing Basic Concepts* is divided into two parts. The first is designed to promote an understanding of the product and quotient properties of logarithms, which play a role in the solution of logarithmic and exponential equations.

We begin by looking at some logarithmic expressions that are obviously true:

$$\log_2 2 = 1 \quad \log_2 4 = 2 \quad \log_2 8 = 3 \quad \log_2 16 = 4 \quad \log_2 32 = 5 \quad \log_2 64 = 6$$

Next, we view the same expressions with their value *understood mentally,* illustrated by the numbers in the background, rather than expressly written.

$$\log_2 2 \quad \log_2 4 \quad \log_2 8 \quad \log_2 16 \quad \log_2 32 \quad \log_2 64$$

This will make the product and quotient properties of equality much easier to "see." Recall the product property states: $\log_b M + \log_b N = \log_b(MN)$ and the quotient property states: $\log_b M - \log_b N = \log_b\left(\frac{M}{N}\right)$. Consider the following.

$$\log_2 4 + \log_2 8 = \log_2 32 \qquad \log_2 64 - \log_2 32 = \log_2 2$$

which is the same as saying

$\log_2 4 + \log_2 8 = \log_2(4 \cdot 8)$ (since $4 \cdot 8 = 32$)

$\log_b M + \log_b N = \log_b(MN)$

which is the same as saying

$\log_2 64 - \log_2 32 = \log_2\left(\frac{64}{32}\right)$ (since $\frac{64}{32} = 2$)

$\log_b M - \log_b N = \log_b\left(\frac{M}{N}\right)$

Exercise 1: Repeat this exercise using logarithms of base 3 and various sums and differences. Answers will vary.

Exercise 2: Use the basic concept behind these exercises to combine these expressions: (a) $\log(x) + \log(x + 3)$, (b) $\ln(x + 2) + \ln(x - 2)$, and (c) $\log(x) - \log(x + 3)$. **a.** $\log(x^2 + 3x)$ **b.** $\ln(x^2 - 4)$ **c.** $\log\frac{x}{x+3}$

The second part is similar to the first, but highlights the power property: $\log_b M^x = x\log_b M$. For instance, knowing that $\log_2 64 = 6$, $\log_2 8 = 3$, and $\log_2 2 = 1$, consider the following:

$\log_2 8$ can be written as $\log_2 2^3$ (since $2^3 = 8$). Applying the power property gives $3 \cdot \log_2 2 = 3$.

$\log_2 64$ can be written as $\log_2 2^6$ (since $2^6 = 64$). Applying the power property gives $6 \cdot \log_2 2 = 6$.

$$\log_b M^x = x\log_b M$$

Exercise 3: Repeat this exercise using logarithms of base 3 and various powers. Answers will vary.

Exercise 4: Use the basic concept behind these exercises to rewrite each expression as a product: (a) $\log 3^x$, (b) $\ln x^5$, and (c) $\ln 2^{3x-1}$. **a.** $x\log 3$ **b.** $5\ln x$ **c.** $(3x - 1)\ln 2$

5.4 Exponential/Logarithmic Equations and Applications

LEARNING OBJECTIVES

In Section 5.4 you will learn how to:

A. Write logarithmic and exponential equations in simplified form

B. Solve exponential equations

C. Solve logarithmic equations

D. Solve applications involving exponential and logarithmic equations

INTRODUCTION

In this section, we'll use the relationships between $f(x) = b^x$ and $g(x) = \log_b x$ to solve equations that arise in applications of exponential and logarithmic functions. As you'll see, these functions have a large variety of significant uses.

POINT OF INTEREST

Using a calculator, we find $\log 6 = 0.7781512504$, meaning $10^{0.7781512504} = 6$. In years past, the number 6 was commonly called the *antilogarithm* of 0.7781512504, or the number whose (base-10) log is 0.7781512504. Prior to the widespread use of calculators, *tables* were used to compute with logarithms, and "finding an antilogarithm" simply meant we were searching the entries of a table for the number whose base b exponent was given. For example, the base-10 antilogarithm of 3 is 1000 since $10^3 = 1000$ and the base-e antilogarithm of 3.36729583 is 29 since $e^{3.36729583} = 29$. Now that calculators can easily produce logarithms and exponentials of any base (to over nine decimal places), we find the term "antilogarithm" gradually fading from common use.

A. Writing Logarithmic and Exponential Equations in Simplified Form

As we noted in Section 5.3, sums and differences of logarithmic terms (with like bases) are combined using the product and quotient properties, respectively. This is a fundamental step in equation solving, as it helps to simplify the equation and assist the solution process.

EXAMPLE 1 Rewrite each equation with a single logarithmic term on one side, as in $\log_b x = k$. Do not attempt to solve.

a. $\log_2 x + \log_2(x + 3) = 4$ **b.** $-\ln 2x = \ln x - \ln(x + 1)$

Solution:

a. $\log_2 x + \log_2(x + 3) = 4$ given

$\log_2[x(x + 3)] = 4$ product property

$\log_2[x^2 + 3x] = 4$ result

b. $-\ln 2x = \ln x - \ln(x + 1)$ given

$-\ln 2x = \ln \frac{x}{x + 1}$ quotient property

$0 = \ln \frac{x}{x + 1} + \ln 2x$ set equal to zero

$0 = \ln\left[\left(\frac{x}{x + 1}\right)\left(\frac{2x}{1}\right)\right]$ product property

$0 = \ln\left(\frac{2x^2}{x + 1}\right)$ result

NOW TRY EXERCISES 7 THROUGH 12

EXAMPLE 2 Rewrite each equation with a single exponential term on one side, as in $b^x = k$. Do not solve.

a. $400e^{0.21x} + 325 = 1225$ **b.** $e^{x-1}(e^{3x}) = e^{2x}$

Solution:

a.
$$400e^{0.21x} + 325 = 1225 \quad \text{given}$$
$$400e^{0.21x} = 900 \quad \text{subtract 325}$$
$$e^{0.21x} = 2.25 \quad \text{divide by 400}$$

b.
$$e^{x-1}(e^{3x}) = e^{2x} \quad \text{given}$$
$$e^{4x-1} = e^{2x} \quad \text{product property}$$
$$\frac{e^{4x-1}}{e^{2x}} = 1 \quad \text{divide}$$
$$e^{(4x-1)-2x} = 1 \quad \text{quotient property}$$
$$e^{2x-1} = 1 \quad \text{result}$$

NOW TRY EXERCISES 13 THROUGH 18

CAUTION

One of the most common mistakes in solving exponential and logarithmic equations is to apply the inverse function too early—before the equation is simplified. Unless the equation can be written with like bases on both sides, always try to isolate a single logarithmic or exponential term prior to applying the inverse function.

B. Solving Exponential Equations

An exponential equation is one where at least one term has a variable exponent. If an exponential equation can be written with a single term on each side where both have the same base, the equation is most readily solved using the *uniqueness property* as in Section 5.1. If not, we solve a **base-*b* exponential** equation by applying a **base-*b* logarithm** using properties I through IV from Section 5.2. These properties can be applied for any base but are particularly effective when the exponential is 10 or e, since calculators are programmed with these bases. Consider the following illustrations:

base 10
$$10^x = k \quad \text{base-10 exponential}$$
$$\log_{10}10^x = \log k \quad \text{apply base-10 logarithms}$$
$$x = \log k \quad \text{property III; find } \log k \text{ using a calculator}$$

base *e*
$$e^x = k \quad \text{base } e \text{ exponential}$$
$$\ln e^x = \ln k \quad \text{apply base } e \text{ logarithms}$$
$$x = \ln k \quad \text{property III; find } \ln k \text{ using a calculator}$$

For exponential bases other than 10 or e, we apply either base and use the power property of logarithms to solve for x: $\log_b k^x = x \log_b k$.

neither 10 nor *e*
$$b^x = k \quad \text{base-}b\text{ exponential}$$
$$\log b^x = \log k \quad \text{apply either logarithm to both sides (we chose base 10)}$$
$$x \log b = \log k \quad \text{power property}$$
$$x = \frac{\log k}{\log b} \quad \text{solution (divide both sides by } \log b)$$

The main ideas are summarized here.

SOLVING EXPONENTIAL EQUATIONS

For any real numbers b, x, and k, where $b > 0$ and $b \neq 1$,

1. If $10^x = k$,	2. If $e^x = k$,	3. $b^x = k$
$\log 10^x = \log k$	$\ln e^x = \ln k$	$x \log b = \log k$
$\Rightarrow x = \log k$	$\Rightarrow x = \ln k$	$\Rightarrow x = \dfrac{\log k}{\log b}$

EXAMPLE 3 Solve each exponential equation and check your answer.

a. $3e^{x+1} - 5 = 7$ **b.** $4^{3x} - 1 = 8$

Solution: **a.**

$$3e^{x+1} - 5 = 7 \quad \text{given}$$
$$3e^{x+1} = 12 \quad \text{add 5}$$
$$e^{x+1} = 4 \quad \text{divide by 3}$$

Since the left-hand side is a base-*e* exponential, we apply the base-*e* logarithm.

$$\ln e^{x+1} = \ln 4 \quad \text{apply base-}e\text{ logarithms}$$
$$x + 1 = \ln 4 \quad \text{property III}$$
$$x = \ln 4 - 1 \quad \text{solve for } x \text{ (exact form)}$$
$$\approx 0.38629 \quad \text{approximate form (to five decimal places)}$$

Check:

$$3e^{x+1} - 5 = 7 \quad \text{original equation}$$
$$3e^{0.38629+1} - 5 = 7 \quad \text{substitute 0.38629 for } x$$
$$3e^{1.38629} - 5 = 7 \quad \text{simplify exponent}$$
$$3(4) - 5 = 7 ✓ \quad e^{1.38629} \approx 4$$

GRAPHICAL SUPPORT

We can verify solutions to exponential equations using the same methods as for other equations. For Example 3(a), enter $3e^{x+1} - 5$ as Y_1 and 7 as Y_2 on the Y= screen of your graphing calculator. Using the 2nd TRACE **(CALC) 5: Intersect** option, we find the graphs intersect at $x = 0.38629436$, the solution we found in Example 3(a).

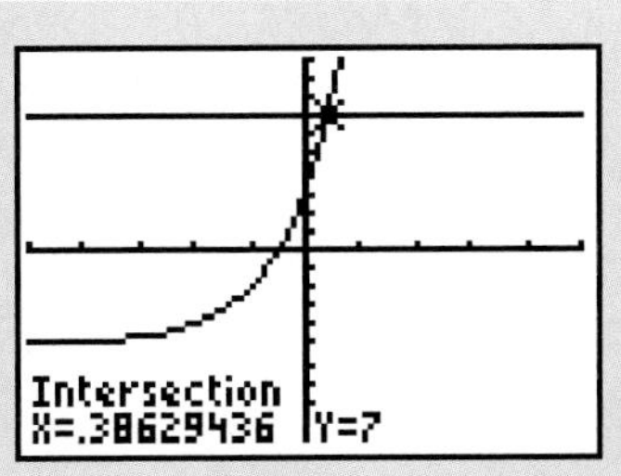

b.

$$4^{3x} - 1 = 8 \quad \text{given}$$
$$4^{3x} = 9 \quad \text{add 1 to both sides}$$

The left-hand side is neither base 10 or base *e*, so we chose base *e* to solve:

$$\ln 4^{3x} = \ln 9 \quad \text{apply base-}e\text{ logarithms}$$
$$3x \ln 4 = \ln 9 \quad \text{power property}$$

$$3x = \frac{\ln 9}{\ln 4}$$ divide by ln 4

$$x = \frac{\ln 9}{3 \cdot \ln 4}$$ solve for x (exact form)

$$\approx 0.52832$$ approximate form (to five decimal places)

Check:

$$4^{3x} - 1 = 8$$ original equation

$$4^{3(0.52832)} - 1 = 8$$ substitute 0.52832 for x

$$4^{1.58496} - 1 = 8$$ simplify exponent

$$9 - 1 \approx 8 ✓$$ $4^{1.58496} \approx 9$

NOW TRY EXERCISES 19 THROUGH 54

In some cases, there may be exponential terms with unlike bases on *both sides* of the equation. As you apply the solution process to these equations, be sure to distinguish between constant terms like $\ln 5$ and variable terms like $x \ln 5$. As with all equations, the goal is to isolate the variable terms on one side of the equation, with all constant terms on the other.

EXAMPLE 4 Solve the exponential equation: $5^{x+1} = 6^{2x}$.

Solution:

$$5^{x+1} = 6^{2x}$$ given

To begin, we take the natural log (or base-10 log) of both sides:

$$\ln 5^{x+1} = \ln 6^{2x}$$ apply base-e logarithms

$$(x + 1)\ln 5 = 2x \ln 6$$ power property

$$x \ln 5 + \ln 5 = 2x \ln 6$$ distribute

$$\ln 5 = 2x \ln 6 - x \ln 5$$ variable terms to one side

$$\ln 5 = x(2 \ln 6 - \ln 5)$$ factor out x

$$\frac{\ln 5}{2 \ln 6 - \ln 5} = x$$ solve for x (exact form)

$$0.81528 = x$$ approximate form

The check is left to the student.

NOW TRY EXERCISES 55 THROUGH 58

In many important applications of exponential functions, the exponential term appears as part of a quotient. In this case we simply "clear denominators" and attempt to isolate the exponential term as before.

EXAMPLE 5 Solve the exponential equation: $\frac{258}{1 + 20e^{-0.009t}} = 192$

Solution

$$\frac{258}{1 + 20e^{-0.009t}} = 192$$ given

$$258 = 192(1 + 20e^{-0.009t})$$ clear denominators

$$1.34375 = 1 + 20e^{-0.009t}$$ divide by 192

$$0.0171875 = e^{-0.009t}$$ subtract 1, divide by 20

$$\ln 0.0171875 = -0.009t$$ apply base-e logarithms

$$\frac{\ln 0.0171875}{-0.009} = t$$ solve for t (exact form)

$$451.51 \approx t$$ approximate form

NOW TRY EXERCISES 59 THROUGH 62

C. Solving Logarithmic Equations

As with exponential functions, the fact that logarithmic functions are one-to-one enables us to quickly solve equations that can be rewritten with a single logarithmic term on each side (assuming both have a like base, as in Section 5.1). In particular we have

LOGARITHMIC EQUATIONS WITH LIKE BASES: THE UNIQUENESS PROPERTY

For all real numbers b, m, and n where $b > 0$ and $b \neq 1$,

If $\log_b m = \log_b n$, then $m = n$

If $m = n$, then $\log_b m = \log_b n$

Equal bases imply equal arguments.

EXAMPLE 6 Solve each equation using the uniqueness property of logarithms.

a. $\log(x + 2) = \log 7 + \log x$ **b.** $\log_3 87 - \log_3 x = \log_3 29$

Solution:

a. $\log(x + 2) = \log 7 + \log x$

$$\log(x + 2) = \log 7x$$ properties of logarithms

$$x + 2 = 7x$$ uniqueness property

$$2 = 6x$$ solve for x

$$\frac{1}{3} = x$$ result

b. $\log_3 87 - \log_3 x = \log_3 29$

$$\log_3 \frac{87}{x} = \log_3 29$$

$$\frac{87}{x} = 29$$

$$87 = 29x$$

$$3 = x$$

NOW TRY EXERCISES 63 THROUGH 68

If the equation results in a single logarithmic term, the uniqueness property cannot be used and we solve by isolating this term on one side and applying a **base-*b* exponential** (exponentiate both sides) as illustrated here:

$$\log_b x = k$$ exponential equation

$$b^{\log_b x} = b^k$$ exponentiate both sides (using base b)

$$x = b^k$$ property IV (find b^k using a calculator)

Note the end result is simply the exponential form of the equation, and we will actually view the solution process in this way.

SOLVING LOGARITHMIC EQUATIONS

For any algebraic expression X and real numbers b and k, where $b > 0$ and $b \neq 1$,

1. If $\log X = k$, $X = 10^k$
2. If $\ln X = k$, $X = e^k$
3. If $\log_b X = k$, $X = b^k$

As we saw in our study of rational and radical equations, when the domain of a function is something other than all real numbers, **extraneous roots** sometimes arise. Logarithmic equations can also produce such roots, and checking all results is a good practice. See Example 7(b).

EXAMPLE 7 Solve each logarithmic equation and check the solutions.

a. $\ln(x + 7) - \ln 5 = 1.4$ **b.** $\log(x + 12) - \log x = \log(x + 9)$

Solution: **a.** $\ln(x + 7) - \ln 5 = 1.4$ given

Bases are alike → combine terms and write equation in exponential form (uniqueness property cannot be applied).

$$\ln\left(\frac{x+7}{5}\right) = 1.4 \quad \text{quotient property}$$

$$\left(\frac{x+7}{5}\right) = e^{1.4} \quad \text{exponential form}$$

$$x + 7 = 5e^{1.4} \quad \text{clear denominator}$$

$$x = 5e^{1.4} - 7 \quad \text{solve for } x \text{ (exact form)}$$

$$\approx 13.27600 \quad \text{approximate form (to five decimal places)}$$

Check $x = 13.276$:

$$\ln(x + 7) - \ln 5 = 1.4 \quad \text{original equation}$$

$$\ln(13.276 + 7) - \ln 5 = 1.4 \quad \text{substitute 13.27600 for } x$$

$$\ln 20.276 - \ln 5 = 1.4 \quad \text{simplify}$$

$$1.4 = 1.4 \checkmark \quad \text{result checks}$$

WORTHY OF NOTE

If all digits of the answer given by your calculator are used, the calculator will generally produce "exact" answers when they are checked. Try using the solution $x = 13.27599983$ for Example 7(a).

b. $\log(x + 12) - \log x = \log(x + 9)$ given

Left-hand side can be simplified → write the equation with a *single logarithmic term* on each side and solve using the uniqueness property.

$$\log\left(\frac{x+12}{x}\right) = \log(x + 9) \quad \text{quotient property}$$

$$\frac{x+12}{x} = x + 9 \quad \text{uniqueness property}$$

$$x + 12 = x^2 + 9x \quad \text{clear denominator}$$

$$0 = x^2 + 8x - 12 \quad \text{set equal to 0}$$

The quadratic formula gives solutions $x = -4 \pm 2\sqrt{7}$. The solution $x = -4 + 2\sqrt{7}$ ($x \approx 1.29150$) checks, but when $-4 - 2\sqrt{7}$ ($x \approx -9.29150$) is substituted into the original equation, we get $\log 2.7085 - \log(-9.2915) = \log(-0.2915)$ and two of the three terms do not represent real numbers. The "solution" $x = -4 - 2\sqrt{7}$ is an extraneous root.

NOW TRY EXERCISES 69 THROUGH 94

GRAPHICAL SUPPORT

Logarithmic equations can also be checked using the intersection of graphs method. For Example 7(b), we first enter $\log(x + 12) - \log x$ as Y_1 and $\log(x + 9)$ as Y_2 on the Y= screen. Using 2nd TRACE (CALC) **5:Intersect,** we find the graphs intersect at $x = 1.2915026$, and that *this is the only solution* (knowing the graph's basic shape, we conclude they cannot intersect again).

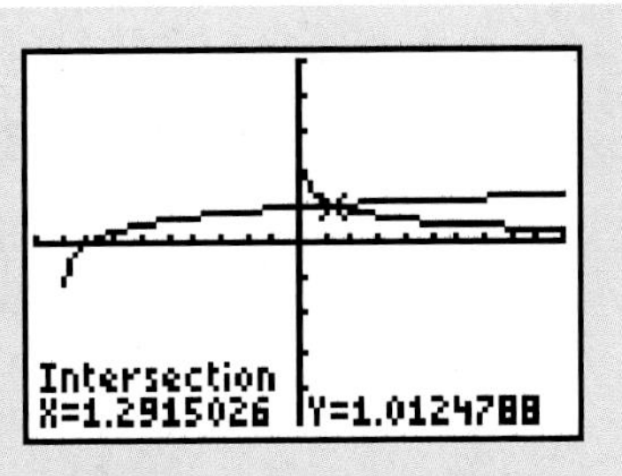

D. Applications of Exponential and Logarithmic Functions

Applications of exponential and logarithmic functions take many different forms and it would be impossible to illustrate them all. As you work through the exercises, try to adopt a "big picture" approach, applying the general principles illustrated in this section to the various applications. Some may look familiar and may have been introduced in previous sections. The difference here is that we now have the ability to *solve for unknowns* as well as to evaluate the relationships.

Newton's law of cooling relates the temperature of a given object to the constant temperature of a surrounding medium. One form of this relationship is $T = T_1 + (T_0 - T_1)e^{-kh}$, where T_0 is the initial temperature of the object, T_1 is the temperature of the surrounding medium, and T is the temperature after h hours (k is a constant that depends on the materials involved).

EXAMPLE 8 If a can of soft drink is taken from a 50°F cooler and placed in a room where the temperature is 75°F, how long will it take the drink to warm to 70°F? Assume $k = 0.95$ and answer in hours and minutes.

Solution:

$T = T_1 + (T_0 - T_1)e^{-kh}$ — given

$70 = 75 + (50 - 75)e^{-0.95h}$ — substitute 50 for T_0, 75 for T_1, 70 for T, and 0.95 for k

$-5 = -25e^{-0.95h}$ — simplify

$0.2 = e^{-0.95h}$ — divide by -25

$\ln 0.2 = \ln e^{-0.95h}$ — apply base-e logarithms

$\ln 0.2 = -0.95h$ — $\ln e^k = k$

$\frac{\ln 0.2}{-0.95} = h$ — solve for h

$1.69 \approx h$ — result

The can of soda will warm to a temperature of 70°F in approximately 1 hour and 41 min ($0.69 \times 60 \approx 41$).

NOW TRY EXERCISES 97 AND 98

TECHNOLOGY HIGHLIGHT

Using a Graphing Calculator to Explore Exponential/Logarithmic Equations

The keystrokes shown apply to a TI-84 Plus model. Please consult our Internet site or your manual for other models.

Even with the new equation-solving abilities in Section 5.4, there remain a large number of exponential and logarithmic equations that are very difficult to solve using inverse functions and manual methods alone. One example would be the equation $e^{(x-3)} + 1 = 5\ln(x-1) + 2$, in which both exponential and logarithmic functions occur. For equations of this nature, graphing technology remains our best tool. To solve $e^{(x-3)} + 1 = 5\ln(x-1) + 2$, enter the left-hand member as Y_1 and the right-hand member as Y_2 on the **Y=** screen of your graphing calculator. Based on what we know about the graphs of $y = e^x$ and $y = \ln x$, it is likely that solutions (points of intersection) will occur on the standard screen. Graph both by pressing **ZOOM** 6 (see Figure 5.14). From the graphs and our knowledge of the basic functions, it is apparent the equation has two solutions (the graphs have two points of intersection). Recall that to find the intersections, we use the **2nd** **TRACE** **(CALC) 5: intersect** option. Press **ENTER** to identify the first graph, then **ENTER** once again to identify (select) the second graph. The smaller solution seems to be near $x = 2$, so we enter a "2" when the calculator asks for a guess (Figure 5.15). After a moment, the calculator determines the smaller root is approximately $x = 1.8735744$ (Figure 5.16). Repeating these keystrokes using a guess of "5" reveals the second solution is about $x = 5.0838288$. Recall that the TI-84 Plus will temporarily store the last calculated solution as the variable x, accessed using the **X,T,θ,n** key. This will enable a quick check of the solution by simply entering the original expressions on the home screen, as shown in Figure 5.17 Use these ideas to solve these equations.

Figure 5.14

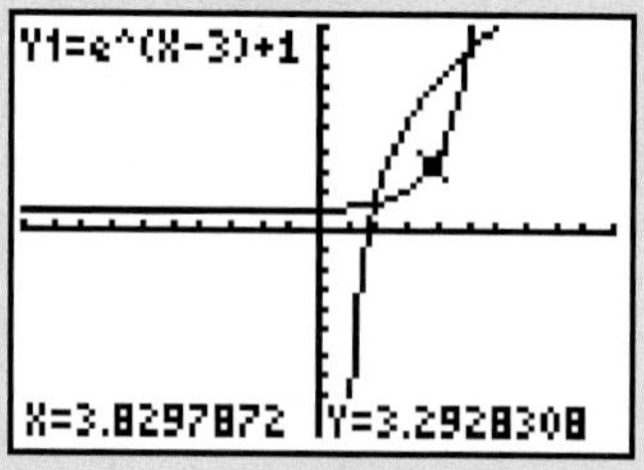

Y1=e^(X-3)+1
X=3.8297872 Y=3.2928308

Figure 5.15

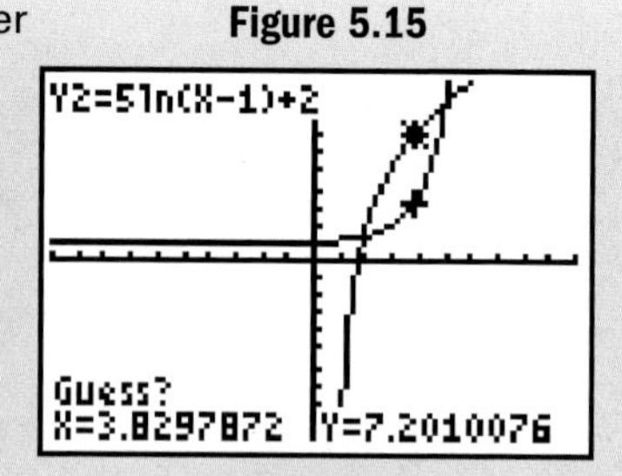

Figure 5.16

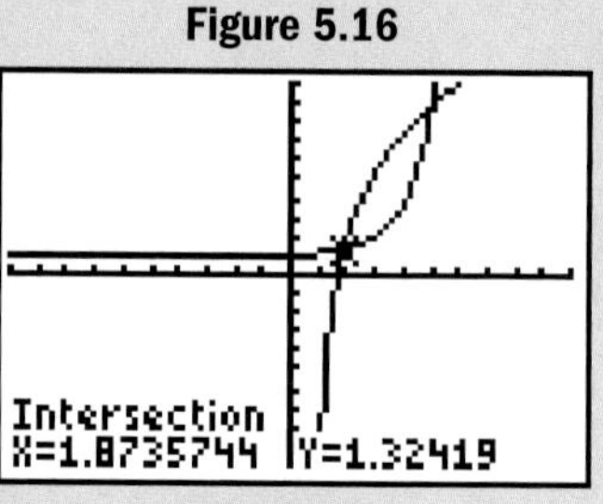

Figure 5.17

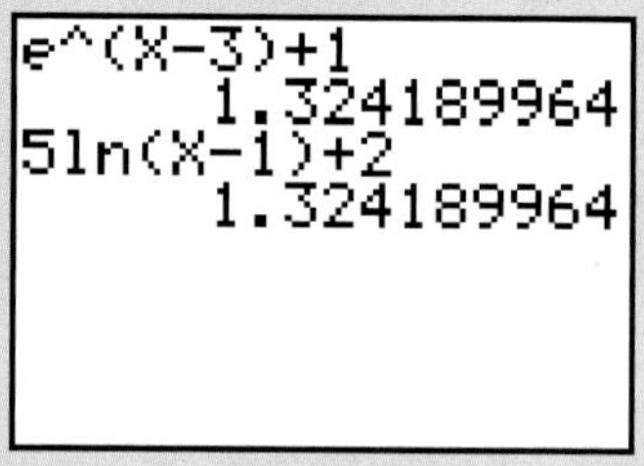

e^(X-3)+1
1.324189964
5ln(X-1)+2
1.324189964

Note the solution to Exercise 4 can be completed/verified without the aid of a calculator (use a u-substitution and the quadratic form).

Exercise 1: $5\log(2x + 3) = e^{2x} - 6$ 1.1311893, −1.467671

Exercise 2: $x^2 + 2 = 3\ln(x + 2)$ −0.0506028, 1.2329626

Exercise 3: $-4\log(x - 3) + 2 = \ln x$ 4.3556075

Exercise 4: $[\ln(2x)]^2 + 3\ln(2x) - 4 = 0$ $\frac{e}{2} \approx 1.3591409, \frac{1}{2e^4} \approx 0.00915782$

5.4 EXERCISES

CONCEPTS AND VOCABULARY

Fill in each blank with the appropriate word or phrase. Carefully reread the section if needed.

1. For $e^{-0.02x+1} = 10$, the solution process is most efficient if we apply a base __e__ logarithm to both sides.

2. To solve $\ln x - \ln(x + 3) = 0$, we can combine terms using the __quotient__ property, or add $\ln(x + 3)$ to both sides and use the __uniqueness__ property.

37. $x = \log 97$; $x \approx 1.9868$
38. $x = \log 12{,}089$; $x \approx 4.0824$
39. $x = \frac{\log 879}{2}$; $x \approx 1.4720$
40. $x = \frac{\log 12{,}089}{3}$; $x \approx 1.3608$
41. $x = \log 879 - 3$; $x \approx -0.0560$
42. $x = \log 4589 + 2$; $x \approx 5.6617$
43. $x = \ln 389$; $x \approx 5.9636$
44. $x = \ln 25$; $x \approx 3.2189$
45. $x = \frac{\ln 1389}{2}$; $x \approx 3.6182$
46. $x = \frac{\ln 2507}{3}$; $x \approx 2.6089$
47. $x = \ln 257 - 1$; $x \approx 4.5491$
48. $x = \ln 589 + 3$; $x \approx 9.3784$
49. $x = 4\ln\left(\frac{5}{2}\right)$; $x \approx 3.6652$
50. $x = \frac{25\ln 4}{2}$; $x \approx 17.3287$
51. $x = \frac{\ln 231}{\ln 7} - 2$; $x \approx 0.7968$
52. $x = \frac{\ln 3589}{\ln 6} - 2$; $x \approx 2.5685$
53. $x = \frac{\ln 128{,}965}{3\ln 5} + \frac{2}{3}$; $x \approx 3.1038$
54. $x = \frac{\ln 78{,}462}{5\ln 9} + \frac{3}{5}$; $x \approx 1.6259$
55. $x = \frac{\ln 2}{\ln 3 - \ln 2}$; $x \approx 1.7095$
56. $x = \frac{\ln 4}{2\ln 4 - \ln 7}$; $x \approx 1.6769$
57. $x = \frac{\ln 9 - \ln 5}{2\ln 5 - \ln 9}$; $x \approx 0.5753$
58. $x = \frac{\ln 5 + 3\ln 2}{\ln 2 + \ln 5}$; $x \approx 1.6021$
59. $t = \frac{-50}{3}\ln\left(\frac{37}{150}\right)$; $t \approx 23.3286$
60. $t = -25\ln\left(\frac{5}{32}\right)$; $t \approx 46.4074$
61. $t = \frac{100}{29}\ln(236)$; $t \approx 18.8408$
62. $t = \frac{-20}{7}\ln\left(\frac{27}{980}\right)$; $t \approx 10.2620$

3. Since logarithmic functions are not defined for all real numbers, we should check all "solutions" for extraneous roots.

4. In the equation $3\ln 5 + x\ln 5 = 2x\ln 6$, $x\ln 5$ is a variable term and should be added to both sides so the variable x can be factored out.

5. Solve the equation here, giving a step-by-step discussion of the solution process: $\ln(4x + 3) + \ln(2) = 3.2$ 2.316566275

6. Describe the difference between *evaluating* the equation below given $x = 9.7$ and *solving* the equation given $y = 9.7$: $y = 3\log_2(x - 1.7) - 2.3$. Answers will vary.

DEVELOPING YOUR SKILLS

Write each equation in the simplified form $\log_b x = k$ (logarithmic term = constant). Do not solve.

7. $3\ln(x + 4) + 7 = 13$ $\ln(x + 4) = 2$

8. $-6 = 2\log_3(x - 5) - 10$ $\log_3(x - 5) = 2$

9. $\log(x + 2) + \log x = 4$ $\log(x^2 + 2x) = 4$

10. $2 = \ln\left(\frac{2}{x} + 3\right) + \ln x$ $\ln(2 + 3x) = 2$

11. $2\log_2(x) + \log_2(x - 1) = 2$ $\log_2(x^3 - x^2) = 2$

12. $\ln(x - 2) - \ln x = -\ln(3x)$ $\ln(3x - 6) = 0$

Write each equation in the simplified form $b^x = k$ (exponential term = constant). Do not solve.

13. $4e^{x-2} + 5 = 69$ $e^{x-2} = 16$

14. $2 - 3e^{0.4x} = -7$ $e^{0.4x} = 3$

15. $250e^{0.05x+1} + 175 = 512.5$ $e^{0.05x+1} = 1.35$

16. $-150 = 290.8 - 190e^{-0.75x}$ $e^{-0.75x} = 2.32$

17. $3^x(3^{2x-1}) = 81$ $3^{3x-1} = 81$

18. $(4^{3x+5})4^{-x} = 64$ $4^{2x+5} = 64$

Solve each equation two ways: by equating bases and using the uniqueness properties, and by applying a base-10 or base-e logarithm and using the power property of logarithms.

19. $2^x = 128$ $x = 7$

20. $3^x = 243$ $x = 5$

21. $5^{3x} = 3125$ $x = \frac{5}{3}$

22. $4^{3x} = 1024$ $x = \frac{5}{3}$

23. $5^{x+1} = 625$ $x = 3$

24. $6^{x-1} = 216$ $x = 4$

25. $\left(\frac{1}{2}\right)^{n-1} = \frac{1}{256}$ $n = 9$

26. $\left(\frac{1}{3}\right)^{n-1} = \frac{1}{729}$ $n = 7$

27. $\frac{1}{625} = \left(\frac{1}{5}\right)^{n-1}$ $n = 5$

28. $\frac{1}{216} = \left(\frac{1}{6}\right)^{n-1}$ $n = 4$

29. $\frac{128}{2187} = \left(\frac{2}{3}\right)^{n-1}$ $n = 8$

30. $\frac{729}{64} = \left(\frac{3}{2}\right)^{n-1}$ $n = 7$

31. $\frac{16}{625} = \left(\frac{2}{5}\right)^{n-1}$ $n = 5$

32. $\frac{729}{4096} = \left(\frac{3}{4}\right)^{n-1}$ $n = 7$

33. $\frac{5}{243} = 5\left(\frac{1}{3}\right)^{n-1}$ $n = 6$

34. $\frac{5}{32} = 10\left(\frac{1}{2}\right)^{n-1}$ $n = 7$

35. $\frac{56}{125} = 7\left(\frac{2}{5}\right)^{n-1}$ $n = 4$

36. $\frac{243}{64} = 16\left(\frac{3}{4}\right)^{n-1}$ $n = 6$

Solve using the method of your choice. Answer in exact form and approximate form rounded to four decimal places.

37. $10^x = 97$

38. $10^x = 12{,}089$

39. $879 = 10^{2x}$

40. $10^{3x} = 12{,}089$

41. $879 = 10^{x+3}$

42. $4589 = 10^{x-2}$

43. $e^x = 389$

44. $e^x = 25$

45. $e^{2x} = 1389$

46. $e^{3x} = 2507$

47. $e^{x+1} = 257$

48. $e^{x-3} = 589$

49. $2e^{0.25x} = 5$

50. $3e^{0.08x} = 12$

51. $7^{x+2} = 231$

52. $6^{x+2} = 3589$

53. $5^{3x-2} = 128{,}965$

54. $9^{5x-3} = 78{,}462$

55. $2^{x+1} = 3^x$

56. $7^x = 4^{2x-1}$

57. $5^{2x+1} = 9^{x+1}$

58. $\left(\frac{1}{5}\right)^{x-1} = \left(\frac{1}{2}\right)^{3-x}$

59. $\frac{87}{1 + 3e^{-0.06t}} = 50$

60. $\frac{39}{1 + 4e^{-0.04t}} = 24$

61. $160 = \frac{200}{1 + 59e^{-0.29t}}$

62. $98 = \frac{152}{1 + 20e^{-0.35t}}$

Solve each equation using the uniqueness property of logarithms.

63. $\log(5x + 2) = \log 2$ $x = 0$

64. $\log(2x - 3) = \log 3$ $x = 3$

65. $\log_4(x + 2) - \log_4 3 = \log_4(x - 1)$ $x = \frac{5}{2}$

66. $\log_3(x + 6) - \log_3 x = \log_3 5$ $x = \frac{3}{2}$

67. $\ln(8x - 4) = \ln 2 + \ln x$ $x = \frac{2}{3}$

68. $\ln(x - 1) + \ln 6 = \ln(3x)$ $x = 2$

Solve each equation by converting to exponential form.

69. $\log(3x - 1) = 2$ $x = 33$

70. $\log(2x + 3) = 2$ $x = \frac{97}{2}$

71. $\log_5(x + 7) = 3$ $x = 118$

72. $\log_3(x - 1) = 2$ $x = 10$

73. $\ln(x + 7) = 2$ $x = e^2 - 7$

74. $\ln(x - 2) = 3$ $x = e^3 + 2$

75. $-2 = \log(2x - 1)$ $x = \frac{101}{200}$

76. $-3 = \log(1 + x)$ $x = -0.999$

77. $\log(2x) - 5 = -3$ $x = 50$

78. $\log(3x) + 7 = 8$ $x = \frac{10}{3}$

79. $-2\ln(x + 1) = -6$ $x = e^3 - 1$

80. $-3\ln(x - 3) = -9$ $x = e^3 + 3$

Solve each logarithmic equation using any appropriate method. Clearly identify any extraneous roots. If there are no solutions, so state.

81. $\log(2x - 1) + \log 5 = 1$

82. $\log(x - 7) + \log 3 = 2$

83. $\log_2(9) + \log_2(x + 3) = 3$

84. $\log_3(x - 4) + \log_3(7) = 2$

85. $\ln(x + 7) + \ln 9 = 2$

86. $\ln 5 + \ln(x - 2) = 1$

87. $\log(x + 8) + \log x = \log(x + 18)$

88. $\log(x + 14) - \log x = \log(x + 6)$

89. $\ln(2x + 1) = 3 + \ln 6$

90. $\ln 21 = 1 + \ln(x - 2)$

91. $\log(-x - 1) = \log(5x) - \log x$

92. $\log(1 - x) + \log x = \log(x + 4)$

93. $\ln(2t + 7) = \ln 3 - \ln(t + 1)$

94. $\ln(5 - r) - \ln 6 = \ln(r + 2)$

81. $x = \frac{3}{2}$
82. $x = \frac{121}{3}$
83. $x = \frac{-19}{9}$
84. $x = \frac{37}{7}$
85. $x = \dfrac{e^2 - 63}{9}$
86. $x = \dfrac{e + 10}{5}$
87. $x = 2$; -9 is extraneous
88. $x = 2$; -7 is extraneous
89. $x = 3e^3 - \frac{1}{2}$; $x \approx 59.75661077$
90. $x = 21e^{-1} + 2$; $x \approx 9.725468265$
91. no solution
92. no solution
93. $t = -\frac{1}{2}$; -4 is extraneous
94. $r = -1$

WORKING WITH FORMULAS

95. Half-life of a radioactive substance: $A = A_0\left(\frac{1}{2}\right)^{\frac{t}{h}}$

The **half-life** of radioactive material is the amount of time required for one-half of an initial amount of the substance to vanish due to the decay. The amount of material remaining can be determined using the formula shown, where t represents elapsed time, h is the half-life of the material, A_0 is the initial amount, and A represents the amount remaining. The sodium isotope $^{24}_{11}Na$ has a half-life of 15 hr. If 500 g were initially present, how much is left after 60 hr? (For more on this formula, see page 527 in Section 5.5). 31.25 g

96. Forensics—estimating time of death: $h = -3.9\ln\left(\frac{T - t}{98.6 - t}\right)$

Under certain conditions, a forensic expert can determine the approximate time of death for a person found recently expired using the formula shown, where T is the body temperature when it was found, t is the (constant) temperature of the room where the person was found, and h is the number of hours since death. If the body was discovered at 9:00 A.M. in a 73°F air-conditioned room, and had a temperature of 86.2°F, at approximately what time did the person expire? 6:25 A.M.

APPLICATIONS

Newton's law of cooling was discussed in Example 8 of this section: $T = T_1 + (T_0 - T_1)e^{-kh}$, where T_0 is the initial temperature of the object, T_1 is the temperature of the surrounding medium, and T is the temperature after elapsed time h in hours (k is a constant that depends on the materials involved).

97. Cooling time: If a can of soft drink at a room temperature of 75°F is placed in a 40°F refrigerator, how long will it take the drink to cool to 45°F? Assume $k = 0.61$ and answer in hours and minutes. 3 hr 11 min

98. Cooling time: Suppose that the temperature in Esconabe, Michigan, was 47°F when a 5°F arctic cold front moved over the state. How long would it take a puddle of water to begin freezing over? (Water freezes at 32°F.) Assume $k = 0.9$ and answer in minutes. 30 min

Use the *barometric equation* $h = (30T + 8000)\ln\left(\frac{P_0}{P}\right)$ for Exercises 99 and 100.

99. Altitude and pressure: Determine the atmospheric pressure at the summit of Mount McKinley (in Alaska), a height of 6194 m. Assume the temperature at the summit is −10°C. 34 cmHg

100. Altitude and pressure: A plane is flying at an altitude of 10,029 m. If the barometric pressure is 22 cm of mercury, what is the temperature at this altitude (to the nearest degree)? 3°C

101. Investment growth: Use the compound interest formula $A = P\left(1 + \frac{r}{n}\right)^{nt}$ to determine how long it would take \$2500 to grow to \$6000 if the annual rate is 8% and interest was compounded monthly. 11.0 yrs

102. Radioactive half-lives: Use the formula discussed in Exercise 95 to find the half-life of polonium, if 1000 g of the substance decayed to 125 g in 420 days. 140 days

103. Advertising and sales: An advertising agency determines the number of items sold is related to the amount spent on advertising by the equation $N(A) = 1500 + 315 \ln A$, where A represents the advertising budget and $N(A)$ gives the number of sales. If a company wants to generate 5000 sales, how much money should be set aside for advertising? Round to the nearest dollar. \$66,910

104. Automobile depreciation: The amount of time required for a certain new car to depreciate to a given value can be determined using the formula $T(v_c) = 5 \ln\left(\frac{v_n}{v_c}\right)$, where v_c represents the current value of the car, v_n represents the value of the car when new, and $T(v_c)$ gives the elapsed time in years. A new car is purchased for \$28,500. Find the current value 3 yr later. \$15,641.13

105. Spaceship velocity: In space travel, the change in the velocity of a spaceship V_s (in km/sec) depends on the mass of the ship M_s (in tons), the mass of the fuel that has been burned M_f (in tons), and the escape velocity of the exhaust V_e (in km/sec). Disregarding frictional forces, these are related by the equation $V_s = V_e \ln\left(\frac{M_s}{M_s - M_f}\right)$. Find the mass of the fuel that has been burned when $V_s = 6$ km/sec, if the escape velocity of the exhaust is 8 km/sec and the ship's mass is 100 tons. 52.76 tons

106. Carbon-14 dating: After the death of an organism, it no longer absorbs the natural radioactive element known as "carbon-14" (^{14}C) from our atmosphere. Scientists theorize that the age of the organism (now a fossil) can be estimated by measuring the amount of ^{14}C that remains in the fossil, since the half-life of carbon-14 is known. One version of the formula used is $T = -7978 \ln x$, where x is the percentage of ^{14}C that remains in the fossil and T is the time in years since the organism died. If an archeologist claimed the bones of a recently discovered skeleton were 9800 years old, what percent of ^{14}C did she determine remained in the bones? 29.3%

WRITING, RESEARCH, AND DECISION MAKING

107. Virtually all ocean life depends on microscopic plants called *phytoplankton.* These plants can only thrive in what is called the *photic zone* of the ocean, or the top layer of ocean, where there is sufficient light for photosynthesis to take place. The depth of this zone depends on various factors, and is measured using an exponential formula called the Beer-Lambert law. Do some research on this mathematical model and write a report on how it is used. Include several examples and a discussion of the factors that most affect the depth of the photic zone. Answers will vary.

108. In 1798, the English economist Thomas Malthus wrote a paper called "Essay on the Principle of Population," in which he forecast that human populations would grow exponentially, while the supply of food would only grow linearly. This dire predication had a huge impact on the social and economic thinking of the day. Do some research on Thomas Malthus and investigate the mathematical models he used to predict the growth of the food supply versus population growth. Why did his predictions have such an impact? Why were his predictions never realized? Answers will vary.

109. Match each equation with the most appropriate solution strategy, and justify/discuss why.

a.	$e^{x+1} = 25$	d	apply base-10 logarithm to both sides
b.	$\log(2x + 3) = \log 53$	e	rewrite and apply uniqueness property for exponentials
c.	$\log(x^2 - 3x) = 2$	b	apply uniqueness property for logarithms
d.	$10^{2x} = 97$	f	apply either base-10 or base-e logarithm
e.	$2^{5x-3} = 32$	a	apply base-e logarithm
f.	$7^{x+2} = 23$	c	write in exponential form

EXTENDING THE CONCEPT

Solve the following equations.

110. $2e^{2x} - 7e^x = 15$ $x = 1.609438$

111. $3e^{2x} - 4e^x - 7 = -3$ $x = 0.69319718$

112. $\log_2(x + 5) = \log_4(21x + 1)$ $x = 3$

113. Solve by exponentiating both sides to an appropriate base: $\log_3 3^{2x} = \log_3 5$. $x = \frac{\log_3 5}{2}$

114. Use the algebraic method from Section 3.2 to find the inverse function for $f(x) = 2^{x+1}$. $f^{-1}(x) = \frac{\ln x}{\ln 2} - 1$

Show that $f(x)$ and $g(x)$ are inverse functions by composing the functions and using logarithmic properties.

115. $f(x) = 3^{x-2}$; $g(x) = \log_3 x + 2$

116. $f(x) = e^{x-1}$; $g(x) = \ln(x) + 1$

117. Show $y = 2^x$ is equivalent to $y = e^{x \ln 2}$.

118. Show $y = b^x$ is equivalent to $y = e^{rx}$, where $r = \ln b$.

115. $(f \circ g)(x) = 3^{(\log_3 x + 2) - 2} = 3^{\log_3 x} = x$; $(g \circ f)(x) = \log_3(3^{x-2}) + 2 = x - 2 + 2 = x$
116. $(f \circ g)(x) = e^{[\ln x + 1] - 1} = e^{\ln x} = x$; $(g \circ f)(x) = \ln e^{x-1} + 1 = x - 1 + 1 = x$
117. $y = e^{x \ln 2} = e^{\ln 2^x} = 2^x$; $y = 2^x = \ln y = x \ln 2$, $e^{\ln y} = e^{x \ln 2} \Rightarrow y = e^{x \ln 2}$
118. $y = b^x$, $\ln y = x \ln b$, $e^{\ln y} = e^{x \ln b}$, $y = e^{xr}$ for $r = \ln b$

MAINTAINING YOUR SKILLS

119. (2.4) Match the graph shown with its correct equation, without actually graphing the function.

a. $y = x^2 + 4x - 5$ **b.** $y = -x^2 - 4x + 5$

c. $y = -x^2 + 4x + 5$ **d.** $y = x^2 - 4x - 5$

120. (R.3) Determine the value of the following expression in exact form (without using a calculator):
$\sqrt{(3.2 \times 10^{-23})(2.45 \times 10^{33})}$

121. (3.3) State the domain and range of the functions

a. $y = \sqrt{2x + 3}$ **b.** $y = |x + 2| - 3$

122. (4.6) Graph the function $r(x) = \frac{x^2 - 4}{x - 1}$. Label all intercepts and asymptotes.

123. (4.3) Use synthetic division and the RRT to find all zeroes (real/complex) of $f(x) = x^3 + x + 10$.

124. (3.6) Suppose the maximum load (in tons) that can be supported by a cylindrical post varies directly with its diameter raised to the fourth power and inversely as the square of its height. A post 8 ft high and 2 ft in diameter can support 6 tons. How many tons can be supported by a post 12 ft high and 3 ft in diameter?

119. b
120. 280,000
121. a. $x \in [-\frac{3}{2}, \infty)$, $y \in [0, \infty)$
b. $x \in (-\infty, \infty)$, $y \in [-3, \infty)$
122.

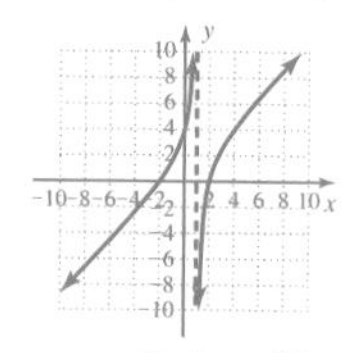

123. $x = -2, 1 \pm 2i$
124. 13.5 tons

5.5 Applications from Business, Finance, and Physical Science

LEARNING OBJECTIVES

In Section 5.5 you will study applications of:

A. Interest compounded *n* times per year

B. Interest compounded continuously

C. Exponential growth and decay

D. Annuities and amortization

INTRODUCTION

Would you pay $950,000 for a home worth only $250,000? Surprisingly, when a conventional mortgage is repaid over 30 years, this is not at all rare. Over time, the accumulated interest on the mortgage is easily more than two or three times the original value of the house. In this section we explore how interest is paid or charged, and look at other applications of exponential and logarithmic functions from business and finance, and the physical and social sciences.

POINT OF INTEREST

One common application of exponential functions is the calculation of interest. Interest is an amount of money paid *by* you for the use of money that is borrowed, or paid *to* you for money that you invest. The custom of charging or paying interest is very ancient, and there are references to this practice that date back as far as 2000 B.C. in ancient Babylon. In this section, we investigate some of the more common ways interest is charged or paid—applications that require the use of exponential and logarithmic functions.

A. Simple and Compound Interest

Simple interest is an amount of interest that is computed only once during the lifetime of an investment (or loan). In the world of finance, the initial deposit or base amount is referred to as the **principal *p*,** the **interest rate *r*** is given as a percentage and is usually stated as an annual rate, with the term of the investment or loan most often given as *time* ***t*** in years. Simple interest is merely an application of the basic percent equation, with the additional element of time coming into play. The **simple interest formula** is *interest* = *principal* × *rate* × *time*, or $I = prt$. To find the total amount A that has accumulated (for deposits) or is due (for loans) after t years, we merely add the accumulated interest to the initial principal: $A = p + prt$ or $A = p(1 + rt)$ after factoring.

WORTHY OF NOTE

If a loan is kept for only a certain number of months, weeks, or days, the time t should be stated as a fractional part of a year so the time period for the rate (years) matches the time period over which the loan is repaid.

SIMPLE INTEREST FORMULA

If principal p is deposited or borrowed at interest rate r for a period of t years, the simple interest on this account will be

$$I = prt$$

The total amount A accumulated or due after this period will be:

$$A = p + prt \quad \text{or} \quad A = p(1 + rt)$$

EXAMPLE 1 Many finance companies offer what have become known as PayDay Loans—a small $50 loan to help people get by until payday, usually no longer than 2 weeks. If the cost of this service is $12.50, determine the annual rate of interest charged by these companies.

Solution: The interest charge is $12.50, the initial principal is $50.00 and the time period is 2 weeks or $\frac{2}{52} = \frac{1}{26}$ of a year. The simple interest formula yields

$I = prt$ simple interest formula

$12.50 = 50r\left(\frac{1}{26}\right)$ substitute $12.50 for I, $50.00 for p, and $\frac{1}{26}$ for t

$6.5 = r$ result

The annual interest rate on these loans is a whopping 650%!

NOW TRY EXERCISES 7 THROUGH 14

Compound Interest

Many financial institutions pay **compound interest** on deposits they receive, which is interest paid on previously accumulated interest. The most common compounding periods are yearly, semiannually (two times per year), quarterly (four times per year), monthly (12 times per year), and daily (365 times per year). Applications of compound interest typically involve exponential functions. For convenience, consider $1000 in principal, deposited at 8% for 3 yr. The simple interest calculation shows $240 in interest is earned and there will be $1240 in the account: $A = 1000[1 + (0.08)(3)] = \$1240$. If the interest is *compounded each year* ($t = 1$) instead of once at the start of the three-year period, the interest calculation shows

$A_1 = 1000(1 + 0.08) = 1080$ in the account at the end of year 1,

$A_2 = 1080(1 + 0.08) = 1166.40$ in the account at the end of year 2,

$A_3 = 1166.40(1 + 0.08) \approx 1259.71$ in the account at the end of year 3.

The account has earned an additional $19.71 interest. More importantly, notice that we're multiplying by $(1 + 0.08)$ each compounding period, meaning results can be computed more efficiently by simply applying the factor $(1 + 0.08)^t$ to the initial principal p. For example:

$$A_3 = 1000(1 + 0.08)^3 \approx \$1259.71.$$

In general, for interest compounded yearly the **accumulated value equation** is $A = p(1 + r)^t$. Notice that solving this equation for p will tell us the amount we need to deposit *now,* in order to accumulate A dollars in t years: $p = \frac{A}{(1 + r)^t}$. This is called the **present value equation.**

INTEREST COMPOUNDED ANNUALLY

If a principal p is deposited at interest rate r and compounded yearly for a period of t years, the **accumulated value** is

$$A = p(1 + r)^t$$

If an accumulated value A is desired after t years, and the money is deposited at interest rate r and compounded yearly, the *present value* is

$$p = \frac{A}{(1 + r)^t}$$

EXAMPLE 2 An initial deposit of \$1000 is made into an account paying 6% compounded yearly. How long will it take for the money to double?

Solution: Using the formula for interest compounded yearly we have

$$A = p(1 + r)^t \quad \text{given}$$
$$2000 = 1000(1 + 0.06)^t \quad \text{substitute 2000 for } A\text{, 1000 for } p\text{, and 0.06 for } r$$
$$2 = 1.06^t \quad \text{isolate variable term}$$
$$\ln 2 = t \ln 1.06 \quad \text{apply base-}e\text{ logarithms}$$
$$\frac{\ln 2}{\ln 1.06} = t \quad \text{solve for } t$$
$$11.9 \approx t \quad \text{approximate form}$$

The money will double in just under 12 years.

NOW TRY EXERCISES 15 THROUGH 20

When interest is compounded more than once a year, say monthly, the bank will divide the interest rate by 12 (the number of compoundings) to maintain a constant yearly rate, but then pays you interest 12 times per year (interest is compounded). The net effect is an increased gain in the interest you earn, and the final compound interest formula takes this form:

$$\text{total amount} = \text{principal}\left(1 + \frac{\text{interest rate}}{\text{number of compoundings per year}}\right)^{(\text{number of years} \times \text{number of compoundings per year})}$$

COMPOUNDED INTEREST FORMULA

If principal p is deposited at interest rate r and compounded n times per year for a period of t years, the *accumulated value* will be:

$$A = p\left(1 + \frac{r}{n}\right)^{nt}$$

EXAMPLE 3 Macalyn won \$150,000 in the Missouri lottery and decides to invest the money for retirement in 20 yr. Of all the options available here, which one will produce the most money for retirement?

a. A certificate of deposit paying 5.4% compounded yearly.

b. A money market certificate paying 5.35% compounded semiannually.

c. A bank account paying 5.25% compounded monthly.

d. A bond issue paying 5.2% compounded daily.

Solution:

a. $A = \$150{,}000\left(1 + \frac{0.054}{1}\right)^{(20\times1)}$

$\approx \$429{,}440.97$

b. $A = \$150{,}000\left(1 + \frac{0.0535}{2}\right)^{(20\times2)}$

$\approx \$431{,}200.96$

c. $A = \$150{,}000\left(1 + \frac{0.0525}{4}\right)^{(20\times4)}$

$\approx \$425{,}729.59$

d. $A = \$150{,}000\left(1 + \frac{0.052}{365}\right)^{(20\times365)}$

$\approx \$424{,}351.12$

The best choice is (b), semiannual compounding at 5.35% for 20 yr.

NOW TRY EXERCISES 21 THROUGH 24

B. Interest Compounded Continuously

It seems natural to wonder what happens to the interest accumulation as n (the number of compounding periods) becomes very large. It appears the interest rate becomes very small (because we're dividing by n), but the exponent becomes very large (since we're multiplying by n). To see the result of this interplay more clearly, it will help to rewrite the compound interest formula $A = p\left(1 + \frac{r}{n}\right)^{nt}$ using the substitution $n = xr$. This gives $\frac{r}{n} = \frac{1}{x}$, and by direct substitution $\left(xr \text{ for } n \text{ and } \frac{1}{x} \text{ for } \frac{r}{n}\right)$ we obtain the form $A = p\left[\left(1 + \frac{1}{x}\right)^{x}\right]^{rt}$ by regrouping. This allows for a more careful study of the "denominator versus exponent" relationship using $\left(1 + \frac{1}{x}\right)^{x}$, *the same expression we used in Section 5.4 to define the number e.* Once again, note what happens as $x \to \infty$ (meaning the number of compounding periods increase without bound).

x	1	10	100	1000	10,000	100,000	1,000,000
$\left(1 + \frac{1}{x}\right)^{x}$	2	2.56374	2.70481	2.71692	2.71815	2.71827	2.71828

As before, we have as $x \to \infty$, $\left(1 + \frac{1}{x}\right)^{x} \to e$. The net result of this investigation is a formula for **interest compounded continuously,** derived by replacing $\left(1 + \frac{1}{x}\right)^{x}$ with the number e in the formula for compound interest, where $A = p\left[\left(1 + \frac{1}{x}\right)^{x}\right]^{rt}$ becomes $A = p[e]^{rt}$.

> **INTEREST COMPOUNDED CONTINUOUSLY**
>
> If a principal p is deposited at interest rate r and compounded continuously for a period of t years, the *accumulated value* will be
>
> $A = pe^{rt}$

EXAMPLE 4 Jaimin has \$10,000 to invest and wants to have at least \$25,000 in the account in 10 yr for his daughter's college education fund. If the account pays interest compounded continuously, what interest rate is required?

Solution: In this case, $P = \$10{,}000$, $A = \$25{,}000$, and $t = 10$.

$$A = pe^{rt} \quad \text{given}$$
$$25{,}000 = 10{,}000e^{10r} \quad \text{substitute 25,000 for } A\text{, 10,000 for } p\text{, and 10 for } t$$
$$2.5 = e^{10r} \quad \text{isolate variable term}$$
$$\ln 2.5 = 10r \ln e \quad \text{apply base-}e\text{ logarithms } (\ln e = 1)$$
$$\frac{\ln 2.5}{10} = r \quad \text{solve for } r$$
$$0.092 \approx r \quad \text{approximate form}$$

Jaimin will need an interest rate of about 9.2% to meet his goal.

NOW TRY EXERCISES 25 THROUGH 34

GRAPHICAL SUPPORT

To check the result from Example 4, use $Y_1 = 10{,}000e^{10x}$ and $Y_2 = 25{,}000$, then look for their point of intersection. We need only set an appropriate window size to ensure the answer will appear in the viewing window. Since 25,000 is the goal, $y \in [0, 30{,}000]$ seems reasonable for y. Although 12% interest ($x = 0.12$) is too good to be true, $x \in [0, 0.12]$ leaves a nice frame for the x-values. Verify that the calculator's answer is equal to $\frac{\ln 2.5}{10}$.

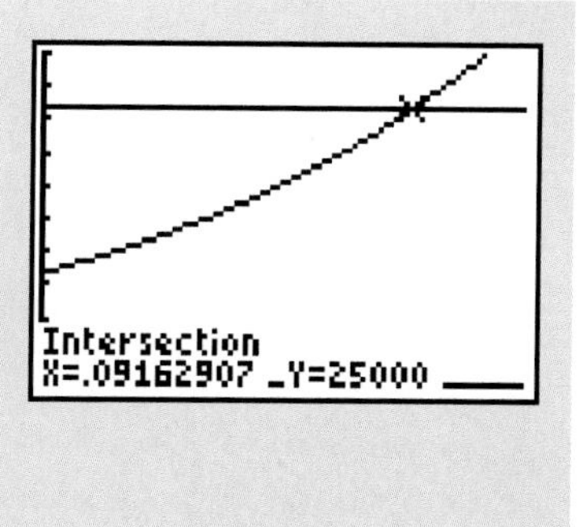

There are a number of interesting applications in the exercise set (see Exercises 37 through 46).

WORTHY OF NOTE

Notice the formula for exponential growth is virtually identical to the formula for interest compounded continuously. In fact, both are based on the same principles. If we let $A(t)$ represent the amount in an account after t years and A_0 represent the initial deposit (instead of P), we have: $A(t) = A_0e^{rt}$ versus $Q(t) = Q_0e^{rt}$ and the two can hardly be distinguished.

C. Applications Involving Exponential Growth and Decay

Closely related to the formula for interest compounded continuously are applications of **exponential growth** and **exponential decay.** If Q (quantity) and t (time) are variables, then Q grows exponentially as a function of t if $Q(t) = Q_0e^{rt}$ for the positive constants Q_0 and r. Careful studies have shown that population growth, whether it be humans, bats, or bacteria, can be modeled by these "base-e" exponential growth functions. If $Q(t) = Q_0e^{-rt}$, then we say Q decreases or **decays exponentially** over time. The constant r determines how rapidly a quantity grows or decays and is known as the **growth rate** or **decay rate** constant. Graphs of exponential growth and decay functions are shown here for arbitrary Q_0 and r. Note the graph of $Q(t) = Q_0e^{-rt}$ (Figure 5.19) is simply a reflection across the y-axis of $Q(t) = Q_0e^{rt}$ (Figure 5.18).

Figure 5.18

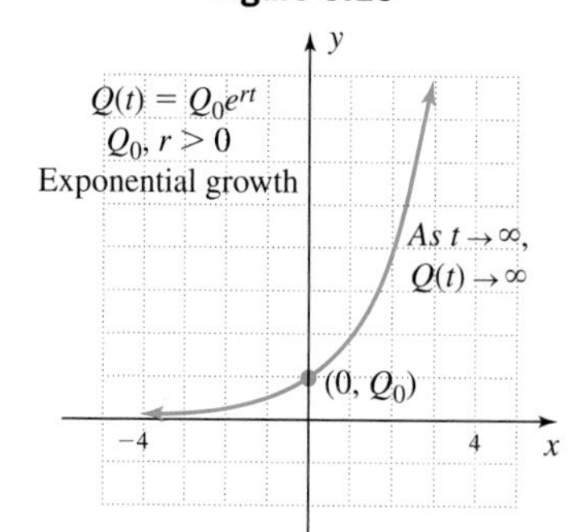

Figure 5.19

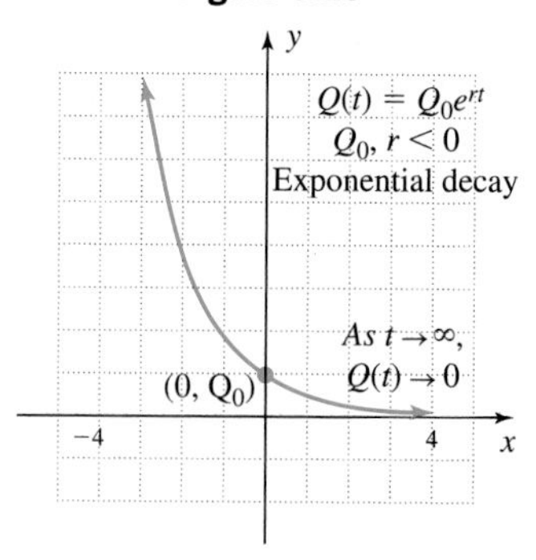

EXAMPLE 5 Because fruit flies multiply very quickly, they are often used in a study of genetics. Given the necessary space and food supply, a certain population of fruit flies is known to double every 12 days. If there were 100 flies to begin, find (a) the growth rate r and (b) the number of days until the population reaches 2000 flies.

Solution: **a.** Using the formula for exponential growth with $Q_0 = 100$, $t = 12$, and $Q(t) = 200$, we can solve for the growth rate r.

$$Q(t) = Q_0e^{rt} \quad \text{exponential growth function}$$
$$200 = 100e^{12r} \quad \text{substitute 200 for } Q(t)\text{, 100 for } Q_0\text{, and 12 for } t$$
$$2 = e^{12r} \quad \text{isolate variable term}$$
$$\ln 2 = 12r\ln e \quad \text{apply base-}e\text{ logarithms } (\ln e = 1)$$
$$\frac{\ln 2}{12} = r \quad \text{solve for } r$$
$$0.05776 \approx r \quad \text{approximate form}$$

The growth rate is approximately 5.78%.

b. To find the number of days until the fly population reaches 2000 we substitute 0.05776 for r in the exponential growth function.

$$Q(t) = Q_0e^{rt} \quad \text{exponential growth function}$$
$$2000 = 100e^{0.05776t} \quad \text{substitute 2000 for } Q(t)\text{, 100 for } Q_0\text{, and 0.05776 for } r$$
$$20 = e^{0.05776t} \quad \text{isolate variable term}$$
$$\ln 20 = 0.05776t\ln e \quad \text{apply base-}e\text{ logarithms } (\ln e = 1)$$
$$\frac{\ln 20}{0.05776} = t \quad \text{solve for } t$$
$$51.87 \approx t \quad \text{approximate form}$$

The fruit fly population will reach 2000 on day 51.

NOW TRY EXERCISES 47 AND 48

WORTHY OF NOTE

Many population growth models assume an unlimited supply of resources, nutrients, and room for growth. When this is not the case, a logistic growth model often results. See Section 5.6.

Perhaps the best known examples of exponential decay involve radioactivity. Ever since the end of World War II, common citizens have been aware of the existence of **radioactive elements** and the power of atomic energy. Today, hundreds of additional applications have been found for radioactive materials, from areas as diverse as biological research, radiology, medicine, and archeology. Radioactive elements decay of their own accord by emitting radiation. The rate of decay is measured using the **half-life** of the substance, which is the time required for a mass of radioactive material to decay until only one-half of its original mass remains. This half-life is used to find the rate of decay r, first mentioned in Section 5.3. In general, we have

$$Q(t) = Q_0 e^{-rt} \quad \text{exponential decay function}$$

$$\frac{1}{2} Q_0 = Q_0 e^{-rt} \quad \text{substitute } \tfrac{1}{2} Q_0 \text{ for } Q(t)$$

$$\frac{1}{2} = \frac{1}{e^{rt}} \quad \text{divide by } Q_0\text{; use negative exponents to rewrite expression}$$

$$2 = e^{rt} \quad \text{property of ratios}$$

$$\ln 2 = rt \ln e \quad \text{apply base-}e\text{ logarithms } (\ln e = 1)$$

$$\frac{\ln 2}{t} = r \quad \text{solve for } r$$

RADIOACTIVE RATE OF DECAY

If t represents the half-life of a radioactive substance per unit time, the nominal rate of decay per a like unit of time is given by

$$r = \frac{\ln 2}{t}$$

The rate of decay for known radioactive elements varies a great deal. For example, the element carbon-14 has a half-life of about 5730 yr, while the element lead-211 has a half-life of only about 3.5 min. Radioactive elements can be detected in extremely small amounts. If a drug is "labeled" (mixed with) a radioactive element and injected into a living organism, its passage through the organism can be traced and information on the health of internal organs can be obtained.

EXAMPLE 6 The radioactive element potassium-42 is often used in biological experiments, since it has a half-life of only about 12.4 hr and desired results can be measured accurately and experiments repeated if necessary. How much of a 2-g sample will remain after 18 hr and 45 min?

Solution: To begin we must find the nominal rate of decay r and use this value in the exponential decay function.

$$r = \frac{\ln 2}{t} \quad \text{radioactive rate of decay}$$

$$r = \frac{\ln 2}{12.4} \quad \text{substitute 12.4 for } t$$

$$r \approx 0.056 \quad \text{result}$$

The rate of decay is approximately 5.6%. To determine how much of the sample remains after 18.75 hr, we use $r = 0.056$ in the decay function and evaluate it at $t = 18.75$.

$$Q(t) = Q_0 e^{-rt} \quad \text{exponential decay function}$$

$$Q(18.75) = 2e^{(-0.056)(18.75)} \quad \text{substitute 2 for } Q_0\text{, 0.056 for } r\text{, and 18.75 for } t$$

$$Q(18.75) \approx 0.7 \quad \text{evaluate}$$

After 18 hr and 45 min, only 0.7 g of potassium-42 will remain.

NOW TRY EXERCISES 49 THROUGH 52

WORTHY OF NOTE

In the case of regularly scheduled loan payments, an amortization schedule is used. The word **amortize** comes from the French word *amortir,* which literally means "to kill off." To amortize a loan means to pay it off, with the schedule of payments often computed using an annuity formula.

D. Applications Involving Annuities and Amortization

Our previous calculations for simple and compound interest involved a single deposit (the principal) that accumulated interest over time. Many savings and investment plans involve a regular schedule of deposits (monthly, quarterly, or annual deposits) over the life of the investment. Such an investment plan is called an **annuity.**

Similar to our work with compound interest, formulas exist for the *accumulated value* of an annuity and the *periodic payment required* to meet future goals. Suppose that for 4 yr, \$100 is deposited annually into an account paying 8% compounded yearly. Using the compound interest formula we can track the total amount A in the account:

$$A = 100 + 100(1.08)^1 + 100(1.08)^2 + 100(1.08)^3$$

WORTHY OF NOTE

It is often assumed that the first payment into an annuity is made *at the end of a compounding period,* and hence earns no interest. This is why the first \$100 deposit is not multiplied by the interest factor. These terms are actually the terms of a **geometric sequence,** which we will study later in Section 8.3.

To develop an annuity formula, we multiply the annuity equation by 1.08, then subtract the original equation. This leaves only the first and last terms, since the other (interior) terms sum to zero:

$$1.08A = 100(1.08) + 100(1.08)^2 + 100(1.08)^3 + 100(1.08)^4 \quad \text{multiply by 1.08}$$
$$-A = -[100 + 100(1.08)^1 + 100(1.08)^2 + 100(1.08)^3] \quad \text{original equation}$$
$$1.08A - A = 100(1.08)^4 - 100 \quad \text{subtract ("interior terms" sum to zero)}$$
$$0.08A = 100[(1.08)^4 - 1] \quad \text{factor out 100}$$
$$A = \frac{100[(1.08)^4 - 1]}{0.08} \quad \text{solve for } A$$

This result can be generalized for any periodic payment p, interest rate r, number of compounding periods n, and number of years t. This would give $A = \dfrac{p\left[\left(1 + \frac{r}{n}\right)^{nt} - 1\right]}{\frac{r}{n}}$.

The formula can be made less formidable using $R = \frac{r}{n}$, where R is the interest rate per compounding period.

ACCUMULATED VALUE OF AN ANNUITY

If a periodic payment P is deposited n times per year at an *annual interest rate* r with interest compounded n times per year for t years, the accumulated value is given by

$$A = \frac{P}{R}[(1 + R)^{nt} - 1], \text{ where } R = \frac{r}{n}$$

This is also referred to as the **future value** of the account.

EXAMPLE 7 Since he was a young child, Fitisemanu's parents have been depositing \$50 each month into an annuity that pays 6% annually and is compounded monthly. If the account is now worth \$9875, how long has it been open?

Solution: In this case $p = 50$, $r = 0.06$, $n = 12$, $R = 0.005$, and $A = 9875$. The formula gives

$$A = \frac{P}{R}[(1 + R)^{nt} - 1] \quad \text{future value formula}$$

$$9875 = \frac{50}{0.005}[(1.005)^{(12)(t)} - 1] \quad \text{substitute 9875 for } A\text{, 50 for } p\text{, 0.005 for } R\text{, and 12 for } n$$

$$1.9875 = 1.005^{12t} \quad \text{simplify and isolate variable term}$$

$$\ln(1.9875) = 12t(\ln 1.005) \quad \text{apply base-}e\text{ logarithms}$$

$$\frac{\ln(1.9875)}{12\ln(1.005)} = t \quad \text{solve for } t$$

$$11.5 \approx t \quad \text{approximate form}$$

The account has been open approximately 11.5 yr.

NOW TRY EXERCISES 53 THROUGH 56

The periodic payment required to meet a future goal or obligation can be computed by solving for P in the previous formula: $P = \frac{AR}{[(1 + R)^{nt} - 1]}$. In this form, P is referred to as a **sinking fund.**

EXAMPLE 8 Sheila is determined to stay out of debt and decides to save \$20,000 to pay cash for a new car in 4 yr. The best investment vehicle she can find pays 9% compounded monthly. If \$300 is the most she can invest each month, can she meet her "4-yr" goal?

Solution: Here we have $P = 300$, $A = 20{,}000$, $r = 0.09$, $n = 12$, and $R = 0.0075$. The sinking fund formula gives

$$P = \frac{AR}{[(1 + R)^{nt} - 1]} \quad \text{sinking fund}$$

$$300 = \frac{(20{,}000)(0.0075)}{(1.0075)^{12t} - 1} \quad \text{substitute 300 for } P\text{, 20,000 for } A\text{, 0.0075 for } R\text{, and 12 for } n$$

$$300(1.0075^{12t} - 1) = 150 \quad \text{multiply in numerator and clear denominators}$$

$$1.0075^{12t} = 1.5 \quad \text{isolate variable term}$$

$$12t\ln(1.0075) = \ln 1.5 \quad \text{apply base-}e\text{ logarithms}$$

$$t = \frac{\ln(1.5)}{12\ln(1.0075)} \quad \text{solve for } t$$

$$\approx 4.5 \quad \text{approximate form}$$

No. She is close, but misses her original 4-yr goal.

NOW TRY EXERCISES 57 AND 58

For Example 8, we could have substituted 4 for t and left P unknown, to see if a payment of \$300 per month would be sufficient. You can verify the result would be $P \approx \$347.70$, which is what Sheila would need to invest to meet her 4-yr goal exactly.

Using a graphing calculator allows for various other investigations, as demonstrated in the following *Technology Highlight.*

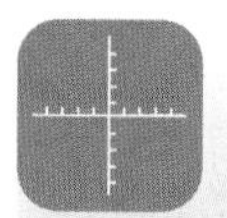

TECHNOLOGY HIGHLIGHT

Using a Graphing Calculator to Explore Compound Interest

The keystrokes shown apply to a TI-84 Plus model. Please consult our Internet site or your manual for other models.

The graphing calculator is an excellent tool for exploring mathematical relationships, particularly when many variables work simultaneously to produce a single result. For example, the formula $A = P\left(1 + \frac{r}{n}\right)^{nt}$ has five different unknowns: the total amount A, initial principal P, interest rate r, compounding periods per year n, and number of years t. In Example 2, we asked how long it would take $1000 to double if it were compounded yearly at 6% ($n = 1$, $r = 0.06$). What if we deposited $5000 instead of $1000? Compounded daily instead of quarterly? Or invested at 12% rather than 10%? There are many ways a graphing calculator can be used to answer such questions. In this exercise, we make use of the calculator's "alpha constants." The TI-84 Plus can use any of the 26 letters of the English alphabet (and even a few other symbols) to store constant values. One advantage is we can use them to write a formula using these constants on the Y= screen, then change any constant from the home screen to see how other values are affected. On the TI-84 Plus, these alpha constants are shown in green and are accessed by pressing the ALPHA key followed by the key with the letter desired. Suppose we wanted to study the relationship between an interest rate r and the time t required for a deposit to double. Using Y_1 in place of A as output variable, and x in place of t, enter $A = P\left(1 + \frac{r}{n}\right)^{nt}$ as Y_1 on the Y= screen (Figure 5.20). To assign initial values to the constants P, r, and n we use the STO▸ and ALPHA keys. Let's start with a deposit of $1000 at 7% interest compounded monthly. The keystrokes are: 1000 STO▸ ALPHA 8 ENTER, 0.07 STO▸ ALPHA X ENTER, and 12 STO▸ ALPHA LOG ENTER (Figure 5.21). After setting an appropriate window size (perhaps Xmax = 15 and Ymax = 3000), we can begin investigating how the interest rate affects this growth. It will help to enter $Y_2 = 2000$ to easily check "doubling time." Graph both functions and use the intersection of graphs method to find the doubling time. This produces the result in Figure 5.22, where we note it will take about 9.9 yr to double under these conditions. Return to the home screen (2nd MODE), change the interest rate to 10%, and graph the functions again. This time the point of intersection is just less than 7 (yr). Experiment with other rates and compounding periods to explore further.

Figure 5.20

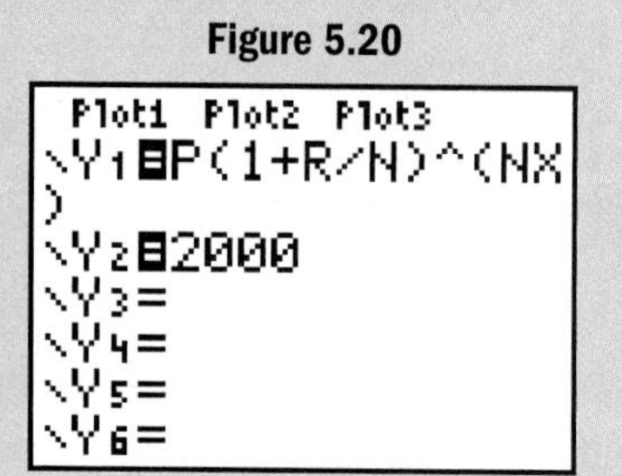

Figure 5.21

Figure 5.22

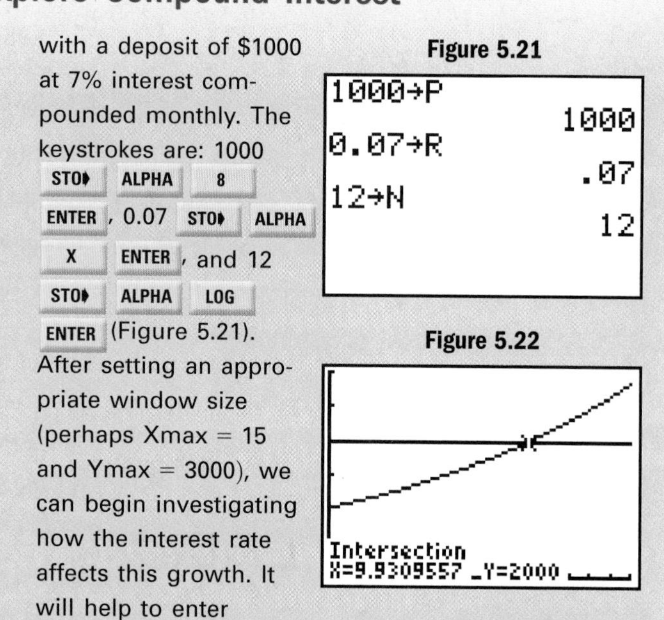

Exercise 1: With $P = \$1000$, and $r = 0.08$, investigate the "doubling time" for interest compounded quarterly, monthly, daily, and hourly. 8.75; 8.69; 8.665; 8.664

Exercise 2: With $P = \$1000$, investigate "doubling time" for rates of 6%, 8%, 10%, and 12%, and $n = 4$, $n = 12$, and $n = 365$. Which had a more significant impact, more compounding periods, or a greater interest rate? 6%: 11.64; 11.58; 11.55, 8%: 8.75; 8.69; 8.67, 10%: 7.02; 6.96; 6.93, 12%: 5.86; 5.81; 5.78; a greater interest rate has more impact

Exercise 3: Will a larger principal cause the money to double faster? Investigate and find out. no

5.5 EXERCISES

CONCEPTS AND VOCABULARY

Fill in each blank with the appropriate word or phrase. Carefully reread the section if needed.

1. Compound interest is interest paid to you on previously accumulated interest.

2. The formula for interest compounded continuously is $A = pe^{rt}$, where e is approximately 2.72.

3. Given constants Q_0 and r, and that Q decays exponentially as a function of t, the equation model is $Q(t) =$ Q_0e^{-rt}.

4. Investment plans calling for regularly scheduled deposits are called annuities. The annuity formula gives the future value of the account.

5. Explain/describe the difference between the future value and present value of an annuity. Include an example. Answers will vary.

6. Describe/explain how you would find the rate of growth r, given that a population of ants grew from 250 to 3000 in 6 weeks. Answers will vary.

DEVELOPING YOUR SKILLS

For simple interest accounts, the interest earned or due depends on the principal p, interest rate r, and the time t in years according to the formula $I = prt$.

7. Find p given $I = \$229.50$, $r = 6.25\%$, and $t = 9$ months. $4896

8. Find r given $I = \$1928.75$, $p = \$8500$, and $t = 3.75$ yr. 6.05%

9. Larry came up a little short one month at bill-paying time and had to take out a title loan on his car at Check Casher's, Inc. He borrowed \$260, and 3 weeks later he paid off the note for \$297.50. What was the annual interest rate on this title loan? (*Hint:* How much *interest* was charged?) 250%

10. Angela has \$750 in a passbook savings account that pays 2.5% simple interest. How long will it take the account balance to hit the \$1000 mark at this rate of interest, if she makes no further deposits? (*Hint:* How much *interest* will be paid?) 13.3 yr

For simple interest accounts, the amount A accumulated or due depends on the principal p, interest rate r, and the time t in years according to the formula $A = p(1 + rt)$.

11. Find p given $A = \$2500$, $r = 6.25\%$, and $t = 31$ months. $2152.47

12. Find r given $A = \$15{,}800$, $p = \$10{,}000$, and $t = 3.75$ yr. 15.47%

13. Olivette Custom Auto Service borrowed \$120,000 at 4.75% simple interest to expand their facility from three service bays to four. If they repaid \$149,925, what was the term of the loan? 5.25 yr

14. Healthy U sells nutritional supplements and borrows \$50,000 to expand their product line. When the note is due 3 yr later, they repay the lender \$62,500. If it was a simple interest note, what was the annual interest rate? 8.33%

For accounts where interest is compounded annually, the amount A accumulated or due depends on the principal p, interest rate r, and the time t in years according to the formula $A = p(1 + r)^t$.

15. Find t given $A = \$48{,}428$, $p = \$38{,}000$, and $r = 6.25\%$. 4 yr

16. Find p given $A = \$30{,}146$, $r = 5.3\%$, and $t = 7$ yr. $21,000.57

17. How long would it take \$1525 to triple if invested at 7.1%? 16 yr

18. What interest rate will ensure a \$747.26 deposit will be worth \$1000 in 5 yr? 6%

For accounts where interest is compounded annually, the principal P needed to ensure an amount A has been accumulated in the time period t when deposited at interest rate r is given by the formula $P = \dfrac{A}{(1 + r)^t}$.

19. The Stringers need to make a $10,000 balloon payment in 5 yr. How much should be invested now at 5.75%, so that the money will be available? $7561.33

20. Morgan is 8 yr old. If her mother wants to have $25,000 for Morgan's first year of college (in 10 yr), how much should be invested now if the account pays a 6.375% fixed rate? $13,475.48

For compound interest accounts, the amount A accumulated or due depends on the principal p, interest rate r, number of compoundings per year n, and the time t in years according to the formula $A = p\left(1 + \dfrac{r}{n}\right)^{nt}$.

21. Find t given $A = \$129{,}500$, $p = \$90{,}000$, and $r = 7.125\%$ compounded weekly. 5 yr

22. Find r given $A = \$95{,}375$, $p = \$65{,}750$, and $t = 15$ yr with interest compounded monthly. 2.48%

23. How long would it take a $5000 deposit to double, if invested at a 9.25% rate and compounded daily? 7.5 yr

24. What principal should be deposited at 8.375% compounded monthly to ensure the account will be worth $20,000 in 10 yr? $8681.04

For accounts where interest is compound continuously, the amount A accumulated or due depends on the principal p, interest rate r, and the time t in years according to the formula $A = pe^{rt}$.

25. Find t given $A = \$2500$, $p = \$1750$, and $r = 4.5\%$. 7.9 yr

26. Find r given $A = \$325{,}000$, $p = \$250{,}000$, and $t = 10$ yr. 2.62%

27. How long would it take $5000 to double if it is invested at 9.25%. Compare the result to Exercise 23. 7.5 yr

28. What principal should be deposited at 8.375% to ensure the account will be worth $20,000 in 10 yr? Compare the result to Exercise 24. $8655.82

Solve for the indicated unknowns.

29. $A = p + prt$
- **a.** solve for t
- **b.** solve for p

30. $A = p(1 + r)^t$
- **a.** solve for t
- **b.** solve for r

31. $A = p\left(1 + \dfrac{r}{n}\right)^{nt}$
- **a.** solve for r
- **b.** solve for t

32. $A = pe^{rt}$
- **a.** solve for p
- **b.** solve for r

33. $Q(t) = Q_0e^{rt}$
- **a.** solve for Q_0
- **b.** solve for t

34. $p = \dfrac{AR}{[(1 + R)^{nt} - 1]}$
- **a.** solve for A
- **b.** solve for n

29. a. $t = \dfrac{A - p}{pr}$

b. $p = \dfrac{A}{1 + rt}$

30. a. $t = \dfrac{\ln\left(\frac{A}{p}\right)}{\ln(1 + r)}$

b. $r = \sqrt[t]{\dfrac{A}{p}} - 1$

31. a. $r = n\left(\sqrt[nt]{\dfrac{A}{p}} - 1\right)$

b. $t = \dfrac{\ln\left(\frac{A}{p}\right)}{n\ln\left(1 + \frac{r}{n}\right)}$

32. a. $p = \dfrac{A}{e^{rt}}$

b. $r = \dfrac{\ln\left(\frac{A}{p}\right)}{t}$

33. a. $Q_0 = \dfrac{Q}{e^{rt}}$

b. $t = \dfrac{\ln\left(\frac{Q}{Q_0}\right)}{r}$

34. a. $A = \dfrac{p[(1 + R)^{nt} - 1]}{R}$

b. $n = \dfrac{\ln\left(\frac{AR}{p} + 1\right)}{t\ln(1 + R)}$

WORKING WITH FORMULAS

35. Amount of a mortgage payment: $P = \dfrac{AR}{1 - (1 + R)^{-nt}}$

The mortgage payment required to pay off (or amortize) a loan is given by the formula shown, where P is the payment amount, A is the original amount of the loan, t is the time in years, r is the annual interest rate, n is the number of payments per year, and $R = \dfrac{r}{n}$. Find the *monthly payment* required to amortize a $125,000 home, if the interest rate is 5.5%/year and the home is financed over 30 yr. $709.74

36. Total interest paid on a home mortgage: $I = \left[\dfrac{prt}{1 - \left(\dfrac{1}{1 + 0.08\bar{3}r}\right)^{12t}}\right] - p$

The total interest I paid in t years on a home mortgage of p dollars is given by the formula shown, where r is the interest rate on the loan. If the original mortgage was \$198,000 at an interest rate of 6.5%, (a) how much interest has been paid in 10 yr? (b) Use a table of values to determine how many years it will take for the interest paid to exceed the amount of the original mortgage. a. \$71,789.99 b. 25 yr

APPLICATIONS

37. Simple interest: The owner of Paul's Pawn Shop loans Larry \$200.00 using his Toro riding mower as collateral. Thirteen weeks later Larry comes back to get his mower out of pawn and pays Paul \$240.00. What was the annual simple interest rate on this loan? 80%

38. Simple interest: To open business in a new strip mall, Laurie's Custom Card Shoppe borrows \$50,000 from a group of investors at 4.55% simple interest. Business booms and blossoms, enabling Laurie to repay the loan fairly quickly. If Laurie repays \$62,500, how long did it take? $5\frac{1}{2}$ yr

39. Compound interest: As a curiosity, David decides to invest \$10 in an account paying 10% interest compounded 10 times per year for 10 yr. Is that enough time for the \$10 to triple in value? no

40. Compound interest: As a follow-up experiment (see Exercise 39), David invests \$10 in an account paying 12% interest compounded 10 times per year for 10 yr, and another \$10 in an account paying 10% interest compounded 12 times per year for 10 yr. Which produces the better investment—more compounding periods or a higher interest rate? higher interest rate

41. Compound interest: Due to demand, Donovan's Dairy (Wisconsin, USA) plans to double its size in 4 yr and will need \$250,000 to begin development. If they invest \$175,000 in an account that pays 8.75% compounded semiannually, (a) will there be sufficient funds to break ground in 4 yr? (b) If not, use a table to find the *minimum interest rate* that will allow the dairy to meet its 4-yr goal. no 9.12%

42. Compound interest: To celebrate the birth of a new daughter, Helyn invests 6000 Swiss francs in a college savings plan to pay for her daughter's first year of college in 18 yr. She estimates that 25,000 francs will be needed. If the account pays 7.2% compounded daily, (a) will she meet her investment goal? (b) if not, use a table to find the *minimum rate of interest* that will enable her to meet this 18-yr goal. no 7.93%

43. Interest compounded continuously: Valance wants to build an addition to his home outside Madrid (Spain) so he can watch over and help his parents in their old age. He hopes to have 20,000 euros put aside for this purpose within 5 yr. If he invests 12,500 euros in an account paying 8.6% interest compounded continuously, (a) will he meet his investment goal? (b) If not, find the *minimum rate of interest* that will enable him to meet this 5-yr goal. no 9.4%

44. Interest compounded continuously: Minh-Ho just inherited her father's farm near Mito (Japan), which badly needs a new barn. The estimated cost of the barn is 8,465,000 yen and she would like to begin construction in 4 yr. If she invests 6,250,000 yen in an account paying 6.5% interest compounded continuously, (a) will she meet her investment goal? (b) If not, find the *minimum rate of interest* that will enable her to meet this 4-yr goal. no 7.58%

45. Interest compounded continuously: William and Mary buy a small cottage in Dovershire (England), where they hope to move after retiring in 7 yr. The cottage needs about 20,000 euros worth of improvements to make it the retirement home they desire. If they invest 12,000 euros in an account paying 5.5% interest compounded continuously, (a) will they have enough to make the repairs? (b) If not, find the *minimum amount they need to deposit* that will enable them to meet this goal in 7 yr. no approx 13,609 euros

46. **Interest compounded continuously:** After living in Oslo (Norway) for 20 years, Zirkcyt and Shybrt decide to move inland to help operate the family ski resort. They hope to make the move in 6 yr, after they have put aside 140,000 kroner. If they invest 85,000 kroner in an account paying 6.9% interest compounded continuously, (a) will they meet their 140,000 kroner goal? (b) If not, find the *minimum amount they need to deposit* that will allow them to meet this goal in 6 yr. no approx 92,540 kroner

47. **Exponential growth:** As part of a lab experiment, Luamata needs to grow a culture of 200,000 bacteria, which are known to double in number in 12 hr. If he begins with 1000 bacteria, (a) find the growth rate r and (b) find how many hours it takes for the culture to produce the 200,000 bacteria. **a.** 5.78% **b.** 91.67 hr

48. **Exponential growth:** After the wolf population was decimated due to overhunting, the rabbit population in the Boluhti Game Reserve began to double every 6 months. If there were an estimated 120 rabbits to begin, (a) find the growth rate r and (b) find the number of months required for the population to reach 2500. **a.** 11.55% **b.** 26.29 mo

49. **Radioactive decay:** The radioactive element iodine-131 has a half-life of 8 days and is often used to help diagnose patients with thyroid problems. If a certain thyroid procedure requires 0.5 g and is scheduled to take place in 3 days, what is the minimum amount that must be on hand now (to the nearest hundredth of a gram)? 0.65 g

50. **Radioactive decay:** The radioactive element sodium-24 has a half-life of 15 hr and is used to help locate obstructions in blood flow. If the procedure requires 0.75 g and is scheduled to take place in 2 days (48 hr), what minimum amount must be on hand *now* (to the nearest hundredth of a gram)? 6.89 g

51. **Radioactive decay:** The radioactive element americium-241 has a half-life of 432 yr and although extremely small amounts are used (about 0.0002 g), it is the most vital component of standard household smoke detectors. How many years will it take a 10-g mass of americium-241 to decay to 2.7 g? 818 yr

52. **Radioactive decay:** Carbon-14 is a radioactive compound that occurs naturally in all living organisms, with the amount in the organism constantly renewed. After death, no new carbon-14 is acquired and the amount in the organism begins to decay exponentially. If the half-life of carbon-14 is 5700 yr, how old is a mummy having only 30% of the normal amount of carbon-14? about 9901 yr

Ordinary annuities: If a periodic payment p is deposited n times per year, with annual interest rate r also compounded n times per year for t years, the future value of the account is given by $A = \frac{p[(1 + R)^{nt} - 1]}{R}$, where $R = \frac{r}{n}$ (i.e., if the rate is 9% compounded monthly, $R = \frac{0.09}{12} = 0.0075$).

53. How long would it take Jasmine to save \$10,000 if she deposits \$90/month at an annual rate of 7.75% compounded monthly? about 7 yr

54. What quarterly investment amount is required to ensure that Larry can save \$4700 in 4 yr at an annual rate of 8.5% compounded quarterly? \$250.00

55. **Saving for college:** At the birth of their first child, Latasha and Terrance opened an annuity account and have been depositing \$50 per month in the account ever since. If the account is now worth \$30,000 and the interest on the account is 6.2% compounded monthly, how old is the child? 23 yr

56. **Saving for a bequest:** When Cherie (Brandon's first granddaughter) was born, he purchased an annuity account for her and stipulated that she should receive the funds (in trust, if necessary) upon his death. The quarterly annuity payments were \$250 and interest on the account was 7.6% compounded quarterly. The account balance of \$17,500 was recently given to Cherie. How much longer did Brandon live? 11 yr

57. **Saving for a down payment:** Tae-Hon is tired of renting and decides that within the next 5 yr he must save \$22,500 for the down payment on a home. He finds an investment company that offers

8.5% interest compounded monthly and begins depositing \$250 each month in the account. (a) Is this monthly amount sufficient to help him meet his 5 yr goal? (b) If not, find the *minimum amount he needs to deposit each month* that will allow him to meet his goal in 5 yr. **a.** no **b.** \$302.25

58. Saving to open a business: Madeline feels trapped in her current job and decides to save \$75,000 over the next 7 yr to open up a Harley Davidson franchise. To this end, she invests \$145 every week in an account paying $7\frac{1}{2}\%$ interest compounded weekly. (a) Is this weekly amount sufficient to help her meet the seven-year goal? (b) If not, find the *minimum amount she needs to deposit each week* that will allow her to meet this goal in 7 yr? **a.** no **b.** \$156.81

WRITING, RESEARCH, AND DECISION MAKING

59. \$12,488,769.67; answers will vary
60. higher; compounded quarterly at 6.75%

59. Many claim that inheritance taxes are put in place simply to prevent a massive accumulation of wealth by a select few. Suppose that in 1890, your great-grandfather deposited \$10,000 in an account paying 6.2% compounded continuously. If the account were to pass to you untaxed, what would it be worth in 2005? Do some research on the inheritance tax laws in your state. In particular, what amounts can be inherited untaxed (i.e., before the inheritance tax kicks in)?

60. One way to compare investment possibilities is to compute what is called the **effective rate of interest.** This is the yearly simple interest rate ($t = 1$) that would generate the same amount of interest as the stated compound interest rate. (a) Would you expect the effective rate to be higher or lower than the stated compound interest rate? (b) Do some research to find a formula that will compute the effective rate of interest, and use it to compare an investment that is compounded monthly at 6.5% with one that is compounded quarterly at 6.75%. Which is the better investment?

61. Willard Libby, an American chemist, won the 1960 Nobel Prize in Physical Chemistry for his discovery and development of radiocarbon dating. Do some research on how radiocarbon dating is used and write a short report. Include several examples and discuss/illustrate how the concepts in this section are needed for the method to work effectively. Answers will vary.

EXTENDING THE CONCEPT

62. If you have not already completed Exercise 42, please do so. For *this* exercise, *solve the compound interest equation for r* to find the exact rate of interest that will allow Helyn to meet her 18-yr goal. 7.93%

63. If you have not already completed Exercise 55, please do so. Suppose the final balance of the account was \$35,100 with interest again being compounded monthly. For *this* exercise, use a graphing calculator to find r, the exact rate of interest the account would have been earning. 7.2%

66. 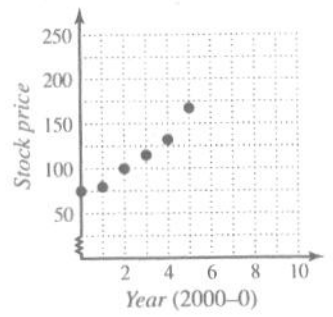

64. Suppose the decay of radioactive elements was measured in terms of a one-fourth life (instead of a half-life). What would the conversion formula be to convert from "fourth-life" to the decay rate r? Polonium-210 has a half-life of 140 days. What is its " fourth-life"? What is its decay rate r? If 20 g of polonium-210 are initially present, how many days until less than 1 g remains? $r = \frac{\ln 4}{t}$; 280 days; ≈ 0.00495; 605 days

MAINTAINING YOUR SKILLS

Exercise 66

Year (2000 = 0)	Stock Price
0	76
1	80
2	98
3	112
4	130
5	170

65. (2.1) In an effort to boost tourism, a trolley car is being built to carry sightseers from a strip mall to the top of Mt. Vernon, 1580-m high. Approximately how long will the trolley cables be? 2548.8 m

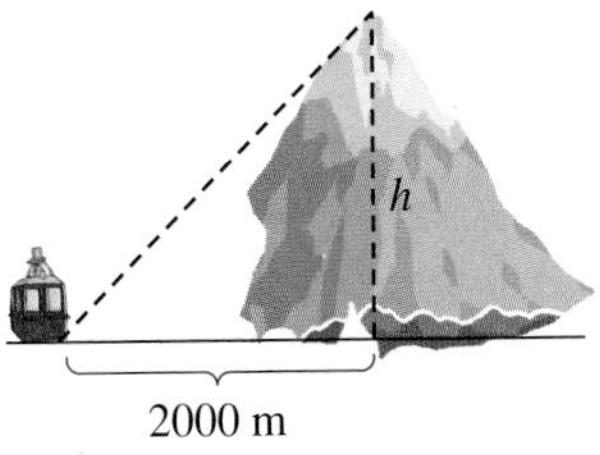

66. (2.6) The table shown gives the average price for a share of stock in IBN, a large firm researching sources of alternative energy. Draw a scatter-plot of the data and decide on an appropriate form of regression. Then use a calculator to find a regression equation. If this rate of growth continues, what will a share of stock be worth in 2010? $y = 3.2x^2 + 2.0x + 76.4$; \$416.40

69. a. $f(x) = x^3$, $f(x) = x$, $f(x) = \sqrt{x}$, $f(x) = \sqrt[3]{x}$, $f(x) = \frac{1}{x}$
b. $f(x) = |x|$, $f(x) = x^2$, $f(x) = \frac{1}{x^2}$
c. $f(x) = x$, $f(x) = x^3$, $f(x) = \sqrt{x}$, $f(x) = \sqrt[3]{x}$
d. $f(x) = \frac{1}{x}$, $f(x) = \frac{1}{x^2}$

67. (2.2) Is the following relation a function? If not, state how the definition of a function is violated. yes

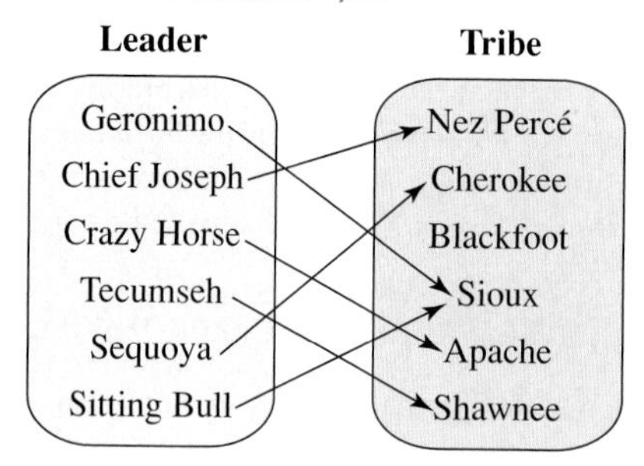

69. (2.4/3.8) Name the toolbox functions that are (a) one-to-one, (b) even, (c) increasing for $x \in \mathbb{R}$, and (d) asymptotic.

68. (4.3) A polynomial is known to have the zeroes $x = 3$, $x = -1$, and $x = 1 + 2i$. Find the equation of the polynomial, given it has degree 4 and a y-intercept of $(0, -15)$.
$P(x) = x^4 - 4x^3 + 6x^2 - 4x - 15$

70. a. $4x^3 + 6x^2 - 8x - 15$ b. $x + \frac{9}{2} + \frac{37}{2(2x - 3)}$
c. $8x^2 - 12x + 5$ d. $4x^2 + 12x + 7$
e. $\frac{x + 3}{2}$ f. $\frac{-3}{2}$

70. (3.1) Given $f(x) = 2x^2 + 6x + 5$ and $g(x) = 2x - 3$, find (a) $(f \cdot g)(x)$, (b) $\left(\frac{f}{g}\right)(x)$, (c) $(f \circ g)(x)$, (d) $(g \circ f)(x)$, (e) $g^{-1}(x)$, and (f) $(f + g)(-\frac{1}{2})$.

5.6 Exponential, Logarithmic, and Logistic Regression Models

LEARNING OBJECTIVES

In Section 5.6 you will learn how to:

A. Choose an appropriate form of regression using context

B. Use a graphing calculator to obtain exponential, logarithmic, and logistic regression models

C. Use a regression model to answer questions and solve problems

D. Determine when a logistics model is appropriate and apply logistics models to a set of data

INTRODUCTION

The basic concepts involved in calculating a regression equation were presented in Section 2.6. In this section, we apply the regression concept to data sets that are best modeled by power, exponential, logarithmic, or logistic functions. All data sets, while contextual and accurate, *have been carefully chosen* to provide a maximum focus on regression fundamentals and the mathematical concepts that follow. In reality, data sets are often not so "well-behaved" and many require sophisticated statistical tests before any conclusions can be drawn.

POINT OF INTEREST

Many will likely remember the 2000 presidential election and the debacle regarding the vote count in Florida. The table here shows the number of votes by which Bush led Gore beginning on November 7 ($t = 0$ days) and ending on December 8 ($t = 31$ days), according to the national media. The data are graphed in Figures 5.23, 5.24, and 5.25 with three different forms of regression applied. You are asked to analyze each graph in this context in Exercise 63.

t (days since 11/7)	N (number of votes in Bush lead)
0	1725
11	930
19	537
31	193

[Source: *USA Today*]

Figure 5.23

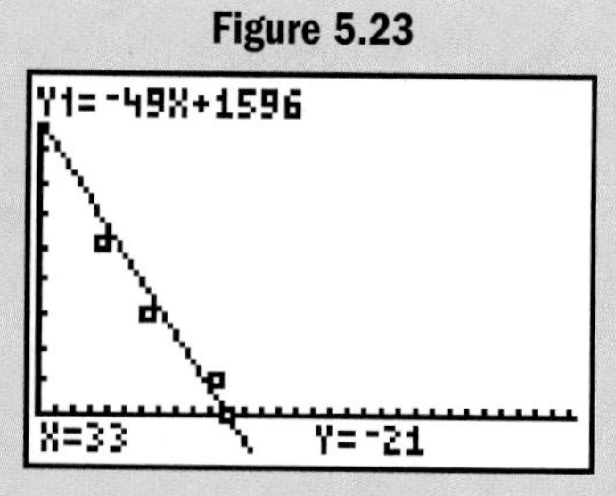

Linear regression

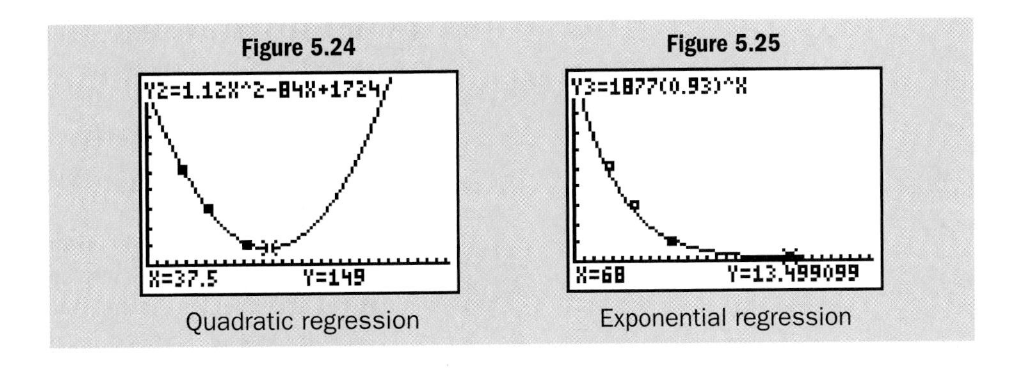

Quadratic regression

Exponential regression

A. Choosing an Appropriate Form of Regression

WORTHY OF NOTE

Recall that a "best-fit" equation model is one that minimizes the vertical distance between all data points and the graph of the proposed model.

Most graphing calculators have the ability to perform 8 to 10 different forms of regression, and selecting which of these to use is a critical issue. When various forms are applied to a given data set, some are easily discounted due to a poor fit, others may fit very well for only a portion of the data, while still others may compete for being the "best-fit" equation. In a statistical study of regression, an in-depth look at the correlation coefficient (r), the coefficient of determination (r^2 or R^2), and a study of **residuals** are used to help make an appropriate choice. For our purposes, the correct or best choice will generally depend on two things: (1) how well the graph appears to fit the scatter-plot, and (2) the context or situation that generated the data, coupled with a dose of common sense.

WORTHY OF NOTE

For more information on the use of residuals, see the *Calculator Exploration and Discovery* feature from Chapter 3.

As we've noted previously, the final choice of regression can rarely be based on the scatter-plot alone, although relying on the basic characteristics and end behavior of certain graphs can be helpful. With an awareness of the toolbox functions, polynomial graphs, and applications of exponential and logarithmic functions, the context of the data can aid a decision.

EXAMPLE 1 Suppose a set of data is generated from each context given. Use common sense, previous experience, or your own knowledge base to state whether a linear, quadratic, logarithmic, exponential, or power regression might be most appropriate. Justify your answers.

a. population growth of the United States since 1800

b. the distance covered by a jogger running at a constant speed

c. height of a baseball t seconds after it's thrown

d. the time it takes for a cup of hot coffee to cool to room temperature

Solution:

a. From examples in Section 5.5 and elsewhere, we've seen that animal and human populations tend to grow exponentially over time. Here, an exponential model is likely most appropriate.

b. Since the jogger is moving at a constant speed, the rate-of-change $\frac{\Delta\text{distance}}{\Delta\text{time}}$ is constant and a linear model would be most appropriate.

c. As seen in numerous places throughout the text, the height of a projectile is modeled by the equation $h(t) = -16t^2 + vt + k$, where $h(t)$ is the height after t seconds. Here, a quadratic model would be most appropriate.

d. Many have had the experience of pouring a cup of hot chocolate, coffee, or tea, only to leave it on the counter as they turn their attention to other things. The hot drink seems to cool quickly at first, then slowly approach room temperature. This experience, perhaps coupled with our awareness of *Newton's Law of Cooling,* shows a logarithmic or exponential model might be appropriate here.

NOW TRY EXERCISES 7 THROUGH 20

B. Exponential and Logarithmic Regression Models

We now focus our attention on regression models that involve exponential and logarithmic functions. Recall the process of developing a regression equation involves these five stages: (1) clearing old data; (2) entering new data; (3) displaying the data; (4) calculating the regression equation; (5) displaying and using the regression graph and equation.

EXAMPLE 2 The number of centenarians (people who are 100 years of age or older) has been climbing steadily over the last half century. The table shows the number of centenarians (per million population) for selected years. Use the data and a graphing calculator to draw the scatter-plot, then use the scatter-plot and context to decide on an appropriate form of regression.

Source: Data from 2004 *Statistical Abstract of the United States,* Table 14; various other years

Year "t" (1950 → 0)	Number "N" (per million)
0	16
10	18
20	25
30	74
40	115
50	262

Figure 5.26

L1	L2	L3 2
0	16	------
10	18	
20	25	
30	74	
40	115	
50	262	
------	------	

L2(6) =262

Figure 5.27

Solution: After clearing any existing data in the data lists, enter the input values (years since 1950) in L1 and the output values (number of centenarians per million population) in L2 (Figure 5.26). For the viewing window, scale the x-axis (years since 1950) from –10 to 70 and the y-axis (number per million) from –50 to 500 to comfortably fit the data and allow room for the coordinates to be shown at the bottom of the screen (Figure 5.27). The scatter-plot rules out a linear model. While a quadratic

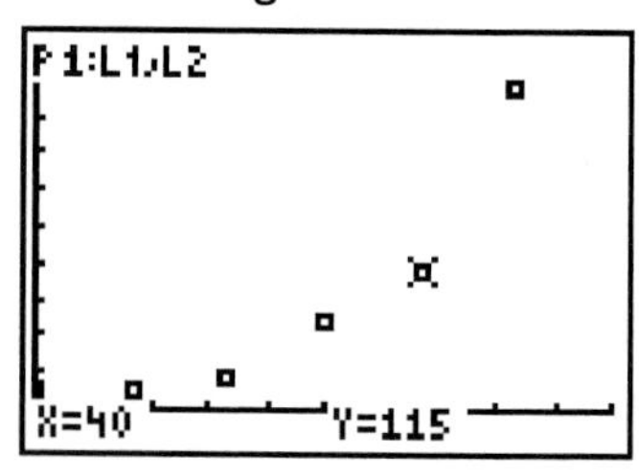

WORTHY OF NOTE

The regression equation can be sent directly to the Y= screen, or pasted to the Y= screen from the home screen. To learn how, see the *Technology Highlight* that follows Example 7.

model may fit the data, we expect that the correct model should exhibit asymptotic behavior since extremely few people lived to be 100 years of age prior to dramatic advances in hygiene, diet, and medical care. This would lead us toward an exponential equation model. The keystrokes STAT ▶ brings up the **CALC** menu, with **ExpReg** (exponential regression) being option "0." The option can be selected by simply pressing "0" and ENTER, or by using the up arrow ▲ or down arrow ▼ to scroll to **0:ExpReg.** The exponential model seems to fit the data very well and gives a high correlation coefficient (Figures 5.28 and 5.29). To four decimal places the equation model is $y = (11.5090)1.0607^x$.

Figure 5.28

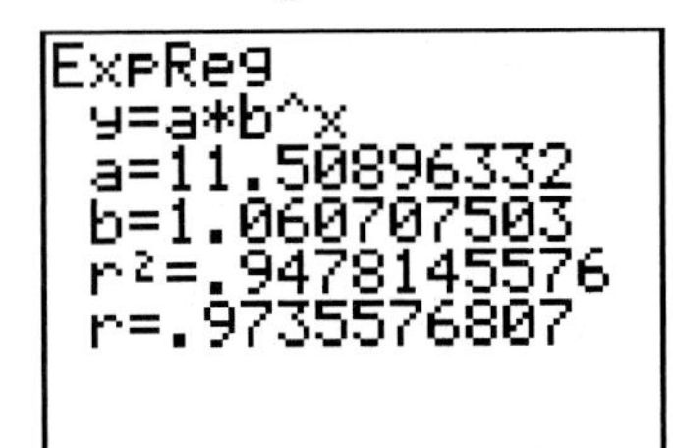

Figure 5.29

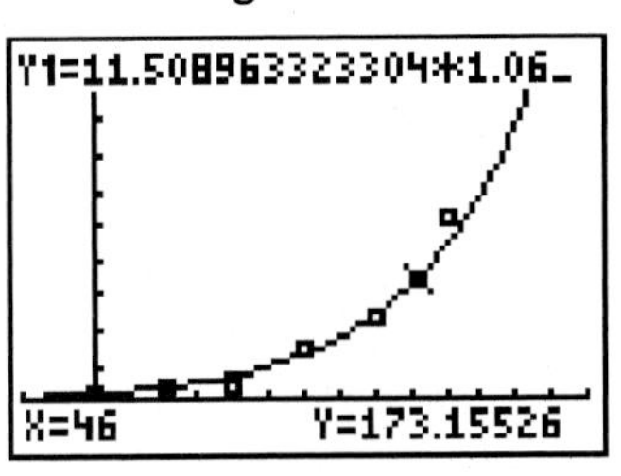

NOW TRY EXERCISES 21 AND 22

Given a general exponential function $y = ab^x$, the growth rate constant (discussed in Section 5.3) can be determined by using properties of logarithms to rewrite the relation in the form $y = ae^{kx}$. This is done by setting them equal to each other and solving for k:

$$
\begin{aligned}
ab^x &= ae^{kx} && \text{set equations equal} \\
b^x &= e^{kx} && \text{divide by } a \\
\ln b^x &= \ln e^{kx} && \text{take natural logs} \\
x \ln b &= kx \ln e && \text{power property} \\
\ln b &= k && \text{solve for } k \ (\ln e = 1)
\end{aligned}
$$

This shows that $b^x = e^{kx}$ for $k = \ln b$.

EXAMPLE 3 Identify the growth rate constant for the equation model from Example 2, then write the equation as a base e exponential function.

Solution: From Example 2 we have $y = (11.5090)1.0607^x$, with $b = 1.0607$, and $k = \ln 1.0607 \approx 0.0589$. The growth rate is about 5.9%. The corresponding base e function is $y = (11.5090)e^{0.0589x}$.

NOW TRY EXERCISES 23 AND 24

For applications involving exponential and logarithmic functions, it helps to remember that while both basic functions are increasing, a logarithmic function increases at a much slower rate. Consider Example 4.

EXAMPLE 4 One measure used in studies related to infant growth, nutrition, and development, is the relation between the circumference of a child's head and their age. The table to the right shows the average circumference of a female child's head for ages 0 to 36 months. Use the data and a graphing calculator to draw the scatter-plot, then use the scatter-plot and context to decide on an appropriate form of regression.

Source: *National Center for Health Statistics*

Age *a* (months)	Circumference *C* (cm)
0	34.8
6	43.0
12	45.2
18	46.5
24	47.5
30	48.2
36	48.6

Solution: After clearing any existing data, enter the child's age (in months) as L1 and the circumference of the head (in cm) as L2. For the viewing window, scale the x-axis from -5 to 50 and the y-axis from 25 to 60 to comfortably fit the data (Figure 5.30). The scatter-plot again rules out a linear model, and the context rules out a polynomial model due to end-behavior. As we expect the circumference of the head to continue increasing slightly for many more months, it appears a logarithmic model may be the best fit. Note that since $\ln(0)$ is undefined, $a = 0.1$ was used to represent the age at birth (rather than $a = 0$), prior to running the regression. The **LnReg** (logarithmic regression) option is option 9, and the keystrokes STAT ▶ **(CALC) 9:LinReg** ENTER gives the equation shown in Figure 5.31, which returns a very high correlation coefficient and fits the data very well (Figure 5.32).

Figure 5.30

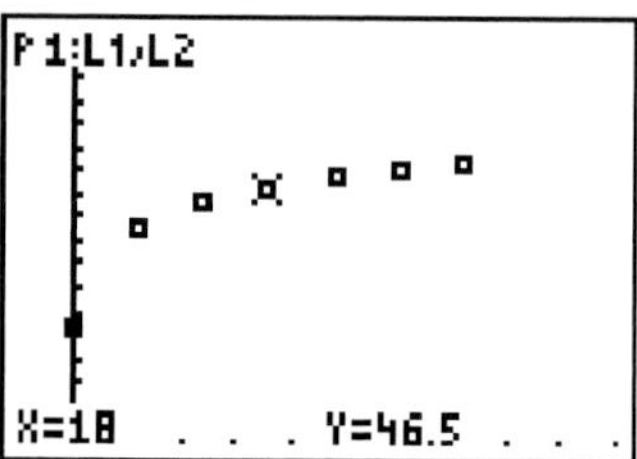

Figure 5.31

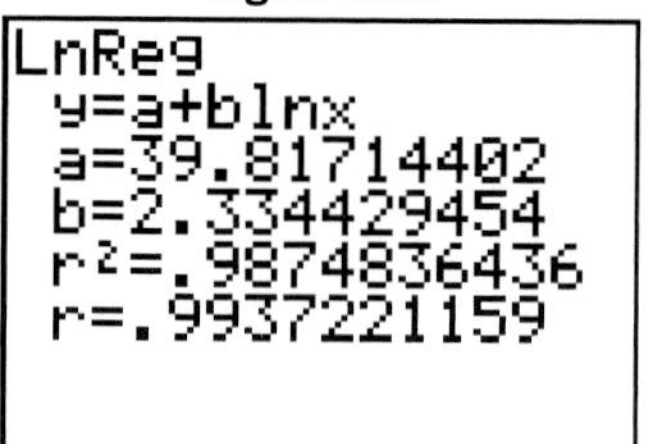

Figure 5.32

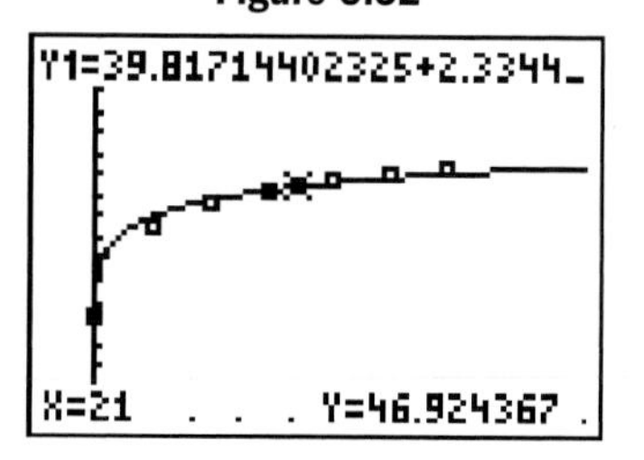

NOW TRY EXERCISES 25 AND 26

For more on the correlation coefficient and its use, see Exercise 62.

C. Applications of Regression

Once a model for the data has been obtained, it is generally put to two specific uses. First, it can be used to **extrapolate** or predict future values or occurrences. When using extrapolation, values are projected *beyond* the given set of data. Second, the equation model can be used to **interpolate** or approximate values that occur *between* those given in the data set. There is a wide variety of applications in the Exercise set. See Exercises 44 through 55.

EXAMPLE 5 Use the regression equation from Example 2 to answer the following questions:

a. Approximately how many centenarians (per million) were there in 1995?

b. Approximately how many centenarians (per million) will there be in 2010?

c. In what year will there be approximately 300 centenarians per million?

Solution: **a.** Writing the function using function notation gives $f(x) = 11.509(1.061)^x$. For the year 1995 we have $x = 45$, so we find the value of $f(45)$:

$$f(x) = 11.509(1.061)^x \quad \text{regression equation}$$
$$f(45) = 11.509(1.061)^{45} \quad \text{substitute } x = 45$$
$$= 11.509(14.3612511) \quad \text{evaluate exponential first (order of operations)}$$
$$= 165.2836389 \quad \text{result}$$

In 1995, there were approximately 165 centenarians per million population.

b. For the year 2010 we have $x = 60$, so we find the value of $f(60)$:

$$f(x) = 11.509(1.061)^x \quad \text{regression equation}$$
$$f(60) = 11.509(1.061)^{60} \quad \text{substitute } x = 60$$
$$= 11.509(34.90784461) \quad \text{evaluate exponential first (order of operations)}$$
$$= 401.7543836 \quad \text{result}$$

In 2010, there will be approximately 402 centenarians per million population.

c. For Part (c) we are given the *number of centenarians* [the output value $f(x)$] and are asked to find the year x in which this number (300) occurs. So we substitute $f(x) = 300$ and solve for x:

$$f(x) = 11.509(1.061)^x \quad \text{regression equation}$$
$$300 = 11.509(1.061)^x \quad \text{substitute } f(x) = 300$$
$$\frac{300}{11.509} = (1.061)^x \quad \text{divide both sides by 11.509}$$
$$\ln\left(\frac{300}{11.509}\right) = \ln(1.061)^x \quad \text{take the base-}e\text{ (or base-10) logarithm of both sides}$$
$$\ln\left(\frac{300}{11.509}\right) = x\ln 1.061 \quad \text{apply power property of logarithms: } \log_b k^x = x \log_b k$$
$$\frac{\ln\left(\frac{300}{11.509}\right)}{\ln 1.061} = x \quad \text{solve for } x \text{ (exact form): divide both sides by ln(1.601)}$$
$$55.06756851 = x \quad \text{approximate form}$$

In the year 2005 (the 55th year after 1950), there will be approximately 300 centenarians per million population.

NOW TRY EXERCISES 56 AND 57

CAUTION

Be very careful when you evaluate expressions like $\frac{\ln\left(\frac{300}{11.509}\right)}{\ln 1.061}$ from Example 5. It is best to compute the result in stages, evaluating the numerator first: $\ln\left(\frac{300}{11.509}\right) = 3.260653137$, then dividing by the denominator: $3.260653137 \div \ln 1.061 = 55.06756851$.

EXAMPLE 6 Use the regression equation from Example 4 to answer the following questions:

a. What is the average circumference of a female child's head, if the child is 21 months old?

b. According to the equation model, what will the average circumference be when the child turns $3\frac{1}{2}$ years old?

c. If the circumference of the child's head is 46.9 cm, about how old is the child?

Solution:

a. Using function notation we have $C(a) \approx 39.8171 + 2.3344\ln(a)$. Substituting 21 for a gives:

$$C(21) \approx 39.8171 + 2.3344\ln(21) \quad \text{substitute 21 for } x$$
$$\approx 46.9 \quad \text{result}$$

The circumference is approximately 46.9 cm.

b. Substituting 3.5 yr $\times$ 12 = 42 months for a gives:

$$C(42) \approx 39.8171 + 2.3344\ln(42) \quad \text{substitute 42 for } x$$
$$\approx 48.5 \quad \text{result}$$

The circumference will be approximately 48.5 cm.

c. For Part (*c*) we're given the circumference C and are asked to find the age "a" in which this circumference (46.9) occurs. Substituting 46.9 for $C(a)$ we obtain:

$$46.9 = 39.8171 + 2.3344\ln(a) \quad \text{substitute 1866 for } f(x)$$
$$\frac{7.0829}{2.3344} = \ln(a) \quad \text{subtract 39.8171, then divide by 2.3344}$$
$$e^{\frac{7.0859}{2.3344}} = a \quad \text{write in exponential form}$$
$$20.8 = a \quad \text{result}$$

The child must be about 21 months old.

NOW TRY EXERCISES 58 AND 59

D. Logistics Equations and Regression Models

Many population growth models assume an unlimited supply of resources, nutrients, and room for growth, resulting in an exponential growth model. When resources become scarce or room for further expansion is limited, the result is often a **logistic growth model.** At first, growth is very rapid (like an exponential function), but this growth begins to taper off and slow down as nutrients are used up, living space becomes restricted, or due to other factors. Surprisingly, this type of growth can take many forms, including population growth, the spread of a disease, the growth of a tumor, or the spread of a stain in fabric. Specific logistic equations were encountered in Section 5.4. The general equation model for logistic growth is

LOGISTIC GROWTH EQUATION

Given constants a, b, and c, the logistic growth $P(t)$ of a population depends on time t according to the model

$$P(t) = \frac{c}{1 + ae^{-bt}}$$

The constant c is called the **carrying capacity** of the population, in that as $t \to \infty$, $P(t) \to c$. In words, as the elapsed time becomes very large, growth will approach (but not exceed) c.

EXAMPLE 7 Yeast cultures have a number of applications that are a great benefit to civilization and have been an object of study for centuries. A certain strain of yeast is grown in a lab, with its population checked at 2-hr intervals, and the data gathered are given in the table. Use the data to draw a scatter-plot, and decide on an appropriate form of regression. If a logistic regression is the best model, attempt to estimate the capacity coefficient c prior to using your calculator to find the regression equation. How close were you to the actual value?

Elapsed Time (hours)	Population (100s)
2	20
4	50
6	122
8	260
10	450
12	570
14	630
16	650

Solution: After clearing the data lists, enter the input values (elapsed time) in L1 and the output values (population) in L2. For the viewing window, scale the t-axis from 0 to 20 and the P-axis from 0 to 700 to comfortably fit the data. From the context and scatter-plot, it's apparent the data are

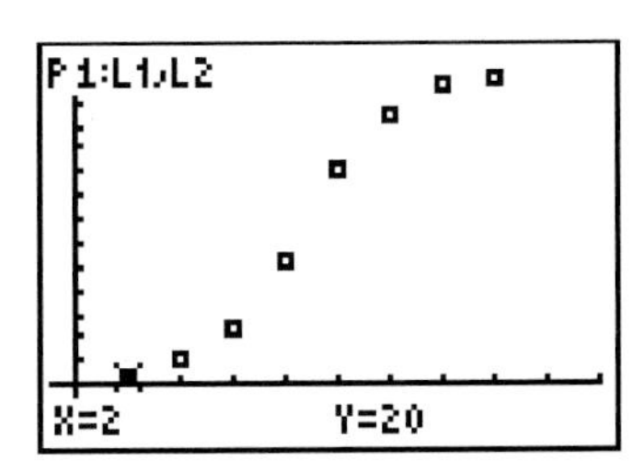

best modeled by a logistic function. Noting that Ymax = 700 and the data seem to level off near the top of the window, a good estimate for c would be about 675. Using logistic regression on the home screen (option **B:Logistic**), we obtain the equation

$$Y_1 = \frac{663}{1 + 123.9e^{-0.553x}} \text{ (rounded).}$$

NOW TRY EXERCISES 60 AND 61

TECHNOLOGY HIGHLIGHT

Pasting the Regression Equation to the Y = Screen

The keystrokes shown apply to a TI-84 Plus model. Please consult our Internet site or your manual for other models.

Once a regression equation has been calculated, the entire equation can be transferred directly from the home screen to the Y= screen of a graphing calculator. The regression curve and scatter-plot can both be viewed by simply pressing GRAPH, assuming an appropriate window has been set. On the TI-84 Plus, this feature is accessed using the VARS key. Using the regression from Example 7, we begin by placing the cursor at Y_1 on the Y= screen (CLEAR the old equation and leave the cursor in the empty slot). Pressing the VARS key gives the screen shown in Figure 5.33, where we select option **5:Statistics** and press ENTER. This brings up the screen shown in Figure 5.34, where we select menu option **EQ,** then option **1:EQ** (for equation). Pressing the ENTER key at this point will paste the regression equation directly to the current location of the cursor, which we left at Y_1 on the Y= screen. We can now use the equation to further investigate the population of the yeast culture.

Figure 5.33

```
VARS Y-VARS
1:Window...
2:Zoom...
3:GDB...
4:Picture...
5:Statistics...
6:Table...
7:String...
```

Figure 5.34

```
XY Σ EQ TEST PTS
1:RegEQ
2:a
3:b
4:c
5:d
6:e
7↓r
```

As an alternative, you can add the argument "Y_1" to the logistic regression command on the home screen. The sequence **Logistic Y_1** ENTER will calculate the equation and automatically place the result in Y_1.

Exercise 1: Use the ideas from this *Technology Highlight* to paste the equation from Examples 4 and 6 directly into Y_1. Then use the calculator to recompute the answers to questions (a), (b), and (c) of Example 6. Were the answers close?
a. $c \approx 46.92$ cm **b.** $c \approx 48.54$ cm **c.** $a \approx 20.78$ mo yes, very close

5.6 EXERCISES

CONCEPTS AND VOCABULARY

Fill in each blank with the appropriate word or phrase. Carefully reread the section if needed.

1. The type of regression used often depends on (a) whether a particular graph appears to fit the data and (b) the context or situation that generated the data.
2. The final choice of regression can rarely be based on the scatter-plot alone. Relying on the basic characteristics and end-behavior of certain graphs can be helpful.
3. To extrapolate means to use the data to predict values beyond the given data.
4. To interpolate means to use the data to predict values between the given data.

5. 1. clear old data
2. enter new data
3. display the data
4. calculate the regression equation
5. display and use the regression graph and equation

6. $f(x) = x$, $f(x) = |x|$, $f(x) = x^2$, $f(x) = x^3$, $f(x) = \sqrt[3]{x}$, $f(x) = \sqrt{x}$

7. e
8. c
9. a
10. b
11. d
12. f

5. List the five steps used to find a regression equation using a calculator. Discuss possible errors that can occur if the first step is skipped. After the new data have been entered, what precautionary step should always be included? Answers will vary.

6. Consider the eight toolbox functions and the exponential and logarithmic functions. How many of these satisfy the condition as $x \to \infty$, $y \to \infty$? For those that satisfy this condition, discuss/explain how you would choose between them judging from the scatter-plot alone. Answers will vary.

DEVELOPING YOUR SKILLS

Match each scatter-plot given with one of the following: (a) likely linear, (b) likely quadratic, (c) likely exponential, (d) likely logarithmic, (e) likely logistic, or (f) none of these.

7.

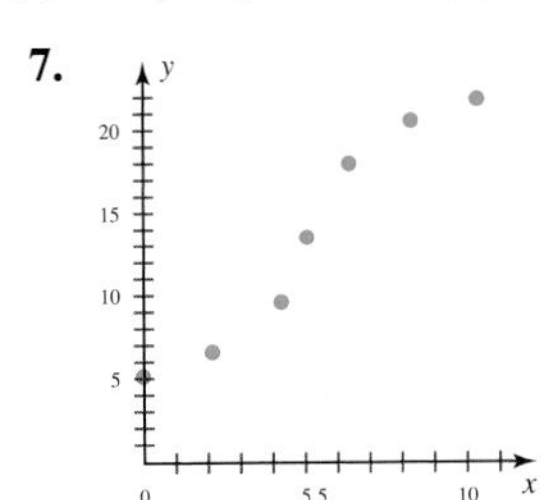

8.

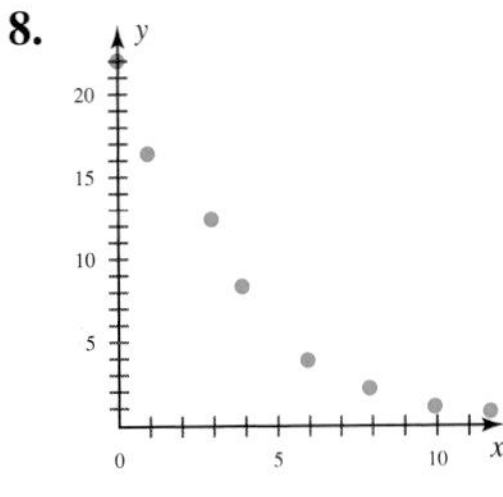

9.

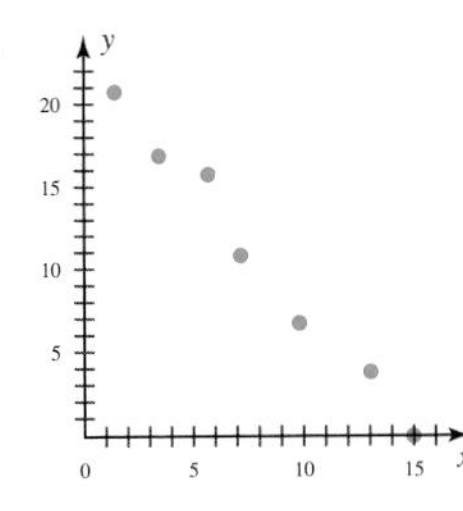

10.

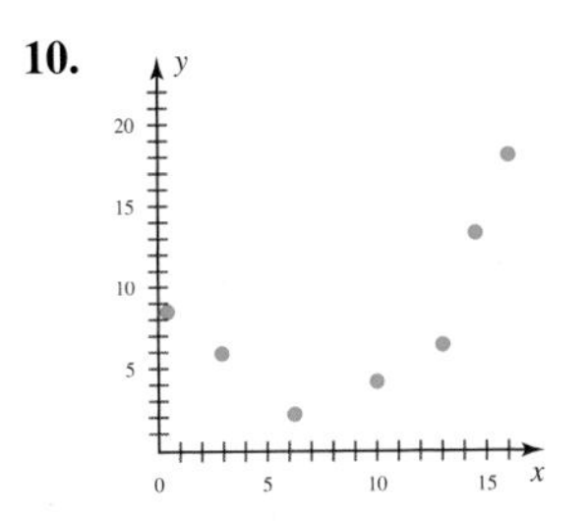

11.

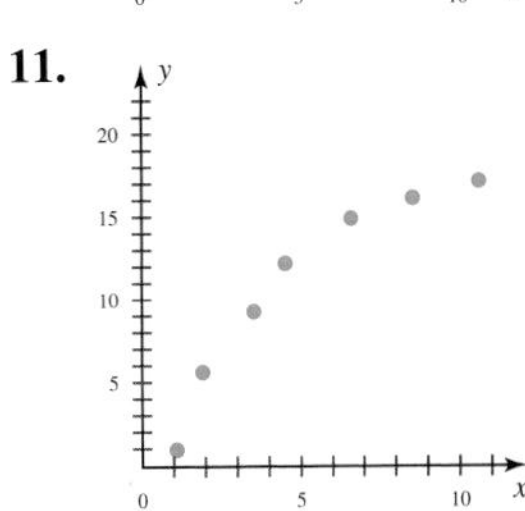

12.

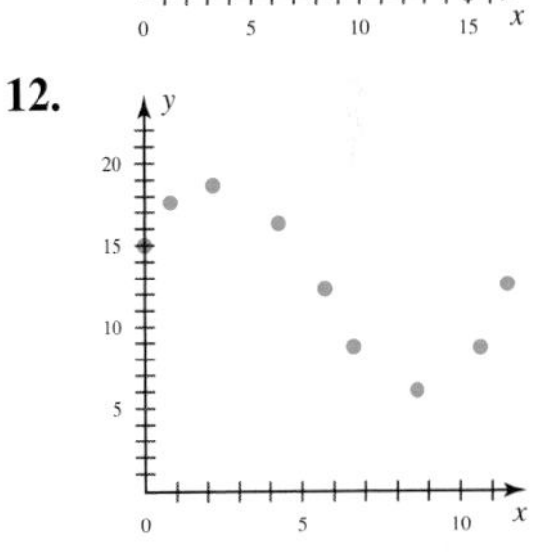

For Exercises 13 to 20, suppose a set of data is generated from the context indicated. Use common sense, previous experience, or your own knowledge base to state whether a linear, quadratic, logarithmic, exponential, power, or logistic regression might be most appropriate. Justify your answers.

13. total revenue and number of units sold linear

14. page count in a book and total number of words linear

15. years on the job and annual salary exponential

16. population growth with unlimited resources exponential

17. population growth with limited resources logistic

18. elapsed time and the height of a projectile quadratic

19. the cost of a gallon of milk over time exponential

20. elapsed time and radioactive decay exponential

Discuss why an exponential model could be an appropriate form of regression for each data set, then find the regression equation.

21. As time increases, the amount of radioactive material decreases but will never truly reach 0 or become negative. Exponential with $b < 1$ and $k > 0$ is the best choice.

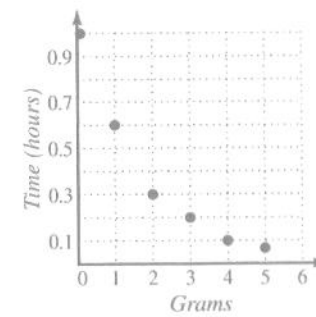

$y \approx (1.042)0.5626^x$

22. Populations usually grow exponentially or logistically. This growth does not appear to taper off. Exponential with $b > 1$ and $k > 0$ is the best choice.

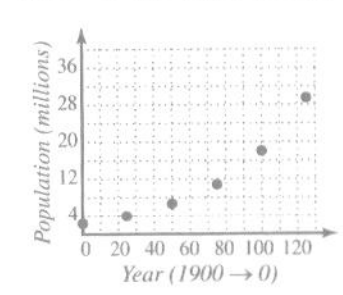

$y \approx (2.6550)1.1754^x$

21. Radioactive Studies

Time in Hours	Grams of Material
0.1	1.0
1	0.6
2	0.3
3	0.2
4	0.1
5	0.06

22. Rabbit Population

Month	Population (in hundreds)
0	2.5
3	5.0
6	6.1
9	12.3
12	17.8
15	30.2

23. Identify the growth rate constant for the equation from Exercise 21. $k \approx -0.5752$, about -57.5%

24. Identify the growth rate constant for the equation from Exercise 22. $k \approx 0.1616$, about 16.2%

Discuss why a logarithmic model could be an appropriate form of regression for each data set, then find the regression equation.

25. Sales will increase rapidly, then level off as the market is saturated with ads and advertising becomes less effective, possibly modeled by a logarithmic function. $y \approx 120.4938 + 217.2705\ln(x)$

26. In a productive mine, we expect that initially, the diamonds may be nearer the surface and more plentiful, becoming more scarce and harder to find as time goes on. A logarithmic model seems to fit this description. $y \approx 454.7845 + 1087.8962\ln(x)$

25. Total number of sales compared to the amount spent on advertising

Advertising costs ($1000s)	Total number of sales
1	125
5	437
10	652
15	710
20	770
25	848
30	858
35	864

26. Cumulative weight of diamonds extracted from a diamond mine

Time (months)	Weight (carats)
1	500
3	1748
6	2263
9	2610
12	3158
15	3501
18	3689
21	3810

The applications in this section require solving equations similar to those that follow. Solve each equation.

27. $96.35 = (9.4)1.6^x$ 4.95

28. $(3.7)2.9^x = 1253.93$ 5.47

29. $(-10.04)1.046^x = -396.58$ 81.74

30. $12{,}110 = (193.76)0.912^x$ -44.89

31. $4.8x^{2.5} = 468.75$ 6.25

32. $4375 = 1.4x^{-1.25}$ 0.0016

33. $2.103x^{0.6} = 56.781$ 243

34. $52 = 63.9 - 6.8\ln x$ 5.75

35. $498.53 + 2.3\ln x = 2595.9$ $e^{911.9}$

36. $49.05 + 12.7\ln x = 258.6$ 14,650,719.43

37. $9 = 68.76 - 7.2\ln x$ 4023.87

38. $52 = \dfrac{67}{1 + 20e^{-0.62x}}$ 6.84

39. $\dfrac{975}{1 + 82.3e^{-0.423x}} = 890$ 15.98

40. $1080 = \dfrac{1100}{1 + 37.2e^{-0.812x}}$ 9.37

41. $5 = \dfrac{8}{1 + 9.3e^{-0.65x}}$ 4.22

WORKING WITH FORMULAS

42. Learning curve: $P(t) = 5.9 + 12.6\ln(t)$

The number of toy planes a new employee can assemble from its component parts depends on how long the employee has been working on the assembly line. This is modeled by the function shown, where t represents the number of days and $P(t)$ is the number of planes the worker is able to assemble. How many planes is an employee making after 5 days on the job? 26 How long will it take until the employee is able to assemble 35 planes per day? 10 days

43. Bicycle sales since 1920: $N(t) = 0.325(1.057)^t$

Despite the common use of automobiles and motorcycles, bicycle sales have continued to grow as a means of transportation as well as a form of recreation. The number of bicycles sold each year (in millions) can be approximated by the formula shown, where t is the number of years after 1920 ($1920 \to 0$). According to this model, in what year did bicycle sales exceed 10 million? 1981

Source: 1976/1992 *Statistical Abstract of the United States,* Tables 406/395; various other years

APPLICATIONS

Answer the questions using the given data and the related regression equation. All extrapolations assume the mathematical model will continue to represent future trends.

44.
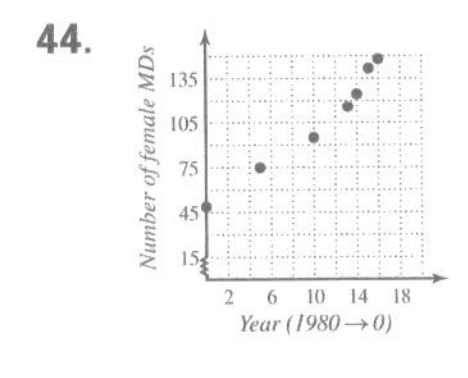

44. Female physicians: The number of females practicing medicine as MDs is given in the table for selected years. Use the data to draw a scatter-plot, then use the context and scatter-plot to find the best regression equation (see Exercise 50). $y = 50.21(1.07)^x$

Source: 2004 *Statistical Abstract of the United States,* Table 149, various other years

Exercise 44

Year (1980 → 0)	Number (in 1000s)
0	48.7
5	74.8
10	96.1
13	117.2
14	124.9
15	140.1
16	148.3

Exercise 45

Year (1900 → 0)	Number (per capita/ per year)
0	38
20	180
40	260
60	590
80	1250
97	2325

45.
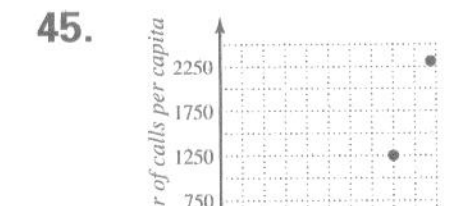

45. Telephone use: The number of telephone calls per capita has been rising dramatically since the invention of the telephone in 1876. The table given shows the number of phone calls per capita per year for selected years. Use the data to draw a scatter-plot, then use the context and scatter-plot to find the best regression equation (also see Exercise 51). $y = 53.24(1.04)^x$

Source: Data from 1997/2000 *Statistical Abstract of the United States,* Tables 922/917

46.

46. Milk production: With milk production becoming a big business, the number of family farms with milk-producing cows has been decreasing as larger corporations take over. The number of farms that keep milk cows for commercial production is given in the table for selected years. Use the data to draw a scatter-plot, then use the context and scatter-plot to find the best regression equation (also see Exercise 52). $y = 346.79(0.94)^x$

Source: 2000 *Statistical Abstract of the United States,* Table 1141

Exercise 46

Year (1980 → 0)	Number (in 1000s)
0	334
5	269
10	193
15	140
17	124
18	117
19	111

Exercise 47

Time (seconds)	Height of Froth (in.)
0	0.90
2	0.65
4	0.40
6	0.21
8	0.15
10	0.12
12	0.08

47.

47. Froth height—carbonated beverages: The height of the froth on carbonated drinks and other beverages can be manipulated by the ingredients used in making the beverage and lends itself very well to the modeling process. The data in the table given shows the froth height of a certain beverage as a function of time, after the froth has reached a maximum height. Use the data to draw a scatter-plot, then use the context and scatter-plot to find the best regression equation (see Exercise 53). $y = 0.89(0.81)^x$

48.

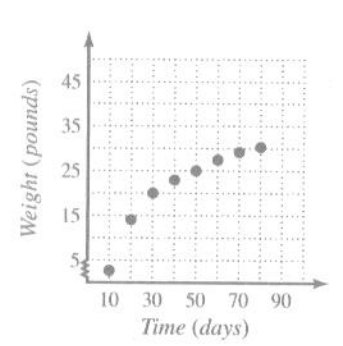

48. Weight loss: Harold needed to lose weight and started on a new diet and exercise regimen. The number of pounds he's lost since the diet began is given in the table. Draw the scatter-plot, then use a calculator to linearize the data and determine whether an exponential or logarithmic equation model is more appropriate. Finally, find the equation model (see Exercise 54). $y = -27.4 + 13.5 \ln x$

Exercise 48

Time (days)	Pounds Lost
10	2
20	14
30	20
40	23
50	25.5
60	27.6
70	29.2
80	30.7

Exercise 49

Time (months)	Ounces Mined
5	275
10	1890
15	2610
20	3158
25	3501
30	3789
35	4109
40	4309

49.

49. Depletion of resources: The longer an area is mined for gold, the more difficult and expensive it gets to obtain. The cumulative total of the ounces of gold produced by a particular mine is shown in the table. Draw the scatter-plot, and use the graph and the context of the application to determine whether an exponential or logarithmic equation model is more appropriate. Finally, find the equation model (see Exercise 55). $y = -2635.6 + 1904.8 \ln x$

50. Use the regression equation from Exercise 44 to answer the following questions:

a. What was the approximate number of female MDs in 1988? 86,270

b. Approximately how many female MDs will there be in 2005? 272,511

c. In what year did the number of female MDs exceed 100,000? 1990

51. Use the regression equation from Exercise 45 to answer the following questions:

a. What was the approximate number of calls per capita in 1970? 829

b. Approximately how many calls per capita will there be in 2005? 3271

c. In what year did the number of calls per capita exceed 1800? 1989

52. Use the regression equation from Exercise 46 to answer the following questions:

a. What was the approximate number of farms with milk cows in 1993? 155,142

b. Approximately how many farms will have milk cows (for commercial production) in 2004? 78,548

c. In what year did the number of farms with milk cows drop below 150,000? 1993

53. Use the regression equation from Exercise 47 to answer the following questions:

a. What was the approximate height of the froth after 6.5 sec? 0.23 in.

b. How long does it take for the height of the froth to reach one-half of its maximum height? 3.2 sec

c. According to the model, how many seconds until the froth height is 0.02 in.? 18 sec

54. Use the regression equation from Exercise 48 to answer the following questions:

a. What was Harold's total weight loss after 15 days? 9.2 lb

b. Approximately how many days did it take to lose a total of 18 lb? 29 days

c. According to the model, what is the projected weight loss for 100 days? 34.8 lb

55. Use the regression equation from Exercise 49 to answer the following questions:

a. What was the total number of ounces mined after 18 months? 2870 oz

b. About how many months did it take to mine a total of 4000 oz? 32.6 months

c. According to the model, what is the projected total after 50 months? 4816 oz

56. 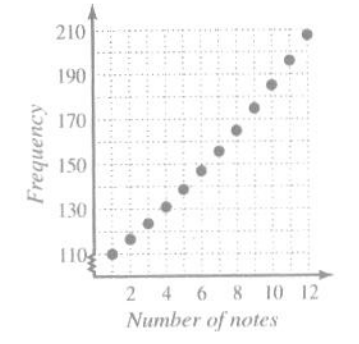

57. 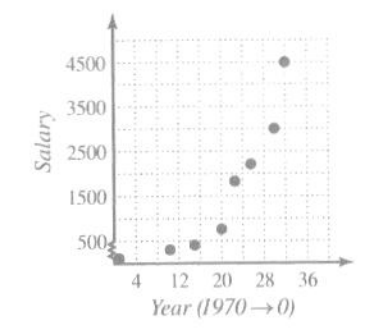

58. 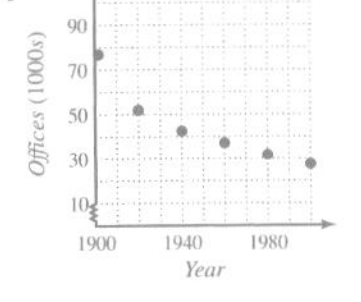

59. 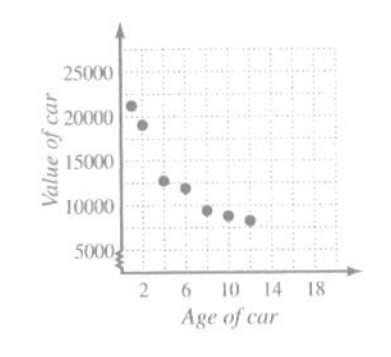

56. Musical notes: The table shown gives the frequency (vibrations per second) for each of the 12 notes in a selected octave from the standard chromatic scale. Use the data to draw a scatter-plot, then use the context and scatter-plot to find the best regression equation. $y = (103.83)1.0595^x$

a. What is the frequency of the "A" note that is an octave higher than the one shown? [*Hint:* The names repeat every 12 notes (one octave) so this would be the 13th note in this sequence.] 220

b. If the frequency is 370.00, what note is being played? The 22nd note, or F#

c. What pattern do you notice for the F#'s in each octave (the 10th, 22nd, 34th, and 46th notes in sequence)? Does the pattern hold for all notes? frequency doubles, yes

Number	Note	Frequency
1	A	110.00
2	A#	116.54
3	B	123.48
4	C	130.82
5	C#	138.60
6	D	146.84
7	D#	155.56
8	E	164.82
9	F	174.62
10	F#	185.00
11	G	196.00
12	G#	207.66

57. Basketball salaries: In 1970, the average player salary for a professional basketball player was about $43,000. Since that time, player salaries have risen dramatically. The average player salary for a professional player is given in the table for selected years. Use the data to draw a scatter-plot, then use the context and scatter-plot to find the best regression equation. $y = 39.86(1.16)^x$

Source: 1998 *Wall Street Journal Almanac,* p. 985; National Basketball Association Data

a. What was the approximate salary for a player in 1988? $576,000

b. What is the projection for the average salary in 2005? $7,187,000

c. In what year did the average salary exceed $1,000,000? 1992

Year (1970 → 0)	Salary (1000s)
0	43
10	171
15	325
20	750
25	1900
27	2300
30	3500
32	4500

58. Number of U.S. post offices: Due in large part to the ease of travel and increased use of telephones, email, and instant messaging, the number of post offices in the United States has been on the decline since the turn of the century. The data given show the number of post offices (in thousands) for selected years. Use the data to draw a scatter-plot, then use the context and scatter-plot to find the best regression equation. $y = 78.8 - 10.3 \ln x$

Source: 1985/2000 *Statistical Abstract of the United States,* Tables 918/1112

a. Approximately how many post offices were there in 1915? 51,000

b. In what year did the number of post offices drop below 34,000? 1977

c. According to the model, how many post offices will there be in the year 2010? 30,400

Exercise 58

Year (1900 → 0)	Offices (1000s)
1	77
20	52
40	43
60	37
80	32
100	28

59. Automobile value: While it is well known that most cars decrease in value over time, what is the best equation model for this decline? Use the data given to draw a scatter-plot, then use the context and scatter-plot to find the best regression equation. Linearize the data if needed. $y = 19{,}943.7 - 5231.4 \ln x$

a. What was the car's value after 7.5 yr? $9402.94

b. Exactly how old is the car if its current value is $8150? 9.5 yr

c. Using the model, how old is the car when value ≤ $3000? 25.5 yr

Exercise 59

Age of Car	Value of Car
1	19,500
2	16,950
4	12,420
6	11,350
8	8375
10	7935
12	6900

Exercise 60

Days after Outbreak	Cumulative Total
0	100
14	560
21	870
35	1390
56	1660
70	1710
84	1750

60. **Spread of disease:** Estimates of the cumulative number of SARS (Sudden Acute Respiratory Syndrome) cases reported in Hong Kong during the spring of 2003 are shown in the table, with day 0 corresponding to February 20. Use the data to draw a scatter-plot, then use the context and scatter-plot to find the best regression equation. $y = \dfrac{1719}{1 + 10.2e^{-0.11x}}$

Source: Center for Disease Control @ www.cdc.gov/ncidod/EID/vol9no12

a. What was the approximate number of SARS cases by day 25? 1041

b. About what day did the number of SARS cases exceed 1500? day 38

c. Using this model, what is the projected number of SARS cases on the 84th day? Why does this differ from the value in the table? 1717 The model only approximates the data.

60.
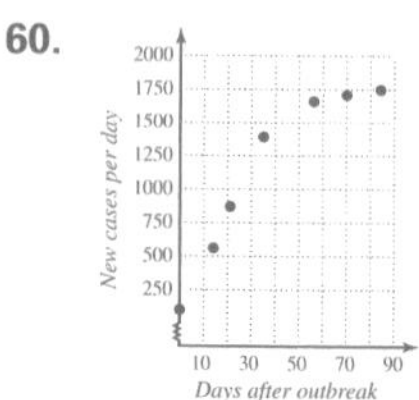

61. **Cable television subscribers:** The percentage of American households having cable television is given in the table to the right for select years from 1976 to 2004. Use the data to draw a scatter-plot, then use the context and scatter-plot to find the best regression equation. $y = \dfrac{69.99}{1 + 4.00e^{-0.22x}}$

Source: Data pooled from the 2001 *New York Times Almanac,* page 393; 2004 *Statistical Abstract of the United States,* Table 1120; various other years

Year (1976 → 0)	Percentage with Cable TV
0	16
4	22.6
8	43.7
12	53.8
16	61.5
20	66.7
24	68
28	70

a. Approximately what percentage of households had cable TV in 1990? 59.1%

b. In what year did the percentage having cable TV exceed 50%? 1986

c. Using this model, what projected percentage of households will have cable in 2008? 69.7%

61.
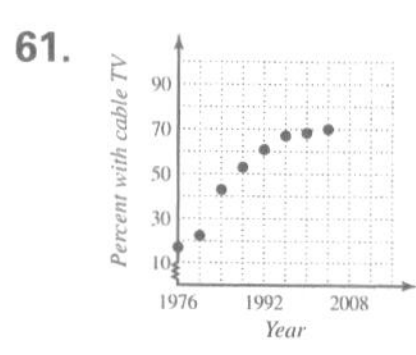

WRITING, RESEARCH, AND DECISION MAKING

62. Although correlation coefficients can be very helpful, other factors must also be considered when selecting the most appropriate equation model for a set of data. To see why, use the data given to draw a scatter-plot. Does the scatter-plot clearly suggest a particular form of regression? Use the STAT feature of your graphing calculator to (a) find a linear regression equation and note its correlation coefficient, and (b) find an exponential regression equation and note its correlation coefficient. What do you notice? Without knowing the context of the data, would you be able to tell which model might be more suitable?

Input	Output
1	1
2	3.5
3	11
4	9.8
5	15
6	27.5

62. No.
a. $y = 4.737x - 5.280$, $r \approx 0.939$
b. $y = (0.914)1.813^x$, $r \approx 0.939$
The correlation coefficients are equal.
No.

63. Carefully read the *Point of Interest* paragraph at the beginning of this section, then study the data and the three related graphs. The graphs show the result of applying linear regression, quadratic regression, and exponential regression respectively, to the data.

a. If the linear regression is correct, what is the significance of the x-intercept? What would be the outcome of the election?

b. If the quadratic regression is correct, what is the significance of the vertex? What would be the outcome of the election?

c. If the exponential regression is correct, what is the significance of the asymptote? What would be the outcome of the election?

d. Which of the three regressions would most merit the involvement of the Supreme Court of the United States? Why?

63. a. Bush & Gore are tied; Gore wins.
b. Bush has his smallest lead; Bush would win.
c. Bush would always have a lead. Bush would win.
d. exponential; outcome would be too close to call

EXTENDING THE CONCEPT

64. The regression fundamentals applied in this section and elsewhere can be extended to other forms as well. The table shown gives the time required for the first six planets to make one complete revolution around the Sun (in years), along with the average orbital radius of the planet (in astronomical units, 1 AU = 92.96 million miles). Use a graphing calculator to draw the scatter-plot, then use the scatter-plot, the context and any previous experience to decide whether a polynomial, exponential, logarithmic or power regression is most appropriate. Then find the regression equation and use it to: (a) estimate the average orbital radius of Saturn, given it orbits the Sun every 29.46 years, and (b) estimate how many years it takes Uranus to orbit the Sun, given it has an average orbital radius of 19.2 AU. power regression; **a.** 9.5 AU; **b.** 84.8 yr

Planet	Years	Radius
Mercury	0.24	0.39
Venus	0.62	0.72
Earth	1.00	1.00
Mars	1.88	1.52
Jupiter	11.86	5.20

65A. Using the points (2, 2) and (7, 7), find the exponential regression equation and the logarithmic regression equation that contains these two points. Enter the functions as Y_1 and Y_2 respectively and graph the points and both equations on the standard screen (zoom **6:ZStandard**). Do you notice anything? Enter the function $Y_3 = x$ and re-graph the functions using window size $x \in [0, 10]$ and $y \in [0, 15]$ to obtain an approximately square viewing window. Name two ways we can verify that Y_1 and Y_2 are inverse functions.

65A. $Y_1 \approx (1.2117)1.2847^x$,
$Y_2 \approx -0.7665 + 3.9912\ln(x)$,
1. symmetry about $y = x$;
2. for (a, b) on Y_1 (b, a) is on Y_2.

65B. Verify that the functions Y_1 and Y_2 from Exercise 65A are inverse functions by: (a) rewriting Y_1 as a base e exponential function (see Example 3), and (b) using the algebraic method to find the inverse function for the general form $y = a + b\ln(x)$, then using the corresponding values from Y_2 in the inverse function to reconcile and verify that results match.

65B. **a.** $y = 1.2117214e^{0.2505524714x}$
b. $a + b\ln(x) = y$
$a + b\ln(y) = x$
$\ln(y) = \dfrac{x - a}{b}$
$y = e^{\frac{x-a}{b}} = e^{\frac{x}{b}}e^{\frac{-a}{b}}$
Substituting $a = -0.7664737786$ and $b = 3.991178001$ at this point gives the result from (a).

MAINTAINING YOUR SKILLS

66. (4.3) State the domain of the function, then write it in lowest terms:

$$h(x) = \frac{x^2 - 6x + 5}{x^3 - 4x^2 - 7x + 10}\quad \frac{1}{x+2}$$

66. $D: x \in (-\infty, -2) \cup (-2, 1) \cup (1, 5) \cup (5, \infty)$

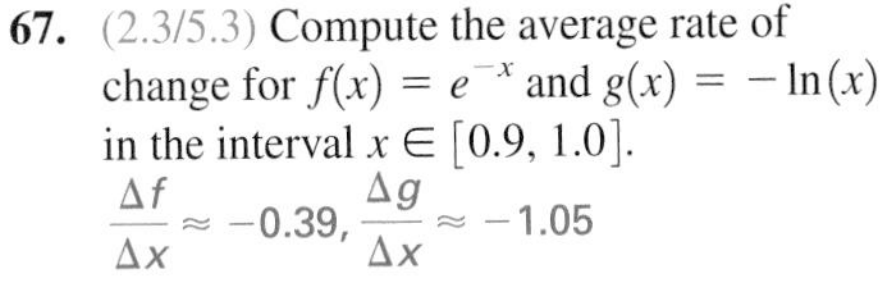

67. (2.3/5.3) Compute the average rate of change for $f(x) = e^{-x}$ and $g(x) = -\ln(x)$ in the interval $x \in [0.9, 1.0]$.
$\dfrac{\Delta f}{\Delta x} \approx -0.39$, $\dfrac{\Delta g}{\Delta x} \approx -1.05$

68. (3.4) Graph the function by completing the square. Clearly label the vertex and all intercepts: $p(x) = x^2 + 5x + 1$.

68.

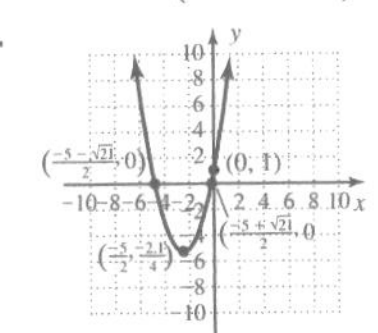

69. (3.7) Find a linear function that will make $p(x)$ continuous. $y = -\frac{3}{2}x + 7; 2 \le x < 4$

$$p(x) = \begin{cases} x^2 & -2 \le x < 2 \\ ?? & ? \le x < ? \\ \sqrt{x-4} + 1 & x \ge 4 \end{cases}$$

70. (3.8) For the graph of $f(x)$ given, estimate max/min values to the nearest tenth and state intervals where $f(x)\uparrow$ and $f(x)\downarrow$.

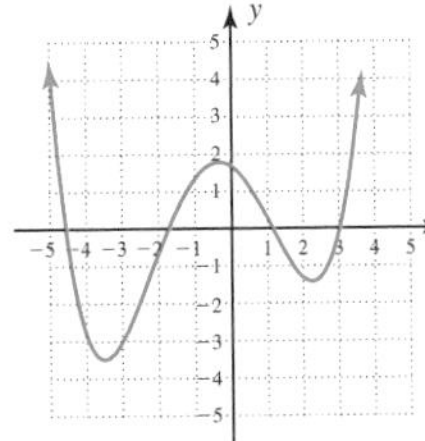

70. max: $(-0.4, 1.8)$
min: $(-3.5, -3.5)$, $(2.3, -1.4)$
$f(x)\uparrow$: $x \in (-3.5, -0.4) \cup (2.3, \infty)$
$f(x)\downarrow$: $x \in (-\infty, -3.5) \cup (-0.4, 2.3)$

71. (3.3) The graph of $f(x) = x^{\frac{2}{3}}$ is given. Use it to sketch the graph of $F(x) = (x - 2)^{\frac{2}{3}} + 3$, and use the graph to state the domain and range of F.
$D: x \in (-\infty, \infty)$ $R: y \in [3, \infty)$

71.

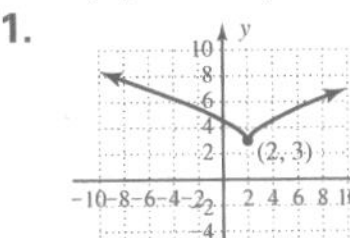

SUMMARY AND CONCEPT REVIEW

SECTION 5.1 Exponential Functions

KEY CONCEPTS

- An exponential function is one of the form $f(x) = ab^x$, where $b > 0, b \neq 1$, and a, b, and x are real numbers.
- For functions of the form $f(x) = b^x$, $b > 0$ and $b \neq 1$, we have:
  - one-to-one function
  - y-intercept $(0, 1)$
  - domain: $x \in \mathbb{R}$
  - range: $y \in (0, \infty)$
  - increasing if $b > 1$
  - decreasing if $0 < b < 1$
  - asymptotic to x-axis
- The graph of $y = b^{x \pm h} \pm k$ is a translation of the basic graph of $y = b^x$, horizontally h units opposite the sign and vertically k units in the same direction as the sign.
- To solve exponential equations with like bases, we use the fact that b^x represents a unique number. In other words, if $b^m = b^n$, then $m = n$ (equal bases imply equal exponents). This is referred to as the Uniqueness Property.
- All previous properties of exponents also apply to exponential functions.

EXERCISES

Graph each function using *transformations of the basic function,* then strategically plotting a few points to check your work and round out the graph. Draw and label the asymptote.

1. $y = 2^x + 3$ **2.** $y = 2^{-x} - 1$ **3.** $y = -3^x - 2$

Solve the exponential equations using the uniqueness property.

4. $3^{2x-1} = 27$ 2 **5.** $4^x = \frac{1}{16}$ -2 **6.** $3^x \cdot 27^{x+1} = 81$ $\frac{1}{4}$

7. A ballast machine is purchased new for $142,000 by the AT & SF Railroad. The machine loses 15% of its value each year and must be replaced when its value drops below $20,000. How many years will the machine be in service? 12.1 yr

SECTION 5.2 Logarithms and Logarithmic Functions

KEY CONCEPTS

- A logarithm is an exponent. For $b > 0$, $b \neq 1$, and $x > 0$, $y = \log_b x$ means $b^y = x$ and $b^{\log_b x} = x$.
- A logarithmic *function* is defined as $f(x) = \log_b x$, where $b > 0$, $b \neq 1$, and x is a positive real number. $y = \log_{10} x = \log x$ is called the common logarithmic function.
- For logarithmic functions of the form $f(x) = \log_b x$, $b > 0$ and $b \neq 1$, we have:
  - one-to-one function
  - x-intercept $(1, 0)$
  - domain: $x \in (0, \infty)$
  - range: $y \in \mathbb{R}$
  - increasing if $b > 1$
  - decreasing if $0 < b < 1$
  - asymptotic to y-axis
- The graph of $y = \log_b(x \pm h) \pm k$ is a translation of the basic graph of $y = \log_b x$, horizontally h units opposite the sign and vertically k units in the same direction as the sign.
- To solve logarithmic equations with like bases, we can use the fact that $\log_b x$ is a unique number. In other words, if $\log_b m = \log_b n$, then $m = n$ (equal bases imply equal arguments).

EXERCISES

Write each expression in *exponential* form.

8. $\log_3 9 = 2$ $3^2 = 9$ **9.** $\log_5 \frac{1}{125} = -3$ $5^{-3} = \frac{1}{125}$ **10.** $\log_2 16 = 4$ $2^4 = 16$

17. 18.

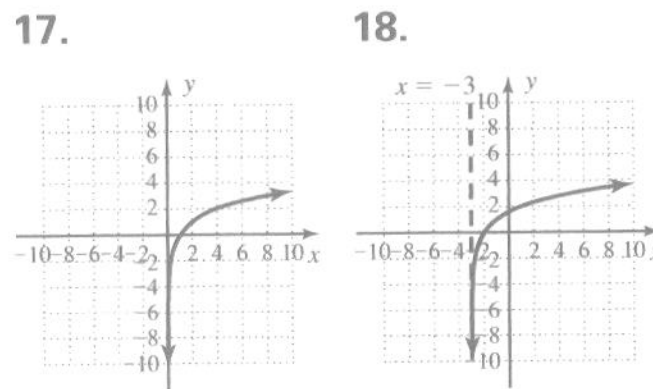

19.

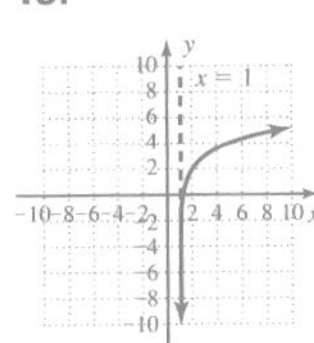

Write each expression in *logarithmic* form.

11. $5^2 = 25$ $\log_5 25 = 2$ **12.** $2^{-3} = \frac{1}{8}$ $\log_2(\frac{1}{8}) = -3$ **13.** $3^4 = 81$ $\log_3 81 = 4$

Solve for x.

14. $\log_2 32 = x$ 5 **15.** $\log_x 16 = 2$ 4 **16.** $\log(x - 3) = 1$ 13

Graph each function using *transformations of the basic function,* then strategically plotting a few points to check your work and round out the graph. Draw and label the asymptote.

17. $f(x) = \log_2 x$ **18.** $f(x) = \log_2(x + 3)$ **19.** $f(x) = 2 + \log_2(x - 1)$

20. The magnitude of an earthquake is given by $M(I) = \log\frac{I}{I_0}$, where I is the intensity and I_0 is the reference intensity. (a) Find $M(I)$ given $I = 62{,}000I_0$ and (b) find the intensity I given $M(I) = 7.3$. **a.** 4.79; **b.** $10^{7.3}\,I_0$

SECTION 5.3 The Exponential Function and Natural Logarithms

KEY CONCEPTS

- The natural exponential function is $y = e^x$, where $e \approx 2.71828$.
- The natural logarithmic function is $y = \log_e x$, most often written in abbreviated form as $y = \ln x$.
- Base-e logarithms can be found using a scientific or graphing calculator and the LN key.
- To evaluate logs with bases other than 10 or e, we use the change-of-base formula:
  $$\log_b x = \frac{\log_d x}{\log_d b}.$$
- Since a logarithm is an exponent, it has properties that parallel those of exponents.
  - Product Property: like base and multiplication, add exponents: $b^n b^m = b^{n+m}$
  - Quotient Property: like base and division, subtract exponents: $\log_b(\frac{M}{N}) = \log_b M - \log_b N$
  - Power Property: exponent raised to a power, multiply exponents: $\log_b(k)^x = x\log_b(k)$
- The logarithmic properties can be used to expand an expression, as in $\log(2x) = \log 2 + \log x$, or to contract an expression (combine like terms), as in $\ln(2x) - \ln(x + 3) = \ln\frac{2x}{x + 3}$.

EXERCISES

21. Solve each equation.

a. $\ln(x) = 32$ **b.** $e^x = 9.8$ **c.** $e^x = \sqrt{7}$ **d.** $\ln(x) = 2.38$

22. Evaluate using the change-of-base formula. Answer in exact form and approximate form, rounding to thousandths.

a. $\log_6 45$ **b.** $\log_3 128$ **c.** $\log_2 108$ **d.** $\log_5 200$

23. Use the product or quotient property of logarithms to write each sum or difference as a single term.

a. $\ln 7 + \ln 6$ **b.** $\log_9 2 + \log_9 15$

c. $\ln(x + 3) - \ln(x - 1)$ **d.** $\log x + \log(x + 1)$

24. Use the power property of logarithms to rewrite each term as a product.

a. $\log_5 9^2$ **b.** $\log_7 4^2$ **c.** $\ln 5^{2x-1}$ **d.** $\ln 10^{3x+2}$

25. Evaluate the following logarithmic expressions using only properties of logarithms (without the aid of a calculator or the change-of-base formula), given $\log_5 2 \approx 0.43$ and $\log_5 3 \approx 0.68$.

a. $\log_5 \frac{2}{3}$ **b.** $\log_5 \frac{3}{2}$ **c.** $\log_5 12$ **d.** $\log_5 18$

21. a. e^{32}
b. $\ln 9.8$
c. $\ln\sqrt{7} = \frac{1}{2}\ln 7$
d. $e^{2.38}$

22. a. $\frac{\log 45}{\log 6} \approx 2.125$
b. $\frac{\log 128}{\log 3} \approx 4.417$
c. $\frac{\log 108}{\log 2} \approx 6.755$
d. $\frac{\log 200}{\log 5} \approx 3.292$

23. a. $\ln 42$
b. $\log_9 30$
c. $\ln\left(\frac{x + 3}{x - 1}\right)$
d. $\log(x^2 + x)$

24. a. $2\log_5 9$
b. $2\log_7 4$
c. $(2x - 1)\ln 5$
d. $(3x + 2)\ln 10$

25. a. ≈ -0.25
b. ≈ 0.25
c. ≈ 1.54
d. ≈ 1.80

26. a. $4 \ln x + \frac{1}{2} \ln y$
b. $\frac{1}{3} \ln p + \ln q$
c. $\frac{5}{3} \log x + \frac{4}{3} \log y - \frac{5}{2} \log x - \frac{3}{2} \log y$
d. $\log 4 + \frac{5}{3} \log p + \frac{4}{3} \log q - \frac{3}{2} \log p - \log q$

26. Use the properties of logarithms to write the following expressions as a sum or difference of simple logarithmic terms.

a. $\ln(x^4\sqrt{y})$ **b.** $\ln(\sqrt[3]{pq})$ **c.** $\log\left(\frac{\sqrt[3]{x^5y^4}}{\sqrt{x^5y^3}}\right)$ **d.** $\log\left(\frac{4\sqrt[3]{p^5q^4}}{\sqrt{p^3q^2}}\right)$

27. The rate of decay for radioactive material is related to its half-life by the formula $R(h) = \frac{\ln 2}{h}$, where h represents the half-life of the material and $R(h)$ is the rate of decay expressed as a decimal. The element radon-222 has a half-life of approximately 3.9 days. (a) Find its rate of decay to the nearest 100th of a percent. (b) Find the half-life of thorium-234 if its rate of decay is 2.89% per day. **a.** 17.77% **b.** 23.98 days

SECTION 5.4 Exponential/Logarithmic Equations and Applications

KEY CONCEPTS

- To solve an exponential or logarithmic equation, first simplify by combining like terms if possible.
- The way logarithms are defined gives rise to four useful properties: For any base b where $b > 0, b \neq 1$,
  - $\log_b b = 1$ (since $b^1 = b$)
  - $\log_b b^x = x$ (since $b^x = b^x$)
  - $\log_b 1 = 0$ (since $b^0 = 1$)
  - $b^{\log_b x} = x$
- If the equation can be written with like bases on both sides, solve using the uniqueness property.
- If a single logarithmic or exponential term can be isolated on one side, then for any base b:
  - If $b^x = k$, then $x = \frac{\log k}{\log b}$.
  - If $\log_b x = k$, then $x = b^k$.

EXERCISES

Solve each equation.

28. $2^x = 7$ $\frac{\ln 7}{\ln 2}$

29. $3^{x+1} = 5$ $\frac{\ln 5}{\ln 3} - 1$

30. $4^{x-2} = 3^x$ $\frac{2 \ln 4}{\ln 4 - \ln 3}$

31. $\log_5(x + 1) = 2$ 24

32. $\log x + \log(x - 3) = 1$ 5; −2 is extraneous

33. $\log_{25}(x - 2) - \log_{25}(x + 3) = \frac{1}{2}$ no solution

34. The *barometric equation* $H = (30T + 8000)\ln\left(\frac{P_0}{P}\right)$ relates the altitude H to atmospheric pressure P, where $P_0 = 76$ cm of mercury. Find the atmospheric pressure at the summit of Mount Pico de Orizaba (Mexico), whose summit is 5657 m. Assume the temperature at the summit is $T = 12°$C. 38.63 cmHg

SECTION 5.5 Applications from Investment, Finance, and Physical Science

KEY CONCEPTS

- Simple interest: $I = prt$; p is the initial principal, r is the interest rate per year, and t is the time in years.
- Amount in an account after t years: $A = p + prt$ or $A = p(1 + rt)$.

- Interest compounded n times per year: $A = p\left(1 + \frac{r}{n}\right)^{nt}$; p is the initial principal, r is the interest rate per year, t is the time in years, and n is the times per year interest is compounded.
- Interest compounded continuously: $A = pe^{rt}$; p is the initial principal, r is the interest rate per year, and t is the time in years. The base e is the exponential constant $e \approx 2.71828$.
- Closely related to the formula for interest compounded continuously are the more general formulas for exponential growth and decay, $Q(t) = Q_0e^{rt}$ and $Q(t) = Q_0e^{-rt}$, respectively.
- If a loan or savings plan calls for a regular schedule of deposits, the plan is called an annuity.
- For periodic payment P, deposited or paid n times per year, at annual interest rate r, with interest compounded or calculated n times per year for t years, and $R = \frac{r}{n}$:
  - The accumulated value of the account is $A = \frac{P}{R}[(1 + R)^{nt} - 1]$.
  - The payment required to meet a future goal is $P = \frac{AR}{[(1 + R)^{nt} - 1]}$.
  - The payment required to amortize an amount A is $P = \frac{AR}{1 - (1 + R)^{-nt}}$.

EXERCISES

Solve each application.

35. Jeffery borrows \$600.00 from his dad, who decides it's best to charge him interest. Three months later Jeff repays the loan plus interest, a total of \$627.75. What was the annual interest rate on the loan? 18.5%

36. To save money for her first car, Cheryl invests the \$7500 she inherited in an account paying 7.8% interest compounded monthly. She hopes to buy the car in 6 yr and needs \$12,000. Is this possible? Almost, she needs \$42.15 more.

37. Eighty prairie dogs are released in a wilderness area in an effort to repopulate the species. Five years later a statistical survey reveals the population has reached 1250 dogs. Assuming the growth was exponential, approximate the growth rate to the nearest tenth of a percent. 55.0%

38. To save up for the vacation of a lifetime, Al-Harwi decides to save \$15,000 over the next 4 yr. For this purpose he invests \$260 every month in an account paying $7\frac{1}{2}$% interest compounded monthly. (a) Is this monthly amount sufficient to meet the four-year goal? (b) If not, find the *minimum amount he needs to deposit each month* that will allow him to meet this goal in 4 yr. **a.** no **b.** \$268.93

SECTION 5.6 Exponential, Logarithmic, and Logistic Regression Models

KEY CONCEPTS

- The choice of regression models generally depends on: (a) whether the graph appears to fit the data, (b) the context or situation that generated the data, and (c) certain tests applied to the data.
- The regression equation can be used to *extrapolate* or predict future values or occurrences. When using extrapolation, values are projected *beyond* the given set of data.

- The regression equation can be used to *interpolate* or approximate intermediate values. When using interpolation, the values occur *between* those given in the data set.

EXERCISES

39. a. logistic
b. logistic; growth rate exceeds population growth
c. exponential: 44.6 million, logistic: 56.1 million exponential: 347.3, 1253 million; logistic, 179.3, 201.0 million; projections from the exponential equation are excessive

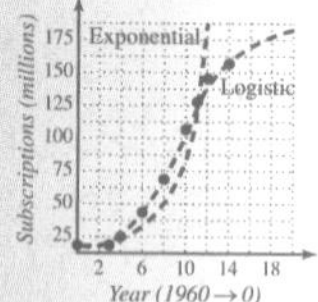

39. The tremendous surge in cell phone subscriptions that began in the early nineties has continued unabated into the new century. The total number of subscriptions is shown in the table for selected years, with 1990 → 0 and the number of subscriptions in millions. Draw a scatter-plot of the data and complete the following.

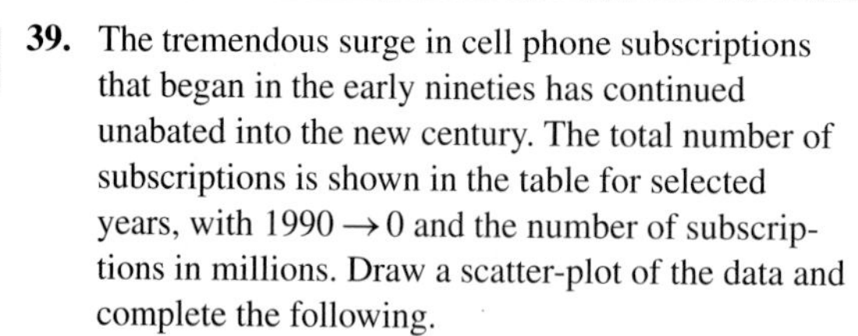

Source: 2000/2004 *Statistical Abstract of the United States,* Tables 919 and 1144

Year (1990 → 0)	Subscriptions (millions)
0	5.3
3	16.0
4	24.1
6	44.0
8	69.2
10	109.5
11	128.4
12	140.0
13	158.7

a. Run both an exponential regression and a logistic regression on the data, and graph them on the same screen with the scatter-plot. Which seems to "fit" the data better?

b. Considering the context of the data and the current population of the United States (approximately 291 million in 2003), which equation model seems more likely to accurately predict the number of cell phone subscriptions in future years? Why?

c. Use both regression equations to approximate the number of subscriptions in 1997. How many subscriptions does each project for 2005? 2010? What do you notice?

40.

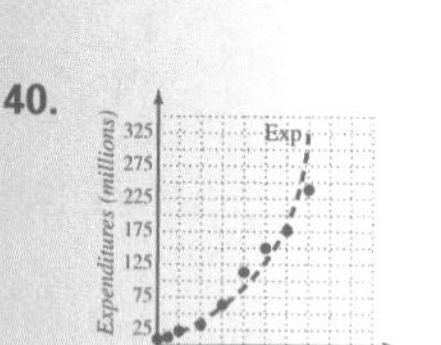

40. The development of new products, improved health care, greater scientific achievement, and other advances is fueled by huge investments in research and development (R & D). Since 1960, total expenditures in the United States have shown a distinct pattern of growth, and the data is given in the table for selected years from 1960 to 1999. Use the data to draw a scatter-plot and complete the following:

Source: 2004 *Statistical Abstract of the United States,* Table 978

Year (1960 → 0)	Expenditures (billion \$)
0	13.7
5	20.3
10	26.3
15	35.7
20	63.3
25	114.7
30	152.0
35	183.2
39	247.0

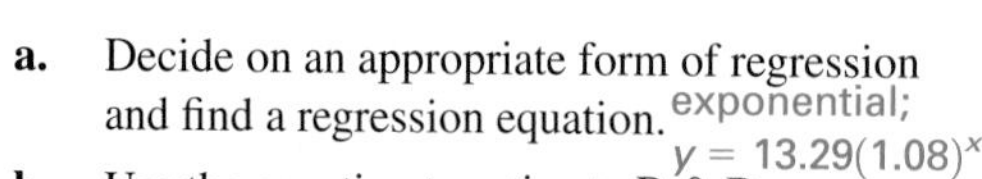

a. Decide on an appropriate form of regression and find a regression equation. exponential; $y = 13.29(1.08)^x$

b. Use the equation to estimate R & D expenditures in 1992. 156.0 billion dollars

c. If current trends continue, how much will be spent on R & D in 2005? 424.2 billion dollars

MIXED REVIEW

1. Evaluate each expression using the change-of-base formula.

a. $\log_2 30$ $\frac{\log 30}{\log 2} \approx 4.9069$ b. $\log_{0.25} 8$ -1.5 c. $\log_8 2$ $\frac{1}{3}$

2. Solve each equation using the uniqueness property.

a. $10^{4x-5} = 1000$ 2 b. $5^{3x-1} = \sqrt{5}$ $\frac{1}{2}$ c. $2^x \cdot 2^{0.5x} = 64$ 4

3. Use the power property of logarithms to rewrite each term as a product.

a. $\log_{10} 20^2$ $2\log_{10} 20$ b. $\log 10^{0.05x}$ $0.05x$ c. $\ln 2^{x-3}$ $(x-3)\ln 2$

4.

5.

6.

7.

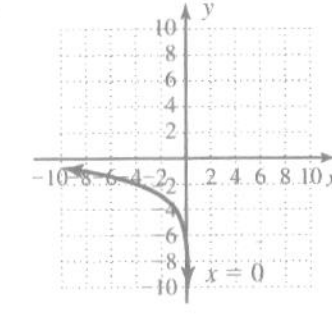

15. $\frac{9}{4} + \frac{\sqrt{129}}{4}$

Graph each of the following functions by shifting the basic function, then strategically plotting a few points to check your work and round out the graph. Graph and label the asymptote.

4. $y = -e^x + 15$

5. $y = 5 \cdot 2^{-x}$

6. $y = \ln(x + 5) + 7$

7. $y = \log_2(-x) - 4$

8. Use the properties of logarithms to write the following expressions as a sum or difference of simple logarithmic terms.

a. $\ln\left(\frac{x^3}{2y}\right)$ $3\ln x - \ln 2 - \ln y$

b. $\log\left(10a\sqrt[3]{a^2b}\right)$ $1 + \log a + \frac{2}{3}\log a + \frac{1}{3}\log b$

c. $\log_2\left(\frac{8x^4\sqrt{x}}{3\sqrt{y}}\right)$ $3 + \frac{9}{2}\log_2(x) - \log_2 3 - \frac{1}{2}\log_2(y)$

9. Write the following expressions in exponential form.

a. $\log_5 625 = 4$ $5^4 = 625$

b. $\ln 0.15x = 0.45$ $e^{0.45} = 0.15x$

c. $\log(0.1 \times 10^8) = 7$ $10^7 = 0.1 \times 10^8$

10. Write the following expressions in logarithmic form.

a. $343^{1/3} = 7$ $\log_{343} 7 = \frac{1}{3}$

b. $256^{3/4} = 64$ $\log_{256} 64 = \frac{3}{4}$

c. $2^{-3} = \frac{1}{8}$ $\log_2(\frac{1}{8}) = -3$

Solve the following equations. State answers in exact form.

11. $\log_2 128 = x$ 7

12. $\log_5(4x + 7) = 0$ $\frac{-3}{2}$

13. $10^{x-4} = 200$ $6 + \log 2$

14. $e^{x+1} = 3^x$ $x = \frac{1}{\ln 3 - 1}$

15. $\log_2(2x - 5) + \log_2(x - 2) = 4$

16. $\log(3x - 4) - \log(x - 2) = 1$ $\frac{16}{7}$

Solve each application.

17. The magnitude of an earthquake is given by $M(I) = \log\left(\frac{I}{I_0}\right)$, where I is the intensity of the quake and I_0 is the reference intensity 2×10^{11} (energy released from the smallest detectable quake). On October 23, 2004, the Niigata region of Japan was hit by an earthquake that registered 6.5 on the Richter scale. Find the intensity of this earthquake by solving the following equation for I: $6.5 = \log\left(\frac{I}{2 \times 10^{11}}\right)$. $I = 6.3 \times 10^{17}$

18. Serene is planning to buy a house. She has $6500 to invest in a certificate of deposit that compounds interest quarterly at an annual rate of 4.4%. (a) Find how long it will take for this account to grow to the $12,500 she will need for a 10% down payment for a $125,000 house. Round to the nearest tenth of a year. 14.9 yr (b) Suppose instead of investing an initial $6500, Serene deposits $500 a quarter in an account paying 4% each quarter. Find how long it will take for this account to grow to $12,500. Round to the nearest tenth of a year. 5.6 yr

Exercise 19

19. British artist Simon Thomas designs sculptures he calls hypercones. These sculptures involve rings of exponentially decreasing radii rotated through space. For one sculpture, the radii follow the model $r(n) = 2(0.8)^n$, where n counts the rings (outer-most first) and $r(n)$ is radii in meters. Find the radii of the six largest rings in the sculpture. Round to the nearest hundredth of a meter. 1.6 m, 1.28 m, 1.02 m, 0.82 m, 0.66 m, 0.52 m

Source: http://www.plus.maths.org/issue8/features/art/

20. The following data report the increase in health care cost per employee for an employer from 1999 to 2004. Use the data to make a scatter-plot, then determine which type of regression should be applied and find a regression equation. Then answer the following questions. $y = 0.55x + 3.69$

Year (1999 → 0)	Employer Average Health Care Cost per Employee (1000s)
0	3.8
1	4.2
2	4.7
3	5.2
4	5.9
5	6.5

a. Use the model to find the projected cost in 2005. 6.99

b. If this pattern of data continued, when would the projected cost exceed $10,000? 2010

20.

Average health care cost

Year (1999 → 0)

PRACTICE TEST

1. Write the expression $\log_3 81 = 4$ in exponential form. $3^4 = 81$

2. Write the expression $25^{1/2} = 5$ in logarithmic form. $\log_{25} 5 = \frac{1}{2}$

3. Write the expression $\log_b\left(\frac{\sqrt{x^5}y^3}{z}\right)$ as a sum or difference of logarithmic terms. $\frac{5}{2}\log_b x + 3\log_b y - \log_b z$

4. Write the expression $\log_b m + \left(\frac{3}{2}\right)\log_b n - \frac{1}{2}\log_b p$ as a single logarithm. $\log_b \frac{m\sqrt{n^3}}{\sqrt{p}}$

Solve for x by writing each expression using the same base.

5. $5^{x-7} = 125$ $x = 10$

6. $2 \cdot 4^{3x} = \frac{8^x}{16}$ $x = \frac{-5}{3}$

Given $\log_a 3 \approx 0.48$ and $\log_a 5 \approx 1.72$, evaluate the following without the use of a calculator:

7. $\log_a 45$ 2.68

8. $\log_a 0.6$ -1.24

Graph using transformations of the parent function. Verify answers using a graphing calculator.

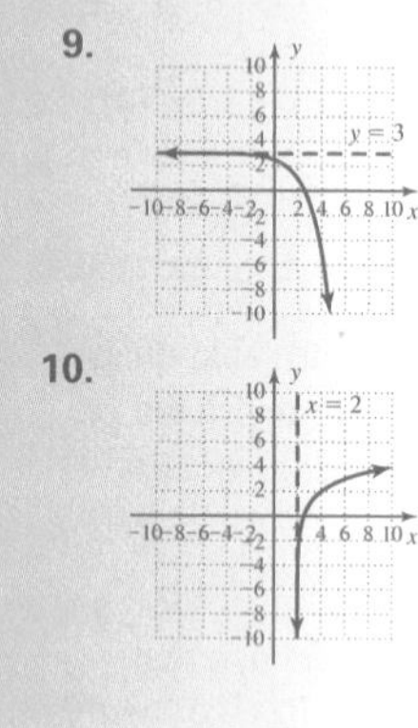

9. $g(x) = -2^{x-1} + 3$

10. $h(x) = \log_2(x - 2) + 1$

Use the change-of-base formula to evaluate. Verify results using a calculator.

11. $\log_3 100$ 4.19

12. $\log_6 0.235$ -0.81

Solve each equation.

13. $3^{x-1} = 89$ $x = 1 + \frac{\ln 89}{\ln 3}$

14. $\log_5 x + \log_5(x + 4) = 1$ $x = 1$; $x = -5$ is extraneous

15. A copier is purchased new for \$8000. The machine loses 18% of its value each year and must be replaced when its value drops below \$3000. How many years will the machine be in service? $\approx$5 yr

16. How long would it take \$1000 to double if invested at 8% annual interest compounded daily? $\approx$8.7 yr

17. The number of ounces of unrefined platinum drawn from a mine is modeled by $Q(t) = -2600 + 1900\ln(t)$, where $Q(t)$ represents the number of ounces mined in t months. How many months did it take for the number of ounces mined to exceed 3000? 19.1 months

18. Septashi can invest his savings in an account paying 7% compounded semi-annually, or in an account paying 6.8% compounded daily. Which is the better investment? 7% compounded semi-annually

19. Jacob decides to save \$4000 over the next 5 yr so that he can present his wife with a new diamond ring for their 20th anniversary. He invests \$50 every month in an account paying $8\frac{1}{4}$% interest compounded monthly. (a) Is this amount sufficient to meet the 5-yr goal? no (b) If not, find the *minimum amount he needs to save monthly* that will enable him to meet this goal. \$54.09

20. Using time-lapse photography, the growth of a stain is tracked in 0.2 second intervals, as a small amount of liquid is dropped on various fabrics. Use the data given to draw a scatter-plot and decide on an appropriate regression model. Precisely how long, to the nearest hundredth of a second, did it take the stain to reach a size of 15 mm? logistic; $y = \frac{39.1156}{1 + 314.6617e^{-5.9483x}}$; 0.89 sec

Exercise 20

Time (sec)	Size (mm)
0.2	0.39
0.4	1.27
0.6	3.90
0.8	10.60
1.0	21.50
1.2	31.30
1.4	36.30
1.6	38.10
1.8	39.00

20.

Calculator Exploration and Discovery

Investigating Logistic Equations

The keystrokes shown apply to a TI-84 Plus model. Please consult our Internet site or your manual for other models.

As we saw in Section 5.6, logistics models have the form $P(t) = \frac{c}{1 + ae^{-bt}}$, where a, b, and c are constants and $P(t)$ represents the population at time t. For populations modeled by a logistics curve (sometimes called an "S" curve) growth is very rapid at first (like an exponential function), but this growth begins to slow down and level off due to various factors. This *Calculator Exploration and Discovery* is designed to investigate the effects that a, b, and c have on the resulting graph.

I. From our earlier observation, as t becomes larger and larger, the term ae^{-bt} becomes smaller and smaller (approaching 0) because it is a decreasing function: as $t \to \infty$, $ae^{-bt} \to 0$. If we allow that the term eventually becomes so small it can be disregarded, what remains is $P(t) = \frac{c}{1}$ or c. This is why c is called the capacity constant and the population can get no larger than c. In Figure 5.35, the graph of $P(t) = \frac{1000}{1 + 50e^{-1x}}$ ($a = 50$, $b = 1$, and $c = 1000$) is shown using a lighter line, while the graph of $P(t) = \frac{750}{1 + 50e^{-1x}}$ ($a = 50$, $b = 1$, and $c = 750$), is given in bold. The window size is indicated in Figure 5.36.

Figure 5.35

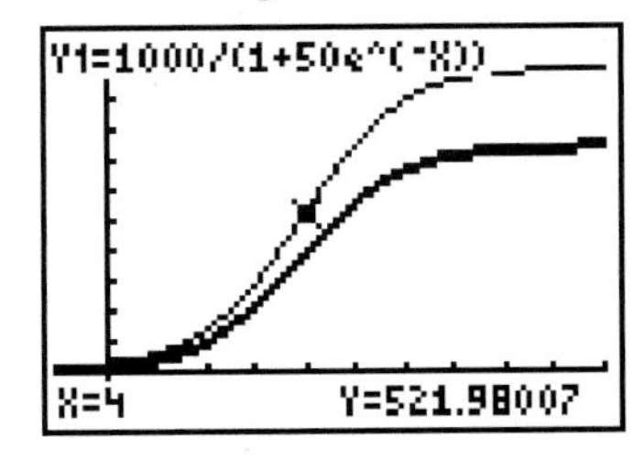

Figure 5.36

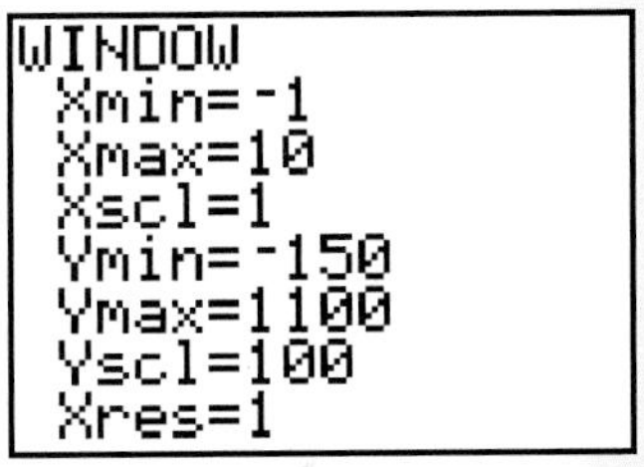

WINDOW
Xmin=-1
Xmax=10
Xscl=1
Ymin=-150
Ymax=1100
Yscl=100
Xres=1

Also note that if a is held constant, smaller values of c cause the "interior" of the S curve to grow at a slower rate than larger values, a concept studied in some detail in a Calculus I class.

II. If $t = 0$, $ae^{-bt} = ae^0 = a$, and we note the ratio $P(0) = \frac{c}{1 + a}$ represents the *initial population*. This also means for constant values of c, larger values of a make the ratio $\frac{c}{1 + a}$ smaller; while smaller values of a make the ratio $\frac{c}{1 + a}$ larger. From this we conclude that a primarily affects the initial population. For the screens shown next, $P(t) = \frac{1000}{1 + 50e^{-1x}}$ (from I. above) is graphed using a lighter line. For comparison, the graph of $P(t) = \frac{1000}{1 + 5e^{-1x}}$ ($a = 5$, $b = 1$, and $c = 1000$) is shown in bold in Figure 5.37, while the graph of $P(t) = \frac{1000}{1 + 500e^{-1x}}$ ($a = 500$, $b = 1$, and $c = 1000$) is shown in bold in Figure 5.38.

Figure 5.37

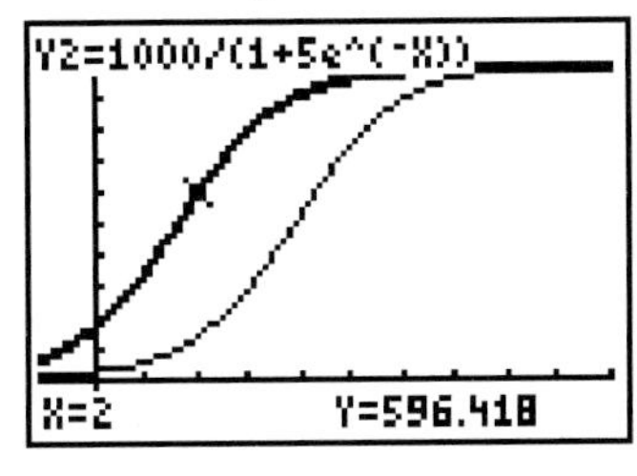

Figure 5.38

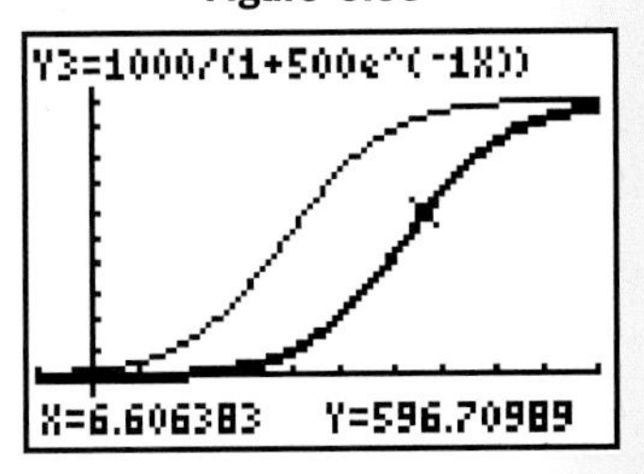

Note that changes in a appear to have no effect on the rate of growth in the interior of the S curve.

III. As for the value of b, we might expect that it affects the rate of growth in much the same way as the growth rate r does for exponential functions $Q(t) = Q_0e^{-rt}$. Sure enough, we note from the graphs shown that b has no effect on the initial value or the eventual capacity, but causes the population to approach this capacity more quickly for larger values of b, and more slowly for smaller values of b. For the screens shown, $P(t) = \dfrac{1000}{1 + 50e^{-1x}}$ ($a = 50$, $b = 1$, and $c = 1000$) is graphed using a lighter line. For comparison, the graph of $P(t) = \dfrac{1000}{1 + 50e^{-1.2x}}$ ($a = 50$, $b = 1.2$, and $c = 1000$) is shown in bold in Figure 5.39, while the graph of $P(t) = \dfrac{1000}{1 + 50e^{-0.8x}}$ ($a = 50$, $b = 0.8$, and $c = 1000$) is shown in bold in Figure 5.40.

Figure 5.39

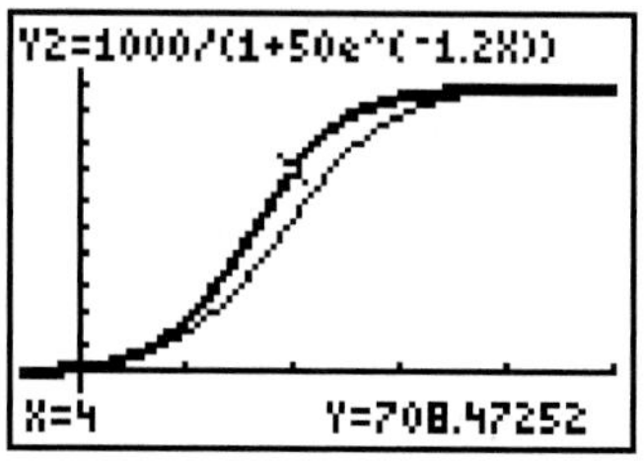

Y2=1000/(1+50e^(-1.2X))
X=4 Y=708.47252

Figure 5.40

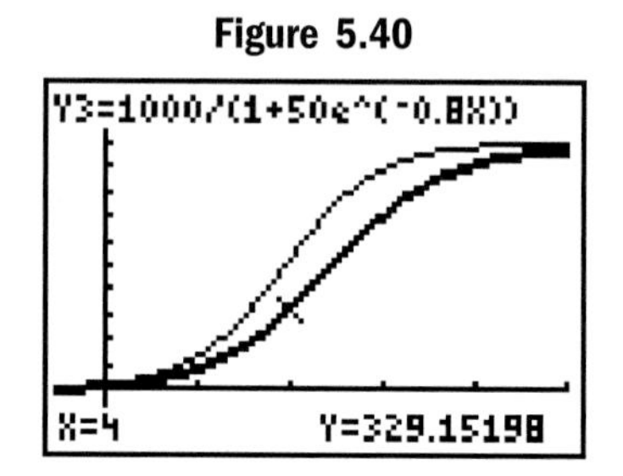

The following exercises are based on the population of an ant colony, modeled by the logistic function $P(t) = \dfrac{2500}{1 + 25e^{-0.5x}}$. Respond to Exercises 1 through 6 without the use of a calculator.

1. Identify the values of a, b, and c for this logistics curve. $a = 25$ $b = 0.5$ $c = 2500$

2. What was the approximate initial population of the colony? 96 ants

3. Which gives a larger initial population: (a) $c = 2500$ and $a = 25$ or (b) $c = 3000$ and $a = 15$? b

4. What is the maximum population capacity for this colony? 2500

5. Would the population of the colony surpass 2000 more quickly if $b = 0.6$ or if $b = 0.4$? $b = 0.6$

6. Which causes a slower population growth: (a) $c = 2000$ and $a = 25$ or (b) $c = 3000$ and $a = 25$? a

7. Verify your responses to Exercises 2 through 6 using a graphing calculator.

STRENGTHENING CORE SKILLS

More on Solving Exponential and Logarithmic Equations

In order to more effectively solve exponential and logarithmic equations, it might help to see how the general process of equation solving applies to a wide range of equation types. Consider the following sequence of equations:

$$2x + 3 = 11 \qquad 2|x| + 3 = 11$$
$$2\sqrt{x} + 3 = 11 \qquad 2e^x + 3 = 11$$
$$2x^2 + 3 = 11 \qquad 2\log x + 3 = 11$$

Each of these equations can be represented generally as $2f(x) + 3 = 11$, where we note that the process of equation solving applies identically to all: first subtract 3, then divide by 2, and finally apply the appropriate inverse function. If we do this symbolically, we have

$$\begin{aligned} 2f(x) + 3 &= 11 && \text{general equation} \\ 2f(x) &= 8 && \text{subtract 3} \\ f(x) &= 4 && \text{divide by 2} \\ f^{-1}[f(x)] &= f^{-1}(4) && \text{apply inverse function} \\ x &= f^{-1}(4) && \text{solution} \end{aligned}$$

The importance of this result cannot be overstated, as it tells us that every solution in the previous sequence is represented by $x = f^{-1}(4)$, where f^{-1} is the inverse function appropriate to the equation. This serves as a dramatic reminder that we must *isolate the term or factor that contains the unknown* before attempting to apply the inverse function or solution process. If we isolate an exponential term with either base 10 or base e, the inverse function is a logarithm of base 10 or base e, respectively. This can easily be seen using the power property of logarithms and the fact that $\ln e = 1$ and $\log 10 = 1$:

ILLUSTRATION 1 Solve for x: **a.** $2e^x + 3 = 11$ **b.** $(2)10^x + 3 = 11$

Solution:

a.

$$\begin{aligned} 2e^x + 3 &= 11 && \text{original equation} \\ 2e^x &= 8 && \text{subtract 3} \\ e^x &= 4 && \text{divide by 2} \\ \ln e^x &= \ln 4 && \text{apply } f^{-1}(x) \\ x \ln e &= \ln 4 && \text{power property} \\ x &= \ln 4 && \text{solution} \end{aligned}$$

b.

$$\begin{aligned} (2)10^x + 3 &= 11 \\ (2)10^x &= 8 \\ 10^x &= 4 \\ \log 10^x &= \log 4 \\ x \log 10 &= \log 4 \\ x &= \log 4 \end{aligned}$$

If we isolate an exponential term with base other than 10 or e, the same process is applied using *either* base 10 or base e, with the additional step of dividing by $\log k$ or $\ln k$ to solve for x:

ILLUSTRATION 2 Solve for x: **a.** $(2)5^x + 3 = 11$ **b.** $(2)1.23^x + 3 = 11$

Solution:

a.

$$\begin{aligned} (2)5^x + 3 &= 11 && \text{original equation} \\ (2)5^x &= 8 && \text{subtract 3} \\ 5^x &= 4 && \text{divide by 2} \\ \ln 5^x &= \ln 4 && \text{apply } f^{-1}(x) \\ x \ln 5 &= \ln 4 && \text{power property} \\ x &= \frac{\ln 4}{\ln 5} && \text{solution (exact form)} \end{aligned}$$

b.

$$\begin{aligned} (2)1.23^x + 3 &= 11 \\ (2)1.23^x &= 8 \\ 1.23^x &= 4 \\ \log 1.23^x &= \log 4 \\ x \log 1.23 &= \log 4 \\ x &= \frac{\log 4}{\log 1.23} \end{aligned}$$

In many applications, the coefficients and bases of an exponential equation are not "nice looking" values. In other words, they are often noninteger or even irrational. Regardless, the solution process remains the same and we should try to see that all such equations belong to the same family.

ILLUSTRATION 3 For the St. Louis Cardinals, player salaries on the opening day roster for the 2000–2001 season were closely modeled by the exponential equation $y = (12.268)0.866^x$, where x represents a player's ordinal rank (1st, 2nd, 3rd, etc.) on the salary schedule and y approximates the player's actual salary in millions. According to this model, how many players make more than $5,000,000?

Source: Sports Illustrated at www.sportsillustrated.com/baseball/mlb/news/2001/04/09

Solution:

$y = (12.268)0.866^x$		original model
$5 = (12.268)0.866^x$		substitute $y = 5$
$0.4075644 \approx 0.866^x$		divide by 12.268
$\ln(0.4075644) \approx \ln 0.866^x$		apply natural log to both sides
$\ln(0.4075644) \approx x \ln 0.866$		power property
$\dfrac{\ln 0.4075644}{\ln 0.866} \approx x$		solve for x (divide by ln 0.866)
$6.24 \approx x$		simplify

According to this model, six or seven players make more than \$5,000,000.

Exercise 1: For the New York Yankees, player salaries on the opening day roster for the 2000–2001 season were closely modeled by the exponential equation $y = (21.303)0.842^x$, where x represents a player's ordinal rank (1st, 2nd, 3rd, etc.) on the salary schedule and y approximates the player's actual salary in millions. According to this model, how many players make more than \$5 million?

Source: Sports Illustrated at www.sportsillustrated.com/baseball/mlb/news/2001/04/09

between eight and nine players

CUMULATIVE REVIEW CHAPTERS 1-5

Use the quadratic formula to solve for x.

1. $x^2 - 4x + 53 = 0$ $x = 2 \pm 7i$

2. $6x^2 + 19x = 36$ $x = \frac{4}{3}, x = \frac{-9}{2}$

3. Use substitution to show that $4 + 5i$ is is a zero of $f(x) = x^2 - 8x + 41$.

4. Graph using transformations of a basic function: $y = 2\sqrt{x + 2} - 3$.

5. Find $(f \circ g)(x)$ and $(g \circ f)(x)$ and comment on what you notice: $f(x) = x^3 - 2$; $g(x) = \sqrt[3]{x + 2}$.

6. State the domain of $h(x)$ in interval notation: $h(x) = \dfrac{\sqrt{x+3}}{x^2 + 6x + 8}$. $x \in [-3, -2) \cup (-2, \infty)$

7. According to the 2002 *National Vital Statistics Report* (Vol. 50, No. 5, page 19) there were 3100 sets of triplets born in the United States in 1991, and 6740 sets of triplets born in 1999. Assuming the relationship (year, sets of triplets) is linear: (a) find the equation of the line, (b) explain the meaning of the slope in this context, and (c) use the equation to estimate the number of sets born in 1996, and to project the number of sets that will be born in 2007 if this trend continues.

8. State the following geometric formulas:

a. area of a circle $A = \pi r^2$

b. Pythagorean theorem $a^2 + b^2 = c^2$

c. perimeter of a rectangle $P = 2L + 2W$

d. area of a trapezoid $A = \frac{h}{2}(B + b)$

9. Graph the following piecewise-defined function and state its domain, range, and intervals where it is increasing and decreasing.

$$h(x) = \begin{cases} -4 & -10 \le x < -2 \\ -x^2 & -2 \le x < 3 \\ 3x - 18 & x \ge 3 \end{cases}$$

10. Solve the inequality and write the solution in interval notation: $\dfrac{2x+1}{x-3} \ge 0$. $x \in (-\infty, -\frac{1}{2}] \cup (3, +\infty)$

11. Use the rational roots theorem to find all zeroes of $f(x) = x^4 - 3x^3 - 12x^2 + 52x - 48$. $x = 3$, $x = 2$ (multiplicity 2); $x = -4$

12. Given $f(c) = \frac{9}{5}c + 32$, find k, where $k = f(25)$. Then find the inverse function using the algebraic method, and verify that $f^{-1}(k) = 25$.

13. Solve the formula $V = \frac{1}{2}\pi b^2 a$ (the volume of a paraboloid) for the variable b. $\sqrt{\dfrac{2V}{\pi a}} = b$

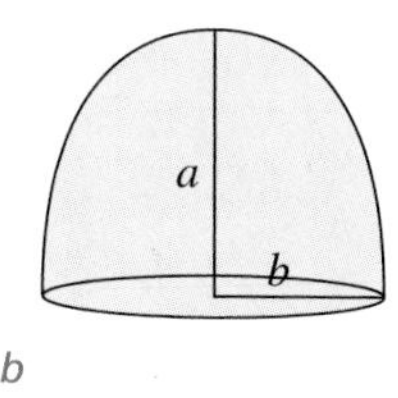

Answers:

3. $(4 + 5i)^2 - 8(4 + 5i) + 41 = 0$
$-9 + 40i - 32 - 40i + 41 = 0$
$0 = 0$ ✓

4.

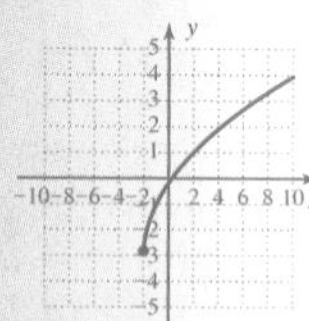

5. $f(g(x)) = x$
$g(f(x)) = x$
Since $(f \circ g)(x) = (g \circ f)(x)$, they are inverse functions.

7. a. $T(t) = 455t + 2645$ $(1991 \rightarrow \textit{year } 1)$

b. $\frac{\Delta T}{\Delta t} = \frac{455}{1}$, triple births increase by 455 each year

c. $T(6) = 5375$ sets of triplets, $T(17) = 10{,}380$ sets of triplets

9.
$D: x \in [-10, \infty)$, $R: y \in [-9, \infty)$
$h(x)\uparrow: x \in (-2, 0) \cup (3, \infty)$
$h(x)\downarrow: x \in (0, 3)$

12. $k = 77$
$f^{-1}(c) = \frac{5}{9}(f - 32)$
$f^{-1}(77) = \frac{5}{9}(77 - 32)$
$= 25$ ✓

14. Use the *Guidelines for Graphing* to graph the polynomial $p(x) = x^3 - 4x^2 + x + 6$.

15. Use the *Guidelines for Graphing* to graph the rational function $r(x) = \frac{5x^2}{x^2 + 4}$.

16. Solve for x: $10 = -2e^{-0.05x} + 25$.

17. Solve for x: $\ln(x + 3) + \ln(x - 2) = \ln(24)$.

18. Once in orbit, satellites are often powered by radioactive isotopes. From the natural process of radioactive decay, the power output declines over a period of time. For an initial amount of 50 g, suppose the power output is modeled by the function $p(t) = 50e^{-0.002t}$, where $p(t)$ is the power output in watts, t days after the satellite has been put into service. (a) Approximately how much power remains 6 *months* later? (b) How many *years* until only one-fourth of the original power remains?

After reading a report from The National Center for Health Statistics regarding the growth of children from age 0 to 36 months, Maryann decides to track the relationships (length in inches, weight in pounds) and (age in months, circumference of head in centimeters) for her newborn child, a beautiful baby girl—Morgan.

19. For the (length, weight) data given: (a) draw a scatter-plot, decide on an appropriate form of regression, and find a regression equation; (b) use the equation to find Morgan's weight when she reaches a height (length) of 39 in.; and (c) determine her length when she attains a weight of 28 lb.

Exercise 19

Length (inches)	Weight (pounds)
17.5	5.50
21	10.75
25.5	16.25
28.5	19.00
33	25.25

Exercise 20

Age (months)	Circumference (centimeters)
1	38.0
6	44.0
12	46.5
18	48.0
21	48.3

20. For the (age, circumference) data given (a) draw a scatter-plot, decide on an appropriate form of regression, and find a regression equation; (b) use the equation to find the circumference of Morgan's head when she reaches an age of 27 months; and (c) determine the age when the circumference of her head reaches 50 cm.

14.

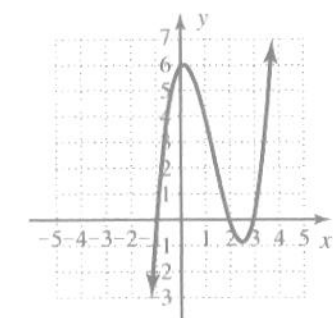

15.

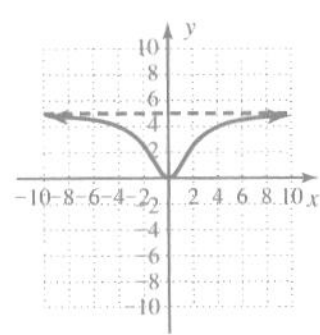

16. $x \approx -40.298$

17. $x = 5$, $x = -6$ is an extraneous root

18. a. $P(183) \approx 34.7$ W

b. $P(t) = \frac{50}{4}$

$t \approx 693$ days

Approx. 1 yr 11 months

19.

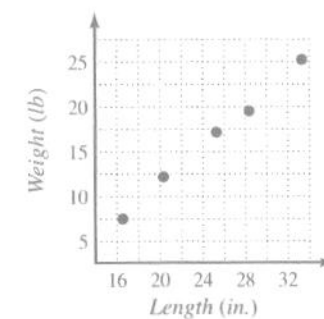

a. linear
$W = 1.24L - 15.83$

b. 32.5 lb

c. 35.3 in.

20.

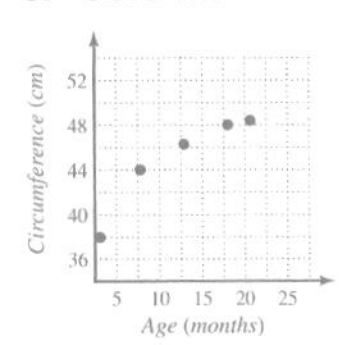

a. $C(a) \approx 37.9694 + 3.4229\ln(a)$

b. about 49.3 cm

c. about 34 mo

Chapter 6

Systems of Equations and Inequalities

Chapter Outline

6.1 Linear Systems in Two Variables with Applications 566

6.2 Linear Systems in Three Variables with Applications 577

6.3 Systems of Inequalities and Linear Programming 590

6.4 Systems and Absolute Value Equations and Inequalities 604

6.5 Solving Linear Systems Using Matrices and Row Operations 616

6.6 The Algebra of Matrices 627

6.7 Solving Linear Systems Using Matrix Equations 640

6.8 Matrix Applications: Cramer's Rule, Partial Fractions, and More 655

Preview

A delivery service operates a fleet of trucks in a large metropolitan area. How can the company maximize the number of packages it delivers, given the trucks are limited by both the weight and volume of packages they carry? The answer can be found using a technique called **linear programming,** which involves solving a **system of linear inequalities.** Other applications of systems include the timing of traffic lights, the optimal routing of phone calls over a network, and business decisions that affect a company's revenue. We begin by considering a system of two linear equations with two variables, also called a **system of simultaneous equations.**

6.1 Linear Systems in Two Variables with Applications

LEARNING OBJECTIVES

In Section 6.1 you will learn how to:

- **A. Verify ordered pair solutions**
- **B. Solve linear systems by graphing**
- **C. Solve linear systems by substitution**
- **D. Solve linear systems by elimination**
- **E. Recognize inconsistent systems (no solutions) and dependent systems (infinitely many solutions)**
- **F. Use a system of equations to mathematically model and solve applications**

INTRODUCTION

In earlier chapters we used linear equations in two variables to model a number of real-world situations. Graphing these equations gave us a visual picture of how the variables were related, and helped us investigate and explore the relationship. Frequently two (or more) equations are used to model more than one relationship between two variables, leading to a **linear system of two equations in two unknowns,** often called a **2 × 2 system** (two-by-two system).

POINT OF INTEREST

Without the aid of a computer, larger systems of equations (many equations with many unknowns) are difficult to solve. In 1946, engineers at the University of Pennsylvania built the first digital computer controlled by vacuum tubes. This computer contained about 17,500 tubes and filled a large room, and was first used by the U.S. Army to solve artillery problems during World War II. Today's computers can quickly solve very large systems, like those used by the federal government for economic forecasts.

A. Solutions to a System of Equations

A **system of equations** is a set of two or more equations to be considered simultaneously. They arise very naturally in a number of contexts, when two or more equation models can be formed using different measures of the same variable quantity. For example, consider an amusement park that brought in \$3660 in revenue one day by charging \$9.00 for adults and \$5.00 for children and selling 560 total tickets. Using a for adult and c for children, we could write one equation modeling the number of tickets sold—$a + c = 560$—and a second modeling the amount of revenue brought in—$9a + 5c = 3660$. To show that we're considering both equations simultaneously as a model of the situation, a large "left brace" is used: $\begin{cases} a + c = 560 \\ 9a + 5c = 3660 \end{cases}$. Note both equations in the system are linear and, by careful inspection, that they have different slopes. If two lines with different slopes are graphed on a coordinate grid, they will intersect at some point. Since every point on a line satisfies the equation of that line, this point of intersection must satisfy both equations simultaneously and is called the **solution to the system.** In general, solutions to 2 × 2 systems of equations are ordered pairs that satisfy all equations in the system.

EXAMPLE 1 Verify that (215, 345) is a solution to $\begin{cases} a + c = 560 \\ 9a + 5c = 3660 \end{cases}$.

Solution: Substitute 215 for a and 345 for c in each equation.

$$a + c = 560 \quad \text{first equation} \qquad 9a + 5c = 3660 \quad \text{second equation}$$

$$(215) + (345) = 560 \qquad 9(215) + 5(345) = 3660$$

$$560 = 560 ✓ \qquad 3660 = 3660 ✓$$

Since (215, 345) satisfies both equations, it is a solution for the system and we find the amusement park sold 215 adult and 345 youth tickets.

NOW TRY EXERCISES 7 THROUGH 18

B. Solving Systems Graphically

To **solve a system of equations** means we apply various methods in an attempt to find ordered pair solutions. For a 2×2 system, solutions occur where two lines intersect, since every point on a line must satisfy the equation of that line. It seems reasonable that we first investigate finding solutions by *graphing.* Any method for graphing the lines can be used, but to keep important concepts fresh, the slope-intercept method is used here.

EXAMPLE 2 Solve the system by graphing: $\begin{cases} 4x - 3y = 9 \\ -2x + y = -5 \end{cases}$.

Solution: First we write each equation in slope-intercept form (solve for y):

$$\begin{cases} 4x - 3y = 9 \\ -2x + y = -5 \end{cases} \longrightarrow \begin{cases} y = \frac{4}{3}x - 3 \\ y = 2x - 5 \end{cases}$$

For the first line, $\frac{\Delta y}{\Delta x} = \frac{4}{3}$ with y-intercept $(0, -3)$. The second equation yields $\frac{\Delta y}{\Delta x} = \frac{2}{1}$ with $(0, -5)$ as the y-intercept. Both are then graphed on the grid as shown. The point of intersection appears to be $(3, 1)$, and checking this point in both equations gives

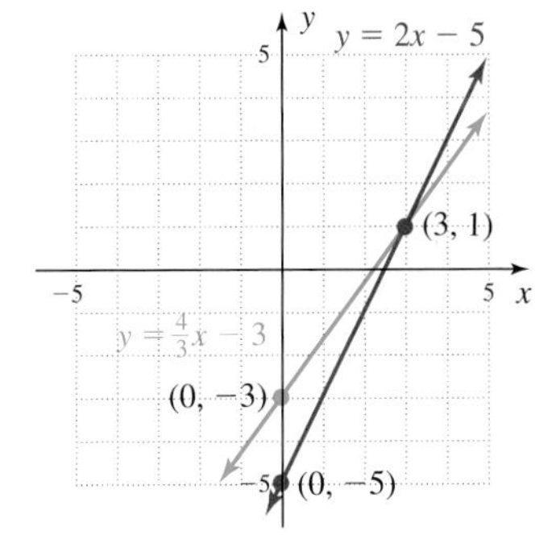

$$4x - 3y = 9 \qquad\qquad -2x + y = -5$$

$$4(3) - 3(1) = 9 \quad \text{substitute 3 for } x \text{ and 1 for } y \quad -2(3) + (1) = -5$$

$$9 = 9 \checkmark \qquad\qquad -5 = -5 \checkmark$$

This shows $(3, 1)$ is the solution to the system.

NOW TRY EXERCISES 19 THROUGH 22

C. Solving Systems by Substitution

While a graphical approach best illustrates *why* the solution must be an ordered pair, it does have one obvious drawback—noninteger solutions are difficult to spot. The ordered pair $\left(\frac{2}{5}, \frac{12}{5}\right)$ is the solution to $\begin{cases} 4x + y = 4 \\ y = x + 2 \end{cases}$, but this would be difficult to "pinpoint" as a precise location on a hand-drawn graph. To overcome this limitation, we next consider a nongraphical method known as **substitution.** The method involves converting a system of two equations in two variables into a single equation in one variable by using an appropriate replacement for one of the variables. For $\begin{cases} 4x + y = 4 \\ y = x + 2 \end{cases}$, the second equation says "y is two more than x." This means ordered pairs such as $(4, 6)$, $(0, 2)$, and $(-7, -5)$ are solutions. We reason that *all* points on this line are related this way, *including the point where this line intersects the other.* For this reason, we can substitute $x + 2$ directly for y in the first equation, obtaining a single equation in x.

EXAMPLE 3 Solve using substitution: $\begin{cases} 4x + y = 4 \\ y = x + 2 \end{cases}$.

Solution: Since $y = x + 2$, we can replace y with $x + 2$ in the first equation.

$$\begin{aligned} 4x + y &= 4 && \text{first equation} \\ 4x + (x + 2) &= 4 && \text{substitute } x + 2 \text{ for } y \\ 5x + 2 &= 4 && \text{simplify} \\ x &= \frac{2}{5} && \text{result} \end{aligned}$$

The x-coordinate is $\frac{2}{5}$. To find the y-coordinate, substitute $\frac{2}{5}$ for x into either of the original equations. Substituting in the second equation gives

$$\begin{aligned} y &= x + 2 && \text{second equation} \\ &= \frac{2}{5} + 2 && \text{substitute } \frac{2}{5} \text{ for } x \\ &= \frac{12}{5} && \frac{2}{1} = \frac{10}{5}, \frac{10}{5} + \frac{2}{5} = \frac{12}{5} \end{aligned}$$

The solution to the system is $\left(\frac{2}{5}, \frac{12}{5}\right)$. Verify by substituting $\frac{2}{5}$ for x and $\frac{12}{5}$ for y into both equations.

NOW TRY EXERCISES 23 THROUGH 32

If neither equation allows an immediate substitution, we first solve for one of the variables, either x or y, and *then* substitute. After all, the equation $x - y = -2$ can be written $x = y - 2$ or $y = x + 2$ and still describe the same relationship between x and y. The substitution method is summarized here.

SOLVING A SYSTEM OF EQUATIONS USING SUBSTITUTION

1. Solve one of the equations for x in terms of y or y in terms of x.
2. Substitute for the appropriate variable in the *other* equation.
3. Solve the resulting equation for the remaining variable. (This will give you one coordinate of the ordered pair solution.)
4. Substitute the value from step 3 back into either of the original equations to determine the value of the other coordinate.
5. Write the answer as an ordered pair and check.

D. Solving Systems Using Elimination

Now consider the system $\begin{cases} 2x - 3y = 5 \\ 6y + 5x = 4 \end{cases}$, where solving for any one of the variables will result in fractional values. The substitution method can still be used, but often the **elimination method** is more efficient. In addition, many future topics are based on this method. The elimination method takes its name from what happens when you add certain equations in a system (by adding the like terms from each). If the coefficients of either x or y are additive inverses—they sum to zero and are *eliminated.* When neither of the variable terms meet this condition, we multiply one or both equations by some non-zero constant to "match up" the coefficients, so that an elimination will take place. Multiplying both sides of an equation by a constant term (other than zero) does not change the solution set. The ordered pairs that satisfy $2x - 3y = 7$ will also satisfy $4x - 6y = 14$, the equation obtained by multiplying both sides by 2, since they are **equivalent equations.** Before beginning a solution using elimination, check to make sure the equations are written in the **standard form** $Ax + By = C$, so that like terms will appear above/below each other. Throughout this chapter, we will use R1 to represent the equation in *row 1* of the system, R2 to represent the equation in *row 2*, and so on. These designations are used to help describe and document the steps being used to solve a system, as in Example 4.

EXAMPLE 4 Solve using elimination: $\begin{cases} 2x - 3y = 7 \\ 6y + 5x = 4 \end{cases}$

Solution: The second equation is not in standard form, so we re-write the system as $\begin{cases} 2x - 3y = 7 \\ 5x + 6y = 4 \end{cases}$. If we "add the equations" now, we would get $7x + 3y = 11$, with neither variable eliminated. However, if we multiply *both sides* of the first equation by 2, the y-coefficients will be additive inverses. The sum then results in an equation with x as the only unknown.

$$\begin{array}{c} 2\text{R1} \\ + \\ \text{R2} \\ \text{sum} \end{array} \begin{cases} 4x - 6y = 14 \\ 5x + 6y = 4 \end{cases}$$
$$9x + 0y = 18$$
$$x = 2 \quad \text{solve for } x$$

Substituting 2 for x back into either of the original equations yields $y = -1$. The ordered pair solution is $(2, -1)$. Verify using the original equations.

NOW TRY EXERCISES 33 THROUGH 38

The elimination method is summarized here. If either equation has fraction or decimal coefficients, we can "clear" them by using an appropriate multiplier.

SOLVING A SYSTEM OF EQUATIONS USING ELIMINATION

1. Be sure each equation is in standard form: $Ax + By = C$.
2. If desired or needed, clear fractions or decimals using a multiplier.
3. Multiply one or both equations by a number that will create coefficients of x (or y) that are additive inverses.
4. Combine the two equations using vertical addition.
5. Solve the resulting equation for the remaining variable.
6. Substitute this x- or y-value back into either of the original equations and solve for the other unknown.
7. Write the answer as an ordered pair and check the solution in both original equations.

EXAMPLE 5 Solve the system using elimination: $\begin{cases} \frac{5}{8}x - \frac{3}{4}y = \frac{1}{4} \\ \frac{1}{2}x = \frac{2}{3}y + 1 \end{cases}$.

Solution: Multiplying the first equation by 8 and the second equation by 6 will clear the fractions from each. We symbolize this using 8R1 and 6R2 respectively, and the result is $\begin{cases} 5x - 6y = 2 \\ 3x - 4y = 6 \end{cases}$. The x-terms can be eliminated if we now use 3R1 and −5R2.

$$\begin{array}{r} 3R1 \\ + \\ -5R2 \end{array} \begin{cases} 15x - 18y = 6 \\ -15x + 20y = -30 \end{cases}$$
$$\text{sum} \quad 0x + 2y = -24$$
$$y = -12 \quad \text{solve for } y$$

Substituting $y = -12$ in either of the original equations yields $x = -14$, and the solution is $(-14, -12)$. Verify by substituting in both equations.

NOW TRY EXERCISES 39 THROUGH 44

CAUTION

Be sure to multiply *all* terms of the equation when using a constant multiplier. Also, note that for Example 5, we could have eliminated the y-terms using 2R1 with −3R2.

E. Inconsistent and Dependent Systems

A system having *at least one* solution is called a **consistent system.** As seen in Example 2, if the lines have different slopes, they intersect at a single point. In some sense, the lines are *independent* of each other and the system is called a **consistent/independent system.** If the lines have equal slopes *and* the same y-intercept, they are identical or **coincident lines.** Since one is right atop the other, they *intersect at all points,* with an infinite number of solutions. In a sense, one equation *depends* on the other and the system is called a **consistent/dependent system.** Using either substitution or elimination on a consistent/dependent system results in the elimination of all variable terms and leaves a statement that is *always true,* such as $0 = 0$ or some other simple equality. Finally, if the lines have equal slopes *but different y-intercepts,* they are parallel and the system will have no solution. A system with no solutions is called an **inconsistent system.** Whether using substitution or elimination, an "inconsistent system" produces an "inconsistent answer," such as $12 = 0$ or some other false statement. In other words, all variable terms are once again eliminated, but the remaining statement is *false.* A summary of the three possibilities is given graphically here.

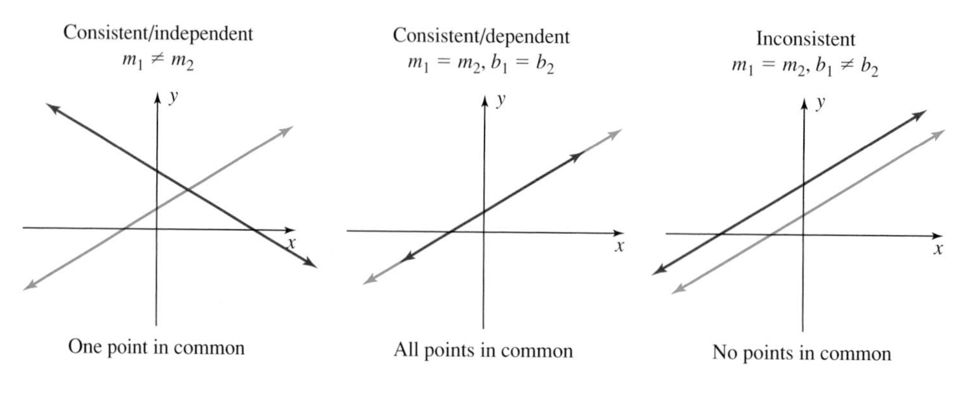

EXAMPLE 6 Solve using elimination: $\begin{cases} 3x + 4y = 12 \\ 6x = 24 - 8y \end{cases}$.

Solution: Writing the system in standard form gives $\begin{cases} 3x + 4y = 12 \\ 6x + 8y = 24 \end{cases}$.

By applying $-2R1$, we can eliminate the variable x:

$$\begin{array}{r} -2R1 \\ + \\ R2 \\ \hline \text{sum} \end{array} \begin{cases} -6x - 8y = -24 \\ 6x + 8y = 24 \end{cases}$$

$$0x + 0y = 0 \quad \text{variables are eliminated}$$

$$0 = 0 \quad \text{true statement}$$

Although we didn't expect it, both variables were eliminated and the final statement is true ($0 = 0$). This indicates the system is consistent/dependent. The graph given here shows the lines are coincident. Writing both equations in slope-intercept form verifies they represent the same line.

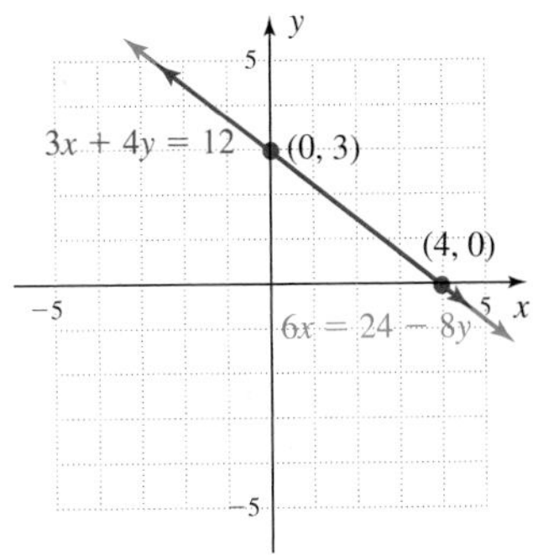

$$\begin{cases} 3x + 4y = 12 \\ 6x + 8y = 24 \end{cases} \longrightarrow \begin{cases} 4y = -3x + 12 \\ 8y = -6x + 24 \end{cases} \longrightarrow \begin{cases} y = -\dfrac{3}{4}x + 3 \\ y = -\dfrac{3}{4}x + 3 \end{cases}$$

NOW TRY EXERCISES 45 THROUGH 56

The solutions of a consistent/dependent system are often written in set notation as the set of ordered pairs (x, y), where y is a specified function of x. For Example 6 the solution would be $\{(x, y) \mid y = -\frac{3}{4}x + 3\}$. Using an ordered pair with an arbitrary variable is also common: $\left(a, \dfrac{-3a}{4} + 3\right)$.

F. Systems and Modeling

In previous chapters we solved numerous real-world problems by writing all unknowns in terms of a single variable. Many situations are easier to model and solve using a system of equations and more than one variable. In these cases, carefully read the description of how two quantities are related, then model each relationship given using an equation in two variables. These relationships can be given in many different ways and the exercise set contains a wide variety of applications. We end this section with a **mixture** application. Although they appear in many different forms (coin problems, metal alloys, investments, merchandising, and so on), mixture problems all have a similar theme. Generally one equation will be related to quantity (how much of each quantity is being combined) and one equation will be related to value (what is the value of each item being combined).

EXAMPLE 7 A jeweler is commissioned to create a piece of artwork that will weigh 14 oz and consist of 75% gold. She has on hand two alloys that are 60% and 80% gold, respectively. How much of each should she use?

Solution: Let x represent ounces of the 60% alloy and y represent ounces of the 80% alloy. The first equation must be $x + y = 14$, since the piece of art must weigh exactly 14 oz (this is the *quantity* equation). The x ounces are 60% gold, the y ounces are 80% gold, and the 14 oz are 75% gold. This gives the *value* equation: $0.6x + 0.8y = 0.75(14)$. The system is $\begin{cases} x + y = 14 \\ 6x + 8y = 105 \end{cases}$ (after clearing decimals). As an estimation tool, note that if equal amounts of the 60% and 80% alloys were used (7 oz each), the result would be a 70% alloy (halfway in between). Since a 75% alloy is needed, more of the 80% gold will be used. Solving for y in the first equation gives $y = 14 - x$. Substituting $14 - x$ for y in the second equation gives

$$\begin{aligned} 6x + 8y &= 105 && \text{second equation} \\ 6x + 8(14 - x) &= 105 && \text{substitute } 14 - x \text{ for } y \\ -2x + 112 &= 105 && \text{simplify} \\ x &= \frac{7}{2} && \text{solve for } x \end{aligned}$$

Substituting $\frac{7}{2}$ for x in the first equation gives $y = \frac{21}{2}$. She should use 3.5 oz of the 60% alloy and 10.5 oz of the 80% alloy.

NOW TRY EXERCISES 63 THROUGH 74

TECHNOLOGY HIGHLIGHT

Solving Systems Graphically

The keystrokes shown apply to a TI-84 Plus model. Please consult your manual or our Internet site for other models.

When used with care, graphing calculators offer an accurate way to solve linear systems and to check the solution(s) obtained by hand. Begin by solving for y in both equations, carefully enter them on the Y= screen as Y_1 and Y_2. Then press ZOOM 6 to graph these equations on the standard screen (in this section, both equations will appear on the standard screen and window size is not an issue (see Section 2.1—*Technology Highlight*). To have the calculator compute the point of intersection, press 2nd CALC and select option **5: intersect** by pressing the number 5 or scrolling down to this option.

Figure 6.1

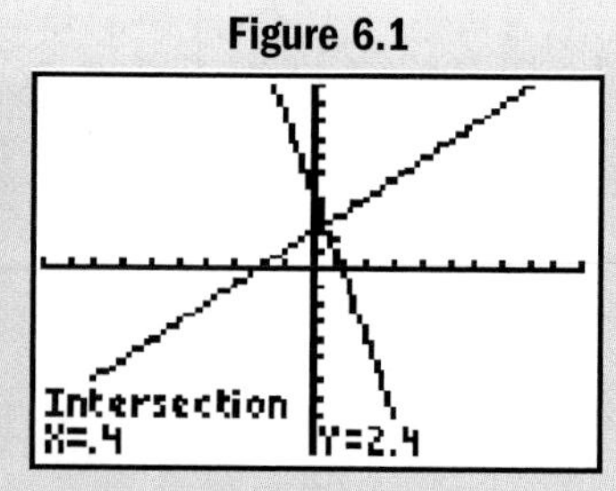

Solution appears at bottom.

Then press ENTER *three* times: The first " ENTER " selects Y_1, the second " ENTER " selects Y_2, and the third " ENTER " bypasses the "GUESS" option (this option is most often used if the graphs from the system intersect at more than one point). The x- and y-coordinates of the solution will appear at the bottom of the screen (see Figure 6.1). We illustrate by confirming the solution to Example 3: $\begin{cases} 4x + y = 4 \\ y = x + 2 \end{cases}$, which we found was $\left(\frac{2}{5}, \frac{12}{5}\right)$.

1. Solve for y in both equations:
$$\begin{cases} y = -4x + 4 \\ y = x + 2 \end{cases}$$
2. Enter the equations as
$Y_1 = -4x + 4$
$Y_2 = x + 2$
3. Press ZOOM 6 to graph the equations.
4. Press 2nd CALC 5 ENTER ENTER ENTER to have the calculator compute the point of intersection.

The coordinates of the point of intersection appear as decimal fractions at the bottom of the screen. The calculator automatically registers the x-coordinate as its most recent entry, and can be converted to a standard fraction by going to the home screen and pressing MATH 1: ▶Frac ENTER .

Note: You can get an *approximate solution* by tracing along either line toward the point of intersection using the TRACE key and the left or right arrows. If you suspect the solution is integer-valued, you can check the nearest integer-valued ordered pair in the original equations.

Solve each system graphically using a graphing calculator.

Exercise 1: $\begin{cases} 3x - y = -7 \\ y + 5x = -1 \end{cases}$ $(-1, 4)$

Exercise 2: $\begin{cases} 2x - 3y = 3 \\ 6 = 8x - 3y \end{cases}$ $\left(\frac{1}{2}, \frac{-2}{3}\right)$

6.1 EXERCISES

CONCEPTS AND VOCABULARY

Fill in the blank with the appropriate word or phrase. Carefully reread the section if needed.

1. Systems that have no solution are called inconsistent systems.

2. Systems having at least one solution are called consistent systems.

3. If the lines in a system intersect at a single point, the system is said to be consistent and independent.

4. If the lines in a system are coincident, the system is referred to as consistent and dependent.

5. The given systems are equivalent. How do we obtain the second system from the first?

$$\begin{cases} \frac{2}{3}x + \frac{1}{2}y = \frac{5}{3} \\ 0.2x + 0.4y = 1 \end{cases} \qquad \begin{cases} 4x + 3y = 10 \\ 2x + 4y = 10 \end{cases}$$

Multiply the first equation by 6 and the second equation by 10.

6. For $\begin{cases} 2x + 5y = 8 \\ 3x + 4y = 5 \end{cases}$, which solution method would be more efficient, substitution or elimination? Discuss/explain why. Answers will vary.

DEVELOPING YOUR SKILLS

Show the lines in each system would intersect in a single point by writing the equations in slope-intercept form.

7. $\begin{cases} 7x - 4y = 24 \\ 4x + 3y = 15 \end{cases}$ $y = \frac{7}{4}x - 6,$ $y = \frac{-4}{3}x + 5$

8. $\begin{cases} 0.3x - 0.4y = 2 \\ 0.5x + 0.2y = -4 \end{cases}$ $y = \frac{3}{4}x - 5,$ $y = \frac{-5}{2}x - 20$

An ordered pair is a solution to an equation if it makes the equation true. Given the graph shown here, determine which equation(s) have the indicated point as a solution. If the point satisfies more than one equation, write the system for which it is a solution.

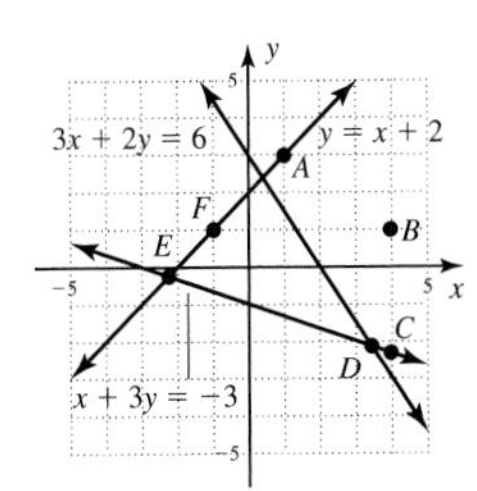

9. *A* $y = x + 2$

10. *B* none

11. *C* $x + 3y = -3$

12. *D* $x + 3y = -3,$ $3x + 2y = 6$

13. *E* $y = x + 2,$ $x + 3y = -3$

14. *F* $y = x + 2$

19. **20.**

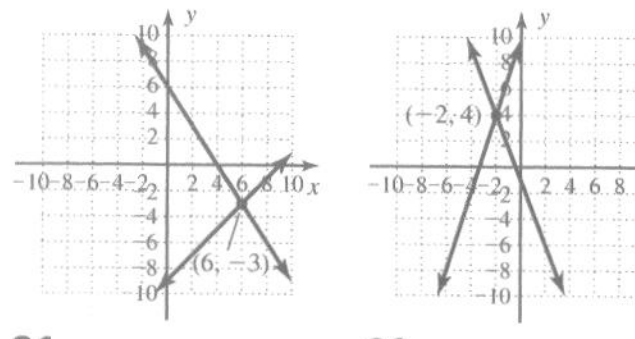

21. **22.**

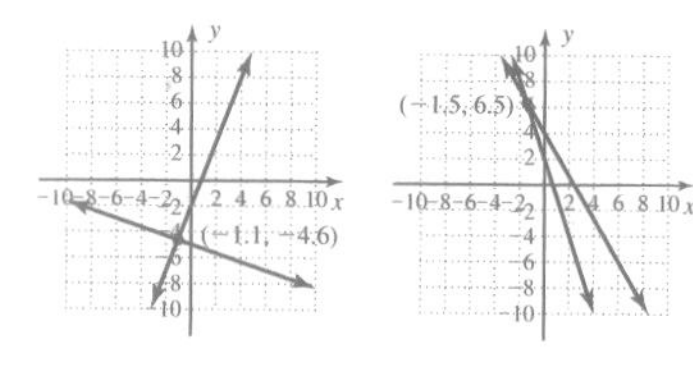

23. $(-4, 1)$
24. $(3, 1)$
25. $(3, -5)$
26. $(4, 2)$
27. second equation, y, $(4, -3)$
28. second equation, x, $(5, 2)$
29. second equation, x, $(10, -1)$
30. first equation, y, $(3, 5)$
31. second equation, x, $\left(\frac{5}{2}, \frac{7}{4}\right)$
32. second equation, y, $\left(\frac{5}{6}, \frac{2}{3}\right)$
33. $(3, -1)$
34. $(2, 2)$
35. $(-2, -3)$
36. $(5, 2)$
37. $\left(\frac{11}{2}, 2\right)$
38. $\left(1, \frac{7}{2}\right)$
39. $(-2, 3)$
40. $(7, -2)$
41. $(-3, 4)$
42. $(0.5, 2)$
43. $(-6, 12)$
44. $(4, -15)$
45. $(2, 8)$; consistent/independent
46. $(6, -2)$; consistent/independent
47. $\varnothing$; inconsistent
48. $\varnothing$; inconsistent
49. $\{(x, y) | 6x - 22 = -y\}$; consistent/dependent
50. $\{(x, y) | 9x - 5y = -15\}$; consistent/dependent
51. $(4, 1)$; consistent/independent
52. $(5, -2)$; consistent/independent
53. $(-3, -4)$; consistent/independent
54. $(2, 1)$; consistent/independent
55. $\left(\frac{-1}{2}, \frac{4}{3}\right)$; consistent/independent
56. $\left(\frac{1}{3}, \frac{-3}{2}\right)$; consistent/independent

Substitute the x- and y-values indicated by the ordered pair to determine if it solves the system.

15. $\begin{cases} 3x + y = 11 \\ -5x + y = -13; \end{cases}$ $(3, 2)$ yes

16. $\begin{cases} 3x + 7y = -4 \\ 7x + 8y = -21; \end{cases}$ $(-6, 2)$ no

17. $\begin{cases} 8x - 24y = -17 \\ 12x + 30y = 2; \end{cases}$ $\left(-\frac{7}{8}, \frac{5}{12}\right)$ yes

18. $\begin{cases} 4x + 15y = 7 \\ 8x + 21y = 11; \end{cases}$ $\left(\frac{1}{2}, \frac{1}{3}\right)$ yes

Solve each system by *graphing*. If the solution does not appear to be a lattice point, estimate the solution to the nearest tenth (indicate that your solution is an estimate).

19. $\begin{cases} 3x + 2y = 12 \\ x - y = 9 \end{cases}$

20. $\begin{cases} 5x + 2y = -2 \\ -3x + y = 10 \end{cases}$

21. $\begin{cases} 5x - 2y = 4 \\ x + 3y = -15 \end{cases}$

22. $\begin{cases} 3x + y = 2 \\ 5x + 3y = 12 \end{cases}$

Solve each system using *substitution*. Write solutions as an ordered pair.

23. $\begin{cases} x = 5y - 9 \\ x - 2y = -6 \end{cases}$

24. $\begin{cases} 4x - 5y = 7 \\ 2x - 5 = y \end{cases}$

25. $\begin{cases} y = \frac{2}{3}x - 7 \\ 3x - 2y = 19 \end{cases}$

26. $\begin{cases} 2x - y = 6 \\ y = \frac{3}{4}x - 1 \end{cases}$

Identify the equation and variable that makes the substitution method easiest to use. Then solve the system.

27. $\begin{cases} 3x - 4y = 24 \\ 5x + y = 17 \end{cases}$

28. $\begin{cases} 3x + 2y = 19 \\ x - 4y = -3 \end{cases}$

29. $\begin{cases} 0.7x + 2y = 5 \\ x - 1.4y = 11.4 \end{cases}$

30. $\begin{cases} 0.8x + y = 7.4 \\ 0.6x + 1.5y = 9.3 \end{cases}$

31. $\begin{cases} 5x - 6y = 2 \\ x + 2y = 6 \end{cases}$

32. $\begin{cases} 2x + 5y = 5 \\ 8x - y = 6 \end{cases}$

Solve using *elimination*. In some cases, the system must first be written in standard form.

33. $\begin{cases} 2x - 4y = 10 \\ 3x + 4y = 5 \end{cases}$

34. $\begin{cases} -x + 5y = 8 \\ x + 2y = 6 \end{cases}$

35. $\begin{cases} 4x - 3y = 1 \\ 3y = -5x - 19 \end{cases}$

36. $\begin{cases} 5y - 3x = -5 \\ 3x + 2y = 19 \end{cases}$

37. $\begin{cases} 2x = -3y + 17 \\ 4x - 5y = 12 \end{cases}$

38. $\begin{cases} 2y = 5x + 2 \\ -4x = 17 - 6y \end{cases}$

39. $\begin{cases} 0.5x + 0.4y = 0.2 \\ 0.3y = 1.3 + 0.2x \end{cases}$

40. $\begin{cases} 0.2x + 0.3y = 0.8 \\ 0.3x + 0.4y = 1.3 \end{cases}$

41. $\begin{cases} 0.32m - 0.12n = -1.44 \\ -0.24m + 0.08n = 1.04 \end{cases}$

42. $\begin{cases} 0.06g - 0.35h = -0.67 \\ -0.12g + 0.25h = 0.44 \end{cases}$

43. $\begin{cases} -\frac{1}{6}u + \frac{1}{4}v = 4 \\ \frac{1}{2}u - \frac{2}{3}v = -11 \end{cases}$

44. $\begin{cases} \frac{3}{4}x + \frac{1}{3}y = -2 \\ \frac{3}{2}x + \frac{1}{5}y = 3 \end{cases}$

Solve using any method and identify the system as consistent or inconsistent.

45. $\begin{cases} 4x + \frac{3}{4}y = 14 \\ -9x + \frac{5}{8}y = -13 \end{cases}$

46. $\begin{cases} \frac{2}{3}x + y = 2 \\ 2y = \frac{5}{6}x - 9 \end{cases}$

47. $\begin{cases} 0.2y = 0.3x + 4 \\ 0.6x - 0.4y = -1 \end{cases}$

48. $\begin{cases} 1.2x + 0.4y = 5 \\ 0.5y = -1.5x + 2 \end{cases}$

49. $\begin{cases} 6x - 22 = -y \\ 3x + \frac{1}{2}y = 11 \end{cases}$

50. $\begin{cases} 15 - 5y = -9x \\ -3x + \frac{5}{3}y = 5 \end{cases}$

51. $\begin{cases} -10x + 35y = -5 \\ y = 0.25x \end{cases}$

52. $\begin{cases} 2x + 3y = 4 \\ x = -2.5y \end{cases}$

53. $\begin{cases} 7a + b = -25 \\ 2a - 5b = 14 \end{cases}$

54. $\begin{cases} -2m + 3n = -1 \\ 5m - 6n = 4 \end{cases}$

55. $\begin{cases} 4a = 2 - 3b \\ 6b + 2a = 7 \end{cases}$

56. $\begin{cases} 3p - 2q = 4 \\ 9p + 4q = -3 \end{cases}$

57. $\left(-2, \frac{5}{2}\right)$
58. $\left(\frac{3}{4}, -6\right)$
59. $(2, -1)$
60. $(1, 2)$

The substitution method can be used for like variables *or for like expressions.* Solve the following systems, *using the expression* common to both equations (do not solve for x or y alone).

57. $\begin{cases} 2x + 4y = 6 \\ x + 12 = 4y \end{cases}$ **58.** $\begin{cases} 8x = 3y + 24 \\ 8x - 5y = 36 \end{cases}$ **59.** $\begin{cases} 5x - 11y = 21 \\ 11y = 5 - 8x \end{cases}$ **60.** $\begin{cases} -6x = 5y - 16 \\ 5y - 6x = 4 \end{cases}$

WORKING WITH FORMULAS

61. Uniform motion with current: $\begin{cases} (R + C)T_1 = D_1 \\ (R - C)T_2 = D_2 \end{cases}$

The formula shown can be used to solve uniform motion problems involving a *current,* where D represents distance traveled, R is the rate of the object with no current, C is the speed of the current, and T is the time. Chan-Li rows 9 mi up river (against the current) in 3 hr. It only took him 1 hr to row 5 mi downstream (with the current). How fast was the current? How fast can he row in still water? 1 mph 4 mph

62. Fahrenheit and Celsius temperatures: $\begin{cases} y = \frac{9}{5}x + 32 & °F \\ y = \frac{5}{9}(x - 32) & °C \end{cases}$

Many people are familiar with temperature measurement in degrees Celsius and degrees Fahrenheit, but few realize that the equations are linear and there is one temperature at which the two scales agree. Solve the system using the method of your choice and find this temperature. $-40°C = -40°F$

APPLICATIONS

Although some of the following applications are contrived, they offer excellent opportunities for further skill and concept development, and lead to applications that are more substantial and meaningful. Solve each using a linear system. Be sure to clearly indicate what each variable represents.

63. Manufacturing cost: The cost of manufacturing a certain product is found by adding the fixed cost of \$12, to \$4 times the number of units produced. The cost equation is $y = 4x + 12$. If the item sells for \$8, the revenue equation is $y = 8x$. The graph of both equations is shown.

Number of units produced and sold

a. Estimate the break-even point (cost = revenue) from the graph. (3, 24)

b. Solve $\begin{cases} y = 4x + 12 \\ y = 8x \end{cases}$ and explain the connection to part (a). (3, 24); these are the equations of the two lines.

c. If 8 units are made and sold what is the profit (profit = revenue − cost)? \$20

64. Supply and demand: The demand for a product depends on its price (low price creates high demand). The supply of a product also depends on price, with more companies willing to *supply* the product if the price is high (this is called the *law of supply and demand*). When supply and demand are equal we have *equilibrium,* with both the buyer and seller satisfied with the price. The given graph illustrates the supply/demand graphs for a popular gas grill.

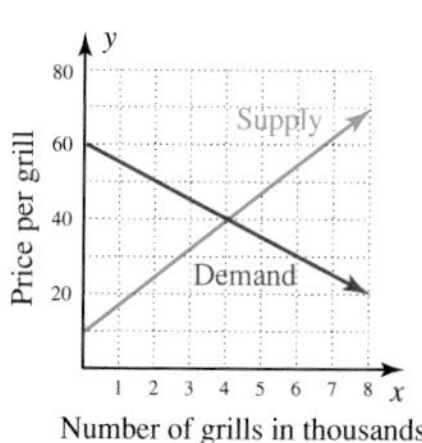

Number of grills in thousands

a. Estimate the "equilibrium point" from the graph (Supply = Demand). (4, 40)

b. Solve the system $\begin{cases} y = 7.5x + 10 \\ y = -5x + 60 \end{cases}$ and explain the connection to Part (a). (4, 40); these are the equations of the supply and demand lines.

c. If $x > 5000$, which is greater—supply or demand? What happens to the price if supply > demand? supply; it goes down

Descriptive Translation

65. If you sum the year that the Declaration of Independence was signed and the year that the Civil War ended, you get 3641. There are 89 yr that separate the two events. What year was the Declaration signed? What year did the Civil War end? 1776; 1865

66. When it was first constructed in 1889, the Eiffel Tower in Paris, France, was the tallest structure in the world. In 1975, the CN Tower in Toronto, Canada, became the world's tallest structure. The CN Tower is 153 ft less than twice the height of the Eiffel Tower, and the sum of their heights is 2799 ft. How tall is each tower? Eiffel Tower is 984 ft; CN Tower is 1815 ft

Geometry

67. **Football field size:** The rectangular field used for college football games has a perimeter of 1040 ft (including end zones). The length of the field is 40 ft more than twice the width. What are the dimensions of a college football field? 160 ft wide and 360 ft long

68. **Concentric rectangles:** As part of an art project, two concentric rectangles (one inside the other) are to be formed using gold wire along their perimeters. The perimeter of the larger rectangle is twelve centimeters less than twice the perimeter of the smaller rectangle. If a 144-cm piece of gold wire is cut into two pieces for this purpose, what is the perimeter of each rectangle? 52 cm and 92 cm

Mixture

69. At a recent production of *A Comedy of Errors,* the Community Theater brought in a total of $30,495 in revenue. If adult tickets were $9 and children's tickets were $6.50, how many tickets of each type were sold if 3800 tickets in all were sold? 2318 adult tickets; 1482 child tickets

70. A dietician needs to mix 10 gal of milk that is $2\frac{1}{2}\%$ milk fat for the day's rounds. He has some milk that is 4% milk fat and some that is $1\frac{1}{2}\%$ milk fat. How much of each should be used? 6 gallons of $1\frac{1}{2}\%$ milk fat; 4 gallons of 4% milk fat

Uniform Motion

71. As part of a training exercise, recruits must cover 48 mi with a combination of running and biking. The average recruit can make the trip in 4 hr. If recruits bike at a rate of 15 mph and run at a rate of 7 mph, how long do they bike? How far do they bike? 2.5 hr; 37.5 mi

72. Anthony and Cleopatra are 200 ft apart when they simultaneously see each other across a large field. They begin running toward each other, longing for that close embrace. If Anthony runs at 14 ft/sec and Cleopatra runs at 11 ft/sec, in how many seconds will they meet? 8 sec

Investment

73. A wealthy alumni donated $10,000 to his alma mater. The college used the funds to make a loan to a science major at 7% interest and a loan to a nursing student at 6% interest. That year the college earned $635 in interest. How much was loaned to each student? nursing student $6500; science major $3500

74. A total of $12,000 is invested in two municipal bonds, one paying 10.5% and the other 12% simple interest. Last year the annual interest earned on the two investments was $1335. How much was invested at each rate? $7000 invested at 10.5%; $5000 invested at 12%

WRITING, RESEARCH, AND DECISION MAKING

Creating your own system of equations can be a fun and challenging exercise. Use the information gained from the research indicated in each problem to create a problem that uses linear systems.

75. Use an almanac, encyclopedia, or the Internet to find the number of electoral votes (based on the 2000 census) for the state of Texas and the number for the state of Illinois. One equation can involve the total number of votes between the two states, a second equation can involve how many more electoral votes Texas has than Illinois. Create an application problem based on this information and ask a classmate to solve it. Answers will vary.

76. Answer using observations only—no calculations. Is the given system consistent/independent, consistent/dependent, or inconsistent? Explain/discuss your answer. $\begin{cases} y = 5x + 2 \\ y = 5.01x + 1.9 \end{cases}$
Different slopes so they cannot be the same line or parallel lines.

EXTENDING THE CONCEPT

77. Federal income tax reform has been a hot political topic for many years. Suppose tax plan A calls for a flat tax of 20% tax on all income (no deductions or loopholes). Tax plan B requires taxpayers to pay \$5000 plus 10% of all income. For what income level do both plans require the same tax? \$50,000

78. Suppose a certain amount of money was invested at 6% per year, and another amount at 8.5% per year, with a total return of \$1250. If the amounts invested at each rate were switched, the yearly income would have been \$1375. To the nearest whole dollar, how much was invested at each rate? \$6552 at 8.5%; \$11,551 at 6%

79.

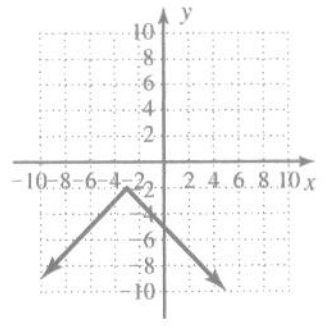

82.

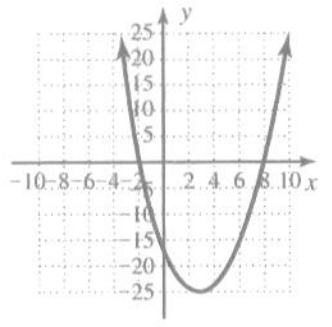

84.

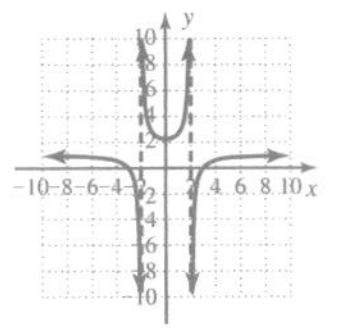

MAINTAINING YOUR SKILLS

79. (3.3) Given the parent function $f(x) = |x|$, sketch the graph of $F(x) = -|x + 3| - 2$.

80. (4.3) Use the RRT to write the polynomial in completely factored form: $3x^4 - 19x^3 + 15x^2 + 27x - 10 = 0$.
$(x - 5)(x - 2)(x + 1)(3x - 1) = 0$

81. (5.4) Solve for x (rounded to the nearest thousandth): $33 = 77.5e^{-0.0052x} - 8.37$
$x \approx 120.716$

82. (3.4) Graph $y = x^2 - 6x - 16$ by completing the square and state the interval where $f(x) \leq 0$. $x \in [-2, 8]$

83. (1.4) Find the sum, difference, product, and quotient of $1 + 3i$ and $1 - 3i$.
$2, 6i, 10, \frac{-4}{5} + \frac{3}{5}i$

84. (4.5) Graph the rational function $h(x) = \dfrac{x^2 - 9}{x^2 - 4}$.

6.2 Linear Systems in Three Variables with Applications

LEARNING OBJECTIVES

In Section 6.2 you will learn how to:

- **A. Visualize a solution in three dimensions**
- **B. Check ordered triple solutions**
- **C. Solve linear systems in three variables**
- **D. Recognize inconsistent and dependent systems**
- **E. Use a system of three equations in three variables to model and solve applications**

INTRODUCTION

The transition from 2×2 systems to systems having three equations in three variables **(3×3 systems)** requires a fair amount of "visual gymnastics" along with good organizational skills. Although the techniques we use are identical and similar results are obtained, the third equation and variable give us more to keep track of, and we must work more carefully toward a solution.

POINT OF INTEREST

In 1545, Girolamo Cardano wrote in his work *Ars Magna* (Great Art), "The first power refers to a line, the square [second power] to a surface, the cube [third power] to a solid, and it would be fatuous indeed for us to progress beyond for the reason that it is contrary to nature." However, as mathematics progressed it was realized that many problems could not be solved in three dimensions—the fourth required "dimension" could be time. Three coordinates are needed to fix the location of a point in space, with the fourth dimension modeling the time it took to reach this location. It was this space-time relationship that aided Albert Einstein's (1879–1955) development of the theory of relativity in the early 1900s.

A. Visualizing Solutions in Three Dimensions

The solution to an equation in one variable is the single number that satisfies the equation. For $x + 1 = 3$, the solution is $x = 2$ and its graph is a single *point* on the number line, a **one-dimensional graph.** The solution to an equation in two variables, such as $x + y = 3$, is an ordered pair (x, y) that satisfies the equation. When we graph this solution set, the result is a *line* on the xy-coordinate grid, a **two-dimensional graph.** The solutions to an equation in three variables, such as $x + y + z = 6$, are the **ordered triples** (x, y, z) that satisfy the equation. When we graph this solution set, the result is a **plane** in **space,** a *graph in three dimensions.* Recall a plane is a flat surface having infinite length and width, but no depth. Although more difficult to visualize, we can attempt to graph this plane using the intercept method and the result is shown in Figure 6.2. For graphs in three dimensions, the xy-plane is parallel to the ground (the y-axis points to the right) and z is the **vertical axis.** To find an additional point on this plane, we use any three numbers whose sum is 6, such as (2, 3, 1). Move 2 units along the x-axis, 3 units parallel to the y-axis, and 1 unit parallel to the z-axis, as shown in Figure 6.3.

WORTHY OF NOTE

We can visualize the location of a point in space by considering a large rectangular box 2 ft long × 3 ft wide × 1 ft tall, placed snugly in the corner of a room. The floor is the xy-plane, one wall is the xz-plane, and the other wall is the yz-plane. The z-axis is formed where the two walls meet and the corner of the room is the origin (0, 0, 0). To find the corner of the box located at (2, 3, 1), first locate the point (2, 3) in the xy-plane (the floor), then move up 1 ft.

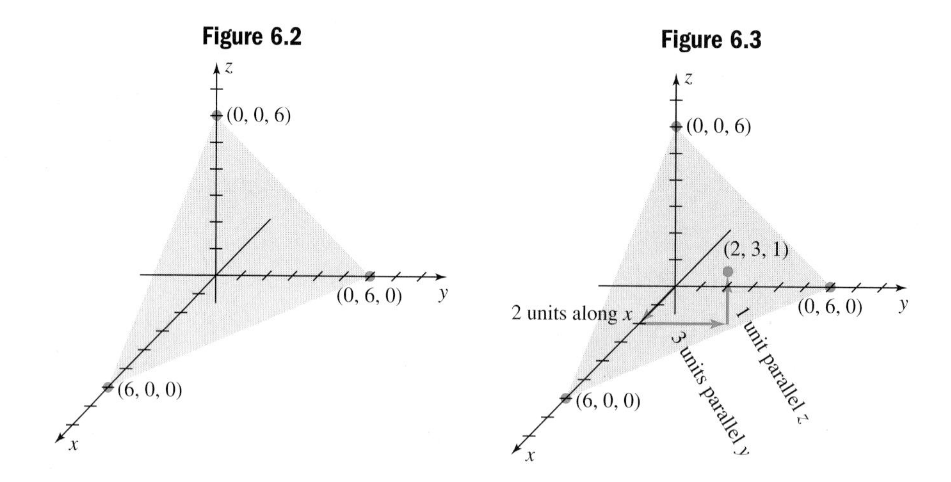

EXAMPLE 1 Use a guess-and-check method to find four additional points on the plane determined by $x + y + z = 6$.

Solution: We can begin by letting $x = 0$, then use any combination of y and z that sum to 6. Two examples are (0, 2, 4) and (0, 5, 1). We could also select any two values for x and y, then determine a value for z that ensures a sum of 6. Two examples are $(-2, 9, -1)$ and $(8, -3, 1)$.

NOW TRY EXERCISES 7 THROUGH 10

B. Solutions to a System of Three Equations in Three Variables

When solving a linear system of three equations in three variables, remember each equation represents a plane in space. These planes can meet in different configurations (Figures 6.4 to 6.7), creating various possibilities for a solution set. The system could have a **unique solution** (x, y, z) if the planes meet at a single point, as shown in Figure 6.4. This solution would satisfy all three equations simultaneously. Alternatively, the system could have an infinite number of solutions. If the planes intersect in a line, as shown in Figure 6.5, the system has **linear dependence** and the coordinates (x, y, z) of the solution set can be written in terms of (depend on) a single variable. If the planes intersect at all points, the system has **coincident dependence** (see Figure 6.6). This indicates the equations of the system differ by only a constant multiple—they are all "disguised forms" of the *same equation*. The solution set is any ordered triple (x, y, z) satisfying this equation. Finally, the system may have no solutions. This can happen a number of different ways, most notably if the planes intersect as shown in Figure 6.7 (other possibilities are discussed in the exercises). In the case of "no solutions," an ordered triple may satisfy none of the equations, only one of the equations, only two of the equations, but not all three equations.

Figure 6.4

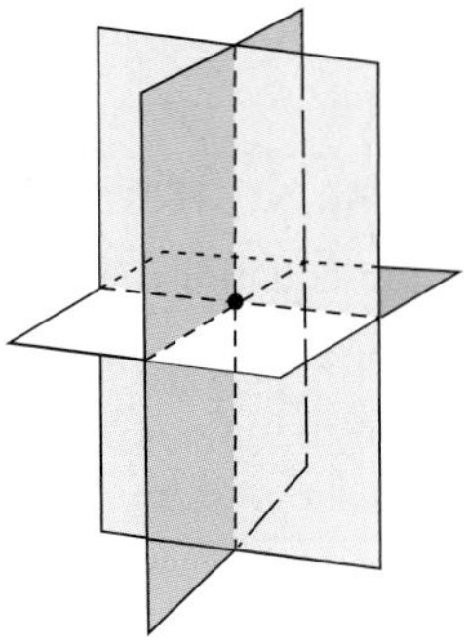

Unique solution

Figure 6.5

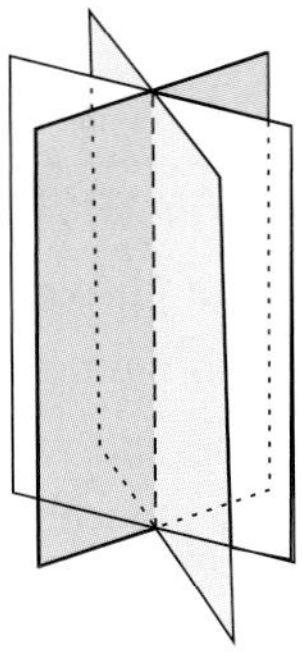

Linear dependence

Figure 6.6

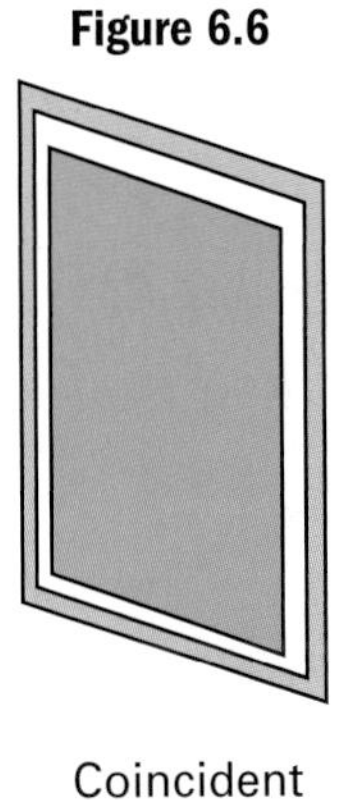

Coincident dependence

Figure 6.7

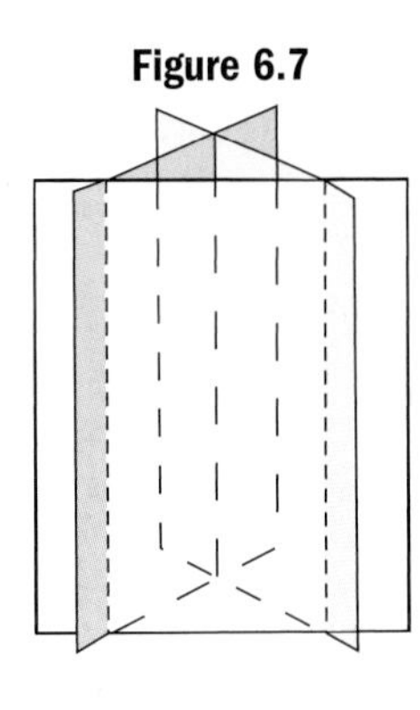

No solutions

EXAMPLE 2 Determine if the ordered triple $(1, -2, 3)$ is a solution to the systems shown.

a. $\begin{cases} x + 4y - z = -10 \\ 2x + 5y + 8z = 4 \\ x - 2y - 3z = -4 \end{cases}$ **b.** $\begin{cases} 3x + 2y - z = -4 \\ 2x - 3y - 2z = 2 \\ x - y + 2z = 9 \end{cases}$

Solution: Substitute 1 for x, -2 for y, and 3 for z in the first system.

a. $\begin{cases} x + 4y - z = -10 \\ 2x + 5y + 8z = 4 \\ x - 2y - 3z = -4 \end{cases} \longrightarrow \begin{cases} (1) + 4(-2) - (3) = -10 \\ 2(1) + 5(-2) + 8(3) = 4 \\ (1) - 2(-2) - 3(3) = -4 \end{cases} \longrightarrow \begin{cases} -10 = -10 \quad \text{true} \\ 16 = 4 \quad \text{false} \\ -4 = -4 \quad \text{true} \end{cases}$

No, the ordered triple $(1, -2, 3)$ is not a solution to the first system.
Now use the same substitutions in the second system.

b. $\begin{cases} 3x + 2y - z = -4 \\ 2x - 3y - 2z = 2 \\ x - y + 2z = 9 \end{cases} \longrightarrow \begin{cases} 3(1) + 2(-2) - (3) = -4 \\ 2(1) - 3(-2) - 2(3) = 2 \\ (1) - (-2) + 2(3) = 9 \end{cases} \longrightarrow \begin{cases} -4 = -4 \quad \text{true} \\ 2 = 2 \quad \text{true} \\ 9 = 9 \quad \text{true} \end{cases}$

The ordered triple $(1, -2, 3)$ is a solution to the second system only.

NOW TRY EXERCISES 11 AND 12

C. Solving Systems of Three Equations in Three Variables Using Elimination

Two systems of equations are **equivalent** if they have the same solution set. The systems $\begin{cases} 2x + y - 2z = -7 \\ x + y + z = -1 \\ -2y - z = -3 \end{cases}$ and $\begin{cases} 2x + y - 2z = -7 \\ y + 4z = 5 \\ z = 1 \end{cases}$ are equivalent, since it can be shown by substitution that both have the unique solution $(-3, 1, 1)$. In addition, it is evident that the second system can be solved more easily, since R2 and R3 have fewer variables than the first system. In the simpler system, mentally substituting $z = 1$ into R2 immediately gives $y = 1$, and these values can be back-substituted into the first equation to find that $x = -3$. This observation guides us to a general approach for solving larger systems—we would like to *eliminate variables in the second and third equations, until we obtain an equivalent system that can easily be solved by back-substitution.* Although there will be many different choices as to where we begin the process, a systematic approach will reduce the number of errors and increase your confidence in solving larger systems. The approach can always be modified as you gain more experience. To begin, let's look at three things that transform a given system into an equivalent system.

TRANSFORMATIONS OF A LINEAR SYSTEM

1. You can change the order the equations are written.
2. You can multiply both sides of an equation by any nonzero constant.
3. You can add two equations in the system and use the result to replace any other equation in the system.

We'll begin by actually solving the system $\begin{cases} 2x + y - 2z = -7 \\ x + y + z = -1 \\ -2y - z = -3 \end{cases}$ from the beginning of this subsection, using the elimination method and these three ideas.

SOLVING A 3 × 3 SYSTEM USING ELIMINATION

1. If the x-term in any equation has a coefficient of 1, interchange equations (if necessary) so this equation becomes R1. Otherwise look for coefficients of 1 on the y- or z-terms.
2. Name R1 the *SOURCE* equation and use it to *TARGET* and eliminate the designated variable in R2 and R3.
3. Use the results as the *new* R2 and R3. The original R1 along with the new R2 and R3 form an equivalent system that contains a 2×2 **subsystem** of the original.
4. Solve the 2×2 subsystem using elimination and keep the result as the new R3. The new R3, along with R1 and R2 from step 2, form an equivalent system that we can solve using back-substitution.

While not absolutely needed for the elimination process, there are two reasons for wanting the coefficient of x to be "1" in R1. First, it makes the elimination method more efficient since we can easily see what to use as a multiplier. Second, it lays the foundation for developing other methods of solving larger systems. If no equation has an x-coefficient of 1, we simply use the y- or z-variable instead (see Example 6). Since solutions to 3×3 systems generally are worked out in stages, we will sometimes track the transformations used by writing them *between* the original system and the equivalent system, rather than to the left as we did in Section 6.1.

EXAMPLE 3

Solve using elimination: $\begin{cases} 2x + y - 2z = -7 \\ x + y + z = -1 \\ -2y - z = -3 \end{cases}$.

Solution:

1. If the x-term in any equation has a coefficient of 1, interchange equations so this equation becomes R1.

$$\begin{cases} 2x + y - 2z = -7 \\ x + y + z = -1 \\ -2y - z = -3 \end{cases} \xrightarrow{R2 \leftrightarrow R1} \begin{cases} x + y + z = -1 \\ 2x + y - 2z = -7 \\ -2y - z = -3 \end{cases}$$

2. Name R1 as the *SOURCE* equation and use it to *TARGET* and eliminate the x-term in R2 and R3.

$$\begin{array}{r} SOURCE: \\ TARGET\ x: \\ TARGET\ x: \end{array} \begin{cases} x + y + z = -1 & \text{R1} \\ 2x + y - 2z = -7 & \text{R2} \\ -2y - z = -3 & \text{R3} \end{cases}$$

Note that R3 has no x-term, so the only elimination needed is the x-term from R2. Using $-2\text{R1} + \text{R2}$ will eliminate this term:

$$\begin{array}{lrl} -2\text{R1} & -2x - 2y - 2z = & 2 \\ + & & \\ \text{R2} & 2x + \ y - 2z = & -7 \\ \hline & 0x - 1y - 4z = & -5 \quad \text{sum} \\ & y + 4z = & 5 \quad \text{simplify} \end{array}$$

The new R2 is $y + 4z = 5$.

3. Use the results as the new R2 and R3. The original R1 along with the new R2 and R3 form an equivalent system that contains a 2×2 **subsystem** of the larger system. Note: The notation $-2\text{R1} + \text{R2} \rightarrow \text{R2}$ is used to show that the result of $-2\text{R1} + \text{R2}$ becomes the new R2.

$$\begin{array}{r} SOURCE: \\ TARGET\ x: \\ TARGET\ x: \end{array} \begin{cases} x + y + z = -1 \\ 2x + y - 2z = -7 \\ -2y - z = -3 \end{cases} \xrightarrow[\text{R3} \rightarrow \text{R3}]{-2\text{R1} + \text{R2} \rightarrow \text{R2}} \begin{cases} x + y + z = -1 \\ y + 4z = 5 \\ -2y - z = -3 \end{cases} \begin{array}{l} \text{new} \\ \text{equivalent} \\ \text{system} \end{array}$$

4. Solve the 2×2 subsystem for either y or z using elimination, and keep the result as a *new* R3. We choose to eliminate y using $2\text{R2} + \text{R3}$:

$$\begin{array}{lrl} 2\text{R2} & 2y + 8z = & 10 \\ + & & \\ \text{R3} & -2y - \ z = & -3 \\ \hline & 0y + 7z = & 7 \quad \text{sum} \\ & z = & 1 \quad \text{simplify} \end{array}$$

The new R3 is $z = 1$.

$$\begin{cases} x + y + z = -1 \\ y + 4z = 5 \\ -2y - z = -3 \end{cases} \xrightarrow{2\text{R2} + \text{R3} \rightarrow \text{R3}} \begin{cases} x + y + z = -1 \\ y + 4z = 5 \\ z = 1 \end{cases} \begin{array}{l} \text{new} \\ \text{equivalent} \\ \text{system} \end{array}$$

The new R3, along with the original R1 and the R2 from step 2 form an equivalent system that can be solved using back-substitution. Substituting $z = 1$ in R2 yields $y = 1$. Substituting $z = 1$ and $y = 1$ in R1 yields $x = -3$. The solution to the system is $(-3, 1, 1)$.

NOW TRY EXERCISES 13 THROUGH 24

D. Inconsistent and Dependent Systems

As mentioned, it is possible for a 3×3 system to have no solutions or an infinite number of solutions. Similar to the 2×2 case, an inconsistent system (no solutions) will produce inconsistent results, ending with a statement such as $0 = -3$ or some other **contradiction.**

EXAMPLE 4

Solve using elimination: $\begin{cases} 2x + y - 3z = -3 \\ 3x - 2y + 4z = 2 \\ 4x + 2y - 6z = -7 \end{cases}$.

Solution:

1. This system has no equation where the coefficient of x is 1.
2. We can still use R1 as the *SOURCE* equation, but this time we'll use the variable y since it *does* have coefficient 1. Targeting y in R2 and R3 we have

$$\begin{array}{r} SOURCE: \\ TARGET\ y: \\ TARGET\ y: \end{array} \begin{cases} 2x + y - 3z = -3 & \text{R1} \\ 3x - 2y + 4z = 2 & \text{R2} \\ 4x + 2y - 6z = -7 & \text{R3} \end{cases}$$

Using $2\text{R1} + \text{R2}$ eliminates the y-term from R2, yielding a new R2: $7x - 2z = -4$.

Using $-2\text{R1} + \text{R3}$ eliminates the y-term from R3, but a contradiction results:

$$\begin{array}{lrl} -2\text{R1} & -4x - 2y + 6z = & 6 \\ + & & \\ \underline{\text{R3}} & \underline{4x + 2y - 6z} = & \underline{-7} \\ & 0x + 0y + 0z = & -1 \quad \text{sum} \\ & 0 = & -1 \quad \text{result} \end{array}$$

We conclude the system is inconsistent. The answer is the empty set $\varnothing$, and we need work no further.

NOW TRY EXERCISES 25 THROUGH 27

Unlike our work with systems having only two variables, a system in three variables can have two forms of dependence—*linear dependence* or *coincident dependence.* To help understand linear dependence, consider a system of two equations in three variables: $\begin{cases} -2x + 3y - z = 5 \\ x - 3y + 2z = -1 \end{cases}$. Each of these equations represents a plane, and unless the planes are parallel, their intersection will be a line (see Figure 6.5). In three-dimensional space, lines do not have a simple equation like they do in two dimensions. To state the solution set, we name a generalized ordered triple (x, y, z), where two of the variables are written in terms of the third variable, now called a **parameter.** The relationships named will identify a point on the line for any given value of the parameter.

EXAMPLE 5 Solve using elimination: $\begin{cases} -2x + 3y - z = 5 \\ x - 3y + 2z = -1 \end{cases}$.

Solution: Using R1 + R2 eliminates the y-term from R2, yielding $-x + z = 4$. This means (x, y, z) will satisfy both equations only when $x = z - 4$ (the x-coordinate must be 4 less than the z-coordinate). Since x is already written in terms of z, we choose z as our parameter and substitute $z - 4$ for x *in either equation* to find how y is related to z. Using R2 we have: $(z - 4) - 3y + 2z = -1$, which yields $y = z - 1$ (verify). This means the y-coordinate of the solution must be 1 less than z. Parameterized solutions are most often written as an ordered triple and here the solution is $(x, y, z) = (z - 4, z - 1, z)$.

NOW TRY EXERCISES 28 THROUGH 30

Both 2×2 and 3×3 systems are called **square systems,** meaning there are exactly as many equations as there are variables. A system of linear equations cannot have a unique solution unless there are at least as many equations as there are variables in the system. The system in Example 5 was nonsquare, having two equations but three variables.

EXAMPLE 6 Solve using elimination: $\begin{cases} 3x - 2y + z = -1 \\ 2x + y - z = 5 \\ 10x - 2y = 8 \end{cases}$.

Solution:

1. This system has no equation where the coefficient of x is 1.
2. We can still use R1 as the *SOURCE,* but this time we'll *TARGET* z in R2 and R3.

$$\begin{array}{rl} \textit{SOURCE:} & \begin{cases} 3x - 2y + z = -1 & \text{R1} \\ 2x + y - z = 5 & \text{R2} \\ 10x - 2y = 8 & \text{R3} \end{cases} \\ \textit{TARGET } z\text{:} & \\ \textit{TARGET } z\text{:} & \end{array}$$

3. Using R1 + R2 eliminates the z-term from R2, yielding a new R2: $5x - y = 4$. There is no z-term in R3.
4. $$\begin{cases} 3x - 2y + z = -1 \\ 2x + y - z = 5 \\ 10x - 2y = 8 \end{cases} \xrightarrow[\text{R3} \rightarrow \text{R3}]{\text{R1} + \text{R2} \rightarrow \text{R2}} \begin{cases} 3x - 2y + z = -1 \\ 5x - y = 4 \\ 10x - 2y = 8 \end{cases}$$

We next solve the 2×2 subsystem. Using $-2\text{R2} + \text{R3}$ eliminates the y term in R3, but also all other terms:

$$\begin{array}{lrcl} -2\text{R2} & -10x + 2y & = & -8 \\ + & & & \\ \underline{\text{R3}} & \underline{10x - 2y} & = & \underline{8} \\ & 0x + 0y & = & 0 \quad \text{sum} \\ & 0 & = & 0 \quad \text{identity} \end{array}$$

Since R3 is the same as 2R2, the system is linearly dependent and equivalent to $\begin{cases} 3x - 2y + z = -1 \\ 5x - y = 4 \end{cases}$. We can solve for y in R2 to write y in terms of x: $y = 5x - 4$. Substituting $5x - 4$ for y in R1 allows us to write z in terms of x:

$3x - 2y \quad + z = -1$	R1
$3x - 2(5x - 4) + z = -1$	substitute $5x - 4$ for y
$3x - 10x + 8 \quad + z = -1$	distribute
$-7x + z = -9$	simplify
$z = 7x - 9$	solve for z

The general solution is $(x, 5x - 4, 7x - 9)$.

NOW TRY EXERCISES 31 THROUGH 34

For Example 6, two solutions would be $(0, -4, -9)$ for $x = 0$ and $(2, 6, 5)$ for $x = 2$. Solutions to linearly dependent systems can actually be written in terms of either x, y, or z, depending on which variable is eliminated in the first step and the variable we elect to solve for afterward.

For **coincident dependence** the equations in a system differ by only a constant multiple. After applying the elimination process—all variables are eliminated from the "target" equations, leaving statements that are always true (such as $2 = 2$ or some other). See Exercises 35 and 36.

E. 3 × 3 Systems and Modeling

Applications of 3×3 systems are simply an extension of our work with 2×2 systems. Once again, the applications come in a variety of forms and range from the contrived (simply made-up to help us learn the fundamentals) to those that are more substantial and meaningful. In the world of business and finance, systems can be used to diversify investments or spread out liabilities, a financial strategy hinted at in Example 7.

EXAMPLE 7 A small business borrowed \$225,000 from three different lenders to expand their product line. The interest rates were 5%, 6%, and 7%. Find how much was borrowed at each rate if the annual interest came to \$13,000 and twice as much was borrowed at the 5% rate than was borrowed at the 7% rate.

Solution: Let x, y, and z represent the amount borrowed at 5%, 6%, and 7%, respectively. This means our first equation is $x + y + z = 225$ (in thousands). The second equation is determined by the total interest paid, which was \$13,000: $0.05x + 0.06y + 0.07z = 13$. The third is found by carefully reading the problem.

"twice as much was borrowed at the 5% rate than was borrowed at the 7% rate": $\xrightarrow{\text{translation}} x = 2z$

"twice as much"

@7% @5%

These equations form the system: $\begin{cases} x + y + z = 225 \\ 0.05x + 0.06y + 0.07z = 13. \\ x = 2z \end{cases}$

The first equation already has a coefficient of 1, so this is our *SOURCE*. Written in standard form:

$$\begin{cases} SOURCE: & x + y + z = 225 & R1 \\ TARGET\ x: & 5x + 6y + 7z = 1300 & R2 \text{ (multiplied by 100)} \\ TARGET\ x: & x - 2z = 0 & R3 \end{cases}$$

Using $-5R1 + R2$ will eliminate the x term in R2, while $-R1 + R3$ will eliminate the x-term in R3. The new R2 is $y + 2z = 175$ and the new R3 is $y + 3z = 225$, yielding the system $\begin{cases} x + y + z = 225 \\ y + 2z = 175 \\ y + 3z = 225 \end{cases}$.

Solving the 2×2 subsystem using $-R2 + R3$ yields $z = 50$, so \$50,000 was borrowed at the 7% rate. Back-substitution shows \$75,000 was borrowed at 6% and \$100,000 at 5%. The solution is the ordered triple $(x, y, z) \rightarrow (100, 75, 50)$.

NOW TRY EXERCISES 54 THROUGH 60

TECHNOLOGY HIGHLIGHT

More on Parameterized Solutions

The keystrokes shown apply to a TI-84 Plus model. Please consult your manual or our Internet site for other models.

For linearly dependent systems, a graphing calculator can be used to both find and check multiple possibilities from the parametric solution. This is done by assigning the chosen parameter to Y_1, then using Y_2 and Y_3 to form the other coordinates of the solution. We can then build the equations in the system using Y_1, Y_2, and Y_3 in place of x, y, and z. The system from Example 6 is $\begin{cases} 3x - 2y + z = -1 \\ 2x + y - z = 5 \\ 10x - 2y = 8 \end{cases}$, which we found had solutions of the form $(x, 5x - 4, 7x - 9)$. We first form the solution using $Y_1 = X$, $Y_2 = 5Y_1 - 4$ (for y), and $Y_3 = 7Y_1 - 9$ (for z). Then we form the equations in the system using $Y_4 = 3Y_1 - 2Y_2 + Y_3$, $Y_5 = 2Y_1 + Y_2 - Y_3$, and $Y_6 = 10Y_1 - 2Y_2$ (see Figure 6.8). After setting up the table (set on **AUTO**), solutions can be found by enabling only Y_1, Y_2, and Y_3, which gives values of x, y, and z, respectively (see Figure 6.9—use the right arrow ▶ to view Y_3). By enabling Y_4, Y_5, and Y_6 you can verify that for any value of the parameter, the first equation is equal to -1, the second is equal to 5, and the third is equal to 8 (see Figure 6.10—use the right arrow ▶ to view Y_6).

Figure 6.8

```
Plot1 Plot2 Plot3
\Y1=X
\Y2=5Y1-4
\Y3=7Y1-9
\Y4=3Y1-2Y2+Y3
\Y5=2Y1+Y2-Y3
\Y6=10Y1-2Y2
\Y7=
```

Figure 6.9

X	Y1	Y2
-3	-3	-19
-2	-2	-14
-1	-1	-9
0	0	-4
1	1	1
2	2	6
3	3	11

X= -3

Figure 6.10

X	Y4	Y5
-3	-1	5
-2	-1	5
-1	-1	5
0	-1	5
1	-1	5
2	-1	5
3	-1	5

X= -3

Exercise 1: Use the ideas from this Technology Highlight to (a) find four specific solutions to Example 5, (b) check multiple variations of the solution given, and (c) determine if $(-9, -6, -5)$, $(-2, 1, 2)$, and $(6, 2, 4)$ are solutions. **a.** Answers will vary. **c.** $(-9, -6, -5)$, and $(-2, 1, 2)$, are solutions and $(6, 2, 4)$ is not a solution.

6.2 EXERCISES

CONCEPTS AND VOCABULARY

Fill in the blank with the appropriate word or phrase. Carefully reread the section if needed.

1. The solution to an equation in three variables is an ordered triple.

2. The graph of the solutions to an equation in three variables is a(n) plane.

3. Systems that have the same solution set are called equivalent systems.

4. If a 3×3 system is linearly dependent, the ordered triple solutions can be written in terms of a single variable called a(n) parameter.

5. Find a value of z that makes the ordered triple $(2, -5, z)$ a solution to $2x + y + z = 4$. Discuss/explain how this is accomplished. 5, substitute and solve for remaining variable

6. Explain the difference between linear dependence and coincident dependence, and describe how the equations are related. Answers will vary.

DEVELOPING YOUR SKILLS

Find any four ordered triples that satisfy the equation given.

7. $x + 2y + z = 9$ Answers will vary.

8. $3x + y - z = 8$ Answers will vary.

9. $-x + y + 2z = -6$ Answers will vary.

10. $2x - y + 3z = -12$ Answers will vary.

Determine if the given ordered triple is a solution of the system.

11. $\begin{cases} x + y - 2z = -1 \\ 4x - y + 3z = 3 \\ 3x + 2y - z = 4 \end{cases}$; $(0, 3, 2)$ yes

12. $\begin{cases} 2x + 3y + z = 2 \\ 5x - 2y - z = 4 \\ x - y - 2z = -3 \end{cases}$; $(1, -1, 3)$ no

Solve each system using elimination and back-substitution.

13. $\begin{cases} x - y - 2z = -10 \\ x - z = 1 \\ z = 4 \end{cases}$ (5, 7, 4)

14. $\begin{cases} x + y + 2z = -1 \\ 4x - y = 3 \\ 3x = 6 \end{cases}$ (2, 5, −4)

15. $\begin{cases} x + 3y + 2z = 16 \\ -2y + 3z = 1 \\ 8y - 13z = -7 \end{cases}$ (−2, 4, 3)

16. $\begin{cases} -x + y + 5z = 1 \\ 4x + y = 1 \\ -3x - 2y = 8 \end{cases}$ (2, −7, 2)

17. $\begin{cases} 2x - y + 4z = -7 \\ x + 2y - 5z = 13 \\ y - 4z = 9 \end{cases}$ (1, 1, −2)

18. $\begin{cases} 2x + 3y + 4z = -18 \\ x - 2y + z = 4 \\ 4x + z = -19 \end{cases}$ (−5, −4, 1)

Solve each system using the elimination method.

19. $\begin{cases} -x + y + 2z = -10 \\ x + y - z = 7 \\ 2x + y + z = 5 \end{cases}$ (4, 0, −3)

20. $\begin{cases} x + y - 2z = -1 \\ 4x - y + 3z = 3 \\ 3x + 2y - z = 4 \end{cases}$ (0, 3, 2)

21. $\begin{cases} 3x + y - 2z = 3 \\ x - 2y + 3z = 10 \\ 4x - 8y + 5z = 5 \end{cases}$ (3, 4, 5)

22. $\begin{cases} 2x - 3y + 2z = 0 \\ 3x - 4y + z = -20 \\ x + 2y - z = 16 \end{cases}$ (5, 12, 13)

23. $\begin{cases} 3x - y + z = 6 \\ 2x + 2y - z = 5 \\ x - 2y + 2z = 7 \end{cases}$ (1, 6, 9)

24. $\begin{cases} 2x - 3y - 2z = 7 \\ x - y + 2z = -5 \\ 3x + 2y - z = 11 \end{cases}$ (2, 1, −3)

Solve using the elimination method. If a system is inconsistent or dependent, so state. For systems with linear dependence, write the answer as an ordered triple in terms of a parameter.

25. no solution, inconsistent
26. no solution, inconsistent
27. no solution, inconsistent
28. $(-z + 2, -2z - 1, z)$, other solutions possible
29. $\left(-\frac{5}{3}z - \frac{2}{3}, -z - 2, z\right)$, other solutions possible
30. $\left(-\frac{1}{4}z + 4, \frac{9}{8}z + \frac{5}{2}, z\right)$, other solutions possible

25. $\begin{cases} 3x + y + 2z = 3 \\ x - 2y + 3z = 1 \\ 4x - 8y + 12z = 7 \end{cases}$ **26.** $\begin{cases} 2x - y + 3z = 8 \\ 3x - 4y + z = 4 \\ -4x + 2y - 6z = 5 \end{cases}$ **27.** $\begin{cases} 4x + y + 3z = 3 \\ x - 2y + 3z = 1 \\ 2x - 4y + 6z = 5 \end{cases}$

28. $\begin{cases} 4x - y + 2z = 9 \\ 3x + y + 5z = 5 \end{cases}$ **29.** $\begin{cases} 6x - 3y + 7z = 2 \\ 3x - 4y + z = 6 \end{cases}$ **30.** $\begin{cases} 2x - 4y + 5z = -2 \\ 3x - 2y + 3z = 7 \end{cases}$

Solve using elimination. If the system is linearly dependent, state the general solution in terms of a parameter. Different forms of the solution are possible.

31. $\begin{cases} 3x - 4y + 5z = 5 \\ -x + 2y - 3z = -3 \\ 3x - 2y + z = 1 \end{cases}$ $(z - 1, 2z - 2, z)$ **32.** $\begin{cases} 5x - 3y + 2z = 4 \\ -9x + 5y - 4z = -12 \\ -3x + y - 2z = -12 \end{cases}$ $(-z + 8, -z + 12, z)$

33. $\begin{cases} x + 2y - 3z = 1 \\ 3x + 5y - 8z = 7 \\ x + y - 2z = 5 \end{cases}$ $(z + 9, z - 4, z)$ **34.** $\begin{cases} -2x + 3y - 5z = 3 \\ 5x - 7y + 12z = -8 \\ x - y + 2z = -2 \end{cases}$ $(-z - 3, z - 1, z)$

Solve using elimination. If the system has coincident dependence, state the solution in set notation.

35. $\begin{cases} -0.2x + 1.2y - 2.4z = -1 \\ 0.5x - 3y + 6z = 2.5 \\ x - 6y + 12z = 5 \end{cases}$ **36.** $\begin{cases} 6x - 3y + 9z = 21 \\ 4x - 2y + 6z = 14 \\ -2x + y - 3z = -7 \end{cases}$

35. $\{(x, y, z) | x - 6y + 12z = 5\}$
36. $\left\{(x, y, z) | x - \frac{1}{2}y + \frac{3}{2}z = \frac{7}{2}\right\}$
37. $(1, 1, 2)$
38. $(3, 0, -2)$
39. $\left\{(x, y, z) | x - \frac{5}{2}y - 2z = 3\right\}$
40. $\left(-1, \frac{-3}{2}, 2\right)$
41. $\left(2, 1, \frac{-1}{3}\right)$
42. $(-z - 17, -z - 4, z)$
43. $(z + 5, z - 2, z)$
44. $(12, 6, 4)$
45. $(18, -6, 10)$
46. $(1, -5, -6)$
47. $\left(\frac{11}{3}, \frac{10}{3}, \frac{7}{3}\right)$
48. $(1, -2, 3)$
49. $(1, -2, 3)$
50. $\left(\frac{1}{5}, \frac{1}{2}, -2\right)$
51. $\left(\frac{1}{2}, \frac{1}{3}, 3\right)$

Solve using the elimination method. If a system is inconsistent or dependent, so state. For systems with linear dependence, write the answer in terms of a parameter. For coincident dependence, state the solution in set notation.

37. $\begin{cases} x + 2y - z = 1 \\ x + z = 3 \\ 2x - y + z = 3 \end{cases}$ **38.** $\begin{cases} 3x + 5y - z = 11 \\ 2x + y - 3z = 12 \\ y + 2z = -4 \end{cases}$ **39.** $\begin{cases} 2x - 5y - 4z = 6 \\ x - 2.5y - 2z = 3 \\ -3x + 7.5y + 6z = -9 \end{cases}$

40. $\begin{cases} x - 2y + 2z = 6 \\ 2x - 6y + 3z = 13 \\ 3x + 4y - z = -11 \end{cases}$ **41.** $\begin{cases} 4x - 5y - 6z = 5 \\ 2x - 3y + 3z = 0 \\ x + 2y - 3z = 5 \end{cases}$ **42.** $\begin{cases} x - 5y - 4z = 3 \\ 2x - 9y - 7z = 2 \\ 3x - 14y - 11z = 5 \end{cases}$

43. $\begin{cases} 2x + 3y - 5z = 4 \\ x + y - 2z = 3 \\ x + 3y - 4z = -1 \end{cases}$ **44.** $\begin{cases} \frac{1}{6}x + \frac{1}{3}y - \frac{1}{2}z = 2 \\ \frac{3}{4}x - \frac{1}{3}y + \frac{1}{2}z = 9 \\ \frac{1}{2}x - y + \frac{1}{2}z = 2 \end{cases}$ **45.** $\begin{cases} \frac{x}{2} + \frac{y}{3} - \frac{z}{2} = 2 \\ \frac{2x}{3} - y - z = 8 \\ \frac{x}{6} + 2y + \frac{3z}{2} = 6 \end{cases}$

In Section 6.8, you will learn to decompose or rewrite a rational expression as a sum of "partial fractions." This skill often leads to systems similar to those that follow. Solve using elimination.

46. $\begin{cases} -2A - B - 3C = 21 \\ B - C = 1 \\ A + B = -4 \end{cases}$ **47.** $\begin{cases} -A + 3B + 2C = 11 \\ 2B + C = 9 \\ B + 2C = 8 \end{cases}$ **48.** $\begin{cases} A + 2C = 7 \\ 2A - 3B = 8 \\ 3A + 6B - 8C = -33 \end{cases}$

49. $\begin{cases} A - 2B = 5 \\ B + 3C = 7 \\ 2A - B - C = 1 \end{cases}$ **50.** $\begin{cases} C = -2 \\ 5A - 2C = 5 \\ -4B - 9C = 16 \end{cases}$ **51.** $\begin{cases} C = 3 \\ 2A + 3C = 10 \\ 3B - 4C = -11 \end{cases}$

WORKING WITH FORMULAS

52. Dimensions of a rectangular solid:

$$\begin{cases} 2w + 2h = P_1 \\ 2l + 2w = P_2 \\ 2l + 2h = P_3 \end{cases}$$

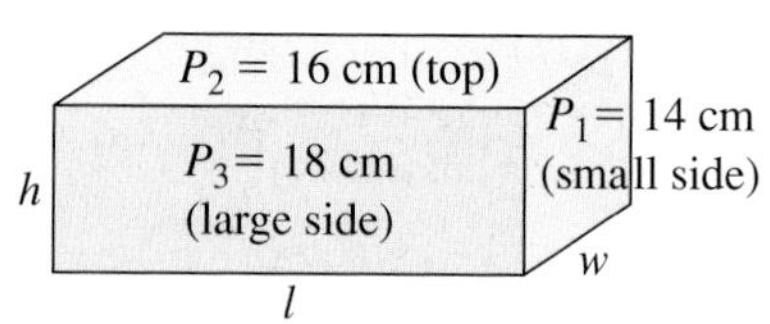

Using the formula shown, the dimensions of a rectangular solid can be found if the perimeter of the three distinct faces are known. Find the dimensions of the solid shown. (5 cm, 3 cm, 4 cm)

53. Distance from a point (x, y, z) to the plane $Ax + By + Cz = D$: $\left|\dfrac{Ax + By + Cz - D}{\sqrt{A^2 + B^2 + C^2}}\right|$

The perpendicular distance from a given point (x, y, z) to the plane defined by $Ax + By + Cy = D$ is given by the formula shown. Consider the plane given in Figure 6.2 $(x + y + z = 6)$. What is the distance from this plane to the point $(3, 4, 5)$? $\approx$3.464 units

APPLICATIONS

Solve the following applications using the information given to create a 3×3 system. Note that some equations may have only two of the three variables used to create the system.

Descriptive Translation

54. Major wars: The United States has fought three major wars in modern times: World War II, the Korean War, and the Vietnam War. If you sum the years that each conflict ended, the result is 5871. The Vietnam War ended 20 years after the Korean War and 28 years after World War II. In what year did each end? World War II, 1945; Korean, 1953; Vietnam, 1973

55. Animal gestation periods: The average gestation period (in days) of an elephant, rhinoceros, and camel sum to 1520 days. The gestation period of a rhino is 58 days longer than that of a camel. Twice the camel's gestation period decreased by 162 gives the gestation period of an elephant. What is the gestation period of each?

55. elephant, 650 days; rhino, 464 days; camel, 406 days

56. Moments in U.S. history: If you sum the year the Declaration of Independence was signed, the year the 13th Amendment to the Constitution abolished slavery, and the year the Civil Rights Act was signed, the total would be 5605. Ninety-nine years separate the 13th Amendment and the Civil Rights Act. The Civil Rights Act was signed 188 years after the Declaration of Independence. What year was each enacted?

56. Declaration of Independence 1776, 13th Amendment 1865; Civil Rights Act 1964

Mixtures

57. Chemical mixtures: A chemist mixes three different solutions with concentrations of 20%, 30%, and 45% glucose to obtain 10 L of a 38% glucose solution. If the amount of 30% solution used is 1 L more than twice the amount of 20% solution used, find the amount of each solution used. 1 L 20% solution; 3 L 30% solution; 6 L 45% solution

58. Value of gold coins: As part of a promotion, a local bank invites its customers to view a large sack full of \$5, \$10, and \$20 gold pieces, promising to give the sack to the first person able to state the number of coins for each denomination. Customers are told there are exactly 250 coins and with a total face value of \$1875. If there are also seven times as many \$5 gold pieces as \$20 gold pieces, how many of each denomination are there?

58. 175 \$5 gold pieces; 50 \$10 gold pieces; 25 \$20 gold pieces

Investment/Finance and Simple Interest Problems

59. Investing the winnings: After winning \$280,000 in the lottery, Maurika decided to place the money in three different investments: a certificate of deposit paying 4%, a money market certificate paying 5%, and some Aa bonds paying 7%. After 1 yr she earned \$15,400 in interest. Find how much was invested at each rate if \$20,000 more was invested at 7% than at 5%. \$80,000 at 4%; \$90,000 at 5%; \$110,000 at 7%

60. Purchase at auction: At an auction, a wealthy collector paid \$7,000,000 for three paintings: a Monet, a Picasso, and a van Gogh. The Monet cost \$800,000 more than the Picasso. The price of the van Gogh was \$200,000 more than twice the price of the Monet. What was the price of each painting? Monet \$1,900,000; Picasso \$1,100,000; van Gogh \$4,000,000

WRITING, RESEARCH, AND DECISION MAKING

61. Use an almanac, encyclopedia, or the Internet to find the lengths of the three longest rivers in South America, then create your own 3×3 system using the information. One equation can involve their combined length. Other equations can be created by forming relationships *between* their lengths (twice as long, 54 mi less than three times as long, etc.). Ask a classmate to solve the system. Answers will vary.

62. Use an almanac, encyclopedia, or the Internet to find the number of electoral votes (based on the 2000 census) owned by the state of Texas, the state of California, and the state of Illinois, then create your own 3×3 system using the information. One equation can involve the total number of votes for the three states. Other equations can involve how many more electoral votes Texas has than Illinois, or the difference in the number between California and Texas. Ask a classmate to solve the system. Answers will vary.

EXTENDING THE CONCEPT

Exercise 65

A B G H F C D E

63. The system $\begin{cases} x - 2y - z = 2 \\ x - 2y + kz = 5 \\ 2x - 4y + 4z = 10 \end{cases}$ is inconsistent if $k = \underline{-1 \text{ or } \frac{1}{2}}$, and dependent if $k = \underline{2}$.

64. One form of the equation of a circle is $x^2 + y^2 + Dx + Ey + F = 0$. Find the equation of the circle through the points $(2, -1)$, $(4, -3)$, and $(2, -5)$. $x^2 + y^2 - 4x + 6y + 9 = 0$

65. The lengths of each side of the squares *A*, *B*, *C*, *D*, *E*, *F*, *G*, *H*, and *I* (the smallest square) shown are whole numbers. Square *B* has sides of 15 cm and square *G* has sides of 7 cm. What are the dimensions of square *D*? b

a. 9 cm **b.** 10 cm **c.** 11 cm **d.** 12 cm **e.** 13 cm

MAINTAINING YOUR SKILLS

66. (2.5) If $p(x) = 2x^2 - x - 3$, in what intervals is $p(x) \leq 0$? $[-1, \frac{3}{2}]$

67. (3.2) Prove that if $f(x) = 2x^3 - 3$, then $f^{-1}(x) = \sqrt[3]{\frac{x+3}{2}}$. $f(f^{-1}(x)) = x$; $f^{-1}(f(x)) = x$

68. (4.4) Graph the polynomial defined by $f(x) = x^4 - 5x^2 + 4$.

69. (5.4) Solve the logarithmic equation: $\log(x + 2) + \log x = \log 3$ $x = 1$

70. (4.2) Use synthetic division to determine if the values given are zeroes of $f(x) = 2x^3 - 5x^2 - x + 6$:

a. $x = -1$ yes **b.** $x = 1$ no **c.** $x = 1.5$ yes **d.** $x = 2$ yes

71. (3.8) Analyze the graph of *g* shown. Clearly state the domain and range, the zeroes of *g*, intervals where $g(x) > 0$, intervals where $g(x) < 0$, local maximums or minimums, and intervals where the function is increasing or decreasing. Assume each tick mark is one unit and estimate endpoints to the nearest tenths.

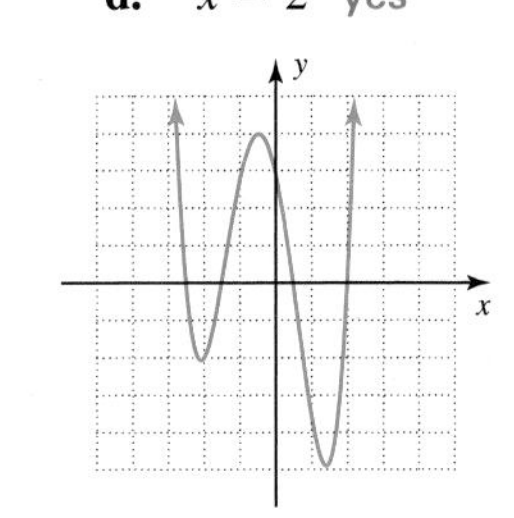

68.

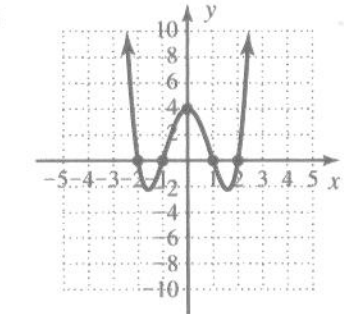

71.
$D: x \in (-\infty, \infty)$
$R: y \in [-5, \infty)$
zeroes: $x = -2.5$, $x = -1.5$, $x = 0.5$, $x = 2$
$g(x) > 0: x \in (-\infty, -2.5) \cup (-1.5, 0.5) \cup (2, \infty)$
$g(x) < 0: x \in (-2.5, -1.5) \cup (0.5, 2)$
max: $(-0.5, 4)$
min: $(-2, -2)$, $(1.5, -5)$
$f(x)\uparrow: x \in (-2, -0.5) \cup (1.5, \infty)$
$f(x)\downarrow: x \in (-\infty, -2) \cup (-0.5, 1.5)$

6.3 Systems of Inequalities and Linear Programming

LEARNING OBJECTIVES

In Section 6.3 you will learn how to:

A. Solve a linear inequality in two variables

B. Solve a system of linear inequalities

C. Solve applications using a system of linear inequalities

D. Solve applications using linear programming

INTRODUCTION

While systems of linear *equations* have an unlimited number of applications, there are many situations that can only be modeled using a system of linear *inequalities.* This is because important decisions in business and industry are usually based on a large number of limitations or constraints, and there may be various ways these constraints can be satisfied.

POINT OF INTEREST

If set $A = \{-6, -5, -4, -3, -2, -1, 0, 1\}$ and set $B = \{-1, 0, 1, 2, 3, 4, 5, 6\}$, the elements common to both sets is called their **intersection** and denoted as $A \cap B$. In 1880, John Venn (1834–1923), building on some ideas developed earlier by Leonhard Euler, introduced a method of diagramming relationships between sets. These are now commonly called **Venn diagrams.** The intersection $A \cap B$ is shown to the right, where we see that $A \cap B = \{-1, 0, 1\}$. The solution to a system of linear inequalities employs a similar idea, in that the solution is a region of the coordinate plane where the individual inequalities overlap.

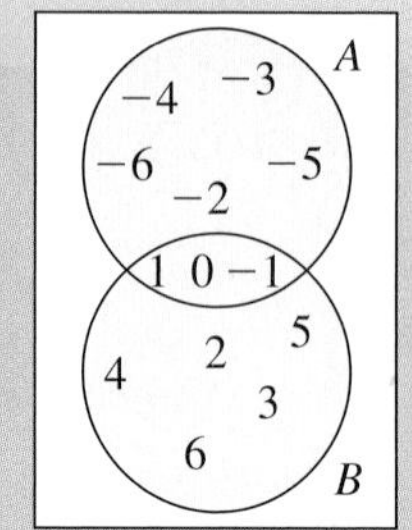

A. Linear Inequalities in Two Variables

A linear equation in two variables is any equation that can be written in the form $Ax + By = C$, where A and B are not both zero. A **linear inequality** in two variables is similarly defined, with the "=" sign replaced by the "<," ">," "≤," or "≥" symbols:

$$Ax + By < C \qquad Ax + By > C$$
$$Ax + By \le C \qquad Ax + By \ge C$$

The process for solving a linear inequality in two variables has many similarities with the one variable case. For one variable, we graph the *boundary* *point* on a number line, decide whether the endpoint is *included* or *excluded,* and *shade the appropriate half line.* For $x + 1 \le 3$, we have the solution $x \le 2$ with the endpoint included and the line shaded to the left (Figure 6.11):

Figure 6.11

−∞ ∞

−3 −2 −1 0 1 2 3

Interval notation: $x \in (-\infty, 2]$

Figure 6.12

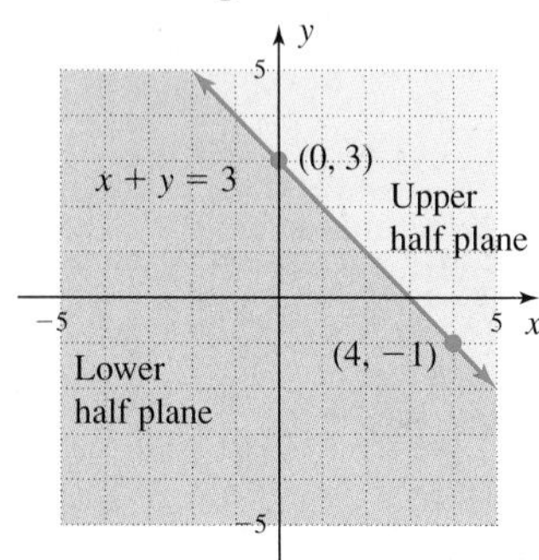

For linear inequalities in two variables, we graph a *boundary* *line,* decide whether the boundary line is *included* or *excluded,* and *shade the appropriate half plane.* For $x + y \le 3$, the boundary line $x + y = 3$ is graphed in Figure 6.12. Note it divides the coordinate plane into two regions called **half planes,** and it forms the **boundary** between the two regions. If the boundary is **included** in the solution set, we graph it using a *solid line.* If the boundary is **excluded,** a *dashed line* is used. Recall that solutions to a linear equation are ordered pairs that make the equation true. We use a similar idea to find or verify solutions to linear inequalities, noting that if any one point in a half plane makes the inequality true, all points in that half plane will satisfy the inequality.

EXAMPLE 1 Determine whether the ordered pairs are solutions to $-x + 2y < 2$: **a.** $(4, -3)$ **b.** $(-2, 1)$

Solution:

a. Substitute 4 for x and -3 for y: $-(4) + 2(-3) < 2$ substitute 4 for x, -3 for y

$-10 < 2$ true

$(4, -3)$ is a solution.

b. Substitute -2 for x and 1 for y: $-(-2) + 2(1) < 2$ substitute -2 for x, 1 for y

$4 < 2$ false

$(-2, 1)$ is not a solution.

NOW TRY EXERCISES 7 THROUGH 10

WORTHY OF NOTE

This relationship is often called the **trichotomy axiom** or the "*three-part truth.*" Given any two quantities, they are either equal to each other, or the first is less than the second, or the first is greater than the second.

Earlier we graphed linear equations by plotting a small number of ordered pairs or by solving for y and using the slope-intercept method. The line represented all ordered pairs that made the equation true, meaning *the left-hand expression was equal to the right-hand expression.* To graph linear inequalities, we reason that if the line represents all ordered pairs that make the expressions *equal,* then any point *not on that line* must make the expressions *unequal*—either greater than or less than (see the *Worthy of Note*). These ordered pair solutions must lie in one of the half planes formed by the line, which we shade to indicate the **solution region.** Note this implies the boundary line for any inequality *is determined by the related equation,* temporarily replacing the inequality symbol with an "=" sign.

EXAMPLE 2 Solve the inequality $-x + 2y \leq 2$.

Solution: The related equation and boundary line is $-x + 2y = 2$. Since the inequality is inclusive (less than *or equal to*), we graph a solid line. Using the intercepts, we graph the line through $(0, 1)$ and $(-2, 0)$ shown in Figure 6.13. To determine the solution region and which side to shade, we select $(0, 0)$ as a test point, which results in a true statement: $-(0) + 2(0) < 2$✓. Since $(0, 0)$ is in the "lower" half plane, we shade this side of the boundary, indicating all points in this region will satisfy the inequality (see Figure 6.14).

Figure 6.13 **Figure 6.14**

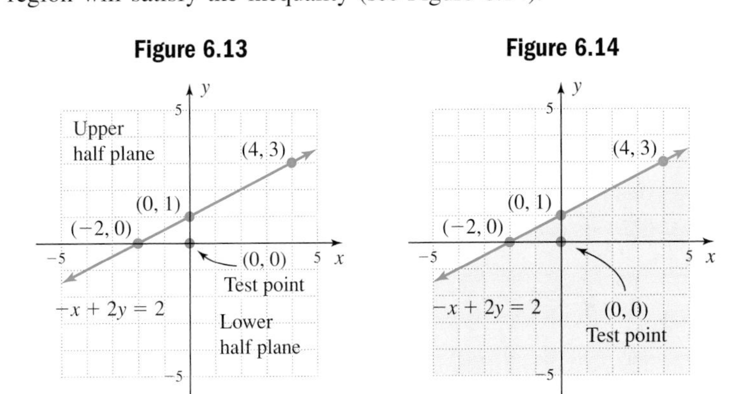

NOW TRY EXERCISES 11 THROUGH 14

The same solution would be obtained if we first solve for y and graph the boundary line using the slope-intercept method. However, this approach does offer a distinct advantage—test points are no longer necessary since solutions to "less than" inequalities will always appear below the boundary line and solutions to "greater than" inequalities appear above the line. Written in slope-intercept form, the inequality from Example 2 is $y \leq \frac{1}{2}x + 1$. Note that (0, 0) still results in a true statement, but the "less than or equal to" symbol now indicates directly that solutions will be found in the lower half plane. This observation leads to our general approach for solving linear inequalities:

SOLVING A LINEAR INEQUALITY

1. Graph the boundary line by solving for y and using the slope-intercept form.
   - Use a solid line if the boundary is included in the solution set.
   - Use a dashed line if the boundary is excluded from the solution set.
2. For "greater than" inequalities shade the upper half plane. For "less than" inequalities shade the lower half plane.
3. Select a test point from the shaded solution region and substitute the x- and y-values into the original inequality to verify the correct region is shaded.

B. Solving Systems of Linear Inequalities

To solve a **system of inequalities,** we apply the procedure outlined to all inequalities in the system, and note the ordered pairs that satisfy *all inequalities simultaneously.* In other words, we find *the intersection of all solution regions* (where they overlap), which then represents the solution for the system. In the case of vertical boundary lines, the designations *"above"* or *"below" the line* cannot be applied, and instead we simply note that for any vertical line $x = k$, points with x-coordinates larger than k will occur to the right.

EXAMPLE 3 Solve the system of inequalities: $\begin{cases} 2x + y \geq 4 \\ x - y < 2 \end{cases}$.

Solution: Solving for y, we obtain $y \geq -2x + 4$ and $y > x - 2$. The line $y = -2x + 4$ will be a solid boundary line (included), while $y = x - 2$ will be dashed (not included). Both inequalities are "greater than" and the upper half plane is shaded for each. The regions overlap and form the solution region (the lavender region shown). This sequence of events is illustrated here:

Shade above $y = -2x + 4$ (in blue)

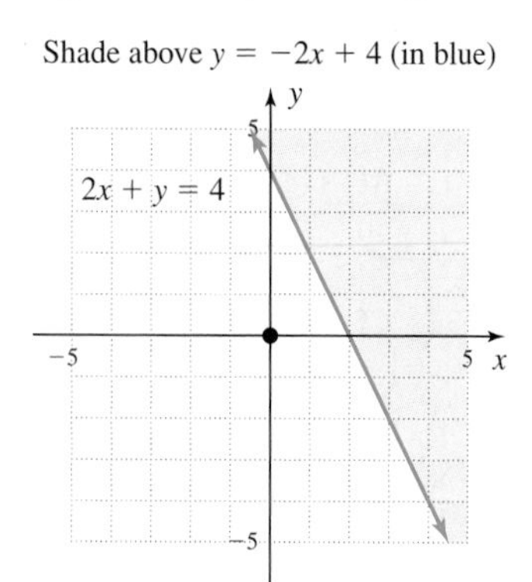

Shade above $y = x - 2$ (in pink)

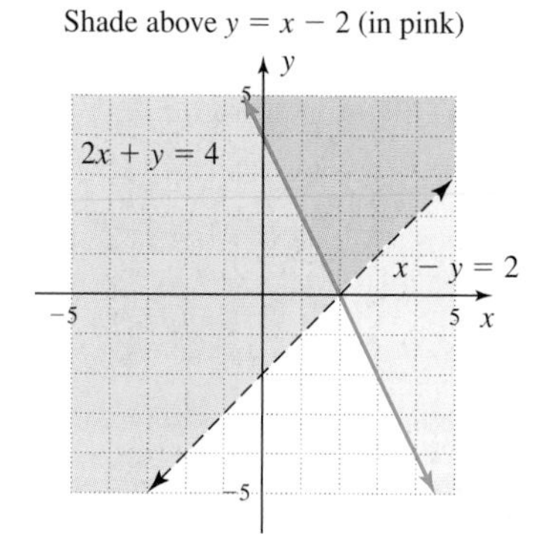

Overlapping region

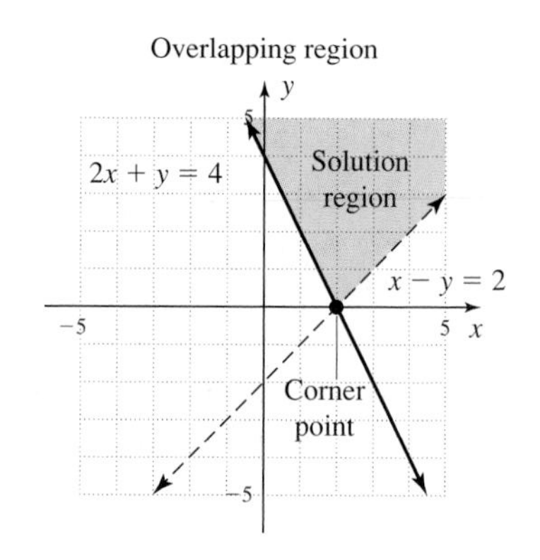

The solutions are all ordered pairs found in this region and its included boundaries. The point of intersection (2, 0) is called a **corner point** or **vertex** of the solution region. To verify the result, test the point (2, 3) from inside the region, (5, −2) from outside the region, and the vertex (2, 0) in both inequalities.

NOW TRY EXERCISES 15 THROUGH 42

Figure 6.15

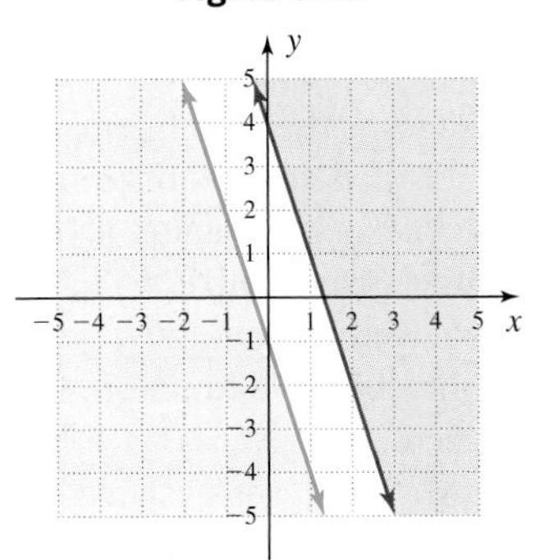

In Example 3 we saw that vertices of a solution region occur where boundary lines intersect. If the point of intersection is not easily found from the graph, we can find it by solving a linear system using the two lines. In this case, the system is $\begin{cases} 2x + y = 4 \\ x - y = 2 \end{cases}$ and solving by elimination gives $3x = 6$, $x = 2$, and $(2, 0)$ as the point of intersection [note $(2, 0)$ is actually *not a solution* since one of the boundary lines is not included]. It is also worth noting that a system of inequalities may have no solutions. For $\begin{cases} y \geq -3x + 4 \\ y \leq -3x - 1 \end{cases}$, the boundary lines are parallel and we would be shading above the line with the greater y-intercept, and below the line with the smaller y-intercept. As shown in Figure 6.15 the regions do not overlap.

C. Applications of Systems of Linear Inequalities

Systems of inequalities give us a way to model the decision-making process when there are certain ***constraints*** that must be satisfied. A constraint is a fact or consideration that somehow limits or governs possible solutions, like the number of acres a farmer plants—which may be limited by time, size of land, government regulation, and so on.

EXAMPLE 4 As part of their retirement planning, James and Lily decide to invest up to \$30,000 in two separate investment vehicles. The first is a bond issue paying 9% and the second is a money market certificate paying 5%. A financial adviser suggests they invest at least \$10,000 in the certificate and not more than \$15,000 in the bond issue. What various amounts can be invested in each?

Solution: Consider the ordered pairs (B, M) where B represents the money invested in bonds and M the money invested in the certificate. Since they plan to invest no more than \$30,000, the investment constraint would be $B + M \leq 30$ (in thousands). Following the adviser's recommendations, the constraints on each investment would be $B \leq 15$ and $M \geq 10$. Since they cannot invest less than zero dollars, the last two constraints are $B \geq 0$ and $M \geq 0$.

The resulting system is $\begin{cases} B + M \leq 30 \\ B \leq 15 \\ M \geq 10 \\ B \geq 0 \\ M \geq 0 \end{cases}$ shown in the figure,

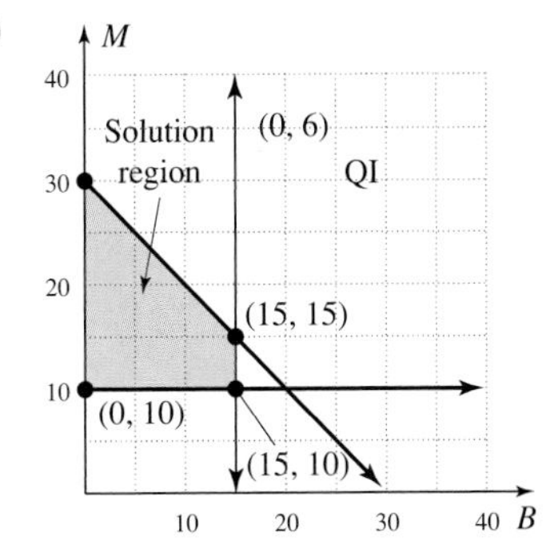

and indicates solutions will be in the first quadrant. There is a vertical boundary line at $B = 15$ with shading to the left (less than) and a horizontal boundary line at $M = 10$ with shading above (greater than). After graphing $M = 30 - B$, we see the solution region is a quadrilateral with vertices at $(0, 10)$, $(0, 30)$, $(15, 10)$, and $(15, 15)$, as shown.

NOW TRY EXERCISES 53 AND 54

D. Linear Programming

To become as profitable as possible, corporations look for ways to maximize their revenue and minimize their costs, while keeping up with delivery schedules and product demand. In order to operate at peak efficiency, plant managers must find ways to maximize productivity, while minimizing related costs and considering employee welfare, union agreements, and other factors. Problems where the goal is to **maximize** or **minimize** the value of a given quantity under certain **constraints** or restrictions are called programming problems. The quantity we seek to maximize or minimize is called the **objective function.** For situations where *linear* programming is used, the objective function is given as a linear function in two variables and is denoted $f(x, y)$. A function in two variables is evaluated in much the same way as a single variable function. To evaluate $f(x, y) = 2x + 3y$ at the point (4, 5), we substitute 4 for x and 5 for y: $f(4, 5) = 2(4) + 3(5) = 23$.

EXAMPLE 5 Determine which of the following ordered pairs maximizes the value of $f(x, y) = 5x + 4y$: (0, 6), (5, 0), (0, 0), or (4, 2).

Solution: Organizing our work in table form gives

Given Point	Evaluate $f(x, y) = 5x + 4y$	Result
(0, 6)	$f(0, 6) = 5(0) + 4(6)$	$f(0, 6) = 24$
(5, 0)	$f(5, 0) = 5(5) + 4(0)$	$f(5, 0) = 25$
(0, 0)	$f(0, 0) = 5(0) + 4(0)$	$f(0, 0) = 0$
(4, 2)	$f(4, 2) = 5(4) + 4(2)$	$f(4, 2) = 28$

The function $f(x, y) = 5x + 4y$ is maximized at (4, 2).

NOW TRY EXERCISES 43 THROUGH 46

When the objective is stated as a linear function in two variables and the constraints are expressed as a system of linear inequalities, we have what is called a **linear programming** problem. The systems of inequalities solved earlier produced a solution region that was either **bounded** (as in Example 4) or **unbounded** (as in Example 3). We interpret the word *bounded* to mean we can enclose the solution region within a circle of appropriate size. If we cannot draw a circle around the region because it extends indefinitely in some direction, the region is said to be *unbounded.* In this study, we will consider only situations that produce a bounded solution region, meaning the region will have three or more vertices. Under these conditions, it can be shown that solution(s) to a linear programming problem *must occur at one of the corner points of the solution region,* also called the **feasible region.**

EXAMPLE 6 Find the maximum value of the objective function $f(x, y) = 2x + y$

given the constraints shown in the system $\begin{cases} x + y \le 4 \\ 3x + y \le 6 \\ x \ge 0 \\ y \ge 0 \end{cases}$.

Solution: Begin by noting that the solutions must be in QI, since we have $x \geq 0$ and $y \geq 0$. Graph the boundary lines $y = -x + 4$ and $y = -3x + 6$, shading the lower half plane in each case since they are "less than" inequalities. This produces the feasible region shown in lavender. There are four corner points to this region: (0, 0), (0, 4), (2, 0), and (1, 3). Three of these points are intercepts and can be found quickly. The point (1, 3) was found by solving the system $\begin{cases} x + y = 4 \\ 3x + y = 6 \end{cases}$. Knowing that the objective function will be maximized at one of the corner points, we test these points in the objective function, using a table to organize our work.

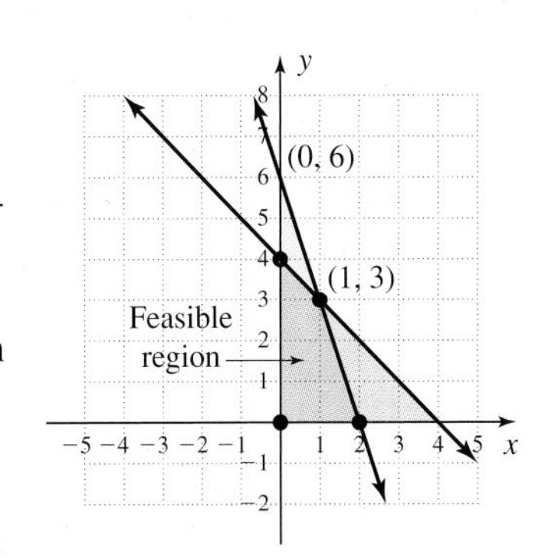

Corner Point	Objective Function $f(x, y) = 2x + y$	Result
(0, 0)	$f(0, 0) = 2(0) + (0)$	0
(0, 4)	$f(0, 4) = 2(0) + (4)$	4
(2, 0)	$f(2, 0) = 2(2) + (0)$	4
(1, 3)	$f(1, 3) = 2(1) + (3)$	5

The objective function $f(x, y) = 2x + y$ is maximized at (1, 3).

NOW TRY EXERCISES 47 THROUGH 50

Figure 6.16

To help understand why solutions must occur at a vertex, consider that the objective function $f(x, y)$ is to be maximized using only (x, y) ordered pairs from the feasible region. If we let K represent this maximum value, the function from Example 6 becomes $K = 2x + y$ or $y = -2x + K$, which is a line with slope -2 and y-intercept K. The table in Example 6 suggests that K should range from 0 to 5 and graphing this line for $K = 1$, $K = 3$, and $K = 5$ produces the family of parallel lines shown in Figure 6.16. Note that values of K larger than 5 will cause the line to miss the solution region, and the maximum value of 5 occurs where the line intersects the feasible region at the vertex (1, 3). These observations lead to the following principles, which we offer without a formal proof.

LINEAR PROGRAMMING SOLUTIONS

1. If the feasible region is convex and bounded, a maximum and a minimum value exist.
2. If a unique solution exists, it will occur at a vertex of the feasible region.
3. If more than one solution exists, at least one of them occurs at a vertex of the feasible region.
4. If the feasible region is unbounded, a linear programming problem may have no solutions.

By *convex* we mean that for any two points in the feasible region, the line segment between them is also in the region (Figure 6.17).

Figure 6.17

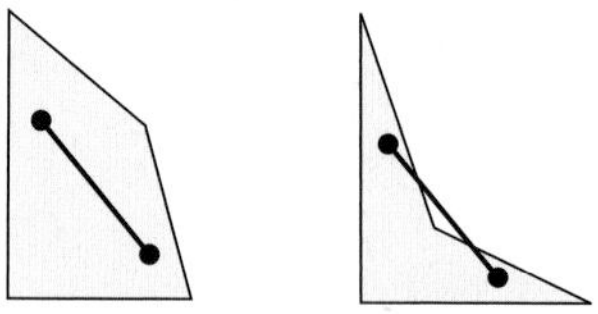

Convex   Not convex

Solving linear programming problems depends in large part on two things: (1) identifying the **objective** and the **decision variables,** and (2) using the decision variables to write the *objective function* and **constraint inequalities.** This brings us to our five-step approach for solving linear programming applications.

SOLVING LINEAR PROGRAMMING APPLICATIONS

1. Identify the main objective and the decision variables (descriptive variables may help).
2. Write the objective function in terms of these variables.
3. Organize all information in a table, with the *decision variables* and *constraints* heading up the columns, and their *components* leading each row. Complete the table using the information given, and write the constraint inequalities using the decision variables, constraints, and the domain.
4. Graph the constraint inequalities and determine the feasible region.
5. Identify all corner points of the feasible region and test these points in the objective function to determine the optimal solution(s).

EXAMPLE 7 The owner of a snack food business wants to create two nut mixes for the holiday season. The regular mix will have 14 oz of peanuts and 4 oz of cashews, while the deluxe mix will have 12 oz of peanuts and 6 oz of cashews. The owner estimates he will make a profit of \$3 on the regular mixes and \$4 on the deluxe mixes. How many of each should be made in order to maximize profit, if only 840 oz of peanuts and 360 oz of cashews are available?

Solution: Our *objective* is to maximize profit, and the *decision variables* could be r to represent the regular mixes sold, and d for the number of deluxe mixes. This gives $P(r, d) = \$3r + \$4d$ as our *objective function.* The information is organized in Table 6.1, using the variables r and d to head each column:

Table 6.1

$$P(r, d) = \$3r \quad + \quad \$4d$$

	Regular, r	Deluxe, d	Constraints: Total Ounces Available
Peanuts	14	12	840
Cashews	4	6	360

Since the mixes are composed of peanuts and cashews, these lead the rows in the table. Reading the table from left to right along the "peanut" row and the "cashew" row, gives the constraint inequalities

$14r + 12d \le 840$ and $4r + 6d \le 360$. Realizing that we cannot make a negative number of mixes, the remaining constraints are $r \ge 0$ and $d \ge 0$. The complete system of constraint inequalities is

$$\begin{cases} 14r + 12d \le 840 \\ 4r + 6d \le 360 \\ r \ge 0 \\ d \ge 0 \end{cases}$$

Note once again that the solutions must be in QI, since $r \ge 0$ and $d \ge 0$. Graphing $d \le -\frac{7}{6}r + 70$ and $d \le -\frac{2}{3}r + 60$ produces the feasible region shown in lavender. The four corner points are (0, 0), (60, 0), (0, 60), and $(20, 46.\overline{6})$. Three of these points are intercepts and can be read from a table of values or the graph itself. The point $(20, 46\frac{2}{3})$ was found by solving the system $\begin{cases} 14r + 12d = 840 \\ 4r + 6d = 360 \end{cases}$.

Since we can't sell a fractional part of a box, we actually must check the point (20, 46) since the constraints prevent us from completing 47 deluxe mixes, and the point (19, 47) since a 47th box can be made if one fewer regular mixes is made (although neither is actually a "corner point"). Knowing the objective function will be maximized at one of these points, we test them in the objective function (Table 6.2).

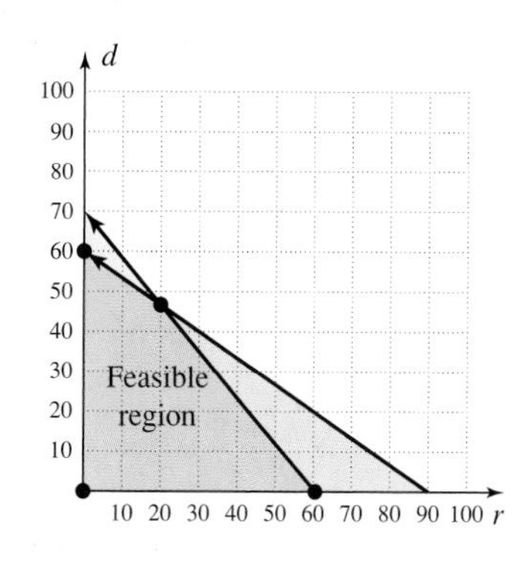

Table 6.2

Corner Point	Objective Function $P(r, d) = \$3r + \$4d$	Result
(0, 0)	$P(0, 0) = \$3(0) + \$4(0)$	$P(0, 0) = 0$
(60, 0)	$P(60, 0) = \$3(60) + \$4(0)$	$P(60, 0) = \$180.00$
(0, 60)	$P(0, 60) = \$3(0) + \$4(60)$	$P(0, 60) = \$240.00$
(20, 46)	$P(20, 46) = \$3(20) + \$4(46)$	$P(20, 46) = \$244.00$
(19, 47)	$P(19, 47) = \$3(19) + \$4(47)$	$P(19, 47) = \$245.00$

Profit will be maximized if 19 boxes of the regular mix and 47 boxes of the deluxe mix are made and sold.

NOW TRY EXERCISES 55 THROUGH 58

Linear programming can also be used to minimize an objective function, as in Example 8.

EXAMPLE 8 A beverage producer needs to minimize shipping costs from its two primary plants in Kansas City (KC) and St. Louis (STL). All wholesale orders within the state are shipped from one of these plants. An outlet in Macon orders 200 cases of soft drinks on the same day an order for 240 cases comes from Springfield. The plant in KC has

300 cases ready to ship and the plant in STL has 200 cases. The cost of shipping each case to Macon is \$0.50 from KC and \$0.70 from STL. The cost of shipping each case to Springfield is \$0.60 from KC and \$0.65 from STL. How many cases should be shipped from each warehouse to minimize costs?

Solution: Our *objective* is to minimize costs, which depends on the number of cases shipped from each plant. To begin we use the following assignments:

$A \rightarrow$ cases shipped from KC to Macon

$B \rightarrow$ cases shipped from KC to Springfield

$C \rightarrow$ cases shipped from STL to Macon

$D \rightarrow$ cases shipped from STL to Springfield

This gives the objective function for total cost as $T = 0.5A + 0.6B + 0.7C + 0.65D$, an equation in *four* variables. But since Macon ordered 200 cases and Springfield 240, we know $A + C = 200$ and $B + D = 240$. This enables us to substitute $C = 200 - A$ and $D = 240 - B$, giving this objective function:

$$\begin{aligned} T(A, B) &= 0.5A + 0.6B + 0.7(200 - A) + 0.65(240 - B) \\ &= 0.5A + 0.6B + 140 - 0.7A + 156 - 0.65B \\ &= 296 - 0.2A - 0.05B \end{aligned}$$

The constraints involving the KC plant are $A + B \leq 300$ with $A \geq 0$ and $B \geq 0$. The constraints involving the STL plant are $C + D \leq 200$ with $C \geq 0$ and $D \geq 0$, but substituting $C = 200 - A$ and $D = 240 - B$ shows these constraints become $A + B \geq 240$ with $A \leq 200$ and $B \leq 240$ (substitute and verify). The following system and solution region are obtained.

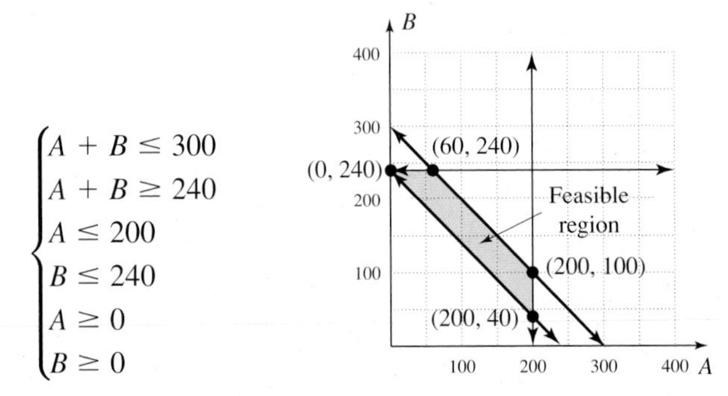

To find the minimum cost, we check each vertex in the objective function.

Vertices	Objective Function $T(A, B) = 296 - 0.2A - 0.05B$	Result
(0, 240)	$P(0, 240) = 296 - 0.2(0) - 0.05(240)$	\$284
(60, 240)	$P(60, 240) = 296 - 0.2(60) - 0.05(240)$	\$272
(200, 100)	$P(200, 100) = 296 - 0.2(200) - 0.05(100)$	\$251
(200, 40)	$P(200, 40) = 296 - 0.2(200) - 0.05(40)$	\$254

The minimum cost occurs when $A = 200$ and $B = 100$, meaning the producer should ship the following quantities:

$A \rightarrow$ cases shipped from KC to Macon $= 200$
$B \rightarrow$ cases shipped from KC to Springfield $= 100$
$C \rightarrow$ cases shipped from STL to Macon $= 0$
$D \rightarrow$ cases shipped from STL to Springfield $= 140$

NOW TRY EXERCISES 59 AND 60

TECHNOLOGY HIGHLIGHT

Systems of Linear Inequalities

The keystrokes shown apply to a TI-84 Plus model. Please consult your manual or our Internet site for other models.

Solving systems of linear inequalities on the TI-84 Plus requires only a simple combination of skills and keystrokes learned previously. This process involves three steps, which are performed on both equations: (1) enter the related equations in Y_1 and Y_2 (solve for y in each equation) to create the boundary lines, (2) graph both lines and test the resulting half planes, and (3) shade the appropriate half plane. As a final note, many real-world applications of linear inequalities preclude the use of negative numbers, so we **set Xmin = 0 and Ymin = 0 for the WINDOW size.** Xmax and Ymax will depend on the equations given.

We illustrate by solving the system $\begin{cases} 3x + 2y < 14 \\ x + 2y < 8 \end{cases}$.

1. *Enter the related equations.* For $3x + 2y = 14$, we have $y = -1.5 + 7$. For $x + 2y = 8$, we have $y = -0.5x + 4$. Enter these as Y_1 and Y_2 on the Y= screen.
2. *Graph the boundary lines.* Note the x- and y-intercepts of both lines are less than 10, so we can graph them using a friendly window where $x \in [0, 9.4]$ and $y \in [0, 6.2]$. After setting the window, press GRAPH to graph the lines.
3. *Shade the appropriate half plane.* Testing (0, 0) in the first inequality results in a true statement, so we have the calculator shade below the line using the ◣ feature located to the far left of Y_1. Simply overlay the diagonal line and press ENTER repeatedly until the ◣ symbol appears (Figure 6.18). Test (0, 0) in the second inequality. A true statement again results so we shade below this line as well. After pressing the GRAPH key, the calculator draws both lines and shades the appropriate regions (Figure 6.19). Note the calculator is programmed to use two different kinds of shading. This makes it easy to identify the solution region—it will be the "checker- board area" where the horizontal lines cross the vertical lines. As a final check, you should navigate the position marker into the solution region and test a few points in both equations.

Figure 6.18

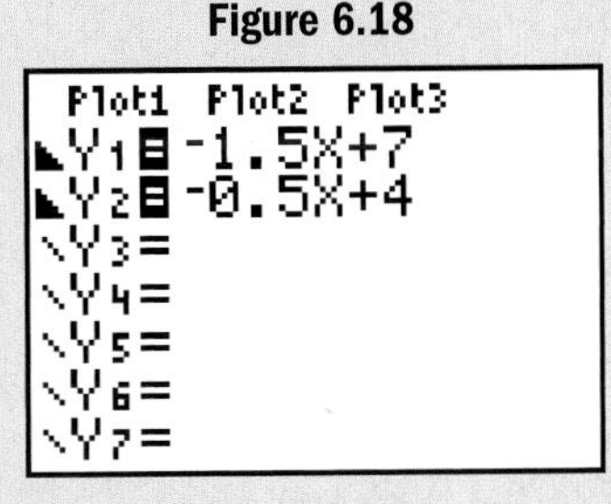

Figure 6.19

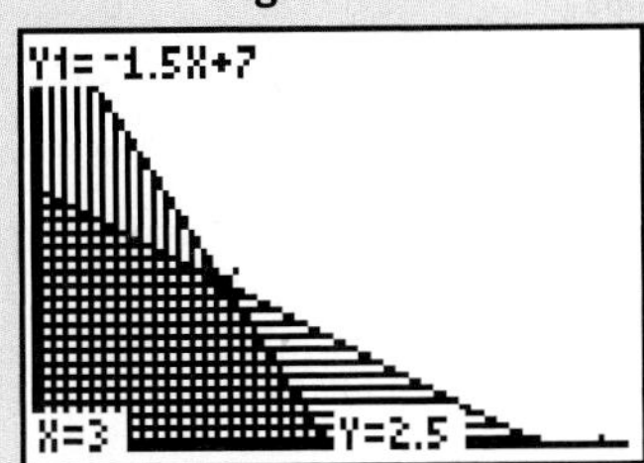

Use these ideas to solve the following systems of linear inequalities. Assume all solutions lie in Quadrant I.

Exercise 1: $\begin{cases} y + 2x < 8 \\ y + x < 6 \end{cases}$ Exercise 2: $\begin{cases} 3x + y < 8 \\ x + y < 4 \end{cases}$ Exercise 3: $\begin{cases} -4x - y > -9 \\ -3x - y > -7 \end{cases}$

Exercise 1.

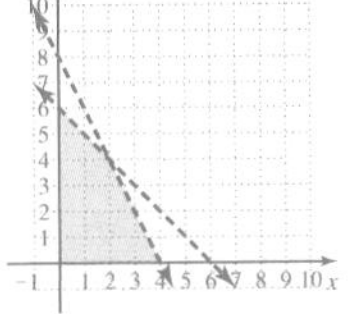

Exercise 2.

Exercise 3.

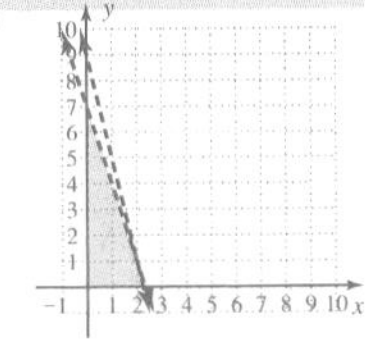

6.3 EXERCISES

5. The feasible region may be bordered by three or more oblique lines, with two of them intersecting outside and away from the feasible region.
7. No, No, No, No
8. No, Yes, No, No
9. No, Yes, Yes, No
10. No, Yes, Yes, No

11. 12.

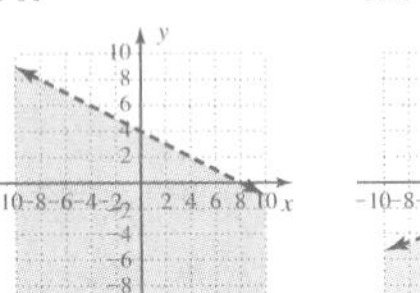

13. 14.

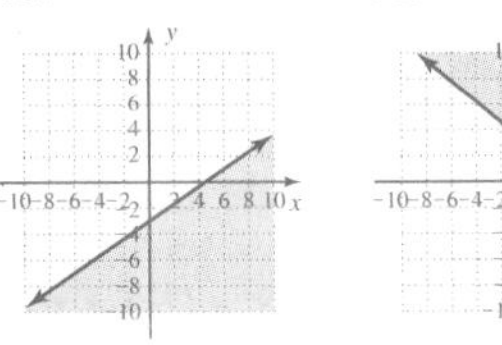

17. 18.

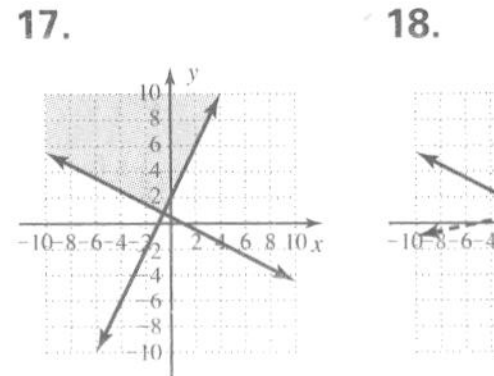

19. 20.

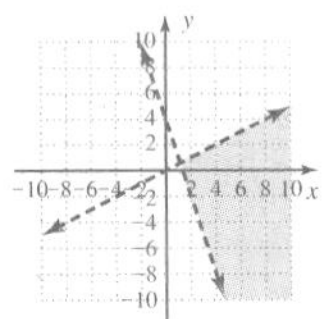

21. 22.

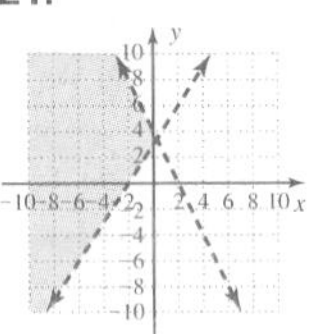

23. 24.

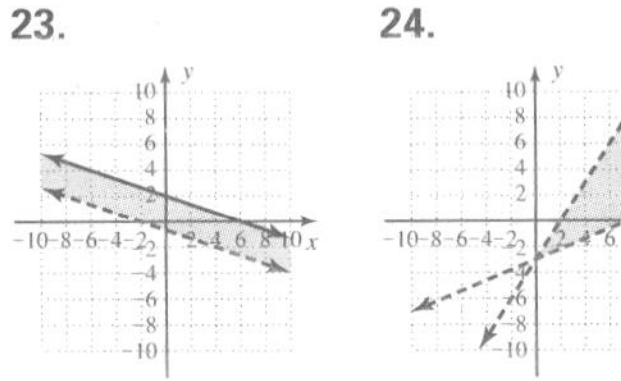

Additional answers can be found in the Instructor Answer Appendix.

CONCEPTS AND VOCABULARY

Fill in the blank with the appropriate word or phrase. Carefully reread the section if needed.

1. Any line $y = mx + b$ drawn in the coordinate plane divides the plane into two regions called half planes.

2. For the line $y = mx + b$ drawn in the coordinate plane, solutions to $y > mx + b$ are found in the region above the line.

3. The overlapping region of two or more linear inequalities in a system is called the solution region.

4. If a linear programming problem has a unique solution (x, y), it must be a vertex of the feasible region.

5. Suppose two boundary lines in a system of linear inequalities intersect, but the point of intersection is not a vertex of the feasible region. Describe how this is possible.

6. Describe the conditions necessary for a linear programming problem to have multiple solutions. (*Hint:* Consider the diagram in Figure 6.16, and the slope of the line from the objective function.)
Objective function and one of the boundary lines may be collinear.

DEVELOPING YOUR SKILLS

Determine whether the ordered pairs given are solutions.

7. $2x + y > 3$; $(0, 0)$, $(3, -5)$, $(-3, -4)$, $(-3, 9)$

8. $3x - y > 5$; $(0, 0)$, $(4, -1)$, $(-1, -5)$, $(1, -2)$

9. $4x - 2y \le -8$; $(0, 0)$, $(-3, 5)$, $(-3, -2)$, $(-1, 1)$

10. $3x + 5y \ge 15$; $(0, 0)$, $(3, 5)$, $(-1, 6)$, $(7, -3)$

Solve the linear inequalities by shading the appropriate half plane.

11. $x + 2y < 8$ **12.** $x - 3y > 6$ **13.** $2x - 3y \ge 9$ **14.** $4x + 5y \ge 15$

Determine whether the ordered pairs given are solutions to the accompanying system.

15. $\begin{cases} 5y - x \ge 10 \\ 5y + 2x \le -5; (-2, 1), (-5, -4), (-6, 2), (-8, 2.2) \end{cases}$ No, No, No, Yes

16. $\begin{cases} 8y + 7x \ge 56 \\ 3y - 4x \ge -12 \\ y \ge 4;\ 1, 5), (4, 6), (8, 5), (5, 3) \end{cases}$ No, Yes, No, No

Solve each system of inequalities by graphing the solution region. Verify the solution using a test point.

17. $\begin{cases} x + 2y \ge 1 \\ 2x - y \le -2 \end{cases}$ **18.** $\begin{cases} -x + 5y < 5 \\ x + 2y \ge 1 \end{cases}$ **19.** $\begin{cases} 3x + y > 4 \\ x > 2y \end{cases}$ **20.** $\begin{cases} 3x \le 2y \\ y \ge 4x + 3 \end{cases}$

21. $\begin{cases} 2x + y < 4 \\ 2y > 3x + 6 \end{cases}$ **22.** $\begin{cases} x - 2y < -7 \\ 2x + y > 5 \end{cases}$ **23.** $\begin{cases} x > -3y - 2 \\ x + 3y \le 6 \end{cases}$ **24.** $\begin{cases} 2x - 5y < 15 \\ 3x - 2y > 6 \end{cases}$

25. $\begin{cases} 5x + 4y \ge 20 \\ x - 1 \ge y \end{cases}$ **26.** $\begin{cases} 10x - 4y \le 20 \\ 5x - 2y > -1 \end{cases}$ **27.** $\begin{cases} 0.2x > -0.3y - 1 \\ 0.3x + 0.5y \le 0.6 \end{cases}$

28. $\begin{cases} x > -0.4y - 2.2 \\ x + 0.9y \le -1.2 \end{cases}$ **29.** $\begin{cases} y \le \frac{3}{2}x \\ 4y \ge 6x - 12 \end{cases}$ **30.** $\begin{cases} 3x + 4y > 12 \\ y < \frac{2}{3}x \end{cases}$

31.

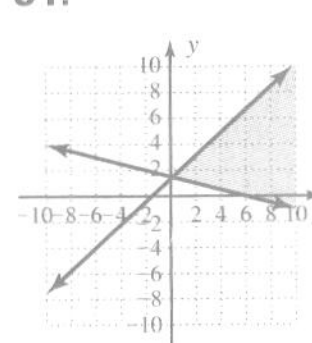

32.

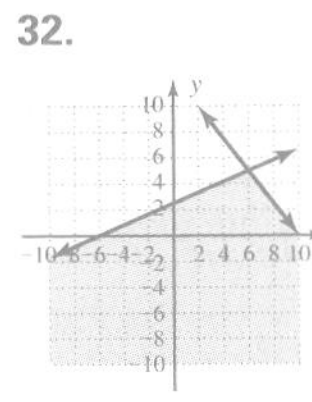

33.

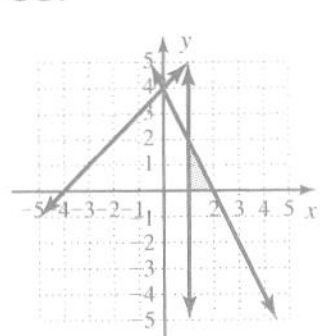

34.

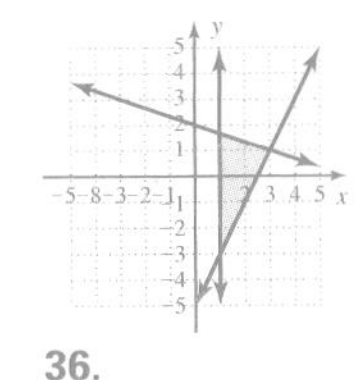

35.

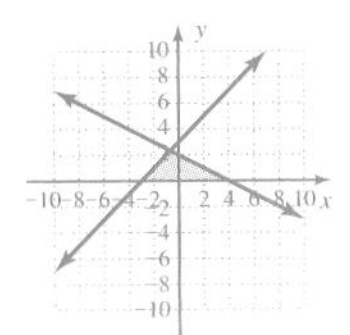

36.

37.

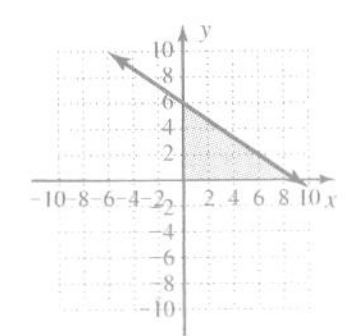

38.

39. $\begin{cases} y - x \le 1 \\ x + y > 3 \end{cases}$ 40. $\begin{cases} y - x \ge 1 \\ x + y > 3 \end{cases}$

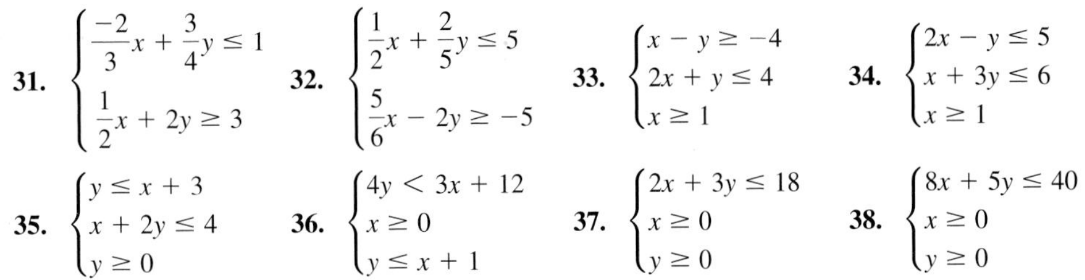

31. $\begin{cases} \dfrac{-2}{3}x + \dfrac{3}{4}y \le 1 \\ \dfrac{1}{2}x + 2y \ge 3 \end{cases}$ 32. $\begin{cases} \dfrac{1}{2}x + \dfrac{2}{5}y \le 5 \\ \dfrac{5}{6}x - 2y \ge -5 \end{cases}$ 33. $\begin{cases} x - y \ge -4 \\ 2x + y \le 4 \\ x \ge 1 \end{cases}$ 34. $\begin{cases} 2x - y \le 5 \\ x + 3y \le 6 \\ x \ge 1 \end{cases}$

35. $\begin{cases} y \le x + 3 \\ x + 2y \le 4 \\ y \ge 0 \end{cases}$ 36. $\begin{cases} 4y < 3x + 12 \\ x \ge 0 \\ y \le x + 1 \end{cases}$ 37. $\begin{cases} 2x + 3y \le 18 \\ x \ge 0 \\ y \ge 0 \end{cases}$ 38. $\begin{cases} 8x + 5y \le 40 \\ x \ge 0 \\ y \ge 0 \end{cases}$

Use the equations given to write the system of linear inequalities represented by each graph.

39.

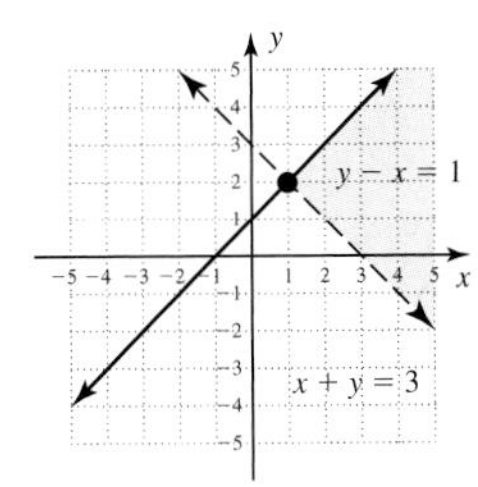

40.

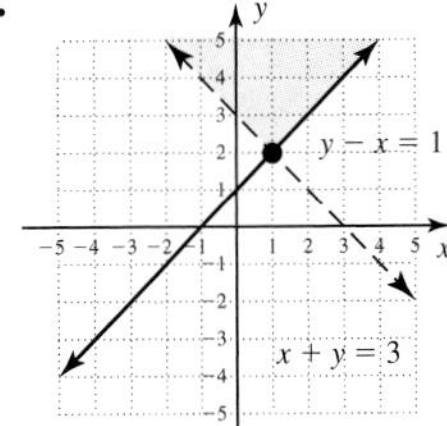

41.

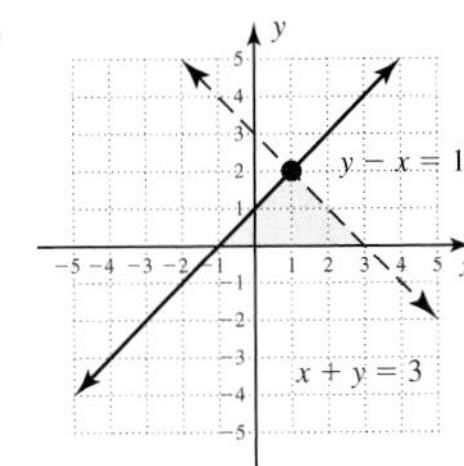

41. $\begin{cases} y - x \le 1 \\ x + y < 3 \\ y \ge 0 \end{cases}$

42.

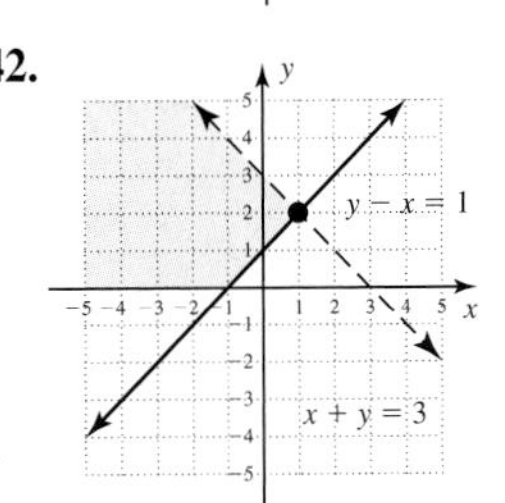

42. $\begin{cases} y - x \ge 1 \\ x + y < 3 \\ y \ge 0 \end{cases}$

Determine which of the ordered pairs given produces the maximum value of $f(x, y)$.

43. $f(x, y) = 12x + 10y$; (0, 0), (0, 8.5), (7, 0), (5, 3) (5, 3)

44. $f(x, y) = 50x + 45y$; (0, 0), (0, 21), (15, 0), (7.5, 12.5) (0, 21)

Determine which of the ordered pairs given produces the minimum value of $f(x, y)$.

45. $f(x, y) = 8x + 15y$; (0, 20), (35, 0), (5, 15), (12, 11) (12, 11)

46. $f(x, y) = 75x + 80y$; (0, 9), (10, 0), (4, 5), (5, 4) (5, 4)

Find the *maximum* value of the objective function $f(x, y) = 8x + 5y$ given the constraints shown.

47. $\begin{cases} x + 2y \le 6 \\ 3x + y \le 8 \\ x \ge 0 \\ y \ge 0 \end{cases}$ (2, 2)

48. $\begin{cases} 2x + y \le 7 \\ x + 2y \le 5 \\ x \ge 0 \\ y \ge 0 \end{cases}$ (3, 1)

Find the *minimum* value of the objective function $f(x, y) = 36x + 40y$ given the constraints shown.

49. $\begin{cases} 3x + 2y \ge 18 \\ 3x + 4y \ge 24 \\ x \ge 0 \\ y \ge 0 \end{cases}$ (4, 3)

50. $\begin{cases} 2x + y \ge 10 \\ x + 4y \ge 3 \\ x \ge 2 \\ y \ge 0 \end{cases}$ (5, 0)

WORKING WITH FORMULAS

Area Formulas

51. The area of a triangle is usually given as $A = \frac{1}{2}BH$, where B and H represent the base and height, respectively. The area of a rectangle can be stated as $A = BH$. If the base of both the triangle and rectangle is equal to 20 in., what are the possible values for H if the triangle must have an area *greater than* 50 in^2 and the rectangle must have an area *less than* 200 in^2? $5 < H < 10$

Volume Formulas

52. The volume of a cone is $V = \frac{1}{3}\pi r^2 h$, where r is the radius of the base and h is the height. The volume of a cylinder is $V = \pi r^2 h$. If the radius of both the cone and cylinder is equal to 10 cm, what are the possible values for h if the cone must have a volume *greater than* 200 cm^2 and the volume of the cylinder must be *less than* 850 cm^2? $\frac{6}{\pi} < h < \frac{17}{2\pi}$

APPLICATIONS

Write a system of linear inequalities that models the information given, then solve.

53. Gifts to grandchildren: Grandpa Faires is considering how to divide a \$50,000 gift between his two grandchildren, Julius and Anthony. After weighing their respective positions in life and family responsibilities, he decides he must bequeath at least \$20,000 to Julius, but no more than \$25,000 to Anthony. Determine the possible ways that Grandpa can divide the \$50,000.

54. Guns versus butter: Every year, governments around the world have to make the decision as to how much of their revenue must be spent on national defense and domestic improvements (guns versus butter). Suppose total revenue for these two needs was \$120 billion, and a government decides they need to spend at least \$42 billion on butter and no more than \$80 billion on defense. Determine the possible amounts that can go toward each need.

Solve the following linear programming problems.

55. Land/crop allocation: A farmer has 500 acres of land on which to plant corn and soybeans. During the last few years, market prices have been stable and the farmer anticipates a profit of \$900 per acre on the corn harvest and \$800 per acre on the soybeans. The farmer must take into account the time it takes to plant and harvest each crop, which is 3 hr/acre for corn and 2 hr/acre for soybeans. If the farmer has at most 1300 hr to plant, care for, and harvest each crop, how many acres of each crop should be planted in order to maximize profits?

56. Coffee blends: The owner of a coffee shop has decided to introduce two new blends of coffee in order to attract new customers—a *Deluxe Blend* and a *Savory Blend.* Each pound of the deluxe blend contains 30% Colombian and 20% Arabian coffee, while each pound of the savory blend contains 35% Colombian and 15% Arabian coffee (the remainder of each is made up of cheap and plentiful domestic varieties). The profit on the deluxe blend will be \$1.25 per pound, while the profit on the savory blend will be \$1.40 per pound. How many pounds of each should the owner make in order to maximize profit, if only 455 pounds of Colombian coffee and 250 pounds of Arabian coffee are currently available?

57. Manufacturing screws: A machine shop manufactures two types of screws—sheet metal screws and wood screws, using three different machines. Machine Moe can make a sheet metal screw in 20 sec and a wood screw in 5 sec. Machine Larry can make a sheet metal screw in 5 sec and a wood screw in 20 sec. Machine Curly, the newest machine (nyuk, nyuk) can make a sheet metal screw in 15 sec and a wood screw in 15 sec. (Shemp couldn't get a job because he failed the math portion of the employment exam.) Each machine can operate for only 3 hr each day before shutting down for maintenance. If sheet metal screws sell for 10 cents and wood screws sell for 12 cents, how many of each type should the machines be programmed to make in order to maximize revenue? (*Hint:* Standardize time units.)

58. Hauling hazardous waste: A waste disposal company is contracted to haul away some hazardous waste material. A full container of liquid waste weighs 800 lb and has a volume of 20 ft^3. A full container of solid waste weighs 600 lb and has a volume of 30 ft^3. The trucks used can carry at most 10 tons (20,000 lb) and have a carrying volume of 800 ft^3. If the trucking company makes \$300 for disposing of liquid waste and \$400 for disposing of solid waste, what is the maximum revenue per truck that can be generated?

59. Minimizing shipping costs: An oil company is trying to minimize shipping costs from its two primary refineries in Tulsa, Oklahoma, and Houston, Texas. All orders within the region are shipped from one of these two refineries. An order for 220,000 gallons comes in from a location

53.

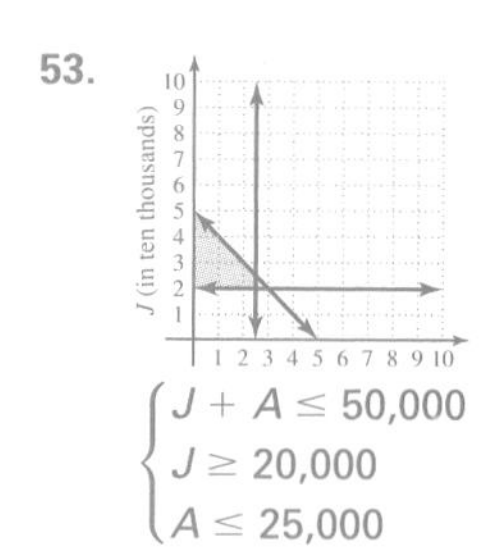

$\begin{cases} J + A \le 50{,}000 \\ J \ge 20{,}000 \\ A \le 25{,}000 \end{cases}$

54.

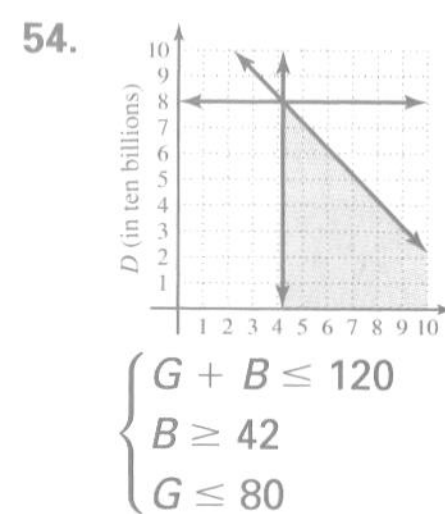

$\begin{cases} G + B \le 120 \\ B \ge 42 \\ G \le 80 \end{cases}$

55. 300 acres of corn; 200 acres of soybeans
56. 770 lb of deluxe and 640 lb of savory
57. 240 sheet metal screws; 480 wood screws
58. \$11,000
59. 220,000 gallons from Tulsa to Colorado; 100,000 gallons from Tulsa to Mississippi; 0 thousand gallons from Houston to Colorado; 150,000 gallons from Houston to Mississippi

in Colorado, and another for 250,000 gallons from a location in Mississippi. The Tulsa refinery has 320,000 gallons ready to ship, while the Houston refinery has 240,000 gallons. The cost of transporting each gallon to Colorado is \$0.05 from Tulsa and \$0.075 from Houston. The cost of transporting each gallon to Mississippi is \$0.06 from Tulsa and \$0.065 from Houston. How many gallons should be distributed from each refinery to minimize the cost of filling both orders?

60. **Minimizing transportation costs:** Robert's Las Vegas Tours needs to drive 375 people and 19,450 lb of luggage from Salt Lake City, Utah, to Las Vegas, Nevada, and can charter buses from two companies. The buses from company X carry 45 passengers and 2750 lb of luggage at a cost of \$1250 per trip. Company Y offers buses that carry 60 passengers and 2800 lb of luggage at a cost of \$1350 per trip. How many buses should be chartered from each company in order for Robert to minimize the cost?

WRITING, RESEARCH, AND DECISION MAKING

61. Is it possible for a 2×2 system of linear inequalities to have no solution? Justify your answer and include a graph.

62. Is it possible to have the entire rectangular grid as a solution to a system of linear inequalities? If so, give an example. If not, explain why not.

63. Graph the feasible region formed by the system $\begin{cases} x \geq 0 \\ y \geq 0 \\ y \leq 3 \\ x \leq 3 \end{cases}$. How would you describe this region? Select random points within the region or on any boundary line and evaluate the objective function $f(x, y) = 4.5x + 7.2y$. At what point (x, y) will this function be maximized? How does this relate to optimal solutions to a linear programing problem?

EXTENDING THE CONCEPT

64. I'm thinking of a three-digit number. Read backward, the new number is smaller than the original number, with a difference of 693. The digits sum to 17. What is the number?

65. Find the maximum value of the objective function $f(x, y) = 22x + 15y$ given the constraints $\begin{cases} 2x + 5y \leq 24 \\ 3x + 4y \leq 29 \\ x + 6y \leq 26 \\ x \geq 0 \\ y \geq 0 \end{cases}$.

MAINTAINING YOUR SKILLS

66. (1.3/1.4) Find all solutions (real and complex) by factoring: $x^3 - 5x^2 + 3x - 15 = 0$.

67. (4.7) Solve the rational inequality. Write your answer in interval notation. $\frac{x + 2}{x^2 - 9} > 0$

68. (1.1) Yolanda receives a \$10,000 inheritance from a distant aunt and decides to invest part of the money at 11% and the rest at 8%. If the total interest from the two investments for one year is \$995, how much did she invest at each rate?

69. (3.6) The resistance to current flow in copper wire varies directly as its length and inversely as the square of its diameter. A wire 8 m long with a 0.004-m diameter has a resistance of 1500 Ω. Find the resistance in a wire of like material that is 2.7 m long with a 0.005-m diameter.

70. (5.4) Solve for x: $-350 = 211e^{-0.025x} - 450$.

71. (5.6) Evaluate at $x = 8$, 10, and 12: $f(x) = \frac{250}{1 + 52e^{-0.75x}}$.

Answers

60. 3 buses from company X; 4 buses from company Y
61. Yes, answers will vary.
62. No, answers will vary.
63. (3, 3); optimal solutions occur at vertices

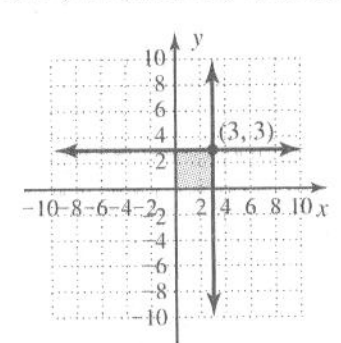

64. 881
65. max: $212.\overline{6}$ at $(\frac{29}{3}, 0)$
66. $x = 5, \pm\sqrt{3}\,i$
67. $x \in (-3, -2) \cup (3, \infty)$
68. \$6500 at 11%; \$3500 at 8%
69. 324 Ω
70. $x \approx 29.87$
71. ≈221.46, 243.01, 248.41

6.4 Systems and Absolute Value Equations and Inequalities

LEARNING OBJECTIVES

In Section 6.4 you will learn how to:

A. Solve absolute value equations graphically using a system

B. Solve absolute value inequalities graphically using a system

C. Use a procedural method for solving absolute value equations and inequalities

INTRODUCTION

Solving absolute value equations and inequalities is a much more conceptual endeavor than solving other types. This is because the endpoints that help form solution sets cannot be found using only the familiar properties of equality and inequality. These must be combined with the definition of absolute value to accurately name a solution. When a system approach is used, we get a visual picture of the solution set and a better understanding of *why* there are different possibilities, perhaps making fundamental ideas easier to retain and apply in context.

POINT OF INTEREST

The term "absolute value" and the vertical bars used as notation are of a very recent origin, having been in use only for the last 50 to 75 years. It is thought that the term came into use because its etymology (word history) can be traced to the Latin words *absolutus* and *absolvere,* the latter being a compound of "*ab,*" meaning away from or off, and "*solvere,*" meaning to free or loosen. Hence the term *absolute value* means a value that is *free* or *away from* any sign. *Source: The Words of Mathematics,* by Steven Schwartzman, Mathematical Association of America © 1994

A. Systems, Graphing, and Absolute Value Equations

The graph of $f(x) = |x|$ was first introduced in Section 2.2, where we noted it was a "V"-shaped graph with a vertex at (0, 0). Using transformations of the basic graph enables us to quickly sketch most any absolute value function. For $f(x) = |x - 2|$ we would simply draw the graph of $f(x) = |x|$ shifted 2 units to the right.

Consider the equation $4|x - 2| - 5 = 7$, or $|x - 2| = 3$ in simplified form. Since the two expressions forming this equation can be treated as separate functions, we can write the equation in system form as $\begin{cases} y_1 = |x - 2| \\ y_2 = 3 \end{cases}$. Solving this system graphically reveals some important information regarding absolute value equations, in particular that *two solutions are possible.*

EXAMPLE 1 Solve by graphing: $\begin{cases} y_1 = |x - 2| \\ y_2 = 3 \end{cases}$

Solution: The graph of $y = |x - 2|$ is a "V" function shifted 2 units right. The graph of $y = 3$ is a horizontal line through (0, 3). The graphs are shown in the figure, where we note they intersect at $(-1, 3)$ and $(5, 3)$. This means that $x = -1$ and $x = 5$ are solutions to the original equation given: $4|x - 2| - 5 = 7$. Verify by substitution.

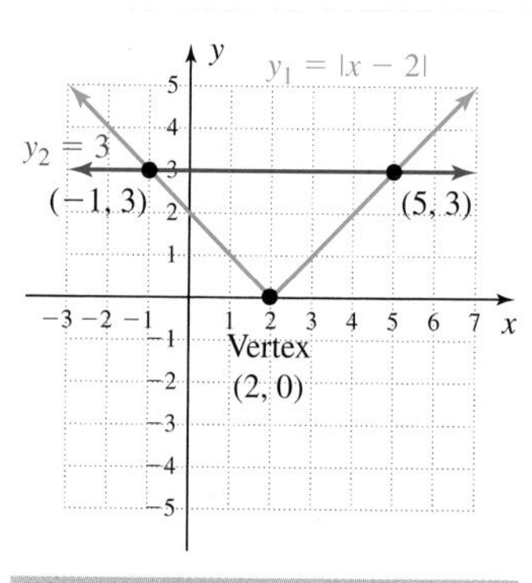

NOW TRY EXERCISES 7 THROUGH 18

From our previous work with systems, we know two graphs can intersect in different ways, or perhaps not intersect at all. Attempting to solve $3|x + 1| + 8 = 2$ leads to $|x + 1| = -2$ in simplified form (verify), which has no solution since $|x + 1|$ is always positive. Graphically, the line $y = -2$ is below the graph of $y = |x + 1|$ so they never intersect.

B. Solving Absolute Value Inequalities Using Systems and Graphing

The solution of absolute value *inequalities* can also be understood graphically. Since the intersection of two graphs indicates where the functions are equal, *the functions are unequal* (greater than or less than) *at all other points.* The process involves graphing the related equations and finding the interval(s) where output values from one function are greater than (graph is above) or less than (graph is below) the other.

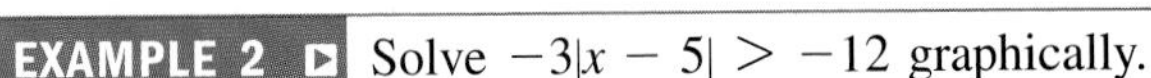

EXAMPLE 2 Solve $-3|x - 5| > -12$ graphically.

Solution: After dividing by -3, we have $|x - 5| < 4$ as the simplified form, and the related system is $\begin{cases} y_1 = |x - 5| \\ y_2 = 4 \end{cases}$. First graph $y = |x - 5|$, the graph of $y = |x|$ shifted 5 units right. The x-intercept is (5, 0) and the y-intercept is (0, 5). The graph of $y = 4$ is a horizontal line through (0, 4). The system is shown in the figure, where we note the graphs intersect at (1, 4) and (9, 4). Observe the graph of $y = |x - 5|$ is ***below the graph*** of $y = 4$ when x is between 1 and 9. This shows $|x - 5| < 4$ in this interval, and the solution set of the original inequality is $x > 1$ *and* $x < 9$. In interval notation we have $x \in (1, 9)$ with the endpoints excluded due to the strict inequality. Check using any point from this interval in the original inequality.

When x is between 1 and 9, $|x - 5|$ is less than 4.

NOW TRY EXERCISES 19 THROUGH 24

Graphing the simplified form of a "less than" absolute value inequality will show the solution set is always a **continuous interval.** As the next example illustrates, graphing the simplified form of a "greater than" absolute value inequality gives a solution set made up of two **disjoint intervals** (meaning the solution cannot be written as a single, continuous interval).

EXAMPLE 3 Solve $\frac{|3x + 2|}{-4} \le -1$ by graphing.

Solution: The simplified form is $|3x + 2| \ge 4$ and the related system is $\begin{cases} y_1 = |3x + 2| \\ y_2 = 4 \end{cases}$. First graph $y = |3x + 2|$. Solving $3x + 2 = 0$ gives $x = -\frac{2}{3}$, which is the vertex and x-intercept. The result will be a "V"-shaped graph with vertex at $\left(-\frac{2}{3}, 0\right)$ and y-intercept of (0, 2).

The graph of $y = 4$ is a horizontal line through (0, 4). The system is shown here, and we note the graphs intersect at $(-2, 4)$ and $\left(\frac{2}{3}, 4\right)$ using symmetry. The graph of $y = |3x + 2|$ is above that of $y = 4$ when $x < -2$ *or* $x > \frac{2}{3}$. The solution *cannot be written as a single interval,* since solutions occur in two disjoint parts. In interval notation we have: $x \in (-\infty, -2] \cup [\frac{2}{3}, \infty)$. Check using any point from these intervals.

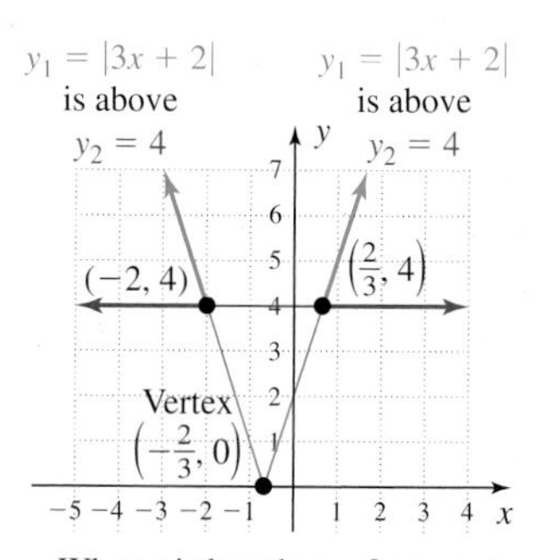

When x is less than -2 or greater than $\frac{2}{3}$, $|3x + 2|$ is greater than 4.

NOW TRY EXERCISES 25 THROUGH 29

C. Procedures for Solving Absolute Value Equations and Inequalities

Example 3 shows one of the limitations of graphical solutions, namely, that noninteger points of intersection may be hard to read. And while solving absolute value equations/inequalities graphically is a great way to build concepts and make connections, it can be too time consuming for practical use. Fortunately, the basic ideas can be summarized using the standard definition of absolute value given in Section R.1: $|x| = \begin{cases} -x \text{ if } x < 0 \\ x \text{ if } x \geq 0 \end{cases}$, along with the connectors "and" and "or" from Section 1.2.

Absolute Value Equations

Absolute value is a measure of a number's distance from zero. This means $|x| = 4$ will have two solutions, since there are two numbers that are 4 units from zero, -4 and 4.

This basic idea can be expanded to include situations where the quantity within the absolute value bars is an algebraic expression and suggests the following property.

ABSOLUTE VALUE EQUATIONS

If k is a positive number or zero and X represents a variable or an algebraic expression, the equation $|X| = k$ is equivalent to $X = -k$ *or* $X = k$. If $k < 0$, the equation $|X| = k$ has no solutions.

EXAMPLE 4 Solve $-5|2m - 7| + 2 = -13$.

Solution: Begin by writing the equation in simplified form.

$$-5|2m - 7| + 2 = -13 \quad \text{original equation}$$
$$-5|2m - 7| = -15 \quad \text{subtract 2}$$
$$|2m - 7| = 3 \quad \text{divide by } -5 \text{ (simplified form)}$$

If we consider $2m - 7$ as the variable expression "X" in the preceding property, we have:

$$2m - 7 = -3 \quad or \quad 2m - 7 = 3 \quad \text{apply properly}$$
$$2m = 4 \quad or \quad 2m = 10 \quad \text{add 7}$$
$$m = 2 \quad or \quad m = 5 \quad \text{result}$$

This equation has the two solutions noted: $\{2, 5\}$.

NOW TRY EXERCISES 30 THROUGH 35

The fact that $|2m - 7| = 3$ has two solutions is reinforced by our awareness of the related graphs. One is a "V"-shaped graph with a vertex at (3.5, 0), the other is a horizontal line through (0, 3) and the graphs must intersect at two points. We follow up with Example 5, where the expression within absolute value bars is nonlinear.

EXAMPLE 5 Solve the equation $|x^2 - 3x - 11| - 3 = 4$.

Solution: Write the equation in simplified form.

$$|x^2 - 3x - 11| = 7 \quad \text{simplified form}$$

Applying the preceding property leads to

$$\begin{array}{lcll} x^2 - 3x - 11 = -7 & or & x^2 - 3x - 11 = 7 & \text{apply property} \\ x^2 - 3x - 4 = 0 & or & x^2 - 3x - 18 = 0 & \text{set equal to zero} \\ (x - 4)(x + 1) = 0 & or & (x - 6)(x + 3) = 0 & \text{solve by factoring} \\ x = 4 \text{ or } x = -1 & or & x = 6 \text{ or } x = -3 & \text{result} \end{array}$$

This equation has the four solutions noted: $\{-3, -1, 4, 6\}$.

NOW TRY EXERCISES 36 THROUGH 47

GRAPHICAL SUPPORT

The fact that there are *four solutions* in Example 5 is easily verified using the graphs of $y_1 = |x^2 - 3x - 11|$ and $y_2 = 7$. For $y = x^2 - 3x - 11$, the portion of the graph that was below the x-axis (negative) has been reflected upward, and the line $y = 7$ intersects the graph in four places.

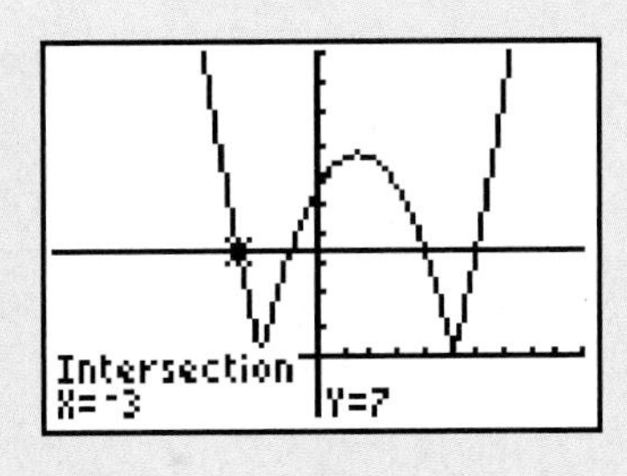

"Less Than" Absolute Value Inequalities

Absolute value inequalities can also be solved using the inherent properties of absolute value, including the concept of distance. The inequality $|x| \le 4$ asks for all numbers x whose distance from zero is **less than or equal to 4.** The solutions are found where $x \ge -4$ and $x \le 4$, which can be written as the single interval $-4 \le x \le 4$ (see Figure 6.20).

Figure 6.20

Distance from zero is less than 4

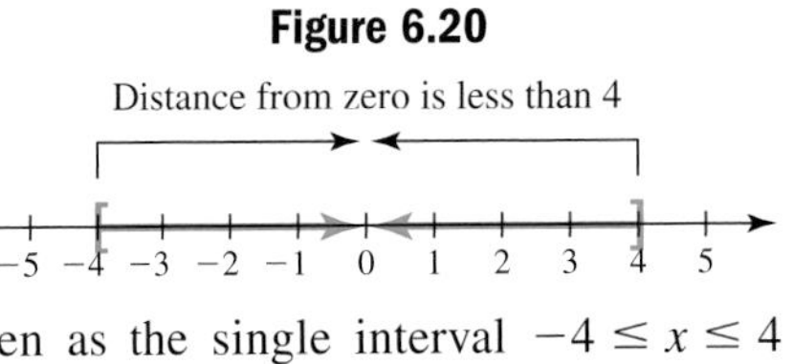

Expanding this concept to include algebraic expressions within the absolute value bars gives the following property:

SOLVING "LESS THAN" ABSOLUTE VALUE INEQUALITIES

If k is a positive number or zero, and X represents a variable or an algebraic expression, the inequality $|X| \le k$ is equivalent to $X \ge -k$ *and* $X \le k$, which *can be written* as a single interval (the interval is continuous). Similar results are obtained if a strict inequality is used. If $k < 0$, the inequality has no solutions.

EXAMPLE 6 Solve the inequality $-2\left|3 - \frac{x}{5}\right| - 1 \geq -3$. Write the answer in interval notation.

Solution:

$$-2\left|3 - \frac{x}{5}\right| - 1 \geq -3 \quad \text{original inequality}$$

$$-2\left|3 - \frac{x}{5}\right| \geq -2 \quad \text{add 1}$$

$$\left|3 - \frac{x}{5}\right| \leq 1 \quad \text{divide by } -2, \textit{ reverse symbol} \text{ (simplest form)}$$

Considering $3 - \frac{x}{5}$ as the "X" in the preceding property:

$$3 - \frac{x}{5} \geq -1 \quad \textit{and} \quad 3 - \frac{x}{5} \leq 1 \quad \text{apply property}$$

$$-\frac{x}{5} \geq -4 \quad \textit{and} \quad -\frac{x}{5} \leq -2 \quad \text{subtract 3}$$

$$x \leq 20 \quad \textit{and} \quad x \geq 10 \quad \text{multiply by } -5, \textit{ reverse symbol}$$

Solutions are written in the order they occur on the number line, giving the solution $\{x|10 \leq x \leq 20\}$. In interval notation: $x \in [10, 20]$.

NOW TRY EXERCISES 48 THROUGH 53

"Greater Than" Absolute Value Inequalities

For "greater than" inequalities, consider $|x| \geq 4$. Now we're asked to find all numbers x whose distance from zero is **greater than or equal to 4.** As Figure 6.21 shows, solutions are found to the left of -4 or in the interval to the right of 4 (including these endpoints). The fact that the intervals are disjoint is reflected in how we write the solution: $x \leq -4$ *or* $x \geq 4$.

Figure 6.21

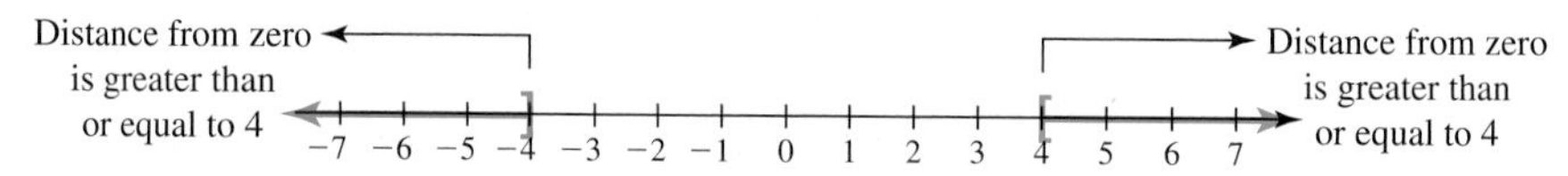

In general we have the following property:

> **SOLVING "GREATER THAN" ABSOLUTE VALUE INEQUALITIES**
> If k is a positive number or zero, and X represents a variable or an algebraic expression, the absolute value inequality $|X| \geq k$ is equivalent to $X \leq -k$ *or* $X \geq k$, which *cannot be written* as a single interval (the intervals are disjoint). Similar results are obtained if a strict inequality is used. If $k < 0$, the solution set is $x \in (-\infty, \infty)$.

EXAMPLE 7 Solve $3|7 - 5w| - 9 > 15$. Write the answer in interval notation.

Solution:

$$3|7 - 5w| - 9 > 15 \quad \text{original inequality}$$

$$|7 - 5w| > 8 \quad \text{add 9, then divide by 3 (simplest form)}$$

Consider $7 - 5w$ as the "X" in the preceding property:

$7 - 5w < -8$	*or*	$7 - 5w > 8$	apply property
$-5w < -15$	*or*	$-5w > 1$	subtract 7
$w > 3$	*or*	$w < -0.2$	divide by -5, *reverse symbol*

The solution is $\{w|w < -0.2 \text{ or } w > 3\}$, or in interval notation: $w \in (-\infty, -0.2) \cup (3, \infty)$.

NOW TRY EXERCISES 54 THROUGH 62

GRAPHICAL SUPPORT

From our earlier work, we can visualize the solution to Example 7 by noting that for $|7 - 5w| > 8$, the graph of $Y_1 = |7 - 5w|$ is above the graph of $Y_2 = 8$ only in the left- or right-hand intervals from their points of intersection, as shown.

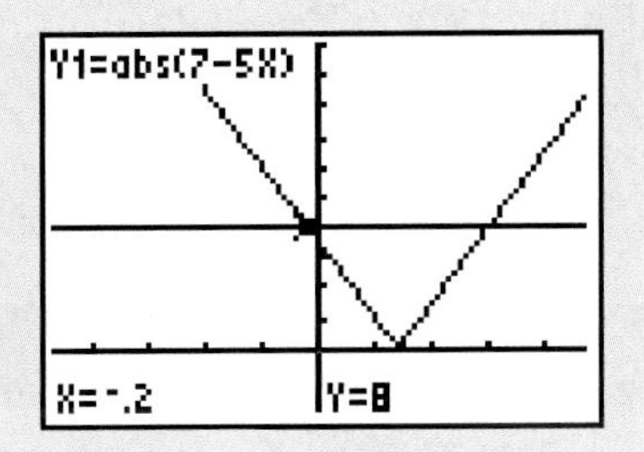

Absolute value inequalities are a fundamental part of the transition from algebra to calculus, playing a lead role in concepts related to the limit of a function. In fact, you'll find the direction/approach notation, domain, continuity, and many other ideas highlighted in this text form a strong bridge to exciting studies in higher mathematics. For a small glimpse, see Exercise 81.

TECHNOLOGY HIGHLIGHT

Visualizing Absolute Value Equations and Inequalities

The keystrokes shown apply to a TI-84 Plus model. Please consult your manual or our Internet site for other models.

In Section 1.1 we noted that an equation is simply a statement that two expressions are equal, and the solution to an equation was an input value that made the left-hand expression equal to the right-hand expression. This realization enables us to further explore what it means to solve an equation graphically, because we can then *treat each expression as a separate function,* create a system, then graph the system and find the points of intersection (if they exist). For $-2|x - 1| + 3 = -2$, we can enter the expression $-2|x - 1| + 3$ (without simplifying) as Y_1 on the Y= screen, and -2 as Y_2. Using the **Zoom4:ZDecimal** feature will produce the graph shown in Figure 6.22. We can find the points of intersection by TRACE ing (since we're using a friendly window) or using the 2nd TRACE **(CALC) 5:Intersect** feature (Section 6.1 *Technology Highlight*), or using a TABLE. Here we elect to use the **5:Intersect** option and we find the solutions are $(-1.5, -2)$ and $(3.5, -2)$. See Figure 6.23.

Figure 6.22

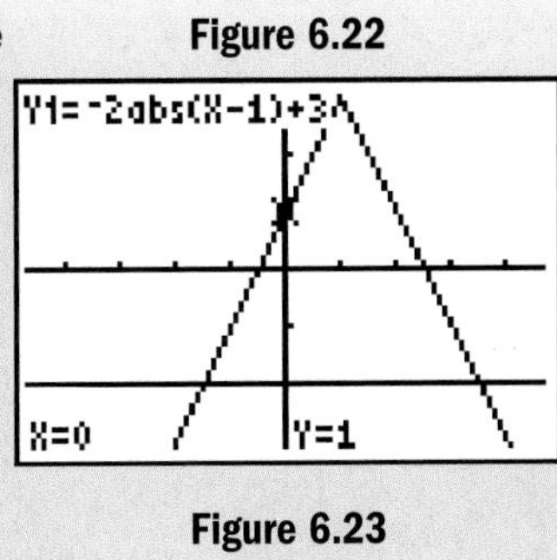

Figure 6.23

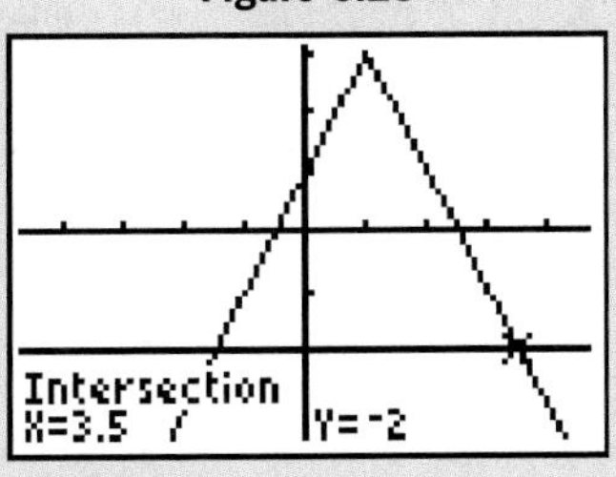

Placing the TABLE in **Auto** **Ask** mode and setting ΔTbl $= 0.5$ on the **2nd** **WINDOW** **(TblSet)** screen (since we know the intersection points), will help us solve any of the related inequalities, $-2|x - 1| + 3 > -2$, $-2|x - 1| + 3 < -2$, or any of the others. Scrolling up or down verifies the expressions are equal at $x = -1.5$ and 3.5, and that $Y_1 > Y_2$ anywhere in between (Figure 6.24). Use these ideas to solve the equations and inequalities given. Then verify the solutions by solving with paper and pencil.

Figure 6.24

X	Y1	Y2
-1.5	-2	-2
-1	-1	-2
-.5	0	-2
0	1	-2
.5	2	-2
1	3	-2
1.5	2	-2

X=-1.5

Exercise 1: $-|x + 3| + 3 = -1$ $x = 1, x = -7$

Exercise 2: $\left|x + \frac{3}{5}\right| - \frac{7}{10} = \frac{9}{5}$ $x = 1.9, x = -3.1$

Exercise 3: $-2.5|x - 2| + 8.6 \geq -1.4$ $-2 \leq x \leq 6$

Exercise 4: $3.2|x - 2.9| + 1.6 \leq 4.5$ $1.99375 \leq x \leq 3.80625$

6.4 EXERCISES

CONCEPTS AND VOCABULARY

Fill in the blank with the appropriate word or phrase. Carefully reread the section if needed.

1. When multiplying or dividing by a negative quantity, we <u>reverse</u> the inequality to maintain the correct relationship between the left- and right-hand expressions.
2. To write an absolute value equation or inequality in simplified form, we <u>isolate</u> the absolute value expression on one side of the equality or inequality.
3. The absolute value equation $|2x + 3| = 7$ is true when $2x + 3 =$ <u>7</u> or when $2x + 3 =$ <u>−7</u>.
4. The absolute value inequality $|3x - 6| < 12$ is true when $3x - 6 <$ <u>12</u> and $3x - 6 >$ <u>−12</u>.

Describe each solution set (assume $k > 0$). Justify your answer and include a diagram.

5. $|ax + b| < -k$
no solution; answers will vary
6. $|ax + b| > -k$
all real numbers; answers will vary

DEVELOPING YOUR SKILLS

Write each absolute value equation or inequality in simplified form. Do not solve.

7. $5|3m + 7| - 2.8 < 13.7$ $|3m + 7| < 3.3$
8. $7|2t + 5| + 6.3 < 11.2$ $|2t + 5| < 0.7$
9. $-2|p| - 3 = -4$ $|p| = 0.5$
10. $-3|n| - 5 = -8$ $|n| = 1$
11. $-3\left|\frac{x}{2} + 4\right| - 1 < -4$
12. $-2\left|3 - \frac{v}{3}\right| + 1 < -5$ $\left|3 - \frac{v}{3}\right| > 3$

Solve each absolute value equation by graphing the related system.

13. $-2|x + 3| - 4 = -14$
14. $-3|2x - 3| + 7 = -14$ $x = -2, x = 5$
15. $3|x + 5| + 6 = -3$
16. $4|3x - 1| + 8 = -4$ no solution
17. $-2|3x| + 5 = -19$
18. $-5|2x| - 4 = -14$ $x = 1, x = -1$

Solve each inequality by graphing the related system. Write solutions in interval notation.

19. $|x - 2| \leq 7$
20. $|x + 1| \leq 3$ $x \in [-4, 2]$
21. $-3|x| - 2 > 4$
22. $-2|x| + 3 > 7$ no solution
23. $\frac{|2x + 5|}{3} + 8 < 9$
24. $\frac{|3x - 2|}{2} + 6 < 8$ $x \in \left(-\frac{2}{3}, 2\right)$
25. $|x + 3| > 7$ $x \in (-\infty, -10) \cup (4, \infty)$
26. $|x - 1| > 5$ $x \in (-\infty, -4) \cup (6, \infty)$
27. $-2|x| - 5 \leq -11$ $x \in (-\infty, -3] \cup [3, \infty)$

11. $\left|\frac{x}{2} + 4\right| > 1$

13. $x = 2, x = -8$

15. no solution

17.

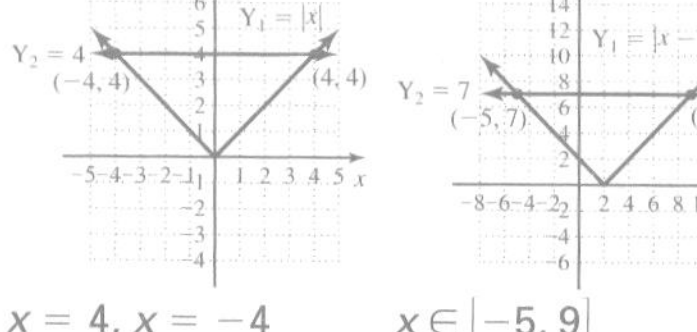

$x = 4, x = -4$

19. $x \in [-5, 9]$

21.

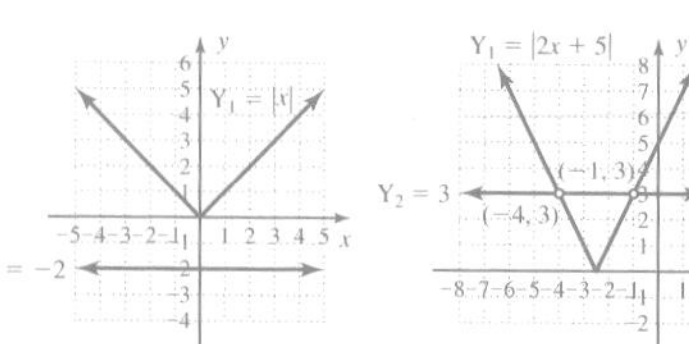

no solution

23. $x \in (-4, -1)$

25.

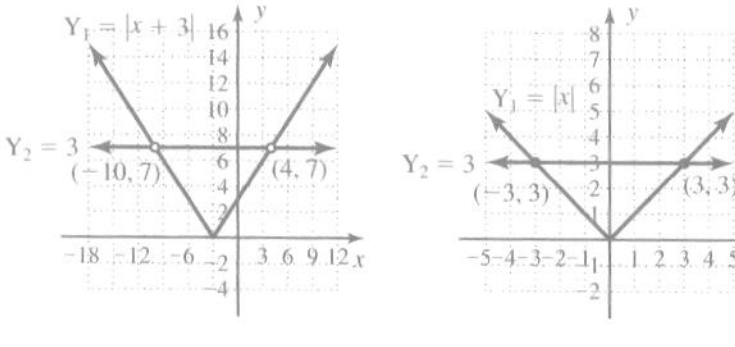

27.

28. $Y_1 < Y_2$
29. $Y_2 \geq Y_1$
30. $x = 1, x = 9$
31. $x = -9, x = 3$
32. $n = -6, n = 1$
33. $n = \frac{-17}{2}, n = \frac{3}{2}$
34. $x = \frac{-11}{2}, x = \frac{-3}{2}$
35. $n = -6, n = -2$
36. $x = 7, x = -5, x = 5, x = -3$
37. $x = 7, x = -2, x = 6, x = -1$
38. $x = -7, x = 4, x = -5, x = 2$
39. $x = -9, x = 5, x = -7, x = 3$
40. $x = \frac{7}{2}, x = -3, x = \frac{3}{2}, x = -1$
41. $x = \frac{-10}{3}, x = 4, x = \frac{-4}{3}, x = 2$
42. $x = -5.2, x = 2.8$
43. $n = 1.8, n = 9.8$
44. $n = \frac{1}{6}, n = \frac{13}{6}$
45. $x = -3.5, x = 1$
46. $x = 3.5, x = 11.5$
47. $n = -16.2, n = -2.2$
48. $x \in (-5, -3)$
49. $m \in [-1, 5]$
50. $m \in \left(\frac{8}{3}, \frac{14}{3}\right)$
51. $n \in \left(\frac{-9}{2}, \frac{3}{2}\right)$
52. no solution
53. no solution
54. $x \in \left(-\infty, \frac{3}{7}\right] \cup [1, \infty)$
55. $x \in \left(-\infty, \frac{-9}{2}\right] \cup \left[\frac{-5}{2}, \infty\right)$
56. $x \in (-\infty, \infty)$
57. $p \in (-\infty, \infty)$
58. $x \in (-\infty, 0) \cup (5, \infty)$
59. $x \in \left(-\infty, \frac{5}{8}\right) \cup \left(\frac{7}{8}, \infty\right)$
60. no solution
61. $q \in (-\infty, -1] \cup \left[\frac{13}{3}, \infty\right)$
62. $p \in (-\infty, -18.5) \cup (11.5, \infty)$
63. $x \in [-1, 3]$
64. $x \in [-1, 4]$
65. $x \in (-\infty, -3] \cup [3, \infty)$
66. $x \in [-2, 2]$
67. $x \in (-\infty, -4] \cup [-2, 2] \cup [4, \infty)$
68. $x \in (-3, -1) \cup (1, 3)$
69. 45 in. $\leq d \leq$ 51 in.
70. $41.775 < h < 58.225$; no
71. $2{,}325{,}000 \leq B \leq 2{,}575{,}000$

Write the inequality represented by the given graphs and solution set.

28.

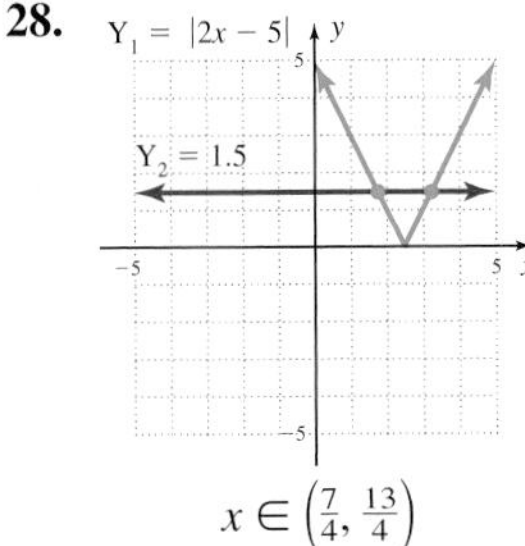

$x \in \left(\frac{7}{4}, \frac{13}{4}\right)$

29.

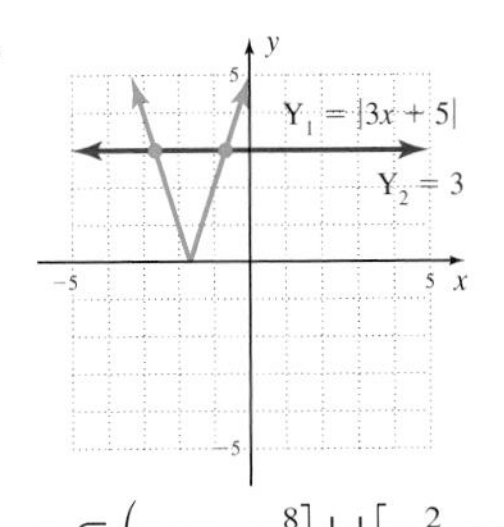

$x \in \left(-\infty, -\frac{8}{3}\right] \cup \left[-\frac{2}{3}, \infty\right)$

Solve the absolute value equations.

30. $|x - 5| + 8 = 12$
31. $|x + 3| + 4 = 10$
32. $|2n + 5| + 3 = 10$
33. $|2n + 7| - 1 = 9$
34. $-2|2x + 7| - 1 = -9$
35. $-2|n + 4| + 3 = -1$
36. $|x^2 - 2x - 25| = 10$
37. $|x^2 - 5x - 10| = 4$
38. $|x^2 + 3x - 19| = 9$
39. $|x^2 + 4x - 33| = 12$
40. $|2x^2 - x - 12| = 9$
41. $|3x^2 - 2x - 24| = 16$
42. $2.5|x + 1.2| - 3.6 = 6.4$
43. $1.2|n - 5.8| + 6.7 = 11.5$
44. $\left|\frac{3}{5}n - \frac{7}{10}\right| + \frac{4}{15} = \frac{13}{15}$
45. $\left|\frac{2}{3}x + \frac{5}{6}\right| - \frac{7}{12} = \frac{11}{12}$
46. $8.7|x - 7.5| - 26.6 = 8.2$
47. $5.3|n + 9.2| + 6.7 = 43.8$

Solve the absolute value inequalities. Write solutions in interval notation.

48. $3|x + 4| + 5 < 8$
49. $5|m - 2| - 7 \leq 8$
50. $|3m - 11| + 6 < 9$
51. $|2n + 3| - 5 < 1$
52. $|4 - 3x| + 12 < 7$
53. $|2 - 7x| + 7 \leq 4$
54. $3|5 - 7x| + 9 \geq 15$
55. $5|2x + 7| + 1 \geq 11$
56. $|4x - 9| + 6 \geq 4$
57. $|5p - 3| + 8 > 6$
58. $4|5 - 2x| - 9 > 11$
59. $10|3 - 4x| - 8 > -3$
60. $-4.1|2p + 7| - 12.3 > 16.4$
61. $-3.9|3q - 5| + 8.7 \leq -22.5$
62. $0.9|2p + 7| - 16.11 > 10.89$
63. $|x^2 - 2x| \leq 3$
64. $|x^2 - 3x| \leq 4$
65. $|x^2 - 4| \geq 5$
66. $|x^2 - 1| \leq 3$
67. $|x^2 - 10| \geq 6$
68. $|x^2 - 5| < 4$

WORKING WITH FORMULAS

69. Spring oscillation $|d - x| \leq L$

A weight attached to a spring hangs at rest a distance of x in. off the ground. If the weight is pulled down (stretched) a distance of L in. and released, the weight begins to bounce and its distance d off the ground at any time satisfies the given inequality. If x equals 4 ft and the spring is stretched 3 in. and released, solve the inequality to find between what distances from the ground the weight will oscillate.

70. A "fair" coin $\left|\frac{h - 50}{5}\right| < 1.645$

If we flip a coin 100 times, we expect "heads" to come up about 50 times if the coin is "fair." In a study of probability, it can be shown that the number of heads h that appears in such an experiment must satisfy the given inequality to be considered "fair." Solve this inequality for h. If you flipped a coin 100 times and obtained 40 heads, is the coin "fair"?

APPLICATIONS

71. Barrels of oil transported: The number of barrels of oil B that a tanker must carry to make a profit is described by the inequality $|B - 2{,}450{,}000| \leq 125{,}000$. Find the range of values for B that represent a profitable level.

72. $300 < d < 400$
73. lowest \$41,342; highest \$65,330
74. $32,500 \le h \le 37,600$; yes
75. $6.5 < d < 13.5$
76. $d < 210$ or $d > 578$
77. $|W - 14| \le 0.1$; $W \in [13.9, 14.1]$; weight must be at least 13.9 oz but no more than 14.1 oz

72. Optimal fishing depth: When deep-sea fishing, the optimal depths d (in feet) for catching a certain type of fish satisfy the inequality $28|d - 350| - 1400 < 0$. Find the range of depths that offer the best fishing.

73. Faculty salaries: At a private college, the range of faculty salaries (in dollars) is given by the inequality $|s - 53{,}336| \le 11{,}994$. Find the highest and lowest salaries on campus.

74. Altitude of jet stream: To take advantage of the jet stream, an airplane must fly at a height h (in feet) that satisfies the inequality $|h - 35{,}050| \le 2550$. If the pilot flies at 34,000 ft will she be in the jet stream?

75. Distance from shore: For the best fishing grounds, a trawler should lower its nets within d mi from shore, given by $2|d - 10| < 7$. Find the range of distances that offers the best fishing.

76. Submarine depth: The sonar operator on a submarine detects an old World War II submarine net and must decide to detour over or under the net. The computer gives him a depth model $|d - 394| - 20 > 164$ where d is the depth in feet that represents safe passage. At what depth should the submarine travel to go under or over the net?

WRITING, RESEARCH, AND DECISION MAKING

77. The machines that fill boxes of breakfast cereal are programmed to fill each box within a certain tolerance. If the box is overfilled, the company loses money. If it is underfilled, it is considered unsuitable for sale. Suppose that boxes marked "14 ounces" of cereal must be filled to within 0.1 oz. Write this relationship as an absolute value inequality, then solve the inequality and explain what your answer means. Let W represent weight.

x	y
−4	4
−2	2
0	0
2	2
4	4

78. Graph the function $y = \sqrt{x^2}$ by completing the given table and plotting the points. Do you recognize the graph? Can you write $y = \sqrt{x^2}$ as another function? Write the function $y = \sqrt{(x - 3)^2}$ in the same alternative form. What can you conclude?

78. $y = |x|$; $y = |x - 3|$
79. a. $x = 4$
b. $x \in [\frac{4}{3}, 4]$
c. $x = 0$
d. $x \in (-\infty, \frac{3}{5}]$
80. $x \in [-2, -1] \cup [0, 1]$
81. $x \in (-1.005, -0.995)$
82. $m = \frac{-7}{4}$; $(0, 2)$; $(4, -5)$
83. D: $x \in (-\infty, \infty)$
R: $y \in [0, 3] \cup [6, \infty)$
84. $f(x) = (x - 4)(x - 2)(x + 1)$; $x = 4, x = 2, x = -1$
85. $x = \dfrac{\ln 2.5 - 1}{-0.025}$; $x \approx 3.35$

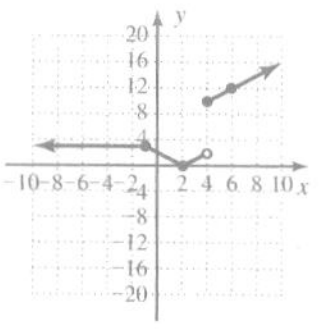

EXTENDING THE CONCEPT

79. Determine the value or values (if any) that will make the equation or inequality true.

a. $|x| + x = 8$ **b.** $|x - 2| \le \dfrac{x}{2}$ **c.** $x - |x| = x + |x|$ **d.** $|x + 3| \ge 6x$

80. Solve the inequality $|x^2 + x - 1| \le 1$. Answer in interval notation.

81. For $f(x) = 2x + 3$, $L = 1$, and $\epsilon = 0.01$, what values of x satisfy $|f(x) - L| < \epsilon$?

MAINTAINING YOUR SKILLS

82. (2.3) Given $7x + 4y = 8$, find $m = \dfrac{\Delta y}{\Delta x}$, then locate the y-intercept and use this slope to locate another point on the line.

83. (3.7) Graph the piecewise-defined function

$$f(x) = \begin{cases} 3 & x \le -1 \\ |x - 2| & -1 < x < 4 \\ x + 6 & x \ge 4 \end{cases}$$

and state its domain and range.

84. (4.3) Use the rational roots theorem to write $f(x) = x^3 - 5x^2 + 2x + 8$ in completely factored form, then solve $f(x) = 0$.

85. (5.1) Given $f(x) = -150e^{-0.025x+1}$, what is the value of x if $f(x) = -375$? Answer in exact and approximate form.

86.

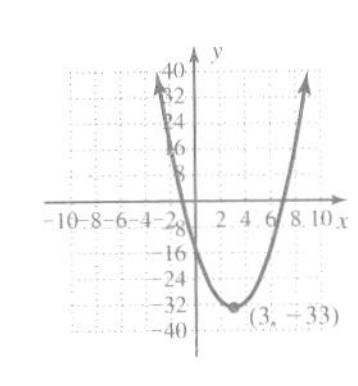

86. (3.4) Graph the quadratic function by completing the square: $f(x) = 2x^2 - 12x - 15$.

87. (3.1) Given $f(x) = x^3 + 3x^2 - 10x$ and $g(x) = x - 2$, find:

a. $(f - g)(x)$ **b.** $(f \cdot g)(x)$ **c.** $\left(\frac{f}{g}\right)(x)$ **d.** $(f \circ g)(x)$

a. $x^3 + 3x^2 - 11x + 2$ b. $x^4 + x^3 - 16x^2 + 20x$ c. $x^2 + 5x$ d. $x^3 - 3x^2 - 10x + 24$

MID-CHAPTER CHECK

1. (1, 1); consistent
2. (5, 3); consistent
3. 20 oz
4. no
5. The second equation is a multiple of the first equation.
6. (1, 2, 3)
7. (1, 2, 3)
8. no solution
9. Mozart = 8 yr; Morphy = 13 yr; Pascal = 16 yr
10. 2 table candles, 9 holiday candles

1. Solve using the substitution method. State whether the system is consistent, inconsistent, or dependent. $\begin{cases} x - 3y = -2 \\ 2x + y = 3 \end{cases}$

2. Solve the system using elimination. State whether the system is consistent, inconsistent, or dependent. $\begin{cases} x - 3y = -4 \\ 2x + y = 13 \end{cases}$

3. Solve using a system of linear equations and any method you choose: How many ounces of a 40% acid, should be mixed with 10 ounces of a 64% acid, to obtain a 48% acid solution?

4. Determine whether the ordered triplet is a solution to the system.

$$\begin{cases} 5x + 2y - 4z = 22 \\ 2x - 3y + z = -1; \quad (2, 0, -3) \\ 3x - 6y + z = 2 \end{cases}$$

5. The system shown here is a dependent system. Without solving, state why.

$$\begin{cases} x + 2y - 3z = 3 \\ 2x + 4y - 6z = 6 \\ x - 2y + 5z = -1 \end{cases}$$

6. Solve the system of equations:

$$\begin{cases} x + 2y - 3z = -4 \\ 2y + z = 7 \\ 5y - 2z = 4 \end{cases}$$

7. Solve using elimination:

$$\begin{cases} 2x + 3y - 4z = -4 \\ x - 2y + z = 0 \\ -3x - 2y + 2z = -1 \end{cases}$$

8. Solve using the method of your choice: $-2|x - 3.5| + 29 \geq 34$.

9. If you add Mozart's age when he wrote his first symphony, with the age of American chess player Paul Morphy when he began dominating the international chess scene, and the age of Blaise Pascal when he formulated his well-known *Essai pour les coniques* (Essay on Conics), the sum is 37. At the time of each event, Paul Morphy's age was three years less than twice Mozart's, and Pascal was three years older than Morphy. Set up a system of equations and find the age of each.

10. Solve using linear programming: Dave and Karen make table candles and holiday candles and sell them out of their home. Dave works 10 min and Karen works 30 min on each table candle, while Karen works 40 min and Dave works 20 min on each holiday candle. Dave can work at most 3 hr and 20 min (200 min) per day on the home business, while Karen can work at most 7 hr. If table candles sell for \$4 and holiday candles sell for \$6, how many of each should be made to maximize their revenue?

REINFORCING BASIC CONCEPTS

Window Size and Graphing Technology

Since most substantial applications involve noninteger values, technology can play an important role in applying mathematical models. But with its use comes a heavy responsibility to use it carefully. A very real effort must be made to determine the best approach and to secure a reasonable estimate. This is the only way to guard against (the inevitable) keystroke errors, or ensure a window size that properly displays the results.

Rationale

On October 1, 1999, the newspaper *USA TODAY* ran an article titled, "Bad Math added up to Doomed Mars Craft." The article told of how a \$125,000,000.00 spacecraft was lost, apparently because the team of scientists that *plotted the course* for the craft used U.S. units of measurement, while the team of scientists *guiding* the craft were using metric units. NASA's space chief was later quoted, "The problem here was not the error, it was the failure of . . . the checks and balances in our process to detect the error." No matter how powerful the technology, always try to begin your problem-solving efforts with an estimate.

Estimation and Viewing Windows

In applying graphing technology to linear systems and problem solving, our initial concern is setting the size of the viewing window. For example, if we tried solving the system

$$\begin{cases} y = \dfrac{(15x + 127.6)}{5} \\ y = -2.9x + 52 \end{cases}$$

using the standard window of a graphing calculator (ZOOM **6:ZStandard** on the TI-84 Plus), the "graph" shown in Figure 6.25 would appear. Note the screen displays only one of the lines, with the other being "out of the viewing window." To prevent a chaotic search for a better window, we can preset the size of the window using a reasoned estimate. Begin by exploring the **context** of the problem, asking questions about the range of possibilities: How fast can a human run? How much does a new car cost? What is a reasonable price for a ticket? What is the total available to invest? There is no calculating involved in these estimates, they simply rely on "horse sense" and human experience. In many applied problems, the input and output values must be positive—which means the solution will appear in the first quadrant, narrowing the possibilities considerably.

Figure 6.25

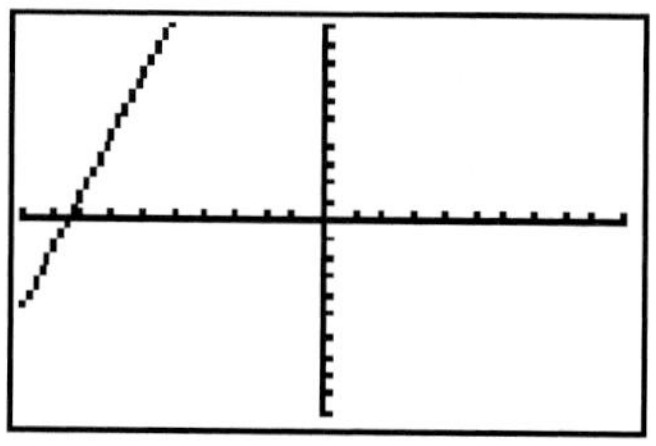

Example 1: "Erin just filled both her boat and Blazer with gas, at a total cost of \$125.97. She purchased 35.7 gallons of premium for her boat and 15.3 gal of regular for her Blazer. Premium gasoline cost \$0.10 per gallon more than regular. What was the cost per gallon of each grade of gasoline?" Use this information to set the viewing window of your graphing calculator, in preparation for solving the problem using a system and graphing technology.

Solution: Asking how much *you* paid for gas the last time you filled up should serve as a fair estimate. Certainly (in 2005) a cost of \$4.00 or more per gallon in the United States is too high, and a cost of \$1.50 per gallon or less would be too low. Also, we can estimate a solution by assuming that both kinds of gasoline cost the same. This would mean 51 gal were purchased for about \$126, and a quick division would place the estimate at

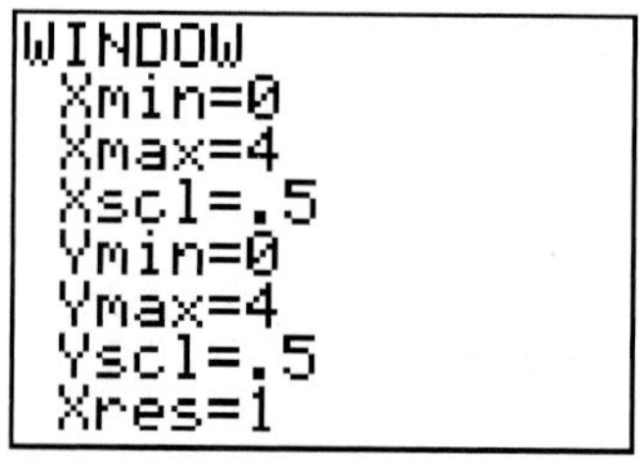

WINDOW
Xmin=0
Xmax=4
Xscl=.5
Ymin=0
Ymax=4
Yscl=.5
Xres=1

near $\frac{104}{51} \approx \$2.47$ per gallon. A good viewing window would be restricted to the first quadrant (since cost > 0) with maximum values of Xmax = 4 and Ymax = 4.

Example 2: "Sharon just won a \$12,500 cash prize and placed the money in two investments. She invests \$4520 in a tax-exempt bond, even though it pays 3.45% less interest than the second account, where she invests the remaining prize money (\$7980). After 1 yr she has earned \$1056.56 in interest. What is the interest rate of each investment?" Use this information to set the viewing window of your calculator, in preparation for solving the problem using a system of equations and graphing technology.

Solution: Reason that a 13% to 14% return on a nonspeculative investment would be "on the high side," so we expect the answer to be no higher. If we assume that all the money went into a single investment, a quick division shows a rate of return of approximately $\frac{1060}{12{,}500} \approx 0.085$ or 8.5%. A good viewing window would be restricted to the first quadrant (since values must be positive) with maximum values of Xmax = 0.14 and Ymax = 0.14.

```
WINDOW
 Xmin=0
 Xmax=.14
 Xscl=.01
 Ymin=0
 Ymax=.14
 Yscl=.01
 Xres=1
```

Solving a 2 × 2 System Using Graphing Technology

In Section 6.1 we concentrated on building a system of equations from the context of an application. Once an estimate has been determined and a window size set, solving a linear system of equations by graphing becomes only a matter of proper input and interpretation of results. The process is summarized here:

SOLVING A 2 × 2 SYSTEM USING GRAPHING TECHNOLOGY

1. Build a system of equations from a careful reading of the problem.
2. Estimate a solution, then use the estimate and context of the exercise to manually set an appropriate window size.
3. Solve both equations for the same variable, using it as the dependent or "y" variable in the system. Enter both equations on the Y= screen.
4. Graph the lines to see if the solution appears in the viewing window (that both lines appear and do intersect). Adjust the window as needed.
5. Have the calculator find this point of intersection using the 2nd CALC **5:Intersect** feature.

Exercise 1: Solve Examples 1 and 2 here using graphing technology.

Exercise 2: Re-solve Exercises 69 and 70 from Section 6.1 using graphing technology.

Ex 1 $\begin{cases} 15.3R + 35.7P = 125.97 \\ P = R + 0.10 \end{cases}$
Premium: \$2.50/gal
Regular: \$2.40/gal

Ex 2 $\begin{cases} 4520x + 7980y = 1056.56 \\ y = x + 0.0345 \end{cases}$
Bond: 6.25%
Second account: 9.7%

6.5 Solving Linear Systems Using Matrices and Row Operations

LEARNING OBJECTIVES

In Section 6.5 you will learn how to:

A. State the size of a matrix and identify entries in a specified row and column

B. Form the augmented matrix of a system of equations

C. Solve a system of equations using row operations

D. Recognize inconsistent and dependent systems

E. Model and solve applications using linear systems

INTRODUCTION

While the current methods of substitution and elimination can be applied to systems of any size, they become somewhat cumbersome and time-consuming for those larger than 3×3. In point of fact, most applications in operations research, business, government, and industry involve extremely large systems, and to solve them efficiently, a procedure that streamlines the solution process is needed. Matrix methods make this process much more routine. In addition, these methods are easily programmable, offering the ability to solve large systems in an instant.

POINT OF INTEREST

Although the Japanese mathematician Seki Kowa (1642–1708) apparently used matrices to solve systems of equations years before, credit for the development of matrix methods is often attributed to Arthur Cayley (1821–1895). Cayley got the idea of matrices by noticing the coefficient patterns in equations and used their matrix form as "a convenient mode of expression."

A. Introduction to Matrices

In general terms, a **matrix** is simply a rectangular arrangement of numbers, called the **entries** of the matrix. **Matrices** (plural of matrix) are denoted by enclosing the entries between a left and right bracket, and named using a capital letter, such as $A = \begin{bmatrix} 1 & -3 & 2 \\ 5 & 1 & -1 \end{bmatrix}$ and $B = \begin{bmatrix} 2 & -1 & 3 \\ 4 & 6 & -2 \\ 1 & 0 & -1 \end{bmatrix}$. They occur in many different sizes as defined by the number of **rows** and **columns** each has, with the number of rows always given first. Matrix A is said to be a 2×3 (two by three) matrix, since it has two rows and three columns (*columns* refer to those entries in a vertical stack, much like the vertical columns that support a heavy roof). Matrix B is a 3×3 (three by three) matrix.

EXAMPLE 1A Determine the size of each matrix and identify the entry located in the second row and second column.

a. $B = \begin{bmatrix} 3 & -2 \\ 1 & 5 \\ -4 & 3 \end{bmatrix}$ **b.** $C = \begin{bmatrix} 0.2 & -0.5 & 0.7 \\ -1 & 0.3 & 1 \\ 2.1 & -0.1 & 0.6 \end{bmatrix}$

Solution: **a.** Matrix B is 3×2. The row 2, column 2 entry is 5.

b. Matrix C is 3×3. The row 2, column 2 entry is 0.3.

If a matrix has the same number of rows and columns, it's called a **square matrix.** From Example 1(a), matrix C is a square matrix, while matrix B is not. For square matrices, the values on a diagonal line *from the upper left to the lower right* are called the **diagonal entries** and are said to be **on the diagonal** of the matrix. When solving systems using matrices, much of our focus is on these diagonal entries.

EXAMPLE 1B Name the diagonal entries of each matrix.

a. $A = \begin{bmatrix} 1 & 4 \\ -2 & -3 \end{bmatrix}$ **b.** $C = \begin{bmatrix} 0.2 & -0.5 & 0.7 \\ -1 & 0.3 & 1 \\ 2.1 & -0.1 & 0.6 \end{bmatrix}$

Solution: **a.** The diagonal entries for matrix A are 1 and -3.

b. For matrix C, the diagonal entries are 0.2, 0.3, and 0.6.

NOW TRY EXERCISES 7 THROUGH 9

B. The Augmented Matrix of a System of Equations

A matrix derived from a system of equations is called an **augmented matrix.** It is created by augmenting or joining the **coefficient matrix,** formed by the variable coefficients, with the **matrix of constants** into a single matrix. The coefficient matrix for the system $\begin{cases} 2x + 3y - z = 1 \\ x + z = 2 \\ x - 3y + 4z = 5 \end{cases}$ is $\begin{bmatrix} 2 & 3 & -1 \\ 1 & 0 & 1 \\ 1 & -3 & 4 \end{bmatrix}$, and the matrix of constants is $\begin{bmatrix} 1 \\ 2 \\ 5 \end{bmatrix}$. These two are joined to form the augmented matrix, with a dotted line often used to separate the two as shown here: $\left[\begin{array}{ccc|c} 2 & 3 & -1 & 1 \\ 1 & 0 & 1 & 2 \\ 1 & -3 & 4 & 5 \end{array}\right]$. It's important to note the use of a zero placeholder for the y-variable in the second row of the matrix, signifying there is no y-variable in the corresponding equation.

EXAMPLE 2 Form the augmented matrix for each system, and name the diagonal entries of each matrix of coefficients.

a. $\begin{cases} \frac{1}{2}x + y = -7 \\ x + \frac{2}{3}y + \frac{5}{6}z = \frac{11}{12} \\ -2y - z = -3 \end{cases}$ **b.** $\begin{cases} x + 4y - z = -10 \\ 2x + 5y + 8z = 4 \\ x - 2y - 3z = -7 \end{cases}$

Solution: **a.** $\begin{cases} \frac{1}{2}x + y = -7 \\ x + \frac{2}{3}y + \frac{5}{6}z = \frac{11}{12} \\ -2y - z = -3 \end{cases} \longrightarrow \left[\begin{array}{ccc|c} \frac{1}{2} & 1 & 0 & -7 \\ 1 & \frac{2}{3} & \frac{5}{6} & \frac{11}{12} \\ 0 & -2 & -1 & -3 \end{array}\right]$

Diagonal entries: $\frac{1}{2}$, $\frac{2}{3}$, and -1.

b. $\begin{cases} x + 4y - z = -10 \\ 2x + 5y + 8z = 4 \\ x - 2y - 3z = -7 \end{cases} \longrightarrow \left[\begin{array}{ccc|c} 1 & 4 & -1 & -10 \\ 2 & 5 & 8 & 4 \\ 1 & -2 & -3 & -7 \end{array}\right]$

Diagonal entries: 1, 5, and -3.

NOW TRY EXERCISES 10 THROUGH 12

This process can easily be reversed to write a system of equations from a given augmented matrix.

EXAMPLE 3 Write the system of equations corresponding to each matrix.

a. $\left[\begin{array}{ccc|c} 1 & 4 & -1 & 4 \\ 2 & 5 & 8 & 15 \\ 1 & 3 & -3 & 1 \end{array}\right]$ **b.** $\left[\begin{array}{ccc|c} 1 & 4 & -1 & 4 \\ 0 & -3 & 10 & 7 \\ 0 & -1 & -2 & -3 \end{array}\right]$

c. $\left[\begin{array}{ccc|c} 1 & 4 & -1 & -10 \\ 0 & -3 & 10 & 7 \\ 0 & 0 & 16 & 16 \end{array}\right]$

Solution: **a.** $\left[\begin{array}{ccc|c} 1 & 4 & -1 & 4 \\ 2 & 5 & 8 & 15 \\ 1 & 3 & -3 & 1 \end{array}\right] \longrightarrow \begin{cases} 1x + 4y - 1z = 4 \\ 2x + 5y + 8z = 15 \\ 1x + 3y - 3z = 1 \end{cases}$

b. $\left[\begin{array}{ccc|c} 1 & 4 & -1 & 4 \\ 0 & -3 & 10 & 7 \\ 0 & -1 & -2 & -3 \end{array}\right] \longrightarrow \begin{cases} 1x + 4y - 1z = 4 \\ 0x - 3y + 10z = 7 \\ 0x - 1y - 2z = -3 \end{cases}$

c. $\left[\begin{array}{ccc|c} 1 & 4 & -1 & -10 \\ 0 & -3 & 10 & 7 \\ 0 & 0 & 16 & 16 \end{array}\right] \longrightarrow \begin{cases} 1x + 4y - 1z = -10 \\ 0x - 3y + 10z = 7 \\ 0x + 0y + 16z = 16 \end{cases}$

NOW TRY EXERCISES 13 THROUGH 18

C. Solving a System Using Matrices

Matrix solutions to a system of equations can take many forms. In this section, we'll solve systems by **triangularizing the augmented matrix,** using the same operations we applied in previous sections. In this context, the operations are referred to as **elementary row operations.**

ELEMENTARY ROW OPERATIONS

1. Any two rows in a matrix can be interchanged.
2. The elements of any row can be multiplied by a nonzero constant.
3. Any two rows can be added together, and the sum used to replace one of the rows.

A matrix is said to be in **triangular form** when all of the entries below the diagonal are zero. For example, the matrix $\left[\begin{array}{ccc|c} 1 & 4 & -1 & -10 \\ 0 & -3 & 10 & 7 \\ 0 & 0 & 1 & 1 \end{array}\right]$ is in triangular form: $\left[\begin{array}{ccc|c} 1 & 4 & -1 & -10 \\ 0 & -3 & 10 & 7 \\ 0 & 0 & 1 & 1 \end{array}\right]$. This form simply models the results of the elimination method, meaning a matrix written in triangular form can also be used to solve the system. We'll illustrate by solving $\begin{cases} 1x + 4y - 1z = 4 \\ 2x + 5y + 8z = 15 \\ 1x + 3y - 3z = 1 \end{cases}$ here, using elimination to the left, and *row* *operations on the augmented matrix* to the right. As before, R1 represents the first equation in the system and the first row of the matrix, R2 represents equation 2 and row 2, and so on. The calculations involved are shown for the first stage only.

Elimination (System of Equations)

$$\begin{cases} x + 4y - z = 4 \\ 2x + 5y + 8z = 15 \\ x + 3y - 3z = 1 \end{cases}$$

Row Operations (Augmented Matrix)

$$\left[\begin{array}{ccc|c} 1 & 4 & -1 & 4 \\ 2 & 5 & 8 & 15 \\ 1 & 3 & -3 & 1 \end{array}\right]$$

To eliminate the x-term in R2 we use $-2\text{R1} + \text{R2} \rightarrow \text{R2}$. For R3 the operations would be $-1\text{R1} + \text{R3} \rightarrow \text{R3}$. Identical operations are performed on the matrix, which begins the process of triangularizing the matrix.

$$\begin{array}{cr} -2\text{R1} & -2x - 8y + 2z = -8 \\ + & \\ \text{R2} & 2x + 5y + 8z = 15 \\ \hline \text{New R2} & -3y + 10z = 7 \end{array} \qquad \begin{array}{crrrr} -2\text{R1} & -2 & -8 & 2 & -8 \\ + & & & & \\ \text{R2} & 2 & 5 & 8 & 15 \\ \hline \text{New R2} & 0 & -3 & 10 & 7 \end{array}$$

$$\begin{array}{cr} -1\text{R1} & -1x - 4y + 1z = -4 \\ + & \\ \text{R3} & 1x + 3y - 3z = 1 \\ \hline \text{New R3} & -1y - 2z = -3 \end{array} \qquad \begin{array}{crrrr} -1\text{R1} & -1 & -4 & 1 & -4 \\ + & & & & \\ \text{R3} & 1 & 3 & -3 & 1 \\ \hline \text{New R3} & 0 & -1 & -2 & -3 \end{array}$$

WORTHY OF NOTE

Example 4 actually illustrates a method called **Gaussian elimination** (Carl Friedrich Gauss; 1777–1855) in which the final matrix is written in what is called **row-echelon form.** It's also possible to solve a system entirely using only the augmented matrix. That is, we could continue using row operations until a complete solution is found, without having to back-substitute. This is done by transforming the matrix of coefficients into one where the diagonal entries are 1's, with zeroes for all other entries. A 3 × 3 example is shown here. The process is then referred to as **Gauss-Jordan elimination** (Wilhelm Jordan; 1842–1899) with the form shown being one example of **reduced row-echelon form.** For more information on these topics, see Appendix III and the Chapter 6 *Technology Extension on the World Wide Web.*

Reduced Row-Echelon Form

$$\left[\begin{array}{ccc|c} 1 & 0 & 0 & a \\ 0 & 1 & 0 & b \\ 0 & 0 & 1 & c \end{array}\right]$$

As always, we should look for opportunities to simplify any equation in the system (and any row in the matrix). Note that -1R3 will make the coefficients and related matrix entries positive. Here is the new system and matrix.

New System

$$\begin{cases} 1x + 4y - 1z = 4 \\ \quad -3y + 10z = 7 \\ \quad 1y + 2z = 3 \end{cases}$$

New Matrix

$$\left[\begin{array}{ccc|c} 1 & 4 & -1 & 4 \\ 0 & -3 & 10 & 7 \\ 0 & 1 & 2 & 3 \end{array}\right]$$

On the left, we finish by solving the 2 × 2 subsystem, eliminating the y-term from R3. On the right, we eliminate the corresponding entry (third row, second column) to triangularize the matrix. This is accomplished using $\text{R2} + 3\text{R3}$, with the result becoming the new R3.

R2 + 3R3 → R3 **R2 + 3R3 → R3**

$$\begin{cases} 1x + 4y - 1z = 4 \\ \quad -3y + 10z = 7 \\ \quad 16z = 16 \end{cases} \qquad \left[\begin{array}{ccc|c} 1 & 4 & -1 & 4 \\ 0 & -3 & 10 & 7 \\ 0 & 0 & 16 & 16 \end{array}\right]$$

Dividing R3 by 16 gives $z = 1$ in the system and entries of 0 0 1 1 in the augmented matrix. Completing the solution by back-substitution in the system gives the ordered triple (1, 1, 1).

SOLVING SYSTEMS BY TRIANGULARIZING THE AUGMENTED MATRIX

1. Write the system as an augmented matrix.
2. Use row operations to obtain zeroes below the first entry of the diagonal.
3. Use row operations to obtain zeroes below the second entry of the diagonal.
4. Continue until the matrix is triangularized (all entries below the diagonal are zero).
5. Divide to obtain a "1" in the final entry of the diagonal (if it is nonzero), then convert the matrix back into equation form and complete the solution using back-substitution.

Note: At each stage, look for opportunities to simplify row entries using multiplication or division. Also, to begin the process any equation with an x-coefficient of 1 can be made R1 by interchanging the equations.

EXAMPLE 4

Solve by triangularizing the augmented matrix: $\begin{cases} 2x + y - 2z = -7 \\ x + y + z = -1 \\ -2y - z = -3 \end{cases}$

Solution:

$$\begin{cases} 2x + y - 2z = -7 \\ x + y + z = -1 \\ -2y - z = -3 \end{cases} \quad \text{R1} \leftrightarrow \text{R2} \quad \begin{cases} x + y + z = -1 \\ 2x + y - 2z = -7 \\ -2y - z = -3 \end{cases} \quad \text{matrix form} \rightarrow \left[\begin{array}{ccc|c} 1 & 1 & 1 & -1 \\ 2 & 1 & -2 & -7 \\ 0 & -2 & -1 & -3 \end{array}\right]$$

$$\left[\begin{array}{ccc|c} 1 & 1 & 1 & -1 \\ 2 & 1 & -2 & -7 \\ 0 & -2 & -1 & -3 \end{array}\right] \quad -2\text{R1} + \text{R2} \rightarrow \text{R2} \quad \left[\begin{array}{ccc|c} 1 & 1 & 1 & -1 \\ 0 & -1 & -4 & -5 \\ 0 & -2 & -1 & -3 \end{array}\right] \quad -1\text{R2} \rightarrow \text{R2} \quad \left[\begin{array}{ccc|c} 1 & 1 & 1 & -1 \\ 0 & 1 & 4 & 5 \\ 0 & -2 & -1 & -3 \end{array}\right]$$

$$\left[\begin{array}{ccc|c} 1 & 1 & 1 & -1 \\ 0 & 1 & 4 & 5 \\ 0 & -2 & -1 & -3 \end{array}\right] \quad 2\text{R2} + \text{R3} \rightarrow \text{R3} \quad \left[\begin{array}{ccc|c} 1 & 1 & 1 & -1 \\ 0 & 1 & 4 & 5 \\ 0 & 0 & 7 & 7 \end{array}\right] \quad \frac{\text{R3}}{7} \rightarrow \text{R3} \quad \left[\begin{array}{ccc|c} 1 & 1 & 1 & -1 \\ 0 & 1 & 4 & 5 \\ 0 & 0 & 1 & 1 \end{array}\right]$$

Converting the augmented matrix back into equation form yields

$\begin{cases} x + y + z = -1 \\ y + 4z = 5 \\ z = 1 \end{cases}$. Back-substitution gives the solution $(-3, 1, 1)$.

NOW TRY EXERCISES 19 THROUGH 36

Here is an additional example of a solution using matrices and row reduction.

D. Inconsistent and Dependent Systems

Due to the strong link between a linear system and its augmented matrix, inconsistent and dependent systems can be recognized just as in Sections 6.1 and 6.2. An inconsistent system will yield an inconsistent or contradictory statement such as $0 = -12$, meaning all entries in a row of the matrix of coefficients are zero, but the constant is not. A dependent system will yield an identity statement such as $0 = 0$, meaning all entries in an entire row of the matrix are zero.

EXAMPLE 5 Solve the system using the augmented matrix: $\begin{cases} x + y - 5z = 3 \\ x - 2z = 1 \\ 2x - y - z = 0 \end{cases}$.

Solution:

$$\begin{cases} x + y - 5z = 3 \\ x - 2z = 1 \\ 2x - y - z = 0 \end{cases} \quad \text{standard form} \rightarrow \quad \begin{cases} x + y - 5z = 3 \\ x + 0y - 2z = 1 \\ 2x - y - z = 0 \end{cases} \quad \text{matrix form} \rightarrow \quad \left[\begin{array}{ccc|c} 1 & 1 & -5 & 3 \\ 1 & 0 & -2 & 1 \\ 2 & -1 & -1 & 0 \end{array}\right]$$

$$\left[\begin{array}{ccc|c} 1 & 1 & -5 & 3 \\ 1 & 0 & -2 & 1 \\ 2 & -1 & -1 & 0 \end{array}\right] \quad -1R1 + R2 \rightarrow R2 \quad \left[\begin{array}{ccc|c} 1 & 1 & -5 & 3 \\ 0 & -1 & 3 & -2 \\ 2 & -1 & -1 & 0 \end{array}\right] \quad -2R1 + R3 \rightarrow R3 \quad \left[\begin{array}{ccc|c} 1 & 1 & -5 & 3 \\ 0 & -1 & 3 & -2 \\ 0 & -3 & 9 & -6 \end{array}\right]$$

$$\left[\begin{array}{ccc|c} 1 & 1 & -5 & 3 \\ 0 & -1 & 3 & -2 \\ 0 & -3 & 9 & -6 \end{array}\right] \quad \frac{R3}{-3} \rightarrow R3 \quad \left[\begin{array}{ccc|c} 1 & 1 & -5 & 3 \\ 0 & -1 & 3 & -2 \\ 0 & 1 & -3 & 2 \end{array}\right] \quad R2 + R3 \rightarrow R3 \quad \left[\begin{array}{ccc|c} 1 & 1 & -5 & 3 \\ 0 & -1 & 3 & -2 \\ 0 & 0 & 0 & 0 \end{array}\right]$$

Since all entries in the last row are zeroes and it's the only row of zeroes, we conclude the system is linearly dependent and equivalent to $\begin{cases} x + y - 5z = 3 \\ y - 3z = 2 \end{cases}$. As before, we demonstrate this dependence by writing the (x, y, z) solution in terms of a parameter. Solving for y in R2 gives y in terms of z: $y = 3z + 2$. Substituting this relation into R1 enables us to write x in terms of z:

$$\begin{aligned} x + y - 5z &= 3 && \text{R1} \\ x + 3z + 2 - 5z &= 3 && \text{substitue } 3z + 2 \text{ for } y \\ x - 2z &= 1 && \text{simplify} \\ x &= 2z + 1 && \text{solve for } x \end{aligned}$$

As written, the solutions all depend on z: $x = 2z + 1$, $y = 3z + 2$, and $z = z$. Selecting z as the parameter (or some other "neutral" variable), we write the solution as $(2z + 1, 3z + 2, z)$. Two of the infinite number of solutions would be $(1, 2, 0)$ for $z = 0$, and $(-1, -1, -1)$ for $z = -1$. Test these triples in the original equations.

NOW TRY EXERCISES 37 THROUGH 45

E. Solving Applications Using Matrices

As in other areas, solving applications using systems relies heavily on the ability to mathematically model information given verbally or in context. As you work through the exercises, read each problem carefully, looking for relationships that yield a system of two equations in two variables or three equations in three variables.

EXAMPLE 6 A museum purchases a famous painting, a ruby tiara, and a rare coin for its collection, spending a total of \$30,000. One year later, the painting has tripled in value, while the tiara and the coin have doubled in value, giving the items a total worth of \$75,000. Find the purchase price of each if the original price of the painting was \$1000 more than twice the coin.

Solution: Let P represent the price of the painting, T the tiara, and C the coin.

Total spent was \$30,000: $\rightarrow P + T + C = 30{,}000$

One year later: $\rightarrow 3P + 2T + 2C = 75{,}000$

Value of painting versus coin: $\rightarrow P = 2C + 1000$

$$\begin{cases} P + T + C = 30000 \\ 3P + 2T + 2C = 75000 \\ P = 2C + 1000 \end{cases} \text{standard form} \rightarrow \begin{cases} 1P + 1T + 1C = 30000 \\ 3P + 2T + 2C = 75000 \\ 1P + 0T - 2C = 1000 \end{cases} \text{matrix form} \rightarrow \left[\begin{array}{ccc|c} 1 & 1 & 1 & 30000 \\ 3 & 2 & 2 & 75000 \\ 1 & 0 & -2 & 1000 \end{array}\right]$$

$$\left[\begin{array}{ccc|c} 1 & 1 & 1 & 30000 \\ 3 & 2 & 2 & 75000 \\ 1 & 0 & -2 & 1000 \end{array}\right] \begin{array}{c} -3R1 + R2 \rightarrow R2 \\ \\ -1R1 + R3 \rightarrow R3 \end{array} \left[\begin{array}{ccc|c} 1 & 1 & 1 & 30000 \\ 0 & -1 & -1 & -15000 \\ 0 & -1 & -3 & -29000 \end{array}\right] \begin{array}{c} -1R2 \rightarrow R2 \\ \\ -1R3 \rightarrow R3 \end{array} \left[\begin{array}{ccc|c} 1 & 1 & 1 & 30000 \\ 0 & 1 & 1 & 15000 \\ 0 & 1 & 3 & 29000 \end{array}\right]$$

$$\left[\begin{array}{ccc|c} 1 & 1 & 1 & 30000 \\ 0 & 1 & 1 & 15000 \\ 0 & 1 & 3 & 29000 \end{array}\right] -1R2 + R3 \rightarrow R3 \left[\begin{array}{ccc|c} 1 & 1 & 1 & 30000 \\ 0 & 1 & 1 & 15000 \\ 0 & 0 & 2 & 14000 \end{array}\right] \frac{R3}{2} \rightarrow R3 \left[\begin{array}{ccc|c} 1 & 1 & 1 & 30000 \\ 0 & 1 & 1 & 15000 \\ 0 & 0 & 1 & 7000 \end{array}\right]$$

From R3 of the triangularized form, $C = \$7000$ directly. Since R2 represents $T + C = 15{,}000$, we find the tiara was purchased for $T = \$8000$. Substituting these values into the first equation shows the painting was purchased for \$15,000. The solution is (15,000, 8000, 7000).

NOW TRY EXERCISES 48 THROUGH 55

TECHNOLOGY HIGHLIGHT

Solving Systems Using Matrices and Calculating Technology

The keystrokes shown apply to a TI-84 Plus model. Please consult your manual or our Internet site for other models.

Graphing calculators offer a very efficient way to solve systems of equations using matrices. Once the system has been written in matrix form, it can easily be entered and solved by asking the calculator to instantly perform the row operations needed to produce a diagonal of ones, and zeroes for all other entries (one example of *reduced row-echelon form*—see Appendix III). We illustrate here using the TI-84 Plus. Pressing the 2nd X^{-1} (MATRIX) gives the screen shown in Figure 6.26, where we begin by selecting the **EDIT** option (push the right arrow ▶ twice). Pressing ENTER places you on a screen where you can EDIT matrix *A*, changing the size as needed, then inputting the entries of the matrix. Using the 3 × 4 matrix from Example 4, we press 3 and ENTER, then 4 and ENTER, giving the screen shown in Figure 6.27.

Figure 6.26

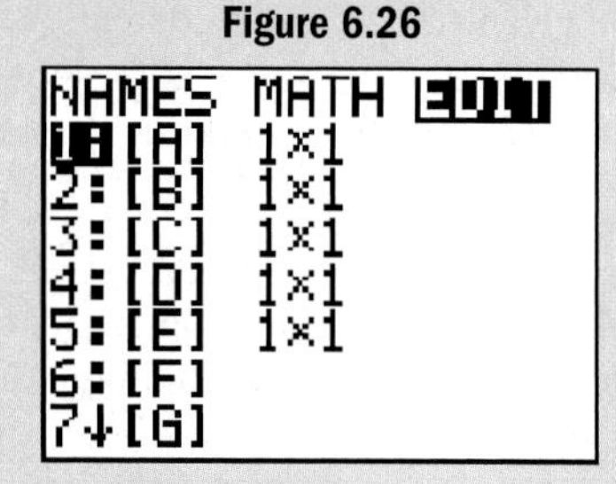

Figure 6.27

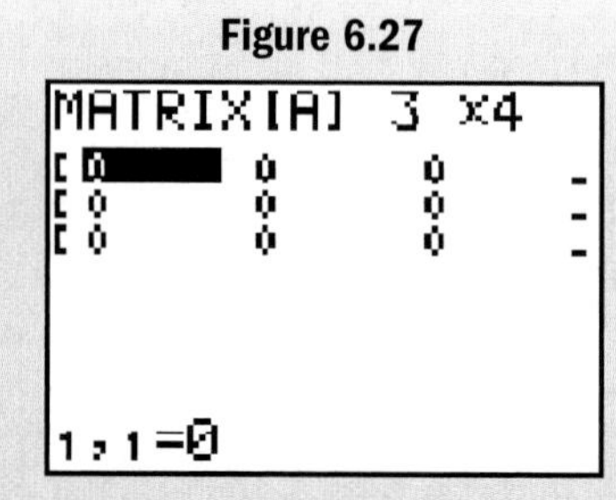

The dash marks to the right indicate that there is a fourth column that cannot be seen, but that comes into view as you enter the elements of the matrix. Begin entering the first row of the matrix, which has entries {1, 1, 1, −1}. Press ENTER after each entry and the cursor automatically goes to the next position in the matrix (note that the TI-84 Plus automatically shifts left and right to allow all four columns to be entered). After entering the second row {2, 1, −2, −7} and the third row {0, −2, −1, −3}, the completed matrix should look like the one shown in Figure 6.28 (the matrix is currently shifted to the right, showing the fourth column). Again note that we have created the matrix named [A], which is a 3×4 matrix. As stated earlier, we solve the system by having the calculator write the matrix in reduced echelon form or **rref** as the TI-84 Plus calls it. To do this we must get back to the home screen by pressing **2nd** **MODE** **(QUIT)**. Press the **CLEAR** key for a clean home screen. To access the **rref** function, press **2nd** **X⁻¹** **(MATRIX)** and the right arrow **▶** to reach the **MATH** option, then scroll upward (or downward) until you get to **B:rref**. Pressing **ENTER** places this function on the home screen, where we must tell it to perform the **rref** operation on matrix [A]. Pressing **2nd** **X⁻¹** **(MATRIX)** once again allows us to select a matrix (notice that matrix **NAMES** is automatically highlighted. Press **ENTER** since we want matrix [A], and matrix [A] is placed on the home screen as the object of the **rref** function. After pressing **ENTER** the calculator quickly computes the row-echelon form and displays it on the screen in Figure 6.29. The solution is easily read as $x = -3$, $y = 1$, and $z = 1$, as we found in Example 4. Use these ideas to complete the following.

Figure 6.28

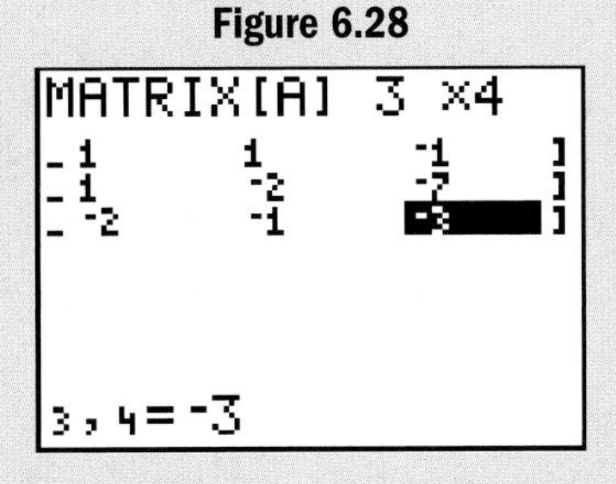

Figure 6.29

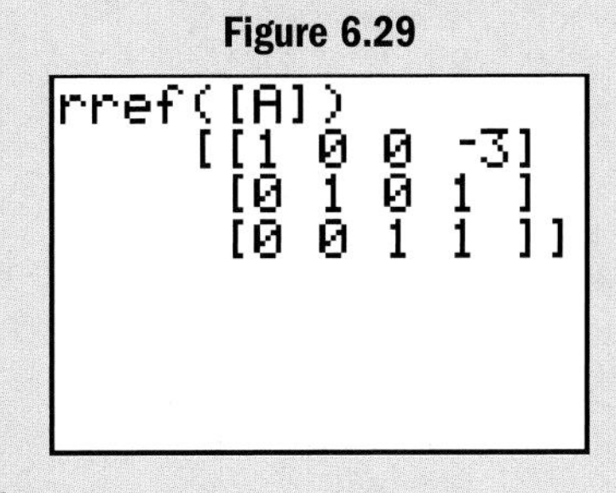

Exercise 1: Use this method to solve the 2×2 system from Exercise 30. (10, 12)

Exercise 2: Use this method to solve the 3×3 system from Exercise 32. (2, 1, −3)

6.5 EXERCISES

CONCEPTS AND VOCABULARY

Fill in the blank with the appropriate word or phrase. Carefully reread the section if needed.

1. A matrix with the same number of rows and columns is called a(n) square matrix.

2. When the coefficient matrix is used with the matrix of constants, the result is a(n) augmented matrix.

3. Matrix $A = \begin{bmatrix} 2 & 4 & -3 \\ 1 & -2 & 1 \end{bmatrix}$ is a 2 by 3 matrix. The entry in the second row and third column is 1.

4. Given matrix B shown here, the diagonal entries are 1, 5, and 1.
$$\begin{bmatrix} 1 & 4 & 3 \\ -1 & 5 & 2 \\ 3 & -2 & 1 \end{bmatrix}$$

5. The notation $-2R1 + R2 \rightarrow R2$ indicates that an equivalent matrix is formed by performing what operations/replacements?

6. Describe how to tell an inconsistent system apart from a dependent system when solving using matrix methods (row reduction).

5. Multiply R1 by −2 and add that result to R2.
6. If all entries in a row are zero, we conclude the system is dependent. If all entries in a row are zero except for the constant, the system is inconsistent.

DEVELOPING YOUR SKILLS

Determine the size of each matrix and identify the second row and second column entry.

7. $\begin{bmatrix} 1 & 0 \\ 2.1 & 1 \\ -3 & 5.8 \end{bmatrix}$ 3×2, 1

8. $\begin{bmatrix} 1 & 0 & 4 \\ 1 & 3 & -7 \\ 5 & -1 & 2 \end{bmatrix}$ 3×3, 3

9. $\begin{bmatrix} 1 & 0 & 4 & 2 \\ 1 & 3 & -7 & -3 \\ 5 & -1 & 2 & 9 \end{bmatrix}$ 3×4, 3

Form the augmented matrix, then name the diagonal entries of the coefficient matrix.

10. $\begin{cases} 2x - 3y - 2z = 7 \\ x - y + 2z = -5 \\ 3x + 2y - z = 11 \end{cases}$

11. $\begin{cases} x + 2y - z = 1 \\ x + z = 3 \\ 2x - y + z = 3 \end{cases}$

12. $\begin{cases} 2x + 3y + z = 5 \\ 2y - z = 7 \\ x - y - 2z = 5 \end{cases}$

Write the system of equations for each matrix. Then use back-substitution to find its solution.

13. $\left[\begin{array}{cc|c} 1 & 4 & 5 \\ 0 & 1 & \frac{1}{2} \end{array}\right]$

14. $\left[\begin{array}{cc|c} 1 & -5 & -15 \\ 0 & -1 & -2 \end{array}\right]$

15. $\left[\begin{array}{ccc|c} 1 & 2 & -1 & 0 \\ 0 & 1 & 2 & 2 \\ 0 & 0 & 1 & 3 \end{array}\right]$

16. $\left[\begin{array}{ccc|c} 1 & 0 & 7 & -5 \\ 0 & 1 & -5 & 15 \\ 0 & 0 & 1 & -26 \end{array}\right]$

17. $\left[\begin{array}{ccc|c} 1 & 3 & -4 & 29 \\ 0 & 1 & -\frac{3}{2} & \frac{21}{2} \\ 0 & 0 & 1 & 3 \end{array}\right]$

18. $\left[\begin{array}{ccc|c} 1 & 2 & -1 & 3 \\ 0 & 1 & \frac{1}{6} & \frac{2}{3} \\ 0 & 0 & 1 & \frac{22}{7} \end{array}\right]$

Perform the indicated row operation(s) and write the new matrix.

19. $\left[\begin{array}{cc|c} \frac{1}{2} & -3 & -1 \\ -5 & 2 & 4 \end{array}\right]$ $2R1 \to R1$, $5R1 + R2 \to R2$

20. $\left[\begin{array}{cc|c} 7 & 4 & 3 \\ 4 & -8 & 12 \end{array}\right]$ $\frac{1}{4}R2 \to R2$, $R1 \leftrightarrow R2$

21. $\left[\begin{array}{ccc|c} -2 & 1 & 0 & 4 \\ 5 & 8 & 3 & -5 \\ 1 & -3 & 3 & 2 \end{array}\right]$ $R1 \leftrightarrow R3$, $-5R1 + R2 \to R2$

22. $\left[\begin{array}{ccc|c} -3 & 2 & 0 & 0 \\ 1 & 1 & 2 & 6 \\ 4 & 1 & -3 & 2 \end{array}\right]$ $R1 \leftrightarrow R2$, $-4R1 + R3 \to R3$

23. $\left[\begin{array}{ccc|c} 3 & 1 & 1 & 8 \\ 6 & -1 & -1 & 10 \\ 4 & -2 & -3 & 22 \end{array}\right]$ $-2R1 + R2 \to R2$; $-4R1 + 3R3 \to R3$

24. $\left[\begin{array}{ccc|c} 2 & 1 & -1 & -3 \\ 3 & 1 & 1 & 0 \\ 4 & 3 & 2 & 3 \end{array}\right]$ $-3R1 + 2R2 \to R2$; $-2R1 + R3 \to R3$

What row operations would produce zeroes beneath the first entry in the diagonal?

25. $\left[\begin{array}{ccc|c} 1 & 3 & 0 & 2 \\ -2 & 4 & 1 & 1 \\ 3 & -1 & -2 & 9 \end{array}\right]$

26. $\left[\begin{array}{ccc|c} 1 & 1 & -4 & -3 \\ 3 & 0 & 1 & 5 \\ -5 & 3 & 2 & 3 \end{array}\right]$

27. $\left[\begin{array}{ccc|c} 1 & 2 & 0 & 10 \\ 5 & 1 & 2 & 6 \\ -4 & 3 & -3 & 2 \end{array}\right]$

Solve each system by triangularizing the matrix and using back-substitution. Simplify by clearing fractions or decimals before beginning.

28. $\begin{cases} 2y = 5x + 4 \\ -5x = 2 - 4y \end{cases}$ $\left(\frac{-6}{5}, -1\right)$

29. $\begin{cases} 0.15g - 0.35h = -0.5 \\ -0.12g + 0.25h = 0.1 \end{cases}$ (20, 10)

30. $\begin{cases} -\frac{1}{5}u + \frac{1}{4}v = 1 \\ \frac{1}{10}u + \frac{1}{2}v = 7 \end{cases}$ (10, 12)

31. $\begin{cases} x - 2y + 2z = 7 \\ 2x + 2y - z = 5 \\ 3x - y + z = 6 \end{cases}$ (1, 6, 9)

32. $\begin{cases} 2x - 3y - 2z = 7 \\ x - y + 2z = -5 \\ 3x + 2y - z = 11 \end{cases}$ (2, 1, −3)

33. $\begin{cases} x + 2y - z = 1 \\ x + z = 3 \\ 2x - y + z = 3 \end{cases}$ (1, 1, 2)

34. $\begin{cases} 2x + 3y + z = 5 \\ 2y - z = 7 \\ x - y - 2z = 5 \end{cases}$ (1, 2, −3)

35. $\begin{cases} -x + y + 2z = 2 \\ x + y - z = 1 \\ 2x + y + z = 4 \end{cases}$ (1, 1, 1)

36. $\begin{cases} x + y - 2z = -1 \\ 4x - y + 3z = 3 \\ 3x + 2y - z = 4 \end{cases}$ (0, 3, 2)

Solve each system by triangularizing the matrix and using back-substitution. If the system is linearly dependent, give the solution in terms of a parameter.

37. $\begin{cases} 4x - 8y + 8z = 24 \\ 2x - 6y + 3z = 13 \\ 3x + 4y - z = -11 \end{cases}$

38. $\begin{cases} -x + 2y + 3z = -6 \\ x - y + 2z = -4 \\ 3x + y - z = 2 \end{cases}$

39. $\begin{cases} y - 4z = -16 \\ x + 3y + 5z = 20 \\ 3x - 2y + 9z = 36 \end{cases}$

40. $\begin{cases} 2x - y + 3z = 1 \\ 4x - 2y + 6z = 2 \\ 10x - 5y + 15z = 5 \end{cases}$

41. $\begin{cases} 3x - 4y + 2z = -2 \\ \frac{3}{2}x - 2y + z = -1 \\ -6x + 8y - 4z = 4 \end{cases}$

42. $\begin{cases} 3x + y + 2z = 3 \\ x - 2y + 3z = 1 \\ 4x - 8y + 12z = 7 \end{cases}$

---

10. $\left[\begin{array}{ccc|c} 2 & -3 & -2 & 7 \\ 1 & -1 & 2 & -5 \\ 3 & 2 & -1 & 11 \end{array}\right]$; diagonal entries 2, −1, −1

11. $\left[\begin{array}{ccc|c} 1 & 2 & -1 & 1 \\ 1 & 0 & 1 & 3 \\ 2 & -1 & 1 & 3 \end{array}\right]$; diagonal entries 1, 0, 1

12. $\left[\begin{array}{ccc|c} 2 & 3 & 1 & 5 \\ 0 & 2 & -1 & 7 \\ 1 & -1 & -2 & 5 \end{array}\right]$; diagonal entries 2, 2, −2

13. $\begin{cases} x + 4y = 5 \\ y = \frac{1}{2} \end{cases} \to \left(3, \frac{1}{2}\right)$

14. $\begin{cases} x - 5y = -15 \\ -y = -2 \end{cases} \to (-5, 2)$

15. $\begin{cases} x + 2y - z = 0 \\ y + 2z = 2 \\ z = 3 \end{cases} \to (11, -4, 3)$

16. $\begin{cases} x + 7z = -5 \\ y - 5z = 15 \\ z = -26 \end{cases} \to (177, -115, -26)$

17. $\begin{cases} x + 3y - 4z = 29 \\ y - \frac{3}{2}z = \frac{21}{2} \\ z = 3 \end{cases} \to (-4, 15, 3)$

18. $\begin{cases} x + 2y - z = 3 \\ y + \frac{1}{6}z = \frac{2}{3} \\ z = \frac{22}{7} \end{cases} \to \left(\frac{41}{7}, \frac{1}{7}, \frac{22}{7}\right)$

19. $\left[\begin{array}{cc|c} 1 & -6 & -2 \\ 0 & -28 & -6 \end{array}\right]$

20. $\left[\begin{array}{cc|c} 1 & -2 & 3 \\ 7 & 4 & 3 \end{array}\right]$

21. $\left[\begin{array}{ccc|c} 1 & -3 & 3 & 2 \\ 0 & 23 & -12 & -15 \\ -2 & 1 & 0 & 4 \end{array}\right]$

22. $\left[\begin{array}{ccc|c} 1 & 1 & 2 & 6 \\ -3 & 2 & 0 & 0 \\ 0 & -3 & -11 & -22 \end{array}\right]$

23. $\left[\begin{array}{ccc|c} 3 & 1 & 1 & 8 \\ 0 & -3 & -3 & -6 \\ 0 & -10 & -13 & 34 \end{array}\right]$

24. $\left[\begin{array}{ccc|c} 2 & 1 & -1 & -3 \\ 0 & -1 & 5 & 9 \\ 0 & 1 & 4 & 9 \end{array}\right]$

25. $2R_1 + R_2 \to R_2$, $-3R_1 + R_3 \to R_3$

26. $-3R_1 + R_2 \to R_2$, $5R_1 + R_3 \to R_3$

27. $-5R_1 + R_2 \to R_2$, $4R_1 + R_3 \to R_3$

37. $\left(-1, \frac{-3}{2}, 2\right)$

38. (0, 0, −2)

39. (0, 0, 4)

40. dependent $\{(x, y, z) | 2x - y + 3z = 1\}$

41. dependent $\{(x, y, z) | 3x - 4y + 2z = -2\}$

42. no solution

43. no solution
44. $(x, \frac{1}{3}x, 4 - \frac{5}{3}x)$
45. $(-\frac{5}{4}z - 3, \frac{1}{8}z - \frac{1}{2}, z)$

43. $\begin{cases} 2x - y + 3z = 1 \\ 2y + 6z = 2 \\ x - \frac{1}{2}y + \frac{3}{2}z = 5 \end{cases}$

44. $\begin{cases} x + 2y + z = 4 \\ 3x - 4y + z = 4 \\ 6x - 8y + 2z = 8 \end{cases}$

45. $\begin{cases} -2x + 4y - 3z = 4 \\ 5x - 6y + 7z = -12 \\ x + 2y + z = -4 \end{cases}$

WORKING WITH FORMULAS

Area of a triangle in the plane: $A = \pm\frac{1}{2}(x_1y_2 - x_2y_1 + x_2y_3 - x_3y_2 + x_3y_1 - x_1y_3)$

The area of a triangle in the plane is given by the formula shown, where the vertices of the triangle are located at the points (x_1, y_1), (x_2, y_2), and (x_3, y_3), and the sign is chosen to ensure a positive value.

46. Find the area of a triangle whose vertices are $(-1, -3)$, $(5, 2)$, and $(1, 8)$. 28 units^2

47. Find the area of a triangle whose vertices are $(6, -2)$, $(-5, 4)$, and $(-1, 7)$. 28.5 units^2

APPLICATIONS

Model each problem using a system of linear equations. Then solve using the augmented matrix.

Descriptive Translation

48. LA to STL, 1600 mi; STL to CIN, 310 mi; CIN to NY, 570 mi

48. The distance (via air travel) from Los Angeles (LA), California, to Saint Louis (STL), Missouri, to Cincinnati (CIN), Ohio, to New York City (NYC), New York, is approximately 2480 mi. Find the distances between each city if the distance from LA to STL is 50 mi more than five times the distance between STL and CIN and 110 mi less than three times the distance from CIN to NYC.

49. The Chicago Bulls won their third straight NBA championship when John Paxson hit a 3-point shot with 3.9 sec left in the deciding game against the Phoenix Suns. Determine the final score if it was two consecutive integers whose sum was 197. 98 to 99

50. Moe is lecturing Larry and Curly once again (Moe, Larry, and Curly of *The Three Stooges* fame) claiming he is twice as smart as Larry and three times as smart as Curly. If he is correct and the sum of their IQs is 165, what is the IQ of each stooge? Moe 90; Larry 45, Curly 30

51. A collector of rare books buys a handwritten, autographed copy of Edgar Allan Poe's *Annabel Lee,* an original advance copy of L. Frank Baum's *The Wonderful Wizard of Oz,* and a first print copy of *The Caine Mutiny* by Herman Wouk, paying a total of \$100,000. Find the cost of each one, given that the cost of *Annabel Lee* and twice the cost of *The Caine Mutiny* sum to the price paid for *The Wonderful Wizard of Oz,* and *The Caine Mutiny* cost twice as much as *Annabel Lee.* Poe, \$12,500; Baum, \$62,500; Wouk, \$25,000

Geometry

52. A right triangle has a hypotenuse of 39 m. If the perimeter is 90 m, and the longer leg is 6 m longer than twice the shortest, find the dimensions of the triangle. 15 m, 36 m, 39 m

53. In triangle *ABC*, the sum of angles *A* and *C* is equal to three times angle *B*. Angle *C* is 10 degrees more than twice angle *B*. Find the measure of each angle. $A = 35°$, $B = 45°$, $C = 100°$

Investment and Finance

54. Suppose \$10,000 is invested in three different investment vehicles paying 5%, 7%, and 9% annual interest. Find the amount invested at each rate if the interest earned after 1 yr is \$760 and the amount invested at 9% is equal to the sum of the amounts invested at 5% and 7%. \$2000 at 5%; \$3000 at 7%; \$5000 at 9%

55. The trustee of a union's pension fund has invested the funds in three ways: a savings fund paying 4% annual interest, a money market fund paying 7%, and government bonds paying 8%.

Find the amount invested in each if the interest earned after one year is \$0.178 million and the amount in government bonds is \$0.3 million more than twice the amount in money market funds. The total amount invested is \$2.5 million dollars. \$.4 million at 4%; \$.6 million at 7%; \$1.5 million at 8%

WRITING, RESEARCH, AND DECISION MAKING

56. In a free-market economy, the supply and demand for many products is directly related to the price of the product. The relationship between supply (amount produced by companies) and demand (amount purchased by consumers) can sometimes be modeled using a system of linear equations. Using an encyclopedia, business textbook, or the Internet, locate some background information on the *law of supply and demand,* and find out what is meant by an equilibrium point. Prepare a report on what you find. Try to include at least one example involving a linear system. Answers will vary.

57. In previous sections, we noted that one condition for a 3 × 3 system to be dependent was for the third equation to be a linear combination of the other two. To test this, write any two equations using the same three variables, then form a third equation by performing some combination of elementary row operations. Solve the resulting 3 × 3 system. What do you notice? Answers will vary.

EXTENDING THE CONCEPT

58. Given the drawing shown, use a system of equations and the matrix method to find the measure of the angles labeled as x and y. Recall that vertical angles are equal and that the sum of the angles in a triangle is 180°. $x = 84°$; $y = 25°$

71°

y x

$(x - 59)°$

59. The system given here has a solution of $(1, -2, 3)$. Find the value of a and b.

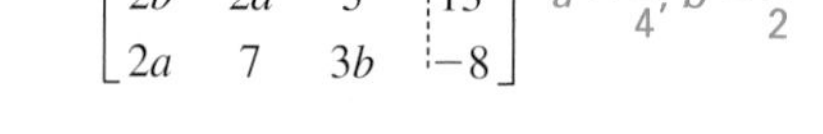

$$\begin{bmatrix} 1 & a & b & \vdots & 1 \\ 2b & 2a & 5 & \vdots & 13 \\ 2a & 7 & 3b & \vdots & -8 \end{bmatrix}$$

$a = \frac{3}{4}$, $b = \frac{1}{2}$

MAINTAINING YOUR SKILLS

60. (3.1) Given $f(x) = x^3 - 8$ and $g(x) = x - 2$, find $f + g$, $f - g$, fg, and $\frac{f}{g}$. $x^3 + x - 10$; $x^3 - x - 6$; $x^4 - 2x^3 - 8x + 16$; $x^2 + 2x + 4$

61. (4.3) Given $x = 2$ is a zero of $h(x) = x^4 - x^2 - 12$, find all zeroes of h, real and complex. $x = \pm 2$; $x = \pm\sqrt{3}\,i$

62. (2.4) Sketch the graph of $f(x) = \frac{1}{x}$ and $g(x) = \frac{1}{x^2}$ without plotting any points.

62.

63. (3.2) Given $p(x) = x^3 - 8$, find $p^{-1}(x)$, then show that $(p \circ p^{-1})(x) = x = (p^{-1} \circ p)(x)$. $p^{-1}(x) = \sqrt[3]{x + 8}$

64. (2.7/5.6) The percentage of adults in the United States who smoke has been gradually declining for the last four decades. The table shows the percentage of adults who were smokers for the years indicated. Use a scatter-plot and your calculator to decide if a linear or exponential model is more appropriate, then graph the scatter-plot and equation model on the same grid. According to the model, in what year will the percentage of adults who smoke drop below 20%? $P(t) = (47.1)0.982^t$; 2007

Source: 1996 *Statistical Abstract of the United States,* Table 222; various other years

64.

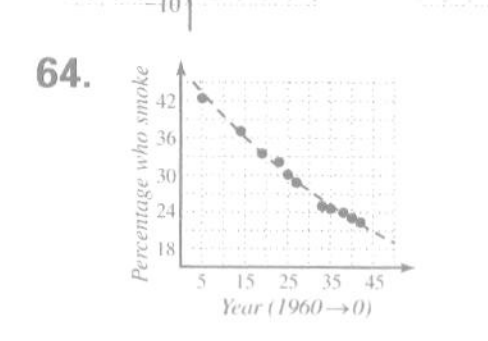

Exercise 64

Year 1960 → 0	Percentage Who Smoke
5	42.4
14	37.1
19	33.5
23	32.1
25	30.1
27	28.8
33	25.0
35	24.6
38	24.0
40	23.1
42	22.4

65. (5.4) If a set amount of money d is deposited regularly (daily, weekly, monthly, etc.) n times per year at a fixed interest rate r, the amount of money M accumulated in t years is given by the formula shown. If a parent deposits \$250 per month for 18 yr at 4.6% beginning when her first child was born, how much has been accumulated to help pay for college expenses? \$83,811.95

$$M = \frac{d\left[\left(1 + \frac{r}{n}\right)^{nt} - 1\right]}{\frac{r}{n}}$$

6.6 The Algebra of Matrices

LEARNING OBJECTIVES

In Section 6.6 you will learn how to:

A. Determine when two matrices are equal

B. Add and subtract matrices

C. Compute the product of two matrices

INTRODUCTION

Matrices serve a much wider purpose than just a convenient method for solving systems. To understand their broader application, we need to know more about matrix theory, the various ways matrices can be combined, and some of their more practical uses. The common operations of addition, subtraction, multiplication, and division are all defined for matrices, as are other operations. Practical applications of matrix theory can be found in the social sciences, inventory management, genetics, operations research, engineering, and many other fields.

POINT OF INTEREST

The word "matrix" shares the same etymology (word history) as the word "mother." Its primary definition actually centers around a "womb-like" state, from which something else is developed. In a metaphorical sense, matrices can be very "womb-like," since they are indeed used to develop or generate many different mathematical results.

A. Equality of Matrices

To effectively study matrix algebra, we first give them a more general definition, or "algebratize" our current view. For the general matrix A, all entries will be denoted using the lowercase letter "a," with their position in the matrix designated by the dual subscript a_{ij}. The letter "i" gives the *row* and the letter "j" gives the *column* of the entry's location. A general $m \times n$ matrix is written:

$$A = \begin{array}{r} \\ \text{row } 1 \rightarrow \\ \text{row } 2 \rightarrow \\ \text{row } 3 \rightarrow \\ \\ \text{row } i \rightarrow \\ \\ \text{row } m \rightarrow \end{array} \begin{array}{c} \begin{array}{ccccccc} \text{col } 1 & \text{col } 2 & \text{col } 3 & & \text{col } j & & \text{col } n \end{array} \\ \begin{bmatrix} a_{11} & a_{12} & a_{13} & \cdots & a_{1j} & \cdots & a_{1n} \\ a_{21} & a_{22} & a_{23} & \cdots & a_{2j} & \cdots & a_{2n} \\ a_{31} & a_{32} & a_{33} & \cdots & a_{3j} & \cdots & a_{3n} \\ \vdots & \vdots & \vdots & & \vdots & & \vdots \\ a_{i1} & a_{i2} & a_{i3} & \cdots & a_{ij} & \cdots & a_{in} \\ \vdots & \vdots & \vdots & & \vdots & & \vdots \\ a_{m1} & a_{m2} & a_{m3} & \cdots & a_{mj} & \cdots & a_{mn} \end{bmatrix} \end{array}$$

a_{ij} is a general matrix element

The size of a matrix is also referred to as its **order,** and we say the order of matrix A is $m \times n$. Note that diagonal entries have the same row and column number, a_{ij} where $i = j$. Also, where the general entry of matrix A is a_{ij}, the general entry of matrix B is b_{ij}, of matrix C is c_{ij}, and so on.

EXAMPLE 1 State the order of each matrix and name the entries corresponding to a_{22} and a_{31}.

a. $A = \begin{bmatrix} 1 & 4 \\ -2 & -3 \end{bmatrix}$ **b.** $B = \begin{bmatrix} 3 & -2 \\ 1 & 5 \\ -4 & 3 \end{bmatrix}$ **c.** $C = \begin{bmatrix} 0.2 & -0.5 & 0.7 \\ -1 & 0.3 & 1 \\ 2.1 & -0.1 & 0.6 \end{bmatrix}$

Solution:

a. matrix A: order 2×2. Entry $a_{22} = -3$ (the row 2, column 2 entry is -3). There is no a_{31} entry (matrix is only 2×2).

b. matrix B: order 3×2. Entry $b_{22} = 5$, entry $b_{31} = -4$.

c. matrix C: order 3×3. Entry $c_{22} = 0.3$, entry $c_{31} = 2.1$.

NOW TRY EXERCISES 7 THROUGH 12

Equality of Matrices

Two matrices are equal if they have the same order and their corresponding entries are equal. In symbols, this means that $A = B$ if $a_{ij} = b_{ij}$ for all i and j.

EXAMPLE 2 Determine whether the following statements are true, false, or conditional. If false, explain why. If conditional, state values that will make the statement true.

a. $\begin{bmatrix} 1 & 4 \\ -2 & -3 \end{bmatrix} = \begin{bmatrix} -3 & -2 \\ 4 & 1 \end{bmatrix}$ b. $\begin{bmatrix} 3 & -2 \\ 1 & 5 \\ -4 & 3 \end{bmatrix} = \begin{bmatrix} 3 & -2 & 1 \\ 5 & -4 & 3 \end{bmatrix}$

c. $\begin{bmatrix} 1 & 4 \\ -2 & -3 \end{bmatrix} = \begin{bmatrix} a-2 & 2b \\ c & -3 \end{bmatrix}$

Solution:

a. $\begin{bmatrix} 1 & 4 \\ -2 & -3 \end{bmatrix} = \begin{bmatrix} -3 & -2 \\ 4 & 1 \end{bmatrix}$ is false. The matrices have the same order and the same entries, but corresponding entries are not equal.

b. $\begin{bmatrix} 3 & -2 \\ 1 & 5 \\ -4 & 3 \end{bmatrix} = \begin{bmatrix} 3 & -2 & 1 \\ 5 & -4 & 3 \end{bmatrix}$ is false. Their orders are not equal.

c. $\begin{bmatrix} 1 & 4 \\ -2 & -3 \end{bmatrix} = \begin{bmatrix} a-2 & 2b \\ c & -3 \end{bmatrix}$ is conditional. The statement is true when $a = 3$, $b = 2$, $c = -2$, and is false otherwise.

NOW TRY EXERCISES 13 THROUGH 16

B. Addition and Subtraction of Matrices

A sum or difference of matrices is found by combining the corresponding entries. This limits the operations to matrices of like orders, otherwise some entries in one matrix would have no "corresponding entries" in the other. This also means the result is a new matrix of *like order,* whose entries are the corresponding sums or differences.

ADDITION AND SUBTRACTION OF MATRICES

Given matrices A, B, C, and D with like orders.
The sum $A + B$ is a matrix C, where $[c_{ij}] = [a_{ij} + b_{ij}]$.
The difference $A - B$ is a matrix D, where $[d_{ij}] = [a_{ij} - b_{ij}]$.

EXAMPLE 3 Compute the sum or difference of the matrices indicated.

$$A = \begin{bmatrix} 2 & 3 \\ 1 & 0 \\ 1 & -3 \end{bmatrix} \quad B = \begin{bmatrix} -3 & 2 & -1 \\ -5 & 4 & 3 \end{bmatrix} \quad C = \begin{bmatrix} 3 & -2 \\ 1 & 5 \\ -4 & 3 \end{bmatrix}$$

a. $A + C$ **b.** $A + B$ **c.** $C - A$

Solution:

a. $A + C = \begin{bmatrix} 2 & 3 \\ 1 & 0 \\ 1 & -3 \end{bmatrix} + \begin{bmatrix} 3 & -2 \\ 1 & 5 \\ -4 & 3 \end{bmatrix}$ sum of A and C

$= \begin{bmatrix} 2+3 & 3+(-2) \\ 1+1 & 0+5 \\ 1+(-4) & -3+3 \end{bmatrix} = \begin{bmatrix} 5 & 1 \\ 2 & 5 \\ -3 & 0 \end{bmatrix}$ add corresponding entries

b. $A + B = \begin{bmatrix} 2 & 3 \\ 1 & 0 \\ 1 & -3 \end{bmatrix} + \begin{bmatrix} -3 & 2 & -1 \\ -5 & 4 & 3 \end{bmatrix}$ Addition and subtraction are not defined for matrices of unlike order.

c. $C - A = \begin{bmatrix} 3 & -2 \\ 1 & 5 \\ -4 & 3 \end{bmatrix} - \begin{bmatrix} 2 & 3 \\ 1 & 0 \\ 1 & -3 \end{bmatrix}$ difference of C and A

$= \begin{bmatrix} 3-2 & -2-3 \\ 1-1 & 5-0 \\ -4-1 & 3-(-3) \end{bmatrix} = \begin{bmatrix} 1 & -5 \\ 0 & 5 \\ -5 & 6 \end{bmatrix}$ subtract corresponding entries

NOW TRY EXERCISES 17 THROUGH 20

C. Matrices and Multiplication

The algebraic terms $2a$ and ab have counterparts in matrix algebra. The product $2A$ represents a constant times a matrix and is called **scalar multiplication.** The product AB represents the product of two matrices.

Scalar Multiplication

The product of a scalar times a matrix is defined by taking the product of the constant with *each entry* in the matrix, forming a new matrix of like size. In symbols, for any real number k and matrix A, $kA = [ka_{ij}]$. Similar to standard algebraic practice, $-A$ represents the scalar product $-1 \cdot A$ and any subtraction can be rewritten as an algebraic sum: $A - B = A + (-B)$. Note that for any matrix A, the sum $A + (-A)$ will yield the **zero matrix Z,** a matrix whose entries are all zeroes. Matrix $-A$ is the **additive inverse** for A, and Z is the **additive identity.**

EXAMPLE 4

Given $A = \begin{bmatrix} 4 & 3 \\ \frac{1}{2} & 1 \\ 0 & -3 \end{bmatrix}$ and $B = \begin{bmatrix} 3 & -2 \\ 0 & 6 \\ -4 & 0.4 \end{bmatrix}$, compute the following:

a. $\frac{1}{2}B$ **b.** $-4A - \frac{1}{2}B$

Solution: **a.** $\frac{1}{2}B = \left(\frac{1}{2}\right)\begin{bmatrix} 3 & -2 \\ 0 & 6 \\ -4 & 0.4 \end{bmatrix}$

$$= \begin{bmatrix} \left(\frac{1}{2}\right)3 & \left(\frac{1}{2}\right)(-2) \\ \left(\frac{1}{2}\right)0 & \left(\frac{1}{2}\right)6 \\ \left(\frac{1}{2}\right)(-4) & \left(\frac{1}{2}\right)0.4 \end{bmatrix} = \begin{bmatrix} \frac{3}{2} & -1 \\ 0 & 3 \\ -2 & 0.2 \end{bmatrix}$$

b. $-4A - \frac{1}{2}B = -4A + \left(-\frac{1}{2}\right)B$ rewrite using algebraic addition

$$= \begin{bmatrix} -16 & -12 \\ -2 & -4 \\ 0 & 12 \end{bmatrix} + \begin{bmatrix} -\frac{3}{2} & 1 \\ 0 & -3 \\ 2 & -0.2 \end{bmatrix}$$ from Part (a)

$$= \begin{bmatrix} -16 - \frac{3}{2} & -12 + 1 \\ -2 + 0 & -4 - 3 \\ 0 + 2 & 12 - 0.2 \end{bmatrix} = \begin{bmatrix} -\frac{35}{2} & -11 \\ -2 & -7 \\ 2 & 11.8 \end{bmatrix}$$ result

NOW TRY EXERCISES 21 THROUGH 24

Matrix Multiplication

When it comes to the *product of two matrices,* it's helpful to know in advance the method was defined to facilitate solving **matrix equations.** Consider the 3×3 system $\begin{cases} x + 4y - z = 10 \\ 2x + 5y - 3z = 7 \\ 8x + y - 2z = 11 \end{cases}$. When this system is written as a matrix equation, it is possible to solve for the variables x, y, and z *simultaneously,* instead of individually as when using row-reduction. To accomplish this, we need x, y, and z in their own **matrix of variables,** so that with appropriate operations we can isolate the variables on one side, giving all three solutions at once. To begin, we use the 3×1 matrix of constants for the right-hand side, and rewrite the left-hand side as the product of the coefficient matrix times a 3×1 matrix of variables. For the system given we have: $\begin{bmatrix} 1 & 4 & -1 \\ 2 & 5 & -3 \\ 8 & 1 & -2 \end{bmatrix} \cdot \begin{bmatrix} x \\ y \\ z \end{bmatrix} = \begin{bmatrix} 10 \\ 7 \\ 11 \end{bmatrix}$. With this setup, can you determine how matrix multiplication must be defined to obtain the original system? While the answer *may* be apparent, the following illustration will help. Consider a cable company offering three different levels of Internet service: Bronze—fast, Silver—very fast, and Gold—lightning fast. Table 6.3 shows the number and types of programs sold to households and businesses for the week. Each program has an incentive package consisting of a rebate and a certain number of free weeks, as shown in Table 6.4.

Table 6.3 Matrix A

	Bronze	Silver	Gold
Homes	40	20	**25**
Businesses	10	15	45

Table 6.4 Matrix B

	Rebate	Free Weeks
Bronze	$15	2
Silver	$25	4
Gold	**$35**	6

To compute the amount of rebate money the cable company paid to households for the week, we would take the first row (R1) in Table 6.3 and multiply by the corresponding entries (bronze with bronze, silver with silver, and so on) in the first column (C1) of Table 6.4. In matrix form, we have $\begin{bmatrix} 40 & 20 & \mathbf{25} \end{bmatrix} \cdot \begin{bmatrix} 15 \\ 25 \\ \mathbf{35} \end{bmatrix} = 40 \cdot 15 + 20 \cdot 25 + \mathbf{25} \cdot \mathbf{35} = \$\mathbf{1975}$. Using R1 of Table 6.3 with C2 from Table 6.4 gives the number of free weeks awarded to homes: $\begin{bmatrix} 40 & 20 & \mathbf{25} \end{bmatrix} \cdot \begin{bmatrix} 2 \\ 4 \\ \mathbf{6} \end{bmatrix} = 40 \cdot 2 + 20 \cdot 4 + \mathbf{25} \cdot \mathbf{6} = 310$.

Using the second row (R2) of Table 6.3 with the two columns from Table 6.4 will give the amount of rebate money and the number of free weeks, respectively, awarded to business customers. When all computations are complete, the result is a product matrix P with order 2×2. This is because the product of R1 from matrix A, with C1 from matrix B, *gives the entry in position* P_{11} *of the product matrix:* $\begin{bmatrix} 40 & 20 & 25 \\ 10 & 15 & 45 \end{bmatrix} \cdot \begin{bmatrix} 15 & 2 \\ 25 & 4 \\ 35 & 6 \end{bmatrix} = \begin{bmatrix} \mathbf{1975} & \mathbf{310} \\ 2100 & 350 \end{bmatrix}$.

Likewise, the product R1 · C2 will give entry P_{12}, the product of R2 with C1 will give P_{21}, and so on. This "row × column" multiplication can be generalized, and leads to the following conclusions: (1) for matrix multiplication to be defined, the rows of the first matrix must have the same number of entries as the columns of the second; and (2) the resulting product will have the same number of rows as the first and the same number of columns as the second. Given $m \times n$ matrix A and $p \times q$ matrix B,

A $(m \times n)$	B $(p \times q)$	A $(m \times n)$	B $(p \times q)$
matrix multiplication is possible only when $n = p$		result will be a $m \times q$ matrix	

In more formal terms, we have the following definition of matrix multiplication.

MATRIX MULTIPLICATION
Given the $m \times n$ matrix $A = [a_{ij}]$ and the $p \times q$ matrix $B = [b_{ij}]$.
If $n = p$, then matrix multiplication is possible and the product AB is an $m \times q$ matrix $P = [p_{ij}]$, where p_{ij} is product of the ith row of A with the jth column of B.

In Example 5, two of the matrix products [parts (a) and (e)] are shown in full detail, with the first entry of the product matrix color-coded.

EXAMPLE 5 Given the matrices A through E shown here, compute the following products:

a. AB **b.** AC **c.** CA **d.** DE **e.** ED

$$A = \begin{bmatrix} -2 & 1 \\ 3 & 4 \end{bmatrix} \quad B = \begin{bmatrix} 4 & 3 \\ 6 & 1 \end{bmatrix} \quad C = \begin{bmatrix} -2 & -1 \\ 3 & 0 \\ 1 & 2 \end{bmatrix} \quad D = \begin{bmatrix} -2 & 1 & 3 \\ 1 & 0 & 2 \\ 4 & 1 & -1 \end{bmatrix} \quad E = \begin{bmatrix} 2 & 5 & 1 \\ 4 & -1 & 1 \\ 0 & 3 & -2 \end{bmatrix}$$

Solution:

a. $AB = \begin{bmatrix} -2 & 1 \\ 3 & 4 \end{bmatrix}\begin{bmatrix} 4 & 3 \\ 6 & 1 \end{bmatrix} = \begin{bmatrix} -2 & -5 \\ 36 & 13 \end{bmatrix}$

Computation: $\begin{bmatrix} (-2)(4) + (1)(6) & (-2)(3) + (1)(1) \\ (3)(4) + (4)(6) & (3)(3) + (4)(1) \end{bmatrix}$

A (2 × 2) B (2 × 2): multiplication is possible since 2 = 2; result will be a 2 × 2 matrix

b. $AC = \begin{bmatrix} -2 & 1 \\ 3 & 4 \end{bmatrix}\begin{bmatrix} -2 & -1 \\ 3 & 0 \\ 1 & 2 \end{bmatrix}$

A (2 × 2) C (3 × 2): multiplication is not possible since 2 ≠ 3

c. $CA = \begin{bmatrix} -2 & -1 \\ 3 & 0 \\ 1 & 2 \end{bmatrix}\begin{bmatrix} -2 & 1 \\ 3 & 4 \end{bmatrix} = \begin{bmatrix} 1 & -6 \\ -6 & 3 \\ 4 & 9 \end{bmatrix}$

C (3 × 2) A (2 × 2): multiplication is possible since 2 = 2; result will be a 3 × 2 matrix

d. $DE = \begin{bmatrix} -2 & 1 & 3 \\ 1 & 0 & 2 \\ 4 & 1 & -1 \end{bmatrix}\begin{bmatrix} 2 & 5 & 1 \\ 4 & -1 & 1 \\ 0 & 3 & -2 \end{bmatrix} = \begin{bmatrix} 0 & -2 & -7 \\ 2 & 11 & -3 \\ 12 & 16 & 7 \end{bmatrix}$

D (3 × 3) E (3 × 3): multiplication is possible since 3 = 3; result will be a 3 × 3 matrix

Computation: $\begin{bmatrix} (-2)(2) + (1)(4) + (3)(0) & (-2)(5) + (1)(-1) + (3)(3) & (-2)(1) + (1)(1) + (3)(-2) \\ (1)(2) + (0)(4) + (2)(0) & (1)(5) + (0)(-1) + (2)(3) & (1)(1) + (0)(1) + (2)(-2) \\ (4)(2) + (1)(4) + (-1)(0) & (4)(5) + (1)(-1) + (-1)(3) & (4)(1) + (1)(1) + (-1)(-2) \end{bmatrix}$

e. $ED = \begin{bmatrix} 2 & 5 & 1 \\ 4 & -1 & 1 \\ 0 & 3 & -2 \end{bmatrix}\begin{bmatrix} -2 & 1 & 3 \\ 1 & 0 & 2 \\ 4 & 1 & -1 \end{bmatrix} = \begin{bmatrix} 5 & 3 & 15 \\ -5 & 5 & 9 \\ -5 & -2 & 8 \end{bmatrix}$

E (3 × 3) D (3 × 3): multiplication is possible since 3 = 3; result will be a 3 × 3 matrix

NOW TRY EXERCISES 25 THROUGH 36

In addition to seeing how matrix multiplication works, Example 5 shows that, in general, matrix multiplication is not commutative. Parts (d) and (e) show $DE \neq ED$ since we get different results, and Parts (b) and (c) show $AC \neq CA$, since AC is not defined while CA *is*.

Operations on matrices can be a laborious process for larger matrices and for matrices with noninteger or large entries. For these, we turn to available technology for assistance, shifting our focus from a meticulous computation of entries, to carefully entering each matrix into the calculator, double-checking each entry, and appraising results to see if they're reasonable.

EXAMPLE 6 Use a calculator to compute the difference $A - B$ for the matrices given.

$$A = \begin{bmatrix} \frac{2}{11} & -0.5 & \frac{6}{5} \\ 0.9 & \frac{3}{4} & -4 \\ 0 & 6 & -\frac{5}{12} \end{bmatrix} \qquad B = \begin{bmatrix} \frac{1}{6} & \frac{-7}{10} & 0.75 \\ \frac{11}{25} & 0 & -5 \\ -4 & \frac{-5}{9} & \frac{-5}{12} \end{bmatrix}$$

Solution: The entries for matrix A are shown in Figure 6.30. After entering matrix B, exit to the home screen [2nd MODE **(QUIT)**], call up matrix A, press the − (subtract) key, then call up matrix B and press ENTER. The calculator quickly finds the difference and displays the results shown in Figure 6.31. The last line on the screen shows the result can be stored for future use in a new matrix C by pressing the STO▸ key, calling up matrix C, and pressing ENTER.

Figure 6.30

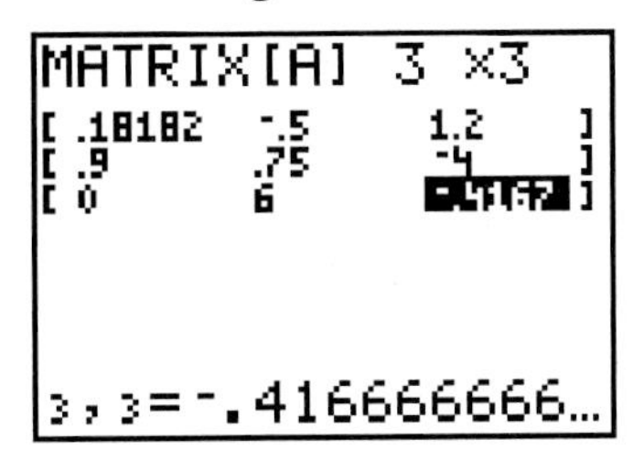

Figure 6.31

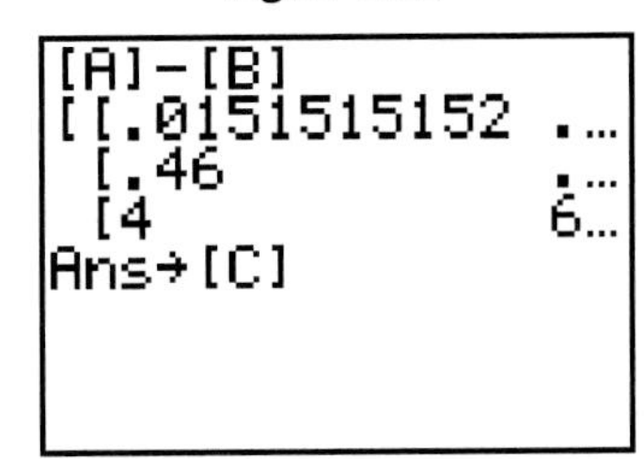

NOW TRY EXERCISES 37 THROUGH 40

Figure 6.32

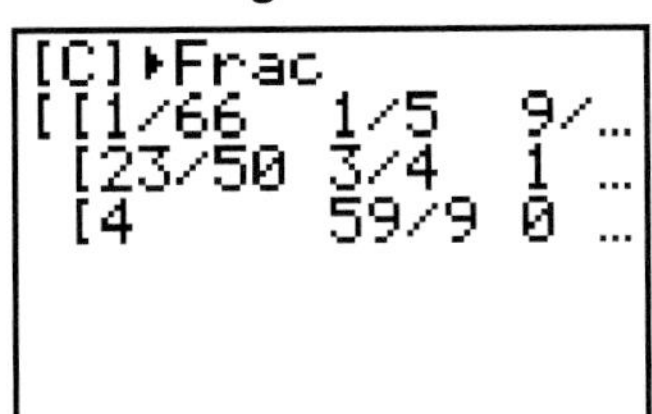

In Figure 6.31 from Example 6, the dots to the right on the calculator screen indicate additional digits or matrix columns that can't fit on the display, as often happens with larger matrices or decimal numbers. Sometimes, converting entries to fraction form will provide a display that's easier to read. Here, this is done by calling up the matrix C, and using the MATH **1: ▶ Frac** option. After pressing ENTER, all entries are converted to fractions in simplest form (where possible), as in Figure 6.32. The third column can be viewed by pressing the right arrow.

EXAMPLE 7 Use your calculator to compute the product AB.

$$A = \begin{bmatrix} 2 & -3 & 0 \\ -1 & 5 & 4 \\ 6 & 0 & 2 \\ 3 & 2 & -1 \end{bmatrix} \quad B = \begin{bmatrix} \frac{1}{2} & -0.7 & 1 \\ 0.5 & 3.2 & -3 \\ -2 & \frac{3}{4} & 4 \end{bmatrix}$$

Solution: Carefully enter matrices A and B into your calculator, then press 2nd MODE (QUIT) to get to the home screen. Use [A][B] ENTER, and the calculator finds the product shown in Figure 6.33.

A (4×3)	B (3×3)	A (4×3)	B (3×3)
multiplication is possible since 3 = 3		result will be a 4 × 3 matrix	

$$AB = \begin{bmatrix} 2 & -3 & 0 \\ -1 & 5 & 4 \\ 6 & 0 & 2 \\ 3 & 2 & -1 \end{bmatrix} \begin{bmatrix} \frac{1}{2} & -0.7 & 1 \\ 0.5 & 3.2 & -3 \\ -2 & \frac{3}{4} & 4 \end{bmatrix}$$

Figure 6.33

```
[A][B]
 [[-.5  -11   11]
  [-6   19.7  0 ]
  [-1   -2.7  14]
  [4.5  3.55  -7]]
```

NOW TRY EXERCISES 41 THROUGH 52

Properties of Matrix Multiplication

Earlier, Example 5 demonstrated that matrix multiplication is not commutative, but what about some of the other properties? Here is a group of properties that *do* hold for matrices. You are asked to check these properties in the exercise set using various matrices. See Exercises 53 through 56.

PROPERTIES OF MATRIX MULTIPLICATION

Given matrices A, B, and C for which the products are defined:

I. $A(BC) = (AB)C$ — matrix multiplication is associative

II. $A(B + C) = AB + BC$ — matrix multiplication is distributive from the left

III. $(B + C)A = BA + CA$ — matrix multiplication is distributive from the right

IV. $k(A + B) = kA + kB$ — a constant k can be distributed over multiplication

We close this section with an application of matrix multiplication. There are many other interesting applications in the exercise set.

EXAMPLE 8 In a certain country, the likelihood that a volunteer will join a particular branch of the military depends on their age. This information is stored in matrix B (Table 6.5). The number of males and females from each age group that are projected to join the military this year is stored in matrix A (Table 6.6). (a) Compute the product AB and discuss/interpret (in context) what is indicated by the entries P_{11}, P_{13}, and P_{24} of the product matrix. (b) How many males are expected to join the Navy this year?

Table 6.5 Matrix *B*

B	Likelihood of Joining			
Age Group	Army	Navy	Air Force	Marines
18–19	0.42	0.28	0.17	0.13
20–21	0.38	0.26	0.27	0.09
22–23	0.33	0.25	0.35	0.07

Table 6.6 Matrix *A*

A	Age Groups		
Sex	18–19	20–21	22–23
Female	1000	1500	500
Male	2500	3000	2000

Solution: a. Matrix A has order 2×3 and matrix B has order 3×4. The product matrix P can be found and is a 2×4 matrix. Carefully enter the matrices in your calculator. Figure 6.34 shows the entries of matrix B. Return to the home screen [2nd MODE (QUIT)], use $[A][B]$ ENTER, and the calculator finds the product matrix, shown in Figure 6.35. Pressing the right arrow shows the complete product matrix is $P = \begin{bmatrix} 1155 & 795 & 750 & 300 \\ 2850 & 1980 & 1935 & 735 \end{bmatrix}$. The entry P_{11} comes from the product of R1 from A with C1 from B, and indicates that for the year, 1155 females are projected to join the Army. In like manner, entry P_{13} shows that 750 females are projected to join the Air Force. Entry P_{24} indicates that 735 males are projected to join the Marines.

Figure 6.34

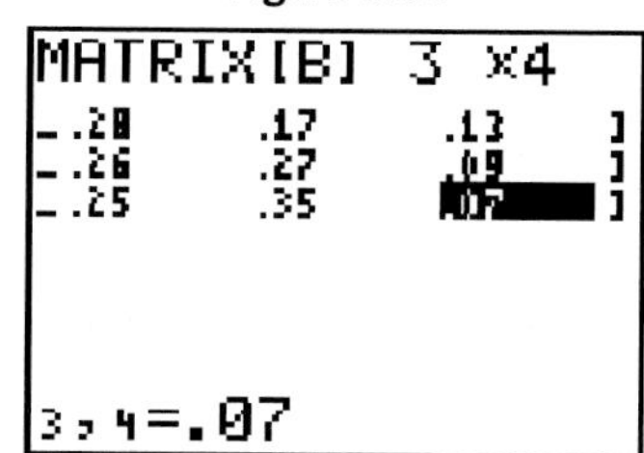

Figure 6.35

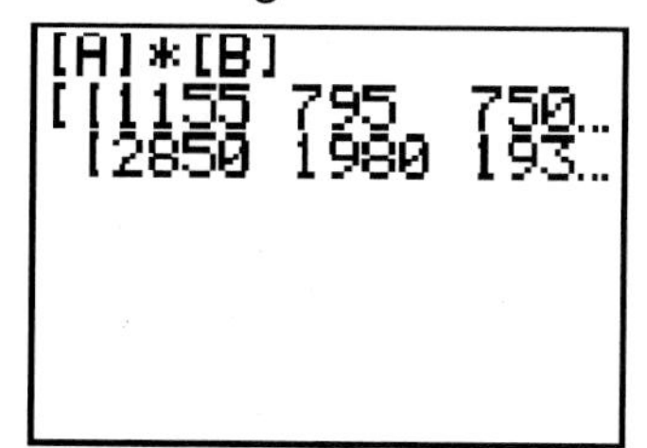

b. The product R2 (males) · C2 (Navy) gives $P_{22} = 1980$, meaning 1980 males will join the Navy.

NOW TRY EXERCISES 59 THROUGH 66

6.6 EXERCISES

CONCEPTS AND VOCABULARY

Fill in the blank with the appropriate word or phrase. Carefully reread the section if needed.

1. Two matrices are equal if they are like size and the corresponding entries are equal. In symbols, $A = B$ if $\underline{a_{ij}} = \underline{b_{ij}}$.

2. The sum of two matrices (of like size) is found by adding the corresponding entries. In symbols, $A + B = \underline{[a_{ij} + b_{ij}]}$.

3. The product of a constant times a matrix is called <u>scalar</u> multiplication.

4. The size of a matrix is also referred to as its <u>order</u>. The order of $A = \begin{bmatrix} 1 & 2 & 3 \\ 4 & 5 & 6 \end{bmatrix}$ is <u>2×3</u>.

5. Give two reasons why matrix multiplication is generally not commutative. Include several examples using matrices of various sizes. Answers will vary.

6. Discuss the conditions under which matrix multiplication is defined. Include several examples using matrices of various sizes. Answers will vary.

DEVELOPING YOUR SKILLS

State the order of each matrix and name the entries in positions a_{12} and a_{23} if they exist. Then name the position a_{ij} of the **5** in each.

7. $\begin{bmatrix} 1 & -3 \\ \mathbf{5} & -7 \end{bmatrix}$

8. $\begin{bmatrix} 19 \\ -11 \\ \mathbf{5} \end{bmatrix}$

9. $\begin{bmatrix} 2 & -3 & 0.5 \\ 0 & \mathbf{5} & 6 \end{bmatrix}$

10. $\begin{bmatrix} 2 & 0.4 \\ -0.1 & \mathbf{5} \\ 0.3 & -3 \end{bmatrix}$

11. $\begin{bmatrix} -2 & 1 & -7 \\ 0 & 8 & 1 \\ \mathbf{5} & -1 & 4 \end{bmatrix}$

12. $\begin{bmatrix} 89 & 55 & 34 & 21 \\ 13 & 8 & \mathbf{5} & 3 \\ 2 & 1 & 1 & 0 \end{bmatrix}$

Determine if the following statements are true, false, or conditional. If false, explain why. If conditional, find values of a, b, c, p, q, and r that will make the statement true.

13. $\begin{bmatrix} \sqrt{1} & \sqrt{4} & \sqrt{8} \\ \sqrt{16} & \sqrt{32} & \sqrt{64} \end{bmatrix} = \begin{bmatrix} 1 & 2 & 2\sqrt{2} \\ 4 & 4\sqrt{2} & 8 \end{bmatrix}$

14. $\begin{bmatrix} \frac{3}{2} & \frac{-7}{5} & \frac{13}{10} \\ \frac{-1}{2} & \frac{-2}{5} & \frac{1}{3} \end{bmatrix} = \begin{bmatrix} 1.5 & -1.4 & 1.3 \\ -0.5 & -0.4 & 0.\overline{3} \end{bmatrix}$

7. 2×2, $a_{12} = -3$, $a_{21} = 5$
8. 3×1, $a_{31} = 5$
9. 2×3, $a_{12} = -3$, $a_{23} = 6$, $a_{22} = 5$
10. 3×2, $a_{12} = 0.4$, $a_{22} = 5$
11. 3×3, $a_{12} = 1$, $a_{23} = 1$, $a_{31} = 5$
12. 3×4, $a_{12} = 55$, $a_{23} = 5$
13. true
14. true

15. conditional, $c = -2$, $a = -4$, $b = 3$

16. conditional, $p = 3$, $q = -7$, $r = 4$

17. $\begin{bmatrix} 10 & 0 \\ 0 & 10 \end{bmatrix}$ 18. $\begin{bmatrix} 0 & 0 & 0 \\ 0 & 0 & 0 \\ 0 & 0 & 0 \end{bmatrix}$

19. different orders, sum not possible

20. $\begin{bmatrix} 0 & 2 & 0 \\ 0 & 2 & -2 \\ -4 & -3 & 7 \end{bmatrix}$

21. $\begin{bmatrix} 20 & -15 \\ -25 & -10 \end{bmatrix}$

22. $\begin{bmatrix} -1 & 2 & 0 \\ 0 & 1 & -2 \\ -4 & -3 & 6 \end{bmatrix}$

23. $\begin{bmatrix} \frac{-5}{2} & -1 & 0 \\ 0 & \frac{-7}{2} & 1 \\ 2 & \frac{3}{2} & -6 \end{bmatrix}$

24. $\begin{bmatrix} 2 & -1 & 3 \\ 4 & 0 & -2 \end{bmatrix}$

25. $\begin{bmatrix} 1 & -2 & 0 \\ 0 & -1 & 2 \\ 4 & 3 & -6 \end{bmatrix}$

26. $\begin{bmatrix} 1 & -2 & 0 \\ 0 & -1 & 2 \\ 4 & 3 & -6 \end{bmatrix}$

27. $\begin{bmatrix} 1 & 0 \\ 0 & 1 \end{bmatrix}$ 28. $\begin{bmatrix} 1 & 0 \\ 0 & 1 \end{bmatrix}$

29. $\begin{bmatrix} 6 & -3 & 9 \\ 12 & 0 & -6 \end{bmatrix}$

30. matrix mult not possible

31. $\begin{bmatrix} 12 & -24 & 90 \\ -6 & 15 & -57 \end{bmatrix}$

32. $\begin{bmatrix} 0 \\ -7 \\ 29 \end{bmatrix}$ 33. $\begin{bmatrix} 79 & -30 \\ -50 & 19 \end{bmatrix}$

34. matrix mult not possible

35. $\begin{bmatrix} 42 & 18 & -60 \\ -12 & -42 & 36 \end{bmatrix}$

36. matrix mult not possible

37. $\begin{bmatrix} 0.71 & 0.65 \\ 1.78 & 3.55 \end{bmatrix}$

38. $\begin{bmatrix} -4.84 & 3.93 \\ 2.95 & 8.91 \end{bmatrix}$

39. $\begin{bmatrix} 1 & -1.25 & 0.25 \\ -0.5 & -0.63 & 2.13 \\ 3.75 & 3.69 & -5.94 \end{bmatrix}$

40. $\begin{bmatrix} -1 & 2.75 & 0.25 \\ -0.5 & 1.38 & -1.88 \\ -4.25 & -2.31 & 6.06 \end{bmatrix}$

Additional answers can be found in the Instructor Answer Appendix.

15. $\begin{bmatrix} -2 & 3 & a \\ 2b & -5 & 4 \\ 0 & -9 & 3c \end{bmatrix} = \begin{bmatrix} c & 3 & -4 \\ 6 & -5 & -a \\ 0 & -3b & -6 \end{bmatrix}$

16. $\begin{bmatrix} 2p+1 & -5 & 9 \\ 1 & 12 & 0 \\ q+5 & 9 & -2r \end{bmatrix} = \begin{bmatrix} 7 & -5 & 2-q \\ 1 & 3r & 0 \\ -2 & 3p & -8 \end{bmatrix}$

For matrices A through H as given, perform the indicated operation(s), if possible. Do not use a calculator. If an operation cannot be completed, state why.

$A = \begin{bmatrix} 2 & 3 \\ 5 & 8 \end{bmatrix}$ $B = \begin{bmatrix} 2 \\ 1 \\ -3 \end{bmatrix}$ $C = \begin{bmatrix} 2 & 0.5 \\ 0.2 & 5 \\ -1 & 3 \end{bmatrix}$

$D = \begin{bmatrix} 1 & 0 & 0 \\ 0 & 1 & 0 \\ 0 & 0 & 1 \end{bmatrix}$ $E = \begin{bmatrix} 1 & -2 & 0 \\ 0 & -1 & 2 \\ 4 & 3 & -6 \end{bmatrix}$ $F = \begin{bmatrix} 6 & -3 & 9 \\ 12 & 0 & -6 \end{bmatrix}$

$G = \begin{bmatrix} -1 & 2 & 0 \\ 0 & 1 & -2 \\ -4 & -3 & 6 \end{bmatrix}$ $H = \begin{bmatrix} 8 & -3 \\ -5 & 2 \end{bmatrix}$

17. $A + H$ **18.** $E + G$ **19.** $F + H$ **20.** $G + D$

21. $3H - 2A$ **22.** $2E + 3G$ **23.** $\frac{1}{2}E - 3D$ **24.** $F - \frac{2}{3}F$

25. ED **26.** DE **27.** AH **28.** HA

29. FD **30.** FH **31.** HF **32.** EB

33. H^2 **34.** F^2 **35.** FE **36.** EF

For matrices A through H as given, use a calculator to perform the indicated operation(s), if possible. If an operation cannot be completed, state why.

$A = \begin{bmatrix} -5 & 4 \\ 3 & 9 \end{bmatrix}$ $B = \begin{bmatrix} 1 & 0 \\ 0 & 1 \end{bmatrix}$ $C = \begin{bmatrix} \frac{\sqrt{3}}{2} & \frac{\sqrt{3}}{3} \\ \sqrt{3} & 2\sqrt{3} \end{bmatrix}$

$D = \begin{bmatrix} 1 & 0 & 0 \\ 0 & 1 & 0 \\ 0 & 0 & 1 \end{bmatrix}$ $E = \begin{bmatrix} 1 & -2 & 0 \\ 0 & -1 & 2 \\ 4 & 3 & -6 \end{bmatrix}$ $F = \begin{bmatrix} -0.52 & 0.002 & 1.032 \\ 1.021 & -1.27 & 0.019 \end{bmatrix}$

$G = \begin{bmatrix} 0 & \frac{3}{4} & \frac{1}{4} \\ \frac{-1}{2} & \frac{3}{8} & \frac{1}{8} \\ \frac{-1}{4} & \frac{11}{16} & \frac{1}{16} \end{bmatrix}$ $H = \begin{bmatrix} \frac{-3}{19} & \frac{4}{57} \\ \frac{1}{19} & \frac{5}{57} \end{bmatrix}$

37. $C + H$ **38.** $A - H$ **39.** $E + G$ **40.** $G - E$

41. AH **42.** HA **43.** EG **44.** GE

45. HB **46.** BH **47.** DG **48.** GD

49. C^2 **50.** E^2 **51.** FG **52.** AF

For Exercises 53 through 56, use a calculator and matrices A, B, and C to verify:

$A = \begin{bmatrix} -1 & 3 & 5 \\ 2 & 7 & -1 \\ 4 & 0 & 6 \end{bmatrix}$ $B = \begin{bmatrix} 0.3 & -0.4 & 1.2 \\ -2.5 & 2 & 0.9 \\ 1 & -0.5 & 0.2 \end{bmatrix}$ $C = \begin{bmatrix} 45 & -1 & 3 \\ -6 & 10 & -15 \\ 21 & -28 & 36 \end{bmatrix}$

53. Matrix multiplication is not generally commutative: (a) $AB \neq BA$, (b) $AC \neq CA$, and (c) $BC \neq CB$. verified

54. Matrix multiplication is distributive from the left: $A(B + C) = AB + AC$. verified

55. Matrix multiplication is distributive from the right: $(B + C)A = BA + CA$. verified

56. Matrix multiplication is associative: $(AB)C = A(BC)$. verified

WORKING WITH FORMULAS

$$\begin{bmatrix} 2 & 2 \\ W & 0 \end{bmatrix} \cdot \begin{bmatrix} L \\ W \end{bmatrix} = \begin{bmatrix} \textbf{Perimeter} \\ \textbf{Area} \end{bmatrix}$$

The perimeter and area of a rectangle can be simultaneously calculated using the matrix formula shown, where L represents the length and W represents the width of the rectangle. Use the matrix formula and your calculator to find the perimeter and area of the rectangles shown, then check using $P = 2L + 2W$ and $A = LW$.

57. 6.374 cm; 4.35 cm

58. 5.02 cm; 3.75 cm

APPLICATIONS

59. Custom T's designs and sells specialty T-shirts and sweatshirts, with plants in Verdi and Minsk. The company offers this apparel in three quality levels: standard, deluxe, and premium. Last fall the Verdi office produced 3820 standard, 2460 deluxe, and 1540 premium T-shirts, along with 1960 standard, 1240 deluxe, and 920 premium sweatshirts. The Minsk office produced 4220 standard, 2960 deluxe, and 1640 premium T-shirts, along with 2960 standard, 3240 deluxe, and 820 premium sweatshirts in the same time period.

a. Write a 3×2 "production matrix" for each plant [$V \rightarrow$ Verdi, $M \rightarrow$ Minsk], with a *T-shirt* column, a *sweatshirt* column, and three rows showing how many of the different types of apparel were manufactured.

b. Use the matrices from Part (a) to determine how many more or less articles of clothing were produced by Minsk than Verdi.

c. Use scalar multiplication to find how many shirts of each type will be made at Verdi and Minsk next fall, if each is expecting a 4% increase in business.

d. What will be Custom T's total production next fall (from both plants), for each type of apparel?

60. Terry's Tire Store sells automobile and truck tires through three retail outlets. Sales at the Cahokia store for the months of January, February, and March broke down as follows: 350, 420, and 530 auto tires and 220, 180, and 140 truck tires. The Shady Oak branch sold 430, 560, and 690 auto tires and 280, 320, and 220 truck tires during the same 3 months. Sales figures for the downtown store were 864, 980, and 1236 auto tires and 535, 542, and 332 truck tires.

a. Write a 2×3 "sales matrix" for each store [$C \rightarrow$ Cahokia, $S \rightarrow$ Shady Oak, $D \rightarrow$ Downtown], with *January, February,* and *March* columns, and two rows showing the sales of auto and truck tires respectively.

b. Use the matrices from Part (a) to determine how many more or fewer tires of each type the downtown store sold (each month) over the other two stores combined.

c. Market trends indicate that for the same three months in the following year, the Cahokia store will likely experience a 10% increase in sales, the Shady Oak store a 3% decrease, with sales at the downtown store remaining level (no change). What will be the combined monthly sales from all three stores next year, for each type of tire?

57. $P = 21.448$ cm; $A = 27.7269$ cm^2

58. $P = 17.54$ cm; $A = 18.825$ cm^2

59. a. $V \rightarrow \begin{matrix} & T & S \\ S & 3820 & 1960 \\ D & 2460 & 1240 \\ P & 1540 & 920 \end{matrix}$ $\quad M \rightarrow \begin{matrix} & T & S \\ S & 4220 & 2960 \\ D & 2960 & 3240 \\ P & 1640 & 820 \end{matrix}$

b. 3900 more by Minsk

c. $V \rightarrow \begin{bmatrix} 3972.8 & 2038.4 \\ 2558.4 & 1289.6 \\ 1601.6 & 956.8 \end{bmatrix}$ $\quad M \rightarrow \begin{bmatrix} 4388.8 & 3078.4 \\ 3078.4 & 3369.6 \\ 1705.6 & 852.8 \end{bmatrix}$

d. $\begin{bmatrix} 8361.6 & 5116.8 \\ 5636.8 & 4659.2 \\ 3307.2 & 1809.6 \end{bmatrix}$

60. a. $C \rightarrow \begin{matrix} & J & F & M \\ A & 350 & 420 & 530 \\ T & 220 & 180 & 140 \end{matrix}$ $\quad S \rightarrow \begin{matrix} & J & F & M \\ A & 430 & 560 & 690 \\ T & 280 & 320 & 220 \end{matrix}$ $\quad D \rightarrow \begin{matrix} & J & F & M \\ A & 864 & 980 & 1236 \\ T & 535 & 542 & 332 \end{matrix}$

b. $\begin{bmatrix} 84 & 0 & 16 \\ 35 & 42 & -28 \end{bmatrix}$

c. $\begin{bmatrix} 1666.1 & 1985.2 & 2488.3 \\ 1048.6 & 1050.4 & 699.4 \end{bmatrix}$

61. [22,000 19,000 23,500 14,000]; *total profit*
North: $22,000
South: $19,000
East: $23,500
West: $14,000

62. [23.5 20 31]; he will donate $23.50 for Mon, he will donate $20 for Tue, and he will donate $31 for Wed.

63. a. $108.20
b. $101
c.
$$\begin{matrix} \text{Science} \\ \text{Math} \end{matrix}\begin{bmatrix} 100 & 101 & 119 \\ 108.2 & 107 & 129.5 \end{bmatrix}$$
First row, total cost for science from each restaurant; Second row, total cost for math from each restaurant.

64. a. $50.70
b. $44.75
c.
$$\begin{matrix} \\ \text{Home} \\ \text{Comm} \\ \text{Prof} \end{matrix}\begin{matrix} \text{Coast} & \text{Midwest} \\ \begin{bmatrix} 32.65 & 28.85 \\ 50.7 & 44.75 \\ 60.25 & 53.125 \end{bmatrix} \end{matrix}$$
Each entry is the cost to package, transport, and install the three types of pool tables from either the coast or midwest.

65. a. 10.3
b. 19.5
c.
$$\begin{matrix} \\ \text{Female} \\ \text{Male} \end{matrix}\begin{matrix} \text{Spanish} & \text{Chess} & \text{Writing} \\ \begin{bmatrix} 32.4 & 10.3 & 21.3 \\ 29.9 & 9.6 & 19.5 \end{bmatrix}' \end{matrix}$$
the number of females in the writing club.

61. **Home improvements:** Dream-Makers Home Improvements specializes in replacement windows, replacement doors, and new siding. During the peak season, the number of contracts that came from various parts of the city (North, South, East, and West) are shown in matrix C. The average profit per contract is shown in matrix P. Compute the product PC and discuss what each entry of the product matrix represents.

$$\begin{matrix} \\ \text{Windows} \\ \text{Doors} \\ \text{Siding} \end{matrix}\begin{matrix} \text{N} & \text{S} & \text{E} & \text{W} \\ \begin{bmatrix} 9 & 6 & 5 & 4 \\ 7 & 5 & 7 & 6 \\ 2 & 3 & 5 & 2 \end{bmatrix} \end{matrix} = C$$

$$\begin{matrix} \text{Windows} & \text{Doors} & \text{Siding} \\ [1500 & 500 & 2500] \end{matrix} = P$$

62. **Classical music:** Station 90.7—*The Home of Classical Music*— is having their annual fund drive. Being a loyal listener, Mitchell decides that for the next 3 days he will donate money according to his favorite composers, by the number of times their music comes on the air: $3 for every piece by Mozart (M), $2.50 for every piece by Beethoven (B), and $2 for every piece by Vivaldi (V). This information is displayed in matrix D. The number of pieces he heard from each composer is displayed in matrix C. Compute the product DC and discuss what each entry of the product matrix represents.

$$\begin{matrix} \\ M \\ B \\ V \end{matrix}\begin{matrix} \text{Mon.} & \text{Tue.} & \text{Wed.} \\ \begin{bmatrix} 4 & 3 & 5 \\ 3 & 2 & 4 \\ 2 & 3 & 3 \end{bmatrix} \end{matrix} = C$$

$$\begin{matrix} \text{M} & \text{B} & \text{V} \\ [3 & 2.5 & 2] \end{matrix} = D$$

63. **Pizza and salad:** The science department and math department of a local college are at a pre-semester retreat, and decide to have pizza, salads, and soft drinks for lunch. The quantity of food ordered by each department is shown in matrix Q. The cost of the food order is shown in matrix C using the published prices from three popular restaurants: Pizza Home (PH), Papa Jeff's (PJ), and Dynamos (D).

$$\begin{matrix} \\ \text{Science} \\ \text{Math} \end{matrix}\begin{matrix} \text{Pizza} & \text{Salad} & \text{Drink} \\ \begin{bmatrix} 8 & 12 & 20 \\ 10 & 8 & 18 \end{bmatrix} \end{matrix} = Q$$

$$\begin{matrix} \\ \text{Pizza} \\ \text{Salad} \\ \text{Drink} \end{matrix}\begin{matrix} \text{PH} & \text{PJ} & \text{D} \\ \begin{bmatrix} 8 & 7.5 & 10 \\ 1.5 & 1.75 & 2 \\ 0.90 & 1 & 0.75 \end{bmatrix} \end{matrix} = C$$

a. What is the total cost to the math department if the food is ordered from Pizza Home?

b. What is the total cost to the science department if the food is ordered from Papa Jeff's?

c. Compute the product QC and discuss the meaning of each entry in the product matrix.

64. **Manufacturing pool tables:** Cue Ball Incorporated makes three types of pool tables, for homes, commercial use, and professional use. The amount of time required to pack, load, and install each is summarized in matrix T, with all times in hours. The costs of these components are summarized in matrix C for two of its warehouses, one on the west coast and the other in the midwest.

$$\begin{matrix} \\ \text{Home} \\ \text{Comm} \\ \text{Prof} \end{matrix}\begin{matrix} \text{Pack} & \text{Load} & \text{Install} \\ \begin{bmatrix} 1 & 0.2 & 1.5 \\ 1.5 & 0.5 & 2.2 \\ 1.75 & 0.75 & 2.5 \end{bmatrix} \end{matrix} = T$$

a. What is the cost to package, load, and install a commercial pool table from the coastal warehouse?

b. What is the cost to package, load, and install a commercial pool table from the warehouse in the midwest?

$$\begin{matrix} \\ \text{Pack} \\ \text{Load} \\ \text{Install} \end{matrix}\begin{matrix} \text{Coast} & \text{Midwest} \\ \begin{bmatrix} 10 & 8 \\ 12 & 10.5 \\ 13.5 & 12.5 \end{bmatrix} \end{matrix} = C$$

c. Compute the product TC and discuss the meaning of each entry in the product matrix.

65. **Joining a club:** Each school year, among the students planning to join a club, the likelihood a student joins a particular club depends on their class standing. This information is stored in matrix C. The number of males and females from each class that are projected to join a club each year is stored in matrix J. Compute the product JC and use the result to answer the following:

$$\begin{matrix} \\ \text{Female} \\ \text{Male} \end{matrix}\begin{matrix} \text{Fresh} & \text{Soph} & \text{Junior} \\ \begin{bmatrix} 25 & 18 & 21 \\ 22 & 19 & 18 \end{bmatrix} \end{matrix} = J$$

$$\begin{matrix} \\ \text{Fresh} \\ \text{Soph} \\ \text{Junior} \end{matrix}\begin{matrix} \text{Spanish} & \text{Chess} & \text{Writing} \\ \begin{bmatrix} 0.6 & 0.1 & 0.3 \\ 0.5 & 0.2 & 0.3 \\ 0.4 & 0.2 & 0.4 \end{bmatrix} \end{matrix} = C$$

a. Approximately how many females joined the chess club?

b. Approximately how many males joined the writing club?

c. What does the entry P_{13} of the product matrix tells us?

66. a. \$11,900
b. \$11,750
c.

	hand	machine
med	4350	2850
large	7200	4700
x-large	7100	4650

\$4350 is the revenue from the medium, hand-finished shirts.

66. Designer shirts: The SweatShirt Shoppe sells three types of designs on its products: stenciled (S), embossed (E), and applique (A). The quantity of each size sold is shown in matrix Q. The retail price of each sweatshirt depends on its size and whether it was finished by hand or machine. Retail prices are shown in matrix C. Assuming all stock is sold,

$$\begin{array}{c c} & \begin{array}{ccc} S & E & A \end{array} \\ \begin{array}{r} \text{med} \\ \text{large} \\ \text{x-large} \end{array} & \begin{bmatrix} 30 & 30 & 15 \\ 60 & 50 & 20 \\ 50 & 40 & 30 \end{bmatrix} = Q \end{array}$$

a. How much revenue was generated by the large sweatshirts?

b. How much revenue was generated by the extra-large sweatshirts?

c. What does the entry P_{11} of the product matrix QC tell us?

$$\begin{array}{c c} & \begin{array}{cc} \text{Hand} & \text{Machine} \end{array} \\ \begin{array}{r} S \\ E \\ A \end{array} & \begin{bmatrix} 40 & 25 \\ 60 & 40 \\ 90 & 60 \end{bmatrix} = C \end{array}$$

WRITING, RESEARCH, AND DECISION MAKING

67. no; no; A and B must be square matrices
68. A and B must be square matrices
69. $\begin{bmatrix} 2^{n-1} & 0 & 2^{n-1} \\ 2^n - 1 & 1 & 2^n - 1 \\ 2^{n-1} & 0 & 2^{n-1} \end{bmatrix}$

67. In a study of operations on polynomials, you likely noted that a binomial square can be computed using the template $(A + B)^2 = A^2 + 2AB + B^2$. Is the same true for operations on matrices? In other words, if A and B represent matrices where all needed products are defined, does the same relationship hold? Investigate this question using matrices of your own choosing, and justify your response. What conditions are necessary for the pattern to hold?

68. Use the results from Exercise 67 to investigate the relationship $(A + B)(A - B) = A^2 - B^2$ in the light of matrix multiplication.

69. For the matrix A shown to the right, use your calculator to compute A^2, A^3, A^4, and A^5. Do you notice a pattern? Try to write a "matrix formula" for A^n, where n is a positive integer, then use your formula to find A^6. Check results using a calculator.

$$A = \begin{bmatrix} 1 & 0 & 1 \\ 1 & 1 & 1 \\ 1 & 0 & 1 \end{bmatrix}$$

EXTENDING THE CONCEPT

70. Even powers generate identity matrix
71. $a = 2, b = 1, c = -3, d = -2$

70. The matrix $M = \begin{bmatrix} 2 & 1 \\ -3 & -2 \end{bmatrix}$ has some very interesting properties. Compute the powers M^2, M^3, M^4, and M^5, then discuss what you find. Try to find/create another 2×2 matrix that has similar properties.

71. For the "matrix equation" $\begin{bmatrix} 2 & 1 \\ -3 & -2 \end{bmatrix} \cdot \begin{bmatrix} a & b \\ c & d \end{bmatrix} = \begin{bmatrix} 1 & 0 \\ 0 & 1 \end{bmatrix}$, use matrix multiplication and two systems of equations to find the entries a, b, c, and d that make the equation true.

MAINTAINING YOUR SKILLS

72. $(-1, 1, -2)$
73. $x \in [-3, -1] \cup [1, 3]$
74. ≈ 4.39
75. $x^2 + 2x - 5$
76. a. $x = \pm\frac{3}{2}$
b. $x = 6$
c. $x = 2, x = -1 \pm \sqrt{3}i$
d. $x = 0, x = -9$
77.

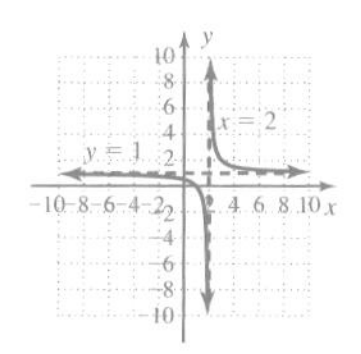

72. (6.2) Solve the system using elimination.

$$\begin{cases} x + 2y - z = 3 \\ -2x - y + 3z = -5 \\ 5x + 3y - 2z = 2 \end{cases}$$

73. (4.7) Use the graph of $f(x)$ given to solve $f(x) \geq 0$.

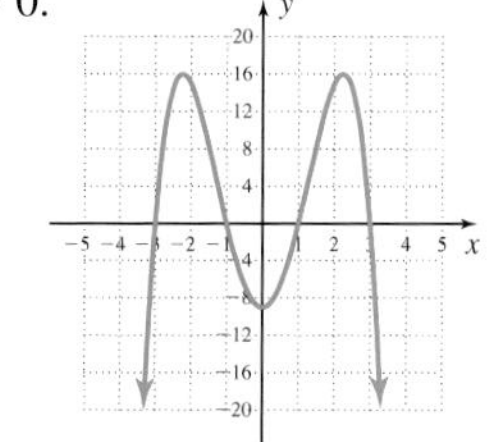

74. (5.3) Evaluate using the change-of-base formula, then check using exponentiation.

$$\log_2 21$$

75. (4.1) Find the quotient using synthetic division, then check using multiplication.

$$\frac{x^3 - 9x + 10}{x - 2}$$

76. (1.3/1.4) Solve each of the following by factoring. Find all roots, real and complex.

a. $4x^2 - 9 = 0$ **b.** $x^2 - 12x + 36 = 0$

c. $x^3 - 8 = 0$ **d.** $x^2 + 9x = 0$

77. (3.5) Graph using a transformation and state the equations of the asymptotes.

$$h(x) = \frac{1}{x - 2} + 1$$

6.7 Solving Linear Systems Using Matrix Equations

LEARNING OBJECTIVES

In Section 6.7 you will learn how to:

A. Recognize an identity matrix for multiplication

B. Find the inverse of a square matrix

C. Solve systems using matrix equations

D. Use determinants to find whether a matrix is invertible

INTRODUCTION

There is a close parallel between solving simple linear equations and solving linear systems using matrix equations. The likeness is shown here in the form of a *specific* equation, a *general* equation, and a *matrix* equation (A, B, X, and I represent matrices). In each case, a multiplicative inverse is applied to obtain the identity, which then yields solution form.

$$\begin{aligned} 3x &= 7 \\ \left(\frac{1}{3}\right)(3x) &= \left(\frac{1}{3}\right)7 \\ 1x &= \left(\frac{1}{3}\right)7 \\ x &= \left(\frac{1}{3}\right)7 \end{aligned} \qquad \begin{aligned} ax &= b \\ \left(\frac{1}{a}\right)(ax) &= \left(\frac{1}{a}\right)b \\ 1x &= \left(\frac{1}{a}\right)b \\ x &= \left(\frac{1}{a}\right)b \end{aligned} \qquad \begin{aligned} AX &= B \\ \left(\frac{1}{A}\right)(AX) &= \left(\frac{1}{A}\right)B \\ IX &= \left(\frac{1}{A}\right)B \\ X &= \left(\frac{1}{A}\right)B \end{aligned}$$

The only real difference is a notational one, in that we prefer using A^{-1} to represent the inverse of matrix A, just as a^{-1} is the multiplicative inverse of the real number a: $a^{-1}(a) = a(a^{-1}) = 1$. Here is the last line of each equation, written using this notation.

$$x = 3^{-1}(7) \qquad x = a^{-1}b \qquad X = A^{-1}B$$

POINT OF INTEREST

The applications of matrix algebra extend in many directions, and are not limited to solving systems of equations. Matrices play an important part in the study of linear transformations, invariants, determinants, Markov chains, cryptography, routing problems, and many other areas.

A. Multiplication and Identity Matrices

From the properties of real numbers, 1 is the identity for multiplication since $n \cdot 1 = 1 \cdot n = n$. A similar identity exists for matrix multiplication. Consider the 2×2 matrix $A = \begin{bmatrix} 1 & 4 \\ -2 & 3 \end{bmatrix}$. While matrix multiplication is not *generally* commutative, if we can find a matrix B where $AB = BA = A$, then B is a prime candidate for the identity matrix, which is denoted I. For the products AB and BA to be possible and have the same order as A, B must also be a 2×2 matrix. Using the arbitrary matrix $B = \begin{bmatrix} a & b \\ c & d \end{bmatrix}$, we can write the equation $AB = A$ as $\begin{bmatrix} 1 & 4 \\ -2 & 3 \end{bmatrix}\begin{bmatrix} a & b \\ c & d \end{bmatrix} = \begin{bmatrix} 1 & 4 \\ -2 & 3 \end{bmatrix}$, and solve for the entries of B using the equality of matrices and systems of equations.

EXAMPLE 1A For $\begin{bmatrix} 1 & 4 \\ -2 & 3 \end{bmatrix}\begin{bmatrix} a & b \\ c & d \end{bmatrix} = \begin{bmatrix} 1 & 4 \\ -2 & 3 \end{bmatrix}$, use matrix multiplication, the equality of matrices, and systems of equations to find the value of a, b, c, and d.

Solution: The product on the left gives $\begin{bmatrix} \mathbf{a + 4c} & b + 4d \\ -2a + 3c & -2b + 3d \end{bmatrix} = \begin{bmatrix} \mathbf{1} & 4 \\ -2 & 3 \end{bmatrix}$.
Since corresponding entries must be equal (shown by matching colors), we can find a, b, c, and d by solving the systems $\begin{cases} \mathbf{a + 4c = 1} \\ -2a + 3c = -2 \end{cases}$ and $\begin{cases} b + 4d = 4 \\ -2b + 3d = 3 \end{cases}$. For the first system, 2R1 + R2 shows $a = 1$ and $c = 0$. Using 2R1 + R2 for the second shows $b = 0$ and $d = 1$. It appears $\begin{bmatrix} 1 & 0 \\ 0 & 1 \end{bmatrix}$ is a candidate for the identity matrix.

Before we name B as the identity matrix, we must show that $AB = BA = A$.

EXAMPLE 1B Given $A = \begin{bmatrix} 1 & 4 \\ -2 & -3 \end{bmatrix}$ and $B = \begin{bmatrix} 1 & 0 \\ 0 & 1 \end{bmatrix}$, determine if $AB = A$ and $BA = A$.

Solution:

$$AB = \begin{bmatrix} 1 & 4 \\ -2 & -3 \end{bmatrix}\begin{bmatrix} 1 & 0 \\ 0 & 1 \end{bmatrix} = \begin{bmatrix} 1(1) + 4(0) & 1(0) + 4(1) \\ -2(1) + (-3)(0) & -2(0) + (-3)(1) \end{bmatrix} = \begin{bmatrix} 1 & 4 \\ -2 & -3 \end{bmatrix} \checkmark$$

$$BA = \begin{bmatrix} 1 & 0 \\ 0 & 1 \end{bmatrix}\begin{bmatrix} 1 & 4 \\ -2 & -3 \end{bmatrix} = \begin{bmatrix} 1(1) + 0(-2) & 1(4) + 0(-3) \\ 0(1) + 1(-2) & 0(4) + 1(-3) \end{bmatrix} = \begin{bmatrix} 1 & 4 \\ -2 & -3 \end{bmatrix} \checkmark$$

Since $AB = A = BA$, B is the identity matrix I.

NOW TRY EXERCISES 7 THROUGH 10

By replacing the entries of $A = \begin{bmatrix} 1 & 4 \\ -2 & -3 \end{bmatrix}$ with the entries of the general matrix $\begin{bmatrix} a_{11} & a_{12} \\ a_{21} & a_{22} \end{bmatrix}$, we can show that $I = \begin{bmatrix} 1 & 0 \\ 0 & 1 \end{bmatrix}$ is the identity for *all* 2×2 matrices. In considering the identity for larger matrices, we find that only *square matrices* have inverses, since $AI = IA$ is the primary requirement (the multiplication must be possible in both directions). This is commonly referred to as *multiplication from the right* and *multiplication from the left.* Using the same procedure as above we can show $\begin{bmatrix} 1 & 0 & 0 \\ 0 & 1 & 0 \\ 0 & 0 & 1 \end{bmatrix}$ is the identity for 3×3 matrices (denoted I_3) and further extend the idea to verify that the $n \times n$ identity matrix I_n consists of 1s down the main diagonal and 0s for all other entries. Also, the identity I_n for a square matrix is unique.

As in Section 6.6, a graphing calculator can be used to investigate operations on matrices and matrix properties. For the 3×3 matrix $A = \begin{bmatrix} 2 & 5 & 1 \\ 4 & -1 & 1 \\ 0 & 3 & -2 \end{bmatrix}$ and the matrix $I_3 = \begin{bmatrix} 1 & 0 & 0 \\ 0 & 1 & 0 \\ 0 & 0 & 1 \end{bmatrix}$, a calculator will confirm that $AI_3 = A = I_3A$, since I_3 is the

Figure 6.36

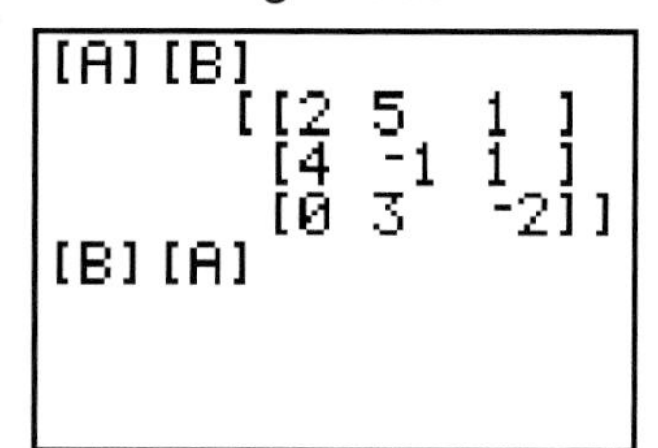

multiplicative identity for 3×3 matrices. Carefully enter A into your calculator as matrix A, and I_3 as matrix B. Figure 6.36 shows $AB = A$ and after pressing ENTER, the calculator will verify $BA = A$, although the screen cannot display the result without scrolling. See Exercises 11 through 14.

B. The Inverse of a Matrix

Again from the properties of real numbers, we know the multiplicative inverse for a is $a^{-1} = \frac{1}{a}$ $(a \neq 0)$, since the products $a \cdot a^{-1}$ and $a^{-1} \cdot a$ yield the identity 1. To show that a similar inverse exists for matrices, consider the square matrix $A = \begin{bmatrix} 6 & 5 \\ 2 & 2 \end{bmatrix}$ and an arbitrary matrix $B = \begin{bmatrix} a & b \\ c & d \end{bmatrix}$. If we can find a matrix B, where $AB = BA = I$, then B is a prime candidate for the inverse matrix of A, which is denoted A^{-1}. Proceeding as in Examples 1A and 1B gives the result shown in Example 2.

EXAMPLE 2 For $\begin{bmatrix} 6 & 5 \\ 2 & 2 \end{bmatrix}\begin{bmatrix} a & b \\ c & d \end{bmatrix} = \begin{bmatrix} 1 & 0 \\ 0 & 1 \end{bmatrix}$, use matrix multiplication, the equality of matrices and systems of equations to find the entries of $B = \begin{bmatrix} a & b \\ c & d \end{bmatrix}$.

Solution: The product on the left gives $\begin{bmatrix} \mathbf{6a + 5c} & 6b + 5d \\ 2a + 2c & 2b + 2d \end{bmatrix} = \begin{bmatrix} \mathbf{1} & 0 \\ 0 & 1 \end{bmatrix}$. Since corresponding entries must be equal (shown by matching colors), we find the values of a, b, c, and d by solving the systems $\begin{cases} \mathbf{6a + 5c = 1} \\ 2a + 2c = 0 \end{cases}$ and $\begin{cases} 6b + 5d = 0 \\ 2b + 2d = 1 \end{cases}$. Using $-3R2 + R1$ for the first system shows $a = 1$ and $c = -1$, while $-3R2 + R1$ for the second system shows $b = -2.5$ and $d = 3$. Matrix $B = \begin{bmatrix} a & b \\ c & d \end{bmatrix} = \begin{bmatrix} 1 & -2.5 \\ -1 & 3 \end{bmatrix}$ is the prime candidate for A^{-1}.

NOW TRY EXERCISES 15 THROUGH 18

To determine if the inverse of matrix A has been found, we test whether multiplication from the right and multiplication from the left yields the matrix I: $AB = BA = I$.

EXAMPLE 3 For the matrices $A = \begin{bmatrix} 6 & 5 \\ 2 & 2 \end{bmatrix}$ and $B = \begin{bmatrix} 1 & -2.5 \\ -1 & 3 \end{bmatrix}$ from Example 2, determine if $AB = BA = I$.

$$AB = \begin{bmatrix} 6 & 5 \\ 2 & 2 \end{bmatrix}\begin{bmatrix} 1 & -2.5 \\ -1 & 3 \end{bmatrix} \qquad BA = \begin{bmatrix} 1 & -2.5 \\ -1 & 3 \end{bmatrix}\begin{bmatrix} 6 & 5 \\ 2 & 2 \end{bmatrix}$$

$$= \begin{bmatrix} 6(1) + 5(-1) & 6(-2.5) + 5(3) \\ 2(1) + 2(-1) & 2(-2.5) + 2(3) \end{bmatrix} \qquad = \begin{bmatrix} 1(6) + (-2.5)(2) & 1(5) + (-2.5)(2) \\ -1(6) + 3(2) & -1(5) + 3(2) \end{bmatrix}$$

$$= \begin{bmatrix} 1 & 0 \\ 0 & 1 \end{bmatrix} \checkmark \qquad = \begin{bmatrix} 1 & 0 \\ 0 & 1 \end{bmatrix} \checkmark$$

Since $AB = BA = I$, we conclude $B = A^{-1}$.

NOW TRY EXERCISES 19 THROUGH 22

These observations guide us to the following definition of an inverse matrix.

THE INVERSE OF A MATRIX

Given an $n \times n$ matrix A. If there exists an $n \times n$ matrix A^{-1} such that $AA^{-1} = A^{-1}A = I_n$, then A^{-1} is the inverse of matrix A.

We will soon discover that while only square matrices have inverses, not every square matrix has an inverse. If an inverse exists, the matrix is said to be **invertible.** For 2×2 matrices that are invertible, a simple formula exists for computing an inverse.

THE INVERSE OF A 2 × 2 MATRIX

If matrix $A = \begin{bmatrix} a & b \\ c & d \end{bmatrix}$, then $A^{-1} = \frac{1}{ad - bc}\begin{bmatrix} d & -b \\ -c & a \end{bmatrix}$.

To "test" the formula, again consider the matrix $A = \begin{bmatrix} 6 & 5 \\ 2 & 2 \end{bmatrix}$, where $a = 6$, $b = 5$, $c = 2$, and $d = 2$:

$$A^{-1} = \frac{1}{(6)(2) - (5)(2)}\begin{bmatrix} 2 & -5 \\ -2 & 6 \end{bmatrix}$$

$$= \frac{1}{2}\begin{bmatrix} 2 & -5 \\ -2 & 6 \end{bmatrix} = \begin{bmatrix} 1 & -2.5 \\ -1 & 3 \end{bmatrix} \checkmark$$

See Exercises 59 through 62 for more practice with this formula.

WORTHY OF NOTE

For more on the **equivalent matrix method,** see Exercises 75 and 76. The **augmented matrix method** for finding an inverse is discussed in the *Strengthening Core Skills* feature at the end of the chapter.

Almost without exception, real-world applications involve much larger matrices, with entries that are not integer-valued. Although the methods already employed can be extended to find the inverse of these larger matrices, the process becomes very tedious and too time consuming to be useful. For practical reasons, we will rely on a calculator to produce larger inverse matrices. This is done by: (1) carefully entering a square matrix A into the calculator, (2) returning to the home screen, (3) calling up matrix A and pressing the X^{-1} key and ENTER to find A^{-1}. In the context of matrices, calculators are programmed to compute an inverse matrix, rather than to somehow find a reciprocal. See Exercises 23 through 26.

C. Solving Systems Using Matrix Equations

As we saw in Section 6.6, one reason matrix multiplication has its row × column definition is to assist in writing a linear system of equations as a matrix equation. The equation consists of the matrix of constants B on the right, and a product of the coefficient matrix A with the matrix of variables X on the left: $AX = B$. For $\begin{cases} x + 4y - z = 10 \\ 2x + 5y - 3z = 7, \\ 8x + y - 2z = 11 \end{cases}$ the matrix equation is $\begin{bmatrix} 1 & 4 & -1 \\ 2 & 5 & -3 \\ 8 & 1 & -2 \end{bmatrix}\begin{bmatrix} x \\ y \\ z \end{bmatrix} = \begin{bmatrix} 10 \\ 7 \\ 11 \end{bmatrix}$. Note the product on the left indeed yields the original system.

Once written as a matrix equation, the system can be solved using an inverse matrix and the sequence outlined in the introduction. If A represents the matrix of coefficients, X the matrix of variables, B the matrix of constants, and I the appropriate identity, the sequence is

(1) $AX = B$ matrix equation

(2) $A^{-1}(AX) = A^{-1}B$ multiply from the left by the inverse of A

(3) $(A^{-1}A)X = A^{-1}B$ associative property

(4) $IX = A^{-1}B$ $A^{-1}A = I$

(5) $X = A^{-1}B$ $IX = X$

Lines 1 through 5 illustrate the steps that make the method work. In actual practice, after carefully entering the matrices, only step 5 is used when solving matrix equations using technology. Once matrix A is entered, the calculator will automatically *find* and *use* A^{-1} in the equation $X = A^{-1}B$.

EXAMPLE 4 Use a calculator and a matrix equation to solve the system

$$\begin{cases} x + 4y - z = 10 \\ 2x + 5y - 3z = 7. \\ 8x + y - 2z = 11 \end{cases}$$

Solution: As before, the matrix equation is $\begin{bmatrix} 1 & 4 & -1 \\ 2 & 5 & -3 \\ 8 & 1 & -2 \end{bmatrix}\begin{bmatrix} x \\ y \\ z \end{bmatrix} = \begin{bmatrix} 10 \\ 7 \\ 11 \end{bmatrix}$.

Carefully enter (and double-check) the matrix of coefficients as matrix A in your calculator, and the matrix of constants as matrix B. The product $A^{-1}B$ shows the solution is $x = 2$, $y = 3$, $z = 4$. Verify by substitution.

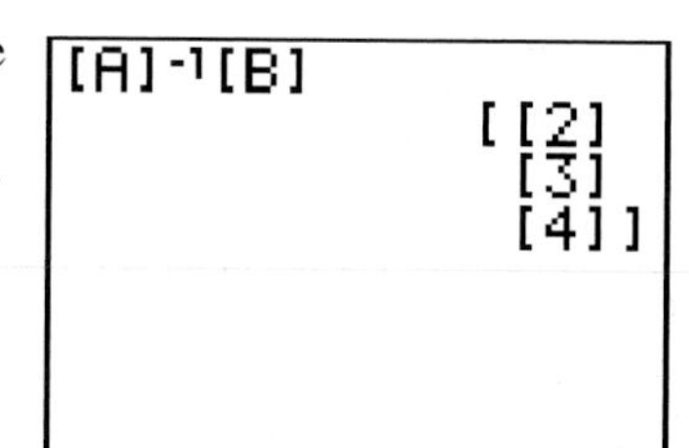

NOW TRY EXERCISES 27 THROUGH 44

The matrix equation method does have a few shortcomings. Consider the system whose corresponding matrix equation is $AX = B$: $\begin{bmatrix} 4 & -10 \\ -2 & 5 \end{bmatrix}\begin{bmatrix} x \\ y \end{bmatrix} = \begin{bmatrix} -8 \\ 13 \end{bmatrix}$. After entering the matrix of coefficients A and matrix of constants B, attempting to compute $A^{-1}B$ results

Figure 6.37

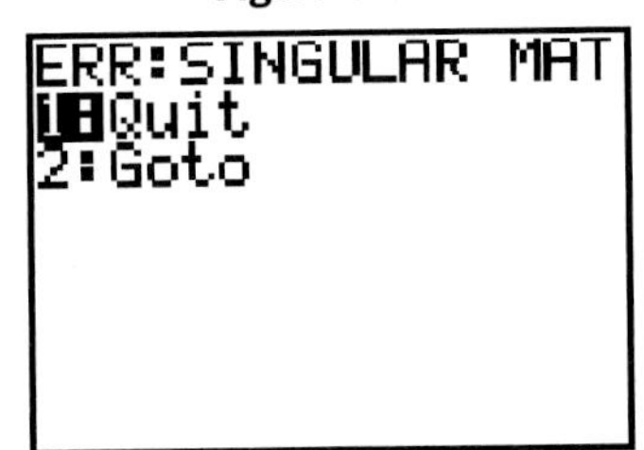

in the error message shown in Figure 6.37. Apparently the calculator is unable to return a solution due to something called a **"singular matrix."** To investigate further, we attempt to find A^{-1} for $\begin{bmatrix} 4 & -10 \\ -2 & 5 \end{bmatrix}$ using the formula for a 2×2 matrix. For $A = \begin{bmatrix} 4 & -10 \\ -2 & 5 \end{bmatrix}$, we have $a = 4$, $b = -10$, $c = -2$, and $d = 5$:

$$A^{-1} = \frac{1}{ad - bc}\begin{bmatrix} d & -b \\ -c & a \end{bmatrix} = \frac{1}{(4)(5) - (-10)(-2)}\begin{bmatrix} 5 & 10 \\ 2 & 4 \end{bmatrix}$$

$$= \frac{1}{0}\begin{bmatrix} 5 & 10 \\ 2 & 4 \end{bmatrix}.$$

We are unable to proceed further (since division by zero is undefined) and conclude that matrix A has no inverse. A matrix having no inverse is said to be **singular** or **non-invertible.** Solving systems using matrix equations is only possible when the matrix of coefficients is **nonsingular.** Otherwise, we must use row reduction and a parameter (as needed) to state a solution.

D. Determinants and Singular Matrices

As a practical matter, it becomes important to know ahead of time whether a particular matrix has an inverse. To help with this, we introduce one additional operation on a square matrix, that of calculating its **determinant.** Every square matrix has a real number associated with it called its determinant. For a 1×1 matrix the determinant is the entry itself. For a 2×2 matrix $A = \begin{bmatrix} a_{11} & a_{12} \\ a_{21} & a_{22} \end{bmatrix}$, the determinant, written as det(A) and denoted by vertical bars as $|A|$, is computed as *a difference of diagonal products* beginning with the upper-left entry: $\det(A) = \begin{vmatrix} a_{11} & a_{12} \\ a_{21} & a_{22} \end{vmatrix} = a_{11}a_{22} - a_{21}a_{12}$.

THE DETERMINANT OF A 2 × 2 MATRIX

Given any 2×2 matrix $A = \begin{bmatrix} a_{11} & a_{12} \\ a_{21} & a_{22} \end{bmatrix}$,

$$\det(A) = |A| = a_{11}a_{22} - a_{21}a_{12}$$

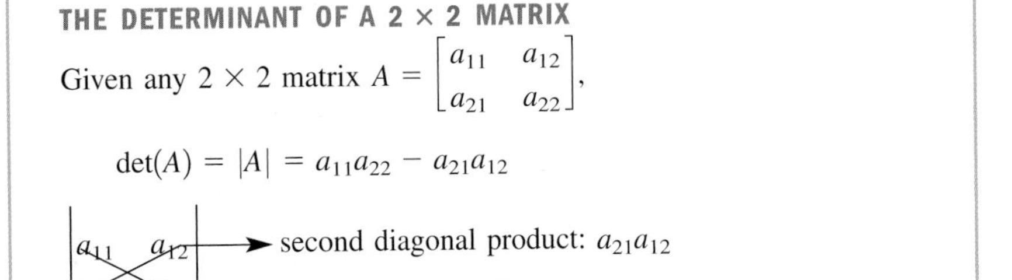

second diagonal product: $a_{21}a_{12}$

first diagonal product: $a_{11}a_{22}$

difference of diagonal products: $a_{11}a_{22} - a_{21}a_{12}$

EXAMPLE 5 Compute the determinant of each matrix given.

a. $A = \begin{bmatrix} 1 & 2 \\ 3 & 5 \end{bmatrix}$ **b.** $B = \begin{bmatrix} 3 & 2 \\ 1 & -6 \end{bmatrix}$

c. $C = \begin{bmatrix} 5 & 2 & 1 \\ -1 & -3 & 4 \end{bmatrix}$ **d.** $D = \begin{bmatrix} 4 & -10 \\ -2 & 5 \end{bmatrix}$

Solution: a. $\det(A) = \begin{vmatrix} 1 & 2 \\ 3 & 5 \end{vmatrix} = (1)(5) - (3)(2) = -1$

b. $\det(B) = \begin{vmatrix} 3 & 2 \\ 1 & -6 \end{vmatrix} = (3)(-6) - (1)(2) = -20$

c. Determinants are only defined for square matrices.

d. $\det(D) = \begin{vmatrix} 4 & -10 \\ -2 & 5 \end{vmatrix} = (4)(5) - (-2)(-10) = 20 - 20 = 0$

NOW TRY EXERCISES 45 THROUGH 48

Notice from Example 5(d), the determinant of $\begin{bmatrix} 4 & -10 \\ -2 & 5 \end{bmatrix}$ was zero, and this is the same matrix we earlier found had no inverse. This observation can be extended to larger matrices and offers the connection we seek between a given matrix, its inverse, and matrix equations.

SINGULAR MATRICES

If $\det(A) = 0$, the inverse matrix *does not exist* and A is said to be *singular* or *noninvertible.*

In summary, inverses exist only for square matrices, but not every square matrix has an inverse. If the determinant of a square matrix is zero, an inverse does not exist and the method of matrix equations cannot be used to solve the system.

WORTHY OF NOTE

For the determinant of a general $n \times n$ matrix using **cofactors,** see Appendix III.

To use the determinant test for a 3×3 system, we need to compute a 3×3 determinant. Although the computation can be a bit tedious, learning how it's done gives insight into how determinants of even larger matrices are computed. This insight contributes to a better understanding of other matrix applications, and gives a certain justification for scalar multiplication and 2×2 determinants. At first glance, our experience with 2×2 determinants appears to be of little help. However, every entry in a 3×3 matrix is associated with a smaller 2×2 matrix, formed by *deleting the row and column* of that entry and using those entries that remain. These 2×2's are called the **associated minor matrices** or simply the **minors.** Using a general matrix of coefficients, we'll identify the minors associated with the entries in the first row.

$$\begin{bmatrix} a_{11} & a_{12} & a_{13} \\ a_{21} & a_{22} & a_{23} \\ a_{31} & a_{32} & a_{33} \end{bmatrix} \quad \begin{bmatrix} a_{11} & a_{12} & a_{13} \\ a_{21} & a_{22} & a_{23} \\ a_{31} & a_{32} & a_{33} \end{bmatrix} \quad \begin{bmatrix} a_{11} & a_{12} & a_{13} \\ a_{21} & a_{22} & a_{23} \\ a_{31} & a_{32} & a_{33} \end{bmatrix}$$

Entry: a_{11} associated minor $\begin{bmatrix} a_{22} & a_{23} \\ a_{32} & a_{33} \end{bmatrix}$

Entry: a_{12} associated minor $\begin{bmatrix} a_{21} & a_{23} \\ a_{31} & a_{33} \end{bmatrix}$

Entry: a_{13} associated minor $\begin{bmatrix} a_{21} & a_{22} \\ a_{31} & a_{32} \end{bmatrix}$

To illustrate, we'll consider the system shown next, and (1) form the matrix of coefficients, (2) identify the minor matrices associated with the entries in the first row, and (3) compute the determinant of each *minor.*

$$\begin{cases} 2x + 3y - z = 1 \\ x - 4y + 2z = -3 \\ 3x + y = -1 \end{cases}$$

(1) Matrix of coefficients: $\begin{bmatrix} 2 & 3 & -1 \\ 1 & -4 & 2 \\ 3 & 1 & 0 \end{bmatrix}$

(2) $\begin{bmatrix} \textcircled{2} & 3 & -1 \\ 1 & -4 & 2 \\ 3 & 1 & 0 \end{bmatrix}$ $\begin{bmatrix} 2 & \textcircled{3} & -1 \\ 1 & -4 & 2 \\ 3 & 1 & 0 \end{bmatrix}$ $\begin{bmatrix} 2 & 3 & \textcircled{-1} \\ 1 & -4 & 2 \\ 3 & 1 & 0 \end{bmatrix}$

Entry a_{11}: 2 associated minor	**Entry a_{12}: 3 associated minor**	**Entry a_{13}: 1 associated minor**
$\begin{bmatrix} -4 & 2 \\ 1 & 0 \end{bmatrix}$	$\begin{bmatrix} 1 & 2 \\ 3 & 0 \end{bmatrix}$	$\begin{bmatrix} 1 & -4 \\ 3 & 1 \end{bmatrix}$
(3) **Determinant of minor**	**Determinant of minor**	**Determinant of minor**
$(-4)(0) - (1)(2) = -2$	$(1)(0) - (3)(2) = -6$	$(1)(1) - (3)(-4) = 13$

For computing a 3×3 determinant, we'll illustrate a technique called **expansion by minors.** Each term in the expansion is formed by the product of an entry from a specified row (or column) with the determinant of the associated minor. The signs used between the terms of the expansion depend on the row or column used, according to a specified pattern. The determinant of a matrix is unique and the expansion can be done using *any* row or column. For this reason, it's helpful to select the row or column having the most zero, positive, and/or smaller entries.

COMPUTING THE DETERMINANT OF A 3 × 3 MATRIX USING EXPANSION BY MINORS

For the 3×3 matrix M shown, det(M) is a unique number that can be found using expansion by minors, which is computed as follows:

Matrix M

$$\begin{bmatrix} a_{11} & a_{21} & a_{31} \\ a_{21} & a_{22} & a_{32} \\ a_{31} & a_{23} & a_{33} \end{bmatrix}$$

1. The *terms of the expansion* are formed by the product of the entries from a specified row (or column) with the determinant of the associated minor matrix. See the following illustration, which uses the entries in row 1.
2. The *signs used between terms* of the expansion depend on the row or column chosen according to the *sign chart* shown.

Sign Chart

$$\begin{bmatrix} + & - & + \\ - & + & - \\ + & - & + \end{bmatrix}$$

$$\det(M) = \begin{vmatrix} a_{11} & a_{12} & a_{13} \\ a_{21} & a_{22} & a_{23} \\ a_{31} & a_{32} & a_{33} \end{vmatrix}$$

$$= +a_{11}\begin{vmatrix} a_{22} & a_{23} \\ a_{32} & a_{33} \end{vmatrix} - a_{12}\begin{vmatrix} a_{21} & a_{23} \\ a_{31} & a_{33} \end{vmatrix} + a_{13}\begin{vmatrix} a_{21} & a_{22} \\ a_{31} & a_{32} \end{vmatrix}$$

EXAMPLE 6 Compute the determinant of $M = \begin{bmatrix} 2 & 1 & -3 \\ 1 & -1 & 0 \\ -2 & 1 & 4 \end{bmatrix}$.

Solution: Since the second row has the "smallest" entries as well as a zero entry, we compute the determinant using this row. According to the sign chart, the signs of the expansion will be negative–positive–negative, so the expansion is

$$\begin{aligned}\det(M) &= -(1)\begin{vmatrix}1 & -3\\ 1 & 4\end{vmatrix} + (-1)\begin{vmatrix}2 & -3\\ -2 & 4\end{vmatrix} - (0)\begin{vmatrix}2 & 1\\ -2 & 1\end{vmatrix}\\ &= -1(4+3) + (-1)(8-6) + (0)(2+2)\\ &= \quad -7 \quad + \quad (-2) \quad + \quad 0\\ &= -9 \rightarrow \text{The value of } \det(M) \text{ is } -9.\end{aligned}$$

NOW TRY EXERCISES 49 THROUGH 54

Try computing the determinant of M two more times, using a different row or column each time. Since the determinant of a given matrix is unique, you should obtain the same result.

EXAMPLE 7 Given the system shown here, (1) form the matrix equation $AX = B$; (2) compute the determinant of the coefficient matrix and determine if you can proceed; and (3) if so, solve the system using a matrix equation.

$$\begin{cases} 2x + 1y - 3z = 11\\ 1x - 1y = 1\\ -2x + 1y + 4z = -8\end{cases}$$

Solution:

1. Form the matrix equation $AX = B$:

$$\begin{bmatrix}2 & 1 & -3\\ 1 & -1 & 0\\ -2 & 1 & 4\end{bmatrix}\begin{bmatrix}x\\ y\\ z\end{bmatrix} = \begin{bmatrix}11\\ 1\\ -8\end{bmatrix}$$

2. Since $\det(A)$ is nonzero (from Example 6), we proceed with solving the system.
3. $X = A^{-1}B$

$$X = \begin{bmatrix}\frac{4}{9} & \frac{7}{9} & \frac{1}{3}\\ \frac{4}{9} & -\frac{2}{9} & \frac{1}{3}\\ \frac{1}{9} & \frac{4}{9} & \frac{1}{3}\end{bmatrix}\begin{bmatrix}11\\ 1\\ -8\end{bmatrix} = \begin{bmatrix}3\\ 2\\ -1\end{bmatrix}$$

calculator finds and uses A^{-1} in one step

The solution is the ordered triple $(3, 2, -1)$

NOW TRY EXERCISES 55 THROUGH 58

We close this section with an application involving a 4×4 system.

EXAMPLE 8 At a local theater, four sizes of soft drinks are sold: extra-large, 32 oz @ \$2.25; large, 24 oz @ \$1.90; medium, 16 oz @ \$1.50; and small, 12 oz @ \$1.20/each. As part of a "free guest pass" promotion, the manager asks employees to try and determine the number of each size sold, given the following information: (1) the total revenue rung under the soft-drink key was \$719.80; (2) there were 9096 oz of soft drink sold; (3) there was a total of 394 soft drinks sold; and (4) the number of large and small drinks sold was 12 more than the number of extra-large and medium drinks sold. Your best friend works at the theater, and asks for your help. Write a system of equations that models this information, then solve the system using a matrix equation.

Solution: If we let x, l, m, and s represent the number of extra-large, large, medium, and small soft drinks sold, the following system is produced:

$$\begin{cases} 2.25x + 1.90l + 1.50m + 1.20s = 719.8 & \text{revenue} \\ 32x + 24l + 16m + 12s = 9096 & \text{ounces sold} \\ x + l + m + s = 394 & \text{quantity sold} \\ l + s = x + m + 12 & \text{relationship between amounts sold} \end{cases}$$

When written as a matrix equation the system becomes:

$$\begin{bmatrix} 2.25 & 1.9 & 1.5 & 1.2 \\ 32 & 24 & 16 & 12 \\ 1 & 1 & 1 & 1 \\ -1 & 1 & -1 & 1 \end{bmatrix} \begin{bmatrix} x \\ l \\ m \\ s \end{bmatrix} = \begin{bmatrix} 719.8 \\ 9096 \\ 394 \\ 12 \end{bmatrix}$$

To solve, carefully enter the matrix of coefficients as matrix A, and the matrix of constants as matrix B, then compute $A^{-1}B = X$ [verify $\det(A) \neq 0$]. This gives a solution of $(x, l, m, s) = (112, 151, 79, 52)$, meaning 112 extra-large, 151 large, 79 medium, and 52 small soft drinks were sold.

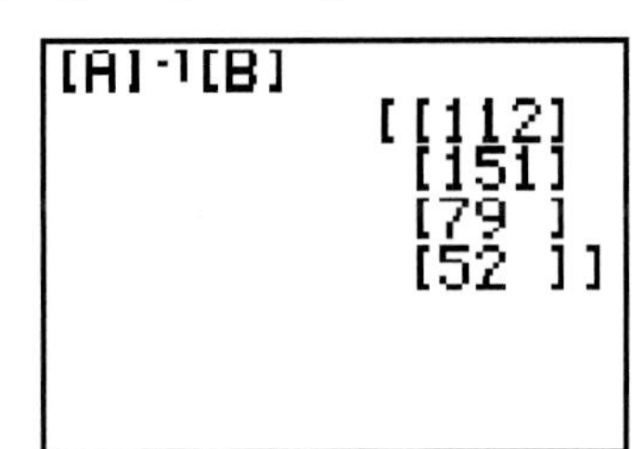

NOW TRY EXERCISES 63 TO 72

6.7 EXERCISES

CONCEPTS AND VOCABULARY

Fill in the blank with the appropriate word or phrase. Carefully reread the section if needed.

1. The $n \times n$ identity matrix I_n, consists of 1's down the main diagonal and zeroes for all other entries.

2. Given square matrices A and B of like size, B is the inverse of A if $AB = BA = I$. Notationally we write $B = A^{-1}$.

3. The product of a square matrix A and its inverse A^{-1} yields the identity matrix.

4. If the determinant of a matrix is zero, the matrix is said to be singular or noninvertible, meaning no inverse exists.

5. Explain why inverses exist only for square matrices, then discuss why some square matrices do not have an inverse. Illustrate each point with an example. Answers will vary.

6. What is the connection between the determinant of a 2×2 matrix and the formula for finding its inverse? Use the connection to create a 2×2 matrix that is invertible, and another that is not. Answers will vary.

DEVELOPING YOUR SKILLS

Use matrix multiplication, equality of matrices, and the arbitrary matrix given to show that $\begin{bmatrix} a & b \\ c & d \end{bmatrix} = \begin{bmatrix} 1 & 0 \\ 0 & 1 \end{bmatrix}$.

7. verified
8. verified
9. verified
10. verified
11. verified
12. verified
13. verified
14. verified

15. $\begin{bmatrix} \frac{1}{9} & \frac{2}{9} \\ \frac{-1}{9} & \frac{5}{18} \end{bmatrix}$
16. $\begin{bmatrix} 1 & \frac{-5}{4} \\ 0 & \frac{-1}{4} \end{bmatrix}$

17. $\begin{bmatrix} -5 & 1.5 \\ -2 & 0.5 \end{bmatrix}$
18. $\begin{bmatrix} \frac{-2}{5} & \frac{1}{5} \\ \frac{1}{2} & 1 \end{bmatrix}$

23. $\begin{bmatrix} \frac{-2}{39} & \frac{1}{13} & \frac{10}{39} \\ \frac{1}{3} & 0 & \frac{1}{3} \\ \frac{-4}{39} & \frac{2}{13} & \frac{-19}{39} \end{bmatrix}$

24. $\begin{bmatrix} \frac{22}{5} & \frac{-4}{3} & \frac{-6}{5} \\ -8 & \frac{10}{3} & 4 \\ 4 & 0 & -2 \end{bmatrix}$

25. $\begin{bmatrix} \frac{-9}{80} & \frac{31}{400} & \frac{27}{400} \\ \frac{1}{80} & \frac{41}{400} & \frac{-3}{400} \\ \frac{-1}{20} & \frac{-1}{100} & \frac{-17}{100} \end{bmatrix}$

26. $\begin{bmatrix} 1 & \frac{-3}{8} & 0 & \frac{3}{8} \\ 1 & \frac{-3}{8} & -1 & \frac{3}{8} \\ 2 & \frac{3}{4} & -2 & \frac{-3}{4} \\ 1 & \frac{-3}{8} & -1 & \frac{-5}{8} \end{bmatrix}$

27. $\begin{bmatrix} 2 & -3 \\ -5 & 7 \end{bmatrix}\begin{bmatrix} x \\ y \end{bmatrix} = \begin{bmatrix} 9 \\ 8 \end{bmatrix}$

28. $\begin{bmatrix} 0.5 & -0.6 \\ -0.7 & 0.4 \end{bmatrix}\begin{bmatrix} x \\ y \end{bmatrix} = \begin{bmatrix} 0.6 \\ -0.375 \end{bmatrix}$

29. $\begin{bmatrix} 1 & 2 & -1 \\ 1 & 0 & 1 \\ 2 & -1 & 1 \end{bmatrix}\begin{bmatrix} x \\ y \\ z \end{bmatrix} = \begin{bmatrix} 1 \\ 3 \\ 3 \end{bmatrix}$

30. $\begin{bmatrix} 2 & -3 & -2 \\ \frac{1}{4} & -\frac{2}{5} & \frac{3}{4} \\ -2 & 1.3 & -3 \end{bmatrix}\begin{bmatrix} x \\ y \\ z \end{bmatrix} = \begin{bmatrix} 4 \\ \frac{-1}{3} \\ 5 \end{bmatrix}$

31. $\begin{bmatrix} -2 & 1 & -4 & 5 \\ 2 & -5 & 1 & -3 \\ -3 & 1 & 6 & 1 \\ 1 & 4 & -5 & 1 \end{bmatrix}\begin{bmatrix} w \\ x \\ y \\ z \end{bmatrix} = \begin{bmatrix} -3 \\ 4 \\ 1 \\ -9 \end{bmatrix}$

32. $\begin{bmatrix} 1.5 & 2.1 & -0.4 & 1 \\ 0.2 & -2.6 & 1 & 0 \\ 0 & 3.2 & 0 & 1 \\ 1.6 & 4 & -5 & 2.6 \end{bmatrix}\begin{bmatrix} w \\ x \\ y \\ z \end{bmatrix} = \begin{bmatrix} 1 \\ 5.8 \\ 2.7 \\ -1.8 \end{bmatrix}$

7. $A = \begin{bmatrix} 2 & 5 \\ -3 & -7 \end{bmatrix}\begin{bmatrix} a & b \\ c & d \end{bmatrix} = \begin{bmatrix} 2 & 5 \\ -3 & -7 \end{bmatrix}$

8. $A = \begin{bmatrix} 9 & -7 \\ -5 & 4 \end{bmatrix}\begin{bmatrix} a & b \\ c & d \end{bmatrix} = \begin{bmatrix} 9 & -7 \\ -5 & 4 \end{bmatrix}$

9. $A = \begin{bmatrix} 0.4 & 0.6 \\ 0.3 & 0.2 \end{bmatrix}\begin{bmatrix} a & b \\ c & d \end{bmatrix} = \begin{bmatrix} 0.4 & 0.6 \\ 0.3 & 0.2 \end{bmatrix}$

10. $A = \begin{bmatrix} \frac{1}{2} & \frac{1}{4} \\ \frac{1}{3} & \frac{1}{8} \end{bmatrix}\begin{bmatrix} a & b \\ c & d \end{bmatrix} = \begin{bmatrix} \frac{1}{2} & \frac{1}{4} \\ \frac{1}{3} & \frac{1}{8} \end{bmatrix}$

For $I_2 = \begin{bmatrix} 1 & 0 \\ 0 & 1 \end{bmatrix}$, $I_3 = \begin{bmatrix} 1 & 0 & 0 \\ 0 & 1 & 0 \\ 0 & 0 & 1 \end{bmatrix}$, and $I_4 = \begin{bmatrix} 1 & 0 & 0 & 0 \\ 0 & 1 & 0 & 0 \\ 0 & 0 & 1 & 0 \\ 0 & 0 & 0 & 1 \end{bmatrix}$, show $AI = IA = A$ for the matrices of like size. Use a calculator for Exercise 14.

11. $\begin{bmatrix} -3 & 8 \\ -4 & 10 \end{bmatrix}$

12. $\begin{bmatrix} 0.5 & -0.2 \\ -0.7 & 0.3 \end{bmatrix}$

13. $\begin{bmatrix} -4 & 1 & 6 \\ 9 & 5 & 3 \\ 0 & -2 & 1 \end{bmatrix}$

14. $\begin{bmatrix} 9 & 1 & 3 & -1 \\ 2 & 0 & -5 & 3 \\ 4 & 6 & 1 & 0 \\ 0 & -2 & 4 & 1 \end{bmatrix}$

Find the inverse of each 2×2 matrix using matrix multiplication, equality of matrices, and a system of equations.

15. $\begin{bmatrix} 5 & -4 \\ 2 & 2 \end{bmatrix}$

16. $\begin{bmatrix} 1 & -5 \\ 0 & -4 \end{bmatrix}$

17. $\begin{bmatrix} 1 & -3 \\ 4 & -10 \end{bmatrix}$

18. $\begin{bmatrix} -2 & 0.4 \\ 1 & 0.8 \end{bmatrix}$

Demonstrate that $B = A^{-1}$, by showing $AB = BA = I$. Do not use a calculator.

19. $A = \begin{bmatrix} 1 & 5 \\ -2 & -9 \end{bmatrix}$, $B = \begin{bmatrix} -9 & -5 \\ 2 & 1 \end{bmatrix}$ Verified

20. $A = \begin{bmatrix} -2 & -6 \\ 4 & 11 \end{bmatrix}$, $B = \begin{bmatrix} 5.5 & 3 \\ -2 & -1 \end{bmatrix}$ Verified

21. $A = \begin{bmatrix} 4 & -5 \\ 0 & 2 \end{bmatrix}$, $B = \begin{bmatrix} \frac{1}{4} & \frac{5}{8} \\ 0 & \frac{1}{2} \end{bmatrix}$ Verified

22. $A = \begin{bmatrix} -2 & 5 \\ 3 & -4 \end{bmatrix}$, $B = \begin{bmatrix} \frac{4}{7} & \frac{5}{7} \\ \frac{3}{7} & \frac{2}{7} \end{bmatrix}$ Verified

Use a calculator to find $A^{-1} = B$, then confirm the inverse by showing $AB = BA = I$.

23. $A = \begin{bmatrix} -2 & 3 & 1 \\ 5 & 2 & 4 \\ 2 & 0 & -1 \end{bmatrix}$

24. $A = \begin{bmatrix} 0.5 & 0.2 & 0.1 \\ 0 & 0.3 & 0.6 \\ 1 & 0.4 & -0.3 \end{bmatrix}$

25. $A = \begin{bmatrix} -7 & 5 & -3 \\ 1 & 9 & 0 \\ 2 & -2 & -5 \end{bmatrix}$

26. $A = \frac{1}{12}\begin{bmatrix} 12 & -6 & 3 & 0 \\ 0 & -4 & 8 & -12 \\ 12 & -12 & 0 & 0 \\ 0 & 12 & 0 & -12 \end{bmatrix}$

Write each system in the form of a matrix equation. Do not solve.

27. $\begin{cases} 2x - 3y = 9 \\ -5x + 7y = 8 \end{cases}$

28. $\begin{cases} 0.5x - 0.6y = 0.6 \\ -0.7x + 0.4y = -0.375 \end{cases}$

29. $\begin{cases} x + 2y - z = 1 \\ x + z = 3 \\ 2x - y + z = 3 \end{cases}$

30. $\begin{cases} 2x - 3y - 2z = 4 \\ \frac{1}{4}x - \frac{2}{5}y + \frac{3}{4}z = \frac{-1}{3} \\ -2x + 1.3y - 3z = 5 \end{cases}$

31. $\begin{cases} -2w + x - 4y + 5z = -3 \\ 2w - 5x + y - 3z = 4 \\ -3w + x + 6y + z = 1 \\ w + 4x - 5y + z = -9 \end{cases}$

32. $\begin{cases} 1.5w + 2.1x - 0.4y + z = 1 \\ 0.2w - 2.6x + y = 5.8 \\ 3.2x + z = 2.7 \\ 1.6w + 4x - 5y + 2.6z = -1.8 \end{cases}$

Write each system as a matrix equation and solve (if possible) using inverse matrices and your calculator. If the coefficient matrix is singular, write *no solution.*

33. $\begin{cases} 0.05x - 3.2y = -15.8 \\ 0.02x + 2.4y = 12.08 \end{cases}$ (4, 5)

34. $\begin{cases} 0.3x + 1.1y = 3.5 \\ -0.5x - 2.9y = -10.1 \end{cases}$ (−3, 4)

35. $\begin{cases} \frac{-1}{6}u + \frac{1}{4}v = 1 \\ \frac{1}{2}u - \frac{2}{3}v = -2 \end{cases}$ (12, 12)

36. $\begin{cases} \sqrt{2}a + \sqrt{3}b = \sqrt{5} \\ \sqrt{6}a + 3b = \sqrt{7} \end{cases}$ no solution

37. $\begin{cases} \frac{-1}{8}a + \frac{3}{5}b = \frac{5}{6} \\ \frac{5}{16}a - \frac{3}{2}b = \frac{-4}{5} \end{cases}$ no solution

38. $\begin{cases} 3\sqrt{2}a + 2\sqrt{3}b = 12 \\ 5\sqrt{2}a - 3\sqrt{3}b = 1 \end{cases}$ $(\sqrt{2}, \sqrt{3})$

39. $\begin{cases} 0.2x - 1.6y + 2z = -1.9 \\ -0.4x - y + 0.6z = -1 \\ 0.8x + 3.2y - 0.4z = 0.2 \end{cases}$ (1.5, −0.5, −1.5)

40. $\begin{cases} 1.7x + 2.3y - 2z = 41.5 \\ 1.4x - 0.9y + 1.6z = -10 \\ -0.8x + 1.8y - 0.5z = 16.5 \end{cases}$ (5, 10, −5)

41. $\begin{cases} x - 2y + 2z = 6 \\ 2x - 1.5y + 1.8z = 2.8 \\ \frac{-2}{3}x + \frac{1}{2}y - \frac{3}{5}z = -\frac{11}{30} \end{cases}$ no solution

42. $\begin{cases} 4x - 5y - 6z = 5 \\ \frac{1}{8}x - \frac{3}{5}y + \frac{5}{4}z = \frac{-2}{3} \\ -0.5x + 2.4y - 5z = 5 \end{cases}$ no solution

43. $\begin{cases} -2w + 3x - 4y + 5z = -3 \\ 0.2w - 2.6x + y - 0.4z = 2.4 \\ -3w + 3.2x + 2.8y + z = 6.1 \\ 1.6w + 4x - 5y + 2.6z = -9.8 \end{cases}$ (−1, −0.5, 1.5, 0.5)

44. $\begin{cases} 2w - 5x + 3y - 4z = 7 \\ 1.6w + 4.2y - 1.8z = 5.4 \\ 3w + 6.7x - 9y + 4z = -8.5 \\ 0.7x - 0.9z = 0.9 \end{cases}$ (0.5, 0, $\frac{2}{3}$, −1)

Compute the determinant of each matrix and state whether an inverse matrix exists. Do not use a calculator.

45. $\begin{bmatrix} 4 & -7 \\ 3 & -5 \end{bmatrix}$ 1, yes

46. $\begin{bmatrix} 0.6 & 0.3 \\ 0.4 & 0.5 \end{bmatrix}$ 0.18, yes

47. $\begin{bmatrix} 1.2 & -0.8 \\ 0.3 & -0.2 \end{bmatrix}$ 0, no

48. $\begin{bmatrix} -2 & 6 \\ -3 & 9 \end{bmatrix}$ 0, no

Compute the determinant of each matrix without using a calculator. If the determinant is zero, write *singular matrix.*

49. $A = \begin{bmatrix} 1 & 0 & -2 \\ 0 & -1 & -1 \\ 2 & 1 & -4 \end{bmatrix}$ 1

50. $B = \begin{bmatrix} -2 & 2 & 1 \\ 0 & -1 & 2 \\ 4 & -4 & 0 \end{bmatrix}$ 4

51. $C = \begin{bmatrix} -2 & 3 & 4 \\ 0 & 6 & 2 \\ 1 & -1.5 & -2 \end{bmatrix}$ singular matrix

52. $D = \begin{bmatrix} 1 & 2 & -0.8 \\ 2.5 & 5 & -2 \\ 3 & 0 & -2.5 \end{bmatrix}$ singular matrix

Use a calculator to compute the determinant of each matrix. If the determinant is zero, write *singular matrix.* If the determinant is nonzero, find A^{-1} and store the result as matrix B (**STO▸** **2nd** **X^{-1}** **2:[B]** **ENTER**). Then verify each inverse by showing $AB = BA = I$.

53. $A = \begin{bmatrix} 1 & 0 & 3 & -4 \\ 2 & 5 & 0 & 1 \\ 8 & 15 & 6 & -5 \\ 0 & 8 & -4 & 1 \end{bmatrix}$ singular matrix

54. $M = \begin{bmatrix} 1 & 2 & 1 & 1 \\ 0 & 1 & -3 & 2 \\ -1 & 0 & 2 & -3 \\ 2 & -1 & 1 & 4 \end{bmatrix}$

54. $\begin{bmatrix} \frac{1}{2} & \frac{-9}{4} & -3 & \frac{-5}{4} \\ \frac{1}{3} & \frac{1}{3} & \frac{1}{3} & 0 \\ 0 & \frac{1}{2} & 1 & \frac{1}{2} \\ \frac{-1}{6} & \frac{13}{12} & \frac{4}{3} & \frac{3}{4} \end{bmatrix}$

For each system shown, form the matrix equation $AX = B$; compute the determinant of the coefficient matrix and determine if you can proceed; and if possible, solve the system using the matrix equation.

55. $\begin{cases} x - 2y + 2z = 7 \\ 2x + 2y - z = 5 \\ 3x - y + z = 6 \end{cases}$ det(A) = −5; (1, 6, 9)

56. $\begin{cases} 2x - 3y - 2z = 7 \\ x - y + 2z = -5 \\ 3x + 2y - z = 11 \end{cases}$ det(A) = −37; (2, 1, −3)

57. $\begin{cases} x - 3y + 4z = -1 \\ 4x - y + 5z = 7 \\ 3x + 2y + z = -3 \end{cases}$ det(A) = 0

58. $\begin{cases} 5x - 2y + z = 1 \\ 3x - 4y + 9z = -2 \\ 4x - 3y + 5z = 6 \end{cases}$ det(A) = 0

WORKING WITH FORMULAS

The inverse of a 2 × 2 matrix: $A = \begin{bmatrix} a & b \\ c & d \end{bmatrix} \rightarrow A^{-1} = \frac{1}{ad - bc} \cdot \begin{bmatrix} d & -b \\ -c & a \end{bmatrix}$

The inverse of a 2 × 2 matrix can be found using the formula shown, as long as $ad - bc \neq 0$. Use the formula to find inverses for the matrices here, then verify by showing $A \cdot A^{-1} = A \cdot A^{-1} = I$.

59. $A = \begin{bmatrix} 3 & -5 \\ 2 & 1 \end{bmatrix}$

60. $B = \begin{bmatrix} 2 & 3 \\ -5 & -4 \end{bmatrix}$

61. $C = \begin{bmatrix} 0.3 & -0.4 \\ -0.6 & 0.8 \end{bmatrix}$ singular

62. $\begin{bmatrix} 0.2 & 0.3 \\ -0.4 & -0.6 \end{bmatrix}$ singular

59. $A^{-1} = \begin{bmatrix} \frac{1}{13} & \frac{5}{13} \\ \frac{-2}{13} & \frac{3}{13} \end{bmatrix}$

60. $B^{-1} = \begin{bmatrix} \frac{-4}{7} & \frac{-3}{7} \\ \frac{5}{7} & \frac{2}{7} \end{bmatrix}$

APPLICATIONS

Solve each application using a matrix equation.

Descriptive Translation

63. **Convenience store sales:** The local Moto-Mart sells four different sizes of Slushies—behemoth, 60 oz @ \$2.59; gargantuan, 48 oz @ \$2.29; mammoth, 36 oz @ \$1.99, and jumbo, 24 oz @ \$1.59. As part of a promotion, the owner offers free gas to any customer who can tell how many of each size were sold last week, given the following information: (1) The total revenue for the Slushies was \$402.29; (2) 7884 ounces were sold; (3) a total of 191 Slushies were sold; and (4) the number of behemoth Slushies sold was one more than the number of jumbo. How many of each size were sold?

63. 31 behemoth, 52 gargantuan, 78 mammoth, 30 jumbo

64. **Cartoon characters:** In America, four of the most beloved cartoon characters are Foghorn Leghorn, Elmer Fudd, Bugs Bunny, and Tweety Bird. Suppose that Bugs Bunny is four times as tall as Tweety Bird. Elmer Fudd is as tall as the combined height of Bugs Bunny and Tweety Bird. Foghorn Leghorn is 20 cm taller than the combined height of Elmer Fudd and Tweety Bird. The combined height of all four characters is 500 cm. How tall is each one?

64. Foghorn 200 cm; Elmer 150 cm; Tweety 30 cm; Bugs 120 cm

65. **Rolling Stones music:** One of the most prolific and popular rock-and-roll bands of all time is the Rolling Stones. Four of their many great hits include: *Jumpin' Jack Flash, Tumbling Dice, You Can't Always Get What You Want,* and *Wild Horses.* The total playing time of all four songs is 20.75 min. The combined playing time of *Jumpin' Jack Flash* and *Tumbling Dice* equals that of *You Can't Always Get What You Want. Wild Horses* is 2 min longer than *Jumpin' Jack Flash,* and *You Can't Always Get What You Want* is twice as long as *Tumbling Dice.* Find the playing time of each song.

65. 3.75 min; 3.75 min; 7.5 min; 5.75 min

66. Mozart wrote some of vocal music's most memorable arias in his operas, including *Tamino's Aria, Papageno's Aria,* the *Champagne Aria,* and the *Catalogue Aria.* The total playing time of all four arias is 14.3 min. *Papageno's Aria* is 3 min shorter than the *Catalogue Aria.* The *Champagne Aria* is 2.7 min shorter than *Tamino's Aria.* The combined time of *Tamino's Aria* and *Papageno's Aria* is five times that of the *Champagne Aria.* Find the playing time of all four arias.

66. 4.1 min; 2.9 min; 1.4 min; 5.9 min

Manufacturing

67. **Resource allocation:** Time Pieces Inc. manufactures four different types of grandfather clocks. Each clock requires these four

Dept.	Clock A	Clock B	Clock C	Clock D
Assemble	2.2	2.5	2.75	3
Install	1.2	1.4	1.8	2
Test	0.2	0.25	0.3	0.5
Pack	0.5	0.55	0.75	1.0

67. 30 of clock A; 20 of clock B; 40 of clock C; 12 of clock D

stages: (1) assembly, (2) installing the clockworks, (3) inspection and testing, and (4) packaging for delivery. The time required for each stage is shown in the table, for each of the four clock types. At the end of busy week, the owner determines that personnel on the assembly line worked for 262 hours, the installation crews for 160 hours, the testing department for 29 hours, and the packaging department for 68 hours. How many clocks of each type were made?

68. 25 small; 15 medium; 30 large; 18 x-large

68. Resource allocation: Figurines Inc. makes and sells four sizes of metal figurines, mostly historical figures and celebrities. Each figurine goes through four stages of development: (1) casting, (2) trimming, (3) finishing, and (4) painting. The time required for each stage is shown in the table, for each of the four sizes. At the end of busy week, the manager finds that the casting department put in 62 hr, and the trimming department worked for 93.5 hr, with the polishing and painting departments logging 138 hr and 358 hr respectively. How many figurines of each type were made?

Dept.	Small	Medium	Large	X-Large
Casting	0.5	0.6	0.75	1
Trimming	0.8	0.9	1.1	1.5
Polishing	1.2	1.4	1.7	2
Painting	2.5	3.5	4.5	6

Curve Fitting

69. $y = x^3 + 2x^2 - 9x - 10$
70. $y = 2x^3 - 10x + 1$

69. Cubic fit: Find a cubic function of the form $y = ax^3 + bx^2 + cx + d$ such that $(-4, -6)$, $(-1, 0)$, $(1, -16)$, and $(3, 8)$ are on the graph of the function.

70. Cubic fit: Find a cubic function of the form $y = ax^3 + bx^2 + cx + d$ such that $(-2, 5)$, $(0, 1)$, $(2, -3)$, and $(3, 25)$ are on the graph of the function.

Nutrition

71. 2 oz Food I, 1 oz Food II, 4 oz Food III
72. 7 oz Food I; 2 oz Food II; 8 oz Food III

71. Animal diets: A zoo dietician needs to create a specialized diet that regulates an animal's intake of fat, carbohydrates, and protein during a meal. The table given shows three different foods and the amount of these nutrients (in grams) that each food provides. How many ounces of each should the dietician recommend to supply 20 g of fat, 30 g of carbohydrates, and 44 g of protein?

Nutrient	Food I	Food II	Food III
Fat	2	4	3
Carb.	4	2	5
Protein	5	6	7

72. Training diet: A physical trainer is designing a workout diet for one of her clients, and wants to supply him with 24 g of fat, 244 g of carbohydrates, and 40 g of protein for the noontime meal. The table given shows three different foods and the amount of these nutrients (in grams) that each food provides. How many ounces of each should the trainer recommend?

Nutrient	Food I	Food II	Food III
Fat	2	5	0
Carb.	10	15	18
Protein	2	10	0.75

WRITING, RESEARCH, AND DECISION MAKING

73. Answers will vary.

73. Some matrix applications require that you solve a matrix equation of the form $AX + B = C$, where A, B, and C are matrices with the appropriate number of rows and columns and A^{-1} exists. Investigate the solution process for such equations using $A = \begin{bmatrix} 2 & 3 \\ -5 & -4 \end{bmatrix}$, $B = \begin{bmatrix} 4 \\ 9 \end{bmatrix}$, $C = \begin{bmatrix} 12 \\ -4 \end{bmatrix}$, and $X = \begin{bmatrix} x \\ y \end{bmatrix}$, then solve $AX + B = C$ for X *symbolically* (using A^{-1}, I, and so on).

74. Both coefficient matrices are singular, even though one system has no solutions while the other has infinitely many.

75. $A^{-1} = \begin{bmatrix} 1 & 2 & -1 \\ 0 & 2 & -1 \\ -1 & -5 & 3 \end{bmatrix}$

76. $A^{-1} = \begin{bmatrix} 5 & 3 & 1 \\ 3 & 4 & 1 \\ 4 & 3 & 1 \end{bmatrix}$

74. The system $\begin{cases} y = -2x + 3 \\ y = -2x + 1 \end{cases}$ is inconsistent (lines are parallel) and $\begin{cases} y = -2x + 3 \\ 2y = -4x + 6 \end{cases}$ is dependent (one equation is a multiple of the other). Write the systems as matrix equations and attempt to solve using an inverse matrix. What do you notice? Would you be able to tell which was inconsistent and which was dependent using only the inverse matrix approach? Now attempt to solve each system using row reduction. How do you distinguish the inconsistent system from the dependent system using the results of row reduction?

The "equivalent matrices and systems of equations" method used in Example 2 can also be applied to a 3 × 3 matrix. Use this method to find the inverse matrix A^{-1} for each matrix A given. Verify the inverse is correct by showing $A^{-1}A = AA^{-1} = I$.

75. $A = \begin{bmatrix} 1 & -1 & 0 \\ 1 & 2 & 1 \\ 2 & 3 & 2 \end{bmatrix}$

76. $A = \begin{bmatrix} 1 & 0 & -1 \\ 1 & 1 & -2 \\ -7 & -3 & 11 \end{bmatrix}$

EXTENDING THE CONCEPT

77. It is possible for the matrix of coefficients to be singular, yet for solutions to exist. If the system is dependent instead of inconsistent, there may be infinitely many solutions and the solution set must be written using a parameter. Try solving the exercise given here using matrix equations. If this is not possible, discuss why, then solve using elimination. If the system is dependent, find at least *two* sets of three fractions that fit the criteria. Answers will vary.

The sum of the two smaller fractions equals the larger, the larger less the smaller equals the "middle" fraction, and four times the smaller fraction equals the sum of the other two.

78. Many solutions are possible.

78. Find 2 × 2 nonzero matrices A and B whose product gives the zero matrix $\begin{bmatrix} 0 & 0 \\ 0 & 0 \end{bmatrix}$.

80.

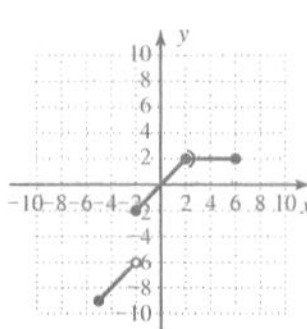

$D: x \in [-5, 6]$
$R: y \in [-9, -6) \cup [-2, 2]$

81. $x = 6, x = 3, x = -2$

82. $x = \frac{-9}{2}, x = \frac{-1}{2}$

83. a. shifts right 2
b. shifts down 2

84. $y \approx (3.705)\sqrt{2}^x$, about 118,570

MAINTAINING YOUR SKILLS

79. (3.1/3.2) Given $f(x) = x^3 + 2$ and $g(x) = \sqrt[3]{x - 2}$, find $(f \circ g)(x)$ and $(g \circ f)(x)$. What can you conclude? They are inverses.

80. (3.7) Graph the piecewise function shown and state its domain and range.
$\begin{cases} x - 4 & -5 \le x < -2 \\ x & -2 \le x \le 2 \\ 2 & 2 < x \le 6 \end{cases}$

81. (4.3) Solve using the rational roots theorem: $x^3 - 7x^2 = -36$

82. (6.4) Solve the absolute value inequality: $-3|2x + 5| - 7 = -19$.

83. (5.2) Match each equation to its related graph. Justify your answers.

$y = \log_2(x - 2)$ $\qquad$ $y = \log_2 x - 2$

a.

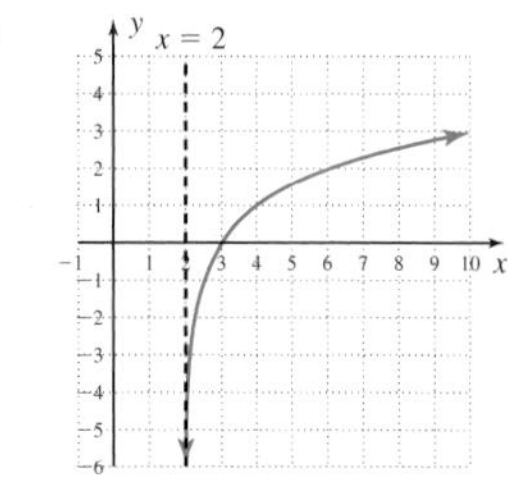

b.

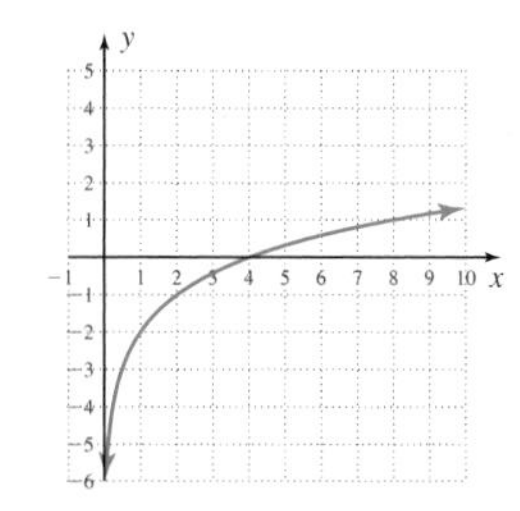

Exercise 84

Year 2000 → 0	Population (1000s)
1	5.24
3	10.48
5	20.96
7	41.92 (est.)
9	83.84 (est.)

84. (5.6) The population growth of a certain city is given in the table. Find an exponential regression equation for the data and use it to estimate the population of the city in the year 2010.

6.8 Matrix Applications: Cramer's Rule, Partial Fractions, and More

LEARNING OBJECTIVES

In Section 6.8 you will learn how to:

A. Solve a system using determinants and Cramer's Rule

B. Use determinants in applications involving geometry in the coordinate plane

C. Decompose a rational expression into partial fractions

INTRODUCTION

In addition to their use in solving systems, matrices can be used to accomplish such diverse things as finding the volume of a three-dimensional solid or establishing certain geometrical relationships in the coordinate plane. Numerous uses are also found in higher mathematics, such as checking whether solutions to a differential equation are linearly independent.

POINT OF INTEREST

Although problems suggesting the use of determinants already existed, it was Gottfried von Leibniz (1646–1716) who formalized their study and offered a written notation for them in 1693. However, the name "determinant" was not applied to the topic or method until 1812 by Augustin-Louis Cauchy (1789–1857), and the vertical bar notation in common use today was not used until 1841, after being introduced by Arthur Cayley.

A. Solving Systems Using Determinants and Cramer's Rule

In addition to their use in identifying singular matrices, determinants can actually be used to *develop a formula approach* for the solution of a system. Consider the following illustration, in which we solve a *general* 2×2 system by modeling the process after a *specific* 2×2 system. With a view toward a solution involving determinants, the coefficients of x are written as a_{11} and a_{21} in the general system, and the coefficients of y are a_{12} and a_{22}.

Specific System

$$\begin{cases} 2x + 5y = 9 \\ 3x + 4y = 10 \end{cases}$$

eliminate the x-term in R2

$-3\text{R1} + 2\text{R2}$

sums to zero

$$\begin{cases} -3 \cdot 2x - 3 \cdot 5y = -3 \cdot 9 \\ 2 \cdot 3x + 2 \cdot 4y = 2 \cdot 10 \end{cases}$$

$$2 \cdot 4y - 3 \cdot 5y = 2 \cdot 10 - 3 \cdot 9$$

General System

$$\begin{cases} a_{11}x + a_{12}y = c_1 \\ a_{21}x + a_{22}y = c_2 \end{cases}$$

eliminate the x-term in R2

$-a_{21}\text{R1} + a_{11}\text{R2}$

sums to zero

$$\begin{cases} -a_{21}a_{11}x - a_{21}a_{12}y = -a_{21}c_1 \\ a_{11}a_{21}x + a_{11}a_{22}y = a_{11}c_2 \end{cases}$$

$$a_{11}a_{22}y - a_{21}a_{12}y = a_{11}c_2 - a_{21}c_1$$

Notice the x-terms sum to zero in both systems. We are deliberately leaving the solution on the left unsimplified to show the pattern developing on the right. Next we solve for y.

Factor Out y

$$(2 \cdot 4 - 3 \cdot 5)y = 2 \cdot 10 - 3 \cdot 9$$

$$y = \frac{2 \cdot 10 - 3 \cdot 9}{2 \cdot 4 - 3 \cdot 5}$$

Factor Out y

$$(a_{11}a_{22} - a_{21}a_{12})y = a_{11}c_2 - a_{21}c_1$$

$$y = \frac{a_{11}c_2 - a_{21}c_1}{a_{11}a_{22} - a_{21}a_{12}}$$

On the left we find $y = \frac{-7}{-7} = 1$ and back-substitution shows $x = 2$. But more importantly, on the right we obtain a formula for the y-value of *any* 2×2 system: $y = \frac{a_{11}c_2 - a_{21}c_1}{a_{11}a_{22} - a_{21}a_{12}}$. If we had chosen to solve for x, the "formula" solution would be

$x = \dfrac{a_{22}c_1 - a_{12}c_2}{a_{11}a_{22} - a_{21}a_{12}}$. Note these formulas are defined only if $a_{11}a_{22} - a_{21}a_{12} \neq 0$. You may have already noticed, but this denominator is the *determinant of the matrix of coefficients* $\begin{bmatrix} a_{11} & a_{12} \\ a_{21} & a_{22} \end{bmatrix}$ from the previous section! Since the numerator is also a difference of two products, we investigate the possibility that it too can be expressed as a determinant. Working backward, we're able to reconstruct the numerator for x in determinant form as $\begin{bmatrix} c_1 & a_{12} \\ c_2 & a_{22} \end{bmatrix}$, where it is apparent this matrix was formed by *replacing the coefficients of the x-variables with the constant terms.*

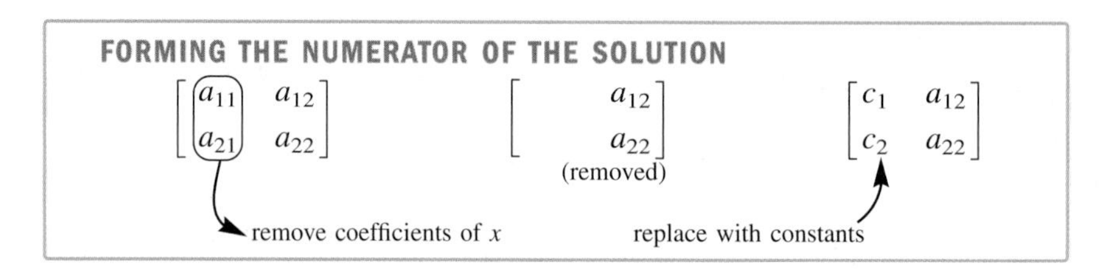

In a similar fashion, the numerator for y can be written in determinant form as $\begin{bmatrix} a_{11} & c_1 \\ a_{21} & c_2 \end{bmatrix}$, or the determinant formed by *replacing the coefficients of the y-variables with the constant terms.* If we use the notation D_y for this determinant, D_x for the determinant where x coefficients were replaced by the constants, and D as the determinant for the matrix of coefficients—the solutions can be written as shown next, with the result known as **Cramer's rule.**

CRAMER'S RULE APPLIED TO 2 × 2 SYSTEMS

Given a 2 × 2 system of linear equations

$$\begin{cases} a_{11} + a_{12} = c_1 \\ a_{21} + a_{22} = c_2 \end{cases}$$

the solution is the ordered pair (x, y), where

$$x = \frac{D_x}{D} = \frac{\begin{bmatrix} c_1 & a_{12} \\ c_2 & a_{22} \end{bmatrix}}{\begin{bmatrix} a_{11} & a_{12} \\ a_{21} & a_{22} \end{bmatrix}} \quad \text{and} \quad y = \frac{D_y}{D} = \frac{\begin{bmatrix} a_{11} & c_1 \\ a_{21} & c_2 \end{bmatrix}}{\begin{bmatrix} a_{11} & a_{12} \\ a_{21} & a_{22} \end{bmatrix}}$$

provided $D \neq 0$.

EXAMPLE 1 Use Cramer's rule to solve the system $\begin{cases} 2x - 5y = 9 \\ -3x + 4y = -10 \end{cases}$.

Solution: For $x = \dfrac{D_x}{D}$ and $y = \dfrac{D_y}{D}$, begin by finding the value of D, D_x, and D_y.

$$D = \begin{vmatrix} 2 & -5 \\ -3 & 4 \end{vmatrix} \qquad D_x = \begin{vmatrix} 9 & -5 \\ -10 & 4 \end{vmatrix} \qquad D_y = \begin{vmatrix} 2 & 9 \\ -3 & -10 \end{vmatrix}$$

$$(2)(4) - (-3)(-5) \qquad (9)(4) - (-10)(-5) \qquad (2)(-10) - (-3)(9)$$

$$= -7 \qquad = -14 \qquad = 7$$

This gives $x = \frac{D_x}{D} = \frac{-14}{-7} = 2$ and $y = \frac{D_y}{D} = \frac{7}{-7} = -1$. The solution is $(2, -1)$. Check by substituting these values into the original equations.

NOW TRY EXERCISES 7 THROUGH 14

Regardless of the method used to solve a system, always be aware that a consistent, inconsistent, or dependent system is possible. The system $\begin{cases} y - 2x = -3 \\ 4x + 6 = 2y \end{cases}$ yields $\begin{cases} -2x + y = -3 \\ 4x - 2y = -6 \end{cases}$ in standard form, with $D = \begin{vmatrix} -2 & 1 \\ 4 & -2 \end{vmatrix} = (-2)(-2) - (4)(1) = 0$. We stop here since Cramer's rule cannot be applied, knowing the system is either inconsistent or dependent. To find out which, we write the equations in function form (solve for y). The result is $\begin{cases} y = 2x - 3 \\ y = 2x + 3 \end{cases}$, showing the system consists of two parallel lines and has no solutions.

Cramer's Rule for 3 × 3 Systems

Cramer's rule can be extended to a 3 × 3 system of linear equations, using the same pattern as for 2 × 2 systems. Given the general 3 × 3 system,

$$\begin{cases} a_{11}x + a_{12}y + a_{13}z = c_1 \\ a_{21}x + a_{22}y + a_{23}z = c_2 \\ a_{31}x + a_{32}y + a_{33}z = c_3 \end{cases}$$

the solutions are $x = \frac{D_x}{D}$, $y = \frac{D_y}{D}$, and $z = \frac{D_z}{D}$, where D_x, D_y, and D_z are again formed by replacing the coefficient of the indicated variable with the constants, and D is the determinant of the matrix of coefficients.

CRAMER'S RULE APPLIED TO 3 × 3 SYSTEMS

Given a 3 × 3 system of linear equations

$$\begin{cases} a_{11}x + a_{12}y + a_{13}z = c_1 \\ a_{21}x + a_{22}y + a_{23}z = c_2 \\ a_{31}x + a_{32}y + a_{33}z = c_3 \end{cases}$$

The solution is an ordered triple (x, y, z), where

$$x = \frac{D_x}{D} = \frac{\begin{vmatrix} c_1 & a_{12} & a_{13} \\ c_2 & a_{22} & a_{23} \\ c_3 & a_{32} & a_{33} \end{vmatrix}}{\begin{vmatrix} a_{11} & a_{12} & a_{13} \\ a_{21} & a_{22} & a_{23} \\ a_{31} & a_{32} & a_{33} \end{vmatrix}} \qquad y = \frac{D_y}{D} = \frac{\begin{vmatrix} a_{11} & c_1 & a_{13} \\ a_{21} & c_2 & a_{23} \\ a_{31} & c_3 & a_{33} \end{vmatrix}}{\begin{vmatrix} a_{11} & a_{12} & a_{13} \\ a_{21} & a_{22} & a_{23} \\ a_{31} & a_{32} & a_{33} \end{vmatrix}}$$

$$z = \frac{D_z}{D} = \frac{\begin{vmatrix} a_{11} & a_{12} & c_1 \\ a_{21} & a_{22} & c_2 \\ a_{31} & a_{32} & c_3 \end{vmatrix}}{\begin{vmatrix} a_{11} & a_{12} & a_{13} \\ a_{21} & a_{22} & a_{23} \\ a_{31} & a_{32} & a_{33} \end{vmatrix}}, \text{ provided } D \neq 0.$$

EXAMPLE 2 Solve using Cramer's rule: $\begin{cases} x - 2y + 3z = -1 \\ -2x + y - 5z = 1 \\ 3x + 3y + 4z = 2 \end{cases}$

Solution: Begin by computing the determinant of the matrix of coefficients, to ensure that Cramer's rule can be applied. Using the third row, we have

$$D = \begin{vmatrix} 1 & -2 & 3 \\ -2 & 1 & -5 \\ 3 & 3 & 4 \end{vmatrix} = +3\begin{vmatrix} -2 & 3 \\ 1 & -5 \end{vmatrix} - 3\begin{vmatrix} 1 & 3 \\ -2 & -5 \end{vmatrix} + 4\begin{vmatrix} 1 & -2 \\ -2 & 1 \end{vmatrix}$$

$$= 3(7) - 3(1) + 4(-3) = 6$$

Since $D \neq 0$ we continue, electing to compute the remaining determinants using a calculator.

$$D_x = \begin{vmatrix} -1 & -2 & 3 \\ 1 & 1 & -5 \\ 2 & 3 & 4 \end{vmatrix} = 12 \qquad D_y = \begin{vmatrix} 1 & -1 & 3 \\ -2 & 1 & -5 \\ 3 & 2 & 4 \end{vmatrix} = 0 \qquad D_z = \begin{vmatrix} 1 & -2 & -1 \\ -2 & 1 & 1 \\ 3 & 3 & 2 \end{vmatrix} = -6$$

The solution is $x = \frac{D_x}{D} = \frac{12}{6} = 2$, $y = \frac{D_y}{D} = \frac{0}{6} = 0$, and $z = \frac{D_z}{D} = \frac{-6}{6} = -1$, or $(2, 0, -1)$ in triple form. Check this solution in the original equations.

NOW TRY EXERCISES 15 THROUGH 22

B. Determinants, Geometry, and the Coordinate Plane

As mentioned in the introduction, the use of determinants extends far beyond solving systems of equations. Here, we'll demonstrate how determinants can be used to find the area of a triangle whose vertices are given as three points in the coordinate plane.

THE AREA OF A TRIANGLE IN THE *xy*-PLANE

Given a triangle with vertices at (x_1, y_1), (x_2, y_2), and (x_3, y_3).

The area is given by the absolute value of one-half the determinant of T, where $T = \begin{bmatrix} x_1 & y_1 & 1 \\ x_2 & y_2 & 1 \\ x_3 & y_3 & 1 \end{bmatrix}$: Area $= \left|\frac{|T|}{2}\right|$

EXAMPLE 3 Find the area of a triangle with vertices at (3, 1), (−2, 3), and (1, 7).

Solution: Applying the formula just given, we have

$$\text{Area} = \left|\frac{\begin{vmatrix} x_1 & y_1 & 1 \\ x_2 & y_2 & 1 \\ x_3 & y_3 & 1 \end{vmatrix}}{2}\right| = \left|\frac{\begin{vmatrix} 3 & 1 & 1 \\ -2 & 3 & 1 \\ 1 & 7 & 1 \end{vmatrix}}{2}\right|$$

$$= \left|\frac{-26}{2}\right| \qquad |T| = -26$$

$$= 13 \qquad |-13| = 13$$

The area of this triangle is 13 units^2.

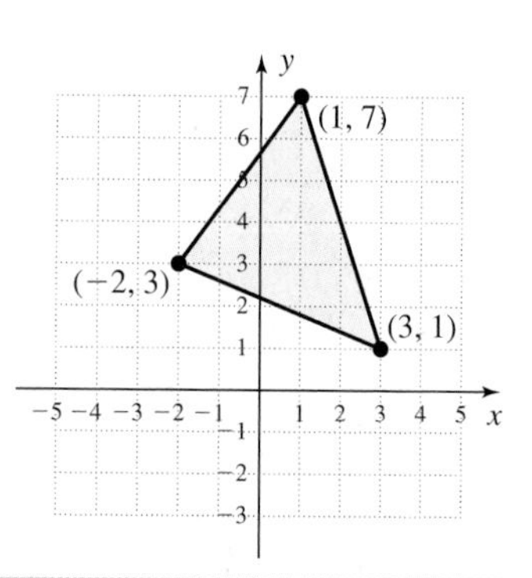

NOW TRY EXERCISES 25 THROUGH 30

As an extension of this determinant formula, what if the three points were collinear? After a moment, it may occur to you that the formula would give an area of 0 units^2, since no triangle could be formed. This gives rise to a **test for collinear points.**

TEST FOR COLLINEAR POINTS

Three points (x_1, y_1), (x_2, y_2), and (x_3, y_3) are collinear (lie on the same line) if the determinant of A is zero, where $|A| = \begin{vmatrix} x_1 & y_1 & 1 \\ x_2 & y_2 & 1 \\ x_3 & y_3 & 1 \end{vmatrix}$.

See Exercises 31 through 36.

C. Rational Expressions and Partial Fractions

One application widely used in higher mathematics involves rewriting a rational expression as a sum of **partial fractions.** The addition of rational expressions is widely taught in courses prior to college algebra, and here we seek to reverse the process. To begin, we make these two observations:

$$(1)\quad \frac{3}{x-1} + \frac{5}{x^2-2x+1} = \frac{3}{x-1} + \frac{5}{(x-1)(x-1)}$$
$$= \frac{3(x-1)}{(x-1)(x-1)} + \frac{5}{(x-1)(x-1)}$$
$$= \frac{(3x-3)+5}{(x-1)(x-1)}$$
$$= \frac{3x+2}{(x-1)^2}$$

Notice that while the new denominator is the repeated factor $(x-1)^2$, *both* $(x-1)$ and $(x-1)^2$ were denominators in the original sum. Assuming we didn't have the original sum to look at, reversing the process would require us to write

$$\frac{3x+2}{(x-1)^2} = \frac{A}{x-1} + \frac{B}{(x-1)^2},$$

where A and B represent some constant term. Multiplying both sides by $(x-1)^2$ to clear denominators gives $3x + 2 = A(x-1) + B$ or $3x + 2 = Ax - A + B$. Equating the corresponding parts of each expression (terms of like degree) gives $A = 3$ and $-A + B = 2 \rightarrow B = 5$, as indicated in the original sum. Note that for denominators with repeated *linear* factors (as above), we use a single constant in the numerator of each partial fraction to ensure we obtain unique solutions.

$$(2)\quad \frac{4}{x} + \frac{2x+3}{x^2+1} = \frac{4(x^2+1)}{x(x^2+1)} + \frac{(2x+3)x}{(x^2+1)x}$$
$$= \frac{4(x^2+1)+(2x+3)x}{x(x^2+1)}$$
$$= \frac{6x^2+3x+4}{x^3+x}$$

In factored form the new denominator has a linear factor, and a quadratic factor that is prime. To be in proper form, the numerator of the linear factor must be some constant A, and the numerator of the quadratic factor *will be of the form* $BX + C$. Once again without knowing the original sum in advance, reversing the process would require us to write

$$\frac{6x^2 + 3x + 4}{x(x^2 + 1)} = \frac{A}{x} + \frac{Bx + C}{x^2 + 1},$$

and we could duplicate the process above using a system of equations to find A, B, and C.

Using these observations, we can begin to formulate a general method for decomposing rational expressions. A solid understanding relies on these three things:

1. The expression $\frac{P(x)}{Q(x)}$ is in proper form when the degree of P is less than the degree of Q.
2. Every polynomial with real coefficients can be factored into a product of linear factors (that are not necessarily unique) and quadratic factors that are prime (cannot be factored further using integers).
3. As noted, if $(ax + k)^n$ is a factor of Q, then $(ax + k)^n, (ax + k)^{n-1}, \ldots, (ax + k)^1$ all have the *potential* to occur as denominators in the decomposition. The same can be said for any prime quadratic factors that are repeated.

The general method for writing a rational expression as a sum or difference of its partial fractions follows.

DECOMPOSING A RATIONAL EXPRESSION $\frac{P(x)}{Q(x)}$ INTO PARTIAL FRACTIONS

For $\frac{P(x)}{Q(x)}$ a rational expression and constants $A, B, C, D, \ldots,$

1. If the degree of P is greater than or equal to the degree of Q, find the quotient and remainder using polynomial division. Only the remainder portion need be decomposed into partial fractions.
2. Factor Q completely into linear factors and quadratic factors that are prime.
3. For *each* linear factor $(ax + b)^n$ of Q, the decomposed form will include
$$\frac{A}{(ax + b)^1} + \frac{B}{(ax + b)^2} + \cdots + \frac{N}{(ax + b)^n}$$
4. For *each* quadratic factor $(ax^2 + bx + c)^n$ of Q, the decomposed form will include
$$\frac{Ax + B}{(ax^2 + bx + c)^1} + \frac{Cx + D}{(ax^2 + bx + c)^2} + \cdots + \frac{Mx + N}{(ax^2 + bx + c)^n}$$
5. Set $\frac{P(x)}{Q(x)}$ equal to its potential decomposed form and multiply both sides by $Q(x)$ to clear denominators.
6. Multiply out the right-hand side, collect like terms, and equate corresponding terms from the right-hand and left-hand sides. The result can be written as a system of equations and solved using any method.

EXAMPLE 5 Use matrix equations and a graphing calculator to decompose the given expression into partial fractions: $\dfrac{12x^3 + 62x^2 + 102x + 56}{x^4 + 6x^3 + 12x^2 + 8x}$.

Solution: The degree of the numerator is less than that of the denominator, so we begin by factoring the denominator. After removing the common factor of x and applying the rational roots theorem with synthetic division ($c = -2$), the completely factored form is $\dfrac{12x^3 + 62x^2 + 102x + 56}{x(x + 2)^3}$. The decomposed form will be

$$\frac{12x^3 + 62x^2 + 102x + 56}{x(x + 2)^3} = \frac{A}{x} + \frac{B}{(x + 2)^1} + \frac{C}{(x + 2)^2} + \frac{D}{(x + 2)^3}.$$

Clearing denominators and simplifying yields

$$12x^3 + 62x^2 + 102 + 56 = A(x + 2)^3 + Bx(x + 2)^2 + Cx(x + 2) + Dx \qquad \text{clear denominators}$$

After expanding the powers on the right, grouping like terms, and factoring, we have

$$12x^3 + 62x^2 + 102x + 56 = (A + B)x^3 + (6A + 4B + C)x^2 + (12A + 4B + 2C + D)x + 8A$$

By equating the coefficients of like terms, the following system and matrix equation are obtained:

$$\begin{cases} A + B = 12 \\ 6A + 4B + C = 62 \\ 12A + 4B + 2C + D = 102 \\ 8A = 56 \end{cases} \rightarrow \begin{bmatrix} 1 & 1 & 0 & 0 \\ 6 & 4 & 1 & 0 \\ 12 & 4 & 2 & 1 \\ 8 & 0 & 0 & 0 \end{bmatrix} \begin{bmatrix} A \\ B \\ C \\ D \end{bmatrix} = \begin{bmatrix} 12 \\ 62 \\ 102 \\ 56 \end{bmatrix}.$$

After carefully entering the matrices F (coefficients) and G (constants), we obtain the solution $A = 7$, $B = 5$, $C = 0$, and $D = -2$ as shown in the figure. The decomposed form is

```
[F]-1[G]
              [[7 ]
               [5 ]
               [0 ]
               [-2]]
```

$$\frac{12x^3 + 62x^2 + 102x + 56}{x(x + 2)^3} = \frac{7}{x} + \frac{5}{(x + 2)^1} + \frac{-2}{(x + 2)^3}.$$

NOW TRY EXERCISES 37 THROUGH 58

As a final note, if the degree of the numerator is *greater than* the degree of the denominator, divide using long division and apply the preceding methods to the remainder polynomial. For instance, you can check that $\dfrac{3x^3 + 6x^2 + 5x - 7}{x^2 + 2x + 1} = 3x + \dfrac{2x - 7}{(x + 1)^2}$, and decomposing the remainder polynomial gives a final result of $3x + \dfrac{2}{x + 1} - \dfrac{9}{(x + 1)^2}$.

6.8 EXERCISES

CONCEPTS AND VOCABULARY

Fill in the blank with the appropriate word or phrase. Carefully reread the section if needed.

1. The determinant $\begin{vmatrix} a_{11} & a_{12} \\ a_{21} & a_{22} \end{vmatrix}$ is evaluated as: $a_{11}a_{22} - a_{21}a_{12}$.

2. Cramer's rule uses a ratio of determinants to solve for the unknowns in a system.

3. Given the matrix of coefficients D, the matrix D_x is formed by replacing the coefficients of x with the constant terms.

4. The three points (x_1, y_1), (x_2, y_2), and (x_3, y_3) are collinear if $|T| = \begin{vmatrix} x_1 & y_1 & 1 \\ x_2 & y_2 & 1 \\ x_3 & y_3 & 1 \end{vmatrix}$ has a value of 0.

5. Discuss/explain the process of writing $\dfrac{8x - 3}{x^2 - x}$ as a sum of partial fractions. Answers will vary.

6. Discuss/explain why Cramer's rule cannot be applied if $D = 0$. Use an example to illustrate. Answers will vary.

DEVELOPING YOUR SKILLS

Write the determinants D, D_x, and D_y for the systems given. Do not solve.

7. $\begin{cases} 2x + 5y = 7 \\ -3x + 4y = 1 \end{cases}$

8. $\begin{cases} -x + 5y = 12 \\ 3x - 2y = -8 \end{cases}$

Solve each system of equations using Cramer's rule, if possible. Do not use a calculator.

9. $\begin{cases} 4x + y = -11 \\ 3x - 5y = -60 \end{cases}$ $(-5, 9)$

10. $\begin{cases} x = -2y - 11 \\ y = 2x - 13 \end{cases}$ $(3, -7)$

11. $\begin{cases} \frac{x}{8} + \frac{y}{4} = 1 \\ \frac{y}{5} = \frac{x}{2} + 6 \end{cases}$ $\left(\frac{-26}{3}, \frac{25}{3}\right)$

12. $\begin{cases} \frac{2}{3}x - \frac{3}{8}y = \frac{7}{5} \\ \frac{5}{6}x + \frac{3}{4}y = \frac{11}{10} \end{cases}$ $\left(\frac{9}{5}, \frac{-8}{15}\right)$

13. $\begin{cases} 0.6x - 0.3y = 8 \\ 0.8x - 0.4y = -3 \end{cases}$ no solution

14. $\begin{cases} -2.5x + 6y = -1.5 \\ 0.5x - 1.2y = 3.6 \end{cases}$ no solution

Write the determinants D, D_x, D_y, and D_z for the systems given, then determine if a solution using Cramer's rule is possible by computing the value of D without the use of a calculator (do not solve the system). Try to determine how the system from Part (a) is related to the system in Part (b).

15. a. $\begin{cases} 4x - y + 2z = -5 \\ -3x + 2y - z = 8 \\ x - 5y + 3z = -3 \end{cases}$ **b.** $\begin{cases} 4x - y + 2z = -5 \\ -3x + 2y - z = 8 \\ x + y + z = -3 \end{cases}$

16. a. $\begin{cases} 2x + 3z = -2 \\ -x + 5y + z = 12 \\ 3x - 2y + z = -8 \end{cases}$ **b.** $\begin{cases} 2x + 3z = -2 \\ -x + 5y + z = 12 \\ x + 5y + 4z = -8 \end{cases}$

Use Cramer's rule to solve each system of equations.

17. $\begin{cases} x + 2y + 5z = 10 \\ 3x + 4y - z = 10 \\ x - y - z = -2 \end{cases}$

18. $\begin{cases} x + 3y + 5z = 6 \\ 2x - 4y + 6z = 14 \\ 9x - 6y + 3z = 3 \end{cases}$

19. $\begin{cases} y + 2z = 1 \\ 4x - 5y + 8z = -8 \\ 8x - 9z = 9 \end{cases}$

---

7. $D = \begin{vmatrix} 2 & 5 \\ -3 & 4 \end{vmatrix}$; $D_x = \begin{vmatrix} 7 & 5 \\ 1 & 4 \end{vmatrix}$; $D_y = \begin{vmatrix} 2 & 7 \\ -3 & 1 \end{vmatrix}$

8. $D = \begin{vmatrix} -1 & 5 \\ 3 & -2 \end{vmatrix}$; $D_x = \begin{vmatrix} 12 & 5 \\ -8 & -2 \end{vmatrix}$; $D_y = \begin{vmatrix} -1 & 12 \\ 3 & -8 \end{vmatrix}$

15. a. $D = \begin{vmatrix} 4 & -1 & 2 \\ -3 & 2 & -1 \\ 1 & -5 & 3 \end{vmatrix}$ $D_x = \begin{vmatrix} -5 & -1 & 2 \\ 8 & 2 & -1 \\ -3 & -5 & 3 \end{vmatrix}$ $D_y = \begin{vmatrix} 4 & -5 & 2 \\ -3 & 8 & -1 \\ 1 & -3 & 3 \end{vmatrix}$ $D_z = \begin{vmatrix} 4 & -1 & -5 \\ -3 & 2 & 8 \\ 1 & -5 & -3 \end{vmatrix}$
$D = 22$, solutions possible
b. $D = 0$, Cramer's rule cannot be used.

16. a. $D = \begin{vmatrix} 2 & 0 & 3 \\ -1 & 5 & 1 \\ 3 & -2 & 1 \end{vmatrix}$, $D_x = \begin{vmatrix} -2 & 0 & 3 \\ 12 & 5 & 1 \\ -8 & -2 & 1 \end{vmatrix}$, $D_y = \begin{vmatrix} 2 & -2 & 3 \\ -1 & 12 & 1 \\ 3 & -8 & 1 \end{vmatrix}$, $D_z = \begin{vmatrix} 2 & 0 & -2 \\ -1 & 5 & 12 \\ 3 & -2 & -8 \end{vmatrix}$
$D = -25$, solutions possible
b. $D = 0$, Cramer's rule cannot be used

17. $(1, 2, 1)$ 18. $(-1, -1, 2)$

19. $\left(\frac{3}{4}, \frac{5}{3}, \frac{-1}{3}\right)$

20. $\left(\frac{14}{5}, \frac{13}{5}, \frac{2}{5}\right)$
21. $(0, -1, 2, -3)$
22. $(1, -1, 2, -2)$

20. $\begin{cases} x + 2y + 5z = 10 \\ 3x - z = 8 \\ -y - z = -3 \end{cases}$

21. $\begin{cases} w + 2x - 3y = -8 \\ x - 3y + 5z = -22 \\ 4w - 5x = 5 \\ -y + 3z = -11 \end{cases}$

22. $\begin{cases} w - 2x + 3y - z = 11 \\ 3w - 2y + 6z = -13 \\ 2x + 4y - 5z = 16 \\ 3x - 4z = 5 \end{cases}$

WORKING WITH FORMULAS

Area of a Norman window: $A = \begin{vmatrix} L & r^2 \\ -\frac{\pi}{2} & W \end{vmatrix}$

The determinant formula shown can be used to find the area of a Norman window with length L and width W. Use the formula to find the area of the following windows.

23.

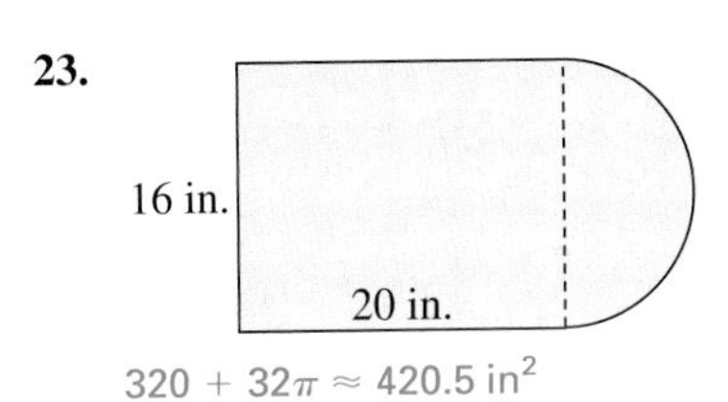

$320 + 32\pi \approx 420.5$ in^2

24.

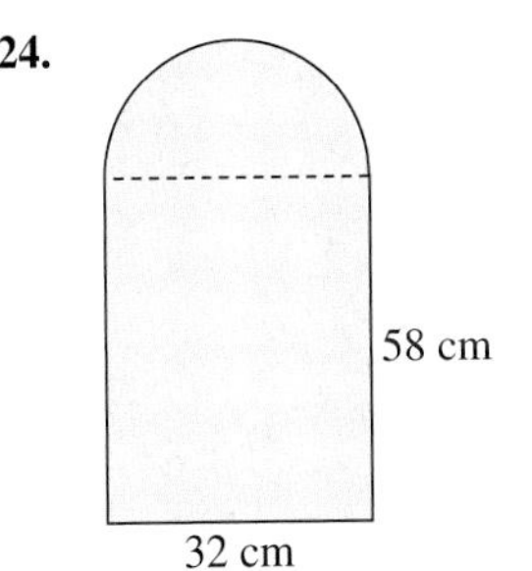

$1856 + 128\pi \approx 2258.1$ cm^2

APPLICATIONS

Geometric Applications

Find the area of the triangle with the vertices given. Assume area units are cm^2.

25. $(2, 1), (3, 7)$, and $(5, 3)$ 8 cm^2

26. $(-2, 3), (-3, -4)$, and $(-6, 1)$ 13 cm^2

Find the area of the parallelogram with vertices given. Assume area units are ft^2.

27. $(-4, 2), (-6, -1), (3, -1)$, and $(5, 2)$ 27 ft^2

28. $(-5, -6), (5, 0), (5, 4)$, and $(-5, -2)$ 40 ft^2

The volume of a triangular pyramid is given by the formula $V = \frac{1}{3}Bh$, where B represents the area of the triangular base and h is the height of the pyramid. Find the volume of a triangular pyramid whose height is given and whose base has the coordinates shown. Assume area units are m^2.

29. $h = 6$ m; vertices $(3, 5), (-4, 2)$, and $(-1, 6)$ 19 m^3

30. $h = 7.5$ m; vertices $(-2, 3), (-3, -4)$, and $(-6, 1)$ 32.5 m^3

Determine if the following sets of points are collinear.

31. $(1, 5), (-2, -1)$, and $(4, 11)$ yes

32. $(1, 1), (3, -5)$, and $(-2, 9)$ no

33. $(-2.5, 5.2), (1.2, -5.6)$, and $(2.2, -8.5)$ no

34. $(-0.5, 1.25), (-2.8, 3.75)$, and $(3, 6.25)$ no

For each linear equation given, substitute the first two points to verify they are solutions. Then use the test for collinear points to determine if the third point is also a solution.

35. $2x - 3y = 7$; $(2, -1), (-1.3, -3.2)$, $(-3.1, -4.4)$ yes yes yes

36. $5x + 2y = 4$; $(2, -3), (3.5, -6.75)$, $(-2.7, 8.75)$ yes yes yes

Decomposition of Rational Expressions

These exercises are designed solely to reinforce the various possibilities for decomposing a rational expression. All are proper fractions whose denominators are completely factored. Set up the partial fraction decomposition using appropriate numerators, but *do not solve.*

37. $\dfrac{3x+2}{(x+3)(x-2)}$

38. $\dfrac{-4x+1}{(x-2)(x-5)}$

39. $\dfrac{3x^2-2x+5}{(x-1)(x+2)(x-3)}$

40. $\dfrac{-2x^2+3x-4}{(x+3)(x+1)(x-2)}$

41. $\dfrac{x^2+5}{x(x-3)(x+1)}$

42. $\dfrac{x^2-7}{(x+4)(x-2)x}$

43. $\dfrac{x^2+x-1}{x^2(x+2)}$

44. $\dfrac{x^2-3x+5}{(x-3)(x+2)^2}$

45. $\dfrac{x^3+3x-2}{(x+1)(x^2+2)^2}$

46. $\dfrac{2x^3+3x^2-4x+1}{x(x^2-3)^2}$

Decompose each rational expression into partial fractions.

47. $\dfrac{4-x}{x^2+x}$

48. $\dfrac{3x+13}{x^2+5x+6}$

49. $\dfrac{2x-27}{2x^2+x-15}$

50. $\dfrac{-11x+6}{5x^2-4x-12}$

51. $\dfrac{8x^2-3x-7}{x^3-x}$

52. $\dfrac{x^2+24x-12}{x^3-4x}$

53. $\dfrac{3x^2+7x-1}{x^3+2x^2+x}$

54. $\dfrac{-2x^2-7x+28}{x^3-4x^2+4x}$

55. $\dfrac{3x^2+10x+4}{8-x^3}$

56. $\dfrac{2x^2-14x-7}{x^3-2x^2+5x-10}$

57. $\dfrac{x^4-x^2-2x+1}{x^5-2x^3+x}$

58. $\dfrac{-3x^4+13x^2+x-12}{x^5-4x^3+4x}$

Write a linear system that models each application. Then solve using Cramer's rule.

59. **Return on investments:** If \$15,000 is invested at a certain interest rate and \$25,000 is invested at another interest rate, the total return was \$2900. If the investments were reversed the return would be \$2700. What was the interest rate paid on each investment?

60. **Cost of fruit:** Many years ago, two pounds of apples, 2 lb of kiwi, and 10 lb of pears cost \$3.26. Three pounds of apples, 2 lb of kiwi, and 7 lb of pears cost \$2.98. Two pounds of apples, 3 lb of kiwi, and 6 lb of pears cost \$2.89. Find the cost of a pound of each fruit.

61. **Seasonal business:** The owner of Scott's Ski Shoppe has noticed a distinct pattern to the amount of money the store takes in during the course of each year. Business is brisk in the winter and summer for snow and water skiing, but much slower in the spring and fall. Given the data shown, find an approximate fit polynomial that will help him predict the company's revenue for any month of the year. Use the model to approximate the company's revenue for May ($x = 5$) and July ($x = 7$). Is more revenue earned in April or November of each year?

Exercise 61

Month	Revenue (1000s)
Jan. (1)	16.0
Feb. (2)	7.0
Mar. (3)	3.5
Jun. (6)	7.5
Aug. (8)	8.0
Oct. (10)	4.5
Dec. (12)	8.5

62. **Employee productivity:** A company president notices distinct patterns in the productivity of an assembly line. In an effort to understand these fluctuations she orders a study of the production output during an 8-hr shift. Given the data shown, find an approximate fit polynomial that will help her predict the assembly line's output at any point during a shift. Use the model to predict the output level at 12:00 noon ($x = 4$). Why do you think output is low at this time?

Exercise 62

Hours 8:00 A.M. → 0	Output (100s)
0	0
1	4.6
3	2.3
5	3.6
7	6.3

37. $\frac{A}{x+3}+\frac{B}{x-2}$
38. $\frac{A}{x-2}+\frac{B}{x-5}$
39. $\frac{A}{x-1}+\frac{B}{x+2}+\frac{C}{x-3}$
40. $\frac{A}{x+3}+\frac{B}{x+1}+\frac{C}{x-2}$
41. $\frac{A}{x}+\frac{B}{x-3}+\frac{C}{x+1}$
42. $\frac{A}{x+4}+\frac{B}{x-2}+\frac{C}{x}$
43. $\frac{A}{x}+\frac{B}{x^2}+\frac{C}{x+2}$
44. $\frac{A}{x-3}+\frac{B}{x+2}+\frac{C}{(x+2)^2}$
45. $\frac{A}{x+1}+\frac{Bx+C}{x^2+2}+\frac{Dx+E}{(x^2+2)^2}$
46. $\frac{A}{x}+\frac{Bx+C}{x^2-3}+\frac{Dx+E}{(x^2-3)^2}$
47. $\frac{4}{x}-\frac{5}{x+1}$
48. $\frac{7}{x+2}-\frac{4}{x+3}$
49. $\frac{-4}{2x-5}+\frac{3}{x+3}$
50. $\frac{-6}{5x+6}-\frac{1}{x-2}$
51. $\frac{7}{x}+\frac{2}{x+1}-\frac{1}{x-1}$
52. $\frac{3}{x}-\frac{7}{x+2}+\frac{5}{x-2}$
53. $\frac{-1}{x}+\frac{4}{x+1}+\frac{5}{(x+1)^2}$
54. $\frac{7}{x}-\frac{9}{x-2}+\frac{3}{(x-2)^2}$
55. $\frac{3}{2-x}-\frac{4}{4+2x+x^2}$
56. $\frac{-3}{x-2}+\frac{5x-4}{x^2+5}$
57. $\frac{1}{x}+\frac{x-2}{(x^2-1)^2}$
58. $\frac{-3}{x}+\frac{x+1}{(x^2-2)^2}$
59. $\begin{cases}15{,}000x+25{,}000y=2900\\25{,}000x+15{,}000y=2700\end{cases}$; 6%, 8%
60. $\begin{cases}2x+2y+10z=3.26\\3x+2y+7z=2.98\\2x+3y+6z=2.89\end{cases}$; apples, 29¢/lb; kiwi, 39¢/lb; pears, 19¢/lb

Additional answers can be found in the Instructor Answer Appendix.

Estimate production levels at 10:00 A.M. ($x = 2$) and 2:00 P.M. ($x = 6$). Why do you think production levels are high at these times?

WRITING, RESEARCH, AND DECISION MAKING

63. Solve the given system four different ways: (1) elimination, (2) row reduction, (3) Cramer's rule, and (4) using a matrix equation. Which method seems to be the least error-prone? Which method seems most efficient (takes the least time)? Discuss the advantages and drawbacks of each method. Answers will vary.

$$\begin{cases} x + 3y + 5z = 6 \\ 2x - 4y + 6z = 14 \\ 9x - 6y + 3z = 3 \end{cases}$$

64. Using an encyclopedia, the Internet, or the resources of your local library, research the history and evolution of matrices and determinants. How did they come about? Who is credited with their development? Why is the vertical bar notation used? Write a short summary of what you find. Answers will vary.

EXTENDING THE CONCEPT

65. Find the area of the pentagon whose vertices are: $(-5, -5)$, $(5, -5)$, $(8, 6)$, $(-8, 6)$, and $(0, 12.5)$.

66. The polynomial form for the equation of a circle is $x^2 + y^2 + Dx + Ey + F = 0$. Find the equation of the circle that contains the points $(-1, 7)$, $(2, 8)$, and $(5, -1)$.

MAINTAINING YOUR SKILLS

67. (1.4) Compute the sum, difference, product, and quotient of the following complex numbers: $2 - \sqrt{3}i$ and $2 + \sqrt{3}i$.

68. (4.4) Graph the polynomial using information about end behavior, y-intercept, x-intercept(s), and midinterval points: $f(x) = x^3 - 2x^2 - 7x + 6$.

69. (5.1/5.4) Solve the equation $3^{2x-1} = 9^{2-x}$ two ways. First using logarithms, then by equating the bases and using properties of equality.

70. (3.6) The volume of a pyramid varies jointly as its altitude and the area of its base. If a pyramid with an altitude of 10 m and a base of area 120 m^2 has a volume of 400 m^3, find the volume of a pyramid with an altitude of 90 m and a base having an area of 150 m^2.

71. (3.3) Which is the graph (left or right) of $g(x) = -|x + 1| + 3$? Justify your answer.

72. (4.4) Which is the graph (left or right) of a degree 3 polynomial? Justify your answer.

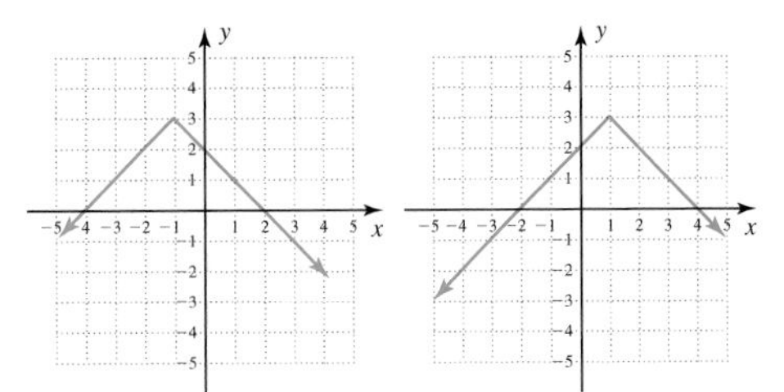

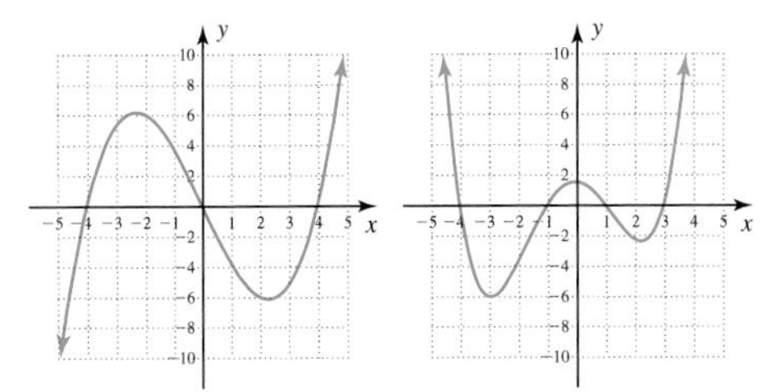

65. $A = 195$ units^2
66. $x^2 + y^2 - 4x - 6y - 12 = 0$
67. $4, -2\sqrt{3}\,i, 7, \frac{1}{7} - \frac{4\sqrt{3}}{7}i$
68.
69. $3^{2x-1} = 3^{4-2x}$; $x = 1.25$
70. 4500 m^3
71. left graph; graph shifts left 1
72. left graph; right shows four roots

SUMMARY AND CONCEPT REVIEW

SECTION 6.1 Linear Systems in Two Variables with Applications

KEY CONCEPTS

- A *solution* to a linear system in two variables is any ordered pair (x, y) that makes all equations in the system true.
- Since every point on the graph of a line satisfies the equation of that line, a point where two lines intersect must satisfy both equations and is a solution of the system.
- A system with at least one solution is called a *consistent system.*
- If the lines have different slopes, there is a unique solution to the system (they intersect at a single point). The system is called a *consistent* and *independent system.*
- If the lines have equal slopes and the same y-intercept, they form identical or *coincident lines.* Since one line is right atop the other, they intersect at all points with an infinite number of solutions. The system is called a *consistent* and *dependent system.*
- If the lines have equal slopes but different y-intercepts, they will never intersect. The system has no solution and is called an *inconsistent system.*

EXERCISES

Solve each system by graphing. If the solution is not a lattice point on the graph (x- and y-values both integers), indicate your solution is an estimate. If the system is inconsistent or dependent, so state.

1. $\begin{cases} 3x - 2y = 4 \\ -x + 3y = 8 \end{cases}$ **2.** $\begin{cases} 0.2x + 0.5y = -1.4 \\ x - 0.3y = 1.4 \end{cases}$ **3.** $\begin{cases} 2x + y = 2 \\ x - 2y = 4 \end{cases}$

Solve using substitution. Indicate whether each system is consistent, inconsistent, or dependent. Write unique solutions as an ordered pair.

4. $\begin{cases} y = 5 - x \\ 2x + 2y = 13 \end{cases}$ **5.** $\begin{cases} x + y = 4 \\ 0.4x + 0.3y = 1.7 \end{cases}$ **6.** $\begin{cases} x - 2y = 3 \\ x - 4y = -1 \end{cases}$

Solve using elimination. Indicate whether each system is consistent, inconsistent, or dependent. Write unique solutions as an ordered pair.

7. $\begin{cases} 2x - 4y = 10 \\ 3x + 4y = 5 \end{cases}$ **8.** $\begin{cases} -x + 5y = 8 \\ x + 2y = 6 \end{cases}$ **9.** $\begin{cases} 2x = 3y + 6 \\ 2.4x + 3.6y = 6 \end{cases}$

10. When it was first constructed in 1968, the John Hancock building in Chicago, Ilinois, was the tallest structure in the world. In 1985, the Sears Tower in Chicago became the world's tallest structure. The Sears Tower is 323 ft taller than the John Hancock Building, and the sum of their heights is 2577 ft. How tall is each structure?

1. $(4, 4)$

2.

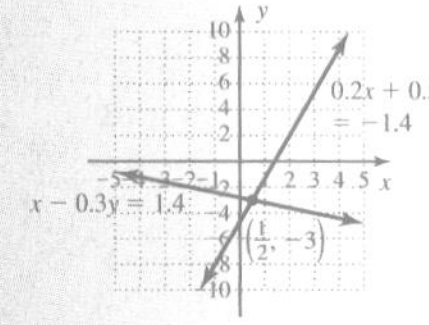

$\left(\frac{1}{2}, -3\right)$

3.

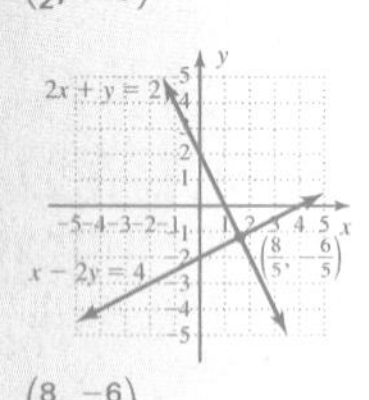

$\left(\frac{8}{5}, \frac{-6}{5}\right)$

4. no solution; inconsistent
5. $(5, -1)$; consistent
6. $(7, 2)$; consistent
7. $(3, -1)$; consistent
8. $(2, 2)$; consistent
9. $\left(\frac{11}{4}, \frac{-1}{6}\right)$; consistent
10. Sears Tower is 1450 ft; Hancock Building is 1127 ft.

SECTION 6.2 Linear Systems in Three Variables with Applications

KEY CONCEPTS

- The graph of a linear equation in three variables is a *plane.*
- A linear system in three variables has the following possible solution sets:
  - If the planes intersect at a point, the system could have one *unique solution* (x, y, z).
  - If the planes intersect at a line, the system has *linear dependence* and the solution (x, y, z) can be written as linear combinations of a single variable (*a parameter*).

- If the planes are *coincident,* the equations in the system differ by a constant multiple, meaning they are all "disguised forms" of the *same equation.* The solutions have *coincident dependence,* and the solution set can be represented by any one of the equations.
- In all other cases, the system has *no solutions* and is an inconsistent system.

EXERCISES

Solve using elimination. If a system is inconsistent or dependent, so state. For systems with linear dependence, give the answer as an ordered triple using a parameter.

11. $\begin{cases} x + y - 2z = -1 \\ 4x - y + 3z = 3 \\ 3x + 2y - z = 4 \end{cases}$

12. $\begin{cases} -x + y + 2z = 2 \\ x + y - z = 1 \\ 2x + y + z = 4 \end{cases}$

13. $\begin{cases} 3x + y + 2z = 3 \\ x - 2y + 3z = 1 \\ 4x - 8y + 12z = 7 \end{cases}$

Solve using a system of three equations in three variables.

14. A large coin jar is full of nickels, dimes, and quarters. There are 217 coins in all. The number of nickels is 12 more than the number of quarters. The value of the dimes is $4.90 more than the value of the nickels. How many coins of each type are in the bank?

15. After winning $280,000 in the lottery, Maurika decided to place the money in three different investments: a certificate of deposit paying 4%, a money market certificate paying 5%, and some Aa bonds paying 7%. After one year she earned $15,400 in interest. Find how much was invested at each rate if $20,000 more was invested at 7% than at 5%.

11. (0, 3, 2)
12. (1, 1, 1)
13. no solution, inconsistent
14. 72 nickels, 85 dimes, 60 quarters
15. $80,000 at 4% $90,000 at 5% $110,000 at 7%

SECTION 6.3 Systems of Inequalities and Linear Programming

KEY CONCEPTS

- To solve a *system of inequalities,* we find the intersecting or *overlapping regions* of the solution regions from the individual inequalities. The common area is called the *feasible region.*
- The process known as *linear programming* seeks to *maximize* or *minimize* the value of a given quantity under certain *constraints* or restrictions.
- The quantity we attempt to maximize or minimize is called the *objective function.*
- The solution(s) to a linear programming problem *occur at one of the corner points of the feasible region.*
- The process of solving a linear programming application contains these six steps:
  - Identify the main objective and the decision variables.
  - Write the objective function in terms of these variables.
  - Organize all information in a table, using the decision variables and constraints.
  - Fill in the table with the information given and write the constraint inequalities.
  - Graph the constraint inequalities and determine the feasible region.
  - Identify all corner points of the feasible region and test these points in the objective function.

16.

17.

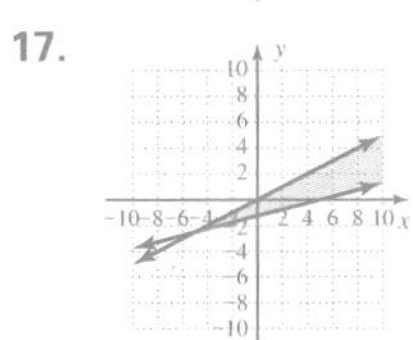

18.

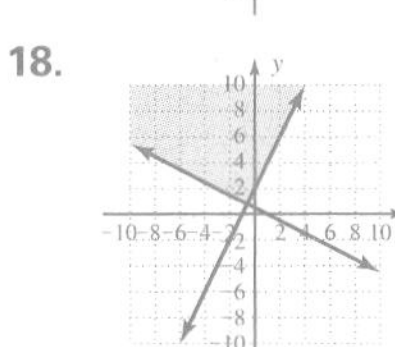

EXERCISES

Graph the solution region for each system of linear inequalities and verify the solution using a test point.

16. $\begin{cases} -x - y > -2 \\ -x + y < -4 \end{cases}$

17. $\begin{cases} x - 4y \le 5 \\ -x + 2y \le 0 \end{cases}$

18. $\begin{cases} x + 2y \ge 1 \\ 2x - y \le -2 \end{cases}$

19. 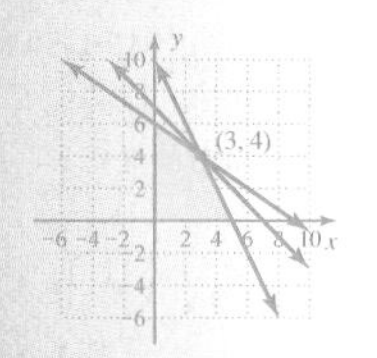

Maximum of 270 occurs at both (0, 6) and (3, 4).

20. 50 cows, 425 chickens

19. Carefully graph the feasible region for the system of inequalities shown, then maximize the objective function: $f(x, y) = 30x + 45y$ $\begin{cases} x + y \leq 7 \\ 2x + y \leq 10 \\ 2x + 3y \leq 18 \\ x \geq 0 \\ y \geq 0 \end{cases}$

20. After retiring, Oliver and Lisa Douglas buy and work a small farm (near Hooterville) that consists mostly of milk cows and egg-laying chickens. Although the price of a commodity is rarely stable, suppose that milk sales bring in an average of \$85 per cow and egg sales an average of \$50 per chicken over a period of time. During this time period, the new ranchers estimate that care and feeding of the animals took about 3 hr per cow and 2 hr per chicken, while maintaining the related equipment took 2 hr per cow and 1 hr per chicken. How many animals of each type should be maintained in order to maximize profits, if at most 1000 hr can be spent on care and feeding, and at most 525 hr on equipment maintenance?

SECTION 6.4 Systems and Absolute Value Equations and Inequalities

KEY CONCEPTS

- For absolute value equations and inequalities, the solution process begins by writing the equation in simplified form, with the absolute value quantity isolated on one side.
- The graph of an absolute value function of the form $|ax + b|$ is "V"-shaped, with its vertex at the x-intercept and its branches formed by $y = ax + b$ and $y = -(ax + b)$.
- The equation $|ax + b| = k$ can be written in system form as $\begin{cases} y = |ax + b| \\ y = k \end{cases}$, with solutions occurring where the graphs intersect.
- The inequalities $|ax + b| < k$ and $|ax + b| > k$ can be solved using the related system of equations, by identifying intervals where the absolute value function is above (greater than) or below (less than) the line $y = k$, according to the original inequality.
- Absolute value equations and inequalities can be solved using the definition of absolute value as a distance.
  - *absolute value equations:* If k is a positive number or zero, the absolute value equation $|ax + b| = k$ is equivalent to $ax + b = -k$ or $ax + b = k$.
  - *absolute value inequalities (less than):* If k is a positive number, the absolute value inequality $|ax + b| < k$ is equivalent to $-k < ax + b < k$.
  - *absolute value inequalities (greater than):* If k is a positive number, the absolute value inequality $|ax + b| > k$ is equivalent to $ax + b < -k$ or $ax + b > k$.
- These properties also apply to less than or equal to ($\leq$) and greater than or equal to ($\geq$) inequalities.

21.

22.

23.

24. $x = -4, x = -1$

25. $x \in (-\infty, -6) \cup (2, \infty)$

26. $x \in (-\infty, \infty)$

EXERCISES

Sketch the graph of the absolute value function by finding the x-intercept (vertex) and then plotting at least two points on each side of the vertex.

21. $f(x) = |x - 3|$ **22.** $g(x) = 2|x + 2|$ **23.** $h(x) = |-2x + 3|$

Solve the equation or inequality using a system of equations. Write solutions in interval notation.

24. $\dfrac{|2x + 5|}{3} + 8 = 9$ **25.** $-3|x + 2| - 2 < -14$ **26.** $\left|\dfrac{x}{2} - 9\right| \geq -7$

27. $x \in [-2, 6]$
28. no solution
29. $x \in (-10, 41)$

Solve the equation or inequality using any method. Write solutions in interval notation.

27. $5|m - 2| - 12 \leq 8$ **28.** $\dfrac{|3x - 2|}{2} + 6 = 4$ **29.** $|0.2x - 3.1| - 1.9 < 3.2$

SECTION 6.5 Solving Linear Systems Using Matrices and Row Operations

KEY CONCEPTS

- A *matrix* is a rectangular arrangement of numbers. An $m \times n$ matrix has m rows and n columns.
- The matrix derived from a system of linear equations is called the *augmented matrix.* It is created by augmenting the *coefficient matrix* (formed by the variable coefficients) with the *matrix of constants.*
- One matrix method for solving systems involves the augmented matrix and *row-reduction.*
  - If possible, interchange equations so that the coefficient of x is a "1" in R1.
  - Write the system in augmented matrix form (coefficient matrix with matrix of constants).
  - Use row operations to obtain zeroes below the first entry of the diagonal.
  - Use row operations to obtain zeroes below the second entry of the diagonal.
  - Continue until the matrix is triangularized (all entries below the diagonal are zero).
  - Convert the augmented matrix back into equation form and solve for z.

EXERCISES

Solve by triangularizing the matrix. Use a calculator for Exercise 32.

30. $\begin{cases} x - 2y = 6 \\ 4x - 3y = 4 \end{cases}$ $(-2, -4)$

31. $\begin{cases} x - 2y + 2z = 7 \\ 2x + 2y - z = 5 \\ 3x - y + z = 6 \end{cases}$ $(1, 6, 9)$

32. $\begin{cases} 2w + x + 2y - 3z = -19 \\ w - 2x - y + 4z = 15 \\ x + 2y - z = 1 \\ 3w - 2x - 5z = -60 \end{cases}$ $(-2, 7, 1, 8)$

SECTION 6.6 The Algebra of Matrices

KEY CONCEPTS

- The entries of a matrix are denoted a_{ij}, where i gives the row and j gives the column of its location.
- The $m \times n$ size of a matrix is also referred to as its *order.*
- Two matrices A and B are equal if corresponding entries are equal: $A = B$ if $a_{ij} = b_{ij}$.
- The sum or difference of two matrices is found by combining corresponding entries: $A + B = [a_{ij} + b_{ij}]$.
- The *identity matrix* for addition is an $m \times n$ matrix whose entries are all zeroes.
- The product of a constant times a matrix is called *scalar multiplication,* and is found by taking the product of the scalar with each entry, forming a new matrix of like size. For matrix A: $kA = ka_{ij}$.
- Matrix multiplication is performed as row entry $\times$ column entry according to the following procedure: For an $m \times n$ matrix $A = [a_{ij}]$ and a $p \times q$ matrix $B = [b_{ij}]$, matrix multiplication is possible if $n = p$, and the result will be an $m \times q$ matrix P. In symbols $A \cdot B = [p_{ij}]$, where p_{ij} is product of the ith row of A with the jth column of B.

33. $\begin{bmatrix} -7.25 & 5.25 \\ 0.875 & -2.875 \end{bmatrix}$

34. $\begin{bmatrix} -6.75 & 6.75 \\ 1.125 & -1.125 \end{bmatrix}$

35. not possible

36. $\begin{bmatrix} -2 & -6 \\ -1 & -7 \end{bmatrix}$

37. $\begin{bmatrix} 1 & 0 \\ 0 & 1 \end{bmatrix}$

38. $\begin{bmatrix} 1 & 0 & 4 \\ 5.5 & -1 & -1 \\ 10 & -2.9 & 7 \end{bmatrix}$

39. $\begin{bmatrix} 3 & -6 & -4 \\ -4.5 & 3 & -1 \\ -2 & 3.1 & 3 \end{bmatrix}$

- When technology is used to perform operations on matrices, the focus shifts from a meticulous computation of new entries, to carefully entering each matrix into the calculator, double checking that each entry is correct, and appraising the results to see if they are reasonable.

EXERCISES

Compute the operations indicated below (if possible), using the following matrices.

$A = \begin{bmatrix} \frac{-1}{4} & \frac{-3}{4} \\ \frac{-1}{8} & \frac{-7}{8} \end{bmatrix}$ $B = \begin{bmatrix} -7 & 6 \\ 1 & -2 \end{bmatrix}$ $C = \begin{bmatrix} -1 & 3 & 4 \\ 5 & -2 & 0 \\ 6 & -3 & 2 \end{bmatrix}$ $D = \begin{bmatrix} 2 & -3 & 0 \\ 0.5 & 1 & -1 \\ 4 & 0.1 & 5 \end{bmatrix}$

33. $A + B$ **34.** $B - A$ **35.** $C - B$ **36.** $8A$ **37.** BA

38. $C + D$ **39.** $D - C$ **40.** BC **41.** $-4D$ **42.** CD

40. not possible 41. $\begin{bmatrix} -8 & 12 & 0 \\ -2 & -4 & 4 \\ -16 & -0.4 & -20 \end{bmatrix}$ 42. $\begin{bmatrix} 15.5 & 6.4 & 17 \\ 9 & -17 & 2 \\ 18.5 & -20.8 & 13 \end{bmatrix}$

SECTION 6.7 Solving Linear Systems Using Matrix Equations

KEY CONCEPTS

- The *identity matrix I* for multiplication has 1s on the main diagonal and 0s for all other entries. For any $n \times n$ matrix A, the identity matrix is also $n \times n$, where $AI = IA = A$.
- For an $n \times n$ (square) matrix A, the *inverse matrix* for multiplication is a matrix B such that $AB = BA = I$. Only square matrices have an inverse. For matrix A the inverse is denoted A^{-1}.
- Any $n \times n$ system of equations can be written as a matrix equation and solved (if solutions exist) using an inverse matrix. For $\begin{cases} 2x + 3y - z = 5 \\ -3x + 2y + 4z = 13 \\ x - 5y + 2z = -3 \end{cases}$, the matrix equation is $\begin{bmatrix} 2 & 3 & -1 \\ -3 & 2 & 4 \\ 1 & -5 & 2 \end{bmatrix} \cdot \begin{bmatrix} x \\ y \\ z \end{bmatrix} = \begin{bmatrix} 5 \\ 13 \\ -3 \end{bmatrix}$.
- Every square matrix has a real number associated with it called its determinant. For a 2×2 matrix $A = \begin{vmatrix} a_{11} & a_{12} \\ a_{21} & a_{22} \end{vmatrix}$, the determinant $|A|$ is the difference of diagonal products: $a_{11}a_{22} - a_{21}a_{12}$.
- If the determinant of a matrix is zero, the matrix is said to be *singular* or *non-invertible*.
- For matrix equations, if the coefficient matrix is non-invertible, the system has no unique solution.

EXERCISES

Complete Exercises 43 through 45 using the following matrices:

$A = \begin{bmatrix} 1 & 0 \\ 0 & 1 \end{bmatrix}$ $B = \begin{bmatrix} 0.2 & 0.2 \\ -0.6 & 0.4 \end{bmatrix}$ $C = \begin{bmatrix} 2 & -1 \\ 3 & 1 \end{bmatrix}$ $D = \begin{bmatrix} 10 & -6 \\ -15 & 9 \end{bmatrix}$

43. Exactly one of the matrices given is singular. Compute each determinant to identify it. *D*

44. Show that $AB = BA = B$. What can you conclude about the matrix A? It's an identity.

45. Show that $BC = CB = I$. What can you conclude about the matrix C? It's the inverse of *B*.

Complete Exercises 46 through 49 using these matrices:

$E = \begin{bmatrix} 1 & -2 & 3 \\ -2 & 1 & -5 \\ -1 & -1 & -2 \end{bmatrix}$ $F = \begin{bmatrix} 1 & -1 & 1 \\ 0 & 1 & 0 \\ -2 & 1 & -1 \end{bmatrix}$ $G = \begin{bmatrix} 1 & 0 & 0 \\ 0 & 1 & 0 \\ 0 & 0 & 1 \end{bmatrix}$ $H = \begin{bmatrix} -1 & 0 & -1 \\ 0 & 1 & 0 \\ 2 & 1 & 1 \end{bmatrix}$

46. Exactly one of the matrices above is singular. Use a calculator to determine which one. *E*

47. Show that $GF = FG = F$. What can you conclude about the matrix G? It's an identity matrix.

48. Show that $FH = HF = I$. What can you conclude about the matrix H? It's the inverse of F.

49. Verify that $EH \neq HE$ and $EF \neq FE$ (Exercise 48). matrix multiplication is not generally commutative

Solve manually using a matrix equation.

50. $\begin{cases} 2x - 5y = 14 \\ -3y + 4x = -14 \end{cases}$ $(-8, -6)$

Solve using a matrix equation and your calculator.

51. $\begin{cases} 0.5x - 2.2y + 3z = -8 \\ -0.6x - y + 2z = -7.2 \\ x + 1.5y - 0.2z = 2.6 \end{cases}$ $(2, 0, -3)$

SECTION 6.8 Matrix Applications: Cramer's Rule, Partial Fractions, and More

KEY CONCEPTS

- Cramer's rule uses a ratio of determinants to solve systems of equations (if they exist).
- The determinant of the 2×2 matrix $\begin{vmatrix} a_{11} & a_{12} \\ a_{21} & a_{22} \end{vmatrix}$ is $a_{11}a_{22} - a_{21}a_{12}$.
- To compute the value of 3×3 and larger determinants, a calculator is generally used.
- Determinants can be used to find the area of a triangle in the plane if the vertices of the triangle are known, and as a test to see if three points are collinear.
- A system of equations can be used to write a rational expression as a sum of its partial fractions.

EXERCISES

Solve using Cramer's rule.

52. $\begin{cases} 5x + 6y = 8 \\ 10x - 2y = -9 \end{cases}$

53. $\begin{cases} 2x + y = -2 \\ -x + y + 5z = 12 \\ 3x - 2y + z = -8 \end{cases}$

54. $\begin{cases} 2x + y - z = -1 \\ x - 2y + z = 5 \\ 3x - y + 2z = 8 \end{cases}$

55. Find the area of a triangle whose vertices have the coordinates $(6, 1)$, $(-1, -6)$, and $(-6, 2)$.

56. Find the partial fraction decomposition for $\dfrac{7x^2 - 5x + 17}{x^3 - 2x^2 + 3x - 6}$.

52. $\left(\frac{-19}{35}, \frac{25}{14}\right)$

53. $\left(\frac{-37}{19}, \frac{36}{19}, \frac{31}{19}\right)$

54. $(1, -1, 2)$

55. $\frac{91}{2}$ units^2

56. $\frac{5}{x-2} + \frac{2x-1}{x^2+3}$

MIXED REVIEW

1. Write the equations in each system in slope-intercept form, then state whether the system is consistent/independent, consistent/dependent, or inconsistent. Do not solve.

a. $\begin{cases} -3x + 5y = 10 \\ 6x + 20 = 10y \end{cases}$

b. $\begin{cases} 4x - 3y = 9 \\ -2x + 5y = -10 \end{cases}$

c. $\begin{cases} x - 3y = 9 \\ -6y + 2x = 10 \end{cases}$

2. Solve by graphing.
$\begin{cases} x - 2y = 6 \\ -2x + y = -9 \end{cases}$

3. Solve using a substitution.
$\begin{cases} 2x + 3y = 5 \\ -x + 5y = 17 \end{cases}$ $(-2, 3)$

4. Solve using elimination.
$\begin{cases} 7x - 4y = -5 \\ 3x + 2y = 9 \end{cases}$ $(1, 3)$

Additional answers can be found in the Instructor Answer Appendix.

Solve using elimination.

5. $\begin{cases} x + 2y - 3z = -4 \\ -3x + 4y + z = 1 \\ 2x - 6y + z = 1 \end{cases}$ $(1, \frac{1}{2}, 2)$

6. $\begin{cases} 0.1x - 0.2y + z = 1.7 \\ 0.3x + y - 0.1z = 3.6 \\ -0.2x - 0.1y + 0.2z = -1.7 \end{cases}$ $(9, 1, 1)$

Solve using row operations to triangularize the matrix.

7. $\begin{cases} \frac{1}{2}x + \frac{2}{3}y = 3 \\ \frac{-2}{5}x - \frac{1}{4}y = 1 \end{cases}$ $(-10, 12)$

8. $\begin{cases} -2x + y - 4z = -11 \\ x + 3y - z = -4 \\ 3x - 2y + z = 7 \end{cases}$ $(1, -1, 2)$

Solve the absolute value inequalities using the method of your choice.

9. $-2|x + 3| - 7 \geq -15$ $x \in [-7, 1]$

10. $-\frac{3}{2}|x - 2| + 3 \leq -9$ $x \in (-\infty, -6] \cup [10, \infty)$

Compute as indicated for

$$A = \begin{bmatrix} 2 & -1 \\ 0 & 3 \end{bmatrix} \quad B = \begin{bmatrix} 1 & -2 \\ 3 & 0 \\ 2 & 4 \end{bmatrix} \quad C = \begin{bmatrix} 1 & -4 & 2 \\ -2 & 0 & -1 \end{bmatrix} \quad D = \begin{bmatrix} 3 & 0 & 1 \\ -1 & 2 & 0 \\ 1 & 1 & -4 \end{bmatrix}$$

11. a. $-2AC$ **b.** CD

12. a. BA **b.** $CB - 4A$

13. Solve using a matrix equation:
$\begin{cases} -x - 2z = 5 \\ 2y + z = -4 \\ -x + 2y = 3 \end{cases}$ $(-9, -3, 2)$

14. Use a matrix equation and a calculator to solve:
$\begin{cases} w + \frac{1}{2}x + \frac{2}{3}y - z = -3 \\ \frac{3}{4}x - y + \frac{5}{8}z = \frac{41}{8} \\ \frac{2}{5}w - x - \frac{3}{10}z = -\frac{27}{10} \\ w + 2x - 3y + 4z = 16 \end{cases}$ $(-1, 2, -3, 1)$

15. Solve using Cramer's rule:
$\begin{cases} -x + 5y - 2z = 1 \\ 2x + 3y - z = 3 \\ 3x - y + 3z = -2 \end{cases}$ $\left(\frac{33}{31}, \frac{-10}{31}, \frac{-57}{31}\right)$

16. Decompose the expression into partial fractions: $\dfrac{x^2 + 1}{x^3 - 3x + 2}$

17. Graph the solution region for the system of inequalities. $\begin{cases} 4x + 2y \leq 14 \\ 2x + 3y \leq 15 \\ y \geq 0 \\ x \geq 0 \end{cases}$

18. Maximize $P(x, y) = 2.5x + 3.75y$, given $\begin{cases} x + y \leq 8 \\ x + 2y \leq 14 \\ 4x + 3y \leq 30 \\ x, y \geq 0 \end{cases}$

19. It's the end of another big day at the circus, and the clowns are putting away their riding equipment—a motley collection of unicycles, bicycles, and tricycles. As she loads them into the storage shed, Trixie counts 21 cycles in all with a total of 40 wheels. In addition, she notes the number of bicycles is one fewer than twice the number of tricycles. How many cycles of each type do the clowns use?

20. A local fitness center is offering incentives in an effort to boost membership. If you buy a year's membership (Y), you receive a \$50 rebate and six tickets to a St. Louis Cardinals home game. For a half-year membership (H), you receive a \$30 rebate and four tickets to a Cardinals home game. For a monthly trial membership (M), you receive a \$10 rebate and

---

11. a. $\begin{bmatrix} -8 & 16 & -10 \\ 12 & 0 & 6 \end{bmatrix}$
b. $\begin{bmatrix} 9 & -6 & -7 \\ -7 & -1 & 2 \end{bmatrix}$

12. a. $\begin{bmatrix} 2 & -7 \\ 6 & -3 \\ 4 & 10 \end{bmatrix}$
b. $\begin{bmatrix} -15 & 10 \\ -4 & -12 \end{bmatrix}$

16. $\dfrac{4}{9(x-1)} + \dfrac{2}{3(x-1)^2} + \dfrac{5}{9(x+2)}$

17.

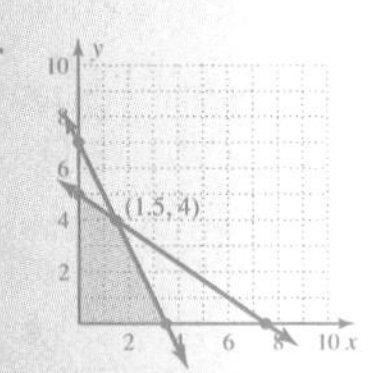

18.

(x, y)	$P(x, y) = 2.5x + 3.75y$
(0, 0)	0
(0, 7)	26.25
(7.5, 0)	18.75
(2, 6)	27.5
(6, 2)	22.5

19. 7 unicycles; 9 bicycles; 5 tricycles

20.

	Y	H	M
M	40	52	70
W	50	44	60

	R	T
Y	50	6
H	30	4
M	10	2

$$\begin{bmatrix} 40 & 52 & 70 \\ 50 & 44 & 60 \end{bmatrix} \begin{bmatrix} 50 & 6 \\ 30 & 4 \\ 10 & 2 \end{bmatrix} = \begin{bmatrix} 4260 & 588 \\ 4420 & 596 \end{bmatrix}$$

a. \$4260
b. 596

two tickets to a Cardinals home game. During the last month, male clients purchased 40 one-year, 52 half-year, and 70 monthly memberships, while female clients purchased 50 one-year, 44 half-year, and 60 monthly memberships. Write the number of sales of each type to males and females as a 2×3 matrix, and the amount of the rebates and number of Cards tickets awarded per type of membership as a 3×2 matrix. Use these matrices to determine (a) the total amount of rebate money paid to males and (b) the number of Cardinals tickets awarded to females.

PRACTICE TEST

Solve each system and state whether the system is consistent, inconsistent, or dependent.

1. Solve graphically: $\begin{cases} 3x + 2y = 12 \\ -x + 4y = 10 \end{cases}$

2. Solve using substitution: $\begin{cases} 3x - y = 2 \\ -7x + 4y = -6 \end{cases}$

3. Solve using elimination: $\begin{cases} 5x + 8y = 1 \\ 3x + 7y = 5 \end{cases}$

4. Solve using elimination: $\begin{cases} x + 2y - z = -4 \\ 2x - 3y + 5z = 27 \\ -5x + y - 4z = -27 \end{cases}$

5. Given matrices A and B, compute:

   a. $A - B$ b. $\frac{2}{5}B$ c. AB
   d. A^{-1} e. $|A|$

   $A = \begin{bmatrix} -3 & -2 \\ 5 & 4 \end{bmatrix}$ $B = \begin{bmatrix} 3 & 3 \\ -3 & -5 \end{bmatrix}$

6. Given matrices C and D, use a calculator to find:

   a. $C - D$ b. $-0.6D$ c. DC
   d. D^{-1} e. $|D|$

   $C = \begin{bmatrix} 0.5 & 0 & 0.2 \\ 0.4 & -0.5 & 0 \\ 0.1 & -0.4 & -0.1 \end{bmatrix}$

   $D = \begin{bmatrix} 0.5 & 0.1 & 0.2 \\ -0.1 & 0.1 & 0 \\ 0.3 & 0.4 & 0.8 \end{bmatrix}$

7. Solve using matrices and row reduction: $\begin{cases} 4x - 5y - 6z = 5 \\ 2x - 3y + 3z = 0 \\ x + 2y - 3z = 5 \end{cases}$

8. Solve using a calculator and Cramer's rule: $\begin{cases} 2x + 3y + z = 3 \\ x - 2y - z = 4 \\ x - y - 2z = -1 \end{cases}$

9. Solve using matrix equation and your calculator: $\begin{cases} 2x - 5y = 11 \\ 4x + 7y = 4 \end{cases}$

10. Solve using matrix equation and your calculator: $\begin{cases} x - 2y + 2z = 7 \\ 2x + 2y - z = 5 \\ 3x - y + z = 6 \end{cases}$

Create a system of equations to model each exercise, then solve using the method of your choice.

11. The perimeter of a "legal-size" paper is 114.3 cm. The length of the paper is 7.62 cm less than twice the width. Find the dimensions of a legal-size sheet of paper.

12. The island nations of Tahiti and Tonga have a combined land area of 692 mi^2. Tahiti's land area is 112 mi^2 more than Tonga's. What is the land area of each island group?

13. Many years ago, two cans of corn (C), 3 cans of green beans (B), and 1 can of peas (P) cost \$1.39. Three cans of C, 2 of B, and 2 of P cost \$1.73. One can of C, 4 of B, and 3 of P cost \$1.92. What is the price of a single can of C, B, and P?

1.

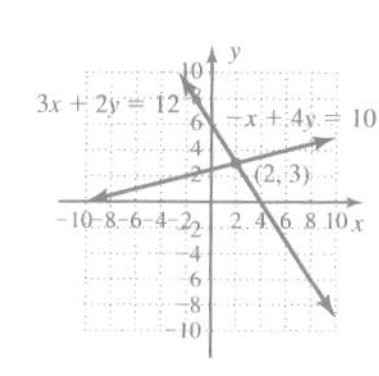

(2, 3)

2. $\left(\frac{2}{5}, \frac{-4}{5}\right)$
3. $(-3, 2)$
4. $(2, -1, 4)$
5. a. $\begin{bmatrix} -6 & -5 \\ 8 & 9 \end{bmatrix}$ b. $\begin{bmatrix} 1.2 & 1.2 \\ -1.2 & -2 \end{bmatrix}$
   c. $\begin{bmatrix} -3 & 1 \\ 3 & -5 \end{bmatrix}$ d. $\begin{bmatrix} -2 & -1 \\ 2.5 & 1.5 \end{bmatrix}$
   e. -2
6. a. $\begin{bmatrix} 0 & -0.1 & 0 \\ 0.5 & -0.6 & 0 \\ -0.2 & -0.8 & -0.9 \end{bmatrix}$
   b. $\begin{bmatrix} -0.3 & -0.06 & -0.12 \\ 0.06 & -0.06 & 0 \\ -0.18 & -0.24 & -0.48 \end{bmatrix}$
   c. $\begin{bmatrix} 0.31 & -0.13 & 0.08 \\ -0.01 & -0.05 & -0.02 \\ 0.39 & -0.52 & -0.02 \end{bmatrix}$
   d. $\begin{bmatrix} \frac{40}{17} & 0 & \frac{-10}{17} \\ \frac{40}{17} & 10 & \frac{-10}{17} \\ \frac{-35}{17} & -5 & \frac{30}{17} \end{bmatrix}$
   e. $\frac{17}{500}$
7. $\left(2, 1, \frac{-1}{3}\right)$
8. $(3, -2, 3)$
9. $\left(\frac{97}{34}, \frac{-18}{17}\right)$
10. $(1, 6, 9)$
11. 21.59 cm by 35.56 cm
12. Tahiti 402 mi^2; Tonga 290 mi^2
13. Corn 25¢
    Beans 20¢
    Peas 29¢

14. $15,000 at 7%
$8000 at 5%
$7000 at 9%

15. $x \in (-\infty, -1) \cup (5, \infty)$

16. $x \in (-\infty, -7] \cup [1, \infty)$

17.

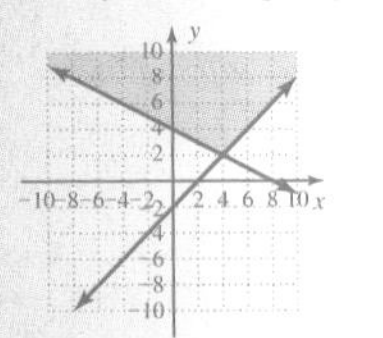

18. (5, 0)

19.

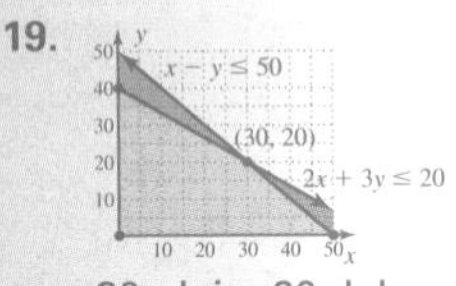

30 plain; 20 deluxe

20. $\dfrac{1}{x-3} + \dfrac{3x+2}{x^2+3x+9}$

14. After inheriting $30,000 from a rich aunt, David decides to place the money in three different investments: a savings account paying 5%, a bond account paying 7%, and a stock account paying 9%. After 1 yr he earned $2080 in interest. Find how much was invested at each rate if $8000 less was invested at 9% than at 7%.

15. Solve using a system:
$\frac{-2}{3}|x - 2| + 7 < 5$

16. Solve analytically:
$5|x + 3| - 12 \geq 8$

17. Solve the system of inequalities by graphing. $\begin{cases} x - y \leq 2 \\ x + 2y \geq 8 \end{cases}$

18. Maximize the objective function:
$P = 50x - 12y$
$\begin{cases} x + 2y \leq 8 \\ 8x + 5y \geq 40 \\ x, y \geq 0 \end{cases}$

Solve the linear programming problem.

19. A company manufactures two types of T-shirts, a plain T-shirt and a deluxe monogrammed T-shirt. To produce a plain shirt requires 1 hr of working time on machine A and 2 hr on machine B. To produce a deluxe shirt requires 1 hr on machine A and 3 hr on machine B. Machine A is available for at most 50 hr/week, while machine B is available for at most 120 hr/week. If a plain shirt can be sold at a profit of $4.25 each and a deluxe shirt can be sold at a profit of $5.00 each, how many of each should be manufactured to maximize the profit?

20. Decompose the expression into partial fractions: $\dfrac{4x^2 - 4x + 3}{x^3 - 27}$.

CALCULATOR EXPLORATION AND DISCOVERY

Optimal Solutions and Linear Programming

In Section 6.3 we learned, *"If a linear programming problem has a unique solution, it must occur at a vertex."* Although we explored the reason why using the feasible region and a family of lines from the objective function, it sometimes helps to have a good "old fashioned," point-by-point verification to support the facts. In this exercise, we'll use a graphing calculator to explore various areas of the feasible region, repeatedly evaluating the objective function to see where the maximal values (optimal solutions) seem to "congregate." If all goes as expected, ordered pairs nearest to a vertex should give relatively larger values. To demonstrate, we'll use Example 6 from Section 6.3, stated below.

Example 6: Find the maximum value of the objective function $f(x, y) = 2x + y$ given the constraints shown in the system $\begin{cases} x + y \leq 4 \\ 3x + y \leq 6 \\ x \geq 0 \\ y \geq 0 \end{cases}$.

Solution: Begin by noting the solutions must be in QI, since $x \geq 0$ and $y \geq 0$. Graph the boundary lines $y = -x + 4$ and $y = -3x + 6$, shading the lower half plane in each case since they are "less than" inequalities. This produces the feasible region shown in lavender. There are four corner points to this region: (0, 0), (0, 4), (2, 0), and (1, 3), the latter found by solving the system: $\begin{cases} x + y = 4 \\ 3x + y = 6 \end{cases}$.

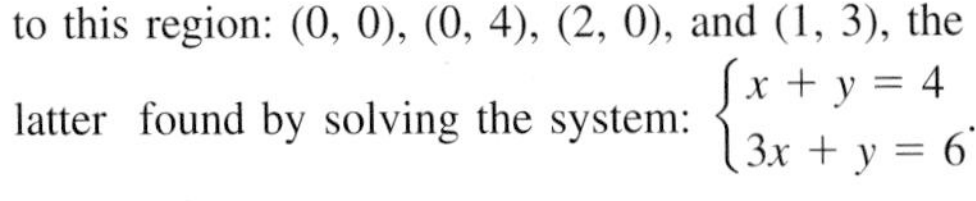

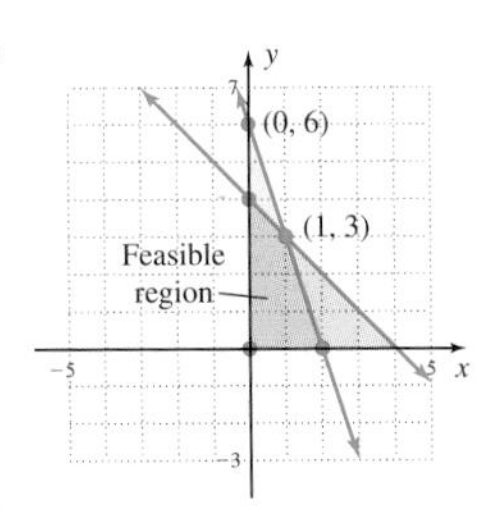

Figure 6.38

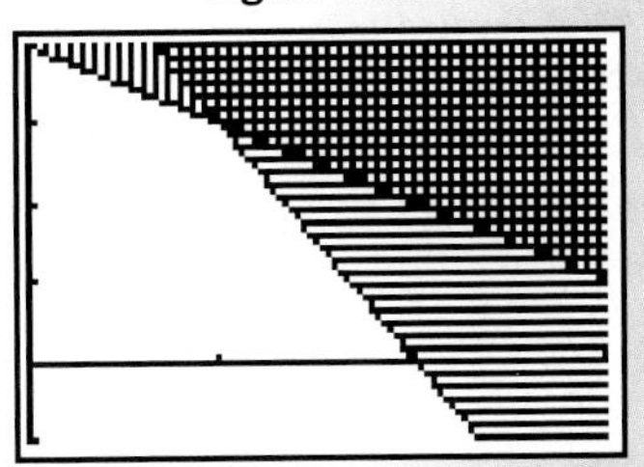

Figure 6.39

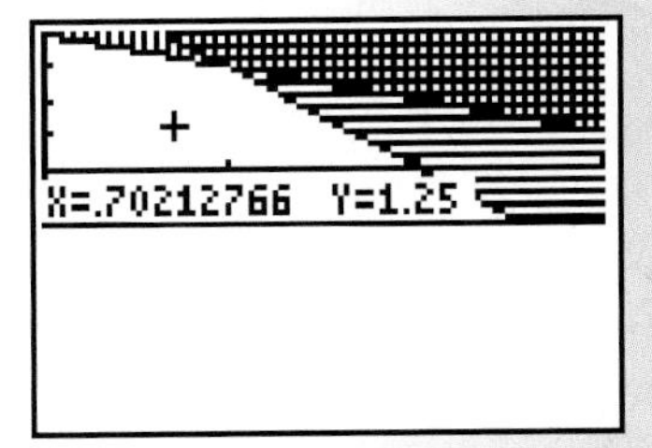

Figure 6.40

Figure 6.41

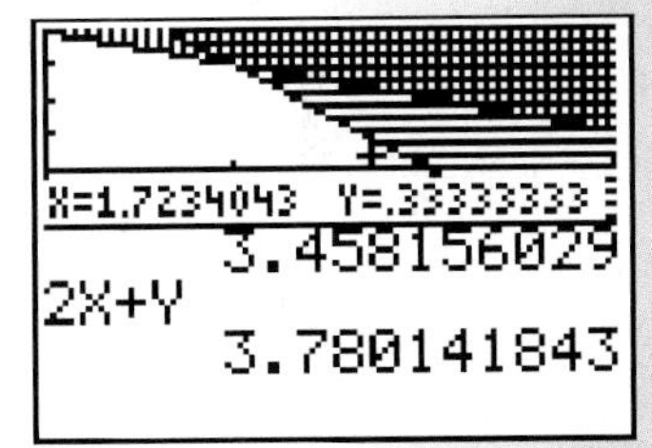

Figure 6.42

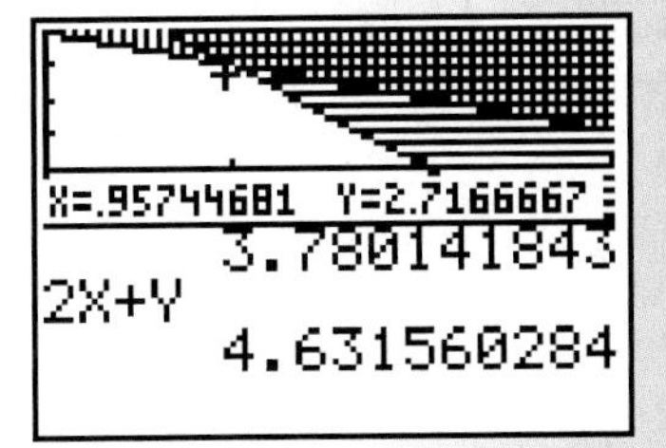

To explore this feasible region in terms of the objective function $f(x, y) = 2x + y$, enter the boundary lines $Y_1 = -x + 4$ and $Y_2 = -3x + 6$ on the **Y=** screen. However, instead of shading below the lines to show the feasible region (using the ◣ feature to the extreme left), we shade above both lines (using the ◥ feature) so that the feasible region remains clear. Setting the window size at $x \in [0, 3]$ and $y \in [-1.5, 4]$ produces Figure 6.38. Using YMin $= -1.5$ will leave a blank area just below QI that allows you to cleanly explore the feasible region as the x- and y-values are displayed. Next we place the calculator in "split-screen" mode so that we can view the graph and the home screen simultaneously. Press the **MODE** key and notice the second-to-last line reads **Full Horiz G-T.** The **Full** (screen) mode is the default operating mode. The **Horiz** mode splits the screen horizontally, placing the graph directly above a shorter home screen. The **G-T** mode splits the screen vertically, with the graph to the left and a table to the right. Highlight **Horiz,** then press **ENTER** and **GRAPH** to have the calculator reset the screen in this mode. The TI-84 Plus has a free-moving cursor (a cursor you can move around without actually tracing a curve). Pressing the left ◀ or right ▶ arrow brings it into view (Figure 6.39). A useful feature of the free-moving cursor is that it automatically stores the current X value as the variable X (**X,T,θ,n** or **ALPHA** **STO▸**) and the current Y value as the variable Y (**ALPHA** 1), which allows us to evaluate the objective function $f(x, y) = 2x + y$ right on the home screen as we explore various corners of the feasible region. To access the graph and free-moving cursor you must press **GRAPH** each time, and to access the home screen you must press **2nd** **MODE** (**QUIT**) each time. Begin by moving the cursor to the upper-left corner of the region, near the y-intercept [we stopped at $(\sim 0.0957, 3.2\overline{6})$]. Once you have the cursor "tucked up into the corner," press **2nd** **MODE** (**QUIT**) to get to the home screen, then enter the objective function: 2X + Y. Pressing **ENTER** automatically evaluates the function for the values indicated by the cursor's location (Figure 6.40). It appears the value of the objective function for points (x, y) in this corner are close to 4, and it's no accident that at the corner point (0, 4) the maximum value is in fact 4. Now let's explore the area in the lower-right of the feasible region. Press **GRAPH** and move the cursor using the arrow keys until it is "tucked" over into the lower-right corner [we stopped at $(\sim 1.72, 0.\overline{3})$]. Pressing **2nd** **ENTER** recalls 2X + Y and pressing **ENTER** re-evaluates the objective function at these new X and Y values (Figure 6.41). It appears that values of the objective function are also close to 4 in this corner, and it's again no accident that *at* the corner point (2, 0) the maximum value *is* 4. Finally, press **GRAPH** to explore the region in the upper-right corner, where the lines intersect. Move the cursor to this vicinity, locate it very near the point of intersection [we stopped at $(\sim 0.957, 2.71\overline{6})$] and return to the home screen and evaluate (Figure 6.42). The value of the objective function is near 5 in this corner of the region, and at the corner point (1, 3) the maximum

Exercise 1: Answers will vary.
Exercise 2: (0, 5)
Exercise 3: (5, 1)

value is 5. This investigation can be repeated for any feasible region and for any number of points within the region, and serves to support the statement that, "*If a linear programming problem has a unique solution, it must occur at a vertex.*"

Exercise 1: Use the ideas discussed here to explore the solution to Example 7 from Section 6.3.

Exercise 2: The feasible region for the system given to the right has five corner points. Use the ideas here to explore the area near each corner point of the feasible region to determine which point is the likely candidate to produce the *maximum* value of the objective function $f(x, y) = -3.5x + y$. Then solve the linear programming problem to verify your guess.

$$\begin{cases} 8x + 3y \leq 30 \\ 5x + 4y \leq 23 \\ x + 2y \leq 10 \\ x, y \geq 0 \end{cases}$$

Exercise 3: The feasible region for the system given to the right has four corner points. Use the ideas here to explore the area near each corner point of the feasible region to determine which point is the likely candidate to produce the *minimum* value of the objective function $f(x, y) = 2x + 4y$. Then solve the linear programming problem to verify your guess.

$$\begin{cases} 2x + 2y \leq 15 \\ x + y \geq 6 \\ x + 4y \geq 9 \\ x, y \geq 0 \end{cases}$$

STRENGTHENING CORE SKILLS

Augmented Matrices and Matrix Inverses

The formula for finding the inverse of a 2×2 matrix has its roots in the more general method of computing the inverse of an $n \times n$ matrix. This involves augmenting a square matrix M with its corresponding identity I_n on the right (forming an $n \times 2n$ matrix), and using row operations to *transform M into the identity*. In some sense, as the original matrix is transformed, the "identity part" keeps track of the operations we used to convert M and we can use the results to "get back home," so to speak. We'll illustrate with the 2×2 matrix from Section 6.7, Example 3, where we found that $\begin{bmatrix} 1 & -2.5 \\ -1 & 3 \end{bmatrix}$ was the inverse matrix for $\begin{bmatrix} 6 & 5 \\ 2 & 2 \end{bmatrix}$. We begin by augmenting $\begin{bmatrix} 6 & 5 \\ 2 & 2 \end{bmatrix}$ with the 2×2 identity.

$$\left[\begin{array}{cc|cc} 6 & 5 & 1 & 0 \\ 2 & 2 & 0 & 1 \end{array}\right] \; -2R1 + 6R2 \longrightarrow R2 \left[\begin{array}{cc|cc} 6 & 5 & 1 & 0 \\ 0 & 2 & -2 & 6 \end{array}\right] \; \frac{R2}{2} \longrightarrow R2 \left[\begin{array}{cc|cc} 6 & 5 & 1 & 0 \\ 0 & 1 & -1 & 3 \end{array}\right]$$

$$\left[\begin{array}{cc|cc} 6 & 5 & 1 & 0 \\ 0 & 1 & -1 & 3 \end{array}\right] \; -5R2 + R1 \longrightarrow R1 \left[\begin{array}{cc|cc} 6 & 0 & 6 & -15 \\ 0 & 1 & -1 & 3 \end{array}\right] \; \frac{R1}{6} \longrightarrow R1 \left[\begin{array}{cc|cc} 1 & 0 & 1 & -2.5 \\ 0 & 1 & -1 & 3 \end{array}\right]$$

As you can see, the identity is automatically transformed into the inverse matrix when this method is applied. Performing similar row operations on the general matrix $\begin{bmatrix} a & b \\ c & d \end{bmatrix}$ results in the formula given earlier.

$$\left[\begin{array}{cc|cc} a & b & 1 & 0 \\ c & d & 0 & 1 \end{array}\right] \; -cR1 + aR2 \longrightarrow R2 \left[\begin{array}{cc|cc} a & b & 1 & 0 \\ 0 & ad - bc & -c & a \end{array}\right]$$

$$\frac{R2}{ad - bc} \longrightarrow R2 \left[\begin{array}{cc|cc} a & b & 1 & 0 \\ 0 & 1 & \dfrac{-c}{ad - bc} & \dfrac{a}{ad - bc} \end{array}\right]$$

$$\left[\begin{array}{cc|cc} a & b & 1 & 0 \\ 0 & 1 & \dfrac{-c}{ad - bc} & \dfrac{a}{ad - bc} \end{array}\right] \; -bR2 + R1 \longrightarrow R1 \left[\begin{array}{cc|cc} a & 0 & \dfrac{bc}{ad - bc} + 1 & \dfrac{-ba}{ad - bc} \\ 0 & 1 & \dfrac{-c}{ad - bc} & \dfrac{a}{ad - bc} \end{array}\right]$$

Finding a common denominator for $\frac{bc}{ad - bc} + 1$ and combining like terms gives $\frac{bc}{ad - bc} + \frac{ad - bc}{ad - bc} = \frac{ad}{ad - bc}$.

$$\left[\begin{array}{cc|cc} a & 0 & \frac{ad}{ad-bc} & \frac{-ba}{ad-bc} \\ 0 & 1 & \frac{-c}{ad-bc} & \frac{a}{ad-bc} \end{array}\right] \begin{array}{c} \frac{R1}{a} \longrightarrow R1 \end{array} \left[\begin{array}{cc|cc} 1 & 0 & \frac{d}{ad-bc} & \frac{-b}{ad-bc} \\ 0 & 1 & \frac{-c}{ad-bc} & \frac{a}{ad-bc} \end{array}\right].$$

$$\text{This shows } A^{-1} = \begin{bmatrix} \frac{d}{ad-bc} & \frac{-b}{ad-bc} \\ \frac{-c}{ad-bc} & \frac{a}{ad-bc} \end{bmatrix}$$

and factoring out $\frac{1}{ad - bc}$ produces the familiar formula. As you might imagine, attempting this on a general 3×3 matrix is problematic at best, and instead we simply apply the procedure to any given 3×3 matrix. Here we'll use the augmented matrix method to find A^{-1}, given $A = \begin{bmatrix} 2 & 1 & 0 \\ -1 & 3 & -2 \\ 3 & -1 & 2 \end{bmatrix}$.

$$\left[\begin{array}{ccc|ccc} 2 & 1 & 0 & 1 & 0 & 0 \\ -1 & 3 & -2 & 0 & 1 & 0 \\ 3 & -1 & 2 & 0 & 0 & 1 \end{array}\right] \begin{array}{c} R1 + 2R2 \longrightarrow R2 \\ \\ -3R1 + 2R3 \longrightarrow R3 \end{array} \left[\begin{array}{ccc|ccc} 2 & 1 & 0 & 1 & 0 & 0 \\ 0 & 7 & -4 & 1 & 2 & 0 \\ 0 & -5 & 4 & -3 & 0 & 2 \end{array}\right]$$

$$\begin{array}{c} R2 - 7R1 \longrightarrow R1 \\ \\ 5R2 + 7R3 \longrightarrow R3 \end{array} \left[\begin{array}{ccc|ccc} -14 & 0 & -4 & -6 & 2 & 0 \\ 0 & 7 & -4 & 1 & 2 & 0 \\ 0 & 0 & 8 & -16 & 10 & 14 \end{array}\right]$$

$$\frac{R3}{8} \longrightarrow R3 \left[\begin{array}{ccc|ccc} -14 & 0 & -4 & -6 & 2 & 0 \\ 0 & 7 & -4 & 1 & 2 & 0 \\ 0 & 0 & 1 & -2 & 1.25 & 1.75 \end{array}\right]$$

$$\begin{array}{c} 4R3 + R2 \longrightarrow R2 \\ \\ 4R3 + R1 \longrightarrow R1 \end{array} \left[\begin{array}{ccc|ccc} -14 & 0 & 0 & -14 & 7 & 7 \\ 0 & 7 & 0 & -7 & 7 & 7 \\ 0 & 0 & 1 & -2 & 1.25 & 1.75 \end{array}\right]$$

Using $\frac{R1}{-14} \longrightarrow R1$ and $\frac{R2}{7} \longrightarrow R2$ produces the inverse matrix $A^{-1} = \begin{bmatrix} 1 & -0.5 & -0.5 \\ -1 & 1 & 1 \\ -2 & 1.25 & 1.75 \end{bmatrix}$.

To verify, we show $AA^{-1} = I$:

$$\begin{bmatrix} 2 & 1 & 0 \\ -1 & 3 & -2 \\ 3 & -1 & 2 \end{bmatrix} \begin{bmatrix} 1 & -0.5 & -0.5 \\ -1 & 1 & 1 \\ -2 & 1.25 & 1.75 \end{bmatrix} = \begin{bmatrix} 1 & 0 & 0 \\ 0 & 1 & 0 \\ 0 & 0 & 1 \end{bmatrix} \checkmark.$$

Exercise 1: Use the preceding inverse and a matrix equation to solve the system

$$\begin{cases} 2x + y = -2 \\ -x + 3y - 2z = -15. \\ 3x - y + 2z = 9 \end{cases}$$ (1, −4, 1)

CUMULATIVE REVIEW CHAPTERS 1–6

Graph each of the following. Include x- and y-intercepts and other important features of each graph.

1. $y = \frac{2}{3}x + 2$
2. $f(x) = |x - 2| + 3$
3. $g(x) = \sqrt{x - 3} + 1$
4. $h(x) = \dfrac{1}{x - 1} + 2$
5. $g(x) = (x - 3)(x + 1)(x + 4)$
6. $y = 2^x + 3$
7. Determine the following for the graph shown to the right. Write all answers in interval notation:
   a. domain
   b. range
   c. interval(s) where $f(x)$ is increasing or decreasing
   d. interval(s) where $f(x)$ is constant
   e. location of any maximum or minimum value(s)
   f. interval(s) where $f(x)$ is positive or negative
   g. the average rate of change using $(-4, 0)$ and $(-2, 3.5)$.

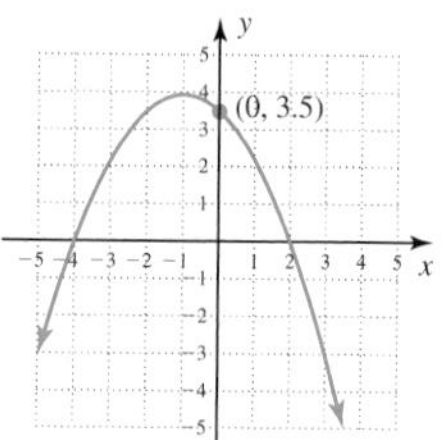

8. Suppose the cost of making a rubber ball is given by $C(x) = 3x + 10$, where x is the number of balls in hundreds. If the revenue from the sale of these balls is given by $R(x) = -x^2 + 123x - 1990$, find the profit function (Profit = Revenue − Cost). How many balls should be produced and sold to obtain the maximum profit? What is this maximum profit?
9. Find all zeroes (real or complex): $g(v) = v^3 - 9v^2 + 2v - 18$.
10. A polynomial has roots $x = 2$ and $x = 2 \pm 3i$. Find the polynomial and write it in standard form.

Given $f(x) = 2x - 5$ and $g(x) = 3x^2 + 2x$ find:

11. $(g - f)(x)$
12. $(fg)(-2)$
13. $(g \circ f)(2)$
14. Calculate the difference quotient for $f(x) = x^2 - 3x$.
15. Use the rational roots theorem to factor the polynomial completely: $x^4 - 6x^3 - 13x^2 + 24x + 36$.

Solve each inequality. Write your answer using interval notation.

16. $x^2 - 3x - 10 < 0$
17. $\dfrac{x - 2}{x + 3} \le 3$
18. Solve each equation.
   a. $\sqrt{x} - 2 = \sqrt{3x + 4}$
   b. $x^{\frac{3}{2}} + 8 = 0$
   c. $2|n + 4| + 3 = 13$
   d. $x^2 - 6x + 13 = 0$
   e. $x^{-2} - 3x^{-1} - 40 = 0$
   f. $4 \cdot 2^{x+1} = \dfrac{1}{8}$
   g. $3^{x-2} = 7$
   h. $\log_3 81 = x$
   i. $\log_3 x + \log_3 (x - 2) = 1$
19. Solve using matrices and row-reduction.
$$\begin{cases} 4x + 3y = 13 \\ -9x + 5y = 6 \end{cases}$$
20. Solve using your calculator and a matrix equation.
$$\begin{cases} x + 2y - z = 0 \\ 2x - 5y + 4z = 6 \\ -x + 3y - 4z = -5 \end{cases}$$

Answers

1. 2.

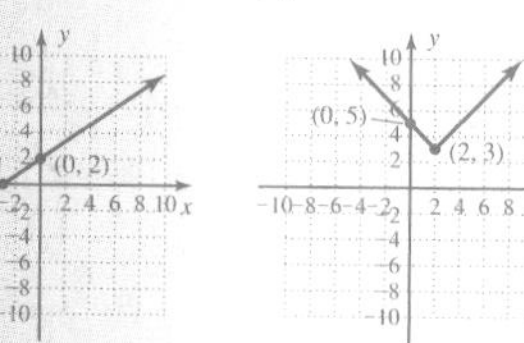

3. 4.

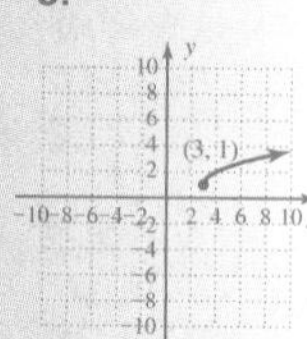

5. 6.

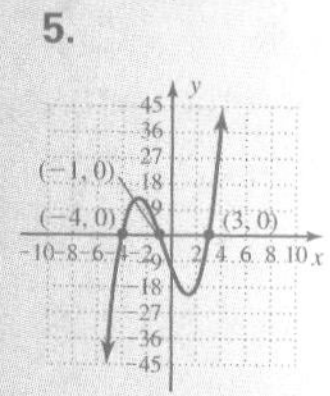

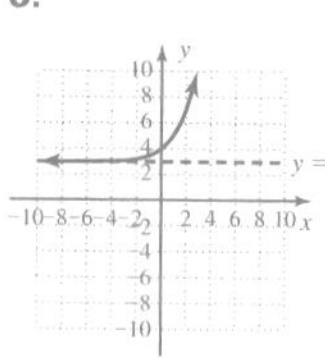

7. a. $D: x \in (-\infty, \infty)$
   b. $R: y \in (-\infty, 4]$
   c. $f(x)\uparrow: x \in (-\infty, 1)$; $f(x)\downarrow: x \in (-1, \infty)$
   d. n/a
   e. max: $(-1, 4)$
   f. $f(x) > 0: x \in (-4, 2)$; $f(x) < 0: x \in (-\infty, -4) \cup (2, \infty)$
   g. $\dfrac{\Delta y}{\Delta x} = \dfrac{7}{4}$
8. $P(x) = -x^2 + 120x - 2000$; 6000; \$1600
9. $x = 9$, $\pm\sqrt{2}\,i$
10. $P(x) = x^3 - 6x^2 + 21x - 26$
11. $3x^2 + 5$
12. -72
13. 1
14. $2x + h - 3$
15. $(x - 2)(x + 2)(x - 3 - 3\sqrt{2})(x - 3 + 3\sqrt{2})$
16. $x \in (-2, 5)$
17. $x \in (-\infty, \frac{-11}{2}] \cup [-3, \infty)$
18. a. no solution
   b. no solution
   c. $n = 1, n = -9$
   d. $x = 3 \pm 2i$
   e. $x = \frac{1}{8}, x = \frac{-1}{5}$
   f. $x = -6$
   g. $x = \dfrac{\ln 7}{\ln 3} + 2$
   h. $x = 4$
   i. $x = 3$
19. $(1, 3)$
20. $(1, 0, 1)$

21. ≈ 9.7 yr

22.

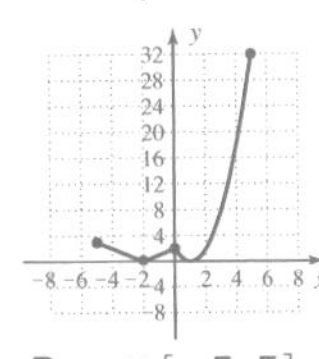

D: $x \in [-5, 5]$
R: $y \in [0, 32]$

23. $\dfrac{3}{x+3} + \dfrac{1}{x-3}$

24.

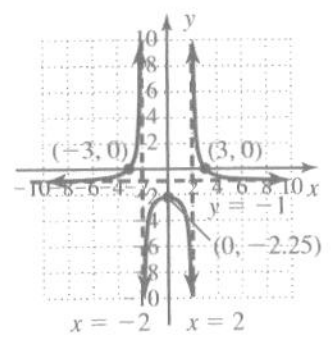

25. a.

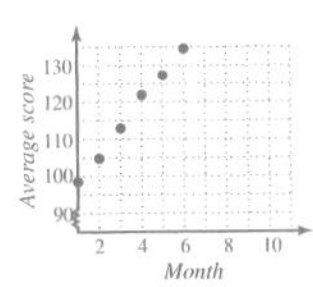

b. $y = 8.14x + 86.67$
c. increase of 8 pts per month
d. 9th month

21. If a person invests \$5000 at 9% compounded continuously, how long until the account grows to \$12,000?

22. Graph the piecewise function shown and state its domain and range.

$$\begin{cases} |x+2| & -5 \le x \le 0 \\ 2(x-1)^2 & 0 < x \le 5 \end{cases}$$

23. Decompose the rational expression into a sum of partial fractions: $\dfrac{4x-6}{x^2-9}$.

24. Graph $h(x) = \dfrac{9-x^2}{x^2-4}$. Give the coordinates of all intercepts and the equation of all asymptotes.

25. Chance's skill at bowling is slowly improving with practice. The table shown gives his average score each month for the past 6 months. Assuming the relationship is linear, use a graphing calculator to:

a. Draw a scatter-plot.

b. Calculate the line of best fit.

c. Round the slope value to the nearest whole and explain its meaning in this context.

d. Predict the month when Chance's average score will exceed 159.

Month	Avg. Score
Jan. (1)	95
Feb. (2)	102
Mar. (3)	111
Apr. (4)	121
May. (5)	127
Jun. (6)	135

Chapter 7

Conic Sections and Nonlinear Systems

Chapter Outline

7.1 The Circle and the Ellipse 682

7.2 The Hyperbola 694

7.3 Nonlinear Systems of Equations and Inequalities 706

7.4 Foci and the Analytic Ellipse and Hyperbola 716

7.5 The Analytic Parabola 729

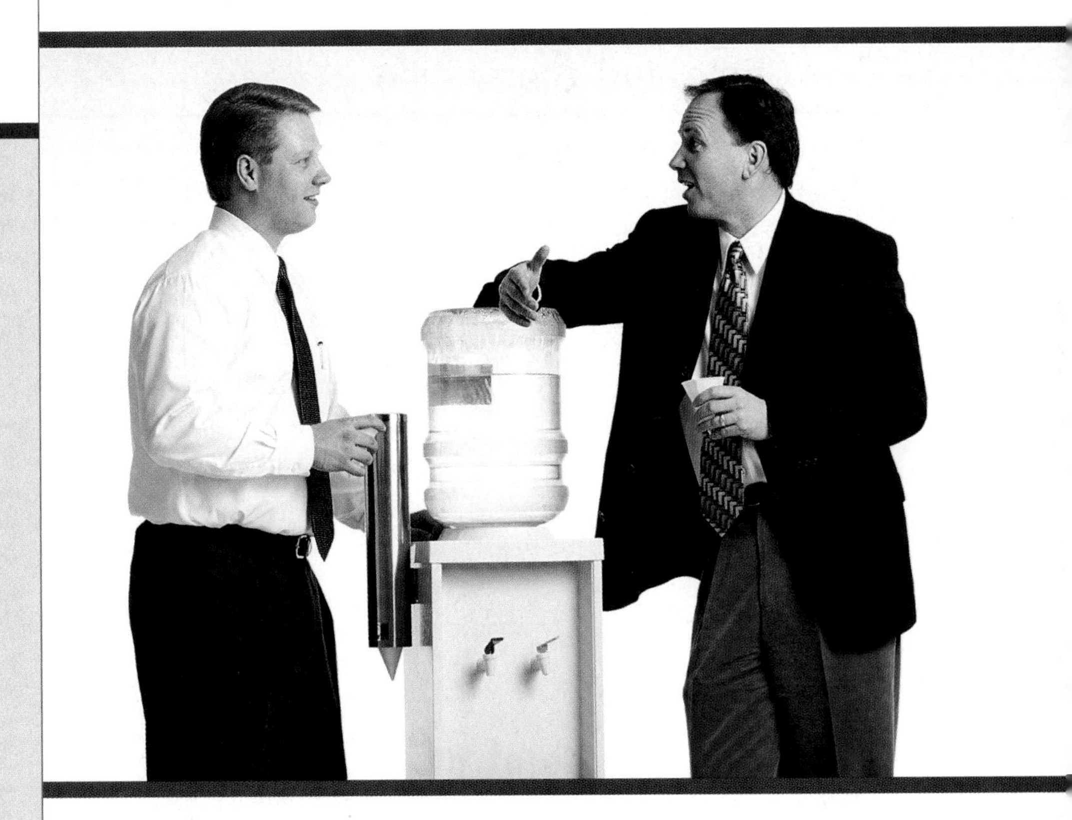

Preview

In this chapter we study a family of curves called the **conic sections.** In common use, a cone brings to mind the cone-shaped paper cups at a water cooler (see photo). The point of the cone is called a **vertex** and the sheet of paper forming the sides is called a **nappe.** Mathematically speaking, a cone can have two nappes that meet at a vertex, and extend infinitely in both directions (Figure 7.1). The conic sections are so named because all curves in the family can be formed by looking at a *section* of the *cone,* or more precisely—the intersection

Figure 7.1

of a plane with a **right circular cone.*** The intersection can be manipulated to form a point or a line, but generally the term *conic* refers to a **circle, ellipse, hyperbola,** or **parabola,** which can also be formed. Each conic section can be represented by a second-degree equation in two variables. In this chapter, we study each of the curves in turn, and expand on some of the ideas from Chapter 6 to solve **nonlinear systems.**

7.1 The Circle and the Ellipse

LEARNING OBJECTIVES

In Section 7.1 you will learn how to:

A. Identify the center-shifted form, polynomial form, and standard form of the equation of a circle and graph central and noncentral circles

B. Identify the center-shifted form, polynomial form, and standard form of the equation of an ellipse and graph central and noncentral ellipses

INTRODUCTION

Consider the different ways a plane can intersect a right circular cone. If the plane is perpendicular to the axis of the cone and does not contain the vertex, a circle is formed (Figure 7.2). The size of the circle depends on the distance of the plane from the vertex. If the plane is slightly tilted from perpendicular, an ellipse is formed (Figure 7.3). In this section we introduce the equation of each conic.

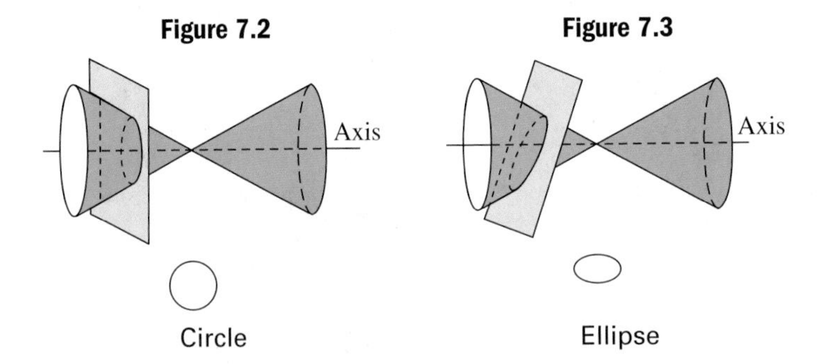

POINT OF INTEREST

Suppose a satellite is orbiting the Earth at an altitude of 200 mi. If the satellite maintains a velocity of approximately 4.8 mi/sec, the orbit will be circular. If the velocity of the satellite is greater or less than 4.8 mi/sec, the orbit becomes elliptical—unless velocity is so small that the satellite returns to Earth, or the velocity is so great that the satellite escapes Earth's gravity.

A. The Equation of a Circle

A circle can be defined as the set of all points in a plane that are a *fixed distance* called the **radius,** from a *fixed point* called the **center.** Since the definition involves *distance,* we can construct the general equation of a circle using the distance formula. Assume the center has coordinates (h, k), and let (x, y) represent any point on the graph. Since the distance between these points must be r, the distance formula yields: $\sqrt{(x - h)^2 + (y - k)^2} = r$.

*A line through the vertex (called the axis) is perpendicular to a circular base.

Center-Shifted Form

Squaring both sides gives $(x - h)^2 + (y - k)^2 = r^2$, sometimes called the **center-shifted form** since the location of the center changes or shifts for different values of h and k.

> **THE EQUATION OF A CIRCLE IN CENTER-SHIFTED FORM**
> An equation written in the form $(x - h)^2 + (y - k)^2 = r^2$ represents a circle of radius r with center at (h, k).

Note the equation contains a sum of squared terms, with radius $\sqrt{r^2} = r$ $(r^2 > 0)$. If $h = 0$ and $k = 0$, the circle is centered at $(0, 0)$ and the graph is a **central circle** with equation $x^2 + y^2 = r^2$. At other values for h or k, the circle shifts horizontally h units and vertically k units with no change in the radius. Note this implies that shifts will be "opposite the sign."

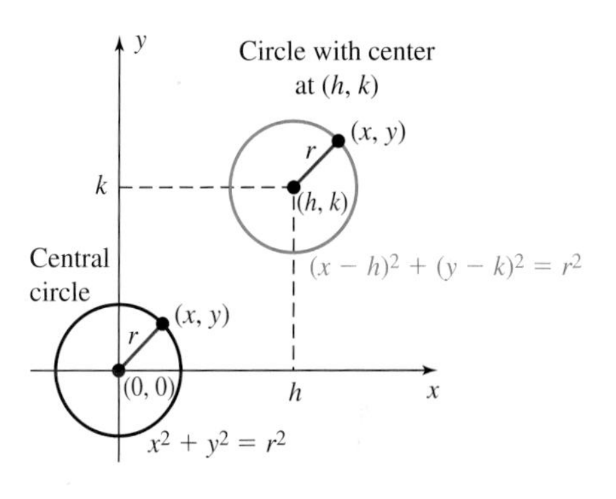

EXAMPLE 1 Find the equation of a circle with center $(0, -1)$ and radius 4, then sketch the graph.

Solution: Since the center is at $(0, -1)$ we have $\mathbf{h} = \mathbf{0}$, $k = -1$, and $r = 4$. Using the center-shifted form $(x - h)^2 + (y - k)^2 = r^2$ we obtain

$$(x - \mathbf{0})^2 + [y - (-1)]^2 = 4^2 \quad \text{substitute 0 for } h, -1 \text{ for } k, \text{ and 4 for } r$$

$$x^2 + (y + 1)^2 = 16 \quad \text{simplify}$$

To graph the circle, it's easiest to begin at $(0, -1)$ and count $r = 4$ units in each horizontal direction, and $r = 4$ units in each vertical direction, knowing the radius is four in *any* direction. Neatly complete the circle by freehand drawing or using a compass. The graph shown is obtained.

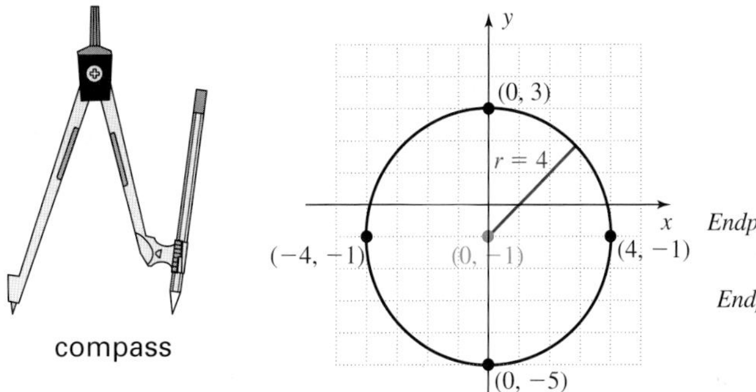

Circle

Center: $(0, -1)$
Radius: $r = 4$
Diameter: $2r = 8$

Endpoints of horizonal diameter
$(-4, -1)$ and $(4, -1)$

Endpoints of vertical diameter
$(0, 3)$ and $(0, -5)$

NOW TRY EXERCISES 7 THROUGH 24

EXAMPLE 2 Graph the circle represented by $(x - 2)^2 + (y + 3)^2 = 12$. Clearly state the center, radius, diameter, and the coordinates of the endpoints of the horizontal and vertical diameters.

Solution: Comparing the given equation with the center-shifted form, we find the center is at $(2, -3)$ and the radius is $r = 2\sqrt{3} \approx 3.5$.

$$(x - h)^2 + (y - k)^2 = r^2 \quad \text{center-shifted form}$$
$$\downarrow \qquad \downarrow \qquad \downarrow$$
$$(x - 2)^2 + (y + 3)^2 = 12 \quad \text{given equation}$$
$$-h = -2 \qquad -k = 3 \qquad r^2 = 12$$
$$h = 2 \qquad k = -3 \qquad r = \sqrt{12} = 2\sqrt{3} \quad \text{radius must be positive}$$

Plot the center $(2, -3)$ and count approximately 3.5 units in the horizontal and vertical directions. Complete the circle by freehand drawing or by using a compass. The graph shown is obtained.

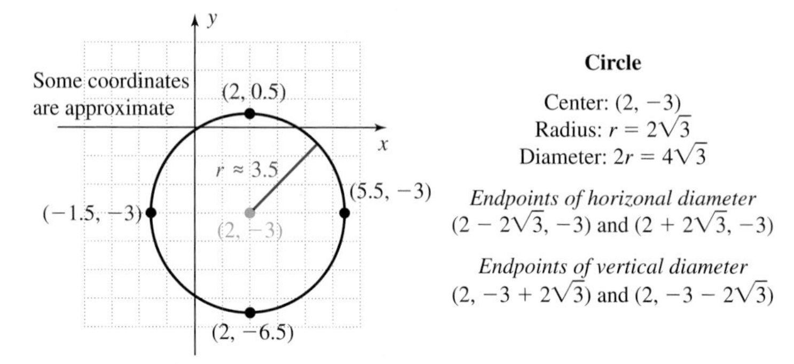

NOW TRY EXERCISES 25 THROUGH 30

Polynomial Form

In Example 2, note the equation is composed of binomial squares in both x and y. Expanding the binomials and collecting like terms places the equation of the circle in **polynomial form:**

$$(x - 2)^2 + (y + 3)^2 = 12 \quad \text{equation in center-shifted form}$$
$$x^2 - 4x + 4 + y^2 + 6y + 9 = 12 \quad \text{expand binomials}$$
$$x^2 + y^2 - 4x + 6y + 1 = 0 \quad \text{combine like terms—polynomial form}$$

For future reference, observe the polynomial form contains a *sum* of second degree terms in x and y, and that *both terms have the same coefficient* (in this case, "1").

Since this form of the equation was derived by squaring binomials, it seems reasonable to assume we can go back to center-shifted form by creating binomial squares in x and y. This is accomplished by *completing the square*.

WORTHY OF NOTE

After writing the equation in center-shifted form, it is possible to end up with a constant that is zero or negative. In the first case, the graph is a single point. In the second case, no graph is possible since roots of the equation will be complex numbers. These are called *degenerate cases*.

EXAMPLE 3 Find the center and radius of the circle whose equation is given, then sketch its graph: $x^2 + y^2 + 2x - 4y - 4 = 0$. Clearly state the center, radius, diameter, and the coordinates of the endpoints of the horizontal and vertical diameters.

Solution: To find the center and radius, we complete the square in both x and y.

$$x^2 + y^2 + 2x - 4y - 4 = 0 \quad \text{given equation}$$
$$(x^2 + 2x + \underline{\quad}) + (y^2 - 4y + \underline{\quad}) = 4 \quad \text{group } x\text{-terms and } y\text{-terms; add 4}$$
$$(x^2 + 2x + 1) + (y^2 - 4y + 4) = 4 + 1 + 4$$

adds 1 to left side — adds 4 to left side — add 1 + 4 to right side

$$(x + 1)^2 + (y - 2)^2 = 9 \quad \text{factor and simplify}$$

The center is at $(-1, 2)$ and the radius is $r = \sqrt{9} = 3$.

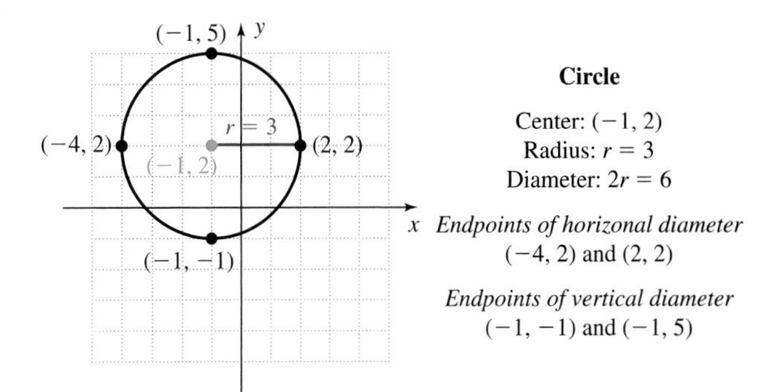

NOW TRY EXERCISES 31 THROUGH 36

Standard Form

The equation of a circle in **standard form** provides a useful link to some of the other conic sections, and is obtained by *setting the equation equal to 1*. In the case of a circle, this means we simply divide by r^2.

$$(x - h)^2 + (y - k)^2 = r^2 \quad \text{center-shifted form}$$

$$\frac{(x - h)^2}{r^2} + \frac{(y - k)^2}{r^2} = 1 \quad \text{divide by } r^2\text{— standard form}$$

In this form, the value of r in each denominator gives the *horizontal* and *vertical* distances, respectively, from the center to the graph. This is not so important in the case of a circle, since this distance is the same in *any* direction. But for other conics, these *horizontal* and *vertical* distances are *not* the same, making the standard form a valuable tool for graphing. To distinguish the horizontal from the vertical distance, r^2 is replaced by a^2 in the "x-term" (horizontal distance), and by b^2 in the "y-term" (vertical distance).

THE EQUATION OF A CIRCLE IN STANDARD FORM

Given $\frac{(x - h)^2}{a^2} + \frac{(y - k)^2}{b^2} = 1$.

If $a = b$ the equation represents the graph of a circle with center at (h, k) and radius $r = a = b$.

- $|a|$ gives the horizontal distance from center to graph.
- $|b|$ gives the vertical distance from center to graph.

For $x^2 + y^2 + 8x - 4 = 0$, we have $(x^2 + 8x + 16) + y^2 = 4 + 16$ or $(x + 4)^2 + y^2 = 20$ in center-shifted form. For standard form, we divide by 20:

$$\frac{(x + 4)^2}{20} + \frac{y^2}{20} = 1 \quad \text{or} \quad \frac{(x + 4)^2}{(2\sqrt{5})^2} + \frac{y^2}{(2\sqrt{5})^2} = 1,$$

where we find $a = b = 2\sqrt{5}$, as expected. See exercises 37 through 42.

B. The Equation of an Ellipse

From our previous work, it seems reasonable to ask, "What happens to the graph when $a \neq b$?" To answer, consider $\frac{(x + 4)^2}{(4)^2} + \frac{y^2}{(3)^2} = 1$, the same equation as before but with $a = 4$ and $b = 3$. The center of the curve is still at $(-4, 0)$ since $h = -4$ and $k = 0$ remain unchanged. For $y = 0$, $(x + 4)^2 = 16$, which gives $x = 0$ and $x = -8$ by inspection: $(0 + 4)^2 = 16$✓ and $(-8 + 4)^2 = 16$✓. This shows the horizontal distance from the center to the graph is $a = 4$, and the points $(-8, 0)$ and $(0, 0)$ are on the graph (see Figure 7.4). For $x = -4$, we have $y^2 = 9$, giving $y = -3$ and $y = 3$ by inspection and showing the vertical distance from the center to the graph is $b = 3$, with points $(-4, 3)$ and $(-4, -3)$ also on the graph. Using this information to sketch the curve reveals the "circle" is elongated and has become an **ellipse.**

Figure 7.4

For an ellipse, the longest distance across the graph is called the **major axis,** with the endpoints of the major axis called **vertices.** The segment perpendicular to and bisecting the major axis with its endpoints on the ellipse is called the **minor axis.** If $a > b$, the major axis is horizontal (parallel to the x-axis) with length $2a$, and the minor axis is vertical with length $2b$. If $b > a$, the major axis is vertical (parallel to the y-axis) with length $2b$, and the minor axis is horizontal with length $2a$ (see Figure 7.5).

Figure 7.5

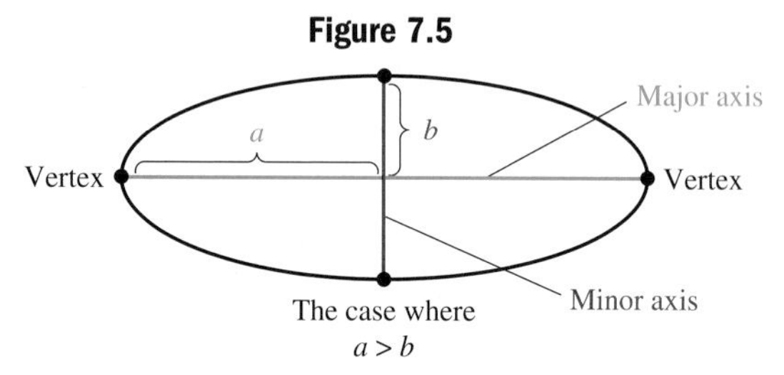

Standard Form

Generalizing this observation we obtain the equation of an ellipse in standard form:

> **THE EQUATION OF AN ELLIPSE IN STANDARD FORM**
>
> Given $\dfrac{(x - h)^2}{a^2} + \dfrac{(y - k)^2}{b^2} = 1$.
>
> If $a \neq b$ the equation represents the graph of an ellipse with center at (h, k).
>
> - $|a|$ gives the horizontal distance from center to graph.
> - $|b|$ gives the vertical distance from center to graph.

EXAMPLE 4 Sketch the graph of $\dfrac{(x - 2)^2}{25} + \dfrac{(y + 1)^2}{4} = 1$.

Solution: Noting $a \neq b$, we have an ellipse with center $(h, k) = (2, -1)$. The horizontal distance from the center to the graph is $a = 5$, and the vertical distance from the center to the graph is $b = 2$. The graph is shown here.

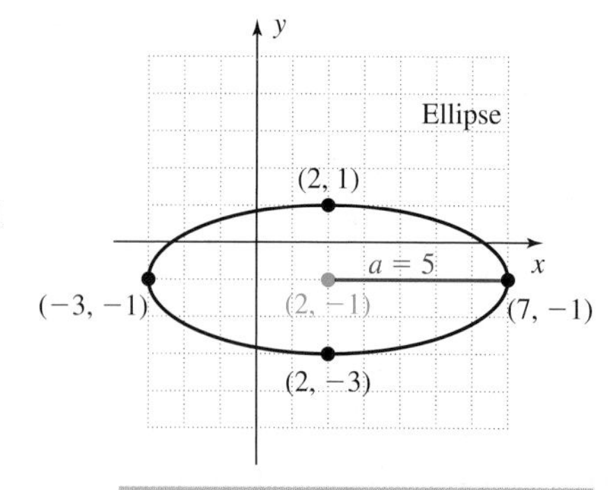

NOW TRY EXERCISES 43 THROUGH 48

Center-Shifted Form

To obtain the center-shifted form of the ellipse from Example 4, we clear the denominators using the least common multiple of 25 and 4, which is 100. This gives

$$(100)\frac{(x-2)^2}{25} + (100)\frac{(y+1)^2}{4} = 1(100)$$
$$4(x-2)^2 + 25(y+1)^2 = 100$$

In this form, we can still identify the center as $(2, -1)$, but now the distinguishing characteristic is that the coefficients of the binomial squares are not equal. The binomials can be expanded to obtain the polynomial form, yielding $4x^2 + 25y^2 - 16x + 50y - 59 = 0$. We still note a sum of second degree terms with unequal coefficients. In general, we have:

> **THE EQUATION OF AN ELLIPSE IN CENTER-SHIFTED FORM**
> An equation written in the form $A(x-h)^2 + B(y-k)^2 = F$, with $A, B, F > 0$, represents an ellipse with center at (h, k).

Note the equation contains a sum of squared terms in x and y. If $A = B$, the equation becomes that of a circle.

EXAMPLE 5 Identify each equation as that of a circle or an ellipse, then find the radius (circles) or the value of a and b (ellipses).

a. $4(x+1)^2 + 4(y-3)^2 = 64$

b. $25(x-2)^2 + 9(y-2)^2 = 225$

c. $9(x-1)^2 + 16(y+2)^2 = 144$

d. $5(x+1)^2 + 5(y+1)^2 = 90$

Solution:

a. Since $A = B$ and A, B, and F have the same sign, this is the equation of a circle. The center is at $(-1, 3)$ and division by 4 (to obtain center-shifted form) shows $r^2 = 16$ so $r = 4$.

b. Since $A \neq B$ and A, B, and F have the same sign, this is the equation of an ellipse. The center is at $(2, 2)$ and division by 225 (to obtain standard form) shows $a = 3$ and $b = 5$.

c. Since $A \neq B$ and A, B, and F have the same sign, the equation is that of an ellipse. The center is at $(1, -2)$ and division by 144 shows $a = 4$ and $b = 3$.

d. Since $A = B$ and A, B, and F have the same sign, this is the equation of a circle. The center is at $(-1, -1)$ and division by 5 shows $r^2 = 18$, so $r = 3\sqrt{2}$.

NOW TRY EXERCISES 49 THROUGH 54

EXAMPLE 6 For $25x^2 + 4y^2 = 100$, (a) write the equation in standard form and identify the center and the value of a and b, (b) identify the major and minor axes and name the vertices, and (c) sketch the graph.

Solution: The coefficients of x^2 and y^2 are unequal, and 25, 4, and 100 have like signs. The equation represents an ellipse with center at (0, 0). To obtain standard form:

a.

$$25x^2 + 4y^2 = 100 \quad \text{given equation}$$

$$\frac{25x^2}{100} + \frac{4y^2}{100} = 1 \quad \text{divide by 100}$$

$$\frac{x^2}{4} + \frac{y^2}{25} = 1 \quad \text{standard form}$$

$$\frac{x^2}{2^2} + \frac{y^2}{5^2} = 1 \quad \text{write denominators in squared form}$$

b. The result shows $a = 2$ and $b = 5$, indicating the major axis will be vertical and the minor axis will be horizontal. With the center at the origin, the x-intercepts will be (2, 0) and (−2, 0), with the vertices (and y-intercepts) at (0, 5) and (0, −5).

c. Plotting these intercepts and sketching the ellipse results in the graph shown.

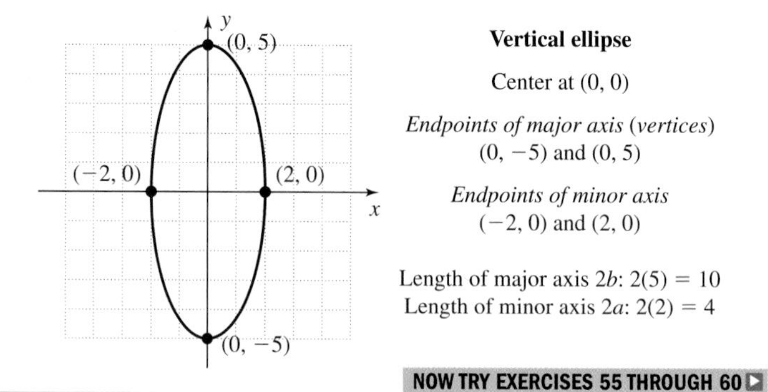

NOW TRY EXERCISES 55 THROUGH 60

If the equation is given in polynomial form, we complete the square in both x and y, then write the equation in standard form to sketch the graph. Figure 7.6 illustrates how the central ellipse and the shifted ellipse are related.

While the center-shifted form is helpful for *identifying* ellipses and provides an intermediate link between the polynomial form and standard form, it is not very helpful when it comes to *graphing* an ellipse. For graphing, the standard form is more useful since it immediately tells us the *distance from the center to the graph in the horizontal and vertical directions.*

Figure 7.6

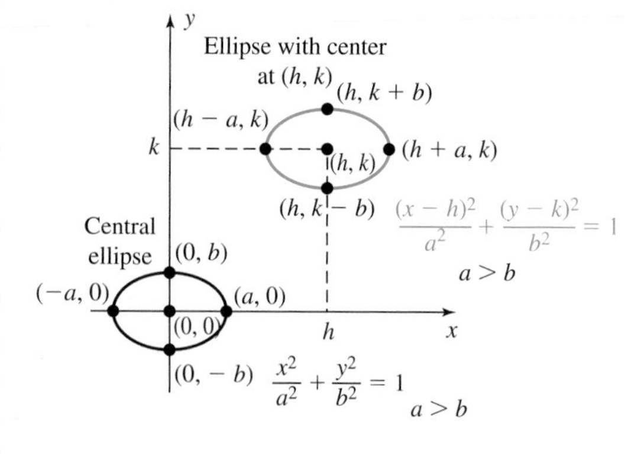

EXAMPLE 7 Sketch the graph of $25x^2 + 4y^2 + 150x - 16y + 141 = 0$.

Solution: The coefficients of x^2 and y^2 are unequal and have like signs, and we assume the equation represents an ellipse but wait until we have the center-shifted form to be certain.

$$25x^2 + 4y^2 + 150x - 16y + 141 = 0 \quad \text{given equation (polynomial form)}$$
$$25x^2 + 150x + 4y^2 - 16y = -141 \quad \text{group like terms; subtract 141}$$
$$25(x^2 + 6x + \_\_) + 4(y^2 - 4y + \_\_) = -141 \quad \text{factor out leading coefficient from each group}$$
$$25(x^2 + 6x + 9) + 4(y^2 - 4y + 4) = -141 + 225 + 16 \quad \text{complete the square; add 225 + 16 to right}$$

adds 25(9) = 225 adds 4(4) = 16

$$25(x + 3)^2 + 4(y - 2)^2 = 100 \quad \text{center-shifted form (an ellipse)}$$
$$\frac{25(x + 3)^2}{100} + \frac{4(y - 2)^2}{100} = \frac{100}{100} \quad \text{divide both sides by 100}$$
$$\frac{(x + 3)^2}{4} + \frac{(y - 2)^2}{25} = 1 \quad \text{simplify (standard form)}$$
$$\frac{(x + 3)^2}{2^2} + \frac{(y - 2)^2}{5^2} = 1 \quad \text{write denominators in squared form}$$

The result is a vertical ellipse with center at $(-3, 2)$, with $a = 2$ and $b = 5$. The vertices are a vertical distance of 5 units from center, and the endpoints of the minor axis are a horizontal distance of 2 units from center.

Note this is the same ellipse as in Example 6, but shifted 3 units left and 2 up.

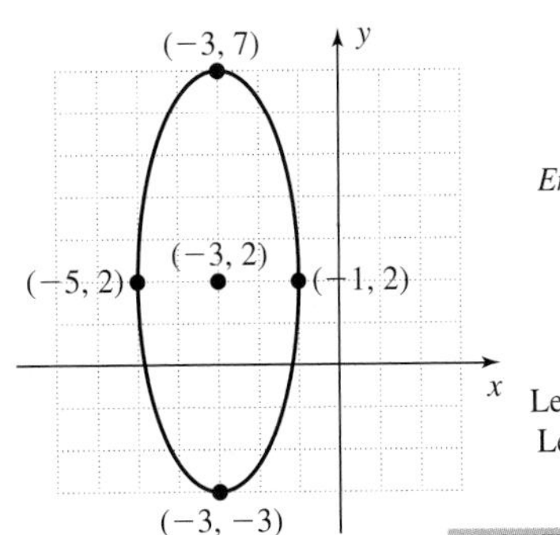

Vertical ellipse

Center at $(-3, 2)$

Endpoints of major axis (vertices) $(-3, -3)$ and $(-3, 7)$

Endpoints of minor axis $(-5, 2)$ and $(-1, 2)$

Length of major axis $2b$: $2(5) = 10$
Length of minor axis $2a$: $2(2) = 4$

NOW TRY EXERCISES 61 THROUGH 68

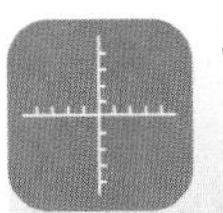

TECHNOLOGY HIGHLIGHT

Using a Graphing Calculator to Study Circles and Ellipses

The keystrokes shown apply to a TI-84 Plus model. Please consult your manual or our Internet site for other models.

When using a graphing calculator to study the conic sections, it is important to keep two things in mind. First, most graphing calculators are only capable of graphing *functions,* which means we must modify the equations of those conic sections that are relations (the circle, ellipse, hyperbola, and horizontal parabola) before they can be graphed using this technology. Second, most standard viewing windows have the x- and y-values preset at $[-10, 10]$ even though the calculator screen may be 94 pixels wide and 64 pixels high. This tends to compress the y-values and give a skewed image of the graph. Consider the *relation* $x^2 + y^2 = 25$. From our work in this section, we know this is the equation of a circle centered at (0, 0) with radius $r = 5$. To enable the calculator to graph this relation, we must define it in two pieces, each of which is a *function,* by solving for y:

$$x^2 + y^2 = 25 \quad \text{original equation}$$
$$y^2 = 25 - x^2 \quad \text{isolate } y^2$$
$$y = \pm\sqrt{25 - x^2} \quad \text{solve for } y$$

Note that we can separate this result into two parts, each of which is a function, enabling the calculator to draw the graph: $Y_1 = \sqrt{25 - x^2}$ gives the "upper half" of the circle, and $Y_2 = -\sqrt{25 - x^2}$, which gives the "lower half." Enter these on the Y= screen (note that $Y_2 = -Y_1$ can be used instead of reentering the entire expression: VARS ▶ ENTER). But if we graph Y_1 and Y_2 on the standard screen, the result appears more elliptical than circular (Figure 7.7). One way to fix this (there are other ways), is to use the ZOOM **5:ZSquare** option, which places the tick marks equally spaced on both axes, instead of trying to force both to display points from -10 to 10. Accessing this option by using the keystrokes ZOOM 5 gives the final result shown (Figure 7.8). Although it is a much improved graph, the circle does not appear "closed" as the calculator lacks sufficient pixels to show the proper curvature. A second alternative is to manually set a friendly window (see the Technology Highlight, Section 2.5). Using Xmin = −9.4, Xmax = 9.4, Ymin = −6.2, and Ymax = 6.2 will generate a better graph, which we can use to study the relation more closely. Note that we can jump between the upper and lower halves of the circle using the up ▲ or down ▼ arrows.

Figure 7.7

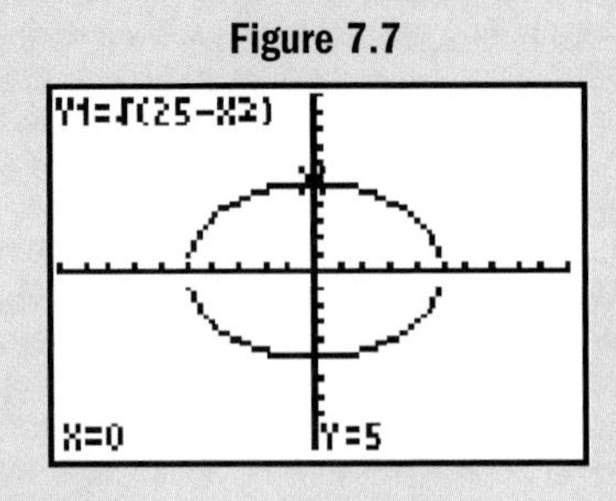

Figure 7.8

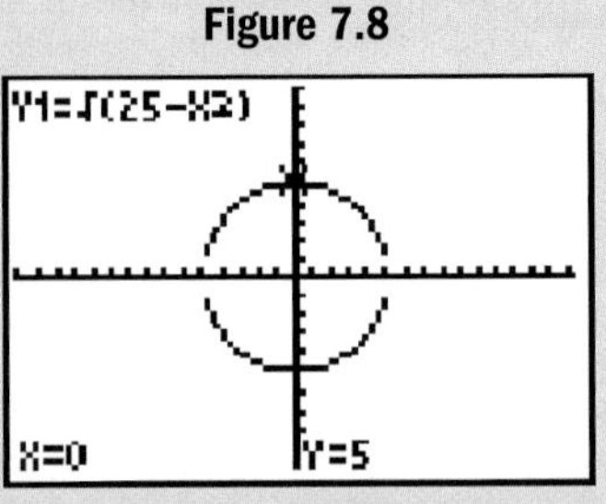

Exercise 1: Use these ideas to graph the circle defined by $x^2 + y^2 = 36$ using a friendly window. Then use the TRACE feature to find the value of y when $x = 3.6$. Now find the value of y when $x = 4.8$. Explain why the values seem "interchangeable." $y = \pm 4.8$; $y = \pm 3.6$

Exercise 2: Use these ideas to graph the ellipse defined by $4x^2 + y^2 = 36$ using a friendly window. Then use the TRACE feature to find the value of the x- and y-intercepts. $(-3, 0), (3, 0), (0, 6), (0, -6)$

7.1 EXERCISES

CONCEPTS AND VOCABULARY

Fill in the blank with the appropriate word or phrase. Carefully reread the section if needed.

1. A circle is defined to be the set of all points an equal distance, called the radius, from a given point, called the center.

2. For $x^2 + y^2 = r^2$, the center of the circle is at (0, 0) and the length of the radius is r. The graph is called a(n) central circle.

3. To write the equation $x^2 + y^2 - 6x = 7$ in standard form, complete the square in x, then set the equation equal to 1.

4. The longest distance across an ellipse is called the major axis and the endpoints are called vertices.

5. Explain/discuss how the relations $a > b$, $a = b$, and $a < b$ affect the graph of a conic section with equation $\frac{(x-h)^2}{a^2} + \frac{(y-k)^2}{b^2} = 1$. Include several illustrative examples. Answers will vary.

6. Compare/contrast the center-shifted, polynomial, and standard forms of the equations discussed in this section. Discuss times when one form may be more useful than another (include examples). Answers will vary.

DEVELOPING YOUR SKILLS

Find the equation of a circle satisfying the conditions given, then sketch its graph.

7. center (0, 0), radius 3 $x^2 + y^2 = 9$

8. center (0, 0), radius 6 $x^2 + y^2 = 36$

9. center (5, 0), radius $\sqrt{3}$ $(x - 5)^2 + y^2 = 3$

10. center (0, 4), radius $\sqrt{5}$ $x^2 + (y - 4)^2 = 5$

7. 8.

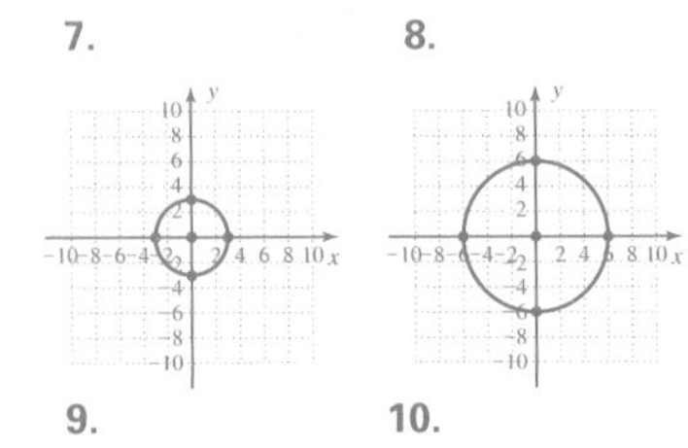

9. 10.

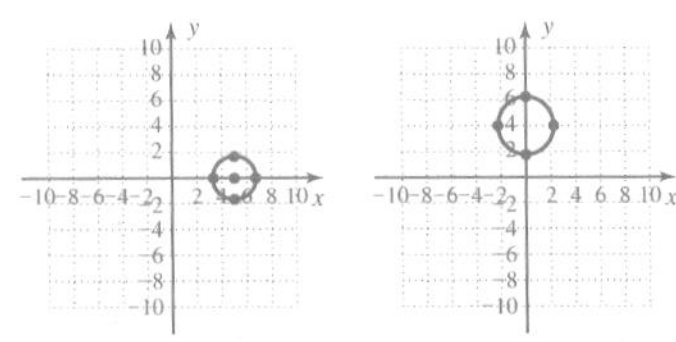

11.

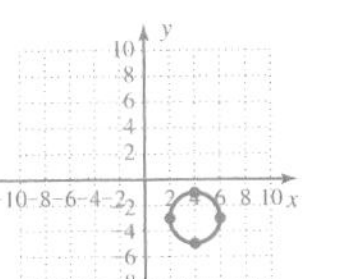

12.

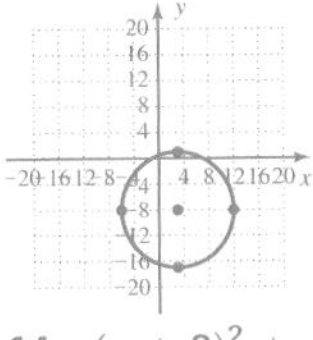

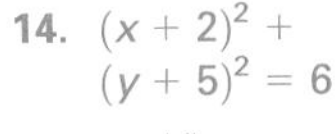

13. $(x + 7)^2 + (y + 4)^2 = 7$

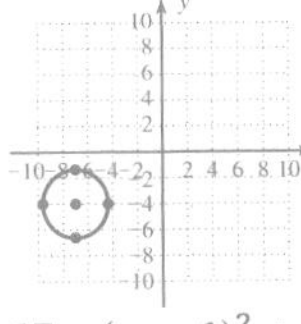

14. $(x + 2)^2 + (y + 5)^2 = 6$

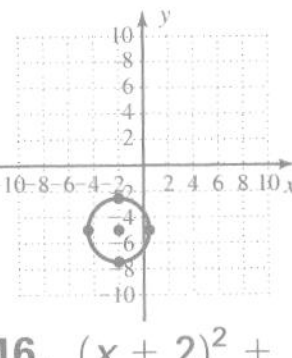

15. $(x - 1)^2 + (y + 2)^2 = 12$

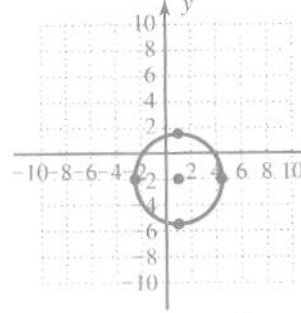

16. $(x + 2)^2 + (y - 3)^2 = 18$

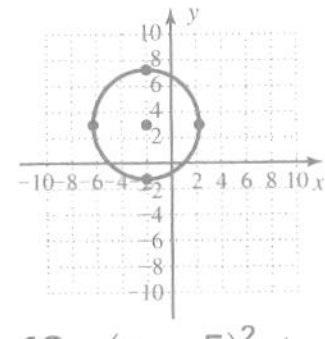

17. $(x - 4)^2 + (y - 5)^2 = 12$

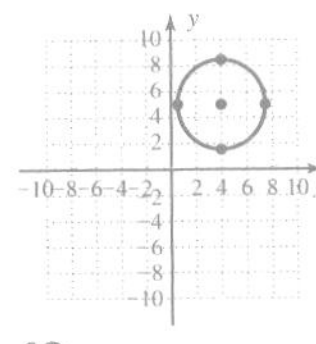

18. $(x - 5)^2 + (y - 1)^2 = 20$

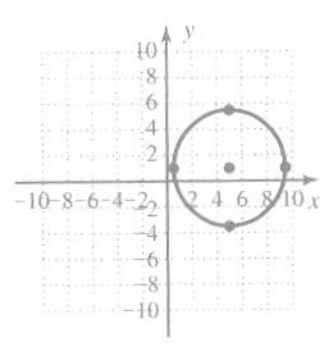

19. 20.

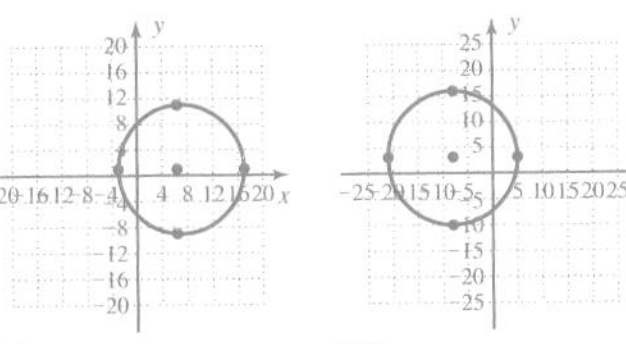

21. 22.

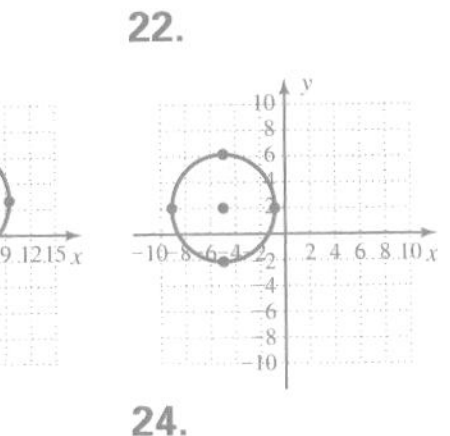

23. 24.

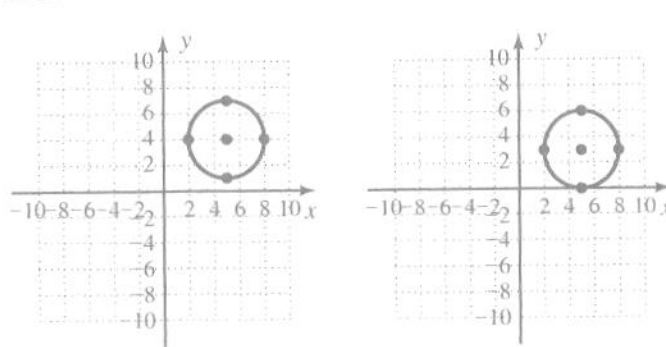

Additional answers can be found in the Instructor Answer Appendix.

11. center $(4, -3)$, radius 2 $(x - 4)^2 + (y + 3)^2 = 4$

12. center $(3, -8)$, radius 9 $(x - 3)^2 + (y + 8)^2 = 81$

13. center $(-7, -4)$, radius $\sqrt{7}$

14. center $(-2, -5)$, radius $\sqrt{6}$

15. center $(1, -2)$, radius $2\sqrt{3}$

16. center $(-2, 3)$, radius $3\sqrt{2}$

17. center $(4, 5)$, diameter $4\sqrt{3}$

18. center $(5, 1)$, diameter $4\sqrt{5}$

19. center at $(7, 1)$, graph contains the point $(1, -7)$ $(x - 7)^2 + (y - 1)^2 = 100$

20. center at $(-8, 3)$, graph contains the point $(-3, 15)$ $(x + 8)^2 + (y - 3)^2 = 169$

21. center at $(3, 4)$, graph contains the point $(7, 9)$ $(x - 3)^2 + (y - 4)^2 = 41$

22. center at $(-5, 2)$, graph contains the point $(-1, 3)$ $(x + 5)^2 + (y - 2)^2 = 17$

23. diameter has endpoints $(5, 1)$ and $(5, 7)$ $(x - 5)^2 + (y - 4)^2 = 9$

24. diameter has endpoints $(2, 3)$ and $(8, 3)$ $(x - 5)^2 + (y - 3)^2 = 9$

Identify the center and radius of each circle, then graph. Also state the domain and range of the relation.

25. $(x - 2)^2 + (y - 3)^2 = 4$ $(2, 3), r = 2, x \in [0, 4], y \in [1, 5]$

26. $(x - 5)^2 + (y - 1)^2 = 9$ $(5, 1), r = 3, x \in [2, 8], y \in [-2, 4]$

27. $(x + 1)^2 + (y - 2)^2 = 12$

28. $(x - 7)^2 + (y + 4)^2 = 20$

29. $(x + 4)^2 + y^2 = 81$ $(-4, 0), r = 9, x \in [-13, 5], y \in [-9, 9]$

30. $x^2 + (y - 3)^2 = 49$ $(0, 3), r = 7, x \in [-7, 7], y \in [-4, 10]$

Write each equation in center-shifted form to find the center and radius of the circle. Then sketch the graph.

31. $x^2 + y^2 - 10x - 12y + 4 = 0$ $(x - 5)^2 + (y - 6)^2 = 57, (5, 6), r = \sqrt{57}$

32. $x^2 + y^2 + 6x - 8y - 6 = 0$ $(x + 3)^2 + (y - 4)^2 = 31, (-3, 4), r = \sqrt{31}$

33. $x^2 + y^2 - 10x + 4y + 4 = 0$

34. $x^2 + y^2 + 6x + 4y + 12 = 0$

35. $x^2 + y^2 + 6y - 5 = 0$ $x^2 + (y + 3)^2 = 14, (0, -3), r = \sqrt{14}$

36. $x^2 + y^2 - 8x + 12 = 0$ $(x - 4)^2 + y^2 = 4, (4, 0), r = 2$

Write each equation in center-shifted form to find the center and radius of the circle, and sketch its graph. Then use the equation in standard form (set equal to 1) to identify the values of a and b. What do you notice about a, b, and r?

37. $x^2 + y^2 + 4x + 10y + 18 = 0$

38. $x^2 + y^2 - 8x - 14y - 47 = 0$

39. $x^2 + y^2 + 14x + 12 = 0$

40. $x^2 + y^2 - 22y - 5 = 0$

41. $2x^2 + 2y^2 - 12x + 20y + 4 = 0$ $(x - 3)^2 + (y + 5)^2 = 32, (3, -5), r = 4\sqrt{2}, a = 4\sqrt{2}, b = 4\sqrt{2}$; they are equal

42. $3x^2 + 3y^2 - 24x + 18y + 3 = 0$ $(x - 4)^2 + (y + 3)^2 = 24, (4, -3); r = 2\sqrt{6}; a = 2\sqrt{6}; b = 2\sqrt{6}$; they are equal

Sketch the graph of each ellipse.

43. $\dfrac{(x - 1)^2}{9} + \dfrac{(y - 2)^2}{16} = 1$

44. $\dfrac{(x - 3)^2}{4} + \dfrac{(y - 1)^2}{25} = 1$

45. $\dfrac{(x - 2)^2}{25} + \dfrac{(y + 3)^2}{4} = 1$

46. $\dfrac{(x + 5)^2}{1} + \dfrac{(y - 2)^2}{16} = 1$

47. $\dfrac{(x + 1)^2}{16} + \dfrac{(y + 2)^2}{9} = 1$

48. $\dfrac{(x + 1)^2}{36} + \dfrac{(y + 3)^2}{9} = 1$

Identify each equation as that of an ellipse or circle, then sketch its graph.

49. $(x + 1)^2 + 4(y - 2)^2 = 16$ ellipse

50. $9(x - 2)^2 + (y + 3)^2 = 36$ ellipse

51. $2(x - 2)^2 + 2(y + 4)^2 = 18$ circle

52. $(x - 6)^2 + y^2 = 49$ circle

53. $4(x - 1)^2 + 9(y - 4)^2 = 36$ ellipse

54. $25(x - 3)^2 + 4(y + 2)^2 = 100$ ellipse

For each exercise, (a) write the equation in standard form then identify the center and the values of a and b, (b) state the coordinates of the vertices and the coordinates of the endpoints of the minor axis, and (c) sketch the graph.

55. $x^2 + 4y^2 = 16$

56. $9x^2 + y^2 = 36$

57. $16x^2 + 9y^2 = 144$

58. $25x^2 + 9y^2 = 225$

59. $2x^2 + 5y^2 = 10$

60. $3x^2 + 7y^2 = 21$

61. $x^2 + \frac{(y+3)^2}{4} = 1$

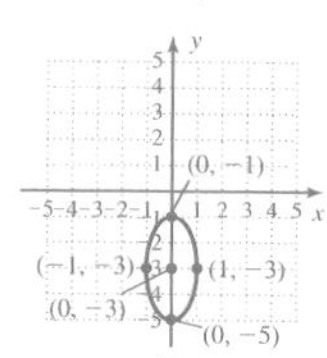

62. $\frac{(x+4)^2}{9} + \frac{y^2}{3} = 1$

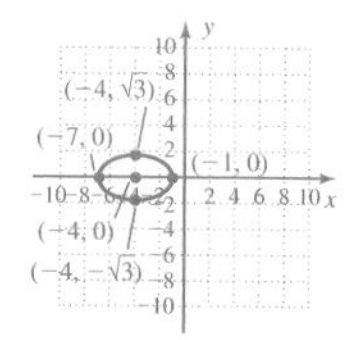

63. $\frac{(x+2)^2}{16} + \frac{(y-1)^2}{4} = 1$

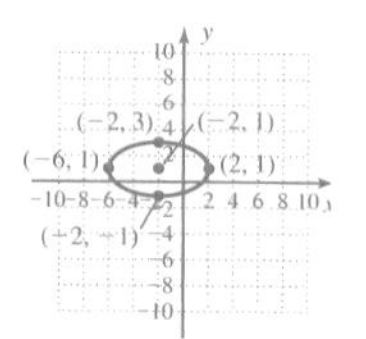

64. $\frac{(x+2)^2}{12} + \frac{(y-4)^2}{36} = 1$

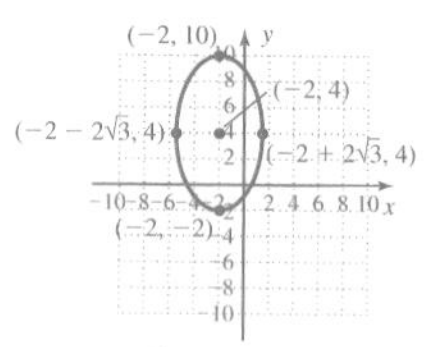

65. $\frac{(x-3)^2}{4} + \frac{(y+5)^2}{10} = 1$

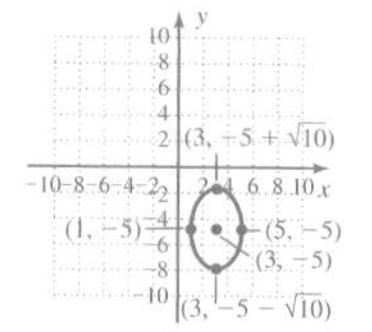

66. $\frac{(x-2)^2}{9} + \frac{(y+1)^2}{4} = 1$

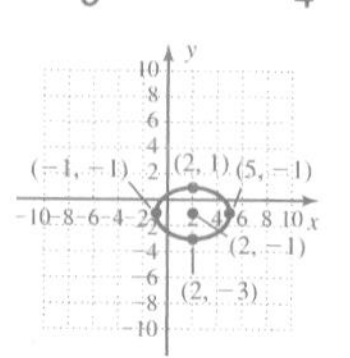

Additional answers can be found in the Instructor Answer Appendix.

Complete the square in both x and y to write each equation in standard form. Then draw a complete graph of the relation and identify all important features.

61. $4x^2 + y^2 + 6y + 5 = 0$

62. $x^2 + 3y^2 + 8x + 7 = 0$

63. $x^2 + 4y^2 - 8y + 4x - 8 = 0$

64. $3x^2 + y^2 - 8y + 12x - 8 = 0$

65. $5x^2 + 2y^2 + 20y - 30x + 75 = 0$

66. $4x^2 + 9y^2 - 16x + 18y - 11 = 0$

67. $2x^2 + 5y^2 - 12x + 20y - 12 = 0$

68. $6x^2 + 3y^2 - 24x + 18y - 3 = 0$

WORKING WITH FORMULAS

69. Area of an inscribed square: $A = 2r^2$

The area of a square inscribed in a circle is found by using the formula given where r is the radius of the circle. Find the area of the inscribed square shown. $A = 50$ units^2

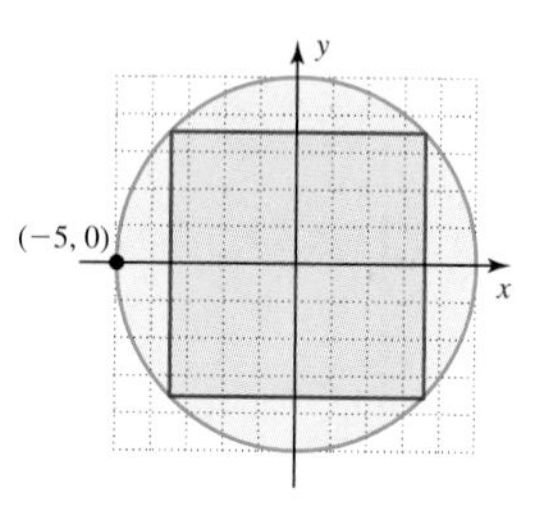

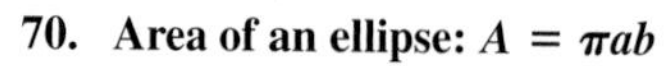

70. Area of an ellipse: $A = \pi ab$

The area of an ellipse is given by the formula shown, where a is the distance from the center to the graph in the horizontal direction and b is the distance from center to graph in the vertical direction. Find the area of the ellipse defined by $16x^2 + 9y^2 = 144$. $A = 12\pi$ units^2

APPLICATIONS

71. Radar detection: The radar on a luxury liner has a range of 25 nautical miles in any direction. If this ship is located at coordinates (5, 12) can the radar pick up its sister ship located at coordinates (15, 36)? Assume coordinates indicate a location in nautical miles from (0, 0). No

72. Earthquake movement: If the epicenter (point of origin) of an earthquake was located at map coordinates (3, 7) and the quake could be felt up to 12 mi away, would a person located at (13, 1) have felt the quake? Assume coordinates indicate a location in statute miles from (0, 0). Yes

73. Inscribed circle: Find the equation for both the red and blue circles, then find the area of the region shaded in blue. Red: $(x-2)^2 + (y-2)^2 = 4$; Blue: $(x-2)^2 + y^2 = 16$; Area blue $= 12\pi$ units^2

Exercise 73

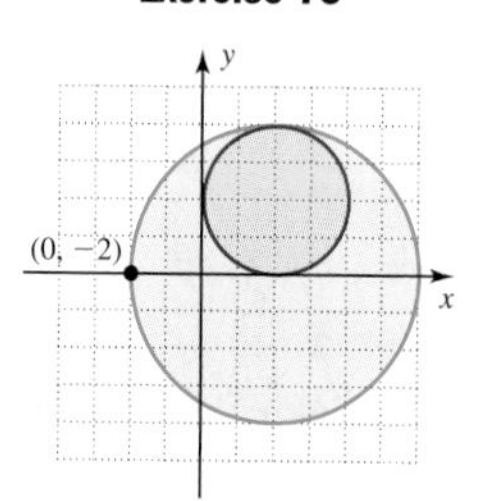

Exercise 74

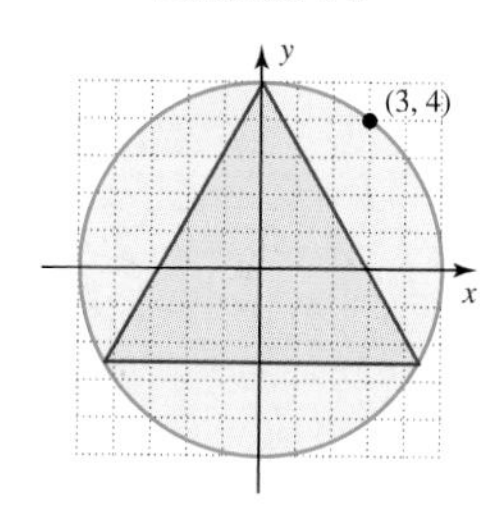

74. Inscribed triangle: The area of an equilateral triangle inscribed in a circle is given by the formula $A = \frac{3\sqrt{3}}{4}r^2$, where r is the radius of the circle. Find the area of the equilateral triangle shown. $\frac{75\sqrt{3}}{4}$ units^2

75. Radio broadcast range: Two radio stations may not use the same frequency if their broadcast areas *overlap*. Suppose station KXRQ has a broadcast area bounded by $x^2 + y^2 + 8x - 6y = 0$ and WLRT has a broadcast area bounded by $x^2 + y^2 - 10x + 4y = 0$. Graph the circle representing each broadcast area on the same grid to determine if both stations may broadcast on the same frequency. No, distance between centers is less than sum of radii.

76. Radio broadcast range: The emergency radio broadcast system is designed to alert the population by relaying an emergency signal to all points of the country. A signal is sent from a station whose broadcast area is bounded by $x^2 + y^2 = 2500$ (x and y in miles) and the signal is picked up and relayed by a transmitter with range $(x - 20)^2 + (y - 30)^2 = 900$. Graph the circle representing each broadcast area on the same grid to determine the greatest distance from the original station that this signal can be received. Be sure to scale the axes appropriately. ≈ 66 mi

As a planet orbits around the Sun, it traces out an ellipse. If the center of the ellipse were placed at (0, 0) on a coordinate grid, the Sun would be actually off-centered (located at a point called the *focus* of the ellipse). Use this information and the graphs provided to complete Exercises 77 and 78.

77. **Orbit of Mercury:** The approximate orbit of the planet Mercury is shown in the figure given. Find an equation that models this orbit. $\frac{x^2}{36^2} + \frac{y^2}{(35.25)^2} = 1$

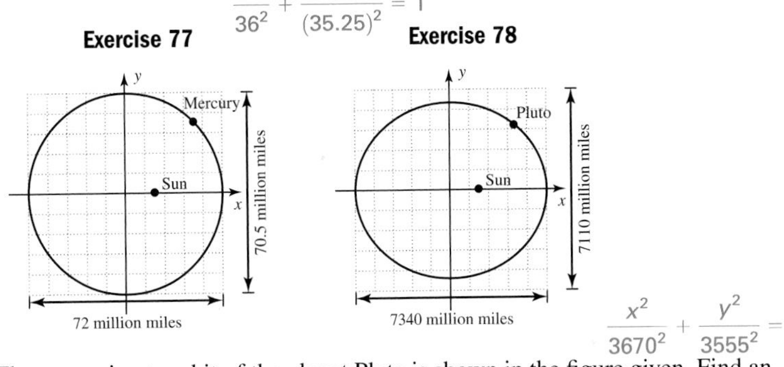

$\frac{x^2}{3670^2} + \frac{y^2}{3555^2} = 1$

78. **Orbit of Pluto:** The approximate orbit of the planet Pluto is shown in the figure given. Find an equation that models this orbit.

79. **Race track area:** Suppose the *Coronado 500* is a car race that is run on an elliptical track. The track is bounded by two ellipses with equations of $4x^2 + 9y^2 = 900$ and $9x^2 + 25y^2 = 900$, where x and y are in hundreds of yards. Use the formula given in Exercise 70 to find the area of the race track. 9000π yd^2

80. **Area of a border:** The tablecloth for a large oval table is elliptical in shape. It is designed with two concentric ellipses (one within the other), as shown in the figure. The equation of the outer ellipse is $9x^2 + 25y^2 = 225$, and the equation of the inner ellipse is $4x^2 + 16y^2 = 64$ with x and y in feet. Use the formula given in Exercise 70 to find the area of the border of the tablecloth. 7π ft^2

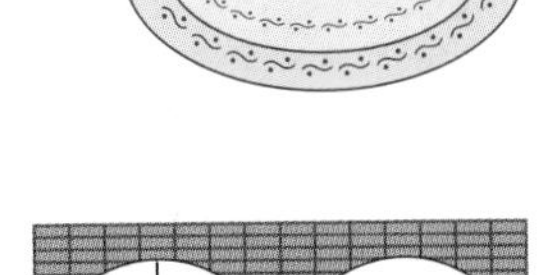

81. **Elliptical arches:** In some situations, bridges are built using uniform elliptical archways, as shown in the figure. Find the equation of the ellipse forming each arch if it has a total width of 30 ft and a maximum center height (above level ground) of 8 ft. What is the height of a point 9 ft to the right of the center of the ellipse? $\frac{x^2}{15^2} + \frac{y^2}{8^2} = 1$, 6.4 ft

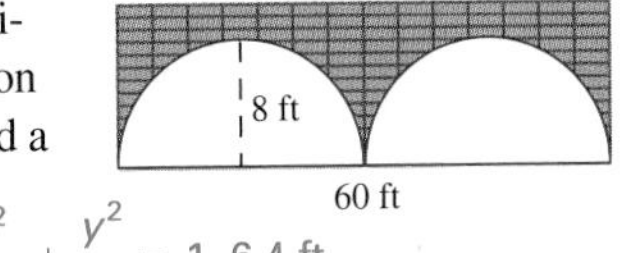

82. **Elliptical arches:** An elliptical arch bridge is built across a one-lane highway. The arch is 20 ft across and has a maximum center height of 12 ft. Will a farm truck hauling a load 10 ft wide with a clearance height of 11 ft be able to go through the bridge without damage? (*Hint:* See Exercise 81.) no

WRITING, RESEARCH, AND DECISION MAKING

83. In Exercises 77 and 78, we saw that the orbits of the planets around the Sun are elliptical in shape. The maximum distance from a planet to the Sun is called the *aphelion,* and the maximum distance from the center of the orbit is called the *semimajor axis.* Use a reference book to find the aphelion of the planet Mars and the length of its semimajor axis.

83. Answers may vary; aphelion: 155 million miles semi-major axis: 142 million miles

84. You have likely heard that Pluto is the farthest planet from the Sun, but may be surprised to hear that this is not always true. Do some reading and research on the orbits of the outer planets and try to determine why. How often does the phenomenon occur? How long does it last? Answers will vary.

85. Attempt to find the x- and y-intercepts for the relation defined by $-9x^2 + 16y^2 = 144$. What happens? Why is this *not* the equation of an ellipse? x^2 and y^2 must have the same sign.

88. a. $xy(1 + 4x)(3 - 2y)$
b. $2x(2x - 5y)^2$
c. $c(11a + 16b)(11a - 16b)$

90.
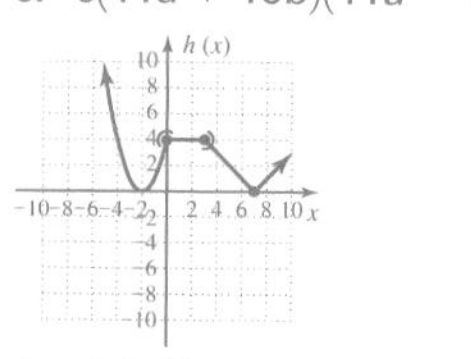

91. $1 - 2i\sqrt{3}$; complex roots must occur in conjugate pairs

92. a. logistic
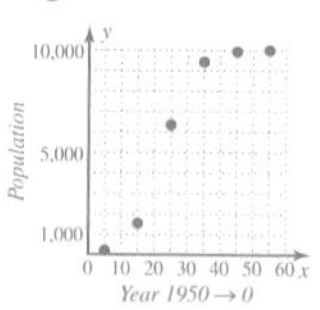

b. $f(x) = \dfrac{10,007}{1 + 159e^{-0.225x}}$
c. $x \approx 29$; 1979

93. a. $D: x \in (-\infty, \infty)$
$R: y \in (-\infty, 4]$
b. $f(x) \geq 0$ for $x \in [-3, -1] \cup [1, 5]$
c. max: $(-2, 2)$ $(3, 4)$
min: $(0, -1)$
d. $f(x)\uparrow$ for $x \in (-\infty, -2) \cup (0, 3)$; $f(x)\downarrow$ for $x \in (-2, 0) \cup (3, \infty)$

EXTENDING THE CONCEPT

86. A circle centered at (3, 4) is tangent to the y-axis. Find all values of y that satisfy $(1, y)$ for this circle. $y = 4 \pm \sqrt{5}$

87. Find the equation of the ellipse that passes through the four points $(-1, 1)$, $(5, 1)$, $(2, 3)$ and $(2, -1)$. $\frac{(x-2)^2}{9} + \frac{(y-1)^2}{4} = 1$

MAINTAINING YOUR SKILLS

88. (R.4) Factor the following expressions:

a. $3xy + 12x^2y - 2xy^2 - 8x^2y^2$

b. $8x^3 - 40x^2y + 50xy^2$

c. $121a^2c - 256cb^2$

89. (4.2) Use synthetic division and the remainder theorem to determine which value of x is a zero of $f(x) = x^4 - 4x^3 - 7x^2 + 22x + 24$.
$x = 1$ $x = 2$ $x = 3$ $x = -4$
$x = 3$ is a zero

90. (3.7) Graph the piecewise-defined function:

$$h(x) = \begin{cases} (x+2)^2 & x < 0 \\ 4 & 0 \leq x \leq 3. \\ |x - 7| & x > 3 \end{cases}$$

91. (4.3) Use substitution to verify that $z = 1 + 2i\sqrt{3}$ is a zero of $f(x) = x^3 - 4x^2 + 17x - 26$. What other complex number must also be a solution? Why?

92. (5.6) For the table of values given:

a. Draw the scatter-plot and select an appropriate form of regression.

b. Find a regression function $f(x)$.

c. Use the equation to solve $f(x) = 8100$.

Year 1950 → 0	Population
5	175
15	1550
25	6350
35	9425
45	9950
55	10,000

93. (3.8) For the graph of $f(x)$ given:

a. State the domain and range.

b. Name intervals where $f(x) \geq 0$.

c. Name the local maximums and/or minimums.

d. Name intervals where $f(x)\uparrow$ and $f(x)\downarrow$.

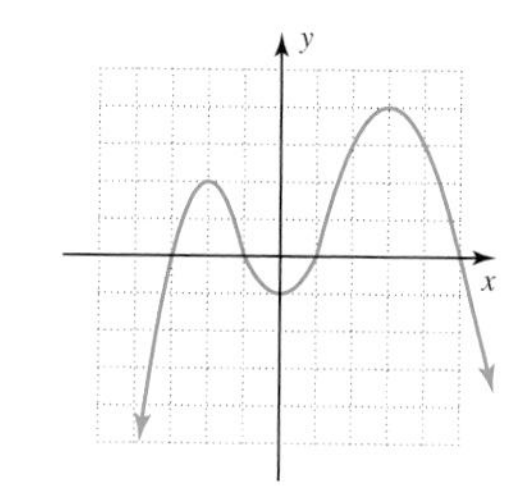

7.2 The Hyperbola

LEARNING OBJECTIVES

In Section 7.2 you will learn how to:

A. Identify the center-shifted form, polynomial form, and standard form of the equation of a hyperbola and graph central and noncentral hyperbolas

B. Distinguish between the equations of a circle, ellipse, and hyperbola

INTRODUCTION

As shown in Figure 7.9, a hyperbola is a conic section formed by a plane that cuts both nappes of a right circular cone (the plane need not be parallel to the axis). A hyperbola has two symmetric parts called **branches,** which open in opposite directions. Although the branches appear to resemble parabolas, we will soon discover they are actually a very different curve.

Figure 7.9

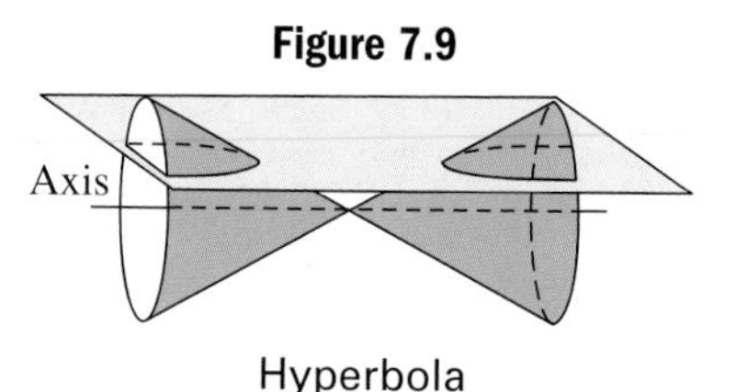

Hyperbola

POINT OF INTEREST

By comparing the orbits of a number of earlier comets, the British astronomer Edmond Halley (1656–1742) showed the great comet of 1682 to be the same as those that had appeared in 1607 and 1531, and successfully predicted the return of the comet in 1759. Earlier appearances of Halley's comet have now been identified from records dating as early as 240 B.C. Most comets have elliptical orbits, and the periods (the time they take to orbit the Sun) of about 200 comets have been calculated. Comets that have a very high velocity and a large mass cannot be captured by the Sun's gravitational field, and the path of these comets trace out one branch of a hyperbola.

A. The Center-Shifted, Polynomial, and Standard Form of the Equation of a Hyperbola

In Section 7.1, we noted the equation $A(x - h)^2 + B(y - k)^2 = F\ (A, B, F > 0)$, could be used to describe the equation of both a circle and an ellipse. If $A = B$, the equation is that of a circle, if $A \neq B$, the equation represents an ellipse. Both cases contain a *sum* of second-degree terms. Perhaps driven by curiosity, we might wonder what happens if the equation has a *difference* of second-degree terms. As you'll see, the result is noteworthy.

Center-Shifted Form

Consider the equation $9x^2 - 16y^2 = 144$. It appears the graph will be centered at (0, 0) since no shifts are applied (h and k are both zero). Using the intercept method to graph this equation reveals an entirely new curve, called a *hyperbola.*

EXAMPLE 1 Graph the equation $9x^2 - 16y^2 = 144$ using intercepts and the center-shifted form.

Solution:

$$9x^2 - 16y^2 = 144 \quad \text{given}$$
$$9(0)^2 - 16y^2 = 144 \quad \text{substitute 0 for } x$$
$$-16y^2 = 144 \quad \text{simplify}$$
$$y^2 = -9 \quad \text{divide by } -16$$

Since y^2 can never be negative, we conclude that the graph has *no y-intercepts.* Substituting $y = 0$ to find the x-intercepts gives

$$9x^2 - 16y^2 = 144 \quad \text{given}$$
$$9x^2 - 16(0)^2 = 144 \quad \text{substitute 0 for } y$$
$$9x^2 = 144 \quad \text{simplify}$$
$$x^2 = 16 \quad \text{divide by 9}$$
$$x = \sqrt{16} \text{ and } x = -\sqrt{16} \quad \text{square root property}$$
$$x = 4 \quad \text{and} \quad x = -4 \quad \text{simplify}$$
$$(4, 0) \quad \text{and} \quad (-4, 0) \quad x\text{-intercepts}$$

Knowing the graph has no y-intercepts, we select inputs greater than 4 and less than -4 to help sketch the graph. Using $x = -5$ and $x = 5$ yields

$9x^2 - 16y^2 = 144$	given	$9x^2 - 16y^2 = 144$
$9(5)^2 - 16y^2 = 144$	substitute for x	$9(-5)^2 - 16y^2 = 144$
$9(25) - 16y^2 = 144$	$5^2 = (-5)^2 = 25$	$9(25) - 16y^2 = 144$
$225 - 16y^2 = 144$	simplify	$225 - 16y^2 = 144$
$-16y^2 = -81$	subtract 225	$-16y^2 = -81$
$y^2 = \frac{81}{16}$	divide by -16	$y^2 = \frac{81}{16}$
$y = \frac{9}{4}$ $\quad y = -\frac{9}{4}$	square root property	$y = \frac{9}{4}$ $\quad y = -\frac{9}{4}$
$y = 2.25$ $\quad y = -2.25$	decimal form	$y = 2.25$ $\quad y = -2.25$
$(5, 2.25)$ $\quad (5, -2.25)$	ordered pairs	$(-5, 2.25)$ $\quad (-5, -2.25)$

Plotting these points and connecting them with a smooth curve, while *knowing there are no y-intercepts,* produces the graph in the figure. The point at the origin (in blue) is not a part of the graph, and is given only to indicate the "center" of the hyperbola.

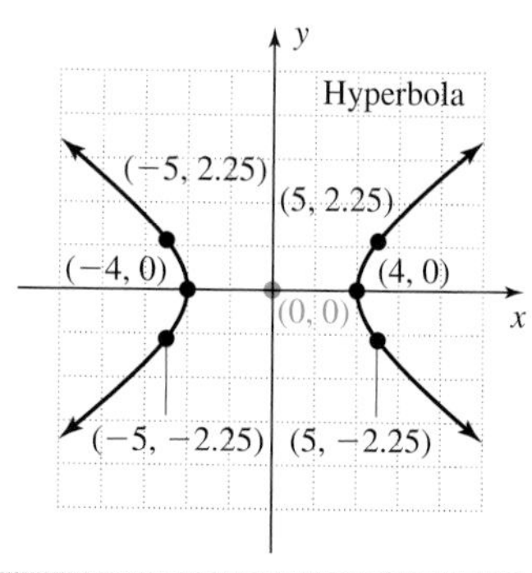

NOW TRY EXERCISES 7 THROUGH 22

Since the hyperbola crosses a horizontal axis, it is referred to as a **horizontal hyperbola.** The points $(-4, 0)$ and $(4, 0)$ are called **vertices,** and the **center** of the hyperbola is always the point halfway between them. If the center is at the origin, we have a **central hyperbola.** The line passing through the center and both vertices is called the **transverse axis** (vertices are always on the transverse axis), and the line passing through the center and perpendicular to this axis is called the **conjugate axis** (see Figure 7.10).

In Example 1, the coefficient of the term containing x^2 was positive and we were subtracting the term containing y^2. If the y^2-term is positive and we subtract the term containing x^2, the result is a vertical hyperbola (Figure 7.11).

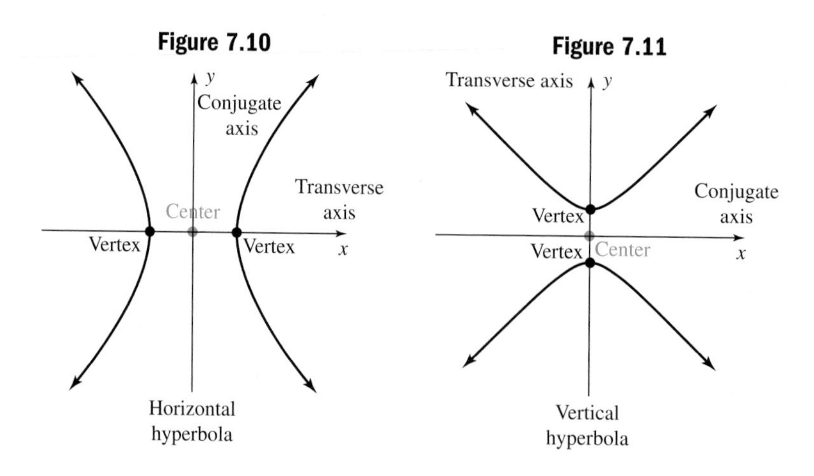

These observations lead us to the equation of a hyperbola in center-shifted form.

THE EQUATION OF A HYPERBOLA IN CENTER-SHIFTED FORM (A, B, $F > 0$)

The equation	The equation
$A(x - h)^2 - B(y - k)^2 = F$	$B(y - k)^2 - A(x - h)^2 = F$
represents a *horizontal hyperbola* with center at (h, k):	represents a *vertical hyperbola* with center at (h, k):
transverse axis $y = k$, conjugate axis $x = h$.	transverse axis $x = h$, conjugate axis $y = k$.

Note each equation contains a *difference* of squared terms in x and y ($A \neq B$ is not a requirement for hyperbolas).

EXAMPLE 2 For the hyperbola shown, state the location of the vertices and the equation of the transverse axis. Then identify the location of the center and the equation of the conjugate axis.

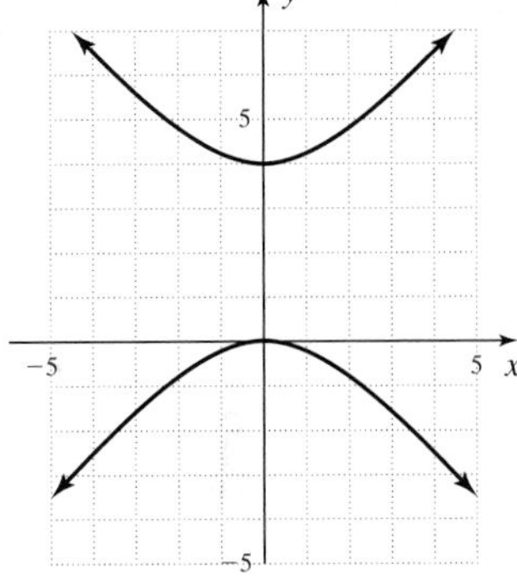

Solution: By inspection we locate the vertices at (0, 0) and (0, 4). The equation of the transverse axis is $x = 0$. The center is halfway between the vertices at (0, 2), meaning the equation of the conjugate axis is $y = 2$.

NOW TRY EXERCISES 23 THROUGH 26

Standard Form

As with the ellipse, the center-shifted form of the equation is helpful for *identifying* hyperbolas, but not very helpful when it comes to *graphing* a hyperbola (since we still must go through the laborious process of finding additional points). For graphing, standard form is once again preferred, with the equation being set equal to 1. Consider the hyperbola $9x^2 - 16y^2 = 144$ from Example 1. To write the equation in standard form, we divide by 144 and obtain $\frac{x^2}{4^2} - \frac{y^2}{3^2} = 1$. By comparing the standard form to the graph, we note $a = 4$ represents the distance from center to vertices, similar to the way we used a previously. But since the graph has no y-intercepts, what could $b = 3$ represent? The answer lies in the fact that branches of a hyperbola are **asymptotic,** meaning they will approach and become very close to imaginary lines that can be used to sketch the graph. As it turns out, the slope of the asymptotic lines are given by the ratios $\frac{b}{a}$ and $-\frac{b}{a}$, with the related equations being $y = \frac{b}{a}x$ and $y = -\frac{b}{a}x$ since this is a central hyperbola (see Exercise 70). The graph from Example 1 is repeated in Figure 7.12, with the asymptotes drawn.

Figure 7.12

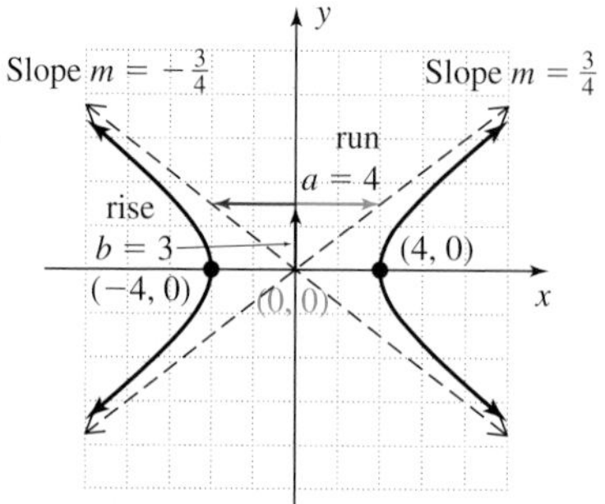

Slope method

Figure 7.13

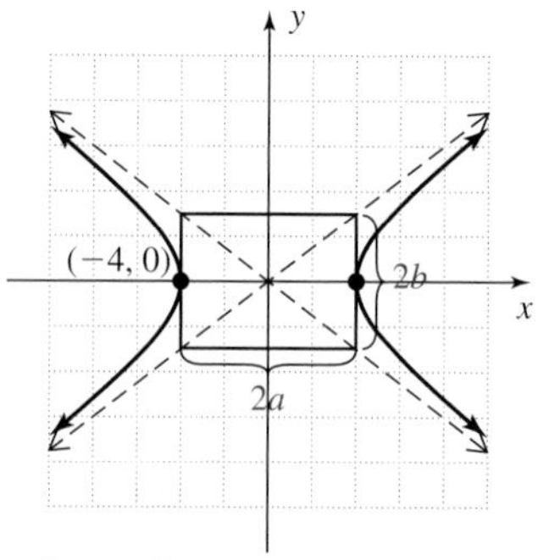

Central rectangle method

A second method of drawing the asymptotes involves drawing a **central rectangle** with dimensions $2a$ by $2b$, as shown in Figure 7.13. The asymptotes will be the *extended diagonals* of this rectangle. This brings us to the equation of a hyperbola in standard form.

THE EQUATION OF A HYPERBOLA IN STANDARD FORM

The equation

$$\frac{(x-h)^2}{a^2} - \frac{(y-k)^2}{b^2} = 1$$

represents a *horizontal* hyperbola with *transverse* axis $y = k$ and *conjugate* axis $x = h$.

The equation

$$\frac{(y-k)^2}{b^2} - \frac{(x-h)^2}{a^2} = 1$$

represents a *vertical* hyperbola with *transverse* axis $x = h$ and *conjugate* axis $y = k$.

The center of each hyperbola is (h, k). The asymptotes can be drawn by starting at the center (h, k) and counting slopes of $m = \pm\frac{b}{a}$.

As an alternative, a rectangle of dimensions $2a$ by $2b$ centered at (h, k) can be drawn. The asymptotes are the extended diagonals of this rectangle.

EXAMPLE 3 Sketch the graph of $16(x-2)^2 - 9(y-1)^2 = 144$. Include the center, vertices, and asymptotes.

Solution: Begin by noting a difference of the second-degree terms, with the x^2-term occurring first. This means we'll be graphing a horizontal hyperbola whose center is at (2, 1). Continue by writing the equation in standard form.

$$16(x-2)^2 - 9(y-1)^2 = 144 \quad \text{given equation}$$

$$\frac{16(x-2)^2}{144} - \frac{9(y-1)^2}{144} = \frac{144}{144} \quad \text{divide by 144}$$

$$\frac{(x-2)^2}{9} - \frac{(y-1)^2}{16} = 1 \quad \text{simplify}$$

$$\frac{(x-2)^2}{3^2} - \frac{(y-1)^2}{4^2} = 1 \quad \text{write denominators in squared form}$$

Since $a = 3$, the vertices are a horizontal distance of 3 units from the center (2, 1), giving $(2 + 3, 1) \to (5, 1)$ and $(2 - 3, 1) \to (-1, 1)$. After plotting the center and vertices, we can begin at the center and count off slopes of $m = \pm\frac{b}{a} = \pm\frac{4}{3}$, or draw a rectangle with dimensions $2(3) = 6$ (horizontal dimension) by $2(4) = 8$ (vertical dimension) centered at $(2, 1)$ to sketch the asymptotes. The complete graph is shown here.

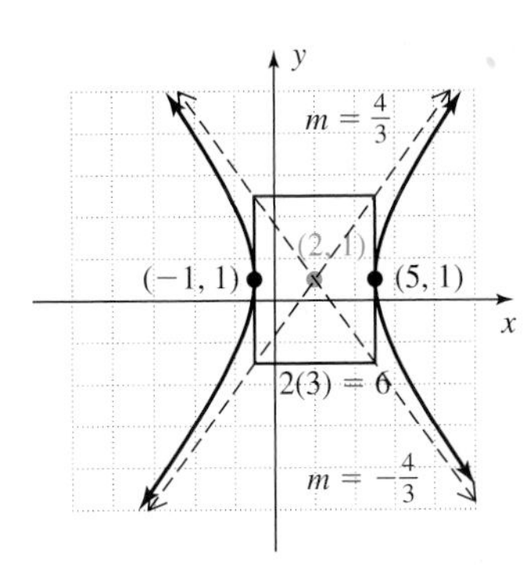

Horizontal hyperbola

Center at (2, 1)

Vertices at (−1, 1) and (5, 1)

Transverse axis: $y = 1$
Conjugate axis: $x = 2$

Width of rectangle (horizontal dimension and distance between vertices) $2a = 2(3) = 6$

Length of rectangle (vertical dimension) $2b = 2(4) = 8$

NOW TRY EXERCISES 27 THROUGH 38

Polynomial Form

If the equation is given in polynomial form, complete the square in x and y to write the equation in standard form.

EXAMPLE 4 Graph the equation $9y^2 - x^2 + 54y + 4x + 68 = 0$.

Solution: Since the y^2-term occurs first, we assume the equation represents a vertical hyperbola, but wait for the center-shifted form to be sure (see Exercise 68).

$$9y^2 - x^2 + 54y + 4x + 68 = 0 \quad \text{given}$$

$$9y^2 - x^2 + 54y + 4x = -68 \quad \text{group like terms; subtract 68}$$

$$9(y^2 + 6y + \_\_) - 1(x^2 - 4x + \_\_) = -68 \quad \text{factor out 9 from } y\text{-terms and } -1 \text{ from } x\text{-terms}$$

$$9(y^2 + 6y + 9) - 1(x^2 - 4x + 4) = -68 + 81 + (-4) \quad \text{complete the square}$$

adds $9(9) = 81$; adds $-1(4) = -4$ — add $81 + (-4)$ to right

$$9(y+3)^2 - 1(x-2)^2 = 9 \quad \text{center-shifted form} \to \text{vertical hyperbola}$$

$$\frac{(y+3)^2}{1} - \frac{(x-2)^2}{9} = 1 \quad \text{divide by 9 (standard form)}$$

$$\frac{(y+3)^2}{1^2} - \frac{(x-2)^2}{3^2} = 1 \quad \text{write denominators in squared form}$$

The center of the hyperbola is $(2, -3)$, with $a = 3$, $b = 1$, and a transverse axis of $x = 2$. The vertices are at $(2, -3 + 1)$ and $(2, -3 - 1)$, $\rightarrow$ $(2, -2)$ and $(2, -4)$. After plotting the center and vertices, we draw a rectangle centered at $(2, -3)$ with a horizontal "width" of $2(3) = 6$ and a vertical "length" of $2(1) = 2$ to sketch the asymptotes. The completed graph is given in the figure.

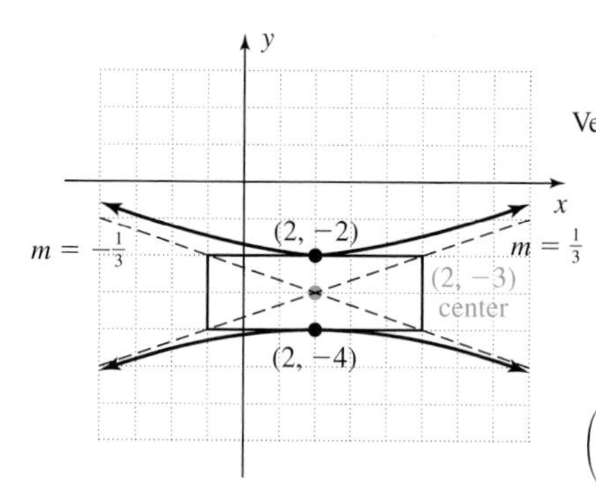

Vertical hyperbola

Center at $(2, -3)$
Vertices at $(2, -2)$ and $(2, -4)$

Transverse axis: $x = 2$
Conjugate axis: $y = -3$

Width of rectangle (horizontal dimension)
$2a = 2(3) = 6$

Length of rectangle (vertical dimension and distance between vertices)
$2b = 2(1) = 2$

NOW TRY EXERCISES 39 THROUGH 48

B. Distinguishing Between the Equation of a Circle, Ellipse, and Hyperbola

So far we've explored numerous graphs of circles, ellipses, and hyperbolas. In Example 5 we'll attempt to identify a given conic section from its equation alone (without graphing the equation). As you've seen, the corresponding equations have unique characteristics that can help distinguish one from the other.

EXAMPLE 5 Identify each equation as that of a circle, ellipse, or hyperbola. Justify your choice and name the center, but do not draw the graphs.

a. $y^2 = 36 + 9x^2$ **b.** $4x^2 = 16 - 4y^2$

c. $x^2 = 225 - 25y^2$ **d.** $25x^2 = 100 + 4y^2$

e. $3(x - 2)^2 + 4(y + 3)^2 = 12$

f. $4(x + 5)^2 = 36 + 9(y - 4)^2$

Solution:

a. Writing the equation in center-shifted form gives $y^2 - 9x^2 = 36$. Since the equation contains a difference of second-degree terms, it is the equation of a (vertical) hyperbola. The center is at $(0, 0)$.

b. In center-shifted form we have $4x^2 + 4y^2 = 16$. Dividing by 4 gives $x^2 + y^2 = 4$. The equation represents a circle of radius 2, with the center at $(0, 0)$.

c. Writing the equation as $x^2 + 25y^2 = 225$ we note a sum of second-degree terms with unequal coefficients. The equation is that of an ellipse, with the center at $(0, 0)$.

d. In center-shifted form we have $25x^2 - 4y^2 = 100$ and the equation contains a difference of second-degree terms. The equation represents a central (horizontal) hyperbola, whose center is at $(0, 0)$.

e. The equation is already in center-shifted form and contains a sum of second-degree terms with unequal coefficients. This is the equation of an ellipse with the center at $(2, -3)$.

f. Writing the equation in center-shifted form gives $4(x + 5)^2 - 9(y - 4)^2 = 36$. With a difference of second-degree terms the equation represents a horizontal hyperbola with center $(-5, 4)$.

NOW TRY EXERCISES 49 THROUGH 60

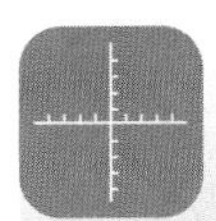

TECHNOLOGY HIGHLIGHT

Using a Graphing Calculator to Study Hyperbolas

The keystrokes shown apply to a TI-84 Plus model. Please consult your manual or our Internet site for other models.

As with the circle and ellipse, the hyperbola must also be defined in two pieces in order to use a graphing calculator to study its graph. Consider the *relation* $4x^2 - 9y^2 = 36$. From our work in this section, we know this is the equation of a horizontal hyperbola centered at (0, 0). Solving for y gives:

$$4x^2 - 9y^2 = 36 \quad \text{original equation}$$
$$-9y^2 = 36 - 4x^2 \quad \text{isolate } y^2\text{-term}$$
$$y^2 = \frac{36 - 4x^2}{-9} \quad \text{divide by } -9$$
$$y = \pm\sqrt{\frac{36 - 4x^2}{-9}} \quad \text{solve for } y$$

We can again separate this result into two parts: $Y_1 = \sqrt{\frac{36 - 4x^2}{-9}}$ gives the "upper half" of the hyperbola, and $Y_2 = -\sqrt{\frac{36 - 4x^2}{-9}}$ gives the "lower half."

Entering these on the Y= screen, graphing them on the standard screen, and pressing the TRACE key gives the graph shown in Figure 7.14 (this time the standard screen gives a fairly nice graph of the function, even though the y-values are still compressed). Note the location of the cursor at $x = 0$, but no y-value is displayed.

Figure 7.14

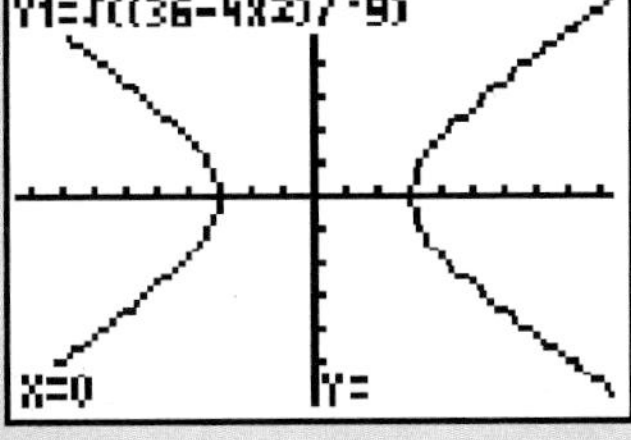

This is because the hyperbola is not defined at $x = 0$. Press the right arrow key ▶ and walk the cursor to the right until the y-values begin appearing. In fact, they begin to appear at (3, 0), which is one of the vertices of the hyperbola. We could also graph the asymptotes for this hyperbola ($y = \pm\frac{2}{3}x$) by entering the lines as Y_3 and Y_4 on the Y= screen. The resulting graph is shown in Figure 7.15 (the TRACE key has been pushed and the down arrow used to highlight Y_2). Use these ideas and the features of your graphing calculator to complete the following exercises.

Figure 7.15

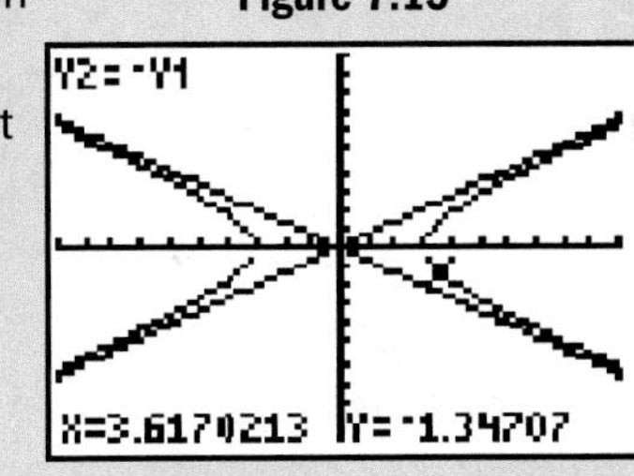

Exercise 1: Graph the hyperbola defined by $25y^2 - 4x^2 = 100$ using a friendly window. What are the coordinates of the vertices of this hyperbola? (0, 2), (0, −2) Use the TRACE feature to find the value(s) of y when $x = 4$. Determine (from the graph) the value(s) of y when $x = -4$, then verify your response using the TABLE feature of your calculator. $y = \pm 2.5612497$

Exercise 2: Graph the hyperbola defined by $9x^2 - 16y^2 = 144$ using the standard window. Then determine the equations of the asymptotes and graph these as well. Why do the asymptotes of this hyperbola intersect at the origin? When will the asymptotes of a hyperbola *not* intersect at the origin? $y = \frac{3}{4}x,\ y = \frac{-3}{4}x$

That is the center. Anytime the center is not at the origin.

7.2 EXERCISES

7.

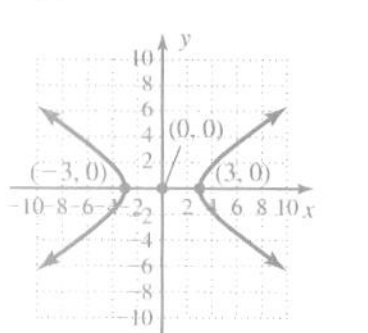

8.

9.

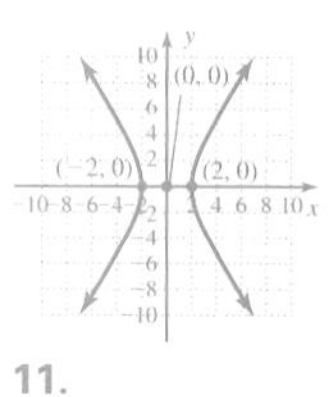

10.

11.

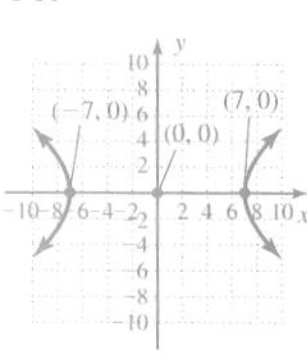

12.

13. 14.

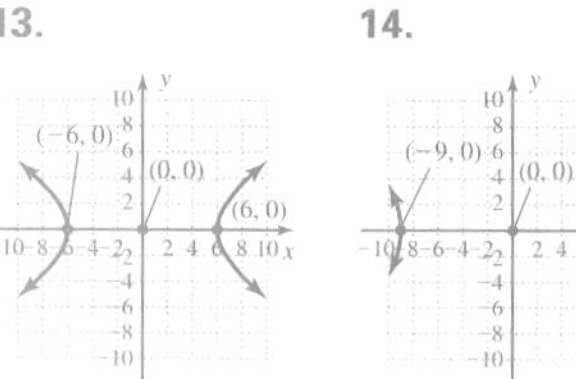

15.

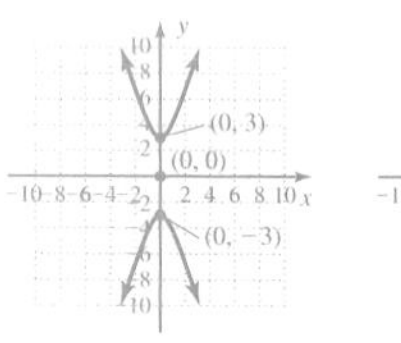

16.

17.

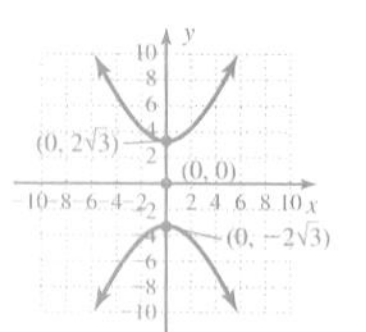

18.

19.

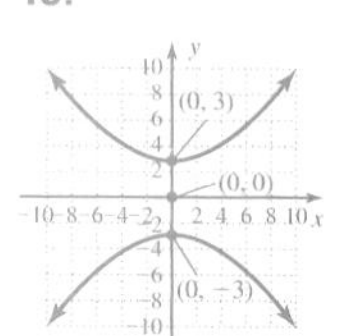

20.

21.

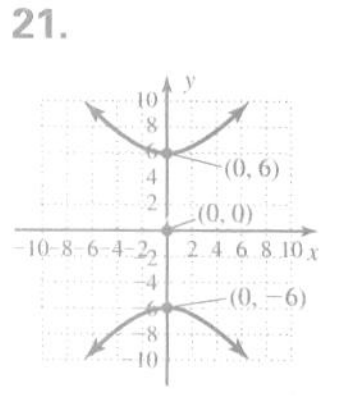

22.

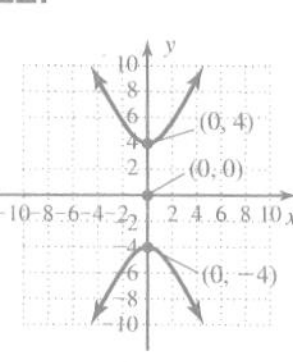

CONCEPTS AND VOCABULARY

Fill in the blank with the appropriate word or phrase. Carefully reread the section if needed.

1. The line that passes through the vertices of a hyperbola is called the transverse axis.

2. The conjugate axis is perpendicular to the transverse axis and contains the center of the hyperbola.

3. The center of a hyperbola is located midway between the vertices.

4. The center of the hyperbola defined by $\frac{(x-2)^2}{4^2} - \frac{(y-3)^2}{5^2} = 1$ is at (2, 3).

5. Compare/contrast the two methods used to find the asymptotes of a hyperbola. Include an example illustrating both methods. Answers will vary.

6. Explore/explain why $A(x-h)^2 - B(y-k)^2 = F$ results in a hyperbola regardless of whether $A = B$ or $A \neq B$. Illustrate with an example. Answers will vary.

DEVELOPING YOUR SKILLS

Graph each hyperbola. Label the center, vertices, and any additional points used.

7. $\frac{x^2}{9} - \frac{y^2}{4} = 1$

8. $\frac{x^2}{16} - \frac{y^2}{9} = 1$

9. $\frac{x^2}{4} - \frac{y^2}{9} = 1$

10. $\frac{x^2}{25} - \frac{y^2}{16} = 1$

11. $\frac{x^2}{49} - \frac{y^2}{16} = 1$

12. $\frac{x^2}{25} - \frac{y^2}{9} = 1$

13. $\frac{x^2}{36} - \frac{y^2}{16} = 1$

14. $\frac{x^2}{81} - \frac{y^2}{16} = 1$

15. $\frac{y^2}{9} - \frac{x^2}{1} = 1$

16. $\frac{y^2}{1} - \frac{x^2}{4} = 1$

17. $\frac{y^2}{12} - \frac{x^2}{4} = 1$

18. $\frac{y^2}{9} - \frac{x^2}{18} = 1$

19. $\frac{y^2}{9} - \frac{x^2}{9} = 1$

20. $\frac{y^2}{4} - \frac{x^2}{4} = 1$

21. $\frac{y^2}{36} - \frac{x^2}{25} = 1$

22. $\frac{y^2}{16} - \frac{x^2}{4} = 1$

For the graphs given, state the location of the vertices and the equation of the transverse axis. Then identify the location of the center and the equation of the conjugate axis. Assume all coordinates are lattice points.

23. (−4, −2), (2, −2), y = −2, (−1, −2), x = −1

24. (−3, 3), (−3, −1), x = −3, (−3, 1), y = 1

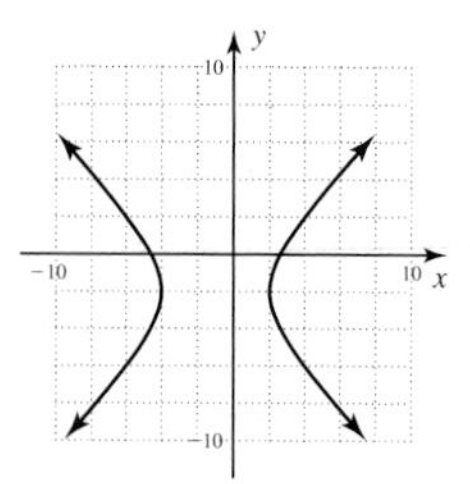

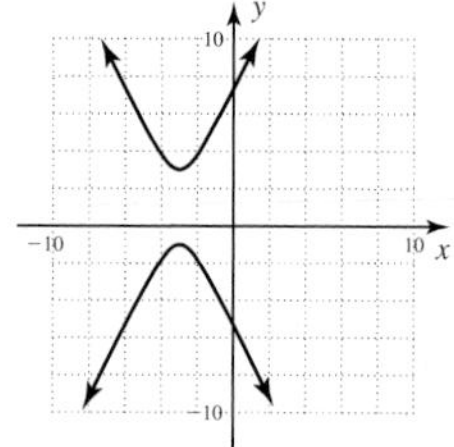

25. (4, 1), (4, −3), x = 4, (4, −1), y = −1

26. (−1, 2), (3, 2), y = 2, (1, 2), x = 1

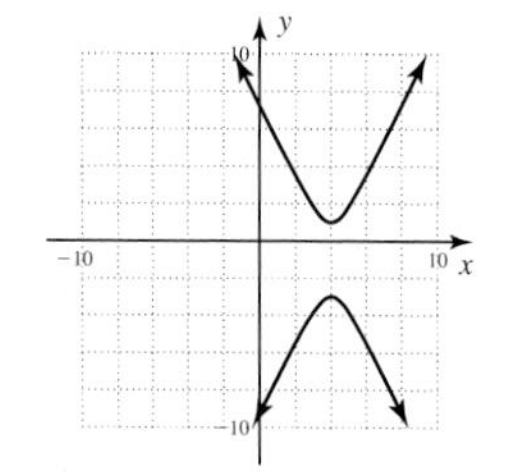

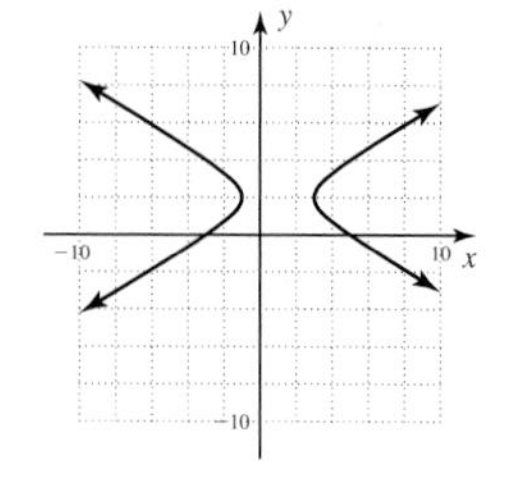

27. 28.

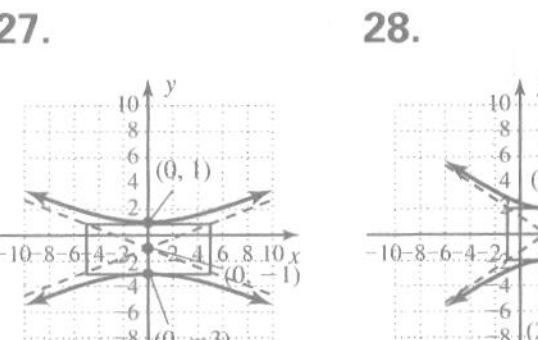

29. 30.

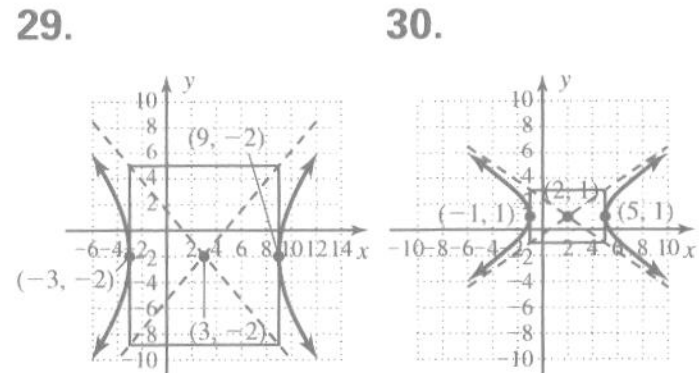

31. 32.

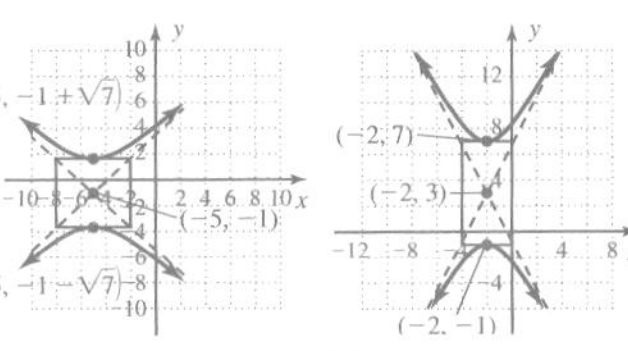

33. 34.

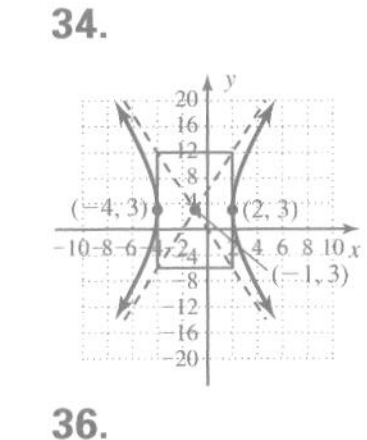

35. 36.

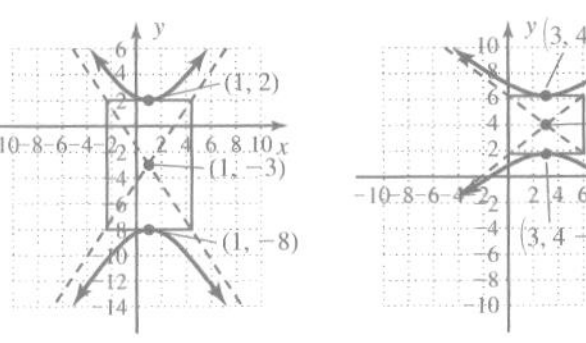

37. 38.

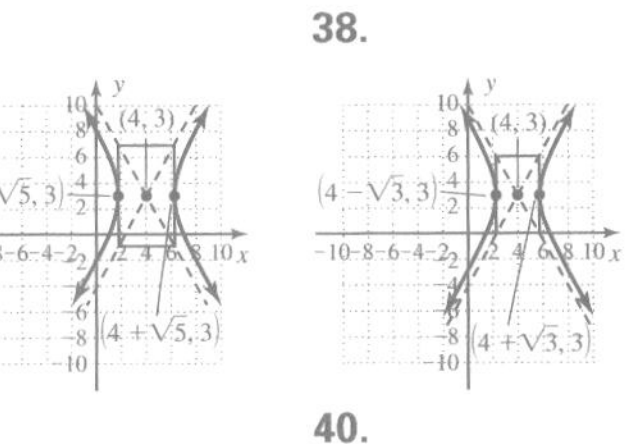

39. 40.

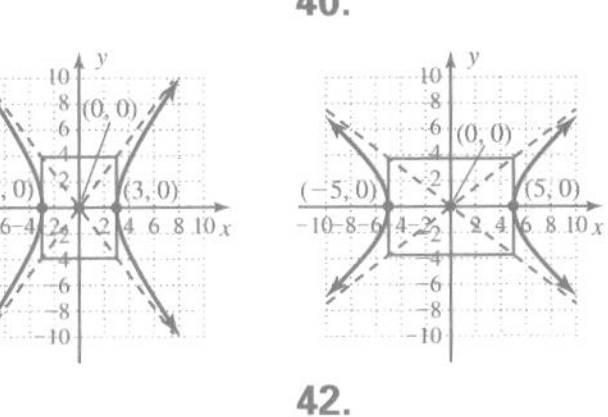

41. 42.

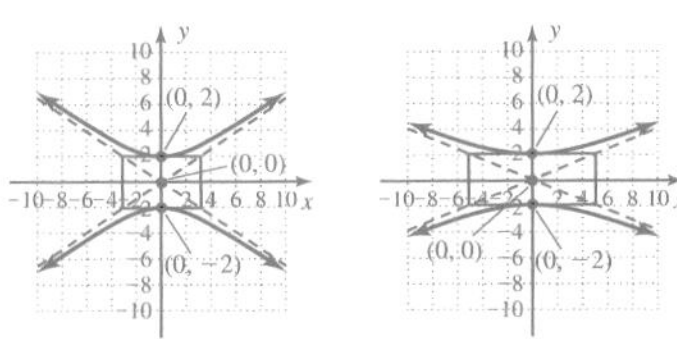

Additional answers can be found in the Instructor Answer Appendix.

Sketch a complete graph of each equation, including the asymptotes. Be sure to identify the center and vertices.

27. $\frac{(y+1)^2}{4} - \frac{x^2}{25} = 1$

28. $\frac{y^2}{4} - \frac{(x-2)^2}{9} = 1$

29. $\frac{(x-3)^2}{36} - \frac{(y+2)^2}{49} = 1$

30. $\frac{(x-2)^2}{9} - \frac{(y-1)^2}{4} = 1$

31. $\frac{(y+1)^2}{7} - \frac{(x+5)^2}{9} = 1$

32. $\frac{(y-3)^2}{16} - \frac{(x+2)^2}{5} = 1$

33. $(x-2)^2 - 4(y+1)^2 = 16$

34. $9(x+1)^2 - (y-3)^2 = 81$

35. $2(y+3)^2 - 5(x-1)^2 = 50$

36. $9(y-4)^2 - 5(x-3)^2 = 45$

37. $12(x-4)^2 - 5(y-3)^2 = 60$

38. $8(x-4)^2 - 3(y-3)^2 = 24$

39. $16x^2 - 9y^2 = 144$

40. $16x^2 - 25y^2 = 400$

41. $9y^2 - 4x^2 = 36$

42. $25y^2 - 4x^2 = 100$

43. $12x^2 - 9y^2 = 72$

44. $36x^2 - 20y^2 = 180$

45. $4x^2 - y^2 + 40x - 4y + 60 = 0$

46. $x^2 - 4y^2 - 12x - 16y + 16 = 0$

47. $x^2 - 4y^2 - 24y - 4x - 36 = 0$

48. $-9x^2 + 4y^2 - 18x - 24y - 9 = 0$

Classify each equation as that of a circle, ellipse, or hyperbola. Justify your response.

49. $-4x^2 - 4y^2 = -24$ circle

50. $9y^2 = -4x^2 + 36$ ellipse

51. $x^2 + y^2 = 2x + 4y + 4$ circle

52. $x^2 = y^2 + 6y - 7$ hyperbola

53. $2x^2 - 4y^2 = 8$ hyperbola

54. $36x^2 + 25y^2 = 900$ ellipse

55. $x^2 + 5 = 2y^2$ hyperbola

56. $x + y^2 = 3x^2 + 9$ hyperbola

57. $2x^2 = -2y^2 + x + 20$ circle

58. $2y^2 + 3 = 6x^2 + 8$ hyperbola

59. $16x^2 + 5y^2 - 3x + 4y = 538$ ellipse

60. $9x^2 + 9y^2 - 9x + 12y + 4 = 0$ circle

WORKING WITH FORMULAS

61. **Equation of a semi-hyperbola:** $y = \sqrt{\frac{36 - 4x^2}{-9}}$

a. $y = \frac{2}{3}\sqrt{x^2 - 9}$ b. $x \in (-\infty, -3] \cup [3, \infty)$ c. $y = \frac{-2}{3}\sqrt{x^2 - 9}$

The "upper half" of a certain hyperbola is given by the equation shown. (a) Simplify the radicand, (b) state the domain of the expression, and (c) enter the expression as Y_1 on a graphing calculator and graph. What is the equation for the "lower half" of this hyperbola?

62. **Focal chord of a hyperbola:** $L = \frac{2b^2}{a}$

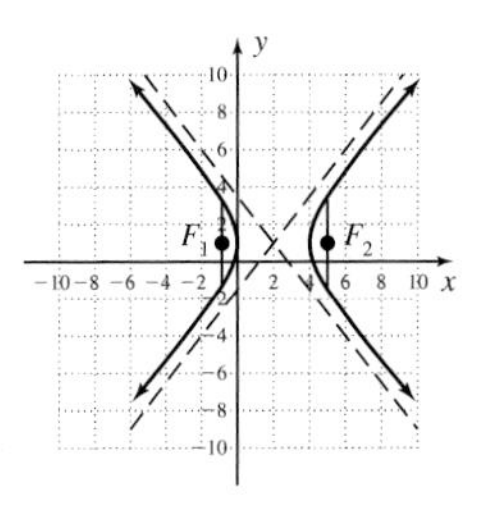

The focal chords of a hyperbola are line segments parallel to the conjugate axis with endpoints on the hyperbola, and containing certain points F_1 and F_2 called the *foci* (see grid). The length of the chord is given by the formula shown. Use the formula to find its length for the hyperbola indicated, then compare the calculated value to their estimated length taken from the graph: $\frac{(x-2)^2}{4} - \frac{(y-1)^2}{5} = 1$. 5

APPLICATIONS

63. **Stunt pilots:** At an air show, a stunt plane dives along a hyperbolic path whose vertex is directly over the grandstands. If the plane's flight path can be modeled by the hyperbola $25y^2 - 1600x^2 = 40{,}000$, what is the minimum altitude of the plane as it passes over the stands? Assume x and y are in yards. 40 yd

64. Flying clubs: To test their skill as pilots, the members of a flight club attempt to drop sandbags on a target placed in an open field, by diving along a hyperbolic path whose vertex is directly over the target area. If the flight path of the plane flown by the club's president is modeled by $9y^2 - 16x^2 = 14{,}400$, what is the minimum altitude of her plane as it passes over the target? Assume x and y are in feet. 40 ft

65. Nuclear cooling towers: The natural draft cooling towers for nuclear power stations are called *hyperboloids of one sheet.* The perpendicular cross sections of these hyperboloids form two branches of a hyperbola. Suppose the central cross section of one such tower is modeled by the hyperbola $1600x^2 - 400(y - 50)^2 = 640{,}000$. What is the minimum distance between the sides of the tower? Assume x and y are in feet. 40 ft

66. Charged particles: It has been shown that when like particles with a common charge are hurled at each other, they deflect and travel along paths that are hyperbolic. Suppose the path of two such particles is modeled by the hyperbola $x^2 - 9y^2 = 36$. What is the minimum distance between the particles as they approach each other? Assume x and y are in microns. 12 microns

WRITING, RESEARCH, AND DECISION MAKING

67. Hyperbolas such as $x^2 - y^2 = 9$ and $y^2 - x^2 = 9$ are referred to as **conjugate hyperbolas.** Graph both on the same grid and discuss why the name is appropriate. Compare and contrast the equations and graphs, then give the equations of two other conjugate hyperbolas. Answers will vary.

68. a. $\dfrac{(x-4)^2}{\frac{1}{4}} - (y-2)^2 = 0$

b. $(x-2)^2 + \dfrac{(y-4)^2}{\frac{1}{5}} = 0$

68. Referring to the introduction and Figure 7.9, it is possible for the plane to intersect only the vertex of the cone or to be tangent to the sides. These are called **degenerate cases** of a conic section. Many times we're unable to tell if the equation represents a degenerate case until it's written in center-shifted or standard form. For example, write the following equations in standard form and comment.

a. $4x^2 - 32x - y^2 + 4y + 60 = 0$ **b.** $x^2 - 4x + 5y^2 - 40y + 84 = 0$

69. LORAN is a long distance radio-navigation system for ships and aircraft, developed and deployed extensively during World War II. Using any of the resources available to you, determine how this system uses hyperbolas to pinpoint the location of a ship or aircraft. Submit a detailed report that includes diagrams and examples of how LORAN works. Answers will vary.

70. For a greater understanding as to *why* the branches of a hyperbola are asymptotic, solve the basic equation $\dfrac{x^2}{a^2} - \dfrac{y^2}{b^2} = 1$ for y, then consider what happens as $x \to \infty$ (recall from our earlier work that $x^2 - k \approx x^2$ for large x). $y = \pm\sqrt{\dfrac{b^2}{a^2}x^2 - b^2}$; as $x \to \infty$, $y \to \pm\sqrt{\dfrac{b^2}{a^2}x^2} = \pm\dfrac{b}{a}x$

EXTENDING THE CONCEPT

71. Find the equation of the circle that shares the same center as the hyperbola given, if the vertices of the hyperbola are on the circle: $9(x - 2)^2 - 25(y - 3)^2 = 225$. $(x-2)^2 + (y-3)^2 = 25$

72. $\dfrac{(x-2)^2}{25} + \dfrac{(y-3)^2}{9} = 1$

72. Find the equation of the ellipse that shares the same center as the hyperbola given, if the length of the minor axis is equal to the height of the central rectangle and the hyperbola and ellipse share the same vertices: $9(x - 2)^2 - 25(y - 3)^2 = 225$.

73. Which has a greater area: (a) The central rectangle of the hyperbola given by $(x - 5)^2 - (y + 4)^2 = 57$, (b) the circle given by $(x - 5)^2 + (y + 4)^2 = 57$; or (c) the ellipse given by $9(x - 5)^2 + 10(y + 4)^2 = 570$? a

MAINTAINING YOUR SKILLS

74. $y = \pm\dfrac{3}{2}\sqrt{3 + 2x - x^2} + 1$

75.

74. (7.1) Solve the equation for y:
$$\frac{(x-1)^2}{4} + \frac{(y-1)^2}{9} = 1$$

75. (3.7) Graph the piecewise-defined function:
$$f(x) = \begin{cases} 4 - x^2 & -2 \le x < 3 \\ 5 & x \ge 3 \end{cases}$$

76. (4.2) Use synthetic division and the remainder theorem to determine if $x = 2$ is a zero of $g(x) = x^5 - 5x^4 + 4x^3 + 16x^2 - 32x + 16$. If yes, find its multiplicity. yes , 3

77. (R.7) Find the area and perimeter of the isosceles trapezoid in cm.

48 in.

54 cm

200 cm

$A = 8691.84\text{ cm}^2$
$P \approx 455.19\text{ cm}$

78. (1.4) The number $z = 1 + i\sqrt{2}$ is a solution to two out of the three equations given. Which two? b and c

a. $x^4 + 4 = 0$ **b.** $x^3 - 6x^2 + 11x - 12 = 0$ **c.** $x^2 - 2x + 3 = 0$

79. (6.3) A government-approved company is licensed to haul toxic waste. Each container of solid waste weighs 800 lb and has a volume of 100 ft^3. Each container of liquid waste weighs 1000 lb and is 60 ft^3 in volume. The revenue from hauling solid waste is \$300 per container, while the revenue from liquid waste is \$375 per container. The truck used by this company has a weight capacity of 39.8 tons and a volume capacity of 6960 ft^3. What combination of solid and liquid weight containers will produce the maximum revenue? 42 solid, 46 liquid

MID-CHAPTER CHECK

1.

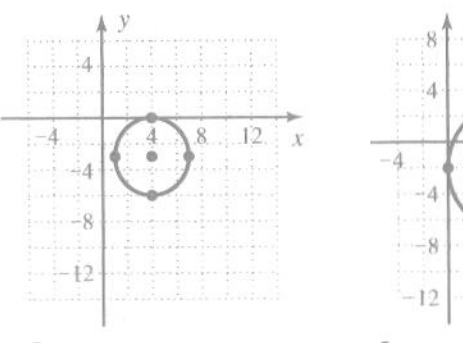

2.

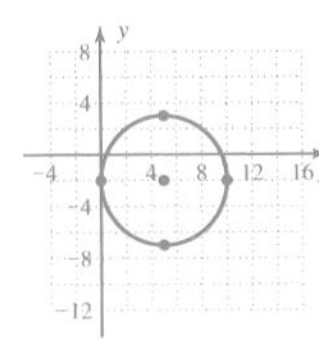

3.

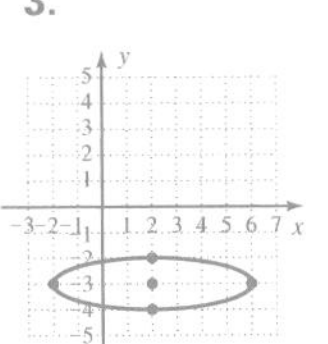

4.

5.

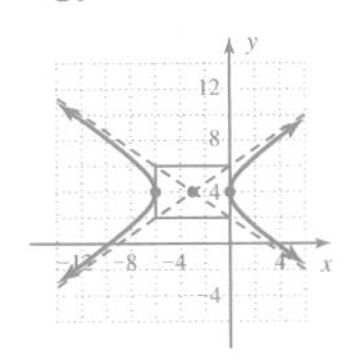

6.

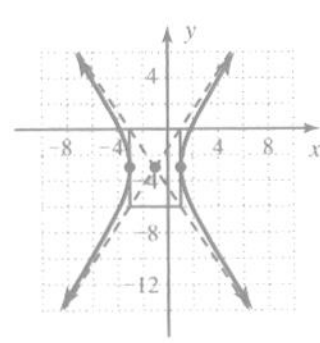

Additional answers can be found in the Instructor Answer Appendix.

Sketch the graph of each conic section.

1. $(x - 4)^2 + (y + 3)^2 = 9$

2. $x^2 + y^2 - 10x + 4y + 4 = 0$

3. $\dfrac{(x - 2)^2}{16} + \dfrac{(y + 3)^2}{1} = 1$

4. $9x^2 + 4y^2 + 18x - 24y + 9 = 0$

5. $\dfrac{(x + 3)^2}{9} - \dfrac{(y - 4)^2}{4} = 1$

6. $9x^2 - 4y^2 + 18x - 24y - 63 = 0$

7. Find the equation of each relation and state its domain and range [7(c) is review].

a.

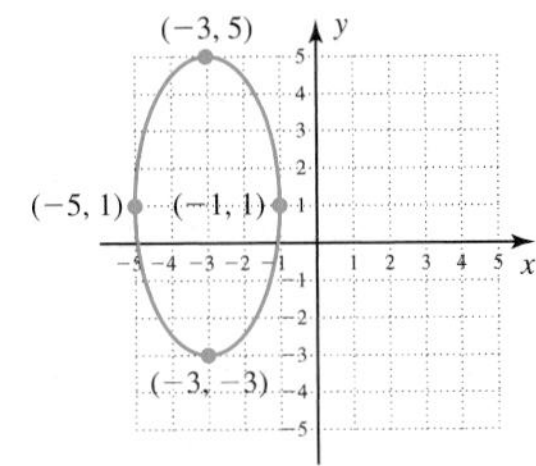

b.

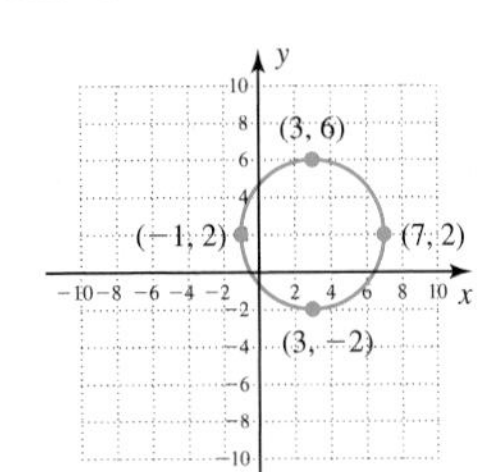

c.

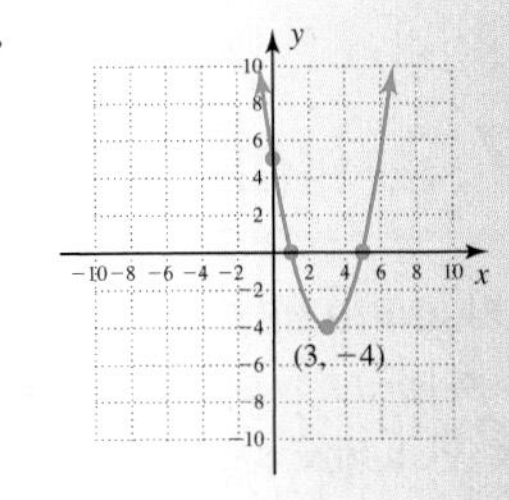

8. Find an equation for the circle with center at $(-2, 5)$, given $(0, 3)$ is a point on the circle. $(x + 2)^2 + (y - 5)^2 = 8$

9. Find the equation of the ellipse (in standard form) if the vertices are $(-4, 0)$ and $(4, 0)$ and the minor axis has a length of 4 units. $\dfrac{x^2}{16} + \dfrac{y^2}{4} = 1$

10. The radio signal emanating from a tall radio tower spreads evenly in all directions with a range of 50 mi. If the tower is located at coordinates (20, 30) and my home is at coordinates (10, 78), will I be able to pick up this station on my home radio? Assume coordinates are in miles. yes, distance $d \approx 49$ mi

REINFORCING BASIC CONCEPTS

More on Completing the Square

From our work so far in Chapter 7, we realize the process of *completing the square* has much greater use than simply as a tool for working with quadratic equations. It is a valuable tool in the application of the conic sections, as well as other areas. The purpose of this *Reinforcing Basic Concepts* is to strengthen the ability and confidence needed to apply the process correctly. This is important since many conic equations include cases where a and b are fractions or irrational numbers. No matter what the context, (1) *the process begins with a coefficient of 1.* Consider $20x^2 + 120x + 27y^2 - 54y + 192 = 0$. We recognize this as the equation of an ellipse, since the coefficients of the squared terms are positive and unequal. To study or graph this ellipse, we use the center-shifted form to identify a, b, and c. Grouping the like-variable terms gives $(20x^2 + 120x + \_\_) + (27y^2 - 54y + \_\_) + 192 = 0$, or $20(x^2 + 6x + \_\_) + 27(y^2 - 2y + \_\_) + 192 = 0$ after factoring. (2) *Use* $\left(\frac{1}{2} \cdot \textit{linear coefficient}\right)^2$ to complete the square. For this example the result is $\left(\frac{1}{2} \cdot 6\right)^2 = 9$ and $\left(\frac{1}{2} \cdot -2\right)^2 = 1$, respectively, and these numbers are inserted into their related group: $20(x^2 + 6x + 9) + 27(y^2 - 2y + 1) + 192 = 0$. Due to the distributive property, we have in effect added $20 \cdot 9 = 180$ and $27 \cdot 1 = 27$ (for a total of 207) to the left side of the equation. This brings us to the third step: (3) *keep the equation in balance.* Since the left side was increased by 207, we must also increase the right side by 207: $20(x^2 + 6x + 9) + 27(y^2 - 2y + 1) + 192 = 207$. The quantities in parenthesis now can be factored as binomial squares: $20(x + 3)^2 + 27(y - 1)^2 + 192 = 207$. Subtracting 192 at any time during this process gives $20(x + 3)^2 + 27(y - 1)^2 = 15$. Division by 15 and simplifying gives $\frac{4(x + 3)^2}{3} + \frac{9(y - 1)^2}{5} = 1$. Note the coefficient of each binomial square is not 1, even after setting the equation equal to 1. In the *Strengthening Core Skills* feature of this chapter, we'll look at how to write equations of this type in standard form to obtain the values of a and b. For now, practice completing the square using these exercises.

Exercise 1: $100x^2 - 400x - 18y^2 - 108y + 230 = 0$ $\quad \frac{25(x-2)^2}{2} - \frac{9(y+3)^2}{4} = 1$

Exercise 2: $28x^2 - 56x + 48y^2 + 192y + 195 = 0$ $\quad \frac{28(x-1)^2}{25} + \frac{48(y+2)^2}{25} = 1$

7.3 Nonlinear Systems of Equations and Inequalities

LEARNING OBJECTIVES

In Section 7.3 you will learn how to:

A. Visualize possible solutions

B. Solve nonlinear systems using substitution

C. Solve nonlinear systems using elimination

D. Solve nonlinear systems using a calculator

E. Solve nonlinear systems of inequalities

INTRODUCTION

Recall that linear equations can be written as $Ax + By = C$, noting particularly that the exponent on both x and y is 1. Equations where the variables have exponents other than 1 or that are transcendental (like logarithmic and exponential equations) are called nonlinear equations. A *nonlinear system* of equations has at least one nonlinear equation, and numerous possibilities exist for the solution set.

POINT OF INTEREST

LORAN is an abbreviation of the phrase *lo*ng *ra*nge *n*avigation, a navigational system created during World War II. It is a system that allows navigators in a ship or aircraft to establish their position by computing the *difference* in the time it takes two radio signals from synchronized transmitters (spaced a significant distance apart) to reach them. The location of all points where signals from the two transmitters are separated by a given time interval is modeled by a hyperbola, sometimes called a *loran line.* The intersection of two loran lines gives the location of the ship (see Exercise 69 from Section 7.2). Today, the location of ships and airplanes is found with great accuracy using GPS tracking systems.

A. Possible Solutions for a Nonlinear System

When solving nonlinear systems, it is often helpful to *visualize* the graphs represented by the equations in the system. This can help determine various possibilities for their intersection and further assist the solution process. Consider Example 1.

EXAMPLE 1 Identify each equation in the system as the equation of a line, parabola, circle, ellipse, or hyperbola. Then determine the number of solutions possible by considering the different ways the graphs might intersect: $\begin{cases} 4x^2 + 9y^2 = 36 \\ 2x + 3y = 6 \end{cases}$. Finally, solve the system by graphing.

Solution: The first equation contains a sum of second-degree terms with unequal coefficients, and we recognize this as a central ellipse. The second equation is obviously linear. This means the system may have no solution, one solution, or two solutions, as shown in Figure 7.16. The graph of the system is shown in Figure 7.17 and the two points of intersection appear to be (3, 0) and (0, 2). After checking these in the original equations we find that both satisfy the system.

Figure 7.16

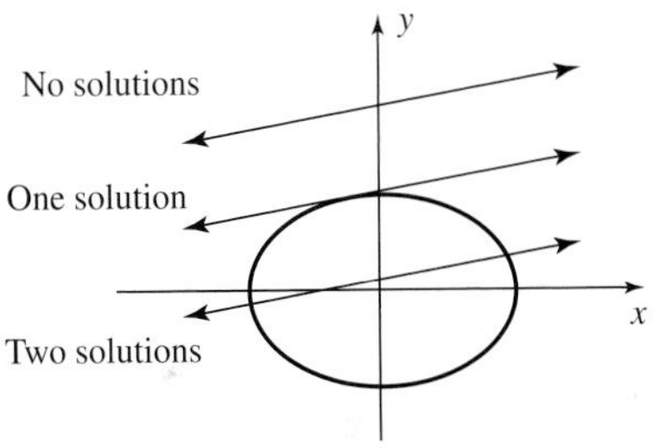

y
No solutions
One solution
Two solutions
x

Figure 7.17

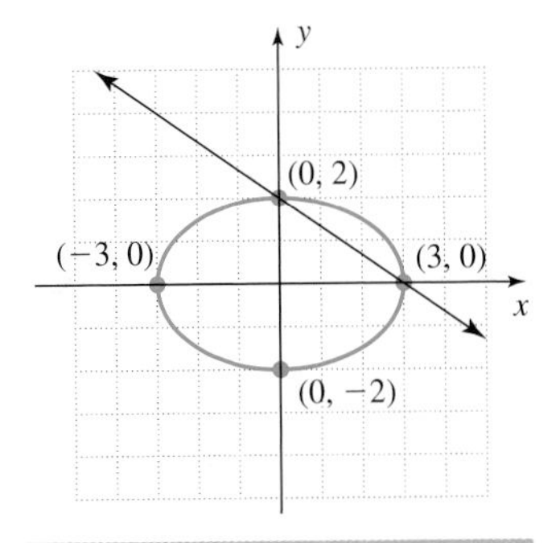

NOW TRY EXERCISES 7 THROUGH 12

B. Solving a Nonlinear System by Substitution

Since graphical methods at best offer an estimate for the solution (points of intersection may not be lattice points), we more often turn to the algebraic methods developed earlier in Chapter 6. Recall the substitution method involves solving one of the equations for a variable or expression that can be substituted in the other equation to eliminate one of the variables.

EXAMPLE 2 Solve the system using substitution: $\begin{cases} y = x^2 - 2x - 3 \\ 2x - y = 7 \end{cases}$.

Solution: The first equation contains a single second-degree term in x, and is the equation of a vertical parabola. The second equation is linear. Since the first equation is already written with y in terms of x, we can substitute $x^2 - 2x - 3$ for y in the second equation to obtain

$$\begin{aligned} 2x - y &= 7 && \text{second equation} \\ 2x - (x^2 - 2x - 3) &= 7 && \text{substitute } x^2 - 2x - 3 \text{ for } y \\ 2x - x^2 + 2x + 3 &= 7 && \text{distribute} \\ -x^2 + 4x + 3 &= 7 && \text{simplify} \\ x^2 - 4x + 4 &= 0 && \text{set equal to zero} \\ (x - 2)^2 &= 0 && \text{factor} \end{aligned}$$

$x = 2$ is a repeated root.

Since the second equation is simpler than the first, we substitute 2 for x in this equation and find $y = -3$. The system has only one (repeated) solution at $(2, -3)$, which is supported by the graph of the system shown in the figure.

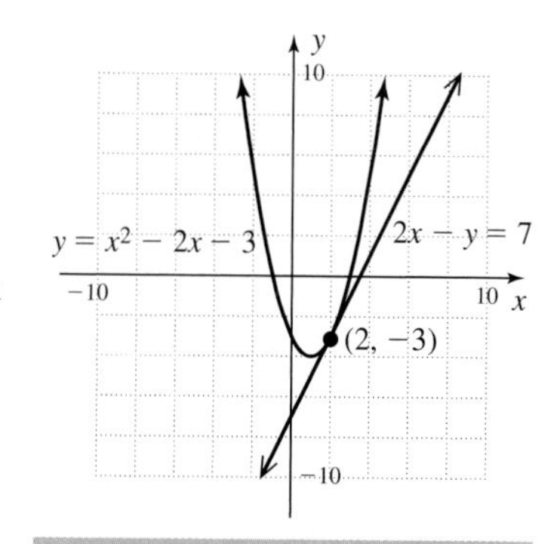

WORTHY OF NOTE

Similar to our earlier work with repeated roots of polynomials, note the graph of $y = x^2 - 2x - 3$ "bounces" off (is tangent to) the line $y = 2x - 7$.

NOW TRY EXERCISES 13 THROUGH 18

C. Solving Nonlinear Systems by Elimination

When both equations in the system have second-degree terms with like variables, it is generally easier to use the elimination method, rather than substitution. As in Chapter 6, watch for systems that have no solutions.

EXAMPLE 3 Solve the system using elimination: $\begin{cases} 2y^2 - 5x^2 = 13 \\ 3x^2 + 4y^2 = 39 \end{cases}$.

Solution: The first equation contains a *difference of second-degree terms* and is the equation of a central hyperbola. The second has a *sum of second-degree terms* with unequal coefficients, and represents a central ellipse. By mentally visualizing the possibilities, there could be zero, one, two, three, or four points of intersection (see Example 1). However, with both centered at (0, 0), we find there can only be zero, two, or four solutions. After writing the system so that x- and y-terms are in the same order, we find that multiplying the first equation by -2 will help eliminate the variable y:

$$\begin{cases} 10x^2 - 4y^2 = -26 & \text{rewrite first equation and multiply by } -2 \\ 3x^2 + 4y^2 = 39 & \text{original second equation} \end{cases}$$

$$13x^2 + 0 = 13 \quad \text{add}$$

$$x^2 = 1 \quad \text{divide by 13}$$
$$x = -1 \quad \text{or} \quad x = 1 \quad \text{square root property}$$

Substituting $x = 1$ and $x = -1$ into the second equation we obtain:

$$\begin{aligned} 3(1)^2 + 4y^2 &= 39 \\ 3 + 4y^2 &= 39 \\ 4y^2 &= 36 \\ y^2 &= 9 \\ y = -3 \quad &\text{or} \quad y = 3 \end{aligned} \qquad \begin{aligned} 3(-1)^2 + 4y^2 &= 39 \\ 3 + 4y^2 &= 39 \\ 4y^2 &= 36 \\ y^2 &= 9 \\ y = -3 \quad &\text{or} \quad y = 3 \end{aligned}$$

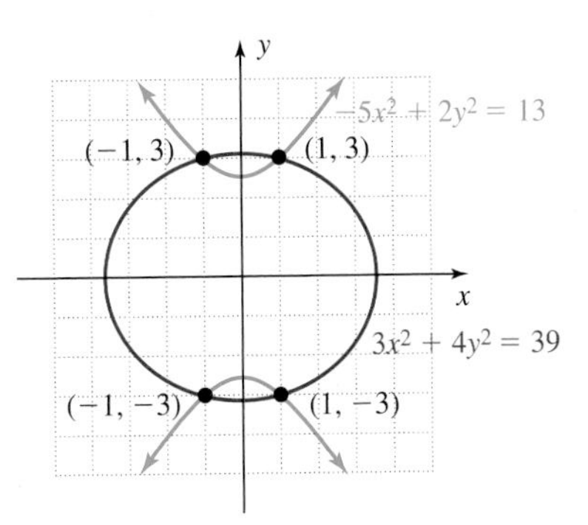

Since -1 and 1 each generated *two outputs,* there are a total of *four* ordered pair solutions: $(1, -3)$, $(1, 3)$, $(-1, -3)$, and $(-1, 3)$. Once again, the graph shown supports these calculations.

NOW TRY EXERCISES 19 THROUGH 24

Nonlinear systems may involve other relations as well, including power, polynomial, logarithmic, or exponential functions. These are solved using the same methods.

EXAMPLE 4 Solve the system using the method of your choice:
$$\begin{cases} y = -\log(x + 7) + 2 \\ y = \log(x + 4) + 1 \end{cases}.$$

Solution: Since both equations have y written in terms of x, substitution appears to be the better choice. The result is a logarithmic equation, which we can solve using the techniques from Chapter 5.

$\log(x + 4) + 1 = -\log(x + 7) + 2$	substitute $\log(x + 4) + 1$ for y in first equation
$\log(x + 4) + \log(x + 7) = 1$	add $\log(x + 7)$; subtract 1
$\log(x + 4)(x + 7) = 1$	product property of logarithms
$(x + 4)(x + 7) = 10^1$	exponential form
$x^2 + 11x + 18 = 0$	eliminate parentheses and set equal to zero
$(x + 9)(x + 2) = 0$	factor
$x + 9 = 0$ or $x + 2 = 0$	zero factor theorem
$x = -9$ or $x = -2$	possible solutions

By inspection, we see that $x = -9$ is not a solution, since $\log(-9 + 4)$ and $-\log(-9 + 7)$ are not real numbers. Substituting -2 for x in the second equation we find one form of the (exact) solution is $(-2, \log 2 + 1)$. If we substitute -2 for x in the first equation the exact solution is $(-2, -\log 5 + 2)$. Use a calculator to verify the answers are equivalent and approximately $(-2, 1.3)$.

NOW TRY EXERCISES 25 THROUGH 36

D. Using a Graphing Calculator to Solve Nonlinear Systems

To solve nonlinear systems using a graphing calculator, be aware that most calculators can only graph *functions* and not relations like the conic sections. We can work around this limitation by writing the equation in two parts, each of which is a function. In the following discussion, all keystrokes are illustrated using a TI-84 Plus calculator. For assistance with other models, please consult your manual or go to www.mhhe.com/coburn.

Consider the equation of a circle centered at (0, 0) with radius $r = 3$: $x^2 + y^2 = 9$. Solving for y, to write the equation in function form, gives

$$x^2 + y^2 = 9 \quad \text{given}$$
$$y^2 = 9 - x^2 \quad \text{subtract } x^2$$
$$Y_1 = -\sqrt{9 - x^2} \text{ or } Y_2 = \sqrt{9 - x^2} \quad \text{square root property}$$

Figure 7.18

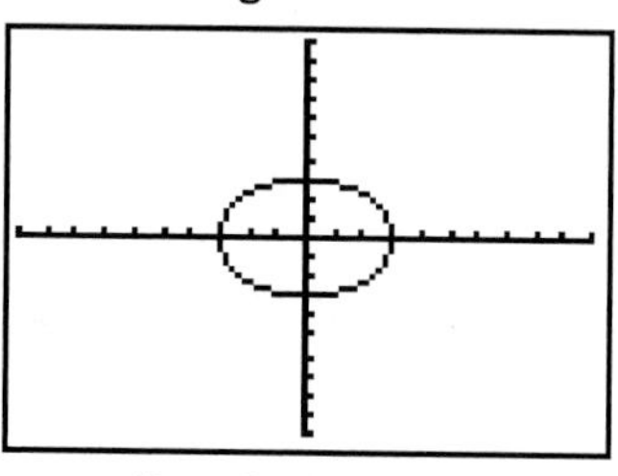

Standard window

The equations of two **semicircles** are the result. The negative radical gives the lower half of the circle and the positive radical gives the upper half. These can be entered as Y_1 and Y_2 on the Y= screen of a graphing calculator to produce a full circle. However, due to the limitations of most calculators, the graph will not appear circular in the standard window (Figure 7.18), and we usually need to apply various ZOOM options to produce a better graph. See Figures 7.19 through 7.21.

Figure 7.19

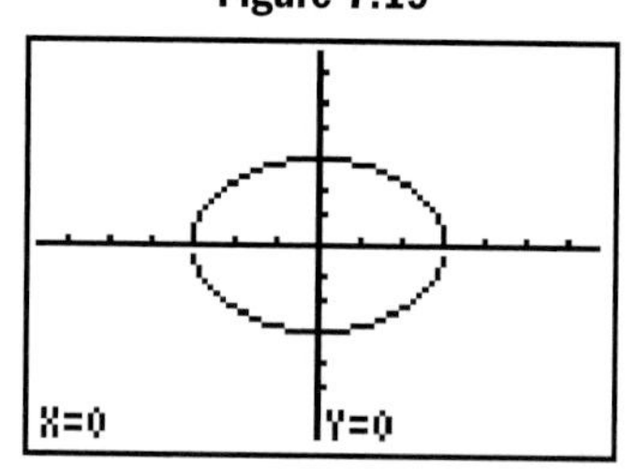

ZOOM 2: Zoom In; the graph still appears elliptical.

Figure 7.20

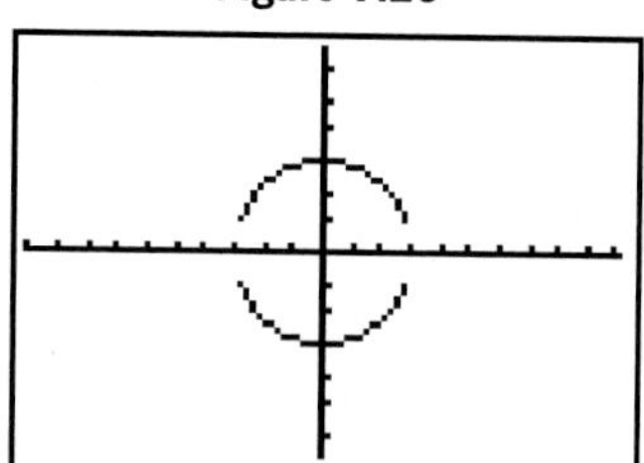

ZOOM 5: ZSquare; the graph appears circular, but broken.

Figure 7.21

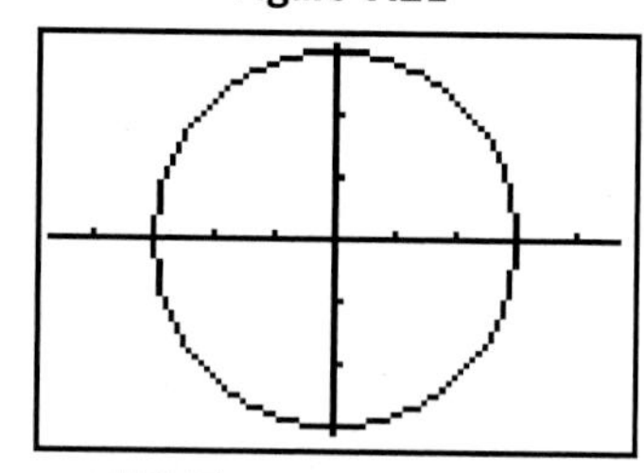

ZOOM 4: ZDecimal; the graph appears nicely circular.

The equations of an ellipse and a hyperbola can likewise be entered. However, as you use your calculator to solve systems of nonlinear equations—you must remember that they are entered in two parts and you may have to jump from Y_1 to Y_2 using the up arrow ▲ and down arrow ▼ keys. We'll illustrate by re-solving the system from Example 3.

EXAMPLE 5 Solve the system $\begin{cases} 2y^2 - 5x^2 = 13 \\ 3x^2 + 4y^2 = 39 \end{cases}$ using a graphing calculator.

Solution: **1.** Solving for y in the first equation gives $y = \sqrt{\dfrac{5x^2 + 13}{2}}$ and $y = -\sqrt{\dfrac{5x^2 + 13}{2}}$.

Solving for y in the second gives $y = \sqrt{\dfrac{39 - 3x^2}{4}}$ and $y = -\sqrt{\dfrac{39 - 3x^2}{4}}$.

2. Enter these as Y_1 and Y_2 for the first equation, with Y_3 and Y_4 for the second respectively:

$$Y_1 = \sqrt{\frac{5x^2 + 13}{2}},\ Y_2 = -\sqrt{\frac{5x^2 + 13}{2}};\ Y_3 = \sqrt{\frac{39 - 3x^2}{4}},\ Y_4 = -\sqrt{\frac{39 - 3x^2}{4}}.$$

Note you could also use $Y_2 = -Y_1$ and $Y_4 = -Y_3$.

3. Press ZOOM 6 to graph the equations on the standard screen (Figure 7.22), then apply any of the ZOOM options to obtain the best graph possible. ZOOM **4:ZDecimal** (Figure 7.23) *almost* gives a good view, but too much of the hyperbola is "missing." Instead of applying other ZOOM options, we simply adjust the window using **Ymin = −5** and **Ymax = 5** (Figure 7.24).

Figure 7.22

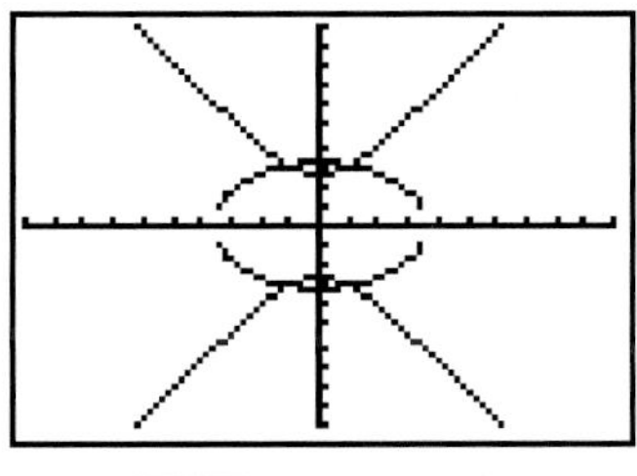

ZOOM 6: ZStandard; graph appears broken.

Figure 7.23

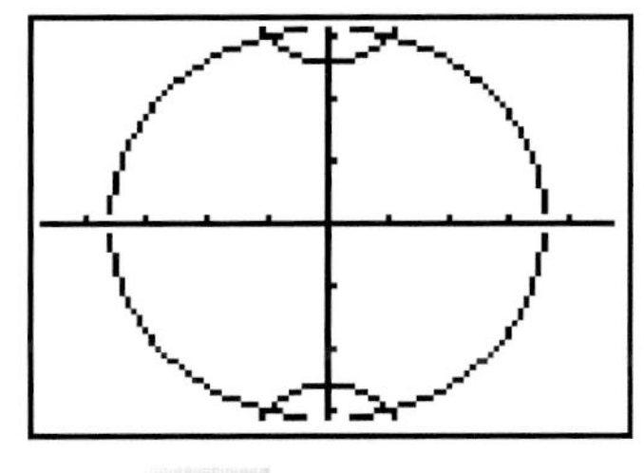

ZOOM 4: ZDecimal; hyperbola is cut off.

Figure 7.24

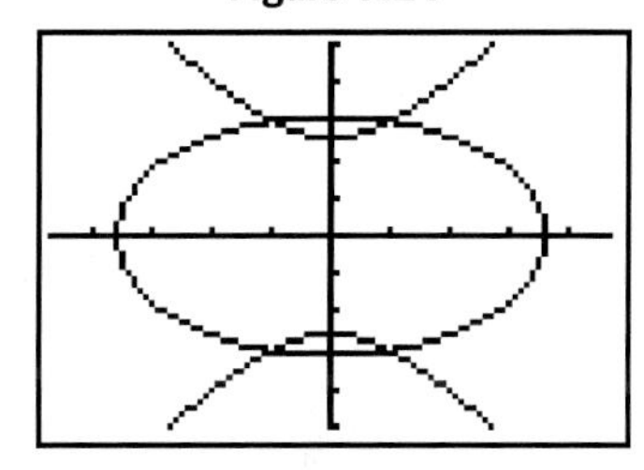

Ymin = −5, Ymax = 5, others as is; both graphs are plainly visible.

4. Press the TRACE key, then use the arrow keys to walk the cursor near any point of intersection. Use 2nd CALC **5:intersect** to begin identifying points of intersection. Be sure to *correctly identify each curve* for the calculator, because it will "run down the list" in the order the functions were entered. In other words, after identifying Y_1 by pressing ENTER, the calculator jumps to Y_2, *which does not intersect* Y_1! Use the down arrow ▼ to skip Y_2, then press ENTER to identify Y_3. Press ENTER once again (skip "Guess") to have the calculator find this point of intersection. Figure 7.25 shows that one solution is at (1, 3). Using this process to locate the other points of intersection (or using the symmetry of the graphs) reveals the remaining solutions are (−1, 3), (−1, −3), and (1, −3). Check the solutions in the original equations.

Figure 7.25

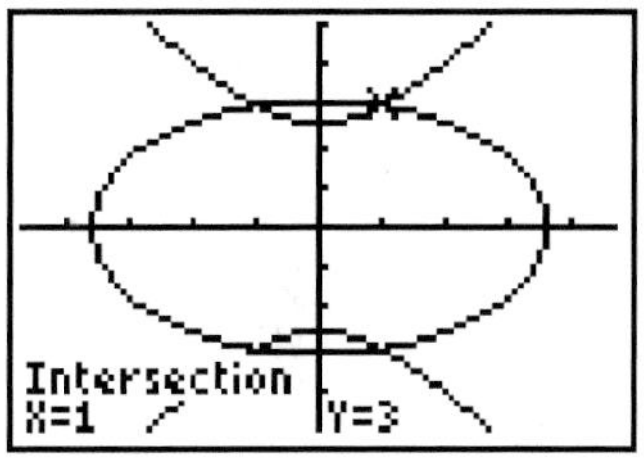

Intersection
X=1 Y=3

NOW TRY EXERCISES 37 THROUGH 42

E. Systems and Nonlinear Inequalities

Figure 7.26

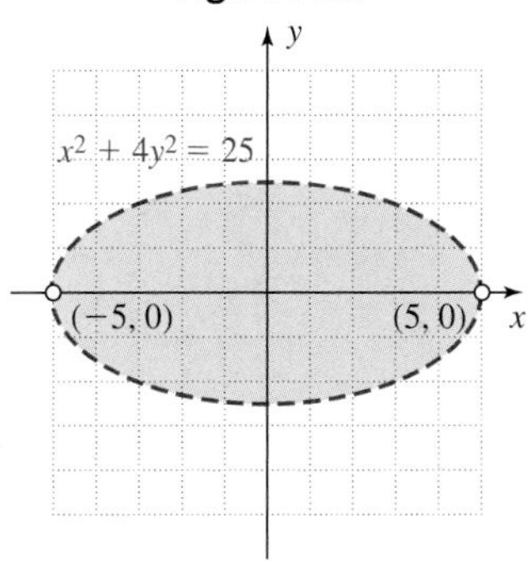

As with our previous work with inequalities in two variables (Sections 6.3 and 6.4), nonlinear inequalities can be solved by graphing the boundary given by the related equation, and checking the regions that result using a test point. For example, the inequality $x^2 + 4y^2 < 25$ is solved by graphing $x^2 + 4y^2 = 25$ [a central ellipse with vertices at (−5, 0) and (5, 0)], deciding if the boundary is included or excluded (in this case it is not), and using a test point from either the "outside" or "inside" region formed. The test point (0, 0) results in a true statement since $(0)^2 + 4(0)^2 < 25$, so the inside of the ellipse is shaded to indicate the solution region (Figure 7.26). For a ***system* of nonlinear inequalities,** we identify regions where the solution set for each inequality overlap, paying special attention to points of intersection.

EXAMPLE 6 Solve the system: $\begin{cases} x^2 + 4y^2 < 25 \\ -x + 4y \geq 5 \end{cases}$.

Solution: We recognize the first inequality from Figure 7.26, an ellipse with vertices at $(-5, 0)$ and $(5, 0)$, and a solution region in the interior. The second inequality is linear and after solving for x in the related equation we use a substitution to find points of intersection (if they exist). For $-x + 4y = 5$, we have $x = 4y - 5$ and substitute $4y - 5$ for x in $x^2 + 4y^2 = 25$:

$$\begin{aligned} x^2 + 4y^2 &= 25 && \text{given} \\ (4y - 5)^2 + 4y^2 &= 25 && \text{substitute } 4y - 5 \text{ for } x \\ 20y^2 - 40y + 25 &= 25 && \text{expand and simplify} \\ y^2 - 2y &= 0 && \text{subtract 25; divide by 20} \\ y(y - 2) &= 0 && \text{factor} \\ y = 0 \quad \text{or} \quad y &= 2 && \text{result} \end{aligned}$$

Back-substitution shows the graphs intersect at $(-5, 0)$ and $(3, 2)$. Graphing a line through these points and using $(0, 0)$ as a test point shows the upper half plane is the solution region for the linear inequality $[-(0) + 4(0) \geq 5$ is *false*$]$. The overlapping (solution) region for *both* inequalities is the elliptical sector shown. Note the points of intersection are graphed using "open dots," (see figure) since points on the graph of the ellipse are excluded from the solution set.

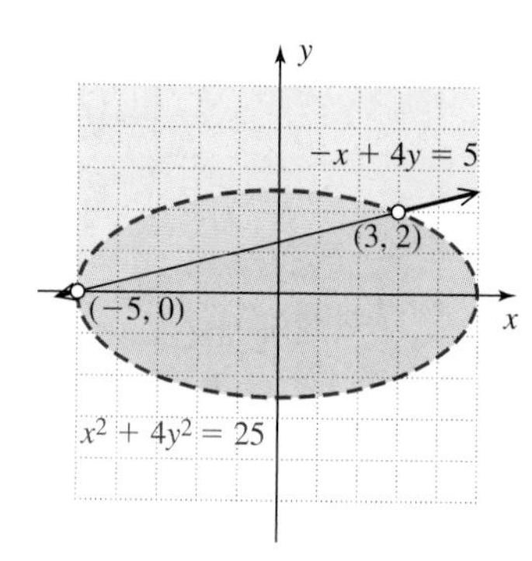

NOW TRY EXERCISES 43 THROUGH 50

7.3 EXERCISES

CONCEPTS AND VOCABULARY

1. Draw sketches showing the different ways each pair of relations can intersect and give one, two, three, and/or four points of intersection. If a given number of intersections is not possible, so state.

   a. circle and line **b.** parabola and line **c.** circle and parabola

   d. circle and hyperbola **e.** hyperbola and ellipse **f.** circle and ellipse

2. By inspection only, identify the systems having *no solutions* and justify your choices.

   a. $\begin{cases} y^2 - x^2 = 16 \\ x^2 + y^2 = 9 \end{cases}$ **b.** $\begin{cases} y = x^2 + 4 \\ x^2 + 4y^2 = 4 \end{cases}$ **c.** $\begin{cases} y = x + 1 \\ 3x^2 + 4y^2 = 12 \end{cases}$

   a and b; circle between hyperbola and parabola above ellipse

3. The solution to a system of nonlinear inequalities is a(n) region of the plane where the solutions for each individual inequality overlap.

4. When both equations in the system have at least one second -degree term, it is generally easier to use the elimination method to find a solution.

Additional answers can be found in the Instructor Answer Appendix.

5. Suppose a nonlinear system contained a central hyperbola and an exponential function. Are three solutions possible? Are four solutions possible? Explain/discuss. Answers will vary.

6. Solve the system twice, once using elimination, then again using substitution. Compare/contrast each process and comment on which is more efficient in this case: $\begin{cases} 4x^2 + y^2 = 25 \\ 2x^2 + y = 5 \end{cases}$. Answers will vary.

DEVELOPING YOUR SKILLS

Identify each equation in the system as that of a line, parabola, circle, ellipse, or hyperbola, and solve the system by graphing.

7. $\begin{cases} x^2 + y = 6 \\ x + y = 4 \end{cases}$ first: parabola; second: line

8. $\begin{cases} -2x + y = 4 \\ 4x^2 + y^2 = 16 \end{cases}$ first: line; second: ellipse

9. $\begin{cases} y - x^2 = -1 \\ 4x^2 + y^2 = 100 \end{cases}$ first: parabola; second: ellipse

10. $\begin{cases} x^2 + y^2 = 25 \\ x^2 + y = 13 \end{cases}$ first: circle; second: parabola

11. $\begin{cases} x^2 - y^2 = 9 \\ x^2 + y^2 = 41 \end{cases}$ first: hyperbola; second: circle

12. $\begin{cases} 4x^2 - y^2 = 36 \\ y^2 + 9x^2 = 289 \end{cases}$ first: hyperbola; second: ellipse

Solve using substitution. [*Hint:* Substitute for x^2 (not y) in Exercises 17 and 18.]

13. $\begin{cases} x^2 + y^2 = 25 \\ y - x = 1 \end{cases}$ $(-4, -3), (3, 4)$

14. $\begin{cases} x + 7y = 50 \\ x^2 + y^2 = 100 \end{cases}$ $(8, 6), (-6, 8)$

15. $\begin{cases} x^2 + 4y^2 = 25 \\ x + 2y = 7 \end{cases}$ $\left(4, \frac{3}{2}\right), (3, 2)$

16. $\begin{cases} x^2 - 2y^2 = 8 \\ x + y = 6 \end{cases}$ $(20, -14), (4, 2)$

17. $\begin{cases} x^2 + y = 13 \\ 9x^2 - y^2 = 81 \end{cases}$ $(\sqrt{10}, 3), (-\sqrt{10}, 3), (5, -12), (-5, -12)$

18. $\begin{cases} y - x^2 = -10 \\ 4x^2 + y^2 = 100 \end{cases}$ $(0, -10), (4, 6), (-4, 6)$

Solve using elimination.

19. $\begin{cases} x^2 + y^2 = 25 \\ 2x^2 - 3y^2 = 5 \end{cases}$ $(4, 3), (4, -3), (-4, 3), (-4, -3)$

20. $\begin{cases} y^2 - x^2 = 12 \\ x^2 + y^2 = 20 \end{cases}$ $(2, 4), (2, -4), (-2, 4), (-2, -4)$

21. $\begin{cases} x^2 - y = 4 \\ x^2 - y^2 = 16 \end{cases}$ no solutions

22. $\begin{cases} 4x^2 + y^2 = 13 \\ x^2 + y^2 = 1 \end{cases}$ no solutions

23. $\begin{cases} 5x^2 - 2y^2 = 75 \\ 2x^2 + 3y^2 = 125 \end{cases}$ $(5, -5), (5, 5), (-5, 5), (-5, -5)$

24. $\begin{cases} 3x^2 - 7y^2 = 20 \\ 4x^2 + 9y^2 = 45 \end{cases}$ $(3, 1), (3, -1), (-3, 1), (-3, -1)$

Solve using the method of your choice.

25. $\begin{cases} y = \log x + 5 \\ y = 6 - \log(x - 3) \end{cases}$ $(5, \log 5 + 5)$

26. $\begin{cases} y = \log(x + 4) + 1 \\ y = 2 - \log(x + 7) \end{cases}$ $(-2, \log 2 + 1)$

27. $\begin{cases} y = 4^{x+3} \\ y - 2^{x^2+3x} = 0 \end{cases}$ $(-3, 1), (2, 1024)$

28. $\begin{cases} y - 3^{x^2+2x} = 0 \\ y = 9^{x+2} \end{cases}$

29. $\begin{cases} x^3 - y = 2x \\ y - 5x = -6 \end{cases}$

30. $\begin{cases} y - x^3 = -2 \\ y + 4 = 3x \end{cases}$

31. $\begin{cases} x^2 + 2y^2 = 17 \\ y + x^2 = 11 \end{cases}$

32. $\begin{cases} x^2 - 3y^2 = 6 \\ y - 3x^2 = -26 \end{cases}$

33. $\begin{cases} 3y^2 - 5x^2 = 7 \\ xy = 6 \end{cases}$

34. $\begin{cases} x^2 + 4y^2 = 20 \\ xy = 4 \end{cases}$

35. $\begin{cases} 2y^2 + xy - 7 = x^2 \\ x - 2y = 5 \end{cases}$

36. $\begin{cases} x^2 + 3y - xy = 27 \\ y - x = -2 \end{cases}$

Solve each system using a graphing calculator. Round solutions to hundredths (as needed).

37. $\begin{cases} 3x^2 + 4y^2 = 35 \\ 5y^2 - x^2 = 1 \end{cases}$

38. $\begin{cases} 5x^2 - 2y^2 = 30 \\ y + 2x = x^2 - 3 \end{cases}$

39. $\begin{cases} y = 2^x - 3 \\ y + 2x^2 = 9 \end{cases}$

40. $\begin{cases} y = -2\log(x + 8) \\ y + x^3 = 4x - 2 \end{cases}$

41. $\begin{cases} y = \dfrac{1}{(x - 3)^2} + 2 \\ (x - 3)^2 + y^2 = 10 \end{cases}$

42. $\begin{cases} y^2 - x^2 = 5 \\ y = \dfrac{1}{x - 1} + 2 \end{cases}$

Solve each system of inequalities.

43. $\begin{cases} y - x^2 \geq 1 \\ x + y \leq 3 \end{cases}$

44. $\begin{cases} x^2 + y^2 \leq 25 \\ x + 2y \leq 5 \end{cases}$

45. $\begin{cases} x^2 + y^2 > 9 \\ 25x^2 + 16y^2 \leq 400 \end{cases}$

7. $(-1, 5)$, $(2, 2)$

8. $(-2, 0)$, $(0, 4)$

9. $(-3, 8)$, $(3, 8)$

10. $(-3, 4)$, $(3, 4)$, $(-4, -3)$, $(4, -3)$

11. $(-5, 4)$, $(5, 4)$, $(-5, -4)$, $(5, -4)$

12. $(-5, 8)$, $(5, 8)$, $(-5, -8)$, $(5, -8)$

28. $(2, 6561), (-2, 1)$
29. $(-3, -21), (1, -1), (2, 4)$
30. $(-2, -10), (1, -1)$
31. $(3, 2), (-3, 2), \left(\frac{5\sqrt{2}}{2}, \frac{-3}{2}\right), \left(-\frac{5\sqrt{2}}{2}, \frac{-3}{2}\right)$
32. $(3, 1), (-3, 1), \left(\sqrt{\frac{226}{27}}, \frac{-8}{9}\right), \left(-\sqrt{\frac{226}{27}}, \frac{-8}{9}\right)$
33. $(2, 3), (-2, -3)$
34. $(-2, -2)$ $(2, 2), (4, 1), (-4, -1)$
35. $\left(\frac{11}{15}, \frac{-32}{15}\right)$
36. $\left(\frac{33}{5}, \frac{23}{5}\right)$
37. $(3, 1.41), (-3, 1.41), (3, -1.41), (-3, -1.41)$
38. $(4, 5), (2.6, -1.4)$
39. $(-2.43, -2.81), (2, 1)$
40. $(0.05, -1.81), (2, -2), (-2.05, -1.55)$
41. $(0.72, 2.19), (2, 3), (4, 3), (5.28, 2.19)$
42. $(0.77, -2.37), (2, 3)$
43.
44.
45.

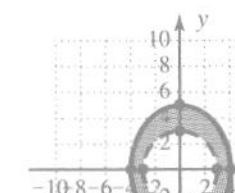

46. no solution

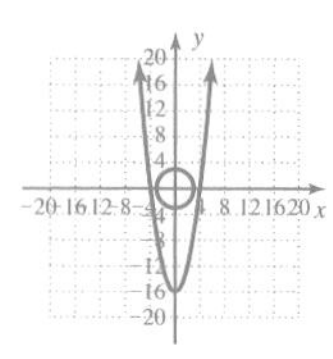

47. 48.

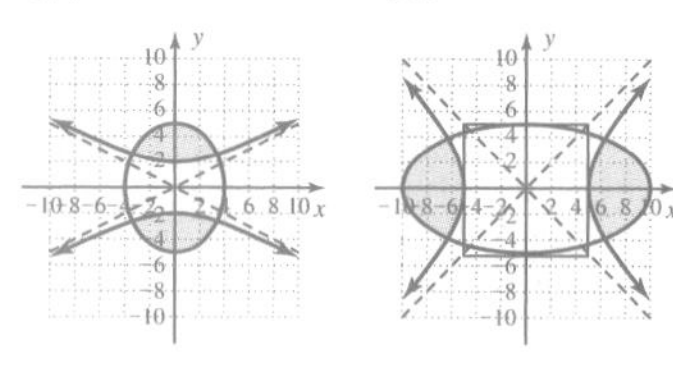

49. no solution 50.

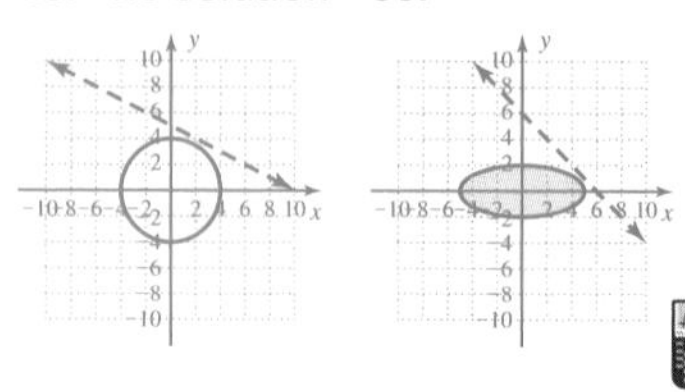

46. $\begin{cases} y - x^2 \leq -16 \\ y^2 + x^2 < 9 \end{cases}$

47. $\begin{cases} 4y^2 - x^2 \geq 16 \\ 25x^2 + 16y^2 \leq 400 \end{cases}$

48. $\begin{cases} 4y^2 + x^2 \leq 100 \\ x^2 - y^2 \geq 25 \end{cases}$

49. $\begin{cases} x^2 + y^2 \leq 16 \\ x + 2y > 10 \end{cases}$

50. $\begin{cases} 25y^2 + 4x^2 \leq 100 \\ x + y < 6 \end{cases}$

WORKING WITH FORMULAS

51. Tunnel clearance: $h = b\sqrt{1 - \left(\frac{d}{a}\right)^2}$

The maximum rectangular clearance allowed by an elliptical tunnel can be found using the formula shown, where $\left\{\frac{x^2}{a^2} + \frac{y^2}{b^2} = 1\right.$ models the tunnel's elliptical cross section and h is the height of the tunnel at a distance d from the center. If $a = 50$ and $b = 30$, find the maximum clearance at distances of $d = 20$, 30, and 40 ft from center. $h \approx 27.5$ ft; $h = 24$ ft; $h = 18$ ft

52. Manufacturing cylindrical vents: $\begin{cases} A = 2\pi rh \\ V = \pi r^2 h \end{cases}$

In the manufacture of cylindrical vents, a rectangular piece of sheet metal is rolled, riveted, and sealed to form the vent. The radius and height required to form a vent with a specified volume, using a piece of sheet metal with a given area, can be found by solving the system shown. Use the system to find the radius and height if the volume required is 4071 cm^3 and the area of the rectangular piece is 2714 cm^2. $r = 3$ cm, $h \approx 143.98$ cm

APPLICATIONS

Solve by setting up and solving a system of nonlinear equations.

53. Dimensions of a flag: A large American flag has an area of 85 m^2 and a perimeter of 37 m. Find the dimensions of the flag. 8.5 m $\times$ 10 m

54. Dimensions of a sail: The sail on a boat is a right triangle with a perimeter of 36 ft and a hypotenuse of 15 ft. Find the height and width of the sail. 9 ft, 12 ft

55. Dimensions of a tract: The area of a rectangular tract of land is 45 km^2. The length of a diagonal is $\sqrt{106}$ km. Find the dimensions of the tract. 5 km, 9 km

56. Dimensions of a deck: A rectangular deck has an area of 192 ft^2 and the length of the diagonal is 20 ft. Find the dimensions of the deck. 12 ft, 16 ft

57. Dimensions of a trailer: The surface area of a rectangular trailer with square ends is 928 ft^2. If the sum of all edges of the trailer is 164 ft, find its dimensions. $8 \times 8 \times 25$ ft

58. Dimensions of a cylindrical tank: The surface area of a closed cylindrical tank is 192π m^2. Find the dimensions of the tank if the volume is 320π m^3 and the radius is as small as possible. $r = 4$ m, $h = 20$ m

Market equilibrium: In a free-enterprise (supply and demand) economy, the amount buyers are willing to pay for an item and the number of these items manufacturers are willing to produce depend on the price of the item. As the price increases, demand for the item decreases since buyers are less willing to pay the higher price. On the other hand, an increase in price increases the supply of the item since manufacturers are now more willing to supply it. When the **supply and demand curves** are graphed, their point of intersection is called the **market equilibrium** for the item.

59. Suppose the monthly market demand D (in ten-thousands of gallons) for a new synthetic oil is related to the price P in dollars by the equation $10P^2 + 6D = 144$. For the market price P, assume the amount D that manufacturers are willing to supply is modeled by $8P^2 - 8P - 4D = 12$. (a) What is the minimum price at which manufacturers are willing to begin supplying the oil? (b) Use this information to create a system of nonlinear equations, then solve the system to find the market equilibrium price (per gallon) and the quantity of oil supplied and sold at this price. $1.83; $3 90,000 gal $\begin{cases} 10P^2 + 6D = 144 \\ 8P^2 - 8P - 4D = 12 \end{cases}$

60. The weekly demand D for organically grown carrots (in thousands of pounds) is related to the price per pound P by the equation $8P^2 + 4D = 84$. At this market price, the amount that growers are willing to supply is modeled by the equation $8P^2 + 6P - 2D = 48$. (a) What is the minimum price at which growers are willing to supply the organically grown carrots? (b) Use this information to create a system of nonlinear equations, then solve the system to find the market equilibrium price (per pound) and the quantity of carrots supplied and sold at this price. $2.11 8500 lb $2.50 $\begin{cases} 8P^2 + 4D = 84 \\ 8P^2 + 6P - 2D = 48 \end{cases}$

WRITING, RESEARCH, AND DECISION MAKING

61. The area of a vertical parabolic segment is given by $A = \frac{2}{3}BH$, where B is the length of the horizontal base of the segment and H is the height from the base to the vertex. Investigate how this formula can be used to find the *area* of the solution region for the general system of inequalities shown. $\begin{cases} y \geq x^2 - bx + c \\ y \leq c + bx - x^2 \end{cases}$

(*Hint:* Begin by investigating with $b = 6$ and $c = 8$, then use other values and try to generalize what you find.) Answers will vary.

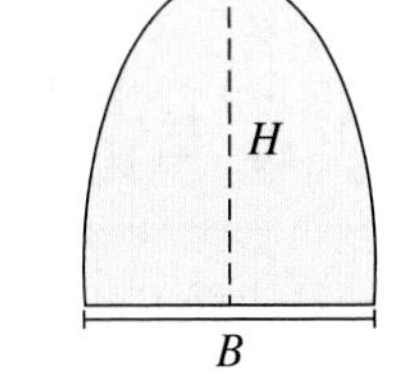

62. For what values of r will the volume of a sphere be numerically equal to its surface area? For what values of r will the volume of a cylinder be numerically equal to its lateral surface area? Can a similar relationship be found for the volume and lateral surface area of a cone? Why or why not? $r = 3$; $r = 2$; yes, $r = \frac{3h}{\sqrt{h^2 - 9}}$ where $h > 3$ is the height of the cone

EXTENDING THE CONCEPT

63. Find the area of the parallelogram formed by joining the points where the hyperbola $xy = 4$ and the ellipse $x^2 + 4y^2 = 20$ intersect. 12 units^2

64. Solve the nonlinear system: $\begin{cases} y = 3(4^x) - 8 \\ y = 4^{2x} - 2(4^x) - 4 \end{cases}$ (1, 4) (0, −5)

65. A rectangular fish tank has a bottom and four sides made out of glass. Use a system of equations to help find the dimensions of the tank if the height is 18 in., surface area is 4806 in^2, the tank must hold 108 gal (1 gal = 231 in^3), and all three dimensions are integers. 18 in. by 18 in. by 77 in.

MAINTAINING YOUR SKILLS

66. (1.3) Solve by factoring:

a. $2x^2 + 5x - 63 = 0$ $x = -7, x = \frac{9}{2}$

b. $4x^2 - 121 = 0$ $x = \frac{-11}{2}, x = \frac{11}{2}$

c. $2x^3 - 3x^2 - 8x + 12 = 0$ $x = 2, x = -2, x = \frac{3}{2}$

67. (1.3) Solve each equation:

a. $3x^2 + 4x - 12 = 0$ $x = \frac{-2 \pm 2\sqrt{10}}{3}$

b. $\sqrt{3x + 1} - \sqrt{2x} = 1$ $x = 0, x = 8$

c. $\frac{1}{x + 2} + \frac{3}{x^2 + 5x + 6} = \frac{2}{x + 3}$ $x = 2$

69. a. b.

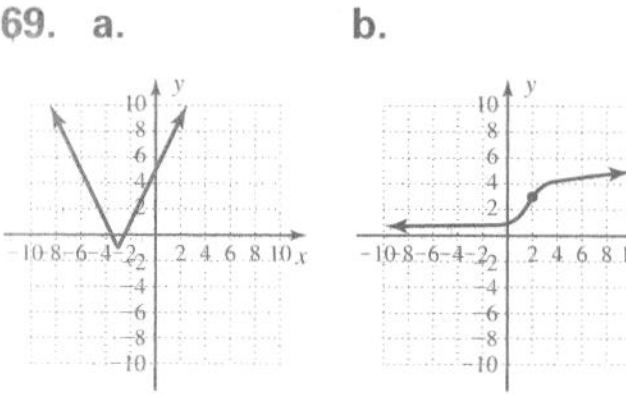

c. d.

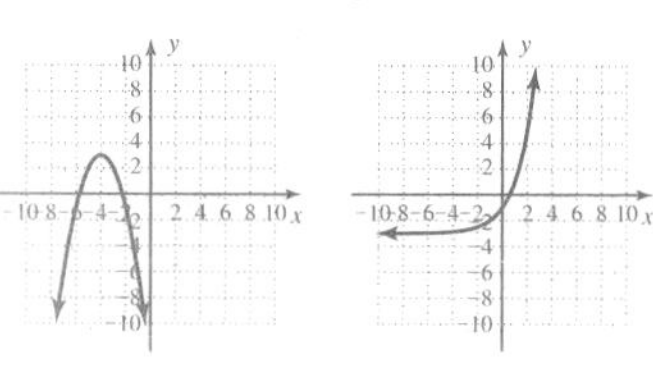

68. (4.3) Use the rational roots theorem to help find all solutions, real and complex: $x^4 + x^3 + 10x^2 + 12x - 24 = 0$
$x = -2, x = 1, \pm 2\sqrt{3}i$

69. (3.3/5.1) Sketch each transformation:

a. $y = 2|x + 3| - 1$

b. $y = \sqrt[3]{x - 2} + 3$

c. $y = -(x + 4)^2 + 3$

d. $y = 2^{x+1} - 3$

70. (6.2) Solve using any method. As an investment for retirement, Donovan bought three properties for a total of \$250,000. Ten years later, the first property had doubled in value, the second property had tripled in value, and the third property was worth \$10,000 less than when he bought it, for a current value of \$485,000. Find the original purchase price if he paid \$20,000 more for the first property than he did for the second. \$95,000, \$75,000, \$80,000

71. (2.3) In 2001, a small business purchased a copier for \$4500. In 2004, the value of the copier had decreased to \$3300. Assuming the depreciation is linear: (a) find the rate-of-change $m = \frac{\Delta\text{value}}{\Delta\text{time}}$ and discuss its meaning in this context; (b) find the depreciation equation; and (c) use the equation to predict the copier's value in 2008. (d) If the copier is traded in for a new model when its value is less than \$700, how long will the company use this copier?
$m = \frac{-400}{1}$, the copier depreciates by \$400 a year. $y = -400x + 4500$ \$1700 9.5 yr

7.4 Foci and the Analytic Ellipse and Hyperbola

LEARNING OBJECTIVES

In Section 7.4 you will learn how to:

A. Locate the foci of an ellipse and use the foci and other features to construct the equation of an ellipse

B. Locate the foci of a hyperbola and use the foci and other features to construct the equation of a hyperbola

C. Solve applications involving foci

INTRODUCTION

Previously, we developed equations for the ellipse and hyperbola by looking at changes in the center-shifted form of the equation of a circle. While this development sheds some light on their equations and related graphs, it limited our ability to use these conics in some significant ways. In this section we develop the equation of the ellipse and hyperbola from their analytic definition.

POINT OF INTEREST

Until the time of Johannes Kepler (1571–1630) astronomers assumed, for philosophical and aesthetic reasons, that all heavenly bodies moved in circular orbits. However, Kepler noted that the careful planetary observations of Tycho Brahe (1546–1601) could not be explained or predicted by such orbits. After years of careful study and searching, Kepler discovered that planetary orbits are actually elliptical, a result he published in his book *New Astronomy* in 1609. This is now referred to as Kepler's first law. Additional discoveries followed soon after. Kepler's second law states that a line joining the planet and the Sun sweeps out equal areas in equal intervals of time, meaning a planet moves slower near its aphelion, and very fast near its perihelion.

A. The Foci of an Ellipse

The Museum of Science and Industry in Chicago, Illinois (http://www.msichicago.org), has a permanent exhibit called the *Whispering Gallery.* The construction of the room is based on some of the reflective properties of an ellipse. If two people stand at designated points in the room and one of them whispers very softly, the other person can hear the whisper quite clearly—even though they are over 40 feet apart! The point at which each person stands is called a **focus** of the ellipse (together they are called the **foci**). This reflective property also applies to light and radiation, giving the ellipse some powerful applications in science, medicine, acoustics, and other areas. To understand and appreciate these applications, we introduce the analytic definition of an ellipse:

WORTHY OF NOTE

You can easily draw an ellipse that satisfies the definition. Press two pushpins (these form the foci of the ellipse) halfway down into a piece of heavy cardboard about 6 in. apart. Take an 8-in. piece of string and loop each end around the pins. Use a pencil to draw the string taut and keep it taut as you move the pencil in a circular motion—and the result is an ellipse! A different length of string or a different distance between the foci will produce a different ellipse.

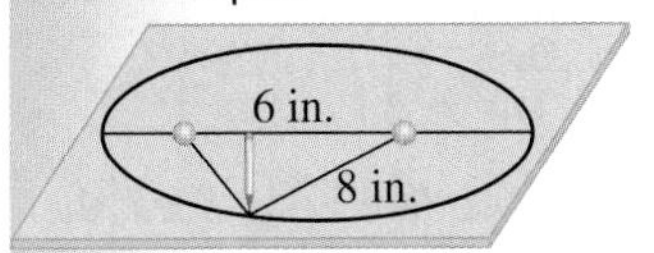

DEFINITION OF AN ELLIPSE

Given any two fixed points F_1 and F_2 in a plane, an ellipse is defined to be the set of all points $P(x, y)$ such that the distance $|F_1P|$ added to the distance $|F_2P|$ remains constant. In symbols,

$$|F_1P| + |F_2P| = k$$

The fixed points F_1 and F_2 are called the *foci* of the ellipse, and the points $P(x, y)$ are points on the graph of the ellipse.

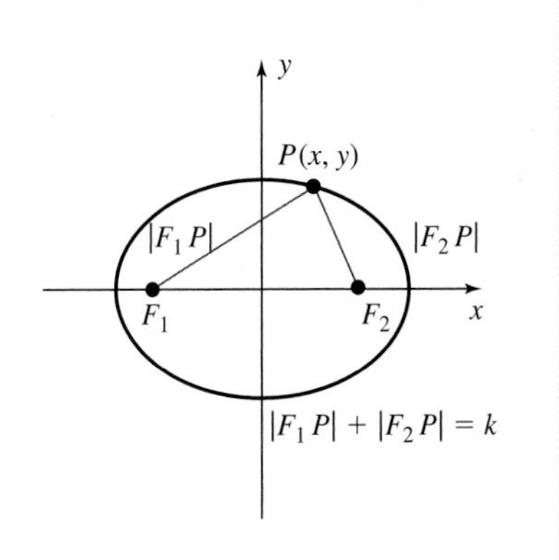

The equation of an ellipse is obtained by combining the definition just given with the distance formula. Consider the general ellipse shown in Figure 7.27 (for calculating ease we use a central ellipse). Note the vertices have coordinates $(-a, 0)$ and $(a, 0)$, and the endpoints of the minor axis have coordinates $(0, -b)$ and $(0, b)$ as before. It is customary to assign foci the coordinates $F_1 \rightarrow (-c, 0)$ and $F_2 \rightarrow (c, 0)$. We can calculate the distance between $(c, 0)$ and any point $P(x, y)$ on the ellipse using the distance formula: $\sqrt{(x - c)^2 + (y - 0)^2}$. Likewise the distance between $(-c, 0)$ and any point (x, y) is $\sqrt{(x + c)^2 + (y - 0)^2}$. According to the definition, the sum must be constant: $\sqrt{(x - c)^2 + (y - 0)^2} + \sqrt{(x + c)^2 + (y - 0)^2} = k$.

Figure 7.27

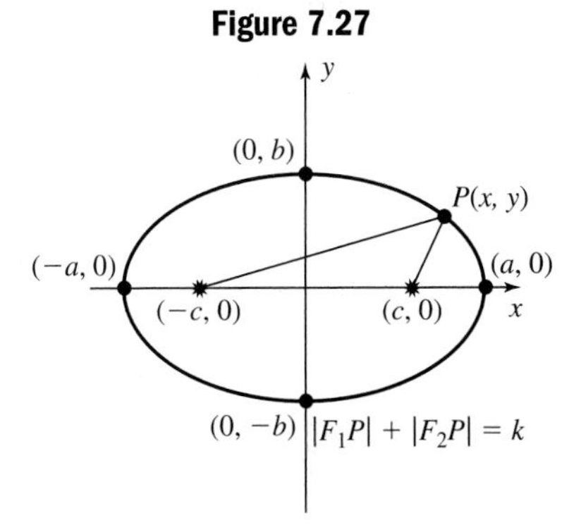

EXAMPLE 1 Use the definition of an ellipse and the diagram given to determine the "length of the string" used to form this ellipse (see *Worthy of Note*). Note that $a = 5$, $b = 3$, and $c = 4$.

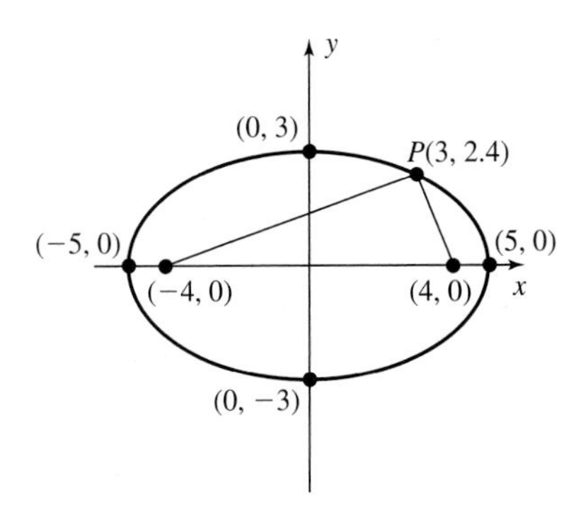

WORTHY OF NOTE

Note that if the foci are coincident (both at the origin) the "ellipse" will actually be a circle with radius $\frac{k}{2}$; $\sqrt{x^2 + y^2} + \sqrt{x^2 + y^2} = k$ leads to $x^2 + y^2 = \frac{k^2}{4}$. In Example 1 we found $k = 10$, giving $\frac{10}{2} = 5$, and if we used the "string" to draw the circle the pencil would be 5 units from the center creating a circle of radius 5.

Solution:

$$\sqrt{(x - c)^2 + (y - 0)^2} + \sqrt{(x + c)^2 + (y - 0)^2} = k \quad \text{given}$$
$$\sqrt{(3 - 4)^2 + (2.4 - 0)^2} + \sqrt{(3 + 4)^2 + (2.4 - 0)^2} = k \quad \text{substitute}$$
$$\sqrt{(-1)^2 + 2.4^2} + \sqrt{7^2 + 2.4^2} = k \quad \text{add}$$
$$\sqrt{6.76} + \sqrt{54.76} = k \quad \text{simplify radicals}$$
$$2.6 + 7.4 = k \quad \text{compute square roots}$$
$$10 = k \quad \text{result}$$

The "length of string" used to form this ellipse is 10 units long.

NOW TRY EXERCISES 7 THROUGH 10

In Example 1, the length of the string could also be found by moving the point P to the location of a vertex, then using the symmetry of the ellipse. This helps to show the constant k is equal to $2a$ *regardless of the distance between foci.* When $P(x, y)$ is coincident with vertex $(a, 0)$, the length of the "string" is identical to the length of the major axis, since the overlapping part of the string from $(c, 0)$ to $(a, 0)$ is the same length as from $(-a, 0)$ to $(-c, 0)$.

The result is an equation with two radicals, similar to those in Section 1.3.

$$\sqrt{(x - c)^2 + (y - 0)^2} + \sqrt{(x + c)^2 + (y - 0)^2} = 2a$$

To simplify this equation, we isolate one of the radicals and square both sides, then isolate the resulting radical expression and square again. The details are given in Appendix IV, and the result is very close to the standard form we saw in Section 7.1: $\frac{x^2}{a^2} + \frac{y^2}{a^2 - c^2} = 1$.

By comparing the standard form $\frac{x^2}{a^2} + \frac{y^2}{b^2} = 1$ with $\frac{x^2}{a^2} + \frac{y^2}{a^2 - c^2} = 1$, we might suspect that $b^2 = a^2 - c^2$, and this is indeed the case. Note from Example 1 the relationship yields

$$b^2 = a^2 - c^2$$
$$3^2 = 5^2 - 4^2$$
$$9 = 25 - 16 \checkmark$$

Additionally, when we consider that $(0, b)$ is a point on the ellipse, the distance from $(0, b)$ to $(c, 0)$ *must be equal to* a due to symmetry (the "constant distance" used to form the ellipse is always $2a$). See Figure 7.28. The Pythagorean theorem (with a as the hypotenuse) gives $b^2 + c^2 = a^2$ or $b^2 = c^2 - a^2$.

While the equation $\frac{x^2}{a^2} + \frac{y^2}{b^2} = 1$ is identical to the one obtained in the Section 7.3, we now have the ability to *locate the foci of any ellipse*—an important step toward using the ellipse in practical applications. Because we're often asked to find the location of the foci, it is best to remember the relationship as $c^2 = |a^2 - b^2|$, with the absolute value bars used to allow for a vertical major axis. Also note that for an ellipse $c < a$ (*major axis horizontal*) or $c < b$ (*major axis vertical*).

Figure 7.28

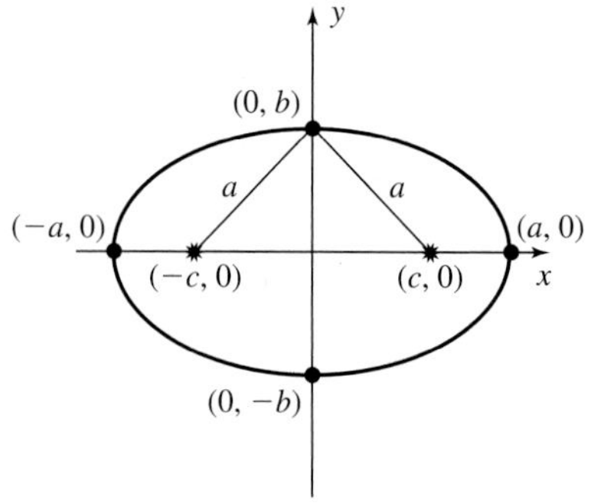

EXAMPLE 2 For the ellipse defined by $25x^2 + 9y^2 - 100x - 54y - 44 = 0$, find the coordinates of the center, vertices, foci, and endpoints of the minor axis. Then sketch the graph.

$$25x^2 + 9y^2 - 100x - 54y - 44 = 0 \quad \text{given}$$

$$25x^2 - 100x + 9y^2 - 54y = 44 \quad \text{group terms; add 44}$$

$$25(x^2 - 4x + \_\_) + 9(y^2 - 6y + \_\_) = 44 \quad \text{factor out lead coefficients}$$

$$25(x^2 - 4x + 4) + 9(y^2 - 6y + 9) = 44 + 100 + 81 \quad \text{complete the square}$$

adds $25(4) = 100$ adds $9(9) = 81$ add $100 + 81$ to right-hand side

$$25(x - 2)^2 + 9(y - 3)^2 = 225 \quad \text{center-shifted form}$$

$$\frac{25(x - 2)^2}{225} + \frac{9(y - 3)^2}{225} = \frac{225}{225} \quad \text{divide by 225}$$

$$\frac{(x - 2)^2}{9} + \frac{(y - 3)^2}{25} = 1 \quad \text{simplify (standard form)}$$

$$\frac{(x - 2)^2}{3^2} + \frac{(y - 3)^2}{5^2} = 1 \quad \text{write denominators in squared form}$$

The result shows a vertical ellipse with $a = 3$ and $b = 5$. The center of the ellipse is at (2, 3). The vertices are a vertical distance of 5 units from center at (2, 8) and (2, −2). The endpoints of the minor axis are a horizontal distance of 3 units from center at (−1, 3) and (5, 3). To locate the foci, we use the foci formula for an ellipse: $c^2 = |a^2 - b^2|$, giving $c^2 = |3^2 - 5^2| = 16$. The result indicates the foci "✷" are located a vertical distance of 4 units from center at (2, 7) and (2, −1).

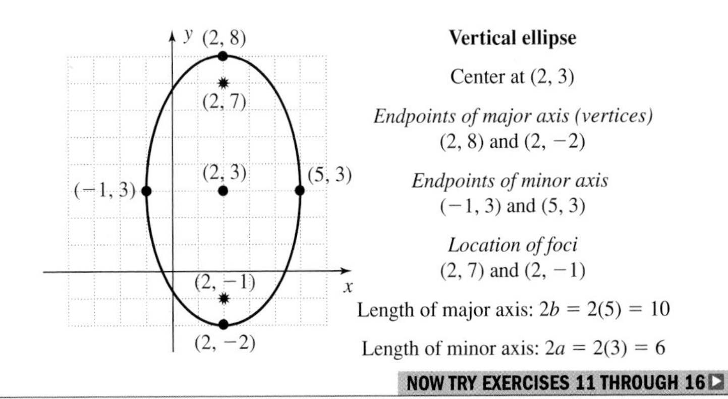

NOW TRY EXERCISES 11 THROUGH 16

For future reference, remember the foci of an ellipse always occur on the major axis, with $a > c$ and $a^2 > c^2$ for a horizontal ellipse. This makes it easier to remember the **foci formula** for ellipses: $c^2 = |a^2 - b^2|$. Since a^2 is larger, it must be decreased by b^2 to equal c^2.

If any two of the values for a, b, and c are known, the relationship between them can be used to construct the equation of the ellipse.

LOOKING AHEAD

For the hyperbola, we'll find that $c > a$, and the formula for the foci of a hyperbola will be $c^2 = a^2 + b^2$.

EXAMPLE 3 Find the equation of the ellipse (in standard form) that has foci at $(0, -2)$ and $(0, 2)$, with a minor axis 6 units in length.

Solution: Since the foci must be on the major axis, we know this is a vertical and central ellipse with $c = 2$ and $c^2 = 4$. The minor axis has a length of $2a = 6$ units, meaning $a = 3$ and $a^2 = 9$. To find b^2, use the foci equation and solve:

$c^2 = |a^2 - b^2|$ foci equation (ellipse)

$4 = |9 - b^2|$ substitute

$-4 = 9 - b^2 \qquad 4 = 9 - b^2$ solve

$b^2 = 13 \qquad b^2 = 5$ result

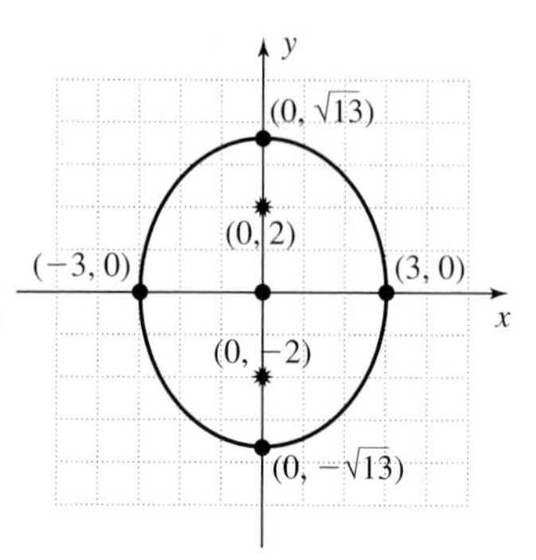

Since we know b^2 must be greater than a^2 (the major axis is always longer), $b^2 = 5$ can be discarded. The standard form is $\frac{x^2}{3^2} + \frac{y^2}{(\sqrt{13})^2} = 1$.

NOW TRY EXERCISES 17 THROUGH 20

B. The Foci of a Hyperbola

Like the ellipse, the foci of a hyperbola play an important part in their application. A long distance radio navigation system (called LORAN for short), can be used to determine the location of ships and airplanes and is based on the characteristics of a hyperbola (see Exercises 55 and 56). Hyperbolic mirrors are also used in some telescopes, and have the property that a beam of light directed at one focus will be reflected to the second focus. To understand and appreciate these applications, we use the analytic definition of a hyperbola:

DEFINITION OF A HYPERBOLA

Given any two fixed points F_1 and F_2 in a plane, a hyperbola is defined to be the set of all points $P(x, y)$ such that the distance $|F_2P|$ subtracted from the distance $|F_1P|$ is a positive constant. In symbols,

$$|F_1P| - |F_2P| = k,\ k > 0$$

The fixed points F_1 and F_2 are called the foci of the hyperbola, and the points $P(x, y)$ are points on the graph of the hyperbola.

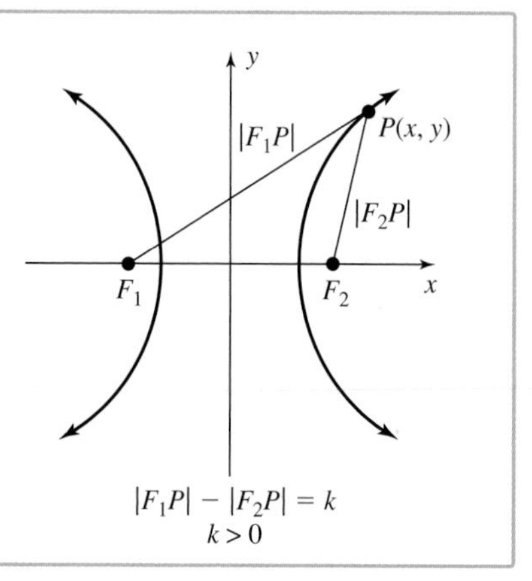

Figure 7.29

The general equation of a hyperbola is obtained by combining the foci definition with the distance formula. Consider the hyperbola shown in Figure 7.29 (for calculating ease we use a central hyperbola). Note the vertices have coordinates $(-a, 0)$ and $(a, 0)$ and as before, we assign foci the coordinates $F_1 \to (-c, 0)$ and $F_2 \to (c, 0)$. We can calculate the distance between $(c, 0)$ and any point $P(x, y)$ on the hyperbola using the distance formula: $\sqrt{(x - c)^2 + (y - 0)^2}$. Likewise the distance between $(-c, 0)$ and any point $P(x, y)$ is $\sqrt{(x + c)^2 + (y - 0)^2}$. According to the definition the *difference* must equal a positive constant: $\left|\sqrt{(x+c)^2+(y-0)^2}-\sqrt{(x-c)^2+(y-0)^2}\right| = k$. The absolute value is used to allow the point $P(x, y)$ to be on either branch of the hyperbola.

EXAMPLE 4 Use the definition of a hyperbola and the diagram given to determine the "constant" and "positive" length used to form the hyperbola. Note that $a = 4$, $b = 3$, and $c = 5$.

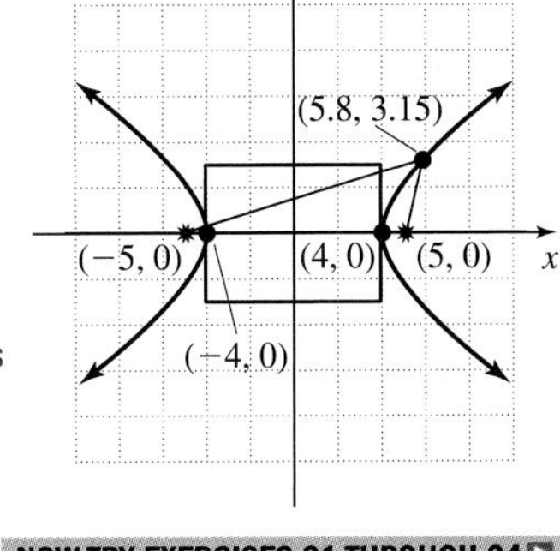

Solution:

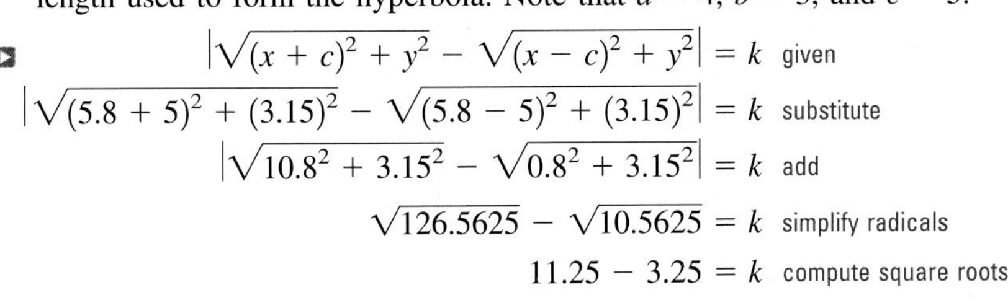

$$\left|\sqrt{(x+c)^2+y^2}-\sqrt{(x-c)^2+y^2}\right| = k \quad \text{given}$$
$$\left|\sqrt{(5.8+5)^2+(3.15)^2}-\sqrt{(5.8-5)^2+(3.15)^2}\right| = k \quad \text{substitute}$$
$$\left|\sqrt{10.8^2+3.15^2}-\sqrt{0.8^2+3.15^2}\right| = k \quad \text{add}$$
$$\sqrt{126.5625}-\sqrt{10.5625} = k \quad \text{simplify radicals}$$
$$11.25 - 3.25 = k \quad \text{compute square roots}$$
$$8 = k \quad \text{result}$$

The positive constant used to form this hyperbola is 8.

NOW TRY EXERCISES 21 THROUGH 24

To actually derive the general equation, it helps to note the constant distance k is again equal to $2a$ (as seen in Example 4), *regardless of the location of the foci.* When the point $P(x, y)$ is directly over the vertex $(a, 0)$, we can take the distance from F_1 to $(a, 0)$ and subtract the distance from F_2 to $(a, 0)$ and end up with a distance identical to that between the vertices themselves, which is $2a$. The result is again an equation with two radicals.

$$\left|\sqrt{(x+c)^2+y^2}-\sqrt{(x-c)^2+y^2}\right| = 2a$$

To develop the equation, assume the first distance is greater than the second and drop the absolute value bars. To solve, we would isolate one of the radicals and square both sides, then isolate the resulting radical expression and square again. The complete derivation is given in Appendix IV, and the result is very close to the standard form we saw in Section 7.2.

$$\frac{x^2}{a^2}-\frac{y^2}{c^2-a^2} = 1$$

By comparing the standard form $\frac{x^2}{a^2}-\frac{y^2}{b^2} = 1$ with $\frac{x^2}{a^2}-\frac{y^2}{c^2-a^2} = 1$, we suspect that $b^2 = c^2 - a^2$, and this is once again the case. From Example 4 the relationship yields:

$$b^2 = c^2 - a^2$$
$$3^2 = 5^2 - 4^2$$
$$9 = 25 - 16 \checkmark$$

We now have the ability to *find the foci of any hyperbola*—and can use this information in many significant applications. Since the location of the foci play such an important role, it is best to remember the relationship as $c^2 = a^2 + b^2$ (called the **foci formula** for hyperbolas), noting that for a hyperbola, $c > a$ and $c^2 > a^2$ (also $c > b$ and $c^2 > b^2$). Recall the asymptotes of any hyperbola can be found by counting off the slope ratio $\frac{\text{rise}}{\text{run}} = \pm\frac{b}{a}$ beginning at the center, or by drawing a central rectangle of dimensions $2a$ by $2b$ and using the extended diagonals of the rectangle.

EXAMPLE 5 For the hyperbola defined by $7x^2 - 9y^2 - 14x + 72y - 200 = 0$, find the coordinates of the center, vertices, foci, and the dimensions of the central rectangle. Then sketch the graph.

Solution:

$$7x^2 - 9y^2 - 14x + 72y - 200 = 0 \quad \text{given}$$

$$7x^2 - 14x - 9y^2 + 72y = 200 \quad \text{group terms; add 200}$$

$$7(x^2 - 2x + \_\_) - 9(y^2 - 8y + \_\_) = 200 \quad \text{factor out lead coefficients}$$

$$7(x^2 - 2x + 1) - 9(y^2 - 8y + 16) = 200 + 7 + (-144) \quad \text{complete the square}$$

adds $7(1) = 7$; adds $-9(16) = -144$; $\rightarrow$ add $7 + (-144)$ to right-hand side

$$7(x - 1)^2 - 9(y - 4)^2 = 63 \quad \text{center-shifted form}$$

$$\frac{(x - 1)^2}{9} - \frac{(y - 4)^2}{7} = 1 \quad \text{divide by 63 and simplify}$$

$$\frac{(x - 1)^2}{3^2} - \frac{(y - 4)^2}{(\sqrt{7})^2} = 1 \quad \text{write denominators in squared form}$$

From the result we find this is a horizontal hyperbola with $a = 3$ and $a^2 = 9$ and $b = \sqrt{7}$ and $b^2 = 7$. The center of the hyperbola is at $(1, 4)$. The vertices are a horizontal distance of 3 units from center at $(-2, 4)$ and $(4, 4)$. To locate the foci, we use the foci formula for a hyperbola: $c^2 = a^2 + b^2$. This yields $c^2 = 16$, showing the foci are located a horizontal distance of 4 units from center at $(-3, 4)$ and $(5, 4)$. The central rectangle is $2\sqrt{7} \approx 5.29$ by $2(3) = 6$. Draw the rectangle and sketch the asymptotes using the extended diagonals. The completed graph is shown in the figure.

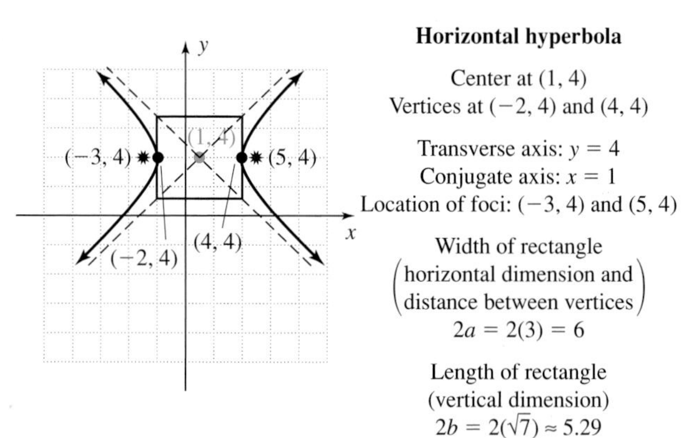

NOW TRY EXERCISES 25 THROUGH 30

As with the ellipse, if any two of the values for a, b, and c are known, the relationship between them can be used to construct the equation of the hyperbola. See Exercises 31 through 34.

C. Applications Involving Foci

Applications involving the foci of a conic section can take various forms. In many cases, only partial information about the ellipse or hyperbola is available and the ideas from Example 3 must be used to "fill in the gaps." In other applications, we must rewrite a known or given equation to find information related to the values of a, b, and c.

EXAMPLE 6 In Washington, D.C., there is a park called the *Ellipse* located between the White House and the Washington Monument. The park is surrounded by a path that forms an ellipse with the length of the major axis being about 1502 ft and the minor axis having a length of 1280 ft. Suppose the park manager wants to install water fountains at the location of the foci. Find the distance between the fountains rounded to the nearest foot.

Solution: Assume the center of the park has the coordinates (0, 0) and that the ellipse is horizontal. Since the major axis has length $2a = 1502$, we know $a = 751$ and $a^2 = 564{,}001$. The minor axis has length $2b = 1280$, meaning $b = 640$ and $b^2 = 409{,}600$. To find c, use the foci equation:

$$\begin{aligned} c^2 &= a^2 - b^2 \\ &= 564{,}001 - 409{,}600 \\ &= 154{,}401 \\ c \approx -393 &\text{ and } c \approx 393 \end{aligned}$$

The distance between the water fountains would be $2(393) = 786$ ft.

NOW TRY EXERCISES 49 THROUGH 52

EXAMPLE 7 As mentioned in the *Point of Interest,* in Section 7.2, comets with a large mass and high velocity cannot be captured by the Sun's gravity, but are slung around the Sun in a hyperbolic path with the Sun at one focus. If the path illustrated to the right is modeled by the equation $2116x^2 - 400y^2 = 846{,}400$, how close did the comet get to the Sun? Assume units are in millions of miles and round to the nearest million.

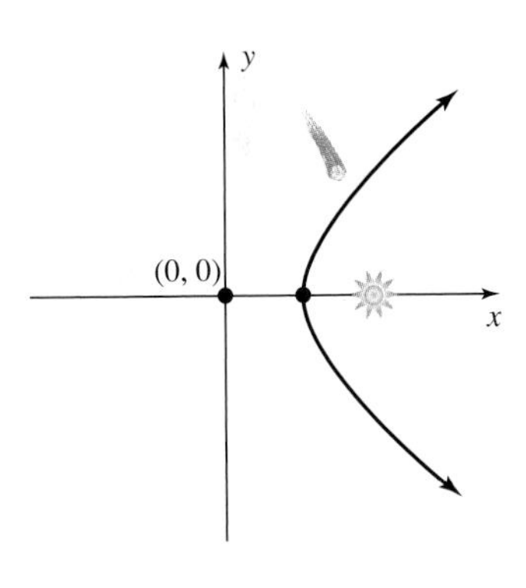

Solution: We are essentially asked to find the distance between a vertex and focus. Begin by writing the equation in standard form:

$$2116x^2 - 400y^2 = 846{,}400 \quad \text{given}$$

$$\frac{x^2}{400} - \frac{y^2}{2116} = 1 \quad \text{divide by 846,400}$$

$$\frac{x^2}{20^2} - \frac{y^2}{46^2} = 1 \quad \text{write denominators in squared form}$$

This is a horizontal hyperbola with $a = 20$ and $a^2 = 400$ and $b = 46$ and $b^2 = 2116$. Use the foci formula for a hyperbola to find c^2 and c.

$$\begin{aligned} c^2 &= a^2 + b^2 \\ c^2 &= 400 + 2116 \\ c^2 &= 2516 \\ c \approx -50 &\text{ and } c \approx 50 \end{aligned}$$

Since $a = 20$ and $|c| \approx 50$, the comet came within $50 - 20 = 30$ million miles of the Sun.

NOW TRY EXERCISES 53 AND 54

TECHNOLOGY HIGHLIGHT

Graphing Calculators and the Definition of a Conic

The keystrokes shown apply to a TI-84 Plus model. Please consult your manual or our Internet site for other models.

Recall that if F_1 and F_2 are the foci of an ellipse and $P(x, y)$ is a point on the graph of the ellipse, then the distance from P to F_1 plus the distance from P to F_2 must be equal to some constant k regardless of the point chosen.

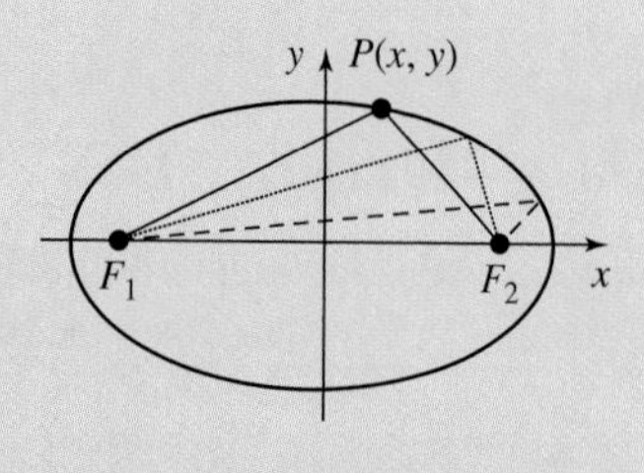

In this *Technology Highlight* we'll use lists L1 through L5, an x-axis ellipse, and the distance formula to check this definition for a select number of points on the ellipse (see the *Technology Highlight* from Sections 1.1 and 2.1 for any needed review in working with lists). To begin, write the equation in standard form, clearly identify the values of a, b, and c, then solve for y and enter the positive root as Y_1 on the Y= screen (only the upper half of the ellipse is used for this exercise). For $4x^2 + 9y^2 = 36$, this leads to $\frac{x^2}{9} + \frac{y^2}{4} = 1$, giving $a = 3$, $b = 2$, and $c = \sqrt{9 - 4} = \sqrt{5}$. Solving for y and simplifying yields $y = \frac{2}{3}\sqrt{9 - x^2}$, which we enter as Y_1. Since the domain of this relation is $x \in [-3, 3]$, we enter the integers from this interval in L1, and the related y-values in L2, using $L2 = Y_1(L1)$. Be sure the cursor is in the header of L2 as you begin [see Figure 7.30(a)]. Next we'll calculate the distance between the points on the ellipse stored as $(x, y) \rightarrow (L1, L2)$, and the foci located at $F_1 = (-\sqrt{5}, 0)$ and $F_2 = (\sqrt{5}, 0)$. For the distance between $(\sqrt{5}, 0)$ and (L1, L2) we'll use $L3 = \sqrt{(L1 - \sqrt{5})^2 + (L2 - 0)^2}$ [see Figure 7.30(b)]. For the distance between $(-\sqrt{5}, 0)$ and (L1, L2) we'll use $L4 = \sqrt{(L1 + \sqrt{5})^2 + (L2 - 0)^2}$. Figure 7.31(a) shows the results of these calculations. Finally, we compute the sum of these two distances using $L5 = L3 + L4$, noting that for all points the sum is equal to 6 [Figure 7.31(b)], which is identical to $2a = 2(3)$.

Figure 7.30(a)

L1	L2	L3
-3	------	------
-2		
-1		
0		
1		
2		
3		

L2 =Y1(L1)

Figure 7.30(b)

L1	L2	L3
-3	0	------
-2	1.4907	
-1	1.8856	
0	2	
1	1.8856	
2	1.4907	
3	0	

L3 =...−√(5))²−L2²)

Figure 7.31(a)

L2	L3	L4
0	5.2361	.76393
1.4907	4.4907	1.5093
1.8856	3.7454	2.2546
2	3	3
1.8856	2.2546	3.7454
1.4907	1.5093	4.4907
0	.76393	5.2361

L2(1)=0

Figure 7.31(b)

L3	L4	L5
5.2361	.76393	6
4.4907	1.5093	6
3.7454	2.2546	6
3	3	6
2.2546	3.7454	6
1.5093	4.4907	6
.76393	5.2361	6

L5(1)=6

Exercise 1: Rework the exercise using the ellipse $4x^2 + 25y^2 = 100$. What do you notice? The sum of corresponding distances is $k = 10$, or $2a = 2(5) = 10$.

Exercise 2: Modify the exercise so that it checks the definition of a hyperbola. Use the hyperbola $9x^2 - 16y^2 = 144$ for verification. Verified. The difference of corresponding distances is $k = 8$, or $2a = 2(4) = 8$.

7.4 EXERCISES

CONCEPTS AND VOCABULARY

Fill in the blank with the appropriate word or phrase. Carefully reread the section if needed.

1. For an ellipse, the relationship between a, b, and c is given by the foci equation $c^2 = |a^2 - b^2|$, since $c < a$ or $c < b$.

2. For a hyperbola, the relationship between a, b, and c is given by the foci equation $c^2 = a^2 + b^2$, since $c > a$ and $c > b$.

11. a. (2, 1)
b. (−3, 1) and (7, 1)
c. $(2 - \sqrt{21}, 1)$ and $(2 + \sqrt{21}, 1)$
d. (2, 3) and (2, −1)
e.

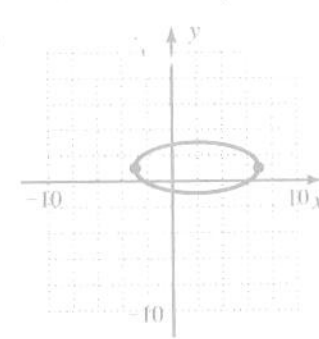

12. a. (3, 2)
b. (−1, 2) and (7, 2)
c. $(3 - \sqrt{7}, 2)$ and $(3 + \sqrt{7}, 2)$
d. (3, 5) and (3, −1)
e.

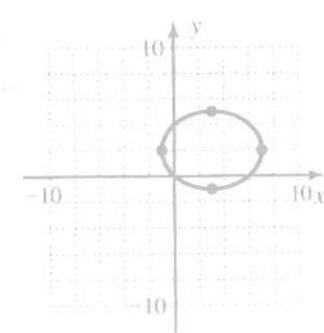

13. a. (4, −3)
b. (4, 2) and (4, −8)
c. (4, 0) and (4, −6)
d. (0, −3) and (8, −3)
e.

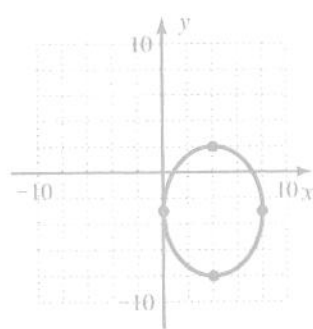

14. a. (−2, 5)
b. (−2, 12) and (−2, −2)
c. $(-2, 5 + 3\sqrt{5})$ and $(-2, 5 - 3\sqrt{5})$
d. (−4, 5) and (0, 5)
e.

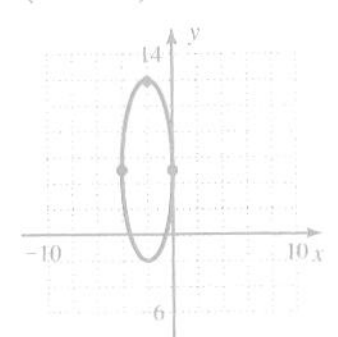

15. a. (−2, −2)
b. (−5, −2) and (1, −2)
c. $(-2 + \sqrt{3}, -2)$ and $(-2 - \sqrt{3}, -2)$
d. $(-2, -2 + \sqrt{6})$ and $(-2, -2 - \sqrt{6})$
e.

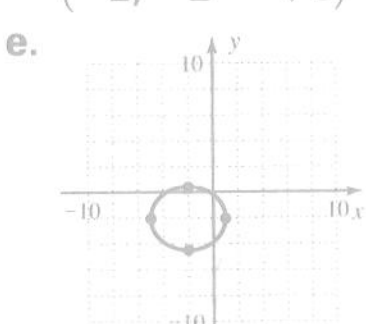

Additional answers can be found in the Instructor Answer Appendix.

3. For a horizontal hyperbola, the length of the transverse axis is $\underline{2a}$ and the length of the conjugate axis is $\underline{2b}$.

4. For a vertical ellipse, the length of the minor axis is $\underline{2a}$ and the length of the major axis is $\underline{2b}$.

5. Suppose foci are located at (−2, 5) and (−2, −3). Discuss/explain the conditions necessary for the graph to be a hyperbola. Answers will vary.

6. Suppose foci are located at (−3, 2) and (5, 2). Discuss/explain the conditions necessary for the graph to be an ellipse. Answers will vary.

DEVELOPING YOUR SKILLS

Use the definition of an ellipse to find the length of the major axes (figures are not drawn to scale).

7.

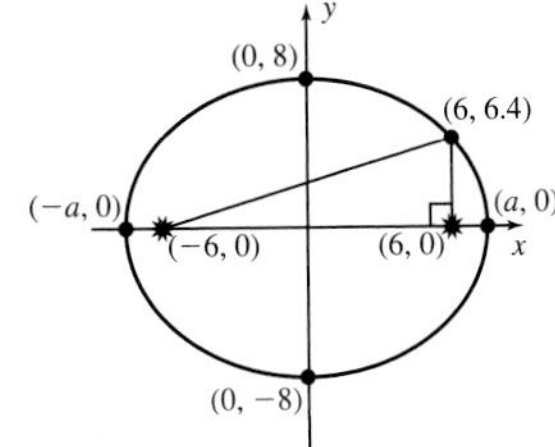

20

8.

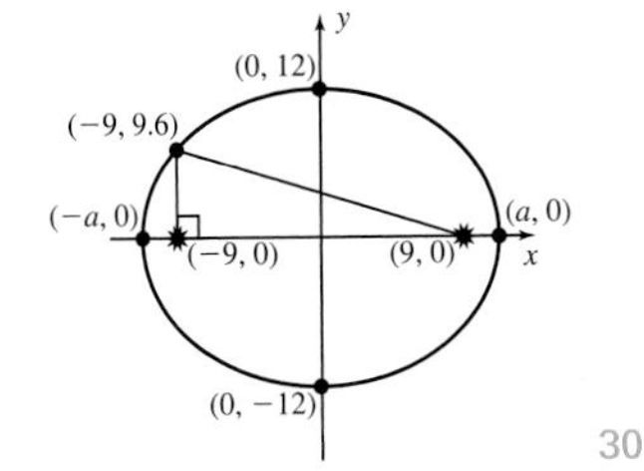

30

9.

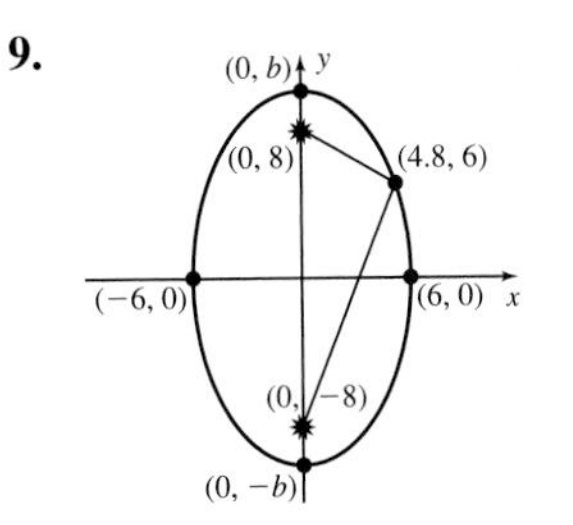

20

10.

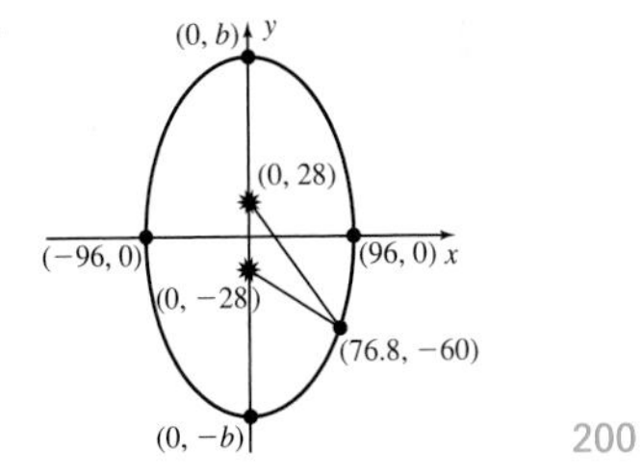

200

Find the coordinates of the (a) center, (b) vertices, (c) foci, and (d) endpoints of the minor axis. Then (e) sketch the graph.

11. $4x^2 + 25y^2 - 16x - 50y - 59 = 0$

12. $9x^2 + 16y^2 - 54x - 64y + 1 = 0$

13. $25x^2 + 16y^2 - 200x + 96y + 144 = 0$

14. $49x^2 + 4y^2 + 196x - 40y + 100 = 0$

15. $6x^2 + 24x + 9y^2 + 36y + 6 = 0$

16. $5x^2 - 50x + 2y^2 - 12y + 93 = 0$

Find the equation of an ellipse (in standard form) that satisfies the following conditions:

17. vertices at (−6, 0) and (6, 0); foci at (−4, 0) and (4, 0) $\quad \frac{x^2}{36} + \frac{y^2}{20} = 1$

18. vertices at (−8, 0) and (8, 0); foci at (−5, 0) and (5, 0) $\quad \frac{x^2}{64} + \frac{y^2}{39} = 1$

19. foci at (0, −4) and (0, 4); length of minor axis: 6 units $\quad \frac{x^2}{9} + \frac{y^2}{25} = 1$

20. foci at (−6, 0) and (6, 0); length of minor axis: 8 units $\quad \frac{x^2}{52} + \frac{y^2}{16} = 1$

Use the definition of a hyperbola to find the distance between the vertices and the dimensions of the rectangle centered at (*h*, *k*). Figures are not drawn to scale. Note that Exercises 23 and 24 are *vertical hyperbolas*.

21.

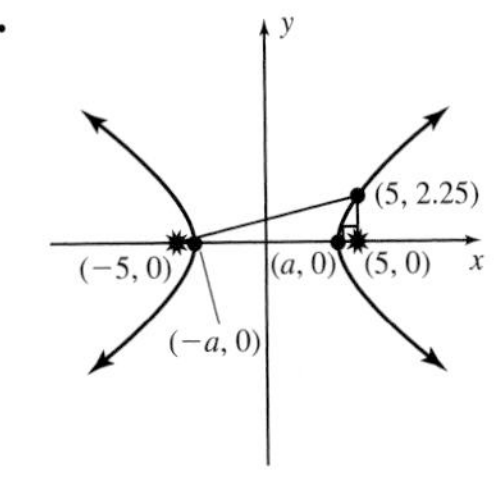

8, 2a = 8, 2b = 6

22.

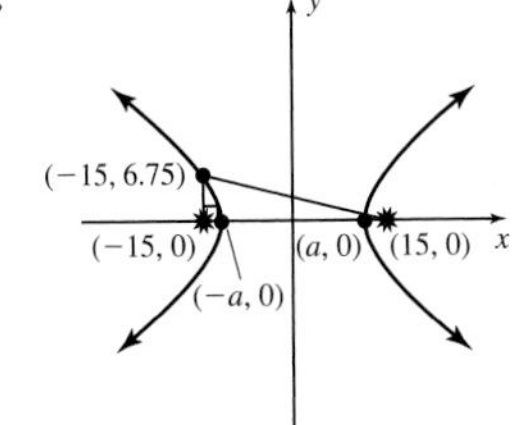

24, 2a = 24, 2b = 18

25. a. (3, 4)
b. (0, 4) and (6, 4)
c. $(3 - \sqrt{13}, 4)$ and $(3 + \sqrt{13}, 4)$
d. $2a = 6, 2b = 4$
e.

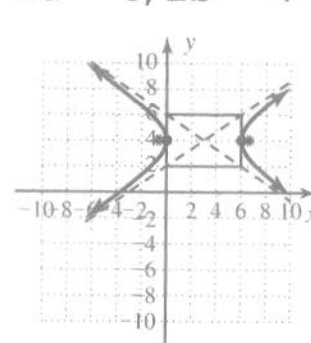

26. a. (5, 2)
b. (−1, 2) and (11, 2)
c. $(5 - 2\sqrt{10}, 2)$ and $(5 + 2\sqrt{10}, 2)$
d. $2a = 12, 2b = 4$
e.

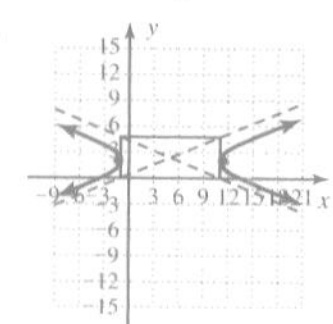

27. a. (0, 3)
b. (−2, 3) and (2, 3)
c. $(-2\sqrt{5}, 3)$ and $(2\sqrt{5}, 3)$
d. $2a = 4,\ 2b = 8$
e.

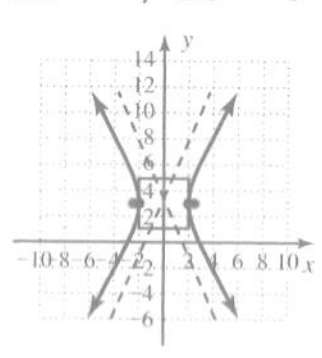

33. $\frac{y^2}{9} - \frac{x^2}{9} = 1$
34. $\frac{x^2}{20} - \frac{y^2}{16} = 1$
35. $(-3\sqrt{5}, 0), (3\sqrt{5}, 0)$
36. $(-\sqrt{5}, 0), (\sqrt{5}, 0)$
37. $(0, 4), (0, -4)$
38. $(0, 2\sqrt{5}), (0, -2\sqrt{5})$
39. $(-\sqrt{6}, 0), (\sqrt{6}, 0)$
40. $(2\sqrt{3}, 0), (-2\sqrt{3}, 0)$
41. $(-\sqrt{13}, 0), (\sqrt{13}, 0)$
42. $(-\sqrt{41}, 0)\ (\sqrt{41}, 0)$
43. $(0, \sqrt{61}), (0, -\sqrt{61})$
44. $(0, 2\sqrt{5}), (0, -2\sqrt{5})$
45. $(-2\sqrt{15}, 0), (2\sqrt{15}, 0)$
46. $(-2\sqrt{15}, 0), (2\sqrt{15}, 0)$
47. $\frac{x^2}{16} + \frac{y^2}{36} = 1$
48. 223.9 million miles

Additional answers can be found in the Instructor Answer Appendix.

23.

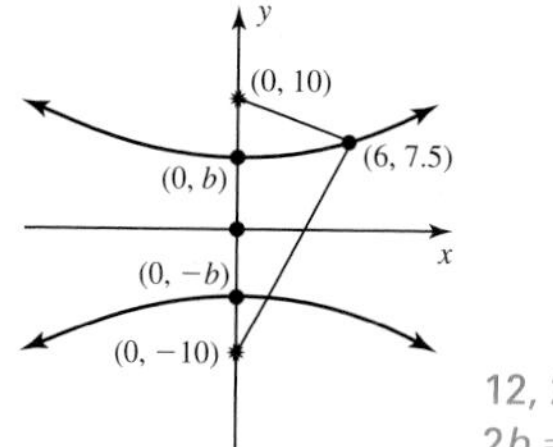

12, $2a = 16$, $2b = 12$

24.

(0, 13)
(−9, 6.25)
(0, b)
(0, −b)
(0, −13)

10, $2a = 24$, $2b = 10$

Find and list the coordinates of the (a) center, (b) vertices, (c) foci, and (d) dimensions of the central rectangle. Then (e) sketch the graph, including the asymptotes.

25. $4x^2 - 9y^2 - 24x + 72y - 144 = 0$

26. $4x^2 - 36y^2 - 40x + 144y - 188 = 0$

27. $16x^2 - 4y^2 + 24y - 100 = 0$

28. $81x^2 - 162x - 4y^2 - 243 = 0$

29. $9x^2 - 3y^2 - 54x - 12y + 33 = 0$

30. $10x^2 + 60x - 5y^2 + 20y - 20 = 0$

Find the equation of the hyperbola (in standard form) that satisfies the following conditions:

31. vertices at $(-6, 0)$ and $(6, 0)$; foci at $(-8, 0)$ and $(8, 0)$

32. vertices at $(-4, 0)$ and $(4, 0)$; foci at $(-6, 0)$ and $(6, 0)$

33. foci at $(0, -3\sqrt{2})$ and $(0, 3\sqrt{2})$; length of conjugate axis: 6 units

34. foci at $(-6, 0)$ and $(6, 0)$; length of conjugate axis: 8 units

Find the coordinates of the foci for the conic sections defined by the equations given. Note that both ellipses and hyperbolas are represented.

35. $\frac{x^2}{49} + \frac{y^2}{4} = 1$

36. $\frac{x^2}{9} + \frac{y^2}{4} = 1$

37. $\frac{x^2}{9} + \frac{y^2}{25} = 1$

38. $\frac{x^2}{16} + \frac{y^2}{36} = 1$

39. $\frac{x^2}{18} + \frac{y^2}{12} = 1$

40. $\frac{x^2}{20} + \frac{y^2}{8} = 1$

41. $\frac{x^2}{4} - \frac{y^2}{9} = 1$

42. $\frac{x^2}{25} - \frac{y^2}{16} = 1$

43. $\frac{y^2}{36} - \frac{x^2}{25} = 1$

44. $\frac{y^2}{16} - \frac{x^2}{4} = 1$

45. $\frac{x^2}{28} - \frac{y^2}{32} = 1$

46. $\frac{x^2}{40} - \frac{y^2}{20} = 1$

WORKING WITH FORMULAS

47. The eccentricity of a conic section: $e = \frac{c}{a}$

In lay terms, the eccentricity of a conic section is a measure of its "roundness," or more exactly, how much the conic section deviates from being "round." A circle has an eccentricity of $e = 0$, since it is perfectly round. Ellipses have an eccentricity between zero and one, or $0 < e < 1$. An ellipse with $e = 0.16$ is closer to circular than one where $e = 0.72$. Here, c represents the distance from the center of the ellipse to its focus, and a represents the length of the semimajor axis (the distance from center to either vertex). Determine which of the following ellipses is closest to being circular: $\frac{x^2}{9} + \frac{y^2}{25} = 1$ or $\frac{x^2}{16} + \frac{y^2}{36} = 1$.

48. The perimeter of an ellipse: $P \approx 2\pi\sqrt{\frac{a^2 + b^2}{2}}$

The perimeter of an ellipse can be *approximated* by the formula shown, where a represents the length of the semimajor axis and b represents the length of the semiminor axis. Estimate the perimeter of the orbit of the planet Mercury, defined by the equation $\frac{x^2}{1296} + \frac{y^2}{1243} = 1$.

APPLICATIONS

49. $\sqrt{7} \approx 2.65$ ft; 2.25 ft

49. Decorative fireplaces: A bricklayer intends to build an elliptical fireplace 3 ft high and 8 ft wide, with two glass doors that open at the middle. The hinges to these doors are to be screwed onto a spine that is perpendicular to the hearth and goes through the foci of the ellipse. How far from center will the spines be located? What is the height of the spine?

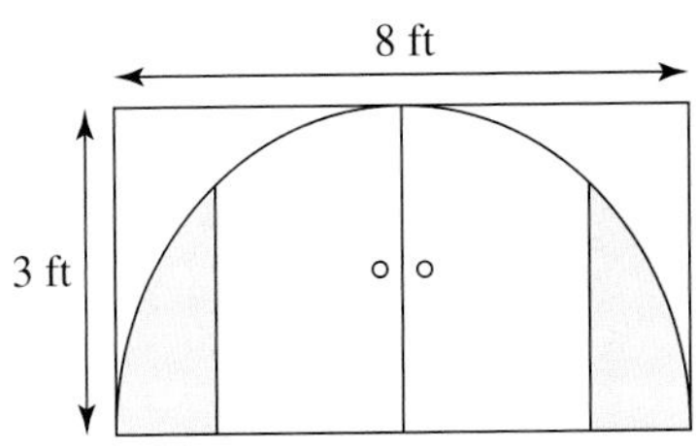

50. Decorative gardens: A retired math teacher decides to present her husband with a beautiful elliptical garden to help celebrate their 50th anniversary. The ellipse is to be 8 m long and 5 m across, with decorative fountains located at the foci. To the nearest hundredth of a meter, how far from the center of the ellipse should the fountain be? How far apart are the fountains? 3.12 m; 6.24 m

51. Attracting attention to art: As part of an art show, a gallery owner asks a student from the local university to design a unique exhibit that will highlight one of the more significant pieces in the collection, an ancient sculpture. The student decides to create an elliptical showroom with reflective walls, with a rotating laser light on a stand at one foci, and the sculpture placed at the other foci on a stand of equal height. The laser light then points continually at the sculpture as it rotates. If the elliptical room is 24 ft long and 16 ft wide, how far from the center of the ellipse should the stands be located (round to the nearest tenth of a foot)? How far apart are the stands? 8.9 ft; 17.9 ft

52. Medical procedures: The medical procedure called *lithotripsy* is a noninvasive medical procedure that is used to break up kidney and bladder stones in the body. A machine called a *lithotripter* uses its three-dimensional semi-elliptical shape and the foci properties of an ellipse to concentrate shock waves generated at one focus on a kidney stone located at the other focus (see diagram—not drawn to scale). If the lithotripter has a length (semimajor axis) of 16 cm and a radius (semiminor axis) of 10 cm, how far from the vertex should a kidney stone be located for the best result? Round to the nearest hundredth. 28.49 cm

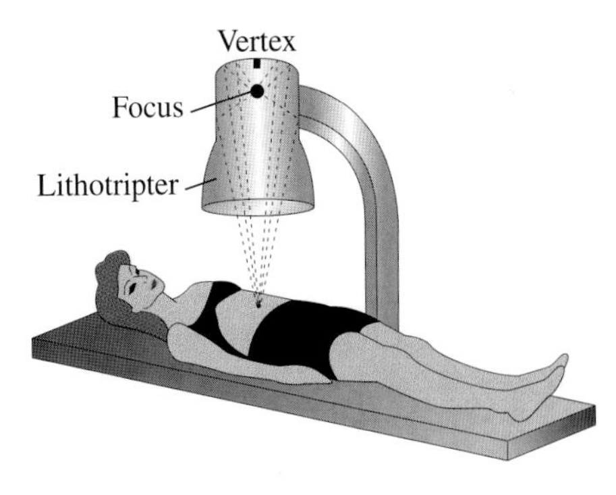

53. $a \approx 142$ millon miles
$b \approx 141$ million miles
orbit time ≈ 686 days

53. Planetary orbits: Except for small variations, a planet's orbit around the Sun is elliptical with the Sun at one foci. The aphelion (maximum distance from the Sun) of the planet Mars is approximately 156 million miles, while the perihelion (minimum distance from the sun) of Mars is about 128 million miles. Use this information to find the lengths of the semimajor and semiminor axes, rounded to the nearest million. If Mars has an orbital velocity of 54,000 mph (1.296 million miles per day), how many days does it take Mars to orbit the Sun? (*Hint:* Use the formula from Exercise 48.)

54. $a \approx 890$ million miles
$b \approx 889$ million miles
orbit time $\approx 10{,}748$ days; 29.4 yrs

54. Planetary orbits: The aphelion (maximum distance from the Sun) of the planet Saturn is approximately 940 million miles, while the perihelion (minimum distance from the Sun) of Saturn is about 840 million miles. Use this information to find the lengths of the semimajor and semiminor axes, rounded to the nearest million. If Saturn has an orbital velocity of 21,650 mph (about 0.52 million miles per day), how many days does it take Saturn to orbit the Sun? How many years?

55. $\frac{x^2}{225} - \frac{y^2}{2275} = 1$, about (24.1, 60) or (−24.1, 60)

55. Locating a ship using radar: Under certain conditions, the properties of a hyperbola can be used to help locate the position of a ship. Suppose two radio stations are located 100 km apart along a straight shoreline. A ship is sailing parallel to the shore and is 60 km out to sea. The ship sends out a distress call that is picked up by the closer station in 0.4 milliseconds (msec—one-thousandth of a second), while it takes 0.5 msec to reach the station that is farther away.

Radio waves travel at a speed of approximately 300 km/msec. Use this information to find the equation of a hyperbola that will help you find the location of the ship, then find the coordinates of the ship. (*Hint:* Draw the hyperbola on a coordinate system with the radio stations on the x-axis at the foci, then use the definition of a hyperbola.)

56. $\frac{x^2}{900} - \frac{y^2}{700} = 1$, about $(-37.6, 20)$ or $(37.6, 20)$

56. Locating a plane using radar: Two radio stations are located 80 km apart along a straight shoreline, when a "mayday" call (a plea for immediate help) is received from a plane that is about to ditch in the ocean (attempt a water landing). The plane was flying at low altitude, parallel to the shoreline, and 20 km out when it ran into trouble. The plane's distress call is picked up by the closer station in 0.1 msec, while it takes 0.3 msec to reach the other. Use this information to construct the equation of a hyperbola that will help you find the location of the ditched plane, then find the coordinates of the plane. Also see Exercise 55.

WRITING, RESEARCH, AND DECISION MAKING

57. When graphing the conic sections, it is often helpful to use what is called a **focal chord,** as it gives additional points on the graph with very little effort. A focal chord is a line segment through a focus (perpendicular to the major or transverse axis), with the endpoints on the graph. For ellipses and hyperbolas, the length of the focal chord is given by $L = \frac{2b^2}{a}$, where a is a vertex. The focus will always be the midpoint of this line segment. Find the length of the focal chord for the hyperbola $\frac{x^2}{4} - \frac{y^2}{9} = 1$ and the coordinates of the endpoints. Verify (by substituting into the equation) that these endpoints are indeed points on the graph, then use them to help complete the graph. $L = 9$ units; verified

58. (4, 3); 5 units; same distance; formulas are identical

58. Using graph paper, draw a complete and careful graph of the hyperbola $9x^2 - 16y^2 = 144$. Be particularly sure that the central rectangle is carefully drawn and the foci are accurately located. What are the coordinates of the upper-right corner of this central rectangle? Use the Pythagorean theorem to find the distance from (0, 0) to this corner point. How does this distance compare to the distance from the center to the focus? What does this relationship have to do with the formula for finding the focus? Explain and discuss what you find.

EXTENDING THE CONCEPT

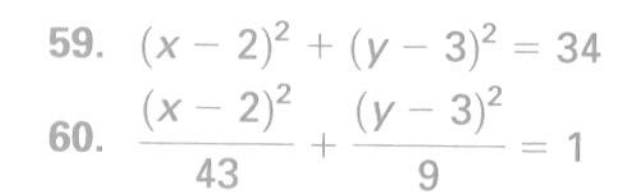
59. $(x - 2)^2 + (y - 3)^2 = 34$

60. $\frac{(x - 2)^2}{43} + \frac{(y - 3)^2}{9} = 1$

59. Find the equation of the circle shown, given the equation of the hyperbola.

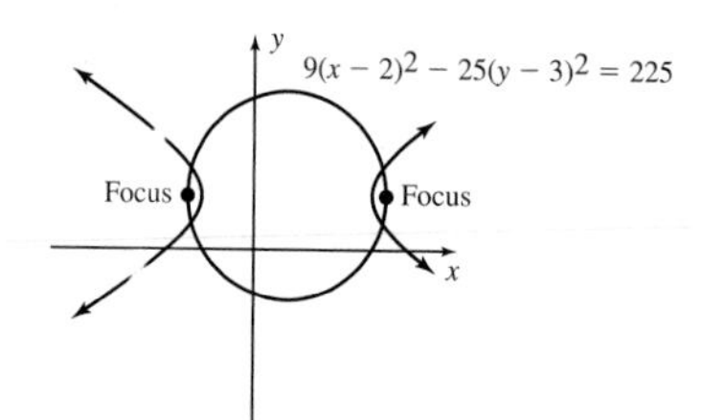

60. Find the equation of the ellipse shown, given the equation of the hyperbola and (2, 0) is on the graph of the ellipse. The hyperbola and ellipse share the same foci.

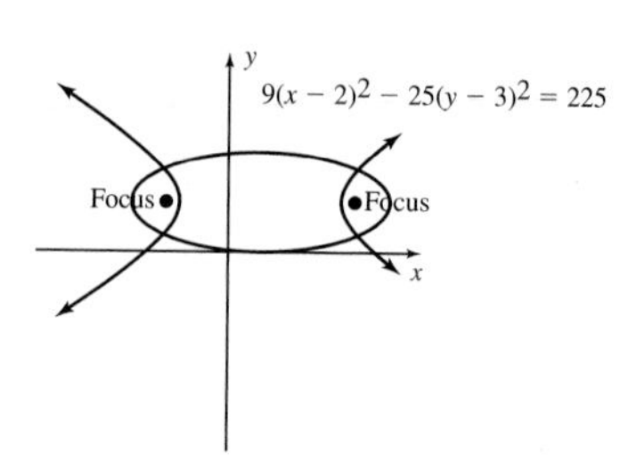

61. Verify that for the horizontal ellipse $\frac{x^2}{a^2} + \frac{y^2}{b^2} = 1$, the length of the focal chord is $\frac{2b^2}{a}$. Also see Exercise 57. verified

62. Verify that for a central hyperbola, a circle that circumscribes the central rectangle must also go through both foci. verified

64. $\dfrac{(30 + 10\sqrt{3}) + (30 - 10\sqrt{3})i,\ (3 - \sqrt{3}) + (3 + \sqrt{3})i}{10}$

65. a. b.

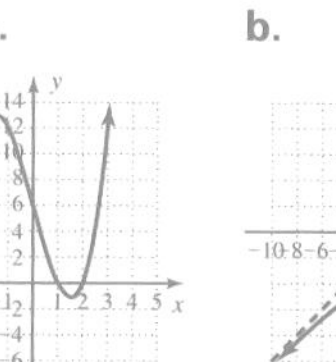

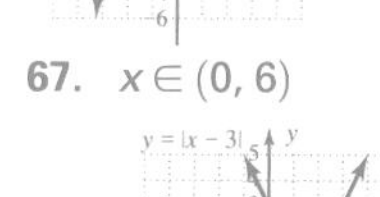

67. $x \in (0, 6)$

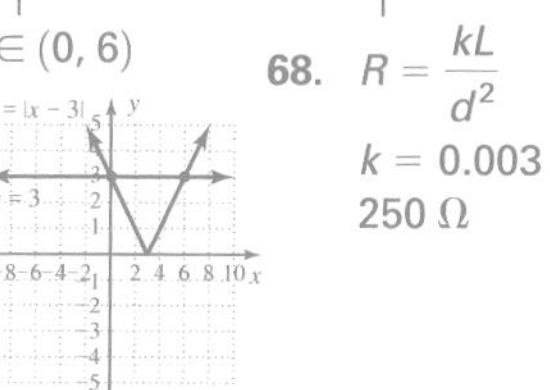

68. $R = \dfrac{kL}{d^2}$
$k = 0.003$
250 Ω

MAINTAINING YOUR SKILLS

63. (5.3) Evaluate the expression using the change-of-base formula: $\log_3 20$. $\dfrac{\log 20}{\log 3} \approx 2.73$

64. (1.4) Compute the product z_1z_2 and quotient $\dfrac{z_1}{z_2}$ given $z_1 = 2\sqrt{3} + 2\sqrt{3}i$; $z_2 = 5\sqrt{3} - 5i$.

65. (4.4/4.5) Graph the functions given:
a. $f(x) = x^3 - 7x + 6$
b. $h(x) = \dfrac{x^3}{x^2 - 4}$

66. (5.5) How long would it take \$2000 to triple if deposited at 6.5% compounded continuously? 16.9 yr

67. (6.4) Solve the absolute value inequality (a) graphically and (b) analytically: $-2|x - 3| + 10 > 4$.

68. (3.6) The resistance to current flow in an electrical wire varies directly as the length L of the wire and inversely as the square of its diameter d. (a) Write the equation of variation, (b) find the constant of variation if a wire 2 m long with diameter $d = 0.005$ m has a resistance of 240 Ω, and (c) find the resistance in a similar wire 3 m long and 0.006 m in diameter.

7.5 The Analytic Parabola

LEARNING OBJECTIVES

In Section 7.5 you will learn how to:

A. Graph parabolas with a horizontal axis of symmetry

B. Identify and use the focus-directrix form of the equation of a parabola

INTRODUCTION

Earlier we saw the graph of a quadratic function was a parabola. Parabolas are actually the fourth and final member of the family of conic sections, and like the others, the graph can be obtained by observing the intersection of a plane and a cone. If the plane is parallel to one **element** of the cone (shown as a dark line in Figure 7.32), the intersection of the plane with one nappe forms a parabola. In this section we develop the equation of a parabola from its analytic definition, opening a new realm of applications that extends far beyond those involving only zeroes and extreme values.

Figure 7.32

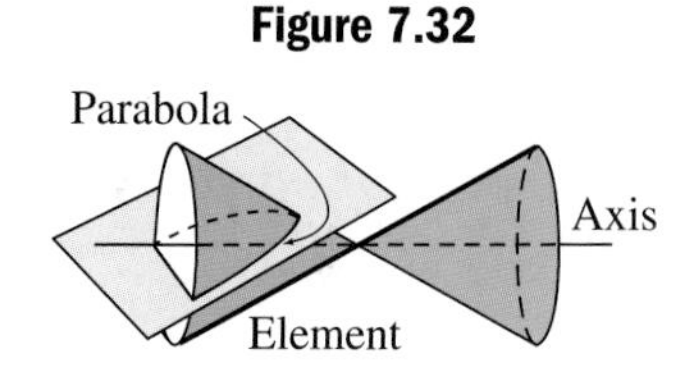

POINT OF INTEREST

Parabolas have a reflective property that is similar to that of ellipses. For an ellipse, light or sound emanating at one focus is reflected to the second. For a parabola, light or sound emanating from the focus reflects in a path that is parallel to the parabola's axis. This makes parabolas singularly valuable for lighting—where rays of light from a light source at the focus can be directed—and for telescopes—where parallel rays of light from objects far out in space are brought together and observed at the focus of the telescope's parabolic mirror.

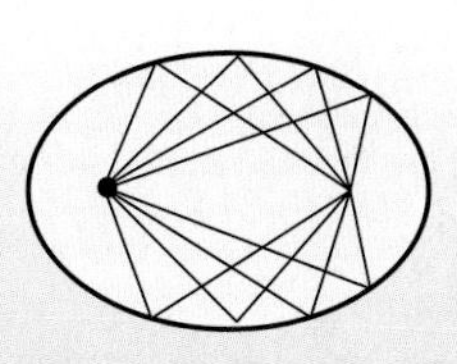

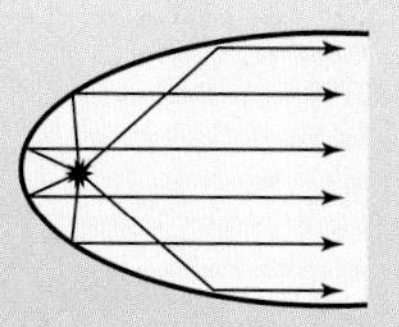

A. Parabolas with a Horizontal Axis

An introductory study of parabolas generally involves those with a vertical axis, defined by the equation $y = ax^2 + bx + c$. Unlike the previous conic sections, this equation has *only one second-degree (squared) term in x* and defines a function rather than a relation. As a review, recall that to graph this function, a five-step method can be used, as outlined and shown in Figure 7.33.

Figure 7.33

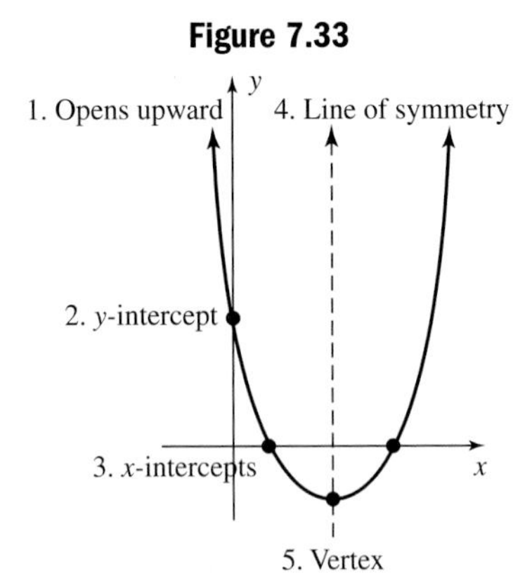

Vertical Parabolas

1. *Determine the concavity:*
   Concave up if $a > 0$, down if $a < 0$.
2. *Find the y-intercept:* The y-intercept is the ordered pair $(0, c)$.
3. *Find the x-intercepts (if they exist):* Solve the related equation $0 = ax^2 + bx + c$ by factoring or using the quadratic formula.
4. *Graph the line of symmetry:* The line of symmetry is the vertical line $x = \frac{-b}{2a}$.
5. *Find the coordinates of the vertex and determine the maximum/minimum value:*

   Since the axis of symmetry will contain the vertex, it has coordinates $\left(\frac{-b}{2a}, f\left(\frac{-b}{2a}\right)\right)$. If $a > 0$, the y-coordinate of the vertex is the minimum value of the function. If $a < 0$, the y-coordinate of the vertex is the maximum value. See Exercises 7 through 12.

Horizontal Parabolas

Similar to our study of horizontal and vertical hyperbolas, the graph of a parabola can open *to the right or left,* as well as up or down. After interchanging the variables x and y in the standard equation, we obtain the parabola $x = ay^2 + by + c$, which opens to the right if $a > 0$ and to the left if $a < 0$. This equation can also be written in shifted form as $x = a(y - k)^2 + h$, where (h, k) is the vertex of the parabola. However, this time the axis of symmetry is the horizontal line $y = k$ and factoring or the quadratic formula is used to find the *y-intercepts* (if they exist). It is important to note that although the graph is still a parabola—*it is not the graph of a function!*

EXAMPLE 1 Graph the relation whose equation is $x = y^2 + 3y - 4$, then state the domain and range of the relation.

Solution: Since the equation has a single squared term in y, the graph will be a horizontal parabola. With $a > 0$ $(a = 1)$, the parabola opens to the right. The x-intercept is $(-4, 0)$. Factoring shows the y-intercepts are $y = -4$ and $y = 1$. The axis of symmetry is $y = \frac{-3}{2} = -1.5$, and substituting this value into the original equation gives $x = -6.25$. The coordinates of the vertex are $(-6.25, -1.5)$. Using horizontal and vertical boundary lines we find the domain for this relation is $x \in [-6.25, \infty)$ and the range is $y \in (-\infty, \infty)$. The graph is shown.

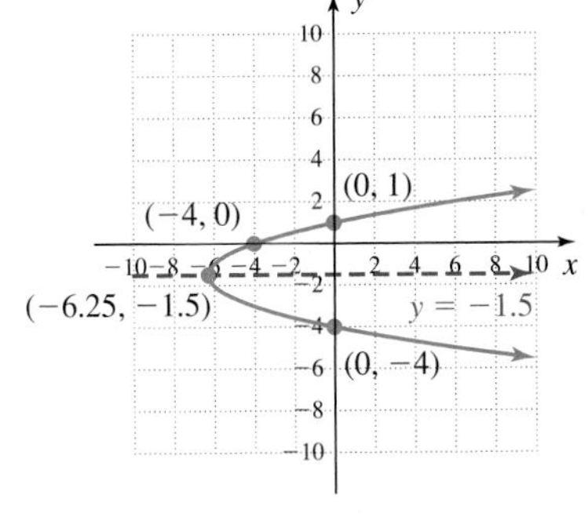

NOW TRY EXERCISES 13 THROUGH 18

The characteristics of horizontal parabolas are summarized here:

> **HORIZONTAL PARABOLAS**
>
> For a second-degree equation of the form $x = ay^2 + by + c$,
>
> 1. The graph is a parabola that opens right if $a > 0$, left if $a < 0$.
> 2. The x-intercept is $(c, 0)$.
> 3. The y-intercept(s) can be found by substituting $x = 0$, then solving by factoring or using the quadratic formula.
> 4. The axis of symmetry is $y = \frac{-b}{2a}$.
> 5. The vertex (h, k) can be found by completing the square and writing the equation in *shifted form* as $x = a(y - k)^2 + h$.

EXAMPLE 2 Graph by completing the square: $x = -2y^2 - 8y - 9$.

Solution: Using the original equation, we note the graph will be a horizontal parabola opening to the left ($a = -2$) and have an x-intercept of $(-9, 0)$. Completing the square gives $x = -2(y^2 + 4y + 4) - 9 + 8$, so $x = -2(y + 2)^2 - 1$. The vertex is at $(-1, -2)$ and $y = -2$ is the axis of symmetry. This means there are no y-intercepts, a fact that comes to light when we attempt to solve the equation after substituting 0 for x:

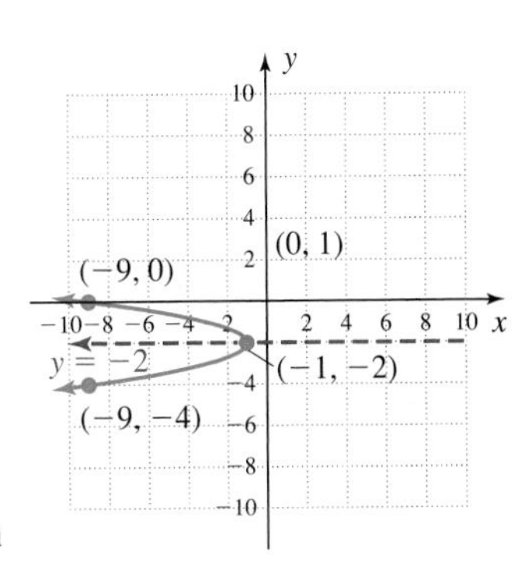

$$-2(y + 2)^2 - 1 = 0 \quad \text{substitute 0 for } x$$

$$(y + 2)^2 = -\frac{1}{2} \quad \text{isolate squared term}$$

The equation has no real roots and there are no y-intercepts. Using symmetry, the point $(-9, -4)$ is also on the graph. Using these points we obtain the graph shown.

NOW TRY EXERCISES 19 THROUGH 36

B. The Focus-Directrix Form of the Equation of a Parabola

As with the ellipse and hyperbola, many significant applications of the parabola rely on its analytical definition rather than its shifted form. From the construction of radio telescopes to the manufacture of flashlights, the location of the focus of a parabola is critical. To understand these and other applications, we introduce the analytic definition of a parabola.

DEFINITION OF A PARABOLA

Given a fixed point F and fixed line D in the plane, a parabola is defined as the set of all points $P(x, y)$ in the plane such that the distance from F to P is equal to the perpendicular distance from line D to P. In symbols, $|FP| = |PP_d|$, where P_d is a point on line D. The fixed point F is called the **focus** of the parabola, and the fixed line D is called the **directrix.**

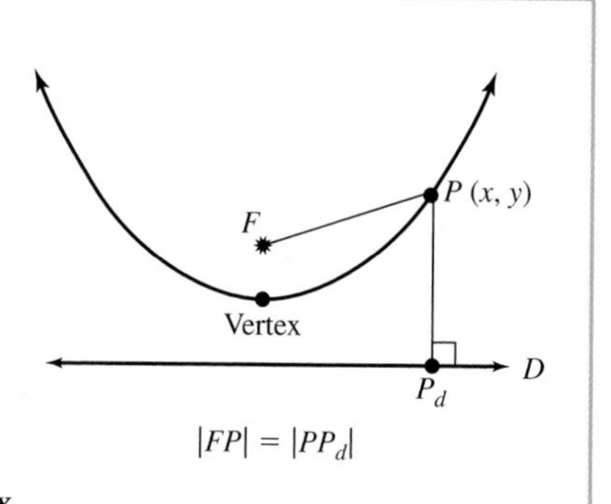

The equation of a parabola can be obtained by combining this definition with the distance formula. With no loss of generality, we can assume the parabola shown in the definition box is oriented in the plane with the vertex at (0, 0) and the focus at $(0, p)$. For the parabola, we use the coordinates $(0, p)$ for the focus, since we already designated $(0, c)$ as the y-intercept. As the diagram in Figure 7.34 indicates, this gives the directrix an equation of $y = -p$ and the point P_d coordinates of $(x, -p)$.

Figure 7.34

Using $|FP| = |PP_d|$, the distance formula yields

$$\sqrt{(x-0)^2 + (y-p)^2} = \sqrt{(x-x)^2 + (y+p)^2} \quad \text{from definition}$$

$$(x-0)^2 + (y-p)^2 = (x-x)^2 + (y+p)^2 \quad \text{square both sides}$$

$$x^2 + y^2 - 2py + p^2 = 0 + y^2 + 2py + p^2 \quad \text{simplify; expand binomials}$$

$$x^2 - 2py = 2py \quad \text{subtract } p^2 \text{ and } y^2$$

$$x^2 = 4py \quad \text{isolate } x^2$$

The resulting equation is called the **focus-directrix form** of a *vertical parabola* with center at (0, 0). If we had begun by orienting the parabola so it opened to the right, we would have obtained the equation of a *horizontal parabola* with center (0, 0): $y^2 = 4px$.

THE EQUATION OF A PARABOLA IN FOCUS-DIRECTRIX FORM WITH VERTEX (0, 0)

Vertical Parabola	Horizontal Parabola
$x^2 = 4py$	$y^2 = 4px$
focus $(0, p)$, directrix: $y = -p$	focus at $(p, 0)$, directrix: $x = -p$
If $p > 0$, concave up.	If $p > 0$, parabola opens right.
If $p < 0$, concave down.	If $p < 0$, parabola opens left.

For a parabola, note there is only one second-degree term.

EXAMPLE 3 Find the vertex, focus, and directrix for the parabola defined by the equation $x^2 = -12y$. Then sketch the graph, including the focus and directrix.

Solution: Since the x-term is squared and no shifts have been applied, the graph will be a vertical parabola with a vertex of (0, 0). Use a direct comparison between the given equation and the focus-directrix form to determine the value of p:

$$x^2 = -12y \quad \text{given equation}$$
$$\downarrow$$
$$x^2 = 4py \quad \text{focus-directrix form}$$

This shows:

$$4p = -12$$
$$p = -3$$

Since $p = -3$ ($p < 0$), the parabola is concave down, with the focus at $(0, -3)$ and directrix $y = 3$. To complete the graph we need a few additional points. Since 36 (6^2) is divisible by 12, we can use inputs of $x = 6$ and $x = -6$, giving the points $(6, -3)$ and $(-6, -3)$. Note the axis of symmetry is $x = 0$. The graph is shown.

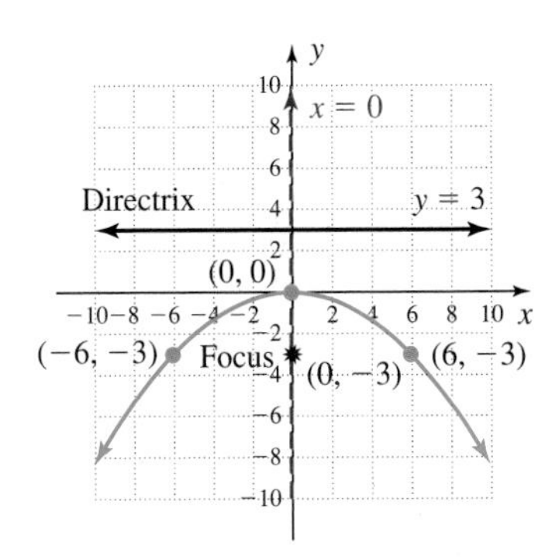

NOW TRY EXERCISES 37 THROUGH 42

Figure 7.35

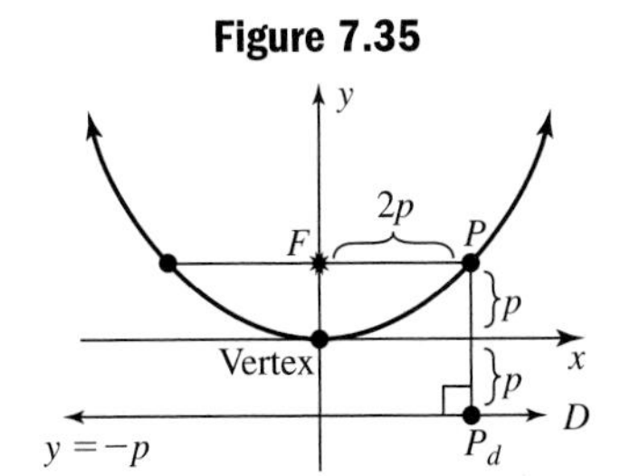

As an alternative to calculating additional points to sketch the graph, we can use what is called the **focal chord** of the parabola. Similar to the ellipse and hyperbola, the focal chord is a line segment that contains the focus, is parallel to the directrix, and has its endpoints on the graph. Using the definition of a parabola and the diagram in Figure 7.35, we see the distance $|PP_d|$ is $2p$. Since $|PP_d| = |FP|$, a line segment parallel to the directrix from the focus to the graph will also have a length of $|2p|$, and the focal chord of any parabola has a total length of $|4p|$. Note that in Example 3, the points we happened to choose were actually the end points of the focal chord.

EXAMPLE 4 Find the vertex, focus, and directrix for the parabola defined by the equation $y^2 = 8x$. Then sketch the graph, including the focus and directrix.

Solution: Since the y-term is squared and no shifts have been applied, the graph will be a horizontal parabola with a vertex of (0, 0). Use a direct comparison between the given equation and the focus-directrix form to determine the value of p:

$$y^2 = 8x \quad \text{given}$$
$$y^2 = 4px \quad \text{focus-directrix form}$$

The result shows:

$$4p = 8$$
$$p = 2$$

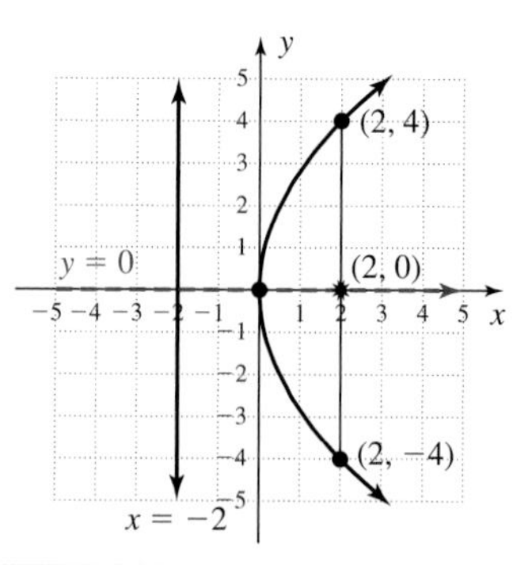

Since $p > 0$, the parabola opens to the right, and has a focus at (2, 0) with directrix $x = -2$. The vertical distance from the focus to the graph is $|2p| = 4$ units, so (2, 4) and (2, −4) are on the graph. The axis of symmetry is $y = 0$. See the figure.

NOW TRY EXERCISES 43 THROUGH 48

Recall that when we graph a relation using transformations and shifts of a basic graph, all features of the graph are likewise shifted. These shifts apply to the focus and directrix of a parabola, as well as to the vertex and axis of symmetry.

EXAMPLE 5 Find the vertex, focus, and directrix for the parabola whose equation is given, then sketch the graph, including the focus and directrix: $x^2 - 6x + 12y - 15 = 0$.

Solution: Since only the x-term is squared, the graph will be a vertical parabola. To find the concavity, vertex, focus, and directrix, we complete the square in x and use a direct comparison between the shifted form and the focus-directrix form

$x^2 - 6x + 12y - 15 = 0$	given equation
$x^2 - 6x + \_\_ = -12y + 15$	complete the square in x
$x^2 - 6x + 9 = -12y + 24$	add 9
$(x - 3)^2 = -12(y - 2)$	factor

Notice the parabola has been shifted 3 right and 2 up, so *all features of the parabola will likewise be shifted.* Since we have $4p = -12$ (the coefficient of the linear term), we know $p = -3$ ($p < 0$) and the parabola is concave down. If the parabola were in standard position, the vertex would be at (0, 0), the focus at (0, −3) and the directrix a horizontal line at $y = 3$. But since the parabola is shifted 3 right and 2 up, we add 3 to all x-values and 2 to all y-values to locate the features of the shifted parabola. The vertex is at $(0 + 3, 0 + 2) = (3, 2)$. The focus is $(0 + 3, -3 + 2) = (3, -1)$ and the directrix is $y = 3 + 2 = 5$. Finally, the horizontal distance from the focus to the graph is $|2p| = 6$ units (since $|4p| = 12$), giving us the additional points $(-3, -1)$ and $(9, -1)$. See the figure.

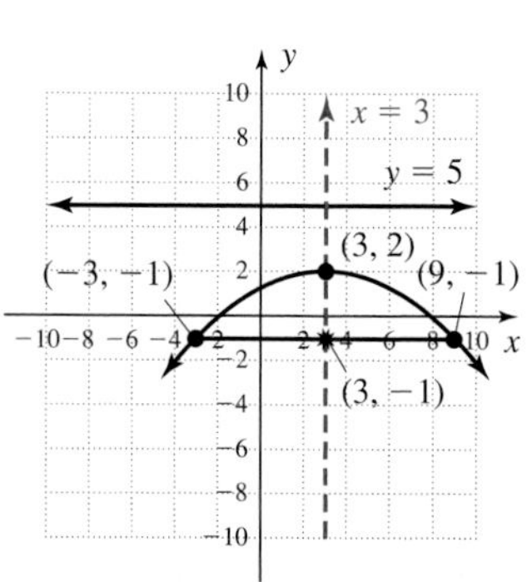

NOW TRY EXERCISES 49 THROUGH 60

Here is just one of the many ways the analytic definition of a parabola can be applied. There are several others in the Exercise Set.

EXAMPLE 6 The diagram shows the cross section of a radio antenna dish. Engineers have located a point on the cross section that is 0.75 m above and 6 m to the right of the vertex. At what coordinates should the engineers build the focus of the antenna?

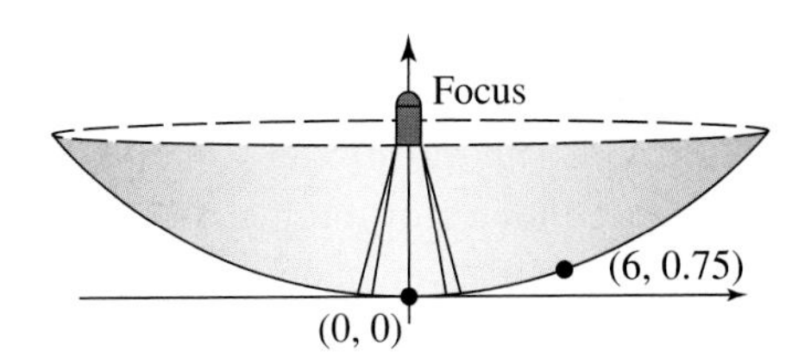

Solution: By inspection we see this is a vertical parabola with center at (0, 0). This means its equation must be of the form $x^2 = 4py$. Because we know (6, 0.75) is a point on this graph, we can substitute (6, 0.75) in this equation and solve for p:

$$x^2 = 4py \quad \text{equation for vertical parabola, vertex at (0, 0)}$$
$$(6)^2 = 4p(0.75) \quad \text{substitute } x = 6 \text{ and } y = 0.75$$
$$36 = 3p \quad \text{simplify}$$
$$p = 12 \quad \text{result}$$

With $p = 12$, we see that the focus must be located at (0, 12), or 12 m directly above the vertex.

NOW TRY EXERCISES 63 THROUGH 68

Note that in many cases, the focus of a parabolic dish may be taller than the rim of the dish.

TECHNOLOGY HIGHLIGHT

The Focus of a Parabola Given in Standard Form

The keystrokes shown apply to a TI-84 Plus model. Please consult our Internet site or your manual for other models.

In this *Technology Highlight,* we attempt to verify that for *any* parabola, the distance from the focus to a point on the graph is equal to the distance from this point to the directrix. While our website contains a TI-84 Plus program that accomplishes this very nicely, we'll use a more deliberate and rudimentary approach here. The quadratic function $y = ax^2 + bx + c$ and its graphs are studied extensively in developmental courses, but usually no mention is made of the parabola's focus and directrix. Generally, when $a \geq 1$, the focus of a parabola is very near its vertex. To see why, write this function in focus-directrix form by completing the square. For convenience, assume $c = 0$:

$$y = ax^2 + bx \quad \text{quadratic function; } c = 0$$
$$= a\left(x^2 + \frac{b}{a}x + \_\_\right) \quad \text{factor out } a$$
$$= a\left(x^2 + \frac{b}{a}x + \frac{b^2}{4a^2}\right) - \frac{b^2}{4a} \quad \text{complete the square; } a\left(\frac{b^2}{4a^2}\right) = \frac{b^2}{4a}$$
$$\frac{1}{a}\left(y + \frac{b^2}{4a}\right) = \left(x - \frac{b}{2a}\right)^2 \quad \text{focus-directrix form}$$

Regardless of the coefficients chosen, this development shows that $4p = \frac{1}{a}$ so $p = \frac{1}{4a}$, so the larger the lead coefficient, the smaller p becomes.

Consider the function $y = (x - 2)^2 + 1$, which is a parabola, concave up, with a vertex of (2, 1). Enter this function as Y_1 on the Y= screen of your graphing calculator. Since $a = 1$, the focus of the parabola *in standard position* would be $\left(0, \frac{1}{4}\right)$, but this parabola is shifted 2 right and 1 up, so the focus is actually at (2, 1.25). The directrix is $y = 0.75$ ($y = 1 - 0.25$). Store the coordinates of the focus (2, 1.25) in locations A and B, respectively, and the value of the directrix in location D. To find the distance between the focus (A, B) and a point $(x, f(x))$ on the graph we use $\sqrt{(X - A)^2 + (Y_1(X) - B)^2}$ (the distance formula), entering this expression as Y_2. To find the distance between $(x, f(x))$ and (x, D) we enter $\sqrt{(Y_1(X) - D)^2}$ as Y_3 (only one addend under the radical since $(X - X)^2 = 0$). Deactivate Y_1, leaving Y_2 and Y_3 active (see Figure 7.36). To verify that these distances are identical, go to the TABLE (use 2nd GRAPH) and compare the entries in Y_2 with those in Y_3. For the following functions, locate the focus and use the ideas from this *Technology Highlight* to verify the definition of a parabola.

Figure 7.36

```
Plot1 Plot2 Plot3
\Y1=(X-2)²+1
\Y2=√((X-A)²+(Y1(X)-B)²)
\Y3=√(Y1(X)-D)²
\Y4=
\Y5=
\Y6=
```

Exercise 1: $y = (x + 3)^2 - 2$ $\left(-3, -\frac{7}{4}\right)$; verified

Exercise 2: $y = -2(x - 1)^2 + 5$ $\left(1, \frac{39}{8}\right)$; verified

Exercise 3: $y = \frac{1}{2}(x - 4)^2 + 2$ $\left(4, \frac{5}{2}\right)$; verified

Exercise 4: $y = x^2 - 4x + 7$ $\left(2, \frac{13}{4}\right)$; verified

7.5 EXERCISES

CONCEPTS AND VOCABULARY

Fill in the blank with the appropriate word or phrase. Carefully reread the section if needed.

1. The equation $x = ay^2 + by + c$ is that of a(n) horizontal parabola, opening to the right if $a > 0$ and to the left if $a < 0$.

2. If point P is on the graph of a parabola with directrix D, the distance from P to line D is equal to the distance between P and the focus of the parabola.

3. Given $y^2 = 4px$, the focus is at $(p, 0)$ and the equation of the directrix is $x = -p$.

4. Given $x^2 = -16y$, the value of p is -4 and the coordinates of the focus are $(0, -4)$.

5. Discuss/explain how to find the vertex, directrix, and focus from the equation $(x - h)^2 = 4p(y - k)$. Answers will vary.

6. If a horizontal parabola has a vertex of $(2, -3)$ with $a > 0$, what can you say about the y-intercepts? Will the graph always have an x-intercept? Explain. Answers will vary.

7. $x \in (-\infty, \infty)$, $y \in [-4, \infty)$
8. $x \in (-\infty, \infty)$, $y \in [-4, \infty)$

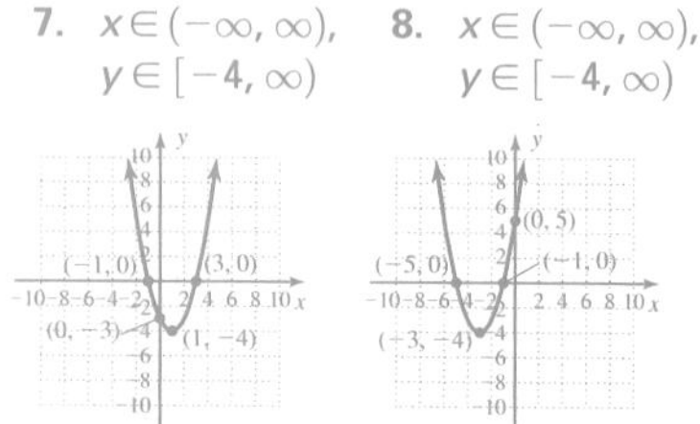

9. $x \in (-\infty, \infty)$, $y \in [-18, \infty)$
10. $x \in (-\infty, \infty)$, $y \in [-27, \infty)$

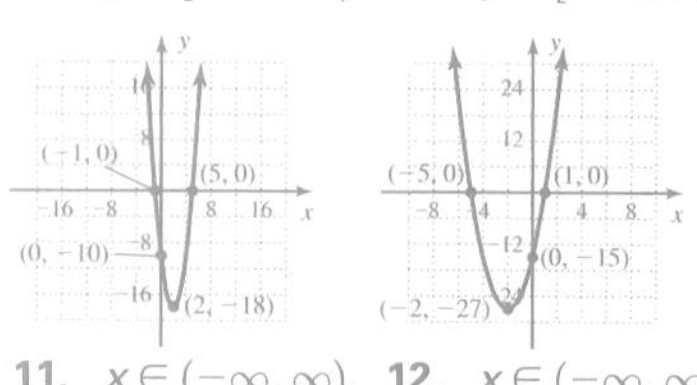

11. $x \in (-\infty, \infty)$, $y \in [-10.125, \infty)$
12. $x \in (-\infty, \infty)$, $y \in [-3.125, \infty)$

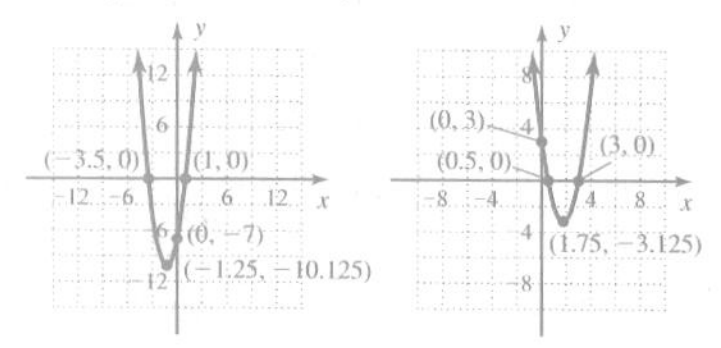

DEVELOPING YOUR SKILLS

Find the x- and y-intercepts (if they exist) and the vertex of the parabola. Then sketch the graph by using symmetry and a few additional points or completing the square and shifting a parent function. Scale the axes as needed to comfortably fit the graph and state the domain and range.

7. $y = x^2 - 2x - 3$
8. $y = x^2 + 6x + 5$
9. $y = 2x^2 - 8x - 10$
10. $y = 3x^2 + 12x - 15$
11. $y = 2x^2 + 5x - 7$
12. $y = 2x^2 - 7x + 3$

13. $x \in [-4, \infty)$, $y \in (-\infty, \infty)$ 14. $x \in [-16, \infty)$, $y \in (-\infty, \infty)$

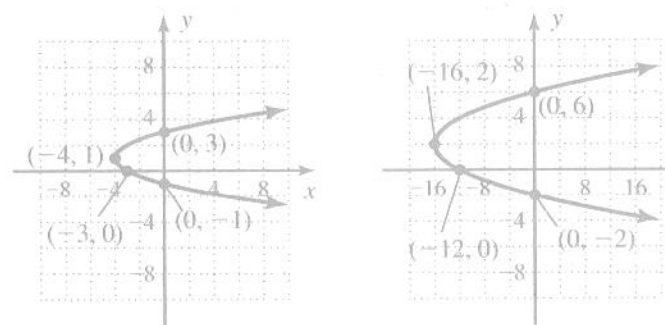

15. $x \in (-\infty, 16]$, $y \in (-\infty, \infty)$ 16. $x \in (-\infty, 4]$, $y \in (-\infty, \infty)$

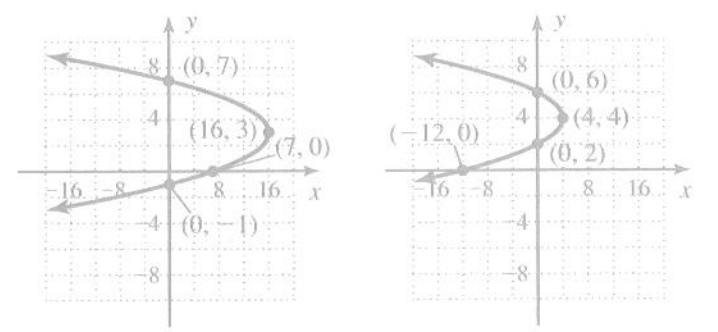

17. $x \in (-\infty, 0]$, $y \in (-\infty, \infty)$ 18. $x \in (-\infty, 0]$, $y \in (-\infty, \infty)$

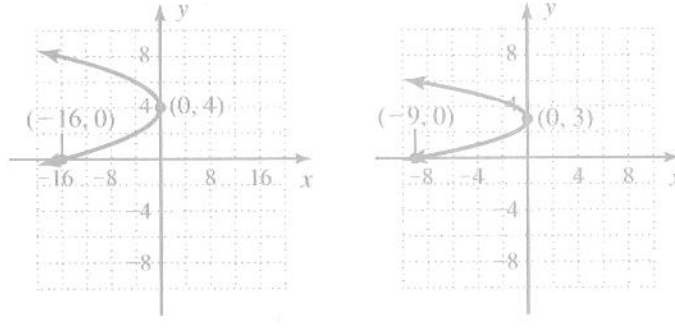

19. $x \in [-9, \infty)$, $y \in (-\infty, \infty)$ 20. $x \in [-16, \infty)$, $y \in (-\infty, \infty)$

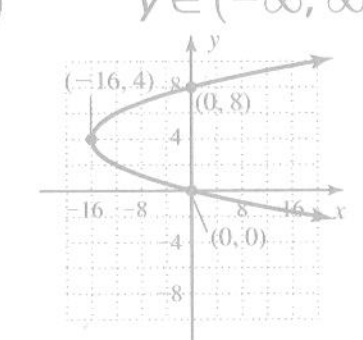

21. $x \in [-4, \infty)$, $y \in (-\infty, \infty)$ 22. $x \in [-9, \infty)$, $y \in (-\infty, \infty)$

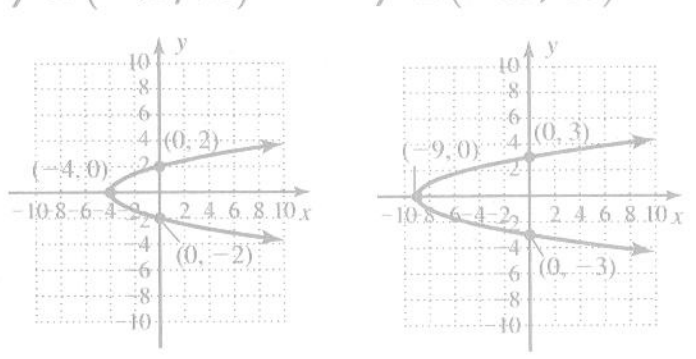

23. $x \in (-\infty, 0]$, $y \in (-\infty, \infty)$ 24. $x \in (-\infty, 0]$, $y \in (-\infty, \infty)$

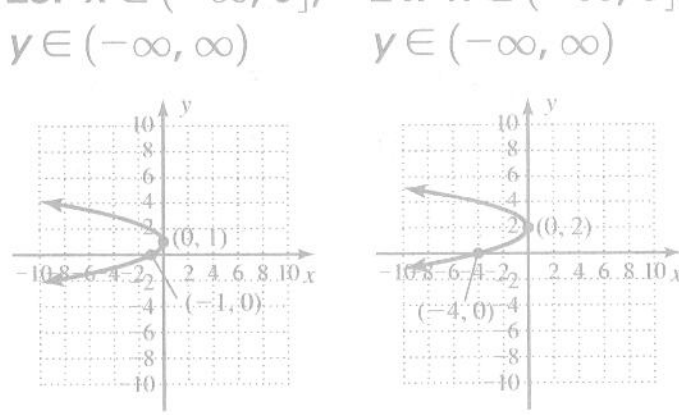

Additional answers can be found in the Instructor Answer Appendix.

Find the x- and y-intercepts (if they exist) and the vertex of the graph. Then sketch the graph using symmetry and a few additional points (scale the axes as needed). Finally, state the domain and range of the relation.

13. $x = y^2 - 2y - 3$ **14.** $x = y^2 - 4y - 12$ **15.** $x = -y^2 + 6y + 7$

16. $x = -y^2 + 8y - 12$ **17.** $x = -y^2 + 8y - 16$ **18.** $x = -y^2 + 6y - 9$

Sketch using symmetry and shifts of a basic function. Be sure to find the x- and y-intercepts (if they exist) and the vertex of the graph, the state the domain and range of the relation.

19. $x = y^2 - 6y$ **20.** $x = y^2 - 8y$ **21.** $x = y^2 - 4$

22. $x = y^2 - 9$ **23.** $x = -y^2 + 2y - 1$ **24.** $x = -y^2 + 4y - 4$

25. $x = y^2 + y - 6$ **26.** $x = y^2 + 4y - 5$ **27.** $x = y^2 - 10y + 4$

28. $x = y^2 + 12y - 5$ **29.** $x = 3 - 8y - 2y^2$ **30.** $x = 2 - 12y + 3y^2$

31. $y = (x - 2)^2 + 3$ **32.** $y = (x + 2)^2 - 4$ **33.** $x = (y - 3)^2 + 2$

34. $x = (y + 1)^2 - 4$ **35.** $x = 2(y - 3)^2 + 1$ **36.** $x = -2(y + 3)^2 - 5$

Find the vertex, focus, and directrix for the parabolas defined by the equations given, then use this information to sketch a complete graph (illustrate and name these features). For Exercises 43 to 60, also include the focal chord.

37. $x^2 = 8y$ **38.** $x^2 = 16y$ **39.** $x^2 = -24y$ **40.** $x^2 = -20y$

41. $x^2 = 6y$ **42.** $x^2 = 18y$ **43.** $y^2 = -4x$ **44.** $y^2 = -12x$

45. $y^2 = 18x$ **46.** $y^2 = 20x$ **47.** $y^2 = -10x$ **48.** $y^2 = -14x$

49. $x^2 - 8x - 8y + 16 = 0$ **50.** $x^2 - 10x - 12y + 25 = 0$ **51.** $x^2 - 14x - 24y + 1 = 0$

52. $x^2 - 10x - 12y + 1 = 0$ **53.** $3x^2 - 24x - 12y + 12 = 0$ **54.** $2x^2 - 8x - 16y - 24 = 0$

55. $y^2 - 12y - 20x + 36 = 0$ **56.** $y^2 - 6y - 16x + 9 = 0$ **57.** $y^2 - 6y + 4x + 1 = 0$

58. $y^2 - 2y + 8x + 9 = 0$ **59.** $2y^2 - 20y + 8x + 2 = 0$ **60.** $3y^2 - 18y + 12x + 3 = 0$

WORKING WITH FORMULAS

61. The area of a right parabolic segment: $A = \frac{2}{3}ab$

A *right parabolic segment* is that part of a parabola formed by a line perpendicular to its axis, which cuts the parabola. The area of this segment is given by the formula shown, where b is the length of the chord cutting the parabola and a is the perpendicular distance from the vertex to this chord. What is the area of the parabolic segment shown in the figure? 16 units^2

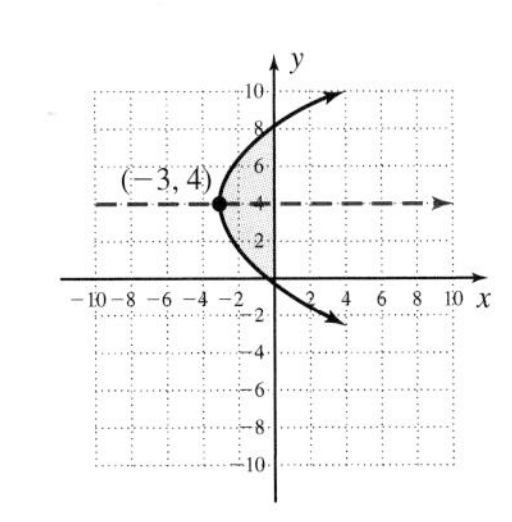

62. The arc length of a right parabolic segment:

$$\frac{1}{2}\sqrt{b^2 + 16a^2} + \frac{b^2}{8a}\ln\left(\frac{4a + \sqrt{b^2 + 16a^2}}{b}\right)$$

Although a fairly simple concept, finding the length of the parabolic arc traversed by a projectile requires a good deal of computation. To find the length of the arc ABC shown, we use the formula given where a is the maximum height attained by the projectile and b is the horizontal distance it traveled. Suppose a baseball thrown from centerfield reaches a maximum height of 20 ft and traverses an arc length of 340 ft. Will the ball reach the catcher 310 ft away without bouncing? yes

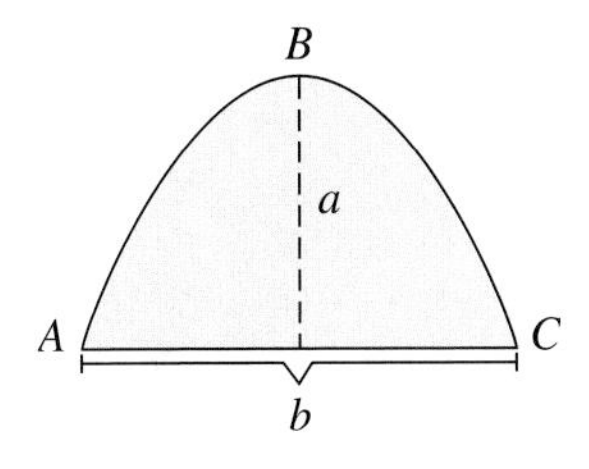

63.

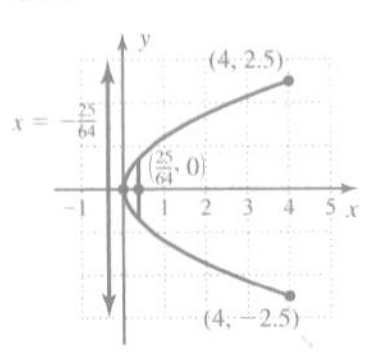

64.

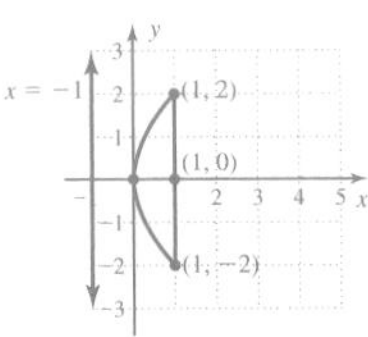

Exercise 66

APPLICATIONS

63. Parabolic car headlights: The cross section of a typical car headlight can be modeled by an equation similar to $25x = 16y^2$, where x and y are in inches and $x \in [0, 4]$. Use this information to graph the relation for the indicated domain.

64. Parabolic flashlights: The cross section of a typical flashlight reflector can be modeled by an equation similar to $4x = y^2$, where x and y are in centimeters and $x \in [0, 2.25]$. Use this information to graph the relation for the indicated domain.

65. Parabolic sound receivers: Sound technicians at professional sports events often use parabolic receivers (see the diagram) as they move along the sidelines. If a two-dimensional cross section of the receiver is modeled by the equation $y^2 = 54x$, and is 36 in. in *diameter*, how deep is the parabolic receiver? What is the location of the focus? [*Hint:* Graph the parabola on the coordinate grid (scale the axes).] 6 in.; (13.5, 0)

Exercise 65

66. Parabolic sound receivers: Private investigators will often use a smaller and less expensive parabolic receiver (see Exercise 65) to gather information for their clients. If a two-dimensional cross section of the receiver is modeled by the equation $y^2 = 24x$, and the receiver is 12 in. in *diameter*, how deep is the parabolic dish? What is the location of the focus? 1.5 in.; (6, 0)

67. Parabolic radio wave receivers: The program known as S.E.T.I. (Search for Extra-Terrestrial Intelligence) identifies a group of scientists using radio telescopes to look for radio signals from possible intelligent species in outer space. The radio telescopes are actually parabolic dishes that vary in size from a few feet to hundreds of feet in diameter. If a particular radio telescope is 100 ft in diameter and has a cross section modeled by the equation $x^2 = 167y$, how deep is the parabolic dish? What is the location of the focus? [*Hint:* Graph the parabola on the coordinate grid (scale the axes).] 14.97 ft, (0, 41.75)

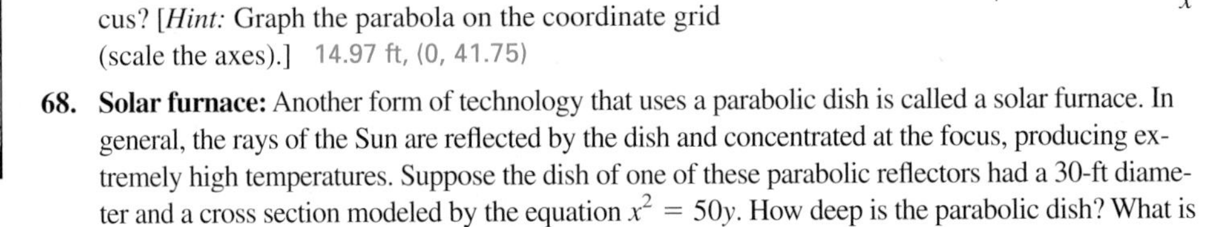

68. Solar furnace: Another form of technology that uses a parabolic dish is called a solar furnace. In general, the rays of the Sun are reflected by the dish and concentrated at the focus, producing extremely high temperatures. Suppose the dish of one of these parabolic reflectors had a 30-ft diameter and a cross section modeled by the equation $x^2 = 50y$. How deep is the parabolic dish? What is the location of the focus? 4.5 ft, (0, 12.5)

WRITING, RESEARCH, AND DECISION MAKING

69. Although no mention of a parabola's focus and directrix was made in Chapters 2 and 3, the quadratic function and its graph were studied extensively. Generally, when $a \geq 1$, the focus of a parabola is very near its vertex. Complete the square of the function $y = 2x^2 - 8x$ and write the result in the form $(x - h)^2 = 4p(y - k)$. What is the value of p? What are the coordinates of the vertex? $(x - 2)^2 = \frac{1}{2}(y + 8)$; $p = \frac{1}{8}$; $(2, -8)$

70. Have someone in your class bring an inexpensive flashlight to class, one where the bulb assembly can easily be removed from the body of the flashlight. As carefully as you can, measure the diameter and depth of the parabolic reflector in millimeters. Use these measurements to draw a cross section of the parabolic reflector on a coordinate grid, with the vertex of the parabola at (0, 0). The parabola will have an equation of the form $y = ax^2$, where a point (x, y) can be determined from the graph (the endpoints of the diameter). Use these values of x and y to find the value of a, then use the equation to locate the focus of this parabolic reflector. How closely does your answer seem to fit the location of the filament of the lightbulb when it is held in place? Answers will vary.

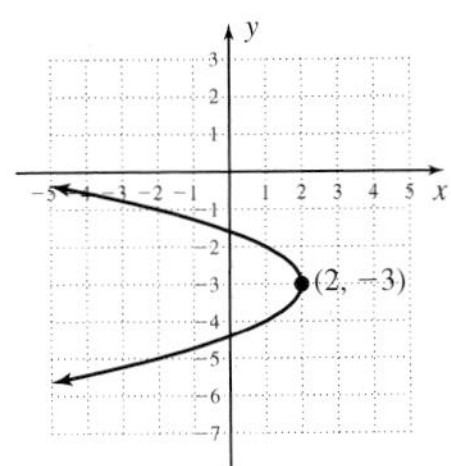

71. Match the graph with the correct equation. Then write a paragraph explaining how you made your choice. b

a. $y = (x - 2)^2 - 3$

b. $x = -(y + 3)^2 + 2$

c. $x = (y - 3)^2 - 2$

d. $y = -(x - 3)^2 + 2$

72. With a diameter of 1000 ft, the radio telescope located at Arecibo, Puerto Rico, is the largest in the world. Do some research on this remarkable telescope, and be sure to include information on the unique way the telescope was manufactured and the ingenious way the focus is held in place. Answers will vary.

EXTENDING THE CONCEPT

73. Find the equation and area of the circle whose center is the vertex of the parabola $y = -x^2 - 6x - 5$ and that intersects the parabola at the two points $(-1, 0)$ and $(-5, 0)$. $(x + 3)^2 + (y - 4)^2 = 20$; 20π units^2

Exercise 74

(−3, 5)
(0, 4)
(−6, −3)

74. In Exercise 61, a formula was given for the area of a right parabolic segment. The area of an *oblique* parabolic segment (the line segment cutting the parabola is *not* perpendicular to the axis) is more complex, as it involves locating the point where a line parallel to this segment is tangent (touches at only one point) to the parabola. The formula is $A = \frac{4}{3}T$, where T represents the area of the triangle formed by the endpoints of the segment and this point of tangency. What is the area of the parabolic segment shown (assuming the lines are parallel)? See Section 6.8, Example 3. 18 units^2

MAINTAINING YOUR SKILLS

75. (5.6) After gold was discovered in the nearby hills, the population of Goldsbury experienced rapid growth, as shown in the table. Use the data to draw a scatter-plot, then find an appropriate exponential equation to model this growth. According to the model: (a) What was the population of Goldsbury in 1995? (b) In what year did the population exceed 150,000? (c) Goldsbury will need dramatic improvements to its infrastructure when the population reaches 250,000 residents. If this growth rate continues, what year will this occur?

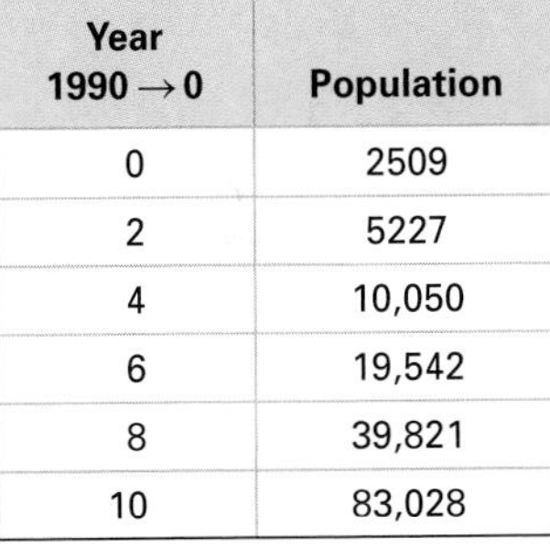

Year $1990 \to 0$	Population
0	2509
2	5227
4	10,050
6	19,542
8	39,821
10	83,028

a. $y = 2527.4(1.414)^x$; 14,286 b. 2001 c. 2003

76. (6.7) Construct a system of three equations in three variables using the general equation $y = ax^2 + bx + c$ and the points $(-3, 3)$, $(0, 6)$, and $(1, -1)$. Then use a matrix equation and your calculator to find the equation of the parabola containing these points. $a = -2$, $b = -5$, $c = 6$

77. (4.4) Use the function $f(x) = x^5 + 2x^4 + 17x^3 + 34x^2 - 18x - 36$ to comment and give illustrations of the tools available for working with polynomials: (a) synthetic division, (b) rational roots theorem, (c) the remainder and factor theorems, (d) the test for $x = -1$ and $x = 1$, (e) the upper/lower bounds property, (f) Descartes's rule of signs, and (g) roots of multiplicity (bounces, cuts, alternating intervals). Answers will vary.

78. (4.3) Find all roots (real and complex) to the equation $x^6 - 64 = 0$. (*Hint:* Begin by factoring the expression as the difference of two perfect squares.) $-2, 2, 1 + \sqrt{3}i, 1 - \sqrt{3}i, -1 + \sqrt{3}i, -1 - \sqrt{3}i$

79. (1.1) Solve the equation and comment. $3 - (x + 2) + 4x = 2(x - 1) + x + 1$ no solution

80. (3.8) What are the characteristics of an *even* function? symmetric with respect to the y-axis
What are the characteristics of an *odd* function? symmetric with respect to the origin

SUMMARY AND CONCEPT REVIEW

SECTION 7.1 The Circle and the Ellipse

KEY CONCEPTS

- The equation of a circle centered at (h, k) with radius r is $(x - h)^2 + (y - k)^2 = r^2$.
- After dividing both sides by r^2, we obtain the standard form $\frac{(x-h)^2}{r^2} + \frac{(y-k)^2}{r^2} = 1$, showing the horizontal and vertical distance from center to graph is r.
- The equation of an ellipse in center-shifted form is $A(x - h)^2 + B(y - k)^2 = F$, where $A \neq B$.
- After dividing both sides by F and simplifying, we obtain the standard form $\frac{(x-h)^2}{a^2} + \frac{(y-k)^2}{b^2} = 1$. The center of the ellipse is (h, k) with horizontal distance a and vertical distance b from center to graph.
- For an ellipse, note the *sum* of second-degree terms with $a \neq b$.

EXERCISES

Sketch the graph of each equation in Exercises 1 through 5.

1. $x^2 + y^2 = 16$

2. $x^2 + 4y^2 = 36$

3. $9x^2 + y^2 - 18x - 27 = 0$

4. $x^2 + y^2 + 6x + 4y + 12 = 0$

5. $\frac{(x+3)^2}{16} + \frac{(y-2)^2}{9} = 1$

6. Find the equation of a circle if $(-4, 5)$ and $(2, -3)$ are endpoints of the diameter. $(x + 1)^2 + (y - 1)^2 = 25$

SECTION 7.2 The Hyperbola

KEY CONCEPTS

- The equation of a *horizontal* hyperbola in center-shifted form is $A(x - h)^2 - B(y - k)^2 = F$.
- After dividing both sides by F and simplifying, we obtain the standard form $\frac{(x-h)^2}{a^2} - \frac{(y-k)^2}{b^2} = 1$. The center of the hyperbola is (h, k) with horizontal distance a from center to vertices and vertical distance b from center to midpoint of one side of the central rectangle.
- The equation of a vertical hyperbola in center-shifted form is $B(y - k)^2 - A(x - h)^2 = F$.
- For a hyperbola, note the *difference* of second-degree terms.

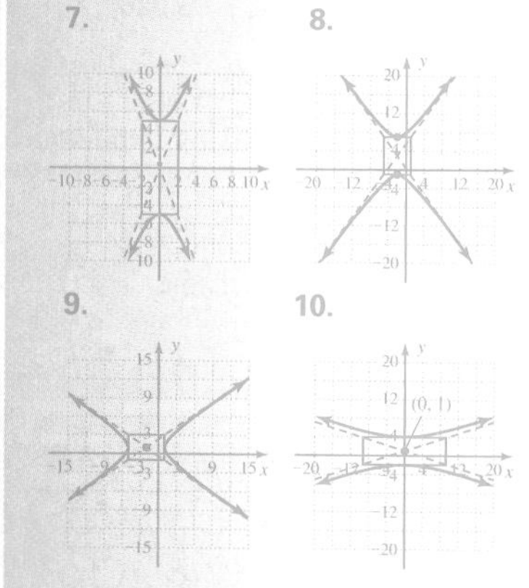

EXERCISES

Sketch the graph of each equation, indicating the center, vertices, and asymptotes. For Exercise 12, also give the equation of the hyperbola in standard form.

7. $4y^2 - 25x^2 = 100$

8. $\frac{(y-3)^2}{16} - \frac{(x+2)^2}{9} = 1$

9. $\frac{(x+2)^2}{9} - \frac{(y-1)^2}{4} = 1$

10. $9y^2 - x^2 - 18y - 72 = 0$

11. $x^2 - 4y^2 - 12x - 8y + 16 = 0$

12. vertices at $(-3, 0)$ and $(3, 0)$, asymptotes of $y = \pm\frac{4}{3}x$

Additional answers can be found in the Instructor Answer Appendix.

SECTION 7.3 Nonlinear Systems of Equations and Inequalities

15.

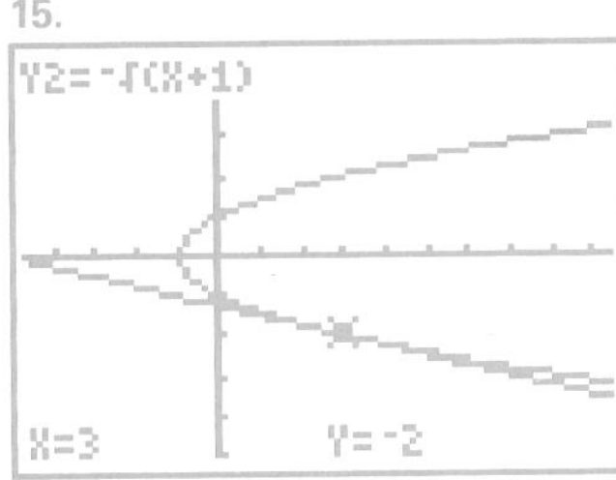

16.

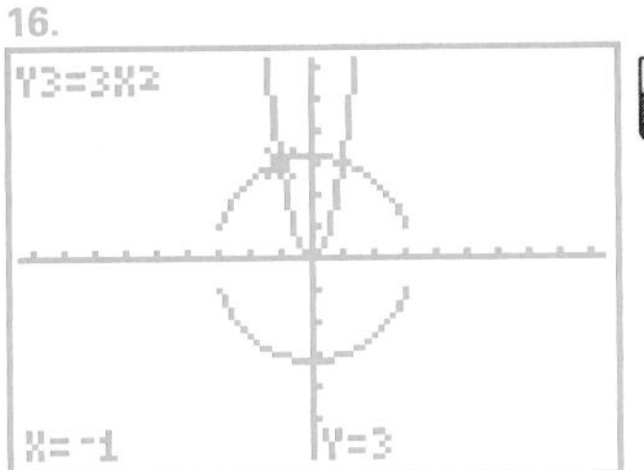

Additional answers can be found in the Instructor Answer Appendix.

KEY CONCEPTS

- Nonlinear systems of equations can be solved using substitution, elimination, or graphing technology.
- Identify the graphs of the equations in the system to determine the number of possible solutions.
- The solution to nonlinear systems of inequalities is a region of the plane where the solutions for individual inequalities in the system overlap.

EXERCISES

Solve Exercises 13 and 14 each using substitution or elimination, solve Exercises 15 and 16 using a graphing calculator. Identify the graph of each relation in the system before you begin.

13. $\begin{cases} x^2 + y^2 = 25 \\ y - x = -1 \end{cases}$ circle, line, (4, 3), (−3, −4)

14. $\begin{cases} x^2 - y^2 = 5 \\ x^2 + y^2 = 13 \end{cases}$ hyperbola, circle, (3, 2), (3, −2), (−3, 2), (−3, −2)

15. $\begin{cases} x = y^2 - 1 \\ x + 4y = -5 \end{cases}$ parabola, line, (3, −2)

16. $\begin{cases} x^2 + y^2 = 10 \\ y - 3x^2 = 0 \end{cases}$ circle, parabola, (1, 3), (−1, 3)

17. $\begin{cases} y \le x^2 - 2 \\ x^2 + 4y^2 \le 16 \end{cases}$

18. $\begin{cases} 4y^2 - 9x^2 \ge 36 \\ x^2 + y^2 \le 48 \end{cases}$

SECTION 7.4 Foci and the Analytic Ellipse and Hyperbola

KEY CONCEPTS

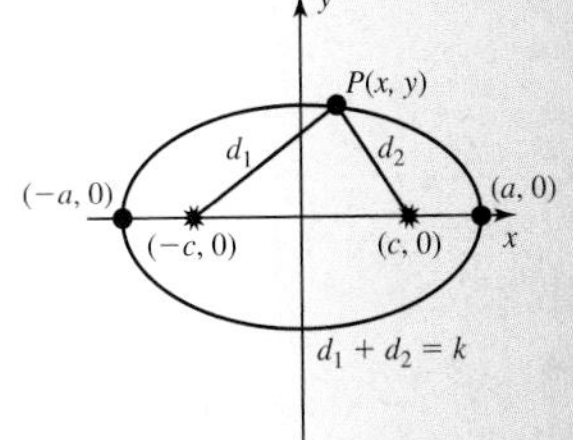

- Given any two fixed points F_1 and F_2 in a plane, an ellipse is defined to be the set of all points $P(x, y)$ such that the distance from the first focus to point P, plus the distance from the second focus to P, remains constant.
- For an ellipse, the distance from the center to one of the vertices is greater than the distance from the center to one of the foci.
- To find the foci of an ellipse: $c^2 = |a^2 - b^2|$ (since $a > c$ or $b > c$).
- Given any two fixed points F_1 and F_2 in a plane, a hyperbola is defined to be the set of all points $P(x, y)$ such that the distance from the first focus to point P, less the distance from the second focus to point P, remains constant.
- For a hyperbola, the distance from center to one of the vertices is less than the distance from center to one of the foci.
- To find the foci of a hyperbola: $c^2 = a^2 + b^2$ (since $c > a$ or $c > b$).

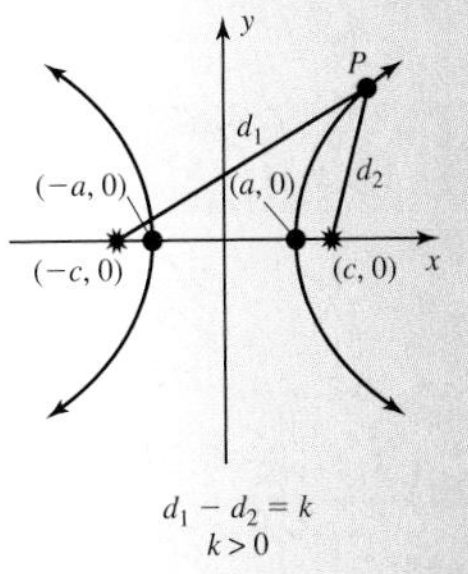

19. 20.

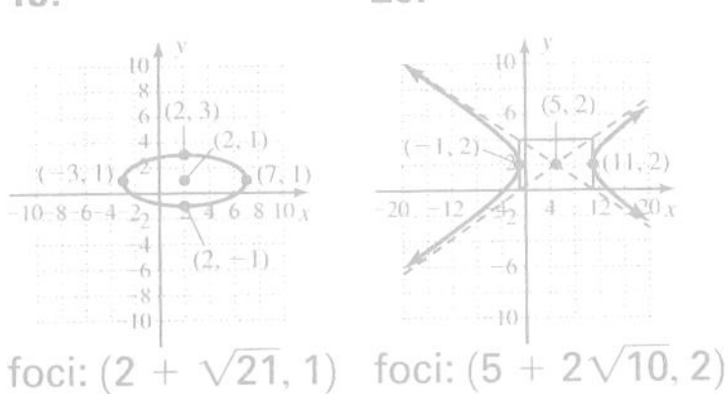

foci: $(2 + \sqrt{21}, 1)$, $(2 - \sqrt{21}, 1)$ foci: $(5 + 2\sqrt{10}, 2)$, $(5 - 2\sqrt{10}, 2)$

EXERCISES

Sketch each graph, noting all special features.

19. $4x^2 + 25y^2 - 16x - 50y - 59 = 0$

20. $4x^2 - 36y^2 - 40x + 144y - 188 = 0$

21. Find the equation of the ellipse given,
    a. vertices at $(-13, 0)$ and $(13, 0)$; foci at $(-12, 0)$ and $(12, 0)$ $\frac{x^2}{169} + \frac{y^2}{25} = 1$
    b. foci at $(0, -16)$ and $(0, 16)$; length of major axis: 40 units $\frac{x^2}{144} + \frac{y^2}{400} = 1$

22. Find the equation of the hyperbola given:
    a. vertices at $(-15, 0)$ and $(15, 0)$; foci at $(-17, 0)$ and $(17, 0)$ $\frac{x^2}{225} - \frac{y^2}{64} = 1$
    b. foci at $(0, -5)$ and $(0, 5)$; vertical length of central rectangle: 8 units $\frac{y^2}{16} - \frac{x^2}{9} = 1$

SECTION 7.5 The Analytic Parabola

KEY CONCEPTS

- Horizontal parabolas have equations of the form $x = ay^2 + by + c$.
- A horizontal parabola will open to the right if $a > 0$ and to the left if $a < 0$. The axis of symmetry is $y = \frac{-b}{2a}$, with the vertex (h, k) found by evaluating at $y = \frac{-b}{2a}$ or by completing the square and writing the equation in shifted form: $x = a(y - k)^2 + h$.
- Given a fixed point F and fixed line D in the plane, a parabola is defined to be the set of all points $P(x, y)$ such that the distance from point F to point P is equal to the perpendicular distance from point P to line D.
- The equation $x^2 = 4py$ describes a vertical parabola, concave up if $p > 0$ and concave down if $p < 0$.
- The equation $y^2 = 4px$ describes a horizontal parabola, opening to the right if $p > 0$, and opening to the left if $p < 0$.
- The focal chord of a parabola is a line segment that contains the focus and is parallel the directrix, with its endpoints on the graph. It has a total length of $|4p|$, meaning the distance from the focus to a point of the graph is $|2p|$. It is commonly used to assist in drawing a graph of the parabola.

23. 24.

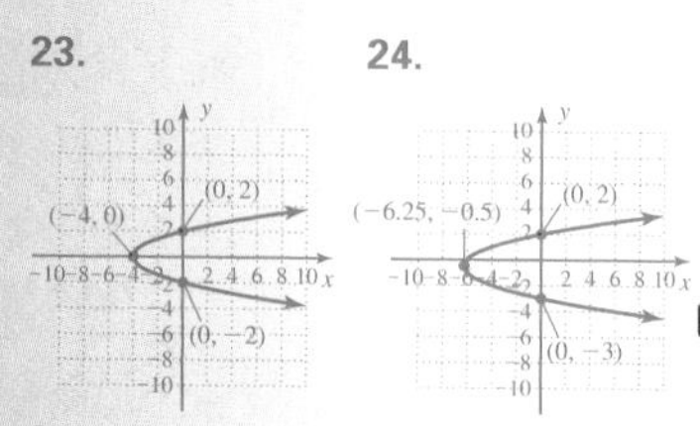

25. 26.

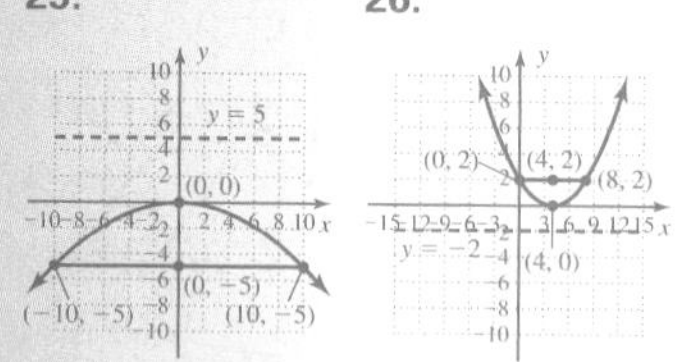

EXERCISES

For Exercises 23 and 24, find the vertex and x- and y-intercepts if they exist. Then sketch the graph using symmetry and a few points or by completing the square and shifting a parent function.

23. $x = y^2 - 4$

24. $x = y^2 + y - 6$

For Exercises 25 and 26, find the vertex, focus, and directrix for each parabola. Then sketch the graph using the vertex, focus, and focal chord. Also graph the directrix.

25. $x^2 = -20y$

26. $x^2 - 8x - 8y + 16 = 0$

MIXED REVIEW

1. 2.

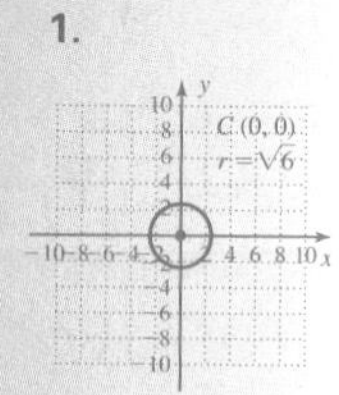

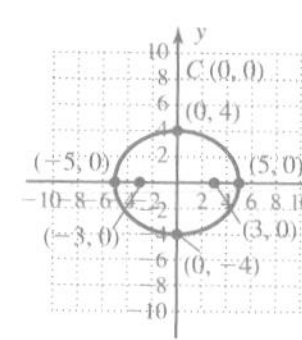

3. 4.

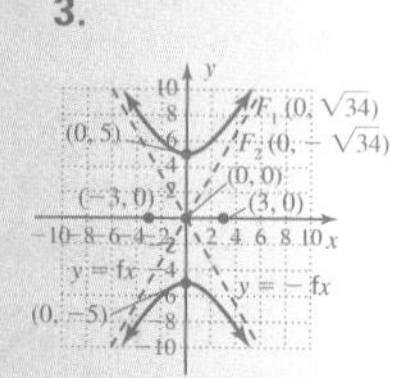

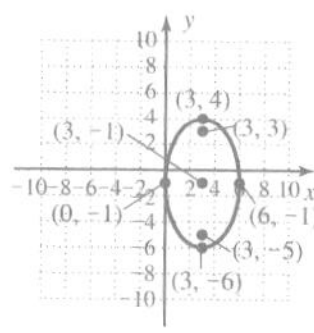
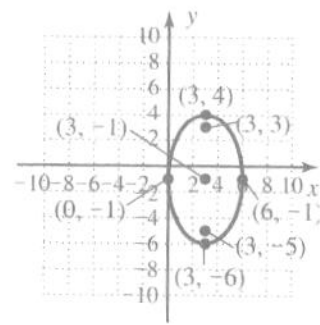

Additional answers can be found in the Instructor Answer Appendix.

For Exercises 1 through 21, graph the conic section and label/draw the center, vertices, directrix, foci, focal chords, asymptotes, and other important features as these apply to a particular equation and conic section.

1. $9x^2 + 9y^2 = 54$

2. $16x^2 + 25y^2 = 400$

3. $9y^2 - 25x^2 = 225$

4. $\frac{(x - 3)^2}{9} + \frac{(y + 1)^2}{25} = 1$

5. $\frac{(y - 2)^2}{25} - \frac{(x + 3)^2}{16} = 1$

6. $4(x - 1)^2 - 36(y + 2)^2 = 144$

7. $16(x + 2)^2 + 4(y - 1)^2 = 64$

8. $49(x + 2)^2 + (y - 3)^2 = 49$

9. $x^2 + y^2 - 8x + 12y + 16 = 0$

10. $y = 2x^2 - 10x + 15$

11. $y = 2x^2 - 8x - 10$

12. $x = -y^2 - 8y - 11$

13. $x = -y^2 + 2y + 3$

14. $x = (y + 2)^2 - 3$

15. $x = y^2 - 9$

16. $x = y^2 - y - 6$

17. $x^2 = -24y$

18. $x^2 - 8x - 4y + 16 = 0$

19. $4x^2 + 16y^2 - 12x - 48y - 19 = 0$

20. $4x^2 - 25y^2 - 24x + 150y - 289 = 0$

21. $x^2 + y^2 + 8x + 12y + 2 = 0$

22. Solve using elimination: $\begin{cases} 4x^2 - y^2 = -9 \\ x^2 + 3y^2 = 79 \end{cases}$ $(2, 5), (2, -5)$ $(-2, 5), (-2, -5)$

23. Verify the closed figure with vertices $P_1(-4, -9)$, $P_2(6, 1)$, and $P_3(-6, 3)$ is an isosceles triangle. $d(P_1P_3) = d(P_2P_3)$

24. A theorem from Euclidean geometry states: *A line drawn from the vertex of an isosceles triangle to the midpoint of the base is the perpendicular bisector of the base.* Verify the theorem is true for the isosceles triangle given in Exercise 23. verified $m(P_1P_2) = (1, -4) = P_4$; slope $(P_1P_2) = 1$; slope $(P_3P_4) = -1$

25. **Planetary orbits:** Except for small variations, a planet's orbit around the Sun is elliptical, with the Sun at one focus. The *perihelion* or minimum distance from the planet Mercury to the Sun is about 46 million kilometers. Its *aphelion* or maximum distance from the Sun is approximately 70 million kilometers. Assume this orbit is a central ellipse on the coordinate grid. (a) Find the coordinates of the Sun; (b) find the length $2a$ of the major axis, and the length $2b$ of the minor axis; and (c) determine the equation model for the orbit of Mercury in standard form. a. $(12, 0)$ b. $2a = 116$; $2b \approx 113.5$ c. $\frac{x^2}{3364} + \frac{y^2}{3220} = 1$

PRACTICE TEST

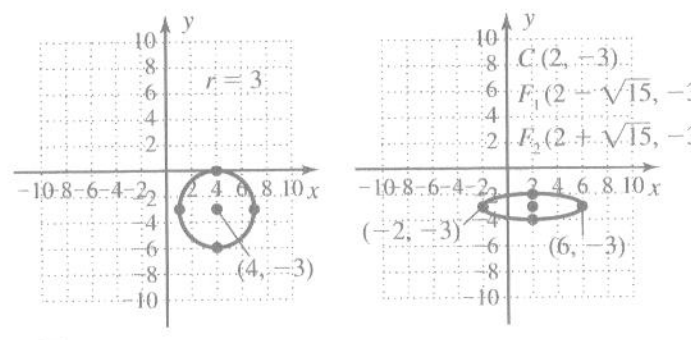

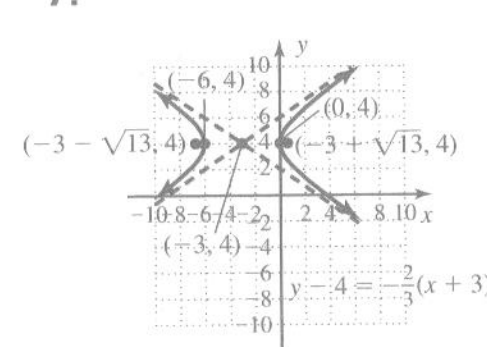

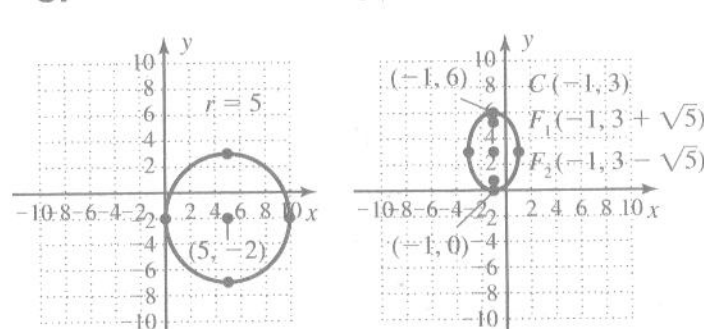

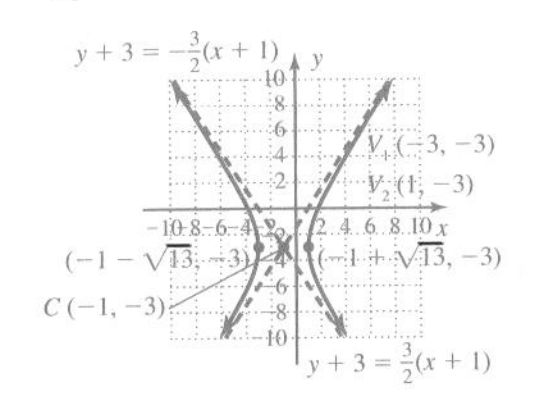

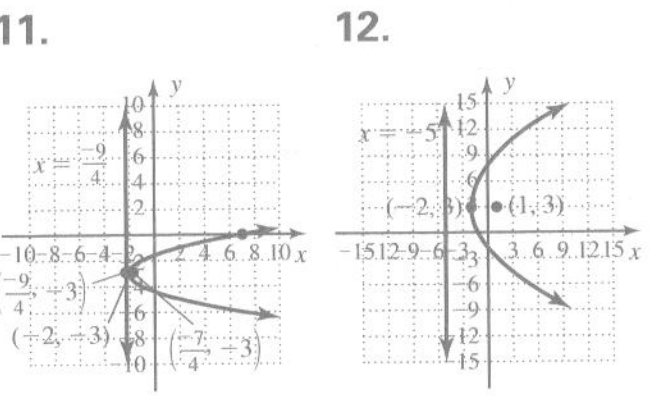

By inspection only (no graphing, completing the square, etc.), match each equation to its correct description.

1. $x^2 + y^2 - 6x + 4y + 9 = 0$ __c__

2. $4y^2 + x^2 - 4x + 8y + 20 = 0$ __d__

3. $y - x^2 - 4x + 20 = 0$ __a__

4. $x^2 - 4y^2 - 4x + 12y + 20 = 0$ __b__

a. Parabola b. Hyperbola c. Circle d. Ellipse

Graph each conic section, and label the center, vertices, foci, focal chords, asymptotes, and other important features where applicable.

5. $(x - 4)^2 + (y + 3)^2 = 9$

6. $\frac{(x - 2)^2}{16} + \frac{(y + 3)^2}{1} = 1$

7. $\frac{(x + 3)^2}{9} - \frac{(y - 4)^2}{4} = 1$

8. $x^2 + y^2 - 10x + 4y + 4 = 0$

9. $9x^2 + 4y^2 + 18x - 24y + 9 = 0$

10. $9x^2 - 4y^2 + 18x - 24y - 63 = 0$

11. $x = (y + 3)^2 - 2$

12. $y^2 - 6y - 12x - 15 = 0$

Solve each nonlinear system using the technique of your choice.

13. a. $\begin{cases} 4x^2 - y^2 = 16 \\ y - x = 2 \end{cases}$ $(\frac{10}{3}, \frac{16}{3}), (-2, 0)$

b. $\begin{cases} 4y^2 - x^2 = 4 \\ x^2 + y^2 = 4 \end{cases}$ $(\frac{2\sqrt{15}}{5}, \frac{2\sqrt{10}}{5}), (\frac{2\sqrt{15}}{5}, \frac{-2\sqrt{10}}{5}), (\frac{-2\sqrt{15}}{5}, \frac{2\sqrt{10}}{5}), (\frac{-2\sqrt{15}}{5}, \frac{-2\sqrt{10}}{5})$

14. A support bracket on the frame of a large ship is a steel right triangle with a hypotenuse of 25 ft and a perimeter of 60 ft. Find the lengths of the other sides using a system of nonlinear equations. 15 ft, 20 ft

15. Find an equation for the circle whose center is at $(-2, 5)$ and whose graph goes through the point $(0, 3)$. $(x + 2)^2 + (y - 5)^2 = 8$

16. Find the equation of the ellipse (in standard form) with vertices at $(-4, 0)$ and $(4, 0)$ with foci located at $(-2, 0)$ and $(2, 0)$. $\frac{x^2}{16} + \frac{y^2}{12} = 1$

17. The orbit of Mars around the Sun is elliptical, with the Sun at one focus. When the orbit is expressed as a central ellipse on the coordinate grid, its equation is $\frac{x^2}{(141.65)^2} + \frac{y^2}{(141.03)^2} = 1$, with a and b in millions of miles. Use this information to find the *aphelion* of Mars (distance from the Sun at its farthest point), and the *perihelion* of Mars (distance from the Sun at its closest point). 154.89 million miles; 128.41 million miles

18. $y = (x - 1)^2 - 4$; $D: x \in (-\infty, \infty)$, $R: y \in [-4, \infty)$; focus: $\left(1, \frac{-15}{4}\right)$
19. $(x - 1)^2 + (y - 1)^2 = 25$; $D: x \in [-4, 6]$, $R: y \in [-4, 6]$
20. $\frac{(x + 3)^2}{9} + \frac{y^2}{36} = 1$; $D: x \in [-6, 0]$, $R: y \in [-6, 6]$; foci: $(-3, -3\sqrt{3})$, $(-3, 3\sqrt{3})$

Determine the equation of each relation and state its domain and range. For the parabola and the ellipse, also give the location of the foci.

18.

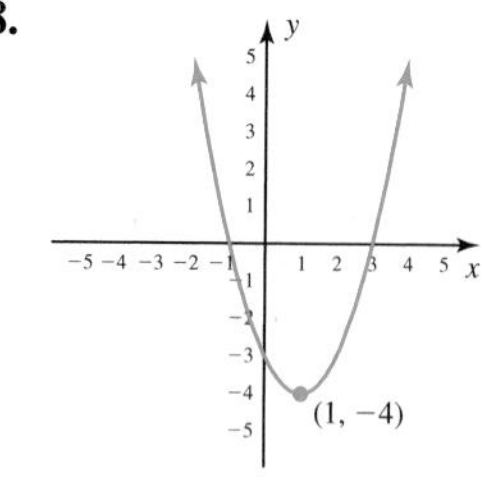

19.

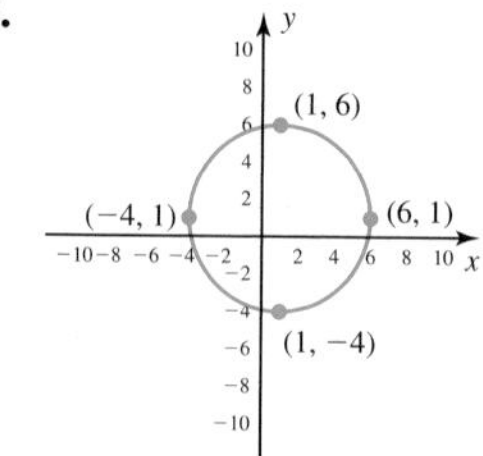

20.

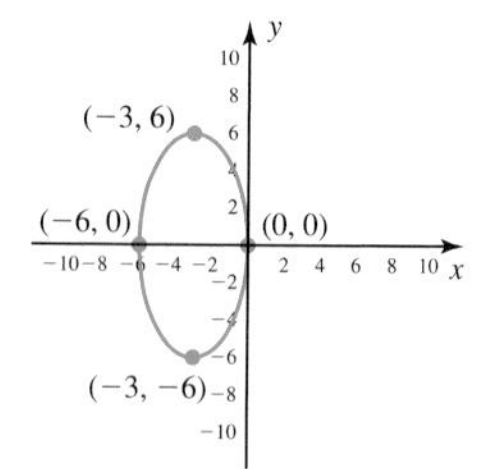

CALCULATOR EXPLORATION AND DISCOVERY

Elongation and Eccentricity

Technically speaking, a circle is an ellipse with both foci at the center. As the distance between foci increases, the ellipse becomes more elongated. We saw other instances of elongation in stretches and compressions of parabolic graphs, and in hyperbolic graphs where the asymptotic slopes varied depending on the values a and b. The measure used to quantify this elongation is called the *eccentricity* e, and is determined by the ratio $e = \frac{c}{a}$. For this *Exploration and Discovery,* we'll use the **repeat graph** feature of a graphing calculator to explore the eccentricity of the graph of a conic. The "repeat graph" feature enables you to graph a family of curves by enclosing changes in a parameter in braces "{ }." For instance, entering $y = \{-2, -1, 0, 1, 2\}x + 3$ as Y_1 on the Y= screen will automatically graph these five lines:

$$y = -2x + 3 \qquad y = -x + 3 \qquad y = 3 \qquad y = x + 3 \qquad y = 2x + 3$$

We'll use this feature to graph a family of ellipses, observing the result and calculating the eccentricity for each curve in the family. The standard form is $\frac{x^2}{a^2} + \frac{y^2}{b^2} = 1$, which we solve for y and enter as Y_1 (see *Technology Highlight* from Section 7.1). After simplification the result is $y = b\sqrt{1 - \frac{x^2}{a^2}}$, but for this investigation we'll use the constant $b = 2$ and vary the parameter a using the values $a = 2, 4, 6$, and 8. The result is $y = 2\sqrt{1 - \frac{x^2}{\{4, 16, 36, 64\}}}$. Note from Figure 7.37 that we've set $Y_2 = -Y_1$ to graph the lower half of the ellipse. Using the "friendly window" shown (Figure 7.38) gives the result shown in Figure 7.39, where we see the ellipse is increasingly elongated in the horizontal direction (note when $a = 2$ the result is a circle since $a = b$). Using $a = 2, 4, 6$, and 8 with $b = 2$ in the foci formula $c = \sqrt{a^2 - b^2}$ gives $c = 0, 2\sqrt{3}, 4\sqrt{2}$, and $2\sqrt{15}$, respectively, with these eccentricities: $e = \frac{0}{2}, \frac{2\sqrt{3}}{4}, \frac{4\sqrt{2}}{6}$, and $\frac{2\sqrt{15}}{8}$. While difficult to see in radical form, we find that the eccentricity of an ellipse always satisfies the inequality $0 < e < 1$ (excluding the circle =

Figure 7.37

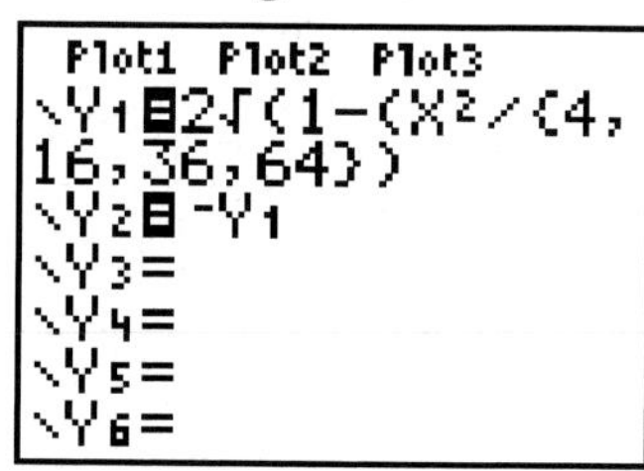

Figure 7.38

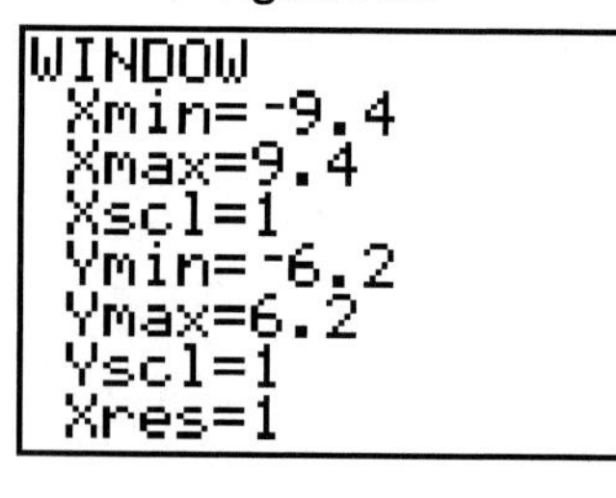

ellipse case). To two decimal places, the values are $e = 0, 0.87, 0.94$, and 0.97, respectively. It's interesting to note how the $e = \frac{c}{a}$ definition of eccentricity relates to our everyday use of the word "eccentric." A normal or "noneccentric" person is thought to be well-rounded, and sure enough $e = 0$ produces a well-rounded figure—a circle. A person who is highly eccentric is thought to be far from the norm, deviating greatly from the center, and greater values of e produce very elongated ellipses.

Figure 7.39

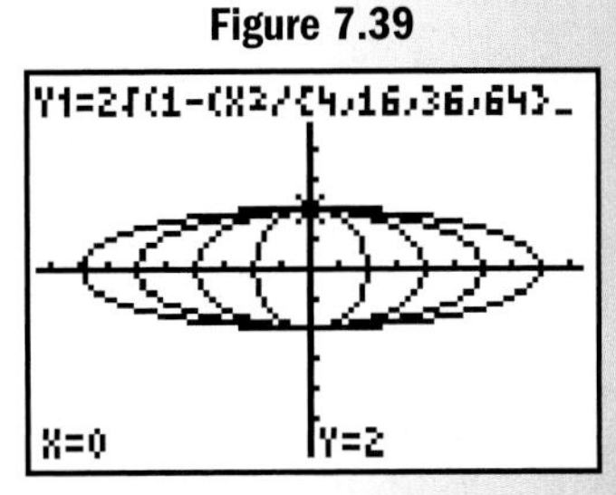

Exercise 1: Perform a similar exploration using a family of *hyperbolas.* What do you notice about the eccentricity? $e > 1$

Exercise 2: Perform a similar exploration using a family of *parabolas.* What do you notice about the eccentricity? $e = 1$

STRENGTHENING CORE SKILLS

Ellipses and Hyperbolas with Rational/Irrational Values of *a* and *b*

Using the process known as completing the square, we were able to convert from the polynomial form of a conic section to the standard form. However, for some equations, values of a and b are somewhat difficult to identify, since the coefficients are not factors. Consider the equation $20x^2 + 120x + 27y^2 - 54y + 192 = 0$, the equation of an ellipse.

$$20x^2 + 120x + 27y^2 - 54y + 192 = 0 \qquad \text{original equation}$$

$$20(x^2 + 6x + \underline{\quad\quad}) + 27(y^2 - 2y + \underline{\quad\quad}) = -192 \qquad \text{subtract 192, begin process}$$

$$20(x^2 + 6x + 9) + 27(y^2 - 2y + 1) = -192 + 27 + 180 \qquad \text{complete the square in } x \text{ and } y$$

$$20(x + 3)^2 + 27(y - 1)^2 = 15 \qquad \text{factor and simplify}$$

$$\frac{4(x + 3)^2}{3} + \frac{9(y - 1)^2}{5} = 1 \qquad \text{standard form}$$

Unfortunately, we cannot easily identify the values of a and b, since the coefficients of each binomial square were not "1." In these cases, we can write the equation in standard form by using a simple property of fractions—the numerator and denominator of any fraction can be divided by the same quantity to obtain an equivalent fraction. Although the result may look odd, it can nevertheless be applied here, giving a result of $\frac{(x + 3)^2}{\frac{3}{4}} + \frac{(y - 1)^2}{\frac{5}{9}} = 1$. We can now identify a and b by writing these denominators in squared form, which gives the following expression: $\frac{(x + 3)^2}{\left(\frac{\sqrt{3}}{2}\right)^2} + \frac{(y - 1)^2}{\left(\frac{\sqrt{5}}{3}\right)^2} = 1$. The values of a and b are now easily seen as $a = \frac{\sqrt{3}}{2} \approx 0.866$ and $b = \frac{\sqrt{5}}{3} \approx 0.745$. Use this idea to complete the following exercises.

Exercise 1: Identify the values of a and b by writing the equation $100x^2 - 400x - 18y^2 - 108y + 230 = 0$ in standard form.

Exercise 2: Identify the values of a and b by writing the equation $28x^2 - 56x + 48y^2 + 192y + 195 = 0$ in standard form.

1. $\frac{(x - 2)^2}{\left(\frac{\sqrt{2}}{5}\right)^2} - \frac{(y + 3)^2}{\left(\frac{2}{3}\right)^2} = 1$

   $a = \frac{\sqrt{2}}{5}, b = \frac{2}{3}$

2. $\frac{(x - 1)^2}{\left(\frac{5\sqrt{7}}{14}\right)^2} + \frac{(y + 2)^2}{\left(\frac{5\sqrt{3}}{12}\right)^2} = 1$

   $a = \frac{5\sqrt{17}}{14}, b = \frac{5\sqrt{3}}{12}$

3. 4.

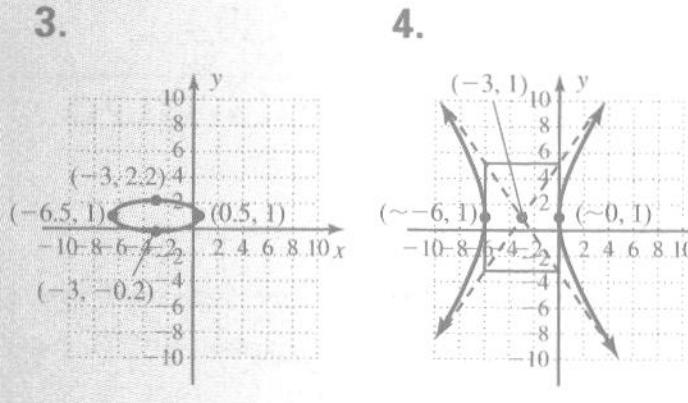

Exercise 3: Write the equation in standard form, then identify the values of a and b and use them to graph the ellipse.

$$\frac{4(x+3)^2}{49} + \frac{25(y-1)^2}{36} = 1$$

$\dfrac{(x+3)^2}{\left(\frac{7}{2}\right)^2} + \dfrac{(y-1)^2}{\left(\frac{6}{5}\right)^2} = 1;\ a = \dfrac{7}{2},\ b = \dfrac{6}{5}$

Exercise 4: Write the equation in standard form, then identify the values of a and b and use them to graph the hyperbola.

$$\frac{9(x+3)^2}{80} - \frac{4(y-1)^2}{81} = 1$$

$\dfrac{(x+3)^2}{\left(\frac{4\sqrt{5}}{3}\right)^2} - \dfrac{(y-1)^2}{\left(\frac{9}{2}\right)^2} = 1;\ a = \dfrac{4\sqrt{5}}{3} \approx 3,\ b = \dfrac{9}{2}$

Cumulative Review Chapters 1–7

10. 11.

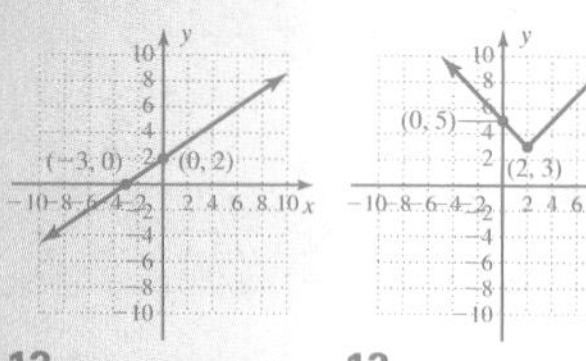

12. 13.

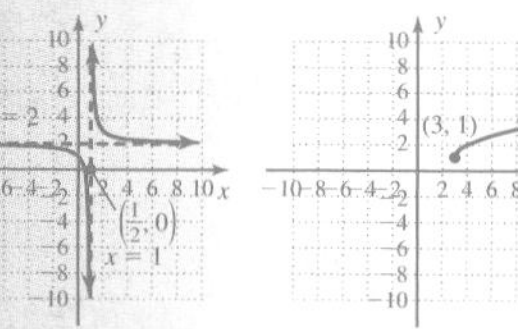

14. a. b.

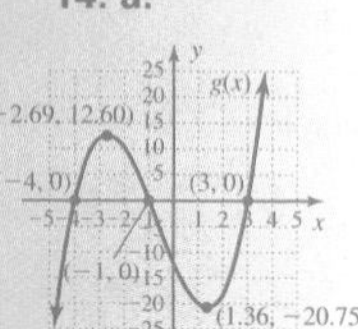

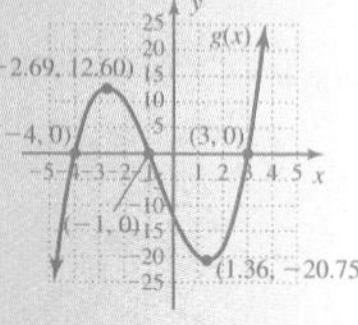

15. 16.

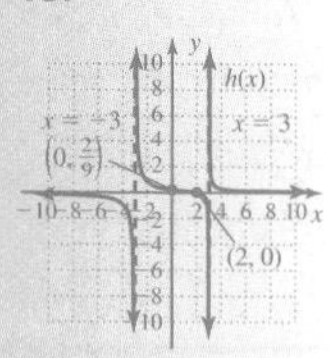

17. 18.

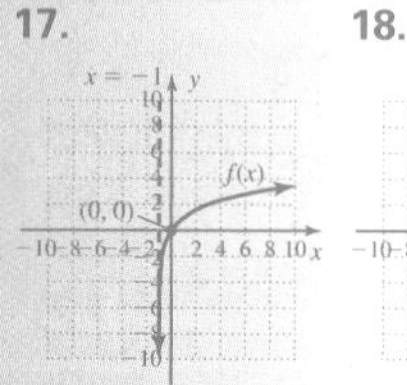

19. 20.

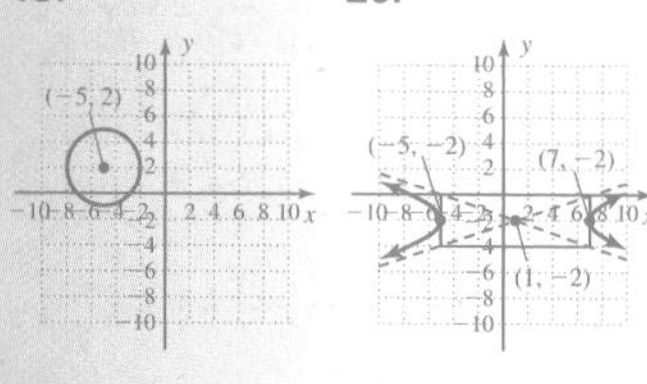

Solve each equation.

1. $x^3 - 2x^2 + 4x - 8 = 0$ $x = 2, x = \pm 2i$
2. $2|n + 4| + 3 = 13$ $n = 1, n = -9$
3. $\sqrt{x+2} - 2 = \sqrt{3x+4}$ no solution
4. $x^{\frac{2}{3}} + 8 = 0$ no solution
5. $x^2 - 6x + 13 = 0$ $x = 3 \pm 2i$
6. $4 \cdot 2^{x+1} = \frac{1}{8}$ $x = -6$
7. $3^{x-2} = 7$ $x = 2 + \frac{\ln 7}{\ln 3}$
8. $\log_3 81 = x$ $x = 4$
9. $\log_3 x + \log_3(x - 2) = 1$ $x = 3$

Graph each function. Include vertices, x- and y-intercepts, asymptotes, and other features.

10. $y = \frac{2}{3}x + 2$
11. $y = |x - 2| + 3$
12. $y = \dfrac{1}{x-1} + 2$
13. $y = \sqrt{x - 3} + 1$
14. a. $g(x) = (x - 3)(x + 1)(x + 4)$ b. $f(x) = x^4 + x^3 - 13x^2 - x + 12$
15. $h(x) = \dfrac{x-2}{x^2 - 9}$
16. $q(x) = 2^x + 3$
17. $f(x) = \log_2(x + 1)$
18. $x = y^2 + 4y + 7$
19. $x^2 + y^2 + 10x - 4y + 20 = 0$
20. $4(x - 1)^2 - 36(y + 2)^2 = 144$
21. Determine the following for the indicated graph (write all answers in interval notation): (a) the domain, (b) the range, (c) interval(s) where $f(x)$ is increasing or decreasing, (d) interval(s) where $f(x)$ is constant, (e) location of any maximum or minimum value(s), (f) interval(s) where $f(x)$ is positive, and (g) interval(s) where $f(x)$ is negative.

a. $x \in (-\infty, \infty)$ b. $y \in (-\infty, 4]$ c. $f(x)\uparrow: x \in (-\infty, -1), f(x)\downarrow: x \in (-1, \infty)$ d. none e. max: $(-1, 4)$ f. $f(x) > 0: x \in (-4, 2)$ g. $f(x) < 0: x \in (-\infty, -4) \cup (2, \infty)$

Solve each system of equations.

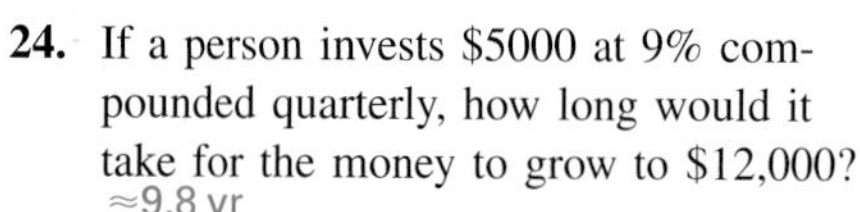

22. $\begin{cases} 4x + 3y = 13 \\ -9y + 5z = 19 \\ x - 4z = -4 \end{cases}$ $(4, -1, 2)$

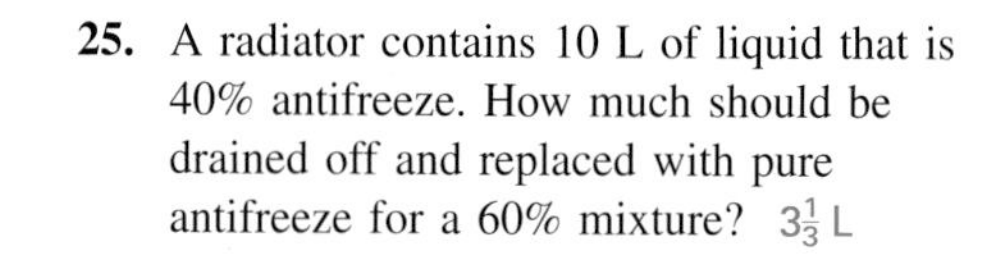

23. $\begin{cases} x^2 + y^2 = 25 \\ 64x^2 + 12y^2 = 768 \end{cases}$ $(3, 4), (3, -4), (-3, 4), (-3, -4)$

24. If a person invests \$5000 at 9% compounded quarterly, how long would it take for the money to grow to \$12,000? ≈9.8 yr

25. A radiator contains 10 L of liquid that is 40% antifreeze. How much should be drained off and replaced with pure antifreeze for a 60% mixture? $3\frac{1}{3}$ L

Chapter 8

Additional Topics in Algebra

Chapter Outline

8.1 Sequences and Series 748

8.2 Arithmetic Sequences 759

8.3 Geometric Sequences 768

8.4 Mathematical Induction 781

8.5 Counting Techniques 792

8.6 Introduction to Probability 807

8.7 The Binomial Theorem 823

Preview

One important way humans relate to the real world is the recognition and study of patterns. Sometimes the patterns don't directly involve numbers, as when a historian studies human events and their relationship to each other—a *sequence of events*—or when a geologist uses *sequence stratigraphy* to study patterns of change in the Earth's surface. But many patterns do involve numbers, and for many reasons, these patterns of numbers are also called *sequences.* In the early part of this chapter we'll investigate how sequences are used as mathematical models. Later in the chapter we'll study the many ways that additional patterns and probabilities are used in such diverse fields as politics, manufacturing, gambling, opinion polls, and others.

8.1 Sequences and Series

LEARNING OBJECTIVES

In Section 8.1 you will learn how to:

A. Write out the terms of a sequence given the general or *n*th term

B. Work with recursive sequences and sequences involving a factorial

C. Determine the general term of a sequence

D. Find the partial sum of a series

E. Use summation notation to write and evaluate series

F. Use sequences to solve applied problems

INTRODUCTION

A *sequence* can be thought of as a pattern of numbers listed in a prescribed order. A *series* is the sum of the numbers in a sequence. Sequences and series come in countless varieties, and we'll introduce some general forms here. In following sections we'll focus on two special types: arithmetic and geometric sequences. These are used in a number of different fields, with a wide variety of significant applications.

POINT OF INTEREST

Sequences come in many different forms. There are infinite sequences: 1, 3, 5, 7, 9, . . . and finite sequences: 0, $0.1\overline{6}$, $0.\overline{3}$, 0.5, $0.\overline{6}$, $0.8\overline{3}$, 1. There are increasing sequences: 1, 4, 7, 10, . . . ; decreasing sequences: $\frac{13}{4}, \frac{23}{8}, \frac{5}{2}, \frac{17}{8}, \ldots$; and alternating sequences: −324, 108, −36, 12, ... Sequences can be constructed using addition: 1, 5, 9, 13, (add four to obtain the next term); multiplication: 729, 81, 9, 1, $\frac{1}{9}$, (multiply by $\frac{1}{9}$ to obtain the next term); or a combination of operations. A large part of mathematical investigation and inquiry has been devoted to discovering patterns that occur naturally in the world around us, as well as to purely numerical patterns that have led to some very important mathematical results.

A. Finding the Terms of a Sequence Given the General Term

Suppose a person had \$10,000 to invest, and decided to place the money in government bonds that guarantee an annual return of 7%. From our work in Chapter 5, we know the amount of money in the account after x years can be modeled by the function $f(x) = 10{,}000(1.07)^x$. If you reinvest your earnings each year, the amount in the account would be (rounded to the nearest dollar):

Year:	$f(1)$	$f(2)$	$f(3)$	$f(4)$	$f(5)\ldots$
	↓	↓	↓	↓	↓
Value:	\$10,700	\$11,449	\$12,250	\$13,108	\$14,026 . . .

Note the relationship (year, value) is a function that pairs 1 with \$10,700, 2 with \$11,449, 3 with \$12,250 and so on. This is an example of a **sequence.** To distinguish sequences from other algebraic functions, we commonly name the functions a instead of f, use the variable n instead of x, and employ a subscript notation. The function $f(x) = 10{,}000(1.07)^x$ would then be written $a_n = 10{,}000(1.07)^n$. Using this notation the first five terms are:

Year:	a_1	a_2	a_3	a_4	a_5
	↓	↓	↓	↓	↓
Value:	\$10,700	\$11,449	\$12,250	\$13,108	\$14,026

WORTHY OF NOTE

Just as f, g, and h are the most frequent names used for functions, a, b, and c are the most frequent names used for sequences.

The values a_1, a_2, a_3, a_4, . . . are called the **terms** of the sequence. If the account were closed after a certain number of years (for example, after the fifth year) we have a **finite sequence.** If we let the investment grow indefinitely, the result is called an

infinite sequence. The expression a_n that defines the sequence is called the **general** or ***n*th term** and the terms immediately preceding it are called the $(n-1)$st term, the $(n-2)$nd term, and so on.

WORTHY OF NOTE

Although our working definition limits the domain of a finite sequence to the natural numbers 1 through n, sequences can actually start with any natural number. For instance, the sequence $a_n = \frac{2}{n-1}$ must start at $n = 2$ to avoid division by zero. In addition, we will sometimes use a_0 to indicate a preliminary or inaugural element. For the preceding illustration, $a_0 = \$10{,}000$ would indicate the amount of money initially held, prior to investing it.

SEQUENCES

A *finite sequence* is a function a_n whose domain is the set of natural numbers from 1 to n: $\{1, 2, 3, 4, \ldots, k, k+1, \ldots n\}$. The terms of the sequence are labeled $a_1, a_2, a_3, a_4, \ldots, a_k, a_{k+1}, \ldots, a_{n-1}, a_n$. The expression a_n, which defines the sequence, is called the *nth term.* An *infinite sequence* is a function whose domain is the set of all natural numbers: $\{1, 2, 3, 4, 5, \ldots\}$.

EXAMPLE 1A For $a_n = \frac{n+1}{n^2}$, find a_1, a_3, a_6, and a_7.

Solution:

$a_1 = \frac{1+1}{1^2} = 2$ $\qquad a_3 = \frac{3+1}{3^2} = \frac{4}{9}$

$a_6 = \frac{6+1}{6^2} = \frac{7}{36}$ $\qquad a_7 = \frac{7+1}{7^2} = \frac{8}{49}$

WORTHY OF NOTE

When the terms of a sequence *alternate in sign* as in Example 1B, we call it an **alternating sequence.**

EXAMPLE 1B Find the first four terms of the sequence $a_n = (-1)^n 2^n$. Write the terms of the sequence as a list.

Solution:

$a_1 = (-1)^1 2^1 = -2$ $\qquad a_2 = (-1)^2 2^2 = 4$

$a_3 = (-1)^3 2^3 = -8$ $\qquad a_4 = (-1)^4 2^4 = 16$

The sequence can be written $-2, 4, -8, 16, \ldots,$ or more generally as $-2, 4, -8, 16, \ldots, (-1)^n 2^n, \ldots$ to show how each term was generated.

NOW TRY EXERCISES 7 THROUGH 32

B. Recursive Sequences and Factorial Notation

Sometimes the formula defining a sequence uses the preceding term or terms to generate those that follow. These are called **recursive sequences** and are particularly useful in writing computer programs. Because of how they are defined, recursive sequences must give an inaugural term or **seed element,** to begin the recursion process.

Perhaps the most famous recursive sequence is associated with the work of Leonardo of Pisa (A.D. 1180–1250), better known to history as *Fibonacci.* In fact, it is commonly called the Fibonacci sequence.

WORTHY OF NOTE

One application of the Fibonacci sequence involves the Fibonacci spiral, found in the growth of many ferns and the spiral shell of many mollusks.

EXAMPLE 2 Write out the first eight terms of the recursive sequence defined by $c_1 = 1$, $c_2 = 1$, and $c_n = c_{n-1} + c_{n-2}$.

Solution: The first two terms are given, so we begin with $n = 3$.

$$\begin{aligned} c_3 &= c_{3-1} + c_{3-2} \\ &= c_2 + c_1 \\ &= 1 + 1 \\ &= 2 \end{aligned} \qquad \begin{aligned} c_4 &= c_{4-1} + c_{4-2} \\ &= c_3 + c_2 \\ &= 2 + 1 \\ &= 3 \end{aligned} \qquad \begin{aligned} c_5 &= c_{5-1} + c_{5-2} \\ &= c_4 + c_3 \\ &= 3 + 2 \\ &= 5 \end{aligned}$$

At this point we can simply use the fact that each successive term is simply the sum of the preceding two, and find that $c_6 = 3 + 5 = 8$, $c_7 = 5 + 8 = 13$, and $c_8 = 13 + 8 = 21$. The first eight terms are 1, 1, 2, 3, 5, 8, 13, and 21.

NOW TRY EXERCISES 33 THROUGH 38

Sequences can also be defined using a **factorial,** which is the product of a given natural number with all those that precede it. The expression **5!** is read, "five factorial," and is evaluated as: $5! = 5 \cdot 4 \cdot 3 \cdot 2 \cdot 1 = 120$.

FACTORIALS

For any natural number n,

$$n! = n \cdot (n - 1) \cdot (n - 2) \cdot \cdots \cdot 3 \cdot 2 \cdot 1$$

Rewriting a factorial in equivalent forms often makes it easier to simplify certain expressions. For example, we can rewrite $5! = 5 \cdot 4!$ or $5! = 5 \cdot 4 \cdot 3!$. Consider Example 3.

EXAMPLE 3 Simplify by writing the numerator in an equivalent form.

a. $\dfrac{9!}{7!}$ **b.** $\dfrac{11!}{8!2!}$ **c.** $\dfrac{6!}{3!5!}$

Solution:

a. $\dfrac{9!}{7!} = \dfrac{9 \cdot 8 \cdot \cancel{7!}}{\cancel{7!}} = 9 \cdot 8 = 72$

b. $\dfrac{11!}{8!2!} = \dfrac{11 \cdot 10 \cdot 9 \cdot \cancel{8!}}{\cancel{8!}2!} = \dfrac{990}{2} = 495$

c. $\dfrac{6!}{3!5!} = \dfrac{6 \cdot \cancel{5!}}{3!\cancel{5!}} = \dfrac{6}{6} = 1$

NOW TRY EXERCISES 39 THROUGH 44

EXAMPLE 4 Find the third term of each sequence.

a. $a_n = \dfrac{n!}{2^n}$ **b.** $c_n = \dfrac{(-1)^n(2n - 1)!}{n!}$

Solution:

a. $a_3 = \dfrac{3!}{2^3} = \dfrac{6}{8} = \dfrac{3}{4}$

b. $c_3 = \dfrac{(-1)^3[2(3) - 1]!}{3!} = \dfrac{(-1)(5)!}{3!} = \dfrac{(-1)[5 \cdot 4 \cdot 3!]}{3!} = -20$

NOW TRY EXERCISES 45 THROUGH 50

C. Finding the *n*th Term of a Sequence

When given a finite sequence of numbers, we can sometimes predict the next number in the sequence based on a recognizable pattern. For instance, given 4, 7, 10, 13, . . . we can anticipate that 16 is the next term (since all differ by 3) and perhaps even recognize the terms are being generated by $a_n = 3n + 1$. Look for such patterns in Example 5.

EXAMPLE 5 For each sequence, determine the next term and the general term.

a. 2, 4, 8, 16, . . . **b.** $1, \frac{1}{2}, \frac{1}{3}, \frac{1}{4}, \ldots$ **c.** $-3, 9, -27, 81, \ldots$

Solution: **a.** The terms appear to be consecutive powers of 2, meaning the next term will be $2^5 = 32$. The general term is $a_n = 2^n$.

b. These terms are fractions whose denominators are consecutive integers. The next term will be $\frac{1}{5}$ and the general term is $a_n = \frac{1}{n}$.

c. The terms alternate in sign and are consecutive powers of 3, so the next term is -243. For the *n*th term, the factor $(-1)^n$ will produce alternating signs and the *n*th term is $a_n = (-1)^n 3^n$.

NOW TRY EXERCISES 51 THROUGH 58

D. Series and Partial Sums

Sometimes the terms of a sequence are dictated by context rather than a formula. Consider the stacking of large pipes in a storage yard. If there are 10 pipes in the bottom row, then 9 pipes, then 8 (see Figure 8.1), how many pipes are in the stack if there is a single pipe at the top? The sequence generated is 10, 9, 8, . . . , 3, 2, 1 and to answer the question we would have to *compute the sum of all terms in the sequence.* When the terms of a finite sequence are added, the result is called a **finite series.**

Figure 8.1

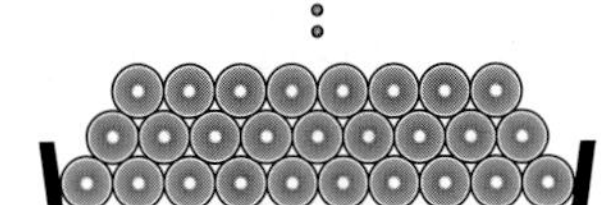

FINITE SERIES

Given the sequence $a_1, a_2, a_3, a_4, \ldots, a_n$, the sum of the terms is called a **finite series** or **partial sum** and is denoted S_n:

$$S_n = a_1 + a_2 + a_3 + a_4 + \cdots + a_n$$

EXAMPLE 6 Given $a_n = 2n$, find the value of (a) S_4 and (b) S_7.

Solution: Since we eventually need the sum of the first seven terms (for Part b), begin by writing out these terms: 2, 4, 6, 8, 10, 12, and 14.

a.
$$\begin{aligned} S_4 &= a_1 + a_2 + a_3 + a_4 \\ &= 2 + 4 + 6 + 8 \\ &= 20 \end{aligned}$$

b.
$$\begin{aligned} S_7 &= a_1 + a_2 + a_3 + a_4 + a_5 + a_6 + a_7 \\ &= 2 + 4 + 6 + 8 + 10 + 12 + 14 \\ &= 56 \end{aligned}$$

NOW TRY EXERCISES 59 THROUGH 64

E. Summation Notation

When the general term of a sequence is known, the Greek letter *sigma* Σ can be used to write the related series as a formula. For instance, to indicate the sum of the first four terms of $a_n = 3n + 1$, we write $\sum_{i=1}^{4}(3i + 1)$. This result is called **summation** or **sigma notation** and the letter i is called the **index of summation.** The letters j, k, l, and m are also used as index numbers, and the summation need not start at 1.

EXAMPLE 7 Compute each sum: (a) $\sum_{i=0}^{4}(3i + 1)$, (b) $\sum_{j=1}^{5}\frac{1}{j}$, and (c) $\sum_{k=3}^{6}(-1)^k k^2$.

Solution:

a.
$$\sum_{i=0}^{4}(3i + 1) = (3 \cdot 0 + 1) + (3 \cdot 1 + 1) + (3 \cdot 2 + 1) + (3 \cdot 3 + 1) + (3 \cdot 4 + 1)$$
$$= 1 + 4 + 7 + 10 + 13 = 35$$

b.
$$\sum_{j=1}^{5}\frac{1}{j} = \frac{1}{1} + \frac{1}{2} + \frac{1}{3} + \frac{1}{4} + \frac{1}{5}$$
$$= \frac{60}{60} + \frac{30}{60} + \frac{20}{60} + \frac{15}{60} + \frac{12}{60} = \frac{137}{60}$$

c.
$$\sum_{k=3}^{6}(-1)^k k^2 = (-1)^3 \cdot 3^2 + (-1)^4 \cdot 4^2 + (-1)^5 \cdot 5^2 + (-1)^6 \cdot 6^2$$
$$= -9 + 16 + -25 + 36 = 18$$

NOW TRY EXERCISES 65 THROUGH 76

If a definite pattern is noted in a given series expansion, this process can be reversed, with the expanded form being expressed in summation notation using the nth term.

EXAMPLE 8 Write each of the following sums in summation (sigma) notation.

a. $1 + 3 + 5 + 7 + 9$ **b.** $6 + 9 + 12 + \cdots$

Solution:

a. The series has five terms and each term is an odd number, or 1 less than a multiple of 2. The general term is $a_n = 2n - 1$, and the series is $\sum_{n=1}^{5}(2n - 1)$.

b. This is an infinite sum whose terms are multiples of 3. The general term is $a_n = 3n$, but the series starts at 2 and not 1. The series is $\sum_{j=2}^{\infty} 3j$.

NOW TRY EXERCISES 77 THROUGH 86

Since the commutative and associative laws hold for the addition of real numbers, summations have the following properties:

PROPERTIES OF SUMMATION

Given any real number c and natural number n,

(I) $\sum_{i=1}^{n} c = cn$

If you add a constant c "n" times the result is cn.

(II) $\sum_{i=1}^{n} ca_i = c\sum_{i=1}^{n} a_i$

A constant can be factored out of a sum.

(III) $\sum_{i=1}^{n} (a_i \pm b_i) = \sum_{i=1}^{n} a_i \pm \sum_{i=1}^{n} b_i$

A summation can be distributed to two (or more) sequences.

(IV) $\sum_{i=1}^{m} a_i + \sum_{i=m+1}^{n} a_i = \sum_{i=1}^{n} a_i;\ 1 \le m < n$

A summation is cumulative and can be written as a sum of smaller parts.

The verification of property II depends solely on the distributive property.

Proof:

$$\sum_{i=1}^{n} ca_i = ca_1 + ca_2 + ca_3 + \cdots + ca_n \quad \text{expand sum}$$

$$= c(a_1 + a_2 + a_3 + \cdots + a_n) \quad \text{factor out } c$$

$$= c\sum_{i=1}^{n} a_i \quad \text{write series in summation form}$$

The verification of properties III and IV simply uses the commutative and associative properties. You are asked to prove property III in Exercise 100.

EXAMPLE 9 Write the series in summation notation: $\frac{1}{60} + \frac{2}{15} + \frac{9}{20} + \frac{16}{15} + \frac{25}{12} + \frac{18}{5}$.

Solution: All denominators are factors of 60 and by factoring out $\frac{1}{60}$ we obtain: $\frac{1}{60}(1 + 8 + 27 + 64 + 125 + 216)$. The remaining terms are perfect cubes, giving a series form of $\frac{1}{60}\sum_{i=1}^{6} i^3$.

NOW TRY EXERCISES 87 THROUGH 90

F. Applications of Sequences

To solve applications of sequences, (1) identify where the sequence begins (the initial term) and (2) write out the first few terms to help identify the nth term.

EXAMPLE 10 Lorenzo already owned 1420 shares of stock when his company began offering employees the opportunity to purchase 175 discounted shares per year. If he made no purchases other than these discounted shares each year, how many shares will he have 9 yr later? If this continued for the 25 yr he will work for the company, how many shares will he have at retirement?

Solution: To begin, it helps to simply write out the first few terms of the sequence. Since he already had 1420 shares before the company made this offer, we let $a_0 = 1420$ be the inaugural element, showing $a_1 = 1595$ (after 1 year, he owns 1595 shares). The first few terms are 1595, 1770, 1945, 2120, and so on. This supports a general term of $a_n = 1595 + 175(n - 1)$.

After 9 years	**After 25 years**
$a_9 = 1595 + 175(8)$	$a_{25} = 1595 + 175(24)$
$= 2995$	$= 5795$

After 9 yr he would have 2995 shares. Upon retirement he would own 5795 shares of company stock.

NOW TRY EXERCISES 93 THROUGH 98

TECHNOLOGY HIGHLIGHT

Using a Graphing Calculator to Generate Sequences and Series

The keystrokes shown apply to a TI-84 Plus model. Please consult your manual or our Internet site for other models.

To support a study of sequences and series, we can use a graphing calculator to generate the desired terms. This can be done either on the home screen or directly into the LIST feature of the calculator (see Section 8.2 *Technology Highlight*). On the TI-84 Plus this is accomplished using the **"seq("** and **"sum("** commands, which are accessed using the keystrokes 2nd STAT (**LIST**) and the screen shown in Figure 8.2. The **"seq("** feature is option 5 under the **OPS** submenu (press ▶ 5) and the **"sum("** feature is option 5 under the **MATH** submenu (press ◀ 5). Although "n" is the preferred variable for sequences and series, most calculators make it most convenient to continue using x.

Figure 8.2

```
NAMES OPS MATH
1:SortA(
2:SortD(
3:dim(
4:Fill(
5:seq(
6:cumSum(
7↓ΔList(
```

EXAMPLE 1: Use a graphing calculator to generate the first four terms of the sequence $a_n = n^2 + 1$, and to find the sum of these four terms.

Solution: CLEAR the home screen and press 2nd STAT ▶ 5 to place the **"seq("** feature on the home screen. The **"seq("** command requires four inputs: a_n (the nth term), variable used (must be specified since the calculator can work with any letter), which term you want the sequence to begin with (often $n = 1$), and the last term of the sequence. For this example the screen would read, "**seq**($x^2 + 1, x, 1, 4$)," with the result being the four terms shown in Figure 8.3. To find the *sum* of these terms, we simply precede the **seq**($x^2 + 1, x, 1, 4$) command by **"sum(."** This can be done by placing the **"sum("** feature on the home screen and recalling the previous ANS (2nd (–)) or by retyping the command. Both methods are shown in Figure 8.4.

Figure 8.3

```
seq(X²+1,X,1,4)
         {2 5 10 17}
```

Figure 8.4

```
seq(X²+1,X,1,4)
         {2 5 10 17}
sum(Ans)
                  34
sum(seq(X²+1,X,1
,4)
                  34
```

Each of the following sequences have some interesting properties or mathematical connections. Use your graphing calculator to generate the first 10 terms of each sequence and the sum of the first 10 terms. As a second stage of the investigation, generate the first 20 terms of each sequence and the sum of the first 20 terms (your calculator takes longer to generate 20 terms than 10). What conclusion (if any) can you reach about the sum of each sequence? Note that if the terms of the sequence are (rational) decimal values, you can display the sequence in fraction form using the MATH 1: ▶Frac option.

Exercise 1: $a_n = \frac{1}{3^n}$ sum → 0.5

Exercise 2: $a_n = \frac{2}{n(n+1)}$ sum → 2

Exercise 3: $a_n = \frac{1}{(2n-1)(2n+1)}$ sum → 0.5

8.1 EXERCISES

5. formula defining the sequence uses the preceding term(s); answers will vary.
6. terms alternate in sign; answers will vary.
7. 1, 3, 5, 7; $a_8 = 15$; $a_{12} = 23$
8. 5, 7, 9, 11; $a_8 = 19$; $a_{12} = 27$
9. 0, 9, 24, 45; $a_8 = 189$; $a_{12} = 429$
10. −10, 4, 42, 116; $a_8 = 1012$; $a_{12} = 3444$
11. −1, 2, −3, 4; $a_8 = 8$; $a_{12} = 12$
12. $-1, \frac{1}{2}, -\frac{1}{3}, \frac{1}{4}$; $a_8 = \frac{1}{8}$; $a_{12} = \frac{1}{12}$
13. $\frac{1}{2}, \frac{2}{3}, \frac{3}{4}, \frac{4}{5}$; $a_8 = \frac{8}{9}$; $a_{12} = \frac{12}{13}$
14. $2, \frac{9}{4}, \frac{64}{27}, \frac{625}{256}$; $a_8 = \left(\frac{9}{8}\right)^8$; $a_{12} = \left(\frac{13}{12}\right)^{12}$
15. $\frac{1}{2}, \frac{1}{4}, \frac{1}{8}, \frac{1}{16}$; $a_8 = \frac{1}{256}$; $a_{12} = \frac{1}{4096}$
16. $\frac{2}{3}, \frac{4}{9}, \frac{8}{27}, \frac{16}{81}$; $a_8 = \left(\frac{2}{3}\right)^8$; $a_{12} = \left(\frac{2}{3}\right)^{12}$
17. $1, \frac{1}{2}, \frac{1}{3}, \frac{1}{4}$; $a_8 = \frac{1}{8}$; $a_{12} = \frac{1}{12}$
18. $1, \frac{1}{4}, \frac{1}{9}, \frac{1}{16}$; $a_8 = \frac{1}{64}$; $a_{12} = \frac{1}{144}$
19. $\frac{-1}{2}, \frac{1}{6}, \frac{-1}{12}, \frac{1}{20}$; $a_8 = \frac{1}{72}$; $a_{12} = \frac{1}{156}$
20. $1, -\frac{1}{7}, \frac{1}{17}, -\frac{1}{31}$; $a_8 = -\frac{1}{127}$; $a_{12} = -\frac{1}{287}$

CONCEPTS AND VOCABULARY

Fill in the blank with the appropriate word or phrase. Carefully reread the section if needed.

1. A sequence is a(n) __pattern__ of numbers listed in a specific __order__.
2. A series is the __sum__ of the numbers from a given sequence.
3. When each term of a sequence is larger than the preceding term, the sequence is said to be __increasing__.
4. When each term of a sequence is smaller than the preceding term, the sequence is said to be __decreasing__.
5. Describe the characteristics of a recursive sequence and give one example.
6. Describe the characteristics of an alternating sequence and give one example.

DEVELOPING YOUR SKILLS

Find the first four terms, then find the 8th and 12th term for each nth term given.

7. $a_n = 2n - 1$
8. $a_n = 2n + 3$
9. $a_n = 3n^2 - 3$
10. $a_n = 2n^3 - 12$
11. $a_n = (-1)^n n$
12. $a_n = \frac{(-1)^n}{n}$
13. $a_n = \frac{n}{n+1}$
14. $a_n = \left(1 + \frac{1}{n}\right)^n$
15. $a_n = \left(\frac{1}{2}\right)^n$
16. $a_n = \left(\frac{2}{3}\right)^n$
17. $a_n = \frac{1}{n}$
18. $a_n = \frac{1}{n^2}$
19. $a_n = \frac{(-1)^n}{n(n+1)}$
20. $a_n = \frac{(-1)^{n+1}}{2n^2 - 1}$
21. $a_n = (-1)^n 2^n$ −2, 4, −8, 16; $a_8 = 256$; $a_{12} = 4096$
22. $a_n = (-1)^n 2^{-n}$ $\frac{-1}{2}, \frac{1}{4}, \frac{-1}{8}, \frac{1}{16}$; $a_8 = \frac{1}{256}$; $a_{12} = \frac{1}{4096}$

Find the indicated term for each sequence.

23. $a_n = n^2 - 2$; a_9 79
24. $a_n = (n - 2)^2$; a_9 49
25. $a_n = \frac{(-1)^{n+1}}{n}$; a_5 $\frac{1}{5}$
26. $a_n = \frac{(-1)^{n+1}}{2n-1}$; a_5 $\frac{1}{9}$
27. $a_n = 2\left(\frac{1}{2}\right)^{n-1}$; a_7 $\frac{1}{32}$
28. $a_n = 3\left(\frac{1}{3}\right)^{n-1}$; a_7 $\frac{1}{243}$

29. $a_n = \left(1 + \frac{1}{n}\right)^n$; a_{10} $\left(\frac{11}{10}\right)^{10}$

30. $a_n = \left(n + \frac{1}{n}\right)^n$; a_9 $\left(\frac{82}{9}\right)^9$

31. $a_n = \frac{1}{n(2n + 1)}$; a_4 $\frac{1}{36}$

32. $a_n = \frac{1}{(2n - 1)(2n + 1)}$; a_5 $\frac{1}{99}$

Find the first five terms of each recursive sequence.

33. $\begin{cases} a_1 = 2 \\ a_n = 5a_{n-1} - 3 \end{cases}$ 2, 7, 32, 157, 782

34. $\begin{cases} a_1 = 3 \\ a_n = 2a_{n-1} - 3 \end{cases}$ 3, 3, 3, 3, 3

35. $\begin{cases} a_1 = -1 \\ a_n = (a_{n-1})^2 + 3 \end{cases}$ −1, 4, 19, 364, 132,499

36. $\begin{cases} a_1 = -2 \\ a_n = a_{n-1} - 16 \end{cases}$ −2, −18, −34, −50, −66

37. $\begin{cases} c_1 = 64, c_2 = 32 \\ c_n = \dfrac{c_{n-2} - c_{n-1}}{2} \end{cases}$ 64, 32, 16, 8, 4

38. $\begin{cases} c_1 = 1, c_2 = 2 \\ c_n = c_{n-1} + (c_{n-2})^2 \end{cases}$ 1, 2, 3, 7, 16

Simplify each factorial expression.

39. $\frac{8!}{5!}$ 336

40. $\frac{12!}{10!}$ 132

41. $\frac{9!}{7!2!}$ 36

42. $\frac{6!}{3!3!}$ 20

43. $\frac{8!}{2!6!}$ 28

44. $\frac{10!}{3!7!}$ 120

Write out the first four terms in each sequence.

45. $a_n = \frac{n!}{(n + 1)!}$ $\frac{1}{2}, \frac{1}{3}, \frac{1}{4}, \frac{1}{5}$

46. $a_n = \frac{n!}{(n + 3)!}$ $\frac{1}{24}, \frac{1}{60}, \frac{1}{120}, \frac{1}{210}$

47. $a_n = \frac{(n + 1)!}{(3n)!}$ $\frac{1}{3}, \frac{1}{120}, \frac{1}{15{,}120}, \frac{1}{3{,}991{,}680}$

48. $a_n = \frac{(n + 3)!}{(2n)!}$ $12, 5, 1, \frac{1}{8}$

49. $a_n = \frac{n^n}{n!}$ $1, 2, \frac{9}{2}, \frac{32}{3}$

50. $a_n = \frac{2^n}{n!}$ $2, 2, \frac{4}{3}, \frac{2}{3}$

Predict the next term and find the general term a_n for each sequence. Answers are not necessarily unique.

51. 2, 4, 8, 16, 32, . . . $64, a_n = 2^n$

52. 3, 9, 27, 81, 243, . . . $729, a_n = 3^n$

53. 1, 4, 9, 16, 25, . . . $36, a_n = n^2$

54. 1, 8, 27, 64, 125, . . . $216, a_n = n^3$

55. −3, 4, 11, 18, 25, . . . $32, a_n = 7n - 10$

56. 2, −1, −4, −7, −10, . . . $-13, a_n = -3n + 5$

57. $\frac{1}{2}, \frac{1}{4}, \frac{1}{8}, \frac{1}{16}, \frac{1}{32}, \ldots$ $\frac{1}{64}, a_n = \frac{1}{2^n}$

58. $\frac{1}{2}, \frac{2}{3}, \frac{3}{4}, \frac{4}{5}, \frac{5}{6}, \ldots$ $\frac{6}{7}, a_n = \frac{n}{n + 1}$

Find the indicated partial sum for each sequence.

59. $a_n = n$; S_5 15

60. $a_n = n^2$; S_7 140

61. $a_n = 2n - 1$; S_8 64

62. $a_n = 3n - 1$; S_6 57

63. $a_n = \frac{1}{n}$; S_5 $\frac{137}{60}$

64. $a_n = \frac{n}{n + 1}$; S_4 $\frac{163}{60}$

Expand and evaluate each series.

65. $\sum_{i=1}^{4} (3i - 5)$ 10

66. $\sum_{i=1}^{5} (2i - 3)$ 15

67. $\sum_{k=1}^{5} (2k^2 - 3)$ 95

68. $\sum_{k=1}^{5} (k^2 + 1)$ 60

69. $\sum_{k=1}^{7} (-1)^k k$ −4

70. $\sum_{k=1}^{5} (-1)^k 2^k$ −22

71. $\sum_{i=1}^{4} \frac{i^2}{2}$ 15

72. $\sum_{i=2}^{4} i^2$ 29

73. $\sum_{j=3}^{7} 2j$ 50

74. $\sum_{j=3}^{7} \frac{j}{2^j}$ $\frac{119}{128}$

75. $\sum_{k=3}^{8} \frac{(-1)^k}{k(k - 2)}$ $\frac{-27}{112}$

76. $\sum_{k=2}^{6} \frac{(-1)^{k+1}}{k^2 - 1}$ $\frac{-11}{42}$

Write each sum using sigma notation. Answers are not necessarily unique.

77. $4 + 8 + 12 + 16 + 20$

78. $5 + 10 + 15 + 20 + 25$

79. $-1 + 4 - 9 + 16 - 25 + 36$

80. $1 - 8 + 27 - 64 + 125 - 216$

For the given general term a_n, write the indicated sum using sigma notation.

81. $a_n = n + 3$; S_5

82. $a_n = \frac{n^2 - 1}{n + 1}$; S_4

83. $a_n = \frac{n^2}{3}$; third partial sum

84. $a_n = 2n - 1$; sixth partial sum

85. $a_n = \frac{n}{2^n}$; sum for $n = 3$ to 7

86. $a_n = n^2$; sum for $n = 2$ to 6

Write each series in sigma notation. Simplify the series (if possible) before you begin.

87. $\frac{1}{20} + \frac{1}{10} + \frac{3}{20} + \frac{1}{5} + \frac{1}{4}$

88. $\frac{1}{30} + \frac{1}{20} + \frac{1}{15} + \frac{1}{12} + \frac{1}{10}$

89. $\frac{1}{8} + \frac{4}{27} + \frac{9}{64} + \frac{16}{125} + \cdots$

90. $\frac{1}{2} + \frac{2}{3} + \frac{3}{4} + \frac{4}{5} + \frac{5}{6} + \cdots$

WORKING WITH FORMULAS

91. Sum of $a_n = 3n - 2$: $S_n = \frac{n(3n - 1)}{2}$

The sum of the first n terms of the sequence defined by $a_n = 3n - 2 = 1, 4, 7, 10, \ldots, (3n - 2), \ldots$ is given by the formula shown. Find S_5 using the formula, then verify by direct calculation. 35

92. Sum of $a_n = 3n - 1$: $S_n = \frac{n(3n + 1)}{2}$

The sum of the first n terms of the sequence defined by $a_n = 3n - 1 = 2, 5, 8, 11, \ldots, (3n - 1), \ldots$ is given by the formula shown. Find S_8 using the formula, then verify by direct calculation. Observing the results of Exercises 91 and 92, can you now state the sum formula for $a_n = 3n - 0$? 100 $S_n = \frac{n(3n + 3)}{2}$

APPLICATIONS

Use the information given in each exercise to determine the nth term a_n for the sequence described. Then use the nth term to list the specified number of terms.

93. Blue-book value: Steve's car has a blue-book value of \$6000. Each year it loses 20% of its value (its value each year is 80% of the year before). List the value of Steve's car for the next 5 yr. (Hint: For $a_1 = 6000$, we need the *next* five terms.)

94. Effects of inflation: Suppose inflation (an increase in value) will average 4% for the next 5 yr. List the growing cost (year by year) of a DVD that costs \$15 right now. (Hint: For $a_1 = 15$, we need the *next* five terms.)

95. Wage increases: Latisha gets \$5.20 an hour for filling candy machines for Archtown Vending. Each year she receives a \$0.50 hourly raise. List Latisha's wage for the first 5 yr. How much will she make in the fifth year if she works 8 hr per day for 240 working days?

96. Average birth weight: The average birth weight of a certain animal species is 900 g, with the baby gaining 125 g each day for the first 10 days. List the infant's weight for the first 10 days. How much does the infant weigh on the 10th day?

77. $\sum_{n=1}^{5}(4n)$

78. $\sum_{n=1}^{5} 5n$

79. $\sum_{n=1}^{6}(-1)^n n^2$

80. $\sum_{n=1}^{6}(-1)^{n+1} n^3$

81. $\sum_{n=1}^{5}(n + 3)$

82. $\sum_{n=1}^{4}\frac{n^2 - 1}{n + 1}$

83. $\sum_{n=1}^{3}\frac{n^2}{3}$

84. $\sum_{n=1}^{6}(2n - 1)$

85. $\sum_{n=3}^{7}\frac{n}{2^n}$

86. $\sum_{n=2}^{6} n^2$

87. $\frac{1}{20}\sum_{n=1}^{5} n$

88. $\frac{1}{60}\sum_{n=1}^{5}(n + 1)$

89. $\sum_{n=1}^{\infty}\frac{n^2}{(n + 1)^3}$

90. $\sum_{n=1}^{\infty}\frac{n}{n + 1}$

93. $a_n = 6000(0.8)^{n-1}$; 6000, 4800, 3840, 3072, 2457.6, 1966.1

94. $a_n = 15(1.04)^{n-1}$; 15, 15.60, 16.22, 16.87, 17.55, 18.25

95. 5.20, 5.70, 6.20, 6.70, 7.20, \$13,824

96. 900, 1025, 1150, 1275, 1400, 1525, 1650, 1775, 1900, 2025, 2025 g

97. **Stocking a lake:** A local fishery stocks a large lake with 1500 bass and then adds an additional 100 mature bass per month until the lake nears maximum capacity. If the bass population grows at a rate of 5% per month through natural reproduction, the number of bass in the pond after n months is given by the recursive sequence $b_0 = 1500$, $b_n = 1.05b_{n-1} + 100$. How many bass will be in the lake after 6 months? ≈2690

98. **Species preservation:** The Interior Department introduces 50 wolves (male and female) into a large wildlife area in an effort to preserve the species. Each year about 12 additional adult wolves are added from capture and relocation programs. If the wolf population grows at a rate of 10% per year through natural reproduction, the number of wolves in the area after n years is given by the recursive sequence $w_0 = 50$, $w_n = 1.10w_{n-1} + 12$. How many wolves are in the wildlife area after 6 years? ≈181

WRITING, RESEARCH, AND DECISION MAKING

99. One of the most fascinating sequences known is called the *Hailstone sequence*,

$$a_n = \begin{cases} \dfrac{a_{n-1}}{2} & \text{if } a_{n-1} \text{ is even} \\ 3(a_{n-1}) + 1 & \text{if } a_{n-1} \text{ is odd} \end{cases}$$

because the numbers in the sequence move "up and down" like hailstones in a cloud. Just as the hailstone will eventually become so heavy it falls to earth, the numbers in this sequence eventually "fall to earth" (metaphorically speaking), ending in the sequence 4, 2, 1, regardless of the inaugural term a_0 chosen. Study this sequence using $a_0 = 5$, $a_0 = 12$, and $a_0 = 11$. What do you notice? Why do the sequences have a different number of terms? Find seed numbers that give longer and shorter sequences. Answers will vary.

100. Verify that a summation may be distributed to two (or more) sequences. That is, verify that the following statement is true: $\sum_{i=1}^{n}(a_i \pm b_i) = \sum_{i=1}^{n} a_i \pm \sum_{i=1}^{n} b_i$. verified

EXTENDING THE CONCEPT

Surprisingly, some of the most celebrated numbers in mathematics can be represented or approximated by a series expansion. Use your calculator to find the partial sums for $n = 4$, $n = 8$, and $n = 12$ for the summations given, and attempt to name the number the summation approximates:

101. $\sum_{k=0}^{n} \frac{1}{k!}$ approaches e

102. $\sum_{k=1}^{n} \frac{1}{3^k}$ approaches $\frac{1}{2}$

103. $\sum_{k=1}^{n} \frac{1}{2^k}$ approaches 1

MAINTAINING YOUR SKILLS

104. (5.2) Write $\log_3 \frac{1}{81} = -x$ in exponential form, then solve by equating bases.

105. (2.4) Set up the difference quotient for $f(x) = \sqrt{x}$, then rationalize the numerator.

106. (7.3) Solve the nonlinear system.

$$\begin{cases} x^2 + y^2 = 9 \\ 9y^2 - 4x^2 = 16 \end{cases}$$

107. (6.7) Solve the system using a matrix equation.

$$\begin{cases} 25x + y - 2z = -14 \\ 2x - y + z = 40 \\ -7x + 3y - z = -13 \end{cases}$$

108. (3.3) Use a transformation to sketch the graph of $f(x) = -|x - 2| + 3$.

109. (7.1) Find an equation for a circle centered at $(2, -3)$ with a radius of 7 units.

104. $3^{-x} = \frac{1}{81}$; $x = 4$

105. $\frac{\sqrt{x+h} - \sqrt{x}}{h}$; $\frac{1}{\sqrt{x+h} + \sqrt{x}}$

106. $(\sqrt{5}, -2)$, $(\sqrt{5}, 2)$, $(-\sqrt{5}, -2)$, $(-\sqrt{5}, 2)$

107. (3, 21, 55)

108.

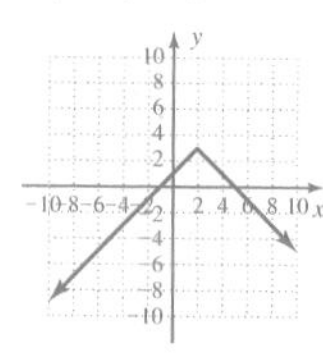

109. $(x - 2)^2 + (y + 3)^2 = 49$

8.2 Arithmetic Sequences

LEARNING OBJECTIVES

In Section 8.2 you will learn how to:

A. Identify an arithmetic sequence and its common difference

B. Find the *n*th term of an arithmetic sequence

C. Find the *n*th partial sum of an arithmetic sequence

D. Solve applications involving arithmetic sequences

INTRODUCTION

Similar to the way polynomials fall into certain groups or families (linear, quadratic, cubic, etc.), sequences and series with common characteristics are likewise grouped. In this section, we focus on sequences where each successive term is generated by adding a constant value, as in the sequence 1, 8, 15, 22, 29, . . . , where 7 is added to a given term in order to produce the next term.

POINT OF INTEREST

Carl Friedrich Gauss (1777–1855) was an outstanding mathematician, astronomer, and physicist. A popular story is told of the young man as an elementary school student, being asked to leave the room for being disruptive and as punishment, being required to compute the sum of the first 100 positive integers. To the teacher's amazement, he returned shortly with the correct answer. The method he used is illustrated in this section.

A. Identifying an Arithmetic Sequence and Finding the Common Difference

An **arithmetic sequence** is one where each successive term is found by adding a fixed constant to the preceding term. For instance 3, 7, 11, 15, . . . is an arithmetic sequence, since adding 4 to any given term produces the next term. This also means if you take the difference of any two consecutive terms, the result will be 4 and in fact, 4 is called the **common difference *d*** for this sequence. Using the notation developed earlier, we can write $d = a_{k+1} - a_k$, where a_k represents any term of the sequence and a_{k+1} represents the term that follows a_k.

ARITHMETIC SEQUENCES

Given a sequence $a_1, a_2, a_3, \ldots, a_k, a_{k+1}, \ldots, a_n$, where $k, n \in \mathbb{N}$ and $k < n$, if there exists a common difference d such that $a_{k+1} - a_k = d$, then the sequence is an *arithmetic sequence.*

The difference of successive terms can be rewritten as $a_{k+1} = a_k + d$ (for $k \geq 1$) to highlight that each following term is found by adding d to the previous term.

EXAMPLE 1 Determine if the given sequence is arithmetic.

a. 2, 5, 8, 11, . . .　　**b.** $\frac{1}{2}, \frac{5}{6}, \frac{4}{3}, \frac{7}{6}, \ldots$

Solution:

a. Begin by looking for a common difference $d = a_{k+1} - a_k$. Checking each pair of consecutive terms we have

$$5 - 2 = 3 \qquad 8 - 5 = 3 \qquad 11 - 8 = 3 \quad \text{and so on.}$$

This is an arithmetic sequence with common difference $d = 3$.

b. Checking each pair of consecutive terms yields

$$\frac{5}{6} - \frac{1}{2} = \frac{5}{6} - \frac{3}{6} = \frac{2}{6} = \frac{1}{3} \qquad \frac{4}{3} - \frac{5}{6} = \frac{8}{6} - \frac{5}{6} = \frac{3}{6} = \frac{1}{2}$$

Since the difference is not constant, this cannot be an arithmetic sequence.

NOW TRY EXERCISES 7 THROUGH 18

EXAMPLE 2 Write the first five terms of the arithmetic sequence, given the first term a_1 and the common difference d.

a. $a_1 = 12$ and $d = -4$ **b.** $a_1 = \frac{1}{2}$ and $d = \frac{1}{3}$

Solution:

a. $a_1 = 12$ and $d = -4$. Starting at $a_1 = 12$, add -4 to each new term to generate the sequence: $12, 8, 4, 0, -4, \ldots$.

b. $a_1 = \frac{1}{2}$ and $d = \frac{1}{3}$. Starting at $a_1 = \frac{1}{2}$, add $\frac{1}{3}$ to each new term to generate the sequence: $\frac{1}{2}, \frac{5}{6}, \frac{7}{6}, \frac{3}{2}, \frac{11}{6}, \ldots$.

NOW TRY EXERCISES 19 THROUGH 30

B. Finding the *n*th Term of an Arithmetic Sequence

If the values a_1 and d from an arithmetic sequence are known, we could generate the terms of the sequence by adding *multiples of d to the first term,* instead of adding d to each new term. For example, we can generate the sequence 3, 8, 13, 18, 23 by adding multiples of 5 to the first term $a_1 = 3$:

$$\begin{aligned} 3 &= 3 + 0d & a_1 &= a_1 + 0d \\ 8 &= 3 + (1)5 & a_2 &= a_1 + 1d \\ 13 &= 3 + (2)5 & a_3 &= a_1 + 2d \\ 18 &= 3 + (3)5 & a_4 &= a_1 + 3d \\ 23 &= 3 + (4)5 & a_5 &= a_1 + 4d \end{aligned}$$

current term (a_5), initial term (a_1), coefficient of common difference (4)

It's helpful to note the coefficient of d is always 1 less than the subscript of the current term (as shown): $5 - 1 = 4$. This observation leads us to the following formula for the nth term.

THE *n*TH TERM OF AN ARITHMETIC SEQUENCE

The nth term of an *arithmetic sequence* is given by

$$a_n = a_1 + (n - 1)d$$

where d is the common difference.

If the term a_1 is unknown or not given, the nth term can be written

$$a_n = a_k + (n - k)d \text{ where } n = k + (n - k)$$

(the subscript of the given term and coefficient of d sum to n).

EXAMPLE 3 Find the 24th term of the sequence 0.1, 0.4, 0.7, 1, ...

Solution: Instead of creating all terms up to the 24th, we determine the constant d and use the nth term formula. By inspection we note $a_1 = 0.1$ and $d = 0.3$.

$$\begin{aligned} a_n &= a_1 + (n-1)d && n\text{th term formula} \\ &= 0.1 + (n-1)0.3 && \text{substitute 1 for } a_1 \text{ and 3 for } d \\ &= 0.1 + 0.3n - 0.3 && \text{eliminate parentheses} \\ &= 0.3n - 0.2 && \text{simplify} \end{aligned}$$

To find the 24th term we substitute 24 for n:

$$\begin{aligned} a_{24} &= 0.3(24) - 0.2 && \text{substitute 24 for } n \\ &= 7.0 && \text{result} \end{aligned}$$

NOW TRY EXERCISES 31 THROUGH 42

EXAMPLE 4 Find the number of terms in the arithmetic sequence 2, −5, −12, −19, . . . , −411.

Solution: By inspection we see that $a_1 = 2$ and $d = -7$. As before,

$$\begin{aligned} a_n &= a_1 + (n-1)d && n\text{th term formula} \\ &= 2 + (n-1)(-7) && \text{substitute 2 for } a_1 \text{ and } -7 \text{ for } d \\ &= -7n + 9 && \text{simplify} \end{aligned}$$

Although we don't know the number of terms in the sequence, we *do* know the last or nth term is −411. Substituting −411 for a_n gives

$$\begin{aligned} -411 &= -7n + 9 && \text{substitute } -411 \text{ for } a_n \\ 60 &= n && \text{solve for } n \end{aligned}$$

There are 60 terms in this sequence.

NOW TRY EXERCISES 43 THROUGH 50

EXAMPLE 5 Given an arithmetic sequence where $a_6 = 0.55$ and $a_{13} = 0.9$, find the common difference d and the value of a_1.

Solution: At first it seems that not enough information is given, but recall we can express a_{13} as the sum of any earlier term and the appropriate multiple of d. Since a_6 is known, we write $a_{13} = a_6 + 7d$ (note $13 - 6 = 7$ as required).

$$\begin{aligned} a_{13} &= a_6 + 7d && n\text{th term formula using } a_6 \\ 0.9 &= 0.55 + 7d && \text{substitute 0.9 for } a_{13} \text{ and 0.55 for } a_6 \\ 0.35 &= 7d && \text{subtract 0.55} \\ d &= 0.05 && \text{solve for } d \end{aligned}$$

Having found d, we can now solve for a_1.

$a_{13} = a_1 + 12d$	nth term formula for $n = 13$
$0.9 = a_1 + 12(0.05)$	substitute 0.9 for a_{13} and 0.05 for d
$0.9 = a_1 + 0.6$	simplify
$a_1 = 0.3$	solve for a_1

The first term is $a_1 = 0.3$ and the common difference is $d = 0.05$.

NOW TRY EXERCISES 51 THROUGH 56

C. Finding the *n*th Partial Sum of an Arithmetic Sequence

Using sequences and series to solve applications often requires computing the sum of a given number of terms. As we develop a formula for these sums, we encounter the method Gauss might have used to add the positive integers from 1 to 100 (see the *Point of Interest*). Consider the sequence $a_1, a_2, a_3, a_4, \ldots, a_n$ with common difference d. Use S_n to represent the sum of the first n terms and write the original series, then the series in reverse order underneath. Since one row increases at the same rate the other decreases, the sum of each column remains constant, and for simplicity's sake we choose $a_1 + a_n$ to represent this sum.

$$\begin{array}{rcccccccccccccc} S_n = & a_1 & + & a_2 & + & a_3 & + \cdots + & a_{n-2} & + & a_{n-1} & + & a_n \\ S_n = & a_n & + & a_{n-1} & + & a_{n-2} & + \cdots + & a_3 & + & a_2 & + & a_1 \\ \hline 2S_n = & (a_1 + a_n) & + & (a_1 + a_n) & + & (a_1 + a_n) & + \cdots + & (a_1 + a_n) & + & (a_1 + a_n) & + & (a_1 + a_n) \end{array}$$

add columns vertically

Since there are n columns, we end up with $2S_n = n(a_1 + a_n)$, and solving for S_n gives the formula for the first n terms of an arithmetic sequence.

THE *n*TH PARTIAL SUM OF AN ARITHMETIC SEQUENCE

Given an arithmetic sequence with first term a_1, the nth partial sum is given by

$$S_n = n\left(\frac{a_1 + a_n}{2}\right).$$

In words: The sum of an arithmetic sequence is the number of terms times the average of the first and last term.

EXAMPLE 6 Find the sum: $\sum_{k=1}^{75}(2k - 1)$ (the sum of the first 75 positive, odd integers).

Solution: The initial terms of the sequence are 1, 3, 5, . . . and we note $a_1 = 1$, $d = 2$, and $n = 75$. To use the sum formula, we need the value of $a_n = a_{75}$. The nth term formula shows $a_{75} = a_1 + 74d = 1 + 74(2)$, so $a_{75} = 149$.

$S_n = \dfrac{n(a_1 + a_n)}{2}$	sum formula
$S_{75} = \dfrac{75(a_1 + a_{75})}{2}$	substitute 75 for n

$$= \frac{75(1 + 149)}{2} \quad \text{substitute 1 for } a_1\text{, 149 for } a_{75}$$

$$= 5625 \quad \text{result}$$

The sum of the first 75 positive, odd integers is 5625.

NOW TRY EXERCISES 57 THROUGH 62

By substituting the nth term formula directly into the formula for partial sums, we're able to find a partial sum without actually having to find the nth term:

$$S_n = \frac{n(a_1 + a_n)}{2} \quad \text{sum formula}$$

$$= \frac{n(a_1 + [a_1 + (n - 1)d])}{2} \quad \text{substitute } a_1 + (n-1)d \text{ for } a_n$$

$$= \frac{n}{2}[2a_1 + (n - 1)d] \quad \textit{alternative formula for the } n\textit{th partial sum}$$

See Exercises 63 through 68 for more on this alternative formula.

EXAMPLE 7 Find the number of terms in the sequence $2 + (-3) + (-8) + (-13) + \cdots + (-308)$.

Solution: This is a partial sum with $a_1 = 2$, $d = -5$, and $a_n = -308$ (the last term in the sequence). Determine n using the nth term formula:

$$a_n = a_1 + (n - 1)d \quad n\text{th term formula}$$

$$-308 = 2 + (n - 1)(-5) \quad \text{substitute 2 for } a_1, -5 \text{ for } d, -308 \text{ for } a_n$$

$$-308 = 2 + (-5n) + 5 \quad \text{simplify}$$

$$-315 = -5n \quad \text{subtract 7}$$

$$n = 63 \quad \text{solve for } n$$

There are 63 terms in this sequence.

NOW TRY EXERCISES 69 THROUGH 78

Figure 8.5

spiral fern

D. Applications

In the evolution of certain plants and shelled animals, sequences and series seem to have been one of nature's favorite tools (see Figures 8.5 and 8.6). Sequences and series also provide a good mathematical model for a variety of other situations as well.

Figure 8.6

nautilus

EXAMPLE 8 Cox Auditorium is an amphitheater that has 40 seats in the first row, 42 seats in the second row, 44 in the third, and so on. If there are 75 rows in the auditorium, what is the auditorium's seating capacity?

Solution: The number of seats in each row gives the terms of an arithmetic sequence with $a_1 = 40$, $d = 2$, and $n = 75$. To find the seating capacity, we need to find the total number of seats, which is the sum of this arithmetic sequence. Since the value of a_{75} is unknown, we opt for the alternative formula $S_n = \frac{n}{2}[2a_1 + (n - 1)d]$.

$$S_n = \frac{n}{2}[2a_1 + (n - 1)d] \quad \text{sum formula}$$

$$S_{75} = \frac{75}{2}[2(40) + (75 - 1)(2)] \quad \text{substitute 40 for } a_1\text{, 2 for } d \text{ and 75 for } n$$

$$= \frac{75}{2}(228) \quad \text{simplify}$$

$$= 8550 \quad \text{result}$$

The seating capacity for Cox Auditorium is 8550.

NOW TRY EXERCISES 81 THROUGH 92

TECHNOLOGY HIGHLIGHT

Using a Graphing Calculator to Study Arithmetic Sequences and Series

The keystrokes shown apply to a TI-84 Plus model. Please consult your manual or our Internet site for other models.

It is often helpful to have the terms of a sequence entered into the LIST feature of a calculator, so that we can take advantage of the other operations and statistical measures available and have a clear view of numerous terms in the sequence. This is done by executing the **"seq("** command in the heading of a specified list. There are many ways to use this combination of features. Here we'll use a sequence to enter the first 15 natural numbers in L1, then use the entries of L1 to generate the first 15 terms of $a_n = (-1)^n(2n - 1)$ into L2. On the TI-84 Plus, we begin by clearing all lists using STAT **4:CLRLIST**(L1, L2) on the home screen. Then access the LISTs using STAT ENTER, press ▲ and move the cursor to the heading of List 1, then input **seq(X, X, 1, 15)** and press ENTER. This will automatically fill List 1 with the first 15 natural numbers (Figure 8.7). Now move the cursor to the heading of List 2. We will be entering the sequence $a_n = (-1)^n(2n - 1)$, but using L1 as the input variable instead of n or x. In other words, the entry will be $L2 = (-1)^{L1}(2L1 - 1)$. The screen shown in Figure 8.7 displays this input just prior to pressing ENTER. The resulting output is displayed in Figure 8.8. We can now view all terms of the sequence by scrolling through the list, using STAT **CALC 1:1–Var Stats L2** to find the sum of the series, or using other available features.

Figure 8.7

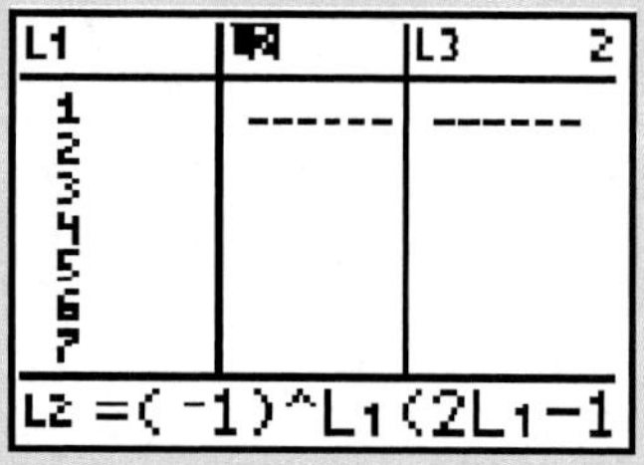

Figure 8.8

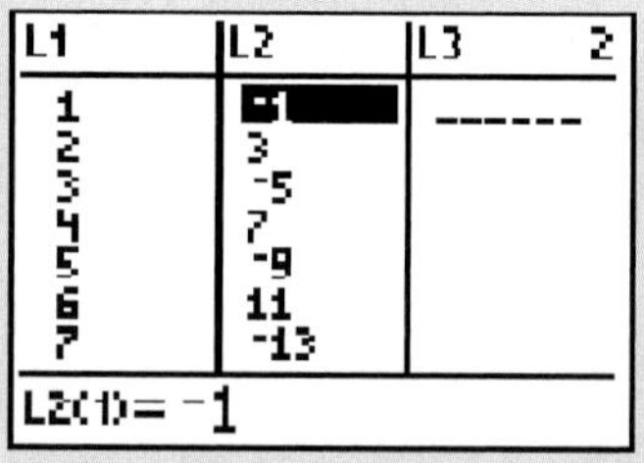

Exercise 1: Enter the natural numbers 1 through 15 in L1, and use L1 to generate the first 15 terms of $a_n = 0.5 + (n - 1)0.25$. Then use STAT **CALC 1:1–Var Stats** to find the sum of these 15 terms. 33.75

Exercise 2: Enter the natural numbers 4 through 24 in L1, and use L1 to generate the terms a_4 through a_{24} of $a_n = \frac{2}{3} + (n - 1)\frac{5}{6}$. Identify the terms a_8, a_{11}, and a_{15} (be careful—we started with the fourth term!). How many terms are in this sequence? Use STAT **CALC 1:1–Var Stats** to find the sum of these terms. $\frac{13}{2}, 9, \frac{37}{3}$; 21; 241.5

8.2 EXERCISES

CONCEPTS AND VOCABULARY

Fill in the blank with the appropriate word or phrase. Carefully reread the section if needed.

1. Consecutive terms in an arithmetic sequence differ by a constant called the common difference.

2. The sum of the first n terms of an arithmetic sequence is called the nth partial sum.

3. The formula for the nth partial sum of an arithmetic sequence is $S_n = \frac{n(a_1 + a_n)}{2}$, where a_n is the nth term.

4. The nth term formula for an arithmetic sequence is $a_n = a_1 + (n - 1)d$, where a_1 is the first term and d is the common difference.

5. Discuss how the terms of an arithmetic sequence can be written in various ways using the relationship $a_n = a_k + (n - k)d$. Answers will vary.

6. Describe how the formula for the nth partial sum was derived, and illustrate its application using a sequence from the exercise set. Answers will vary.

DEVELOPING YOUR SKILLS

Determine if the sequence given is arithmetic. If yes, name the common difference. If not, try to determine the pattern that forms the sequence.

7. $-5, -2, 1, 4, 7, 10, \ldots$

8. $1, -2, -5, -8, -11, -14, \ldots$

9. $0.5, 3, 5.5, 8, 10.5, \ldots$

10. $1.2, 3.5, 5.8, 8.1, 10.4, \ldots$

11. $2, 3, 5, 7, 11, 13, 17, \ldots$

12. $1, 4, 8, 13, 19, 26, 34, \ldots$

13. $\frac{1}{24}, \frac{1}{12}, \frac{1}{8}, \frac{1}{6}, \frac{5}{24}, \ldots$

14. $\frac{1}{12}, \frac{1}{15}, \frac{1}{20}, \frac{1}{30}, \frac{1}{60}, \ldots$

15. $1, 4, 9, 16, 25, 36, \ldots$

16. $-125, -64, -27, -8, -1, \ldots$

17. $\pi, \frac{5\pi}{6}, \frac{2\pi}{3}, \frac{\pi}{2}, \frac{\pi}{3}, \frac{\pi}{6}, \ldots$

18. $\pi, \frac{7\pi}{8}, \frac{3\pi}{4}, \frac{5\pi}{8}, \frac{\pi}{2}, \ldots$

Write the first four terms of the arithmetic sequence with the given first term and common difference.

19. $a_1 = 2, d = 3$

20. $a_1 = 8, d = 3$

21. $a_1 = 7, d = -2$

22. $a_1 = 60, d = -12$

23. $a_1 = 0.3, d = 0.03$

24. $a_1 = 0.5, d = 0.25$

25. $a_1 = \frac{3}{2}, d = \frac{1}{2}$

26. $a_1 = \frac{1}{5}, d = \frac{1}{10}$

27. $a_1 = \frac{3}{4}, d = -\frac{1}{8}$

28. $a_1 = \frac{1}{6}, d = -\frac{1}{3}$

29. $a_1 = -2, d = -3$

30. $a_1 = -4, d = -4$

Identify the first term and the common difference, then write the expression for the general term a_n and use it to find the 6th, 10th, and 12th terms of the sequence.

31. $2, 7, 12, 17, \ldots$

32. $7, 4, 1, -2, -5, \ldots$

33. $5.10, 5.25, 5.40, \ldots$

34. $9.75, 9.40, 9.05, \ldots$

35. $\frac{3}{2}, \frac{9}{4}, 3, \frac{15}{4}, \ldots$

36. $\frac{5}{7}, \frac{3}{14}, -\frac{2}{7}, -\frac{11}{14}, \ldots$

Find the indicated term using the information given.

37. $a_1 = 5, d = 4$; find a_{15} 61

38. $a_1 = 9, d = -2$; find a_{17} −23

39. $a_1 = \frac{3}{2}, d = -\frac{1}{12}$; find a_7 1

40. $a_1 = \frac{12}{25}, d = -\frac{1}{10}$; find a_9 $\frac{-8}{25}$

41. $a_1 = -0.025, d = 0.05$; find a_{50} 2.425

42. $a_1 = 3.125, d = -0.25$; find a_{20} −1.625

Find the number of terms in each sequence.

43. $a_1 = 2, a_n = -22, d = -3$ 9

44. $a_1 = 4, a_n = 42, d = 2$ 20

45. $a_1 = 0.4, a_n = 10.9, d = 0.25$ 43

46. $a_1 = -0.3, a_n = -36, d = -2.1$ 18

7. arithmetic; $d = 3$
8. arithmetic; $d = -3$
9. arithmetic; $d = 2.5$
10. arithmetic; $d = 2.3$
11. not arithmetic; all prime
12. not arithmetic; each difference increases by 1
13. arithmetic; $d = \frac{1}{24}$
14. arithmetic; $d = -\frac{1}{60}$
15. not arithmetic; $a_n = n^2$
16. not arithmetic; $a_n = (n - 6)^3$
17. arithmetic; $d = \frac{-\pi}{6}$
18. arithmetic; $d = -\frac{\pi}{8}$
19. 2, 5, 8, 11
20. 8, 11, 14, 17
21. 7, 5, 3, 1
22. 60, 48, 36, 24
23. 0.3, 0.33, 0.36, 0.39
24. 0.5, 0.75, 1, 1.25
25. $\frac{3}{2}, 2, \frac{5}{2}, 3$
26. $\frac{1}{5}, \frac{3}{10}, \frac{2}{5}, \frac{1}{2}$
27. $\frac{3}{4}, \frac{5}{8}, \frac{1}{2}, \frac{3}{8}$
28. $\frac{1}{6}, \frac{-1}{6}, \frac{-1}{2}, \frac{-5}{6}$
29. $-2, -5, -8, -11$
30. $-4, -8, -12, -16$
31. $a_1 = 2, d = 5, a_n = 5n - 3$, $a_6 = 27, a_{10} = 47, a_{12} = 57$
32. $a_1 = 7, d = -3, a_n = -3n + 10$, $a_6 = -8, a_{10} = -20, a_{12} = -26$
33. $a_1 = 5.10, d = 0.15$, $a_n = 0.15n + 4.95, a_6 = 5.85$, $a_{10} = 6.45, a_{12} = 6.75$
34. $a_1 = 9.75, d = -0.35$, $a_n = -0.35n + 10.10$ $a_6 = 8.00, a_{10} = 6.60, a_{12} = 5.90$
35. $a_1 = \frac{3}{2}, d = \frac{3}{4}, a_n = \frac{3}{4}n + \frac{3}{4}$, $a_6 = \frac{21}{4}, a_{10} = \frac{33}{4}, a_{12} = \frac{39}{4}$
36. $a_1 = \frac{5}{7}, d = -\frac{1}{2}$; $a_n = \frac{5}{7} + (n - 1)(-\frac{1}{2})$, $a_6 = \frac{-25}{14}; a_{10} = \frac{-53}{14}; a_{12} = \frac{-67}{14}$

47. $-3, -0.5, 2, 4.5, 7, \ldots, 47$ 21

48. $-3.4, -1.1, 1.2, 3.5, \ldots, 38$ 19

49. $\frac{1}{12}, \frac{1}{8}, \frac{1}{6}, \frac{5}{24}, \frac{1}{4}, \ldots, \frac{9}{8}$ 26

50. $\frac{1}{12}, \frac{1}{15}, \frac{1}{20}, \frac{1}{30}, \ldots, -\frac{1}{4}$ 21

Find the common difference d and the value of a_1 using the information given.

51. $a_3 = 7, a_7 = 19$ $d = 3, a_1 = 1$

52. $a_5 = -17, a_{11} = -2$ $d = \frac{5}{2}, a_1 = -27$

53. $a_2 = 1.025, a_{26} = 10.125$ $d = \frac{91}{240}, a_1 = \frac{31}{48}$

54. $a_6 = -12.9, a_{30} = -5.4$ $d = \frac{5}{16}, a_1 = \frac{-1157}{80}$

55. $a_{10} = \frac{13}{18}, a_{24} = \frac{27}{2}$ $d = \frac{115}{126}, a_1 = \frac{-472}{63}$

56. $a_4 = \frac{5}{4}, a_8 = \frac{9}{4}$ $d = \frac{1}{4}, a_1 = \frac{1}{2}$

Evaluate each sum. For Exercises 61 and 62, use the summation properties from Section 8.1.

57. $\sum_{n=1}^{30}(3n - 4)$ 1275

58. $\sum_{n=1}^{29}(4n - 1)$ 1711

59. $\sum_{n=1}^{37}\left(\frac{3}{4}n + 2\right)$ 601.25

60. $\sum_{n=1}^{20}\left(\frac{5}{2}n - 3\right)$ 465

61. $\sum_{n=4}^{15}(3 - 5n)$ -534

62. $\sum_{n=7}^{20}(7 - 2n)$ -280

Use the alternative formula for the nth partial sum to compute the sums indicated.

63. The sum S_{15} for the sequence $-12 + (-9.5) + (-7) + (-4.5) + \cdots$ 82.5

64. The sum S_{20} for the sequence $\frac{9}{2} + \frac{7}{2} + \frac{5}{2} + \frac{3}{2} + \cdots$ -100

65. The sum S_{30} for the sequence $0.003 + 0.173 + 0.343 + 0.513 + \cdots$ 74.04

66. The sum S_{50} for the sequence $(-2) + (-7) + (-12) + (-17) + \cdots$ -6225

67. The sum S_{20} for the sequence $\sqrt{2} + 2\sqrt{2} + 3\sqrt{2} + 4\sqrt{2} + \cdots$ $210\sqrt{2}$

68. The sum S_{10} for the sequence $12\sqrt{3} + 10\sqrt{3} + 8\sqrt{3} + 6\sqrt{3} + \cdots$ $30\sqrt{3}$

Find the number of terms in each series and then find the sum.

69. $2 + 5 + 8 + 11 + \cdots + 74$

70. $3 + 7 + 11 + 15 + \cdots + 119$

71. $4 + 9 + 14 + 19 + \cdots + 154$

72. $6 + 9 + 12 + 15 + \cdots + 93$

73. $7 + 10 + 13 + 16 + \cdots + 100$

74. $20 + 17 + 14 + 11 + \cdots + (-97)$

75. $(-13) + (-9) + (-5) + (-1) + \cdots + 183$

76. $(-2) + (-5) + (-8) + (-11) + \cdots + (-80)$

77. $\frac{1}{2} + \frac{3}{4} + 1 + \frac{5}{4} + \cdots + \frac{13}{4}$

78. $\frac{19}{3} + \frac{17}{3} + \frac{15}{3} + \cdots + \frac{-7}{3}$

69. $n = 25, S_{25} = 950$
70. $n = 30, S_{30} = 1830$
71. $n = 31, S_{31} = 2449$
72. $n = 30, S_{30} = 1485$
73. $n = 32, S_{32} = 1712$
74. $n = 40, S_{40} = -1540$
75. $n = 50, S_{50} = 4250$
76. $n = 27, S_{27} = -1107$
77. $n = 12, S_{12} = \frac{45}{2}$
78. $n = 14, S_{14} = 28$

WORKING WITH FORMULAS

79. Sum of the first n natural numbers: $S_n = \frac{n(n+1)}{2}$

The sum of the first n natural numbers can be found using the formula shown, where n represents the number of terms in the sum. Verify the formula by adding the first six natural numbers by hand, and then evaluating S_6. Then find the sum of the first 75 natural numbers. $S_6 = 21$; $S_{75} = 2850$

80. Sum of the squares of the first n natural numbers: $S_n = \frac{n(n+1)(2n+1)}{6}$

If the first n natural numbers are squared, the sum of these squares can be found using the formula shown, where n represents the number of terms in the sum. Verify the formula by computing the sum of the squares of the first six natural numbers by hand, and then evaluating S_6. Then find the sum of the squares of the first 20 natural numbers: $(1^2 + 2^2 + 3^2 + \cdots + 20^2)$. $S_6 = 91$; $S_{20} = 2870$

APPLICATIONS

81. Find the sum of the first 30 multiples of 2. 930

82. Find the sum of the first 5 multiples of $\frac{2}{3}$. 10

83. Savings account balance: Not trusting a bank, Chance Calvert's grandpa has placed \$2500 in a cookie jar for his favorite grandson. Each month Chance removes \$85 for a truck payment. Find the amount in the cookie jar 19 months later. (Hint: For $a_1 = 2500$, the amount remaining *19 mo* later will be what term of the sequence?) \$885

84. Stud screws and drywall: Mitchell Benjamin's mom buys two boxes of stud screws (350 in each box) to frame out their basement for drywall. Each stud requires five screws. How many screws remain after 125 studs have been put up? (Hint: For $a_1 = 700$, the amount remaining *after 125 studs* will be what term of the sequence?) 75 screws

85. Record wind speed: The highest winds ever recorded on Earth were measured at 318 mph inside a tornado that occurred in the state of Oklahoma during May of 1999. Suppose that as the winds began to die down, it was noticed that they declined by about 13 mph each hour. How long before the winds had decreased to 6 mph? (Hint: For $a_1 = 318$, *x hrs later* will be what term of the sequence?) 24 hr

86. Temperature fluctuation: At 5 P.M. in Coldwater, the temperature was a chilly 36°F. If the temperature decreased by 3°F every half-hour for the next 7 hr, at what time did the temperature hit 0°F? at 11 P.M.

87. Arc of a baby swing: When Mackenzie's baby swing is started, the first swing (one way) is a 30-in. arc. As the swing slows down, each successive arc is $\frac{3}{2}$ in. less than the previous one. Find (a) the length of the tenth swing and (b) how far Mackenzie has traveled during the ten swings. 16.5 in.; 232.5 in.

88. Computer animations: The animation on a new computer game initially allows the hero of the game to jump a (screen) distance of 10 in. over booby traps and obstacles. Each successive jump is limited to $\frac{3}{4}$ in. less than the previous one. Find (a) the length of the seventh jump and (b) the total distance covered after seven jumps. 5.5 in.; 54.25 in.

89. Seating capacity: The Fox Theater creates a "theater in the round" when it shows any of Shakespeare's plays. The first row has 80 seats, the second row has 88, the third row has 96, and so on. How many seats are in the 10th row? If there is room for 25 rows, how many chairs will be needed to set up the theater? 152; 4400

90. Marching formations: During the Rose Bowl parade and football game, many high school bands perform and display their musical and marching talents. San Marino High performs at halftime and uses a formation with 12 marchers in the first row, 14 in the second, 16 in the third, and so on for 16 rows. How many marchers are in the last row? How many band members are in the performance? 42, 432

91. Sales goals: *At the time that I was newly hired, 100 sales per month was what I required. Each following month—the last plus 20 more, as I work for the goal of top sales award. When 2500 sales are thusly made, it's Tahiti, Hawaii, and pina coladas in the shade.* How many sales were made by this person after the seventh month? 220 What were the total sales after the 12th month? 2520 Was the goal of 2500 total sales met after the 12th month? yes

92. Bequests to charity: *At the time our mother left this Earth, she gave $9000 to her children of birth. This we kept and each year added $3000 more, as a lasting memorial from the children she bore. When $42,000 is thusly attained, all goes to charity that her memory be maintained.* What was the balance in the sixth year? In what year was the goal of $42,000 met?
$a_6 = 27{,}000$; $a_{11} = 42{,}000$; the goal was met in 11 yr.

WRITING, RESEARCH, AND DECISION MAKING

93. Investigate the similarities (and differences) between the linear function $f(x) = 3 + 2(x - 1)$ and the arithmetic sequence $a_n = 3 + 2(n - 1)$. Make specific comments regarding the domain and graph of each, including any connections regarding slopes and intercepts. Although the graphs cannot be the same (why?), what change in a_n would cause it to more closely approximate the graph of $f(x)$? Answers will vary.

94. From a study of numerical analysis, a function is known to be linear if its "first differences" (differences between each output) are constant. Likewise, a function is known to be quadratic if its "first differences" form an *arithmetic sequence.* Use this information to determine if the following sets of output come from a linear or quadratic function:

a. 19, 11.8, 4.6, −2.6, −9.8, −17, −24.2, . . . linear function

b. −10.31, −10.94, −11.99, −13.46, −15.35, . . . quadratic

EXTENDING THE CONCEPT

95. From elementary geometry it is known that the interior angles of a triangle sum to 180°, the interior angles of a quadrilateral sum to 360°, the interior angles of a pentagon sum to 540°, and so on. Use the pattern created by the relationship between the number of sides to the number of angles to develop a formula for the sum of the interior angles of an n-sided polygon. The interior angles of a decagon (10 sides) sum to how many degrees? $180(n-2)$, 1440°

96. From a study of numerical analysis, a function is known to be cubic if its "second differences" form an arithmetic sequence. For second differences, calculate and list the difference of each output value, then calculate and list the difference between each of *these* values (see Exercise 94).

a. Verify this property using $f(x) = x^3 - 9x - 8$ for $x \in [-4, 4]$. Compute second differences by making a list of first differences, then computing the differences between consecutive terms in this list. verified

b. Use the property to determine if the following data came from a cubic function: $(-4, -35), (-3, -4), (-2, 9), (-1, 10), (0, 5), (1, 0)$ yes

MAINTAINING YOUR SKILLS

97. (5.4) Solve for t: $2530 = 500e^{0.45t}$ $t \approx 3.6$

98. (3.4) Graph by completing the square. Label all important features: $y = x^2 - 2x - 3$.

99. (6.5) Solve using matrices and row reduction: $\begin{cases} 2x - 3y = -1 \\ -4x + 9y = 4 \end{cases}$ $\left(\frac{1}{2}, \frac{2}{3}\right)$

100. (4.3) Verify that $x = 3 + 2i$ is a solution to $x^3 - 7x^2 + 19x - 13 = 0$. verified

101. (2.3) In 2000, the deer population was 972. By 2005 it had grown to 1217. Assuming the growth is linear, find the function that models this data and use it to estimate the deer population in 2008. $f(x) = 49x + 972$; 1364

102. (3.6) Given y varies inversely with x and jointly with w. If $y = 4$ when $x = 15$ and $w = 52.5$, find the value of y if $x = 32$ and $w = 208$. $y = \frac{52}{7}$

8.3 Geometric Sequences

LEARNING OBJECTIVES

In Section 8.3 you will learn how to:

A. Identify a geometric sequence and its common ratio

B. Find the nth term of a geometric sequence

C. Find the nth partial sum of a geometric sequence

D. Find the sum of an infinite geometric series

E. Solve application problems involving geometric sequences and series

INTRODUCTION

Recall that arithmetic sequences are those where each term is found by *adding* a constant value to the preceding term. In this section we consider **geometric sequences,** where each term is found by *multiplying* the preceding term by a constant value. Geometric sequences have many interesting applications, as do **geometric series.**

POINT OF INTEREST

In 1761, Robert Wallace (1697–1771) published a work entitled *Various Prospects of Mankind,* in which he put forth the idea that any economic progress made by mankind would eventually be overwhelmed by the unchecked growth of a population. Acknowledging the work of Wallace, Thomas Robert Malthus (1766–1834) published his *Essay on the Principle of Population* in 1798, predicting that population growth would eventually outstrip the food supply, reducing mankind to subsistence-level living conditions. His contentions were based on the idea that population grows at a *geometric* rate, while the food supply tends to grow at an *arithmetic* rate. Mathematically his argument was sound, as any (increasing) geometric sequence will eventually grow beyond an arithmetic sequence, but he failed to consider that new technologies could expand agricultural production many times over, and his predictions never materialized.

A. Geometric Sequences

A geometric sequence is one where each successive term is found by multiplying the preceding term by a fixed constant. Consider growth of a bacteria population, where a single cell splits in two every hour over a 24-hr period. Beginning with a single bacterium ($a_0 = 1$), after 1 hr there are 2, after 2 hr there are 4, and so on. Writing the number of bacteria as a sequence we have:

hours:	a_1	a_2	a_3	a_4	a_5	...
	↓	↓	↓	↓	↓	
bacteria:	2	4	8	16	32	...

The sequence 2, 4, 8, 16, 32, . . . is a geometric sequence since each term is found by multiplying the previous term by the constant factor 2. This also means that the ratio of any two consecutive terms must be 2 and in fact, 2 is called the **common ratio *r*** for this sequence. Using the notation from Section 8.1 we can write $r = \frac{a_{k+1}}{a_k}$, where a_k represents any term of the sequence and a_{k+1} represents the term that follows a_k.

EXAMPLE 1 Determine if the given sequence is geometric.

a. 1, 0.5, 0.25, 0.125, . . . **b.** $\frac{1}{8}, \frac{1}{4}, \frac{3}{4}, 3, 15, \ldots$

Solution: Apply the definition to check for a common ratio $r = \frac{a_{k+1}}{a_k}$.

a. For 1, 0.5, 0.25, 0.125, . . . , the ratio of consecutive terms gives

$$\frac{0.5}{1} = 0.5, \qquad \frac{0.25}{0.5} = 0.5, \qquad \frac{0.125}{0.25} = 0.5, \qquad \text{and so on.}$$

This is a geometric sequence with common ratio $r = 0.5$.

b. For $\frac{1}{8}, \frac{1}{4}, \frac{3}{4}, 3, 15, \ldots$, we have:

$$\frac{1}{4} \div \frac{1}{8} = \frac{1}{4} \cdot \frac{8}{1} = 2 \qquad \frac{3}{4} \div \frac{1}{4} = \frac{3}{4} \cdot \frac{4}{1} = 3 \qquad 3 \div \frac{3}{4} = \frac{3}{1} \cdot \frac{4}{3} = 4 \qquad \text{and so on.}$$

Since the ratio is not constant, this is not a geometric sequence.

NOW TRY EXERCISES 7 THROUGH 24

EXAMPLE 2 Write the first five terms of the geometric sequence, given the first term $a_1 = -16$ and the common ratio $r = 0.25$.

Solution: Given $a_1 = -16$ and $r = 0.25$. Starting at $a_1 = -16$, multiply each term by 0.25 to generate the sequence.

$$a_2 = -16 \cdot 0.25 = -4 \qquad a_3 = -4 \cdot 0.25 = -1$$
$$a_4 = -1 \cdot 0.25 = -0.25 \qquad a_5 = -0.25 \cdot 0.25 = -0.0625$$

The first five terms of this sequence are -16, -4, -1, -0.25, and -0.0625.

NOW TRY EXERCISES 25 THROUGH 38

B. Find the *n*th Term of a Geometric Sequence

If the values a_1 and r from a geometric sequence are known, we could generate the terms of the sequence by applying *additional factors of r to the first term,* instead of multiplying each new term by r. If $a_1 = 3$ and $r = 2$, we simply begin at a_1, and continue applying additional factors of r for each successive term.

$$3 = 3 \cdot 2^0 \qquad a_1 = a_1r^0$$
$$6 = 3 \cdot 2^1 \qquad a_2 = a_1r^1$$
$$12 = 3 \cdot 2^2 \qquad a_3 = a_1r^2$$
$$24 = 3 \cdot 2^3 \qquad a_4 = a_1r^3$$
$$48 = 3 \cdot 2^4 \qquad a_5 = a_1r^4$$

current term — initial term — exponent on common ratio

From this pattern, we note the exponent on r is always 1 less than the subscript of the current term: $5 - 1 = 4$, which leads us to the formula for the nth term of a geometric sequence.

THE *n*TH TERM OF A GEOMETRIC SEQUENCE

The nth term of a *geometric sequence* is given by

$$a_n = a_1r^{n-1}$$

where r is the common ratio.

If the term a_1 is unknown or not given, the nth term can be written

$$a_n = a_kr^{n-k} \text{ where } n = k + (n - k)$$

(the subscript on the given term and the exponent on r sum to n).

EXAMPLE 3 Find the 10th term of the sequence 3, -6, 12, -24, ...

Solution: Instead of writing out all 10 terms, we determine the constant ratio r and use the nth term formula. By inspection we note that $a_1 = 3$ and $r = -2$.

$$a_n = a_1r^{n-1} \qquad \text{nth term formula}$$
$$= 3(-2)^{n-1} \qquad \text{substitute 3 for } a_1 \text{ and } -2 \text{ for } r$$

To find the 10th term we substitute $n = 10$:

$$a_{10} = 3(-2)^{10-1} \qquad \text{substitute 10 for } n$$
$$= 3(-2)^9 = -1536 \qquad \text{simplify}$$

NOW TRY EXERCISES 39 THROUGH 46

EXAMPLE 4 Find the number of terms in the geometric sequence 4, 2, 1, . . . , $\frac{1}{64}$.

Solution: Observing that $a_1 = 4$ and $r = \frac{1}{2}$, we have

$$a_n = a_1 r^{n-1} \quad \text{nth term formula}$$
$$= 4\left(\frac{1}{2}\right)^{n-1} \quad \text{substitute 4 for } a_1 \text{ and } \frac{1}{2} \text{ for } r$$

Although we don't know the number of terms in the sequence, we *do* know the last or nth term is $\frac{1}{64}$. Substituting $a_n = \frac{1}{64}$ gives

$$\frac{1}{64} = 4\left(\frac{1}{2}\right)^{n-1} \quad \text{substitute } \frac{1}{64} \text{ for } a_n$$
$$\frac{1}{256} = \left(\frac{1}{2}\right)^{n-1} \quad \text{divide by 4 } \left(\text{multiply by } \frac{1}{4}\right)$$

From our work in Chapter 5, we try to write both sides as exponentials with a like base, or apply logarithms. Since $256 = 2^8$, we equate bases.

$$\left(\frac{1}{2}\right)^8 = \left(\frac{1}{2}\right)^{n-1} \quad \text{write } \frac{1}{256} \text{ as } \left(\frac{1}{2}\right)^8$$
$$\rightarrow 8 = n - 1 \quad \text{like bases imply that exponents must be equal}$$
$$9 = n \quad \text{solve for } n$$

This shows $\frac{1}{64}$ is the ninth term, and there are nine terms in the sequence.

NOW TRY EXERCISES 47 THROUGH 58

EXAMPLE 5 Given a geometric sequence where $a_4 = 0.075$ and $a_7 = 0.009375$, find the common ratio r and the value of a_1.

Solution: Since a_1 is not known, we express a_7 as the product of a known term and the appropriate number of common ratios: $a_7 = a_4 r^3$ $(7 - 4 = 3$, as required).

$$a_7 = a_4 \cdot r^3 \quad \text{nth term formula where } a_1 \text{ is unknown}$$
$$0.009375 = 0.075r^3 \quad \text{substitute 0.009375 for } a_7 \text{ and 0.075 for } a_4$$
$$0.125 = r^3 \quad \text{divide by 0.075}$$
$$r = 0.5 \quad \text{solve for } r$$

Having found r, we can now solve for a_1.

$$a_7 = a_1 r^6 \quad \text{nth term formula}$$
$$0.009375 = a_1(0.5)^6 \quad \text{substitute 0.009375 for } a_7 \text{ and 0.5 for } r$$
$$0.009375 = a_1(0.015625) \quad \text{simplify}$$
$$a_1 = 0.6 \quad \text{solve for } a_1$$

The first term is $a_1 = 0.6$ and the common ratio is $r = 0.5$.

NOW TRY EXERCISES 59 THROUGH 64

C. Find the *n*th Partial Sum of a Geometric Sequence

As with arithmetic series, applications of geometric series often involve computing a sum of consecutive terms. We can adapt the method for finding the sum of an arithmetic sequence to develop a formula for adding the first n terms of a geometric sequence.

For the nth term $a_n = a_1r^{n-1}$, we have $S_n = a_1 + a_1r + a_1r^2 + a_1r^3 + \cdots + a_1r^{n-1}$. Multiply S_n by $-r$, then add the original series.

$$\begin{aligned} -rS_n &= -a_1r + (-a_1r^2) + (-a_1r^3) + \cdots + (-a_1r^{n-1}) + (-a_1r^n) \\ +\ S_n &= a_1 + a_1r + a_1r^2 + \cdots + a_1r^{n-2} + a_1r^{n-1} \\ \hline S_n - rS_n &= a_1 + 0 + 0 + 0 + 0 + 0 + (-a_1r^n) \end{aligned}$$

We then have $S_n - rS_n = a_1 - a_1r^n$, and can now solve for S_n:

$$\begin{aligned} S_n - rS_n &= a_1 - a_1r^n && \text{difference of } S_n \text{ and } rS_n \\ S_n(1 - r) &= a_1 - a_1r^n && \text{factor out } S_n \\ S_n &= \frac{a_1 - a_1r^n}{1 - r} && \text{solve for } S_n \end{aligned}$$

The result is a formula for the nth partial sum of a geometric sequence.

THE *n*TH PARTIAL SUM OF A GEOMETRIC SEQUENCE

Given a geometric sequence with first term a_1 and common ratio r, the nth partial sum (the sum of the first n terms) is

$$S_n = \frac{a_1 - a_1r^n}{1 - r} = \frac{a_1(1 - r^n)}{1 - r},\ r \neq 1$$

In words: The sum of a geometric sequence is the difference of the first term and $(n + 1)$th term, divided by 1 minus the common ratio.

EXAMPLE 6 Find the sum: $\sum_{i=1}^{9} 3^i$ (the first nine powers of 3).

Solution: The initial terms of this series are $3 + 9 + 27 + \cdots$, and we note $a_1 = 3$, $r = 3$, and $n = 9$. We could find the first nine terms and add, but using the partial sum formula will be much quicker.

$$\begin{aligned} S_n &= \frac{a_1(1 - r^n)}{1 - r} && \text{sum formula} \\ S_9 &= \frac{3(1 - 3^9)}{1 - 3} && \text{substitute 3 for } a_1\text{, 9 for } n\text{, and 3 for } r \\ &= \frac{3(-19{,}682)}{-2} && \text{simplify} \\ &= 29{,}523 && \text{result} \end{aligned}$$

NOW TRY EXERCISES 65 THROUGH 82

EXAMPLE 7 Find the 11th partial sum of the geometric sequence whose third term is 8 and whose sixth term is 1.

Solution: In order to use the sum formula, we need to know both a_1 and r. Using $a_3 = 8$ and $a_6 = 1$ in the nth term formula, we find

$$a_6 = a_3r^3 \quad n\text{th term formula where } a_1 \text{ is unknown}$$
$$1 = 8r^3 \quad \text{substitute 1 for } a_6 \text{ and 8 for } a_3$$
$$\frac{1}{8} = r^3 \quad \text{divide by 8}$$
$$r = \frac{1}{2} \quad \text{solve for } r$$

Having found r, we can now solve for a_1.

$$a_6 = a_1r^5 \quad n\text{th term formula}$$
$$1 = a_1\left(\frac{1}{2}\right)^5 \quad \text{substitute 1 for } a_6 \text{ and } \frac{1}{2} \text{ for } r$$
$$1 = a_1\frac{1}{32} \quad \left(\frac{1}{2}\right)^5 = 32$$
$$32 = a_1 \quad \text{solve for } a_1$$

With $a_1 = 32$ and $r = \frac{1}{2}$, we can now use the sum formula:

$$S_n = \frac{a_1(1 - r^n)}{1 - r} \quad \text{sum formula}$$
$$S_{11} = \frac{32\left(1 - \left(\frac{1}{2}\right)^{11}\right)}{1 - \frac{1}{2}} \quad \text{substitute 11 for } n\text{, 32 for } a_1\text{, and } \tfrac{1}{2} \text{ for } r$$
$$= \frac{32\left(1 - \frac{1}{2048}\right)}{\frac{1}{2}} \quad \text{simplify}$$
$$= 64\left(\frac{2047}{2048}\right) \quad \text{simplify in parentheses } \left(\frac{2048}{2048} = 1\right)\text{, invert and multiply}$$
$$= \frac{2047}{32}$$

The sum of the first eleven terms of this sequence is $\frac{2047}{32}$.

NOW TRY EXERCISES 83 THROUGH 88

D. The Sum of an Infinite Geometric Series

To this point we've considered only partial sums of a geometric series. While it is impossible to add an infinite number of these terms, some of these "infinite sums" appear to have a limiting value. The sum appears to get ever closer to this value but never exceeds it—much like the asymptotic behavior of some graphs. We will define the sum of this **infinite geometric series** to be this limiting value, if it exists. Consider the illustration in Figure 8.9, where a standard sheet of typing paper is cut in half. One of the halves is again cut in half and the process is continued indefinitely, as shown. Notice the "halves" create an infinite sequence $\frac{1}{2}, \frac{1}{4}, \frac{1}{8}, \frac{1}{16}, \frac{1}{32}, \cdots$ with $a_1 = \frac{1}{2}$ and $r = \frac{1}{2}$. The corresponding infinite series is $\frac{1}{2} + \frac{1}{4} + \frac{1}{8} + \frac{1}{16} + \frac{1}{32} + \cdots + \frac{1}{2^n} + \cdots$.

Figure 8.9

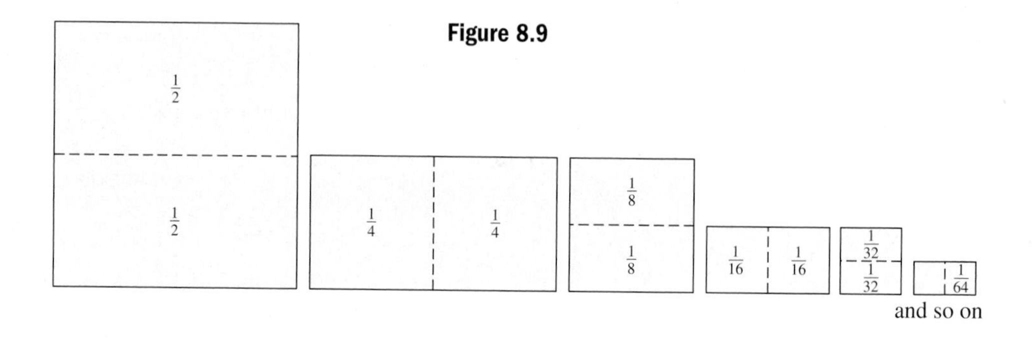

Figure 8.10

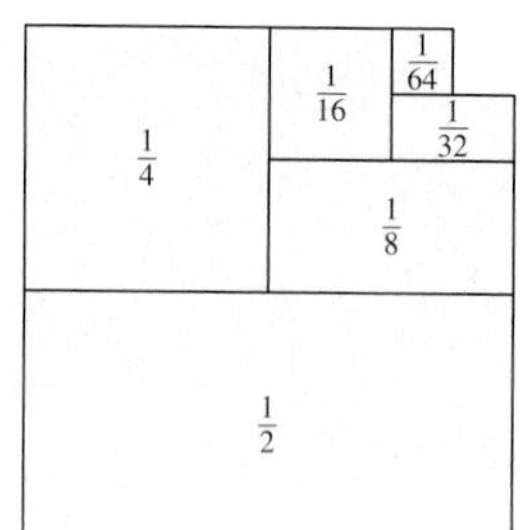

If we arrange one of the halves from each stage as shown in Figure 8.10, we would be rebuilding the original sheet of paper. As we add more and more of these halves together, we get closer and closer to the size of the original sheet. We gain an intuitive sense that this series must add to 1, because the *pieces* of the original sheet of paper must add to 1 whole sheet. To explore this idea further, consider what happens to $\left(\frac{1}{2}\right)^n$ as n becomes large.

$$n = 4{:}\left(\frac{1}{2}\right)^4 = 0.0625 \qquad n = 8{:}\left(\frac{1}{2}\right)^8 \approx 0.004 \qquad n = 12{:}\left(\frac{1}{2}\right)^{12} \approx 0.0002$$

Further exploration with a calculator seems to support the idea that as $n \to \infty$, $\left(\frac{1}{2}\right)^n \to 0$, although a definitive proof is left for a future course. In fact, it can be shown that for any $|r| < 1$, r^n becomes very close to zero as n becomes large. In symbols: as $n \to \infty$, $r^n \to 0$. Returning to $S_n = \frac{a_1 - a_1 r^n}{1 - r} = \frac{a_1}{1 - r} - \frac{a_1 r^n}{1 - r}$, note that if $|r| < 1$ and "we sum an infinite number of terms," the numerator of the subtrahend (second term) becomes zero, leaving only the minuend (first term). In other words, the limiting value (represented by S_∞) is $S_\infty = \frac{a_1}{1 - r}$.

> **INFINITE GEOMETRIC SERIES**
>
> Given a geometric sequence with first term a_1 and $|r| < 1$, the sum of the related infinite series is given by
>
> $$S_\infty = \frac{a_1}{1 - r}; r \neq 1$$
>
> If $|r| > 1$, no finite sum exists.
>
> In words: the sum of an infinite series is the first term divided by 1 minus the common ratio.

EXAMPLE 8 Find the limiting value of each infinite geometric series (if it exists).

a. $1 + 2 + 4 + 8 + \cdots$ **b.** $3 + 2 + \frac{4}{3} + \frac{8}{9} + \cdots$

c. $0.185 + 0.000185 + 0.000000185 + \cdots$

Solution: Begin by determining if the infinite series is geometric. If so, use $S_\infty = \frac{a_1}{1 - r}$.

a. Since $r = 2$ (by inspection), a finite sum does not exist.

b. Using the ratio of consecutive terms we find $r = \frac{2}{3}$ and the infinite sum exists. With $a_1 = 3$, we have

$$S_\infty = \frac{3}{1 - \frac{2}{3}} = \frac{3}{\frac{1}{3}} = 9$$

c. This series is equivalent to the repeating decimal $0.185185185\ldots = 0.\overline{185}$. The common ratio is $r = \frac{0.185185}{0.185} = 0.001$ and the infinite sum exists.

$$S_\infty = \frac{0.185}{1 - 0.001} = \frac{5}{27}.$$

NOW TRY EXERCISES 89 THROUGH 104

E. Applications Involving Geometric Sequences and Series

Here are a few of the ways these ideas can be put to use.

EXAMPLE 9 A pendulum is any object attached to a fixed point and allowed to swing freely under the influence of gravity. Suppose each swing is 0.9 the length of the previous one. Gradually the swings become shorter and shorter and at some point the pendulum will appear to have stopped (although *theoretically* it never does).

a. How far does the pendulum travel on its eighth swing, if the first swing was 2 m?

b. What is the total distance traveled by the pendulum for these 8 swings?

c. How many swings until the length of each swing falls below 0.5 m?

d. What total distance does the pendulum travel before coming to rest?

Solution: **a.** The lengths of each swing form the terms of a geometric sequence with $a_1 = 2$ m and $r = 0.9$. The first few terms are 2, 1.8, 1.62, 1.458, and so on. For the 8th term we have:

$$a_n = a_1 r^{n-1} \quad \text{nth term formula}$$
$$a_8 = 2(0.9)^{8-1} \quad \text{substitute 8 for } n\text{, 2 for } a_1\text{, and 0.9 for } r$$
$$\approx 0.956$$

The pendulum travels about 0.956 m on its 8th swing.

b. For the total distanced travelled after 8 swings, we compute the value of S_8.

$$S_n = \frac{a_1(1 - r^n)}{1 - r} \quad n\text{th partial sum formula}$$

$$S_8 = \frac{2(1 - 0.9^8)}{1 - 0.9} \quad \text{substitute 2 for } a_1\text{, 0.9 for } r \text{ and 8 for } n$$

$$\approx 11.4$$

The pendulum has traveled about 11.4 meters by the end of the 8th swing.

c. To find the number of swings until the length of each swing is less than 0.5 m, we solve for n in the equation $0.5 = 2(0.9)^{n-1}$. This yields:

$$0.25 = (0.9)^{n-1} \quad \text{divide by 2}$$

$$\ln 0.5 = (n - 1)\ln 0.9 \quad \text{take the natural log, apply power property}$$

$$\frac{\ln 0.5}{\ln 0.9} + 1 = n \quad \text{solve for } n \text{ (exact form)}$$

$$14.16 \approx n \quad \text{solve for } n \text{ (approximate form)}$$

After the 14th swing, each successive swing will be less than 0.5 m.

d. For the total distance travelled before coming to rest, we consider the related infinite geometric series, with $a_1 = 2$ and $r = 0.9$.

$$S_\infty = \frac{a_1}{1 - r} \quad \text{infinite sum formula}$$

$$S_\infty = \frac{2}{1 - 0.9} \quad \text{substitute 2 for } a_1 \text{ and 0.9 for } r$$

$$= 20 \quad \text{result}$$

The pendulum would travel 20 m before coming to rest.

NOW TRY EXERCISES 107 THROUGH 121

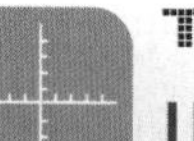

TECHNOLOGY HIGHLIGHT

Using a Graphing Calculator to Study Geometric Sequences and Series

The keystrokes shown apply to a TI-84 Plus model. Please consult your manual or our Internet site for other models.

In the *Technology Highlight* from Section 8.2, we worked with sequences and series in **function mode** using the (blue) 2nd functions L1 through L6 located above the digits 1 through 6. Most graphing calculators have a **sequence mode** that enables you to work with sequences using graphs and tables, in much the same way as you work with functions $f(x)$. To place the calculator in sequence mode, press MODE and use the down ▼ arrow to get to the fourth line: **Func Par Pol Seq**. Use the right arrow ▶ to highlight **Seq** and press ENTER. The calculator is in sequence mode when pressing the Y= key gives the screen shown in Figure 8.11, and then pushing the X,T,θ,n key produces "n" on the home screen (instead of "X").

Figure 8.11

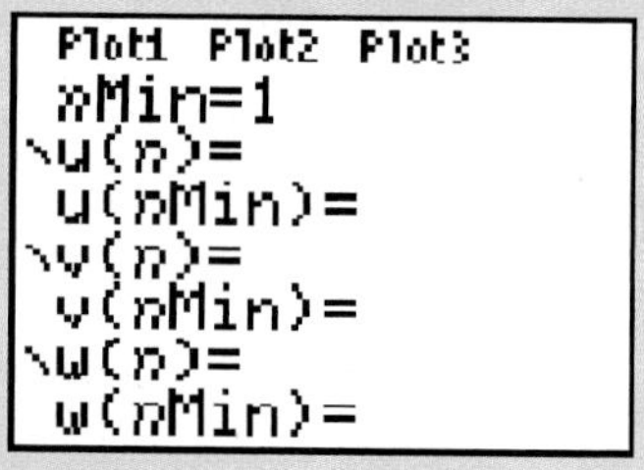

```
Plot1 Plot2 Plot3
nMin=1
\u(n)=
 u(nMin)=
\v(n)=
 v(nMin)=
\w(n)=
 w(nMin)=
```

Experiment with this feature by defining the function **u(n)** $= 2 \cdot (0.5)^{n-1}$ and using graphs and tables. Here we'll use a defined variable to study sequences and series on the home screen. This is done using the (blue) 2nd functions **u**, **v**, and **w** above the digits 7, 8, and 9 (notice the *variable locations* **u**, **v**, *and* **w** *are green*). Once defined, **u**, **v**, and **w** can be used to generate any specific term or a finite sequence. They can also be combined with the **"sum("** operation to find the sum of n terms [access using 2nd STAT **MATH 5:sum(...]**.

EXAMPLE 1 Using your calculator, generate the first five terms of $a_n = 2 \cdot (0.5)^{n-1}$ on the home screen, find the sum of these five terms, then generate the 8th term.

Solution: Begin by CLEAR ing the home screen. In **Seq** mode, define a sequence "**u(n)**," by placing the nth term in quotes "$2 \cdot (0.5)^{n-1}$" [ALPHA +], then store the result in **u:** STO→ 2nd 7 .
To display the first five terms, we supply three operators to **u** in the form **u(***start, stop, increment***)**, where *start* is the first term we want evaluated, *stop* is the last term we want evaluated, and *increment* is what we want the calculator to "count by" (similar to ΔTbl on the 2nd WINDOW TBLSET menu). For this exercise we used **u**(1, 5, 1), which produces the outputs {2, 1, 0.5, 0.25, and 0.125} as seen in Figure 8.12. To find the sum of these terms we can (1) use "sum(ANS)" (recall the previous result using 2nd (–)); (2) recall **u**(1, 5, 1), place the cursor in position over **u** and insert the "**sum(**" operation; or (3) simply retype sum (**u**(1, 5, 1)) and press ENTER . The first option is used in Figure 8.12. To generate the eighth term of the sequence (or any other) simply enter **u**(8) as shown and press ENTER .

Figure 8.12

```
"2*0.5^(n-1)"→u
                Done
u(1,5,1)
{2 1 .5 .25 .12...
sum(Ans)
               3.875
u(8)
```

Exercise 1: Using your calculator, (a) generate the first five terms of $a_n = 3 \cdot (0.2)^{n-1}$ on the home screen; (b) find the sum of these five terms, then seven terms, then 10 terms. What do you notice about the results from part (b)? Use the formula $S_\infty = \frac{a_1}{1 - r}$ to verify your "suspicions." 3, 0.6, 0.12, 0.024, 0.0048; 3.7488, 3.749952, 3.749999616 $\rightarrow 3.75$ $S_\infty = 3.75$

Exercise 2: Using your calculator, (a) generate the first five terms of $a_n = 2000 \cdot (1.05)^{n-1}$ on the home screen; (b) generate the eighth term; and (c) compute the ratios $u(8)/u(7)$ and $u(9)/u(8)$. What is the significance of the results from part (c)? geometric sequence, $r = 1.05$
a. 2000, 2100, 2205, 2315.25, 2431.0125 **b.** 2814.200845
c. 1.05, 1.05

8.3 EXERCISES

CONCEPTS AND VOCABULARY

Fill in the blank with the appropriate word or phrase. Carefully reread the section if needed.

1. In a geometric sequence, each successive term is found by <u>multiplying</u> the preceding term by a fixed value r.

2. In a geometric sequence, the common ratio r can be found by computing the <u>quotient</u> of any two consecutive terms.

3. The nth term of a geometric sequence is given by $a_n =$ <u>$a_1 r^{n-1}$</u>, for any $n \geq 1$.

4. For the general sequence $a_1, a_2, a_3, \ldots, a_k, \ldots, a_n$ the fifth partial sum is given by $S_5 =$ <u>$\frac{a_1(1 - r^5)}{1 - r}$</u>.

5. Describe/discuss how the formula for the nth partial sum is related to the formula for the sum of an infinite geometric series. Answers will vary.

6. Describe the difference(s) between an arithmetic and a geometric sequence. How can a student prevent confusion between the formulas? Answers will vary.

10. $r = \frac{-1}{4}$
11. $a_n = n^2 + 1$
12. $a_n = 4n - 17$
15. not geometric; ratio of terms decreases by 1
16. not geometric; ratio of terms decreases by 1
18. $r = \frac{-2}{3}$
23. not geometric; $a_n = \dfrac{240}{n!}$
25. 5, 10, 20, 40
26. 2, −8, 32, −128
27. $-6, 3, \frac{-3}{2}, \frac{3}{4}$
28. $\frac{2}{3}, \frac{2}{15}, \frac{2}{75}, \frac{2}{375}$
29. $4, 4\sqrt{3}, 12, 12\sqrt{3}$
30. $\sqrt{5}, 5, 5\sqrt{5}, 25$
33. $-\frac{3}{8}$ 34. $-\frac{16}{81}$ 35. $\frac{25}{4}$
39. $a_1 = \frac{1}{27}, r = -3,$ $a_n = \frac{1}{27}(-3)^{n-1}, a_6 = -9,$ $a_{10} = -729, a_{12} = -6561$
40. $a_1 = -\frac{7}{8}; r = -2,$ $a_n = -\frac{7}{8}(-2)^{n-1};$ $a_6 = 28; a_{10} = 448;$ $a_{12} = 1792$
41. $a_1 = 729, r = \frac{1}{3},$ $a_n = 729(\frac{1}{3})^{n-1},$ $a_6 = 3, a_{10} = \frac{1}{27}, a_{12} = \frac{1}{243}$
42. $a_1 = 625, r = \frac{1}{5},$ $a_n = 625(\frac{1}{5})^{n-1}, a_6 = \frac{1}{5},$ $a_{10} = \frac{1}{3125}, a_{12} = \frac{1}{78125}$
43. $a_1 = \frac{1}{2}, r = \sqrt{2},$ $a_n = \frac{1}{2}(\sqrt{2})^{n-1}, a_6 = 2\sqrt{2},$ $a_{10} = 8\sqrt{2}, a_{12} = 16\sqrt{2}$
44. $a_1 = 36\sqrt{3}, r = \dfrac{1}{\sqrt{3}},$ $a_n = 36\sqrt{3}\left(\dfrac{1}{\sqrt{3}}\right)^{n-1}, a_6 = 4,$ $a_{10} = \dfrac{4}{9}, a_{12} = \dfrac{4}{27}$
45. $a_1 = 0.2, r = 0.4,$ $a_n = 0.2(0.4)^{n-1}$ $a_6 = 0.002048,$ $a_{10} = 0.0000524288,$ $a_{12} = 0.000008388608$
46. $a_1 = 0.5, r = -0.7,$ $a_n = 0.5(-0.7)^{n-1},$ $a_6 = -0.084035,$ $a_{10} = -0.0201768035,$ $a_{12} = -0.0098866337$
67. $\frac{3872}{27} \approx 143.41$
68. $\frac{765}{32} \approx 23.91$
69. 257.375
70. $\frac{11{,}605}{512} \approx 22.67$
71. 728
72. 10,922
73. 10.625

DEVELOPING YOUR SKILLS

Determine if the sequence given is geometric. If yes, name the common ratio. If not, try to determine the pattern that forms the sequence.

7. 4, 8, 16, 32, . . . $r = 2$
8. 2, 6, 18, 54, 162, . . . $r = 3$
9. 3, −6, 12, −24, 48, . . . $r = -2$
10. 128, −32, 8, −2, . . .
11. 2, 5, 10, 17, 26, . . .
12. −13, −9, −5, −1, 3, . . .
13. 3, 0.3, 0.03, 0.003, . . . $r = 0.1$
14. 12, 0.12, 0.0012, 0.000012, . . . $r = 0.01$
15. −1, 3, −12, 60, −360, . . .
16. $-\frac{2}{3}, 2, -8, 40, -240, \ldots$
17. $25, 10, 4, \frac{8}{5}, \ldots$ $r = \frac{2}{5}$
18. $-36, 24, -16, \frac{32}{3}, \ldots$
19. $\frac{1}{2}, \frac{1}{4}, \frac{1}{8}, \frac{1}{16}, \ldots$ $r = \frac{1}{2}$
20. $\frac{2}{3}, \frac{4}{9}, \frac{8}{27}, \frac{16}{81}, \ldots$ $r = \frac{2}{3}$
21. $3, \dfrac{12}{x}, \dfrac{48}{x^2}, \dfrac{192}{x^3}, \ldots$ $r = \frac{4}{x}$
22. $5, \dfrac{10}{a}, \dfrac{20}{a^2}, \dfrac{40}{a^3}, \ldots$ $r = \frac{2}{a}$
23. 240, 120, 40, 10, 2, . . .
24. −120, −60, −20, −5, −1, . . . not geometric, $a_n = -120\left(\dfrac{1}{n!}\right)$

Write the first four terms of the sequence, given a_1 and r.

25. $a_1 = 5, r = 2$
26. $a_1 = 2, r = -4$
27. $a_1 = -6, r = -\frac{1}{2}$
28. $a_1 = \frac{2}{3}, r = \frac{1}{5}$
29. $a_1 = 4, r = \sqrt{3}$
30. $a_1 = \sqrt{5}, r = \sqrt{5}$
31. $a_1 = 0.1, r = 0.1$ 0.1, 0.01, 0.001, 0.0001
32. $a_1 = 0.024, r = 0.01$ 0.024, 0.00024, 0.0000024, 0.000000024

Find the indicated term for each sequence.

33. $a_1 = -24, r = \frac{1}{2}$; find a_7
34. $a_1 = 48, r = -\frac{1}{3}$; find a_6
35. $a_1 = -\frac{1}{20}, r = -5$; find a_4
36. $a_1 = \frac{3}{20}, r = 4$; find a_5 $\frac{192}{5}$
37. $a_1 = 2, r = \sqrt{2}$; find a_7 16
38. $a_1 = \sqrt{3}, r = \sqrt{3}$; find a_8 81

Identify a_1 and r, then write the expression for the nth term $a_n = a_1r^{n-1}$ and use it to find a_6, a_{10}, and a_{12}.

39. $\frac{1}{27}, -\frac{1}{9}, \frac{1}{3}, -1, 3, \ldots$
40. $-\frac{7}{8}, \frac{7}{4}, -\frac{7}{2}, 7, -14, \ldots$
41. 729, 243, 81, 27, 9, . . .
42. 625, 125, 25, 5, 1, . . .
43. $\frac{1}{2}, \frac{\sqrt{2}}{2}, 1, \sqrt{2}, 2, \ldots$
44. $36\sqrt{3}, 36, 12\sqrt{3}, 12, 4\sqrt{3}, \ldots$
45. 0.2, 0.08, 0.032, 0.0128, . . .
46. 0.5, −0.35, 0.245, −0.1715, . . .

Find the number of terms in each sequence.

47. $a_1 = 9, a_n = 729, r = 3$ 5
48. $a_1 = 1, a_n = -128, r = -2$ 8
49. $a_1 = 16, a_n = \frac{1}{64}, r = \frac{1}{2}$ 11
50. $a_1 = 4, a_n = \frac{1}{512}, r = \frac{1}{2}$ 12
51. $a_1 = -1, a_n = -1296, r = \sqrt{6}$ 9
52. $a_1 = 2, a_n = 1458, r = -\sqrt{3}$ 13
53. 2, −6, 18, −54, . . . , −4374 8
54. 3, −6, 12, −24, . . . , −6144 12
55. $64, 32\sqrt{2}, 32, 16\sqrt{2}, \ldots, 1$ 13
56. $243, 81\sqrt{3}, 81, 27\sqrt{3}, \ldots, 1$ 11
57. $\frac{3}{8}, -\frac{3}{4}, \frac{3}{2}, -3, \ldots, 96$ 9
58. $-\frac{5}{27}, \frac{5}{9}, -\frac{5}{3}, -5, \ldots, -135$ 7

Find the common ratio r and the value of a_1 using the information given.

59. $a_3 = 324, a_7 = 64$ $r = \frac{2}{3}, a_1 = 729$
60. $a_5 = 6, a_9 = 486$ $r = 3, a_1 = \frac{2}{27}$
61. $a_4 = \frac{4}{9}, a_8 = \frac{9}{4}$ $r = \frac{3}{2}, a_1 = \frac{32}{243}$
62. $a_2 = \frac{16}{81}, a_5 = \frac{2}{3}$ $r = \frac{3}{2}, a_1 = \frac{32}{243}$
63. $a_4 = \frac{32}{3}, a_8 = 54$ $r = \frac{3}{2}, a_1 = \frac{256}{81}$
64. $a_3 = \frac{16}{25}, a_7 = 25$ $r = \frac{5}{2}, a_1 = \frac{64}{625}$

Find the indicated sum. For Exercises 81 and 82, use the summation properties from Section 8.1.

65. $a_1 = 8, r = -2$; find S_{12} −10,920
66. $a_1 = 2, r = -3$; find S_8 −3280
67. $a_1 = 96, r = \frac{1}{3}$; find S_5
68. $a_1 = 12, r = \frac{1}{2}$; find S_8
69. $a_1 = 8, r = \frac{3}{2}$; find S_7
70. $a_1 = -1, r = -\frac{3}{2}$; find S_{10}
71. $2 + 6 + 18 + \cdots$; find S_6
72. $2 + 8 + 32 + \cdots$; find S_7
73. $16 - 8 + 4 - \cdots$; find S_8

74. -6560
75. ≈ 1.60
76. $\frac{547}{18} \approx 30.39$

74. $4 - 12 + 36 - \cdots$; find S_8 **75.** $\frac{4}{3} + \frac{2}{9} + \frac{1}{27} + \cdots$; find S_9 **76.** $\frac{1}{18} - \frac{1}{6} + \frac{1}{2} - \cdots$; find S_7

77. $\sum_{j=1}^{5} 4^j$ 1364 **78.** $\sum_{k=1}^{10} 2^k$ 2046 **79.** $\sum_{k=1}^{8} 5\left(\frac{2}{3}\right)^{k-1}$ $\frac{31{,}525}{2187} \approx 14.41$

80. $\sum_{j=1}^{7} 3\left(\frac{1}{5}\right)^{j-1}$ ≈ 3.75 **81.** $\sum_{i=4}^{10} 9\left(-\frac{1}{2}\right)^{i-1}$ $\frac{-387}{512} \approx -0.76$ **82.** $\sum_{i=3}^{8} 5\left(-\frac{1}{4}\right)^{i-1}$ ≈ 0.25

Find the indicated partial sum using the information given. Write all results in simplest form.

83. $a_2 = -5, a_5 = \frac{1}{25}$; find S_5 $\frac{521}{25}$ **84.** $a_3 = 1, a_6 = -27$; find S_6 $\frac{-182}{9}$ **85.** $a_3 = \frac{4}{9}, a_7 = \frac{9}{64}$; find S_6 $\frac{3367}{1296}$

86. $a_2 = \frac{16}{81}, a_5 = \frac{2}{3}$; find S_8 $\frac{6305}{972}$ **87.** $a_3 = 2\sqrt{2}, a_6 = 8$; find S_7 $14 + 15\sqrt{2}$ **88.** $a_2 = 3, a_5 = 9\sqrt{3}$; find S_7 $39 + 40\sqrt{3}$

Determine whether the infinite geometric series has a finite sum. If so, find the limiting value.

89. $3 + 6 + 12 + 24 + \cdots$ no **90.** $4 + 8 + 16 + 32 + \cdots$ no **91.** $9 + 3 + 1 + \cdots$ $\frac{27}{2}$

92. $36 + 24 + 16 + \cdots$ 108 **93.** $25 + 10 + 4 + \frac{8}{5} + \cdots$ $\frac{125}{3}$ **94.** $10 + 2 + \frac{2}{5} + \frac{2}{25} + \cdots$ $\frac{25}{2}$

95. $6 + 3 + \frac{3}{2} + \frac{3}{4} + \cdots$ 12 **96.** $-49 + (-7) + \left(-\frac{1}{7}\right) + \cdots$ $\frac{-343}{6}$ **97.** $6 - 3 + \frac{3}{2} - \frac{3}{4} + \cdots$ 4

98. $10 - 5 + \frac{5}{2} - \frac{5}{4} + \cdots$ $\frac{20}{3}$ **99.** $3 + 0.3 + 0.03 + 0.003 + \cdots$ $\frac{10}{3}$

100. $0.63 + 0.0063 + 0.000063 + \cdots$ $\frac{7}{11}$ **101.** $\sum_{k=1}^{\infty} \frac{3}{4}\left(\frac{2}{3}\right)^k$ $\frac{3}{2}$

102. $\sum_{i=1}^{\infty} 5\left(\frac{1}{2}\right)^i$ 5 **103.** $\sum_{j=1}^{\infty} 9\left(-\frac{2}{3}\right)^j$ $-\frac{18}{5}$ **104.** $\sum_{k=1}^{\infty} 12\left(\frac{4}{3}\right)^k$ no

WORKING WITH FORMULAS

105. Sum of the cubes of the first n natural numbers: $S_n = \dfrac{n^2(n+1)^2}{4}$

Compute $1^3 + 2^3 + 3^3 + \cdots + 8^3$ using the formula given. Then confirm the result by direct calculation. 1296

106. Student loan payment: $A_n = P(1 + r)^n$

If P dollars is borrowed at an annual interest rate r with interest compounded annually, the amount of money to be paid back after n years is given by the indicated formula. Find the total amount of money that the student must repay to clear the loan, if $8000 is borrowed at 4.5% interest and the loan is paid back in 10 yr. $12,423.76

APPLICATIONS

107. Pendulum movement: On each swing, a pendulum travels only 80% as far as it did on the previous swing. If the first swing is 24 ft, how far does the pendulum travel on the 7th swing? What total distance is traveled before the pendulum comes to rest? about 6.3 ft; 120 ft

108. Pendulum movement: Ernesto is swinging to and fro on the swing set at a local park. Using his legs and body, he pumps each swing until reaching a maximum height, then suddenly relaxes until the swing comes to a stop. With each swing, Ernesto travels 75% as far as he did on the previous swing. If the first arc (or swing) is 30 ft, find the distance Ernesto travels on the 5th arc. What total distance will he travel before coming to rest. about 9.5 ft; 120 ft

109. Cost of an education: A college education today costs an average of $50,000. Determine the cost of a college education 10 years from now, if inflation holds steady at 3% per year. approx. $67,196

110. Effects of inflation: In 1960, a loaf of bread cost $0.25. If inflation averaged 3% per year since then, how much would a loaf of bread cost in the year 2010, 50 years later? approx. $1.10

111. Depreciation: A certain new SUV depreciates in value about 20% per year (meaning it holds 80% of its value each year). If the SUV is purchased for $26,000, how much is it worth 5 years later? How many years until its value is less than $2792? $8520; 10 yr

112. Depreciation: A new photocopier under heavy use will depreciate about 25% per year (meaning it holds 75% of its value each year). If the copier is purchased for $7000, how much is it worth 4 yr later? How many years until its value is less than $1246? $2215; 6 yr

113. Equipment aging: Tests have shown that the pumping power of a heavy-duty oil pump decreases by 3% per month. If the pump can move 160 gallons per minute (gpm) new, how many gpm can the pump move 8 months later? If the pumping rate falls below 118 gpm, the pump must be replaced. How many months until this pump is replaced? 125.4 gpm; 10 months

114. Equipment aging: At the local mill, a certain type of saw blade can saw approximately 2 log-feet/sec when it is new. As time goes on the blade becomes worn, and loses 6% of its cutting speed each week. How many log-feet/sec can the saw blade cut after 6 weeks? If the cutting speed falls below 1.2 log-feet/sec, the blade must be replaced. During what week of operation will this blade be replaced? 1.38 log ft; 8th week

115. Population growth: At the beginning of the year 2000, the population of the United States was approximately 277 million. If the population is growing at a rate of 2.3% per year, what will the population be in 2010, 10 yr later? about 347.7 million

116. Population growth: The population of the Zeta Colony on Mars is 1000 people. Determine the population of the Colony 20 yr from now, if the population is growing at a constant rate of 5% per year. about 2653 people

117. 51,200 bacteria; 12 half-hours later (6 hr)
118. 3504 people; 7 two-month periods (14 months)

117. Population growth: A biologist finds that the population of a certain type of bacteria doubles *each half-hour.* If an initial culture has 50 bacteria, what is the population after 5 hours? How long will it take for the number of bacteria to reach 204,800?

118. Population growth: Suppose the population of a "boom town" in the old west doubled *every two months* after gold was discovered. If the initial population was 219, what was the population 8 months later? How many months until the population exceeds 28,000?

119. Elastic rebound: Megan discovers that a rubber ball dropped from a height of 20 m rebounds four-fifths of the distance it has previously fallen. How high does it rebound on the 7th bounce? How far does the ball travel before coming to rest? 4.2 m; 180 m

120. Elastic rebound: The screen saver on my computer is programmed to send a colored ball vertically down the middle of the screen so that it rebounds 95% of the distance it last traversed. If the ball always begins at the top and the screen is 36 cm tall, how high does the ball bounce after its 8th rebound? How far does the ball travel before coming to rest (and a new screen saver starts)? 23.9 cm; 1404 cm

121. Creating a vacuum: To create a vacuum, a hand pump is used to remove the air from an air-tight cube with a volume of 462 in^3. With each stroke of the pump, two-fifths of the air that remains in the cube is removed. How much air remains inside after the 5th stroke? How many strokes are required to remove all but 12.9 in^3 of the air? 35.9 in^3; 7 strokes

WRITING, RESEARCH, AND DECISION MAKING

122. As part of a science experiment, identical rubber balls are dropped from a certain height on these surfaces: slate, cement, and asphalt. When dropped on slate, the ball rebounds 80% of the height from which it last fell. On concrete the figure is 75% and on asphalt the figure is 70%. The ball is dropped from 130 m on the slate, 175 m on the cement, and 200 m on the asphalt. Which ball has traveled the shortest total distance at the time of the fourth bounce? Which ball will travel farthest before coming to rest? dropped on slate; dropped on cement

123. Consider the following situation. A person is hired at a salary of $40,000 per year, with a guaranteed raise of $1750 per year. At the same time, inflation is running about 4% per year. How many years until this person's salary is overtaken and eaten up by the actual cost of living? 6 yr

124. A standard piece of typing paper is approximately 0.001 in. thick. Suppose you were able to fold this piece of paper in half 26 times. How thick would the result be? (a) As tall as a hare, (b) as tall as a hen, (c) as tall as a horse, (d) as tall as a house, or (e) over 1 mi high? Find the actual height by computing the 27th term of a geometric sequence. Discuss what you find. e; ≈67,109 in. This is about 1.06 mi.

125. $a_n = 1000(1.05)^{n-1}$; after 6 yr $\rightarrow a_7$ is needed: $a_7 = 1340.10$; amount in the account after 6 yr; amount in the account after 7 yr; a_n generates the terms of the sequence before any interest is applied, while $A(t)$ gives the amount in the account after interest has been applied. Here $a_7 = A(6)$.

131.

132.

133. $p(50) \approx 2562.1$
$p(75) \approx 3615.6$
$p(100) \approx 4035.1$
$p(150) \approx 4189.1$

134. $\frac{-x-2}{x^2+3} + \frac{1}{x+1}$

125. Consider the geometric sequence formed by depositing \$1000 at 5% annual interest for 6 years. Write the nth term formula for this sequence and use it to find the value of the account after 6 years. Next consider the formula $A(t) = P(1 + r)^t$ from Chapter 5, with $P = \$1000$ and $r = 0.05$. What does $A(6)$ represent? What does $A(7)$ represent? Discuss/explain the relationship between the functions a_n and $A(t)$.

EXTENDING THE CONCEPT

126. Given $a_n = a_1 r^{n-1}$, expressions for a_{n-1} and a_{n+1} can be developed by noting the relationship between the subscripts and exponents. Specifically, we have $a_{n-1} = a_1 r^{n-2}$ and $a_{n+1} = a_1 r^n$, respectively. Use this idea to write the sum formula for S_{n+1}. $\frac{a_1 - a_1 r^{n+1}}{1 - r}$

127. Find an alternative formula for the sum $S_n = \sum_{k=1}^{n} \log k$, that does not use the sigma notation. $S_n = \log n!$

128. Verify the following statements:

a. If $a_1, a_2, a_3, \ldots, a_n$ is a geometric sequence with r and a_1 greater than zero, then $\log a_1, \log a_2, \log a_3, \ldots, \log a_n$ is an arithmetic sequence. verified

b. If $a_1, a_2, a_3, \ldots, a_n$ is an arithmetic sequence, then $10^{a_1}, 10^{a_2}, \ldots, 10^{a_n}$, is a geometric sequence. verified

MAINTAINING YOUR SKILLS

129. (1.5) Find the zeroes of f using the quadratic formula: $f(x) = x^2 + 5x + 9$. $x = \frac{-5}{2} \pm \frac{\sqrt{11}}{2}i$

130. (1.3) Solve for x:
$$\frac{3}{x^2 - 3x - 10} - \frac{4}{x - 5} = \frac{1}{x + 2}$$ $x = 0$

131. (4.6) Graph the rational function:
$$h(x) = \frac{x^2}{x - 1}$$

132. (3.3) Graph $f(x) = \frac{1}{(x-2)^2} + 1$ using transformations of the parent function.

133. (5.6) Given the logistics function shown, find $p(50)$, $p(75)$, $p(100)$, and $p(150)$:
$$p(t) = \frac{4200}{1 + 10e^{-0.055t}}$$

134. (6.8) Decompose the expression into partial fractions: $\frac{-3x + 1}{x^3 + x^2 + 3x + 3}$

8.4 Mathematical Induction

LEARNING OBJECTIVES

In Section 8.5 you will learn how to:

A. Use subscript notation to evaluate and compose functions

B. Apply the principle of mathematical induction to sum formulas involving natural numbers

C. Apply the principle of mathematical induction to general statements involving natural numbers

INTRODUCTION

Since middle school (or even before) we have accepted that, "The product of two negative numbers is a positive number." But have you ever been asked to *prove* it? It's not as easy as it seems. We may think of several patterns that yield the result, analogies that indicate its truth, or even number line illustrations that lead us to believe the statement. But most of us have never seen a *proof* (see www.mhhe.com/coburn). In this section we introduce one of mathematics' most powerful tools for proving a statement, called **proof by induction.**

POINT OF INTEREST

The word *induction* stems from the Latin word "in," whose meaning is identical to its current use, and a form of the Latin word *ducere,* which means "to lead." Reasoning by induction uses specific examples "to lead you in" to a more general pattern. The word may also imply using this pattern to again "lead you" to a conclusive proof.

A. Subscript Notation and Composition of Functions

One of the challenges in understanding a proof by induction is working with the notation. Earlier in the chapter we introduced subscript notation as an alternative to function notation, since it is more commonly used when the functions are defined by a sequence. But regardless of the notation used, the functions can still be simplified, evaluated, composed, and even graphed. Consider the function $f(x) = 3x^2 - 1$ and the sequence defined by $a_n = 3n^2 - 1$. Both can be evaluated and graphed, with the only difference being that $f(x)$ is continuous with domain $x \in \mathbb{R}$, while a_n is discrete (made up of distinct points) with domain $n \in \mathbb{N}$.

Many applications require the composition of functions, and the notation for sequences and series can also be used for this purpose.

EXAMPLE 1 For $f(x) = 3x^2 - 1$ and $a_n = 3n^2 - 1$, find $f(k + 1)$ and a_{k+1}.

Solution:

$$\begin{aligned} f(k+1) &= 3(k+1)^2 - 1 \\ &= 3(k^2 + 2k + 1) - 1 \\ &= 3k^2 + 6k + 2 \end{aligned} \qquad \begin{aligned} a_{k+1} &= 3(k+1)^2 - 1 \\ &= 3(k^2 + 2k + 1) - 1 \\ &= 3k^2 + 6k + 2 \end{aligned}$$

NOW TRY EXERCISES 7 THROUGH 18

No matter which notation is used, every occurrence of the input variable is replaced by the new value or expression indicated by the composition.

B. Mathematical Induction Applied to Sums

Consider the sum of odd numbers $1 + 3 + 5 + 7 + 9 + 11 + 13 + \cdots$. The sum of the first four terms is $1 + 3 + 5 + 7 = 16$, or $S_4 = 16$. If we now add a_5 (the next term in line), would we get the same answer as if we had simply computed S_5? Common sense would say, "Yes!" since $S_5 = 1 + 3 + 5 + 7 + 9 = 25$ and $S_4 + a_5 = 16 + 9 = 25$✓. In diagram form, we have

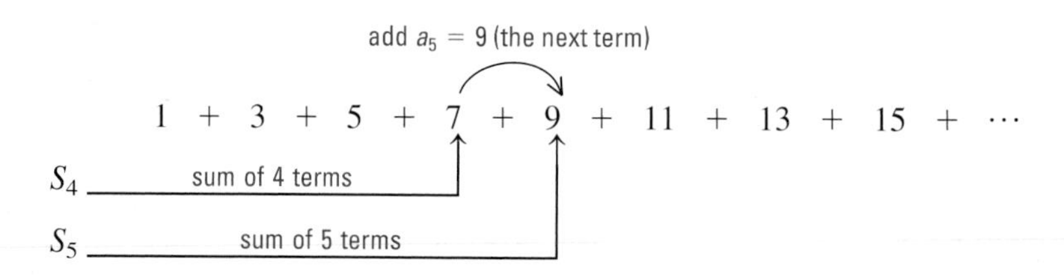

Our goal is to develop this same degree of clarity in the *notational scheme* of things. For a given series, if we find the kth partial sum S_k (shown next) and then add the next term a_{k+1}, would we get the same answer if we had simply computed S_{k+1}? In other words, is $S_k + a_{k+1} = S_{k+1}$ true?

add next term a_{k+1}

$$a_1 + a_2 + a_3 + \cdots + a_{k-1} + a_k + a_{k+1} + \cdots + a_{n-1} + a_n$$

S_k — sum of k terms

S_{k+1} — sum of $k + 1$ terms

Now, let's return to the sum $1 + 3 + 5 + 7 + \cdots + 2n - 1$. This is an arithmetic series with $a_1 = 1$, $d = 2$, and nth term $a_n = 2n - 1$. Using the sum formula for an arithmetic sequence, an alternative formula for *this sum* can be established.

$$S_n = \frac{n(a_1 + a_n)}{2} \quad \text{summation formula for an arithmetic sequence}$$

$$= \frac{n(1 + 2n - 1)}{2} \quad \text{substitute 1 for } a_1 \text{ and } 2n-1 \text{ for } a_n$$

$$= \frac{n(2n)}{2} \quad \text{simplify}$$

$$= n^2 \quad \text{result}$$

This shows that the sum of the first n positive odd integers is given by $S_n = n^2$. As a check we compute $S_5 = 1 + 3 + 5 + 7 + 9 = 25$ and compare to $S_5 = 5^2 = 25$✓. We also note $S_6 = 6^2 = 36$, and $S_5 + a_6 = 25 + 11 = 36$, showing $S_6 = S_5 + a_6$. For more on this relationship, see Exercises 19 through 24.

While it may seem simplistic now, showing $S_5 + a_6 = S_6$ and $S_k + a_{k+1} = S_{k+1}$ (in general) is a critical component of a proof by induction. Unfortunately, general summation formulas for many sequences cannot be established from known formulas. In addition, just because a formula works for the first few values of n, we cannot assume that it will hold true for *all* values of n (there are infinitely many). As an illustration, the formula $a_n = n^2 - n + 41$ yields a prime number for *every natural number n from 1 to 40,* but fails to yield a prime for $n = 41$! This helps demonstrate the need for a more conclusive proof, particularly when a relationship appears to be true, and can be "verified" in a finite number of cases, but whether it is true in *all* cases remains in question.

WORTHY OF NOTE

No matter how distant the city or how many relay stations are involved, if the generating plant is working and the kth station relays to the $(k + 1)$st station, the city will get its power.

Proof by induction is based on a relatively simple idea. To help understand how it works, consider n relay stations that are used to transport electricity from a generating plant to a distant city. If we know the generating plant is operating, and if we assume that the kth relay station (any station in the series) is making the transfer to the $(k + 1)$st station (the next station in the series), then we're sure the city will have electricity.

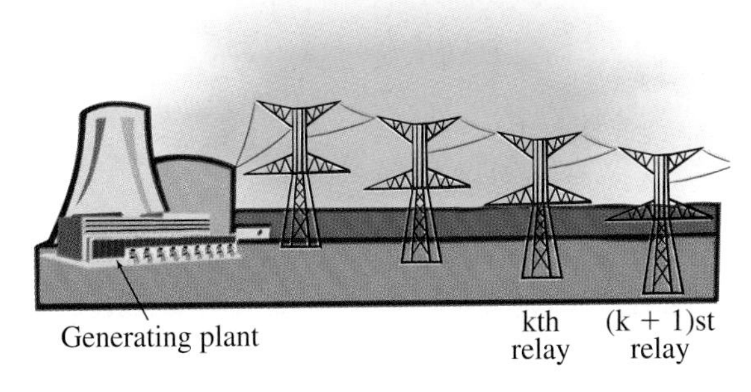

This idea can be applied mathematically as follows. Consider the statement, "The sum of the first n positive, even integers is $n^2 + n$." In other words, $2 + 4 + 6 + 8 + \cdots + 2n = n^2 + n$. We can certainly verify the statement for the first few even numbers:

The first even number is 2 and . . . $(1)^2 + 1 = 2$

The sum of the first *two* even numbers is $2 + 4 = 6$ and . . . $(2)^2 + 2 = 6$

The sum of the first *three* even numbers is $2 + 4 + 6 = 12$ and . . . $(3)^2 + 3 = 12$

The sum of the first *four* even numbers is $2 + 4 + 6 + 8 = 20$ and . . . $(4)^2 + 4 = 20$

While we could continue this process for a very long time (or even use a computer), *no finite number of checks can prove a statement is universally true.* To prove the statement true for *all* positive integers, we use a reasoning similar to that applied in the relay stations example. If we are sure the formula works for $n = 1$ (the generating station is operating), and assume that if the formula is true for $n = k$, it must also be true for $n = k + 1$ [the kth relay station is transferring electricity to the $(k + 1)$st station], then the statement is true for all n (the city will get its electricity). The case where $n = 1$ is called the **base case** of an inductive proof, and the assumption that the formula is true for $n = k$ is called the **induction hypothesis.** When the induction hypothesis is applied to a sum formula, we attempt to show that $S_k + a_{k+1} = S_{k+1}$. Since k and $k + 1$ are arbitrary, the statement must be true for all n.

MATHEMATICAL INDUCTION APPLIED TO SUMS

Let S_n be a sum formula involving positive integers.

If 1. S_1 is true, and

2. the truth of S_k implies that S_{k+1} is true,

then S_n must be true for all positive integers n.

WORTHY OF NOTE

To satisfy our finite minds, it might help to show that S_n is true for the first few cases, prior to extending the ideas to the infinite case.

Both parts 1 and 2 must be verified for the proof to be complete. Since the process requires the terms S_k, a_{k+1}, and S_{k+1}, we will usually compute these first.

EXAMPLE 2 Use induction to prove that *the sum of the first n perfect squares is:*

$$1 + 4 + 9 + 16 + 25 + \cdots + n^2 = \frac{n(n+1)(2n+1)}{6}.$$

Solution: Given $a_n = n^2$ and $S_n = \dfrac{n(n+1)(2n+1)}{6}$, the needed components are . . .

For $a_n = n^2$: $a_k = k^2$ and $a_{k+1} = (k+1)^2$

For $S_n = \dfrac{n(n+1)(2n+1)}{6}$: $S_k = \dfrac{k(k+1)(2k+1)}{6}$ and $S_{k+1} = \dfrac{(k+1)(k+2)(2k+3)}{6}$

1. Show S_n is true for $n = 1$.

$$S_n = \frac{n(n+1)(2n+1)}{6} \quad \text{sum formula}$$

$$S_1 = \frac{1(2)(3)}{6} \quad \textit{base case: } n = 1$$

$$= 1\checkmark \quad \text{result checks, the first term is 1}$$

2. Assume S_k is true,

$$1 + 4 + 9 + 16 + \cdots + k^2 = \frac{k(k+1)(2k+1)}{6} \quad \text{induction hypothesis: } S_k \text{ is true}$$

and use it to show the truth of S_{k+1} follows. That is,

$$\underbrace{1 + 4 + 9 + 16 + \cdots + k^2}_{S_k} + \underbrace{(k+1)^2}_{a_{k+1}} = \underbrace{\frac{(k+1)(k+2)(2k+3)}{6}}_{S_{k+1}}$$

Working with the left-hand side, we have

$$\underbrace{1 + 4 + 9 + 16 + \cdots + k^2}+ (k + 1)^2$$

$$= \frac{k(k+1)(2k+1)}{6} + (k+1)^2 \quad \text{use the induction hypothesis: } 1 + 4 + 9 + 16 + 25 + \cdots + k^2 = \frac{k(k+1)(2k+1)}{6}$$

$$= \frac{k(k+1)(2k+1) + 6(k+1)^2}{6} \quad \text{common denominator}$$

$$= \frac{(k+1)[k(2k+1) + 6(k+1)]}{6} \quad \text{factor out } k + 1$$

$$= \frac{(k+1)[2k^2 + 7k + 6]}{6} \quad \text{multiply and combine terms}$$

$$= \frac{(k+1)(k+2)(2k+3)}{6} \quad \text{factor the trinomial, result is } S_{k+1} ✓$$

Since the truth of S_{k+1} follows from S_k, the formula is true for all n.

NOW TRY EXERCISES 27 THROUGH 38

LOOKING AHEAD 

With all this emphasis on proof by induction, it may help to consider the fundamental role it plays in connecting the finite to the infinite. In Section 3.3, (Example 9), we observed the area under a curve actually has an intriguing relationship to distance, rate, and time. Having a formula that gives the "sum of an infinite number of terms," enables us to compute the area under more complicated graphs using an infinite number of rectangles, trapezoids, or other shapes with simple formulas for area. See the *Reinforcing Basic Concepts* feature on page 790.

C. The General Principle of Mathematical Induction

Proof by induction can be used to verify many other kinds of relationships involving a natural number n. In this regard, the basic principles remain the same but are stated more broadly. Rather than having S_n represent a sum, we take it to represent *any statement or relationship* we might wish to verify. This broadens the scope of the proof and makes it more widely applicable, while maintaining its value to the sum formulas verified earlier.

THE GENERAL PRINCIPLE OF MATHEMATICAL INDUCTION

Let S_n be a statement involving positive integers.

If 1. S_1 is true, and

2. the truth of S_k implies that S_{k+1} is also true

then S_n must be true for all positive integers n.

EXAMPLE 3 Use the general principle of mathematical induction to show the statement S_n is true for all positive integers n. S_n: $2^n \geq n + 1$

Solution: The statement S_n is defined as $2^n \geq n + 1$. This means that S_k is represented by $2^k \geq k + 1$ and S_{k+1} by $2^{k+1} \geq k + 2$.

1. Show S_n is true for $n = 1$:

S_n: $2^n \geq n + 1$ given statement

S_1: $2^1 \geq 1 + 1$ base case: $n = 1$

$2 \geq 2$ ✓ true

Although not a part of the formal proof, a table of values can help to illustrate the relationship we're trying to establish. It *appears* that the statement is true.

n	1	2	3	4	5
2^n	2	4	8	16	32
$n + 1$	2	3	4	5	6

2. Assume that S_k is true,

S_k: $2^k \geq k + 1$ induction hypothesis

and use it to show that the truth of S_{k+1}. That is,

S_{k+1}: $2^{k+1} \geq k + 2$.

Begin by working with the left-hand side of the inequality, 2^{k+1}.

$$2^{k+1} = 2(2^k) \quad \text{properties of exponents}$$
$$\geq 2(\mathbf{k + 1}) \quad \textit{induction hypothesis:} \text{ substitute } k + 1 \text{ for } 2^k \text{ (symbol changes since } k + 1 \text{ is less than } 2^k)$$
$$\geq 2k + 2 \quad \text{distribute}$$

Since k is a positive integer, $2k + 2 \geq k + 2$,

showing $2^{k+1} \geq k + 2$.

Since the truth of S_{k+1} follows from S_k, the formula is true for all n.

WORTHY OF NOTE

Note there is no reference to a_n, a_k, or a_{k+1} in the statement of the general principle of mathematical induction.

NOW TRY EXERCISES 39 THROUGH 42

EXAMPLE 4 Let S_n be the statement, "*$4^n - 1$ is divisible by 3 for all positive integers n.*" Use mathematical induction to prove that S_n is true.

Solution: If a number is evenly divisible by three, it can be written as the product of 3 and some positive integer we will call p.

1. Show S_n is true for $n = 1$:

$$S_n: \quad 4^n - 1 = 3p \qquad 4^n - 1 = 3p,\ p \in \mathbb{Z}$$
$$S_1: \quad 4^{(1)} - 1 = 3p \qquad \text{substitute 1 for } n$$
$$3 = 3p\checkmark \qquad \text{statement is true for } n = 1$$

2. Assume that S_k is true . . .

S_k: $4^k - 1 = 3p$ induction hypothesis

and use it to show the truth of S_{k+1}. That is,

S_{k+1}: $4^{k+1} - 1 = 3q$ for $q \in \mathbb{Z}$ is also true.

Beginning with the left-hand side we have:

$$4^{k+1} - 1 = 4 \cdot 4^k - 1 \qquad \text{properties of exponents}$$
$$= 4 \cdot (\mathbf{3p + 1}) - 1 \qquad \text{induction hypothesis}$$
$$= 12p + 3 \qquad \text{distribute and simplify}$$
$$= 3(4p + 1) = 3q \qquad \text{factor}$$

The last step shows $4^{k+1} - 1$ is divisible by 3. Since the original statement is true for $n = 1$, and the truth of S_k implies the truth of S_{k+1}, the statement, "*$4^n - 1$ is divisible by 3*" is true for all positive integers n.

NOW TRY EXERCISES 43 THROUGH 47

We close this section with some final notes. Although the base step of a proof by induction seems trivial, both the base step and the induction hypothesis are necessary parts of the proof. For example, the statement $\frac{1}{3^n} < \frac{1}{3n}$ is false for $n = 1$, but true for all other positive integers. Finally, for a fixed natural number p, some statements are false for all $n < p$, but true for all $n \geq p$. By modifying the base case to begin at p, we can use the induction hypothesis to prove the statement is true for all n greater than p. For example, $3^n > n^3$ is false for $n < 4$, but true for all $n \geq 4$. See Exercise 49.

8.4 EXERCISES

CONCEPTS AND VOCABULARY

Fill in the blank with the appropriate word or phrase. Carefully reread the section if needed.

1. No __finite__ number of verifications can prove a statement __universally__ true.

2. Showing a statement is true for $n = 1$ is called the __base__ __case__ of an inductive proof.

3. Assuming that a statement/formula is true for $n = k$ is called the __induction hypothesis__.

4. The graph of a sequence is __discrete__, meaning it is made up of distinct points.

5. Explain the equation $S_k + a_{k+1} = S_{k+1}$. Begin by saying, "Since the kth term is arbitrary . . ." (continue from here).
Answers will vary.

6. Discuss the similarities and differences between mathematical induction applied to sums and the general principle of mathematical induction.
Answers will vary.

DEVELOPING YOUR SKILLS

For the given nth term a_n, find a_4, a_5, a_k, and a_{k+1}.

7. $a_n = 10n - 6$ **8.** $a_n = 6n - 4$ **9.** $a_n = n$

10. $a_n = 7n$ **11.** $a_n = 2^{n-1}$ **12.** $a_n = 2(3^{n-1})$

For the given sum formula S_n, find S_4, S_5, S_k, and S_{k+1}.

13. $S_n = n(5n - 1)$ **14.** $S_n = n(3n - 1)$ **15.** $S_n = \frac{n(n+1)}{2}$

16. $S_n = \frac{7n(n+1)}{2}$ **17.** $S_n = 2^n - 1$ **18.** $S_n = 3^n - 1$

Verify that $S_4 + a_5 = S_5$ for each exercise. These are identical to Exercises 13 through 18.

19. $a_n = 10n - 6$; $S_n = n(5n - 1)$ verified

20. $a_n = 6n - 4$; $S_n = n(3n - 1)$ verified

21. $a_n = n$; $S_n = \frac{n(n+1)}{2}$ verified

22. $a_n = 7n$; $S_n = \frac{7n(n+1)}{2}$ verified

23. $a_n = 2^{n-1}$; $S_n = 2^n - 1$ verified

24. $a_n = 2(3^{n-1})$; $S_n = 3^n - 1$ verified

7. $a_4 = 34$, $a_5 = 44$, $a_k = 10k - 6$, $a_{k+1} = 10k + 4$
8. $a_4 = 20$, $a_5 = 26$, $a_k = 6k - 4$, $a_{k+1} = 6k + 2$
9. $a_4 = 4$, $a_5 = 5$, $a_k = k$, $a_{k+1} = k + 1$
10. $a_4 = 28$, $a_5 = 35$, $a_k = 7k$, $a_{k+1} = 7k + 7$
11. $a_4 = 8$, $a_5 = 16$, $a_k = 2^{k-1}$, $a_{k+1} = 2^k$
12. $a_4 = 54$, $a_5 = 162$, $a_k = 2(3^{k-1})$, $a_{k+1} = 2(3^k)$
13. $S_4 = 76$, $S_5 = 120$, $S_k = k(5k - 1)$, $S_{k+1} = (k + 1)(5k + 4)$
14. $S_4 = 44$, $S_5 = 70$, $S_k = k(3k - 1)$, $S_{k+1} = (k + 1)(3k + 2)$
15. $S_4 = 10$, $S_5 = 15$, $S_k = \frac{k(k+1)}{2}$, $S_{k+1} = \frac{(k+1)(k+2)}{2}$
16. $S_4 = 70$, $S_5 = 105$, $S_k = \frac{7k(k+1)}{2}$, $S_{k+1} = \frac{7(k+1)(k+2)}{2}$
17. $S_4 = 15$, $S_5 = 31$, $S_k = 2^k - 1$, $S_{k+1} = 2^{k+1} - 1$
18. $S_4 = 80$, $S_5 = 242$, $S_k = 3^k - 1$, $S_{k+1} = 3^{k+1} - 1$

25. verified*
26. verified
27. verified
28. verified
29. verified
30. verified
31. verified*
32. verified
33. verified
34. verified
35. verified*
36. verified
37. verified*
38. verified
39. verified*
40. verified
41. verified
42. verified
43. verified*
44. verified
45. verified
46. verified
47. verified*

*complete proof shown in solution appendix

WORKING WITH FORMULAS

25. Sum of the first n cubes (alternative form): $(1 + 2 + 3 + 4 + \cdots + n)^2$

Earlier we noted the formula for the sum of the first n cubes was $\frac{n^2(n+1)^2}{4}$. An alternative is given by the formula shown. Use a proof by induction to verify the formula. That is, prove that for $a_n = n^3$, $1 + 8 + 27 + 64 + \cdots + n^3 = (1 + 2 + 3 + 4 + \cdots + n)^2$.

26. Powers of the imaginary unit: $i^{n+4} = i^n$, where $i = \sqrt{-1}$

Use a proof by induction to prove that powers of the imaginary unit are cyclic. That is, that they cycle through the numbers i, -1, $-i$, and 1 for consecutive powers.

APPLICATIONS

Use mathematical induction to prove the indicated sum formula is true for all natural numbers n.

27. $2 + 4 + 6 + 8 + 10 + \cdots + 2n$; $a_n = 2n$, $S_n = n(n + 1)$

28. $3 + 7 + 11 + 15 + 19 + \cdots + (4n - 1)$; $a_n = 4n - 1$, $S_n = n(2n + 1)$

29. $5 + 10 + 15 + 20 + 25 + \cdots + 5n$; $a_n = 5n$, $S_n = \frac{5n(n+1)}{2}$

30. $1 + 4 + 7 + 10 + 13 + \cdots + (3n - 2)$; $a_n = 3n - 2$, $S_n = \frac{n(3n-1)}{2}$

31. $5 + 9 + 13 + 17 + \cdots + (4n + 1)$; $a_n = 4n + 1$, $S_n = n(2n + 3)$

32. $4 + 12 + 20 + 28 + 36 + \cdots + (8n - 4)$; $a_n = 8n - 4$, $S_n = 4n^2$

33. $3 + 9 + 27 + 81 + 243 + \cdots + 3^n$; $a_n = 3^n$, $S_n = \frac{3(3^n - 1)}{2}$

34. $5 + 25 + 125 + 625 + \cdots + 5^n$; $a_n = 5^n$, $S_n = \frac{5(5^n - 1)}{4}$

35. $2 + 4 + 8 + 16 + 32 + 64 + \cdots + 2^n$; $a_n = 2^n$, $S_n = 2^{n+1} - 2$

36. $1 + 8 + 27 + 64 + 125 + 216 + \cdots + n^3$; $a_n = n^3$, $S_n = \frac{n^2(n+1)^2}{4}$

37. $\frac{1}{1(3)} + \frac{1}{3(5)} + \frac{1}{5(7)} + \cdots + \frac{1}{(2n-1)(2n+1)}$; $a_n = \frac{1}{(2n-1)(2n+1)}$, $S_n = \frac{n}{2n+1}$

38. $\frac{1}{1(2)} + \frac{1}{2(3)} + \frac{1}{3(4)} + \cdots + \frac{1}{n(n+1)}$; $a_n = \frac{1}{n(n+1)}$, $S_n = \frac{n}{n+1}$

Use the principle of mathematical induction to prove that each statement is true for all natural numbers n.

39. $3^n \geq 2n + 1$

40. $2^n \geq n + 1$

41. $3 \cdot 4^{n-1} \leq 4^n - 1$

42. $4 \cdot 5^{n-1} \leq 5^n - 1$

43. $n^2 - 7n$ is divisible by 2

44. $n^3 - n + 3$ is divisible by 3

45. $n^3 + 3n^2 + 2n$ is divisible by 3

46. $5^n - 1$ is divisible by 4

47. $6^n - 1$ is divisible by 5

WRITING, RESEARCH, AND DECISION MAKING

48. Answers will vary.
49. The relation cannot be verified for all n; $n = 8$.

48. When using proof by induction, it's important to remember that both parts of the proof are necessary. For example, it is possible that the first part is true, while the second part is false: $\frac{1}{n} \geq \frac{2}{n+1}$ is true for $n = 1$, but false for all other natural numbers n. It is also possible that the second part is true, while the first part is false: $\frac{1}{\sqrt{1}} + \frac{1}{\sqrt{2}} + \frac{1}{\sqrt{3}} + \frac{1}{\sqrt{4}} + \cdots + \frac{1}{\sqrt{n}} > \sqrt{n}$ is false for $n = 1$, but true for all other natural numbers n. Do some research or experimentation to find other relationships where Part 1 of the inductive proof is true while Part 2 is false and vice versa.

49. At first glance it may appear the inequality $3n^2 - 1 > 4(n - 1)^2$ is true. Try to "prove" the statement using an inductive proof. What happens? What is the first natural number for which the statement is false?

Additional answers can be found in the Instructor Answer Appendix.

50. correlation coefficient $r = 1$
52. verified
53. $A + B = \begin{bmatrix} 1 & 1 \\ 7 & 4 \end{bmatrix}$, $A - B = \begin{bmatrix} -3 & 3 \\ -1 & -2 \end{bmatrix}$, $2A - 3B = \begin{bmatrix} -8 & 7 \\ -6 & -7 \end{bmatrix}$, $AB = \begin{bmatrix} 6 & 7 \\ 10 & 0 \end{bmatrix}$, $BA = \begin{bmatrix} -5 & 3 \\ 5 & 11 \end{bmatrix}$, $B^{-1} = \begin{bmatrix} 0.3 & 0.1 \\ -0.4 & 0.2 \end{bmatrix}$
54. $\left(\frac{-4 + \sqrt{26}}{2}, 0\right)$ $\left(\frac{-4 - \sqrt{26}}{2}, 0\right)$
55. $D: x \in (-\infty, \infty)$ $R: y \in [-2, \infty)$
56. $(x - 4)^2 + (y - 3)^2 = 25$
57. $x = \frac{1 + \ln 4}{2}$
58. $x = 2, x = -2, (0, \frac{1}{4}), (1, 0)$

Additional answers can be found in the Instructor Answer Appendix.

50. You may have noticed that the sum formula for the first n integers was *quadratic,* and the formula for the first n integer squares was *cubic.* Is the formula for the first n integer cubes, if it exists, a quartic (degree four) function? Use your calculator to run a quartic regression on the first five perfect cubes (enter 1 through 5 in L_1 and the cumulative sums in L_2). What did you find? Use proof by induction to show that the sum of the first n cubes is:

$$1 + 8 + 27 + \cdots + n^3 = \frac{n^4 + 2n^3 + n^2}{4} = \frac{n^2(n + 1)^2}{4}.$$

EXTENDING THE CONCEPT

51. Use mathematical induction to prove that $\frac{x^n - 1}{x - 1} = (1 + x + x^2 + x^3 + \cdots + x^{n-1})$.

52. Use mathematical induction to prove that for $1^4 + 2^4 + 3^4 + \cdots + n^4$, where $a_n = n^4$,

$$S_n = \frac{n(n + 1)(2n + 1)(3n^2 + 3n - 1)}{30}.$$

MAINTAINING YOUR SKILLS

53. (6.6) Given the matrices $A = \begin{bmatrix} -1 & 2 \\ 3 & 1 \end{bmatrix}$ and $B = \begin{bmatrix} 2 & -1 \\ 4 & 3 \end{bmatrix}$, find $A + B$, $A - B$, $2A - 3B$, AB, BA, and B^{-1}.

54. (3.4) The parabola defined by $g(x) = 2x^2 + 8x - 5$ has a vertex at $(-2, -13)$. Find the x-intercepts using the vertex/root formula.

55. (3.8) State the domain and range of the piecewise function shown here.

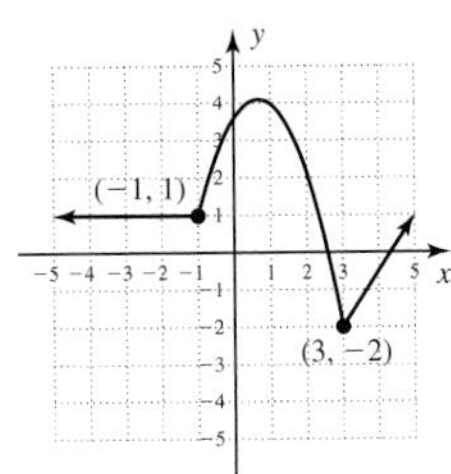

56. (7.1) State the equation of the circle whose graph is shown here.

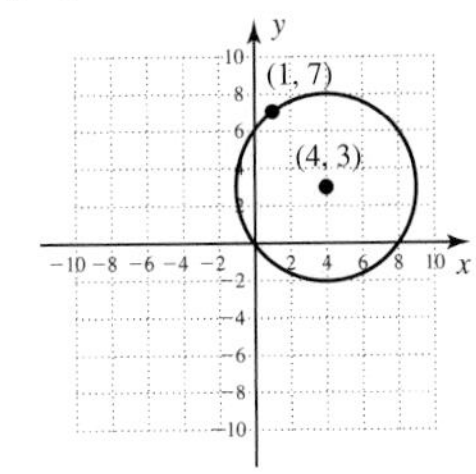

57. (5.4) Solve: $3e^{(2x-1)} + 5 = 17$. Answer in exact form.

58. (4.6) Give the equations of all asymptotes and coordinates of all intercepts for $h(x) = \frac{x^3 - 1}{x^2 - 4}$.

MID-CHAPTER CHECK

1. 3, 10, 17, $a_9 = 59$
2. 4, 7, 12, $a_9 = 84$
3. −1, 3, −5, $a_9 = -17$
4. 360
5. $\sum_{k=1}^{6}(3k - 2)$

In Exercises 1 to 3, the nth term is given. Write the first three terms of each sequence and find a_9.

1. $a_n = 7n - 4$ **2.** $a_n = n^2 + 3$ **3.** $a_n = (-1)^n(2n - 1)$

4. Evaluate the sum $\sum_{n=1}^{4} 3^{n+1}$

5. Rewrite using sigma notation.

$1 + 4 + 7 + 10 + 13 + 16$

Match each formula to its correct description.

6. $S_n = \frac{n(a_1 + a_n)}{2}$ d

7. $a_n = a_1r^{n-1}$ e

a. sum of an infinite geometric series

b. nth term formula for an arithmetic series

8. $S_\infty = \dfrac{a_1}{1 - r}$ a

9. $a_n = a_1 + (n - 1)d$ b

10. $S_n = \dfrac{a_1(1 - r^n)}{1 - r}$ c

c. sum of a finite geometric series

d. summation formula for an arithmetic series

e. nth term formula for a geometric series

11. a. $a_1 = 2$, $d = 3$, $a_n = 3n - 1$
b. $a_1 = \frac{3}{2}$, $d = \frac{3}{4}$, $a_n = \frac{3}{4}n + \frac{3}{4}$

11. Identify a_1 and the common difference d. Then find an expression for the general term a_n.

a. 2, 5, 8, 11, . . . **b.** $\frac{3}{2}, \frac{9}{4}, 3, \frac{15}{4}, \ldots$

12. $n = 25$, $S_{25} = 950$
13. $n = 16$, $S_{16} = 128$
14. $S_{10} = -5$
15. $S_{10} = \dfrac{-29{,}524}{27}$
16. a. $a_1 = 2$, $r = 3$, $a_n = 2(3)^{n-1}$
b. $a_1 = \frac{1}{2}$, $r = \frac{1}{2}$, $a_n = (\frac{1}{2})^n$
17. $n = 8$, $S_8 = \frac{1640}{27}$
18. $\frac{-343}{6}$
19. 1785
20. ≈ 4.5 ft; ≈ 127.9 ft

Find the number of terms in each series and then find the sum.

12. $2 + 5 + 8 + 11 + \cdots + 74$

13. $\frac{1}{2} + \frac{3}{2} + \frac{5}{2} + \frac{7}{2} + \cdots + \frac{31}{2}$

14. For an arithmetic series, $a_3 = -8$ and $a_7 = 4$. Find S_{10}.

15. For a geometric series, $a_3 = -81$ and $a_7 = -1$. Find S_{10}.

16. Identify a_1 and the common ratio r. Then find an expression for the general term a_n.

a. 2, 6, 18, 54, . . . **b.** $\frac{1}{2}, \frac{1}{4}, \frac{1}{8}, \frac{1}{16}, \ldots$

17. Find the number of terms in the series then compute the sum.
$\frac{1}{54} + \frac{1}{18} + \frac{1}{6} + \cdots + \frac{81}{2}$

18. Find the infinite sum (if it exists).
$-49 + (-7) + (-1) + (-\frac{1}{7}) + \cdots$

19. Barrels of toxic waste are stacked at a storage facility in pyramid form, with 60 barrels in the first row, 59 in the second row, and so on, until there are 10 barrels in the top row. How many barrels are in the storage facility?

20. As part of a conditioning regimen, a drill sergeant orders her platoon to do 25 continuous standing broad jumps. The best of these recruits was able to jump 96% of the distance from the previous jump, with a first jump distance of 8 ft. Use a sequence/series to determine the distance the recruit jumped on the 15th try, and the total distance traveled by the recruit after all 25 jumps.

Reinforcing Basic Concepts

Applications of Summation

Summation properties and formulas have far reaching and powerful applications in a continuing study of mathematics. Here we'll take a closer look at selected properties and a few of the ways they can be applied. The properties of summation play a huge role in the development of key ideas in a first semester calculus course, and the following summation formulas are an integral part of these ideas. The first three formulas were verified in Section 8.4, while proof of the fourth was part of Exercise 50 on page 789.

(1) $\displaystyle\sum_{i=1}^{n} c = cn$ (2) $\displaystyle\sum_{i=1}^{n} i = \frac{n(n+1)}{2}$

(3) $\displaystyle\sum_{i=1}^{n} i^2 = \frac{n(n+1)(2n+1)}{6}$ (4) $\displaystyle\sum_{i=1}^{n} i^3 = \frac{n^2(n+1)^2}{4}$

To see the various ways they can be applied consider the following.

EXAMPLE 1 Over several years, the owner of Morgan's LawnCare has noticed that the company's monthly profits (in thousands) can be approximated by the sequence $a_n = 0.0625n^3 - 1.25n^2 + 6n$, with the points plotted in Figure 8.13 (the continuous graph is shown for effect only). Find the company's approximate annual profit.

Figure 8.13

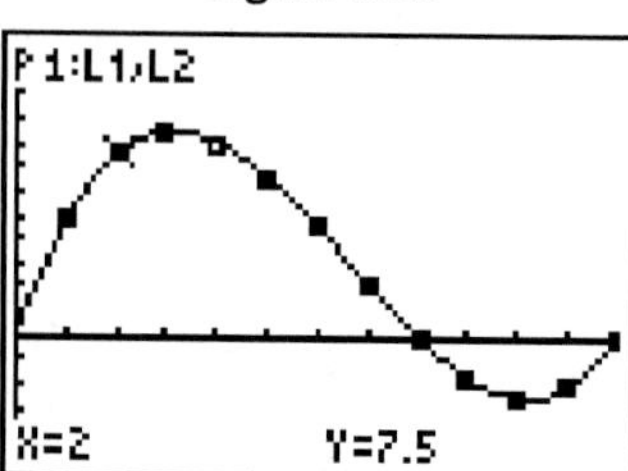

Solution:

a. The most obvious approach would be to simply compute terms a_1 through a_{12} (January through December) and find their sum. Using the "**sum(**" and "**seq(**" features of a graphing calculator (covered in Section 8.1), we input: **sum(seq(Y1, X, 1, 12),** as seen in Figure 8.14. After pressing **ENTER** we find that annual profits are \$35,750.

Figure 8.14

```
sum(seq(Y1,X,1,1
2)
           35.75
```

Figure 8.15

```
sum(seq(Y1,X,1,8
)
              42
sum(seq(Y1,X,9,1
2)
           -6.25
```

b. As an alternative, we could add the amount of profit earned by the company in the first eight months, then add the amount the company lost (or broke even) during the last four months. In other words, we could apply Summation Property IV (page 753): $\sum_{i=1}^{12} a_n = \sum_{i=1}^{8} a_n + \sum_{i=9}^{12} a_n$. Using the "**sum(**" and "**seq(**" features produces the values shown in Figure 8.15, and gives the same result as Part (A): $\sum_{i=1}^{8} a_n + \sum_{i=9}^{12} a_n = 42 + (-6.25) = 35.75$ or \$35,750.

c. As a third alternative, we could use summation properties along with the appropriate summation formulas, and compute the result manually. Note the function is now written in terms of "i."

Distribute summations and factor out constants (Properties II and III):

$$\sum_{i=1}^{12}(0.0625i^3 - 1.25i^2 + 6i) = 0.0625\sum_{i=1}^{12} i^3 - 1.25\sum_{i=1}^{12} i^2 + 6\sum_{i=1}^{12} i$$

Replace each summation with the appropriate summation formula, substituting 12 for n:

$$= 0.0625\left[\frac{n^2(n+1)^2}{4}\right] - 1.25\left[\frac{n(n+1)(2n+1)}{6}\right] + 6\left[\frac{n(n+1)}{2}\right]$$

$$= 0.0625\left[\frac{(12)^2(13)^2}{4}\right] - 1.25\left[\frac{(12)(13)(25)}{6}\right] + 6\left[\frac{(12)(13)}{2}\right]$$

$$= 0.0625(6084) - 1.25(650) + 6(78) \text{ or } 35.75$$

As we expected, the result shows profit was \$35,750. As you consider and apply these ideas, bear in mind that while some approaches seem "easier" than others, all have great value, are applied in different ways at different times, and are necessary to adequately develop key concepts in future classes.

Exercises:

1. Repeat (a), (b), and (c) from Example 1 if the profit sequence is $a_n = 0.125x^3 - 2.5x^2 + 12x$.

   \$71,500

2. Change the index so the sum begins at $j = 1$, then compute.

   a. $\sum_{j=10}^{20}[2j - 1]$ **b.** $\sum_{j=7}^{13} j^2$

   a. 319 b. 728

8.5 Counting Techniques

LEARNING OBJECTIVES

In Section 8.5 you will learn how to:

A. Count possibilities using lists and tree diagrams

B. Count possibilities using the fundamental principle of counting

C. Quick-count distinguishable permutations

D. Quick-count nondistinguishable permutations

E. Quick-count using combinations

INTRODUCTION

How long would it take to estimate the number of fans sitting shoulder-to-shoulder at a sold-out basketball game? Well, it depends. You could actually begin counting 1, 2, 3, 4, 5, . . . , which would take a very long time, or you could try to simplify the process by counting the number of fans in the first row and multiplying by the number of rows. Techniques for the "quick-counting" the objects in a set or various subsets of a large set play an important role in a study of probability.

POINT OF INTEREST

In the game of "Clue"® (Parker Brothers), a crime is committed in one of nine rooms, with one of six implements, by one of six people. Using the ideas discussed in this section, we can systematically count the number of possible ways to combine a suspect, implement, and room, and hence the number of ways the crime might be committed. Without the ability to count in an organized fashion, counting these possibilities could take a very long time.

A. Counting by Listing and Tree Diagrams

The most elementary way to count the possibilities of a desired **outcome** is to create an **ordered list.** Consider the simple spinner shown in Figure 8.16, which is divided into three equal parts. What are the different possible outcomes for two spins, spin 1 followed by spin 2? We might begin by listing A as the first possible outcome, for no reason other than it's the first letter of the alphabet. For the second spin the possibilities are again A, B, or C, giving the list AA, AB, and AC. Similarly, if B is spun first the possibilities are BA, BB, and BC. Finally, a first spin of C gives CA, CB, and CC, showing there are nine possibilities. You might imagine, however, that if the spinner had the five letters A, B, C, D, and E or if we requested the possibilities for three spins, the listing method would quickly become tedious and ineffective. As an alternative, we can organize the possibilities using a **tree diagram.** As the name implies, each choice or possibility appears as the branch of a tree, with the total possibilities being equal to the number of (unique) paths from the beginning point to the end of a branch. Figure 8.17 shows how the spinner exercise would appear (possibilities for two spins). Moving from top to bottom we can trace the nine paths shown in the previous listing.

Figure 8.16

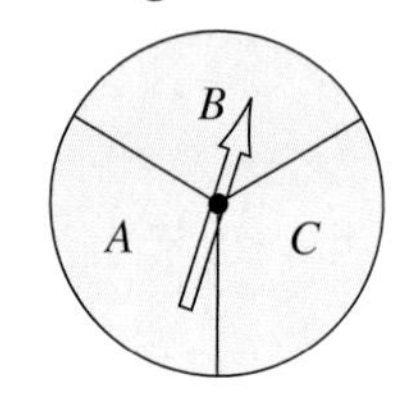

Figure 8.17

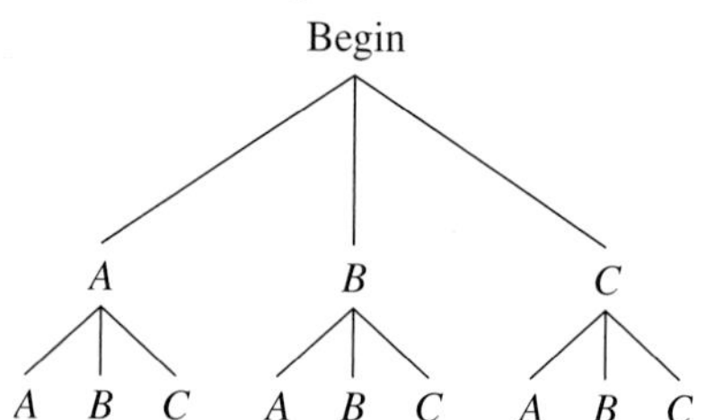

EXAMPLE 1 A basketball player is fouled and awarded three free throws. Let H represent the possibility of a hit (basket is made), and M the possibility of a miss. Determine the possible outcomes for the three shots using a tree diagram.

Solution: Each shot has two possibilities, hit (H) or miss (M), so the tree will branch in two directions at each level. As illustrated in the figure, there are a total of eight possibilities: HHH, HHM, HMH, HMM, MHH, MHM, MMH, and MMM.

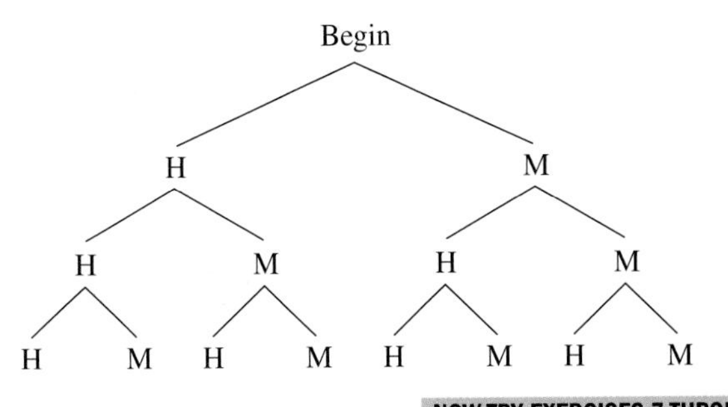

NOW TRY EXERCISES 7 THROUGH 10

WORTHY OF NOTE

Sample spaces may vary depending on how we define the experiment, and for simplicity's sake we consider only those experiments having outcomes that are equally likely.

To assist our discussion, an **experiment** is any task that can be done repeatedly and has a well-defined set of possible outcomes. Each repetition of the experiment is called a **trial.** A **sample outcome** is any potential outcome of a trial, and a **sample space** is a set of all possible outcomes.

In our first illustration, the *experiment* was spinning a spinner, there were *three sample outcomes* (A, B, or C), the experiment had *two trials* (spin 1 and spin 2), and there were *nine* elements in the *sample space.* Note that after the first trial, each of the three sample outcomes will again have three possibilities (A, B, and C). For two trials, we have two factors of three giving $3^2 = 9$ possibilities, while three trials yields a sample space with $3^3 = 27$ possibilities. In general, we have

> **A "QUICK-COUNTING" FORMULA FOR THE SAMPLE SPACE OF AN EXPERIMENT**
>
> If an experiment has N sample outcomes that are equally likely and the experiment is repeated t times, the number of elements in the sample space is given by the N^t.

EXAMPLE 2 Many combination locks have the digits 0 through 39 arranged along a circular dial. Opening the lock requires stopping at a sequence of three numbers within this range, going counterclockwise to the first number, clockwise to the second, and counterclockwise to the third. How many three number combinations are possible?

Solution: There are 40 sample outcomes ($N = 40$) in this experiment, and three trials ($t = 3$). The number of possible combinations is identical to the number of elements in the sample space. The quick-counting formula gives $40^3 = 64{,}000$ possible combinations.

NOW TRY EXERCISES 11 AND 12

B. Fundamental Principle of Counting

Some security systems, license plates, and telephone numbers exclude certain numbers. For example, phone numbers cannot begin with 0 or 1 because these are reserved for operator assistance, long distance, and international calls. Constructing a three-digit area code is like filling in three blanks $\underset{\text{digit}}{\underline{\quad}}\ \underset{\text{digit}}{\underline{\quad}}\ \underset{\text{digit}}{\underline{\quad}}$ with three digits. Since the area code must start with a number between 2 and 9, there are eight choices for the first blank. Since there are 10 choices for the second digit and 10 choices for the third, there are $8 \cdot 10 \cdot 10 = 800$ possibilities in the sample space.

EXAMPLE 3 A digital security system requires that you enter a four digit PIN (personal identification number), using only the digits 1 through 9. How many codes are possible if (a) repetition of digits is allowed and (b) repetition is not allowed?

Solution:

a. Consider filling in the four blanks $\underset{\text{digit}}{\underline{\quad}}\ \underset{\text{digit}}{\underline{\quad}}\ \underset{\text{digit}}{\underline{\quad}}\ \underset{\text{digit}}{\underline{\quad}}$ with the number of ways the digit can be chosen. If repetition is allowed, the experiment is similar to that of Example 2 and there are $N^t = 9^4 = 6561$ possible PINs.

b. If repetition is not allowed, there are only eight possible choices for the second digit of the PIN, then seven for the third, and six for the fourth. The number of possible PIN numbers decreases to $9 \cdot 8 \cdot 7 \cdot 6 = 3024$.

NOW TRY EXERCISES 13 THROUGH 16

Given *any* experiment involving a sequence of tasks, if the first task can be completed in p possible ways, the second task has q possibilities, and the third task has r possibilities, a tree diagram will show the sample space for task_1–task_2–task_3 is $p \cdot q \cdot r$. Even though the examples we've considered to this point have varied a great deal, this idea was fundamental to counting all possibilities in a sample space and is, in fact, known as the **fundamental principle of counting** (FPC).

FUNDAMENTAL PRINCIPLE OF COUNTING

Given any experiment with three defined tasks, if there are p possibilities for the first task, q possibilities for the second, and r possibilities for the third, the total number of ways the experiment can be completed is $p \cdot q \cdot r$. This fundamental principle can be extended to include any number of tasks.

EXAMPLE 4 How many five-digit numbers can be formed using only the odd digits 1, 3, 5, 7, and 9, if (a) no repetitions are allowed, (b) repetitions are allowed, and (c) repetitions are not allowed, with the number being greater than 40,000 and divisible by 5?

Solution: Essentially we're asked for the number of ways to fill in five blanks by considering the number of possibilities for each blank.

a. If no repetitions are allowed we have five choices for the first blank, four for the second, and so on. There are $5! = 120$ numbers that can be formed.

b. If repetitions are allowed there are five choices for all five blanks: so there are $5^5 = 3125$ different numbers that can be formed.

c. For the number to be divisible by 5, the last digit must be a 5, which can happen in only one way. To be greater than 40,000 the first digit must be 5 or more, giving only the two choices 7 or 9, since the 5 is already being used. The remaining three digits can be chosen in any order from those that remain, giving

$\underset{\text{choices}}{\underline{\ 2\ }}\ \underset{\text{choices}}{\underline{\ 3\ }}\ \underset{\text{choices}}{\underline{\ 2\ }}\ \underset{\text{choices}}{\underline{\ 1\ }}\ \underset{\text{choices}}{\underline{\ 1\ }}$ or $2 \cdot 3 \cdot 2 \cdot 1 \cdot 1 = 12$ such numbers that can be formed.

NOW TRY EXERCISES 17 THROUGH 20

EXAMPLE 5 Adrienne, Bob, Carol, Dax, Earlene, and Fabian bought tickets to see *The Marriage of Figaro.* Assuming they sat together in a row of six seats, how many different seating arrangements are possible if (a) Bob and Carol are sweethearts and must sit together? (b) Bob and Carol are enemies and must not sit together?

Solution:

a. Since a restriction has been placed on the seating arrangement, it will help to divide the experiment into a sequence of tasks: *task 1:* they sit together; *task 2:* either Bob is on the left or Bob is on the right; and *task 3:* the other four are seated. Bob and Carol can sit together in five different ways, as shown in Figure 8.18, so there are five possibilities for task 1. There are two ways they can be side-by-side: Bob on the left and Carol on the right, as shown, or Carol on the left and Bob on the right. The remaining four people can be seated randomly, so task 3 has $4! = 24$ possibilities. Under these conditions they can be seated $5 \cdot 2 \cdot 4! = 240$ ways.

b. This is similar to Part (a), but now we have to count the number of ways they can be separated by *at least one seat: task 1:* Bob and Carol are in nonadjacent seats; *task 2:* either Bob is on the left or Bob is on the right; and *task 3:* the other four are seated. For task 1, be careful to note there is no multiplication involved, just a simple counting. If Bob sits in seat 1, there are four nonadjacent seats, as shown in Figure 8.19. If Bob sits in seat 2, there are three nonadjacent seats, and so on. This gives $4 + 3 + 2 + 1 = 10$ possibilities for Bob and Carol not sitting together. Task 2 and task 3 have the same number of possibilities as in Part (a), giving $10 \cdot 2 \cdot 4! = 480$ possible seating arrangements.

WORTHY OF NOTE

We could also reason that since there are $6! = 720$ random seating arrangements and 240 of them consist of Bob and Carol sitting together [Example 5(a)], that the remaining 480 must consist of Bob and Carol *not* sitting together. More will be said about this type of reasoning later in Section 8.6.

Figure 8.18

Seat 1	Seat 2	Seat 3	Seat 4	Seat 5	Seat 6
Bob	Carol				
	Bob	Carol			
		Bob	Carol		
			Bob	Carol	
				Bob	Carol

Figure 8.19

Seat 1	Seat 2	Seat 3	Seat 4	Seat 5	Seat 6
Bob		Carol			
Bob			Carol		
Bob				Carol	
Bob					Carol

NOW TRY EXERCISES 21 THROUGH 28

C. Distinguishable Permutations

In the game of Scrabble® (Milton Bradley), players attempt to form words by rearranging letters. Suppose a player has the letters P, S, T, and O at the end of the game. These letters could be rearranged or *permuted* to form the words POTS, SPOT, TOPS, or STOP. These arrangements are called permutations of the four letters. A permutation is any new arrangement, listing, or sequence of objects obtained by changing an existing order. A **distinguishable permutation** is a permutation that produces a result different from the original. For example, a distinguishable permutation of the digits in the number 1989 is 8199.

Example 5 considered six people, six seats, and the various ways they could be seated. But what if there were fewer seats than people? By the FPC, six people could sit in four seats in $6 \cdot 5 \cdot 4 \cdot 3 = 360$ different arrangements, six people could sit in three seats in $6 \cdot 5 \cdot 4 = 120$ different arrangements, and so on. These rearrangements are called distinguishable permutations. You may have noticed that for six people and six seats, we used all six factors of 6!, while for six people and four seats we used the first four, six people and three seats required only the first three, and so on. Generally, for n people and r seats, the first r factors of $n!$ will be used. The notation and formula for *distinguishable permutations of n objects taken r at a time* is ${}_nP_r = \frac{n!}{(n-r)!}$. By defining $0! = 1$, the formula includes the case where all n objects are selected, which of course results in ${}_nP_n = \frac{n!}{(n-n)!} = \frac{n!}{0!} = \frac{n!}{1} = n!$.

DISTINGUISHABLE PERMUTATIONS: UNIQUE ELEMENTS

If r objects are selected from a set containing n unique elements $(r \leq n)$ and placed in an ordered arrangement, the number of distinguishable permutations is

$${}_nP_r = \frac{n!}{(n-r)!} \quad \text{or} \quad {}_nP_r = n(n-1)(n-2)\cdots(n-r+1)$$

Both yield the first r factors of n!

EXAMPLE 6 Compute each value of ${}_nP_r$ using the methods just described.

a. ${}_7P_4$ **b.** ${}_{10}P_3$

Solution: Begin by evaluating each expression using the formula ${}_nP_r = \frac{n!}{(n-r)!}$, noting that the third line (in bold) gives r factors of $n!$.

a. $${}_7P_4 = \frac{7!}{(7-4)!} = \frac{7 \cdot 6 \cdot 5 \cdot 4 \cdot 3!}{3!} = \mathbf{7 \cdot 6 \cdot 5 \cdot 4} = 840$$

b. $${}_{10}P_3 = \frac{10!}{(10-3)!} = \frac{10 \cdot 9 \cdot 8 \cdot 7!}{7!} = \mathbf{10 \cdot 9 \cdot 8} = 720$$

NOW TRY EXERCISES 29 THROUGH 36

EXAMPLE 7 A local chapter of the Outdoor Club needs to elect a chair, vice-chair, and treasurer from among its 12 members. How many different ways can these three offices be filled?

Solution: The result will be an ordered arrangement of 3 people taken from a group of 12. There are ${}_{12}P_3 = 12 \cdot 11 \cdot 10 = 1320$ different ways the offices can be filled.

NOW TRY EXERCISES 37 THROUGH 40

EXAMPLE 8 As part of a sorority's initiation process, the nine new inductees must participate in a 1-mi race. Assuming there are no ties, how many finishes are possible if it is well known that Molasses Mary will finish last and Lightning Louise will finish first?

Solution: To help understand the situation, we can diagram the possibilities for finishing first through fifth. Since Louise will finish first, this slot can be filled in only one way, by Louise herself. The same goes for Mary and her fifth-place finish: $\underset{\text{1st}}{\underline{\text{Louise}}}\ \underset{\text{2nd}}{\underline{\quad\quad}}\ \underset{\text{3rd}}{\underline{\quad\quad}}\ \underset{\text{4th}}{\underline{\quad\quad}}\ \underset{\text{5th}}{\underline{\text{Mary}}}$. The remaining three slots can be filled in ${}_7P_3 = 7 \cdot 6 \cdot 5$ different ways, indicating that under these conditions, there are $1 \cdot 7 \cdot 6 \cdot 5 \cdot 1 = 210$ different ways to finish.

NOW TRY EXERCISES 41 AND 42

D. Nondistinguishable Permutations

As the name implies, certain permutations are nondistinguishable, meaning you cannot tell one apart from another. Such is the case when the original set contains elements or sample outcomes that are identical. Consider a family with four children, Lyddell, Morgan, Michael, and Mitchell, who are at the photo studio for a family picture. Michael and Mitchell are identical twins and cannot be told apart. In how many ways can they be lined up for the picture? Since this is an ordered arrangement of four children taken from a group of four, there are ${}_4P_4 = 24$ ways to line them up. A few of them are

Lyddell	Morgan	Michael	Mitchell	Lyddell	Morgan	Mitchell	Michael
Lyddell	Michael	Morgan	Mitchell	Lyddell	Mitchell	Morgan	Michael
Michael	Lyddell	Morgan	Mitchell	Mitchell	Lyddell	Morgan	Michael

But of these six arrangements, half will appear to be the same picture, since the difference between Michael and Mitchell cannot be distinguished. In fact, of the 24 total permutations, every picture where Michael and Mitchell have switched places will be nondistinguishable. To find the distinguishable permutations, we need to take the total permutations (${}_4P_4$) *and divide by* 2!, *the number of ways the twins can be permuted:* $\frac{{}_4P_4}{(2)!} = \frac{24}{2} = 12$ distinguishable pictures.

These ideas can be generalized and stated in the following way.

NONDISTINGUISHABLE PERMUTATIONS: NONUNIQUE ELEMENTS

In a set containing n elements where one element is repeated p times, another is repeated q times, and another is repeated r times $(p + q + r \leq n)$, the number of nondistinguishable permutations is given by

$$\frac{{}_nP_n}{p!q!r!} = \frac{n!}{p!q!r!}$$

The idea can be extended to include any number of repeated elements.

WORTHY OF NOTE

If a Scrabble player is able to play all seven letters in one turn, he or she "bingos" and is awarded 50 extra points. The player in Example 9 did just that. Can you determine what word was played?

EXAMPLE 9 A Scrabble player has the seven letters S, A, O, O, T, T, and T in his rack. How many distinguishable arrangements can be formed as he attempts to play a word?

Solution: Essentially the exercise asks for the number of distinguishable permutations of the seven letters, given T is repeated three times and O is repeated twice. There are $\frac{{}_7P_7}{3!2!} = 420$ distinguishable permutations.

NOW TRY EXERCISES 43 THROUGH 54

E. Combinations

Similar to nondistinguishable permutations, there are other times the total number of permutations must be reduced to quick-count the elements of a desired subset. Consider a vending machine that offers a variety of 40¢ snacks. If you have a quarter (Q), dime (D), and nickel (N), the machine wouldn't care about the order the coins were deposited. Even though QDN, QND, DQN, DNQ, NQD, and NDQ give the ${}_3P_3 = 6$ possible permutations, the machine considers them as equal and will vend your snack. Using sets, this is similar to saying the set $A = \{X, Y, Z\}$ has only one subset with three elements, since $\{X, Z, Y\}$, $\{Y, X, Z\}$, $\{Y, Z, X\}$, and so on, all represent the same set. Similarly, there are six, two-letter permutations of X, Y, and Z (${}_3P_2 = 6$): XY, XZ, YX, YZ, ZX, and ZY, but only three two-letter subsets: $\{X, Y\}$, $\{X, Z\}$ and $\{Y, Z\}$. When permutations having the same elements are considered identical, the result is the number of possible **combinations** and is denoted ${}_nC_r$. Since the r objects can be selected in $r!$ ways, we divide ${}_nP_r$ by $r!$ to "quick-count" the number of possibilities: ${}_nC_r = \frac{{}_nP_r}{r!}$, which can be thought of as *the first r factors of n!, divided by r!*. Take special note that when r objects are selected from a set with n elements and the order they're listed is unimportant (because you end up with the same subset), the result is a *combination* not a permutation.

COMBINATIONS

The number of combinations of n objects taken r at a time is given by

$${}_nC_r = \frac{{}_nP_r}{r!}.$$

EXAMPLE 10 Compute each value of ${}_nC_r$ given.

a. ${}_7C_4$ b. ${}_8C_3$ c. ${}_5C_2$

Solution:

a. ${}_7C_4 = \frac{7 \cdot 6 \cdot 5 \cdot 4}{4!} = 35$

b. ${}_8C_3 = \frac{8 \cdot 7 \cdot 6}{3!} = 56$

c. ${}_5C_2 = \frac{5 \cdot 4}{2!} = 10$

NOW TRY EXERCISES 55 THROUGH 64

EXAMPLE 11 A city lottery is getting ready to draw five numbers from 1 through 9 to determine the winner(s) for its "Little-Lottery" game. If a player picks the same five numbers, they win. In how many ways can the winning numbers be drawn?

Solution: Since the winning numbers can be drawn in any order, we have a combination of 9 things taken 5 at a time. The three numbers can be drawn in ${}_9C_5 = \frac{9 \cdot 8 \cdot 7 \cdot 6 \cdot 5}{5!} = 126$ ways.

NOW TRY EXERCISES 65 THROUGH 66

Somewhat surprisingly, there are many situations where the order things are listed is not important. Such situations include:

- The formation of committees, since the order people volunteer is unimportant
- Card games with a standard deck, since the order cards are dealt is unimportant
- Playing BINGO, since the order the numbers are called is unimportant

When the order in which people or objects are selected from a group is unimportant, the number of possibilities is a *combination,* not a permutation.

Another way to tell the difference between permutations and combinations is the following memory device: *P*ermutations have *P*riority or *P*recedence; in other words, the *P*osition of each element matters. By contrast, a *C*ombination is like a *C*ommittee of

WORTHY OF NOTE

By substituting $\frac{n!}{(n-r)!}$ for ${}_nP_r$ in the formula for combinations, we find an alternative method for computing ${}_nC_r$ is $\frac{n!}{r!(n-r)!}$. This form is used more extensively later in this chapter.

Colleagues or Collection of Commoners; all members have equal rank. For permutations, *a-b-c* is different from *b-a-c*. For combinations, *a-b-c* is the same as *b-a-c*.

EXAMPLE 12 The Sociology Department of Caskill Community College has 12 dedicated faculty members. (a) In how many ways can a three-member textbook selection committee be formed? (b) If the department is in need of a Department Chair, Curriculum Chair, and Technology Chair, in how many ways can the positions be filled?

Solution:

a. Since textbook selection depends on a *C*ommittee of *C*olleagues, the order members are chosen is not important. This is a *C*ombination of 12 people taken 3 at a time, and there are ${}_{12}C_3 = 220$ ways the committee can be formed.

b. Since those selected will have *P*osition or *P*riority, this is a *P*ermutation of 12 people taken 3 at a time, giving ${}_{12}P_3 = 1320$ ways the positions can be filled.

NOW TRY EXERCISES 67 THROUGH 78

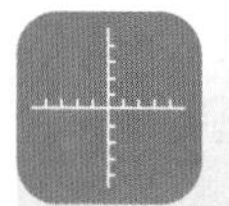

TECHNOLOGY HIGHLIGHT

Calculating Permutations and Combinations Using a Calculator

The keystrokes shown apply to a TI-84 Plus model. Please consult your manual or our Internet site for other models.

We now know the ${}_nP_r$ function introduced in this section gives the permutations of n objects taken r at a time. Both the ${}_nP_r$ and ${}_nC_r$ functions are accessed using the MATH key and the **PRB** submenu (see Figure 8.20a). To compute the permutations of 12 objects taken 9 at a time (${}_{12}P_9$), clear the home screen and enter a 12, then press MATH ◄ **2:${}_nP_r$** to access the ${}_nP_r$ operation, which is automatically pasted on the home screen next to the 12. Now enter a 9, press ENTER and a result of 79833600 is displayed (Figure 8.20b).

Figure 8.20(a)

```
MATH NUM CPX PRB
1:rand
2:nPr
3:nCr
4:!
5:randInt(
6:randNorm(
7:randBin(
```

Figure 8.20(b)

```
12 nPr 9
          79833600
12 nCr 9
               220
220*9!
          79833600
```

Repeat the sequence to compute the value of ${}_{12}C_9$ (MATH ◄ **3:${}_nC_r$**). Note that the value of ${}_{12}P_9$ is much larger than ${}_{12}C_9$ and that they differ by a factor of 9! since ${}_nC_r = \frac{{}_nP_r}{r!}$. Use these features to respond to the following:

Exercise 1: The Department of Humanities has nine faculty members who must serve on at least one committee per semester. How many different committees can be formed that have (a) two members, 36 (b) three members, 84 (c) four members, 126 and (d) five members? 126

Exercise 2: A certain state places 45 ping-pong balls numbered 1 through 45 in a container, then draws out five to form the winning lottery numbers. How many different ways can the five numbers be picked? 1,221,759

Exercise 3: Dairy King maintains six different toppings at a self-service counter, so that customers can top their ice cream sundaes with as many as they like. How many different sundaes can be created if a customer were to select any three ingredients? 20

8.5 EXERCISES

CONCEPTS AND VOCABULARY

Fill in the blank with the appropriate word or phrase. Carefully reread the section if needed.

1. A(n) experiment is any task that can be repeated and has a(n) well-defined set of possible outcomes.

2. If an experiment has N equally likely outcomes and is repeated t times, the number of elements in the sample space is given by N^t.

3. When unique elements of a set are rearranged, the result is called a(n) distinguishable permutation.

4. If some elements of a group are identical, certain rearrangements are identical and the result is a(n) nondistinguishable permutation.

5. A three-digit number is formed from digits 1 to 9. Explain how forming the number with repetition differs from forming it without repetition. Answers will vary.

6. Discuss/explain the difference between a permutation and a combination. Try to think of new ways to help remember the distinction. Answers will vary.

DEVELOPING YOUR SKILLS

Exercise 7

7. For the spinner shown here, (a) draw a tree diagram illustrating all possible outcomes for two spins and (b) create an ordered list showing all possible outcomes for two spins.

Exercise 8

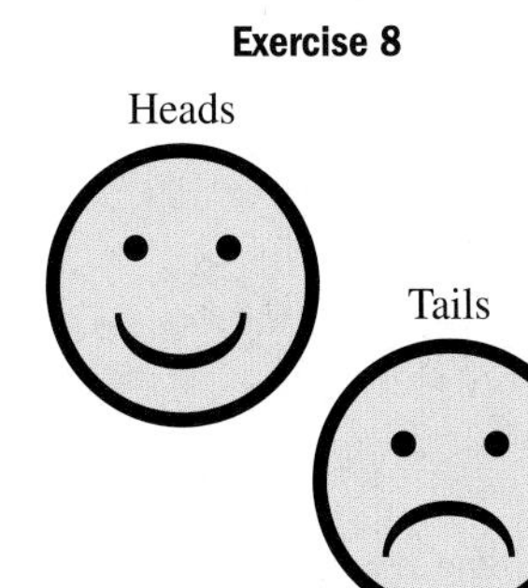

8. For the fair coin shown here, (a) draw a tree diagram illustrating all possible outcomes for four flips and (b) create an ordered list showing the possible outcomes for four flips.

9. A fair coin is flipped five times. If you extend the tree diagram from Exercise 8, how many elements are in the sample space? 32

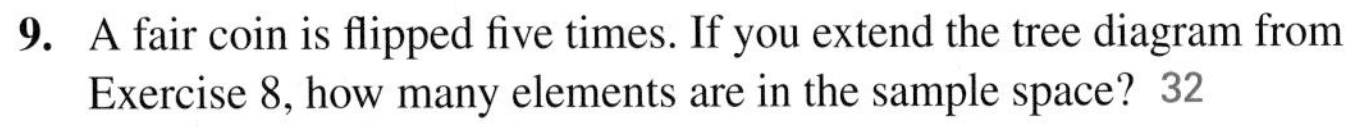

10. A spinner has the two equally likely outcomes A or B and is spun four times. How is this experiment related to the one in Exercise 8? How many elements are in the sample space? 16

11. An inexpensive lock uses the numbers 0 to 24 for a three-number combination. How many different combinations are possible? 15,625

12. Grades at a local college consist of A, B, C, D, F, and W. If four classes are taken, how many different report cards are possible? 1296

License plates. In a certain (English-speaking) country, license plates for automobiles consist of two letters followed by one of four symbols (■, ◆, ◎, or ●), followed by three digits. How many license plates are possible if

13. Repetition is allowed? 2,704,000

14. Repetition is not allowed? 1,872,000

15. A remote access door opener requires a five-digit (1–9) sequence. How many sequences are possible if (a) repetition is allowed? (b) repetition is not allowed? 59,049; 15,120

16. An instructor is qualified to teach Math 020, 030, 140, and 160. How many different four-course schedules are possible if (a) repetition is allowed? (b) repetition is not allowed? 256; 24

Use the fundamental principle of counting and other quick-counting techniques to respond.

17. **Menu items:** At Joe's Diner, the manager is offering a dinner special that consists of one choice of entree (chicken, beef, soy meat, or pork), two vegetable servings (corn, carrots, green beans, peas, broccoli, or okra), and one choice of pasta, rice, or potatoes. How many different meals are possible? 360 if double veggies are not allowed, 432 if double veggies are allowed.

Additional answers can be found in the Instructor Answer Appendix.

18. Getting dressed: A frugal businessman has five shirts, seven ties, four pairs of dress pants, and three pairs of dress shoes. Assuming that all possible arrangements are appealing, how many different shirt-tie-pants-shoes outfits are possible? 420

19. Number combinations: How many four-digit numbers can be formed using the even digits 0, 2, 4, 6, 8, if (a) no repetitions are allowed; (b) repetitions are allowed; (c) repetitions are not allowed and the number must be less than 6000 and divisible by 10. 120 625 12

20. Number combinations: If I was born in March, April, or May, after the 19th but before the 30th, and after 1949 but before 1981, how many different MM–DD–YYYY dates are possible for my birthday? 930

Seating arrangements: William, Xayden, York, and Zelda decide to sit together at the movies. How many ways can they be seated if

21. They sit in random order? 24

22. York must sit next to Zelda? 12

23. York and Zelda must be on the outside? 4

24. William must have the aisle seat? 6

Course schedule: A college student is trying to set her schedule for the next semester and is planning to take five classes: English, Art, Math, Fitness, and Science. How many different schedules are possible if

25. The classes can be taken in any order. 120

26. She wants her science class to immediately follow her math class. 24

27. She wants her English class to be first and her Fitness class to be last. 6

28. She can't decide on the best order and simply takes the classes in alphabetical order. 1

Find the value of $_nP_r$ in two ways: (a) compute r factors of $n!$ and (b) use the formula $_nP_r = \frac{n!}{(n-r)!}$.

29. $_{10}P_3$ 720

30. $_{12}P_2$ 132

31. $_9P_4$ 3024

32. $_5P_3$ 60

33. $_8P_7$ 40,320

34. $_8P_1$ 8

Determine the number of three-letter permutations of the letters given, then use an organized list to write them all out. How many of them are actually words or common names?

35. T, R, and A 6; 3

36. P, M, and A 6; 3

37. The regional manager for an office supply store needs to replace the manager and assistant manager at the downtown store. In how many ways can this be done if she selects the personnel from a group of 10 qualified applicants? 90

38. The local chapter of Mu Alpha Theta will soon be electing a president, vice-president, and treasurer. In how many ways can the positions be filled if the chapter has 15 members? 2730

39. The local school board is going to select a principal, vice-principal, and assistant vice-principal from a pool of eight qualified candidates. In how many ways can this be done? 336

40. From a pool of 32 applicants, a board of directors must select a president, vice-president, labor relations liaison, and a director of personnel for the company's day-to-day operations. Assuming all applicants are qualified and willing to take on any of these positions, how many ways can this be done? 863,040

41. A hugely popular chess tournament now has six finalists. Assuming there are no ties, (a) in how many ways can the finalists place in the final round? (b) In how many ways can they finish first, second, and third? (c) In how many ways can they finish if it's sure that Roberta Fischer is going to win the tournament and that Geraldine Kasparov will come in sixth? 720; 120; 24

42. A field of 10 horses has just left the paddock area and is heading for the gate. Assuming there are no ties in the big race, (a) in how many ways can the horses place in the race? (b) In how many ways can they finish in the win, place, or show positions? (c) In how many ways can they finish if it's sure that John Henry III is going to win, Seattle Slew III will come in second (place), and either Dumb Luck II or Calamity Jane I will come in tenth? 3,628,800; 720; 10,080

Assuming all multiple births are identical and the children cannot be told apart, how many distinguishable photographs can be taken of a family of six, if they stand in a single row and there is

43. one set of twins 360

44. one set of triplets 120

45. one set of twins and one set of triplets 60

46. one set of quadruplets 30

47. How many distinguishable numbers can be made by rearranging the digits of 105,001? 60

48. How many distinguishable numbers can be made by rearranging the digits in the palindrome 1,234,321? 630

How many distinguishable permutations can be formed from the letters of the given word?

49. logic 120

50. leave 60

51. lotto 30

52. levee 20

A Scrabble player (see Example 9) has the six letters shown remaining in her rack. How many distinguishable, six-letter permutations can be formed? (If all six letters are played, what was the word?)

53. A, A, A, N, N, B 60, BANANA

54. D, D, D, N, A, E 120, ADDEND

Find the value of ${}_nC_r$: (a) using ${}_nC_r = \frac{{}_nP_r}{r!}$ (r factors of $n!$ over $r!$) and (b) using ${}_nC_r = \frac{n!}{r!(n-r)!}$.

55. ${}_9C_4$ 126

56. ${}_{10}C_3$ 120

57. ${}_8C_5$ 56

58. ${}_6C_3$ 20

59. ${}_6C_6$ 1

60. ${}_6C_0$ 1

Use a calculator to verify that each pair of combinations is equal.

61. ${}_9C_4$, ${}_9C_5$ verified

62. ${}_{10}C_3$, ${}_{10}C_7$ verified

63. ${}_8C_5$, ${}_8C_3$ verified

64. ${}_7C_2$, ${}_7C_5$ verified

65. A platoon leader needs to send four soldiers to do some reconnaissance work. There are 12 soldiers in the platoon and each soldier is assigned a number between 1 and 12. The numbers 1 through 12 are placed in a helmet and drawn randomly. If a soldier's number is drawn, then that soldier goes on the mission. In how many ways can the reconnaissance team be chosen? 495

66. Seven colored balls (red, indigo, violet, yellow, green, blue, and orange) are placed in a bag and three are then withdrawn. In how many ways can the three colored balls be drawn? 35

67. When the company's switchboard operators went on strike, the company president asked for three volunteers from among the managerial ranks to temporarily take their place. In how many ways can the three volunteers "step forward," if there are 14 managers and assistant managers in all? 364

68. Becky has identified 12 books she wants to read this year and decides to take four with her to read while on vacation. She chooses *Pastwatch* by Orson Scott Card for sure, then decides to randomly choose any three of the remaining books. In how many ways can she select the four books she'll end up taking? 165

69. A new garage band has built up their repertoire to 10 excellent songs that really rock. Next month they'll be playing in a *Battle of the Bands* contest, with the winner getting some guaranteed gigs at the city's most popular hot spots. In how many ways can the band select 5 of their 10 songs to play at the contest? 252

70. Pierre de Guirré is an award-winning chef and has just developed 12 delectable, new main-course recipes for his restaurant. In how many ways can he select three of the recipes to be entered in an international culinary competition? 220

For each exercise, determine whether a permutation, a combination, counting principles, or a determination of the number of subsets is the most appropriate tool for obtaining a solution, then solve. Some exercises can be completed using more than one method.

71. In how many ways can eight second-grade children line up for lunch? 40,320

72. If you flip a fair coin five times, how many different outcomes are possible? 32

73. Eight sprinters are competing for the gold, silver, and bronze metals. In how many ways can the metals be awarded? 336

74. Motorcycle license plates are made using two letters followed by three numbers. How many plates can be made if repetition of letters (only) is allowed? 486,720

75. A committee of five students is chosen from a class of 20 to attend a seminar. How many different ways can this be done? 15,504

76. If onions, cheese, pickles, and tomatoes are available to dress a hamburger, how many different hamburgers can be made? 16

77. A caterer offers eight kinds of fruit to make various fruit trays. How many different trays can be made using four different fruits? 70

78. Eighteen females try out for the basketball team, but the coach can only place 15 on her roster. How many different teams can be formed? 816

WORKING WITH FORMULAS

Stirling's Formula: $n! \approx \sqrt{2\pi} \cdot (n^{n+0.5}) \cdot e^{-n}$

79. Values of $n!$ grow very quickly as n gets larger (13! is already in the billions). For some applications, scientists find it useful to use the approximation for $n!$ shown, called Stirling's Formula.

a. Compute the value of 7! on your calculator, then use Stirling's Formula with $n = 7$. By what percent does the approximate value differ from the true value? $\approx$1.2%

b. Compute the value of 10! on your calculator, then use Stirling's Formula with $n = 10$. By what percent does the approximate value differ from the true value? $\approx$0.83%

80. Factorial formulas: For whole numbers n and k, where $n > k$,

$$\frac{n!}{(n-k)!} = n(n-1)(n-2)\cdots(n-k+1)$$

a. Verify the formula for $n = 7$ and $k = 5$. verified

b. Verify the formula for $n = 9$ and $k = 6$. verified

APPLICATIONS

81. Yahtzee: In the game of "Yahtzee"® (Milton Bradley) five dice are rolled simultaneously on the first turn in an attempt to obtain various arrangements (worth various point values). How many different arrangements are possible? 7776

82. Twister: In the game of "Twister"® (Milton Bradley) a simple spinner is divided into four quadrants designated Left Foot (LF), Right Hand (RH), Right Foot (RF), and Left Hand (LH), with four different color possibilities in each quadrant (red, green, yellow, blue). Determine the number of possible outcomes for three spins. 4096

83. Clue: As mentioned in the *Point of Interest,* in the game of "Clue"® (Parker Brothers) a crime is committed in one of nine rooms, with one of six implements, by one of six people. In how many different ways can the crime be committed? 324

BINGO is a game played on a 5-by-5 grid with the letters of the word B-I-N-G-O heading the five columns. Numbers in the "B" column range from 1 to 15, in the "I" column they range from 16 to 30, "N" from 31 to 45, "G" from 46 to 60, and "O" from 61 to 75. Numbers such as "B-12" and "G-51" are drawn and called out, until a player completes a prescribed "BINGO" pattern—usually five in a row horizontally, vertically, or diagonally. There is one "free-space" in the middle of the grid in the "N" column.

B	I	N	G	O
01	17	45	49	73
03	16	36	57	62
10	22	FREE	52	75
07	26	33	59	61
14	19	41	47	71
♦ LUCKY CARDS ♦				

84. How many different ways can the "B" column be filled? 360,360

85. How many different ways can the "N" column be filled? 32,760

86. How many different ways can any one of the diagonals be filled? 50,625

87. How many different BINGO cards can be formed (as in a black-out)? 1,474,200

Phone numbers in North America have 10 digits: a three-digit area code, a three-digit exchange number, and the four final digits that make each phone number unique. Neither area codes nor exchange numbers can start with 0 or 1. Prior to 1994 the second digit of the area code *had to be* a 0 or 1. Sixteen area codes are reserved for special services (such as 911 and 411). In 1994, the last area code was used up and the rules were changed to allow the digits 2 through 9 as the middle digit in area codes.

88. How many different area codes were possible prior to 1994? 144

89. How many different exchange numbers were possible prior to 1994? 800

90. How many different phone numbers were possible *prior to* 1994? 1,152,000,000

91. How many different phone numbers were possible *after* 1994? 6,272,000,000

Aircraft N-numbers: In the United States, private aircraft are identified by an "N-Number," which is generally the letter "N" followed by five characters and includes these restrictions: (1) the N-Number can consist of five digits, four digits followed by one letter, or three digits followed by two letters; (2) the first digit cannot be a zero; (3) to avoid confusion with the numbers zero and one, the letters O and I cannot be used; and (4) repetition of digits and letters is allowed. How many unique N-Numbers can be formed

92. that have four digits and one letter? 216,000

93. that have three digits and two letters? 518,400

94. that have five digits? 90,000

95. that have three digits, two letters with no repetitions of any kind allowed? 357,696

Seating arrangements: Eight people would like to be seated. Assuming some will have to stand, in how many ways can the seats be filled if the number of seats available is

96. eight 40,320

97. five 6720

98. three 336

99. one 8

Seating arrangements: In how many different ways can eight people (six students and two teachers) sit in a row of eight seats if

100. the teachers must sit on the ends 1440

101. the teachers must sit together 10,080

Television station programming: A television station needs to fill eight half-hour slots for its Tuesday evening schedule with eight programs. In how many ways can this be done if

102. there are no constraints 40,320

103. *Seinfeld* must have the 8:00 P.M. slot 5040

104. *Seinfeld* must have the 8:00 P.M. slot and *The Drew Carey Show* must be shown at 6:00 P.M. 720

105. *Friends* can be aired at 7:00 or 9:00 P.M. and *Everybody Loves Raymond* can be aired at 6:00 or 8:00 P.M. 2880

Scholarship awards: Fifteen students at Roosevelt Community College have applied for six available scholarship awards. How many ways can the awards be given if

106. there are six different awards given to six different students 3,603,600

107. there are six identical awards given to six different students 5005

Committee composition: The local city council has 10 members and is trying to decide if they want to be governed by a committee of three people or by a president, vice-president, and secretary.

108. If they are to be governed by committee, how many unique committees can be formed? 120

109. How many different president, vice-president, and secretary possibilities are there? 720

110. **Team rosters:** A soccer team has three goalies, eight defensive players, and eight forwards on its roster. How many different starting line-ups can be formed (one goalie, three defensive players, and three forwards)? 9408

111. **e-mail addresses:** A business wants to standardize the e-mail addresses of its employees. To make them easier to remember and use, they consist of two letters and two digits (followed by @esmtb.com), with zero being excluded from use as the first digit and no repetition of letters or digits allowed. Will this provide enough unique addresses for their 53,000 employees worldwide? 52,650, no

WRITING, RESEARCH, AND DECISION MAKING

112. Compute the combinations ${}_7C_7, {}_7C_6, {}_7C_5, \ldots, {}_7C_0$ by hand and notice the pattern that develops. Repeat the exercise for ${}_6C_6, {}_6C_5, {}_6C_4, \ldots, {}_6C_0$ to see if the pattern holds. In Exercises 61 to 64 you used a calculator to verify that for specific values of n and r, ${}_nC_r = {}_nC_{n-r}$ for $r \le n$. Rework Exercises 61 to 64 by hand and use the patterns observed here to explore why this relationship is true. Comment on what you find.

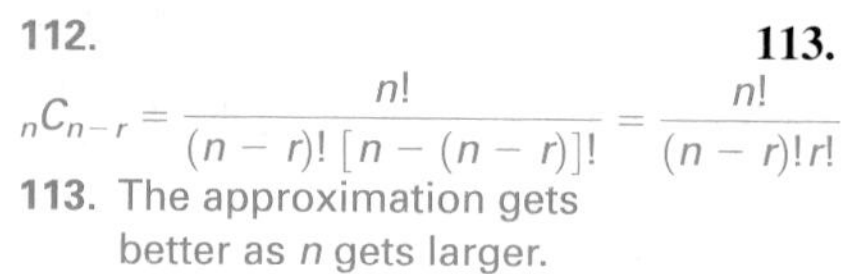

113. In Exercise 79, we learned that an approximation for $n!$ can be found using Stirling's Formula: $n! \approx \sqrt{2\pi}(n^{n+0.5})e^{-n}$. As with other approximations, mathematicians are very interested in whether the approximation gets better or worse for larger values of n (does their ratio get closer to 1 or farther from 1). Use your calculator to investigate and answer the question.

EXTENDING THE CONCEPT

114. Verify that the following equations are true, then generalize the patterns and relationships noted to create your own equation. Afterward, write each of the four factors from Part (a) (the two combinations on each side) in expanded form and discuss/explain why the two sides are equal.

a. ${}_{10}C_3 \cdot {}_7C_2 = {}_{10}C_2 \cdot {}_8C_5$ $\frac{10!}{2!3!5!}$

b. ${}_9C_3 \cdot {}_6C_2 = {}_9C_2 \cdot {}_7C_4$ $\frac{9!}{2!3!4!}$

c. ${}_{11}C_4 \cdot {}_7C_5 = {}_{11}C_5 \cdot {}_6C_4$ $\frac{11!}{4!5!2!}$

d. ${}_8C_3 \cdot {}_5C_2 = {}_8C_2 \cdot {}_6C_3$ $\frac{8!}{2!3!3!}$

EXTENDING THE CONCEPT

115. **Tic-Tac-Toe:** In the game *Tic-Tac-Toe,* players alternately write an "X" or an "O" in one of nine squares on a 3 × 3 grid. If either player gets three in a row horizontally, vertically, or diagonally, that player wins. If all nine squares are played with neither person winning, the game is a draw. Assuming "X" always goes first,

a. How many different "boards" are possible if the game ends after five plays? 1440

b. How many different "boards" are possible if the game ends after six plays? 5328

116. A circular permutation is one where the objects are arranged in a circle rather than a line. In this case, these four permutations of the letters in the word "MATH" would be considered the same: M-A-T-H, A-T-H-M, T-H-M-A, H-M-A-T → (M, H, T, A arranged in a circle). Normally, there would be $4! = 24$ permutations. How many *circular* permutations are there? 6

MAINTAINING YOUR SKILLS

117. (4.6) Graph the rational function. Clearly label all asymptotes and intercepts:

$$h(x) = \frac{x^3 - x}{x^2 - 4}.$$

118. (6.3) Solve the given system of linear inequalities by graphing. Shade the feasible region.

$$\begin{cases} 2x + y < 6 \\ x + 2y < 6 \\ x \geq 0 \\ y \geq 0 \end{cases}$$

119. (8.2) For the series $1 + 5 + 9 + 13 + \cdots + 197$, state the nth term formula then find the 35th term and the sum of the first 35 terms.

120. (6.6) Given matrices A and B shown, use a calculator to find $A + B$, AB, and A^{-1}.

$$A = \begin{bmatrix} 1 & 0 & 3 \\ -2 & 5 & 1 \\ 2 & 1 & 4 \end{bmatrix}$$

$$B = \begin{bmatrix} 0.5 & 0.2 & -7 \\ -9 & 0.1 & 8 \\ 1.2 & 0 & 6 \end{bmatrix}$$

121. (7.2) Graph the hyperbola that is defined by $\frac{(x-2)^2}{4} - \frac{(y+3)^2}{9} = 1$.

122. (8.4) Use the principle of mathematical induction to prove that $\frac{1}{2} + \frac{1}{6} + \frac{1}{12} + \cdots + \frac{1}{n(n+1)} = \frac{n}{n+1}$.

117.

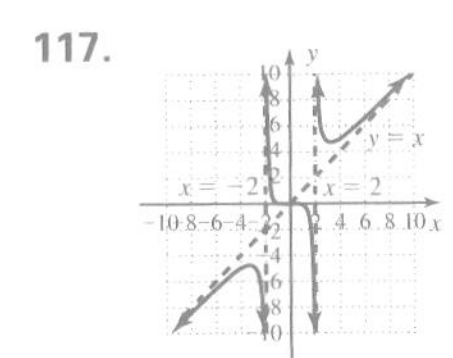

118.

119. $a_n = 1 + (n-1)4$; 137; 2415

120. $A + B = \begin{bmatrix} 1.5 & 0.2 & -4 \\ -11 & 5.1 & 9 \\ 3.2 & 1 & 10 \end{bmatrix}$, $AB = \begin{bmatrix} 4.1 & 0.2 & 11 \\ -44.8 & 0.1 & 60 \\ -3.2 & 0.5 & 18 \end{bmatrix}$, $A^{-1} = \begin{bmatrix} \frac{-19}{17} & \frac{-3}{17} & \frac{15}{17} \\ \frac{-10}{17} & \frac{2}{17} & \frac{7}{17} \\ \frac{12}{17} & \frac{1}{17} & \frac{-5}{17} \end{bmatrix}$

121.

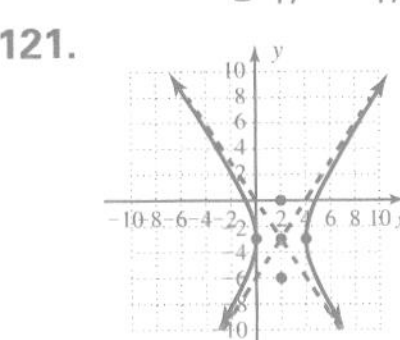

8.6 Introduction to Probability

LEARNING OBJECTIVES

In Section 8.6 you will learn how to:

- **A.** Define an event on a sample space
- **B.** Compute elementary probabilities
- **C.** Use certain properties of probability
- **D.** Compute probabilities using quick-counting techniques
- **E.** Compute probabilities involving mutually exclusive events
- **F.** Compute probabilities involving nonexclusive events

INTRODUCTION

There are few areas of mathematics that give us a better view of the world than **probability** and **statistics.** Unlike statistics, which seeks to analyze and interpret data, probability (for our purposes) attempts to use observations and data to make statements concerning the likelihood of future events. Such predictions of what *might* happen have found widespread application in such diverse fields as politics, manufacturing, gambling, opinion polls, product life, and many others. In this section we develop the basic elements of probability.

POINT OF INTEREST

Probability has a long and colorful history, originating in games of chance that can be traced back to ancient times. Modern probability theory is of a much more recent vintage, and its foundation stems from questions placed to the French mathematicians Blaise Pascal (1623–1662) and Pierre de Fermat (1601–1665) by gamblers desiring to know the "fair value" of bets on games of chance. After

Pascal and Fermat decided that these were questions worthy of their mathematical talents, it wasn't long until the study of probability became formalized and axiomatic, with its uses and applications finding their way into many areas outside of gaming.

A. Defining an Event

WORTHY OF NOTE

As mentioned in Section 8.5, our study of probability will involve only those sample spaces with events that are equally likely.

In Section 8.5 we defined the following terms: experiment and sample outcome. Flipping a coin twice in succession is an *experiment,* and two sample outcomes are HH and HT. An **event** E is *any designated set of sample outcomes,* and is a subset of the sample space. One event might be E_1: (two heads occur), another possibility is E_2: (at least one tail occurs).

EXAMPLE 1 Consider the experiment of rolling one standard, six-sided die (plural is dice). State the sample space S and define any two events relative to S.

Solution: S is the set of all possible outcomes, so $S = \{1, 2, 3, 4, 5, 6\}$. Two possible events are E_1: (a 5 is rolled) and E_2: (an even number is rolled).

NOW TRY EXERCISES 7 THROUGH 10

B. Elementary Probability

When rolling the die, we know the result can be any of the six equally likely outcomes in the sample space, so the chance of E_1:(a five is rolled) is $\frac{1}{6}$. Since three of the elements in S are even numbers, the chance of E_2:(an even number is rolled) is $\frac{3}{6} = \frac{1}{2}$. This suggests the following definition.

WORTHY OF NOTE

Probabilities can be written in fraction form, decimal form, or as a percent. For $P(E_2)$ from Example 1, the probability is $\frac{1}{2}$, 0.5, or 50%.

THE PROBABILITY OF AN EVENT *E*

Given S is a sample space of equally likely events and E is an event relative to S, the probability of E, written $P(E)$, is computed as

$$P(E) = \frac{n(E)}{n(S)}$$

where $n(E)$ represents the number of elements in E, and $n(S)$ represents the number of elements in S.

A standard deck of playing cards consists of 52 cards divided in four groups or *suits.* There are 13 hearts (♥), 13 diamonds (♦), 13 spades (♠), and 13 clubs (♣). As you can see in the illustration, each of the 13 cards in a suit is labeled 2, 3, 4, 5, 6, 7, 8, 9, 10, J, Q, K, and A. Also notice that 26 of the cards are red (hearts and diamonds) and 26 are black (spades and clubs).

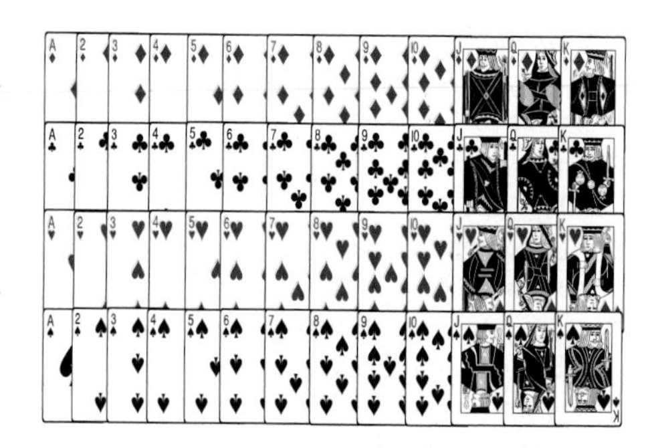

EXAMPLE 2 A single card is drawn from a well-shuffled deck. Define S and state the probability of any single outcome. Then define E as *a King is drawn* and find $P(E)$.

Solution: Sample space: $S = \{\text{the 52 cards}\}$. There are 52 equally likely outcomes, so the probability of any one outcome is $\frac{1}{52}$. Since S has four Kings,

$$P(E) = \frac{n(E)}{n(S)} = \frac{4}{52} \text{ or about } 0.077.$$

NOW TRY EXERCISES 11 THROUGH 14

EXAMPLE 3 A family of five has two girls and three boys named Sophie, Maria, Albert, Isaac, and Pythagorus. Their ages are 21, 19, 15, 13, and 9, respectively. One is to be selected randomly. Find the probability a teenager is chosen.

Solution: The sample space is $S = \{9, 13, 15, 19, 21\}$. Three of the five are teenagers, meaning the probability is $\frac{3}{5}$, 0.6, or 60%.

NOW TRY EXERCISES 15 AND 16

C. Properties of Probability

A study of probability necessarily includes recognizing some basic and fundamental properties. For example, when a fair die is rolled, what is $P(E)$ if E is defined as *a 1, 2, 3, 4, 5, or 6 is rolled?* The event E will occur 100% of the time, since 1, 2, 3, 4, 5, 6 are the only possibilities. In symbols we write P(outcome is in the sample space) or simply $P(S) = 1$ (100% written as a decimal number).

What percent of the time will a result *not* in the sample space occur? Since the die has only the six sides numbered 1 through 6, the probability of rolling something else is zero. In symbols, P(outcome is not in sample space) $= 0$ or simply $P({\sim}S) = 0$.

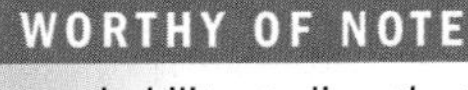

WORTHY OF NOTE

In probability studies, the tilde "~" acts as a negation symbol. For any event E defined on the sample space, $\sim E$ means the Event does not occur.

PROPERTIES OF PROBABILITY

Given sample space S and any event E defined relative to S.

1. $P(S) = 1$ 2. $P({\sim}S) = 0$ 3. $0 \le P(E) \le 1$

EXAMPLE 4 A game is played using a spinner like the one shown. Determine the probability of the following "events:"

E_1: A nine is spun. E_2: An integer greater than 0 and less than 9 is spun.

Solution: The sample space consists of eight equally likely outcomes.

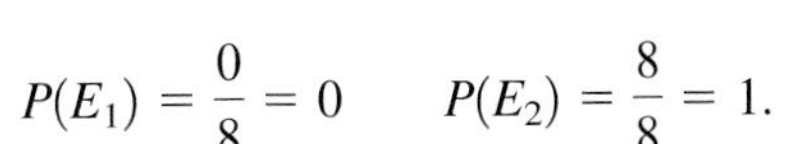

$$P(E_1) = \frac{0}{8} = 0 \qquad P(E_2) = \frac{8}{8} = 1.$$

Note the first outcome is not really an "event," since it is not in the sample space and cannot occur, while E_2 contains the entire sample space and must occur.

NOW TRY EXERCISES 17 AND 18

Because we know $P(S) = 1$ and all sample outcomes are equally likely, the probabilities of all simple events defined on the sample space must sum to 1. For the experiment of rolling a fair die, the sample space has six outcomes that are equally likely. Note that $P(1) = P(2) = P(3) = P(4) = P(5) = P(6) = \frac{1}{6}$, and $\frac{1}{6} + \frac{1}{6} + \frac{1}{6} + \frac{1}{6} + \frac{1}{6} + \frac{1}{6} = 1$

PROBABILITY AND SAMPLE OUTCOMES

Given a sample space S with n equally likely sample outcomes $s_1, s_2, s_3, \ldots, s_n$.

$$\sum_{i=1}^{n} P(s_i) = P(s_1) + P(s_2) + P(s_3) + \cdots + P(s_n) = 1$$

The **complement** of an event E is the set of sample outcomes in S not contained in E. Symbolically, $\sim E$ is the complement of E.

PROBABILITY AND COMPLEMENTARY EVENTS

Given sample space S and any event E defined relative to S, the complement of E, written $\sim E$, is the set of all outcomes not in E and:

1. $P(E) = 1 - P(\sim E)$
2. $P(E) + P(\sim E) = 1$

EXAMPLE 5 Use complementary events to answer the following questions:

a. A single card is drawn from a well-shuffled deck. What is the probability that it is not a diamond?

b. A single letter is picked at random from the letters in the word "divisibility." What is the probability it is not an "i"?

Solution:

a. Since there are 13 diamonds in a standard 52-card deck, there are 39 nondiamonds: $P(\sim D) = 1 - P(D) = 1 - \frac{13}{52} = \frac{39}{52} = 0.75$.

b. Of the 12 letters in d-i-v-i-s-i-b-i-l-i-t-y, 5 are "i's." This means $P(\sim \text{i}) = 1 - P(\text{i})$, or $1 - \frac{5}{12} = \frac{7}{12}$. The probability of choosing a letter other than i is $0.58\overline{3}$.

NOW TRY EXERCISES 19 THROUGH 22

Notice the second property in the preceding box was obtained by adding $P(\sim E)$ to both sides of the first. Using a Venn diagram (see the *Point of Interest* from Section 6.3) we can give a visual interpretation of complementary events. The rectangle in Figure 8.21 represents the entire sample space of an experiment and the circular area represents all sample outcomes belonging to a defined event E. This implies that the area within the rectangle but outside the circle must represent $\sim E$, the complement of E. Notice again that $(E) + (\sim E)$ gives S, and $P(E) + P(\sim E) = 1$ (the entire rectangle).

Figure 8.21

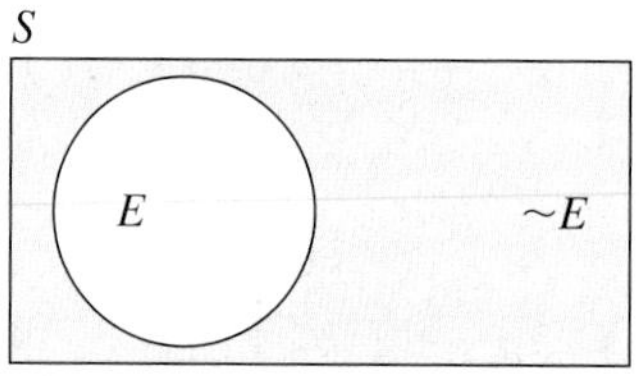

EXAMPLE 6 Inter-Island Waterways has just opened hydrofoil service between several islands. The hydrofoil is powered by two engines, one forward and one aft, and will operate if either of its two engines are functioning. Due to testing and past experience, the company knows the probability of the aft engine failing is $P(\text{aft engine fails}) = 0.05$, the probability of the forward engine failing is $P(\text{forward engine fails}) = 0.03$, and the probability that both fail is $P(\text{both engines simultaneously fail}) = 0.012$. What is the probability the hydrofoil completes its next trip?

Solution: Although the answer *seems* complicated at first, note that P(trip is completed) and P(both engines simultaneously fail) are complements.

$$P(\text{trip is completed}) = 1 - P(\text{both engines simultaneously fail})$$
$$= 1 - 0.012$$
$$= 0.988$$

There is close to a 99% probability that each trip will be completed safely.

NOW TRY EXERCISES 23 AND 24

The chart in Figure 8.22 shows all 36 possible outcomes (the sample space) from the experiment of rolling two fair dice.

Figure 8.22

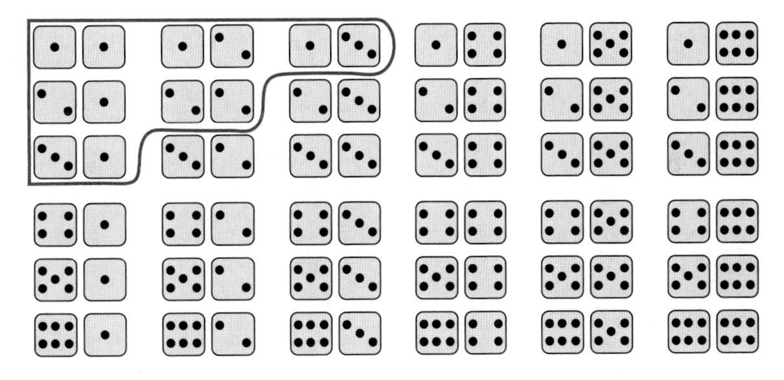

EXAMPLE 7 Two fair dice are rolled. What is the probability that the sum of both dice is greater than or equal to 5, $P(\text{sum} \geq 5)$?

Solution: See Figure 8.22. For $P(\text{sum} \geq 5)$ it's easier to use complements: $P(\text{sum} \geq 5) = 1 - P(\text{sum} < 5)$, which gives $1 - \frac{6}{36} = 1 - \frac{1}{6} = \frac{5}{6} = 0.8\overline{3}$.

NOW TRY EXERCISES 25 AND 26

D. Probability and Quick-Counting

Quick-counting techniques were introduced earlier to help count the number of elements in a large or more complex sample space, and the number of sample outcomes in an event.

EXAMPLE 8A Five cards are drawn from a shuffled 52-card deck. Which event has the greater probability E_1:(*all five cards are face cards*) or E_2:(*all five cards are hearts*)?

Solution: The sample space for both events consists of all five-card groups that can be formed from the 52 cards or ${}_{52}C_5$. For E_1 we are to select five face cards from the 12 that are available (three from each suit), or ${}_{12}C_5$. The probability of five face cards is $\frac{n(E)}{n(S)} = \frac{{}_{12}C_5}{{}_{52}C_5}$, which gives $\frac{792}{2{,}598{,}960} \approx 0.0003$. For E_2 we are to select five hearts from the

13 available, or ${}_{13}C_5$. The probability of five hearts is $\frac{n(E)}{n(S)} = \frac{{}_{13}C_5}{{}_{52}C_5}$, which is $\frac{1287}{2,598,960} \approx 0.0005$. There is a slightly greater probability of drawing five hearts.

EXAMPLE 8B Of the 42 seniors at Jacoby High School, 23 are female and 19 are male. A group of five students is to be selected at random to attend a conference in Reno, Nevada. What is the probability the group will have exactly three females?

Solution: The sample space consists of all five-person groups that can be formed from the 42 seniors or ${}_{42}C_5$. The event consists of selecting 3 females from the 23 available (${}_{23}C_3$) and 2 males from the 19 available (${}_{19}C_2$). Using the fundamental principle of counting $n(E) = {}_{23}C_3 \cdot {}_{19}C_2$ and the probability the group has three females is $\frac{n(E)}{n(S)} = \frac{{}_{23}C_3 \cdot {}_{19}C_2}{{}_{42}C_5}$, which gives $\frac{302,841}{850,668} \approx 0.356$. There is approximately a 35.6% probability the group will have exactly three females.

NOW TRY EXERCISES 27 THROUGH 34

E. Probability and Mutually Exclusive Events

Two events that have no common outcomes are called **mutually exclusive** events (one excludes the other and vice versa). For example in rolling one die, E_1:(*a 2 is rolled*) and E_2:(*an odd number is rolled*) are mutually exclusive, since 2 is not an odd number. Given this fact, what is the probability of E_3:(*a 2 is rolled* or *an odd number is rolled*)? Intuitively we know the probability must be greater than $\frac{1}{2}$ since $P(E_2) = \frac{1}{2}$ all by itself. Recall that the conjunctive "or" implies the union of two disjoint intervals or sets, in this case $E_1 \cup E_2$ (which is the same as E_3 here). Since E_1 and E_2 have no common elements, $n(E_1 \cup E_2) = n(E_1) + n(E_2)$, giving the following sequence:

$$P(E_1 \cup E_2) = \frac{n(E_1 \cup E_2)}{n(S)} \quad \text{definition of probability}$$

$$= \frac{n(E_1) + n(E_2)}{n(S)} \quad \text{substitute } n(E_1) + n(E_2) \text{ for } n(E_1 \cup E_2)$$

$$= \frac{n(E_1)}{n(S)} + \frac{n(E_2)}{n(S)} \quad \text{property of rational expressions}$$

$$= P(E_1) + P(E_2) \quad \text{definition of probability}$$

PROBABILITY AND MUTUALLY EXCLUSIVE EVENTS

Given sample space S and mutually exclusive events E_1 and E_2 defined relative to S, the probability of E_1 *or* E_2 is given by

$$P(E_1 \cup E_2) = P(E_1) + P(E_2)$$

EXAMPLE 9 Pilialoha has written a program that will simulate drawing one card from a well-shuffled deck of 52 cards. Define the events E_1:(a number card is drawn), E_2:(a Jack is drawn), and E_3:(a number card or Jack is drawn). After several thousand trials the program gives a value of 0.76846 for $P(E_3) = P(E_1 \cup E_2)$. By what percent does this differ from the computed theoretical value?

Solution: E_1 and E_2 are mutually exclusive events, so $P(E_1 \cup E_2) = P(E_1) + P(E_2)$.

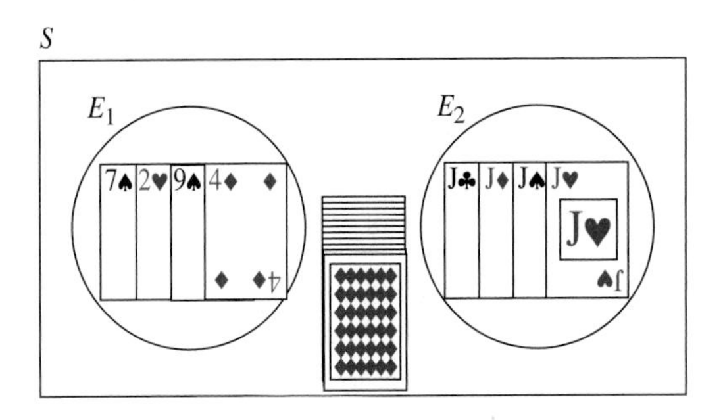

There are 36 number cards and four Jacks in a standard deck, so $P(E_1) = \frac{36}{52}$ and $P(E_2) = \frac{4}{52}$. The value of $P(E_3) = P(E_1 \cup E_2)$ is $\frac{36}{52} + \frac{4}{52} = \frac{40}{52} \approx 0.76923$. The results differ by approximately one-tenth of 1%.

NOW TRY EXERCISES 35 AND 36

F. Probability and Nonexclusive Events

WORTHY OF NOTE

This can be verified by simply counting the elements involved: $n(E_1) = 13$ and $n(E_2) = 12$ so $n(E_1) + n(E_2) = 25$. However, there are only 22 total possibilities—the J♣, Q♣, and K♣ got counted twice.

Sometimes the way events are defined causes them to share sample outcomes. Using a standard deck of playing cards once again, if we define the events E_1:(a club is drawn) and E_2:(a face card is drawn), they share the outcomes J♣, Q♣, and K♣ as shown in Figure 8.23. This overlapping region is the intersection of the events, or $E_1 \cap E_2$. If we compute $n(E_1 \cup E_2)$ as $n(E_1) + n(E_2)$ as before, this intersecting region gets counted *twice*!

Figure 8.23

In cases where the events are **nonexclusive** (not mutually exclusive), we maintain the correct count by subtracting one of the two intersections, obtaining $n(E_1 \cup E_2) = n(E_1) + n(E_2) - n(E_1 \cap E_2)$. This leads to the following calculation for nonexclusive events:

$$P(E_1 \cup E_2) = \frac{n(E_1) + n(E_2) - n(E_1 \cap E_2)}{n(S)} \quad \text{definition of probability}$$

$$= \frac{n(E_1)}{n(S)} + \frac{n(E_1)}{n(S)} - \frac{n(E_1 \cap E_2)}{n(S)} \quad \text{property of rational expressions}$$

$$= P(E_1) + P(E_2) - P(E_1 \cap E_2) \quad \text{definition of probability}$$

PROBABILITY AND NONEXCLUSIVE EVENTS

Given sample space S and *nonexclusive events* E_1 and E_2 defined relative to S, the probability of E_1 *or* E_2 is given by

$$P(E_1 \cup E_2) = P(E_1) + P(E_2) - P(E_1 \cap E_2)$$

EXAMPLE 10A What is the probability that a club or a face card is drawn from a standard deck of 52 well-shuffled cards?

Solution: As before, define the events E_1:(a club is drawn) and E_2:(a face card is drawn). Since there are 13 clubs and 12 face cards, $P(E_1) = \frac{13}{52}$ and $P(E_2) = \frac{12}{52}$. But three of the face cards are clubs, so $P(E_1 \cap E_2) = \frac{3}{52}$. This leads to

$$\begin{aligned} P(E_1 \cup E_2) &= P(E_1) + P(E_2) - P(E_1 \cap E_2) && \text{nonexclusive events} \\ &= \frac{13}{52} + \frac{12}{52} - \frac{3}{52} && \text{substitute} \\ &= \frac{22}{52} \approx 0.423 && \text{combine terms} \end{aligned}$$

EXAMPLE 10B A survey of 100 voters was taken to gather information on critical issues and the demographic information collected is shown in the table. One out of the 100 voters is to be drawn at random to be interviewed on the 5 o'Clock News. What is the probability the person is a woman or a Republican?

	Women	Men	Totals
Republican	17	20	37
Democrat	22	17	39
Independent	8	7	15
Green Party	4	1	5
Tax Reform	2	2	4
Totals	53	47	100

Solution: Since there are 53 women and 37 Republicans, $P(W) = 0.53$ and $P(R) = 0.37$. The table shows 17 people are both female and Republican so $P(W \cap R) = 0.17$.

$$\begin{aligned} P(W \cup R) &= P(W) + P(R) - P(W \cap R) && \text{nonexclusive events} \\ &= 0.53 + 0.37 - 0.17 && \text{substitute} \\ &= 0.73 && \text{combine} \end{aligned}$$

There is a 73% probability the person is a woman or a Republican.

NOW TRY EXERCISES 37 THROUGH 50

TECHNOLOGY HIGHLIGHT

Principles of Quick-Counting, Combinations, and Probability

The keystrokes shown apply to a TI-84 Plus model. Please consult your manual or our Internet site for other models.

You likely noticed that retrieving the ${}_nC_r$ operation from Example 8 required a lot of "fancy finger work," since the operation is contained in a submenu and requires an argument both before ${}_nC_r$ (the number n) and after ${}_nC_r$ (the value of r). There are many ways to make this process more efficient. Here we'll illustrate how to use the ${}_nC_r$ operation as a function. In many other cases, a simple program can be employed.

At this point you are likely using the Y= screen and tables (TBLSET , 2nd GRAPH **TABLE,** and so on) with relative ease. When probability calculations require a repeated use of permutations and combinations, these features can make the work much more efficient and help to explore the various patterns they generate. Let's observe the various possibilities for choosing r children from a group of six children ($n = 6$). To begin, set the TBLSET to **AUTO,** then press Y= and enter 6 ${}_nC_r$ X as Y_1 (Figure 8.24). Access the **TABLE** (2nd GRAPH) and note that the calculator automatically computes the value of ${}_6C_0$, ${}_6C_1$, ${}_6C_2$, . . . , ${}_6C_6$ (Figure 8.25) Also note the pattern of outputs is symmetric. If we wanted to investigate calculations similar to those required in Example 8B (${}_{23}C_3 \cdot {}_{19}C_2$), we can enter $Y_1 = 23$ ${}_nC_r$ X, $Y_2 = 19$ ${}_nC_r$(X − 1), and $Y_3 = Y_1 \cdot Y_2$, or any variation of these depending on what the situation calls for. Use these ideas to work the following exercises.

Figure 8.24

```
Plot1 Plot2 Plot3
\Y1■6 nCr X
\Y2=
\Y3=
\Y4=
\Y5=
\Y6=
\Y7=
```

Figure 8.25

X	Y_1
0	1
1	6
2	15
3	20
4	15
5	6
6	1

X=0

Exercise 1: Use your calculator to display the values of ${}_5C_0$, ${}_5C_1$, . . . , ${}_5C_5$. Is the result a pattern similar to that for ${}_6C_0$, ${}_6C_1$, ${}_6C_2$, . . . , ${}_6C_6$? Repeat for ${}_7C_r$. Why are the "middle values" repeated for $n = 7$ and $n = 5$, but not for $n = 6$? Because of the symmetry of ${}_nC_r$

Exercise 2: A committee consists of 10 Republicans and eight Democrats. To maintain a simple majority on an important subcommittee that is to be formed, there must be one more Republican than Democrat. The subcommittee can have as few as three (2 R's and 1 D) and as many as nine (five R's and four D's) members. Use your calculator and the preceding ideas to explore the various possibilities for forming such a committee. Specifically, in how many ways can a committee of four Republicans and three Democrats be formed? 11,760

8.6 EXERCISES

CONCEPTS AND VOCABULARY

Fill in the blank with the appropriate word or phrase. Carefully reread the section if needed.

1. Given a sample space S and an event E defined relative to S, $P(E) = \frac{n(E)}{n(S)}$.

2. In elementary probability, we consider all events in the sample space to be equally likely.

3. Given a sample space S and an event E defined relative to S: $\underline{0} \le P(E) \le \underline{1}$, $P(S) = \underline{1}$, and $P(\sim S) = \underline{0}$.

4. The complement of an event E is the set of sample outcomes in S, which are not contained in E.

5. Discuss/explain the difference between mutually exclusive events and nonmutually exclusive events. Give an example of each. Answers will vary.

6. A single die is rolled. With no calculations, explain why the probability of rolling an even number is greater than rolling a number greater than four. Answers will vary.

DEVELOPING YOUR SKILLS

State the sample space S and the probability of a single outcome. Then define any two events E relative to S (many answers possible).

Exercise 8

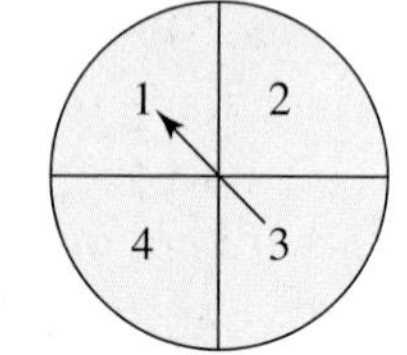

7. Two fair coins (heads and tails) are flipped. $S = \{HH, HT, TH, TT\}, \frac{1}{4}$

8. The simple spinner shown is spun. $S = \{1, 2, 3, 4\}, \frac{1}{4}$

9. The head coaches for six little league teams (the Patriots, Cougars, Angels, Sharks, Eagles, and Stars) have gathered to discuss new changes in the rule book. One of them is randomly chosen to ask the first question.

10. Experts on the planets Mercury, Venus, Mars, Jupiter, Saturn, Uranus, Neptune, and Pluto have gathered at a space exploration conference. One group of experts is selected at random to speak first.

9. S = {coach of Patriots, Cougars, Angels, Sharks, Eagles, Stars}, $\frac{1}{6}$

10. S = {experts for Mercury, Venus, Mars, Jupiter, Saturn, Uranus, Neptune, Pluto}, $\frac{1}{8}$

Find $P(E)$ for the events E_n defined.

11. Nine index cards numbered 1 through 9 are shuffled and placed in an envelope, then one of the cards is randomly drawn. Define event E as *the number drawn is even.* $P(E) = \frac{4}{9}$

12. Eight flash cards used for studying basic geometric shapes are shuffled and one of the cards is drawn at random. The eight cards include information on circles, squares, rectangles, kites, trapezoids, parallelograms, pentagons, and triangles. Define event E as *a quadrilateral is drawn.* $P(\text{quad}) = \frac{5}{8}$

13. One card is drawn at random from a standard deck of 52 cards. What is the probability of

   a. drawing a Jack $\frac{1}{13}$
   b. drawing a spade $\frac{1}{4}$
   c. drawing a black card $\frac{1}{2}$
   d. drawing a red three $\frac{1}{26}$

14. Pinochle is a card game played with a deck of 48 cards consisting of 2 Aces, 2 Kings, 2 Queens, 2 Jacks, 2 Tens, and 2 Nines in each of the four standard suits [hearts (♥), diamonds (♦), spades (♠), and clubs (♣)]. If one card is drawn at random from this deck, what is the probability of

   a. drawing an Ace $\frac{1}{6}$
   b. drawing a club $\frac{1}{4}$
   c. drawing a red card $\frac{1}{2}$
   d. drawing a face card (Jack, Queen, King) $\frac{1}{2}$

15. A group of finalists on a game show consists of three males and five females. Hank has a score of 520 points, with Harry and Hester having 490 and 475 points, respectively. Madeline has 532 points, with Mackenzie, Morgan, Maggie, and Melanie having 495, 480, 472, and 470 points, respectively. One of the contestants is randomly selected to start the final round. Define E_1 as *Hester is chosen,* E_2 as *a female is chosen,* and E_3 as *a contestant with less than 500 points is chosen.* Find the probability of each event. $P(E_1) = \frac{1}{8}, P(E_2) = \frac{5}{8}, P(E_3) = \frac{3}{4}$

16. Soccer coach Maddox needs to fill the last spot on his starting roster for the opening day of the season and has to choose between three forwards and five defenders. The forwards have jersey numbers 5, 12, and 17, while the defenders have jersey numbers 7, 10, 11, 14, and 18. Define E_1 as *a forward is chosen,* E_2 as *a defender is chosen,* and E_3 as *a player whose jersey number is greater than 10 is chosen.* Find the probability of each event. $P(E_1) = \frac{3}{8}, P(E_2) = \frac{5}{8}, P(E_3) = \frac{5}{8}$

17. A game is played using a spinner like the one shown. For each spin,

   a. What is the probability the arrow lands in a shaded region? $\frac{3}{4}$
   b. What is the probability your spin is less than 5? 1
   c. What is the probability you spin a 2? $\frac{1}{4}$
   d. What is the probability the arrow lands on a prime number? $\frac{1}{2}$

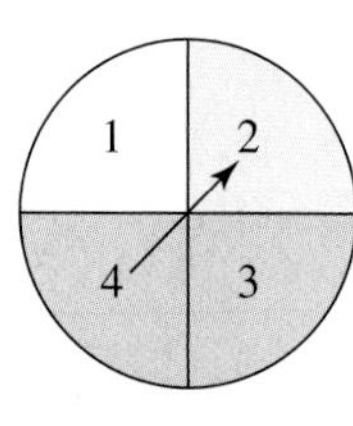

18. A game is played using a spinner like the one shown here. For each spin,

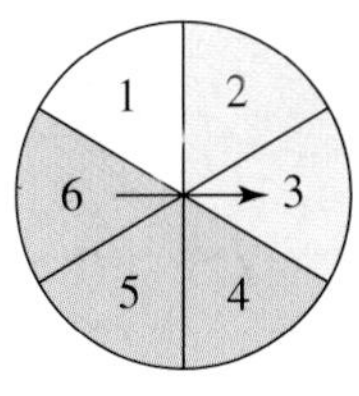

a. What is the probability the arrow lands in a lightly shaded region? $\frac{1}{3}$

b. What is the probability your spin is greater than 2? $\frac{2}{3}$

c. What is the probability the arrow lands in a shaded region? $\frac{5}{6}$

d. What is the probability you spin a 5? $\frac{1}{6}$

Use the complementary events to complete Exercises 19 through 22.

19. One card is drawn from a standard deck of 52. What is the probability it is not a club? $\frac{3}{4}$

20. Four standard dice are rolled. What is the probability the sum is less than 24? $\frac{1295}{1296}$

21. A single digit is randomly selected from among the digits of 10!. What is the probability the digit is not a 2? $\frac{6}{7}$

22. A corporation will be moving its offices to Los Angeles, Miami, Atlanta, Dallas, or Phoenix. If the site is randomly selected, what is the probability Dallas is not chosen? $\frac{4}{5}$

23. A large manufacturing plant can remain at full production as long as one of its two generators is functioning. Due to past experience and the age difference between the systems, the plant manager estimates the probability of the main generator failing is 0.05, the probability of the secondary generator failing is 0.01, and the probability of both failing is 0.009. What is the probability the plant remains in full production today? 0.991

24. A fire station gets an emergency call from a shopping mall in the mid-afternoon. From a study of traffic patterns, Chief Nozawa knows the probability the most direct route is clogged with traffic is 0.07, while the probability of the secondary route being clogged is 0.05. The probability both are clogged is 0.02. What is the probability they can respond to the call unimpeded using one of these routes? 0.98

25. Two fair dice are rolled (see Figure 8.22). What is the probability of

a. a sum less than four $\frac{1}{12}$

b. a sum less than eleven $\frac{11}{12}$

c. the sum is not nine $\frac{8}{9}$

d. a roll is not a "double" (both dice the same) $\frac{5}{6}$

"Double-six" dominos is a game played with the 28 numbered tiles shown in the diagram.

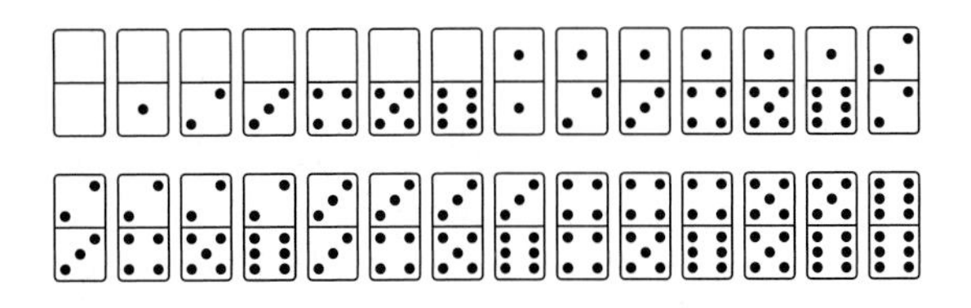

26. The 28 dominos are placed in a bag, shuffled, and then one domino is randomly drawn. What is the probability the total number of dots on the domino

a. is three or less $\frac{1}{7}$

b. is greater than three $\frac{6}{7}$

c. does not have a blank half $\frac{3}{4}$

d. is not a "double" (both sides the same) $\frac{3}{4}$

Find $P(E)$ given the values for $n(E)$ and $n(S)$ shown.

27. $n(E) = {}_6C_3 \cdot {}_4C_2; n(S) = {}_{10}C_5$ $\frac{10}{21}$

28. $n(E) = {}_{12}C_9 \cdot {}_8C_7; n(S) = {}_{20}C_{16}$ $\frac{352}{969}$

29. $n(E) = {}_9C_6 \cdot {}_5C_3; n(S) = {}_{14}C_9$ $\frac{60}{143}$

30. $n(E) = {}_7C_6 \cdot {}_3C_2; n(S) = {}_{10}C_8$ $\frac{7}{15}$

31. Five cards are drawn from a well-shuffled, standard deck of 52 cards. Which has the greater probability (a) all five cards are red or (b) all five cards are numbered cards? How much greater? b, about 12%

32. Five cards are drawn from a well-shuffled pinochle deck of 48 cards (see Exercise 14). Which has the greater probability (a) all five cards are face cards (King, Queen, or Jack) or (b) all five cards are black? How much greater? probabilities are equal, 0

33. A dietetics class has 24 students. Of these, 9 are vegetarians and 15 are not. The instructor receives enough funding to send six students to a conference. If the students are selected randomly, what is the probability the group will have

a. exactly two vegetarians 0.3651
b. exactly four nonvegetarians 0.3651
c. at least three vegetarians 0.3969

34. A large law firm has a support staff of 15 employees: 6 paralegals and 9 legal assistants. Due to recent changes in the law, the firm wants to send five of them to a forum on the new changes. If the selection is done randomly, what is the probability the group will have

a. exactly three paralegals 0.2398
b. exactly two legal assistants 0.2398
c. at least two paralegals 0.7063

35. Two fair dice are rolled. What is the probability of rolling

a. boxcars (a sum of 12) or snake eyes (a sum of 2) $\frac{1}{18}$
b. a sum of 7 or a sum of 11 $\frac{2}{9}$
c. an even-numbered sum or a prime sum $\frac{8}{9}$
d. an odd-numbered sum or a sum that is a multiple of 4 $\frac{3}{4}$
e. a sum of 15 or a multiple of 12 $\frac{1}{36}$
f. a sum that is a prime number $\frac{5}{12}$

36. *Eight Ball* is a game played on a pool table with 15 balls numbered 1 through 15 and a cue ball that is solid white. Of the 15 numbered balls, 8 are a solid (nonwhite) color and numbered 1 through 8, and seven are striped balls numbered 9 through 15. All 16 balls are placed in a large leather bag and mixed, then one is drawn out. Consider the cue ball as "0." What is the probability of drawing

a. a striped ball $\frac{7}{16}$
b. a solid-colored ball $\frac{9}{16}$
c. a polka-dotted ball 0
d. the cue ball $\frac{1}{16}$
e. the cue ball or the eight ball $\frac{1}{8}$
f. a striped ball or a number less than five $\frac{3}{4}$
g. a solid color or a number greater than 12 $\frac{3}{4}$
h. an odd number or a number divisible by 4 $\frac{3}{4}$

Find the probability indicated using the information given.

37. Given $P(E_1) = 0.7$, $P(E_2) = 0.5$, and $P(E_1 \cap E_2) = 0.3$, compute $P(E_1 \cup E_2)$. 0.9

38. Given $P(E_1) = 0.6$, $P(E_2) = 0.3$, and $P(E_1 \cap E_2) = 0.2$, compute $P(E_1 \cup E_2)$. 0.7

39. Given $P(E_1) = \frac{3}{8}$, $P(E_2) = \frac{3}{4}$, and $P(E_1 \cup E_2) = \frac{15}{18}$; compute $P(E_1 \cap E_2)$. $\frac{7}{24}$

40. Given $P(E_1) = \frac{1}{2}$, $P(E_2) = \frac{3}{5}$, and $P(E_1 \cup E_2) = \frac{17}{20}$; compute $P(E_1 \cap E_2)$. $\frac{1}{4}$

41. Given $P(E_1 \cup E_2) = 0.72$, $P(E_2) = 0.56$, and $P(E_1 \cap E_2) = 0.43$; compute $P(E_1)$. 0.59

42. Given $P(E_1 \cup E_2) = 0.85$, $P(E_1) = 0.4$, and $P(E_1 \cap E_2) = 0.21$; compute $P(E_2)$. 0.66

43. Two fair dice are rolled. What is the probability the sum of the dice is

a. a multiple of 3 and an odd number $\frac{1}{6}$
b. a sum greater than 5 and a 3 on one die $\frac{7}{36}$
c. an even number and a number greater than 9 $\frac{1}{9}$
d. an odd number and a number less than 10 $\frac{4}{9}$

44. The fifteen numbered pool balls (no cueball—see Exercise 36) are placed in a large bowl and mixed, then one is drawn out. What is the probability of drawing

a. the eight ball $\frac{1}{15}$
b. a number greater than fifteen 0
c. an even number $\frac{7}{15}$
d. a multiple of three $\frac{1}{3}$
e. a solid color and an even number $\frac{4}{15}$
f. a striped ball and an odd number $\frac{4}{15}$
g. an even number and a number divisible by three $\frac{2}{15}$
h. an odd number and a number divisible by 4 0

45. A survey of 50 veterans was taken to gather information on their service career and what life is like out of the military. A breakdown of those surveyed is shown in the table. One out of the 50 will be selected at random for an interview and a biographical sketch. What is the probability the person chosen is

	Women	Men	Totals
Private	6	9	15
Corporal	10	8	18
Sergeant	4	5	9
Lieutenant	2	1	3
Captain	2	3	5
Totals	24	26	50

a. a woman and a sergeant $\frac{2}{25}$

b. a man and a private $\frac{9}{50}$

c. a private and a sergeant 0

d. a woman and an officer $\frac{2}{25}$

e. a person in the military 1

46. Referring to Exercise 45, what is the probability the person chosen is

a. a woman or a sergeant $\frac{29}{50}$

b. a man or a private $\frac{16}{25}$

c. a woman or a man 1

d. a woman or an officer $\frac{14}{25}$

e. a captain or a lieutenant $\frac{4}{25}$

A computer is asked to randomly generate a three-digit number. What is the probability the

47. ten's digit is odd or the one's digit is even $\frac{3}{4}$

48. first digit is prime and the number is a multiple of 10 $\frac{4}{45}$

A computer is asked to randomly generate a four-digit number. What is the probability the number is

49. at least 4000 or a multiple of 5 $\frac{11}{15}$

50. less than 7000 and an odd number $\frac{1}{3}$

WORKING WITH FORMULAS

51. Games involving a fair spinner (with numbers 1 through 4): $P(n) = (\frac{1}{4})^n$

Games that involve moving pieces around a board using a fair spinner are fairly common. If a fair spinner has the numbers 1 through 4, the probability that any one number is spun n times in succession is given by the formula shown, where n represents the number of spins. What is the probability (a) the first player spins a two? (b) all four players spin a two? (c) Discuss the graph of $P(n)$ and explain the connection between the graph and the probability of consistently spinning a two. $\frac{1}{4}$; $\frac{1}{256}$; answers will vary.

52. Games involving a fair coin (heads and tails): $P(n) = (\frac{1}{2})^n$

When a fair coin is flipped, the probability that heads (or tails) is flipped n times in a row is given by the formula shown, where n represents the number of flips. What is the probability (a) the first flip is a heads? (b) the first four flips are heads? (c) Discuss the graph of $P(n)$ and explain the connection between the graph and the probability of consistently flipping heads. $\frac{1}{2}$; $\frac{1}{16}$; answers will vary.

APPLICATIONS

53. To improve customer service, a company tracks the number of minutes a caller is "on hold" and waiting for a customer service representative. The table shows the probability that a caller will wait m minutes. Based on the table, what is the probability a caller waits

a. at least two minutes 0.33

b. less than two minutes 0.67

c. four minutes or less 1

d. over four minutes 0

e. less than two or more than four minutes 0.67

f. three or more minutes 0.08

Exercise 53

Wait Time (minutes m)	Probability
0	0.07
$0 < m < 1$	0.28
$1 \le m < 2$	0.32
$2 \le m < 3$	0.25
$3 \le m < 4$	0.08

Exercise 54

Number of Computers	Probability
0	9%
1	51%
2	28%
3	9%
4	3%

54. To study the impact of technology on American families, a researcher first determines the probability that a family has n computers at home. Based on the table, what is the probability a home

a. has at least one computer 91% **b.** has two or more computers 40%
c. has less than four computers 97% **d.** has five computers 0%
e. has one, two, or three computers 88% **f.** does not have two computers 72%

Jolene is an experienced markswoman and is able to hit a 10 in. by 20 in. target 100% of the time at a range of 100 yd. Assuming the probability she hits a target is related to its area, what is the probability she hits the shaded portions shown?

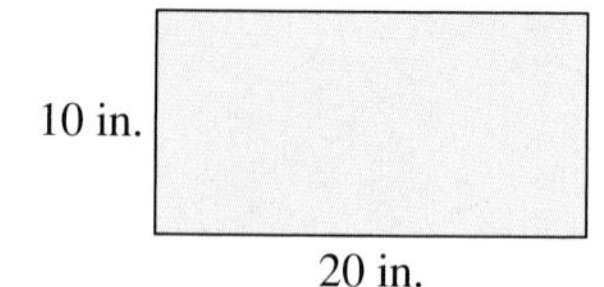

55. a. $\frac{1}{2}$
b. $\frac{1}{2}$
c. 0.2165

55. a. isosceles triangle **b.** right triangle **c.** equilateral triangle

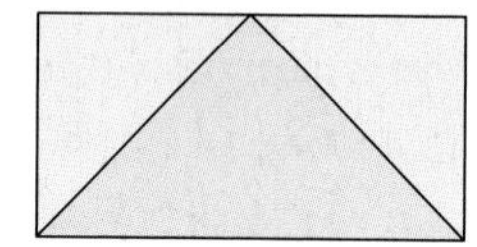
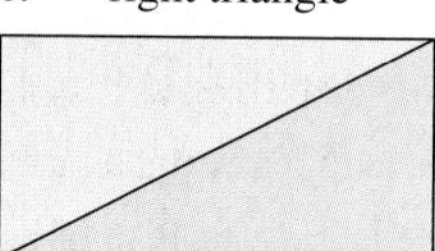
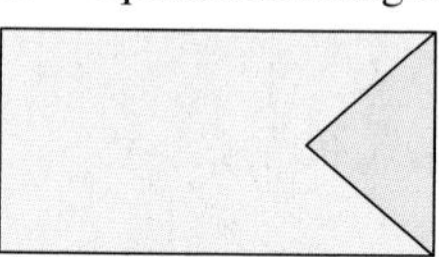

56. a. $\frac{1}{2}$
b. $\frac{\pi}{8}$
c. $\frac{3}{4}$

56. a. square **b.** circle **c.** isosceles trapezoid with $b = \frac{B}{2}$

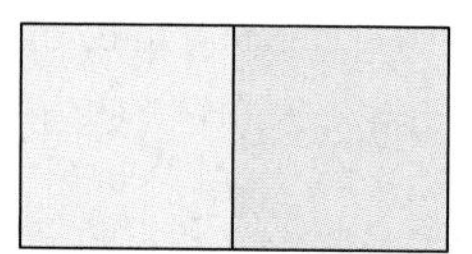
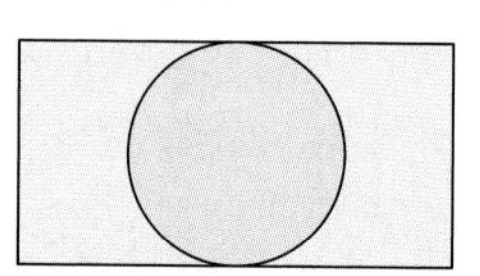
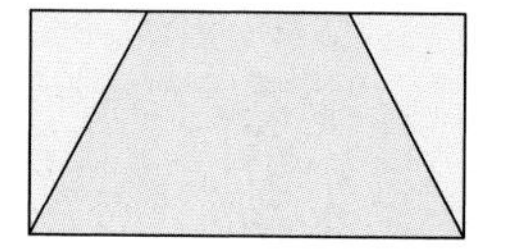

57. A circular dartboard has a total radius of 8 in., with circular bands that are 2 in. wide, as shown. You are skilled enough to hit this board 100% of the time so you always score at least two points each time you throw a dart. Assuming the probabilities are related to area, on the next dart that you throw what is the probability you

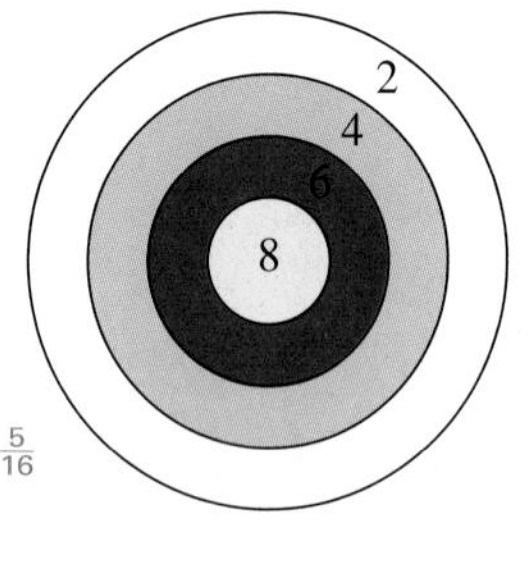

a. score at least a 4? $\frac{9}{16}$ **b.** score at least a 6? $\frac{1}{4}$
c. hit the bull's-eye? $\frac{1}{16}$ **d.** score exactly 4 points? $\frac{5}{16}$

Exercise 58

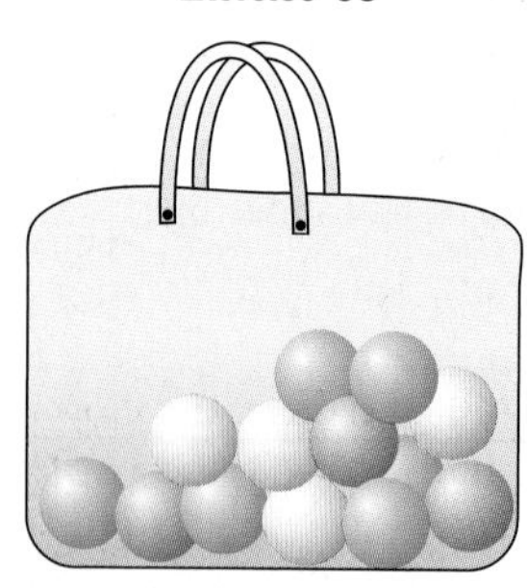

58. Three red balls, six blue balls, and four white balls are placed in a bag. What is the probability the first ball you draw out is

a. red $\frac{3}{13}$ **b.** blue $\frac{6}{13}$ **c.** not white $\frac{9}{13}$
d. purple 0 **e.** red or white $\frac{7}{13}$ **f.** red and white 0

59. Three red balls, six blue balls, and four white balls are placed in a bag, then two are drawn out and placed in a rack. What is the probability the balls drawn are

a. first red, second blue $\frac{3}{26}$ **b.** first blue, second red $\frac{3}{26}$ **c.** both white $\frac{1}{13}$
d. first blue, second not red $\frac{9}{26}$ **e.** first white, second not blue $\frac{2}{13}$ **f.** first not red, second not blue $\frac{11}{26}$

60. Consider the 210 discrete points found in the first and second quadrants where $-10 \leq x \leq 10$, $1 \leq y \leq 10$, and x and y are integers. The coordinates of these points are written on a slip of paper and placed in a box. One of the slips is then randomly drawn. What is the probability the point (x, y) drawn

a. is on the graph of $y = |x|$ $\frac{2}{21}$ **b.** is on the graph of $y = 2|x|$ $\frac{1}{21}$
c. is on the graph of $y = 0.5|x|$ $\frac{1}{21}$ **d.** has coordinates $(x, y > -2)$ 1
e. has coordinates $(x \leq 5, y > -2)$ $\frac{16}{21}$ **f.** is between the branches of $y = x^2$ $\frac{1}{5}$

61. Your instructor surprises you with a True/False quiz for which you are totally unprepared and must guess randomly. What is the probability you pass the quiz with an 80% or better if there are

a. three questions $\frac{1}{8}$ **b.** four questions $\frac{1}{16}$ **c.** five questions $\frac{3}{16}$

62. A robot is sent out to disarm a timed explosive device by randomly changing some switches from a neutral position to a *positive flow* or *negative flow* position. The problem is, the switches are independent and unmarked, and it is unknown which direction is positive and which direction is negative. The bomb is harmless if a majority of the switches yield a positive flow. All switches must be thrown. What is the probability the device is disarmed if there are

a. three switches $\frac{1}{2}$ **b.** four switches $\frac{5}{16}$ **c.** five switches $\frac{1}{2}$

63. A survey of 100 retirees was taken to gather information concerning how they viewed the Vietnam War back in the early 1970s. A breakdown of those surveyed is shown in the table. One out of the hundred will be selected at random for a personal, taped interview. What is the probability the person chosen had a

Career	Support	Opposed	T
Military	9	3	12
Medical	8	16	24
Legal	15	12	27
Business	18	6	24
Academics	3	10	13
Totals	53	47	100

a. career of any kind and opposed the war $\frac{47}{100}$

b. medical career and supported the war $\frac{2}{25}$

c. military career and opposed the war $\frac{3}{100}$

d. legal or business career and opposed the war $\frac{9}{50}$

e. academic or medical career and supported the war $\frac{11}{100}$

64. Referring to Exercise 63, what is the probability the person chosen

a. had a career of any kind or opposed the war 1

b. had a medical career or supported the war 0.69

c. supported the war or had a military career 0.56

d. had a medical or a legal career 0.51

e. supported or opposed the war 1

65. The Board of Directors for a large hospital has 15 members. There are six doctors of nephrology (kidneys), five doctors of gastroenterology (stomach and intestines), and four doctors of endocrinology (hormones and glands). Eight of them will be selected to visit the nation's premier hospitals on a 3-week, expenses-paid tour. What is the probability the group of eight selected consists of exactly

a. four nephrologists and four gastroenterologists $\frac{5}{429}$

b. three endocrinologists and five nephrologists $\frac{8}{2145}$

66. A support group for hodophobics (an irrational fear of travel) has 32 members. There are 15 aviophobics (fear of air travel), eight siderodrophobics (fear of train travel), and nine thalassophobics (fear of ocean travel) in the group. Twelve of them will be randomly selected to participate in a new therapy. What is the probability the group of 12 selected consists of exactly

a. two aviophobics, six siderodrophobics, and four thalassophobics ≈ 0.002

b. five thalassophobics, four aviophobics, and three siderodrophobics ≈ 0.043

67. A trained chimpanzee is given a box containing eight wooden cubes with the letters p, a, r, a, l, l, e, l printed on them (one letter per block). Assuming the chimp can't read or spell, what is the probability he draws the eight blocks in order and actually forms the word "parallel?" $\frac{1}{3360}$

68. A number is called a "perfect number" if the sum of its proper factors is equal to the number itself. Six is the first perfect number since the sum of its proper factors is six: $1 + 2 + 3 = 6$. Twenty-eight is the second since: $1 + 2 + 4 + 7 + 14 = 28$. A young child is given a box containing eight wooden blocks with the following numbers (one per block) printed on them: four 3's, two 5's, one 0, and one 6. What is the probability she draws the eight blocks in order and forms the fifth perfect number: 33,550,336? $\frac{1}{840}$

WRITING, RESEARCH, AND DECISION MAKING

69. The function $f(x) = \left(\frac{1}{2}\right)^x$ gives the probability that x number of flips will all result in heads (or tails). Compute the probability that 20 flips results in *20 heads in a row,* then use the Internet or some other resource to find the probability of winning a state lottery. Which is more likely to happen (which has the greater probability)? Were you surprised? $\frac{1}{1,048,576}$; answers will vary; 20 heads in a row.

70. It is well known that the cost of auto insurance varies depending on age, gender, driving record, and other factors. Insurance for young men is more expensive than for young women. Insurance for teens costs more than insurance for adults. Either by Internet research, phone calls, or personal interviews, gather information regarding the probabilities (likelihood of an accident) an insurance company assigns to various groups in order to set their rates. Summarize how these probabilities impact the cost of a person's insurance. Answers will vary.

EXTENDING THE CONCEPT

71. The concept of mutually exclusive and nonexclusive events has a very nice geometric interpretation. Consider the Venn diagram shown, which uses squares rather than circles to represent an event. Notice the area of the sample space is $(20)(36) = 720\text{ cm}^2$ and the area of each square is $(12)(12) = 144\text{ cm}^2$. In this case, the probability of an event can be viewed as $P(E) = \frac{\text{area of } E}{\text{area of } S}$, which means $P(E_1) = \frac{144}{720} = P(E_2)$ or 0.2. As drawn, the two squares are mutually exclusive (no overlap). The squares are still mutually exclusive if we draw them side by side to form a rectangle, with the rectangle showing the combined probability of E_1 or E_2 is still $P(E_1 \cup E_2) = \frac{\text{area of rectangle}}{\text{area of } S}$ or $\frac{288}{720} = 0.4$. *If the squares overlap by 4 cm,* they are no longer mutually exclusive and the dimensions of the rectangle becomes 20 by 12, giving a ratio of $\frac{\text{area of rectangle}}{\text{area of } S} = \frac{240}{720} = 0.\overline{3}$. This is exactly the value obtained using the area concept and $P(E_1 \cup E_2) = P(E_1) + P(E_2) - P(E_1 \cap E_2)$, where $P(E_1 \cup E_2) = \frac{144}{720} + \frac{144}{720} - \frac{48}{720} = \frac{240}{720}$ or $0.\overline{3}$. Repeat this investigation using two 10×10 in. squares on a 12×24 in. sample space, with the squares overlapping 3 in. Include the diagrams. $P(E_1) = P(E_2) = \frac{100}{288}$; $P(E_1 \cup E_2) = \frac{170}{288}$

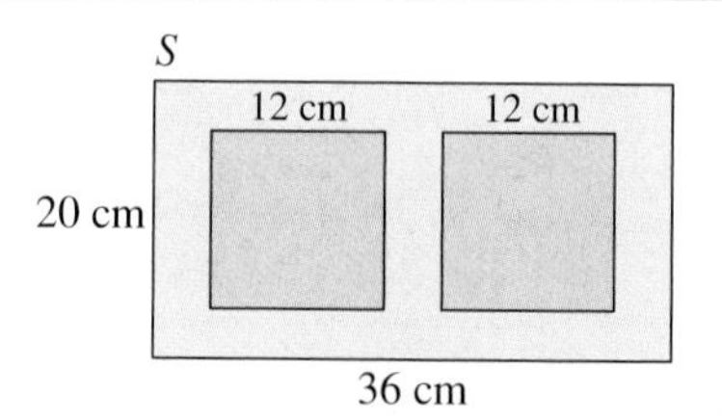

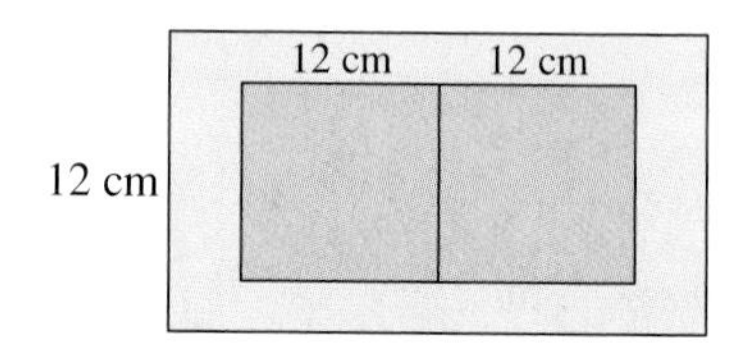

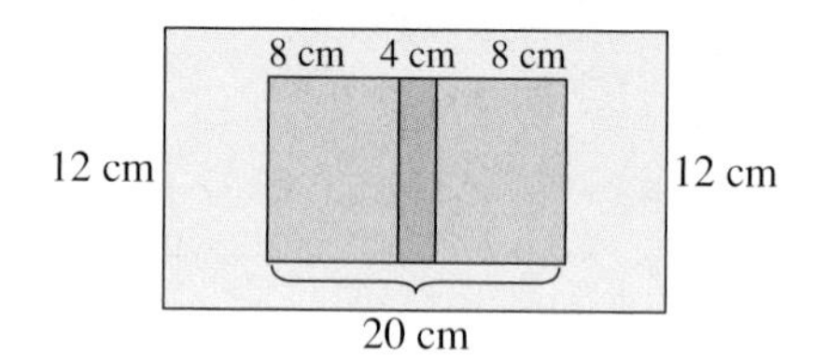

72. Recall that a function is a relation in which each element of the domain is paired with only one element from the range. Is the relation defined by $C(x) = {}_nC_x$ (n is a constant) a function? To investigate, plot the points generated by $C(x) = {}_6C_x$ for $x = 0$ to $x = 6$ and answer the following questions:

a. Is the resulting graph continuous or discrete (made up of distinct points)? discrete

b. Does the resulting graph pass the vertical line test? yes

c. Discuss the features of the relation and its graph, including the domain, range, maximum or minimum values, and symmetries observed. Answers will vary.

MAINTAINING YOUR SKILLS

73. (6.5) Solve the system using matrices and row reduction: $\begin{cases} x - 2y + 3z = 10 \\ 2x + y - z = 18 \\ 3x - 2y + z = 26 \end{cases}$ (9, 1, 1)

74. (5.3) Complete the following logarithmic properties:

$\log_b b = \underline{1}$ $\quad \log_b 1 = \underline{0}$

$\log_b b^n = \underline{n}$ $\quad b^{\log_b n} = \underline{n}$

75.

77.

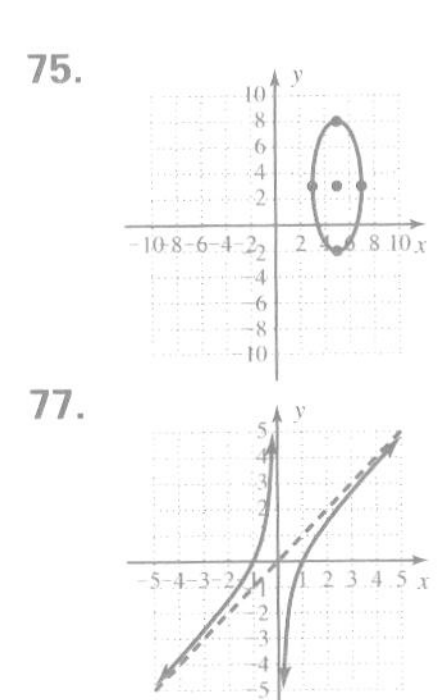

75. (7.1) Graph the ellipse defined by $\frac{(x-5)^2}{4} + \frac{(y-3)^2}{25} = 1$

76. (8.1) Compute the first six terms of the recursive sequence defined by $a_1 = 64$ and $a_{n+1} = \frac{a_n}{-2} + 128$. 64, 96, 80, 88, 84, 86

77. (4.6/4.7) Solve the inequality by graphing the function and labeling the appropriate interval(s): $\frac{x^2 - 1}{x} \geq 0$. $x \in [-1, 0) \cup [1, \infty)$

78. (8.3) A rubber ball is dropped from a height of 25 ft onto a hard surface. With each bounce, it rebounds 60% of the height from which it last fell. Use sequences/series to find (a) the height of the sixth bounce, (b) the total distance traveled up to the sixth bounce, and (c) the distance the ball will travel before coming to rest. a. ≈1.17 ft b. ≈94.17 ft c. 100 ft

8.7 The Binomial Theorem

LEARNING OBJECTIVES

In Section 8.7 you will learn how to:

A. Use Pascal's triangle to find $(a + b)^n$

B. Find binomial coefficients using $\binom{n}{k}$ notation

C. Use the binomial theorem to find $(a + b)^n$

D. Find a specific term of a binomial expansion

INTRODUCTION

In Section R.3 we defined a binomial as a polynomial with two terms. This limited us to terms with real number coefficients and whole number powers on variables. In this section, we will loosely regard a binomial as the sum or difference of *any* two terms. Hence $3x^2 - y^4$, $\sqrt{x} + 4$, $x + \frac{1}{x}$, and $-\frac{1}{2} + \frac{\sqrt{3}}{2}i$ are all "binomials." Our goal is to develop an ability to raise a binomial to any natural number power, with the results having important applications in genetics, probability, polynomial theory, and other areas. The tool used for this purpose is called the *binomial theorem*.

POINT OF INTEREST

Blaise Pascal, French philosopher, mathematician, and physicist, is considered one of the great intellects in the history of western civilization. In 1642 he invented one of the first mechanical calculators to aid his father, a tax official, in his calculations. Twelve years later (1654), in response to questions posed regarding the games of chance that were popular at the time, Pascal and his good friend Pierre de Fermat laid the foundation for the mathematical theory of probability. In 1665, his treatise entitled *Traité du trinagle artihmétique* was posthumously published, forever fixing his name to the table of numbers we now know as Pascal's triangle—although there is evidence the table was known to the Chinese and Arab cultures as early as 1050.

A. Binomial Powers and Pascal's Triangle

A good deal of mathematical understanding comes from a study of patterns. One area where the study of patterns has been particularly fruitful is **Pascal's triangle** (Figure 8.26), named after the French scientist Blaise Pascal although the triangle was well known before his time. It begins with a "1" at the vertex of the triangle, with 1's extending diagonally downward to the left and right as shown. The entries on the interior of the triangle are found by adding the two entries directly above and to the left and right of each new position.

There are a variety of patterns hidden within the triangle. In this section we'll use the *horizontal rows* of the triangle to help us raise a binomial to various powers. To begin, recall that $(a + b)^0 = 1$ and $(a + b)^1 = 1a + 1b$ (unit coefficients are included for emphasis). In our earlier work, we saw that a binomial square (a binomial raised to the second power) always followed the pattern $(a + b)^2 = 1a^2 + 2ab + 1b^2$. Observe the overall pattern that is developing as we include $(a + b)^3$:

Figure 8.26

1	First row
1 1	Second row
1 2 1	Third row
1 3 3 1	Fourth row
1 ? 6 ? 1	
1 ? ? ? ? 1	
and so on	

$$\begin{array}{lcr} (a+b)^0 & 1 & \text{row 1} \\ (a+b)^1 & 1a + 1b & \text{row 2} \\ (a+b)^2 & 1a^2 + 2ab + 1b^2 & \text{row 3} \\ (a+b)^3 & 1a^3 + 3a^2b + 3ab^2 + 1b^3 & \text{row 4} \end{array}$$

Apparently the coefficients of $(a + b)^n$ will occur in row $n + 1$ of Pascal's triangle. Also observe that in each term of the expansion, the exponent of the first term *a decreases by 1* as the exponent on the second term *b increases by 1*, keeping the degree of each term constant (recall the degree of a term with more than one variable is the sum of the exponents).

$$\begin{array}{ccccccc} 1a^3b^0 & + & 3a^2b^1 & + & 3a^1b^2 & + & 1a^0b^3 \\ 3+0 & & 2+1 & & 1+2 & & 0+3 \\ \text{degree 3} & & \text{degree 3} & & \text{degree 3} & & \text{degree 3} \end{array}$$

These observations help us to quickly expand a binomial power.

EXAMPLE 1 Use Pascal's triangle and the patterns noted to expand $\left(x + \frac{1}{2}\right)^4$.

Solution: Working step-by-step we have

1. The coefficients will be in the fifth row of Pascal's triangle.

   1 4 6 4 1

2. The exponents on x begin at 4 and *decrease,* while the exponents on $\frac{1}{2}$ begin at 0 and *increase.*

$$1x^4\left(\frac{1}{2}\right)^0 + 4x^3\left(\frac{1}{2}\right)^1 + 6x^2\left(\frac{1}{2}\right)^2 + 4x^1\left(\frac{1}{2}\right)^3 + 1x^0\left(\frac{1}{2}\right)^4$$

3. Simplify each term.

   The result is $x^4 + 2x^3 + \frac{3}{2}x^2 + \frac{1}{2}x + \frac{1}{16}$.

NOW TRY EXERCISES 7 THROUGH 12

If the exercise involves a difference rather than a sum, we simply rewrite the expression using algebraic addition and proceed as before. In other words, to write the expansion for $(3 - 2i)^5$, first rewrite it as $[3 + (-2i)]^5$.

EXPANDING BINOMIAL POWERS $(a + b)^n$

1. The coefficients will be in row $n + 1$ of Pascal's triangle.
2. The exponents on *the first term* begin at n and *decrease*, while the exponents on *the second term* begin at 0 and *increase*.
3. For any binomial difference $(a - b)^n$, rewrite the base as $[a + (-b)]^n$ using algebraic addition and proceed as before, then simplify each term.

B. Binomial Coefficients and Factorials

Pascal's triangle can easily be used to find the coefficients of $(a + b)^n$, as long as the exponent is relatively small. If we needed to expand $(a + b)^{25}$, writing out the first 26 rows of the triangle would be rather tedious. To overcome this limitation, we introduce a *formula* for the binomial coefficients that will enable us to find the coefficients of any expansion.

WORTHY OF NOTE

Once again, the coefficients for $(a + b)^n$ are found in the $(n + 1)$st row of Pascal's triangle, since we started the triangle at $(a + b)^0$, which gave the "1" at the vertex.

THE BINOMIAL COEFFICIENTS

For natural numbers n and r where $n \geq r$, the expression $\binom{n}{r}$, read "n choose r," is called the **binomial coefficient** and evaluated as:

$$\binom{n}{r} = \frac{n!}{r!(n-r)!}$$

In Example 1, we found the coefficients of $(a + b)^4$ using the fifth or $(n + 1)$st row of Pascal's triangle. In Example 2, these coefficients are found using the formula for binomial coefficients.

EXAMPLE 2 Evaluate $\binom{n}{r} = \frac{n!}{r!(n-r)!}$ as indicated:

a. $\binom{4}{1}$ **b.** $\binom{4}{2}$ **c.** $\binom{4}{3}$

Solution:

a. $\binom{4}{1} = \frac{4!}{1!(4-1)!} = \frac{4 \cdot 3!}{1!3!} = 4$

b. $\binom{4}{2} = \frac{4!}{2!(4-2)!} = \frac{4 \cdot 3 \cdot 2!}{2!2!} = \frac{4 \cdot 3}{2} = 6$

c. $\binom{4}{3} = \frac{4!}{3!(4-3)!} = \frac{4 \cdot 3!}{3!1!} = 4$

NOW TRY EXERCISES 13 THROUGH 20

Note $\binom{4}{1} = 4$, $\binom{4}{2} = 6$, and $\binom{4}{3} = 4$ give the *interior entries* in the fifth row of Pascal's triangle: 1 4 6 4 1. For consistency and symmetry, we define $0! = 1$, which enables the formula to generate all entries of the triangle, including the "1's."

$$\binom{4}{0} = \frac{4!}{0!(4-0)!} \quad \text{apply formula}$$
$$= \frac{4!}{1 \cdot 4!} = 1 \quad 0! = 1$$

$$\binom{4}{4} = \frac{4!}{4!(4-4)!} = \frac{4!}{4! \cdot 0!} \quad \text{apply formula}$$
$$= \frac{4!}{4! \cdot 1} = 1 \quad 0! = 1$$

The formula for $\binom{n}{r}$ with $0 \le r \le n$ now gives all coefficients in the $(n + 1)$st row. For $n = 5$, we have

$$\begin{matrix} \binom{5}{0} & \binom{5}{1} & \binom{5}{2} & \binom{5}{3} & \binom{5}{4} & \binom{5}{5} \\ 1 & 5 & 10 & 10 & 5 & 1 \end{matrix}$$

EXAMPLE 3 Evaluate the binomial coefficients:

a. $\binom{7}{2}$ **b.** $\binom{9}{0}$ **c.** $\binom{6}{6}$

Solution:

a. $\binom{7}{2} = \frac{7!}{2!(7-2)!} = \frac{7 \cdot 6 \cdot 5!}{2!5!} = 21$

b. $\binom{9}{0} = \frac{9!}{0!(9-0)!} = \frac{9!}{9!} = 1$

c. $\binom{6}{6} = \frac{6!}{6!(6-6)!} = \frac{6!}{6!} = 1$

NOW TRY EXERCISES 21 THROUGH 24

You may have noticed that the formula for $\binom{n}{r}$ is identical to that of ${}_nC_r$, and both yield like results for given values of n and r (see Exercise 53). For future use, it will help to commit the following to memory: $\binom{n}{0} = 1$, $\binom{n}{1} = n$, $\binom{n}{n-1} = n$, and $\binom{n}{n} = 1$.

C. The Binomial Theorem

Using $\binom{n}{r}$ notation and the observations made regarding binomial powers, we can now state the **binomial theorem.**

BINOMIAL THEOREM

For any binomial $(a + b)$ and natural number n,

$$(a+b)^n = \binom{n}{0}a^nb^0 + \binom{n}{1}a^{n-1}b^1 + \binom{n}{2}a^{n-2}b^2 + \cdots + \binom{n}{n-1}a^1b^{n-1} + \binom{n}{n}a^0b^n$$

The theorem can also be stated in summation form as:

$$(a+b)^n = \sum_{r=0}^{n}\binom{n}{r}a^{n-r}b^r$$

The expansion actually looks overly impressive in this form, and it helps to summarize the process in words, as we did earlier. The exponents on the first term a begin at n and decrease, while the exponents on the second term b begin at 0 and increase, keeping the degree of each term constant. The $\binom{n}{r}$ notation simply gives the coefficients of each term. As a final note, observe that the r in $\binom{n}{r}$ gives the exponent on b (the second factor) for each term of the binomial.

EXAMPLE 4 Expand $(a + b)^6$ using the binomial theorem.

Solution:

$$(a + b)^6 = \binom{6}{0}a^6b^0 + \binom{6}{1}a^5b^1 + \binom{6}{2}a^4b^2 + \binom{6}{3}a^3b^3 + \binom{6}{4}a^2b^4 + \binom{6}{5}a^1b^5 + \binom{6}{6}a^0b^6$$

$$= \frac{6!}{0!6!}a^6 + \frac{6!}{1!5!}a^5b^1 + \frac{6!}{2!4!}a^4b^2 + \frac{6!}{3!3!}a^3b^3 + \frac{6!}{4!2!}a^2b^4 + \frac{6!}{5!1!}a^1b^5 + \frac{6!}{6!0!}b^6$$

$$= 1a^6 + 6a^5b + 15a^4b^2 + 20a^3b^3 + 15a^2b^4 + 6ab^5 + 1b^6$$

NOW TRY EXERCISES 25 THROUGH 32

EXAMPLE 5 Find the first three terms of $(2x + y^2)^{10}$.

Solution: Use the binomial theorem with $a = 2x$, $b = y^2$, and $n = 10$.

$$(2x + y^2)^{10} = \binom{10}{0}(2x)^{10}(y^2)^0 + \binom{10}{1}(2x)^9(y^2)^1 + \binom{10}{2}(2x)^8(y^2)^2 + \cdots$$ first three terms

$$= (1)1024x^{10} + (10)512x^9y^2 + \frac{10!}{2!8!}256x^8y^4 + \cdots$$ $\binom{10}{0} = 1, \binom{10}{1} = 10$

$$= 1024x^{10} + 5120x^9y^2 + (45)256x^8y^4 + \cdots$$ $\frac{10!}{2!8!} = 45$

$$= 1024x^{10} + 5120x^9y^2 + 11{,}520x^8y^4 + \cdots$$ result

NOW TRY EXERCISES 33 THROUGH 36

D. Finding a Specific Term of the Binomial Expansion

In some applications of the binomial theorem, our main interest is a *specific term* of the expansion, rather than the expansion as a whole. To find a specified term, it helps to consider that the expansion of $(a + b)^n$ has $n + 1$ terms: $(a + b)^0$ has one term, $(a + b)^1$ has two terms, $(a + b)^2$ has three terms, and so on. Because the notation $\binom{n}{r}$ always begins at $r = 0$ for the first term, the value of r will be *1 less than the term we are seeking.* In other words, for the seventh term of $(a + b)^9$, we use $r = 6$. This means the coefficient is $\binom{9}{6}$, the exponent on b is 6 (same as r), and the exponent on a is 3 (exponents must sum to 9). The fifth term is $\binom{9}{6}a^3b^6$. In general, we have

> **THE kTH TERM OF A BINOMIAL EXPANSION**
> For the binomial expansion $(a + b)^n$, the kth term is given by
> $$\binom{n}{r}a^{n-r}b^r, \text{ where } r = k - 1.$$

EXAMPLE 6 Find the eighth term in the expansion of $(x + 2y)^{12}$.

Solution: By comparing $(x + 2y)^{12}$ to $(a + b)^n$ we have $a = x$, $b = 2y$, and $n = 12$. Since we want the eighth term, $k = 8 \rightarrow r = 7$. The eighth term of the expansion is

$$\binom{12}{7}x^5(2y)^7 = \frac{12!}{7!5!}128x^5y^7$$ $2^7 = 128$

$$= (792)(128x^5y^7)$$ $\binom{12}{7} = 792$

$$= 101{,}376x^5y^7$$ result

NOW TRY EXERCISES 37 THROUGH 42

One application of the Binomial Theorem involves a **binomial experiment** and **binomial probability.** For binomial probabilities, the following must be true: (1) The experiment must have only two possible outcomes, typically called success and failure, and (2) If the experiment has n trials, the probability of success must be constant for all n trials. If the probability of success for each trial is p, the formula $\binom{n}{k}(1-p)^{n-k}p^k$ gives the probability that exactly k trials will be successful.

BINOMIAL PROBABILITY

Given a binomial experiment with n trials, where the probability for success in each trial is p. The probability that exactly k trials are successful is given by $\binom{n}{k}(1-p)^{n-k}p^k$

EXAMPLE 7 Paula Rodrigues has a free-throw shooting average of 85%. On the last play of the game, with her team behind by 3 points, she is fouled at the three-point line, and is awarded two additional free throws via technical fouls on the opposing coach (a total of 5 free-throws). What is the probability she makes *at least 3* (meaning they at least tie the game)?

Solution: Here we have $p = 0.85$, $1 - p = 0.15$ and $n = 5$. The key idea is to recognize the phrase *at least three* means "3 or 4 or 5." So $P(\text{at least } 3) = P(3 \cup 4 \cup 5)$.

$$\begin{aligned}
P(\text{at least } 3) &= P(3 \cup 4 \cup 5) && \text{"or" implies a union} \\
&= P(3) + P(4) + (5) && \text{sum of probabilities} \\
&= \binom{5}{3}(0.15)^2(0.85)^3 + \binom{5}{4}(0.15)^1(0.85)^4 + \binom{5}{5}(0.15)^0(0.85)^5 \\
&\approx 0.1382 + 0.3915 + 0.4437 \\
&= 0.9734
\end{aligned}$$

Paula's team has an excellent chance ($\approx 97.3\%$) of at least tying the game.

NOW TRY EXERCISES 43 THROUGH 46

TECHNOLOGY HIGHLIGHT

Binomial Coefficients, Pascal's Triangle, and $_nC_r$

The keystrokes shown apply to a TI-84 Plus model. Please consult your manual or our Internet site for other models.

The TI-84 Plus can easily generate any row of Pascal's triangle on the home screen, using the $_nC_r$ function and the calculator's ability to display multiple outputs for various values of r. This ability can sometimes be very helpful in applications of the binomial theorem. As in the *Technology Highlight* from Section 3.6, we use the "{" and "}" symbols that are the

2nd functions for the parentheses keys. For the sixth row of Pascal's triangle, we enter 5 $_nC_r$ {0, 1, 2, 3, 4, 5} on the home screen as shown in Figure 8.27, then press ENTER. The calculator neatly outputs the coefficients in the form {1 5 10 10 5 1}. If there are more outputs than the calculator can display, as when computing the seventh row of the triangle, the right arrow key ▶ is used to bring the remaining entries into view. Use these ideas to complete the following exercises.

Figure 8.27

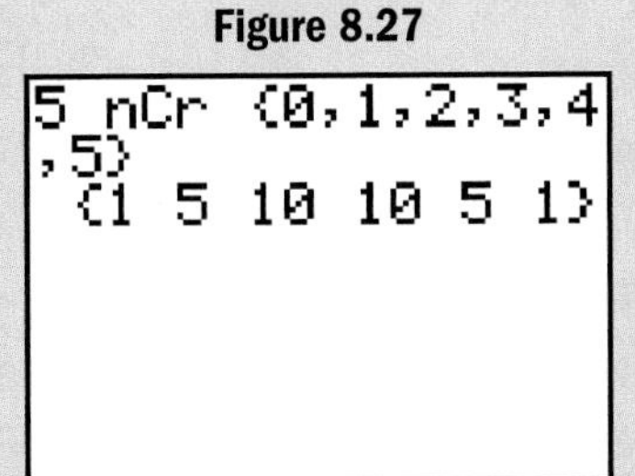

Exercise 1: Write the coefficients in the seventh row of Pascal's triangle, then write the binomial expansion for $(a + b)^6$. {1 6 15 20 15 6 1} $a^6 + 6a^5b^1 + 15a^4b^2 + 20a^3b^3 + 15a^2b^4 + 6a^1b^5 + b^6$

Exercise 2: Write the first three coefficients in the tenth row of Pascal's triangle, then use the result and your knowledge of the binomial theorem to write the first three terms in the expansion of $(a + 2b)^9$. {1 9 36 . . .} $a^9 + 9a^8(2b) + 36a^7(2b)^2 = a^9 + 18a^8b + 144a^7b^2$

8.7 EXERCISES

CONCEPTS AND VOCABULARY

Fill in the blank with the appropriate word or phrase. Carefully reread the section if needed.

1. In any binomial expansion, there is always <u>one</u> more term than the power applied.

2. In all terms in the expanded form of $(a + b)^n$, the exponents on a and b must sum to <u>n</u>.

3. To expand a binomial *difference* such as $(a - 2b)^5$, we rewrite the binomial as <u>$(a + (-2b))^5$</u> and proceed as before.

4. In a binomial experiment with n trials, the probability there are exactly k successes is given by the formula <u>$\binom{n}{k}(1 - p)^{n-k}p^k$</u>.

5. Discuss why the expansion of $(a + b)^n$ has $n + 1$ terms. Answers will vary.

6. For any defined binomial experiment, discuss the relationships between the phrases, "exactly k success," and "at least k successes. Answers will vary.

DEVELOPING YOUR SKILLS

Use *Pascal's triangle* and the patterns explored to write each expansion.

7. $(x + y)^5$ **8.** $(a + b)^6$ **9.** $(2x + 3)^4$ **10.** $\left(x^2 + \frac{1}{3}\right)^3$

11. $(p - q)^7$ **12.** $\left(\frac{1}{2}m - n\right)^4$

Evaluate each of the following

13. $\binom{7}{4}$ 35 **14.** $\binom{8}{2}$ 28 **15.** $\binom{5}{3}$ 10 **16.** $\binom{9}{5}$ 126

17. $\binom{20}{17}$ 1140 **18.** $\binom{30}{26}$ 27,405 **19.** $\binom{40}{3}$ 9880 **20.** $\binom{45}{3}$ 14,190

21. $\binom{6}{0}$ 1 **22.** $\binom{5}{0}$ 1 **23.** $\binom{15}{15}$ 1 **24.** $\binom{10}{10}$ 1

Use the *binomial theorem* to expand the following expressions. Write the general form first, then simplify.

25. $(c + d)^5$ **26.** $(v + w)^4$ **27.** $(a - b)^6$ **28.** $(x - y)^7$

29. $(2x - 3)^4$ **30.** $(a - 2b)^5$ **31.** $(1 - 2i)^3$ **32.** $(2 + \sqrt{3}i)^5$

7. $x^5 + 5x^4y + 10x^3y^2 + 10x^2y^3 + 5xy^4 + y^5$
8. $a^6 + 6a^5b^1 + 15a^4b^2 + 20a^3b^3 + 15a^2b^4 + 6a^1b^5 + b^6$
9. $16x^4 + 96x^3 + 216x^2 + 216x + 81$
10. $x^6 + x^4 + \frac{1}{3}x^2 + \frac{1}{27}$
11. $p^7 - 7p^6q + 21p^5q^2 - 35p^4q^3 + 35p^3q^4 - 21p^2q^5 + 7pq^6 - q^7$
12. $\frac{1}{16}m^4 - \frac{1}{2}m^3n + \frac{3}{2}m^2n^2 - 2mn^3 + n^4$
25. $c^5 + 5c^4d + 10c^3d^2 + 10c^2d^3 + 5cd^4 + d^5$
26. $v^4 + 4v^3w + 6v^2w^2 + 4vw^3 + w^4$
27. $a^6 - 6a^5b + 15a^4b^2 - 20a^3b^3 + 15a^2b^4 - 6ab^5 + b^6$
28. $x^7 - 7x^6y + 21x^5y^2 - 35x^4y^3 + 35x^3y^4 - 21x^2y^5 + 7xy^6 - y^7$
29. $16x^4 - 96x^3 + 216x^2 - 216x + 81$
30. $a^5 - 10a^4b + 40a^3b^2 - 80a^2b^3 + 80ab^4 - 32b^5$
31. $-11 + 2i$
32. $-118 - 31\sqrt{3}i$

33. $x^9 + 18x^8y + 144x^7y^2 + \cdots$
34. $6561p^8 + 17{,}496p^7q + 20{,}412p^6q^2$
35. $v^{24} - 6v^{22}w + \frac{33}{2}v^{20}w^2$
36. $\frac{1}{1024}a^{10} - \frac{5}{256}a^9b^2 + \frac{45}{256}a^8b^4$
37. $35x^4y^3$
38. $15m^2n^4$
39. $1792p^2$
40. $-39{,}405{,}366a^5$
41. $264x^2y^{10}$
42. $10{,}206n^4m^5$
47. $16x^4 - 160x^3 + 600x^2 - 1000x + 625$
48. $73 + 88\sqrt{3}i$

Use the *binomial theorem* to write the first three terms.

33. $(x + 2y)^9$ **34.** $(3p + q)^8$ **35.** $\left(v^2 - \frac{1}{2}w\right)^{12}$ **36.** $\left(\frac{1}{2}a - b^2\right)^{10}$

Find the indicated term for each binomial expansion.

37. $(x + y)^7$; 4th term **38.** $(m + n)^6$; 5th term **39.** $(p - 2)^8$; 7th term

40. $(a - 3)^{14}$; 10th term **41.** $(2x + y)^{12}$; 11th term **42.** $(3n + m)^9$; 6th term

43. Batting averages: Tony Gwen (San Diego Padres) had a lifetime batting average of 0.347, ranking him as one of the greatest hitters of all time. Suppose he came to bat 5 times in any given game.

a. What is the probability that he will get exactly 3 hits? $\approx 17.8\%$

b. What is the probability that he will get at least 3 hits ? $\approx 23.0\%$

44. Pollution testing: Erin suspects that a nearby iron smelter is contaminating the drinking water over a large area. A statistical study reveals that 83% of the wells in this area are likely contaminated. If the figure is accurate, find the probability that if another 10 wells are tested,

a. exactly 8 are contaminated $\approx 29.3\%$ **b.** at least 8 are contaminated $\approx 76.6\%$

45. Late rental returns: The manager of Victor's DVD Rentals knows that 6% of all DVDs rented are returned *late*. Of the 8 videos rented in the last hour, what is the probability that

a. exactly 5 are returned on time $\approx 0.88\%$ **b.** exactly 6 are returned on time $\approx 6.9\%$

c. at least 6 are returned on time $\approx 99.0\%$ **d.** none of them will be returned late $\approx 61.0\%$

46. Opinion polls: From past experience, a research firm knows that 20% of telephone respondents will agree to answer an opinion poll. If 20 people are contacted by phone, what is the probability that

a. exactly 18 refuse to be polled $\approx 13.7\%$ **b.** exactly 19 refuse to be polled $\approx 5.8\%$

c. at least 18 refuse to be polled $\approx 79.4\%$ **d.** none of them agree to be polled $\approx 1.2\%$

Expand each binomial in three different ways: (a) by direct multiplication; (b) using Pascal's triangle and the observed patterns; and (c) using the binomial theorem. Verify that the results are identical.

47. $(5 - 2x)^4$ **48.** $(1 - 2i\sqrt{3})^4$

WORKING WITH FORMULAS

49. Binomial probability: $P(k) = \binom{n}{k}\left(\frac{1}{2}\right)^k\left(\frac{1}{2}\right)^{n-k}$

The theoretical probability of getting exactly k heads in n flips of a fair coin is given by the formula above. What is the probability that you would get 5 heads in 10 flips of the coin? ≈ 0.25

50. Binomial probability: $P(k) = \binom{n}{k}\left(\frac{1}{5}\right)^k\left(\frac{4}{5}\right)^{n-k}$

A multiple choice test has five options per question. The probability of guessing correctly k times out of n questions is found using the formula shown. What is the probability a person scores a 70% by guessing randomly (7 out of 10 questions correct)? ≈ 0.0008

WRITING, RESEARCH, AND DECISION MAKING

51. Prior to calculators and computers, the binomial theorem was frequently used to approximate the value of compound interest given by the expression $\left(1 + \frac{r}{n}\right)^{nt}$ by expanding the first three terms. For example, if the interest rate were 8% ($r = 0.08$) and the interest was compounded quarterly ($n = 4$) for 5 yr ($t = 5$), we have $\left(1 + \frac{0.08}{4}\right)^{(4)(5)} = (1 + 0.02)^{20}$. The first three terms of the expansion give a value of: $1 + 20(0.02) + 190(0.0004) = 1.476$.

a. Calculate the percent error: $\%\text{error} = \frac{\text{approximate value}}{\text{actual value}}$ 99.33%

b. What is the percent error if only two terms are used. 94.22%

52. If you sum the entries in each row of Pascal's triangle, a pattern emerges. Find a formula that generalizes the result for any row of the triangle, and use it to find the sum of the entries in the 12th row of the triangle. 2^{n-1}, 2048

53. The $\binom{n}{r}$ notation produces exactly the same values as the ${}_nC_r$ notation, indicating a connection between the terms of a binomial expansion and the combination of n objects taken r at a time. Use some out-of-class resources (the Internet, probability texts, and so on) to explore and discuss this connection. Answers will vary.

EXTENDING THE CONCEPT

54. Show that $\binom{n}{k} = \binom{n}{n-k}$ for $n = 6$ and $k \le 6$. $\binom{6}{6} = \binom{6}{0} = 1$; $\binom{6}{5} = \binom{6}{1} = 6$; $\binom{6}{4} = \binom{6}{2} = 15$

55. Use Pascal's triangle to show $(3 - 2i)^4 = -119 - 120i$. verified

56. The *derived polynomial* of $f(x)$ is $f(x + h)$ or the original polynomial evaluated at $x + h$. Use Pascal's triangle or the binomial theorem to find the derived polynomial for $f(x) = x^3 + 3x^2 + 5x - 11$. Simplify the result completely.
$x^3 + 3x^2h + 3xh^2 + h^3 + 3x^2 + 6xh + 3h^2 + 5x + 5h - 11$

MAINTAINING YOUR SKILLS

57. (3.7) Graph the function shown and find $f(3)$: $f(x) = \begin{cases} x + 2 & x \le 2 \\ (x-4)^2 & x > 2 \end{cases}$

57. $f(3) = 1$

58. (4.4) Show that $x = -1 + i$ is a solution to $x^4 + 2x^3 - x^2 - 6x - 6 = 0$. verified

59. (4.4) Graph the function $g(x) = x^3 - x^2 - 6x$. Clearly indicate all intercepts and intervals where $g(x) > 0$.
$g(x) > 0$: $x \in (-2, 0) \cup (3, \infty)$

59. (−2, 0), (0, 0), (3, 0)

60. (5.1) If \$2500 is deposited for 10 years at 6% compounded continuously, how much is in the account at the end of the 10 yr? \$4555.30

61. (2.6) During a catch-and-release program, the girth measurements (circumference) of numerous catfish are taken and compared to the fishes' length. The data gathered are shown below.

a. Draw the scatter-plot.

b. Does the association appear to be linear? yes

c. Find an approximate line of fit using two representative points, and use the equation to predict the length of a catfish that has a thirty-five centimeter girth. $l(g) = 2.57g - 8.10$; ≈82 cm

61. a. Length (cm) vs. Girth (cm)

Girth (cm)	Length (cm)	Girth (cm)	Length (cm)
9	15	18	38
12	24	21	43
15	30	24	56

62. (3.6) A rectangular wooden beam is laid across a ditch. The maximum safe load the beam can support varies jointly as the width of the beam and the square of its depth, and inversely as its length. Write the variation equation. $S = k\frac{wd^2}{L}$

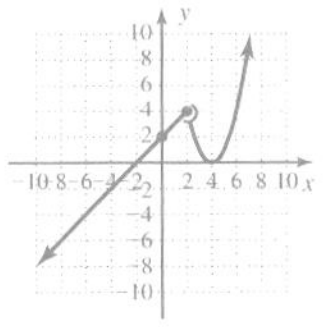

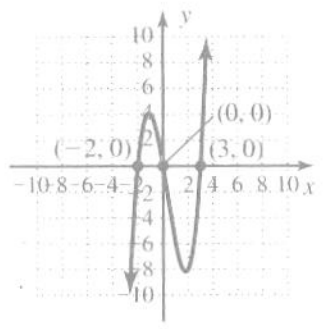

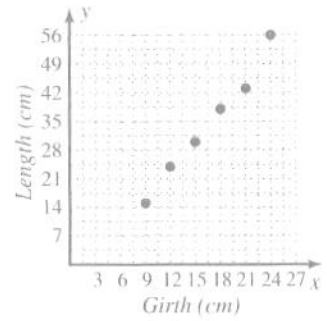

SUMMARY AND CONCEPT REVIEW

SECTION 8.1 Sequences and Series

KEY CONCEPTS

- A *finite sequence* is a function a_n whose domain is the set of natural numbers from 1 to n.
- The terms of the sequence are labeled $a_1, a_2, a_3, \ldots, a_{k-1}, a_k, a_{k+1}, \ldots, a_{n-2}, a_{n-1}, a_n$.
- The expression a_n, which defines the sequence (generates the terms in order), is called the nth term.

- An *infinite sequence* is a function whose domain is the set of *all* natural numbers: $\{1, 2, 3, 4, 5, \ldots\}$.
- When each term of a sequence is larger than the preceding term, it is called an increasing sequence.
- When each term of a sequence is smaller than the preceding term, it is called a decreasing sequence.
- When successive terms of a sequence alternate in sign, it is called an alternating sequence.
- When the terms of a sequence are generated using previous term(s), it is called a recursive sequence.
- Sequences are sometimes defined using factorials, which are the product of a given natural number with all natural numbers that precede it: $n! = n \cdot (n-1) \cdot (n-2) \cdot \cdots \cdot 3 \cdot 2 \cdot 1$.
- Given the sequence: $a_1, a_2, a_3, a_4, \ldots, a_n$ the sum of the terms is called a finite series and is denoted S_n.
- $S_n = a_1 + a_2 + a_3 + a_4 + \cdots + a_n$. The sum of the first n terms is called a partial sum or finite series.
- A series can also be written using sigma notation. The expression $\sum_{i=1}^{k} a_i = a_1 + a_2 + a_3 + \cdots + a_k$.
- When sigma notation is used, the letter "i" is called the index of summation.

EXERCISES

Write the first four terms that are defined and the value of a_{10}.

1. $a_n = 5n - 4$ 1, 6, 11, 16; $a_{10} = 46$

2. $a_n = \dfrac{n+1}{n^2-1}$ not defined, 1, $\frac{1}{2}, \frac{1}{3}, \frac{1}{4}$; $a_{10} = \frac{1}{9}$

Find the general term a_n for each sequence, and the value of a_6.

3. 1, 16, 81, 256, . . . $a_n = n^4$; $a_6 = 1296$

4. $-17, -14, -11, -8, \ldots$ $a_n = -17 + (n-1)(3)$; $a_6 = -2$

Find the eighth partial sum (S_8).

5. $\frac{1}{2}, \frac{1}{4}, \frac{1}{8}, \ldots$ $\frac{255}{256}$

6. $-21, -19, -17, \ldots$ -112

Evaluate each sum.

7. $\sum_{n=1}^{7} n^2$ 140

8. $\sum_{n=1}^{5} (3n-2)$ 35

Write the first five terms that are defined.

9. $a_n = \dfrac{n!}{(n-2)!}$ not defined, 2, 6, 12, 20, 30

10. $\begin{cases} a_1 = \frac{1}{2} \\ a_{n+1} = 2a_n - \frac{1}{4} \end{cases}$ $\frac{1}{2}, \frac{3}{4}, \frac{5}{4}, \frac{9}{4}, \frac{17}{4}$

Write as a single summation and evaluate.

11. $\sum_{n=1}^{7} n^2 + \sum_{n=1}^{7} (3n-2)$ $\sum_{n=1}^{7} (n^2 + 3n - 2)$; 210

SECTION 8.2 Arithmetic Sequences

KEY CONCEPTS

- In an arithmetic sequence, successive terms are found by adding a fixed constant to the preceding term.
- In other words, if there exists a number d, called the common difference, such that $a_{k+1} - a_k = d$, then the sequence is arithmetic. Alternatively we can write $a_{k+1} = a_k + d$ for $k \geq 1$.
- The nth term a_n of an arithmetic sequence is given by $a_n = a_1 + (n-1)d$, where a_1 is the first term and a_n represents the general term of a finite sequence.

- If the initial term is unknown or is not a_1, the nth term can be written $a_n = a_k + (n - k)d$, where the subscript of the term a_k and the coefficient of d sum to n.
- For an arithmetic sequence with first term a_1, the nth partial sum (the sum of the first n terms) is given by $S_n = \frac{n(a_1 + a_n)}{2}$. If the nth term is unknown, the alternative formula $S_n = \frac{n}{2}[2a_1 + (n - 1)d]$ can be used.

EXERCISES

Find the general term (a_n) for each arithmetic sequence. Then find the indicated term.

12. 2, 5, 8, 11, . . . ; find a_{40}
$a_n = 2 + 3(n - 1)$; 119

13. 3, 1, −1, −3, . . . ; find a_{35}
$a_n = 3 + (-2)(n - 1)$; −65

Find the sum of each series.

14. $-1 + 3 + 7 + 11 + \cdots + 75$ 740

15. $1 + 4 + 7 + 10 + \cdots + 88$ 1335

16. $3 + 6 + 9 + 12 + \cdots; S_{20}$ 630

17. $1 + \frac{3}{4} + \frac{1}{2} + \frac{1}{4} + \cdots; S_{15}$ −11.25

18. $\sum_{n=1}^{25} (3n - 4)$ 875

19. $\sum_{n=1}^{40} (4n - 1)$ 3240

SECTION 8.3 Geometric Sequences

KEY CONCEPTS

- In a geometric sequence, successive terms are found by multiplying the preceding term by a nonzero constant.
- In other words, if there exists a number r, called the common ratio, such that $\frac{a_{k+1}}{a_k} = r$, then the sequence is geometric. Alternatively, we can write $a_{k+1} = a_k r$ for $r \geq 1$.
- The nth term a_n of a geometric sequence is given by $a_n = a_1 r^{n-1}$, where a_1 is the first term and a_n represents the general term of a finite sequence.
- If the initial term is unknown or is not a_1, the nth term can be written $a_n = a_k r^{n-k}$, where the subscript of the term a_k and the exponent on r sum to n.
- For an arithmetic sequence with first term a_1, the nth partial sum (the sum of the first n terms) is given by $S_n = \frac{a_1(1 - r^n)}{1 - r}$.
- If $|r| < 1$, the value of r^n becomes very small for large values of n: As $n \to \infty$, $r^n \to 0$. This means the sum of an infinite geometric series can be represented by $S_\infty = \frac{a_1}{1 - r}$.

EXERCISES

Find the indicated term for each geometric sequence.

20. $a_1 = 5, r = 3$; find a_7
3645

21. $a_1 = 4, r = \sqrt{2}$; find a_7
32

22. $a_1 = \sqrt{7}, r = \sqrt{7}$; find a_8
2401

Find the indicated sum, if it exists.

23. $16 - 8 + 4 - \cdots$ find S_7 10.75

24. $2 + 6 + 18 + \cdots$; find S_8 6560

25. $\frac{4}{5} + \frac{2}{5} + \frac{1}{5} + \frac{1}{10} + \cdots$; find S_{12} $\frac{819}{512}$

26. $4 + 8 + 12 + 24 + \cdots$ does not exist

27. $5 + 0.5 + 0.05 + 0.005 + \cdots$ $\frac{50}{9}$

28. $6 - 3 + \frac{3}{2} - \frac{3}{4} + \cdots$ 4

29. $\sum_{n=1}^{8} 5\left(\frac{2}{3}\right)^n$ $\frac{63{,}050}{6561}$

30. $\sum_{n=1}^{\infty} 12\left(\frac{4}{3}\right)^n$ does not exist

31. $\sum_{n=1}^{\infty} 5\left(\frac{1}{2}\right)^n$ 5

32. Charlene began to work for Grayson Natural Gas in January of 1990 with an annual salary of \$26,000. Her contract calls for a \$1220 raise each year. Use a sequence/series to compute her salary after nine years, and her total earnings up to and including that year. (Hint: For $a_1 = 26{,}000$, her salary after 9 *yrs* will be what term of the sequence?) $a_9 = \$36{,}980$; $S_9 = \$314{,}900$

33. Sumpter reservoir contains 121,500 ft^3 of water and is being drained in the following way. Each day one-third of the water is *drained* (and not replaced). Use a sequence/series to compute how much water *remains in the pond* after 7 days. $\approx 7111.1\ ft^3$

34. Credit-hours taught at Cody Community College have been increasing at 7% per year since it opened in 2000 and taught 1225 credit-hours. For the new faculty, the college needs to predict the number of credit-hours that will be taught in 2009. Use a sequence/series to compute the credit-hours for 2009 and to find the total number of credit hours taught through the 2009 school year. $a_9 \approx 2105$ credit hrs; $S_9 \approx 14{,}673$ credit hours

SECTION 8.4 Mathematical Induction

KEY CONCEPTS

- Functions written in subscript notation can be evaluated, graphed, and composed with other functions.
- A sum formula involving only natural numbers n as inputs can be proven valid using a proof by induction. Given that S_n represents a sum formula involving natural numbers, if (1) S_1 is true and (2) $S_k + a_{k+1} = S_{k+1}$, then S_n must be true for all natural numbers.
- Proof by induction can also be used to validate other relationships, using a more general statement of the principle. The new statement has a much wider scope, but can still be applied to sums. Let S_n be a statement involving the natural numbers n. If (1) S_1 is true (S_n for $n = 1$) and (2) the truth of S_k implies that S_{k+1} is also true, then S_n must be true for all natural numbers n.
- Both parts 1 and 2 are necessary for a proof by induction.

EXERCISES

Use the mathematical induction to prove the indicated sum formula is true for all natural numbers n.

35. $1 + 2 + 3 + 4 + 5 + \cdots + n$; $a_n = n$ and $S_n = \dfrac{n(n+1)}{2}$. verified

36. $1 + 4 + 9 + 16 + 25 + 36 + \cdots + n^2$; $a_n = n^2$ and $S_n = \dfrac{n(n+1)(2n+1)}{6}$. verified

Use the principle of mathematical induction to prove that each statement is true for all natural numbers n.

37. $4^n \geq 3n + 1$ verified

38. $6 \cdot 7^{n-1} \leq 7^n - 1$ verified

39. $3^n - 1$ is divisible by 2 verified

SECTION 8.5 Counting Techniques

KEY CONCEPTS

- An **experiment** is any task that can be repeated and has a well-defined set of possible outcomes.
- Each repetition of an experiment is called a trial.
- Any potential outcome of an experiment is called a sample outcome.
- The set of all sample outcomes is called the sample space.
- An experiment with N (equally likely) sample outcomes that is repeated t times, has a sample space with N^t elements.
- If a sample outcome can be used more than once, the counting is said to be with repetition. If a sample outcome can be used only once the counting is said to be without repetition.

See additional answers in Instructor Answer Appendix.

- The fundamental principle of counting states: If there are p possibilities for a first task, q possibilities for the second, and r possibilities for the third, the total number of ways the experiment can be completed is pqr. This fundamental principle can be extended to include any number of tasks.
- If the elements of a sample space have precedence or priority (order or rank is important), the number of elements is counted using a permutation, denoted ${}_nP_r$ and read, "the distinguishable permutations of n objects taken r at a time."
- To expand ${}_nP_r$, we can write out the first r factors of $n!$ or use the formula ${}_nP_r = \frac{n!}{(n-r)!}$.
- If any of the sample outcomes are identical, certain permutations will be nondistinguishable. In a set containing n elements where one element is repeated p times, another is repeated q times, and another r times $(p + q + r \leq n)$, the number of distinguishable permutations is given by $\frac{{}_nP_n}{p!q!r!} = \frac{n!}{p!q!r!}$.
- If the elements of a set have no rank, order, or precedence (as in a committee of colleagues—order/rank is not important), permutations with the same elements are considered identical. The result is the number of combinations, given by ${}_nC_r = \frac{{}_nP_r}{r!}$ (the first r factors of $n!$ divided by $r!$).

EXERCISES

40. Three slips of paper with the letters A, B, and C are placed in a box and randomly drawn one at a time. Show all possible ways they can be drawn using a tree diagram.

41. The combination for a certain bicycle lock consists of three digits. How many combinations are possible if (a) repetition of digits is not allowed and (b) repetition of digits is allowed. 720; 1000

42. Jethro has three work shirts, four pairs of work pants, and two pairs of work shoes. How many different ways can he dress himself (shirt, pants, shoes) for a day's work? 24

43. From a field of 12 contestants in a pet show, three cats are chosen at random to be photographed for a publicity poster. In how many different ways can the cats be chosen? 220

44. How many subsets can be formed from the elements of this set: {■, ⊡, □, ■, ▣}? 32

45. Compute the following values by hand, showing all work:

a. $7!$ 5040 **b.** ${}_7P_4$ 840 **c.** ${}_7C_4$ 35

46. Six horses are competing in a race at the McClintock Ranch. Assuming there are no ties, (a) how many different ways can the horses finish the race? (b) How many different ways can the horses finish first, second, and third place? (c) How many finishes are possible if it is well known that Nellie-the-Nag will finish last and Sea Biscuit will finish first?

46. a. 720 b. 120 c. 24

47. How many distinguishable permutations can be formed from the letters in the word "tomorrow?" 3360

48. Quality Construction Company has 12 equally talented employees. (a) How many ways can a three-member crew be formed to complete a small job? (b) If the company is in need of a Foreman, Assistant Foreman, and Crew Chief, in how many ways can the positions be filled? a. 220 b. 1320

SECTION 8.6 Introduction to Probability

KEY CONCEPTS

- An event E is any designated set of sample outcomes.
- Given S is a sample space of equally likely sample outcomes and E is an event relative to S, the probability of E, written $P(E)$, is computed as $P(E) = \frac{n(E)}{n(S)}$, where $n(E)$ represents the number of elements in E, and $n(S)$ represents the number of elements in S.
- The complement of an event E is the set of sample outcomes in S, but not in E and is denoted ~E.

Additional answers can be found in the Instructor Answer Appendix.

- Given sample space S and any event E defined relative to S, these properties of probability hold: (1) $P(\sim S) = 0$, (2) $0 \le P(E) \le 1$, (3) $P(S) = 1$, (4) $P(E) = 1 - P(\sim E)$, and (5) $P(E) + P(\sim E) = 1$.
- Two events that have no outcomes in common are said to be mutually exclusive.
- If two events are mutually exclusive, $P(E_1 \text{ or } E_2) \rightarrow P(E_1 \cup E_2) = P(E_1) + P(E_2)$.
- If two events are not mutually exclusive, $P(E_1 \text{ or } E_2) \rightarrow P(E_1 \cup E_2) = P(E_1) + P(E_2) - P(E_1 \cap E_2)$

EXERCISES

49. One card is drawn from a standard deck. What is the probability the card is a ten or a face card? $\frac{4}{13}$

50. One card is drawn from a standard deck. What is the probability the card is a Queen or a face card? $\frac{3}{13}$

51. One die is rolled. What is the probability the result is not a three? $\frac{5}{6}$

52. Given $P(E_1) = \frac{3}{8}$, $P(E_2) = \frac{3}{4}$, and $P(E_1 \cup E_2) = \frac{5}{6}$, compute $P(E_1 \cap E_2)$. $\frac{7}{24}$

53. Find $P(E)$ given that $n(E) = {}_7C_4 \cdot {}_5C_3$ and $n(S) = {}_{12}C_7$. $\frac{175}{396}$

Wait (days d)	Probability
0	0.002
$0 < d < 10$	0.07
$10 \le d < 20$	0.32
$20 \le d < 30$	0.43
$30 \le d < 40$	0.178

54. To determine if more physicians should be hired, a medical clinic tracks the number of days between a patient's request for an appointment and the actual appointment date. The table to the left shows the probability that a patient must wait "d" days. Based on the table, what is the probability a patient must wait

a. at least 20 days 0.608
b. less than 20 days 0.392
c. 40 days or less 1
d. over 40 days 0
e. less than 40 and more than 10 days 0.928
f. 30 or more days 0.178

SECTION 8.7 The Binomial Theorem

KEY CONCEPTS

- To expand $(a + b)^n$ for n of "moderate size," we can use Pascal's triangle and observed patterns.
- For any natural numbers n and r, where $n \ge r$, the expression $\binom{n}{r}$ (read "n choose r") is called the *binomial coefficient* and evaluated as $\binom{n}{r} = \frac{n!}{r!(n-r)!}$. The expression $\binom{n}{r}$ is equivalent to ${}_nC_r$.
- If n is large, it is more efficient to expand using the binomial coefficients and binomial theorem.
- When simplifying $\binom{n}{r}$ to compute binomial coefficients, it helps to recognize that $8! = 8 \cdot 7 \cdot 6 \cdot 5 \cdot 4!$; $8! = 8 \cdot 7 \cdot 6 \cdot 5!$; $8! = 8 \cdot 7 \cdot 6!$; and so on.
- The following binomial coefficients are useful/common and should be committed to memory:
$$\binom{n}{0} = 1 \qquad \binom{n}{1} = n \qquad \binom{n}{n-1} = n \qquad \binom{n}{n} = 1$$
- To apply these ideas more generally, we define $0! = 1$: for example, $\binom{n}{n} = \frac{n!}{n!(n-n)!} = \frac{1}{0!} = \frac{1}{1} = 1$.
- The binomial theorem: $(a+b)^n = \binom{n}{0}a^nb^0 + \binom{n}{1}a^{n-1}b^1 + \binom{n}{2}a^{n-2}b^2 + \cdots + \binom{n}{n-1}a^1b^{n-1} + \binom{n}{n}a^0b^n$.
- The kth term of $(a + b)^n$ can be found using the formula $\binom{n}{r}a^{n-r}b^r$, where $r = k - 1$, making it unnecessary to expand the binomial until a specific term is reached.

EXERCISES

55. Evaluate each of the following:

a. $\binom{7}{5}$ 21 b. $\binom{8}{3}$ 56

56. Use Pascal's triangle to expand the binomials:

a. $(x - y)^4$ $x^4 - 4x^3y + 6x^2y^2 - 4xy^3 + y^4$ b. $(1 + 2i)^5$ $41 - 38i$

Use the binomial theorem to:

57. Write the first four terms of

a. $(a + \sqrt{3})^8$ b. $(5a + 2b)^7$

57. a. $a^8 + 8\sqrt{3}a^7 + 84a^6 + 168\sqrt{3}a^5$
b. $78{,}125a^7 + 218{,}750a^6b + 262{,}500a^5b^2 + 175{,}000a^4b^3$

58. Find the indicated term of each expansion.

a. $(x + 2y)^7$; fourth $280x^4y^3$ b. $(2a - b)^{14}$; 10th $-64{,}064a^5b^9$

MIXED REVIEW

1. Identify each sequence as arithmetic, geometric, or neither. If neither, try to identify the pattern that forms the sequence.

a. 120, 163, 206, 249, . . .
b. 4, 4, 4, 4, 4, 4, . . .
c. 1, 2, 6, 24, 120, 720, 5040, . . .
d. 2.00, 1.95, 1.90, 1.85, . . .
e. $\frac{5}{8}, \frac{5}{64}, \frac{5}{512}, \frac{5}{4096}, \ldots$
f. −5.5, 6.05, −6.655, 7.3205, . . .
g. $0.\overline{1}, 0.\overline{2}, 0.\overline{3}, 0.\overline{4}, \ldots$
h. 525, 551.25, 578.8125, . . .
i. $\frac{1}{2}, \frac{1}{4}, \frac{1}{6}, \frac{1}{8}, \ldots$

1. a. arithmetic b. $a_n = 4$ c. $a_n = n!$ d. arithmetic e. geometric f. geometric g. arithmetic h. geometric i. $a_n = \frac{1}{2n}$

2. Compute by hand (show your work).

a. 10! 3,628,800
b. $\frac{10!}{6!}$ 5040
c. ${}_{10}P_4$ 5040
d. ${}_{10}P_9$ 3,628,800
e. ${}_{10}C_6$ 210
f. ${}_{10}C_4$ 210

3. The call letters for a television station must consist of four letters and begin with either a K or a W. How many distinct call letters are possible if repeating any letter is not allowed? 27,600

4. Given $a_1 = 9$ and $r = \frac{1}{3}$, write out the first five terms and the 15th term.

4. 9, 3, 1, $\frac{1}{3}$, $\frac{1}{9}$; $a_{15} = \frac{1}{531{,}441}$

5. Given $a_1 = 0.1$ and $r = 5$, write out the first five terms and the 20th term.

5. 0.1, 0.5, 2.5, 12.5, 62.5; $a_{20} = 1{,}907{,}348{,}632{,}812$

6. One card is drawn from a well-shuffled deck of standard cards. What is the probability the card is a Queen or an Ace? $\frac{2}{13}$

7. Two fair dice are rolled. What is the probability the result is not doubles (doubles = same number on both die)? $\frac{5}{6}$

8. A house in a Boston suburb cost $185,000 in 1985. Each year its value increased by 8%. If this appreciation were to continue, find the value of the house in 2005 and 2015 using a sequence. $862,277.07; $1,861,591.52

9. Evaluate each sum using summation formulas.

a. $\sum_{n=1}^{\infty}\left(\frac{2}{3}\right)^n$ 2
b. $\sum_{n=1}^{10}(9 + 2n)$ 200
c. $\sum_{n=1}^{5}12n + \sum_{n=1}^{5}(-5) + \sum_{n=1}^{5}n^2$ 210

10. Expand each binomial using the binomial theorem. Simplify each term.

a. $(2x + 5)^5$
b. $(1 - 2i)^4$

10. a. $32x^5 + 400x^4 + 2000x^3 + 5000x^2 + 6250x + 3125$
b. $-7 + 24i$

11. For $(a + b)^n$, determine:

a. the first three terms for $n = 20$
b. the last three terms for $n = 20$
c. the fifth term for $n = 35$
d. the fifth term for $n = 35$, $a = 0.2$, and $b = 0.8$

11. a. $a^{20} + 20a^{19}b + 190a^{18}b^2$
b. $190a^2b^{18} + 20ab^{19} + b^{20}$
c. $52{,}360a^{31}b^4$
d. 4.6×10^{-18}

12. On average, bears older than 3 yr old increase their weight by 0.87% per day from July to November. If a bear weighed 110 kg on June 30th: (a) identify the type of sequence that gives the bear's weight each day; (b) find the general term for the sequence; and (c) find the bear's weight on July 1, July 2, July 3, July 31, August 31, and September 30.

12. a. geometric
b. $a_n = 110(1.0087)^n$
c. 110.96, 111.92, 112.90, 143.88, 188.21, 244.06

13. Use a proof by induction to show that $3 + 6 + 9 + \cdots + 3n = \frac{3n(n + 1)}{2}$.

13. verified

14. a. 6720;
b. $8^5 = 32{,}768$

14. The owner of an arts and crafts store makes specialty key rings by placing five colored beads on a nylon cord and tying it to the ring that will hold the keys. If there are eight different colors to choose from, (a) how many distinguishable key rings are possible if no colors are repeated? (b) How many distinguishable key rings are possible if a repetition of colors is allowed?

15. Donell bought 15 raffle tickets from the Inner City Children's Music School, and five tickets from the Arbor Day Everyday raffle. The Music School sold a total of 2000 tickets and the Arbor Day foundation sold 550 tickets. For E_1: Donell wins the Music School raffle and E_2: Donell wins the Arbor Day raffle, find $P(E_1 \text{ or } E_2)$. 0.01659

Find the sum if it exists.

16. $\frac{1}{3} + \frac{2}{3} + 1 + \frac{4}{3} + \cdots + \frac{20}{3}$ $S_n = 70$

17. $0.36 + 0.0036 + 0.000036 + 0.00000036 + \cdots$ $\frac{4}{11}$

Find the first five terms of the sequences in Exercises 18 and 19.

18. $a_n = \frac{12!}{(12-n)!}$ 12; 132; 1320; 11,880; 95,040

19. $\begin{cases} a_1 = 10 \\ a_{n+1} = a_n(\frac{1}{5}) \end{cases}$ 10, 2, $\frac{2}{5}$, $\frac{2}{25}$, $\frac{2}{125}$

20. A random survey of 200 college students produces the data shown. One student from this group is randomly chosen for an interview. Use the data to find

a. P(student works more than 10 hr) 0.84

b. P(student takes less than 13 credit-hours) 0.72

c. P(student works more than 20 hr and takes more than 12 credit-hours) 0.1

d. P(student works between 11 and 20 hr or takes 6 to 12 credit-hours) 0.795

	0 to 10 hr	11 to 20 hr	More than 20 hr	Total
1 to 5 credits	3	7	10	20
6 to 12 credits	21	55	48	124
13 or more credits	8	28	20	56
Total:	32	90	78	200

PRACTICE TEST

1. a. $\frac{1}{2}, \frac{4}{5}, 1, \frac{8}{7}$; $a_8 = \frac{16}{11}$, $a_{12} = \frac{8}{5}$
b. 6, 12, 20, 30; $a_8 = 90$, $a_{12} = 182$
c. $3, 2\sqrt{2}, \sqrt{7}, \sqrt{6}$; $a_8 = \sqrt{2}$, $a_{12} = i\sqrt{2}$
2. a. 165
b. $\frac{311}{420}$
c. $\frac{-2343}{512}$
d. 7
3. a. $a_1 = 7, d = -3$, $a_n = 10 - 3n$
b. $a_1 = -8, d = 2$, $a_n = 2n - 10$
c. $a_1 = 4, r = -2$, $a_n = 4(-2)^{n-1}$
d. $a_1 = 10, r = \frac{2}{5}$, $a_n = 10(\frac{2}{5})^{n-1}$

1. The general term of a sequence is given. Find the first four terms, the 8th term, and the 12th term.

a. $a_n = \frac{2n}{n+3}$ **b.** $a_n = \frac{(n+2)!}{n!}$ **c.** $a_n = \begin{cases} a_1 = 3 \\ a_{n+1} = \sqrt{(a_n)^2 - 1} \end{cases}$

2. Expand each series and evaluate.

a. $\sum_{k=2}^{6}(2k^2 - 3)$ **b.** $\sum_{j=2}^{6}(-1)^j\left(\frac{j}{j+1}\right)$ **c.** $\sum_{j=1}^{5}(-2)\left(\frac{3}{4}\right)^j$ **d.** $\sum_{k=1}^{\infty}7\left(\frac{1}{2}\right)^k$

3. Identify the first term and the common difference or common ratio. Then find the general term a_n.

a. 7, 4, 1, −2, . . . **b.** −8, −6, −4, −2, . . .

c. 4, −8, 16, −32, . . . **d.** 10, 4, $\frac{8}{5}$, $\frac{16}{25}$, . . .

4. Find the indicated value for each sequence.

a. $a_1 = 4$, $d = 5$; find a_{40} 199 **b.** $a_1 = 2$, $a_n = -22$, $d = -3$; find n 9

c. $a_1 = 24$, $r = \frac{1}{2}$; find a_6 $\frac{3}{4}$ **d.** $a_1 = -2$, $a_n = 486$, $r = -3$; find n 6

5. Find the sum of each series.

a. $7 + 10 + 13 + \cdots + 100$ 1712 **b.** $\sum_{k=1}^{37}(3k + 2)$ 2183

c. $4 - 12 + 36 - 108 + \cdots$; find S_7 2188 **d.** $6 + 3 + \frac{3}{2} + \frac{3}{4} + \cdots$ 12

6. Each swing of a pendulum (in one direction) is 95% of the previous one. If the first swing is 12 ft; (a) find the length of the seventh swing and (b) determine the distance traveled by the pendulum for the first seven swings. a. ≈8.82 ft b. ≈72.4 ft

7. A rare coin that cost \$3000 appreciates in value 7% per year. Find the value after 12 yr. \$6756.57

8. A car that costs \$50,000 decreases in value by 15% per year. Find the value of the car after 5 yr. \$22,185.27

9. Use mathematical induction to prove that for $a_n = 5n - 3$, the sum formula $S_n = \frac{5n^2 - n}{2}$ is true for all natural numbers n. verified

10. Use the principle of mathematical induction to prove that S_n: $2 \cdot 3^{n-1} \le 3^n - 1$ is true for all natural numbers n. verified

11. Three colored balls (Aqua, Brown, and Creme) are to be drawn without replacement from a bag. List all possible ways they can be drawn using (a) a tree diagram and (b) an organized list. ABC, ACB, BAC, BCA, CAB, CBA

12. Suppose that license plates for motorcycles must consist of three numbers followed by two letters. How many license plates are possible if zero and "Z" cannot be used and no repetition is allowed? 302,400

13. How many subsets can be formed from the elements in this set: {🖂, 🖂, 🖂, 🖂, 🖂, 🖂}. 64

14. Compute the following values by hand, showing all work: (a) 6!, (b) $_6P_3$ (c) $_6C_3$ 720 120 20

15. An English major has built a collection of rare books that includes two identical copies of *The Canterbury Tales* (Chaucer), three identical copies of *Romeo and Juliet* (Shakespeare), four identical copies of *Faustus* (Marlowe), and four identical copies of *The Faerie Queen* (Spenser). If these books are to be arranged on a shelf, how many distinguishable permutations are possible? 900,900

16. A company specializes in marketing various *cornucopia* (traditionally a curved horn overflowing with fruit, vegetables, gourds, and ears of grain) for Thanksgiving table settings. The company has seven fruit, six vegetable, five gourd, and four grain varieties available. If two from each group (without repetition) are used to fill the horn, how many different cornucopia are possible? 302,400

17. Use Pascal's triangle to expand/simplify:

a. $(x - 2y)^4$ b. $(1 + i)^4$

18. Use the binomial theorem to write the first three terms of (a) $(x + \sqrt{2})^{10}$ and (b) $(a - 2b^3)^8$.

17. a. $x^4 - 8x^3y + 24x^2y^2 - 32xy^3 + 16y^4$
b. -4
18. a. $x^{10} + 10\sqrt{2}\,x^9 + 90x^8$
b. $a^8 - 16a^7b^3 + 112a^6b^6$

19. Michael and Mitchell are attempting to make a nonstop, 100-mi trip on a tandem bicycle. The probability that Michael cannot continue pedaling for the entire trip is 0.02. The probability that Mitchell cannot continue pedaling for the entire trip is 0.018. The probability that neither one can pedal the entire trip is 0.011. What is the probability that they complete the trip? 0.989

20. The spinner shown is spun once. What is the probability of spinning

a. a striped wedge $\frac{1}{4}$
b. a shaded wedge $\frac{5}{12}$
c. a clear wedge $\frac{1}{3}$
d. an even number $\frac{1}{2}$
e. a two or an odd number $\frac{7}{12}$
f. a number greater than nine $\frac{1}{4}$
g. a shaded wedge *or* a number greater than 12 $\frac{5}{12}$
h. a shaded wedge *and* a number greater than 12 0

21. To improve customer service, a cable company tracks the number of days a customer must wait until their cable service is installed. The table shows the probability that a customer must wait d days. Based on the table, what is the probability a customer waits

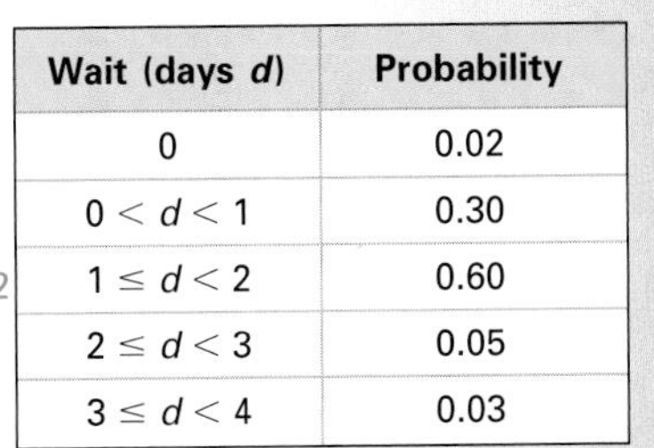

Wait (days d)	Probability
0	0.02
$0 < d < 1$	0.30
$1 \le d < 2$	0.60
$2 \le d < 3$	0.05
$3 \le d < 4$	0.03

a. at least 2 days 0.08
b. less than 2 days 0.92
c. 4 days or less 1
d. over 4 days 0
e. less than 2 or at least 3 days 0.95
f. three or more days 0.03

Additional answers can be found in the Instructor Answer Appendix.

22. An experienced archer can hit the rectangular target shown 100% of the time at a range of 75 m. Assuming the probability the target is hit is related to its area, what is the probability the archer hits within the

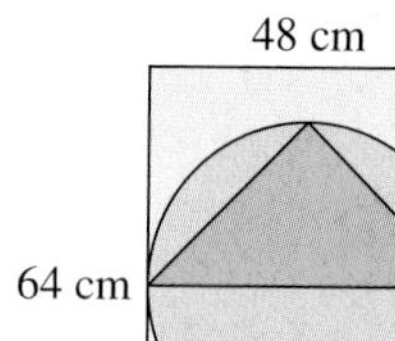

a. triangle 0.1875
b. circle 0.589
c. circle but outside the triangle 0.4015
d. lower half-circle 0.2945
e. rectangle but outside the circle 0.4110
f. lower half-rectangle, outside the circle 0.2055

23. A survey of 100 union workers was taken to register concerns to be raised at the next bargaining session. A breakdown of those surveyed is shown in the table to the right. One out of the hundred will be selected at random for a personal interview. What is the probability the person chosen is a

Expertise Level	Women	Men	Total
Apprentice	16	18	34
Technician	15	13	28
Craftsman	9	9	18
Journeyman	7	6	13
Contractor	3	4	7
Totals	50	50	100

a. woman or a craftsman $\frac{59}{100}$
b. man or a contractor $\frac{53}{100}$
c. man and a technician $\frac{13}{100}$
d. journeyman or an apprentice $\frac{47}{100}$

24. Cheddar is a 12-year-old male box turtle. Provolone is an 8-year-old female box turtle. The probability that Cheddar will live another 8 yr is 0.85. The probability that Provolone will live another 8 yr is 0.95. Find the probability that

a. both turtles live for another 8 yr 0.8075
b. neither turtle lives for another 8 yr 0.0075
c. at least one of them will live another 8 yr 0.9925

25. Use a proof by induction to show that the sum of the first n natural numbers is $\frac{n(n+1)}{2}$. That is, prove $1 + 2 + 3 + \cdots + n = \frac{n(n+1)}{2}$. verified

CALCULATOR EXPLORATION AND DISCOVERY

Infinite Series, Finite Results

Although there were many earlier flirtations with infinite processes, it may have been the paradoxes of Zeno of Elea (~450 B.C.) that crystallized certain questions that simultaneously frustrated and fascinated early mathematicians. The first paradox, called the dichotomy paradox, can be summarized by the following question: How can one ever finish a race, seeing that one-half the distance must first be traversed, then one-half the remaining distance, then one-half the distance that then remains, and so on an infinite number of times? Although we easily accept that races can be finished, the subtleties involved in this question stymied mathematicians for centuries and were not satisfactorily resolved until the eighteenth century. In modern notation, Zeno's first paradox says $\frac{1}{2} + \frac{1}{4} + \frac{1}{8} + \frac{1}{16} + \cdots < 1$. This is a geometric series with $a_1 = \frac{1}{2}$ and $r = \frac{1}{2}$.

EXAMPLE 1 For the geometric sequence with $a_1 = \frac{1}{2}$ and $r = \frac{1}{2}$, the nth term is $a_n = \frac{1}{2^n}$. Use the "**sum(**" and "**seq(**" features of your calculator to compute S_5, S_{10}, and S_{15} (see *Technology Highlight* from Section 8.1). Does the sum appear to be approaching some "limiting value"? If so, what is this value? Now compute S_{20}, S_{25}, and S_{30}. Does there still appear to be a limit to the sum? What happens when you have the calculator compute S_{35}?

Solution: CLEAR the calculator and enter sum(seq(0.5^X, X, 1, 5)) on the home screen. Pressing ENTER gives $S_5 = 0.96875$ (Figure 8.28). Press 2nd ENTER to recall the expression and overwrite the 5, changing it to a 10. Pressing ENTER shows $S_{10} = 0.9990234375$. Repeating these steps gives $S_{15} = 0.9999694824$, and it seems that "1" may be a limiting value. Our conjecture receives further support as S_{20}, S_{25}, and S_{30} are closer and closer to 1, but do not exceed it (Figure 8.29). At S_{35}, the number of digits needed to display the exact result exceeds the internal capacity of the calculator, and the calculator returns a "1."

Figure 8.28

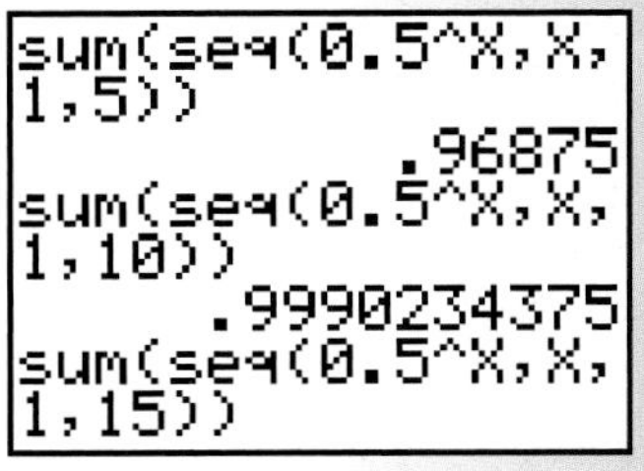

```
sum(seq(0.5^X,X,
1,5))
           .96875
sum(seq(0.5^X,X,
1,10))
      .9990234375
sum(seq(0.5^X,X,
1,15))
```

Figure 8.29

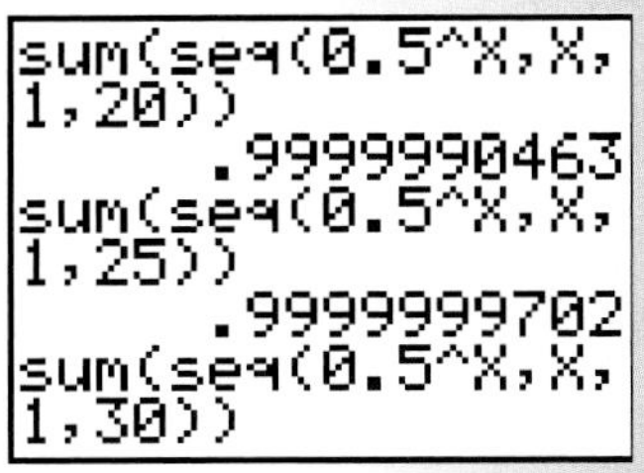

```
sum(seq(0.5^X,X,
1,20))
       .9999990463
sum(seq(0.5^X,X,
1,25))
       .9999999702
sum(seq(0.5^X,X,
1,30))
```

Note that the sum of additional terms will create a longer string of 9's. That the sum of an infinite number of these terms **IS 1** can be understood by converting the repeating decimal $0.\overline{9}$ to its fractional form (as shown). For $x = 0.\overline{9}$, $10x = 9.\overline{9}$ and it follows that:

$$\begin{aligned} 10x &= 9.\overline{9} \\ -x &= -0.\overline{9} \\ \hline 9x &= 9 \\ x &= 1 \end{aligned}$$

For a geometric sequence, the result of an infinite sum can be verified using $S_\infty = \dfrac{a_1}{1 - r}$. However, there are many nongeometric, infinite series that also have a limiting value. In some cases these require many, many more terms before the limiting value can be observed.

For Exercises 1, 2, and 3, use a calculator to write the first five terms and the techniques illustrated here to find S_5, S_{10}, and S_{15}. Decide if the sum appears to be approaching some limiting value, then compute S_{20} and S_{25}. Do these sums support your conjecture? For Exercise 4, repeat the investigation using S_{50}, S_{100}, S_{150}, S_{200}, and S_{250}.

Exercise 1: $a_1 = \frac{1}{3}$ and $r = \frac{1}{3}$ $\frac{1}{2}$

Exercise 2: $a_1 = 0.2$ and $r = 0.2$ $\frac{1}{4}$

Exercise 3: $a_n = \dfrac{1}{(n - 1)!}$ e

Exercise 4: $a_n = \dfrac{2}{n(n + 1)}$ 1.9608, 1.9802, 1.9868, 1.9900, 1.9920

Additional Insight: Zeno's first paradox can also be "resolved" by observing that the "half-steps" needed to complete the race require increasingly shorter (infinitesimally short) amounts of time. Eventually the race is complete.

STRENGTHENING CORE SKILLS

Probability, Quick-Counting, and Card Games

There are few areas in the college mathematics curriculum that help develop a student's reasoning powers like probability. This is likely due to the logical and organized thought processes that are often required to develop a solution. Even the best problem solvers sometimes stumble here. After carefully analyzing a situation and understanding the question completely, we still have several decisions to make, such as should a permutation or a combination be used? Is the fundamental principle of counting (FCP) involved? What is the sample space? Are the events mutually exclusive? Questions like these cannot be answered using superficial thinking or by haphazardly applying formulas. This *Strengthening Core Skills* is designed to help you apply elements of probability to situations where such decisions are routinely called for—determining the probability of being dealt a certain hand of five cards from a well-shuffled deck. The card game known as *Five Card Stud* is often played for fun and relaxation, using toothpicks, beans, or pocket change as players attempt to develop a winning "hand" from the five cards dealt. The various "hands" are given here with the higher value hands listed first (e.g., a full house is a better/higher hand than a flush).

Five Card Hand	Description	Probability of Being Dealt
royal flush	five cards of the same suit in sequence from Ace to 10	0.000 001 540
straight flush	any five cards of the same suit in sequence (exclude royal)	0.000 013 900
four of a kind	four cards of the same rank, any fifth card	
full house	three cards of the same rank, with one pair	
flush	five cards of the same suit, no sequence required	0.001 970
straight	five cards in sequence, regardless of suit	
three of a kind	three cards of the same rank, any two other cards	
two pairs	two cards of the one rank, two of another rank, one other card	0.047 500
one pair	two cards of the same rank, any three others	0.422 600

Let's consider the probability of being dealt a few of these hands. Surprisingly, some of the computations require only a direct application of the FCP, others require using combinations and the FCP jointly, while still others further require some additional logic (see Illustration 2 and elsewhere). Instead of studying each hand by order of value, it will help to organize our investigation into two parts: (1) the hands that are based on suit (the flushes) and (2) the hands that are based on rank (four-of-a-kind, pairs, etc.). For each computation, we consider the sample space to be five cards dealt from a deck of 52, or ${}_{52}C_5$.

(1) The hands that are based on suit (the flushes): A flush consists of five cards in the same suit, a straight consists of five cards in sequence. Let's consider that an Ace can be used as either a high card (as in 10, J, Q, K, A) or a low card (as in A, 2, 3, 4, 5). Since the dominant characteristic of a flush is its *suit,* we first consider choosing one suit of the four, then the number of ways that the straight can be formed (if needed).

ILLUSTRATION 1 What is the probability of being dealt a royal flush?

Solution: For a royal flush, all cards must be of one suit. Since there are four suits, it can be chosen in ${}_4C_1$ ways. A royal flush must have the cards A, K, Q, J, and 10 and once the suit has been decided, it can happen in only this (one) way or ${}_1C_1$. This means P (royal flush) $=$

$$\frac{{}_4C_1 \cdot {}_1C_1}{{}_{52}C_5} \approx 0.000\,001\,540.$$

ILLUSTRATION 2 What is the probability of being dealt a straight flush?

Solution: Once again all cards must be of one suit, which can be chosen in ${}_4C_1$ ways. A straight flush is any five cards in sequence and once the suit has been decided, this can happen in 10 ways (Ace on down, King on down, Queen on down, and so on). By the FCP, there are ${}_4C_1 \cdot {}_{10}C_1 = 40$ ways this can happen, but *four of these will be royal flushes that are of a higher* *value* and must be subtracted from this total. So in the intended context we have $P(\text{straight flush}) = \dfrac{{}_4C_1 \cdot {}_{10}C_1 - 4}{{}_{52}C_5} \approx 0.000\,013\,900$

ILLUSTRATION 3 What is the probability of being dealt a flush?

Solution: All cards must be of the same suit, again chosen in ${}_4C_1$ ways. But once the suit is chosen, any five cards from the suit (13 cards) will do, which can be done in ${}_{13}C_5$ ways. This means the event can occur in ${}_4C_1 \cdot {}_{13}C_5$ ways, but *this again includes hands of a higher value*—namely, the 40 straight flushes from Illustrations 1 and 2. This indicates $P(\text{flush}) = \dfrac{{}_4C_1 \cdot {}_{13}C_5 - 40}{{}_{52}C_5} \approx 0.001\,970.$

(2) The hands that are based on rank (four-of-a-kind, pairs, etc.): We illustrate the process of finding probabilities that are based on rank using the computations for one pair and for two pairs. Surprisingly, these are more involved that those for the other hands. Since the dominant characteristic here is the chosen *rank,* we first consider finding the number of ways that rank can be chosen, then the number of ways the "matches" (pairs, triples, and so on) can be chosen from each rank. Since we must end up with five cards, we then consider the number of ways the remaining cards can be dealt.

ILLUSTRATION 4 What is the probability of being dealt two pairs?

Solution: For two pairs (which must be of different ranks) we begin by choosing two ranks of the 13 possible, or ${}_{13}C_2$. We then consider that two cards must be chosen from each of these ranks, ${}_4C_2$ from the first rank and ${}_4C_2$ from the second. Since the fifth card cannot match either of the ranks chosen (or we would have a full house), its rank must be chosen from the 11 that remain $({}_{11}C_1)$ and can be any one of the four cards in this rank $({}_4C_1)$. The number of possibilities in the event is ${}_{13}C_2 \cdot {}_4C_2 \cdot {}_4C_2 \cdot {}_{11}C_1 \cdot {}_4C_1$. Therefore $P(\text{two pair}) = \dfrac{{}_{13}C_2 \cdot {}_4C_2 \cdot {}_4C_2 \cdot {}_{11}C_1 \cdot {}_4C_1}{{}_{52}C_5} \approx 0.047\,500.$

ILLUSTRATION 5 What is the probability of being dealt one pair?

Solution: We begin by choosing 1 rank from the 13 possible, giving ${}_{13}C_1$. We can then choose the two cards from the selected rank in ${}_4C_2$ ways. Similar to Illustration 4, the remaining three cards must be chosen from the 12 ranks that remain (so we end up with only one pair) giving ${}_{12}C_3$ possibilities, and once each (different) rank is selected we must chose one of the four cards $({}_4C_1)$. The number of possibilities in this event is ${}_{13}C_1 \cdot {}_4C_2 \cdot {}_{12}C_3 \cdot {}_4C_1 \cdot {}_4C_1 \cdot {}_4C_1$, so $P(\text{one pair}) = \dfrac{{}_{13}C_1 \cdot {}_4C_2 \cdot {}_{12}C_3 \cdot {}_4C_1 \cdot {}_4C_1 \cdot {}_4C_1}{{}_{52}C_5} \approx 0.422\,600.$

Using the insights gained from Illustrations 1 through 5, compute the probability of being dealt:

Exercise 1: A hand with four-of-a-kind. 0.000 240

Exercise 2: A full house. 0.001 441

Exercise 3: A hand with three-of-a-kind. 0.021 128

Exercise 4: Four of the same suit (almost a flush) and one other card. 0.042 920

Exercise 5: A straight. [*Hint:* Of the 13 ranks, only 10 can lead to a (five-card) straight, and any royal or straight flushes should be removed from the count. 0.003 925

CUMULATIVE REVIEW CHAPTERS 1–8

1. a. 23 cards are assembled each hour.
   b. $y = 23x - 155$
   c. 184 cards
   d. ≈6:45 A.M.
2. verified
3. $x = \frac{-5 \pm \sqrt{109}}{6}$; $x \approx 0.91$; $x \approx -2.57$
4. 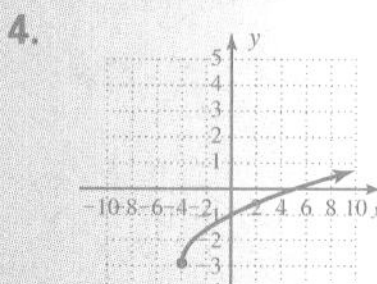
5. $Y = \frac{kVW}{X}$; $Y = \frac{3VW}{2X}$
6. D: $x \in [-3, 3]$
   R: $y \in [-2, -2] \cup (-1, 2) \cup [4, 9]$
7. verified; reflections across $y = x$
8. a. $x = 0$ b. $x \in (-1, 0)$
   c. $x \in (-\infty, -1) \cup (0, \infty)$
   d. $x \in (-\infty, -1) \cup (-1, 1)$
   e. $x \in (1, \infty)$ f. $y = 3$ at $(1, 3)$
   g. none
   h. $x \approx -2.3, 0.4, 2$
   i. $x \approx 0.25$
   j. does not exist
   k. $-\infty$ l. 0
   m. $x \in (-\infty, 1) \cup (-1, \infty)$
9. a. $4x + 2h - 3$
   b. $\frac{-1}{(x + h - 2)(x - 2)}$
10. 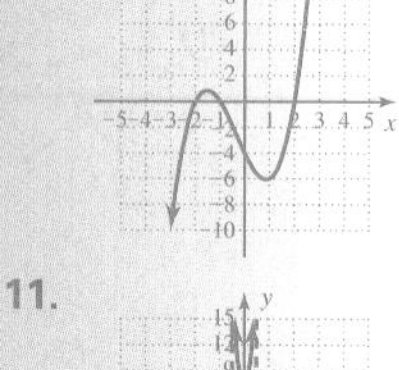
11. 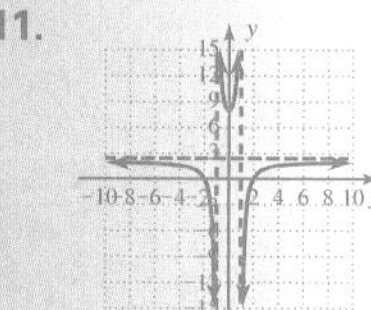
12. a. $\log_{10} x = y$
    b. $\log_3 \frac{1}{81} = -4$

Additional answers can be found in the Instructor Answer Appendix.

1. Robot Moe is assembling memory cards for computers. At 9:00 A.M., 52 cards had been assembled. At 11:00 A.M., a total of 98 had been made. Assuming the production rate is linear

a. Find the slope of this line and explain what it means in this context.

b. Find a linear equation model for this data.

c. Determine how many boards Moe can assemble in an eight-hour day.

d. Determine the approximate time that Moe began work this morning.

2. Verify by direct substitution that $x = 2 + i$ is a solution to $x^2 - 4x + 5 = 0$.

3. Solve using the quadratic formula: $3x^2 + 5x - 7 = 0$. State your answer in exact and approximate form.

4. Sketch the graph of $y = \sqrt{x + 4} - 3$ using transformations of a parent function. Plot the x- and y-intercepts.

5. Write a variation equation and find the value of k: Y varies inversely with X and jointly with V and W. Y is equal to 10 when $X = 9$, $V = 5$, and $W = 12$.

6. Graph the piecewise function and state the domain and range.

$$y = \begin{cases} -2 & -3 \le x \le -1 \\ x & -1 < x < 2 \\ x^2 & 2 \le x \le 3 \end{cases}$$

7. Verify that $f(x) = x^3 - 5$ and $g(x) = \sqrt[3]{x + 5}$ are inverse functions. How are the graphs of f and g related?

8. For the graph of $g(x)$ shown, state where

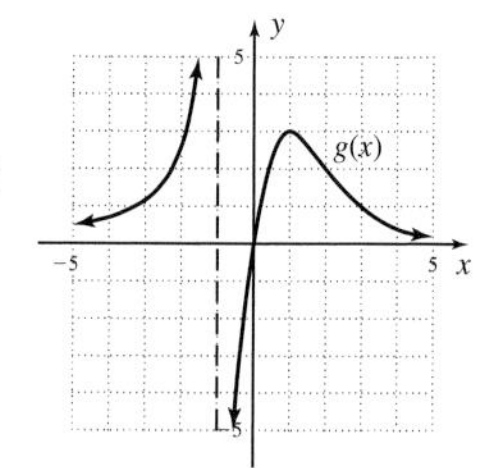

a. $g(x) = 0$ **b.** $g(x) < 0$ **c.** $g(x) > 0$ **d.** $g(x)\uparrow$
e. $g(x)\downarrow$ **f.** local max **g.** local min **h.** $g(x) = 2$
i. $g(4)$ **j.** $g(-1)$ **k.** as $x \to -1^+$, $g(x) \to$ ____
l. as $x \to \infty$, $g(x) \to$ ____ **m.** the domain of $g(x)$

9. Compute the difference quotient for each function given.

a. $f(x) = 2x^2 - 3x$ **b.** $h(x) = \frac{1}{x - 2}$

10. Graph the polynomial function given. Clearly indicate all intercepts.
$f(x) = x^3 + x^2 - 4x - 4$

11. Graph the rational function $h(x) = \frac{2x^2 - 8}{x^2 - 1}$. Clearly indicate all asymptotes and intercepts.

12. Write each expression in logarithmic form:

a. $x = 10^y$ **b.** $\frac{1}{81} = 3^{-4}$

13. Write each expression in exponential form:

a. $3 = \log_x(125)$ **b.** $\ln(2x - 1) = 5$

14. What interest rate is required to ensure that \$2000 will double in 10 yr if interest is compounded continuously?

15. Solve for x.

a. $e^{2x-1} = 217$

b. $\log(3x - 2) + 1 = 4$

16. Solve using matrices and row reduction:

$$\begin{cases} 2a + 3b - 6c = 15 \\ 4a - 6b + 5c = 35 \\ 3a + 2b - 5c = 24 \end{cases}$$

17. Solve using a calculator and inverse matrices.

$$\begin{cases} 0.7x + 1.2y - 3.2z = -32.5 \\ 1.5x - 2.7y + 0.8z = -7.5 \\ 2.8x + 1.9y - 2.1z = 1.5 \end{cases}$$

18. Find the equation of the hyperbola with foci at $(-6, 0)$ and $(6, 0)$ and vertices at $(-4, 0)$ and $(4, 0)$.

19. Write $x^2 + 4y^2 - 24y + 6x + 29 = 0$ by completing the square, then identify the center, vertices, and foci.

20. Use properties of sequences to determine a_{20} and S_{20}.

a. 262144, 65536, 16384, 4096, . . .

b. $\frac{7}{8}, \frac{27}{40}, \frac{19}{40}, \frac{11}{40}, \ldots$

21. 1333

23. a. ≈7.0%
b. ≈91.9%
c. ≈98.9%
d. $\binom{12}{0}(0.04)^{12}(0.96)^{0}$; virtually nil

24. verified

25. $y = -9.75x^2 + 128.4x - 285.15$
a. July
b. ≈136
c. 134
d. early October to late February

21. Empty 55-gal drums are stacked at a storage facility in the form of a pyramid with 52 barrels in the first row, 51 barrels in the second row, and so on, until there are 10 barrels in the top row. Use properties of sequences to determine how many barrels are in this stack.

22. Three \$20 bills, six \$10 bills, and four \$5 bills are placed in a box, then two bills are drawn out and placed in a savings account. What is the probability the bills drawn are

a. first \$20, second \$10 $\frac{3}{26}$ **b.** first \$10, second \$20 $\frac{3}{26}$

c. both \$5 $\frac{1}{13}$ **d.** first \$5, second not \$20 $\frac{3}{13}$

e. first \$5, second not \$10 $\frac{2}{13}$ **f.** first not \$20, second \$20 $\frac{5}{26}$

23. The manager of Tom's Tool and Equipment Rentals knows that 4% of all tools rented are returned late. Of the 12 tools rented in the last hour, what is the probability that

a. exactly ten will be returned on time **b.** at least eleven will be returned on time

c. at least ten will be returned on time **d.** none of them will be returned on time

24. Use a proof by induction to show $3 + 7 + 11 + 15 + \cdots + (4n - 1) = n(2n + 1)$ for all natural numbers n.

25. A park ranger tracks the number of campers at a popular park from March ($m = 3$) to September ($m = 9$) and collects the following data (month, number of campers): (3, 12), (5, 114), (7, 135), and (9, 81). Assuming the data is quadratic, draw a scatter-plot and use your calculator to obtain a parabola of best fit. (a) What month had the maximum number of campers? (b) What was this maximum number? (c) How many campers might be expected in June? (d) Based on your model, what month(s) is the park apparently closed to campers (number of campers is zero or negative)?

Appendix I

U.S. Standard Units and the Metric System

U.S. Standard				Metric Standard			
DISTANCE				DISTANCE			
1 mile	=	1760	yards	1 kilometer	=	1000	meters
1 mile	=	5280	feet	1 hectometer	=	100	meters
1 yard	=	3	feet	1 dekameter	=	10	meters
1 yard	=	36	inches	**1 meter**	**=**	**1**	**meter**
1 foot	=	12	inches	1 decimeter	=	0.1	meter
				1 centimeter	=	0.01	meter
				1 millimeter	=	0.001	meter
WEIGHT				WEIGHT			
1 ton	=	2000	pounds	1 kilogram	=	1000	grams
1 pound	=	16	ounces	1 hectogram	=	100	grams
				1 dekagram	=	10	grams
				1 gram	**=**	**1**	**gram**
				1 decigram	=	0.1	gram
				1 centigram	=	0.01	gram
				1 milligram	=	0.001	gram
CAPACITY or VOLUME				CAPACITY or VOLUME			
1 gallon	=	4	quarts	1 kiloliter	=	1000	liters
1 quart	=	2	pints	1 hectoliter	=	100	liters
1 pint	=	2	cups	1 dekaliter	=	10	liters
1 cup	=	8	ounces	**1 liter**	**=**	**1**	**liter**
1 tablespoon	=	3	teaspoons	1 deciliter	=	0.1	liter
				1 centiliter	=	0.01	liter
				1 milliliter	=	0.001	liter

Conversion Factors: U.S. to Metric				Conversion Factors: Metric to U.S.			
DISTANCE				DISTANCE			
1 inch	=	2.540	centimeters	1 centimeter	=	0.394	inch
1 foot	=	0.305	meter	1 meter	=	3.281	feet
1 yard	=	0.914	meter	1 meter	=	1.094	yards
1 mile	=	1.609	kilometers	**1 kilometer**	**=**	**0.621**	**mile**
WEIGHT				WEIGHT			
1 ounce	=	28.349	grams	1 gram	=	0.035	ounce
1 pound	=	0.454	kilogram	1 dekagram	=	0.353	ounce
1 ton	=	907.185	kilograms	**1 kilogram**	**=**	**2.205**	**pounds**
CAPACITY or VOLUME				CAPACITY or VOLUME			
1 ounce	=	29.573	milliliters	1 milliliter	=	0.034	ounce
1 pint	=	0.473	liters	1 liter	=	2.114	pints
1 quart	=	0.946	liter	1 liter	=	1.057	quarts
1 gallon	=	3.785	liters	**1 liter**	**=**	**0.264**	**gallon**

Appendix II

Rational Expressions and the Least Common Denominator

FINDING THE LEAST COMMON DENOMINATOR (LCD)

1. Factor each denominator, using exponents for repeated factors.
2. List all *unique* factors that occur in *any denominator.*
3. Use the largest exponent that appears on each factor in the list.
4. The LCD is the product of these factors.

EXAMPLE Find the LCD for each pair of rational expressions:

a. $\frac{4}{18x^2}, \frac{7}{24x}$ **b.** $\frac{5x}{x^2-9}, \frac{2}{x^2+6x+9}$ **c.** $\frac{2y}{y-3}, \frac{5y^2}{y+1}$

Solution: **a.** $\frac{4}{18x^2} = \frac{5}{2 \cdot 3^2 \cdot x^2}$ $\quad \frac{7}{24x} = \frac{7}{2^3 \cdot 3 \cdot x}$ factor

$2, 3, x$ unique factors: 2, 3, and x

$2^3, 3^2, x^2$ use largest exponent

LCD: $72x^2$ LCD is the product of these factors

b. $\frac{5x}{x^2-9} = \frac{5x}{(x-3)(x+3)}$ $\quad \frac{2}{x^2+6x+9} = \frac{2}{(x+3)^2}$ factor

$(x-3), (x+3)$ unique factors: $(x-3), (x+3)$

$(x-3)^1, (x+3)^2$ use largest exponent

LCD: $(x-3)^1(x+3)^2$ LCD is product of these factors

c. $\frac{2y}{y-3}$ $\quad \frac{5y^2}{y+1}$ denominators are prime

$(y-3), (y+1)$ unique factors: $(y-3), (y+1)$

$(y-3)^1, (y+1)^1$ use largest exponent

LCD: $(y-3)(y+1)$ LCD is product of these factors

It is perhaps more intuitive to focus on the words *common denominator,* and simply supply the factors needed by either expression (using the fundamental property of fractions) to match up the denominators. For instance, to find a common denominator and build equivalent expressions for $\frac{2x}{x^2-3x-10}$ and $\frac{3}{x^2-25}$, we note the factored forms are $\frac{2x}{(x-5)(x+2)}$ and $\frac{3}{(x-5)(x+5)}$, with the first denominator lacking a factor of $(x+5)$ and the second lacking a factor of $(x+2)$. After supplying these factors we obtain $\frac{2x(x+5)}{(x-5)(x+2)(x+5)}$ and $\frac{3(x+2)}{(x-5)(x+5)(x+2)}$.

Reduced Row-Echelon Form and More on Matrices

A matrix is in reduced row-echelon form if it satisfies the following conditions:

1. All null rows (zeroes for all entries) occur at the bottom of the matrix.
2. The first entry of any nonzero row must be a "1."
3. For any two consecutive, nonzero rows, the leading "1" in the higher row is to the left of the 1 in the lower row.
4. Every *column* with a leading 1 has zeroes for all other entries in the column.

Matrices A through D are in reduced row-echelon form.

$$A = \begin{bmatrix} 0 & 1 & 0 & 0 & 0 & 5 \\ 0 & 0 & 0 & 1 & 0 & 3 \\ 0 & 0 & 0 & 0 & 1 & 2 \end{bmatrix} \quad B = \begin{bmatrix} 1 & 0 & 0 & 5 \\ 0 & 1 & 0 & 3 \\ 0 & 0 & 0 & 0 \end{bmatrix} \quad C = \begin{bmatrix} 1 & 0 & 0 & 5 \\ 0 & 1 & 3 & -2 \\ 0 & 0 & 0 & 0 \end{bmatrix} \quad D = \begin{bmatrix} 1 & 0 & 0 & 5 \\ 0 & 1 & 0 & 2 \\ 0 & 0 & 1 & -2 \end{bmatrix}$$

Where *Gaussian elimination* places a matrix in *row-echelon form* (satisfying the first three conditions), *Gauss-Jordan elimination* is an example of *reduced row-echelon form*. To obtain this form, continue applying row operations to the matrix until the fourth condition above is also satisfied. For a 3×3 system having a unique solution, the diagonal entries of the coefficient matrix will be 1's, with 0's for all other entries. To illustrate, we'll extend Example 4 from Section 6.5 until reduced row-echelon form is obtained.

Example 4 (Section 6.5): Solve using Gauss-Jordan elimination: $\begin{cases} 2x + y - 2z = -7 \\ x + y + z = -1 \\ -2y - z = -3 \end{cases}$.

$$\begin{cases} 2x + y - 2z = -7 \\ x + y + z = -1 \\ -2y - z = -3 \end{cases} \quad \text{R1} \leftrightarrow \text{R2} \quad \begin{cases} x + y + z = -1 \\ 2x + y - 2z = -7 \\ -2y - z = -3 \end{cases} \quad \text{matrix form} \rightarrow \quad \left[\begin{array}{ccc|c} 1 & 1 & 1 & -1 \\ 2 & 1 & -2 & -7 \\ 0 & -2 & -1 & -3 \end{array}\right]$$

$$\left[\begin{array}{ccc|c} 1 & 1 & 1 & -1 \\ 2 & 1 & -2 & -7 \\ 0 & -2 & -1 & -3 \end{array}\right] \quad -2\text{R1} + \text{R2} \rightarrow \text{R2} \quad \left[\begin{array}{ccc|c} 1 & 1 & 1 & -1 \\ 0 & -1 & -4 & -5 \\ 0 & -2 & -1 & -3 \end{array}\right] \quad -1\text{R2} \quad \left[\begin{array}{ccc|c} 1 & 1 & 1 & -1 \\ 0 & 1 & 4 & 5 \\ 0 & -2 & -1 & -3 \end{array}\right]$$

$$\left[\begin{array}{ccc|c} 1 & 1 & 1 & -1 \\ 0 & 1 & 4 & 5 \\ 0 & -2 & -1 & -3 \end{array}\right] \quad 2\text{R2} + \text{R3} \rightarrow \text{R3} \quad \left[\begin{array}{ccc|c} 1 & 1 & 1 & -1 \\ 0 & 1 & 4 & 5 \\ 0 & 0 & 7 & 7 \end{array}\right] \quad \frac{\text{R3}}{7} \rightarrow \text{R3} \quad \left[\begin{array}{ccc|c} 1 & 1 & 1 & -1 \\ 0 & 1 & 4 & 5 \\ 0 & 0 & 1 & 1 \end{array}\right]$$

$$\left[\begin{array}{ccc|c} 1 & 1 & 1 & -1 \\ 0 & 1 & 4 & 5 \\ 0 & 0 & 1 & 1 \end{array}\right] \quad -\text{R2} + \text{R1} \rightarrow \text{R1} \quad \left[\begin{array}{ccc|c} 1 & 0 & -3 & -6 \\ 0 & 1 & 4 & 5 \\ 0 & 0 & 1 & 1 \end{array}\right] \quad \begin{array}{c} 3\text{R3} + \text{R1} \rightarrow \text{R1} \\ -4\text{R3} + \text{R2} \rightarrow \text{R2} \end{array} \quad \left[\begin{array}{ccc|c} 1 & 0 & 0 & -3 \\ 0 & 1 & 0 & 1 \\ 0 & 0 & 1 & 1 \end{array}\right]$$

The final matrix is in reduced row-echelon form and gives the solution $(-3, 1, 1)$, just as obtained in Section 6.5.

The Determinant of a General Matrix

To compute the determinant of a general square matrix, we introduce the idea of a **cofactor.** For the $n \times n$ matrix A, $A_{ij} = (-1)^{i+j}|M_{ij}|$ is the cofactor of matrix element a_{ij}, where $|M_{ij}|$ represents the determinant of the corresponding minor matrix. Note that $i + j$ is the sum of the row and column numbers of the entry, and if this sum is even, $(-1)^{i+j} = 1$, while if the sum is odd, $(-1)^{i+j} = -1$ (this is how the sign table for a 3×3 determinant was generated). To compute the determinant of an $n \times n$ matrix, multiply each element in any row or column by its cofactor and add. The result is a tier-like process in which the determinant of a larger matrix requires computing the determinant of smaller matrices. In the case of a 4×4 matrix, each of the minor matrices will be size 3×3, whose determinant then requires the computation of other 2×2 determinants. In the following illustration, two entries in the first row are zero for convenience. For matrix

$$A = \begin{bmatrix} -2 & 0 & 3 & 0 \\ 1 & 2 & 0 & -2 \\ 3 & -1 & 4 & 1 \\ 0 & -3 & 2 & 1 \end{bmatrix}, \text{ we have: } \det(A) = -2 \cdot (-1)^{1+1} \begin{vmatrix} 2 & 0 & -2 \\ -1 & 4 & 1 \\ -3 & 2 & 1 \end{vmatrix} + (3) \cdot (-1)^{1+3} \begin{vmatrix} 1 & 2 & -2 \\ 3 & -1 & 1 \\ 0 & -3 & 1 \end{vmatrix}.$$

Computing the first cofactor gives -16, the second cofactor is 14. This gives:

$$\begin{aligned} \det(A) &= -2(-16) + 3(14) \\ &= 74 \end{aligned}$$

Appendix IV

Deriving the Equation of a Conic

The Equation of an Ellipse

In Section 7.4, the equation $\sqrt{(x + c)^2 + y^2} + \sqrt{(x - c)^2 + y^2} = 2a$ was developed using the distance formula and the definition of an ellipse. To find the standard form of the equation, we treat this result as a radical equation, isolating one of the radicals and squaring both sides.

$$\sqrt{(x + c)^2 + y^2} = 2a - \sqrt{(x - c)^2 + y^2} \quad \text{isolate one radical}$$

$$(x + c)^2 + y^2 = 4a^2 - 4a\sqrt{(x - c)^2 + y^2} + (x - c)^2 + y^2 \quad \text{square both sides}$$

We continue by simplifying the equation, isolating the remaining radical, and squaring again.

$$x^2 + 2xc + c^2 + y^2 = 4a^2 - 4a\sqrt{(x - c)^2 + y^2} + x^2 - 2xc + c^2 + y^2 \quad \text{expand binomials}$$

$$4xc = 4a^2 - 4a\sqrt{(x - c)^2 + y^2} \quad \text{simplify}$$

$$a\sqrt{(x - c)^2 + y^2} = a^2 - xc \quad \text{isolate radical; divide by 4}$$

$$a^2[(x - c)^2 + y^2] = a^4 - 2a^2xc + x^2c^2 \quad \text{square both sides}$$

$$a^2x^2 - 2a^2xc + a^2c^2 + a^2y^2 = a^4 - 2a^2xc + x^2c^2 \quad \text{expand and distribute } a^2 \text{ on the left}$$

$$a^2x^2 - x^2c^2 + a^2y^2 = a^4 - a^2c^2 \quad \text{simplify}$$

$$x^2(a^2 - c^2) + a^2y^2 = a^2(a^2 - c^2) \quad \text{factor}$$

$$\frac{x^2}{a^2} + \frac{y^2}{a^2 - c^2} = 1 \quad \text{divide by } a^2(a^2 - c^2)$$

Since $a > c$, we know $a^2 > c^2$ and $a^2 - c^2 > 0$. For convenience, we let $b^2 = a^2 - c^2$ and it also follows that $a^2 > b^2$ and $a > b$ (since $c > 0$). Substituting b^2 for $a^2 - c^2$ we obtain the standard form of the equation of an ellipse (major axis horizontal, since we stipulated $a > b$): $\frac{x^2}{a^2} + \frac{y^2}{b^2} = 1$. Note once again the x-intercepts are $(\pm a, 0)$, while the y-intercepts are $(0, \pm b)$.

The Equation of a Hyperbola

Also in Section 7.4, the equation $\sqrt{(x + c)^2 + y^2} - \sqrt{(x - c)^2 + y^2} = 2a$ was developed using the distance formula and the definition of a hyperbola. To find the standard form of this equation, we apply the same procedures as before.

$$\sqrt{(x + c)^2 + y^2} = 2a + \sqrt{(x - c)^2 + y^2} \quad \text{isolate one radical}$$

$$(x + c)^2 + y^2 = 4a^2 + 4a\sqrt{(x - c)^2 + y^2} + (x - c)^2 + y^2 \quad \text{square both sides}$$

$$x^2 + 2xc + c^2 + y^2 = 4a^2 + 4a\sqrt{(x - c)^2 + y^2} + x^2 - 2xc + c^2 + y^2 \quad \text{expand binomials}$$

$$4xc = 4a^2 + 4a\sqrt{(x - c)^2 + y^2} \quad \text{simplify}$$

$$xc - a^2 = a\sqrt{(x - c)^2 + y^2} \quad \text{isolate radical; divide by 4}$$

$$x^2c^2 - 2a^2xc + a^4 = a^2[(x - c)^2 + y^2] \quad \text{square both sides}$$

$$x^2c^2 - 2a^2xc + a^4 = a^2x^2 - 2a^2xc + a^2c^2 + a^2y^2 \quad \text{expand and distribute } a^2 \text{ on the right}$$

$$x^2c^2 - a^2x^2 - a^2y^2 = a^2c^2 - a^4 \quad \text{simplify}$$

$$x^2(c^2 - a^2) - a^2y^2 = a^2(c^2 - a^2) \quad \text{factor}$$

$$\frac{x^2}{a^2} - \frac{y^2}{c^2 - a^2} = 1 \quad \text{divide by } a^2(c^2 - a^2)$$

From the definition of a hyperbola we have $0 < a < c$, showing $c^2 > a^2$ and $c^2 - a^2 > 0$. For convenience, we let $b^2 = c^2 - a^2$ and substitute to obtain the standard form of the equation of a hyperbola (transverse axis horizontal): $\frac{x^2}{a^2} - \frac{y^2}{b^2} = 1$. Note the x-intercepts are $(0, \pm a)$ and there are no y-intercepts.

Student Answer Appendix

CHAPTER R

Exercises R.1, pp. 10–13

1. subset of, element of **3.** positive, negative, 7, −7, principal
5. Order of operations requires multiplication before addition.
7. a. $\{1, 2, 3, 4, 5\}$ **b.** $\{\ \}$ **9.** True **11.** True **13.** True
15. $1.\overline{3}$

−1 0 1 2 3 4

17. $2.\overline{5}$

−1 0 1 2 3 4

19. ≈ 2.65

−2 −1 0 1 2 3

21. ≈ 1.73

0 1 2 3 4 5

23. a. i. $\{8, 7, 6\}$ **ii.** $\{8, 7, 6\}$ **iii.** $\{-1, 8, 7, 6\}$
iv. $\{-1, 8, 0.75, \frac{9}{2}, 5.\overline{6}, 7, \frac{3}{5}, 6\}$ **v.** $\{\ \}$ **vi.** $\{-1, 8, 0.75, \frac{9}{2}, 5.\overline{6}, 7, \frac{3}{5}, 6\}$
b. $\{-1, \frac{3}{5}, 0.75, \frac{9}{2}, 5.\overline{6}, 6, 7, 8\}$
c.

$\frac{3}{5}$ 0.75 $\frac{9}{2}$ $5.\overline{6}$

−2 −1 0 1 2 3 4 5 6 7 8

25. a. i. $\{\sqrt{49}, 2, 6, 4\}$ **ii.** $\{\sqrt{49}, 2, 6, 0, 4\}$
iii. $\{-5, \sqrt{49}, 2, -3, 6, -1, 0, 4\}$ **iv.** $\{-5, \sqrt{49}, 2, -3, 6, -1, 0, 4\}$
v. $\{\sqrt{3}, \pi\}$ **vi.** $\{-5, \sqrt{49}, 2, -3, 6, -1, \sqrt{3}, 0, 4, \pi\}$
b. $\{-5, -3, -1, 0, \sqrt{3}, 2, \pi, 4, 6, \sqrt{49}\}$
c.

$\sqrt{3}$ π $\sqrt{49}$

−6 −5 −4 −3 −2 −1 0 1 2 3 4 5 6 7

27. False; not all real numbers are irrational. **29.** False; not all rational numbers are integers. **31.** False; $\sqrt{25} = 5$ is not irrational.
33. c IV **35.** a VI **37.** d III **39.** Let a represent Kylie's age: $a \geq 6$.
41. Let n represent the number of incorrect words: $n \leq 2$. **43.** 2.75
45. −4 **47.** $\frac{1}{2}$ **49.** $\frac{3}{4}$ **51.** 10 **53.** negative **55.** $-n$ **57.** −8, 2
59. undefined, since $12 \div 0 = k$ implies $k \cdot 0 = 12$ **61.** undefined, since $7 \div 0 = k$ implies $k \cdot 0 = 7$ **63. a.** positive **b.** negative **c.** negative
d. negative **65.** $-\frac{11}{6}$ **67.** −2 **69.** $9^2 = 81$ is closest **71.** 7
73. −2.185 **75.** $4\frac{1}{3}$ **77.** $-\frac{29}{12}$ or $-2\frac{5}{12}$ **79.** 0 **81.** −5 **83.** $-\frac{1}{10}$
85. $-\frac{7}{8}$ **87.** −17 **89.** $\frac{-11}{12}$ **91.** 64 **93.** 4489.70 **95.** $D \approx 4.3$ cm
97. 32°F **99.** 179°F **101.** Tsu Ch'ung-chih: $\frac{355}{113}$ **103.** negative

Exercises R.2, pp. 19–21

1. constant **3.** coefficient **5.** $-5 + 5 = 0$, $-5 \cdot (-\frac{1}{5}) = 1$ **7.** two; 3 and −5 **9.** two; 2 and $\frac{1}{4}$ **11.** three; −2, 1, and −5 **13.** one; −1
15. $n - 7$ **17.** $n + 4$ **19.** $(n - 5)^2$ **21.** $2n - 13$ **23.** $n^2 + 2n$
25. $\frac{2}{3}n - 5$ **27.** $3(n + 5) - 7$ **29.** Let w represent the width. Then $2w$ represents twice the width and $2w - 3$ represents three meters less than twice the width. **31.** Let b represent the speed of the bus. Then $b + 15$ represents 15 mph more than the speed of the bus. **33.** 14
35. 19 **37.** 0 **39.** 16 **41.** −36 **43.** 51 **45.** 2 **47.** 144
49. $\frac{-41}{5}$ **51.** 24

53.

x	Output
−3	14
−2	6
−1	0
0	−4
1	−6
2	−6
3	−4

−1 has an output of 0.

55.

x	Output
−3	−18
−2	−15
−1	−12
0	−9
1	−6
2	−3
3	0

3 has an output of 0.

57.

x	Output
−3	−5
−2	8
−1	9
0	4
1	−1
2	0
3	13

2 has an output of 0.

59. a. $7 + (-5) = 2$ **b.** $n + (-2)$ **c.** $a + (-4.2) + 13.6 = a + 9.4$
d. $x + 7 - 7 = x$ **61. a.** 3.2 **b.** $\frac{5}{6}$ **63.** $-5x + 13$ **65.** $-\frac{2}{15}p + 6$
67. $-2a$ **69.** $\frac{17}{12}x$ **71.** $-2a^2 + 2a$ **73.** $6x^2 - 3x$ **75.** $2a + 3b + 2c$
77. $\frac{29}{8}n + \frac{38}{5}$ **79.** $7a^2 - 13a - 5$ **81.** $-6b$ **83.** $17g + 9h$
85. $9x^2 - 3x$ **87.** $-0.9y - 11.4x$ **89.** 10 ohms **91.** $\frac{1}{2}j$ **93.** $2w + 3$
95. $c + 22$; 37¢ **97.** $25t + 43.50$; \$81 **99. a.** positive odd integer

Exercises R.3, pp. 31–34

1. power **3.** $20x$, 0 **5. a.** has addition of unlike terms **b.** is multiplication **7.** $6p^5q^4$ **9.** $-16a^5b^3$ **11.** $14x^7y^7$ **13.** $216p^3q^6$
15. $32.768\,h^3k^6$ **17.** $\frac{p^2}{4q^2}$ **19.** $49c^{14}d^4$ **21.** $\frac{9}{16}x^6y^2$ **23.** $\frac{9}{4}x^3y^2$
25. a. $V = 27x^6$ **b.** 1728 units^3 **27.** $3w^3$ **29.** $-3ab$ **31.** $\frac{27}{8}$
33. $2h^3$ **35.** $\frac{4p^8}{q^6}$ **37.** $\frac{8x^6}{27y^9}$ **39.** $\frac{25m^4n^6}{4r^8}$ **41.** $\frac{25p^2q^2}{4}$ **43.** $\frac{3p^2}{-4q^2}$
45. $\frac{5}{3h^7}$ **47.** $\frac{1}{a^3}$ **49.** $\frac{a^{12}}{b^4c^8}$ **51.** $\frac{-12}{5x^4}$ **53.** $\frac{-2b^7}{27a^9c^3}$ **55.** 2 **57.** $\frac{7}{10}$
59. $\frac{13}{9}$ **61.** −4 **63.** 1.45×10^{10} **65.** \$4,800,000,000 **67.** 26,571.4 hr; 1107.1 days **69.** polynomial, none of these, degree 3 **71.** nonpolynomial because exponents are not whole numbers, NA, NA **73.** polynomial, binomial, degree 3 **75.** $-w^3 - 3w^2 + 7w + 8.2$; −1
77. $c^3 + 2c^2 - 3c + 6$; 1 **79.** $\frac{-2}{3}x^2 + 12$; $\frac{-2}{3}$ **81.** $3p^3 - 3p^2 - 12$
83. $7.85b^2 - 0.6b - 1.9$ **85.** $\frac{1}{4}x^2 - 8x + 6$
87. $q^6 + q^5 - q^4 + 2q^3 - q^2 - 2q$ **89.** $-3x^3 + 3x^2 + 18x$
91. $3r^2 - 11r + 10$ **93.** $x^3 - 27$ **95.** $b^3 - b^2 - 34b - 56$
97. $21v^2 - 47v + 20$ **99.** $9 - m^2$ **101.** $p^2 + 1.1p - 9$
103. $x^2 + \frac{3}{4}x + \frac{1}{8}$ **105.** $m^2 - \frac{9}{16}$ **107.** $6x^2 + 11xy - 10y^2$
109. $12c^2 + 23cd + 5d^2$ **111.** $2x^4 - x^2 - 15$ **113.** $4m + 3$; $16m^2 - 9$ **115.** $7x + 10$; $49x^2 - 100$ **117.** $6 - 5k$; $36 - 25k^2$
119. $ab^2 - c$; $a^2b^4 - c^2$ **121.** $x^2 + 8x + 16$ **123.** $16g^2 + 24g + 9$
125. $16p^2 - 24pq + 9q^2$ **127.** $4m^2 + 12mn + 9n^2$
129. $xy + 2x - 3y - 6$ **131.** $k^3 + 3k^2 - 28k - 60$ **133. a.** 340 mg
b. 292.5 mg **c.** Less, amount is decreasing **d.** after 5 hr
135. $F = kPQd^{-2}$ **137.** $5x^{-3} + 3x^{-2} + 2x^{-1} + 4$ **139.** \$15
141. 6

Exercises R.4, pp. 41–44

1. product **3.** binomial, conjugate **5.** Answers will vary.
7. a. $-17(x^2 - 3)$ **b.** $7b(3b^2 - 2b + 8)$ **c.** $-3a^2(a^2 + 2a - 3)$

9. a. $(a+2)(2a+3)$ **b.** $(b^2+3)(3b+2)$ **c.** $(n+7)(4m-11)$ **11. a.** $(3q+2)(3q^2+5)$ **b.** $(h-12)(h^4-3)$ **c.** $(k^2-7)(k^3-5)$ **13. a.** $(b-7)(b+2)$ **b.** $(a-9)(a+5)$ **c.** $(n-4)(n-5)$ **15. a.** $(3p+2)(p-5)$ **b.** $(4q-5)(q+3)$ **c.** $(5u+3)(2u-5)$ **17. a.** $(2s+5)(2s-5)$ **b.** $(3x+7)(3x-7)$ **c.** $2(5x+6)(5x-6)$ **d.** $(11h+12)(11h-12)$ **19. a.** $(a-3)^2$ **b.** $(b+5)^2$ **c.** $(2m-5)^2$ **d.** $(3n-7)^2$ **21. a.** $(2p-3)(4p^2+6p+9)$ **b.** $(m+\frac{1}{2})(m^2-\frac{1}{2}m+\frac{1}{4})$ **c.** $(g-0.3)(g^2+0.3g+0.09)$ **d.** $-2t(t-3)(t^2+3t+9)$ **23. a.** $(x+3)(x-3)(x+1)(x-1)$ **b.** $(x^2+9)(x^2+4)$ **c.** $(x-2)(x^2+2x+4)(x+1)(x^2-x+1)$ **25. a.** $(n+1)(n-1)$ **b.** $(n-1)(n^2+n+1)$ **c.** $(n+1)(n^2-n+1)$ **d.** $7x(2x+1)(2x-1)$ **27.** $(a+5)(a+2)$ **29.** $(x-2)(x-10)$ **31.** $(8+3m)(8-3m)$ **33.** $(r-3)(r-6)$ **35.** $(2h+3)(h+2)$ **37.** $(3k-4)^2$ **39.** prime **41.** $4m(m+5)(m-2)$ **43.** $(a+5)(a-12)$ **45.** $(2x-5)(4x^2+10x+25)$ **47.** $(m-4)(4m-3)$ **49.** $(x-5)(x+3)(x-3)$ **51. a.** H **b.** E **c.** C **d.** F **e.** B **f.** A **g.** I **h.** D **i.** G **53.** $H=-33$ **55.** $S=\pi r^2(2+h)$, $S=75\pi\approx 235.62\text{ cm}^2$ **57. a.** 3 in. **b.** 5 in. **c.** $V=2(5)(7)=70\text{ ft}^3$

59. $L=L_0\sqrt{\left(1+\frac{v}{c}\right)\left(1-\frac{v}{c}\right)}$
$L=12\sqrt{(1+0.75)(1-0.75)}$
$=3\sqrt{7}$ in. $\approx$ 7.94 in.

61. a. $\frac{1}{8}(4x^4+x^3-6x^2+32)$ **b.** $\frac{1}{18}(12b^5-3b^3+8b^2-18)$ **63.** $2x(16x-27)(6x+5)$

Exercises R.5, pp. 50–54

1. 1, −1 **3.** common factor **5.** F **7. a.** $-\frac{1}{3}$ **b.** $\frac{x+3}{2x(x-2)}$ **9. a.** simplified **b.** $\frac{a-4}{a-7}$ **11. a.** −1 **b.** −1 **13. a.** $-3ab^9$ **b.** $\frac{x+3}{9}$ **c.** $-1(y+3)$ **d.** $\frac{-1}{m}$ **15. a.** $\frac{2n+3}{n}$ **b.** $\frac{3x+5}{2x+3}$ **c.** $x+2$ **d.** $n-2$ **17.** $\frac{(a-2)(a+1)}{(a+3)(a+2)}$ **19.** 1 **21.** $\frac{(p-4)^2}{p^2}$ **23.** $\frac{-15}{4}$ **25.** $\frac{3}{2}$ **27.** $\frac{8(a-7)}{a-5}$ **29.** $\frac{y}{x}$ **31.** $\frac{m}{m-4}$ **33.** $\frac{y+3}{3y(y+4)}$ **35.** $\frac{x+0.3}{x-0.2}$ **37.** $\frac{3(a^2+3a+9)}{2}$ **39.** $\frac{2n+1}{n}$ **41.** $\frac{3+20x}{8x^2}$ **43.** $\frac{14y-x}{8x^2y^4}$ **45.** $\frac{2}{p+6}$ **47.** $\frac{-3m-16}{(m+4)(m-4)}$ **49.** $\frac{-5m+37}{m-7}$ **51.** $\frac{-y+11}{(y+6)(y-5)}$ **53.** $\frac{2a-5}{(a+4)(a-5)}$ **55.** $\frac{1}{y+1}$ **57.** $\frac{m^2-6m+21}{(m+3)^2(m-3)}$ **59.** $\frac{y^2+26y-1}{(5y+1)(y+3)(y-2)}$ **61. a.** $\frac{1}{p^2}-\frac{5}{p};\frac{1-5p}{p^2}$ **b.** $\frac{1}{x^2}+\frac{2}{x^3};\frac{x+2}{x^3}$ **63.** $\frac{4a}{20+a}$ **65.** $p-1$ **67.** $\frac{x}{9x-12}$ **69.** $\frac{2}{-y-31}$ **71. a.** $\frac{1+\frac{3}{m}}{1-\frac{3}{m}};\frac{m+3}{m-3}$ **b.** $\frac{1+\frac{2}{x^2}}{1-\frac{2}{x^2}};\frac{x^2+2}{x^2-2}$ **73. a.** \$300 million, \$2550 million **b.** It would require many resources. **c.** No **75.** $\frac{f_2+f_1}{f_1f_2}$ **77.** $\frac{-a}{x(x+h)}$ **79.** $\frac{-(2x+h)}{2x^2(x+h)^2}$

81.

Day	Price
0	10
1	16.67
2	32.76
3	47.40
4	53.51
5	52.86
6	49.25
7	44.91
8	40.75
9	37.04
10	33.81

Price rises rapidly for first four days, then begins a gradual decrease. Yes, on the 35th day of trading.

83. $t=8$ weeks **85.** (b); others equal 2 **87.** $\frac{6}{23}$; $\frac{ac}{ad+bc}$

Exercises R.6, pp. 64–67

1. even **3.** $\left(16^{\frac{1}{4}}\right)^3$ **5.** Answers will vary. **7. a.** 9 **b.** 10 **9. a.** $7|p|$ **b.** $|x-3|$ **c.** $9m^2$ **d.** $|x-3|$ **11. a.** 4 **b.** $-5x$ **c.** $6z^4$ **d.** $\frac{v}{-2}$ **13. a.** 2 **b.** not a real number **c.** $3x^2$ **d.** $-3x$ **e.** $k-3$ **f.** $|h+2|$ **15. a.** −5 **b.** $-3|n^3|$ **c.** not a real number **d.** $\frac{7|v^5|}{6}$ **17. a.** 4 **b.** $\frac{64}{125}$ **c.** $\frac{125}{8}$ **d.** $\frac{9p^4}{4q^2}$ **19. a.** −1728 **b.** not a real number **c.** $\frac{1}{9}$ **d.** $\frac{-256}{81x^4}$ **21. a.** $\frac{32n^{10}}{p^2}$ **b.** $\frac{1}{2y^{\frac{1}{4}}}=\frac{y^{\frac{3}{4}}}{2y}$ **23. a.** $3m\sqrt{2}$ **b.** $10pq^2\sqrt[3]{q}$ **c.** $\frac{3}{2}mn\sqrt[3]{n^2}$ **d.** $4pq^3\sqrt{2p}$ **e.** $-3+\sqrt{7}$ **f.** $\frac{9}{2}-\sqrt{2}$ **25. a.** $15a^2$ **b.** $-4b\sqrt{b}$ **c.** $\frac{x^4\sqrt{y}}{3}$ **d.** $3u^2v\sqrt[3]{v}$ **27. a.** $2m^2$ **b.** $3n$ **c.** $\frac{3\sqrt{5}}{4x}$ **d.** $\frac{18\sqrt[3]{3}}{z^3}$ **29. a.** $2x^2y^3$ **b.** $x^2\sqrt[4]{x}$ **c.** $\sqrt[12]{b}$ **d.** $\frac{1}{\sqrt[6]{6}}=\frac{\sqrt[6]{6^5}}{6}$ **31. a.** $9\sqrt{2}$ **b.** $14\sqrt{3}$ **c.** $16\sqrt{2m}$ **d.** $-5\sqrt{7p}$ **33. a.** $-x\sqrt[3]{2x}$ **b.** $2-\sqrt{3x}+3\sqrt{5}$ **c.** $6x\sqrt{2x}+5\sqrt{2}-\sqrt{7x}+3\sqrt{3}$ **35. a.** 98 **b.** $\sqrt{15}+\sqrt{21}$ **c.** n^2-5 **d.** $39-12\sqrt{3}$ **37. a.** −19 **b.** $\sqrt{10}+\sqrt{65}-2\sqrt{7}-\sqrt{182}$ **c.** $12\sqrt{5}+2\sqrt{14}+36\sqrt{15}+6\sqrt{42}$ **39.** Answers will vary. **41.** Answers will vary. **43. a.** $\frac{\sqrt{3}}{2}$ **b.** $\frac{2\sqrt{15x}}{9x^2}$ **c.** $\frac{3\sqrt{6b}}{10b}$ **d.** $\frac{\sqrt[3]{2p^2}}{2p}$ **45. a.** $-12+4\sqrt{11}$; 1.27 **b.** $\frac{6\sqrt{x}+6\sqrt{2}}{x-2}$ **47. a.** $\sqrt{30}-2\sqrt{5}-3\sqrt{3}+3\sqrt{2}$; 0.05 **b.** $\frac{7+7\sqrt{2}+\sqrt{6}+2\sqrt{3}}{-3}$; −7.60 **49.** 8.33 ft **51.** 23.9 m **53.** $\frac{1}{\sqrt{x+2}+\sqrt{x}}$ **55. a.** 365.02 days **b.** 688.69 days **c.** 87.91 days **57. a.** 36 mph **b.** 46.5 mph **59.** $12\pi\sqrt{34}\approx 219.82\text{ m}^2$ **61. a.** $(x+\sqrt{5})(x-\sqrt{5})$ **b.** $(n+\sqrt{19})(n-\sqrt{19})$ **63.** Because $m^4\geq 0$ for $m\in\mathbb{R}$ **65.** 3

Practice Test, pp. 68–69

1. a. True **b.** True **c.** False; $\sqrt{2}$ cannot be expressed as a ratio of two integers. **d.** True **2. a.** 11 **b.** −5 **c.** not a real number **d.** 20 **3. a.** $\frac{9}{8}$ **b.** $\frac{-7}{6}$ **c.** 0.5 **d.** −4.6 **4. a.** $\frac{28}{3}$ **b.** 0.9 **c.** 4 **d.** −7

5. ≈4439.28 **6. a.** 0 **b.** undefined **7. a.** 3; −2, 6, 5 **b.** 2; $\frac{1}{3}$, 1
8. a. −13 **b.** ≈7.29 **9. a.** $x^3 - (2x - 9)$ **b.** $2n - 3\left(\frac{n}{2}\right)^2$
10. a. Let r represent Earth's radius. Then $11r - 119$ represents Jupiter's radius. **b.** Let e represent this year's earnings. Then $4e + 1.2$ represents last year's earnings. **11. a.** $9v^2 + 3v - 7$ **b.** $-7b + 8$ **c.** $6x + x^2$
12. a. $(3x + 4)(3x - 4)$ **b.** $v(2v - 3)^2$ **c.** $(x + 5)(x + 3)(x - 3)$
13. a. $5b^3$ **b.** $4a^{12}b^{12}$ **c.** $\frac{m^6}{8n^3}$ **d.** $\frac{25}{4}p^2q^2$ **14. a.** $-4ab$
b. $6.4 \times 10^{-2} = 0.064$ **c.** $\frac{a^{12}}{b^4c^8}$ **d.** −6 **15. a.** $9x^4 - 25y^2$
b. $4a^2 + 12ab + 9b^2$ **16. a.** $7a^4 - 5a^3 + 8a^2 - 3a - 18$
b. $-7x^4 + 4x^2 + 5x$ **17. a.** −1 **b.** $\frac{2 + n}{2 - n}$ **c.** $x - 3$ **d.** $\frac{x - 5}{3x - 2}$
e. $\frac{x - 5}{3x + 1}$ **f.** $\frac{3(m + 7)}{5(m + 4)(m - 3)}$ **18. a.** $|x + 11|$ **b.** $\frac{-2}{3v}$ **c.** $\frac{64}{125}$
d. $-\frac{1}{2} + \frac{\sqrt{2}}{2}$ **e.** $11\sqrt{10}$ **f.** $x^2 - 5$ **g.** $\frac{\sqrt{10x}}{5x}$ **h.** $2(\sqrt{6} + \sqrt{2})$
19. $-0.5x^2 + 10x + 1200$; **a.** 10 decreases of 0.50 or $5.00
b. Maximum revenue is $1250. **20.** 58 cm

CHAPTER 1

Exercises 1.1, pp. 80–83

1. identity, unknown **3.** literal, two **5.** Answers will vary. **7.** linear
9. linear **11.** nonlinear; two variables are multiplied together
13. $\frac{9}{10}$ **15.** $\frac{6}{5}$ **17.** $\frac{-3}{8}$ **19.** 12 **21.** −56 **23.** $\frac{20}{21}$ **25.** $\frac{-27}{4}$
27. $\frac{-123}{19}$ **29.** −1 **31.** contradiction **33.** conditional; $x = -1.1$
35. identity **37.** $R = \frac{I}{PT}$ **39.** $r = \frac{C}{2\pi}$ **41.** $R = \frac{W}{I^2}$ **43.** $h = \frac{3V}{4\pi r^2}$
45. $A = 6s^2$ **47.** $P = \frac{2(S - B)}{S}$ **49.** $y = \frac{-Ax}{B} + \frac{C}{B}$
51. $y = \frac{-20}{9}x + \frac{16}{3}$ **53.** $y = \frac{-4}{5}x - 5$ **55.** $a = 3; b = 2; c = -19;$ $x = -7$ **57.** $a = -6; b = 1; c = 33; x = \frac{-16}{3}$ **59.** $a = 2; b = -13;$ $c = -27; x = -7$ **61.** $h = 17$ cm **63.** 510 ft **65.** 56 in. **67.** 3084 ft
69. 48; 50 **71.** 5; 7 **73.** 3 P.M. **75.** 36 min **77.** 4 quarts; 50% O.J.
79. 16 lb; $1.80 lb **81.** 6 gal **83.** 16 lb **85.** about 1.4 million
87. $106\frac{2}{3}$ oz of 15% acid; $93\frac{1}{3}$ oz of 45% acid **89.** 69 **91.** $-3, 1, -\frac{1}{3}, 7$
93. commutative property **95.** $x^3 - 3x^2 + 7x - 9$

Exercises 1.2, pp. 90–94

1. set, interval **3.** intersection, union **5.** Answers will vary.
7. $w \geq 45$ **9.** $250 < T < 450$

11. 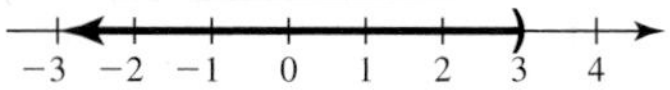

13. 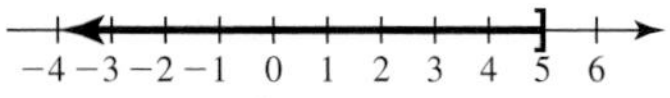

15. 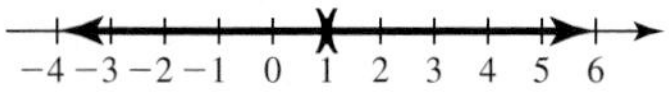

17. 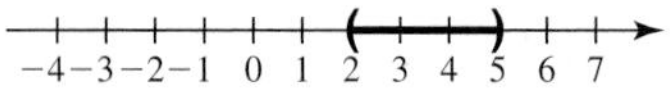

19. $\{a \mid a \geq 2\}$;
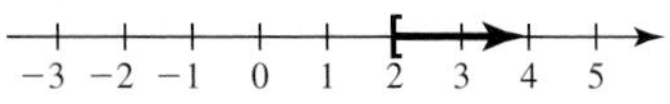
; $a \in [2, \infty)$

21. $\{n \mid n \geq 1\}$;

; $n \in [1, \infty)$

23. $\{x \mid x < \frac{-32}{5}\}$;
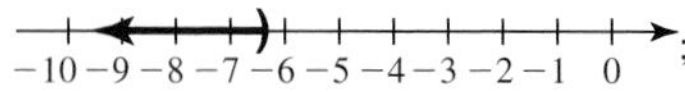
; $x \in (-\infty, \frac{-32}{5})$

25. $\{y \mid 2 < y < 5\}$;
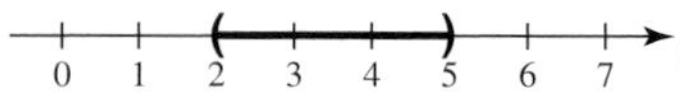
; $y \in (2, 5)$

27. $\{m \mid 2 < m \leq 6\}$;
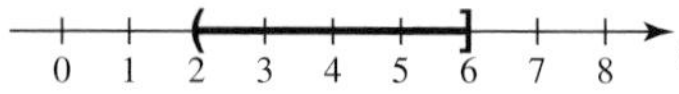
; $m \in (2, 6]$

29. $\{m \mid \frac{4}{3} \leq m < \frac{11}{6}\}$;
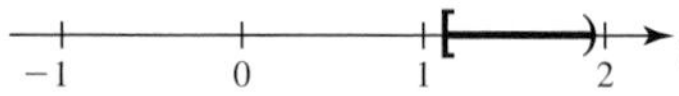
; $m \in [\frac{4}{3}, \frac{11}{6})$

31. $\{x \mid x \geq -2\}$; $[-2, \infty)$ **33.** $\{x \mid -2 \leq x \leq 1\}$; $[-2, 1]$
35. {2}; {−3, −2, −1, 0, 1, 2, 3, 4, 6, 8}
37. { }; {−3, −2, −1, 0, 1, 2, 3, 4, 5, 6, 7} **39.** {4, 6}; {2, 4, 5, 6, 7, 8}

41. 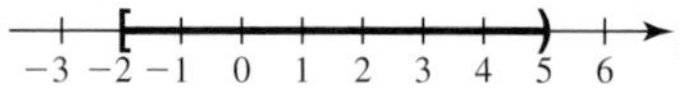
; $x \in [-2, 5)$

43. 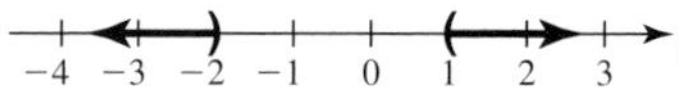
; $x \in (-\infty, -2) \cup (1, \infty)$

45. no solution

47. $x \in (-\infty, \infty)$;
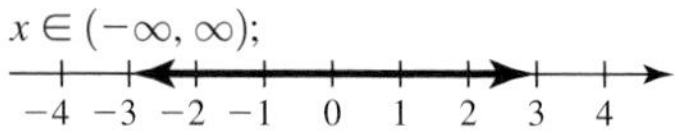

49. $x \in [-5, 0]$;
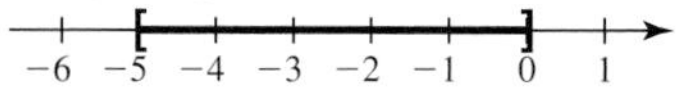

51. $x \in (\frac{-1}{3}, \frac{-1}{4})$;
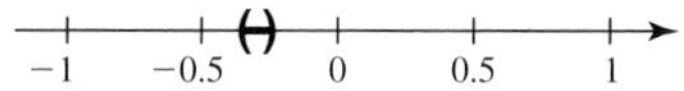

53. $x \in (-\infty, \infty)$;
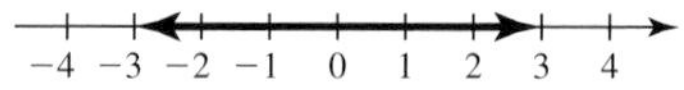

55. $x \in [-4, 1)$;
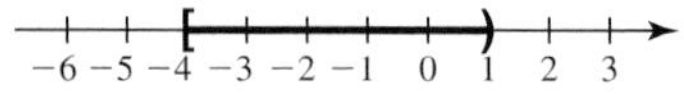

57. $x \in [-1.4, 0.8]$;
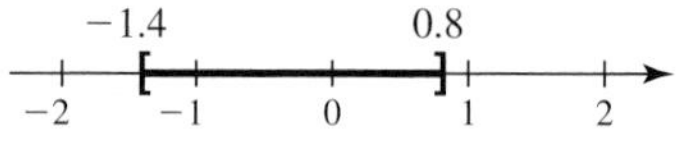

59. $x \in [-16, 8)$;
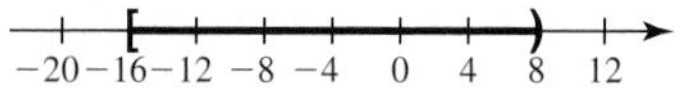

61. $m \in (-\infty, 0) \cup (0, \infty)$ **63.** $y \in (-\infty, -7) \cup (-7, \infty)$
65. $a \in (-\infty, \frac{1}{2}) \cup (\frac{1}{2}, \infty)$ **67.** $x \in (-\infty, 4) \cup (4, \infty)$ **69.** $x \in [2, \infty)$
71. $n \in [4, \infty)$ **73.** $b \in [\frac{4}{3}, \infty)$ **75.** $y \in (-\infty, 2]$ **77.** < **79.** <
81. < **83.** > **85.** 177.34 lb or less **87.** $x \geq 97\%$
89. 7.2°C < T < 29.4°C **91.** $b \geq \$2000$ **93.** $h \geq 8$ in.
95. Alaska −80°F ≤ T ≤ 100°F; Hawaii 12°F ≤ T ≤ 100°F; Alaska; Answers will vary.

97. 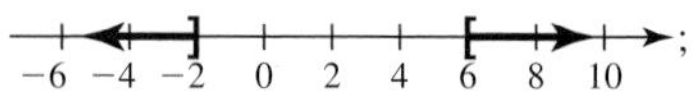
; $x \in (-\infty, -2] \cup [6, \infty)$

99. $2n - 8$ **101.** $420 + 30.625\pi$ cm^3 **103.** $x = 4$

Exercises 1.3, pp. 101–105

1. excluded **3.** extraneous **5.** Answers will vary. **7.** $x = 5$ or $x = -3$ **9.** $m = 4$ **11.** $p = 0$ or $p = 2$ **13.** $h = 0$ or $h = \frac{-1}{2}$
15. $a = 3$ or $a = -3$ **17.** $g = -9$ **19.** $m = -5$ or $m = -3$ or $m = 3$

21. $c = -3$ or $c = 15$ **23.** $r = 8$ or $r = -3$ **25.** $t = -13$ or $t = 2$ **27.** $x = 5$ or $x = -3$ **29.** $w = -\frac{1}{2}$ or $w = 3$ **31.** $x = 4$ or $x = -1$ or $x = 5$ or $x = -2$ **33.** $x = 1$ **35.** $a = 10$ **37.** $y = 12$ **39.** $x = 3$; $x = 7$ is extraneous **41.** $n = 7$ **43.** $a = -1$ **45.** $f = \dfrac{f_1 f_2}{f_1 + f_2}$ **47.** $r = \dfrac{E - IR}{I}$ **49.** $h = \dfrac{3V}{\pi r^2}$ **51.** $r^3 = \dfrac{3V}{4\pi}$ **53. a.** $x = \frac{14}{3}$ **b.** $x = \frac{8}{3}$ **c.** $m = 40$ **55. a.** $m = 3$ **b.** $x = 5$ **c.** $m = -64$ **d.** $x = -16$ **57. a.** $x = 2, 18$ **b.** $x = 11$ **59.** $x = -32$ **61.** $x = 9$ **63.** $x = \frac{16}{25}$ **65.** $x = -27, 125$ **67.** $x = 1, 4$ **69.** $x = -3, -2, 1, 2$ **71.** $x = -1, \frac{1}{4}$ **73.** $x = \pm\frac{1}{3}, \pm\frac{1}{2}$ **75.** $x = -4, 45$ **77.** $x = -6$; $x = \frac{-74}{9}$ is extraneous **79.** $S = 12\pi\sqrt{34}\text{ m}^2$ **81.** 7, 9, 11 **83.** $n = 5$ **85.** 11 in. by 13 in. **87.** $a = 6$ cm **89.** either \$50 or \$30 **91. a.** -32 ft **b.** 11 sec **c.** pebble is at canyon's rim **93.** $11\frac{2}{3}$ hr or 11 hr 40 min **95.** 140 mi **97.** $P \approx 52.1\%$ **99. a.** 36 million mi **b.** 67 million mi **c.** 93 million mi **d.** 142 million mi **e.** 484 million mi **f.** 887 million mi **101.** The constant "3" was not multiplied by the LCM; $3x(x + 3) - 8x = x + 3$; $x = -1, 1$ **103.** Answers will vary. **105. a.** $x = \frac{3}{4}, 5$ **b.** $x = -7, -\frac{2}{3}$ **107.** 1.7 hr **109.** $-7, -5, -3$ **111.** $3\frac{1}{2}$

Mid-Chapter Check, p. 106

1. a. $x = \frac{-2}{3}, \frac{2}{3}$ **b.** $x = \frac{3}{2}$ **c.** $x = -2, 5, 0$ **d.** $x = \frac{-1}{3}, \frac{1}{3}$ **e.** $x = -16, 3$ **f.** $x = \frac{3}{4}, 3$ **g.** $x = 2, \frac{-\sqrt{5}}{2}, \frac{\sqrt{5}}{2}$ **h.** $x = -1, \frac{3}{2}$ **2.** $x = 3, -1$ **3.** $x = \frac{7}{4}$ **4.** $x = -2$ **5.** $x = 9$ **6.** $v_0 = \dfrac{H + 16t^2}{t}$ **7.** $r = \sqrt{\dfrac{S}{\pi(2 + y)}}$

8. a. $x \geq 1$ or $x \leq -2$;

−4 −3 −2 −1 0 1 2 3 4

b. $16 < x \leq 19$;

0 15 16 17 18 19 20

9. $x = 0, 7$ **10. a.** at $t = 1$, $H = 80$ ft **b.** at $H = 140$, $t = \frac{5}{2}$ or $\frac{7}{2}$ sec

Reinforcing Basic Concepts, pp. 106–107

1. $x(x + 4) = 0$; $x = 0, -4$ **2.** $x(x + 7) = 0$; $x = 0, -7$ **3.** $x(x - 5) = 0$; $x = 0, 5$ **4.** $x(x - 2) = 0$; $x = 0, 2$ **5.** $x(x + \frac{1}{2}) = 0$; $x = 0, -\frac{1}{2}$ **6.** $x(x + \frac{2}{5}) = 0$; $x = 0, -\frac{2}{5}$ **7.** $x(x - \frac{2}{3}) = 0$; $x = 0, \frac{2}{3}$ **8.** $x(x - \frac{5}{6}) = 0$; $x = 0, \frac{5}{6}$ **9.** $(x + 9)(x - 9) = 0$; $x = -9, 9$ **10.** $(x + 11)(x - 11) = 0$; $x = -11, 11$ **11.** $(x + \sqrt{7})(x - \sqrt{7}) = 0$; $x = -\sqrt{7}, \sqrt{7}$ **12.** $(x + \sqrt{31})(x - \sqrt{31}) = 0$; $x = -\sqrt{31}, \sqrt{31}$ **13.** $(x + 7)(x - 7) = 0$; $x = -7, 7$ **14.** $(x + \sqrt{13})(x - \sqrt{13}) = 0$; $x = -\sqrt{13}, \sqrt{13}$ **15.** $(x + \sqrt{21})(x - \sqrt{21}) = 0$; $x = -\sqrt{21}, \sqrt{21}$ **16.** $(x + 4)(x - 4) = 0$; $x = -4, 4$ **17.** $(x + 9)(x - 5) = 0$; $x = -9, 5$ **18.** $(x - 9)(x - 4) = 0$; $x = 9, 4$ **19.** $(x + 8)(x + 2) = 0$; $x = -8, -2$ **20.** $(x - 11)(x + 4) = 0$; $x = 11, -4$ **21.** $(x + 8)(x - 2) = 0$; $x = -8, 2$ **22.** $(x - 17)(x - 3) = 0$; $x = 17, 3$ **23.** $(x + 1)(x + 7) = 0$; $x = -1, -7$ **24.** $(x - 9)(x + 3) = 0$; $x = 9, -3$

Exercises 1.4, pp. 113–117

1. $3 - 2i$ **3.** $2, 3\sqrt{2}$ **5.** (b) is correct. **7. a.** $4i$ **b.** $7i$ **c.** $3\sqrt{3}$ **d.** $6\sqrt{2}$ **9. a.** $-3i\sqrt{2}$ **b.** $-5i\sqrt{2}$ **c.** $15i$ **d.** $6i$ **11. a.** $i\sqrt{19}$ **b.** $i\sqrt{31}$ **c.** $\dfrac{2\sqrt{3}}{5}i$ **d.** $\dfrac{3\sqrt{2}}{8}i$ **13. a.** $1 + i$; $a = 1, b = 1$ **b.** $2 + \sqrt{3}\,i$; $a = 2, b = \sqrt{3}$ **15. a.** $4 + 2i$; $a = 4, b = 2$ **b.** $2 - \sqrt{2}\,i$; $a = 2, b = -\sqrt{2}$ **17. a.** $5 + 0i$; $a = 5, b = 0$ **b.** $0 + 3i$; $a = 0, b = 3$ **19. a.** $18i$; $a = 0, b = 18$ **b.** $\dfrac{\sqrt{2}}{2}i$; $a = 0, b = \dfrac{\sqrt{2}}{2}$ **21. a.** $4 + 5\sqrt{2}\,i$; $a = 4, b = 5\sqrt{2}$ **b.** $-5 + 3\sqrt{3}\,i$; $a = -5, b = 3\sqrt{3}$ **23.** $\dfrac{7}{4} + \dfrac{7\sqrt{2}}{8}i$; $a = \dfrac{7}{4}, b = \dfrac{7\sqrt{2}}{8}$ **b.** $\dfrac{1}{2} + \dfrac{\sqrt{10}}{2}i$; $a = \dfrac{1}{2}, b = \dfrac{\sqrt{10}}{2}$ **25. a.** $19 + i$ **b.** $2 - 4i$ **c.** $9 + 10\sqrt{3}\,i$ **27. a.** $-3 + 2i$ **b.** 8 **c.** $2 - 8i$ **29. a.** $2.7 + 0.2i$ **b.** $15 + \dfrac{1}{12}i$ **c.** $-2 - \dfrac{1}{8}i$ **31. a.** 15 **b.** 16 **33. a.** $-21 - 35i$ **b.** $-42 - 18i$ **35. a.** $-12 - 5i$ **b.** $1 + 5i$ **37. a.** $4 - 5i$; 41 **b.** $3 + i\sqrt{2}$; 11 **39. a.** $-7i$; 49 **b.** $\frac{1}{2} + \frac{2}{3}i$; $\frac{25}{36}$ **41. a.** 41 **b.** 74 **43. a.** 11 **b.** $\frac{17}{36}$ **45. a.** $-5 + 12i$ **b.** $-7 - 24i$ **47. a.** $-21 - 20i$ **b.** $7 + 6\sqrt{2}\,i$ **49.** no **51.** yes **53.** yes **55.** yes **57.** yes **59.** Answers will vary. **61. a.** 1 **b.** -1 **c.** $-i$ **d.** i **63. a.** $\frac{2}{7}i$ **b.** $\frac{-4}{5}i$ **65. a.** $\frac{21}{13} - \frac{14}{13}i$ **b.** $\frac{-10}{13} - \frac{15}{13}i$ **67. a.** $1 - \frac{3}{4}i$ **b.** $-1 - \frac{2}{3}i$ **69. a.** $\sqrt{13}$ **b.** $\sqrt{41}$ **c.** $\sqrt{11}$ **71.** $A + B = 10$ $AB = 40$ **73.** $7 - 5i\ \Omega$ **75.** $25 + 5i$ V **77.** $\frac{7}{4} + i\ \Omega$ **79. a.** $(x + 3i)(x - 3i)$, **b.** verified **81.** 1 **83. a.** $3 + 4i$ **b.** $3 - 2i$ **c.** $\dfrac{3\sqrt{2}}{2} + \dfrac{\sqrt{2}}{2}i$

85. a. Six is not a rational number—False. **b.** The rational numbers are a subset of the reals—True. **c.** 103 is an element of the set {3, 4, 5 . . .}—True. **d.** The real numbers are not a subset of the complex—False. **87.** $\dfrac{x - 2}{x - 5}$; $x \neq -5, 2, 5$ **89. a.** $(x + 2)(x - 2)(x^2 + 4)$ **b.** $(n - 3)(n^2 + 3n + 9)$ **c.** $(x - 1)^2(x + 1)$ **d.** $m(2n - 3m)^2$

Exercises 1.5, pp. 126–129

1. descending, 0 **3.** quadratic, 1 **5.** $x = \pm\frac{\sqrt{5}}{2}$, The square root property is easier. **7.** $a = -1$; $b = 2$; $c = -15$ **9.** not quadratic **11.** $a = \frac{1}{4}$; $b = -6$; $c = 0$ **13.** $a = 2$; $b = 0$; $c = 7$ **15.** not quadratic **17.** $a = 1$; $b = -1$; $c = -5$ **19.** $m = \pm 4$ **21.** $y = \pm 2\sqrt{7}$; $y \approx \pm 5.29$ **23.** no real solutions **25.** $x = \pm\frac{\sqrt{21}}{4}$; $x \approx \pm 1.15$ **27.** $n = 9$; $n = -3$ **29.** $w = -5 \pm \sqrt{3}$; $w \approx -3.27$ or $w \approx -6.73$ **31.** no real solutions **33.** $m = 2 \pm \frac{3\sqrt{2}}{7}$; $m \approx 2.61$ or $m \approx 1.39$ **35.** 9; $(x + 3)^2$ **37.** $\frac{9}{4}$; $(n + \frac{3}{2})^2$ **39.** $\frac{1}{9}$; $(p + \frac{1}{3})^2$ **41.** $x = -1$ or $x = -5$ **43.** $p = 3 \pm \sqrt{6}$; $p \approx 5.45$ or $p \approx 0.55$ **45.** $p = -3 \pm \sqrt{5}$; $p \approx -0.76$ or $p \approx -5.24$ **47.** $m = \frac{-3}{2} \pm \frac{\sqrt{13}}{2}$; $m \approx 0.30$ or $m \approx -3.30$ **49.** $n = \frac{5}{2} \pm \frac{3\sqrt{5}}{2}$; $n \approx 5.85$ or $n \approx -0.85$ **51.** $x = \frac{1}{2}$ or $x = -4$ **53.** $n = 3$ or $n = \frac{-3}{2}$ **55.** $p = \frac{3}{8} \pm \frac{\sqrt{41}}{8}$; $p \approx 1.18$ or $p \approx -0.43$ **57.** $m = \frac{7}{2} \pm \frac{\sqrt{33}}{2}$; $m \approx 6.37$ or $m \approx 0.63$ **59.** $x = 6$ or $x = -3$ **61.** $m = \pm\frac{5}{2}$ **63.** $n = \frac{2 \pm \sqrt{5}}{2}$; $n \approx 2.12$ or $n \approx -0.12$ **65.** $w = \frac{2}{3}$ or $w = \frac{-1}{2}$ **67.** $m = \frac{3}{2} \pm \frac{\sqrt{6}}{2}i$; $m \approx 1.5 \pm 1.22i$ **69.** $n = \pm\frac{3}{2}$ **71.** $w = \frac{-4}{5}$ or $w = 2$ **73.** $a = \frac{1}{6} \pm \frac{\sqrt{23}}{6}i$; $a \approx 0.16 \pm 0.80i$ **75.** $p = \frac{3 \pm 2\sqrt{6}}{5}$; $p \approx 1.58$ or $p \approx -0.38$ **77.** $w = \frac{1 \pm \sqrt{21}}{10}$; $w \approx 0.56$ or $w \approx -0.36$ **79.** $a = \frac{3}{4} \pm \frac{\sqrt{31}}{4}i$; $a \approx 0.75 \pm 1.39i$ **81.** $p = 1 \pm \frac{3\sqrt{2}}{2}i$; $p \approx 1 \pm 2.12i$ **83.** $w = \frac{-1}{3} \pm \frac{\sqrt{2}}{3}$; $w \approx 0.14$ or $w \approx -0.80$ **85.** $a = \frac{-6 \pm 3\sqrt{2}}{2}$; $a \approx -0.88$ or $a \approx -5.12$ **87.** $p = \frac{4 \pm \sqrt{394}}{6}$; $p \approx 3.97$ or $p \approx -2.64$ **89.** $w = -5$ or $w = 2$ **91.** $a = \frac{1 \pm \sqrt{57}}{4}$; $a \approx 2.14$ or $a \approx -1.64$ **93.** $n = 4$ or $n = -1$ **95.** two rational **97.** two complex **99.** two rational **101.** two complex **103.** two irrational **105.** one repeated **107.** $x = \frac{3}{2} \pm \frac{1}{2}i$ **109.** $x = -\frac{1}{2} \pm \frac{i\sqrt{3}}{2}$ **111.** $x = \dfrac{5}{4} \pm \dfrac{3i\sqrt{7}}{4}$ **113.** $t = \dfrac{v \pm \sqrt{v^2 - 64h}}{32}$ **115.** $t = \dfrac{6 + \sqrt{138}}{2}$ sec, $t \approx 8.87$ sec **117.** 30,000 ovens **119.** 36 ft, 78 ft **121. a.** $P = -x^2 + 120x - 2000$ **b.** 10,000 **123.** $t = 2.5$ sec, 6.5 sec **125.** $x = -2i$; $x = 5i$ **127.** $x = \dfrac{-3}{4}i$; $x = 2i$ **129.** $x = -1 - i$; $x = -13 - i$ **131.** $k = \pm 4$

133. Answers will vary. **135.** $x = 2$ or $x = 3$ or $x = -5$
137. a. $x = 9$ or $x = -4$ **b.** $x = \pm\frac{5}{2}$ **c.** $x = -6, -2, 2$
139. 700 \$30 tickets; 200 \$20 tickets **141.** $2 - i\sqrt{2}$

Summary and Concept Review Exercises, pp. 129–134

1. a. linear **b.** nonlinear, variable as divisor **c.** nonlinear, exponent on g will be 2 **2.** $b = 6$ **3.** $n = 4$ **4.** $m = -1$ **5.** $x = \frac{1}{6}$ **6.** no solution
7. $g = 10$ **8.** $h = \frac{V}{\pi r^2}$ **9.** $L = \frac{P - 2W}{2}$ **10.** $x = \frac{c - b}{a}$
11. $y = \frac{2}{3}x - 2$ **12.** 4 gal of 20% sugar, 8 gal of 50% sugar
13. $12 + \frac{9}{8}\pi$ ft² **14.** $\frac{3}{4}$ hr **15.** $a \ge 35$ **16.** $a < 2$ **17.** $s \le 65$
18. $c \ge 1200$ **19.** $(5, \infty)$ **20.** $(-10, \infty)$ **21.** $(-\infty, 2]$ **22.** $(-9, 9]$
23. $(-6, \infty)$ **24.** $(-\infty, \frac{-8}{5}) \cup (2.3, \infty)$ **25. a.** $(-\infty, 3) \cup (3, \infty)$ **b.** $(-\infty, \frac{3}{2}) \cup (\frac{3}{2}, \infty)$ **c.** $[-5, \infty)$ **d.** $(-\infty, 6]$ **26.** $x \ge 96\%$
27. a. $x = -3$ or $x = 5$ or $x = -1$ or $x = 4$ **b.** $x = 0$ or $x = \frac{-5}{2}$ or $x = 9$ or $x = \frac{1}{2}$ **28. a.** $x = -9$ or $x = 2$ **b.** $n = -9$ or $n = -3$ **c.** $z = \frac{3}{2}$ or $z = -1$ **d.** $r = 0$ or $r = 4$ or $r = -1$ **e.** $b = 0$ or $b = -3$ or $b = 3$ **f.** $a = \frac{3}{2}$ or $a = -2$ or $a = 2$ **29.** $x = \frac{-1}{2}$
30. $h = 4$ **31.** $n = -1$ **32.** $x = -3; x = 3$ **33.** $x = -4; x = 5$
34. $x = -1$ **35.** 0 and 1; 5 and 6 **36.** width, 6 in.; length, 9 in.
37. 1 sec, 244 ft, 8 sec **38.** \$24 per load; \$42 per load **39.** $6\sqrt{2}\,i$
40. $24\sqrt{3}\,i$ **41** $-2 + \sqrt{2}\,i$ **42.** $3\sqrt{2}\,i$ **43.** i **44.** $21 + 20i$
45. $-2 + i$ **46.** $-5 + 7i$ **47.** 13 **48.** $-20 - 12i$
49. $(5i)^2 - 9 = -34$ $\quad(-5i)^2 - 9 = -34$
$25i^2 - 9 = -34$ $\quad 25i^2 - 9 = -34$
$-25 - 9 = -34$✓ $\quad -25 - 9 = -34$✓
50. $(2 + i\sqrt{5})^2 - 4(2 + i\sqrt{5}) + 9 = 0$ $\quad (2 - i\sqrt{5})^2 - 4(2 - i\sqrt{5}) + 9 = 0$
$4 + 4i\sqrt{5} + 5i^2 - 8 - 4i\sqrt{5} + 9 = 0$ $\quad 4 - 4i\sqrt{5} + 5i^2 - 8 + 4i\sqrt{5} + 9 = 0$
$5 + (-5) = 0$✓ $\quad 5 + (-5) = 0$✓
51. a. $2x^2 + 3 = 0$; $a = 2, b = 0, c = 3$ **b.** not quadratic **c.** $x^2 - 8x - 99 = 0$; $a = 1, b = -8, c = -99$ **d.** $x^2 + 16 = 0$; $a = 1, b = 0, c = 16$ **52. a.** $x = \pm 3$ **b.** $x = 2 \pm \sqrt{5}$ **c.** $x = \pm\sqrt{5}\,i$ **d.** $x = \pm 5$ **53. a.** $x = 3$ or $x = -5$ **b.** $x = -8$ or $x = 2$ **c.** $x = 1 \pm \frac{\sqrt{10}}{2}$; $x \approx 2.58$ or $x \approx -0.58$ **d.** $x = 2$ or $x = \frac{1}{3}$ **54. a.** $x = 2 \pm \sqrt{5}\,i$ $x \approx 2 \pm 2.24i$ **b.** $x = \frac{3 \pm \sqrt{2}}{2}$; $x \approx 2.21$ or $x \approx 0.79$ **c.** . $= \frac{3}{2} \pm \frac{1}{2}i$ **55.** 1.3 sec, 4.66 sec, 6 sec **56.** 0.8 sec, 3.2 sec, 5 sec **57.** \$3.75; 3000 **58.** 6 hr

Mixed Review, p. 134

1. a. $x \in (8, \infty)$ **b.** $x \in (-\infty, \frac{-4}{3}) \cup (\frac{-4}{3}, \infty)$
3. a. $x(x + 2)(x + 8)$ **b.** $-2(m - 9)(m + 3)$ **c.** $2(3z + 5)(3z - 5)$ **d.** $(v + 2)(v + 3)(v - 3)$ **5.** $y = \frac{-3}{4}x - 3$ **7. a.** $x = -2$ **b.** $n = 5$
9. $x = 7, 11$ **11.** $x = -\sqrt{6}, \sqrt{6}$ **13.** $x = \frac{4}{5}$ **15.** $x = \pm\sqrt{5}, \pm i\sqrt{5}$
17. a. $v = 6$ **b.** $x = -5; x = 4$ **19.** 6′10″

Practice Test, p. 135

1. $x = 27$ **2.** $C = \frac{P}{1 + k}$ **3.** $x = 2$ **4.** 30 gal **5.** $x < -30$
6. $-5 \le x < 4$ **7.** $x \in \mathbb{R}$ **8.** Jacques needs at least a 177
9. $z = -3, 10$ **10.** $x = \pm\frac{5}{2}$ **11.** $x = \frac{2}{3}, 6$ **12.** $x = -2, \frac{-3}{2}, \frac{3}{2}$
13. \$4.50/tin, 90 tins **14.** $\frac{-4}{3} + \frac{i\sqrt{5}}{3}$ **15.** $-i$ **16. a.** 1 **b.** $i\sqrt{3}$ **c.** 1 **17.** $\frac{-3}{5} + \frac{6}{5}i$ **18.** 34 **19.** $x = 1 \pm \sqrt{3}\,i$
20. $(2 - 3i)^2 - 4(2 - 3i) + 13 = 0$
$-5 - 12i - 8 + 12i + 13 = 0$
$0 = 0$✓
21. $x = 5 \pm \frac{\sqrt{2}}{2}$ **22.** $x = \frac{5}{4} \pm \frac{i\sqrt{7}}{4}$ **23.** $x = 1 \pm \frac{\sqrt{3}}{3}$ **24.** $x = 1 \pm 3i$
25. a. $t = 5$ (May) **b.** $t = 9$ (September) **c.** July; \$3000 more

Strengthening Core Skills, pp. 137–138

Exercise 1: $\frac{7}{2} + (-1) = \frac{5}{2} = -\frac{b}{a}$✓
$\frac{7}{2} \cdot (-1) = \frac{-7}{2} = \frac{c}{a}$✓

Exercise 2: $\frac{2 + 3\sqrt{2}}{2} + \frac{2 - 3\sqrt{2}}{2} = \frac{4}{2} = \frac{-b}{a}$✓
$\frac{2 + 3\sqrt{2}}{2} \cdot \frac{2 - 3\sqrt{2}}{2} = \frac{-14}{4} = \frac{-7}{2} = \frac{c}{a}$✓

Exercise 3: $(5 + 2\sqrt{3}i) + (5 - 2\sqrt{3}i) = 10 = \frac{-b}{a}$✓
$(5 + 2\sqrt{3}i)(5 - 2\sqrt{3}i) = 25 + 12 = 37 = \frac{c}{a}$✓

CHAPTER 2

Exercises 2.1, pp. 150–155

1. lattice **3.** $y = 0, x = 0$ **5.** $m = \frac{15}{2}$, Answers will vary.

7.

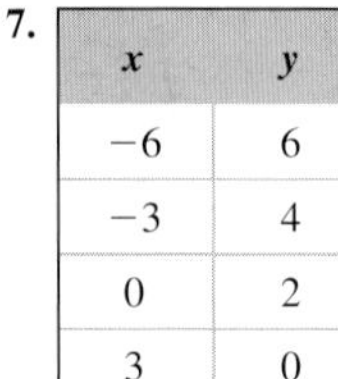

x	y
−6	6
−3	4
0	2
3	0

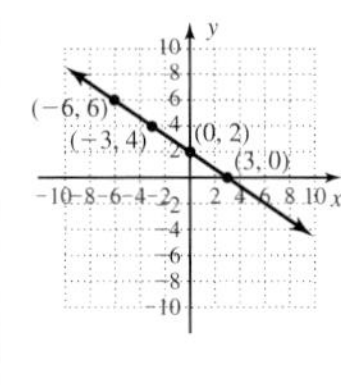

9.

x	y
−2	1
0	4
2	7
4	10

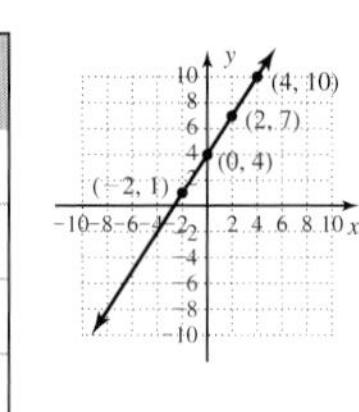

11.

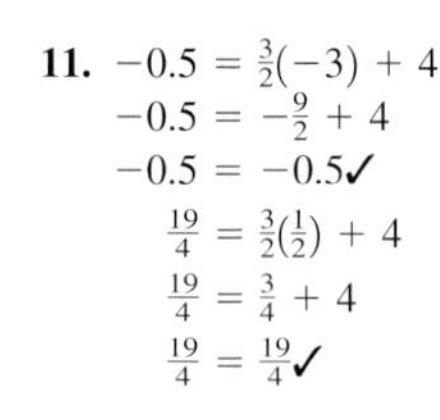

$-0.5 = \frac{3}{2}(-3) + 4$
$-0.5 = -\frac{9}{2} + 4$
$-0.5 = -0.5$✓
$\frac{19}{4} = \frac{3}{2}(\frac{1}{2}) + 4$
$\frac{19}{4} = \frac{3}{4} + 4$
$\frac{19}{4} = \frac{19}{4}$✓

13.

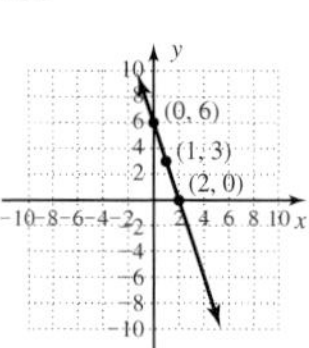

15.

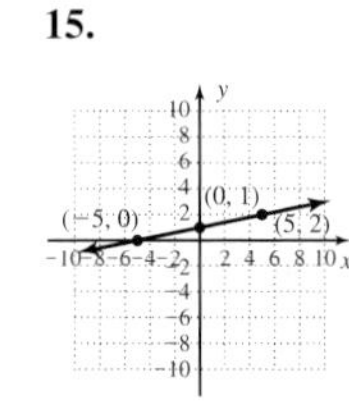

17.

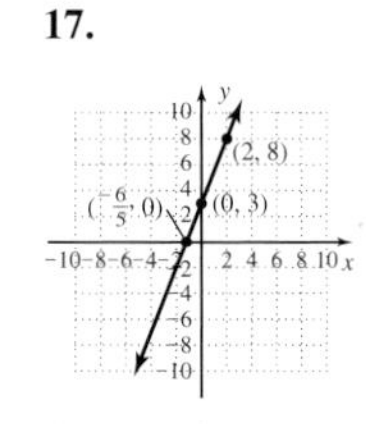

19.

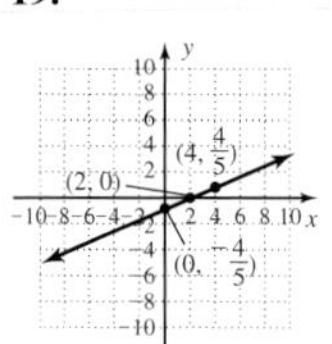

21.

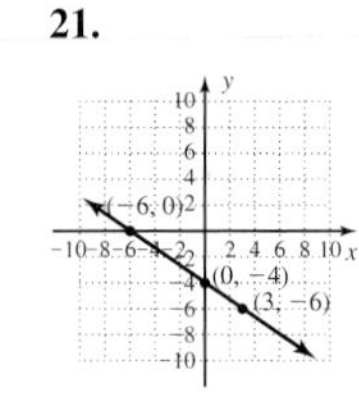

23.

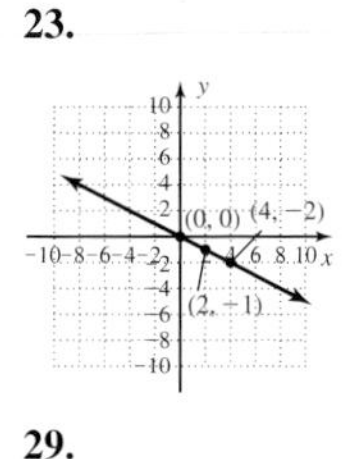

25.

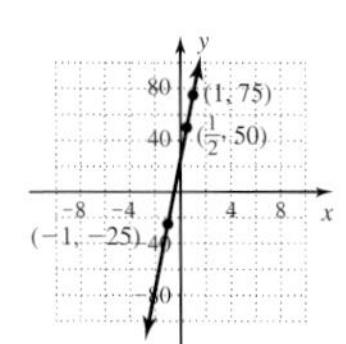

27.

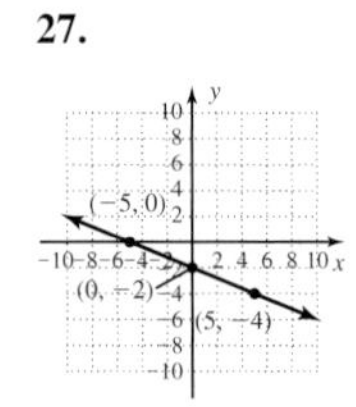

29.

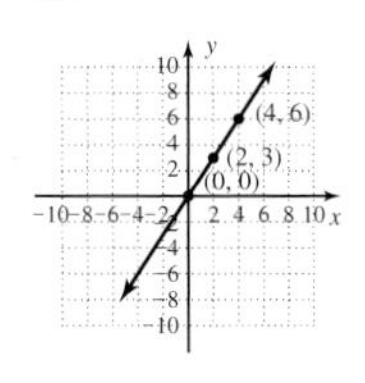

31.

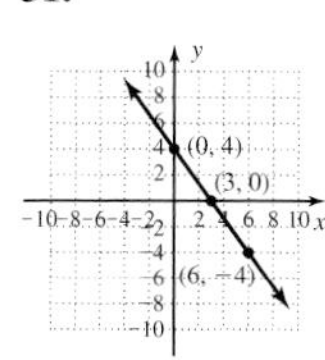

33.

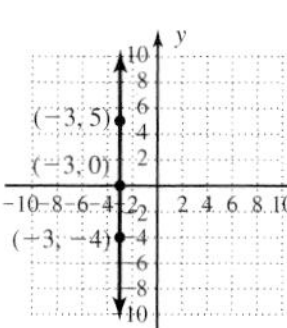

35.

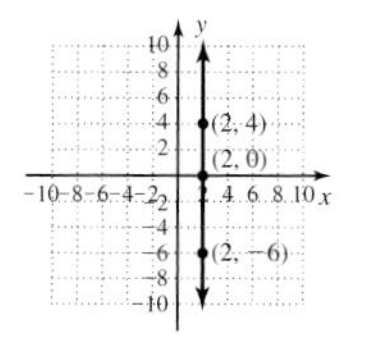

37. L_1: $x = 2$; L_2: $y = 4$; point of intersection (2, 4) **39. a.** $m = 125$, cost increased \$125,000 per 1000 sq ft **b.** \$375,000 **41. a.** $m = 22.5$, distance increases 22.5 mph **b.** about 186 mi **43. a.** $m = \frac{23}{6}$, a person weighs 23 lb more for each additional 6 in. in height **b.** 3.8

45. $m = 1$; (2, 4) and (1, 3)

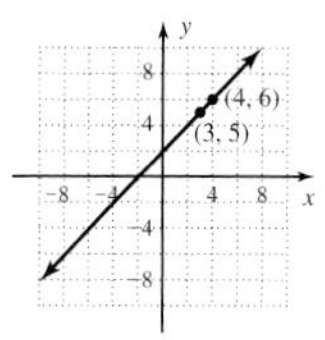

47. $m = \frac{4}{3}$; (5, −1) and (1, −9)

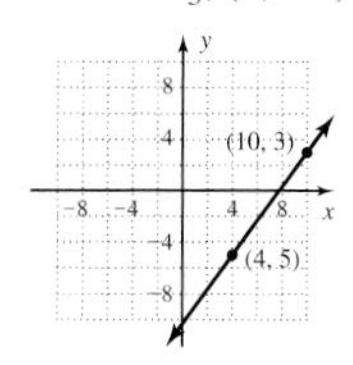

49. $m = \frac{15}{2} = \frac{7.5}{1}$; (2, −0.5) and (4, 14.5)

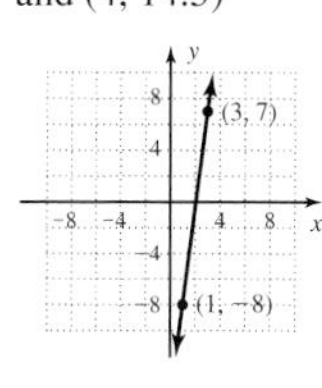

51. $m = 0$; (6, −8) and (6, 3)

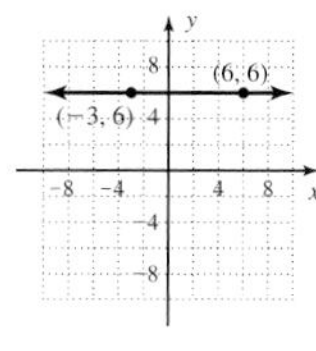

53. parallel **55.** neither **57.** parallel **59.** not a right triangle **61.** not a right triangle **63.** right triangle **65.** (3,1) **67.** (−0.7, −0.3) **69.** $(\frac{1}{20}, \frac{1}{24})$ **71.** (0, −1) **73.** (−1, 0) **75.** $2\sqrt{34}$ **77.** 10 **79. a.** 76.4 yr **b.** 2010 **81. a.** \$3500 **b.** 5 yr **83. a.** \$2349 **b.** 2011 **85. a.** 23% **b.** 2005 **87.** Answers will vary. **89.** e **91.** $x = 9, x = -2$ **93.** 12 gal **95.** $x \neq 5, x \neq -3$; $x = 7$ or $x = -4$

Exercises 2.2, pp. 167–174

1. first **3.** range **5.** Answers will vary.

7.

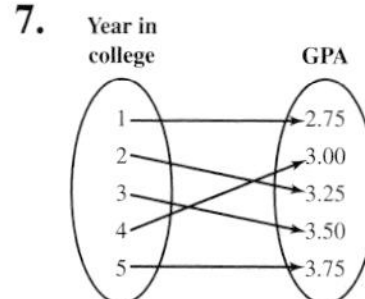

$D = \{1, 2, 3, 4, 5\}$
$R = \{2.75, 3.00, 3.25, 3.50, 3.75\}$

9. $D = \{1, 3, 5, 7, 9\}$; $R = \{2, 4, 6, 8, 10\}$ **11.** $D = \{4, -1, 2, -3\}$; $R = \{0, 5, 4, 2, 3\}$

13.

x	y
−6	5
−3	3
0	1
3	−1
6	−3
8	$\frac{-13}{3}$

15.

x	y
−2	0
0	2, −2
1	3, −3
3	5, −5
6	8, −8
7	9, −9

17.

x	y
−3	8
−2	3
0	−1
2	3
3	8
4	15

19.

x	y
−4	3
−3	4
0	5
2	$\sqrt{21}$
3	4
4	3

21.

x	y
10	3, −3
5	2, −2
4	$\sqrt{3}, -\sqrt{3}$
2	1, −1
1.25	0.5, −0.5
1	0

23.

x	y
−9	−2
−2	−1
−1	0
0	1
4	$\sqrt[3]{5}$
7	2

25. function **27.** Not a function. The Shaq is paired with two heights. **29.** Not a function; 4 is paired with 2 and −5. **31.** function **33.** function **35.** Not a function; −2 is paired with 3 and −4. **37.** function **39.** function **41.** Not a function; 0 is paired with 4 and −4. **43.** function **45.** Not a function; −3 is paired with −2 and 2. **47.** function **49.** function, $x \in [-4, -5]$ $y \in [-2, 3]$ **51.** function, $x \in [-4, \infty)$ $y \in [-4, \infty)$ **53.** function, $x \in [-4, 4]$, $y \in [-5, -1]$ **55.** function, $x \in (-\infty, \infty)$, $y \in (-\infty, \infty)$ **57.** Not a function, $x \in [-3, 5]$, $y \in [-3, 3]$ **59.** Not a function, $x \in (-\infty, 3]$ $y \in (-\infty, \infty)$ **61.** $x \in (-\infty, 5) \cup (5, \infty)$ **63.** $x \in [\frac{-5}{3}, \infty)$ **65.** $x \in (-\infty, -5) \cup (-5, 5) \cup (5, \infty)$ **67.** $x \in (-\infty, -3\sqrt{2}) \cup (-3\sqrt{2}, 3\sqrt{2}) \cup (3\sqrt{2}, \infty)$ **69.** $x \in (-\infty, \infty)$ **71.** $x \in (-\infty, \infty)$ **73.** $x \in (-\infty, \infty)$ **75.** $x \in (-\infty, -2) \cup (-2, 5) \cup (5, \infty)$ **77.** $x \in [2, \frac{5}{2}) \cup (\frac{5}{2}, \infty)$ **79.** $f(-6) = 0, f(\frac{3}{2}) = \frac{15}{4}, f(2c) = c + 3, f(c + 2) = \frac{1}{2}c + 4$ **81.** $f(-6) = 132, f(\frac{3}{2}) = \frac{3}{4}, f(2c) = 12c^2 - 8c, f(c + 2) = 3c^2 + 8c + 4$ **83.** $h(3) = 1, h(\frac{-2}{3}) = \frac{-9}{2}, h(3a) = \frac{1}{a}, h(a - 1) = \frac{3}{a - 1}$ **85.** $h(3) = 5, h(\frac{-2}{3}) = -5, h(3a) = -5$ if $a < 0$ or 5 if $a > 0$, $h(a - 1) = -5$ if $a < 1$ or 5 if $a > 1$ **87.** $g(0.4) = 0.8\pi, g(\frac{9}{4}) = \frac{9}{2}\pi, g(h) = 2\pi h, g(h + 3) = 2\pi h + 6\pi$ **89.** $g(0.4) = 0.16\pi, g(\frac{9}{4}) = \frac{81}{16}\pi$, $g(h) = \pi h^2, g(h + 3) = \pi(h + 3)^2$ **91.** $p(0.5) = 2, p(\frac{9}{4}) = \frac{\sqrt{30}}{2}, p(a) = \sqrt{2a + 3}, p(a + 3) = \sqrt{2a + 9}$ **93.** $p(0.5) = -17, p(\frac{9}{4}) = \frac{163}{81}, p(a) = \frac{3a^2 - 5}{a^2}, p(a + 3) = \frac{3a^2 + 18a + 22}{a^2 + 6a + 9}$ **95. a.** $D = \{-1, 0, 1, 2, 3, 4, 5\}$ **b.** 1 **c.** −1 **d.** $R = \{-2, -1, 0, 1, 2, 3, 4\}$ **97. a.** $x \in [-5, 5]$ **b.** −2 **c.** −2 **d.** $y \in [-3, 4]$ **99. a.** $x \in [-3, \infty)$ **b.** 2 **c.** 0 **d.** $y \in (-\infty, 4]$ **101. a.** 186.5 lb **b.** 37 lb **103. a.** $N(g) = 23g$ **b.** $g \in [0, 15]$; $N \in [0, 345]$ **105. a.** $[0, \infty)$ **b.** ≈244 units^3 **c.** 8×6 **107. a.** $c(t) = 12.50t + 19.50$ **b.** \$63.25 **c.** ≈8 hr **d.** $t \in [0, 10.44]$; $c \in [0, 150]$ **109. a.** Yes. Each x is paired with exactly one y. **b.** 9 P.M. **c.** $3\frac{1}{2}$ m **d.** 5 P.M. and 1 A.M. **111.** Answers will vary. **113. a.** Son, 72.5 sec **b.** 10 m **c.** 45 sec **d.** 3

115.

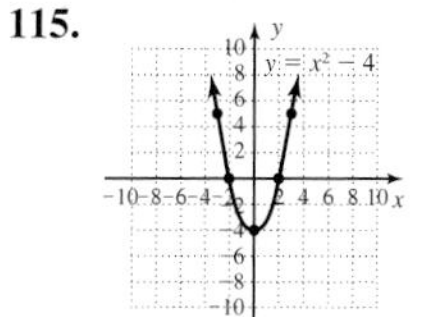

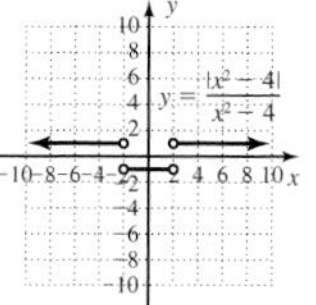

117. a. $19\sqrt{6}$ **b.** 1 **119. a.** $(x - 3)(x - 5)(x + 5)$ **b.** $(2x + 3)(x - 8)$ **c.** $(2x - 5)(4x^2 + 10x + 25)$ **121. a.** $x^6y^2z^4$ **b.** $\frac{27}{8}$

Exercises 2.3, pp. 182–188

1. $-\frac{7}{4}$, (0, 3) **3.** 2.5. **5.** Answers will vary.

7. $y = \frac{-4}{5}x + 2$

x	y
−5	6
−2	$\frac{18}{5}$
0	2
1	$\frac{6}{5}$
3	$\frac{-2}{5}$

9. $y = 2x + 7$

x	y
−5	−3
−2	3
0	7
1	9
3	13

11. $y = \frac{-5}{3}x - 5$

x	y
−5	$\frac{10}{3}$
−2	$\frac{-5}{3}$
0	−5
1	$\frac{-20}{3}$
3	−10

13. $f(x) = 2x - 3$, new coeff. 2, constant −3 **15.** $f(x) = \frac{-5}{3}x - 7$, new coeff. $\frac{-5}{3}$, constant −7 **17.** $f(x) = \frac{-35}{6}x - 4$, new coeff. $\frac{-35}{6}$, constant −4

19.

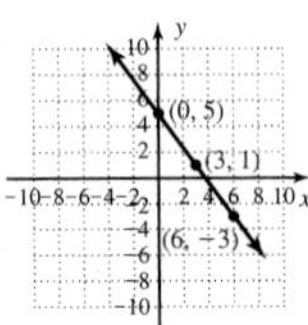

21.

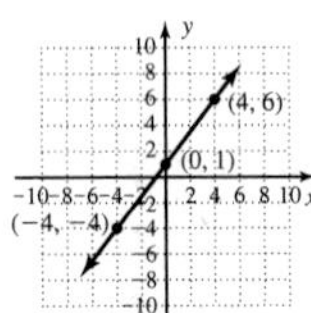

23.

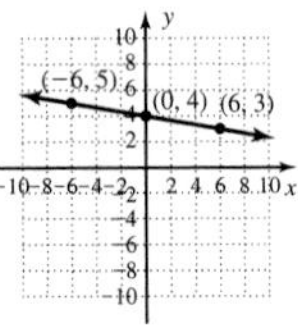

25. a. $\frac{-3}{4}$ **b.** $f(x) = \frac{-3}{4}x + 3$ **c.** The coeff. of x is the slope and the constant is the y-intercept. **27. a.** $\frac{2}{5}$ **b.** $f(x) = \frac{2}{5}x - 2$ **c.** The coeff. of x is the slope and the constant is the y-intercept. **29. a.** $\frac{4}{5}$ **b.** $f(x) = \frac{4}{5}x + 3$ **c.** The coeff. of x is the slope and the constant is the y-intercept. **31.** $y = \frac{-2}{3}x + 2$, $m = \frac{-2}{3}$, y-intercept (0, 2) **33.** $y = \frac{-5}{4}x + 5$, $m = \frac{-5}{4}$, y-intercept (0, 5) **35.** $y = \frac{1}{3}x$, $m = \frac{1}{3}$, y-intercept (0, 0) **37.** $y = \frac{-3}{4}x + 3$, $m = \frac{-3}{4}$, y-intercept (0, 3) **39.** $y = \frac{2}{3}x + 1$ **41.** $y = 3x + 3$ **43.** $y = 3x + 2$ **45.** $f(x) = 250x + 500$ **47.** $f(x) = \frac{75}{2}x + 150$ **49.** $f(x) = 2x - 13$ **51.** $y = \frac{-4}{5}x + 4$ **53.** $y = \frac{5}{3}x - 5$ **55.**

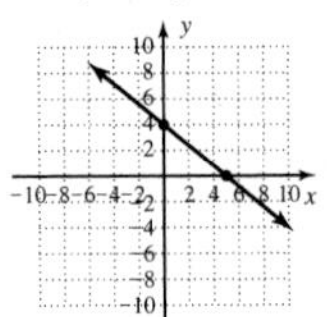

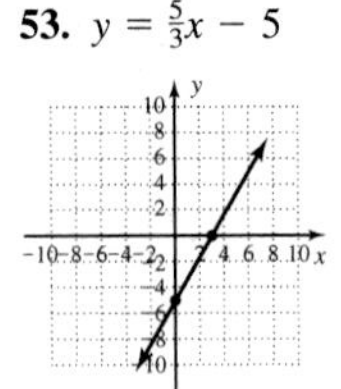

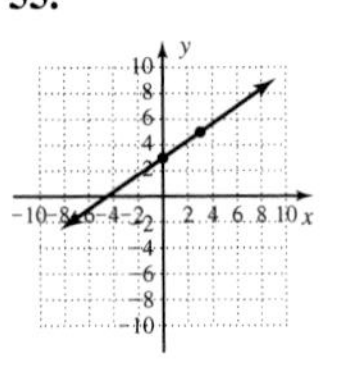

57. **59.** **61.**

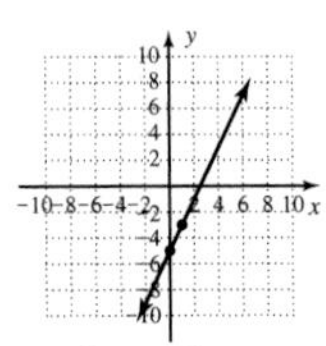

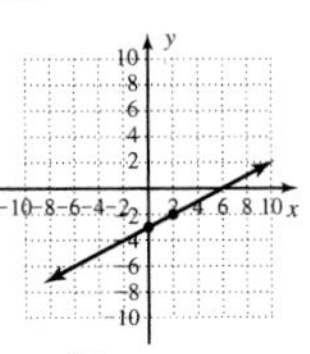

63. $y = \frac{2}{5}x - \frac{16}{5}$ **65.** $y = \frac{-5}{3}x + \frac{22}{3}$ **67.** $y = \frac{-12}{5}x - \frac{29}{5}$ **69.** perpendicular **71.** neither **73.** neither **75.** $f(x) = 2x - 9$ **77.** $f(x) = \frac{3}{8}x - \frac{41}{8}$ **79.** $f(x) = 0.5x - 4$

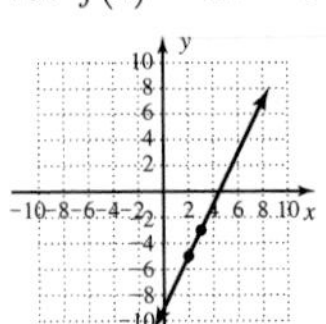

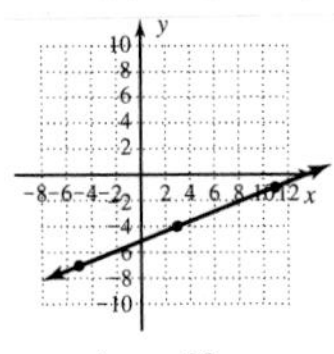

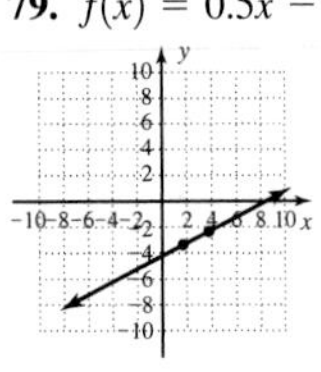

81. a. $y = \frac{-3}{4}x - \frac{5}{2}$ **b.** $y = \frac{4}{3}x - \frac{20}{3}$ **83. a.** $y = \frac{4}{9}x + \frac{31}{9}$ **b.** $y = \frac{-9}{4}x + \frac{3}{4}$ **85. a.** $y = \frac{-1}{2}x - 2$ **b.** $y = 2x - 2$ **87.** $y = \frac{6}{5}x - \frac{14}{5}$; For each 5000 additional sales, income rises \$6000. **89.** $y = -20x + 110$; For every hour of television, a student's final grade falls 20%. **91.** $y = \frac{35}{2}x + \frac{5}{4}$; Every 2 in. of rainfall increases the number of cattle raised per acre by 35. **93.** C **95.** A **97.** B **99.** D **101. a.** $m = \frac{-3}{4}$, y-intercept (0, 2) **b.** $m = \frac{-2}{5}$, y-intercept (0, −3) **c.** $m = \frac{5}{6}$, y-intercept (0, 2) **d.** $m = \frac{5}{3}$, y-intercept (0, 3) **103. a.** As the temperature increases 5°C, the velocity of sound waves increases 3 m/s. At a temperature of 0°C, the velocity is 331 m/s. **b.** 343 m/s **c.** 50°C **105. a.** $V(t) = \frac{20}{3}t + 150$ **b.** Every 3 yr the value of the coin increases by \$20; the initial value was \$150. **107. a.** $N(t) = 7x + 9$ **b.** Every 1 yr the number of homes with Internet access increases by 7 million. **c.** 1993 **109. a.** \$223.33 **b.** 15 years, in 2013 **c.** 3 yr **111. a.** 86 million **b.** 13 yr **c.** 2010 **113. a.** $P(t) = 58{,}000t + 740{,}000$ **b.** Each year, the prison population increases by 58,000. **c.** 1,726,000 **115.** Answers will vary. **117.** (1) d, (2) a, (3) c, (4) b, (5) f, (6) h **119.** $x = \frac{5 \pm 2\sqrt{13}}{3}$; $x \approx -0.74$ or $x \approx 4.07$ **121.** 113.10 yd^2 **123.** (−4, 5)

Mid-Chapter Check, pp. 188–189

1. **2.** $\frac{-18}{7}$

3. positive, loss is decreasing (profit is increasing); $m = \frac{3}{2}$, yes; $\frac{1.5}{1}$, each year Data.com's loss decreases by 1.5 million.

4. $y = \frac{3}{2}x + \frac{5}{2}$

5. $x = -3$; no; Input −3 is paired with more than one output. **6.** $y = \frac{-4}{3}x + 4$; yes; Each input is paired with only one output. **7. a.** 0 **b.** $x \in [-3, 5]$ **c.** 3.5 **d.** $y \in [-4, 5]$ **8.** from $x = 1$ to $x = 2$; steeper line → greater slope **9.** $F(p) = \frac{3}{4}p + \frac{5}{4}$; For every 4000 pheasants, the fox population increases by 300; 1625. **10. a.** $x \in \{-3, -2, -1, 0, 1, 2, 3, 4\}$ $y \in \{-3, -2, -1, 0, 1, 2, 3, 4\}$ **b.** $x \in [-3, 4]$ $y \in [-3, 4]$ **c.** $x \in (-\infty, \infty)$ $y \in (-\infty, \infty)$

Reinforcing Basic Concepts, pp. 189–190

1. a. $\frac{1}{3}$, increasing **b.** $y - 5 = \frac{1}{3}(x - 0)$ **c.** $y = \frac{1}{3}x + 5$ **d.** $x - 3y = -15$ **e.** (0, 5), (−15, 0)

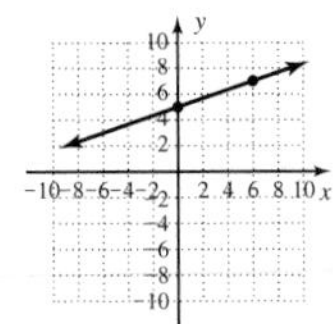

2. a. $\frac{-7}{3}$, decreasing **b.** $y - 9 = \frac{-7}{3}(x - 0)$ **c.** $y = \frac{-7}{3}x + 9$ **d.** $7x + 3y = 27$ **e.** (0, 9), $(\frac{27}{7}, 0)$

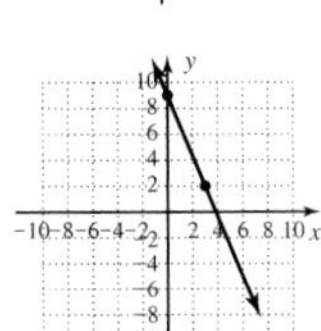

3. a. $\frac{1}{2}$, increasing **b.** $y - 2 = \frac{1}{2}(x - 3)$ **c.** $y = \frac{1}{2}x + \frac{1}{2}$ **d.** $x - 2y = -1$ **e.** $(0, \frac{1}{2})$, (−1, 0)

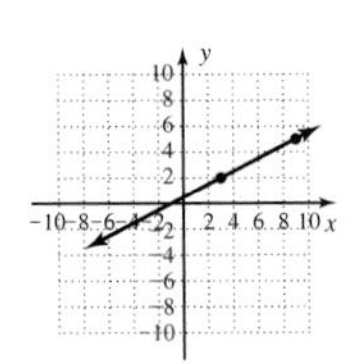

4. a. $\frac{3}{4}$, increasing **b.** $y + 4 = \frac{3}{4}(x + 5)$ **c.** $y = \frac{3}{4}x - \frac{1}{4}$ **d.** $3x - 4y = 1$ **e.** $(0, \frac{-1}{4}), (\frac{1}{3}, 0)$

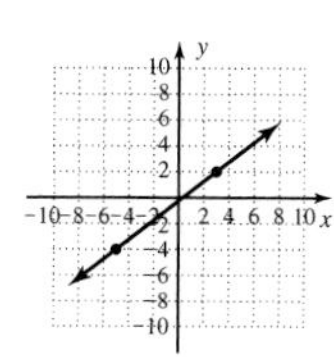

5. a. $\frac{-3}{4}$, decreasing **b.** $y - 5 = \frac{-3}{4}(x + 2)$ **c.** $y = \frac{-3}{4}x + \frac{7}{2}$ **d.** $3x + 4y = 14$ **e.** $(0, \frac{7}{2}), (\frac{14}{3}, 0)$

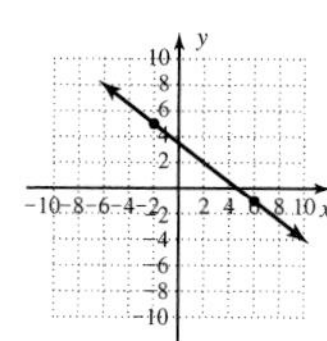

6. a. $\frac{-1}{2}$, decreasing **b.** $y + 7 = \frac{-1}{2}(x - 2)$ **c.** $y = \frac{-1}{2}x - 6$ **d.** $x + 2y = -12$ **e.** $(0, -6), (-12, 0)$

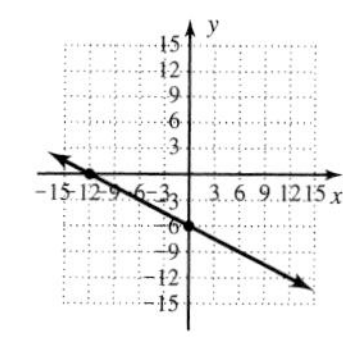

Exercises 2.4, pp. 200–205

1. parabola **3.** point, inflection **5.** Answers will vary.

7.

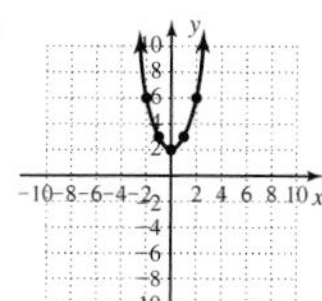

9.

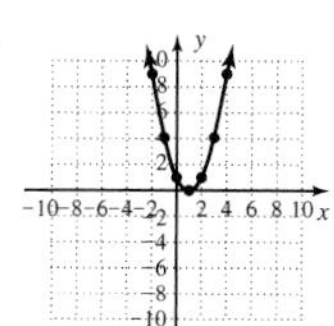

11. a. up/up, $(-2, -4)$, $x = -2$, $(0, 0)$, $(-4, 0)$, $(0, 0)$ **b.** $x \in (-\infty, \infty)$, $y \in [-4, \infty)$ **13. a.** up/up, $(1, -4)$, $x = 1$, $(-1, 0)$, $(3, 0)$, $(0, -3)$ **b.** $x \in (-\infty, \infty)$, $y \in [-4, \infty)$ **15. a.** up/up, $(2, -9)$, $x = 2$, $(-1, 0)$, $(5, 0)$, $(0, -5)$ **b.** $x \in (-\infty, \infty)$, $y \in [-9, \infty)$

17.

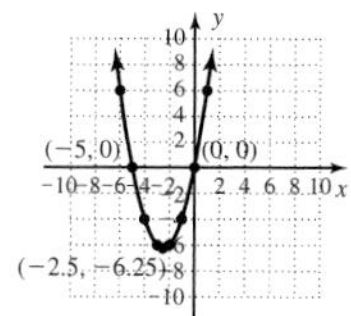

19.

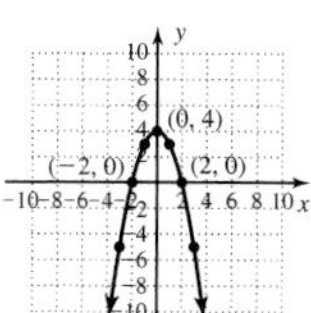

21.

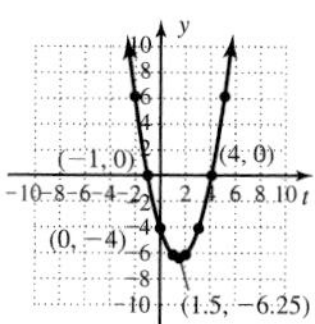

23.

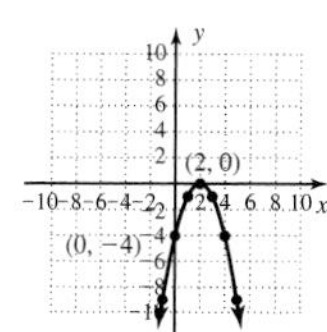

25.

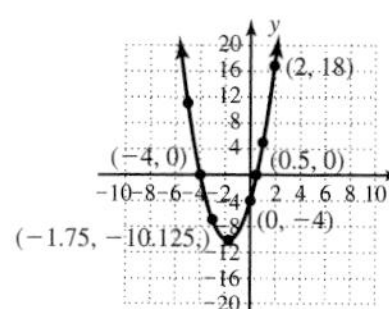

27.

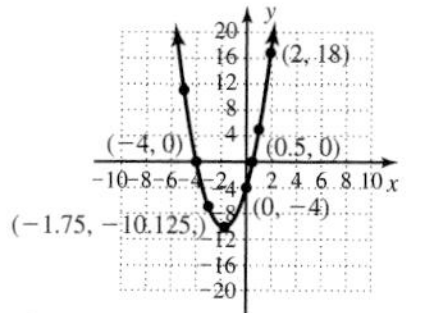

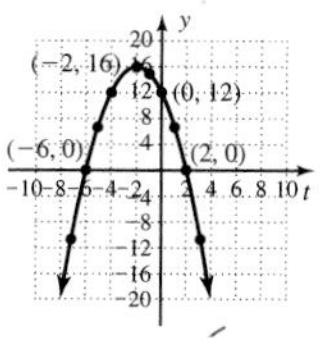

29.

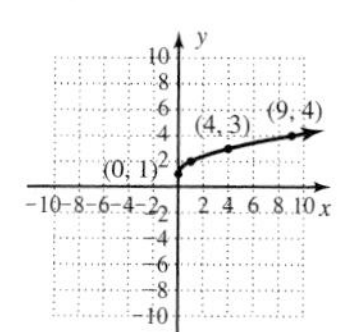

31.

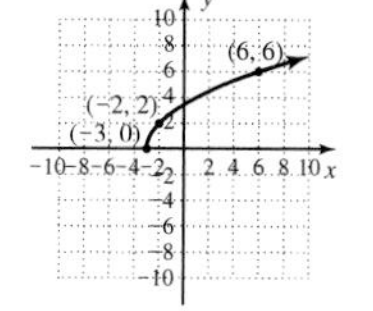

33.

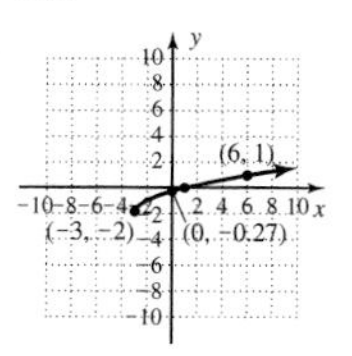

35.

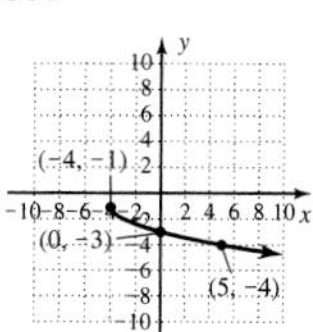

37.

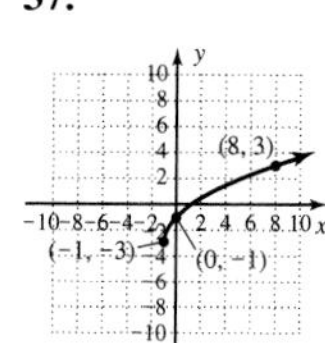

39.

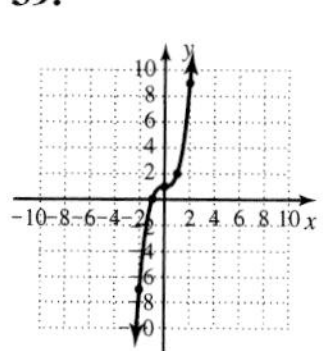

41.

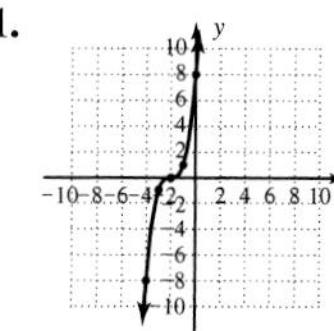

43. a. up on left, down on right; $(1, 0)$, $(0, 1)$ **b.** $x \in (-\infty, \infty)$ $y \in (-\infty, \infty)$ **c.** $(1, 0)$ **45. a.** down on left, up on right; $(-4, 0)$, $(-1, 0)$, $(1, 0)$, $(0, -4)$ **b.** $x \in (-\infty, \infty)$ $y \in (-\infty, \infty)$ **c.** $(-1.3, 2.1)$ **47. a.** up on left, down on right; $(-1, 0)$, $(2, 0)$, $(4, 0)$, $(0, -8)$ **b.** $x \in (-\infty, \infty)$ $y \in (-\infty, \infty)$ **c.** $(1.7, -2.1)$

49. **51.** 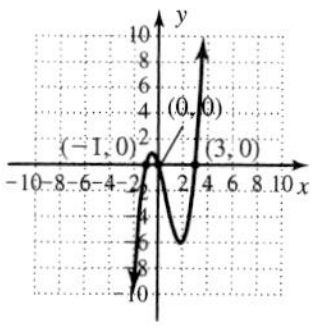 **53.**

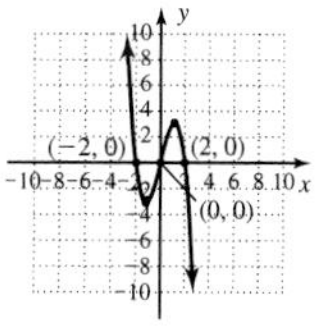

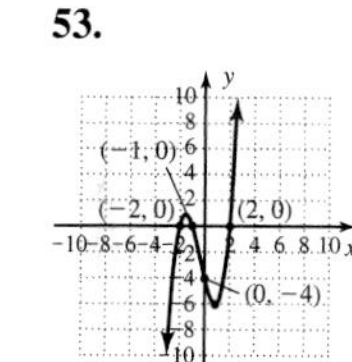

55.

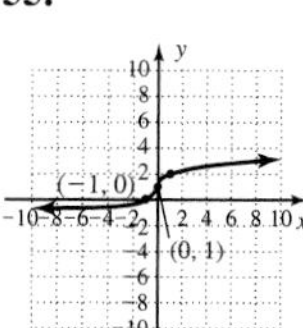

57.

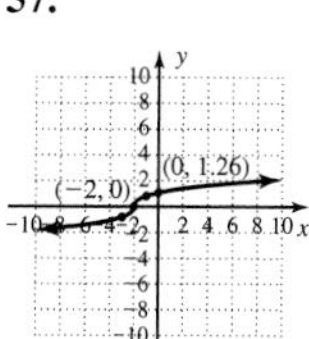

59.

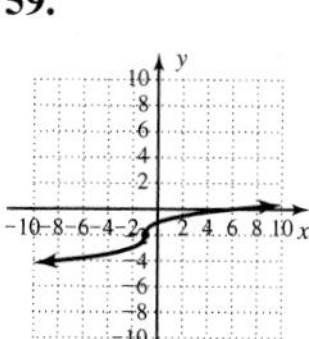

61.

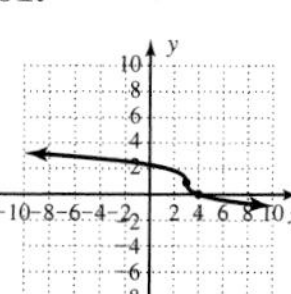

63.

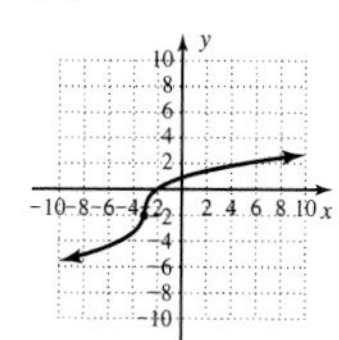

65. c **67.** a **69.** k **71.** d **73.** i **75.** l **77.** 40 ft/sec, 110.25 ft **79. a.** 7 **b.** 7 **c.** They are the same. **d.** Slopes are equal.

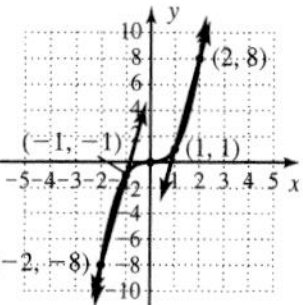

81. a. 176 ft **b.** 320 ft **c.** 144 ft/sec **d.** -144 ft/sec; The arrow is going down. **83. a.** 17.89 ft/sec; 25.30 ft/sec **b.** 30.98 ft/sec; 35.78 ft/sec **c.** Between 5 and 10 **d.** 1.482 ft/sec, 0.96 ft/sec **85. a.** They are the same. **b.** Slope $(\frac{2}{3})$ is constant for a line. **c.** $\frac{2}{3}$, Every change of 1 in x results in $\frac{2}{3}$ unit change in y. **87.** 62,500 ft^2

89.

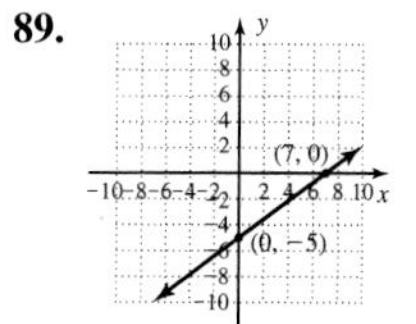

91. $x \in (-6, \infty)$

93. No, $x = 2$ is paired with $y = 2$ and $y = -2$.

Exercises 2.5, pp. 212–216

1. zeroes **3.** $x \in \mathbb{R}$ **5.** $x \in (-3, 0) \cup (3, \infty)$ **7.** $x \in (\frac{2}{3}, \infty)$
9. $x \in [8, \infty)$ **11.** no solution **13.** $x \in (-\infty, \frac{5}{2})$ **15.** $x \in (-\infty, \frac{4}{3})$
17. $x \in [8, \infty)$ **19.** $x \in [4, \infty)$ **21.** $x \in [-3, 3]$ **23.** no solution
25. $x \in (-\infty, \infty)$ **27.** $x \in (-\infty, 3)$ **29.** $x \in (-\infty, -3]$
31. no solution **33.** $x \in (-\infty, -2) \cup (3, \infty)$ **35.** $x \in [-4, 3]$
37. no solution **39.** $x \in (-\infty, -4) \cup (-1, 2)$ **41.** $x \in [-4, -3]$
43. $x \in (-\infty, -3] \cup \{2\}$ **45.** $x \in (0, 4)$ **47.** $x \in (-\infty, -5] \cup [1, \infty)$
49. $x \in (-1, \frac{7}{2})$ **51.** $x \in [-\sqrt{7}, \sqrt{7}]$
53. $x \in (-\infty, \frac{-3-\sqrt{33}}{2}] \cup [\frac{-3+\sqrt{33}}{2}, \infty)$ **55.** $x \in (-\infty, -1] \cup [\frac{5}{3}, \infty)$
57. $x \in (-\infty, \infty)$ **59.** no solution **61.** $x \in (-\infty, \infty)$
63. no solution **65.** $x \in (-\infty, \infty)$ **67.** $x \in (-\infty, \infty)$
69. $x \in (-\infty, -3] \cup [5, \infty)$ **71.** $x \in (\frac{-3}{2}, 5)$
73. $x \in [-2\sqrt{2}, 0] \cup [2\sqrt{2}, \infty)$ **75.** $x \in [-4, 4]$ **77.** $x \in (-\infty, \frac{4}{3}]$
79. $x \in (-\infty, -5] \cup [5, \infty)$ **81.** $x \in (-\infty, 0] \cup [5, \infty)$
83. $x \in (-\infty, -3] \cup [5, \infty)$ **85.** $x \in (-\infty, \infty)$ **87.** $x \in [-3, 0] \cup [5, \infty)$
89. April–September; December–March, October–December
91. a. 2 sec **b.** 2 sec **c.** 5 ft **d.** 10 ft **93.** $x \in \{-2\} \cup [4, \infty)$
95. $x \in [-3, 0] \cup [3, \infty)$ **97. a.** $6\frac{1}{2}$ **b.** $\frac{m^5}{27n^9}$ **99.** $\frac{3}{4}$
101. $4, 6i, 13, -\frac{5}{13} + \frac{12}{13}i$

Exercises 2.6, pp. 224–232

1. scatter-plot **3.** linear **5.** Answers will vary. **7.** positive
9. cannot be determined **11.** positive **13. a.** linear **b.** positive
15. a, d, c, b **a.** positive **b.** negative **c.** negative **d.** positive
17. a. **b.** positive **c.** strong
19. a. **b.** negative, moderate **c.** $y = -0.4x + 82.8$
21. a. **b.** positive, strong **c.** $y = 0.5x + 30.2$
23. a. **b.** positive, strong **c.** $y = 2.4x + 69.4$, 74,200, 103,000 **25. a.** $h(t) = -14.5t^2 + 90t$ **b.** $v = 90$ ft/sec **c.** Venus
27. a. **b.** linear **c.** positive **d.** $y = 0.96x + 1.55$, 63.95 in.
29. a. **b.** linear **c.** positive **d.** $y = 9.55x + 70.42$; about 232,800 The number of applications, since the line has a greater slope.
31. a.

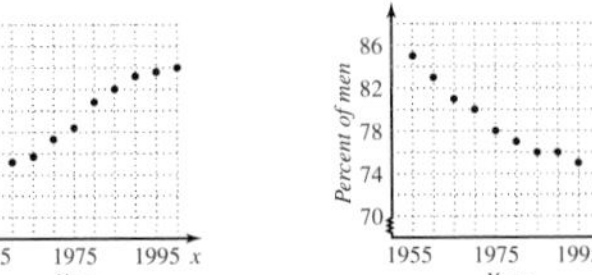

b. men: linear **c.** negative **d.** yes, |slope| is greater
b. women: linear **c.** positive
33. a. **b.** strong **c.** $y = 0.07x^2 - 2.02x + 21.77$
35. a. **b.** strong **c.** $y = 0.04x^2 - 11.32x + 807.88$
37. a. linear **b.** $y = 108.2x + 330.2$, strong **c.** \$1736.8 billion; about \$2170 billion
39. (b), since there is a recognizable and fixed correspondence between the independent and dependent variables
41.

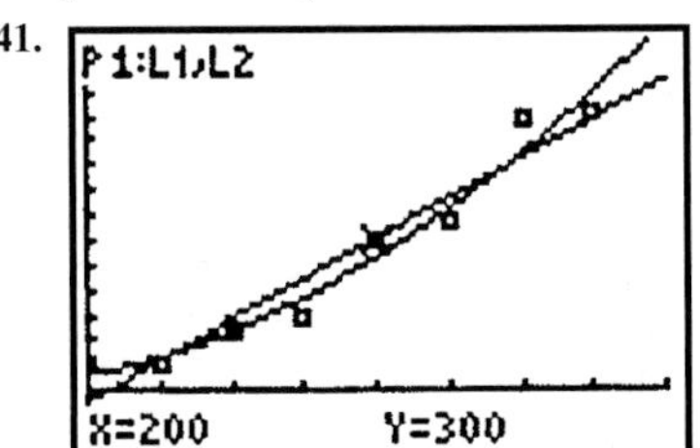

Very strong; virtually equal; context and goodness of fit.

43. Answers will vary
45. $66 + 9\pi$ cm, $40.5\pi + 432$ cm^2 **47.** $w = \frac{-7}{10} \pm \frac{\sqrt{51}}{10}i$
49. $\frac{3}{2}, \pm\sqrt{7}i$

Summary and Concept Review, pp. 232–238

1. a. $\frac{-5}{9}$, (14, −7) **b.** $\frac{1}{3}$, (0,3)

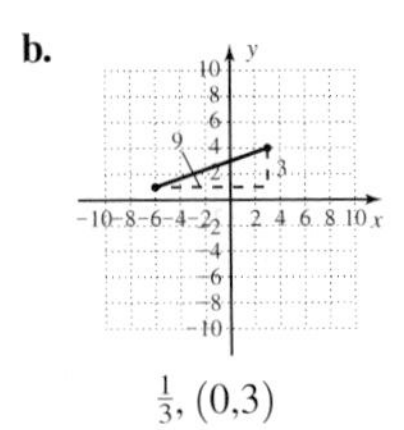

2. a. parallel **b.** perpendicular

3. a.

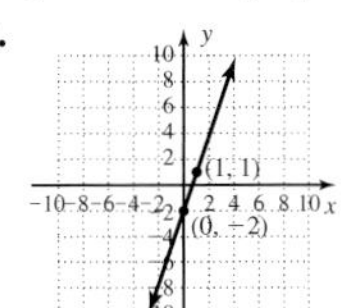

b.

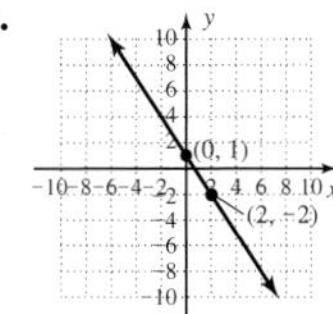

4. a.

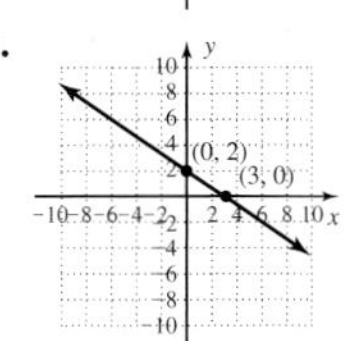

b.

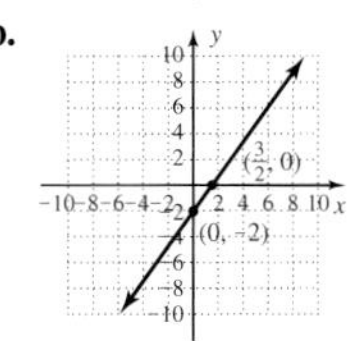

5. a. vertical **b.** horizontal **c.** neither

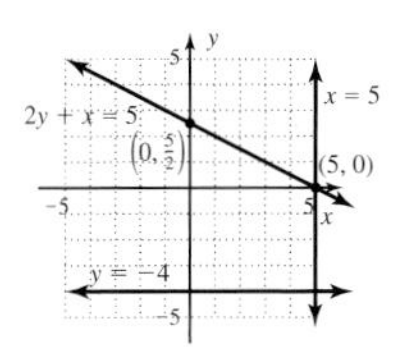

6. yes **7.** $m = \frac{2}{3}$, y-intercept $(0, 2)$, when the rodent population increases by 2000, the hawk population increases by 300.

8. $(\frac{1}{2}, 0)$, $\sqrt{149}$

9.

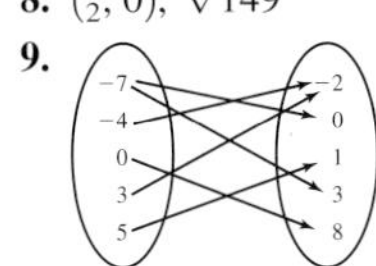

$x \in \{-7, -4, 0, 3, 5\}$ $y \in \{-2, 0, 1, 3, 8\}$

10.

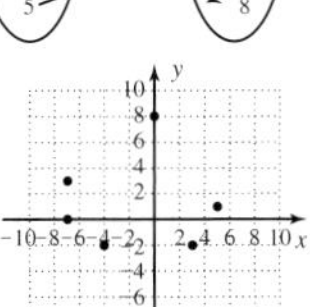

No, −7 is paired with 3 and 0.

11. a. $x \in [-\frac{5}{4}, \infty)$ **b.** $x \in (-\infty, -2) \cup (-2, 3) \cup (3, \infty)$

12. $\frac{26}{9}$; $18a^2 - 9a$; $2a^2 - 7a + 5$

13.

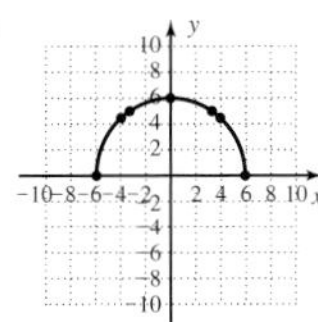

$x \in [-6, 6]$
$y \in [0, 6]$

Yes, passes the vertical line test

14. yes **15. I. a.** $D = \{-1, 0, 1, 2, 3, 4, 5\}$ $R = \{-2, -1, 0, 1, 2, 3, 4\}$ **b.** 1 **c.** 2 **II. a.** $x \in [-5, 4]$ $y \in [-5, 4]$ **b.** −3 **c.** −2 **III. a.** $x \in [-3, \infty)$ $y \in [-4, \infty)$ **b.** −1 **c.** −3 or 3

16. a. $y = \frac{-4}{3}x + 4$, $m = \frac{-4}{3}$, y-intercept $(0, 4)$ **b.** $y = \frac{5}{3}x - 5$, $m = \frac{5}{3}$, y-intercept $(0, -5)$

17. a.

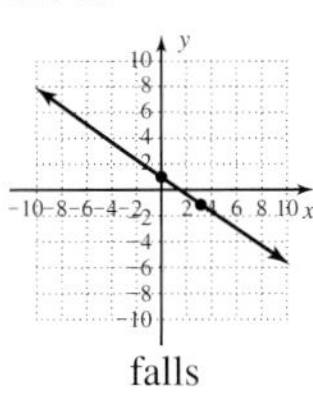

falls

b.

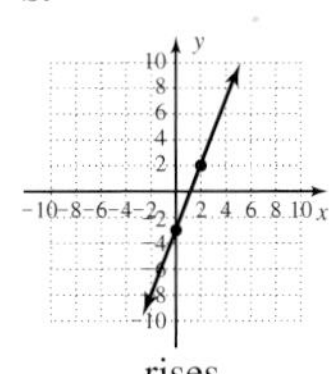

rises

18. a.

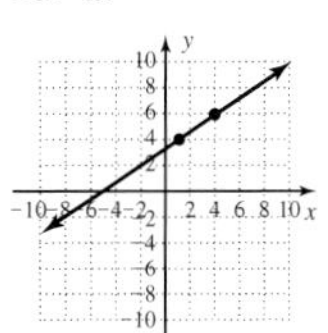

b.

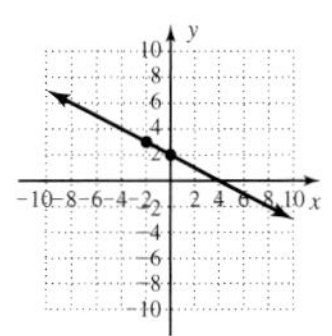

19. $y = 5, x = -2; y = 5$ **20.** $y = \frac{-3}{4}x + \frac{11}{4}$ **21.** $f(x) = \frac{4}{3}x$

22. $m = \frac{2}{5}$, y-intercept $(0, 2)$, $y = \frac{2}{5}x + 2$; When the rabbit population increases by 500, the wolf population increases by 200.

23. a. $y = \frac{-15}{2}x + 105$ **b.** $(14, 0), (0, 105)$ **c.** $f(x) = \frac{-15}{2}x + 105$ **d.** $f(20) = -45, x = 12$

24.

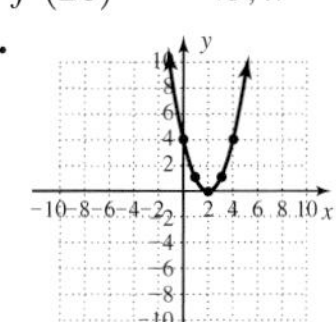

25.

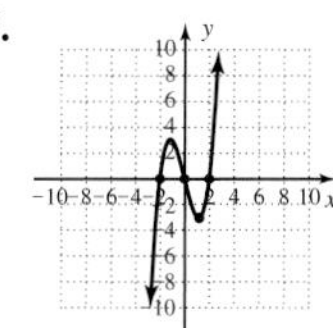

26.

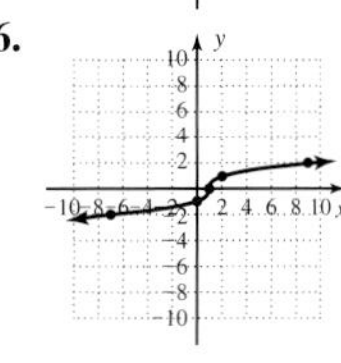

27.

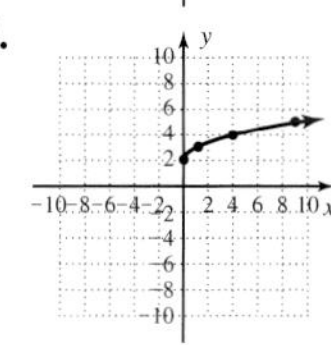

28. a. $(-4, 0), (1, 0), (0, -3)$ **b.** up/up **c.** $(-1.5, -4)$

29. a. $(-3, 0), (0, 2)$ **b.** up on the right **c.** $(-4, -2)$

30. a. $(-2, 0), (1, 0), (4, 0), (0, -3)$ **b.** up on left, down on right **c.** $(1, 0)$ **31. a.** $(-1.5, 0), (2, 0), (0, -1.5)$ **b.** up/up **c.** $(\frac{1}{2}, -2)$

32. a. $(-2, 0), (0, -2)$ **b.** up on left, down on right **c.** $(-1, -1)$

33. a. $(1, 0), (0, -2)$ **b.** down on left, up on right **c.** not applicable

34.

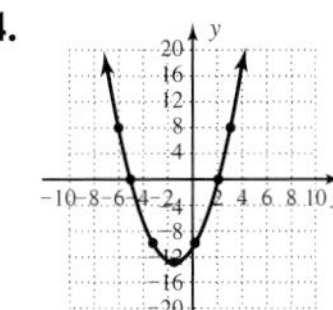

35.

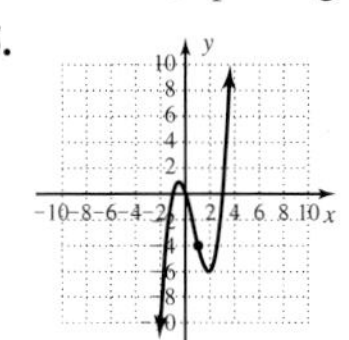

36.

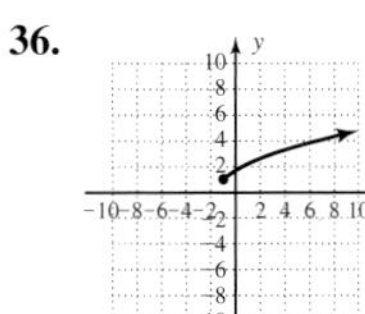

37.

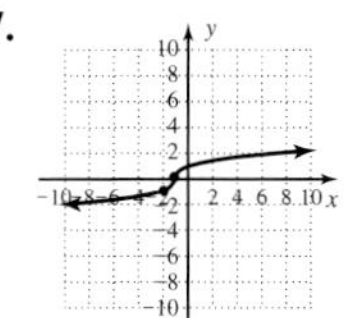

38. $w = 10$ to $w = 15$; $\frac{7}{100}$; $\frac{19}{100}$ **39.** $x \in (-4, 1)$

40. $x \in (-\infty, -4) \cup (3, \infty)$ **41.** $x \in (-\infty, -2]$ **42.** $x \in (-\infty, \infty)$

43. $x \in (\frac{2}{3}, \infty)$ **44.** no solution **45.** $x \in (0, 5)$ **46.** $x \in [-5, 1]$

47. $x \in (-\infty, -2) \cup (2, \infty)$ **48.** $x \in (-\infty, -1) \cup (0, 1)$

49. $x \in (-\infty, 0] \cup [5, \infty)$ **50.** $x \in [-1, 0] \cup [1, \infty)$

51. a.

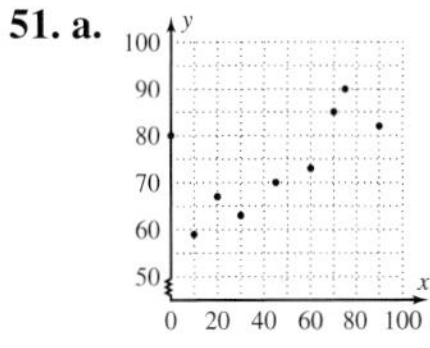

b. linear **c.** yes **d.** positive

52. $y = 0.35x + 56.10$ **53.** 98

Mixed Review, pp. 238–240

1. $y = \frac{-5}{3}x - 3$ **3. a.** $x \in (-\infty, -5) \cup (-5, 5) \cup (5, +\infty)$ **b.** $x \in \left[\frac{5}{3}, \infty\right)$ **5.** $y = \frac{-3}{4}x + 1$ **7.** $d = \sqrt{65} \approx 8.06$ units; midpoint: $\left(1, \frac{1}{2}\right)$

9.

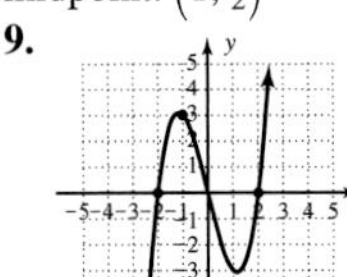

11.

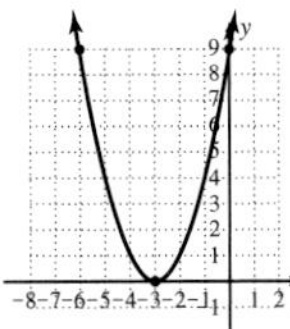

13.

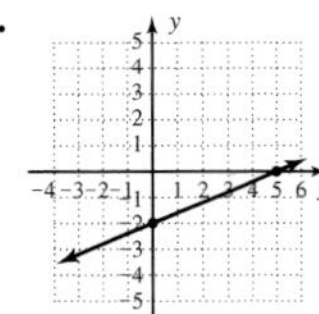

15. $x \in [-4, -3]$ **17.** $x \in (-\infty, -3] \cup [3, \infty)$

19. a.

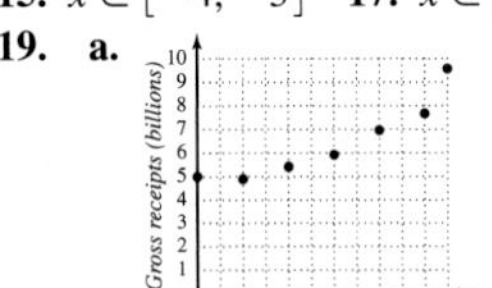

quadratic **b.** $g(t) = 0.0357t^2 - 0.0602t + 4.9795$ **c.** $g(15) \approx 12.1$ billion (2005); $g(18) \approx 15.5$ billion (2008)

Practice Test, pp. 240–241

1. a. a and c are nonfunctions, they do not pass the vertical line test

2.

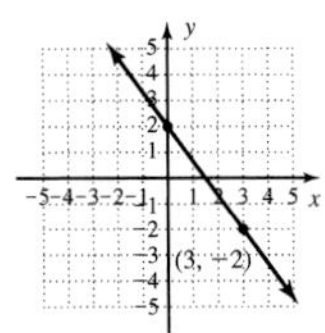

3. neither **4.** $y = \frac{-2}{3}x$ **5. a.** yes **b.** no **6.** $a(t) = \frac{20}{3}t + \frac{20}{3}$; 80 mph **7. a.** (7.5, 1.5), **b.** ≈ 61.27 mi **8.** L_1: $x = -3$ L_2: $y = 4$ **9. a.** $x \in \{-4, -2, 0, 2, 4, 6\}$ $y \in \{-2, -1, 0, 1, 2, 3\}$ **b.** $x \in [-2, 6]$ $y \in [1, 4]$ **10. a.** 300 **b.** 30 **c.** $W(h) = \frac{25}{2}h$ **d.** wages are \$12.50 per hr **e.** $h \in [0, 40]$; $w \in [0, 500]$

11. 4; $-5 - 7\sqrt{2}$; $8 - 33i$ **12.** $x = 3 \pm \frac{\sqrt{6}}{3}i$ **13.** $x = 3, \frac{1}{2}$

14. $x \in (-\infty, \frac{10}{3})$ **15.** $x \in (-\infty, -7] \cup [5, \infty)$ **16. I. a.** square root **b.** $x \in [-4, \infty)$, $y \in [-3, \infty)$ **c.** $(-2, 0), (0, 1)$ **d.** up on right **e.** $x \in (-2, \infty)$ **f.** $x \in [-4, -2)$ **II. a.** cubic **b.** $x \in (-\infty, \infty)$ $y \in (-\infty, \infty)$ **c.** $(2, 0), (0, -1)$ **d.** down on left, up on right **e.** $x \in (2, \infty)$ **f.** $x \in (-\infty, 2)$ **III. a.** absolute value **b.** $x \in (-\infty, \infty)$ $y \in (-\infty, 4]$ **c.** $(-1, 0), (3, 0), (0, 2)$ **d.** down/down **e.** $x \in (-1, 3)$ **f.** $x \in (-\infty, -1) \cup (3, \infty)$ **IV. a.** quadratic **b.** $x \in (-\infty, \infty)$; $y \in [-5.5, \infty)$ **c.** $(0, 0), (5, 0)$ **d.** up/up **e.** $x \in (-\infty, 0) \cup (5, \infty)$ **f.** $x \in (0, 5)$

17.

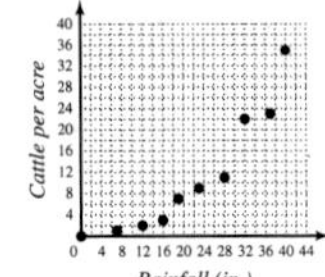

18. yes; positive **19.** ≈ 53 cattle per acre **20. a.** no; graph is less steep **b.** $\frac{\Delta S}{\Delta t} = 25$ for [5, 6] $\frac{\Delta S}{\Delta t} = 29$ for [6, 7]

Strengthening Core Skills, pp. 243–244

1. between 6 and 7

x	$-x^2$	$3x$	24	$-x^2 + 3x + 24$
0	0	0	24	24
1	−1	3	24	26
2	−4	6	24	26
3	−9	9	24	24
4	−16	12	24	20
5	−25	15	24	14
6	−36	18	24	6
7	−49	21	24	−4
8	−64	24	24	−16

2. For $x > 3$; answers will vary; down/down; $x \in (-3, 3)$

x	x^2	−9	$x^2 - 9$
0	0	−9	−9
1	1	−9	−8
2	4	−9	−5
3	9	−9	0
4	16	−9	7
5	25	−9	16
6	36	−9	27
7	49	−9	40

x	9	$-x^2$	$9 - x^2$
0	9	0	9
1	9	−1	8
2	9	−4	5
3	9	−9	0
4	9	−16	−7
5	9	−25	−16
6	9	−36	−27
7	9	−49	−40

3. Answers will vary.

Cumulative Review pp. 244–245

1. $2n - 5 = n + 3$ **3. a.** $\frac{-3n^2m}{2}$ **b.** 15.3×10^{-3} **c.** $\frac{b^6c^6}{8a^3}$ **d.** $3\frac{1}{2}$

5. a. $\frac{x - 7}{(x - 5)(x + 2)}$ **b.** $\frac{b^2 - 4ac}{4a^2}$ **7.** $x = 1$

9. $-3 < x < 2$ **11. a.** $-21 + 20i$ **b.** $\frac{-3}{5} - \frac{4}{5}i$

13. $x = -5 \pm \frac{\sqrt{2}}{2}$; $x \approx -5.707$, $x \approx -4.293$

15. $W = 31$ cm, $L = 47$ cm

17.

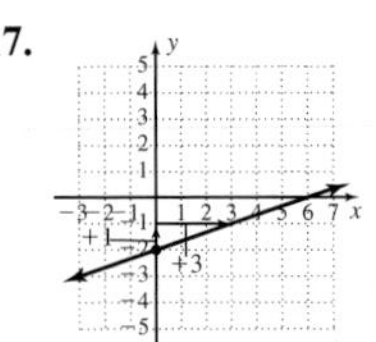

19. $m_1 = \frac{1}{2}$, $m_2 = -2$, $\Rightarrow y = -2x + 4$

CHAPTER 3

Exercises 3.1, pp. 256–261

1. $(f + g)(x)$, $A \cap B$ **3.** intersection, $g(x)$ **5.** Answers will vary. **7.** $h(x) = 3x^2 + 4x - 3$; $x \in (-\infty, \infty)$ **9.** -46 **11. a.** -151 **b.** not defined **c.** $x \in [\frac{5}{4}, \infty)$ **13.** $h(x) = \sqrt{x} - 2$; $h(x)$ shifts $f(x)$ down 2 units.

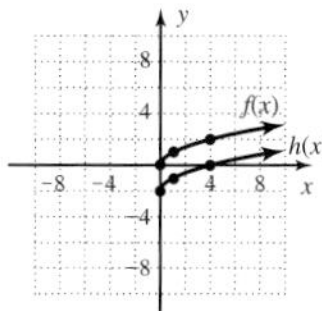

15. $h(x) = 6x^3 - x^2 - 10x - 4$; $x \in (-\infty, \infty)$ **17. a.** $H(x) = \sqrt{(x + 5)(2 - x)}$, **b.** $2\sqrt{3}$; not defined **c.** $[-5, 2]$ **19.** $h(x) = x^2 - 6x$; $x \in (-\infty, 1) \cup (1, \infty)$ **21. a.** $H(x) = \dfrac{2x - 3}{\sqrt{x^2 - x - 6}}$ **b.** $H(-2)$ is not defined; $H(5) = \dfrac{\sqrt{14}}{2}$ **c.** $x \in (-\infty, -2) \cup (3, \infty)$ **23.** $h(x) = \dfrac{x + 1}{x - 5}$; $x \in (-\infty, 5) \cup (5, \infty)$ **25.** $h(x) = \dfrac{x - 5}{\sqrt{x - 2}}$; $x \in (2, \infty)$ **27.** $h(x) = \dfrac{x^2 - 9}{\sqrt{x + 1}}$; $x \in (-1, \infty)$ **29.** $h(x) = x - 4$; $x \in (-\infty, -4) \cup (-4, \infty)$ **31.** $h(x) = x^2 - 2$; $x \in (-\infty, -4) \cup (-4, \infty)$ **33.** $h(x) = \dfrac{3x + 6}{x - 3}$; $x \in (-\infty, -2) \cup (-2, 3) \cup (3, \infty)$ **35.** sum: $3x + 1$, $x \in (-\infty, \infty)$; difference: $x + 5$, $x \in (-\infty, \infty)$; product: $2x^2 - x - 6$, $x \in (-\infty, \infty)$; quotient: $\dfrac{2x + 3}{x - 2}$, $x \in (-\infty, 2) \cup (2, \infty)$ **37.** sum: $x^2 + 3x + 5$, $x \in (-\infty, \infty)$; difference: $x^2 - 3x + 9$, $x \in (-\infty, \infty)$; product: $3x^3 - 2x^2 + 21x - 14$, $x \in (-\infty, \infty)$; quotient: $\dfrac{x^2 + 7}{3x - 2}$, $x \in \left(-\infty, \dfrac{2}{3}\right) \cup \left(\dfrac{2}{3}, \infty\right)$ **39.** sum: $x^2 + 3x - 4$, $x \in (-\infty, \infty)$; difference: $x^2 + x - 2$, $x \in (-\infty, \infty)$; product: $x^3 + x^2 - 5x + 3$, $x \in (-\infty, \infty)$; quotient: $x + 3$, $x \in (-\infty, 1) \cup (1, \infty)$ **41.** sum: $3x + 1 + \sqrt{x - 3}$, $x \in [3, \infty)$; difference: $3x + 1 - \sqrt{x - 3}$, $x \in [3, \infty)$; product: $(3x + 1)\sqrt{x - 3}$, $x \in [3, \infty)$; quotient: $\dfrac{3x + 1}{\sqrt{x - 3}}$, $x \in (3, \infty)$ **43.** sum: $2x^2 + \sqrt{x + 1}$, $x \in [-1, \infty)$; difference: $2x^2 - \sqrt{x + 1}$, $x \in [-1, \infty)$; product: $2x^2\sqrt{x + 1}$, $x \in [-1, \infty)$; quotient: $\dfrac{2x^2}{\sqrt{x + 1}}$, $x \in (-1, \infty)$ **45.** sum: $\dfrac{7x - 11}{(x - 3)(x + 2)}$, $x \in (-\infty, -2) \cup (-2, 3) \cup (3, \infty)$; difference: $\dfrac{-3x + 19}{(x - 3)(x + 2)}$, $x \in (-\infty, -2) \cup (-2, 3) \cup (3, \infty)$; product: $\dfrac{10}{x^2 - x - 6}$, $x \in (-\infty, -2) \cup (-2, 3) \cup (3, \infty)$; quotient: $\dfrac{2x + 4}{5x - 15}$, $x \in (-\infty, -2) \cup (-2, 3) \cup (3, \infty)$ **47.** 0; 0; $a^2 - 5a - 14$; $a^2 - 9a$ **49. a.** $h(x) = \sqrt{2x - 2}$ **b.** $H(x) = 2\sqrt{x + 3} - 5$ **c.** D of $h(x)$: $x \in [1, \infty)$; D of $H(x)$: $x \in [-3, \infty)$ **51. a.** $h(x) = \dfrac{10}{5 + 3x}$ **b.** $H(x) = \dfrac{5x + 15}{2x}$ **c.** D of $h(x)$: $\{x \mid x \in \mathbb{R}, x \neq 0, x \neq \frac{-5}{3}\}$; D of $H(x)$: $\{x \mid x \in \mathbb{R}, x \neq -3, x \neq 0\}$ **53. a.** $h(x) = x^2 + x - 2$ **b.** $H(x) = x^2 - 3x + 2$ **c.** D of $h(x)$: $x \in (-\infty, \infty)$ D of $H(x)$: $x \in (-\infty, \infty)$ **55. a.** $h(x) = x^2 + 7x + 8$ **b.** $H(x) = x^2 + x - 1$ **c.** D of $h(x)$: $x \in (-\infty, \infty)$ D of $H(x)$: $x \in (-\infty, \infty)$ **57. a.** $h(x) = \sqrt{3x + 1}$ **b.** $H(x) = 3\sqrt{x - 3} + 4$ **c.** D of $h(x)$: $x \in [-\frac{1}{3}, \infty)$ D of $H(x)$: $x \in [3, \infty)$

59. a. $h(x) = |-3x + 1| - 5$ **b.** $H(x) = -3|x| + 16$ **c.** D of $h(x)$: $(-\infty, \infty)$ D of $H(x)$: $(-\infty, \infty)$ **61. a.** $h(x) = 4x - 20$ **b.** $H(x) = \dfrac{x}{4 - 5x}$ **c.** D of $h(x)$: $\{x \mid x \in \mathbb{R}, x \neq 5\}$ D of $H(x)$: $\{x \mid x \in \mathbb{R}, x \neq 0, x \neq \frac{4}{5}\}$ **63. a.** 41 **b.** 41 **65.** $A = 2\pi r(20 + r)$; $f(r) = 2\pi r$, $g(r) = 20 + r$; $A(5) = 250\pi$ units^2 **67. a.** 4 **b.** 0 **c.** 2 **d.** 3 **e.** $\frac{2}{3}$ **f.** 6 **g.** -3 **h.** 1 **i.** 1 **j.** undefined **k.** 8 **l.** -6 **69. a.** $p(n) = 11.45n - 0.1n^2$ **b.** \$123 **c.** \$327 **d.** $C(115) > R(115)$ **71.** $h(x) = x - 2.5$; 10.5 **73. a.** 4160 **b.** 45,344 **c.** $M(x) = 453.44x$; yes **75. a.** 6 ft **b.** 36π ft^2 **c.** $A(t) = 9\pi t^2$; yes **77.** Answers may vary. **79. a.** 1995 to 1996; 1999 to 2004 **b.** 30; 1995 **c.** 20 seats; 1997 **d.** The total number in the senate (50); the number of additional seats held by the majority. **81.** Answers will vary. **83. a.** 3321 **b.** 212 **85. a.** **b.** **c.**

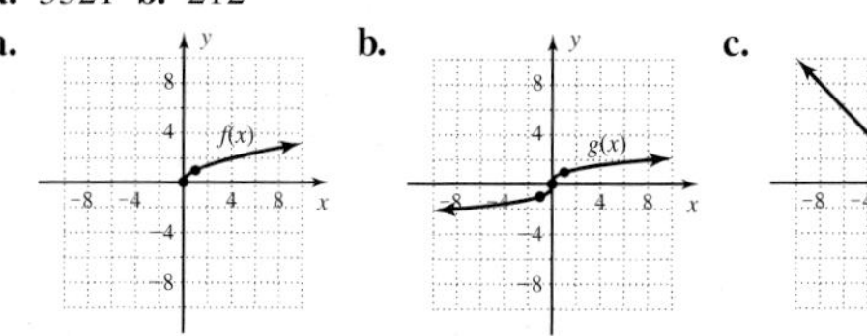

87. D of $f(x)$: $x \in [-2, 2]$ D of $g(x)$: $x \in (-\infty, -2] \cup [2, \infty)$ **89. a.** volume of a cone **b.** volume of a sphere

Exercises 3.2, pp. 268–273

1. second, one **3.** $(-11, -2), (-5, 0), (1, 2), (19, 4)$ **5.** False, answers will vary. **7.** one-to-one **9.** one-to-one **11.** one-to-one **13.** one-to-one **15.** one-to-one **17.** not one-to-one, $y = 7$ is paired with $x = -2$ and $x = 2$ **19.** one-to-one **21.** one-to-one **23.** not one-to-one; $p(t) > 5$, corresponds to two x-values **25.** one-to-one **27.** one-to-one **29.** $f^{-1}(x) = \{(1, -2), (4, -1), (5, 0), (9, 2), (15, 5)\}$ **31.** $v^{-1}(x) = \{(3, -4), (2, -3), (1, 0), (0, 5), (-1, 12), (-2, 21), (-3, 32)\}$ **33.** $f^{-1}(x) = x - 5$ **35.** $p^{-1}(x) = \dfrac{-5}{4}x$ **37.** $f^{-1}(x) = \dfrac{x - 3}{4}$ **39.** $Y_1^{-1} = x^3 + 4$ **41.** $f^{-1}(x) = \dfrac{x - 7}{2}$ **43.** $f^{-1}(x) = x^2 + 2$; $x \geq 0$ **45.** $f^{-1}(x) = \sqrt{x - 3}$; $x \geq 3$ **47.** $f^{-1}(x) = \sqrt[3]{x - 1}$ **49.** $(f \circ g)(x) = x$, $(g \circ f)(x) = x$ **51.** $(f \circ g)(x) = x$, $(g \circ f)(x) = x$ **53.** $(f \circ g)(x) = x$, $(g \circ f)(x) = x$ **55.** $(f \circ g)(x) = x$, $(g \circ f)(x) = x$ **57.** $f^{-1}(x) = \dfrac{x + 5}{3}$ **59.** $f^{-1}(x) = 2x + 5$ **61.** $f^{-1}(x) = 2x + 6$ **63.** $f^{-1}(x) = \sqrt[3]{x - 3}$ **65.** $f^{-1}(x) = \dfrac{x^3 - 1}{2}$ **67.** $f^{-1}(x) = 2\sqrt[3]{x} + 1$ **69.** $f^{-1}(x) = \dfrac{x^2 - 2}{3}$, $x \geq 0$ **71.** $p^{-1}(x) = \dfrac{x^2}{4} + 3$; $x \geq 0$ **73.** $v^{-1}(x) = \sqrt{x - 3}$ **75.**

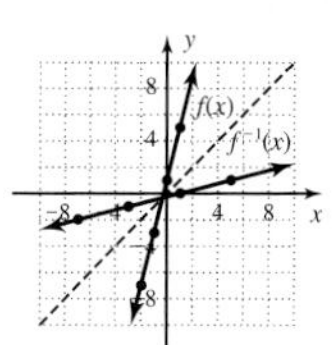

77. **79.** **81.**

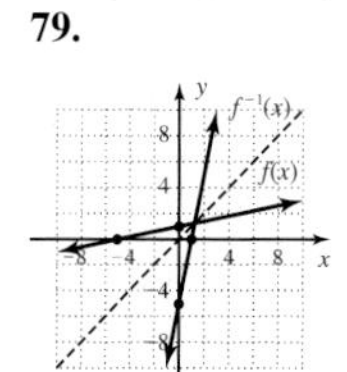

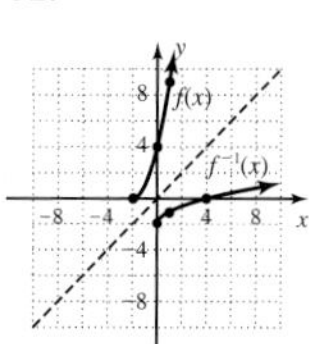

83. $D: x \in [0, \infty), R: y \in [-2, \infty);$
$D: x \in [-2, \infty), R: y \in [0, \infty)$

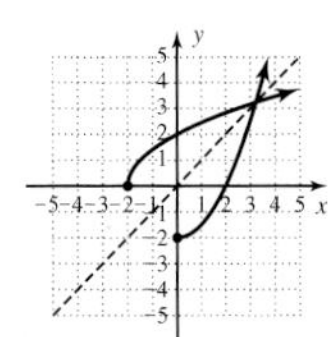

85. $D: x \in (0, \infty), R: y \in (-\infty, \infty);$
$D: x \in (-\infty, \infty), R: y \in (0, \infty)$

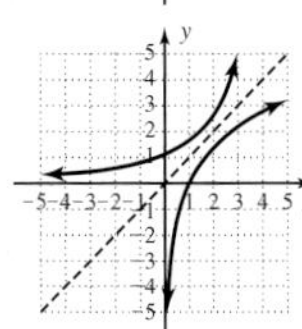

87. $D: x \in (-\infty, 4], R: y \in (-\infty, 4];$
$D: x \in (-\infty, 4], R: y \in (-\infty, 4]$

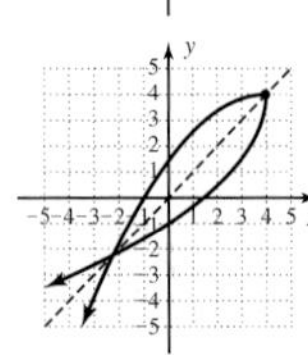

89. a. 31.5 cm **b.** The result is 80 cm. It gives the distance of the projector from the screen. **91. a.** −63.5°F **b.** $f^{-1}(x) = \frac{-2}{7}(x - 59)$; it is 35 **c.** 22,000 ft **93. a.** 144 ft **b.** $f^{-1}(x) = \frac{\sqrt{x}}{4}$, 3 sec, the original input for $f(x)$ **c.** 7 sec **95. a.** 28,260 ft^3 **b.** $f^{-1}(h) = \sqrt[3]{\frac{3h}{\pi}}$, 30 ft, the original input for $f(h)$ **c.** 9 ft **97. a.** 5 cm **b.** $f^{-1}(x) = \sqrt[3]{\frac{x}{\pi}}$, $f^{-1}(392.5) \approx 5$; same as original input for $f(x)$ **c.** $f^{-1}(x)$
99. a. verified **b.**

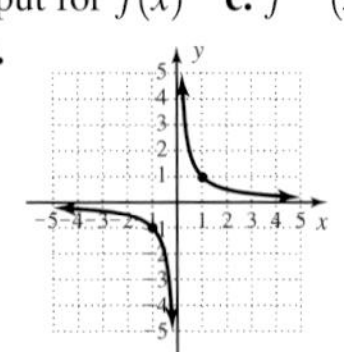

c. (1, 1) and (−1, −1); x and y coordinates are identical on $f(x) = x$
101. a. 0 **b.** 3 **c.** 81 **d.** 3 **103.** ≈0.472, ≈0.365; rate of change is greater in [1, 2] due to shape of the one-wing function.
105. $x = 2 \pm 3\sqrt{5}, x \approx 8.71, x \approx -4.71$
107. a.

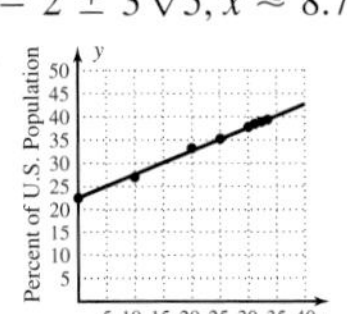

b. $P(t) = 0.51t + 22.51$, very strong
c. 2005: 40.4%, 2010: 43%

Exercises 3.3, pp. 283–288

1. stretch, compression **3.** (−5, −9), up **5.** Answers will vary.
7. square root function; y-int (0, 2); x-int (−3, 0); node (−4, −2); up on right **9.** cubic function; y-int (0, −2); x-int (−2, 0); pivot (−1, −1); up, down
11.

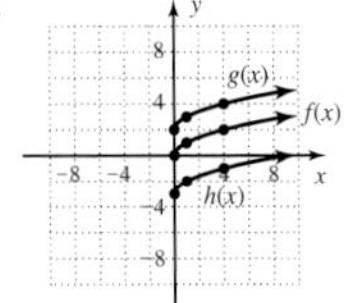

13. **15.**

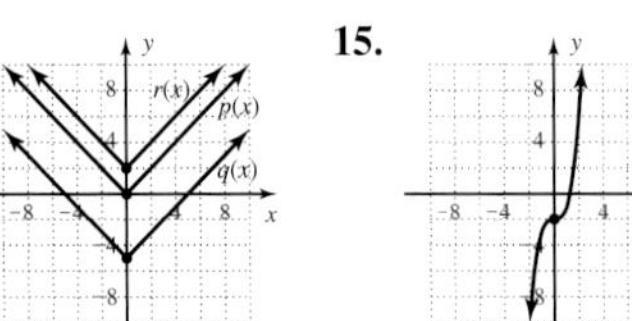

17.

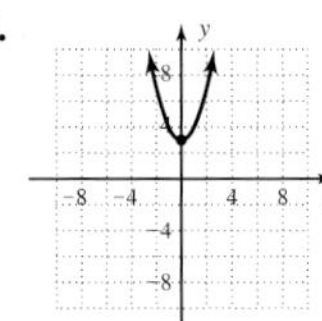

19.

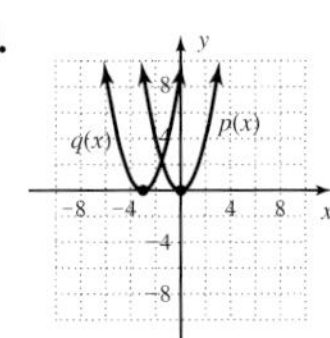

21.

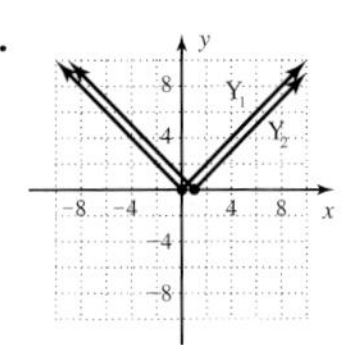

23.

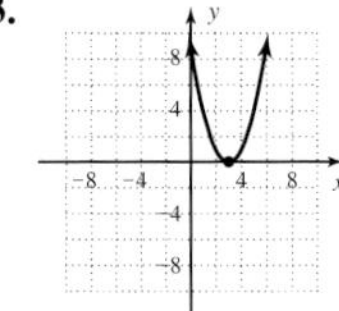

25.

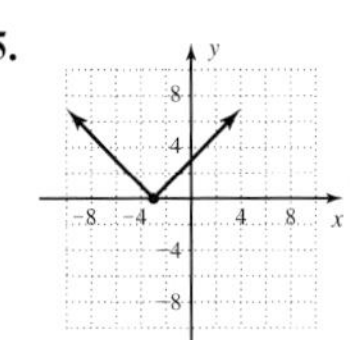

27.

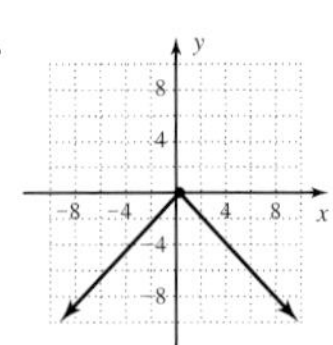

29.

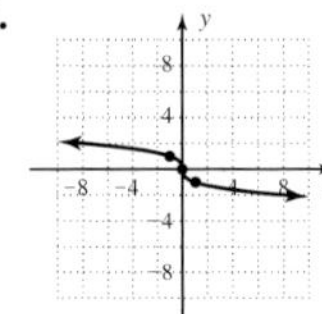

31.

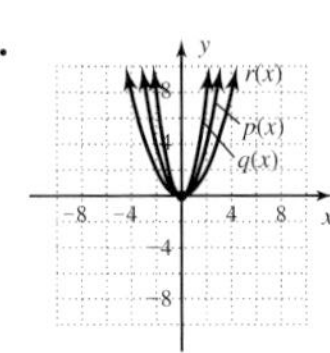

33.

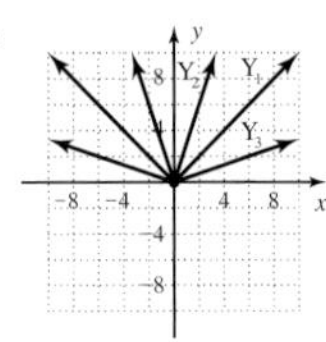

35.

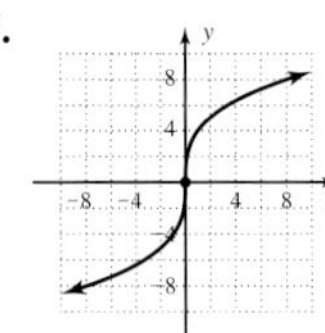

37.

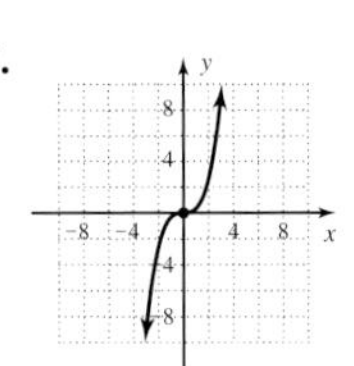

39. g **41.** i **43.** e **45.** j **47.** l **49.** c
51. left 2, down 1

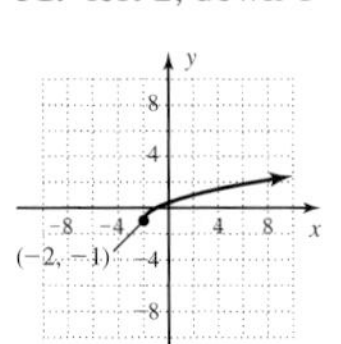

53. reflected across x-axis; left 3, down 2

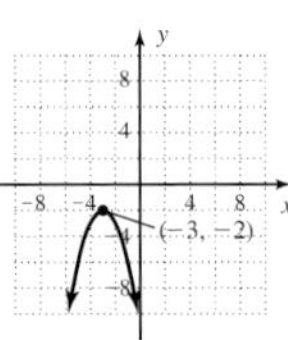

55. left 3, down 1

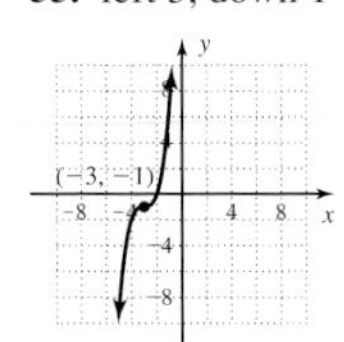

57. left 1, down 2

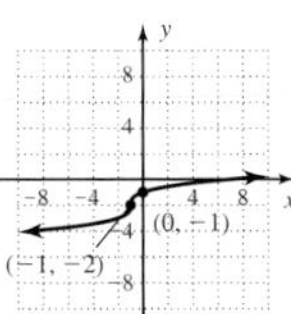

59. reflected across x-axis, left 3, down 2

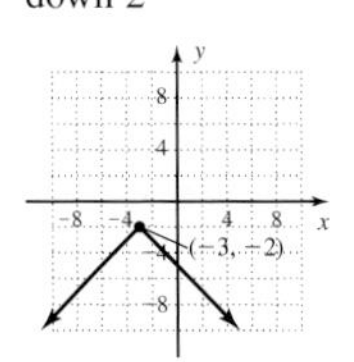

61. stretched vertically; reflected across x-axis, left 1, down 3

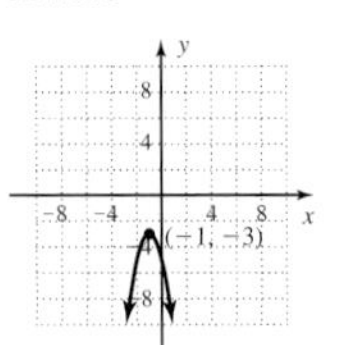

63. reflected across x-axis, left 2, down 1, compressed vertically

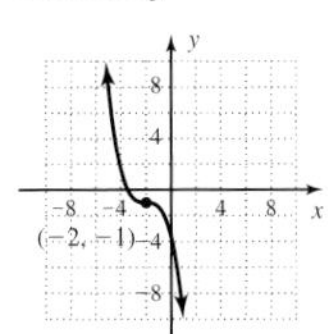

65. reflected across x-axis, right 4, down 3, stretched vertically

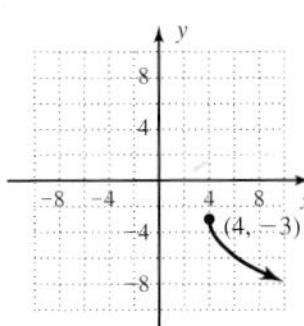

67. right 3, up 1, compressed vertically

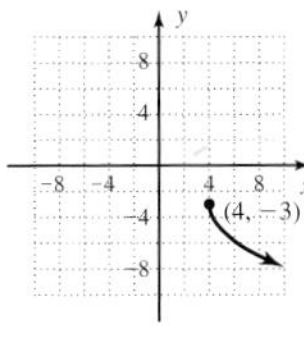

69. a.

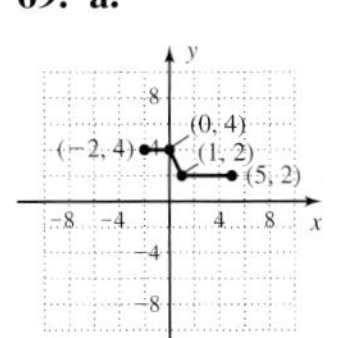

b.

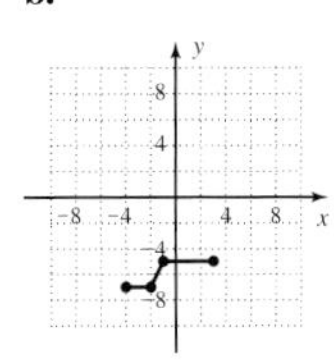

c.

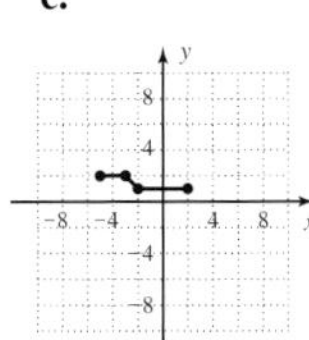

d.

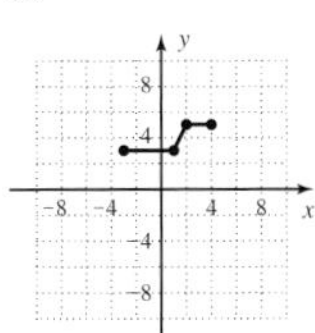

71. a.

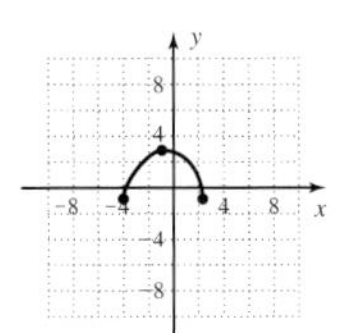

b.

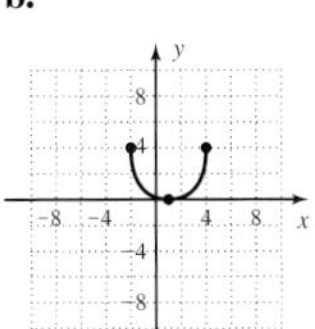

c.

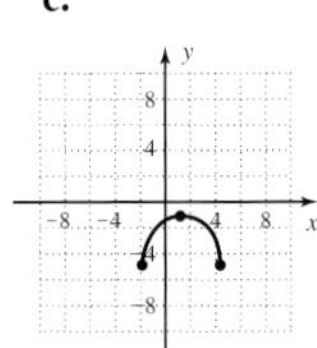

d.

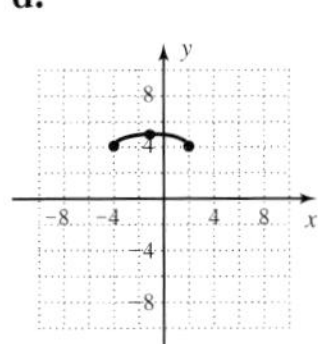

73. $A = 27$ units^2

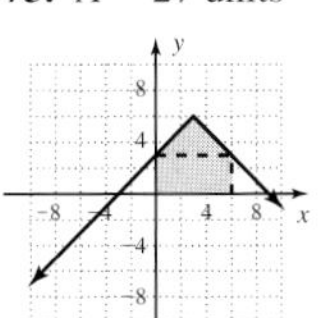

75. $A = 22\frac{2}{3}$ units^2

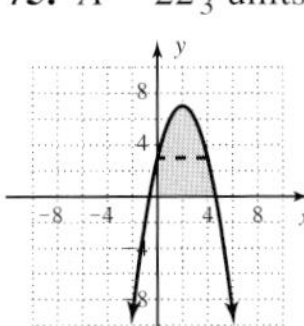

77. $A = \frac{9}{2}\pi + 12$ units^2

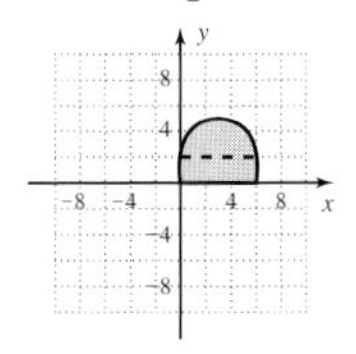

79. ≈ 4.2, 70 units^3, 65.4 units^3, yes

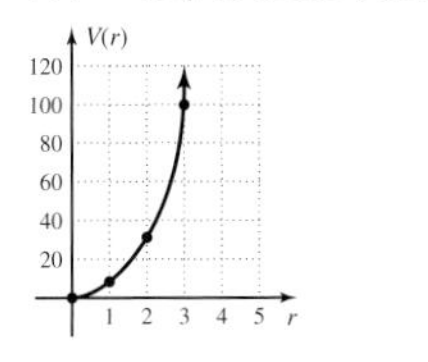

81. compressed vertically, 2.25 sec

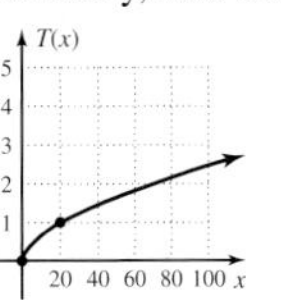

83. compressed vertically, 216 W

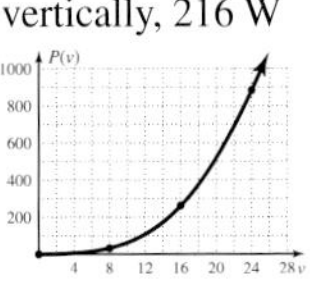

85. vertical stretch by a factor of 4; 10 ft/sec; 12.5 ft

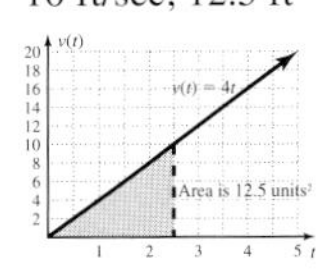

87. $x \in (0, 4)$ yes, $x \in (4, \infty)$ yes

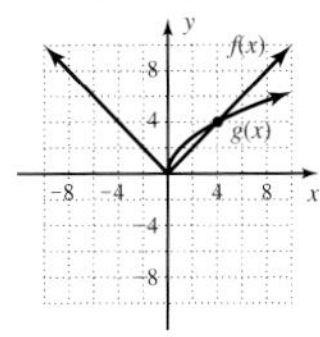

89. The result are identical; $2(x - 3) = 2x - 6$ via the distributive property

91. Any points in Quadrants III and IV will reflect in x-axis and move to Quadrants I and II

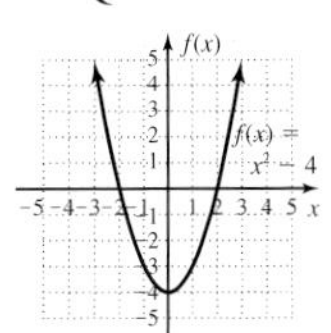

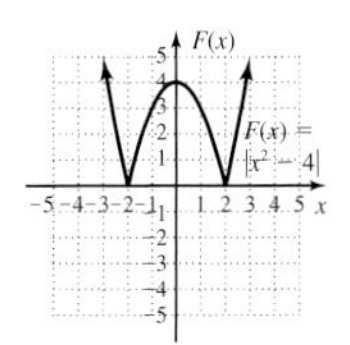

93. $x = 2, x = -1 \pm \sqrt{3}i$ **95.** $d = 29, m = \frac{-21}{20}$ **97.** $x = -5$

Exercises 3.4, pp. 295–299

1. $\frac{25}{2}$ **3.** vertex **5.** Answers will vary.

7. left 2, down 9

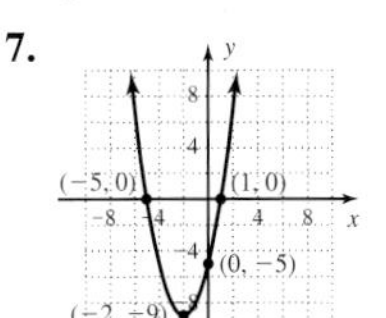

9. reflected across x-axis; right 1, up 4

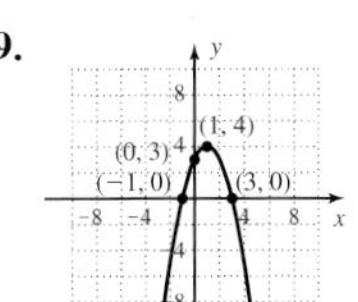
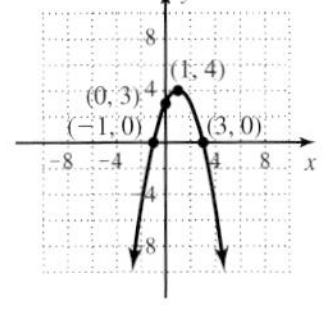

11. right $\frac{5}{2}$, down $\frac{17}{4}$

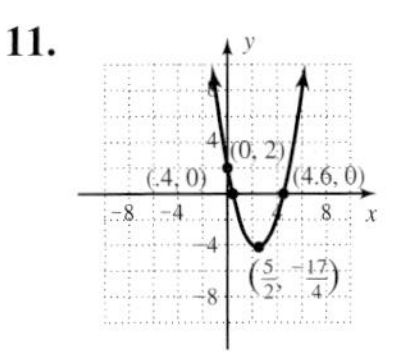

13. stretched vertically; left 1, down 8

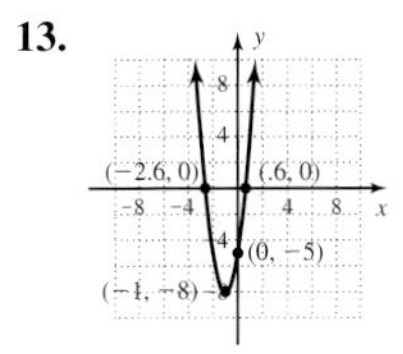

15. 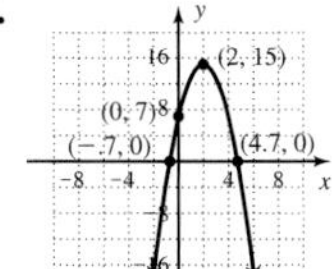

stretched vertically; reflected across x-axis; right 2, up 15

17. 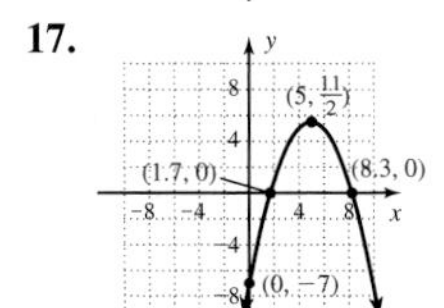

compressed vertically; reflected across x-axis; right 5, up $\frac{11}{2}$

19. 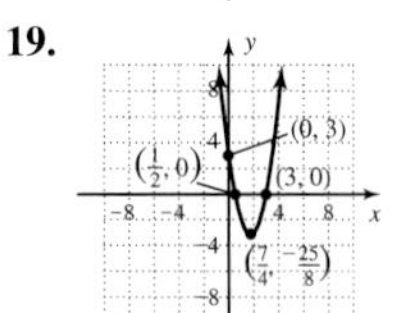

stretched vertically; right $\frac{7}{4}$, down $\frac{-25}{8}$

21. 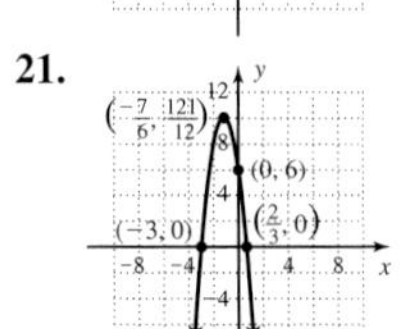

stretched vertically; reflected across x-axis; left $\frac{7}{6}$, up $\frac{121}{12}$

23. 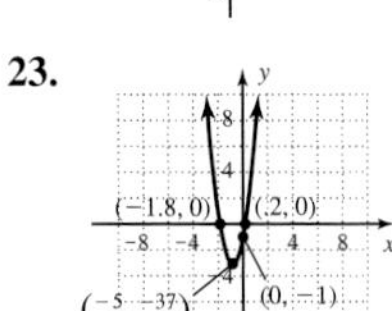 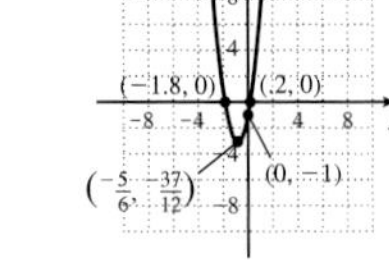

stretched vertically; left $\frac{5}{6}$, down $\frac{37}{12}$

25. $x = -3 \pm \sqrt{5}$ **27.** $x = -4 \pm \frac{\sqrt{14}}{2}$ **29.** $x = -2.7, x = 1.3$

31. 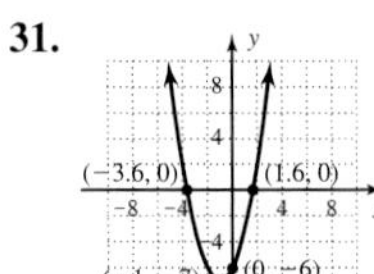 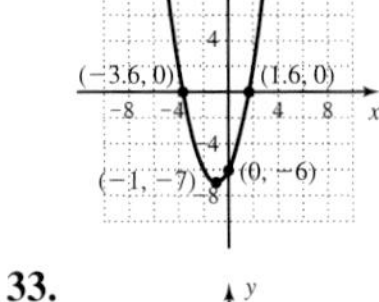

left 1, down 7

33. 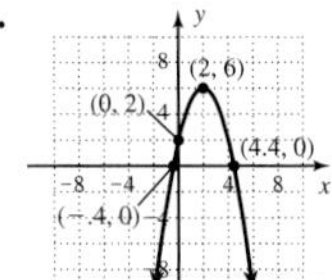

reflected across x-axis; right 2, up 6

35. 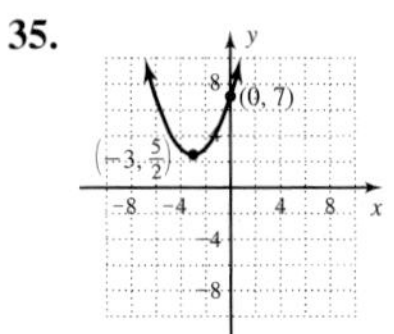

compressed vertically; left 3, up $\frac{5}{2}$

37. 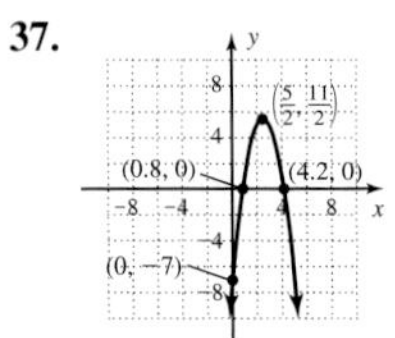

stretched vertically; reflected across x-axis, right $\frac{5}{2}$, up $\frac{11}{2}$

39. 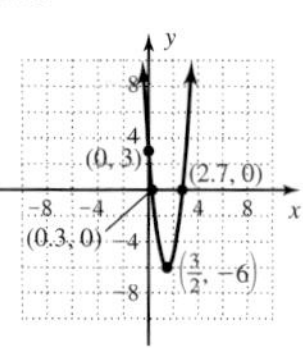

stretched vertically; right $\frac{3}{2}$, down 6

41. 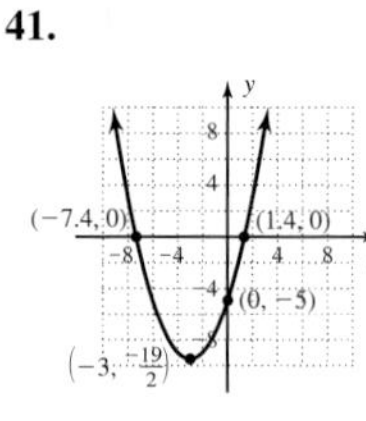

compressed vertically; left 3, down $\frac{19}{2}$

43. $f(x) = -(x-2)^2$ **45.** $p(x) = 1.5\sqrt{x+3}$ **47.** $f(x) = \frac{4}{5}|x+4|$ **49.** 55 **51. a.** (0, −66,000); when no cars are produced, there is a loss of \$66,000. **b.** (20, 0), (330, 0); no profit will be made if less than 20 or more than 330 cars are produced. **c.** 175 **d.** \$240,250 **53. a.** 6 mi **b.** 3600 ft **c.** 3200 ft **d.** 12 mi **55. a.** (0, −3300); if no appliances are sold, the loss will be \$3300. **b.** (20, 0), (330, 0); if less than 20 or more than 330 appliances are made and sold, there will be no profit. **c.** $0 \le x \le 200$ **d.** 175, \$12,012.50 **57. a.** 288 ft

b. 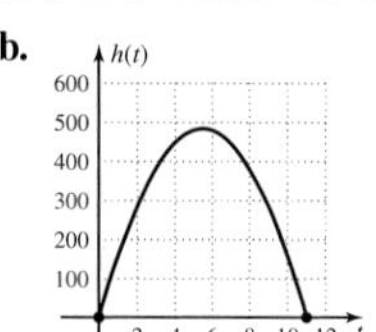

c. 484 ft; 5.5 sec **d.** 11 sec **59.** 6000; \$3200

61. Answers will vary. **63.** $f(x) = x^2 - 4x + 13$ **65.** $y = 81$ **67.** $m = \frac{4}{3}$, y-intercept (0, 3) **69.** $(f \circ g)(x) = x, (g \circ f)(x) = x$

Mid-Chapter Check, p. 299

1. a. 31 **b.** $6x^3 - x^2 - 15x$ **2. a.** $x \in (-\infty, 0) \cup (0, 5) \cup (5, \infty)$ **b.** −6 **3. a.** 40 m **b.** 15 m **c.** 11.25 m, 15 m **d.** yes **4. a.** 3 **b.** −10 **c.** 3 **d.** −3

5. a. 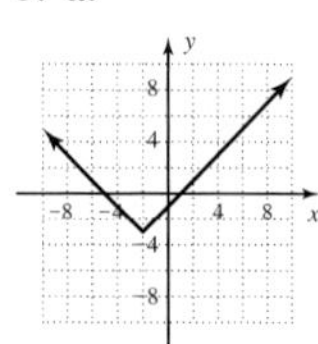

b. 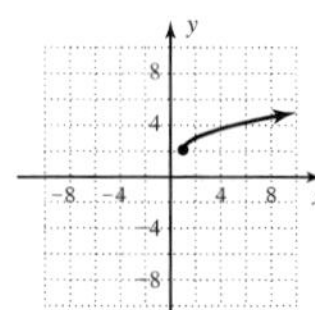

c. **d.** 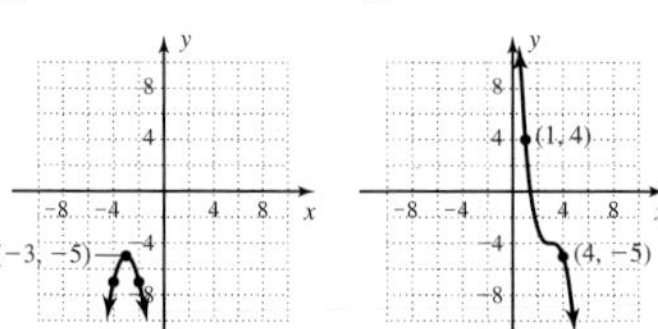

6. $f^{-1}(x) = x^2 + 3$, D: $x \in [0, \infty)$; R: $y \in [3, \infty)$; verified

7. 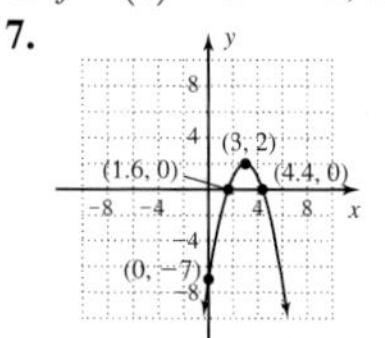

8. 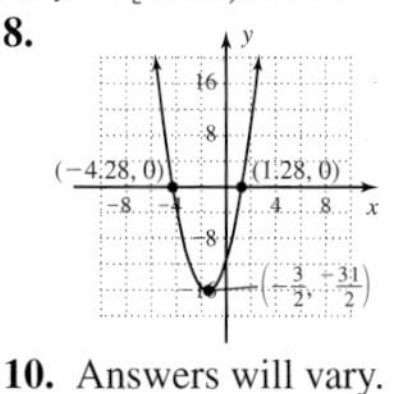

9. $(-\frac{5}{4}, \frac{-81}{8})$, (0, 7), (−3.5, 0), (1, 0)

10. Answers will vary.

Reinforcing Basic Concepts, p. 300

1. $h(x) = x^2 - 28;\ x = 4 \pm 2\sqrt{7}$ **2.** $h(x) = x^2 + 1;\ x = -2 \pm i$
3. $h(x) = 2x^2 - \frac{3}{2};\ x = \dfrac{5}{2} \pm \dfrac{\sqrt{3}}{2}$ **4.** $h(x) = x^3 - 9x - 3$
5. $h(x) = x^3 - 10x - 5$

Exercises 3.5, pp. 307–312

1. $x \to -\infty, y \to 2^-$ **3.** vertical, $y = 2$ **5.** In the reciprocal quadratic function, all range values are positive. **7.** as $x \to \infty, y \to 2^-$, as $x \to -\infty, y \to 2^+$, as $x \to -3^-, y \to \infty$, as $x \to -3^+, y \to -\infty$, $D: x \in (-\infty, -3) \cup (-3, \infty)$, $R: y \in (-\infty, 2) \cup (2, \infty)$
9. as $x \to \infty, y \to -2^+$, as $x \to -\infty, y \to -2^+$, as $x \to 1^-, y \to \infty$, as $x \to 1^+, y \to \infty$, $D: x \in (-\infty, 1) \cup (1, \infty)$, $R: y \in (-2, \infty)$
11. $y = -2, x = -1$ $f(x) = \dfrac{1}{(x+1)^2} - 2$ **13.** $y = -2, x = -1$ $f(x) = \dfrac{1}{x+1} - 2$ **15.** $y = -5, x = -2$; $f(x) = \dfrac{1}{(x+2)^2} - 5$
17. $\to -2^+$ **19.** $\to -\infty$ **21.** $-1, \pm\infty$
23. down 1, $x \in (-\infty, 0) \cup (0, \infty)$, $y \in (-\infty, -1) \cup (-1, \infty)$

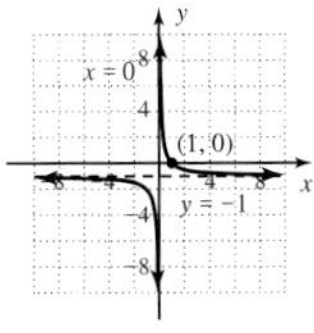

25. left 2, $x \in (-\infty, -2) \cup (-2, \infty)$, $y \in (-\infty, 0) \cup (0, \infty)$

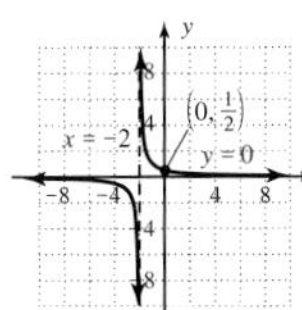

27. reflected across x-axis, right 2, $x \in (-\infty, 2) \cup (2, \infty)$, $y \in (-\infty, 0) \cup (0, \infty)$

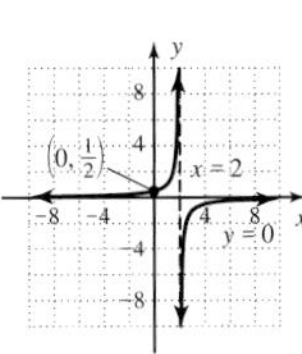

29. left 2, down 1, $x \in (-\infty, -2) \cup (-2, \infty)$, $y \in (-\infty, -1) \cup (-1, \infty)$

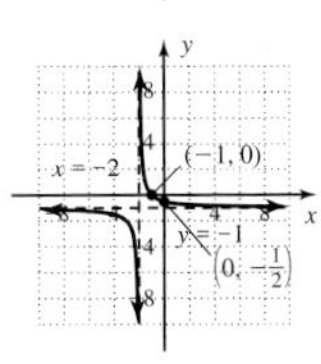

31. right 1, $x \in (-\infty, 1) \cup (1, \infty)$, $y \in (0, \infty)$

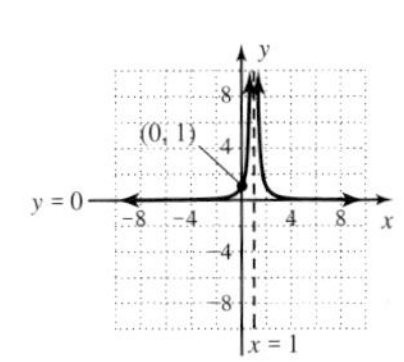

33. reflected across x-axis, left 2, $x \in (-\infty, -2) \cup (-2, \infty)$, $y \in (-\infty, 0)$

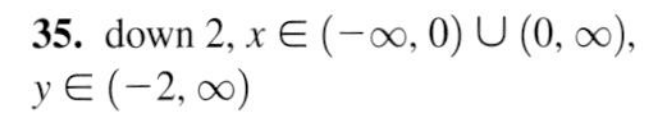
35. down 2, $x \in (-\infty, 0) \cup (0, \infty)$, $y \in (-2, \infty)$

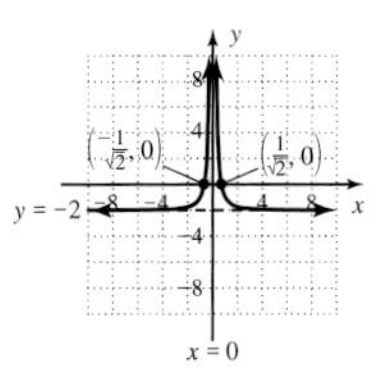

37. left 2, up 1, $x \in (-\infty, -2) \cup (-2, \infty)$, $y \in (1, \infty)$

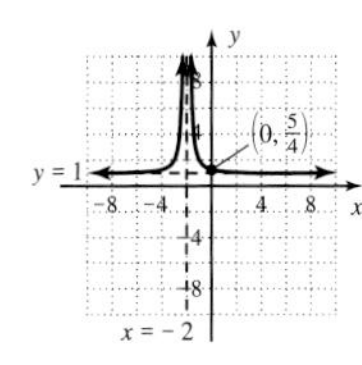

39. left 4, up 3, $x \in (-\infty, -4) \cup (-4, \infty)$, $y \in (-\infty, 3) \cup (3, \infty)$

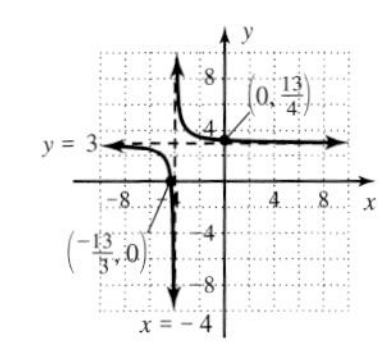

41. reflected across x-axis, right 1, down 3, $x \in (-\infty, 1) \cup (1, \infty)$, $y \in (-\infty, -3) \cup (-3, \infty)$

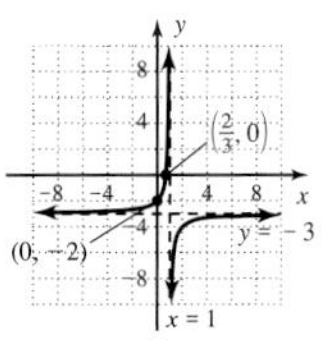

43. reflected across x-axis, right 2, up 3, $x \in (-\infty, 2) \cup (2, \infty)$, $y \in (-\infty, 3)$
$\left(\dfrac{6 - \sqrt{3}}{3}, 0\right), \left(\dfrac{6 + \sqrt{3}}{3}, 0\right)$

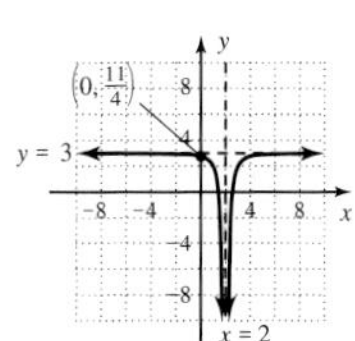

45. reflected across x-axis, left 2, up 3, $x \in (-\infty, -2) \cup (-2, \infty)$, $y \in (-\infty, 3)$
$\left(\dfrac{-6 - \sqrt{3}}{3}, 0\right), \left(\dfrac{-6 + \sqrt{3}}{3}, 0\right)$

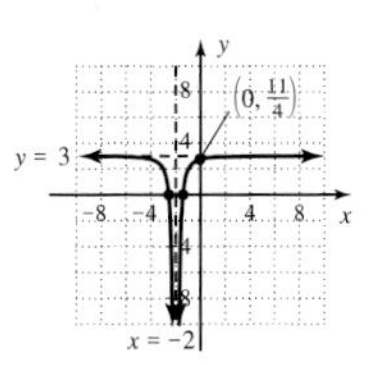

47. F becomes very small; $y = \dfrac{1}{x^2}$

49. a. It decreases; 75, 25, 15 **b.** It approaches 0. **c.** as p decreases, D becomes very large; as $p \to 0, D \to \infty$

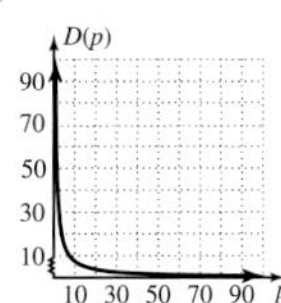

51. a. It decreases; 100, 25, $11.\overline{1}$. **b.** toward the light source **c.** Intensity becomes large; as $d \to 0, I \to \infty$

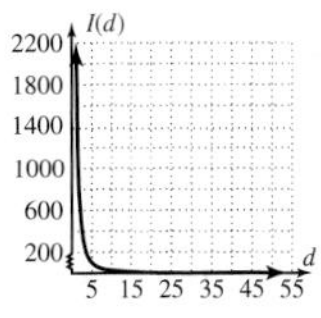

53. a. \$20,000, \$80,000, \$320,000; cost increases dramatically

b.

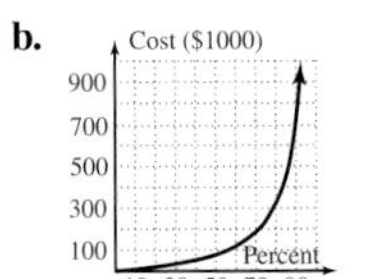

c. as $p \to 100^-$; $C \to \infty$

55. a. 5 hr; about 0.28 **b.** $-0.019, -0.005$; As the number of hours increases, the rate of change decreases. **c.** $h \to \infty$, $C \to 0^+$, horizontal asymptote **57.** $f^{-1}(x) = f(x)$; Answers will vary. **59.** Answers will vary. **61.** $\frac{-16}{3}, \frac{3}{4}$ $(3x + 16)(4x - 3) = 0$ **63.** $-5 - i\sqrt{3}$ **65.** verified

Exercises 3.6, pp. 321–326

1. constant **3.** reciprocal quadratic **5.** Answers will vary. **7.** $d = kr$ **9.** $F = ka$ **11.** $y = 0.025x$

x	y
500	12.5
650	16.25
750	18.75

13. \$321.30; the hourly wage; $k = \$9.18$/hr

15. a. $k = \frac{192}{47}$ $S = \frac{192}{47}h$ **b.**

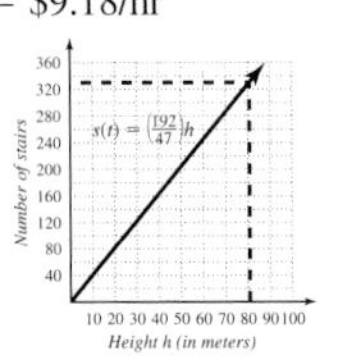

c. 330 stairs

d. $S = 331$; yes

17. $A = kS^2$ **19.** $P = kc^2$

21.

q	p
45	226.8
55	338.8
70	548.8

$k = 0.112$; $p = 0.112\,q^2$

23. $k = 6$, $A = 6s^2$; about 55,303,776 m^2

25. a. $k = 16$ $d = 16t^2$ **b.**

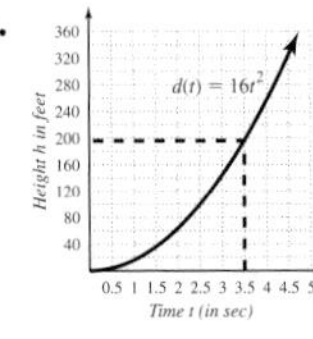

c. about 3.5 sec

d. 3.5 sec; yes

e. 2.75 sec

27. $F = \frac{k}{d^2}$ **29.** $S = \frac{k}{L}$ **31.** $Y = \frac{12{,}321}{Z^2}$

Z	Y
37	9
74	2.25
111	1

33. $w = \frac{3{,}072{,}000{,}000}{r^2}$; 48 kg **35.** $I = krt$ **37.** $A = kh(B + b)$

39. $V = ktr^2$ **41.** $C = \frac{6.75R}{S^2}$

R	S	C
120	6	22.5
200	12.5	8.64
350	15	10.5

43. $E = 0.5\,mv^2$; 612.50 J

45. cube root family; answers will vary; 0.054 or 5.4%

Amount A	Rate R
1.0	0.000
1.05	0.016
1.10	0.032
1.15	0.048
1.20	0.063
1.25	0.077

47. $T = \frac{48}{V}$; 32 volunteers **49.** $M = \frac{1}{6}E$; 41.7 kg

51. $D = 21.6\sqrt{S}$; 144.9 ft **53.** $C = 8.5LD$; \$76.50

55. $C = (4.4 \times 10^{-4})\frac{p_1p_2}{d^2}$; about 223 calls

57. a. Scatter-plot shows data are obviously nonlinear; decreasing to increasing pattern rules out a power function.

b. $p(t) = 0.0148t^2 - 0.9175t + 19.5601$; 6.5%; 6.3%

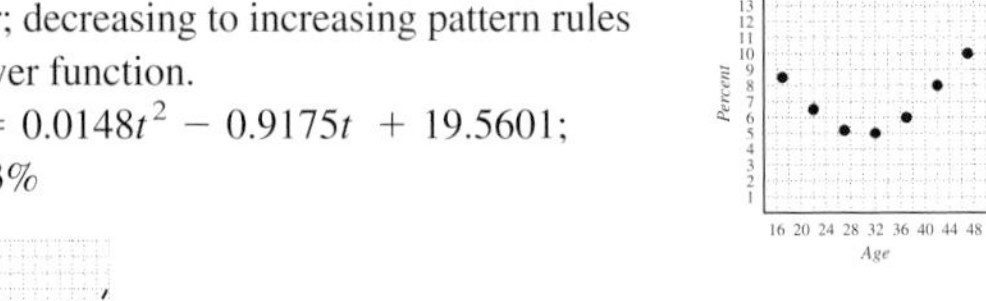

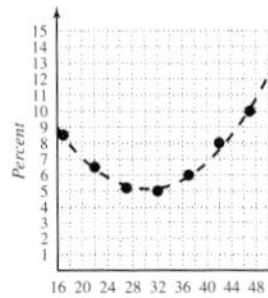

59. Answers will vary. **61.** 6.67×10^{-7} **63.** $\frac{9y^2}{4x^2}$ **65.** yes **67.** 60 gal

Exercises 3.7, pp. 335–340

1. continuous **3.** smooth **5.** Answers will vary.

7. a. $f(x) = \begin{cases} x^2 - 6x + 10 & 0 \le x \le 5 \\ \frac{3}{2}x - \frac{5}{2} & 5 < x \le 9 \end{cases}$ **b.** $y \in [1, 11]$

9. a. $f(t) = \begin{cases} -t^2 + 6t & 0 \le t \le 5 \\ 5 & t > 5 \end{cases}$ **b.** $y \in [0, 9]$

11. $-2, 2, 0, \frac{1}{2}, 2.999, 5$ **13.** $5, 5, 0, -4, 5, 11$

15. a.

Year (0 → 1950)	Percent
5	7.33
15	14.13
25	14.93
35	22.65
45	41.55
55	60.45

b. Each piece gives a slightly different value due to rounding of coefficients in each model. At $t = 30$, we use the "first" piece: $P(30) = 13.08$.

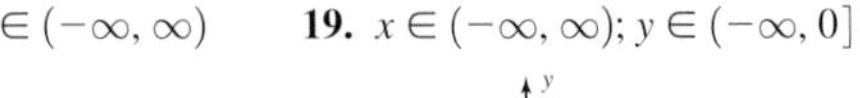

17. $x \in (-\infty, \infty)$; $y \in (-\infty, \infty)$

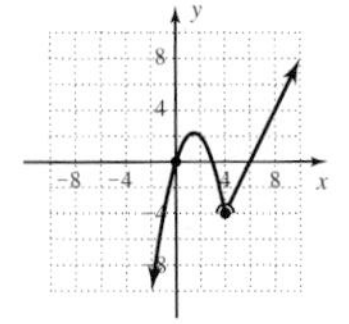

19. $x \in (-\infty, \infty)$; $y \in (-\infty, 0]$

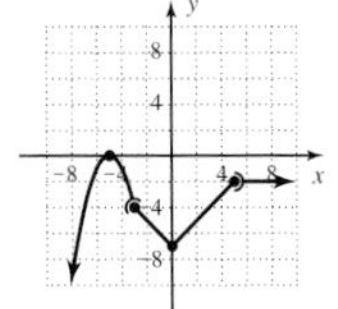

21. $x \in (-\infty, 9);\ y \in [2, \infty)$

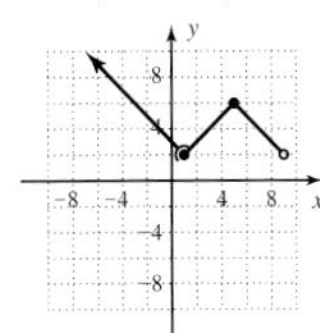

23. $x \in (-\infty, \infty);\ y \in [0, \infty)$

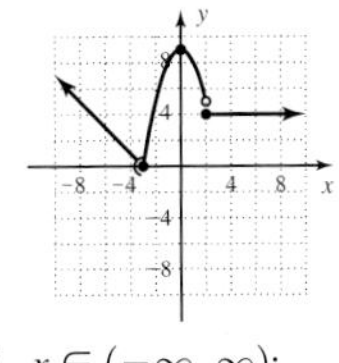

25. $x \in (-\infty, \infty)$;
$y \in (-\infty, -6) \cup (-6, \infty)$;
discontinuity at $x = -3$,
redefine $f(x) = -6$ at $x = -3$

27. $x \in (-\infty, \infty)$;
$y \in [0.75, \infty) \cup \{-4\}$;
discontinuity at $x = 1$,
redefine $f(x) = 3$ at $x = 1$

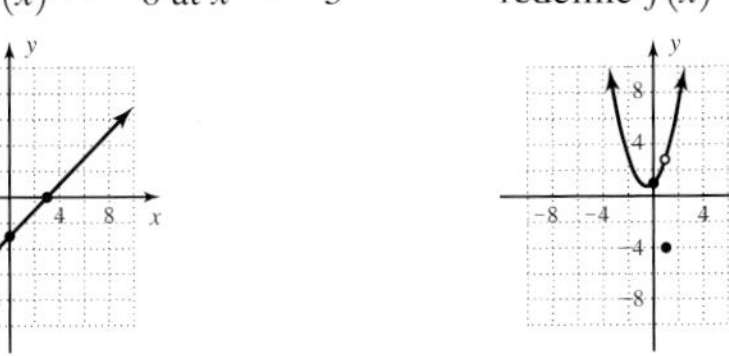

29. Graph is discontinuous at $x = 0$; $f(x) = 1$ for $x > 0$; $f(x) = -1$ for $x < 0$.

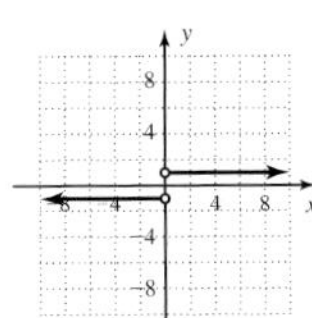

31. $C(p) = \begin{cases} 0.09p & 0 \le p \le 1000 \\ 0.18p - 90 & p > 1000 \end{cases}$;

$126

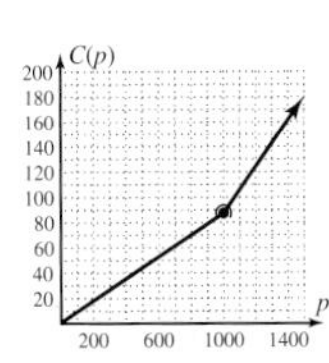

33. $C(t) = \begin{cases} 0.75t & 0 \le t \le 25 \\ 1.5t - 18.75 & t > 25 \end{cases}$;

$48.75

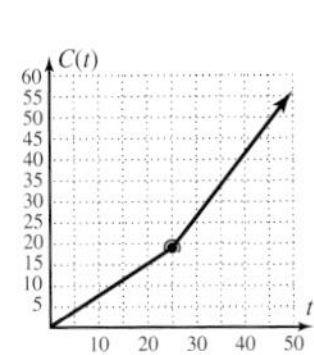

35. $S(t) = \begin{cases} -1.35t^2 + 31.9t + 152; & 0 \le t \le 12 \\ 2.5t^2 - 80.6t + 950; & 12 < t \le 22 \end{cases}$

$498 billion, $653 billion, $782 billion

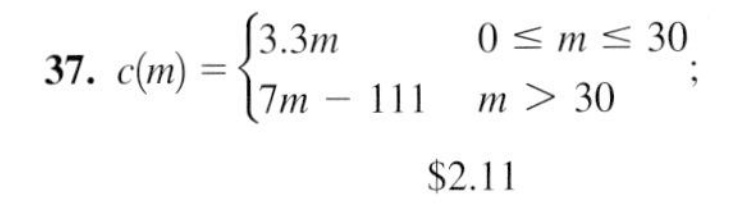

37. $c(m) = \begin{cases} 3.3m & 0 \le m \le 30 \\ 7m - 111 & m > 30 \end{cases}$;

$2.11

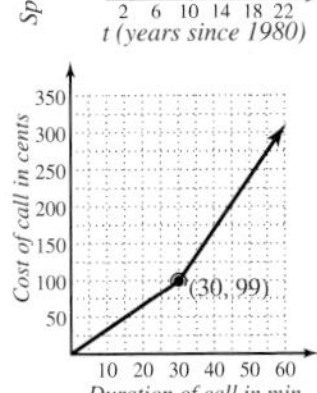

39. $C(a) = \begin{cases} 0 & a < 2 \\ 2 & 2 \le a < 13 \\ 5 & 13 \le a < 20 \\ 7 & 20 \le a < 65 \\ 5 & a \ge 65 \end{cases}$ $38

41. yes; $h(x) = \begin{cases} 5 & x \le -3 \\ -2x - 1 & -3 < x < 2 \\ -5 & x \ge 2 \end{cases}$

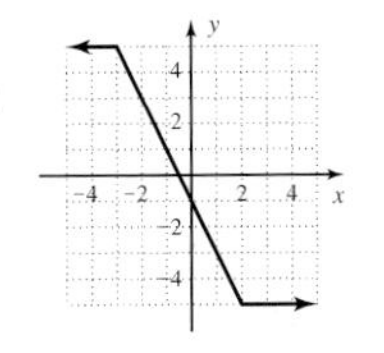

43. Answers will vary. **45.** $h(x) = 4x - 3,\ 1 \le x \le 3$
47. $x = -7, x = 4$ **49.** $y = 2\sqrt{x + 4} - 1$ **51. a.** 25 **b.** $6x^2 + 1$

Exercises 3.8, pp. 350–357

1. cut, linear, bounce **3.** increasing **5.** Answers will vary.
7. $D: x \in (-\infty, 3) \cup (3, \infty),\ R: y \in (-\infty, -4) \cup (-4, \infty)$
9. $D: x \in (-\infty, \infty),\ R: y \in (-\infty, 5]$ **11.** $D: x \in (-\infty, 2]$, $R: y \in (-\infty, 3]$ **13.**

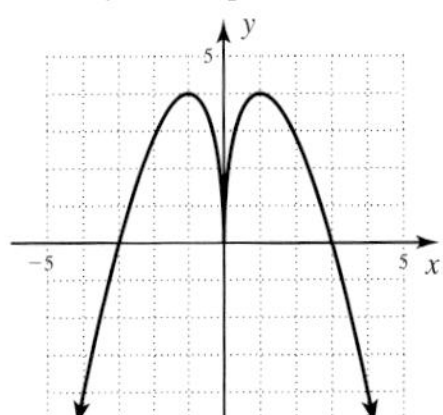

15. even **17.** not even **19.**

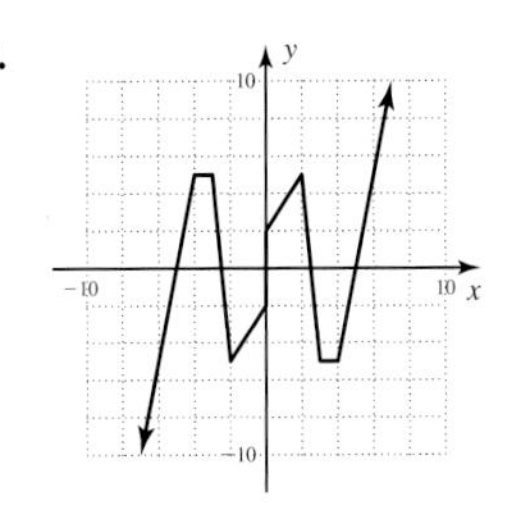

21. odd **23.** not odd **25.** $x \in [-1, 1] \cup [3, \infty)$
27. $x \in (-\infty, -1) \cup (-1, 1) \cup (1, \infty)$ **29.** $f(x)\uparrow: (1, 4)$
$f(x)\downarrow: (-2, 1) \cup (4, \infty)$ constant: $(-\infty, -2)$ **31.** $V(x)\uparrow: (-3, 1) \cup (4, 6)$
$V(x)\downarrow: (-\infty, -3) \cup (1, 4)$ constant: none **33. a.** $x \in (-\infty, \infty)$, $y \in (-\infty, 5]$ **b.** (1, 0), (3, 0) **c.** $H(x) \ge 0: x \in [1, 3]$, $H(x) \le 0: x \in (-\infty, 1] \cup [3, \infty)$ **d.** $H(x)\uparrow: x \in (-\infty, 2)$, $H(x)\downarrow: x \in (2, \infty)$ **e.** max: (2, 5) **f.** none
35. a. $x \in (-\infty, \infty), y \in (-\infty, \infty)$ **b.** (−1, 0)
c. $q(x) \ge 0: x \in (-\infty, -1]$, $q(x) \le 0: x \in [-1, \infty)$ **d.** $q(x)\uparrow$: none
$q(x)\downarrow: x \in (-\infty, \infty)$ **e.** none **f.** none
37. a. $x \in (-\infty, \infty), y \in (-\infty, \infty)$ **b.** (−1, 0), (5, 0)
c. $g(x) \ge 0: x \in [-1, \infty)$, $g(x) \le 0: x \in (-\infty, -1] \cup \{5\}$
d. $g(x)\uparrow: x \in (-\infty, 1) \cup (5, \infty)$, $g(x)\downarrow: x \in (1, 5)$ **e.** max: (1, 6), min: (5, 0) **f.** none **39. a.** $x \in (-\infty, \infty), y \in (-\infty, \infty)$
b. (−2, 0), (4, 0), (8, 0) **c.** $q(x) \ge 0: x \in \{-2\} \cup [4, \infty)$,
$q(x) \le 0: x \in (-\infty, 4) \cup \{8\}$ **d.** $q(x)\uparrow: x \in (-\infty, -2) \cup (1, 6) \cup (8, \infty)$, $q(x)\downarrow: x \in (-2, 1) \cup (6, 8)$ **e.** max: (−2, 0), (6, 2) min: (1, −5), (8, 0) **f.** none **41. a.** $x \in (-\infty, \infty), y \in (-\infty, 3]$
b. (0, 0), (2, 0) **c.** $Y_2 \ge 0: x \in [0, 2]$, $Y_2 \le 0: x \in (-\infty, 0] \cup [2, \infty)$
d. $Y_2\uparrow: x \in (-\infty, 1)$, $Y_2\downarrow: x \in (1, \infty)$
e. max: (1, 3) **f.** none **43. a.** $x \in (-\infty, 2) \cup (2, \infty), y \in (-\infty, 4)$
b. (1, 0), (3, 0) **c.** $Y_2 \ge 0: x \in (-\infty, 1] \cup [3, \infty)$,
$Y_2 \le 0: x \in [1, 2) \cup (2, 3]$ **d.** $Y_2\uparrow: x \in (2, \infty)$, $Y_2\downarrow: x \in (-\infty, 2)$
e. none **f.** $x = 2, y = 4$ **45. a.** $x \in (-\infty, \infty), y \in \{-4\} \cup [-3, \infty)$
b. (−4, 0) **c.** $H(x) \ge 0: x \in (-\infty, -4] \cup (-2, 0) \cup (0, 2)$,
$H(x) \le 0: x \in [-4, -2] \cup [3, \infty)$ **d.** $H(x)\uparrow: x \in (-2, 0)$,
$H(x)\downarrow: x \in (-\infty, -2) \cup (0, 2)$, $H(x)$ constant: $x \in (3, \infty)$ **e.** none

f. $x = 0$ **47.** $\frac{\Delta y}{\Delta x} = 2x - 4 + h$; $[0.00, 0.01]: \frac{\Delta y}{\Delta x} = -3.9$;

$[3.00, 3.01]: \frac{\Delta y}{\Delta x} \approx 2.0$; Answers will vary.

49. $\frac{\Delta y}{\Delta x} = \frac{1}{\sqrt{x + h} + \sqrt{x}}$ $[1.00, 1.01]: \frac{\Delta y}{\Delta x} \approx 0.5$;

$[4.00, 4.01]: \frac{\Delta y}{\Delta x} \approx .025$; Answers will vary.

51. $y = \sin x$ **a.** $y \in [-1, 1]$ **b.** $(-180, 0), (0, 0), (180, 0), (360, 0)$ **c.** $f(x)\uparrow: x \in (-90, 90) \cup (270, 360)$, $f(x)\downarrow: x \in (-180, -90) \cup (90, 270)$ **d.** min: $(-90, -1)$ and $(270, -1)$, max: $(90, 1)$ **e.** odd
$y = \cos x$
a. $y \in [-1, 1]$ **b.** $(-90, 0), (90, 0), (270, 0)$ **c.** $f(x)\uparrow: x \in (-180, 0) \cup (180, 360)$, $f(x)\downarrow: x \in (0, 180)$ **d.** min: $(-180, -1)$ and $(180, -1)$, max: $(0, 1)$ and $(360, 1)$ **e.** even **53. a.** $t \in (0, 1) \cup (3, 4) \cup (7, 10)$ **b.** $t \in (4, 7)$ **c.** $t \in (1, 3)$ **d.** $(4, 12), (10, 16)$ **e.** $(7, -4)$ **f.** $t \in (0, 6) \cup (8, 10)$ **g.** $t \in (6, 8)$ **h.** $(6, 0), (8, 0)$ **55.** $h(x) = |x^2 - 4| - 5$ **a.** $x \in (-\infty, \infty), y \in [-5, \infty)$ **b.** $(-3, 0), (3, 0)$ **c.** $h(x) \geq 0: x \in (-\infty, -3] \cup [3, \infty)$; $h(x) \leq 0: x \in [-3, 3]$ **d.** $h(x)\uparrow: x \in (-2, 0) \cup (2, \infty)$; $h(x)\downarrow: x \in (-\infty, -2) \cup (0, 2)$ **e.** max: $(0, -1)$; min: $(-2, -5), (2, -5)$ **f.** none **57. a.** $t \in [75, 102]$, $D \in [-300, 230]$ **b.** $D(t)\uparrow: t \in (76, 77) \cup (83, 84) \cup (86, 87) \cup (92, 100)$; $D(t)\downarrow: t \in (75, 76) \cup (77, 83) \cup (84, 86) \cup (89, 92) \cup (100, 102)$ $D(t)$ constant: $t \in (87, 89)$ **c.** max: $(75, -40), (77, -50), (84, -170), (100, 240)$; min: $(76, -70), (83, -210), (86, -220), (92, -300), (102, -140)$ **d.** increase: 96 to 97 or 99 to 100; decrease: 101 to 102 **59. a.** zeroes: $(-9, 0), (-3, 0), (6, 0)$; min: $(-6, -6), (6, 0)$; max: $(3, 6)$

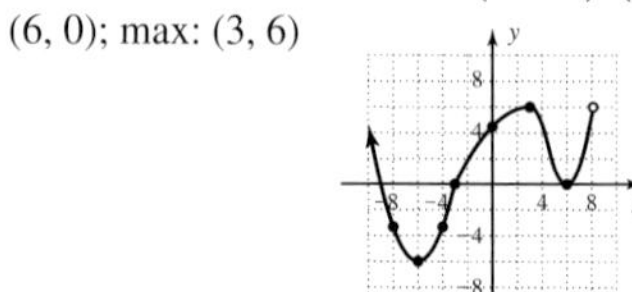

61. Answers will vary. **63.**

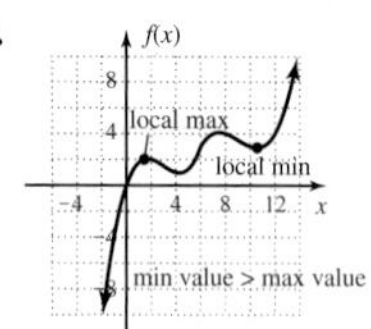

Answers will vary.

65. $x = -2, x = 10$ **67. a.** $\frac{1}{64}$ **b.** $\frac{4}{9}$ **69.** $y = |x - 1| - 2$

Summary and Concept Review, pp. 358–364

1. $a^2 + 7a - 2$ **2.** 147 **3.** $x \in (-\infty, \frac{2}{3}) \cup (\frac{2}{3}, \infty)$ **4.** $4x^2 + 8x - 3$ **5.** 99 **6.** x, x **7.** $f(x) = \sqrt{x} + 1$; $g(x) = 3x - 2$ **8.** $f(x) = 3 - |x|$; $g(x) = x^2 - 1$ **9.** $f(x) = x^2 - 3x - 10$; $g(x) = x^{\frac{1}{3}}$ **10.** $A(t) = \pi(2t + 3)^2$ **11.** no **12.** no **13.** yes **14.** $f^{-1}(x) = \dfrac{x - 2}{-3}$ **15.** $f^{-1}(x) = \sqrt{x + 2}$ **16.** $f^{-1}(x) = x^2 + 1$; $x \geq 0$ **17.** $f(x)$: $D: x \in [-4, \infty)$, $R: y \in [0, \infty)$; $f^{-1}(x)$: $D: x \in [0, \infty)$, $R: y \in [-4, \infty)$ **18.** $f(x)$: $D: x \in (-\infty, \infty)$, $R: y \in (-\infty, \infty)$; $f^{-1}(x)$: $D: (-\infty, \infty)$, $R: y \in (-\infty, \infty)$ **19.** $f(x)$: $D: x \in (-\infty, \infty)$, $R: y \in (0, \infty)$; $f^{-1}(x)$: $D: x \in (0, \infty)$, $R: y \in (-\infty, \infty)$ **20. a.** \$3.05 **b.** $f^{-1}(t) = \dfrac{t - 2}{0.15}$, $f^{-1}(3.05) = 7$ **c.** 12 days

21. quadratic **22.** absolute value **23.** cubic

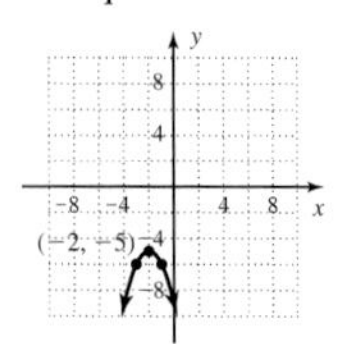

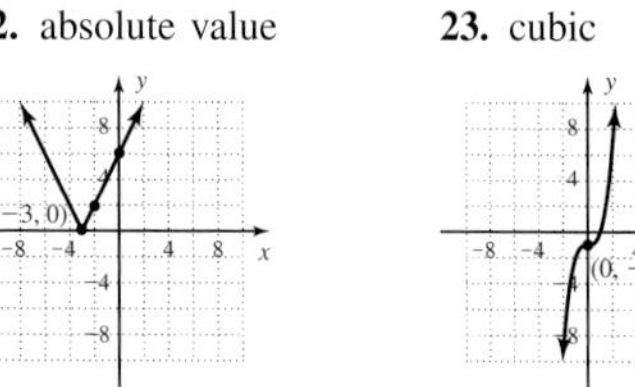

24. square root **25.** cube root **26.** linear

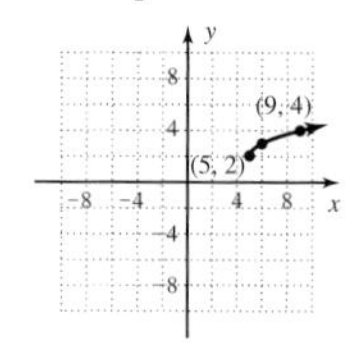

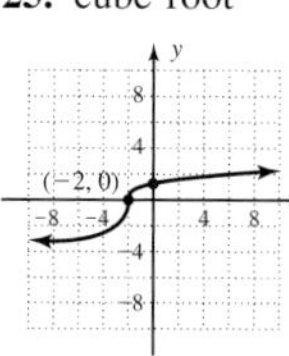

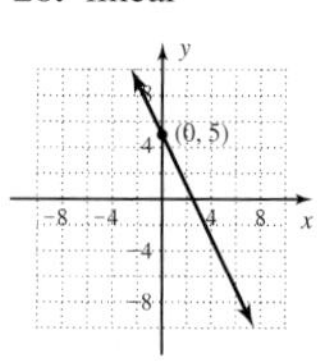

27. a. **b.** **c.**

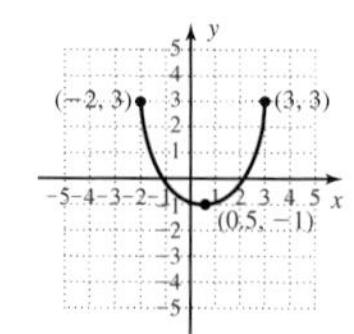

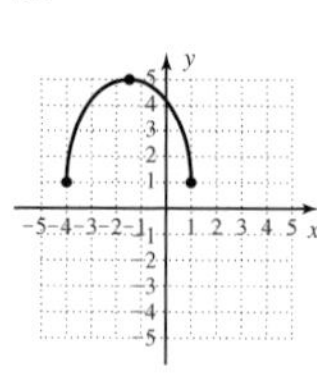

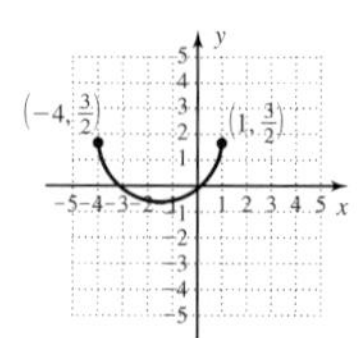

28. $g(x) = -2|x + 1| + 4$

29. **30.** **31.**

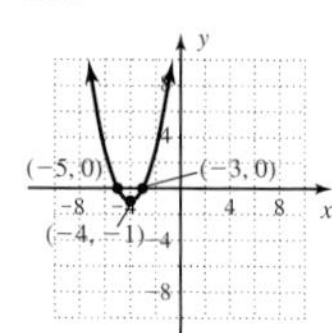

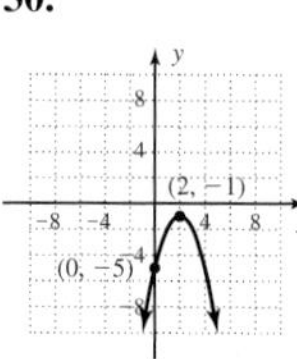

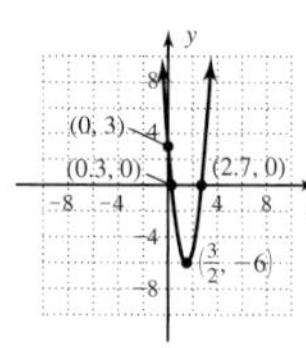

32. $y = -|x - 2| + 6$ **33.** $y = \frac{5}{2}\sqrt{x + 4} - 4$ **34.** $y = \frac{1}{9}(x + 3)^3 - 1$ **35. a.** 2 ft **b.** 130 ft **c.** 4 sec **d.** 146 ft; 3 sec

36. **37.**

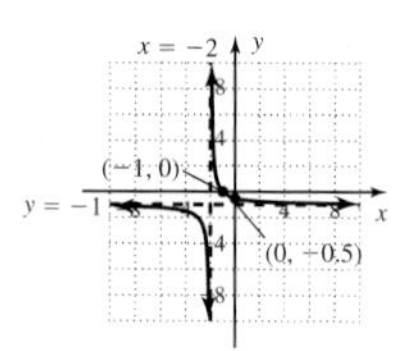

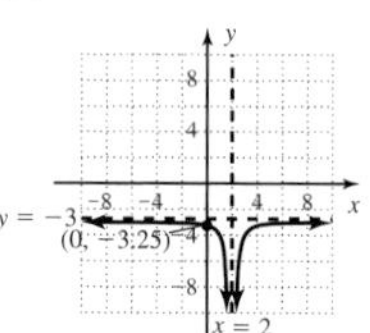

38. a. ≈\$32,143; \$75,000; \$175,000; \$675,000; cost increases dramatically

b.

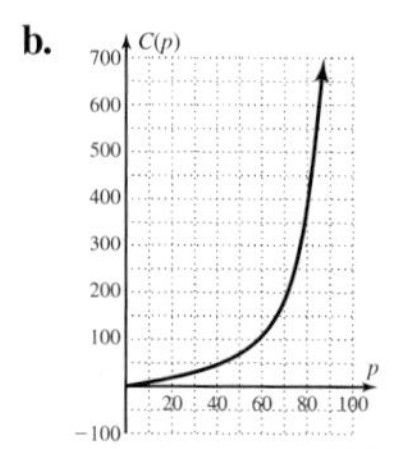

c. as $p \to 100$, $C \to \infty$

39. $k = 17.5$; $y = 17.5\sqrt[3]{x}$ **40.** $k = 0.72$; $z = \dfrac{0.72v}{w^2}$

x	y
216	105
0.343	12.25
729	157.5

v	w	z
196	7	2.88
38.75	1.25	17.856
24	0.6	48

41. 4.5 sec **42. a.** $f(x) = \begin{cases} 5 & x \leq -3 \\ -x + 1 & -3 < x \leq 3 \\ 3\sqrt{x - 3} - 1 & x > 3 \end{cases}$

b. $R: y \in [-2, \infty)$ **43.** $D: x \in (-\infty, \infty)$, $R: y \in (-\infty, -8) \cup (-8, \infty)$, discontinuity at $x = -3$; define $h(x) = -8$ at $x = -3$

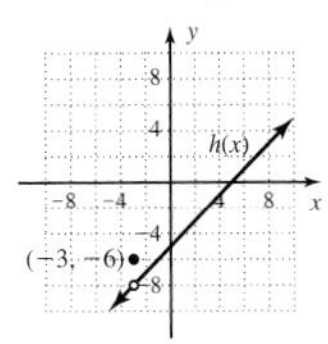

44. $-4, -4, -4.5, -4.99, 3\sqrt{3} - 9, 3\sqrt{3.5} - 9$
45. $D: x \in (-\infty, \infty)$, $R: y \in [-4, \infty)$

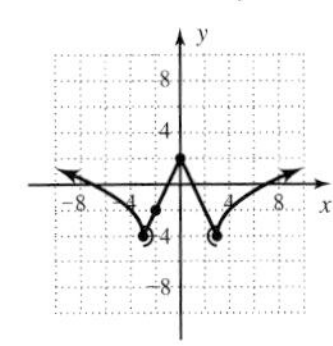

46. $T(x) = \begin{cases} 11x^2 - 197.4x + 1737.3 & 8 \le x \le 11 \\ 17x + 708.67 & x > 11 \end{cases}$

47. a. $D: x \in (-\infty, \infty)$, $R\ y \in [2, \infty)$ **b.** 38, 18, 6, 3, 2 **c.** none **d.** $f(x) < 0$: none, $f(x) > 0, x \in (-\infty, \infty)$ **e.** min: (3, 2) **f.** $f(x)\uparrow: x \in (3, \infty)$, $f(x)\downarrow: x \in (-\infty, 3)$

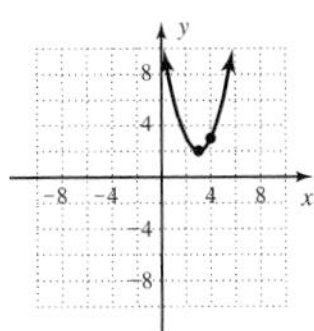

48. a. $D: x \in (-\infty, \infty)$, $R: y \in (-\infty, 3]$ **b.** $-2, 0, 2, 3, 2$ **c.** $(-1, 0)$, $(5, 0)$ **d.** $f(x) < 0: x \in (-\infty, -1) \cup (5, \infty)$; $f(x) > 0: x \in (-1, 5)$ **e.** max: (2, 3) **f.** $f(x)\uparrow: x \in (-\infty, 2)$, $f(x)\downarrow: x \in (2, \infty)$

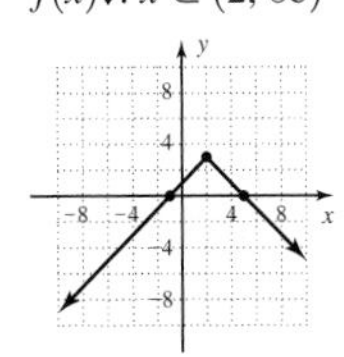

49. a. $D: x \in (-\infty, -2) \cup (-2, \infty)$, $R: y \in (-\infty, -3) \cup (-3, \infty)$ **b.** $-4, -2, -2.\bar{6}, -2.75, -2.8$ **c.** $(\frac{-5}{3}, 0)$ **d.** $f(x) < 0: x \in (-\infty, -2) \cup (\frac{-5}{3}, \infty)$, $f(x) > 0: x \in (-2, \frac{-5}{3})$ **e.** none **f.** $f(x)\uparrow$: none, $f(x)\downarrow: x \in (-\infty, -2) \cup (-2, \infty)$

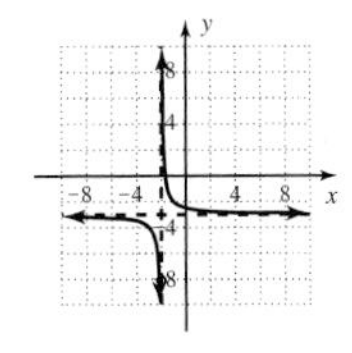

50. $D: x \in (-\infty, \infty)$, $R: y \in [-5, \infty)$, $f(x)\uparrow: x \in (2, \infty)$, $f(x)\downarrow: x \in (-\infty, 2)$, $f(x) > 0: x \in (-\infty, -1) \cup (5, \infty)$, $f(x) < 0: x \in (-1, 5)$ **51.** $D: x \in [-3, \infty)$, $R: y \in (-\infty, 0)$, $f(x)\uparrow$: none, $f(x)\downarrow: x \in (-3, \infty)$, $f(x) > 0$: none, $f(x) < 0: x \in (-3, \infty)$ **52.** $D: x \in (-\infty, \infty)$, $R: y \in (-\infty, \infty)$, $f(x)\uparrow: x \in (-\infty, -3) \cup (1, \infty)$, $f(x)\downarrow: x \in (-3, 1)$, $f(x) > 0: x \in (-5, -1) \cup (4, \infty)$, $f(x) < 0: x \in (-\infty, -5) \cup (-1, 4)$

53. zeroes: $(-6, 0)$, $(0, 0)$, $(6, 0)$, $(9, 0)$, min: $(-3, -8)$, $(7.5, -2)$, max: $(-6, 0)$, $(3, 4)$

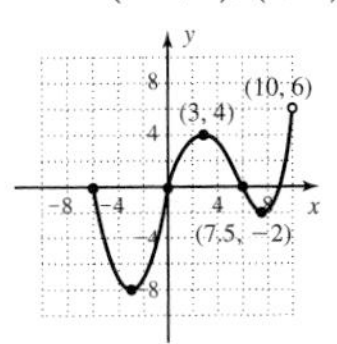

Mixed Review, pp. 365–366

1. $k = 5.4$; $y = \dfrac{5.4}{x^2}$

x	y
1	5.4
3	0.6
10	0.054

3. $\frac{-10}{3}$ **5.** $\frac{9}{8}$ **7.** $\{x|x \in \mathbb{R}, x \neq 1 \pm \sqrt{3}\}$ **9.** $f(x) = \sqrt[3]{x}$; $g(x) = x^2 + 5$ (others possible)
11. cube root family

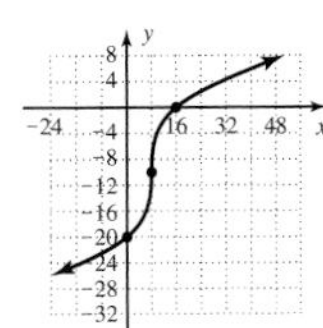

13. $D: x \in (-\infty, \infty)$; $R: y \in (-\infty, 7]$; $f(x)\uparrow: x \in (-\infty, -3) \cup (0, 3)$; $f(x)\downarrow: x \in (-3, 0) \cup (3, \infty)$; $f(x)$ constant: none
$f(x) > 0: x \in (-4, -1) \cup (1, 5)$;
$f(x) < 0: x \in (-\infty, -4) \cup (-1, 1) \cup (5, \infty)$; max: $(-3, 4)$ and $(3, 7)$; min: $(0, -2)$
15.

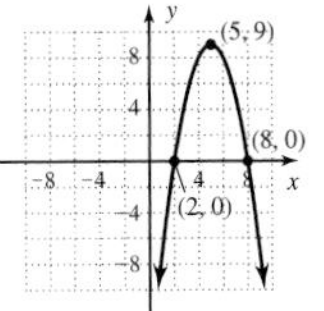

17. zeroes: (2, 0), (10, 0); max: (15, 10), min: (5, −10)

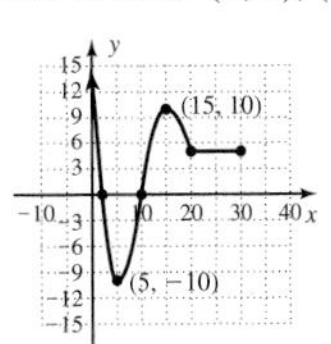

19. $f(x) = -2x^2 + x + 3$

Practice Test, pp. 366–367

1. 42 **2.** $4a^2 + 4a - 2$ **3.** $x \in [0, \sqrt{3}) \cup (\sqrt{3}, \infty)$ **4.** $f^{-1}(x) = \dfrac{x - 1}{2}$; $g^{-1}(x) = \sqrt{x + 3}$

5.

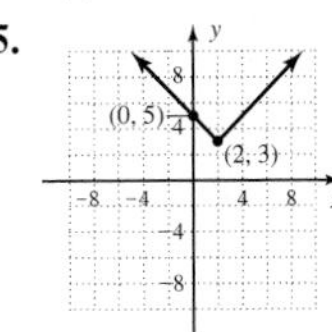

6.

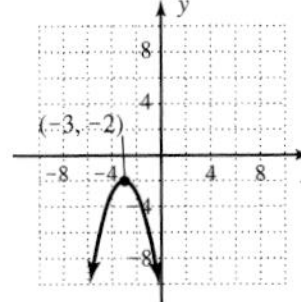

7.

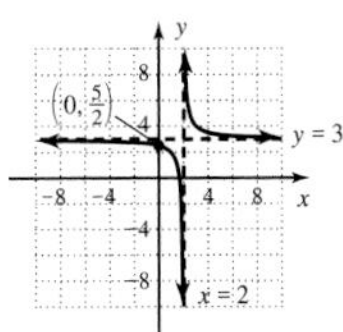

8.

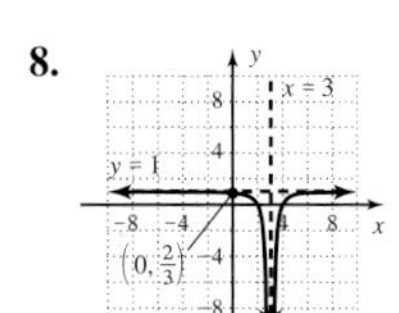

9.

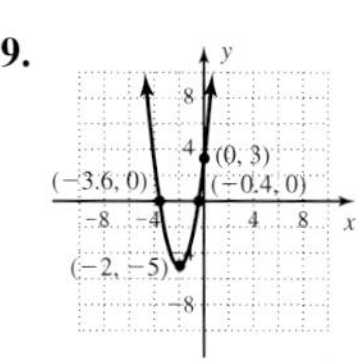

10.

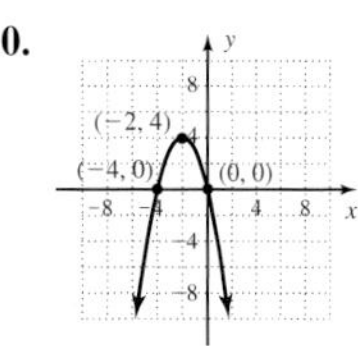

11. 1; −3; 1

12.

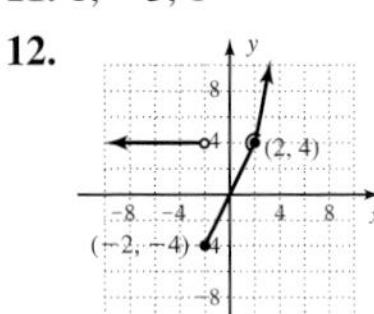

13. a. $D: x \in (-\infty, \infty)$; $R: y \in (-\infty, 4]$ **b.** $f(-1) = 4$ **c.** $(-4, 0)$ and $(2, 0)$ **d.** $f(x) < 0: x \in (-\infty, -4) \cup (2, \infty)$; $f(x) > 0: x \in (-4, 2)$ **e.** max: $(-1, 4)$; min: none **f.** $f(x)\uparrow: x \in (-\infty, -1)$; $f(x)\downarrow: x \in (-1, \infty)$ **g.** $f(x) = -\frac{4}{9}(x + 1)^2 + 4$ **14. a.** $D: x \in (-4, \infty)$; $R: y \in (-3, \infty)$ **b.** $f(-1) \approx 2.2$ **c.** $(-3, 0)$ **d.** $f(x) < 0: x \in (-4, -3)$; $f(x) > 0: x \in (-3, \infty)$ **e.** max: none; min $(-4, -3)$ **f.** $f(x)\uparrow: x \in (-3, \infty)$; $f(x)\downarrow$: none **g.** $f(x) = 3\sqrt{x + 4} - 3$

15. a. $V(t) = \frac{4}{3}\pi(\sqrt{t})^3$ **b.** 36π in^3

16. a.

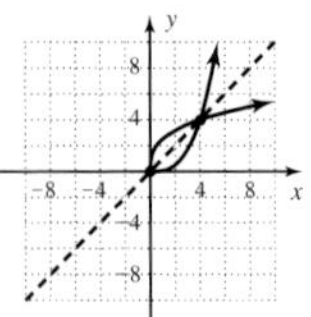

b.

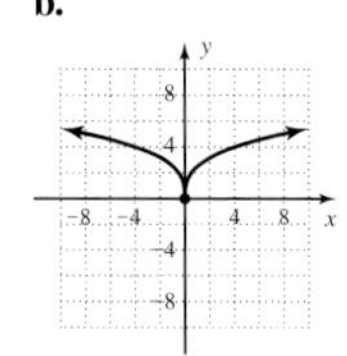

c.

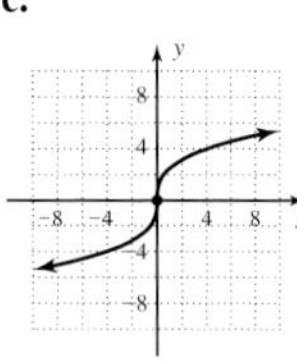

17. a. Yes, passes vertical line test. **b.** Yes, passes horizontal line test. **c.** odd **d.** $(6, 4), (-4, -2)$ **18. a.** $g(x)$ **b.** $g(x)$ **19. a.** 40 ft, 48 ft **b.** 49 ft **c.** 14 sec **20.** 520 lb

Strengthening Core Skills, pp. 369–371

1.

2.

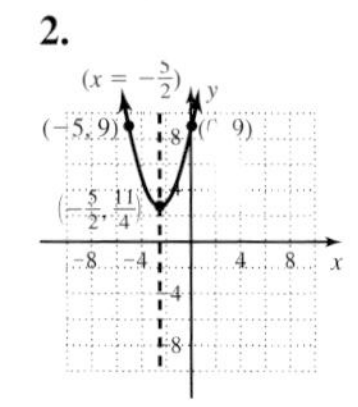

3.

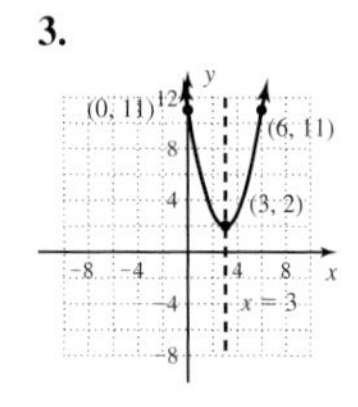

4.

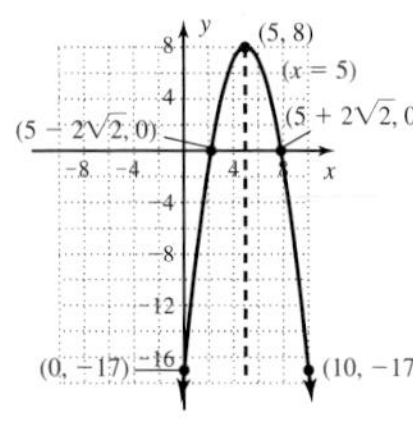

5.

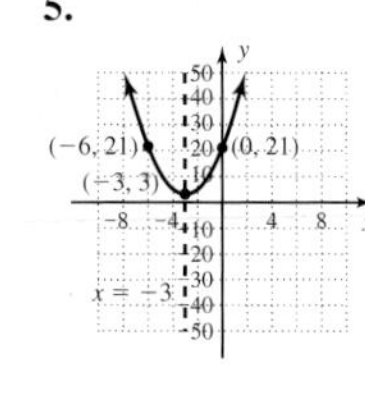

6.

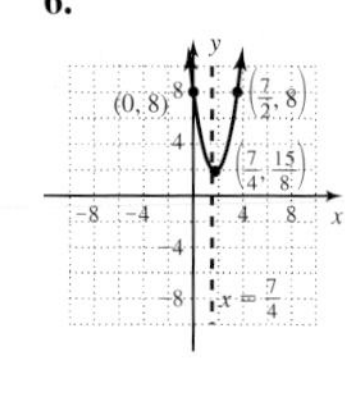

Cumulative Review, pp. 371–372

1. $x^2 + 2$ **3.** 29.45 cm **5.** $x = \dfrac{-7 \pm \sqrt{89}}{4}$ **7. a.** $\dfrac{-1}{3}$ **b.** $\dfrac{3}{5}$

9. $y = \frac{1}{2}x + \frac{7}{2}$

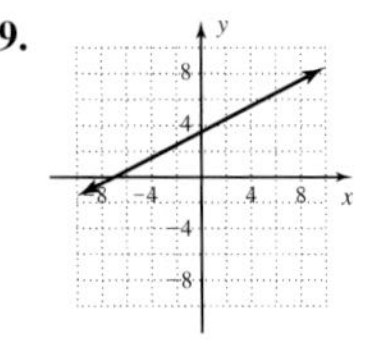

11.

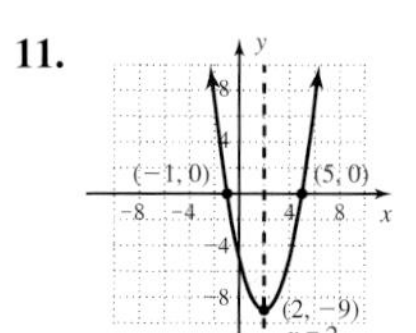

13. $(f \cdot g)(x) = 3x^3 - 12x^2 + 12x$; $\left(\dfrac{f}{g}\right)(x) = 3x, x \neq 2$; $(g \circ f)(-2) = 22$ **15. a.** $(0, \frac{1}{2})$, $(1, 0)$

b.

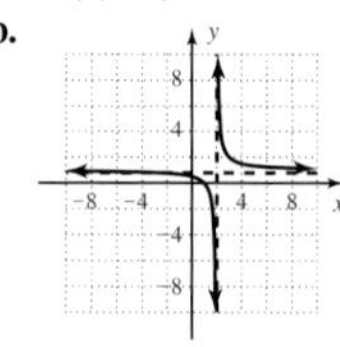

17. a. $f(x)$ **b.** $g(x)$ **19.** $y = 0.42x + 0.81$, about 43 ppsi

CHAPTER 4

Exercises 4.1, pp. 380–383

1. $x^3 + 2x^2 + 3x - 4$ **3.** $x^2 + 3x + 6$
5. $x^3 + 2x^2 + 3x - 4 = (x^2 + 3x + 6)(x - 1) + 2$ **7.** $x - 7$
9. $2r + 3$ **11.** $x^2 - 3x - 10$ **13.** $4n^2 - 2n + 1$ **15.** $3b^2 - 3b - 2$
17. (1) $9b^2 - 24b + 16 = (3b - 4)(3b - 4)$
(2) $\dfrac{9b^2 - 24b + 16}{3b - 4} = 3b - 4$
19. (1) $2n^3 - n^2 - 19n + 4 = (2n^2 - 7n + 2)(n + 3) - 2$
(2) $\dfrac{2n^3 - n^2 - 19n + 4}{n + 3} = 2n^2 - 7n + 2 + \dfrac{-2}{n + 3}$
21. (1) $g^4 - 15g^2 + 10g + 24 = (g^3 - 4g^2 + g + 6)(g + 4)$
(2) $\dfrac{g^4 - 15g^2 + 10g + 24}{g + 4} = g^3 - 4g^2 + g + 6$
23. (1) $(x^4 - 16x^2 - 5x - 24) = (x^3 - 4x^2 - 5)(x + 4) - 4$
(2) $\dfrac{x^4 - 16x^2 - 5x - 24}{x + 4} = x^3 - 4x^2 - 5 - \dfrac{4}{x + 4}$

25. $\begin{array}{r|rrrr} -2 & 1 & 5 & -1 & -14 \\ & & -2 & -6 & 14 \\ \hline & 1 & 3 & -7 & 0 \end{array}$ **27.** $\begin{array}{r|rrrr} -7 & 1 & 12 & 34 & -7 \\ & & -7 & -35 & 7 \\ \hline & 1 & 5 & -1 & 0 \end{array}$

29. $x^3 + 3x^2 - 8x - 13 = (x^2 + 2x - 10)(x + 1) - 3$; $x^2 + 2x - 10 + \frac{-3}{x + 1}$ **31.** $x^3 - 15x + 12 = (x^2 + 3x - 6)(x - 3) - 6$; $x^2 + 3x - 6 + \frac{-6}{x - 3}$ **33.** $(x - 9)(x + 7)$ **35.** $(x + 2)(x - 1)(x + 3)$
37. $(x - 1)(x - 2)(x + 6)$ **39.** $(x + 3)(x - 1)(x - 2)$
41. $(x + 2)(x - 3)(x - 6)$ **43.** $(x + 5)(x + 4)(x - 6)$
45. $(x + 2)(x + 1)(x - 1)(x - 3)$ **47.** $B = \dfrac{2A}{h} - \dfrac{\cancel{h}b}{\cancel{h}} = \dfrac{2A}{h} - b$
49. $k = -6$ **51.** $k = -16$ **53.** $k = -1$ **55.** $k = -8$
57. a. $5x + 18$ **b.** $x = 18$ **c.** 60×108 **59.** Answers will vary.
61. $k = \frac{1}{6}$ **63.** $k = 2$ **65.** 4×10^8 **67.** $x = 6$, $x = 2$ **69.** $x \in [\frac{-3}{2}, \infty)$

Exercises 4.2, pp. 389–393

1. linear, $P(k)$, remainder **3.** $a - bi$, complex conjugate **5.** Answers will vary.

7. $\begin{array}{r|rrrr} -3 & 1 & 2 & -5 & -6 \\ & & -3 & 3 & 6 \\ \hline & 1 & -1 & -2 & 0 \end{array}$ **9.** $\begin{array}{r|rrrr} 2 & 1 & 0 & -7 & 6 \\ & & 2 & 4 & -6 \\ \hline & 1 & 2 & -3 & 0 \end{array}$

11. $\begin{array}{r|rrrr} 4 & 1 & -6 & 0 & 32 \\ & & 4 & -8 & -32 \\ \hline & 1 & -2 & -8 & 0 \end{array}$

13. a. -30 **b.** 12 **15. a.** -2 **b.** -22 **17. a.** -1 **b.** 3
19. a. -5 **b.** 0 **21. a.** -10 **b.** 0 **23. a.** $\frac{-7}{8}$ **b.** $\frac{-32}{27}$
25. $P(x) = (x + 2)(x - 3)(x + 5), P(x) = x^3 + 4x^2 - 11x - 30$
27. $P(x) = (x + 2)(x - \sqrt{3})(x + \sqrt{3}), P(x) = x^3 + 2x^2 - 3x - 6$
29. $P(x) = (x + 5)(x - 2\sqrt{3})(x + 2\sqrt{3}), P(x) = x^3 + 5x^2 - 12x - 60$
31. $P(x) = (x - 1)(x + 2)(x - \sqrt{10})(x + \sqrt{10}),$
$P(x) = x^4 + x^3 - 12x^2 - 10x + 20$ **33.** $P(3i) = 0$ **35.** $P(-2i) = 0$
37. $P(1 + 2i) = 0$ **39.** $P(x) = x^3 - x^2 + 8x + 10$
41. $P(x) = x^4 - 2x^3 + 2x^2 + 6x - 15$
43. $P(x) = x^4 - 4x^2 + 12x - 9$ **45.** $P(x) = x^4 - 4x^3 + 2x^2 + 4x - 8$
47. $P(x) = x^3 - 4x^2 + 9x - 36$ **49.** $P(x) = x^3 - 9x^2 + 33x - 65$
51. $P(x) = x^3 - 8x^2 + 23x - 22$ **53.** $P(x) = x^4 + 7x^2 + 12$
55. $P(x) = x^4 - x^3 - 3x^2 + 17x - 30$
57. $P(x) = x^4 - 6x^3 + 15x^2 - 26x + 6$ **59.** $x = -3$; multiplicity one; $x = 3$; multiplicity two; degree 3 **61.** $x = 2$; multiplicity three; degree 3
63. $x = 3$; multiplicity three; $x = -3$; multiplicity one; degree 4
65. $x = -4$; multiplicity two; $x = 3$; multiplicity two; $x = -3$; multiplicity one; degree 5 **67.** 4-in. squares; 16 in. × 10 in. × 4 in.
69. $P(x) = x^3 - 2x^2 + 9x - 18$ **71.** $P(x) = x^4 - 4x^3 + 13x^2 - 36x + 36$ **73.** $P(x) = x^3 - 3x^2 + 7x - 5$
75. $P(x) = (x - 1)^3(x - 1 - 2i)(x - 1 + 2i)$
77. $P(x) = (x + 3)^2(x - 1 - \sqrt{2})(x - 1 + \sqrt{2})(x - 5i)(x + 5i)$
79.

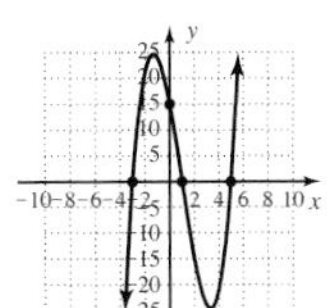

81.

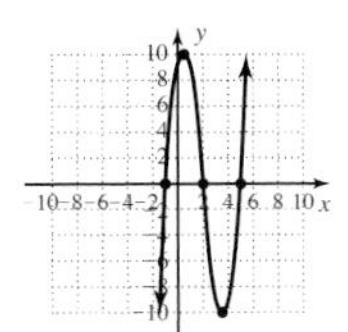

83.

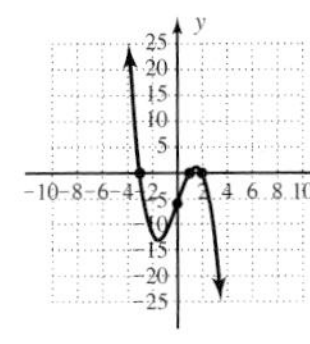

85. a. week 10, 22.5 thousand **b.** one week before closing, 36 thousand **c.** week 9 **87. a.** $P(x) = x^2 - 4x + 13$ **b.** $P(x) = x^2 - 2x + 3$
89. $k = 22$ **91.** $S_3 = 36$; $S_5 = 225$ **93.**

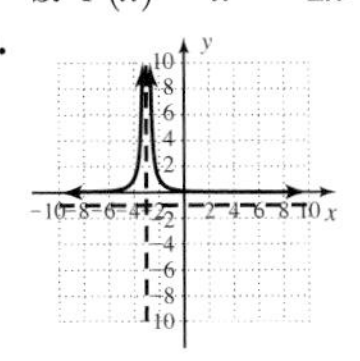

95. $x = 0, x = \frac{3}{2}, x = -1$ **97.** -1.9

Exercises 4.3, pp. 402–406

1. coefficients **3.** coefficients, sum, 0, root **5.** b; 4 is not a factor of 6
7. $C(x) = (x - 4i)(x + 3)(x - 2), x = -3, x = 2$
9. $C(x) = (x - 3i)(x - 1 - 2i)(x - 1 + 2i), x = 1 \pm 2i$
11. $C(x) = (x - 6i)(x - 1 - \sqrt{3}\,i)(x - 1 + \sqrt{3}\,i), x = 1 \pm \sqrt{3}\,i$
13. $C(x) = (x - 2 + i)(x - 3i)(x + i), x = 3i, x = -i$
15. a. yes **b.** no **c.** yes
17. a. no **b.** yes **c.** yes
19. $\frac{\{\pm1, \pm15, \pm3, \pm5\}}{\{\pm1, \pm4, \pm2\}}$
21. $\frac{\{\pm1, \pm15, \pm3, \pm5\}}{\{\pm1, \pm2\}}$ **23.** $\frac{\{\pm1, \pm28, \pm2, \pm14, \pm4, \pm7\}}{\{\pm1, \pm6, \pm2, \pm3\}}$
25. $\frac{\{\pm1, \pm3\}}{\{\pm1, \pm32, \pm2, \pm16, \pm4, \pm8\}}$
27. $(x + 4)(x - 1)(x - 3), x = -4, 1, 3$
29. $(x + 3)(x + 2)(x - 5), x = -3, -2, 5$
31. $(x + 3)(x - 1)(x - 4), x = -3, 1, 4$
33. $(x + 2)(x - 3)(x - 5), x = -2, 3, 5$
35. $(x + 4)(x + 1)(x - 2)(x - 3), x = -4, -1, 2, 3$
37. $(x + 7)(x + 2)(x + 1)(x - 3), x = -7, -2, -1, 3$
39. $(2x + 3)(2x - 1)(x - 1); x = -\frac{3}{2}, \frac{1}{2}, 1$
41. $(2x + 3)^2(x - 1); x = -\frac{3}{2}, 1$
43. $(x + 2)(x - 1)(2x - 5); x = -2, 1, \frac{5}{2}$
45. $(x + 1)(2x + 1)(x - \sqrt{5})(x + \sqrt{5}); x = -1, -\frac{1}{2}, \sqrt{5}, -\sqrt{5}$
47. $(x - 1)(3x - 2)(x - 2i)(x + 2i); x = 1, \frac{2}{3}, -2i, -2i$
49. $x = 1, 2, 3, \frac{-3}{2}$ **51.** $x = -2, 1, \frac{-2}{3}$ **53.** $x = -2, \frac{-3}{2}, 1, 4$
55. $x = -1, 2, 3, \frac{5}{3}$ **57.** $x = 1, 2, -3, \pm\sqrt{7}i$ **59.** $x = -2, \frac{2}{3}, 1, \pm\sqrt{3}\,i$
61. $x = 1, 2, 4, -2$ **63.** $x = -3, 1, \pm\sqrt{2}$ **65.** $x = -1, \frac{3}{2}, \pm\sqrt{3}\,i$
67. $x = \frac{1}{2}, 1, 2, \pm\sqrt{3}\,i$ **69.** possible roots: $\{\pm1, \pm8, \pm2, \pm4\}$; neither -1 nor 1 is a root; 3 or 1 positive roots, 1 negative root; roots must lie between -2 and 2 **71.** possible roots: $\{\pm1, \pm2\}$; -1 is a root; 2 or 0 positive roots, 3 or 1 negative roots; roots must lie between -3 and 2 **73.** possible roots: $\{\pm1, \pm12, \pm2, \pm6, \pm3, \pm4\}$; $x = 1$ and $x = -1$ are roots; 4, 2, or 0 positive roots, 1 negative root; roots must lie between -1 and 4 **75.** possible roots: $\frac{\{\pm1, \pm20, \pm2, \pm10, \pm4, \pm5\}}{\{\pm1, \pm2\}}$; $x = 1$ is a root; 1 positive root, 1 negative root; roots must lie between -2 and 1
77. $(x - 4)(2x - 3)(2x + 3); x = 4, \frac{3}{2}, -\frac{3}{2}$
79. $(2x + 1)(3x - 2)(x - 12); x = -\frac{1}{2}, \frac{2}{3}, 12$
81. $(x - 2)(2x - 1)(2x + 1)(x + 12); x = 2, \frac{1}{2}, -\frac{1}{2}, -12$
83. $x = -1, \frac{-1}{3}, \frac{2}{3}$ **85.** $x = -2, -1, 1, 5$
87. $x = -5, -2, \frac{3}{2}$ **89.** $x = 3, 7, \frac{1}{4}$ **91.** $x = -2$, 1 multiplicity 2, $\frac{-2}{3}$
93. $t = 1, \frac{3}{4}, \frac{-1}{8}$ **95. a.** 5 **b.** 13 **c.** 2 **97.** yes **99.** yes
101. a. 4 cm × 4 cm × 4 cm **b.** 5 cm × 5 cm × 5 cm
103. length 10 in., width 5 in., height 3 in. **105.** 1994, 1998, 2002, 5 yr
107. $x = -4, 1, 3, \pm i$ **109. a.** $w = 1, 1 \pm \sqrt{3}\,i$ **b.** $w = 2, 2 \pm 3i$
111. 2, 1, 4

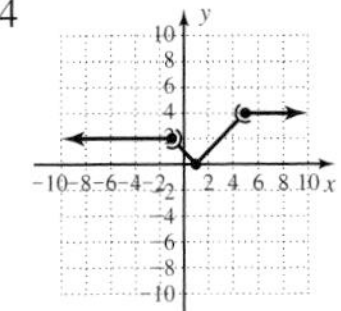

113. a. $f(x) \geq 0$: $x \in [-4, -1] \cup [1, \infty)$ **b.** max: $(-3, 4)$; min: $(0, -2)$
c. $f(x)\uparrow$: $x \in (-\infty, -3) \cup (0, \infty)$, $f(x)\downarrow$: $x \in (-3, 0)$
115. $\{(22, -3), (13, -2), (6, -1), (1, 0), (-2, 1), (-3, 2)\}$;
D: $\{22, 13, 6, 1, -2, -3\}$; R: $\{-3, -2, -1, 0, 1, 2\}$

Exercises 4.4, pp. 415–420

1. quartic **3.** bounce, flatter **5.** Answers will vary. **7.** up/down
9. down/down **11.** down/up; $(0, -2)$ **13.** down/down; $(0, -6)$
15. up/down; $(0, -6)$ **17.** c **19.** e **21.** f
23. degree 6; up/up; $(0, -12)$ **25.** degree 5; up/down; $(0, -24)$
27. degree 6; up/up; $(0, -192)$ **29.** degree 5; up/down; $(0, 2)$
31. a. even **b.** -3 odd, -1 even, 3 odd **c.** $f(x) = (x + 3)(x + 1)^2(x - 3)$ **33. a.** even **b.** -3 odd, -1 odd, 2 odd, 4 odd
c. $f(x) = (x + 3)(x + 1)(x - 2)(x - 4)$ **35. a.** odd **b.** -1 even, 3 odd
c. $f(x) = -(x + 1)^2(x - 3)$ **37.** b **39.** e **41.** c

43.

45.

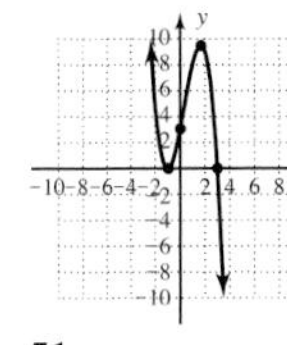

47.

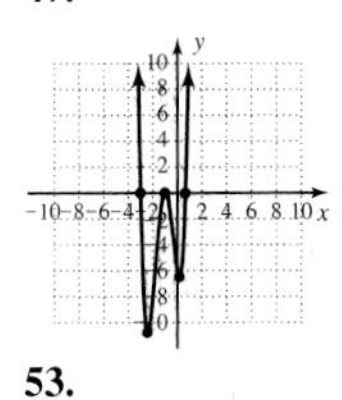

49.

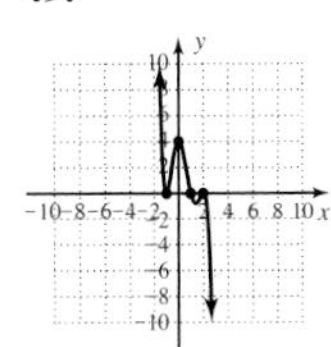

51.

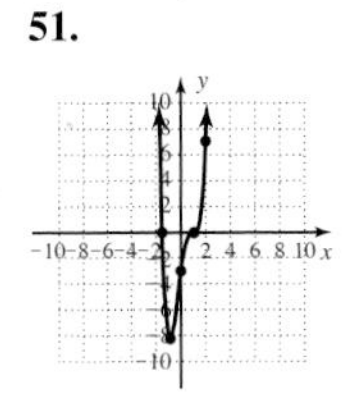

53.

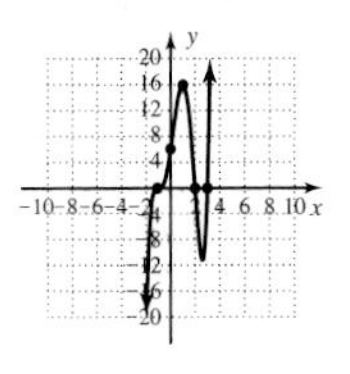

55.

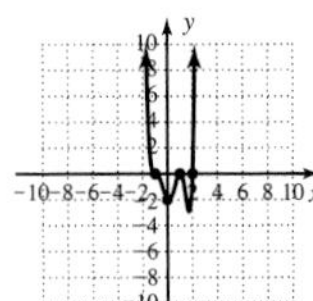
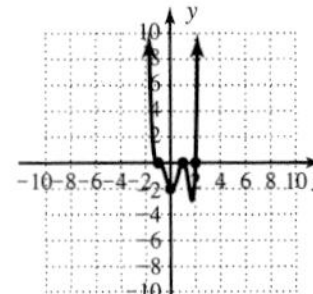

57. $P(x) = x^4 - 2x^3 - 13x^2 + 14x + 24$

59.

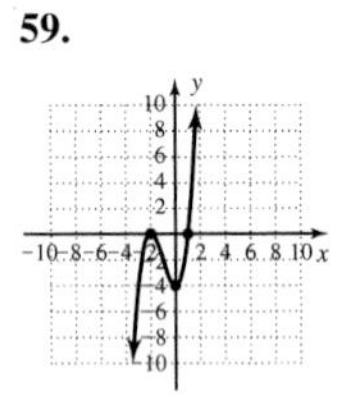

61.

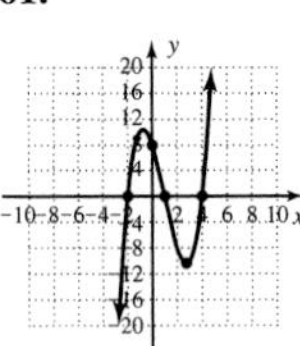

63.

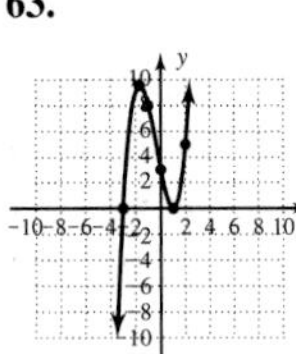

65.

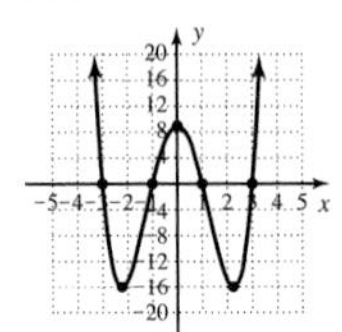

67.

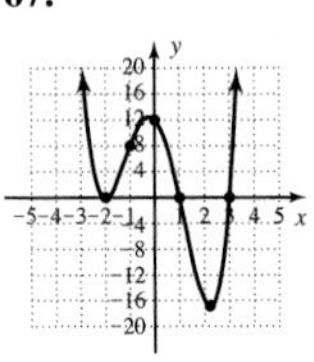

69.

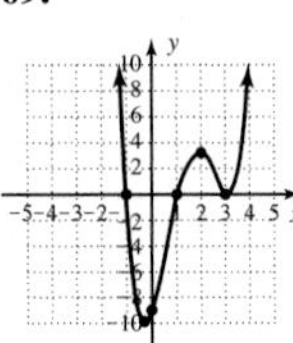

71.

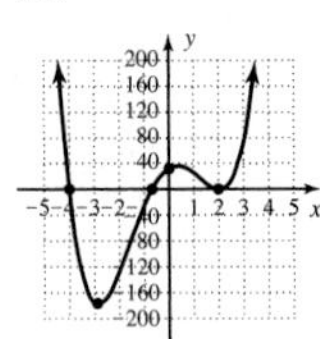

73.

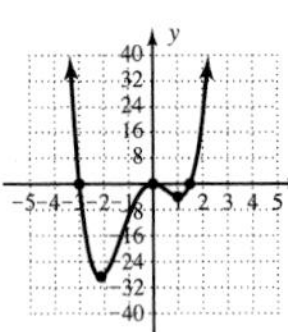

75.

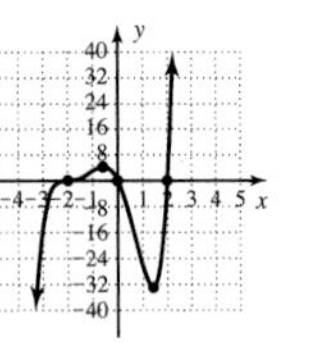

77.

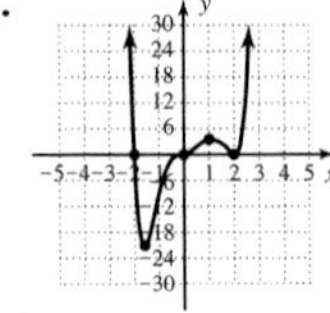

79. $h(x) = (x + 4)(x - \sqrt{3})(x + \sqrt{3})(x - \sqrt{3}i)(x + \sqrt{3}i)$
81. $f(x) = (x + \frac{5}{2})(x - \sqrt{2})(x + \sqrt{2})(x - \sqrt{3})(x + \sqrt{3})$
83. $P(x) = \frac{1}{6}(x + 4)(x - 1)(x - 3)$
85. a. 3 **b.** 5 **c.** $B(x) = \frac{1}{4}x(x - 4)(x - 9)$, −\$80,000
87. End behavior precludes extended use.
89. a. $f(x) \to \infty$, $f(x) \to -\infty$ **b.** $g(x) \to \infty$, $g(x) \to \infty$; $x^4 \geq 0$ for all x
91. verified **93.** $h(x) = \dfrac{1 - 2x}{x^2}$; D: $x \in \{x \mid x \neq 0\}$, R: $y \in \{y \geq -1\}$,
$H(x) = \dfrac{1}{x^2 - 2x}$, D: $x \in \{x \mid x \neq 0, x \neq 2\}$, R: $y \in \{y \mid y \neq 0\}$ **95.** yes
97. $\approx 2827.43\ \text{m}^2$

Mid-Chapter Check, p. 420

1. (1) $x^3 + 8x^2 + 7x - 14 = (x^2 + 6x - 5)(x + 2) - 4$
(2) $\dfrac{x^3 + 8x^2 + 7x - 14}{x + 2} = x^2 + 6x - 5 - \dfrac{4}{x + 2}$
2. $f(x) = (2x + 3)(x + 1)(x - 1)(x - 2)$ **3.** $f(-2) = 7$
4. $f(x) = x^3 - 2x + 4$ **5.** $g(2) = -8$ and $g(3) = 5$ have opposite signs
6. $f(x) = (x - 2)(x + 1)(x + 2)(x + 4)$ **7.** $x = -2, x = 1, x = -1 \pm 3i$

8.

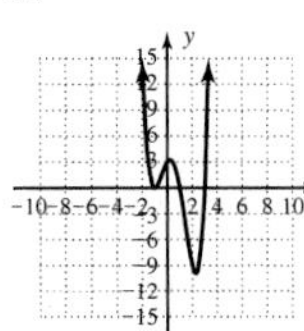

9.

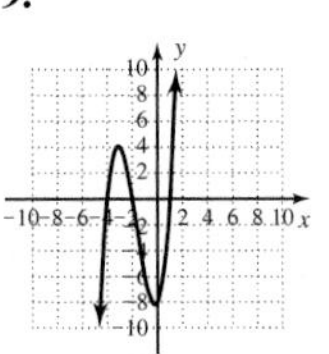

10. a. degree 4; three turning points **b.** 2 sec
c. $A(t) = t^4 - 10t^3 + 32t^2 - 38t + 15$, $A(2) = 3$, altitude is 300 ft above hard-deck; $A(4) = -9$, altitude is 900 ft below hard-deck

Reinforcing Basic Concepts, p. 421

1. 1.532 **2.** −2.152, 1.765

Exercises 4.5, pp. 431–437

1. all real, zeroes **3.** denominator, numerator **5.** $x = 15$ **7.** $x = 3$, $\{x \mid x \in \mathbb{R}, x \neq 3\}$ **9.** $x = 3$, $x = -3$, $\{x \mid x \in \mathbb{R}, x \neq 3, x \neq -3\}$
11. $x = \frac{-5}{2}$, $x = 1$, $\{x \mid x \in \mathbb{R}, x \neq \frac{-5}{2}, x \neq 1\}$ **13.** No V.A., $\{x \mid x \in \mathbb{R}\}$
15. $x = 3$, yes; $x = -2$, yes **17.** $x = 3$, no **19.** $x = 2$, yes; $x = -2$, no **21.** (0, 0) cut, (3, 0) cut **23.** (−4, 0) cut, (0, 4)
25. (0, 0) cut, (3, 0) bounce **27.** $y = 0$, crosses at $(\frac{3}{2}, 0)$
29. $y = 4$, crosses at $(-\frac{21}{4}, 4)$ **31.** $y = 3$

33. $\dfrac{2}{x - 1} + 3$

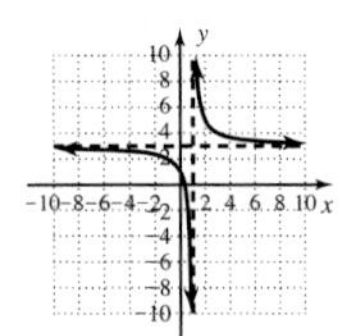

35. $\dfrac{2}{x + 3} + 1$

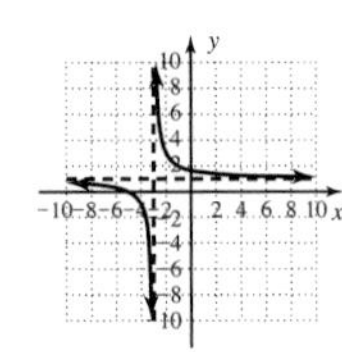

37. $\dfrac{-1}{(x - 3)^2} + 1$

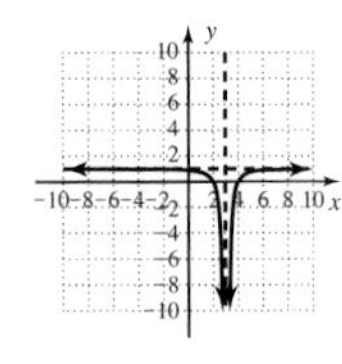

39.

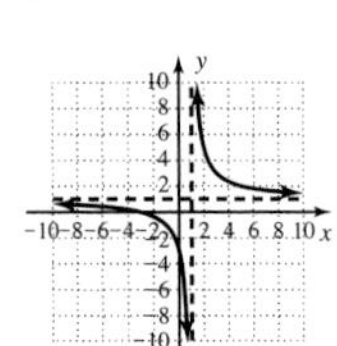

41.

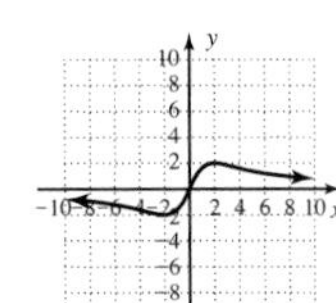

43.

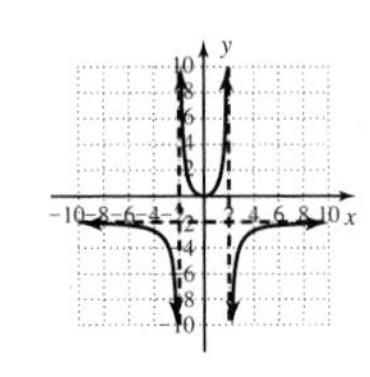

45.

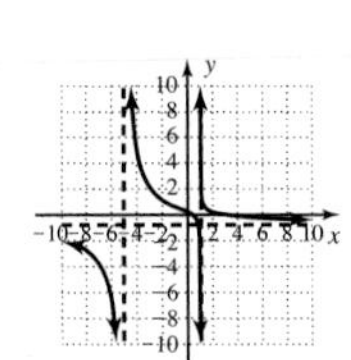

47.

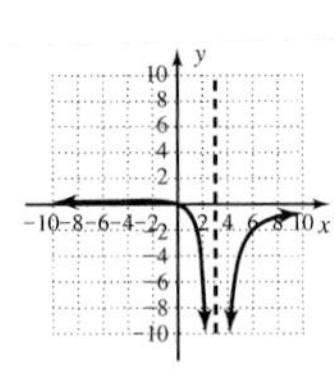

49.

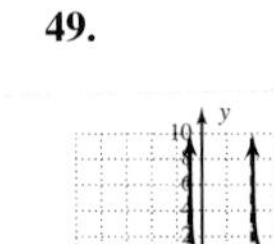

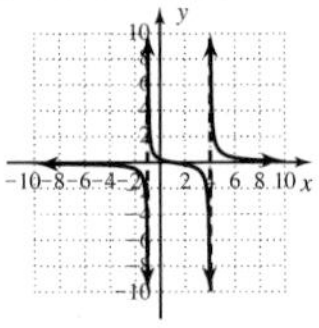

51.

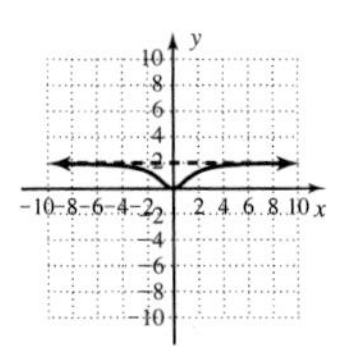

53.

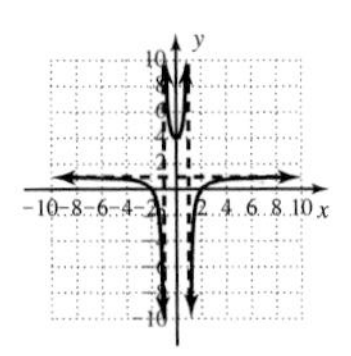

55.

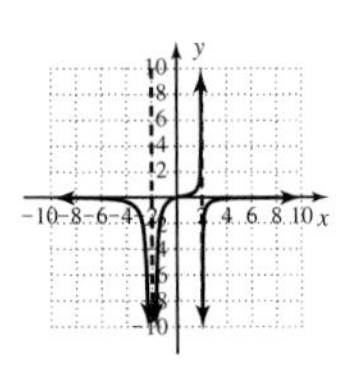

57. $f(x) = \dfrac{(x-4)(x+1)}{(x+2)(x-3)}$ **59.** $f(x) = \dfrac{x^2-4}{9-x^2}$

61. a. population density approaches zero far from town
b. 10 mi, 20 mi **c.** 4.5 mi, 704

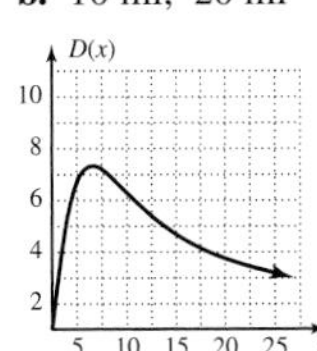

63. a. 2; 10 **b.** 10; 20
c. on average, 6 words will be remembered for life

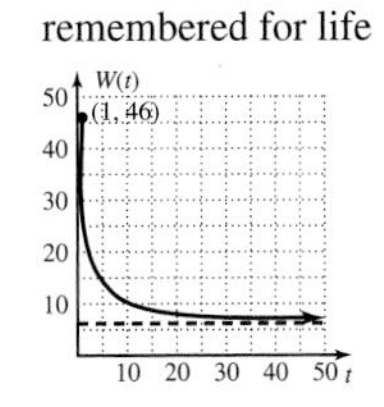

65. a.

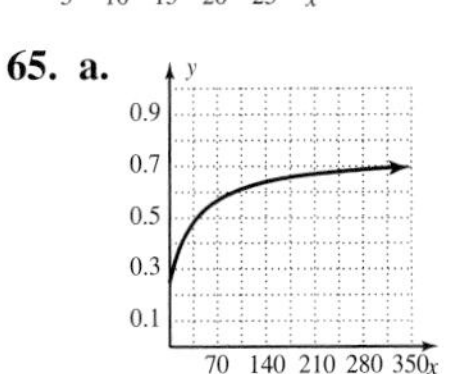

b. 35%; 62.5%; 160 gal
c. 160 gal; 200 gal
d. 70%; 75%

67. a. \$225; \$175 **b.** 2000 heaters **c.** 4000
d. The horizontal asymptote at $y = 125$ means the average cost approaches \$125 as monthly production gets very large. Due to limitations on production (maximum of 5000 heaters) the average cost will never fall below $A(5000) = 135$. **69. a.** 5 **b.** 18 **c.** The horizontal asymptote at $y = 95$ means her average grade will approach 95 as the number of tests taken increases; no **d.** 6 **71. a.** 16.0 28.7 65.8 277.8 **b.** 12.7, 37.1, 212.0 **c.** (A) 22.4, 40.2, 92.1, 388.9 (B) 17.8, 51.9, 296.8; answers will vary
73. $f(x) = \dfrac{(x-5)(x-2)(x+2)}{(x-5)(x-3)(x+3)}$; vertical asymptotes at $x = -3$,

$x = 3$, hole at $\left(5, \dfrac{21}{16}\right)$; $f(x) = \begin{cases} f(x) & x \neq 5 \\ \dfrac{21}{16} & x = 5 \end{cases}$

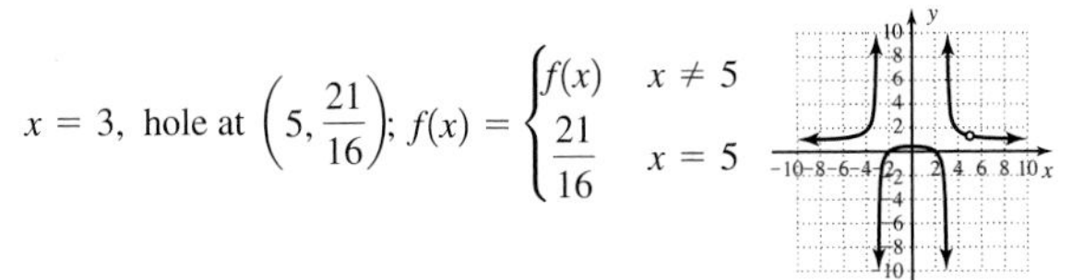

75. $\mathbb{C}: a + bi$, where $a, b \in \mathbb{R}$, $i = \sqrt{-1}$,
$\mathbb{Q}: \left\{\dfrac{a}{b} \middle| a \in \mathbb{Z}, b \in \mathbb{Z}; b \neq 0\right\}$, $\mathbb{Z}: \{\ldots -2, -1, 0, 1, 2 \ldots\}$

77. no, $f(x)$ is not one-to-one

79. 39, $\frac{3}{2}$, 1

Exercises 4.6, pp. 445–451

1. nonremovable **3.** two **5.** Answers will vary.

7. $F(x) = \begin{cases} \dfrac{x^2-4}{x+2} & x \neq -2 \\ -4 & x = -2 \end{cases}$

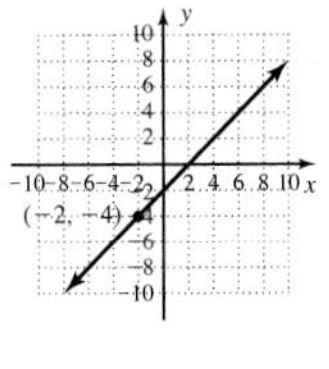

9. $G(x) = \begin{cases} \dfrac{x^2-2x-3}{x+1} & x \neq -1 \\ -4 & x = -1 \end{cases}$

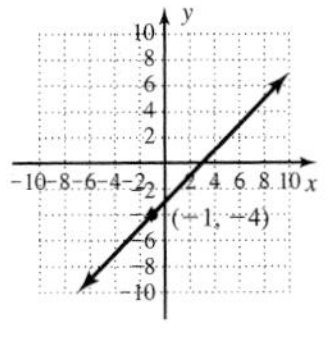

11. $H(x) = \begin{cases} \dfrac{3x-2x^2}{2x-3} & x \neq \dfrac{3}{2} \\ \dfrac{-3}{2} & x = \dfrac{3}{2} \end{cases}$

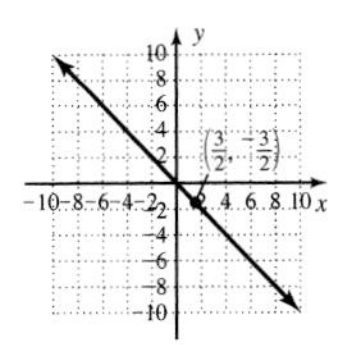

13. $p(x) = \begin{cases} \dfrac{x^3-8}{x-2} & x \neq 2 \\ 12 & x = 2 \end{cases}$

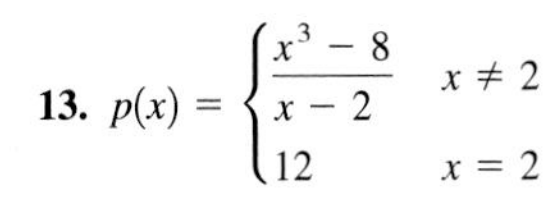

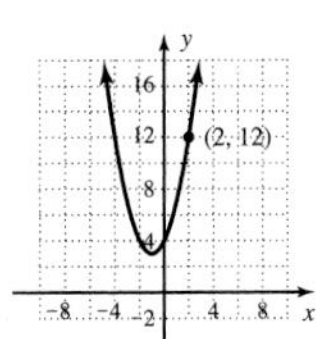

15. $Q(x) = \begin{cases} x^2 - x - 6 & x \neq -1 \\ -4 & x = -1 \end{cases}$

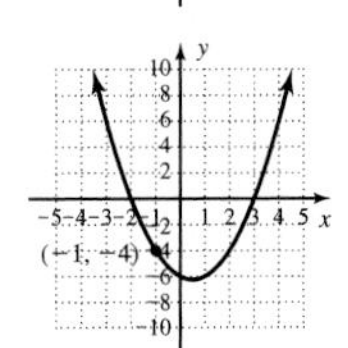

17. $r(x) = \begin{cases} \dfrac{x^3 + 3x^2 - x^{-3}}{x^2 + 2x - 3} & x \neq -3, x \neq 1 \\ -2 & x = -3 \\ 2 & x = 1 \end{cases}$

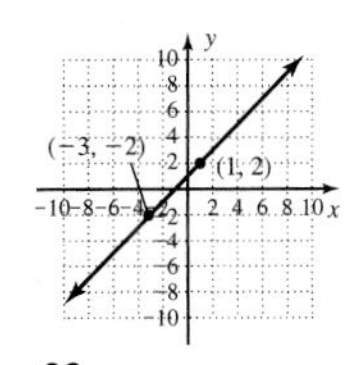

19.

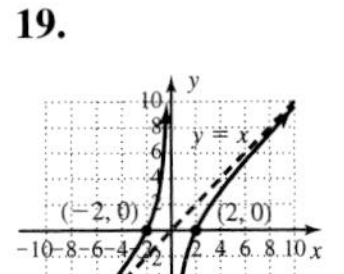

21.

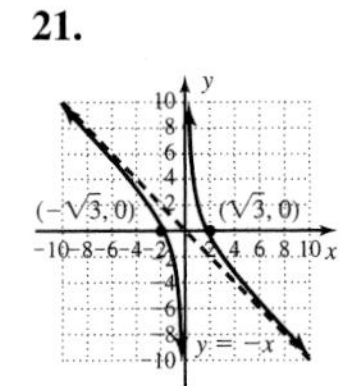

23.

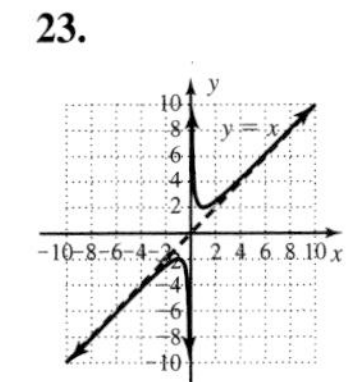

25.

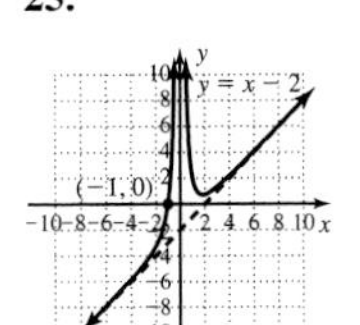

27.

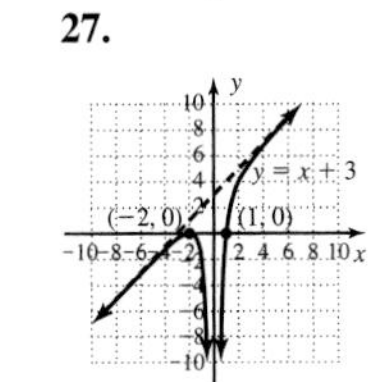

29.

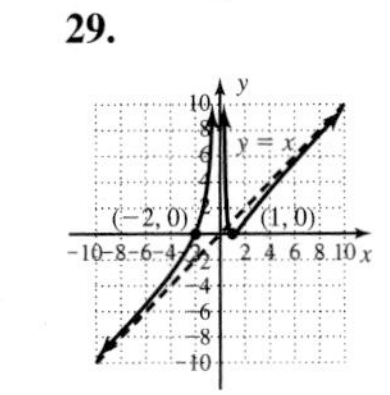

31.

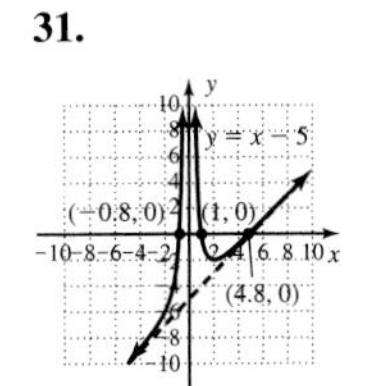

33.

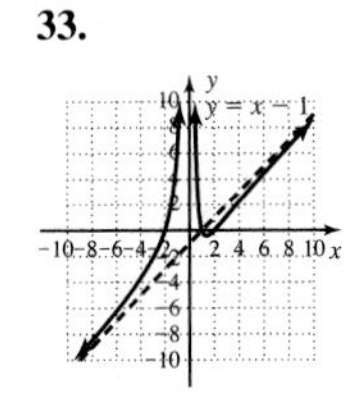

35.

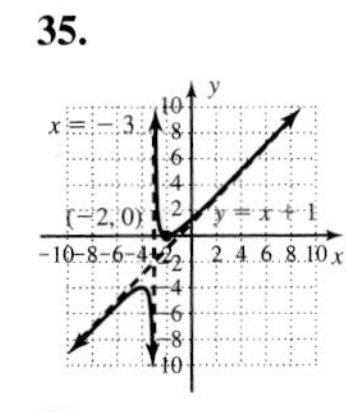

37.

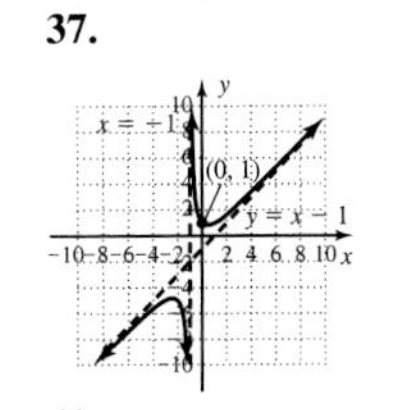

39.

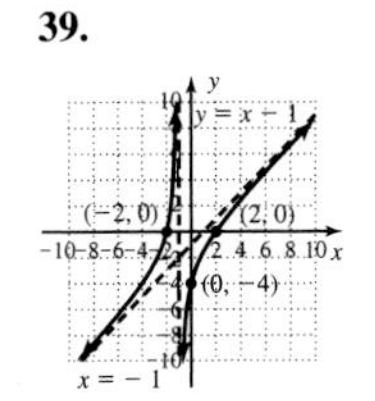

41.

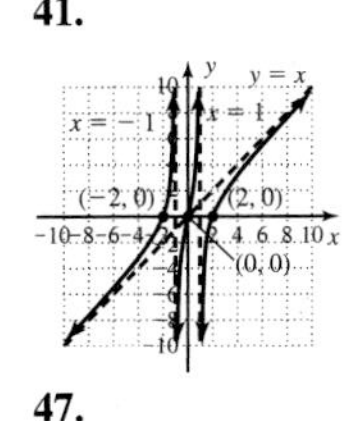

43.

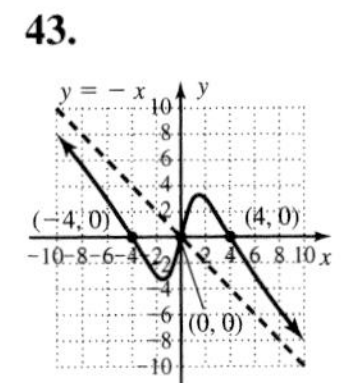

45.

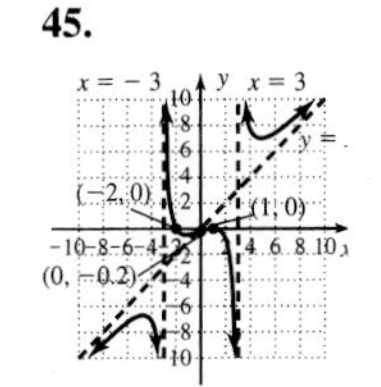

47.

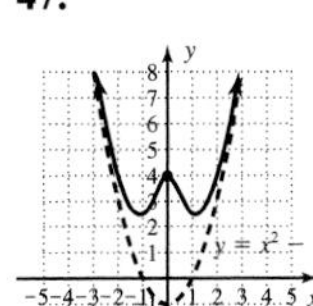

49.

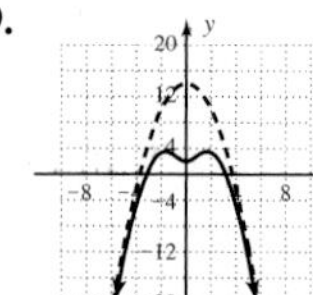

51. 119.1 **53. a.** $a = 5, y = 3a$ **b.** 60.5 **c.** 10

55. a. $A(x) = \dfrac{4x^2 + 53x + 250}{x}$, $x = 0$, $g(x) = 4x + 53$
b. cost: \$307, \$372, \$445, Avg. cost: \$307, \$186, \$148.33
c. 8, \$116.25 **d.**

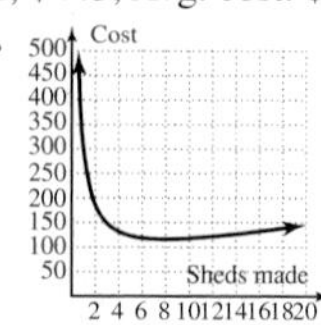

57. a. $S(x, y) = 2x^2 + 4xy$; $V(x, y) = x^2y$ **b.** $S(x) = \dfrac{2x^3 + 48}{x}$
c. $S(x)$ is asymptotic to $y = 2x^2$ **d.** $x = 2$ ft 3.5 in.; $y = 2$ ft 3.5 in.
59. a. $A(x, y) = xy$; $R(x, y) = (x - 2.5)(y - 2)$
b. $y = \dfrac{2x + 55}{x - 2.5}$, $A(x) = \dfrac{2x^2 + 55x}{x - 2.5}$
c. $A(x)$ is asymptotic to $y = 2x + 55$
d. $x \approx 11.16$ in.; $y = 8.93$ in.
61. a. $h = \dfrac{V}{\pi r^2}$ **b.** $S = 2\pi r^2 + \dfrac{2V}{r}$ **c.** $S = \dfrac{2\pi r^3 + 2V}{r}$
d. $r \approx 5.76$ cm; $h \approx 11.52$ cm; $S \approx 625.13$ cm^3 **63.** Answers will vary. **65. a.** $m = \dfrac{k}{h - a}$; $y = \dfrac{k(x - a)}{h - a}$ **b.** $\left(0, \dfrac{-ka}{h - a}\right)$, $(a, 0)$
c. $A = \dfrac{1}{2}\left(\dfrac{-ka^2}{h - a}\right)$ **d.** base $a = 2h$; height $y = 2k$; triangle is isosceles; $A = \frac{1}{2}(8)(8) = 32$ units^2; $A = \dfrac{1}{2}\left[\dfrac{4a^2}{a - 4}\right]$ has a minimum at (8, 32)
67. $x \in (-\infty, \frac{4 - \sqrt{10}}{2}) \cup (\frac{4 + \sqrt{10}}{2}, \infty)$; $x \in (-\infty, 2)$ **69.** $2 + i$
71. a. $b^2 - 4ac > 0$, with $b^2 - 4ac$ a perfect square **b.** $b^2 - 4ac > 0$, but not a perfect square **c.** $b^2 - 4ac = 0$ **d.** none **e.** none
f. $b^2 - 4ac < 0$

Exercises 4.7, pp. 458–464

1. vertical, multiplicity **3.** 0 **5.** Answers will vary. **7.** $x \in (-3, 5)$
9. $x \in [4, \infty)$ **11.** $x \in [-2, 4]$ **13.** $x \in (-2 - \sqrt{3}, -2 + \sqrt{3})$
15. $x \in (-\infty, -3] \cup \{1\}$ **17.** $x \in (-3, 1) \cup (2, \infty)$
19. $x \in (-\infty, -3) \cup (-1, 1) \cup (3, \infty)$
21. $x \in (-\infty, -2) \cup (-2, 1) \cup (3, \infty)$ **23.** $x \in [-1, 1] \cup \{3\}$
25. $x \in [-3, 2)$ **27.** $x \in (-\infty, -2) \cup (-2, -1)$
29. $x \in (-\infty, -2) \cup [2, 3)$ **31.** $x \in (-\infty, -5) \cup (0, 1) \cup (2, \infty)$
33. $x \in (-4, -2] \cup (1, 2] \cup (3, \infty)$ **35.** $x \in (-7, -3) \cup (2, \infty)$
37. $x \in (-\infty, -2] \cup (0, 2)$ **39.** $x \in (-\infty, -17) \cup (-2, 1) \cup (7, \infty)$
41. $x \in (-3, \frac{-7}{4}] \cup (2, \infty)$ **43.** $x \in (-2, \infty)$ **45.** $x \in (-1, \infty)$
47. $(-\infty, -3) \cup (3, \infty)$ **49.** b **51.** b **53. a.** verified
b. $D = -(p + 3)^2(p + \frac{3}{4})$, $p = -3, q = -2$; $p = \frac{-3}{4}, q = \frac{1}{4}$
c. $(-\infty, -3) \cup (-3, \frac{-3}{4})$ **d.** verified **55.** $d(x) = k(x^3 - 192x + 1024)$
a. $x \in (5, 8]$ **b.** 320 units **c.** $x \in [0, 3)$ **d.** 2 ft
57. $x \in [-3, 1] \cup [\frac{5}{2}, \infty)$ **59.** $x \in (-5, -2] \cup (7, \infty)$ **61. a.** verified
b. horizontal: $r_2 = 20$, as r_1 increases, r_2 decreases to maintain $R = 40$, vertical: $r_1 = 20$, as r_1 decreases, r_2 increases to maintain $R = 40$
c. $r_1 \in (20, 40)$ **63. a.** $(0°, 30°)$ **b.** $(20°, \infty)$ **c.** $(50°, \infty)$
65. a. $n \geq 4$ **b.** $n \leq 9$ **c.** 13 **67. a.** yes **b.** the method of this section
c. lose critical values **69.** $x(x + 2)(x - 1)^2 > 0$; $\dfrac{x(x + 2)}{(x - 1)^2} > 0$
71. $f'(x) > 0$ for $x \in (-2, 1) \cup (4, \infty)$
73.

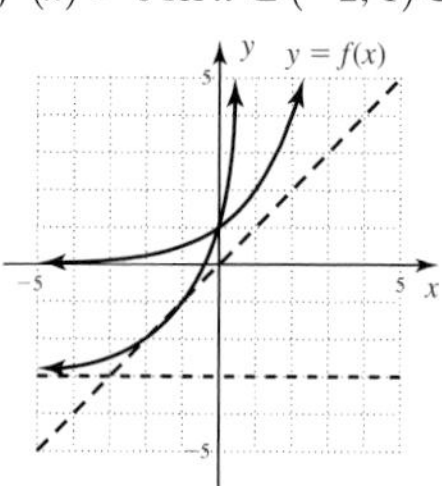

75.

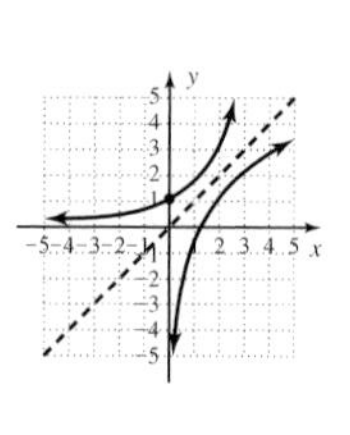

77. $x = 0$

Summary and Concept Review, pp. 464–468

1. $q(x) = x^2 + 6x + 7$; $R = 8$ **2.** $q(x) = x + 1$; $R = 3x - 4$

3.
$$\begin{array}{r|rrrrr} -7 & 2 & 13 & -6 & 9 & 14 \\ & & -14 & 7 & -7 & -14 \\ \hline & 2 & -1 & 1 & 2 & 0 \end{array}$$
Since $R = 0$, -7 is a root and $x + 7$ is a factor

4. $x^3 - 4x + 5 = (x - 2)(x^2 + 2x) + 5$;
$\dfrac{x^3 - 4x + 5}{x - 2} = x^2 + 2x + \dfrac{5}{x - 2}$ **5.** $(x + 4)(x + 1)(x - 3)$

6. $k = 16$ **7.**
$$\begin{array}{r|rrrr} \frac{1}{2} & 4 & 8 & -3 & -1 \\ & & 2 & 5 & 1 \\ \hline & 4 & 10 & 2 & 0 \end{array}$$
Since $R = 0$, $\frac{1}{2}$ is a root and $\left(x - \frac{1}{2}\right)$ is a factor.

8.
$$\begin{array}{r|rrrr} 3i & 1 & -2 & 9 & -18 \\ & & 3i & -9 - 6i & 18 \\ \hline & 1 & -2 + 3i & -6i & 0 \end{array}$$
Since $R = 0$, $3i$ is a root and $(x - 3i)$ is a factor.

9.
$$\begin{array}{r|rrrr} -7 & 1 & 9 & 13 & -10 \\ & & -7 & -14 & 7 \\ \hline & 1 & 2 & -1 & -3 \end{array}$$
$P(-7) = -3$

10. $P(x) = x^3 - x^2 - 5x + 5$
11. $C(x) = x^4 - 2x^3 + 5x^2 - 8x + 4$ **12. a.** $C(0) = 350$ customers, more at 2 P.M., 170 **b.** Busier at 1 P.M. $760 > 710$
13. $x = i, x = 4i$ **14.** Zeroes are in $[1, 2]$ and $[4, 5]$.
15. No, -2 is not since last row does not alternate in sign. Yes, -3 is a lower bound. **16.** $g(x)$ has one variation in sign $\longrightarrow$ 1 pos root; $g(-x)$ has three variations in sign $\longrightarrow$ 3 or 1 neg root

Pos	Neg	Complex
1	3	0
1	1	2

A grapher shows the second row is correct.

17. $P(x) = (x - 4)(x + 1)(2x + 3)$; $x = 4, -1, \frac{-3}{2}$ **18.** The possibilities are ± 1 and ± 3, none give a zero remainder. Therefore, h has no rational roots. **19.** degree 5; up/down; $(0, -4)$ **20.** degree 4; up/up; $(0, 8)$
21.

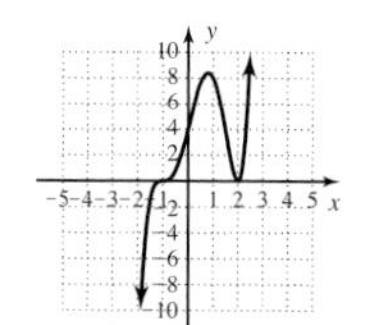

22.

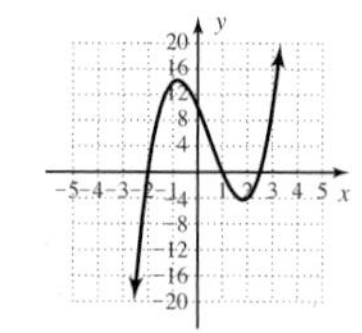

23.

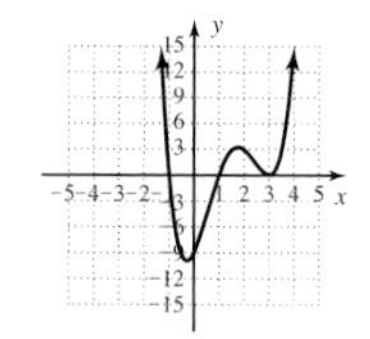

24. a. even **b.** $x = -2$, odd; $x = -1$, even; $x = 1$, odd
c. deg 6: $P(x) = (x + 2)(x + 1)^2(x - 1)^3$
25. a. $\{x | x \in \mathbb{R}; x \neq -1, 4\}$ **b.** HA: $y = 1$; VA: $x = -1, x = 4$
c. $r(0) = \frac{9}{4}$ (y-intercept); $x = -3, 3$ (x-intercepts) **d.** $r(1) = \frac{4}{3}$

26. $h(x) = 3 + \dfrac{1}{x - 2}$ **27.** **28.**

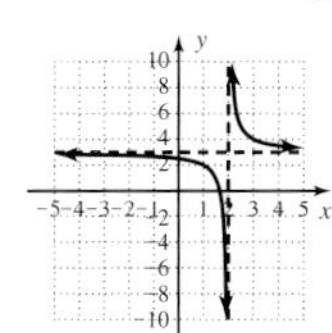

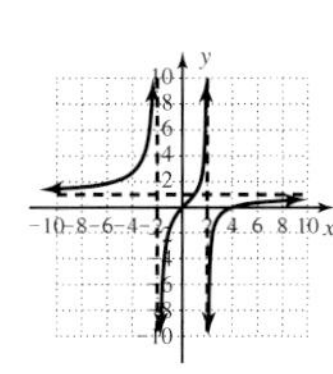

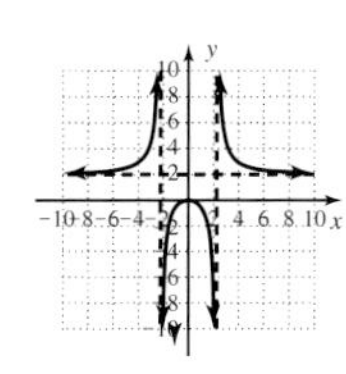

29. $r(x) = \dfrac{x^2 - x - 12}{x^2 - x - 6}$; $r(0) = 2$

30. a. $y = 15$; as $|x| \to \infty$, $A(x) \to 15^+$. As production increases, average cost decreases and approaches 15. **b.** $x > 2000$

31. $H(x) = \begin{cases} \dfrac{x^2 - 3x - 4}{x + 1} & x \neq -1 \\ -5 & x = -1 \end{cases}$

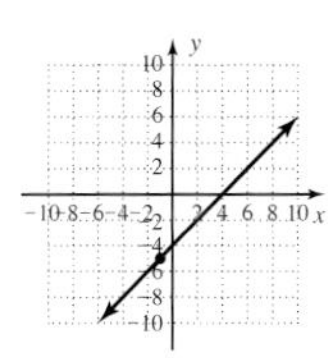

32. **33.**

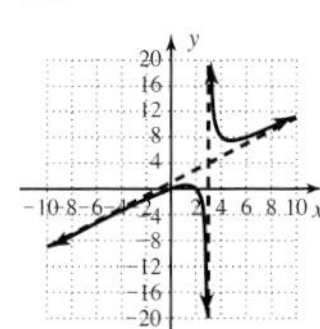

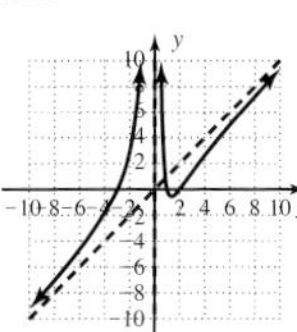

34. factored form $(x + 4)(x - 1)(x - 2) > 0$

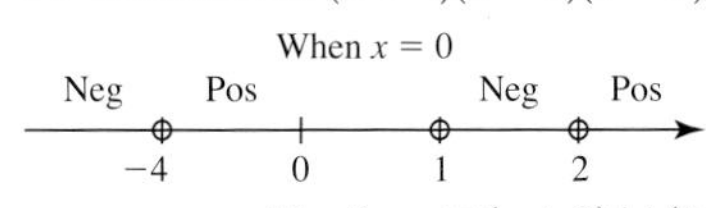

outputs are positive for $x \in (-4, 1) \cup (2, \infty)$

35. $\dfrac{x^2 - 3x - 10}{x - 2} = \dfrac{(x - 5)(x + 2)}{x - 2} \geq 0$

When $x = 0$
Neg Pos Neg Pos
-2 0 2 5

outputs are positive or zero for $x \in [-2, 2) \cup [5, \infty)$

36. $\dfrac{(x + 2)(x - 1)}{x(x - 2)} \leq 0$

When $x = -1$
Pos Neg Pos Neg Pos
-2 -1 0 1 2

outputs are negative or zero for $x \in [-2, 0) \cup [1, 2)$

Mixed Review, p. 469

1. $q(x) = x^2 - 5$; $R = 8$ **3.** a, c, d **5.** $k = 6$
7. $x^3 - 3x^2 + 25x - 75$ **9.** $x = 9$; $x = \frac{8}{3}$
11. **13.** **15.**

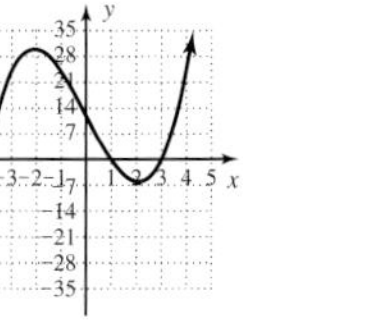

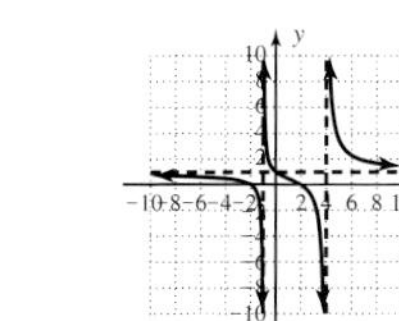

17. **19.** $x \in (-2, 0) \cup [6, \infty)$

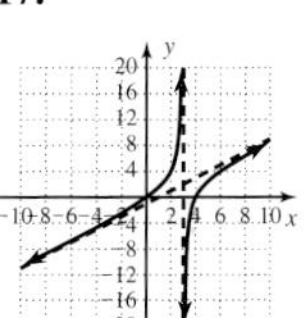

Practice Test, pp. 470–471

1. $x - 5 + \dfrac{14x + 3}{x^2 + 2x + 1}$ **2.** $x^2 + 2x - 9 + \dfrac{-2}{x + 2}$ **3.** $k = -35$

4.
$$\begin{array}{r|rrrrr} -3 & 1 & 0 & -15 & -10 & 24 \\ & & -3 & 9 & 18 & -24 \\ \hline & 1 & -3 & -6 & 8 & 0 \end{array} \quad R = 0 \checkmark$$

5. -1

6. $P(x) = x^3 - 2x^2 + 9x - 18$ **7.** $Q(x) = (x - 2)^2(x - 1)^2(x + 1)$, 2 mult 2, 1 mult 2, -1 mult 1 **8.** $C(x) = (x - 3i)(x + 4)(x + 2)$

9. a. $\dfrac{\pm 1, \pm 18, \pm 2, \pm 9, \pm 3, \pm 6}{\pm 1}$ **b.** 1 positive root, 3 or 1 negative roots; 2 or 0 complex roots **c.** $C(x) = (x + 2)(x - 1)(x - 3i)(x + 3i)$

10. a. 1992, 1994, 1998 **b.** 4 yr **c.** surplus of \$2.5 million

11. **12.** **13.**

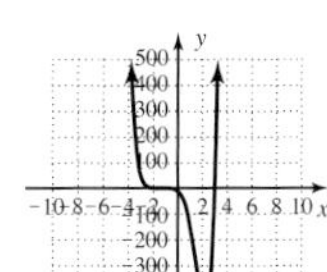

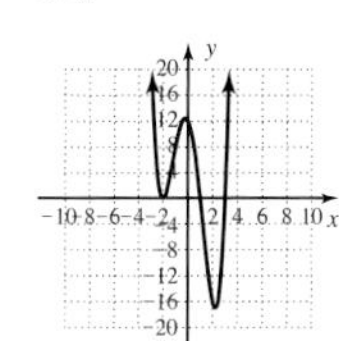

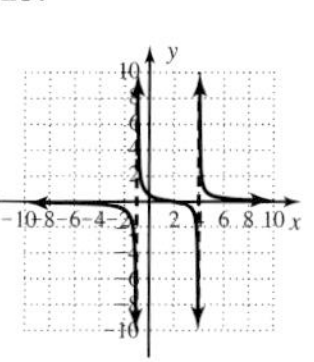

14. a. removal of 100% of the contaminants **b.** \$1,700,000, \$500,000, \$3,000,000 **c.** 88%

15. **16.**

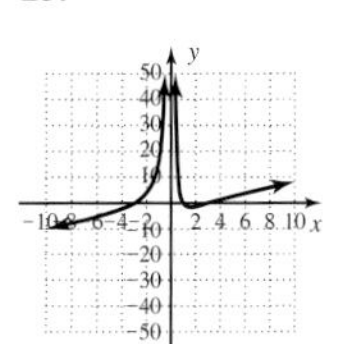

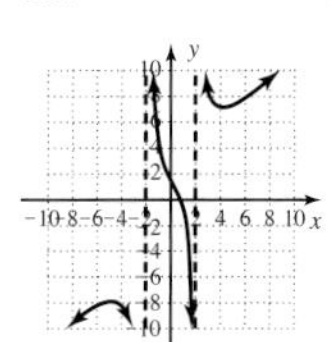

17. 800 **18.** $x \in (-\infty, -3] \cup [-1, 4]$ **19.** $x \in (-\infty, -4) \cup (0, 2)$

20. a.

b. $h = -\sqrt[3]{55}$; no **c.** 28.6% 29.6% **d.** 12 hr **e.** 4 hr, 43.7%
f. The chemical will eventually disappear from the bloodstream.

Strengthening Core Skills, p. 473

1. $x \in (-\infty, 3]$ **2.** $x \in (-2, -1) \cup (2, \infty)$ **3.** $x \in (-\infty, -4) \cup (1, 3)$
4. $x \in [-2, \infty)$ **5.** $x \in (-\infty, -2) \cup (2, \infty)$
6. $x \in [-3, 1] \cup [3, \infty)$

Cumulative Review pp. 473–474

1. $R = \dfrac{R_1 R_2}{R_1 + R_2}$ **3. a.** $(x - 1)(x^2 + x + 1)$ **b.** $(x - 3)(x + 2)(x - 2)$

5. all reals **7.** verified **9.** $y = \dfrac{11}{60}x + \dfrac{1009}{60}$; 39 min, driving time increases 11 min every 60 days

11. $y = 1.18x^2 - 10.99x + 4.6$; **13.** $f^{-1}(x) = \dfrac{x^3 + 3}{2}$,

15.

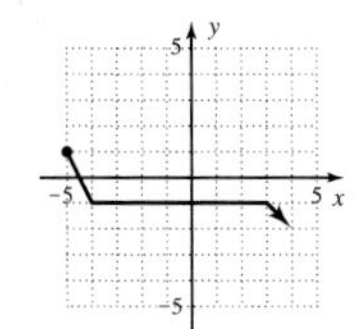

17. $X = 63$ **19.**

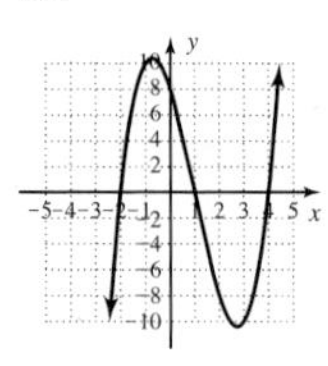

CHAPTER 5

Exercises 5.1, pp. 482–485

1. b^x, b, b, x **3.** a, 1 **5.** False; for $|b| < 1$ and $x_2 > x_1$, $b^{x_2} < b^{x_1}$ so function is decreasing **7.** 40,000; 5000; 20,000; 27,589.162 **9.** 500; 1.581; 2.321; 221.168 **11.** 10,000; 1975.309; 1487.206; 1316.872

13.

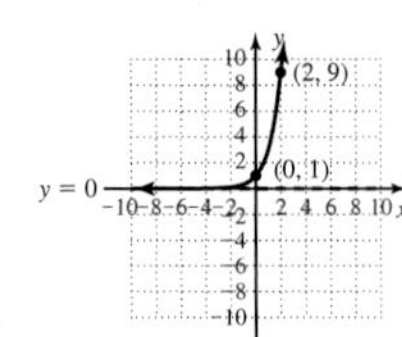

increasing

15.

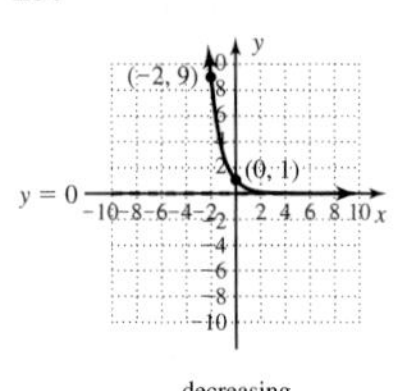

decreasing

17. up 2

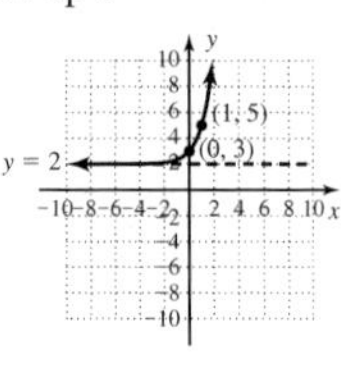

19. left 3

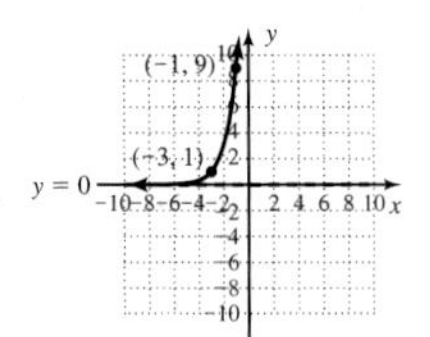

21. reflect across y-axis

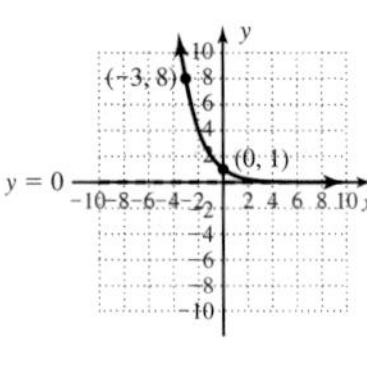

23. reflect across y-axis, up 3

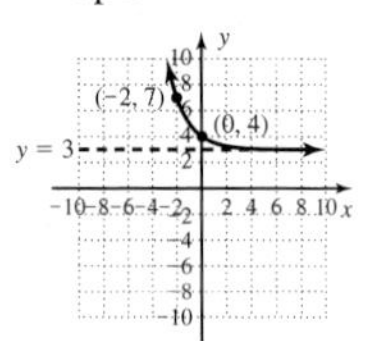

25. left 1, down 3

27. up 1

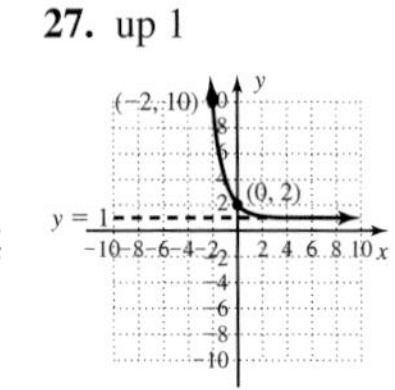

29. right 2

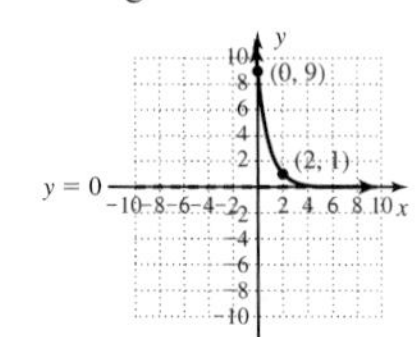

31. down 2

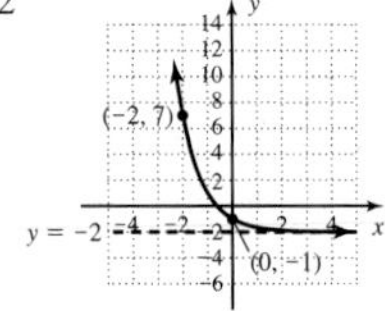

33. e **35.** a **37.** b **39.** 3 **41.** $\frac{3}{2}$ **43.** $\frac{-1}{3}$ **45.** 4 **47.** −3 **49.** 3 **51.** 2 **53.** −2 **55. a.** 1732, 3000, 5196, 9000 **b.** yes **c.** as $t \to \infty$, $P \to \infty$ **d.**

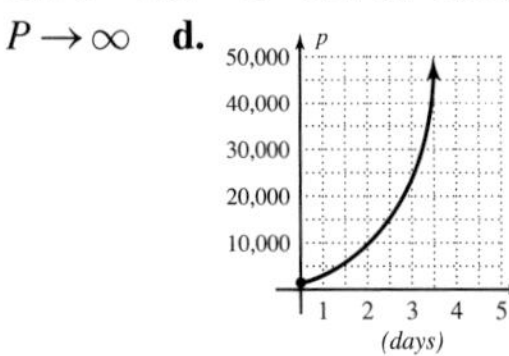

57. a. ≈\$12,875 **b.** ≈\$152 **59. a.** \$100,000 **b.** 3 yr **61. a.** ≈\$86,806 **b.** 3 yr **63. a.** \$40 million **b.** 7 yr **65.** ≈\$32,578 **67. a.** 8 g **b.** 48 min **69.** 9.5×10^{-7}; answers will vary **71.** $\frac{1}{5}$

73. 5, $\frac{-7}{9}$, $2a^2 - 3a$, $2a^2 + 4ah + 2h^2 - 3a - 3h$ **75.** $y = 1|x - 2| + 1$ **77. a.** $\emptyset$ **b.** $\{-5, 6\}$

Exercises 5.2, pp. 491–494

1. $\log_b x$, b, b, greater **3.** (1, 0), 0 **5.** 5; answers will vary **7.** $2^3 = 8$ **9.** $7^{-1} = \frac{1}{7}$ **11.** $9^0 = 1$ **13.** $8^{\frac{1}{3}} = 2$ **15.** $2^1 = 2$ **17.** $7^2 = 49$ **19.** $10^2 = 100$ **21.** $10^{-1} = 0.1$ **23.** $\log_4 64 = 3$ **25.** $\log_3 \frac{1}{9} = -2$ **27.** $\log_9 1 = 0$ **29.** $\log_{\frac{1}{3}} 27 = -3$ **31.** $\log 1000 = 3$ **33.** $\log \frac{1}{100} = -2$ **35.** $\log_4 8 = \frac{3}{2}$ **37.** $\log_4 \frac{1}{8} = \frac{-3}{2}$ **39.** 2 **41.** 5 **43.** −2 **45.** 1 **47.** 1 **49.** $\frac{1}{2}$ **51.** 25 **53.** 6 **55.** 6 **57.** 2 **59.** $\frac{1}{125}$ **61.** $\frac{5}{3}$

63. shift up 3

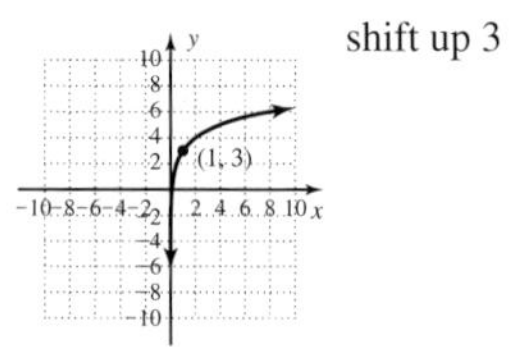

65. shift right 2, up 3

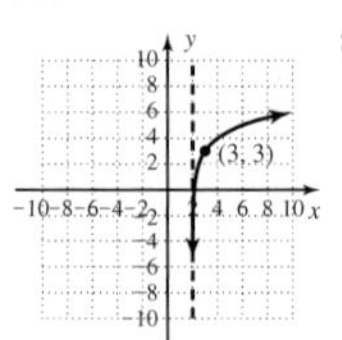

67. shift left 1

69. reflect across x-axis, shift left 1

71. $x \in (-\infty, -1) \cup (3, \infty)$ **73.** $x \in \left(\frac{3}{2}, \infty\right)$ **75.** $x \in (-3, 3)$ **77.** ≈ 2.2430 **79.** ≈ 5.1071 **81.** ≈ −0.6990 **83.** pH ≈ 4.1; acid **85. a.** ≈2.4 **b.** ≈1.2 **87. a.** ≈4.7 **b.** ≈4.9 **89. a.** 20 dB **b.** 120 dB **91. a.** 95% **b.** 67% **c.** 39% **93.** ≈4.3; acid **95.** Answers will vary. **a.** 0 dB **b.** 90 dB **c.** 15 dB **d.** 120 dB **e.** 100 dB **f.** 120 dB **97.** Answers will vary. **99.** $\left(\frac{1}{2}\right)^y = x$, $2^{-y} = x$, $-y = \log_2 x$, $y = -\log_2 x$

101. a. **b.**

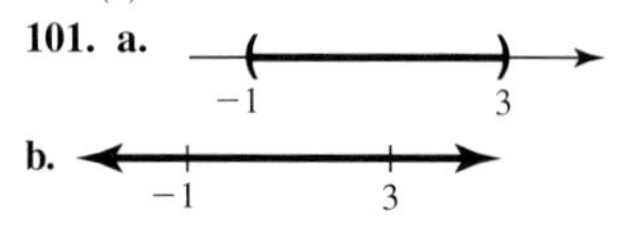

103. a. $(x - 2)(x^2 + 2x + 4)$ **b.** $(a + 7)(a - 7)$ **c.** $(n - 5)(n - 5)$ **d.** $(2b - 3)(b - 2)$

105. $x \in (-\infty, -5)$; $f(x) = (x + 5)(x - 4)^2 = x^3 - 3x^2 - 24x + 80$

Exercises 5.3, pp. 502–507

1. $(1 + \frac{1}{x})^x \to \infty$ **3.** ln x, exponent, e **5.** $\log_2 9 > 3$; $\log_3 26 < 3$ **7.** 2.718282 **9.** 7.389056 **11.** 4.481689 **13.** 4.113250

15.

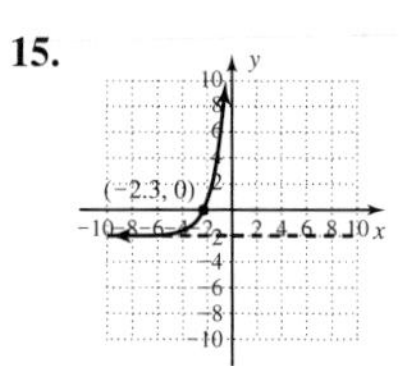

17.

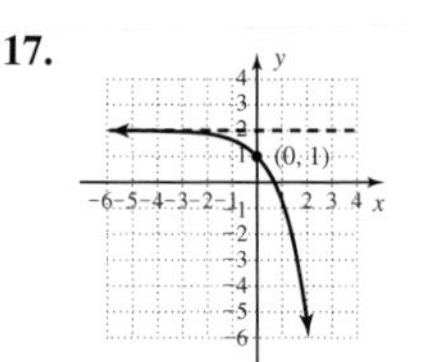

19.

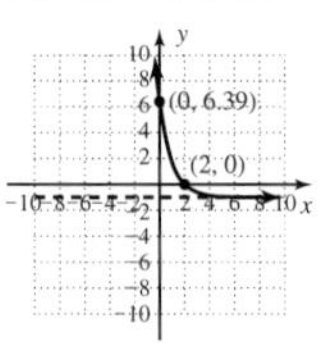

21. 3.912023 **23.** −0.693147 **25.** 5.416100 **27.** 0.346574 **29.** $x = e$; $x \approx 2.718$ **31.** $x = e^{-1.961}$; $x \approx 0.141$ **33.** $x = \pm\sqrt{e^{2.4}}$; $x \approx \pm 3.320$ **35.** $x = -4.2$; $x = -4.2$ **37.** $x = 0$; $x = 0$ **39.** $x = \ln 7.389$; $x \approx 1.99999$ **41.** $x = \dfrac{5 \ln 1.396}{2}$; $x \approx 0.83403$ **43.** not a real number

45. $\ln(2x^2 - 14x)$ **47.** $\log(x^2 - 1)$ **49.** $\log_3 4$ **51.** $\log\left(\dfrac{x}{x + 1}\right)$

53. $\ln\left(\frac{x-5}{x}\right)$ **55.** $\ln(x-2)$ **57.** $\log_2 42$ **59.** $\log_5(x-2)$
61. $(x+2)\log 8$ **63.** $(2x-1)\ln 5$ **65.** $\frac{1}{2}\log 22$ **67.** $4\log_5 3$
69. $3\log a + \log b$ **71.** $\ln x + \frac{1}{4}\ln y$ **73.** $2\ln x - \ln y$
75. $\frac{1}{2}[\log(x-2) - \log x]$ **77.** $\ln 7 + \ln x + \frac{1}{2}\ln(3-4x) - \ln 2 - 3\ln(x-1)$ **79.** $\frac{\ln 60}{\ln 7}$; 2.104076884 **81.** $\frac{\ln 152}{\ln 5}$; 3.121512475
83. $\frac{\log 1.73205}{\log 3}$; 0.499999576 **85.** $\frac{\log 0.125}{\log 0.5}$; 3 **87.** $f(x) = \frac{\log(x)}{\log(3)}$; $f(5) \approx 1.4650$; $f(15) \approx 2.4650$; $f(45) \approx 3.4650$; outputs increase by 1; $f(3^3 \cdot 5) = 4.465$ **89.** $h(x) = \frac{\log(x)}{\log(9)}$; $h(2) \approx 0.3155$; $h(4) \approx 0.6309$; $h(8) \approx 0.9464$; outputs are multiples of 0.3155; $h(2^4) = 4(0.3155) \approx 1.2619$ **91. a.** 7.19 mi **b.** 3.45 mi **c.** 1.72 mi; losing altitude
93. a. 4833.5 ft, 7492.1 ft, 14,434.4 ft **b.** 29,032.8 ft
95. No; $234,612.01 **97. a.** ≈36.6 yr **b.** ≈20 yr **c.** ≈13.7 yr **d.** ≈11%. **99. a.** 6 hr **b.** 18% **101.** ≈17,255 yr **103. a.** ~1 **b.** ~0.5 **c.** ~0.33 **d.** ~0.25 **105.** $\frac{\Delta y}{\Delta x} \approx 20$, $f(3) \approx 20$; $\frac{\Delta y}{\Delta x} \approx 7.39$; $f(2) \approx 7.39$; apparently $\frac{\Delta y}{\Delta x} = f(x)$; $f(4) \approx 54.6$; $\frac{\Delta y}{\Delta x} \approx 54.6$
107. a. $D_f: x \in (0, \infty)$; $R_f: y \in \mathbb{R}$; $D_g: x \in [0, \infty)$; $R_y: y \in [-1, \infty)$ **b.** $(1, 0)$ **c.** $-0.16, -30.42, -374.38$; as $x \to \infty$, the difference between the two functions increases **109.** Answers will vary.
111. verified; 1.000, 2.303, 0.693 **113. a.** $\log_3 4 + \log_3 5 = 2.7269$ **b.** $\log_3 4 - \log_3 5 = -0.2031$ **c.** $2\log_3 5 = 2.9300$

115.

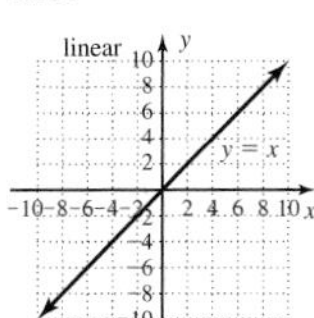

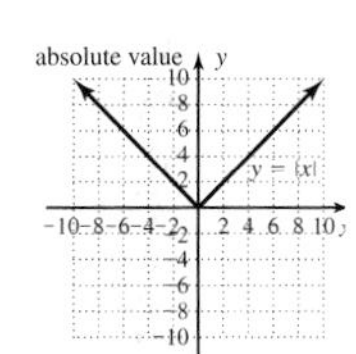

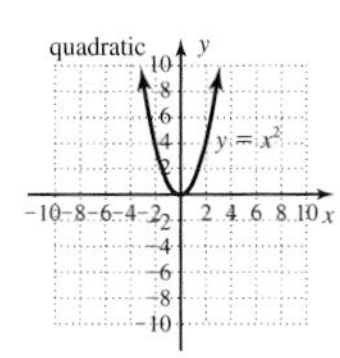

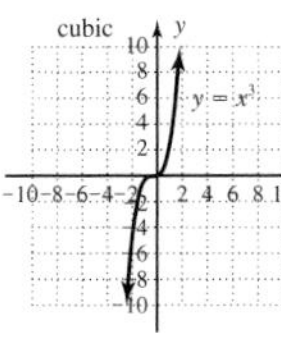

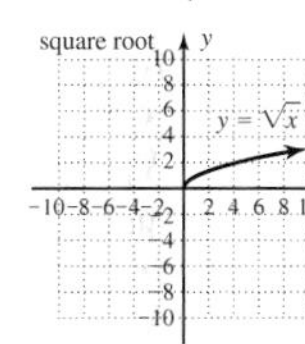

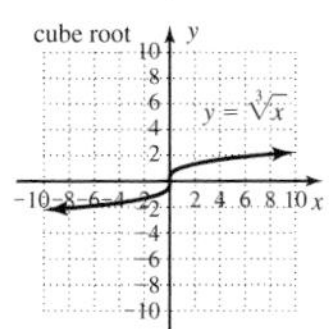

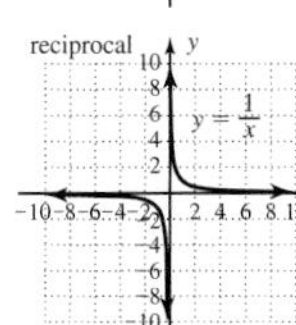

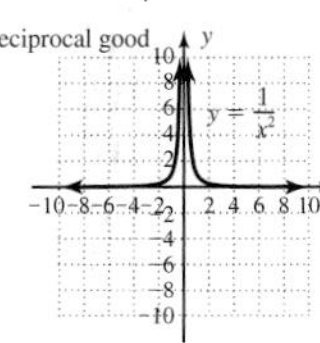

117.

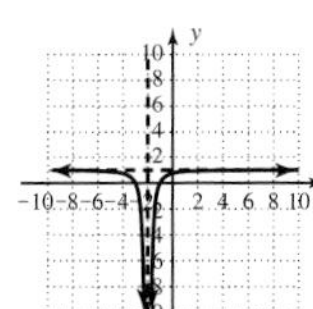

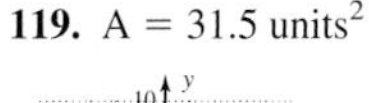
119. A = 31.5 units^2

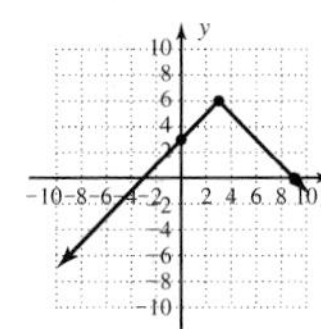

Mid-Chapter Check, p. 507

1. a. $\frac{2}{3} = \log_{27} 9$ **b.** $\frac{5}{4} = \log_{81} 243$ **2. a.** $8^{\frac{5}{3}} = 32$ **b.** $1296^{0.25} = 6$
3. a. $x = 5$ **b.** $b = \frac{5}{4}$ **4. a.** $x = 3$ **b.** $b = 5$ **5. a.** $71,191.41 **b.** 6 yr **6. a.** 2.385606274 **b.** 4.93286932 **7. a.** $\frac{\log 19}{\log 3} \approx 2.68$ **b.** $\frac{\log 60}{\log 2} \approx 5.91$ **8. a.** 2.48% **b.** 12 yr **9. a.** $\log(4x^2 - 9)$ **b.** $\log\left(\frac{1}{x-5}\right)$ **10. a.** 2.0104 **b.** 0.3562 **c.** 1.6542 **d.** 3.1937

Reinforcing Basic Concepts, p. 508

1. Answers will vary. **2. a.** $\log(x^2 + 3x)$ **b.** $\ln(x^2 - 4)$ **c.** $\log\frac{x}{x+3}$ **3.** Answers will vary. **4. a.** $x\log 3$ **b.** $5\ln x$ **c.** $(3x-1)\ln 2$

Exercises 5.4 pp. 516–520

1. e **3.** extraneous **5.** 2.316566275 **7.** $\ln(x+4) = 2$
9. $\log(x^2 + 2x) = 4$ **11.** $\log_2(x^3 - x^2) = 2$ **13.** $e^{x-2} = 16$
15. $e^{0.05x+1} = 1.35$ **17.** $3^{3x-1} = 81$ **19.** $x = 7$ **21.** $x = \frac{5}{3}$
23. $x = 3$ **25.** $n = 9$ **27.** $n = 5$ **29.** $n = 8$ **31.** $n = 5$ **33.** $n = 6$
35. $n = 4$ **37.** $x = \log 97$; $x \approx 1.9868$ **39.** $x = \frac{\log 879}{2}$; $x \approx 1.4720$
41. $x = \log 879 - 3$; $x \approx -0.0560$ **43.** $x = \ln 389$; $x \approx 5.9636$
45. $x = \frac{\ln 1389}{2}$; $x \approx 3.6182$ **47.** $x = \ln 257 - 1$; $x \approx 4.5491$
49. $x = 4\ln\left(\frac{5}{2}\right)$; $x \approx 3.6652$ **51.** $x = \frac{\ln 231}{\ln 7} - 2$; $x \approx 0.7968$
53. $x = \frac{\ln 128{,}965}{3\ln 5} + \frac{2}{3}$; $x \approx 3.1038$ **55.** $x = \frac{\ln 2}{\ln 3 - \ln 2}$; $x \approx 1.7095$
57. $x = \frac{\ln 9 - \ln 5}{2\ln 5 - \ln 9}$; $x \approx 0.5753$ **59.** $t = \frac{-50}{3}\ln\left(\frac{37}{150}\right)$; $t \approx 23.3286$
61. $t = \frac{100}{29}\ln(236)$; $t \approx 18.8408$ **63.** $x = 0$ **65.** $x = \frac{5}{2}$ **67.** $x = \frac{2}{3}$
69. $x = 33$ **71.** $x = 118$ **73.** $x = e^2 - 7$ **75.** $x = \frac{101}{200}$ **77.** $x = 50$
79. $x = e^3 - 1$ **81.** $x = \frac{3}{2}$ **83.** $x = \frac{-19}{9}$ **85.** $x = \frac{e^2 - 63}{9}$
87. $x = 2$; -9 is extraneous **89.** $x = 3e^3 - \frac{1}{2}$; $x \approx 59.75661077$
91. no solution **93.** $t = -\frac{1}{2}$; -4 is extraneous **95.** 31.25 g
97. 3 hr 11 min **99.** 34 cmHg **101.** 11.0 yrs **103.** $66,910
105. 52.76 tons **107.** Answers will vary. **109. a.** d **b.** e **c.** b **d.** f **e.** a **f.** c **111.** $x = 0.69319718$ **113.** $x = \frac{\log_3 5}{2}$
115. $(f \circ g)(x) = 3^{(\log_3 x + 2) - 2} = 3^{\log_3 x} = x$; $(g \circ f)(x) = \log_3(3^{x-2}) + 2 = x - 2 + 2 = x$ **117.** $y = e^{x\ln 2} = e^{\ln 2^x} = 2^x$; $y = 2^x = \ln y = x\ln 2$, $e^{\ln y} = e^{x\ln 2} \Rightarrow y = e^{x\ln 2}$ **119.** b
121. a. $x \in [-\frac{3}{2}, \infty)$, $y \in [0, \infty)$ **b.** $x \in (-\infty, \infty)$, $y \in [-3, \infty)$
123. $x = -2, 1 \pm 2i$

Exercises 5.5, pp. 531–536

1. Compound **3.** $Q_0 e^{-rt}$ **5.** Answers will vary. **7.** $4896 **9.** 250%
11. $2152.47 **13.** 5.25 yr **15.** 4 yr **17.** 16 yr **19.** $7561.33
21. 5 yr **23.** 7.5 yr **25.** 7.9 yr **27.** 7.5 yr **29. a.** $t = \frac{A - p}{pr}$
b. $p = \frac{A}{1 + rt}$ **31. a.** $r = n\left(\sqrt[nt]{\frac{A}{p}} - 1\right)$
b. $t = \frac{\ln\left(\frac{A}{p}\right)}{n\ln\left(1 + \frac{r}{n}\right)}$ **33. a.** $Q_0 = \frac{Q}{e^{rt}}$ **b.** $t = \frac{\ln\left(\frac{Q}{Q_0}\right)}{r}$

35. \$709.74 **37.** 80% **39.** no **41.** no, 9.12% **43.** no, 9.4% **45.** approx 13,609 euros **47. a.** 5.78% **b.** 91.67 hr **49.** 0.65 g **51.** 818 yr **53.** about 7 yr **55.** 23 yr **57. a.** no **b.** \$302.25 **59.** \$12,488,769.67; answers will vary **61.** Answers will vary. **63.** 7.2% **65.** 2548.8 m **67.** yes **69. a.** $f(x) = x^3$, $f(x) = x$, $f(x) = \sqrt{x}$, $f(x) = \sqrt[3]{x}$, $f(x) = \frac{1}{x}$ **b.** $f(x) = |x|$, $f(x) = x^2$, $f(x) = \dfrac{1}{x^2}$ **c.** $f(x) = x$, $f(x) = x^3$, $f(x) = \sqrt{x}$, $f(x) = \sqrt[3]{x}$ **d.** $f(x) = \dfrac{1}{x}$, $f(x) = \dfrac{1}{x^2}$

Exercises 5.6, pp. 544–551

1. data, context, situation **3.** beyond **5.** Answers will vary. 1. clear old data 2. enter new data 3. display the data 4. calculate the regression equation 5. display and use the regression graph and equation **7.** e **9.** a **11.** d **13.** linear **15.** exponential **17.** logistic **19.** exponential

21. As time increases, the amount of radioactive material decreases but will never truly reach 0 or become negative. Exponential with $b < 1$ and $k > 0$ is the best choice.

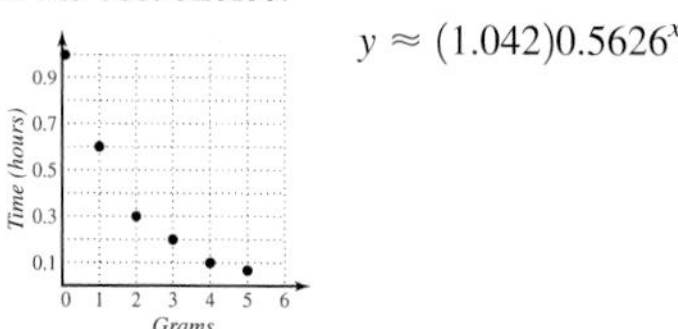

$y \approx (1.042)0.5626^x$

23. $k \approx -0.5752$, about -57.5% **25.** Sales will increase rapidly, then level off as the market is saturated with ads and advertising becomes less effective, possibly modeled by a logarithmic function. $y \approx 120.4938 + 217.2705 \ln(x)$ **27.** 4.95 **29.** 81.74 **31.** 6.25 **33.** 243 **35.** $e^{911.9}$ **37.** 4023.87 **39.** 15.98 **41.** 4.22 **43.** 1981

45. $y = 53.24(1.04)^x$

47. $y = 0.89(0.81)^x$

49. $y = -2635.6 + 1904.8 \ln x$

51. a. 829 **b.** 3271 **c.** 1989 **53. a.** 0.23 in. **b.** 3.2 sec **c.** 18 sec **55. a.** 2870 oz **b.** 32.6 months **c.** 4816 oz

57. $y = 39.86(1.16)^x$ **a.** \$576,000 **b.** \$7,187,000 **c.** 1992

59. $y = 19{,}943.7 - 5231.4 \ln x$ **a.** \$9402.94 **b.** 9.5 yr **c.** 25.5 yr

61.

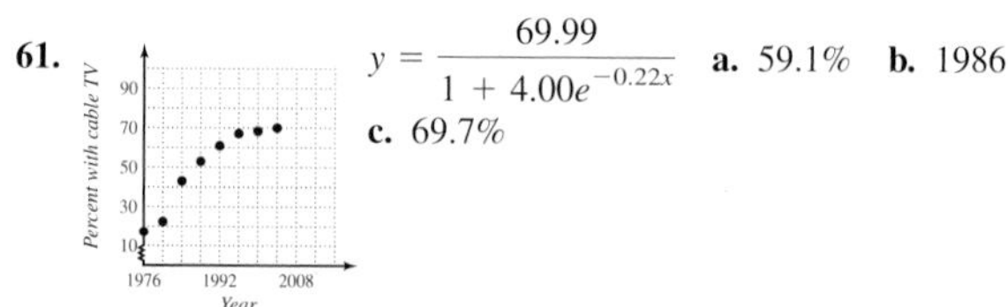

$y = \dfrac{69.99}{1 + 4.00e^{-0.22x}}$ **a.** 59.1% **b.** 1986 **c.** 69.7%

63. a. Bush & Gore are tied; Gore wins. **b.** Bush has his smallest lead; Bush would win. **c.** Bush would always have a lead. Bush would win. **d.** exponential; outcome would be too close to call

65A. $Y_1 \approx (1.2117)1.2847^x$, $Y_2 \approx -0.7665 + 3.9912 \ln(x)$, 1. symmetry about $y = x$; 2. for (a, b) on Y_1 (b, a) is on Y_2.

65B. a. $y = 1.2117214e^{0.2505524714x}$ **b.** $a + b \ln(x) = y$, $a + b \ln(y) = x$, $\ln(y) = \dfrac{x - a}{b}$, $y = e^{\frac{x-a}{b}} = e^{\frac{x}{b}}e^{\frac{-a}{b}}$

Substituting $a = -0.7664737786$ and $b = 3.991178001$ at this point gives the result from (a).

67. $\dfrac{\Delta f}{\Delta x} \approx -0.39$, $\dfrac{\Delta g}{\Delta x} \approx -1.05$ **69.** $y = -\frac{3}{2}x + 7$; $2 \le x < 4$

71. $D: x \in (-\infty, \infty)$, $R: y \in [3, \infty)$

Summary and Concept Review, pp. 552–556

1.

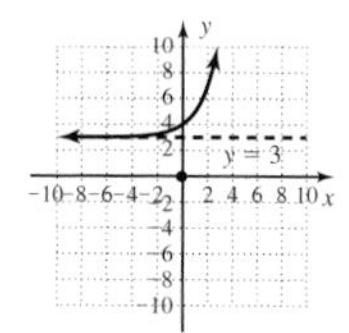

2.

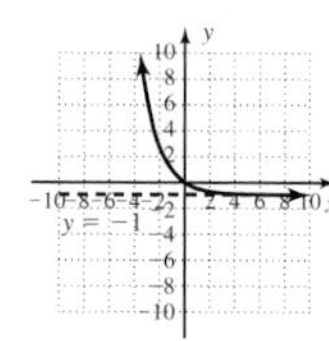

3.

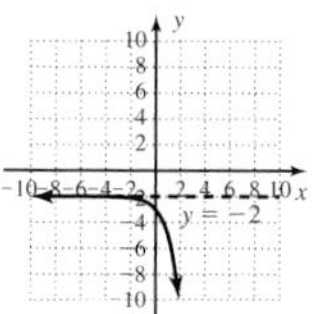

4. 2 **5.** −2 **6.** $\frac{1}{4}$ **7.** 12.1 yr **8.** $3^2 = 9$ **9.** $5^{-3} = \frac{1}{125}$ **10.** $2^4 = 16$ **11.** $\log_5 25 = 2$ **12.** $\log_2(\frac{1}{8}) = -3$ **13.** $\log_3 81 = 4$ **14.** 5 **15.** 4 **16.** 13

17.

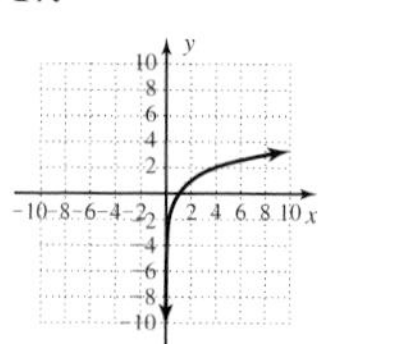

18.

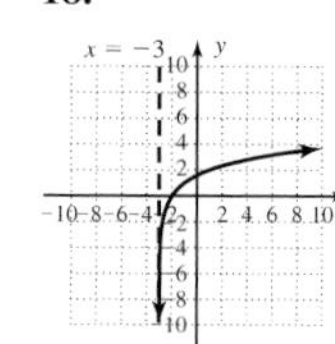

19.

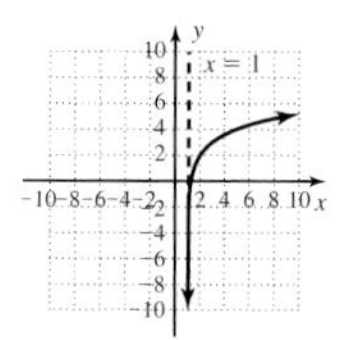

20. a. 4.79 **b.** $10^{7.3} I_0$ **21. a.** e^{32} **b.** ln 9.8 **c.** $\ln \sqrt{7} = \frac{1}{2} \ln 7$ **d.** $e^{2.38}$ **22. a.** $\dfrac{\log 45}{\log 6} \approx 2.125$ **b.** $\dfrac{\log 128}{\log 3} \approx 4.417$ **c.** $\dfrac{\log 108}{\log 2} \approx 6.755$ **d.** $\dfrac{\log 200}{\log 5} \approx 3.292$ **23. a.** ln 42 **b.** $\log_9 30$ **c.** $\ln\left(\dfrac{x+3}{x-1}\right)$ **d.** $\log(x^2 + x)$ **24. a.** $2 \log_5 9$ **b.** $2 \log_7 4$

c. $(2x - 1)\ln 5$ **d.** $(3x + 2)\ln 10$ **25. a.** ≈ -0.25 **b.** ≈ 0.25 **c.** ≈ 1.54 **d.** ≈ 1.80 **26. a.** $4\ln x + \frac{1}{2}\ln y$ **b.** $\frac{1}{3}\ln p + \ln q$ **c.** $\frac{5}{3}\log x + \frac{4}{3}\log y - \frac{5}{2}\log x - \frac{3}{2}\log y$ **d.** $\log 4 + \frac{5}{3}\log p + \frac{4}{3}\log q - \frac{3}{2}\log p - \log q$ **27. a.** 17.77% **b.** 23.98 days **28.** $\frac{\ln 7}{\ln 2}$ **29.** $\frac{\ln 5}{\ln 3} - 1$ **30.** $\frac{2\ln 4}{\ln 4 - \ln 3}$ **31.** 24 **32.** 5; -2 is extraneous **33.** no solution **34.** 38.63 cmHg **35.** 18.5% **36.** Almost, she needs \$42.15 more. **37.** 55.0% **38. a.** no **b.** \$268.93 **39. a.** logistic **b.** logistic; growth rate exceeds population growth **c.** exponential: 44.6 million, logistic: 56.1 million, exponential: 347.3, 1253 million; logistic, 179.3, 201.0 million; projections from the exponential equation are excessive

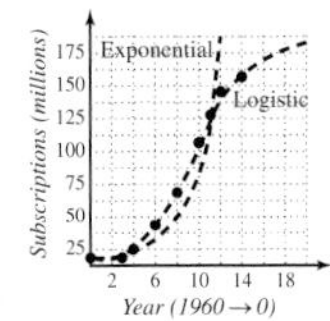

40. a. exponential; $y = 13.29(1.08)^x$ **b.** 156 billion dollars **c.** 424.2 billion dollars

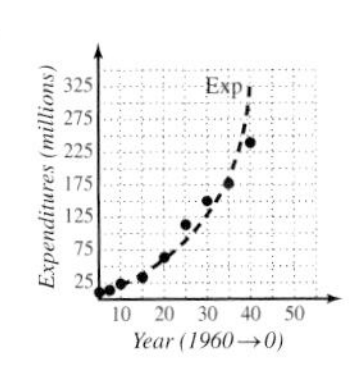

Mixed Review, pp. 556–557

1. a. $\frac{\log 30}{\log 2} \approx 4.9069$ **b.** -1.5 **c.** $\frac{1}{3}$ **3. a.** $2\log_{10}20$ **b.** $0.05x$ **c.** $(x - 3)\ln 2$ **5.**

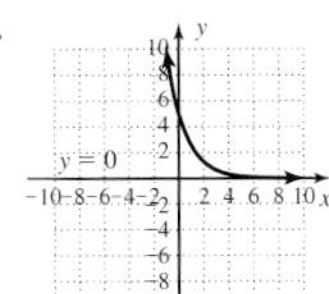

7.

9. a. $5^4 = 625$ **b.** $e^{0.45} = 0.15x$ **c.** $10^7 = 0.1 \times 10^8$ **11.** 7 **13.** $6 + \log 2$ **15.** $\frac{9}{4} + \frac{\sqrt{129}}{4}$ **17.** $I = 6.3 \times 10^{17}$ **19.** 1.6 m, 1.28 m, 1.02 m, 0.82 m, 0.66 m, 0.52 m

Practice Test, p. 558

1. $3^4 = 81$ **2.** $\log_{25}5 = \frac{1}{2}$ **3.** $\frac{5}{2}\log_b x + 3\log_b y - \log_b z$ **4.** $\log_b \frac{m\sqrt{n^3}}{\sqrt{p}}$ **5.** $x = 10$ **6.** $x = \frac{-5}{3}$ **7.** 2.68 **8.** -1.24

9.

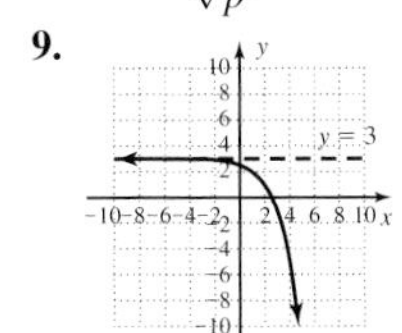

10.

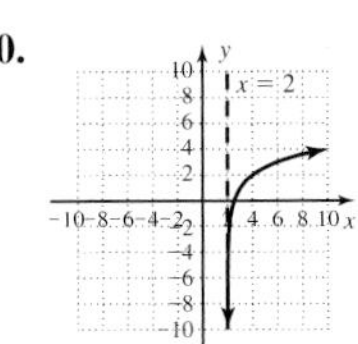

11. 4.19 **12.** -0.81 **13.** $x = 1 + \frac{\ln 89}{\ln 3}$ **14.** $x = 1$; $x = -5$ is extraneous **15.** ≈ 5 yr **16.** ≈ 8.7 yr **17.** 19.1 months **18.** 7% compounded semi-annually **19.** no, \$54.09 **20.** logistic; $y = \frac{39.1156}{1 + 314.6617e^{-5.9483x}}$; 0.89 sec

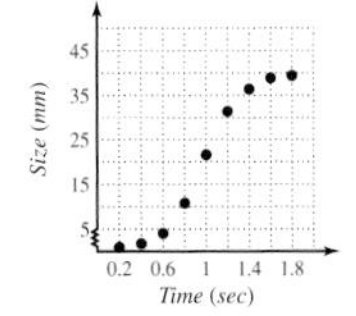

Strengthening Core Skills, pp. 560–562

1. between eight and nine players

Cumulative Review, pp. 562–563

1. $x = 2 \pm 7i$ **3.** $(4 + 5i)^2 - 8(4 + 5i) + 41 = 0, -9 + 40i - 32 - 40i + 41 = 0, 0 = 0$ ✓ **5.** $f(g(x)) = x$, $g(f(x)) = x$ Since $(f \circ g)(x) = (g \circ f)(x)$, they are inverse functions. **7. a.** $T(t) = 455t + 2645$ (1991 → *year* 1) **b.** $\frac{\Delta T}{\Delta t} = \frac{455}{1}$, triple births increase by 455 each year **c.** $T(6) = 5375$ sets of triplets, $T(17) = 10{,}380$ sets of triplets

9.

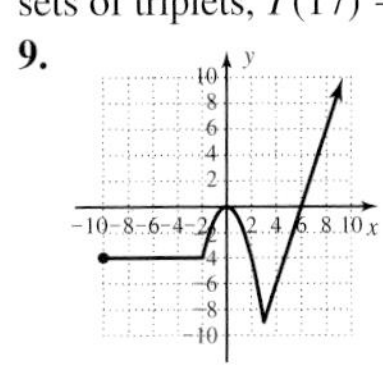

$D: x \in [-10, \infty)$, $R: y \in [-9, \infty)$
$h(x)\uparrow: x \in (-2, 0) \cup (3, \infty)$
$h(x)\downarrow: x \in (0, 3)$

11. $x = 3$, $x = 2$ (multiplicity 2); $x = -4$ **13.** $\sqrt{\frac{2V}{\pi a}} = b$

15.

17. $x = 5$, $x = -6$ is an extraneous root

19.

a. linear; $W = 1.24L - 15.83$ **b.** 32.5 lb **c.** 35.3 in.

CHAPTER 6

Exercises 6.1, pp. 573–577

1. inconsistent **3.** consistent, independent **5.** Multiply the first equation by 6 and the second equation by 10. **7.** $y = \frac{7}{4}x - 6$, $y = \frac{-4}{3}x + 5$ **9.** $y = x + 2$ **11.** $x + 3y = -3$ **13.** $y = x + 2$, $x + 3y = -3$ **15.** yes **17.** yes

19.

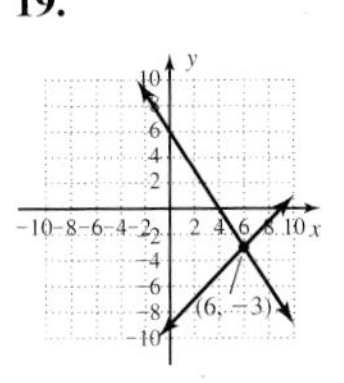

21.

23. $(-4, 1)$ **25.** $(3, -5)$ **27.** second equation, y, $(4, -3)$ **29.** second equation, x, $(10, -1)$ **31.** second equation, x, $\left(\frac{5}{2}, \frac{7}{4}\right)$ **33.** $(3, -1)$ **35.** $(-2, -3)$ **37.** $\left(\frac{11}{2}, 2\right)$ **39.** $(-2, 3)$ **41.** $(-3, 4)$ **43.** $(-6, 12)$ **45.** $(2, 8)$; consistent/independent **47.** $\varnothing$; inconsistent **49.** $\{(x, y)|6x - 22 = -y\}$; consistent/dependent **51.** $(4, 1)$; consistent/independent **53.** $(-3, -4)$; consistent/independent **55.** $\left(\frac{-1}{2}, \frac{4}{3}\right)$; consistent/independent **57.** $\left(-2, \frac{5}{2}\right)$ **59.** $(2, -1)$ **61.** 1 mph, 4 mph **63. a.** $(3, 24)$ **b.** $(3, 24)$; these are the equations of the two lines. **c.** \$20 **65.** 1776; 1865 **67.** 160 ft wide and 360 ft long

69. 2318 adult tickets; 1482 child tickets **71.** 2.5 hr; 37.5 mi
73. nursing student \$6500; science major \$3500 **75.** Answers will vary.
77. \$50,000
79.

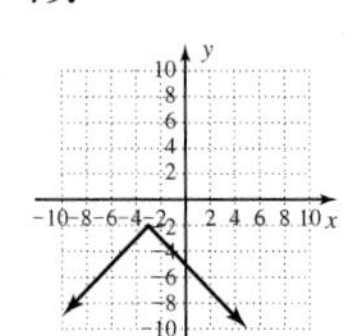

81. $x \approx 120.716$ **83.** $2, 6i, 10, \frac{-4}{5} + \frac{3}{5}i$

Exercises 6.2, pp. 586–589

1. triple **3.** equivalent, systems **5.** 5, substitute and solve for remaining variable **7.** Answers will vary. **9.** Answers will vary. **11.** yes
13. $(5, 7, 4)$ **15.** $(-2, 4, 3)$ **17.** $(1, 1, -2)$ **19.** $(4, 0, -3)$
21. $(3, 4, 5)$ **23.** $(1, 6, 9)$ **25.** no solution, inconsistent
27. no solution, inconsistent **29.** $\left(-\frac{5}{3}z - \frac{2}{3}, -z - 2, z\right)$, other solutions possible **31.** $(z - 1, 2z - 2, z)$ **33.** $(z + 9, z - 4, z)$
35. $\{(x, y, z) | x - 6y + 12z = 5\}$ **37.** $(1, 1, 2)$
39. $\left\{(x, y, z) | x - \frac{5}{2}y - 2z = 3\right\}$ **41.** $\left(2, 1, \frac{-1}{3}\right)$
43. $(z + 5, z - 2, z)$ **45.** $(18, -6, 10)$ **47.** $\left(\frac{11}{3}, \frac{10}{3}, \frac{7}{3}\right)$
49. $(1, -2, 3)$ **51.** $\left(\frac{1}{2}, \frac{1}{3}, 3\right)$ **53.** ≈ 3.464 units
55. elephant, 650 days; rhino, 464 days; camel, 406 days **57.** 1 L 20% solution; 3 L 30% solution; 6 L 45% solution **59.** \$80,000 at 4%; \$90,000 at 5%; \$110,000 at 7% **61.** Answers will vary.
63. -1 or $\frac{1}{2}, 2$ **65.** b **67.** $f(f^{-1}(x)) = x$; $f^{-1}(f(x)) = x$ **69.** $x = 1$
71. $D: x \in (-\infty, \infty)$, $R: y \in [-5, \infty)$, zeroes: $x = -2.5, x = -1.5, x = 0.5, x = 2$,
$g(x) > 0: x \in (-\infty, -2.5) \cup (-1.5, 0.5) \cup (2, \infty)$,
$g(x) < 0: x \in (-2.5, -1.5) \cup (0.5, 2)$, max: $(-0.5, 4)$,
min: $(-2, -2), (1.5, -5)$, $f(x)\uparrow: x \in (-2, -0.5) \cup (1.5, \infty)$,
$f(x)\downarrow: x \in (-\infty, -2) \cup (-0.5, 1.5)$

Exercises 6.3, pp. 600–603

1. half, planes **3.** solution **5.** The feasible region may be bordered by three or more oblique lines, with two of them intersecting outside and away from the feasible region. **7.** No, No, No, No **9.** No, Yes, Yes, No
11.

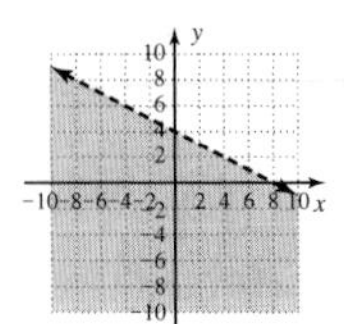

13.

15. No, No, No, Yes

17.

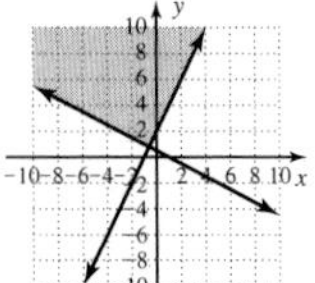

19.

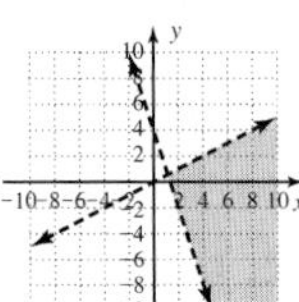

21.

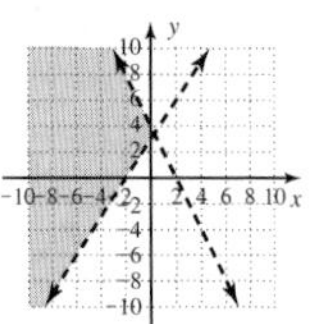

23.

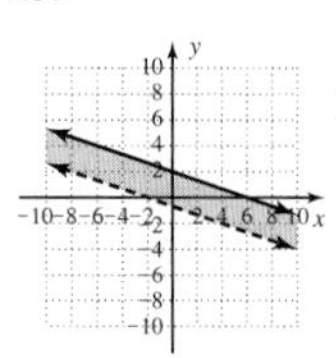

25.

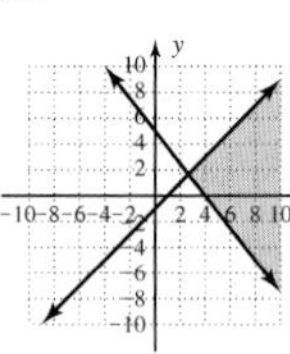

27.

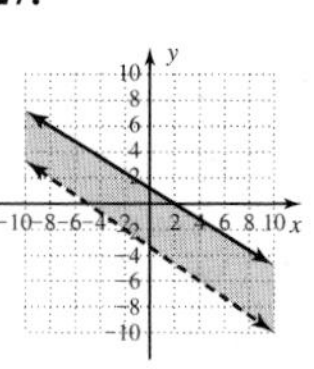

29.

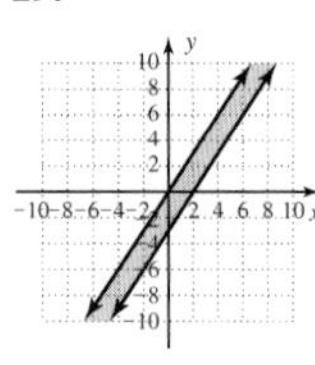

31.

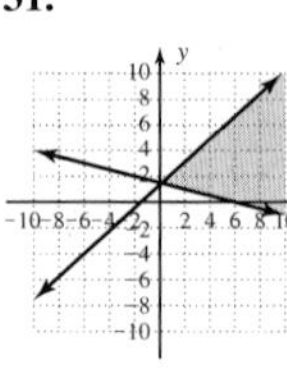

33.

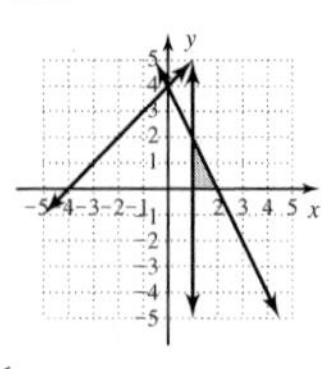

35.

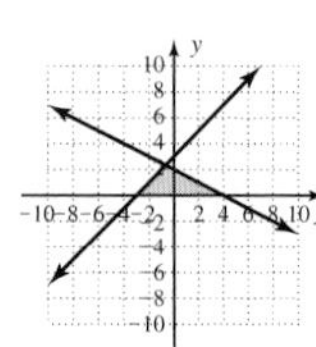

37.

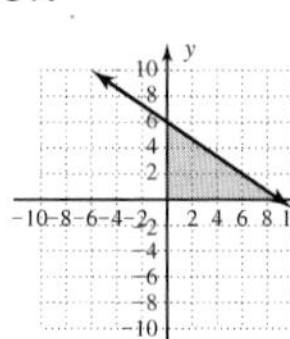

39. $\begin{cases} y - x \le 1 \\ x + y > 3 \end{cases}$

41. $\begin{cases} y - x \le 1 \\ x + y < 3 \\ y \ge 0 \end{cases}$ **43.** $(5, 3)$ **45.** $(12, 11)$ **47.** $(2, 2)$

49. $(4, 3)$ **51.** $5 < H < 10$
53.

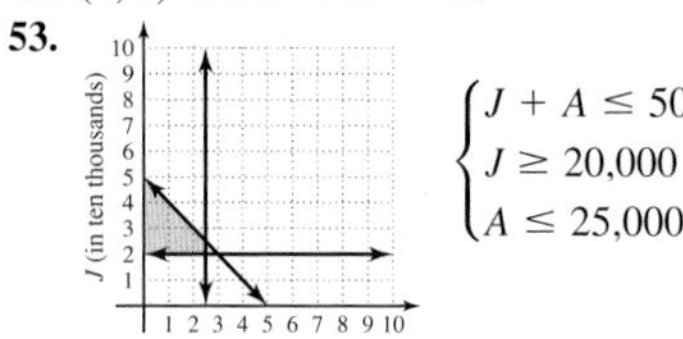

$\begin{cases} J + A \le 50{,}000 \\ J \ge 20{,}000 \\ A \le 25{,}000 \end{cases}$

55. 300 acres of corn; 200 acres of soybeans **57.** 240 sheet metal screws; 480 wood screws **59.** 220,000 gallons from Tulsa to Colorado; 100,000 gallons from Tulsa to Mississippi; 0 thousand gallons from Houston to Colorado; 150,000 gallons from Houston to Mississippi
61. Yes, answers will vary.
63. $(3, 3)$; optimal solutions occur at vertices

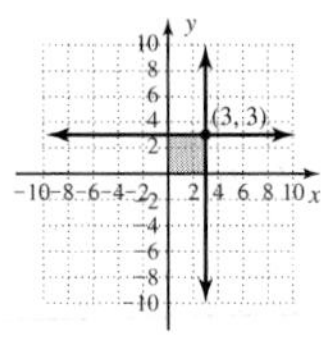

65. max: $212.\overline{6}$ at $\left(\frac{29}{3}, 0\right)$ **67.** $x \in (-3, -2) \cup (3, \infty)$ **69.** 324 Ω
71. $\approx 221.46, 243.01, 248.41$

Exercises 6.4, pp. 610–613

1. reverse **3.** 7, −7 **5.** no solution; answers will vary
7. $|3m + 7| < 3.3$ **9.** $|p| = 0.5$ **11.** $\left|\frac{x}{2} + 4\right| > 1$
The graphs for 13, 15, 17, 19, 21, 23, 25 and 27 appear on page SA-45.
13. $x = 2, x = -8$ **15.** no solution **17.** $x = 4, x = -4$
19. $x \in [-5, 9]$ **21.** no solution **23.** $x \in (-4, -1)$
25. $x \in (-\infty, -10) \cup (4, \infty)$ **27.** $x \in (-\infty, -3] \cup [3, \infty)$
29. $Y_2 \ge Y_1$ **31.** $x = -9, x = 3$ **33.** $n = \frac{-17}{2}, n = \frac{3}{2}$
35. $n = -6, n = -2$ **37.** $x = 7, x = -2, x = 6, x = -1$
39. $x = -9, x = 5, x = -7, x = 3$ **41.** $x = \frac{-10}{3}, x = 4, x = \frac{-4}{3}, x = 2$

43. $n = 1.8, n = 9.8$ **45.** $x = -3.5, x = 1$ **47.** $n = -16.2, n = -2.2$ **49.** $m \in [-1, 5]$ **51.** $n \in \left(\frac{-9}{2}, \frac{3}{2}\right)$ **53.** no solution **55.** $x \in \left(-\infty, \frac{-9}{2}\right] \cup \left[\frac{-5}{2}, \infty\right)$ **57.** $p \in (-\infty, \infty)$ **59.** $x \in \left(-\infty, \frac{5}{8}\right) \cup \left(\frac{7}{8}, \infty\right)$ **61.** $q \in (-\infty, -1] \cup \left[\frac{13}{3}, \infty\right)$ **63.** $x \in [-1, 3]$ **65.** $x \in (-\infty, -3] \cup [3, \infty)$ **67.** $x \in (-\infty, -4] \cup [-2, 2] \cup [4, \infty)$ **69.** 45 in. $\le d \le$ 51 in. **71.** $2{,}325{,}000 \le B \le 2{,}575{,}000$ **73.** lowest \$41,342; highest \$65,330 **75.** $6.5 < d < 13.5$ **77.** $|W - 14| \le 0.1$; $W \in [13.9, 14.1]$; weight must be at least 13.9 oz but no more than 14.1 oz **79. a.** $x = 4$ **b.** $x \in \left[\frac{4}{3}, 4\right]$ **c.** $x = 0$ **d.** $x \in \left(-\infty, \frac{3}{5}\right]$ **81.** $x \in (-1.005, -0.995)$

83.

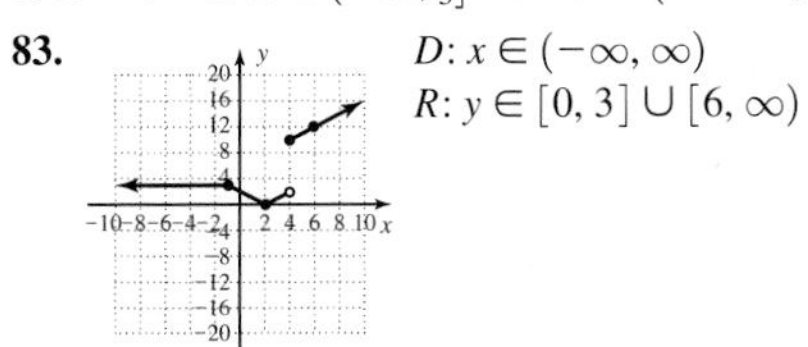

$D: x \in (-\infty, \infty)$
$R: y \in [0, 3] \cup [6, \infty)$

85. $x = \dfrac{\ln 2.5 - 1}{-0.025}$; $x \approx 3.35$ **87. a.** $x^3 + 3x^2 - 11x + 2$ **b.** $x^4 + x^3 - 16x^2 + 20x$ **c.** $x^2 + 5x$ **d.** $x^3 - 3x^2 - 10x + 24$

Mid-Chapter Check, p. 613

1. (1, 1); consistent **2.** (5, 3); consistent **3.** 20 oz **4.** no **5.** The second equation is a multiple of the first equation. **6.** (1, 2, 3) **7.** (1, 2, 3) **8.** no solution **9.** Mozart = 8 yr; Morphy = 13 yr; Pascal = 16 yr **10.** 2 table candles, 9 holiday candles

Reinforcing Basic Concepts, pp. 613–615

Ex 1 $\begin{cases} 15.3R + 35.7P = 125.97 \\ P = R + 0.10 \end{cases}$, Premium: \$2.50/gal, Regular: \$2.40/gal

Ex 2 $\begin{cases} 4520x + 7980y = 1056.56 \\ y = x + 0.0345 \end{cases}$, Bond: 6.25%, Second account: 9.7%

Exercises 6.5, pp. 623–626

1. square **3.** 2, 3, 1 **5.** Multiply R1 by −2 and add that result to R2. **7.** 3×2, 1 **9.** 3×4, 3 **11.** $\left[\begin{array}{ccc|c} 1 & 2 & -1 & 1 \\ 1 & 0 & 1 & 3 \\ 2 & -1 & 1 & 3 \end{array}\right]$; diagonal entries 1, 0, 1

13. $\begin{cases} x + 4y = 5 \\ y = \frac{1}{2} \end{cases} \rightarrow \left(3, \frac{1}{2}\right)$ **15.** $\begin{cases} x + 2y - z = 0 \\ y + 2z = 2 \\ z = 3 \end{cases} \rightarrow (11, -4, 3)$

17. $\begin{cases} x + 3y - 4z = 29 \\ y - \frac{3}{2}z = \frac{21}{2} \\ z = 3 \end{cases} \rightarrow (-4, 15, 3)$ **19.** $\left[\begin{array}{cc|c} 1 & -6 & -2 \\ 0 & -28 & -6 \end{array}\right]$

21. $\left[\begin{array}{ccc|c} 1 & -3 & 3 & 2 \\ 0 & 23 & -12 & -15 \\ -2 & 1 & 0 & 4 \end{array}\right]$ **23.** $\left[\begin{array}{ccc|c} 3 & 1 & 1 & 8 \\ 0 & -3 & -3 & -6 \\ 0 & -10 & -13 & 34 \end{array}\right]$

25. $2R_1 + R_2 \rightarrow R_2$, $-3R_1 + R_3 \rightarrow R_3$ **27.** $-5R_1 + R_2 \rightarrow R_2$, $4R_1 + R_3 \rightarrow R_3$ **29.** (20, 10) **31.** (1, 6, 9) **33.** (1, 1, 2) **35.** (1, 1, 1) **37.** $\left(-1, \frac{-3}{2}, 2\right)$ **39.** (0, 0, 4) **41.** dependent $\{(x, y, z) | 3x - 4y + 2z = -2\}$ **43.** no solution **45.** $\left(-\frac{5}{4}z - 3, \frac{1}{8}z - \frac{1}{2}, z\right)$ **47.** 28.5 units^2 **49.** 98 to 99 **51.** Poe, \$12,500; Baum, \$62,500; Wouk, \$25,000 **53.** $A = 35°$, $B = 45°$, $C = 100°$ **55.** \$.4 million at 4%; \$.6 million at 7%; \$1.5 million at 8% **57.** Answers will vary. **59.** $a = \frac{3}{4}, b = \frac{1}{2}$ **61.** $x = \pm 2$; $x = \pm\sqrt{3}\,i$ **63.** $p^{-1}(x) = \sqrt[3]{x + 8}$ **65.** \$83,811.95

Exercises 6.6, pp. 635–639

1. a_{ij}, b_{ij} **3.** scalar **5.** Answers will vary. **7.** $2 \times 2, a_{12} = -3, a_{21} = 5$ **9.** $2 \times 3, a_{12} = -3, a_{23} = 6, a_{22} = 5$ **11.** $3 \times 3, a_{12} = 1, a_{23} = 1, a_{31} = 5$ **13.** true **15.** conditional, $c = -2$, $a = -4$, $b = 3$

17. $\begin{bmatrix} 10 & 0 \\ 0 & 10 \end{bmatrix}$ **19.** different order, sum not possible **21.** $\begin{bmatrix} 20 & -15 \\ -25 & -10 \end{bmatrix}$

23. $\begin{bmatrix} \frac{-5}{2} & -1 & 0 \\ 0 & \frac{-7}{2} & 1 \\ 2 & \frac{3}{2} & -6 \end{bmatrix}$ **25.** $\begin{bmatrix} 1 & -2 & 0 \\ 0 & -1 & 2 \\ 4 & 3 & -6 \end{bmatrix}$ **27.** $\begin{bmatrix} 1 & 0 \\ 0 & 1 \end{bmatrix}$

29. $\begin{bmatrix} 6 & -3 & 9 \\ 12 & 0 & -6 \end{bmatrix}$ **31.** $\begin{bmatrix} 12 & -24 & 90 \\ -6 & 15 & -57 \end{bmatrix}$ **33.** $\begin{bmatrix} 79 & -30 \\ -50 & 19 \end{bmatrix}$

35. $\begin{bmatrix} 42 & 18 & -60 \\ -12 & -42 & 36 \end{bmatrix}$ **37.** $\begin{bmatrix} 0.71 & 0.65 \\ 1.78 & 3.55 \end{bmatrix}$

39. $\begin{bmatrix} 1 & -1.25 & 0.25 \\ -0.5 & -0.63 & 2.13 \\ 3.75 & 3.69 & -5.94 \end{bmatrix}$ **41.** $\begin{bmatrix} 1 & 0 \\ 0 & 1 \end{bmatrix}$ **43.** $\begin{bmatrix} 1 & 0 & 0 \\ 0 & 1 & 0 \\ 0 & 0 & 1 \end{bmatrix}$

45. $\begin{bmatrix} \frac{-3}{19} & \frac{4}{57} \\ \frac{1}{19} & \frac{5}{57} \end{bmatrix}$ **47.** $\begin{bmatrix} 0 & \frac{3}{4} & \frac{1}{4} \\ \frac{-1}{2} & \frac{3}{8} & \frac{1}{8} \\ \frac{-1}{4} & \frac{11}{16} & \frac{1}{16} \end{bmatrix}$ **49.** $\begin{bmatrix} 1.75 & 2.5 \\ 7.5 & 13 \end{bmatrix}$

51. $\begin{bmatrix} -0.26 & 0.32 & -0.07 \\ 0.63 & 0.30 & 0.10 \end{bmatrix}$ **53.** verified **55.** verified

57. $P = 21.448$ cm; $A = 27.7269$ cm^2

59. a. $V \rightarrow \begin{array}{c} \\ S \\ D \\ P \end{array}\begin{array}{cc} T & S \\ \left[\begin{array}{cc} 3820 & 1960 \\ 2460 & 1240 \\ 1540 & 920 \end{array}\right] \end{array}$, $M \rightarrow \begin{array}{c} \\ S \\ D \\ P \end{array}\begin{array}{cc} T & S \\ \left[\begin{array}{cc} 4220 & 2960 \\ 2960 & 3240 \\ 1640 & 820 \end{array}\right] \end{array}$

b. 3900 more by Minsk

c. $V \rightarrow \begin{bmatrix} 3972.8 & 2038.4 \\ 2558.4 & 1289.6 \\ 1601.6 & 956.8 \end{bmatrix}$, $M \rightarrow \begin{bmatrix} 4388.8 & 3078.4 \\ 3078.4 & 3369.6 \\ 1705.6 & 852.8 \end{bmatrix}$

d. $\begin{bmatrix} 8361.6 & 5116.8 \\ 5636.8 & 4659.2 \\ 3307.2 & 1809.6 \end{bmatrix}$

61. [22,000 19,000 23,500 14,000]; *total profit*
North: \$22,000
South: \$19,000
East: \$23,500
West: \$14,000

63. a. \$108.20 **b.** \$101 **c.** $\begin{array}{r} \text{Science} \\ \text{Math} \end{array}\begin{bmatrix} 100 & 101 & 119 \\ 108.2 & 107 & 129.5 \end{bmatrix}$, First row, total cost for science from each restaurant; Second row, total cost for math from each restaurant. **65. a.** 10.3 **b.** 19.5

c. $\begin{array}{r} \\ \text{Female} \\ \text{Male} \end{array}\begin{array}{ccc} \text{Spanish} & \text{Chess} & \text{Writing} \\ \left[\begin{array}{ccc} 32.4 & 10.3 & 21.3 \\ 29.9 & 9.6 & 19.5 \end{array}\right] \end{array}$,

the number of females in the writing club.

67. no; no; A and B must be square matrices

69. $\begin{bmatrix} 2^{n-1} & 0 & 2^{n-1} \\ 2^n - 1 & 1 & 2^n - 1 \\ 2^{n-1} & 0 & 2^{n-1} \end{bmatrix}$ **71.** $a = 2, b = 1, c = -3, d = -2$

73. $x \in [-3, -1] \cup [1, 3]$ **75.** $x^2 + 2x - 5$ **77.**

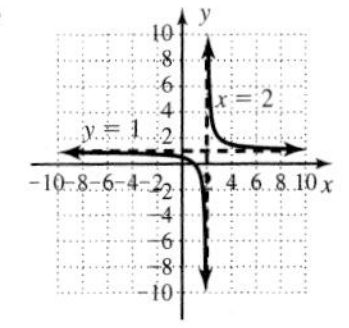

Exercises 6.7, pp. 649-654

1. main, diagonal, zeroes **3.** identity **5.** Answers will vary.

7. verified **9.** verified **11.** verified **13.** verified **15.** $\begin{bmatrix} \frac{1}{9} & \frac{2}{9} \\ \frac{-1}{9} & \frac{5}{18} \end{bmatrix}$

17. $\begin{bmatrix} -5 & 1.5 \\ -2 & 0.5 \end{bmatrix}$ **19.** verified **21.** verified **23.** $\begin{bmatrix} \frac{-2}{39} & \frac{1}{13} & \frac{10}{39} \\ \frac{1}{3} & 0 & \frac{1}{3} \\ \frac{-4}{39} & \frac{2}{13} & \frac{-19}{39} \end{bmatrix}$

25. $\begin{bmatrix} \frac{-9}{80} & \frac{31}{400} & \frac{27}{400} \\ \frac{1}{80} & \frac{41}{400} & \frac{-3}{400} \\ \frac{-1}{20} & \frac{-1}{100} & \frac{-17}{100} \end{bmatrix}$ **27.** $\begin{bmatrix} 2 & -3 \\ -5 & 7 \end{bmatrix}\begin{bmatrix} x \\ y \end{bmatrix} = \begin{bmatrix} 9 \\ 8 \end{bmatrix}$

29. $\begin{bmatrix} 1 & 2 & -1 \\ 1 & 0 & 1 \\ 2 & -1 & 1 \end{bmatrix}\begin{bmatrix} x \\ y \\ z \end{bmatrix} = \begin{bmatrix} 1 \\ 3 \\ 3 \end{bmatrix}$ **31.** $\begin{bmatrix} -2 & 1 & -4 & 5 \\ 2 & -5 & 1 & -3 \\ -3 & 1 & 6 & 1 \\ 1 & 4 & -5 & 1 \end{bmatrix}\begin{bmatrix} w \\ x \\ y \\ z \end{bmatrix} = \begin{bmatrix} -3 \\ 4 \\ 1 \\ -9 \end{bmatrix}$

33. (4, 5) **35.** (12, 12) **37.** no solution **39.** (1.5, −0.5, −1.5) **41.** no solution **43.** (−1, −0.5, 1.5, 0.5) **45.** 1, yes **47.** 0, no **49.** 1 **51.** singular matrix **53.** singular matrix **55.** $\det(A) = -5$; (1, 6, 9)

57. $\det(A) = 0$ **59.** $A^{-1} = \begin{bmatrix} \frac{1}{13} & \frac{5}{13} \\ \frac{-2}{13} & \frac{3}{13} \end{bmatrix}$ **61.** singular

63. 31 behemoth, 52 gargantuan, 78 mammoth, 30 jumbo **65.** 3.75 min; 3.75 min; 7.5 min; 5.75 min **67.** 30 of clock A; 20 of clock B; 40 of clock C; 12 of clock D **69.** $y = x^3 + 2x^2 - 9x - 10$ **71.** 2 oz Food I, 1 oz Food II, 4 oz Food III **73.** Answers will vary.

75. $A^{-1} = \begin{bmatrix} 1 & 2 & -1 \\ 0 & 2 & -1 \\ -1 & -5 & 3 \end{bmatrix}$ **77.** Answers will vary.

79. They are inverses. **81.** $x = 6, x = 3, x = -2$ **83. a.** shifts right 2 **b.** shifts down 2

Exercises 6.8, pp. 662–665

1. $a_{11}a_{22} - a_{21}a_{12}$ **3.** constant **5.** Answers will vary.

7. $D = \begin{vmatrix} 2 & 5 \\ -3 & 4 \end{vmatrix}; D_x = \begin{vmatrix} 7 & 5 \\ 1 & 4 \end{vmatrix}; D_y = \begin{vmatrix} 2 & 7 \\ -3 & 1 \end{vmatrix}$ **9.** (−5, 9)

11. $\left(\frac{-26}{3}, \frac{25}{3}\right)$ **13.** no solution **15. a.** $D = \begin{vmatrix} 4 & -1 & 2 \\ -3 & 2 & -1 \\ 1 & -5 & 3 \end{vmatrix}$,

$D_x = \begin{vmatrix} -5 & -1 & 2 \\ 8 & 2 & -1 \\ -3 & -5 & 3 \end{vmatrix}$ $D_y = \begin{vmatrix} 4 & -5 & 2 \\ -3 & 8 & -1 \\ 1 & -3 & 3 \end{vmatrix}$,

$D_z = \begin{vmatrix} 4 & -1 & -5 \\ -3 & 2 & 8 \\ 1 & -5 & -3 \end{vmatrix}$, $D = 22$, solutions possible

b. $D = 0$, Cramer's rule cannot be used. **17.** (1, 2, 1) **19.** $\left(\frac{3}{4}, \frac{5}{3}, \frac{-1}{3}\right)$

21. (0, −1, 2, −3) **23.** $320 + 32\pi \approx 420.5$ in² **25.** 8 cm² **27.** 27 ft²

29. 19 m³ **31.** yes **33.** no **35.** yes yes yes **37.** $\frac{A}{x+3} + \frac{B}{x-2}$

39. $\frac{A}{x-1} + \frac{B}{x+2} + \frac{C}{x-3}$ **41.** $\frac{A}{x} + \frac{B}{x-3} + \frac{C}{x+1}$

43. $\frac{A}{x} + \frac{B}{x^2} + \frac{C}{x+2}$ **45.** $\frac{A}{x+1} + \frac{Bx+C}{x^2+2} + \frac{Dx+E}{(x^2+2)^2}$

47. $\frac{4}{x} - \frac{5}{x+1}$ **49.** $\frac{-4}{2x-5} + \frac{3}{x+3}$ **51.** $\frac{7}{x} + \frac{2}{x+1} - \frac{1}{x-1}$

53. $\frac{-1}{x} + \frac{4}{x+1} + \frac{5}{(x+1)^2}$ **55.** $\frac{3}{2-x} - \frac{4}{4+2x+x^2}$

57. $\frac{1}{x} + \frac{x-2}{(x^2-1)^2}$ **59.** $\begin{cases} 15{,}000x + 25{,}000y = 2900 \\ 25{,}000x + 15{,}000y = 2700 \end{cases}$; 6%, 8%

61. $y \approx 0.0257x^4 - 0.7150x^3 + 6.7852x^2 - 25.1022x + 35.1111$, May: \$5900; July: \$8300; November $(4.7 > 4.1)$ **63.** Answers will vary.

65. $A = 195$ units² **67.** 4, $-2\sqrt{3}\,i$, 7, $\frac{1}{7} - \frac{4\sqrt{3}}{7}i$

69. $3^{2x-1} = 3^{4-2x}$; $x = 1.25$ **71.** left graph; graph shifts left 1

Summary and Concept Review, pp. 666–671

The graphs for 1, 2 and 3 appear on page SA-45.
1. (4, 4) **2.** $\left(\frac{1}{2}, -3\right)$ **3.** $\left(\frac{8}{5}, \frac{-6}{5}\right)$ **4.** no solution; inconsistent **5.** (5, −1); consistent **6.** (7, 2); consistent **7.** (3, −1); consistent **8.** (2, 2); consistent **9.** $\left(\frac{11}{4}, \frac{-1}{6}\right)$; consistent **10.** Sears Tower is 1450 ft; Hancock Building is 1127 ft. **11.** (0, 3, 2) **12.** (1, 1, 1) **13.** no solution, inconsistent **14.** 72 nickels, 85 dimes, 60 quarters **15.** \$80,000 at 4%, \$90,000 at 5%, \$110,000 at 7%

16.

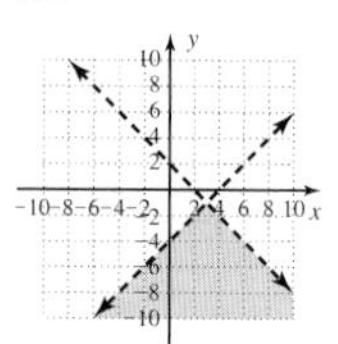

17.

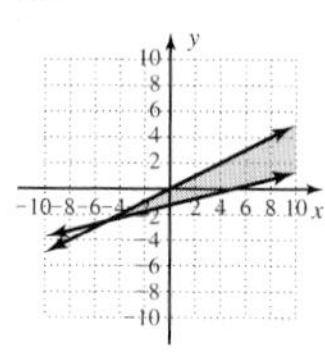

18.

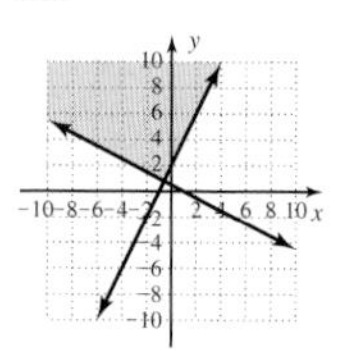

19. Maximum of 270 occurs at both (0, 6) and (3, 4).

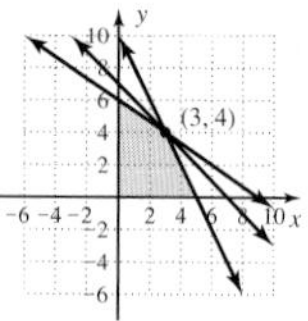

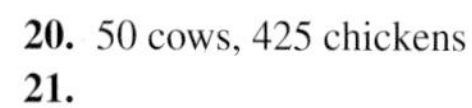

20. 50 cows, 425 chickens

21.

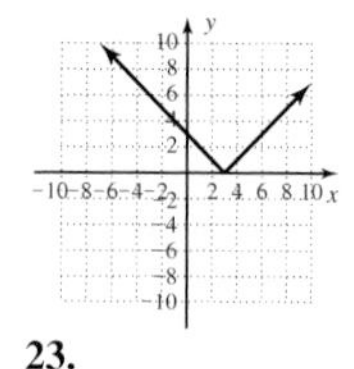

22.

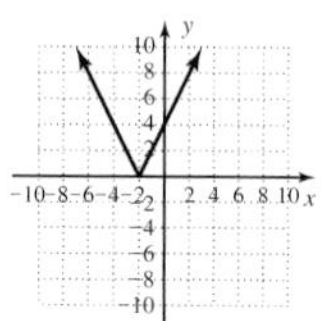

23.

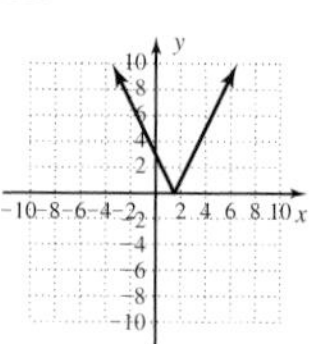

24. $x = -4, x = -1$ **25.** $x \in (-\infty, -6) \cup (2, \infty)$ **26.** $x \in (-\infty, \infty)$ **27.** $x \in [-2, 6]$ **28.** no solution **29.** $x \in (-10, 41)$ **30.** (−2, −4) **31.** (1, 6, 9) **32.** (−2, 7, 1, 8)

33. $\begin{bmatrix} -7.25 & 5.25 \\ 0.875 & -2.875 \end{bmatrix}$ **34.** $\begin{bmatrix} -6.75 & 6.75 \\ 1.125 & -1.125 \end{bmatrix}$ **35.** not possible

36. $\begin{bmatrix} -2 & -6 \\ -1 & -7 \end{bmatrix}$ **37.** $\begin{bmatrix} 1 & 0 \\ 0 & 1 \end{bmatrix}$ **38.** $\begin{bmatrix} 1 & 0 & 4 \\ 5.5 & -1 & -1 \\ 10 & -2.9 & 7 \end{bmatrix}$

39. $\begin{bmatrix} 3 & -6 & -4 \\ -4.5 & 3 & -1 \\ -2 & 3.1 & 3 \end{bmatrix}$ **40.** not possible **41.** $\begin{bmatrix} -8 & 12 & 0 \\ -2 & -4 & 4 \\ -16 & -0.4 & -20 \end{bmatrix}$

42. $\begin{bmatrix} 15.5 & 6.4 & 17 \\ 9 & -17 & 2 \\ 18.5 & -20.8 & 13 \end{bmatrix}$ **43.** D **44.** It's an identity.

45. It's the inverse of B. **46.** E **47.** It's an identity matrix **48.** It's the inverse of F. **49.** verified; matrix multiplication is not generally commutative

50. (−8, −6) **51.** (2, 0, −3) **52.** $\left(\frac{-19}{35}, \frac{25}{14}\right)$ **53.** $\left(\frac{-37}{19}, \frac{36}{19}, \frac{31}{19}\right)$

54. (1, −1, 2) **55.** $\frac{91}{2}$ units² **56.** $\frac{5}{x-2} + \frac{2x-1}{x^2+3}$

Mixed Review, pp. 671–673

1. a. $\begin{cases} y = \frac{3}{5}x + 2 \\ y = \frac{3}{5}x + 2; \end{cases}$ consistent/dependent

b. $\begin{cases} y = \frac{4}{3}x - 3 \\ y = \frac{2}{5}x - 2; \end{cases}$ consistent/independent **c.** $\begin{cases} y = \frac{1}{3}x - 3 \\ y = \frac{1}{3}x - \frac{5}{3}; \end{cases}$ inconsistent

3. $(-2, 3)$ **5.** $(1, \frac{1}{2}, 2)$ **7.** $(-10, 12)$ **9.** $x \in [-7, 1]$

11. a. $\begin{bmatrix} -8 & 16 & -10 \\ 12 & 0 & 6 \end{bmatrix}$ **b.** $\begin{bmatrix} 9 & -6 & -7 \\ -7 & -1 & 2 \end{bmatrix}$ **13.** $(-9, -3, 2)$

15. $(\frac{33}{31}, \frac{-10}{31}, \frac{-57}{31})$ **17.** **19.** 7 unicycles; 9 bicycles; 5 tricycles

Practice Test, pp. 673–674

The graphs for 1 and 19 appear on page SA-45.

1. $(2, 3)$ **2.** $\left(\frac{2}{5}, \frac{-4}{5}\right)$ **3.** $(-3, 2)$ **4.** $(2, -1, 4)$ **5. a.** $\begin{bmatrix} -6 & -5 \\ 8 & 9 \end{bmatrix}$

b. $\begin{bmatrix} 1.2 & 1.2 \\ -1.2 & -2 \end{bmatrix}$ **c.** $\begin{bmatrix} -3 & 1 \\ 3 & -5 \end{bmatrix}$ **d.** $\begin{bmatrix} -2 & -1 \\ 2.5 & 1.5 \end{bmatrix}$ **e.** -2

6. a. $\begin{bmatrix} 0 & -0.1 & 0 \\ 0.5 & -0.6 & 0 \\ -0.2 & -0.8 & -0.9 \end{bmatrix}$ **b.** $\begin{bmatrix} -0.3 & -0.06 & -0.12 \\ 0.06 & -0.06 & 0 \\ -0.18 & -0.24 & -0.48 \end{bmatrix}$

c. $\begin{bmatrix} 0.31 & -0.13 & 0.08 \\ -0.01 & -0.05 & -0.02 \\ 0.39 & -0.52 & -0.02 \end{bmatrix}$ **d.** $\begin{bmatrix} \frac{40}{17} & 0 & \frac{-10}{17} \\ \frac{40}{17} & 10 & \frac{-10}{17} \\ \frac{-35}{17} & -5 & \frac{30}{17} \end{bmatrix}$ **e.** $\frac{17}{500}$

7. $\left(2, 1, \frac{-1}{3}\right)$ **8.** $(3, -2, 3)$ **9.** $\left(\frac{97}{34}, \frac{-18}{17}\right)$ **10.** $(1, 6, 9)$

11. 21.59 cm by 35.56 cm **12.** Tahiti 402 mi^2; Tonga 290 mi^2

13. Corn 25¢, Beans 20¢, Peas 29¢ **14.** \$15,000 at 7%, \$8000 at 5%, \$7000 at 9% **15.** $x \in (-\infty, -1) \cup (5, \infty)$

16. $x \in (-\infty, -7] \cup [1, \infty)$ **17.**

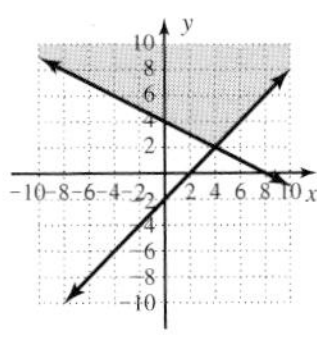

18. $(5, 0)$

19. 30 plain; 20 deluxe **20.** $\dfrac{1}{x - 3} + \dfrac{3x + 2}{x^2 + 3x + 9}$

Strengthening Core Skills, pp. 676–677

Exercise 1: $(1, -4, 1)$

Cumulative Review Chapters 1–6

1.

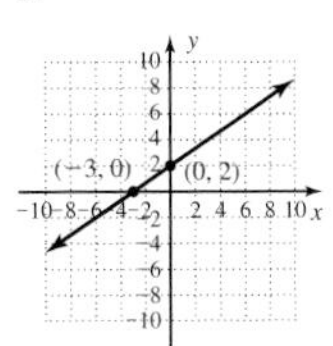

3.

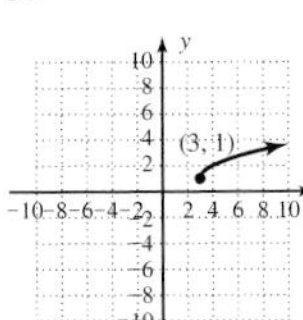

5.

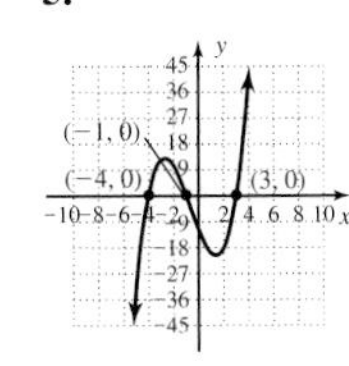

7. a. $D: x \in (-\infty, \infty)$ **b.** $R: y \in (-\infty, 4)$ **c.** $f(x)\uparrow: x \in (-\infty, 1)$, $f(x)\downarrow: x \in (-1, \infty)$ **d.** n/a **e.** max: $(-1, 4)$

f. $f(x) > 0$: $x \in (-4, 2)$, $f(x) < 0$: $x \in (-\infty, -4) \cup (2, \infty)$

g. $\dfrac{\Delta y}{\Delta x} = \dfrac{7}{4}$ **9.** $x = 9$, $\pm\sqrt{2}\,i$ **11.** $3x^2 + 5$ **13.** 1

15. $(x - 2)(x + 2)(x - 3 - 3\sqrt{2})(x - 3 + 3\sqrt{2})$

17. $x \in (-\infty, \frac{-11}{2}] \cup [-3, \infty)$ **19.** $(1, 3)$ **21.** ≈ 9.7 yr

23. $\dfrac{3}{x + 3} + \dfrac{1}{x - 3}$ **25. a.**

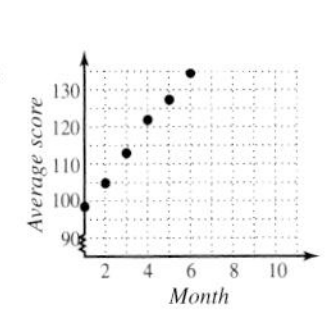

b. $y = 8.14x + 86.67$ **c.** increase of 8 pts per month **d.** 9th month

CHAPTER 7

Exercises 7.1, pp. 690–694

1. radius, center **3.** complete, square, 1 **5.** Answers will vary.

7. $x^2 + y^2 = 9$ **9.** $(x - 5)^2 + y^2 = 3$ **11.** $(x - 4)^2 + (y + 3)^2 = 4$

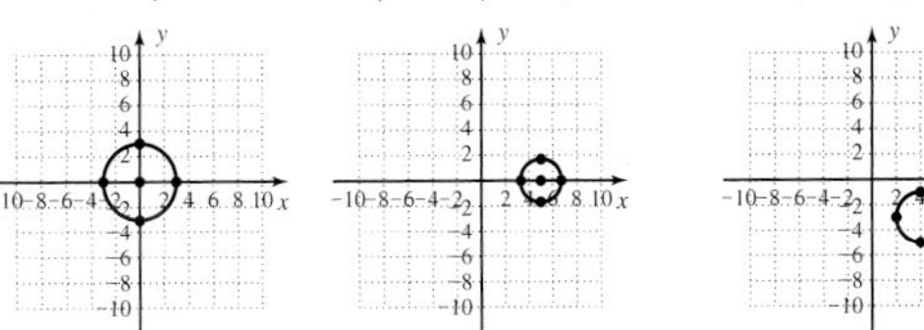

13. $(x + 7)^2 + (y + 4)^2 = 7$ **15.** $(x - 1)^2 + (y + 2)^2 = 12$

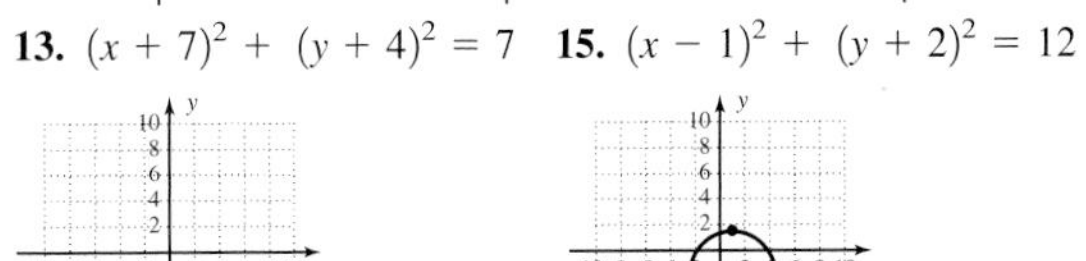

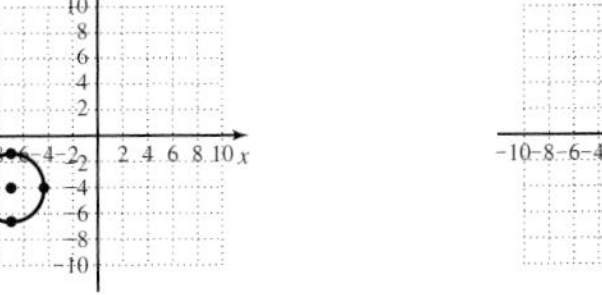

17. $(x - 4)^2 + (y - 5)^2 = 12$ **19.** $(x - 7)^2 + (y - 1)^2 = 100$

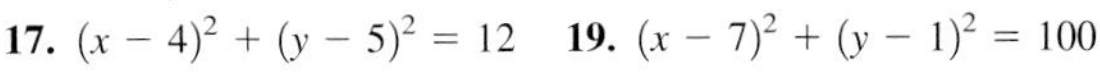

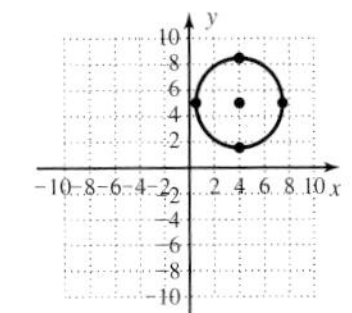

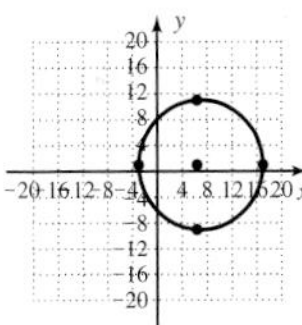

21. $(x - 3)^2 + (y - 4)^2 = 41$ **23.** $(x - 5)^2 + (y - 4)^2 = 9$

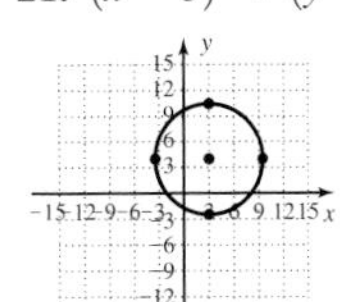

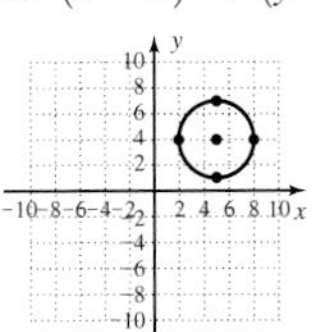

25. $(2, 3)$, $r = 2$, $x \in [0, 4]$, $y \in [1, 5]$

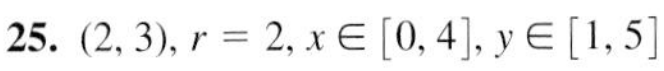

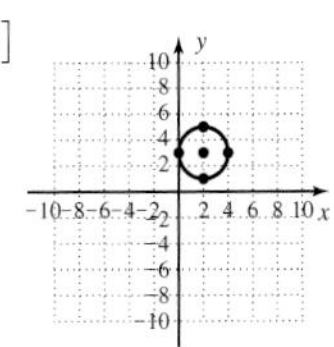

27. $(-1, 2)$, $r = 2\sqrt{3}$, $x \in [-1 - 2\sqrt{3}, -1 + 2\sqrt{3}]$, $y \in [2 - 2\sqrt{3}, 2 + 2\sqrt{3}]$

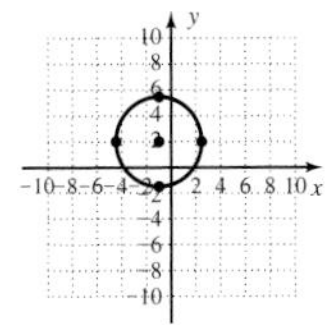

29. $(-4, 0), r = 9, x \in [-13, 5], y \in [-9, 9]$

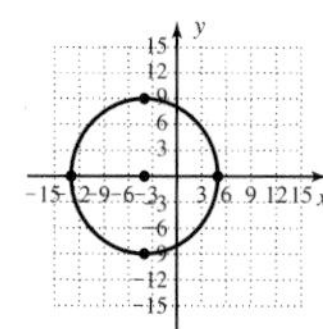

31. $(x - 5)^2 + (y - 6)^2 = 57, (5, 6), r = \sqrt{57}$

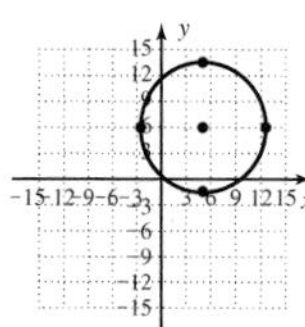

33. $(x - 5)^2 + (y + 2)^2 = 25, (5, -2), r = 5$

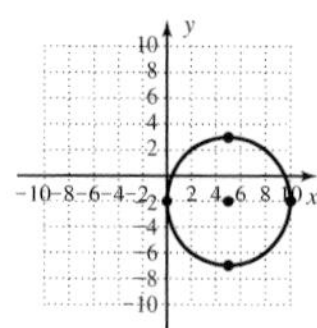

35. $x^2 + (y + 3)^2 = 14, (0, -3), r = \sqrt{14}$

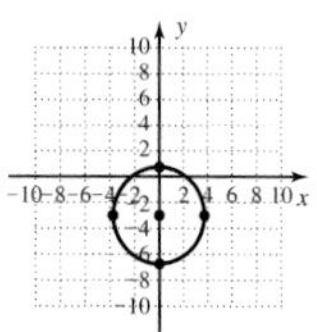

37. $(x + 2)^2 + (y + 5)^2 = 11, (-2, -5),$
$r = \sqrt{11}, a = \sqrt{11}, b = \sqrt{11}$; they are equal

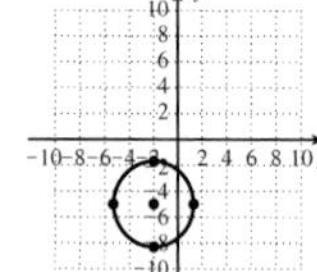

39. $(x + 7)^2 + y^2 = 37, (-7, 0), r = \sqrt{37},$
$a = \sqrt{37}, b = \sqrt{37}$; they are equal

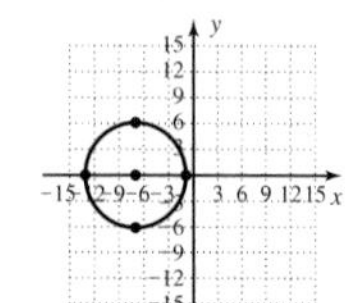

41. $(x - 3)^2 + (y + 5)^2 = 32, (3, -5), r = 4\sqrt{2},$
$a = 4\sqrt{2}, b = 4\sqrt{2}$; they are equal

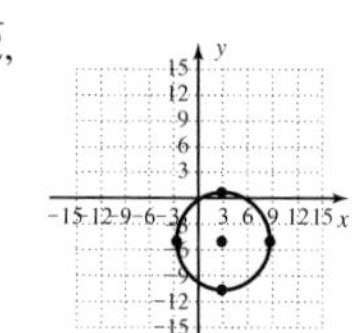

43.

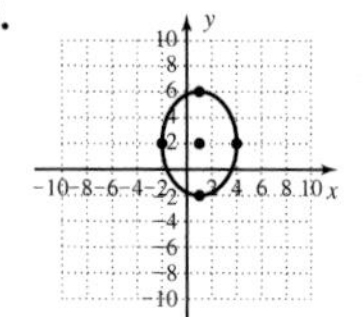

45.

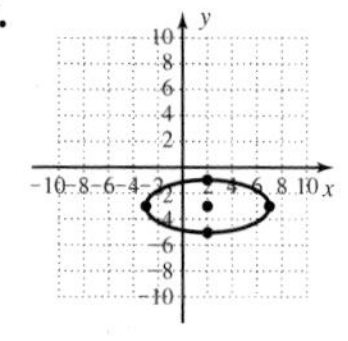

47.

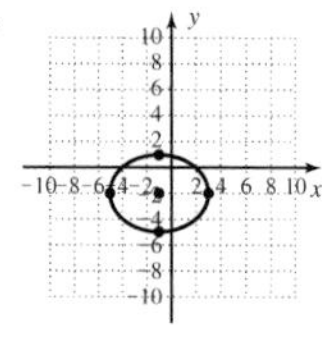

49. ellipse

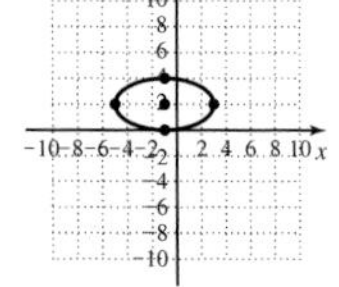

51. circle

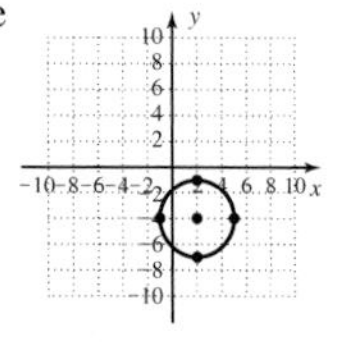

53. ellipse

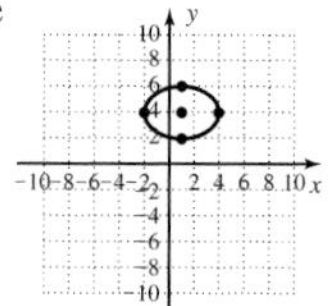

55. a. $\frac{x^2}{16} + \frac{y^2}{4} = 1, (0, 0), a = 4, b = 2$
b. $(-4, 0), (4, 0), (0, -2), (0, 2)$ **c.**

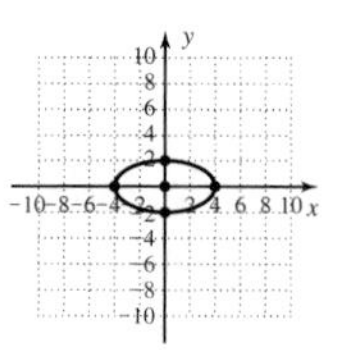

57. a. $\frac{x^2}{9} + \frac{y^2}{16} = 1, (0, 0), a = 3, b = 4$
b. $(0, -4), (0, 4)\ (-3, 0)\ (3, 0)$ **c.**

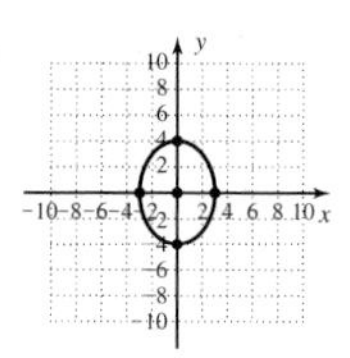

59. a. $\frac{x^2}{5} + \frac{y^2}{2} = 1, (0, 0), a = \sqrt{5}, b = \sqrt{2}$
b. $(-\sqrt{5}, 0), (\sqrt{5}, 0), (0, -\sqrt{2}), (0, \sqrt{2})$ **c.**

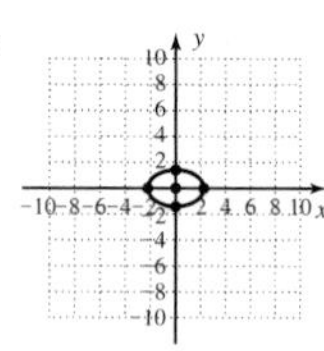

61. $x^2 + \frac{(y + 3)^2}{4} = 1$

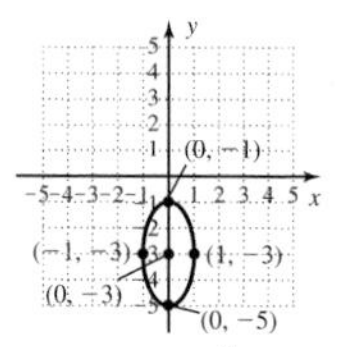

63. $\frac{(x + 2)^2}{16} + \frac{(y - 1)^2}{4} = 1$

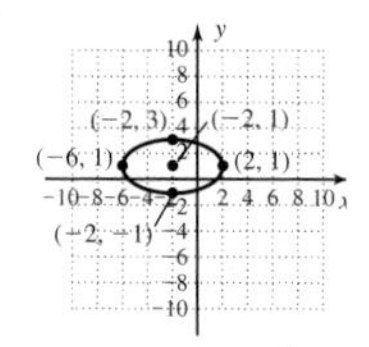

65. $\frac{(x - 3)^2}{4} + \frac{(y + 5)^2}{10} = 1$

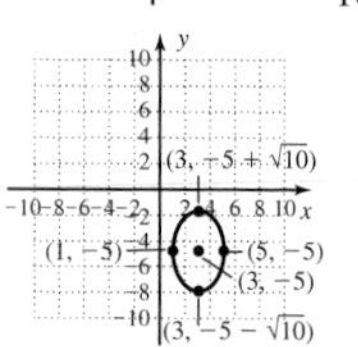

67. $\frac{(x - 3)^2}{25} + \frac{(y + 2)^2}{10} = 1$

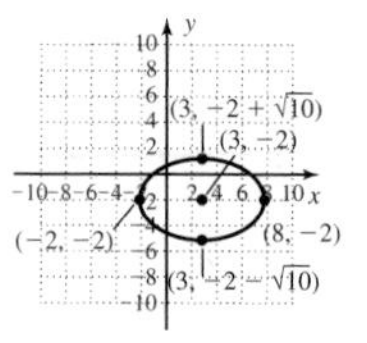

69. $A = 50 \text{ units}^2$ **71.** No **73.** Red: $(x - 2)^2 + (y - 2)^2 = 4$;
Blue: $(x - 2)^2 + y^2 = 16$; Area blue $= 12\pi \text{ units}^2$

75.

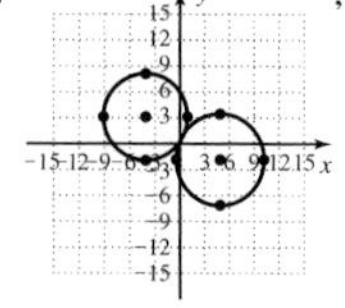

, no; distance between centers is less than sum of radii.

77. $\frac{x^2}{36^2} + \frac{y^2}{(35.25)^2} = 1$ **79.** $9000\pi \text{ yd}^2$

81. $\dfrac{x^2}{15^2} + \dfrac{y^2}{8^2} = 1$, 6.4 ft
83. Answers may vary; aphelion: 155 million miles semimajor axis: 142 million miles
85. x^2 and y^2 must have the same sign. **87.** $\dfrac{(x-2)^2}{9} + \dfrac{(y-1)^2}{4} = 1$
89. $x = 3$ is a zero **91.** $1 - 2i\sqrt{3}$; complex roots must occur in conjugate pairs **93. a.** D: $x \in (-\infty, \infty)$, R: $y \in (-\infty, 4]$ **b.** $f(x) \geq 0$ for $x \in [-3, -1] \cup [1, 5]$ **c.** max: $(-2, 2)$, $(3, 4)$; min: $(0, -1)$
d. $f(x)\uparrow$ for $x \in (-\infty, -2) \cup (0, 3)$; $f(x)\downarrow$ for $x \in (-2, 0) \cup (3, \infty)$

Exercises 7.2, pp. 702–705

1. transverse **3.** midway **5.** Answers will vary.

7.
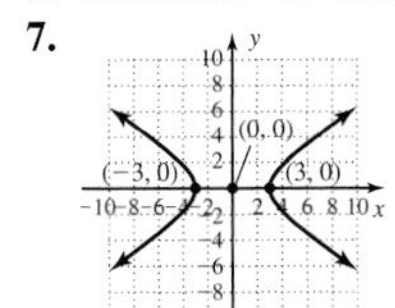

9.
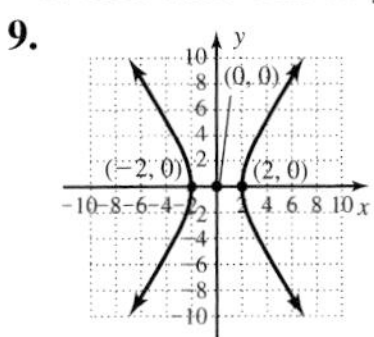

11.
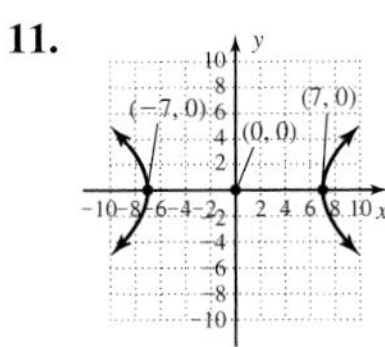

13.
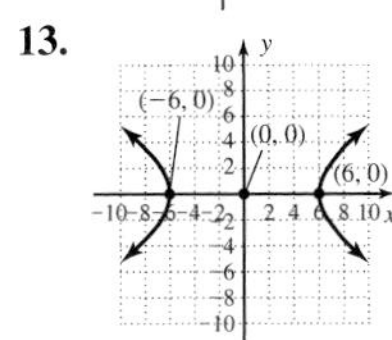

15.
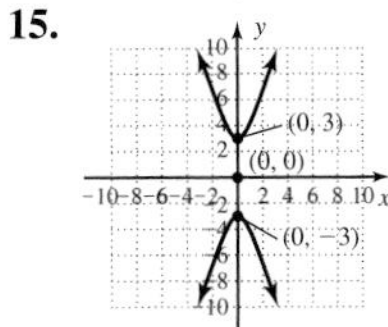

17.
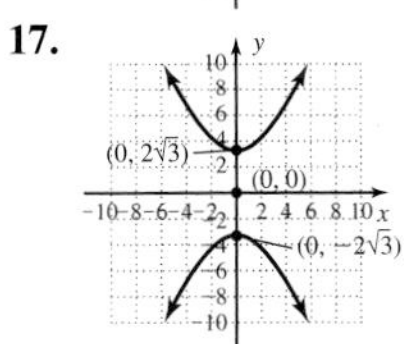

19.
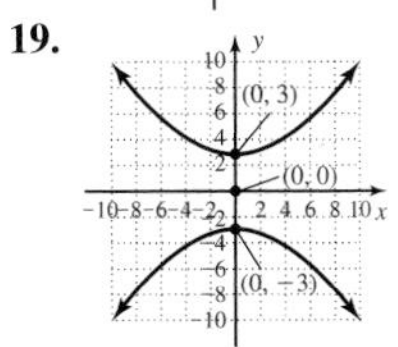

21.
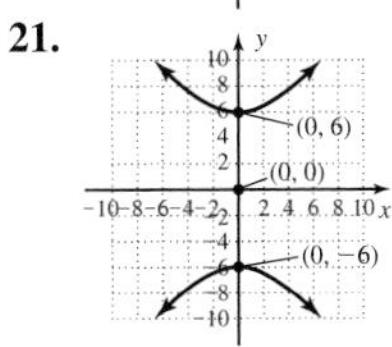

23. $(-4, -2)$, $(2, -2)$, $y = -2$, $(-1, -2)$, $x = -1$
25. $(4, 1)$, $(4, -3)$, $x = 4$, $(4, -1)$, $y = -1$

27.
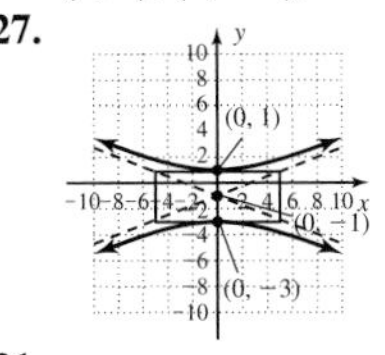

29.
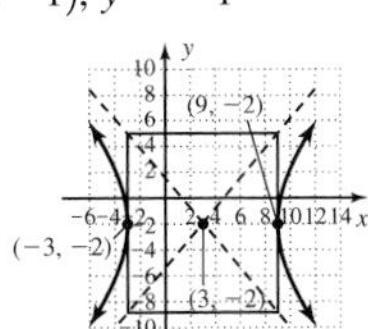

31.
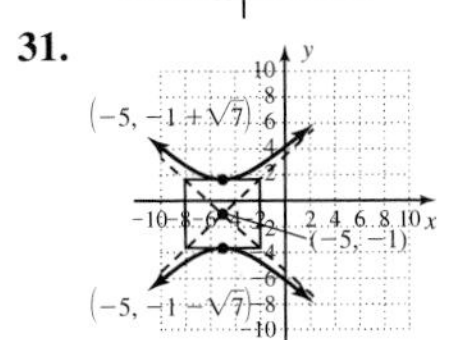

33.
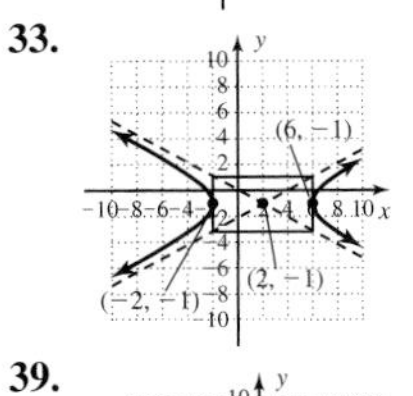

35.
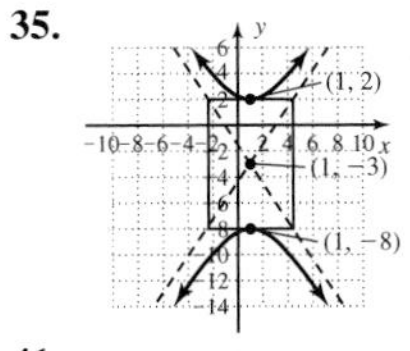

37.
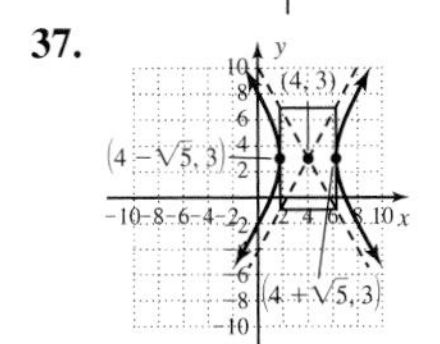

39.
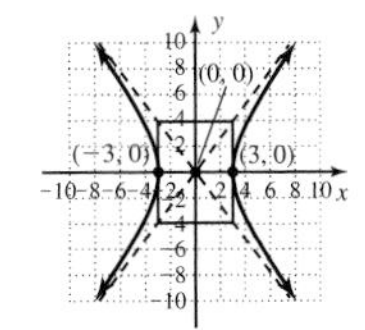

41.
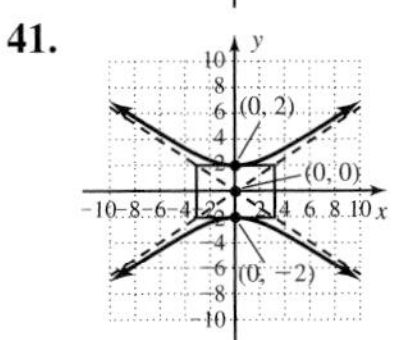

43.
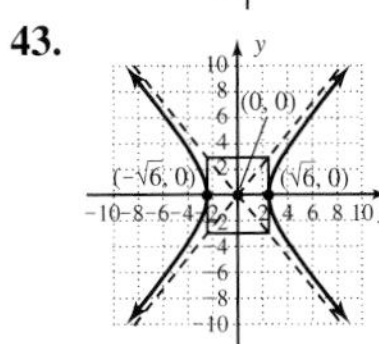

45.
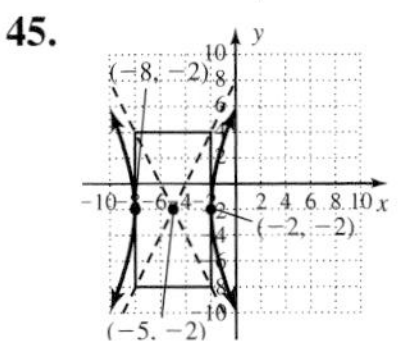

47.
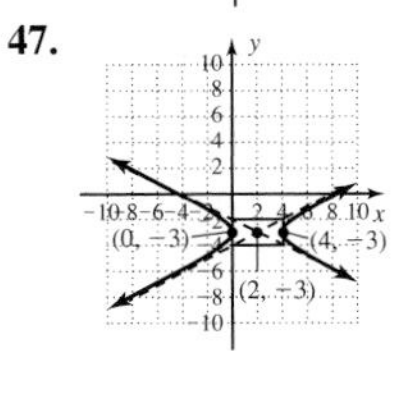

49. circle **51.** circle **53.** hyperbola **55.** hyperbola **57.** circle **59.** ellipse **61. a.** $y = \frac{2}{3}\sqrt{x^2 - 9}$ **b.** $x \in (-\infty, -3] \cup [3, \infty)$ **c.** $y = \frac{-2}{3}\sqrt{x^2 - 9}$ **63.** 40 yd **65.** 40 ft **67.** Answers will vary. **69.** Answers will vary. **71.** $(x-2)^2 + (y-3)^2 = 25$ **73.** a

75.
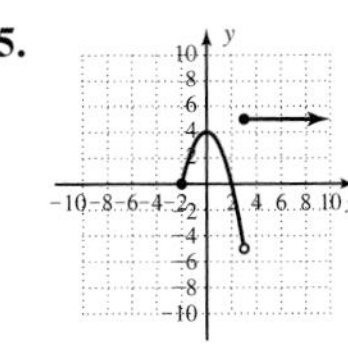

77. $A = 8691.84 \text{ cm}^2$, $P = 455.19$ cm
79. 42 solid, 46 liquid

Mid-Chapter Check, p. 705

1.
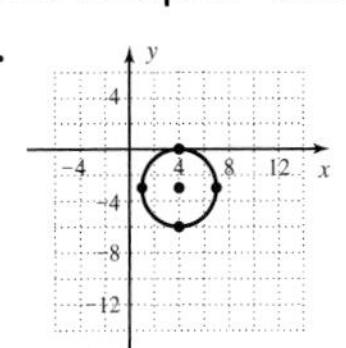

2.
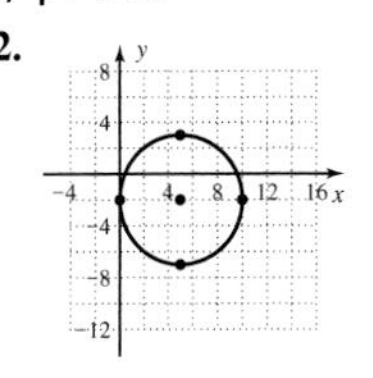

3.
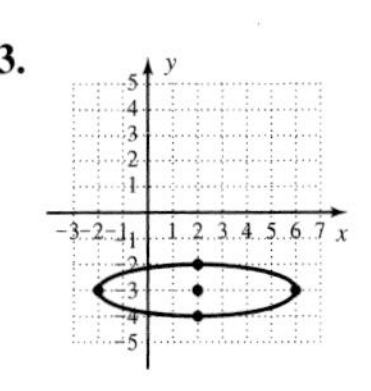

4.
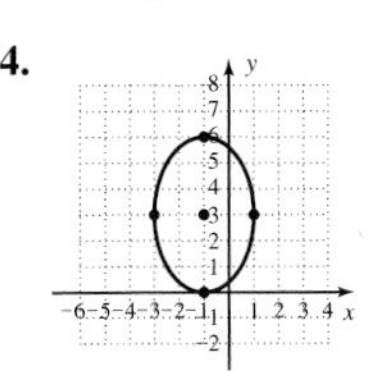

5.
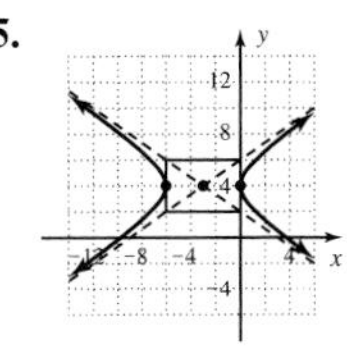

6.
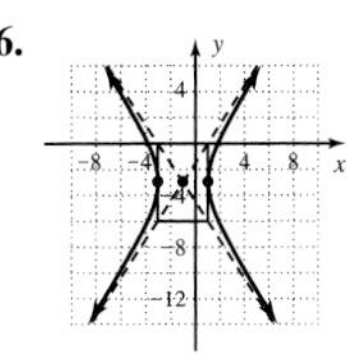

7. a. $\dfrac{(x+3)^2}{4} + \dfrac{(y-1)^2}{16} = 1$; D: $x \in [-5, -1]$; R: $y \in [-3, 5]$
b. $(x-3)^2 + (y-2)^2 = 16$; D: $x \in [-1, 7]$; R: $y \in [-2, 6]$
c. $y = (x-3)^2 - 4$; D: $x \in (-\infty, \infty)$ R: $y \in (-4, \infty)$
8. $(x+2)^2 + (y-5)^2 = 8$ **9.** $\dfrac{x^2}{16} + \dfrac{y^2}{4} = 1$ **10.** yes, distance $d \approx 49$ mi

Reinforcing Basic Concepts, p. 706

Exercise 1: $\dfrac{25(x-2)^2}{2} - \dfrac{9(y+3)^2}{4} = 1$

Exercise 2: $\dfrac{28(x-1)^2}{25} + \dfrac{48(y+2)^2}{25} = 1$

Exercises 7.3, pp. 712–716

1. a. 3 or 4 not possible

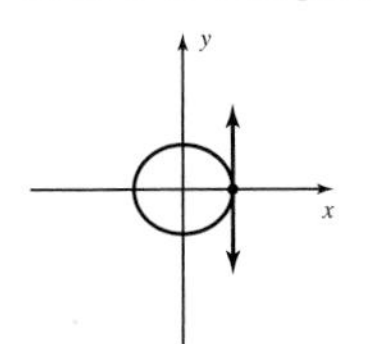

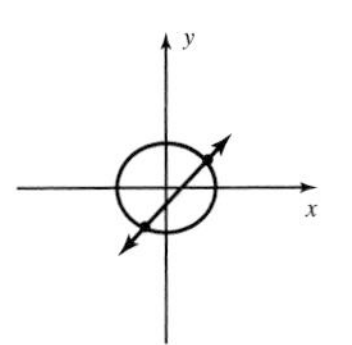

b. 3 or 4 not possible

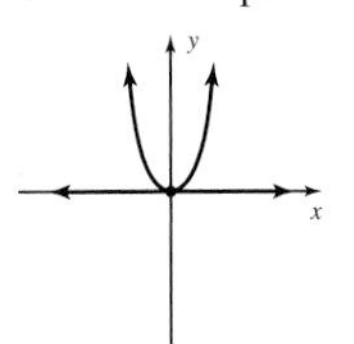

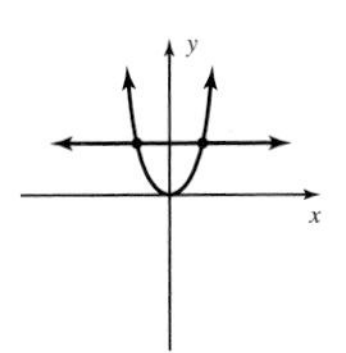

c.

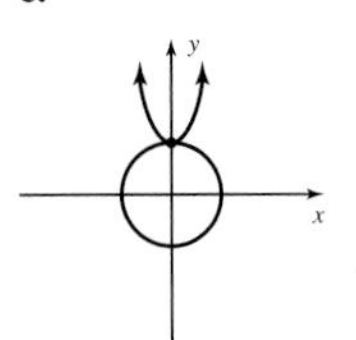

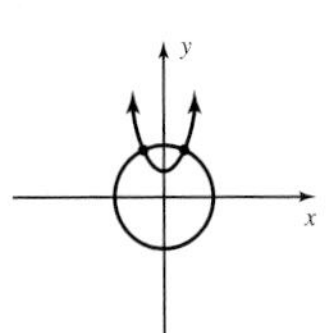

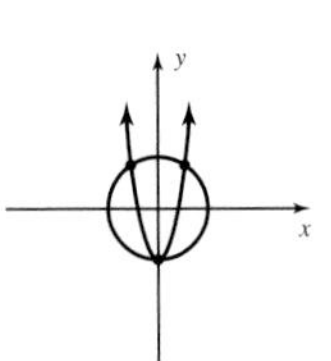

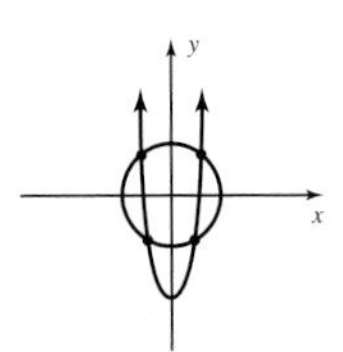

d.

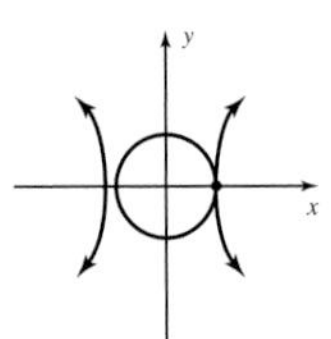

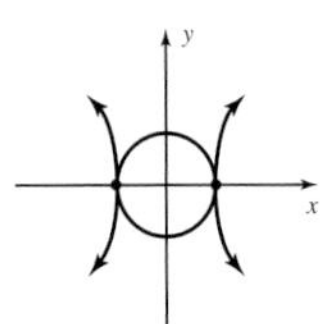

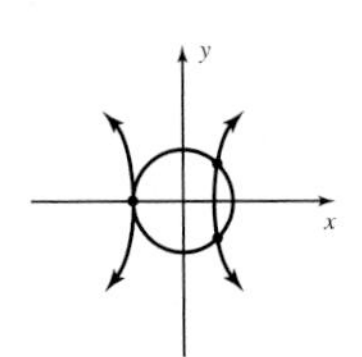

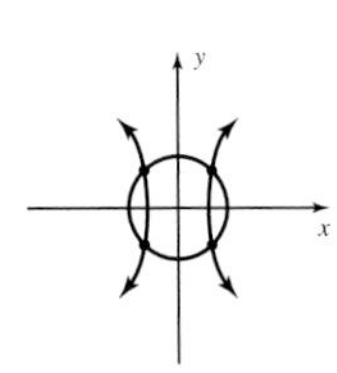

e.

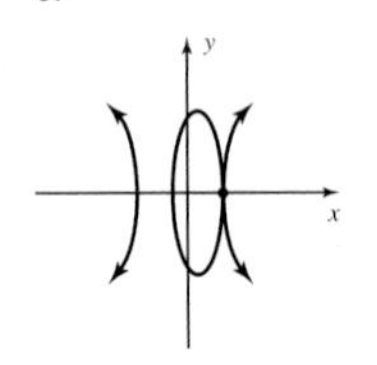

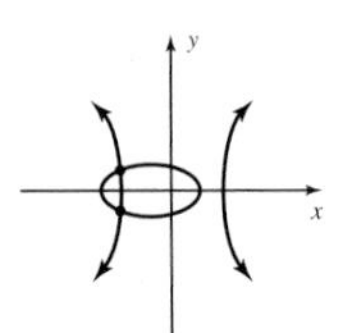

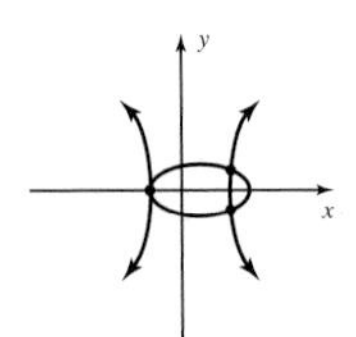

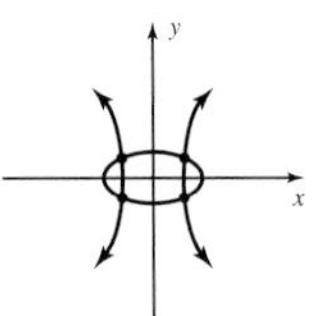

f.

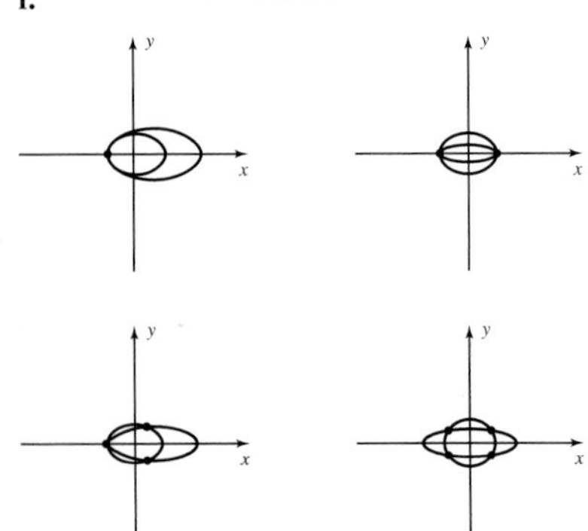

3. region, solutions **5.** Answers will vary.
7. first: parabola; second: line **9.** first: parabola; second: ellipse

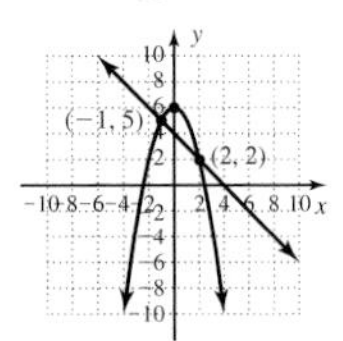

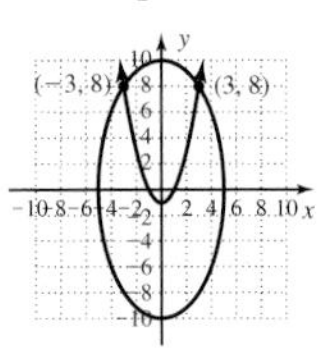

11. first: hyperbola; second: circle **13.** $(-4, -3), (3, 4)$

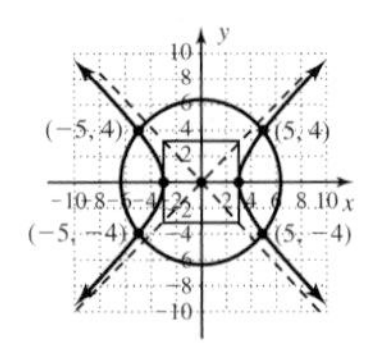

15. $\left(4, \frac{3}{2}\right), (3, 2)$ **17.** $(\sqrt{10}, 3), (-\sqrt{10}, 3), (5, -12), (-5, -12)$
19. $(4, 3), (4, -3), (-4, 3), (-4, -3)$ **21.** no solutions
23. $(5, -5), (5, 5), (-5, 5), (-5, -5)$ **25.** $(5, \log 5 + 5)$
27. $(-3, 1), (2, 1024)$ **29.** $(-3, -21), (1, -1), (2, 4)$
31. $(3, 2), (-3, 2), \left(\frac{5\sqrt{2}}{2}, \frac{-3}{2}\right), \left(-\frac{5\sqrt{2}}{2}, \frac{-3}{2}\right)$ **33.** $(2, 3), (-2, -3)$
35. $\left(\frac{11}{15}, \frac{-32}{15}\right)$ **37.** $(3, 1.41), (-3, 1.41), (3, -1.41), (-3, -1.41)$
39. $(-2.43, -2.81), (2, 1)$ **41.** $(0.72, 2.19), (2, 3), (4, 3), (5.28, 2.19)$

43.

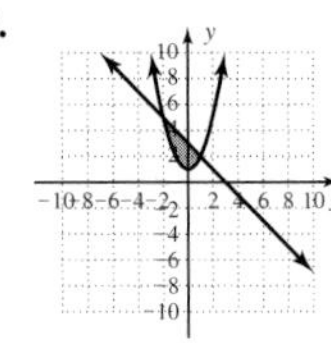

45.

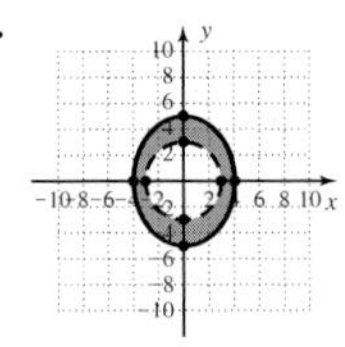

47.

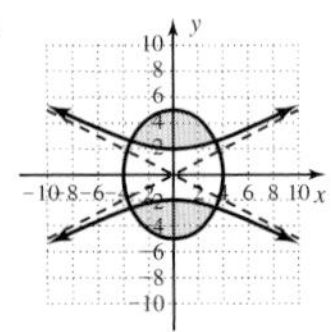

49.

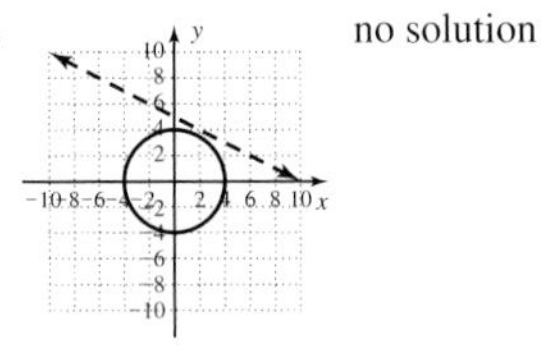

no solution

51. $h \approx 27.5$ ft; $h = 24$ ft; $h = 18$ ft **53.** 8.5 m × 10 m
55. 5 km, 9 km **57.** 8 × 8 × 25 ft
59. \$1.83; \$3, 90,000 gal $\begin{cases} 10P^2 + 6D = 144 \\ 8P^2 - 8P - 4D = 12 \end{cases}$
61. Answers will vary. **63.** 12 units^2 **65.** 18 in. by 18 in. by 77 in.
67. a. $x = \frac{-2 \pm 2\sqrt{10}}{3}$ **b.** $x = 0, x = 8$ **c.** $x = 2$

69. a.

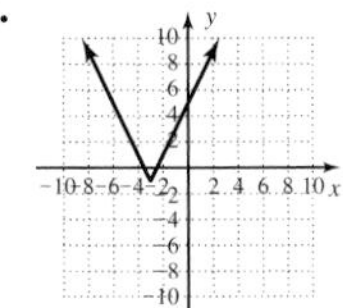

b.

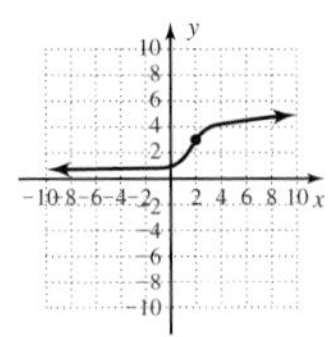

c.

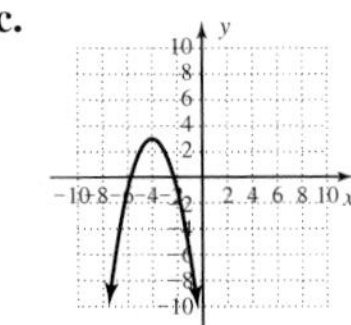

d.

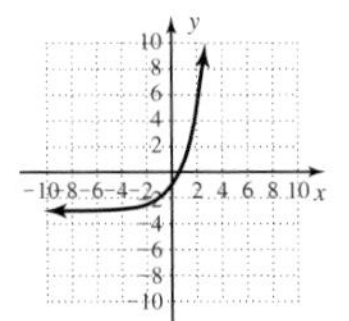

71. a. $m = \frac{-400}{1}$, the copier depreciates by \$400 a year. **b.** $y = -400x + 4500$ **c.** \$1700 **d.** 9.5 yr

Exercises 7.4, pp. 724–729

1. $c^2 = |a^2 - b^2|$ **3.** $2a, 2b$ **5.** Answers will vary **7.** 20 **9.** 20
11. a. $(2, 1)$ **b.** $(-3, 1)$ and $(7, 1)$ **c.** $(2 - \sqrt{21}, 1)$ and $(2 + \sqrt{21}, 1)$ **d.** $(2, 3)$ and $(2, -1)$ **e.**

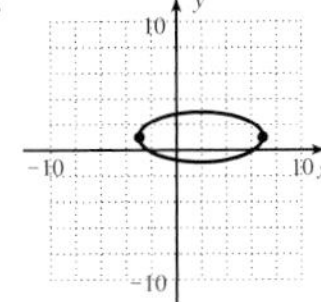

13. a. $(4, -3)$ **b.** $(4, 2)$ and $(4, -8)$ **c.** $(4, 0)$ and $(4, -6)$ **d.** $(0, -3)$ and $(8, -3)$ **e.**

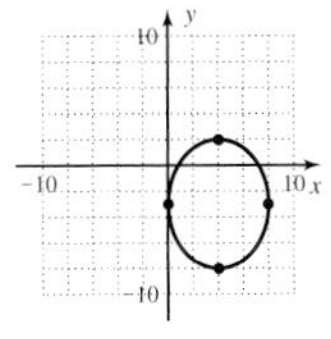

15. a. $(-2, -2)$ **b.** $(-5, -2)$ and $(1, -2)$ **c.** $(-2 + \sqrt{3}, -2)$ and $(-2 - \sqrt{3}, -2)$ **d.** $(-2, -2 + \sqrt{6})$ and $(-2, -2 - \sqrt{6})$
e.

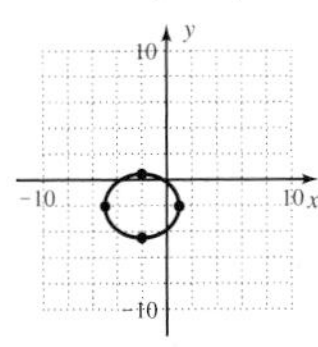

17. $\frac{x^2}{36} + \frac{y^2}{20} = 1$ **19.** $\frac{x^2}{9} + \frac{y^2}{25} = 1$
21. 8, $2a = 8$, $2b = 6$ **23.** 12, $2a = 16$, $2b = 12$
25. a. $(3, 4)$ **b.** $(0, 4)$ and $(6, 4)$ **c.** $(3 - \sqrt{13}, 4)$ and $(3 + \sqrt{13}, 4)$ **d.** $2a = 6$, $2b = 4$ **e.**

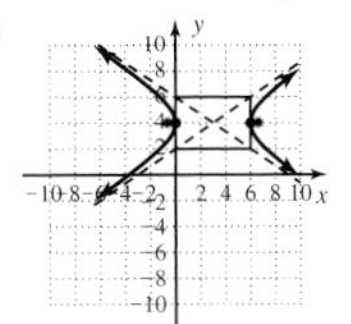

27. a. $(0, 3)$ **b.** $(-2, 3)$ and $(2, 3)$ **c.** $(-2\sqrt{5}, 3)$ and $(2\sqrt{5}, 3)$ **d.** $2a = 4$, $2b = 8$ **e.**

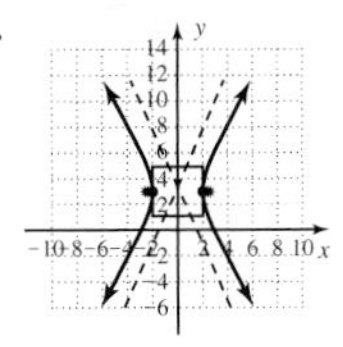

29. a. $(3, -2)$ **b.** $(1, -2)$ and $(5, -2)$ **c.** $(-1, -2)$ and $(7, -2)$ **d.** $2a = 4$, $2b = 4\sqrt{3}$ **e.**

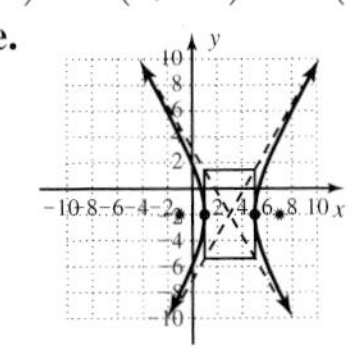

31. $\frac{x^2}{36} - \frac{y^2}{28} = 1$ **33.** $\frac{y^2}{9} - \frac{x^2}{9} = 1$ **35.** $(-3\sqrt{5}, 0), (3\sqrt{5}, 0)$
37. $(0, 4), (0, -4)$ **39.** $(-\sqrt{6}, 0), (\sqrt{6}, 0)$ **41.** $(-\sqrt{13}, 0), (\sqrt{13}, 0)$
43. $(0, \sqrt{61}), (0, -\sqrt{61})$ **45.** $(-2\sqrt{15}, 0), (2\sqrt{15}, 0)$
47. $\frac{x^2}{16} + \frac{y^2}{36} = 1$ **49.** $\sqrt{7} \approx 2.65$ ft; 2.25 ft **51.** 8.9 ft; 17.9 ft
53. $a \approx 142$ million miles, $b \approx 141$ million miles, orbit time ≈ 686 days
55. $\frac{x^2}{225} - \frac{y^2}{2275} = 1$ about (24.1, 60) or (224.1, 60) **57.** $L = 9$ units; verified **59.** $(x - 2)^2 + (y - 3)^2 = 34$ **61.** verified **63.** $\frac{\log 20}{\log 3} \approx 2.73$

65. a.

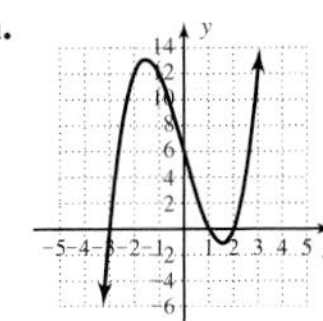

b.

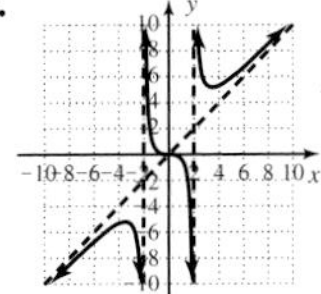

67. $x \in (0, 6)$

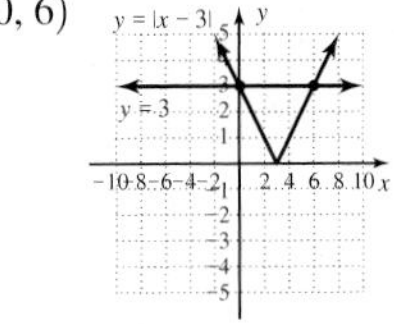

Exercises 7.5, pp. 736–739

1. horizontal, right, $a < 0$ **3.** $(p, 0)$, $x = -p$ **5.** Answers will vary.
7. $x \in (-\infty, \infty)$, $y \in [-4, \infty)$ **9.** $x \in (-\infty, \infty)$, $y \in [-18, \infty)$

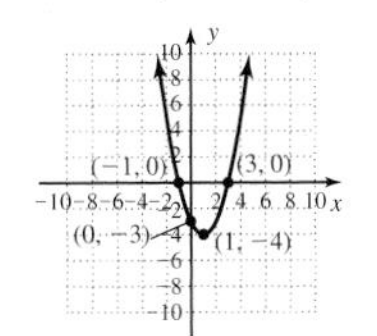

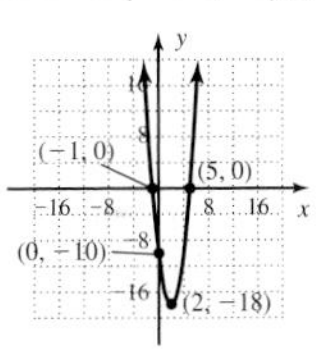

11. $x \in (-\infty, \infty)$, $y \in [-10.125, \infty)$ **13.** $x \in [-4, \infty)$, $y \in (-\infty, \infty)$

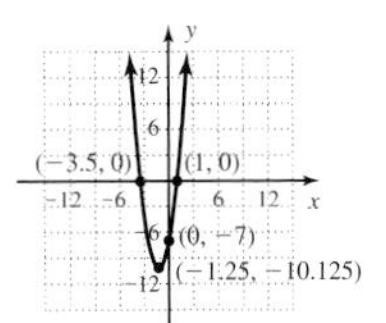

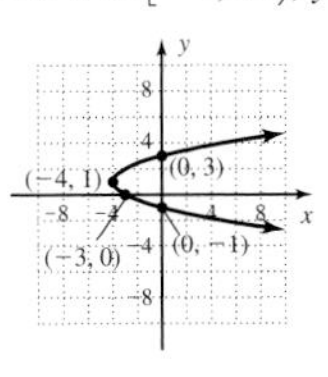

15. $x \in (-\infty, 16]$, $y \in (-\infty, \infty)$ **17.** $x \in (-\infty, 0]$, $y \in (-\infty, \infty)$

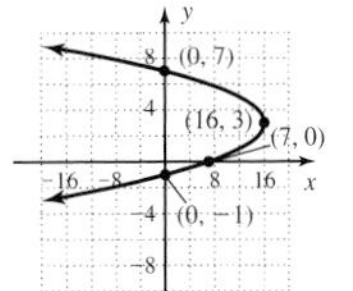

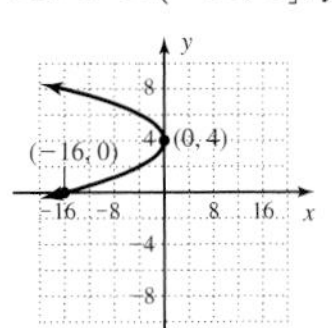

19. $x \in [-9, \infty)$, $y \in (-\infty, \infty)$ **21.** $x \in [-4, \infty)$, $y \in (-\infty, \infty)$

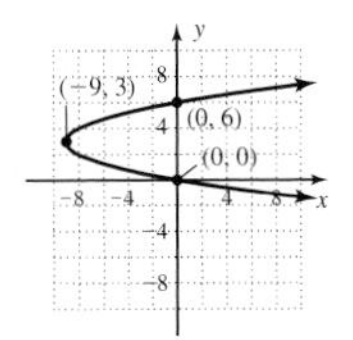

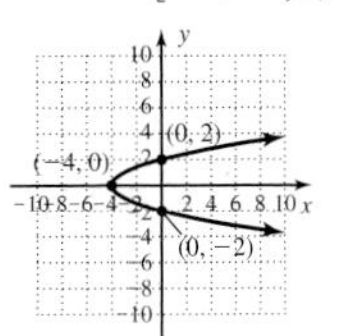

23. $x \in (-\infty, 0], y \in (-\infty, \infty)$

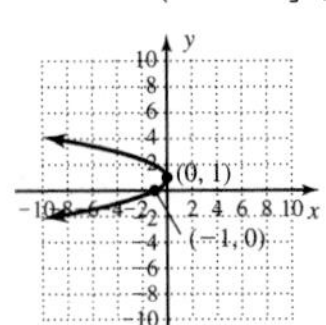

25. $x \in [-6.25, \infty), y \in (-\infty, \infty)$

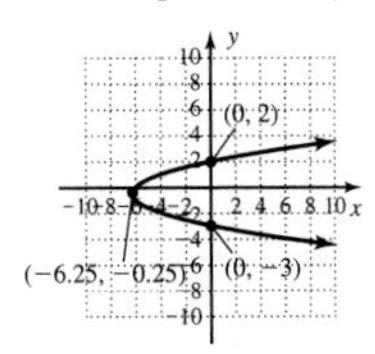

27. $x \in [-21, \infty), y \in (-\infty, \infty)$

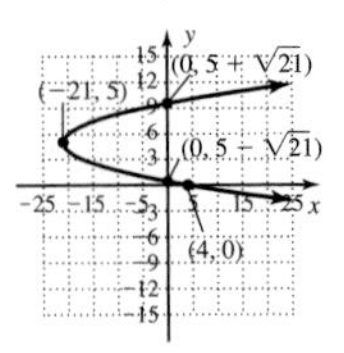

29. $x \in (-\infty, 11], y \in (-\infty, \infty)$

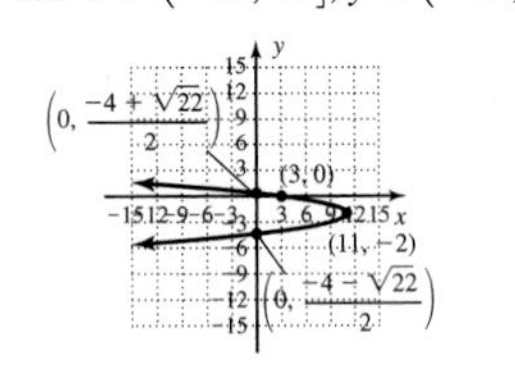

31. $x \in (-\infty, \infty), y \in [3, \infty)$

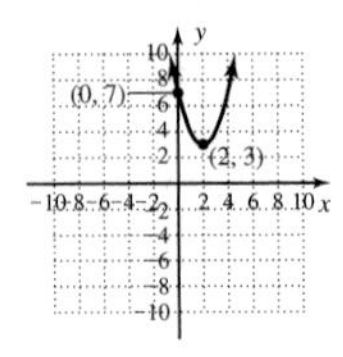

33. $x \in [2, \infty), y \in (-\infty, \infty)$

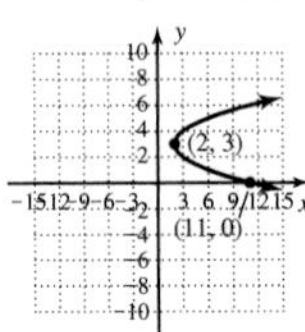

35. $x \in [1, \infty), y \in (-\infty, \infty)$

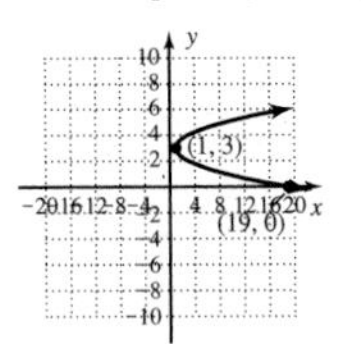

37.

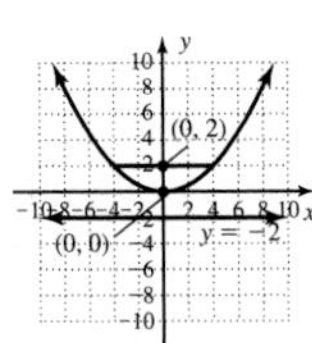

39.

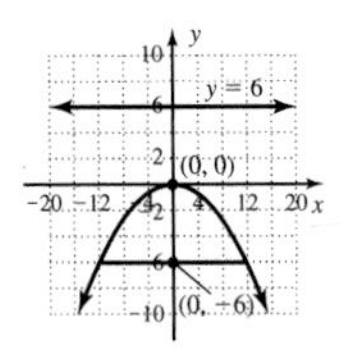

41.

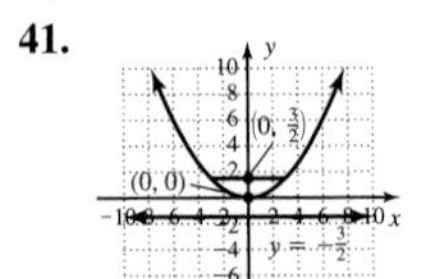

43.

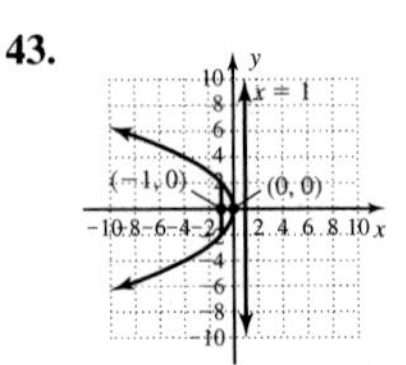

45.

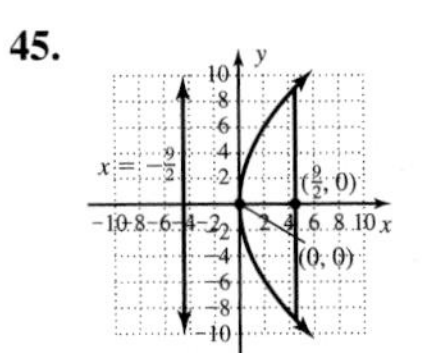

47.

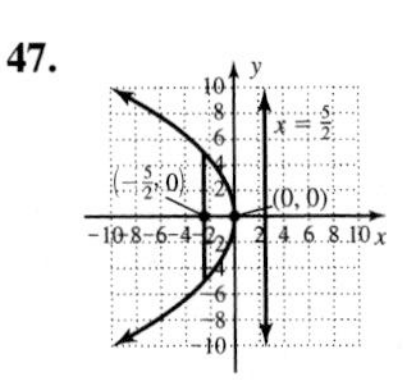

49.

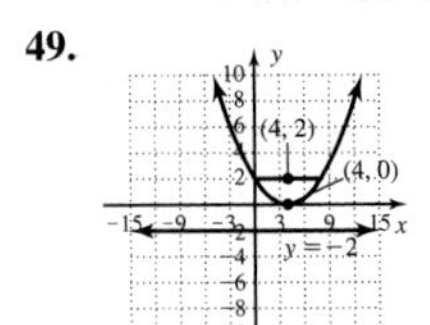

51.

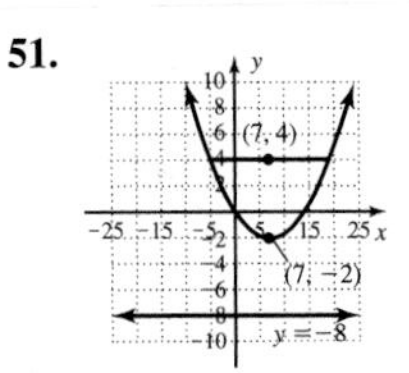

53.

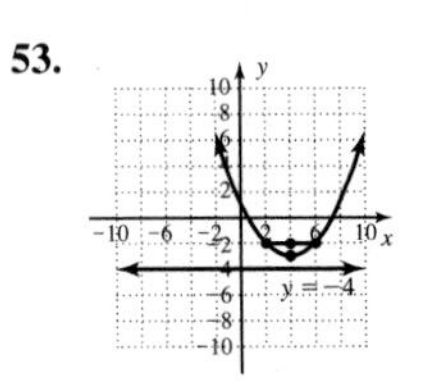

55.

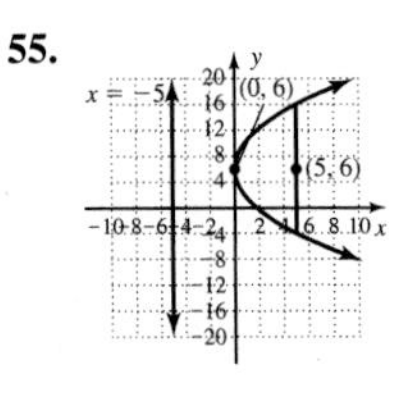

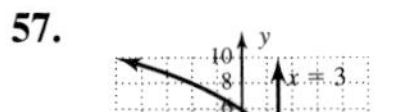

57.

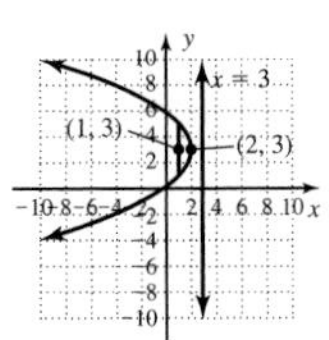

59.

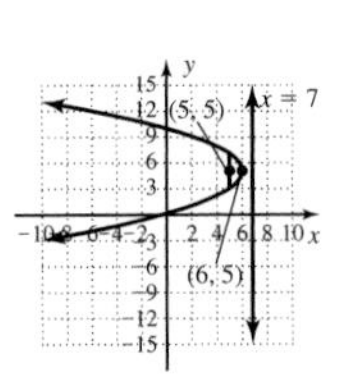

61. 16 units^2 **63.**

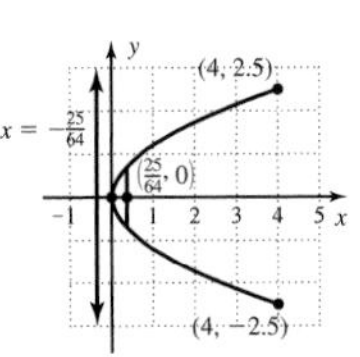

65. 6 in.; (13.5, 0)

67. 14.97 ft, (0, 41.75) **69.** $(x - 2)^2 = \frac{1}{2}(y + 8); p = \frac{1}{8}; (2, -8)$
71. b **73.** $(x + 3)^2 + (y - 4)^2 = 20; 20\pi$ units^2
75. a. $y = 2527.4(1.414)^x$; 14,286 **b.** 2001 **c.** 2003
77. Answers will vary. **79.** no solution

Summary and Concept Review, pp. 740–742

1.

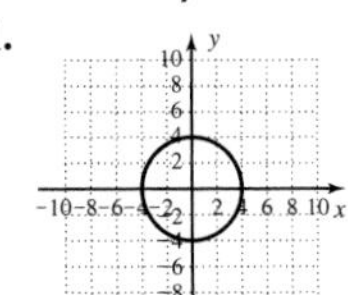

2.

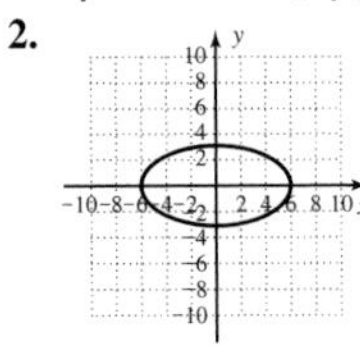

3.

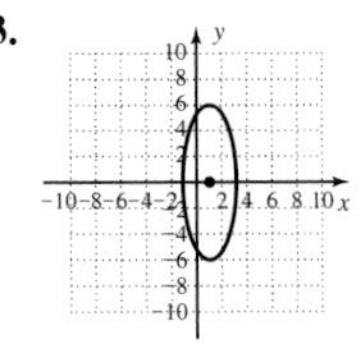

4.

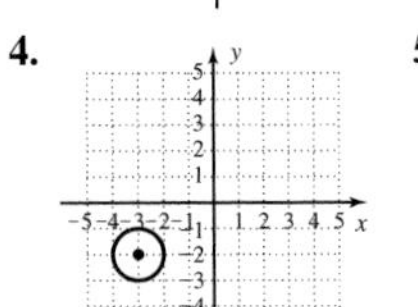

5.

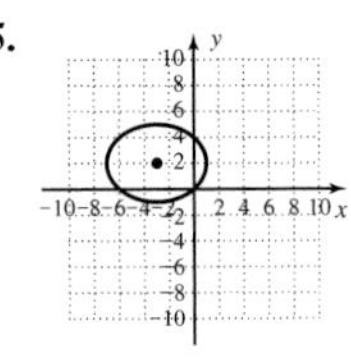

6. $(x + 1)^2 + (y - 1)^2 = 25$

7.

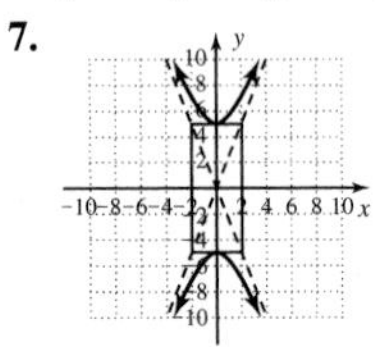

8.

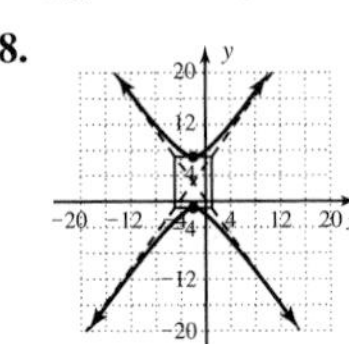

9.

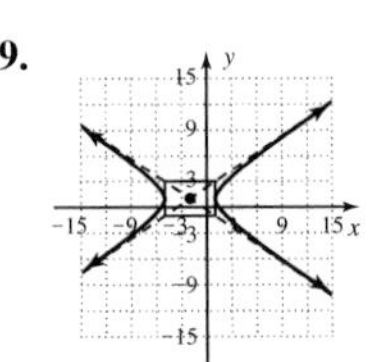

10.

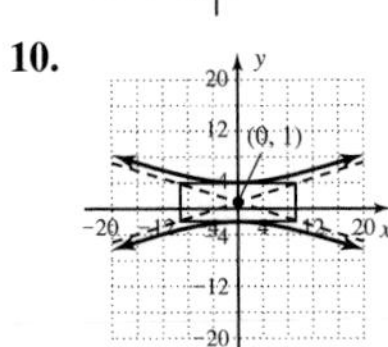

11.

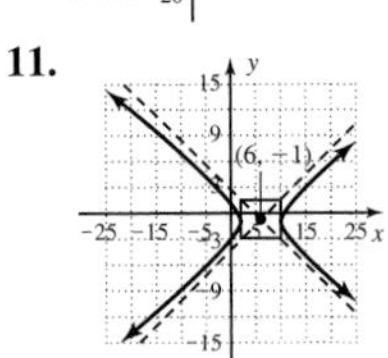

12.

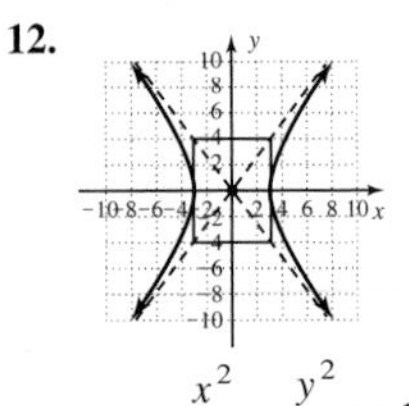

$$\frac{x^2}{9} - \frac{y^2}{16} = 1$$

13. circle, line, (4, 3), (−3, −4)
14. hyperbola, circle, (3, 2), (3, −2),(−3, 2),(−3, −2)
15. parabola, line, (3, −2)

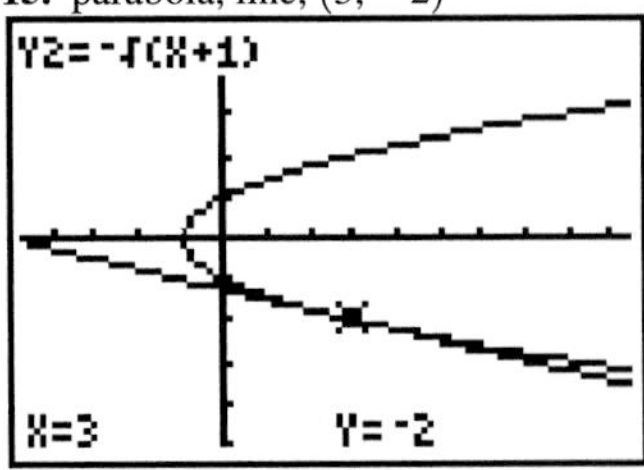

16. circle, parabola, (1, 3), (−1, 3)

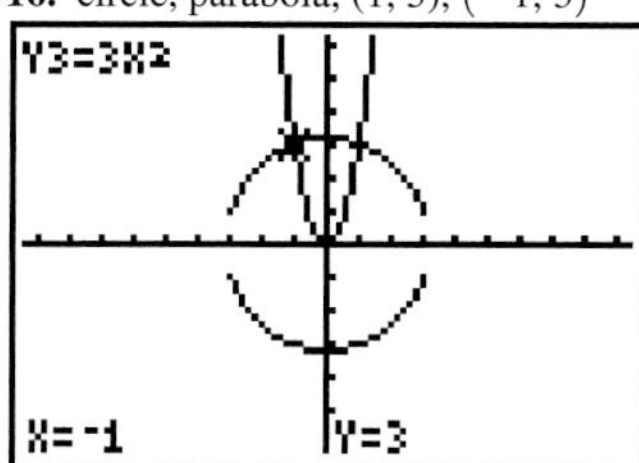

17.

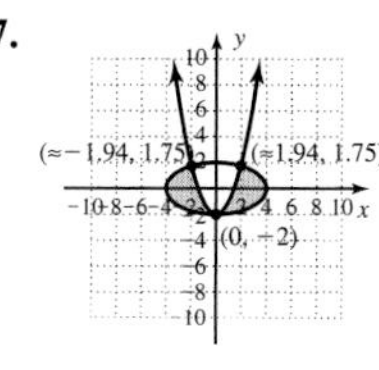

18.

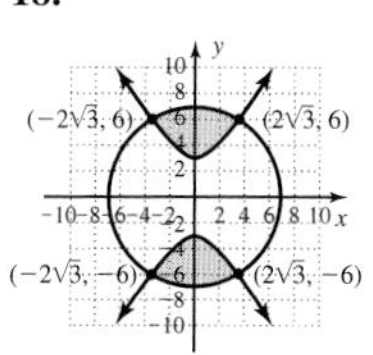

19.

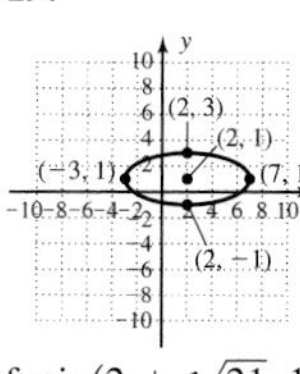

foci: $(2 + \sqrt{21}, 1)$ $(2 - \sqrt{21}, 1)$

20. foci: $(5 + 2\sqrt{10}, 2)$ $(5 - 2\sqrt{10}, 2)$

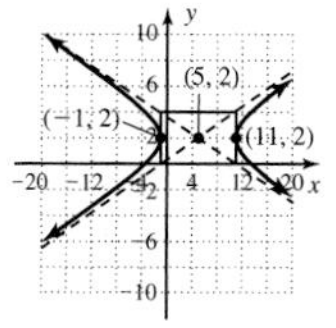

21. a. $\frac{x^2}{169} + \frac{y^2}{25} = 1$ **b.** $\frac{x^2}{144} + \frac{y^2}{400} = 1$

22. a. $\frac{x^2}{225} - \frac{y^2}{64} = 1$ **b.** $\frac{y^2}{16} - \frac{x^2}{9} = 1$

23.

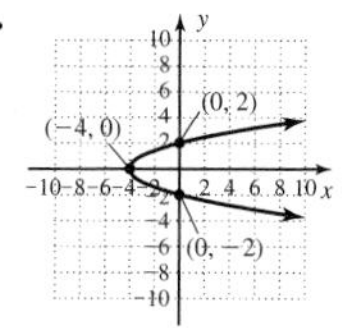

24.

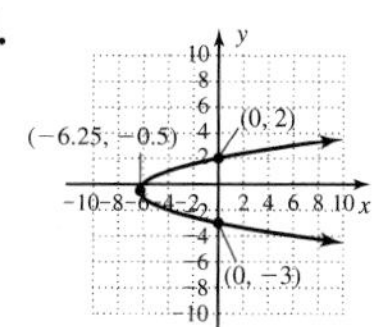

25.

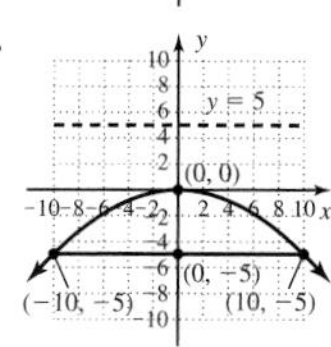

26.

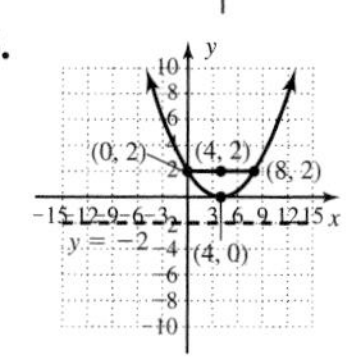

Mixed Review, pp. 742–743

1.

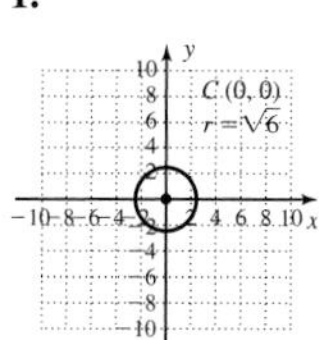

3.

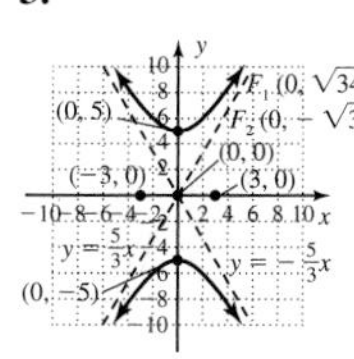

5.

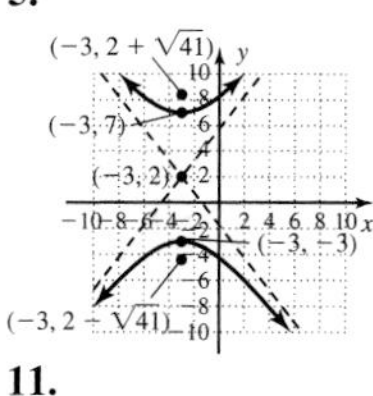

7.

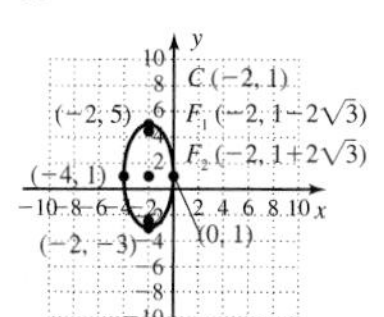

9.

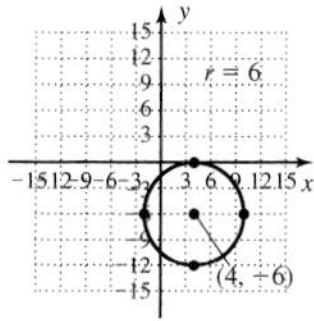

11.

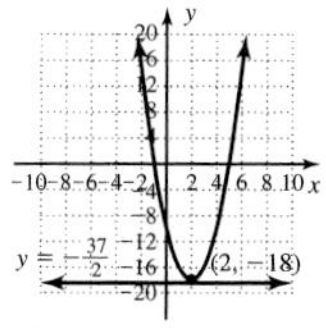

13.

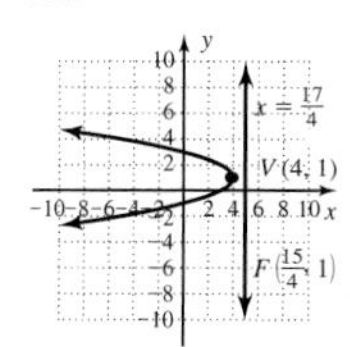

15.

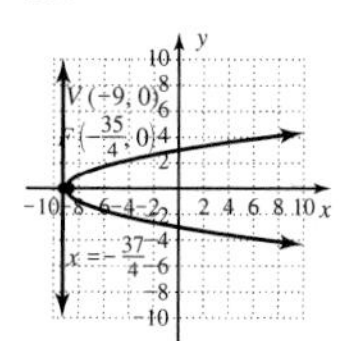

17.

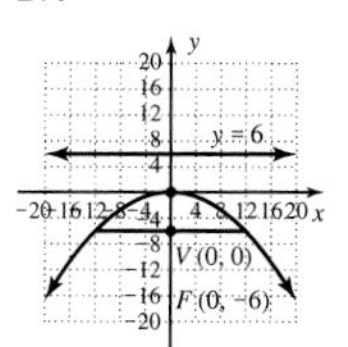

19.

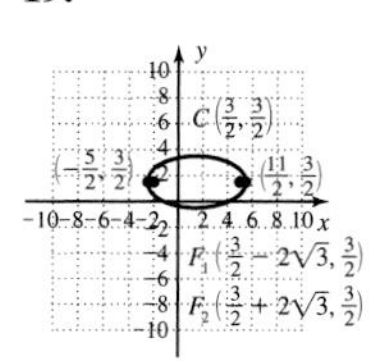

21.

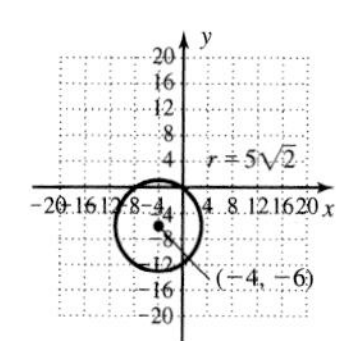

23. $d(P_1P_3) = d(P_2P_3)$; $m(P_1P_2) = (1, -4) = P_4$; slope $(P_1P_2) = 1$; slope $(P_3P_4) = -1$ **25. a.** (12, 0) **b.** $2a = 116$; $2b \approx 113.5$

c. $\frac{x^2}{3364} + \frac{y^2}{3220} = 1$

Practice Test, pp. 743–744

1. c **2.** d **3.** a **4.** b

5.

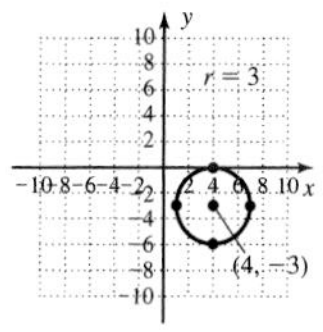

6.

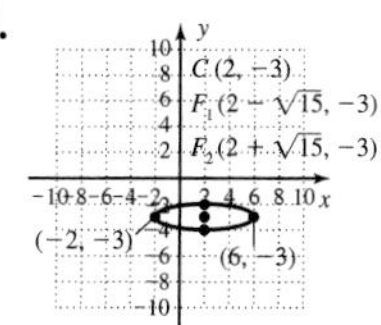

7.

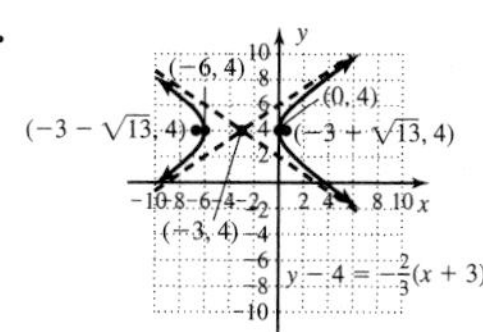

8.

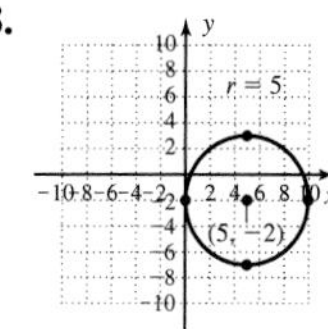

9.

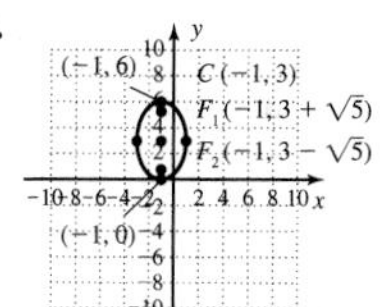

10.

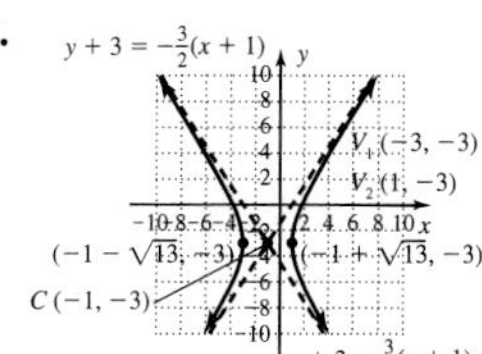

11.

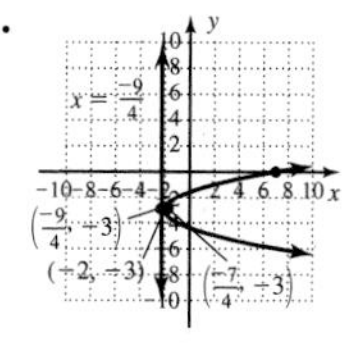

12.

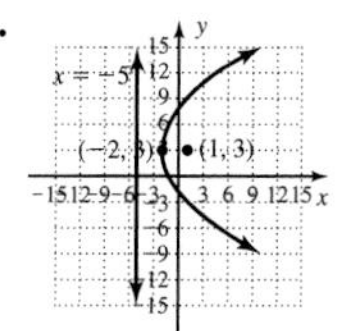

13. a. $(\frac{10}{3}, \frac{16}{3})$, $(-2, 0)$ **b.** $(\frac{2\sqrt{15}}{5}, \frac{2\sqrt{10}}{5})$, $(\frac{2\sqrt{15}}{5}, \frac{-2\sqrt{10}}{5})$; $(\frac{-2\sqrt{15}}{5}, \frac{2\sqrt{10}}{5})$, $(\frac{-2\sqrt{15}}{5}, \frac{-2\sqrt{10}}{5})$ **14.** 15 ft, 20 ft

15. $(x + 2)^2 + (y - 5)^2 = 8$ **16.** $\frac{x^2}{16} + \frac{y^2}{12} = 1$

17. 154.89 million miles; 128.41 million miles

18. $y = (x - 1)^2 - 4$; $D: x \in (-\infty, \infty)$, $R: y \in [-4, \infty)$; focus: $\left(1, \frac{-15}{4}\right)$

19. $(x - 1)^2 + (y - 1)^2 = 25$; $D: x \in [-4, 6]$, $R: y \in [-4, 6]$

20. $\dfrac{(x+3)^2}{9} + \dfrac{y^2}{36} = 1$; $D: x \in [-6, 0]$, $R: y \in [-6, 6]$; foci: $(-3, -3\sqrt{3}), (-3, 3\sqrt{3})$

Strengthening Core Skills, pp. 745–746

Exercise 1. $\dfrac{(x-2)^2}{\left(\frac{\sqrt{2}}{5}\right)^2} - \dfrac{(y+3)^2}{\left(\frac{2}{3}\right)^2} = 1$; $a = \dfrac{\sqrt{2}}{5}$, $b = \dfrac{2}{3}$

Exercise 2. $\dfrac{(x-1)^2}{\left(\frac{5\sqrt{7}}{14}\right)^2} + \dfrac{(y+2)^2}{\left(\frac{5\sqrt{3}}{12}\right)^2} = 1$; $a = \dfrac{5\sqrt{17}}{14}$, $b = \dfrac{5\sqrt{3}}{12}$

Exercise 3. $\dfrac{(x+3)^2}{\left(\frac{7}{2}\right)^2} + \dfrac{(y-1)^2}{\left(\frac{6}{5}\right)^2} = 1$; $a = \dfrac{7}{2}$, $b = \dfrac{6}{5}$

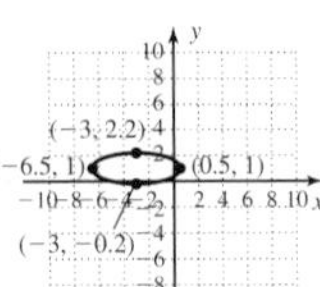

Exercise 4. $\dfrac{(x+3)^2}{\left(\frac{4\sqrt{5}}{3}\right)^2} - \dfrac{(y-1)^2}{\left(\frac{9}{2}\right)^2} = 1$; $a = \dfrac{4\sqrt{5}}{3} \approx 3$; $b = \dfrac{9}{2}$

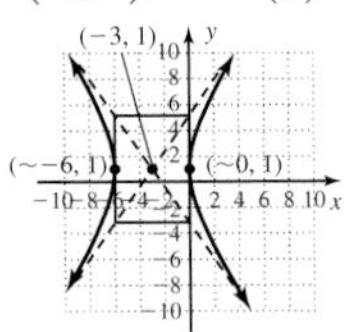

Cumulative Review Chapters 1–7, p. 746

1. $x = 2, x = \pm 2i$ **3.** no solution **5.** $x = 3 \pm 2i$ **7.** $x = 2 + \dfrac{\ln 7}{\ln 3}$

9. $x = 3$ **11.**

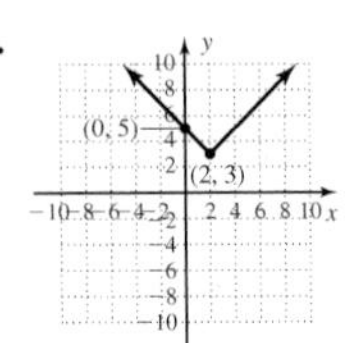

13.

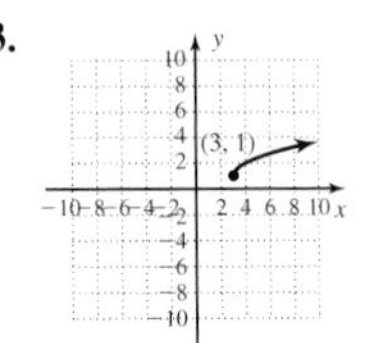

15. **17.** **19.**

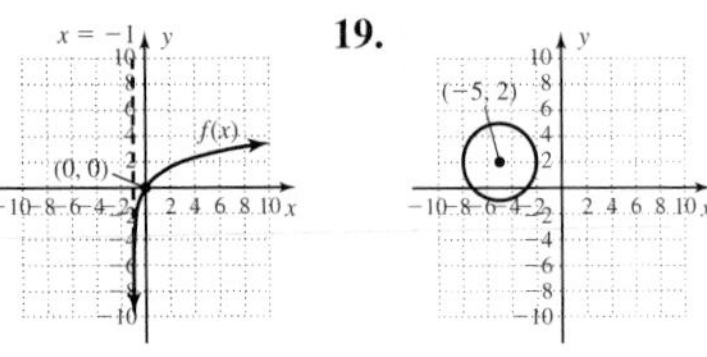

21. a. $x \in (-\infty, \infty)$ **b.** $y \in (-\infty, 4]$ **c.** $f(x)\uparrow: x \in (-\infty, -1)$, $f(x)\downarrow: x \in (-1, \infty)$ **d.** none **e.** max: $(-1, 4)$ **f.** $f(x) > 0: x \in (-4, 2)$ **g.** $f(x) < 0: x \in (-\infty, -4) \cup (2, \infty)$

23. $(3, 4), (3, -4), (-3, 4), (-3, -4)$ **25.** $3\frac{1}{3}$ L

CHAPTER 8

Exercises 8.1, pp. 755–758

1. pattern, order **3.** increasing **5.** formula defining the sequence uses the preceding term(s); answers will vary. **7.** $1, 3, 5, 7$; $a_8 = 15$; $a_{12} = 23$ **9.** $0, 9, 24, 45$; $a_8 = 189$; $a_{12} = 429$ **11.** $-1, 2, -3, 4$; $a_8 = 8$; $a_{12} = 12$ **13.** $\frac{1}{2}, \frac{2}{3}, \frac{3}{4}, \frac{4}{5}$; $a_8 = \frac{8}{9}$; $a_{12} = \frac{12}{13}$ **15.** $\frac{1}{2}, \frac{1}{4}, \frac{1}{8}, \frac{1}{16}$; $a_8 = \frac{1}{256}$; $a_{12} = \frac{1}{4096}$ **17.** $1, \frac{1}{2}, \frac{1}{3}, \frac{1}{4}$; $a_8 = \frac{1}{8}$; $a_{12} = \frac{1}{12}$ **19.** $\frac{-1}{2}, \frac{1}{6}, \frac{-1}{12}, \frac{1}{20}$; $a_8 = \frac{1}{72}$; $a_{12} = \frac{1}{156}$ **21.** $-2, 4, -8, 16$; $a_8 = 256$; $a_{12} = 4096$ **23.** 79 **25.** $\frac{1}{5}$ **27.** $\frac{1}{32}$ **29.** $\left(\frac{11}{10}\right)^{10}$ **31.** $\frac{1}{36}$ **33.** 2, 7, 32, 157, 782 **35.** −1, 4, 19, 364, 132, 499 **37.** 64, 32, 16, 8, 4 **39.** 336 **41.** 36 **43.** 28 **45.** $\frac{1}{2}, \frac{1}{3}, \frac{1}{4}, \frac{1}{5}$ **47.** $\frac{1}{3}, \frac{1}{120}, \frac{1}{15{,}120}, \frac{1}{3{,}991{,}680}$ **49.** $1, 2, \frac{9}{2}, \frac{32}{3}$ **51.** $64, a_n = 2^n$ **53.** $36, a_n = n^2$ **55.** $32, a_n = 7n - 10$ **57.** $\frac{1}{64}, a_n = \frac{1}{2^n}$ **59.** 15 **61.** 64 **63.** $\frac{137}{60}$ **65.** 10 **67.** 95 **69.** −4 **71.** 15 **73.** 50 **75.** $\frac{-27}{112}$ **77.** $\sum_{n=1}^{5}(4n)$ **79.** $\sum_{n=1}^{6}(-1)^n n^2$ **81.** $\sum_{n=1}^{5}(n+3)$ **83.** $\sum_{n=1}^{3}\frac{n^2}{3}$ **85.** $\sum_{n=3}^{7}\frac{n}{2^n}$ **87.** $\frac{1}{20}\sum_{n=1}^{5} n$ **89.** $\sum_{n=1}^{\infty}\frac{n^2}{(n+1)^3}$ **91.** 35 **93.** $a_n = 6000(0.8)^{n-1}$; 6000, 4800, 3840, 3072, 2457.6, 1966.1 **95.** 5.20, 5.70, 6.20, 6.70, 7.20, \$13,824 **97.** ≈ 2690 **99.** Answers will vary. **101.** approaches e **103.** approaches 1 **105.** $\dfrac{\sqrt{x+h} - \sqrt{x}}{h}$; $\dfrac{1}{\sqrt{x+h} + \sqrt{x}}$ **107.** (3,21,55) **109.** $(x-2)^2 + (y+3)^2 = 49$

Exercises 8.2, pp. 765–768

1. common, difference **3.** $\dfrac{n(a_1 + a_n)}{2}$, nth **5.** Answers will vary. **7.** arithmetic; $d = 3$ **9.** arithmetic; $d = 2.5$ **11.** not arithmetic; all prime **13.** arithmetic; $d = \frac{1}{24}$ **15.** not arithmetic; $a_n = n^2$ **17.** arithmetic; $d = \frac{-\pi}{6}$ **19.** 2, 5, 8, 11 **21.** 7, 5, 3, 1 **23.** 0.3, 0.33, 0.36, 0.39 **25.** $\frac{3}{2}, 2, \frac{5}{2}, 3$ **27.** $\frac{3}{4}, \frac{5}{8}, \frac{1}{2}, \frac{3}{8}$ **29.** −2, −5, −8, −11 **31.** $a_1 = 2, d = 5, a_n = 5n - 3, a_6 = 27, a_{10} = 47, a_{12} = 57$ **33.** $a_1 = 5.10, d = 0.15, a_n = 0.15n + 4.95, a_6 = 5.85, a_{10} = 6.45, a_{12} = 6.75$ **35.** $a_1 = \frac{3}{2}, d = \frac{3}{4}, a_n = \frac{3}{4}n + \frac{3}{4}, a_6 = \frac{21}{4}, a_{10} = \frac{33}{4}, a_{12} = \frac{39}{4}$ **37.** 61 **39.** 1 **41.** 2.425 **43.** 9 **45.** 43 **47.** 21 **49.** 26 **51.** $d = 3, a_1 = 1$ **53.** $d = \frac{91}{240}, a_1 = \frac{31}{48}$ **55.** $d = \frac{115}{126}, a_1 = \frac{-472}{63}$ **57.** 1275 **59.** 601.25 **61.** −534 **63.** 82.5 **65.** 74.04 **67.** $210\sqrt{2}$ **69.** $n = 25, S_{25} = 950$ **71.** $n = 31, S_{31} = 2449$ **73.** $n = 32, S_{32} = 1712$ **75.** $n = 50, S_{50} = 4250$ **77.** $n = 12, S_{12} = \frac{45}{2}$ **79.** $S_6 = 21$; $S_{75} = 2850$ **81.** 930, 10 **83.** \$885 **85.** 24 hr **87.** 16.5 in.; 232.5 in. **89.** 152; 4400 **91.** 220, 2520, yes **93.** Answers will vary. **95.** $180(n-2)$, 1440° **97.** $t \approx 3.6$ **99.** $\left(\frac{1}{2}, \frac{2}{3}\right)$ **101.** $f(x) = 49x + 972$; 1364

Exercises 8.3, pp. 777–781

1. multiplying **3.** $a_1 r^{n-1}$ **5.** Answers will vary. **7.** $r = 2$ **9.** $r = -2$ **11.** $a_n = n^2 + 1$ **13.** $r = 0.1$ **15.** not geometric; ratio of terms decreases by 1 **17.** $r = \frac{2}{5}$ **19.** $r = \frac{1}{2}$ **21.** $r = \frac{4}{x}$ **23.** not geometric; $a_n = \dfrac{240}{n!}$ **25.** 5, 10, 20, 40 **27.** $-6, 3, \frac{-3}{2}, \frac{3}{4}$ **29.** $4, 4\sqrt{3}, 12, 12\sqrt{3}$ **31.** 0.1, 0.01, 0.001, 0.0001 **33.** $-\frac{3}{8}$ **35.** $\frac{25}{4}$ **37.** 16 **39.** $a_1 = \frac{1}{27}, r = -3, a_n = \frac{1}{27}(-3)^{n-1}, a_6 = -9, a_{10} = -729, a_{12} = -6561$ **41.** $a_1 = 729, r = \frac{1}{3}, a_n = 729(\frac{1}{3})^{n-1}, a_6 = 3, a_{10} = \frac{1}{27}, a_{12} = \frac{1}{243}$ **43.** $a_1 = \frac{1}{2}, r = \sqrt{2}, a_n = \frac{1}{2}(\sqrt{2})^{n-1}, a_6 = 2\sqrt{2}, a_{10} = 8\sqrt{2}, a_{12} = 16\sqrt{2}$ **45.** $a_1 = 0.2, r = 0.4, a_n = 0.2(0.4)^{n-1}, a_6 = 0.002048, a_{10} = 0.0000524288, a_{12} = 0.000008388608$ **47.** 5 **49.** 11 **51.** 9 **53.** 8 **55.** 13 **57.** 9 **59.** $r = \frac{2}{3}; a_1 = 729$ **61.** $r = \frac{3}{2}, a_1 = \frac{32}{243}$ **63.** $r = \frac{3}{2}, a_1 = \frac{256}{81}$ **65.** −10,920 **67.** $\frac{3872}{27} \approx 143.41$

69. 257.375 **71.** 728 **73.** 10.625 **75.** ≈ 1.60 **77.** 1364
79. $\frac{31{,}525}{2187} \approx 14.41$ **81.** $\frac{-387}{512} \approx -0.76$ **83.** $\frac{521}{25}$ **85.** $\frac{3367}{1296}$
87. $14 + 15\sqrt{2}$ **89.** no **91.** $\frac{27}{2}$ **93.** $\frac{125}{3}$ **95.** 12 **97.** 4 **99.** $\frac{10}{3}$
101. $\frac{3}{2}$ **103.** $-\frac{18}{5}$ **105.** 1296 **107.** about 6.3 ft; 120 ft **109.** approx. \$67,196 **111.** \$8520; 10 yr **113.** 125.4 gpm; 10 months **115.** about 347.7 million **117.** 51,200 bacteria; 12 half-hours later (6 hr)
119. 4.2 m; 180 m **121.** 35.9 in^3; 7 strokes **123.** 6 yr
125. $a_n = 1000(1.05)^{n-1}$; after 6 yr $\rightarrow a_7$ is needed: $a_7 = 1340.10$; amount in the account after 6 yr, amount in the account after 7 yr; a_n generates the terms of the sequence before any interest is applied, while $A(t)$ gives the amount in the account after interest has been applied. Here $a_7 = A(6)$. **127.** $S_n = \log n!$ **129.** $x = \frac{-5}{2} \pm \frac{\sqrt{11}}{2}i$

131.

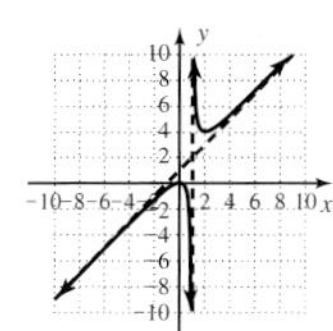

133. $p(50) \approx 2562.1$, $p(75) \approx 3615.6$, $p(100) \approx 4035.1$, $p(150) \approx 4189.1$

Exercises 8.4, pp. 787–789 (Only selected proofs are shown.)

1. finite, universally **3.** induction, hypothesis **5.** Answers will vary.
7. $a_4 = 34, a_5 = 44, a_k = 10k - 6, a_{k+1} = 10k + 4$
9. $a_4 = 4, a_5 = 5, a_k = k, a_{k+1} = k + 1$
11. $a_4 = 8, a_5 = 16, a_k = 2^{k-1}, a_{k+1} = 2^k$
13. $S_4 = 76, S_5 = 120, S_k = k(5k - 1), S_{k+1} = (k + 1)(5k + 4)$
15. $S_4 = 10, S_5 = 15, S_k = \dfrac{k(k+1)}{2}, S_{k+1} = \dfrac{(k+1)(k+2)}{2}$
17. $S_4 = 15, S_5 = 31, S_k = 2^k - 1, S_{k+1} = 2^{k+1} - 1$
19. verified **21.** verified **23.** verified

25. (1) Show S_n is true for $n = 1$: $S_1 = (1)^2 = 1$✓
(2) Assume S_k is true:
$1 + 8 + 27 + \cdots + k^3 = (1 + 2 + 3 + \cdots + k)^2$
Use it to show the truth of S_{k+1}:
$1 + 8 + 27 + \cdots + k^3 + (k+1)^3 = [1 + 2 + 3 + \cdots + k + (k+1)]^2$
left-hand side: $(1 + 8 + 27 + \cdots + k^3) + (k+1)^2(k+1)$
$= (1 + 8 + 27 + \cdots + k^3) + k(k+1)^2 + (k+1)^2$
$= (1 + 8 + 27 + \cdots + k^3) + k(k+1)(k+1) + (k+1)^2$
$= (1 + 8 + 27 + \cdots + k^3) + 2\dfrac{k(k+1)}{2}(k+1) + (k+1)^2$
$= \mathbf{(1 + 2 + 3 + \cdots + k)^2} + 2(1 + 2 + 3 + \cdots + k)(k+1) + (k+1)^2$
$= [(1 + 2 + 3 + \cdots + k) + (k+1)]^2$

27. verified **29.** verified

31. (1) Show S_n is true for $n = 1$: $S_1 = 1[2(1) + 3)] = 5$✓
(2) Assume S_k is true:
$5 + 9 + 13 + \cdots + (4k + 1) = k(2k + 3)$
Use it to show the truth of S_{k+1}:
$5 + 9 + 13 + \cdots + (4k + 1) + [4(k+1) + 1] = (k+1)[(2(k+1) + 3)]$
$5 + 9 + 13 + \cdots + (4k + 1) + (4k + 5) = (k+1)(2k + 5)$
left-hand side: $\mathbf{k(2k + 3)} + (4k + 5) = 2k^2 + 3k + 4k + 5$
$= 2k^2 + 7k + 5 = (2k + 5)(k + 1)$ **33.** verified

35. (1) Show S_n is true for $n = 1$: $S_1 = 2^{1+1} - 2 = 2^2 - 2 = 2$✓
(2) Assume S_k is true:
$2 + 4 + 8 + \cdots + 2^k = 2^{k+1} - 2$
Use it to show the truth of S_{k+1}:
$2 + 4 + 8 + \cdots + 2^k + 2^{k+1} = 2^{(k+1)+1} - 2$
left-hand side: $\mathbf{2^{k+1} - 2} + 2^{k+1} = 2(2^{k+1}) - 2$
$= 2^{(k+1)+1} - 2$

37. (1) Show S_n is true for $n = 1$: $S_1 = \dfrac{1}{2(1) + 1} = \dfrac{1}{3}$✓
(2) Assume S_k is true:
$\dfrac{1}{1(3)} + \dfrac{1}{3(5)} + \cdots + \dfrac{1}{(2k-1)(2k+1)} = \dfrac{k}{2k+1}$
Use it to show the truth of S_{k+1}:
$\dfrac{1}{1(3)} + \dfrac{1}{3(5)} + \cdots + \dfrac{1}{(2k-1)(2k+1)} + \dfrac{1}{(2k+1)(2k+3)} = \dfrac{k+1}{2k+3}$
left-hand side: $\mathbf{\dfrac{k}{2k+1}} + \dfrac{1}{(2k+1)(2k+3)}$
$= \dfrac{k(2k+3)}{(2k+1)(2k+3)} + \dfrac{1}{(2k+1)(2k+3)}$
$= \dfrac{2k^2 + 3k + 1}{(2k+1)(2k+3)} = \dfrac{(2k+1)(k+1)}{(2k+1)(2k+3)}$
$= \dfrac{k+1}{2k+3}$

39. (1) Show S_n is true for $n = 1$: S_1: $3^1 \geq 2(1) + 1$✓
(2) Assume S_k is true: $3^k \geq 2k + 1$
Use it to show the truth of S_{k+1}:
$3^{k+1} \geq 2(k+1) + 1 = 2k + 3$
left-hand side: $3^{k+1} = 3(3^k)$
$\geq 3\mathbf{(2k + 1)} = 6k + 3$
Since k is a positive integer, $6k + 3 \geq 2k + 3$ showing
$3^{k+1} \geq 2k + 3$

41. verified

43. (1) Show S_n is true for $n = 1$: S_1: $1^2 - 7 = -6$ or $2(-3)$✓
(2) Assume S_k is true: $k^2 - 7k = 2p$ for $p \in \mathbb{Z}$
Use it to show the truth of S_{k+1}:
$(k+1)^2 - 7(k+1) = 2q$ for $q \in \mathbb{Z}$
left-hand side: $k^2 + 2k + 1 - 7k - 7$
$= k^2 - 7k + 2k - 6$
$= \mathbf{2p} + 2k - 6$
$= 2(p + k - 3) = 2q$ is divisible by 2

45. verified

47. (1) Show S_n is true for $n = 1$: S_1: $6^1 - 1 = 5$✓
(2) Assume S_k is true: $6^k - 1 = 5p$ for $p \in \mathbb{Z}$
Use it to show the truth of S_{k+1}:
$6^{k+1} - 1 = 5q$ for $q \in \mathbb{Z}$
left-hand side: $6^{k+1} - 1 = 6 \cdot 6^k - 1$
$= 6(6^k) - 1$
$= 6\mathbf{(5p + 1)} - 1$
$= 30p + 5$
$= 5(6p + 1) = 5q$ is divisible by 5

49. The relation cannot be verified for all n; $n = 8$.

51. (1) Show S_n is true for $n = 1$. $\dfrac{x^1 - 1}{x - 1} = 1$✓ result checks
(2) Assume S_k is true. $\dfrac{x^k - 1}{x - 1} = 1 + x + x^2 + \cdots + x^{k-1}$ and use it to show the truth of S_{k+1} follows: $\dfrac{x(x^k - 1)}{x - 1} = x(1 + x + x^2 + \cdots + x^{k-1})$
$\dfrac{x^{k+1} - x}{x - 1} = x + x^2 + x^3 + \cdots + x^k$
$\dfrac{x^{k+1} - x}{x - 1} + \dfrac{x - 1}{x - 1} = (x + x^2 + x^3 + \cdots + x^k) + 1$
$\dfrac{x^{k+1} - 1}{x - 1} = 1 + x + x^2 + x^3 + \cdots + x^k$✓
Since the steps are reversible, the truth of S_{k+1} follows from S_k and the formula is true for all n.

53. $A + B = \begin{bmatrix} 1 & 1 \\ 7 & 4 \end{bmatrix}, A - B = \begin{bmatrix} -3 & 3 \\ -1 & -2 \end{bmatrix},$
$2A - 3B = \begin{bmatrix} -8 & 7 \\ -6 & -7 \end{bmatrix}, AB = \begin{bmatrix} 6 & 7 \\ 10 & 0 \end{bmatrix},$
$BA = \begin{bmatrix} -5 & 3 \\ 5 & 11 \end{bmatrix}, B^{-1} = \begin{bmatrix} 0.3 & 0.1 \\ -0.4 & 0.2 \end{bmatrix}$

55. $D: x \in (-\infty, \infty), R: y \in [-2, \infty)$ **57.** $x = \dfrac{1 + \ln 4}{2}$

Mid-Chapter Check, pp. 789–790

1. 3, 10, 17, $a_9 = 59$ **2.** 4, 7, 12, $a_9 = 84$ **3.** $-1, 3, -5, a_9 = -17$
4. 360 **5.** $\sum_{k=1}^{6}(3k - 2)$ **6.** d **7.** e **8.** a **9.** b **10.** c
11. a. $a_1 = 2, d = 3, a_n = 3n - 1$ **b.** $a_1 = \frac{3}{2}, d = \frac{3}{4}, a_n = \frac{3}{4}n + \frac{3}{4}$
12. $n = 25, S_{25} = 950$ **13.** $n = 16, S_{16} = 128$ **14.** $S_{10} = -5$
15. $S_{10} = \dfrac{-29{,}524}{27}$ **16. a.** $a_1 = 2, r = 3, a_n = 2(3)^{n-1}$
b. $a_1 = \frac{1}{2}, r = \frac{1}{2}, a_n = \left(\frac{1}{2}\right)^n$ **17.** $n = 8, S_8 = \frac{1640}{27}$ **18.** $\dfrac{-343}{6}$
19. 1785 **20.** $\approx$4.5 ft; $\approx$127.9 ft

Reinforcing Basic Concepts, pp. 790–791

1. \$71,500 **2. a.** 319 **b.** 728

Exercises 8.5, pp. 801–807

1. experiment, well-defined **3.** distinguishable **5.** Answers will vary.
7. a.

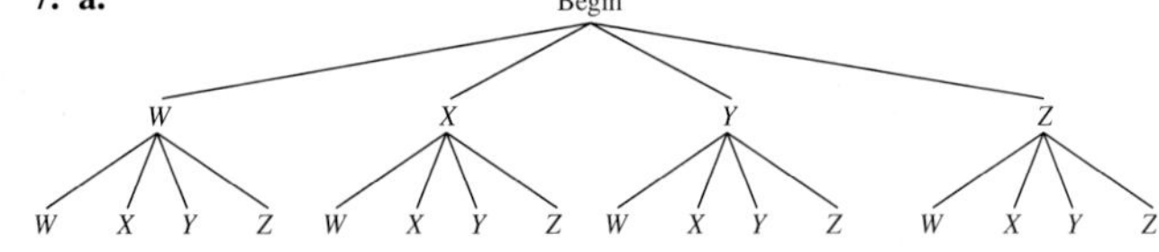

b. *WW, WX, WY, WZ, XW, XX, XY, XZ, YW, YX, YY, YZ, ZW, ZX, ZY, ZZ*
9. 32 **11.** 15,625 **13.** 2,704,000 **15. a.** 59,049 **b.** 15,120
17. 360 if double veggies are not allowed, 432 if double veggies are allowed.
19. a. 120 **b.** 625 **c.** 12 **21.** 24 **23.** 4 **25.** 120 **27.** 6 **29.** 720
31. 3024 **33.** 40,320 **35.** 6; 3 **37.** 90 **39.** 336 **41.** 720; 120; 24
43. 360 **45.** 60 **47.** 60 **49.** 120 **51.** 30 **53.** 60, BANANA
55. 126 **57.** 56 **59.** 1 **61.** verified **63.** verified **65.** 495 **67.** 364
69. 252 **71.** 40,320 **73.** 336 **75.** 15,504 **77.** 70 **79. a.** $\approx$1.2%
b. $\approx$0.83% **81.** 7776 **83.** 324 **85.** 32,760 **87.** 1,474,200
89. 800 **91.** 6,272,000,000 **93.** 518,400 **95.** 357,696 **97.** 6720
99. 8 **101.** 10,080 **103.** 5040 **105.** 2880 **107.** 5005 **109.** 720
111. 52,650, no **113.** The approximation gets better as *n* gets larger.
115. a. 1440 **b.** 5328
117.

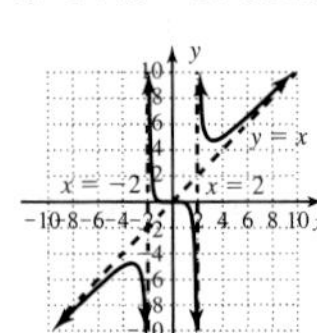

119. $a_n = 1 + (n - 1)4$; 137; 2415

121.

Exercises 8.6, pp. 815–823

1. $n(E)$ **3.** 0, 1, 1, 0 **5.** Answers will vary. **7.** $S = \{$HH, HT, TH, TT$\}$, $\frac{1}{4}$ **9.** $S = \{$coach of Patriots, Cougars, Angels, Sharks, Eagles, Stars$\}$, $\frac{1}{6}$
11. $P(E) = \frac{4}{9}$ **13. a.** $\frac{1}{13}$ **b.** $\frac{1}{4}$ **c.** $\frac{1}{2}$ **d.** $\frac{1}{26}$ **15.** $P(E_1) = \frac{1}{8}, P(E_2) = \frac{5}{8}, P(E_3) = \frac{3}{4}$ **17. a.** $\frac{3}{4}$ **b.** 1 **c.** $\frac{1}{4}$ **d.** $\frac{1}{2}$ **19.** $\frac{3}{4}$ **21.** $\frac{6}{7}$ **23.** 0.991
25. a. $\frac{1}{12}$ **b.** $\frac{11}{12}$ **c.** $\frac{8}{9}$ **d.** $\frac{5}{6}$ **27.** $\frac{10}{21}$ **29.** $\frac{60}{143}$ **31.** b, about 12% **33. a.** 0.3651 **b.** 0.3651 **c.** 0.3969 **35. a.** $\frac{1}{18}$ **b.** $\frac{2}{9}$ **c.** $\frac{8}{9}$
d. $\frac{3}{4}$ **e.** $\frac{1}{36}$ **f.** $\frac{5}{12}$ **37.** 0.9 **39.** $\frac{7}{24}$ **41.** 0.59 **43. a.** $\frac{1}{6}$ **b.** $\frac{7}{36}$ **c.** $\frac{1}{9}$
d. $\frac{4}{9}$ **45. a.** $\frac{2}{25}$ **b.** $\frac{9}{50}$ **c.** 0 **d.** $\frac{2}{25}$ **e.** 1 **47.** $\frac{3}{4}$ **49.** $\frac{11}{15}$
51. $\frac{1}{4}$; $\frac{1}{256}$; answers will vary. **53. a.** 0.33 **b.** 0.67 **c.** 1 **d.** 0
e. 0.67 **f.** 0.08 **55. a.** $\frac{1}{2}$ **b.** $\frac{1}{2}$ **c.** 0.2165 **57. a.** $\frac{9}{16}$ **b.** $\frac{1}{4}$ **c.** $\frac{1}{16}$
d. $\frac{5}{16}$ **59. a.** $\frac{3}{26}$ **b.** $\frac{3}{26}$ **c.** $\frac{1}{13}$ **d.** $\frac{9}{26}$ **e.** $\frac{2}{13}$ **f.** $\frac{11}{26}$ **61. a.** $\frac{1}{8}$ **b.** $\frac{1}{16}$
c. $\frac{3}{16}$ **63. a.** $\frac{47}{100}$ **b.** $\frac{2}{25}$ **c.** $\frac{3}{100}$ **d.** $\frac{9}{50}$ **e.** $\frac{11}{100}$ **65. a.** $\frac{5}{429}$
b. $\frac{8}{2145}$ **67.** $\frac{1}{3360}$ **69.** $\frac{1}{1{,}048{,}576}$; answers will vary; 20 heads in a row.
71. $P(E_1) = P(E_2) = \frac{100}{288}$; $P(E_1 \cup E_2) = \frac{170}{288}$ **73.** (9, 1, 1)
75.

77.

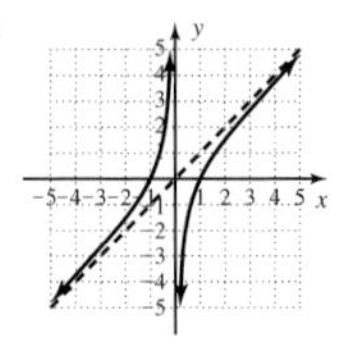

$x \in [-1, 0) \cup [1, \infty)$

Exercises 8.7, pp. 829–831

1. one **3.** $(a + (-2b))^5$ **5.** Answers will vary.
7. $x^5 + 5x^4y + 10x^3y^2 + 10x^2y^3 + 5xy^4 + y^5$
9. $16x^4 + 96x^3 + 216x^2 + 216x + 81$
11. $p^7 - 7p^6q + 21p^5q^2 - 35p^4q^3 + 35p^3q^4 - 21p^2q^5 + 7pq^6 - q^7$
13. 35 **15.** 10 **17.** 1140 **19.** 9880 **21.** 1 **23.** 1
25. $c^5 + 5c^4d + 10c^3d^2 + 10c^2d^3 + 5cd^4 + d^5$
27. $a^6 - 6a^5b + 15a^4b^2 - 20a^3b^3 + 15a^2b^4 - 6ab^5 + b^6$
29. $16x^4 - 96x^3 + 216x^2 - 216x + 81$ **31.** $-11 + 2i$
33. $x^9 + 18x^8y + 144x^7y^2 + \cdots$ **35.** $v^{24} - 6v^{22}w + \frac{33}{2}v^{20}w^2$
37. $35x^4y^3$ **39.** $1792p^2$ **41.** $264x^2y^{10}$ **43. a.** $\approx$17.8% **b.** $\approx$23.0%
45. a. $\approx$0.88% **b.** $\approx$6.9% **c.** $\approx$99.0% **d.** $\approx$61.0%
47 $16x^4 - 160x^3 + 600x^2 - 1000x + 625$ **49.** $\approx$0.25 **51. a.** 99.33%
b. 94.22% **53.** Answers will vary. **55.** verified
57.

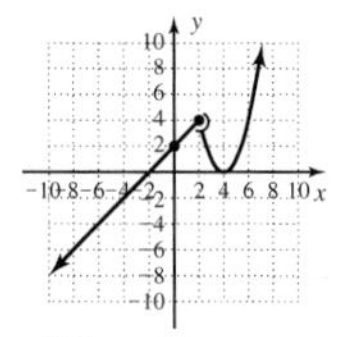

$f(3) = 1$

59. $g(x) > 0: x \in (-2, 0) \cup (3, \infty)$

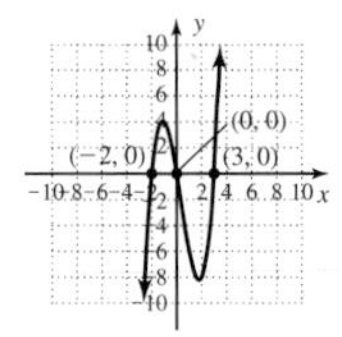

61. a. **b.** yes **c.** $l(g) = 2.57g - 8.10$; $\approx$82 cm

Summary and Concept Review, pp. 831–837

1. 1, 6, 11, 16; $a_{10} = 46$ **2.** not defined, 1, $\frac{1}{2}, \frac{1}{3}, \frac{1}{4}$; $a_{10} = \frac{1}{9}$
3. $a_n = n^4$; $a_6 = 1296$ **4.** $a_n = -17 + (n - 1)(3)$; $a_6 = -2$ **5.** $\frac{255}{256}$
6. -112 **7.** 140 **8.** 35 **9.** not defined, 2, 6, 12, 20, 30

10. $\frac{1}{2}, \frac{3}{4}, \frac{5}{4}, \frac{9}{4}, \frac{17}{4}$ **11.** $\sum_{n=1}^{7}(n^2 + 3n - 2)$; 210
12. $a_n = 2 + 3(n - 1)$; 119 **13.** $a_n = 3 + (-2)(n - 1)$; -65
14. 740 **15.** 1335 **16.** 630 **17.** -11.25 **18.** 875 **19.** 3240
20. 3645 **21.** 32 **22.** 2401 **23.** 10.75 **24.** 6560 **25.** $\frac{819}{512}$
26. does not exist **27.** $\frac{50}{9}$ **28.** 4 **29.** $\frac{63{,}050}{6561}$ **30.** does not exist **31.** 5
32. $a_9 = \$36{,}980$; $S_9 = \$314{,}900$ **33.** $\approx 7111.1 \text{ ft}^3$
34. $a_9 \approx 2105$ credit hours; $S_9 \approx 14{,}673$ credit hours

35. (1) Show S_n is true for $n = 1$: $S_1 = \dfrac{1(1 + 1)}{2} = 1$✓

(2) Assume S_k is true:

$$1 + 2 + 3 + \cdots + k = \frac{k(k + 1)}{2}$$

Use it to show the truth of S_{k+1}:

$$1 + 2 + 3 + \cdots + k + (k + 1) = \frac{(k + 1)(k + 2)}{2}$$

left-hand side: $1 + 2 + 3 + \cdots + k + (k + 1)$

$$= \frac{\mathbf{k(k + 1)}}{\mathbf{2}} + \frac{2(k + 1)}{2} = \frac{k(k + 1) + 2(k + 1)}{2}$$

$$= \frac{k^2 + 3k + 2}{2} = \frac{(k + 1)(k + 2)}{2}$$

36. (1) Show S_n is true for $n = 1$: $S_1 = \dfrac{1[2(1) + 1](1 + 1)}{6} = 1$✓

(2) Assume S_k is true:

$$1 + 4 + 9 + \cdots + k^2 = \frac{k(2k + 1)(k + 1)}{6}$$

Use it to show the truth of S_{k+1}:

$$1 + 4 + 9 + \cdots + k^2 + (k + 1)^2 = \frac{(k + 1)(2k + 3)(k + 2)}{6}$$

left-hand side: $1 + 4 + 9 + \cdots + k^2 + (k + 1)^2$

$$= \frac{\mathbf{k(k + 1)(2k + 1)}}{\mathbf{6}} + \frac{6(k + 1)^2}{6} = \frac{(k + 1)[(2k^2 + k + 6k + 6]}{6}$$

$$= \frac{(k + 1)(2k^2 + 7k + 6)}{6} = \frac{(k + 1)(2k + 3)(k + 2)}{6}$$

37. (1) Show S_n is true for $n = 1$: S_1: $4^1 \geq 3(1) + 1$✓
(2) Assume S_k is true: $4^k \geq 3k + 1$
Use it to show the truth of S_{k+1}:
$4^{k+1} \geq 3(k + 1) + 1 = 3k + 4$
left-hand side: $4^{k+1} = 4(4^k)$
$\geq 4(\mathbf{3k + 1}) = 12k + 4$
Since k is a positive integer, $12k + 4 \geq 3k + 4$ showing
$4^{k+1} \geq 3k + 4$

38. (1) Show S_n is true for $n = 1$: S_1: $6 \cdot 7^{1-1} \leq 7^1 - 1$✓
(2) Assume S_k is true: $6 \cdot 7^{k-1} \leq 7^k - 1$
Use it to show the truth of S_{k+1}:
$6 \cdot 7^k \leq 7^{k+1} - 1$
left-hand side: $6 \cdot 7^k = 7 \cdot 6 \cdot 7^{k-1}$
$\leq 7 \cdot \mathbf{7^k - 1}$
$\leq 7^{k+1} - 1$

39. (1) Show S_n is true for $n = 1$: S_1: $3^1 - 1 = 2$ or $2(1)$✓
(2) Assume S_k is true: $3^k - 1 = 2p$ for $p \in \mathbb{Z}$
Use it to show the truth of S_{k+1}:
$3^{k+1} - 1 = 2q$ for $q \in \mathbb{Z}$
left-hand side: $3^{k+1} - 1 = 3 \cdot 3^k - 1$
$= 3 \cdot \mathbf{2p}$
$= 2(3p) = 2q$ is divisible by 2

40. 6 ways

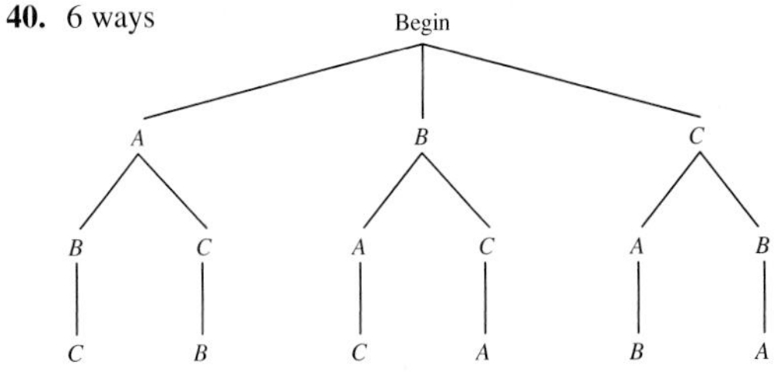

41. 720; 1000 **42.** 24 **43.** 220 **44.** 32 **45. a.** 5040 **b.** 840 **c.** 35
46. a. 720 **b.** 120 **c.** 24 **47.** 3360 **48. a.** 220 **b.** 1320 **49.** $\frac{4}{13}$
50. $\frac{3}{13}$ **51.** $\frac{5}{6}$ **52.** $\frac{7}{24}$ **53.** $\frac{175}{396}$ **54. a.** 0.608 **b.** 0.392 **c.** 1 **d.** 0
e. 0.928 **f.** 0.178 **55. a.** 21 **b.** 56
56. a. $x^4 - 4x^3y + 6x^2y^2 - 4xy^3 + y^4$ **b.** $41 - 38i$
57. a. $a^8 + 8\sqrt{3}a^7 + 84a^6 + 168\sqrt{3}a^5$ **b.** $78{,}125a^7 + 218{,}750a^6b + 262{,}500a^5b^2 + 175{,}000a^4b^3$ **58. a.** $280x^4y^3$ **b.** $-64{,}064a^5b^9$

Mixed Review, pp. 837–838

1. a. arithmetic **b.** $a_n = 4$ **c.** $a_n = n!$ **d.** arithmetic **e.** geometric
f. geometric **g.** arithmetic **h.** geometric **i.** $a_n = \dfrac{1}{2n}$ **3.** 27,600
5. 0.1, 0.5, 2.5, 12.5, 62.5; $a_{20} = 1{,}907{,}348{,}632{,}812$ **7.** $\frac{5}{6}$
9. a. 2 **b.** 200 **c.** 210 **11. a.** $a^{20} + 20a^{19}b + 190a^{18}b^2$
b. $190a^2b^{18} + 20ab^{19} + b^{20}$ **c.** $52{,}360a^{31}b^4$ **d.** 4.6×10^{-18}
13. verified **15.** 0.01659 **17.** $\frac{4}{11}$ **19.** 10, 2, $\frac{2}{5}$, $\frac{2}{25}$, $\frac{2}{125}$

Practice Test, pp. 838–840

1. a. $\frac{1}{2}, \frac{4}{5}, 1, \frac{8}{7}$; $a_8 = \frac{16}{11}$, $a_{12} = \frac{8}{5}$ **b.** 6, 12, 20, 30; $a_8 = 90$, $a_{12} = 182$
c. $3, 2\sqrt{2}, \sqrt{7}, \sqrt{6}$; $a_8 = \sqrt{2}$, $a_{12} = i\sqrt{2}$ **2. a.** 165 **b.** $\frac{311}{420}$ **c.** $\frac{-2343}{512}$
d. 7 **3. a.** $a_1 = 7, d = -3, a_n = 10 - 3n$ **b.** $a_1 = -8, d = 2,$
$a_n = 2n - 10$ **c.** $a_1 = 4, r = -2, a_n = 4(-2)^{n-1}$ **d.** $a_1 = 10, r = \frac{2}{5},$
$a_n = 10(\frac{2}{5})^{n-1}$ **4. a.** 199 **b.** 9 **c.** $\frac{3}{4}$ **d.** 6 **5. a.** 1712 **b.** 2183
c. 2188 **d.** 12 **6. a.** ≈ 8.82 ft **b.** ≈ 72.4 ft **7.** \$6756.57
8. \$22,185.27 **9.** verified **10.** verified **11.** ABC, ACB, BAC, BCA, CAB, CBA

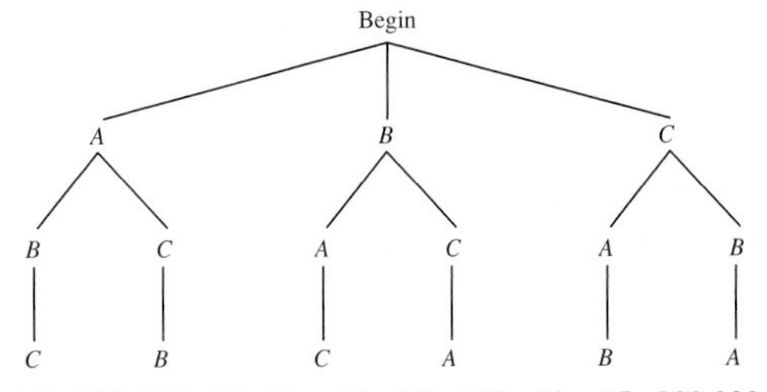

12. 302,400 **13.** 64 **14.** 720, 120, 20 **15.** 900,900 **16.** 302,400
17. a. $x^4 - 8x^3y + 24x^2y^2 - 32xy^3 + 16y^4$ **b.** -4
18. a. $x^{10} + 10\sqrt{2}x^9 + 90x^8$ **b.** $a^8 - 16a^7b^3 + 112a^6b^6$
19. 0.989 **20. a.** $\frac{1}{4}$ **b.** $\frac{7}{12}$ **c.** $\frac{1}{3}$ **d.** $\frac{1}{2}$ **e.** $\frac{7}{12}$ **f.** $\frac{1}{4}$ **g.** $\frac{3}{4}$ **h.** 0
21. a. 0.08 **b.** 0.92 **c.** 1 **d.** 0 **e.** 0.95 **f.** 0.03 **22. a.** 0.1875
b. 0.589 **c.** 0.4015 **d.** 0.2945 **e.** 0.4110 **f.** 0.2055 **23. a.** $\frac{59}{100}$
b. $\frac{53}{100}$ **c.** $\frac{13}{100}$ **d.** $\frac{47}{100}$ **24. a.** 0.8075 **b.** 0.0075 **c.** 0.9925
25. verified

Strengthening Core Skills, pp. 841–843

Exercise 1: 0.000 240 **Exercise 2:** 0.001 441 **Exercise 3:** 0.021 128
Exercise 4: 0.042 920 **Exercise 5:** 0.003 925

Cumulative Review Chapters 1–8, pp. 844–845

1. a. 23 cards are assembled each hour. **b.** $y = 23x + 52$ **c.** 236 cards **d.** ≈6:45 A.M. **3.** $x = \dfrac{-5 \pm \sqrt{109}}{6}$; $x \approx 0.91$; $x \approx -2.57$ **5.** $Y = \dfrac{KVW}{X}$; $Y = \dfrac{3VW}{2X}$ **7.** verified; reflections across $y = x$ **9. a.** $4x + 2h - 3$ **b.** $\dfrac{-1}{(x + h - 2)(x - 2)}$ **11.**

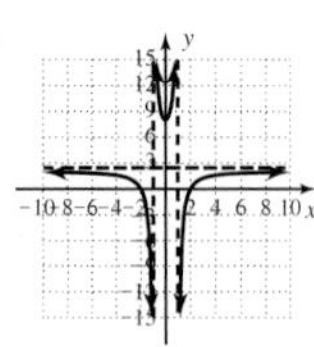

13. a. $x^3 = 125$ **b.** $e^5 = 2x - 1$ **15. a.** $x \approx 3.19$ **b.** $x = 334$ **17.** (5, 10, 15) **19.** $(-3, 3)$; $(-7, 3)$, $(1, 3)$; $(-3 - 2\sqrt{3}, 3)$, $(-3 + 2\sqrt{3}, 3)$ **21.** 1333 **23. a.** ≈7.0% **b.** ≈91.9% **c.** ≈98.9% **d.** $\binom{12}{0}(0.04)^{12}(0.96)^0$; virtually nil **25.** $y = -9.75x^2 + 128.4x - 285.15$ **a.** July **b.** ≈136 **c.** 134 **d.** early October to late February

CHAPTER 6

Exercises 6.4, pp. 610–613

13. $Y_1 = |x + 3|$, $Y_2 = 5$; $(-8, 5)$, $(2, 5)$

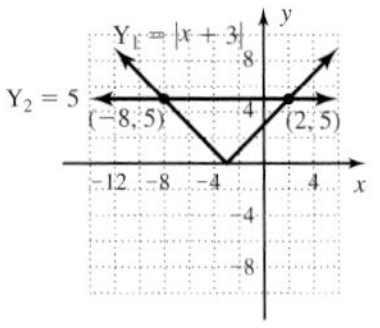

15.

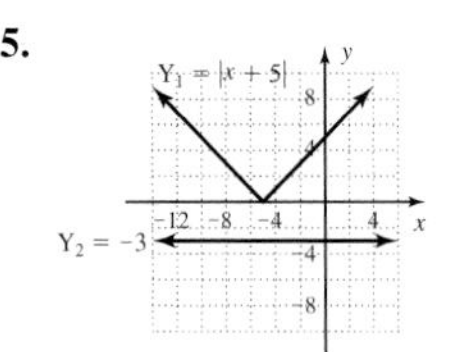

17. $Y_1 = |x|$, $Y_2 = 4$; $(-4, 4)$, $(4, 4)$

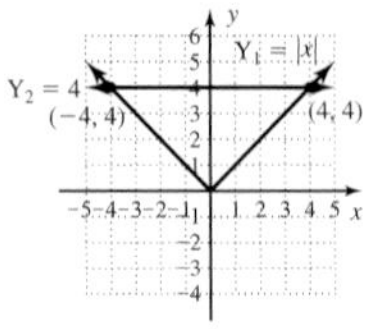

19.

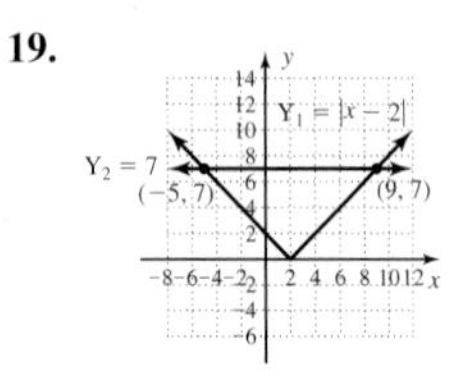

21.

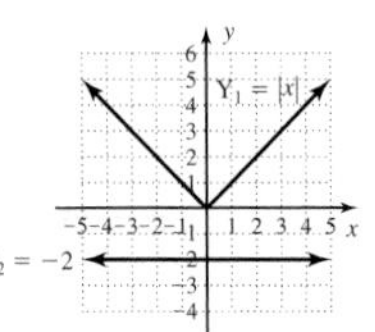

23.

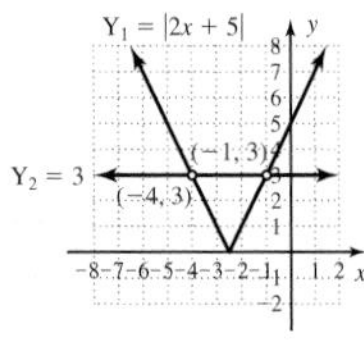

25.

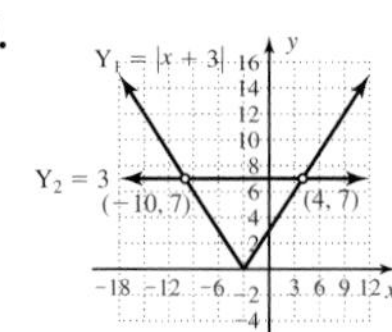

27.

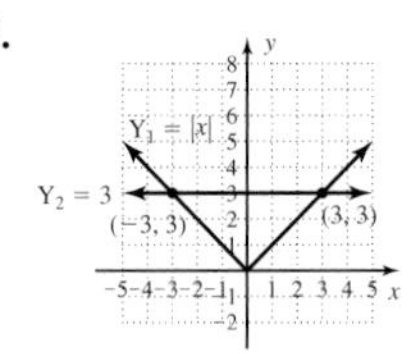

Summary and Concept Review, pp. 666–671

1.

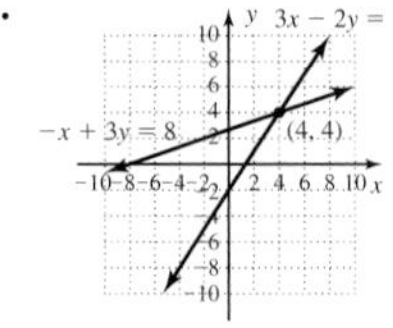

2.

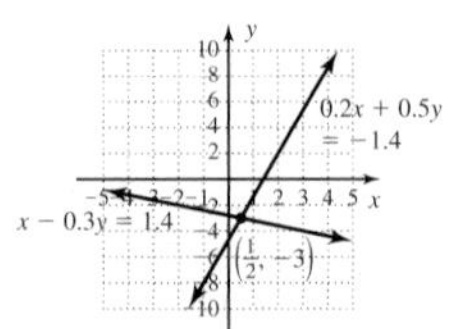

3. $2x + y = 2$, $x - 2y = 4$; $\left(\frac{8}{5}, -\frac{6}{5}\right)$

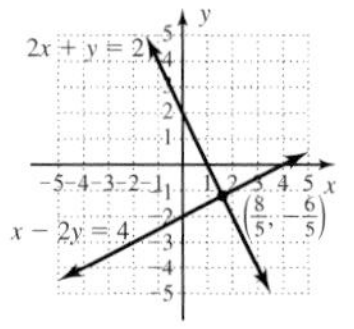

Practice Test, pp. 673–674

1. $3x + 2y = 12$, $-x + 4y = 10$; (2, 3)

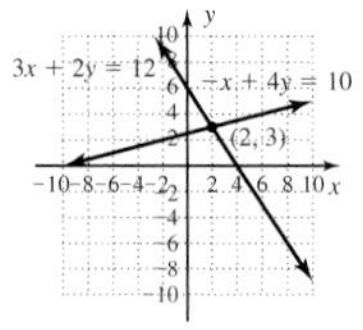

19.

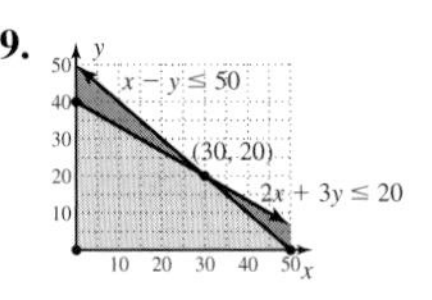

Instructor Answer Appendix

CHAPTER R

Exercises R.1

25. a. i. $\{\sqrt{49}, 2, 6, 4\}$ **ii.** $\{\sqrt{49}, 2, 6, 0, 4\}$
iii. $\{-5, \sqrt{49}, 2, -3, 6, -1, 0, 4\}$ **iv.** $\{-5, \sqrt{49}, 2, -3, 6, -1, 0, 4\}$
v. $\{\sqrt{3}, \pi\}$ **vi.** $\{-5, \sqrt{49}, 2, -3, 6, -1, \sqrt{3}, 0, 4, \pi\}$
b. $\{-5, -3, -1, 0, \sqrt{3}, 2, \pi, 4, 6, \sqrt{49}\}$
c.

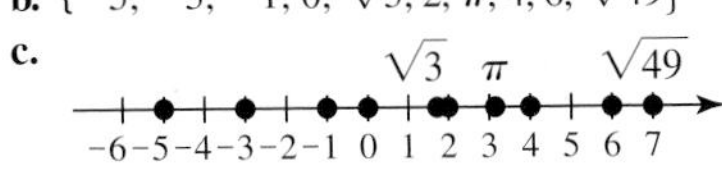

26. a. i. $\{5, \sqrt{64}\}$ **ii.** $\{5, \sqrt{64}\}$ **iii.** $\{-8, 5, \sqrt{64}\}$
iv. $\{-8, 5, -2\frac{3}{5}, 1.75, -0.6, \frac{7}{2}, \sqrt{64}\}$ **v.** $\{-\sqrt{2}, \pi\}$
vi. $\{-8, 5, -2\frac{3}{5}, 1.75, -\sqrt{2}, -0.6, \pi, \frac{7}{2}, \sqrt{64}\}$
b. $\{-8, -2\frac{3}{5}, -\sqrt{2}, -0.6, 1.75, \pi, \frac{7}{2}, 5, \sqrt{64}\}$
c.

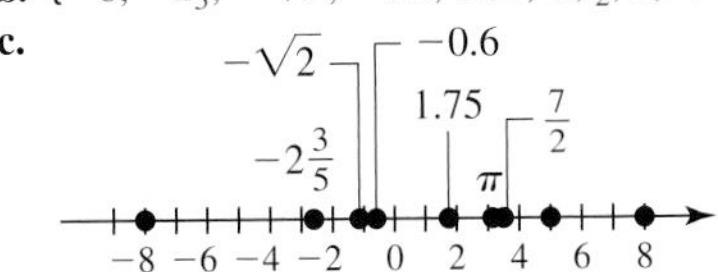

CHAPTER 1

Exercises 1.2

27. $\{m|2 < m \leq 6\}$;

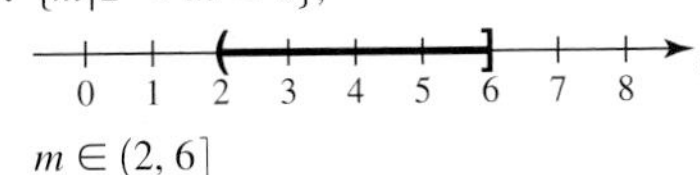

$m \in (2, 6]$

28. $\{x|-3 < x \leq \frac{5}{2}\}$;

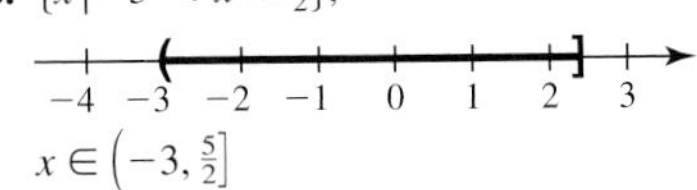

$x \in \left(-3, \frac{5}{2}\right]$

29. $\{m|\frac{4}{3} \leq m < \frac{11}{6}\}$;

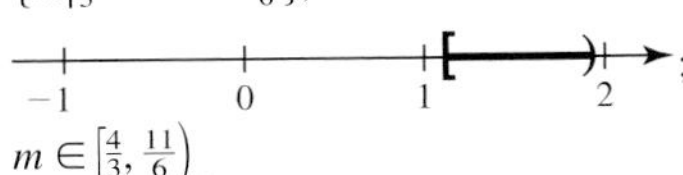

$m \in \left[\frac{4}{3}, \frac{11}{6}\right)$

30. $\{x|\frac{-10}{3} < x \leq -1\}$;

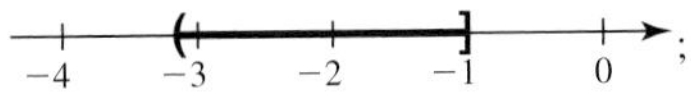

$x \in \left(\frac{-10}{3}, -1\right]$

41. 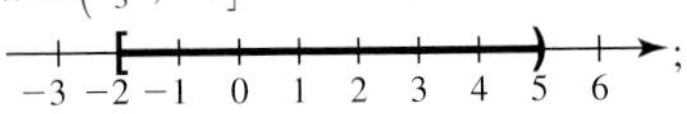 $x \in [-2, 5)$

42. 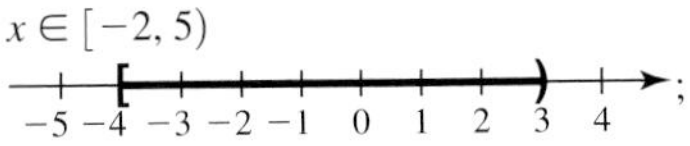 $x \in [-4, 3)$

43. 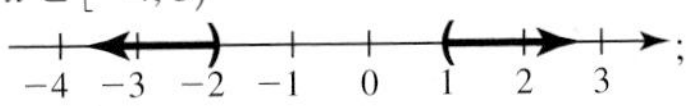 $x \in (-\infty, -2) \cup (1, \infty)$

44. 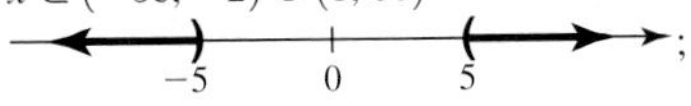 $x \in (-\infty, -5) \cup (5, \infty)$

47. $x \in (-\infty, \infty)$;

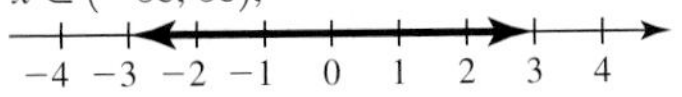

48. $x \in (-\infty, 2]$;

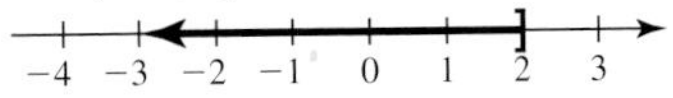

49. $x \in [-5, 0]$;

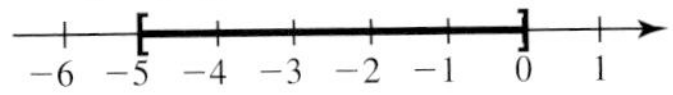

50. $x \in [-4, 0]$;

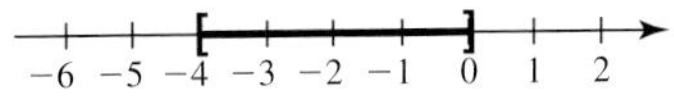

Exercises 1.5

92. $z = \frac{-5 \pm \sqrt{105}}{20}$; $z \approx 0.26$ or $z \approx -0.76$ **95.** two rational **96.** two rational **97.** two complex **98.** two complex **99.** two rational **100.** two rational **101.** two complex **102.** two complex **103.** two irrational **104.** two irrational **105.** one repeated **106.** one repeated **107.** $x = \frac{3}{2} \pm \frac{1}{2}i$ **108.** $x = \frac{5}{2} \pm \frac{3}{2}i$ **109.** $x = -\frac{1}{2} \pm \frac{i\sqrt{3}}{2}$ **110.** $x = -1 \pm 3i\sqrt{2}$ **111.** $x = \frac{5}{4} \pm \frac{3i\sqrt{7}}{4}$ **112.** $x = \frac{2}{5} \pm \frac{i\sqrt{11}}{5}$

CHAPTER 2

Exercises 2.1

25.

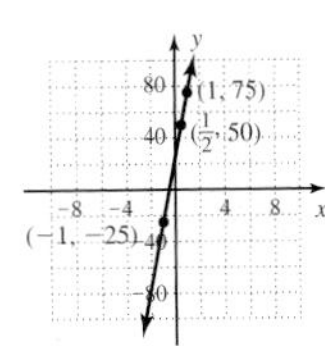

26.

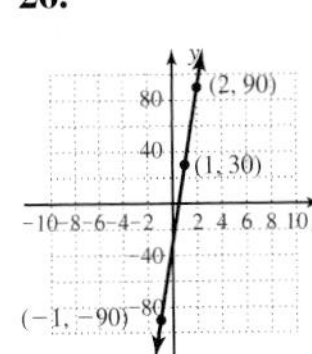

27.

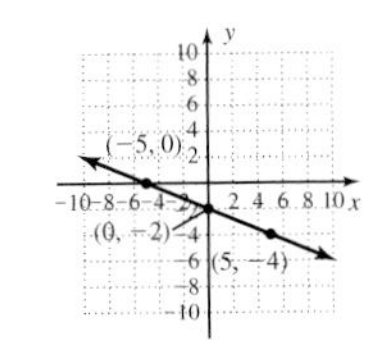

28.

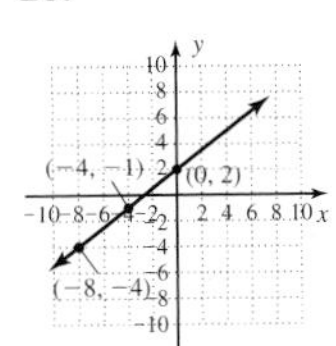

29.

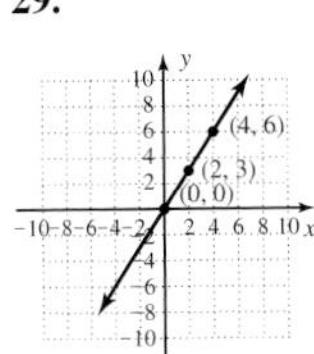

30.

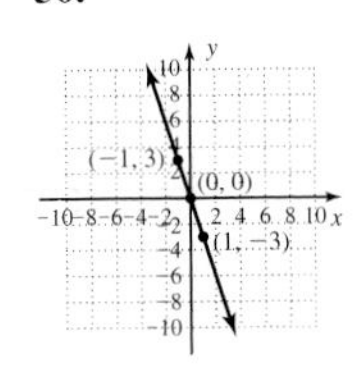

31.

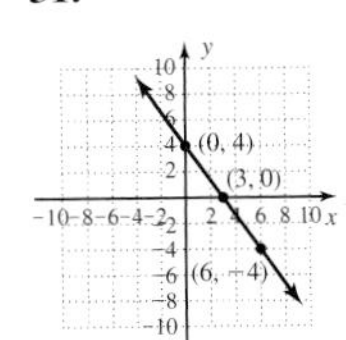

32.

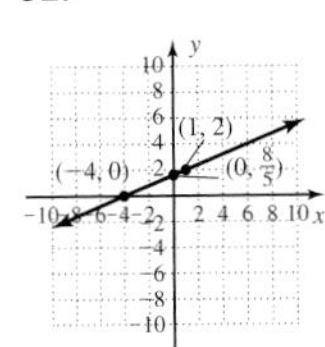

33.

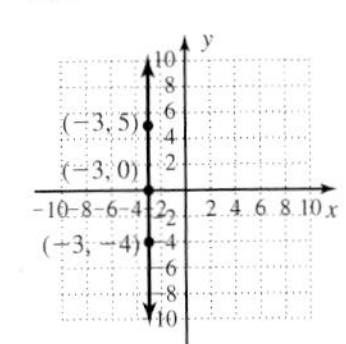

34.

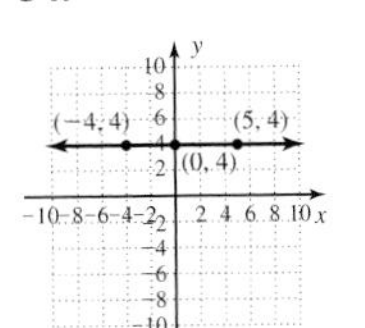

35.

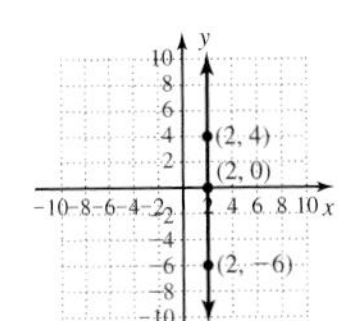

36.

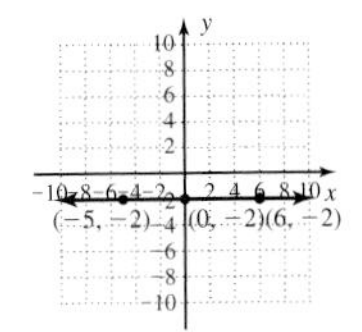

Exercises 2.2

94. $p(0.5) = 14$, $p(\frac{9}{4}) = \frac{70}{27}$, $p(a) = \dfrac{2a^2 + 3}{a^2}$, $p(a + 3) = \dfrac{2a^2 + 12a + 21}{a^2 + 6a + 9}$
95. a. $D = \{-1, 0, 1, 2, 3, 4, 5\}$ **b.** 1 **c.** -1 **d.** $R = \{-2, -1, 0, 1, 2, 3, 4\}$
96. a. $D = \{-5, -4, -3, -2, -1, 0, 1, 2, 3, 4, 5\}$ **b.** 0 **c.** -4 **d.** $R = \{-1, 0, 1, 2, 3, 4, 5\}$
97. a. $x \in [-5, 5]$ **b.** -2 **c.** -2 **d.** $y \in [-3, 4]$
98. a. $x \in [-3, 5]$ **b.** -4 **c.** -1 **d.** $y \in [-4, 5]$
99. a. $x \in [-3, \infty)$ **b.** 2 **c.** 0 **d.** $y \in (-\infty, 4]$
100. a. $x \in [-5, \infty)$ **b.** -3 **c.** -4 **d.** $y \in [-4, \infty)$

Exercises 2.3

35. $y = \frac{1}{3}x$, $m = \frac{1}{3}$, y-intercept $(0, 0)$ **36.** $y = \frac{-2}{5}x$, $m = \frac{-2}{5}$, y-intercept $(0, 0)$ **37.** $y = \frac{-3}{4}x + 3$, $m = \frac{-3}{4}$, y-intercept $(0, 3)$
38. $y = \frac{3}{5}x - 4$, $m = \frac{3}{5}$, y-intercept $(0, -4)$

51. $y = \frac{-4}{5}x + 4$

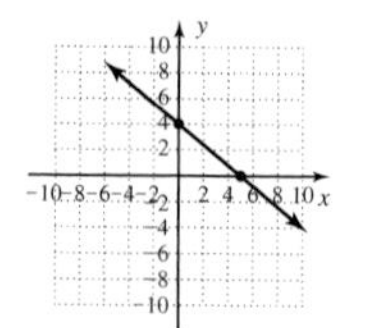

52. $y = \frac{1}{2}x + 2$

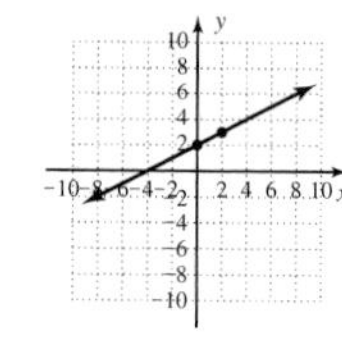

53. $y = \frac{5}{3}x - 5$

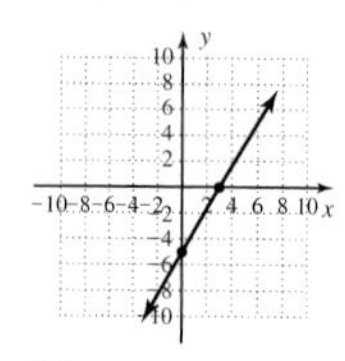

54. $y = \frac{-2}{5}x + 2$

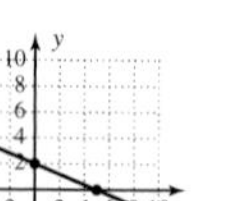

55.

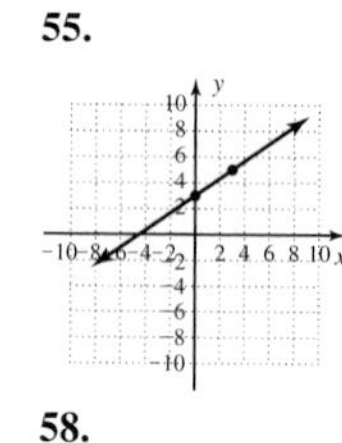

56.

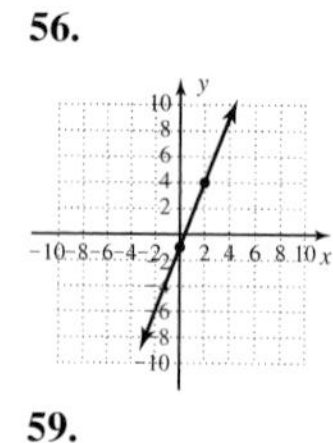

57.

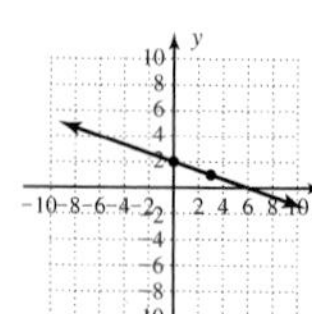

58.

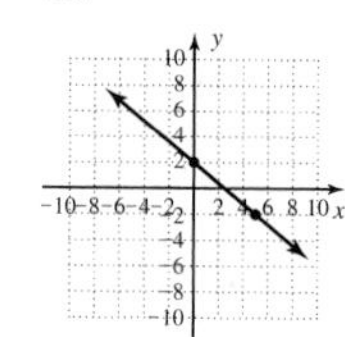

59.

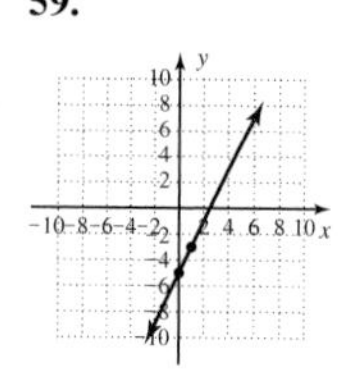

60.

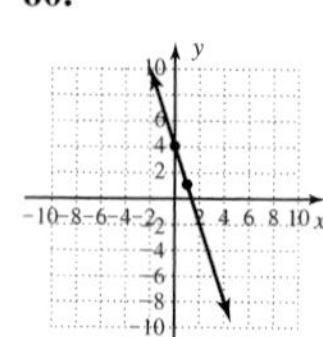

61.

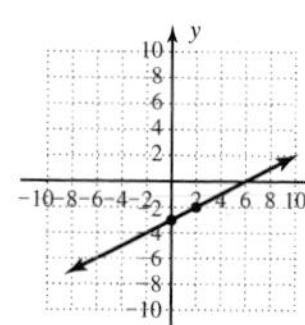

62.

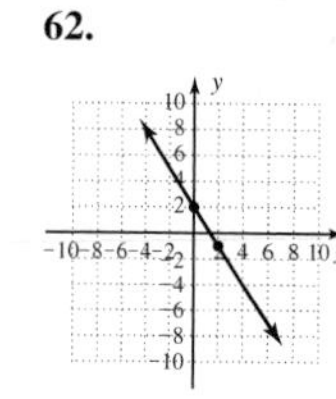

87. $y = \frac{6}{5}x - \frac{14}{5}$; For each 5000 additional sales, income rises \$6000.
88. $y = \frac{-3}{2}x + \frac{27}{2}$; Every two years, 30,000 typewriters are no longer in service. **89.** $y = -20x + 110$; For every hour of television, a student's final grade falls 20%.

Mid-Chapter Check

6. $y = \frac{-4}{3}x + 4$; yes; Each input is paired with only one output.
7. a. 0 **b.** $x \in [-3, 5]$ **c.** 3.5 **d.** $y \in [-4, 5]$

Reinforcing Basic Concepts

1. a. $\frac{1}{3}$, increasing **b.** $y - 5 = \frac{1}{3}(x - 0)$
c. $y = \frac{1}{3}x + 5$ **d.** $x - 3y = -15$
e. $(0, 5)$, $(-15, 0)$

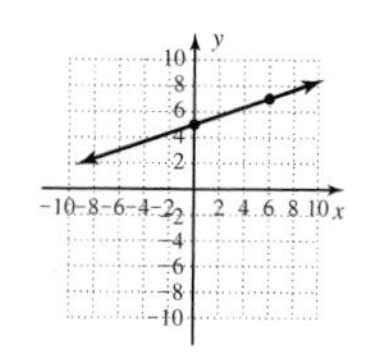

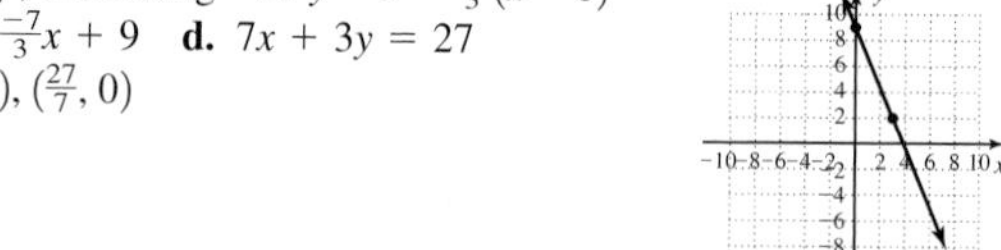

2. a. $\frac{-7}{3}$, decreasing **b.** $y - 9 = \frac{-7}{3}(x - 0)$
c. $y = \frac{-7}{3}x + 9$ **d.** $7x + 3y = 27$
e. $(0, 9)$, $(\frac{27}{7}, 0)$

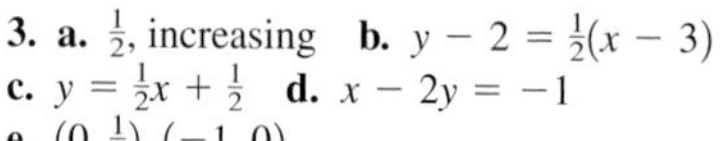

3. a. $\frac{1}{2}$, increasing **b.** $y - 2 = \frac{1}{2}(x - 3)$
c. $y = \frac{1}{2}x + \frac{1}{2}$ **d.** $x - 2y = -1$
e. $(0, \frac{1}{2})$, $(-1, 0)$

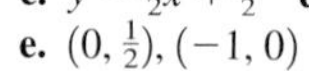

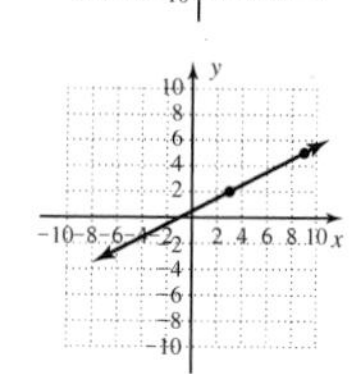

4. a. $\frac{3}{4}$, increasing **b.** $y + 4 = \frac{3}{4}(x + 5)$
c. $y = \frac{3}{4}x - \frac{1}{4}$ **d.** $3x - 4y = 1$
e. $(0, \frac{-1}{4})$, $(\frac{1}{3}, 0)$

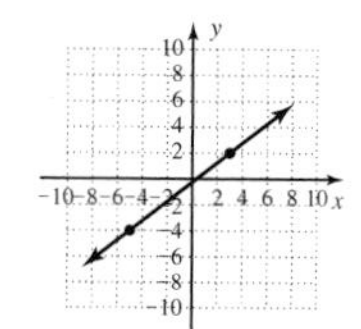

5. a. $\frac{-3}{4}$, decreasing **b.** $y - 5 = \frac{-3}{4}(x + 2)$
c. $y = \frac{-3}{4}x + \frac{7}{2}$ **d.** $3x + 4y = 14$
e. $(0, \frac{7}{2})$, $(\frac{14}{3}, 0)$

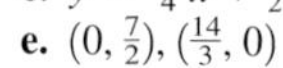

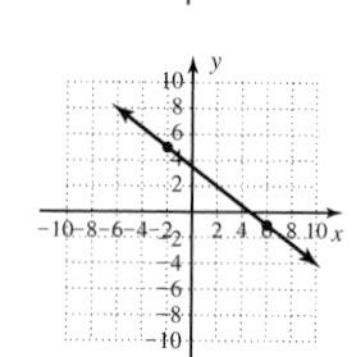

6. a. $\frac{-1}{2}$, decreasing **b.** $y + 7 = \frac{-1}{2}(x - 2)$
c. $y = \frac{-1}{2}x - 6$ **d.** $x + 2y = -12$
e. $(0, -6)$, $(-12, 0)$

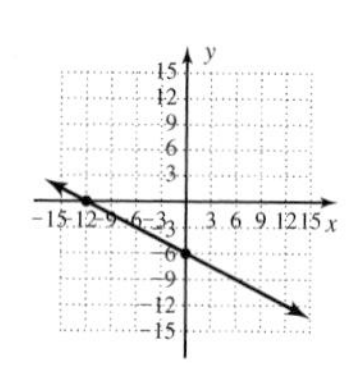

Exercises 2.4

23.

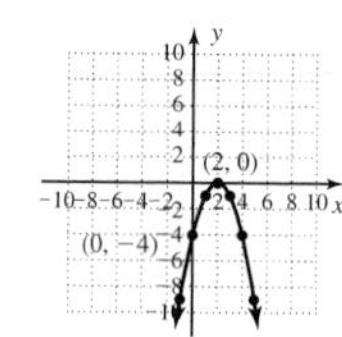

24.

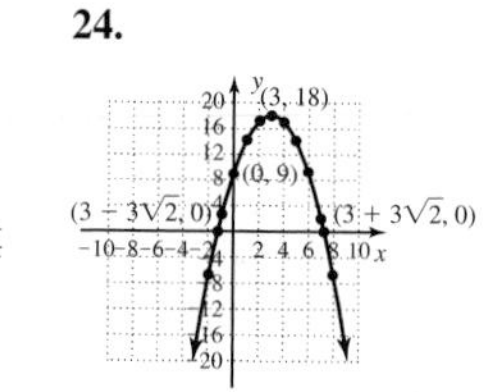

25.

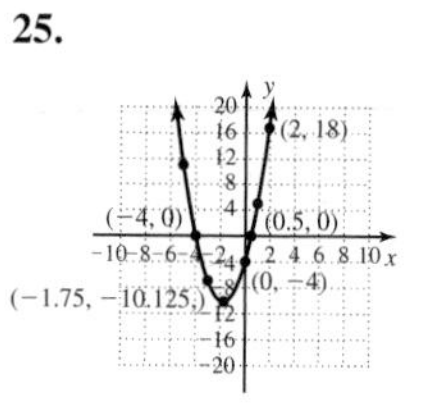

26.

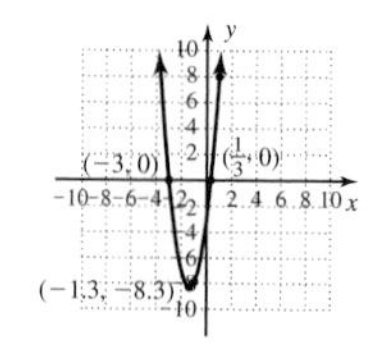

27.

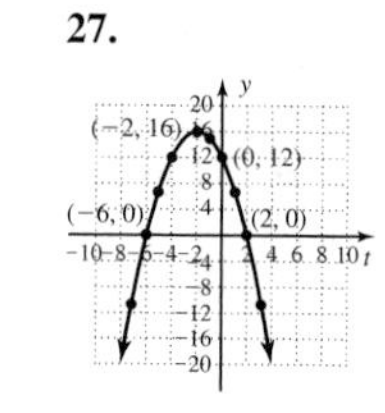

28.

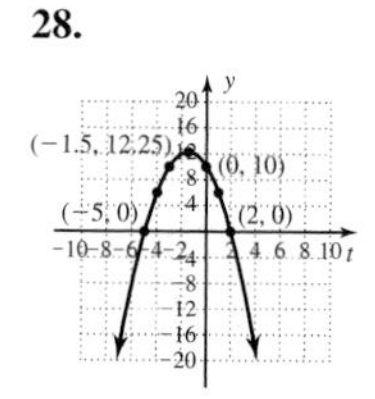

29.

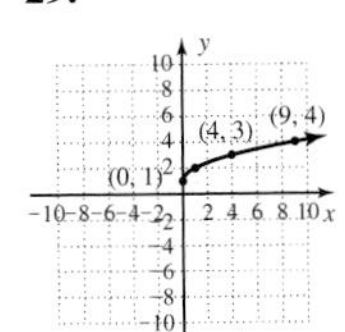

30.

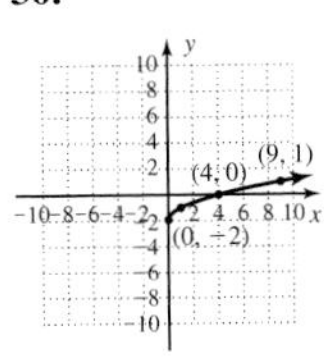

31.

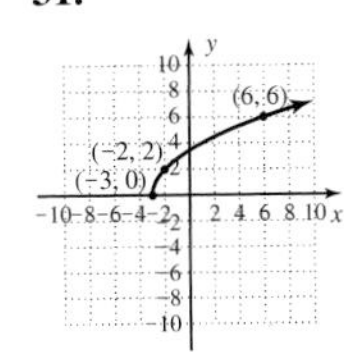

32.

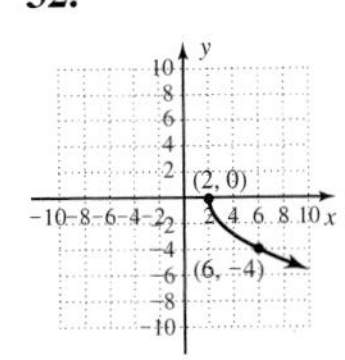

33.

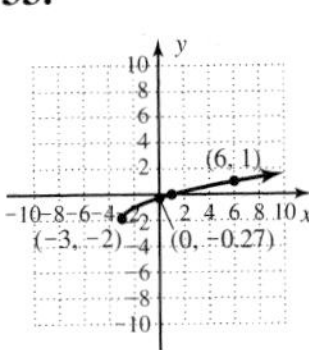

34.

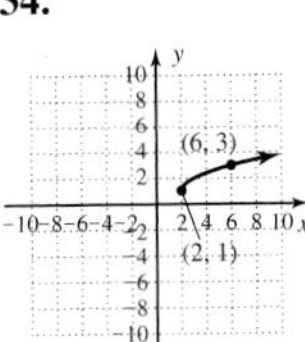

35.

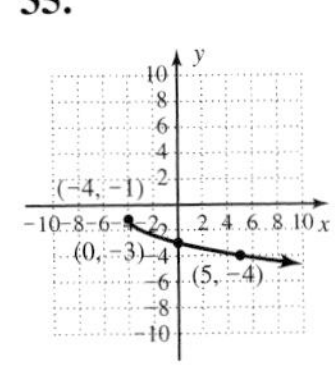

36.

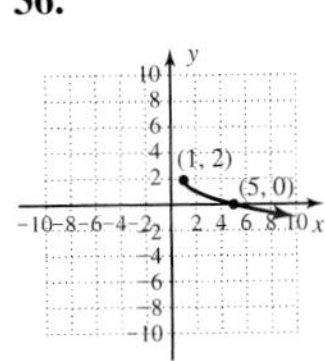

37.

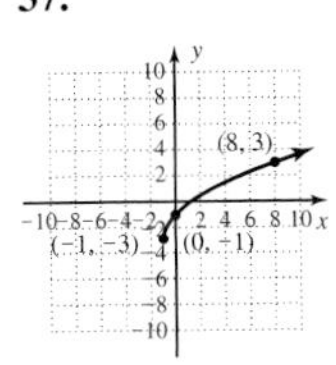

38.

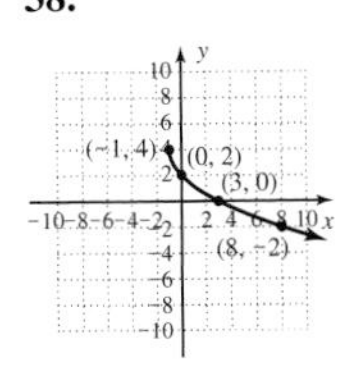

39.

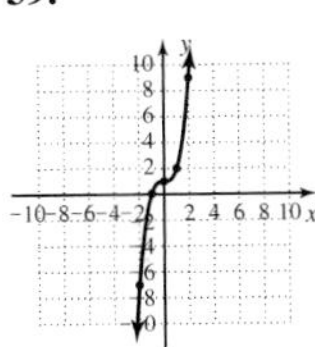

40.

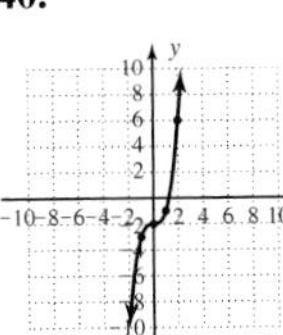

41.

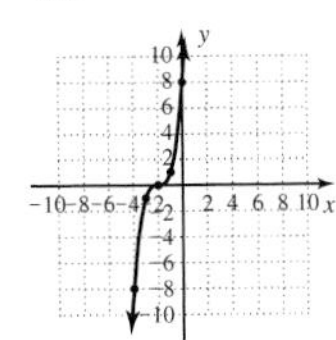

42.

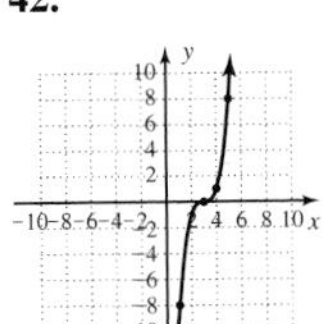

Strengthening Core Skills

Exercise 2

x	x^2	-9	$x^2 - 9$
0	0	−9	−9
1	1	−9	−8
2	4	−9	−5
3	9	−9	0
4	16	−9	7
5	25	−9	16
6	36	−9	27
7	49	−9	40

x	9	x^2	$9 - x^2$
0	9	0	9
1	9	−1	8
2	9	−4	5
3	9	−9	0
4	9	−16	−7
5	9	−25	−16
6	9	−36	−27
7	9	−49	−40

CHAPTER 3

Exercises 3.1

26. $h(x) = \dfrac{x+1}{\sqrt{x+3}};\ x \in (-3, \infty)$

27. $h(x) = \dfrac{x^2-9}{\sqrt{x+1}};\ x \in (-1, \infty)$

28. $h(x) = \dfrac{x^2-1}{\sqrt{x-3}};\ x \in (3, \infty)$

29. $h(x) = x - 4;\ x \in (-\infty, -4) \cup (-4, \infty)$

30. $h(x) = x + 7;\ x \in (-\infty, 7) \cup (7, \infty)$

31. $h(x) = x^2 - 2;\ x \in (-\infty, -4) \cup (-4, \infty)$

32. $h(x) = x^2 + 2;\ x \in (-\infty, 5) \cup (5, \infty)$

33. $h(x) = \dfrac{3x+6}{x-3};\ x \in (-\infty, -2) \cup (-2, 3) \cup (3, \infty)$

34. $h(x) = \dfrac{2x-4}{x+1};\ x \in (-\infty, -1) \cup (-1, 2) \cup (2, \infty)$

35. sum: $3x + 1,\ x \in (-\infty, \infty)$; difference: $x + 5,\ x \in (-\infty, \infty)$;
product: $2x^2 - x - 6,\ x \in (-\infty, \infty)$;
quotient: $\dfrac{2x+3}{x-2},\ x \in (-\infty, 2) \cup (2, \infty)$

36. sum: $3x - 8,\ x \in (-\infty, \infty)$; difference: $-x - 2,\ x \in (-\infty, \infty)$;
product: $2x^2 - 13x + 15,\ x \in (-\infty, \infty)$;
quotient: $\dfrac{x-5}{2x-3},\ x \in \left(-\infty, \dfrac{3}{2}\right) \cup \left(\dfrac{3}{2}, \infty\right)$

37. sum: $x^2 + 3x + 5,\ x \in (-\infty, \infty)$;
difference: $x^2 - 3x + 9,\ x \in (-\infty, \infty)$;
product: $3x^3 - 2x^2 + 21x - 14,\ x \in (-\infty, \infty)$;
quotient: $\dfrac{x^2+7}{3x-2},\ x \in \left(-\infty, \dfrac{2}{3}\right) \cup \left(\dfrac{2}{3}, \infty\right)$

38. sum: $x^2 - 2x + 4,\ x \in (-\infty, \infty)$;
difference: $x^2 - 4x - 4,\ x \in (-\infty, \infty)$;
product: $x^3 + x^2 - 12x,\ x \in (-\infty, \infty)$;
quotient: $\dfrac{x^2-3x}{x+4},\ x \in (-\infty, -4) \cup (-4, \infty)$

39. sum: $x^2 + 3x - 4,\ x \in (-\infty, \infty)$;
difference: $x^2 + x - 2,\ x \in (-\infty, \infty)$;
product: $x^3 + x^2 - 5x + 3,\ x \in (-\infty, \infty)$;
quotient: $x + 3,\ x \in (-\infty, 1) \cup (1, \infty)$

40. sum: $x^2 - x - 12,\ x \in (-\infty, \infty)$;
difference: $x^2 - 3x - 18,\ x \in (-\infty, \infty)$;
product: $x^3 + x^2 - 21x - 45,\ x \in (-\infty, \infty)$;
quotient: $x - 5,\ x \in (-\infty, -3) \cup (-3, \infty)$

41. sum: $3x + 1 + \sqrt{x-3},\ x \in [3, \infty)$;
difference: $3x + 1 - \sqrt{x-3},\ x \in [3, \infty)$;
product: $(3x + 1)\sqrt{x-3},\ x \in [3, \infty)$;
quotient: $\dfrac{3x+1}{\sqrt{x-3}},\ x \in (3, \infty)$

42. sum: $x + 2 + \sqrt{x+6},\ x \in [-6, \infty)$;
difference: $x + 2 - \sqrt{x-6},\ x \in [-6, \infty)$;
product: $(x + 2)\sqrt{x+6},\ x \in [-6, \infty)$;
quotient: $\dfrac{x+2}{\sqrt{x+6}},\ x \in (-6, \infty)$

43. sum: $2x^2 + \sqrt{x+1},\ x \in [-1, \infty)$;
difference: $2x^2 - \sqrt{x+1},\ x \in [-1, \infty)$;
product: $2x^2\sqrt{x+1},\ x \in [-1, \infty)$;
quotient: $\dfrac{2x^2}{\sqrt{x+1}},\ x \in (-1, \infty)$

44. sum: $x^2 + 2 + \sqrt{x - 5}, x \in [5, \infty)$;
difference: $x^2 + 2 - \sqrt{x - 5}, x \in [5, \infty)$;
product: $(x^2 + 2)\sqrt{x - 5}, x \in [5, \infty)$;
quotient: $\dfrac{x^2 + 2}{\sqrt{x - 5}}, x \in (5, \infty)$

45. sum: $\dfrac{7x - 11}{(x - 3)(x + 2)}, x \in (-\infty, -2) \cup (-2, 3) \cup (3, \infty)$;
difference: $\dfrac{-3x + 19}{(x - 3)(x + 2)}, x \in (-\infty, -2) \cup (-2, 3) \cup (3, \infty)$;
product: $\dfrac{10}{x^2 - x - 6}, x \in (-\infty, -2) \cup (-2, 3) \cup (3, \infty)$;
quotient: $\dfrac{2x + 4}{5x - 15}, x \in (-\infty, -2) \cup (-2, 3) \cup (3, \infty)$

46. sum: $\dfrac{5x + 17}{(x - 3)(x + 5)}, x \in (-\infty, -5) \cup (-5, 3) \cup (3, \infty)$;
difference: $\dfrac{3x + 23}{(x - 3)(x + 5)}, x \in (-\infty, -5) \cup (-5, 3) \cup (3, \infty)$;
product: $\dfrac{4}{x^2 + 2x - 15}, x \in (-\infty, -5) \cup (-5, 3) \cup (3, \infty)$;
quotient: $\dfrac{4x + 20}{x - 3}, x \in (-\infty, -5) \cup (-5, 3) \cup (3, \infty)$

Exercises 3.2

77.

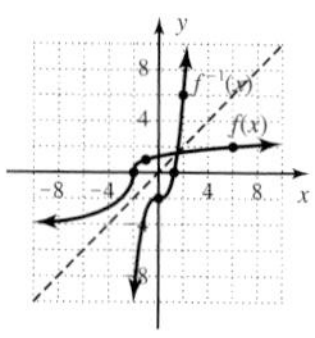

78.

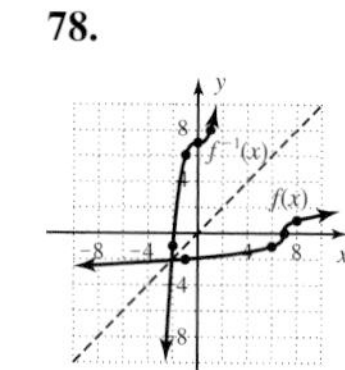

79.

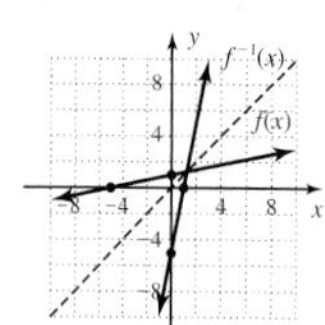

80.

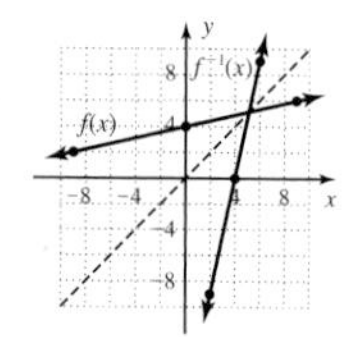

81.

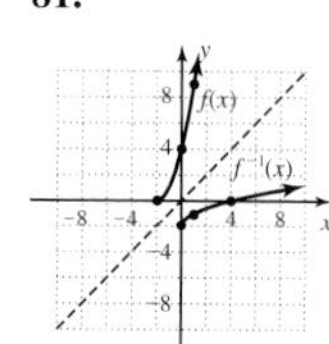

82.

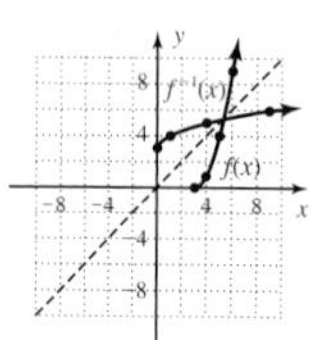

83. $D: x \in [0, \infty), R: y \in [-2, \infty); D: x \in [-2, \infty), R: y \in [0, \infty)$

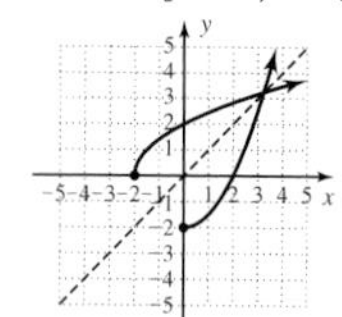

84. $D: x \in (-\infty, \infty), R: y \in (-\infty, \infty)$;
$D: x \in (-\infty, \infty), R: y \in (-\infty, \infty)$

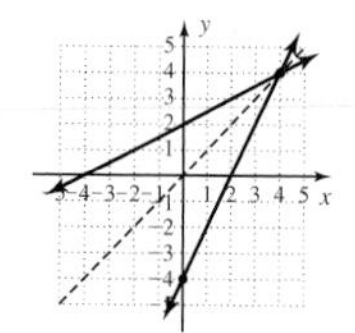

85. $D: x \in (0, \infty), R: y \in (-\infty, \infty); D: x \in (-\infty, \infty), R: y \in (0, \infty)$

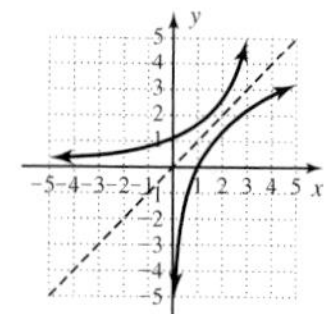

Exercises 3.3

27.

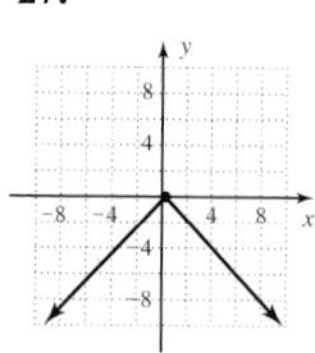

28.

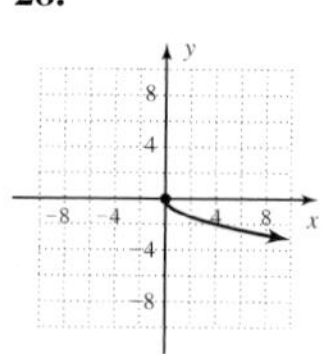

29.

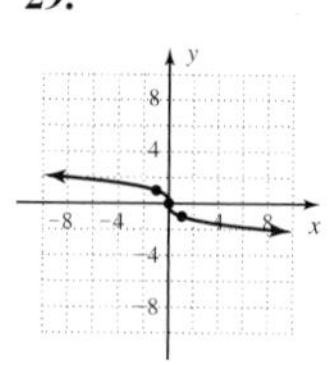

30.

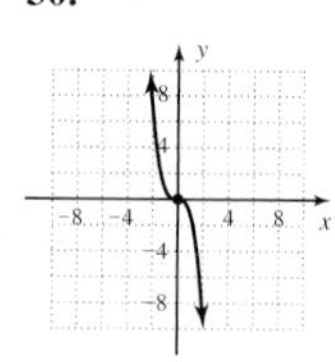

31.

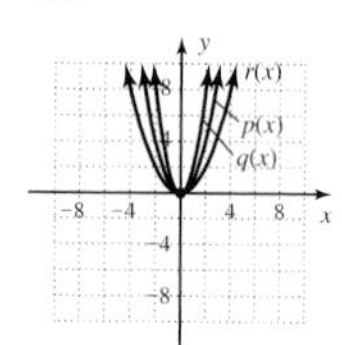

32.

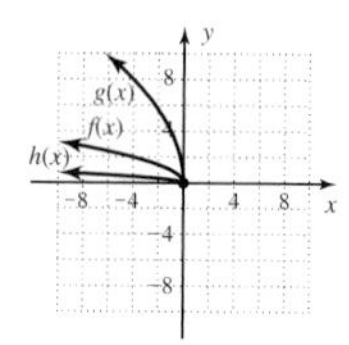

33.

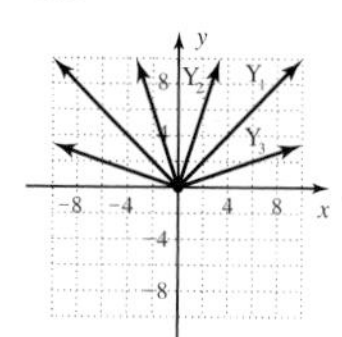

34.

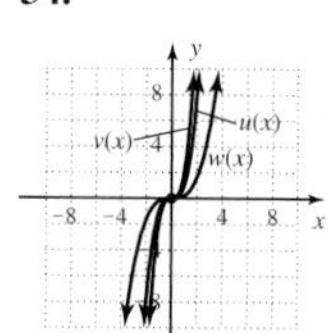

35.

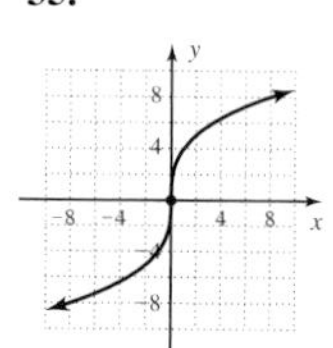

36.

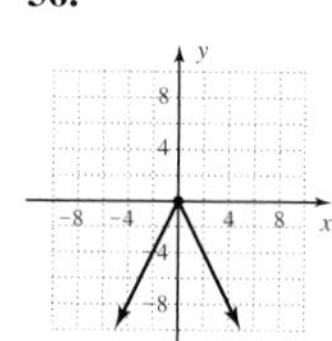

37.

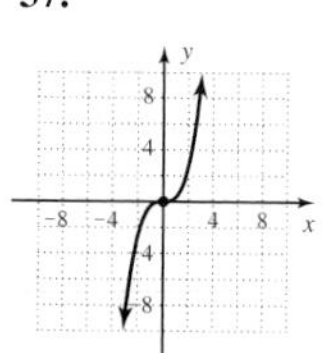

38.

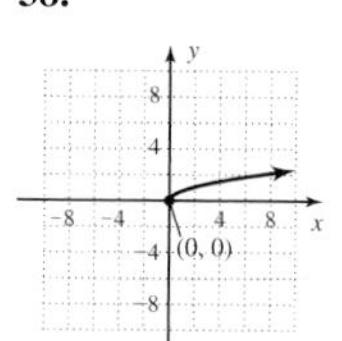

63. reflected across x-axis, left 2, down 1, compressed vertically

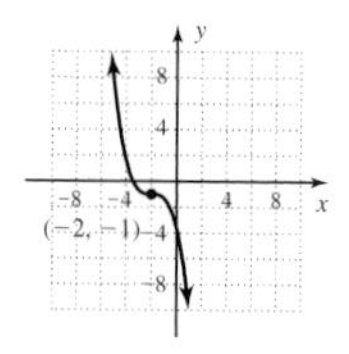

64. left 1, up 2, stretched vertically

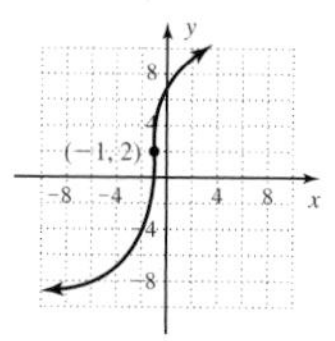

65. reflected across x-axis, right 4, down 3, stretched vertically

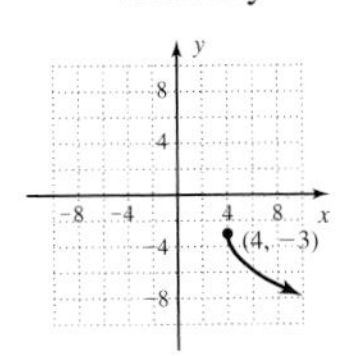

66. reflected across y-axis, right 2, down 1, stretched vertically

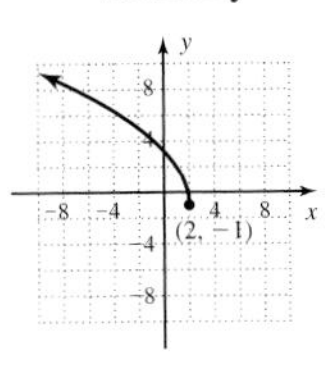

67. right 3, up 1, compressed vertically

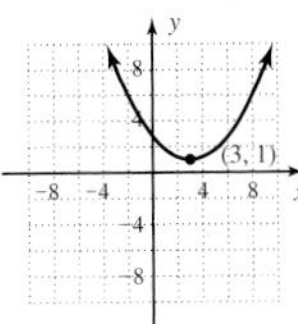

68. reflected across x-axis, right 3, up 4, stretched vertically

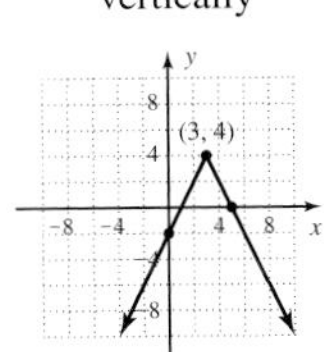

69. a.

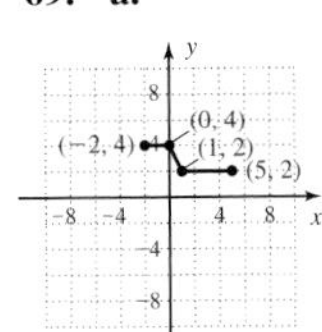

b.

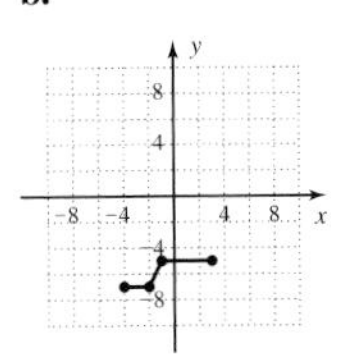

c.

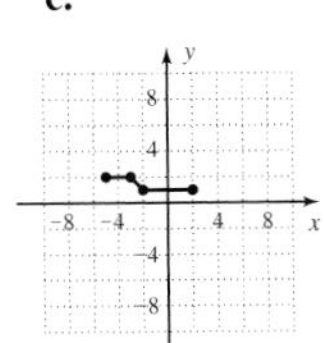

d.

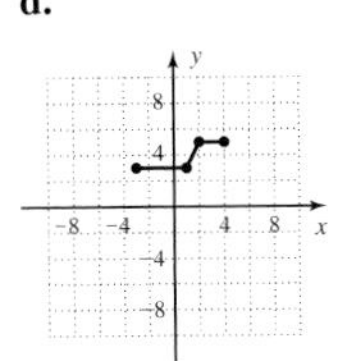

70. a.

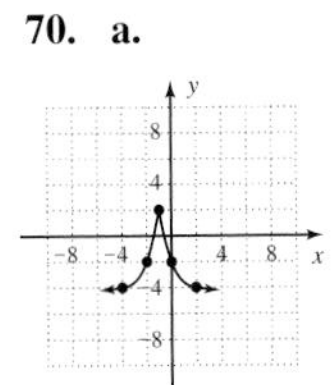

b.

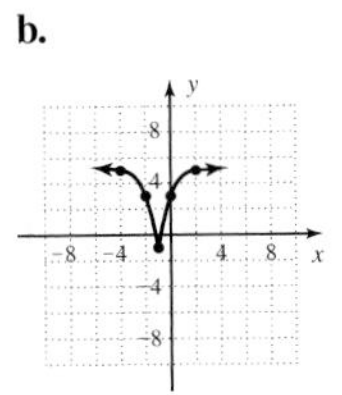

c.

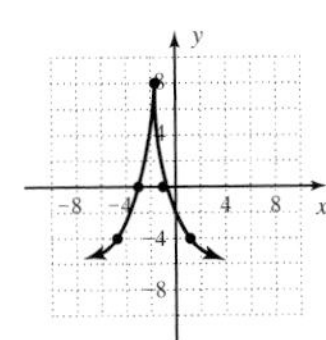

d.

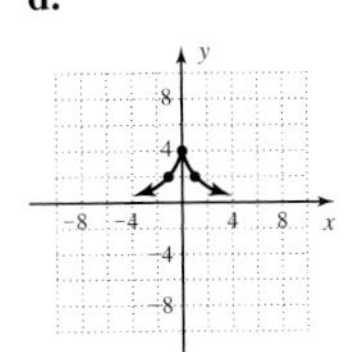

79.

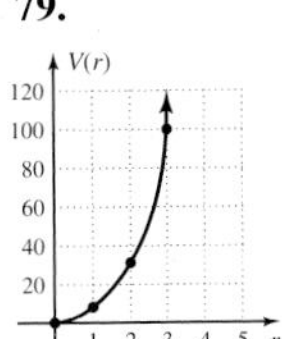

80.

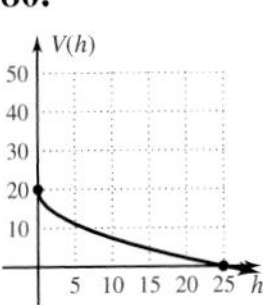

81.

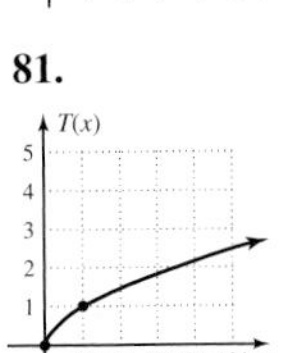

82.

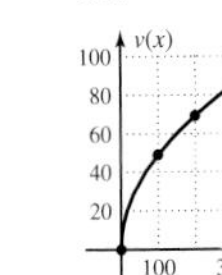

Exercises 3.4

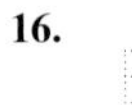

16. reflected across x-axis; right 2; up 5; stretched vertically

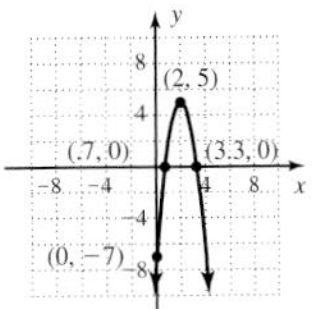

17. reflected across x-axis; right 5, up $\frac{11}{2}$; compressed vertically

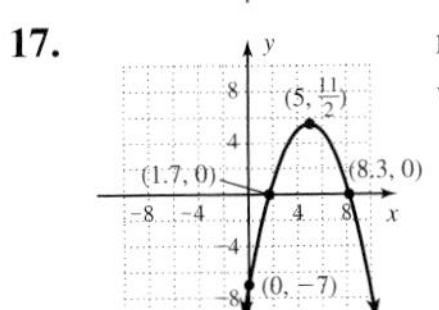

18. reflected across x-axis; right 3, up 8; compressed vertically

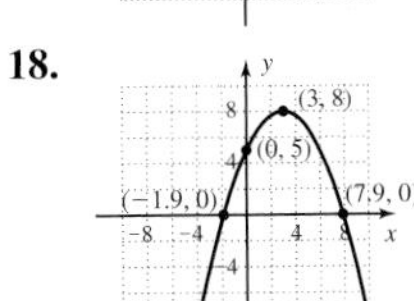
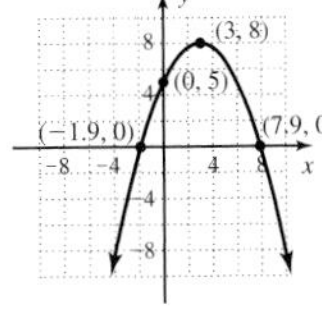

19. right $\frac{7}{4}$, down $\frac{-25}{8}$, stretched vertically

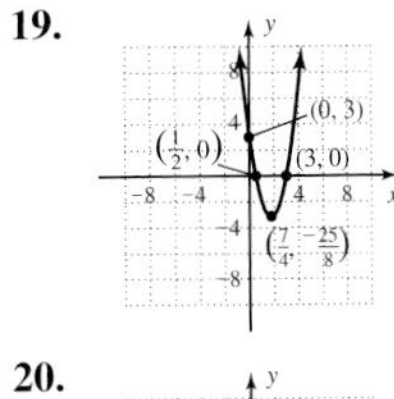

20. right $\frac{9}{8}$, down $\frac{49}{16}$; stretched vertically

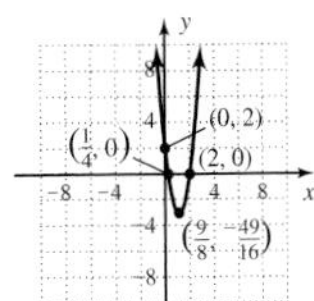

21. reflected across x-axis; left $\frac{7}{6}$, up $\frac{121}{12}$; stretched vertically

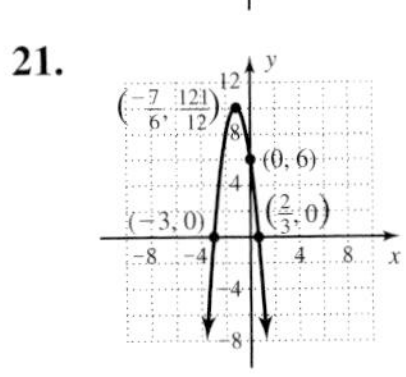

22. 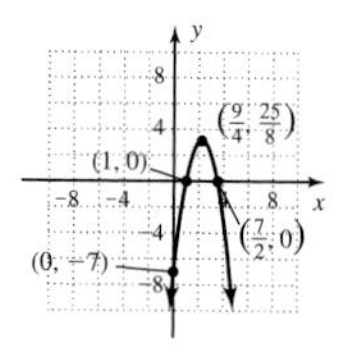

reflected across x-axis; right $\frac{9}{4}$, up $\frac{25}{8}$; stretched vertically

23. 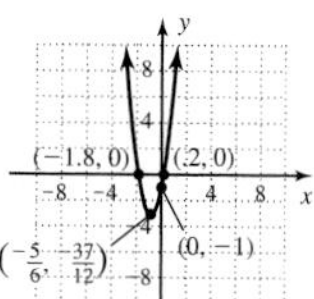

left $\frac{5}{6}$, down $\frac{37}{12}$; stretched vertically

24. 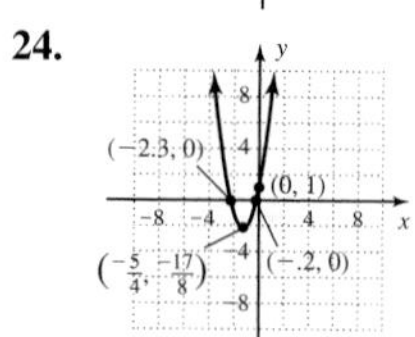

left $\frac{5}{4}$, down $\frac{17}{8}$; stretched vertically

31. 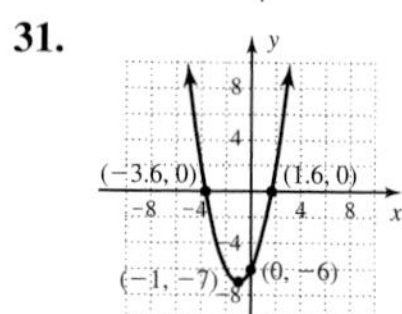

left 1, down 7

32. 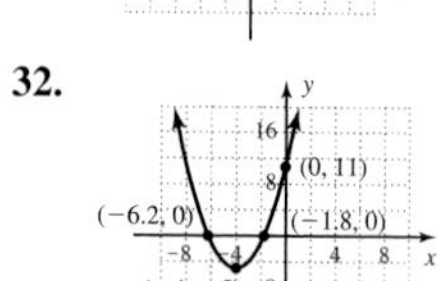 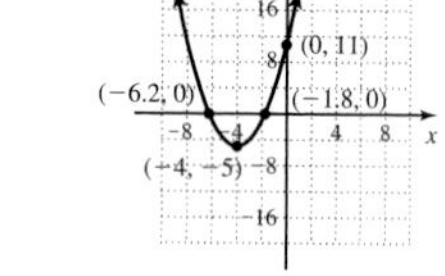

left 4, down 5

33. 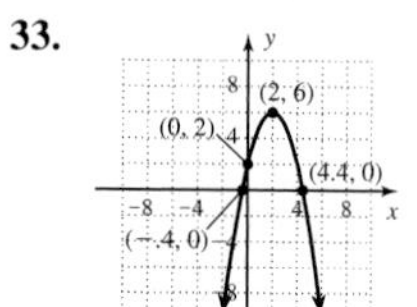

reflected across x-axis; right 2, up 6

34. 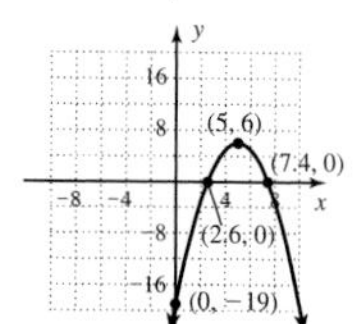

reflected across x-axis; right 5, up 6;

35. 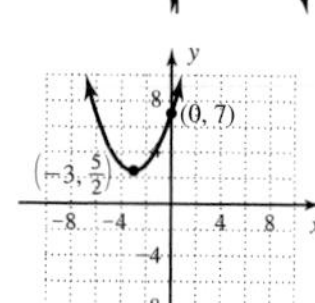

left 3, up $\frac{5}{2}$; compressed vertically

36. 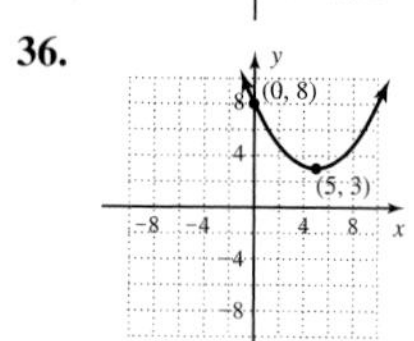

right 5, up 3; compressed vertically

37. 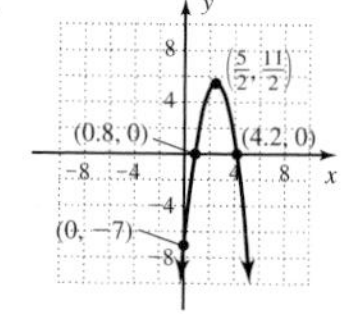

reflected across x-axis; right $\frac{5}{2}$, up $\frac{11}{2}$; stretched vertically

38. 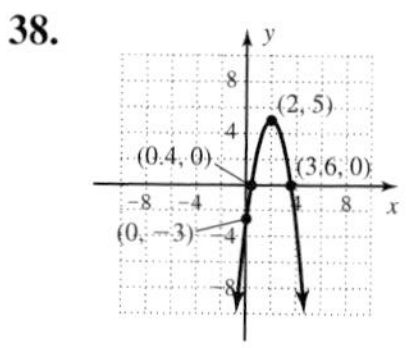

reflected across x-axis; right 2, up 5; stretched vertically

39. 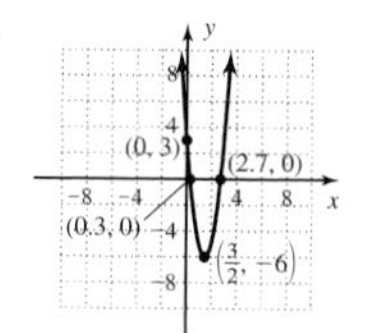

right $\frac{3}{2}$, down 6; stretched vertically

40. 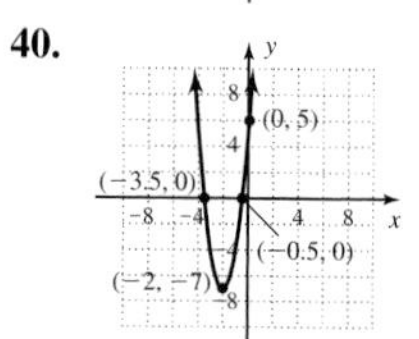

left 2, down 7; stretched vertically

41. 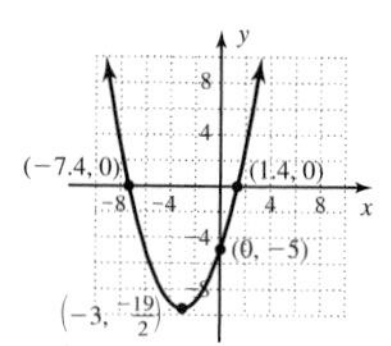

left 3, down $\frac{19}{2}$; compressed vertically

42. 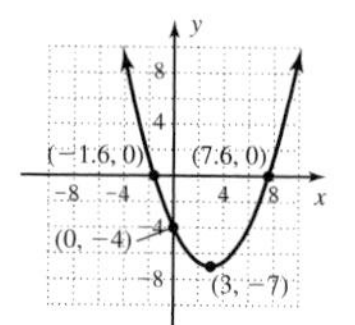

right 3, down 7; compressed vertically

Exercises 3.5

30. right 3, up 2,
$x \in (-\infty, 3) \cup (3, \infty)$,
$y \in (-\infty, 2) \cup (2, \infty)$

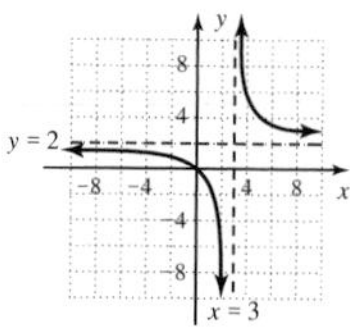

31. right 1, $x \in (-\infty, 1) \cup (1, \infty)$,
$y \in (0, \infty)$

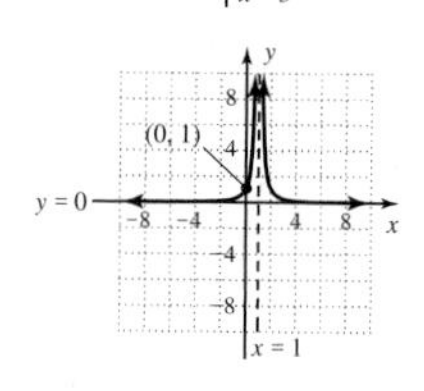

32. left 5, $x \in (-\infty, -5) \cup (-5, \infty)$, $y \in (0, \infty)$

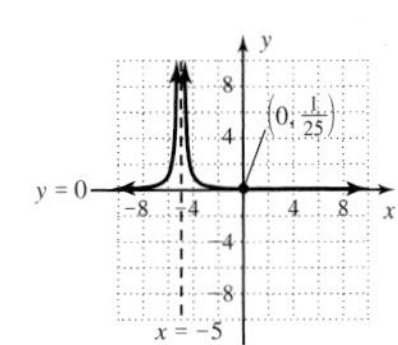

33. reflected across x-axis, left 2, $x \in (-\infty, -2) \cup (-2, \infty)$, $y \in (-\infty, 0)$

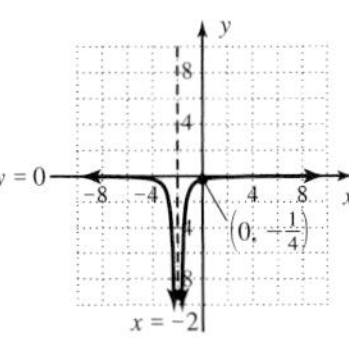

34. reflected across x-axis, down 2, $x \in (-\infty, 0) \cup (0, \infty)$, $y \in (-\infty, -2)$

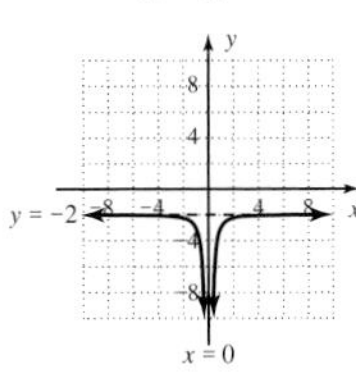

35. down 2, $x \in (-\infty, 0) \cup (0, \infty)$, $y \in (-2, \infty)$

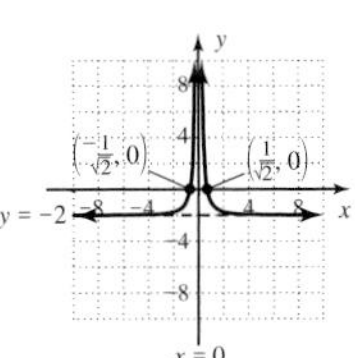

36. up 3, $x \in (-\infty, 0) \cup (0, \infty)$, $y \in (3, \infty)$

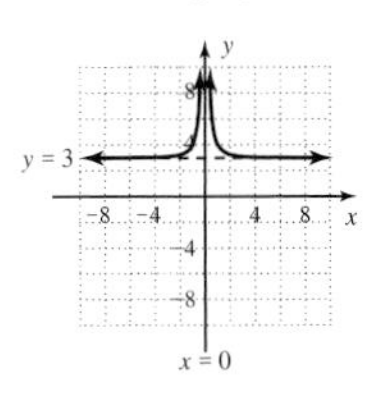

37. left 2, up 1, $x \in (-\infty, -2) \cup (-2, \infty)$, $y \in (1, \infty)$

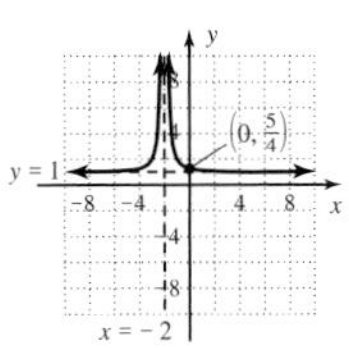

38. right 1, down 2, $x \in (-\infty, 1) \cup (1, \infty)$, $y \in (-2, \infty)$ $\left(\dfrac{2-\sqrt{2}}{2}, 0\right), \left(\dfrac{2+\sqrt{2}}{2}, 0\right)$

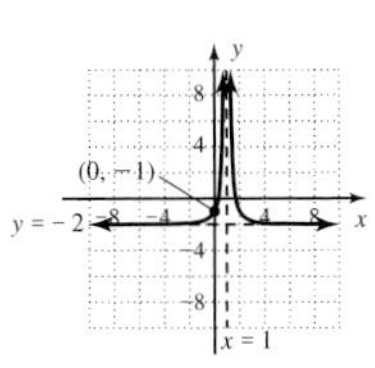

39. left 4, up 3, $x \in (-\infty, -4) \cup (-4, \infty)$, $y \in (-\infty, 3) \cup (3, \infty)$

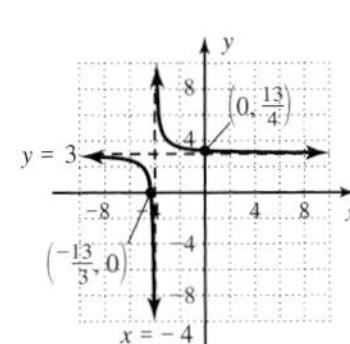

40. left 2, down 2, $x \in (-\infty, -2) \cup (-2, \infty)$, $y \in (-\infty, -2) \cup (-2, \infty)$

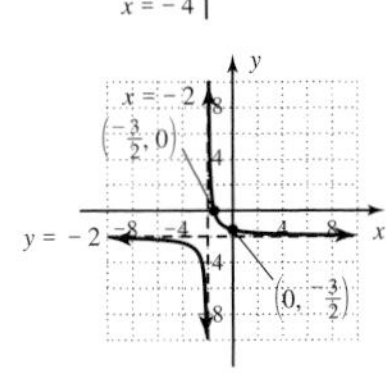

41. reflected across x-axis, right 1, down 3, $x \in (-\infty, 1) \cup (1, \infty)$, $y \in (-\infty, -3) \cup (-3, \infty)$

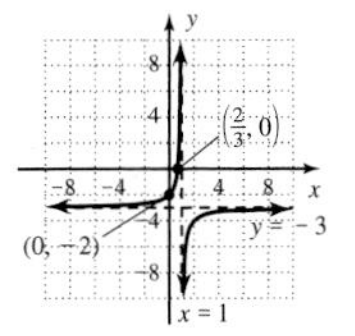

42. reflected across x-axis, left 3, down 1, $x \in (-\infty, -3) \cup (-3, \infty)$, $y \in (-\infty, -1) \cup (-1, \infty)$

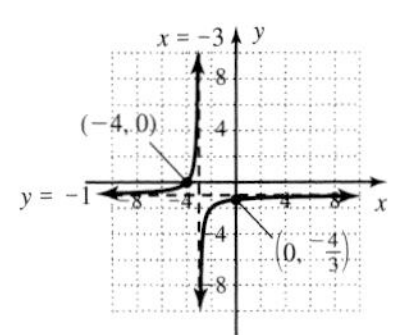

43. reflected across x-axis, right 2, up 3, $x \in (-\infty, 2) \cup (2, \infty)$, $y \in (-\infty, 3)$ $\left(\dfrac{6-\sqrt{3}}{3}, 0\right), \left(\dfrac{6+\sqrt{3}}{3}, 0\right)$

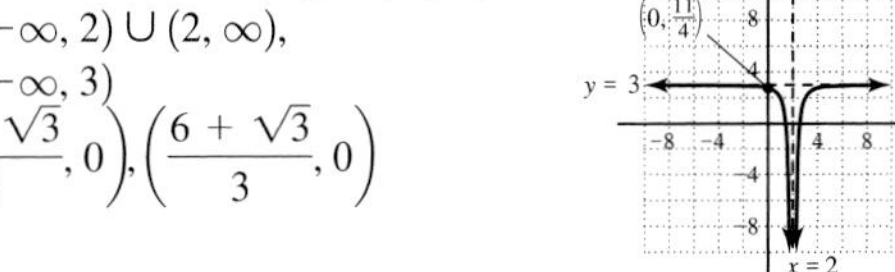

44. reflected across x-axis, left 1, down 2, $x \in (-\infty, -1) \cup (-1, \infty)$, $y \in (-\infty, -2)$

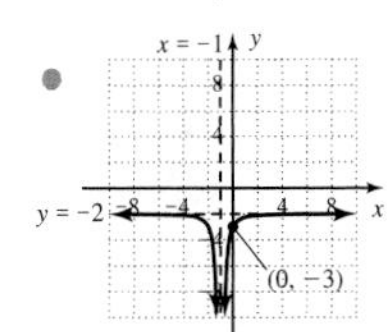

45. reflected across x-axis, left 2, up 3, $x \in (-\infty, -2) \cup (-2, \infty)$, $y \in (-\infty, 3)$ $\left(\dfrac{-6-\sqrt{3}}{3}, 0\right), \left(\dfrac{-6+\sqrt{3}}{3}, 0\right)$

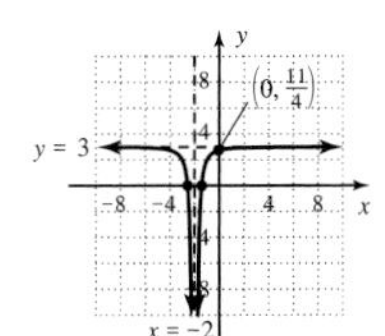

46. reflected across x-axis, right 5, down 2, $x \in (-\infty, 5) \cup (5, \infty)$, $y \in (-\infty, -2)$

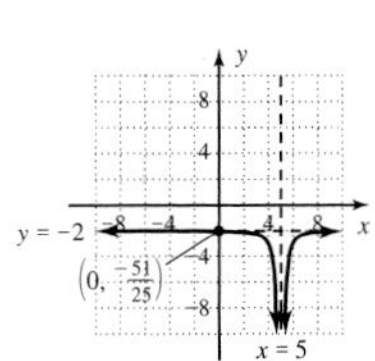

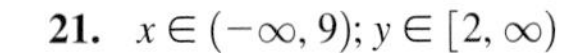

Exercises 3.7

21. $x \in (-\infty, 9)$; $y \in [2, \infty)$

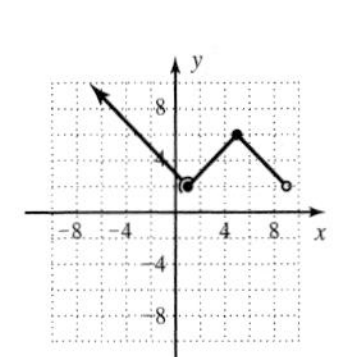

22. $x \in (-\infty, 6]$; $y \in (-\infty, 7]$

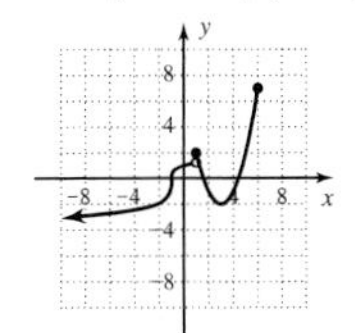

23. $x \in (-\infty, \infty)$; $y \in [0, \infty)$

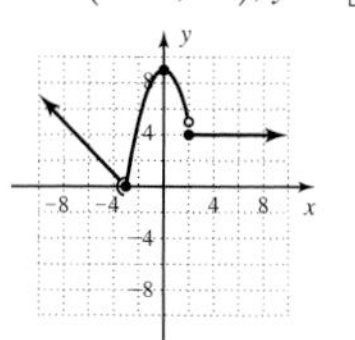

24. $x \in (-\infty, \infty)$; $y \in [0, \infty)$

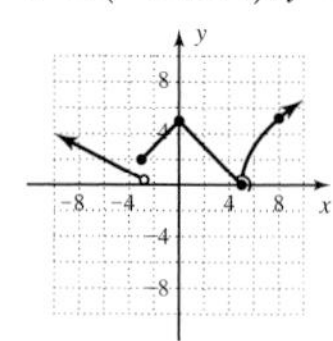

25. $x \in (-\infty, \infty)$; $y \in (-\infty, -6) \cup (-6, \infty)$; discontinuity at $x = -3$, redefine $f(x) = -6$ at $x = -3$

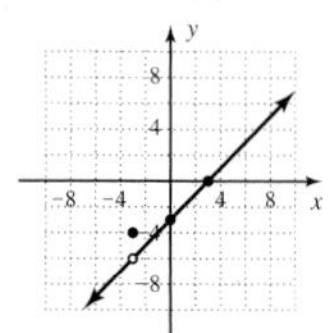

26. $x \in (-\infty, \infty)$; $y \in (-\infty, 7) \cup (7, \infty)$; discontinuity at $x = 5$, redefine $f(x) = 7$ at $x = 5$

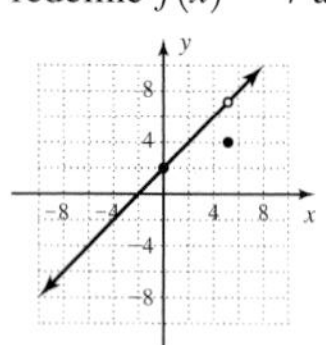

27. $x \in (-\infty, \infty)$; $y \in [0.75, \infty)$; discontinuity at $x = 1$, redefine $f(x) = 3$ at $x = 1$

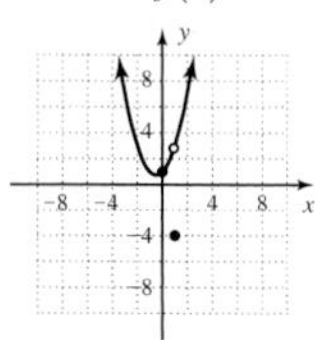

28. $x \in (-\infty, \infty)$; $y \in (-\infty, 1]$; discontinuity at $x = -2$, redefine $f(x) = -8$ at $x = -2$

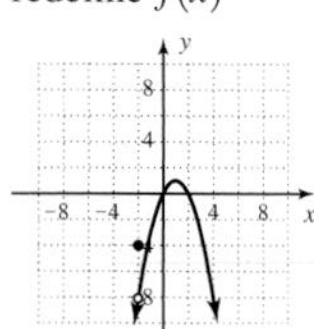

Exercises 3.8

40. **a.** $x \in [-4, \infty)$, $y \in (-\infty, 3]$
b. $(-4, 0)$, $(2, 0)$
c. $Y_1 \geq 0$: $x \in [-4, 2]$
$Y_1 \leq 0$: $x \in [-2, \infty)$
d. $Y_1 \uparrow$: $x \in (-4, -2)$
$Y_1 \downarrow$: $x \in (-2, \infty)$
e. min: $(-4, 0)$, max: $(-2, 3)$
f. none

41. **a.** $x \in (-\infty, \infty)$, $y \in (-\infty, 3]$
b. $(0, 0)$, $(2, 0)$
c. $Y_2 \geq 0$: $x \in [0, 2]$
$Y_2 \leq 0$: $x \in (-\infty, 0] \cup [2, \infty)$
d. $Y_2 \uparrow$: $x \in (-\infty, 1)$
$Y_2 \downarrow$: $x \in (1, \infty)$
e. max: $(1, 3)$
f. none

42. **a.** $x \in (-\infty, 2) \cup (2, \infty)$, $y \in (-\infty, -1) \cup (-1, \infty)$
b. $(3, 0)$
c. $Y_1 \geq 0$: $x \in (2, 3]$
$Y_1 \leq 0$: $x \in (-\infty, 2) \cup [3, \infty)$
d. $Y_1 \uparrow$: none
$Y_1 \downarrow$: $x \in (-\infty, 2) \cup (2, \infty)$
e. none
f. $x = 2$, $y = -1$

43. **a.** $x \in (-\infty, 2) \cup (2, \infty)$, $y \in (-\infty, 4)$
b. $(1, 0)$, $(3, 0)$
c. $Y_2 \geq 0$: $x \in (-\infty, 1] \cup [3, \infty)$
$Y_2 \leq 0$: $x \in [1, 2) \cup (2, 3]$
d. $Y_2 \uparrow$: $x \in (2, \infty)$
$Y_2 \downarrow$: $x \in (-\infty, 2)$
e. none
f. $x = 2$, $y = 4$

51. $y = \sin x$
a. $y \in [-1, 1]$
b. $(-180, 0)$, $(0, 0)$, $(180, 0)$, $(360, 0)$
c. $f(x)\uparrow$: $x \in (-90, 90) \cup (270, 360)$,
$f(x)\downarrow$: $x \in (-180, -90) \cup (90, 270)$
d. min: $(-90, -1)$ and $(270, -1)$, max: $(90, 1)$
e. odd

$y = \cos x$
a. $y \in [-1, 1]$
b. $(-90, 0)$, $(90, 0)$, $(270, 0)$
c. $f(x)\uparrow$: $x \in (-180, 0) \cup (180, 360)$, $f(x)\downarrow$: $x \in (0, 180)$
d. min: $(-180, -1)$ and $(180, -1)$, max: $(0, 1)$ and $(360, 1)$
e. even

Strengthening Core Skills

4.

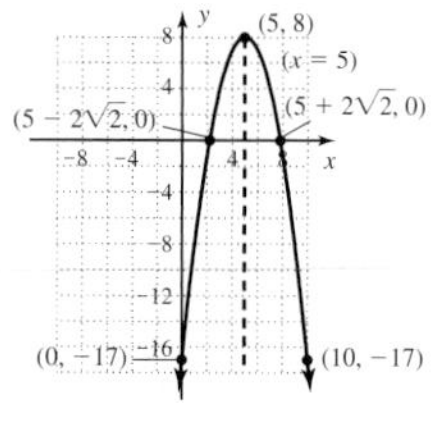

5.

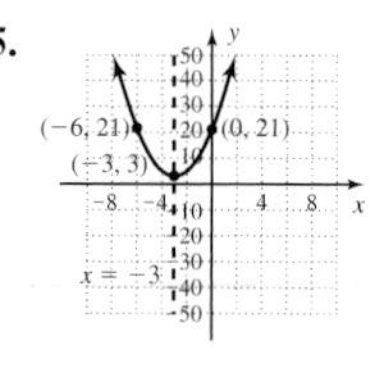

6.

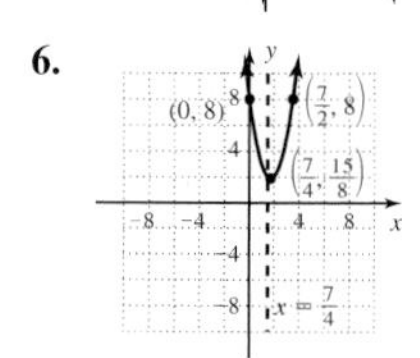

Cumulative Review

9. 

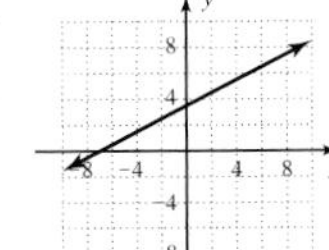

$y = \frac{1}{2}x + \frac{7}{2}$ **10.** no

11.

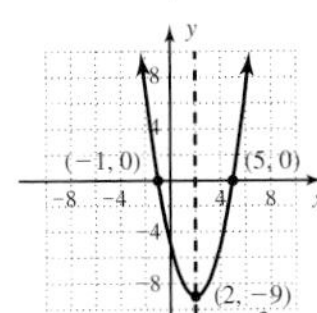

12. $x \in [1, 6]$

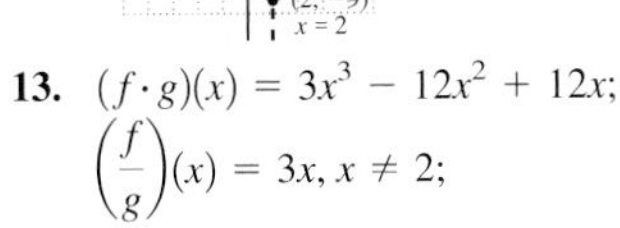

13. $(f \cdot g)(x) = 3x^3 - 12x^2 + 12x;$

$\left(\frac{f}{g}\right)(x) = 3x, x \neq 2;$

$(g \circ f)(-2) = 22$

14. $f^{-1}(x) = \dfrac{5x + 20}{3}$ **15. a.** $(0, \frac{1}{2}), (1, 0)$ **b.**

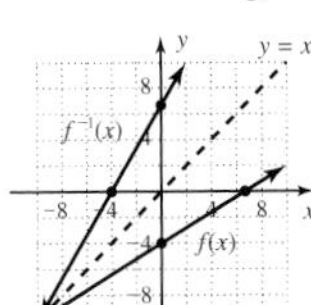

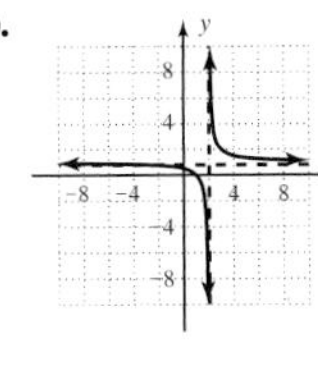

CHAPTER 4

Exercises 4.3

35. $(x + 4)(x + 1)(x - 2)(x - 3), x = -4, -1, 2, 3$
36. $(x + 4)(x + 2)(x - 1)(x - 5), x = -4, -2, 1, 5$
37. $(x + 7)(x + 2)(x + 1)(x - 3), x = -7, -2, -1, 3$
38. $(x + 6)(x + 2)(x - 1)(x - 3),$
$x = -6, -2, 1, 3$
39. $(2x + 3)(2x - 1)(x - 1); x = -\frac{3}{2}, \frac{1}{2}, 1$
40. $(x - 1)(3x + 1)(3x + 2); x = 1, -\frac{1}{3}, -\frac{2}{3}$
41. $(2x + 3)^2(x - 1); x = -\frac{3}{2}, 1$
42. $(x - 1)(3x + 2)^2; x = 1, -\frac{2}{3}$
43. $(x + 2)(x - 1)(2x - 5); x = -2, 1, \frac{5}{2}$
44. $(3x - 2)(x - 1)(x - 3); x = \frac{2}{3}, 1, 3$
45. $(x + 1)(2x + 1)(x - \sqrt{5})(x + \sqrt{5});$
$x = -1, -\frac{1}{2}, \sqrt{5}, -\sqrt{5}$
46. $(x + 1)(3x - 2)(x - \sqrt{3})(x + \sqrt{3});$
$x = -1, \frac{2}{3}, -\sqrt{3}, \sqrt{3}$
47. $(x - 1)(3x - 2)(x - 2i)(x + 2i); x = 1, \frac{2}{3}, 2i, -2i$
48. $(x - 1)(2x + 1)(x - 3i)(x + 3i); x = 1, -\frac{1}{2}, 3i, -3i$
77. $(x - 4)(2x - 3)(2x + 3); x = 4, \frac{3}{2}, -\frac{3}{2}$
78. $(2x + 1)(3x - 4)(x - 6); x = -\frac{1}{2}, \frac{4}{3}, 6$
79. $(2x + 1)(3x - 2)(x - 12); x = -\frac{1}{2}, \frac{2}{3}, 12$
80. $(x + 14)(2x + 3)(2x - 1); x = -14, -\frac{3}{2}, \frac{1}{2}$
81. $(x - 2)(2x - 1)(2x + 1)(x + 12);$
$x = 2, \frac{1}{2}, -\frac{1}{2}, -12$
82. $(2x + 3)(x - 12)(2x^2 + 1);$
$x = -\frac{3}{2}, 12, \frac{\sqrt{2}}{2}i, -\frac{\sqrt{2}}{2}i$
83. $x = -1, \frac{-1}{3}, \frac{2}{3}$
84. $x = \frac{-1}{4}, 1, \frac{3}{2}$
85. $x = -2, -1, 1, 5$
86. $x = 3, 1, -1, \frac{-5}{2}$
87. $x = -5, -2, \frac{3}{2}$
88. $x = -4, 5, \frac{1}{3}$
89. $x = 3, 7, \frac{1}{4}$
90. $x = 3, 5, \frac{1}{6}$
91. $x = -2$, 1 multiplicity 2, $\frac{-2}{3}$
92. $x = \frac{-2}{5}$, 1 multiplicity 2, 2
93. $t = 1, \frac{3}{4}, \frac{-1}{8}$
94. $n = -1, \frac{-3}{8}, \frac{2}{3}$
95. a. 5 **b.** 13 **c.** 2
96. $2 + i, 2 + 3i, \frac{\sqrt{6}}{2} + \frac{\sqrt{2}}{2}i$

Exercises 4.4

59.

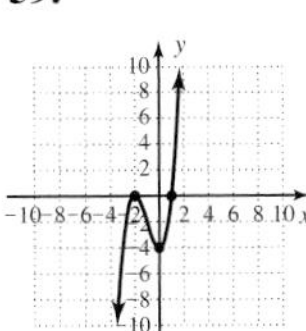

60.

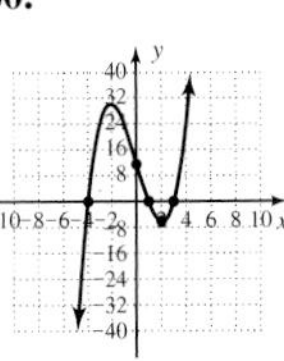

61.

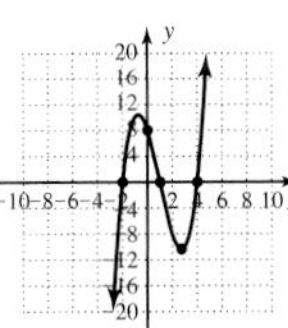

62.

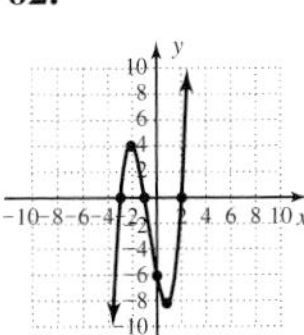

63.

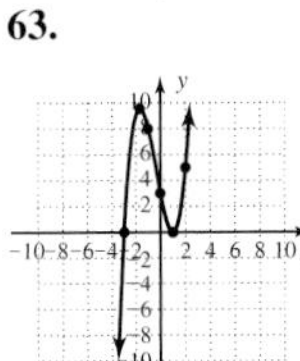

64.

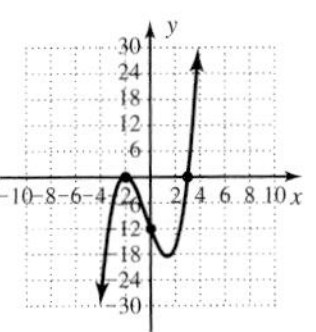

81. $f(x) = (x + \frac{5}{2})(x - \sqrt{2})$
$(x + \sqrt{2})(x - \sqrt{3})(x + \sqrt{3})$
82. $g(x) = (x + \frac{2}{3})(x - \sqrt{2})$
$(x + \sqrt{2})(x - \sqrt{6})(x + \sqrt{6})$
83. $P(x) = \frac{1}{6}(x + 4)(x - 1)(x - 3)$
84. $P(x) = \frac{1}{12}(x + 3)(x - 1)^2(x - 4)$
85. a. 3
b. 5
c. $B(x) = \frac{1}{4}x(x - 4)(x - 9)$, −$80,000
86. a. 4
b. 3
c. $P(x) = \frac{3}{4}(x - 1)(x - 2)(x - 4)(x - 6)$; 4.5 in, −9 in

Exercises 4.5

55.

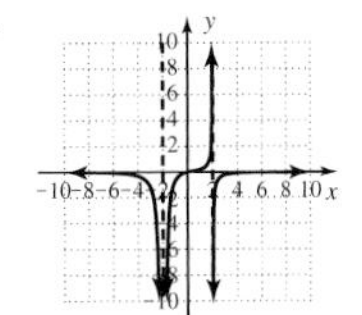

56.

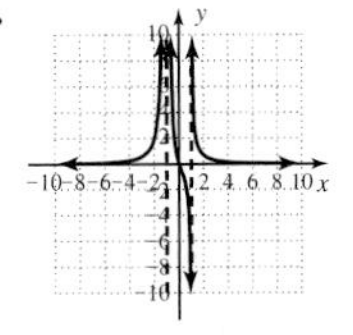

57. $f(x) = \dfrac{(x - 4)(x + 1)}{(x + 2)(x - 3)}$ **58.** $f(x) = \dfrac{5x}{(x + 3)^2(x - 3)}$

59. $f(x) = \dfrac{x^2 - 4}{9 - x^2}$ **60.** $f(x) = \dfrac{(x - 1)^2}{(x - 2)^2}$

61.

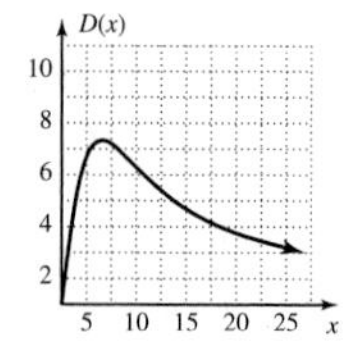

Exercises 4.6

13. $P(x) = \begin{cases} \dfrac{x^3 - 8}{x - 2} & x \neq 2 \\ 12 & x = 2 \end{cases}$

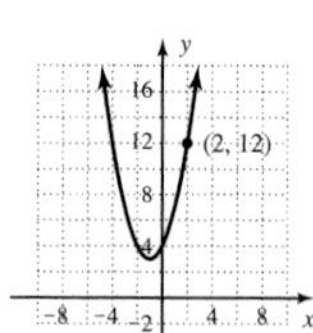

14. $P(x) = \begin{cases} \dfrac{8x^3 - 1}{2x - 1} & x \neq \dfrac{1}{2} \\ 3 & x = \dfrac{1}{2} \end{cases}$

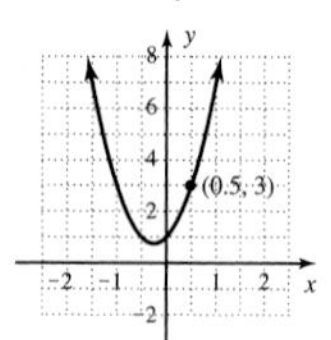

15. $q(x) = \begin{cases} \dfrac{x^3 - 7x - 6}{x + 1} & x \neq -1 \\ -4 & x = -1 \end{cases}$

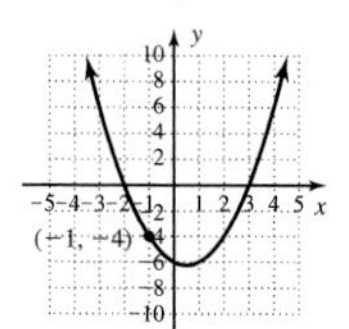

16. $q(x) = \begin{cases} \dfrac{x^3 - 3x + 2}{x + 2} & x \neq -2 \\ 9 & x = -2 \end{cases}$

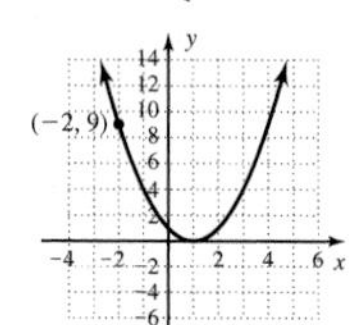

17. $R(x) = \begin{cases} \dfrac{x^3 + 3x^2 - x - 3}{x^2 + 2x - 3} & x \neq -3, x \neq 1 \\ -2 & x = -3 \\ 2 & x = 1 \end{cases}$

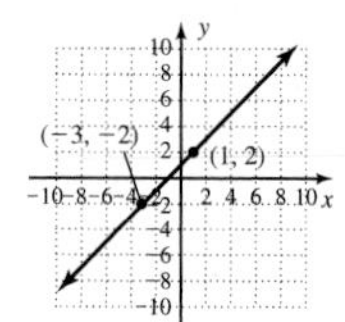

18. $R(x) = \begin{cases} \dfrac{x^3 - 2x^2 - 4x + 8}{x^2 - 4} & x \neq -2, x \neq 2 \\ -4 & x = -2 \\ 0 & x = 2 \end{cases}$

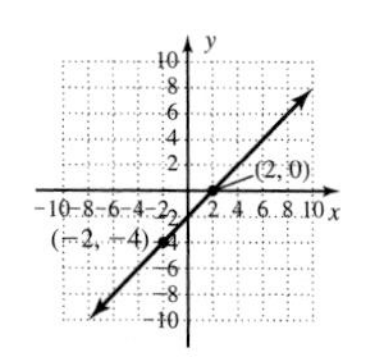

19.

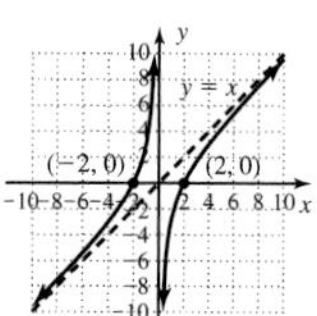

20.

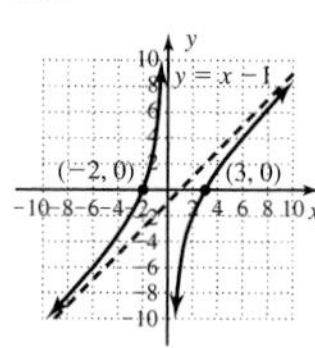

21.

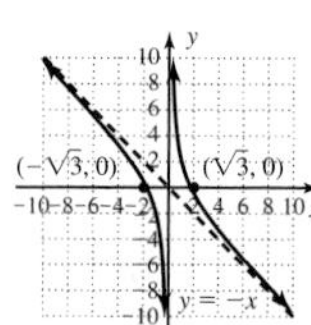

22.

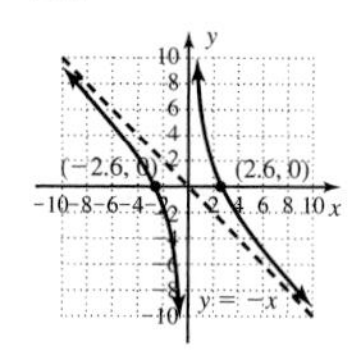

23.

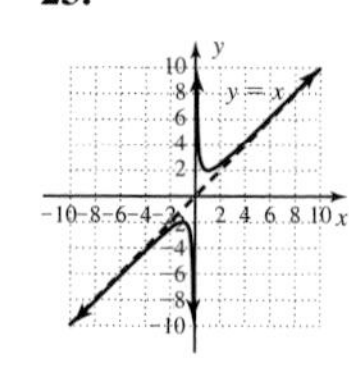

24.

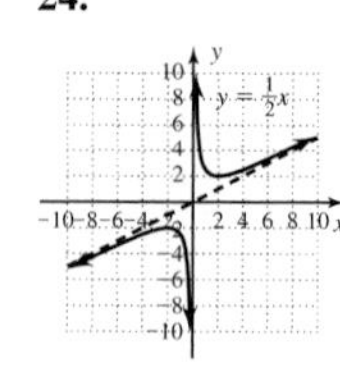

25.

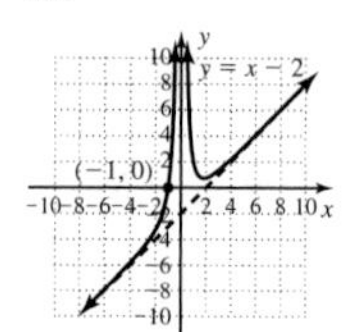

26.

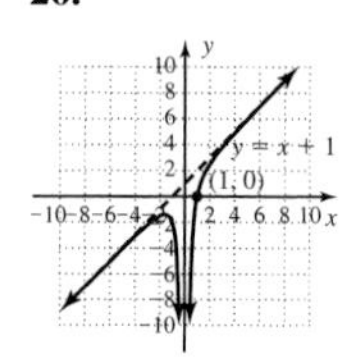

27.

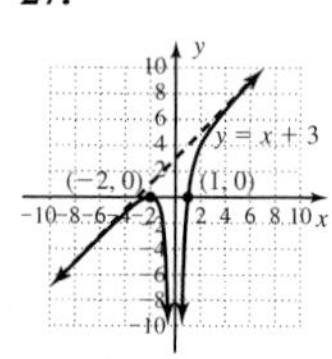

28.

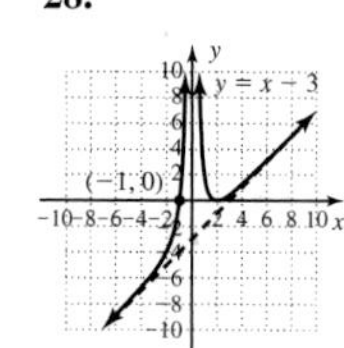

29.

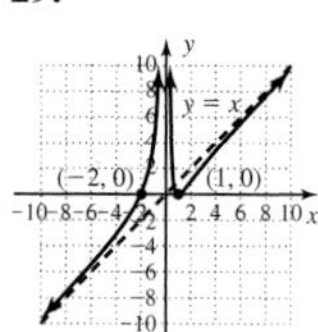

30.

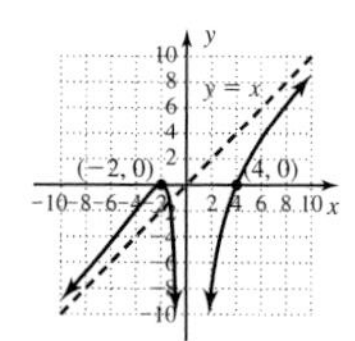

31.

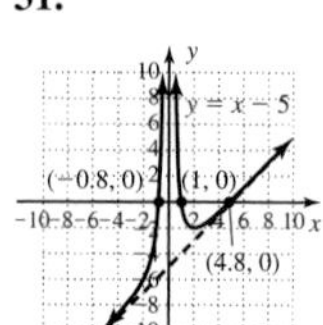

32.

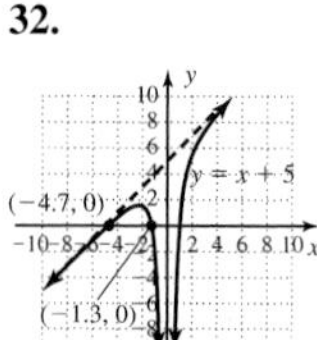

33.

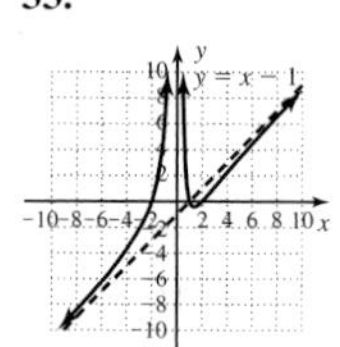

34.

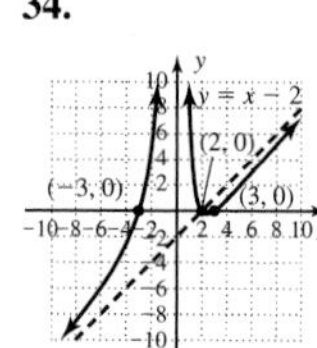

35.

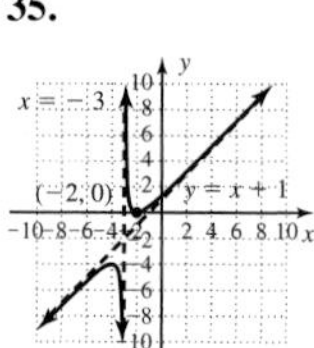

36.

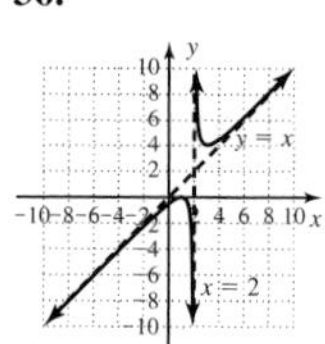

37.

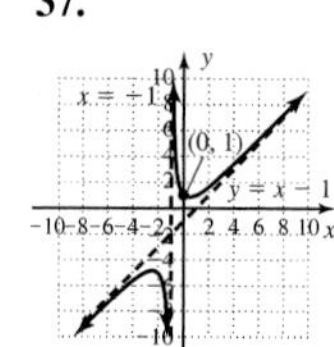

38.

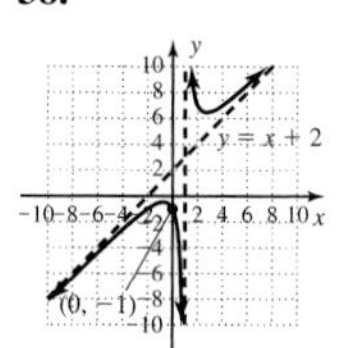

39.

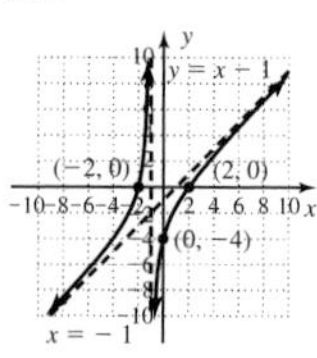

40.

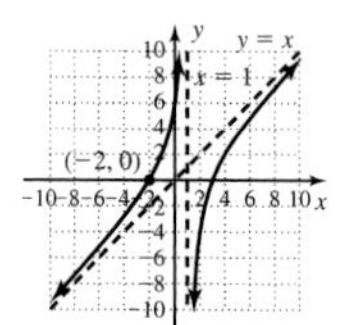

41.

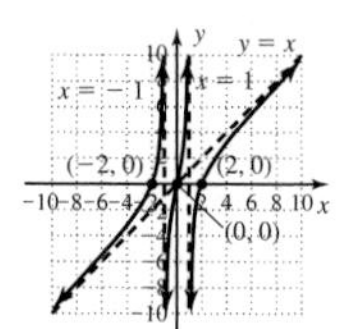

42.

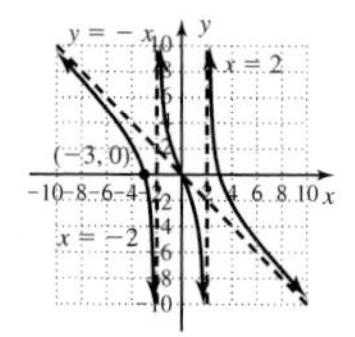

43.

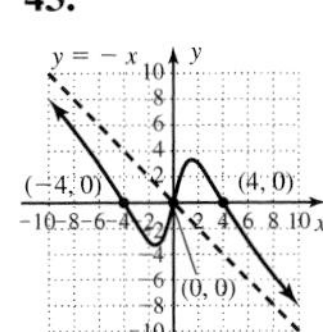

44.

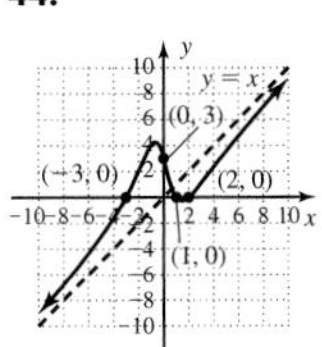

45.

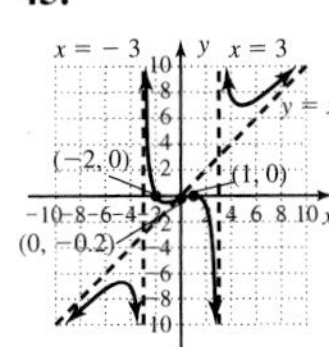

46.

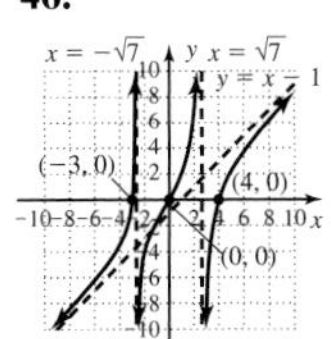

47.

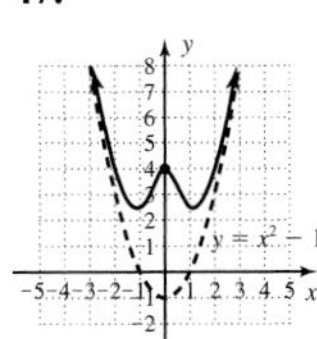

48.

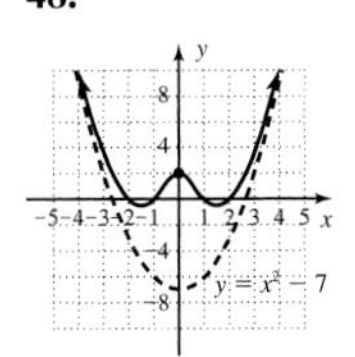

49.

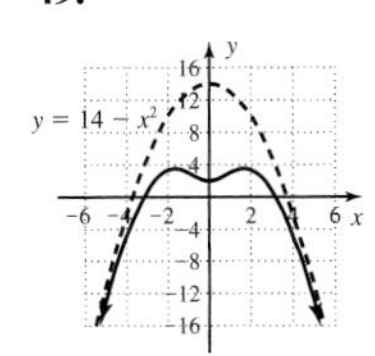

50.

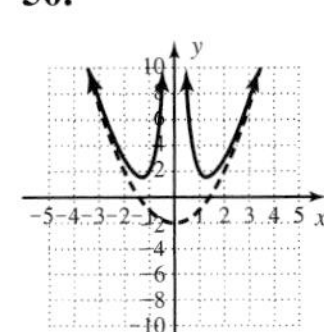

Cumulative Review

6.

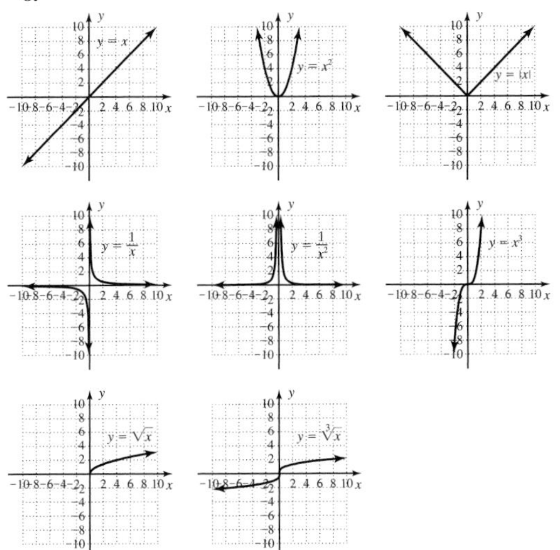

CHAPTER 5

Exercises 5.1

21. reflect across y-axis

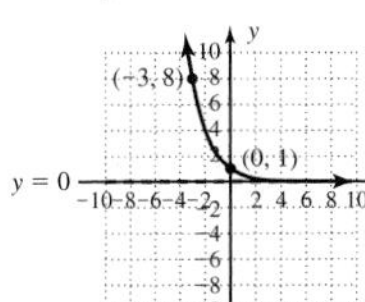

22. reflect across y-axis

23. reflect across y-axis, up 3

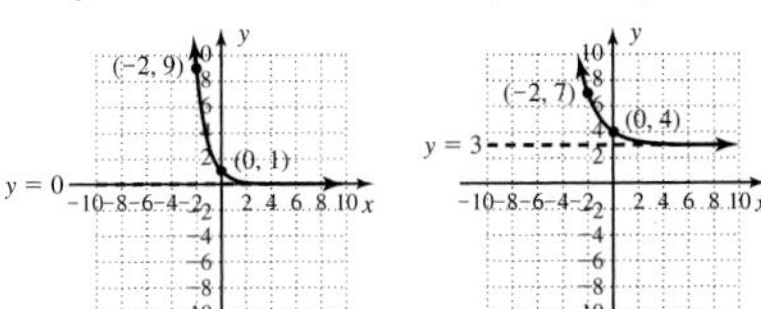

24. reflect across y-axis, down 2

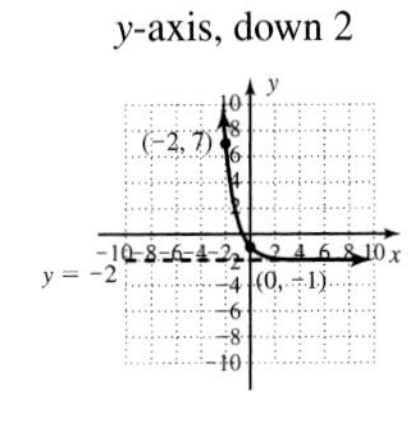

25. left 1, down 3

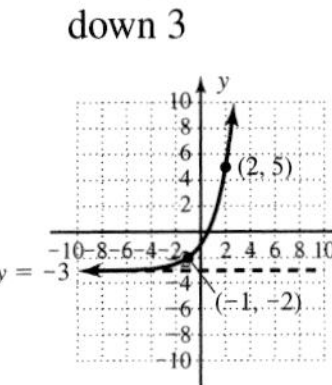

26. right 2, up 1

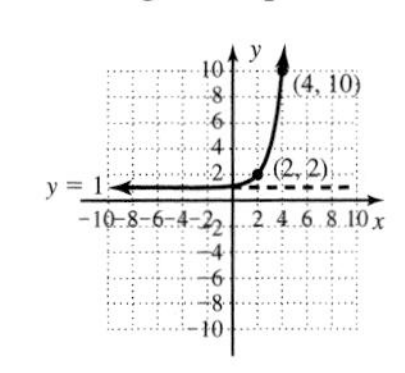

27. up 1

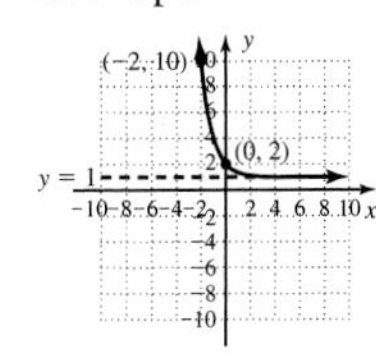

28. down 4

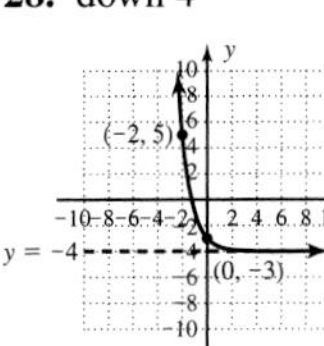

29. right 2

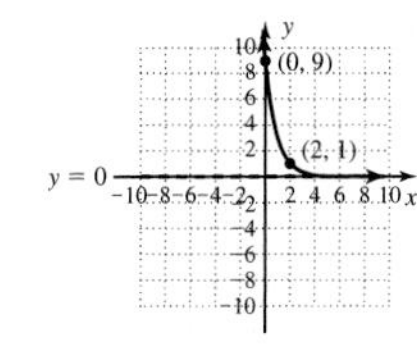

30. left 2

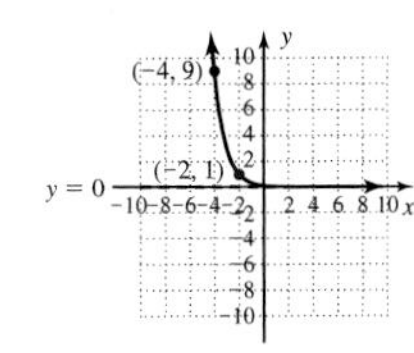

31. down 2

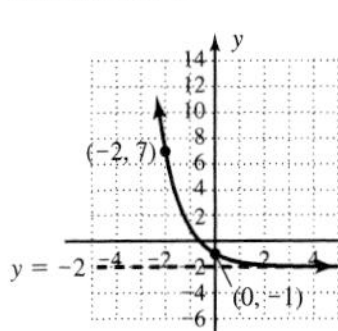

32. up 2

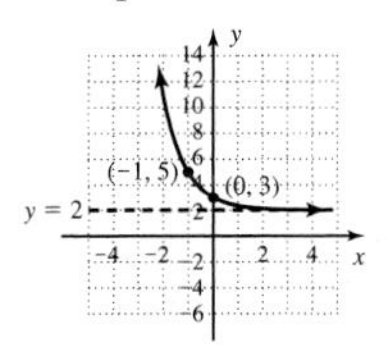

Exercises 5.3

50. $\log_6 3$ **51.** $\log\left(\frac{x}{x+1}\right)$ **52.** $\log\left(\frac{x-2}{x}\right)$ **53.** $\ln\left(\frac{x-5}{x}\right)$

54. $\ln\left(\frac{x+3}{x-1}\right)$ **55.** $\ln(x-2)$ **56.** $\ln(x-5)$ **57.** $\log_2 42$

58. $\log_9 30$ **59.** $\log_5(x-2)$ **60.** $\log_3(3x+5)$

115.

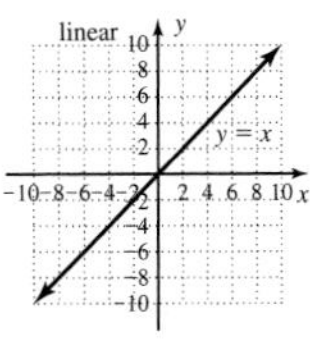

absolute value

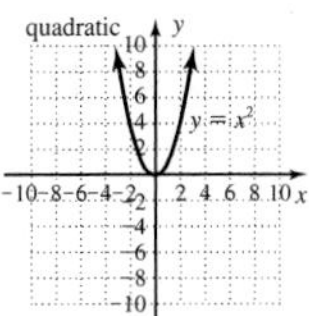

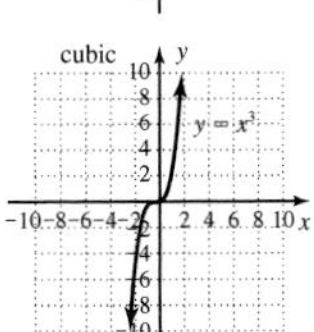

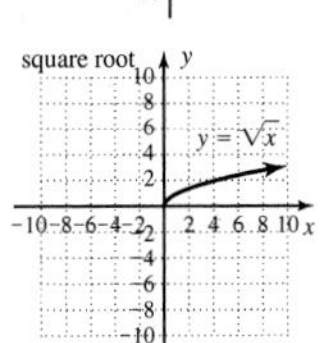

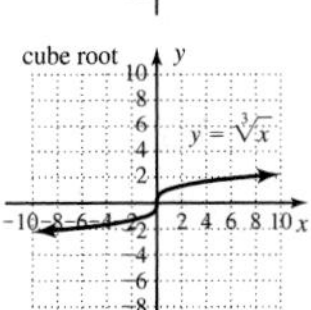

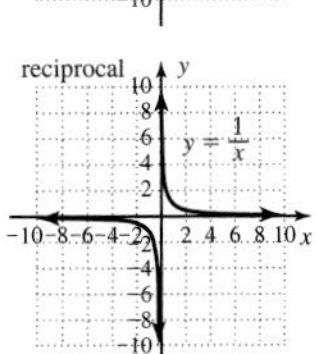

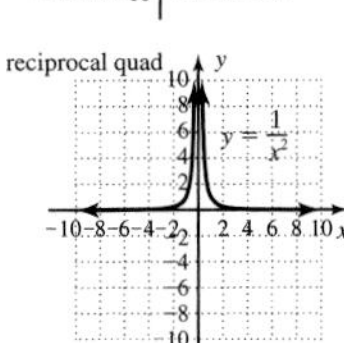

116.

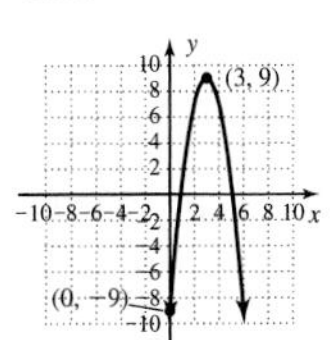

117.

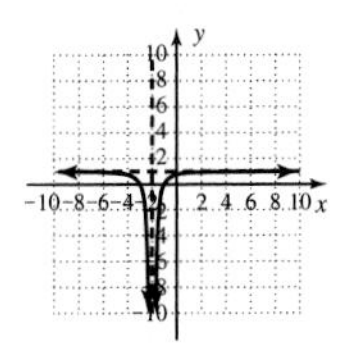

118. $D: x \in [-4, 10]$; $R: y \in [0, 8] \cup \{-2\}$

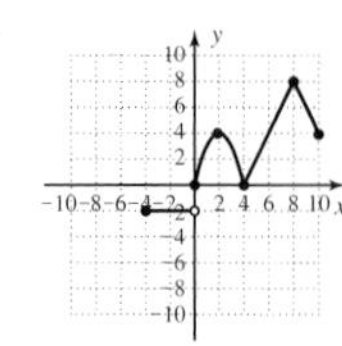

119. $A = 31.5$ units^2

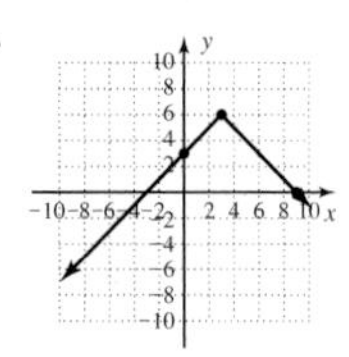

120. $y = 510.8x - 14.9$;

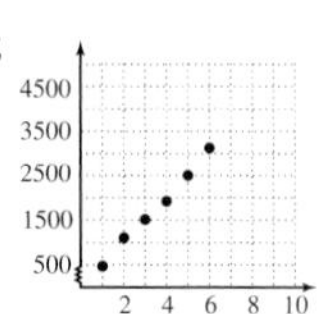

a. \$510.80;
b. \$6114.70

CHAPTER 6

Exercises 6.3

25.

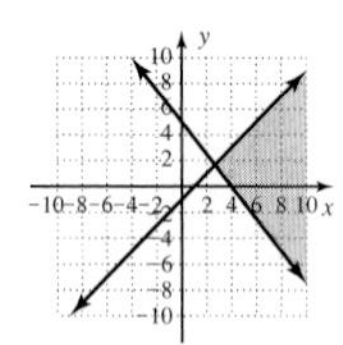

26.

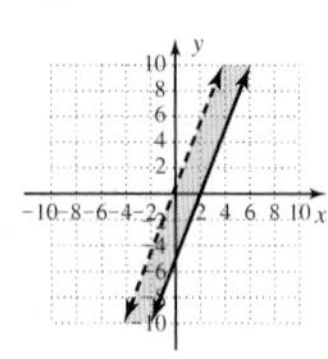

27.

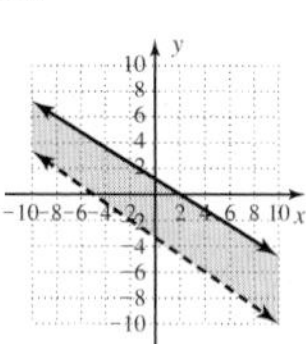

28.

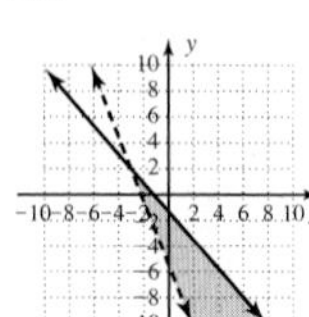

29.

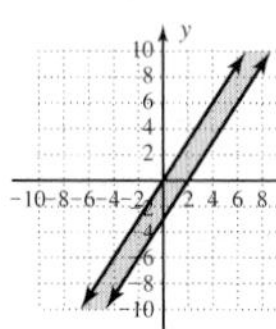

30.

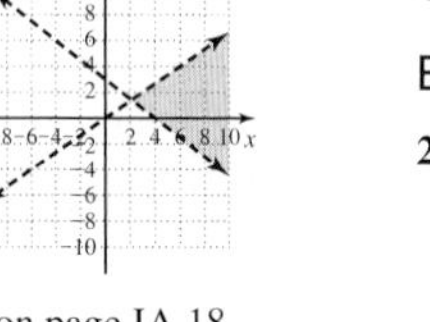

The graphs for 13, 15, 17, 19, 21, 23, 25 and 27 appear on page IA 18.

Exercises 6.6

41. $\begin{bmatrix} 1 & 0 \\ 0 & 1 \end{bmatrix}$ **42.** $\begin{bmatrix} 1 & 0 \\ 0 & 1 \end{bmatrix}$

43. $\begin{bmatrix} 1 & 0 & 0 \\ 0 & 1 & 0 \\ 0 & 0 & 1 \end{bmatrix}$ **44.** $\begin{bmatrix} 1 & 0 & 0 \\ 0 & 1 & 0 \\ 0 & 0 & 1 \end{bmatrix}$

45. $\begin{bmatrix} \frac{-3}{19} & \frac{4}{57} \\ \frac{1}{19} & \frac{5}{57} \end{bmatrix}$ **46.** $\begin{bmatrix} \frac{-3}{19} & \frac{4}{57} \\ \frac{1}{19} & \frac{5}{57} \end{bmatrix}$

47. $\begin{bmatrix} 0 & \frac{3}{4} & \frac{1}{4} \\ \frac{-1}{2} & \frac{3}{8} & \frac{1}{8} \\ \frac{-1}{4} & \frac{11}{16} & \frac{1}{16} \end{bmatrix}$ **48.** $\begin{bmatrix} 0 & \frac{3}{4} & \frac{1}{4} \\ \frac{-1}{2} & \frac{3}{8} & \frac{1}{8} \\ \frac{-1}{4} & \frac{11}{16} & \frac{1}{16} \end{bmatrix}$

49. $\begin{bmatrix} 1.75 & 2.5 \\ 7.5 & 13 \end{bmatrix}$ **50.** $\begin{bmatrix} 1 & 0 & -4 \\ 8 & 7 & -14 \\ -20 & -29 & 42 \end{bmatrix}$

51. $\begin{bmatrix} -0.26 & 0.32 & -0.07 \\ 0.63 & 0.30 & 0.10 \end{bmatrix}$ **52.** $\begin{bmatrix} 6.68 & -5.09 & -5.08 \\ 7.63 & -11.42 & 3.27 \end{bmatrix}$

Exercises 6.8

61. $y \approx 0.0257x^4 - 0.7150x^3 + 6.7852x^2 - 25.1022x + 35.1111$
May: \$5900
July: \$8300
November ($4.7 > 4.1$)

62. $y \approx -0.0742x^4 + 1.1408x^3 - 5.5158x^2 + 9.0492x$
196 units
397 units
598 units
Answers will vary.

Mixed Review

1. a. $\begin{cases} y = \frac{3}{5}x + 2 \\ y = \frac{3}{5}x + 2; \end{cases}$
consistent/dependent

b. $\begin{cases} y = \frac{4}{3}x - 3 \\ y = \frac{2}{5}x - 2; \end{cases}$
consistent/independent

c. $\begin{cases} y = \frac{1}{3}x - 3 \\ y = \frac{1}{3}x - \frac{5}{3}; \end{cases}$
inconsistent

2.

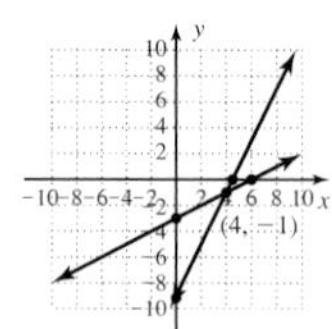

CHAPTER 7

Exercises 7.1

25.

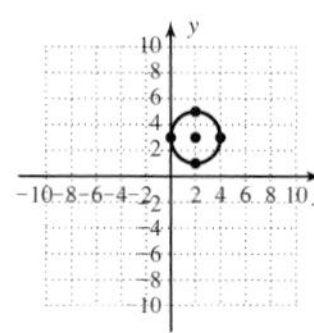

26.

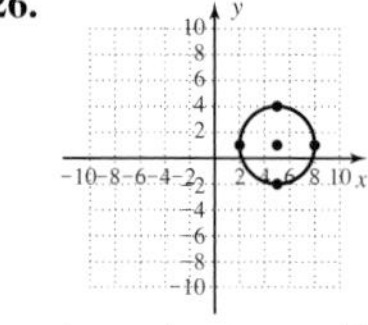

27. $(-1, 2)$, $r = 2\sqrt{3}$,
$x \in [-1 - 2\sqrt{3}, -1 + 2\sqrt{3}]$,
$y \in [2 - 2\sqrt{3}, 2 + 2\sqrt{3}]$

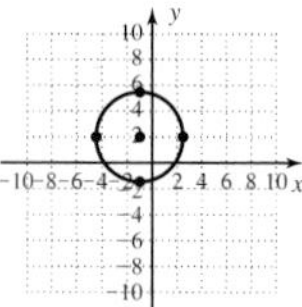

28. $(7, -4)$, $r = 2\sqrt{5}$,
$x \in [7 - 2\sqrt{5}, 7 + 2\sqrt{5}]$,
$y \in [-4 - 2\sqrt{5}, -4 + 2\sqrt{5}]$

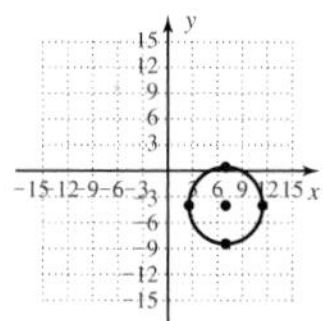

29.

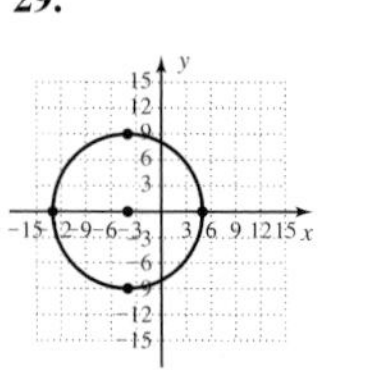

30.

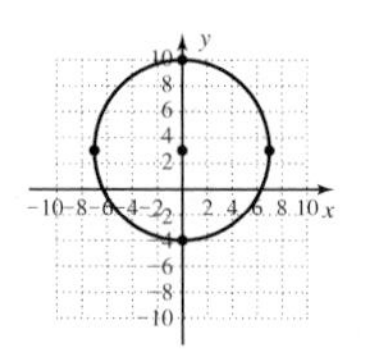

31.

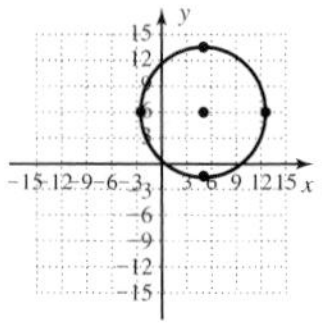

32.

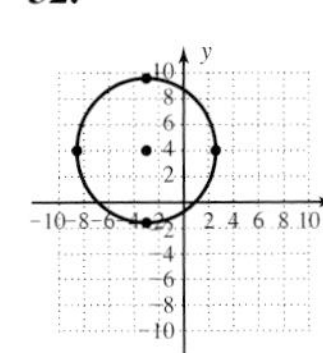

33. $(x - 5)^2 + (y + 2)^2 = 25$, $(5, -2)$, $r = 5$

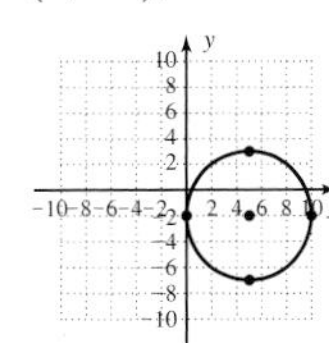

34. $(x + 3)^2 + (y + 2)^2 = 1$, $(-3, -2)$, $r = 1$

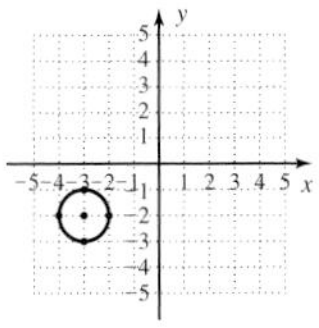

35.

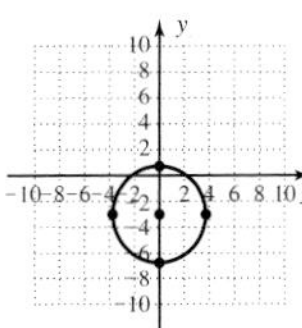

36.

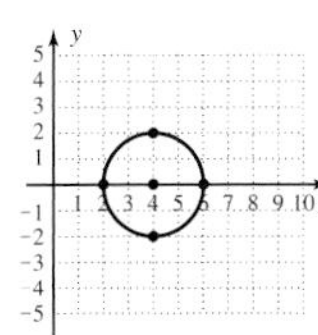

37. $(x + 2)^2 + (y + 5)^2 = 11$, $(-2, -5)$, $r = \sqrt{11}$, $a = \sqrt{11}$, $b = \sqrt{11}$; they are equal

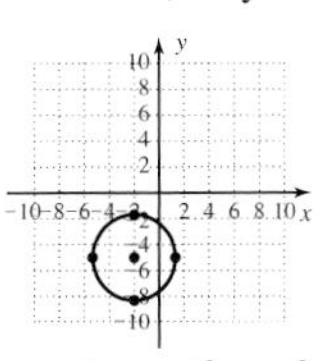

38. $(x - 4)^2 + (y - 7)^2 = 112$, $(4, 7)$, $r = 4\sqrt{7}$, $a = 4\sqrt{7}$, $b = 4\sqrt{7}$; they are equal

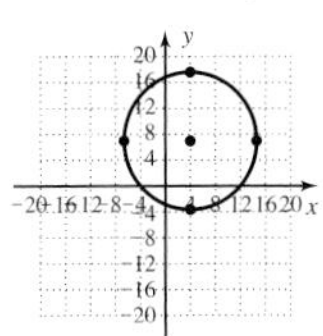

39. $(x + 7)^2 + y^2 = 37$, $(-7, 0)$, $r = \sqrt{37}$, $a = \sqrt{37}$, $b = \sqrt{37}$; they are equal

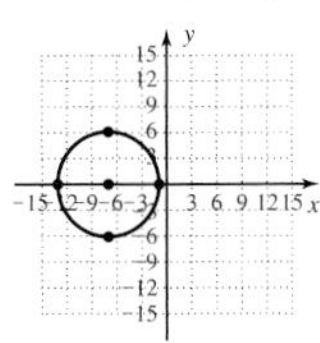

40. $x^2 + (y - 11)^2 = 126$, $(0, 11)$, $r = 3\sqrt{14}$, $a = 3\sqrt{14}$, $b = 3\sqrt{14}$; they are equal

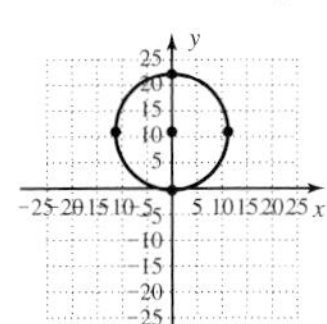

41.

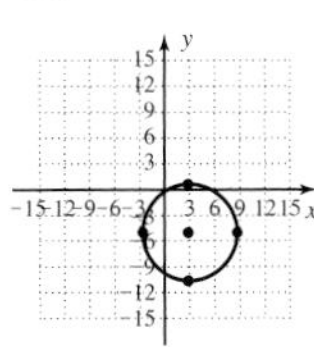

42.

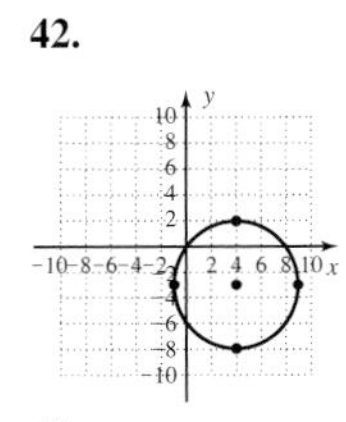

43.

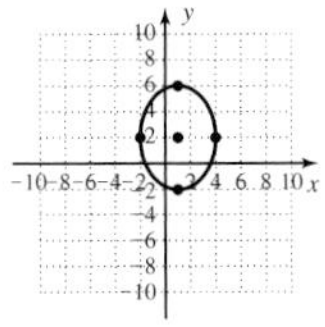

44.

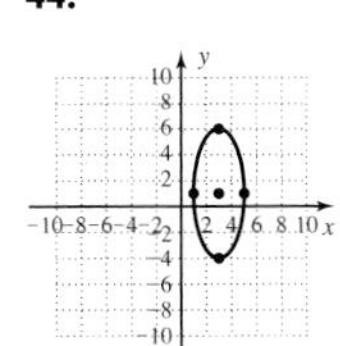

45.

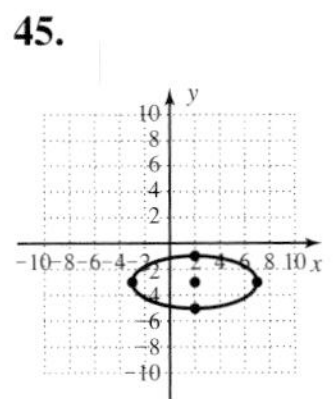

46.

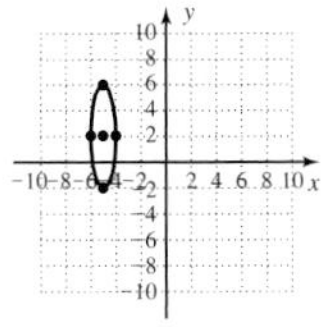

47.

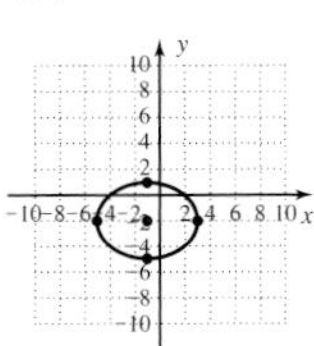

48.

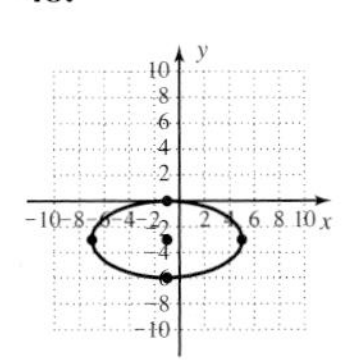

49.

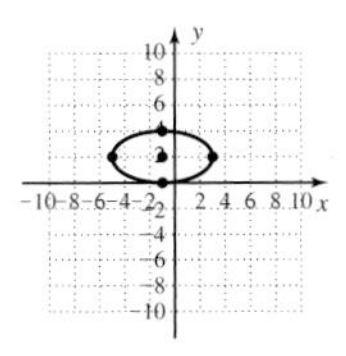

50.

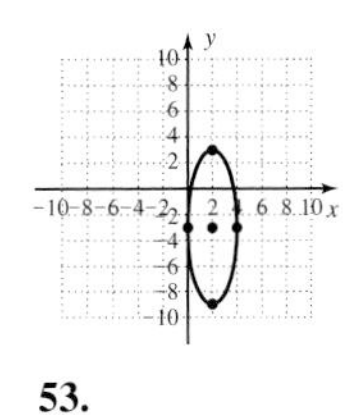

51.

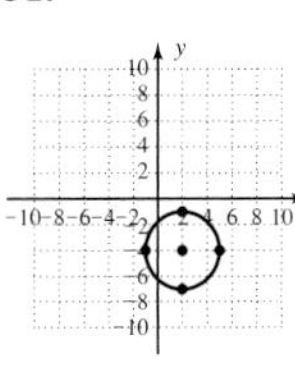

52.

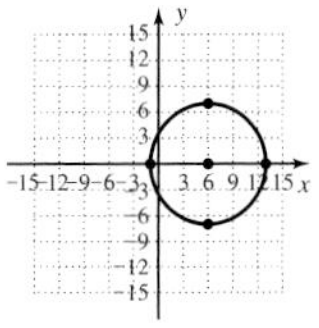

53.

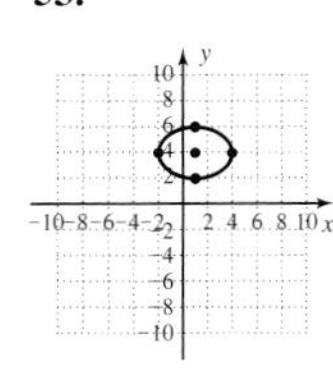

54.

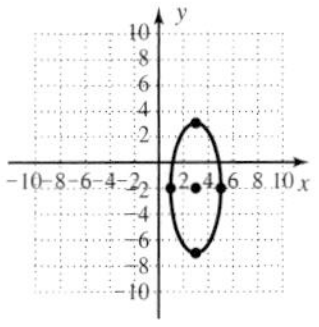

55. a. $\frac{x^2}{16} + \frac{y^2}{4} = 1$, $(0, 0)$, $a = 4$, $b = 2$
b. $(-4, 0)$, $(4, 0)$, $(0, -2)$, $(0, 2)$ **c.**

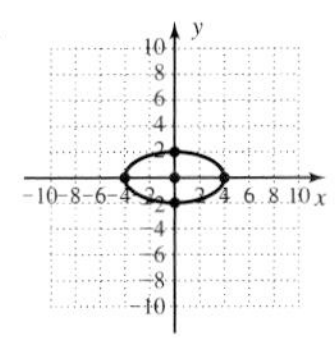

56. a. $\frac{x^2}{4} + \frac{y^2}{36} = 1$, $(0, 0)$, $a = 2$, $b = 6$
b. $(0, -6)$, $(0, 6)$, $(-2, 0)$, $(2, 0)$ **c.**

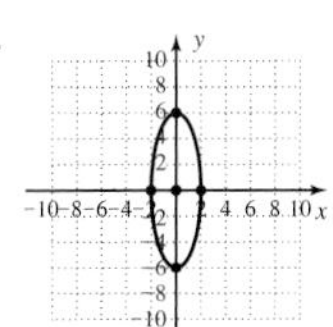

57. a. $\frac{x^2}{9} + \frac{y^2}{16} = 1$, $(0, 0)$, $a = 3$, $b = 4$
b. $(0, -4)$, $(0, 4)$ $(-3, 0)$ $(3, 0)$ **c.**

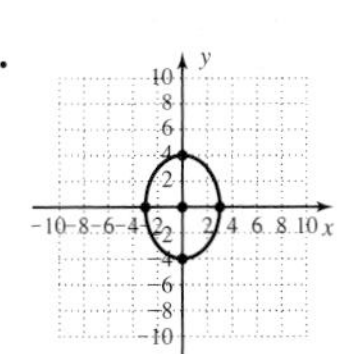

58. a. $\frac{x^2}{9} + \frac{y^2}{25} = 1$, $(0, 0)$, $a = 3$, $b = 5$
b. $(0, -5)$, $(0, 5)$, $(-3, 0)$, $(3, 0)$ **c.**

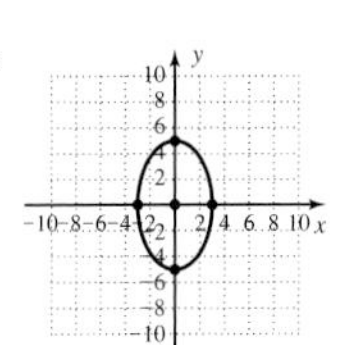

59. a. $\frac{x^2}{5} + \frac{y^2}{2} = 1$, $(0, 0)$, $a = \sqrt{5}$, $b = \sqrt{2}$
b. $(-\sqrt{5}, 0)$, $(\sqrt{5}, 0)$, $(0, -\sqrt{2})$, $(0, \sqrt{2})$ **c.**

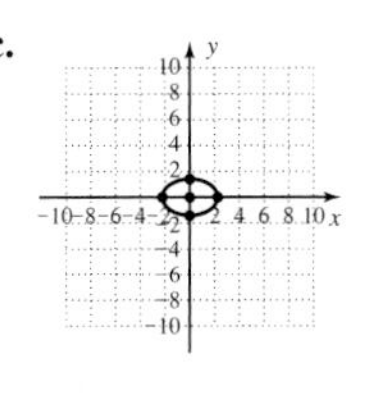

60. a. $\frac{x^2}{7} + \frac{y^2}{3} = 1$, $(0, 0)$, $a = \sqrt{7}$, $b = \sqrt{3}$
b. $(-\sqrt{7}, 0)$, $(\sqrt{7}, 0)$, $(0, -\sqrt{3})$, $(0, \sqrt{3})$ **c.**

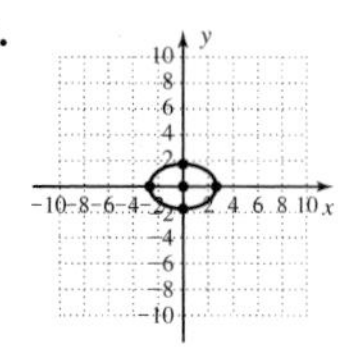

67. $\frac{(x-3)^2}{25} + \frac{(y+2)^2}{10} = 1$

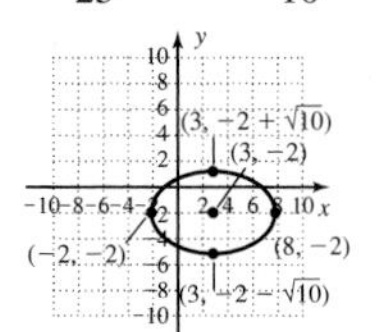

68. $\frac{(x-2)^2}{9} + \frac{(y+3)^2}{18} = 1$

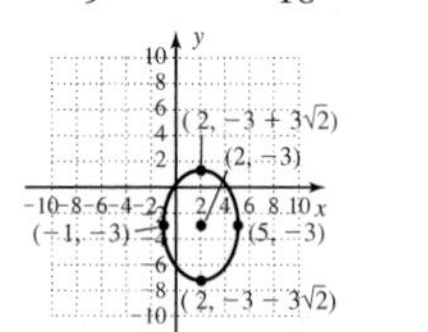

75.

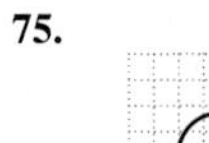

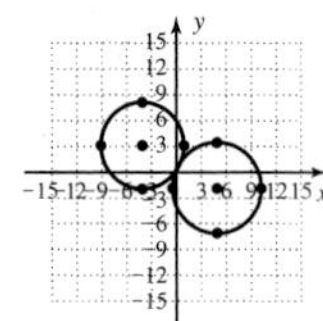

76.

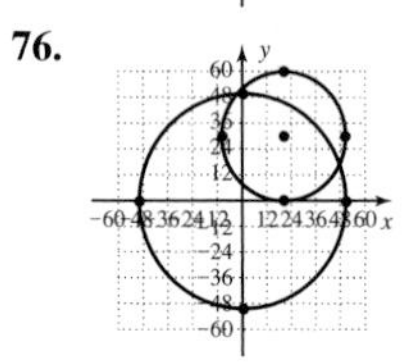

Exercises 7.2

43.

44.

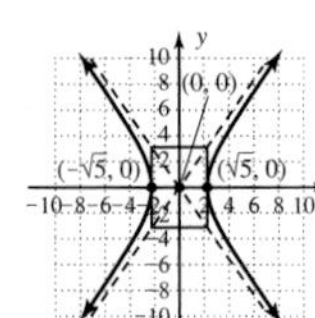

45.

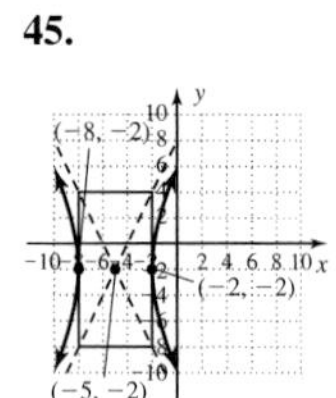

46.

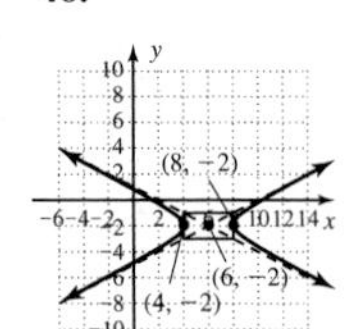

47.

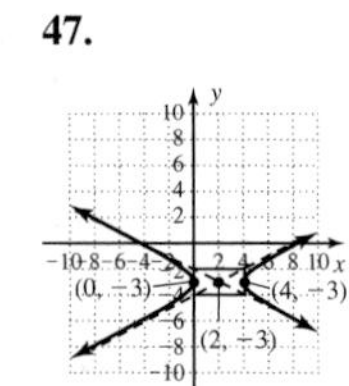

48.

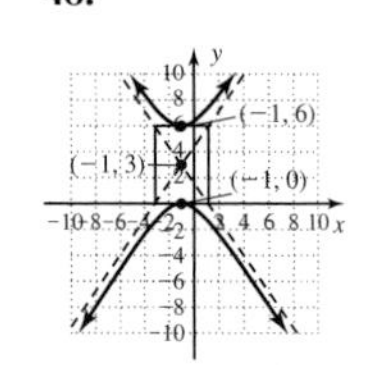

Mid-Chapter Check

7. a. $\frac{(x+3)^2}{4} + \frac{(y-1)^2}{16} = 1$; D: $x \in [-5, -1]$; R: $y \in [-3, 5]$
b. $(x-3)^2 + (y-2)^2 = 16$; D: $x \in [-1, 7]$; R: $y \in [-2, 6]$
c. $y = (x-3)^2 - 4$ D: $x \in (-\infty, \infty)$; R: $y \in (-4, \infty)$

Exercises 7.3

1. a. 3 or 4 not possible

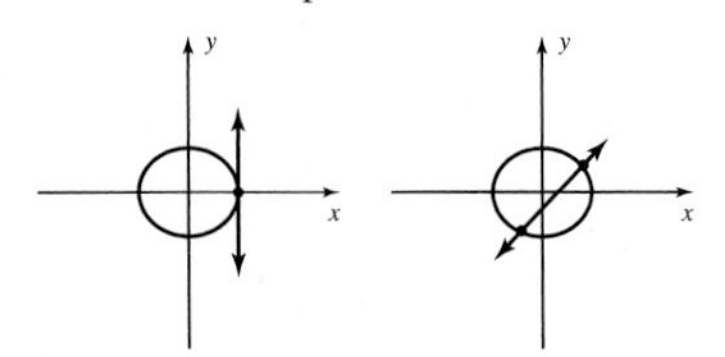

b. 3 or 4 not possible

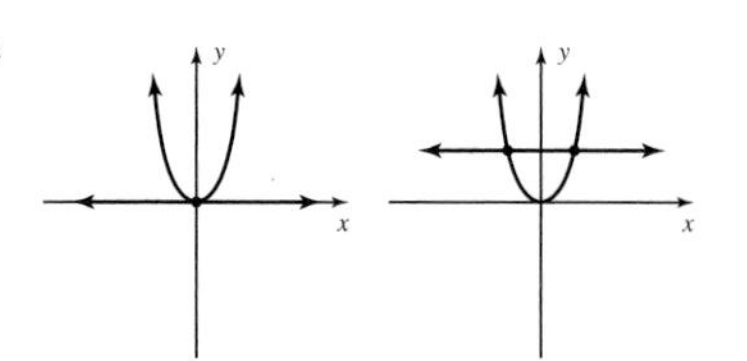

c.

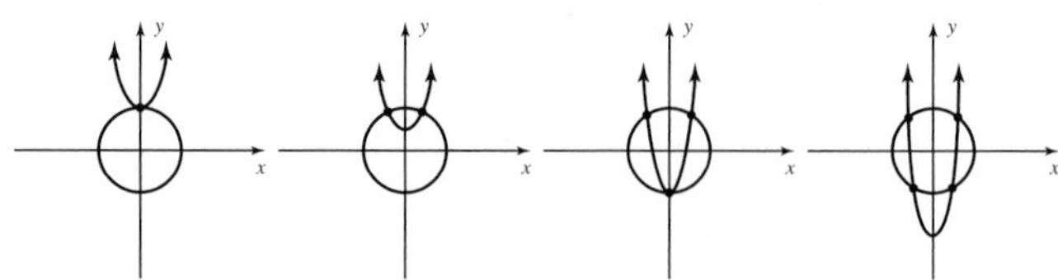

d.

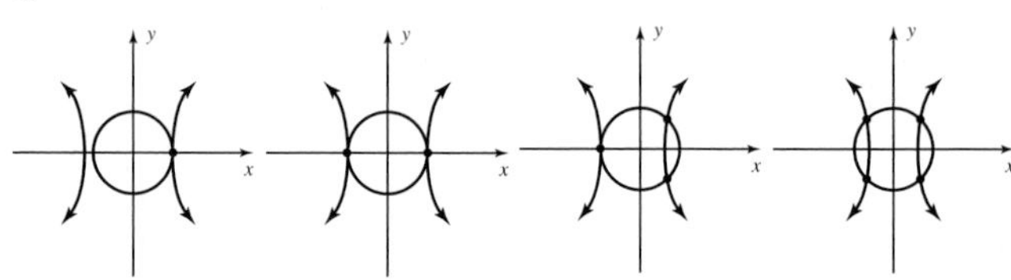

e.

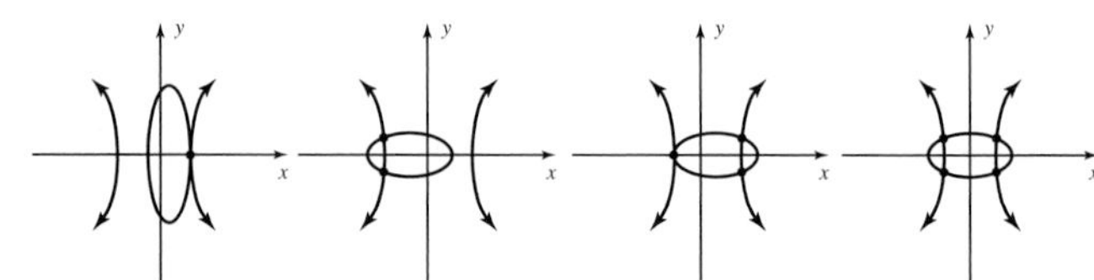

f.

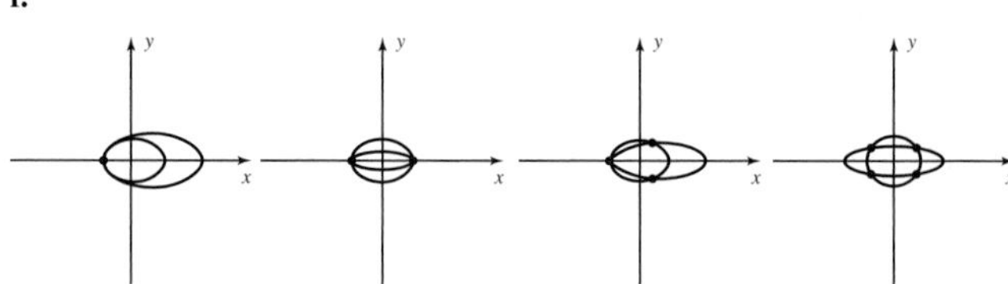

Exercises 7.4

16. a. $(5, 3)$ **b.** $(5, 8)$ and $(5, -2)$ **c.** $(5, 3 + \sqrt{15})$ and $(5, 3 - \sqrt{15})$
d. $(5 - \sqrt{10}, 3)$ and $(5 + \sqrt{10}, 3)$ **e.**

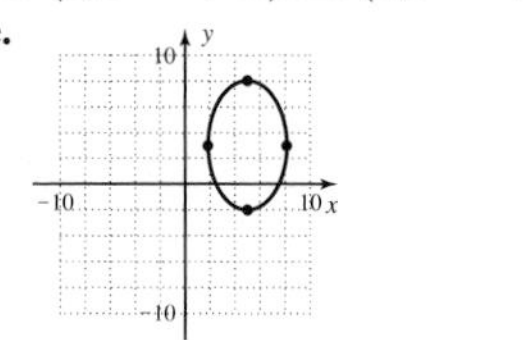

28. a. $(1, 0)$ **b.** $(-1, 0)$ and $(3, 0)$ **c.** $(1 - \sqrt{85}, 0)$ and $(1 + \sqrt{85}, 0)$
d. $2a = 4$, $2b = 18$ **e.**

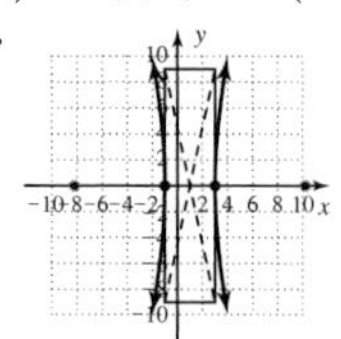

29. a. $(3, -2)$ **b.** $(1, -2)$ and $(5, -2)$ **c.** $(-1, -2)$ and $(7, -2)$
d. $2a = 4$, $2b = 4\sqrt{3}$ **e.**

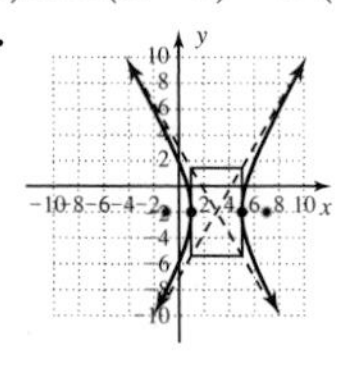

30. a. $(-3, 2)$ **b.** $(-6, 2)$ and $(0, 2)$ **c.** $(-3 - 3\sqrt{3}, 2)$ and $(-3 + 3\sqrt{3}, 2)$ **d.** $2a = 6, 2b = 6\sqrt{2}$ **e.**

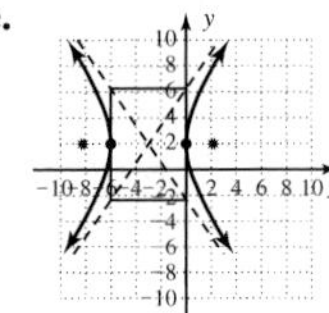

31. $\frac{x^2}{36} - \frac{y^2}{28} = 1$ **32.** $\frac{x^2}{16} - \frac{y^2}{20} = 1$

Exercises 7.5

25. $x \in [-6.25, \infty), y \in (-\infty, \infty)$

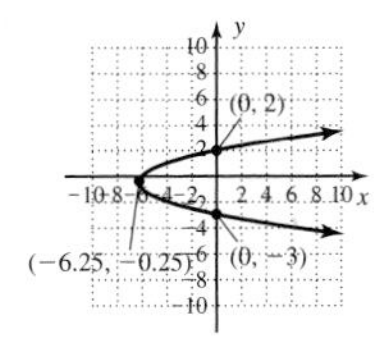

26. $x \in [-9, \infty), y \in (-\infty, \infty)$

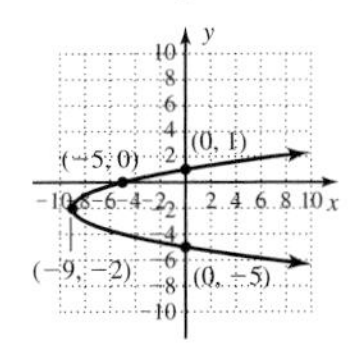

27. $x \in [-21, \infty), y \in (-\infty, \infty)$

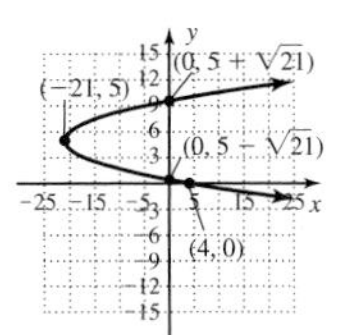

28. $x \in [-41, \infty), y \in (-\infty, \infty)$

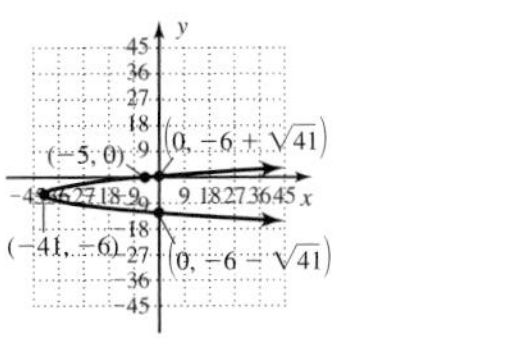

29. $x \in (-\infty, 11], y \in (-\infty, \infty)$

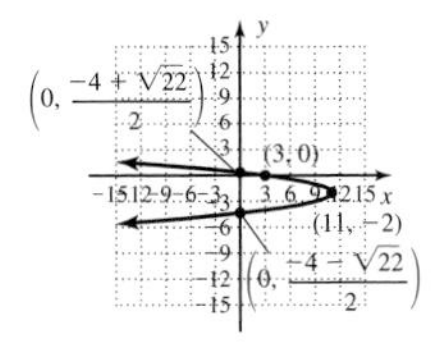

30. $x \in [-10, \infty), y \in (-\infty, \infty)$

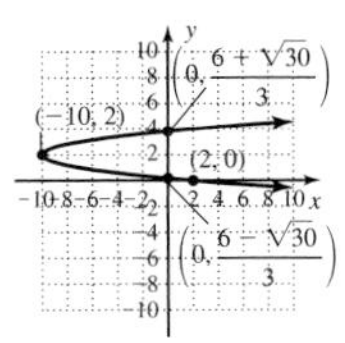

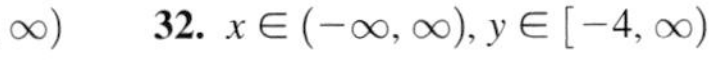

31. $x \in (-\infty, \infty), y \in [3, \infty)$

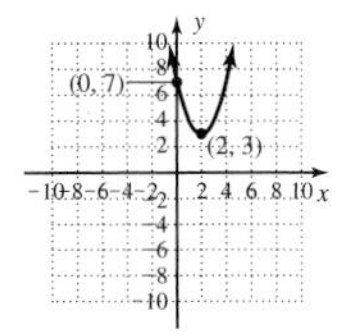

32. $x \in (-\infty, \infty), y \in [-4, \infty)$

33. $x \in [2, \infty), y \in (-\infty, \infty)$

34. $x \in [-4, \infty), y \in (-\infty, \infty)$

35. $x \in [1, \infty), y \in (-\infty, \infty)$

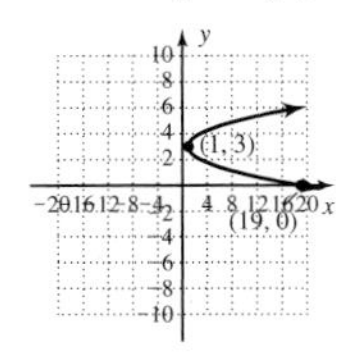

36. $x \in (-\infty, -5], y \in (-\infty, \infty)$

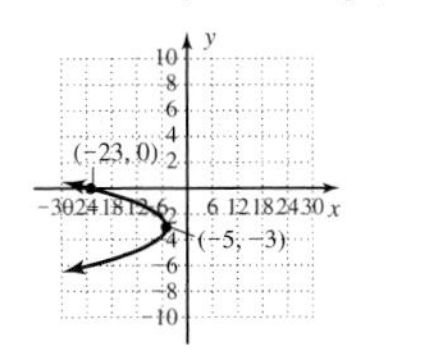

37.

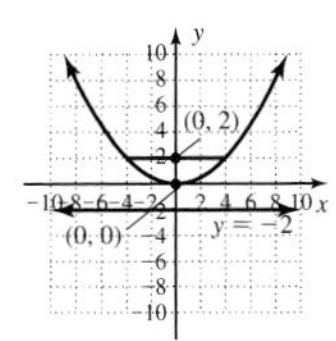

38.

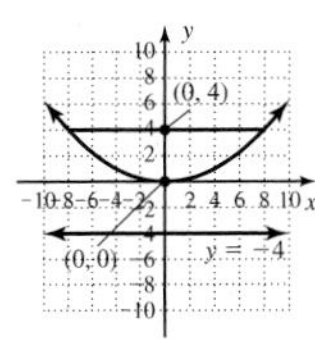

39.

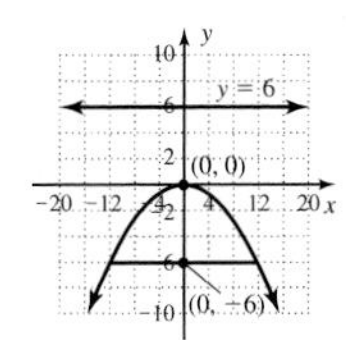

40.

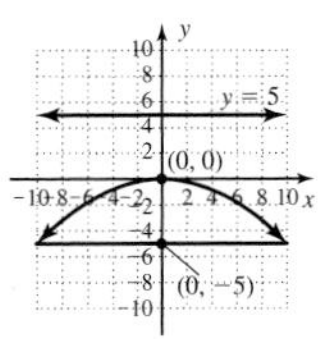

41.

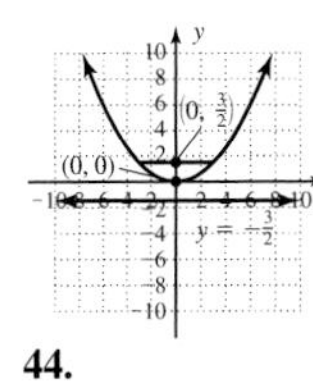

42.

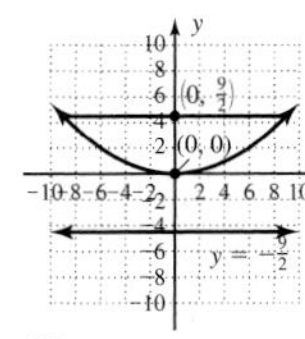

43.

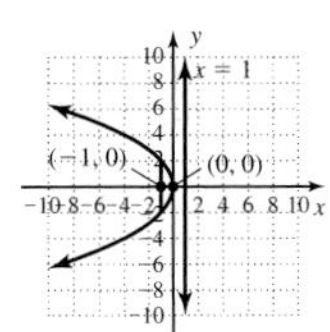

44.

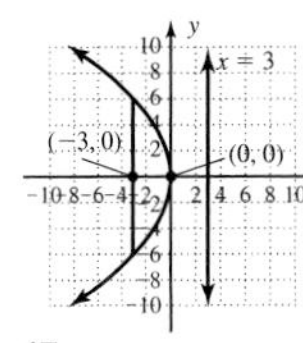

45.

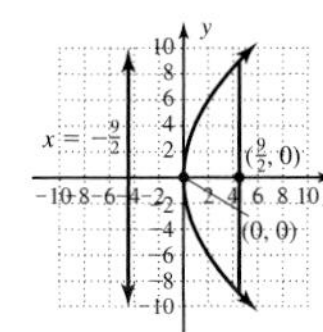

46.

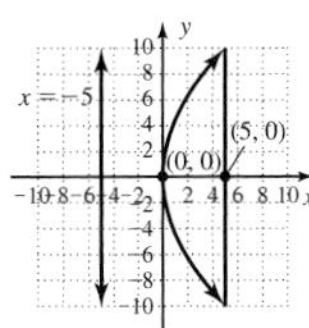

47.

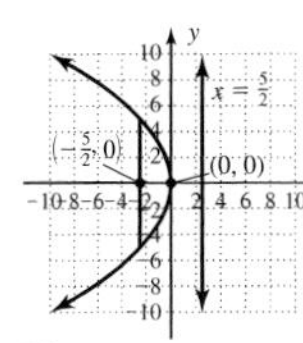

48.

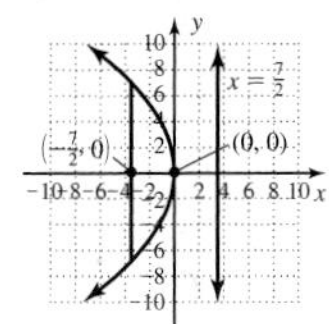

49.

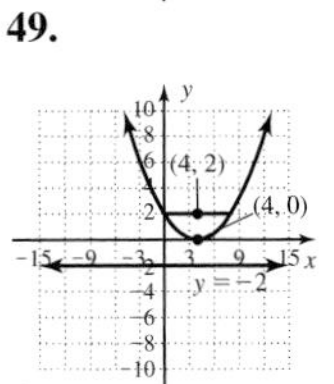

50.

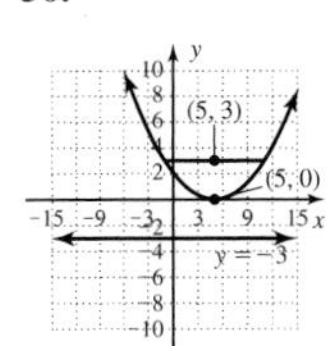

51.

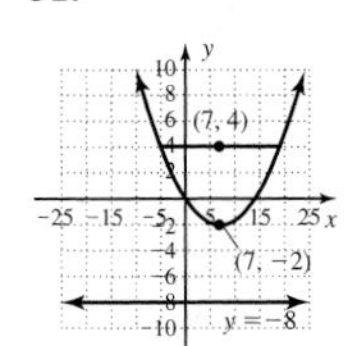

52.

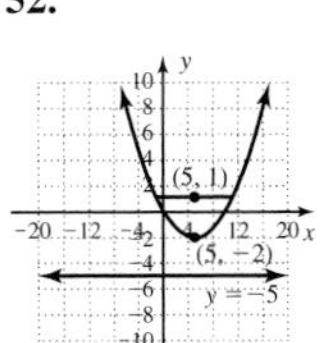

53.

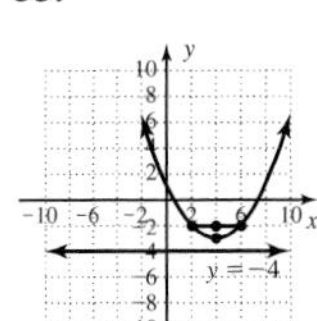

54.

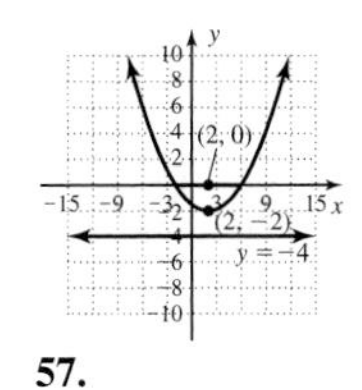

55.

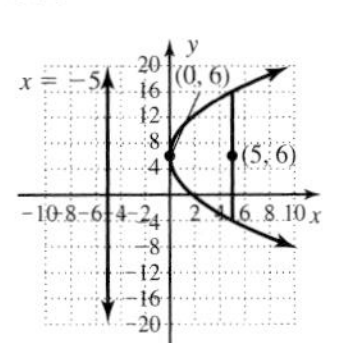

56.

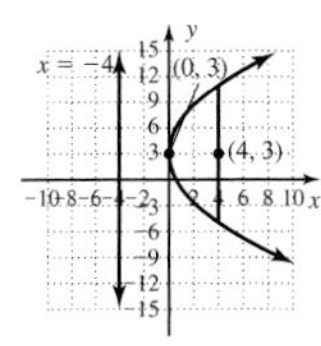

57.

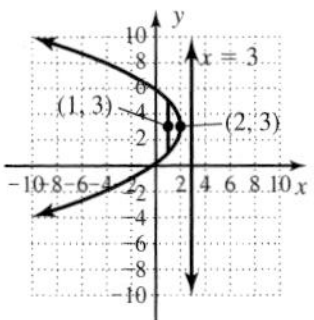

58.

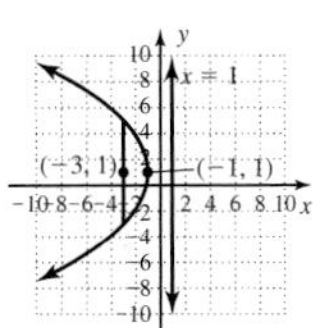

59.

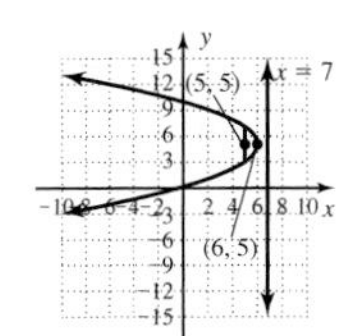

60.

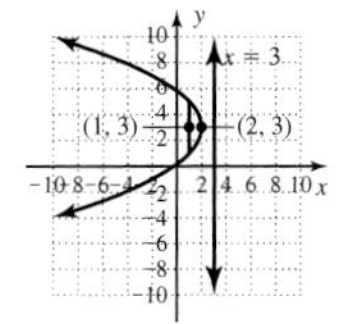

Summary and Concept Review

Section 7.2

11.

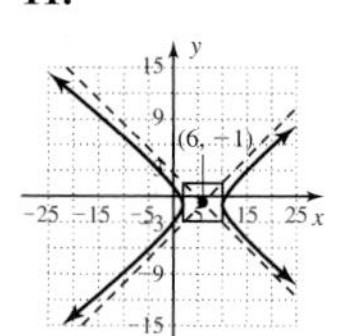

12.

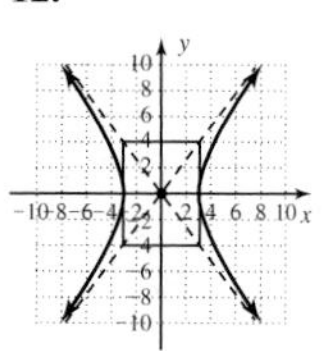

$$\frac{x^2}{9} - \frac{y^2}{16} = 1$$

Section 7.3

17.

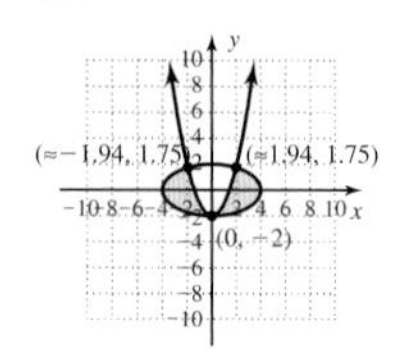

18.

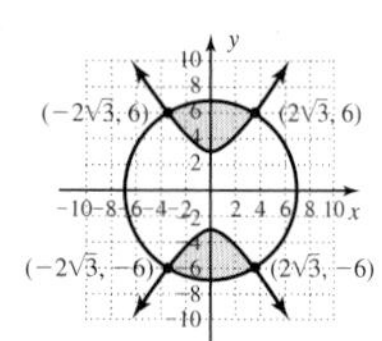

Mixed Review

5.

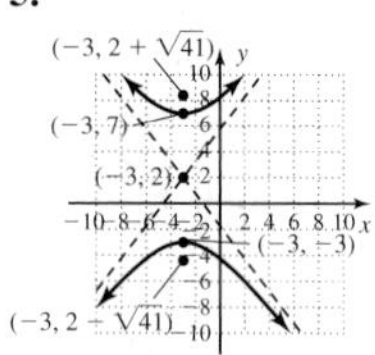

6.

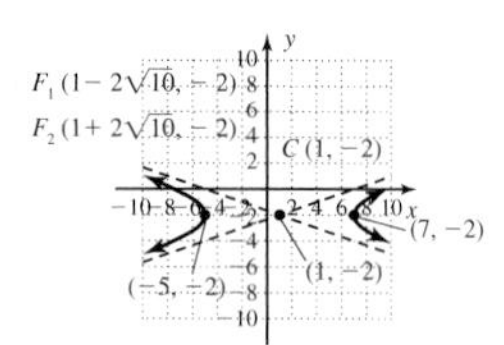

7.

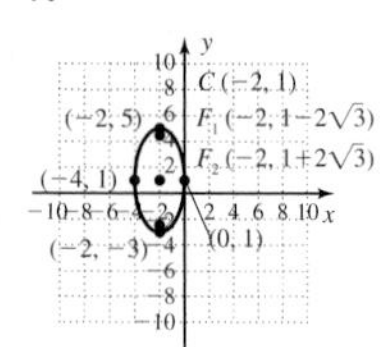

8.

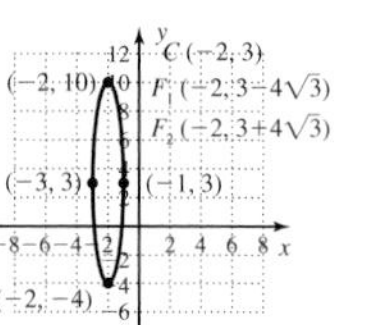
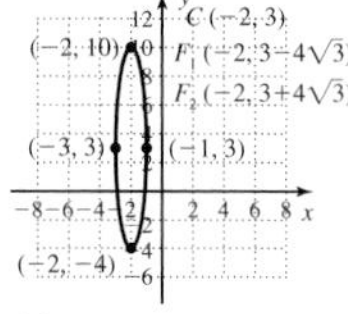

9.

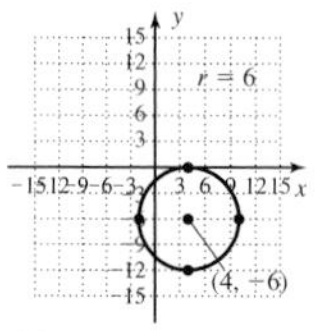

10.

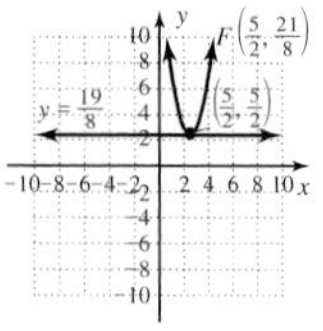

11.

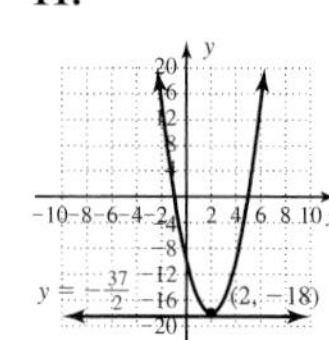
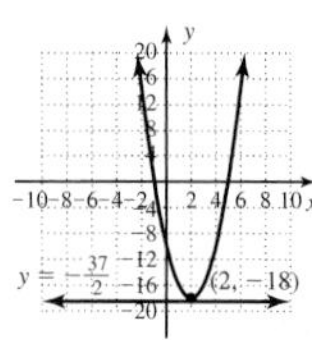

12.

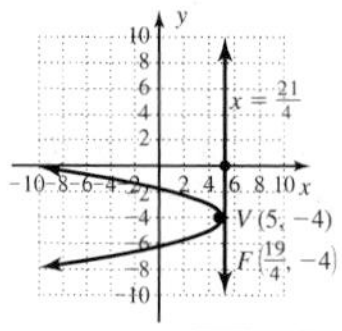

13.

14.

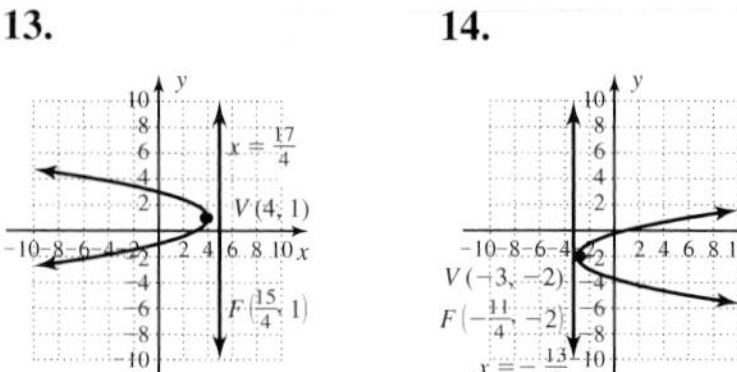

15.

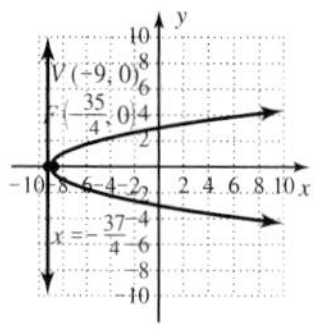

16.

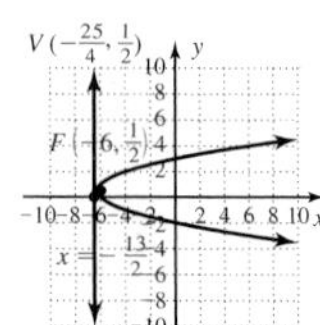

17.

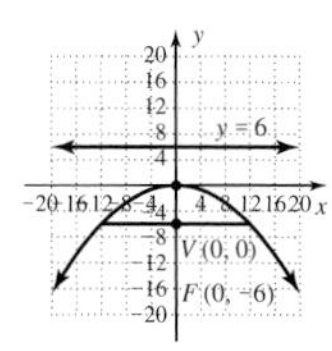

18.

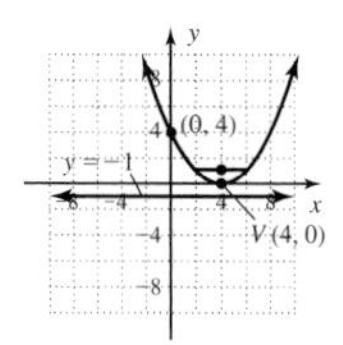

19.

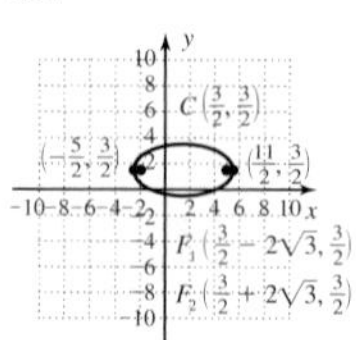

20.

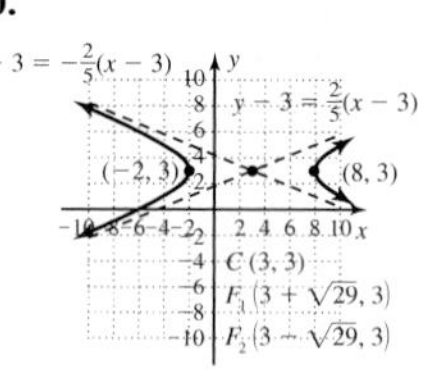

21.

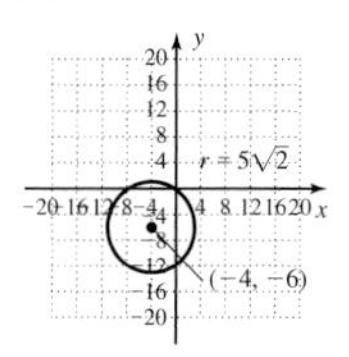

CHAPTER 8

Exercises 8.4 (Only selected proofs are shown.)

25. (1) Show S_n is true for $n = 1$: $S_1 = (1)^2 = 1$✓

(2) Assume S_k is true:

$1 + 8 + 27 + \cdots + k^3 = (1 + 2 + 3 + \cdots + k)^2$

Use it to show the truth of S_{k+1}:

$1 + 8 + 27 + \cdots + k^3 + (k + 1)^3 = [1 + 2 + 3 + \cdots + k + (k + 1)]^2$

left-hand side: $(1 + 8 + 27 + \cdots + k^3) + (k + 1)^2(k + 1)$

$= (1 + 8 + 27 + \cdots + k^3) + k(k + 1)^2 + (k + 1)^2$

$= (1 + 8 + 27 + \cdots + k^3) + k(k + 1)(k + 1) + (k + 1)^2$

$= (1 + 8 + 27 + \cdots + k^3) + 2\dfrac{k(k+1)}{2}(k + 1) + (k + 1)^2$

$= \mathbf{(1 + 2 + 3 + \cdots + k)^2} + 2(1 + 2 + 3 + \cdots + k)(k + 1) + (k + 1)^2$

$= [(1 + 2 + 3 + \cdots + k) + (k + 1)]^2$

31. (1) Show S_n is true for $n = 1$: $S_1 = 1[2(1) + 3)] = 5$✓

(2) Assume S_k is true:

$5 + 9 + 13 + \cdots + (4k + 1) = k(2k + 3)$

Use it to show the truth of S_{k+1}:

$5 + 9 + 13 + \cdots + (4k + 1) + [4(k + 1) + 1] = (k + 1)[(2(k + 1) + 3)]$

$5 + 9 + 13 + \cdots + (4k + 1) + (4k + 5) = (k + 1)(2k + 5)$

left-hand side: $\mathbf{k(2k + 3)} + (4k + 5) = 2k^2 + 3k + 4k + 5$

$= 2k^2 + 7k + 5 = (2k + 5)(k + 1)$

35. (1) Show S_n is true for $n = 1$: $S_1 = 2^{1+1} - 2 = 2^2 - 2 = 2$✓

(2) Assume S_k is true:

$2 + 4 + 8 + \cdots + 2^k = 2^{k+1} - 2$

Use it to show the truth of S_{k+1}:

$2 + 4 + 8 + \cdots + 2^k + 2^{k+1} = 2^{(k+1)+1} - 2$

left-hand side: $\mathbf{2^{k+1} - 2} + 2^{k+1} = 2(2^{k+1}) - 2$

$= 2^{(k+1)+1} - 2$

37. (1) Show S_n is true for $n = 1$: $S_1 = \dfrac{1}{2(1) + 1} = \dfrac{1}{3}$✓

(2) Assume S_k is true:

$\dfrac{1}{1(3)} + \dfrac{1}{3(5)} + \cdots + \dfrac{1}{(2k - 1)(2k + 1)} = \dfrac{k}{2k + 1}$

Use it to show the truth of S_{k+1}:

$\dfrac{1}{1(3)} + \dfrac{1}{3(5)} + \cdots + \dfrac{1}{(2k - 1)(2k + 1)} + \dfrac{1}{(2k + 1)(2k + 3)} = \dfrac{k + 1}{2k + 3}$

left-hand side: $\mathbf{\dfrac{k}{2k + 1}} + \dfrac{1}{(2k + 1)(2k + 3)}$

$= \dfrac{k(2k + 3)}{(2k + 1)(2k + 3)} + \dfrac{1}{(2k + 1)(2k + 3)}$

$= \dfrac{2k^2 + 3k + 1}{(2k + 1)(2k + 3)} = \dfrac{(2k + 1)(k + 1)}{(2k + 1)(2k + 3)}$

$= \dfrac{k + 1}{2k + 3}$

39. (1) Show S_n is true for $n = 1$: S_1: $3^1 \geq 2(1) + 1$✓

(2) Assume S_k is true: $3^k \geq 2k + 1$

Use it to show the truth of S_{k+1}:

$3^{k+1} \geq 2(k + 1) + 1 = 2k + 3$

left-hand side: $3^{k+1} = 3(3^k)$

$\geq 3(\mathbf{2k + 1}) = 6k + 3$

Since k is a positive integer, $6k + 3 \geq 2k + 3$ showing
$3^{k+1} \geq 2k + 3$

43. (1) Show S_n is true for $n = 1$: S_1: $1^2 - 7 = -6$ or $2(-3)$✓

(2) Assume S_k is true: $k^2 - 7k = 2p$ for $p \in \mathbb{Z}$

Use it to show the truth of S_{k+1}:
$(k+1)^2 - 7(k+1) = 2q$ for $q \in \mathbb{Z}$

left-hand side: $k^2 + 2k + 1 - 7k - 7$
$= k^2 - 7k + 2k - 6$
$= \mathbf{2p} + 2k - 6$
$= 2(p + k - 3) = 2q$ is divisible by 2

47. (1) Show S_n is true for $n = 1$: S_1: $6^1 - 1 = 5$✓

(2) Assume S_k is true: $6^k - 1 = 5p$ for $p \in \mathbb{Z}$

Use it to show the truth of S_{k+1}:
$6^{k+1} - 1 = 5q$ for $q \in \mathbb{Z}$

left-hand side: $6^{k+1} - 1 = 6 \cdot 6^k - 1$
$= 6(6^k) - 1$
$= 6(\mathbf{5p + 1}) - 1$
$= 30p + 5$
$= 5(6p + 1) = 5q$ is divisible by 5

51. (1) Show S_n is true for $n = 1$.

$\frac{x^1 - 1}{x - 1} = 1$✓ result checks

(2) Assume S_k is true.

$\frac{x^k - 1}{x - 1} = 1 + x + x^2 + \cdots + x^{k-1}$

and use it to show the truth of S_{k+1} follows.

$\frac{x(x^k - 1)}{x - 1} = x(1 + x + x^2 + \cdots + x^{k-1})$

$\frac{x^{k+1} - x}{x - 1} = x + x^2 + x^3 + \cdots + x^k$

$\frac{x^{k+1} - x}{x - 1} + \frac{x - 1}{x - 1} = (x + x^2 + x^3 + \cdots + x^k) + 1$

$\frac{x^{k+1} - 1}{x - 1} = 1 + x + x^2 + x^3 + \cdots + x^k$✓

Since the steps are reversible, the truth of S_{k+1} follows from S_k and the formula is true for all n.

Exercises 8.5

7. a. 16 possible

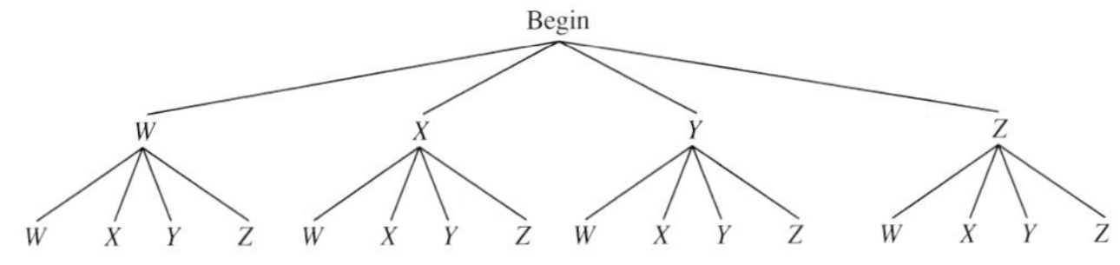

b. *WW, WX, WY, WZ, XW, XX, XY, XZ, YW, YX, YY, YZ, ZW, ZX, ZY, ZZ*

8. a. 16 possible

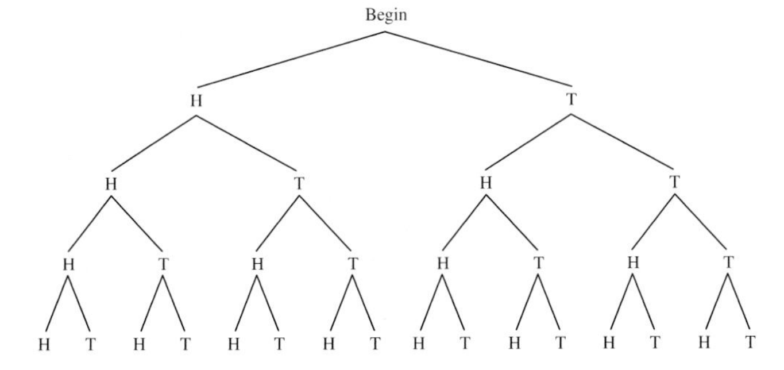

b. HHHH, HHHT, HHTH, HHTT, HTHH, HTHT, HTTH, HTTT, THHH, THHT, THTH, THTT, TTHH, TTHT, TTTH, TTTT

Summary and Concept Review

35. (1) Show S_n is true for $n = 1$: $S_1 = \frac{1(1+1)}{2} = 1$✓

(2) Assume S_k is true:

$1 + 2 + 3 + \cdots + k = \frac{k(k+1)}{2}$

Use it to show the truth of S_{k+1}:

$1 + 2 + 3 + \cdots + k + (k+1) = \frac{(k+1)(k+2)}{2}$

left-hand side: $1 + 2 + 3 + \cdots + k + (k+1)$

$= \frac{\mathbf{k(k+1)}}{\mathbf{2}} + \frac{2(k+1)}{2} = \frac{k(k+1) + 2(k+1)}{2}$

$= \frac{k^2 + 3k + 2}{2} = \frac{(k+1)(k+2)}{2}$

36. (1) Show S_n is true for $n = 1$: $S_1 = \frac{1[2(1) + 1](1 + 1)}{6} = 1$✓

(2) Assume S_k is true:

$1 + 4 + 9 + \cdots + k^2 = \frac{k(2k+1)(k+1)}{6}$

Use it to show the truth of S_{k+1}:

$1 + 4 + 9 + \cdots + k^2 + (k+1)^2 = \frac{(k+1)(2k+3)(k+2)}{6}$

left-hand side: $1 + 4 + 9 + \cdots + k^2 + (k+1)^2$

$= \frac{\mathbf{k(k+1)(2k+1)}}{\mathbf{6}} + \frac{6(k+1)^2}{6} = \frac{(k+1)[(2k^2 + k + 6k + 6]}{6}$

$= \frac{(k+1)(2k^2 + 7k + 6)}{6} = \frac{(k+1)(2k+3)(k+2)}{6}$

37. (1) Show S_n is true for $n = 1$: S_1: $4^1 \geq 3(1) + 1$✓

(2) Assume S_k is true: $4^k \geq 3k + 1$

Use it to show the truth of S_{k+1}:
$4^{k+1} \geq 3(k+1) + 1 = 3k + 4$

left-hand side: $4^{k+1} = 4(4^k)$
$\geq 4(\mathbf{3k + 1}) = 12k + 4$

Since k is a positive integer, $12k + 4 \geq 3k + 4$ showing
$4^{k+1} \geq 3k + 4$

38. (1) Show S_n is true for $n = 1$: S_1: $6 \cdot 7^{1-1} \leq 7^1 - 1$✓

(2) Assume S_k is true: $6 \cdot 7^{k-1} \leq 7^k - 1$

Use it to show the truth of S_{k+1}:
$6 \cdot 7^k \leq 7^{k+1} - 1$

left-hand side: $6 \cdot 7^k = 7 \cdot 6 \cdot 7^{k-1}$
$\leq 7 \cdot \mathbf{7^k - 1}$
$\leq 7^{k+1} - 1$

39. (1) Show S_n is true for $n = 1$: S_1: $3^1 - 1 = 2$ or $2(1)$✓

(2) Assume S_k is true: $3^k - 1 = 2p$ for $p \in \mathbb{Z}$

Use it to show the truth of S_{k+1}:
$3^{k+1} - 1 = 2q$ for $q \in \mathbb{Z}$

left-hand side: $3^{k+1} - 1 = 3 \cdot 3^k - 1$
$= 3 \cdot \mathbf{2p}$
$= 2(3p) = 2q$ is divisible by 2

40. 6 Ways

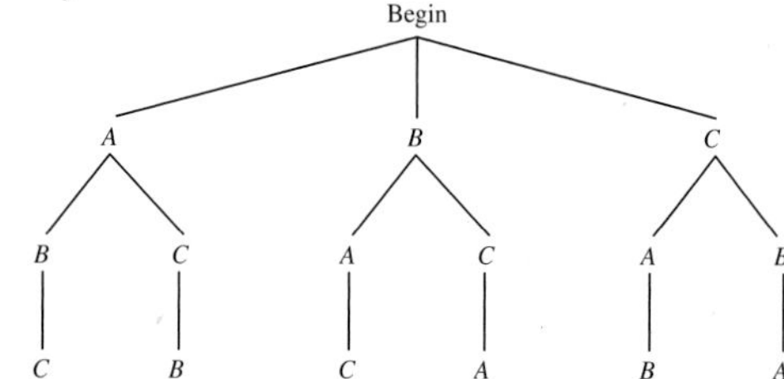

Practice Test

11.

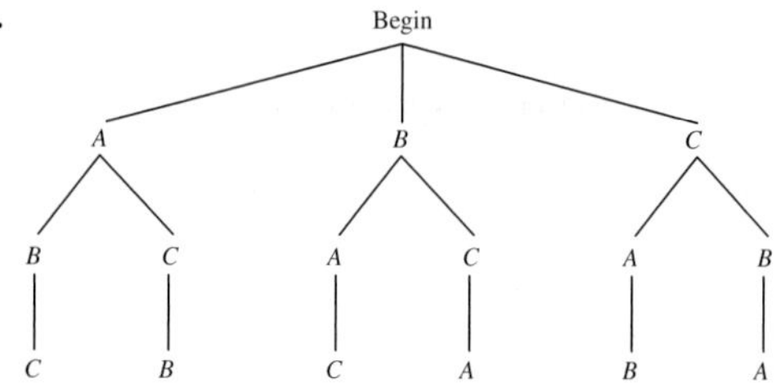

Cumulative Review

13. a. $x^3 = 125$ **b.** $e^5 = 2x - 1$ **14.** 6.93% **15. a.** $x \approx 3.19$ **b.** $x = 334$ **16.** (9, 1, 1) **17.** (5, 10, 15) **18.** $\frac{x^2}{16} - \frac{y^2}{20} = 1$

19. $(-3, 3)$; $(-7, 3)$, $(1, 3)$; $(-3 - 2\sqrt{3}, 3)$, $(-3 + 2\sqrt{3}, 3)$

20. a. $a_{20} = \frac{1}{1{,}048{,}576} \approx 0.00000095$; $S_{20} = 349525.\overline{3}$

b. $a_{20} = -\frac{117}{40}$; $S_{20} = -20.5$

Index

A

Abscissa, 312
Absolute value
  definition of, 6
  of real numbers, 6–7
  use of term, 604
Absolute value equations, 606
  on graphing calculator, 609–610
  procedures to solve, 606–609
  systems and, 604–605, 668
Absolute value function, 338
Absolute value inequalities
  greater than, 608–609
  less than, 607–608
  procedures to solve, 606–609
  systems and graphing to solve, 605–606, 668
Absolute value relation, 161
Accumulated value, 522, 528
Addition
  associative property of, 17
  commutative property of, 16
  of complex numbers, 109–110
  distributive property of multiplication over, 18
  of fractions, 47
  of functions, 248–249
  of matrices, 628–629, 669
  of polynomials, 27–28
  of radical expressions, 60
  of rational expressions, 47–48
Additive identity, 17, 629
Additive inverse, 17, 629
Additive property
  of equality, 72–73, 85
  of inequality, 85
Algebra
  fundamental theorem of, 393–394, 397
  of matrices, 627–635, 669–670
  rhetorical, 13
Algebraic expressions, 14, 72
  evaluation of, 16
  simplification of, 18–19
Algebraic fractions. *See* Rational expressions
Algebraic functions, 94
Algebraic method, 264–265
Algebraic terms, 13
Allowable values, 88–89
Alternating sequences, 749
Altitude, 504
Amortization, 528–529
Annuities, 528–529, 534
Antilogarithm, 509
Approximate form, 4
Approximate solutions, 119
Arc length, 737
Area. *See also* Surface area
  of circular sector, 309
  of cone, 102
  under curve, 280–283
  of ellipse, 692
  of rectangle, 637
  of right parabolic segment, 737
  of square, 692
  of trapezoid, 381
  of triangle, 447, 658–659
Arithmetic sequences, 759, 832–833
  applications of, 763–764
  finding common difference of, 759–760
  finding nth partial sum of, 762–763
  finding nth term of, 760–762
Associated minor matrices, 646, 647
Associations
  linear, 218
  negative, 217
  nonlinear, 218
  positive, 217
Associative properties, 17
Asymptotes
  horizontal, 304–306, 361, 426–427
  nonlinear, 440, 442
  oblique, 438, 440, 441
  vertical, 304–306, 362, 422–424
Asymptotic discontinuities, 330
Augmented matrices
  matrix inverses and, 676–677
  of system of equations, 617–618
  triangularizating, 618–620
Augmented matrix method, 643
Average distance, 147
Average rate of change, 198–199
Axis of symmetry, 191, 192

B

Base, 7
Base case, of inductive proof, 784
Base functions, 369–371
Bernoulli, Johann, 155
Best fit, 537
  line of, 221
  parabola of, 223
Binomial coefficients, 825–826
  on graphing calculator, 828–829
Binomial conjugates, 29–30
Binomial expansion, 827–828
Binomial powers, 823–825
Binomial probability formula, 830
Binomials, 27
  cube of, 115
  F-O-I-L method for multiplying, 29
  product of, 28
  quotients of, 374
  square of, 30–31
Binomial theorem, 823, 826–827, 836
  applications of, 827–828
  binomial coefficients and, 825–826, 828–829
  Pascal's triangle and, 823–825
Bisection, 421
Body mass index formula, 92
Boundary, 84, 590, 594
Boyle, Robert, 44
Boyle's law, 44, 45, 317
Brahe, Tycho, 716
Branches, of absolute value function, of hyperbola, 694
Briggs, Henry, 489

C

Calculators. *See also* Graphing calculators
  exponentials on, 509
  logarithms on, 489–490, 509
Cantor, Georg, 155
Carbon-14 dating, 505, 519
Cardano, Girolomo, 107, 383, 577
Carrying capacity, 543
Cartesian coordinate system, 140
Cauchy, Augustin-Louis, 655
Cayley, Arthur, 616, 655
Celsius, converted to Fahrenheit, 172
Center, of circle, 682
Center-shifted form
  of equation of circle, 683
  of equation of ellipse, 687–688
  of equation of hyperbola, 695–697
Change-of-base formula, 499–500
  on graphing calculator, 502
Charles, Jacques, 44

Charles's law, 45
Circles, 740
center, 683
equation of, 682–685, 700–701
on graphing calculator, 689–690
Clark's rule, 12
Closed form solutions, 119
Coefficients, 13
binomial, 825–826
correlation, 219
factor theorem and, 385–386
lead, 27
Cofactors, A-5
Coincident dependence, 579, 584
Coincident lines, 570
Collinear points test, 659
Combinations, 798–800
on graphing calculator, 815
Combined variation, 317–318
Comets, 694–695
Common binomial factors, 36
Common difference, for sequence, 759
Common factors
rational functions and, 438–439
reduction of, 45
Common logarithms
applications of, 490–491
calculators and, 489–490
Common ratio, 769
Commutative properties
of addition, 16–17
of multiplication, 16–17
Complementary events, 810
Completely factored form, 36
Completing the square, 117, 706
graphing quadratic functions by, 288–290
solving quadratic equations by, 119–120, 288
Complex conjugates, 110–111, 123
product of, 111
roots of multiplicity and, 386–388
Complex conjugates theorem, 386–388
proof of, 386–387
Complex numbers, 109, 132
absolute value of, 115, 404
addition and subtraction of, 109–110
division of, 112
on graphing calculator, 113
historical background of, 107, 108
identifying and simplifying, 109
multiplication of, 110–112
square root of, 404
standard form of, 109
Complex roots
on graphing calculator, 471–472
of polynomials, 385–386, 389
Composition
of functions, 251–254, 358
transformations via, 300
Compound annual growth, 258
Compound fractions, 48–49
Compound inequalities, 86
method to solve, 86–88
Compound interest, 522–523
on graphing calculator, 530
Compound interest formula, 480, 523, 524
Compressions, 278, 360
Concave down, 198
Concave up, 191
Concavity, 191, 192
Conditional equations, 73
Cones, lateral surface area of, 102
Conic, 682. *See also* Circles; Ellipses; Hyperbolas; Parabolas
deriving equation of, A-6–A-7
on graphing calculator, 744–745
graphing calculators and, 724
Conic sections, 681, 682
working with, 354
Conjugates
binomial, 29–30
complex, 110–111, 123, 386–388
Consecutive integers, 77
Consistent/dependent system of equations, 570
Consistent/independent system of equations, 570
Consistent system of equations, 570
Constant of variation, 313
Constant of proportion, 313
Constant terms, 13
Constraint inequalities, 596
Continuous graphs, 198, 327
Continuous intervals, 605
Continuously compounded interest formula, 523–525
Contradictions, 73–74
Conversion
Celsius–Fahrenheit, 172
U.S. Standard Units—Metric System, A-1–A-2
Coordinate grid, 141
Coordinate plane, 141
Coordinates, 3, 461
Correlation, 219
Correlation coefficient, 219
Counting, 834–835
combinations and, 798–800
distinguishable permutations and, 796–797
fundamental principle of, 794–796, 835
graphing calculator and, 800
by listing and tree diagrams, 792–794
nondistinguishable permutations and, 797–798
Cramer's Rule, 656–658, 671
Cube root function, 197–198
Cube roots
definition for, 9
notation for, 8
simplification of, 55–56
Cubes
sum of cubes of n natural numbers, 779
sum and difference of two perfect, 39–40
Cubic equations, 460
Cubing function, 195–197
Curves, area under, 280–283
Cylinders
surface area of, 81, 258
volume of, 43
Cylindrical shell volume formula, 43
Cylindrical vents, 714

D

Data analysis, 237
toolbox functions and, 318–320
Data sets, regression and, 536
Decay functions, 478
Decay rate, 525, 527
Decimals, 4
Decision variables, 596
Decomposition
of composite functions, 255
of rational expressions, 660–661
of rational terms, 14
Decreasing function, 198
Degenerate cases, 684
Degrees, of monomials, 26–27
Demand curves, 714
Demographics, 476
Denominators, rationalizing, 61–62
Dependent systems, 570–571, 582–584, 620–621
Depreciation, 505
Depreciation equation, 180
Descartes, René, 108, 140, 155, 190, 288
Descartes rule of signs, 399–401, 421
Descriptive translation, 98, 100
Descriptive variables, 5, 6, 15
Determinants
to find area of triangle, 658–659
of general matrix, A-5
of minor matrices, 647
of singular matrices, 645–649
to solve systems, 655–658
Dichotomy paradox, 840
Difference quotient, 348–349
Direction/approach notation, 302
Direct proportion, 313

Directrix, 732
Direct variation, 313, 362
  toolbox functions and, 313–316
Dirichlet, Lejeune, 155
Discontinuities
  asymptotic, 330
  jump or nonremovable, 331, 332
  nonremovable, 438
  removable, 332
Discriminant
  on graphing calculator, 125
  of quadratic formula, 122–123
Disjoint intervals, 605
Distance formula, 148
Distributive property of multiplication over addition, 18
Dividends, 374
Division
  of complex numbers, 112
  of functions, 248–249
  of polynomials, 250, 374–380, 464
  of radical expressions, 60–62
  of rational expressions, 46, 47
  synthetic, 364, 375–379
  zero and, 7
Divisor, 374
Domain, 88–89
  effective, 328
  of functions, 255–256, 264
  on graphing calculator, 307
  implied, 163
  of logarithmic functions, 488–489
  of piecewise-defined functions, 327–328
  of rational functions, 422–423
  of relation, 156
  vertical boundary lines and, 161–162
Drug absorption, 505

E

Earthquake intensity, 490, 493
East/west reflection, 278
Eccentricity, 744–745
Einstein, Albert, 577
Electrical resistance, 21
Elementary row operations, 618
Elements, 2
Elimination
  analysis of function, 568–570
  Gaussian, 619, A-4
  Gauss-Jordan, 619, A-4
  to solve linear systems, 579–581
  to solve nonlinear systems, 708–709
Ellipses, 740, 744
  center-shifted form of, 687–688
  definition of, 717
  equation of, 685–686, 700–701, 741, A-6
  focal chord of, 728
  foci formula for, 719
  foci of, 716–720
  on graphing calculator, 689–690
  perimeter of, 726
  with rational/irrational values, 745
  reflective property of, 729
  standard form of equation of, 686
Elongation, 744–745
Empty sets. *See* Null set
End behavior, 191, 192, 243
  of polynomial graphs, 407–408
  of rational graphs, 302
  of rational inequalities, 455
Endpoints, 6, 84
Equality
  additive property of, 72
  of matrices, 627–628
  multiplicative property of, 72, 73
  power property of, 96, 97
  square root property of, 118–119
Equations, 72. *See also* Systems of linear equations; Systems of linear equations in three variables; Systems of linear equations in two variables; Systems of nonlinear equations; *specific types of equations*
  accumulated value, 522
  of circle, 682–685
  conditional, 73
  of conic, A-6–A-7
  of ellipse, 685–689
  equivalent, 569
  exponential, 481, 486–488, 554, 560–562
  families of, 72, 75
  linear, 72–78, 129–130, 140–149
  literal, 49–50, 63, 74–75
  logarithmic, 509–510, 513–514, 554, 560–562
  logistics, 543–544
  matrix, 640–649
  method to solve, 72, 205
  polynomial, 94–95, 131
  present value, 522
  quadratic, 94, 118–122, 125, 133, 288
  radical, 94, 96–98
  rational, 94
  regression, 181, 367–368
  solution form for, 73
Equivalent equations, 569
Equivalent matrix method, 643
Euler, Leonhard, 108, 155, 590
Even functions, 342
Events, 808
  complement of, 810
  exclusive, 812–813
  nonexclusive, 813–814
Exact form, 4
Exact form solutions, 119
Expansion of minors, 647–648
Experiments, 793, 834
Exponential decay, 525
Exponential equations
  base-*b*, 510, 513
  on graphing calculator, 481
  logarithmic form and, 486–488
  method to solve, 510–512, 560–562
  in simplified form, 509–510
Exponential form, 7–8
Exponential functions, 476, 552, 554
  applications of, 480–481, 515, 521–529
  base *e*, 495–496
  evaluation of, 476–477
  graphs of, 477–479
  method to solve, 479–480
  natural, 495
  power functions versus, 477
Exponential growth, 525
Exponential growth formula, 525, 526
Exponential notation, 7–8, 22
Exponential terms, 22
Exponents, 7
  on graphing calculator, 491
  properties of, 22–25
  quotient property of, 24
  rational, 56–58, 97–98, 100
  zero and negative numbers as, 24–25
Extraneous roots, 96, 514
Extrapolation, 180, 540
Extreme values, 293–294, 412

F

Factorable polynomials, 378–379
  rational roots theorem and, 398
Factorial formulas, 804
Factorial notation, 750
Factors/factoring, 35
  common binomial, 36
  greatest common, 35
  by grouping, 36
  nested, 43
  of special forms, 38–41
Factor theorem, 384–385, 465
  complex numbers and coefficients and, 385–386
  proof of, 385
Fahrenheit, converting Celsius to, 172
Falling bodies, 203
False positive, 72
Feasible region, 594
Fermat, Pierre de, 807–808, 823
Fibonacci, 749

Fibonacci sequence, 749
Fibonacci spiral, 749
Finite sequences, 748, 749, 831
Finite series, 748, 751
Fish length to weight relationship formula, 66
Flowcharts
 decision making process in, 35
 factoring in, 40, 41
Fluid motion, formula for, 286
Focal chord, 728
 of ellipse, 728
 of hyperbola, 703
 of parabola, 733
Foci, 741
 applications of, 722–723
 of ellipse, 716–720
 of hyperbola, 720–722
Focus-directrix form of equation of parabola, 732–735
F-O-I-L method, 29
Folium of Descartes, 461
Force, 324
Formulas, 74
 absolute value of complex number, 115, 404
 altitude of airplane, 504
 amount of mortgage payment, 532
 arc length, 737
 area, 309, 381, 447, 601, 625, 663, 692, 737
 binomial cubes, 115
 binomial probability, 830
 body mass index, 92
 Celsius to Fahrenheit conversion, 172
 central semicircle equation, 214
 change-of-base, 499–500, 502
 compound annual growth, 258
 compound interest, 480, 523, 524
 conic sections, 354, 726
 continuously compounded interest, 523–525
 coordinates for folium of Descartes, 461
 cost of removing pollutants, 434
 cylindrical shell volume, 43
 cylindrical vents, 714
 dimensions of rectangular solid, 588
 discriminant of reduced cubic equation, 460
 distance, 148, 588
 electrical resistance, 21
 exponential growth, 525
 factorial, 804
 fluid motion, 286
 focal chord of hyperbola, 703
 force between charged particles, 324
 forensics—estimating time of death, 518
 games, 483, 819
 general linear equation, 185
 gravitational attraction, 309
 growth of bacteria population, 483
 half-life of radioactive substance, 518
 height of projected image, 271
 height of projectile, 127, 228
 human life expectancy, 153
 ideal weight for males, 171
 intercept/intercept form of linear equation, 185
 interest earnings, 153
 inverse of matrices, 652
 lateral surface area of cone, 102
 learning curve, 546
 length to weight relationship, 66
 medication in bloodstream, 33
 midpoint, 147–148
 mortgage payment, 34
 nested factoring, 43
 painted area on canvas, 102–103
 perimeter of ellipse, 726
 perpendicular distance from point to line, 214
 pH level, 492–493
 population density, 433
 power of wind-driven generator, 203
 powers of imaginary unit, 788
 radius of sphere, 271
 required interest rate, 324
 root tests for quartic polynomials, 417
 sand dune function, 338
 semi-hyperbola equation, 703
 simple interest, 521
 simplification of, 49–50
 slope, 144–145, 189
 spring oscillation, 611
 square root of complex numbers, 404
 Stirling's formula, 804
 student loan payment, 757
 sum of cubes, 779
 sum of first *n* cubes, 788
 sum of first *n* natural numbers, 766
 sum of first *n* positive integers, 296
 sum of first *n* squares, 766
 sum of *n*th terms of sequence, 757
 surface area of cylinder, 81, 127, 258, 447
 surface area of rectangular box with square ends, 296
 temperature measurement, 575
 time required to double, 493, 504
 timing falling object, 66
 total interest paid on home mortgage, 533
 tunnel clearance, 714
 uniform motion with current, 575
 velocity of falling bodies, 203
 vertex, 291
 vertex/intercept, 291–292
 volume, 21, 228, 286, 381, 390, 601
Fractions
 addition and subtraction of, 47
 compound, 48–49
 least common multiple to clear, 73
 major, 48
 minor, 48
 partial, 659–661
 simplifying compound, 48–49
Frustum, volume of, 228
Function families, 190
 on graphing calculator, 282–283
Function form, 174
 linear equations and, 174–175
 slope of line and, 175–176
Function inequalities
 applications of, 210–211
 solution to, 209–210
Functions, 174, 233–234, 237, 267
 analyzing graphs of, 341–350, 364
 base, 369–371
 composite, 253–255
 composition of, 251–253, 358
 continuous, 329–330
 cube root, 197–198
 cubing, 195–197
 decay, 478
 discontinuous, 330–331
 domain of, 255–256, 264
 even, 342
 exponential, 476–481, 495–496, 552, 553
 finding equations of, 292–293
 on graphing calculator, 253–254
 growth, 478
 historical background of, 155
 importance of, 155
 intervals and increasing or decreasing, 345–346
 intervals and positive or negative, 344–345
 inverse, 261, 263–268, 359
 linear, 176–180, 206–208, 235
 logarithmic, 485, 487–489, 552
 maximum and minimum value of, 347–348
 monotonically increasing, 478
 notation for, 163–165
 objective, 594, 597–599
 odd, 343–344
 one-to-one, 261–263, 359
 piecewise-defined, 327–335, 363
 polynomial, 393–402
 products and quotients of, 249–251
 quadratic, 192–193, 236, 288–294, 361, 479

rational, 301–307, 422–431, 438–445, 468
reciprocal, 301–303, 305
as relations, 159–160
slope-intercept form to graph, 176–179
square root, 194–195
step, 332–333
study of, 312
sum and difference of, 248–249
transcendental, 485
trigonometric, 354
vertical line test for, 160, 161
zeroes of, 207, 242
Fundamental principle of counting (FPC), 794–796, 835
Fundamental property of rational expressions, 45
Fundamental theorem of algebra, 393–394, 397
proof of, 393
Future value, 528–529
regression and, 540

G

Galileo Galilei, 54
Gaunt, John, 216
Gauss, Carl Friedrich, 107, 108, 393, 394, 619, 759
Gaussian elimination, 619, A-4
Gauss-Jordan elimination, 619, A-4
General linear equations, 185
General matrix, A-5
General solutions for family of equations, 75
Geometric sequences, 528, 768, 769, 833
applications of, 775–776
finding *n*th partial sum of, 772–773
finding *n*th term of, 770–771
on graphing calculator, 776–777, 840–841
Geometric series, 773–775, 840–841
Geometry, 98–99
Girard, Albert, 107, 383
Graphing calculator features
function mode, 776
LISTs, 166
repeat graph, 744
sequence mode, 776
STATPLOT, 166
TRACE, 211–212
Graphing calculators
calculating permutations and combinations on, 800
change-of-base formula on, 502
complex numbers on, 113
complex polynomials on, 401–402
composite functions on, 253–254
compound interest on, 530
domain of function on, 255–256
equation of line on, 181–182
estimating irrational roots using, 294
evaluating expressions and looking for patterns using, 136–137
exponential equations on, 511, 516
exponential functions on, 481
exponents on, 491
factors of polynomials on, 379–380
graphing lines on, 178
guide to, 79
intermediate value theorem and, 397
inverse functions on, 267–268
joint variation on, 320
linear equations on, 149–150
linear programming on, 674–676
linear regression on, 221–222
logarithms on, 489–491, 502, 515, 516
logistic equations on, 559–560
matrics on, 622–623, 633, 635
maximums and minimums on, 350
parabolas on, 735–736
parameterized solutions on, 585
piecewise-defined functions on, 334–335
polynomial graphs on, 414
polynomial inequalities on, 458
polynomials with complex roots on, 389
probability on, 815
quadratic equations and discriminant on, 125
quadratic functions on, 200
quadratic regression on, 223–224
rational exponents on, 100
rational functions on, 307, 455
rational graphs on, 431
rational inequalities on, 458
regression on, 537–540, 544
removable discontinuities on, 444–445
roots on, 471–472
sequences on, 754–755, 764, 776–777
series on, 754–755, 776–777, 840–841
solve nonlinear systems on, 710–711
split screen viewing on, 397
summation applications on, 790–791
systems of equations on, 572–573, 604–605
systems of inequalities on, 599
2×2 systems on, 615
unions, intersections, and inequalities on, 90
window size on, 613–615
zeroes of function on, 242
Graphs, 3, 232–234
analysis of function, 341–350, 364
applications of linear, 149
boundaries of, 142
continuous, 198, 327
of continuous functions, 329–330
of discontinuous functions, 330–331
distance formula and, 148
of exponential functions, 477–479
finding equations of functions from, 292–293
of function and its inverse, 266–267
horizontal and vertical lines on, 142–143, 145–146
importance of reading, 164–165
intercept method for, 141–142
of linear equations, 140–149, 232–233
of linear systems, 567, 572–573, 578
one-dimensional, 3, 578
one-wing, 194
origin of term for, 407
parallel and perpendicular lines on, 146–147
of piecewise-defined functions, 329–332
of polynomial functions, 412–414, 466
quadratic, 369–371
of quadratic functions, 190–193, 288–292
of rational functions, 301–306, 422–431, 467
of relations, 156–158
slope and rates of change on, 143–146
smooth, 198, 327
to solve absolute value equations, 604–605
to solve absolute value inequalities, 605–606
to solve systems of equations, 567
stretches and compressions in, 278
symmetry and, 342–344
transformations of, 274–282
translation of, 274
trigonometric, 354
two-dimensional, 578
Gravitational attraction, 309
Greater than, 5, 6
Greatest common factors (GCF), 35
Grid lines, 141
Grouping, factoring by, 36
Grouping symbols, with functions, 248
Growth functions, 478
Growth rate, 525

H

Half-life of radioactive substances, 484, 505, 518, 526–527
Half planes, 141, 590
Halley, Edmond, 694
Hannibal, 341
Harriott, Thomas, 83

Horizontal asymptotes, 304–306, 361
 of rational functions, 426–427
Horizontal boundary lines, 162
Horizontal change, 144
Horizontal lines
 applications for, 142–143
 slope of, 146
Horizontal line test, 262
Horizontal parabola, 158
Horizontal propeller, 198
Horizontal reflections, 277–278
Horizontal shifts, 276, 360
Horizontal translations, 275–276
Human life expectancy, 153
Hyperbolas, 694, 740, 741
 analysis of, 354
 center-shifted form of equation of, 695–697
 definition of, 720–721
 equation of, 700–701, A-6
 focal chord of, 703
 foci formula for, 721
 foci of, 720–722
 polynomial form of equation of, 699–700
 with rational/irrational values, 745
 standard form of equation of, 697–698
Hypotenuse, 62, 63

I

Ideal gas law, 45
Identities
 additive, 17, 629
 linear equations and, 73
 multiplicative, 17
Identity element, 17
Identity matrices, 640–642, 670
Imaginary numbers
 historical background of, 107
 identifying and simplifying, 108–109
Implied domain, 163
Inclusion, of endpoints, 84
Inconsistent systems of equations, 570, 582–584, 620–621
Increasing function, 198
Index
 of radicals, 8
 of summation, 752
Induction, 781–787, 834. *See also* Mathematical induction
Induction hypothesis, 784
Inequalities, 237. *See also* Linear inequalities; Systems of inequalities
 absolute value, 605–609
 additive property of, 85
 applications of, 88–89
 compound, 86–88
 constraint, 596
 function, 209–211
 on graphing calculator, 90
 joint, 87
 linear in one variable, 83–89, 130
 method to solve, 205–209
 multiplicative property of, 85
 nonstrict, 6
 polynomial, 451–454, 457, 458, 468
 push principle to solve, 472–473
 quadratic, 208–209
 rational, 451, 454–458, 468
 solution sets and, 84
 strict, 6
 symbols for, 5–6, 84
Infinite sequences, 749, 832
Infinite series
 geometric, 773–775
 on graphing calculator, 840–841
Inflection, point of, 196
Input/output table, 140
Input value, 16
Integers, 3, 77
Intercept method, 142
Interest
 compound, 522–524, 555
 continuously compounded, 524–525, 555
 simple, 521, 554
Interest earnings, 153, 324
Interest rate *r,* 521
Intermediate value theorem (IVT), 396, 397
Interpolation, 180, 540
Intersection, 86
 on graphing calculator, 90
 of sets, 590
Interval notation, 84
Intervals
 continuous, 605
 disjoint, 605
 where function is increasing or decreasing, 345–346
 where function is positive or negative, 344–345
Interval tests, solving function inequalities using, 209–210
Inverse, of matrices, 642–643, 676–677
Inverse element, 17
Inverse functions, 261, 263–264, 267, 359, 486
 algebraic method to find, 264–265
 on graphing calculator, 267–268
 graphs of, 266–267
 solving exponential and logarithmic equations using, 510
Inverse operation, 55
Inverse variation, 316–317, 362
Irrational numbers, 4
Irrational roots, 294
Irreducible polynomials, 395

J

Joint variation, 317–318
 on graphing calculator, 320
Jump discontinuity, 331, 332

K

Kepler, Johannes, 716
Kepler's third law of planetary motion, 67
Kowa, Seki, 616

L

Lagrance, Joseph-Louis, 155
Lateral surface area of cone, 102
Lattice points, 141
Lead coefficients, 27
Learning curve, 546
Least common denominator (LCD), A-3
Least common multiple (LCM), 73
Lebesque, Henri, 1
Leonardo of Pisa, 749
Less than, 5, 6
Liebniz, Gottfried von, 655
Lift capacity formula, 92
Like terms, 18
Line
 equation of, 181–182
 slope of, 175–176
Linear association, 218
Linear dependence, 578
Linear depreciation, 180
Linear equation model, 220–221
Linear equations, 72, 129–130, 140. *See also* Equations
 addition and multiplication properties to solve, 72–73
 forms of, 189–190
 function form and, 174–175
 general, 185
 general solution for family of, 75
 graphs of, 140–149
 identities and contradictions and, 73–74
 intercept/intercept form of, 185
 method for solving, 72, 73
 in point-slope form, 179
 problem-solving guide for, 75–78
 specified variables and literal equations and, 74–75
 standard form of, 72, 189
 in two variables, 140–149
Linear factorization theorem, 394, 395
 corollary to, 396
 proof of, 394–395

Linear functions, 235. *See also* Functions
family of, 179–180
inequalities and, 206–208
slope-intercept form to graph, 176–179
Linear inequalities, 83, 130. *See also* Inequalities; Systems of inequalities
applications of, 88–89
compound, 86–88
method to solve, 85–86
in one variable, 83–89, 130
solutions sets and, 84
in two variables, 590–592
Linear programming, 565, 667
on graphic calculators, 674–676
objective function and, 597–599
principles of, 595
solutions to problems in, 596–597
systems of linear inequalities and, 594–599, 667
Linear regression, 221–222
Linear systems. *See* Systems of linear equations; Systems of linear equations in three variables; Systems of linear equations in two variables
Line of best fit, 221–222
Lines
characteristics of, 189
coincident, 570
horizontal, 142–143
horizontal boundary, 162
parallel, 146
perpendicular, 146–147
secant, 184, 199
vertical, 142–143
vertical boundary, 161–162
Line segment, 147
LISTs feature, 166
Literal equations, 74
method to solve, 74–75
rewriting, 63
simplification of, 49–50
Loan payments, 779
Logarithmic equations, 554
with like bases, 513, 514
method to solve, 513–514, 560–562
in simplified form, 509–510
Logarithmic functions, 485, 487, 552
applications of, 515
graphs of, 488–489
natural, 496–497
use of, 486
Logarithmic tables, 489
Logarithms, 552
applications for, 486, 489–491
base-*b*, 510
change-of-base formula and, 499–500
common, 489–491
on graphing calculator, 489–491, 502, 515
historical background of, 485–486
natural, 500–501, 553
properties of, 488, 497–499, 508
Logistic equations, 543–544
on graphing calculator, 559–560
Logistic growth model, 543
LORAN, 706–707, 720
Lorentz transformation, 44
Lowest terms, 45

M

Magnitudes
of earthquakes, 490, 493
of numbers, 3
of stars, 493
Major fractions, 48
Malthus, Thomas Robert, 768–769
Mapping notation, 156, 159
Market equilibrium, 714
Mathematical induction, 781, 834
applied to sums, 782–785
general principle of, 785–787
proof by, 781–785
Mathematical models, 14–15
Matrices, 616–617
addition and subtraction of, 628–629, 669
applications using, 621–622, 640, 655–661
associated minor, 646, 647
augmented, 617–618, 620–621, 676–677
coefficient, 617
of constants, 617
equality of, 627–628
on graphing calculator, 622–623, 633
historical background of, 616
identity, 640–642
inverse of, 642–643, 676–677
meaning of term, 627
multiplication of, 629–635, 669
noninvertible, 645
reduced row-echelon form and, 619, A-4–A-5
singular, 645–649
solving systems using, 618–620, 622–623, 669
in triangular form, 618–620
Matrix equations, 640
solving systems using, 643–645, 670
Maximum value, 191, 594
of functions, 347–348
on graphing calculator, 350
Measurement
Metric System, A-1–A-2
U. S. Standard Units, A-1–A-2
Medication in bloodstream formula, 33
Metric System, A-1–A-2
Midpoint, of line segment, 147
Midpoint formula, 147–148
Minimum value, 191, 594
of functions, 347–348
on graphing calculator, 350
Minor fractions, 48
Mixture problems
method to solve, 78
systems and, 571–572
Modeling
systems of linear equations in three variables and, 584–585
systems of linear equations in two variables and, 571–572
Monomials, 26–27
product of, 28
Monotonically increasing functions, 478
Mortgage payment formula, 34
Mortgage payments, 532, 533
Multiplication
associative property of, 17
commutative property of, 16
of complex numbers, 110–112
of functions, 249–251
identity matrices and, 640–642, 670
of matrices, 629–635, 669
of polynomials, 28–29
of radical expressions, 60–62
of rational expressions, 46
scalar, 629–630
Multiplicative identity, 17
Multiplicative inverse, 17
Multiplicative property
of equality, 72–73, 85
of inequality, 85
Multiplicity
roots of, 345, 388
vertical asymptotes and, 424
zeroes of, 410–412
Mutually exclusive events, 812–813

N

Napier, John, 485
Nappe, 681
Natural exponential functions, 495
Natural logarithmic functions, 496–497, 553
applications of, 500–501
Natural numbers, 2
Negative association, 217
Negative exponents, 25

Negative numbers, 3
as exponents, 24–25
Negative reciprocals, 147
Nested factoring, 43
Nested factoring formula, 43
Newton's law of cooling, 515
Nondistinguishable permutations, 797–798
Nonexclusive events, 813–814
Noninvertible matrix, 645
Nonlinear association, 218
Nonlinear asymptotes, 440, 442
Nonlinear inequalities, systems of, 711–712, 741
Nonlinear systems of equations. *See* Systems of nonlinear equations
Nonremovable discontinuity, 331, 438
Nonstrict inequality, 6
Nonterminating decimals, 4
North/south reflection, 277
Notation
direction/approach, 302
exponential, 7–8, 22
factorial, 750
function, 163–165
historical background of, 13
inequality, 5–6
interval, 84
mapping, 156
ordinary, 26
scientific, 26
square and cube root, 8, 489
summation, 752–753
Null set, 2
Number line, 3, 84
Number puzzles, 98
Numbers
complex, 107–113, 132
imaginary, 107–109
irrational, 4
magnitude of, 3
natural, 2
negative, 3, 24–25
positive, 3
rational, 3–4, 44
real, 4–9, 16–18
sets of, 2
whole, 2–3
Numerators, rationalizing, 63
Numerical coefficients, 13
Nuñez, Pedro, 374

O

Objective function, 594
Objective variables, 596
Object variables, 74
Oblique asymptotes, 438
rational functions with, 440, 441
Odd functions, 343–344
One-dimensional graphs, 3, 578
One-to-one functions, 261–263, 359
One-wing graph, 194
Opposites, 17
Ordered lists, 792
Ordered pair, 140, 141
Ordered pair form, 156
Ordered triples, 578
Order of operations, 9
Ordinary notation, 26
Ordinate, 312
Origin, 141
symmetry to, 343
Oughtred, William, 83
Output, 16

P

Painted area on canvas formula, 102–103
Parabolas, 191, 729, 742
of best fit, 223
definition of, 732
focus-directrix form of, 732–735
focus of, 735–736
horizontal, 730–731
reflective property of, 729
standard form of, 735–736
vertical, 730
Parallel lines, 146
slope of, 177–178
Parameters, 582
Partial fractions, 659–661
Partial sums
of arithmetic sequence, 762–763
of geometric sequence, 772–773
of infinite geometric series, 774–775
of series, 751, 773–775
Pascal, Blaise, 807–808, 823
Pascal's triangle, 22, 823–825
finding coefficients using, 825–826
on graphing calculator, 828–829
Perfect cubes, 8
sum and difference of two, 39–40
Perfect squares, 8
difference of two, 29
factoring difference of two, 38
Perfect square trinomials, 30, 39
Perimeter, of rectangle, 637
Period, 369
Permutations
distinguishable, 796–797
nondistinguishable, 797–798
Perpendicular distance, 214
Perpendicular lines, 146–147
slope of, 177
pH level, 492–493
Piecewise-defined functions, 327, 363
applications of, 332–334
effective domains of, 327–328
evaluation of, 329
on graphing calculator, 334–335
graphs of, 329–332
Pitch diameter, 12
Placeholder substitution, 40
Plane, in space, 578
Planetary orbits, 743
Point of inflection, 196
Point-slope form, 189
applications of, 179–181
linear equations in, 179
Poiseuille's law, 44
Polynomial equations, 94–95, 131
applications of, 98–100
standard form for, 118
Polynomial expressions, 26–27
Polynomial form
of equation of circle, 684
of equation of ellipses, 685–689
of equation of hyperbolas, 699–700
of equations of parabolas, 729
Polynomial functions
graphs of, 412–414, 466
zeroes of, 393–402, 465–466
Polynomial graphs
end behavior of, 407–408
on graphing calculator, 414
with roots of multiplicity, 409–412
turning points and, 412
Polynomial inequalities, 452–454, 468
applications of, 457
on graphing calculator, 458
steps to solve, 452
study of, 451
Polynomials, 27
addition and subtraction of, 27–28
complex, 401–402
with complex roots, 385–386, 389
degree of, 27
division of, 250, 374–380, 464
factorable, 378–379, 398
factoring quadratic, 36–38
families of, 27
graphical tests for factors of, 379–380
on graphing calculator, 401–402
irreducible, 395
prime, 37
products of, 28–31
quartic, 417
real, 396
Polynomial zeroes theorem, 345, 388
Population density, 433
Positive association, 217
Positive numbers, 3

Power
of imaginary unit, 788
to power property, 23
of wind-driven generator, 203
Power functions, 409, 477
Power property
of equality, 96–98
of exponents, 23
of logarithms, 497
power to, 23
product to, 23
proof of, 498
quotient to, 23
Power regressions, 319
Power to power property, 23
Present value equation, 522
Pressure, 21
Prime polynomials, 37
Principal *p*, 521
Principle roots, 4
Principle square roots, 54
Probability, 807, 835–836
applications of, 841–843
elementary, 808–809
formula for binomial, 830
on graphing calculator, 815
historical background of, 807–808
mutually exclusive events and, 812–813
nonexclusive events and, 813–814
properties of, 809–811
quick-counting and, 811–812
Problem solving, methods for, 75–78
Product property
of exponents, 22
of logarithms, 497
proof of, 498
of radicals, 60
Products, to power property, 23
Projectile motion, 98
Projectiles, 123
height of, 127, 228
Projectile velocity, 198–199
Proof, by induction, 781–785
Push principle, 472–473
Pythagorean theorem, 62–63

Q

Quadrants, 141
Quadratic, 118
Quadratic equations, 94, 133
completing the square to solve, 119–120, 288
on graphing calculator, 125
method to solve, 137–138, 288
quadratic formula to solve, 120–122
square root property of equality and, 118–119
standard form of, 94, 118
Quadratic forms, 40
Quadratic formula, 121
applications of, 123–124
discriminant and, 122–123
solving quadratic equations with, 120–122
Quadratic functions, 236, 479
completing the square to graph, 288–290
extreme values and, 293–294
on graphing calculator, 200
graphs of, 192–193, 290–292, 361
reciprocal, 301, 303, 305
Quadratic graphs
base functions and, 369–371
characteristics of, 190–192
Quadratic inequalities, 208–209
Quadratic polynomials, factoring, 36–38
Quadratic regression, 222–223
Quartic polynomials, 408, 417
Quick-counting techniques, 811–812, 815
Quotient property
of exponents, 24
of logarithms, 497
of radicals, 59
Quotients
difference, 348–349
of polynomials, 374–375
to power property, 23

R

Radical equations, 94
power property of equality and, 96
solution to, 96–98
Radical expressions, 54
addition and subtraction of, 60
multiplication and division of, 60–62
rewriting, 63
simplification of, 54–56, 60
Radicals, 8
product property of, 58–59
quotient property of, 59
Radicand, 8
Radioactive decay, 484, 505, 526, 527
Radius, of circle, 682
Range, 156, 162
Rates of change, 144, 235, 274
difference quotient and, 348–349
method to compute, 501
slope as, 178
Rate of increase, 318
Rational equations, 94
applications of, 100
method to solve, 95–96
Rational exponents, 57–58
on graphing calculator, 100
power property of equality and, 97–98
to simplify radical expressions, 60
use of, 56–57
Rational expressions, 44
addition and substraction of, 47–48
fundamental property of, 45
least common denominator and, A-3
multiplication and division of, 46–47
partial fractions and, 659–661
rewriting, 49–50
in simplest form, 45–46
simplification of compound, 48–49
Rational functions, 301–303, 423, 468
applications of, 306–307, 442–444
common factors and, 438–439
on graphing calculator, 431
graphs of, 422–431, 467
horizontal and vertical asymptotes of, 304–306, 422–423, 426–427
with oblique and nonlinear asymptotes, 439–442
Rational inequalities, 454–458, 468
applications of, 457
on graphing calculator, 458
study of, 451
Rationalizing the denominator, 61–62
Rational numbers, 3–4, 44
Rational roots theorem (RRT), 397–399
Rational terms, 14
Raw data, 219
Real data, 219
Real numbers, 4–5
absolute value of, 6–7
operations on, 7–9
order property of, 5
properties of, 16–18
subsets of, 4–5
Real polynomials, 396
Real roots, 421
Reciprocal functions, 301–303, 305
graph of, 301
quadratic, 301, 303, 305
Reciprocals, 17
Rectangles, 637
Rectangular coordinate system, 141, 232–233
Recursive sequences, 749–750
Reduced row-echelon form, 619, A-4–A-5
Reference intensity, 490–491
Reflections
across *x*-axis, 277
across $y = x$, 277–278
horizontal, 277–278
vertical, 276–277

Regression, 216, 237, 555–556
  applications of, 540–542
  data sets and, 536
  forms of, 537
  on graphing calculator, 537–540, 544
  power, 319
  quadratic, 222–223
Regression equations, 181
  applications of, 541–542
  calculation of, 367–368
Regression line, 221
Regression models, 538–540
  logistics equations and, 543–544
Relations, 156, 233
  absolute value, 161
  functions as, 159–160
  graphs of, 156–158
Remainder, 374
Remainder theorem, 383–384, 465
  complex numbers and coefficients and, 385–386
  proof of, 384
Removable discontinuity, 332
  on graphing calculator, 444–445
Repeated roots, 471–472
Required interest rate, 324
Residuals, 368, 537
Revenue models, 98–100
Rhetorical algebra, 13
Richter values, 490
Right circular cone, 682
Right triangles, 62–63
Rigid transformations, 278
Rise, 144
Roman numerals, 248
Roots
  approximation of real, 421
  complex, 385–386, 389, 471–472
  cube, 8, 9, 55–56
  extraneous, 96, 514
  irrational, 294
  principle, 4
  repeated, 471–472
  square, 4, 8, 9, 54–56
Roots of multiplicity, 345
  attributes of polynomial graphs with, 409–412
  complex conjugates and, 388
Root tests, for quartic polynomials, 417
Row-echelon form, 619, A-4
  reduced, 619, A-4–A-5
Row operations, elementary, 618
Run, 144

S

Sample space, 793
Sand dune function, 338
Scalar multiplication, 629–630
Scale of data, 222
Scatter-plots, 217
  linear/nonlinear association and, 218
  positive/negative association and, 217–218
  strong/weak association and, 219
Scientific notation, 26
Secant lines, 184, 199
Seed element, 749
Semicircle
  equation of central, 214
  graph of, 158
Semi-hyperbola, equation of, 703
Sequences, 747, 748, 831–832. *See also* Series; *specific types of sequences*
  alternating, 749
  applications of, 753–754
  arithmetic, 759–764, 832–833
  finding *n*th term of, 751, 760–762, 770–771
  finding terms of, 748–749
  finite, 748, 749, 831
  forms of, 748
  geometric, 768–777, 833
  on graphing calculator, 754–755
  induction and, 782
  infinite, 749, 832
  partial sums in, 751, 762–763, 772–773
  recursive, 749–750
  summation notation and, 752–753
Sequence stratigraphy, 747
Series, 748, 832. *See also* Sequences
  applications of, 775–776
  finite, 751
  geometric, 773–775
  on graphing calculator, 754–755, 776–777, 840–841
  induction and, 782
  infinite, 773–775, 840–841
  partial sum of, 751, 773–775
  summation notation to evaluate, 752–753
Set notation, 2, 84
Sets
  intersection of, 86
  null, 2
  of numbers, 2
Shifted form, of quadratic functions, 288
Shifts
  horizontal, 276, 360
  vertical, 275, 360
Sigma notation, 752
Simple interest, 521
Simplest form, 45, 61
Singular matrices, 645, 646
  determinants and, 645–649
Sinking fund, 529
Slope, 143
  of horizontal and vertical lines, 146
  importance of understanding, 174
  of parallel lines, 146, 177
  of perpendicular lines, 146–147, 177
  positive and negative, 145
  as rate of change, 178
Slope formula, 144–145, 189
  use of, 348
Slope-intercept form, 176, 189
  graph of line and, 176–179
Slope triangle, 143–144
Smooth graphs, 198, 327
Solution form, 73
Solution sets, 84, 590
Solution to the system, 566
Sound, intensity of, 493
Spheres, 271, 286
Spherical cap, 381
Spring oscillation, 611
Square root function, 194–195
Square root property of equality, 118–119
Square roots, 4
  of complex numbers, 404
  definition for, 9
  notation for, 8, 489
  principle of, 54
  simplification of, 54–56
Squares
  of binomials, 30–31
  perfect, 8, 29, 38
Square systems, 583
Squaring function, 190–191
Standard form
  of complex numbers, 109
  of equation of circle, 685
  of equation of ellipse, 686
  of equation of hyperbola, 697–698
  of linear equations, 69, 72, 189
  of polynomial equations, 118
  of polynomial expressions, 27
  of quadratic equations, 94
Statistics, 807
STATPLOT feature, 166
Step functions, 332–333
Stirling's formula, 804
Strength of correlation, 219
Stretches, 278, 360
Strict inequality, 6
Strong association, 219
Subsets, 2
Substitution
  placeholder, 40
  to solve nonlinear systems, 707–708
  to solve system of equations, 567–568
Subsystems, 580, 581

Subtraction
of complex numbers, 109–110
of fractions, 47
of functions, 248–249
of matrices, 628–629, 669
of polynomials, 27–28
of radical expressions, 60
of rational expressions, 47–48
Summation
applications of, 790–791
notation for, 752–753
properties of, 753
Supply curves, 714
Surface area. *See also* Area
of cone, 102
of cylinder, 81, 127, 258, 447
of rectangular box, 296
Symbols
grouping, 248
inequality, 5–6, 84
Symmetry
axis of, 191, 192
graphs and, 342–344
to origin, 343
Synthetic division, 375–377, 464
factorable polynomials and, 378–379
Systems approach
to solve absolute value equations, 604–605, 668
to solve absolute value inequalities, 605–606, 668
Systems of inequalities, 590–599, 667. *See also* Linear inequalities
applications of, 590, 593
on graphing calculator, 599
linear programming and, 565, 594–599, 667
solution to, 590, 592–593
Systems of linear equations
augmented matrix of, 617–618
determinants and Cramer's Rule to solve, 655–658
inconsistent and dependent, 620–621
matrices and, 616–623, 669
matrix equations to solve, 640–649, 670
Systems of linear equations in three variables, 566, 666–667
elimination to solve, 579–581
graphing to solve, 578
inconsistent and dependent, 582–584
modeling and, 584–585
solutions to, 578–579
visualizing solutions to, 578
Systems of linear equations in two variables, 666
elimination to solve, 568–570
graphing to solve, 567, 572–573
inconsistent and dependent, 570–571
modeling and, 571–572
simultaneous, 565
solutions to, 566
substitution to solve, 567–568
Systems of nonlinear equations, 706, 741
elimination to solve, 708–709
possible solutions for, 707
substitution to solve, 707–708
using graphing calculator to solve, 710–711
Systems of nonlinear inequalities, 711–712

T

Terminating decimals, 4
Test for collinear points, 659
Theory of relativity, 577
3 × 3 systems, 577–585
Tightness of fit, 219
Timing a falling object, 66
Toolbox functions, 190, 236, 359–360
data analysis and, 318–320
direct variation and, 313–316
graphs of, 198–199
inverse variation and, 316–317
joint or combined variation and, 317–318
transformations and, 274
Transcendental functions, 94, 485
Transformations, 360
area under curve and, 280–283
of general function, 279
of graphs, 274
horizontal reflections and, 277–278
horizontal shifts and, 276
of linear systems, 580
nonrigid, 278, 360
of quadratic graphs, 369
rigid, 278, 360
vertical reflection and, 276–277
vertical shifts and, 275
via composition, 300
Translations
of graphs, 274–276
horizontal, 275–276
vertical, 274–275
Trapezoids, area of, 381
Tree diagrams, 792–793
Trial-and-error process, 37
Trials, 793
Triangles
area of, 447, 625
determinants to find area of, 658–659
inscribed, 692
Pascal's, 22, 823–826, 828–829
right, 62–63
slope, 143–144
Triangularizing, of augmented matrix, 618–620
Trichotomy axiom, 591
Trigonometric graphs, 354
Trinomials, 27
perfect square, 30, 39
in quadratic form, 40
Tunnel clearance, 714
Turning points, 412
Two-dimensional graphs, 578
2 × 2 systems, 566–573, 577

U

Unbounded region, 594
Uniform motion problems, 77–78
Union, 86, 90
Uniqueness property, 479–480, 513
Unique solutions, 578
Upper and lower bounds property, 399–401
U. S. Standard Units, A-1–A-2

V

Variables, 5
decision, 596
descriptive, 5, 6
matrix of, 630
object, 74
objective, 596
Variable terms, 13
Variation
constant of, 313
direct, 313–316, 362
inverse, 316–317, 362
joint, 317–318, 320
Variation problems
examples of, 314–316
steps to solve, 314, 362
Velocity
average rate of change applied to projectile, 198–199
of falling body, 203
Venn diagrams, 590
Verbal information, translated into mathematical model, 14–15
Vertex, 161, 191–193, 681
Vertex formula, 291
Vertex/intercept formula, 291–292
use of, 370
Vertical asymptotes, 304–306, 362
multiplicities and, 424
of rational functions, 422–423
Vertical axis, 578
Vertical boundary lines, 161–162
Vertical change, 144
Vertical format, 16
Vertical lines
applications for, 142–143
slope of, 146
Vertical line test for functions, 160, 161
Vertical propeller, 196
Vertical reflections, 276–277
Vertical shifts, 275, 360

Vertical translations, 274–275
Viéte, François, 107
Volume
  of cylinder, 43
  formula for pressure and, 21
  of frustum, 228
  of open box, 390
  of spherical cap, 381
  surface area of cylinder with fixed, 447

W

Weak association, 219
Whole numbers, 2–3
Written information, translation of, 14–15, 76

X

x-axis, 141
  reflections across, 277
x-intercepts, 142, 189
  of rational functions, 424–425

Y

y-axis, 141
  symmetric to, 342
y-intercepts, 142, 189
  of rational functions, 424–425

Z

Zeno of Elea, 840
Zeroes
  division and, 7
  as exponents, 24
  of function, 207, 242
  of multiplicity, 410–412
  of polynomial functions, 393–402, 465–466
Zero factor property, 94–95

The Toolbox and Other Functions

linear

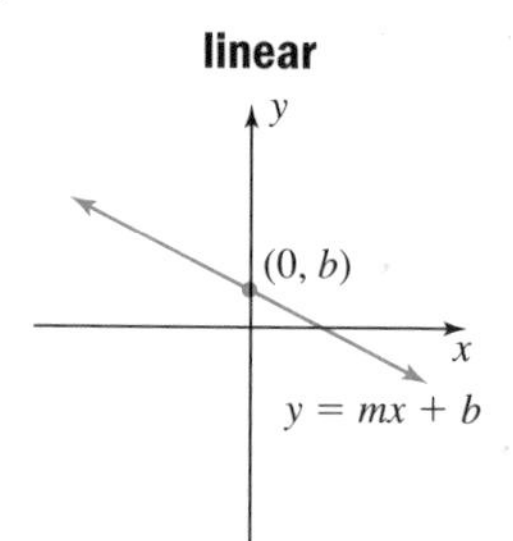

$m < 0, b > 0$

linear

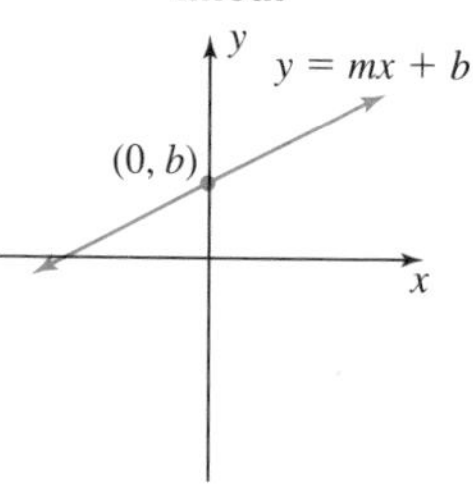

$m > 0, b > 0$

identity

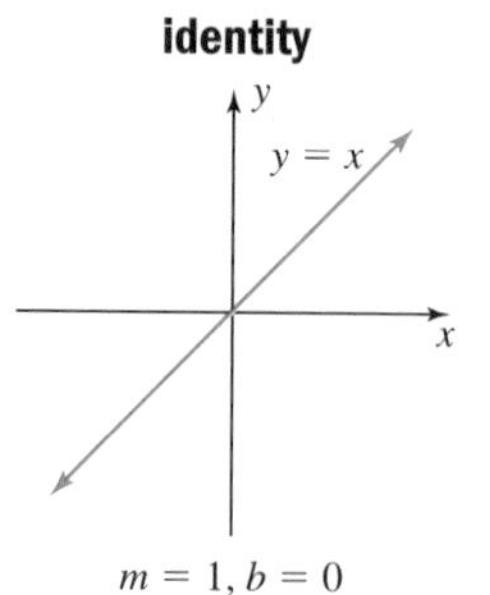

$m = 1, b = 0$

constant

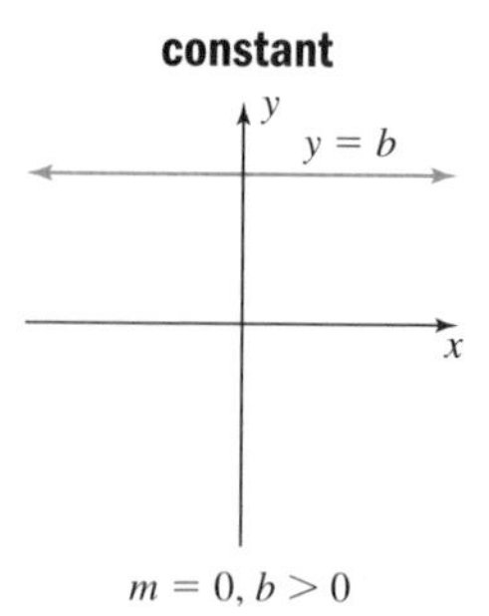

$m = 0, b > 0$

absolute value

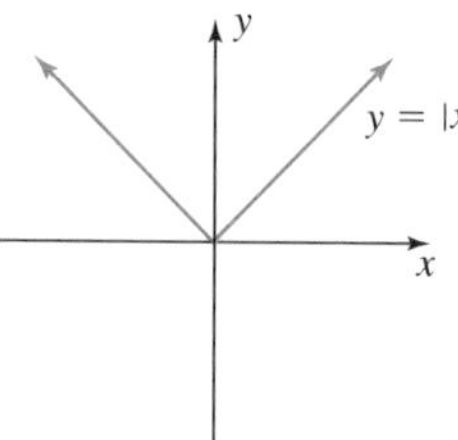
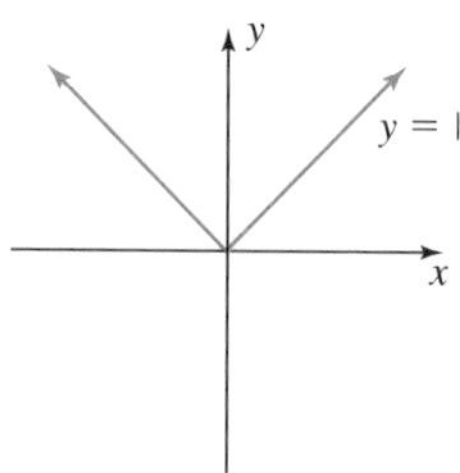

squaring

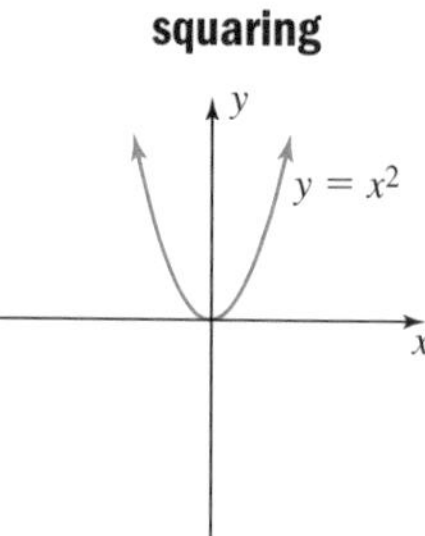

cubing

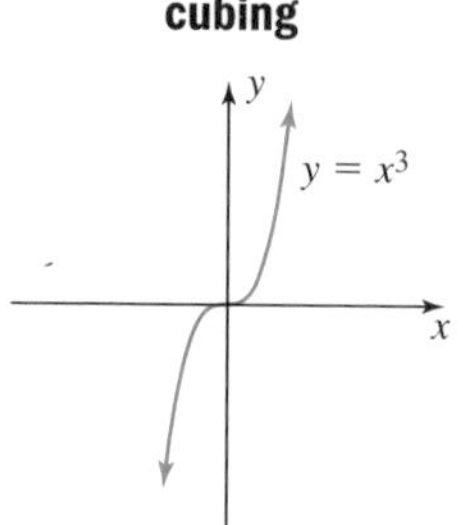

square root

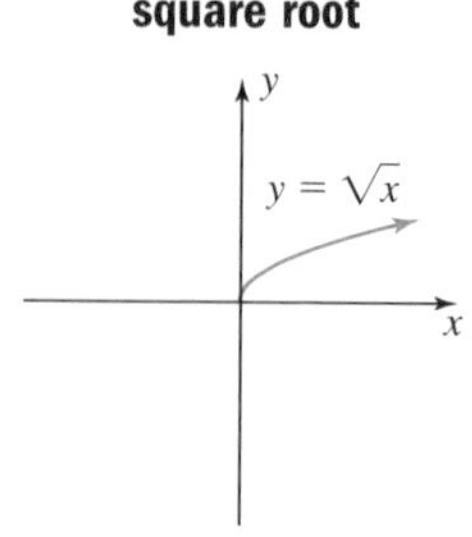

cube root

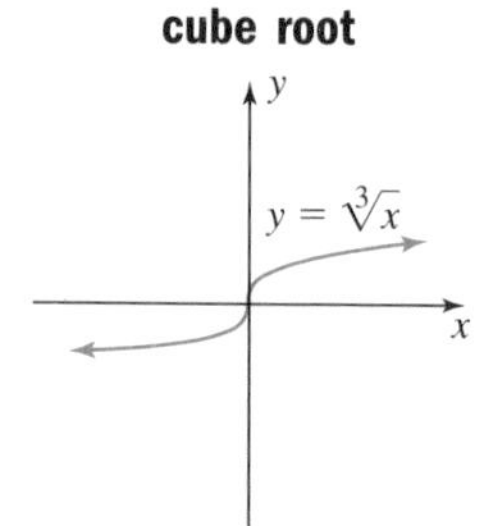

greatest integer

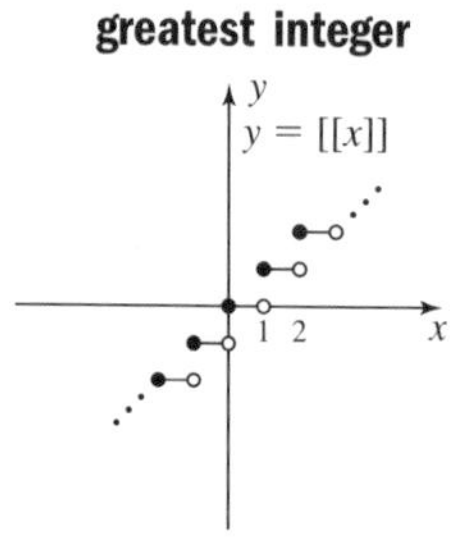

reciprocal

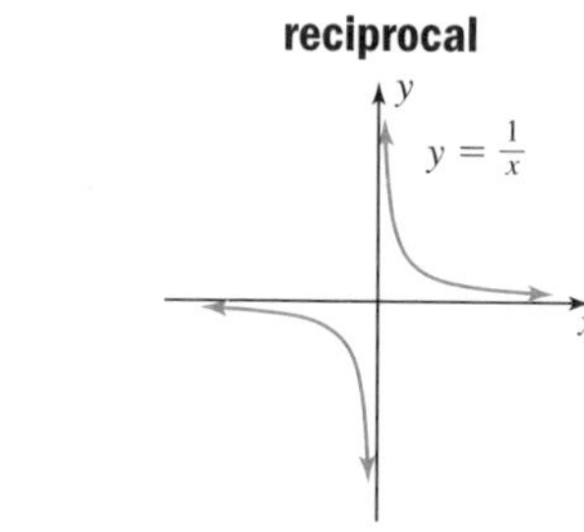

reciprocal quadratic

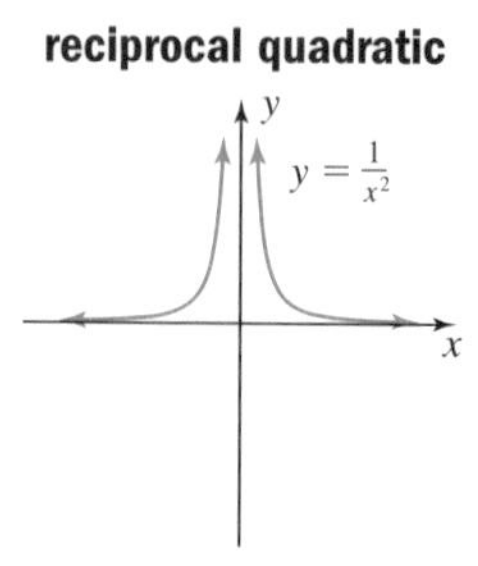

exponential

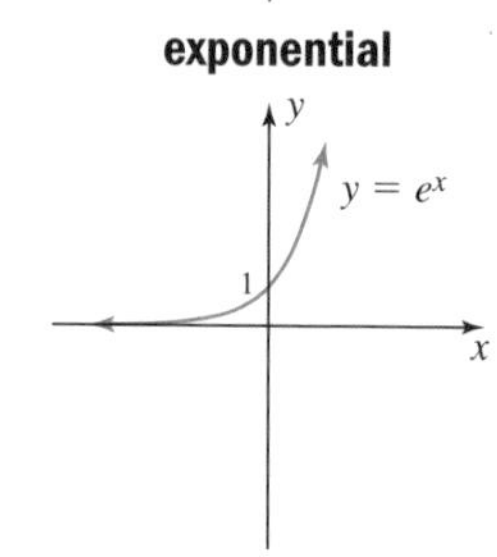

exponential

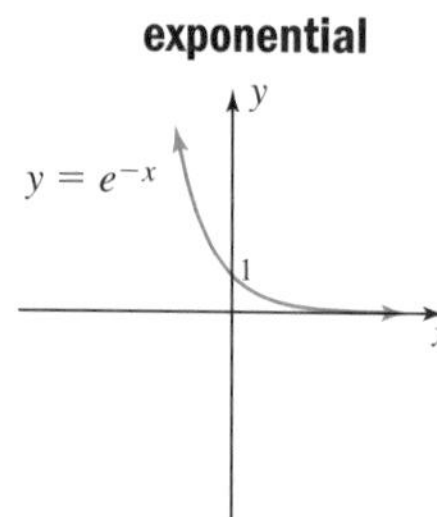

logarithmic

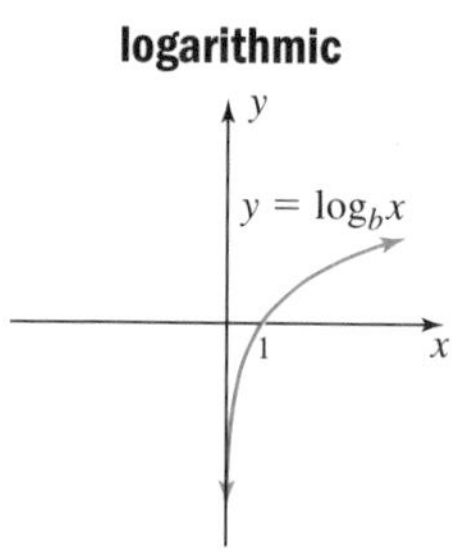

logistic

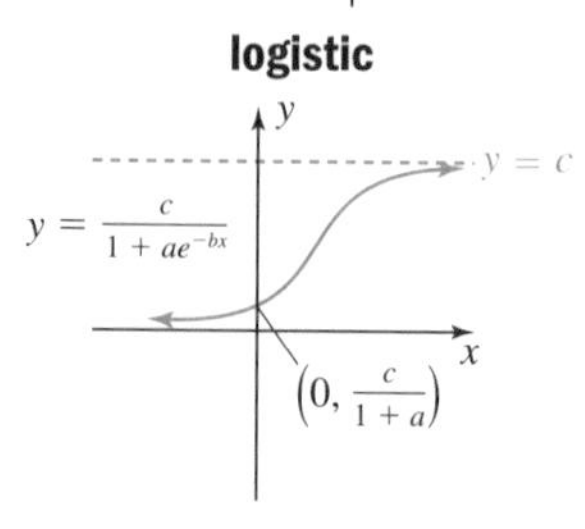

Transformations of Basic Graphs

Given Function	Transformation of Given Function
$y = f(x)$	$y = af(x \pm h) \pm k$

north/south reflections
vertical stretches/compressions

horizontal shift h units, opposite direction of sign

vertical shift k units, same direction as sign

Average Rate of Change of $f(x)$

For linear function models, the average rate of change on the interval $[x_1, x_2]$ is constant, and given by the slope formula: $\frac{\Delta y}{\Delta x} = \frac{y_2 - y_1}{x_2 - x_1}, x_2 \neq x_1$. The average rate of change for many other function models is nonconstant. By writing the slope formula in function form using $y_1 = f(x_1)$ and $y_2 = f(x_2)$, we can compute the average rate of change of other functions:

$$\frac{\Delta y}{\Delta x} = \frac{f(x_2) - f(x_1)}{x_2 - x_1}$$